W9-AAE-699

SECOND EDITION

Industrial Communication Technology Handbook

EDITED BY

Richard Zurawski
ISA Group, San Francisco, California, USA

CRC Press is an imprint of the
Taylor & Francis Group, an **informa** business

CRC Press
Taylor & Francis Group
6000 Broken Sound Parkway NW, Suite 300
Boca Raton, FL 33487-2742

Printed on acid-free paper
Version Date: 20140626

International Standard Book Number-13: 978-1-4822-0732-3 (Hardback)

Library of Congress Cataloging-in-Publication Data

Industrial communication technology handbook / edited by Richard Zurawski. -- Second edition.
pages cm
Includes bibliographical references and index.
ISBN 978-1-4822-0732-3
1. Computer networks. 2. Data transmission systems. 3. Wireless communication systems. I. Zurawski, Richard.

TK5105.5.I48 2015
621.39'81--dc23 2013047692

Visit the Taylor & Francis Web site at
http://www.taylorandfrancis.com

and the CRC Press Web site at
http://www.crcpress.com

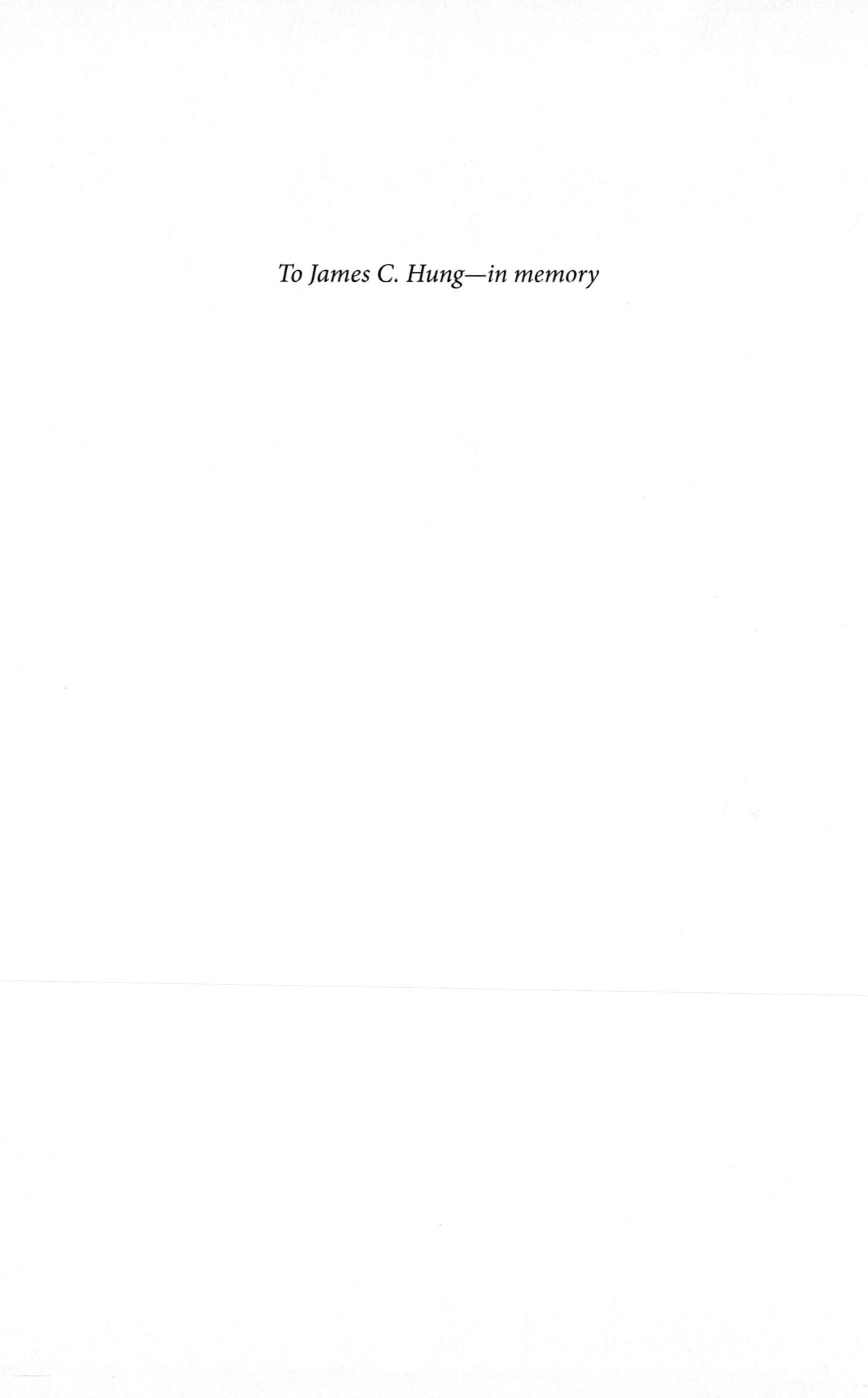

To James C. Hung—in memory

Contents

Preface xiii
Editor xv
Acknowledgments xvii
Contributors xix
International Advisory Board xxv

SECTION I Field Area and Control Networks

1 Fieldbus System Fundamentals 1-1
Thilo Sauter

2 Networked Control Systems for Manufacturing 2-1
James R. Moyne, Dawn M. Tilbury, and Dhananjay Anand

3 Configuration and Management of Networked Embedded Devices 3-1
Wilfried Elmenreich and Andrea Monacchi

4 Smart Transducer Interface Standard for Sensors and Actuators 4-1
Kang B. Lee

5 IO-Link (Single-Drop Digital Communication System) for Sensors and Actuators 5-1
Wolfgang Stripf

6 AS-Interface 6-1
Tilman Schinke

7 HART over Legacy 4–20 mA Signal Base 7-1
Mark Nixon, Wally Pratt, and Eric Rotvold

8 HART Device Networks 8-1
Mark Nixon and Deji Chen

9 Common Industrial Protocol (CIP™) and the Family of CIP Networks 9-1
Viktor Schiffer

10 Modbus Protocol 10-1
Rudy Belliardi and Ralf Neubert

11 PROFIBUS 11-1
Ulrich Jecht, Wolfgang Stripf, and Peter Wenzel

12 PROFINET 12-1
Peter Wenzel

13 Sercos® Automation Bus 13-1
Scott Hibbard, Peter Lutz, and Ronald M. Larsen

14 FOUNDATION Fieldbus 14-1
Salvatore Cavalieri

15 INTERBUS 15-1
Jürgen Jasperneite

SECTION II Industrial Ethernet

16 Switched Ethernet in Automation 16-1
Gunnar Prytz, Per Christian Juel, Rahil Hussain, and Tor Skeie

17 Real-Time Ethernet for Automation Applications 17-1
Max Felser

18 Ethernet for Control Automation Technology 18-1
Gianluca Cena, Stefano Scanzio, Adriano Valenzano, and Claudio Zunino

19 Ethernet POWERLINK 19-1
Federico Tramarin and Stefano Vitturi

20 IEEE 802.1 Audio/Video Bridging and Time-Sensitive Networking 20-1
Wilfried Steiner, Norman Finn, and Matthias Posch

SECTION III Fault-Tolerant Clock Synchronization in Industrial Automation Networks

21 Clock Synchronization in Distributed Systems Using NTP and PTP 21-1
Reinhard Exel, Thilo Sauter, Paolo Ferrari, and Stefano Rinaldi

SECTION IV Accessing Factory Floor Data

22 Linking Factory Floor and the Internet 22-1
Thilo Sauter

23 MTConnect 23-1
Dave Edstrom

24 Standard Message Specification for Industrial Automation Systems 24-1
Karlheinz Schwarz

25 Extending CEA-709 Control Networks over IP Channels 25-1
Dietmar Loy and Stefan Soucek

26 Web Services for Embedded Devices 26-1
Vlado Altmann, Hendrik Bohn, and Frank Golatowski

SECTION V Safety Technologies in Industrial Networks and Network Security

27 PROFIsafe 27-1
Wolfgang Stripf and Herbert Barthel

28 SafetyNET p Protocol 28-1
Marco Cereia, Jochen Streib, and Reinhard Sperrer

29 Security in Industrial Communications 29-1
Thilo Sauter and Albert Treytl

SECTION VI Wireless Industrial Networks

30 Wireless LAN Technology for the Factory Floor 30-1
Andreas Willig

31 WirelessHART 31-1
Alessandra Flammini and Emiliano Sisinni

32 ISA100.11a 32-1
Stig Petersen and Simon Carlsen

33 Comparison of WirelessHART and ISA100.11a for Wireless Instrumentation 33-1
Stig Petersen and Simon Carlsen

34 IEC 62601: Wireless Networks for Industrial Automation–Process Automation (WIA-PA) 34-1
Ivanovitch Silva and Luiz Affonso Guedes

35 Wireless Extensions of Real-Time Industrial Networks 35-1
Gianluca Cena, Adriano Valenzano, and Stefano Vitturi

36 Wireless Sensor Networks for Automation 36-1
Tomas Lennvall, Jan-Erik Frey, and Mikael Gidlund

37 Design and Implementation of a Truly Wireless Real-Time Sensor/Actuator Interface for Discrete Manufacturing Automation 37-1
Guntram Scheible, Dacfey Dzung, Jan Endresen, and Jan-Erik Frey

38 IPv6 over Low-Power Wireless Personal Area Networks (6LoWPAN) and Constrained Application Protocol (CoAP) **38**-1
Guido Moritz and Frank Golatowski

SECTION VII Time-Triggered Communication

39 Concepts of Time-Triggered Communication **39**-1
Roman Obermaisser

40 Time-Triggered Protocol (TTP/C) **40**-1
Roman Obermaisser and Michael Paulitsch

41 Time-Triggered CAN **41**-1
Roland Kammerer and Roman Obermaisser

42 Time-Triggered Ethernet **42**-1
Wilfried Steiner and Michael Paulitsch

SECTION VIII Avionics and Aerospace

43 MIL-STD-1553B Digital Time Division Command/Response Multiplex Data Bus **43**-1
Chris deLong

44 ASCB **44**-1
Michael Paulitsch

45 ARINC Specification 429 Mark 33 Digital Information Transfer System **45**-1
Daniel A. Martinec and Samuel P. Buckwalter

46 ARINC 629 ..... **46**-1
Michael Paulitsch

47 Commercial Standard Digital Bus **47**-1
Lee H. Harrison

48 SAFEbus **48**-1
Michael Paulitsch and Kevin R. Driscoll

49 Dimensioning of Civilian Avionics Networks **49**-1
Jean-Luc Scharbarg and Christian Fraboul

SECTION IX Automotive Communication Technologies

50 In-Vehicle Communication Networks **50**-1
Nicolas Navet and Françoise Simonot-Lion

51 Standardized Basic System Software for Automotive Embedded Applications 51-1
Thomas M. Galla

52 Protocols and Services in Controller Area Networks 52-1
Gianluca Cena and Adriano Valenzano

53 FlexRay Communication Technology 53-1
Roman Nossal-Tueyeni and Dietmar Millinger

54 The LIN Standard 54-1
Antal Rajnák and Anders Kallerdahl

SECTION X Building Automation

55 State of the Art in Smart Homes and Buildings 55-1
Wolfgang Kastner, Lukas Krammer, and Andreas Fernbach

56 Fundamentals of LonWorks®/CEA 709 Networks 56-1
Dietmar Loy and Stefan Soucek

57 BACnet 57-1
Frank Schubert

58 KNX: A Worldwide Standard Protocol for Home and Building Automation 58-1
Michele Ruta, Floriano Scioscia, Giuseppe Loseto, and Eugenio Di Sciascio

59 Future Trends in Smart Homes and Buildings 59-1
Wolfgang Kastner, Markus Jung, and Lukas Krammer

SECTION XI Energy and Power Systems

60 Protocols for Automatic Meter Reading 60-1
Klaas De Craemer, Geert Deconinck, and Matthias Stifter

61 Communication Protocols for Power System Automation 61-1
Peter Palensky, Friederich Kupzog, Thomas Strasser, Matthias Stifter, and Thomas Leber

62 IEC 61850: A Single Standard for the Engineering, Protection, Automation, and Supervision of Power Systems on All Voltage Levels 62-1
Karlheinz Schwarz

63 Fundamentals of the IEC 61400-25 Standard 63-1
Federico Pérez, Elisabet Estévez, Isidro Calvo, Darío Orive, and Marga Marcos

SECTION XII Communication Networks and Services in Railway Applications

64 Communication in Train Control 64-1
Heinz Kantz, Stefan Resch, and Christoph Scherrer

65 Ethernet Extensions for Train Communication Network 65-1
Stephan Rupp and Reiner Gruebmeyer

SECTION XIII Semiconductor Equipment and Materials International

66 Semiconductor Equipment and Materials International Interface and Communication Standards 66-1
Kiah Mok Goh, Lay Siong Goh, Geok Hong Phua, Aik Meng Fong, Yan Guan Lim, Ke Yi, and Oo Tin

SECTION XIV Emerging Protocols and Technologies

67 Brain 67-1
Michael Paulitsch, Brendan Hall, and Kevin R. Driscoll

Index **Index**-1

Preface

The first edition of the *Industrial Communication Technology Handbook* was published almost a decade ago, in 2005. It gave a fairly comprehensive picture of the specialized communication networks used in diverse application areas. Solutions and technologies proposed for and deployed in process automation and on the factory floor dominated the volume. Not surprisingly, the late 1980s and the 1990s were the years when a large number of frequently competing solutions and technologies were introduced by major automation vendors and industry consortia. At the end of the 1990s, the Ethernet emerged as a contender for real-time applications, including safety-critical ones—largely on the factory floor. The Ethernet was also viewed as a potential solution for the vertical integration of functional layers of the industrial automation architectures, as it enabled a seamless data/command flow between the factory floor and upper layers. Another emerging area of research and development embarked upon at that time by the control and automation industry sector was the use of commercial wireless technologies in the automation of plants and on the factory floor.

But plant and factory automation were not the only application areas for specialized communication networks. The automotive industry has been exploring from the mid-1980s the possibility of the use of dedicated networks to automate different car functions and domains, aiming to replace mechanical or hydraulic systems with electrical/electronic ones. Production models released from the beginning of the 1990s integrated a range of networks to support different car functions and domains.

Building automation and control (BAC) is aimed at reducing energy consumption. As early as the mid-1990s, research and development activities commenced in Japan and in the United States to come up with a system to control light and temperature (coupled, particularly close to window areas) in office buildings to save energy and provide "personal comfort." Due to the highly distributed nature of the systems involved, using specialized communication networks was a necessity.

The use of specialized communication networks in avionics was a world of its own. At the time when the first edition was published, any technical publications were seldom available to the broad engineering profession. Most technical details were confined to technical reports, sometimes available for a substantial fee.

The last ten years or so have seen a remarkable success of Ethernet-based solutions adopted, and standardized, for real-time applications—some safety-critical. (The automotive industry began evaluating IEEE 802.1 TSN/AVB for the driver assistance.) The vertical integration of industrial enterprises no longer poses major technical obstacles. This is due to the emergence of Ethernet-based solutions with real-time capabilities. But, perhaps, the most remarkable development affecting control and automation of plants and factories has been the introduction of the WAN- and PAN-based wireless technologies, stand-alone networks, and hybrid wireline-wireless solutions. It was the control and automation industry sector that saw the introduction and deployment on the factory floor of sophisticated large-scale wireless sensor networks, to mention WISA, the ABB proprietary solution. The industrial control and automation sector invested in the development and standardization of new wireless solutions (WirelessHART, ISA 100.11a, and WIA-PA), aiming at process automation to support noncritical monitoring and control functions with the prospect of addressing safety-critical applications in future releases.

The need to manage electrical power systems to conserve energy and optimally distribute electric energy during periods of excessive demand mandated development of new solutions and technologies for automatic meter reading in order to integrate commercial and domestic energy consumption into the energy demand management decision systems.

The rapid evolution of some technologies and development of new ones called for a new edition of this handbook. It is hoped that the new edition will be useful to a broad spectrum of professionals involved in the conception, design and development, standardization, and the use of specialized communication networks. It will also have the potential to be adopted by academic instructors. The industry demand for practical knowledge of specialized communication networks, as well as hands-on exposure to the equipment, is on the rise. Academic institutions engaged in engineering education and vocational training have an important role in adequately preparing future engineering graduates to enter the profession equipped with the practical knowledge sought by the industry.

The book offers a comprehensive treatment of specialized communication networks, with chapters segregated by industry sectors and application domains. The basics of communication networks, and some aspects relevant to industrial applications that are widely available in numerous professional journals, magazines, and technical books, have been omitted to keep the new edition at a reasonable size. Certain technologies covered in the first edition have been omitted in the second edition as they are either no longer maintained or have a small user base.

Half of the contributions to the second edition are from major technology vendors, or industry consortia, involved in the technology creation, maintenance, and standardization—making this edition an authoritative volume. The other half is from governmental and private research establishments—contributors are researchers from leading academic institutions with distinguished publication records in the areas of their expertise.

This new edition, with twice as many chapters as the first one, adds new sections on industrial Ethernet, wireless industrial networks, time-triggered communication, networks in avionics and aerospace, fault-tolerant clock synchronization in industrial automation networks, networks in energy and power systems, and emerging protocols and technologies. The other sections have been updated and expanded substantially.

The second edition of the *Industrial Communication Technology Handbook* is divided into 14 sections consisting of 67 chapters and covers the following topics: field area and control networks, industrial Ethernet, fault-tolerant clock synchronization in industrial automation networks, accessing factory floor data, safety technologies in industrial networks and network security, wireless industrial networks, time-triggered communication, automotive communication technologies, networks in avionics and aerospace, networks in building automation and control, networks for energy and power systems, communication networks and services in railway applications, SEMI, and emerging protocols and technologies.

To ensure the integrity of the content, there is some overlap among selected chapters. Some chapter titles specifically refer to the technology presented in that chapter; others have more general titles, covering a range of technologies. There is a table of contents on the first page of each chapter to guide the reader. IEEE 802.15.1/Bluetooth and IEEE 802.15.4/ZigBee do not have dedicated chapters but are discussed extensively in the chapters on WISA (36 and 37), WirelessHART (31 and 36), and ISA 100 (32 and 36). CAN, which was developed specifically for automotive applications, and is also used in industrial control and automation, is discussed in Section IX, Automotive Communication Technologies, and also in Section VII, Time-Triggered Communication.

Editor

Richard Zurawski, a fellow of the Institute of Electrical Electronics Engineers (USA), is with ISA Group, San Francisco, California. He has over 35 years of academic and industrial experience, including a regular professorial appointment at the University of Tokyo and a full-time R&D advisory position with Kawasaki Electric, Tokyo. He provided consulting services to Kawasaki Electric, Ricoh, and Toshiba Corporations, Japan. He has also participated in a number of Japanese Intelligent Manufacturing Systems programs.

Dr. Zurawski is the editor of two book series: The Industrial Information Technology series and the Embedded Systems series—both published by CRC Press/Taylor & Francis Group. He is the former editor in chief of the *IEEE Transactions on Industrial Informatics* (2007–2010). He served as editor at large of the *IEEE Transactions on Industrial Informatics* (2006) and as associate editor of the *IEEE Transactions on Industrial Electronics* (1994–2005), *Real-Time Systems*, and *The International Journal of Time-Critical Computing Systems* (Kluwer Academic Publishers, 1997–2003).

Dr. Zurawski was a guest editor of a special issue of the prestigious *Proceedings of the IEEE* dedicated to industrial communication systems. He was also invited by the *IEEE Spectrum* to contribute an article on Java technology to "Technology 1999: Analysis and Forecast Issue." He was a guest editor of three special sections in the *IEEE Transactions on Industrial Electronics* on factory automation and factory communication systems.

Dr. Zurawski served as a vice president of the Industrial Electronics Society (1994–1997), chairman of the IES Factory Automation Council (1994–1997), and chairman of the IES Technical Committee on Factory Automation (2005–2010). He was also on the steering committee of the *ASME/IEEE Journal of Microelectromechanical Systems*. In 1996, he received the Anthony J. Hornfeck Service Award from the IEEE Industrial Electronics Society.

Dr. Zurawski was the editor of six major handbooks: *The Industrial Information Technology Handbook*, CRC Press (2004); *The Industrial Communication Technology Handbook*, CRC Press (2005); *Embedded Systems Handbook*, CRC Press/Taylor & Francis Group (2005, 2014); *Integration Technologies for Industrial Automated Systems*, CRC Press/Taylor & Francis Group (2008); *Embedded Systems Design and Verification*, vol. 1, *Embedded Systems Handbook*, 2nd edition (2009) and *Networked Embedded Systems*, vol. 2, *Embedded Systems Handbook*, 2nd edition (2009), CRC Press/Taylor & Francis Group; and *Electrical Engineering Technology and Systems Handbook* (seven volume set, 2015), CRC Press/Taylor & Francis.

Dr. Zurawski received an M.Eng. in electronics, and a Ph.D. in computer science.

Acknowledgments

I would like to thank all contributing authors, particularly authors from industry and industry consortia whose job priorities usually are different from writing chapters in books. I wish to acknowledge the work of Dr. Ronald Schoop of Schneider Electric, who passed away before finishing his contribution. I am grateful to the members of the *Handbook*'s Advisory Board for so generously taking part in the making of this book.

I also wish to thank at CRC Press, Nora Konopka, publisher of this book; Michele Smith, Jessica Vakili, and Richard Tressider, production editor.

Contributors

Vlado Altmann
Institute of Applied Microelectronics and Computer Engineering
University of Rostock
Rostock, Germany

Dhananjay Anand
Information Technology Laboratory
National Institute of Standards and Technology
Gaithersburg, Maryland

Herbert Barthel
Siemens AG
Nürnberg, Germany

Rudy Belliardi
Schneider Electric
Paris, France

Hendrik Bohn
Nedbank Group Ltd.
Johannesburg, South Africa

Samuel P. Buckwalter
ARINC
Annapolis, Maryland

Isidro Calvo
Department of Automatic Control and Systems Engineering
University of the Basque Country
Vitoria-Gasteiz, Spain

Simon Carlsen
Statoil ASA
Harstad, Norway

Salvatore Cavalieri
Department of Electrical, Electronics and Computer Engineering
University of Catania
Catania, Italy

Gianluca Cena
Institute of Electronics, Computer and Telecommunication Engineering
National Research Council of Italy
Torino, Italy

Marco Cereia
Institute of Electronics, Computer and Telecommunication Engineering
National Research Council of Italy
Torino, Italy

Deji Chen
Emerson Process Management
Round Rock, Texas

Geert Deconinck
ESAT/ELECTA
Katholieke Universiteit Leuven
Leuven, Belgium

Klaas De Craemer
ESAT/ELECTA
Katholieke Universiteit Leuven
Leuven, Belgium

Chris deLong
Honeywell Aerospace
Albuquerque, New Mexico

Eugenio Di Sciascio
Dipartimento di Ingegneria Elettrica e dell'Informazione
Politecnico di Bari
Bari, Italy

Kevin R. Driscoll
Honeywell International Inc.
Maple Grove, Minnesota

Dacfey Dzung
ABB Switzerland Ltd.
Baden, Switzerland

Dave Edstrom
Virtual Photons Electrons, LLC
Ashburn, Virginia

Wilfried Elmenreich
Institute of Networked and Embedded Systems
University of Klagenfurt
Klagenfurt, Austria

Jan Endresen
ABB AS Corporate Research
Billingstad, Norway

Elisabet Estévez
System Engineering and Automation Department
University of Jaén
Jaén, Spain

Reinhard Exel
Center for Integrated Sensor Systems
Danube University Krems
Wiener Neustadt, Austria

Max Felser
Bern University of Applied Sciences
Burgdorf, Switzerland

Andreas Fernbach
Institute of Computer Aided Automation
Vienna University of Technology
Vienna, Austria

Paolo Ferrari
Department of Information Engineering
University of Brescia
Brescia, Italy

Norman Finn
Cisco Systems
Milpitas, California

Alessandra Flammini
Department of Information Engineering
University of Brescia
Brescia, Italy

Aik Meng Fong
Singapore Institute of Manufacturing Technology
Singapore

Christian Fraboul
University of Toulouse
Toulouse, France

Jan-Erik Frey
ABB AB System Automation
Västerås, Sweden

Thomas M. Galla
Elektrobit Austria GmbH
Vienna, Austria

Mikael Gidlund
ABB AB Corporate Research
Västerås, Sweden

Kiah Mok Goh
Singapore Institute of Manufacturing Technology
Singapore

Lay Siong Goh
Singapore Institute of Manufacturing Technology
Singapore

Frank Golatowski
Institute of Applied Microelectronics and Computer Engineering
University of Rostock
Rostock, Germany

Reiner Gruebmeyer
Kontron AG
Eching, Germany

Luiz Affonso Guedes
Department of Computer Engineering and Automation
Federal University of Rio Grande do Norte
Natal, Brazil

Brendan Hall
Honeywell International Inc.
Eden Prairie, Minnesota

Lee H. Harrison
Galaxy Scientific Corp.
Egg Harbor Township, New Jersey

Scott Hibbard
Bosch Rexroth Corporation
Hoffman Estates, Illinois

Rahil Hussain
ABB Corporate Research
Billingstad, Norway

Jürgen Jasperneite
Fraunhofer Anwendungszentrum Industrial Automation
Fraunhofer Institut für Optronik, Systemtechnik und Bildauswertung
Lemgo, Germany

Ulrich Jecht
UJ Process Analytics
Baden-Baden, Germany

Per Christian Juel
ABB Corporate Research
Billingstad, Norway

Markus Jung
Institute of Computer Aided Automation
Vienna University of Technology
Vienna, Austria

Anders Kallerdahl
Mocean Laboratories AB
Gothenburg, Sweden

Roland Kammerer
Vienna University of Technology
Vienna, Austria

Heinz Kantz
Thales Austria GmbH
Vienna, Austria

Wolfgang Kastner
Institute of Computer Aided Automation
Vienna University of Technology
Vienna, Austria

Lukas Krammer
Institute of Computer Aided Automation
Vienna University of Technology
Vienna, Austria

Friederich Kupzog
Austrian Institute of Technology
Vienna, Austria

Ronald M. Larsen
Sercos North America
Santa Rosa Beach, Florida

Thomas Leber
Vienna University of Technology
Vienna, Austria

Kang B. Lee
National Institute of Standards and Technology
Gaithersburg, Maryland

Tomas Lennvall
ABB AB Corporate Research
Västerås, Sweden

Yan Guan Lim
Singapore Institute of Manufacturing Technology
Singapore

Giuseppe Loseto
Dipartimento di Ingegneria Elettrica e dell'Informazione
Politecnico di Bari
Bari, Italy

Dietmar Loy
LOYTEC Electronics GmbH
Vienna, Austria

Peter Lutz
Sercos International e.V.
Suessen, Germany

Marga Marcos
Department of Automatic Control and Systems Engineering
University of the Basque Country
Bilbao, Spain

Daniel A. Martinec
ARINC
Annapolis, Maryland

Dietmar Millinger
Technology Consulting
Vienna, Austria

Andrea Monacchi
Institute of Networked and Embedded Systems
University of Klagenfurt
Klagenfurt, Austria

Guido Moritz
SIV AG
Roggentin, Germany

James R. Moyne
Department of Mechanical Engineering
University of Michigan
Ann Arbor, Michigan

Nicolas Navet
University of Luxembourg
Luxembourg, Luxembourg

Ralf Neubert
Schneider Electric
Marktheidenfeld, Germany

Mark Nixon
Emerson Process Management
Round Rock, Texas

Roman Nossal-Tueyeni
Austro Control GmbH
Vienna, Austria

Roman Obermaisser
Embedded Systems
University of Siegen
Netphen, Germany

Darío Orive
Department of Automatic Control and Systems Engineering
University of the Basque Country
Bilbao, Spain

Peter Palensky
Austrian Institute of Technology
Vienna, Austria

Michael Paulitsch
Airbus Group Innovations
Munich, Germany

Federico Pérez
Department of Automatic Control and Systems Engineering
University of the Basque Country
Bilbao, Spain

Stig Petersen
Information and Communication Technology
SINTEF ICT
Trondheim, Norway

Geok Hong Phua
Singapore Institute of Manufacturing Technology
Singapore

Matthias Posch
Vienna University of Technology
Stetten, Austria

Wally Pratt
HART Communication Foundation
Austin, Texas

Gunnar Prytz
ABB Corporate Research
Billingstad, Norway

Antal Rajnák
Mentor Graphics Corp.
Geneva, Switzerland

Stefan Resch
Thales Austria GmbH
Vienna, Austria

Stefano Rinaldi
Department of Information Engineering
University of Brescia
Brescia, Italy

Eric Rotvold
Emerson Process Management
Chanhassen, Minnesota

Stephan Rupp
Duale Hochschule Baden-Württemberg Stuttgart
Stuttgart, Germany

Michele Ruta
Dipartimento di Ingegneria Elettrica e dell'Informazione
Politecnico di Bari
Bari, Italy

Thilo Sauter
Center for Integrated Sensor Systems
Danube University Krems
Wiener Neustadt, Austria

Stefano Scanzio
Institute of Electronics, Computer and Telecommunication Engineering
National Research Council of Italy
Torino, Italy

Jean-Luc Scharbarg
University of Toulouse
Toulouse, France

Guntram Scheible
ABB Automation Products GmbH
Heidelberg, Germany

Christoph Scherrer
Thales Austria GmbH
Vienna, Austria

Viktor Schiffer (Retired)
Rockwell Automation GmbH
Düsseldorf, Germany

Tilman Schinke
AS-International Association e.V.
Gelnhausen, Germany

Frank Schubert
MBS GmbH
Krefeld, Germany

Karlheinz Schwarz
Schwarz Consulting Company
Karlsruhe, Germany

Floriano Scioscia
Dipartimento di Ingegneria Elettrica e dell'Informazione
Politecnico di Bari
Bari, Italy

Ivanovitch Silva
Institute Metropolis Digital
Federal University of Rio Grande do Norte
Natal, Brazil

Françoise Simonot-Lion
Nancy Université
Nancy, France

Emiliano Sisinni
Department of Information Engineering
University of Brescia
Brescia, Italy

Tor Skeie
Simula Research Laboratory
University of Oslo
Oslo, Norway

Stefan Soucek
LOYTEC Electronics GmbH
Vienna, Austria

Reinhard Sperrer
Safety Network International e.V.
Ostfildern, Germany

Wilfried Steiner
TTTech Computertechnik AG
Vienna, Austria

Matthias Stifter
Energy Department
Austrian Institute of Technology
Vienna, Austria

Thomas Strasser
Austrian Institute of Technology
Vienna, Austria

Jochen Streib
Safety Network International e.V.
Ostfildern, Germany

Wolfgang Stripf
IO-Link Community
and
PROFIBUS and PROFINET International
Karlsruhe, Germany

Dawn M. Tilbury
Department of Mechanical Engineering
University of Michigan
Ann Arbor, Michigan

Oo Tin
Singapore Institute of Manufacturing Technology
Singapore

Federico Tramarin
Institute of Electronics, Computer and Telecommunication Engineering
National Research Council of Italy
Padova, Italy

Albert Treytl
Center for Integrated Sensor Systems
Danube University Krems
Wiener Neustadt, Austria

Adriano Valenzano
Institute of Electronics, Computer and Telecommunication Engineering
National Research Council of Italy
Torino, Italy

Stefano Vitturi
Institute of Electronics, Computer and Telecommunication Engineering
National Research Council of Italy
Padova, Italy

Peter Wenzel
PROFIBUS Nutzerorganisation e.V.
and
PROFIBUS and PROFINET International
Karlsruhe, Germany

Andreas Willig
Department of Computer Science and Software Engineering
University of Canterbury
Christchurch, New Zealand

Ke Yi
Singapore Institute of Manufacturing Technology
Singapore

Claudio Zunino
Institute of Electronics, Computer and Telecommunication Engineering
National Research Council of Italy
Torino, Italy

International Advisory Board

I

Field Area and Control Networks

1 **Fieldbus System Fundamentals** *Thilo Sauter* 1-1
Introduction • What Is a Fieldbus? • History • Communication Fundamentals: The OSI Model • Fieldbus Characteristics • Industrial Ethernet: The New Fieldbus • Aspects for Future Evolution • Appendix • References

2 **Networked Control Systems for Manufacturing** *James R. Moyne, Dawn M. Tilbury, and Dhananjay Anand* 2-1
Introduction • Parameterization of Industrial Networks • Differentiation of Industrial Networks • NCS Characterization • Applications for Industrial Networks • Future Trends • Acknowledgments • References

3 **Configuration and Management of Networked Embedded Devices** *Wilfried Elmenreich and Andrea Monacchi* 3-1
Introduction • Concepts and Terms • Requirements for Configuration and Management • Interface Separation • Profiles, Datasheets, and Descriptions • Application Development • Configuration Interfaces • Management Interfaces • Maintenance in Fieldbus Systems • Conclusion • Acknowledgments • References

4 **Smart Transducer Interface Standard for Sensors and Actuators** *Kang B. Lee* 4-1
Introduction • Smart Transducer Model • Networking Smart Transducers • Establishment of the IEEE 1451 Standards • Goals of IEEE 1451 • IEEE 1451 Standards • Example Application of IEEE 1451.2 • Application of IEEE 1451–Based Sensor Network • Summary • Acknowledgments • References

5 **IO-Link (Single-Drop Digital Communication System) for Sensors and Actuators** *Wolfgang Stripf* 5-1
Motivation and Objectives for a New Technology • IO-Link Technology • Abbreviations • References

6 **AS-Interface** *Tilman Schinke* 6-1
AS-Interface Background • AS-Interface Technology • AS-Interface Devices: Sensors/Actuators • AS-Interface Master: Coupling to Other Automation Systems • Functional Safety with AS-Interface Safety • Open System: Interoperability and Certification • References and Further Probing

7 **HART over Legacy 4–20 mA Signal Base** *Mark Nixon, Wally Pratt, and Eric Rotvold* 7-1
Introduction • HART Networks • HART Networks • HART Circuits • In Closing • References

8 **HART Device Networks** *Mark Nixon and Deji Chen* 8-1
Introduction • HART Architecture • HART Communication Stack • System Tools • Planning and Installation • Application Example: Bioreactor • Future Directions • References

9 **Common Industrial Protocol (CIP™) and the Family of CIP Networks** *Viktor Schiffer* 9-1
Introduction • Description of the CIP Networks Library • Network Adaptations of CIP • Benefits of the CIP Family • Application Layer Enhancements • Conformance Testing • Abbreviations • Terminology • References

10 **Modbus Protocol** *Rudy Belliardi and Ralf Neubert* 10-1
Overview • Modbus Protocol • Modbus over Serial Line • Modbus/TCP • Gateways and Similar Devices • Modbus as Part of the CIP Stack, in ODVA • Modbus on Other Stacks • Conformance

11 **PROFIBUS** *Ulrich Jecht, Wolfgang Stripf, and Peter Wenzel* 11-1
Basics • Transmission Technologies • Communication Protocol • Application Profiles • Integration Technologies • Technical Support • Wide Range of Applications • Abbreviations • References

12 **PROFINET** *Peter Wenzel* 12-1
Introduction • PROFINET Basics • Principles of PROFINET Communication • Conformance Classes and Their Functions • Optional Functions • Integration of Fieldbus Systems • Application Profiles • PROFINET for Process Automation • Installation Technology for PROFINET • Technical Support • Abbreviations • References

13 **Sercos® Automation Bus** *Scott Hibbard, Peter Lutz, and Ronald M. Larsen* 13-1
Description • Features and Operation of Sercos III • Features and Operation of Sercos II • Future Technical Advancements • Acknowledgments • References • Sources for More Information

14 **Foundation Fieldbus** *Salvatore Cavalieri* 14-1
Introduction • Technical Description of Foundation Fieldbus • Wireless Solutions for Foundation Fieldbus • Open Systems Implementation • Conclusions • References

15 **INTERBUS** *Jürgen Jasperneite* 15-1
Introduction to Field Communication • INTERBUS Overview • INTERBUS Protocol • Diagnostics • Functional Safety • Interoperability, Certification • Connectivity • IP over INTERBUS • Performance Evaluation • Conclusions • References

1

Fieldbus System Fundamentals

1.1 Introduction 1-1
1.2 What Is a Fieldbus? 1-3
1.3 History 1-5
Roots of Industrial Networks • Evolution of Fieldbuses
1.4 Communication Fundamentals: The OSI Model 1-9
Layer Structure • Communication Services
1.5 Fieldbus Characteristics 1-15
Traffic Characteristics and Requirements • Fieldbus Systems and the OSI Model • Network Topologies • Medium Access Control • Communication Paradigms • Above the OSI Layers: Interoperability and Profiles • Fieldbus Management
1.6 Industrial Ethernet: The New Fieldbus 1-37
1.7 Aspects for Future Evolution 1-41
Appendix 1-43
References 1-45

Thilo Sauter
Danube University Krems

1.1 Introduction

Few developments have changed the face of automation so profoundly as the introduction of networks did. Especially, fieldbus systems—networks devised for the lowest levels of the automation hierarchy—had an enormous influence on the flexibility and performance of modern automation systems in all application areas. However, fieldbus systems were not the result of some *divine spark*; they emerged in a continuous and often cumbersome evolution process. Today, many applications areas are unthinkable without them: not only factory automation, distributed process control, building and home automation, substation automation, and more generally energy distribution, but also in-vehicle networking, railway applications, and avionics. All these fields heavily rely on the availability of appropriate networks accounting for the special demands of the individual application.

But what exactly is a fieldbus? Even after a quarter of a century of fieldbus development, there exists no clear-cut definition for the term. The *definition* given in the IEC 61158 fieldbus standard is more a programmatic declaration or a least common multiple compromise than a concise formulation [1]: "A fieldbus is a digital, serial, multidrop, data bus for communication with industrial control and instrumentation devices such as – but not limited to – transducers, actuators and local controllers." It comprises some important characteristics, but is far from being complete. On the other hand, it is a bit too restrictive.

A more elaborate explanation is given by the Fieldbus Foundation, the user organization supporting one of the major fieldbus systems [2]: "A Fieldbus is a digital, two-way, multidrop communication link among intelligent measurement and control devices. It serves as a Local Area Network (LAN) for advanced process control, remote input/output and high speed factory automation applications." Again, this is a bit restrictive, for it limits the application to process and factory automation, the primary areas where the Foundation Fieldbus is being used.

The lack of a clear definition is mostly due to the complex evolution history of fieldbuses. In most cases, bus systems emerged primarily to break up the conventional star-type point-to-point wiring schemes connecting simple digital and analog input and output devices to central controllers, thereby laying the grounds for the implementation of really distributed systems with more intelligent devices. As was declared in the original mission statement of the IEC work

> the Field Bus will be a serial digital communication standard which can replace present signalling techniques such as 4-20 mA ... so that more information can flow in both directions between intelligent field devices and the higher level control systems over shared communication medium... [3,4].

Still in more recent publications, this aspect is seen as the only raison d'être for fieldbus systems [5], which is however shortsighted and does not do the fieldbus justice.

Today, fieldbus systems comprising communication networks and devices can also be regarded as networked embedded systems. In the 1980s, when their era began, this term was still unknown. Yet, their main features and development stimuli are comparable:

- *Focused solutions.* Fieldbus systems were no general-purpose developments, even if they were said to be. They were always developed with a concrete application field in mind and designed to meet the respective boundary conditions (like not only temporal behavior, efficiency, and reliability, but also cost) in the best possible way.
- *Smart devices.* An essential objective for embedded systems and fieldbuses alike is to bring more intelligence to the field, that is, to the end devices. Like in embedded systems, fieldbus developers also used the technological building blocks available at the time where possible, such as standard microcontrollers to keep costs low. However, if special needs were to be met, also dedicated solutions were devised.
- *Limited resources.* Embedded applications and fieldbus system both share the fundamental problem that resources are limited. No matter what the state of art in microelectronics is, embedded devices (and field devices) are less powerful than standard computers. Communication (sub)systems usually have less available bandwidth than computer networks, and power consumption is an issue.
- *Comprehensive concepts.* Fieldbus systems are not just networks. Communication is only part of a distributed automation concept with comprehensive application software and tool chains. In some advanced cases, fieldbuses were embedded into special frameworks exhibiting many characteristics of distributed operating systems.
- *Distribution.* A network is the prerequisite of distributed systems; many data processing tasks can be removed from a central controller and placed directly in the field devices if they are sufficiently smart and the interface can handle reasonably complex ways of communication.
- *Flexibility and modularity.* A fieldbus installation like any other network can be extended much more easily than a centralized system, provided the limitations of addressing space, cable length, etc., are not exceeded. For the special case of fieldbuses, simplification of the parameterization and configuration of complex field devices is an additional benefit making system setup and commissioning easier.
- *Maintainability.* Monitoring of devices, applying updates, and other maintenance tasks are easier, if at all possible, via a network.

These aspects are not just theoretical contemplations but actual user demands that influenced the development from the beginning on [4]. However, as the application requirements in the various automation domains were quite different, so were the solutions, and that makes it so difficult to find a comprehensive definition. The purpose of this chapter thus is to give an overview of the nature of fieldbus systems. It briefly reviews the evolution from the historical roots up to the tedious standardization efforts. It discusses in detail typical characteristics that distinguish fieldbuses from other types of networks. Current activities aiming at using Ethernet in automation are reviewed, and evolution prospects are given. It appears that fieldbus systems have reached their climax and that future years will bring a mix of Ethernet-/Internet-based solutions and still-to-be-developed, new field-level networks.

1.2 What Is a Fieldbus?

As mentioned in Section 1.1, fieldbus systems have to be seen as an integrative part of a comprehensive automation concept and not as standalone solutions. The name is therefore programmatic and evocative. Interestingly enough, not even the etymology of the term itself is fully clear. The English word *fieldbus* is definitely not the original one. It appeared around 1985 when the fieldbus standardization project within IEC TC65 was launched [4] and seems to be a straightforward literal translation of the German term *Feldbus*, which can be traced back until about 1980 [6]. Indeed, the overwhelming majority of early publications in the area is available only in German. The word itself was coined in process industry and primarily refers to the process field, designating the area in a plant where lots of distributed field devices, mostly sensors and actuators, are in direct contact with the process to be controlled. Slightly after the German expression and sharing its etymological root, the French word *réseau de terrain* (or *réseau d'instrumentation*, instrumentation network) emerged. This term was not specifically targeted at the process industry but refers also to large areas with scattered devices. The connection of such devices to the central control room was traditionally made via point-to-point links, which resulted in a significant and expensive cabling need. The logical idea, powered by the advances of microelectronics in the late 1970s, was to replace this star-like cabling in the field by a party-line, bus-like installation connecting all devices via a shared medium—the fieldbus [7,8].

Given the large dimensions of process automation plants, the benefits of a bus are particularly evident. However, the concept was not undisputed when it was introduced. The fieldbus approach was an ambitious concept: a step toward decentralization, including the preprocessing of data in the field devices, which both increases the quality of process control and reduces the computing burden for the centralized controllers [9]. Along with it came the possibility to configure and parameterize the field devices remotely via the bus. This advanced concept, on the other hand, demanded increased communication between the devices that goes far beyond a simple data exchange. This seemed infeasible to many developers, and still in the mid-1980s, one could read statements like [10] "The idea of the fieldbus concept seems promising. However, with reasonable effort it is not realizable at present."

The alternative and somewhat more conservative approach was the development of the so-called *field multiplexers*, devices that collect process signals in the field, serialize them, and transfer them via one single cable to a remote location where a corresponding device demultiplexes them again [11]. For quite some time, the two concepts competed and coexisted [12], but ultimately the field multiplexers mostly disappeared, except for niches in process automation, where many users still prefer such *Remote I/O* systems despite the advantages of fieldbus solutions [13]. The central field multiplexer concept of sampling I/O points and transferring their values in simple data frames also survived in some fieldbus protocols especially designed for low-level applications.

The desire to cope with the wiring problem getting out of hand in large installations was certainly the main impetus for the development of fieldbus systems. Other obvious and appealing advantages of the concept are modularity, the possibility to easily extend installations, and the possibility to have much more intelligent field devices that can communicate not just for the sake of process data transfer, but also for maintenance and configuration purposes [14,15]. A somewhat different viewpoint that led to different design approaches was to regard bus systems in process control as the spine of distributed real-time systems [16]. While the wiring optimization concepts were in many cases rather simple bottom-up approaches, these distributed real-time ideas resulted in sophisticated and usually well-investigated top-down designs.

An important role in the fieldbus evolution has been played by the so-called automation pyramid. This hierarchical model was defined to structure the information flow required for factory and process automation. The idea was to create a transparent, multilevel network—the basis for computer-integrated manufacturing (CIM). The numbers vary, but typically this model comprised up to five levels, sometimes more [17]. While the networks for the upper levels already existed by the time the pyramid was defined, the field level was still governed by point-to-point connections. Fieldbus systems were therefore developed also with the aim of finally bridging this gap. The actual integration of field-level networks into the rest of the hierarchy was in fact considered in early standardization [4]; for most of the proprietary developments, however, it was never the primary intention.

In the automation pyramid, fieldbuses actually populate two levels: the field level and the cell/process level. For this reason, they are sometimes further differentiated into two classes:

1. Sensor-actuator buses or device buses have very limited capabilities and serve to connect very simple devices with, for example, programmable logic controllers (PLCs). They can be found exclusively on the field level.
2. Fieldbuses connect control equipment like PLCs and PCs as well as more intelligent devices. They are found on the cell level and are closer to computer networks.

This distinction may seem reasonable but is in fact problematic. There are only few fieldbus systems that can immediately be allocated to one of the groups, most of them are being used in both levels. Therefore, it should be preferable to abandon this arbitrary differentiation.

How do fieldbus systems compare to computer networks? The classical distinction of the different network types used in the automation pyramid hinges on the distances the networks span. From top down, the hierarchy starts with global area networks (GANs), which cover long, preferably intercontinental distances, and nowadays mostly use satellite links. On the second level are wide area networks (WANs). They are commonly associated with telephone networks (no matter if analog or digital). Next come the well-known LANs, with Ethernet as the most widely used technology today. They are the classical networks for office automation and cover only short distances. The highest level of the model shown in Figure 1.1 is beyond the scope of the original definition, but is gaining importance with the availability of the Internet. In fact, Internet technology is penetrating all levels of this pyramid all the way down to the process level. And since IP has reached a dominating position as worldwide networking technology, the discussion about how to distinguish GANs, WANs, and LANs is today mostly only of academic interest.

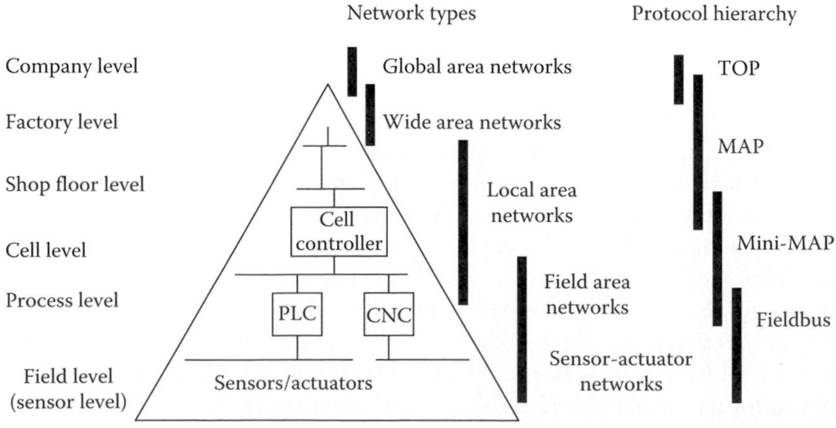

FIGURE 1.1 Hierarchical network levels in automation and protocols originally devised for them.

From GANs to LANs, the classification according to the spatial extension is evident. One step below, on the field level, this criterion fails, because fieldbus systems or field area networks (FANs) can cover even larger distances than LANs. Yet, as LANs and FANs evolved nearly in parallel, some clear distinction between the two network types seemed necessary. As length is inappropriate, the classical border line drawn between LANs and FANs relies mostly on the characteristics of the data transported over these networks. LANs have high data rates and carry large amounts of data in large packets. Timeliness is not a primary concern, and real-time behavior is not required. Fieldbus systems, by contrast, have low data rates. Since they transport mainly process data, the size of the data packets is small, and real-time capabilities are important.

For some time, these distinction criteria between LANs and FANs were sufficient and fairly described the actual situation. Recently, however, drawing the line according to data rates and packet sizes is no longer applicable. In fact, the boundaries between LANs and fieldbus systems have faded. Today, there are fieldbus systems with data rates well above 10 Mbit/s, which is still standard in older LAN installations. In addition, more and more applications require the transmission of video or voice data, which results in large data packets.

On the other hand, Ethernet as *the* LAN technology is becoming more and more popular in automation and is bound to replace some of today's widely used mid-level fieldbus systems. The real-time extensions developed during the last years tackle its most important drawback and will ultimately permit the use of Ethernet also in low-level control applications. However, it seems that industrial Ethernet will not make the lowest-level fieldbuses fully obsolete. They are much better optimized for their specific automation tasks than the general-purpose network Ethernet. But the growing use of Ethernet results in a reduction of the levels in the automation hierarchy. Hence the pyramid gradually turns into a flat structure with at most three, maybe even only two levels.

Consequently, a more appropriate distinction between LANs and FANs should be based on the functionality and the application area of these networks. According to this pragmatic argumentation, a fieldbus is simply a network used in automation, irrespective of topology, data rates, protocols, or real-time requirements. Consequently, it need not be confined to the classical field level; it can be found on higher levels (provided they still exist) as well. A LAN, on the other hand, belongs to the office area. This definition is loose, but mirrors the actual situation. Only one thing seems strange at first: Following this definition, industrial Ethernet changes into a fieldbus, even though many people are inclined to associate it with LANs. However, this is just another evidence that the boundaries between LANs and FANs are fading.

1.3 History

The question of what constitutes a fieldbus is closely linked to the evolution of these industrial networks. The best approach to understanding the essence of the concepts is to review the history and the intentions of the developers. This review will also falsify one of the common errors frequently purported by marketing divisions of automation vendors: that fieldbus systems were a revolutionary invention. They may have revolutionized automation; there is hardly any doubt about it. However, they themselves were only a straightforward evolution that built on preexisting ideas and concepts.

1.3.1 Roots of Industrial Networks

Although the term *fieldbus* appeared only about 30 years ago, the basic idea of field-level networks is much older. Still, the roots of modern fieldbus technology are mixed. Both classical electrical engineering and computer science have contributed their share to the evolution, and we can identify three major sources of influence:

1. Communication engineering with large-scale telephone networks
2. Instrumentation and measurement systems with parallel buses and real-time requirements
3. Computer science with the introduction of high-level protocol design

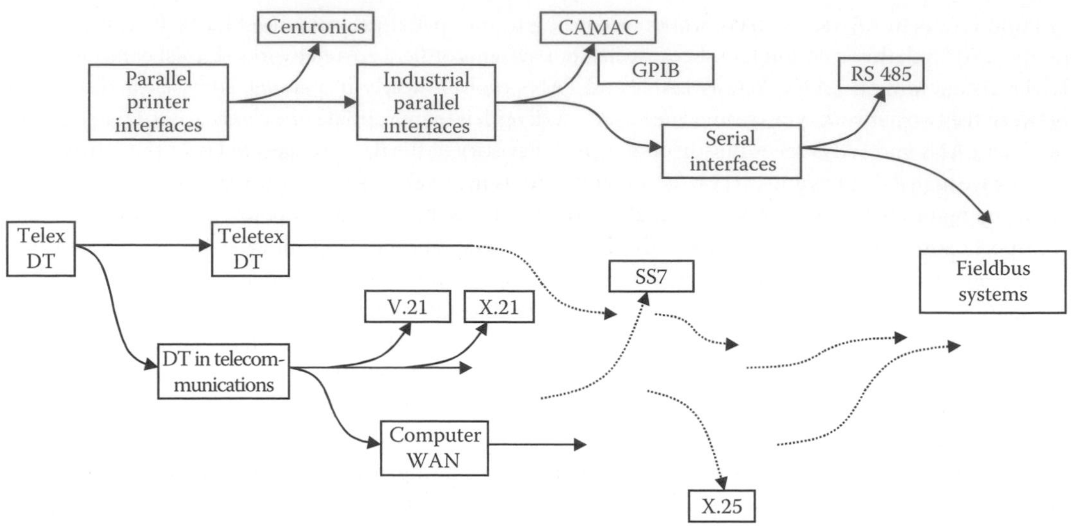

FIGURE 1.2 Historical roots of fieldbus systems.

This early stage is depicted in Figure 1.2. One foundation of automation data transfer has to be seen in the classic telex networks and also in standards for data transmission over telephone lines. Large distances called for serial data transmission, and many of these comparatively early standards still exist, like V.21 (data transmission over telephone lines) and X.21 (data transmission over special data lines). Various protocols have been defined, mostly described in state machine diagrams and rather simple because of the limited computing power of the devices available at that time. Of course, these communication systems have a point-to-point nature and therefore lack the multidrop characteristic of modern fieldbus systems, but nevertheless they were the origin of serial data transmission. Talking about serial data communication, one should notice that the engineers who defined the first protocols often had a different understanding of the expressions *serial* and *parallel* than we have today. For example, the *serial* Interface V.24 transmits the application data serially, but the control data in a parallel way over separate control lines.

In parallel to the development of data transmission in the telecommunication sector, hardware engineers defined interfaces for standalone computer systems to connect peripheral devices such as printers. The basic idea of having standardized interfaces for external devices was soon extended to process control and instrumentation equipment. The particular problems to be solved were the synchronization of spatially distributed measurement devices and the collection of measurement data from multiple devices in large-scale experimental setups. This led to the development of standards like CAMAC (in nuclear science) and GPIB (later also known as IEEE 488). To account for the limited data processing speed and real-time requirements for synchronization, these bus systems had parallel data and control lines, which is also not characteristic for fieldbus systems. However, they were using the typical multidrop structure.

Later on, with higher integration density of integrated circuits and thus increased functionality and processing capability of microcontrollers, devices became smaller and portable. The connectors of parallel bus systems were now too big and clumsy, and alternatives were sought [18]. The underlying idea of developments like I^{2}C [19] was to extend the already existing serial point-to-point connections of computer peripherals (based on the RS 232) to support longer distances and finally also multidrop arrangements. The capability of having a bus structure with more than just two connections together with an increased noise immunity due to differential signal coding eventually made the RS 485 a cornerstone of fieldbus technology up to the present day.

Historically the youngest root of fieldbus systems, but certainly the one that left the deepest mark was the influence of computer science. Its actual contribution was a structured approach to the design of high-level communication systems, contrary to the mostly monolithic design approaches that had been sufficient until then. This change in methodology had been necessitated by the growing number of computers used worldwide and the resulting complexity of communication networks. Conventional telephone networks were no longer sufficient to satisfy the interconnection requirements of modern computer systems. As a consequence, the big communication backbones of the national telephone companies gradually changed from analog to digital systems. This opened the possibility to transfer large amounts of data from one point to another. Together with an improved physical layer, the first really powerful data transmission protocols for WANs were defined, such as X.25 (packet switching) or SS7 (common channel signaling). In parallel to this evolution on the telecommunications sector, LANs were devised for the local interconnection of computers, which soon led to a multitude of solutions. It took nearly a decade until Ethernet and Transmission Control Protocol (TCP)/IP finally gained the dominating position they have today.

1.3.2 Evolution of Fieldbuses

The preceding section gave only a very superficial overview of the roots of networking, which laid the foundations not only of modern computer networks but also of those on the field level. But let us now look more closely at the actual evolution of the fieldbus systems. Here again, we have to consider the different influences of computer science and electrical engineering. First and foremost, the key contribution undoubtedly came from the networking of computer systems, when the ISO/ open system interconnection (OSI) model was introduced [20,21]. This seven-layer reference model was (and still is) the starting point for the development of many complex communication protocols.

The first application of the OSI model to the domain of automation was the definition of manufacturing automation protocol (MAP) in the wake of the CIM idea [22]. MAP was intended to be a framework for the comprehensive control of industrial processes covering all automation levels, and the result of the definition was a powerful and flexible protocol [23]. Its complexity, however, made implementations extremely costly and hardly justifiable for general-purpose use. As a consequence, a tightened version called MiniMAP and using a reduced model based on the OSI layers 1, 2, and 7 was proposed to better address the problems of the lower automation layers [24]. Unfortunately, it did not have the anticipated success either. What did have success was Manufacturing Message Specification (MMS). It defined the cooperation of various automation components by means of abstract objects and services and was later used as a starting point for many other fieldbus definitions [25]. The missing acceptance of MiniMAP as well as the inapplicability of the original MAP/MMS standard to time-critical systems [26] were finally the reasons for the IEC to launch the development of a fieldbus based on the MiniMAP model, but tailored to the needs of the field level. According to the original objectives, the higher levels of the automation hierarchy should be covered by MAP or PROWAY [22].

Independent of this development in computer science, the progress in microelectronics brought forward many different integrated controllers, and new interfaces were needed to interconnect the ICs in an efficient and cheap way. The driving force was the reduction of both the interconnect wires on the printed circuit boards and the number of package pins on the ICs. Consequently, electrical engineers—without the knowledge of the ISO/OSI model or similar architectures—defined simple buses like the I^{2}C. Being interfaces rather than fully fledged bus systems, they have very simple protocols, but they were and still are widely used in various electronic devices.

Long before the *invention* of board-level buses, the demand for a reduction of cabling weight in avionics and space technology had led to the development of the MIL STD 1553 bus, which can be regarded as the first *real* fieldbus. Introduced in 1970, it showed many characteristic properties of modern fieldbus systems: serial transmission of control and data information over the same line, master–slave structure, the possibility to cover longer distances, integrated controllers, and it is still used today. Later on, similar thoughts (reduction of cabling weight and costs) resulted in the development of several bus systems not only in the automotive industry but also in the automation area. A characteristic property of these fieldbuses is that they were defined in the spirit of classical interfaces, with a focus on the lower two protocol layers, and no or nearly no application layer definitions. With time, these definitions were added to make the system applicable to other areas as well. Controller area network (CAN) is a good example of this evolution: For the originally targeted automotive market, the definition of the lowest two OSI layers was sufficient. Even today, automotive applications of CAN typically use only these low-level communication features because they are easy to use and the in-vehicle networks are usually closed. For applications in industrial automation, however, where extensibility and interoperability is an important issue, higher-level functions are important. So, when CAN was found to be interesting also for other application domains, a special application layer was added. The lack of such a layer in the original definition is the reason why there are many different fieldbus systems (like CANopen, Smart Distributed System (SDS), and DeviceNet) using CAN as a low-level interface.

From today's point of view, it can be stated that all fieldbuses that still have some relevance were developed using the *top-down* or *computer science–driven* approach, that is, a proper protocol design with abstract high-level programming interfaces to facilitate usage and integration in complex systems. The fieldbuses that followed the *bottom-up* or *electrical engineering–driven* approach, that is, that were understood as low-level computer interface, did not survive due to their inflexibility and incompatibility with modern software engineering, unless some application layer functions were included in the course of the evolution.

From the early 1980s on, when automation made a great leap forward with PLCs and more intelligent sensors and actuators, something like a gold rush set in. The increasing number of devices used in many application areas called for a reduced cabling, and microelectronics had grown mature enough to support the development of elaborated communication protocols. This was also the birth date for the fieldbus as an individual term. Different application requirements generated different solutions, and from today's point of view, it seems that creating new fieldbus systems was a trendy and fashionable occupation for many companies in the automation business. Those mostly proprietary concepts never had a real future, because the number of produced nodes could never justify the development and maintenance costs. Figure 1.3 depicts the evolution timeline of fieldbus systems and their environment [27]. The list of examples is of course not comprehensive; only systems that still have some significance have been selected. Details about the individual solutions are summarized in the tables in the appendix.

As the development of fieldbus systems was a typical *technology push* activity driven by the device vendors, the users first had to be convinced of the new concepts. Even though the benefits were quite obvious,

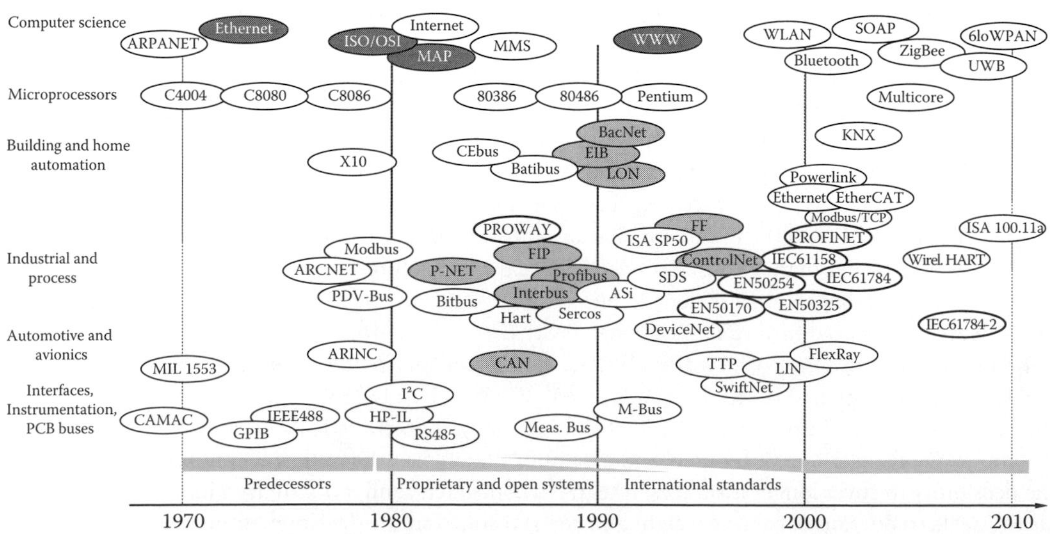

FIGURE 1.3 Milestones of fieldbus evolution and related fields.

the overwhelming number of different systems appalled rather than attracted the customers, who were used to perfectly compatible current-loop or simple digital inputs and outputs as interfaces between field devices and controllers and were reluctant to use new concepts that would bind them to one vendor. What followed was a fierce selection process where not always the fittest survived, but often those with the highest marketing power behind them. Consequently, most of the newly developed systems vanished or remained restricted to small niches. After a few years of struggle and confusion on the user's side, it became apparent that proprietary fieldbus systems would always have only limited success and that more benefit lies in creating *open* specifications so that different vendors may produce compatible devices, which gives the customer back their freedom of choice [8]. As a consequence, user organizations were founded to carry on the definition and promotion of the fieldbus systems independent of individual companies. It was this idea of open systems that finally paved the way for the breakthrough of the fieldbus concept.

The final step to establish the fieldbus in the automation world was international standardization. The basic idea behind it is that a standard establishes a specification in a very rigid and formal way, ruling out the possibility of quick changes. This attaches a notion of reliability and stability to the specification, which in turn secures the trust of the customers and, consequently, also the market position. Furthermore, a standard is vendor-independent, which guarantees openness. Finally, in many countries, standards have a legally binding position, which means that when a standard can be applied (e.g., in connection with a public tender), it *has* to be applied. Hence a standardized system gains a competitive edge over its nonstandardized rivals. This position is typical for, for example, Europe (see [28] for an interesting US-centric comment). It is therefore no wonder that after the race for fieldbus developments, a race for standardization was launched. This was quite easy on a national level, and most of today's relevant fieldbus systems soon became national standards. Troubles started when international solutions were sought. The fieldbus standardization project of IEC, which was started in the technical subcommittee SC65C in 1985, had the ambitious objective of creating one single, universally accepted fieldbus standard for factory and process automation [16,27,29]. Against the backdrop of a quickly evolving market and after 14 years of fierce technical and increasingly political struggles, this goal was abandoned with the multiprotocol standards IEC 61158 and IEC 61784-1 [30–32]. In other application domains, other standards were defined, so that the fieldbus world today consists of a sumptuous collection of well-established approaches.

1.4 Communication Fundamentals: The OSI Model

It has been stated before that the definition of the OSI model was as essential cornerstone for the development of fieldbus systems. The attempts at creating a reference model for data communication arose from the fact that at the time, there were a series of computer networks all of which were incompatible with one another. The expansion of these networks was therefore limited to a specific circle of users. Data transmission from one network to another was possible only with great investment in specialized hardware and software solutions. It was the aim of the OSI model to counteract this development. ISO introduced the concept of an *open system*. Such systems consist of hardware and software components that comply with a given set of standards. These standards guarantee that systems from different manufacturers are compatible with one another and can easily communicate.

To alleviate handling the rather complex task of data communication, it was decided in the committee to partition it into a strictly hierarchical, layered model. All relevant communication functions were counted up and ordered into overlying groups building on one another. On that basis, a reasonable degree of modularization was sought, and seven layers seemed a feasible compromise. In fact, there is no other significance or mystery in the number of layers. The great significance of the OSI model and its value for practical use came about due to the consistent implementation of three essential concepts:

1. *Protocol*. The term *protocol* denotes a set of rules that govern the communication of layers on the same level. If layer N of open system 1 wishes to contact layer N of open system 2, both systems must adhere to specific rules and conventions. Together, these rules make up the protocol of layer N. Layers lying on the same level are also called peer layers.

2. *Service*. This represents any service made available by one layer (called *service provider*) to the layer directly above it (called *service user*). It is important to notice that for the services of a layer, the OSI model defines only their functionality and not how they are actually implemented.
3. *Interface*. There is an interface between every two layers. This clearly defined interface specifies which services are offered by the lower layer to the upper layer, that is, how the service user can access the services of the service provider, what parameters need to be transferred, and what the expected results are.

It is a common misconception to think that the OSI model describes or prescribes actual implementations. It establishes only the effects of each of the layers. Suitable standards were additionally worked out for the layers. These standards are not part of the reference model and were later published in their own right.

1.4.1 Layer Structure

The clear distinction between horizontal and vertical communication was the key to the definition of interconnectable systems. Figure 1.4 shows the layer structure and how the data are passed from one process to another. A system application initiates a transmission via layer 7 of its communication stack. To fulfill this task, layer 7 prepares data for its peer layer on the receiving side, packages this, and requests services of layer 6 to transmit these data. Layer 6 does the same. It prepares data for the peer layer 6′ to fulfill its task and requests services from the next lower layer 5, and so on all the way down to layer 1, which actually transmits the data. On the way down the layers, the data of the application process are augmented by layer-specific data needed to execute the respective protocols. These data are typically address and control information that is mostly combined in a protocol header. In addition, the data may be segmented into individual packets to match the allowed maximum packet size for a given layer. This way, the number of bits being actually transmitted can be significantly larger than the pure user data provided by the application process, and the communication overhead can be substantial. On the receiving side, the peer layers strip all this additional information to recover the user data for the application process. The basic functions of the individual layers are briefly described in the following text.

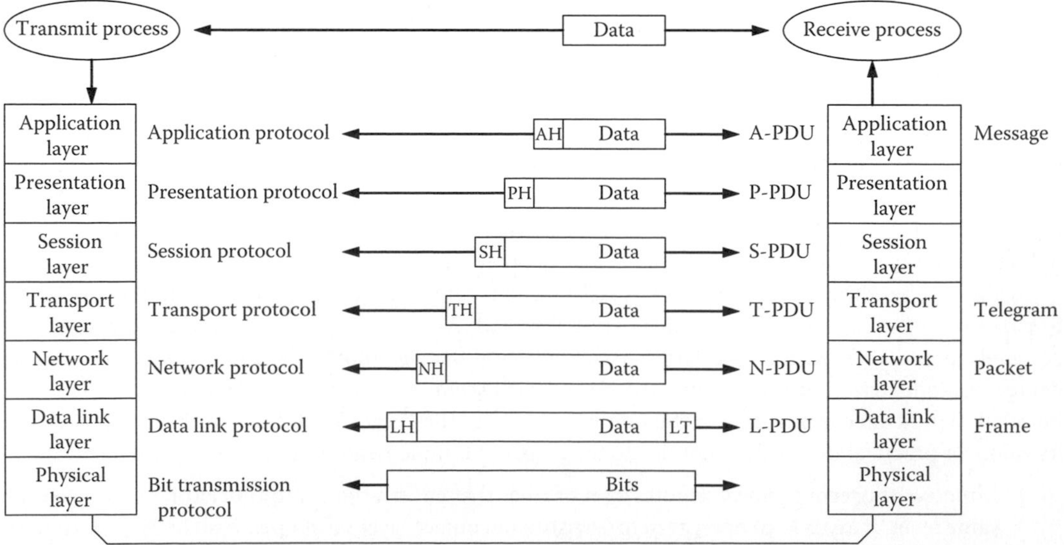

FIGURE 1.4 Layer structure of the OSI model and data frame formation.

1.4.1.1 Layer 1 (Physical Layer)

As the name suggests, this layer describes all mechanical, physical, optical, electrical, and logical properties of the communication system. This includes, for example, definition of the connector type, impedances, transmission frequencies, admissible line lengths, line types, the type of coding of individual bits (such as NRZI: nonreturn to zero inverted, RZ: return to zero, Manchester coding), simultaneous transmission of energy and data, signal energy imitations for intrinsically safe applications, and the like. One of the most widely used standards for layer 1 in fieldbus technology is the RS 485 interface, which naturally forms only a small part of a complete layer-1 definition. Essentially, layer 1 presents to its upper layer all that is needed to transfer a given data frame.

1.4.1.2 Layer 2 (Data Link Layer)

This layer is a pure point-to-point connection with the task of guaranteeing transmission between two network nodes. This firstly involves the formation of the data frame (Figure 1.4), which typically contains a header with control and address information and the actual data. The second task of the data link layer is the coding and checking of the frame (e.g., via CRC: cyclic redundancy check), to allow transmission errors to be detected or even corrected. This also includes the checking of timeouts or verification that the corresponding responses and confirmations are received from the opposite side. The data link layer thus provides the following service to the above lying layer 3: setup of a logic channel to an opposite end device without intermediate nodes or the transmission of a data frame between two end points. In practice, layer 2 becomes too overloaded with functions to allow for a straightforward implementation. Therefore, it is usually subdivided into the logical link control (see IEEE 802.2), which sets up the connection to layer 3 (to which the error detection mechanism is assigned) and the medium access control (MAC) to link to layer 1 (this generally controls who is able to transmit when).

1.4.1.3 Layer 3 (Network Layer)

If there are nodes between the end points of an end-to-end connection, packets must be routed. In layer 3, the paths between origin and destination are established via the specified target addresses. This is easy if the corresponding path lists are available in the nodes. It becomes more complicated when the paths are to be optimized on the basis of various criteria such as cost, quality, load, and delay times if the path conditions change during a transmission, packets need to take different paths due to bandwidth considerations, or the packet size is unsuitable for certain paths. The task of layer 3 is by no means a trivial one, especially when there are various physical transmission media within the network with different transmission speeds. It is also necessary to ensure that congestion does not occur along the paths, which would cause the maximum delay times to be exceeded.

A differentiation is made between connectionless and connection-oriented services. In the case of a connectionless service (datagram service), there is no allocation of fixed channels; every transmitted package must include the complete address and is sent as an independent unit. With a connection-oriented service (virtual circuit service), a virtual channel is made available, which from the point of view of the user offers the advantage that the data packets need not include any addresses. One of the first protocols of this type to be implemented and which is still in use today is the X.25 (ISO 8473). For fieldbus systems, such virtual circuits (and therefore connection-oriented layer-3 services) do not play any practical role. In any case, layer 3 presents to its upper layer a valid path through the network for one individual data packet.

1.4.1.4 Layer 4 (Transport Layer)

The transport layer sets up an end-to-end connection. This means that the receiver does not route the data further, but passes them on to layer 5, already prepared. There are various mechanisms available in layer 4. If the data to be transmitted are too big, layer 4 can split them up and transmit each piece individually. If the transmission times are long or there are a number of possible transmission paths, it is useful to number the split packets. The receiver station must then recombine the individual packets in the right order. On layer 4, connection-oriented and connectionless protocols are available as well. The most popular examples are TCP for connection-oriented and User Datagram Protocol (UDP) for connectionless data transmission. They originated in the Internet world but are becoming increasingly important in the field-level networking domain.

1.4.1.5 Layer 5 (Session Layer)

The main task of layer 5 is to bring together several end devices into a session and to synchronize the *conversation*. This also involves identification or authentication (e.g., password check) and the handling of very large messages. Close cooperation between this layer and the operating system is of vital significance. For this reason, layers 6 and 7 are implemented in fieldbus practice with high transparency for these particular layer-5 functions or a separate channel is created to bypass them. It is also the task of layer 5 to introduce any necessary synchronization markers, so that it knows when to resume after a breakdown in communication.

1.4.1.6 Layer 6 (Presentation Layer)

The presentation layer interprets the incoming data and codes the data to be transmitted. This means that level 6 carries out syntactic and semantic tasks. These include, for example, the meaning of the sequence of bits of a character, to be interpreted as a letter, interpretation of currency as well as physical units, and cryptographic tasks. For encoding, ISO defined the standards Abstract Syntax Notation 1 or Basic Encoding Rules, which are frequently used.

1.4.1.7 Layer 7 (Application Layer)

Layer 7 is a boundary layer (interface to the application) and with that occupies a special position. It forms the interface between the application and the communication unit. In this level, the procedures or protocol processes of various application functions are defined, for calling up data, file transfer, etc. The purpose of layer 7 is a transparent representation of the communication. If, for example, a system accesses databases via the communication unit, layer 7 must be designed so that it does not require any knowledge of the individual tasks of the underlying layers. With that, an efficient communications system allows a database that is distributed among various different locations to be viewed as a single interconnected database. An intelligent switch that transmits information via an OSI communication unit, such as *turn light on*, does not need to know anything about the actual communication protocol; it simply knows the *name* of the lights as well as the functions *turn light on*, *turn light off*, or *dim light by* 40%, etc.

Based on these definitions, it is possible to set up complex communication systems in a structured way. Moreover, the strictly hierarchical layout of the model allows interconnection of heterogeneous systems on different layers. Through the use of repeaters, one can overcome the limitations of a given physical layer. The interconnecting device shares a common data link layer. Bridges interconnect different networks by translating data and protocols on layer 3. Routers link networks on layer 4, whereas gateways (or more precisely, application layer gateways) interconnect entirely different communication systems on the application layer.

1.4.2 Communication Services

Figure 1.5 shows in greater detail how data are exchanged between service user and service provider or between two peer layers. The interface between two neighboring layers is called service access point (SAP). The vertical and horizontal communication is made up of two important units:

1. *Service Data Unit*, SDU. Communication of layer N + 1 with the underlying layer N occurs via its services, or more accurately via the interfaces of these services. For layer N, the transferred data represents pure user data that is passed on to the next lowest layer for further processing.

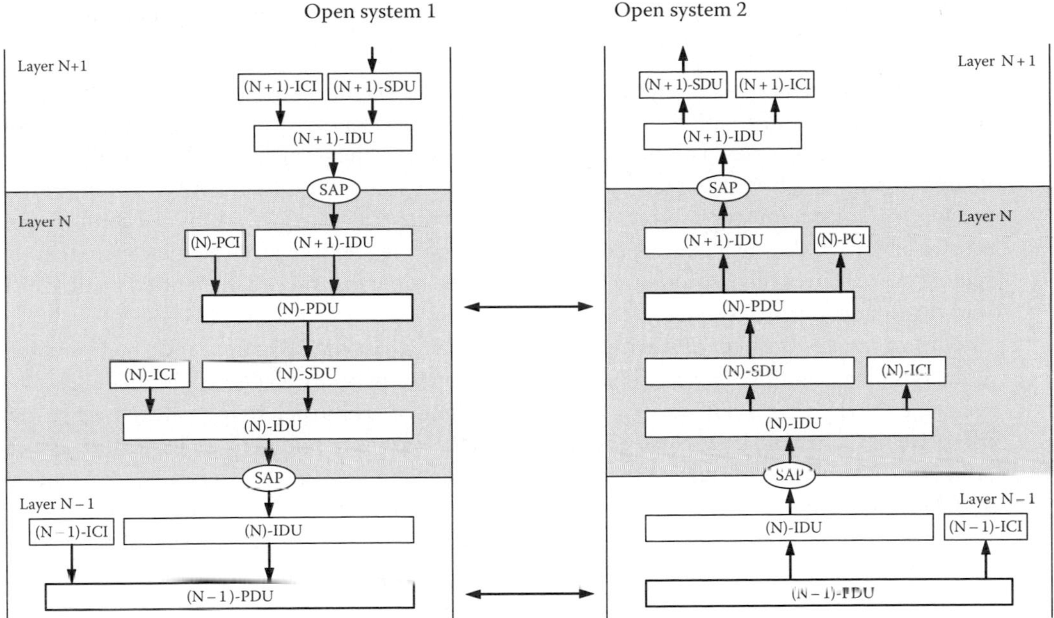

FIGURE 1.5 Communication in the OSI model.

2. *Protocol Data Unit*, PDU. Communication between two peer layers is implemented via so-called protocol data units. These represent the core element of the rule set that can be understood and correctly interpreted only by the peer layers. A PDU consists of the transmitted user data supplemented with parameters of the interface (interface control information, ICI) and unique control information (protocol control information, PCI).

The exact procedure is highlighted in Figure 1.5. The user data of layer N that are to be transmitted to its peer layer are put into an SDU. The interface process of layer N adds an ICI to the message header (the SDU header), whereby the ICI and SDU together form the interface data unit (IDU). The IDU is then transferred via the SAP. In the underlying layer, the ICI is now decoded, and a corresponding process is initiated. The following process (execution of a protocol) should be viewed as a logic process, except in the lowest layer, the physical layer. In the procedure initiated by the ICI, layer N – 1 packs the SDU into a PDU. This involves adding a PCI to the SDU. The assembled PDU is now transmitted, which means that layer N – 1 in turn uses the lower layers and itself now behaves as a service user. Only layer 1 actually physically transmits the information.

In even more detail, the interaction between the individual layers is governed by a series of operations defined in the OSI model. These operations are called service primitives. There are basically four different service primitives: Request, req; indication, ind; confirmation, con; and response, res. Each service primitive that is called up by the service user is termed a request. With the request, layer N receives the order to execute a specific task. The respective task and corresponding data are converted into a corresponding PDU. In accordance with the OSI model, the service provider uses the services of the underlying layer N – 1, in order to carry out its task. This interaction continues until the lowest layer entrusts the data to the physical medium. On the receiver side, the peer layer is activated via an indication (or sometimes even a whole series of indications). After the remote layer N has decoded the PDU and extracted the control information, the user data are passed on to the above lying layer (i.e., the layer directly above or, in the case of the application layer, the actual user process) also by means of an indication. The way back from the receiver side is composed of responses generated by the service users and confirmations issued by the service providers. However, services need not always comprise each of these four service primitives. There are many possible and practically used combinations with varying degrees of reliability, as the following examples show:

1. *Locally confirmed service.* A locally confirmed service (Figure 1.6) comprises a request, an indication, and a confirmation. In this case, layer N of the sender then receives a confirmation from local layer N − 1. Based on the parameters transmitted, the service provider can tell whether its underlying layer was able to accept and process the request accordingly. After a brief preprocessing, this local confirmation is also made known to the service user by means of a confirmation service primitive. As the name suggests, a locally confirmed service gives no guarantee that the remote user receives the information transmitted. The confirmation can be omitted at all, which leads to an unconfirmed service.
2. *Confirmed service.* This type of service also consists of a request, an indication, and a confirmation. With a confirmed service however, the peer layer generates an acknowledgment immediately after receiving the indication (Figure 1.7). The acknowledgment is returned to the sender via a service primitive of the remote layer N − 1 and signaled to the service provider via an indication of local layer N − 1. Note that contrary to the procedure outlined in point 1, the transmitted parameters and data come from the partner side. From the indication, layer N can conclude whether or not the service requested by layer N − 1 was executed without error. If necessary (i.e., if errors have occurred), this service is used again. This depends on the protocol used in layer N. In all cases,

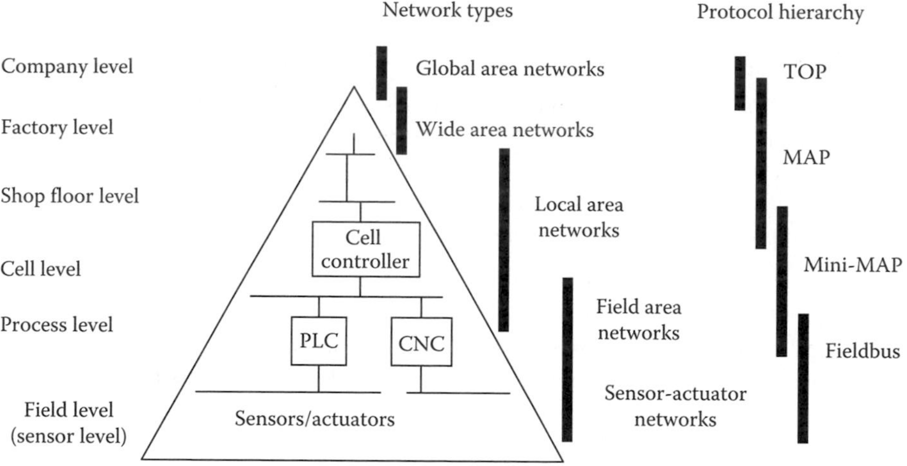

FIGURE 1.6 Locally confirmed service.

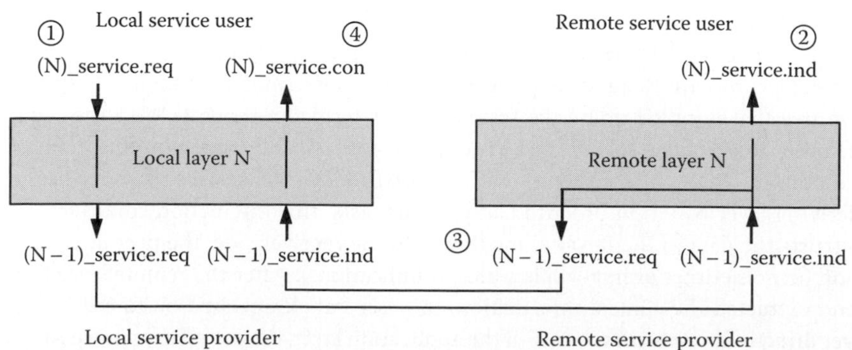

FIGURE 1.7 Confirmed service.

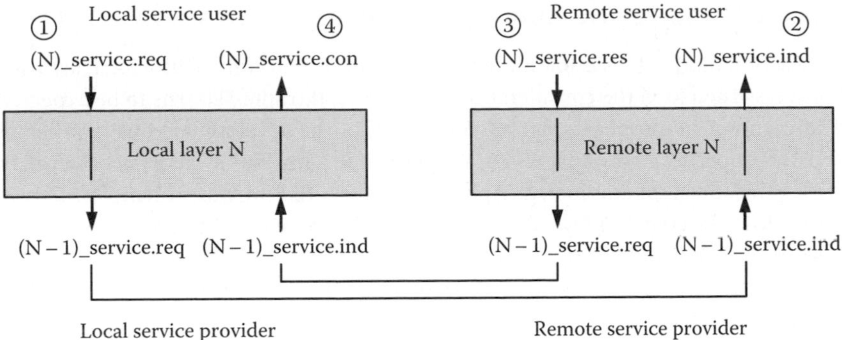

FIGURE 1.8 Answered service.

the service user is informed of the output with a suitable confirmation (positive or negative). From this confirmation, it is possible to derive whether or not the originally requested service has been satisfactorily fulfilled by layer N.

3. *Answered service*. This represents the fully fledged use case of all service primitives (Figure 1.8). Here, after an indication on the partner side, a response is generated by the remote service user. The response is transmitted via a service primitive of layer N – 1 and passed on to the service user from which the original service request came as a confirmation. This mechanism permits data traffic in both directions. Contrary to the other two services, an answered service always consists of request, indication, response, and confirmation.

In short: Request and response are always called up by the service user; the resulting confirmation and indication originate from the corresponding layer. The interaction between the individual layers is best understood if one imagines that the layers lying on top of one another are primarily inactive. The service provider is *awoken* only after a request from the service user. In order to perform this service, it in turn activates the underlying layer by calling up the appropriate service primitive. This interaction continues down to the lowest layer, which then accesses the transmission medium. On the remote side, the indication of a low-lying layer informs the layer lying directly above that a service is to be executed. This indication is then processed in accordance with the service type—it causes either a confirmation or a response.

1.5 Fieldbus Characteristics

The application areas of fieldbus systems are manifold; hence, many different solutions have been developed in the past. Nevertheless, there is one characteristic and common starting point for all those efforts. Just like today's embedded system networks, fieldbus systems were always designed for efficiency, with two main aspects:

1. Efficiency concerning data transfer, meaning that messages are rather short according to the limited size of process data that must be transmitted at a time.
2. Efficiency concerning protocol design and implementation, in the sense that typical field devices do not provide ample computing resources.

These two aspects, together with characteristic application requirements in the individual areas with respect to real-time, topology, and economical constraints, have led to the development of concepts that still are very peculiar of fieldbus systems and present fundamental differences to LANs. In the previous section, the general differences in the context of the OSI model were already sketched. This section will discuss some more peculiarities in greater detail.

1.5.1 Traffic Characteristics and Requirements

The primary function of a fieldbus is to link sensors, actuators, and control units that are used to control a technical process. Therefore, the consideration of the typical traffic patterns to be expected in a given application domain was in most cases the starting point for the development of a new fieldbus. Indeed, the characteristic properties of the various data types inside a fieldbus system differ strongly according to the processes that must be automated. Application areas like manufacturing, process, and building automation pose different *timing* and *consistency* requirements that are not even invariant and consistent within the application areas [16].

As regards timing, there are two essential philosophies to look at the technical process in a black-box form and describe (and thus convey within the network) its behavior. One is a state-based approach focusing on the status of the process defined by its internal state variables together with its inputs and outputs. These variables are continuously sampled and transmitted in discrete-time fashion and form the basis for continuous process control and monitoring (like temperature, pressure, etc.).

The other philosophy is an event-based one, that is, data are transmitted only in case of state changes. This approach is well suited for processes or subprocesses that have a somewhat discrete nature and can be modeled as a state machine. Switches obviously lend themselves to such an interpretation. But also in continuous processes, it may be reasonable to transmit process data only if they exceed certain predefined limits. This naturally requires the implementation of control functions locally in the network nodes rather than in a remote control unit.

As far as consistency is concerned, there are on the one hand process data that are continuously updated and on the other hand parameterization data that are transferred only upon demand. In case of error, the former can easily be reconstructed from historical data via interpolation (or simply be updated by new measurements). The system-wide consistency of configuration data, on the other hand, is an important requirement that cannot be met by mechanisms suitable for process data.

Trying to find an abstract and comprehensive classification scheme for all possible fieldbus traffic patterns, we arrive at the traffic types shown in Figure 1.9 and are described in the following text.

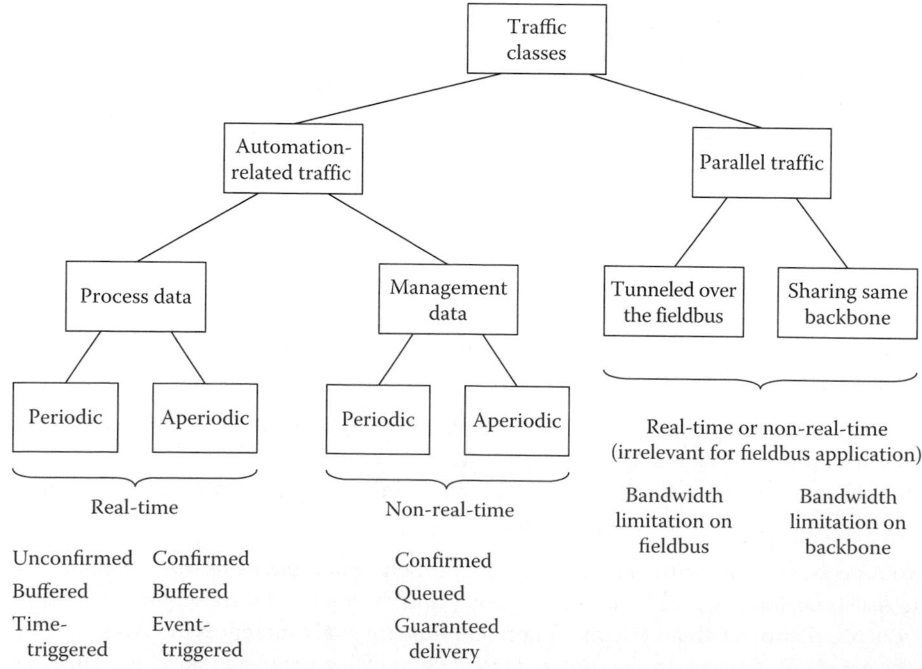

FIGURE 1.9 Traffic classes and typical properties.

In theory, data traffic generation is exclusively a question of the application process and should be seen independently of the underlying communication system. In practice, however, the communication protocol and the services provided therein heavily influence the possible traffic types—it is one of the peculiarities of fieldbus systems that the two cannot be properly disentangled.

1.5.1.1 Process Data

This type is sometimes also called *cyclic* or *identified* traffic because the communication relations must be known once the application is specified. Depending on the characteristics of the process, such data can be periodic or aperiodic. Periodic traffic mostly relates to the state of a process and is typically handled by some sort of time-slot-based communication strategy, where each variable is assigned a dedicated share of the network bandwidth based on the a priori known sampling time or generation rate of the data source. The frequency or data update rate may also be adaptive, so as to dynamically change according to the current state of the process (e.g., alarm conditions may require a more frequent sampling of a state variable).

Aperiodic, *acyclic*, or *spontaneous* traffic is generated on demand in an event-based manner and transmitted according to the availability of free communication bandwidth. If the predominating data transfer scheme is time slot based, there may be spare slots or some idle time in the periodic traffic explicitly reserved for this purpose.

From an application point of view, process data typically are real-time data and require timely delivery. For both periodic and aperiodic data, this means that they should be transmitted (and received) within a limited time window to be meaningful for the purpose of process control. For aperiodic data, there is the additional restriction that the service ought to be reliable, that is, data loss should not occur or at least be detected by some appropriate mechanism.

1.5.1.2 Management Data

This type is also called *parameterization* or *configuration* data and generally refers to all data that are needed to set up and adjust the operation of the automation system as such. Settings of distributed application processes (such as control functions localized in the network nodes), communication and network parameters, and more general all network management data belong to this class. Management data typically are aperiodic as they occur infrequently depending on the state of the complete system. Traditionally, they are therefore often transmitted in some dedicated parameter channels that use a small amount of communication bandwidth left over by the (mostly periodic) process data.

In some cases, management data may also be needed periodically or quasi-periodically, for example, to update session information after a certain time, to exchange authentication information after a given number of messages, or to change communication parameters on a routine basis. Nevertheless, the communication mechanisms for such periodic management messages likely do not differ from the aperiodic ones; they are simply invoked on a periodic basis.

Contrary to process data, management data are mostly not real-time data, which means that timely delivery is not that important. What is more important, however, is guaranteed and correct delivery, which influences the communication services that are used for this type of data. Confirmed services are mandatory.

1.5.1.3 Parallel Traffic

This traffic type does not belong to the applications processes concerned with the actual control of the technical process. Rather, it is generated by independent parallel processes and shares the communication medium. In traditional fieldbus systems, which were closed environments, this type of traffic did not exist. With growing interconnection of automation networks, however, it becomes relevant, either because external traffic (such as IP traffic) may be routed or tunneled through the fieldbus or because fieldbus traffic may be routed through a shared backbone network [33,34]. In particular, this is an issue for industrial Ethernet solutions, which in general foresee some sort of general-purpose IP channel for devices not belonging to the automation system.

From the viewpoint of the automation process, care must be taken that the fieldbus traffic is not negatively influenced. This needs to be done by appropriate quality-of-service arrangements that guarantee reasonable bandwidth and latency to the fieldbus application traffic in case a shared backbone is used. In the other case, the fieldbus network management must ensure that the parallel traffic cannot consume an arbitrary amount of communication resources or block other processes.

1.5.1.4 Implications for the Fieldbus

The various traffic types have also implications for the way communication in fieldbus systems is handled, both from a protocol and from an implementation point of view. Data that are exchanged on a cyclic basis are usually sent via connectionless services. The reason behind this is that for periodically updated process data, it makes no sense to require a confirmation. If a message containing process data gets lost, resending it is not sensible because by the time the data might arrive at the receiver, they are outdated, anyway, and new data might already be available. With respect to practical implementation, the handling of such data, especially at the receiver side, is mostly implemented by means of buffers. In these buffers, the latest data always overwrite older values, even when they are basically implemented in a first in, first out (FIFO) structure. In this case, the *FIFO full* signal is suppressed, so that always the most recent values are in the buffer.

On the contrary, acyclic data need special precautions, irrespective of whether they are related to process variables or management data. Loss of such data is not desirable and might be detrimental to the overall application (e.g., if alarm events are not received). Hence, mechanisms involving some sort of acknowledgment must be used to allow for retransmissions in case of data loss. These mechanisms can be implemented on the lower communication layers or the application level, depending on the basic communication services a fieldbus offers. In terms of actual implementation, such messages are typically handled in queues. As opposed to buffers, messages are not overwritten. If the queue is full, no new messages are accepted until earlier entries have been consumed and deleted. This ensures that no messages are lost before they are processed by the node.

Stimulated by the different ways of looking at and handling data exchange in fieldbus systems (and embedded systems in general), two opposing paradigms have been established in the past and have ignited a long, partly fierce debate: the *time-triggered* and the *event-triggered* paradigm [140]. The time-triggered approach is specifically suited for periodic real-time data, and many fieldbus systems actually use it in one form or another. The event-triggered approach was designed following the idea that only changes in process variables are relevant for transmission. An additional aspect behind it was that such events should be broadcast in the network, so that every node potentially interested in the data can receive them. This makes extension of the network fairly easy by just adding new nodes if the message identifiers they need for their application are known.

1.5.2 Fieldbus Systems and the OSI Model

Like all modern communication systems, fieldbus protocols are essentially modeled according to the ISO/OSI model. However, in most cases, only layers 1, 2, and 7 are actually used [14]. This is in fact a tribute to the lessons learned from the MAP failure, where it was found that a full seven-layer stack requires far too many resources and does not permit an efficient implementation. For this reason, the MiniMAP approach and based on it the IEC fieldbus standard explicitly prescribe a three-layer structure consisting of physical, data link, and application layer. But what about the other layers and the functions defined therein?

The reduced and simplified protocol stack reflects the actual situation found in many automation applications pretty well, anyway. Many fieldbuses are single-segment networks with limited size, and extensions are realized via repeaters or, at most, bridges. Therefore, network and transport layer—which contain routing functionality and end-to-end control—are simply not necessary. The same applies to the upper layers. Originally, fieldbus systems were not meant to be very sophisticated. Fully implemented

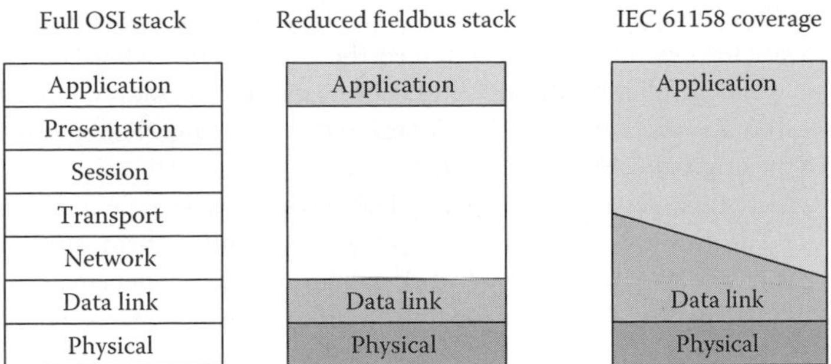

FIGURE 1.10 Layer structure of a typical fieldbus protocol stack as defined by the IEC 61158.

session and presentation layers are therefore not needed, either. Still, certain functions from the layers 3 to 6 might be needed in reduced form. Rudimentary networking aspects could be required or specific coding rules for messages that are better suited for the limited resources available in typical fieldbus nodes. In such cases, these functions are frequently included in the layer 2 or 7. For the IEC 61158 fieldbus standard [1], the rule is that layer 3 and 4 functions can be placed either in layer 2 or layer 7, whereas layer 5 and 6 functionalities are always covered in layer 7 (Figure 1.10).

It would nevertheless be deceptive to think that all fieldbus systems just consist of a physical, data link, and application layer. There are several examples where other layers were explicitly defined. Particularly in the building automation domain, the situation is different. Owing to the possibly high number of nodes, these fieldbus systems must offer the capability of hierarchically structured network topologies, and a reduction to three layers is not sensible. For instance, European Installation Bus (EIB) and KNX use also the network and transport layers to implement routing through the hierarchical network as well as connection-oriented and connectionless end-to-end communication functions. BACnet uses the network layer as well, which is especially important as BACnet was devised as higher-layer protocol to operate on different lower-layer protocols and links such as Ethernet, MS/TP (master–slave/token passing as inexpensive data-link layer protocol based on the RS 485 standard for the physical layer), and LonTalk. For such a heterogeneous approach, a uniform network layer is essential.

The probably most elaborate protocol structure in the fieldbus world is exhibited by LonWorks. Even though it is today chiefly used in building automation, it was in fact designed as a general-purpose control network (LON stands for local operating network) without any particular application area in mind; hence, it resembles much more a LAN than a conventional highly efficient fieldbus. In the LonTalk protocol, all seven OSI layers are defined, even though layer 6 is rather thin in terms of functionality. Specific characteristics are a rich layer 3 that supports a variety of different addressing schemes and advanced routing capabilities, the support of many different physical layers (a common aspect in all building automation networks), and a large number of various communication objects not just for process data exchange and network management, but also for advanced functions like file transfer.

Among the fieldbus systems mainly used in industrial and process automation, ControlNet and P-NET are particular in that they implement also layers 3 and 4. An outstanding characteristic of P-NET is its capacity for multinetwork structures, where the so-called multiport masters can link multiple segments to any arbitrary structure. Layer 3 provides a source routing mechanism (where the path through the network must be defined inside the packet) to manage transmission even in meshed networks. Layer 4 is called service layer and actually contains definitions and processing rules for communication objects that go beyond the usual end-to-end functionality of the OSI transport layer.

An essential part of fieldbus protocol stacks are comprehensive application layers. They are indispensable for open systems and form the basis for interoperability. Powerful application layers offering abstract functionalities to the actual applications, however, require a substantial software implementation effort that can negatively impact the protocol processing time and also the costs for a fieldbus interface. This is why in some cases (like Interbus, PROFIBUS-DP/PA, or CAN), an application layer was originally omitted. While the application areas were often regarded as limited in the beginning, market pressure and the desire for flexibility finally enforced the addition of higher-layer protocols, and the growing performance of controller hardware facilitated their implementation. CAN is a good example for this because a plethora of protocols (like CANopen, SDS, and DeviceNet) appeared in the course of time on the basis of the original CAN layers 1 and 2.

A further vital aspect of any network is an appropriate network management. This includes tasks like incorporating new end devices into an existing network and combining them with other end devices to form functional units. In addition, a modern network management also offers mechanisms for the analysis and diagnosis of systems that are already up and running. Network management was not foreseen in the original OSI model; rather it was put into a dedicated OSI Network Management Framework. It is to be understood as existing in parallel to the OSI layers as it affects all of them:

Physical and data link layers. All end devices must, for example, have the same channel configuration. This involves certain parameters that determine transmission on the underlying medium. In this connection, the applicable bit rate is of particular interest. Network management also concerns the repeaters operating on layer 1 and bridges (implemented in layer 2), because they can also be used to link subnetworks with different channel configurations.

Network layer. Every end device or group of end devices must be provided with a unique address. With the help of intelligent routers, it is possible to form subnetworks and with that reduce the network load. These devices too can be contacted and then configured by the commands made available through the network management (such as loading of new routing tables).

Transport and session layers. These layers are mainly responsible for the service quality that is offered by the underlying network. Corresponding configuration possibilities, which can be carried out with the network management, are therefore closely linked to the term *quality*.

Presentation layer. Common syntax is a basic prerequisite for interoperability. Network management must inform two communicating end devices of the syntax of the matching data types.

Application layer. Here, network management is concerned with the carrying out of tasks that relate to applications. It should be possible not only to load applications and define their specific configurations but also to load and modify tables describing the communication relationships of applications.

Inside fieldbus protocols, network management is traditionally not very highly developed. This stems from the fact that a fieldbus normally is not designed for the setup of large, complex networks. There are again exceptions, especially in building automation, which consequently need to provide more elaborated functions for the setup and maintenance of the network. In most cases, however, the flexibility and functionality of network management are adapted to the functionality and application area of the individual fieldbus. There are systems with comparatively simple (ASi, Interbus, P-NET, and J1939) and rather complex management functions (WorldFIP, CANopen, LonWorks, KNX, and BACnet). The latter systems are typically more flexible in their application range but need more efforts for configuration and commissioning. In any case, network management functions are normally not explicitly present (in parallel to the protocol stack as suggested by the OSI model), but rather directly included in the protocol layers (mostly the application layer).

1.5.3 Network Topologies

One important property of a fieldbus is its topology. Developers of fieldbus systems have been very creative in the selection and definition of the best suited physical layout of the network. Again, this selection was typically influenced by the target application area as well as by the available interface technologies that are used to build the fieldbus. Figure 1.11 shows the most relevant topologies for wired automation networks. It should be noted that the physical layer of a fieldbus has to meet quite demanding requirements like robustness, immunity to electromagnetic disturbances, intrinsic safety

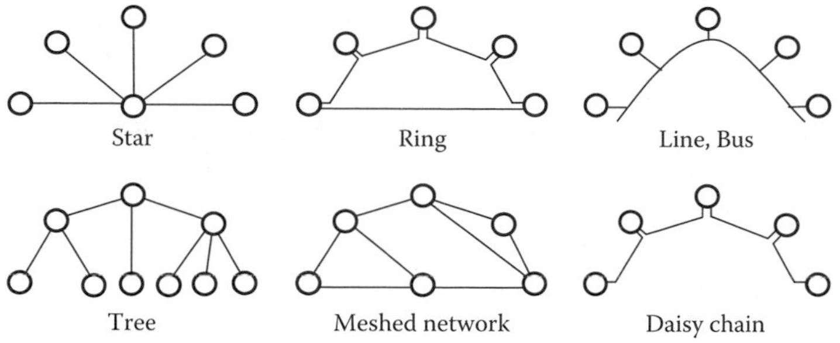

Star
Ring
Line, Bus
Tree
Meshed network
Daisy chain

FIGURE 1.11 Topological network structures typically used in fieldbus systems.

for hazardous areas, or costs. The significance of the physical layer is underpinned by the fact that this area was the first that reached (notably undisputed) consensus in standardization.

The *star* topology was the typical wiring in automation before the introduction of the fieldbus. The PLC was the center, attached to the distributed I/O elements with dedicated lines. The obvious cabling overhead was one of the main reasons to develop serial bus systems. With the adoption of switched Ethernet also for automation purposes, the star topology returned. Today, the central element is the Ethernet switch, and all Ethernet nodes are connected by means of a structured cabling, that is, a dedicated link to each network node. Another application of the star topology is for fieldbus systems using optical fibers as transmission medium. Here, the center is an active star coupler linking the fibers (as in byteflight). Unless the coupler is a fully fledged node (i.e., addressable by the fieldbus protocol), this start topology is logically equivalent to a line structure.

Another simple topological form is the *ring*. Here, each node has two network interfaces (an input and a separate, independent output), and the nodes are arranged one after another in the form of a chain. In its entirety, the ring can be viewed as one large shift register, and it is usually also operated in this way (the most prominent example is INTERBUS). As there is no need to address the nodes explicitly, it is a very fast method to exchange data and also a very deterministic one with low jitter (this is why SERCOS uses this topology for the interconnection of drives). A variant of the ring topology, actually a daisy-chain structure, was introduced with industrial Ethernet. Here, each node contains a small switch, and the nodes are not connected in a star topology, but cascaded like a string of pearls. This layout is used, for example, in PROFINET.

The *line*, often referred to simply as *bus* (although this term is not unique), is the most successful and most commonly used network topology in the fieldbus world. It was the logical and most efficient replacement for the former star-like point-to-point cabling in that one single line should connect all network nodes (just as in the original concepts of Ethernet with the famous yellow cable and other types of coax cables). In many cases, the line topology is based on the RS 485 interface. This is an inexpensive, fully differential, multipoint interface standard using a shielded twisted pair cable with 120 Ohm characteristic impedance. Maximum cable length is 1200 m, and the maximum achievable data rate is 10 Mbit/s. The maximum number of nodes per segment depends on the electrical characteristics of the driver circuits and was originally limited to 32 (termed unit loads). Enhanced transceivers with higher input impedance meanwhile allow up to 256 nodes per segment. Beyond this, repeaters are necessary to regenerate the data signals. Some fieldbus systems (like CAN and all CAN-based systems) use RS 485 transceivers in a modified way (i.e., by essentially applying the data signal not to the actual data input but the output enable used to switch the output to a high-impedance state) to generate asymmetric bit patterns supporting a special form of medium access. A crucial aspect in practice is the proper electrical termination of the bus line in order to avoid signal reflections disturbing the data transfer. Wrong or missing termination is the most frequent problem leading to communication failures. A variant of the line topology used for P-NET therefore requires that the ends of the cable be connected, so as to form a closed loop. Nevertheless, electrically, this structure is still a bus.

The *tree structure* is a composite network structure and characterized by one or more substations being dependent on a root node. Each substation can in turn be a root node for a lower-level segment. In many cases, the actual connections between the stations are regular point-to-point connections or lines. In automation technology, tree structures are the usual way to build hierarchical, relatively complex networks. The root nodes usually have routing capabilities, so that the data traffic can at least partly be confined to individual areas of the network. This however requires caution in the planning phase of the network when defining communication relations between end nodes. Nodes exchanging lots of information should be kept in common subnets to avoid congestion on the backbone links. Tree structures are very common in building automation networks like EIB, LonWorks, or BACnet.

Some fieldbus systems permit *free topologies* where nodes can be connected in a multipoint fashion without restrictions. Essentially, this is a variant of the line topology because the nodes are electrically interconnected. Contrary to the standard line topology, however, there are no stringent limitations concerning line length, the length of stubs, or branches that would cause severe signal degradation. This makes cabling very convenient for the installer, but poses a substantial challenge for the signal processing in the nodes and their transceivers. Fieldbus systems using such free topologies are, for example, LonWorks (where the free topology transceiver is the most widely used because of its robustness even in cases where conventional RS 485 transceivers would be sufficient) or ASi, which was designed for utmost simplicity and ease of handling in harsh industrial environments.

Mesh networks, where multiple paths through the network exist, play only a subordinate role in fieldbus systems because they require appropriate routing strategies to keep messages from circling in the network and causing congestion. LonWorks and P-NET offer the possibility for building meshes.

The general topology used by the fieldbus also influences the way access to the medium is being handled. As a matter of fact, topologies were often selected according to the desired MAC method, or vice versa.

1.5.4 Medium Access Control

Fieldbus designers have been very creative in what concerns allocation of the available communication bandwidth to the individual nodes. On the data link layer, virtually all medium access principles also known from LANs are used, plus many different subtypes, refinements, and hybrid solutions (Figure 1.12). This has to be understood against the backdrop of desired efficiency. Resources in terms of both computing power in the nodes and available data rate and thus bandwidth in the network were very limited at the time most fieldbus systems were invented. Hence, developers wanted to get the most out of what was at hand, and they found that a straightforward application of the medium access strategies used in computer networks were not efficient enough. So, they started designing new, more efficient methods tailored to individual application areas and optimized for the particular type of traffic and communication models they had in mind. In fact, the efficiency or overhead in the sense of how many bits must be transmitted over the network to convey a given amount of user data was a very common parameter to compare (and market) fieldbus systems—even though the significance of such a parameter alone was and is more than questionable.

An important classification for the way the data transfer is controlled is the distinction between single- and multimaster systems. The single-master (or master–slave) approach reflects the tradition of centralized, PLC-based automation systems and is typically used for fieldbus systems in the lowest levels of the automation pyramid where the roles of the nodes in the network can be clearly distributed. Such networks typically have a limited size and a simple, mostly single-segment structure. The master either retrieves data from the slaves in direct request–response communication or explicitly synchronizes time slots that are then used by the slaves to send their data.

The alternative strategy is the multimaster approach, where all nodes are basically equal and must share the communication medium in a democratic and fair way. Such networks can have more complex structures, including hierarchical ones with large numbers of nodes, and are found mostly on

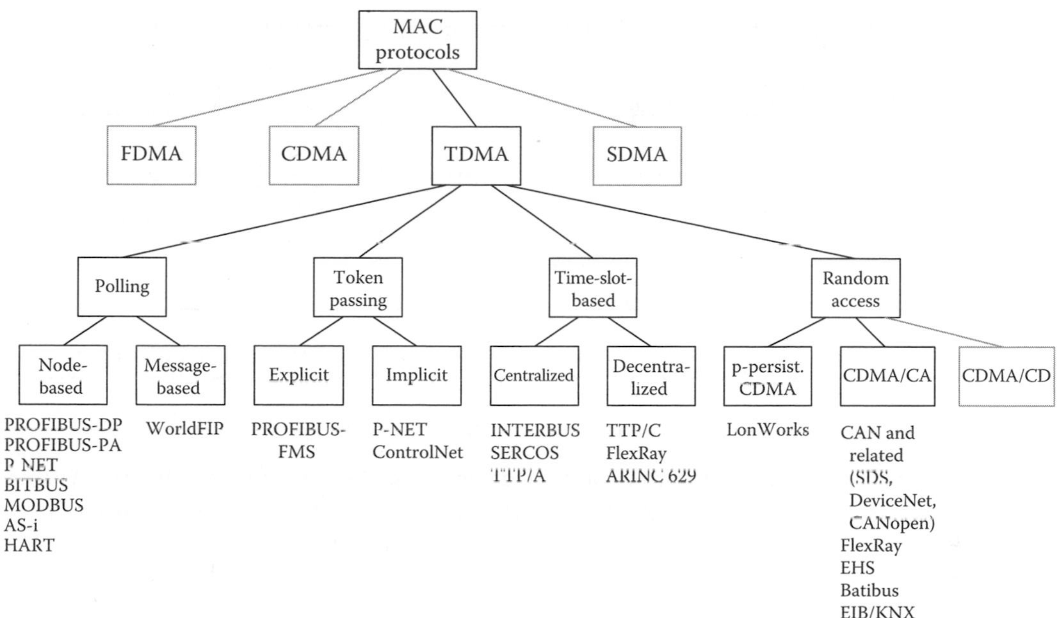

FIGURE 1.12 Medium access control strategies in fieldbus systems and examples.

the middle level of the automation pyramid (the cell level) and in building automation. Historically, multimaster fieldbus systems originated from the wish to have truly distributed systems for process control, and in many cases, such fieldbus systems were more than pure communication networks—they were embedded in comprehensive operating-system-like software frameworks.

As far as the actual MAC strategies are concerned, all fieldbus systems use a time division multiple access (TDMA) strategy in the sense that the bandwidth is shared in the time domain, that is, the network nodes use the communication line sequentially. Other multiplexing methods known in particular from telecommunications, such as frequency division multiple access, code division multiple access, or space division multiple access, play no or only subordinate roles in the fieldbus world. More precisely, medium access is managed either in a centralized fashion by polling or time-slot-based techniques, or it is done in a decentralized way by TP or random access methods. Within these basic methods, which will be briefly reviewed later, various variants and blends are in use.

1.5.4.1 Polling

Polling is a master–slave access scheme and foresees that a slave node is allowed to send data only when explicitly told so by a central master (sometimes also called arbiter or arbitrator). On the network, there is a constant alternation between poll messages from the master to each slave and response messages from the slaves back. In its purest form, polling is strictly cyclic, which means that the master polls the entire list of slaves one by one and then restarts the cycle. The polling rate might be adapted (in theory even dynamically) if individual nodes have more data to transmit than others [35], but this is rarely done in practice.

Evidently, polling is perfectly suited for periodic traffic where all process variables need equidistant sampling (Figure 1.13). Since the traffic is strictly controlled by one node, the cycle time can be kept constant provided that slaves implement antijabber mechanisms so that they cannot block the medium in the case of failure. Likewise, jitter can be kept low. Simple polling is employed in typical sensor–actuator bus systems like ASi, HART, MODBUS, or lower-level fieldbus systems like BITBUS, PROFIBUS-DP, and PROFIBUS-PA. Polling can also be used as an underlying bus access mechanism in a multimaster system, such that the permission to access the slaves is rotated between several master nodes. This strategy is used, for example, by PROFIBUS-FMS or P-NET.

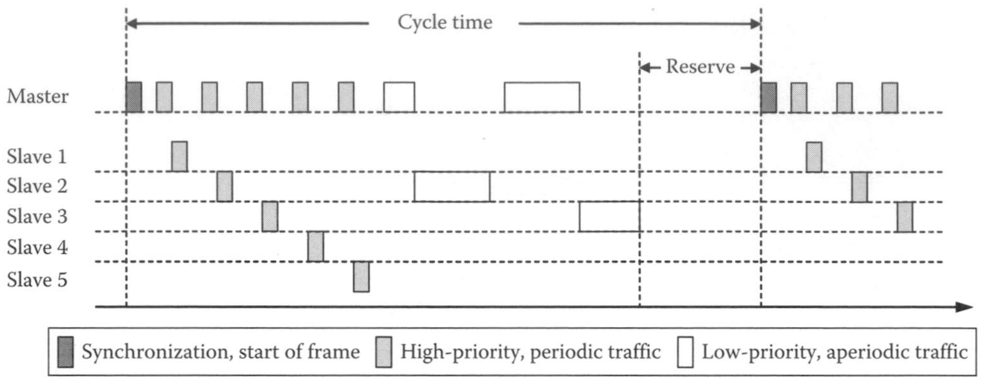

FIGURE 1.13 Bus cycle with cyclic polling and room for aperiodic traffic (PROFIBUS-DP V2).

A distinction between different flavors of polling can be made depending on the way data are addressed. The classical master–slave polling in the examples listed earlier uses explicit node addressing, that is, the communication approach is device-centric. Another possibility is to identify the desired data by name, meaning that the request from the master concerns not a device, but a certain process variable. This is an approach employed by the producer/consumer model in WorldFIP, where the bus arbitrator polls a variable and the node generating (*producing*) it then sends it onto the network, or in LIN, where the master retrieves data from the slaves by polling the message identifiers. The link active scheduler in Foundation Fieldbus performs a similar function by prompting devices to send their scheduled data to the consumers. This polling variant is sometimes called central polling. From the network traffic point of view, the two approaches are equal; each consists of a request–response pair of messages per variable exchange.

A peculiarity of the bus arbitration in WorldFIP, however, is that the polling mechanism accounts for different periodicity requirements of the individual variables. To this end, the polling cycle (here called macrocycle) is compiled at configuration time from several elementary cycles or microcycles in a static schedule to ensure that each variable will be sampled as often as needed under the constraint that the total data rate originating from this scheduled traffic is at any given time lower than the maximum data rate of the bus. The number of elementary cycles a macrocycle must have is essentially defined by the least common multiple of the periodicities divided by the highest common denominator of the periodicities (Figure 1.14).

In strict polling, there is no room for aperiodic traffic, which is a severe drawback for fieldbus systems designed to be more than simple sensor–actuator buses. Therefore, most fieldbus systems foresee some possibility for the master to handle sporadic data exchange for configuration data or process variables that require transfer only on an event basis. The important constraint is that this type of traffic must

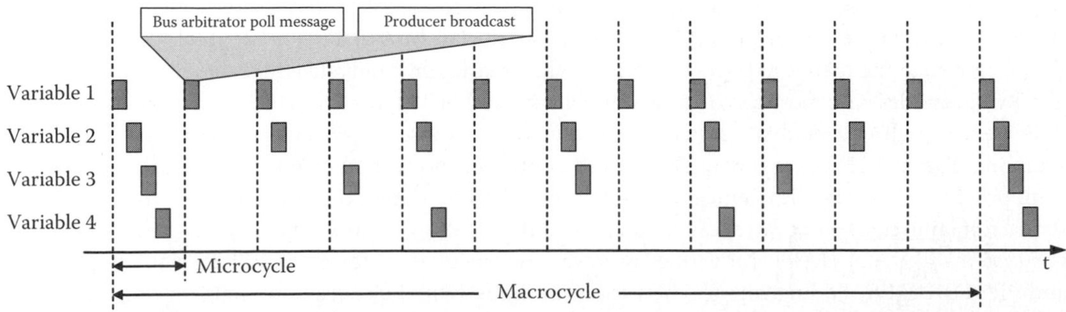

FIGURE 1.14 Bus cycle in WorldFIP.

not interfere with the periodic one. In practice, this is done be reserving some bandwidth in addition to the scheduled periodic messages for a priori unknown aperiodic data transfer. PROFIBUS-DP/PA and many Ethernet-based automation networks (such as PROFINET) use a dedicated portion of the bus cycle after the periodic traffic for aperiodic traffic. In order to keep the cycle period constant, there is usually some spare time after this window to account for possible retransmissions of messages. In WorldFIP, aperiodic traffic can be scheduled in the empty slots of the elementary cycles, so that the macrocycle need not be extended.

Another disadvantage of polling is that slaves cannot become active by themselves. If all of a sudden, a slave—for instance, in the event of an alarm condition—needs to transmit additional data, it first has to signal this alarm condition to the master to initiate an aperiodic request. In most cases (like in WorldFIP), the slaves have the possibility to set a flag in the ordinary cyclic data exchange messages to indicate the need for additional service (in the old CAMAC system, this flag had the descriptive name *look at me*). The master then typically starts a handshake procedure to retrieve these data, possibly by first requesting information about their amount so as to allocate an appropriate time slot for the transfer (also with respect to other pending requests). This basic mechanism can be further extended to allow a communication among slaves. Again, two possibilities exist: The first is to handle it indirectly via the master, which gets the message from the sending slave and forwards it to the addressee (as done in the Measurement Bus). This permits the master to maintain full control over the bus. The alternative is that the master delegates the bus access to the slave for a while until the slave has sent all its data and returns the access right to the master (as in WorldFIP, Foundation Fieldbus, or PROFIBUS-DP-V2). In this case, the master has to implement some timeout mechanism to ensure that the procedure is properly terminated.

1.5.4.2 Token Passing

One way to coordinate medium access between several peer stations in a network is TP. In this method, the right to control the network is represented by a special piece of information called token, which is passed on from node to node. Only a node possessing this token may initiate data transfer. A set of rules ensures that the token is passed in a fair manner and that errors such as lost tokens or duplicate tokens are detected and resolved. Compared with time-slot mechanisms, TP has the advantage that if a node has no data to send, it will hand over the token to the next station immediately, which saves time. Essentially, there are two ways of implementing the token: either explicitly by means of a dedicated short message or implicitly by distributed, synchronized access counters (ACs) in all nodes. TP is often combined with an underlying master–slave mechanism to control a subset of nodes.

The explicit form of TP is employed by PROFIBUS, both FMS and DP/PA (Figure 1.15). Even though PROFIBUS-DP in its basic variant is a single-master system, several DP networks may coexist on the

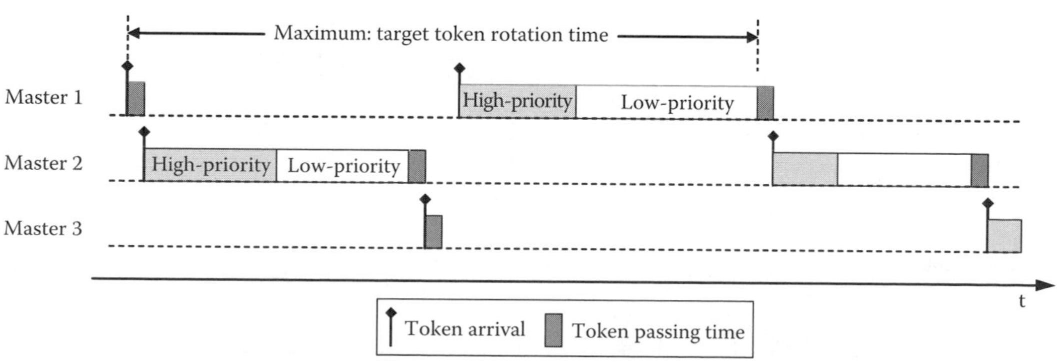

FIGURE 1.15 Token passing in PROFIBUS.

same bus. A crucial point here is the correct setting of a timing parameter called target token rotation time T_{TR}. This parameter indirectly determines how long a master may occupy the bus. When a master receives the token, it starts a timer to measure the real token rotation time. When it receives the token the next time, it is allowed to exchange messages with other masters and slaves only while the timer has not reached the T_{TR}. As soon as this happens, the token must be handed over to the next master in the list. If a master receives the token after the timer is expired, it may send just one high-priority message before having to pass on the token again. The time available for each master therefore also depends on the amount of time all the other masters in the network hold the token, and it might vary from round to round if aperiodic data have to be sent or retransmissions of individual frames are necessary. Periodicity of individual process variables is therefore less exact as with polling, and jitter may be larger. Still, the total time of a token rotation cycle can be upper bounded so as to support real-time requirements. In any case, careful selection of the T_{TR} value must be done based on an a priori analysis of the expected network traffic and is essential to ensure optimal use of the available communication bandwidth.

The implicit form of TP is used by P-NET. This fieldbus uses a hybrid approach together with polling. In order to keep the overhead of the volume of information that needs to be transferred on the bus to the absolute minimum, the token is simulated by two counters, included in every master, and is not actually passed around the bus (Figure 1.16). The *Idle Bus Bit Period Counter* (IC) is incremented every bit period as long as the bus is free and is reset as soon as some node sends data. The *AC* is incremented whenever the IC reaches the values 40, 50, 60 … 360. If the AC status matches the address of a master, this means that the master has the token and is therefore able to access the bus for one single request–response data exchange with some other node. To ensure that the token is not passed on, there are maximum values for the time a master can wait before sending a request to a slave and also for the time a slave may take until it starts its response. If the master has no data to send, it remains silent, and after a further 10 increments, the token is handed over. Once the AC reaches the number of masters contained within the bus system, up to a maximum of 32, it is reset to the value 1. To avoid loss of synchronization due to the freely running system clocks in the nodes during long idle periods, a master has to send at least a synchronization message if the IC value reaches 360. Since the data exchange per master is strictly limited, the token rotation time is upper bounded, even though it is likely not constant.

A similar implicit method is employed by ControlNet. A node possessing the token may transmit one single data frame possibly containing several link layer packets. All nodes monitor the source address of this frame, and at the end of the frame, increment this address in an internal implicit token register. If the value of this register matches the address of a node, it possesses the token and may send its own data. If there are no data to be sent, the node must send an empty frame so that the token can be passed on. This mechanism is used both for scheduled and for unscheduled traffic with two different tokens.

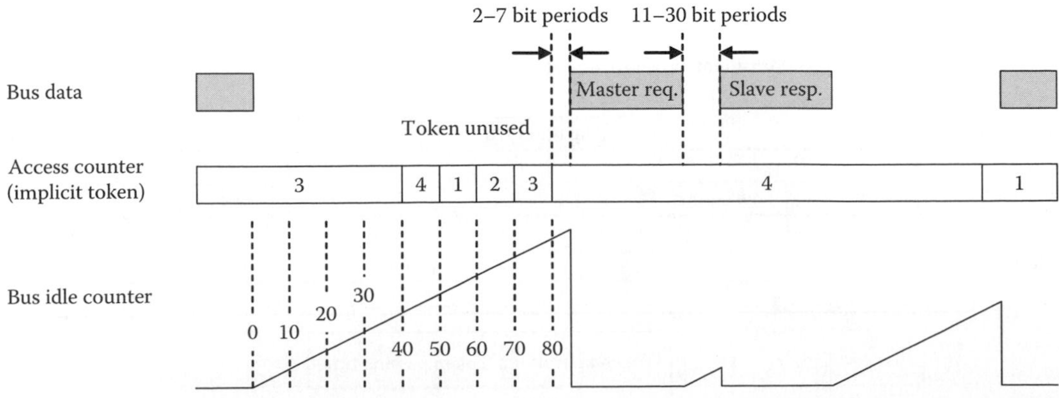

FIGURE 1.16 P-NET polling mechanism.

While every node with scheduled traffic will have a guaranteed bus access every network cycle, the access right for unscheduled traffic changes in a round-robin fashion, that is, the first node to get bus access for unscheduled traffic is rotated every cycle.

1.5.4.3 Time-Slot-Based Access

In time-slot-based methods, the available transmission time on the medium is divided into distinct slots that are assigned to the individual nodes. During its time slot, a node may then access the medium. The difference to cyclic polling strategies—which in fact also come down to partitioning the polling cycle into time windows—is that the nodes are not requested by a central station to send their data; they can do it by themselves. Therefore, time-slot-based methods are mostly also called TDMA (in a narrower sense). The slots need not have the same size; they may be different. Likewise, they may be dynamically distributed to the nodes according to the amount of data they wish to send. This way, aperiodic traffic can be incorporated, whereas TDMA in general favors periodic traffic. A strictly equal distribution of the slots is also called synchronous TDMA and is particularly known from telecommunications technology, where each end device is assigned a time window in which, for example, the correspondingly digitized voice data can be transmitted. There, such systems are referred to as *synchronous bus systems*. The case where time slots are dynamically distributed is called asynchronous TDMA. Contrary to the synchronous version where the position of the time slot within the cycle implicitly identifies the sending nodes, some address information must be sent in the case of asynchronous TDMA.

The fact that nodes may access the medium without prior explicit request does not mean that TDMA methods are generally multimaster or peer-to-peer techniques. Whether they are depends on the way the synchronization of the time slots is achieved. Evidently, if no handshake procedures are implemented for data transfer, all nodes must be properly synchronized so that only one node at a time accesses the bus. This can be done in a centralized or decentralized way.

The simpler version is to have one dedicated bus master sending some sort of synchronization message at the start of the cycle. After that, the nodes exchange their data in their previously assigned time slots. Such a strategy is used, for example, by SERCOS, which is therefore regarded as a master–slave fieldbus (Figure 1.17). In this particular case, the time slots for the slaves are all equal in length, and the master has a larger time window to send its data to the slaves afterward. A similar method is also used by TTP/A. Here, a bus master (apart from having the possibility for simple polling) sends a synchronization message in the first time slot of a round to start the cyclic data transfer from all other nodes. In byteflight, a node (the sync master) sends a periodic synchronization signal to start the data exchange round (in byteflight called slot), and the others initialize slot counters to determine the position of

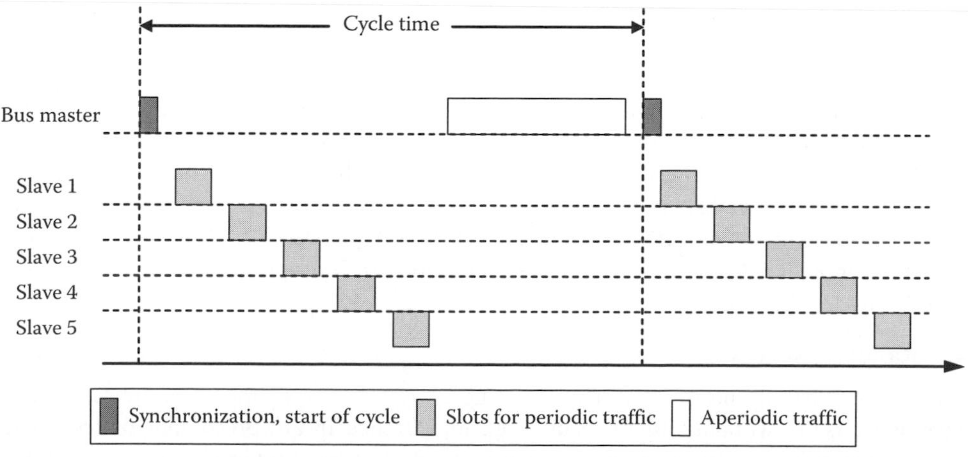

FIGURE 1.17 Centralized TDMA scheme (example SERCOS).

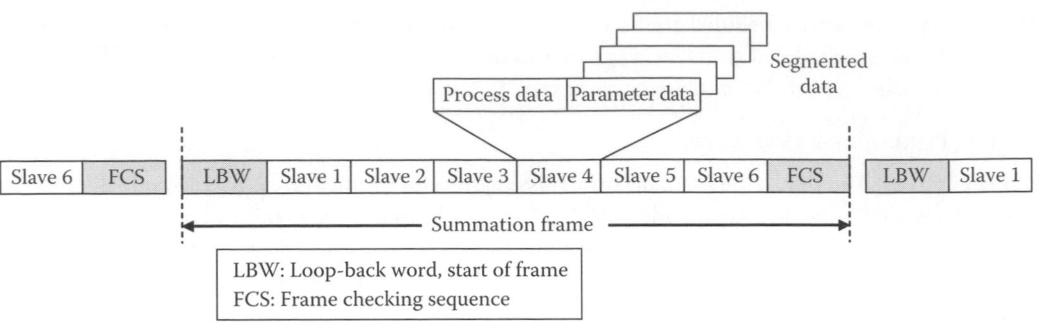

FIGURE 1.18 Summation frame of INTERBUS.

their predefined microslot. If a node has no data to send, the respective microslot is shortened. In this respect, the bus access mechanism resembles an implicit TP method. The spare time of unused microslots becomes available for aperiodic or lower-priority traffic wards the end of the round.

Related to the centralized TDMA approaches is the summation frame protocol of INTERBUS (Figure 1.18). As noted in the previous section, the system is based on a ring structure and acts like a large shift register. Each slave has a predefined slot in the data frame that is selected in accordance with its position in the ring. Output data from the master to the slave and input data from the slave back are exchanged in the respective slot as the frame is shifted through the ring. Thus, one data frame is sufficient for the cyclic updating of all end devices. A clever arrangement of buffers furthermore ensures that despite the ring structure with its inherent delays, all I/Os at the slaves are updated at the same time, so that synchronous operation with respect to the process variables is guaranteed. Aperiodic traffic can be introduced in this rather rigid scheme through a so-called parameter channel. Larger amounts of data are transferred in small packets in this channel without affecting the exchange of cyclic process data.

The second method of managing the time slots is a decentralized one. In this case, there is no dedicated node to initiate the cycle; rather, all devices synchronize themselves either by explicit clock synchronization mechanisms or by a set of timers that settle bus operation down to a stable steady state. Fieldbus systems relying on such distributed mechanisms have a high degree of fault tolerance because there is no single point of failure as in centralized approaches. Therefore, they are well suited for safety-critical applications (such as TTP/C, FlexRay, or ARINC 629), provided that they are designed to have real-time capabilities. Nevertheless, the underlying algorithms are relatively complex, which is why such systems have attracted a lot of scientific interest. The distributed nature of bus access requires proper error containment mechanisms to avoid faulty nodes (babbling idiots) from blocking the medium and jeopardizing real-time behavior.

The real-time properties of time-slot mechanisms also led to the introduction of such approaches in fieldbus systems that originally use different access control methods. As a prominent example, CAN was enhanced in this way by superimposing TDMA structures like in time-triggered CAN (TT-CAN) or flexible time-triggered CAN (FTT-CAN). Last but not least, it should be noted that the boundaries between TDMA and other MAC techniques are sometimes blurred. For instance, the access control method of ControlNet is called Concurrent Time Domain Multiple Access, although it is rather an implicit TP strategy. The Flexible Time Domain Multiple Access of byteflight also has some similarity to TP.

1.5.4.4 Random Access

Random access basically means that a network node tries to access the communication medium whenever it wants to—without limitations imposed, for instance, by any precomputed access schedule. This principle is called carrier sense multiple access (CSMA); it was first implemented in the ALOHA network in 1970 and since then has been modified in various ways. Obviously, it is perfectly suited for spontaneous peer-to-peer communication. Its conceptual simplicity comes at the expense of a severe drawback: collisions inevitably occur when several nodes try to access the network at the same time, even if they first check if the line is idle. The variants of CSMA therefore have one common goal—to deal with these collisions in an effective way without wasting too much communication bandwidth and thereby avoiding excessive communication delays.

The best known CSMA variant is CSMA-CD (collision detection) used in Ethernet, more precisely in its original version with shared medium. Here, collisions are immediately detected by the sending nodes that monitor the bus while sending. After CD, the nodes abort the data transfer and wait for a random time before trying again. In fieldbus systems, this variant is not very common because despite its stochastic component, there is a high probability that collisions remain if the source data rates are too high. In practice, the effectively attainable throughput is limited to values well below 50% of the maximum data rate, which was always deemed insufficient for automation networks. It should be noted, though, that things changed with the introduction of switched Ethernet.

A variant where the backoff time (the time a node waits before a retry) is not just random, but adaptable is called predictive p-persistent CSMA and is used in LonWorks. Here, p denotes the probability that the node will try again in a certain time interval after the collision. This probability can be adapted by each node based on an estimation of its backlog and the monitored network load. In high-load conditions, the nodes reduce their probability for starting a medium access, which in turn reduces the probability for collisions. LonWorks additionally has a number of high-priority *time slots* with differing probabilities for medium access, so that messages with a higher priority still have a better chance to be sent without long delays.

The most widely used CSMA strategy in the fieldbus world, however, uses asymmetric symbols for coding the bits on the bus line, so that when two different bits are sent at a time, one of them wins over the other. In this case, there is no actual collision, bus access is nondestructive, and this is why this strategy is called CSMA-CA (collision avoidance), sometimes also called CSMA-BA (bitwise arbitration). The first fieldbus to use this method was CAN, and this example will be used in the sequel to explain the idea in more detail.

The bus line is designed as an open collector bus so that the low level is *dominant* and the high level remains *recessive*. This means that a "1" sent from an end device can be overwritten by a "0." CAN uses message-based addressing, and after sending a synchronization bit, the nodes write the identifier of their message to the bus bit by bit and at the same time observe the current bus level. If the bits sent and read back differ, arbitration is lost, and the node stops sending. In the end, the remaining node is the one whose message has the lowest identifier (Figure 1.19). The BA method brings with it a major disadvantage: The propagation time of the signals on the line must be short compared with the bit time to yield quasi-simultaneity for all nodes. With the highest bit rate of 1 Mbit/s, this means a maximum bus length of only 40 m. In motor vehicles (the original application field of CAN), this restriction is of lesser importance, but in industrial automation technology, it can lead to a reduction in the data transfer rate.

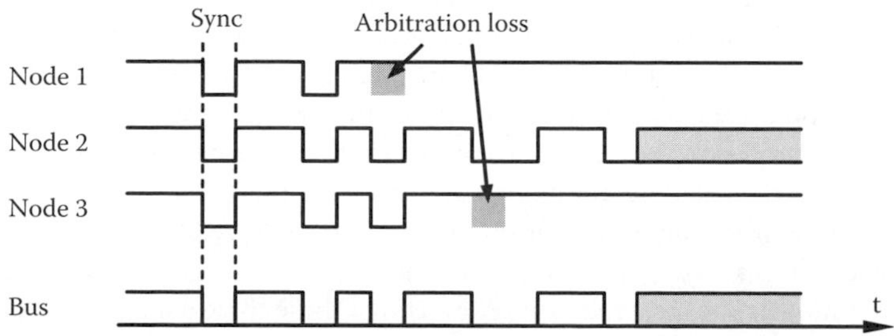

FIGURE1.19 Bitwise arbitration method with CAN.

As with decentralized TDMA, faulty nodes may infinitely block the bus. To prevent this situation, CAN nodes contain error counters that are incremented whenever the node detects transmit or receive errors and are decremented after successful transmit or receive procedures. The counter status reflects the reliability of the node and determines if the node may fully participate in bus traffic or only with certain restrictions. In the extreme case, it is completely excluded from the bus.

BA was so successful that it was used in several other fieldbus systems in similar form. Examples are building automation networks like EIB, BATIBUS, or EHS, as well as in other automotive networks like VAN and FlexRay (for aperiodic traffic). But CAN was also used as a basis for further extension. CAN as such originally only defined layers 1 and 2 in the OSI model. Although this was sufficient for the exchange of short messages within a closed network, it was insufficient for automation technology. For this reason, the CAN-in-Automation user group defined the CAN application layer and then the CANopen protocol. Other protocols for automation technology, also based on CAN, are DeviceNet and SDS. The CAN Kingdom protocol has been specially developed for machine controls and safety-critical applications. These higher-level protocols offer the possibility of exchanging larger volumes of data and of synchronizing end devices. Network management functions solve the problems of node configuration and identifier specification.

CSMA-CA has one inherent problem: Even in the absence of babbling idiots, the highest priority object can practically block the bus, and messages with lower priorities seldom get through. Therefore, quality-of-service guarantees can only be given in a strict sense for the message with the highest priority if no additional measures are taken. A number of solutions have been proposed to overcome this problem. One of them is to limit the frequency of the messages such that after a successful transmission, a node has to wait a certain amount of time before being able to send again. Another possibility is to introduce mechanisms in the upper protocol layers to exchange identifiers cyclically within specific priority classes in addition to restrictions in access frequencies. This is a network management function used, for example, in CANopen. The third method of improving fairness (and thus real-time capabilities) is to superimpose time-slot mechanisms at least for some message and traffic classes. This strategy is being employed by TT-CAN and FTT-CAN.

1.5.5 Communication Paradigms

It was stated before that there are many diverse types of traffic to be considered in fieldbus systems, and there are many different services that a fieldbus should provide to the user application. Apart from basic functions like reading and writing variables, management of objects is required. This includes updates of device configurations, starting and stopping tasks running on nodes, up- or downloading of program, triggering and synchronization of events, establishment and termination of connections between nodes, and access control to node data, to name just of few. In general terms, fieldbus services need to handle objects by identifying them, the actions to be performed on them, as well as the appropriate parameters for the actions. These services must be defined and provided by the upper layers of the protocol stack, in most cases by the application layer.

Communication and cooperation between different application layers and applications in a fieldbus system is based on a basic set of two or three different models [36]. These models again represent different philosophies of the type of information that is exchanged between two or more entities. One approach is to build the cooperation on actions or functions into which a more complex process is decomposed. The responsibility for interpretation of the information is largely concentrated on the sender's side. This is the philosophy behind the client–server model. The other approach is a data-oriented one. Here, not actions but data are exchanged, and the responsibility for their interpretation is with the receiver (which then might take according actions). This is the idea behind the publisher–subscriber model and the producer–consumer model. The most relevant properties of these three are summed up in Table 1.1.

The overview shows that communication paradigms need to be supported by the medium access mechanism of the data link layer to achieve optimal performance.

TABLE 1.1 Properties of Communication Paradigms

	Client–Server Model	Producer–Consumer Model	Publisher–Subscriber Model
Communication relation	Peer-to-peer	Broadcast	Multicast
Communication type	Connection-oriented	Connectionless	Connectionless
Master–slave relation	Monomaster, multimaster	Multimaster	Multimaster
Communication service	Confirmed, unconfirmed, acknowledged	Unconfirmed, acknowledged	Unconfirmed, acknowledged
Application classes	Parameter transfer, cyclic communication	Event notification, alarms, error, synchronization	State changes, event-oriented signal sources (e.g., switches)

1.5.5.1 Client–Server Model

In the client–server paradigm, two entities cooperate in the information exchange. The entity providing a service or data is called server, and the one requesting the service is called client. In general, the server may become active only upon request from the client, which means that this model is better suited for state-based data traffic handled in some scheduled manner. Events, that is, spontaneous traffic, can be handled only if they occur at the client (which may solely initiate communication). If they occur at the server, they cannot be handled in a spontaneous way at all; the server has to postpone the transmission until it receives an appropriate request from the client.

Client–server communication is based on confirmed services with appropriate service primitives (request, indication, response, and confirm) as defined in the OSI model (Figure 1.20a). This is the classical implementation used by many fieldbus systems that derive their application layer protocols essentially from the MMS standard. Examples are PROFIBUS-FMS (DP and PA in reduced form as well), WorldFIP, INTERBUS, or P-NET. The MMS model proposed comprehensive services that are usually only partly implemented in the fieldbus application layers. For the example of PROFIBUS-FMS (in similar form, they exist also in Foundation Fieldbus and WorldFIP), the services comprise the following groups: variable access (read, write), transmission of events, execution of programs, domain management (up- or download of large amounts of data), context management (administration of communication relationships), and object dictionary management (administration of the list of all objects in a device). Depending on the actual service requested, the semantics of the information exchanged obviously differ.

In some cases, and deviating from the standard model described before, a client–server communication can also be built from two unconfirmed services (Figure 1.20b). This means that the response generated by the server is not related to the request by means of the underlying protocols (i.e., it does not generate a confirmation on the local application layer). Rather, the client application has to keep track of the request and needs to relate the indication generated by the response on its own. This mechanism reduces the implementation efforts on the protocol side at the expense of the application complexity. Another—nonstandard—variant of the client–server model may have several responses from the server answering a single request [36]. Such an approach can be beneficial if the service execution on the

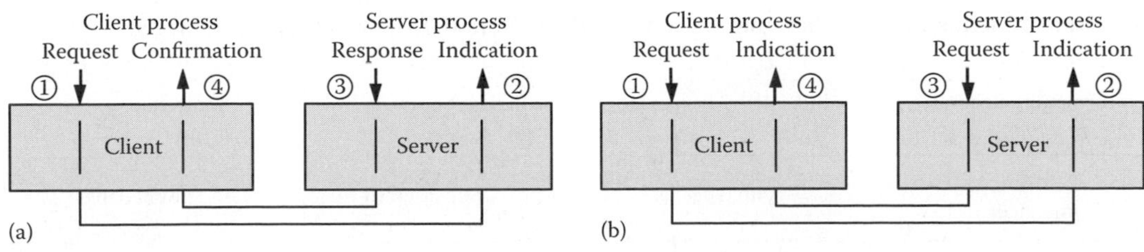

FIGURE 1.20 Communication services used in the client–server model. (a) Normal and (b) Based on unconfirmed services.

server side takes a long time, so that partial results are reasonable or at least indications about the status of the execution. This may help to circumvent problems with, for example, hard-coded timeouts in legacy client applications by means of some keep-alive signal. Somewhat related to this and resembling a publisher–subscriber-type operation is a multiresponse from a server following one single request from the client. This can be useful if strictly periodic data transmissions from the server are required. Contrary to a cyclic client–server polling strategy, which always is affected by network and software delay, the server can control the sampling and transmission period of the data to be sent in a better and more accurate way.

As far as the interrelation of the client–server model and MAC strategies is concerned, it should be noted that basically, a client–server-type communication can be implemented in both mono- and multimaster systems. In the latter cases (which may be based on CSMA, TDMA, or TP), every master can take on the role of a client, whereas in mono-master systems (polling-based or centralized TDMA), this position is reserved for the bus master. Consequently, the client–server paradigm is used mainly for master–slave systems (as represented by PROFIBUS, ASi, MODBUS, P-NET, BITbus, and INTERBUS) and for reliable data transfer on application level (e.g., for parameterization data, file transfer, network, and device management). In particular for management functions, the client–server model is widely used also within fieldbus systems that organize their regular data transfer according to the publisher–subscriber model, such as WorldFIP, EIB, CANopen, DeviceNet, ControlNet, or LonWorks.

1.5.5.2 Publisher–Subscriber Model

The basic idea of the publisher–subscriber model is that certain nodes (the publishers) produce information that they post into the network. The subscribers are typically groups of nodes that listen to information sources. Relating publishers and subscribers can be done at runtime. The producer–consumer model uses very similar mechanisms; the only difference is that broadcast services are being employed instead of multicast communication as in the case of the publisher–subscriber model. However, this distinction is rather subtle and not very relevant in practice.

Depending on how the information exchange is initiated, two different subtypes of the publisher–subscriber paradigm can be distinguished [16]. In the pull model, the publishing action is triggered by a centralized publishing manager (Figure 1.21). Upon receiving the request from the manager, the publisher broadcasts the respective response to the network. Filtering of the message, which determines if the service is a broadcast or multicast, is done locally by the subscribers who listen to the network for message identifiers they subscribe to. It should be noted that the subscriber group may as well consist of just one node, such that the communication relation is effectively a one-to-one relationship.

In terms of communication services, the pull publisher–subscriber model uses a confirmed request/response service for the interaction between manager and publisher. The peculiarity of this service however is that while the request is unicast (i.e., has only one recipient), the response from the publisher is multicast to the publishing manager *and* the subscribers and already contains the information to be published. It is the task of the underlying protocol layers to ensure that the appropriate addressing scheme is used to transform the unicast into a multicast message. For the subscribers, the communication then may look like an unconfirmed service (i.e., they receive only an indication that a subscribed object has been transmitted) although it is actually a confirmed one from the viewpoint of the manager.

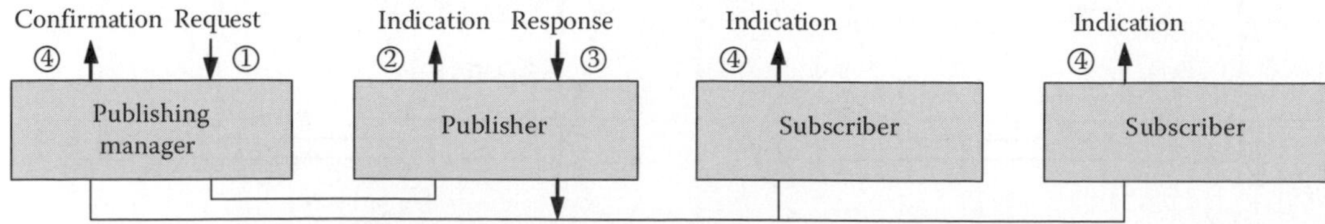

FIGURE 1.21 Pull-type publisher–subscriber model.

Still, there might be subtle differences in the way the request itself is handled. On the one hand, the manager might know the identifier of a variable and requests publication of this specific variable without caring about producer. This method is being used, for example, in WorldFIP. On the other hand, the manager might know the device address of the publisher and send the request directly to this node. This is typically done in fieldbus systems that normally rely on node addressing for data exchange and thus a client–server model. In such fieldbus systems, a publisher–subscriber transaction is the exception rather than the rule, as in PROFIBUS-DP V2. In master–slave systems, this type of communication—although not strictly publisher–subscriber—may also be used to implement a direct communication possibility between slaves under the direct control of the master. This shows that an application layer service for reading data based on the pull model comes very close to a read service implemented according to the client–server model. The essential difference is that the response is directed to a group of receivers instead of only one—the manager.

The second subtype of the publisher–subscriber paradigm is the push model (Figure 1.22) in which the publishers become active by themselves without prior request from a centralized station and distribute their information to the subscribers when they consider it necessary, for example, triggered by the expiration of a timer (as in TDMA systems) or by an external event. In terms of communication services, the publishing of information is implemented by means of unconfirmed (or locally confirmed) services on the publisher's side. Like in the pull variant, the subscribers receive an indication that new data have arrived and do not answer or acknowledge receipt of the message.

A necessary step in either of the two variants is the subscription of the subscribers with the publisher. This is a typical network management of configuration action that can be accomplished statically during system setup and commissioning or dynamically during runtime. As the underlying communication mechanism of the publisher–subscriber model is a multicast to a defined group of nodes, it must be ensured that a node subscribing to a given message or object joins the correct communication group. This is therefore mainly done using client–server-type communication based on confirmed services.

Processes with mostly event based communication can get along very well with publisher–subscriber- or producer–consumer-type communication systems. Depending on the fundamental layer-2 communication methods, real-time requirements may be more (in the case of TDMA methods or centralized polling) or less met (with pure random access methods). The obvious advantage is that all connected devices have direct access to the entire set of information since the broadcasting is based on identification of messages rather than nodes. Reaction times on events can be very short due to the absence of slow polling or token cycles. Generally, publisher–subscriber-type systems (or subsystems) are mostly multimaster systems because every information source (producer) must have the possibility to access the bus. The selection of relevant communication relationships is solely based on message filtering at the subscriber's side. Such filter tables are typically defined during the planning phase of an installation. It is no wonder that this communication model is being widely used in fieldbus systems, as, for instance, WorldFIP, Foundation Fieldbus, CAN, CANopen, DeviceNet, ControlNet, LIN, EIB, or LonWorks.

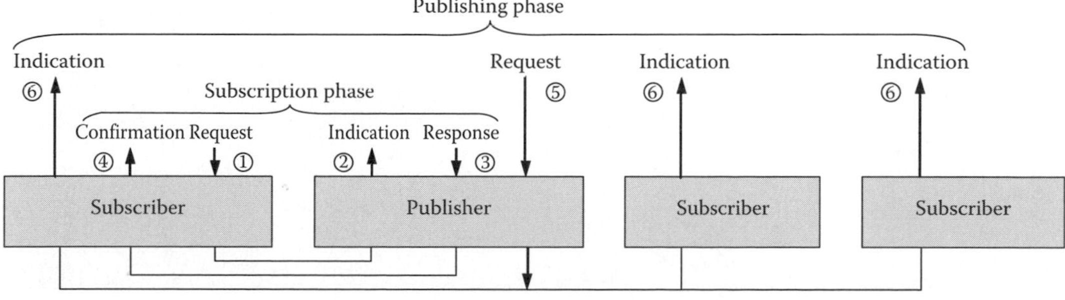

FIGURE 1.22 Push-type publisher–subscriber model.

Still, there is an inherent problem. As both paradigms are message based and therefore connectionless on the application layer, they are not suited for the transmission of sensitive, nonrepetitive data such as parameter and configuration values or commands. Connectionless mechanisms can inform the respective nodes about communication errors on layer 2, but not about errors on application layer. Particularly for fieldbus systems devised for safety-critical applications (such as TTP), additional mechanisms have been developed to enable atomic broad- and multicasts. Atomic in this context means that the communication is successful only if all subscribers receive the distributed information correctly. Only in this case will actions associated with the transaction be executed; otherwise, they are cancelled.

1.5.6 Above the OSI Layers: Interoperability and Profiles

A key point for the acceptance of fieldbus systems was their openness. The availability of publicly available specifications and finally standards made it possible to set up the so-called multivendor systems, where components developed by different manufacturers work together to achieve a common functionality. This possibility was the evidence for interoperability and still is an important argument in fieldbus marketing. The standardization of fieldbuses was originally thought to be sufficient for interoperable systems, but reality quickly showed that it was not. Standards (interoperability standards, to be precise) define the rules that the system components must comply with in order to be able to work with other components, but they often leave room for interpretation, and actual implementations may vary. Certification of the devices is a suitable way to reduce the problems, but by no means a guarantee. Another reason for troubles is that the semantics of data objects are not precisely defined. In fact, there are various degrees of interoperability, as defined by the IEC TC65 SC65C WG7 within the scope of the IEC *Interoperability Definitions* (Figure 1.23) [37]. These definitions relate to application functionalities that can be achieved, that is, the degree to which devices and applications can actually work together.

Incompatible: The two communication partners have no common ground, and their connection, if physically possible, could result in damage. The connectors, voltage levels, and modulation methods can be different.

Compatible: The most fundamental requirement is that both devices use the same communication protocol. They can be connected and may exchange data packets. Still, compliance

Interoperability level / Necessary agreements	Incompatible	Compatible	Inter-connectable	Inter-workable	Inter-operable	Inter-changeable
Temporal behavior						
Functional behavior						
Data semantics						
Definition of the data						
Access to the data						
Protocol (layer 1–7)						

FIGURE 1.23 Application functionalities within the interoperability definitions.

with the protocol (as verified, e.g., by conformity tests according to ISO 9646) alone does not guarantee meaningful cooperation between the applications as they cannot make sense of the received packets.

Interconnectable: The partners have the same methods of accessing the data, and they mutually understand the addressing methods used by the partner. This is important because the exact way in which data are made available in the network is not defined by the protocol layers; they just offer a number of services for data transmission. Being interconnectable, the partners can recognize whether received messages were addressed to them, and they are capable of extracting the data (or encoding them in the opposite direction). Furthermore, they use the application layer services in the same way and have a common understanding of data types used for their applications.

Interworkable: Applications share the same definitions of data and use the same variable types. Therefore, they can correctly exchange data using a common way of coding. As an example, a sensor can send its value in long_int format, and a controller can correctly receive and process this value. Usually, the development tools offer support for this level. If the tools are designed for distributed systems, errors violating interworkability can already be detected in an early development phase.

Interoperable: On this level, the semantics of the data play an important role. Variables take on physical relevance; they do not have just a data type but also a physical unit that needs to be known to the partners. Furthermore, accuracy ranges, meaning, and purpose of variables are defined. If a device produces a larger data set, it is known how the individual data can be distinguished. Interoperable nodes can be used immediately without additional description, their network interfaces are unique, and also the functional behaviors match. It is precisely defined what happens to the data that are sent across the network and which actions or sequence of actions they trigger.

Interchangeable: Even the temporal behavior of two nodes is the same with respect to the needs of the application. Typically, such requirements are maximum reaction times to events or commands. If such constraints are not equal, the overall behavior of the entire network can fundamentally change even if a device replacing another is functionally equivalent.

The problem of interoperability has been disregarded in many cases until recently. In fact, it is not a problem of the fieldbus itself, but of the application. Consequently, it must be tackled beyond the ISO/OSI model. The definition of appropriate *profiles* addresses this problem. A profile defines which variables carry which data, how they are coded, what physical units they have, etc. By virtue of this profile, it is possible to subject devices, which are claimed to satisfy this profile, to a corresponding conformity test. The compatibility of a device is generally also tested when used within a multivendor system.

The creation of profiles originated from the recognition that the definition of the protocol layers alone is not sufficient to allow for the implementation of interoperable products, because there are simply too many degrees of freedom. Therefore, profiles limit the top-level functionality and define specialized subsets for particular application areas [38,39]. Likewise, they specify communication objects, data types, and their encoding. So they can be seen as an additional layer on top of the ISO/OSI model, which is why they have also been called *Layer 8* or *User Layer*. One thing to be kept in mind is that nodes using them literally form islands on a fieldbus, which contradicts the philosophy of an integrated decentralized system. Different profiles may coexist on one fieldbus, but a communication between the device groups is normally very limited or impossible at all.

The concept of profiles has many names. In MMS, they are termed Companion Standards. In P-NET, they are equivalent to the Channel concept. Function blocks in Foundation Fieldbus essentially pursue the same idea. In LonWorks, there are the Standard Network Variable Types, and in EIB, interoperability is achieved by means of the EIB Interworking Standards.

From a systematic viewpoint, profiles can be distinguished into communication, device, and branch. A bus-specific *communication profile* defines the mapping of communication objects onto the services offered by the fieldbus. A *branch profile* specifies common definitions within an application area concerning terms, data types, their coding, and physical meaning. *Device profiles* finally build on communication and branch profiles and describe functionality, interfaces, and in general the behavior of entire device classes such as electric drives, hydraulic valves, or simple sensors and actuators.

The work of defining profiles is scattered among different groups. Communication profiles are usually in the hands of fieldbus user groups. They can provide the in-depth know-how of the manufacturers, which is indispensable for bus-specific definitions. Device and branch profiles are increasingly a topic for independent user groups. For them, the fieldbus is just a means to an end—the efficient communication between devices. What counts more in this respect is the finding and modeling of uniform device structures and parameters for a specific application. This forms the basis for a mapping to a communication system that is generic within a given application context.

The ultimate goal is the definition of fieldbus-independent device profiles. This is an attempt to overcome on a high level the still overwhelming variety of systems. Finally, such profiles are also expected to facilitate the employment of fieldbus systems by the end user who normally is only concerned about the overall functionality of a particular plant—and not about the question which fieldbus to use. The methods used to define data types, indices, default values, coding and meanings, identification data, and device behavior are based on functional abstractions and general modeling techniques [40].

1.5.7 Fieldbus Management

Owing to the different capabilities and application areas of fieldbus systems, the management of a fieldbus shows varying complexity, and its solutions are more or less convenient for the user. It has already been stated that the various fieldbuses offer a wide range of network management services with grossly varying levels of sophistication. Apart from the functional boundary conditions given by the protocols, fieldbus management always strongly relies on the tool support provided by the manufacturers. This significantly adds to inhomogeneity of the fieldbus world in that entirely different control concepts, user interfaces, and implementation platforms are being used. Furthermore, a strict division between communication and application aspects of fieldbus management is usually not drawn.

Typical *communication-related* management functions are network parameter settings like address information, data rate, or timing parameters. These functions are rather low level and implicitly part of all fieldbus protocols. The user can access them via software tools mostly supplied by the device vendor. *Application-related* management functions concern the definition of communication relations, system-wide timing parameters (such as cycle times), priorities, or synchronization. The mechanisms and services offered by the fieldbus systems to support these functions are very diverse and should be integrated in the management framework for the application itself (e.g., the control system using the fieldbus).

As a matter of fact, a common management approach is still not available despite all interoperability achievements, and vendor-specific solutions are preferred. From the user's point of view (which includes not only the end users, but also system integrators), this entails significantly increased costs for the build-up and maintenance of know-how because they must become acquainted with an unmanageable variety of solutions and tools. This situation actually revives one of the big acceptance problems that fieldbus systems originally had among the community of users: the missing interoperability. Communication interoperability (as ensured by the fieldbus standards) is a necessary but not sufficient precondition. For the user, *handling interoperability* of devices from different vendors is equally important [41]. What is needed are harmonized concepts for configuration and management tools. More than that, tools should also be consistent for different aspects of the life cycle of an installation, like planning, configuration, commissioning, test, and diagnosis or maintenance. An increasingly important topic on the overall enterprise level is asset management, which also includes field-level data [42]. Such tools require functionality-oriented abstract views. A major disadvantage of today's tool variety is that they operate in many cases on incompatible data bases, which hampers system integration and is likely to produce consistency problems. More advanced concepts build on unified data sets that present consistent views to the individual tools with well-defined interfaces. The data structures are nevertheless still specific for each fieldbus.

With the increasing importance of LAN and Internet technologies in automation, new approaches for fieldbus management appeared that may be apt to introduce at least a common view at various fieldbuses. All these concepts aim at integrating fieldbus management into existing management applications of the higher-level network, which is nowadays typically IP based. One commonly employed high-level network management protocol is SNMP, which can be used to access also fieldbus data points [43,44]. Another approach involves the use of directory services [45]. These two solutions permit the inclusion of a large number of devices in specialized network management frameworks. An alternative that has become very popular is the use of web technology, specifically HTTP tunneled over the fieldbus, to control device parameters. This trend is supported by the increasing availability of embedded web servers and the use of XML as device description language [46]. The appealing feature of this solution is that no special tools are required and a standard web browser is sufficient. However, web pages are less suitable for the management of complete networks and rather limited to single-device management. Nevertheless, this approach is meanwhile pursued by many manufacturers.

In the context of management and operation frameworks, the unified description of device and system properties becomes of eminent importance. To this end, device description languages were introduced. Over the years, several mutually incompatible languages and dialects were developed [47]. Originally, they were tailored to individual fieldbus systems (e.g., HART DDL, PROFIBUS GSD, CANopen EDS, and FF DDL) and laid the foundation for user-friendly configuration and engineering tools. In recent years, the diversity of description languages is being addressed by the increased usage of universal languages like XML [48], which is also the basis for the electronic device description language standardized in IEC 61804.

For a fieldbus-independent access to the field devices and their data (not necessarily covering the entire life cycle), several solutions have been proposed. They mostly rely on a sort of middleware abstraction layer using object-oriented models. Examples are OPC or Java. Such platforms can ultimately be extended through the definition of suitable application frameworks that permit the embedding of generic or proprietary software components in a unified environment spanning all phases of the life cycle. A popular approach is, for example, the Field Device Tool [47,49]. Typical for all approaches is that they start from a comprehensive, consistent description of the devices and their parameters and filter this data base depending on the life cycle phase. The user therefore gets access only to a specific subset coinciding with his role in the life cycle. The latest trend is to use web services as actual middleware because of their platform independence [50,51] and workflows to describe processes [52]. To gain even more flexibility, ontologies are being used to introduce semantic information about processes [53], and all together leads to the booming field of service-oriented architectures [54–56]. Special attention deserves OPC, which is currently the most widely used solution for data exchange between the field-level and higher-level applications, for example, SCADA systems [57]. Also inside the field level, OPC as interoperable data exchange middleware is a de facto standard [58]. In its most recent version OPC UA (Unified Architecture), web services are used as one technological basis.

1.6 Industrial Ethernet: The New Fieldbus

As stated before, Ethernet has become increasingly popular in automation. And like in the early days of fieldbus systems, this boom is driven mainly by the industry—on an academic level, the use of Ethernet had been discussed decades ago. At that time, Ethernet was considered inappropriate because of its lack of real-time capabilities. With the introduction of switched Ethernet and certain modifications of the protocol, however, these problems have been alleviated. And even if there are still doubts about the predictability of Ethernet, its penetration into the real-time domain will influence the use of fieldbus-based devices and most likely restrict the future use of fieldbus concepts [59]. Today, Ethernet already takes the place of mid-level fieldbus systems, for example, for the connection of PLCs.

Using Ethernet on the field-level one first of all has to overcome the problem of the inherent lack of determinism. For Ethernet as shared medium, and if no hard real-time behavior is requested, various types of traffic smoothing techniques have been proposed [60,61], where real-time packets are given priority over non-real-time ones. The goal is to eliminate contention within each local node and to shape non-real-time traffic so as to reduce the chance for collision with real-time packets from the other nodes. The introduction of switched Ethernet further alleviated the problem [62]. Much research work consequently focused on how to reduce the queuing delays inside the switches [63,64] and on traffic smoothing methods for switched Ethernet [65], even if also switched Ethernet per se is not fully deterministic and leaves room for further research [66].

One of the main arguments used to promote Ethernet on the field level is that because it is the same network technology as in the office world, a straightforward integration is possible, that is, both automation and office domain can in principle be connected to one single enterprise network. A quick look at reality, however, shows that things are different. Ethernet per se is but a solution for the two lower OSI layers, and as fieldbus history already showed, this is not sufficient. Even if the commonly used Internet protocol suite with TCP/UDP/IP is taken into account, only the lower four layers are covered. Consequently, there are several possibilities to get Ethernet or Internet technologies into the domain currently occupied by fieldbus systems:

- Tunneling of a fieldbus protocol over UDP/TCP/IP
- Tunneling of TCP/IP over an existing fieldbus
- Definition of new real-time-enabled protocols
- Limitation of the free medium access in standard Ethernet
- Modifications of the Ethernet hardware to achieve better real-time capabilities

All of these possibilities are actually used in practice (Figure 1.24). In the beginning, all research work carefully avoided any concepts violating the Ethernet standards. Compatibility and conformity were the primary goals. Especially those approaches already contained in the first version of the IEC 61158

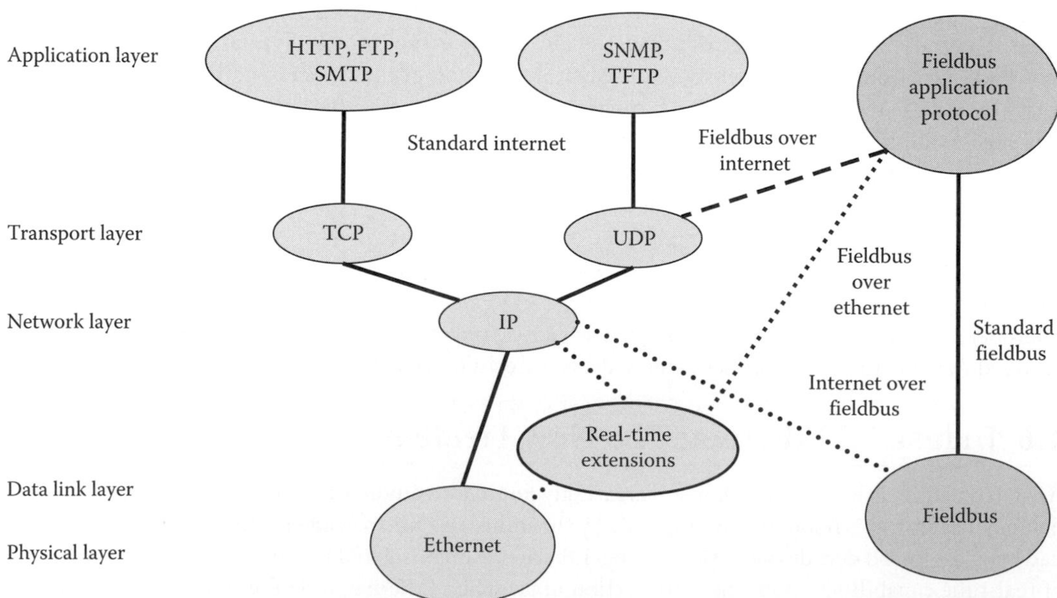

FIGURE 1.24 Structures of Ethernet and Fieldbus combinations.

standard employ existing fieldbus application layer protocols on top of IP-based transport mechanisms (TCP or UDP, respectively, depending on the services needed) that replace the lower fieldbus layers [31]. The following four examples pursue this approach:

1. The high-speed Ethernet variant of Foundation Fieldbus is an application of the existing Fieldbus Foundation's H1 protocol wrapped in UDP/IP packets [67].
2. Ethernet/IP (IP in this case standing for Industrial Protocol) uses the Control and Information Protocol (CIP) already known from ControlNet and DeviceNet [68]. This application layer protocol is sent over TCP or UDP, depending on whether configuration or process data have to be transmitted.
3. Modbus/TCP is based on standard Modbus frames encapsulated in TCP frames. For more stringent real-time requirements, a real-time variant of the publisher–subscriber model was developed by IDA Group (Interface for Distributed Automation, now merged with Modbus Organization), which builds on UDP.
4. P NET on IP defines a way to wrap P-NET messages in UDP packets.

These industrial Ethernet solutions build on Ethernet in its original form, that is, they use the physical and data link layer of ISO/IEC 8802-3 without any modifications. Furthermore, they assume that Ethernet is low loaded or fast Ethernet switching technology is used, in order to get a predictable performance. Switching technology does eliminate collisions, but delays inside the switches and lost packages under heavy load conditions are unavoidable also with switches [69]. This gets worse if switches are used in a multilevel hierarchy and may result in grossly varying communication delays. The real-time capabilities of native Ethernet are therefore limited and must rely on application-level mechanisms controlling the data throughput. For advanced requirements, like drive controls, this is not sufficient. These known limitations of conventional Ethernet stimulated the development of several alternative solutions that were more than just adaptations of ordinary fieldbus systems. These entirely new approaches were originally outside the IEC standardization process, but are now included in the RT Ethernet standard, that is, the second volume of IEC 61784 [70].

The initial and boundary conditions for the standardization work, which started in 2003, were targeted at backward compatibility with existing standards. First of all, real-time Ethernet (RTE) was seen as an extension to the industrial Ethernet solutions already defined in the communication profile families in IEC 61784-1. Furthermore, coexistence with conventional Ethernet is intended. The scope of the working document [71] states that

> the RTE shall not change the overall behavior of an ISO/IEC 8802-3 communication network and their related network components or IEEE 1588, but amend those widely used standards for RTE behaviors. Regular ISO/IEC 8802 3 based applications shall be able to run in parallel to RTE in the same network.

The work program of the RTE working group essentially consists of the definition of a classification scheme with RTE performance classes based on actual application requirements [59,72]. This is a response to market needs that demand scalable solutions for different application domains. One possible classification structure could be based on the reaction time of typical applications:

- A first low-speed class with reaction times around 100 ms. This timing requirement is typical for the case of humans involved in the system observation (10 pictures/s can already be seen as a low-quality movie), for engineering, and for process monitoring. Most processes in process automation and building control fall into this class. This requirement may be fulfilled with a standard system with TCP/IP communication channel without many problems.
- In a second class, the requirement is a reaction time below 10 ms. This is the requirement for most tooling machine control systems like PLCs or PC-based control. To reach this timing behavior, special care has to be taken in the RTE equipment: Sufficient computing resources are needed to handle the TCP/IP protocol in real time or the protocol stack must be simplified and reduced to get these reaction times on simple, cheap resources.

- The third and most demanding class is defined by the requirements of motion control: To synchronize several axes over a network, a time precision well below 1 ms is needed. Current approaches to reach this goal rely on modifications of both protocol medium access and hardware structure of the controllers.

During the last years, a number of industrial solutions appeared that tackled the real-time requirements, mostly on the basis of switched Ethernet. Still, as with fieldbus systems, they were tailored to specific needs. Not even the use of standard Ethernet is really a common denominator, and above the data link layer, the approaches are completely different. Some use standard TCP/UDP/IP mechanisms for transmitting data, maybe enhanced by additional software layers to support both real-time and non-real-time communication, and some use dedicated communication stacks that bypass the entire IP suite. Figure 1.25 sketches the various appearances of the protocol stack. Manifold differences are also possible on the physical layer. Some approaches foresee redundant media (VNET/IP, TCnet), PROFINET-I/O uses dedicated built-in switches to reduce the data transmission jitter [73], and EtherCAT as well as SERCOS III need dedicated controllers [59]. Ethernet Powerlink uses the old shared Ethernet and places a master–slave scheduling system on top of it. Common to many proposed networks is that they employ clock synchronization to support real-time applications. To this end, the standard IEEE 1588 [74], which originally emerged in the instrumentation area, was officially adopted also by IEC. The specific requirements in the automation domain have led to several suggestions for improvement of the standard regarding performance or fault tolerance [75,76] that were taken into account in the subsequent revision.

Given the multitude of approaches, despite the early hope that Ethernet could be the basis for a unique industrial communication solution, the situation has not too much changed compared with the heterogeneity of the traditional fieldbus systems [70]. Interoperability between different industrial Ethernet solutions is not possible in a direct way and must be established on a higher level by means of profiles (which is actually done to allow for cooperation between *old* fieldbuses and *new* Ethernet installations) or middleware layers like OPC. In many cases, higher-layer protocols (in particular, application layer protocols) and at least data objects are compatible with classical fieldbus systems and allow for an interconnection on a relatively high level. From this compatibility viewpoint, four different classes of industrial Ethernet approaches exist:

1. Reuse of higher-layer protocols from the preexisting fieldbus solutions. This applies to MODBUS/TCP, MODBUS/RTPS, high-speed Ethernet (Foundation Fieldbus over UDP/IP), Ethernet/IP (which uses the CIP protocol common to ControlNet and DeviceNet), P-NET on IP, Vnet/IP

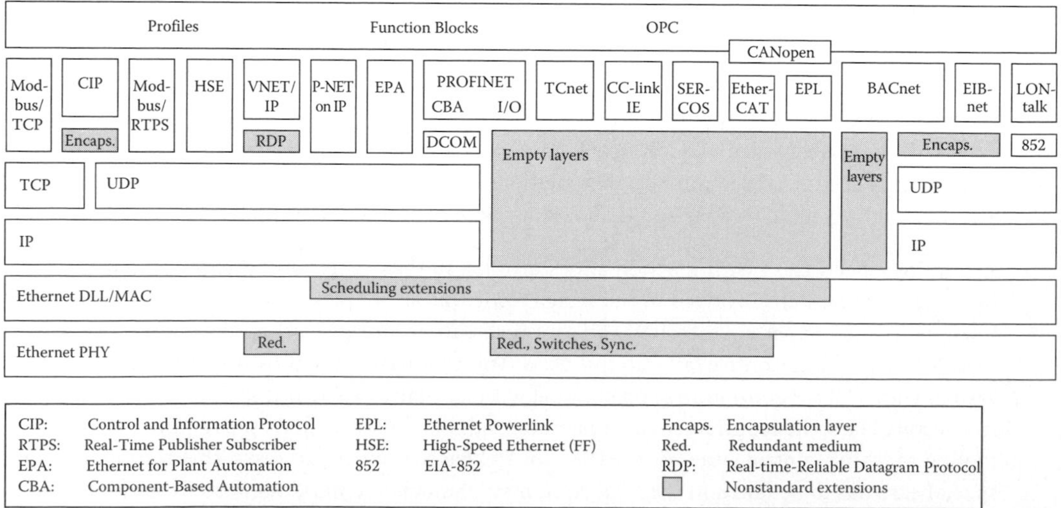

FIGURE 1.25 Protocol architecture of selected real-time Ethernet solutions proposed for IEC 61784-2.

(compatible to Vnet from Yokogawa), SERCOS III (structure of telegrams has been retained from the earlier, fiber-optics-based versions), and CC-Link IE (using the existing CC-Link protocol over Ethernet). Building automation networks take particular advantage of this architecture. BACnet, LonWorks, and EIBnet all use IP-based networks (or plain Ethernet as an alternative for BACnet) as a transport medium for the higher-layer protocols [77].
2. Compatibility of data models and objects with preexisting fieldbuses. An example is PROFINET, where proxy solutions exist to incorporate legacy PROFIBUS devices and networks.
3. Usage of application layer profiles from preexisting fieldbuses without direct protocol compatibility. This is the case for, for example, Ethernet Powerlink and EtherCAT, which use the CANopen application layer to achieve compatibility with widely used device profiles, for example, for drives.
4. Completely new industrial Ethernet developments without backward compatibility because of the lack of older fieldbus solutions. This is the case for the Asian networks Ethernet for Plant Automation and TC-Net.

Given the still large number of solutions, the only conceivable improvement compared to the classical fieldbus technology is that even with all proprietary modifications, Ethernet and, to a large extent, also the IP suite are being recognized as technological basis for the new generation of industrial communication systems. And with a view on vertical integration, the main benefit of industrial Ethernet is that all approaches allow for a standard TCP/UDP/IP communication channel in parallel to *fieldbus* communication. Even the RTE solutions (like PROFINET, Ethernet Powerlink, and EtherCAT) have such a conventional channel for configuration purposes. This is related to the IP-over-fieldbus tunneling concepts discussed earlier, but unlike the classical fieldbus systems, where such IP tunnels were introduced long after the fieldbus development and thus often had to cope with performance problems, they are now an integral part of the system concept right from the beginning on.

The separation of real-time and non-real-time traffic is accomplished on Ethernet MAC level with prioritization or TDMA schemes together with appropriate bandwidth allocation strategies. In such a parallel two-stack model, IP channels are no longer stepchildren of industrial communication, but offer sufficient performance to be used for regular data transfer. While this enables in principle the coexistence of automation and nonautomation applications on industrial Ethernet segments, the mixing of automation and office is not advisable for performance, but more important for security reasons. The value of this standard IP channel is rather to be seen in a simple direct access path to the field devices. Therefore, the currently favored solutions for configuration tools (i.e., XML, SOAP, and more generally web technology) can be used consistently. This again does not mean that industrial Ethernet solutions are interoperable or use the same configuration tools, but at least the basic principles are the same.

Actually, all this could have already been done with traditional fieldbus systems as well, and it certainly would have been done had especially the achievements of the Internet and the WWW been available in the early 1980s. So, what we see today with the rapid evolution of Ethernet in automation can in fact be regarded as a second wave of fieldbus development, which takes into account all the technological achievements of the last decade and exploits them for the field level, thereby making particularly vertical integration significantly easier.

1.7 Aspects for Future Evolution

Fieldbus systems have come a long way from the very first attempts of industrial networking to contemporary highly specialized automation networks. What is currently at hand—even after the selection process during the last two decades—nearly fully covers the complete spectrum of possible applications. Nevertheless, there is enough evolution potential left [17,27]. On the technological side, the communication medium itself allows further innovations. The currently most promising research field for technological evolution is the wireless domain. The benefits are obvious: (1) no failure-prone or costly cabling and (2) high flexibility, even mobility. The problems on the other hand are also obvious: very peculiar properties of the wireless communication channel must be dealt with, such as attenuation, fading, multipath reception, temporarily hidden nodes, the simple access for intruders, and many more [78–81]. Wireless communication options do exist today for several fieldbuses [82–84]. Up to now, they have been used just to replace the conventional data cable. A really efficient use of wireless communication, however, would necessitate an entire redefinition of at least the lower fieldbus protocol layers, while the higher protocol layers might be kept compatible with fieldbus or industrial Ethernet solutions [85–88]. Evaluation of currently available wireless technologies from the computer world with respect to their applicability in automation is a first step in this direction [89,90]. Ultimately, we can expect completely new automation networks optimized for wireless communication, where wired and wireless network segments are seamlessly integrated [91] and maybe only the application layer protocol remains compatible with traditional wired solutions to achieve integration [92–95].

Driven by ever more demanding application areas, communication networks for safety-relevant systems gain importance. In special fields such as x-by-wire for vehicles and avionics, dedicated fieldbus systems were developed that specifically addressed the problem of reliable communication [96]. Examples are TTP, FlexRay, byteflight, or ARINC 629. As this domain is subject to very stringent normative regulations and thus very conservative, it was dominated for a long time (and still is) by point-to-point connections between devices. The first *older* fieldbus system to penetrate this field was the CAN-based Safety Bus [97]. It took a long time and effort for this system to pass the costly certification procedures. Nevertheless, it was finally accepted also by the users, which was by no means obvious in an area concerned with the protection of human life, given that computer networks usually have the psychological disadvantage of being considered unreliable. After this pioneering work, other approaches like the ProfiSafe profile, INTERBUS Safety, ASi Safety, and recently also Ethernet/IPsafety, WorldFIP, and SafetyLON readily followed [98]. All these traditional fieldbus systems that were originally not designed for safety-critical applications implement this particular functionality by means of dedicated profiles. In practice, they wrap additional safety protocols into the normal payload data, which foresee a set of measures to make communication more reliable: sequence numbers, additional CRCs and confirmations, timestamps, heartbeat functions, and timeouts together with safety monitors and built-in test functions for the hardware components detect residual errors and may typically achieve safety integrity level 3 according to IEC 61508, which makes them applicable for fire emergencies, alarming, traffic and transportation applications, or power plant control.

Apart from communication protocol issues, there are two major trends to be noticed. One is the growing complexity of networks and networked systems in general. This is manifested by the increasing integration of fieldbus systems in higher-level, heterogeneous networks and process control systems on the one hand and the massive use of Internet technologies to achieve simplification and possible harmonization of existing solutions on the other. This in fact drastically reduces the number of layers in the traditional network hierarchy, which only reflects the trend toward peer-to-peer networking on a protocol level [99]. The other trend is the still increasing capabilities of embedded devices, the possibility to integrate more computational resources while at the same time reducing energy consumption. Systems on Chip with on-chip memory, network interfaces, and the computing power of a complete industrial PC offer sufficient resources for smart and low-cost sensors and actuators. This evolution is on the one hand the foundation of the current boom of Ethernet in automation. On the other, it will stimulate more research in the already booming field of sensor networks [100–103]. What is likely to be seen in the future are much more ubiquitous Ethernet- and Internet-based concepts, probably optimized to meet special performance requirements on the field level but still compatible with the standards in the management area. At any rate, these concepts and protocols will have to be scalable to allow for seamless integration of low-level, highly specialized sensor/actuator networks tailored to meet the demands of low power consumption, small-footprint implementation, high flexibility, and self-organization. The next evolution step in fieldbus history is just ahead.

Appendix

Tables 1.A.1 through 1.A.4 presented here give an overview of selected fieldbus systems, categorized by application domain. The list is necessarily incomplete, although care has been taken to include all approaches that either exerted a substantial influence on the evolution of the entire field or are significant still today. The year of introduction refers to the public availability of the specification or first products. This year is also the one used in the timeline in Figure 1.3. Note that despite careful research, the information obtained from various sources was frequently inconsistent, so that there may be an uncertainty in the figures. Where respective data could be obtained, the start of the project has been listed as well because there are several cases where the development of the fieldbus took a long time before the first release.

TABLE 1.A.1 Instrumentation and PCB-Level Buses

Fieldbus	Developer (Country)	Introduced in	Standard	Refs.
CAMAC	ESONE (Europe)	1969 (Start of development 1966)	IEEE 583 (1970, 1982, 1994) IEEE 595 (1974, 1982) IEEE 596 (1972, 1982) IEEE 758 (1979)	[104]
GPIB (HP-IB)	Hewlett-Packard (United States)	1974 (Start of development 1965)	ANSI IEEE-488 (1975, 1978) ANSI IEEE-488.2 (1987, 1992) IEC 60625 (1979,1993)	[105,106]
HP IL	Hewlett-Packard (United States)	1980 (Start of development 1976)	—	[18]
I²C	Philips (the Netherlands)	1981	—	[19]
M-Bus	University of Paderborn, TI, Techem (Germany)	1992	EN 1434-3 (1997)	[107]
Measurement bus	Industry consortium (Germany)	1988	DIN 66348-2 (1989) DIN 66348-3 (1996)	

TABLE 1.A.2 Automotive and Aircraft Fieldbuses

Fieldbus	Developer (Country)	Introduced in	Standard	Refs.
ABUS	Volkswagen (Germany)	1987	—	[108]
ARINC	Aeronautical Radio, Inc. (United States)	1978	AEEC ARINC 429 (1978, 1995) AEEC ARINC 629 (1989)	[109]
CAN	Bosch (Germany)	1986 (Start of development 1983), CAL 1992	ISO 11898 (1993, 1995) ISO 11519 (1994)	[110]
Flexay	DaimlerChrysler, BMW (Germany)	2002	—	
J1850	Ford, GM, Chrysler (United States)	1987	SAE J1850 (1994, 2001) ISO 11519-4	[108]
J1939	SAE (United States)	1994	SAE J1939 (1998)	[108]
LIN	Industry consortium	1999	—(open spec)	
MIL 1533	SAE (military and industry consortium, United States)	1970 (Start of development 1968)	MIL-STD-1553 (1973) MIL-STD-1553A (1975) MIL-STD-1553B (1978)	[111]
VAN	Renault, PSA Peugeot-Citroen (France), ISO TC22	1988	ISO 11519-3 (1994)	[108]
SwiftNet	Ship Star Assoc., Boeing (United States)	1997	IEC 61158 (2000)	
TTP	Vienna University of Technology (Austria)	1996	—	[108]

TABLE 1.A.3 Fieldbuses for Industrial and Process Automation and Their Foundations

Fieldbus	Developer (Country)	Introduced in	Standard	Refs.
ARCNET	Datapoint (United States)	1977	ANSI ATA 878 (1999)	[112]
ASi	Industry and university consortium (Germany)	1991	EN 50295-2 (1998, 2002) IEC 62026-2 (2000)	
Bitbus	Intel (United States)	1983	ANSI IEEE 1118 (1990)	
CC-Link	Mitsubishi (Japan)	1996	—(open spec)	
CANopen	CAN in Automation (user group, Germany)	1995 (Start of development 1993)	EN 50325-4 (2002)	[110]
ControlNet	Allen-Bradley (United States)	1996	EN 50170-A3 (2000)	[112]
DeviceNet	Allen-Bradley (United States)	1994	EN 50325-2 (2000)	[119]
FF	Fieldbus Foundation (industry consortium, United States)	1995 (Start of development 1994)	BSI DD 238 (1996) EN 50170-A1 (2000)	[112]
Hart	Rosemount (United States)	1986	—(open spec)	
Interbus-S	Phoenix Contact (Germany)	1987 (Start of development 1983)	DIN 19258 (1993) EN 50254-2 (1998)	[113]
MAP	General Motors (United States)	1982 (Start of development 1980)	MAP 1.0 (1982) MAP 2.0 (1985) MAP 3.0 (1988)	[114]
MMS	ISO TC 184	1986	ISO/IEC 9506 (1988, 2000)	
Modbus	Gould, Modicon (United States)	1979	—(open spec)	
PDV-bus	Industry and university consortium (Germany)	1979 (Start of development 1972)	DIN 19241 (1982)	[115]
P-NET	Proces Data (Denmark)	1983	DS 21906 (1990) EN 50170-1 (1996)	
PROWAY C	IEC TC 65	1986 (Start of development 1975)	ISA S72.01 (1985) IEC 60955 (1989)	[14]
PROFIBUS	Industry and university consortium (Germany)	1989 (Start of development 1984)	FMS: DIN 19245-1—2 (1991) DP: DIN 19245-3 (1993) PA: DIN 19245-4 (1995) FMS/DP: EN 50170-2 (1996) DP: EN 50254-3 (1998) PA: EN 50170-A2 (2000)	
SDS	Honeywell (United States)	1994	EN 50325-3 (2000)	[110]
Sercos	Industry consortium (Germany)	1989 (Start of development 1986)	IEC 61491 (1995) EN 61491 (1998)	
Seriplex	APC, Inc. (United States)	1990	IEC 62026-6 (2000)	[112]
SINEC L2	Siemens (Germany)	1992	—	
SP50 Fieldbus	ISA SP 50 (United States)	1993	ISA SP 50 (1993)	
(World)FIP	Industry and university consortium (France)	1987 (Start of development 1982)	AFNOR NF C46601—7 (1989–1992) EN 50170-3 (1996) DWF: AFNOR NF C46638 (1996 DWF: EN 50254-4 (1998)	[16]

TABLE 1.A.4 Fieldbuses for Building and Home Automation

Fieldbus	Developer (Country)	Introduced in	Standard	Refs.
BACnet	ASHRAE SPC135P (industry consortium, United States)	1991	ANSI/ASHRAE 135 (1995) ENV 1805-1 (1998) ENV 13321-1 (1999) ISO 16484-5 (2003)	
Batibus	Industry consortium (France)	1987	AFNOR NF 46621—3, 9 (1991) ENV 13154-2 (1998)	
CEBus	Industry consortium (United States)	1984	ANSI EIA 600 (1992)	
EHS	Industry consortium (Europe)	1987	ENV 13154-2 (1998)	
EIB	Industry consortium (Germany)	1990	AFNOR NFC 46624—8 (1991) DIN V VDE 0829 (1992) ENV 13154-2 (1998)	[116]
HBS	Industry consortium (Japan)	1986 (Start of development 1981)	EIAJ/REEA ET2101	
LonWorks	Echelon (United States)	1991	ANSI EIA 709 (1999) ENV 13154-2 (1998)	[112,117]
Sigma I	ABB (Germany)	1983	—	
X10	Pico Electronics (United Kingdom)	1970 (Start of development 1975)	—	[117]
KNX	Industry consortium (Europe)	2002	EN 50090 (2003) ISO/IEC 14543-3 (2006)	

References

1. International Electrotechnical Commission, IEC 61158-1, Digital data communications for measurement and control—Fieldbus for use in industrial control systems, Part 1: Introduction, 2003.
2. Fieldbus Foundation, What is fieldbus?, http://www.fieldbus.org/About/FoundationTech/. (accessed August 15, 2013).
3. G. G. Wood, Fieldbus status 1995, *Computing & Control Engineering Journal*, 6, 1995, 251–253.
4. G. G. Wood, Survey of LANs and standards, *Computer Standards & Interfaces*, 6, 1987, 27–36.
5. N. P. Mahalik (ed.), *Fieldbus Technology: Industrial Network Standards for Real-Time Distributed Control*, Springer, Berlin, Germany, 2003..
6. H. Töpfer, W. Kriesel, Zur funktionellen und strukturellen Weiterentwicklung der Automatisierungsanlagentechnik, *Messen Steuern Regeln*, 24, 1981, 183–188.
7. T. Pfeifer, K.-U. Heiler, Ziele und Anwendungen von Feldbussystemen, *Automatisierungstechnische Praxis*, 29(12), 1987, 549–557.
8. H. Steusloff, Zielsetzungen und Lösungsansätze für eine offene Kommunikation in der Feldebene, *Automatisierungstechnik'90*, VDI Berichte 855, DI-Verlag, Düsseldorf, Germany, 1990, pp. 337–357.
9. L. Capetta, A. Mella, F. Russo, Intelligent field devices: User expectations, *IEE Colloquium on Fieldbus Devices: A Changing Future*, London, U.K., 1994, pp. 6/1–6/4.
10. K. Wanser, Entwicklungen der Feldinstallation und ihre Beurteilung, *Automatisierungstechnische Praxis*, 27(5), 1985, 237–240.
11. J. A. H. Pfleger, Anforderungen an Feldmultiplexer, *Automatisierungstechnische Praxis*, 29, 1987, 205–209.
12. H. Junginger, H. Wehlan, Der Feldmultiplexer aus Anwendersicht, *Automatisierungstechnische Praxis*, 31, 1989, 557–564.
13. W. Schmieder, T. Tauchnitz, FuRIOS: Fieldbus and remote I/O—A system comparison, *Automatisierungstechnische Praxis*, 44, 2002, 61–70.

14. P. Pleinevaux, J.-D. Decotignie, Time critical communication networks: Field buses, *IEEE Network*, 2, 1988, 55–63.
15. E. H. Higham, Casting a crystal ball on the future of process instrumentation and process measurements, *IEEE Instrumentation and Measurement Technology Conference* (*IMTC '92*), New York, May 12–14, 1992, pp. 687–691.
16. J. P. Thomesse, Fieldbus technology in industrial automation, *Proceedings of the IEEE*, 93, 2005, 1073–1101.
17. T. Sauter, S. Soucek, W. Kastner, D. Dietrich, The evolution of factory and building automation, *IEEE Industrial Electronics Magazine*, 5(3), 2011, 35–48.
18. R. D. Quick, S. L. Harper, HP-IL: A low-cost digital interface for portable applications, *Hewlett-Packard Journal*, 34, 1983, 3–10.
19. NXP Semiconductor, I^2C-bus specification and user manual, Rev. 5, 2012, http://www.nxp.com/documents/user_manual/UM10204.pdf. (accessed August 15, 2013).
20. H. Zimmermann, OSI reference model: The ISO model of architecture for open system interconnection, *IEEE Transactions on Communications*, 28(4), 1980, 425–432.
21. J. Day, H. Zimmermann, The OSI reference model, *Proceedings of the IEEE*, 71(12), 1983, 1334–1340.
22. D. J. Damsker, Assessment of industrial data network standards, *IEEE Transactions on Energy Conversion*, 3, 1988, 199–204.
23. H. A. Schutz, The role of MAP in factory integration, *IEEE Transactions on Industrial Electronics*, 35(1), 1988, 6–12.
24. B. Armitage, G. Dunlop, D. Hutchison, S. Yu, Fieldbus: An emerging communications standard, *Microprocessors and Microsystems*, 12, 1988, 555–562.
25. S. G. Shanmugham, T. G. Beaumariage, C. A. Roberts, D. A. Rollier, Manufacturing communication: The MMS approach, *Computers and Industrial Engineering*, 28(1), 1995, 1–21.
26. T. Phinney, P. Brett, D. McGovan, Y. Kumeda, FieldBus—Real-time comes to OSI, *International Phoenix Conference on Computers and Communications*, Scottsdale, AZ, March 27–30, 1991, pp. 594–599.
27. T. Sauter, The three generations of field-level networks—Evolution and compatibility issues, *IEEE Transactions on Industrial Electronics*, 57(11), 2010, 3585–3595.
28. Gesmer Updegrove LLP, Government issues and policy, http://www.consortiuminfo.org/government/. (accessed August 15, 2013).
29. P. Leviti, IEC 61158: An offence to technicians?, *IFAC International Conference on Fieldbus Systems and their Applications, FeT 2001*, Nancy, France, November 15–16, 2001, 36pp.
30. T. Sauter, Fieldbus systems: History and evolution, in: R. Zurawski (ed.), *The Industrial Communication Technology Handbook*, CRC Press, Boca Raton, FL, 2005, pp. 7-1–7-39.
31. M. Felser, T. Sauter, The fieldbus war: History or short break between battles?, *IEEE International Workshop on Factory Communication Systems* (*WFCS*), Västerås, Sweden, 2002, pp. 73–80.
32. T. Phinney, Mopping up from bus wars, *World Bus Journal* (*ISA*), 2002, pp. 22–23.
33. S. Soucek, T. Sauter, G. Koller, Effect of delay jitter on quality of control in EIA-852-based networks, *Twenty-Ninth Annual Conference of the IEEE Industrial Electronics Society* (*IECON*), Roanoke, VA, November 2–6, 2003, pp. 1431–1436.
34. S. Soucek, T. Sauter, Quality of service concerns in IP-based control systems, *IEEE Transactions on Industrial Electronics*, 51, 2004, 1249–1258.
35. S. Cavalieri, S. Monforte, A. Corsaro, G. Scapellato, Multicycle polling scheduling algorithms for FieldBus networks, *Real-Time Systems*, 25(2–3), 2003, 157–185.
36. J.-P. Thomesse, Z. Mammeri, L. Vega, Time in distributed systems cooperation and communication models, *Fifth IEEE Computer Society Workshop on Future Trends of Distributed Computing Systems*, Cheju Island, Korea, 1995, pp. 41–49.
37. IEC TC65, Industrial process measurement and control, IEC TC 65/290/DC, Device profile guideline, 2002.

38. M. Wollschlaeger, C. Diedrich, J. Müller, U. Epple, Integration of fieldbus systems into on-line asset management solutions based on fieldbus profile descriptions, *IEEE International Workshop on Factory Communication Systems* (*WFCS*), Västerås, Sweden, 2002, pp. 89–96.
39. R. Frenzel, M. Wollschlaeger, T. Hadlich, C. Diedrich, Tool support for the development of IEC 62390 compliant fieldbus profiles, *IEEE Conference on Emerging Technologies and Factory Automation* (*ETFA*), Bilbao, Spain, September 13–16, 2010, pp. 1–4.
40. R. Simon, P. Neumann, C. Diedrich, M. Riedl, Field devices-models and their realisations, *IEEE International Conference on Industrial Technology* (*ICIT'02*), Bangkok, Thailand, December 11–14, 2002, pp. 307–312.
41. F. Benzi, G. S. Buja, M. Felser, Communication architectures for electrical drives, *IEEE Transactions on Industrial Informatics*, 1(1), 2005, 47–53.
42. S. Theurich, R. Frenzel, M. Wollschlaeger, T. Szczepanski, Web-based asset management for heterogeneous industrial networks, *IEEE International Workshop on Factory Communication Systems* (*WFCS*), Dresden, Germany, May 21–23, 2008, pp. 267–270.
43. M. Knizak, M. Kunes, M. Manninger, T. Sauter, Applying internet management standards to fieldbus systems, *IEEE International Workshop on Factory Communication Systems* (*WFCS*), Barcelona, Spain, October 1–3, 1997, pp. 309–315.
44. T. Sauter, M. Lobashov, How to access factory floor information using Internet technologies and gateways, *IEEE Transactions on Industrial Informatics*, 7(4), 2011, 699–712.
45. M. Wollschlaeger, Integration of VIGO into directory services, *Sixth International P-NET Conference*, Vienna, Austria, May 25–26, 1999.
46. R. P. Pantoni, L. C. Passarini, D. Brandao, Developing and implementing an open and non-proprietary device description for fieldbus devices based on software standards, *IEEE International Symposium on Industrial Electronics* (*ISIE*), Vigo, Spain, June 4–7, 2007, pp. 1887–1892.
47. W. Kastner, F. Kastner-Masilko, EDDL inside FDT/DTM, *IEEE International Workshop on Factory Communication Systems*, Vienna, Austria, 2004, pp. 365–368.
48. M. Wollschlaeger, T. Bangemann, XML based description model as a platform for Web-based maintenance, *IEEE International Conference on Industrial Informatics* (*INDIN*), Berlin, Germany, 2004, pp. 125–130.
49. R. Simon, M. Riedl, C. Diedrich, Integration of field devices using field device tool (FDT) on the basis of electronic device descriptions (EDD), *IEEE International Symposium on Industrial Electronics* (*ISIE*), Rio de Janeiro, Brazil, June 9–11, 2003, pp. 189–194.
50. J. L. M. Lastra, M. Delamer, Semantic web services in factory automation: Fundamental insights and research roadmap, *IEEE Transactions on Industrial Informatics*, 2(1), 2006, 1–11.
51. A. P. Kalogeras, J. V. Gialelis, C. E. Alexakos, M. J. Georgoudakis, S. A. Koubias, Vertical integration of enterprise industrial systems utilizing web services, *IEEE Transactions on Industrial Informatics*, 2(2), 2006, 120–128.
52. A. Talevski, E. Chang, T. S. Dillon, Reconfigurable Web service integration in the extended logistics enterprise, *IEEE Transactions on Industrial Informatics*, 1(2), 2005, 74–84.
53. M. Therani, Ontology development for designing and managing dynamic business process networks, *IEEE Transactions on Industrial Informatics*, 3(2), 2007, 173–185.
54. T. Cucinotta, A. Mancina, G. F. Anastasi, G. Lipari, L. Mangeruca, R. Checcozzo, F. Rusina, A real-time service-oriented architecture for industrial automation, *IEEE Transactions on Industrial Informatics*, 5(3), 2009, 267–277.
55. I. M. Delamer, J. L. M. Lastra, Service-oriented architecture for distributed publish/subscribe middleware in electronics production, *IEEE Transactions on Industrial Informatics*, 2(4), 2006, 281–294.
56. A. Sasa, M. B. Juric, M. Krisper, Service-oriented framework for human task support and automation, *IEEE Transactions on Industrial Informatics*, 4(4), 2008, 292–302.

57. Y. Hongli, L. Feng, Research on OPC UA based on FDT/DTM and EDDL, *Second International Conference on Digital Manufacturing and Automation (ICDMA)*, Zhangjiajie, China, August 5–7, 2011, pp. 992–995.
58. X. Hao, S. Hou, OPC DX and industrial Ethernet glues fieldbus together, *Eighth Control, Automation, Robotics and Vision Conference (ICARCV)*, Kunming, China, December 6–9, 2004, pp. 562–567.
59. M. Felser, Real-time Ethernet—Industry prospective, *Proceedings of the IEEE*, 93, 2005, 1118–1129.
60. S. Kweon, M.-G. Cho, K. G. Shin, Soft real-time communication over Ethernet with adaptive traffic smoothing, *IEEE Transactions on Parallel and Distributed Systems*, 15, 2004, 946–959.
61. L. Lo Bello, G. Kaczynski, O. Mirabella, Improving the real-time behaviour of Ethernet networks using traffic smoothing, *IEEE Transactions on Industrial Informatics*, 1, 2005, 151–161.
62. T. Skeie, S. Johannessen, Ø. Holmeide, Timeliness of real-time IP communication in switched industrial Ethernet networks, *IEEE Transactions on Industrial Informatics*, 2, 2006, 25–39.
63. Y. Song, Time constrained communication over switched Ethernet, *IFAC International Conference on Fieldbus Systems and their Applications (FeT)*, Nancy, France, 2001, pp. 152–159.
64. J. Jasperneite, J. Imtiaz, M. Schumacher, K. Weber, A proposal for a generic real-time Ethernet system, *IEEE Transactions on Industrial Informatics*, 5(2), 2009, 75–85.
65. J. Loeser, H. Haertig, Low-latency hard real-time communication over switched Ethernet, *Sixteenth Euromicro Conference on Real-Time Systems (ECRTS)*, Catania, Italy, June 30–July 2, 2004, pp. 13–22.
66. J. D. Decotignie, Ethernet-based real-time and industrial communications, *Proceedings of the IEEE*, 93, 2005, 1102–1117.
67. S. H. Pee, R. H. Yang, J. Berge, B. Sim, Foundation Fieldbus high speed Ethernet (HSE) implementation, *IEEE International Symposium on Intelligent Control*, Vancouver, British Columbia, Canada, 2002, pp. 777–782.
68. V. Schiffer, The CIP family of fieldbus protocols and its newest member—Ethernet/IP, *Conference on Emerging Technologies and Factory Automation (ETFA 2001)*, Antibes Juan-Les-Pins, France, October 15–18, 2001, pp. 377–384.
69. K.-C. Lee, S. Lee, M. H. Lee, Worst case communication delay of real-time industrial switched Ethernet with multiple levels, *IEEE Transactions on Industrial Electronics*, 53(5), 2006, 1669–1676.
70. J.-D. Decotignie, The many faces of industrial Ethernet, *IEEE Industrial Electronics Magazine*, 3(1), 2009, 8–19.
71. International Electrotechnical Commission, TC65/SC65C, New work item proposal, document number 65C/306/NP, 2003.
72. International Electrotechnical Commission, TC65/SC65C, Meeting minutes, document number 65C/318/INF, 2003.
73. J. Jasperneite, J. Feld, PROFINET: An integration platform for heterogeneous industrial communication systems, *IEEE International Conference on Emerging Technologies and Factory Automation (ETFA)*, Catania, Italy, 2005, pp. 815–822.
74. IEEE Standard 1588, Standard for a precision clock synchronization protocol for networked measurement and control systems, IEEE, 2002.
75. J. Jasperneite, K. Shehab, K. Weber, Enhancements to the time synchronization standard IEEE-1588 for a system of cascaded bridges, *IEEE Workshop on Factory Communication Systems (WFCS)*, Vienna, Austria, 2004, pp. 239–244.
76. G. Gaderer, P. Loschmidt, T. Sauter, Improving fault tolerance in high-precision clock synchronization, *IEEE Transactions on Industrial Informatics*, 6(2), 2010, 206–215.
77. W. Kastner, G. Neugschwandtner, S. Soucek, H. M. Newman, Communication systems for building automation and control, *Proceedings of the IEEE*, 93, 2005, 1178–1203.
78. A. Willig, K. Matheus, A. Wolisz, Wireless technology in industrial networks, *Proceedings of the IEEE*, 93, 2005, 1130–1151.

79. L. Lo Bello, E. Toscano, Coexistence issues of multiple co-located IEEE 802.15.4/ZigBee networks running on adjacent radio channels in industrial environments, *IEEE Transactions on Industrial Informatics*, 5(2), 2009, 157–167.
80. M. Jonsson, K. Kunert, Towards reliable wireless industrial communication with real-time guarantees, *IEEE Transactions on Industrial Informatics*, 5(4), 2009, 429–442.
81. G. Gamba, F. Tramarin, A. Willig, Retransmission strategies for cyclic polling over wireless channels in the presence of interference, *IEEE Transactions on Industrial Informatics*, 6(3), 2010, 405–415.
82. G. Cena, A. Valenzano, S. Vitturi, Hybrid wired/wireless networks for real-time communications, *IEEE Industrial Electronics Magazine*, 2(1), 2008, 8–20.
83. L. Seno, S. Vitturi, C. Zunino, Analysis of Ethernet powerlink wireless extensions based on the IEEE 802.11 WLAN, *IEEE Transactions on Industrial Informatics*, 5(2), 2009, 86–98.
84. G. Cena, L. Seno, A. Valenzano, C. Zunino, On the performance of IEEE 802.11e wireless infrastructures for soft-real-time industrial applications, *IEEE Transactions on Industrial Informatics*, 6(3), 2010, 425–437.
85. S. Vitturi, I. Carreras, D. Miorandi, L. Schenato, A. Sona, Experimental evaluation of an industrial application layer protocol over wireless systems, *IEEE Transactions on Industrial Informatics*, 3(4), 2007, 275–288.
86. J. Kjellsson, A. E. Vallestad, R. Steigmann, D. Dzung, Integration of a wireless I/O interface for PROFIBUS and PROFINET for factory automation, *IEEE Transactions on Industrial Electronics*, 56(10), 2009, 4279–4287.
87. S.-E. Yoo, P. K. Chong, D. Kim, Y. Doh, M.-L. Pham, E. Choi, J. Huh, Guaranteeing real-time services for industrial wireless sensor networks with IEEE 802.15.4, *IEEE Transactions on Industrial Electronics*, 57(11), 2010, 3868–3876.
88. G. Scheible, D. Dzung, J. Endresen, J.-E. Frey, Unplugged but connected—Design and implementation of a truly wireless real-time sensor/actuator interface, *IEEE Industrial Electronics Magazine*, 1(2), Summer 2007, 25–34.
89. G. Anastasi, M. Conti, M. Di Francesco, A comprehensive analysis of the MAC unreliability problem in IEEE 802.15.4 wireless sensor networks, *IEEE Transactions on Industrial Informatics*, 7(1), 2011, 52–65.
90. E. Toscano, L. L. Bello, Multichannel superframe scheduling for IEEE 802.15.4 industrial wireless sensor networks, *IEEE Transactions on Industrial Informatics*, 8(2), 2012, 337–350.
91. T. Sauter, J. Jasperneite, L. Lo Bello, Towards new hybrid networks for industrial automation, *IEEE Conference on Emerging Technologies and Factory Automation* (*ETFA*), Palma de Mallorca, Spain, 2009, pp. 1–8.
92. J. H. Lee, T. Kwon, J. S. Song, Group connectivity model for industrial wireless sensor networks, *IEEE Transactions on Industrial Electronics*, 57(5), 2010, 1835–1844.
93. Y. Ishii, Exploiting backbone routing redundancy in industrial wireless systems, *IEEE Industrial Electronics Magazine*, 56(10), 2009, 4288–4295.
94. M. Baldi, R. Giacomelli, G. Marchetto, Time-driven access and forwarding for industrial wireless multihop networks, *IEEE Transactions on Industrial Informatics*, 5(2), 2009, 99–112.
95. H.-J. Korber, H. Wattar, G. Scholl, Modular wireless real-time sensor/actuator network for factory automation applications, *IEEE Transactions on Industrial Informatics*, 3(2), 2007, 111–119.
96. G. Leen, D. Heffernan, Expanding automotive electronic systems, *IEEE Computer*, 35, 2002, 88–93.
97. R. Piggin, An introduction to safety-related networking, *IEE Computing & Control Engineering*, 15, 2004, 34–39.
98. T. Novak, P. Fischer, M. Holz, M. Kieviet, T. Tamandl, Safe commissioning and maintenance process for a safe fieldbus, *IEEE International Workshop on Factory Communication Systems* (*WFCS*), Dresden, Germany, May 21–23, 2008, pp. 225–232.

99. T. Sauter, The continuing evolution of integration in manufacturing automation, *IEEE Industrial Electronics Magazine*, 1(1), 2007, 10–19.
100. F. De Pellegrini, D. Miorandi, S. Vitturi, A. Zanella, On the use of wireless networks at low level of factory automation systems, *IEEE Transactions on Industrial Informatics*, 2(2), 2006, 129–143.
101. L. Lo Bello, E. Toscano, An adaptive approach to topology management in large and dense real-time wireless sensor networks, *IEEE Transactions on Industrial Informatics*, 5(3), 2009, 314–324.
102. J. Chen, X. Cao, P. Cheng, Y. Xiao, Y. Sun, Distributed collaborative control for industrial automation with wireless sensor and actuator networks, *IEEE Transactions on Industrial Electronics*, 57(12), 2010, 4219–4230.
103. L. Palopoli, R. Passerone, T. Rizano, Scalable offline optimization of industrial wireless sensor networks, *IEEE Transactions on Industrial Informatics*, 7(2), 2011, 328–339.
104. CAMAC, A modular instrumentation system for data handling, EUR4100e, March, 1969.
105. P. Zsolt, GPIB Tutorial, http://www.hit.bme.hu/people/papay/edu/GPIB/tutor.htm. (accessed August 15, 2013).
106. National Instruments, GPIB tutorial, http://nemu.web.psi.ch/doc/manuals/software_manuals/GPIB/GPIB_tutorial.pdf. (accessed August 15, 2013).
107. M-Bus Usergroup, The M-Bus: A documentation, Version 4.8, November 11, 1997, http://www.m-bus.com/mbusdoc/default.html.
108. G. Leen, D. Heffernan, A. Dunne, Digital networks in the automotive vehicle, *IEE Computing and Control Engineering Journal*, 10, 1999, 257–266.
109. N. C. Audsley, A. Grigg, Timing analysis of the ARINC 629 databus for real-time applications, *Microprocessors and Microsystems*, 21, 1997, 55–61.
110. CAN-in-Automation, CAN history, http://www.can-cia.de/can/protocol/history/ (accessed August 15, 2013).
111. Condor Engineering, MIL-STD-1553 Tutorial, http://digilander.libero.it/LeoDaga/Corsi/AD/Documenti/MIL-STD-1553Tutorial.pdf. (accessed August 15, 2013).
112. ER-Soft, The fieldbus comparison chart, http://www.er-soft.com/en/downloads/ER-Soft— -Fieldbus—Comparison— Chart.pdf. (accessed August 15, 2013).
113. Interbus Club, Interbus basics, 2001, http://www.interbus.com/get.php?object = 497. (accessed August 15, 2013).
114. H. Kirrmann, Industrial automation, lecture notes, EPFL, 2004, http://lamspeople.epfl.ch/kirrmann/IA_slides.htm. (accessed August 15, 2013).
115. H. Wölfel, Die Entwicklung der digitalen Prozeßleittechnik—Ein Rückblick (Teil 3), *Automatisierungstechnische Praxis*, 40, 1998, S17–S24.
116. T. Sauter, D. Dietrich, W. Kastner (eds.), *EIB Installation Bus System*, Publicis MCD, Erlangen, Germany, 2001.
117. E. B. Driscoll, The history of X10, http://home.planet.nl/~lhendrix/x10_history.htm. (accessed August 15, 2013).

2

Networked Control Systems for Manufacturing: Parameterization, Differentiation, Evaluation, and Application

2.1 Introduction 2-1
2.2 Parameterization of Industrial Networks 2-4
Speed and Bandwidth • Delay and Jitter • Wired and Wireless QoS Metrics • Network QoS versus System Performance
2.3 Differentiation of Industrial Networks 2-12
Categorization of Networks • Time-Division Multiplexing • RA with Collision Arbitration: CAN (CSMA/AMP) • Ethernet-Based Networks • Impact of Ethernet Application Layer Protocols: OPC • Wireless Networks
2.4 NCS Characterization 2-28
Theoretical Perspective • Experimental Perspective • Analytical Perspective
2.5 Applications for Industrial Networks 2-31
Networks for Control • Networks for Diagnostics • Networks for Safety • Multilevel Factory Networking Example: Reconfigurable Factory Testbed
2.6 Future Trends 2-36
Acknowledgments 2-37
References 2-37

James R. Moyne
University of Michigan

Dawn M. Tilbury
University of Michigan

Dhananjay Anand
National Institute of Standards and Technology

2.1 Introduction

Networks have become an integral part of manufacturing over the past two decades, replacing point-to-point communications at all levels. At lower levels in the factory infrastructure, networks provide higher reliability, visibility, and diagnosability and enable capabilities such as distributed control, diagnostics, safety, and device interoperability. At higher levels, networks can leverage Internet services to enable factory-wide automated scheduling, control, and diagnostics; improve data storage and visibility; and open the door to e-manufacturing.

In general, control networks can replace traditional point-to-point wired systems while providing a number of advantages. Perhaps the simplest but most important advantage is the reduced volume of wiring. Fewer physical potential points of failure, such as connectors and wire harnesses, result in increased reliability. This advantage is further accentuated in wireless networks. Another significant advantage is that networks enable complex distributed control systems to be realized in both horizontal (e.g., peer-to-peer coordinated control among sensors and actuators) and vertical (e.g., machine to cell to system-level control) directions. Other documented advantages of networks include increased capability for troubleshooting and maintenance, enhanced interchangeability and interoperability of devices, improved reconfigurability of control systems, and ease of integration of web service–based capabilities such as the cloud- and applet-based systems [48,93].

The primary application of control networks is supervisory control and data acquisition (SCADA) systems [31,53]. Networked SCADA systems often provide a supervisory-level factory-wide solution for coordination of machine and process diagnostics, along with other factory floor and operations information. However, networks are being used at all levels of the manufacturing hierarchy, loosely defined as device, machine, cell, subsystem, system, factory, and enterprise. Within the manufacturing domain, the application of networks can be further divided into subdomains of *control*, *diagnostics*, and *safety*. Control network operation generally refers to communicating the necessary sensory and actuation information for closed-loop control. The control may be time-critical, such as at a computer numeric controller (CNC) or servo drive level, or event-based, such as at a programmable logic controller (PLC) level. In this subdomain, networks must guarantee a certain level of response time determinism to be effective. Diagnostics network operation usually refers to the communication of sensory information as necessary to deduce the health of a tool, product, or system; this is differentiated from *network diagnostics*, which refers to deducing the health of the network [28,38,40,84]. Systems diagnostics solutions may *close-the-loop* around the diagnostic information to implement control capabilities such as equipment shutdown or continuous process improvement; however, the performance requirements of the system are driven by the data collection, and actuation is usually event based (i.e., not time dependent). An important quality of diagnostics networks is the ability to communicate large amounts of data, with determinism usually less important than in control networks. Issues of data compression and security can also play a large role in diagnostics networks, especially when utilized as a mechanism for communication between user and vendor to support equipment e-diagnostics [22,38,84]. Required data communication rates associated with large volumes of data, along with required data storage, data merging, data quality, and analytics capabilities collectively contribute to the *big data* problem in factories; thus, networks are an integral part of the big data solution [66]. Safety is the newest of the three network subdomains, but is rapidly receiving attention in industry [12,52]. Here, network requirements are often driven by standards, with an emphasis on determinism (guaranteed response time), network reliability, and capability for self-diagnosis [35].

Driven by a desire to minimize cost and maximize interoperability and interchangeability, there continues to be a movement to try to consolidate around a single network technology at different levels of control and across different application domains. For example, Ethernet, which was widely regarded as a high-level-only communication protocol in the past, is now frequently being utilized as a lower-level control network, with flavors of Ethernet actually being used for real-time, high-speed control [15,24,91]. The hierarchical Ethernet network infrastructure has enabled capabilities such as web-based *drill-down* (focused data access) to the sensor level [43,84]. In some situations, wireless technology can be considered as a replacement for wired networks at all levels, primarily to support diagnostics, but also to support control and even safety functionality in specific instances [13,17,101].

This movement toward consolidation, and indeed the technical selection of networks for a particular application, revolves around evaluating and balancing quality of service (QoS) parameters. Multiple components (nodes) are vying for a limited network bandwidth, and they must strike a balance with factors related to the time to deliver information end to end between components. Two parameters that are often involved in this balance are network average speed and determinism; briefly, network speed is a function of the network access time and bit transfer rate, while determinism is a measure of the ability to communicate data consistently from end to end within a guaranteed time. Note that this QoS issue applies to both wired and wireless network applications; however, with wireless, there must be more focus on external factors that can affect the end-to-end transmission performance, reliability, and security.

Network protocols utilize different approaches to provide end-to-end data delivery. The differentiation could be at the lowest physical level (e.g., wired vs. wireless) up through the mechanism at which network access is negotiated, all the way up through application services that are supported. Protocol functionality is commonly described and differentiated utilizing the International Standards Organization–Open Systems Interconnection (ISO–OSI) seven-layer reference model [3]. The seven layers are physical, data link, network, transport, session, presentation, and application.

The network protocol, specifically the media access control (MAC) protocol component, defines the mechanism for delegating this bandwidth in such a way so that the network is *optimized* for a specific type of communication (e.g., large data packet size with low determinism vs. small data packet size with high determinism). Over the past two decades, *bus wars* (referring to sensor bus network technology) resulted in serious technical debates with respect to the optimal MAC approach for different applications [26,60,64].

Over the past 10 years, however, it has become clear that the pervasiveness of Ethernet, especially in domains outside of manufacturing control (e.g., the Internet), will result in its eventual dominance in the manufacturing control domain [19,25,39,70]. This movement has been facilitated in large part by the emergence of switch technology in Ethernet networks, which can increase determinism [59]. Ethernet has become a strong contender in the safety networking for the same reasons. In summary, Ethernet (1) has achieved dominance in the diagnostics domain primarily due to its speed, support for large data sizes, and web friendliness; (2) is achieving dominance in the control domain due mostly to the advent of switched networks and the resulting determinism, and (3) has become a strong contender in the safety domain due to the determinism capability as well as the ability to partition networks, for example, to support diagnostics, control, and safety on the same physical network [53,90].

Even more recently, there has been a strong consideration of wireless as the networking medium at all levels to support all functionalities, though it is not clear to what extent wireless will successfully supplant wired technologies. Research into characterization of wireless technologies in manufacturing environment will help to facilitate this movement to wireless [4,17]. It is the authors' opinion that a large portion of Ethernet solutions, especially at the higher levels, and in the domains of diagnostics and, to a lesser extent, control, will be replaced by wireless over the next decade [57].

The body of research around control networks is very deep and diverse. Networks present not only challenges of timing in control systems, but also opportunities for new control directions enabled by the distribution capabilities of control networks. For example, there has been a significant amount of work on networked control systems (NCSs) [5,7,16]. Despite this rich body of work, one important aspect of control networks remains relatively untouched in the research community, namely, the speed of the devices on the network. Practical application of control networks often reveals that device speeds dominate in determining the system performance to the point that the speed and determinism (network QoS parameters) of the network protocol are irrelevant [47,59,62]. Recent experimental work has shown that actual measured delays can vary significantly from manufacturer specifications [82]. Unfortunately, the academic focus on networks in the analysis of control network systems, often with assumptions of zero delay of devices, has served to further hide the fact that device speed is often the determining factor in assessing the NCS performance.

In light of the strong and diverse academic and industry focus on networks as well as the myriad of network technologies, the prospect of choosing a *best* network technology for a particular environment is ominous. The choice should be governed by a number of factors that balance upfront and recurring costs and performance to an objective function that best fits the particular environment. Thus, the best solution is necessarily application dependent [54]. A methodology is needed that supports the application of theory, experimental results, and analytics to a particular application domain so that the trade-offs can be quantified and a best solution determined [55].

This chapter explores the application of NCSs in the domains of control, diagnostics, and safety in manufacturing [57]. Specifically, Section 2.2 explores the parameterization of networks with respect to balancing QoS capabilities. This parameterization provides a basis for differentiating industrial network types. Section 2.3 introduces common network protocol approaches and differentiates them with respect to functional characteristics. This includes a discussion of all forms of Ethernet being employed in NCSs, from switched Ethernet to real-time Ethernet (RTE) and wireless Ethernet, as well as technologies that are leveraged to support Ethernet in various NCS environments, such as time synchronization. The importance of device performance is also explored. Section 2.4 presents a method for NCS evaluation that includes theoretical, experimental, and analytical components. In Section 2.5, network applications within the domain of manufacturing are explored; these include application subdomains of control, diagnostics, and safety, as well as different levels of control in the factory such as machine level, cell level, and system level. Within this section, an example is presented of a multilevel factory networking solution that supports networked control, diagnostics, and safety. This chapter concludes with a discussion of future trends in industrial networks with a focus on the continued movement to wireless networking technology.

2.2 Parameterization of Industrial Networks

The function of a network is to transmit data from one node to another. Different types of industrial networks use different mechanisms for allocating the bandwidth on the network to individual nodes and for resolving contentions among nodes. Briefly, there are three common mechanisms for allocating bandwidth: time-division multiplexing (TDM), random access (RA) with collision detection (CD), and RA with collision avoidance (CA). In TDM, the access time to the network is allocated in a round-robin fashion among the nodes, either by passing a token (e.g., ControlNet) or having a master poll the slaves (e.g., AS-I). Because the bandwidth is carefully allocated, no collisions will occur. If RA to the network is allowed, collisions can occur if two nodes try to access the network at the same time. The collision can be destructive or nondestructive. With a destructive collision, the data are corrupted, and both nodes must retransmit (e.g., hub-based Ethernet). With a nondestructive collision, one node keeps transmitting and the other backs off (e.g., CAN); in this case, the data are not corrupted. CA mechanisms (e.g., WiFi) use random delay times to minimize the probability that two nodes will try to transmit at the same time, but collisions can still occur. Collisions can be completely avoided if the network is effectively set up in a star configuration with intelligent routing, as is done with switched Ethernet. Network collision arbitration and avoidance mechanisms, and the most common network protocols that use them, will be discussed in detail in Section 2.3.

Although any network protocol can be used to send data, each network protocol has its pros and cons. In addition to the protocol, the type and amount of data to be transmitted are also important when analyzing the network performance: Will the network carry many small packets of data frequently or large packets of data infrequently? Must the data arrive before a given deadline? How many nodes will be competing for the bandwidth, and how will the contention be handled?

The QoS of a network is a multidimensional parameterized measure of how well the network performs its function; the parameter measures include the speed and bandwidth of a network (how much data can be transmitted in a time interval), the delay and jitter associated with data transmission (time for a message to reach its destination and repeatability of this time), and the reliability and security of the network infrastructure [89]. When using networks for control, it is often important to assess determinism as a QoS parameter, specifically evaluating whether message end-to-end communication times can be predicted exactly or approximately, and whether these times can be bounded.

In this section, we will review the basic QoS measures of industrial networks, with a focus on time delays, as they are typically the most important element determining the capabilities of an industrial control system. In Section 2.3, more detailed analysis of the delays for specific networks will be given. In Section 2.4, a methodology will be presented for evaluating the many dimensions of QoS along with other factors as they relate to the particular application environment.

2.2.1 Speed and Bandwidth

The *bandwidth* of an industrial network is given in terms of the number of bits that can be transmitted per second. Industrial networks vary widely in bandwidth, including CAN-based networks, which have a maximum data rate of 1 Mb/s, and Ethernet-based networks, which can support data rates upwards of 10 Gb/s,* although most networks currently used in the manufacturing industry are based on 100 Mb/s or 1 Gb/s Ethernet. DeviceNet, a commonly used network in the manufacturing industry, is based on CAN and has a maximum data rate of 500 kb/s. The *speed* is the inverse of the data rate, thus the time to transmit 1 bit of data over the network is $T_{bit} = 1$ μs for 1 Mb/s CAN and 10 ns for 100 Mb/s Ethernet.

The data rate of a network should be considered together with the packet size and overhead. Data are not just sent across the network one bit at a time. Instead, data are encapsulated into packets, with headers specifying, for example, the source and destination addresses of the packet, and often a checksum for detecting the transmission errors. All industrial networks have a minimum packet size; for example, the minimum packet size is 47 bits for CAN and 64 bytes for Ethernet. A minimum *interframe time* between two packets is required between subsequent messages to ensure that each packet can be distinguished individually; this time is specified by the network protocol and is included herein as part of the frame.

The transmission time for a message on the network can be computed from the network's data rate, the message size, and the distance between two nodes. As most of these quantities can be computed exactly (or approximated closely), transmission time is considered a deterministic parameter in a network system. The transmission time can be written as the sum of the frame time and the propagation time:

$$T_{tx} = T_{frame} + T_{prop}$$

where

T_{frame} is the time required to send the packet across the network
T_{prop} is the time for a message to propagate between any two devices

As the typical transmission speed in a communication medium is 2×10^8 m/s, the propagation time T_{prop} is negligible on a small scale. In the worst case, the propagation delays from one end to the other of the network cable for two typical control networks are $T_{prop} = 67.2$ μs for 2500 m Ethernet,† and $T_{prop} = 1$ μs for 100 m CAN. The propagation delay is not easily characterized because the distance between the source and destination nodes is not constant among different transmissions, but typically it is less than 1 μs (if the devices are less than 100 m apart). Some networks (e.g., Ethernet) are not a single trunk but have multiple links connected by hubs, switches, and/or routers that receive, store, and forward packets from one link to another. The delays associated with these interconnections can dominate propagation delays in a complex network and must also be considered when determining the transmission delays [61].

The frame time, T_{frame}, depends on the size of the data, the overhead, any padding, and the bit time. Let N_{data} be the size of the data in terms of bytes, N_{ovhd} be the number of bytes used as overhead, N_{pad} be the number of bytes used to pad the remaining part of the frame to meet the minimum frame size

* 10 Gb/s solutions are available with fiber-optic cabling.

† Because Ethernet uses Manchester biphase encoding, two bits are transmitted on the network for every bit of data.

requirement, and N_{stuff} be the number of bytes used in a stuffing mechanism (on some protocols).* The frame time can then be expressed by the following equation:

$$T_{frame} = \left[N_{data} + N_{ovhd} + N_{pad} + N_{stuff}\right] \times 8 \times T_{bit} \tag{2.1}$$

In Ref. [45], these values are explicitly described for Ethernet, ControlNet, and DeviceNet protocols.

The effective bandwidth of a control network will depend not only on the physical bandwidth, but also on the efficiency of encoding the data into packets (how much overhead is needed in terms of addressing and padding), how efficiently the network operates in terms of (long or short) interframe times, and whether network time is wasted due to message collisions. For example, to send one bit of data over a 500 kb/s CAN network, a 47 bit message is needed, requiring 94 μs. To send the same one bit of data over 10 Mb/s Ethernet, an 84-byte message is needed (64-byte frame size plus 20 bytes for interframe separation), requiring a 67.2 μs T_{frame}. Thus, even though the raw network speed is 20 times faster for Ethernet, the frame time is only 30% lower than CAN. This example shows that the network speed is only one factor that must be considered when computing the effective bandwidth of a network.

2.2.2 Delay and Jitter

The *time delay* on a network is the total time between the data being available at the source node (e.g., sampled from the environment or computed at the controller) and it being available at the destination node (received and decoded, where the decode level depends on where the delay is evaluated within the end-to-end communication). The *jitter* is the variability in the delay. Many control techniques have been developed for systems with constant time delays [21,78], but variable time delays can be much more difficult to compensate for, especially if the variability is large. Although time delay is an important factor to consider for control systems implemented over industrial networks, it generally is not well defined or studied by standards organizations defining network protocols [94].

To further explain the different components that go into the time delay and jitter on a network, consider the timing diagram in Figure 2.1 showing how messages are sent across a network. The source node A captures (or computes) the data of interest. There is some preprocessing that must be done to encapsulate the data into a message packet and encode it for sending over the network; this time is denoted T_{pre}. If the network is busy, the node may need to wait for some time T_{wait} for the network to become available. This waiting time is a function of the MAC mechanism of the protocol, which is categorized as part of layer 2 of the OSI model. Then, the message is sent across the network, taking time T_{tx} as described in Section 2.2.1. Finally, when the message is received at the destination node B, it must be decoded and postprocessed, taking time T_{post}. Thus, the total time delay can be expressed by the following equation:

$$T_{delay} = T_{pre} + T_{wait} + T_{tx} + T_{post} \tag{2.2}$$

The waiting time T_{wait} can be computed based on the network traffic, how many nodes there are, the relative priority of these nodes and the messages they are sending, and how much data they send. The pre- and postprocessing times T_{pre} and T_{post} depend on the devices. Often the network encoding and decoding are implemented in software or firmware. These times are rarely given as part of device specifications. As they can be the major sources of delay and jitter in a network, a more detailed discussion of these delays is given here.

* The bit-stuffing mechanism in DeviceNet is as follows: if more than 5 bits in a row are "1," then a "0" is added and vice versa. Ethernet uses Manchester biphase encoding and, therefore, does not require bit stuffing.

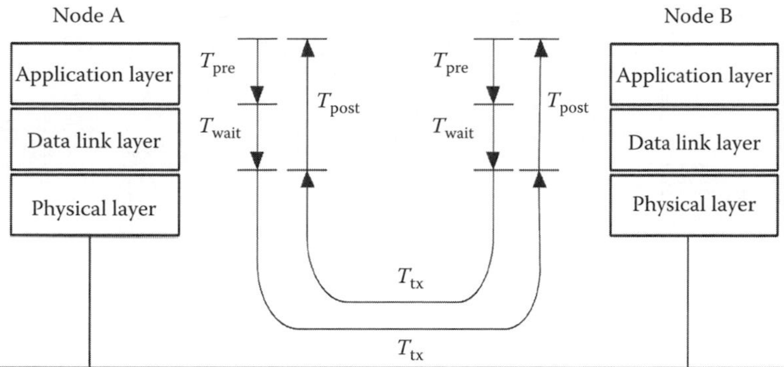

FIGURE 2.1 Timing diagram showing time spent sending a message from a source node to a destination node.

2.2.2.1 Pre- and Postprocessing Times

The preprocessing time at the source node depends on the device software and hardware characteristics. In many cases, it is assumed that the preprocessing time is constant or negligible. However, this assumption is not true in general; in fact, there may be noticeable differences in processing time characteristics between similar devices, and these delays may be significant. The postprocessing time at the destination node is the time taken to decode the network data into the physical data format and output it to the external environment.

In practical applications, it is very difficult to identify each individual timing component. However, a very straightforward experiment can be run with two nodes on the network. The source node A repeatedly requests data from a destination node B and waits until it receives a response before sending another request. Because there are only two nodes on the network, there is never any contention, and thus the waiting time is zero. The request–response frequency is set low enough that no messages are queued up at the sender's buffer. The message traffic on the network is monitored, and each message is time-stamped. The processing time of each request–response pair, that is, $T_{post} + T_{pre}$, can be computed by subtracting the transmission time from the time difference between the request and response messages. Because the time stamps are recorded all at the same location, the problem of time synchronization across the network is avoided.

Figure 2.2 shows the experimentally determined device delays for DeviceNet devices utilizing the aforementioned setup [47,59]. Note that for all devices, the mean delay is significantly longer than the minimum frame time in DeviceNet (94 μs), and the jitter is often significant. The uniform distribution of processing time at some of the devices is due to the fact that they have an internal sampling time, which is mismatched with the request frequency. Hence, the processing time recorded here is the sum of the actual processing time and the waiting time inside the device. The tested devices include photoeyes, input–output terminal blocks, mass flow controllers, and other commercially available DeviceNet devices.

A key point that can be taken from the data presented in Figure 2.2 is that the device processing time can be substantial in the overall calculation of T_{delay}. In fact, this delay often dominates over network delays. Thus, when designing industrial network systems to be used for control, device delay and delay variability should be considered as important factors when choosing the components. In the same manner, controller devices such as off-the-shelf PLCs typically specify scan times and interscan delays on the order of a few milliseconds, thus these controller delays can also dominate over network delays.

2.2.2.2 Waiting Time at Source Nodes

A message may spend time waiting in the queue at the sender's buffer and could be blocked from transmitting by other messages on the network. Depending on the amount of data the source node must send

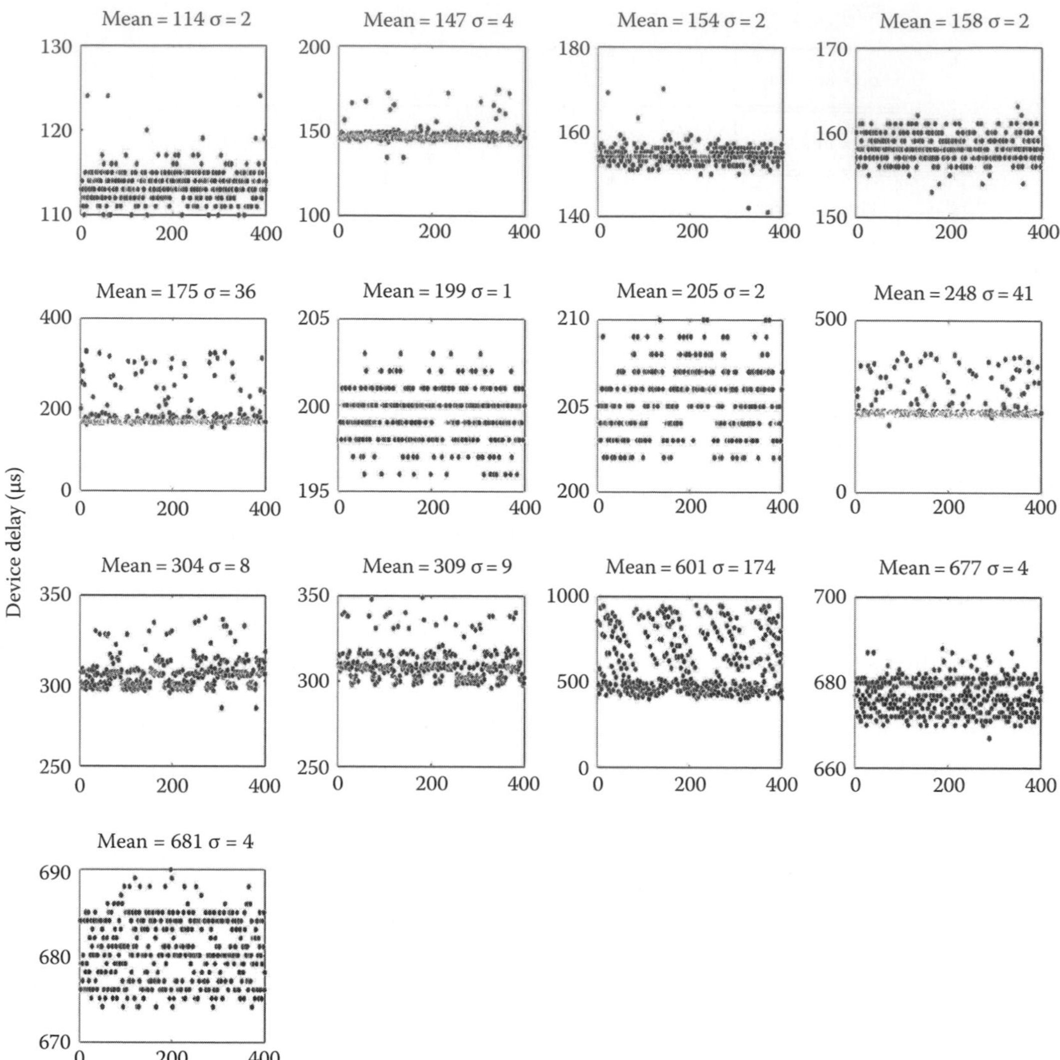

FIGURE 2.2 Device delays for DeviceNet devices in a request–response setup as described in Section 2.2.2.1. Delays are measured with only source and destination node communicating on the network and thus focus only on device delay jitter as described in Section 2.2.2. The stratification of delay times seen in some nodes is due to the fact that the smallest time that can be recorded is 1 μs.

and the traffic on the network, the waiting time may be significant. The main factors affecting waiting time are network protocol, message connection type, and network trafficload.

For control network operation, the message connection type must be specified. In a master–slave (MS) network,* there are three types of message connections: strobe, poll, and change of state (COS)/cyclic. In a *strobe* connection, the master device broadcasts a message to a group of devices, and these devices respond with their current condition. In this case, all devices are considered to sample new information

* In this context, an MS network refers to operation from an end-to-end application layer perspective. Master node applications govern the method by which information is communicated to and from their slave node applications. Note that, as it will be described further in Section 2.3.1, application layer MS behavior does not necessarily require corresponding MS behavior at the MAC layer.

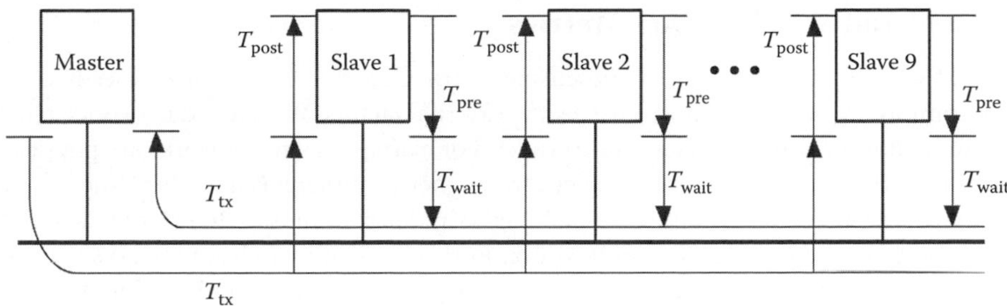

FIGURE 2.3 Waiting time diagram for a strobe message configuration.

at the same time. In a *poll* connection, the master sends individual messages to the polled devices and requests update information from them. Devices respond with new signals only after they have received a poll message. *COS/cyclic* devices send out messages either when their status is changed (COS) or periodically (cyclic). Although the COS/cyclic connection seems most appropriate from a traditional control systems point of view, strobe and poll are commonly used in industrial control networks [20].

For example, consider the strobe message connection in Figure 2.3. If Slave 1 is sending a message, the other eight devices must wait until the network medium is free. In a CAN-based DeviceNet network, it can be expected that Slave 9 will encounter the most waiting time because it has a lower priority on this priority-based network. However, in any network, there will be a nontrivial waiting time after a strobe, depending on the number of devices that will respond to the strobe.

Figure 2.4 shows experimental data of the waiting time of nine identical devices with a strobed message connection on a DeviceNet network; 200 pairs of messages (request and response) were collected. Each symbol denotes the mean, and the distance between the upper and lower bars equals two standard deviations. If these bars are over the limit (maximum or minimum), then the value of the limit is used instead. It can be seen in Figure 2.4 that the average waiting time is proportional to the node number (i.e., priority). Although all these devices have a very low variance of processing time, the devices with the lowest node numbers have a larger variance of waiting time than the others, because the variance of processing time occasionally allows a lower-priority device to access the idle network before a higher-priority one.

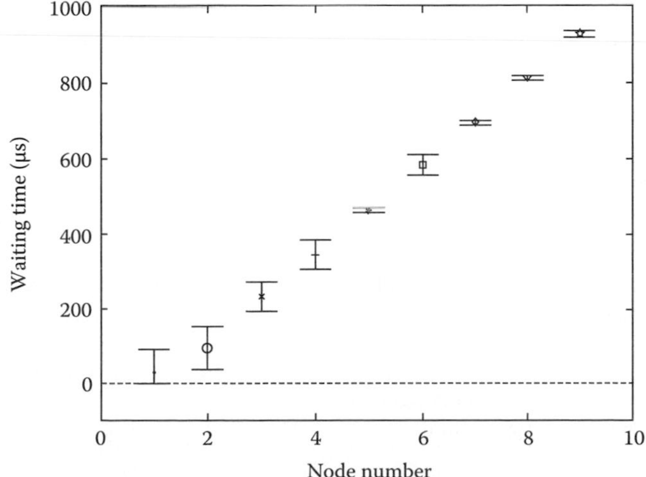

FIGURE 2.4 Nine identical devices with strobed message connection.

2.2.3 Wired and Wireless QoS Metrics

The advent of wireless has added a new dimension to the QoS metric set that focuses heavily on the QoS of the communication medium with respect to external factors. With wired networks, reliability of data transmission is an important factor to consider. For example, some networks are physically more vulnerable than others to data corruption by electromagnetic interference [4]. This issue is much more prevalent in wireless networks as there is a much higher potential exposure to external factors that can reduce the quality of the communication medium. Both wired and wireless networks can use handshaking by sending of acknowledgment (ACK) messages to increase the reliability. If no ACK message is received, the message is resent. These handshaking techniques increase the reliability of a network, but also add to the required overhead and thus decrease the overall effective bandwidth.

Wireless networks also commonly utilize techniques such as frequency hopping and transmission on multiple frequency channels to make the transmission more robust to radio frequency (RF) interference sources such as other industrial wireless networks, cell phones, and manufacturing equipment such as spot welders that generate RF noise [13,101]. These frequency hopping protocols can be quite elaborate and often distinguish the robustness of one wireless technology over another.

Security is another factor that must be considered for both wired and wireless networks. For all networked systems, security is of special concern when networks and operating systems are used that can be vulnerable to Internet-based attacks and viruses [22]. Most industrial fieldbuses were not designed to be highly secure, relying mainly on physical isolation of the network instead of authentication or encryption techniques. When some type of security is provided, the intent is more commonly to prevent accidental misuse of process data than to thwart malicious network attacks [96]. For wireless, the security issue has another dimension in that the transmissions themselves can be intercepted rather easily by a listening device that understands the transmission protocol. The most common solution to this problem is encryption of the transmission through mechanisms such as virtual private network (VPN). While VPN provides for secure keyed *tunnels*, mechanisms such as these also introduce transmission overhead and delay as shown in Figure 2.5. Thus, the *cost of security* must be considered when evaluating the (especially wireless) networked solutions.

Other QoS and related metrics such as device power requirements, distances between nodes, and desire to *not* generate interference (i.e., coexist) with other networks can weigh heavily in the choice of

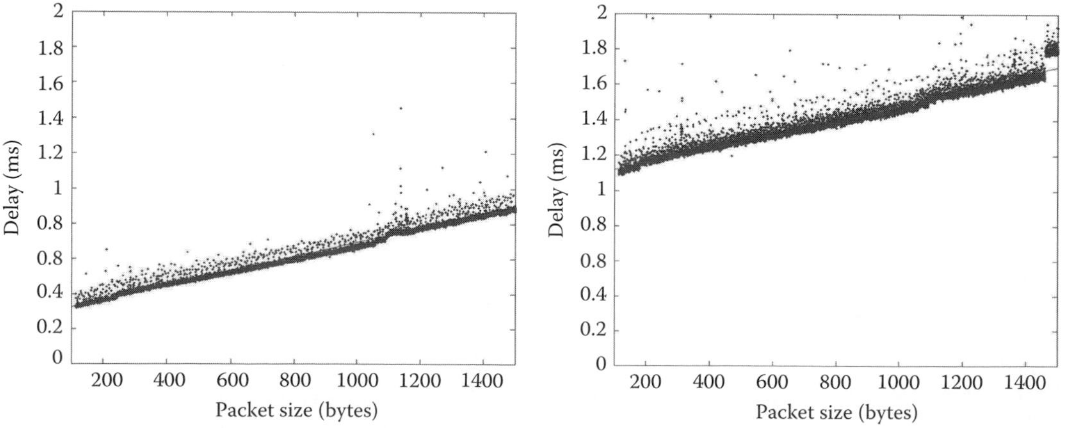

FIGURE 2.5 Example of the impact of VPN security encryption on Ethernet network performance (the figure on the left is the round-trip delay measured for user datagram protocol [UDP]* packet transmission on a 100 Mbps switched Ethernet network; the figure on the right is the delay measured for the same system, but with VPN encryption added).

* UDP is a relatively fast unacknowledged communication service utilized in Ethernet, often utilized as a baseline in defining Ethernet system performance.

a wireless protocol and solution [4,13,17]. The key to evaluating solutions against metrics is to be able to quantify all of the important metrics and evaluate them collectively in the application environment. This process is discussed in Section 2.4.

2.2.4 Network QoS versus System Performance

When a network is used in the feedback loop of a control system, the performance of the system depends not only on the QoS of the network, but also on how the network is used (e.g., sample time, message scheduling, and node prioritization) [47,49]. For example, consider a continuous-time control system that will be implemented with networked communication. Figure 2.6 shows how the control performance varies vs. sampling period in the cases of continuous control, digital control, and networked control. The performance of the continuous control system is independent of the sample time (for a fixed control law). The performance of the digital control system approaches the performance of the continuous-time system as the sampling frequency increases [30]. In an NCS, the performance is worse than the digital control system at low frequencies, due to the extra delay associated with the network (as described in Section 2.2.2). Also, as the sampling frequency increases, the network starts to become saturated, data packets are lost or delayed, and the control performance rapidly degrades. Between these two extremes lies a *sweet spot* where the sample period is optimized to the control and networking environment. Note that, as discussed in Section 2.2.2, the device delay can comprise a significant portion of the end-to-end delay; thus, the optimal sampling period can often be more a function of the device speed than the network speed, especially for faster networks with minimal collisions, such as switched Ethernet.

Typical performance criteria for feedback control systems include overshoot to a step reference, steady-state tracking error, phase margin, or time-averaged tracking error [29]. The performance criteria in Figure 2.6 can be one of these or a combination of them. Due to the interaction of the network and control requirements, the selection of the best sampling period is a compromise. More details on the performance computation and analysis of points A, B, and C in Figure 2.6 can be found in [47], including simulation and experimental results that validate the overall shape of the chart.

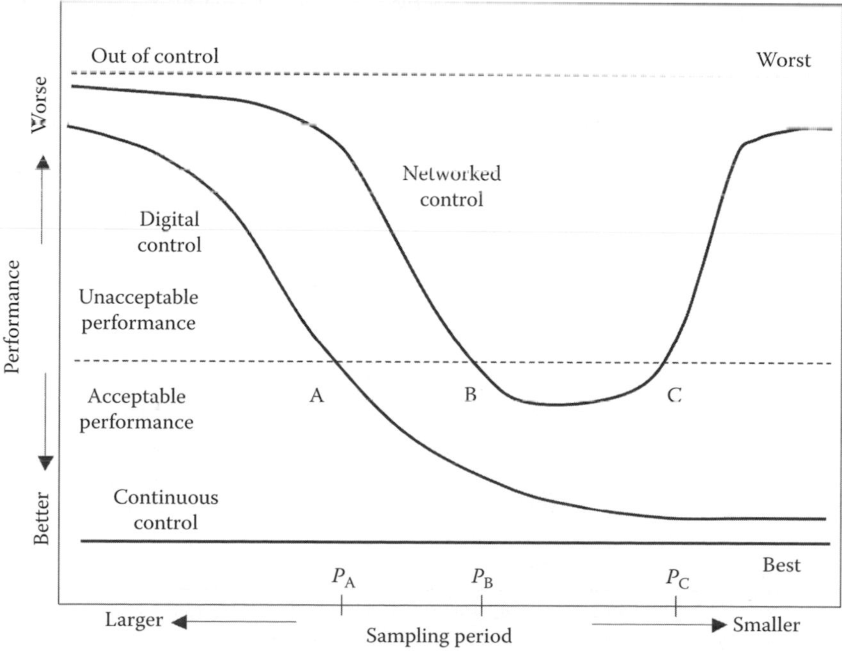

FIGURE 2.6 Performance comparison of continuous control, digital control, and networked control, as a function of sampling frequency.

2.3 Differentiation of Industrial Networks

Networks can be differentiated either by their protocol (at any or all levels of the ISO–OSI seven-layer reference model [3]) or by their primary function (control, diagnostics, and safety). These dimensions of differentiation are somewhat related. In this section, we first define how network protocols are categorized technically with respect to timing and then discuss the different types of protocols that are commonly used in industrial networks. In Section 2.5, we describe how these different types of networks are used for different functions.

2.3.1 Categorization of Networks

When evaluating the network QoS parameters associated with timeliness, determinism, etc., the protocol functionality at the data link layer is the primary differentiator among network protocol types. Specifically, the MAC sublayer protocol within the data link layer describes the protocol for obtaining access to the network. The MAC sublayer thus is responsible for satisfying the time-critical/real-time response requirement over the network and for the quality and reliability of the communication between network nodes [41]. The discussion, categorization, and comparison in this section thus focus on the MAC sublayer protocols.

As noted in Section 2.2, there are three main types of medium access control used in control networks: TDM (such as MS or token-passing [TP]), RA with retransmission when collisions occur (e.g., Hub-based Ethernet and most wireless mechanisms), and RA with prioritization for collision arbitration (e.g., CAN). Implementations can be hybrids of these types; for example, switched Ethernet combines time-division multiplexed (TDM) and RA. Note that, regardless of the MAC mechanism, most network protocols support some form of MS communication at the application level; however, this appearance of TDM at the application level does not necessarily imply the same type of parallel operation at the MAC level. Within each of these three MAC categories, there are numerous network protocols that have been defined and used.

Surveys of the types of control networks used in industry show a wide variety of networks in use; see Table 2.1 and also [31,32,48,65,79,94]. Networks can be classified according to type: RA with CD, CA, or arbitration on message priority (AMP); or TDM using TP or MS.

2.3.2 Time-Division Multiplexing

TDM can be accomplished in one of two ways: MS or TP. In an MS network, a single master polls multiple slaves. Slaves can send data over the network only when requested by the master; there are no

TABLE 2.1 Popular Fieldbuses

Network	Type	Users (%)	Max. Speed	Max. Devices
Ethernet TCP/IP	RA/CD	78	1 Gb/s	1024
Modbus	TDM/MS	48	35 Mb/s	32
DeviceNet	RA/AMP	47	500 kb/s	64
ControlNet	TDM/TP	39	5 Mb/s	99
WiFi (IEEE 802.11b)	RA/CA	35	11 Mb/s	Not specified
Modbus TCP	TDM/MS	34	1 Gb/s	256
PROFIBUS-DP	TDM/MS and TP	27	12 Mb/s	127
AS-I	TDM/MS	17	167 kb/s	31

Sources: Grid Connect, The Grid Connect Fieldbus comparison chart, http://www.synergetic.com/compare.htm; Montague, J., *Control Eng.*, 52(3), 2005.

Note: The maximum speed depends on the physical layer, not the application-level protocol. Note that the totals are more than 100% because most companies use more than one type of bus.

collisions, as the data transmissions are carefully scheduled by the master. ATP network has multiple masters or peers. The token bus protocol (e.g., IEEE 802.4) allows a linear, multidrop, tree-shaped, or segmented topology [100]. The node that currently has the token is allowed to send data. When it has finished sending data, or the maximum token holding time has expired, it *passes* the token to the next logical node on the network. If a node has no message to send, it just passes the token to the successor node. The physical location of the successor is not important because the token is sent to the logical neighbor. Collision of data frames does not occur, as only one node can transmit at a time. Most TP protocols guarantee a maximum time between network accesses for each node, and most also have provisions to regenerate the token if the token holder stops transmitting and does not pass the token to its successor. AS-I, Modbus, and Interbus-S are typical examples of MS networks, while PROFIBUS and ControlNet are typical examples of TP networks. Each peer node in a PROFIBUS network can also behave like a master and communicate with a set of slave nodes during the time it holds the token [74].

TP networks are deterministic because the maximum waiting time before sending a message frame can be characterized by the token rotation time. At high utilizations, TP networks are very efficient and fair. There is no time wasted on collisions, and no single node can monopolize the network. At low utilizations, they are inefficient due to the overhead associated with the TP protocol. Nodes without any data to transmit must still receive and pass the token.

Waiting time in a TDM network can be determined explicitly once the protocol and the traffic to be sent on the network are known. For TP networks, the node with data to send must first wait to receive the token. The time it needs to wait can be computed by adding up the transmission times for all of the messages on nodes ahead of it in the logical ring. For example, in ControlNet, each node holds the token for a minimum of 22.4 μs and a maximum of 827.2 μs.

In MS networks, the master typically polls all slaves every cycle time. Slaves cannot transmit data until they are polled. After they are polled, there is no contention for the network so the waiting time is zero. If new data are available at a slave (e.g., a limit switch trips), the slave must wait until it is polled before it can transmit its information. In many MS networks (such as AS-I), the master will only wait for a response from a slave until a timer has expired. If the slave does not respond within the timeout value for several consecutive polls, it is assumed to have dropped off the network. Also, every cycle time, the master attempts to poll an inactive slave node (in a round-robin fashion) [8]. In this way, new slaves can be added to the network and will be eventually noticed by the master.

2.3.3 RA with Collision Arbitration: CAN (CSMA/AMP)

CAN is a serial communication protocol developed mainly for applications in the automotive industry, also capable of offering good performance in other time critical industrial applications. The CAN protocol is optimized for short messages and uses a carrier sense multiple access (CSMA)/AMP medium access method. Thus, the protocol is message oriented, and each message has a specific priority that is used to arbitrate access to the bus in case of simultaneous transmission. The bit stream of a transmission is synchronized on the start bit, and the arbitration is performed on the following message identifier, in which a logic zero is dominant over a logic one. A node that wants to transmit a message waits until the bus is free and then starts to send the identifier of its message bit by bit. Conflicts for access to the bus are solved during transmission by an arbitration process at the bit level of the arbitration field, which is the initial part of each frame. Hence, if two devices want to send messages at the same time, they first continue to send the message frames and then listen to the network. If one of them receives a bit different from the one it sends out, it loses the right to continue to send its message, and the other wins the arbitration. With this method, an ongoing transmission is never corrupted, and collisions are nondestructive [45].

DeviceNet is an example of a technology based on the CAN specification that has received considerable acceptance in manufacturing applications at the device level. The DeviceNet specification is based on the standard CAN with an additional application and physical layer specification [20,45].

The DeviceNet frame has a total overhead of 47 bits, which includes start of frame, arbitration (11-bit identifier), control, cyclic redundancy check (CRC), ACK, end of frame, and intermission fields. The size of a data field is between 0 and 8 bytes. The DeviceNet protocol uses the arbitration field to provide source and destination addressing as well as message prioritization.

The major disadvantage of CAN compared with the other networks is the slow data rate, limited by the network length. Because of the bit synchronization, the same data must appear at both ends of the network simultaneously. DeviceNet has a maximum data rate of 500 kb/s for a network of 100 m. Thus, the throughput is limited compared with other control networks. CAN is also not suitable for the transmission of messages of large data sizes, although it does support fragmentation of data that is more than 8 bytes into multiple messages.

2.3.4 Ethernet-Based Networks

The proliferation of the Internet has led to the pervasiveness of Ethernet in both homes and businesses. Because of its low cost, widespread availability, and high communication rate, Ethernet has been proposed as the ideal network for industrial automation [19,70]. Previously, some had questioned whether Ethernet would become the de facto standard for automation networks, making all other solutions obsolete [27,88]. The concerns primarily centered around the fact that standard Ethernet (IEEE 802.3) is not a deterministic protocol, and network QoS cannot be guaranteed [19,45]. Collisions can occur on the network, and messages must be retransmitted after random amounts of time. To address this inherent nondeterminism, many different *flavors* of Ethernet were proposed for use in industrial automation. Several of these add layers on top of standard Ethernet or on top of the TCP/IP protocol suite to enable the behavior of Ethernet to be more deterministic [25,27,42]. In this way, the network solutions may no longer be *Ethernet* other than at the physical layer; they may use the same hardware but are not interoperable. As noted in [48], message transmission does not always lead to successful communication: "just because you can make a telephone ring in Shanghai doesn't mean you can speak Mandarin." An effective and accepted solution in recent years has been the utilization of switches to manage the Ethernet bandwidth utilizing a TDM approach among time-critical nodes. Rather than repeat the survey of approaches to industrial Ethernet in [25], in this section, the general MAC protocol of Ethernet is outlined, and the general approaches that are used with Ethernet for industrial purposes are discussed.

Ethernet is an RA network, also often referred to as CSMA. Each node listens to the network and can start transmitting at any time that the network is free. Typically, once the network is clear, a node must wait for a specified amount of time (the interframe time) before sending a message. To reduce collisions on the network, nodes wait an additional random amount of time called the backoff time before they start transmitting. Some types of messages (e.g., MAC layer ACKs) may be sent after a shorter interframe time. Priorities can be implemented by allowing for shorter interframe times for higher-priority traffic. However, if two nodes start sending messages at the exact same time (or if the second node starts transmitting before the first message arrives at the second node), there will be a collision on the network. Collisions in Ethernet are destructive; the data are corrupted, and the messages must be resent.

In the remainder of this section, various flavors and concepts associated with wired Ethernet are discussed.

2.3.4.1 Hub-Based Ethernet (CSMA/CD)

Hub-based Ethernet uses hub(s) to interconnect the devices on a network; this type of Ethernet is common in the office environment. When a packet comes into one hub interface, the hub simply broadcasts the packet to all other hub interfaces. Hence, all of the devices on the same network receive the same packet simultaneously, and message collisions are possible. Collisions are dealt with utilizing the CSMA/CD protocol as specified in the IEEE 802.3 network standard [9,14,92].

This protocol operates as follows: when a node wants to transmit, it listens to the network. If the network is busy, the node waits until the network is idle; otherwise, it can transmit immediately (assuming an interframe delay has elapsed since the last message on the network). If two or more nodes listen to the idle network and decide to transmit simultaneously, the messages of these transmitting nodes collide, and the messages are corrupted. While transmitting, a node must also listen to detect a message collision. On detecting a collision between two or more messages, a transmitting node transmits 32 jam bits and waits a random length of time to retry its transmission. This random time is determined by the standard binary exponential backoff algorithm: The retransmission time is randomly chosen between 0 and (2^i) slot times, where i denotes the ith collision event detected by the node, and one slot time is the minimum time needed for a round-trip transmission. However, after 10 collisions have been reached, the interval is fixed at a maximum of 1023 slots. After 16 collisions, the node stops attempting to transmit and reports failure back to the node microprocessor. Further recovery may be attempted in higher layers [92].

The Ethernet data payload size is between 46 and 1500 bytes. There is a nonzero minimum data size requirement because the standard states that valid frames must be at least 64 bytes long (which includes 18 bytes of overhead). If the data portion of a frame is less than 46 bytes, the pad field is used to fill out the frame to the minimum size.

Several solutions were proposed for using this form of Ethernet in control applications [19]; however, to a large extent, these solutions have been rendered moot with the proliferation of switched Ethernet as described below. On the other hand, many of the same issues reappear with the migration to wireless Ethernet for control.

2.3.4.2 Switched Ethernet (CSMA/CA)

Switched Ethernet utilizes switches to subdivide the network architecture, thereby avoiding collisions, increasing network efficiency, and improving determinism. It is widely used in manufacturing applications. The main difference between switch- and hub-based Ethernet networks is the intelligence of forwarding packets. Hubs simply pass on incoming traffic from any port to all other ports, whereas switches learn the topology of the network and forward packets to the destination port only. In a starlike network layout, every node is connected with a single cable to the switch as a full-duplex point-to-point link. Thus, collisions can no longer occur on any network cable. Switched Ethernet relies on this star cluster layout to achieve this collision-free property.

Switches employ the cut-through or store-and-forward technique to forward packets from one port to another, using per-port buffers for packets waiting to be sent on that port. Switches with cut-through first read the MAC address and then forward the packet to the destination port according to the MAC address of the destination and the forwarding table on the switch. On the other hand, switches with store-and-forward examine the complete packet first. Using the CRC code, the switch will first verify that the frame has been correctly transmitted before forwarding the packet to the destination port. If there is an error, the frame will be discarded. Store-and-forward switches are slower, but will not forward any corrupted packets.

Although there are no message collisions on the networks, congestion may occur inside the switch when one port suddenly receives a large number of packets from the other ports. If the buffers inside the switch overflow, messages will be lost [25]. Three main queuing principles are implemented inside the switch in this case. They are first-in-first-out (FIFO) queue, priority queue, and per-flow queue. The FIFO queue is a traditional method that is fair and simple. However, if the network traffic is heavy, the network QoS for timely and fair delivery cannot be guaranteed. In the priority queuing scheme, the network manager reads some of the data frames to distinguish which queues will be more important. Hence, the packets can be classified into different levels of queues. Queues with high priority will be processed first followed by queues with low priority until the buffer is empty. With the per-flow queuing operation, queues are assigned different levels of priority (or weights). All queues are then processed one by one according to priority; thus, the queues with higher priority will generally have higher performance and could potentially block queues with lower priority [19].

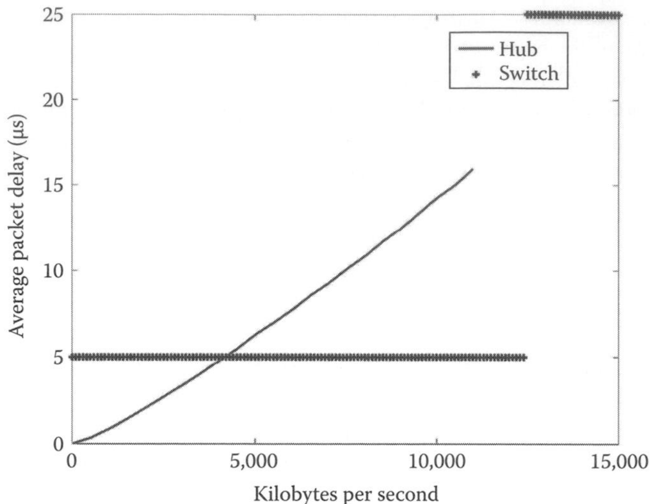

FIGURE 2.7 Packet delay as a function of node traffic for a hub and a switch [61]. Simulation results with baselines (delay magnitudes) computed from experiments.

Thus, although switched Ethernet can avoid the extra delays due to collisions and retransmissions, it can introduce delays associated with buffering and forwarding. This trade-off can be seen in Figure 2.7, which shows the average packet delay as a function of node traffic. The switch delay is small but constant until the buffer saturates and packets must be resent; the hub delay increases more gradually. Examples of timing analysis and performance evaluation of switched Ethernet can be found in [43,61,72,98].

2.3.4.3 Industrial Ethernet

In an effort to package Ethernet to be more suitable for industrial applications, a number of *industrial Ethernet* protocols have emerged. These include EtherNet/IP, Modbus/TCP, and PROFINET. While these protocol specifications vary to some extent at all levels of the OSI model, they all fundamentally utilize or recommend switched Ethernet technology as defined in the previous subsection. Thus, the differences in performance between industrial Ethernet technologies lie more with the devices than the protocols. For example, in an effort to understand the trade-offs between industrial Ethernet technologies, two common industrial Ethernet protocols, EtherNet/IP and PROFINET, were compared in the areas of architecture principles, technologies incorporated, performance, ease of use, diagnostics capabilities, and network management capabilities* [56]. As part of this effort, parallel multilayer switched Ethernet testbeds were developed utilizing each of these technologies, where the network layout was representative of the structure being utilized at a leading automotive manufacturer.

The results indicate that both protocols and protocol devices are fairly similar and are adequate to the task of providing industrial networking capabilities at the PLC level and higher [56]. However, distinct differences were observed, such as those illustrated in Figure 2.8, that indicate additional improvements in device performance may be needed if the solution is to be deployed down to the I/O level.

2.3.4.4 Ethernet for High-Speed and Time-Critical Applications

In the context of NCSs, communication networks that can guarantee the delivery of a transmitted message before a preset operational deadline or within a time interval are called real-time networks. As NCSs are

* Many industrial Ethernet protocols are available, with Modbus/TCP the most widely utilized [32,53]; however, in this instance, the manufacturer was interested in comparing these two industrial Ethernet protocols.

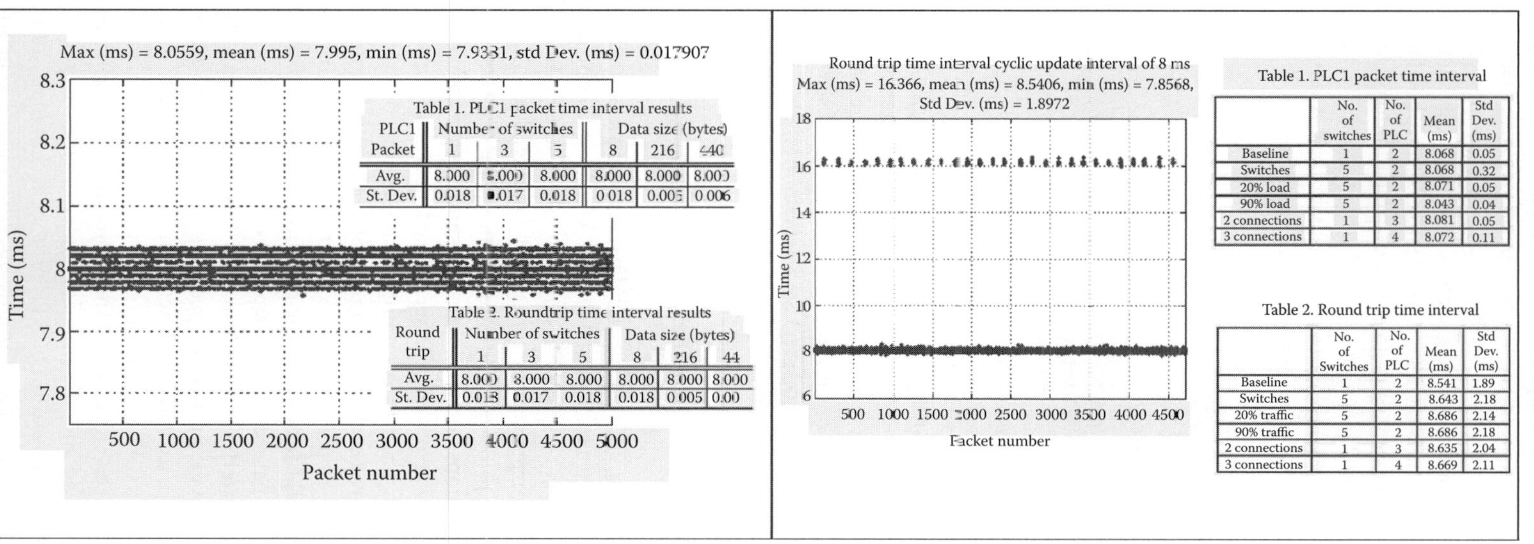

Max (ms) = 8.0559, mean (ms) = 7.995, min (ms) = 7.9331, std Dev. (ms) = 0.017907
8.3
8.2
8.1
8
7.9
7.8
Time (ms)
Table 1. PLC1 packet time interval results
PLC1 Packet | Number of switches 1 | 3 | 5 | Data size (bytes) 8 | 216 | 440
Avg. | 8.000 | 8.000 | 8.000 | 8.000 | 8.000 | 8.000
St. Dev. | 0.018 | 0.017 | 0.018 | 0.018 | 0.003 | 0.006
Table 2. Roundtrip time interval results
Round trip | Number of switches 1 | 3 | 5 | Data size (bytes) 8 | 216 | 44
Avg. | 8.000 | 8.000 | 8.000 | 8.000 | 8.000 | 8.000
St. Dev. | 0.013 | 0.017 | 0.018 | 0.018 | 0.005 | 0.00
500 1000 1500 2000 2500 3000 3500 4000 4500 5000
Packet number
Round trip time interval cyclic update interval of 8 ms
Max (ms) = 16.366, mean (ms) = 8.5406, min (ms) = 7.8568,
Std Dev. (ms) = 1.8972
18
16
14
12
10
8
6
Time (ms)
500 1000 1500 2000 2500 3000 3500 4000 4500
Packet number
Table 1. PLC1 packet time interval
| No. of switches | No. of PLC | Mean (ms) | Std Dev. (ms)
Baseline | 1 | 2 | 8.068 | 0.05
Switches | 5 | 2 | 8.068 | 0.32
20% load | 5 | 2 | 8.071 | 0.05
90% load | 5 | 2 | 8.043 | 0.04
2 connections | 1 | 3 | 8.081 | 0.05
3 connections | 1 | 4 | 8.072 | 0.11
Table 2. Round trip time interval
| No. of Switches | No. of PLC | Mean (ms) | Std Dev. (ms)
Baseline | 1 | 2 | 8.541 | 1.89
Switches | 5 | 2 | 8.643 | 2.18
20% traffic | 5 | 2 | 8.686 | 2.14
90% traffic | 5 | 2 | 8.686 | 2.18
2 connections | 1 | 3 | 8.635 | 2.04
3 connections | 1 | 4 | 8.669 | 2.11

FIGURE 2.8 Sample plots of round-trip timing measurements for PROFINET (5a) and EtherNet/IP (5b). Here, the round-trip times for packets between two PLCs are plotted for a large number of packets to obtain a time distribution. The embedded tables represent the consolidation of a number of these graphs, where the testing environment is modified in terms of switches between sender and receiver, data size transmitted, and network loading (the plots shown represent the *baseline* case).

being used for faster and more time-critical applications, there is growing need to continually reduce the time interval between messages and to specify operational deadlines with greater precision.

Ethernet networks, while growing in popularity for NCSs, are conventionally best-effort networks without real-time guarantees. Network parameters such as bandwidth, latency, and jitter significantly degrade the real-time performance of an NCS. RTE is a collective term applied to all variants of the core Ethernet technology designed to meet varying degrees of real-time requirements.

The IEC 61784-2 specification [36] includes a categorization of existing RTE technologies into 16 *communication profile families* (CPFs) based on their real-time capabilities. According to the specification, real-time capabilities are graded based on nine performance indicators:

1. Average message latency
2. Number of networked nodes
3. Basic network topology
4. Number of switches between end nodes
5. Throughput (for real-time traffic)
6. Throughput (for coexisting non-real-time traffic)
7. Clock synchronization accuracy
8. Jitter or variation in the period of cyclic messages
9. Redundancy or recovery time

As an example, the automotive manufacturing industry currently defines a high-speed automation network as being able to support a 10 ms poll interval with less than 5 ms latency with about 50 connected nodes [77]. Using the IEC 61784-2 specification, the automotive manufacturing requirements call for an RTE in CPF-2. Networked control systems for motion control (e.g., of servo motors or drives) or for the control of highly dynamic chemical processes (e.g., chemical processes used in the manufacture of semiconductor microchips) may require much shorter time intervals with more precise real-time guarantees.

2.3.4.4.1 Real-Time Ethernet

Most RTE protocols attempt to first address the nondeterminism in communication delay introduced by the CSMA-CD network arbitration algorithm. A common approach used to eliminate the need for collision arbitration is to specify a single management node or master for the network and to impose an MS, poll-response schedule on all connected nodes (e.g., EtherNet/IP and PROFINET RT). Many real-time protocols additionally mandate the use of specialized network switches to minimize packet collisions. Modern managed switches are able to significantly improve network determinism and speed by prioritizing Ethernet packets if they are generated by a real-time software application, part of high-priority communication session, or containing time-sensitive data. Switches that are able to analyze each packet across multiple layers of the OSI model and adapt accordingly are colloquially called *Smart Switches*. In order to further ensure that the network delay is deterministic and uniform for all the nodes in the network, some RTE protocols use isochronous communication that have either a passed token or a preset communication time table, for example, EtherCAT, SERCOS, Ethernet Powerlink, TCnet, and CIP motion. EtherCAT, for example, uses isochronous transmissions to provide very high speed (60 μs round-trip time for a network with 1000 nodes) and very low jitter (1 μs variation) between messages [24]. Like many other high-speed automation protocols, EtherCAT requires precise clock synchronization between all connected nodes. Precisely synchronized clocks allow the protocol to more efficiently utilize available bandwidth by reducing the overhead and idle time associated with carrier synchronization. Clock synchronization is so critical to the operation of RTE protocols that many of them include a clock synchronization algorithm. CIP motion, for example, includes a clock synchronization algorithm called CIP Sync [18].

2.3.4.4.2 Precise Clock Synchronization

The need for precise clock synchronization across a network extends beyond the requirements of RTEs. Modern manufacturing processes are increasingly data driven. Precisely, time-stamped measurements from machines enable improved diagnosis, control, and optimization.

NCSs in current manufacturing plants generally operate at periodic intervals of 10 ms or larger [6]. In order to ensure the fidelity of sampled events, the authors in [85] suggest that clock synchronization between networked nodes be accurate within 1 ms. Clock synchronization accuracy required for a system is defined by either the maximum sampling rate within the system or the minimum time difference between monitored events. NCS sampling intervals are continually shrinking with tighter process control requirements. For example, the authors in [71,87] identify fault conditions such as electric arcing in the semiconductor fabrication process that occur over a time frame of 1–100 μs and state that these fast events may also have to be accurately reported and addressed by an NCS in the future [44]. In order to accurately record and respond to events occurring in the order of microseconds, clocks must presumably be synchronized to within a microsecond of accuracy.

For most applications including personal computing and transactions over the Internet, Network Time Protocol (NTP) [50] is used to synchronize clocks. NTP is an application layer Ethernet protocol designed to synchronize clocks over a variable latency network. When used in a local area network, the NTP algorithm has been shown to achieve submillisecond synchronization accuracy [63] between a time-server and several slaves.

The current state-of-the-art Ethernet clock synchronization algorithm is the IEEE 1588 Version 2 Precision Time Protocol (PTP) [23,37]. It provides submicrosecond clock synchronization of nodes across a local area Ethernet network [2]. The several orders of magnitude improvement in the accuracy of PTP over NTP is gained by using specialized network interface hardware to time stamp and decode PTP packets. The need for specialized network interfaces and switches constrains PTP to be used only in industrial local area networks where submicrosecond-level synchronization is required. Typically, these are applications where diagnostic data must be time-stamped with microsecond accuracy or where the sampled data must be reported at the rate of several kiloHertz mandate the use of PTP. Some RTEs such as SERCOS and EtherCAT adopt PTP as their default clock synchronization algorithm [24,83].

Figure 2.9 shows the result of an experiment where PTP and NTP were used to synchronize several slave clocks to a single time-server over a local Ethernet network subject to background network traffic

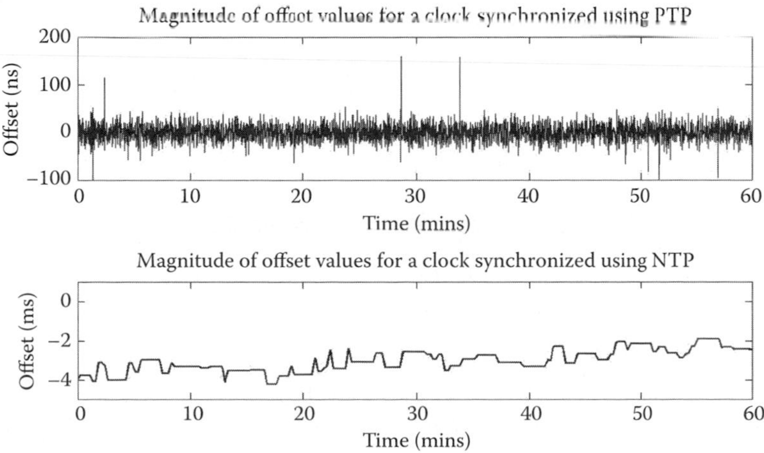

FIGURE 2.9 A comparison of clock offsets derived from the PTP and NTP protocols (note the different time scales on the *y*-axis).

consistent with an automation environment (<50% bandwidth utilization). The figure shows that PTP accuracy is close to five orders of magnitude better for the same network conditions.

2.3.5 Impact of Ethernet Application Layer Protocols: OPC

OPC is an open communication standard that is often used in industry to connect SCADA systems and human–machine interfaces (HMIs) to control systems and fieldbus networks [34,51,86].* It is based on the Microsoft DCOM standard [68] and is the dominant factory-floor application layer protocol utilized for diagnostics and is beginning to be used for sequential control [67]. The main benefit of OPC is that it allows any products that support the standard to share data. Although OPC actually consists of many different communication specifications, its most commonly used form is called data access, which supports both client–server and publisher–subscriber communication models. The server maintains a table of data values, and the client can read or write updates. The overhead associated with OPC (and DCOM in general) is significant, as shown in Figure 2.10. Most of this delay is due to the software implementation of the OPC protocol; OPC was never intended for a real-time environment. However, it is very useful to push data up from the low-level controls to the higher-level supervisors or diagnostic systems. It can also be used to send commands down from the HMIs to the control systems. Its high level of interoperability enables the connection of multiple control systems from different vendors in a unified manner. However, when OPC is used to send control data, the additional delay caused by the higher-level application layer protocol must be considered.

2.3.6 Wireless Networks

The move to wireless has resulted in yet another explosion of protocol technologies due to the fact that the change in the media from a wire to air has resulted in the need for higher level of consideration of many external issues (e.g., noise, coexistence, and security) as well as the opportunity for the development of more elaborate protocols to address these issues (e.g., frequency hopping, dynamic transmission power algorithms, and new CA and backoff-style algorithms) [4,13,17]. Table 2.2 provides a high-level

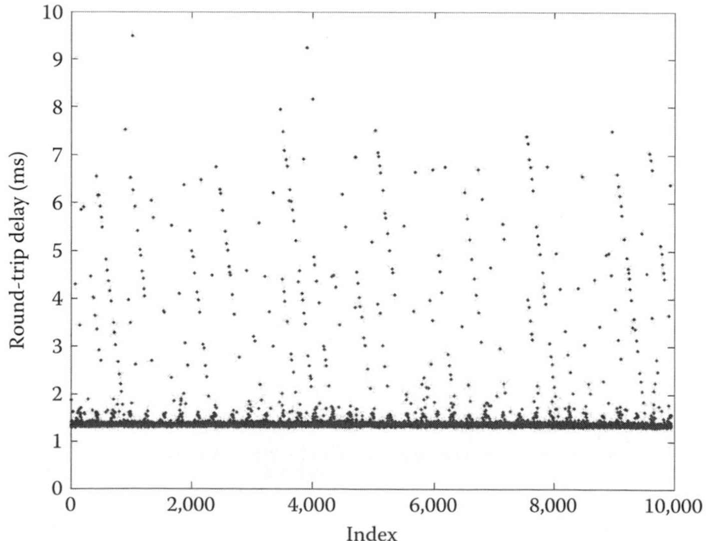

FIGURE 2.10 Illustration of overhead delay associated with OPC over a 100 Mb/s switched Ethernet network. Mean = 1.5 ms with standard deviation = 0.03 ms; for comparison purposes, the mean UDP round-trip time over the same network is 0.33 ms.

* While OPC originally stood for *OLE for Process Control*, the official stance of the OPC Foundation is that OPC is no longer an acronym, and the technology is simply known as OPC.

TABLE 2.2 High-Level Description of Common Wireless Technologies Illustrating Trade-Offs

Characteristic	802.11	Bluetooth	ZigBee
Range	High	Medium, depends on class	Low
Power consumption	High	Medium	Low
Data rate	Very high	High	Low
Optimal data size	High	Medium	Low
Redundancy	Low	Low	High

comparison of three technologies that are utilized in manufacturing that offer somewhat complementary capabilities. From the table, it is clear that, at least at this time, one size does not fit all when it comes to matching a wireless protocol to application requirements. In the remainder of this section, 802.11 and Bluetooth are explored in detail. For additional discussion of ZigBee, see Refs. [13,103].

2.3.6.1 Wireless Ethernet (CSMA/CA)

Wireless Ethernet, based on the IEEE 802.11 standard, can replace wired Ethernet in a transparent way as it implements the two lowest layers of the ISO–OSI model [3,101]. Besides the physical layer, the biggest difference between 802.11 and 802.3 is in the medium access control. Unlike wired Ethernet nodes, wireless stations cannot *hear* a collision. A CA mechanism is used but cannot entirely prevent collisions. Thus, after a packet has been successfully received by its destination node, the receiver sends a short ACK packet back to the original sender. If the sender does not receive an ACK packet, it assumes that the transmission was unsuccessful and retransmits.

The CA mechanism in 802.11 works as follows. If a network node wants to send while the network is busy, it sets its backoff counter to a randomly chosen value. Once the network is idle, the node waits first for an interframe space and then for the backoff time before attempting to send (see Figure 2.11). If another node accesses the network during that time, it must wait again for another idle interval. In this way, the node with the lowest backoff time sends first. Certain messages (e.g., ACK) may start transmitting after a shorter interframe space; thus, they have a higher priority. Collisions may still occur because of the random nature of the backoff time; it is possible for two nodes to have the same backoff time.

Several refinements to the protocol also exist. Nodes may reserve the network either by sending a request to send message or by breaking a large message into many smaller messages (fragmentation); each successive message can be sent after the smallest interframe time. If there is a single master node on the network, the master can poll all the nodes and effectively create a TDM contention-free network.

In addition to time delays, the difference between the theoretical data rate and the practical throughput of a control network should be considered. For example, raw data rates for 802.11 wireless networks range from 11 to 54 Mbits/s. The actual throughput of the network, however, is lower due to both the overhead associated with the interframe spaces, ACK, and other protocol support transmissions, and

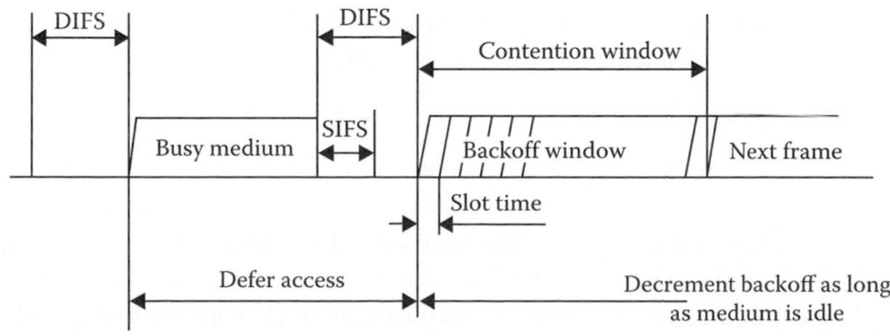

FIGURE 2.11 Timing diagram for wireless Ethernet (IEEE 802.11).

TABLE 2.3 Maximum Throughputs for Different 802.11 Wireless Ethernet Networks

Network Type	802.11a	802.11g	802.11b
Nominal data rate	54	54	11
Theoretical throughput	26.46	17.28	6.49
Measured throughput	23.2	13.6	3.6

Note: All data rates and throughputs are in Mb/s.

TABLE 2.4 Computed Frame Times and Experimentally Measured Delays on Wireless Networks

Network Type	802.11a	802.11g	802.11b
Frame time (UDP), computed	0.011	0.011	0.055
Median delay (UDP), measured	0.346	0.452	1.733
Frame time (OPC), computed	0.080	0.080	0.391
Median delay (OPC), measured	2.335	2.425	3.692

Note: All times in ms.

the actual implementation of the network adapter. Although 802.11a and 802.11g have the same raw data rate, the throughput is lower for 802.11g because its backward compatibility with 802.11b requires that the interframe spaces be as long as they would be on the 802.11b network. Computed and measured throughputs are shown in Table 2.3. The experiments were conducted by continually sending more traffic on the network until a further setpoint increase in traffic resulted in no additional throughput.

Experiments conducted to measure the time delays on wireless networks are summarized in Table 2.4 and Figure 2.12. Data packets were sent from the client to the server and back again, with varying amounts of cross-traffic on the network. The send and receive times on both machines were time-stamped. The packet left the client at time t_a and arrived at the server at time t_b; then left the server at time t_c and arrived at the client at time t_d. The sum of the pre- and postprocessing times and the transmission time on the network for both messages can be computed as (assuming that the two nodes are identical)

$$2 * T_{delay} = 2 * \left(T_{Pre} + T_{wait} + T_{tx} + T_{post}\right) = t_d - t_a - \left(t_c - t_b\right)$$

Note that this measurement does not require that the clocks on the client and the server be synchronized. As the delays at the two nodes can be different, it is this sum of the two delays that is plotted in Figure 2.12 and tabulated in Table 2.4.

Two different types of data packets were considered: user datagram protocol (UDP) and OPC. UDP packets carry only a data load of 50 bytes. OPC requires extra overhead to support an application layer; consequently, the OPC packets contain 512 data bytes (in addition to the overhead). For comparison purposes, the frame times (including the overheads) are computed for different packets.

2.3.6.2 Bluetooth

Bluetooth was originally designed as a wireless replacement for cellular phone headsets; however, its mid-range wireless capabilities in terms of distance and power consumption, combined with its frequency hopping capability for improved noise immunity, have led to its consideration as a wireless solution for lower-level connectivity on the manufacturing floor. Bluetooth operates in the same 2.4 GHz frequency band as 802.11b and g and defines a full OSI communications stack, with the lower levels published as IEEE standard 802.15.1 [11]. Advantages of Bluetooth, which make it ideal for manufacturing, include dynamic connection establishment without need for human interaction, utilization of forward error

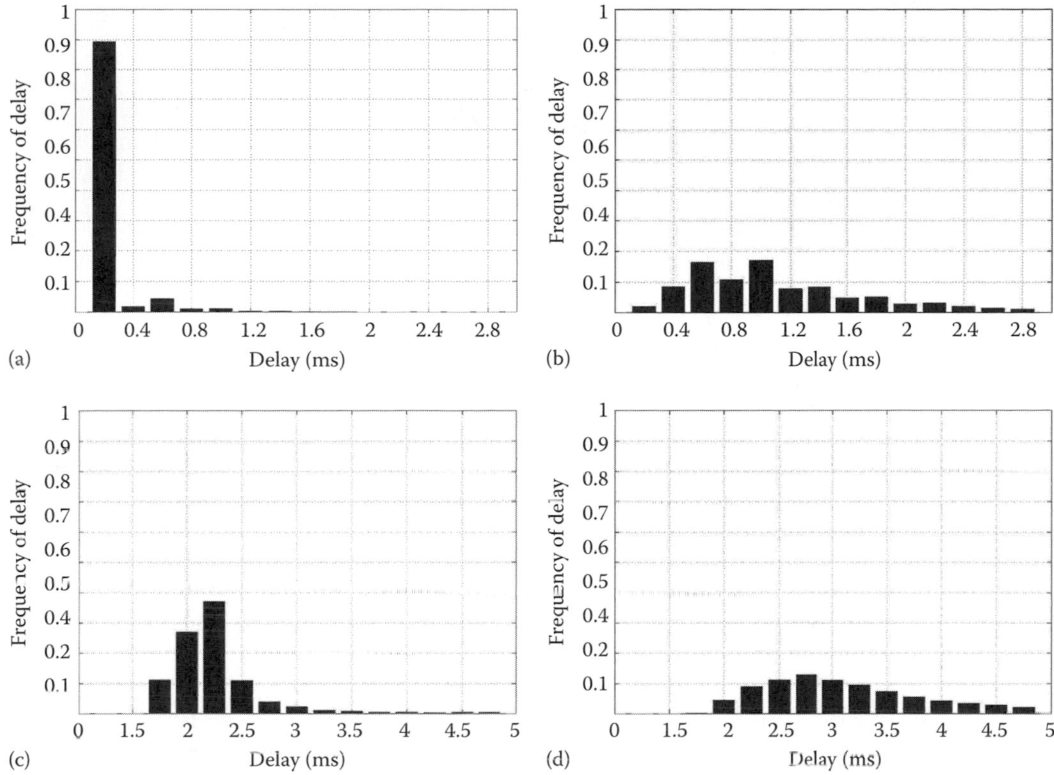

FIGURE 2.12 Distributions of packet delays for different values of cross-traffic throughput on an 802.11a network. (a) UDP delays, 3 Mb/s cross-traffic; (b) UDP delays, 22 Mb/s cross-traffic; (c) OPC delays, 3 Mb/s cross traffic; and (d) OPC delays, 22 Mb/s cross-traffic.

correction (FEC) for delivering the messages without error and without requiring retransmission, and a frequency hopping capability for improved noise immunity and coexistence that includes a capability to *learn* of and avoid areas in the spectrum that are crowded with other transmissions. Drawbacks include loss of efficiency from the FEC capability (a 1 Mbps communications channel can deliver only 721 Kbps) and disruption of 802.11 transmissions (coexistence issue) due to the fact that the technology frequency hops at a faster rate than specified for 802.11b/g [11,13].

Bluetooth applications have also been shown to exhibit nondeterministic response time behavior in the face of ambient interference. While this behavior is not limited to Bluetooth wireless systems, it provides a good example of an issue that must be addressed before common wireless technologies can be utilized effectively in NCSs. As an example, Figure 2.13 shows the results of an experiment to better understand the determinism of a Bluetooth NCS. As shown in Figure 2.13a, the test setup includes two Bluetooth nodes, namely, a controller and an I/O block, and a DeviceNet network hardwired system with a *Packet Sniffer* node capability providing for real-time monitoring and time-stamping of information for performance evaluation. The experiment consists of DeviceNet requests sent out over the Bluetooth network to change the value (i.e., *flip*) of a bit on the I/O block. When a packet is returned indicating that the bit value has been changed (i.e., a *change of state* is observed), a request is then placed to flip it back, etc. The packet interval of DeviceNet transmissions and poll rate of the system are set to be representative of typical DeviceNet operations and to ensure that the packet interval is significantly shorter than the response time of the Bluetooth system. Thus, a practical performance metric of the "number of DeviceNet packets between observed COS occurrences" can be used. Figure 2.13b shows a baseline result with minimal noise interference and ideal distance between the two Bluetooth nodes;

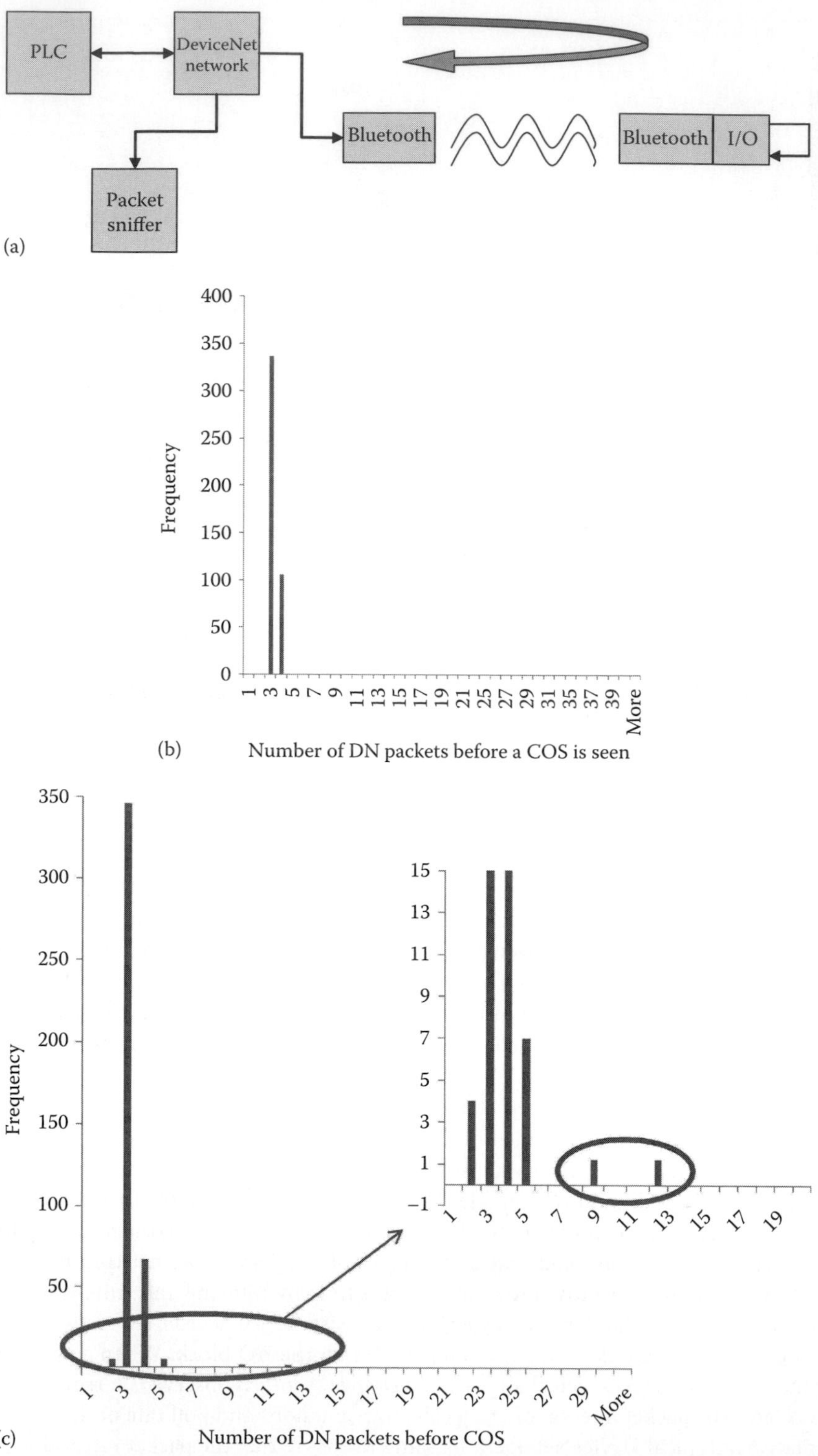

FIGURE 2.13 Experimental characterization of determinism in a Bluetooth system. (a) Bluetooth determinism performance experimental setup, (b) baseline result with minimal channel noise and near ideal distance between nodes, (c) distance test (10 m) is inset expanded view of nondeterminism.

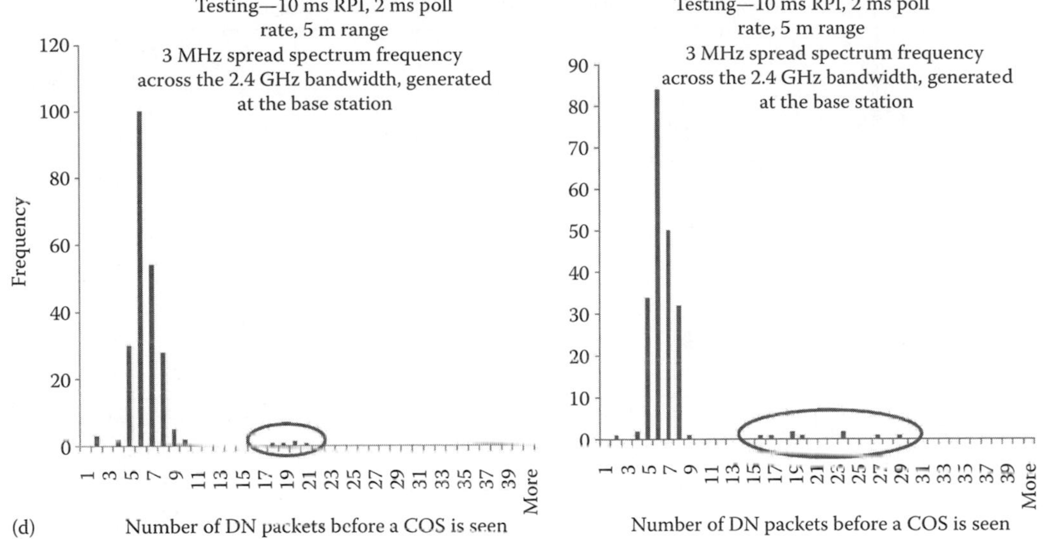

FIGURE 2.13 (continued) Experimental characterization of determinism in a Bluetooth system. (d) interference test, with nondeterminism highlighted.

note the tight and bounded distribution around these packets indicating relatively high determinism. Figure 2.13c shows the same system, but with the distance increased to 10 m and the antennae of the two nodes oriented to be perpendicular to each other. The inset reveals that while average performance does not degrade significantly, there is a significant loss of determinism due to the expanded tail of the distribution. Figure 2.13d illustrates that the same problem can occur when noise interference is introduced (note that the noise levels were purposely set to be below the threshold where the Bluetooth system would invoke frequency hopping to avoid noisy areas of the spectrum).

The results summarized in Figure 2.13 do not represent an exhaustive evaluation of Bluetooth, nor do they illustrate performance issues that are exclusive to Bluetooth. Rather the results illustrate one of the general problems with wireless that must be addressed if wireless is to be utilized in time-critical networked control and safety systems, namely, lack of determinism of the protocol in the face of external factors and the lack of focus on determinism in the design of the protocol as well as products and systems utilizing the protocol. The need for determinism in NCSs is explored further in Section 2.5.

2.3.6.3 Experimental Evaluation of Wireless Performance

To a much greater extent than in wired communication, the physical medium must be considered when characterizing the performance of wireless systems. When operating in license-free wireless bands, the quality of the physical communication channel cannot be guaranteed. Cohabitation with transmitters serving multiple applications must not only be tolerated, but also expected. Further, there are effects resulting from the propagation of radio signals in the conditions inside industrial environments, such as reflections and signal attenuation, which must be tolerated [76]. The four major categories of properties affecting wireless performance are radio propagation path loss, active radio interference, passive channel errors, and channel access errors and constraints. These are discussed in the following subsections.

2.3.6.3.1 Radio Propagation Path Loss

The strength of a received signal is a fundamental property affecting wireless network performance. Most wireless devices have a minimum signal-to-noise ratio (SNR) required to operate. It is therefore important to consider energy lost in the transmission medium when characterizing wireless network performance.

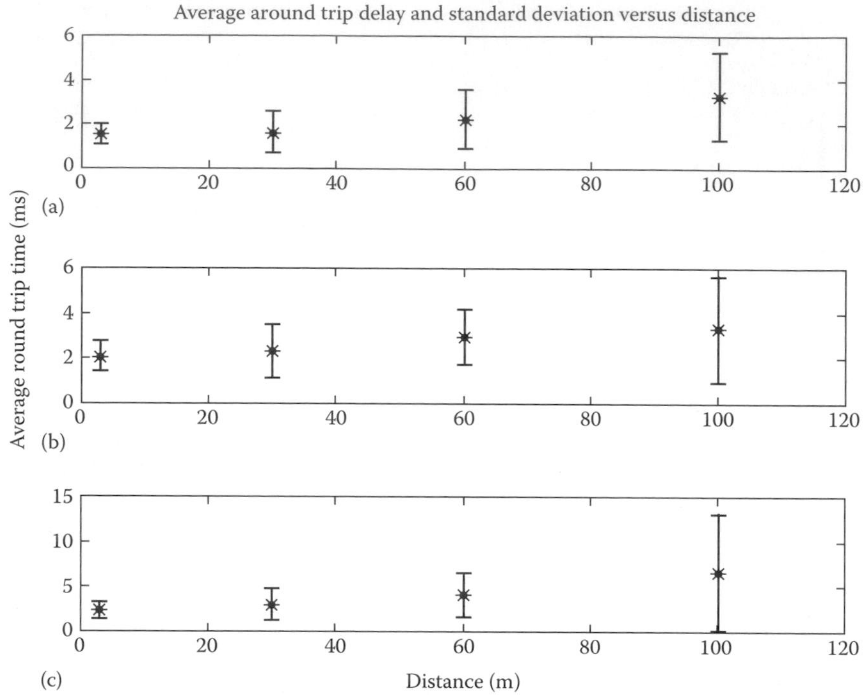

FIGURE 2.14 A comparison of the round-trip delays between an access point and client node placed at increasing distances when using 802.11a (a), 802.11b, (b) and 802.11g (c) respectively.

In a wireless system, increasing the distance between the nodes increases not only the cumulative path loss but also the effects due to reflections and scattering. The result is sporadic deep fades (sudden drops in received signal strength) in the SNR that cause bit transmission errors, which in turn require retransmissions at various levels and consequently greater communication delay.

The distance between two 802.11 devices was varied to study the impact on the round-trip communication delay. Attenuation from walls and floors was not explicitly controlled, but the environment was selected to closely resemble the plant floor of an automotive manufacturing industry. Figure 2.14 shows the results of the experiment for 802.11a, b, and g protocols respectively [4].

2.3.6.3.2 Active Radio Interference

802.11 and Bluetooth ordinarily occupy license-free ISM radio bands, which are shared with microwave ovens, RF identification devices, motion sensors, wireless security cameras, narrowband wireless telephones, and other short-range communication devices. In industrial environments, other sources of radio interference include high energy arc welders, motor drives, and frequency controllers. A significant source of radiation in the 2.45–2.485 GHz frequency range is the commercial microwave oven. Radiation power in a microwave can reach a maximum of 16–33 dBm. An analysis of the impact of microwave oven leakage on 802.11-based WLAN is presented in [102]. A band occupancy study of the 2.4 GHz band [10] shows that while the microwave oven does affect WLAN performance, the most significant source of interference in terms of power and duty cycle is a coexisting WiFi access point. Studies on the spectral footprint of welders and motor controllers conclude that the energy from these devices is limited to frequencies below 1 GHz and pose no significant threat to the communication protocols discussed here [73]. Further, the industrial environment is characterized by sudden and large variations in propagation conditions. The presence of uncoordinated interference sources close to network components induces deep fades and signal distortion that leads to clustered bit errors.

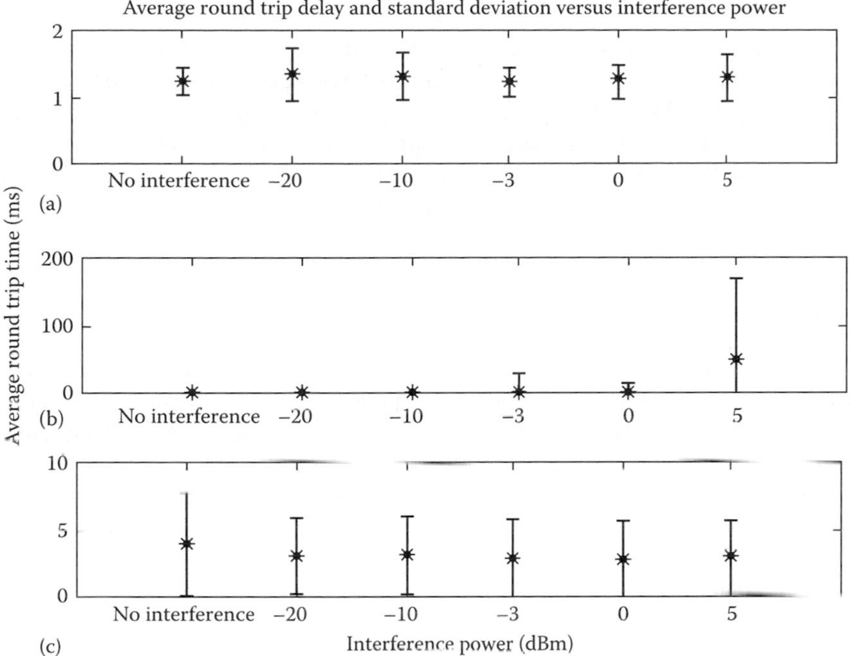

FIGURE 2.15 A comparison of the round-trip delays between an access point and client node subject to increasing interference when using 802.11a (a), 802.11b, (b) and 802.11g (c) respectively.

Figure 2.15 shows the results of an experiment to illustrate the impact of an interfering radio source at close proximity. The interference source was equipped with a band filter and modulation scheme to mimic a rogue 802.11b or 802.11a transmitter (chosen depending on the operating band of interest). The interference power was therefore limited to 20 dBm, and the frequency bandwidth of the signal was also limited to 25 MHz [4].

2.3.6.3.3 *Passive Channel Errors*

Even in conditions where signal strength is above the critical limit and there are no active sources of interference, wireless transmissions are subject to scattering, diffraction, and reflections. In an industrial setting, due to the large size of the plant area and the presence of scattering surfaces, several reflected radio paths with different relative lengths may exist between the transmitter and the receiver. As a result, multiple instances of the same transmitted signal arrive at the receiver dispersed in time. These reflected signals interfere with the original transmission, occasionally causing destructive interference. This leads to a sudden drop in received signal strength called a deep fade. The channel properties on the plant floor are dynamic, and this deep fade condition may rapidly appear and disappear, causing bursts of bit errors and retries during the period of the deep fade.

In an industrial setting, the RMS multipath delay spread can be upwards of 200 ns [73]. In the case of an 802.11b transmission at 11 Mbits/s using single-carrier quadrature phase shift keying, the symbol rate is 11 Msymbols/s. The symbol period is therefore about 91 ns. With a delay spread of 200 ns, it is likely that waveforms corresponding to adjacent symbols will frequently overlap at the receiver, causing intersymbol interference (ISI). Failure to decode the symbols will result in intermittent bit errors.

ISI is less pronounced in Bluetooth transmissions since the symbol rate is much lower (1 Msymbols/s). The use of several subcarriers in the OFDM modulation scheme and the presence of a long guard interval in the 802.11g standards completely eliminate ISI.

2.3.6.3.4 Channel Access Errors and Constraints

Wireless transceivers are half-duplex devices capable of either transmitting or receiving at a particular time. This gives rise to several issues specific to wireless CSMA systems. The *hidden terminal problem* [95] is one where a particular node inaccurately estimates a channel as available when another node is transmitting out of range. The first node transmits over an occupied channel causing interference. Another form of channel access error occurs when spurious radio transmissions interfering with the system will cause all nodes in range to wait indefinitely for the medium to become available, essentially hanging the network.

2.4 NCS Characterization

Determination of an optimal NCS (which includes solutions for control, diagnostics, and/or safety applications) design for a particular manufacturing application requires knowing not only the theory and trade-offs of the NCS protocols being considered (i.e., a theoretical perspective) but also experimental data to augment the theoretical understanding of the NCS capabilities (i.e., experimental perspective) and an analytical study that allows for gauging the relative importance of performance metrics (e.g., end-to-end speed and jitter) to the specific application environment (i.e., analytical perspective). In this section, we explore these three perspectives in detail and show how they can be utilized collectively to provide a methodology for NCS design decision making [55].

2.4.1 Theoretical Perspective

The theoretical information perspective involves a general overview of network operation and the metrics that should be evaluated when assessing the best NCS solution for a particular system. The comparative evaluation of network technologies and implementations is primarily a study of trade-offs. Indeed, the move from point-to-point to networked systems itself is a study of trade-offs. One important aspect of the trade-off study is illustrated with the Lian curve, discussed in Section 2.2.4. This curve illustrates the impact of network congestion on NCS performance in a digital control environment. Finding and maintaining NCS operation in the sweet spot for optimized performance is a study of trade-offs, which is a topic of ongoing research. It is important to note that the term *network* can be used very loosely in this analysis. For example, oftentimes the network congestion is actually observed in the devices themselves as they parse and encode data for end-to-end application communication [47]. The Lian curve phenomena can still be observed in these environments with the network definition extended to include the node processing.

A theoretical understanding of the network choices is a clear prerequisite to an NCS deployment decision. The understanding should extend beyond the network protocol performance to the entire system performance, and an important component of that understanding is the clear definition of performance metrics that are important to the particular application environment. An example of metrics that may be important in automotive manufacturing environment can be found in Ref. [1], which is the output of an industrial network performance workshop; in this case, the top five metrics were *node performance, ease of use and diagnostics tools, cost of security/technology, capability for prioritization*, and *time synchronization*. The exercise of this workshop revealed a few important aspects of the metric selection process, which are important to the analytical application of these metrics in NCS decision making. The first is that many of the metrics such as ease of use or cost of complexity may be difficult to quantify, but these metrics must be incorporated into the quantified decision-making process (i.e., these metrics should not be avoided). Oftentimes, a qualitative metric can be broken down into quantitative submetrics (e.g., ease of use into average time between system crashes plus average time to diagnose a problem) so that it can be compared with other factors; the challenge is understanding the appropriate weighting for these quantifications. The second important aspect is that metrics can often overlap significantly. Care should be taken to break these metrics down into subcomponents so that a nonoverlapping set of subcomponents can be determined. This will avoid any *double counting* in the decision-making process.

2.4.2 Experimental Perspective

Usually, the theoretical knowledge that can be applied to NCS analysis must be augmented with experimental knowledge to complete the performance information base. This need for an experimental perspective is especially true when assessing the contribution of end node performance to the system performance analysis. Node performance should be evaluated in all systems because, as noted in Section 2.2, node performance has been shown to dominate over network performance in many manufacturing system environments. This is especially true in higher-level switched Ethernet systems, where the emphasis is on providing services such as security and self-typing on top of the data delivery mechanism [56,72]. These services provide overhead that contributes to network congestion; in smaller networks, where the node count does not number in the thousands, the degradation in individual node performance due to overhead factors such as these dominates. For example, Figure 2.16 illustrates that, in a particular two-node Ethernet network, node delay accounted for over 90% of the entire end-to-end delay [72]. Figure 2.5 (Section 2.2.3) gives an example of the impact of protocol overhead, in this case VPN to support security.

With respect to both network and node performance, specific network overhead factors that should be considered experimentally include self-typing or higher-level protocols for enabling control systems such as OPC (discussed in Section 2.3.5) and XML,* protocols for security such as VPN, and choices of Ethernet transport layer protocols such as TCP/IP vs. UDP. The impact of these factors is explored in detail in [72].

The wireless NCS application environment demands an additional level of experimental data to provide detailed information about the impact of the medium on NCS performance. In wired systems, the *wire* is generally considered to be a medium that can guarantee end-to-end delivery, provided that the protocol on top of it handles issues of congestion and performance (which is true for most industrial network protocols). There are a number of factors in the various wireless manufacturing environments that can severely compromise wireless NCS performance. For example, a mobile phone or a wire mesh used to strengthen a cement wall can lead to intermittent and difficult-to-diagnose NCS performance issues. Knowing the impact of these issues in advance helps optimize decisions as well as reduce discovery times for NCS performance problems detected during operation [4].

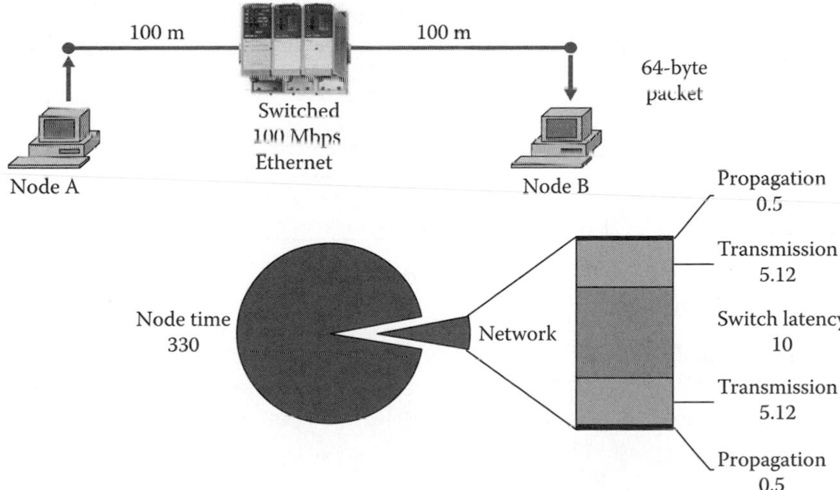

FIGURE 2.16 Illustration of node delay contribution to total delay in a typical two-node Ethernet network (pictured in the image). Time given in microseconds.

* eXtensible Markup Language (XML) is a self-typing language often used to support communications among dissimilar components on an industrial Ethernet network.

2.4.3 Analytical Perspective

While the theoretical and experimental analysis provides an excellent foundation for NCS evaluation, it does not provide a capability to combine the results for differential analysis with respect to the specifics of the application environment. A final analytical step is thus needed where the factors can be weighed and combined to yield a single answer with respect to the best decision. The weighted analysis can be achieved through a simple normalized weighting of factors of concern that are determined through either theoretical or experimental means [54]. The weighting equation takes the form of

$$WCost_{\text{total}} = \frac{W_H}{(W_H + W_E + W_P)} Cost_H + \frac{W_E}{(W_H + W_E + W_P)} Cost_E + \frac{W_P}{(W_H + W_E + W_P)} Cost_P \tag{2.3}$$

where

$Cost_H$, $Cost_E$, and $Cost_P$ are the costs associated with hardware, engineering, and maintenance and performance
W_H, W_E, and W_P are the weights assigned, respectively, to these costs

Each cost factor is made up of a number of contributors, which are themselves weighted in a normalized manner as in

$$Cost_H = \frac{W_1}{\sum_{i=1}^{n} W_i} \cdot Cost_{H1} + \frac{W_2}{\sum_{i=1}^{n} W_i} \cdot Cost_{H2} + \cdots + \frac{W_n}{\sum_{i=1}^{n} W_i} \cdot Cost_{Hn} \tag{2.4}$$

The weights should be chosen to reflect the particular NCS application environment. Equation 2.3 is normalized so that factors can be incrementally added to the formulation as these factors are quantified during the course of understanding the NCS environment. Note that Equations 2.3 and 2.4 assume that the factors are independent; thus, as noted earlier, care should be taken to ensure that factors do not overlap.

Figure 2.17 illustrates this analysis applied to the decision of utilizing (separate) dedicated vs. shared networks to support control and safety NCS functionality. Depending on the prioritization of factors, the best solution can vary.

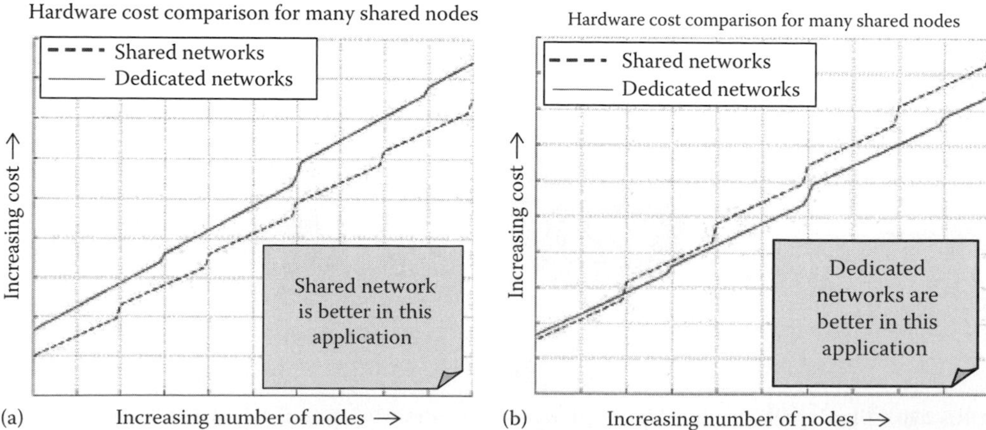

FIGURE 2.17 Illustration of the application of the weighted analytical perspective resulting in decisions of (a) shared or (b) dedicated networks for supporting control and safety functionality, respectively, depending on weights chosen.

It is important to note that manufacturers tend to consistently overemphasize upfront costs such as installation and underemphasize recurring maintenance and performance costs when making network decisions. Utilizing the more formalized approach presented here, this *forgetting* of the recurring costs can be avoided [54].

2.5 Applications for Industrial Networks

In this section, we briefly describe current trends in the use of networks for distributed multilevel control, diagnostics, and safety. There is an enormous amount of data produced in a typical manufacturing system, as thousands of sensors record position, velocity, flow, temperature, and other variables hundreds of times every minute. In addition to this physical information, there are the specifications for the parts that must be produced, the orders for how many parts of each type are needed, and the maintenance schedules for each machine. Generally, the information content can be thought of as supporting a control, diagnostics, or safety function, or some combination of these. To support the aggregate of these functions in a manufacturing environment, separate networks or network partitions are sometimes employed, where each network or partition is dedicated to one or more function types, such as control and diagnostics, with the network protocol chosen that best fits (i.e., balances) the QoS requirements of the function type(s) [90]. In this section, networks for these function types are explored, focusing on the QoS requirements that often govern the network protocol choice.

2.5.1 Networks for Control

Control signals can be divided into two categories: real time and event based. For real-time control, signals must be received within a specified amount of time for correct operation of the system. Examples of real-time signals include continuous feedback values (e.g., position, velocity, and acceleration) for servo systems, temperature and flow in a process system, and limit switches and optical sensors in material flow applications (e.g., conveyors). To support real-time control, networks often must have a high level of determinism; that is, they must be able to guarantee end-to-end communication of a signal within a specified amount of time. Further, QoS of NCSs can be very dependent upon the amount of jitter in the network; thus, for example, fixed determinism is usually preferred over bounded determinism.

Event-based control signals are used by the controller to make decisions, but do not have a time deadline. The system will wait until the signal is received (or a timeout is reached), and then the decision is made. An example of an event-based signal is the completion of a machining operating in a CNC; the part can stay in the machine without any harm to the system until a command is sent to the material handler to retrieve it.

In addition to dividing control signals by their time requirements, the data size that must be transmitted is important. Some control signals are a single bit (e.g., a limit switch) whereas others are very large (e.g., machine vision). Generally speaking, however, and especially with real-time control, data sizes on control networks tend to be relatively small, and high levels of determinism are preferred.

Control networks in a factory are typically divided into multiple levels to correspond to the factory control distributed in a multitier hierarchical fashion. At the lowest level of networked control are device networks, which are usually characterized by smaller numbers of nodes (e.g., less than 64), communicating small data packets at high sample frequencies and with a higher level of determinism. An example of networked control at this level is servo control; here network delay and jitter requirements are very strict. Deterministic networks that support small data packet transmissions, such as CAN-based networks, are very common at this level. Although seemingly nonoptimal for this level of control (large packet sizes), switched Ethernet is becoming common, due to the desire to push Ethernet into all levels in the factory to support web services and greater transparency between control layers. Regardless of the network type, determinism and jitter capabilities for lower-level networked control are enhanced oftentimes by utilizing techniques that minimize the potential for jitter through network contention, such as MS operation, polling techniques, and deadbanding [69].

An intermediate level of network is the cell or subsystem, which includes SCADA. At this level, multiple controllers are connected to the network (instead of devices directly connected to the network). The controllers exchange both information and control signals, but as the cells or subsystems are typically physically decoupled, the timing requirements are not as strict as they are at the lowest levels, or nonexistent if event-driven control is enforced [62]. These networks are also used to download new part programs and updates to the lower-level controllers. TP and Ethernet-based networks are commonly used at this level, with the ability to communicate larger amounts of data and support for network services generally taking precedence over determinism.

Networks at the factory or enterprise level coordinate multiple cells and link the factory-floor control infrastructure to the enterprise-level systems (e.g., part ordering and supply chain integration). Large amounts of data travel over these networks, but the real-time requirements are usually minimal to nonexistent. Ethernet is the most popular choice here primarily because Internet support at this level is usually critical (e.g., cloud services), and Ethernet also brings attractive features to this environment such as support for high data volumes, network services, availability of tools, capability for wide area distribution, and low cost.

Currently, wireless networks are not commonly utilized for control. When they are utilized, it is often because a wired system is impossible or impractical (e.g., a stationary host controller for an autonomous guided vehicle [AGV]). Wireless systems are more applicable to discrete control; issues with the determinism of wireless communications in the face of interference (as explored in Section 2.3.6) usually result in the requirement that wireless time-critical control systems go through extensive and often costly verification prior to deployment.

2.5.2 Networks for Diagnostics

Diagnostic information that is sent over the network often consists of large amounts of data sent infrequently. For example, a tool-monitoring system may capture spindle current at 1 kHz. The entire current trace would be sent to the diagnostic system after each part is cut (if the spindle current is used for real-time control, it could be sent over the network every 1 ms, but this would then be considered control data). The diagnostic system uses this information for higher-level control, such as to schedule a tool change or shut down a tool that is underperforming.

Diagnostics networks are thus usually set up to support large amounts of data with the emphasis on speed over determinism. Ethernet is the dominant network protocol in system diagnostics networks. As with control, diagnostics is often set up in a multitier hierarchical fashion, with different physical layer technology (e.g., wireless, broadband, and fiber-optic) utilized at different levels to support the data volume requirements. Also, a variety of data compression techniques, such as COS reporting and store and forwarding of diagnostic information on a process *run-by-run* basis, are often used in communicating diagnostic information up the layers of the network hierarchy [38,84]. Wireless is an ideal medium for diagnostics because determinism and high levels of data integrity are usually not required, and diagnostics systems can often take advantage of the flexibility afforded by the wireless media. As an example, a personal digital assistant handheld system can be designed to collect diagnostics data wirelessly from a nearby machine and perform system health checks. Maintenance engineers thus could utilize a single portable unit as a diagnostics tool for all machines in a factory.

As noted in Section 2.1, diagnostics networks enable diagnostics of the networked system rather than the network itself (with the latter referred to as network diagnostics). Both types of diagnostics are commonly used in manufacturing systems. Many network protocols have built-in network diagnostics. For example, nodes that are configured to send data only when there is a COS may also send *heartbeat* messages every so often to indicate that they are still on the network.

2.5.3 Networks for Safety

Traditionally, safety interlocks around manufacturing cells have been hardwired using ultrareliable safety relays to ensure that the robots and machines in the cell cannot run if the cell door is open or there is an operator inside the cell. This hardwiring is not easy to reconfigure and can be extremely difficult to troubleshoot if something goes wrong (e.g., a loose connection). Safety networks allow the safety systems to be reconfigurable and significantly improve the ability to troubleshoot. They also allow safety functions to be more easily coordinated across multiple components, for example, shutting down all machines in a cell at the same time and also coordinating *soft shutdown* where appropriate (safe algorithms for gradual shutdown of systems without damage to systems and/or scrapping of product). Further, safety network systems often provide better protection against operators bypassing the safety interlocks and thus making the overall system safer.

Safety networks have the strongest determinism and jitter requirements of all network function types. Safety functions must be guaranteed within a defined time; thus, the network must provide that level of determinism. Further, the network must have a deterministic heartbeat-like capability; if network connectivity fails for any reason, the system must revert to a safe state within the guaranteed defined time; this is often referred to as *fail-to-safe*. CAN-based and switched Ethernet-based networks are popular candidates for networked safety because of their high levels of determinism, the network self-diagnostic mechanisms they can utilize to determine the node and network health, and the ability of the latter to support capabilities such as web services as well as its general dominance in control and diagnostics domains. However, it is important to note that most network protocols, in and of themselves, are adequate to the task of supporting safety networking. This is because safety systems must support a fail-to-safe capability if a communication is not received within a predetermined amount of time. In other words, safety systems must be robust to the possibility of network failure of any network. This is often accomplished by adding functionality at higher levels (e.g., application) of the communication stack to guarantee proper safety functionality [75]. Thus, even wireless safety systems are possible. The key is understanding if the benefit of a more flexible or commonplace network technology, such as wireless, is outweighed by the reduced guaranteed response time of the safety system in the increased occurrence of fail-to-safe events due to network delays.

2.5.4 Multilevel Factory Networking Example: Reconfigurable Factory Testbed

The reconfigurable factory testbed (RFT) at the University of Michigan is a comprehensive platform that enables research, development, education, validation, and transfer of reconfigurable manufacturing system concepts [58]. It consists of both real and virtual machines controlled over a communication network and coordinated through a unified software architecture. The RFT is conceived to be extensible to allow the modular incorporation and integration of additional components (hardware and/or software, real and/or virtual). The hardware components of the RFT include a serial–parallel manufacturing line consisting of two machining cells connected by a conveyor, a suite of communication and control networks, an AGV, and an RFID system. The software components of the RFT include a virtual factory simulation, an open software integration platform and data warehouse, an infrastructure of Web-based HMIs, and a computerized maintenance management system. A schematic of the RFT is shown in Figure 2.18.

The network shown in Figure 2.18 represents a multitier-networked control, diagnostic, and safety network infrastructure that exists on the RFT. The serial–parallel line component of the RFT is the primary component currently being utilized to explore manufacturing networks. With respect to control networks, cell 1 has a DeviceNet network to connect the machines and robot controllers (including the robot gripper and the clamps in the machines); cell 2 uses PROFIBUS for the same purpose. The conveyor system (pallet stops, pallet sensors, motor, and controller) was originally outfitted to communicate via a second DeviceNet network; this was converted to an 802.11 wireless-networked system. The cell-level controllers (including the conveyor controller) communicate with the system level controller (SLC) over Ethernet via OPC and support an event-based control paradigm. The SLC has a wireless network connection with the AGV. All of these control networks are shown in Figure 2.19.

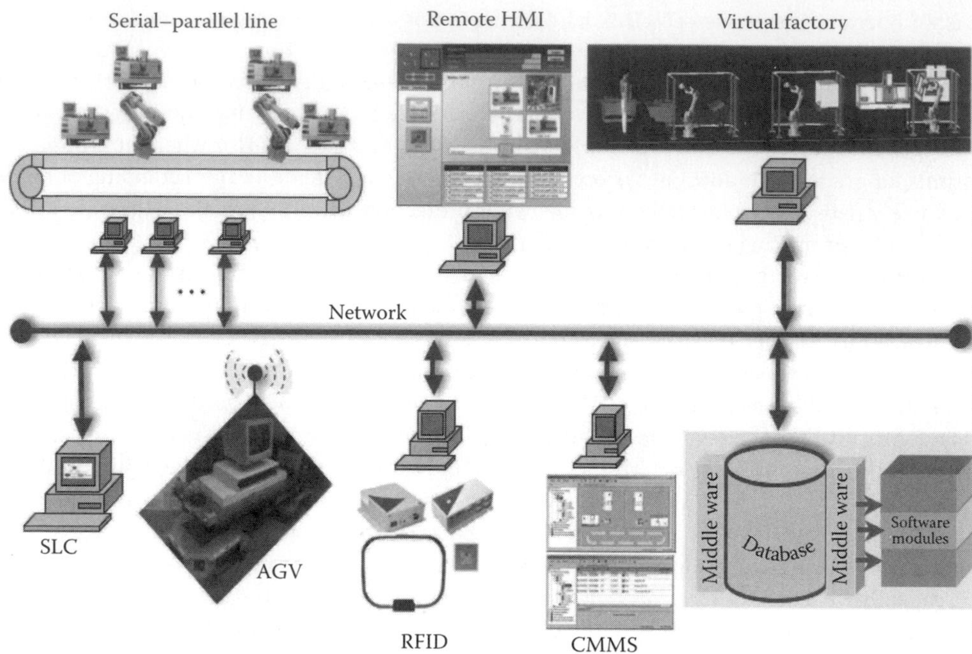

FIGURE 2.18 Reconfigurable factory testbed.

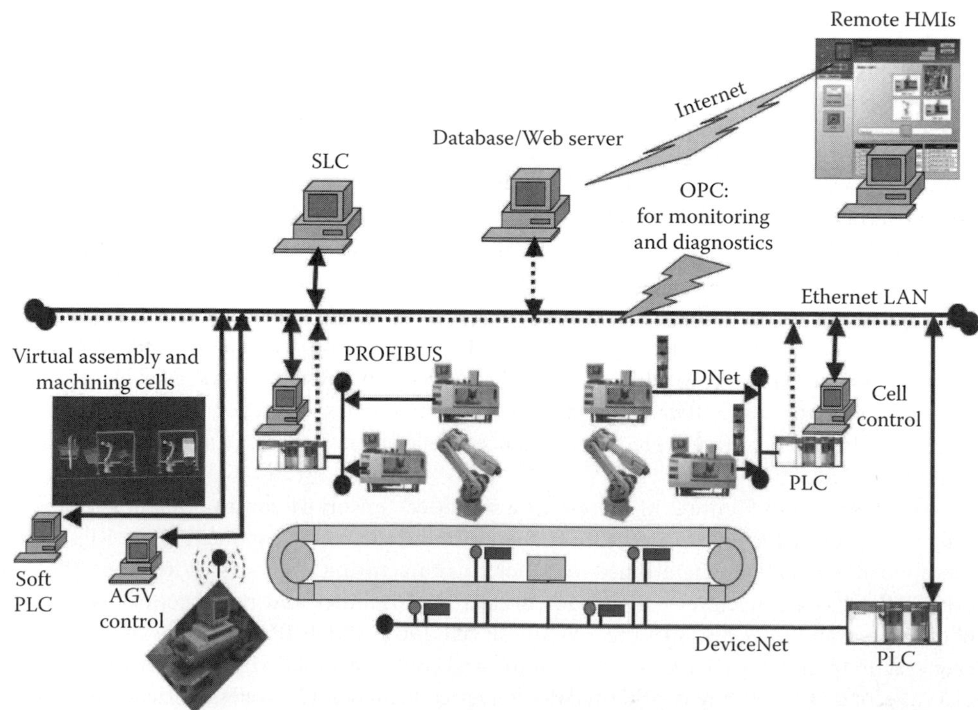

FIGURE 2.19 Networks on the RFT: control networks are indicated by solid lines, and diagnostics networks are indicated by dashed lines.

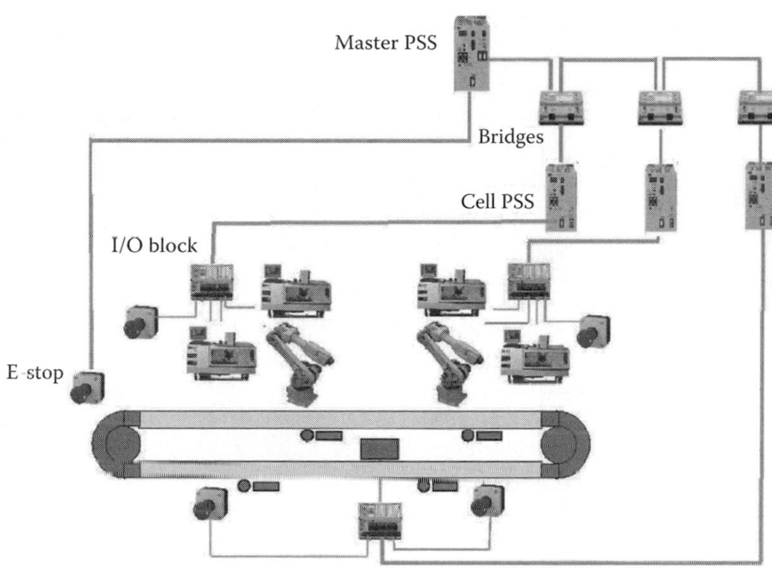

FIGURE 2.20 Safety network implementation on the RFT.

The network infrastructure for collecting the diagnostic data on the RFT uses OPC. For example, for every part that is machined, the spindle current on the machine is sampled at 1 kHz. These time-dense data are directly sampled using LabVIEW* and then stored in the database. Compressed diagnostics data focused on identifying specific features of the current trace are passed to higher levels in the diagnostics network.

Networks for safety are implemented in the serial–parallel line utilizing the SafetyBUS p protocol, as shown in Figure 2.20. As with the control and diagnostics system, the implementation is multitier, corresponding to levels of safety control. Specifically, safety networks are implemented for each of the two cells as well as the conveyor. The safety network interlocks the emergency stops, robot cages, and machine enclosures with the controllers. These three cell-level networks are coordinated through a hierarchy to a high-level master safety network. This implementation not only allows for safety at each cell to be controlled individually but also provides a capability for system-wide safe shutdown. Further, this implementation allows for multitier logic to be utilized for the implementation of safety algorithms.

As manufacturing networks research and development has increasingly shifted toward industrial Ethernet and wireless communications, so too has the focus in RFT networking deployment and analysis. For example, in providing an experimental capability to support NCS design evaluation (see Section 2.4), the wireless performance testbed shown in Figure 2.21 was developed in collaboration with USCAR, a collaboration of automotive manufacturers [97]. Here, a simple parallel point-to-point wireless environment is provided for three common industrial wireless protocols: IEEE 802.11, Bluetooth, and ZigBee. An Ethernet real-time packet *sniffer* is utilized to monitor traffic, thereby determining the performance of the wireless connection. The testbed is set up so that various wireless configurations and interference environments can be evaluated in a laboratory setting. The testbed can thus be utilized as both a design/evaluation and a troubleshooting tool. With respect to the latter, wireless signal generation and analysis equipment are utilized with the testbed to effectively *record* wireless patterns in the field and then *play back* these patterns in the experimental environment for analysis and troubleshooting.

The RFT implementation of multitiered networks for control, diagnostics, and safety provides a rich research environment for exploration into industrial control networks. Specifically, topics that can be

* National Instruments, Austin, TX.

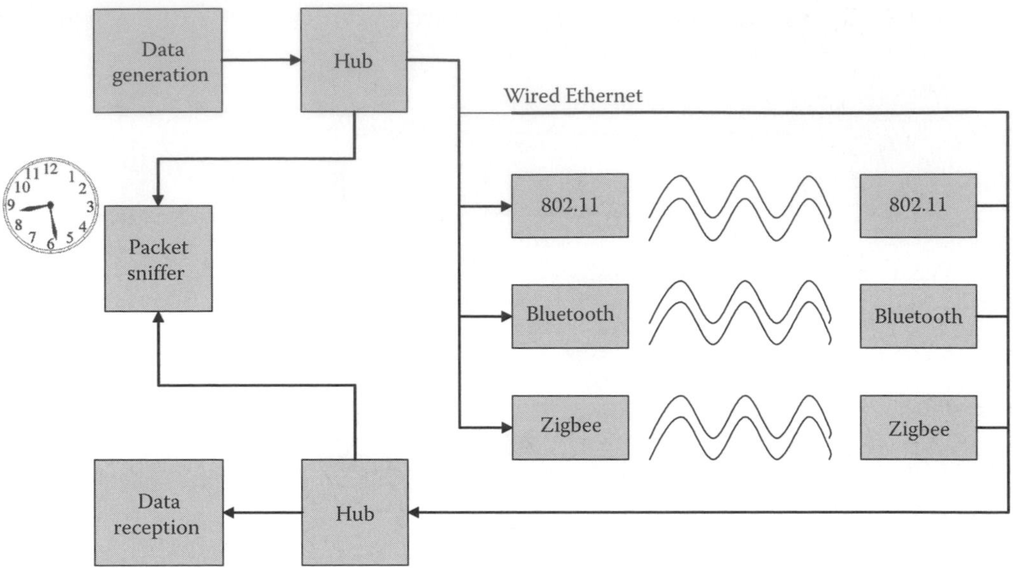

FIGURE 2.21 Experimental setup for determining and investigating the wireless NCS performance.

addressed include (1) coordination of control, diagnostics, and/or safety operation over one, two, or three separate networks; (2) distributed control design and operation over a network; (3) distribution of control and diagnostics in a hierarchical networked system; (4) compression techniques for hierarchical diagnostics systems; (5) remote control safe operation; (6) hierarchical networked safety operation and soft shutdown; (7) heuristics for optimizing control/diagnostics/safety network operation; (8) network standards for manufacturing; (9) best practices for network systems design and operation; as well as (10) seamlessly integrating simulations with physical systems over the network [33,45,46,58,70].

2.6 Future Trends

As the rapid move toward networks in industrial automation continues, the process has now matured to the point that a large portion of the migration is not just toward networks, but from one network technology to another (e.g., the move to switched Ethernet everywhere as a consolidation effort or the move to wireless). The immediate advantages of reduced wiring and improved reliability have been accepted as a fact in industry and are often significant enough by themselves (e.g., return-on-investment—ROI) to justify the move to networked solutions. Further, with the advent of more detailed ROI calculators such as that discussed in Section 2.4, other networking advantages can be quantified. It is expected that diagnostics network adoption will continue to lead the way in the overall network proliferation trend, followed by control and then safety networks, but the ordering is driven by the stricter QoS balancing requirements of the latter, not by any belief of higher ROI of the former. In fact, in gauging the criticality of control and safety with respect to diagnostics, it is conceivable that significantly higher ROI may be achievable in the migration to control and especially safety networking. However, even with diagnostics networks, the possibilities and benefits of e-Diagnostics and (with control) e-Manufacturing are only beginning to be explored.

Looking to the future, the most notable trends appearing in industry are (1) the move to wireless networks at all levels, (2) the domination of switched Ethernet, and (3) the design of systems to include network attributes and thus leverage network advantages and mitigate disadvantages [5,101]. The dominance of Ethernet-based technologies will continue for the most part due to the explosion of web services available. With respect to wireless growth, it is not expected that wireless will replace wired in all applications, but rather the traditional barriers to wireless implementation (e.g., technology, security, and reliability) will continue be lowered, and the trend of incorporating network attributes into systems will better allow for the mitigation of these barriers. This will allow the significant benefits of wireless to be realized, such as further reduction of the volume of wiring needed (although often power is still required), enabling the placement of sensors in difficult locations, and enabling the placement of sensors on moving parts such as on tool tips that rotate at several thousand RPM. Issues with the migration to wireless include interference between multiple wireless networks, security, and reliability and determinism of data transmission as explored in Section 2.3.6.3. The anticipated benefit in a number of domains (including many outside of manufacturing) is driving innovation that should be able to be leveraged to address many of these issues in the manufacturing domain. However, the determinism issue will be difficult to overcome because manufacturing is not a market driver in the typical application domains of most wireless network technologies.

The trend of integration of traditionally separate industrial systems, such as diagnostics and safety, will continue [80,81,90]. For example, a diagnostics system can raise safety alarms when a process variable is outside the expected range of safe operations. Advancements in network partitioning allow the networked safety system operation to leverage networked diagnostics information into improved safety decision making, without the degradation of overall safety network performance. This capability opens the door to a number of opportunities for improving safety, control, and diagnostics in manufacturing systems.

In summary, as the industry moves forward in networked communications, many among the dozens of protocols that have been developed for industrial networks over the last few decades will fall out of favor, but will not die overnight due to the large existing installed base and the long lifetime of manufacturing systems. In addition, new protocols may continue to emerge to address niches, where a unique QoS balance is needed. However, it is expected that Ethernet and wireless will continue to grab larger and larger shares of the industrial network's installed base, driven largely by lower cost through volume, the Internet, higher availability of solutions and tools for these network types (e.g., web-enabled tools), and the unmatched flexibility of wireless. Indeed, it is not unreasonable to expect that, in the next decade, the next major milestone in industrial networking, namely, the *wireless factory*, will be within reach, where diagnostics, control, and safety functions at multiple levels throughout the factory are enabled utilizing wireless technology.

Acknowledgments

The authors would like to acknowledge support from the NSF ERC for reconfigurable manufacturing systems (ERC-RMS) grant EEC95-92125. We would also like to thank the many students of the ERC-RMS who did much of the work on which much of this chapter is based and Feng-Li Lian for his extensive research on NCSs.

References

1. K. Acton, M. Antolovic, N. Kalappa, J. Luntz, J. Moyne, and D. Tilbury, Practical metrics for evaluating network system performance. *UM-ERC/RMS Network Performance Workshop*, Ann Arbor, MI, April 2006. Available at http://erc.engin.umich.edu/publications.
2. J. Amelot, Y. Li-Baboud, C. Vasseur, J. Fletcher, D. M. Anand, and J. Moyne. An IEEE 1588 performance testing dashboard for power industry requirements. In *Proceedings of IEEE ISPCS*, 2011.
3. American National Standards Institute. *OSI Basic Reference Model*. ANSI, New York, 1984. ISO/7498.
4. D. Anand, M. Bhatia, J. Moyne, W. Shahid, and D. Tilbury, *Wireless Test Results Booklet*. University of Michigan, Ann Arbor, MI, 2010. Available at http://www.umich.edu/~tilbury/WirelessTestResults2010.pdf.

5. D. Anand, J. Fletcher, Y. Li-Baboud, and J. Moyne. A practical implementation of distributed system control over an asynchronous Ethernet network using time stamped data. *In Proceedings of IEEE CASE*, 2010.
6. V. Anandarajah, N. Kalappa, R. Sangole, S. Hussaini, Y. Li, and J. R. Moyne. Precise time synchronization in semiconductor manufacturing. In *Proceedings of IEEE ISPCS*, 2007.
7. P. Antsaklis and J. Baillieul (eds.). Special issue on networked control systems. *IEEE Transactions on Automatic Control*, 49(9):1421–1597, September 2004.
8. AS-I Standard, 2005. Available at http://www.as-interface.net.
9. D. Bertsekas and R. Gallager. *Data Networks*, 2nd edn. Prentice-Hall, Englewood Cliffs, NJ, 1992.
10. M. Biggs, A. Henley, and T. Clarkson. Occupancy analysis of the 2.4 GHz ISM band. *IEE Proceedings Communications*, 151:481–488, 2004.
11. Bluetooth Special Interest Group (SIG). Available at https://www.bluetooth.org/.
12. G. Brown, What is the future for safety networking? *Control Engineering Magazine*, February, 2012. Available at http://www.controleng.com/home/single-article/what-is-the-future-for-safety-networking/10bd161b4a9ba727aa747b5d59a3005b.html.
13. D. Caro. *Wireless Networks for Industrial Automation*, 2nd edn. The Instrumentation, Systems and Automation Society, Research Triangle Park, NC, 2005.
14. B. J. Casey. Implementing Ethernet in the industrial environment. In *Proceedings of IEEE Industry Applications Society Annual Meeting*, vol. 2, pp.1469–1477, Seattle, WA, October 1990.
15. G. Cena, L. Seno, A. Valenzano, and S. Vitturi, Performance analysis of Ethernet Powerlink networks for distributed control and automation systems. *Computer Standards and Interfaces*, 31:566–572, 2009.
16. M.-Y. Chow (ed.). Special section on distributed network-based control systems and applications. *IEEE Transactions on Industrial Electronics*, 51(6):1126–1279, December 2004.
17. D. Christin, P. S. Mogre, and M. Hollick, Survey on wireless sensor technologies for industrial automation: The security and quality of service perspectives, *Future Internet*, 2:96–125, 2010.
18. CIP. Integrated Architecture and CIP Sync Configuration Rockwell Automation Technical Report-IA-AT003B-EN-P, June 2011, (http://literature.rockwellautomation.com/idc/groups/literature/documents/at/ia-at003_-en-p.pdf)
19. J.-D. Decotignie. Ethernet-based real-time and industrial communications. *Proceedings of the IEEE*, 93(6):1102–1118, June 2005.
20. DeviceNet specifications, Open DeviceNet Vendor Association. DeviceNet specifications Volume I: DeviceNet communication model and protocol release 2.0. 1997.
21. L. Dugyard and E. I. Verriest. *Stability and Control of Time-Delay Systems*. Springer, New York, 1998.
22. D. Dzung, M. Naedele, T. P. Von Hoff, and M. Crevatin. Security for industrial communication systems. *Proceedings of the IEEE*, 93(6):1152–1177, June 2005.
23. J. C. Eidson. *Measurement, Control, and Communication Using IEEE 1588*, Springer, London, U.K., 2006.
24. EtherCAT 2013. http://www.ethercat.org/.
25. M. Felser. Real-time Ethernet—Industry prospective. *Proceedings of the IEEE*, 93(6):1118–1129, June 2005.
26. M. Felser and T. Sauter. The fieldbus war: History or short break between battles? In *Proceedings of the IEEE International Workshop on Factory Communication Systems (WFCS)*, pp. 73–80, Våsteras, Sweden, August 28, 2002.
27. M. Felser and T. Sauter. Standardization of industrial Ethernet—The next battlefield? In *Proceedings of the IEEE International Workshop on Factory Communication Systems (WFCS)*, pp. 413–421, Vienna, Austria, September 2004.
28. M. Fondl. Network diagnostics for industrial Ethernet. *The Industrial Ethernet Book*, September 16, 2003. Available at http://ethernet.industrial-networking.com.

29. G. F. Franklin, J. D. Powell, and A. Emani-Naeini. *Feedback Control of Dynamic Systems*, 3rd edn. Addison-Wesley, Reading, MA, 1994.
30. G. F. Franklin, J. D. Powell, and M. L. Workman. *Digital Control of Dynamic Systems*, 3rd edn. Addison-Wesley, Reading, MA, 1998.
31. B. Galloway and G. P. Hancke, Introduction to industrial control networks. *IEEE Communications Surveys and Tutorials*, 15(2):860–880, 2013.
32. Grid Connect. The Grid Connect Fieldbus comparison chart. Available at http://www.er-soft.com/files/ER-soft–Fieldbus–Comparison–Chart.pdf.
33. W. S. Harrison, D. M. Tilbury, and C. Yuan. From hardware-in-the-loop to hybrid process simulation: An ontology for the implementation phase of a manufacturing system. *IEEE Transactions on Automation Science and Engineering*, 9(1):96–109, January 2012.
34. D. W. Holley. Understanding and using OPC for maintenance and reliability applications. *IEEE Computing and Control Engineering*, 15(1):28–31, February/March 2004.
35. IEC standard redefines safety systems. *InTech*, 50(7):25–26, July 2003.
36. IEC 61784-2. Industrial communication networks – Profiles – Part 2: Additional fieldbus profiles for real-time networks based on ISO/IEC 8802-3, 2007.
37. IEEE. 1588: Standard for a precision clock synchronization protocol for networked measurement and control systems, 2002. Available at http://www.nist.gov/el/isd/ieee/ieee1588.cfm.
38. International SEMATECH. *Proceedings of the International SEMATECH e-Manufacturing/e-Diagnostics Workshop*, April 2004. Available at http://ismi.sematech.org/emanufacturing/meetings/20040419/index.htm.
39. S.-L. Jämsä-Jounela, Future trends in process automation. *Annual Reviews in Control*, 3:211–220, 2007.
40. H. Kaghazchi and D. Heffernan. Development of a gateway to PROFIBUS for remote diagnostics. In *PROFIBUS International Conference*, Warwickshire, U.K., 2004. Available online at http://www.ul.ie/~arc/techreport.html.
41. S. A. Koubias and G. D. Papadopoulos. Modern fieldbus communication architectures for real-time industrial applications. *Computers in Industry*, 26:243–252, August 1995.
42. L. Larsson. Fourteen industrial Ethernet solutions under the spotlight. *The Industrial Ethernet Book*, (37): March 2007. Available at http://ethernet.industrial-networking.com.
43. K. C. Lee and S. Lee. Performance evaluation of switched Ethernet for networked control systems. In *Proceedings of IEEE Conference of the Industrial Electronics Society*, 4:3170–3175, November 2002.
44. Y. Li-Baboud, X. Zhu, D.M. Anand, S. Hussaini, and J.R. Moyne. Semiconductor manufacturing equipment data acquisition simulation for timing performance analysis. In *Proceedings of IEEE ISPCS*, 2008.
45. F.-L. Lian, J. M. Moyne, and D. M. Tilbury. Performance evaluation of control networks: Ethernet, ControlNet, and DeviceNet. *IEEE Control Systems Magazine*, 21(1):66–83, February 29, 2001.
46. F.-L. Lian, J. M. Moyne, D. M. Tilbury, and P. Otanez. A software toolkit for design and optimization of sensor bus systems in semiconductor manufacturing systems. In *AEC/APC Symposium XIII Proceedings*, Banff, Alberta, Canada, October 2001.
47. F.-L. Lian, J. R. Moyne, and D. M. Tilbury. Network design consideration for distributed control systems. *IEEE Transactions on Control Systems Technology*, 10(2):297–307, March 2002.
48. P. S. Marshall. A comprehensive guide to industrial networks: Part 1. *Sensors Magazine*, 18(6):28–43, June 2001.
49. P. Marti, J. Yepez, M. Velasco, R. Villa, and J. M. Fuertes. Managing quality-of-control in network-based control systems by controller and message scheduling co-design. *IEEE Transactions on Industrial Electronics*, 51(6): 1159–1167, December 2004.
50. D. L. Mills. Internet time synchronization: The network time protocol. *IEEE Transactions on Communications*, 39(10):1482–1493, October 1991.
51. G. A. Mintchell. OPC integrates the factory floor. *Control Engineering*, 48(1):39, January 2001.

52. J. Montague. Safety networks up and running. *Control Engineering*, 51(12):38, December 2004. Available at http://www.controleng.com/article/CA484725.html.
53. J. Montague. Networks busting out all over. *Control Engineering*, 52(3):69, March 2005. Available at http://www.controleng.com/article/CA509788.html.
54. J. Moyne, B. Triden, A. Thomas, K. Schroeder, and D. Tilbury. Cost function and tradeoff analysis of dedicated vs. shared networks for safety and control systems, *ATP International Journal*, 4(2):22–31, September 2006.
55. J. Moyne and D. Tilbury. Determining network control solution designs for manufacturing systems. In *Proceedings of 9th Biennial Conference on Engineering Systems Design and Analysis (ESDA'08)*, Haifa, Israel, July 2008.
56. J. Moyne and D. Tilbury, Performance metrics for industrial Ethernet. *The Industrial Ethernet Book*, 38: April 2007. Available at http://ethernet.industrial-networking.com.
57. J. Moyne and D. Tilbury, The emergence of industrial control networks for manufacturing control, diagnostics and safety data. *IEEE Proceedings, Special Issue on the Emerging Technology of Network Control Systems*, 95(1):29–47, January 2007.
58. J. Moyne, J. Korsakas, and D. M. Tilbury. Reconfigurable factory testbed (RFT): A distributed testbed for reconfigurable manufacturing systems. In *Proceedings of the Japan–U.S.A. Symposium on Flexible Automation*, Denver, CO, July 2004. American Society of Mechanical Engineers (ASME).
59. J. Moyne and F. Lian. Design considerations for a sensor bus system in semiconductor manufacturing. In *International SEMATECH AEC/APC Workshop XII*, September 2000.
60. J. Moyne, N. Najafi, D. Judd, and A. Stock. Analysis of sensor/actuator bus interoperability standard alternatives for semiconductor manufacturing. In *Sensors Expo Conference Proceedings*, Cleveland, OH, September 1994.
61. J. Moyne, P. Otanez, J. Parrott, D. Tilbury, and J. Korsakas. Capabilities and limitations of using Ethernet-based networking technologies in APC and e-Diagnostics applications. In *SEMATECH AEC/APC Symposium XIV Proceedings*, Snowbird, UT, September 2002.
62. J. Moyne, D. Tilbury, and H. Wijaya. An event-driven resource-based approach to high-level reconfigurable logic control and its application to a reconfigurable factory testbed. In *Proceedings of the CIRP International Conference on Reconfigurable Manufacturing Systems*, Ann Arbor, MI, 2005.
63. T. Neagoe, V. Cristea, and L. Banica. NTP versus PTP in computer networks clock synchronization. In *Proceedings of IEEE International Symposium on Industrial Electronics*, 2006.
64. P. Neumann, Communication in industrial automation—What is going on? *Control Engineering Practice*, 15:1332–1347, 2007.
65. Neumann 2010.
66. NIST 2013 http://bigdatawg.nist.gov/home.php.
67. OPC Foundation. OPC data access automation interface standard Version 2.02. 1999. http://www.opcfoundation.org.
68. OPC Foundation. OPC DA 3.00 specification, March 30, 2003. Available at http://www.opcfoundation.org/ua; www.opcfoundation.org/ua.
69. P. G. Otanez, J. R. Moyne, and D. M. Tilbury. Using deadbands to reduce communication in networked control systems. In *Proceedings of the American Control Conference*, pp. 3015–3020, Anchorage, Alaska, May 2002.
70. P. G. Otanez, J. T. Parrott, J. R. Moyne, and D. M. Tilbury. The implications of Ethernet as a control network. In *Proceedings of the Global Powertrain Congress*, Ann Arbor, MI, September 2002.
71. J. Parker, M. Reath, A. F. Krauss, and W. J. Campbell. Monitoring and preventing arc-induced wafer damage in 300 mm manufacturing. In *Proceedings of International Conference on IC Design and Technology*, 2004.
72. J. T. Parrott, J. R. Moyne, and D. M. Tilbury. Experimental determination of network quality of service in Ethernet: UDP, OPC, and VPN. In *Proceedings of the American Control Conference*, Minneapolis, MN, 2006.

73. R. Piggin and D. Brandt. Wireless ethernet for industrial applications. *Assembly Automation*, 26:205–215, 2006.
74. PROFIBUS Standard. IEC 61158 type 3 and IEC 61784. Available at http://www.profibus.com.
75. Railway applications—Communication, signalling and processing systems. Part 1: Safety-related communication in closed transmission systems, 2001. Irish Standard EN 50159-1.
76. T.S. Rappaport. *Wireless Communications: Principles and Practice*, 2nd edn. Prentice Hall PTR, Upper Saddle River, NJ, 2002.
77. M. Read. Wireless Requirements—USCAR plant floor controllers committee. Presented at the University of Michigan Network Performance Workshop, May 2008.
78. J.-P. Richard. Time-delay systems: An overview of some recent advances and open problems. *Automatica*, 39(10):1667–1694, 2003.
79. T. Sauter, The three generations of field-level networks—Evolution and compatibility issues. *IEEE Transactions on Industrial Electronics*, 57(11):3585–3595, November 2010.
80. K. Schroeder, A. Khan, J. Moyne, and D. Tilbury. An application of the direct integration of industrial safety and diagnostics systems. In *Proceedings of the 2008 International Manufacturing Science and Engineering Conference (MSEC2008)*, Evanston, IL, October 2008.
81. C. Scott, G. Law et al. Integrated diagnostics in a process plant having a process control system and a safety system. U.S. Patent 6975966. United States of America, Fisher-Rosemount Systems, Inc., Austin, TX, 2005.
82. L. Seno, F. Tramarin, and S. Vitturi, Performance of industrial communication systems: Real application contexts. *IEEE Industrial Electronics Magazine*, pp. 27–37, June 2012
83. SERCOS 2013. www.sercos.com.
84. A. Shah and A. Raman. Factory network analysis. In *Proceedings of the International SEMATECH e-Diagnostics and EEC Workshop*, July 2003. Available at http://ismi.sematech.org/emanufacturing/meetings/20030718/index.htm.
85. D. Sharma, D. M. Anand, Y. Li-Baboud, and J. R. Moyne. A time synchronization testbed to define and standardize real-time model-based control capabilities in semiconductor manufacturing. In *Proceedings of AEC/APC Symposium*, 2009.
86. B. Shetler. OPC in manufacturing. *Manufacturing Engineering*, 130(6): June 2003.
87. S. Singlevich and K.V.R. Subrahmanyam. Detecting arcing events in semiconductor manufacturing equipment. In *Advanced Semiconductor Manufacturing Conference (ASMC), 23rd Annual SEMI*, vol., pp. 102, 105, May 15–17, 2012
88. P. Sink. Industrial Ethernet: The death knell of fieldbus? *Manufacturing Automation Magazine*, Aurora, Ontario, Canada, April 1999.
89. S. Soucek and T. Sauter. Quality of service concerns in IP-based control systems. *IEEE Transactions on Industrial Electronics*, 51:1249–1258, December 2004.
90. Z. Stank, Bridging safety onto automation networks. *Plant Engineering*, June 2013. Available at http://www.plantengineering.com/single-article/bridging-safety-onto-automation-networks/46dfcf0d84d5d5a1e56010320fce9674.html.
91. Y. Takayanagi and M. Ichikawa, Up-to-date trend of real-time ethernet for industrial automation systems. In *ICROS-SICE International Joint Conference, Fukuoka*, Japan, August 2009.
92. A. S. Tanenbaum. *Computer Networks*, 3rd edn. Prentice-Hall, Upper Saddle River, NJ, 1996.
93. F. Tao, L. Zhang, V. C. Venkatesh, Y. Luo, and Y. Cheng, Cloud manufacturing: A computing and service-oriented manufacturing model. *Proceedings of the Institution of Mechanical Engineers, Part B: Journal of Engineering Manufacture*, 225:1969–1976, 2011.
94. J.-P. Thomesse. Fieldbus technology in industrial automation. *Proceedings of the IEEE*, 93(6):1073–1101, June 2005.
95. F. Tobagi and L. Kleinrock. Packet switching in radio channels: Part II—The hidden terminal problem in carrier sense multiple-access and the busy-tone solution. *IEEE Transactions on Communications*, 23:1417–1433, 1975.

96. A. Treytl, T. Sauter, and C. Schwaiger. Security measures for industrial fieldbus systems—State of the art and solutions for IP-based approaches. In *Proceedings of the IEEE International Workshop on Factory Communication Systems (WFCS)*, pp. 201–209, Vienna, Austria, September 31, 2004.
97. USCAR Web site: www.uscar.org.
98. E. Vonnahme, S. Ruping, and U. Ruckert. Measurements in switched Ethernet networks used for automation systems. In *Proceedings of IEEE International Workshop on Factory Communication Systems*, Porto, Portugal, pp. 231–238, September 2000.
99. J. D. Wheelis. Process control communications: Token bus, CSMA/CD, or token ring? *ISA Transactions*, 32(2):193–198, July 1993.
100. A. Willig, K. Matheus, and A. Wolisz. Wireless technology in industrial networks. *Proceedings of the IEEE*, 93(6):1130–1151, June 2005.
101. Y. Zhao, B. G. Agee, and J. H. Reed. Simulation and measurement of microwave oven leakage for 802.11 WLAN interference management. In *Proceedings of IEEE International Symposium on Microwave, Antenna, Propagation and EMC Technologies*, 2005.
102. ZigBee Alliance. IEEE 802.15. 4, ZigBee standard. 2009. Available at http://www.zigbee.org.

3

Configuration and Management of Networked Embedded Devices

Wilfried Elmenreich
University of Klagenfurt

Andrea Monacchi
University of Klagenfurt

3.1 Introduction 3-2
3.2 Concepts and Terms 3-2
Configuration versus Management • Smart Devices • Plug and Play versus Plug and Participate • State
3.3 Requirements for Configuration and Management 3-4
3.4 Interface Separation 3-5
Interface File System Approach • Service-Oriented Device Architectures • Resource-Oriented Architectures
3.5 Profiles, Datasheets, and Descriptions 3-8
Profiles • Electronic Device Description Language • Field Device Tool/Device Type Manager • Transducer Electronic Datasheet • Interface File System/Smart Transducer Descriptions • Devices Profile for Web Services • Constrained Application Protocol
3.6 Application Development 3-13
3.7 Configuration Interfaces 3-16
Hardware Configuration • Plug and Participate • Service Discovery Mechanisms • Application Configuration and Upload
3.8 Management Interfaces 3-19
Monitoring and Diagnosis • Calibration
3.9 Maintenance in Fieldbus Systems 3-20
3.10 Conclusion 3-22
Acknowledgments 3-22
References 3-22

3.1 Introduction

The advent of embedded microcontrollers and the possibility of equipping microcontrollers with small, fast, and low-cost network interfaces has allowed for the creation of smart distributed systems consisting of a set of networked embedded devices. The main reasons for building a system in a distributed manner are the possibilities of having redundancy and exploiting parallelism. In the context of embedded networks, a distributed system also supports the following applications:

- *Distributed sensing* with smart transducers. A smart transducer is the combination of sensor/actuator, processing element, and network interface. Smart transducers perform local preprocessing of the analog sensor signal and transmit the measurement digitally via the network.
- Systems that integrate *legacy systems*, which cannot be extended or changed due to legal or closed system issues. The unchanged legacy system is thus integrated as a subsystem into a network, establishing a larger overall system.
- *Complexity management* by dividing the overall applications into several subsystems with separate hardware and software. A distributed network of embedded devices also comes with increased complexity for typical configuration and management tasks such as system setup (no hyphen/space), diagnosis, repair, monitoring, calibration, and change. In order to handle this complexity, computer-automated mechanisms can perform tasks such as registering new devices, auto-configuring data flow, and detecting configuration mismatches.

In this chapter, we examine state-of-the art mechanisms for handling these tasks. A considerable number of mature solutions for configuration and management exist in the context of fieldbus systems and service-oriented architectures (SOAs). A fieldbus system is a network for industrial manufacturing plants. The fieldbus system conducts fieldbus nodes comprising sensors, actuators, valves, console lights, switches, and contactors. Challenges for fieldbus systems are interoperability, real-time communication, robustness, support for management, and configuration. Thus, fieldbus systems have evolved a set of interesting concepts for supporting setup, configuration, monitoring, and maintenance of embedded devices connected to a fieldbus. Twenty years ago, a typical process automation plant consisted of various field devices from a half-dozen vendors. Each device had its own setup program with different syntax for the same semantics. The data from the devices often differed in format and in routine to interface each device [1]. Since that time, many fieldbus configuration and management methods have been devised.

The following sections are organized as follows: Section 3.2 gives definitions for the concepts and terms in the context of configuration and management of networked embedded systems. Section 3.3 investigates requirements for configuration and management tasks. Section 3.4 analyzes the necessary interfaces of an intelligent device and proposes meaningful distinctions between interface types. Section 3.5 discusses profiles and other representation mechanisms of system properties for embedded devices. Section 3.6 gives an overview of application development methods and their implications for configuration and management of distributed embedded systems. Section 3.7 examines the initial setup of a system in terms of hardware and software configuration. Section 3.8 deals with approaches for system management, such as application download, diagnosis, and calibration of devices. Section 3.9 presents maintenance methods for reconfiguration, repair, and reintegration of networked devices. Section 3.10 concludes this overview.

3.2 Concepts and Terms

The purpose of this section is to introduce and define some important concepts and terms that are used throughout this chapter.

3.2.1 Configuration versus Management

The term *configuration* is used for a wide range of actions. Part of configuration deals with setting up the hardware infrastructure of a network and its nodes, that is, physically connecting nodes (cabling) and configuring nodes in a network (e.g., by using switches and jumpers). Configuration also involves setting up the network on the logical (i.e., software) level. Therefore, a particular configuration mechanism often depends on the network issues such as topology or underlying communication paradigm. For example, the TTP-Tools [2] are designed for the time-triggered architecture [3]. TTP-Tools provide various design and configuration mechanisms for engineering dependable real-time systems, but it is assumed that the system is based on a time-triggered paradigm [3].

In contrast to configuration, *management* deals with handling of an established system and includes maintenance, diagnosis, monitoring, and debugging. As with configuration, different systems can differ greatly in their support and capabilities in these areas.

Configuration and management are often difficult to separate because procedures such as plug and play (see Section 3.2.3) involve both configuration and management tasks.

3.2.2 Smart Devices

The term *smart* or *intelligent* device was first used in this context by Ko and Fung in [4], meaning "a sensor or actuator device that is equipped with a network interface in order to support an easy integration into a distributed control application."

In the context of networked embedded systems, an *intelligent* device supports its configuration and management by providing its data via a well-defined network interface [5] and/or offering a self-description of its features. The description usually comes in a machine-readable form (e.g., as an XML description) that resides either locally at the embedded device (e.g., IEEE1451.2 [6]) or at a higher network level, referenced by a series number (e.g., Object Management Group [OMG] Smart Transducer Interface [7]).

3.2.3 Plug and Play versus Plug and Participate

Plug and play describes a feature for automatic integration of a newly connected device into a system without user intervention. Although this feature works well for personal computers within an office environment, it is quite difficult to achieve this behavior for automation systems because without user intervention, the system would not be able to ascertain which sensor data should be used and which actuator should be instrumented by a given device. Therefore, in the automation domain, the more correct term *plug and participate* should be used, which describes the initial automatable configuration and integration of a new device. For example, after connecting a new sensor to a network, the sensor could be automatically detected, given a local name, and assigned to a communication slot. The task of a human system integrator is then reduced to deciding on further processing and usage of the sensor data.

3.2.4 State

Zadeh states that "The notion of state of a system at any given time is the information needed to determine the behavior of the system from that time on" [8, p. 3]. In real-time computer systems, we distinguish between the *initialization state* (i-state) and the *history state* (h-state) [9].

The i-state encompasses the static data structure of the computer system, that is, data that are usually located in the static (read-only) memory of the system. The i-state does not change during the execution of a given application, for example, calibration data of a fieldbus node. The h-state is the "dynamic data structure [...] that undergoes change as the computation progresses" [9, p. 91]. Examples of an h-state are the cached results of a sequence of measurements that are used to calculate the current state of a process variable.

The size of the h-state at a given level of abstraction may vary during execution. A good system design will aim at having a *ground state*, that is, when the size of the h-state becomes zero. In a distributed system, this usually requires all tasks to be inactive and no messages to be in transit.

3.3 Requirements for Configuration and Management

The requirements for a configuration and management framework are driven by several factors. We have identified the following points:

- *(Semi)automatic configuration*: The requirement for a plug-and-play-like configuration can be justified by three arguments: Firstly, automatic or semiautomatic configuration saves time and therefore leads to better maintainability and lower costs. Secondly, the necessary qualification of the person who sets up the system may be less if the overall system is easier to configure. Thirdly, the number of configuration faults will decrease because monotonous and error-prone tasks such as looking up configuration parameters in manuals are done by the computer. In most cases, a fully automatic configuration will be possible only if the functionality of the system is reduced to a manageable subset. For more complex applications, consulting a human mind is unavoidable. We thus distinguish two use cases: (1) the automatic setup of simple subsystems and the (2) computer-supported configuration of large distributed systems. The first case mainly concerns systems that require an *automatic* and *autonomous* (i.e., without human intervention) reconfiguration of network and communication participants in order to adapt to different operating environments. Usually, such systems either use very sophisticated (and often costly) negotiation protocols or work only on closely bounded and well-known application domains. The second case is the usual application.
- *Comprehensible interfaces*: In order to minimize errors, all interfaces will be made as comprehensible as possible. This includes the uniform representation of data provided by the interfaces and the capability for selectively restricting an interface to the data required by the interface's user. The comprehensibility of an interface can be expressed by the *mental load* that it puts on the user. Different users need different specialized interfaces, each with minimal mental load. For example, an application developer mostly has a service-centered view of the system. Physical network details and other properties not relevant for the application should be hidden from the developer [10].
- *Uniform data structures*: Configuration and management of distributed embedded systems requires representations of system properties that are usable by software tools. In order to avoid a situation in which each application deals with the required information in its own way, these representations should be generic, highly structured, and exactly specified.
- *Low overhead on embedded system*: Typical embedded hardware is restricted by requirements for cost, size, power consumption, and mechanical robustness. Thus, embedded hardware usually provides far less memory and processing power than average desktop systems. Currently, typical microcontrollers provide about several hundred bytes of RAM and less than 128 kB of Flash ROM. Clocked by an internal oscillator, these microcontrollers provide only a few MIPS of processing power. The local application will consume most of the available resources, and consequently, the designers of configuration and management mechanisms must ensure that there is no excessive overhead on the embedded system nodes. This requirement drives design decisions where configuration data are compressed or even stored on a separate repository outside the embedded network.
- *Use of standard software/hardware*: Computers running standard Windows or Linux operating systems do not provide guaranteed response times for programs, and most hardware interfaces are controlled by the operating system. Lack of a guaranteed response time might violate the special timing requirements of an embedded real-time protocol, and thus, it is often not possible to directly connect a configuration host computer to the fieldbus network without a gateway. Instead, a configuration tool must use some other means of communication, such as standard communication protocols or interfaces such as TCP/IP, RS232, USB, or standard middleware like Common

Object Request Broker Architecture (CORBA). Often, the system uses a dedicated gateway node that routes the configuration and management access to the sensor/actuator nodes; thus, the timing of the embedded network is not disturbed by configuration traffic, and the embedded nodes do not need to implement a USB or CORBA interface and can be kept slim. In order to reduce the complexity of the involved conversion and transformation steps, the interface to and from the embedded node must be comprehensible, structurally simple, and easy to access.

3.4 Interface Separation

If different user groups access a system for different purposes, users should only be provided with an interface that presents information relevant for their respective purposes [11].

Interfaces for different purposes may differ in accessible information and temporal behavior of the access across the interface.

Kopetz, Holzmann, and Elmenreich have identified three interfaces to a transducer node [5]:

The configuration and planning (CP) interface allows the integration and setup of newly connected nodes. The CP interface is used to generate the *glue* in the network that enables network components to interact in the intended manner. The CP interface is not usually time critical.

The diagnostic and management (DM) interface is used for parameterization and calibration of devices and for collecting diagnostic information to support maintenance activities. For example, a remote maintenance console can request diagnostic information from a certain sensor. The DM interface is not usually time critical.

The real-time service (RS) interface is used to communicate the application data, for example, sensor measurements or set values for an actuator. This interface usually has to fulfill timing constraints such as a bounded latency and a small communication jitter. The RS interface has to be configured by the use of the CP (e.g., communication schedules) or DM interface (e.g., calibration data or level monitors).

The TTP/A fieldbus system in [12] uses time-triggered scheduling to provide a deterministic communication scheme for the RS interface. A specified part of the bandwidth is reserved for arbitrary CP and DM activities. Therefore, it is possible to perform CP tasks while the system is in operation without a probe effect on the RS [13].

3.4.1 Interface File System Approach

The concept of an Interface File System (IFS) was introduced in [5]. The IFS provides a unique addressing scheme to all relevant data belonging to nodes in a distributed system. Thus, the IFS maps real-time data, many kinds of configuration data, self-describing information, and internal state reports for diagnostic purposes.

The IFS is organized hierarchically as follows: The *cluster name* addresses a particular fieldbus network. Within the cluster, a specific node is addressed by the *node name*. The IFS of a node is structured into *files* and *records*. Each record is a unit of 4 bytes of data.

The IFS is a generic approach that has been implemented with the TTP/A protocol as a case study for the OMG Smart Transducer Interface. The IFS approach supports the integration and management of heterogeneous fieldbus networks well. The IFS provides the following benefits:

- The IFS establishes a well-defined interface between network communication and local application. The local application uses application programming interface functions to read and write data from/into the IFS. The communication interface accesses the IFS to exchange data across the network.
- The IFS provides transparency on network communication because a task does not have to discriminate between data that are locally provided and data that are communicated via the network.

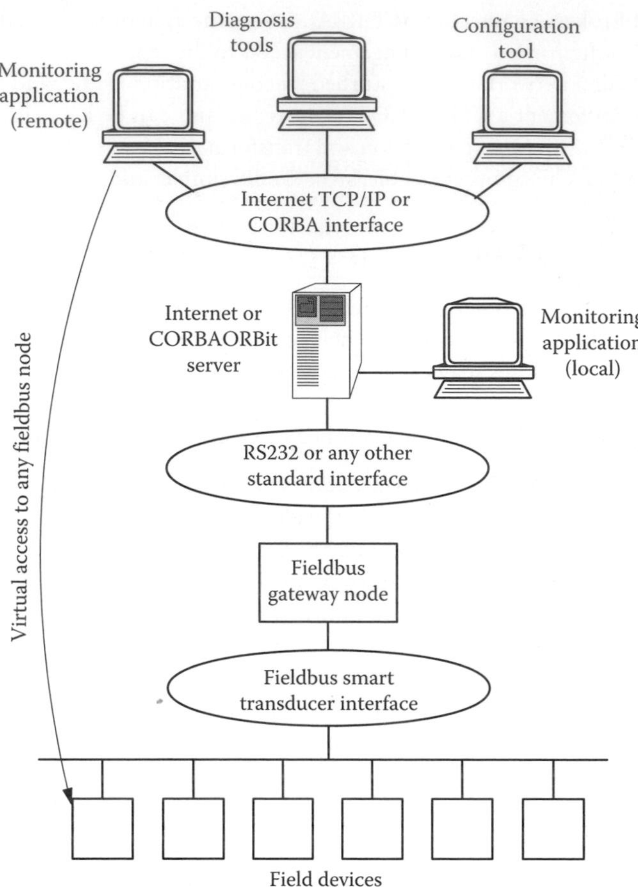

FIGURE 3.1 Architecture for remote configuration and monitoring.

- Because the configuration and management data are mapped into the IFS, the IFS supports access from configuration and management tools from outside the network well. Figure 3.1 depicts an architecture with configuration and management tools that access the IFS of a fieldbus network from the Internet.

The IFS maps different data domains such as RS data, configuration data, and management data using a consistent addressing scheme. In fact, the management interface can be used to define the RS data set dynamically (e.g., to select between a smoothed measurement and a measurement with better dynamics as the output from a sensor). Although real-time guarantees for communication of real-time data are necessary, access to configuration and management data is not time critical. This enables the employment of web-based tools for remote maintenance.

Tools that operate on the IFS have been implemented using CORBA as middleware. CORBA is an object model managed by the OMG that provides transparent communication among remote objects. Objects can be implemented in different programming languages and can run on different platforms. The standardized CORBA protocol IIOP (Internet Inter-ORB Protocol) can be routed over TCP/IP, thus supporting worldwide access to and communication between CORBA objects across the Internet.

Alternatively, it is possible to use web services as management interface for embedded devices. A case study that implements web services on top of the IFS is described in [14].

3.4.2 Service-Oriented Device Architectures

In a network of smart transducers, finding a model that can suit each sensor is necessary for accessing data in the same way and for enabling cheaper development and maintenance of the network.

SOA concerns the organization of self-contained and distributed computing components in services so that complex applications can be developed as compositions of loosely coupled components.

The SOA approach is independent of any technology because architectures specify a common language for the definition of services. The Web Service Description Language (WSDL) provides a means for the specification of functionalities offered by a web service so that the implementation of the service can be separated from the component's interface. Services are described in terms of actions, I/O interface, and constraints on data, as well as communication protocols. In order to interact with other web services, a central service registry may be deployed. The Universal Description, Discovery, and Integration registry is a platform-independent registry that enables components to discover each other. Provider components can publish their WSDL descriptions on the registry, enabling client components to reach the provider components using Simple Object Access Protocol (SOAP) messages. SOAP relies on XML to provide an abstract and structured representation for messages.

Because complex services can be assembled from individual services, interaction in the process needs to be coordinated. Synchronization can be controlled by a central orchestrator or directly managed by the distributed entities according to their roles (service choreography). Such coordination may be hard-coded into the procedures used to access the service. However, this is difficult to apply in networks of numerous distributed services and multiple interleaved business processes. Approaches dealing with this issue are the Business Process Execution Language for Web Services, the Web Services Choreography Interface, and the Web Services Choreography Description Language.

3.4.3 Resource-Oriented Architectures

An interesting characteristic of communication between components concerns the information used for requesting a service. A typical distinction is between remote procedure call (RPC) and the Representation State Transfer (REST) model. Remote procedure call is an interprocess communication mechanism that allows processes to call procedures (which are turned into methods on object-oriented platforms such as Java RMI) on different address spaces. SOAP-based services implement this approach. A module called *stub* manages this mechanism. *Stub* takes care of data serialization (i.e., marshaling) and hides network mechanisms from developers. The main difference between RPC and REST is that in RPC, the sender sends part of its state to the receiver to perform the request and delegates the receiver to carry out the procedure. In contrast, in the REST model, the sender considers the receiver a collection of resources identified by URIs, and only representations of those resources are sent back to the requester. Therefore, REST web services are manipulations of resources (addressable by URI) rather than procedure or method calls to components. REST is an architectural style defining the behavior of web services in terms of client–server interactions. Accordingly, a client requests a resource to a server. The server processes the request and returns a representation of the resource, consisting of a document describing the state of the resource. This requires resources to be identified with a global identifier (e.g., URI) and means that network components use a standard interface and protocol to exchange such representations (e.g., HTTP). Consequently, the user can progress through a web application by navigating links, which represent state transitions where a different state of the application is returned to the user, who is unaware of the actual network infrastructure (e.g., topology). Therefore, REST does not provide any mechanism for device discovery, as users are supposed to start their navigation at an index page and receive a meaningful list of resources available on the site (server), as well as connections to other sites. Due to its resource-centered approach, REST is usually considered a resource-oriented architecture. This approach was first described in Fielding's PhD thesis [15] and during the definition of the HTTP protocol, although a RESTful approach can also be implemented in other application-level protocols (e.g., SOAP). To enable service interoperability, a typical solution adopted by REST designs is to describe the payload of HTTP messages in the XML or JSON (Javascript object notation) format. This means that REST can directly take advantage of HTTP, using a uniform interface to access resources (i.e., GET, POST, PUT, and DELETE operations). On the contrary, SOAP-oriented designs use HTTP only as a means for complex XML messages, describing both the protocol and the data to be exchanged when interacting with the service. Moritz et al. [16] examine possible solutions for IP-based smart cooperating objects. In particular, the author provides a comparison of RESTful architectures with the DPWS and underlines necessary requirements when using REST architectures. Moreover, he claims that DPWS can be easily used to model a RESTful application, but a RESTful design can model only a restricted subset of the DPWS design.

What is missing in RESTful architectures is support for asynchronous messaging and eventing, as well as dynamic discovery of resources and semantics.

The interface of REST web services can be described by means of the Web Application Description Language (WADL), a machine-readable and XML-based description of HTTP-based web applications. Therefore, WADL is the parallel of WSDL for RESTful web services, as WADL focuses only on describing the interface without any support for semantics. Kamilaris et al. [17] propose a service discovery mechanism using WADL to advertise REST web services when joining the network. The approach is then evaluated in a smart home scenario, consisting of a network of smart meters. In WSDL definitions, semantic annotations can be added using the Semantic Annotation for WSDL. Another solution is the OWL-S description language, which uses an RDF ontology to assign a semantic to the web service functionality, although OWL-S does not provide any means to express its relationships to other services [18]. However, as the original RDF specification does not include quantifiers, Verborgh et al. [18] aim at providing a semantic method to express RESTful web services. In particular, because RESTful services strongly depend on their URIs, a descriptive semantic explanation of the relationship between a resource and its URI should be provided. To enable automated agents to consume REST web service, it is necessary to describe the meaning of services and their relationships with other services. RESTdesc is proposed as a valid solution for describing the semantics of RESTful web services according to the functionalities they implement, rather than their input–output interface. RESTdesc allows the semantic description of web services as implications in the Notation 3{n3}* format, which means that services are defined by the pre- and post-conditions needed to execute a request. This gives a reasoner (e.g., EYE {eye}†) the possibility of performing inferences by dynamically evaluating preconditions and applying given axioms, thus providing a flexible solution to handle complex situations. Indeed, the authors suggest that RESTdesc descriptions can be used for automatic context-aware discovery of services, as well as service composition and execution. An example use case illustrating the power of these semantic descriptions is presented in [19]. The authors show that a service composition is a logic entailment binding initial and goal configuration. Accordingly, each service is seen as implication, and service composition is a chain of implications corresponding to a dependency relationship between services. Beyond the formal demonstration, the authors also show how the plan can be implemented by an executor component. Certain rules involve physical data, and thus, these rules are context dependent and are considered only during plan execution.

3.5 Profiles, Datasheets, and Descriptions

In order to support computer-aided configuration, dedicated computer-readable representations of network and node properties are required. This information plays a similar role to that of manuals and datasheets for a computer-based support framework during configuration and management of a system. These computer-readable representations allow the establishment of common rule sets for developing and configuring applications and for accessing devices and system properties (for configuration as well as management functions). In the following section, we will discuss the profile approach as well as several mechanisms following an electronic datasheet approach: Electronic Device Description

* http://www.w3.org/Teamsubmission/n3/

† http://eulersharp.sourceforge.net

Language (EDDL), Field Device Tool/Device Type Manager (FDT/DTM), transducer electronic datasheets (TEDS), smart transducer descriptions (STD) of the IFS, Devices Profile for Web Services (DPWS), and constrained application protocol (CoAP).*

3.5.1 Profiles

Profiles are a widely used mechanism to create interoperability in distributed embedded systems. We distinguish between several types of profiles, namely, application, functional, or device profiles. Heery and Patel propose a very general and short profile definition that we adopt for our discussion. In short, "… profiles are schemata, which consist of data elements drawn from one or more namespaces,[1] combined together by implementers, and optimized for a particular local application" [20].

In many cases, a profile is the result of the joint effort of a group of device vendors in a particular area of application. Usually, a task group is founded that tries to identify reoccurring functions, usage patterns, and properties in the device vendors' domain and then creates strictly formalized specifications according to these identified parts, resulting in the so-called profiles.

More specifically, for each device type, a profile defines exactly what kind of communication objects, variables, and parameters have to be implemented so that a device will conform to the profile. Profiles usually consist of several types of variables and parameters (e.g., process parameters, maintenance parameters, and user-defined parameters) and provide a hierarchical conformance model that allows for the definition of user-defined extensions of a profile. A device profile need not necessarily correspond to a particular physical device, for example, a physical node could consist of multiple *virtual* devices (e.g., multipurpose I/O controller), or a virtual device could be distributed over several physical devices.

Protocols supporting device, functional, and application profiles are CANopen [21], Profibus, and LON [22] (LonMark functional profiles). Figure 3.2 depicts, as an example, the visual specification of a LonMark[2] functional profile for an analog input object. The profile defines a set of network variables (in this example, only mandatory variables are shown) and local configuration parameters (none in this example). The arrow specifies that this profile outputs a digital representation of an analog value, whereas the structure of this output is defined with the standardized (in LON) network variable type `SNVT_lev_percent` (−163.84% to 163.84% of full scale). In addition, a profile also specifies other important properties, such as timing information, valid range, update rate, power-up state, error condition, and behavior (usually as a state diagram).

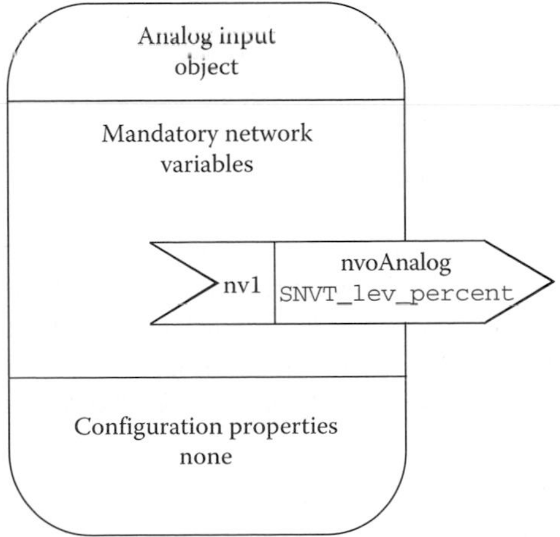

FIGURE 3.2 Functional profile for an analog input.

* http://tools.ietf.org/html/draft-ietf-core-coap-18

Although the approach for creating profiles is comparable in different protocols, the profiles are not always interchangeable between the various fieldbuses, although advancements (at least for process control–related fieldbuses) have been made within the IEC 61158 [23] standard. Block- and class-based concepts such as function blocks, as defined for the Foundation Fieldbus (FF) or the Profibus DP or component classes in IEEE 1451.1 [24] can be considered implementations of the functional profile concept.

3.5.2 Electronic Device Description Language

The EDDL was first used with the HART communication protocol in 1992 [25]. In 2004, EDDL was extended by combining the device description languages of HART, PROFIBUS, and Fieldbus Foundation, resulting in an approved international standard IEC 61804-2 [26]. An additional standard IEC 61804-3 extends EDDL by an enhanced user interface, graphical representations, and persistent data storage [27]. In 2004, the OPC Foundation joined the EDDL Cooperation Team. Subsequently, EDDL is used in the open OPC Unified Architecture.

In EDDL, each fieldbus component is represented by an electronic device descriptor (EDD). An EDD is represented in a text file and is operating system independent. Within the basic control and database server, an EDDL interpreter reads the EDD files corresponding to the devices present at the fieldbus system.

Although EDD files are tested against various fieldbus protocols by the vendors, there is no standardized test process for assuring that an EDD works with every available EDDL interpreter. Another weak point of EDDL is that the functionality that can be described with EDDL is limited by the basic functions required by the IEC 61804-2 standard. Thus, device functionality that cannot be described by these functions is often modeled via additional proprietary plug-ins outside the standard. With fieldbus devices becoming increasingly sophisticated, it will become difficult to adequately describe devices with the current EDDL standard.

3.5.3 Field Device Tool/Device Type Manager

The FDT/DTM is a manufacturer-spanning configuration concept for fieldbus devices [28]. FDT is supported by the FDT group, which currently consists of over 50 members (vendors and users). The FTD group is responsible for the certification of DTMs.

A DTM is an executable software that acts as a device driver. An FDT/DTM configuration system consists of the following parts: an FDT frame application for each distributed control system, a communication DTM (comm-DTM) for each fieldbus system, and a device DTM for each field device type. Device functionality is fully encapsulated into the DTM so FDT/DTM comes with no limits regarding the functionality of more complex future devices. In contrast to EDDL, FDT/DTM comes with higher costs for the device manufacturers because manufacturers have to provide the DTM device drivers. On the other hand, EDDL puts greater strain on system manufacturers.

The FDT software has to be executed on a server separate from the basic control and database server. Currently, DTM can run only on Microsoft Windows, which is a drawback due to Microsoft's upgrading and licensing policies (Windows versions evolve rather fast in comparison to the lifetime of an automation plant, and it is not possible to obtain support and extra licenses for outdated Windows versions).

3.5.4 Transducer Electronic Datasheet

The TEDS was developed to establish a generic electronic datasheet format as part of the smart transducer-related IEEE 1451 standards family. The IEEE 1451.2 [6] standard specifies the TEDS including the digital interface to access datasheets, read sensors, or set actuators.

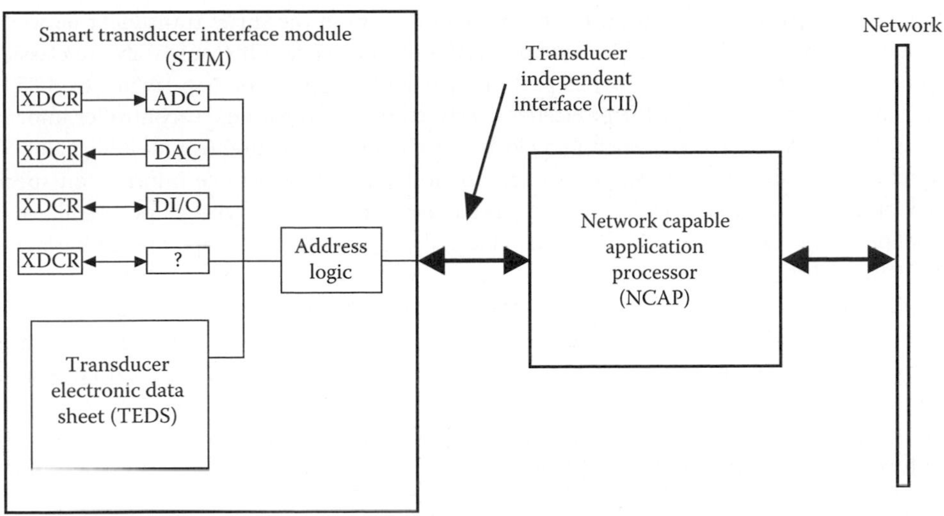

FIGURE 3.3 Smart transducer interface module connected to NCAP.

Figure 3.3 depicts the TEDS in context with the system architecture as defined in the IEEE 1451 standard:

- *Smart transducer interface module* (*STIM*): A STIM contains between 1 and 255 transducers of various predefined types together with their descriptions in the form of the corresponding TEDS.
- *Network capable application processor* (*NCAP*): The NCAP is the interface to the overall network. By providing an appropriate NCAP, the transducer interface is independent of the physical fieldbus protocol.
- *Transducer independent interface* (*TII*): The TII is the interface between the STIM and the NCAP. The TII is specified as an abstract model of a transducer instrumented over 10 digital communication lines.

TEDS describe node-specific properties, such as the structure and temporal properties of devices and transducer data. IEEE 1451 defines self-contained nodes with the TEDS stored in a memory directly located at the node. This causes an overhead on the nodes, so the representation of the configuration information for such a system must be compact. IEEE 1451 achieves this goal by providing a large set of predefined transducer types and modes based on enumerated information, in which identifiers are associated with more detailed prespecified descriptions (similar to error codes). Although the standard defines some possibilities for parameterization of transducer descriptions, this approach restricts the device functionality expressiveness in a similar way as in EDDL.

3.5.5 Interface File System/Smart Transducer Descriptions

STDs as defined in [29] are a method for formally describing properties of devices that follow the CORBA Smart Transducer Interface standard (the descriptions themselves are currently not part of the standard).

The STD uses XML [30] as the primary representation mechanism for all relevant system aspects. Together with related standards, such as XML Schema or XSLT, XML provides advanced structuring, description, representation, and transformation capabilities. XML is becoming the de facto standard for data representation and has extensive support throughout the industry. Some examples where XML has already been used for applications in the fieldbus domain can be found in [31–33].

As the name implies, *STDs* describe the properties of nodes in the smart transducer network. The STD format is used for both describing *static* properties of a device family (comparable to classic datasheets) and devices that are configured as part of a particular application (e.g., if the local STD also contains the local node address). The properties described in STDs cover microcontroller information (e.g., controller vendor, clock frequency, and clock drift), node information (e.g., vendor name, device name/version, and node identifiers), protocol information, and node service information specifying the behavior and the capabilities of a node. In the current approach, a service plays a similar role as a functional profile (see Section 3.5.1) or function block. The functional units align to the interface model of the CORBA STI standard [34].

It is not always possible to store all relevant information outside the node, but by focusing on reducing the amount of required information on the node to the minimum, extensive external meta-information can be used without size constraints. The reference to this external information is the unique combination of series and serial number of the node. The series number is identical for all nodes of the same type. The serial number identifies the instance of a node among all nodes of a series.

The advantages of this approach are twofold:

- Firstly, the overhead at the node is very low. Current low-cost microcontrollers provide internal RAM and EEPROM memory of around 256 bytes. This is not sufficient for storing more than the most basic parts of datasheets according to standards such as the IEEE 1451.2 standard without extra hardware such as external memory. With the proposed description approach, only 8 bytes of memory are required for storing the series and serial numbers.
- Secondly, instead of implicitly representing the node information with many predefined data structures mapped to a compact format, it is possible to have an explicit representation of the information in a well-structured and easy-to-understand way. A typical host computer running the configuration and management tools can easily deal with very extensive generic XML descriptions. Furthermore, XML formats are inherently easy to extend, so the format is open for future extensions of transducer or service types.

3.5.6 Devices Profile for Web Services

The Universal Plug and Play (UPnP) standard is a first step toward the implementation of service-oriented device architectures, and UPnP is already diffused in common consumer electronic devices such as network-attached storage systems and media servers. Other frameworks are commonly used in distributed and service-oriented settings, such as OSGi, Jini, and CORBA. However, these frameworks are usually not applicable to very resource-constrained devices.

The DPWS is a collection of standards that allow embedded devices to support secure web services. DPWS does not rely on a global service registry to store and retrieve service descriptions because DPWS uses dynamic service discovery mechanisms to allow devices to advertise themselves and discover their peers. Moreover, due to metadata description of devices, provided services can be accessed alike. Communication among devices is granted by an event-based mechanism. This provides a way to notify state changes according to a publish/subscribe policy.

To meet constraints of deeply embedded devices, Moritz et al. [35] developed uDPWS, a lightweight version that can be run by 8-bit microcontrollers. In this variant, the authors use a table-driven implementation (consisting in a simple string comparison) to avoid parsing the whole incoming message as XML file.

Since SOAP representations rely on XML, data management introduces a certain overhead. A solution to overcome the heavyweight XML representation is to use a binary format. A potential candidate is the Efficient XML interchange (EXI), recommended by W3C since 2011. EXI provides an alternative compact representation of structured data that can be easily integrated into existing XML-based services by means of special converters. A comparison between different formats is addressed by Sakr [36].

EXI is shown to provide the greatest compression and compactness compared to other representations. Therefore, EXI enhances deeply constrained embedded devices with an efficient way of processing XML information, enabling devices to be integrated in SOAs. An implementation of EXI that can be run by deeply embedded systems is proposed by Kyusakov [37].

3.5.7 Constrained Application Protocol

Using HTTP is the straightforward solution to implement a REST approach. However, as HTTP provides functionalities that can be considered unnecessary in constrained environments, the IETF constrained RESTful environments working group (CoRE) has proposed the CoAP{coap} as an alternative that can be run by deeply embedded systems (e.g., 8-bit microcontrollers). To maintain an easy mapping with HTTP, CoAP uses the same methods (i.e., GET, POST, PUT, and DELETE). However, instead of using TCP, CoAP uses UDP along with a simple retransmission mechanism in which each GET request is associated with a unique identifier. Moreover, CoAP provides an eventing mechanism to the architecture.

In HTTP, data are transferred from a server to a client after a synchronous request. However, in networks of resource-constrained devices, such as sensor nodes, this means that a node should perform polling on another peer in order to monitor the peer's state, thus increasing the network usage and power consumption. To overcome this limitation, CoAP allows a node to subscribe to updates of a certain resource. Therefore, CoAP provides an asynchronous eventing mechanism, in which nodes can become observers of a certain resource and receive notification of status changes. Another characteristic of REST is that REST does not provide any mechanism for device and resource discovery. In HTTP and the web, discovery of resources takes place by accessing an index page listing links to resources on the same or different servers. Therefore, links define relationships between web resources. This means that giving a meaning to links enables machines to automatically calculate how to use resources on a server. For this purpose, a CoRE link format (RFC6690) has been proposed as the format to be used in M2M applications within the CoAP protocol. Accordingly, links are defined by a list of attributes. A resource-type attribute, *rt*, defines the semantic type of a resource, which may be expressed by a URI referencing a specific concept in an ontology.

The interface description attribute, *if*, can be used to describe the service interface, for instance, by URI referencing a machine-readable document such as a WADL definition. Finally, a maximum size attribute, *sz*, can be used to return an indication of the maximum expected size of the resource representation.

According to this approach, a client can contact the server using a GET to a predefined location (i.e., */.well-known/core*), which returns the list of resources exposed by the server and their media type. To provide a basic filtering mechanism, the query can be associated to certain standardized attributes. In particular, *href* filters resources according to their path and type, allowing retrieval of certain Multipurpose Internet Mail Extensions types. Similarly, the attributes *rt* and *if* can be used to select resources according to their application-specific meaning and permitted operations. In addition, it is possible to deploy a resource directory listing links to resources stored on other services. In order to be visible to clients, resource providers can POST their resources to the */.well-known/core* position of the chosen directory node.

3.6 Application Development

In the following section, we examine several application development approaches and how they influence system configuration.

Model-based development is a widely used approach in distributed systems. The basic idea is to create a model of the application that consists of components that are connected via links that represent the communication flow between the components. Different approaches usually differ in what constitutes

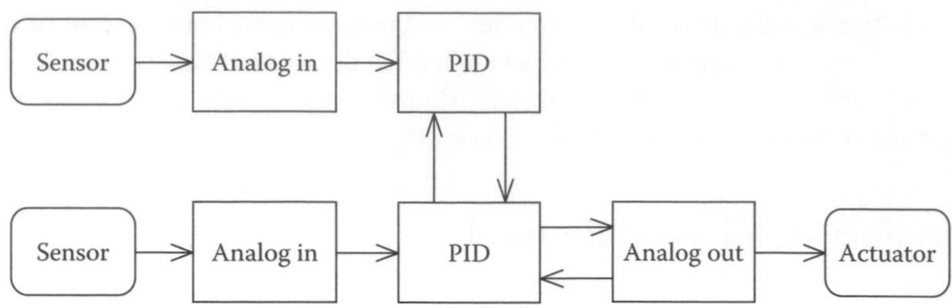

FIGURE 3.4 Example of an application model.

a component (e.g., function blocks, subsystems, services, functional profiles, and physical devices) and the detailed semantics of a link. Many approaches support the recursive definition of components, which allows grouping multiple lower-level components into one higher-level component. Figure 3.4 depicts a model of a typical small control application consisting of two analog inputs receiving values from two sensors, two PIDs, and one analog output controlling an actuator.

The model-based approach is not the only application design approach. Another approach used by multiple fieldbus configuration tools is the ANSI/ISA-88.01-1995 procedural control model [38]. This modeling approach enforces a strictly modular hierarchical organization of the application (see Figure 3.5). There should be little or no interaction between multiple process cells. Interaction between components in a process cell is allowed. To make best use of this approach, the structure of the network site and the application should closely correspond to the hierarchy specified by this model.

The modeling approach conceptually follows the typical hierarchy of process control applications, with multiple locally centralized Programmable Logic Controllers that drive several associated control devices. This eases transition from predecessor systems and improves overall robustness because this approach provides fault containment at the process cell level, the downside being that the coupling between the physical properties of the system and the application is rather tight. An example of a fieldbus protocol that supports this modeling approach is the Profibus PA protocol, which supports this model by providing a universal function block parameter for batch identification [39].

Another design approach is the *two-level design approach* [40], which originated in the domain of safety critical systems. In this approach, the communication between components must be configured before configuring the devices. Although this requires many design decisions to be taken very early in the design process, this approach greatly improves overall composability of the components in the system.

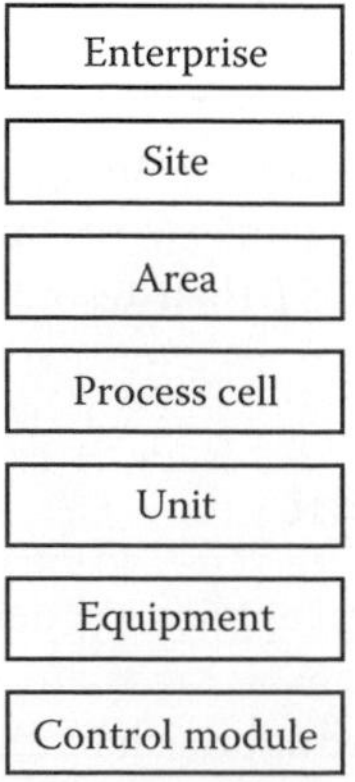

FIGURE 3.5 ANSI/ISA-88.01-1995 hierarchical model.

Abstract application models provide several advantages for application development:

- The modular design of applications helps to deal with complexity by applying a *divide-and-conquer* strategy. Furthermore, modular design supports reuse of application components and physical separation.
- The separation of application logic from physical dependencies allows hardware-independent design that enables application development before hardware is available, as well as easing migration and possibly allowing the reuse (of parts) of applications.

For configuring a physical fieldbus system from such an application model, we must examine (1) how this application model maps to the physical nodes in the network and (2) how information flow is maintained in the network.

In order to map the application model to actual devices, fieldbuses often provide a model for specifying physical devices, for example, in Profibus DP, the physical mapping between function blocks and the physical device is implemented as follows (see Figure 3.6): A physical device can be subdivided in several modules that take the role as *virtual* devices. Each device can have from one (in case of simple functionality) to many slots, which provide the mapping from physical devices to function blocks. A function block is mapped to a slot, whereas slots may have associated physical and transducer blocks. Physical and transducer blocks represent physical properties of a fieldbus device. Parameters of a function block are indexed, and the slot number and parameter index cooperatively define the mapping to data in the device memory.

In contrast, the *FF* follows an object-oriented design philosophy. Thus, all information items related to configuring a device and the application (control strategy) are represented with objects. This includes function blocks, parameters, as well as subelements of parameters. These objects are collected in an object dictionary (OD), whereas each object is assigned an index. This OD defines the mapping to the physical memory on the respective device. In order to understand the methods for controlling the communication flow between the application components, we first examine some recurring important communication properties in fieldbus applications:

- Use of state communication as primary communication mechanism for operating a fieldbus [41]. State communication usually involves cyclically updating the associated application data.
- Support for asynchronous/sporadic communication (event communication) in order to perform management functions and deal with parts of the application that cannot be performed with state communication.

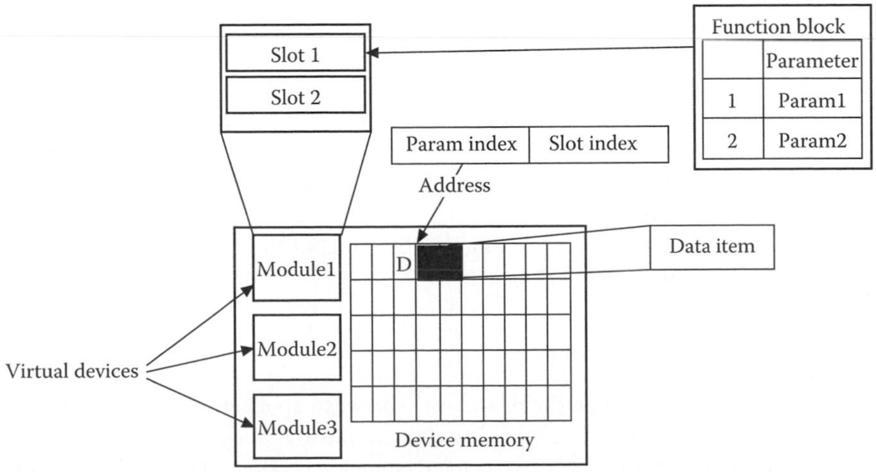

FIGURE 3.6 Mapping of function blocks to physical device in Profibus DP.

A common method to achieve these properties is by scheduling. There are many scheduling approaches with vastly different effects on configuration. The following are some commonly adopted approaches in fieldbus systems:

Multicycle polling: In this approach, communication is controlled by a dedicated node that authorizes other nodes to transmit their data [42]. This dedicated node, typically called the *master node*, polls the other nodes in a time division multiplexing scheme. Thus, the cyclic polling also takes care of bus arbitration. This approach is taken, for example, in WorldFIP, FF, and ControlNet. For configuring the devices in such a network, the master nodes require at least a list of nodes to be polled, that is, in a configuration with a single master, only one node need be configured with time information to control the whole cluster.

Time-triggered: In a time-triggered communication model, the communication schedule is derived from the progression of physical time. This approach requires a predefined collision-free schedule that defines a priori when a device is allowed to broadcast its data and an agreement on a global time, which requires synchronization of the local clocks of all participating devices [3]. Examples of protocols that support time-triggered communication are TTP/A [12], TTP/C [43], and the synchronous part of the Flexray protocol [44]. In order to configure the communication in these systems, the schedules must be downloaded to all nodes in the network.

Event-triggered: Event-triggered communication implements a push model in which the sender decides when to send a message, for example, when a particular value has changed more than a given *delta*. Collisions on the bus are solved by collision detection/retransmission or collision avoidance, that is, bitwise arbitration protocols such as CAN [45]. Event-triggered communication does not depend on scheduling because communication conflicts are either resolved by the protocol at the data link layer (e.g., bit-wise arbitration) or must be resolved by the application.

Scheduling information is usually stored in dedicated data structures that are downloaded to the nodes in the network to be available for use by the network management system functions of the node.

The TTP/A protocol integrates configuration information for application data flow and the communication schedule. Thus, the local communication schedules (named *round descriptor lists*) as well as the interfaces of application services [34] are mapped into the same interfacing mechanism, the *IFS* (see Section 3.4.1).

For representation of the overall system, cluster configuration description (CCD) format was developed, which acts as a central uniform data structure that stores all the information pertinent to the fieldbus system. This information includes the following:

- *Cluster description meta-information*: This description block holds information on the cluster description itself, such as the maintainer, name of the description file, or the version of the CCD format itself.
- *Communication configuration information*: This information includes round sequence lists as well as round descriptor lists, which represent the detailed specification of the communication behavior of the cluster. Additionally, this part of the CCD also includes (partially physical) properties important for communication, such as the UART specification, line driver, and minimum/maximum signal run times.
- *Cluster node information*: This block contains information on the nodes in a cluster, whereas nodes are represented with the STD format.

3.7 Configuration Interfaces

In the previous section, we focused on the relationship between application and configuration. In this section, we examine parts of system configuration that are mostly independent of the application. We briefly look at the physical configuration of networked embedded systems, how nodes are recognized by the configuration system, and how application code is downloaded to nodes.

3.7.1 Hardware Configuration

The hardware configuration encompasses the setup of the plugs and cables of the network. For instance, several fieldbus systems implement means to avoid mistakes, such as connecting a power cable to a sensor input, which would cause permanent damage to the system or even harm people. Moreover, hardware configuration interfaces such as plugs and clamps are often subject to failure in harsh environments, for example, on a machine that induces a lot of vibration.

For hardware configuration, the following approaches can be identified:

- Usage of special jacks and cables that support a tight mechanical connection and avoid mistakes in orientation and polarity by their geometries, for example, the actuator–sensor interface[3] (AS-i) specifies a mechanically coded flat cable that allows the connection of slaves on any position on the cable by using piercing connectors. AS-i uses cables with two wires transporting data and energy via the same line. The piercing connectors support simple connection, safe contacting, and protection up to class IP67.
- Baptizing devices in order to obtain an identifier that allows addressing the newly connected device. This could be done by explicitly assigning an identifier to the device, for example, by setting dip switches or entering a number over a local interface, or implicitly by the cabling topology, for example, devices could be daisy-chained and obtain their name subsequently according to the chain.
- Alternatively, it is possible to assign unique identifiers to nodes in advance. This approach is taken, for example, with Ethernet devices for which the MAC address is a worldwide unique identifier, or in the TTP/A protocol that also uses unique node IDs. However, such a worldwide unique identifier will require many digits making it unpractical to have the number printed somewhere on the device. To overcome this problem, machine-readable identifiers in the form of barcodes or RF tags are used during hardware configuration.
- Simple configuration procedures, which can be carried out and verified by nonexpert personnel.

3.7.2 Plug and Participate

Hardware configuration is intended to be simple, and consequently a networked system should behave intelligently in order to relieve human personnel of error-prone tasks.

In the case of plug and participate, the system runs an integration task that identifies new nodes, obtains information about these nodes, and changes the network configuration in order to include the new nodes in the communication.

Identification of new nodes can be supported with manual baptizing as described in the previous section. Alternatively, it is also possible to automatically search for new nodes and identify them as described in [46].

If there can be different classes of nodes, it is necessary to obtain information on the type of the newly connected nodes. This information will usually be available in the form of an electronic datasheet that can be obtained from the node or from an adequate repository.

The necessary changes of the network configuration for including the new node greatly depend on the employed communication paradigm. In the case of a polling paradigm, only the list of nodes to be polled has to be extended. In the case of a time-triggered paradigm, the schedule has to be changed and updated in all participating nodes. In the case of an event-triggered paradigm, only the new node has to be authorized to send data; however, it is very difficult to predict how a new sender will affect the timing behavior of an event-triggered system. In all three cases, critical timing might be affected due to a change in response time, that is, when the cycle time has to be changed. Thus, in time-critical systems, extensibility must be taken into account during system design, for example, by reserving at first unused bandwidth or including spare communication slots.

3.7.3 Service Discovery Mechanisms

Service discovery defines how network entities can represent and advertise services, so that they can be detected and used by other members. This is a naming problem: assigning each entity a persistent identifier that can be used for retrieving resources. A simple naming system may use broadcasting or multicasting to advertise or retrieve resources (according to push or pull semantics). For instance, this is what happens in the address resolution protocol. Another solution is to use centralized name servers (registries), which can be static nodes or dynamically elected by the community. This approach presents single points of failure that can be improved by organizing nodes in a hierarchy of registries (e.g., domain name service).

Large-scale distributed systems usually use logical overlays, and as such, topologies are set up depending on certain criteria, such as physical distance between nodes. We refer to [47] for a complete overview of resource naming.

Dargie [48] classifies naming systems in three generations: name services (e.g., DNS), directory services (e.g., CORBA and LDAP), and service discovery systems. Although name service offers just mapping between identifiers and resources, directory services enable attribute-based queries because each entry is defined by a set of attributes. However, the author suggests that the first two generations may not be enough in dynamically changing environments such as mobile networks, where entities join and leave the network, thus continuously modifying the network topology. According to Dargie [48], service discovery systems can provide self-configuration and self-healing properties to networks of mobile embedded devices. These systems are indeed able to update the network configuration by discovering new services and detecting failures and disconnected entities (e.g., using leasing on services). Dargie [48] reports on state of the art of service discovery systems. Network size affects bandwidth, so he divides systems into small and large systems. Small systems (e.g., Jini, UPnP, SLP, and Bluetooth) can be implemented as directory based or using multicast. On the other hand, service discovery for large systems (e.g., Ninja SDS, Twine, Jxta, and Ariadne) has to take into account scalability issues. Consequently, systems can be implemented as registries organized in multiple hierarchies or using logical overlays connecting nodes in a peer-to-peer manner.

3.7.4 Application Configuration and Upload

Some frequently recurring applications, such as standard feedback control loops, alert monitoring, and simple control algorithms can often be put in place like building bricks because these applications are generically available (e.g., PID controllers).

For more complex or unorthodox applications, however, it is necessary to implement user-defined applications. These cases require code to be uploaded onto target devices.

About 15 years ago, the most common method of reprogramming a device was to have an EPROM memory chip in a socket that was physically removed from the device, erased under UV radiation, and programmed using a dedicated development system, that is, a PC with a hardware programming device, and then put back into the system.

Today, most memory devices and microcontrollers provide an interface for in-system serial programming of Flash and EEPROM memory. The hardware interface for in-system serial programming usually consists of a connector with four to six pins attached either to an external programming device or directly to the development PC. These programming interfaces are often proprietary to particular processor families, but there also exist some standard interfaces that support a larger variety of devices. For example, the JTAG debugging interface (IEEE Standard 1149.1) also supports the upload of application code.

While the in-system serial programming approach is much more convenient than the socketed EPROM method, both approaches are conceptually reasonably similar because it is still necessary to establish a separate hardware connection to the target system. A more advanced approach for uploading applications is in-programming. In this approach, it is possible to program and configure a device without taking the device out of the distributed target system and without using extra cables and hardware interfaces.

In-system configuration is supported by state-of-the-art flash devices, which can reprogram themselves in part by using a bootloader program. Typically, whenever a new application has to be set up, the programming system sends the node a signal causing it to enter a dedicated upload mode. During the upload phase, the node's service is usually inactive. Failures that lead to misconfiguration must be corrected by locally connecting a programming tool.

Alternatively, application code could be downloaded via the network into the RAM memory at startup. In this case, only the bootloader resides in the persistent memory of the device, and the user-defined application code has to be downloaded at startup. This approach has the advantage of being stateless, so that errors in the system are removed at the next startup, thus engineers could handle many faults by a simple restart of the system. On the other hand, this approach depends on the configuration instance at startup—the system cannot be started if the configuration instance is down. Moreover, the restart time of a system may be considerably longer.

3.8 Management Interfaces

The possibility of performing remote management operations on distributed devices is one of the most important advantages of these systems. Wollschläger states that "in automation systems, engineering functions for administration and optimization of devices are gaining importance in comparison with control functions" [19, p. 89].

Typical management operations are monitoring, diagnosis, or node calibration. Unlike primary fieldbus applications, which often require cyclical, multidrop communication, these management operations usually use a one-to-one (client–server) communication style. For this reason, most fieldbus systems support both communication styles.

A central question is whether and how this management traffic influences the primary application, a problem known as probe effect [13]. System management operations that influence the timing behavior of network communication are critical for typical fieldbus applications (e.g., process control loops) that require exact real-time behavior.

The probe effect can be avoided by reserving a fixed amount of the bandwidth for management operations. For example, in the FF and WorldFIP protocols, the application cycle (macro cycle) is chosen to be longer than strictly required by the application, and the remaining bandwidth is free for management traffic.

In order to avoid collisions within this management traffic window, adequate mechanisms for avoiding or resolving such conflicts must be used (e.g., token-passing between nodes that want to transmit management information and priority-based arbitration).

In TTP/A, management communication is implemented by interleaving real-time data broadcasts (implemented by multipartner rounds) with the so-called master–slave rounds that open a communication channel to individual devices.

If management traffic is directly mingled with application data, such as in CAN, LonWorks, or Profibus PA, care must be taken that this management traffic does not influence the primary control application. This is typically achieved by analyzing network traffic and leaving enough bandwidth headroom. For complex systems and safety-critical systems that require certain guarantees on system behavior, this analysis can become very difficult.

3.8.1 Monitoring and Diagnosis

In order to perform passive monitoring of the communication of the application, it usually suffices to listen on the network. However, the monitoring device must have knowledge of the communication scheme used in the network in order to be able to understand and decode the data traffic. If this scheme is controlled by the physical time, as is the case in time-triggered networks, the monitoring node must also synchronize itself to the network.

Advanced networked embedded devices often have built-in self-diagnostic capabilities and can disclose their own status to the management system. It depends on the capabilities of the system how such information reaches the management framework. Typically, a diagnosis tool or the diagnosis part of the management framework will regularly check the information in the nodes. This method is called *status polling*. In some fieldbus protocols (e.g., FF), devices can also transmit status messages by themselves (*alert reporting*).

In general, restrictions from the implementation of the management interface of a fieldbus protocol also apply to monitoring because in most fieldbus systems, the monitoring traffic is transmitted using the management interface.

For systems that do not provide this separation of management from application information at the protocol level, other means must be taken to ensure that monitoring does not interfere with the fieldbus application. Status polling usually is performed periodically, and thus, it should be straightforward to reserve adequate communication resources during system design so that the control application is not disturbed. In case of alert reporting, the central problem without adequate arbitration and scheduling mechanisms is discerning how to avoid overloading the network in case of *alarm showers*, when many devices want to send their messages at once. It can be very difficult to give timeliness guarantees (e.g., the time between an alarm occurring and the alarm being received by its respective target) in such cases. The typical approach for dealing with this problem (e.g., as taken in CAN) is to provide much bandwidth headroom.

For in-depth diagnosis of devices, it is sometimes also desirable to monitor operation and internals of individual field devices. This temporarily involves greater data traffic, which cannot be easily reserved a priori. Therefore, the management interface must provide some flexibility on the diagnosis data in order to dynamically adjust to the proper level of detail using some kind of *pan and zoom* approach [50].

3.8.2 Calibration

The *calibration* of transducers is an important management function in many fieldbus applications. There is some ambiguity involved concerning the use of this term. Berge strictly distinguishes between *calibration* and *range setting*:

"Calibration is the correction of sensor reading and physical outputs so they match a standard" [39, p. 363]. According to this definition, calibration cannot be performed remotely because the device must be connected to a standardized reference input.

Range setting is used to move the value range of the device so that the resulting value delivers the correctly scaled percentage value. *Range setting* does not require any input and measurement of output, thus *range setting* can be performed remotely. In the HART bus, this operation is called *calibration*, whereas calibration is called *trim*.

Network technology does not influence the way calibration is handled, although information that is required for calibration is stored as part of the properties that describe a device. Such information could be the minimum calibration span limit, this being "the minimum distance between two calibration points within the supported operation range of a device." Additionally, calibration-related information, that is, individual calibration history can be stored in the devices themselves. This information is then remotely available for management tools in order to check the calibration status of devices. Together with the self-diagnosis capabilities of the field devices, this allows performance of a focused and proactive management strategy.

3.9 Maintenance in Fieldbus Systems

Maintenance is the activity of keeping the system in good working order. Typical networked embedded systems (e.g., fieldbuses) provide extensive management features, such as diagnosis and monitoring, which help greatly in maintaining systems. There are several different maintenance schemes that influence the way these steps are executed in detail. Choice of a particular maintenance scheme is usually motivated by the application requirements [39]:

- *Reactive maintenance* is a scheme in which a device is fixed only after it has been found to be broken. Reactive maintenance should be avoided in environments where downtime is costly (such as in factory applications). Thus, designers of such applications will usually choose more active maintenance strategies. Nonetheless, fieldbus systems also provide advantages for this scheme because fieldbus systems support fast detection of faulty devices.
- *Preventive maintenance* is a scheme in which devices are serviced in regular intervals even if devices are working correctly. This strategy prevents unexpected downtime, thus improving availability. Due to the associated costs, this approach will only be taken in safety-related applications such as in aviation, train control, or where unexpected downtime would lead to very high costs.
- *Predictive maintenance* is similar to preventive maintenance, differing in a dynamic service interval that is optimized by using long-time statistics on devices.
- *Proactive maintenance* focuses on devices that are expected to require maintenance.

Maintenance mainly comprises the following steps:

- Recognizing a defective device
- Repairing (replacing) the defective device
- Reintegrating the serviced device

In fieldbus systems, faulty devices will usually be recognized via the network. This is achieved by monitoring the fieldbus nodes and the application or with devices that are capable of sending alerts (also refer to Section 3.8).

After the source of a problem has been found, the responsible node must be serviced. This often means disconnecting the node from the network, thus strategies are required to allow the system to deal with *disconnection of* the node, as well as reconnecting and *reintegrating* the replacement node.

If the whole system has to be powered down for maintenance, the faulty node can simply be replaced, and integration of the replacement node occurs as part of the normal initial startup process. If powering down the whole system is undesirable or even impossible (in the sense of leading to severe consequences, as in the case of safety critical applications), this process becomes more complicated. In this case, we have several options:

- *Implementation of redundancy*: This approach must be taken for safety- or mission-critical devices, where the respective operations must be continued during replacement after a device becomes defective. A detailed presentation of redundancy and fault-tolerant systems can be found in [51].
- *Shut down part of the application*: In the case of factory communication systems that often are organized as multilevel networks and/or use a modular approach, it might be feasible to shut down a local subnetwork (e.g., a local control loop, or a process cell as defined in the ANSI/ISA-88.01-1995 standard).

The replacement node must be configured with individual node data, such as calibration data (these data usually differ between replaced and replacement nodes), and the *state* of a node. The state information can include the following:

- Information accumulated during run-time (the history state of a system): This information must be transferred from the replaced node to the replacement node.
- Timing information so that the node can synchronize with the network: In networks that use a distributed static schedule (e.g., TTP/A), each node must be configured with its part of the global schedule in order to establish a network-wide consistent communication configuration.

One alternative approach for avoiding this transfer of system state is to design (and create) a stateless system. Bauer proposes a generic approach for creating stateless systems from systems with state in [52]. Another possibility is to provide well-defined reintegration points where this state is minimized. Fieldbus applications typically use a cyclical communication style so the start of a cycle is a *natural* reintegration point.

3.10 Conclusion

Configuration and management play an important role for distributed embedded systems. The need for configuration and management in the domain of industrial fieldbus systems has led to interesting mechanisms evolving during the last 20 years. Most of these mechanisms can also be applied in the domain of embedded systems.

The configuration phase can be subdivided into a part that requires local interaction such as connection of hardware and setting dip switches, and a part that can be performed remotely via the fieldbus system. A good design requires the local part to be as simple as possible in order to simplify the interactions of local personnel. The other part should be supported by tools that assist the system integrator in tedious and error-prone tasks such as adjusting parameters according to the datasheet of a device. Examples of systems with such a configuration support are, among others, the IEEE 1451, the FDT, and the EDDL, which all employ machine-readable electronic datasheets in order to support configuration tools.

Management encompasses functions such as monitoring, diagnosis, calibration, and support for maintenance. In contrast to the configuration phase, most management functions are used concurrently to the RS during operation. Some management functions, such as monitoring, may require real-time behavior for themselves. In order to avoid a probe effect on the RS, scheduling of the fieldbus system must be designed to integrate management traffic with real-time traffic.

Acknowledgments

We would like to thank Lizzy Dawes for contributing her English language skills. This work was supported by the Austrian FWF project TTCAR under contract no. P18060-N04, by Lakeside Labs, Austria, by the European Regional Development Fund (ERDF) and the Carinthian Economic Promotion Fund (KWF) under grant KWF 20214-23743-35470 (Project MONERGY: http://www.monergy-project.eu/).

References

1. J. Powell. The "profile" concept in fieldbus technology. Technical article, Siemens Milltronics Process Instruments Inc., Peterborough, Ontario, Canada, 2003.
2. TTTech Computertechnik. Utilizing TTPTools in by-wire prototype projects. White paper, Vienna, Austria, 2002.
3. H. Kopetz and G. Bauer. The time-triggered architecture. *Proceedings of the IEEE*, 91(1):112–126, January 2003.
4. W.H. Ko and C.D. Fung. VLSI and intelligent transducers. *Sensors and Actuators*, 2:239–250, 1982.
5. H. Kopetz, M. Holzmann, and W. Elmenreich. A universal smart transducer interface: TTP/A. *International Journal of Computer System Science & Engineering*, 16(2):71–77, March 2001.
6. Institute of Electrical and Electronics Engineers, Inc. IEEE Std 1451.2-1997. Standard for a smart transducer interface for sensors and actuators—Transducer to micro-processor communication protocols and transducer electronic data sheet (TEDS) formats, New York, September 1997.
7. OMG. Smart transducers interface V1.0. Available specification document number formal/2003-01-01. Object Management Group, Needham, MA, January 2003. Available at: http://doc.omg.org/formal/2003-01-01.

8. L.A. Zadeh. *The Concept of System, Aggregate, and State in System Theory*, Inter-University Electronics Series, vol. 8, pp. 3–42, McGraw-Hill, New York, 1969.
9. H. Kopetz. *Real-Time Systems—Design Principles for Distributed Embedded Applications*, Kluwer Academic Publishers, Boston, MA/Dordrecht/London, U.K., 1997.
10. S. Pitzek and W. Elmenreich. Managing fieldbus systems. In *Proceedings of the Work-in-Progress Session of the 14th Euromicro International Conference*, Vienna, Austria, June 2002.
11. A. Ran and J. Xu. Architecting software with interface objects. In *Proceedings of the Eighth Israeli Conference on Computer-Based Systems and Software Engineering*, Washington, DC, 1997, pp. 30–37.
12. H. Kopetz et al. Specification of the TTP/A protocol. Technical report, Technische Universität Wien, Institut für Technische Informatik, Vienna, Austria, 2002. Version 2.00. Available at: http://www.vmars.tuwien.ac.at/ttpa/.
13. J. Gait. A probe effect in concurrent programs. *Software Practice and Experience*, 16(3):225–233, March 1986.
14. M. Venzke. Spezifikation von interoperablen Webservices mit XQuery. PhD thesis, Technische Universität Hamburg-Harburg, Hamburg-Harburg, Germany, 2003.
15. R.T. Fielding. Architectural styles and the design of network-based software architectures. PhD thesis, 2000. AAI9980887.
16. G. Moritz, E. Zeeb, S. Prüter, F. Golatowski, D. Timmermann, and R. Stoll. Devices profile for web services and the root. In *Proceedings of the Eighth IEEE International Conference on Industrial Informatics (INDIN)*, Shanghai, China, 2010, pp. 584–591.
17. A. Kamilaris, A. Pitsillides, and V. Trifa. The smart home meets the web of things. *International Journal of Ad Hoc and Ubiquitous Computing*, 7(3):145–154, May 2011.
18. R. Verborgh, T. Steiner, D. Van Deursen, J. De Roo, R. Van de Walle, and J.G. Vallés. Description and interaction of RESTful services for automatic discovery and execution. In *Proceedings of the FTRA 2011 International Workshop on Advanced Future Multimedia Services*, Jeju, South Korea, 2011.
19. R. Verborgh, V. Haerinck, T. Steiner, D. Van Deursen, S. Van Hoecke, J. De Roo, R. Van de Walle, and J.G. Vallés. Functional composition of sensor Web APIs. In *Proceedings of the Fifth International Workshop on Semantic Sensor Networks*, Boston, MA, November 2012.
20. R. Heery and M. Patel. Application profiles: Mixing and matching metadata schemas. Ariadne, 25, September 2000. Available at: http://www.ariadne.ac.uk.
21. CAN in Automation e.V. CANopen—Communication Profile for Industrial Systems, 2002. Available at: http://www.can-cia.de/downloads/.
22. D. Loy, D. Dietrich, and H.-J. Schweinzer (eds.). *Open Control Networks*, Kluwer Academic Publishing, Norwell, MA, October 2001.
23. International Electrotechnical Commission (IEC). Digital data communications for measurement and control—Fieldbus for use in industrial control systems—Part 1: Overview and guidance for the IEC 61158 series, April 2003.
24. Institute of Electrical and Electronics Engineers, Inc. Standard for a smart transducer interface for sensors and actuators—Network capable application processor (NCAP) information model, IEEE Std 1451.1-1999. June 1999.
25. Borst Automation. Device description language. The HART book, 9, May 1999. Available at: http://www.thehartbook.com/.
26. International Electrotechnical Commission (IEC). Function Blocks (FB) for process control—Part 2: Specification of FB concept and Electronic Device Description Language (EDDL). IEC Standard 61804-2:2004, 2004.
27. International Electrotechnical Commission (IEC). Function blocks (FB) for process control—Part 3: Electronic Device Description Language (EDDL). IEC 61804-3:2006, 2006.
28. J. Riegert. Field device tool—Mastering diversity & reducing complexity. *PROCESSWest*, pp. 46–48, 2005.

29. S. Pitzek and W. Elmenreich. Configuration and management of a real-time smart transducer network. In *Proceedings of the Ninth IEEE International Conference on Emerging Technologies and Factory Automation*, Lisbon, Portugal, September 2003, vol. 1, pp. 407–414.
30. World Wide Web Consortium (W3C). *Extensible Markup Language (XML) 1.0*, 2nd edn., October 2000. Available at: http://www.w3.org.
31. M. Wollschläger. A framework for fieldbus management using XML descriptions. In *Proceedings on the 2000 IEEE International Workshop on Factory Communication Systems* (*WFCS 2000*), Porto, Portugal, September 2000, pp. 3–10.
32. S. Eberle. XML-basierte Internetanbindung technischer Prozesse. In *Informatik 2000 Neue Horizonte im neuen Jahrhundert*, pp. 356–371, Springer-Verlag, Berlin, Germany, September 2000.
33. D. Bühler. The CANopen Markup Language—Representing fieldbus data with XML. In *Proceedings of the 26th IEEE International Conference of the IEEE Industrial Electronics Society* (*IECON 2000*), Nagoya, Japan, October 2000.
34. W. Elmenreich, S. Pitzek, and M. Schlager. Modeling distributed embedded applications using an interface file system. *The Seventh IEEE International Symposium on Object-Oriented Real-Time Distributed Computing*, Vienna, Austria, 2004.
35. G. Moritz, S. Prüter, D. Timmermann, and F. Golatowski. Real-time service-oriented communication protocols on resource constrained devices. In *International Multiconference on Computer Science and Information Technology* (*IMCSIT*), Wisla, Poland, IEEE, 2008, pp. 695–701.
36. S. Sakr. Xml compression techniques: A survey and comparison. *Journal of Computer and System Sciences*, 75(5):303–322, August 2009.
37. R. Kyusakov. Towards application of service oriented architecture in wireless sensor networks. PhD thesis, Luleå University of Technology, Luleå, Sweden, 2012.
38. ANSI/ISA-88.01. *Batch Control Part 1: Models and Terminology*, ANSI/ISA, Research Triangle Park, NC, December 1995.
39. J. Berge. *Fieldbuses for Process Control: Engineering, Operation, and Maintenance*, The Instrumentation, Systems, and Automation Society, Research Triangle Park, NC, 2002.
40. S. Poledna, H. Angelow, M. Glück, M. Pisecky, I. Smaili, G. Stöger, C. Tanzer, and G. Kroiss. TTP two level design approach: Tool support for composable fault-tolerant real-time systems. *SAE World Congress 2000*, Detroit, MI, March 2000.
41. P. Pleinevaux and J.-D. Decotignie. Time critical communication networks: Field buses. *IEEE Network*, 2(3):55–63, May 1998.
42. S. Cavalieri, S. Monforte, A. Corsaro, and G. Scapellato. Multicycle polling scheduling algorithms for fieldbus networks. *Real-Time Systems*, 25(2–3):157–185, September–October 2003.
43. TTAGroup. *TTP Specification Version 1.1*, TTAGroup, Vienna, Austria, 2003. Available at: http://www.ttagroup.org.
44. T. Führer, F. Hartwich, R. Hugel, and H. Weiler. Flexray—The communication system for future control systems in vehicles. *SAE World Congress 2003*, Detroit, MI, March 2003.
45. BOSCH. *CAN Specification Version 2.0*, Robert Bosch GmbH, Stuttgart, Germany, 1991.
46. W. Elmenreich, W. Haidinger, P. Peti, and L. Schneider. New node integration for master-slave fieldbus networks. In *Proceedings of the 20th IASTED International Conference on Applied Informatics* (*AI 2002*), Innsbruck, Austria, February 2002, pp. 173–178.
47. A.S. Tanenbaum and M. van Steen. *Distributed Systems: Principles and Paradigms*, 2nd edn., Prentice-Hall, Inc., Upper Saddle River, NJ, 2006.
48. W. Dargie. *Context-Aware Computing and Self-Managing Systems*, 1st edn., Chapman & Hall/CRC, Boca Raton, FL, 2009.
49. M. Wollschläger, C. Diedrich, T. Bangemann, J. Müller, and U. Epple. Integration of fieldbus systems into on-line asset management solutions based on fieldbus profile descriptions. In *Proceedings of the Fourth IEEE International Workshop on Factory Communication Systems*, Vasteras, Sweden, August 2002, pp. 89–96.

50. L. Bartram, A. Ho, J. Dill, and F. Henigman. The continuous zoom: A constrained fisheye technique for viewing and navigating large information spaces. In *ACM Symposium on User Interface Software and Technology*, Pittsburgh, PA, 1995, pp. 207–215.
51. S. Poledna. *Fault-Tolerant Real-Time Systems—The Problem of Replica Determinism*, Kluwer Academic Publishers, Boston, MA/Dordrecht, the Netherlands/London, U.K., 1995.
52. G. Bauer. Transparent fault tolerance in a time-triggered architecture. PhD thesis, Technische Universität Wien, Institut für Technische Informatik, Vienna, Austria, 2001.

4

Smart Transducer Interface Standard for Sensors and Actuators

Kang B. Lee
National Institute of Standards and Technology

4.1 Introduction **4**-1
4.2 Smart Transducer Model **4**-2
4.3 Networking Smart Transducers **4**-4
4.4 Establishment of the IEEE 1451 Standards **4**-4
4.5 Goals of IEEE 1451 .... **4**-6
4.6 IEEE 1451 Standards **4**-6
IEEE 1451 Smart Transducer Model • IEEE 1451.0 Common Functionality and Communication Protocols • IEEE P1451.1 Common Network Services • IEEE 1451.2 Transducer Independence Interface • IEEE 1451.3 Distributed Multidrop Systems • IEEE 1451.4 Mixed-Mode Interface • IEEE 1451.5 Wireless Transducer Interface • IEEE 1451.7 Transducers to RFID Interface • Benefits of IEEE 1451
4.7 Example Application of IEEE 1451.2 .... **4**-14
4.8 Application of IEEE 1451–Based Sensor Network **4**-16
4.9 Summary **4**-16
Acknowledgments **4**-16
References **4**-17

4.1 Introduction

Sensors are used in many devices and systems to provide information on the parameters being measured or to identify the states of control. They are good candidates for increased built-in intelligence. Microprocessors can make smart sensors or devices a reality. With this added capability, it is possible for a smart sensor to directly communicate measurements to an instrument or a system. In recent years, the concept of computer networking has gradually migrated into the sensor community. Networking of transducers (sensors or actuators) in a system and communicating transducer information via digital means versus analog cabling facilitate easy distributed measurements and control. In other words, intelligence and control, which were traditionally centralized, are gradually migrating to the sensor level. This can provide flexibility, improve system performance, and ease system installation, upgrade, and maintenance. Thus, the trend in industry is moving toward distributed control with intelligent sensing architecture. These enabling technologies can be applied to aerospace, automotive, industrial automation, military and homeland defenses, manufacturing process control, smart buildings and homes, and smart toys and appliances for consumers.

As examples, (1) in order to reduce the number of personnel to run a naval ship from 400 to less than 100 as required by the reduced-manning program, the US Navy needs tens of thousands of networked sensors per vessel to enhance automation and improve maintenance, and (2) airplane manufacturers would like to network hundreds of sensors for monitoring and characterizing airplane performance.

Sensors are used across industries and are going global [1]. The sensor market is extremely diverse, and it is expected to grow to $252 billion by 2016. The rapid development and emergence of smart sensor and field network technologies have made the networking of smart transducers a very economical and attractive solution for a broad range of measurement and control applications. However, with the existence of a multitude of incompatible networks and protocols, the number of sensor interfaces and amount of hardware and software development efforts required to support this variety of sensor networks are enormous for both sensor producers and users alike. The reason is that a sensor interface customized for a particular network will not necessarily work with another network. It seems that a variety of networks will coexist to serve their specific industries. The sensor manufacturers are uncertain of which network(s) to support and are restrained from full-scale smart sensor product development. Hence, this condition has impeded the widespread adoption of the smart sensor and networking technologies despite a great desire to build and use them. Clearly, either a common sensor network interface standard is needed to help alleviate this problem [2], or many sensor network interfaces have to coexist in a working environment. With the emergence of the Industrial Internet, Internet of Things, Machines to Machines, and Cyber Physical Systems technologies, smart sensors and sensor networks will play a key role in supporting these technologies to achieve their potentials. Standardized smart sensor and sensor network interfaces will help ease the adoption and use of these technologies.

4.2 Smart Transducer Model

In order to develop a sensor interface standard, a smart transducer model should first be defined. As defined in the IEEE Std 1451.2-1997 [3], a smart transducer is a transducer that provides functions beyond those necessary for generating a correct representation of a sensed or controlled quantity. This functionality typically simplifies the integration of the transducer into applications in a networked environment. Thus, let us consider the functional capability of a smart transducer. A smart transducer should have

- Integrated intelligence closer to the point of measurement and control
- Basic computation capability
- Capability to communicate data and information in a standardized digital format to a network

Based on this premise, a smart transducer model is shown in Figure 4.1. It applies to both sensors and actuators. The output of a sensor is conditioned and scaled, then converted to a digital format through an analog-to-digital (A/D) converter. The digitized sensor signal can then be easily processed by a microprocessor using a digital application control algorithm. On the other hand, the output, after being converted to an analog signal via a digital-to-analog (D/A) converter, can then be used to control

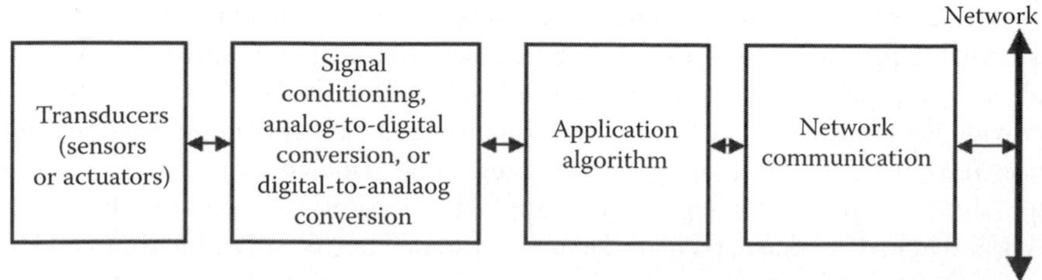

FIGURE 4.1 A smart transducer model.

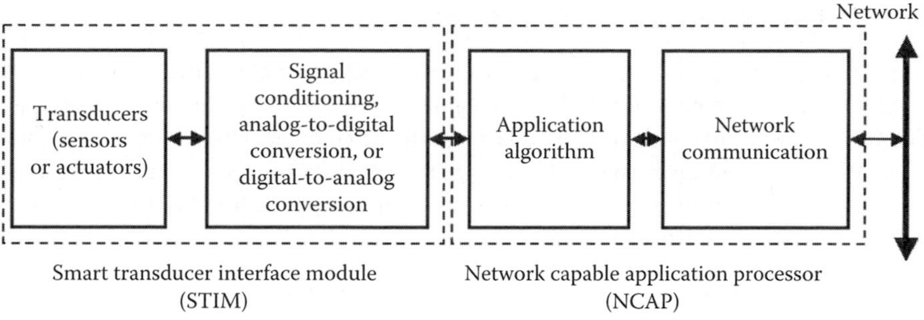

FIGURE 4.2 Functional partitioning.

an actuator. Any of the measured or calculated parameters can be passed on to any device or host in a network by means of network communication protocols.

The different modules of the smart transducer model can be grouped into functional units as shown in Figure 4.2. The transducers and signal conditioning and conversion modules can be grouped into a building block called a smart transducer interface module (STIM). Likewise, the application algorithm and network communication modules can be combined into a single entity called a network capable application processor (NCAP). With this functional partitioning, transducer to network interoperability can be achieved in these manners:

1. STIMs from different sensor manufacturers can *plug and play* with NCAPs from a particular sensor network supplier.
2. STIMs from a sensor manufacturer can *plug and play* with NCAPs supplied by different sensor or field network vendors.
3. STIMs from different manufacturers can be interoperable with NCAPs from different field network suppliers.

Using this partitioning approach, a migration path is provided to those sensor manufacturers who want to build STIMs with their sensors, but do not intend to become field network providers. Similarly, it applies to those sensor network builders who do not want to become sensor manufacturers.

As technology becomes more advanced and microcontrollers become smaller relative to the size of the transducer, integrated networked smart transducers that are economically feasible to implement will emerge in the marketplace. In this case, all the modules are incorporated into a single unit as shown in Figure 4.3. Thus, the interface between the STIM and NCAP is not exposed for external access and separation. The only connection to the integrated transducer is through the network connector. The integrated smart transducer approach simplifies the use of transducers by merely plugging the device into a sensor network.

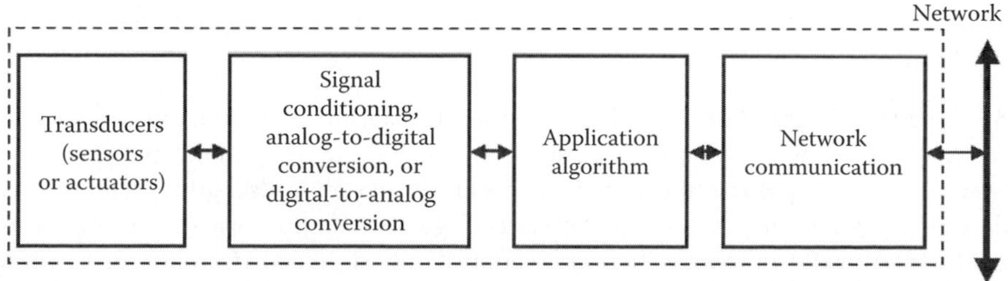

FIGURE 4.3 An integrated networked smart transducer.

4.3 Networking Smart Transducers

Not until recently have sensors been connected to instruments or computer systems by means of a point-to-point or multiplexing scheme. These techniques involve a large amount of cabling, which is very bulky and costly to implement and maintain. With the emergence of computer networking technology, transducer manufacturers and users alike are finding ways to apply this networking technology to their transducers for monitoring, measurement, and control applications [4]. Networking smart sensors provides the following features and benefits:

- Enable peer-to-peer communication and distributed sensing and control
- Significantly lower the total system cost by simplified wiring
- Use prefabricated cables instead of custom laying of cables for ease of installation and maintenance
- Facilitate expansion and reconfiguration
- Allow time-stamping of sensor data
- Enable sharing of sensor measurement and control data
- Provide Internet connectivity, meaning *global* or *anywhere*, access of sensor information

4.4 Establishment of the IEEE 1451 Standards

As discussed earlier, a smart sensor interface standard is needed in industry. In view of this situation, the Technical Committee on Sensor Technology of the Institute of Electrical and Electronics Engineers (IEEE)'s Instrumentation and Measurement Society sponsored a series of projects for establishing a family of IEEE 1451 Standards [5]. These standards specify a set of common interfaces for connecting transducers to instruments, microprocessors, field networks, or the Internet. They cover digital, mixed-mode, distributed multidrop sensor integration with radio frequency identification (RFID), signal treatment, wireless interfaces, network interfaces, and eXtensible Messaging and Presence Protocol (XMPP) to address the needs of different sectors of industry. A key concept in the IEEE 1451 standards is the definition and specification of the Transducer Electronic Data Sheet (TEDS) formats. A TEDS is a standardized method of storing transducer (sensor or actuator) identification and manufacture-related information, such as manufacturer name, sensor types, serial number, and calibration data. The TEDS provides many benefits:

- Enable self-identification of sensors or actuators—a sensor or actuator equipped with the IEEE 1451 TEDS can identify and describe itself to the host or network via the sending of the TEDS.
- Provide long-term self-documentation—the TEDS in the sensor can be updated and stored with information such as location of the sensor, recalibration date, repair record, and many maintenance-related data.
- Reduce human error—automatic transfer of TEDS data to the network or system eliminates the entering of sensor parameters by hands, which could induce errors due to various conditions.
- Ease field installation, upgrade, and maintenance of sensors—this helps to reduce life cycle costs because only lesser skilled persons are needed to perform these tasks by simply doing *plug and play* of sensors.

IEEE 1451, designated as Standard Transducer Interface for Sensors and Actuators, consists of a family of document standards with appropriately named parts, for example, 1451.1 is Common Network Services and 1451.5 is Wireless Communication. This family of standards aims to work with each other to provide a set of seamless interfaces for communicating sensor data/information from devices to networks and the Internet. It works with other application standards like MTConnect for accessing sensor data from machines [6,7]. and Open Geospatial Consortium Sensor Web Enablement sensor data for geospatial applications [8,9]. The IEEE 1451 family reference model is shown in Figure 4.4.

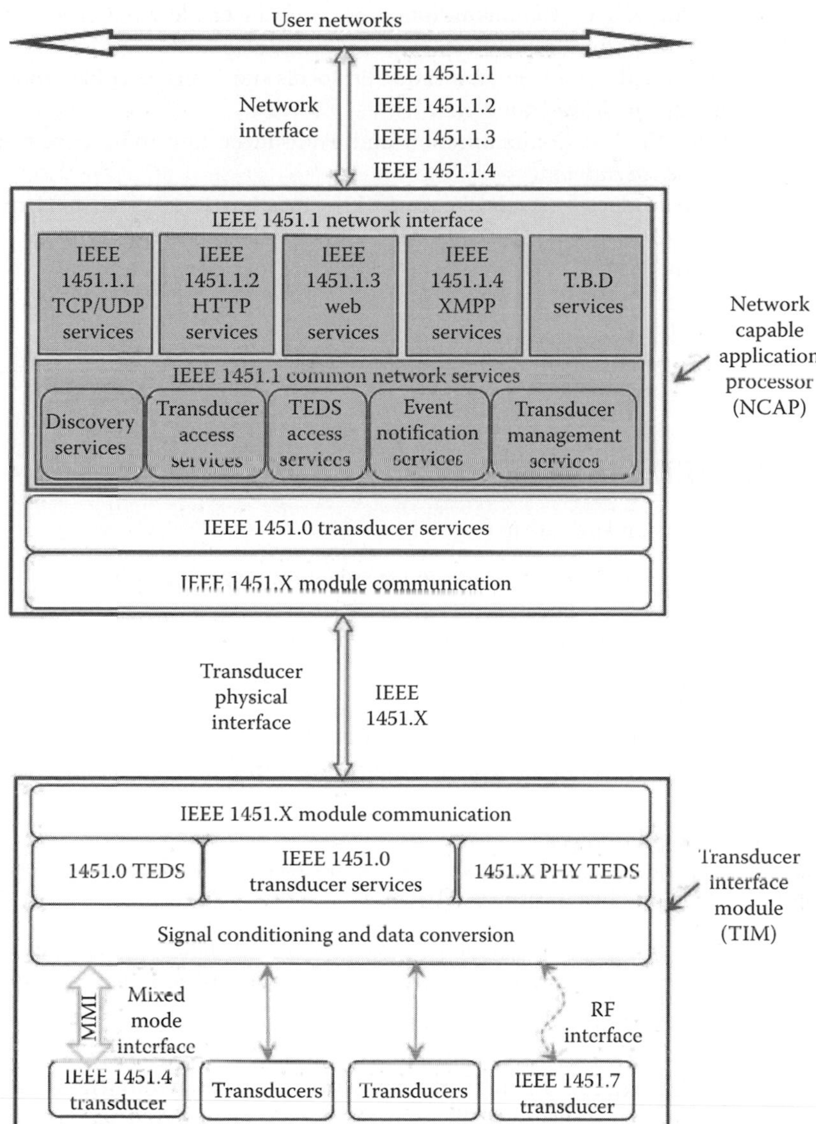

FIGURE 4.4 IEEE 1451 family reference model.

The current status of the family of standards development effort is as follows:

1. IEEE 1451.0-2007, Common Functions, Communication Protocols, and Transducer Electronic Data Sheet (TEDS) Formats—*published standard*
2. IEEE P1451.1,* Common Network Services [6]—*standard being developed*
3. IEEE 1451.2-1997, Transducer to Microprocessor Communication Protocols and Transducer Electronic Data Sheet (TEDS) Formats [3,10]—*published standard*
4. IEEE P1451.2, Serial Point-to-Point Interface—standard being developed. When this standard is completed, it will supersede the IEEE 1451.2-1997

* P1451.1, the "P" designation means P1451.1 is a draft standard development project. Once the draft document is approved as a standard, "P" will be dropped.

5. IEEIEEE 1451.3-2003, Digital Communication and Transducer Electronic Data Sheet (TEDS) Formats for Distributed Multidrop Systems [11]—*published standard*
6. IEEE 1451.4-2004, Mixed-mode Communication Protocols and Transducer Electronic Data Sheet (TEDS) Formats [12]—*published standard*
7. IEEE 1451.5-2005, Wireless Communication and Transducer Electronic Data Sheet (TEDS) Formats [13]—*published standard*
8. IEEE 1451.7-2010, Transducers to Radio Frequency Identification (RFID) Systems Communication Protocols and Transducer Electronic Data Sheet Formats [14]—*published standard*
9. IEEE P1451.1.4, Standard for a Smart Transducer Interface for Sensors, Actuators, and Devices—eXtensible Messaging and Presence Protocol (XMPP) for Networked Device Communication—*standard being developed*
10. IEEE P1451.001, Recommended Practice for Signal Treatment Applied to Smart Transducers—*standard is being developed*

4.5 Goals of IEEE 1451

The goals of the IEEE 1451 standards are to

- Develop network-independent and vendor-independent transducer interfaces
- Define TEDS and standardized data formats
- Support general transducer data, control, timing, configuration, and calibration models
- Allow transducers to be installed, upgraded, replaced, and moved with minimum effort by simply *plug and play*
- Eliminate error-prone, manual entering of data and system configuration steps
- Ease the connection of sensors and actuators to networks or Internet by wireline or wireless means

4.6 IEEE 1451 Standards

4.6.1 IEEE 1451 Smart Transducer Model

The IEEE 1451 smart transducer model parallels the smart transducer model discussed in Figure 4.2. In addition, the IEEE 1451 model includes the TEDS. The model applied to each of the IEEE 1451.X standards is discussed in the following.

4.6.2 IEEE 1451.0 Common Functionality and Communication Protocols

Several standards in the IEEE 1451 family share certain characteristics, but there is no common set of functions, communication protocols, and TEDS formats that facilitate interoperability among these standards. The IEEE 1451.0 standard provides that commonality and simplifies the creation of future standards with different physical layers that will facilitate interoperability in the family. This project defines a set of common functionality for the family of IEEE 1451 smart transducer interface standards. This set of functionality is independent of the physical communications media. It includes the basic functions required to control and manage smart transducers, common communications protocols, and media-independent TEDS formats. The block diagram for IEEE 1451.0 is shown in Figure 4.5. The IEEE 1451.0 defines functional characteristics, but it does not define any physical interface.

The IEEE 1451.0 standard defines different types of TEDS, their data format, and the digital interface and communication protocols between the TIM (Transducer Module in Figure 4.5) and the NCAP. The TIM contains the transducer(s) and the TEDS, which is stored in a nonvolatile memory attached to a transducer. The TEDS contains fields that describe the type, attributes, operation, and calibration of the transducer. The mandatory requirement for the TEDS is less than 300 bytes. The rest of the TEDS specification is optional. A transducer integrated with the TEDS provides a very unique feature that makes

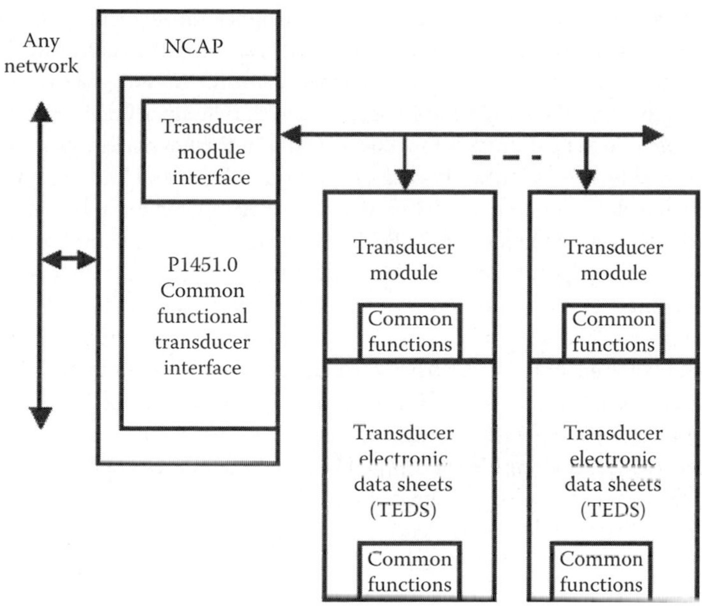

FIGURE 4.5 The block diagram of IEEE 1451.0.

possible the self-describing of transducers to the system or network. Since the manufacture-related data in the TEDS always go with the transducer, and this information is electronically transferred to an NCAP or host, human errors associated with manual entering of sensor parameters into the host are eliminated. Because of this distinctive feature of the TEDS, upgrading transducers with higher accuracy and enhanced capability or replacing transducers for maintenance purpose is simply considered *plug and play*. Many different types of TEDS are defined in the standard, and some of them are mandatory and others are optional. They are listed in Table 4.1.

The TEDS formats are divided into two categories. The first category of TEDS contains data in a machine-readable form, which is intended to be used by the NCAP. The second category of TEDS contains data in a human-readable form. The human-readable TEDS may be represented in multiple languages using different encoding for each language. The Meta TEDS contains the data that describe the TIM. It contains the revision of the standard, the version number of the TEDS, the number of channels in the TIM, and the worst-case timing required to access these channels. This information will allow the NCAP to access the channel information. In addition, the Meta TEDS includes the channel groupings that describe the relationships between channels. Each transducer is represented by a channel. Each channel in the TIM contains a Channel TEDS. The Channel TEDS lists the actual timing parameters for each individual channel. It also lists the type of transducer, the format of the data word being output by the channel, the physical

TABLE 4.1 IEEE 1451.0 TEDS

TEDS Name	TEDS Type	Mandatory/Optional
Meta TEDS	Machine readable	Mandatory
Channel TEDS	Machine readable	Mandatory
Calibration TEDS	Machine readable	Optional
Generic Extension TEDS	Machine readable	Optional
Meta Identification TEDS	Human readable	Optional
Channel Identification TEDS	Human readable	Mandatory
Calibration Identification TEDS	Human readable	Optional
End-User Application-Specific TEDS	Human readable	Mandatory

units (temperature in Kelvin and pressure in Pascal, etc.), the upper and lower range limits, the uncertainty or accuracy, whether or not a calibration TEDS is provided, and where the calibration is to be performed. The Calibration TEDS contains all the necessary information for the sensor data to be converted from the A/D converter raw output into the physical units specified in the Channel TEDS. If actuators are included in the TIM, it also contains the parameters that convert data in the physical units into the proper output format to drive the actuators. It also contains the calibration interval and last calibration date and time. This allows the system to determine when a calibration is needed. A general calibration algorithm is specified in the standard. The Generic Extension TEDS is provided to allow industry groups to provide additional TEDS in a machine-readable format. The Meta Identification TEDS is human-readable data that the system can retrieve from the TIM for display purposes. This TEDS contains fields for the manufacturer's name, the model number and serial number of the TIM, and a date code. The Channel Identification TEDS is similar to the Meta Identification TEDS. When transducers from different manufacturers are built into a TIM, this information will be very useful for the identification of channels. The Channel Identification TEDS provides information about each channel, whereas the Meta Identification TEDS provides information for the TIM. The Calibration Identification TEDS provides details of the calibration in the TIM. This information includes who performed the calibration and what standards were used. The End-User Application-Specific TEDS is not defined in detail by the standard. It allows the user to insert information, such as installation location, time it was installed, or any other desired text.

4.6.3 IEEE P1451.1 Common Network Services

The IEEE P1451.1 defines a set of common network services for smart transducers. These common network services, defined for IEEE 1451 smart transducers, communicate transducer and relevant information through a network via various network communication protocol services providing transducer discovery, transducer and TEDS data access, event notification, and transducer management services. Using this set of common network services, one would be able to access and manage any of the IEEE 1451 sensors and actuators using different popular protocols, such as Hypertext Transfer Protocol and XMPP, as shown in Figure 4.6.

4.6.4 IEEE 1451.2 Transducer Independence Interface

According to the IEEE 1451.2-1999 standard, the STIM (new version of IEEE 1451.2 will use the term TIIM) module can contain a combination of sensors and actuators of up to 255 channels, signal conditioning/processing, A/D converter, D/A converter, and digital logics to support the transducer

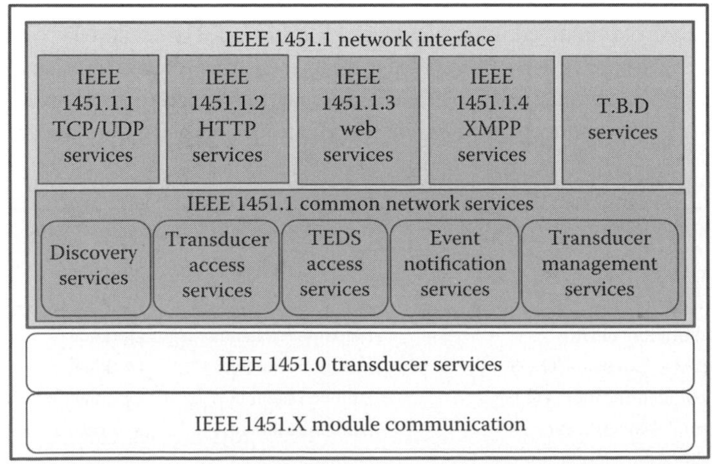

FIGURE 4.6 Network interfaces for smart transducers.

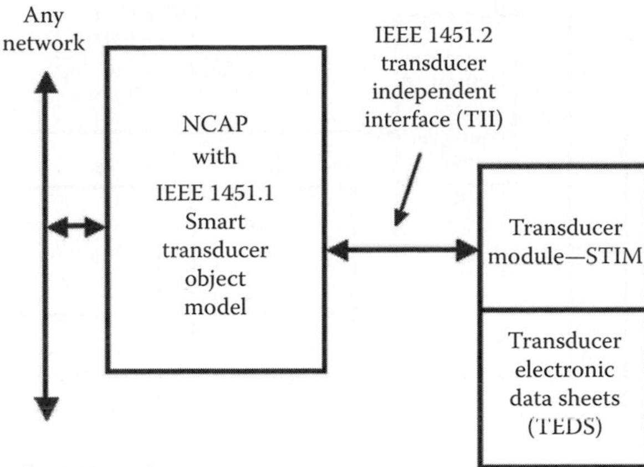

FIGURE 4.7 Communication between NCAP and STIM via transducer independence interface.

TABLE 4.2 Transducer Independence Interface Signal Lines

Line	Logic	Driven by	Function
DIN	Positive logic	NCAP	Address and data transport from NCAP to STIM
DOUT	Positive logic	STIM	Data transport from STIM to NCAP
DCLK	Positive edge	NCAP	Positive-going edge latches data on both DIN and DOUT
NIOE	Active low	NCAP	Signals that the data transport is active and delimits data transport framing
NTRIG	Negative edge	NCAP	Performs triggering function
NACK	Negative edge	STIM	Serves two functions: Trigger acknowledge Data transport acknowledge
NINT	Negative edge	STIM	Used by the STIM to request service from the NCAP
NSDET	Active low	STIM	Used by the NCAP to detect the presence of a STIM
POWER	N/A	NCAP	Nominal 5 V power supply
COMMON	N/A	NCAP	Signal common or ground

independent interface (TII). An NCAP can communicate with one or more STIMs via the transducer independent interface (Figure 4.7).

The transducer independence interface defines an enhanced serial peripheral interface by adding a few extra wires for a total of 10 wires in order to be able to provide a high-speed interface for data capture via handshaking lines—NIOE, NTRIG, NACK, and NINT. The NSDET line is added for the NCAP to detect the presence of an STIM, enabling the plug and play of transducers. The signal lines are given in Table 4.2.

4.6.5 IEEE 1451.3 Distributed Multidrop Systems

The IEEE 1451.3 defines a transducer bus for connecting transducer modules to an NCAP in a distributed multidrop fashion. A block diagram is shown in Figure 4.8. The physical interface for the transducer bus is based on Home Phoneline Networking Alliance (HomePNA) specification. Both power and data run on a twisted pair of wires. Multiple transducer modules, called transducer bus interface modules (TBIMs), can be connected to an NCAP via the bus. Each TBIM contains transducers, signal conditioning/processing, A/D, D/A, and digital logics to support the bus and can accommodate large arrays of transducers for synchronized access at up to 128 Mbps with HomePNA 3.0 and up to 240 Mbps with extensions. The TEDS is defined in the eXtensible Markup Language.

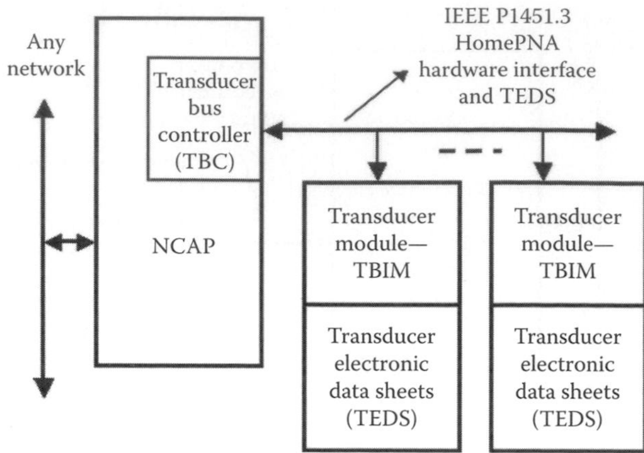

FIGURE 4.8 Block diagram of IEEE 1451.3.

4.6.6 IEEE 1451.4 Mixed-Mode Interface

The IEEE 1451.4 defines a mixed-mode interface (MMI), which is used for connecting transducer modules, mixed-mode transducers (MMT), to an instrument, a computer, or an NCAP. The block diagram of the system is shown in Figure 4.9. The physical transducer interface not only is based on the Maxim/Dallas Semiconductor's one-wire protocol but also supports up to four wires for bridge-type sensors. It is a simple, low-cost connectivity for analog sensors with a very small TEDS—64 bits mandatory and 256 bits optional. The mixed-mode interface supports a digital interface for reading and writing the TEDS by the instrument or NCAP. After the TEDS transaction is completed, the interface switches into analog mode, where the analog sensor signal is sent straight to the instrument and NCAP, which is equipped with A/D to read the sensor data.

4.6.7 IEEE 1451.5 Wireless Transducer Interface

Wireless communication is emerging, and low-cost wireless technology is on the horizon. Wireless communication links could replace the costly cabling for sensor connectivity. It also could greatly reduce sensor installation cost. Industry would like to apply the wireless technology for sensors; however, the wide use of wireless sensor technologies has been hindered by the interoperability problem

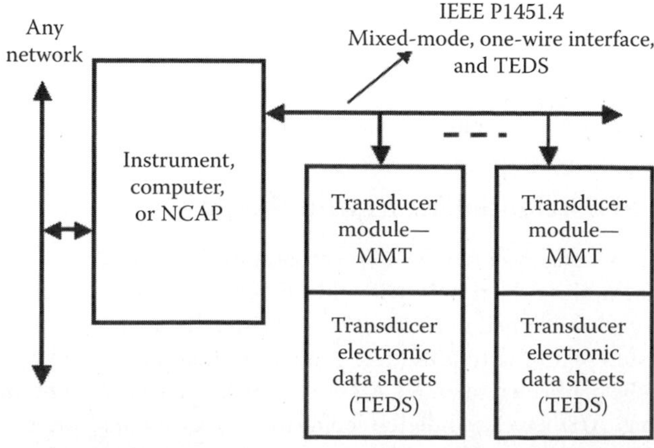

FIGURE 4.9 Block diagram of IEEE 1451.4.

among wireless sensors, equipment, and data. In response to this need, a wireless sensing workshop was held to involve industry to define IEEE 1451.5, which aims to develop a wireless sensor communication interface standard that will leverage existing wireless communication technologies and protocols [15]. The standard defines wireless communication methods and data format for transducers. The standard adopts existing popular wireless communication protocols, such as WiFi (IEEE 802.11), Bluetooth (IEEE 802.15.1), ZigBee (IEEE 802.15.4), and 6LowPAN (IEEE 802.15.4). The 6LowPAN is an Internet Protocol (IP)–driven protocol (IPv6 compatible) that eases the access of transducers from the Internet. A block diagram of IEEE 1451.5 is shown in Figure 4.10. IEEE 1451.5 defines wireless message formats, a data and control model, security model, and physical TEDS that are scalable to meet the needs of low-cost to sophisticated sensor or device manufacturers. Currently, different sensor networks cited earlier do not interoperate with each other due to the use of different hardware and software protocols. Even though ZigBee and 6LowPAN use the same 802.15.4 wireless communication protocol, they cannot interoperate due to the differences in application-level protocol and data format, etc. The IEEE 1451.5 approach is to help achieve data-level interoperability, that is, the sensor data received at user network is the same, whether it is from a Bluetooth-based sensor network or a ZigBee-based sensor network. This concept is illustrated in Figure 4.10.

4.6.8 IEEE 1451.7 Transducers to RFID Interface

RFID technologies are rapidly emerging as a means of tracking products and assets during manufacturing and distribution. For example, airplane manufacturers track their airplane parts from their suppliers, during the assembly process, and throughout the life cycle and maintenance of the airplanes. RFID standards have been developed to meet these needs. Sensors can provide valuable information about the conditions of products, in particular, the perishable kinds, such as fruits, meats, fishes, and drugs for public safety. There is also a great need to provide sensor data as part of the supply chain reporting; however, there is a lack of openly defined standard interfaces between sensors and RFID tags. The purpose of the IEEE 1451.7 standard is to fill the gap by providing such sensor-to-RFID tag interfaces to meet industry needs. The IEEE 1451.7 standard provides standardized sensor data formats and command/reply set for communicating sensor and RFID tag data to the requesters or clients in the network. It also addresses security protocols to provide secure data. The integration of IEEE 1451.7 transducer (sensor and RFID tag) with the IEEE 1451 family of standard as needed is illustrated in Figure 4.11.

4.6.9 Benefits of IEEE 1451

IEEE 1451 defines a set of common transducer interfaces, which will help to lower the cost of designing smart sensors and actuators because designers would have to design only a single set of standardized digital interfaces. Thus, the overall cost to make networked sensors will decrease. Incorporating the TEDS with the sensors will enable self-description of sensors and actuators, eliminating error-prone, manual configuration.

4.6.9.1 Sensor Manufacturers

Sensor manufacturers can benefit from the standard because they have to design only a single standard physical interface. Standard calibration specification and data format can help to design and develop multilevel products based on TEDS with a minimum effort.

4.6.9.2 Application Software Developers

Applications can benefit from the standard as well because standard transducer models for control and data can support and facilitate distributed measurement and control applications. The standard also provides support for multiple languages—good for international developers.

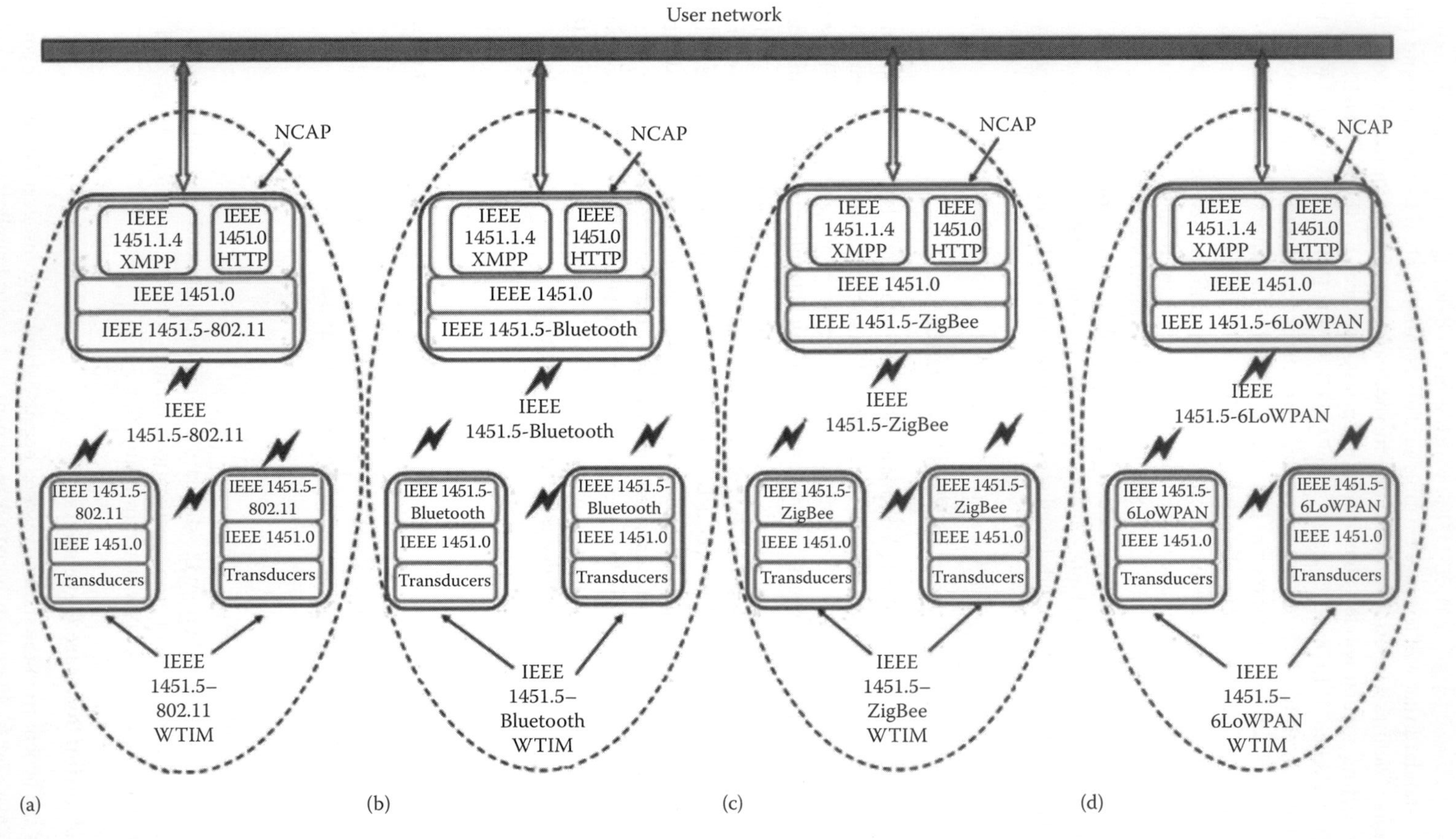

FIGURE 4.10 Interoperable IEEE 1451.5–based wireless sensor networks. (a) IEEE 1451-5-802.11, (b) IEEE 1451.5-Bluetooth WSN, (c) IEEE 1451.5-ZigBee WSN, and (d) IEEE 1451-5-bLOWPAN WSN.

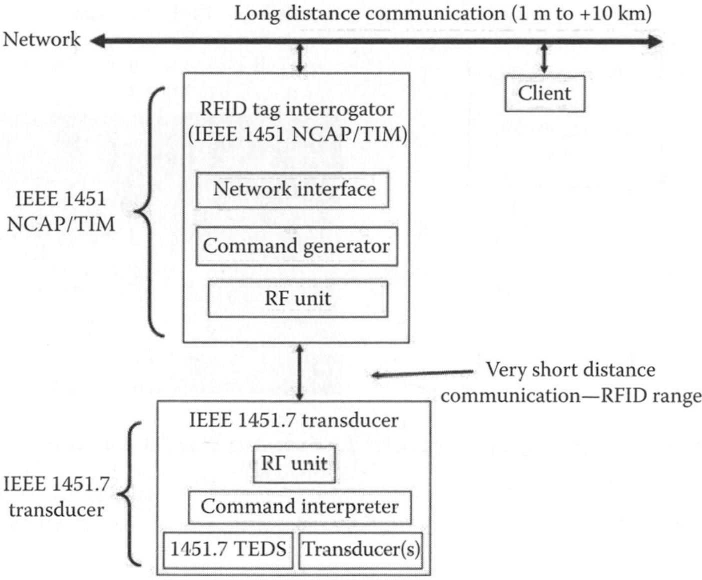

FIGURE 4.11 Integration of an IEEE 1451 NCAP/TIM module with an IEEE 1451.7 transducer in a network.

4.6.9.3 System Integrators

Sensor system integrators can benefit from IEEE 1451 because sensor systems become easier to install, maintain, modify, and upgrade. Quick and efficient transducer replacement results by simply *plug and play*. It can also provide a means to store installation details in the TEDS. Self-documenting of hardware and software is done via the TEDS. Best of all is the ability to choose sensors and networks based on merit.

4.6.9.4 End Users

End users can benefit from a standard interface because sensors will be easy to use by simple *plug and play*. Based on the information provided in the TEDS, software can automatically provide the physical units, readings with significant digits as defined in the TEDS, installation details such as instruction, identification, and location of the sensor.

4.6.9.5 *Plug and Play* of Sensors and Actuators

IEEE 1451 enables *plug and play* of transducers to a network as illustrated in Figure 4.12. In this example, IEEE 1451.4–compatible transducers from different companies are shown to work with a sensor network. IEEE 1451 also enables *plug and play* of transducers to a data acquisition system/instrumentation system

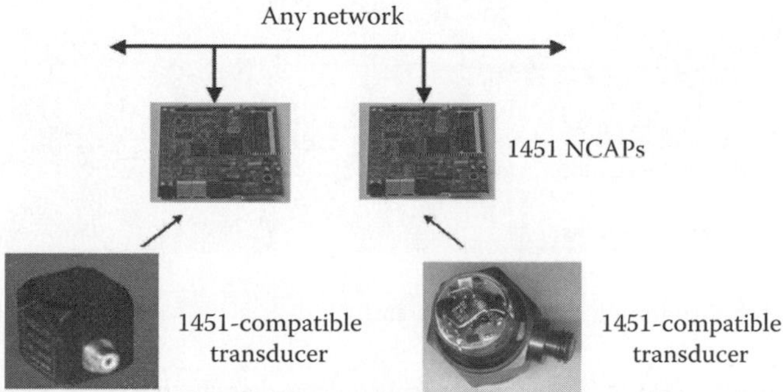

FIGURE 4.12 IEEE 1451 enables *plug and play* of transducers to a network.

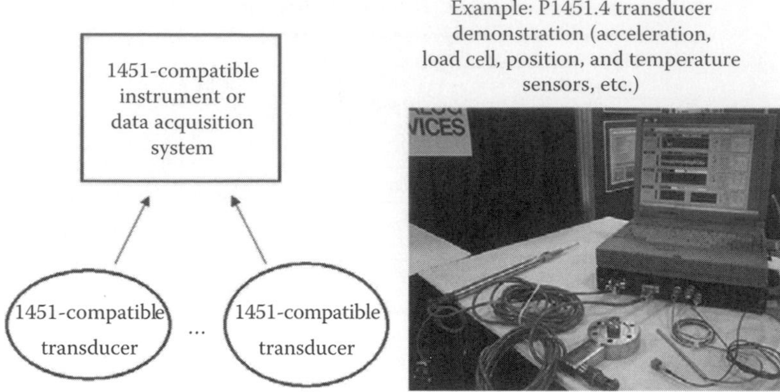

FIGURE 4.13 IEEE 1451 enables *plug and play* of transducers to data acquisition/instrumentation system.

as shown in Figure 4.13. In this example, various IEEE 1451.4–compatible transducers, such as an accelerometer, a thermistor, a load cell, and a linear variable differential transformer are shown to work with a LabView-based system.*

4.7 Example Application of IEEE 1451.2

IEEE 1451–based sensor network consisting of sensors, STIM, and NCAP is designed and built into a cabinet as shown in Figure 4.14. There were a total of four STIM and NCAP network nodes as pointed to in Figure 4.14. Thermistor sensors were used for temperature measurements. They were calibrated in the laboratory to generate IEEE 1451.2–compliance calibration TEDS for all four STIMs and NCAPs. The thermistors were mounted on the spindle motor housing, bearing, and axis drive motors of a three-axis vertical machining center, which is shown in Figure 4.15. Since each NCAP has a built-in micro web server, a custom web page was constructed using the web tool provided with the NCAP.

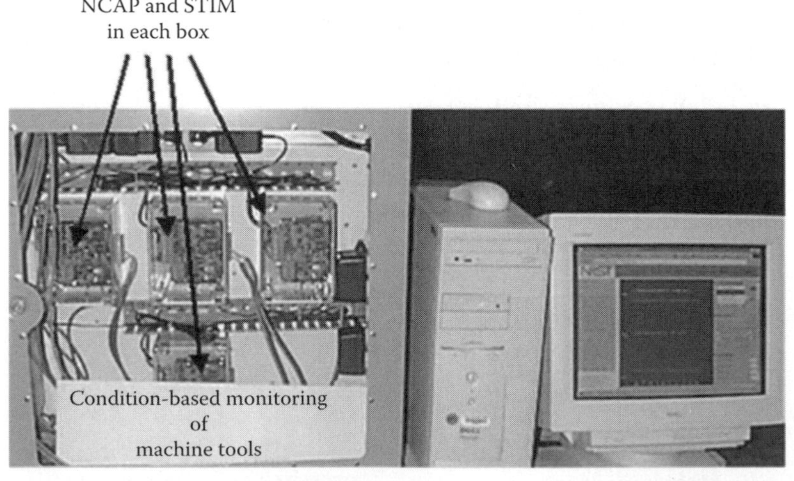

FIGURE 4.14 NCAP-based condition monitoring system.

* Certain commercial products are identified in this chapter in order to describe the system. Such identification does neither imply recommendation or endorsement by the National Institute of Standards and Technology, nor imply that the products identified are necessarily the best or the only ones available for the purpose.

FIGURE 4.15 Three-axis vertical machining center.

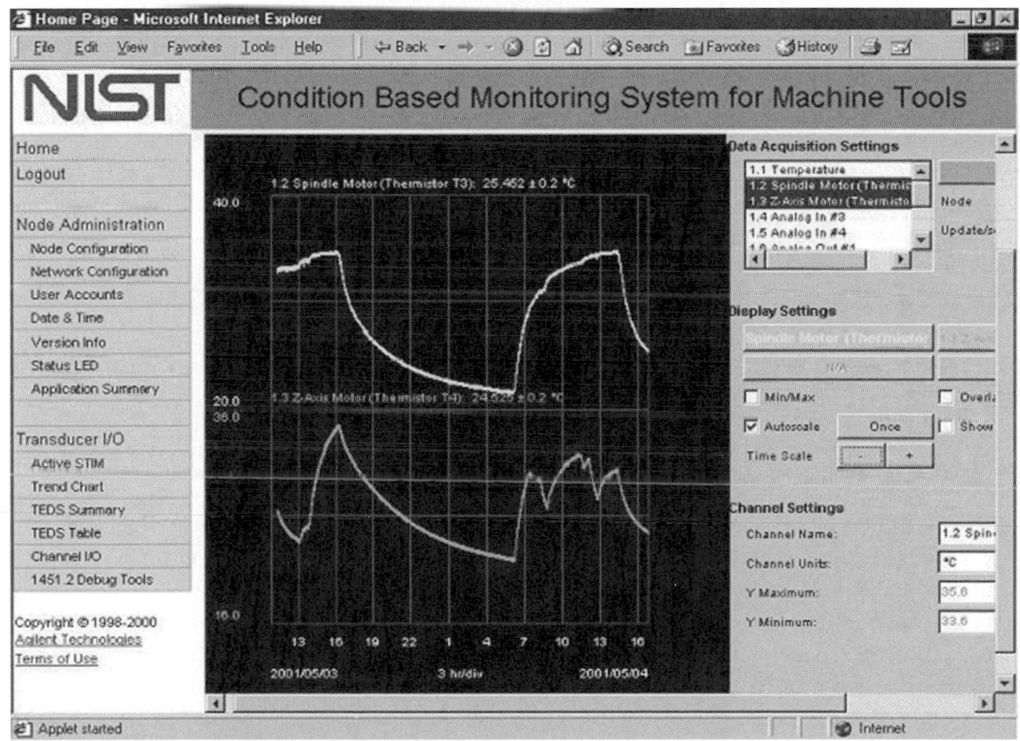

FIGURE 4.16 Temperature trend chart.

Thus, remote monitoring of the machine thermal condition was easily achieved via the Ethernet network and the Internet using a readily available common web browser. The daily trend chart of the temperature of the spindle motor (top trace) and the temperature of the Z-axis drive motor (bottom trace) in the machine is shown in Figure 4.16. The temperature rise tracks the working of the machine during the day, and the temperature fall indicates that the machine is cooling off after the machine shop is closed.

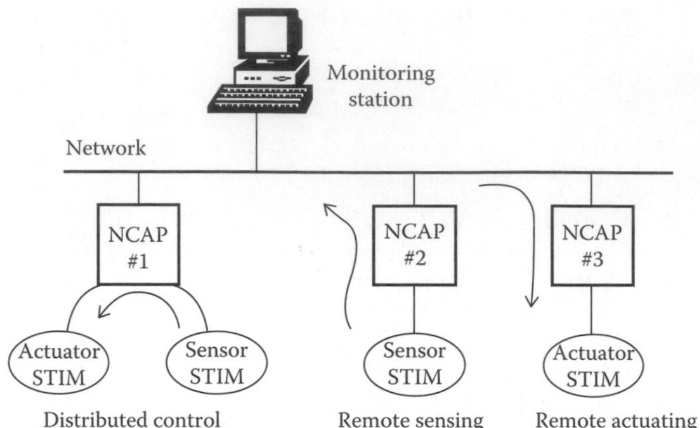

FIGURE 4.17 Application model of IEEE 1451.

4.8 Application of IEEE 1451–Based Sensor Network

A distributed measurement and control system can be easily implemented based on the IEEE 1451 standards [16]. An application model of IEEE 1451 is shown in Figure 4.17. Three NCAP/STIMs are used to illustrate the distributed control, remote sensing or monitoring, and remote actuating. In the first scenario, a sensor and an actuator are connected to the STIM of NCAP #1, and an application software running in the NCAP can perform a locally distributed control function, such as maintaining a constant temperature for a bath. The NCAP reports measurement data, process information, and control status to a remote monitoring station or host. It frees the host from the processor-intensive, closed-loop control operation. In the second scenario, only sensors are connected to NCAP #2, which can perform remote process or condition monitoring functions, such as monitoring the vibration level of a set of bearings in a turbine. In the third scenario, based on the broadcast data received from NCAP #2, NCAP #3 activates an alarm when the vibration level of the bearings exceeds a critical set point. As illustrated in these examples, IEEE 1451–based sensor network can easily facilitate peer-to-peer communications and distributed control functions.

4.9 Summary

The IEEE 1451 smart transducer interface standards are defined to allow a transducer manufacturer to build transducers of various performance capabilities that are interoperable within a networking system. The IEEE 1451 family of standards has provided the common interface and enabling technology for the connectivity of transducers to microprocessors, instrumentation systems, and field networks using wired and wireless means. The standardized TEDS allows the self-description of sensors, which turns out to be a very valuable tool for condition-based maintenance. The expanding Internet market has created a good opportunity for sensor and network manufacturers to exploit web-based and smart sensor technologies. As a result, users will greatly benefit from many innovations and new applications.

Acknowledgments

The author sincerely thanks the IEEE 1451 working groups for the use of the materials in this chapter. Through its program in Smart Machine Tools, the Manufacturing Engineering Laboratory of the National Institute of Standards and Technology has contributed to the development of the IEEE 1451 standards.

References

1. Amos, K., Sensor market goes global, *InTech-The International Journal for Measurement and Control*, pp. 40–43, June 1999.
2. Bryzek, J., Summary report, *Proceedings of the IEEE/NIST First Smart Sensor Interface Standard Workshop*, NIST, Gaithersburg, MD, pp. 5–12, March 31, 1994.
3. IEEE Std 1451.2-1997, *Standard for a Smart Transducer Interface for Sensors and Actuators—Transducer to Microprocessor Communication Protocols and Transducer Electronic Data Sheet (TEDS) Formats*, Institute of Electrical and Electronics Engineers, Inc., Piscataway, NJ, 1997.
4. Eidson, J. and Woods, S., A research prototype of a networked smart sensor system, *Proceedings Sensors Expo Boston*, May 1995, Helmers Publishing. Boston, MA.
5. Lee, K. The proposed smart transducer interface standand *Proceedings of the Instrumentation Measurement Conference (IMTC)*, St. Paul, MN, 1, 129–133A, May 18–21, 1998.
6. Lee, K. B., Song, Y., and Gu P. S., A sensor model for enhancement of manufacturing equipment data interoperability, *Proceedings of ASME 2012 International Mechanical Engineering Congress & Exposition (IMECE 2012)*, Houston, TX, November 9–15, 2012.
7. Lee, K. B., Song, Y., and Gu P. S., Integration of MTConnect and standard-based sensor networks for manufacturing equipment monitoring, *Proceedings of ASME 2012 International Manufacturing Science and Engineering Conference*, Notre Dame, IN, June 4–8, 2012.
8. Lee, K. and Song, E., Integration of IEEE 1451 smart transducers and OGC-SWE using STWS, *IEEE Sensors Applications Symposium*, New Orleans, LA, February 17–19, 2009.
9. O'Reilly, T. C. et al., Instrument interface standards for interoperable ocean sensor networks, *Proceedings for IEEE OCEANS 2009 Conference*, Bremen, Germany, May 11–14, 2009.
10. Woods, S. et al., IEEE-P1451.2 Smart transducer interface module, *Proceedings Sensors Expo*, Helmers Publishing, Philadelphia, PA, pp. 25–38, October 1996.
11. IEEE Std 1451.3-2003, *Standard for a Smart Transducer Interface for Sensors and Actuators—Digital Communication and Transducer Electronic Data Sheet (TEDS) Formats for Distributed Multidrop Systems*, Institute of Electrical and Electronics Engineers, Inc., Piscataway, NJ, 2003.
12. IEEE Std 1451.4-2004, *Standard for a Smart Transducer Interface for Sensors and Actuators—Mixed-mode Communication Protocols and Transducer Electronic Data Sheet (TEDS) Formats*, Institute of Electrical and Electronics Engineers, Inc., Piscataway, NJ, 2004.
13. IEEE Std 1451.5-2005, *Standard for a Smart Transducer Interface for Sensors and Actuators—Wireless Communication and Transducer Electronic Data Sheet (TEDS) Formats*, Institute of Electrical and Electronics Engineers, Inc., Piscataway, NJ, 2005.
14. IEEE Std 1451.7-2010, *Standard for a Smart Transducer Interface for Sensors and Actuators—Transducers to Radio Frequency Identification (RFID) Systems Communication Protocols and Transducer Electronic Data Sheet Formats*, Institute of Electrical and Electronics Engineers, Inc., Piscataway, NJ, 2010.
15. Lee, K. B., Gilsinn, J. D., Schneeman, R. D., Huang, H. M., *First Workshop on Wireless Sensing*, National Institute of Standards and Technology. NISTIR 02-6823, Gaithersburg, MD, February 2002.
16. Lee, K. and Schneeman, R., Distributed measurement and control based on the IEEE 1451 smart transducer interface standards, *Instrumentation and Measurement Technical Conference 1999*, Venice, Italy, May 24–26, 1999.

5

IO-Link (Single-Drop Digital Communication System) for Sensors and Actuators

5.1 Motivation and Objectives for a New Technology 5-1
5.2 IO-Link Technology 5-2
Purpose of Technology • Positioning within the Automation Hierarchy • Wiring, Connectors, and Power • Communication Features of IO-Link • Role of a Master • IO-Link Configuration • Mapping to Fieldbuses and System Integration • Implementation and Engineering Support • Test and Certification • Profiles • Functional Safety • Standardization
Abbreviations 5-8
References 5-9

Wolfgang Stripf
PROFIBUS and PROFINET International

5.1 Motivation and Objectives for a New Technology

The increased use of microcontrollers embedded in low-cost sensors and actuators has provided opportunities for adding diagnosis and configuration data to support increasing application requirements.

The driving force for a new technology called IO-Link™* has been the need of these low cost sensors and actuators to exchange the diagnosis and configuration data with a controller (PC or PLC) using a low-cost digital communication technology while maintaining backward compatibility with the current digital input and digital output (DI/DO) signals.

Another driving force is cost reduction and substitution of error-prone analog transmission such as 0–10 V. Using IO-Link avoids digital/analog conversion on the sensor side and analog/digital conversion on the controller side.

In fieldbus concepts, the IO-Link defines a generic interface for connecting sensors and actuators to a Master† unit, which may be combined with gateway capabilities to become a fieldbus remote I/O node.

Any IO-Link-compliant Device‡ can be attached to any available interface port of the Master. Devices perform physical-to-digital conversion in the Device and then communicate the result directly in a standard format using *coded switching* of the 24 V I/O signaling line, thus removing the need for different DI, DO, AI, AO modules, and a variety of cables.

* IO-Link™ is a trade name of the "IO-Link Community." Use of the registered logos for IO-Link™ requires permission of the "IO-Link Community."

† "Master" (with upper case M) means the IO-Link-compliant counterpart to "Device."

‡ "Device" (with upper case D) means IO-Link-compliant sensor or actuator in this document.

Physical topology is point-to-point from each Device to the Master using three wires over distances up to 20 m. The IO-Link physical interface is backward compatible with the 24 V I/O signaling specified in IEC 61131-2. Transmission rates of 4.8, 38.4, and 230.4 kbit/s are supported.

The Master of the IO-Link interface detects, identifies, and manages Devices plugged into its ports.

Tools allow the association of Devices with their corresponding electronic I/O Device descriptions (IODDs) and their subsequent configuration to match the application requirements.

The IO-Link technology specifies three different levels of diagnostic capabilities: for immediate response by automated needs during the production phase, for medium-term response by operator intervention, or for longer-term commissioning and maintenance via extended diagnosis information.

5.2 IO-Link Technology

5.2.1 Purpose of Technology

Figure 5.1 shows the basic concept of IO-Link.

The single-drop digital communication interface technology for small sensors and actuators IO-Link (also known as SDCI in IEC 61131-9) defines a migration path from the existing DI/DO interfaces for switching 24 V Devices, as defined in IEC 61131-2, toward a point-to-point communication link. Thus, for example, digital I/O modules in existing fieldbus peripherals can be replaced by IO-Link Master modules providing both classic DI/DO interfaces and IO-Link. Analog transmission technology can be replaced by IO-Link combining its robustness, parameterization, and diagnostic features with the saving of digital/analog and analog/digital conversion efforts.

5.2.2 Positioning within the Automation Hierarchy

Figure 5.2 shows the domain of the IO-Link technology within the automation hierarchy.

The IO-Link technology defines a generic interface for connecting sensors and actuators to a Master unit, which may be combined with gateway capabilities to become a fieldbus remote I/O node or via an adapter to personal computers or drives.

Starting point for the design of IO-Link is the classic 24 V DI defined in IEC 61131-2 and output interface (DO) specified in [1]. Thus, IO-Link offers connectivity of classic 24 V sensors (*switching signals)* as a default operational mode. Additional connectivity is provided for actuators when a port has been configured into single-drop communication mode (*coded switching*).

Many sensors and actuators nowadays are already equipped with microcontrollers offering a universal asynchronous receiver transmitter interface that can be extended by the addition of a few hardware components and protocol software to support IO-Link communication. This second operational mode uses *coded switching* of the 24 V I/O signaling line. Once activated, the IO-Link mode supports parameterization, cyclic data exchange, diagnosis reporting, identification and maintenance information,

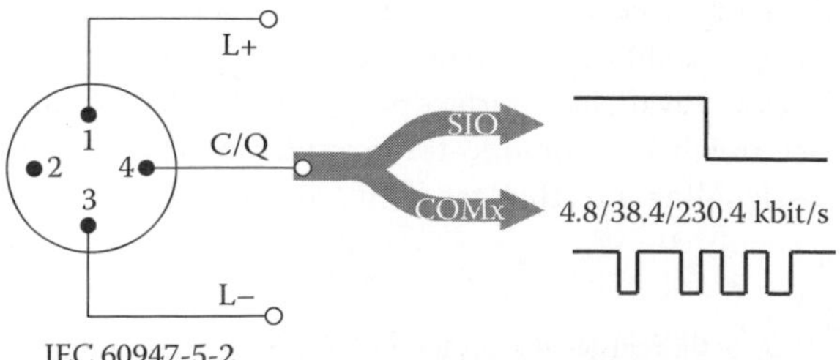

Pin	Signal	Definition	Standard
1	L+	24 V	IEC 61131-2
2	I/Q	Not connected, DI, or DO	IEC 61131-2
3	L−	0 V	IEC 61131-2
4	Q	"Switching signal" DI (SIO)	IEC 61131-2
	C	"Coded switching" (COM1, COM2, COM3)	IO-Link (IEC 61131-9)

FIGURE 5.1 IO-Link compatibility with IEC 61131-2.

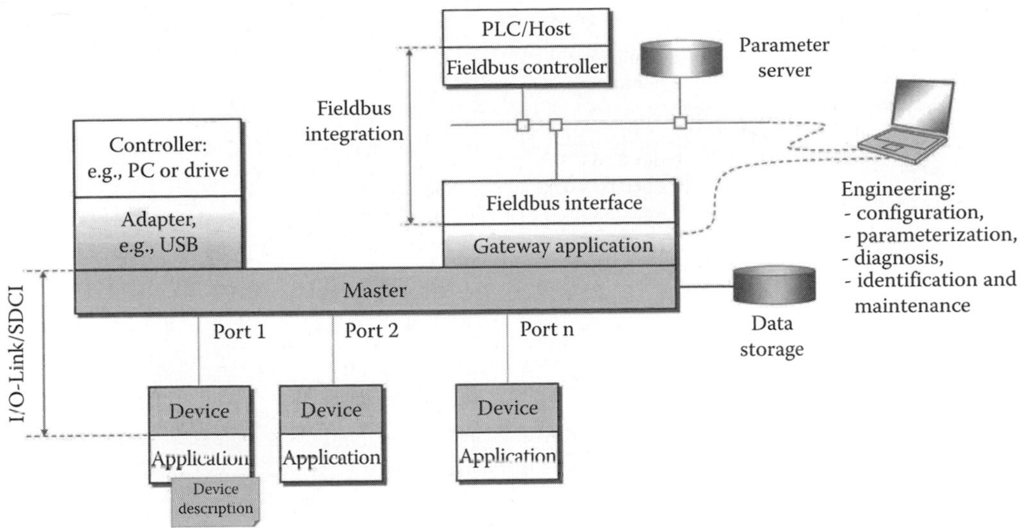

FIGURE 5.2 Domain of the IO-Link technology within the automation hierarchy.

and external parameter storage for Device backup and fast reload of replacement devices. To improve start-up performance, these Devices usually provide nonvolatile storage for parameters.

Configuration and parameterization of Devices are supported through an XML-based device description similar to fieldbus electronic device descriptions (see [2]).

5.2.3 Wiring, Connectors, and Power

The default connection (port class A) comprises four pins (see Figure 5.1). The default wiring for port class A complies with IEC 60947-5-2 and uses only three wires for 24, 0 V, and a signal line. The fourth wire may be used as an additional signal line complying with IEC 61131-2.

Five-pin connections (port class B*) are specified for Devices requiring additional power from an independent 24 V power supply.

Maximum length of cables is 20 m, and shielding is not required.

5.2.4 Communication Features of IO-Link

The generic Device model is shown in Figure 5.3 and explained in the following paragraphs.

A Device may receive process data (PD; out) to control a discrete or continuous automation process or send PD (in) representing its current state or measurement values. The Device usually provides parameters enabling the user to configure its functions to satisfy particular needs. To support this case, a large parameter space is defined with access via an index (0–65, 535; with a predefined organization) and a subindex (0–255).

The first two index entries 0 and 1 are reserved for the direct parameter pages 1 and 2 with a maximum of 16 octets each. Parameter page 1 is mainly dedicated to Master commands such as Device start-up and fallback, retrieval of Device-specific operational and identification information. Parameter page 2 allows for a maximum of 16 octets of Device-specific parameters.

The other indices (2–65, 535) allow access to one record having a maximum size of 232 octets. Subindex 0 specifies transmission of the complete record addressed by the index; other subindices specify transfer of selected data items within the record.

* A port class A Device using the fourth wire is not compatible with a port class B Master.

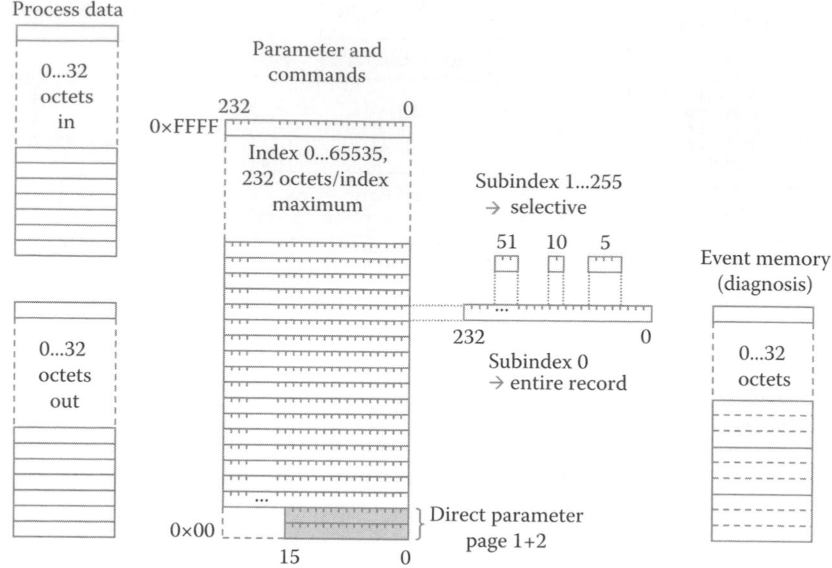

FIGURE 5.3 Generic Device model for IO-Link (Master's view).

Within a record, individual data items may start on any bit offset, and their length may range from 1 bit to 232 octets, but the total number of data items in the record cannot exceed 255. The organization of data items within a record is specified in the IODD (see [2]).

All changes of Device condition that require reporting or intervention are stored within an event memory before transmission. An event flag is then set in the cyclic data exchange to indicate the existence of an event.

Communication between a Master and a Device is point to point and is based on the principle of the Master first sending a request message and then a Device sending a response message (see Figure 5.4).

Both messages together are called an M-sequence. Several M-sequence types are defined to support user requirements for data transmission (see [1]). The first byte of a Master message controls the data transfer direction (read/write). It is followed by the M-sequence type of communication channel and data integrity checking code, and optional bytes with either *PD* and/or *on-request data* (OD).

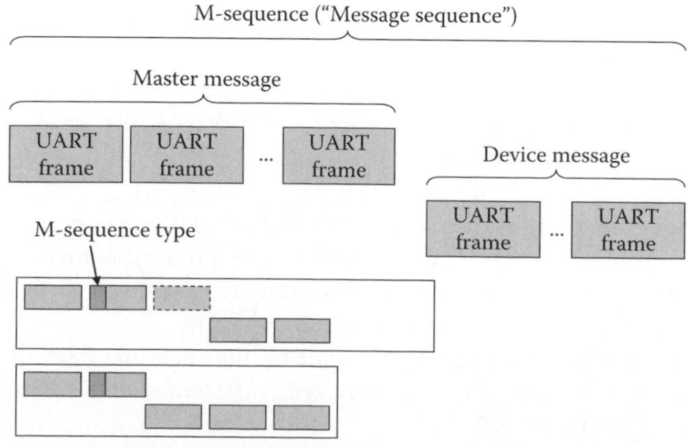

FIGURE 5.4 Principle of message sequences between Master and Device.

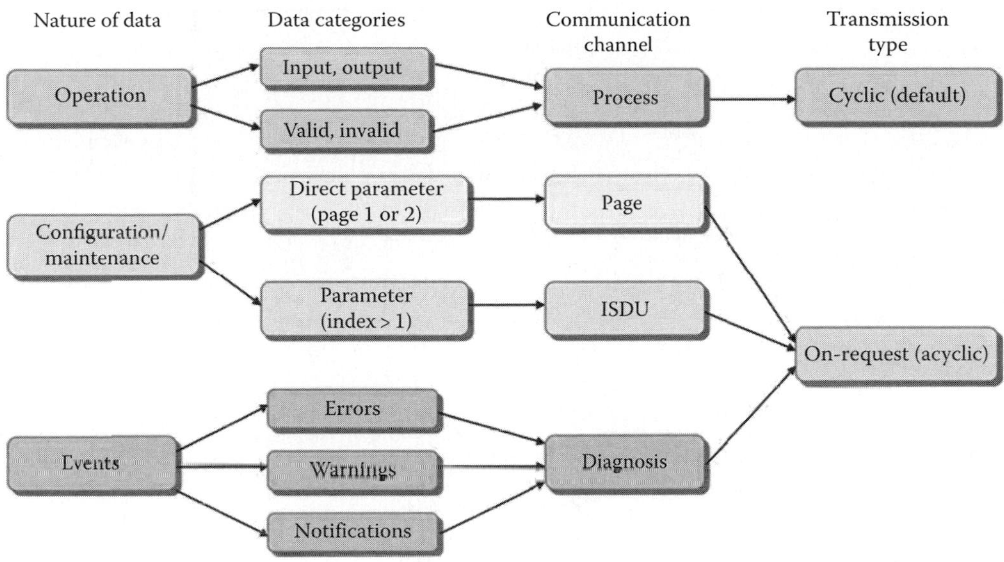

FIGURE 5.5 Relationship between nature of data and transmission types.

Data of various categories are transmitted through separate communication channels within the data link layer (DL), as shown in Figure 5.5.

- Operational data such as Device inputs and outputs are transmitted through a process channel using cyclic transfer. Operational data may also be associated with qualifiers such as valid/invalid.
- Configuration and maintenance parameters are transmitted using acyclic transfers. A page channel is provided for direct access to parameter pages 1 and 2, and an indexed service data unit channel is used for accessing additional parameters and commands via indices.
- Device events are transmitted using acyclic transfers through a diagnostic channel. Device events are reported using three severity levels: error, warning, and notification.

Figure 5.6 shows that each port of a Master has its own DL that interfaces to a common master application layer (AL). Within the AL, the services of the DL are translated into actions on *PD* objects (input/output), *OD* objects (read/write), and events. Master applications include a configuration manager, data storage mechanism, diagnosis unit (DU), OD exchange (ODE), and a PD exchange (PDE).

System management checks identification of the connected Devices and adjusts ports and Devices to match the chosen configuration and the properties of the connected Devices. It controls the state machines in the AL and DL, for example, at start-up.

5.2.5 Role of a Master

A Master accommodates 1 to *n* ports and their associated DLs. During start-up, it changes the ports to the user-selected port modes, which can be INACTIVE, DI, DO, FIXEDMODE, or SCANMODE. If communication is requested, the Master uses a special wake-up current pulse to initiate communication with the Device. The Master then auto-adjusts the transmission rate to COM1, COM2, or COM3 (see [1]) and checks the *personality* of the connected Device, that is, its vendor identification, Device identification, and communication properties.

If there is a mismatch between the Device parameters and the stored parameter set within the Master, the parameters in the Device are overwritten or the stored parameters within the Master are updated depending on configuration.

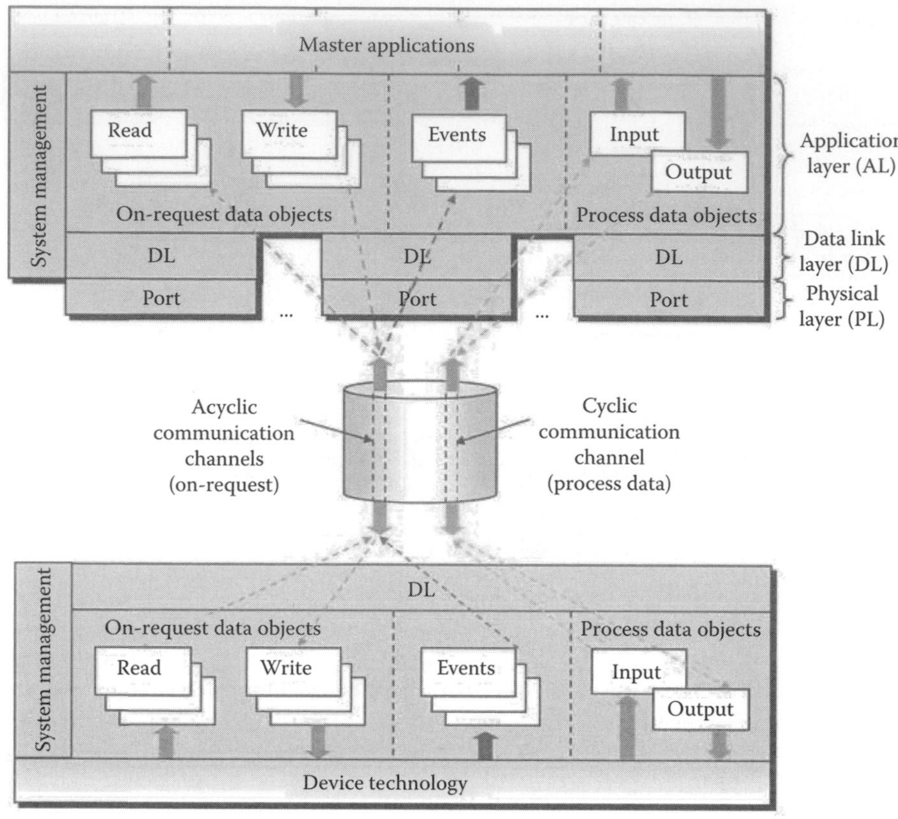

FIGURE 5.6 Object transfer at the application layer (AL) level.

It is also possible to start a device in DI mode, switch to IO-Link communication for configuration and parameterization, and then use a fallback command to switch back to DI mode for normal operation (see Figure 5.7).

Coordination of the ports is also a task of the Master, which the user can configure through the selection of port cycle modes. In *FreeRunning* mode, each port defines its own cycle based on the properties of the connected Device. In *MessageSync* mode, messages sent on the connected ports start at the same time or in a defined staggered manner. In *FixedValue* mode, each port uses a user-defined fixed cycle time (see [1]).

The Master is responsible for the assembling and disassembling of all data from or to the Devices.

The Master provides a data storage area of at least 2048 octets per Device for backup of Device data. The Master may combine these Device data together with all other relevant data for its own operation and make these data available for higher-level applications for Master backup purpose or recipe control.

5.2.6 IO-Link Configuration

Engineering support for a Master is usually provided by a Port and Device Configuration Tool (PDCT) connected directly or indirectly via fieldbus to the Master. The PDCT configures both port properties and Device properties (see parameters shown in Figure 5.3). It combines both an interpreter of the IODD and a configurator (see Figure 5.2). The IODD provides all the necessary properties to establish communication and the necessary parameters and their boundaries to establish the desired function of a sensor or actuator. The PDCT also supports the compilation of the PD for propagation on the fieldbus and vice versa.

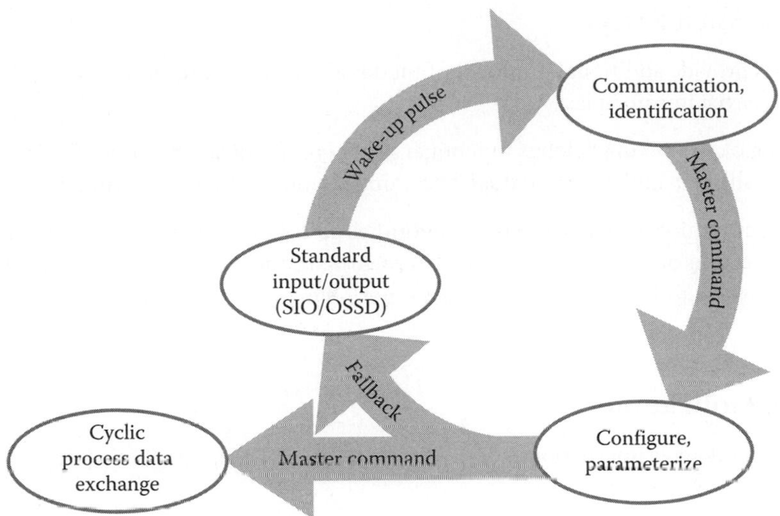

FIGURE 5.7 IO-Link communication states.

5.2.7 Mapping to Fieldbuses and System Integration

Integration of a Master within a fieldbus system, that is, the definition of gateway functions for exchanging data with higher-level entities on a fieldbus, is out of the scope of the IO-Link specification [1]. Examples of these functions include mapping of the PDE, realization of program-controlled parameterization or a remote parameter server, or the propagation of diagnosis information.

The integration of a PDCT into engineering tools of a particular fieldbus is also out of the scope of the IO-Link specification [1].

However, IO-Link integrations (gateways/Master) exist for the most important fieldbuses such as PROFIBUS, PROFINET, DeviceNet, Ethernet/IP, EtherCAT, Powerlink, CC-Link, and ASi. It is the responsibility of the fieldbus organizations to provide corresponding specifications.

5.2.8 Implementation and Engineering Support

A number of application specific integrated circuits and software stacks facilitate the design and implementation of Masters and Devices. Information can be retrieved from www.io-link.com.

5.2.9 Test and Certification

A comprehensive specification for the test of Masters and Devices exist (see [3]), and corresponding test tools can be acquired for both. A manufacturer declaration based on the positive results of the Device tester is required for the usage of the IO-Link word and picture mark. Certification of Devices is not required. Similar rules exist for Masters. The fieldbus organizations are responsible for the quality assurance of the combination of fieldbus gateway and Master.

5.2.10 Profiles

An IO-Link profile specification exists for sensors (see [4]). It describes the common part of a sensor model that should be valid for future Device profiles and a more specific part for the so-called Smart Sensors comprising recommended PD structures, identification objects, binary switching thresholds and hysteresis, best practice handling of quantity measurements with or without associated units, diagnosis objects, and teaching commonalities.

5.2.11 Functional Safety

Many fieldbuses provide additional profiles for functional safety communication (see [5]). For IO-Link, two possibilities exist in principle:

1. Tunneling of an existing fieldbus functional safety communication across IO-Link
2. A specific slim IO-Link functional safety communication profile with a functional safety gateway

The IO-Link community does not prohibit the tunneling solution, which is the responsibility of the corresponding fieldbus organization. In general, for sensor and actuator manufacturers, the second possibility is in favor due to less development and support efforts. A corresponding specification is going to be developed.

5.2.12 Standardization

The IO-Link technology is internationally standardized within IEC 61131-9 (see [6]).

Abbreviations

AI	Analog input
AL	Application layer
AO	Analog output
ASIC	Application-specific integrated circuit
CM	Configuration manager
COM1	IO-Link transmission rates of 4.8 kbit/s
COM2	IO-Link transmission rates of 38.4 kbit/s
COM3	IO-Link transmission rates of 230.4 kbit/s
DI	Digital input
DL	Data link layer
DO	Digital output
DS	Data storage
DU	Diagnosis unit
IEC	International Electrotechnical Commission
I/O	Input/output
IODD	I/O device description
ISDU	Indexed service data unit
OD	On-request data
ODE	On-request data exchange
PC	Personal computer
PD	Process data
PDCT	Port and device configuration
PDE	Process data exchange
PLC	Programmable logic controller
SDCI	Single-drop digital communication interface for small sensors and actuators
SIO	Standard I/O mode according IEC 61131-2
UART	Universal asynchronous receiver transmitter
XML	Extensible mark-up language

References

1. IO-Link Community specification: *IO-Link Interface and System*, V1.1.2, August 2013, downloadable from www.io-link.com. (accessed September 1, 2013).
2. IO-Link Community specification: *IO-Device Description,* V1.1, August 2011, downloadable from www.io-link.com. (accessed September 1, 2013).
3. IO-Link Community specification: *IO-Link Test*, V1.1.2, August 2012, downloadable from www.io-link.com. (accessed September 1, 2013).
4. IO-Link Community specification: *IO-Link Smart Sensor Profile*, V1.0, October 2011, downloadable from www.io-link.com. (accessed September 1, 2013).
5. IEC 61784-3 Industrial communication networks—Profiles—Part 3: Functional safety fieldbuses—General rules and profile definitions.
6. IEC 61131-9 Ed 1: Programmable Controllers—Part 9: Single-drop digital communication interface for small sensors and actuators (SDCI).

6 AS-Interface*

6.1 AS-Interface Background 6-1
6.2 AS-Interface Technology 6-2
AS-Interface System Data in a Nutshell • Data and Energy on One Wire • AS-Interface Electromechanics • AS-Interface Communication System
6.3 AS-Interface Devices: Sensors/Actuators 6-19
Function Request in the Slave • Device Profiles • Extended Profiles: Combined Transaction Types • Device Addressing • Device Coupling to Sensors/Actuators • Structure of the AS-Interface Slave • Device Interface and Diagnostics
6.4 AS-Interface Master: Coupling to Other Automation Systems 6-36
Function Sequence in the Master • Master Levels • Master Calls • Data Fields and Lists on the Sequence Control • Master Analog Part • Master Host Interface: Master Profile
6.5 Functional Safety with AS-Interface Safety 6-47
Transmission Principle • Safety Monitor • Safe Inputs • Safe Outputs • Start-Up of a Safety System: Teaching the Code Table
6.6 Open System: Interoperability and Certification 6-53
References and Further Probing 6-55

Tilman Schinke
AS-International Association e.V.

6.1 AS-Interface Background

AS-Interface (AS-i) (actuator/sensor interface) was defined and developed by a consortium of 11 German and Swiss companies between 1990 and 1994. At this point in time, multiple different fieldbuses had already been developed, but none of them seemed to be suitable to connect simple sensors and actuators. Therefore, it was necessary to devise an entirely new type of a system with the primary goal to replace cable trunks and to minimize installation expenses. Thus, a system was designed based on a simple nonshielded two-conductor cable that was able to transfer data as well as 24 V power.

Since the implementation of the first automation processes using AS-i in 1994, multiple improvements to the system have been introduced. Among others, the number of slaves that can be connected to the system was doubled from 31 to 62, safety signals can now be transmitted via standard AS-i networks, and procedures were introduced that ensure the easy transmission of bytes, 16-bit words, and even longer chains of characters.

However, all these improvements were closely watched concerning their comparability with the current standard. Therefore, the first prototype slaves, developed in 1992, are still operational in all AS-i networks.

* Reorganized and updated based on the publications of the AS-International Association: "AS-Interface die Lösung in der Automation" (2002), "AS-Interface Safety at Work" (2004), and "Automatisieren ist einfach mit AS-Interface" (2009).

AS-i (standardized as EN 50295-2 (1998, 2002) [1] and IEC 62026-2 (2000) [2]) is the fieldbus for the lower levels of the information pyramid in industrial automation technology, with a broad range of applications in factory automation, process automation, building automation, and mobile equipment automation. Today (2014), there are

- More than 27 Mio. installed nodes
- More than 350 members worldwide
- More than 1,800 certified products
- Worldwide over one million safety systems in use
- More than 2,500,000 AS-i safety devices
- Increase of 66% in the field of safety devices

6.2 AS-Interface Technology

6.2.1 AS-Interface System Data in a Nutshell

6.2.2 Data and Energy on One Wire

The AS-i transfers data and energy on a two-conductor cable; the two conductors are usually named AS-i+ and AS-i–. But AS-i– is different to the ground (GND), well known in direct current technology (Table 6.1).

To run an AS-i network, four basic requirements have to be established:

1. Power supply: supply of the outputs of the network with a nominal voltage of 24 V direct current (DC)
2. Safety: secure mains separation (PELV)
3. Balancing of the network: increased noise immunity
4. Data decoupling: separation of power and data

These are the reasons why it is not possible to feed AS-i networks directly with a standard power supply. Special AS-i power supplies or data-decoupling modules for the Power24 technology are required. A manifold range of power supplies are available that suit almost any individual use case, providing current in a range of 1–8A and more.

TABLE 6.1 AS-i System Technical Data

	AS-i
Devices/addresses per system	62
Topology	Line, star, tree, ring
Available voltage	24 V at the outputs
Available current	Up to 8 A
Typical cycle time[a]	3 ms
Slave response time	154 μs
Network length	500 m[b]
Protection class	Up to IP 69K
Amount of I/O using the 8I/8O profile	496 inputs/496 outputs
Standardized analog value transmission	16 bit
Performance level according EN ISO 13849-1	Up to PL_e
SIL according to IEC 61508	Up to SIL3
Hot swap/live insertion	Yes/yes

[a] AS-i uses the master–slave system with cyclic polling.

[b] 500 m by use of repeaters, optional enhancements with termination is not taken into account.

Power Supply

In case of AS-i, the network cable supplies the power. This serves to supply the slaves and a part of the master over the two-conductor cable. And of course, the power for the connected outputs is provided, the common 24 V used in the cabinets. For this reason, the AS-i power supplies provide a DC voltage from 29.5 to 31.6 V at currents up to 8 A, to enable the 24 V at the end of the 100 m AS-i cable. For more details, see 6.2.2.2 AS-Interface cable.

Safety

AS-i is designed as a system for low voltages with safe isolation (protective extralow voltage [PELV]). This means that according to the relevant IEC standards, *safe isolation* is required of the power supply between the supply network and the AS-i network.

Balancing

The power supply is also used to balance the AS-i network. AS-i is operated as a symmetrical, non-grounded system. For optimum noise immunity against symmetrical noise coupling, the AS-i cable needs to be installed as symmetrically as possible. This is accomplished by the balancing circuit shown in Figure 6.1. The shield connection must be grounded at an appropriate point on the machine or system. AS-i allows only this point to be connected to system GND.

Data Decoupling

The data-decoupling network, which is normally located in the same housing as the power supply, consists of two inductors of 50 mH each and two parallel-wired resistors of 39 Ω each. The inductors use differentiation in voltage pulses to convert the current pulses generated by the AS-i transmitters. At the same time, they prevent the AS-i cable from being short-circuited by the power supply for data transmission. (For more information, see 6.2.2.3 Modulation procedure.)

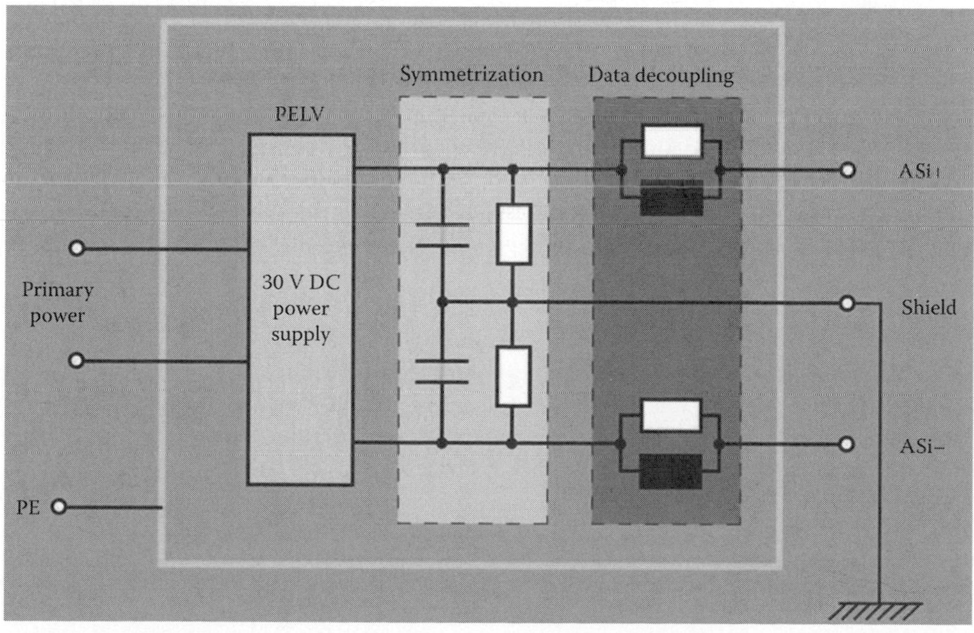

FIGURE 6.1 The AS-i power supply.

6.2.2.1 Power24 Technology

Using Power24 technology, the AS-i network power is supplied by a 24 V power supply. In this case, the data decoupling and balancing has to be done in a data-decoupling module. Sometimes, this module is integrated into the master housing. The use of this technology has the benefit that no additional power supply is necessary. This saves costs but it reduces the functionality as well. The reduced input voltage has consequences, the network length is reduced to 50 m, and, of course, the outputs can provide voltages below 24 V only.

6.2.2.2 AS-Interface Cable

Two different two-conductor unshielded cables were specified as the transmission medium for AS-i. One is a flexible high-voltage cable conforming to CENELEC or DIN VDE 0281, designated *H05VV-F 2 × 1.5*, which is very inexpensive and easily obtainable. The other is an AS-i-specific flat cable with very similar electrical properties but with the additional installation characteristics. The transmission media otherwise found in fieldbus systems (e.g., twisted-pair two-conductor cable, coaxial cable, fiber-optic cable) are not suitable for simultaneous transmission of data and power, complicated to install, or too expensive.

There are two electrical considerations involved in the selection of the suitable transmission medium: the DC resistance for power transmission and the transmission characteristics in the frequency range that is used for communication. At least 2A should be able to be transmitted for powering the slaves.

Assuming

- A cable length of 100 m
- 62 slaves evenly spaced along the cable
- All slaves having the same current requirement on average so that a total current of 2 A flows

the result is a maximum voltage drop between the feed point of the power supply and the terminal point of the last slave that depends on the selected conductor cross section as shown in the Table 6.2.

Since actuators with a permissible voltage tolerance of 24 V + 10%/−15% can be connected on Interface 1 of the slave and since a technically feasible voltage tolerance must be allowed for the power supply, a maximum permissible voltage drop for the cable of 3 V and a conductor cross section of 1.5 mm^2 (AWG 16) was specified for the cables. This solution should cover the vast majority of practical applications. In cases that deviate greatly from the earlier assumptions, a thorough analysis of the power distribution and voltage conditions at the remotest point in the network should be carried out.

Using Power24 technology, the outputs could provide less power only because of the reduced supply current (24 V).

From a purely technical point of view, other conductor cross sections in a range from 0.75 to 2.5 mm^2 (AWG 18 to AWG 14) could be used. This makes the use of special cables possible, such as in applications where motion cables are required or where higher currents are demanded.

The other aspects in selecting the AS-Interface transmission cable have to do with the cable properties in the transmission frequency range. These can be described by the cable models shown in Figure 6.2. Shown are the maximum ratings that must be met by any proposed cable.

TABLE 6.2 DC Resistance and Maximum Voltage Drop as a Function of Conductor Cross Section

Conductor cross section [mm^2]	0.75	1.5	2.5
DC resistance [mΩ/m]	52.0	26.6	16.0
Maximum voltage drop [V]	5.4	2.7	1.6

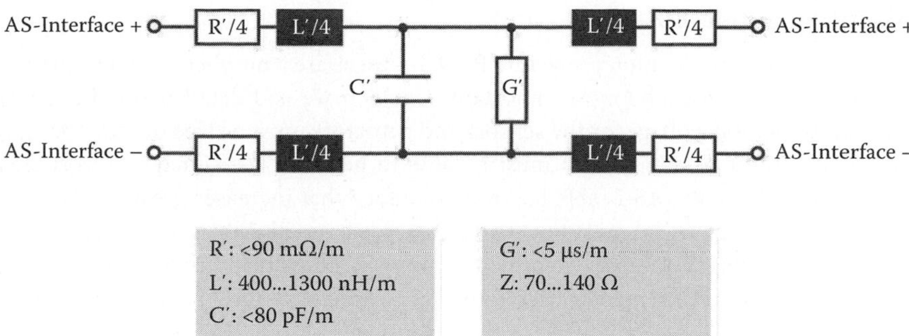

FIGURE 6.2 AS-i cable model and limits.

The transmission properties of the AS-i cable, especially the impedance curve over the frequency, which is critically defining for the transmission behavior, are shown in Figure 6.3. You can see that both lower and higher frequencies up to 167 kHz are strongly attenuated, which has a very positive effect on the behavior of the AS-i system especially when it comes to asymmetrical interference (Figure 6.4).

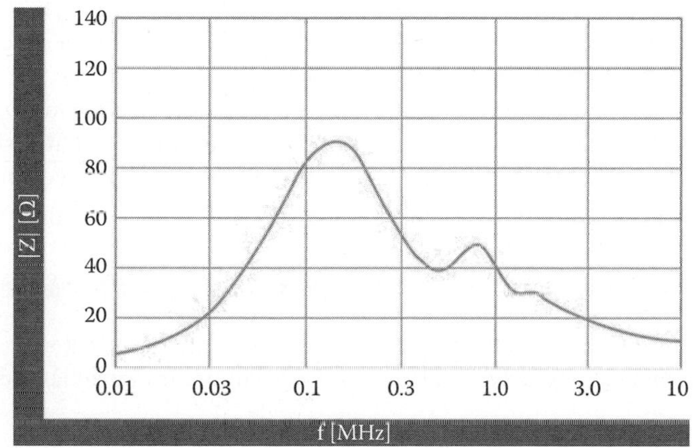

FIGURE 6.3 Transmission properties of the AS-i cable—impedance curve with data decoupling and 30 slaves.

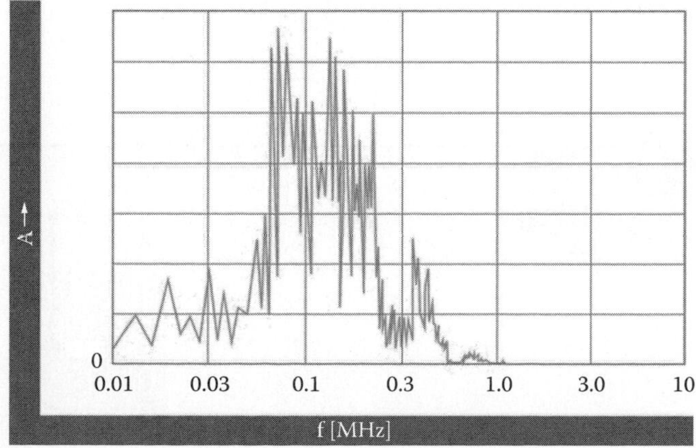

FIGURE 6.4 Transmission properties of the AS-i cable. Frequency spectrum of a typical master poll.

6.2.2.3 Modulation Procedure

When selecting a suitable modulation procedure for AS-i, there are a number of requirements that need to be taken into consideration, the most important of which we will detail here. The message signal superimposed on the supply voltage for the sensors and actuators must be free of DC. The slave's sender (and where possible the master's sender) must be able to generate the signal in a cost-effective and space-saving manner. Since the AS-i cable has an impedance that increases greatly over the frequency (see Figure 6.3), the message signal must be relatively narrowbanded. Nor is it allowed to radiate unacceptable levels of electromagnetic noise. No previously known, and already-used, modulation procedure was able to meet all these requirements. Therefore, it was necessary for AS-i to specify and develop a new procedure.

The result was alternating pulse modulation (APM), a procedure for serial transmission in the base band whose functionality will be explained with reference to Figures 6.5 and 6.6. The send bit sequence is first encoded into a bit sequence that performs a phase change whenever the send signal changes. The result is a send current that, in conjunction with a single inductor in the system, uses differentiation to generate the desired signal voltage level on the AS-i cable.

Each rise in the send current thus results in a negative voltage pulse and each drop to a positive voltage pulse. In this way, it is quite simple to generate signals in slaves that have a higher voltage than their actual supply voltage. This means inductors can be eliminated in the slave, which keeps the electronics integrated in the sensor or actuator small and inexpensive. On the receiver side, these voltage signals are detected on the line and converted back into the send bit sequence. The receiver synchronizes itself with the first detected negative pulse, which it interprets as the start bit of a message.

If the voltage pulses approximate $\sin^2$ pulses, then the requirements for low limit frequency and low noise emission are met at the same time. This is done by means of suitable shaping of the send current pulses, which are generated like the integral of a $\sin^2$ pulse. A frequency spectrum for a typical send pulse sequence is shown in Figure 6.3. A comparison with Figure 6.4 shows that the attenuation curve and the signal spectrum are optimally matched. Using this modulation procedure and the available topologies, bit times of 6 ms are attainable. This allows a gross transmission rate of 167 kbit/s.

The nonterminated cable results in the amplitudes of the AS-i messages that a receiver has to process being able to have the fluctuations shown in Figure 6.7, both within the individual telegrams and between

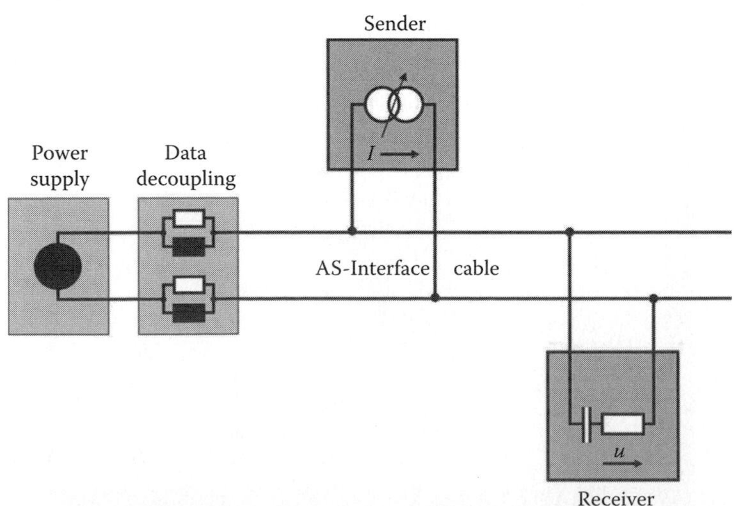

FIGURE 6.5 APM—signal generation.

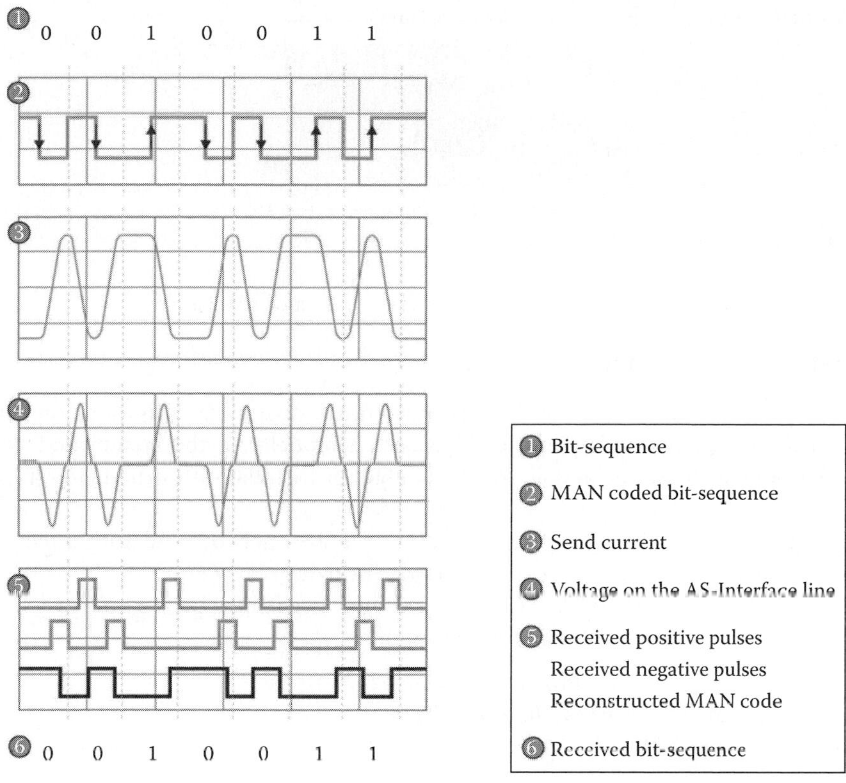

FIGURE 6.6 APM—MAN code.

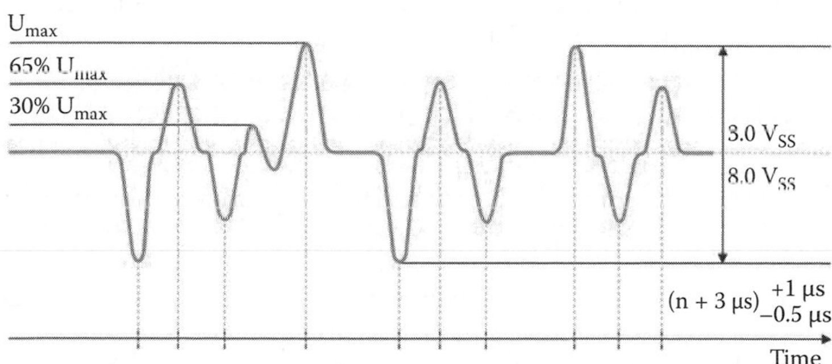

FIGURE 6.7 Slave reply at master receiver.

the telegrams from various slaves. The differences within the message arise from the fact that reflections at the cable end affect higher frequencies (and thereby more densely packed $\sin^2$ pulses) more than lower frequencies. In addition, ringing is not allowed to affect the receiver. Managing these amplitude tolerances ensures that AS-i is extremely robust with respect to any cable topology.

The absolute size of the slave reply depends both on the topology and on the location of the slave. If the slave is located near the power supply, which forms a cable terminator, this slave will generate the reply with the smallest amplitude. Whereas a slave only communicates with a station (the master) that is located at a fixed location along the line, the master must be able to process every slave reply from any location along the cable.

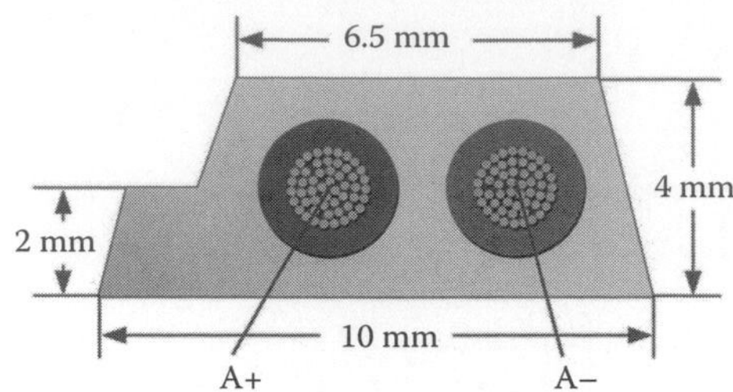

FIGURE 6.8 The yellow AS-i profile cable with a conductor cross section of 1.5 mm².

6.2.3 AS-Interface Electromechanics

The standardized electromechanics of AS-i offer an innovative installation and implementation technology. This technology adds considerably to the simplicity of the system and thus to cost reductions. Figure 6.8 shows the yellow AS-i flat cable for installation in the field. This cable has become the trademark of AS-i.

The cable is available with a *traffic yellow* jacket for the AS-i network and with a *black* jacket for auxiliary power, which can be routed separately to some field devices.

The available jacket materials with their chemical and mechanical properties cover most of the requirements found in industrial automation applications:

- EPR rubber mixture EM3 according to DIN DE 0207 part 21
- TPE YM5 according to DIN VDE 0207 part 5
- PUR according CENELRC HD 22, T 10

The yellow cable has a conductor cross section of 2×1.5 mm² and is equipped with a coding strip ensuring correct polarization and contact on the specified side. The connection is achieved via piercing contact piercers, penetrating the cable jacket and thus establishing a secure contact to the specifically designed fine stranded wire (Figure 6.9). Cutting, stripping, installing ferrules, and attaching a shield can be omitted. Even prelengthening the cable is no longer necessary since the cable can be connected at any arbitrary point. This shows how easy connecting a slave can be—with AS-i!

Many modules and field devices can be directly connected to the yellow flat cable. The product database of AS-i—www.as-interface.net/products—offers a multitude of accessories to establish branches, drop cables, etc., to route and attach the flat cable and cable glands to bring it into junction boxes.

A black profile cable with the same measurements as the yellow cable is used to provide additional power for some field devices or modules with 24 V DC.

In addition to the AS-i flat cable, a two-conductor cable for applications in IP20 environments (in cabinets, distribution boxes, and similar protected environments) was introduced.

The IP20 two-conductor cable has a conductor cross section of 2×0.75 mm² and is mechanical coded with a tab. This two-conductor cable is optimized for connections via piercing technology. For this purpose, each conductor is pressed into a tool, very similar to a fork, while the insulation is cut and a secure contact is achieved. Processes like cutting, stripping, installing ferrules, or attaching a shield are also omitted. This AS-i two-conductor cable is more flexible and easier to handle than the yellow flat cable—due to its single-layer insulation and thus smaller conductor diameter. However, this makes it unsuitable for field installation.

6.2.4 AS-Interface Communication System

6.2.4.1 Access Procedure

Since AS-i was designed to replace star-type two-point connections (traditional cable tree), a bus access procedure was selected, which reproduces this topology and is able to ensure a defined response time

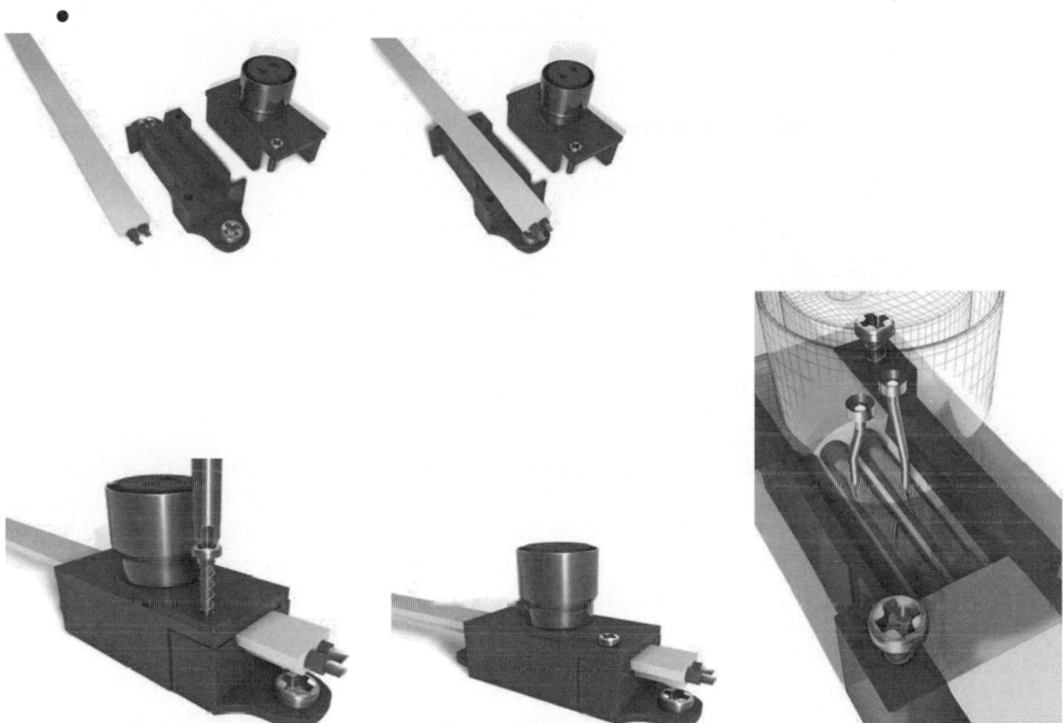

FIGURE 6.9 M12 adapter for the AS-i profile cable with *piercing* technology.

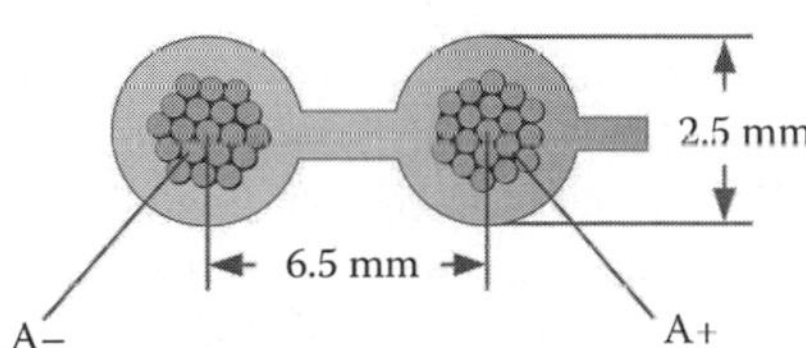

FIGURE 6.10 AS-i two-conductor cable for use in IP 20 applications has a conductor cross section of $2 \times 0.75\ \text{mm}^2$.

(the master–slave access with cyclical polling). The master sends a telegram that is received at a particular slave address, and the slave contacted at this address replies within the provided time (Figure 6.11). From the viewpoint of the transmission system, only one master and a maximum of 62 slaves can be involved at any one time.

The procedure chosen for AS-i allows the construction of very simple and thereby cost-effective slaves while providing at the same time the greatest possible flexibility and integrity. In case of a brief disturbance on the line, for example, the master can repeat telegrams for which it received either no reply or an invalid reply. This means it is not necessary to repeat the entire cycle over again.

The single-master access procedure does not preclude operation with a second (standby) master. The latter can also listen to the data traffic and assume responsibility for communication should the first master fail.

6.2.4.2 AS-Interface Messages

An AS-i message consists, as shown in Figure 6.11, of a master request, a master pause, a slave response, and a slave pause. All master requests are exactly 14 bit times in length, and all slave responses have a length of 7 bit times. A bit time corresponds to a uniform 6 ms.

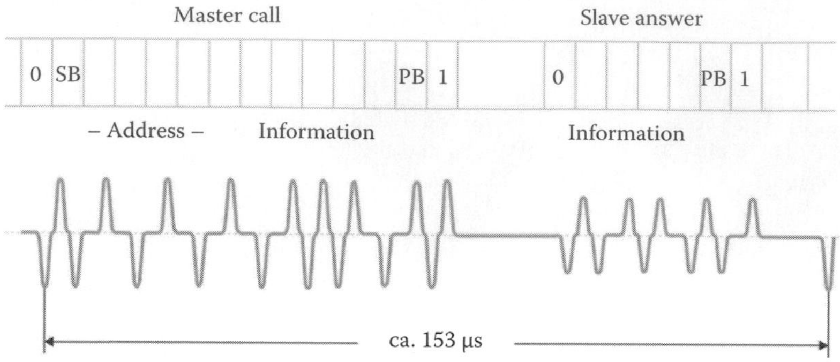

FIGURE 6.11 Structure of an AS-i message.

The master pause is allowed to be at least 2 and a maximum of 10 bit times in length. If the slave is synchronized, it can begin to send its response after as soon as 2 bit times. If it is not synchronized, it requires 2 bit times longer, since it is monitoring the master pause during this time for any additional information before it can accept the poll as valid. If the master, however, has not received the start bit for the slave response after 10 bit times, it can assume that there is no response forthcoming and it can begin with the next request. The pause (slave pause) between the end of the slave response and the next master request should be no more than 1.5 to 2 bit times in length.

The signal sequence shown in Figure 6.11 represents the bit sequence 00101010111011Bin, which means a data request to a standard slave having address 21 and user data EHEX. The slave responds with the bit sequence 0011001Bin, which contains the user data 6HEX.

A master poll of a standard slave is listed in Table 6.3.

The slave response is listed in Table 6.4.

4 bits of user data are contained in a master request, while the slave returns 4 bits of user data in response. AS-i Specification Version 2.1 (1998) created the possibility of connecting 62 slaves to an AS-i network instead of the previous 31. This made it necessary to define an additional address bit without however losing compatibility with the existing systems. For this reason, the following definition of the master request was agreed on for all slaves having the ID code "AHEX," which are operated in the so-called expanded address mode (Table 6.5).

The message shown in Figure 6.11 could also be directed to a slave in expanded addressing mode. In this case, it contains address 21A and user data 6HEX. This means that in a master request to a slave

TABLE 6.3 Master Request

Bit	Reference
Start bit ST	Identifies the beginning of the master request = 0: valid start bit; = 1: not allowed.
Control bit SB	Identifies the type of request = 0: data, parameter request, address assignment = 1 command request.
Address A0–A4	Address of the contacted slave (5 bit).
Information I0–I4	These 5 bits contain in order to the type of request, the information to be transferred to the slave. Details are described by the respective call type.
Parity bit PB	Parity bit the number of all "1" in the master call has to be even.
End bit EB	Identifies the end of the master request = 0: not allowed; = 1: valid end bit.

TABLE 6.4 Slave Response

Bit	Reference
Start bit ST	Identifies the beginning of the slave response = 0: valid start bit; = 1: not allowed.
Information I0–I3	4-bit information.
Parity bit PB	Parity bit the number of all "1" in the slave response has to be even.
End bit EB	Identifies the end of the slave response = 0: not allowed; = 1: valid end bit.

TABLE 6.5 Master Request in Extended Address Mode

Bit	Reference
Start bit ST	Identifies the beginning of the master request = 0: valid start bit; = 1: not allowed.
Control bit SB	Identifies the type of request = 0: data, parameter request, address assignment; = 1: command request.
Address A0–A4 13	Address of the contacted slave (5 bits and select bit).
Information I0–I2, 14	These 4 bits contain in order to the type of request, the information to be transferred to the slave. Details are described by the respective call type.
Parity bit PB	Parity bit the number of all "1" in the master call has to be even.
End bit EB	Identifies the end of the master request = 0: not allowed; = 1: valid end bit.

in this mode, only 3 bits of user data are contained. The response from a slave in expanded addressing mode remained unchanged.

Data Convention

In addition to the described message format, the data convention shown in Figure 6.12 has been established for AS-i. From the point of view of the host system or Interface 3, the value 0 for a data bit means that the sensor, the actuator of the function of the slave, is *turned off* or *inactive*; a value of 1 means *turned on* or *active*.

Since this convention does not apply in the same way to the AS-i telegram level, the master must invert the data bits when outputting (and only then!). One must also note that this definition does not apply to outputting of parameters. Since parameters are system-internal data, the master does not invert them.

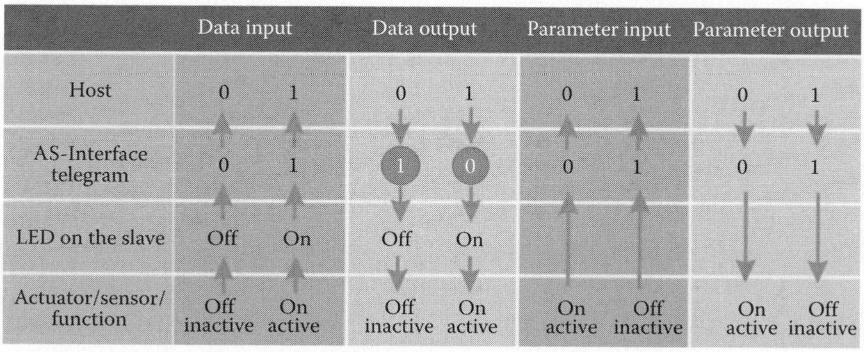

	Data input		Data output		Parameter input		Parameter output	
Host	0	1	0	1	0	1	0	1
AS-Interface telegram	0	1	1	0	0	1	0	1
LED on the slave	Off	On	Off	On				
Actuator/sensor/ function	Off inactive	On active	Off inactive	On active	On active	Off inactive	On active	Off inactive

FIGURE 6.12 Convention for data transmission.

6.2.4.3 Data Safety

No communication system works without any faults. Therefore, it is crucial to be aware of possible interruptions that could affect the fieldbus by falsifying a transmitted message. Consequently, one of the tasks when designing a fieldbus system is to assure that falsified messages can be safely recognized and will not be processed (Figure 6.13).

Of note, even fieldbuses that use optical cables to transmit information are not entirely immune to the danger of potential interruptions. Because optical senders and receivers need to be supplied with electronic energy and interruptions can be transmitted via the supply cable, these systems can also experience bit falsifications (Figure 6.14).

It is easy to see that the probability for an interruption to affect a long message is greater than for a shorter message. A short message can be laid out more fortified against the influence of interruptions.

Falsifications that changed the message during the transmission from sender to receiver need to be recognized on the receiver side. Therefore, the receiver needs to add additional information (proof or safety bits), following a specified algorithm, to the actual information. The receiver has the task to check the received message for potential falsifications and to repeat the calculation via the same algorithm. If the result is correct, the message is determined to be correctly transmitted. Otherwise, it is determined to be lost and a repeat is requested by the sender.

Particularly critical is the unlikely but still possible case that interruptions interfere with a message in a way that several data bits are falsified. In this case, it is possible that a sent message A *close valve*

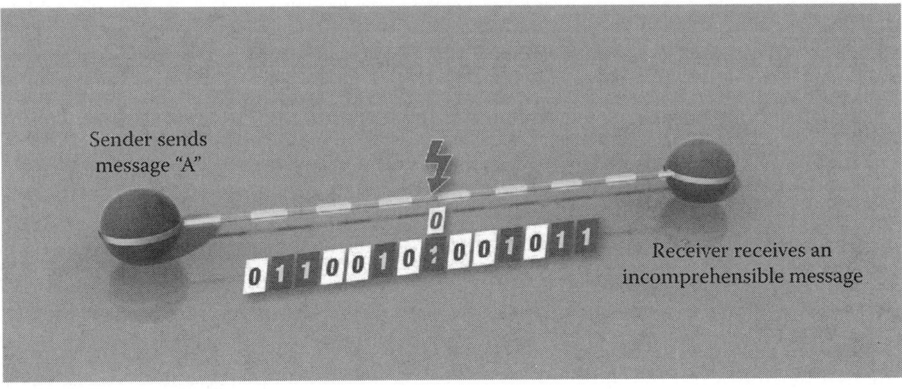

FIGURE 6.13 Loss of a message.

FIGURE 6.14 Falsification of a message.

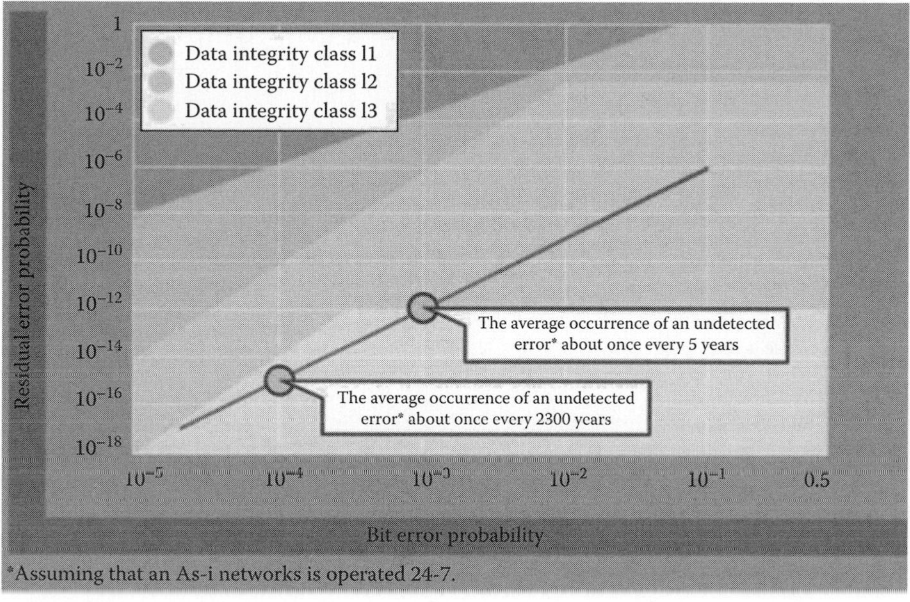

FIGURE 6.15 Data integrity classes according to EN 60870-5.

turns into a valid message B, for instance, *open valve.* This message will be accepted as seemingly correct despite error checking. The statistical probability for the occurrence of such a case is called *residual error probability.* It can create a very dangerous situation for a machine or plant and the human operator and must therefore, under all circumstances, be reduced to an extremely small value.

The standard EN60870-5 defines three data integrity classes that, depending on the kind of controlled process, require the observation of certain residual error probability limits (Figure 6.15). The residual error probability can be used to calculate the average time between two unrecognized falsified messages. Very often, this is the more intuitive value.

AS-Interface Data Integrity and Error Response.

Reliable error recognition is of great importance for faultless communication over the AS-i cable, which is generally unshielded. Error detection is done in the *receive* state.

Data checking in AS-i is accomplished according to a different principle than that used in most of the familiar bus systems. One reason for this lies in the very short telegrams that are exchanged.

The master request contains 11 bits of data to be checked and the slave response 4 bits. To check this conventionally using a hamming distance of 4 (HD = 4), you would have to send an additional 7 bits for the master telegram and 4 bits for the slave telegram. This would drastically reduce the achievable network transmission rate.

So instead, AS-i performs greater checking on the bit transmission itself. This makes use of the knowledge of redundancies in the code and the fixed lengths of the telegrams. As a result, the following errors can be distinguished (Figure 6.16):

- Start bit error
- Alternating error
- Pause error
- Information error
- Parity error
- End bit error
- Telegram length error

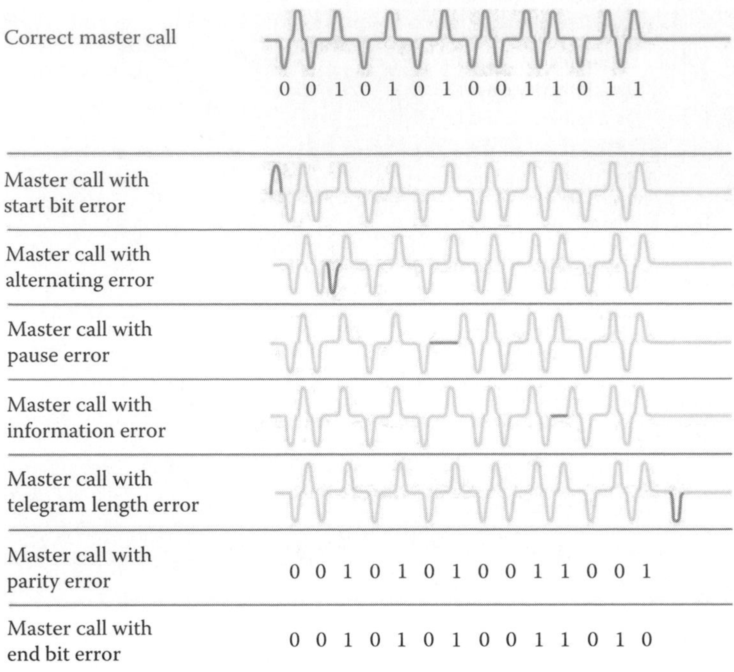

FIGURE 6.16 Error detection on the bit level.

Each master request and each slave response is subjected to these checks. If one of the errors named is detected, the request is considered faulty. The checks can be performed with just a few logical operators, which have little effect on the processing time.

If the slave receives a master request and when checking the received bit sequence detects an error, it goes into *asynchronous* state and waits for the next pause so it can then again read in the next master request in the *receive* state (see Figure 6.22). In this case, the slave does not generate a response telegram, since the error could lie in the address range of the telegram and the telegram could have been meant for a different slave. The master can determine that no slave response has arrived by means of the master pause monitoring and repeat the request if necessary.

If the master request is received without error but does not contain the correct address or contains an unrecognized command code, the slave goes into *synchronous* state and in this case does not generate a slave response.

Only if a command does in fact trigger a response in a slave is a slave response generated and returned to the master as a positive acknowledgment. If this reply is likewise received without error, the message is considered to have been successfully exchanged. It can still happen, however, that the master detects an error in the response and then repeats the request, even though the slave has already correctly executed the command.

An exception to this acknowledgment principle is the broadcast commands (see next section). Here, of course, a slave is not allowed to generate a response. The gross transmission rate of AS-i is 167 kbit/s, which enables a net data rate of maximum 53.3 kbit/s including all pauses necessary for the various functions. The resulting efficiency of the user data transfer can then be calculated as 32%. This is an excellent value compared with other bus systems.

6.2.4.4 System Interfaces

The AS-i system consists, as shown in Figure 6.17, of several hardware components that are connected to each other and to the environment. These components include the transmission system, the slaves, and the master.

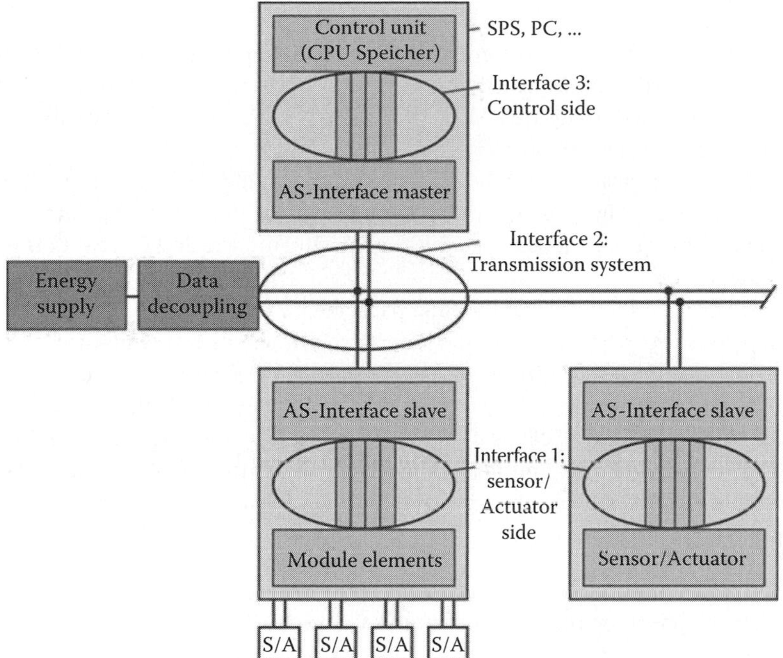

FIGURE 6.17 AS-i components and interfaces.

Transmission System

This system consists of the AS-i two-wire cable that connects the slaves to each other and to the master in freely selectable topologies, the energy (power) supply, and the data decoupling. Both data traffic and the power distribution are handled on this two-wire transmission system.

Slave

The slave uses Interface 1 to create the sensor/actuator side, the connection to the sensor or actuator, where the AS-i is configured so that the slave function can be fully integrated into the sensor/actuator.

Master

The master uses Interface 3, the control side, to create the connection to the host system, which in turn consists of a PLC, a PC, or a coupler to a higher-level fieldbus system. In addition, the master controls the entire data traffic over the transmission system and assumes responsibility for the necessary management functions.

The AS-i standard describes Interface 2 as a transmission system fully and completely. All the conditions necessary for data and power transmission are specified and all the functions for data and parameter transmission, addressing and identification, or status polling of the slaves are defined.

In the standard cited, Interface 1 is described only conceptually, with its specific realization left to the product manufacturer. Profiles, which describe the most common slave types, are used for a more precise description of how the functional interchangeability of the products from various manufacturers is ensured.

Due to the wide variety of structures that a host system can embody, Interface 3 is also described only conceptually. The specific realization is determined by the implementation in a PLC, PC, or fieldbus coupler.

Among the other components of AS-i, which are used only as needed, are repeaters, which allow an extension of the cable length to over 100 m, and GND fault detectors as well as addressing and service devices, which can be used for assigning addresses to slaves and, in case of a fault, provide quick problem diagnostics.

6.2.4.5 Transmission Layer

The transmission layer in the master is responsible for exchanging individual telegrams with the slaves. Along with the *raw* data from the sequence control, it gets data exchange jobs. When a telegram is sent, frames (start and end bit) and parity are added. The correctly received slave reply without parity, or the frame, is returned to the sequence control.

Data exchange between master and slave always consists of a master pause and a slave reply, which is followed by a send pause (Figure 6.18). The master pause is the pause that the slave needs to observe after receiving a master request. The send pause is the time during which the network is in the idle state after the end of a slave reply.

The master specification requires that the send pause be no longer than 2 bit times in undisturbed, normal operation. This ensures that the specified short cycle times can be maintained. When measuring the pause time, note that the last (positive) pulse of the slave reply starts in the middle of the end bit and the first (negative) pulse of the next master request not until the middle of the start bit. The send pause may be extended to a maximum of 500 ms if this does not result in the complete cycle time being longer than 5 ms. This increase in the cycle time for systems having fewer than 62 slaves allows dedicating more processor power for administration functions. The function for data transmission offers the sequence control two transmission methods (Figure 6.19):

One-Time Transmission

Absence of a slave response or receipt of a defective response is immediately reported. The master request is not repeated. This version is used when looking for new slaves.

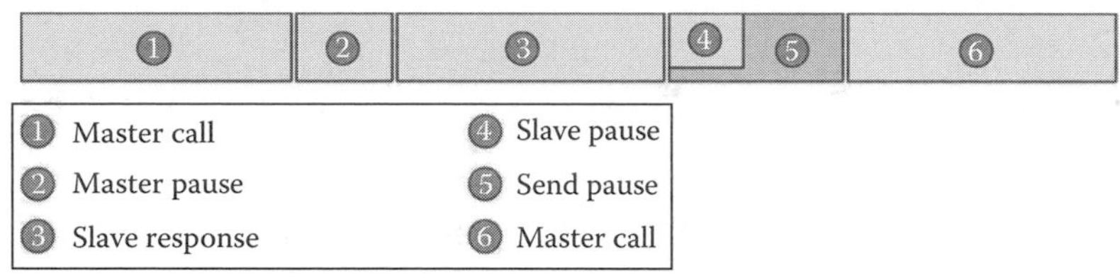

FIGURE 6.18 Timing structure of the AS-i messages.

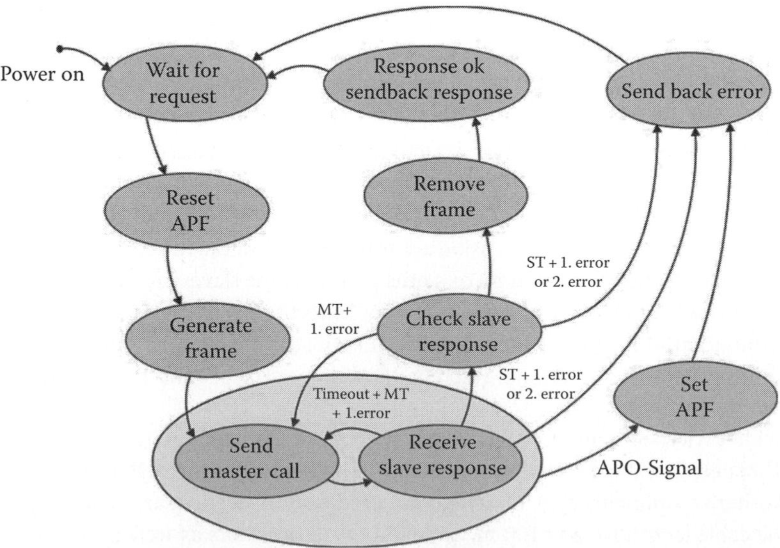

FIGURE 6.19 State diagram of the master transmission level.

Repeatable Transmission

The transmission control repeats the master request one time when a slave response is absent or is defective. This allows single errors to be caught at the lowest level without having to significantly delay the signals from individual slaves.

A transmission is considered successful if after the minimum master pause time has expired and before 10 bit times have passed a slave response begins and is then received without error. The time-out of 10 bit times takes into account the response time of the slave (master pause) of 3–5 bit times and a propagation time on the cable of one bit time. The remaining 4–6 bit times allow for the use of repeaters, which regenerate the signals.

When a slave response is received, Manchester coding, start, end, and parity bits of the incoming telegram are checked. The vast majority of all transmission errors are detected by the Manchester coding, so that only errors in multiple $\sin^2$ pulses of a telegram result in parity errors. When direct repetition of the master request results in a second error, this is reported to the sequence control, which then takes over the error handling.

During transmission of the master request and receipt of the slave response, the transmission control monitors the AS-i power *on* (APO) flag. If within this time the supply voltage on the line is too low, AS-i power fail (APF) is reported to the sequence control.

6.2.4.6 Transmission Phases

At the sequence control level, two procedures work in parallel:

1. The transmission phases for controlling data exchange with the slaves
2. Administration, which carries out the function requests on the master layer and takes care of automatic addressing

The sequence control distinguishes between two operating modes that control the behavior when activating and polling slaves:

1. In the projecting mode, all connected slaves are activated regardless of the nominal configuration and are included in the data exchange. This mode allows start-up of a system without prior projecting.
2. In the protected mode, the master communicates only with projected slaves and thereby allows for automatic checking of the configuration. Only slaves that are noted in the list of projected slaves (LPS) and whose nominal and actual configurations agree are activated. Missing or incorrectly connected slaves result in an error message; automatic addressing is possible.

After the supply voltage has been turned on, the master cycles through the various transmission phases (Figure 6.20). First, initialization takes place in the offline phase. Once initialization is complete, start-up operation begins, in which all connected slaves are detected in the detection phase and are then activated in the activation phase. Following this, the actual data exchange begins in cyclical normal operation.

Initialization

Initialization during the offline phase places the master in the base state. The data in the input data image (IDI) for all slave inputs are set to zero ("0," inactive) and the output data to one. One is the value that is also assumed by the slave outputs after turning on the supply voltage and after a reset. This ensures that the state of the slave outputs does not change when the master is turned on. Since a programmable logic controller sets outputs to zero after a reset, the *write output data* sequence control level function inverts the data that are transferred from the controller. After the supply voltage is turned on, the projected parameters are copied to the parameter field so that at the next activation the preset parameters are used.

The sequence control can be brought from any other state to the offline phase by setting the offline flag. The offline flag thus has the function of resetting the complete network and the master, which corresponds roughly to the *stop* function of many programmable logic controllers.

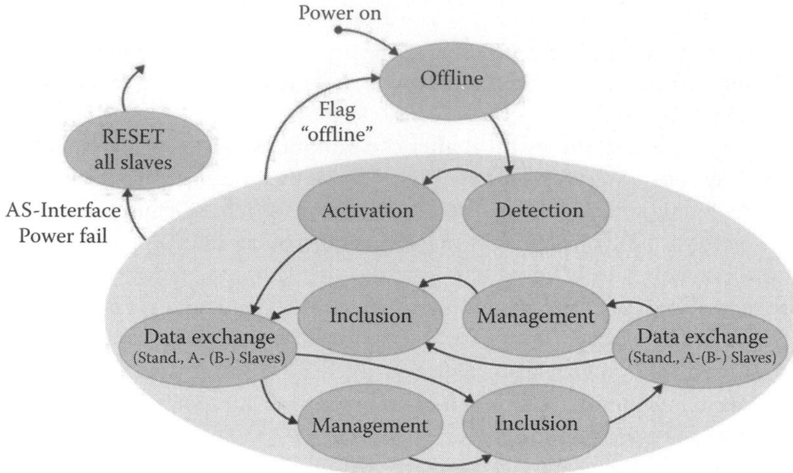

FIGURE 6.20 Transmission level phases.

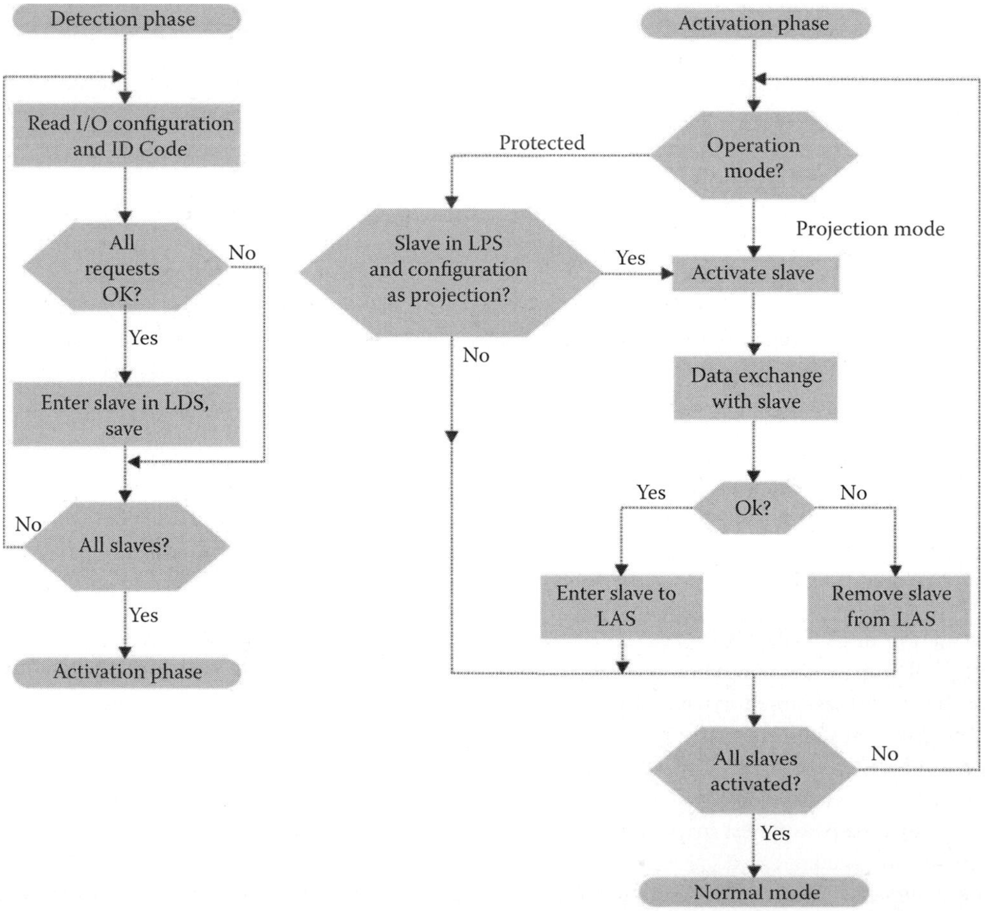

Detection phase
Read I/O configuration and ID Code
All requests OK?
No
Yes
Enter slave in LDS, save
No
All slaves?
Yes
Activation phase

Activation phase
Operation mode?
Protected
Projection mode
Slave in LPS and configuration as projection?
Yes
Activate slave
No
Data exchange with slave
Yes
Ok?
No
Enter slave to LAS
Remove slave from LAS
All slaves activated?
No
Yes
Normal mode

FIGURE 6.21 Start-up operation with detection and activation phase (simplified).

Start-Up

In start-up operation, the sequence control detects all the connected slaves and activates them. Start-up operation consists of the detection phase and the activation phase (inclusion phase); see Figure 6.21.

In the detection phase, the slaves on the AS-i string are searched for by reading the I/O configuration and ID codes. Slaves that respond to all requests are entered in the list of detected slaves (LDS). Their I/O configuration and their ID codes are stored in the configuration data image (CDI). In the activation phase, the previously found slaves are activated by writing the parameter outputs. The IDI is filled with valid values by means of a first data exchange.

In project mode, all detected slaves (with the exception of the zero address) are activated. In protected mode, only the detected slaves that are also listed in the LPS and whose I/O configurations and ID codes agree with the projected configuration data are activated. The master thus exchanges data only with slaves that are intended by the projection. At the end of the activation phase, a check is made to see whether the nominal and actual configurations agree and then the Config_OK flag is set.

Normal Mode

Normal mode is where the actual data exchange with the connected sensors and actuators takes place. A cycle consists of data exchange, management, and inclusion phase.

6.3 AS-Interface Devices: Sensors/Actuators

Communication partners of the master are slaves. They have the task to take the information that is directed to them out of the data stream and to provide their information to the master. To achieve this, the master sends out a telegram with a specific slave address and the slave with this address responds during the time allocated for this process. Thus, every message is acknowledged. After the receipt for the reception of the message was correctly received, the message is deemed correctly sent. The only exception to this rule is the master request *broadcast*. Since it goes out to all slaves at the same time, it cannot be answered.

The duration of the entire message, consisting of the master request and slave reply, is only 152 μs. Therefore, 6500 of these individual messages can be exchanged during 1 s.

Should the master not receive a reply or if the reply cannot be decoded without errors, the message can be repeated. Since the messages are extremely short, isolated repetitions do not exceed the specified time limits. This characteristic contributes considerably to the excellent noise resistance of AS-i.

6.3.1 Function Request in the Slave

The states that the sequence control can assume in the slave and the possible transitions between them are shown in the state diagram Figure 6.22. After a *reset*, the AS-i goes into the *initialize* state.

Here, it performs the following functions:

- Resetting the outputs, the internal registers, and the flags
- Loading the address, the I/O configuration, and the ID codes from the nonvolatile memory
- Switching to the state *asynchronous*

If the sequence control detects errors in reading the nonvolatile memory, it sets Flag S3 in the status register. This could happen, for example, if there is a power loss while the addressing request is being executed, so that the save in nonvolatile memory cannot be completely finished.

In *asynchronous* state, the slave examines the data stream arriving on the line and waits for a pause. In *receive* state, a data telegram is read in starting with the next start bit, and the check described in Section 6.4.5.3 is performed. If there was a receiving error, the slave returns to the *asynchronous* state, where it waits for the next pause. If no errors were detected in the check, it goes into the *decode* state.

In the *decode* state, first, an address comparison is made. Only if the addresses agree is the command analyzed and, where appropriate, a response generated. This is sent to the master in the *send* state.

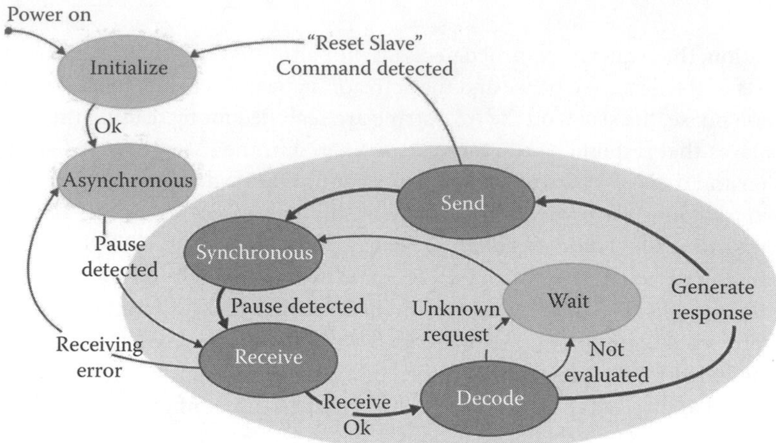

FIGURE 6.22 Main states for sequence control in the slave.

Since communication with the nonvolatile memory can take up a longer time (depending on the request and type of IC up to 500 ms), it runs in the background. It does not disturb data traffic between master and slave, but the master must still take the execution times into account.

After initializing, the slave will execute all commands but not immediately reply to data requests. Only after a valid parameter request has been received is data exchange enabled. This procedure is intended to ensure that the slave is in the desired operating state before the data are output.

6.3.2 Device Profiles

A device profile consists of data and parameters that can be exchanged with a device, as well as the status diagram that provides information about the behavior of the device on the communication network. Device profiles are independent of the characteristic of the fieldbus system being used. In the case of AS-i, these device profiles are called *slave profiles* (in contrast to *master profiles*). A slave profile specifies the following properties of a slave uniquely and bindingly:

- Which data bits carry information and what this information means
- Whether parameters are used and, if yes, what they mean
- Whether the peripheral error bit may be evaluated
- How the slave behaves

An example of a typical status diagram for a simple slave is shown in Figure 6.23. In the case of a slave that, for example, communicates digital data distributed over several cycles, the status diagram is of course more complex. But this is information that has meaning only for the master and slave developer; the user does not have to be concerned about these processes. The unique assignment of a slave to a profile is given by the combination of the I/O configuration, ID code, and extended ID Code 2.

The profile also specifies the following in many cases:

- Which logic levels are defined for the inputs and outputs of the slave module
- Which time delays need to be taken into account for in- and/or output signals
- Which pin configurations need to be observed and what the labeling of the module has to look like

Figure 6.24 shows the pin assignments provided for in the profiles for the most commonly used M12 connector method. These specifications are—as much as possible—compatible with the standards for the connected products. These additional specifications support the interoperability between products made by different manufacturers.

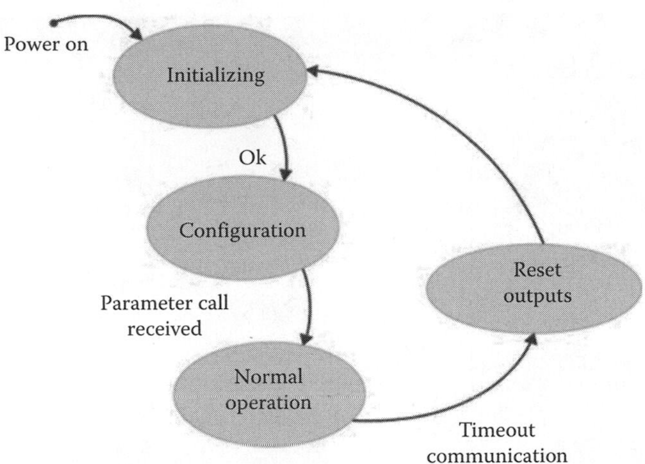

FIGURE 6.23 Status diagram of a slave according to a device profile.

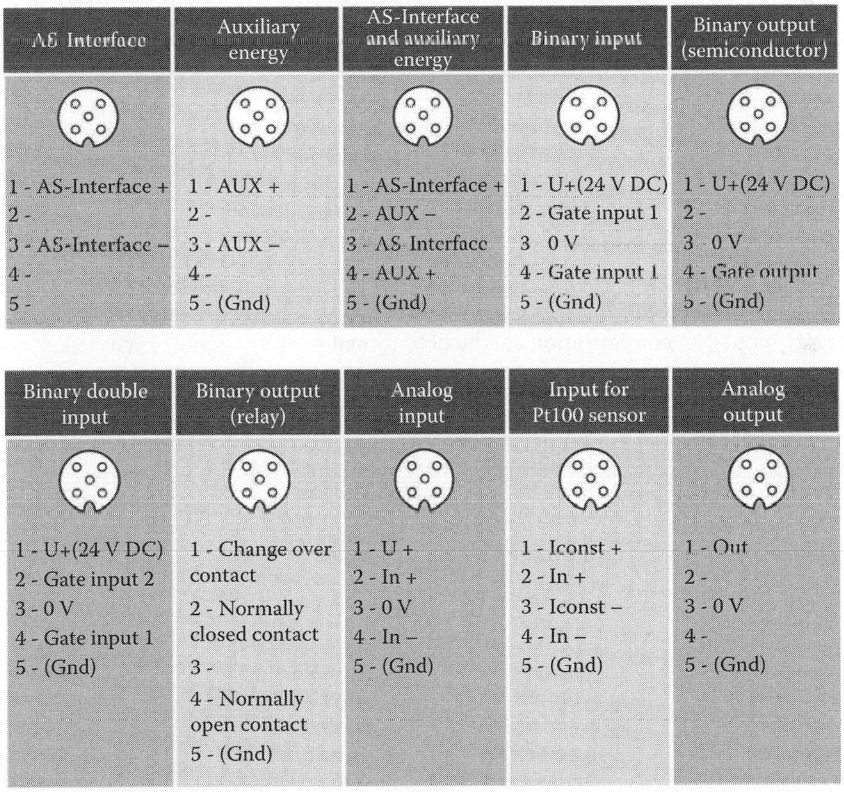

AS-Interface	Auxiliary energy	AS-Interface and auxiliary energy	Binary input	Binary output (semiconductor)
1 - AS-Interface + 2 - 3 - AS-Interface – 4 - 5 -	1 - AUX + 2 - 3 - AUX – 4 - 5 - (Gnd)	1 - AS-Interface + 2 - AUX – 3 - AS-Interface 4 - AUX + 5 - (Gnd)	1 - U+(24 V DC) 2 - Gate input 1 3 - 0 V 4 - Gate input 1 5 - (Gnd)	1 - U+(24 V DC) 2 - 3 - 0 V 4 - Gate output 5 - (Gnd)

Binary double input	Binary output (relay)	Analog input	Input for Pt100 sensor	Analog output
1 - U+(24 V DC) 2 - Gate input 2 3 - 0 V 4 - Gate input 1 5 - (Gnd)	1 - Change over contact 2 - Normally closed contact 3 - 4 - Normally open contact 5 - (Gnd)	1 - U + 2 - In + 3 - 0 V 4 - In – 5 - (Gnd)	1 - Iconst + 2 - In + 3 - Iconst – 4 - In – 5 - (Gnd)	1 - Out 2 - 3 - 0 V 4 - 5 - (Gnd)

FIGURE 6.24 Pin assignment for M12 connectors on AS-i slaves (here female).

In order to not inhibit advances or the offering of specialized modules with too narrowly defined profile definitions, the *free profile* with ID code FHex is provided. Here, essential parts of the device profile definition may be defined between the manufacturer and user.

The overview in Figure 6.25 shows which profiles are currently available. Additional profiles are permanently being reviewed and approved by the technical commission of the AS-International Association.

I/O Code	Slave-profiles	ID-Code 0	1	2	3	4	5	6	7	8	9	A	B	C	D	E	F
0	I, I, I, I	0.0	0.1									0.A	0.B				0.F
1	I, I, I, O	1.0	1.1									1.A					1.F
2	I, I, I, B	2.0										R					2.F
3	I, I, O, O	3.0	3.1									3.A					3.F
4	I, I, B, B	4.0										4.A					4.F
5	I, O, O, O	5.0										5.A					5.F
6	I, B, B, B	6.0										6.A					6.F
7	B, B, B, B	7.0	7.1	7.2	7.3	7.4						7.A	7.B		7.D	7.E	7.F
8	O, O, O, O	8.0	8.1									8.A					8.F
9	O, O, O, I	R	9									9.A					9.F
A	O, O, O, B	A.0	A									R					A.F
B	O, O, I, I	R	B.1									B.A					B.F
C	O, O, B, B	C.0										C.A					C.F
D	O, I, I, I	R	D.1									D.A					D.F
E	O, B, B, B	E.0										E.A					E.F
F	T, T, T, T	For future use															V

FIGURE 6.25 Slave profiles (i, input; O, output; B, bidirectional port; T, tristate; R, reserved).

I/O Configuration

The I/O configuration refers to the data ports on Interface 1 of the slave and is defined as shown in Table 6.6 I/O configuration. Here, *in* means a process input, *out* a process output, I/O a bidirectional behavior of the port, and *tri* a tristate output with no function. The latter state is assumed if during the reset no unambiguous I/O configuration can be determined from the slave's own data memory due to a read error.

TABLE 6.6 I/O Configuration

Code	D0	D1	D2	D3
0_{HEX}	IN	IN	IN	IN
1_{HEX}	IN	IN	IN	OUT
2_{HEX}	IN	IN	IN	I/O
3_{HEX}	IN	IN	OUT	OUT
4_{HEX}	IN	IN	I/O	I/O
5_{HEX}	IN	OUT	OUT	OUT
6_{HEX}	IN	I/O	I/O	I/O
7_{HEX}	I/O	I/O	I/O	I/O
8_{HEX}	OUT	OUT	OUT	OUT
9_{HEX}	OUT	OUT	OUT	IN
A_{HEX}	OUT	OUT	OUT	I/O
B_{HEX}	OUT	OUT	IN	IN
C_{HEX}	OUT	OUT	I/O	I/O
D_{HEX}	OUT	IN	IN	IN
E_{HEX}	OUT	I/O	I/O	I/O
F_{HEX}	TRI	TRI	TRI	TRI

Wherever an output *out* (and no bidirectional behavior) is defined, this means that the corresponding information bit is undefined in the slave response. Likewise, the information bit from the master call remains without meaning, even though it is sent, where an input *in* is defined. Unneeded information can thus be hidden with this I/O configuration. At the same time, the I/O configuration together with the ID codes for identifying slaves is employed. These refer to the slave profile. The I/O configuration is 4 bits long, is determined by the manufacturer, and is stored in the slave, where it cannot be changed.

Actual ID Code

This is used to identify various types of slaves. The ID code *A*, for example, means that the slave works in *extended addressing mode* and interprets master calls differently at times. The ID code *B* indicates a *safety at work* slave, which is designed for sending safety-related signals. The ID code is 4 bits long, is determined by the manufacturer, and is stored in the slave, where it cannot be changed.

Extended ID Code 1

This part of the ID code can be modified by the user. It can be used for identifying slaves that from the manufacturer's point of view are identical but that from the user's perspective are different. One could, for example, assign motor starters having an adjustable overcurrent limit with different ID Code 1 values, so that they cannot be confused when replacing a unit in the system. For the standard slave, this code is 4 bits long, and for a slave with extended addressing mode, 3 bits long.

Extended ID Code 2

This part of the ID code is used for extending the identification possibilities of AS-i slaves. Like the actual ID code, it is 4 bits long, is determined by the manufacturer, and is stored in the slave, where it cannot be changed.

Together with the I/O configuration, the ID codes refer to a device profile. This contains specifications for the meanings of the data and parameter bits that are expected by or sent by the slave, specifications regarding the behavior of the slave, as well as any additional definitions that support the interchange ability of slaves from different manufacturers but having the same function.

The extended ID code is not implemented for slaves conforming to Specification 2. In the case of these slaves, the master expands the missing codes to FHex.

6.3.3 Extended Profiles: Combined Transaction Types

Although most AS-i slaves get by with just a few cyclically exchanged data and acyclically sent parameter bits of information, there are numerous applications where this alone is not sufficient. For these situations, several solutions were created in Version 3.0 of the specification, which will be described in the following text in some detail.

The common denominator of these solutions is that several messages are exchanged one after the other between the master and slave, with the contents of these messages having a defined relationship. This type of communication between master and slave is therefore referred to as *combined transaction types*, abbreviated CTTs. The master recognizes from the profile of a slave that it would like to communicate in one of the defined CTTs and automatically starts with exchange of the necessary information and cyclical data transfer.

For comparison, the data and parameter exchange with a standard slave is shown in Table 6.7. The left half of the illustration shows cyclical data exchange: In each AS-i message, 4 bits of data are sent from the master to the slave and 4 bits of data from the slave back to the master. In the worst case, the slave is contacted every 5 ms with a data telegram, so that the cycle time is maximum 5 ms.

The right half of the illustration shows the acyclic data exchange using the parameter call. Here again, 4 bits of parameter information are sent from the master to the slave. In the new Version 3.0 of the specification, however, the back transfer of 4 bits of parameter data from the slave to the master is also defined, so that these bits may be used for indicating, for example, simple diagnostics information.

Table 6.7 Simple Transaction

	Cyclic Data		Acyclic Data (Parameter)		
	Data Length	Cycle Time		Data Length	Transfer Rate
⇨ Binary output	4 bit	5 ms	➡ Binary output	4 bit	25 Bd
⇦ Binary input	4 bit	5 ms	⬅ Binary input	4 bit	25 Bd

Since only one parameter telegram per AS-i cycle can be sent and, in the worst case, each slave has to be provided with parameter data, a minimum transfer rate of 25 baud (bits/s) in each direction results. Reference to a cycle time does not make sense here, since these data are not sent cyclically.

To exchange cyclical digital data, such as a measured and digitized temperature value, a counter state, a rotational speed set point, or a displacement or angle value, the master has been equipped with the *analog input data image* (AIDI) and the *analog output data image* (AODI), each having a capacity of 124 words (see Section 6.4.4).

For the acyclic data, which are typically diagnostic data, parameter data, or data that identify the slave in detail, the masters have a data channel that is constructed similar to the familiar parameter channel. It allows data packets with lengths up to 32 bytes to be exchanged with the slave.

In the meantime, five different possibilities for byte-by-byte data communication between the master and slave have been specified by AS-International. These are described in greater detail in the following. The data amounts shown in the illustrations are net data, and the cycle times, the maximum times that may occur in the worst case. The prerequisite for using these transaction types is having a master that supports them.

CTT1

Combined transaction type 1 (CTT1) is intended for simple and complex sensors or actuators such as temperature, pressure or flow transmitters, displacement and angle measurement systems, scanners, speed-controlled motors, analog process valves, and displays with up to eight characters or similar field devices. The slave can only be operated in standard addressing mode. It is not possible to change the transfer direction for the cyclical data on the fly.

Whereas slaves having Profile S-7.3 can only receive or send cyclical data and setting of parameter is only possible using the 4-bit parameters (Table 6.8), Profile S-7.4 offers expanded possibilities for communication. Here, you can switch to the parameter channel to query the ID data and diagnostic data for the slave and exchange parameter data records (Table 6.9). The ID data contain among other things the vendor ID and the device ID in code. Furthermore, the operating mode of the slave (analog mode, transparent mode, 4-bit mode) is transmitted.

In analog mode, the available number range of –32768 (8000Hex) to 32767 (7FFFHex) is generally divided into five ranges (Figure 6.26 shows this using the example of a 4–20 mA signal):

1. Error range *wire break*. The measured value is invalid, and the overflow bit is set.
2. Underrange. The measured value is valid but has a greater measuring uncertainty because it is outside the nominal range.
3. Working range. The measured value is acquired at the specified accuracy.
4. Overrange. The measured value is valid but has a greater measuring uncertainty because it is outside the nominal range.
5. Error range *short circuit*. The measured value is invalid, and the overflow bit is set.

The exact limits of the nominal and over-/underflow ranges as well as the error ranges are defined in the vendor's data sheet. The default value, which is substituted in the receiver, for example, when communication is interrupted, is $7FFF_{Hex}$.

In transparent mode, 16 bits of data can be transmitted in any desired format. There is no over- or underflow state of the measured value, and the corresponding bit is not set. The default value, which is substituted in the receiver, for example, when communication is interrupted, is 0000_{Hex}.

TABLE 6.8 CTT1 (Profile S-7.3) for Simple Sensors or Actuators

Profile S-7.3 (Input)

Cyclic Data			Acyclic Data (Parameter)		
	Data Length	Cycle Time		Data Length	Transfer Rate
			➡ Binary output	4 bit	25 Bd
			⬅ Binary input	4 bit	25 Bd
⇦ Digital input	2 byte … 8 byte	70 ms … 280 ms			

Profile S-7.3 (Output)

Cyclic Data			Acyclic Data (Parameter)		
	Data Length	Cycle Time		Data Length	Transfer Rate
			➡ Binary output	4 bit	25 Bd
			⬅ Binary input	4 bit	25 Bd
⇨ Digital output	2 byte … 8 byte	70 ms … 280 ms			

TABLE 6.9 CTT1 (Profile S-7.4) Sensors or Actuators with Need of Parameter Settings

Profile S-7.4 (Input)

Cyclic Data			Acyclic Data (Parameter)		
	Data Length	Cycle Time		Data Length	Transfer Rate
⇨ Binary output[a]	4 bit	5 ms[b]	➡ Binary output	4 bit	25 Bd
⇦ Binary input[a]	4 bit	5 ms[b]	⬅ Binary input	4 bit	25 Bd
			➡ Digital output[a]	2 byte … 32 byte	200 Bd
⇦ Digital input[a]	2 byte … 8 byte	70 ms … 280 ms[b]	⬅ Digital input[a]	2 byte … 32 byte	200 Bd

Profile S-7.4 (Output)

Cyclic Data			Acyclic Data (Parameter)		
	Data Length	Cycle Time		Data Length	Transfer Rate
⇨ Binary output[a]	4 bit	5 ms[b]	➡ Binary output	4 bit	25 Bd
⇦ Binary input[a]	4 bit	5 ms[b]	⬅ Binary input	4 bit	25 Bd
⇨ Digital output[a]	2 byte … 8 byte	70 ms … 280 ms[b]	➡ Digital output[a]	2 byte … 32 byte	200 Bd
			⬅ Digital input[a]	2 byte … 32 byte	200 Bd

[a] selectable.

[b] Cycle times apply as long as no acyclic data is being exchanged.

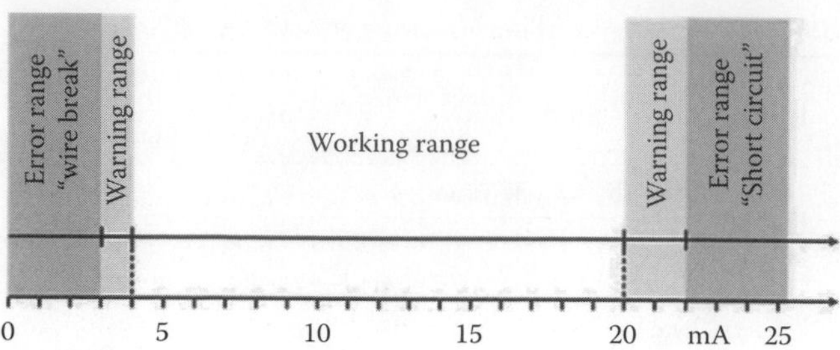

FIGURE 6.26 Warning and working ranges of an analog sensor having a 4–20 mA signal.

In a 4-bit mode, the slave behaves like a standard slave, cyclically exchanging 4 bits of data in both directions. Here, the CTT1 is used only for identifying the slave, for detailed diagnostics, and to set parameter.

The worst-case times and data transfer rates shown in Tables 6.8 and 6.9 can be halved (resp. doubled) through optimal use of the possibilities of the AS-i, namely, when the slave sends back its response to the master call in the same message and does not allow an entire cycle to elapse.

It should be noted that switching between cyclical data communication and parameter communication can take a certain time (several 100 ms) and that during acyclic data exchange no cyclical data can be sent.

CTT2

Combined transaction type 2 (CTT2) is intended for simple and complex field devices such as temperature regulators, displays with more than eight characters, speed-controlled motors with actual value feedback, analog process valves with position feedback, or similar field devices in which digital data need to be sent in both directions (Table 6.10). CTT2 can also serve as an alternative to a HART

TABLE 6.10 CTT2 for Combi Field Devices (Not Shown Here: S-B.A.E)

Profile S-7.5.5					
Cyclic Data			Acyclic Data (Parameter)		
	Data Length	Cycle Time		Data Length	Transfer Rate
⇨ Binary output	2 bit	5 ms	➡ Binary output	4 bit	25 Bd
⇦ Binary input	2 bit	5 ms	⬅ Binary input	4 bit	25 Bd
	1 byte	160 ms		2 byte	
⇨ Digital output	…	…	➡ Digital output	…	33 Bd
	8 byte	720 ms[a]		32 byte	
	1 byte	160 ms		2 byte	
⇦ Digital input	…	…	⬅ Digital input	…	33 Bd
	8 byte	720 ms[a]		32 byte	
Profile S-7.A.5					
Cyclic Data			Acyclic Data (Parameter)		
	Data Length	Cycle Time		Data Length	Transfer Rate
⇨ Binary output	1 bit	10 ms	➡ Binary output	3 bit	10 Bd
⇦ Binary input	2 bit	10 ms	⬅ Binary input	4 bit	13 Bd
	1 byte	320 ms		2 byte	
⇨ Digital output	…	…	➡ Digital output	…	17 Bd
	4 byte	720 ms[a]		32 byte	
	1 byte	320 ms		2 byte	
⇦ Digital input	…	…	⬅ Digital input	…	17 Bd
	4 byte	720 ms[a]		32 byte	

[a] Cycle times apply as long as no acyclic data are being exchanged.

interface to a field device, albeit with some minor limitations. The slave can be operated in standard addressing mode but not in extended addressing mode. Data exchange is full duplex, that is, both binary data and digital input and output data can be sent at the same time. No special switching of the data transfer direction is required.

This transaction type allows use of *combi field devices*. This includes devices that have both binary and digital input and output capability. The binary inputs and outputs can be used, for example, for rapid display of an over- or underlimit violation, and the simultaneously available digital transmission of the precise measured value can be used for diagnostics and maintenance purposes.

The parameter channel can be used in this transaction type as well to transmit the ID data for the slave including vendor ID, product ID, and operating mode. Comprehensive diagnostics and parameter data records can also be exchanged. If acyclic digital data are exchanged, the cyclical digital data traffic is interrupted for the duration of the transmission, whereas the cyclical transmission of the binary data remains unaffected.

The data formats used are the same as for CTT1.

The worst-case times and data transfer rates shown in Table 6.10 can be halved (resp. doubled) through optimal use of the possibilities of the AS-i, namely, when the slave sends back its response to the master call in the same message and does not allow an entire cycle to elapse.

In order to handle the resulting variety of parameters, work is currently underway to incorporate AS-i into the FDT standard. This will result in an OEM neutral platform, which can be used to conveniently set parameters and process remote diagnostics and maintenance.

CTT3

Combined transaction type 3 (CTT3) is intended for binary sensors, actuators, and field devices including keypads with signal light feedback, valve terminals, status indicator columns, and similar devices where binary data need to be transmitted in both directions. Simple analog sensors or actuators can also be operated using CTT3 if the resolution of 8 bits is sufficient. The slave is operated in extended addressing mode (Table 6.11).

Profile S-7.A.7 makes the output data bit *lost* due to the extended addressability available again. This is, however, at the expense of response time, which for this transaction type can be up to maximum 40 ms.

The worst-case times and data transfer rates shown in Table 6.11 can be halved respectively, doubled through optimal use of the possibilities of the AS-i, namely, when the slave sends back its response to the master call in the same message and does not allow an entire cycle to elapse.

If one were to equip an AS-i network with 62 slaves having Profile S-7.A.A, a total of 496 binary inputs and 496 binary outputs would be available, that is, nearly 1000 I/Os in all.

TABLE 6.11 CTT3 for Keypads, Valve Terminals, and Similar Devices

Profile S-7.A.7					
Cyclic Data			Acyclic Data (Parameter)		
	Data Length	Cycle Time		Data Length	Transfer Rate
⇨ Binary output	4 bit	40 ms	➡ Binary output	3 bit	10 Bd
⇦ Binary input	4 bit	10 ms	⬅ Binary input	4 bit	13 Bd
Profile S-7.A.A					
Cyclic Data			Acyclic Data (Parameter)		
	Data Length	Cycle Time		Data Length	Transfer Rate
			➡ Binary output	3 bit	10 Bd
			⬅ Binary input	4 bit	13 Bd
⇨ Digital output	1 byte	80 ms			
⇦ Digital input	1 byte	80 ms			

TABLE 6.12 CTT4 for Single- or Dual-Channel Analog Sensors

Profile S-7.A.8, S-7.A.9 (Input)					
	Cyclic Data		Acyclic Data (Parameter)		
	Data Length	Cyclic Time		Data Length	Transfer Rate
			➡ Binary output	3 bit	10 Bd
			⬅ Binary input	4 bit	13 Bd
⇦ Digital input	2 byte or 4 byte	80 ms 160 ms			

CTT4

Combined transaction type 4 (CTT4) is intended for simple analog sensors such as temperature, pressure or flow transmitters, and similar devices in which digital data only need to be transmitted from the slave to the master. The slave is operated in extended addressing mode (Table 6.12).

This transmission type is an extension of CTT1.

CTT5

Combined transaction type 5 (CTT5) is optimized for fast data transmission. It can be used for fast sensors such as angle encoders, actuators, and field devices that are, for example, components in dynamic digital control loops. 8, 12, or 16 bits of data can be sent full duplex in 5 ms (Table 6.13). This is organized as follows using several slave addresses: The lowest address of 2, 3, or 4 successive addresses is the physical address of the slave. This slave uses the next higher addresses for widening its data channel. In this way, a digital value of up to 16-bit resolution can be sent in full in a single AS-i cycle. The *shadow addresses* may not of course be assigned to other slaves. The slave occupies 2, 3, or 4 standard addresses.

The output value is provided in the output data image (ODI) of the slave with the lowest address and the input value in the corresponding field of the AIDI. The fields of the *shadow slaves* remain empty.

In the case of slaves that use this transaction type, it must be noted that the mechanisms for automatic addressing do not work in the same way as for other slaves. It may therefore be necessary for the user to take other measures if he or she is using slaves with CTT5.

Communication with safe slaves is also part of CTT. Here, however, communication takes place not primarily between the master and the slave but rather between the slave and the safety monitor. This is covered in detail in Section 6.4 of this book.

TABLE 6.13 CTT5 for Fast Analog Sensors, Actuators, and Field Devices

Profile S-6.0.X					
	Cyclic Data		Acyclic Data (Parameter)		
	Data Length	Cyclic Time		Data Length	Transfer Rate
			➡ Binary output	4 bit	25 Bd
			⬅ Binary input	4 bit	25 Bd
⇨ Digital output	1 byte ... 2 byte	5 ms			
⇦ Digital input	1 byte ... 2 byte	5 ms			

In summary, there are now various complementary methods available for simply exchanging analog data, parameter data, ID data, and detailed diagnostic data over AS-i. The task of the product developer is to select the most appropriate procedure for the task at hand from among the variety of possibilities. For the user, this means new types of communication are available that are as simple to use as one has come to expect from AS-i. *Click and go* is now here for the analog world as well.

6.3.4 Device Addressing

The address of every AS-i slave must be set before it is connected to the network. An exception is the replacement of a defective slave in the network during operation (see *automatic addressing*). Addressing is not performed with DIP switches, which are commonly used in other fieldbus systems, but requires the use of an addressing device. The addressing device is connected to the addressing and diagnostics interface, reads the address and the profile of the slave, and enables writing of the target address into the EEPROM of the slave. In general, the addressing device has the approximate size of a multimeter and is battery powered, allowing addressing and diagnostics directly at the location of the slave.

For reasons of compatibility with older versions of the AS-i specifications, two different addressing methods are offered:

1. Standard slaves: with an address between 1 and 31
2. A/B slaves: share the address of a standard slave

For this reason, addresses 1A to 31A as well as 1B to 31B are provided. A/B slaves can be recognized by the ID code "A" (refer to Section 6.3.1.2).

An address can either be used by a standard slave, an A slave, a B slave, or an A and a B slave at the same time.

Every slave address is preset to 0 by the manufacturer. Before an AS-i slave can be connected to the AS-i network, it requires an address between 1A and 31B to enable data exchange with the master.

This address is stored in nonvolatile memory (EEPROM) in the slave. This is the reason why switches to set the address are not used on AS-i. The address can be sent by a master command or the addressing interface on the slave could be used. This interface should really be called *addressing and diagnostics interface*, because it can be used not only to write a new address into the slave but also to thoroughly test all available functions of the slave.

Addressing Interfaces

Due to the large number of very different slaves in the AS-i environment, multiple slave addressing alternatives are available (Figure 6.27):

Alternative 1

Slaves with an M8 or M12 connection to the AS-i network can use this interface to connect to the addressing device and receive their target address. This alternative is usually used on compact IP67 devices designed for harsh environment application.

For addressing, the device has been designed to directly plug into the addressing device. During this process, the addressing device has to supply power to the slave.

Alternative 2

If the slave has a standardized addressing jack, it can be connected to the addressing device using a matching cable to receive its address. This alternative is most commonly used for slaves that are intended for enclosures applications. For IP67 modules, the addressing plug is located behind a cover that needs to be closed thoroughly after use. When connecting the cable to the addressing jack, the slave is separated from the network and connected to the addressing device through the internal contacts in the plug. The addressing device supplies power to the slave as well. The advantage of this alternative is that a slave, already installed and fully wired, does not have to be separated from the network and its peripheral devices.

FIGURE 6.27 Standardized addressing interface.

Alternative 3

Many slaves are equipped with an infrared interface that can be used for addressing. By employing an IR adapter that can be connected to the addressing device instead of the mechanical addressing plug, the connection to the slave can be established and the desired address can be programmed. This alternative is most commonly used on field devices for IP67 environments, since the protections class is not reduced.

Also, this alternative requires no separation between the device and its periphery. The slave recognizes the request to communicate via the IR interface through a special data sequence (*magic sequence*) and interrupts normal communications with the master. During the addressing process, the power for the slave is still supplied by the AS-i network. This can potentially be an advantage for devices with high-power requirements.

Addressing Device

Depending on the device version, an addressing device offers the opportunity to test the connected slave, that is, to control the outputs and read the input information.

Network diagnostics devices are used to search for specific network errors (for instance, electrical connection problems or double addressing). When necessary, these devices can be used to determine the network status, to count and statistically evaluate telegram errors on the network, or to record data traffic in order to identify the causes of rare and sporadically occurring network interruptions.

Some of the diagnostics characteristics mentioned here are already integrated into available masters. For instance, there are masters that permit addressing of individual slaves through the user interface or masters that, for the purpose of statistical evaluation record, collect and store telegram errors over longer periods of time.

Automatic Addressing

Automatic addressing is an important feature of AS-i. It allows a failed sensor or actuator to be replaced without requiring a special procedure for assigning an address to the new one. This speeds up the repair process immeasurably.

If a slave fails (e.g., mechanical damage), it must be replaced by a slave of the same type whose address is set to zero. If exactly one slave has failed, the master can then automatically perform the necessary parameterization of the operating address. The prerequisite for automatic addressing is that the network is projected in the master and it therefore has information about the type and addresses of the correctly operating slaves in the network. Since reassigning of slave addresses is a critical procedure that could also result in errors, some restrictions need to be noted (Figure 6.28, Automatic addressing).

Automatic addressing is possible if in protected mode one projected slave is missing in the LAS and automatic addressing is enabled (Auto_Prog set). This state is indicated by the Auto_Prog_Available flag.

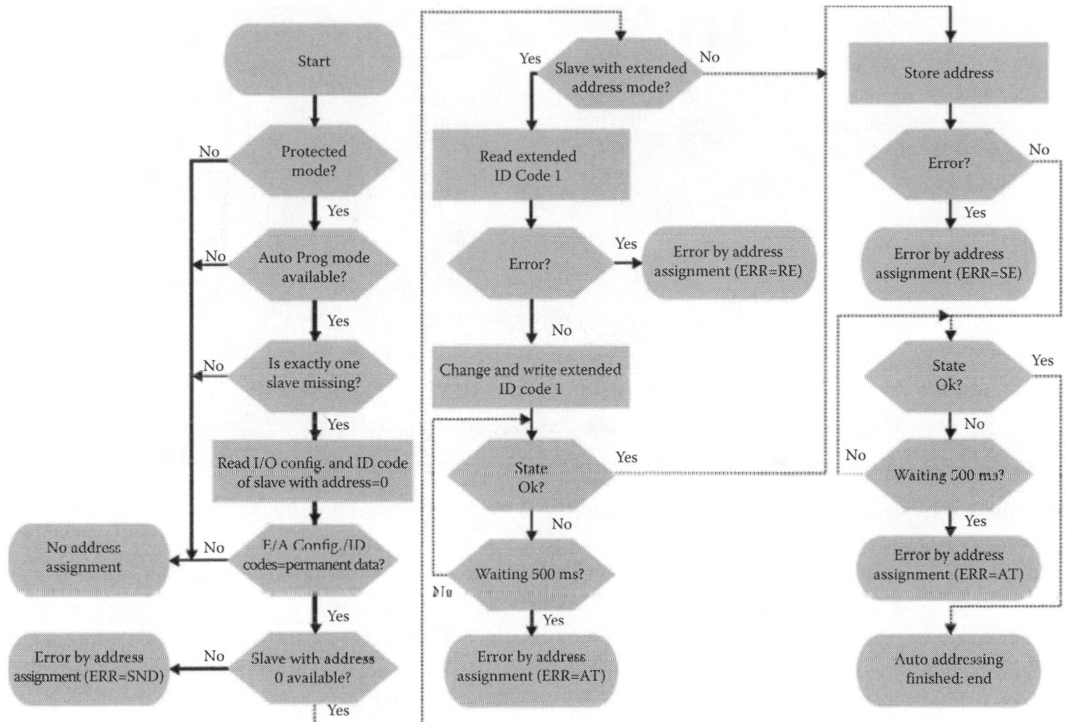

FIGURE 6.28 Automatic addressing.

If all the conditions are met, then when a slave with address zero is connected, the sequence control performs the "Change_operating_address" function to set the slave with address zero to the address of the previously failed slave. Of course, the new slave must be of the same type as the old one (I/O configuration and ID codes correspond to the projected configuration data).

6.3.5 Device Coupling to Sensors/Actuators

During each cycle, 4 data bits are sent back and forth between each slave and the master. Corresponding to these are the voltage levels on the ports of the AS-i slave IC. These ports can be configured for each IC individually as inputs or outputs or as bidirectional ports. The configuration of the ports is specified as *I/O configuration* when the IC is installed in the slave. This means the same AS-i slave IC can control both sensors and actuators.

The use of AS-i can also be expanded to include peripheral elements having more than 4 bits of information, analog devices, or decentralized intelligent devices with low information exchange needs for control. In bidirectional mode, a slave can receive 4 bits of output data per request and send back 4 bits of input data. Multibit transmissions can be realized by distributing the information over multiple cycles. AS-i can thereby be used for all simple sensors and actuators while not being limited to binary peripherals.

There is also a *write parameter* request, which sends 4 bits as parameter values to the slave. This can be used to control special functions of the slave, such as changing the switching sensing distance for optical sensors or a filter function. Parameter requests are acyclical, meaning only one slave can get new parameter values in each AS-i cycle.

Integrated Sensors and Actuators

The AS-i slave IC can be integrated directly into sensors and actuators. This results in a sensor or actuator with *integrated* AS-i (Figure 6.29). All the data and parameter bits are available to this device.

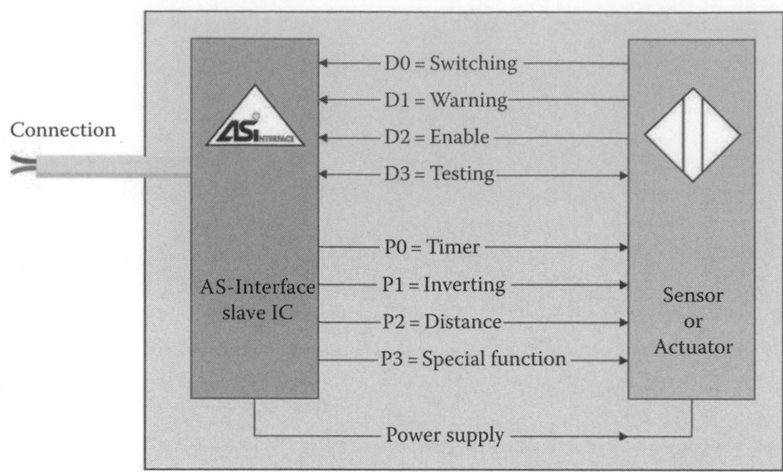

FIGURE 6.29 Sensor or actuator with integrated AS-i.

The communications structure—the AS-i network—is then used not only for transmitting a binary switching bit but also for additional information. Thus, a gradual failure (e.g., contamination of an optical sensor), a spontaneous failure (e.g., a short circuit), an overload on the power supply, and similar cases are reported to the controller and protected against over AS-i. This allows user systems to be constructed in which certain error types cause immediate shutoff in order to save tools and machinery and to prevent longer downtimes. Other error types can generate a maintenance request for preventing any downtime at all (e.g., contamination indicator) or to provide targeted diagnostics to help reduce the downtimes.

From this perspective, the use of *intelligent* peripheral components is especially effective for the user. Only these will allow a diagnosis that includes the sensor or actuator.

Connection by Module Technology

The AS-i IC can be integrated into a module so that conventional sensors and actuators can also be connected to it: *sensor or actuator with external AS-i* (see Figure 6.30). Most of the sensors and actuators currently available on the market can be easily connected to any AS-i network through modules. These modules provide power and data on the device side through sockets or terminals. Shown is a 2I/2O module to which 2 binary sensors and 2 binary actuators are connected. An optional watchdog can be activated and deactivated using Parameter 0. It monitors the communication and disconnects the actuators from power when there is a communication failure.

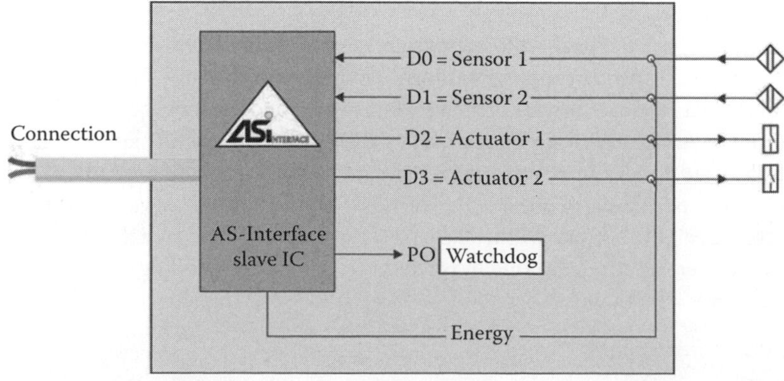

FIGURE 6.30 Module for coupling sensors and actuators without their own slave IC to the AS-i network.

6.3.6 Structure of the AS-Interface Slave

As shown in Figure 6.31, the AS-i slave represents the link between the AS-i transmission system and Interface 1, to which the sensors and actuators are connected. The slave powers the sensors/actuators and handles communication between them and the master. It was clear from the very beginning of the development that the slave, if it was to be integrated directly into the sensor or actuator, needed to be small and compact as well as extremely inexpensive. This can be achieved only through the use of a highly integrated circuit, which will be described at the end of this section.

Regardless of whether the slave is implemented using the integrated circuit or some other way, it will have the structure shown in Figure 6.31 and will have the following connections, whereby the connections for Interface 1 may be physical or logical, and those for Interface 2 must be physically present.

In the supply voltage, the data path is separated from the power path by an inductance. In order to implement it in an IC, this inductance is realized electronically. The supply voltage provides the power for the sensor or actuator on the V_{out} terminal and ensures that the slave remains of sufficiently high resistance in the frequency band of data communication.

In the receiver, the voltage pulses detected on the AS-i cable are filtered, digitized, and written to the receive register. At the same time, the received signal is subjected to various plausibility checks to ensure that no noise pulses have corrupted the master request. Since the signal amplitude can fluctuate over a fairly wide range, the receiver has adaptive response to the trigger level to improve receiving integrity.

In the sender, the information from the send register is encoded and sent out over the AS-i cable as a current pulse sequence according to the modulation procedure described in Section 6.2.2.3. For security reasons, the sender is monitored by *jabber inhibit* logic, which prevents continuous sending and therefore blocking of the AS-i network by a defective slave.

The sequence controller finally decodes the master requests, checks them for errors, carries out the commands encoded in them, and, if appropriate, causes a reply to be sent. The sequence controller

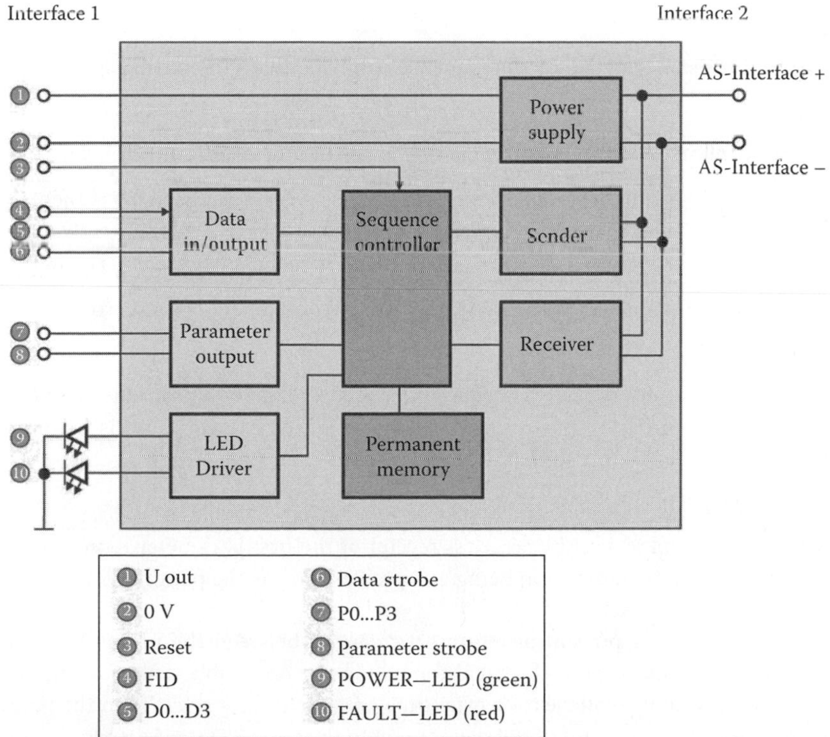

FIGURE 6.31 AS-i Slave block diagram.

also has a memory that is used for recording the slave address and that can store it for an unlimited time without power (nonvolatile).

Address Register

This 5-bit-wide register contains the current slave address. If the address sent in a master request agrees with the address contained in this register, then this slave will reply. After a *reset*, the register is loaded with the address contained in the nonvolatile memory. Its contents can be changed by the master using the commands *delete address* or *address assignment*.

Identification Register

These 4-bit-wide registers contain the I/O configuration shown in Table 6.6 and the ID codes of the slaves. They are permanently stored and are loaded from the nonvolatile memory after a *reset*. They are (with the exception of ID Code 1) fixed at the time the slave is manufactured and cannot be modified.

Data Output Register

The *data output* register is 4 bits wide and contains the data from the last *data request*, which was received without error by the slave. Those bits that are allocated to an output in accordance with the I/O configuration are output on the respective data port, and the information from the other bits is ignored. After a *reset*, the register is loaded with the default value FHex.

Parameter Output Register

This 4-bit-wide register contains the parameters from the last *write parameter* request from the master that was received without error by the slave. The bits are output on the corresponding parameter ports.

Receive Register

The 12-bit-wide receive register contains the last message sent by the master for further processing in the sequence controller.

Send Register

The 5-bit-wide send register is loaded by the sequence controller with the slave response to be sent.

Status Register

The status register contains three flags for indicating certain status conditions or errors:

1. Flag S0 is set during address storage if the new address has not yet been permanently stored.
2. Flag S1 is set if the input FID reports a peripheral error.
3. Flag S3 is set if an error has occurred while reading the address from the permanent memory.

The status register can be evaluated by the master for the purpose of analyzing error status conditions.

Synchronization Flag

If the slave has correctly received a master request, decoded it, and, if appropriate, acknowledged with a reply, the *synchronization* flag is set. In synchronized state, the master pause is monitored after the master request for only one bit time so that the slave response can start after 2 bit times.

Data Exchange Blocked Flag

This flag is set by a *reset* and is reset by errorless receipt of the first parameter request to its own slave address. This prevents data requests from being accepted as long as the parameter ports have not been loaded with the nominal parameters.

This behavior is necessary for preventing *misunderstandings* between the master and slave. It could happen, for example, that due to poor electrical contact on the AS-i cable, a slave—without the master knowing it—receives no supply voltage for a brief time and performs a *reset*. Then the parameter ports are reset and any associated functions of the slave are set to the default state. As a consequence, the slave could respond differently than the master expects.

6.3.7 Device Interface and Diagnostics

As already indicated by Figure 6.31, Interface 1 of the slave has four data ports, which, depending on the selected I/O configuration, can be used either as input ports, as output ports, or for bidirectional communication. A data strobe output is also provided, which signals when output data are present and when input data are expected.

For actuator slaves, it is recommended that the time-out monitor, also called a watchdog and that is integrated in the slave IC, be activated. If, while a timing member is running, no new correctly received *data request* arrives at the address of the slave, the actuator can use the watchdog to place the system in a safe state. Such time-out monitoring allows a variety of error possibilities to be covered, such as hardware faults in the master, interference on the transmission cable, or loss of the address in a slave. Where necessary, an AS-i system can thus be made more secure. A time of 40–100 ms is specified as the time-out period.

In addition to the data ports, which are provided for cyclical data exchange with the master, additional ports are available, which are used for (acyclical) parameter output. Here again, an additional strobe output indicates when a new parameter message has arrived.

The FID input is used to signal peripheral errors. If the slave electronics detect an error (such as an overloaded supply voltage caused by an external short), this input can be used to display the event locally (the red error LED on the slave flashes) and to report it to the master using the status register. The master makes an entry in the list of peripheral errors and sets a collective flag. It is the job of the error handling routines in the master or host system to recognize this situation and take appropriate measures (such as reactivating the slave or reporting the error to the operator). In the case of serious errors, the reset input on the slave IC can also be activated. This will, however, generate a network configuration error.

Finally, a DC voltage is provided to the connected sensor or actuator on the V_{out} port, which generally lies within the tolerance range of 24 V + 10%/−15%.

The new uniform status display on the AS-i slaves (Figure 6.32) supports rapid fault analysis and thus minimizes error times.

Symptom	Possible Reason	Flags in the Master		Indication at the Slave	Indication at the Slave
		Config. error	Periph. fault	Standard	Extended
Normal operation	Everything Ok?	Reset	Reset		
No data exchanges	Master in STOP Mode Slave not in LPS Slave with wrong IO/ID Reset am Slave aktiv	Set	Reset		
No data exchanges because Address = 0	Slave address = 0	Set	Reset		
Peripheral fault	To be defined by the manufacturer	Reset	Set		Alternierend
Serious peripherial error with reset	To be defined by the manufacturer	Set	Undefined		

FIGURE 6.32 Status display on the slave.

6.4 AS-Interface Master: Coupling to Other Automation Systems

The AS-i master establishes the connection between the network and the higher-level control system (Figure 6.33). It independently manages the data traffic on the AS-i cable. Special expertise about baud rates and extensive files with the description of the specific characteristics of the bus devices are not necessary for start-up. Due to the specific design of the AS-i master, the network is mostly self-managed. Additionally, the AS-i master supports *live insertion* of new slaves during operation and allows the exchange of defective slaves during operation (*hot swap*) by using automatic address assignment (see Section 6.3.1.4).

6.4.1 Function Sequence in the Master

In normal operation, after conclusion of the detection and activation phase (which is described in greater detail in Section 6.2.4.6), cyclical communication takes place between the master and all connected slaves. Such a cycle consists of the data exchange phase, the (optional) management phase, and the inclusion phase. The two data exchange phases illustrated in Figure 6.34 differ from each other, only when A and B slaves exist in the network at one address. In the data exchange phase, the master queries all activated slaves one after the other with a data call; the states from the ODI are sent to the slave and the input signals that are contained in the slave response are stored in the IDI. If the transmission level reports an error, the call is repeated in up to three successive cycles. If the third attempt also fails, it is assumed that the corresponding slave is absent or defective. The slave is then removed from all master lists and the IDI is set to zero. This error handling spread out over several cycles is necessary because given these short telegrams a noise burst with extremely high amplitude can interfere with several cycles.

The repetition of the calls also ensures reliable operation under these conditions. Individual noise pulses or interference such as is typically encountered in industrial environments are caught by the immediate repetition already on the transmission level without having a noticeable effect on the time response.

After the data exchange phase is finished, one can transition, if needed, to the management phase. In the management phase, *acyclic* telegrams are sent. Exactly one master call can be performed per cycle.

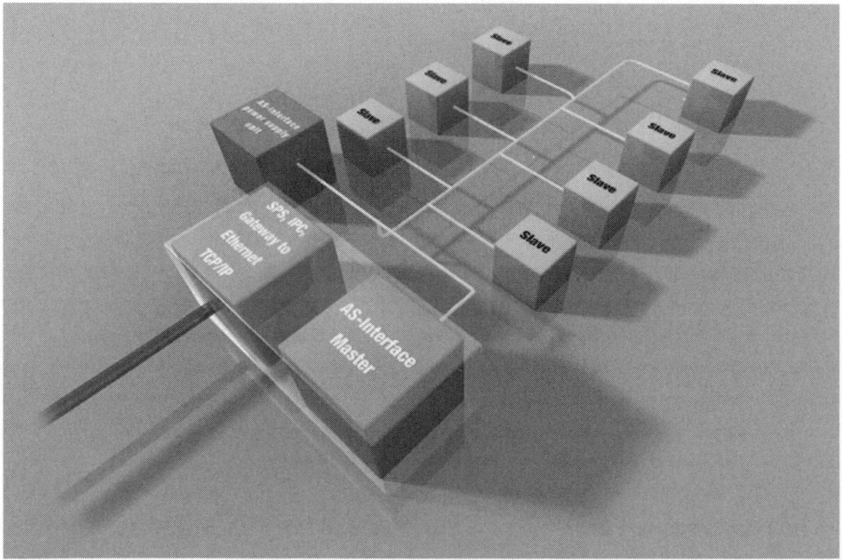

FIGURE 6.33 The AS-i master provides the device data to the control unit.

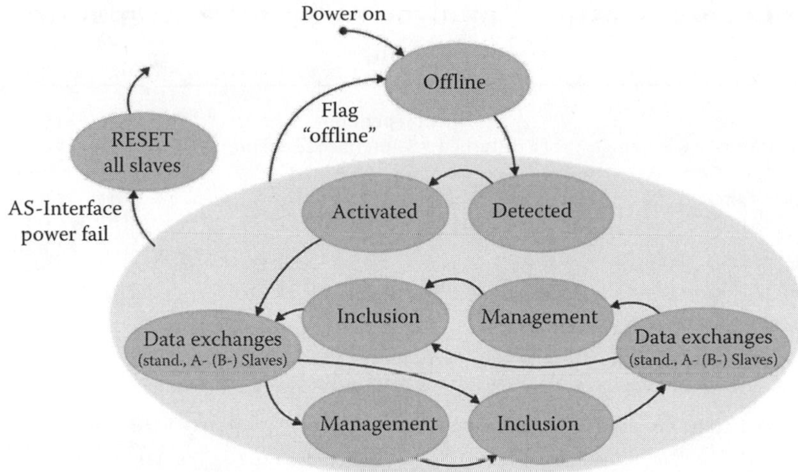

FIGURE 6.34 Main status diagram (simplified) of the master.

With the data necessary for carrying out the functions of the sequence control level (e.g., setting parameters or changing an operating address) can be sent, in part distributed over several cycles. If there is no job, the cycle moves on to the inclusion phase and no telegram is sent.

The following jobs can be executed during the management phase:

- Setting parameter values
- Reading a slave status
- Reading the I/O configuration
- Reading the three ID codes
- Writing the extended ID Code 1
- Resetting the operating address of a slave to 0
- Assigning a new operating address to Slave 0
- Resetting a slave
- Carrying out a broadcast

After the management phase is finished, the inclusion phase is entered, in which at the end of each cycle, newly arriving stations are searched for. During each inclusion phase, a search is made for exactly one slave using the *read status* call. This call is performed from the transmission level without repetition in case of error, since a transmission error at this point is not time critical. The peripheral error information of the slaves is collected for activated slaves in the inclusion phase. This has the advantage that no lengthening of the cycle is required. On the other hand, the querying of all peripheral error information takes several cycles. Already activated slaves also get the *read ID code* call, so that any passive devices connected later in the network (e.g., diagnostic devices) can identify the slaves.

If a slave that is already activated replies, or if there is no slave reply, the next slave is searched for in the next activation phase. If a slave that was previously unknown replies, the complete identification of the slave is determined in the next inclusion phases. Then in the following inclusion phase, the slave is activated depending on the selected operating mode. Activation is accomplished by writing the parameters from the parameter field.

An inclusion or reinclusion of a slave therefore extends over several inclusion phases. The procedure, which here is divided over several cycles with one telegram per cycle, is the same as that used to search for and activate slaves in start-up mode.

If you were to view the sequence of an AS-i cycle described here with an oscilloscope connected to the AS-i line, the screen would look as illustrated in Figure 6.35.

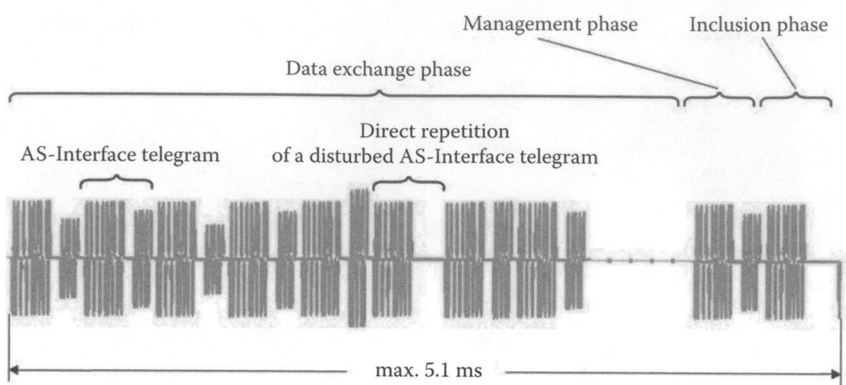

FIGURE 6.35 AS-i cycle.

6.4.2 Master Levels

The specification divides the master into three layers, which describe the master from the AS-i cable to the host interface (e.g., to the PLC) (Figure 6.36).

An important task of all the layers is the handling of errors, or *not normal* operating states, without sending false sensor or actuator signals.

Such states could include, for example, the following:

- Errors in pauses
- Errors in slave responses
- A failed slave
- A slave connected during operation
- An unexpected slave connected

The transmission physics of the master basically has to meet the same requirements as that of a slave. Here is where the actual bit transmission takes place.

The lowest logical layer is the transmission layer, which is responsible for transmitting and receiving individual telegrams. Automatic repetition of telegrams when a slave reply is missing or defective ensures integrity for the higher layers.

The sequence control is just above the transmission layer and passes requests for data transmission to the latter. The sequence control controls the sequence of telegrams. In addition to the actual sequence

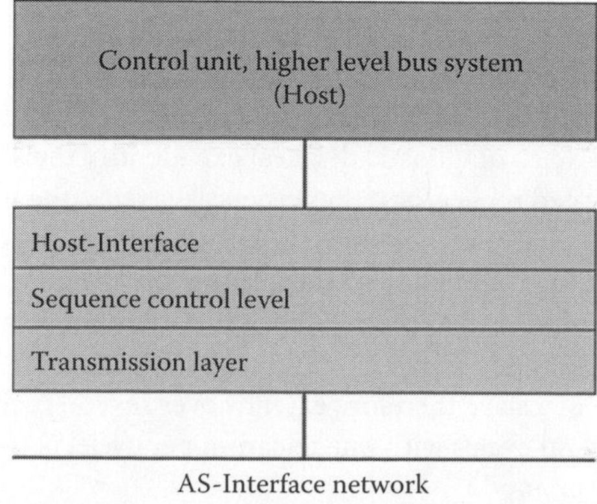

FIGURE 6.36 Logical structure of the AS-i master.

control, the sequence control layer processes functions that are requested by the host through the master layer. The highest layer is the master layer, in which the AS-i functions are adapted to the respective host system. This layer is where profiles are formed, which allow a restriction of the master functions usable by the host.

6.4.3 Master Calls

Table 6.14 summarizes the stipulated AS-i master calls to a standard slave. The following section describes them and the response of the slave in detail.

Read I/O Configuration

The call *read I/O configuration* allows the master to read the set I/O configuration of a slave. This is sent in the slave response to this call and is used together with the calls *read ID code* for unambiguous identification of a slave. The I/O configuration refers to the data ports on Interface 1 of the slave and is defined as shown in Table 6.6. Here, *in* means a process input, *out* a process output, I/O a bidirectional behavior of the port, and *tri* a tristate output with no function. The latter state is assumed if during the reset no unambiguous I/O configuration can be determined from the slave's own data memory due to a read error.

Wherever an output *out* (and no bidirectional behavior) is defined, this means that the corresponding information bit is undefined in the slave response. Likewise, the information bit from the master call remains without meaning, even though it is sent, where an input *in* is defined. Unneeded information can thus be hidden with this I/O configuration. At the same time, the I/O configuration together with the ID codes for identifying slaves is employed. These refer to the slave profile. The I/O configuration is 4 bits long, is determined by the manufacturer, and is stored in the slave, where it cannot be changed.

TABLE 6.14 Overview of the Master Calls to a Standard Slave

	ST	CB	5-Bit Address					5-Bit Information					PB	EB
Data request	0	0	A4	A3	A2	A1	A0	0	D3	D2	D1	D0	PB	1
Parameter request	0	0	A4	A3	A2	A1	A0	1	D3	D2	D1	D0	PB	1
Address assignment	0	0	0	0	0	0	0	1	1	1	1	1	PB	1
Command request: write extended ID Code 1	0	1	0	0	0	0	0	0	|D3	|D2	|D1	|D0	PB	1
Command request: delete address	0	1	A4	A3	A2	A1	A0	0	0	0	0	0	PB	1
Command request: reset slave	0	1	A4	A3	A2	A1	A0	1	1	1	0	0	PB	1
Command request: read I/0 configuration	0	1	A4	A3	A2	A1	A0	1	0	0	0	0	PB	1
Command request: read id code	0	1	A4	A3	A2	A1	A0	1	0	0	0	1	PB	1
Command request: read extended ID Code 1	0	1	A4	A3	A2	A1	A0	1	0	0	1	0	PB	1
Command request: read extended ID Code 2	0	1	A4	A3	A2	A1	A0	1	0	0	1	1	PB	1
Command request: read status	0	1	A4	A3	A2	A1	A0	1	1	1	1	0	PB	1
Broadcast (reset)	0	1	1	1	1	1	1	1	0	1	0	1	PB	1

Read ID Code

The command calls *read ID code* apply to the identification code of the slave. This together with the I/O configuration is used for unique identification of a slave.

Data Call

The master call *data call* (*data exchange*; see [1]) is the most important and frequently used AS-i call. It is used to cyclically send the bit pattern for the data outputs on Interface 1 of the slave accessed by address and then to read the input level and send it back in the slave response.

The data ports can be used in various ways: as outputs to the process, which can be set and cleared by the call, and/or as inputs from the process whose state, initiated by the call, is read in. Determinant here is the slave profile, which is defined by the I/O configuration and the ID codes.

If the data ports are configured as outputs, the last sent value is stored in the slave until it is overwritten by a new value or reset by a reset. Slaves can also be equipped with a watchdog function that resets the outputs when communication with the master is interrupted for longer than 40 ms. After a reset, such as would be caused by turning on the supply voltage, the slave does not send replies to the data call until it has first received a valid parameter call.

A data call to address 00Hex is not possible, as this would be interpreted by the slave as an addressing call.

Parameter Call

The master call *parameter call* is used to send the bit pattern acyclic for the parameter outputs on Interface 1 and then to read the input levels and send them back in the slave response. This parameter port can be used to remotely control certain functions in the slave, such as activating a time function, changing the switching distance of the sensor, or, in the case of a multisensor system, changing from one measuring process to a different one. The last sent value is stored in the slave until it is overwritten by a new value or reset by a reset. Reading the information in on the parameter ports can be used to send simple diagnostics information.

After a reset, the parameter ports are set to a defined state. They are set only after a *parameter call* has been received without error. Only then are *data call* messages properly exchanged.

A standard slave has 4 bits of parameter data available or 3 bits for a slave in extended addressing mode. A parameter call to address 00Hex is not possible, as this would be interpreted by the slave as an addressing call.

Addressing Call

This command allows the master to permanently set the address of the slave that previously had address 00Hex to a new value.

The slave first acknowledges errorless receipt of the command with the reply 6Hex and can then be accessed under the new address. At the same time, the slave starts internally storing the new address in a nonvolatile memory. This procedure can take up to 500 ms and may not be interrupted by a power interruption or a reset command.

The addressing unit uses this call to give the slave a new address over the AS-i line. The master can also use this call to assist in replacement of defective slaves.

Reset Slave

The command *reset AS-i-slave* can be used to reset a slave to its base state. It has the same effect as does the reset after applying power or the reset on the reset input of Interface 1. The slave acknowledges errorless receipt of this command with the reply 6Hex. The reset procedure may take a maximum of 2 ms.

Clear Operating Address

The command *clear operating address* is used for temporary deleting of the slave operating address and is needed in connection with the address assignment, since the *addressing call* can only be carried out

TABLE 6.15 Flags of the Slave's Status Register

Flag	Meaning
S0	*Address volatile* This flag is set when the internal slave routine for permanent storing of the slave address is running.
S1	*Peripheral error* This flag is set when the slave has detected an error in its peripheral. This may indicate, for example, an overload of the sensor supply voltage.
S3	*Nonvolatile memory read error* This flag is set when an error occurs during a reset while reading the nonvolatile memory.

by a slave having operating address 00Hex. If, for example, a slave that used to have address 15Hex needs to be reprogrammed to the new address 09Hex, this can only be done using the following command sequence: *clear operating address* and *addressing call* (09Hex).

The slave acknowledges errorless receipt of *clear operating address* with the reply 0Hex and is then accessible under the new address 00Hex. Clearing the operating addressing by means of this command is not permanent; if after executing this command you need to reload the originally nonvolatile stored address, you can do this using the *reset AS-i-slave* command.

Read Status

The command call *read status* can be used to read out the status register of the contacted slave. Its contents are sent in the slave response to this call. The slave's status register contains three flags (Table 6.15) having the following meanings:

Flag S2 is not yet used and is reserved for future expansions. Simultaneous with the reply to the call, an address comparison is performed by querying the nonvolatile memory. If the address in the address register is not equal to that in the nonvolatile memory, Flag S0 is set. The information from the status register can be used by the master for diagnostic purposes. Version 2.0 slaves do not support the *peripheral error* flag. When the master is communicating with a slave having extended ID Code 2 = FHex, it does not evaluate Flag S1.

Broadcast

Command calls containing 15Hex are defined as *broadcast calls*. They may not be replied to by the slaves. To this extent, they are untypical for normal AS-i data exchange. At the present time, only the broadcast command *reset* is defined. Additional broadcast commands will be implemented in future versions of the specification. Version 2.0 slaves do not support the broadcast function.

Table 6.16 provides an overview of the master calls. This applies to a slave with extended addressing mode:

From Table 6.16, it can be seen that an information bit is used in numerous calls as a *select bit* for distinguishing between the A slave and the B slave. The definition of the select bit is selected such that the A slave behaves just like the standard slave. This enables you to operate *new* slaves in AS-i networks even when they are connected to *old* masters.

6.4.4 Data Fields and Lists on the Sequence Control

The sequence control takes care of initializing, starting, and normal data exchange on the network. The interface to the master level—the interface to the host—consists of several functions that pass along the actual user data, control, and configure the system. The sequence control stores the data required for this in several data fields and Table 6.17 provides an overview of these data.

The data can be categorized into four groups:

1. The user data are information that is exchanged directly with the slaves.
2. The configuration data provide information about the status of the slaves. This makes possible detailed diagnostics of the network.

TABLE 6.16 Overview of the Master Calls to a Slave with Extended Addressing Mode

	ST	CB	5-Bit Address					5-Bit Information					PB	EB
Data request	0	0	A4	A3	A2	A1	A0	0	$\overline{\text{Sel}}$	D2	D1	D0	PB	1
Parameter request	0	0	A4	A3	A2	A1	A0	1	$\overline{\text{Sel}}$	D2	D1	D0	PB	1
Address assignment	0	0	0	0	0	0	0	A4	A3	A2	A1	A0	PB	1
Command request: write extended ID Code 1	0	1	0	0	0	0	0	0	|D3	|D2	|D1	|D0	PB	1
Command request: reset slave	0	1	A4	A3	A2	A1	A0	0	$\overline{\text{Sel}}$	0	0	0	PB	1
Command request: reset slave	0	1	A4	A3	A2	A1	A0	0	$\overline{\text{Sel}}$	1	0	0	PB	1
Command request: read I/O configuration	0	1	A4	A3	A2	A1	A0	0	Sel	0	0	0	PB	1
Command request: read ID code	0	1	A4	A3	A2	A1	A0	0	Sel	0	0	1	PB	1
Command request: read extended ID Code 1	0	1	A4	A3	A2	A1	A0	0	Sel	0	1	0	PB	1
Command request: read extended ID Code 2	0	1	A4	A3	A2	A1	A0	1	Sel	0	1	1	PB	1
Command request: read status	0	1	A4	A3	A2	A1	A0	1	$\overline{\text{Sel}}$	1	1	0	PB	1
Broadcast (reset)	0	1	1	1	1	1	1	1	0	1	0	1	PB	1

3. The projecting data are stored in nonvolatile memory in the master and represent the nominal configuration of the connected network. By comparing the nominal and actual configuration, the master can determine whether there are errors, report them, and, if appropriate, clear them by means of automatic addressing.
4. The flags inform the host (PLC, PC, etc.) about the status of the master and are used to influence it:
   a. Config_OK: Nominal and actual configuration are in agreement. The Config_OK flag enables simple monitoring of the configuration.
   b. LES.0: Slave with address 0 is present.
   c. Auto_Prog: Automatic addressing is enabled.
   d. Auto_Prog_Available: Automatic addressing is possible.
   e. Projecting_active: Projecting mode.
   f. Normal_mode_active: Set in cyclical normal mode.
   g. NOT APO: Low voltage was detected.
   h. Offline_Ready: Offline phase is active and function requests to the sequence control are being processed.
   i. Periphery OK: No slave is reporting a periphery error.

The flags on the master layer are set by the master layer or by the host and affect the function of the master:

- Data_exchange_Active: Enables data exchange between the master and the slaves
- The offline flag: Switches the sequence control to offline phase
- Auto_Address_Enable: Enables automatic addressing

6.4.5 Master Analog Part

The transmission physics of the master describes the electrical coupling to the cable, the analog section of the hardware structure (Figure 6.37). Implementation of this interface has key significance

TABLE 6.17 Data Fields and Lists on the Sequence Control

		Feldgröße
Input data image (IDI)	This array contains the latest actual copies of the received data of all received slaves. Input data of inactive slaves is set to zero (0).	62 × 4 bit = 31 byte
Analog input data image (AIDI)	This array contains the received analog data of the analog inputs from the slaves with profiles 7.3 or 7.4.	31 × 4 × 3 byte = 372 byte
Output data image (ODI)	This array contains the data to be transmitted to the active slaves at the next data exchange.	62 × 4 bit = 31 byte
Analog output data image (AODI)	This array contains the analog data to be transmitted to the outputs of the active slaves with profiles 7.3 or 7.4 at the next data exchange.	31 × 4 × 3 byte = 372 byte
Configuration data image (CDI)	This array contains the input/output configuration and the ID codes (ID, ID1, ID2) of all slaves. The data of inactive slaves are set to the default value 1.	62 × 4 × 4 bit = 124 byte
Permanent configuration data (PCD)	This array contains the projected I/O configurations and ID codes (ID, ID1, ID2) data of all slaves. The data of not projected slaves are set to the default value 1. The data are stored in nonvolatile memory.	62 × 4 × 4 bit = 124 byte
Parameter image (PI)	This array contains the parameter data that have been sent to parameter outputs of all active slaves. The data of inactive slaves are set to the default value FHex.	62 × 4 bit = 31 byte
Permanent parameter (PP)	This array contains the configured parameter data of all slaves. The data are stored in nonvolatile memory, after master power on the data is stored in the array of PI by the sequence control level.	62 × 4 bit = 31 byte
List of detected slaves (LDS)	In this list, one bit is set for each detected slave.	63 bit = 8 byte
List of activated slaves (LAS)	In this list, one bit is set for each active slave.	62 bit = 8 byte
List of projected slaves (LPS)	This list contains all projected slaves that are expected to be in the AS-i network. This list is stored in nonvolatile memory.	62 bit = 8 byte
List of peripheral faults (LPF)	In this list, one bit is set for each slave sending a peripheral fault.	63 bit = 8 byte

for the noise immunity of the system. To prevent excessive signal distortion, the impedance between the interface terminals must, as in the case of the slave, meet certain requirements. Like the slaves, the master is not allowed to disturb the symmetry of the network, since the cable would otherwise become an antenna for radiating noise. The difference in the terminal impedances to GND must therefore be kept as small as possible. In addition, galvanic isolation between the controller and the AS-i network is mandatory.

A comparator checks the DC voltage on the line and uses the APO signal to report when the voltage is sufficient for reliable operation of the slaves at every location along the line.

6.4.6 Master Host Interface: Master Profile

The host interfaces can be designed to be as versatile as the user interfaces. This is due to fact that an AS-i master—as described earlier—can be part of a simple PLC as well as part of a gateway to a higher-level complex fieldbus. The information in the data fields described earlier can be exchanged with the higher-level control system via this interface to the host system.

AS-i gateways are available for all commonly used higher-level fieldbus systems. As mentioned earlier, AS-i is an excellent addition for these systems since it enables the economical collection and

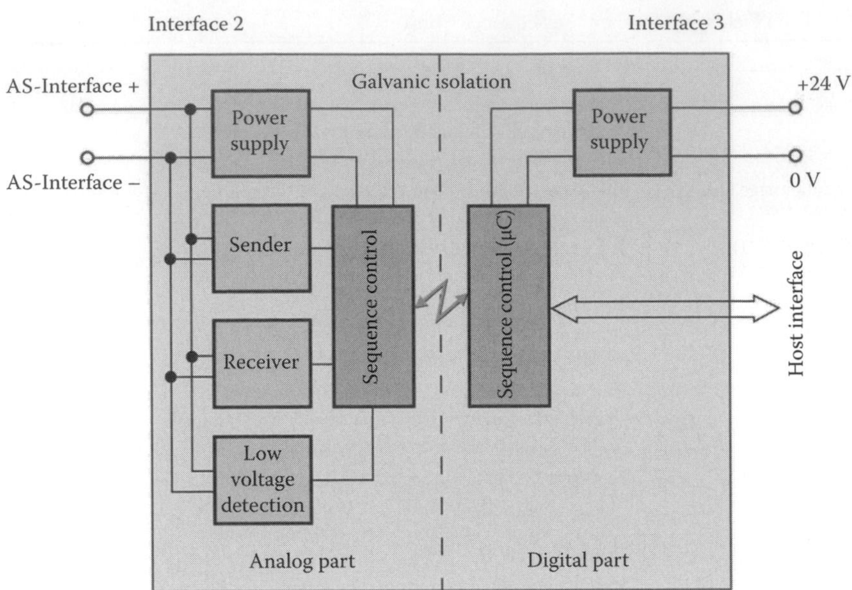

FIGURE 6.37 Hardware structure of the AS-i master.

combination of frequently widely scattered inputs and outputs on a machine or in a plant. Figure 6.38 shows the electronic data sheet (GSD file) of a PROFIBUS/AS-i gateway. The input data field of 152 byte maximum length holds the entire IDI and about half of the AIDI. The ODI and AODI fit similarly. Additionally, 68 bytes are provided for diagnostics.

Due to the fact that AS-i underwent numerous innovations and functional enhancements since its introduction, different master profiles exist that support a variety of different functions. Table 6.18 shows the most current status:

Coupling to Fieldbus Systems

Due to its functionality, AS-i is located directly at the technological process. If you want to make the sensor and actuator data available to the higher levels of the automation hierarchy as well, consistent networking structures are necessary. In the simplest case, this is done by coupling to the PLC in which the master is located. For virtually every commonly available PLC, there are interface modules available for fieldbus systems such as PROFIBUS, InterBus, and DeviceNet. The application program in the PLC is then responsible for providing the data on the fieldbus system, but there is no direct link between the AS-i and the fieldbus (Figure 6.39).

It therefore seemed obvious to provide the AS-i master itself with an interface to a fieldbus system, that is, to create a coupler between AS-i and a fieldbus system. The resulting networking structure (Figure 6.39) offers numerous advantages. The spatial separation of the networked components is increased by the network length of the fieldbus system. The simple cabling of the sensors and actuators using AS-i remains intact. The master interfaces (couplers) can be assigned to multiple PLCs, which make the automation process more flexible. Redundant solutions are then also possible.

The following sections describe a typical coupler between AS-i and a fieldbus, using PROFIBUS as the example. Typical network transitions take place on layers 2 and 3 of the ISO/OSI reference model. In the case of PROFIBUS, layers 1, 2, and 7 are defined, whose functionality differs in the variants DP and FMS. Above this is the application layer interface (ALI), which describes the link between the communication system and the application program. In the case of the AS-i master, such a strict association with the OSI layers is not possible. The sequence control takes over the functionality of multiple layers, with the master layer comparable to the ALI.

```
;=============================================================================
;             GSD file according to DIN 19245 part 3 (PROFIBUS-DP)
;=============================================================================
#Profibus_DP
.....
; Supported DP features
Freeze_Mode_Sipp             = 1       ; supported
Sync_Mode_supp               = 1       ; supported
Auto_Mode_supp               = 1       ; supported
Set_Slave_Add_supp           = 1       ; supported
.....
; Maximum supported sizes
Modular_Station              = 1       ; modular
Max_Module                   = 11
Max_Input_Len                = 152     ; max.   152 Byete Input
Max_Output_Len               = 152     ; max.   152 Byte Output
Max_Data_Len                 = 304     ; max.   304     Byte I/O
Modul_Offset                 = 1
Slave_Family                 = 9@TDF@AS-I       ;      Hauptfamilie Gateway
                                                ;      Unterfamilie AS-I
Max  Diag  Data  Len         = 68
.....
```

FIGURE 6.38 GSD file for a PROFIBUS/AS-i gateway.

TABLE 6.18 Master Profiles and Supported Functions

Master Profile	Standard Address Mode	Safety at Work	Extended Address Mode	Peripheral Error Flag	CTT1	CTT2–CTT5	Extended Diagnostics
M0, M1, M2	Yes	Yes	—	—	—	—	—
M3	Yes	Yes	Yes	Yes	Yes	—	—
M4	Yes	Yes	Yes	Yes	Yes	Yes	Yes

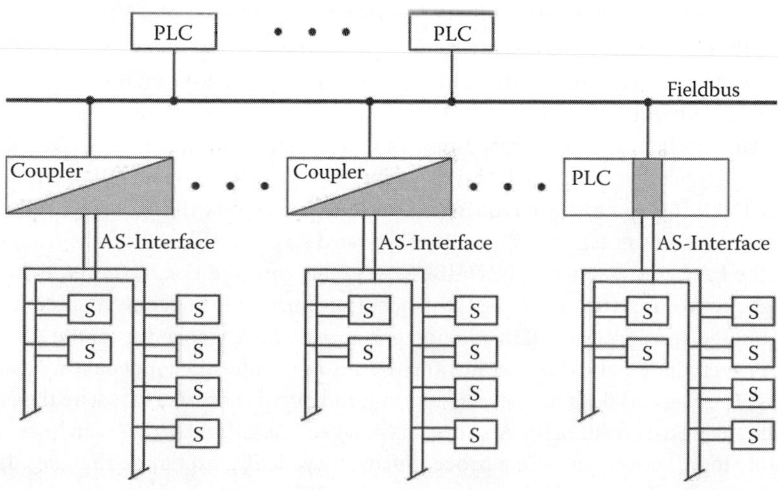

FIGURE 6.39 AS-i in the network.

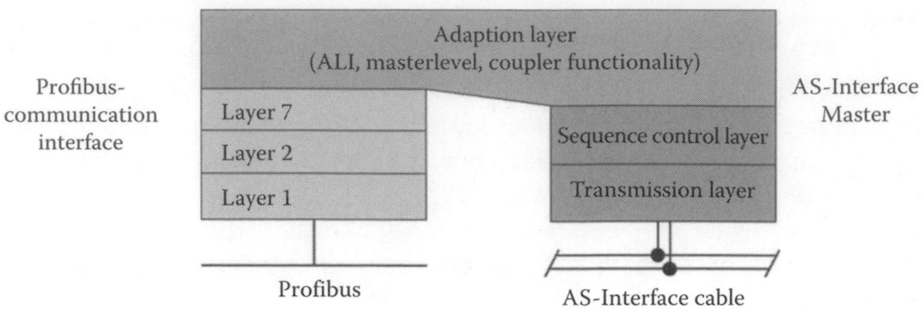

FIGURE 6.40 Structure of the AS-i/PROFIBUS coupler.

Based on the earlier text, the following structure of the coupler with three function units can be outlined (Figure 6.40):

1. The communications interface
2. The AS-i master
3. The adaption layer

For the PROFIBUS interface, a slave implementation is sufficient. The master contains the complete implementation of transmission control and sequence control.

For a PROFIBUS station, AS-i represents the application process. Likewise, from the point of view of an AS-i master, the PROFIBUS is the host control system with which it exchanges data through the master layer. As a result of the strongly differing system properties of AS-i and PROFIBUS, the coupler is preferably located at the application interface. Direct conversion of an AS-i message into a PROFIBUS message and the reverse is not possible. Rather, the logical approach is to combine the data from AS-i and send them in groups over the PROFIBUS. How this grouping is done and which actions are associated with it can be determined only on the application level.

The adaption layer combines the properties of the PROFIBUS ALI, the AS-i master layer, and a coupler process. It is responsible for the following tasks:

At the interface to layer 7, indicated PROFIBUS application services (reading and writing) must be mapped in communication-related operations to the AS-i interface to the sequence control and, if appropriate, a reply sent to layer 7. Likewise, the administration services such as opening and closing a connection have to be processed. In addition, the coupler must itself invoke several services of layer 7, for example, to initialize the PROFIBUS parameters. The way in which AS-i system information is grouped into PROFIBUS telegrams essentially determines the performance capability of the entire data transmission between the user and AS-i. If all the inputs and output data from the AS-i slaves were stored in separate PROFIBUS telegrams, 124 PROFIBUS telegrams would be necessary for a complete exchange of the data; the data throughput would thus drop to an unacceptable level.

By grouping the 4 input data bits for all AS-i slaves into a single telegram, the IDI (cf. Table 6.17) of AS-i can be sent over PROFIBUS in a single request–response sequence. But the AS-i user still does not know whether the data are valid or not. For this, the list of activated slaves (LAS) needs to be read. Reading of the input data and the LAS in two separate PROFIBUS telegrams runs the risk, however, that the state of an AS-i slave changes between the two requests. Consistent transmission is then only assured if input data and the LAS are sent in the same telegram. If in addition one assigns the readable flags of the AS-i master to this telegram, the user gets additional status information over AS-i each time the input data are read. He can then check the Config_OK flag and does not have to take any additional measures to isolate the error unless the flag is not set. The analog data fields of the AS-i master (AIDI and AODI, Table 6.17) can be handled similarly.

Programmable logic controllers use a process map to cyclically exchange the user data with their environment. This procedure can be supported in PROFIBUS with the help of a cyclical connection whereby the communication software automatically takes care of regular data exchange. For users who are not working cyclically (e.g., programming device, control station) and only occasionally access the AS-i data, an acyclical connection is recommended. Data exchange over a cyclical connection takes place in general faster than for an acyclical connection, since multiple procedures can be running concurrently. The AS-i PROFIBUS coupler allows both connection types for accessing the AS-i data.

By opening a connection, an AS-i user opens a channel over which it can then exchange data with the coupler. The user ends proper data exchange by closing the connection, such as when the PLC program is stopped. Connection opening and closing are essential prerequisites for reliable operation of the coupler. If there is no longer a connection between the coupler and the AS-i user, the master also no longer has any valid user data in the coupler that it could exchange with the slaves. It should therefore end the data exchange with the slaves and place them in the base state. Likewise, data exchange in the AS-i system should not be activated at the same time a single connection is opened through which any projecting of the AS-i master needs to take place. The specification for the coupler is, therefore, that data exchange in the AS-i system be activated when the first cyclical connection is opened and deactivated when the last cyclical connection is closed.

6.5 Functional Safety with AS-Interface Safety

In automation procedures affecting the safety of people through dangerous movements, the high basic data transmission reliability for standard input and output data for AS-i is not sufficient by itself. Therefore, additional measures, depending on the level of expected danger of the machine movement, and according to the guidelines for machines of the EU (guideline 2006/42/EG) and to the workplace safety guidelines of the member countries (general framework for regulations *safety and health protection for employees* 89/391/EWG), have to ensure that

- Along the entire transmission path, data cannot be falsified (including through errors in the master or slave electronics)
- Data are not delayed for an inadmissibly long time or get lost entirely
- Data are used from an incorrect source
- It is impossible to wrongly interpret data because of defective repetitions or mix-ups in the sequence

In order to fulfill these requirements, two additional components are added to AS-i:

1. The safe sensor or the safe actuator, respectively
2. The safety monitor

Like any standard AS-i slave, these components can be implemented at arbitrary positions on the network and can be combined with standard components without limitations or restrictions. However, the safe sensor and the safety monitor (as well as the safety monitor and the safe actuator) are in a special communication connection.

AS-i Safety at Work is approved for applications that require safety according to safety integrity level (SIL3) according to IEC(EN)61508 or performance level *e* according to EN ISO 134849-1:2006 (Figure 6.41). *AS-i Safety at Work* offers a guaranteed switch-off reaction time of under 40 ms, independent of the AS-i network configuration. As a result, the switch-off reaction time does not have to be evaluated after changes or expansions of the network.

TUV and trade associations (German BiA) carefully checked and approved the concept and the products *AS-i Safety at Work*. Since its introduction in 2001, Safety at Work has developed into the most successful system in the field of safety technology.

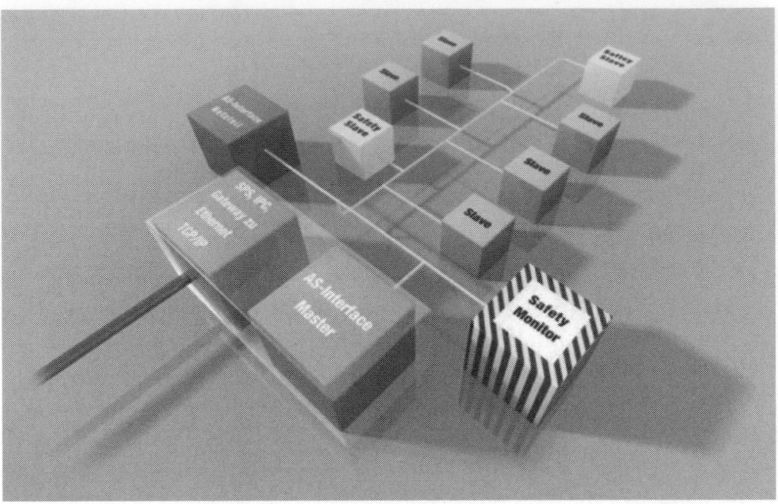

FIGURE 6.41 AS-i network with safety at work components.

6.5.1 Transmission Principle

The Electrical Engineering Committee of the German employers' mutual insurance association (BIA) formed a working group that developed a *proposed basis for testing and certifying bus systems for the transmission of safety-related messages*. This describes, among other things, the faults that can occur when sending messages and measures for overcoming them. For example, in safe bus systems, the message *E-STOP not actuated* often includes a reference to the sender and time information, plus a checksum (e.g., 32-bit CRC) for data integrity. With only seven bits net or four bits of user data per message (slave response), this procedure cannot be used for AS-i.

In AS-i Safety at Work, each slave encodes the information *E-STOP not actuated* into a 32-bit-long code word and sends it as a data frame of 4 bits each per AS-i cycle. Because the transmission takes place in eight sequential telegrams, AS-i refers to this code word as a code sequence. At the receiver, at any desired point in time, a complete code sequence results that is uniquely associated with this slave based on the telegrams received in the last eight AS-i cycles. This is accomplished by means of the rules applied to the encoding. For 32 bits, there are 232 different code sequences, which results in more than 4 billion possibilities. A maximum of 950,000 of these are actually used, or 1 in 4,000. Each safe slave uses a different code sequence from these 950,000, which is stored in the slave when it is manufactured. This procedure is patent protected.

That the process is simpler than it sounds can be seen by looking at the block diagram of a safe slave (Figure 6.42). The code sequence is stored in a 4 × 8 bit memory. With each AS-i telegram, the next four bits from the memory are sent in the slave response. The data strobe signal is used in the slave for advancing to the next value. Actuating the E-STOP interrupts the connection between the memory and the AS-i slave IC. Instead, the value 0000Bin is then present on the inputs of the IC and is transmitted in the slave response.

The receiver of the messages—mostly a safety monitor that listens in on the slave response—has a memory for each connected slave in which the code sequence is also stored. It compares each slave response with the reference value contained in the memory, and in normal operation, the complete code sequence is received after eight slave responses—and thus is the message *E-STOP not activated*. If the slave response contains the value 0000Bin, this corresponds to the message *E-STOP actuated*. If the received message corresponds to neither the expected value in the code sequence or the value 0000Bin, there is a fault or an error. The 8 × 4 bits of the code sequence are chosen so that any error, such as a short circuit between 2 bits or a constant 1 level on an input of the AS-i slave IC, results in discrepancies after the comparison.

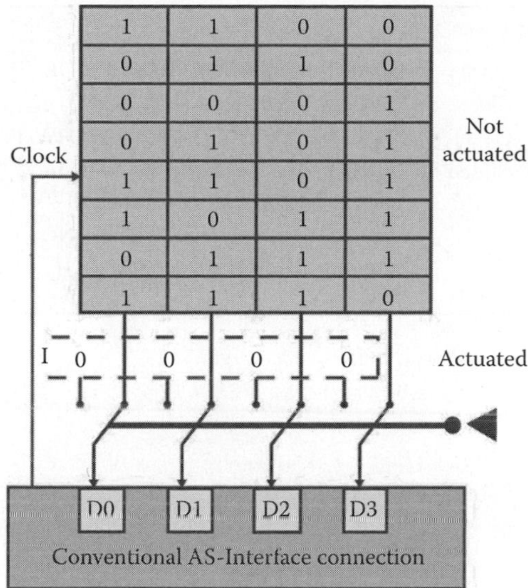

FIGURE 6.42 Block diagram of a save slave.

The error handling for safety performance level e according to EN ISO 13849 requires that a single error shall not result in a loss of function. An E-STOP button is therefore equipped with two normally closed contacts, so that when there is a fault in one contact, the redundant second contact can initiate the standstill. As implemented in the safe slave, this means that the code sequence is divided into two groups of 8 × 2 bits each and each contact affects one respective group. If only one of the contacts opens upon actuation, 2 bits are at zero while the remaining 2 bits of the code sequence are further transmitted. The receiver recognizes this and, in the case of the E-STOP function, generates an error message in addition to the standstill.

Not every deviation from the expected code sequence is caused by a defective slave or sensor. Contacts have mechanical tolerances or bounce. Both when actuating and when unlocking, transition states can arise. The receiver therefore always considers multiple slave responses when determining the state of the sensor. Table 6.19 and Figure 6.43 show the states of a sensor in the process image and the associated state transitions.

These states form the process image and represent the respective current state of a slave or sensor. Only an on-state represents the information *E-STOP not actuated*. A change between two of the first four states shown always involves the transition state, since at least eight slave responses are necessary in order to determine the state of the contacts.

6.5.2 Safety Monitor

In principle, the function of the safety monitor is the same as that of a safety relay. The safety monitor with two output circuits includes two redundantly constructed OSSDs that may be used up to control system of performance level e. Due to additional integrated safe logical elements, the function of the safety monitor extends far beyond that of an individual safety relay. Depending on the application, a safety monitor may replace several conventional safety relays. This reduces the wiring effort significantly and saves space in the switch cabinet.

During the configuration of the safety monitor, the designer must ensure that the safety components match the respective safety functions in order to achieve the required control system category. He must not rely on the safety monitor for this purpose as the monitor can accept configurations that do not guarantee the desired control system performance level. If, for example, a standard slave, that is, a simple non-safety-related switch, is configured as safety component into a safe output circuit, this can neither

Contact 1	Contact 2	Received Valves	State
Open	Open	More than eight times 0000Bin	Off
Open	Closed	More than eight times 00-	Half_01
Closed	Open	More than eight times-00	Half-10
Closed	Closed	More than eight valves of the code sequence	On
		None	No communication
		Others	Transition state, state of the sensor

(a)

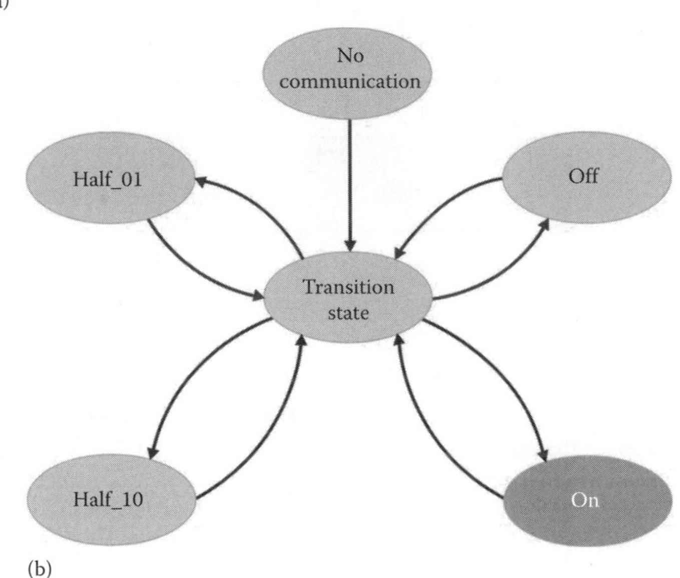

(b)

FIGURE 6.43 (a) States in the process image and (b) state transitions in the process image.

be detected in the safety monitor nor in the configuration software. Only the person who creates or checks the configuration can detect this.

It is important to print out and sign the configuration log included in the safety monitor and ensure that it is attached to the plant documentation. With his/her signature, the designer documents the correctness of the safety configuration. For all further changes or additions to the plant configuration, a new log must be printed, signed, and attached to the plant documentation. Discrepancies can be detected by simply comparing the printed version to the current log stored in the safety monitor.

The safety monitor listens in on the data communication between the master and slaves. A first microcontroller receives both master requests and slave responses and places them in a buffer (Figure 6.44). Every 5 ms, a timer starts evaluation of the telegrams. This task is assumed by two additional microcontrollers that monitor each other. They consider the addressing sequence of the master requests and thus can detect missing slaves, for example. Depending on the user configuration (see the following), they distinguish between safe slaves, for which they create the safe process image, and conventional slaves, whose input and output data they also store. Once all the telegrams from the buffer have been processed, the AS-i-specific part of the safety monitor ends. In the standard AS-i, this corresponds to the interface between the AS-i master and the host (PLC).

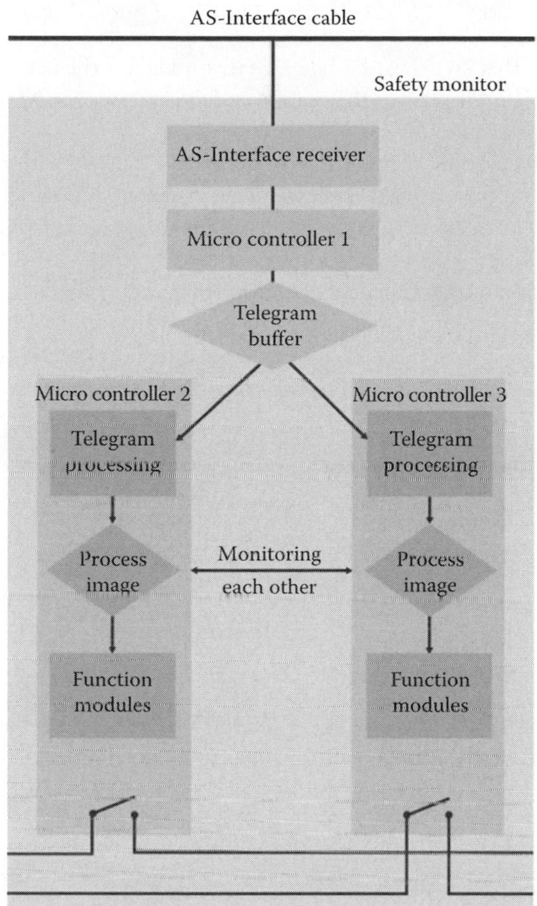

FIGURE 6.44 Structure of the safety monitor.

The next task is to determine from the process image the conditions under which the output relay should be turned on to be able to start a machine—the task of a safety controller. The safety monitor contains software function modules for this purpose that can take on certain monitoring tasks. These function modules are implemented as a state machine. They use the current state (behavior of the slave in the past) and the value in the process image (how do the signals currently look on the slave) to determine the new state. The state is recalculated for each slave every 5 ms. Thus, the implementation of an E-STOP function meets performance level e with a *two-channel forced* function module. It checks whether both contacts in the slave are always in the same state, that is, both open or both closed. Should the safe process image of the associated slave be in one of the two *half* states, it means there is a fault in the E-STOP button. The function module therefore goes into the *slave is defective* state and indicates this, normally by a red flashing LED on the safety monitor. Even if now the process image again displays a proper state for the slave, the error status is not cleared until the operator has acknowledged it. The function module *two-channel dependent* is intended to monitor a safety door with two separate switches. Here, *half* states are allowed, for example, when the operator opens the door just a crack. The function module does not go into an error state but rather into the *test required* state and indicates this normally by a yellow flashing LED on the safety monitor. The operator is thus prompted to open the door fully and close it again to verify correct function of both switches for the safety monitor.

6.5.3 Safe Inputs

Besides the safety monitor that switches the safe signals und starts the actions, the other components connected do have a significant impact to the usability of the system. For AS-i, a huge product range of safe inputs and outputs is available.

This example shows how to connect protective switches to a module with M12 plug connection. For each channel, one or several protective switches may be connected in series but the requirements of the applicable control system category must be observed for this (Figure 6.45). One must also choose a manipulation-proof mounting.

Safe modules usually contain two independently implemented input channels, each of which meets the requirements of control system performance level e according to EN ISO 13849. To assign only a single contact mechanical switch to such safe modules designed for parallel connection, a short-circuit jumper must be installed on the free socket of the emergency-stop connection. This short-circuit jumper must be protected against dirt and falling out.

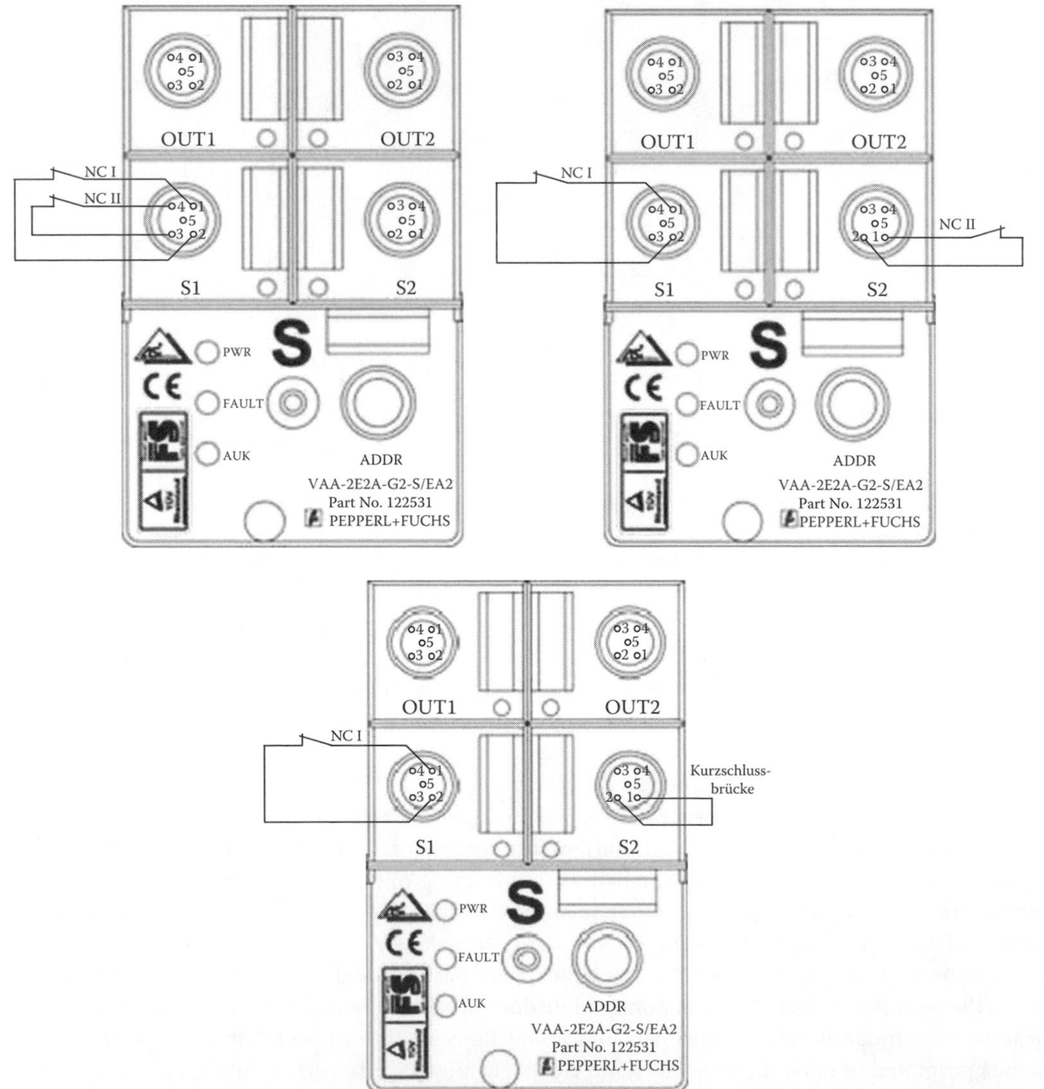

FIGURE 6.45 Connection of a two-channel mechanical switch or of two or one single-channel mechanical switch(es).

Short-Circuit Monitoring

Usually, the inputs of a safe module are monitored for short circuits between each other. For active optoelectronic protective devices with electronic outputs, the short-circuit monitoring is implemented in the sensor. Safe modules without short-circuit monitoring must be used for the connection to the AS-i branch.

6.5.4 Safe Outputs

Besides, the safe outputs integrated to the safety monitor allow the connection to decentralized safe outputs. These might be combined to groups for common switching. Typically, the safe outputs provide external device monitoring to guard the functionality of the relays.

A safe output module has two AS-i addresses: one for switching on operation and one for the safe switch-off. The two addresses need to be different.

The safe address is monitored by the safe output. The safety monitor is sending at this address the known code table. When the safety monitor switches the safe output, it sends the code table 0000 and the safe output will be switching off. Several safe outputs may use the same safe address; this allows groups of safe outputs to be put in the safe state at the same time.

6.5.5 Start-Up of a Safety System: Teaching the Code Table

It is easy to start up an AS-i safety system. It continues the simple plug-and-play philosophy of the system. Nevertheless, you have to do the risk analysis. In the moment you know what protection you need in which position it is all the easy way of AS-i,

- Place the required safety devices with the required performance level in the machinery
- Address the devices and connect them to the AS-i network
- Configure the safety devices in the safety monitor device
- Learn the safe configuration (see next section)
- Print out the documentation by clicking the print button in the configuration software

Learning the Safe Configuration

After completion of the configuration transfer to the AS-i safety monitor connection, the code tables of the slaves to be monitored must be read. Before this, the AS-i bus must be switched on (*on*). This includes all safe slaves. For two-hand buttons, this means that they must be operated simultaneously during the learning of the safe configuration.

To start reading the code tables in the configuration software, click on the command "Teach code tables-..." in the monitor menu. Once the code tables of all safe slaves to be monitored have been read without problem, the preliminary configuration log is transferred to the configuration software.

6.6 Open System: Interoperability and Certification

EN 50295[1] and IEC 62026-2[2] are the standardization basis for AS-i. All AS-i components are developed based on these standards and on the complete specification. The internationally high market acceptance means that the broad AS-i product spectrum is constantly growing. That is another benefit of this open system.

The technical commission of AS-International has developed complete manufacturer-neutral documentation that ensures that the functionality of the system and safe operation in harsh industrial environments can be taken for granted. An important goal of the documentation is interoperability of the various components.

The documentation of the user organization consists of the *complete specification* with profiles and testing specifications. This specification describes the slave and master transmission systems, the minimum requirements and the profiles, as well as everything necessary for a full understanding of the system. The specification also contains notes on technical implementation to simplify the development of new components. The testing specifications allow for verification that the specification has been met. The AS-International Association provides the documentation to its members. This availability of information is doubtless an important factor in the success of the system. Many companies are in this way able to design special solutions and make these individual solutions available for a variety of applications. The AS-i chips from a wide variety of manufacturers are the foundation of the economical product range.

Certification

Certified products are important for the user of a system. Certification guarantees function according to the specification, thus ensuring interoperability of components from different manufacturers in a network. All components in an AS-i network can be subjected to testing for certification. Thus, certification stands for a uniform and high-quality system. Certified products, even from different manufacturers, are guarantees of compatibility and a trouble-free AS-i network (Figure 6.46).

The certification agency is a manufacturer-neutral part of AS-International. Following an application to the certification agency and a type test, a certificate is issued for successfully passing the inspection.

The tests for a certificate can be performed only in a laboratory approved by AS-International in accordance with the AS-i testing specifications and accreditation guidelines. This ensures that the tests can be verified. A list of recognized laboratories is available from the AS-International Association.

The AS-i testing specifications include a detailed description of the certification test. Each testing specification contains precise test instructions to be used in verifying the logical behavior and communication-relevant properties of the AS-i network. The strictly defined measuring procedures and testing methods ensure correct function in the AS-i system.

Noise immunity of the components is subject to strict control and ensures a reliable AS-i network, even under difficult EMC conditions. The most important feature of all tests is to ensure functionality of the system in harsh industrial environments. Symmetry and impedance tests are combined with strict burst tests in *worst-case* topologies to test for EMC compatibility.

The tested products are easily identified by means of the certification sign (Figure 6.46) and associated test number. This logo is a registered trademark of the AS-International Association and may be used exclusively for certified products. The shadow logo together with the test number is a quality mark that is subject to strict controls.

FIGURE 6.46 AS-i certification sign.

References and Further Probing

1. EN 50295-2 (1998, 2002)—Low-voltage switchgear and controlgear—Controller and device interface systems—Actuator Sensor interface (AS-i).
2. IEC 62026-2 (2000)—Low-voltage switchgear and controlgear—Controller-device interfaces (CDIs)—Part 2: Actuator sensor interface (AS-i).
3. http://www.as-interface.net, AS-Interface Academy by AS-International Association.
4. http://www.as-interface.net, AS-Interface Knowledge Base by AS-International Association.
5. R. Becker, B. Müller, A. Schiff, T. Schinke, and H. Walker, AS-Interface the Automation Solution, AS-International Association, 2002.
6. F. Bauder, P. Christiani, D. Grundke, H. Hopp, M. Korff, K. Kern, W. Lehner et al., AS-Interface Safety at Work, AS-International Association, 2004.
7. R. Becker, Automation is easy with AS-Interface, AS-International Association, 2008.

7

HART over Legacy 4–20 mA Signal Base

Mark Nixon
Emerson Process Management

Wally Pratt
HART Communication Foundation

Eric Rotvold
Emerson Process Management

7.1 Introduction 7-1
Evolution of HART • HART Signal • HART System • HART Field Devices • Calibrating HART Field Devices

7.2 HART Networks 7-6
Understanding IO System Trade-Offs • Multidrop Connections • Burst-Mode Communications • Summarizing HART Performance

7.3 HART Networks 7-8
Modulated Analog Signal • HART Physical Layer • HART Network Guidelines • HART Networking Guidelines

7.4 HART Circuits 7-11
HART Communications • Analog Signaling • HART Inputs • HART Outputs • HART Intrinsically Safe Circuits

7.5 In Closing 7-15

References 7-15

7.1 Introduction

7.1.1 Evolution of HART

A Highway Addressable Remote Transducer (HART™) [1] device is a microprocessor-based process transmitter or actuator that supports two-way communications with a host, digitizes the transducer signals, and digitally corrects its process variable (PV) values. The HART digital signal is modulated on to the 4–20 mA signal at a higher frequency and then observed by process control equipment.

Since 4–20 mA transmitters provide a single PV to the control system via the loop current, early HART development focused on the primary PV. In HART, PV includes all of the properties and data items necessary to support the loop current as shown in Figure 7.1. HART commands provide access to all loop current data and properties. Process values associated with PV include the digital value, percent range, and loop current. Properties of the PV include upper- and lower-range values, upper and lower transducer limits, transducer serial number, and minimum span. Status information includes *loop current fixed*, *loop current saturated*, *PV out of limits*, and whether or not the device has malfunctioned.

Initial HART devices focused on improving accuracy and reliability. For example, adding an ambient sensor (e.g., temperature) along with linearization and compensation software dramatically improved accuracy. Today, virtually all HART field devices are multivariable. For example, there are pressure transmitters that provide the sensor temperature, level transmitters that can calculate volume, and flow meters that contain totalizers and perform data logging. The growth in multivariable

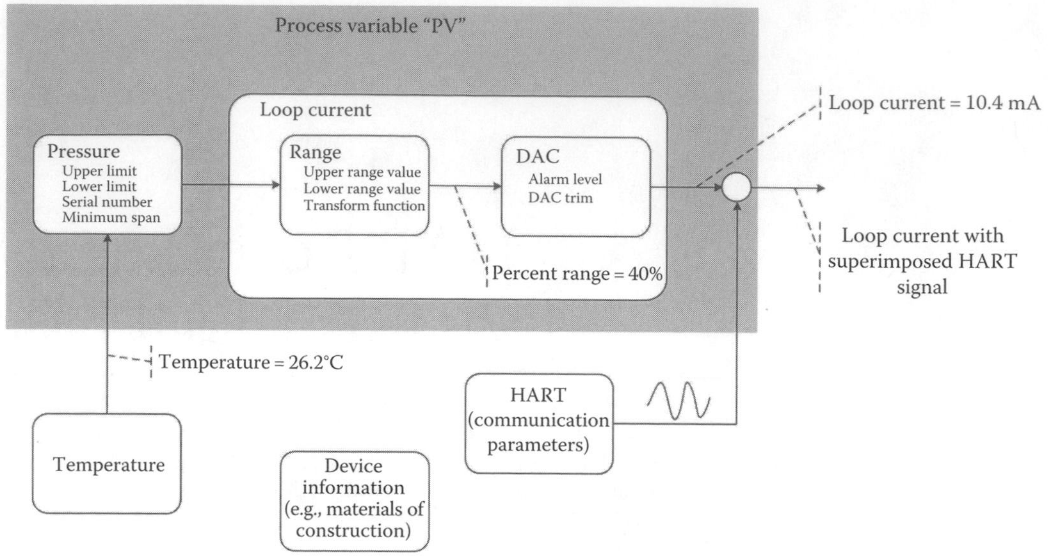

FIGURE 7.1 HART pressure transmitter.

field devices continues to accelerate as microprocessor capabilities increase and power requirements decrease. As shown in Figure 7.5, HART provides access to variables in multivariable field devices and provides mechanisms to map parameters.

7.1.2 HART Signal

All HART field devices support two communication channels: the traditional current loop and HART communication. The current loop occupies the 0–25 Hz band used by all 4–20 mA devices to continuously transmit a single process value. HART communications reside in a higher 500–10,000 Hz band. The earliest HART communication protocol was based on the Bell 202 telephone communication standard and operates using the frequency shift keying (FSK) principle. The FSK digital signal is made up of two frequencies: 1200 and 2200 Hz representing bits 1 and 0, respectively. As shown in Figure 7.2, sine waves of these two frequencies are superimposed on the 4–20 mA analog signal. Because the average value of the FSK signal is always zero, the 4–20 mA analog signal is not affected. The digital communication signal has a response time of approximately two to three data updates per second without interrupting the 4–20 mA analog signal. A minimum loop impedance of 230 Ω is required for communication. Phase shift keying (PSK), a later variation of the physical layer, is also supported. PSK supports the higher speed of 9600 Bd. PSK is not widely used.

AHART field instrument (slave) transmits a HART signal by modulating a high-frequency carrier current of about 1 mA p–p onto its normal output current. This is illustrated in Figure 7.2 for the same 10.4 mA analog signal used in Figure 7.1. The signal must be continuous at the 1200 and 2200 Hz boundaries.

7.1.3 HART System

Figure 7.3 shows a simple HART system setup with a single device connected to the host. In this setup, only one device is connected through the HART interface. As shown in Figure 7.4, it is also possible

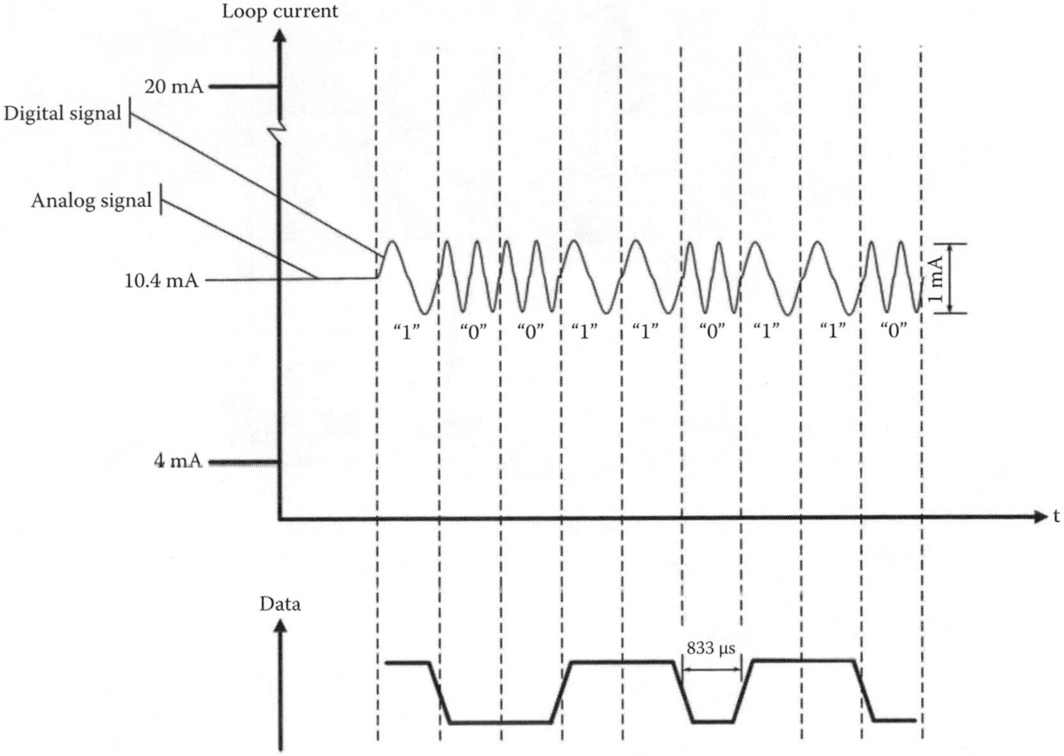

FIGURE 7.2 HART simultaneous analog and digital communication.

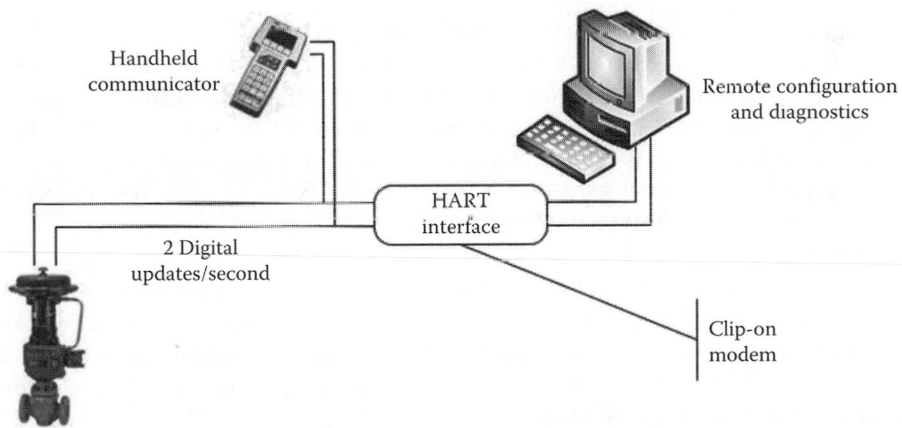

FIGURE 7.3 HART system.

to support more than one device using a multidrop setup. The wire connections, power, and isolation considerations are described later in this chapter.

The multidrop mode of operation requires only a single pair of wires and, if applicable, safety barriers and an auxiliary power supply for up to 15 field devices. All process values are transmitted digitally. In multidrop mode, all field device polling addresses are >0, and the current through each device is fixed to a minimum value (typically 4 mA).

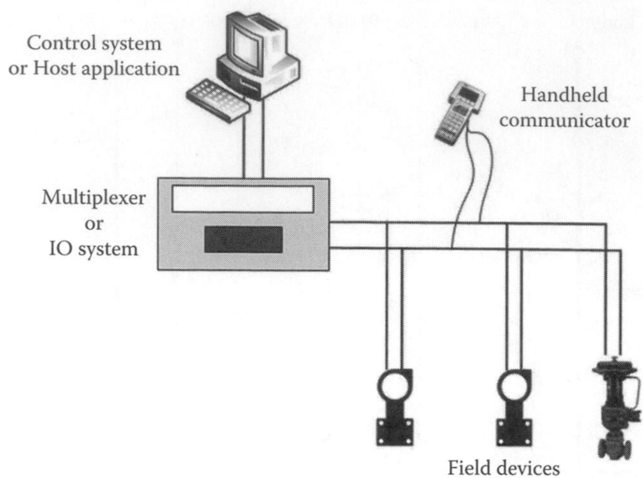

FIGURE 7.4 Multidrop mode of operation. Note: Instrument power is provided by an interface or external power source that is not shown.

7.1.4 HART Field Devices

The minimum requirements, as described in a separate chapter, that all HART field devices must meet include

- Adherence to physical and data link layer requirements
- Support for minimum application layer requirements (e.g., device status information, engineering units, IEEE floating point)
- Support for all Universal Commands

Most HART field devices also support a variety of Common Practice Commands. Whenever a field device supports a Common Practice Command, it must implement that command exactly as specified.

The 4–20 mA current loop communication channel connects the HART field device to the control system. The current loop is always connected to the PV (see the calibration subsection). Furthermore, some devices support more than one current loop connection. When this occurs, the second current loop is connected to the SV and so on. In other words, the dynamic variables connect a HART field device to control system.

A *device variable* directly or indirectly characterizes the connected process. As shown in Figure 7.5, dynamic variables may be mapped to process measurements, calculated parameters, or parameters within the device itself. In this example, device variables 0, 1, and 2 are connected to the process, and device variable 4 is calculated based on device variables 0 and 1. Device variable 3 is within the field device itself (e.g., it may be an onboard temperature sensor).

Multivariable field devices allow the device variable mapped to the PV (i.e., the current loop) to be configured based on application requirements. For example, a level gauge may measure level and calculate volume. In this example, the level gauge allows the user to choose whether the current loop reports level or volume. This allows the data that are most important to the user to be transmitted continuously via the current loop. Simultaneously on the same wire, the other PVs can be accessed using HART communications.

7.1.5 Calibrating HART Field Devices

All field devices need periodic calibration. Calibration of HART devices includes calibrating the digital process value, scaling the process value into a percent of range, and transmitting it on the 4–20 mA

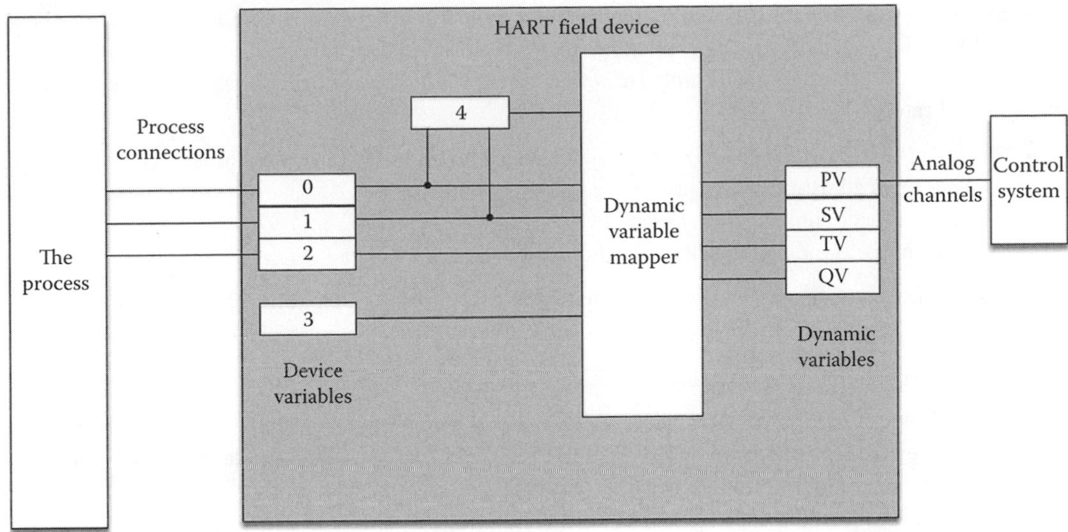

FIGURE 7.5 HART device variables vs. dynamic variables.

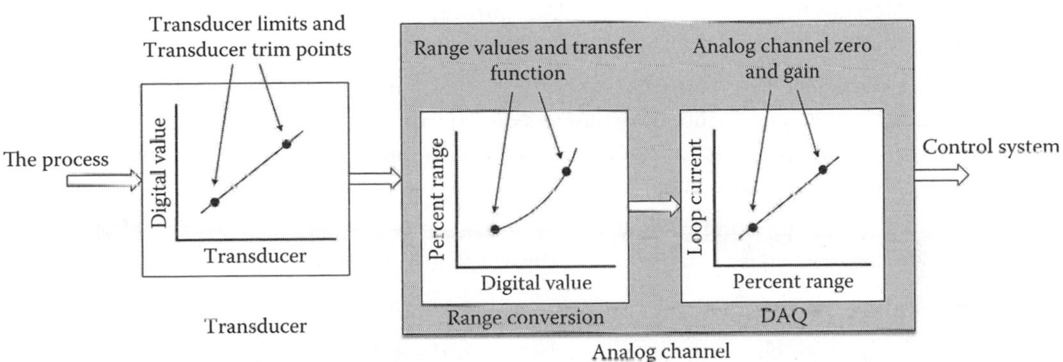

FIGURE 7.6 PV blocks that can be calibrated.

current loop. HART devices use standardized commands to test and calibrate the loop current, rerange a field device, and even perform a transducer trim [2][3].

Figure 7.6 is a block diagram showing the processing steps associated with generating PV and will be used to illustrate the calibration process. Three blocks are used for data acquisition and communication of PV via the current loop. From left to right, these blocks include the transducer block, range block, and DAQ block.

7.1.5.1 Calibrating the Process Variable

The transducer block is responsible for generating a precise digital value representing the connected process. In HART, the transducer block contains properties like the upper and lower transducer limits, transducer trim points, as well as device-specific transducer characterization data. Access to characterization data is usually not available or adjustable in the field. All devices require periodic calibration of the digital value produced by its transducers.

The calibration of the transducer blocks (i.e., transducer trim) consists of supplying a simulated transducer value and comparing the transducer value provided by the field device with the value measured by a traceable reference to determine whether calibration is required. If calibration is required, it is performed using the HART protocol. In general, calibration is performed by providing the field device with correct transducer values: one near the lower limit and another near the upper limit. Using these values, the field device performs internal adjustments to correct its calculations.

There are documenting process calibrators that perform all of these functions. These documenting process calibrators

- Provide the simulated process signal and an accurate, traceable reference value
- Determine if adjustment is required
- Issue the HART commands to complete the calibration process
- Record the as-found and as-left calibration data

Some instrument management systems are capable of automatically uploading the calibration records for trending and later analysis.

7.1.5.2 Scaling the Process Variable

The next two blocks in Figure 7.6 translate the transducer value into a 4–20 mA signal. Traditionally, this has been referred to as the *zero and span* of the device. HART supports standardized commands to rerange a field device and to calibrate the loop current.

In the range block, HART uses the upper- and lower-range values to scale the transducer value into *percent range*. The lower-range value sets the transducer value that produces a 4 mA signal. The upper-range value specifies the transducer value that will produce a 20 mA signal. In many field devices, the upper-range value can be smaller than the lower-range value. For example, a level gauge could be configured to provide 4 mA when the tank is full and a 20 mA value when it is empty. This can be useful, for example, when the level gauge 4–20 mA signal is connected to a proportioning valve.

Note: The differences between ranging and calibrating a device are critical. Considerable confusion can result if the difference between calibrating the PV's digital value and reranging the instrument is not understood. Many hours can be spent attempting to get an *accurate* 4–20 mA signal if the function of the transducer and range blocks is ignored.

The range block can also apply a transfer function (e.g., linear, square root, and cubic spline) to the PV signal. Square root functions are frequently used to approximate a flow from a differential pressure measurement.

7.1.5.3 Producing the Loop Current

The DAQ block is responsible for the actual production of the 4–20 mA signal. HART provides standardized commands to calibrate the loop current. Since the range block already scales the process value, the responsibility of the DAQ block is to ensure 0% produces 4 mA and 100% produces 20 mA. The objective of calibrating this block is to ensure that when the device thinks the loop current is, for example, 7.956 mA, there is exactly 7.956 mA on the wire. The meaning of 7.956 mA, in terms of the process value it represents, has been already established in the previous two blocks.

From a system viewpoint, the purpose of the 4–20 mA current loop is to communicate with the control system. Consequently, the most critical requirement is that the field device and the control system both agree on the value of the current loop. Agreement between the field device and the control system on the milliamp value is much more critical than the absolute accuracy in many applications.

7.2 HART Networks

7.2.1 Understanding IO System Trade-Offs

All communications systems can be characterized by their throughput and latency:

- *Throughput* indicates the maximum number of transactions per second (tps) that can be communicated by the system.
- *Latency* measures the worst-case maximum time between the start of a transaction and the completion of that transaction.

Throughput and latency are affected by the communication protocol requirements and by implementation decisions. Knowledgeable end users will insist on receiving these two statistics when evaluating system components. The HART Application Guide includes a host capability checklist.

The performance (throughput and latency) of IO systems can vary dramatically. This subsection describes the design decisions and architecture that affect HART IO system performance.

HART can provide 2.65 PV updates per second (also referred to as transactions per second or tps). This is a typical value, and actual throughput can be less. For example, slave devices may begin their response later in the allotted transmit time window, or the message might be a HART command containing a longer data field. In addition, hosts have a small time window to begin their message transmission. If that window is missed, the HART network becomes *unsynchronized*, and the host must back off an extra time interval. The resulting extra time lowers performance to 0.88 updates per second (or about three times slower) or less.

7.2.2 Multidrop Connections

All HART-compatible host and slave devices must support multidrop. In multidrop, several HART field devices are connected via the same pair of wires. Latency is increased because several commands may be enqueued and waiting for transmission.

In multidrop networks, throughput remains the same (2–3 tps). However, latency increases and is proportional to the number of device on the loop. Even with the degraded latency, multidrop is suitable for many applications (e.g., temperature monitoring or tank inventory management).

7.2.3 Burst-Mode Communications

Burst-mode allows a field device to be configured to continuously transmit digital process values. All HART hosts must support burst-mode; however, utilization of burst-mode depends on the application. If common multivariable field devices are employed, burst-mode maximizes the update rate of secondary PVs. For example, secondary PVs can be normally updated (using Command 3) 1.8 times per second. With burst-mode, the secondary PVs can be updated 2.4 times per second (about a 30% improvement).

However, configuration of a field device can take somewhat longer. When burst-mode is enabled, the burst-mode device is responsible for token recovery. Consequently, hosts only have network access after a burst message from the field device. This results in a throughput of 0.80 configuration read commands per second. That is about 2.5 times slower than with burst-mode not in use.

7.2.4 Summarizing HART Performance

For optimum performance, I/O systems should include one (hardware or software) modem for each channel. With such an I/O system, using all HART capabilities is practical including continuous status and diagnostics, remote configuration of instruments, acquisition of secondary PVs from multivariable field device, and the deployment of multidrop networks.

Lower-cost I/O systems multiplex HART communications by using one HART modem chip to support several I/O channels. With multiplexed I/O, all transactions must be serialized, and the host must resynchronize its communications each time the HART channel is changed. Consequently, the execution of the last requested command may be delayed by the time required to process all commands already enqueued plus the resynchronization delay time.

Tables 7.1 and 7.2 summarize system performance vs. I/O architecture. While actual performance can vary, these tables provide a basis for evaluating I/O system designs. Multiplexed I/O is clearly the worst performer and is even worse than multidropped I/O. For 8 channel analog I/O (a common format in many systems), multiplexed I/O throughput is 14.5 times worse than when a modem is used for each channel. Clearly, the use of multiplexed systems dramatically degrades system performance.

TABLE 7.1 Summary of Latency for I/O Systems

Number of Channels	Measured in Seconds			
	Point-to-Point	Point-to-Point (Unbuffered)	Multidrop	Multiplexed
1	0.38	1.14	0.38	0.38
4	0.38	1.14	1.51	2.73
8	0.38	1.14	3.02	5.46
16	0.38	1.14	6.04	10.92
32	0.38	1.14	12.08	21.84

TABLE 7.2 Summary of Throughput for I/O Systems

Number of Channels	Measured in Transactions per Second			
	Point-to-Point	Point-to-Point (Unbuffered)	Multidrop	Multiplexed
1	2.65	1.14	2.65	1.47
4	10.60	4.56	2.65	1.47
8	21.19	9.12	2.65	1.47
16	42.38	18.24	2.65	1.47
32	84.77	36.48	2.65	1.47

7.3 HART Networks

7.3.1 Modulated Analog Signal

HART communication is designed to be compatible with existing 4–20 mA systems. HART communication is based on communication techniques long used in telephone and telecommunications, and consequently, HART is very forgiving. For example, there are several ways to calculate HART cable lengths. No matter the technique used, the result is very conservative (i.e., HART will normally be successful on cables 25%–50% longer than calculated.)

HART cable run lengths are a function of the quality of the cable used: the lower the capacitance, the longer the possible cable length. In many cases, the cable runs are determined by intrinsic safety (IS) capacitance limitations before HART limits are reached.

7.3.2 HART Physical Layer

All HART devices support two communications channels: (1) the traditional 4–20 mA signaling and (2) the modulated HART communications. These exist in separate communications bands (see Figure 7.2), and the signals do not interfere with each other. This allows both communications to occur simultaneously. In fact, a significant portion of the physical layer specification is dedicated to defining the filtering and response times to ensure the separation of these bands.

Process control equipment normally contains a low-pass filter to block out noise. Consequently, HART looks like noise to most existing systems and is filtered out. In most cases, this feature allows HART to be used with existing systems and with no special modifications.

All HART devices are required to support the HART FSK physical. The FSK physical layer is based on the Bell 202 modem standard. FSK modulates the digital data into frequencies (*1* is 1200 Hz and *0* is 2200 Hz). FSK is quite robust, can be communicated long distances, and has good noise immunity. A single simple HART modem chip is used by a device to send and receive the HART modulated signals.

7.3.3 HART Network Guidelines

As previously discussed, HART was designed to work in typical 4–20 mA loops. Figure 7.7 illustrates a number of ways to connect devices. Table 7.3 summarizes the characteristics for networks shown in Figure 7.7. As shown in Figure 7.7, devices may be powered from the 4 to 20 mA connection (two wires) or powered from a separate connection (four wires). As an example, the loop with a two-wire transmitter has a current sensor (i.e., one low impedance device), a transmitter controlling the 4–20 mA signal (i.e., only one device varying the analog signal), and a single HART secondary host (when a handheld is connected to the loop). Most control systems use a 250 Ω current sense resistor.

7.3.3.1 HART Two-Wire Connection

HART devices (slaves) transmit by modulating the process 4–20 mA DC loop current with a 1 mA p–p AC current signal. Since the average value of the HART signal is zero, the DC value of the process loop remains unchanged. Receive circuits in a HART slave device amplify, filter, and demodulate the current signal.

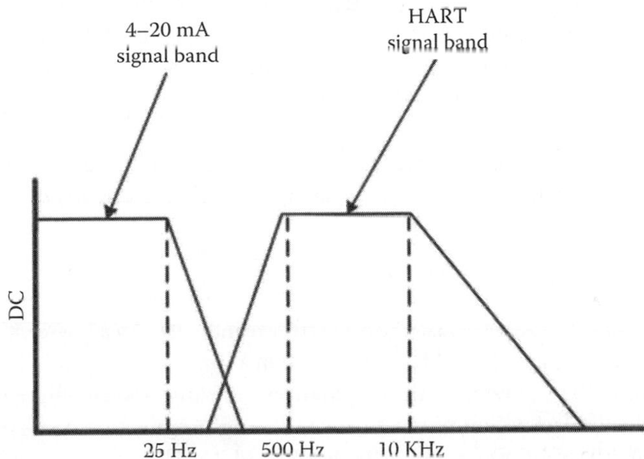

FIGURE 7.7 HART devices and networks.

TABLE 7.3 Characteristics of the HART Devices Shown in Figure 7.7

	Two-Wire Transmitter	Four-Wire Transmitter	Multidrop Network	Control System Current Input	Control System Current Output	Secondary Master	Valve Actuator
Loop-powered	Yes	No	Yes	No	No	No	Yes
4–20 mA signaling	Current sink	Current source	None	Current receiver	Current source	None	Current receiver
4–20 mA impedance	High	High	High	Low	High	Very high	Low
Hart signaling	Current	Current	Current	Voltage	Current	Voltage	Voltage
Hart impedance	High	High	High	Low	High	Medium	Low
Isolated from ground	Yes	No	Yes	No	No	Yes	Yes

7.3.3.2 HART Four-Wire Connection

HART devices (slaves) may also be powered from a separate power source. In these cases, the additional two wires are used to supply power to the device.

7.3.3.3 HART Multidropped Slave Devices

Some current loops use only digital signaling. The field instrument current is fixed at 4 mA or some other convenient value, and only digital communication occurs. Up to 15 such field instruments with unique addresses of values > 0 may be connected.

7.3.3.4 HART Primary and Secondary Master Devices

HART masters transmit by driving the loop with a low impedance voltage. Regardless of whether a master or field device is transmitting, a signal voltage of about 500 mV p–p is developed across the conductors of the current loop (assuming a 500 Ω current sense resistor) and is seen by both devices. Receive circuits in each device filter and demodulate the signal voltage.

In general, a HART primary master is the device that provides the communications between the control system (DCS) and the remote process instruments with the intent to receive process information and perform maintenance operations. A HART network that has a HART master interface integrated into the DCS will usually be configured as a primary master.

It is also possible to connect a second HART interface to a network; this second connection is referred to as a secondary master. An example of a secondary master is a handheld communicator that would be connected directly across a HART network. Such a network may have a primary master. AHART network can have only one primary and one secondary master connected at a time.

7.3.4 HART Networking Guidelines

The HART physical layers are simple. Most of the details are managed by a low-cost modem chip. HART signaling is in a separate frequency band from the 4–20 mA analog signal and, in most cases, does not interfere with existing 4–20 mA systems. The impedances, filtering, and signaling of HART devices are specified to ensure reliable HART communication when using standard 4–20 mA installation practices. HART network guidelines are summarized in Tables 7.4 and 7.5 [5].

TABLE 7.4 HART Network Guidelines

Guideline	Description
1	A network must have at least one, typically only one, low-impedance device. Total loop resistance must be between 170 and 600 Ω.
2	A network must have no more than one device varying the 4–20 mA signal.
3	Only one secondary device is allowed.
4	Cable run lengths of 3000 m for single pair cable and 1500 m for multiconductor cables is typical. Actual length depends on the number of multidropped field devices and the quality of the cable used (refer to Table 7.5).
5	Low-capacitance shielded twisted pair cable is recommended. However, HART has been successfully used over poor-quality, unshielded wiring. Don't replace wiring until you have attempted HART communication.
6	HART is compatible with IS rules, and HART communicates across most IS barriers. In general, zener diode barriers are acceptable as they do not prevent two-way communication. However, isolating barriers must be HART compatible (i.e., some isolating barriers support one-way communication only).

TABLE 7.5 Cable Capacitance

Number of Devices	65 pf/m (m)	95 pf/m (m)	160 pf/m (m)	225 pf/m (m)
1	2769	2000	1292	985
5	2462	1815	1138	892
10	2154	1600	1015	769
15	1846	1415	892	708

Note: Allowable cable lengths for 1.02 mm (#18 AWG) shielded twisted pair cable.

7.4 HART Circuits

7.4.1 HART Communications

Up until this point, the discussion has focused on HART networks. In this section, more details are provided around the actual circuits that may be used with HART. The section concludes with a number of installation examples.

7.4.2 Analog Signaling

HART devices may be connected in a conventional current loop. As shown in Figure 7.8, the HART field device (slave) signals by varying the amount of current flowing through itself.

On the other side of the circuit, the HART master or controller (primary master) detects this current variation by measuring the DC voltage across the current sense resistor. The loop current varies from 4 to 20 mA at frequencies usually under 10 Hz (refer to Figure 7.9).

The HART (digital) signal is superimposed on the 4–20 mA (analog) signal. The master transmits HART signals by applying a voltage signal across the current sense resistor, and it receives a voltage signal by detecting the HART current signal across the sense resistor. Conversely, the slave transmits by modulating the loop current with HART signals and receives HART signals by demodulating the loop current. As discussed earlier in this chapter, the HART signal is at a high-frequency carrier current (*1* is 1200 Hz and *0* is 2200 Hz) of about 1 mA p–p and superimposed on the normal 4–20 mA current.

Depending on the device, the HART signal is modulated by varying the loop current or by directly modulating a voltage onto the loop. Process transmitters vary the loop current 1 mA p–p. On a typical loop,

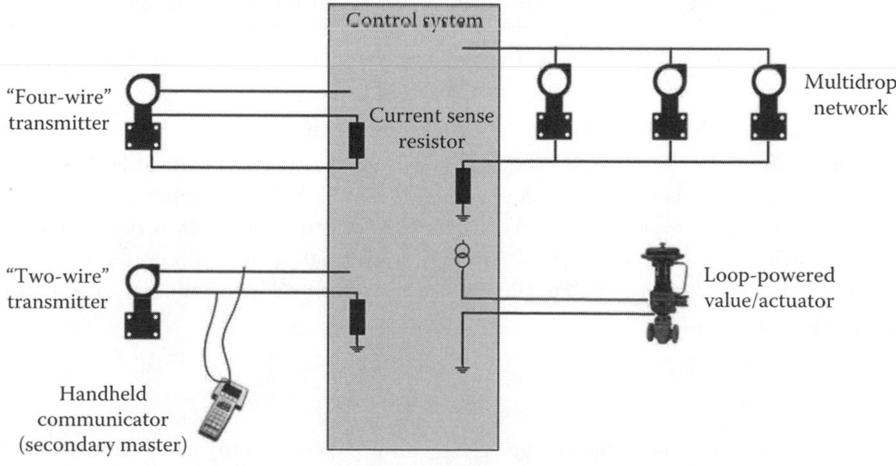

FIGURE 7.8 Current loop with HART signal sources.

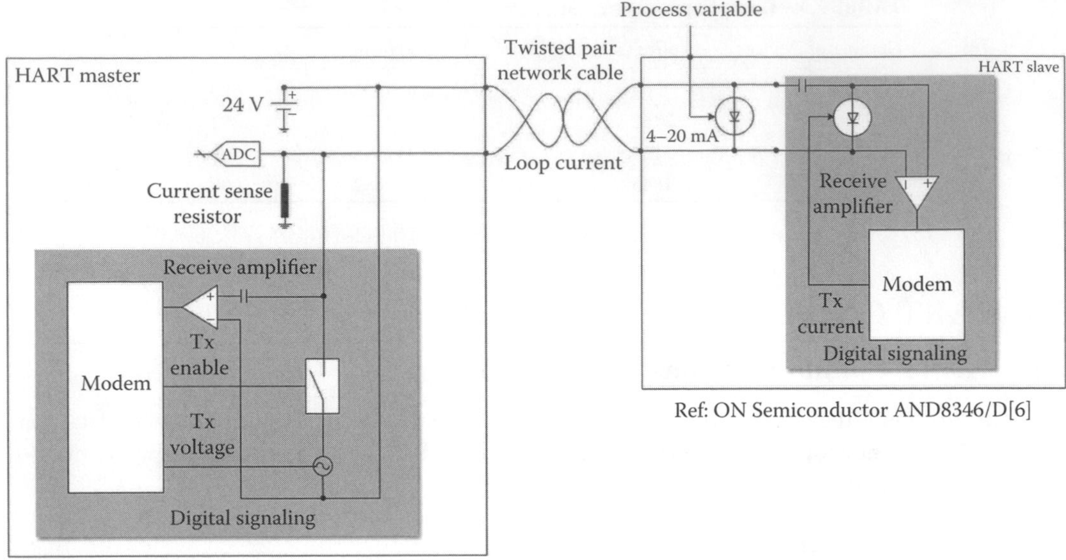

FIGURE 7.9 HART frequency bands.

the current sense resistor is 250 Ω, which produces a 250 mV peak to peak HART signal. Other equipment (e.g., handhelds, DCSs, and valves) modulate a voltage on the loop. For each type, the HCF specifications establish signaling levels, device impedances, and internal filter requirements (refer to Table 7.3) [4].

The slave interface to the network is typically a current regulator. Its output current is controlled by varying a much smaller current into an op-amp summing junction. This junction is a convenient point at which to sum the analog and digital signals, thereby achieving superposition of the digital signal onto the analog. The digital transmit signal (Tx current) is capacitively coupled into the summing junction, to preserve the DC accuracy of the 4–20 mA analog signal.

The HART master interfaces to the network act as a voltage source. The network is driven by a low-output impedance (600 or less) voltage amplifier. The output of the voltage amplifier needs to be high impedance. Typically, a HART master uses a coupling capacitor to ensure that the transmitter circuit meets the output impedance requirements specified in the HART physical layer specification.

7.4.3 HART Inputs

Control systems support from 1 channel per HART modem to as many as 32 channels per HART modem. When multiple channels utilize a single HART modem, the modem must be multiplexed.

To function correctly, the two-wire AI circuits require that 24 VDC be supplied through the bussed field power connection. For four-wire circuits, a four-wire termination block is typically used. Power for four-wire transmitters must be provided from an appropriate external power source. Typical AI specifications are summarized in Table 7.6. Wiring diagrams for two-wire and four-wire AI circuits are shown in Figures 7.10 and 7.11.

7.4.4 HART Outputs

Control systems support from 1 channel per HART modem to as many as 16 channels per HART modem. When multiple channels utilize a single HART modem, the modem must be multiplexed. Typical AO specifications are summarized in Table 7.7. Wiring for AO circuits is shown in Figure 7.12.

TABLE 7.6 AI 4–20 mA Specifications

Item	Specification
Nominal signal range (span)	4–20 mA
Full signal range	1–22.5 mA, with over-range checking
Field circuit per channel	32 mA maximum
Resolution	16 bits
Roll-off frequency	−3 dB at 2.7 Hz, −20.5 dB at one-half the sampling frequency applies to the analog signal
HART scan times	600–800 ms (typical)

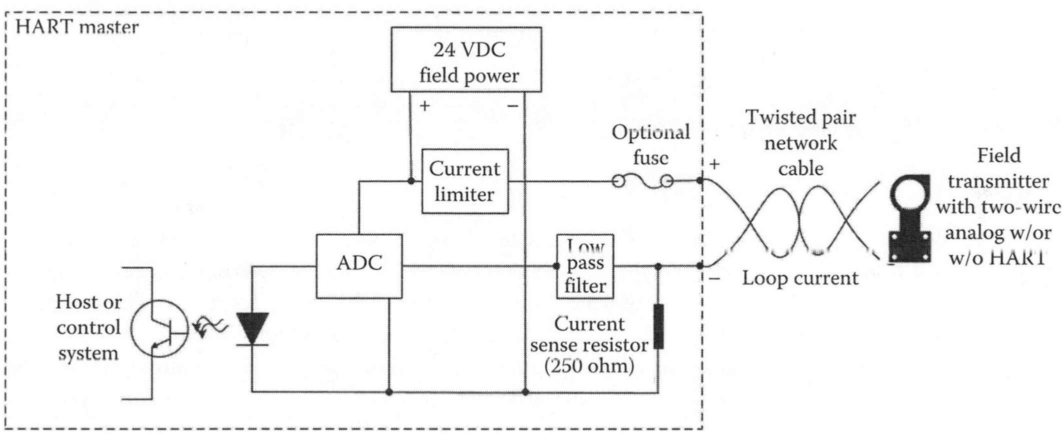

FIGURE 7.10 Wiring diagram for a two-wire AI device. [7]

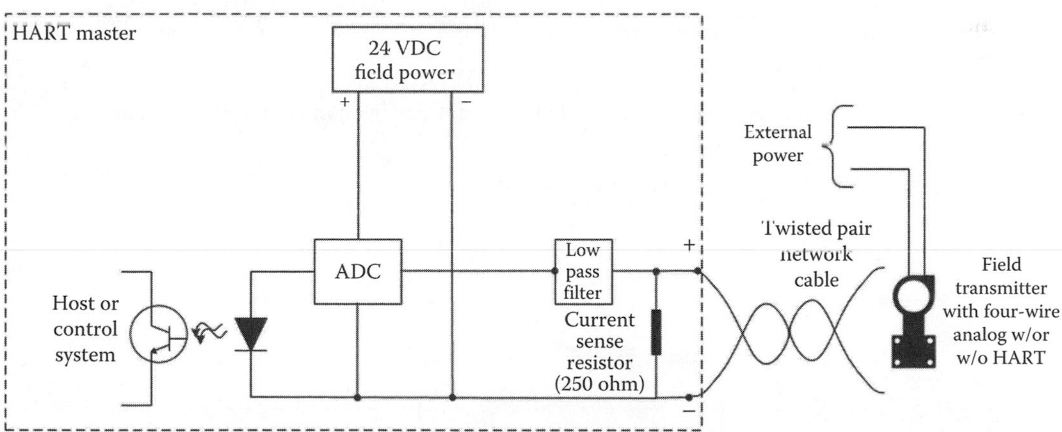

FIGURE 7.11 Wiring diagram for a four wire AI device.

TABLE 7.7 AO 4–20 mA Specifications

Item	Specification
Nominal signal range (span)	4–20 mA
Full signal range	1–23 mA, with over-range checking
Resolution	12–16 bits
HART scan times	600–800 ms (typical)

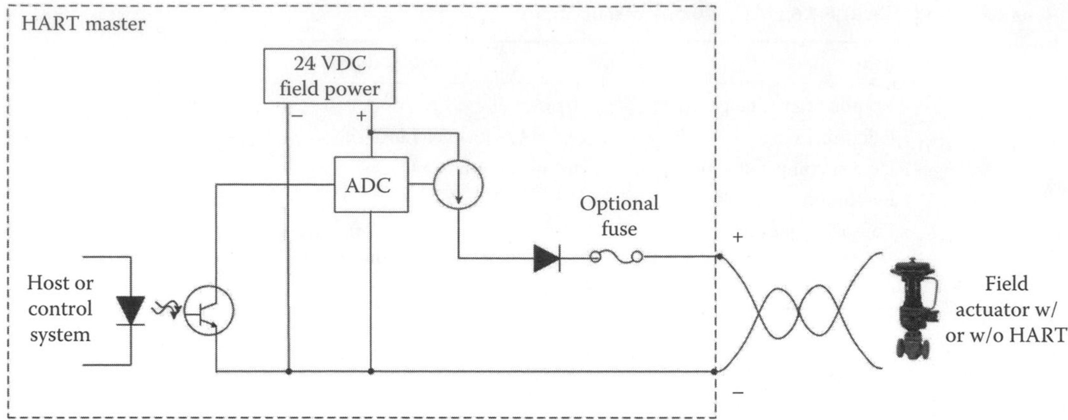

FIGURE 7.12 Wiring diagram for an AO device.

7.4.5 HART Intrinsically Safe Circuits

IS is a method of providing safe operation of electronic process control instrumentation in hazardous areas. IS systems keep the available electrical energy in the system low enough that ignition of the hazardous atmosphere cannot occur. No single field device or wiring is intrinsically safe by itself (except for battery-operated, self-contained devices), but is intrinsically safe only when employed in a properly designed IS system.

HART-communicating devices work well in applications that require IS operation. IS devices (e.g., barriers) are often used with traditional two-wire 4–20 mA instruments to ensure an IS system in hazardous areas. With traditional analog instrumentation, energy to the field can be limited with or without a ground connection by installing one of the following IS devices:

- Shunt-diode (zener) barriers that use a high-quality safety ground connection to bypass excess energy
- Isolators, which do not require a ground connection, that repeat the analog measurement signal across an isolated interface in the safe-side load circuit

A Zener barrier is shown in Figure 7.13. Additional information can be found in the HART planning guide.

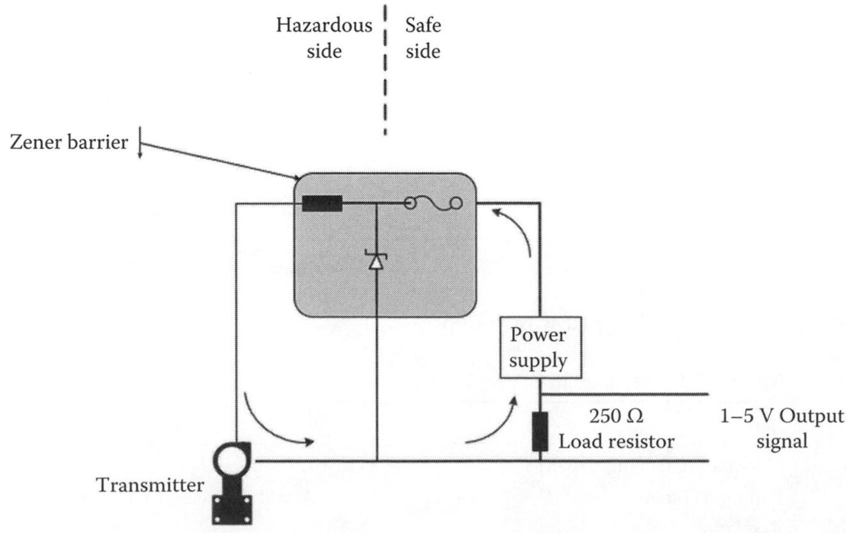

FIGURE 7.13 A 4–20 mA loop with a Zener barrier.

7.5 In Closing

The simplicity of the HART protocol contributes to its popularity and to the low cost of HART compatible equipment. This section demonstrates the power, the value, and the wide range of HART capabilities and applications. Growth of HART capabilities and applications is certain to continue for many more years.

Use this information to understand the capabilities of HART and to evaluate the field devices and host systems you encounter.

References

1. HART Communication Foundation, *HART Field Communications Protocol Specification.* HCF_SPEC-12.
2. Holladay, K.L. Calibrating HART transmitters. ISA Field Calibration Technology Committee. http://www.isa.org/~pmcd/FCTC/CalibrationReferences/CalHart.PDF
3. Holladay, K.L. and Lents, D. Specification for field calibrator interface. ISA Field Calibration Technology Committee. http://www.isa.org/~pmcd/FCTC/FieldCalbrIntf/FieldCalbrIntfIndex.htm
4. HART communication, http://www.hartcomm.org.
5. HART Application Guide—HCF LIT 34, http://www.hartcomm.org.
6. ON Semiconductor AND8346/D Application Note, http://onsemi.com.
7. DeltaV Documentation, http://www2.emersonprocess.com.

8

HART Device Networks

8.1 Introduction 8-1
History • Improved Plant Operations • Reduced Configuration, Installation, and Checkout • Monitoring, Control, and Safety • WirelessHART • Security
8.2 HART Architecture 8-7
Theory of Operation • Communication Modes • HART Network Topologies • HART Commands
8.3 HART Communication Stack 8-15
Overview • Wired Protocol • Wireless Protocol
8.4 System Tools 8-21
Host Interfaces • Hosts with Limitations on Data Handling • Hosts with Pass-Through Messages • Tools Utilizing Device Configuration (DDL) • WirelessHART Tools
8.5 Planning and Installation 8-28
Wired HART • WirelessHART
8.6 Application Example: Bioreactor 8-31
8.7 Future Directions 8-32
References 8-33

Mark Nixon
Emerson Process Management

Deji Chen
Emerson Process Management

8.1 Introduction

8.1.1 History

The HART™ (Highway Addressable Remote Transducer) fieldbus standard has been in existence since the late 1980s. It is a bidirectional communication protocol that provides data access between intelligent field instruments and host systems. It is administrated by the HART Communication Foundation (HCF), located in Austin, Texas [6]. In its initial release, the HART field communication protocol was superimposed on a 4–20 mA signal providing two-way digital communications with smart field instruments without compromising the integrity of the measured data. As shown in Figure 8.1, the HART protocol has evolved from a 4 to 20-mA-based protocol to the current wired and wireless-based technology with extensive features supporting security, unsolicited data transfers, event notifications, block mode transfers, and advanced diagnostics. Diagnostics now include information about the device, the equipment the device is attached to, and in some cases the actual process being monitored.

The earliest HART communication protocol was based on the Bell 202 telephone communication standard and operated using the frequency shift keying (FSK) principle. The digital signal is made up of two frequencies: 1200 and 2200 Hz representing bits 1 and 0, respectively. As shown in Figure 8.2, sine waves of these two frequencies are superimposed on the analog signal cables to provide simultaneous analog and digital communications. Because the average value of the FSK signal is always zero, the 4–20 mA analog signal is not affected. The digital communication signal has a response time of approximately 2–3 data updates/s without interrupting the 4–20 mA analog signal. A minimum loop

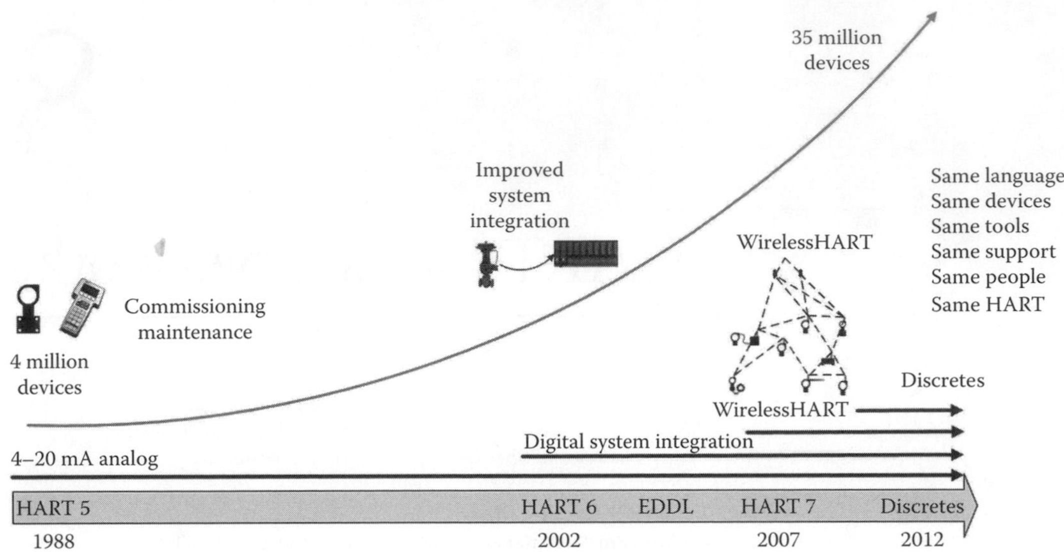

FIGURE 8.1 Evolution of HART.

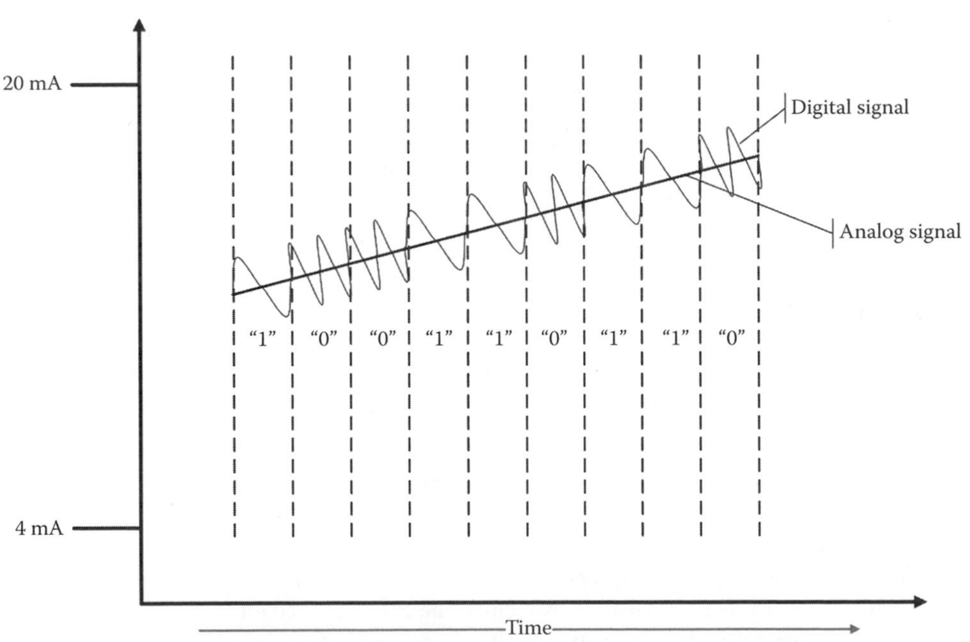

FIGURE 8.2 Wired HART simultaneous analog and digital communication. [7][8]

impedance of 230 Ω is required for communication. Phase shift keying (PSK), a later variation of the physical layer, is also supported. PSK supports the higher speed of 9600 Bd. PSK is not widely used.

Figure 8.3 shows a wired HART system setup. In this setup, only one device is connected to the HART interface. It is also possible to support more than one device using a multidrop setup. In a multidrop deployment, as shown in Figure 8.7, more than one slave device may be connected to the master device on the same HART network. In HART, unlike simpler protocols like Modbus, data are encapsulated together so that key variables always remain in proper context. For example, measured

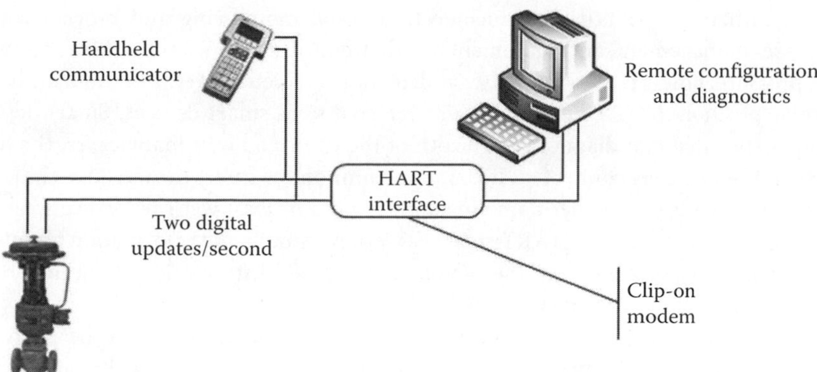

FIGURE 8.3 HART system.

variable, units of measure, and status are always communicated together in the same data packet so that they remain synchronized and in proper relation to each other.

The latest protocol addition to the HART standard is WirelessHART™ [1]. WirelessHART is a secure networking technology operating in the 2.4 GHz ISM radio band. WirelessHART utilizes IEEE 802.15.4–compatible DSSS radios with channel hopping on a packet by packet basis. WirelessHART communication is arbitrated using the WirelessHART network manager to schedule link activity. A given communication link may be dedicated to communication between a network pair or a link may be shared. WirelessHART is shown in Figure 8.4.

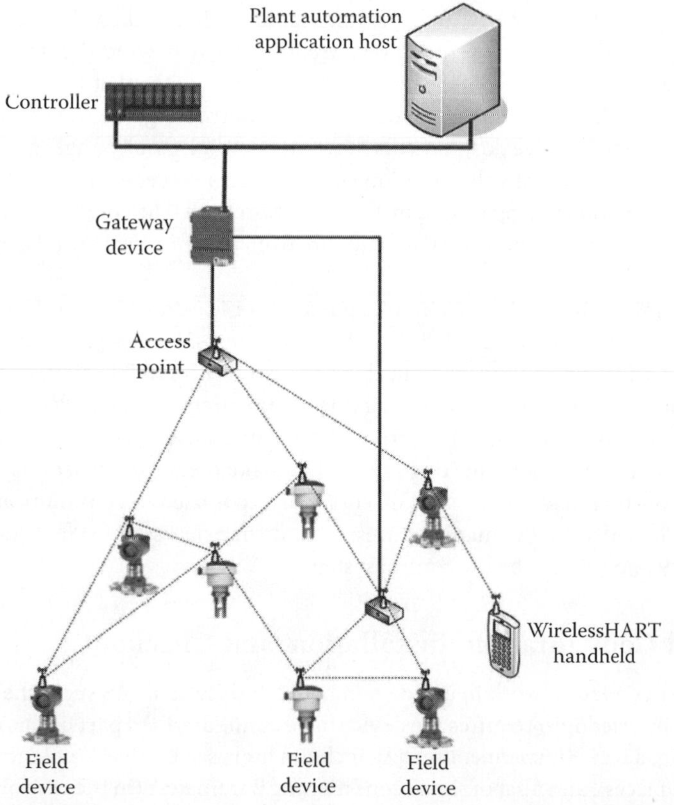

FIGURE 8.4 WirelessHART system.

Applications utilizing HART devices include traditional monitoring and process control, safety applications, asset management, and equipment health monitoring applications. Measurements include temperature, pressure, flow, pH, conductivity, gas detection, discretes, level, vibration, mass flow, energy usage, and valve position. HART devices are also referred to as smart devices. Smart devices include advanced diagnostics that can diagnose the health of the device and, in many cases, the health of the process to which the device is connected. It is not uncommon for smart devices to include diagnostics that can detect plugged lines, burner flame instability, agitator loss, wet gas, orifice wear, leaks, flow cavitation, and excessive vibration. HART smart devices communicate information about the process they are monitoring, information about the overall network, plus information about the process equipment and when they need maintenance.

As the latest addition to the HART family, WirelessHART brings more to the portfolio. Its key factors are security, robustness, availability, and ease of use. To ensure security, all devices joining WirelessHART networks are authenticated and all communications secured. To maintain robustness and high availability, the wireless network employs several redundancy features and is able to adapt to changes in the environment. Ease-of-use includes features such as status on measured values, time stamps, event latching and confirmation, and large data transfer mechanisms. Security, measurement, control, and diagnostics are fully integrated.

8.1.2 Improved Plant Operations

Improved operations include the control systems and the operator's ability to view and adjust the process, interact with equipment, and respond to process conditions. Consider the following scenario in which the pH of a tank is controlled and the pH measurement is oscillating about a setpoint. As a first step in investigating this problem, the operator would likely put the valve target, the actual valve position, and the pH reading and its setpoint on a trend. After reviewing the trended oscillations, the operator may decide to adjust the controller gain. The operator notes that the oscillations remain the same regardless of whether they increase or decrease the controller gain. The operator also notes that the read-back value of actual valve position indicates that the minimum change in valve position is 0.5% and notes that a step in an actual valve position change always precedes the pH change. The operator suspects that the oscillation is being caused by the resolution (stick-slip) of the control valve. To follow-through on this, the operator requests a valve signature from the smart valve and compares what is seen with the actual process.

HART supports this scenario through an integrated set of features. The primary signal, in this case pH, is communicated through the HART 4–20 mA signal. The 4–20 mA signal is fast enough to capture and communicate oscillations as described in this scenario. With HART, digital values are utilized to communicate actual valve position as well as auxiliary measurements. Request/response communications are used to manually adjust the valve output. Block data transfer is used to communicate the valve signature. Periodic and burst data reporting are used to make the most efficient use of communication resources. Event reporting may also be used. Event reporting is used to communicate device alerts and other information included by the manufactures. Both device status and signal status are included as part of Command 9 and utilized by the control system.

8.1.3 Reduced Configuration, Installation, and Checkout

HART devices are designed to work in a wide number of applications. As such, the actual process that a specific device is inserted into requires the device to be configured. As part of this configuration, each device is given a tag, a scale (instrument scaling includes high-scale value, low-scale value, engineering units, and decimal places), and a set of signal conditioning parameters (in the case of a valve, it is important to know which direction the valve goes when the signal valve is increased). The set of parameters that a device supports and the methods available to the device are described in device description language (DDL). Users often make use of standard parameter value templates for different types of devices and use these templates to customize devices for use in their plants (tools such as an AMS 475 field communicator use DDL files to calibrate and configure field devices). In this way, devices can be completely configured off-line at the factory, in the shop, or prior to start-up by downloading the specific configuration into the device as part of the control system setup.

Configuring the control system requires a different set of information: measurements that the device supports and the signal information that is to be associated with those measurements. Signal information includes signal tag (since devices can be multivariable, they can have multiple signal tags associated with them), scaling information, alarm information, linearization, and description. Control system configuration may be downloaded into the control system at various points in the control system development. The HART protocol provides an extensive set of features to support the overall configuration, installation, and checkout procedure. All HART and WirelessHART devices are required to have a maintenance port.

8.1.4 Monitoring, Control, and Safety

Table 8.1 provides a classification for monitor and control applications. As shown in Table 8.1, there are six different classes of sensor and control applications ranging from critical safety (class 0) to condition monitoring and regulatory compliance (classes 4 and 5). HART devices and technology have been deployed for all of these classifications.

To better understand these classifications, it is important to consider the question, *How are measurements being used?* In many cases, the measurements are used by operators to monitor tank levels, emission levels, water quality, equipment health, and a wide range of other things. These measurements are often used to generate reports for the US Food and Drug Administration, US Environmental Protection Agency, and other agencies. These measurements are used by plant personnel to make decisions about the operation of the process, plan maintenance activities, and schedule production runs. These measurements are often directly or indirectly used to validate that the quality requirements of the finished product have been met. These measurements are also often directly or indirectly tied into feedforward and feedback control strategies.

On the control side of the discussion, there are many kinds of final control elements including valves, agitators, blowers, and conveyors. These control elements take a value and perform some action. In the case of an on–off valve, the valve attempts to open or close. In the case of a regulating valve, the read-back of the actual valve position is important in order to determine if there are problems with the final control element itself that may be impacting control performance. In the case of an agitator, it is important to know if the agitator is moving.

TABLE 8.1 Usage Classes

Safety	Class 0: Emergency action	Always critical	Importance of message timeliness increases
Control	Class 1: Closed-loop regulatory control	Often critical	
	Class 2: Closed-loop supervisory control	Usually noncritical	
	Class 3: Open-loop control	Human in the loop	
Monitoring	Class 4: Alerting	Short-term operational consequence (e.g., event-based maintenance)	
	Class 5: Logging and downloading/uploading	No immediate operational consequence (e.g., history collection, sequence-of-events, preventive maintenance)	

Reference ISA100.11a annex C [5].

Devices that take measurements or take final actions based on the controller outputs are generally smart HART devices. HART devices are the most commonly deployed fieldbus devices in the world.

8.1.5 WirelessHART

Protocols such as Foundation Fieldbus, Profibus, and HART are well established in the industrial process control space. Although the cost of installing and maintaining the wiring for these networks is often quite significant, they continue to dominate installations. Many years of experience with wired technologies and procedures for using them are well established. As such, the need to replace these networks with something completely new is relatively low. However, there is also a continued need to improve productivity and safety while at the same time reducing costs. In general, this means more measurements. In many cases, the most effective way to add these measurements is with wireless technology. This has created an opening for WirelessHART, which has emerged as a very significant technology in many plants.

The technology behind WirelessHART includes a combination of device improvements, network technology, and network management. Network management is the key to the operation of the wireless network in that it is used to manage network resources efficiently, schedule communications to meet the requirements of the application, and establish routing to meet reliability and performance goals. Network health reports are used to adapt the network to changing conditions. Network resources are dynamically allocated to cover changes in throughput. Security, reliability, ease of use, long battery life, and support for large numbers of devices are key requirements that can be met by a network manager.

WirelessHART devices can be line-powered or powered by either batteries or energy harvesters. Non-line-powered wireless devices (i.e., battery-powered devices) offer the most flexibility for deployment, but low energy consumption is required to make them practical. While these devices may transmit messages at an interval of every second, more typically, the average will be in increments of 8 s or longer. In general, periodic data updates from these devices will include only a few tens of bytes of data per message. Additionally, some devices may transmit stored files or time-series data on a daily basis that could be tens of kilobytes of data. In all of these cases, the on–off ratio is key to lowering the power required by the non-line-powered device.

8.1.6 Security

While remaining backward compatible, HART evolves continuously to keep up with the latest developments and requirements from the industry. At current release version 7, HART introduced WirelessHART as the first industrial wireless mesh network standard and Discretes for controlling discrete devices. Significantly, WirelessHART has addressed one of the biggest challenges in the process industry these days: plant security. The success gained with WirelessHART provides the guidance for the next generation of secured process plants.

WirelessHART employs robust security measures to protect the network and secure the data at all times. These measures include the latest security technologies to provide the highest levels of protection available. It uses industry standard 128 bit AES encryption algorithm at multiple tiers. The secret network key is used at the data link layer to authenticate each data transmission. At the network layer, each session has a different key to encrypt and authenticate peer-to-peer communication. A different join key is used for each device to encrypt and authenticate during the device join process. In addition, the network manager periodically changes all the keys during the lifetime of the network. WirelessHART also supports access control list (white list) through the quarantine state as part of the joining process.

8.2 HART Architecture

8.2.1 Theory of Operation

HART is a communication protocol. Its purpose is to enable digital data exchange between the field device and the host computer. The host is often represented by a master device or a gateway in the wireless case. Just like with any other network protocol, the HART host first needs to talk to the device, to find out what the device provides, and then makes use of what the device provides. In this section, how HART identifies a device, how HART describes a device's capability, what data are exchanged, and how people could develop products around HART are described.

8.2.1.1 Device Identification

Each HART device has a 38-bit address that consists of the manufacturer ID code, device type code, and device-unique identifier. A unique address is encoded in each device at the time of manufacture. A HART master, for example, the AMS 475, must know the address of a field device in order to communicate successfully with it. A master can learn the address of a slave device by issuing one of two commands that cause the slave device to respond with its address:

1. Command 0, Read Unique Identifier—Command 0 is the preferred method for initiating communication with a slave device because it enables a master to learn the address of each slave device without user interaction. Each polling address (0–15) is probed to learn the unique address for each device.
2. Command 11, Read Unique Identifier by Tag—Command 11 is useful if there are more than 15 devices in the network or if the network devices were not configured with unique polling addresses (multidropping more than 15 devices is possible when the devices are individually powered and isolated.). Command 11 requires the user to specify the tag numbers to be polled.

A WirelessHART device uses this identifier to form its unique long address defined in the wireless network.

8.2.1.2 Electronic Device Description Language

Electronic DDL (EDDL) [2] is a machine-readable language used to describe devices in a common and consistent way. The description describes the device, methods provided by the device, measurement and device parameters that the device supports, configuration information, and, to some extent, the interactions that the users can perform with that device. The description file for a device is called the DD of the device. A DD file provides a picture of all parameters and functions of a device in a standardized language.

HART DDL is used to write the DD that combines all of the information needed by the host application into a single structured file. The DD file identifies which common practice commands are supported as well as the format and structure of all device-specific commands. A device description is shown in Figures 8.14 and 8.15.

DD source files for HART devices resemble files written in the C programming language. DD files are submitted to the HCF for registration in the HCF DD Library. Quality checks are performed on each DD file submitted to ensure specification compliance, to verify that there are no conflicts with DDs already registered, and to verify operation with standard HART hosts. The HCF DD Library is the central location for management and distribution of all HART DDs to facilitate use in host applications such as PCs and handheld terminals.

A DD for a HART field device is roughly equivalent to a printer driver for a computer. DDs eliminate the need for host suppliers to develop and support custom interfaces and drivers. A DD-enabled host application could read and write data according to each device's procedures.

After having identified the device, the host uses the information in the unique identifier to locate the correct DD. The DD is either preloaded into the host, or retrieved from the HCF, or even read out from the device directly. Once the DD is known, meaningful communication may start between the host and the device.

8.2.1.3 Accessing Data

There is considerable variation between device types with regard to what data are available from it. The most common data types are the process variable, a percent of range, and a digital reflection of the analog mA signal or device status. These values are often mapped to the HART protocol primary variable (PV), secondary variable (SV), tertiary variable (TV), and fourth variable (FV). For example, a mass flow meter has derived values that would normally be obtained using standard analog instruments. The PV would be the mass flow value, the SV could be the static pressure, the TV could be the temperature, and the FV could be the digital mA signal reflection. These mappings are device dependent and user selectable. Valves have position and cycle information allowing applications to determine if the valve is functioning properly and when it is time to perform a maintenance cycle.

8.2.1.4 Writing Parameters and Commanding Devices

HART also describes how to write data back to the instrument. For example, the offset and span calibration data may be written. HART also supports commands that may be used to calibrate instruments. For wired devices, all of the communications are carried out over 4–20 mA wiring using the superimposed HART digital signal. For wireless devices, these communications are sent over-the-air using the IEEE 802.15.4 radios.

8.2.1.5 Design Approach

The HCF provides HART specifications that can be used by suppliers to design and build devices, tools, and applications. These HART specifications utilize several design approaches:

- The device description (DD described with DDL).
- HART communicates via HART messages. Using this approach, devices are able to communicate without needing to know specific details about each other.
- Service or protocol structure. The HART specifications include protocol descriptions that describe how to configure, allocate, de-allocate, limit, manage, and diagnose overall device and system resources and behaviors.
- The content of HART messages are HART commands. HART commands are defined in the application layer of the HART protocol.

DDL enables universal host applications. In one place, DDL brings together all the information a host application needs to interoperate with field devices. These applications can access all data and properties of each device, thereby simplifying maintenance, support, and operation of HART-compliant devices. It works well with small handheld hosts and large integrated maintenance systems. It works with embedded applications and applications running on commercial operating systems. DDL's power and portability save costs for host and device vendors alike.

The HART protocol supports message-oriented communication [9]. With message-oriented communications, all communication between applications is based on messages that use well-known descriptions. In the case of HART, these descriptions are HART commands. With this communication pattern, it is not necessary for applications to know internal details about each other. Interaction between applications is accomplished by passing HART commands over a common messaging medium.

HART commands are classified as universal, common practice, and device specific. Universal commands are understood and used by all field devices operating with the HART protocol (device designation, firmware number, etc.). Common practice commands are usually supported by many, but not necessarily all, HART field devices (read variable, set parameter, etc.). Device-specific commands support functions that are unique to each device. Device-specific commands provide access to data about the type and construction of a device as well as information on the maintenance state and start-up. Most field devices support commands of all three classes.

Using messages and commands has several advantages. First, applications can be run in different environments (e.g., one side of the application may be running on a Windows-based host and the other end in an embedded device such as a Rosemount 3051 pressure transmitter). Second, not all devices need to support all commands and services. Third, services allow higher-level applications to be implemented independently of the actual protocol.

8.2.2 Communication Modes

HART was first conceived as a master–slave communication protocol, which means that during normal operation, each slave (field device) communication is initiated by a master communication device. Two masters can connect to each HART loop. The primary master is generally a distributed control system (DCS), programmable logic controller (PLC), or a personal computer (PC). The secondary master can be a handheld terminal or another PC. Slave devices include transmitters, actuators, and controllers that respond to commands from the primary or secondary master.

Most communications between the master and the slave are via request/response communications methods. The master initiates a request and the slave responds. The HART transport layer manages outstanding requests. The content of messages is one or more HART commands.

Later releases of HART added burst mode communications, events and event notifications, and block data transfer. Each of these communication methods is described in the following sections.

8.2.2.1 Request/Response

The HART communication protocol uses request/response messages to access and change parameter values, invoke device methods, configure devices, and, in the case of WirelessHART, manage the wireless network and network devices. As an example, a request/response message pair for reading the PV is shown in Tables 8.2 and 8.3. Command-specific response codes are shown in Table 8.4.

This command is used to read the PV from the device. The HART 4–20 mA loop also transmits a single dynamic variable value using an analog signal. The PV is always associated with the first analog channel of a device. Since the analog signal on the 4–20 mA cable could only convey a single dynamic value, common practice commands are used to define the relationship between the analog signal and the digital PV.

There are many HART commands. The classification and usages of these commands are described later in this chapter.

8.2.2.2 Burst Mode

Burst mode allows the master to instruct the slave device to continuously broadcast a standard HART reply message (e.g., the response of the primary process variable as shown in Table 8.3). The master

TABLE 8.2 Command 1 Read Primary Variable Request

Byte	Format	Description
None		

TABLE 8.3 Command 1 Read Primary Variable Response

Byte	Format	Description
0	Enum	Primary variable units
1–4	Float	Primary variable

TABLE 8.4 Command 1 Read Primary Variable Command-Specific Codes

Code	Class	Description
0	Success	No command-specific errors
6	Error	Device-specific command error
8	Warning	Update failure
16	Error	Access restricted

receives the message at the burst rate until it instructs the slave to stop bursting. Wired devices optionally support burst mode. WirelessHART devices must support burst mode.

There are several supported burst modes. These are summarized in Table 8.5.

The most common burst mode supported is continuous mode. In some versions of WirelessHART devices where higher reporting rates are required, window mode is supported. Window mode is shown in Figure 8.5.

In some applications, particularly when the control strategy must control the process against a narrow constraint, rising and falling modes may be used to adjust reporting rates. On-change is used to trigger a communication when a value changes. On-change is supported by discrete devices.

TABLE 8.5 HART Burst Modes

Burst Code	Mode	Description
0	Continuous	The *Burst Message* is published continuously at (worst case) the minimum update period
1	Window	The *Burst Message* is triggered when the source value deviates more than the specified trigger value
2	Rising	The *Burst Message* is triggered when source value *Rises Above* the specified trigger value
3	Falling	The *Burst Message* is triggered when the source value *Falls Below* the specified trigger value
4	On-change	The burst message is triggered when any value in the message changes

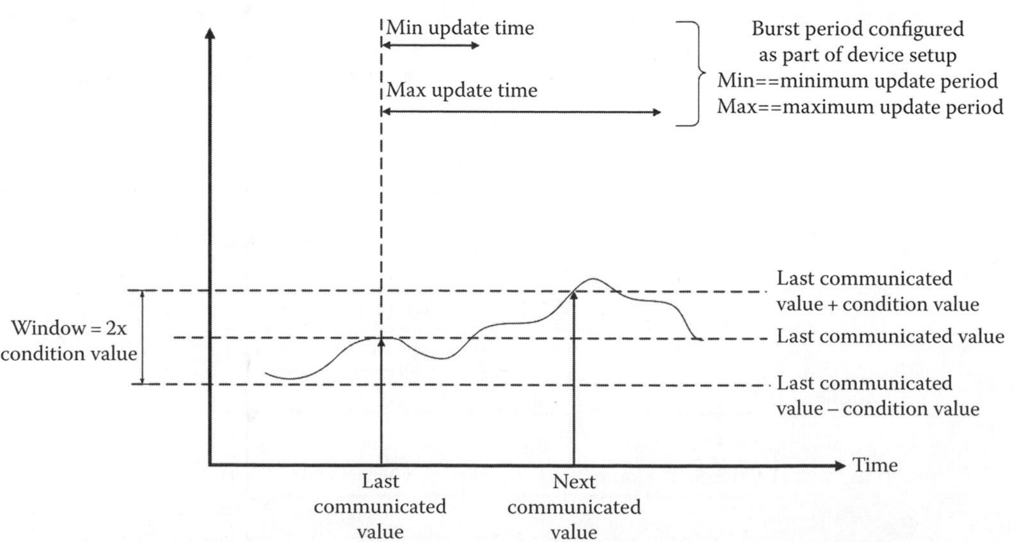

FIGURE 8.5 Window burst mode.

The commands used to setup and manage burst communications are summarized later:

- Command 103: Write burst period
- Command 104: Write burst trigger
- Command 105: Read burst mode configuration
- Command 107: Write burst device variables
- Command 108: Write burst mode command number
- Command 109: Burst mode control

Multiple burst messages may be used if there is a need to communicate many values at different reporting rates.

8.2.2.3 Events and Event Notifications

Event notification publishes changes in the device's status independently from data publishing supported in other burst mode commands. For events, the status included in the device status byte, extended device status byte, and Command 48 may be used. It is possible to specify a limited set of bits that will trigger event notification. To prevent spurious event notifications, a de-bounce interval is configured. This defines the amount of time that a condition must persist before the event notification is time-stamped and sent out. Once an event has been latched, it is transmitted repeatedly at the rate indicated by the retry period until the event has been acknowledged. Event notifications have a low priority but require a time stamp in order to indicate the first time when a notification occurred. Event notification requires and is built upon burst mode operation.

The HART protocol offers two distinct methods to display events: the device status and the Common Practice Command 48. The following commands are used to set up and manage event notification operation:

- Command 115 is used to determine the configuration of the event notification
- Command 116 selects the bits that can trigger an event notification
- Command 117 controls the timing of event notifications
- Command 118 is used to enable or disable event notification
- Command 119 is used to acknowledge the event notification

The latest values for device status, the configuration change counter, and the Command 48 response bytes are always included in Command 119. Command 116 is used to identify the bits that may trigger the event notification. Command 119 returns the status byte received only when the event mask was established using Command 116. The bytes returned in Command 119 reflect the current device status, and Command 48 data are sent regardless of which bits are masked. The time stamp remains the same until an acknowledge is received. If an acknowledge is received, the time stamp is set to the time when the acknowledge was performed.

8.2.2.4 Block Data Transfer

Block data transfer allows devices such as valves, vibration monitors, and analyzers to transfer blocks of information. The block data transfer mechanism is best classified as a transport layer service. The data transfer mechanism is like a pipeline. It establishes a *connection* between the host and the slave device and guarantees the transfer of a stream of data. The mechanism is designed to maximize the utilization of the HART communication bandwidth while performing the transfer.

To access this capability in a field device, the host establishes a connection to a specific *port* in the slave using Command 111. A host opens a port (Command 111), creating a pipeline to the device, and transfers data to or from the field device (Command 112).

The HART protocol includes robust error checking, and the data link layer ensures that a given command will be delivered to the slave device. A master receives a status word indicating whether the command was successful or not. Masters then perform retries to deliver commands when communication errors occur.

TABLE 8.6 Data Transfer Features

Index	Requirement	Feature
1	Maximize data throughput	The maximum number of bytes to be transferred in a transaction is negotiated when the *Port* is opened. The more data transferred in each transaction, the more efficient the block transfer becomes.
2	Slaves and masters have differing communication buffer space	Data are transferred in blocks as large as both the master and the slave support.
3	Transfer must be reliable. No data may be duplicated or lost.	Byte counters are used to track the data transferred. Acknowledgment indicates which byte count is next expected by the recipient.
4	Must be flexible to support differing application needs	Transfer is based on a virtual connection to a *Port*. Different *Ports* support different, well-defined functions. A block of ports are allocated for device-specific requirements.
5	Transfer must be synchronous	Function codes are used to open, close, or reset a port. The port is not closed until the master and the slave both agree the transfer is complete.

Since HART commands are normally stand-alone, implementing the retry mechanism is relatively simple. However, when transferring a block of data, retries become more complex. Since the data are transferred as a stream, the progress of the data transfer must be tracked. Table 8.6 lists features for block data transfer.

8.2.3 HART Network Topologies

Several topologies are supported by the HART specifications. Each of these is described in the following sections.

8.2.3.1 Point-to-Point

In point-to-point mode, the traditional 4–20 mA analog signal is used to communicate one process variable, while additional process variables, configuration parameters, and other device data are transferred digitally using the HART protocol as shown in Figure 8.6. The 4–20 mA analog signal is not affected by

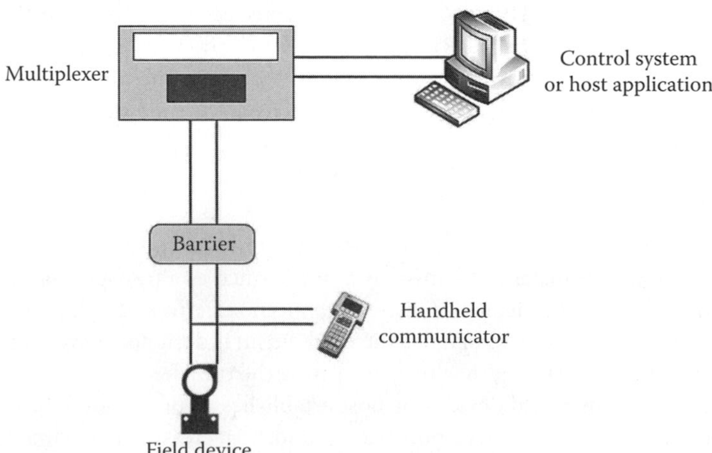

FIGURE 8.6 Point-to-point mode of operation. *Note*: Instrument power is provided by an interface or external power source that is not shown.

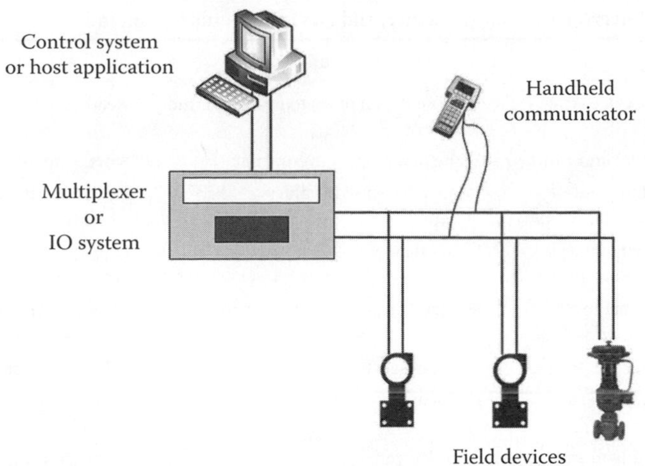

FIGURE 8.7 Multidrop mode of operation. *Note*: Instrument power is provided by an interface or external power source that is not shown.

the HART signal and can be used for control in the standard way. The HART digital signal gives access to SVs and other data that can be used for operations, commissioning, maintenance, and diagnostic purposes.

Point-to-point communications is the most common mode of operation. Most DCS suppliers provide IO with Integrated HART capabilities. In these configurations, the 4–20 mA device signal is directly connected to an IO channel as part of the DCS system.

8.2.3.2 Multidrop

The multidrop mode of operation requires only a single pair of wires and, if applicable, safety barriers and an auxiliary power supply for up to 15 field devices (Figure 8.7). All process values are transmitted digitally. In multidrop mode, all field device polling addresses are >0, and the current through each device is fixed to a minimum value (typically 4 mA).

8.2.3.3 WirelessMesh

WirelessHART extends the application of HART communications by enhancing the HART technology. The topology for WirelessHART is shown in Figure 8.4. It is a mesh network. WirelessHART builds on the wired HART universal, common practice, and device-specific commands. Since the technology is fundamentally HART, existing, previously installed host applications can, using a HART wireless gateway, access wireless-enabled HART field devices and new wireless-only HART field devices.

8.2.4 HART Commands

At the application layer, HART uses commands for data exchanges. The HART command set provides uniform and consistent communication for all field devices. In addition to the full set of commands for process automation, a set of new commands are defined to manage the WirelessHART network. Recently, HCF defined another set of commands to better deal with discrete devices in HART.

TABLE 8.7 HART Universal, Common Practice, and Device-Specific Commands

Universal	Common Practice	Device Specific
Read manufacturer and device type	Read selection of up to four dynamic variables	Read or write low-flow cut-off
Read primary variable (PV) and units	Write damping time constant	Start, stop, or clear totalizer
Read current output and percent of range	Write device range values	Read or write density calibration factor
Read up to four predefined dynamic variables	Calibrate (set zero, set span)	Choose PV (mass, flow, or density)
Read or write 8-character tag, 16-character descriptor, date	Set fixed output current	Read or write materials or construction information
Read or write 32-character message	Perform self-test	Trim sensor calibration
Read device range values, units, and damping time constant	Perform master reset	PID enable
Read or write final assembly number	Trim PV zero	Write PID setpoint
Write polling address	Write PV unit	Valve characterization
	Trim DAC zero and gain	Valve setpoint
	Write transfer function (square root/linear)	Travel limits
	Write sensor serial number	User units
	Read or write dynamic variable assignments	Local display information

8.2.4.1 HART Commands for Process Automation

As shown in Table 8.7, the basic command set includes three classes: universal, common practice, and device specific. Host applications may implement any of the necessary commands for a particular application.

All devices using the HART protocol must recognize and support the universal commands. Universal commands provide access to information useful in normal operations (e.g., read PV and units).

Common practice commands provide functions implemented by many, but not necessarily all, HART communication devices.

Device-specific commands represent functions that are unique to each field device. These commands access setup and calibration information, as well as information about the construction of the device. Information on device-specific commands is available from device manufacturers.

8.2.4.2 WirelessHART Commands

WirelessHART commands build on the same patterns used for device communications. These additional commands are used for gateway communications, network manager communication, and adapter communications. Each of these is summarized in Table 8.8.

8.2.4.3 Discrete Device Commands

The HCF specifications provide support for discrete applications. This support includes

- Discrete field devices supporting, for example, pressure, level, or temperature switches, proximity and limit switches, solenoid valves, motor starters, and simple motion/position control
- Hybrid field devices like level transmitters that include redundant (backup) level switches or positioners with limit switches providing full-open or full-closed valve position
- Discrete adapters that communicate with a connected PLC and allow mini/micro PLCs to be incorporated into HART networks

TABLE 8.8 WirelessHART Commands

Gateway	Network Manager	Devices
Writing networking IDs	Joining	Read wireless device capabilities
Writing network tags	Health reports	Reporting health
Managing device list entries	Report	Supporting networking resources
Managing blacklists and white lists	Path, route, and transport failures	Reporting path, route, and transport failures
Caching published data from devices	Timetable management	Supporting timetables
Managing network constraints	CCA mode management	Supporting routes
Managing stale data settings	Network flow control	Supporting superframes
Supporting host applications	Managing superframes	Supporting links
Supporting active advertising	Managing links	Supporting graphs
Maintaining device lists	Managing graphs	Supporting security keys
Flushing cached device information	Writing security keys	Supporting routes
Time source for network	Monitoring and grooming the network	
Managing device scheduling flags	Managing routes	

Discrete devices communicate process variables and discrete variables using the same communication techniques as used by traditional HART devices. Discrete variables include packed-Boolean and state variables. Discrete values must include the value along with status. In both process variables and discrete variables, the *variables* are really a collection of real-time data and the properties associated with that real-time data. For example, all discrete variables may include a fault behavior in addition to the discrete values. Commands for configuring and communicating discrete devices were added as part of HART 7.4.

8.3 HART Communication Stack

8.3.1 Overview

The HART communication standard has been in existence since the late 1980s. In its initial release, the HART field communication protocol was superimposed on a 4–20 mA signal providing two-way communications with field instruments without compromising the integrity of the analog output. The HART protocol has evolved from the initial 4–20 mA based signal to the current wired and wireless-based technology with extensive features supporting security, unsolicited data transfers, event notifications, block mode transfers, and advanced diagnostics. Figure 8.8 illustrates the architecture of the HART protocol stack according to the OSI seven-layer communication model [4].

The following sections describe the wired and wireless communication layers.

8.3.2 Wired Protocol

HART protocol stack includes four layers: physical layer, data link layer, transport layer, and application layer. As shown in Figure 8.9, the HART protocol allows two masters (primary and secondary) to communicate with slave devices and provide additional operational flexibility. A permanently connected host system may be used simultaneously, while a handheld terminal or PC controller is communicating with a field device.

The HART protocol ensures interoperability among devices through universal commands that enable hosts to easily access and communicate the most common parameters used in field devices. The HART DDL extends interoperability to include information that may be specific to a particular device. DDL enables a single handheld configurator or PC host application to configure and maintain

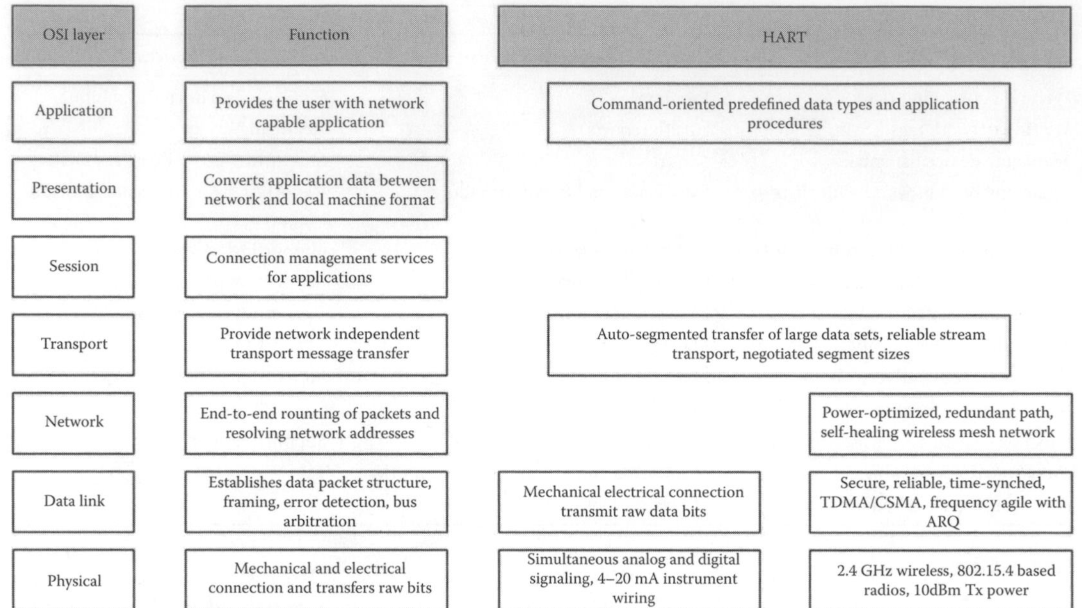

OSI layer
Function
HART
Application
Provides the user with network capable application
Command-oriented predefined data types and application procedures
Presentation
Converts application data between network and local machine format
Session
Connection management services for applications
Transport
Provide network independent transport message transfer
Auto-segmented transfer of large data sets, reliable stream transport, negotiated segment sizes
Network
End-to-end rounting of packets and resolving network addresses
Power-optimized, redundant path, self-healing wireless mesh network
Data link
Establishes data packet structure, framing, error detection, bus arbitration
Mechanical electrical connection transmit raw data bits
Secure, reliable, time-synched, TDMA/CSMA, frequency agile with ARQ
Physical
Mechanical and electrical connection and transfers raw bits
Simultaneous analog and digital signaling, 4–20 mA instrument wiring
2.4 GHz wireless, 802.15.4 based radios, 10dBm Tx power

FIGURE 8.8 HART communication layers.

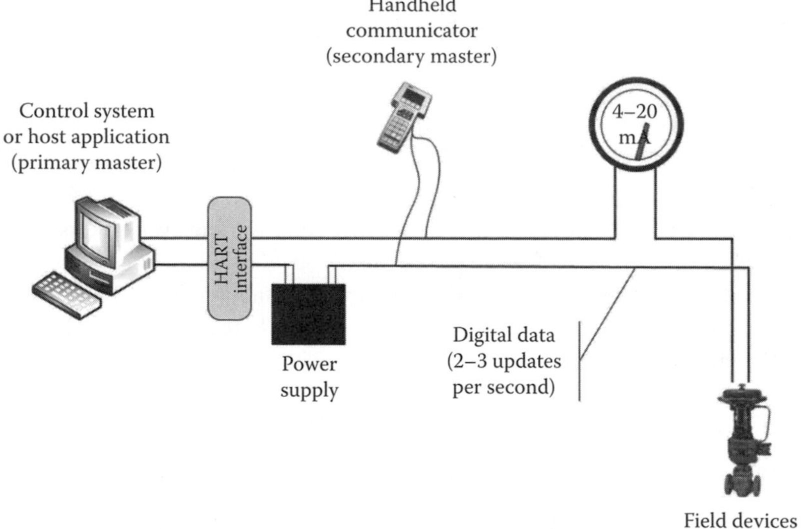

FIGURE 8.9 Multimaster setup.

HART-communicating devices from any manufacturer. The use of common tools for products of different vendors minimizes the amount of equipment and training needed to maintain a plant. The following sections provide additional details on the HART communications stack.

8.3.2.1 Physical Layer

Data transmission between masters and field devices is physically realized by superimposing an encoded digital signal on the 4–20 mA current loop. Since the coding has no mean values, an analog signal transmission taking place at the same time is not affected. This enables the HART protocol to include the existing simplex channel transmitting the current analog signal and an additional half-duplex channel for digital communication in both directions. This is illustrated in Figures 8.2 and 8.3.

The physical layer defines an asynchronous half-duplex interface that operates on the analog current signal line. To encode the bits, the FSK method, based on the Bell 202 communication standard, is used. Digital value 0 is assigned frequency 2200 Hz, and digital value 1 is assigned frequency 1200 Hz. Each individual byte of the layer-2 message is transmitted as 11-bit UART character at a data rate of 1200 bits/s.

The HART specification defines that master devices send voltage signals, while the field devices (slaves) convey their messages using load-independent currents. The current signals are converted to voltage signals at the internal resistance of the receiver (at its load). To ensure a reliable signal reception, the HART protocol specifies the total load of the current loop including the cable resistance to be between 230 and 1100 Ω. Usually, the upper limit is determined by the power output of the power supply unit.

As shown in Figure 8.3, the HART masters are connected in parallel to the field devices. Connecting this way allows handhelds to be connected and disconnected during operation without disrupting the current loop.

HART wiring in the field usually consists of twisted pair cables. If very thin and/or long cables are used, the cable resistance increases and, hence, the total load. As a result, the signal attenuation and distortion increases, while the critical frequency of the transmission network decreases.

If interference signals are a problem, then the wiring must be shielded. The signal loop and the cable shield should be grounded at one common point only. According to the specification, the following configurations work reliably:

- For short distances, simple unshielded 0.2 mm two-wire lines are sufficient.
- For distances of up to 1500 m, individually twisted 0.2 mm two-wire pairs with a common shield over the cable should be used.
- For distances of up to 3000 m, individually twisted 0.5 mm two-wire lines shielded in pairs are required.

HART communication between two or more devices can function properly only when all communication participants are able to interpret the HART signals correctly. To ensure this, both the cables and devices in the current loop that are not part of the HART communication must also fulfill the requirements. Those devices can impede or even prevent the transmission of the data. The reason is that the inputs and outputs of these devices are specified only for the 4–20 mA technology. Since the input and output resistances change with the signal frequency, such devices are likely to short-circuit the higher-frequency HART signals (1200–2200 Hz). To prevent this, the internal resistance must be increased using an additional circuit. An RC low-pass circuit (250 W, 1 mF) performs this function.

8.3.2.2 Data Link Layer

The data link layer provides a reliable, transaction-oriented communication path to and from field devices for digital data transfer. Communications are over twisted pair wire that may be simultaneously carrying the 4–20 mA signals. The data link layer corrects for errors due to noise or other disturbances on the communication links by using error detecting information and an automatic repeat request protocol to request the retransmission of data blocks that may be corrupted.

HART is a master–slave protocol and is loosely organized around the ISO/OSI seven-layer model for communications protocols. The data link layer supports the application layer above it and requires services from the physical layer below it. Furthermore, the data link layer can be divided into two sublayers: the logical link control responsible for addressing, framing, and error detection; and the medium access control that controls the transmission of messages across the physical link.

The data link layer supports long (5 byte) *unique* and short (1 byte) *poll* addresses. Polling addresses may be used only with Command 0. This allows the HART protocol to support both point-to-point and multidropped communication with field devices. If poll (short form) addresses are used, up to 64 slave devices may be multidropped on a single communication link. If unique (long-form) addresses are used, the number of multidropped devices is essentially unlimited and is determined based on the application's required rate of scan of the devices on the communication link.

The data link layer arbitrates access to the field device between a single secondary master device such as a handheld terminal and a single primary master device such as a control or data acquisition system. The data link layer gives equal access to the communication channel to both kinds of masters when they are being simultaneously used. The data link layer does not arbitrate between two secondary or two primary masters that are trying to talk on the same link.

To support the regular transfer of information from field device to master device, the data link layer supports a mode of operation in which field devices periodically broadcast information onto the communication link. A slave device is said to be in *burst mode* when it is providing a synchronous cyclic broadcasting of data, without continuous polling by a master device. No matter how many field devices are on a communication link, only one may be in burst mode.

Information transfer between devices on the communication link is through a defined message format. The entire message is protected by a single parity check product code. Message framing is through a combination of a start of frame delimiter and a message length field.

HART devices supporting burst communication mode send packets with short 75 ms breaks, which can alternately be read by the primary as well as the secondary master. While usually only two transactions per second are possible, the field device can send up to four messages using this method.

The structure of a HART message is shown in Figure 8.10. Each individual byte is sent as an 11-bit UART character equipped with a start, a parity, and a stop bit. In revision 5 and later, the HART protocol provides two message formats, which use different forms of addressing. In addition to the short frame slave address format containing 4 bits, a long frame address format has been introduced as an alternative. This allows more participants to be integrated, while achieving more safety in case of incorrect addressing during transmission failures.

The elements of the HART frame are summarized as follows:

- The *delimiter* is the first field in a HART message. It is used for message framing by indicating the position of the byte count. It also indicates the frame type, which is used for bus arbitration. Three frame types are supported by the HART data link layer:
    - An STX (0x02) indicates a master to field device (i.e., a slave or burst mode device) frame. An STX is the start of a transaction and is normally followed by an ACK.
    - An ACK (0x06) is the slave's or burst mode device's response to an STX.
    - A BACK (0x01) is a burst acknowledge frame periodically transmitted by a burst-mode device. These frames are transmitted without a corresponding STX.

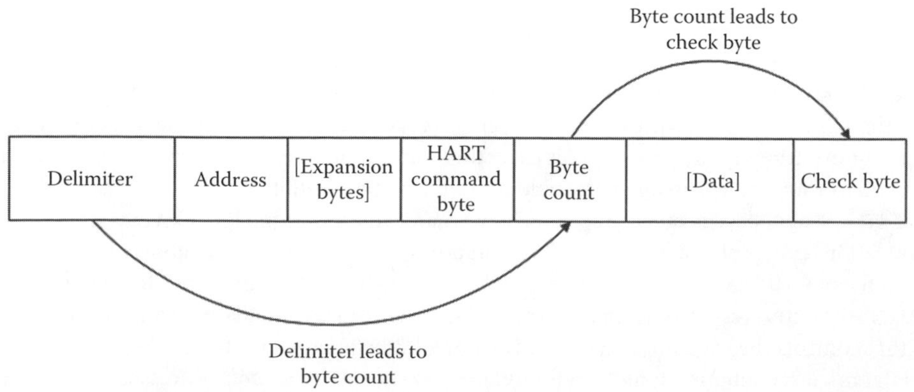

FIGURE 8.10 HART frame structure.

- The *address* field can be short or long. The protocol supports both five (5) byte *unique* addresses and one (1) byte polling addresses. The length of the address field (1 or 5 bytes) is indicated by the delimiter. The address field always includes the master address, the slave address, and (for ACK and BACK frames) whether the slave is in burst mode. The short and long address formats are as follows:
  - The address field of the short frame format contains 1 byte with 1 bit serving to distinguish the two masters and 1 bit to indicate burst-mode messages. For the addressing of the field devices, 4 bits are used (addresses 0–15).
  - The address field of the long frame format contains 5 bytes; hence, the field device is identified using 38 bits.
- The *expansion* bytes are optional. This field is 0–3 bytes long and its length is indicated in the delimiter. The definition of the expansion bytes is controlled by the HCF. If a field device does not know the meaning of all expansion bytes contained in the frame, then the field device must not answer.
- The *command* byte encodes the master commands of the three categories: universal, common practice, and device-specific commands. The significance of these commands depends on the definitions in the application layer.
- The *byte count* character indicates the message length, which is necessary since the number of data bytes per message can vary from 0 to 25. This is the only way to enable the recipient to clearly identify the message and the checksum. The number of bytes depends on the sum of the status and the data bytes.
- The *data* field is optional and consists of an integral number of bytes of application layer data. The data field contains subfields as defined in the command summary specification and contains the information transferred between the host application and the field device. The data can be transmitted as unsigned integers, floating-point numbers, or ASCII-coded character strings. The data format to be used is determined by the command byte. However, not all commands or responses contain data.
- The response message includes two status bytes at the beginning of the data portion of the message. The two status bytes indicate whether the received message was correct and the operational state of the field device. When the field device operates properly, both status bytes are set to logical zero.
- This *check byte* field is 1 byte long. The check byte value is determined by a bitwise exclusive OR of all bytes of a message including the leading delimiter.

The time required to transmit a message results from the bit data rate (1200 Hz) and the number of bits per message. The length of the message varies depending on the message length of 0–25 bytes and the message format. For a payload of 25 bytes, a total of 35 bytes must be transmitted. Since each byte is transmitted as UART character, the following data are obtained:

Bytes per Message	35 (25 Payload Characters + 10 Control Characters)	
Message size	35 characters × 11 bits	385 bits
Usable data rate	25 × 8 bits/385 bits	52%
Time per bit	1/1200 bits/s	0.83 ms
Transaction time	385 × 0.83 ms	0.32 s

8.3.2.3 Transport Layer

Traditionally, the wired HART protocol is considered to support only three layers of the ISO/OSI seven-layer model for communication protocols. The reality is that these are also elements of the transport layer found in the protocol specifications such as block data transfer. The block data transfer mechanism, as is described in Section 8.2.2.4, is best classified as a transport layer service.

The HART transport layer is fully described in HART 7 and is utilized as part of the WirelessHART specifications.

8.3.2.4 Application Layer

The communication routines of HART master devices and operating programs are based on HART commands that are defined in the application layer of the HART protocol. Predefined commands enable the master device to give instructions to a field device or send messages/data. So setpoints, actual values, and parameters can be transmitted and various services for start-up and diagnostics performed. The field devices immediately respond to a request by sending an acknowledgment that can contain requested status reports and/or the data of the field device.

An example command is Command 33. This HART command enables the master to read four transmitter variables of the field device and the corresponding units of measurement with only one command.

HART commands are classified by universal, common practice, and device specific. These classifications were described in Section 8.2.4.

8.3.3 Wireless Protocol

The WirelessHART protocol stack includes five layers: physical layer, data link layer, network layer, transport layer, and application layer. In addition, a centralized network manager is responsible for overall network routing communication scheduling [10][11].

8.3.3.1 Physical Layer

The WirelessHART physical layer is based on the IEEE 802.15.4-2006 2.4 GHz DSSS physical layer, which includes 15 of 16 possible RF channels. WirelessHART fully conforms to IEEE 802.15.4-2006 [13]. Additional physical layers can be easily added in the future as radio technology evolves.

8.3.3.2 Data Link Layer

The WirelessHART data link layer (DLL) is based on a fully compliant IEEE 802.15.4-2006 MAC. The WirelessHART DLL extends the functionality of the MAC by defining a fixed 10 ms timeslot, synchronized frequency hopping, and time division multiple access to provide collision-free and deterministic communications. To manage timeslots, the concept of a *superframe* is introduced that groups a sequence of consecutive timeslots. A superframe is periodic, with the total length of the member slots as the period. All superframes in a WirelessHART network start from the ASN (absolute slot number) 0, the time when the network is first created. Each superframe then repeats itself based on its period. In WirelessHART, a transaction in a timeslot is described by a vector: {frame id, index, type, source address, destination address, channel offset} where frame id identifies the specific superframe; index is the index of the slot in the superframe; type indicates the type of the slot (transmit/receive/idle); source address and destination address are the addresses of the source device and destination device, respectively; and channel offset provides the logical channel to be used in the transaction. To fine-tune the channel usage, WirelessHART introduces the idea of channel blacklisting. Channels affected by consistent interferences could be put in the blacklist. In this way, the network administrator can totally disable the use of those channels in the blacklist. To support channel hopping, each device maintains an active channel table. Due to channel blacklisting, the table may have less than 15 entries.

8.3.3.3 Network Layer

The network layer is responsible for several functions; the most important of which are routing and security within the mesh network. Whereas the DLL moves packets between devices, hop by hop, the network layer moves packets end-to-end within the wireless network. The network layer also includes other features such as route tables and timetables. Route tables are used to route communications along graphs. Timetables are used to allocate communication bandwidth to specific services such as publishing data and transferring blocks of data. Network layer security provides end-to-end data integrity and privacy across the wireless network.

8.3.3.4 Transport Layer

The WirelessHART transport layer provides a reliable, connectionless transport service to the application layer. When selected by the application layer interface, packets sent across the network are acknowledged by the end device so that the originated device can retransmit lost packets.

8.3.3.5 Application Layer

The application layer is HART. Because of this, access to WirelessHART is readily available by most host systems, handhelds, and asset management systems.

One key to these network layers is the network manager. Network manager is a special WirelessHART node that manages the mesh network. It sets the different parameters in each network layer in a comprehensive way so that individual nodes in the network work together as a whole.

8.4 System Tools

We have talked about HART devices, HART hosts, and HART networks. This is not enough. HART technology shines when it is fully integrated with a control system. In this section, we describe ways a control system works with HART. We describe technologies to bridge the gap with legacy control systems. We also introduce system tools that work with HART technology.

8.4.1 Host Interfaces

There are many setups that may be used for connecting HART devices to host systems. Several setups are presented in the following sections.

8.4.1.1 HART Point-to-Point Interface

In this scenario, the HART interface is a point-to-point interface. This topology is most often used for control system interfaces. Instrument power may in some cases be provided by the twisted pair connections. In other cases, external power is required. Figure 8.11 shows a point-to-point scenario without external power.

8.4.1.2 HART Multidrop Interface

In this scenario, the HART interface is provided through an IO system. In some cases, this could be as simple as a modem interfaced to a laptop. This topology could be used to support higher-level applications

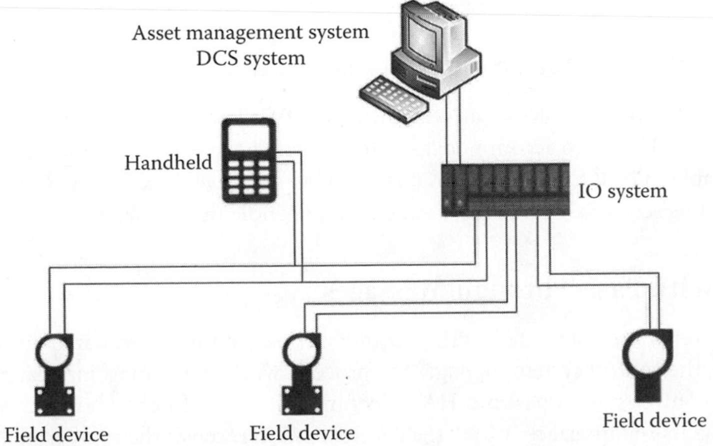

FIGURE 8.11 HART point-to-point interface.

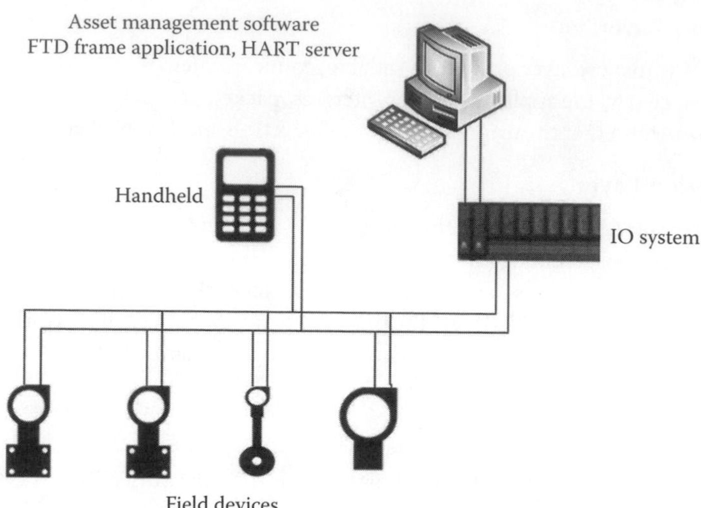

FIGURE 8.12 HART multidrop.

such as an asset management application. The multidrop mode of operation requires only a single pair of wires and, if applicable, safety barriers and an auxiliary power supply for up to 15 field devices. All process values are transmitted digitally. In multidrop mode, all field device polling addresses are >0, and the current through each device is fixed to a minimum value (typically 4 mA). Instrument power is provided by an interface or external power source. This scenario is shown in Figure 8.12 without an external power source.

8.4.1.3 Utilizing an FTA with Legacy Control Systems

Many legacy control systems may not support the HART signal. In these cases, what may be done is to connect the 4–20 mA signal to a field terminal assembly (FTA) and then wire the FTA to both the control system and a separate system for device management and diagnostics. Figure 8.13 is one example. In other cases, the control system may handle HART but not provide support for an asset management system. In this case, commands such as 0–3, 11, and 13 may be handled by the DCS. Other commands, including configuration and diagnostics, are passed to the asset management system.

8.4.2 Hosts with Limitations on Data Handling

All HART-compatible control systems can read the digital PV from a slave device. However, some system architectures may not be able to accommodate textual data (e.g., tag and descriptor fields). In these cases, the controller is able to read the process variable, but may not have direct access to all other data in the HART device. In this scenario, FTA could also be used to handle the textual data.

8.4.3 Hosts with Pass-Through Messages

Some control systems are integrated with a configuration or instrument management application. In these systems, the control system manages the process, while the management application manages the devices. The control system passes a HART command, issued by the management application, to the field device via its I/O interface. When the control system receives the reply from the field device, it sends the reply to the management application. This function is referred to as a pass-through feature of the control system.

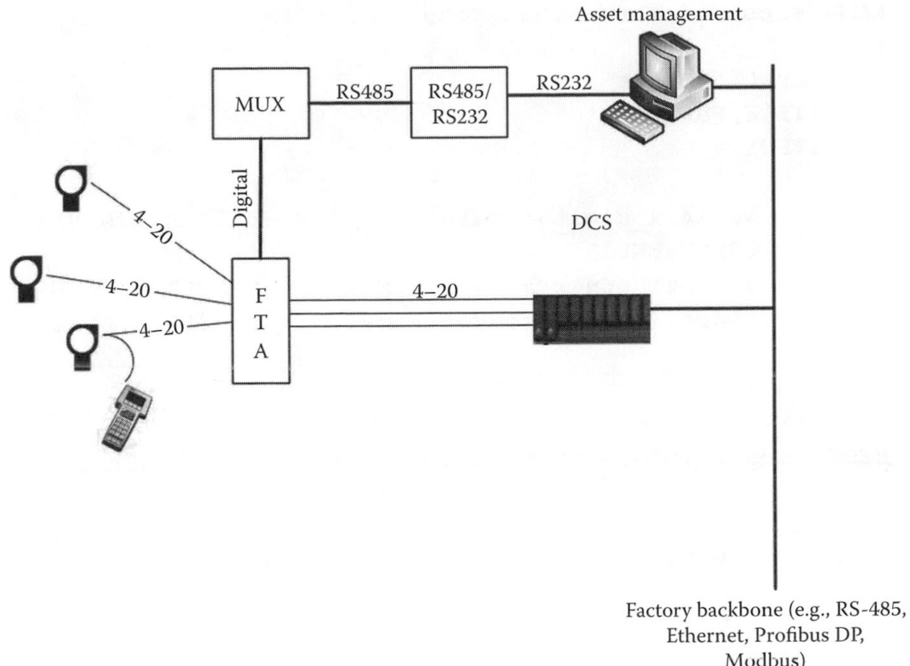

FIGURE 8.13 Using a field terminal assembly.

8.4.4 Tools Utilizing Device Configuration (DDL)

The HART commands are based on the services of the lower layers and enable an open communication between the master and the field devices. This openness and the interchangeability of the devices, independent of the manufacturer, are available only as long as the field devices operate exclusively with the universal and common practice commands and the user does not need more than the simple HART standard notation for the status and fault messages. In practice, this is not often the case since companies differentiate themselves through measurement technology, diagnostics, and advanced applications that are ideally tuned to work with their devices. The HART standard allows for these innovations through the DDL.

DDL allows the manufacture to fully describe device-related information and the special properties of a field device. For this reason, DDL is also used in other fieldbus networks such as foundation fieldbus. It is also currently being utilized as a key part of the emerging FDI standard.

The DDL language is a domain-specific language (DSL). A DSL is a programming language that's targeted at a specific problem; other programming languages such as C++ are more general purpose. A DSL contains the syntax and semantics that model concepts at the same level of abstraction that the problem domain offers. In this case, since the domain is devices, the DSL DDL is designed specifically to support all aspects of device description and management.

With DDL, a generic host system could take advantage of the innovations in a device by reading the device's DD, automatically improving plant performance. An example DD is shown in Figure 8.14.

The item declarations for this layout are shown in Figure 8.15.

The actual display for this DD is shown in Figure 8.16.

8.4.5 WirelessHART Tools

WirelessHART shares the application layer with wired HART. As a consequence, all HART tools currently in use could be used with a WirelessHART device. In some cases, the tools are enhanced to support

```
MENU sensor_information_group
{
    LABEL "Sensor 1";
    STYLE GROUP;
    ITEMS
    {
        sensor1_digital_value           (DISPLAY_VALUE),
        COLUMNBREAK,
        sensor1_upper_sensor_limit      (DISPLAY_VALUE),
        sensor1_lower_sensor_limit      (DISPLAY_VALUE)
    }
}

MENU sensor_information_page
{
    LABEL "Sensor Information";
    STYLE PAGE;
    ITEMS
    {
        sensor_information_group
    }
}

MENU process_values_window
{
    LABEL "Process Variables";
    STYLE WINDOW;
    ITEMS
    {
        sensor_information_page
    }
}
```

FIGURE 8.14 HART device layout described with DDL.

additional features in WirelessHART. For example, WirelessHART adaptors act as master devices for traditional wired HART devices and present them as subdevices on a WirelessHART network. An enhanced HART tool could interact with these subdevices just like with any other HART devices.

In addition, special new tools are available for WirelessHART. In this section, we describe three such tools. Two are exclusively for WirelessHART: Wi-HTest and Wi-Analys. They are used to help develop, configure, and diagnose WirelessHART devices and networks. The third tool is the WirelessHART handheld device, an extension to wired HART handhelds.

8.4.5.1 Wi-HTest

All HART devices must be registered with the HART Foundation. As part of the registration process, a device is thoroughly tested at each of its network layers. The testing system for the HART device is called HTest. In addition, a WirelessHART device is subjected to Wi-HTest, which checks the wireless conformances. The product name for Wi-HTest is HCF_KIT-193™.

```
VARIABLE sensor1_digital_value
{
   HELP "This value represents the process sensor value as measured by this device.";
   LABEL "Process Temperature";
   TYPE FLOAT
   {
       DISPLAY_FORMAT "3.2f";
   }
   CLASS DYNAMIC;
   HANDLING READ;
}

VARIABLE sensor1_digital_units
{
    HELP "Describes the engineering units associated with the digital value.";
    LABEL "Process Temperature Units";
    TYPE ENUMERATED
    {
        { 32,  "degC" },
        { 33,  "degF" },
        { 34,  "degR" },
        { 35,  "Kelvin" }
    }
    HANDLING READ & WRITE;
}

VARIABLE sensor1_upper_sensor_limit
{
   HELP "This is the high point at which the device transducer will become saturated.";
   LABEL "Upper Sensor Limit";
   TYPE FLOAT
   {
       DISPLAY_FORMAT "3.2f";
   }
   HANDLING READ;
}

VARIABLE sensor1_lower_sensor_limit
{
   HELP "This is the low point at which the device transducer will become saturated.";
   LABEL "Lower Sensor Limit";
   TYPE FLOAT
   {
       DISPLAY_FORMAT "3.2f";
   }
   HANDLING READ;
}

UNIT sensor1_relation
{
   sensor1_digital_units:
       sensor1_digital_value,
       sensor1_upper_sensor_limit,
       sensor1_lower_sensor_limit
}
```

FIGURE 8.15 HART DDL item declarations.

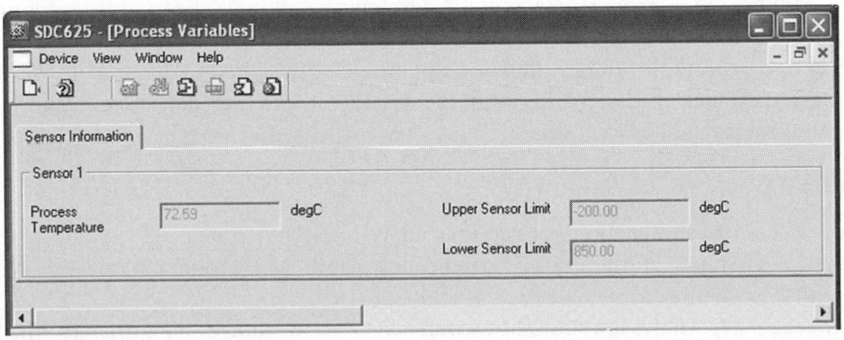

FIGURE 8.16 HART device display built with DDL.

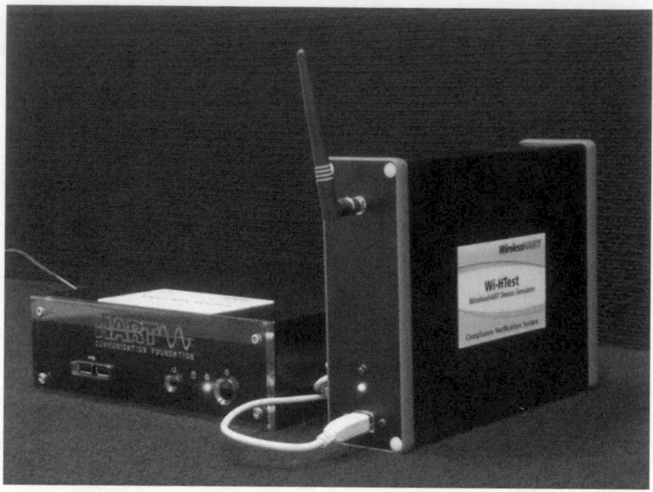

FIGURE 8.17 Wi-HTest tool.

Figure 8.17 shows the actual Wi-HTest tool. It includes two boxes: the tool itself on the left and the access point with radio on the right. The tool box is an embedded computer. The user plugs a mouse, a keyboard, and a monitor into the tool box.

Wi-HTest works one-on-one with a WirelessHART device. They are also connected by wire via the maintenance port. There is a suite of test script software installed in the tool box. Each test script is independently executed and performs a series of interactions with the device. Depending on the focus and complexity, a script checks the device functionality at the physical, data link network, and application layers, or a combination of the layers. The script is written in CINT, an interpretive version of the C language. There is also support for the user to write new test scripts.

A typical script runs in the following sequence:

- Use the maintenance port to set up the device. Wired HART commands are used.
- Send Force-Join Command 771 to the device via the maintenance port; wait for the device to join wirelessly.
- Once the device has joined, configure the device with wireless commands.
- Perform a series of tests, mostly wirelessly, with the device.
- Disconnect the device and end the test.

The usefulness of Wi-HTest goes beyond the conformance test. Companies could purchase it for different uses. It could help in WirelessHART product development. It could be used to configure a device. When used without the maintenance port connection, Wi-HTest could act as a WirelessHART gateway for a single-device network, in which different scripts could be designed to manipulate the device.

The wireless messages exchanged between the Wi-HTest and the device could be captured by Wi-Analys, the tool we shall describe next.

8.4.5.2 Wi-Analys

Wi-Analys is a wireless sniffer tool that captures all WirelessHART messages on all WirelessHART defined physical channels at the 2.4 GHz frequency band. The product name for Wi-Analys is HCF_KIT-190™.

The hardware for Wi-Analys is a single box with an antenna, as shown in Figure 8.18. The box is connected via a USB cable to a Windows PC, on which the software is installed. The software includes a service receiving the streaming messages from the box and a user interface displaying the messages in

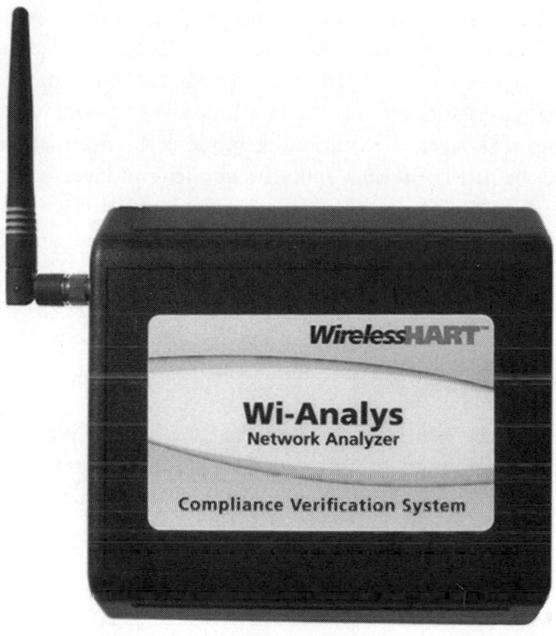

FIGURE 8.18 Wi-Analys tool.

WirelessHART format. The messages are decrypted if the security information is available to the software. The software is intelligent enough to extract some of the keys from the captured messages. Figure 8.19 shows the user interface.

Let's look at the messages displayed in Figure 8.19 column by column. The displayed columns are selected using the check marks in the left-hand pane. The check boxes at the bottom are for message

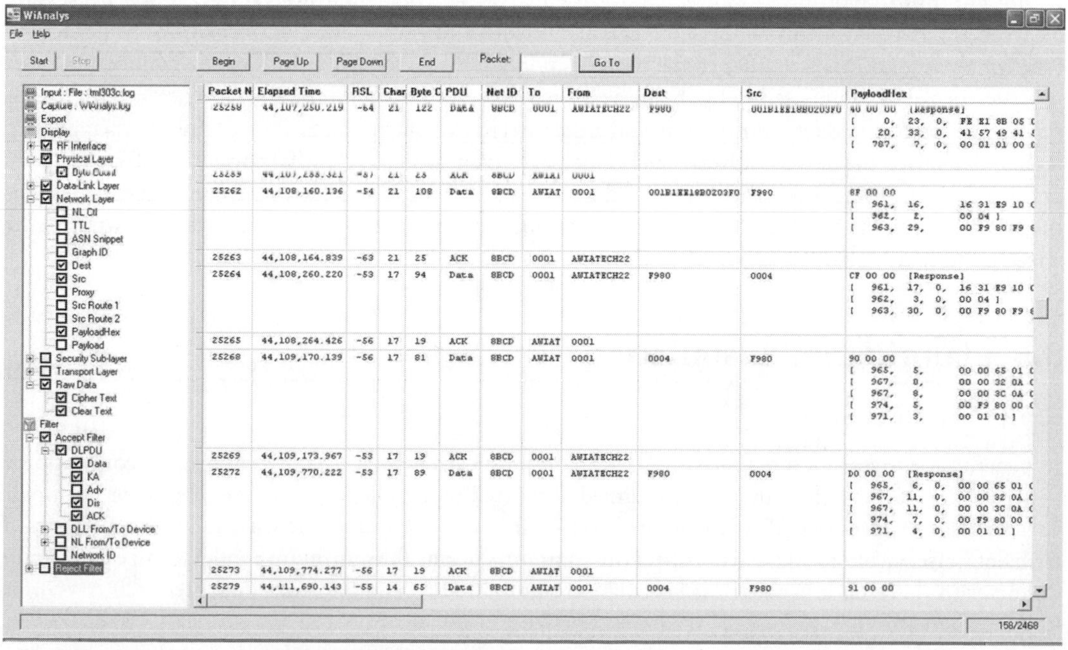

FIGURE 8.19 Wi-Analys screen capture.

filtering. The message in the first row is numbered packet 25258. The elapsed time is the time since the Wi-Analys is powered on until the message is captured. The received signal level is −64. The message is transmitted on physical channel 21. There are 122 bytes in the message. This WirelessHART message type is *data*. The message is sent at the data link layer to node 0x0001 from a node named AWIATECH22. At the network layer, the destination node is the network manager 0xF980, and the source node is unique ID 0x001B1EE18B0203F0. The application layer payload contains three commands, Commands 0, 20, and 787. From this information, we know this is a join request message from a new device with that unique ID to the network manager, and the access point 0x0001 is the proxy node interacting with the device.

Wi-Analys is used together with Wi-HTest in conformance testing. Each Wi-HTest execution log includes a Wi-Analys capture to help postanalyze the test result. As a matter of fact, special wireless messages are sent from Wi-HTest to Wi-Analys for coordination and to better manage the captured messages.

Wi-Analys is not limited to conformance testing. It is a complete network sniffer tool for WirelessHART, just like other commercial sniffer tools for wired or wireless networks such as Ethernet and Wi-Fi network. For any use case for a commercial network sniffer, we have a use case for Wi-Analys.

8.4.5.3 Wireless Handheld

WirelessHART handheld is the natural progression of the wired HART handheld. A HART handheld device is shown in earlier figures in this chapter. It connects by wire to the HART network. A wireless handheld connects without wire to the WirelessHART network. Regardless of how it is connected, the wireless handheld has the same functionality as its wired counterpart.

WirelessHART defines two ways a handheld communicates with the network: talking to a targeted field device or joining the network as a regular device. When joining as a regular device, the handheld could only talk to the network manager and the gateway just like other regular devices. This way the handheld could request the diagnostic and health report information, but it could not talk to an individual device directly.

The handheld could talk to an active device in the network via a preconfigured special session between the two. With this connection, the handheld could gather device information, run diagnostics, calibrate, etc. This is similar to the wired handheld; the difference is in how the connection is set up. The special session is set up by the network manager and preconfigured in the device. The handheld acquires the session information and other network information from the network manager. It then listens for the device's advertisement messages from which the initial communication links are extracted. From these links, the session and related communication bandwidth are established. Once the work is done, the handheld terminates the communication, just like you disconnect the wires of the wired handheld.

8.5 Planning and Installation

8.5.1 Wired HART

In general, the installation practice for HART communicating devices is the same as for conventional 4–20 mA instrumentation. Individually shielded twisted pair cable, either in single-pair or multipair varieties, is the recommended wiring practice. Unshielded cables may be used for short distances if ambient noise and cross talk will not affect the communication. The minimum conductor size is 0.51 mm diameter (#24 AWG) for cable runs less than 1524 m (5000 ft) and 0.81 mm diameter (#20 AWG) for longer distances.

TABLE 8.9 Non-IS Wired HART Cable Lengths

	Cable Capacitance—pf/ft (pf/m)			
	Cable Length—Feet (m)			
Number of Devices	20 pf/ft (65 pf/m)	30 pf/ft (95 pf/m)	50 pf/ft (160 pf/m)	70 pf/ft (225 pf/m)
1	9000 ft (2769 m)	6500 ft (2000 m)	4200 ft (1292 m)	3200 ft (985 m)
5	8000 ft (2462 m)	5900 ft (1815 m)	3700 ft (1138 m)	2900 ft (892 m)
10	7000 ft (2154 m)	5200 ft (1600 m)	3300 ft (1015 m)	2500 ft (769 m)
15	6000 ft (1846 m)	4600 ft (1415 m)	2900 ft (892 m)	2300 ft (708 m)

Allowable cable lengths for 1.02 mm (#18 AWG) shielded twisted pair.

Most installations are well within the 3,000 m (10,000 ft) theoretical limit for HART communication. However, the electrical characteristics of the cable (mostly capacitance) and the combination of connected device ccan affect the maximum allowable cable length of a HART network. Table 8.9 shows the effect of cable capacitance and the number of HART devices on the cable. The table is based on typical installations of HART devices in non-IS (intrinsic safe) environments, that is, no miscellaneous series impedance. Detailed information for determining the maximum cable length for any HART network configuration can be found in the HART physical layer specifications.

HART is often used in IS installations. IS is a method of providing safe operation of electronic process control instrumentation in hazardous areas. IS systems keep the available electrical energy in the system low enough to prevent ignition of the hazardous atmosphere. No single field device or wiring is intrinsically safe by itself (except for battery-operated self-contained devices such as a WirelessHART device), but is intrinsically safe only when employed in a properly designed IS system. Most DCS suppliers provide IS-rated IO that is compatible with HART.

8.5.2 WirelessHART

In many cases, the WirelessHART network may be configured similarly to a wired HART network. In these cases, the gateway is the remote I/O system connecting wireless devices and adaptors to DCSs, PLCs, and other plant automation systems. And the access points are the I/O modules of the gateway. The gateway has one or more access points that connect wireless devices to the gateway. The WirelessHART network starts with the gateway and access points at one end and the field devices at the other. Access points can be geographically dispersed from the gateway electronics and in general should be located near the devices to which they connect.

A key consideration is the number of devices that may be connected, directly or indirectly, to one access point, which is called the access point loading:

$$\text{NumDevices} = \text{Average update period (AUP)} \times 25$$

For example, with an average reporting rate of once per 1 s, 25 devices may be connected with an access point; with an average reporting rate of once per 10 s, 250 devices can be connected with an access point.

This criterion is similar to that used for any traditional I/O: do not crowd the I/O. When in doubt, use more access points. With little additional cost, this increases the number of alternative network paths and makes the network more robust.

Another formula that may be used to estimate the average bandwidth consumed by a WirelessHART network is as follows:

$$\text{Bandwidth Consumed} = \text{NumDevices} \times \left(0.0001\% + \left(\frac{0.02\%}{\text{AUP}}\right)\right)$$

where

0.0001% is used for overhead (network health reports and the like)

0.02% is for data publishing and other network traffic

For example, 100 devices at an average update interval of 1 s will consume 2.01% of total bandwidth; 1500 devices at average update interval of 60 s will consume 0.65% of total bandwidth.

These are very conservative estimates that do not account for the size of the area. For large networks, devices farther apart could use the same channel at the same time. And the bandwidth usages in one place will be less than the total network bandwidth due to the distance fading of the radio energy. With this formula, the user could estimate whether the WirelessHART network will coexist well with other 2.4 GHz networks in the same plant area.

A huge benefit of WirelessHART is its adapter. The adapter can be located anywhere along the current loop from the device to the I/O module. This allows existing HART devices to be adapted to wireless. It also frees the designer from having to locate the device in a better reception area instead of where the device physically should belong.

WirelessHART networks are easier and less costly to install than traditional wired HART systems simply because there are no wires to pull through cable trays. You simply locate where the devices should be mounted in the plant and install them, then install the access points and gateway. All WirelessHART devices are default routers, but to fortify the mesh network, routing devices may be added at strategic locations. After the devices have joined the network, simply configure them, and the wireless mesh forms and communications begin. It is as easy as a traditional HART 4–20 mA installation, with the same tools and know-how. An example installation is shown in Figure 8.20.

An example of a gateway is the Rosemount 1420 Wireless Gateway. The 1420 includes both the security manager and the network manager. The 1420, through a web browser, provides an interface for users to set up security, run diagnostics, and configure information about the wireless mesh. The gateway is

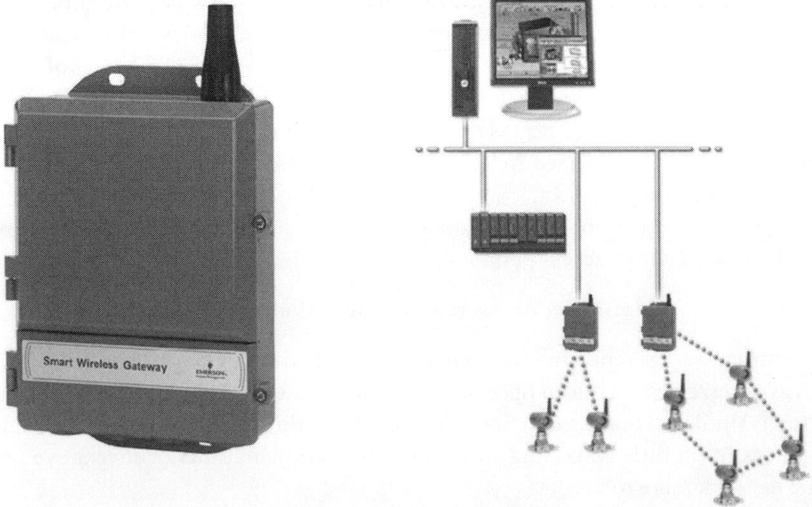

FIGURE 8.20 Connecting WirelessHART to the control system.

also the entry point for host and DCS systems to access wireless device data. The 1420 enables integration with popular communication protocols such as Modbus, OPC, and TCP/IP via Ethernet or serial connections.

In many situations, the gateway connects to the control system through native interfaces and shows up as a part of the overall configuration system. In other cases, industry standard mechanisms such as Modbus and OPC may be used.

8.6 Application Example: Bioreactor

The HART communication protocol enables companies to make sure measurements are as efficient, accurate, and timely as possible. Control and monitoring applications are ideal for a HART point-to-point configuration. The HART network 4–20 mA fast update rates are ideal for pressure and flow measurements. Digital measurements may be used to communicate actual valve position as well as other parameters. Accurate and timely measurement for control of bioreactors is essential in all industries. We use a bioreactor process in Figure 8.21 to show how HART is best suited for such applications.

The measurements, actuators, blocking valves, and scan rates are summarized in Table 8.10.

The primary objective of most field devices will be to provide a process measurement. The frequency at which this information is required by the process automation host is specific to the process equipment and the measurement type (e.g., pressure, temperature, flow, level, and analytical). Thus, as part of the process automation host configuration, the user configures the following information for all network devices that are accessed through the HART host interface:

- Device Tag—which uniquely identifies the device (e.g., HART Tag)
- Measurement value(s) that are to be accessed in the network device
- How often each measurement value is to be communicated to the gateway

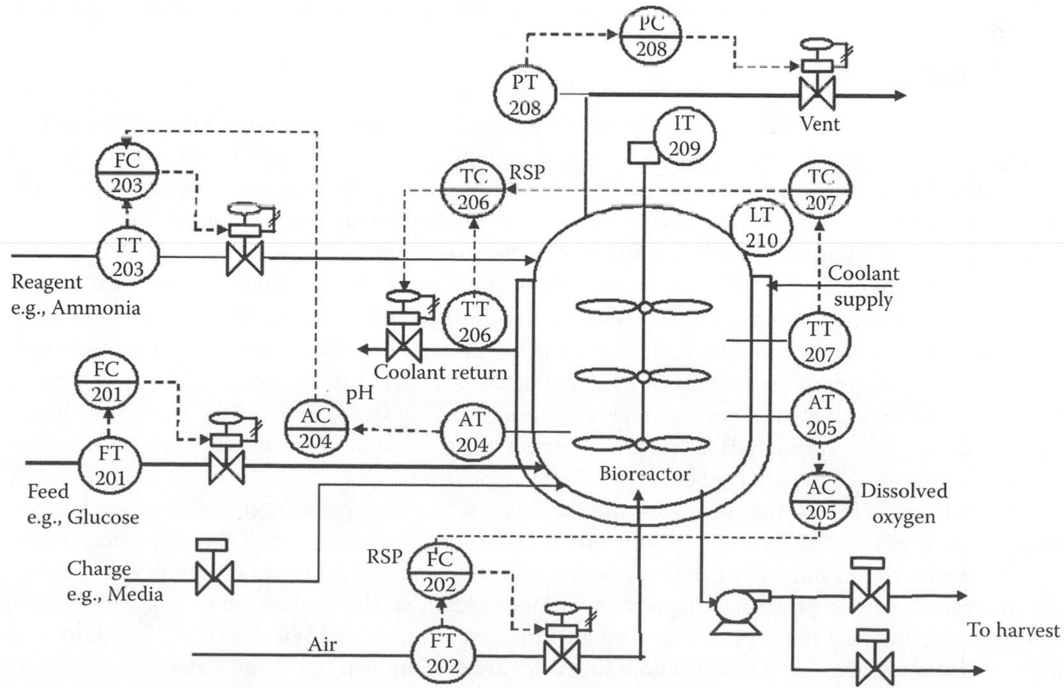

FIGURE 8.21 Bioreactor process.

TABLE 8.10 Instrument and Valve List for Bioreactor

Category	Device	Measurement	Scan Rates (s)
Measurement	C1	Reactor level (LT210)	16
	C2	Feed flow (liquid—FT201)	1
	C3	Reactor gas pressure (PT208)	1
	C4	Reactor temperature (TT207)	4
	C5	Agitator amps (IT209)	8
	C6	Return water temperature (TT206)	16
	C7	Reagent flow (FT203)	1
	C8	Air flow (FT202)	1
	C9	Dissolved oxygen (AT205)	4
	C10	pH (AT204)	4
Regulating valve	A1	Feed flow (FV201)	1
	A2	Reagent flow (FV203)	1
	A3	Coolant flow (FV206)	1
	A4	Vent flow (FV208)	1
	A5	Air flow (FV202)	1
Blocking valve	B1	Charge flow (FZ211)	1
	B2	Harvest flow (FZ212)	1
	B3	Harvest flow (FZ213)	1

Device configuration settings would typically be configured using a HART handheld or asset management system. The control configuration would typically be configured using the control system connected to the bioreactor.

In this example, we assume that all field devices are HART devices, and they are connected to the host using different HART topologies. Note that we could have WirelessHART devices in the mix.

8.7 Future Directions

Wired HART and WirelessHART continue to build on the innovation that was started in the late 1980s. Current advancements follow a well-established foundation in interoperability and design patterns described earlier in this chapter. In this section, we will discuss the following two questions: *What are the business drivers?* and *How will the device infrastructure evolve to support these business drivers?* [11]

Starting with the business drivers, all business performance is based on value that can be generated from its assets. These assets range from people and materials, to intellectual content, to physical properties. Plants are becoming much more integrated with business systems. These plants operate too much tighter requirements, are expected to be able to adjust production schedules in real time to changes in conditions and orders, and are much more regulated. Achieving these objectives requires a much greater understanding of the process, improved understanding of the state of the equipment in the plant, and far better data analysis techniques. The people operating these plants will likely hold degrees and, in many cases, advanced degrees.

This leads to the second question: *How will the device infrastructure evolve to support these business drivers?* The answer to this second question must be considered in several parts. Gaining process insight involves an increased number of measurements, providing more diagnostics on the devices providing the measurements, providing diagnostics on the process that the devices are part of, and moving things online that were in the past done manually. The first release of the WirelessHART standard went a long way toward making it possible both to reach advanced measurements and diagnostics that are already in devices today and to cost effectively measure many things that were difficult to reach in the past. In the first case, many plant infrastructures today are ill equipped to report advanced diagnostics.

Wireless allows these measurements to be communicated on an alternative infrastructure. In other cases, the type of equipment, for example, rotating equipment, made it difficult to take measurements. It is a lot easier to attach devices to this kind of equipment and let the wireless infrastructure take care of the communications. In still other cases where state-of-the-art was manual measurement, wireless makes it cost effective to periodically take these measurements and communicate them. An example of this is equipment health and monitoring. New devices are being designed and built to measure vibration and communicate signal values and diagnostics back to online centralized systems. So what does this mean for HART?

HART today is the workhorse of the industry. There is little evidence to suggest that this will change anytime soon. In this light, the most recent additions such as discrete devices and burst mode enhancements continue to be released for both wired and wireless technologies. Innovation will continue, and both wired and wireless devices will be there to serve users.

References

1. IEC 62591 Ed. 1.0 b:2010, Industrial communication networks—Wireless communication network and communication profiles—Wireless HART™, 2010.
2. IEC 61804-3 Ed. 2.0 b:2010, Function blocks (FB) for process control—Part 3: Electronic device description language (EDDL), 2010.
3. IEEE Std 802.15.4TM:2006, Wireless medium access control (MAC) and physical layer (PHY) specifications for low-rate wireless personal area networks (LR-WPANs), October 2006.
4. ISO 7498-1, Information processing systems—OSI reference model—The basic model, 1996.
5. ISA, http://www.isa.org/.
6. HART communication, http://www.hartcomm.org.
7. HART Application Guide—HCF LIT 34, http://www.hartcomm.org.
8. HART Communications, http://www.samson.de/pdf_en/l452en.pdf.
9. M. Fowler, *Patterns of Enterprise Application Architecture*, Addison-Wesley Professional, Boston, MA, 2002.
10. J. Song, S. Han, A. K. Mok, D. Chen, M. Lucas, M. Nixon, and W. Pratt, WirelessHART: Applying wireless technology in real time industrial process control, *RTAS*, St. Louis, MO, pp. 377–386, 2008.
11. D. Chen, M. Nixon, and A. Mok, *WirelessHART™ Real-Time Network for Industrial Automation*, Springer, New York, 2010.

9

Common Industrial Protocol (CIP™) and the Family of CIP Networks

9.1 Introduction **9**-1
EtherNet/IP: CIP on Ethernet Technology • DeviceNet: CIP on CAN Technology • ControlNet: CIP on Concurrent Time Domain Multiple Access Technology • CompoNet: CIP on Time Division Multiple Access Technology • Functional Safety • Synchronization • Distributed Motion Control • Energy Optimization

9.2 Description of the CIP Networks Library **9**-4
Object Modeling • Services • Messaging Protocol • Communication Objects • Object Library • Device Profiles • Configuration and Electronic Data Sheets • CIP Routing • Data Management • Auxiliary Power Distribution System • Maintenance and Further Development of the Specifications

9.3 Network Adaptations of CIP **9**-22
DeviceNet • ControlNet • EtherNet/IP • CompoNet

9.4 Benefits of the CIP Family **9**-70
Benefits for Device Manufacturers • Benefits for the Users of Devices and Systems

9.5 Application Layer Enhancements **9**-71
CIP Sync and CIP Motion • CIP Safety • Integration of Non-CIP Networks • CIP Energy

9.6 Conformance Testing **9**-92

Abbreviations **9**-94

Terminology **9**-95

References **9**-98

Viktor Schiffer
Rockwell Automation GmbH

9.1 Introduction

Traditionally, networks used in manufacturing enterprises were optimized for performance in specific applications, most commonly for control, information, and safety. While well suited to the functionality for which they were designed, these networks were not developed with a single, coherent enterprise architecture in mind. Since efficiency, reliability, and, ultimately, profitability are generally dependent on having more than one of these capabilities, manufacturers were forced to implement several different networks, none of which communicated innately with the other. As a result, over the course of time, most manufacturing enterprise network environments have been characterized by numerous specialized—and generally incompatible—networks existing in one space.

Today, however, corporate expectations for the manufacturing automation network landscape are dramatically different, thanks to the rapid and ubiquitous adoption of Internet technology. Companies of all sizes, all over the world, are trying to find the best ways to connect the entire enterprise. No longer is control of the manufacturing processes enough: The new manufacturing mandate is to enable users throughout the company to access manufacturing data from any location, at any time, and to integrate these data seamlessly with business information systems.

Due to this adoption and expansion of the use of Internet technologies, a rapidly increasing number of users worldwide have looked to *open* systems as a way to connect their disparate enterprise processes. However, the devices, programs, and processes used at the various layers of the seven-layer Open Systems Interconnection (OSI) [64] model have different options, capabilities, and standards (or lack of). In general, integrating these networks requires extra resources and programming, and even then, gaps between the systems often cannot be fully and seamlessly bridged. Consequently, without a way to seamlessly integrate a network, users compromise their investments and rarely achieve all of the productivity and quality benefits promised by open network technology.

Common application layers are the key to advanced communication and true network integration. ODVA's four best-in-class networks—EtherNet/IP™, DeviceNet™, ControlNet™, and CompoNet™—all are linked by one of the industrial automation's most versatile protocols: the Common Industrial Protocol, known as CIP™. CIP encompasses a comprehensive suite of messages and services for the collection of industrial automation applications—control, safety, energy, synchronization and motion, information, and network management. CIP allows users to integrate these applications with enterprise-level Ethernet networks and the Internet. Supported by hundreds of vendors around the world and truly media independent, CIP provides users with a unified communication architecture throughout the industrial enterprise. CIP allows users to benefit today from the many advantages of open networks and protects their existing automation investments, while providing an extensible and upgradable communication architecture.

With media independence comes choice—the ability to choose the CIP Network best suited for an application. As a single, media-independent platform that is shared by a variety of networking technologies, CIP provides the interoperability and interchangeability that is essential to open networks and open systems. Four network adaptations of CIP are available.

9.1.1 EtherNet/IP: CIP on Ethernet Technology

EtherNet/IP provides users with the network tools to deploy standard Ethernet technology (IEEE 802.3 [63] combined with the TCP/IP Suite) for industrial automation applications while enabling Internet and enterprise connectivity resulting in data anytime and anywhere. EtherNet/IP offers various topology options including a conventional star with standard Ethernet infrastructure devices or Device Level Ring (DLR) with EtherNet/IP devices so enabled. QuickConnect™ functionality allows devices to be exchanged while the network is running.

9.1.2 DeviceNet: CIP on CAN Technology

DeviceNet provides users with a cost-effective network to distribute and manage simple devices throughout their architecture. DeviceNet uses a trunkline/dropline topology and has DC power available on the network cable to simplify installations by providing a single connection point for network communications and device power up to 24 Vdc, 8 A. QuickConnect functionality allows devices to be reconnected quickly while the network is running by using an abbreviated start-up procedure.

9.1.3 ControlNet: CIP on Concurrent Time Domain Multiple Access Technology

ControlNet provides users with the tools to achieve deterministic, high-speed transport of time-critical I/O and peer-to-peer interlocks. ControlNet offers a choice of topology options including trunkline/dropline, star, or tree. Hardware options are also offered for applications requiring intrinsically safe hardware. Redundant network communication is also available.

9.1.4 CompoNet: CIP on Time Division Multiple Access Technology

CompoNet enables users to maximize network throughput for applications needing to transmit small packets of data quickly between controllers, sensors, and actuators. Its simple network connector and cabling scheme reduces overall system cost and time.

In addition to these network implementations, ODVA has published extensions to CIP for critical applications.

9.1.5 Functional Safety

Safety application coverage in CIP provides the ability to mix safety devices and standard devices on the same network or wire for seamless integration and increased flexibility. CIP Safety™ provides fail-safe communication between nodes such as safety I/O blocks, safety interlock switches, safety light curtains, and safety Programmable Logic Controllers (PLCs) in safety applications up to Safety Integrity Level (SIL) 3 according to IEC 61508 standards. CIP Safety has also been adopted by Sercos International. A more detailed description of the CIP Safety extension is given in Section 9.5.2.

9.1.6 Synchronization

Synchronization services in CIP provide the increased control coordination needed for control applications where absolute time synchronization is vital to achieve real-time synchronization between distributed intelligent devices and systems. CIP Sync™ is compliant with IEEE-1588™ standard and allows synchronization accuracy between two devices of better than 100 ns. Real-time synchronization can be achieved over conventional 100 Mbps Ethernet systems with a switch-based architecture. A more detailed description of the CIP Sync extension is given in Section 9.5.1.

9.1.7 Distributed Motion Control

Motion application coverage in CIP eliminates the need for a purpose-built motion-optimized network by allowing high-performance motion control and other devices to be combined on a single EtherNet/IP network. This approach results in a modular and streamlined approach to system design and lowers overall system and training cost. CIP Motion™ achieves real-time deterministic behavior of multiple axes through a common sense of time, allowing for 100 axes to be coordinated with a 1 ms network update to all axes. Clock synchronization between axes of better than 100 ns can be readily achieved, meeting the needs of the most demanding motion control applications.

9.1.8 Energy Optimization

Energy application coverage in CIP provides a family of objects and services for the optimization of energy usage (OEU™) and allows scalability of implementation within the device from basic energy awareness to more advanced functions for control of energy, aggregation, and reporting of energy information or dynamic demand–response. Further, the CIP family of energy objects and services will allow systems to monitor energy usage and manage energy for efficient energy consumption through dynamic control of energy state and analysis of energy information. Protocol-neutral energy attributes allow for

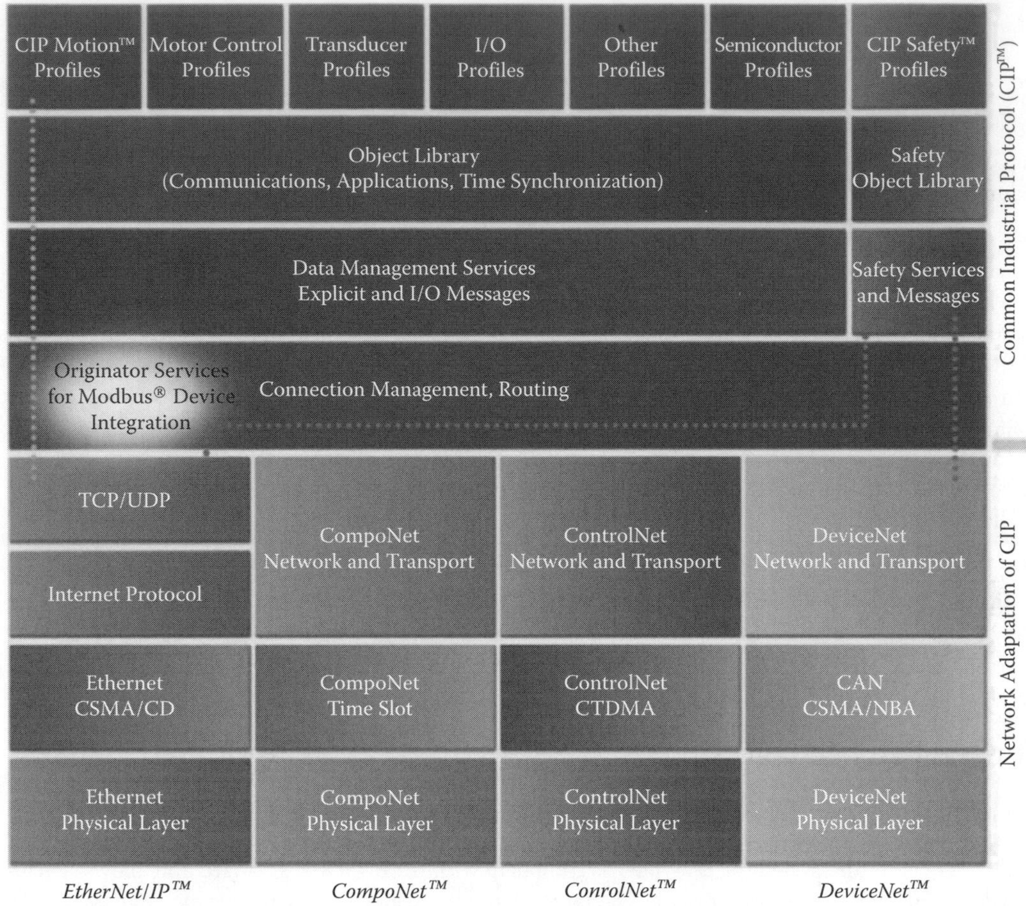

FIGURE 9.1 The Common Industrial Protocol and its network adaptations.

flexibility in the propagation of energy information via multiple protocols to facilitate an e-business model such as capturing energy requirements as a line item on production bills of material or to implement demand–response mechanisms for dynamic energy transactions.

The universal principles of CIP easily lend themselves to possible future implementations on new physical/data link layers. The overall relationship between the four implementations of CIP is shown in Figure 9.1.

9.2 Description of the CIP Networks Library

CIP is a very versatile protocol designed with the automation industry in mind. However, due to its open nature, it can be and has been applied to many more areas. The CIP Networks Library contains several volumes:

- Volume 1 deals with the common aspects of CIP that apply to all of the network adaptations. This volume contains the common object library and the device profile library, along with a general description of the communications model, device configuration, and CIP data management. This volume also defines an auxiliary power distribution system that is common to all adaptations of CIP.
- Volume 2 is the EtherNet/IP Adaptation of CIP, which describes how CIP is adapted to the Ethernet TCP/IP and UDP/IP transportation layers. It also contains any extensions to the material in Volume 1 that are necessary for EtherNet/IP, such as the optional industrial physical layer and connectors.

- Volume 3 is the DeviceNet Adaptation of CIP, which describes how CIP is adapted to the CAN data link layer. It also contains any extensions to the material in Volume 1 that are necessary for DeviceNet.
- Volume 4 is the ControlNet Adaptation of CIP, which describes how CIP is adapted to the ControlNet data link layer. It contains a complete description of the ControlNet data link layer and any extensions to the material in Volume 1 that are necessary for ControlNet.
- Volume 5 is CIP Safety. It contains the information necessary to implement the CIP Safety protocol on CIP Networks.
- Volume 6 is the CompoNet Adaptation of CIP, which describes how CIP is adapted to the CompoNet data link layer. It contains a complete description of the CompoNet data link layer and any extensions to the material in Volume 1 that are necessary for CompoNet.
- Volume 7 is the Integration of Modbus Devices into the CIP Architecture. This volume describes a standard for the integration of Modbus devices into the CIP world.

For brevity, this document will use these volume numbers when referencing the different books in the CIP Networks Library.

Specifications for the CIP Networks referenced earlier, and other documents discussing CIP, are available from ODVA at www.odva.org. It is beyond the scope of this book to fully describe each and every detail of CIP, but key features of the protocol and the auxiliary power distribution system will be discussed, including the following:

- Object modeling
- Services
- Messaging protocol
- Communication objects
- Object library
- Device profiles
- Configuration and electronic data sheets (EDSs)
- Bridging and routing
- Data management
- Auxiliary power distribution system

A few terms used throughout this section are described here to make sure they are well understood; further terms are described in Section 9.9:

- *Client*
  Within a client/server model, the client is the device that sends a request to a server. The client expects a response from the server.
- *Server*
  Within a client/server model, the server is the device that receives a request from a client. The server is expected to give a response to the client.
- *Producer*
  Within the Producer/Consumer model, the producing device places a message on the network for consumption by one or several consumers. Generally, the produced message is not directed to a specific consumer.
- *Consumer*
  Within the Producer/Consumer model, the consumer is one of potentially several consuming devices that picks up a message placed on the network by a producing device.
- *Producer/Consumer Model*
  The Producer/Consumer model is inherently multicast. Nodes on the network determine if they should consume the data in a message based on the connection ID (CID) in the packet. CIP uses the Producer/Consumer model, as opposed to the traditional source/destination message-addressing scheme (see Figure 9.2).

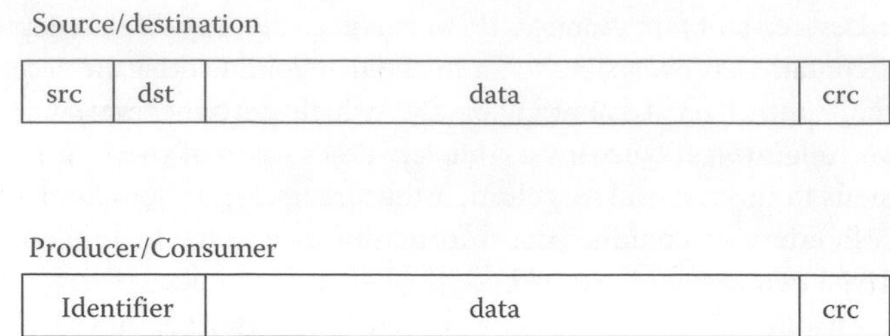

FIGURE 9.2 Source/destination vs. Producer/Consumer model.

- *Explicit Message*
  Explicit Messages contain addressing and service information that directs the receiving device to perform a certain service (action) on a specific part (e.g., an attribute of a given object) of a device.
- *Implicit (I/O) Message*
  Implicit messages do not carry address and/or service information; any consuming nodes already know what to do with the data based on the CID that was assigned when the connection was established. Implicit messages are so named because the meaning of the data is *implied* by the CID. In most cases, they are used to transport I/O data.

Let's have a look at the individual elements of CIP.

9.2.1 Object Modeling

CIP uses abstract object modeling to describe the following:

- The suite of available communication services
- The externally visible behavior of a CIP node
- A common means by which information within CIP products is accessed and exchanged

Every CIP node is modeled as a collection of objects. An object provides an abstract representation of a particular component within a product. Anything not described in object form is not visible through CIP. CIP objects are structured into classes, instances, and attributes.

A class is a set of objects that all represent the same kind of system component. An object instance is the actual representation of a particular object within a class. Each instance of a class not only has the same attributes but also has its own particular set of attribute values. As Figure 9.3 illustrates, multiple object instances within a particular class can reside within a CIP node.

In addition to the instance attributes, an object class may also have class attributes. These are attributes that describe properties of the whole object class, for example, how many instances of this particular

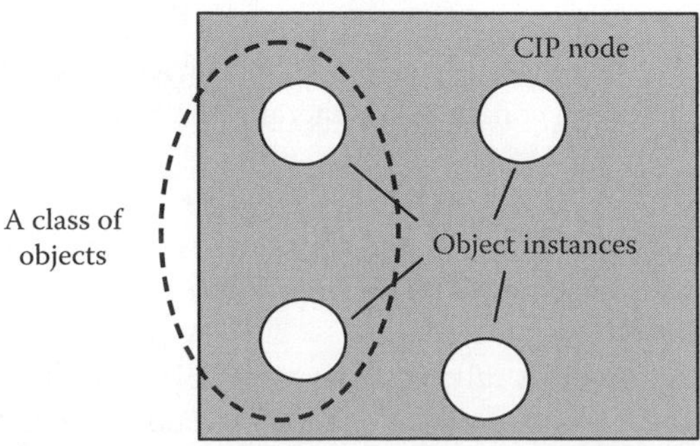

FIGURE 9.3 A class of objects.

object exist. Furthermore, both object instances and the class itself exhibit a certain behavior and allow certain services to be applied to the attributes, instances, or the whole class. All publicly defined objects that are implemented in a device must follow at least the mandatory requirements defined in the various CIP Networks specifications. Vendor-specific objects may also be defined with a set of instances, attributes, and services according to the requirements of the vendor. However, they need to follow certain rules that are also set forth in the specifications.

The objects and their components are addressed by a uniform addressing scheme consisting of the following:

- *Node Address*
  An integer identification value assigned to each node on a CIP Network. On DeviceNet, ControlNet, and CompoNet, this is also called a media access control identifier (MAC ID) and is nothing more than the node number of the device. On EtherNet/IP, the node address is the IP address.
- *Class Identifier* (*Class ID*)
  An integer identification value assigned to each object class accessible from the network.
- *Instance Identifier* (*Instance ID*)
  An integer identification value assigned to an object instance that identifies it among all instances of the same class.
- *Attribute Identifier* (*Attribute ID*)
  An integer identification value assigned to a class or instance attribute.
- *Service Code*
  An integer identification value that denotes an action request that can be directed at a particular object instance or object attribute (see Section 9.2.2).

Object class identifiers are divided into two types of objects: publicly defined objects (ranging from 0x0000 to 0x0063 and 0x00F0 to 0x02FF) and vendor-specific objects (ranging from 0x0064 to 0x00C7 and 0x0300 to 0x04FF). All other class identifiers are reserved for future use. In some cases, for example, within the assembly object class, instance identifiers are divided into two types of instances: publicly defined (ranging from 0x0001 to 0x0063 and 0x00C8 to 0x02FF) and vendor-specific (ranging from 0x0064 to 0x00C7 and 0x0300 to 0x04FF). All other instance identifiers are reserved for future use. Attribute identifiers are divided into two types of attributes: publicly defined (ranging from 0x0000 to 0x0063, 0x0100 to 0x02FF, and 0x0500 to 0x08FF) and vendor-specific (ranging from 0x0064 to 0x00C7, 0x0300 to 0x04FF, and 0x0900 to 0x0CFF). All other attribute identifiers are reserved for future use. While vendor-specific objects can be created with a great deal of flexibility, these objects must adhere to certain rules specified for CIP, for example, they can use whatever instance and attribute IDs the developer wishes, but their class attributes must follow guidelines detailed in Volume 1, Chapter 4, of the CIP Networks Library.

Addressing objects and their attributes can be performed with 8-bit, 16-bit, or 32-bit addresses. In most cases, class and instance addresses are 8- or 16-bit wide, and attribute addresses are only 8-bit wide. Thirty-two-bit addresses are currently reserved for instance addressing only.

Figure 9.4 shows an example of this object-addressing scheme.

9.2.2 Services

Service codes are used to define the action that is requested to take place when an object or parts of an object are addressed through Explicit Messages using the addressing scheme described in Section 9.2.1. Apart from simple read and write functions, a set of CIP services has been defined. These CIP services are common in nature, meaning they may be used in all CIP Networks and they are useful for a variety of objects. Furthermore, there are object-specific service codes that may have a different meaning for the same code, depending on the class of object. Finally, defining vendor-specific

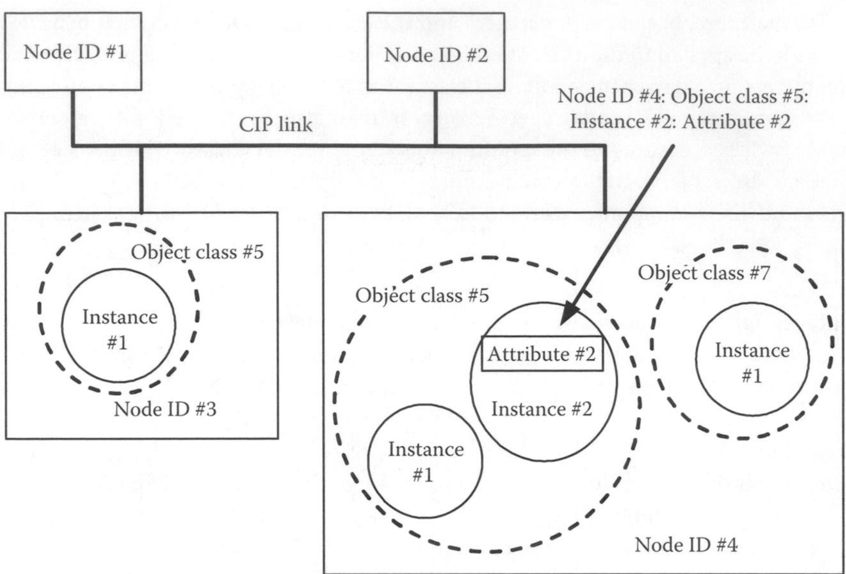

FIGURE 9.4 Object addressing example.

services according to the requirements of the product developer is possible. While this provides a lot of flexibility, the disadvantage of vendor-specific services is that they may not be understood universally. Minimally, vendors provide a description of the public information that their customers will need access to in their literature.

9.2.3 Messaging Protocol

CIP is a connection-based protocol. A CIP connection provides a path between multiple application objects. When a connection is established, the transmissions associated with that connection are assigned a CID (see Figure 9.5). If the connection involves a bidirectional exchange, then two CID values are assigned.

The definition and format of the CID is network dependent. For example, the CID for CIP connections over DeviceNet is based on the CAN identifier field.

Since most messaging on a CIP Network is done through connections, a process has been defined to establish such connections between devices that are not yet connected. This is done through the Unconnected Message Manager (UCMM) function, which is responsible for processing unconnected explicit requests and responses.

Establishing a CIP connection is generally accomplished by sending a UCMM Forward_Open service request message. The Forward_Open is required for all devices that support connections on ControlNet and EtherNet/IP. CompoNet uses a different method described in Section 9.3.4.12. DeviceNet only uses the simplified methods described in Sections 9.3.1.16 and 9.3.1.17.

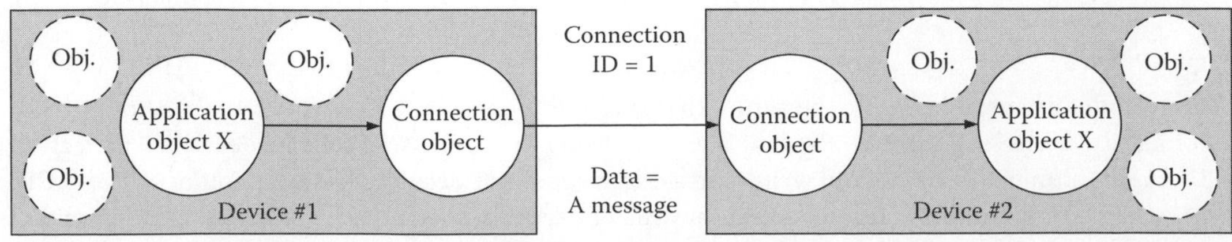

FIGURE 9.5 Connections and Connection IDs.

A Forward_Open request contains all information required to create a connection between the originator and the target device. The resulting data exchange may be unidirectional or bidirectional. In particular, the Forward_Open request contains information on the following:

- Timeout information for this connection
- Network CID for the connection from the originator to the target
- Network CID for the connection from the target to the originator
- Information about the identity of the originator (vendor ID and serial number)
- Maximum data sizes of the messages on this connection
- Whether it will be unicast or multicast
- Trigger mechanisms, for example, cyclic, change of state (COS)
- Electronic key so the target node can verify that it is the proper type of node (optional)
- Connection path for the application object data in the node that will be produced and consumed
- Data Segment containing configuration information for the node (optional)
- Routing information if the connection is to span more than one network (optional)

Some networks, like ControlNet, EtherNet/IP, and CompoNet, also make use of Unconnected Explicit Messaging. DeviceNet uses Unconnected Explicit Messaging only to establish connections.

All connections on a CIP Network can be categorized as I/O connections or explicit messaging connections.

- I/O connections provide dedicated, special-purpose communication paths between a producing application and one or more consuming applications. Application-specific I/O data move through these ports, a process that is often referred to as Implicit Messaging. These messages can be unicast or multicast. These connections are also called *implicit* connections because the meaning of the data is implied by the CID.
- Explicit messaging connections provide generic, multipurpose communication paths between two devices. These connections often are referred to simply as messaging connections. Explicit messages provide typical request/response-oriented network communications. These messages are point to point. They are called *explicit* messages because the data in the request explicitly state what service and object are being requested.

The actual data transmitted in CIP I/O messages are the I/O data in an appropriate format; for example, the data may be prefixed by a Sequence Count Value. This Sequence Count Value can be used to distinguish old data from new, for example, if a message has been resent as a heartbeat in a COS connection. The two states *Run* or *Idle* can be indicated with an I/O message either by prefixing a real time header, as is primarily used for ControlNet and EtherNet/IP, or by sending I/O data (Run) or no I/O data (Idle), a process primarily used for DeviceNet. CompoNet uses a bit within the OUT frame or the TRG frame to indicate the states *Run* and *Idle*. *Run* is the normal operational state of a device with the outputs under the control of the controlling application, while the reaction to receiving an *Idle* event is vendor specific and application-specific. Typically, this means bringing all outputs of the device to a predefined safe *Idle* state (which usually means *off*), that is, de-energized.

Explicit messaging requests contain a service code with path information to the desired object within the target device followed by data (if any). The associated responses repeat the service code followed by status fields followed by data (if any). DeviceNet and CompoNet use a *condensed* format for Explicit Messages in most cases, while ControlNet and EtherNet/IP only use the *full* format.

9.2.4 Communication Objects

CIP communication objects manage and provide the runtime exchange of messages. Communication objects are unique in that they are the focal points for all CIP communication. It therefore makes sense to look at them in more detail.

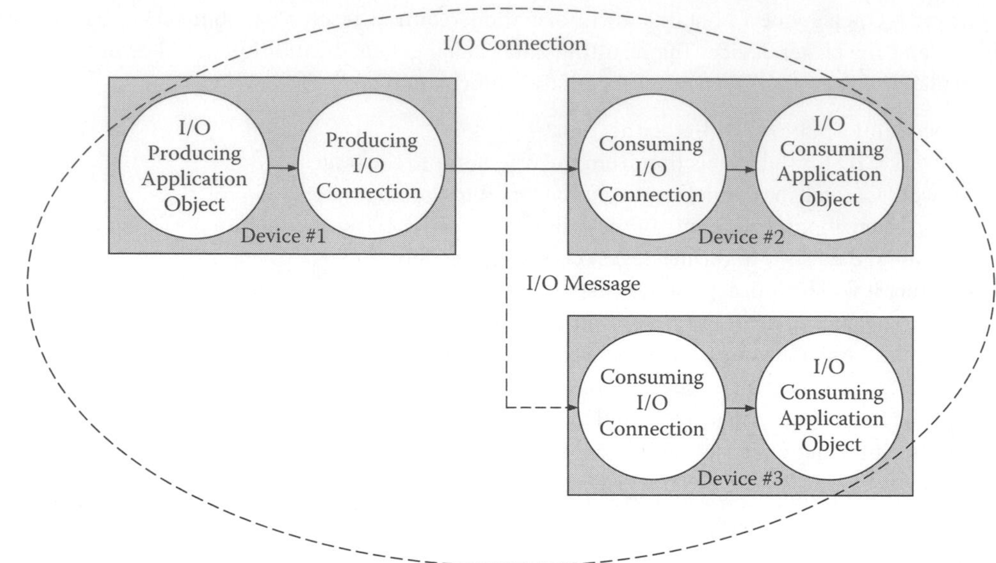

FIGURE 9.6 CIP multicast I/O Connection.

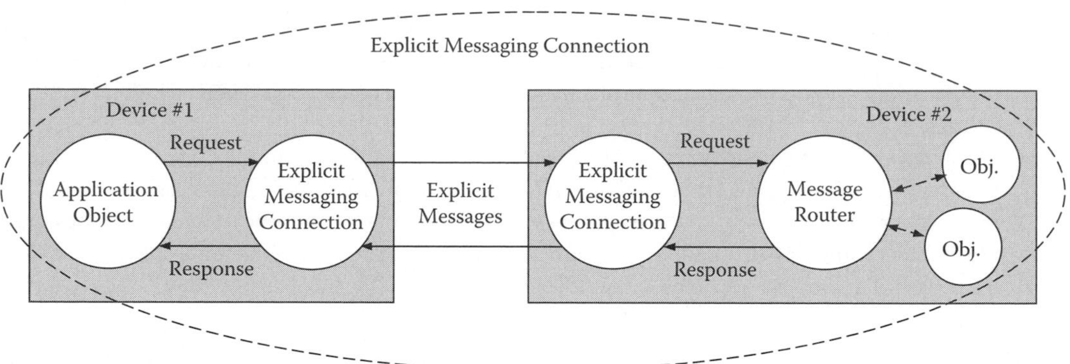

FIGURE 9.7 CIP Explicit Messaging Connection.

Each communication object contains a link producer part, a link consumer part, or both. I/O connections may be either producing or consuming or producing and consuming, while explicit messaging connections are always producing and consuming.

Figures 9.6 and 9.7 show the typical connection arrangement for CIP I/O messaging and CIP explicit messaging. The attribute values in the connection objects define a set of attributes that describe vital parameters of this connection. Note that Explicit Messages are always directed to the Message Router Object.

The attribute values of a connection object specify whether it is an I/O connection or an explicit messaging connection, the maximum size of the data to be exchanged across this connection, and the source and destination of the data. Further attributes define the state and behavior of the connection. Particularly important behaviors include how messages are triggered (from the application, through COS, that is, when data have changed, through cyclic events or by network events) and the timing of the connections (timeout associated with this connection and predefined action if a timeout occurs). CIP allows multiple connections to coexist in a device, although simple devices—for example, simple DeviceNet or CompoNet slaves—typically will only have one or two live connections at any time (only one connection on CompoNet).

9.2.5 Object Library

The CIP family of protocols contains a large collection of commonly defined objects. The overall set of object classes can be subdivided into three types:

1. General-use
2. Application-specific
3. Network-specific

Objects defined in Volume 1 of the CIP Networks Library are available for use on all network adaptations of CIP. Some of these objects may require specific changes or limitations when implemented on some of the network adaptations. These exceptions are noted in the network-specific volume. Therefore, to see a complete picture of a specific network implementation of an object, refer to Chapter 5 in both the Protocol Adaptation Volume and Volume 1.

The following are general use objects (Object IDs in brackets):

Assembly (0x04)	Message Router (0x02)
Acknowledge Handler (0x2B)	Originator Connection List (0x45)
Connection (0x05)	Parameter (0x0F)
Connection Configuration (0xF3)	Parameter Group (0x10)
Connection Manager (0x06)	Port (0xF4)
File (0x37)	Register (0x07)
Identity (0x01)	Selection (0x2E)

The following group of objects is application-specific (Object IDs in brackets):

AC/DC Drive (0x2A)	Position Sensor (0x23)
Analog Group (0x22)	Presence Sensing (0x0E)
Analog Input Group (0x20)	S-Analog Actuator (0x32)
Analog Input Point (0x0A)	S-Analog Sensor (0x31)
Analog Output Group (0x21)	S-Device Supervisor (0x30)
Analog Output Point (0x0B)	S-Gas Calibration (0x34)
Base Energy (0x4E)	S-Partial Pressure (0x38)
Block Sequencer (0x26)	S-Sensor Calibration (0x40)
Command Block (0x27)	S-Single Stage Controller (0x33)
Control Supervisor (0x29)	Safety Analog Input Group (0x4A)
Discrete Group (0x1F)	Safety Analog Input Point (0x49)
Discrete Input Group (0x1D)	Safety Discrete Input Group (0x3E)
Discrete Output Group (0x1E)	Safety Discrete Input Point (0x3D)
Discrete Input Point (0x08)	Safety Discrete Output Group (0x3C)
Discrete Output Point (0x09)	Safety Discrete Output Point (0x3B)
Electrical Energy (0x4F)	Safety Dual Channel Analog Input (0x4B)
Event Log (0x41)	Safety Dual Channel Output (0x3F)
Group (0x12)	Safety Supervisor (0x39)
Motion Device Axis (0x42)	Safety Validator (0x3A)
Motor Data (0x28)	Softstart (0x2D)
Nonelectrical Energy (0x50)	Target Connection List (0x4D)
Overload (0x2C)	Time Sync (0x43)
Position Controller (0x25)	Trip Point (0x35)
Position Controller Supervisor (0x24)	

The last group of objects is network-specific (Object IDs in brackets):

Base Switch (0x51)
CompoNet Link (0xF7)
CompoNet Repeater (0xF8)
ControlNet (0xF0)
ControlNet Keeper (0xF1)
ControlNet Scheduling (0xF2)
Device Level Ring (DLR) (0x47)
DeviceNet (0x03)
Ethernet Link (0xF6)
Modbus (0x44)
Modbus Serial Link (0x46)
Parallel Redundancy Protocol (0x56)
Power Management (0x53)
PRP Nodes Table (0x57)
SERCOS III Link (0x4C)
SNMP (0x52)
QoS (0x48)
RSTP Bridge (0x54)
RSTP Port (0x55)
TCP/IP Interface (0xF5)

The general-use objects can be found in many different devices, while the application-specific objects are typically found only in devices hosting such applications. New objects are added on an ongoing basis by the various ODVA Special Interest Groups (SIGs). As mentioned earlier, there are many vendor-specific objects defined by developers to satisfy needs that may not be met by the existing open objects contained in the published specifications.

Although this looks like a large number of object types, typical devices implement only a subset of these objects. Figure 9.8 shows the object model of such a typical device.

The objects required in a typical device are the following:

- Either a Connection Object or a Connection Manager Object
- An Identity Object
- One or several network-specific link objects (depends on network)
- A Message Router Object (at least its function)

Further objects are added according to the functionality of the device. This enables scalability for each implementation so that simple devices, such as proximity sensors, are not burdened with unnecessary overhead. Developers not only typically use publicly defined objects (see the list given earlier), but can also create their own objects in the vendor-specific areas, for example, Class ID 100–199. However, they are strongly encouraged to work with the SIGs of ODVA to create common definitions for additional objects instead of inventing private ones.

Out of the general-use objects, several are described in more detail in the following text.

9.2.5.1 Identity Object (Class ID: 0x01)

The Identity Object is described in greater detail because, being a relatively simple object, it can serve to illustrate the general principles of CIP objects. In addition, every device must have an Identity Object.

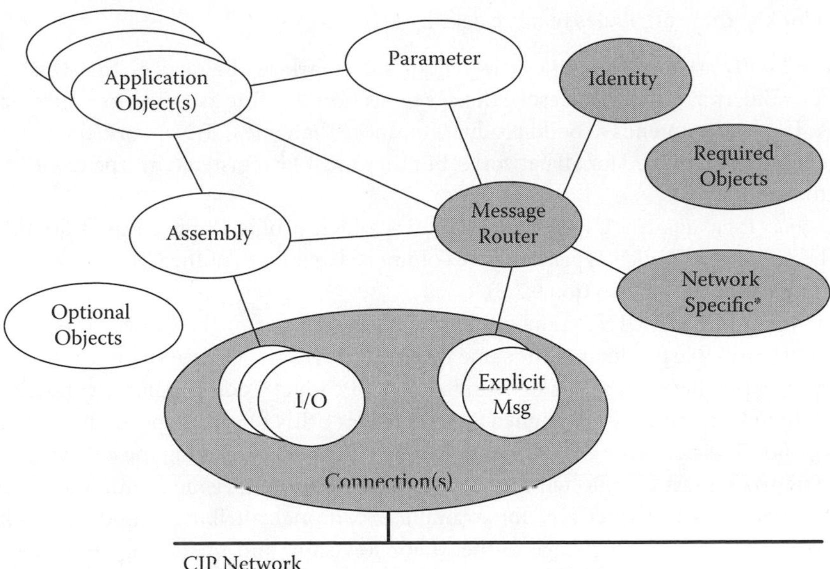

FIGURE 9.8 Typical device object model. *Note*: * DeviceNet Object, ControlNet Object, TCP/IP Interface Object, Ethernet Link Object, CompoNet Link Object, and CompoNet Repeater Object.

The vast majority of devices support only one instance of the Identity Object. Thus, typically, there are no requirements for any class attributes that describe additional class details, for example, how many instances exist in the device. Only instance attributes are required in most cases. These are as follows:

Mandatory Attributes

Vendor ID	Status
Device Type	Serial Number
Product Code	Product Name
Revision	

Optional or Conditional Attributes

State
Configuration Consistency Value
Heartbeat Interval
Active Language
Supported Language List
International Product Name
Semaphore
Assigned_Name
Assigned_Description
Geographic_Location
Modbus Identity Info

Let us have a look at these attributes in more detail:

- The *Vendor ID* attribute identifies the vendor that markets the device. This Unsigned Integer (UINT) value (for Data Type descriptions, see Section 9.2.9) is assigned to a specific vendor by ODVA. If a vendor intends to build products for more than one CIP Network, the same Vendor ID will generally be assigned for all networks, but they must be registered independently with ODVA prior to use.
- The *Device Type*, again a UINT value, specifies which profile has been used for this device. It must be one of the Device Types listed in Volume 1, Chapter 6, of the CIP Networks Library or a vendor-specific type (see Section 9.2.6).
- The *Product Code* is a UINT number defined by the vendor of the device. This code is used to distinguish multiple products of the same Device Type from the same vendor. There is generally a loose association between a Catalog Number and a Product Code, but not necessarily.
- The *Revision* is split into two Unsigned Short Integer (USINT) values specifying a Major Revision and a Minor Revision. Any device change(s) that results in modifying the behavior of the device on the network must be reflected in a change to the Minor Revision at minimum. Any changes in the device's logical interface, for example, additional attributes, modified/additional I/O Assemblies, etc., require a change to the Major Revision, and this change must be reflected in a revised EDS (see Section 9.2.7). Vendor ID, Device Type, Product Code, and Major Revision provide an unambiguous identification of an EDS for this device.
- The *Status* attribute provides information on the status of the device, for example, whether it is owned (controlled by another device) or configured (to something different than the out-of-the-box default), and whether any major or minor faults have occurred.
- The *Serial Number* is used to uniquely identify individual devices in conjunction with the Vendor ID, that is, no two CIP devices from one vendor may carry the same Serial Number. The 32 bits of the Serial Number allow ample space for a subdivision into number ranges that can be used by different divisions of larger companies.
- The *Product Name* attribute allows the vendor to give a meaningful ASCII name string (up to 32 characters) to the device.
- The *State* attribute describes the state of a device in a single UINT value. It is less detailed than the Status attribute.
- The *Configuration Consistency Value* allows a distinction between a device that has been configured and one that has not, or between different configurations in a single device. This helps avoid unnecessary configuration downloads.
- The *Heartbeat Interval* enables the Device Heartbeat Message. This is an unconnected change-of-state message that has a settable background cyclic interval between 1 and 255 s. Currently, this option is defined for use only on DeviceNet.
- The *Supported Language List* and *International Product Name* attributes can be used to describe the product in multiple languages, and the *Active Language* attribute then specifies which of the supported languages is in use.
- The *Semaphore* attribute provides a semaphore for client access synchronization to the entire device.
- The *Assigned_Name*, *Assigned_Description*, and *Geographical_Location* attributes can be used to individualize products by the user of the product.
- The *Modbus Identity Info* attribute can provide identity information in Modbus format to the extent the device supports it.

The services supported by the class and instance attributes are either Get_Attribute_Single (typically implemented in DeviceNet and CompoNet devices) or Get_Attributes_All (typically implemented in ControlNet and EtherNet/IP devices). The only attributes that can be set are as follows: the Heartbeat Interval, the Active Language, the Semaphore, the Assigned_Name, Assigned_Description, and Geographical_Location attributes. The only other service that typically is supported by the Identity Object is the Reset service. This Reset service comes with three different options that can let the device restart in three different ways.

The behavior of the Identity Object is described through a state transition diagram.

9.2.5.2 Parameter Object (Class ID: 0x0F)

This object is described in some detail since it is referred to in Section 9.2.7, Configuration and Electronic Data Sheets. When used, the Parameter Object comes in two types: a complete object and an abbreviated version (Parameter Object Stub). This abbreviated version is used primarily by DeviceNet and CompoNet devices that have only small amounts of memory available. The Parameter Object Stub, in conjunction with the EDS, has roughly the same functionality as the full object (see Section 9.2.7).

The purpose of the Parameter Object is to provide a general means of allowing access to many attributes of the various objects in the device without requiring a tool (such as a handheld terminal) to have any knowledge about specific objects in the device.

The class attributes of the Parameter Object contain information about how many instances exist in this device and a Class Descriptor, indicating, among other properties, whether a full or a stub version is supported. In addition, class attributes tell whether a Configuration Assembly is used and what language is used in the Parameter Object.

The first six Instance Attributes are required for the Object Stub. These are the following:

Parameter Value	The actual parameter
Link Path Size Link Path	These two attributes describe the application object/instance/attribute from which the parameter value was retrieved.
Descriptor	This describes parameter properties, for example, read-only, monitor parameter, etc.
Data Type	This describes the data type (e.g., size and range) using a standard mechanism defined by CIP
Data Size	Data size in bytes

These six attributes allow access, interpretation, and modification of the parameter value, but the remaining attributes make it much easier to understand the meaning of the parameter:

- The next three attributes provide ASCII strings with the name of the parameter, its engineering units, and an associated help text.
- Another three attributes contain the minimum, maximum, and default values of the parameter.
- Four more attributes can link the scaling of the parameter value so that the parameter can be displayed in a more meaningful way, for example, raw value in multiples of 10 mA, scaled value displayed in Amps.
- Another four attributes can link the scaling values to other parameters. This feature allows variable scaling of parameters, for example, percentage scaling to a full range value that is set by another parameter.
- Attribute #21 defines how many decimal places are to be displayed if the parameter value is scaled.
- Finally, the last three attributes are an international language version of the parameter name, its engineering units, and the associated help text.

9.2.5.3 Assembly Object (Class ID: 0x04)

Assembly Objects provide the option of mapping data from the attributes of different instances of various classes into one single attribute (#3) of an Assembly Object. This mapping is generally used for I/O Messages to maximize the efficiency of the control data exchange on the network. Assembly mapping makes the I/O data available in one block; thus, there are fewer Connection Object instances and fewer transmissions on the network. The process data are normally combined from different application objects. An Assembly Object also can be used to configure a device with a single data block, alleviating the need to set individual parameters.

CIP makes a distinction between Input and Output Assemblies. *Input* and *Output* in this context are viewed from the perspective of the controlling element (e.g., a PLC/PAC). An Input Assembly in a device collects data from the input application (e.g., field wiring terminal and proximity sensor) and produces it on the network, where it is consumed by the controlling device and/or operator interface. An Output Assembly in a device consumes data that the controlling element sends to the network and writes that data to the output application (e.g., field wiring terminals and motor speed control). This data mapping is very flexible; even mapping of individual bits is permitted. Assemblies also can be used to transmit a complete set of configurable parameters instead of accessing them individually. These Assemblies are called Configuration Assemblies.

Figure 9.9 shows an example of Assembly mapping. The data from application objects 100 and 101 are mapped in two instances of the Assembly Object. Instance 1 is set up as an Input Assembly for the

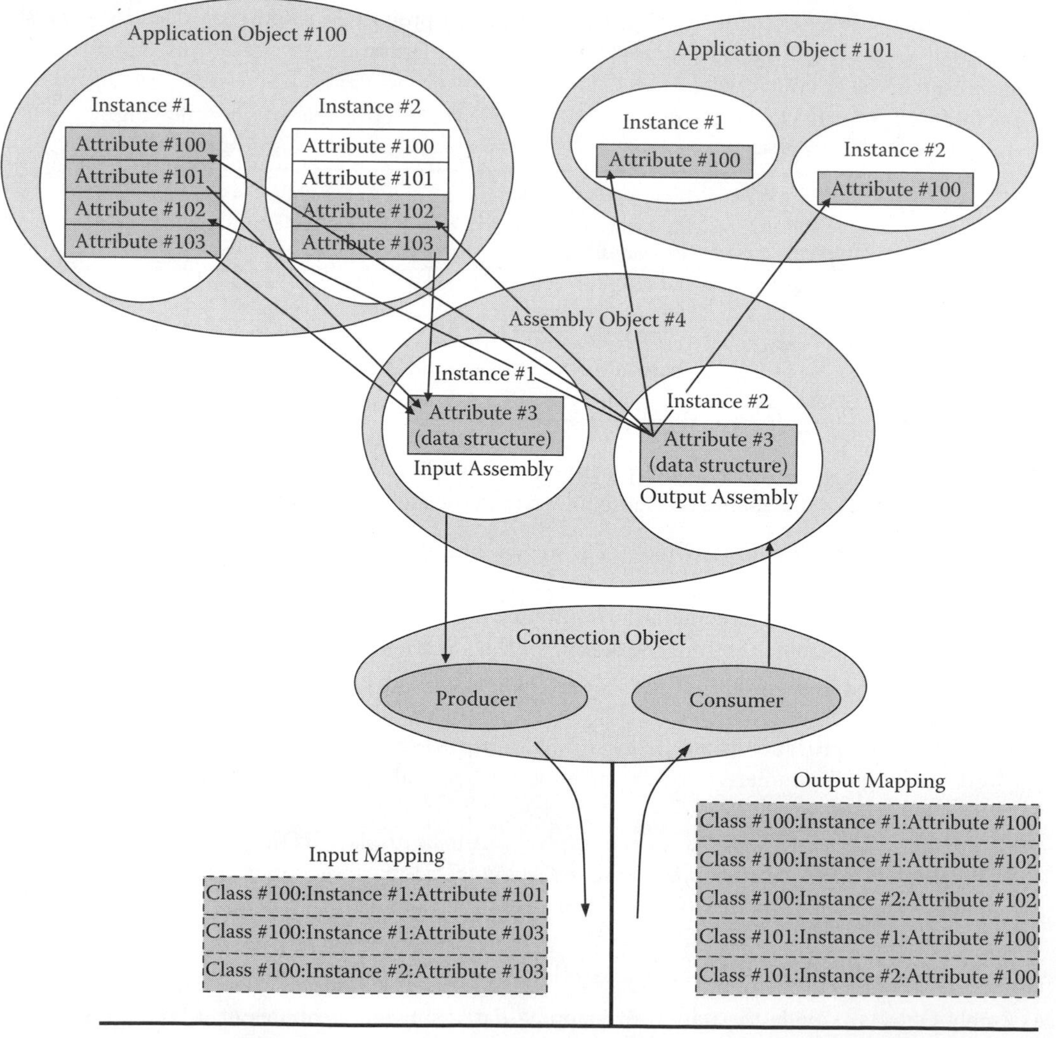

FIGURE 9.9 Example of an Assembly mapping in a typical I/O device.

input data and instance 2 as an Output Assembly for output data. The data block is always accessed via attribute 3 of the relevant Assembly instance. Attributes 1 and 2 contain mapping information.

I/O Assembly mapping is specified for many device profiles in Chapter 6 of Volume 1. Device developers then can choose which Assemblies they support in their products. If none of the publicly defined Assemblies fully represents the functionality of the product, a device vendor may implement additional vendor-specific Assemblies (Instance IDs 100–199).

CIP defines static and dynamic Assembly Objects. Whereas mapping for static Assemblies is permanently programmed in the device (ROM), dynamic Assemblies can be modified and extended through dynamic mapping (RAM). Most simple CIP devices support only static Assembly Objects. Dynamic Assembly Objects may be used in more complex devices, but they are not very common.

9.2.6 Device Profiles

It would be possible to design products using only the definitions of communication networks and objects, but this could easily result in similar products having quite different data structures and behavior. To overcome this situation and to make the application of CIP devices much easier, devices of similar functionality have been grouped into Device Types with associated profiles. Such a CIP profile contains the full description of the object structure and behavior. The following Device Types and associated profiles are defined in Volume 1 (profile numbers in parentheses):

AC Drives (0x02)	Limit Switch (0x04)
CIP Modbus Device (0x28)	Managed Ethernet Switch (0x2C)
CIP Modbus Translator (0x29)	Mass Flow Controller (0x1A)
CIP Motion Drive (0x25)	Motor Overload Device (0x03)
Communications Adapter (0x0C)	Motor Starter (0x16)
CompoNet Repeater (0x26)	Photoelectric Sensor (0x06)
Contactor (0x15)	Pneumatic Valve(s) (0x1B)
ControlNet Physical Layer Component (0x32)	Position Controller (0x10)
Programmable Logic Controller (0x0E)	Process Control Valve (0x1D)
DC Drives (0x13)	Residual Gas Analyzer (0x1E)
DC Power Generator (0x1F)	Resolver (0x09)
Encoder (0x22)	RF Power Generator (0x20)
Enhanced Mass Flow Controller (0x27)	Safety Analog I/O Device (0x2A)
Fluid Flow Controller (0x24)	Safety Discrete I/O Device (0x23)
General Purpose Discrete I/O (0x07)	Softstart Starter (0x17)
Generic Device, keyable (0x2B)	Turbomolecular Vacuum Pump (0x21)
Human Machine Interface (HMI) (0x18)	Vacuum/Pressure Gauge (0x1C)
Inductive Proximity Switch (0x05)	

Device developers must use a Device Type ID to uniquely identify their product. Any device that does not fall into the scope of one of the specialized device profiles must use the Generic Device Profile (0x2B) or a vendor-specific profile. What profile is used and which parts of it are implemented must be described in the user's device documentation.

Every profile consists of a set of objects—some required, some optional—and a behavior associated with that particular type of device. Most profiles also define one or more I/O data formats (Assemblies) that define the makeup of the I/O data. In addition to the commonly defined objects and I/O data Assemblies, vendors can add Objects and Assemblies of their own if their devices provide additional functionality. Furthermore, vendors can create profiles within the vendor-specific profile range. They are then free to define whatever behavior and objects are required for their device as long as they adhere to the general rules for profiles outlined in Chapter 6 of Volume 1 of the CIP Networks Library.

Whenever additional functionality is needed by multiple vendors, ODVA encourages coordinating these new features through SIGs, which can then create new profiles or additions to existing profiles for everybody's use and for the benefit of the device users.

All open (ODVA-defined) profiles carry numbers in the 0x00–0x63 or 0x0100–0x02FF ranges, while vendor-specific profiles carry numbers in the 0x64–0xC7 or 0x0300–0x02FF ranges. All other profile numbers are reserved by CIP.

9.2.7 Configuration and Electronic Data Sheets

CIP provides several options for configuring devices:

- A printed data sheet
- Parameter Objects and Parameter Object Stubs
- An EDS
- A combination of an EDS and Parameter Object Stubs
- A Configuration Assembly combined with any of the methods provided earlier

When using configuration information collected on a printed data sheet, configuration tools can provide only prompts for service, class, instance, and attribute data and relay this information to a device. While this procedure can do the job, it is the least desirable solution since it does not determine the context, content, or format of the data.

Parameter Objects, on the other hand, provide a full description of all configurable data for a device. Since the device itself provides all the necessary information, a configuration tool can gain access to all parameters and maintain a user-friendly interface. However, this method imposes a burden on a device with full parameter information that may be excessive for a small device with limited internal resources. Therefore, an abbreviated version of the Parameter Object, called a Parameter Object Stub, may be used (see Section 9.2.5.2). This option still allows access to the parameter data, but it does not describe any meaning to the data. Parameter Object Stubs in conjunction with a printed data sheet are usable, but certainly not optimal. On the other hand, an EDS supplies all of the information that a full Parameter Object contains, in addition to I/O Connection information, so the EDS provides the full functionality and ease of use of the Parameter Object without imposing an excessive burden on the individual device. In addition, an EDS provides a means for tools to perform offline configuration and to download configuration data to the device at a later time.

An EDS is a simple ASCII text file that can be generated on any ASCII editor. Since the CIP Specification provides a set of rules for the overall design and syntax of an EDS, specialized EDS editing tools, such as ODVA's EZ-EDS, can simplify the creation of EDS files. The main purpose of the EDS is to give information on several aspects of the device's capabilities, the most important ones being the I/O Connections it supports and what parameters for display or configuration exist within the device. It is highly recommended that an EDS describe all supported I/O Connections, as this makes the application of a device in a control system much easier. When it comes to parameters, EDS files should contain the attributes of application objects so that software can provide user access for monitoring and/or configuration purposes.

Let's look at some details of the EDS. First, an EDS is structured into sections, each of which starts with a section name in square brackets []. The first two sections are mandatory for all EDS files.

- [File]: Describes the contents and revision of the file.
- [Device]: Is equivalent to the Identity Object information and is used to match an EDS to a device.
- [Device Classification]: Describes what network the device can be connected to. This section is optional for DeviceNet, required for ControlNet, EtherNet/IP, and CompoNet.
- [ParamClass]: Describes configuration details in addition to class-level attributes of the Parameter Object.

- [Params]: Identifies all configuration parameters in the device, follows the Parameter Object definition. Further details are given later.
- [Groups]: Identifies all parameter groups in the device and lists group name and parameter numbers.
- [Assembly]: Describes the structure of data items.
- [Connection Manager]: Describes connections supported by the device. Typically used in ControlNet and EtherNet/IP.
- [Connection ManagerN]: Same as the [Connection Manager] section, but only for connection entries that do not apply to all CIP ports of the device.
- [Port]: Describes the various network ports a device may have.
- [Capacity]: Specifies the communication capacity of EtherNet/IP and ControlNet devices.
- [Connection Configuration]: This section defines the characteristics of the Connection Configuration Object implemented in this device, if a Connection Configuration Object implementation exists. It is used for EDS-based I/O Scanner configuration.
- [Event Enumeration]: Associates specific event or status codes within a device with an international string.
- [Symbolic Translation]: This section is used to publicize the translation between a Symbolic Segment or an ANSI Extended Symbol Segment encoded EPATH specification to the equivalent ParamN or AssemN entry keywords.
- [Internationalization]: This section allows the representation of all strings within an EDS in multiple languages.
- [Modular]: Describes modular structures inside a device.
- [IO_Info]: Describes I/O connection methods and I/O sizes. Allowed for DeviceNet only.
- [Variant_IO_Info]: Describes multiple IO_Info data sets. Allowed for DeviceNet only.
- [EnumPar]: Enumeration list of parameter choices to present to the user. This is an old enumeration method specified for DeviceNet only.
- [ControlNet Physical Layer]: Describes details of the ControlNet physical layer. Allowed for ControlNet only.
- [CompoNet_Device]: Describes the type of CompoNet device. Allowed for CompoNet only.
- [CompoNet_IO]: Describes the I/O Connection details of CompoNet slaves. Allowed for CompoNet only.
- [Modbus Mapper]: Used to provide a description of individual Modbus items that correspond to a specific CIP object attribute.
- Object Class sections: These sections—one for every object class—can be used to describe all object details, such as instances, attributes, and supported services.

These sections allow a very detailed device description, although only a few of these details are described here. Further reading is available in Refs. [1,2].

A tool with a collection of EDS files will first use the [Device] section to try to match an EDS with each device it finds on a network. Once this is done and a particular combination of device and EDS is chosen, the tool can then display device properties and parameters and allow their modification if the user so chooses. A tool may also display what I/O Connections a device allows and which of these are already in use. EDS-based tools are mainly used for slave or I/O Adapter devices, as I/O Scanner devices typically are too complex to be configured through EDS constructs alone. For those devices, the EDS is used primarily to identify the device and then guide the tool to call a matching configuration applet.

A particular strength of the EDS approach lies in the methodology of parameter configuration. A configuration tool typically takes all of the information that can be supplied by the Parameter Objects and an EDS and displays it in a user-friendly manner. In many cases, this enables the user to configure a device without needing a detailed manual, as the tool presentation of the parameter information and the help text enables decision making for a complete device configuration. This assumes the developer of the product and the EDS file has supplied all required information and any optional information with completeness and accuracy.

A complete description of what can be done with EDS files goes well beyond the scope of this book. Available materials on this topic provide greater detail [1,2].

9.2.8 CIP Routing

CIP defines mechanisms that allow the transmission of messages across multiple networks, provided that the intermediate devices (CIP routers) between the various networks are equipped with the objects and services used in CIP routing. If this is the case, the message will be forwarded from CIP router to CIP router until it has reached its destination node. Here is how it works:

For Unconnected Explicit Messaging, the actual Explicit Message to be executed on the target device is *wrapped up* inside of another Explicit Message service, the so-called Unconnected_Send service (Service Code 0x52 of the Connection Manager Object). This service message contains all the information about the transport mechanism, such as the request timeout (which may be modified as the message moves through each CIP router), the message request path information, and the routing path information.

The first CIP router device that receives an Unconnected_Send message will take its contents and forward it to the next CIP router, as specified within the Route Path section of the message. Before the message is actually sent, the *used* part of the path is removed but is remembered by the CIP router device for the return of the response. The CIP router may subtract some time from the timeout value, thereby reducing the timeout value as it closes in on the destination. This process is executed for every CIP router the message goes through, until the final CIP router is reached. The number of CIP routers an Unconnected_Send may pass through is theoretically limited by the message length.

Once the Unconnected_Send message has arrived at the last CIP router, the Unconnected_Send *wrapper* is removed and the *inner* Explicit Message is sent to the target device, which executes the requested service and generates a response. That response, as received from the target device, is then transported back through all the CIP routers it traversed during its forward journey until it reaches the originating node. It is important to note in this context that the transport mechanism may have been successful in forwarding the message and returning the response, but the response still could contain an indication that the desired service could not be performed successfully in the target network/device. Through this mechanism, the CIP router devices do not need to know anything about the message paths ahead of time so no preconfiguration of the CIP router devices is required. This is often referred to as *seamless routing.*

When a connection (I/O or Explicit) is set up using the Forward_Open service (see Section 9.3.1.14), it may go to a target device on another network. To enable the appropriate setup process, the Forward_Open message may contain a field with path information describing a route to the target device. This is very similar to the Unconnected_Send service described earlier. The routing information is then used to create routed connections within the CIP routing devices between the originator and the target of the message. Once set up, these connections automatically forward any incoming messages for this connection to the proper outgoing port. Again, this is repeated through each CIP router until the message has reached the target node. As with routed Unconnected Explicit Messages, the number of hops is generally limited only by the capabilities of the devices involved. In contrast to routed Unconnected Messages, routed Connected Messages do not carry path information. Since Connected Messages always use the same path for any given connection, the path information that was given to the routing devices during connection setup is held there as long as the connection exists. Again, the CIP routing devices do not have to be preprogrammed; they are self-configured during the connection establishment process.

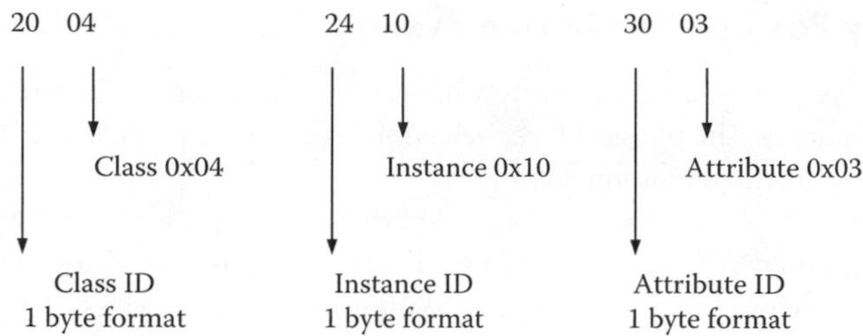

FIGURE 9.10 Logical segment encoding example.

9.2.9 Data Management

The data management part of the CIP Specification describes addressing models for CIP entities and the data structures of the entities themselves.

Entity addressing is done by Segments, which allows flexible usage so that many different types of addressing methods can be accommodated. Two uses of this addressing scheme (Logical Segments and Data Types) are looked at in more detail in the following text.

9.2.9.1 Logical Segments

Logical Segments (first byte = 0x20–0x3F) are addressing Segments that can be used to address objects and their attributes within a device. They are typically structured as follows: [Class ID] [Instance ID] [Attribute ID, if required].

Each element of this structure allows various formats (1 byte, 2 bytes, and 4 bytes). Figure 9.10 shows a typical example of this addressing method.

This type of addressing is commonly used to point to Assemblies, Parameters, and other addressable entities within a device. It is used extensively in EDS files, but also within Explicit Messages, to name just a few application areas.

9.2.9.2 Data Types

Types (first byte = 0xA0–0xDF) can be either structured (first byte = 0xA0–0xA3, 0xA8 or 0xB0) or elementary (first and only byte = 0xC1–0xDE). All other values are reserved. Structured Data Types can be arrays of elementary Data Types or a collection of arrays or elementary Data Types. Of particular importance in the context of this book are elementary Data Types, which are used within EDS files to specify the Data Types of parameters and other entities.

Here is a list of commonly used Data Types:

- 1 bit (encoded into 1 byte)
  - Boolean, BOOL, Type Code 0xC1
- 1 byte
  - Bit string, 8 bits, BYTE, Type Code 0xD1
  - Unsigned 8-bit integer, USINT, Type Code 0xC6
  - Signed 8-bit integer, SINT, Type Code 0xC2
- 2 bytes
  - Bit string, 16 bits, WORD, Type Code 0xD2
  - Unsigned 16-bit integer, UINT, Type Code 0xC7
  - Signed 16-bit integer, INT, Type Code 0xC3
- 4 bytes
  - Bit string, 32 bits, DWORD, Type Code 0xD3
  - Unsigned 32-bit integer, UDINT, Type Code 0xC8
  - Signed 32-bit integer, DINT, Type Code 0xC4

The Data Types in CIP follow the requirements of IEC 61131-3 [3,62].

9.2.10 Auxiliary Power Distribution System

The CIP application layer can be used on a variety of network technologies. Each CIP Network specification consists of two volumes. The physical layer behavior defined on a particular network is described in the appropriate CIP Network adaptation volume.

Chapter 8 of Volume 1 of the CIP Networks Library defines an optional auxiliary power distribution system that is separate and distinct from the physical layer requirements of any of the CIP Networks.

Auxiliary power may be used to provide application power for such devices as Input/Output modules, Emergency Stop circuitry, and other application-specific needs. The cabling system provides 4-wire, two-circuit wiring that supplies 24 V switched and unswitched power. Depending on the cabling selected by the designer, the maximum current ranges from 7 to 10 A. This standard specifies system topologies, cable and connector requirements, and power supply requirements for auxiliary power distribution.

This system is not intended to provide redundant network power for already powered networks such as DeviceNet or CompoNet.

9.2.11 Maintenance and Further Development of the Specifications

ODVA has a set of working groups that maintain the specifications and create protocol extensions, for example, new profiles or functional enhancements such as CIP Sync and CIP Safety. These groups are called Special Interest Groups (SIGs).

The results of these SIGs are written up and presented to the Technical Review Board for approval and then incorporated into the specifications. Only ODVA members can work within the SIGs. These participants have the advantage of advance knowledge of technical changes coming to the specifications. Participation in one or several SIGs is, therefore, highly recommended.

9.3 Network Adaptations of CIP

Four public derivatives of CIP currently exist. There are currently four public adaptations of CIP, each based on different data link layers and transport mechanisms, which maintain the common upper layers of CIP, as illustrated earlier in Figure 9.1.

9.3.1 DeviceNet

9.3.1.1 Introduction

DeviceNet was the first implementation of CIP. As mentioned in Section 9.1, DeviceNet is based on the Controller Area Network (CAN). DeviceNet uses a subset of the CAN protocol (11-bit identifier only, no remote frames). The DeviceNet adaptation of CIP accommodates the 8-byte packet size limitation of the CAN protocol and allows the use of simple devices with minimal processing power. For a more detailed description of the CAN protocol and some of its applications, see Ref. [4].

9.3.1.2 Relationship to Standards

Like other CIP Networks, DeviceNet follows the OSI model, an ISO standard for network communications that is hierarchical in nature. Networks that follow this model define all necessary functions, from physical implementation up to the protocol and methodology to communicate control and information data within and across networks.

Figure 9.11 shows the relationship between CIP and DeviceNet.

The DeviceNet adaptation of CIP is described in Volume 3 of the CIP Networks Library [5]. All other features are based on CIP. For example, DeviceNet is also described in a number of national and international standards [6,7].

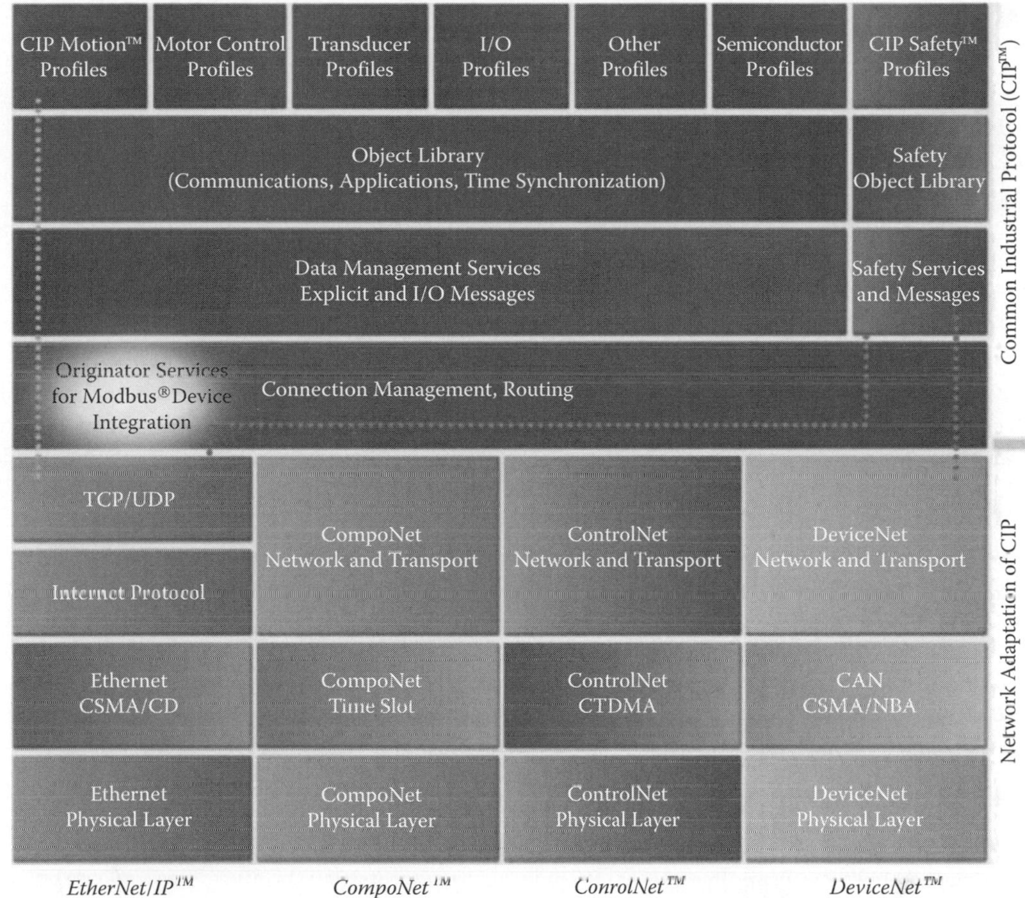

FIGURE 9.11 Relationship between CIP and DeviceNet.

9.3.1.3 DeviceNet Features

DeviceNet is a communication system at the low end (sensors and actuators) of the industrial communication spectrum with the following features:

- Trunkline/dropline configuration
- Support for up to 64 nodes
- Node insertion or removal while the network is up and running
- QuickConnect for devices that are frequently removed from and added to the network, for example, robot tools
- Simultaneous support for both network-powered devices, for example, sensors, and separately powered devices, for example, actuators
- Use of sealed or open-style connectors
- Protection from wiring errors
- Selectable data rates of 125, 250, and 500 kBaud
- Adjustable power configuration to meet individual application needs
- High current capability (up to 16 A per supply)
- Operation with off-the-shelf power supplies
- Power taps that allow the connection of several power supplies from multiple vendors that comply with DeviceNet standards

- Built-in overload protection
- Power available along the bus: both signal and power lines contained in the cable
- Several cables that are suitable for a number of different applications

9.3.1.4 DeviceNet Physical Layer and Relationship to CAN

The physical layer of DeviceNet is an extension of the ISO 11898 standard [8]. This extension defines the following additional details:

- Improved transceiver characteristics that allow the support of up to 64 nodes per network
- Additional circuitry for overvoltage and mis-wiring protection
- Several types of cables for a variety of applications
- Several types of connectors for open (IP 20) and sealed (IP 65/67) devices

The cables described in the CIP Networks Library were designed specifically to meet minimum propagation speed requirements to ensure that they can be used up to the maximum system length. Figure 9.12 shows examples of some of the key characteristics that can be achieved with some of the defined cable types in conjunction with suitable transceiver circuits and proper termination resistors (121 Ω).

ODVA has issued a guideline [9] that gives complete details of how to build the physical layer of a DeviceNet Network; equivalent information can also be found in an IEC standard [10].

Developers of DeviceNet devices can create DeviceNet circuits with or without physical layer isolation (both versions are fully specified). Furthermore, a device may take some or all of its power from the bus, thus avoiding extra power lines for devices that can live on the power supplied through the DeviceNet cable.

All DeviceNet devices must be equipped with one of the connectors described in Volume 3, although hard wiring of a device is allowed, provided the node is removable without severing the trunk.

9.3.1.5 Frame Structure

DeviceNet uses standard CAN frames with an 11-bit identifier; for further details, see [8,11] and Section 9.3.1.15 of this publication.

9.3.1.6 Protocol Adaptations

On the protocol side, there are basically two adaptations of CIP that have been made to better accommodate it to the CAN data frame:

- Shortening CIP Explicit Messages to 8 bytes or less where possible, with the use of message fragmentation for longer messages
- Definition of a Master Slave communications option to minimize the connection establishment overhead (see Section 9.3.1.17).

These two features have been created to allow the use of simple and thus inexpensive microcontrollers. This is particularly important for small, cost-sensitive devices like photoelectric or proximity sensors. As a result of this adaptation, the DeviceNet protocol in its simplest form has been implemented in 8-bit microprocessors with as little as 4 kB of code memory and 175 bytes of RAM.

	Trunk Distance			Drop Length	
Data Rate (kBaud)	Thick Cable (m)	Thin Cable (m)	Flat Cable (m)	Maximum (m)	Cumulative (m)
125	500	100	420	6	156
250	250		200		78
500	100		75		39

FIGURE 9.12 Data Rate vs. Trunk and Drop Length

The message fragmentation mentioned previously comes in two varieties:

1. For I/O Messages, the use of fragmentation is defined by the maximum length of the data to be transmitted through a connection. Any connection that has more than 8 bytes to transmit always uses the fragmentation protocol, even if the actual data to be transmitted is 8 bytes or less, for example, an *Idle* Message.
2. For Explicit Messaging, the use of the fragmentation protocol is indicated in the header of every message, since the actual frame size can vary in length, depending on the content of the Explicit Message. The actual fragmentation protocol is contained in one extra byte within the message that indicates whether the fragment is a start fragment, a middle fragment, or an end fragment. A modulo 64 rolling fragment counter allows very long fragmented messages and is limited in theory only by the maximum Produced or Consumed Connection sizes (65,535 bytes). In reality, the capabilities of the devices limit the message sizes.

9.3.1.7 Indicators and Switches

Indicators and switches are optional on DeviceNet. However, certain DeviceNet users not only require indicators and switches; they also specify what type to use. Many factors must be considered before implementing these devices, including packaging, accessibility, and customer expectations.

Indicators allow the user to determine the state of the device and its network connection(s). Since indicators can be very useful when troubleshooting the operation of a device, manufacturers are advised to incorporate some or all of the indicators described in the DeviceNet specification. While devices may incorporate additional indicators with behavior not described in the specification, any indicators labeled per specification must also follow their specified behavior.

Similarly, devices may be built with or without switches or other directly accessible means for configuration of MAC ID and baud rate. If these switches are used, certain rules apply to how these values are used at power-up and during the operation of the device.

9.3.1.8 Additional Objects

The DeviceNet Specification defines one additional object, the DeviceNet Object.

9.3.1.8.1 DeviceNet Object (Class ID: 0x03)

A DeviceNet Object is required for every DeviceNet port of the device. The instance attributes of this object contain information on how this device uses the DeviceNet port, including the MAC ID of the device and the (expected) baud rate of the DeviceNet network the device is attached to. Both attributes are always expected to be nonvolatile; that is, after a power interruption, the device is expected to try to go online again with the same values that were stored in these attributes before the power interruption. Devices that set these values through switches typically override any stored values at power-up. The DeviceNet Object may also contain information on further aspects associated with its DeviceNet behavior, such as information related to the Master Slave communications status, QuickConnect support (see Section 9.3.1.17.7), and Active Node Table.

9.3.1.9 Network Access

DeviceNet uses the network access mechanisms described in the CAN specification, that is, bitwise arbitration through the CAN Identifier for every frame to be sent. This requires a system design that does not allow multiple uses of any of these identifiers. Since the node number of every device is coded into the CAN Identifier (see Section 9.3.1.15), it is generally sufficient to make sure that none of the node numbers exists more than once on any given network. This is guaranteed through the Network Access algorithm (see Section 9.3.1.10).

9.3.1.10 Going Online

Any device that wants to communicate on DeviceNet must go through a Network Access algorithm before any communication is allowed. The main purpose of this process is to avoid duplicate Node IDs on the same network; a secondary purpose is to announce a node's presence on the link for nodes that maintain an Active Node Table. Every device that is ready to go online sends a Duplicate MAC ID Check Message containing its Port Number, Vendor ID, and Serial Number. If another device is already online with this MAC ID or is in the process of going online with this MAC ID, it responds with a Duplicate MAC ID Response Message that directs the checking device to go offline and not communicate any further.

If two or more devices with the same MAC ID happen to transmit the Duplicate MAC ID Check Message at exactly the same time, all of them will win arbitration at the same time and will proceed with their message. However, since this message has different values (Port Number, Vendor ID, and Serial Number) in the data field, the nodes will detect bit errors and will flag error frames that cause all nodes to discard the frame. This reaction triggers a retransmission of the message by the sending node. While this action may eventually result in a Bus-Off condition for the devices involved, a situation with duplicate Node IDs is safely avoided.

9.3.1.11 Offline Connection Set

The Offline Connection Set is a set of messages created to communicate with devices that have failed to go online (see Section 9.3.1.10), to allow a new MAC ID to be set. At any given point in time, only one offline device and one tool can use the Offline Connection Set; therefore, the first step in its use is to determine if a tool has ownership of the Offline Connection Set. Once a tool has successfully claimed ownership, it can check whether there are any nodes on the network that are in the offline state. If such nodes exist, the tool can then determine their Vendor ID(s) and Serial Number(s). Using this information, which is unique by definition, the tool can then address a specific device that responds by flashing an indicator. Once this identification is complete and the user is certain that communication is established with the intended device, the tool can then send a new MAC address to the device. The target device then restarts the Duplicate MAC ID algorithm and tries to go online with the new MAC Address. More information on this topic can be found in [4,5].

9.3.1.12 DeviceNet Status Indication Messages

There are two optional DeviceNet messages that indicate a status or a status transition of a device. One of them is called *Device Heartbeat* and the other is called *Device Shutdown*. Both messages are transmitted by a UCMM-capable device as an unconnected response message (Message Group 3, Message ID 5) and by a Group 2 Only Server as an unconnected response message (Message Group 2, Message ID 3). These messages are independent of any other communication relationship that may exist with other devices on the network.

The Device Heartbeat Message, sent at a heartbeat interval set in the ID Object, provides a way for a device to indicate its presence on the network and its current *health* condition. The Device Shutdown Message provides a way for a device to indicate that it is in the process of shutting down and going to the offline state.

9.3.1.13 Explicit Messaging

All Explicit Messaging in DeviceNet is done via connections and the associated Connection Object instances. However, these objects first must be set up in the device. This can be done by using the Predefined Master Slave Connection Set to activate a static Connection Object already available in the device or by using the UCMM port of a device, to dynamically set up a Connection Object for Explicit Messaging. The only messages sent to the UCMM are *Open* or *Close* requests that set up or tear down an Explicit Messaging Connection, while the only messages that can be sent to the Master Slave equivalent

Byte Offset	7	6	5	4	3	2	1	0	
	Bit Number								
0	Frag [0]	XID	MAC ID						Message header
1	R/R [0]	Service Code							Message body
2	Class ID								
3	Instance ID								
4	Service data								
...	...								
7	(optional)								

FIGURE 9.13 Nonfragmented Explicit Request Message Format

Byte Offset	7	6	5	4	3	2	1	0	
	Bit Number								
0	Frag [0]	XID	MAC ID						Message header
1	R/R [1]	Service Code							Message body
2	Service data								
...	...								
7	(optional)								

FIGURE 9.14 Nonfragmented Explicit Response Message Format

of the UCMM called the Group 2 Only Unconnected Port are the *Allocate* or *Release* service requests (see Section 9.3.1.17). Explicit Messages always pass via the Message Router Object to the object that is being addressed (refer to Figure 9.8).

As mentioned in Section 9.2.3, Explicit Messages on DeviceNet have a very compact structure to make them fit into the 8-byte frame in most cases. Figure 9.13 shows a typical example of a request message using the 8/8 Message Body Format (8/8 means 1 byte for Class ID and 1 byte for Instance ID).

The consumer of this Explicit Message responds using the format shown in Figure 9.14. The consumer sets the R/R (Request/Response) bit and repeats the Service Code of the request message. Any data transferred with the response is entered in the service data field.

Most messages will use the 8/8 format shown in Figure 9.13, since they need to address Class and Instance IDs only up to 255. If there is a need to address any class/instance combinations above 255, then this is negotiated between the two communication partners during the setup of the connection. Should an error occur, the receiver responds with the Error Response Message. The Service Code for an Error Response message is 0x14, and 2 bytes of error code are included in the service data field to convey more information about the nature of the error. See Refs. [4,5] for further details of message encoding, including the use of fragmentation.

9.3.1.14 I/O Messaging

Since DeviceNet does not use a Real-Time Header or Sequence Count Value like ControlNet and EtherNet/IP do, I/O Messages in DeviceNet have a very compact structure. For I/O data transfers up to 8 bytes long, the data are sent in a nonfragmented message, which uses the entire CAN data field for I/O data. For I/O data transfers longer than 8 bytes, a fragmentation protocol spreads the data over multiple frames. This fragmentation protocol uses 1 byte of the CAN data field to control the fragmentation of the data. See Figure 9.15 and 9.16 for examples of fragmented and nonfragmented I/O messages. I/O Messages without data (i.e., with zero-length data) indicate the *Idle* state of the producing application. Any producing device can do this—master, slave, or peer.

Byte Offset	Bit Number 7	6	5	4	3	2	1	0
0	Process data							
…								
7	(0–8 bytes)							

FIGURE 9.15 Nonfragmented I/O Message Format

Byte Offset	Bit Number 7	6	5	4	3	2	1	0
0	Fragmentation protocol							
1	Process data							
…								
7	(0–7 bytes)							

FIGURE 9.16 Fragmented I/O Message Format

As mentioned, I/O Messages are used to exchange high-priority application and process data via the network, and this communication is based on the Producer/Consumer model. The associated I/O data are always transferred from one producing application object to one or more consuming application objects. This is accomplished using I/O Messages via I/O Messaging Connection Objects (Figure 9.17 shows two consuming applications) that have been preset in the device. This can be done in one of two ways by using

- The Predefined Master Slave Connection Set to activate a static I/O Connection Object already available in the device
- An Explicit Messaging Connection Object already available in the device to dynamically create and set up an appropriate I/O Connection Object

I/O Messages usually pass directly to the data of the assigned application object. The Assembly Object is the most common application object used with I/O Connections. Also refer to Figure 9.8.

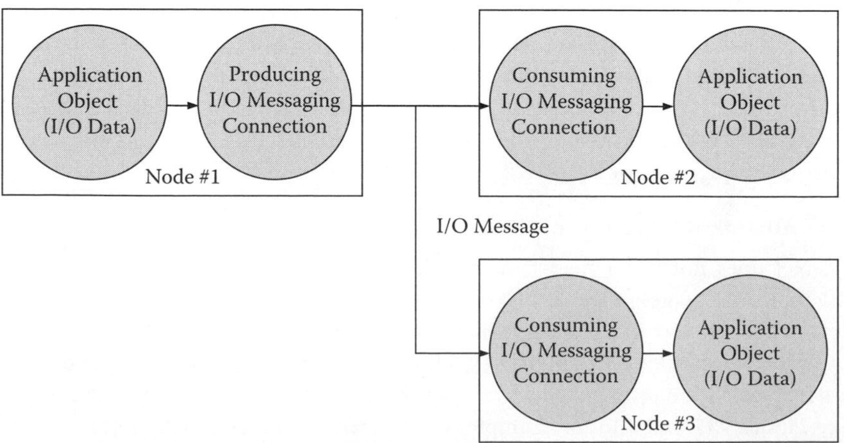

FIGURE 9.17 I/O messaging connections.

9.3.1.15 Using the CAN Identifier

DeviceNet is based on the standard CAN protocol and therefore uses an 11-bit message identifier. A distinction therefore can be made between $2^{11} = 2048$ messages. The 6-bit MAC ID field is sufficient to identify a device because a DeviceNet Network is limited to a maximum of 64 participants.

The overall CAN Identifier range is divided into four Message Groups of varying sizes, as shown in Figure 9.18.

The bitwise arbitration mechanism of CAN determines the priority of messages on DeviceNet. When two nodes transmit simultaneously, the numerically lower CAN Identifier value will win arbitration. The arbitration mechanism is explained in the CAN specification [11]. A detailed description is beyond the scope of this document, but in short, transmitted bits are shifted onto the wire most significant bit first, so a zero in the upper bit positions will take precedence over a one. As is shown in Figure 9.18, Message Group 1 has a zero in bit 10, so it is the highest-priority group. Group 2 is the second highest-priority group because of the zero in bit 9. Group 3 is the next highest priority group, because the CAN IDs contain a one in bits 9 and 10. All valid Group 3 Message IDs are lower numerically than the corresponding bits (8–6) in Group 4 and therefore Group 4 is the lowest priority of all.

In DeviceNet, the CAN Identifier is the CID. This comprises the Message Group ID, the Message ID within this group, and the device's MAC ID, which can be the source or destination address. The definition depends on the Message Group and the Message ID. The significance of the message within the system is defined by the Message Group and Message ID.

The four Message Groups are used as follows:

Message Group 1 is assigned 1024 CAN Identifiers (0x0000–0x03FF), which is 50% of all identifiers available. Up to 16 different Message IDs are available per device (node) within this group. The priority of a message within this group is primarily determined by the Message ID (the significance of the message) and only after that by the source MAC ID (the producing device). If two devices transmit a Message Group 1 message at the same time, then the device with the lower Message ID will always win the arbitration. However, if two devices transmit the same Message ID at the same time on the CAN bus, then the device with the lower MAC ID will win. The messages of Group 1 are, therefore, well suited for the exchange of high-priority process data.

Message Group 2 is assigned 512 identifiers (0x0400–0x05FF). Most of the Message IDs in this group are optionally defined for what is commonly referred to as the Predefined Master Slave Connection Set (see Section 9.3.1.17). One Message ID is reserved for the Duplicate Node ID Check (see Section 9.3.1.10). Priority within Message Group 2 is determined primarily by the MAC ID and, only after that, by the Message ID. This message group was designed so that a CAN controller with an 8-bit mask is able to filter out its Group 2 Messages based on MAC ID. This makes it possible for very low-cost, low-functionality microcontrollers with integral CAN controllers to be suitable for use on DeviceNet.

Connection ID = CAN Identifier (Bits 10:0)											
10	9	8	7	6	5	4	3	2	1	0	Used For
0	Message ID				Source MAC ID						Message Group 1
1	0	MAC ID						Message ID			Message Group 2
1	1	Message ID			Source MAC ID						Message Group 3
1	1	1	1	1	Message ID						Message Group 4
1	1	1	1	1	1	1	×	×	×	×	Invalid CAN Identifiers

FIGURE 9.18 Definition of the Message Groups.

Message Group 3, with 448 CAN Identifiers (0x0600–0x07BF), has a similar structure to Message Group 1; however, it is mainly used for low-priority process data exchange due to the relative priority difference between Groups 1 and 3. In addition, the main use of this group is setting up dynamic Explicit Connections. Seven Message IDs per device are possible, and two of these are reserved for what is commonly referred to as the UCMM port (see Section 9.3.1.16).

Message Group 4, with 48 CAN Identifiers (0x07C0–0x07EF), does not include any MAC IDs, only Message IDs. The messages in this group are used only for network management. Four Message IDs are currently assigned for services of the Offline Connection Set.

The remaining 16 CAN Identifiers (0x07F0–0x07FF) are invalid CAN IDs and thus are not permitted for use in DeviceNet systems.

With this allocation of CAN Identifiers, the unused CAN Identifiers cannot be used by other devices. Therefore, each device has exactly 16 Message IDs in Group 1, 8 Message IDs in Group 2, and 7 Message IDs in Group 3. One advantage of this system is that the CAN Identifiers used in the network can always be clearly assigned to a device. Devices are responsible for managing their own identifiers. This simplifies the design, troubleshooting, and diagnosis of DeviceNet systems, as a central tool that keeps a record of all CAN ID assignments on the network is not needed.

9.3.1.16 Connection Establishment

As described in Sections 9.3.1.12 and 9.3.1.14, messages on DeviceNet are always exchanged in a connection-based manner. Communication objects must be set up for this purpose. These are not initially available when a device is powered on; they first have to be created. There are two ports by which a DeviceNet device can be addressed when first powered on, the UCMM port or the Group 2 Only Unconnected Explicit Request port, which is defined by the Predefined Master Slave Connection Set. Picture these ports as doors to the device. Only one key will unlock each door. The appropriate key for each lock is the CID—that is, the CAN Identifier—of the selected port. Other doors in the device can be opened only if and when the appropriate key is available, and other instances of Connection Objects are set up.

The setting up of communication relationships (i.e., connections) via the UCMM port represents a general procedure that should be adhered to with all DeviceNet devices. Devices that feature the Predefined Master Slave Connection Set and are UCMM Capable are called Group 2 Servers. A Group 2 Server can be addressed by one or more connections from one or more clients.

Since UCMM-capable devices need a good amount of processing power to service multiple communication requests, a simplified communication establishment and I/O data exchange method has been created for low-end devices. This is called the Predefined Master Slave Connection Set (see Section 9.3.1.17). This covers as many as five predefined connections that can be activated (assigned) when accessing the device. The Predefined Master Slave Connection Set represents a subset of the general connection establishment method, and it is limited to pure Master Slave relations. Slave devices that are not UCMM Capable and support only this subset are called Group 2 Only Servers. Only the master that allocates it can address a Group 2 Only Server. All messages received by this device are defined in Message Group 2.

For more details of the connection establishment using UCMM and the Master Slave Connection Set, refer to Refs. [4,5].

9.3.1.17 Predefined Master Slave Connection Set

Establishing a connection via the UCMM port requires a relatively large number of steps that must be completed to allow data exchange via DeviceNet, and the devices must provide resources to administer the dynamic connections. Because every device can set up a connection with every other device and the source MAC ID of the devices is contained in the CID, the CAN Identifier (CID) may have to be filtered via software. This depends on how many connections a device supports and on the type and number of screeners (hardware CAN ID filters) of the CAN chip used in the device's implementation.

While this approach maximizes the use of the multicast, peer-to-peer, and Producer/Consumer capabilities of CAN, it requires a higher-performance CPU and more RAM and ROM resources. These requirements eliminate an entire class of low-cost microcontrollers with internal CAN controllers from consideration, raising the cost of implementation to levels that preclude cost-effective solutions for low-end (e.g., low end-user cost) devices. The Predefined Master Slave Connection Set was designed to minimize message processing and to take advantage of the limited screening capabilities of many CAN controllers. The Predefined Master Slave Connection Set is the way that the vast majority of devices communicate on DeviceNet.

The Predefined Master Slave Connection Set defines an alternate way to establish connections called the Group 2 Only Unconnected Explicit Request Port. This method allows a device to limit the messages received to only those in Group 2 with its own MAC ID. This greatly reduces the amount of packets that a node must deal with. A CAN controller with a single mask and match screener (a so-called BasicCAN screener) can be used in this type of device, which makes it possible to use the low-cost microcontrollers and simple CAN controllers mentioned earlier. Devices that operate in this manner are referred to as Group 2 Only Servers, deriving their name from the fact that they are only required to receive messages in Group 2.

The Predefined Master Slave Connection Set is also used in UCMM-capable devices. Such devices are referred to as Group 2 Servers, deriving their name from the fact that they respond to Group 2 messages but are not limited to just Group 2 messages.

Many of the reasons for defining the Predefined Master Slave Connection Set were due to hardware limitations prevalent when the protocol was first developed. Many of the cost considerations have changed as hardware evolved over time. Today, most devices can be implemented with a UCMM port and still be cost effective. This is the preferred type of device to develop today. For reasons that go beyond the scope of this document, devices that are not capable of UCMM place extra burden on other devices and tools. Except in extremely low-cost situations, UCMM should always be a design goal for DeviceNet products.

The Predefined Master Slave Connection Set provides an interface for a set of up to five preconfigured connection types in a node.

The basis of this model is a 1:n communication structure consisting of one control device and decentralized I/O devices. The central portion of such a system is known as the *Master*, and the decentralized devices are known as *Slaves*. Multiple masters are allowed on the network, but a slave can be allocated to only one master at any time.

The predefined Connection Objects occupy instances 1–5 in the Connection Object (Class ID 0x05, see Section 9.2.4):

- Explicit Messaging Connection
  - Group 2 Explicit Request/Response Message (Instance ID 1)
- I/O Messaging Connections
  - Polled I/O Connection (Instance ID 2)
  - Bit-Strobe I/O Connection (Instance ID 3)
  - COS or Cyclic I/O Connection (Instance ID 4)
  - Multicast Polling I/O Connection (Instance ID 5)

The messages to the slave are defined in Message Group 2, and some of the responses from the slave are contained in Message Group 1. The distribution of CIDs for the Predefined Master Slave Connection Set is defined as shown in Figure 9.19.

Because the CAN ID of most of the messages the master produces contains the destination MAC ID of the slave, it is imperative that only one master talks to any given slave. Therefore, before it can use this Predefined Connection Set, the master must first allocate it with the device. The DeviceNet Object manages this important function in the slave device. It allows only one master to allocate its Predefined Connection Set, thereby preventing duplicate CAN IDs from appearing on the wire.

The two services used are called Allocate_Master Slave_Connection_Set (Service Code 0x4B) and Release_Group_2_Identifier_Set (Service Code 0x4C). These two services always access instance 1 of the DeviceNet Object (Class ID 0x03) (see Figure 9.20).

Connection ID = CAN Identifier (bits 10:0)											
10	9	8	7	6	5	4	3	2	1	0	Used For
0	Group 1 Message ID				Source MAC ID						Group 1 Messages
0	1	1	0	0	Source MAC ID						Slave's I/O Multicast Poll Response
0	1	1	0	1	Source MAC ID						Slave's I/O Change of State or Cyclic Message
0	1	1	1	0	Source MAC ID						Slave's I/O Bit-Strobe Response Message
0	1	1	1	1	Source MAC ID						Slave's I/O Poll Response or COS/Cyclic Ack Message
1	0	MAC ID						Group 2 Message ID			Group 2 Messages
1	0	Source MAC ID						0	0	0	Master's I/O Bit-Strobe Command Message
1	0	Source MAC ID						0	0	1	Master's I/O Multicast Poll Group ID
1	0	Destination MAC ID						0	1	0	Master's Change of State or Cyclic Ack Message
1	0	Source MAC ID						0	1	1	Slave's Explicit/Unconnected Response Messages
1	0	Destination MAC ID						1	0	0	Master's Explicit Request Messages
1	0	Destination MAC ID						1	0	1	Master's I/O Poll Command/COS/Cyclic Message
1	0	Destination MAC ID						1	1	0	Group 2 Only Unconnected Explicit Request Messages

FIGURE 9.19 Connection IDs of the Predefined Master Slave Connection Set

Figure 9.20 shows the Allocate Message with 8-bit Class ID and 8-bit Instance ID, a format that is always used when it is sent as a Group 2 Only Unconnected Message. It also may be sent across an existing connection and in a different format if a format other than 8/8 was agreed during the connection establishment.

The Allocation Choice Byte is used to determine which predefined connections are to be allocated (see Figure 9.21).

The associated connections are activated by setting the appropriate bits. COS and Cyclic are mutually exclusive choices. The COS/Cyclic Connection may be configured as not acknowledged using the acknowledge suppression bit. The individual connection types are described in more detail in the following text.

	Bit Number								
Byte Offset	7	6	5	4	3	2	1	0	
0	Frag [0]	XID	MAC ID						Message header
1	R/R [0]		Service Code [0x4B]						Message body
	Class ID [0x03]								
	Instance ID [0x01]								
2–5	Allocation Choice								
	0	0	Allocator's MAC ID						

FIGURE 9.20 Allocate_Master Slave_Connect_Set Request Message

Bit Number							
7	6	5	4	3	2	1	0
Reserved	Ack suppression	Cyclic	Change of state	Multicast polling	Bit-strobe	Polled	Explicit Message

FIGURE 9.21 Format of the Allocation Choice Byte

The allocator's MAC ID contains the address of the node (master) that wants to assign the Predefined Master Slave Connection Set. Byte 0 of this message differs from the allocator's MAC ID if this service has been passed on to a Group 2 Only Server via a Group 2 Only Client (what is commonly referred to as a *proxy function*).

The slave, if not already claimed, responds with a Success Message. The connections are now in the Configuring State. Setting the Expected_Packet_Rate EPR (Set_Attribute_Single service to attribute 9 in the appropriate Connection Object instance, value in ms) starts the connection's time-monitoring function. The connection then changes into Established State, and I/O Messages begin transferring via this connection.

The allocated connections can be released individually or collectively through the Release_Master Slave_Connection_Set service (Service Code 0x4C), using the same format as that in Figure 9.20, except that the last byte (Allocator's MAC ID) is omitted.

The following is an explanation of the four I/O Connection types in the Predefined Master Slave Connection Set.

9.3.1.17.1 Polled I/O Connection

A Polled I/O Connection is used to implement a classic Master Slave relationship between a control unit and a device. In this setup, a master can transfer data to a slave using the Poll Request and receive data from the slave using the Poll Response. Figure 9.22 shows the exchange of data between one master and three slaves in Polled I/O mode.

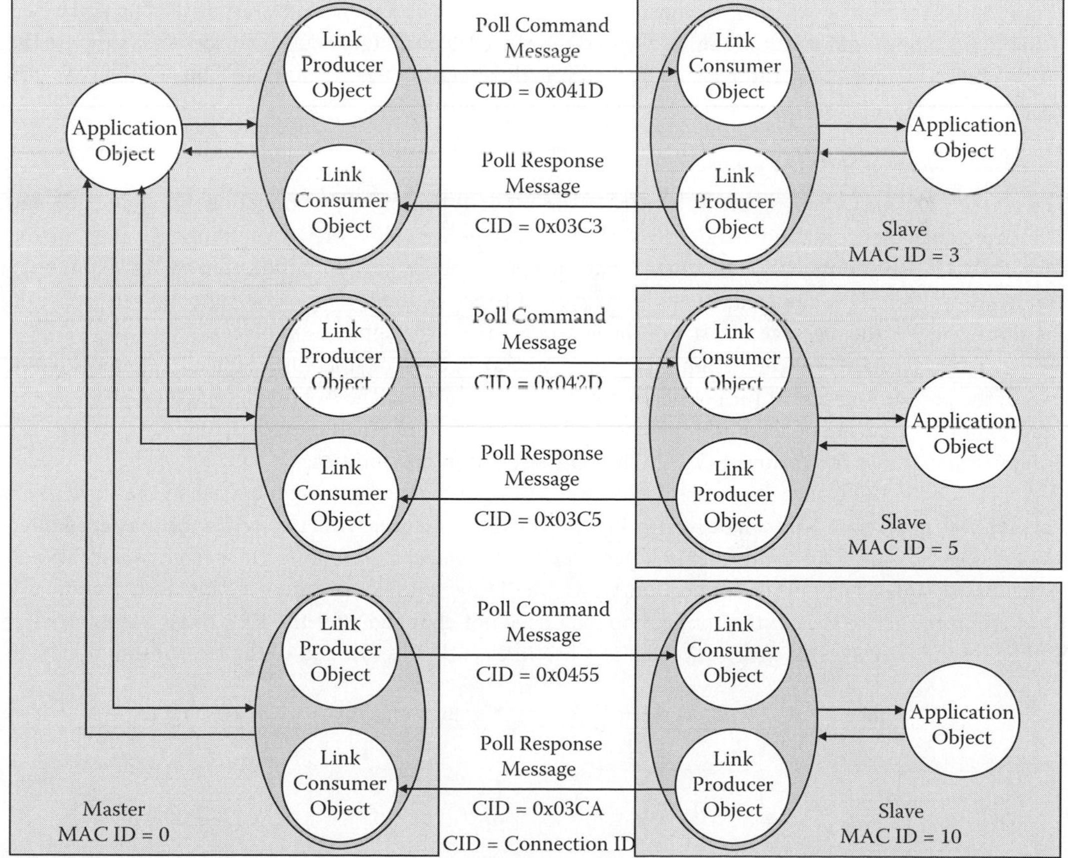

FIGURE 9.22 Polled I/O Connections.

The amount of data transferred in a message between a master and a slave using the Polled I/O Connection can be any length. If the length exceeds 8 bytes, the fragmentation protocol is automatically used. A Polled I/O Connection is always a point-to-point connection between a master and a slave. The slave consumes the Poll Message and sends back an appropriate response (normally, its input data).

The Polled Connection is subject to a time-monitoring function, which can be adjusted, in the device. A Poll Command must have been received within this time (4 × EPR) or else the connection reverts to timeout mode. When a connection times out, the node optionally may go to a preconfigured fault state as set up by the user. A master usually polls all the slaves in a round-robin manner.

A slave's response time to a poll command is not defined in The DeviceNet Specification. While this provides flexibility for slave devices to be tailored to their primary application, it may also exclude the device from use in higher-speed applications.

9.3.1.17.2 Bit-Strobe I/O Connection

The master's transmission on this I/O Connection is the Bit-Strobe Command. Using this command, a master multicasts one message to reach all its slaves allocated for the Bit-Strobe Connection. The frame sent by the master using a Bit-Strobe Command is always 8 bytes or 0 bytes (if Idle). From these 8 bytes, each slave is assigned one bit (see Figure 9.23). Each slave can send back as many as 8 data bytes in its response.

A Bit-Strobe I/O Connection represents a multicast connection between one master and any number of strobe-allocated slaves (see Figure 9.24). Since all devices in a network receive the Bit-Strobe Command at the same time, they can be synchronized by this command. When the Bit-Strobe Command is received, the slave may consume its associated bit and then send a response of up to 8 bytes.

Since this command uses the source MAC ID in the CID (see Figure 9.19), devices that support the Bit-Strobe I/O Connection and have a CAN chip with screening limited to only 8 bits of the CAN ID (11 bits) must perform software screening of the CAN Identifier.

9.3.1.17.3 Change of State/Cyclic I/O Connection

The COS/Cyclic I/O Connection differs from the other types of I/O Connections in that both end points produce their data independently. This can be done in a COS or Cyclic manner. In the former case, the COS I/O Connection recognizes that the application object data indicated by the Produced_Connection_Path have changed. In the latter case, a timer of the Cyclic I/O Connection expires and therefore triggers the message transfer of the latest data from the application object.

A COS/Cyclic I/O Connection can be set up as acknowledged or unacknowledged. When acknowledged, the consuming side of the connection must define a path to the Acknowledge Handler Object to ensure proper handling of acknowledgments and management of any required retries.

Figure 9.25 shows the various COS/Cyclic I/O Connection possibilities.

A COS/Cyclic I/O Connection can also originate from a master, making it appear to the slave like a Polled I/O Connection. This can be seen in Figure 9.19 since the same CID is used for the master's Polled I/O Message as is used for the master's COS/Cyclic I/O Message. COS Connections have two additional behaviors not present in other connection types. The Expected Packet Rate (EPR) is used as a default production trigger so that, if the connection data have not changed after the EPR timer has expired, it will be resent anyway. This *heartbeat*, as it is sometimes called, is utilized so the consuming node can

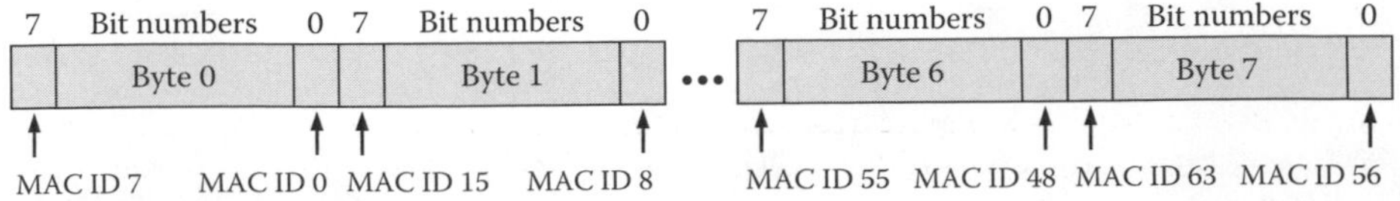

FIGURE 9.23 Data format of the Bit-Strobe I/O Connection.

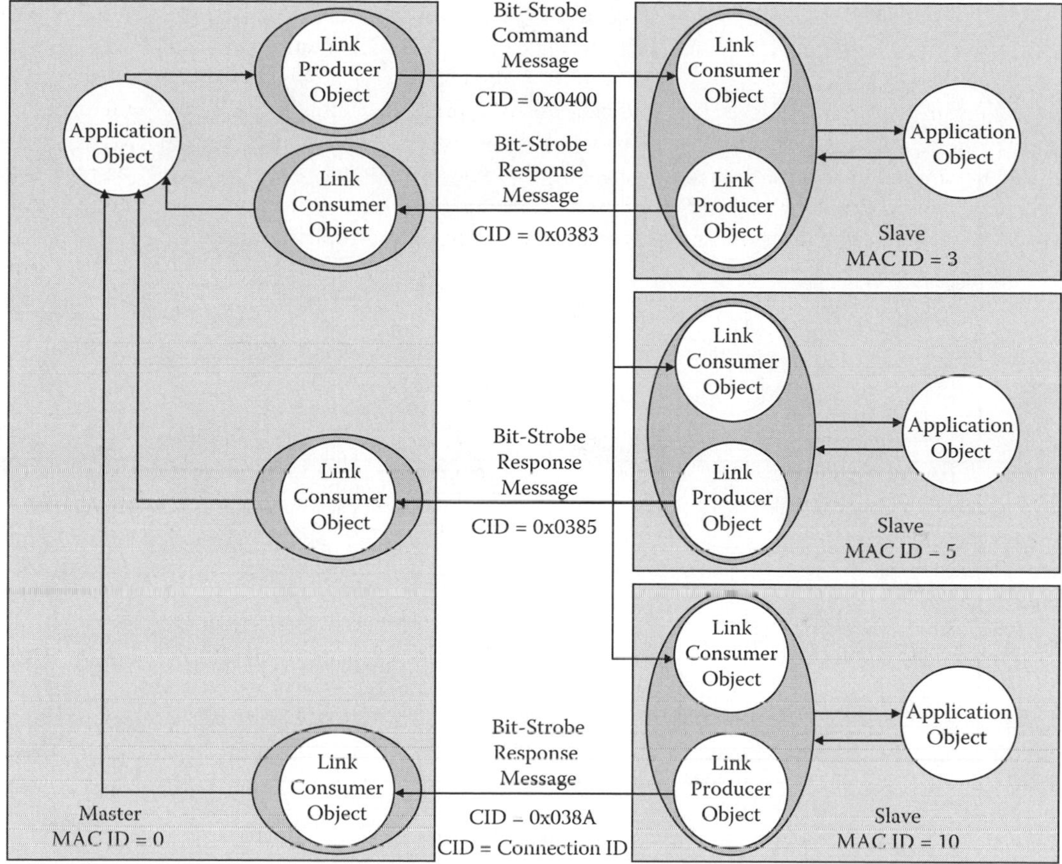

FIGURE 9.24 Bit-Strobe I/O Connections.

know the difference between a node that has gone offline and one whose data have not changed. COS Connections also have a Production Inhibit Timer feature that prevents a node from producing data too often and thus using too much network bandwidth. The production inhibit timer determines the amount of time the node must remain quiet after producing data to the network.

9.3.1.17.4 Multicast Polled I/O Connection

This connection is similar to the regular I/O poll except that all of the slaves belonging to a multicast group consume the same output data from the master. Each slave responds with its own reply data. A unique aspect of this connection is that the master picks the CAN ID from one of the slaves in the multicast group and must then set the consumed CAN ID in each of the other slaves to that same value. If, during runtime, that slave's connection times out, the master must either stop producing its multicast poll command or pick another slave in the group and reset the command CAN ID in all the remaining slaves in the group to that value before sending another Multicast Poll Command.

9.3.1.17.5 I/O Data Sharing

Due to the inherent broadcast nature of all CAN frames, applications can be set up to *listen* to the data produced by other applications. Such a *listen-only* mode is not described in the DeviceNet specification, but some vendors have created products that do exactly that, for example, *shared inputs* in Allen-Bradley I/O Scanners.

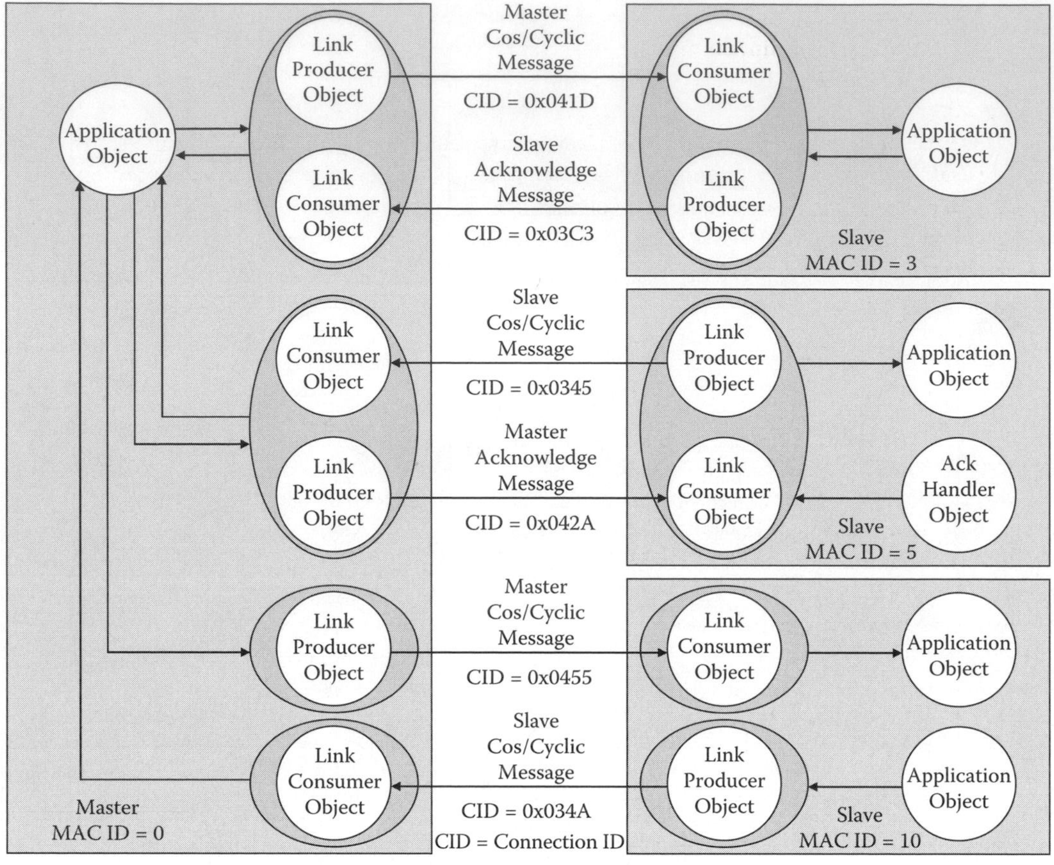

FIGURE 9.25 COS/Cyclic I/O Connections.

9.3.1.17.6 Typical Master Slave Start Sequence

Typically, starting up a DeviceNet Network with an I/O Scanner and a set of slaves is executed as follows:

- All devices run their self-test sequence and then try to go online with the algorithm described in Section 9.3.1.10. Any device that uses an autobaud mechanism to detect the baud rate of a network has to wait with its Duplicate Node ID Message until it has seen enough CAN frames to detect the correct baud rate.
- Once online, slave devices will remain silent, except to defend their MAC ID, until their master allocates them.
- Once online, a master will try to allocate each slave configured into its scan list by running the following sequence of messages:
  - Try to open a connection to the slave using a UCMM Open Message.
  - If successful, the master can then use this connection for further communication with the slave (the device is a Group 2 Server).
  - If not successful, the master will try again after a minimum wait time of 1 s.
  - If unsuccessful again, the master will try to allocate the slave using the Group 2 Only Unconnected Explicit Request Message (at least for Explicit Messaging) after a minimum wait time of 1 s.
  - If successful, the master can then use this connection for further communication with the slave (the device is a Group 2 Only Server).

    - If not successful, the master will try again after a minimum wait time of 1 s.
    - If unsuccessful again, the master will start over with the UCMM Message after a minimum wait time of 1 s. This process will carry on indefinitely or until the master has allocated the slave.
- Once the master has allocated the slave, it may carry out some identity verification to see whether it is safe to start I/O Messaging with the slave. The master also may apply further configuration to the connections it has established, for example, setting the Explicit Messaging Connection to *Deferred Delete*;
- Setting the EPR value(s) brings the I/O Connection(s) to the Established State so that I/O Messaging can commence.

9.3.1.17.7 *QuickConnect Connection Establishment*

DeviceNet also allows an optional method of connection establishment known as QuickConnect. This was designed to provide the same level of protection against duplicate MAC IDs, but to do so in a much shorter time period, allowing connections to be established in a fraction of the time they normally take. This method is useful in applications where nodes are added to an operating network and the time required for establishing connections directly impacts productivity. For example, in robotic applications, the end of arm electronics are often changed out when a new item enters its workspace. These electronics need to be operational very quickly to avoid cycle time delays.

The QuickConnect process includes all the same steps as the typical start-up process, but most of them are done in parallel rather than in sequence. As a result, the device self-check and Duplicate MAC ID Check processes begin immediately, and the node goes online almost simultaneously. A failure of the device self-test or a duplicate MAC ID indication causes the device to remove itself from the bus.

In order for applications to benefit fully from this method, QuickConnect must be implemented in both the master and the slave. This feature is selectable through an EDS entry and, by default, is disabled in nodes that support it.

9.3.1.17.8 *Master Slave Summary*

Device manufacturers can easily support the Predefined Master Slave Connection Set by using simple BasicCAN controllers. Software screening of the CAN Identifier generally is not necessary, which enables the use of low-cost 8-bit controllers. This may represent an advantage as far as the devices are concerned but entails some disadvantages for the system design.

Group 2 Only (i.e., UCMM-incapable) devices permit only one Explicit Messaging Connection between client (master) and server (slave), whereas UCMM-capable devices can maintain Explicit Messaging Connections with more than one client at the same time.

If a device wants to communicate with one of the allocated slaves that does not support UCMM, the master recognizes this situation and sets up a communication relationship with the requestor instead. Any communication between the requestor is then automatically routed via the master. This is called the Proxy function. Since this puts an additional burden on the master and on network bandwidth, it is recommended that slave devices support UCMM.

Although not explicitly defined in The DeviceNet Specification, DeviceNet masters can, under certain conditions, automatically configure their scan lists and/or the devices contained in their scan lists. This functionality simply makes use of the explicit messaging capabilities of masters and slaves that allows the master to read from a slave whatever information is required to start an I/O communication and to download any configurable parameters that have been communicated to the master via EDS. This functionality facilitates the replacement of even complex slave devices without the need for a tool, dramatically reducing system downtime.

9.3.1.18 Device Profiles

DeviceNet devices may utilize any of the device profiles described in the CIP Networks Library. As of the publication date of this book, no DeviceNet-specific profiles have been defined.

9.3.1.19 Configuration

DeviceNet devices typically come with EDSs as described in Section 9.2.7. EDS files for DeviceNet devices can make full use of all EDS features, but they do not necessarily contain all sections. Typical DeviceNet devices contain (apart from the mandatory sections) at least an IO_Info section.

This section specifies which types of Master Slave connections are supported and which one(s) should be enabled as defaults. It also tells which I/O Connections may be used simultaneously.

An EDS also may contain individual parameters and/or a Configuration Assembly with a complete description of all parameters within this assembly. A full description of what can be done in DeviceNet EDS files would go well beyond the scope of this book. For available materials on this topic that go into more detail, see [1,2].

9.3.1.20 Conformance Test

See Chapter 6 of this publication for information on conformance testing.

9.3.1.21 Tools

Tools for DeviceNet networks can be divided into three groups:

1. Physical layer tools are tools (hardware and/or software) that verify the integrity and conformance of the physical layer or monitor the quality of the data transmission.
2. Configuration tools are software tools capable of communicating with individual devices for data monitoring and configuration purposes. They can range from very basic software operating on handheld devices to powerful PC-based software packages used to configure complete networks. Most configuration tools are EDS-based; however, more complex devices like I/O Scanners tend to have their own configuration applets that are only partially based on EDS files. Some of these tools support multiple access paths to the network, for example, via Ethernet and suitable routing devices, and thus allow remote access. High-level tools also actively query the devices on the network to identify them and monitor their *health*.
3. Monitoring tools typically are PC-based software packages that can capture and display CAN frames on the network. A raw CAN frame display may be good enough for some experts, but using a tool that allows both raw CAN display and DeviceNet interpretation of the frames is recommended.

For a typical installation, a configuration tool is all that is needed. However, to ensure that the network is operating reliably, verification with a physical layer tool is highly recommended. Experience shows that the overwhelming majority of DeviceNet network problems are caused by inappropriate physical layer installation. Protocol monitoring tools are used primarily to investigate interoperability problems and to assist during the development process. Turn to the CIP supplier directory on the ODVA website to access a list of vendors that provide tools for DeviceNet.

9.3.1.22 Advice for Developers

Before starting any DeviceNet product development, the following issues should be considered in detail:

- What functionality does the product require today and in future applications?
  - Slave functionality
  - Group 2 Server vs. Group 2 Only Server.
  - Master functionality
  - Combination of Master and Slave functionality

- What are the physical layer requirements? Is IP 65/67 required, or is IP 20 good enough?
- What type of hardware should be chosen for this product?
- What kind of firmware should be used for this product? Will a commercially available communication stack be used?
- Will the development of hardware and/or software be done internally, or will it be designed by an outside company?
- What are the configuration requirements?
- What design and verification tools should be used?
- What kind of configuration software should be used for this product? Will a commercially available software package be used, that is, is an EDS adequate to describe the device or is custom software needed?
- When and where will the product be tested for conformance and interoperability?
- What is an absolute must before my products can be placed on the market (i.e., own the specification, have the company's own Vendor ID, and have the product conformance tested)?

A full discussion of these issues goes well beyond the scope of this publication; for more information, see [12].

9.3.1.23 DeviceNet Summary

Since its introduction in 1994, DeviceNet has been used successfully in tens of millions of nodes in many different applications. It is a de facto standard in many countries, which is reflected in several national and international standards [6,13,7]. Due to its universal communication characteristics, it is one of the most versatile networks for low-end devices. While optimized for devices with small amounts of I/O, it can easily accommodate larger devices as well. Powerful EDS-based configuration tools allow easy commissioning and configuration of even complex devices without the need to consult manuals.

With the introduction of CIP Safety on DeviceNet, many machine-level applications that previously required a set of dedicated networks today can be accommodated on a single DeviceNet network.

Finally, as a member of the CIP family of networks, DeviceNet can be combined into an overall CIP Network structure that allows seamless communication among CIP Networks, as if they were only one network.

9.3.2 ControlNet

9.3.2.1 Introduction

Introduced in 1997, ControlNet is a deterministic digital communications network that provides high-speed transport of time-critical I/O and explicit messaging data—including upload/download of programming and configuration data and peer-to-peer messaging—on a single physical media link. Each device and/or controller is a node on the network.

ControlNet is a Producer/Consumer network that supports multiple communication hierarchies and message prioritization. ControlNet systems offer a single point of connection for configuration and control by supporting both Implicit (I/O) and Explicit Messaging. ControlNet's time-based message scheduling mechanism provides network devices with deterministic and predictable access to the network while preventing network collisions. This scheduling mechanism allows time-critical data, which is required on a periodic, repeatable, and predictable basis, to be produced on a predefined schedule without the loss of efficiency associated with continuously requesting, or *polling*, for the required data.

9.3.2.2 Relationship to Standards

Like other CIP Networks, ControlNet follows the OSI model, an ISO standard for network communications that is hierarchical in nature. Networks that follow this model define all necessary functions, from physical implementation up to the protocol and methodology to communicate control and information data within and across networks.

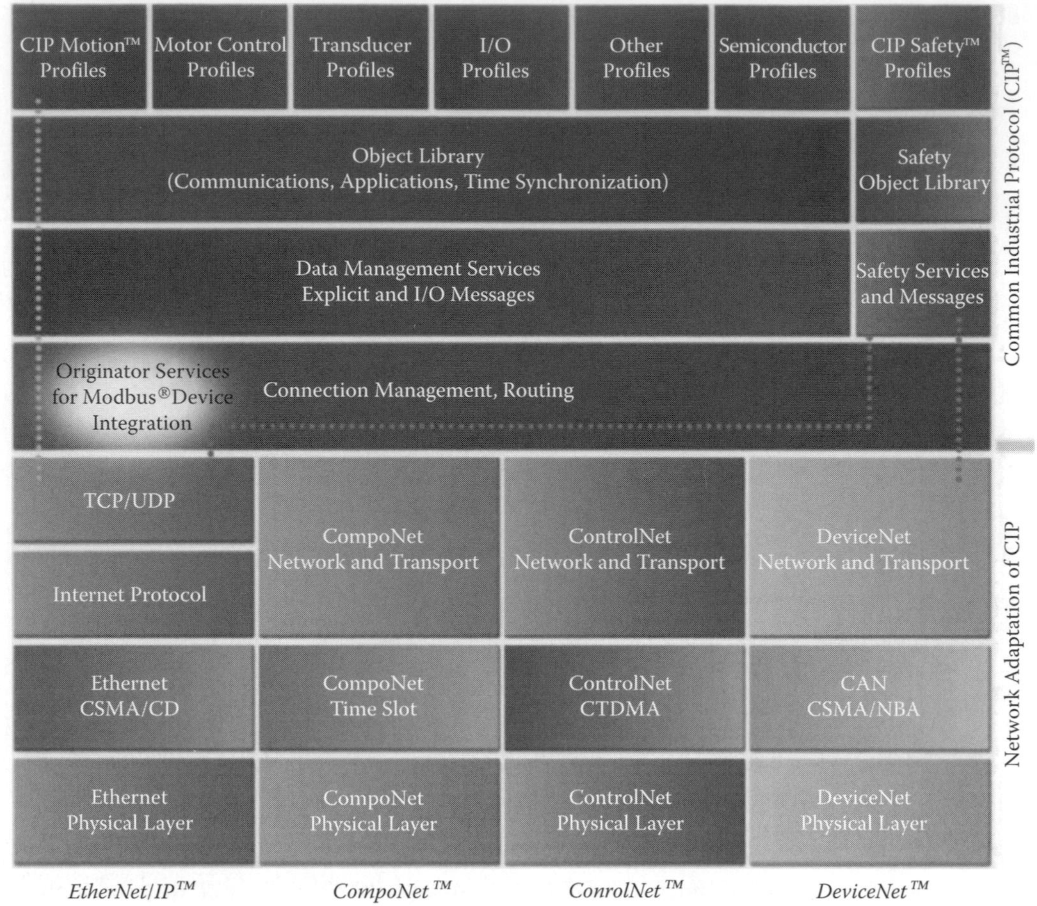

FIGURE 9.26 Relationship between CIP and ControlNet.

Figure 9.26 shows the relationship between CIP and ControlNet.

The ControlNet adaptation of CIP is described in Volume 4 of the CIP Networks Library [14]. All other features are based on CIP. ControlNet is also described in international standards, for example in [15].

9.3.2.3 ControlNet Features

ControlNet is a high-speed deterministic industrial communication system with the following features:

- Trunkline/dropline configuration (copper media), star configuration (optical media)
- Support for media redundancy
- Support for up to 99 nodes
- Node insertion or removal while the network is up and running
- Use of sealed or open-style connectors
- Fixed baud rate (5 Mbaud)

9.3.2.4 ControlNet Physical Layer

The physical layer of ControlNet has been designed specifically for this network; it does not reuse any existing open technology. The basis of the physical layer is a 75 Ω coaxial trunkline (typically of RG-6 type cable) terminated at both ends with 75 Ω terminating resistors. To reduce impedance mismatch, all ControlNet devices are connected to the network through special taps that consist of a coupling

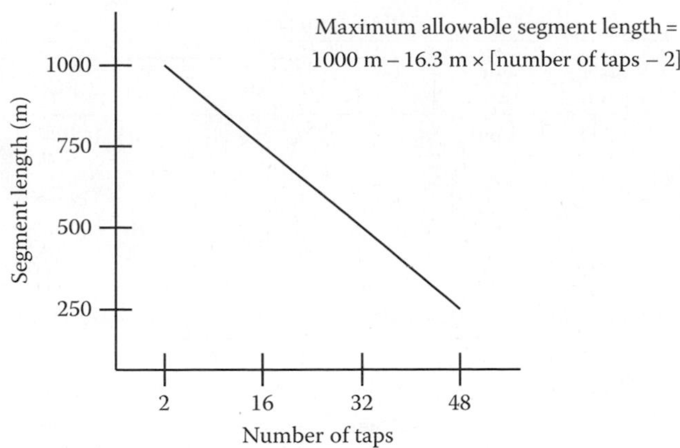

FIGURE 9.27 Coax medium topology limits.

network and a specific length of dropline (1 m). There is no minimum distance requirement between taps, but since every tap introduces some signal attenuation, each tap reduces the maximum length of the trunkline by 16.3 m. This results in a full-length trunkline of 1000 m with only two taps at the ends, while a fully populated physical network with 48 taps allows a trunkline length of 250 m (see Figure 9.27).

This physical layer limitation is addressed by including repeaters that can increase the network size without lowering the speed. Therefore, if a network is to be built with a higher number of nodes (up to 99 nodes are possible) or with a topology that goes beyond the single trunkline limitations, repeaters can be used to extend the bus. It is possible to create any type of topology: tree, star, or linear bus. Even a ring topology is possible using a special type of repeater. Repeaters for fiber-optic media can be used either to further increase the system size or to allow isolation of network segments in harsh EMC environments or for high-voltage applications.

The number of repeaters between any two nodes was initially limited to five, but further technology developments now allow up to 20 repeaters in series. However, regardless of the media technology used, the overall length of a ControlNet system (the distance between any two nodes on the network) is still limited by propagation delay. With currently available media, this translates into approximately 20 km.

To better accommodate industry requirements, ControlNet supports redundant media, allowing bumpless transfer from primary to secondary media or vice versa if one of them should fail or deteriorate. Developers are encouraged to support this redundant media feature in their designs. For cost-sensitive applications, less expensive device variants may then be created by populating one channel only.

Another feature often used in the process industry is the capability of running ControlNet systems into areas with an explosion hazard. ControlNet is fully approved to meet worldwide standards for intrinsic safety (explosion protection).

Copper media uses BNC-type connectors for IP 20-type applications and TNC-type connectors for IP 67 protection. Devices also may implement a Network Access Port. This feature takes advantage of the repeater function of the ControlNet ASICs. It uses an additional connector (RJ-45) with RS 422-based signals that provides easy access to any node on the network for configuration devices.

The signal transmitted on the copper media is a 5-Mbps Manchester-encoded signal with an amplitude of up to 9.5 V (pk-pk) at the transmitter that can be attenuated down to 510 mV (pk-pk) at the receiving end. The specification provides reference transmitting and receiving circuits.

9.3.2.5 Frame Structure

Every frame transmitted on ControlNet has the format of the MAC frame shown in Figure 9.28.

Within every MAC frame, a field of up to 510 bytes is available for transmitting data or messages. This field may be populated with one or several Lpackets (link packets). These Lpackets carry the individual

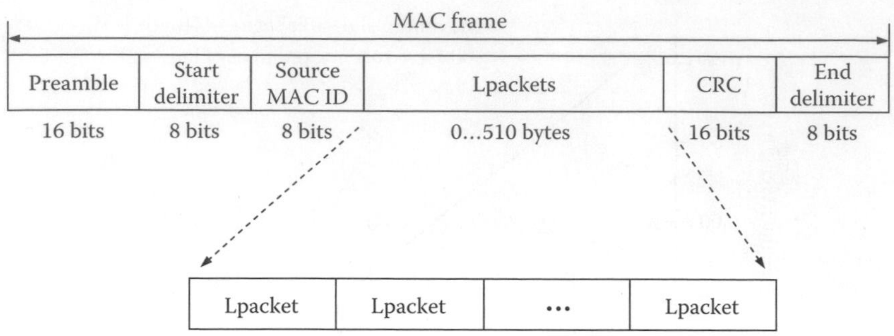

FIGURE 9.28 MAC Frame Format.

CIP messages (I/O or Explicit). Specialized Lpackets are used for network management. Since all nodes always listen to all MAC frames, they have no problem consuming any of the Lpackets in a frame that is unicast, multicast, or broadcast in nature. This feature allows fine-tuned multicasting of small amounts of data to different sets of consumers without much overhead.

There are two types of Lpacket formats: fixed tag and generic tag. Fixed tag Lpackets are used for Unconnected Messaging and network administration packets, while the generic tag Lpackets are used for all Connected Messaging (I/O and Explicit).

Figure 9.29 shows the format of a fixed tag Lpacket. By including the destination MAC ID, this format reflects the fact that these Lpackets are always directed from the requesting device (sending the MAC frame) to the target device (the destination MAC ID). The service byte within a fixed tag Lpacket does not represent the service of an Explicit Message, but a service type on a different level, since the fixed tag Lpacket format can be used for a variety of actions, such as network administration.

Figure 9.30 shows the format of a generic tag Lpacket. The size byte specifies the number of words within the Lpacket, while the control byte gives information on what type of Lpacket this is. The 3-byte CID specifies which connection this Lpacket belongs to. These three bytes are the three lower bytes of the 4-byte CID specified in the Forward_Open message; the uppermost byte is always zero. For a device

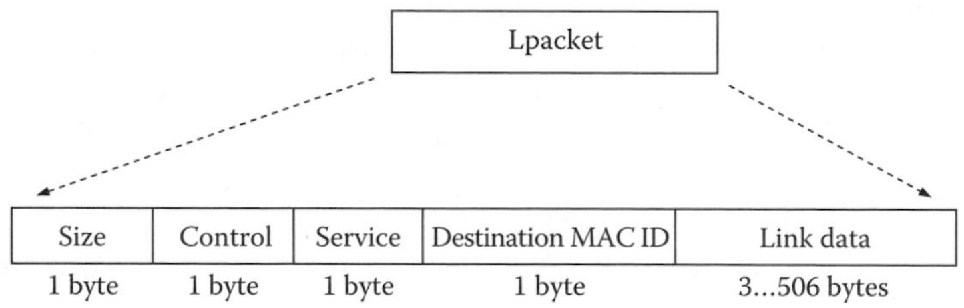

FIGURE 9.29 Fixed tag Lpacket format.

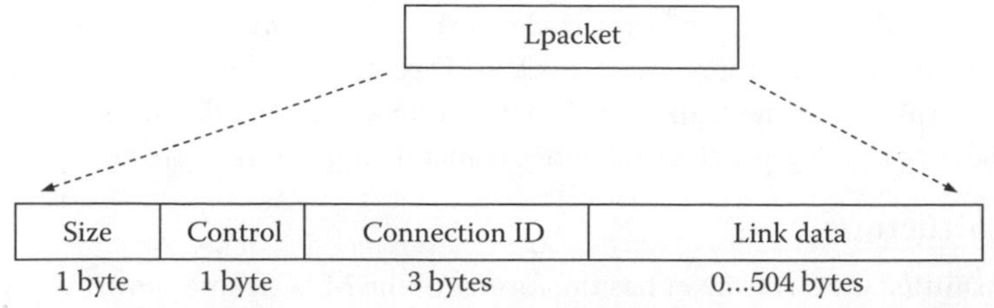

FIGURE 9.30 Generic tag Lpacket format.

that receives the MAC frame, the CID indicates whether to ignore the Lpacket (the device is not part of the connection), to consume the data and forward it to the application (the device is an end point of this connection), or to forward the data to another network (the device acts as a router in a routed connection).

9.3.2.6 Protocol Adaptation

ControlNet can use all features of CIP. The ControlNet frame is big enough that fragmentation is rarely required and thus is only provided by application-specific services that might require it. Since ControlNet is not used in very simple devices, no scaling is required.

9.3.2.7 Indicators and Switches

ControlNet devices must be built with Device Status and Network Status indicators as described in the specification. Devices may have additional indicators that must not carry any of the names of those described in the specification.

Devices may be built with or without switches or other directly accessible means for configuration. If switches for the MAC ID exist, then certain rules apply regarding how these values must be used at power-up and during the operation of the device.

9.3.2.8 Additional Objects

Volume 4 defines three additional objects: the ControlNet Object, the Keeper Object, and the Scheduling Object.

9.3.2.8.1 ControlNet Object (Class ID: 0xF0)

The ControlNet Object contains a host of information about the state of the device's ControlNet interface, among them diagnostic counters, data link and timing parameters, and the MAC ID. A ControlNet Object is required for every physical layer attachment of the device. A redundant channel pair counts as one attachment.

9.3.2.8.2 Keeper Object (Class ID: 0xF1)

The Keeper Object (not required for every device) holds (for the network scheduling software) a copy of the Connection Originator schedule data for all Connection Originator devices using the network. Every ControlNet Network with scheduled I/O traffic must have at least one device with a Keeper Object (typically, a PLC or another Connection Originator). If there are multiple Keeper Objects on a network, they perform negotiations to determine which Keeper is the Master Keeper and which Keeper(s) performs Backup Keeper responsibilities. The Master Keeper is the Keeper actively distributing attributes to the nodes on the network. A Backup Keeper is one that monitors Keeper-related network activity and can transition into the role of Master Keeper should the original Master Keeper become inoperable.

9.3.2.8.3 Scheduling Object (Class ID: 0xF2)

The Scheduling Object is required in every device that can originate an I/O Messaging Connection. Whenever a network scheduling tool accesses a Connection Originator on a ControlNet Network, an instance of the Scheduling Object is created and a set of object-specific services is used to interface with this object. Once the instance is created, the network scheduling tool can then read and write connection data for all connections that originate from this device. After having read the connection data from all Connection Originators, the network scheduling tool can calculate an overall schedule for the ControlNet Network and write these data back to all Connection Originators. The scheduling session is ended by deleting the instance of the Scheduling Object.

9.3.2.9 Network Access

ControlNet's bus access mechanism allows full determinism and repeatability while still maintaining sufficient flexibility for various I/O Message triggers and Explicit Messaging. This bus access mechanism is called Concurrent Time Domain Multiple Access (CTDMA); it is illustrated in Figure 9.31.

The time axis is divided into equal intervals called Network Update Time (NUT). Each NUT is subdivided into the Scheduled Service Time, the Unscheduled Service Time, and the Guardband Time.

Figure 9.32 shows the function of the Scheduled Service. Every node up to, and including, the SMAX node (maximum node number participating in the Scheduled Service) has a chance to send a message within the Scheduled Service. If a particular node has no data to send, it will send a short frame to indicate that it is still alive. If a node fails to send its frame, the next-higher node number will step in after a very short, predetermined waiting time. This process ensures that a node failure will not lead to an interruption of the NUT cycle.

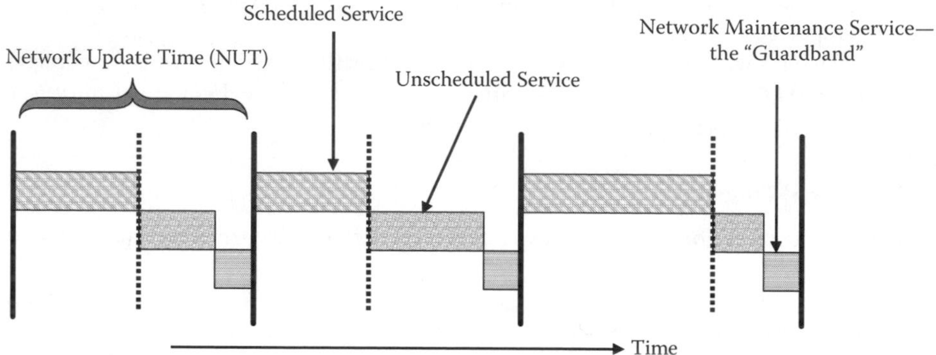

FIGURE 9.31 Media access through CTDMA (Concurrent Time Domain Multiple Access).

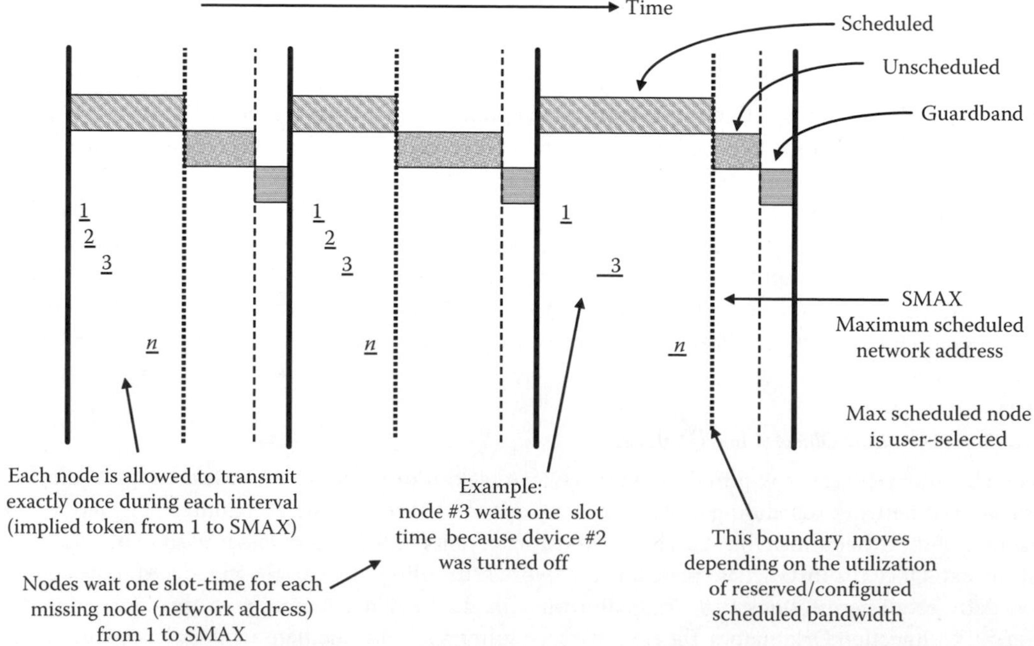

FIGURE 9.32 Scheduled Service.

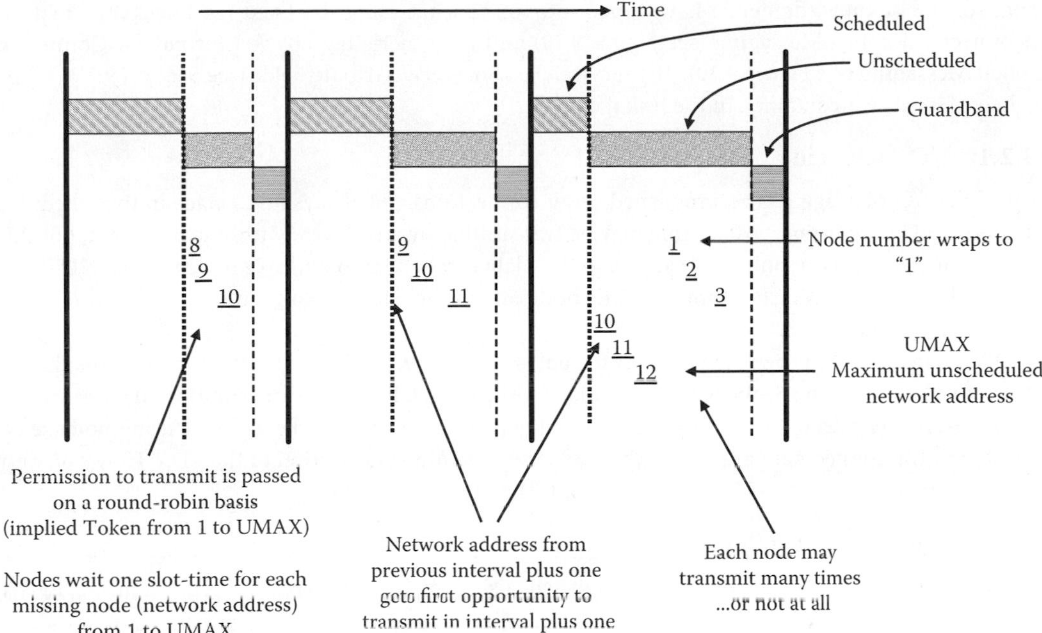

FIGURE 9.33 Unscheduled Service.

Figure 9.33 shows the function of the Unscheduled Service. Since this service is designed for non-time-critical messages, only one node is guaranteed access to the bus during the Unscheduled Service Time. If there is time left, the other nodes (with higher node numbers) will also get a chance to send. As with the Scheduled Service Time, if a node fails to send during its turn, the next node will step in. The node number that is allowed to send first within the Unscheduled Service Time is increased by one in each NUT. This guarantees an equal chance to all nodes. When the node sequencing within a NUT reaches the maximum value, known as UMAX, it wraps around to node 1, and the sequence resumes.

These two service intervals, combined with the Guardband, guarantee determinism and repeatability while still maintaining sufficient freedom to allow for unscheduled message transmissions, for example, for parameterization.

9.3.2.10 Network Start-Up

After power-on, every ControlNet device goes through a process of accessing the ControlNet communication network and learning the current NUT and other timing requirements. This is a fairly complex process typically handled by commercially available ControlNet ASICs. It is beyond the scope of this book to describe the details here.

9.3.2.11 Explicit Messaging

Explicit Messages on ControlNet, unlike those on DeviceNet, can be sent either connected or unconnected; both are transmitted within the unscheduled part of the NUT. Connected Explicit Messaging requires setting up a connection (see Section 9.3.2.13) before messages are exchanged. This means that all resources required for managing the connection are reserved for this purpose as long as the connection exists, which allows more timely responses to message requests. This is very useful when the application requires periodic explicit requests. Most Explicit Messages also can be sent unconnected, but this mechanism makes use of generally limited resources in nodes that sometimes can be highly utilized. For this reason, Unconnected Explicit Messaging should be used only when the application requires very irregular and infrequent request intervals. Every part of an Explicit Message (request, response, and acknowledgments) is wrapped into an Lpacket using the fixed tag Lpacket format for Unconnected Explicit Messaging (see Figure 9.29) and the generic tag Lpacket format for Connected Explicit Messaging (see Figure 9.30). The service/class/instance/attribute fields (see Section 9.2.3) of the Explicit Message are contained in the link data field.

9.3.2.12 I/O Messaging

ControlNet I/O Messaging is accomplished using connections and always takes place in the scheduled part of the NUT. Only one MAC frame may be transmitted by any device within its time slot, but this MAC frame may contain multiple Lpackets so that data can be sent to multiple nodes in one NUT. The individual Lpackets may be consumed by one node only or by multiple nodes if they are set up to consume the same data.

I/O Messages use the generic tag Lpacket format (see Figure 9.30). The link data field contains the I/O data prefixed with a 16-bit Sequence Count Value for the packet. Run/Idle can be indicated within a prefixed Real-Time Header or by sending the data packet (Run) or no data packet (Idle). The method used is indicated in the connection parameters of the Connection Manager section of the EDS. However, only the Real-Time Header method has been used for ControlNet up to now.

9.3.2.13 Connection Establishment

All connections on ControlNet are established using a UCMM Forward_Open message (see Section 9.2.3).

9.3.2.14 Device Classes

Four classes of device functionality are built with CIP. While they are not explicitly defined in the specification, they are useful for distinguishing among several classes of devices. The four classes are described here:

1. The minimal device function is that of an Explicit Message Server, which is used for Explicit Messaging applications only and acts as a target for Unconnected and (optionally) Connected Explicit Messages, for example, for program upload/download, data collection, status monitoring, etc.
2. The next device class is an I/O Server, which adds I/O Messaging Support to an Explicit Message Server device and acts as a target for both Explicit and I/O Messages, for example, simple I/O Devices, Pneumatic Valves, and AC Drives. These devices are also called I/O Adapters.
3. Another device class is an Explicit Message Client, which adds client support to Explicit Message Server applications and acts as a target and as an originator for explicit messaging applications, for example, computer interface cards and HMI devices.
4. The most powerful type of device is an I/O Scanner, which adds I/O Message origination support to the functionality of all the other device classes and which acts as a target and as an originator for Explicit and I/O Messages, for example, PLCs and I/O Scanners.

9.3.2.15 Device Profiles

ControlNet devices may utilize any device profiles described in the CIP Networks Library. As of the publication date of this book, no ControlNet-specific profiles have been defined.

9.3.2.16 Configuration

ControlNet devices typically come with EDSs as described in Section 9.2.6. For EDS-based configuration tools, the EDS should contain a Connection Manager section to describe the details of the connections that can be made into the device. This section basically mirrors the contents of the Forward_Open message that a Connection Originator would send to the device. Multiple connections can be specified within an EDS, then one or more can be chosen by the configuration tool.

An EDS may also contain individual parameters and/or a Configuration Assembly with a complete description of all parameters within this assembly. In many applications, the Configuration Assembly is transmitted as an attachment to the Forward_Open message.

9.3.2.17 Conformance Test

See Section 9.6 of this publication for information on conformance testing.

9.3.2.18 Tools

Tools for ControlNet Networks can be divided into three groups:

1. Physical layer tools are tools (hardware and/or software) that verify the integrity and conformance of the physical layer or monitor the quality of the data transmission.
2. Configuration tools are software tools capable of communicating with individual devices for data monitoring and configuration purposes. Most configuration tools are EDS-based; however, more complex devices like I/O Scanners tend to have their own configuration applets that are only partially based on EDS files. Some of these tools support multiple access paths to the network, for example, via Ethernet and suitable routing devices, and thus allow remote access. High-level tools also actively query the devices on the network to identify them and monitor their health. Configuration tools also may be integrated into other packages like PLC programming software.
3. Monitoring tools typically are PC-based software packages that can capture and display the ControlNet frames on the network. A raw ControlNet frame display may be good enough in some instances, but using a tool that can display both raw ControlNet frames and interpreted frames is recommended.

For a typical installation, a configuration tool is all that is needed. However, to ensure the network is operating reliably, testing with a physical layer tool is highly recommended. Experience shows that the overwhelming majority of ControlNet network problems are caused by inappropriate physical layer installation. Protocol monitoring tools are mainly used to investigate interoperability problems and to assist during the development process.

Turn to the CIP supplier directory on the ODVA website to access a list of vendors that provide tools for ControlNet.

9.3.2.19 Advice for Developers

Before any development of a ControlNet product is started, the following issues should be considered in detail:

- What functionality (Device Classes, see Section 9.3.2.14) does the product require today and in future applications?
  - Explicit Messaging server only.
  - I/O Adapter functionality.
  - Explicit Messaging client.
  - I/O Scanner functionality.
- What are the physical layer requirements? Is IP 65/67 required, or is IP 20 good enough?
- Will the development be based on commercially available hardware components and software packages (recommended) or designed from scratch (possible but costly)?
- What are the configuration requirements?
- What design and verification tools should be used?
- When and where will the product be tested for conformance and interoperability?
- What is an absolute must before products can be placed on the market (own the specification, have the company's own Vendor ID, and have the product conformance tested)?

Rockwell Automation has published a comprehensive developer's handbook that assists vendors in developing products, see [16]. ControlNet chipsets and associated software packages are available from Rockwell Automation. Turn to the ODVA website for a list of companies that can support ControlNet developments.

9.3.2.20 ControlNet Summary

Since its introduction in 1997, ControlNet has been used successfully in millions of nodes in many different applications. It is the network of choice for many high-speed I/O and PLC interlocking applications. Like DeviceNet, ControlNet has become an international standard [15]. Due to its universal communication characteristics, it is one of the most powerful controller-level networks available.

ControlNet's greatest strengths are its media redundancy and its full determinism and repeatability. These strengths make it ideally suited for many applications that require media redundancy and also for many high-speed applications, in which ControlNet maintains full Explicit Messaging capabilities without compromising its real-time behavior.

Finally, as a member of the CIP family of networks, ControlNet can be combined into an overall CIP Network structure that allows seamless communication among CIP Networks, just as if they were only one network.

9.3.3 EtherNet/IP

9.3.3.1 Introduction

Introduced in 2000, EtherNet/IP is another member of the CIP family. Using CIP as its upper-layer protocol, EtherNet/IP extends the application of Ethernet TCP/IP to the plant floor. EtherNet/IP can coexist with any other protocol running on top of the standard TCP/UDP Transport Layer and with other CIP Networks. EtherNet/IP—CIP plus Internet and Ethernet standards—provides a pure, unmodified, standards-based Ethernet solution for interoperability among manufacturing enterprise networks, and it enables Internet and enterprise connectivity anywhere, anytime utilizing commonly available switches. The *IP* in EtherNet/IP stands for the *Industrial Protocol* in CIP; this is not to be confused with *IP* in TCP/IP, which stands for *Internet Protocol*.

Due to the length of Ethernet frames and the typical multimaster structure of Ethernet networks, there are no particular limitations in the EtherNet/IP implementation of CIP. Basically, all that is required is a mechanism to encode CIP messages into Ethernet frames.

9.3.3.2 Relationship to Standards

Like other CIP Networks, EtherNet/IP follows the OSI model, an ISO standard for network communications that is hierarchical in nature. Networks that follow this model define all necessary functions, from physical implementation up to the protocol and methodology to communicate control and information data within and across networks.

Figure 9.34 shows the relationship between CIP and EtherNet/IP.

The EtherNet/IP Adaptation of CIP is described in Volume 2 of the CIP Networks Library. All other features are based on CIP. This volume defines how CIP is adapted for use on Ethernet. An encapsulation mechanism (see Section 9.3.3.11) is defined for EtherNet/IP specifying how I/O and Explicit Messages are carried in Ethernet frames. The well-known TCP/IP protocol is used for encapsulating Explicit Messages, while UDP/IP is used for encapsulating I/O Messages. Since the commonly implemented TCP/IP and UDP/IP protocol stacks are used for encapsulation, many applications will not require extra middleware for this purpose.

Ethernet has its roots in the office computing environment, which is not traditionally concerned with determinism like industrial applications are. However, with the proper selection and configuration of infrastructure devices (see Section 9.3.3.21) using fast data rates with full duplex communications, there

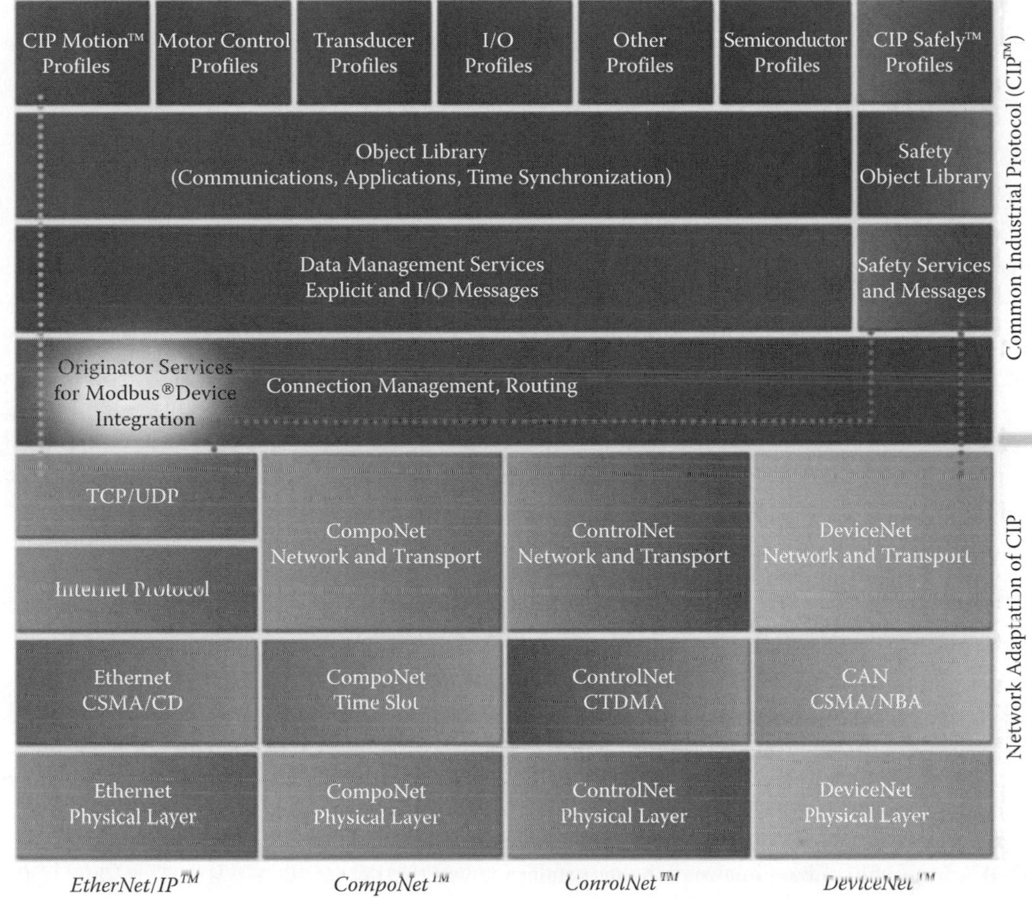

FIGURE 9.34 Relationship between CIP and EtherNet/IP.

will be no collisions or lost packets, giving Ethernet a level of determinism that is more than adequate for use in industrial control applications. Additionally, extensions to CIP like CIP Sync and CIP Motion (see Section 9.5.1) allow EtherNet/IP to be used in highly synchronous and deterministic applications like coordinated drives and motion control.

EtherNet/IP is also described in international standards, that is, the IEC fieldbus standards, see [10,15].

9.3.3.3 EtherNet/IP Features

EtherNet/IP is a communication system built on standard, unmodified Ethernet with the following features:

- Built on and compliant with the relevant Ethernet standards, not just compatible with them
- Fully independent of data rate: 10, 100, 1000 Mbps
- Systems can be built with standard infrastructure
- Virtually unlimited number of nodes in a network
- Networks can be structured into subnets with IP routers
- Full support of communication across subnets since EtherNet/IP uses IP addressing for all communication
- Non-real-time communication and real-time communication can coexist in the same subnet
- Support for coordinated drives and motion control

- Support for DLR that provides single fault tolerance through media redundancy
- QuickConnect for devices that are frequently removed from and added to the network, for example, robot tools
- Coexistence with other upper-layer protocols, such as HTTP, FTP, and VOIP

9.3.3.4 EtherNet/IP Physical Layer

Since EtherNet/IP takes the Ethernet protocol to the factory floor, recommendations are made in Volume 2 [17] regarding grounding, isolation, and cable and connector construction that are designed to make EtherNet/IP successful in a typical factory automation environment. These changes do not affect the actual signaling or interoperability with standard Ethernet products, but simply make devices more suitable for harsher industrial environments. As a result, two levels of performance criteria are defined:

1. The commercial off-the-shelf (COTS) EtherNet/IP Level provides basic Ethernet connectivity. This level includes the well-known RJ-45-type Ethernet connector but specifies topology constraints (e.g., up to 100 m) and cabling requirements through references to specific IEEE, ANSI/TIA/EIA standards. Such devices are typically suited for IP 20 applications.
2. The industrial EtherNet/IP Level goes beyond the COTS Level by specifying minimum environmental, cabling, and connector requirements that include IEC, ANSI/TIA/EIA standards. Connectors required for the industrial EtherNet/IP Level include an enhanced performance RJ-45 connector, a sealed RJ-45 connector, as well as a more compact D-coded M 12-4 connector. The sealed RJ-45 and M 12 connectors can achieve an IP 67 rating.

Cat 5E or Cat 6 shielded or unshielded cables are recommended for EtherNet/IP. The use of shielded cables is specifically recommended in applications where adjacent material, such as metal cable ducts, may have substantial influence on the characteristics of the cable. In accordance with IEEE 802.3, copper media may be used only for distances up to 100 m. Fiber-optic media is recommended for longer distances. Fiber-optic media may also be advisable for applications with very high electromagnetic disturbances or high-voltage potential differences between devices.

ODVA has published a guideline for the installation of Ethernet media, see [18]. This topic is also covered by the international standard IEC 61784-5-2 [10].

9.3.3.5 Frame Structure

EtherNet/IP uses standard Ethernet TCP/IP and UDP/IP frames as defined by international standards [19–22]. Therefore, no further frame details are described here.

9.3.3.6 Protocol Adaptation

EtherNet/IP can use all features of CIP. The Ethernet frame is big enough that fragmentation is rarely required. If it is required, fragmentation is automatically handled by IP fragmentation provided by TCP/IP and UDP/IP. Since EtherNet/IP is not expected to be used in very simple devices, no further scaling is required.

9.3.3.7 Indicators and Switches

EtherNet/IP devices that need to conform to the industrial EtherNet/IP Level must have the two indicators set forth in the specification: Module Status and Network Status. Devices may have additional indicators that must not carry any of the names of those described in the specification.

Devices may be built with or without switches or other manual means for configuration.

9.3.3.8 Additional Objects

Volume 2 defines the following 11 additional objects that are found only on EtherNet/IP devices. Most of these objects are required only when the feature they pertain to is implemented. Exceptions to this are noted where appropriate.

9.3.3.8.1 *TCP/IP Interface Object (Class ID: 0xF5)*

The TCP/IP Interface Object provides a mechanism for configuring a device's TCP/IP network interface. Examples of configurable items include the device's IP address, network mask, and gateway address. Every EtherNet/IP must have at least one instance of this class.

9.3.3.8.2 *Ethernet Link Object (Class ID: 0xF6)*

The Ethernet Link Object maintains configuration parameters, various error counters, and status information for the Ethernet IEEE 802.3 communications interface. Each device has exactly one instance of the Ethernet Link Object for each Ethernet IEEE 802.3 communications interface.

9.3.3.8.3 *Device Level Ring Object (Class ID: 0x47)*

The DLR Object manages all data and behavior associated with the DLR functionality of a device. For further details on DLR, see Section 9.3.3.23.2 of this publication.

9.3.3.8.4 *QoS Object (Class ID: 0x48)*

The Quality of Service (QoS) Object manages all data and behavior associated with the QoS functionality of a device. It includes the settings for DSCP in the IP header and the Frame Prioritization settings for the Ethernet header. If the device supports DLR, then this class must be implemented, too.

9.3.3.8.5 *Base Switch Object (Class ID: 0x51)*

The Base Switch Object provides the CIP application-level interface to basic status information for a Managed Ethernet switch device. Devices shall implement no more than one instance of the Base Switch Object.

9.3.3.8.6 *Simple Network Management Object (Class ID: 0x52)*

The SNMP Object provides parameters used to configure aspects of the SNMP Agent in the device.

9.3.3.8.7 *Power Management Object (Class ID: 0x53)*

The Power Management Object defines a Sleeping state and a Paused state. The method to trigger the transition from the Sleeping state to the Paused state is network adaptation specific.

9.3.3.8.8 *RSTP Bridge Object (Class ID: 0x54)*

The RSTP Bridge Object provides the configuration and diagnostic interface for the RSTP protocol at the bridge level. For further details on the use of RSTP, see Section 9.3.3.24 of this publication.

9.3.3.8.9 *RSTP Port Object (Class ID: 0x55)*

The RSTP Port Object provides a configuration and diagnostic interface for the RSTP protocol at the port level. For further details on the use of RSTP, see Section 9.3.3.24 of this publication.

9.3.3.8.10 *Parallel Redundancy Protocol Object (Class ID: 0x56)*

The Parallel Redundancy Protocol (PRP) Object provides a configuration and diagnostic interface for PRP parameters, if implemented in the product.

9.3.3.8.11 *PRP Nodes Table Object (Class ID: 0x57)*

The PRP Node Table Object keeps the record of all PRP-capable nodes that have been detected on the network.

9.3.3.9 IP Address Assignment

Since the initial development of TCP/IP, numerous methods for configuring a device's IP address have evolved. Not all of these methods are suitable for industrial control devices. In the office environment, for example, it is common for a PC to obtain its IP address via Dynamic Host Configuration Protocol (DHCP), meaning that it can potentially acquire a different address each time the PC reboots. This is acceptable because the PC is typically a client device that only makes requests, so there is no impact if its IP address changes.

However, for an industrial control device that is a target of communication requests, the IP address cannot change at each power-up. A PLC, for example, must be at the same address each time it powers up.

In addition, the only interface common to all EtherNet/IP devices is an Ethernet communications port. Some devices may also have a serial port, a user interface display, hardware switches, or other interfaces, but these are not universally shared across all devices. Since Ethernet is the common interface, the initial IP address should at least be configurable over Ethernet.

The EtherNet/IP Specification, via the TCP/IP Interface Object, defines a number of ways to configure a device's IP address. A device may obtain its IP address via Bootstrap Protocol (BOOTP), via DHCP, or via an explicit Set_Attribute_Single or Set_Attributes_All service. None of these methods is mandated however. As a result, vendors could choose different methods for configuring IP addresses.

From the user's perspective, it is desirable for vendors to support some common mechanism(s) for IP address configuration. The current ODVA recommendations on this subject can be downloaded from the ODVA website [23].

9.3.3.10 Address Conflict Detection

Since IP addresses are often assigned by human interaction or as a default private address by the device manufacturer (e.g., 192.168.1.1), it is not uncommon to find multiple devices on the same network with the same IP address. This situation is undesirable; therefore, duplicate IP address detection and the subsequent address conflict resolution have been defined for EtherNet/IP. The Address Conflict Detection (ACD) mechanism deployed in EtherNet/IP conforms to the IETF RFC 5227 [24]. Support for ACD is optional; however, any EtherNet/IP device that supports it must follow the method described in Volume 2, Appendix F, of The EtherNet/IP Specification.

9.3.3.11 EtherNet/IP Encapsulation

EtherNet/IP is based entirely on existing TCP/IP and UPD/IP technologies and uses them without any modification. TCP/IP is mainly used for the transmission of Explicit Messages while UDP/IP is used mainly for I/O Messaging.

The encapsulation protocol defines two reserved TCP/UDP port numbers. All EtherNet/IP devices accept at least two TCP connections on TCP port number 0xAF12. This port is used for all TCP-based Explicit Messaging, either connected or unconnected. It is also used for the encapsulation protocol commands that are employed when setting up communications between nodes. Some encapsulation commands may also be sent to port 0xAF12 via UDP datagrams.

Port 0x08AE is used by any devices that support EtherNet/IP's I/O messaging over UDP. These messages can be sent either unicast or multicast by taking advantage of the multicast capabilities of IP. Multicast data flow makes more efficient use of the available bandwidth and provides for better data consistency across the system. Being connectionless, UDP is well suited to this purpose as connection management is handled by CIP.

9.3.3.11.1 General Use of the Ethernet Frame

Since EtherNet/IP is completely based on Ethernet with TCP/IP and UDP/IP, all CIP-related messages sent on an EtherNet/IP network are Ethernet frames with an IP header (see Figure 9.35).

The Ethernet header, the IP header, and the TCP or UDP headers are described through international standards (see Section 9.3.3.5); therefore, details of these headers are mentioned only in The EtherNet/IP Specification when it is necessary to understand how they are used to carry CIP.

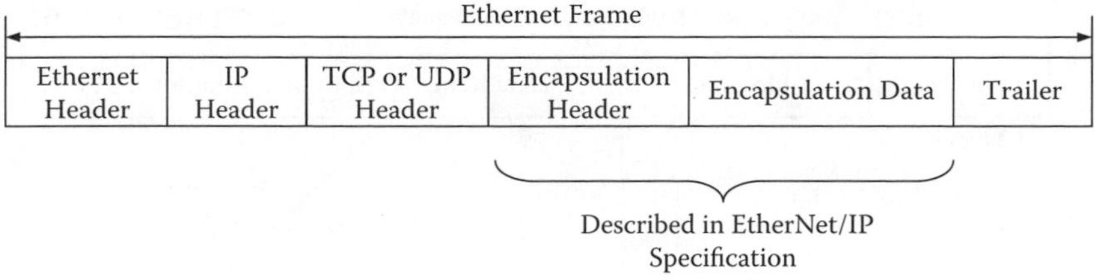

FIGURE 9.35 Relationship between CIP and Ethernet frames.

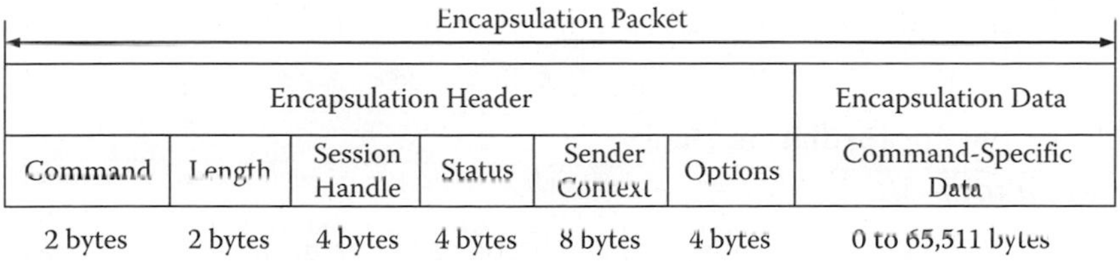

Encapsulation Packet						
Encapsulation Header						Encapsulation Data
Command	Length	Session Handle	Status	Sender Context	Options	Command-Specific Data
2 bytes	2 bytes	4 bytes	4 bytes	8 bytes	4 bytes	0 to 65,511 bytes

FIGURE 9.36 Structure of the encapsulation packet.

The encapsulation header contains a command that determines the meaning of the encapsulation data. Many commands specify the use of the so-called Common Packet Format. I/O Messages sent in UDP frames do not use the encapsulation header, but they still follow the Common Packet Format.

9.3.3.11.2 Encapsulation Header and Encapsulation Commands

The overall encapsulation packet has the structure described in Figure 9.36.

While the description of some of the encapsulation header details would go beyond the scope of this book, the command field requires more attention here. However, only those commands that are needed to understand the EtherNet/IP protocol are described, and their description lists only the main features.

9.3.3.11.2.1 ListIdentity Command The ListIdentity command typically is sent as a broadcast UDP message to tell all EtherNet/IP devices to return a data set with identity information. This command is used by software tools to browse a network.

9.3.3.11.2.2 RegisterSession/UnRegisterSession Commands These two commands are used to open and close an Encapsulation Session between two devices. Once a session is established, it is used to exchange more messages. Only one session may exist between two devices.

The device receiving the RegisterSession request creates a Session Handle that it returns in the RegisterSession reply. This value is used to identify messages sent between the two devices that use this session.

9.3.3.11.2.3 SendRRData/SendUnitData Commands The SendRRData Command is used for Unconnected Explicit Messaging, and the SendUnitData Command is used for Connected Explicit Messaging. The device transmitting the SendRRData request creates a Sender Context value that is returned with the reply. The SendUnitData does not use the Sender Context field.

9.3.3.11.2.4 Common Packet Format The Common Packet Format is a construct that provides a way to structure the Encapsulation Data field for those Encapsulation commands that specify Encapsulation data. The Common Packet Format defines Items that represent different types of information to be exchanged between devices. If the command definition requires it, the Common Packet Format allows packing of multiple Items into one encapsulation frame, as shown in Figure 9.37.

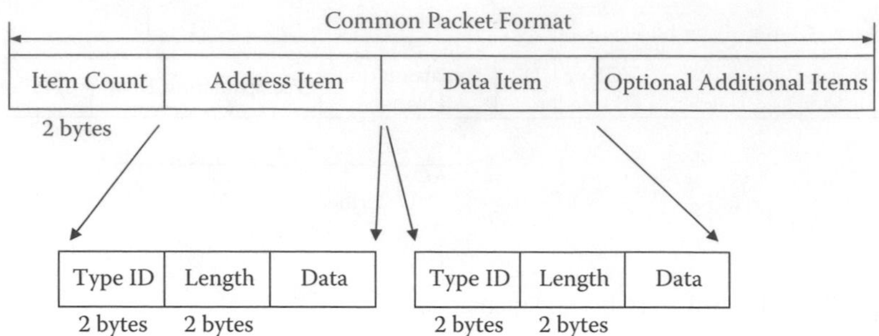

FIGURE 9.37 Example of the Common Packet Format.

9.3.3.12 Use of the Encapsulation Data

9.3.3.12.1 Explicit Messaging

Explicit Messages on EtherNet/IP can be sent either connected or unconnected. Connected Explicit Messaging requires setting up a connection (see Section 9.3.3.13) before messages are exchanged. This means that all resources required for managing the connection are reserved for this purpose as long as the connection exists, which allows for more timely responses to message requests. This is very useful in applications that require periodic explicit requests. Explicit Messages also can be sent unconnected, but this mechanism makes use of generally limited resources in nodes that sometimes can be highly utilized. For this reason, Unconnected Explicit Messaging should be used only when the application requires very irregular and infrequent request intervals. Explicit Messages on EtherNet/IP are sent with a TCP/IP header and use the SendRRData Encapsulation Command (unconnected) or the SendUnitData Encapsulation Command (connected). As an example, the full encapsulation of a UCMM request is shown in Figure 9.38.

The Message Router Request Packet noted in the figure contains the CIP message request or response. This part of the packet follows the general format of Explicit Messages—the Message Router Request/Response Format—defined in Volume 1, Chapter 2, of the CIP Networks Library.

9.3.3.12.2 I/O Messaging

I/O Messages on EtherNet/IP are sent with a UDP/IP header. No encapsulation header is required, but the message still follows the Common Packet Format. See Figure 9.39 for an example.

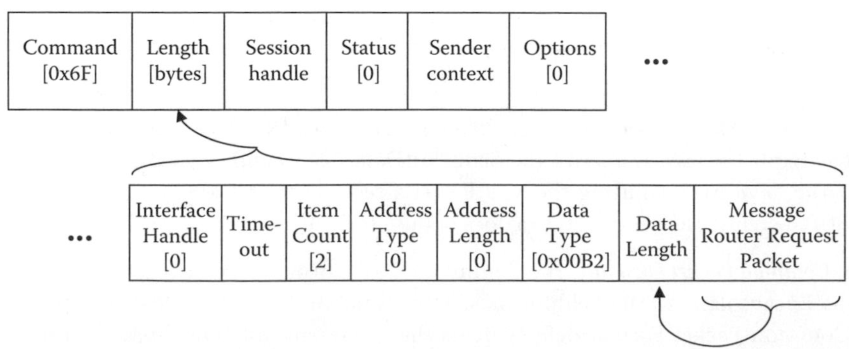

FIGURE 9.38 UCMM request encapsulation.

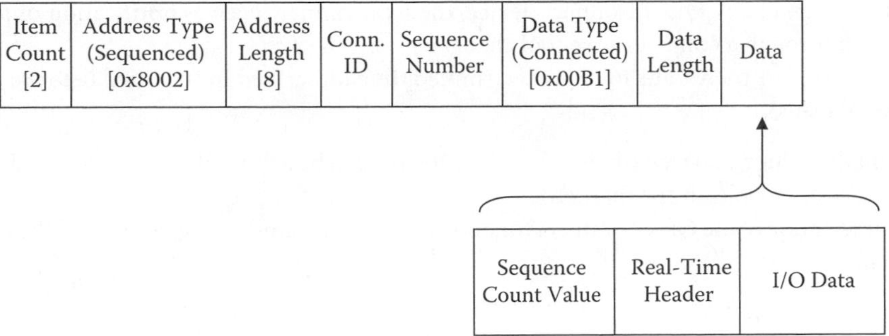

FIGURE 9.39 I/O Message encapsulation.

The data field contains the I/O data prefixed with a 16-bit Sequence Count Value for the packet. I/O data transmission without the Sequence Count Value is possible, but it is used only for CIP Safety connections. Run/Idle can be indicated within a Real-Time Header or by sending the packet with I/O data (Run) or without I/O data (Idle). The method used is indicated in the connection parameters of the Connection Manager section of the EDS. The Real-Time Header method is recommended [25] for use on EtherNet/IP for interoperability reasons, and this is what is shown in Figure 9.39.

I/O Messages from the originator to the target are typically sent as UDP unicast frames, while those sent from the target to the originator can be sent as UDP multicast or unicast frames. Multicast frames allow other EtherNet/IP devices to listen to the input data. To avoid having these UDP multicast frames propagating too widely over the network, the use of switches that support (IGMP) Snooping is highly recommended. IGMP (Internet Group Management Protocol) [26] is a protocol that allows the automatic creation of multicast groups. Using this functionality, the switch will automatically create and maintain a multicast group consisting of the devices that need to consume these multicast messages. Once the multicast groups have been established, the switch will direct such messages only to those devices that have subscribed to the multicast group of that message.

9.3.3.13 Connection Establishment

All connections on EtherNet/IP are established using a UCMM Forward_Open message (see Section 9.2.3).

9.3.3.14 QuickConnect Connection Establishment

While most applications can wait several seconds until a connection is established, there are certain application scenarios that require a device to be operational with only a very short delay after the application of power. Typically, these are devices sitting on an exchangeable tool that a robot would pick up for certain manufacturing steps. In comparison to DeviceNet, EtherNet/IP devices are more complex and larger, and more complex parts of the communication stack are typically implemented in software, so it will take more time to power up a device. The other additional complexity in EtherNet/IP is the TCP layer with a timeout behavior of its own. Furthermore, active infrastructure (switches) may take a long time to reboot. Under consideration of these conditions, the following method was developed for a fast establishment of I/O connections:

- If more than one EtherNet/IP device is mounted onto the exchangeable tool, embedded infrastructure with defined start-up behavior must be used.
- In preparation of a restart, every connection that participates in the QuickConnect application must be shut down using the Forward_Close service before disconnecting the device.
- When the device has responded to the Forward_Close request, it closes the TCP connection with the I/O Scanner.

- At the restart of any QuickConnect device, the I/O Scanner receives notification of power reapplication through a contact in the tool changer.
- The I/O Scanner then waits for a predetermined time (described in the EDS) before a connection is reestablished.

Using this methodology, start-up times of less than 100 ms can be achieved with current technology; the first products are available on the market.

The full description of the QuickConnect functionality can be found in Appendix E of The EtherNet/IP Specification.

9.3.3.15 Device Classes

Four classes of device functionality are built with CIP. While they are not explicitly defined in the specification, they are useful for distinguishing among several classes of devices. The four classes are described here:

1. The minimal device function is that of an Explicit Message Server, which is used for Explicit Messaging applications only and acts as a target for Unconnected and (optionally) Connected Explicit Messages, for example, for program upload/download, data collection, and status monitoring.
2. The next device class is an I/O Server, which adds I/O Messaging Support to an Explicit Message Server device and acts as a target for both Explicit and I/O Messages, for example, simple I/O Devices, Pneumatic Valves, and AC Drives. These devices are also called I/O Adapters.
3. Another device class is an Explicit Message Client, which adds client support to Explicit Message Server applications and acts as a target and as an originator for explicit messaging applications, for example, computer interface cards and HMI devices.
4. The most powerful type of device is an I/O Scanner, which adds I/O Message origination support to the functionality of all the other device classes and which acts as a target and as an originator for Explicit and I/O Messages, for example, PLCs and I/O Scanners.

9.3.3.16 Device Profiles

EtherNet/IP devices may utilize any of the device profiles described in the CIP Networks Library. As of the publication date of this book, no EtherNet/IP-specific device profiles have been defined.

9.3.3.17 Configuration

EtherNet/IP devices typically come with EDSs as described in Section 9.2.7. For EDS-based configuration tools, the EDS should contain a Connection Manager section to describe the details of the connections that can be made into the device. This section basically mirrors what is contained in the Forward_Open message that a Connection Originator would send to the device. Multiple connections can be specified within an EDS that can then be chosen by the configuration tool.

An EDS also may contain individual parameters and/or a Configuration Assembly with a complete description of all parameters within this assembly. In many applications, the Configuration Assembly is transmitted as an attachment to the Forward_Open message.

9.3.3.18 Conformance Test

See Section 9.6 of this publication for information on conformance testing.

9.3.3.19 Requirements for TCP/IP Support

In addition to the various requirements set forth in The EtherNet/IP Specification, all EtherNet/IP hosts are required to have a functional TCP/IP protocol suite and transport mechanism. The minimum host requirements for EtherNet/IP hosts are those covered in RFC 1122 [27], RFC 1123 [28], and RFC 1127 [29] and the subsequent documents that may supersede them. Whenever a feature or protocol is implemented by an EtherNet/IP host, that feature shall be implemented in accordance with the appropriate RFC (Request for Comment) documents, regardless of whether the feature or protocol is considered required or optional by this specification. The Internet and the RFCs are dynamic. There will be changes to the RFCs and to the requirements included in this section as the Internet and The EtherNet/IP Specifications evolve.

All EtherNet/IP devices shall at a minimum support

- Internet Protocol (IP version 4) (RFC 791 [21])
- User Datagram Protocol (UDP) (RFC 768 [20])
- Transmission Control Protocol (TCP) (RFC 793 [22])
- Address Resolution Protocol (RFC 826 [30])
- Internet Control Messaging Protocol (RFC 792 [31])
- Internet Group Management Protocol (IGMP) (RFC 1112 [32] and 2236 [26])
- IEEE 802.3 (Ethernet) as defined in RFC 894 [33]

Although the encapsulation protocol is suitable for use on other networks besides Ethernet that support TCP/IP and products may be implemented on these other networks, conformance testing of EtherNet/IP products is limited to those products on Ethernet. Other suitable networks include

- Point-to-Point Protocol (RFC 1171 [34])
- ARCNET (RFC 1201 [35])
- FDDI (RFC 1103 [36])

9.3.3.20 Coexistence of EtherNet/IP and Other Ethernet-Based Protocols

EtherNet/IP devices are encouraged, but not required, to support other Ethernet-based protocols and applications not specified in The EtherNet/IP Specification. For example, they may support HTTP, Telnet, FTP, etc. The EtherNet/IP protocol makes no requirements with regard to these protocols and applications.

Figure 9.40 shows the relationship between CIP and other typical Ethernet based protocol stacks. Since EtherNet/IP, like many other popular protocols, is based on TCP/IP and UDP/IP, coexistence with many other services and protocols is not a problem, and CIP blends nicely into the set of already existing functions. This means that anyone who is already using some or all of these popular Ethernet services can add CIP without undue burden; the existing services like HTTP or FTP may remain as before, and CIP will become another service on the process layer.

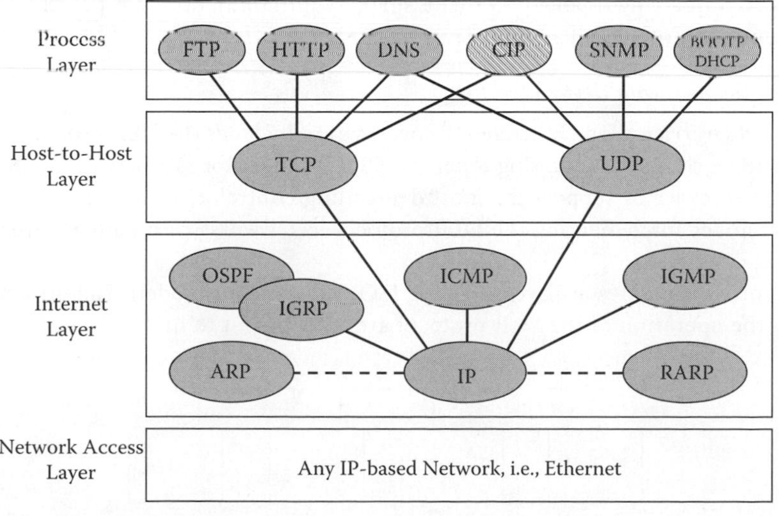

FIGURE 9.40 Relationship of CIP to other typical Ethernet protocols.

9.3.3.21 Ethernet Infrastructure

9.3.3.21.1 Traditional Approach

To apply EtherNet/IP successfully to the automation world, the issue of determinism has to be considered. The inherent principle of the Ethernet bus access mechanism, whereby collisions are detected and nodes back off and try again later, cannot guarantee determinism. While Ethernet in its present form cannot be made strictly deterministic, there are ways to improve this situation.

First, the hubs typically used in many office environments must be replaced by more intelligent switches that will forward only those Ethernet frames intended for nodes connected to these switches. By using wire-speed switching fabric and full duplex switch technology, collisions are completely avoided; instead of colliding, multiple messages sent to the same node at the same time are queued up inside the switch and are then delivered one after another.

As already mentioned in Section 9.3.3.12.2, using switches that support IGMP Snooping is highly recommended.

If EtherNet/IP Networks are to be connected to a general company network, this should always be done through a router. The router keeps the UDP multicast messages from propagating into the company network and ensures that the broadcast or multicast office traffic does not congest the control network. Even though the router separates the two worlds, it can be set up to allow the TCP/IP-based Explicit Messages to pass through so that a configuration tool sitting in a PC in the office environment may be capable of monitoring and configuring devices on the control network.

9.3.3.22 Devices with Multiple Ethernet Ports

Chapter 6 of Volume 2 of the CIP Specifications describes a number of scenarios for devices with multiple Ethernet ports and how these scenarios are to be mapped in the object structure.

9.3.3.23 Ring and Linear Topologies

9.3.3.23.1 Linear Topology

Many end-user applications benefit from connecting devices in a linear or ring topology. With such a topology, end devices typically have two Ethernet ports (with an embedded switch) and are connected in sequence, one device to the next (Figure 9.41).

With linear topology, a failure of one node or a link between two nodes causes nodes on either side of the failure to be unreachable. By using a ring protocol implemented in the end devices, these devices may be configured in a ring topology so that a single-point failure does not prevent communication between the remainder of the functioning devices.

9.3.3.23.2 Ring Topology with Device Level Ring

9.3.3.23.2.1 DLR Overview The EtherNet/IP specification includes the DLR protocol, allowing multiport devices to be connected in a ring topology. DLR provides for fast network fault detection and reconfiguration in order to support the most demanding control applications. For example, a ring network of 50 nodes implementing the DLR protocol has a worst-case fault recovery time of less than 3 ms.

The DLR protocol operates at Layer 2 (in the ISO OSI network model). The presence of the ring topology and the operation of the DLR protocol are transparent to higher-layer protocols, such as

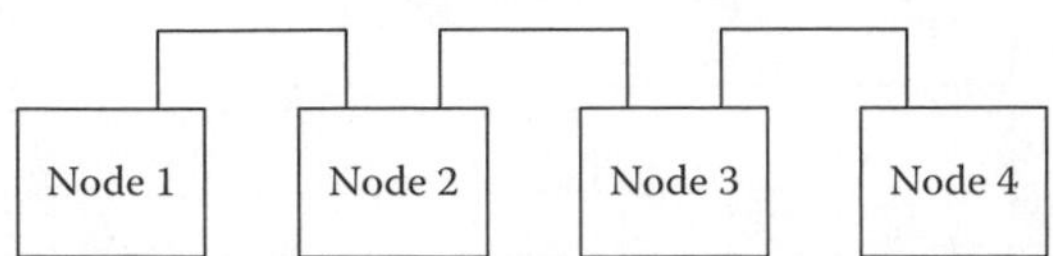

FIGURE 9.41 Linear topology.

TCP/IP and CIP, with the exception of a DLR Object that provides a DLR configuration and diagnostic interface via CIP.

There are several classes of DLR implementation, as described in the following text:

- *Ring Supervisor*
  This class of devices is capable of being a ring supervisor. These devices must implement the required ring supervisor behaviors, including the ability to send and process Beacon frames at the default Beacon interval of 400 μs and may be user-configured for as fast as 100 μs Beacon interval. Ring supervisors must also send Announce frames for those devices that rely on the Announce frame mechanism to detect a change in ring status.
- *Ring Node, Beacon-based*
  This class of devices implements the DLR protocol, but without the ring supervisor capability. The device must be able to process and act on the Beacon frames sent by the ring supervisor. Beacon-based ring nodes must support Beacon rates from 100 μs to 100 ms.
- *Ring Node, Announce-based*
  This class of devices implements the DLR protocol, but without the ring supervisor capability. These devices do not have the capacity to process Beacon frames, so they simply forward Beacon frames received on one port to the other port and instead rely on Announce frames to indicate the ring state. Announce frames are sent at a much slower rate than Beacon frames.

9.3.3.23.2.2 ***Normal DLR Operation*** A DLR network includes at least one node configured to be a ring supervisor and any number of normal ring nodes. The ring supervisor sends special frames (*Beacon* and *Announce* frames) to detect ring fault and ring restoration.

Figure 9.42 illustrates the normal operation of a DLR network. Each node has two Ethernet ports, has implemented an embedded switch, and supports DLR. When a ring node receives a packet on one of its Ethernet ports, it determines whether the packet needs to be received by the ring node itself (e.g., the packet has this node's MAC address as the destination MAC address) or whether the packet should be sent out the other Ethernet port.

The active ring supervisor blocks traffic on one of its ports with the exception of few special frames and does not forward traffic from one port to other. This configuration avoids a network loop, so only one path exists between any two ring nodes during normal operation.

The active ring supervisor transmits a Beacon frame through both of its Ethernet ports once per Beacon interval (400 μs by default). The active ring supervisor also sends Announce frames once per second. The Beacon and Announce frames serve several purposes:

- The presence of Beacon and Announce frames inform ring nodes to transition from linear topology mode to ring topology mode.

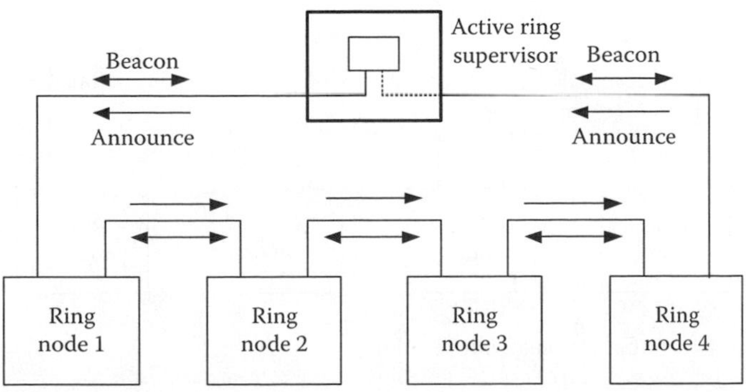

FIGURE 9.42 Normal DLR operation.

- Loss of Beacon frames at the supervisor enables detection of certain types of ring faults. (Note that normal ring nodes are also able to detect and announce ring faults.)
- The Beacon frames carry a precedence value, allowing selection of an active supervisor when multiple ring supervisors are configured.

9.3.3.23.2.3 ***Ring Faults*** Ring faults may include common link failures such as device power failure or media disconnection, or higher-level failures where the physical layer is active but the device has failed.

The ring supervisor detects a ring fault directly via a Link Status message from a ring node or indirectly via loss of Beacon frames. When a ring fault is detected, the active ring supervisor reconfigures the network by unblocking traffic on its previously blocked port and flushing its unicast MAC table. The supervisor immediately sends Beacon and Announce frames with the ring state value indicating that the ring is now faulted.

Ring nodes also flush their unicast MAC tables upon detecting loss of the Beacon in one direction or upon receipt of Beacon or Announce frames with the ring state value indicating the ring fault state. Flushing the unicast MAC tables at both supervisor and ring nodes is necessary for network traffic to reach its intended destination after the network reconfiguration.

Figure 9.43 shows the network configuration after a link failure, with the active ring supervisor passing traffic through both of its ports.

9.3.3.23.2.4 ***Further Information for Developers*** Further details of the DLR protocol operation, including event tables, state diagrams, and other implementation requirements, are included in Chapter 9 of The EtherNet/IP Specification.

9.3.3.24 Use of the Rapid Spanning Tree Protocol

In addition to the DLR protocol, the Rapid Spanning Tree Protocol (RSTP) has also been allowed for use in conjunction with EtherNet/IP.

RSTP was originally designed for networks based on a tree topology where many devices are connected back to an Ethernet switch, which in turn can be connected to other Ethernet switches. RSTP is a mature and widely accepted approach to solve the Ethernet ring recovery issue when one looks at the most current enhancements to the specification. The IEEE Standard 802.1D 2004 edition incorporated RSTP into that part of the standard. Changes were made by the IEEE Standards committee to RSTP, which make it a suitable recovery mechanism for a ring topology for some automation applications.

Chapter 9 of The EtherNet/IP Specification describes the use of RSTP within EtherNet/IP.

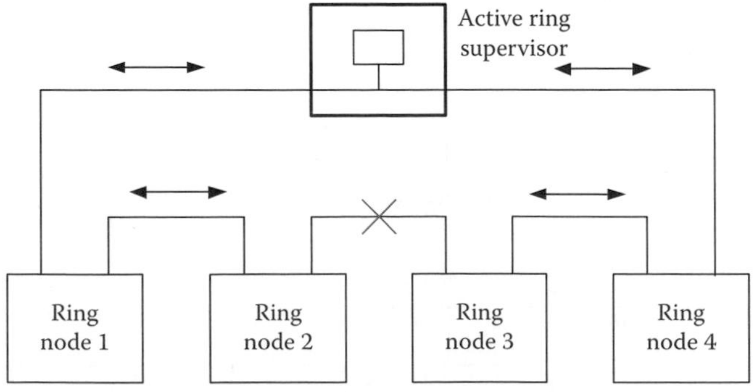

FIGURE 9.43 Network reconfiguration after link failure.

9.3.3.25 Tools

Tools for EtherNet/IP Networks can be divided into four groups:

1. *Physical layer tools* are tools (hardware and/or software) that verify the integrity and conformance of the physical layer or monitor the quality of the data transmission.
2. *Commissioning tools*: are tools that set the IP adress needed for all EtherNet/IP devices (see Section 9.3.3.9). In these cases, a BOOTP/DHCP server tool, such as the free BOOTP/DHCP routine downloadable from the Rockwell Automation website, is required.
3. *Configuration tools* are software tools capable of communicating with individual devices for data monitoring and configuration purposes. Most configuration tools are EDS-based; however, more complex devices like I/O Scanners tend to have their own configuration applets that are only partially based on EDS files. Some of these tools support multiple access paths to the network, for example, via suitable routing devices. High-level tools also actively query the devices on the network to identify them and monitor their health. Configuration tools also may be integrated into other packages like PLC programming software.
4. *Monitoring tools* typically are PC-based software packages (e.g., traffic analyzers or *sniffers*) that can capture and display the Ethernet frames on the network. A raw Ethernet frame display may be good enough in some instances, but using a tool that can both display raw Ethernet frames and provide multiple levels of frame interpretation (IP, TCP/UDP, and EtherNet/IP header interpretation) is recommended. Due to the popularity of Ethernet, a large number of these tools are available, but not all of them support EtherNet/IP decoding.

In a typical installation, only a commissioning tool and a configuration tool are needed. Protocol monitoring tools are used mainly to investigate interoperability problems and to assist during the development process.

Turn to the CIP supplier directory on the ODVA website to access a list of vendors that provide tools for EtherNet/IP.

9.3.3.26 Advice for Developers

Before any development of an EtherNet/IP product is started, the following issues should be considered in detail:

- What functionality (Device Classes, see Section 9.3.2.13) does the product require today and in future applications?
  - Explicit Messaging server only.
  - I/O Adapter functionality.
  - Explicit Messaging client.
  - I/O Scanner functionality.
- What are the physical layer requirements? Is IP 65/67 required, or is IP 20 good enough?
- Will the development be based on commercially available hardware components and software packages (recommended) or designed from scratch (possible but costly)?
- What are the configuration requirements?
- What design and verification tools should be used?
- When and where will the product be tested for conformance and interoperability?
- What is an absolute must before products can be placed on the market (own the specification, have the company's own Vendor ID, and have the product conformance tested)?

Ethernet chipsets and associated base software packages are available from many vendors. For the support of the EtherNet/IP part of development, refer to the ODVA website for a list of companies that can support EtherNet/IP development. An extended description of the development process is available for download from the ODVA website [37].

To help EtherNet/IP developers in creating their products, ODVA runs a series of so-called Implementor Workshops for EtherNet/IP during which various aspects of EtherNet/IP are discussed. These workshops (with a North American and a European series) have created a number of documents with functionality recommendations for EtherNet/IP devices, see [23,25,38–40]. EtherNet/IP devices are then tested against these recommendations during multivendor testing events, called Plugfests. Visit the ODVA web page to learn more about upcoming EtherNet/IP Implementor Workshops and Plugfests.

9.3.3.27 EtherNet/IP Summary

Since its introduction in 2000, EtherNet/IP has shown remarkable growth in many applications that previously used traditional networks. This success (several million nodes installed to date) is largely attributed to the fact that EtherNet/IP uses standard unmodified Ethernet to introduce real-time behavior into the Ethernet domain without sacrificing any of Ethernet's most useful features, such as company-wide access with standard and specialized tools through corporate networks.

A major strength of EtherNet/IP is the fact that it does not require a modified or highly segregated network: standard switches and routers used in the office world can be used for industrial applications without modification. At the same time, all existing transport-level or TCP/UDP/IP-level protocols can continue to be used without any need for special bridging devices. The substantially improved real-time behavior of CIP Sync and the introduction of CIP Safety also allow EtherNet/IP to be used in applications that currently require several dedicated networks.

Finally, as a member of the CIP family of networks, EtherNet/IP Networks can be combined into an overall CIP Network structure that allows seamless communication among CIP Networks, just as if they were only one network.

9.3.4 CompoNet

9.3.4.1 Introduction

CompoNet is a low-level network that provides high-speed communication between higher-level devices such as controllers and simple industrial devices such as sensors and actuators. IEC has published CompoNet as IEC 62026-7 in 2010 [41].

9.3.4.2 Relationship to Standards

Like other CIP Networks, CompoNet follows the OSI model, an ISO standard for network communications that is hierarchical in nature. Networks that follow this model define all necessary functions, from physical implementation up to the protocol and methodology to communicate control and information data within and across networks.

Figure 9.44 shows the relationship between CIP and CompoNet.

The CompoNet adaptation of CIP is described in Volume 6 of the CIP Networks Library [42]. All other features are based on CIP. CompoNet is also described in an international standard [41].

9.3.4.3 CompoNet Features

CompoNet supports both bit-level I/O slaves (BitIN and BitOUT slaves) and byte-level I/O slaves (WordIN and WordOUT slaves) simultaneously in one network. It also supports intelligent repeaters to expand the network flexibly and provide network diagnosis [43].

The main features of CompoNet are as follows:

- Selectable data rates: 4 Mbps/3 Mbps/1.5 Mbps/93.75 kbps.
- Single master network with a large number of slave nodes: 384 slave devices maximum including WordIN: 64; WordOUT: 64; BitIN: 128; BitOUT: 128.

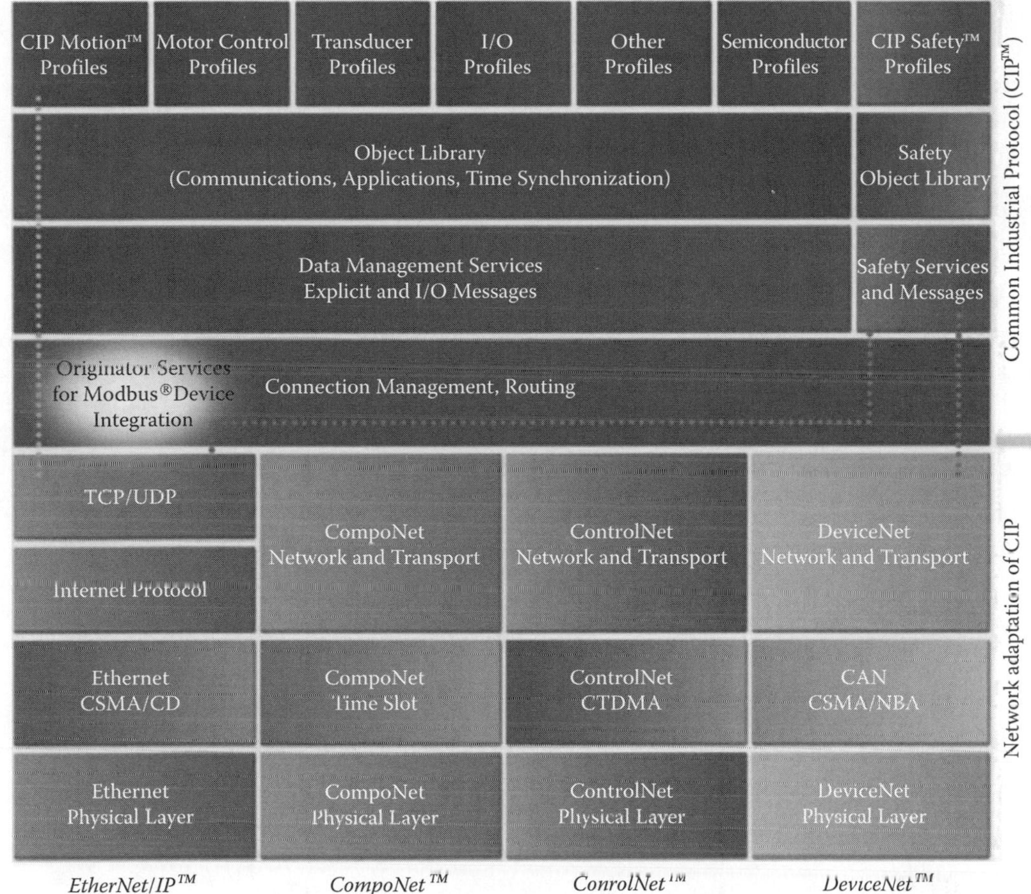

FIGURE 9.44 Relationship between CIP and CompoNet.

- Up to 64 repeaters per network to expand physical covering area and to adapt different cables.
- Up to 32 nodes (slaves and repeaters) per segment.
- I/O capacity: 1280 input/1280 output points.
- Support for flat 4-wire, round 4-wire, and 2-wire cables in bus and branch topologies.
- Maximum of three segment layers. This means up to two repeaters are allowed between any slave and the master.
- 30/30/100/500 m maximum trunk cable distance with respect to data rates, and 150/150/500/2500 m maximum distance between the most distant slaves with repeaters.
- Trunkline/dropline except for 4 Mbps.
- Efficient communication with multicast polling and Time Division Multiple Access (TDMA).

9.3.4.4 CompoNet Physical Layer

The physical layer of CompoNet has been designed specifically for this network; it does not reuse any existing open technology.

CompoNet uses a transformer-coupled transmission method and a Manchester-encoded signal on the wire; the principal circuit of the physical media attachment is shown in Figure 9.45.

Master ports and slave ports use the same physical media attachment with only minor differences in the coupling network. To help vendors design their CompoNet physical media attachment, ODVA

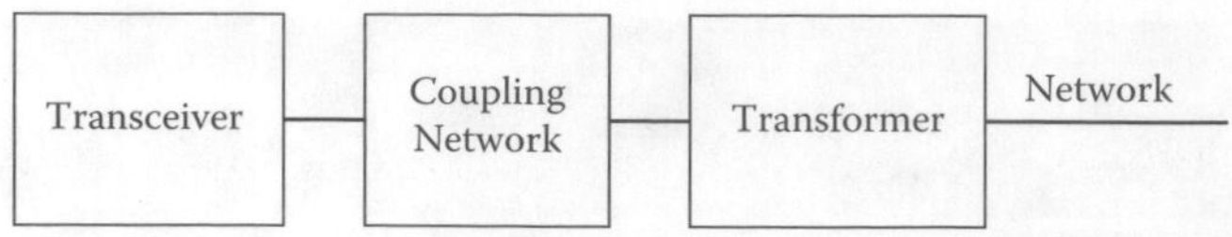

FIGURE 9.45 Physical media attachment of CompoNet.

has published recommended circuits for both masters and slaves in Chapter 8 of The CompoNet Specification [42]. Vendors can also design their own circuits.

9.3.4.5 Frame Structure

A typical message frame is composed of the Preamble, Command Code, Command Code–Dependent Block(s), and Cyclic Redundancy Code (CRC), as shown in Figure 9.46.

All frames use the same Preamble. Two types of CRC generator polynomials, CRC8 (8-bit) and CRC16 (16-bit), are used depending on which frame is used.

There are seven types of frames with varying lengths of command codes as shown in Figure 9.47:

1. *OUT Frame*
   This is a frame from the Master to the Slaves/Repeaters, which delivers OUT data to OUT slaves, specifies the group of slaves/repeaters that should report their status, and synchronizes the slaves/repeaters to start time domain timers of the CN and IN frames. Data are organized in 16-bit words and transmitted, LSB first, word by word in ascending order.
2. *TRG Frame*
   This frame functions like the OUT frame except that it contains no output data; it is a trigger frame sent by the master in place of the OUT frame when it has no outputs to send.
3. *CN Frame*
   This frame is used by slaves/repeaters to report their connection status to the master and notify the master of a request to send an event.
4. *IN Frame*
   This is a frame from input slaves to the master with input data. Data are transmitted, LSB first, word by word in ascending order, if the size is greater than 16 bits.

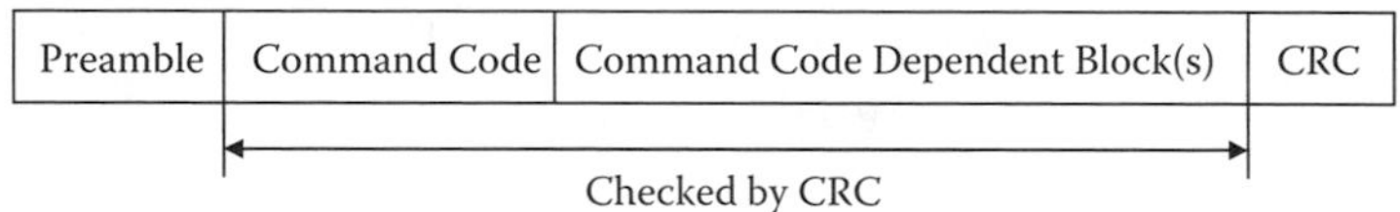

FIGURE 9.46 A general frame.

Command code							Meaning
B0	B1	B2	B3	B4	B5	B6	
0	0	0	1	x	x	x	OUT
0	0	1	1	x	x	x	TRG
0	1	x	x				CN
1	0						IN
1	1	1	x	x	x		A_EVENT
1	1	0	x	x	x		B_EVENT
0	0	0	0	1			BEACON

FIGURE 9.47 Command code in frames.

5. *A_EVENT Frame*
   Any node on the network can send A_EVENT frames, used for acyclic message communications. Data are in 16-bit words and transmitted, LSB first, word by word in ascending order.
6. *B_EVENT Frame*
   This frame is always originated by a master who sets data link parameters or reads the information as described in Section 9.3.4.9.2. Also, it is used to grant a slave/repeater permission to send an A_EVENT request or response.
7. *BEACON Frame*
   This frame specifies the data rate and sends the initial communication parameters to slaves/repeaters.

9.3.4.6 Protocol Adaptation

CompoNet uses connection-based I/O Messaging and Unconnected Explicit Messaging. Thus, every device must have the UCMM Function.

9.3.4.7 Indicators and Switches

CompoNet does not require a product to have indicators. However, if a product includes indicators with any of the legends in Chapter 9 of The CompoNet Specification, they must follow the behavior specified in that chapter.

Chapter 9 of The CompoNet Specification describes how switches are used to set the MAC address and baud rate. The MAC address may also be set via Explicit Messaging.

9.3.4.8 Additional Objects

The CompoNet Specification defines two additional objects: the CompoNet Link Object and the CompoNet Repeater Object.

9.3.4.8.1 CompoNet Link Object (Class ID: 0xF7)

The CompoNet Link Object manages all aspects associated with the CompoNet link, in particular node address and baud rate as well as the switches associated with MAC address and baud rate. Furthermore, this object contains information on the allocation of the I/O communication. Within a master device, this object may also contain information on the slaves that are present on the network.

9.3.4.8.2 CompoNet Repeater Object (Class ID: 0xF8)

Every repeater device on CompoNet must support one instance of the CompoNet Repeater Object. Its main purpose is to monitor the power supply voltage of the subnet it connects to.

9.3.4.9 Network Access

9.3.4.9.1 Network Schedule

In a CompoNet network, the master controls bus communications according to its configuration. A master divides a communication cycle into several time domains or time slots.

CompoNet conducts arbitration under strict time supervision managed by the master. The communication cycle is partitioned into time domains as shown in Figure 9.48. Each node obtains the right to send data to the network within a specified time period after the completion of the OUT time domain.

The first domain of each communication cycle is the OUT time domain. Subsequent domains are the CN time domain, the IN time domain, and the EXTEND time domain.

- *OUT Time Domain*
  The master sends an OUT frame or a TRG frame in this period.
- *CN Time Domain*
  CN frames are sent in this period. The number of CN frames sent in this time domain is determined by the master.

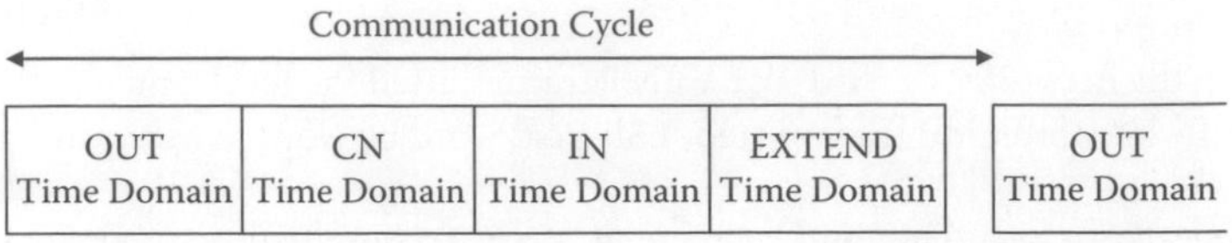

Communication Cycle

OUT Time Domain	CN Time Domain	IN Time Domain	EXTEND Time Domain	OUT Time Domain

FIGURE 9.48 Time domains.

- *IN Time Domain*
  IN frames are sent in this period consecutively by all input-type devices.
- *EXTEND Time Domain*
  The master executes message communications in this period. Event frames, that is, A_EVENT frames and B_EVENT frames, can be sent in this period. BEACON frames shall be sent periodically. The master can send a BEACON before every OUT Time Domain starts or in an idle EXTEND Time Domain.

Figure 9.49 shows the sequence of frames in a communication cycle. The master starts the cycle by sending an OUT frame. The OUT frame is a broadcast message used to send output data to all OUT slaves. Each OUT slave consumes its output data (up to 16 bits) from its offset in the OUT frame. The completion of the OUT frame indicates the end of the OUT Time Domain and triggers slaves and repeaters to start the timers that allow them to correctly participate in the CN Time Domain and the IN Time Domain.

In the CN Time Domain, the slaves or repeaters addressed by the *CN Request MAC ID Mask* field in the OUT frame will transmit their CN frames at the predefined time sequence.

During the IN Time Domain, IN frames are sent by any IN devices that are in the Participated State, at the predefined time sequence (see Figure 9.49). During the IN Time Domain, nodes in the EventOnly Substate do not transmit an IN frame.

During the EXTEND Time Domain, nodes may transmit an event command frame and possibly an immediate acknowledge frame, depending on the event type. These can be sent by the master, slaves, or repeaters. The node designated in the event command frame's Destination MAC ID field will send an event acknowledge frame, if required by the specific event command. CIP explicit messaging is done during this domain.

9.3.4.9.2 Network Access

CompoNet has an algorithm that controls the network access of any of the slaves and repeaters. This is a combination of actions taken by the slave itself and commands from the master using Status Read and Status Write (STW) operation.

Starting from power-up, all devices first need to detect the data rate of the network. Once a device has detected the data rate from the BEACON frames sent by the master, it transitions to the

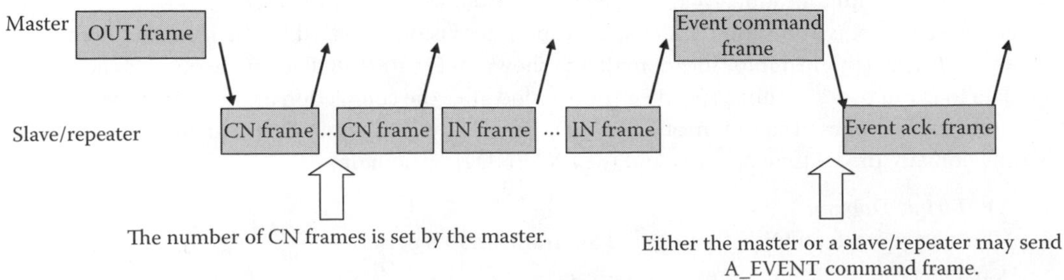

FIGURE 9.49 A typical communication cycle.

Nonparticipated state. The device will stay in this state as long as it receives valid frames within the timeout period (650 ms for 93.75 kbps, 200 ms for the other data rates) and no other event occurs. When the master has determined that it is OK for the slave to transition to the normal operating mode, that is, no duplicate of the slave's node number has been detected, it sends an STW_Run command to the slave and the slave then transitions to the *Participated* state, which is the normal operating state. The slave will leave this state and fall back to the *Nonparticipated* state either when it has experienced a network timeout (no OUT or TRG frames within the timeout period) or when it receives an STW_Standby command from the master.

The detection of duplicate node IDs is also a combination of master and slave actions. The slave will go to the Communication Fault state if told to do so by the master through an STW_Dup command or when its CN Counter overflows due to communication errors caused by duplicate node IDs.

9.3.4.10 Explicit Messaging

CompoNet uses UCMM for explicit messaging; there is no connection-based Explicit Messaging. Explicit Messages are encapsulated into A_EVENT frames as shown in Figure 9.50.

The format of an Explicit Message is defined to have two parts as shown in Figure 9.51: Header and Service Data. The Header is the CompoNet-specific part containing information for flow control, fragmentation protocol, and addresses in a word format. The Service Data part consists of the Request/Response Data as defined in the Message Router Request/Response format in Volume 1, Chapter 2, of the CIP Networks Library.

Two types of Explicit Message formats are defined:

- Compact 1 Octet Class ID and Instance ID (required)
- Expanded—CIP EPATH (optional)

A fragmentation protocol is defined optionally for supporting long data transfers in Explicit Messages.

The SID/Extended SID field is used for reply matching. The client sets the value, and the server echoes it back. Values used are specific to whether the master or the slave is the client in the transaction.

Explicit Messaging transactions are subject to timeouts. The default timeout values can be changed by Explicit Messages.

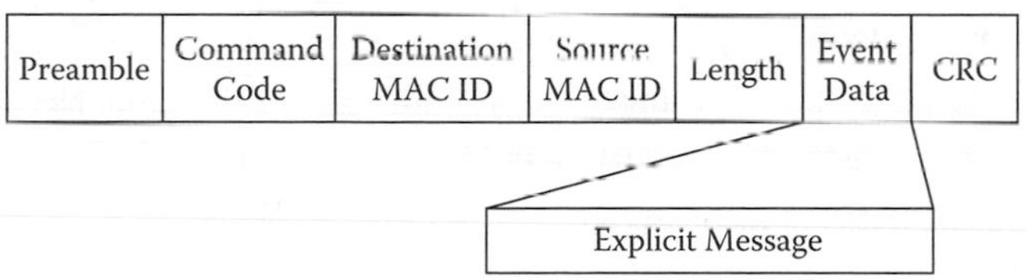

FIGURE 9.50 A_EVENT frame format.

Word offset	B15	...	B00
0 to n	Header: Control Code, Destination MAC ID, Source MAC ID, SID/Extended SID, Size, Service Code, Class ID, Instance ID		
n + 1 to 21	Service Data		

FIGURE 9.51 CompoNet Explicit Message format.

9.3.4.11 I/O Messaging

I/O Messages on CompoNet, like on any other CIP Network, are always exchanged in a connection-based manner. Communication Objects must be set up for this purpose. CompoNet uses a Predefined Allocation service to establish I/O communication between the master and the slaves. Since these I/O connections are not initially available when a device is powered on, they first have to be created. This is done by sending an Allocate Service to the CompoNet Link Object of the slave that is to be allocated. When an I/O Connection is no longer needed, the slave(s) can be released by sending a Release Service to the slave's CompoNet Link Object.

CompoNet uses OUT frames to deliver output data to consuming slaves and to trigger IN frame transmission, IN frames to deliver produced data to the master, and TRG frames to trigger IN frame transmission when the master has no output data to send.

OUT/TRG frames are monitored by I/O slaves to restart a watchdog timer. If an I/O slave has not received any OUT/TRG frames before the watchdog timer expires, it will produce a Timeout event. The timer value of the watchdog is four times the EPR attribute in the I/O connection instance. A watchdog timeout event transitions the connection instance to the *Timeout* state, and the application will be notified.

An OUT frame with I/O-Refresh disabled or a TRG frame indicates *Idle*, and an OUT frame with I/O-Refresh enabled indicates *Run*.

9.3.4.12 I/O Connection Establishment

As described in Section 9.3.4.11, I/O Messages on CompoNet are always exchanged in a connection-based manner. Similar to DeviceNet, a CompoNet slave is allocated by the master by sending an Allocate Service to the CompoNet Link Object of the slave that is to be allocated.

When an I/O Connection is no longer needed, a slave can be released by sending a Release Service to the slave's CompoNet Link Object.

In contrast to DeviceNet, which also uses a Predefined Master Slave Connection Set, CompoNet has only one I/O data exchange mechanism (polled). Therefore, only one type of allocation may take place so there is only a single Allocation Choice bit defined in the Allocate service.

9.3.4.13 Device Profiles

9.3.4.13.1 Bit Slave or Word Slave

All existing profiles must be realized by using the minimum amount of CompoNet communication. The following rules apply to the adaptation of existing profiles:

- CompoNet frame rules must be observed.
- If an existing profile can be realized by a Bit Slave, it must be a Bit Slave.
- An existing profile must be realized with the minimal data length that is feasible.

9.3.4.13.2 Byte Size Differences

CompoNet uses the same definitions as other CIP Network even though it has specific transmission frames. In order to agree with the data size in bytes as used by CIP, CompoNet needs some rules to align a CompoNet I/O frame (which counts I/O size in bits) with CIP objects (which typically count data length in bytes).

For bit-level slaves, the related I/O size shown in the connection object instance is rounded up to 1 byte. For example, a BitIN slave with 1-bit valid input data uses IN frames with 2 bits of data to deliver data on the wire, but the *produced_connection_size* (attribute 7) of its connection instance shall be *1*, which means 1 byte in CIP.

For byte-level slaves, the same sizes shall be used as with other CIP Networks. If the CIP byte size does not match the CompoNet frame size, the size is rounded up to the next even number for the transmission. For example, a device with 3 bytes of application data will have size 3 in its application object (e.g., the Assembly Object) and size of 3 for the *produced_connection_size* (attribute 7) of its connection instance, but the IN frames on the wire have 32 bits of valid data.

9.3.4.14 Configuration

CompoNet devices typically come with EDSs as described in Section 9.2.7.

To support EDS-based configuration, several CompoNet-specific EDS keywords have been added. With these new keywords and most CIP EDS keywords, CompoNet masters can be configured by a tool that can decode CompoNet EDS files. As a minimum (apart from the required sections), CompoNet EDSs should contain the CompoNet-specific sections that describe the I/O Connections available in the slave.

An EDS also may contain individual parameters and/or a Configuration Assembly with a complete description of all parameters within this Assembly.

CompoNet can also be configured by FDT/DTM, which is beyond the scope of CIP [44].

9.3.4.15 Advice for Developers

Before starting any CompoNet product development, the following issues should be considered in detail:

- What functionality does the product require today and in future applications?
  - Slave functionality.
  - Master functionality.
- What are the physical layer requirements? Is IP 65/67 required, or is IP 20 good enough?
- What type of hardware should be chosen for this product?
- What kind of firmware should be used for this product? Will a commercially available communication stack be used?
- Will the development of hardware and/or software be done internally, or will it be designed by an outside company?
- What are the configuration requirements?
- What design and verification tools should be used?
- What kind of configuration software should be used for this product? Will a commercially available software package be used, that is, is an EDS adequate to describe the device or is custom software needed?
- When and where will the product be tested for conformance and interoperability?
- What is an absolute must before my products can be placed on the market (i.e., own the specification, have the company's own Vendor ID, and have the product conformance tested)?

A full discussion of these issues goes well beyond the scope of this publication, see [45,46] instead. ODVA provides a developer's toolkit including working source code for CompoNet slaves and repeaters.

9.3.4.16 Conclusions

CompoNet is a well-adapted CIP Network. It complies with the CIP object modeling, object addressing, as well as the CIP communication model and its configuration rules. It is easy to realize CIP network routing and bridging. Combining with its advantages in aspects of data link and physical layer, CompoNet provides unique solutions with existing CIP resources to vendors and users.

All the changes in automation require consideration of the next generations of networking. In view of these trends and in view of technologies similar to those described earlier, it can be seen how much more successful a network can be if it is from the family of CIP Networks in all levels of hierarchy—instead of being patched together as a result of selecting isolated independent networks specific to each level.

9.4 Benefits of the CIP Family

CIP offers distinct benefits for two groups:

1. Device manufacturers
2. Users of devices and systems

9.4.1 Benefits for Device Manufacturers

For device manufacturers, a major benefit of using CIP is the fact that existing knowledge can be reused from one protocol to another, resulting in lower training costs for development, sales, and support personnel. Manufacturers also can reduce development costs, since certain parts (e.g., parameters and profiles) of the embedded firmware that are the same regardless of the network can be reused from one network to the other. As long as these parts are written in a high-level language, the adaptation is simply a matter of running the right compiler for the new system.

Another important advantage for manufacturers is the easy routing of messages from one system to another. Any routing device can be designed very easily since there is no need to invent a *translation* from one system to another; both systems already speak the same language.

Manufacturers also benefit from working with the same organizations for support and conformance testing.

9.4.2 Benefits for the Users of Devices and Systems

For users of devices and systems, a major benefit of using CIP is the fact that existing knowledge can be reused from one protocol to another, for example, through device profiles, and device behavior is identical from one system to another, resulting in lower training costs. Technical personnel and users do not have to make great changes to adapt an application from one type of CIP Network to another, and the system integrator can choose the CIP Network that is best suited to his or her application without having to sacrifice functionality.

Another important CIP advantage is the ease of bridging and routing between CIP Networks. Moving information among incompatible networks is always difficult and cumbersome since there is seldom a direct translation of all functionality on one network to another. This is where users can reap the full benefits of CIP. Forwarding data and messages from top to bottom and back again is very easy to implement and uses very little system overhead. There is no need to translate from one data structure to another—they are the same. Services and status codes share the same benefit, as these, too, are identical over all CIP Networks. Finally, creating a message that runs through multiple hops of CIP Networks is simply a matter of inserting the full path from the originating to the target device. Not a single line of code or any other configuration is required in the routing devices, resulting in fast and efficient services that are easy to create and maintain. Even though these networks may be used in different parts of the application, messaging from beginning to end really functions as if there is only one network.

Finally, the Producer/Consumer mechanisms used in all CIP Networks provide highly efficient use of transmission bandwidth, resulting in system performance that often is much higher than that of other networks running at higher raw baud rates. With CIP, only the truly important data are transmitted, rather than old data being repeated over and over again.

Planned and future protocol extension will always be integrated in a manner that allows coexistence of *normal* devices with *enhanced* devices like those supporting CIP Sync and/or CIP Safety. Therefore, no strict segmentation into *Standard*, CIP Sync, and CIP Safety networks is required unless there is a compelling reason, for example, unacceptably high response times due to high bus load.

9.5 Application Layer Enhancements

9.5.1 CIP Sync and CIP Motion

9.5.1.1 General Considerations

While CIP Networks [5,14,17,42,47] provide real-time behavior that is appropriate for many applications, a growing number of applications require even tighter control of certain real-time parameters. Let us have a look at some of these parameters:

- *Real-Time*
  This term is being used with a variety of meanings in various contexts. For further use in this section, the following definition is used: A system exhibits real-time behavior when it can react to an external stimulus within a predetermined time. How short or how long this time is depends on the application. Demanding industrial control applications require reactions in the millisecond range while, in some process control applications, a reaction time of several seconds or more is sufficient.
- *Determinism*
  A deterministic system allows for a worst-case scenario (not a prediction or a probability) when deciding on the timing of a specific action. Industrial communication systems may offer determinism to a greater or lesser degree, depending on how they are implemented and used. Networks featuring message transmission at a predetermined point in time, such as ControlNet, Sercos Interface, and Interbus-S, are often said to offer absolute determinism. On the other hand, networks such as Ethernet may become nondeterministic under certain load conditions, specifically, when it is deployed in half-duplex mode with hubs. However, when Ethernet is deployed with full-duplex high-speed switches, it operates in a highly deterministic manner (see Section 9.3.3.21).
- *Reaction Time*
  In an industrial control system, the overall system reaction time is what determines real-time behavior. The communication system is only one of several factors contributing to the overall reaction time. In general, reaction time is the time from an input stimulus to a related output action.
- *Jitter*
  This term defines the time deviation of a certain event from its average occurrence. Some communication systems rely on very little message jitter, while most applications require only that a certain jitter is not exceeded for actions at the borders of the system, such as input sampling jitter and output action jitter.
- *Synchronicity*
  Distributed systems often require certain actions to occur in a coordinated fashion, that is, the actions must take place at a predetermined moment in time, independent of where the action is to take place. A typical application is coordinated motion or electronic gearing. Some of these applications require synchronicity in the microsecond range.
- *Data Throughput*
  This is a system's capability to process a certain amount of data within a specific time span. In communication systems, protocol efficiency, the communication model (e.g., Producer/Consumer), and end-point processing power are most important, while the wire speed sets only the limit of how much raw data can be transmitted across the physical media.

CIP Sync is a communication principle that enables synchronous low-jitter system reactions without the need for low-jitter data transmission. This is of great importance in systems that do not provide absolute deterministic data transmission or where it is desirable for a variety of higher-layer protocols to run in parallel with the application system protocol. The latter situation is characteristic of Ethernet.

Most users of TCP/IP-based Ethernet want to keep using it as before without the need to resort to a highly segregated network segment to run the real-time protocol. The CIP Sync communication principle meets these requirements.

9.5.1.2 Using IEEE 1588 Clock Synchronization

The published IEEE standard 1588—Standard for a Precision Clock Synchronization Protocol for Networked Measurement and Control Systems [48]—lays the foundation for a precise synchronization of real-time clocks in a distributed system.

An IEEE 1588 system consists of a Time Master that distributes its system time to Time Slaves in a tree-like structure. The Time Master may be synchronized with another real-time clock further up in the hierarchy while the Time Slaves may be Time Masters for other devices *below* them. A Time Slave that is Time Master to another set of devices (typically, in another part of the system) is also called a Boundary Clock. The time distribution is done by multicasting a message with the actual time of the master clock. This message originates in a relatively high layer of the communication stack, and, therefore, the actual transmission takes place at a slightly later point in time. Also, the stack processing time varies from one message to another. To compensate for this delay and its jitter, the actual transmission time can be captured in a lower layer of the communication stack, such as noting the *transmit complete* feedback from the communication chip. This update time capture is then distributed in a follow-up message. The average transmission delay also is determined so that the time offset between the master and the slave clock can be compensated.

This protocol has been fully defined for Ethernet UDP/IP systems, and the protocol details for further industrial communication systems will follow. The clock synchronization accuracy that can be achieved with this system depends largely on the precision time capture of the master clock broadcast message. Hardware-assisted time capture systems can reach a synchronization accuracy of 250 ns or less. Some Ethernet chip manufacturers offer integrated IEEE 1588 hardware support.

9.5.1.3 Additional Object

CIP Sync requires the addition of a time synchronization object.

9.5.1.3.1 Time Sync Object (Class ID: 0x43)

The Time Sync Object manages the real-time clock inside a CIP Sync device and provides access to the IEEE 1588 timing information.

9.5.1.4 Fundamentals of CIP Sync Communication

Real-time clocks coordinated through the IEEE 1588 protocol do not, of their own accord, constitute a real-time system yet. Additional details showing how time stamping is used for input sampling and for the coordination of output actions need to be added. Some device profiles need to be extended as well to incorporate time information in their I/O Assemblies.

9.5.1.5 Message Prioritization

Combining these three elements (Sections 9.5.1.2 through 9.5.1.4) in conjunction with a collision-free infrastructure (see Section 9.3.3.21) is sufficient to build a real-time system. However, it is necessary to consider all traffic within the system and to arrange all application messages containing time-critical data in such a way that they are guaranteed to arrive at all consumers in time. When other Ethernet protocols—such as HTTP or FTP, which may have very long frames—need to coexist in the same system, careful configuration may be required. Ethernet frames with up to 1500 bytes of payload (approximately 122 μs long in a 100 Mbps system) can easily congest the system and delay important messages by an undetermined amount of time, possibly too long for the system to function correctly.

This is where message prioritization, known as QoS in the Ethernet world, becomes an important element. EtherNet/IP defines common usage for two standard QoS mechanisms: Differentiated

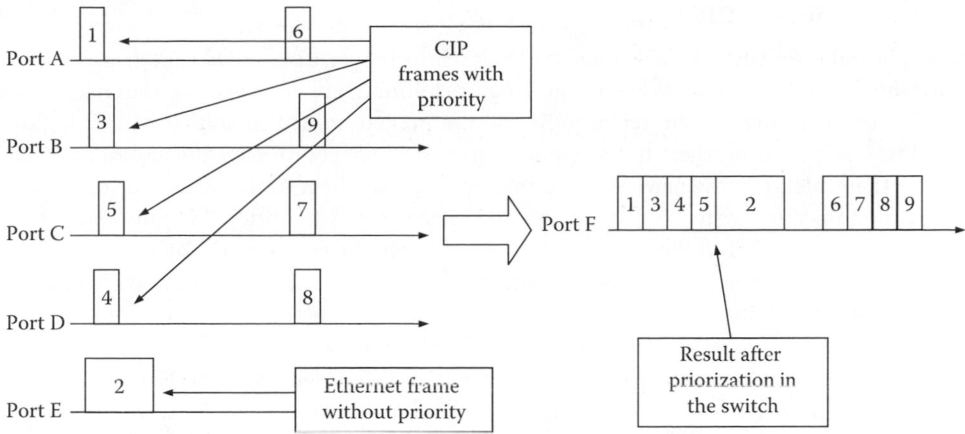

FIGURE 9.52 Ethernet Frame Prioritization.

Services (Diffserv) and IEEE 802.1Q tagged frames. These schemes are supported by many switches available today. QoS allows preferential treatment of Ethernet frames in such a way that those frames with the highest priority will jump the message queues in a switch and will be transmitted first. Messages with high priority will be transmitted while those with lower priority typically have to wait. Suitable priority assignments for all time-critical messages then guarantee their preferential treatment. Standard EtherNet/IP and other Ethernet messages will receive low or no priority and thus have to wait until all higher-priority messages have passed. Once this prioritization scheme is implemented, one full-length frame can be accommodated within each communication cycle consisting of a set of prioritized input (port A through port E) and output (port F) messages. Figure 9.52 illustrates this process.

The overall approach to QoS for EtherNet/IP calls for devices to mark their packets with a priority value, using Diffserv Code Points and/or 802.1D priority values. By explicitly marking packets with a priority value, switches and routers are able to differentiate EtherNet/IP traffic from noncritical traffic as well as differentiate specific EtherNet/IP traffic streams (e.g., IEEE 1588 vs. I/O vs. Explicit Messaging).

The following list summarizes the QoS behavior for EtherNet/IP:

- For CIP transport class 0 and 1 connections (i.e., UDP-based), there is a defined mapping of CIP priorities to 802.1D priorities and DiffServ Code Points.
- For UCMM and CIP transport class 3 connections (i.e., TCP-based), there is a single defined DiffServ Code Point and 802.1D priority value.
- For PTP (IEEE 1588) messages, there are DiffServ Code Points and 802.1D priority values corresponding to the two different types of PTP messages.
- When QoS is implemented, the default behavior is to mark packets with DSCP values. Devices may optionally support sending and receiving 801.1Q frames with the corresponding priority values. If supported, sending tagged frames is disabled by default in order to prevent device interoperability problems. The end user is responsible for enabling the tagged frame behavior and ensuring interoperability between devices.
- The QoS Object provides a means to configure DSCP values and a means to enable/disable sending of 802.1Q tagged frames.
- There are no requirements for devices to mark traffic other than CIP or IEEE 1588, but devices are free to do so.

9.5.1.6 Applications of CIP Sync

Typical applications for CIP Sync are time-stamping sensor inputs, distributed time-triggered outputs, and distributed motion, such as electronic gearing or camming applications. For example, in motion applications, sensors sample their actual positions at a predetermined time, that is, in a highly synchronized way, and transmit them to the application master that coordinates the motion. The application master then calculates the new reference values and sends them to the motion drives. Using CIP Sync, the communication system is not required to have extremely low jitter; it is sufficient to transmit all time-critical messages, and their exact arrival time becomes irrelevant. The assignment of suitable priorities to CIP Sync communication guarantees that all time-critical messages always have the bandwidth they need, and all other traffic automatically is limited to the remaining bandwidth.

As a result of these measures, CIP Sync devices can coexist side by side with other EtherNet/IP devices without any need for network segmentation or special hardware. Even non-EtherNet/IP devices—provided they do not override any of the CIP Sync prioritizations—can be connected without any loss of performance in the CIP Sync application.

9.5.1.7 Expected Performance of CIP Sync Systems

As mentioned, CIP Sync systems can be built to maintain a synchronization accuracy of better than 250 ns, in many cases without the use of Boundary Clocks. The communication cycle and thus the reaction delay to unexpected events are largely governed by the number of CIP Sync devices in a system. Allowing some bandwidth (approximately 40%) for non-CIP Sync messages as described in Section 9.5.1.5, the theoretical limit (close to 100% wire load) for the communication cycle of a CIP Sync system based on a 100-Mbps Ethernet network is around 500 μs for 30 coordinated motion axes, with 32 bytes of data each.

9.5.1.8 CIP Sync Summary

CIP Sync is a natural extension of the EtherNet/IP system into the real-time domain. Unlike many other proposed or existing real-time extensions to other protocols, CIP Sync does not require any strict network segmentation between high-performance, real-time sections and other parts of the communication system. CIP Sync provides the ability to mix parallel TCP/IP-based protocols with industrial communication architectures of any size without compromising performance.

CIP Sync currently has been applied to EtherNet/IP, and an extension to other CIP implementations will follow.

9.5.1.9 CIP Sync and CIP Motion

CIP Motion utilizes CIP Sync to manage real-time motion control. As discussed previously, CIP Sync utilizes the IEEE-1588 "Standard for a Precision Clock Synchronization Protocol for Networked Measurement and Control Systems" [48] to synchronize devices to a very high degree of accuracy. CIP Sync encapsulates the IEEE-1588 services that measure network transmission latencies and correct for infrastructure delays. The result is the ability to synchronize distributed clocks to within hundreds of nanoseconds of accuracy, or less.

Once all the devices in a control system share a synchronized, common understanding of system time, real-time control can be accomplished by including time as a part of the motion information. Unlike the traditional approaches to motion control, the CIP Motion solution doesn't schedule the network to create determinism. Instead, CIP Motion delivers the data and the time stamp for execution as a part of the packet on the network. This allows motion devices to plan and follow positioning path information according to a predetermined execution plan. Since the motion controller and the drives share a common understanding of time, the motion controller can tell the drive where to go—and what time to be there. This direct use of time in the data packet frees the network from the constraint of a rigid data delivery schedule. If data delivery fluctuates slightly on the network, the motion execution is unaffected.

There are many benefits to this approach. Since the network is not *scheduled*, there is flexibility in the amount of data that can be sent back from each device. During runtime, a drive can be reconfigured to send more or less data depending on the needs of the application. In addition, devices can be added or removed from the system because specific time slots are not allocated from the network bandwidth. The motion data packets that move between drives and controllers contain all the relevant information required for real-time motion execution; as long as basic clock synchronization is maintained, time is used as the event for execution—not the data delivery itself.

9.5.1.10 CIP Motion

CIP Motion was added to the CIP Networks Library in 2006, with the addition of the CIP Motion Axis Object and the CIP Motion Device Profile.

9.5.1.10.1 CIP Motion Profile

The CIP application profile used on EtherNet/IP provides a comprehensive set of services and device profiles that provide a wide range of functionality and device support. CIP Motion extends the CIP capability by defining extensions focused on drive control as listed in the following text:

- Torque, velocity, or position control of servo drives and VFDs (Variable Frequency Drives)
- Servo drive and VFD configuration, status, and diagnostic parameters
- Support for feedback-only axes that can provide reference information for camming and line-shafting applications
- Unicast control-to-drive communications
- Multicast peer-to-peer communications (future)

The CIP Motion profile is designed to minimize the differences between servo drive and VFD handling. This facilitates features like common configuration services, common status and diagnostic services, and common application instruction support for servo drives and VFDs making them interchangeable at the application level.

The CIP Motion profile takes advantage of the latest advances in motion control technology to provide a comprehensive, state-of-the-art profile. Extensive use of floating point data eliminates the complexity typically associated with integer math and scaling. The profile focuses on a simplified slave interface, making it easier for drive vendors to develop products that connect to the EtherNet/IP network and utilize the CIP Motion extensions.

9.5.2 CIP Safety

Like other safety protocols based on industry standard networks, CIP Safety adds additional services to transport data with high integrity. Unlike other networks, CIP Safety presents a scalable, network-independent approach to safety network design, one in which the safety services are described in a well-defined layer. Since safety functionality is incorporated into each device—rather than in the network infrastructure—CIP Safety allows both standard and safety devices to operate on the same open network. This capability gives users a choice of network architectures—with or without a safety PLC—for their functional safety networks. This approach also enables safety devices from multiple vendors to communicate seamlessly across standard CIP Networks to other safety devices without requiring difficult-to-manage gateways.

A complete definition of all details of CIP Safety can be found in Volume 5 [49].

9.5.2.1 General Considerations

Hardwired safety systems employ safety relays that are interconnected to provide a safety function. Hardwired systems are difficult to develop and maintain for all but the most basic applications. Furthermore, these systems place significant restrictions on the distance between devices.

Because of these issues, as well as distance and cost considerations, implementing safety services on standard communication networks is highly preferable. The key to developing safety networks is not to create a network that cannot fail, but to create a system where failures in the network cause safety devices to go to a known state. If the user knows which state the system will go to in the event of a failure, they can make their application safe. But this means that significantly more checking and redundant coding information is required.

So, to determine the additional safety requirements, the German Safety Bus committee [50] implemented and later extended an existing railway standard [51]. This committee provided design guidelines to safety network developers to allow their networks and safety devices to be certified according to IEC 61508 [52]. The latest version of this document has been published as GS-ET-26 [53].

Based on these standards, CIP was extended for high-integrity safety services. The result is a scalable, routable, network-independent safety layer that alleviates the need for dedicated safety gateways. Since all safety devices execute the same protocol, independent of the media on which they reside, the user approach is consistent and independent of media or network used.

CIP Safety is an extension to standard CIP that has been approved by TÜV Rheinland for use in IEC 61508 SIL 3 and EN 954-1 Category 4 applications, now ISO 13849-1, performance level e [54]. It extends the model by adding CIP Safety application layer functionality, as shown in Figure 9.53. The additions include several safety-related objects and Safety Device Profiles with specific implementation details of CIP Safety as implemented on DeviceNet, EtherNet/IP, and Sercos.

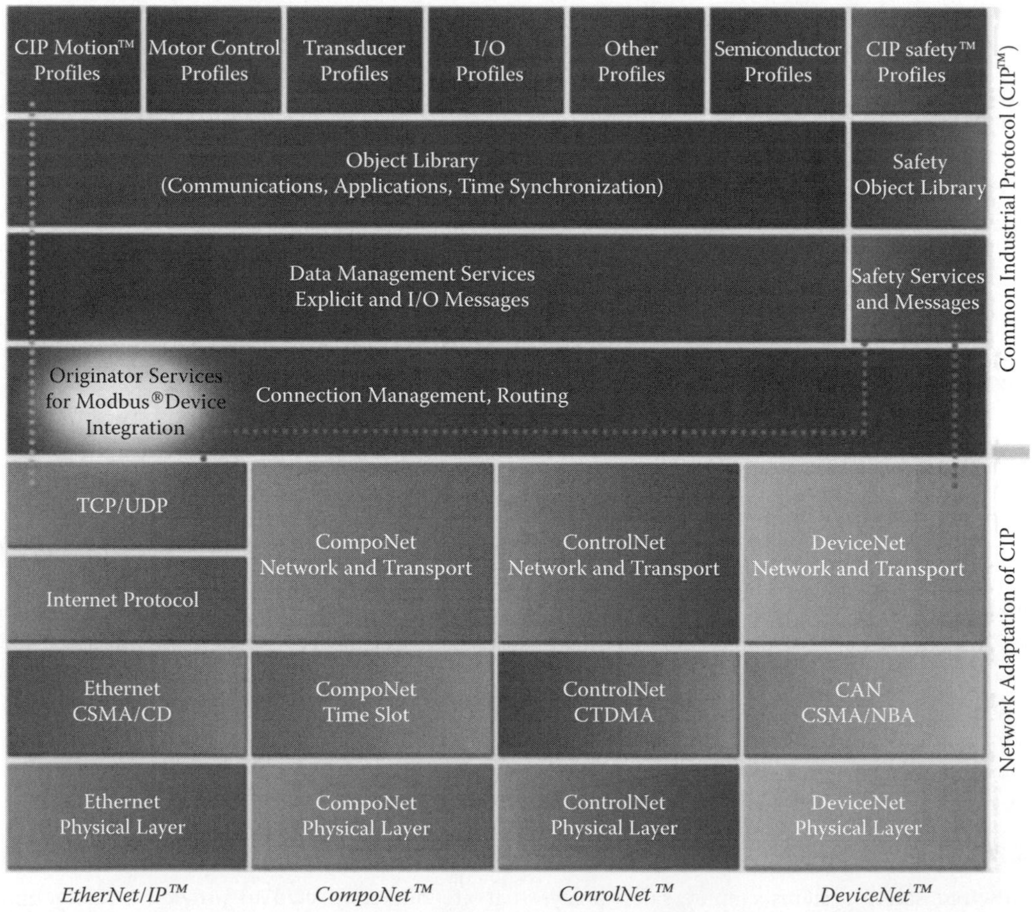

FIGURE 9.53 CIP communication layers, including safety.

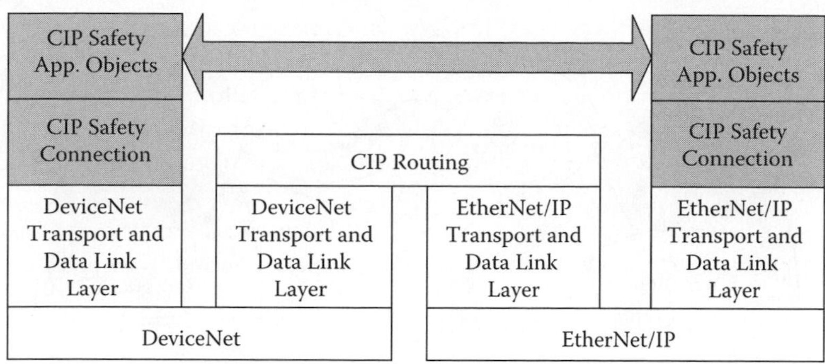

FIGURE 9.54 Routing of safety data.

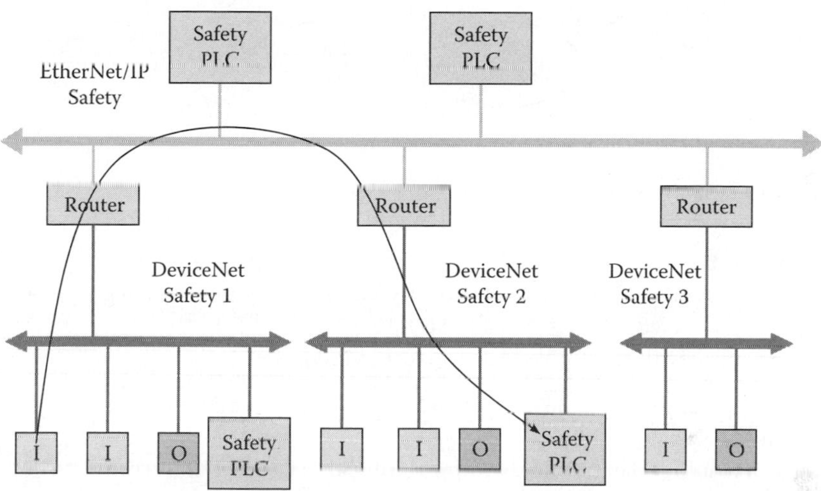

FIGURE 9.55 Network routing.

Because the safety application layer extensions do not rely on the integrity (see Section 9.5.2.3) of the underlying standard CIP as described in Section 9.2 and data link layers as described in Sections 9.3.1 through 9.3.3, single channel (nonredundant) hardware can be used for the data link communication interface. This same partitioning of functionality allows the use of standard routers for safety data, as shown in Figure 9.54. Routing safety messages is possible because the end device is responsible for ensuring the integrity of the data. If an error occurs during data transmission or in the intermediate router, the end device will detect the failure and take appropriate action.

This routing capability allows the creation of safety cells on one network, for example, DeviceNet, with quick reaction times to be interconnected with other cells via a backbone network such as EtherNet/IP for interlocking, as shown in Figure 9.55. Only the safety data that is needed is routed to the required cell, which reduces individual bandwidth requirements. The combination of rapidly responding local safety cells and the intercell routing of safety data allows users to create large safety applications with fast response times. Another benefit of this configuration is the ability to multicast safety messages across multiple networks.

9.5.2.2 Implementation of Safety

As indicated in Figure 9.54, all CIP Safety devices also have underlying standard CIP functionality. The extension to the CIP Safety application layer is specified using a Safety Validator Object. This object is responsible for managing the CIP Safety Connections (standard CIP Connections are managed through

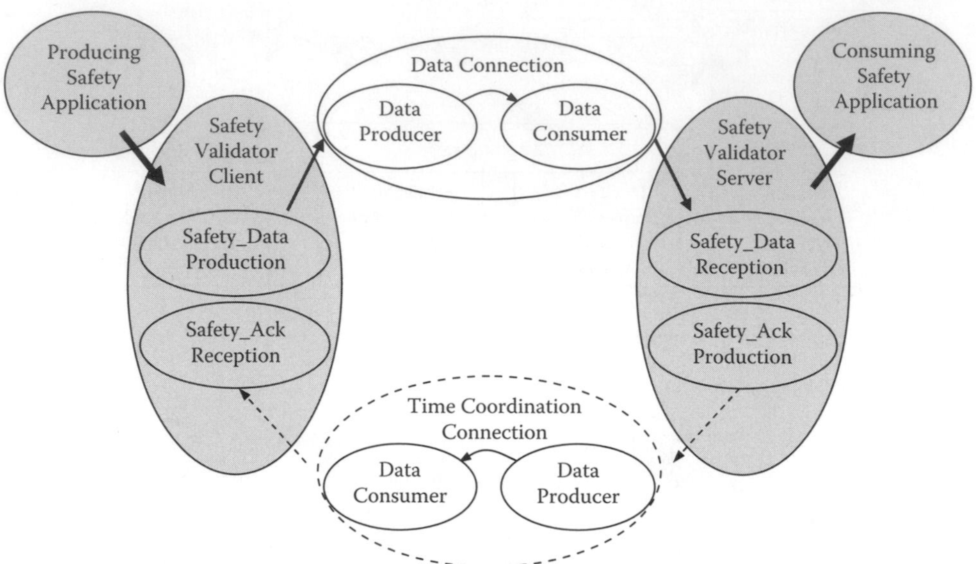

FIGURE 9.56 Relationship of Safety Validators, Unicast Connection.

communication objects) and serves as the interface between the safety application objects and the link layer connections, as shown in Figure 9.56. The Safety Validator ensures the integrity of the safety data transfers by applying the measures described in Section 9.5.2.3.

Functions performed by the Safety Validator Object as follows:

- The producing safety application uses an instance of a Client Validator to produce safety data and ensure time coordination.
- The client uses a link data producer to transmit the data and a link consumer to receive time coordination messages.
- The consuming safety application uses a Server Validator to receive and check data.
- The server uses a link consumer to receive data and a link producer to transmit time coordination messages.

The link producers and consumers have no knowledge of the safety packet and fulfill no safety function. The responsibility for high-integrity transfer and checking of safety data lies within the Safety Validators.

9.5.2.3 Ensuring Integrity

CIP Safety does not prevent communication errors from occurring; rather, it ensures transmission integrity by detecting errors and allowing devices to take appropriate actions. The Safety Validator is responsible for detecting these communication errors. The nine communication errors that must be detected are shown in Figure 9.57, along with the five measures CIP Safety uses to detect these errors [51].

9.5.2.3.1 Time Expectation via Time Stamp

All CIP Safety data are produced with a time stamp that allows Safety Consumers to determine the age of the produced data. This detection measure is superior to the more conventional reception timers. Reception timers can tell how much time has elapsed since a message was last received, but they do not convey any information about the actual age of the data. A time stamp allows transmission, media access/arbitration, queuing, retry, and routing delays to be detected.

	Measures to Detect Communication Errors				
Communication Errors	Time Expectation via Time Stamp	ID for Send and Receive	Safety CRC	Redundancy with Cross-Checking	Diverse Measure
Message Repetition	X		X[a]		
Message Loss	X		X[a]		
Message Insertion	X	X	X[a]		
Incorrect Sequence	X		X[a]		
Message Corruption			X	X	
Message Delay	X				
Coupling of safety and safety data		X			
Coupling of safety and standard data	X	X	X	X	X
Increased age of data in bridge	X				

[a] The Safety CRC provides additional protection for communication errors in fragmented messages.

FIGURE 9.57 Error Detection Measures.

Time is coordinated between producers and consumers using ping requests and ping responses, as shown in Figure 9.58. After a connection is established, the producer generates a ping request, which causes the consumer to respond with its consumer time. The producer will note the time difference between the ping production and the ping response and store this as an offset value. The producer will add this offset value to its producer time for all subsequent data transmissions. This value is transmitted as the time stamp. When the consumer receives a data message, it subtracts its internal clock from the time stamp to determine the data age. If the data age is less than the maximum age allowed, the data are applied; otherwise, the connection goes to the safety state. The device application is notified so that the connection safety state can be reflected accordingly.

The ping request-and-response sequence is repeated periodically to correct for any producer or consumer time base drift.

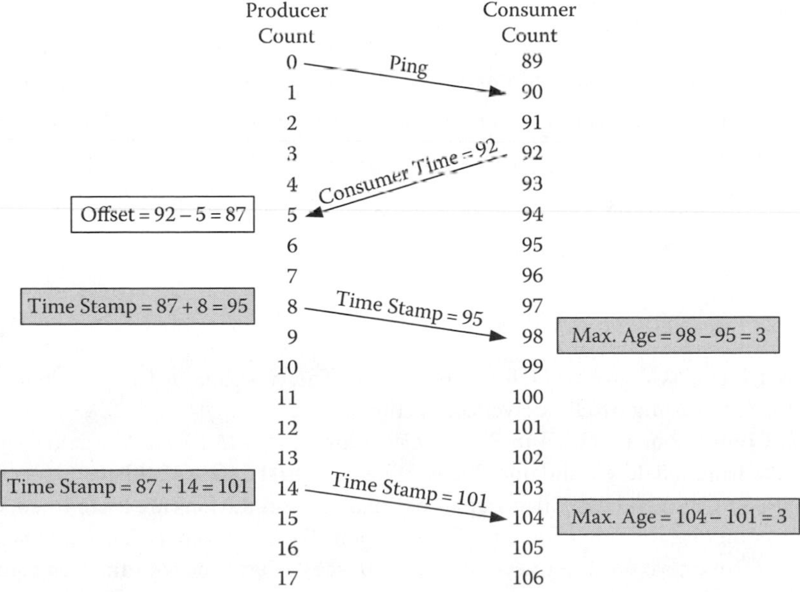

FIGURE 9.58 Time stamp.

9.5.2.3.2 *Production Identifier*

A Production Identifier (PID) is encoded in all data produced by a Safety Connection to ensure that each received message arrives at the correct consumer. The PID is derived from an electronic key, the device Serial Number, and the CIP Connection Serial Number. Any safety device inadvertently receiving a message with the incorrect PID will go to a safety state. Any safety device that does not receive a message within the expected time interval with the correct PID will also go to a safety state. This measure ensures that messages are routed correctly in multinetwork applications.

9.5.2.3.3 *Safety Cyclic Redundancy Code*

All safety transfers on CIP Safety use Safety CRCs to ensure the integrity of the transfer of information. The Safety CRCs serve as the primary means of detecting possible corruption of transmitted data. They provide detection up to a Hamming distance of four for each data transfer section, though the overall Hamming distance coverage is greater for the complete transfer due to the protocol's redundancy. The Safety CRCs are generated in the Safety Producers and checked in the Safety Consumers. Intermediate routing devices do not examine the Safety CRCs. Thus, by employing end-to-end Safety CRCs, the individual data link CRCs are not part of the safety function. This eliminates certification requirements for intermediate devices and helps to ensure that the safety protocol is independent of the network technology. The Safety CRC also provides a strong protection mechanism that allows the detection of underlying data link errors, such as bit stuffing or fragmentation.

While the individual link CRCs are not relied on for safety, they are still enabled. This provides an additional level of protection and noise immunity by allowing data retransmission for transient errors at the local link.

9.5.2.3.4 *Redundancy and Cross-Check*

Data and CRC redundancy with cross-checking provides an additional measure of protection by detecting possible corruption of transmitted data. By effectively increasing the Hamming distance of the protocol, these measures allow long safety data packets—up to 250 bytes—to be transmitted with high integrity. For short packets of 2 bytes or less, data redundancy is not required; however, redundant CRCs are cross-checked to ensure integrity.

9.5.2.3.5 *Diverse Measures for Safety and Standard*

The CIP Safety protocol is present only in safety devices, which prevents standard devices from masquerading as safety devices.

9.5.2.4 Safety Connections

CIP Safety provides two types of Safety Connections:

1. Unicast Connections
2. Multicast Connections

A Unicast Connection, as shown in Figure 9.56, allows a Safety Validator Client to be connected to a Safety Validator Server using two link-layer connections.

A Multicast Connection, as shown in Figure 9.59, allows up to 15 Safety Validator Servers to consume safety data from a Safety Validator Client. When the first Safety Validator Server establishes a connection with a Safety Validator Client, a pair of link layer connections are established, one for data-and-time correction and the other for time coordination. Each new Safety Validator Server uses the existing data-and-time correction connection and establishes a new time coordination connection with the Safety Validator Client.

To optimize throughput on DeviceNet, each Multicast Connection uses three data link connections, as shown in Figure 9.60. CIP Safety implementations on other networks do not require

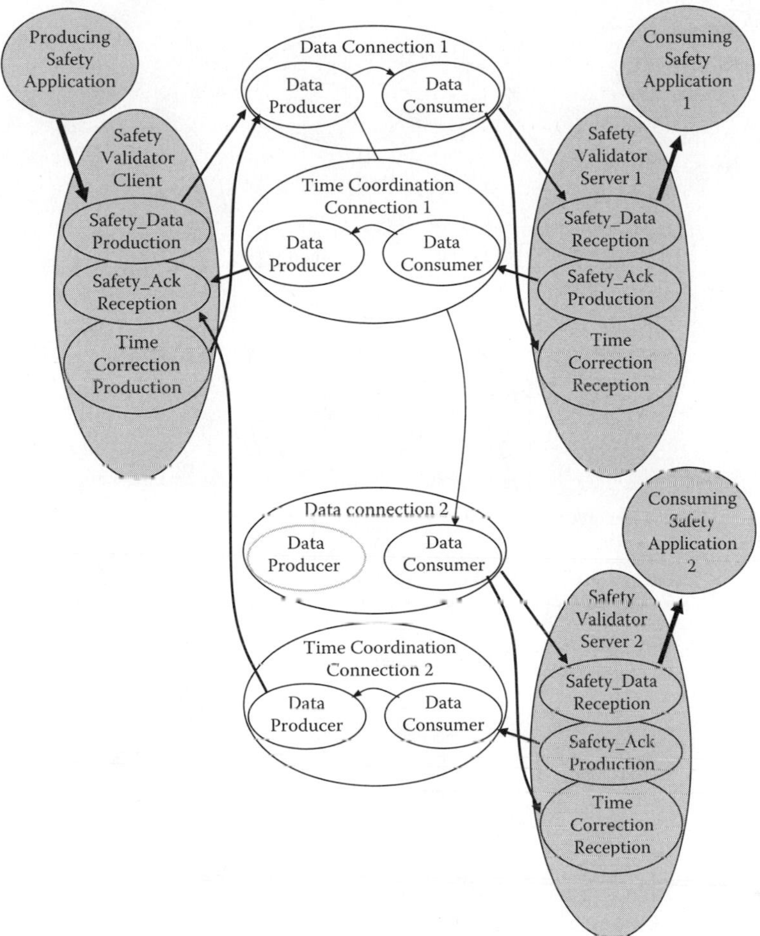

FIGURE 9.59 Multicast Connection.

this split. The data-and-time correction messages are sent on separate connections. This allows short messages to be transmitted on DeviceNet within a single CAN frame and reduces the overall bandwidth, since the time correction and time coordination messages are sent at much slower periodic intervals.

When multicast messages are routed off-link, the router combines the data-and-time correction messages from DeviceNet and separates them when messages reach DeviceNet. Since the safety message contents are unchanged, the router provides no safety function.

9.5.2.5 Message Packet Sections

CIP Safety has four message sections:

1. Data
2. Time stamp
3. Time correction
4. Time coordination

The description of these formats goes beyond the scope of this book. For available materials on this topic that go into more detail see [49,55].

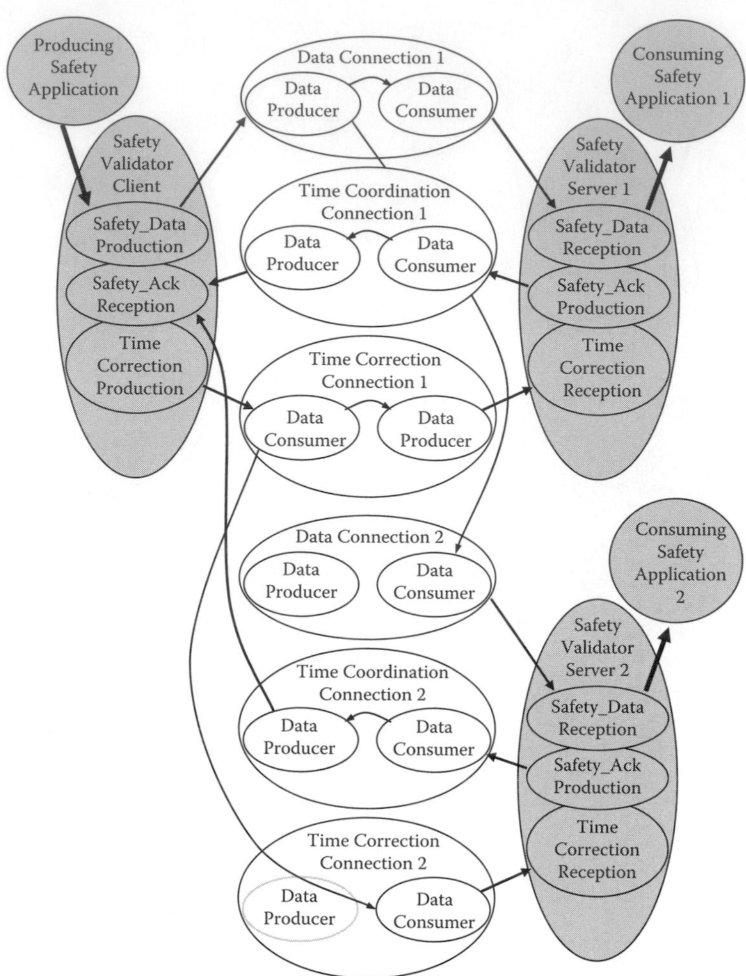

FIGURE 9.60 Multicast Connection on DeviceNet.

9.5.2.6 Configuration

Before safety devices can be used in a safety system, they first must be configured and connections must be established. The process of configuration requires placement of configuration data from a configuration tool in a safety device. There are two possible sequences for configuration:

1. Configuration tool directly to device
2. Via an intermediate device

In the configuration tool-to-device case, as shown in Figure 9.61, the configuration tool writes directly to the device to be configured (1) and (2).

In the case of intermediate device configuration, the tool first writes to an originator (1) and the originator writes to the target using an Originator-to-Target Download (3) or a SafetyOpen service (4). The SafetyOpen service (4) is unique in that it allows a safety connection to be established at the same time that a device is configured.

9.5.2.7 Connection Establishment

CIP provides a connection establishment mechanism using a Forward_Open service that allows producer-to-consumer connections to be established locally or across multiple networks via intermediate

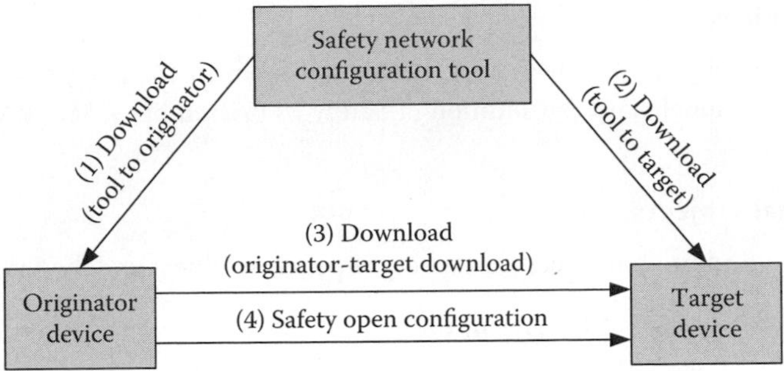

FIGURE 9.61 Configuration transfers.

routers. An extension of the Forward_Open, called the SafetyOpen service, has been created to allow the same multinetwork connections for safety.

There are two types of SafetyOpen requests:

- Type 1: with configuration
- Type 2: without configuration

With the Type 1 SafetyOpen request, configuration and connections are established at the same time, allowing rapid configuration of devices with simple and relatively small configuration data.

With the Type 2 SafetyOpen request, the safety device first must be configured, and the SafetyOpen request then establishes a Safety Connection. This separation of configuration and connection establishment allows the configuration of devices with large and complex configuration data.

In both cases, the SafetyOpen request establishes all underlying link layer connections—across the local network as well as any intermediate networks and routers.

9.5.2.8 Configuration Implementation

CIP Safety provides the following protection measures to ensure configuration integrity:

- Safety Network Number
- Password Protection
- Configuration Ownership
- Configuration Locking

9.5.2.8.1 Safety Network Number

The Safety Network Number provides a unique network identifier for each network in the safety system. The Safety Network Number, combined with the local device address, allows any device in the safety system to be uniquely addressed.

9.5.2.8.2 Password Protection

All safety devices support the use of an optional password. The password mechanism provides an additional protection measure, prohibiting the reconfiguration of a device without the correct password.

9.5.2.8.3 Configuration Ownerships

The owner of a CIP Safety device can be specified and enforced. Each safety device can specify that it be configured only by a selected originator or that the configuration is accomplished by a configuration tool.

9.5.2.8.4 Configuration Locking

Configuration Locking provides the user with a mechanism to ensure that all devices have been verified and tested prior to being used in a safety application.

9.5.2.9 Safety Devices

The relationship of the objects within a safety device is shown in Figure 9.62. Note that CIP Safety extends the CIP object model, with the addition of Safety I/O Assemblies, Safety Validator, and Safety Supervisor Objects.

9.5.2.10 Additional Objects

CIP Safety requires two additional objects, the Safety Supervisor Object and the Safety Validator Object.

9.5.2.10.1 Safety Supervisor Object (Class ID: 0x39)

The Safety Supervisor Object provides a common configuration interface for safety devices. The Safety Supervisor Object centralizes and coordinates application object state behavior and related status information, exception status indications (alarms and warnings) and defines a behavior model which is assumed by objects belonging to safety devices.

9.5.2.10.2 Safety Validator Object (Class ID: 0x3A)

The Safety Validator Object contains the information necessary to coordinate and maintain reliable safety connections between client and server safety applications. The primary role of the Safety Validator Object is to act as a safety transport manager of multiple low-level CIP connections that together form a complete safety connection.

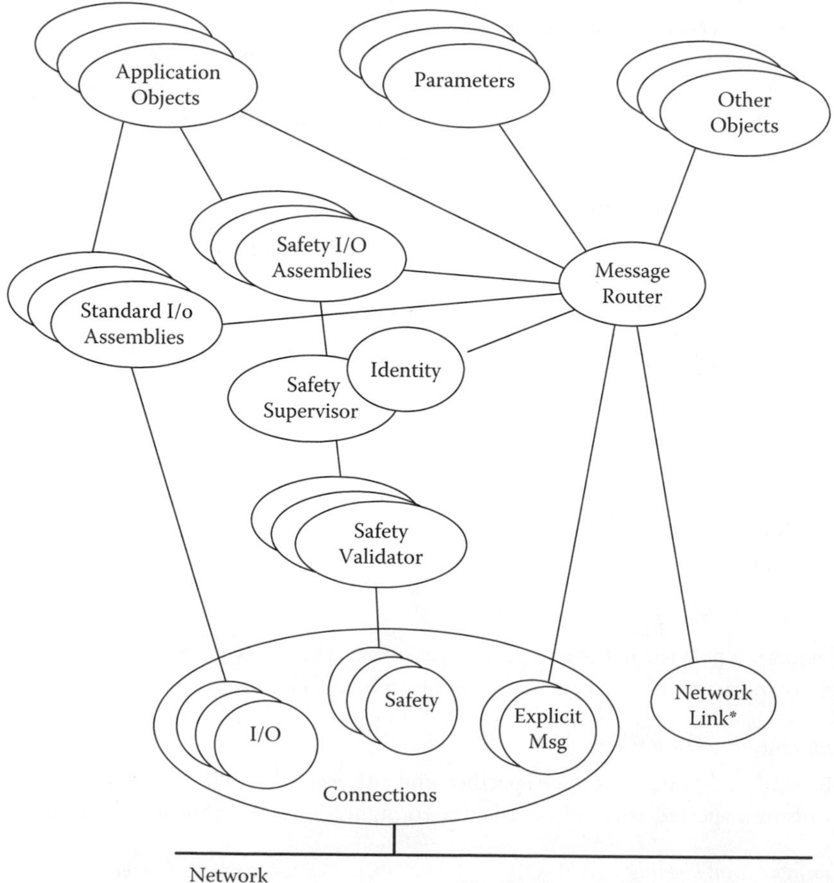

FIGURE 9.62 Safety device objects. *Note*: * DeviceNet, ControlNet, Ethernet, and SERCOS.

9.5.2.11 CIP Safety on Sercos

9.5.2.11.1 What Is Sercos?

Sercos (*SE*rial *R*eal-time *CO*mmunication *S*ystem), the digital drive interface approved as international standard IEC 61491 [56] in 1995, is optimized for high-speed deterministic motion control, where the exact synchronization of multiple drives is required. Sercos has become a globally accepted real-time networking standard for demanding motion control applications over the last decade. Sercos has outstanding technical features like real-time capabilities, high performance, noise immunity, and a very large variety of products and suppliers. Sercos not only defines the protocol structure but also includes an ample variety of profile definitions (parameters and functionalities), which are already successfully used in a large number of applications. Sercos is supported and maintained by Sercos International [57].

The third-generation Sercos (Sercos III) combines the proven mechanisms of Sercos interface with Ethernet's physics and protocol. Typical Sercos III networks use a double-ring structure that provides media redundancy with fast switch over. In addition to the ring structure, a linear structure is also possible (see Figure 9.63). A Sercos ring or line structure consists of one master and multiple slaves—drives, I/O, and sensors. Multiple rings can be used in a network to realize distributed and hierarchical network structures.

The communication is based on a time-slot protocol using fixed and distinct communication cycles. A communication cycle is divided into two channels with a timing control.

In the real-time channel, collective Sercos III telegrams are transferred as broadcast data. This increases the bandwidth and improves the protocol efficiency. The addressing of the Sercos III devices is achieved by predefined addresses or by addresses assigned by the master (remote addressing). Sercos III telegrams are processed on the fly to reduce delay times in a network.

In the non-real-time channel, any non-real-time Ethernet frames can be sent as individual telegrams to any device in the network. The addressing for this is carried out directly via the MAC addresses allocated to the master and slave devices.

9.5.2.11.2 CIP Safety on Sercos

CIP Safety on Sercos extends the functionality of the Sercos III real-time communication system to support both safety-relevant and non-safety relevant data transmission over one single network.

The CIP Safety Stack (CSS) is implemented corresponding to the CIP Safety specification without modifications. Thus, CIP Safety on Sercos uses the safe message format and the CIP Safety objects and services of the CIP Safety specification, as well as the same CRC polynomials and algorithms. For this reason, it is possible for the CIP Safety Adaptation Layer (CSAL) as well as the Sercos III subordinate communication system, including the Sercos Messaging Protocol (SMP), to be viewed as part of the non-safety-relevant transfer (see Figure 9.64).

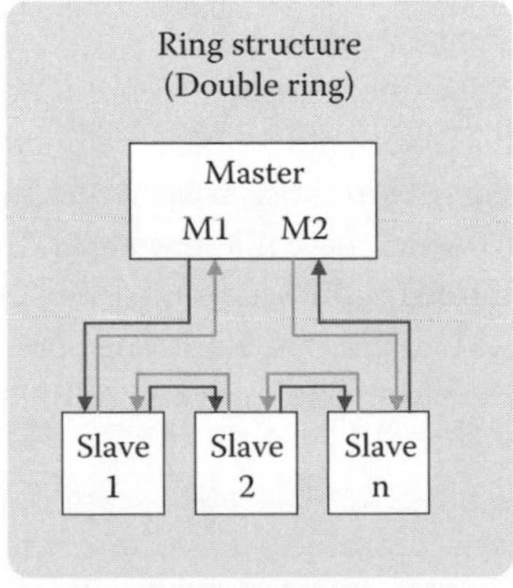

With hardware redundancy

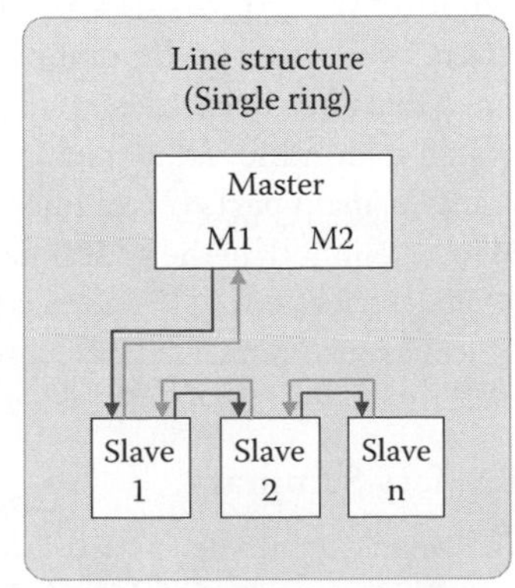

Without hardware redundancy

FIGURE 9.63 Sercos III ring and line topology.

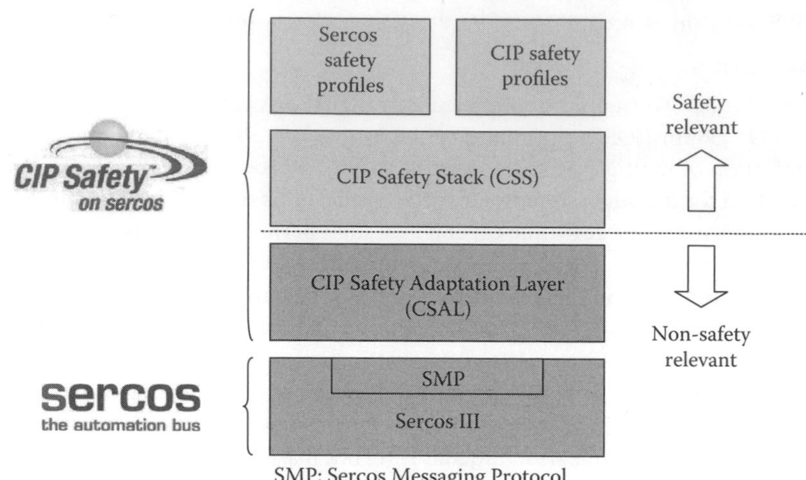

FIGURE 9.64 CIP Safety on Sercos—Structure.

The main task of the CSS in a CIP Safety Device is to create and process CIP Safety messages using cyclic process data and to correspondingly specify communication errors with the help of different error recognition measures. A CIP Safety Device can be a tool, an originator, or a target. In addition to the process data, the CSS must also process configuration data.

Because Sercos III does not provide any transport connections, an adaptation layer is required in order to fill the gaps between CIP Safety and Sercos III. An important task is assigned to the SMP, which is located above the cyclic Sercos III connections. It offers the CSAL services in order to send messages, which can be of any length, from a producer to the consumers. In order to transport these data, the SMP uses data containers with a fixed length in a cyclic connection. Messages that are larger than one data container are split up into several fragments and transferred subsequently. Messages with a higher priority can interrupt the transfer of long fragmented messages with lower priority. This means that an effective multiplexing of several logical communication channels in a single transport container is possible. The SMP is part of the non-safety-relevant transfer. Mechanisms and measures for the data integrity of CIP Safety are not influenced.

With Sercos III and CIP Safety, the implementation of a wide range of topologies is possible. The range stretches from structures with a central safety control to completely decentralized solutions without any safety control. With Sercos III and CIP Safety, it is also possible to route safe data beyond the limits of a Sercos network and also beyond non-safety-relevant participants.

CIP Safety on Sercos has been integrated into the CIP Specification by extending Volume 5 (CIP Safety) and through some minor adjustments to Volume 1 (CIP Common). Since the safety protocol was not changed at all, the main parts of Volume 5 were changed only in those areas where modified wording was needed to accommodate the additional network. However, there is a new appendix in Volume 5 (Appendix G) that describes how CIP Safety is being transported on the Sercos transport layer. Since the CIP Safety protocol as such remained unchanged, most of the adaptation work was done on the Sercos side, which is reflected in the CIP Safety on Sercos specification available from Sercos International [57].

9.5.2.12 CIP Safety Summary

CIP Safety is a scalable, routable, network-independent safety protocol based on extensions to the CIP architecture. This concept can be used in solutions ranging from device-level networks such as DeviceNet to higher-level networks such as EtherNet/IP. Designing network independence into CIP Safety allows multinetwork routing of Safety Connections. Functions such as multinetwork routing and multicast messaging provide a strong foundation that enables users to create the rapidly responding local cells and interconnect remote cells that are required for today's safety applications. CIP Safety's design also enables expansion to future network technologies as they become available.

9.5.3 Integration of Non-CIP Networks

9.5.3.1 Integration of Modbus into CIP

9.5.3.1.1 Overview

With the advent of the Modbus translator, customers can take advantage of EtherNet/IP and Modbus capabilities in the same network. The simplicity of the Modbus protocol is combined with the unique values of EtherNet/IP. The relationship is further strengthened because the Modbus/TCP protocol and EtherNet/IP are both based on standard Ethernet technology. Both protocols can coexist in the same network because they both operate over standard Ethernet. Migration of existing products is enabled with no custom hardware required.

By establishing the Modbus Integration SIG within the ODVA, the seamless connection between the Modbus protocol and EtherNet/IP was enabled. Volume 7 [61], Integration of Modbus Devices into CIP, was developed within the SIG and later approved by the ODVA. Edition 1.0 of Volume 7 supports translation between EtherNet/IP and Modbus/TCP. The Modbus Serial protocol support was added to Volume 7 in Edition 1.1 and approved by the ODVA at the beginning of 2008. Clarifications and updates to Volume 7 are being considered to continually improve the volume.

The Modbus protocol and the EtherNet/IP protocol make up the majority of the installed Ethernet-based device level products to date. Both are widely accepted standards with strong international organizations behind them and solid membership and participation from those organizations. Volume 7 links the two protocols together.

The Modbus Integration SIG was formed in May 2007 with the purpose of creating the Modbus translator. More than 20 different companies belong to the Modbus Integration SIG. The diverse membership of the SIG insured an unbiased specification was developed. The diversity of the SIG membership brought all perspectives into the development of the specification from the Modbus target device developer to the PLC manufacturer to the EtherNet/IP target vendor to the CIP Originator designer. All aspects of the CIP and Modbus networks were and are represented.

The development of Volume 7 came with the requirement that the use of the Modbus translator would not force changes to existing Modbus target devices or the EtherNet/IP target devices. Also a requirement for the development was to minimize the impact to CIP Originators. These requirements were met. The impact to the CIP Originator was focused on the need to support both the Modbus/TCP and Ethernet/IP protocol on the same physical Ethernet port should a customer desire to do so.

The Modbus translator can be implemented in the CIP Originator or as a CIP router; that is, the translator can be in a PLC or as a stand-alone device. The Modbus translator effort and consequently Volume 7 of the CIP suite of protocols were targeted at CIP Originator developers, CIP router developers, and Modbus vendors. The first two audiences are obvious. The Modbus device vendor is targeted so that this vendor can better understand how their device can be more easily integrated into the CIP to Modbus solution.

As the name states, the Modbus translator translates CIP objects and services into Modbus messages and function codes. A CIP Originator communicates with and controls Modbus target devices through the Modbus translator. To the CIP Originator, the Modbus target devices appear as CIP target devices. The Modbus target devices believe that they are being controlled by a Modbus client. The translation is transparent.

The user would place a stand-alone Modbus translator in the user's CIP-based network and then connect Modbus target devices to the Modbus side of the translator. The other side of the translator would be connected to the CIP Originator.

The user also has the option to use a Modbus translator as a module inside the CIP Originator. In this case, the user would talk EtherNet/IP across the backplane to the translator. The Modbus target devices connect externally to the PLC-based translator.

A third approach places the Modbus translator and the Modbus/TCP target devices on the same network as the EtherNet/IP target devices using the shared cabling. The EtherNet/IP traffic destined for the Modbus target devices would be sent to the Modbus translator, and the Modbus translator would send the Modbus/TCP traffic over the same network to the Modbus/TCP target devices. The Modbus/TCP target devices would reply to the Modbus translator and then the translator will translate the Modbus traffic into EtherNet/IP traffic sent to the CIP Originator.

In all three cases, the CIP-to-Modbus translation allows communication using I/O Connections as well as Explicit Messaging. When Explicit Messaging is used, the translator may offer two types of access to the data in the Modbus device, through publicly defined instances for Modbus data in the Assembly Object using Common Services or through the Modbus Object using object-specific services.

Logically, in the OSI model of the TCP/IP stack, the Modbus translation on Ethernet sits above Modbus/TCP application in the application layer (Layer 7) and below the CIP application layers. The Modbus translator on Ethernet is on the same level as the EtherNet/IP encapsulation. The lower layers of the stack are the standard TCP/IP stack layering.

Changes to the CIP Originator were limited. Port types were added for Modbus/TCP and Modbus Serial devices with associated port names and numbering. The Identity Object was extended to include Modbus-specific information, and a Modbus Object was developed and included in the CIP Object Library. Status codes were added to identify errors and status items that are specific to Modbus. Other status codes were modified to better support Modbus and EtherNet/IP together.

All the capabilities of Modbus translator are detailed in Volume 7, *Integration of Modbus Devices into CIP*, within the CIP suite of protocols. The updated Volume 1, Common Industrial Protocol (CIP) Specification, includes support for the Modbus translation as does the updated Volume 2, EtherNet/IP Adaptation of CIP.

Customers now have greater opportunity to improve their automation networks by being able to incorporate Modbus devices into their existing CIP Networks seamlessly, especially in the case of EtherNet/IP and Modbus TCP devices on the same physical network. Customers have a broader range of products available to them and can mix and match features they need from a device without being restricted to a single protocol. Risk to the buyer is reduced by the availability of greater choices to use devices based on EtherNet/IP and Modbus/TCP.

9.5.3.1.2 *Object Extensions*

The definition of the Modbus translation has resulted in extensions of existing objects as well as the creation of new objects.

9.5.3.1.2.1 Extension of Existing Objects The main extension for the Modbus integration is within the Identity Object, which has been enhanced with an attribute containing Modbus-specific identity information plus a translation definition for standard CIP ID information.

A set of Assembly Object and Parameter Object Instances have been defined that mirror Modbus-specific data tables (Holding Registers, Input Registers, Coils, and Discrete Inputs). These instance numbers are now represented by 32-bit numbers (UDINT) to accommodate the four data set ranges represented by 16-bit numbers each.

9.5.3.1.3 *Additional Objects*

Volume 7 defines two additional objects that are found only on Modbus devices, the Modbus Object, and the Modbus Serial Link Object.

9.5.3.1.3.1 ***Modbus Object (Class ID: 0x44)*** The Modbus Object provides an interface to Modbus data existing in a CIP device. Within a CIP to Modbus translator, the Modbus object provides an interface to the data and functions within a target Modbus device.

No Instance Attributes are defined for this object; instead, the Modbus Object defines services that mirror Modbus Function Codes along with their associated address data.

9.5.3.1.3.2 ***Modbus Serial Link Object (Class ID: 0x46)*** The Modbus Serial Link Object is used for configuration of a Modbus serial data communication channel and includes link-specific counters and status information for the port. Each instance of this object represents the client portion of a Modbus serial channel, which allows the Modbus translator to read/write data with external Modbus serial servers.

9.5.3.2 Integration of IO-Link into CIP

IO-Link, standardized as IEC 61131-9, is a communication standard that extends the 24 VDC I/O interface of IEC 61131-2 [3] to allow serial communication between I/O devices (IO-Link Masters) and sensors and actuators (IO-Link Devices).

An ODVA SIG, established in fall 2012, is currently working on the definition of the integration of IO-Link Masters and Devices into CIP. The completion of this work is expected for late 2014/early 2015.

9.5.4 CIP Energy

The optimization of energy usage is a natural expansion of ODVA's application coverage for industrial automation. The objects that support this functionality were added to Volume 1 of the CIP Networks Library in Edition 3.12. The management of energy usage methodology described in the specification defines a set of standard attributes, services, and behaviors that will facilitate the reporting of industrial devices' use of operational energy and the control of industrial devices into and out of nonoperational energy conserving states.

Four objects have been defined as of the publication of this text. A *Base Energy* Application Object standardizes access to the most basic data and services common to the various energy resources used in industry. The Base Energy Object also provides the means to aggregate energy information at various levels of the enterprise and present these data consistently at all levels: from the asset level at the bottom of the energy usage tree, up through the machine and process level to the system level, and ultimately to the production and enterprise domains. Two resource type application objects are defined. An *Electrical Energy Object* provides electrical energy data reporting capabilities and diagnostics for the electrical energy consumers and producers found within the various levels of an industrial facility. A *Nonelectrical Energy Object* provides unified reporting of energy consumption and production of nonelectrical energy data such as gas and steam. A *Power Management Object* provides standardized attributes and services to support the control of devices into and out of paused or sleep states. An overview of these objects in the overall construct of CIP is shown in Figure 9.65.

What follows is a brief introduction to the objects involved in CIP Energy. More information can be found in Refs. [58,59].

9.5.4.1 Additional Objects

Four additional objects have been defined for energy management so far, the Base Energy Object, the Electrical Energy Object, the Nonelectrical Energy Object, and the Power Management Object.

9.5.4.1.1 Base Energy Object (Class ID: 0x4E)

The Base Energy Object acts as an *Energy Supervisor* for energy implementations in CIP. It provides energy mode services and can provide aggregation services for aggregating energy values up through the

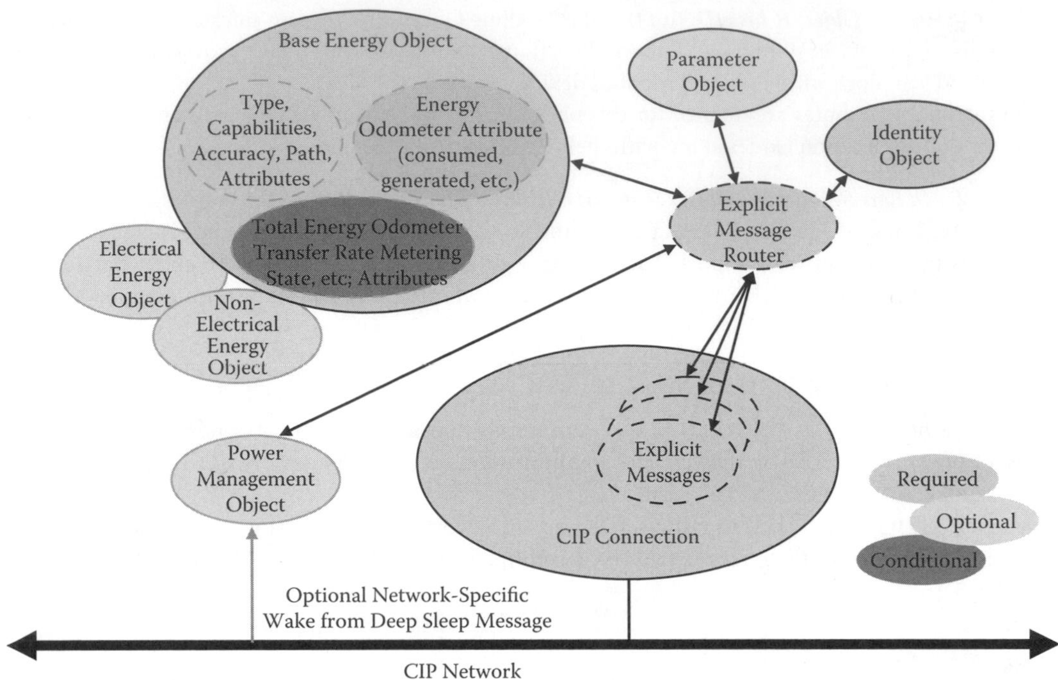

FIGURE 9.65 CIP Energy Objects.

various levels of an industrial facility. It also provides a standard format for reporting energy metering results. The object is independent of the energy type and allows data and functionality specific to the energy type to be integrated into an energy system in a standard way.

Multiple instances of the Base Energy Object may exist in a device. For instance, an electric power monitor may count metering pulse output transitions of a separate metering device. The count of such transitions, represented by a Base Energy Object instance, would reflect the energy consumption measured by the separate device. As another example, a device may act as a proxy for the energy consumed by a number of simple devices, where each device is associated with a separate instance of the Base Energy Object. An instance of the Base Energy Object may exist as a stand-alone instance, or it may exist in conjunction with an instance of either the Electrical Energy Object or the Nonelectrical Energy Object. If an instance of either the Electrical or Nonelectrical Energy Object is implemented in a device, it must be associated with a Base Energy Object instance in the device (i.e., it is a child of the Base Energy Object instance).

The object definition allows for creating five types of devices: Energy Measured, Energy Derived, Energy Proxy, Energy Rate Fixed, and Energy Aggregated devices.

- Energy Measured devices are devices such as a power monitor that measures voltage, current, phase angle, etc., and calculates power and energy.
- Energy Derived devices are devices such as an overload relay that measures motor current, assumes a value of motor voltage, and then derives the value of power and energy.
- Energy Proxy devices are devices such as a controller with discrete outputs that control external devices, each of which has an associated user-provided energy transfer rate that the controller uses to calculate the energy used based on the state of the output.
- Energy Rate Fixed devices are simple devices that report a nominal or user-defined energy value when operating and zero when in a nonoperating state.
- Energy Aggregated devices are devices that can collect energy usage of *child* devices and report them together as an aggregate value.

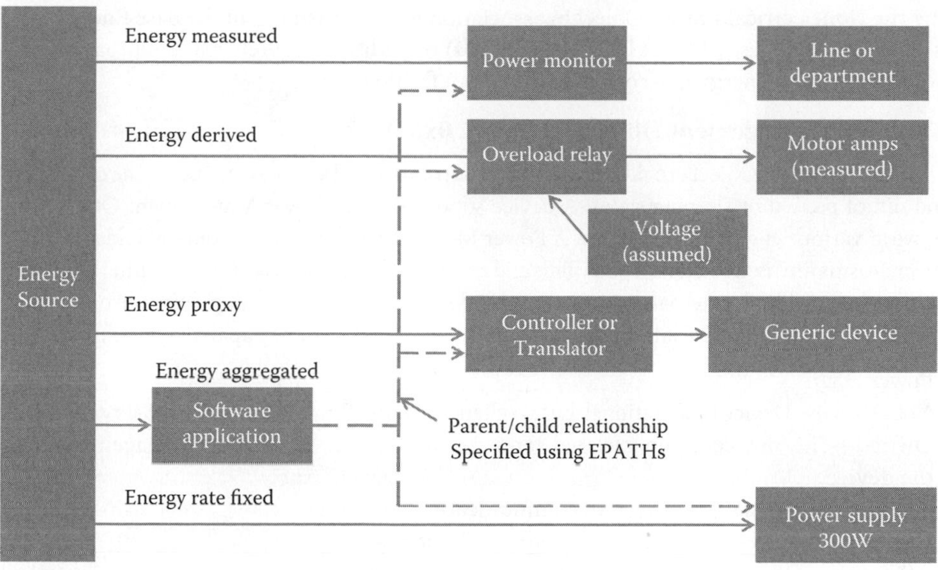

FIGURE 9.66 Types of energy-related devices.

Array [4]	Array [3]	Array [2]	Array [1]	Array [0]
Terawatt-hours (kW h × 10^9)	Gigawatt-hours (kW h × 10^6)	Megawatt-hours (kW h × 10^3)	Kilowatt-hours (kW h)	Watt-hours (kW h × 10^{-3})

FIGURE 9.67 Energy Odometer

Figure 9.66 illustrates these devices.

Energy Odometers are defined to report large amounts of energy usage in a manner similar to a car odometer or a typical power meter on a residential home. The values are in multiples of kilowatt-hours as shown in Figure 9.67:

The Base Energy object may contain information about Electrical and Nonelectrical Energy object instances that it refers to, or it may be a stand-alone instance with no children.

9.5.4.1.2 Electrical Energy Object (Class ID: 0x4F)

The Electrical Energy Object provides unified electrical energy–specific data reporting and diagnostics for the CIP-enabled devices and processes found within the various levels of an industrial facility. Energy management is typically related to the measurement and reporting of a variety of metering results. This object provides for the consistent reporting of electrical energy data. Electrical energy is organized in a separate object to accommodate its alternating and poly-phase characteristics, which result in a collection of attributes that are unique among energy sources.

Using the Electrical Energy Object in association with an instance of the Base Energy Object (via the Associated Base Energy Object Path, Attribute 41) provides a comprehensive approach to reporting usage of electrical energy in a consistent and open fashion.

9.5.4.2 Nonelectrical Energy Object (Class ID: 0x50)

The Nonelectrical Energy Object provides unified nonelectrical energy–specific data reporting and diagnostics for the CIP-enabled devices and processes found within the various levels of an industrial facility. Energy management is typically related to the measurement and reporting of a variety of metering results. This object provides for the consistent reporting of nonelectrical energy data, including, without limitation, natural gas, fuel oil, steam, compressed air, hot water, and chilled water.

Using the Nonelectrical Energy Object in association with an instance of the Base Energy Object (via the Associated Base Energy Object Path, Attribute 41) provides a comprehensive approach to reporting usage of nonelectrical energy in a consistent and open fashion.

9.5.4.3 Power Management Object (Class ID: 0x53)

The Power Management Object provides standardized attributes and services to support the control of devices into and out of paused or sleeping states. A device supporting the Power Management Object can transition between various energy-related states. A Power Management service and optional adaptation-specific sleep mechanisms are used to control entry into and exit from energy-saving states. Within the paused states, a device may have multiple internal energy-saving modes, each with a different power consumption level.

There are six basic energy-related states for CIP Power Management-capable devices:

1. Power Off.
2. Not Owned—Device is operational, but no client *owns* the Power_Management service of the device.
3. Owned—The device is operational, and a client *owns* the Power_Management service of the device.
4. Paused—Energy-saving state; CIP communication continues.
5. Sleeping—Energy-saving state; CIP communication is suspended.
6. Resuming—Device is transitioning from Paused to Owned.

Within a particular power management state, a device's operational capabilities and power levels may remain in transition for a time until the agreed power usage level is attained.

9.6 Conformance Testing

Open specifications, such as those managed by ODVA, both provide vendors with the ability to build products that will interoperate with products from other vendors and allow users to choose products that will interoperate in multivendor systems by ensuring a common network interface for given device types. In order to achieve interoperability of devices from multiple vendors, product compliance with these open specifications is essential.

ODVA drives product compliance with the CIP Network Specifications primarily in two ways. First, each vendor is required to sign a Terms of Usage Agreement for the ODVA technology or technologies, for which they intend to make, have made, sell, or have sold products. In signing this agreement, the vendor agrees to comply with the network technology specification and meet a set of user responsibilities, including the conformance testing of developed and/or sold products. A list of authorized vendors can be found on the ODVA website.

Second, ODVA administers a vendor-independent conformance testing process. The goal of the ODVA conformance testing process is to help to ensure, to the greatest extent practicable, that products implementing ODVA technologies and standards comply with the ODVA specifications and interoperate in multivendor systems.

A cornerstone of this process is the successful completion of the ODVA conformance test at an ODVA-authorized test service provider (TSP). A full list of ODVA-authorized TSPs can be found on the ODVA website. TSPs perform conformance tests that are designed, developed, and managed by ODVA and conduct the tests in accordance with ODVA test requirements and procedures. ODVA TSPs must meet certain standards, including vendor independence, neutrality, and technical competency in networks and testing practices. The ODVA conformance test is typically a composite test comprised of three parts:

1. An automated software test that verifies the function of the network protocol. Depending on the complexity of the device, several thousand messages are transmitted to the device under test (DUT). To ensure a test that is closely adapted to the characteristics of the DUT, the manufacturer must provide a formal description of all relevant CIP features of the DUT.

2. A hardware test that examines the characteristics of the physical layer for conformance. Physical layer tests vary by network and may include product labeling, indicator operation, isolation, connectors, mis-wiring, voltage ranges, timing, etc.
3. An interoperability test that exercises the product using prescribed test scenarios designed to demonstrate the successful interoperability of the product in multivendor systems.

The automated conformance test software is a Windows®-based tool that uses a network interface card in the PC to access the DUT. It is recommended that device developers run this test in their own lab before taking devices to a TSP. The hardware test (where appropriate) and the system interoperability test involve more complex test setups that typically are not available to device developers but are documented in the test plans or other ODVA publications. The vendor of the product may, at its option, observe the test at the TSP.

As of 2013, the range of composite conformance tests available from ODVA included those for DeviceNet, ControlNet, EtherNet/IP, and CompoNet devices or family of devices as listed in the following text:

CIP Safety Nodes

- CIP Safety on DeviceNet
- CIP Safety on EtherNet/IP

CIP Nodes

- CompoNet Master
- CompoNet Slave
- ControlNet I/O Scanner
- ControlNet I/O Adapter
- DeviceNet Node
- DeviceNet Embedded Technologies
- EtherNet/IP Node
- EtherNet/IP Embedded Technologies

Infrastructure Devices

- CompoNet Repeater
- DeviceNet Power Supply

Upon the product's successful completion of the test, the TSP submits the test results to ODVA for review and final approval. Contingent on passing results from the conformance tests and other requirements of ODVA, ODVA issues a Declaration of Conformity for the product. Declarations of Conformity are posted on ODVA's website at www.odva.org.

Adjunct tests are also available from the ODVA headquarters' TSP. These adjunct tests require that the device submitted for additional testing passes the appropriate conformance test first. Adjunct tests include

- DeviceNet Semiconductor Industry (test is in addition to DeviceNet Node test)
- EtherNet/IP performance test

Passing adjunct test results are listed on the Declaration of Conformity for the device or device family.

Products that have received an official Declaration of Conformity from ODVA earn the right to use ODVA's CONFORMANCE TESTED certification marks as appropriate for the network connectivity of the product. (Refer to the ODVA Identity Guidelines on the ODVA website for more information on logo usage.) End users should check the ODVA website under *Product Compliance* for the list of ODVA-issued Declarations of Conformity or look for one of the following marks on a product (Figure 9.68).

DeviceNet	DeviceNet CONFORMANCE TESTED®
EtherNet/IP conformance tested	EtherNet/IP CONFORMANCE TESTED™
CompoNet™	CompoNet CONFORMANCE TESTED™
ControlNet	ControlNet CONFORMANCE TESTED™
CIP Safety™ on DeviceNet™	CIP Safety on DeviceNet CONFORMANCE TESTED™
CIP Safety™ on EtherNet/IP™	CIP Safety on EtherNet/IP CONFORMANCE TESTED™

FIGURE 9.68 Conformance test marks.

Abbreviations

For the purposes of this chapter, the following abbreviations apply:

Abbreviation	Meaning
ASCII	American Standard Code for Information Interchange
CIP	The Common Industrial Protocol defined by Volume 1 of the CIP Networks Library
CID	Connection Identifier
EPR	Expected Packet Rate
ISO	International Standards Organization
MAC ID	Media Access Control Identifier (another name for Network Address)
OSI	Open Systems Interconnection (see ISO 7498)
UCMM	Unconnected Message Manager
CRC	Cyclic Redundancy Check
LED	Light Emitting Diode
MAC	Media Access Control sublayer
NAP	Network Access Port
NUT	Network Update Time
RG-6	Standard for coaxial cable
SMAX	MAC ID of the maximum scheduled node
UMAX	MAC ID of the maximum unscheduled node
FTP	File Transfer Protocol. An Internet application that uses TCP reliable packet transfer to move files between different nodes
RFC	Request For Comments (RFCs)—This document series, which was launched in 1969, describes the Internet suite of protocols and related experiments. Not all (in fact, very few) RFCs describe the Internet Standards, but all Internet Standards are written up as RFCs. The RFC series is unusual in that the proposed protocols are forwarded by the Internet research and development community, acting on their own behalf, as opposed to the formally reviewed and standardized protocols that are promoted by organizations such as CCITT and ANSI [60].
TCP	Transmission Control Protocol (TCP)—An Internet Standard transport layer protocol defined in STD 7, RFC 793. It is connection oriented and stream oriented, as opposed to UDP. See also: connection-oriented, stream-oriented, User Datagram Protocol [60].
UDP	User Datagram Protocol (UDP)—An Internet Standard transport layer protocol defined in STD 6, RFC 768. It is a connectionless protocol that adds a level of reliability and multiplexing to IP. See also: connectionless, Transmission Control Protocol [60].

Terminology

For the purposes of this document, the following definitions apply:

Term	**Definition**
Allocate	In the DeviceNet context, this is the process of reserving resources of the Predefined Master Slave Connection Set in a DeviceNet node. It is associated with services of a similar name, of the DeviceNet Object Class (Class ID 0x03).
Application	Typically refers to the application layer of the ISO-OSI model. The application layer is the part of the product that performs application-specific functions. Typically, application objects that provide the desired behavior are associated with the application.
Attribute	A description of an externally visible characteristic or feature of an object. The attributes of an object contain information about variable portions of an object. Typically, they provide status information or govern the operation of an object. Attributes also may affect the behavior of an object. Attributes are divided into Class Attributes and Instance Attributes.
Behavior	Indication of how the object responds to particular events. Its description includes the relationship between attribute values and services.
Bit	A unit of information consisting of a 1 or a 0. This is the smallest data unit that can be transmitted.
Broadcast	A message that is sent to all nodes on a network. It also refers to the property of a network where all nodes listen to all messages transmitted for purposes of determining bus access/priority.
Byte	A sequence of 8 bits that is treated as a single unit.
Class	A set of objects, all of which represent a similar system component. A class is a generalization of the object, a template for defining variables and methods. All objects in a class are identical in form and behavior, but they may contain different attribute values.
Object-specific service	A service defined by a particular object class to perform a required function that is not performed by any common services. A class-specific service is unique to the object class that defines it.
Client	1. An object that uses the services of another (server) object to perform a task. 2. An initiator of a message to which a server reacts.
Connection	A logical binding between two application objects. These application objects may be the same or different devices.
Connection Identifier	Identifier assigned to a transmission that is associated with a particular connection between producers and consumers that identifies a specific piece of application information.
Connection path	Is made up of a byte stream that defines the application object to which a connection instance applies.
Consumer	A node that is receiving (i.e., consumes) data from a producer.
Consuming application	The application that consumes data.
CRC error	Error that occurs when the cyclic redundancy check (CRC) value does not match the value generated by the transmitter.
Cyclic	Term used to describe events that repeat in a regular and repetitive manner.
Datagram	A transmitted message.
Device	A physical hardware connection to the link. A device may contain more than one node.

Device Profile	A collection of device-dependent information and functionality providing consistency between similar devices of the same device type.
Drop or drop line	The cable that connects one or more nodes to a trunk cable, usually accomplished using a tap.
Encapsulation	The technique used by layered protocols in which a layer adds header information to the protocol data unit (PDU) from the layer above for purposes of carrying one protocol within another.
Ethernet	A standard for LANs, initially developed by Xerox, and later refined by Digital, Intel, and Xerox (DIX). In its original form, all hosts are connected to a coaxial cable, where they contend for network access using a Carrier Sense Multiple Access with Collision Detection (CSMA/CD) paradigm. See also IEEE 802.3, Local Area Network [60].
Expected Packet Rate	A misnomer, the Expected Packet Rate (EPR) is basic interval at which a connection transmits its data.
Fixed tag	A two-byte field in a ControlNet Lpacket that identifies unconnected or station management services the node is expected to perform. The first byte is the specific service code, and the second byte contains the MAC ID of the destination node.
Frame	See MAC Frame.
Generic tag	A three-byte field in a ControlNet Lpacket that serves as the Connection Identifier (CID). It is associated with a specific piece of application information.
Guardband	It is the portion of ControlNet bandwidth that is allocated for the transmission of the moderator frame.
Instance	The actual physical presentation of an object within a class. Identifies one of potentially many objects within the same object class.
Keeper	Object responsible for holding and distributing the Connection Originator schedule data for all Connection Originator devices on a ControlNet Network.
Link or Data Link	Refers to the Data Link layer of the ISO/OSI model.
Lpacket	On ControlNet, the Lpacket (or link packet) is a portion of the MAC Frame where application information that contains a size, control byte, tag, and link data is transmitted. There may be one or more Lpackets in a single MAC Frame.
MAC frame	A collection of MAC symbols transmitted on the network medium that contains the required message formatting/framing necessary to pass a message to another node. For example, a ControlNet MAC Frame consists of a preamble, start delimiter, source MAC ID, Lpackets, CRC, and end delimiter.
Message Router	The object within a node that distributes Explicit Message requests to the appropriate application objects.
Multicast	A packet that is sent to multiple nodes on the network.
Network	A series of nodes connected by some type of communication medium. The connection paths between any pair of nodes can include repeaters and bridges.
Network Access Port	On ControlNet, this is an alternate physical layer connection point on a permanent node that allows a temporary node to be connected to the link. The temporary node has its own network address, but simply shares the permanent node's physical layer connection to the network.
Network address	An integer identification value assigned to each node on a CIP Network.

Network status indicators — Indicators (i.e., LEDs) on a node indicating the status of the Physical and Data Link Layers.

Network Update Time — Repetitive time interval on a ControlNet Network that is used to subdivide the network bandwidth. It determines the fastest rate that real-time data can be transferred on the network.

Node — A connection to a link that requires a single MAC ID.

Object —
1. An abstract representation of a particular component within a product. Objects can be composed of any or all of the following components:
   (a) Data (information which changes with time)
   (b) Configuration (parameters for behavior)
   (c) Methods (things that can be done using data and configuration)
2. A collection of related data (in the form of variables) and methods (procedures) for operating on that data that have clearly defined interface and behavior

Object specific service — A service defined by a particular object class to perform a required function that is not performed by one of the common services. An object-specific service is unique to the object class that defines it.

Originator — The client responsible for establishing a connection path to the target.

Point-to-point — A one-to-one data exchange relationship between two, and only two nodes

Port — A CIP port is the abstraction for a physical network connection to a CIP device. A CIP device has one port for each network connection. Within the EtherNet/IP specific context, a TCP or UDP port is a transport layer demultiplexing value. Each application has a unique port number associated with it [60].

Producer — A node that is responsible for transmitting data.

Redundant media — A system using more than one medium to help prevent communication failures.

Repeater — Two-port active Physical Layer device that reconstructs and retransmits all traffic on one segment to another segment.

Scheduled — On ControlNet, these are data transfers that occur in a deterministic and repeatable manner, on preconfigured NUT-based intervals.

Segment — This term has two uses within CIP. With respect to cable topology, a segment is a length of cable connected via taps with terminators at each end; a segment has no active components and does not include repeaters. With respect to explicit messaging, segments (Logical Segments, Port Segments, etc.) are used in Explicit Messages to describe various addressing elements of devices such as Class IDs, Attribute IDs, ports, Connection Points, etc.

Serial Number — A unique 32-bit integer assigned by each manufacturer to every device. The number needs to be unique only with respect to the manufacturer.

Server — A device or object that provides services to another device (the client).

Service — Operation or function that an object performs upon request from another object.

Tap — Point of attachment on the trunk where one or more drop lines are attached.

Target — The end node to which a connection is established.

Terminator — A resistor placed at the physical extreme ends of trunk segments to prevent transmission reflections from occurring.

Transceiver — The physical component within a node that provides transmission and reception of signals onto and off of the medium.

Trunk or trunkline — The main bus or central part of a cable system, typically terminated at each end by a termination resistor.

Unconnected Message Manager	The function within a node that transmits and receives Unconnected Explicit Messages.
Unscheduled	On ControlNet, this refers to data transfers that use the unscheduled portion of the NUT.

References

1. V. Schiffer, Modular EDSs and other EDS enhancements for DeviceNet, *Proceedings of the Ninth International CAN Conference 2003*, Munich, Germany, 2003.
2. V. Schiffer, Device configuration using electronic data sheets, *ODVA 2003 Conference and Ninth Annual Meeting*, Ann Arbor, MI.
3. IEC 61131-2:2007, Programmable controllers—Part 2: Equipment requirements and tests.
4. R. Hofmann, C. Schlegel, J. Stolberg, and S. Weiher, *Controller Area Network—Basics, Protocols, Chips and Application*. IXXAT Automation, Weingarten, Germany, 2001.
5. CIP Networks Library, Volume 3, DeviceNet Adaptation of CIP, Edition 1.13, Volume 3, April 2013, ODVA, Inc.
6. IEC 62026-3:2008, Low-voltage switchgear and controlgear—Controller-device interfaces (CDIs)—Part 3: DeviceNet.
7. Chinese national standard: GB/T 18858:2003, Low-voltage switchgear and controlgear controller-device interface.
8. ISO 11898:1993, Road vehicles—Interchange of digital information—Controller area network (CAN) for high-speed communication.
9. Planning and Installation Manual, DeviceNet Cable System, Publication PUB00027, 2003, downloadable from ODVA website (http://www.odva.org/).
10. IEC 61784-5-2:2010, Industrial communication networks—Profiles—Part 5-2: Installation of fieldbuses—Installation profiles for CPF 2.
11. Robert Bosch GmbH, *CAN Specifications—Controller Area Network, Version 2.0*, Robert Bosch GmbH, Stuttgart, Germany, 1991. http://www.semiconductors.bosch.de/pdf/can2spec.pdf.
12. V. Schiffer, R. Romito, DeviceNet Development Considerations, 2000.
13. EN 62026-3:2008, Low-voltage switchgear and controlgear—Controller-device interfaces (CDIs)—Part 3: DeviceNet.
14. CIP Networks Library, Volume 4, ControlNet Adaptation of CIP, Edition 1.8, April 2013, ODVA, Inc.
15. IEC 61158:2000, Digital data communications for measurement and control—Network for use in industrial control systems.
16. ControlNet Product Developer's Guide, Rockwell Automation, Publication 9220-6.5.1, downloadable from http://literature.rockwellautomation.com.
17. CIP Networks Library, Volume 2, EtherNet/IP Adaptation of CIP, Edition 1.15, April 2013, ODVA, Inc.
18. EtherNet/IP Media Planning and Installation Manual, Publication PUB00148, downloadable from ODVA website (http://www.odva.org/).
19. IEEE 802.3:2002, Information Technology—Telecommunication & Information Exchange between Systems—LAN/MAN—Specific Requirements—Part 3: Carrier Sense Multiple Access with Collision Detection (CSMA/CD) Access Method and Physical Layer Specifications 2002.
20. RFC 768, User Datagram Protocol, 1980.
21. RFC 791, Internet Protocol, 1981.
22. RFC 793, Transmission Control Protocol, 1981.
23. Recommended IP Addressing Methods for EtherNet/IP Devices, Publication PUB00028, 2003, downloadable from ODVA website (http://www.odva.org/).
24. RFC 5227—IPv4 Address Conflict Detection. 2008, downloadable from http://www.faqs.org/rfcs/.
25. Recommended Functionality for EtherNet/IP Devices, Publication PUB00070, 2013, downloadable from ODVA website (http://www.odva.org/).

26. RFC 2236, Internet Group Management Protocol, Version 2, 1997.
27. RFC 1122, Requirements for Internet Hosts—Communication Layers, 1989.
28. RFC 1123, Requirements for Internet Hosts—Application and Support, 1989.
29. RFC 1127, Perspective on the Host Requirements RFCs, 1989.
30. RFC 826, Ethernet Address Resolution Protocol: Or converting network protocol addresses to 48.bit Ethernet address for transmission on Ethernet hardware, 1982.
31. RFC 792, Internet Control Message Protocol, 1981.
32. RFC 1112, Host extensions for IP multicasting, 1989.
33. RFC 894, Standard for the transmission of IP datagrams over Ethernet networks, 1984.
34. RFC 1171, Point-to-Point Protocol for the transmission of multi-protocol datagrams over Point-to-Point links, 1990.
35. RFC 1201, Transmitting IP traffic over ARCNET networks, 1991.
36. RFC 1103, Proposed standard for the transmission of IP datagrams over FDDI Networks, 1989.
37. EtherNet/IP Quick Start for Vendors Handbook—A Guide for EtherNet/IP Developers, Publication PUB00213, 2013, downloadable from ODVA website (http://www.odva.org/).
38. EtherNet/IP Interoperability Test Procedures, Publication PUB00095, downloadable from ODVA website (http://www.odva.org/).
39. Performance Test Terminology for EtherNet/IP Devices, Publication PUB00080, 2005, downloadable from ODVA website (http://www.odva.org/).
40. Performance Test Methodology for EtherNet/IP Devices, Publication PUB00081, 2005, downloadable from ODVA website (http://www.odva.org/).
41. IEC 62026-7:2010, Low-voltage switchgear and controlgear—Controller-device interfaces (CDIs)—Part 7: CompoNet.
42. CIP Networks Library, Volume 6, CompoNet Adaptation of CIP, Edition 1.7, April 2010, ODVA, Inc.
43. H. Harada, CompoNet: Innovations for high performance sensor–actuator applications, *CIP Networks Conference & 12th Annual Meeting, Presented Papers*, Englewood, CO, 2007, downloadable from ODVA website (http://www.odva.org/).
44. FDT Group Homepage. http://www.fdtgroup.org/.
45. H. Ri, Overview of CompoNet developer's toolkit, Presented at the *ODVA 2007 CIP Networks Conference and 12th Annual Meeting*, Englewood, Colorado, 2007, downloadable from ODVA website (http://www.odva.org/).
46. H. Ri, Design considerations for CompoNet processors, Presented at the *ODVA 2007 CIP Networks Conference and 12th Annual Meeting*, Englewood, Colorado, 2007, downloadable from ODVA website (http://www.odva.org/).
47. CIP Networks Library, Volume 1, *Common Industrial Protocol*, Edition 3.14, April 2013, ODVA, Inc.
48. IEEE 1588-2008, Standard for a precision clock synchronization protocol for networked measurement and control systems.
49. CIP Networks Library, Volume 5, CIP Safety, Edition 2.7, April 2013, ODVA, Inc.
50. BG Fachausschuß Elektrotechnik, Draft proposal test and certification guideline, safety bus systems, May 28, 2000.
51. EN 50159-1:2001, Railway applications, communication, signaling and processing systems.
52. IEC 61508:1998, Functional safety of electrical/electronic/programmable electronic safety-related systems.
53. GS-ET-26, Grundsatz für die Prüfung und Zertifizierung von "Bussystemen für die Übertragung sicherheitsrelevanter Nachrichten," Fachausschuss "Elektrotechnik" Prüf- und Zertifizierungsstelle im BG-PRÜFZERT, 2002, downloadable from http://www.bgetem.de/.
54. ISO 13849-1:2006, Safety of machinery—Safety-related parts of control systems—Part 1: General principles for design.
55. D.A. Vasko, S.R. Nair, CIP safety: Safety networking for the future, *Proceedings of the Ninth International CAN Conference*, Munich, Germany, 2003. http://www.cancia.org/.

56. IEC 61491:1995, Electrical equipment of industrial machines—Serial data link for real-time communications between controls and drives.
57. Sercos International, http://www.sercos.de/EUROPE-English.15.0.html.
58. ODVA Technology at a Glance, Technical Approach to Optimization of Energy Usage, Publication PUB00265, 2012, available through ODVA website (http://www.odva.org/).
59. ODVA white paper, Optimization of Energy Usage (OEU™): ODVA's Vision of Energy Optimization for the Industrial Consumer, Publication PUB00246, 2011, downloadable from ODVA website (http://www.odva.org/).
60. RFC 1392, 1993, Internet Users' Glossary.
61. CIP Networks Library, Volume 7, Integration of Modbus Devices into the CIP Architecture, Edition 1.6, April 2013, ODVA, Inc.
62. IEC 61131-3:2003, Programmable controllers—Part 3: Programming languages.
63. IEEE 802.3:2000, ISO/IEC 8802-3:2000, IEEE Standard for Information technology—Local and metropolitan area networks—Part 3: Carrier sense multiple access with collision detection (CSMA/CD) access method and physical layer specification.
64. ISO/IEC 7498-1:1994, ISO/IEC Standard—Information technology—Open Systems Interconnection—Basic Reference Model.

10

Modbus Protocol*

10.1 Overview 10-1
History • Premise • General • Features
10.2 Modbus Protocol 10-4
Protocol in the Abstract • Client/Server in General • Application Protocol Data Unit • Client/Server Interactions • Data Types • Data Models • Function Code Categories • Function Codes • Exception Codes • Transmission Order
10.3 Modbus over Serial Line 10-12
General • Modbus Serial Line PDU • Modbus Transmission Modes • RTU Transmission Mode • ASCII Transmission Mode
10.4 Modbus/TCP 10-20
General • TCP/IP Encapsulation • Role of the Transaction ID • Assigned Port • Protocol ID • Unit ID and Gateways • TCP as a Streaming Protocol • Connection Establishment and Management • State Machines • Modbus/TCP Flowcharts • [Informative] TCP Interfaces and Parameterization
10.5 Gateways and Similar Devices 10-28
General • Interpretation • Mappings • Tunneling and Bridging
10.6 Modbus as Part of the CIP Stack, in ODVA 10-29
10.7 Modbus on Other Stacks 10-34
10.8 Conformance 10-34

Rudy Belliardi
Schneider Electric

Ralf Neubert
Schneider Electric

10.1 Overview

10.1.1 History

The Modbus™ protocol was created in 1978 by Modicon Inc. as a simple way for communicating control data between controllers and sensors using an RS232 port.

The protocol became widely adopted, quickly reaching the status of de facto standard in the industrial automation field. Today, the Modbus protocol is the single, most supported protocol among automation devices.

Schneider Electric transferred the specifications for Modbus and Modbus/TCP (Modbus over TCP/IP) to Modbus.org. See the web-based community www.Modbus.org.

10.1.2 Premise

The material presented in this chapter does not contain all the details of the Modbus protocol. The authoritative specifications are publicly available and can be found at the web-based community www.Modbus.org.

10.1.3 General

Modbus is an application layer messaging, protocol placed at OSI layer 7, for client/server communication between devices connected on different types of buses or networks. Figure 10.1 shows the messaging between one client and one server and the conventional names given to the messages at the source and the destination.

Modbus is currently transported using any of the following underlying layers:

- RS232, RS422, RS485
- TCP/IP
- Modbus Plus, a token-passing network
- Many other stacks over a variety of media (e.g., fiber, radio, cellular)

This is shown in Figure 10.2.

A network example is illustrated in Figure 10.3; gateways are used to connect the underlying layers, HMIs are human–machine interface stations, and PLCs are programmable logic controllers.

The Modbus messaging/transaction takes place exchanging application protocol data units ([APDUs] described later). While the APDU is the same for all underlying layers, the client/server mechanism employed by Modbus takes full advantage of the different possibilities offered by the currently used underlying layer (one to one, one to many, error checking, pipelining, etc.).

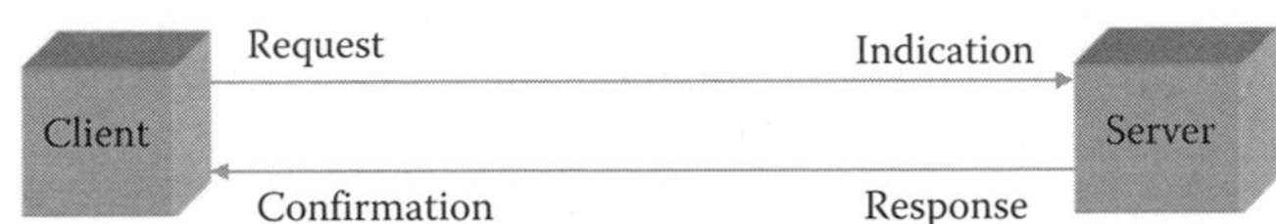

FIGURE 10.1 Client–server messaging.

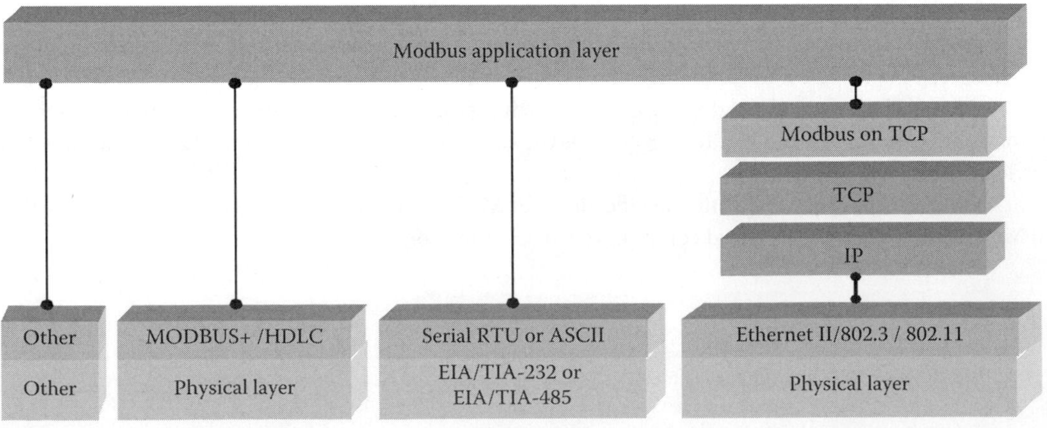

FIGURE 10.2 Modbus communication stacks.

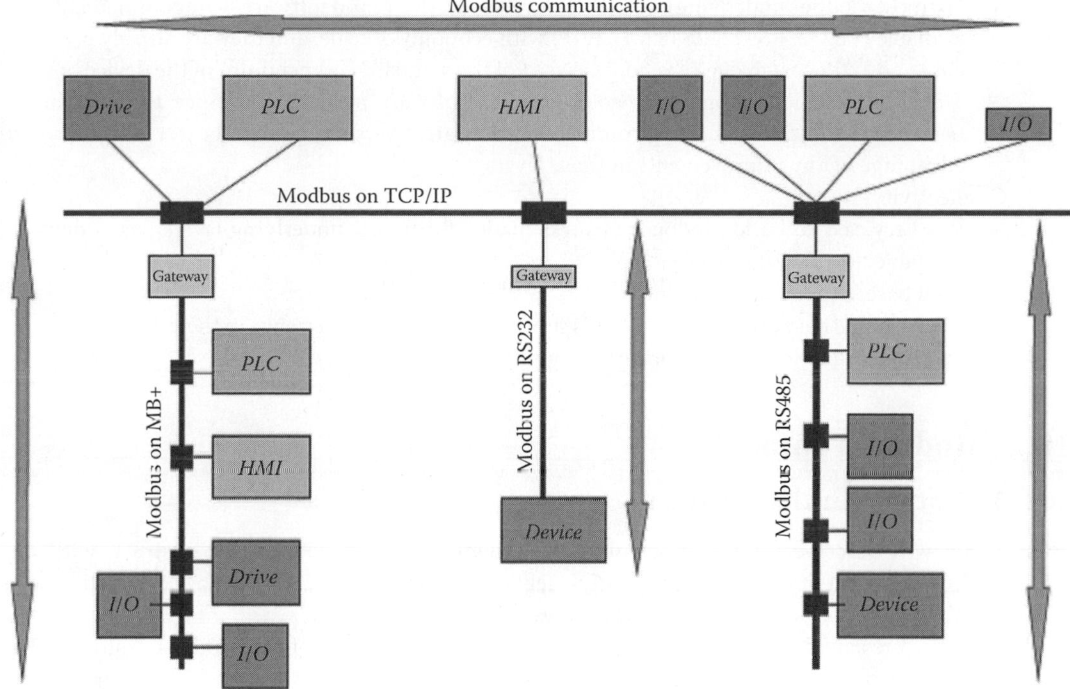

FIGURE 10.3 Example of Modbus network.

10.1.4 Features

The popularity of Modbus stems from a commitment to simplicity, recognizing that industrial automation applications are the most diverse and that there are advantages in delegating the handling of diversity to the applications.

Some features are as follows:

- Simple to implement
  - Can be implemented in days
- Small footprint
  - Can run on any computer system, CPU, or microcontroller, allowing the networking of simple devices
- Scalable in complexity, scope, and range
  - Complexity: A device that has only a simple purpose needs only to implement the needed message types.
  - Scope: From real-time sensing to production data.
  - Range: A collection of devices using Modbus/TCP to communicate can range up to 10,000 or more on a single switched Ethernet network.
- It is simple to administer and enhance
  - There is no need to use complex configuration tools when adding a new node to a network.
- Open and low cost
  - The specifications are freely downloadable from the www.Modbus.org website.
  - In addition to being a de facto standard, it has been standardized by the International Electrotechnical Commission (IEC), the Standardization Administration of China (SAC), and relevant parts by the Semiconductor Equipment and Materials International (SEMI) association.

    - For most Modbus underlying layers, these layers—hardware and software—are commercially available (RS232, RS422, RS485, TCP/IP) with economy of scale, and they are already present at no extra cost to device vendors as part of the standard compendium of the devices.
    - There is no vendor-proprietary equipment or software needed, and since there is no need to modify the underlying commercial standard layers, the protocol can readily take advantage of any improvements in those layers.
- Connectivity
    - It is very easy to build Modbus networks made of different underlying layers, via widely available flexible gateways.
- Installed base, experience, and tools
    - The installed base of Modbus devices is substantial, so is the experience in their deployment and the availability of monitoring/debugging tools.

10.2 Modbus Protocol

10.2.1 Protocol in the Abstract

This section will describe the full protocol and its client/server behavior in the abstract, with no particular instantiation of the layers. The most widely used instantiations, Modbus over serial line and Modbus/TCP, will be described in subsequent sections.

The material presented in this section does not contain all the details of the Modbus protocol. The authoritative specifications are publicly available and can be found at the web-based community www.Modbus.org.

10.2.2 Client/Server in General

The transfer of information between a Modbus client and a Modbus server is initiated when the client sends a request to the server to transfer information, to execute a command, or to perform one of many other possible functions.

After the server receives the request, it executes the command and/or prepares the required data. The server then responds to the client by acknowledging that the command is complete and/or providing the requested data.

The system response time is limited by two main factors: the time required for the client to send the request/receive the response and the ability of the server to answer within a specific amount of time.

A device may implement a Modbus client service, a Modbus server service, or both, depending on the requirements of the device and in some cases on the particular Modbus underlying layer (e.g., in Modbus Plus, both services are always present at the same Modbus address/node, and the client or server role is directed by a token-passing mechanism).

Depending on the underlying layer, a client may be able to initiate Modbus messaging requests to one or more servers, and a server may be able to respond to requests received from one or more clients. These possibilities will be detailed later when the different transport mechanism/underlying layers will be discussed; the distinction between the various cases is not relevant at this time, since the Modbus application protocol is the same in all cases.

A typical HMI or SCADA application implements a client service to initiate communications with PLCs and other devices for information gathering. An I/O device implements a server service so that other devices can read and write its I/O values. Because this I/O device does not need to initiate communications, it does not implement a client service.

A PLC typically implements both client and server services so that it can initiate communications to other PLCs and I/O devices and can respond to requests from other PLCs, SCADA, HMIs, and other devices.

Figure 10.4 illustrates the preceding discussion.

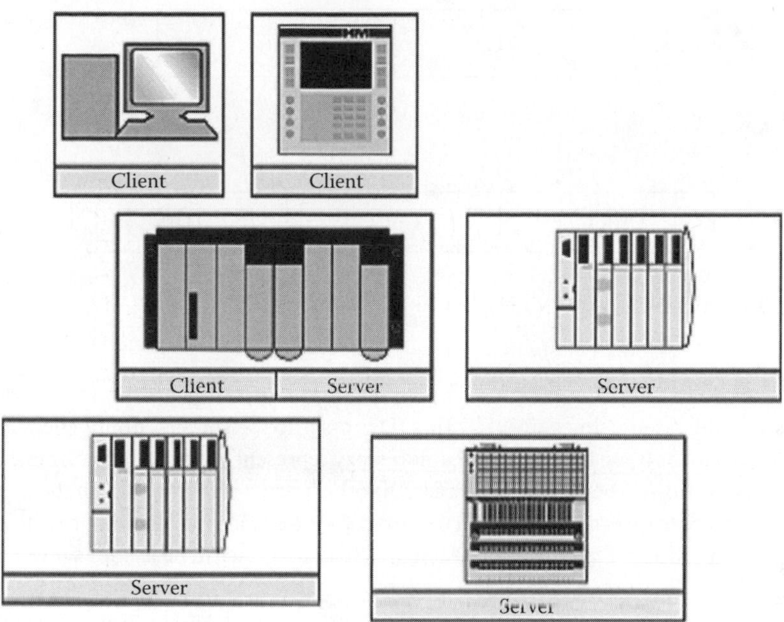

FIGURE 10.4 Clients, servers, and their possible colocation.

10.2.2.1 Modbus Client Service

A device that implements the Modbus client service can initiate Modbus messaging requests to another device that implements a Modbus server. These requests allow the client to transact data with and/or send commands to the remote device.

10.2.2.2 Modbus Server Service

A device that implements the Modbus server service can respond to requests from any Modbus client. The Modbus server service allows a device to make all its internal and I/O data available to remote devices for both reading and writing and allows for the execution of other commands.

10.2.2.3 Specialized Client and Server Services

There are many devices on the market that package together useful Modbus functionality and make it available via easy configuration. One of them is a sophisticated client known as *scanner*; it allows to configure and execute repetitively read/writes of Modbus networked field devices, with handling of retries, all in an optimized fashion. These devices are applications of the protocol and provide higher-level services.

10.2.2.4 Client and Server Messages

The messages exchanged by Modbus client and server services are defined by a structure called APDU, described in the next clause.

10.2.3 Application Protocol Data Unit

10.2.3.1 Structure

While the Modbus application protocol payload, also known as Modbus protocol data unit (PDU), is made of two fields, the code and the data fields, as shown in Figure 10.5, it is convenient to describe the protocol considering at the same time also the unit identifier (ID) field. The aggregate structure is called Modbus APDU, and it is shown in Figure 10.6.

The reason for considering the APDU is that it is common across the various Modbus transports.

FIGURE 10.5 Modbus PDU format.

Unit ID	Code	Data

FIGURE 10.6 Modbus APDU format.

10.2.3.2 Unit ID Field

Server devices are addressed using a unit ID. This ID is assigned to servers, not to clients.

Depending on the underlying transport, the unit ID may represent the full address or just a part of a segmented address leading to a server. In the latter case, the other parts are represented in the transport mechanism. The address needs to be unique across all the servers addressable by a client. In case the unit ID is just a part of a segmented address, then the unit ID assigned to a server has to be unique within that segment.

The unit ID field size is 1 octet.

Not all unit ID values represent addresses; some values of the unit ID have a special meaning, as described in Table 10.1.

A request with a unit ID field set to broadcast is sent to all the servers in the address space of that unit ID. A broadcast request does not have a response. The broadcast mode of communication will be detailed later.

In all allowed cases but the broadcast—which does not have a response—the unit ID field value in the response is the unit ID value of the addressed server (i.e., the unit ID is echoed back).

A unit ID value of 255 is a flag indicating that the server address is being fully specified by the transport. This flag is used in two situations:

1. When the server is already fully identified via the addressing in the transport layer and there is no interest in assigning to it a dedicated unit ID
2. When the server, already fully identified via the addressing in the transport layer, is actually a gateway and the service is being requested locally to the gateway itself

Both situations will be clearer when discussing transport layers and gateways.

10.2.3.3 Code Field

The code field accommodates the encoding of the service asked to the server when in a request and signals the success or failure of the requested service in a response (excluding the broadcast case that does not have a response).

Function codes are the encodings used to indicate the requested services; they have selected values in the 0x01–0x7F range; 0x00 is invalid and 0x80–0xFF are reserved for the exception mechanism, as described in the following. The function code encodings will be discussed later.

The success of a function code processing is signaled by echoing the function code value of the request in the same code field in the response.

TABLE 10.1 Unit ID Field

Broadcast	Server Addressing	Reserved	Address Fully Determined by Next Layer
0	From 1 to 247	From 248 to 254	255

TABLE 10.2 Code Field

Request	Normal Response (Response to Successful Processing)	Exception Response (Response to Unsuccessful Processing)
Function code	Echo function code in request	Exception = OR-ing of 0x80 with function code in request

The failure of a function code processing is signaled by returning the OR-ing of 0x80 with the function code value of the request in the same code field in the response. In this case, the returned value is called an exception.

The code field size is 1 octet. Exceptions are easy to detect as the code field value will have its high bit turned on.

Table 10.2 summarizes the usage of the code field.

10.2.3.4 Data Field

This field contains additional function information when in a request and the results of the function processing when in a response. For successful processing, the contents will be whatever has been requested by the function. For unsuccessful processing, the content will be an exception code.

The data field size is from 0 to 252 octets, depending on the value of the code field.

The maximum size of the data field is dictated by the size inherited from the first Modbus implementation on a serial line network, with an RS485 frame of 256 octets made of unit ID field (1 octet) + code field (1 octet) + data field + cyclical redundancy check (CRC) (2 octets).

10.2.4 Client/Server Interactions

10.2.4.1 General

From a network point of view, the interactions can be of two types:

1. Broadcast (unconfirmed)
2. Unicast (confirmed)

Interactions can succeed or fail due to many reasons. There is a distinction between failures due to communication and failures due to processing; in addition, the failures due to processing return exceptions.

The client issues a request and starts an application-specified time-out.

If the interaction is a broadcast, then the time-out is the so-called turnaround delay and the following will happen:

- No response is expected; the client simply waits for the specified time-out before issuing any other request.

If the interaction is a unicast, then the time-out is the so-called response time-out and one of the following will happen:

- There can be a communication failure, the server never receives the request, and the client eventually times out while waiting for the response.
- The server receives the request, processes it (successfully or not), and issues the response, but then there is a communication failure and the client eventually times out while waiting for the response.
- The server receives the request, processes it successfully, and issues the response, and the response is received within the time-out.
- The server receives the request, processes it unsuccessfully, and issues the response, and the response is received within the time-out.

Figure 10.7 shows a broadcast on a network having three servers. Only a subset of the possible actions can be performed via a broadcast, as detailed later, since there is no response.

Figure 10.8 shows a unicast with successful processing, returning a normal response.

Figure 10.9 shows a unicast with unsuccessful processing, returning an exception response.

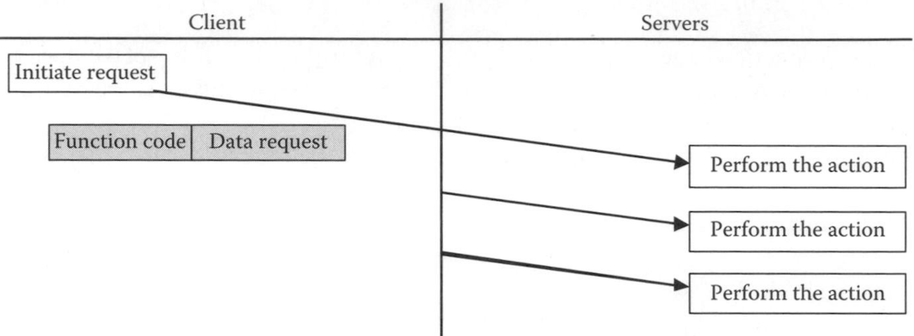

FIGURE 10.7 Broadcast.

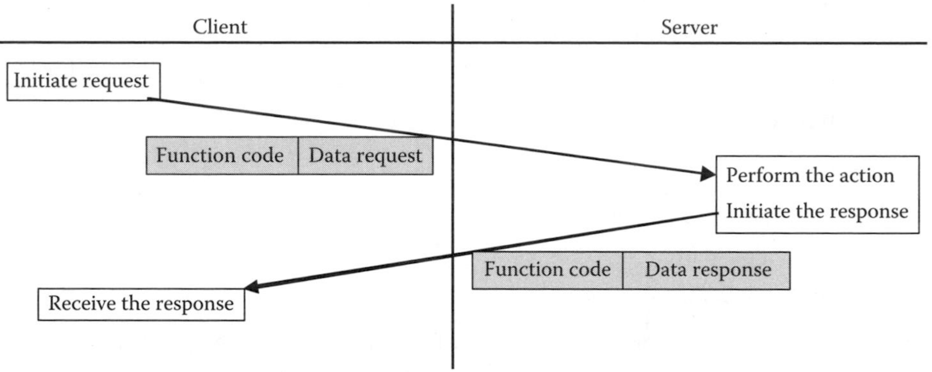

FIGURE 10.8 Unicast with normal response.

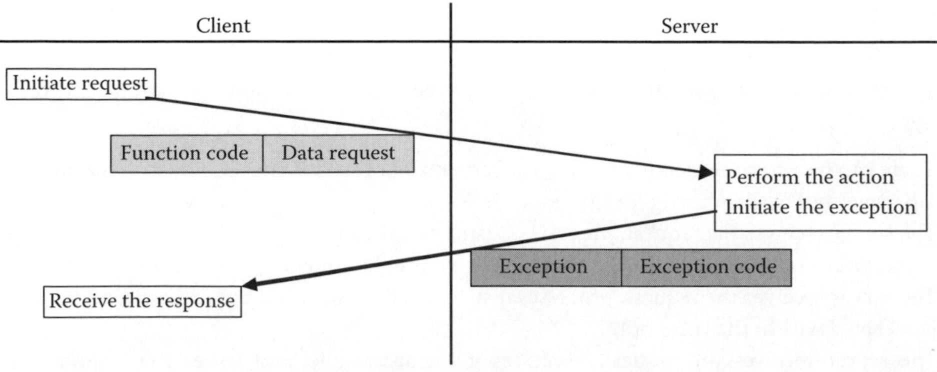

FIGURE 10.9 Unicast with exception response.

10.2.4.2 State Machines

From a client and server individual point of view, the aforementioned is illustrated using state machine diagrams, drawn following the UML standard notation as shown in Figure 10.10, with the following semantics: When a *trigger* event occurs in a system being in *State_A*, that system will transit into *State_B* only if the *guard condition* is true; upon transiting, the specified action is performed.

The client state diagram is shown in Figure 10.11 and the server state diagram is shown in Figure 10.12. With some underlying layers, the state diagrams get a little more complicated, as there are more possibilities (transaction processing in Modbus/TCP), but the protocol essentials are the same as represented here.

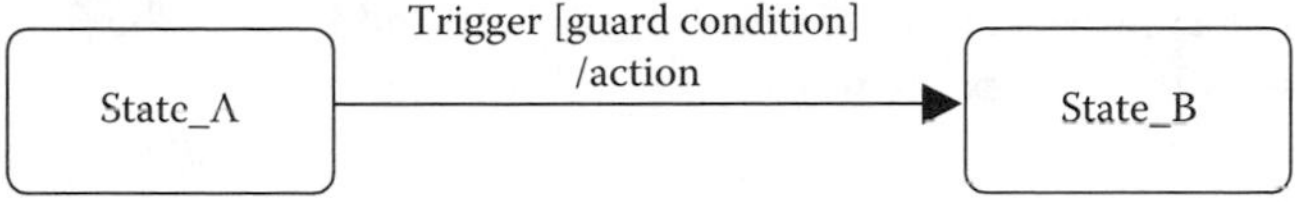

FIGURE 10.10 Syntax of state diagram representation.

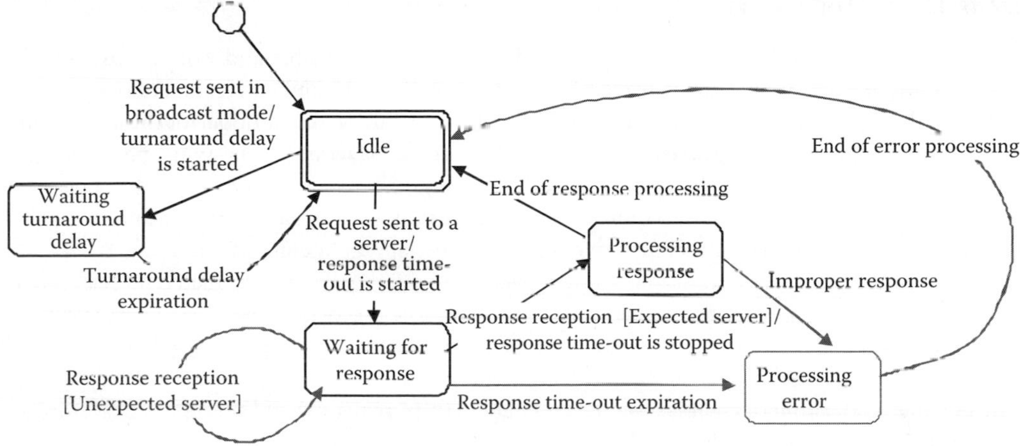

FIGURE 10.11 Client state diagram.

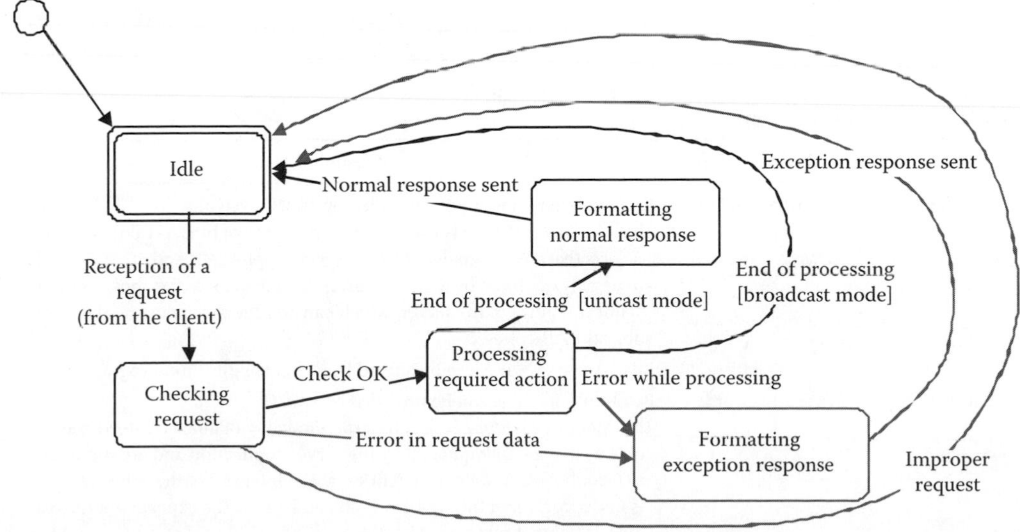

FIGURE 10.12 Server state diagram.

10.2.5 Data Types

10.2.5.1 Discrete, Coil, Input Register, and Holding Register

The most used Modbus data types are the discrete, coil, input register, and holding register, described in Table 10.3.

10.2.5.2 Record and File

The Modbus record and file data types are less known and used only with few function codes.

The Modbus record data type, from an application user (client user) point of view, is a set of contiguous registers of a specified type, characterized by the address of the first register and by the quantity of registers; in the context of this definition, the registers involved have also been called references.

The Modbus file data type is an organization of records, characterized by an unsigned number.

10.2.6 Data Models

The distinctions between inputs and outputs, and between bit-addressable and word-addressable data items, do not imply any application behavior. It is perfectly acceptable, and customary, to regard all four tables as overlaying one another, if this is the most natural interpretation on the target machine in question.

Strictly speaking, the data types refer to what is exchanged *on the wire*, and the actual physical or logical model on the device is up to the application, even if often they may map to real tables.

Figures 10.13 and 10.14 give some common but by no means exhaustive interpretations when using actual tables, respectively, as distinct memory tables and overlapping memory tables.

Discretes, coils, input registers, and holding registers are often collectively called data references or data items. For each of the aforementioned data item types, the protocol allows individual selection of 65,536 data items, and the operations of read or write of those items are in quantities dependent on the service function code.

The possible association of client/server data references (bits, registers) and actual physical storage or logical meaning within devices is a local matter.

The meaning of the content of any data reference is entirely up to the application.

The client/server data reference addresses, used in client/server service function codes, are unsigned integers starting at 0.

TABLE 10.3 Modbus Primitive Data Types

Modbus Data Type	Raw Data Type	Comment
Discrete	Single bit	Read-only, its value can be provided by an I/O system. This Modbus data type is useful in the modeling of binary-valued real objects that are manipulated by the server application and are supposed to be only observed by the client user. The integrity of the above contract is under control of the server, which can confine the exposure of the real objects to discretes.
Coil	Single bit	Read–write, its value can be altered by a client application program.
Input register	16-bit word	Read-only, its value can be provided by an I/O system. This Modbus data type is useful in the modeling of analog-valued real objects that are manipulated by the server application and are supposed to be only observed by the client user. The integrity of the above contract is under control of the server, which can confine the exposure of the real objects to input registers.
Holding register	16-bit word	Read–write, its value can be altered by a client application program.

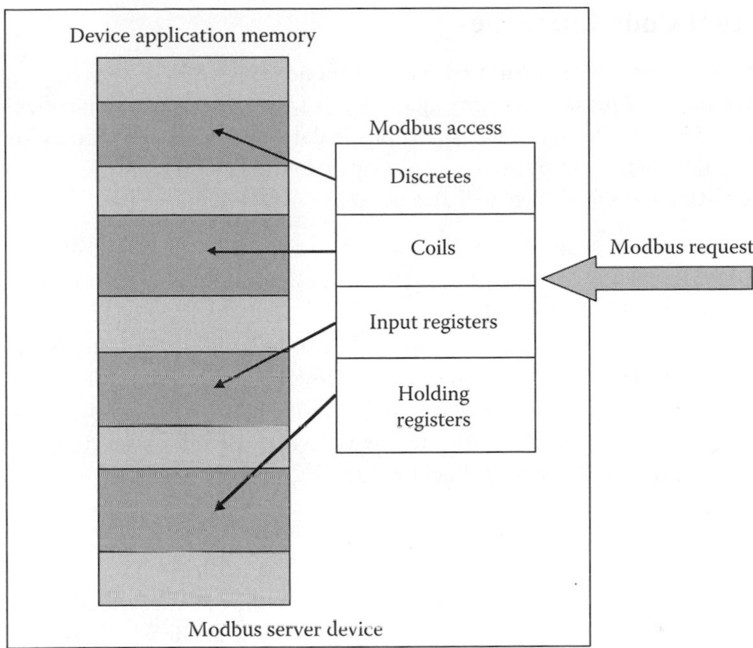

FIGURE 10.13 Interpretation as distinct tables.

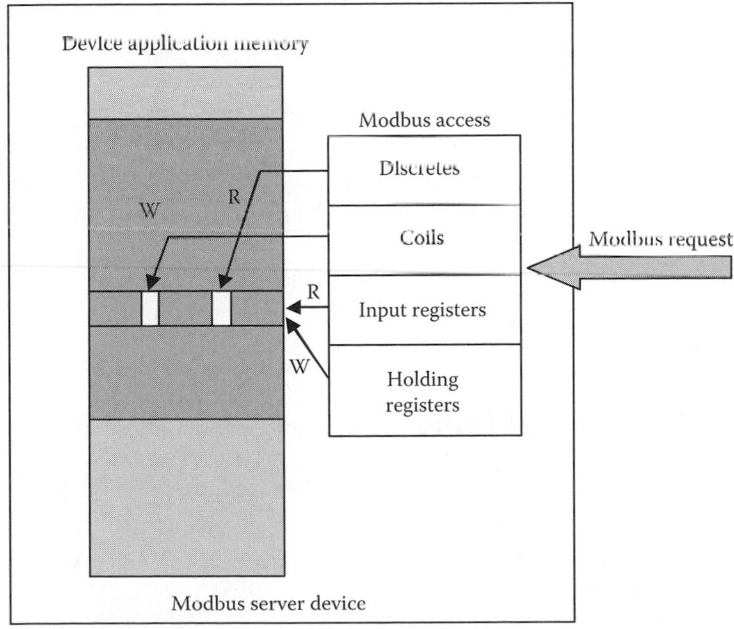

FIGURE 10.14 Interpretation as overlapping tables.

10.2.7 Function Code Categories

Client/server service identifiers are commonly called function codes (FCs).

Function codes are encodings of services requested to a server. Some function codes are further specialized by means of a subcode, specified as part of the data field. These encodings are partitioned in three categories, and since the subdivision may propagate to the subcodes, for the sake of completeness, despite being part of the data field, they will also be mentioned here:

Publicly assigned function codes. These function codes are either assigned to a standard service or reserved for future assignment. The function codes currently assigned to a standard service will be listed in an upcoming section.

User-definable function codes. These function codes can be used for experimentation in a controlled laboratory environment. *They must not be used in an open environment*. There are two ranges: FC 65 (0x41) to 72 (0x48) included and 100 (0x64) to 110 (0x6E) included.

Reserved function codes. These function codes are currently used by some companies for legacy products and are not available for public use.

10.2.8 Function Codes

The function codes publicly assigned to a standard service are the ones described in Table 10.4. Details can be found in the authoritative specifications publicly available at the web-based community www.Modbus.org.

The Modbus protocol does not mandate the presence of any particular subset of these function codes. A Modbus implementation can have any subset.

The function codes publicly assigned to a standard service are listed in Table 10.4.

10.2.9 Exception Codes

Exception codes provide the data field content of exception responses. See Table 10.5.

10.2.10 Transmission Order

Data representation *on the wire*: Modbus uses a *big-endian* convention for addresses and data items. This means that when a numerical quantity larger than a single octet is transmitted, the most significant octet is sent first. An example is given as follows:

Register size, 16 bits; value, 0x1234; then the first octet sent is 0x12 and the second is 0x34.

Note that the remote terminal unit's (RTU) CRC wants CRC low and CRC high (CRC high is the last octet of the RTU message frame) (see Figure 10.23).

10.3 Modbus over Serial Line

10.3.1 General

This section describes the Modbus protocol over serial line, that is, how the Modbus application protocol, at OSI layer 7, is realized over OSI layers 2 and 1, where layer 1 is a serial line, EIA/TIA-485 (RS485) or EIA/TIA-232 E (RS232).

The material presented in this section does not contain all the details of the Modbus protocol. The authoritative specifications are publicly available and can be found at the web-based community www.Modbus.org.

Most of the discussion of this section will be on layer 2 and its support for the client/server behavior described in the previous section. This stack is shown in Table 10.6.

TABLE 10.4 Function Codes Publicly Assigned to a Standard Service

Function Code	Description
FC 2 (0x02)	Read discretes.
FC 1 (0x01)	Read coils.
FC 5 (0x05)	Write single coil.
FC 15 (0x0F)	Write multiple coils.
FC 5 (0x05) with unit ID = 0	Broadcast write single coil.
FC 15 (0x0F) with unit ID = 0	Broadcast write multiple coils.
FC 4 (0x04)	Read input registers.
FC 3 (0x03)	Read holding registers.
FC 6 (0x06)	Write single holding register.
FC 16 (0x10)	Write multiple holding registers.
FC 22 (0x16)	Mask write holding register.
FC 23 (0x17)	Read/write holding registers.
FC 24 (0x18)	Read FIFO.
FC 6 (0x06) with unit ID = 0	Broadcast write single holding register.
FC 16 (0x10) with unit ID = 0	Broadcast write multiple holding registers.
FC 20 (0x14)	Read file record.
FC 21 (0x15)	Write file record.
FC/subcode 43 (0x2B)/14 (0x0E)	Read device identification.
FC 7 (0x07)	Read exception status.
FC 8 (0x08)	Diagnostics.
FC 11 (0x0B)	Get com event counter.
FC 12 (0x0C)	Get com event log.
FC 17 (0x11)	Report server ID.
FC/subcode 43 (0x2B)/13 (0x0D)	CANopen general reference request and response PDU. Please see Note 4.

Note 1: Function code assignments are managed by the Modbus.org industrial consortium.

Note 2: The following function codes, while publicly assigned and described by the Modbus.org specifications, are not covered by the IEC Modbus standardization: FC 7 (0x07, read exception status), FC 8 (0x08, diagnostics), FC 11 (0x0B, get com event counter), FC 12 (0x0C, get com event log), and FC 17 (0x11, report server ID). The applicability of some of these function codes may depend on the underlying layer; please refer to the Modbus.org specifications.

Note 3: The following function codes and function code/subcodes are reserved. FC 8/19 (0x08/0x13), FC 8/21-255 (0x08/0x15-0xFFFF), FC 9 (0x09), FC 10 (0x0A), FC 13 (0x0D), FC 14 (0x0E), FC 41 (0x29), FC 42 (0x2A), FC 43/0-12 (0x2B/0x00-0x0C), FC 43/15-255 (0x2B/0x0F-0xFF), FC 90 (0x5A), FC 91 (0x5B), FC 125 (0x7D), FC 126 (0x7E), and FC 127 (0x7F).

Note 4: The function code FC 43/13 (0x2B/0x0D) is CANopen general reference request and response PDU. Subcode 13 of FC 43 is a Modbus assigned number licensed to CAN in Automation (CiA) for the CANopen General Reference. Please refer to the Modbus.org website or the CiA website for a copy and terms of use that cover function code 43/13. It is not covered by the IEC Modbus standardization.

The network is a serial bus, with only one client at a time on it, by configuration.* An RS485 network is shown in Figure 10.15 and an RS232 network is shown in Figure 10.16.

The addressing of a server by a client is fully accomplished using the unit ID previously described in Section 10.2.3.2. There can be up to 247 servers on the network; each must have a unique unit ID between 1 and 247. RS232 is a point-to-point network; the unit ID, even if not needed, is still required.

The messaging is performed exchanging APDUs (Figure 10.6, but augmented with error checking as will be shown shortly).

The state machines described in Section 10.2.4.2 are in effect, with a specialization of the *improper response* and *improper request*.

* In the RS485 case, there could be more than one client on the bus, but something must ensure, at the application above the Modbus protocol, that only one is active at a time, across a complete client/server interaction, the synchronization between the clients may be complicated.

TABLE 10.5 Exception Codes

Encoding	Name	Description
0x01	Illegal function	The function code received in the request is not an allowable action for the server. This may be because the function code is only applicable to newer devices and was not implemented in the unit selected. It could also indicate that the server is in the wrong state to process a request of this type, for example, because it is not configured and it is being asked to return register values.
0x02	Illegal data address	The data address received in the request is not an allowable address for the server. More specifically, the combination of reference number and transfer length is invalid. For example, for a controller with 100 registers, a request with offset (data address) 96 and length 4 would succeed, and a request with offset 96 and length 5 would generate exception code 0x02.
0x03	Illegal data value	A value contained in the request data field is not an allowable value for server. This indicates a fault in the structure of the remainder of a complex request, such as that the implied length is incorrect. It specifically does *not* mean, for example, that a data item submitted for storage in a register has a value outside the expectation of the application program, since the client/server protocol is unaware of the significance of any particular value for any particular register.
0x04	Server device failure	An unrecoverable error occurred, while the server was attempting to perform the requested action.
0x05	Acknowledge	The server accepted the service invocation but the service requires a relatively long time to execute. The server therefore returns only an acknowledgment of the service invocation receipt. This response is returned to prevent a time-out error from occurring in the client.
0x06	Server busy	The server was unable to accept the request. The client application has the responsibility of deciding if and when to resend the request.
0x08	Memory parity error	For specialized use in conjunction with function codes 20 (0x14) and 21 (0x15), to indicate that the extended file area failed to pass a consistency check. For example, the server attempted to read record file but detected a memory parity error. The client can retry the request, but service may be required on the server device.
0x0A	Gateway path unavailable	For specialized use in conjunction with gateways, hubs, switches, and similar network devices, to indicate that the gateway was unable to allocate an internal communication path from the input port to the output port for processing the request. This usually means that the gateway is misconfigured or overloaded.
0x0B	Gateway target device failed to respond	For specialized use in conjunction with gateways, hubs, switches, and similar network devices, to indicate that no response was obtained from the target device. This usually means that the device is not present on the network.

Note 1: The following exception codes are reserved: 0x00, 0x07, 0x09, and 0x0C-0xFF.

Note 2: The exception code table is as from the Modbus.org Modbus protocol specification. Some exception code descriptions date back to the origin of the protocol. The exception codes 0x02 and 0x03 have historically been interpreted in different ways, to the point that Modbus.org decided to accept them interchangeably when running the conformance test. Going forward, the advice is as follows: An exception response should use exception code 0x02 when the Modbus request was correct, data-wise, but not for this server configuration. An exception response should use exception code 0x03 when the Modbus request was wrong, data-wise, for any server (and indeed the problem could have been detected by the client, i.e., a smarter client would not have produced such a request, irrespective of the server configuration).

TABLE 10.6 OSI Layers and Modbus over Serial Line

Layer	ISO/OSI Model	
7	Application	Modbus application protocol
6	Presentation	Empty
5	Session	Empty
4	Transport	Empty
3	Network	Empty
2	Data link	Modbus over serial line
1	Physical	EIA/TIA-485 or EIA/TIA-232

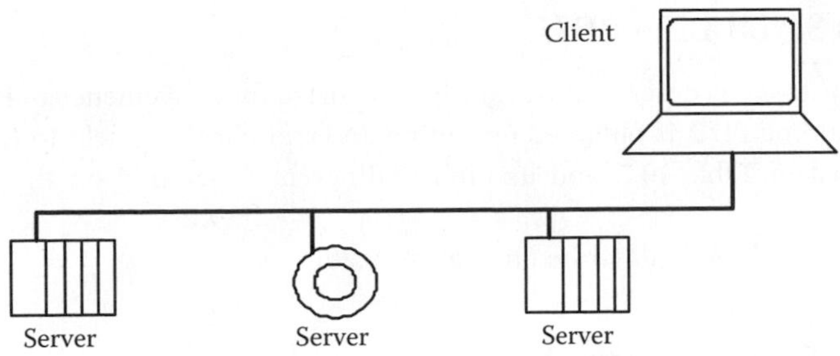

FIGURE 10.15 A serial bus (RS485 shown).

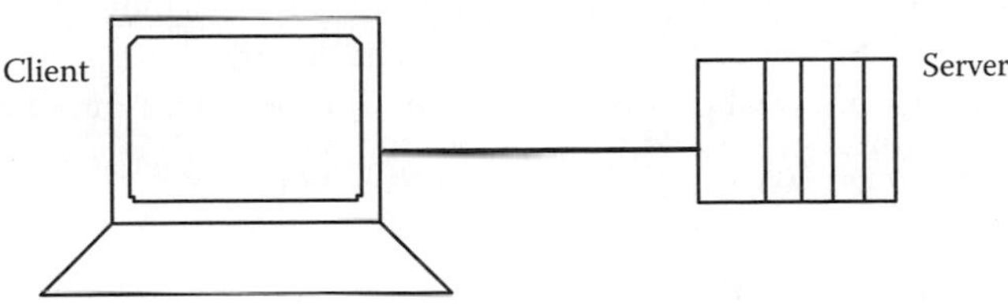

FIGURE 10.16 A serial bus (RS232 shown).

For the client state machine, the *improper response* is a response with a frame error, and such a response is just dropped.

For the server state machine, the *improper request* is either a frame error or a request not addressed to the server, and in both cases, the request is dropped.

The unicast interaction is illustrated in Figure 10.17.

The broadcast interaction, obtained with unit ID = 0 for selected function codes as from Table 10.4, is illustrated in Figure 10.18.

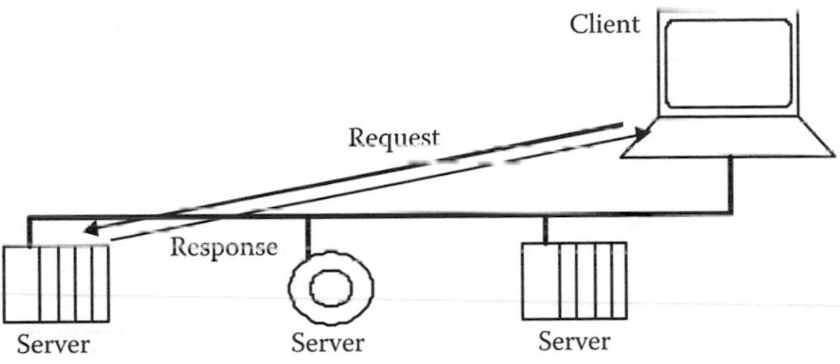

FIGURE 10.17 Unicast on serial bus (RS485 shown).

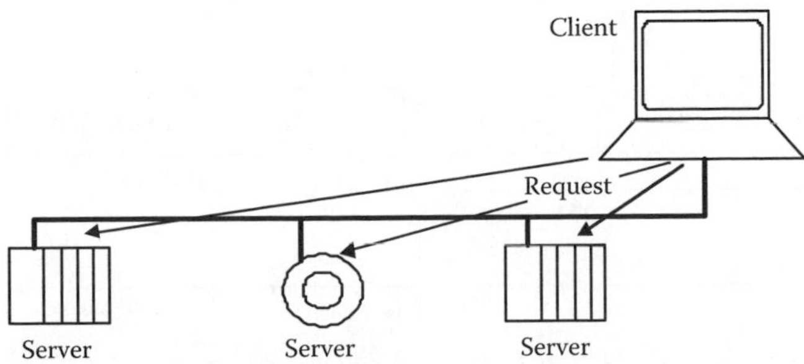

FIGURE 10.18 Broadcast on serial bus (RS485 shown).

10.3.2 Modbus Serial Line PDU

The APDU for client/server as described previously is reported for convenience in Figure 10.19.

The Modbus serial line PDU is obtained by adding an error checking field to the APDU. The error checking is described in Table 10.7, and its kind is different depending on the transmission mode (described next).

The Modbus serial line PDU is described in Figure 10.20.

10.3.3 Modbus Transmission Modes

Modbus over serial line uses one of two distinct transmission modes: remote terminal unit (RTU) and American Standard Code for Information Interchange (ASCII). RTU is more efficient but slightly more complicated to implement; ASCII is very simple. They will be described in upcoming sections.

The transmission mode (and serial port parameters) must be the same for all devices on a Modbus serial line.

10.3.4 RTU Transmission Mode

10.3.4.1 General

When devices communicate on a Modbus serial line using the RTU mode, each character on the wire contains an application message octet. The main advantage of this mode is that its greater density allows better data throughput than ASCII mode for the same baud rate. Each message *must* be transmitted as a continuous stream of characters.

This mode is more efficient than ASCII since each application message octet needs only one character on the wire.

The RTU transmission mode identifies the transmission element boundaries based on time.

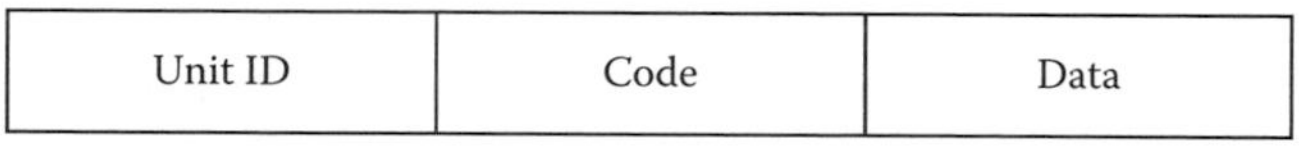

Unit ID	Code	Data

FIGURE 10.19 APDU format.

TABLE 10.7 Error Checking Field

Parameter	Length	Description
Redundancy checking	2 octets	CRC or LRC, depending on the transmission mode, RTU and ASCII, respectively

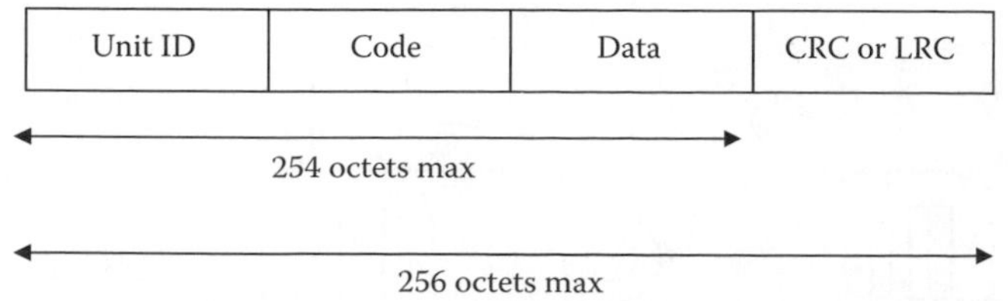

FIGURE 10.20 Modbus serial line PDU.

The format for each character on the wire (11 bits) in RTU mode is as follows:

Coding System
8-bit binary
Two hexadecimal characters (0–9, A–F) contained in the 8-bit field of the character on the wire.
Bits per Character on the Wire
1 start bit
8 data bits, least significant bit (LSB) sent first
1 bit for even/odd parity, no bit for no parity
1 stop bit if parity is used; 2 bits if no parity
Even parity is required, other modes (odd parity, no parity) may also be used. In order to ensure a maximum compatibility with other products, it is *recommended* to support also the no parity mode. The default parity mode *must* be even parity. The format with parity checking is shown in Figure 10.21.
Error Check Field
CRC
How Characters on the Wire Are Transmitted Serially
Each character on the wire is sent in this order (left to right): LSB to most significant bit (MSB)

Devices *may* accept by configuration both even or odd parity checking and no parity checking. If no parity is implemented, an additional stop bit is transmitted to fill out the character on-the-wire frame. The format without parity checking is shown in Figure 10.22.

Figure 10.23 shows the RTU message frame, complete of CRC (CRC low, CRC high).

The maximum size of the Modbus RTU frame is 256 octets.

10.3.4.2 Modbus Message RTU Framing

A Modbus message is placed by the transmitting device into a frame that has a known beginning and ending point. This allows devices that receive a new frame to begin at the start of the message and to know when the message is completed. Partial messages *must* be detected and errors *must* be set as a result.

With parity checking

Start	1	2	3	4	5	6	7	8	Par	Stop

FIGURE 10.21 Bit sequence in RTU mode (with parity).

Without parity checking

Start	1	2	3	4	5	6	7	8	Stop	Stop

FIGURE 10.22 Bit sequence in RTU mode (no parity case).

Unit ID	Code	Data	CRC
1 octet	1 octet	0 up to 252 octets	2 octets CRC Low, CRC Hi

FIGURE 10.23 RTU Modbus message.

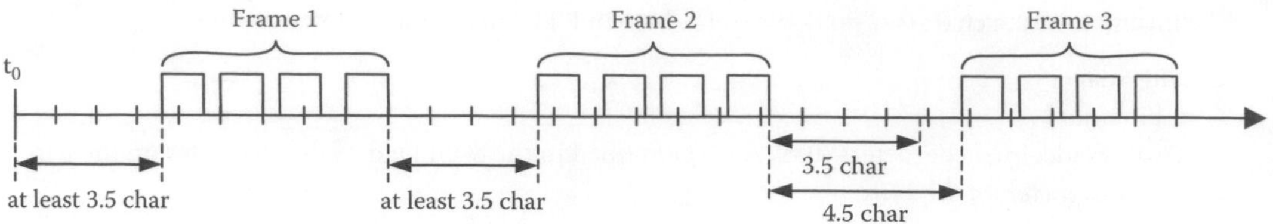

FIGURE 10.24 RTU message frames and separation.

Modbus message

Start	Address	Function	Data	CRC check	End
≥3.5 char	8 bits	8 bits	N × 8 bits	16 bits	≥3.5 char

FIGURE 10.25 RTU message frame.

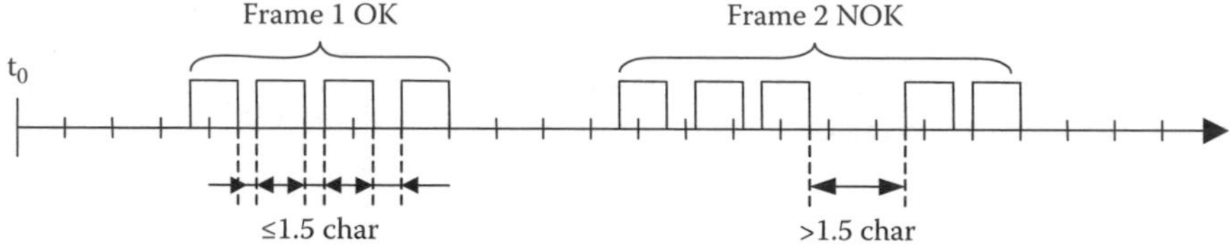

FIGURE 10.26 Good frame and bad frame.

In the RTU mode, message frames are separated by a silent interval of *at least* 3.5 character (on the wire) times. In the following sections, this time interval is called t3.5, and it is shown in Figures 10.24 and 10.25.

The entire message frame *must* be transmitted as a continuous stream of characters.

If a silent interval of more than 1.5 character (on the wire) times occurs between two characters on the wire, the message frame is declared incomplete and should be discarded by the receiver. See Figure 10.26.

Please consult the *Modbus over serial line* specification at www.Modbus.org for more details on timing, on various checks, and on the CRC.

10.3.5 ASCII Transmission Mode

10.3.5.1 General

When devices communicate on a Modbus serial line using the ASCII mode, each application message octet is sent as two ASCII characters on the wire, that is, it will require two characters on the wire. This mode is used when the capabilities of the device do not allow the conformance with RTU mode requirements regarding timers' management.

This mode is less efficient than RTU since each application message octet needs two characters on the wire.

Example: The application message octet 0x5B is encoded as two characters for the wire: 0x35 and 0x42 (0x35 = *5* and 0x42 = *B* in ASCII).

While there are time-outs involved, the ASCII transmission mode identifies the transmission element boundaries based on character values.

The format for each character on the wire (10 bits) in ASCII mode is as follows:

Coding System

Hexadecimal, ASCII characters 0–9, A–F

One hexadecimal character contained encoded 7-bit ASCII character in the 7-bit field of the character on the wire.

Bits per Character on the Wire

1 start bit

7 data bits, LSB sent first

1 bit for even/odd parity; no bit for no parity

1 stop bit if parity is used; 2 bits if no parity

Even parity is required; other modes (odd parity, no parity) may also be used. In order to ensure a maximum compatibility with other products, it is *recommended* to support also no parity mode. The default parity mode *must* be even parity. The format with parity checking is shown in Figure 10.27.

Error Check Field

Longitudinal redundancy check (LRC)

How Characters on the Wire Are Transmitted Serially

Each character on the wire is sent in this order (left to right): LSB to MSB

Devices *may* accept by configuration both even or odd parity checking and no parity checking. If no parity is implemented, an additional stop bit is transmitted to fill out the character on-the-wire frame. The format without parity checking is shown in Figure 10.28.

10.3.5.2 Modbus Message ASCII Framing

A Modbus message is placed by the transmitting device into a frame that has a known beginning and ending point. This allows devices that receive a new frame to begin at the start of the message and to know when the message is completed. Partial messages *must* be detected and errors *must* be set as a result.

In ASCII mode, a message is delimited by specific ASCII characters as start of frames and end of frames. A message *must* start with a *colon* (:) character (ASCII 3A hex) and end with a *carriage return–line feed* (CRLF) pair (ASCII 0D and 0A hex).

With parity checking

Start	1	2	3	4	5	6	7	Par	Stop

FIGURE 10.27 Bit sequence in ASCII mode (with parity).

Without parity checking

Start	1	2	3	4	5	6	7	Stop	Stop

FIGURE 10.28 Bit sequence in ASCII mode (no parity case).

Start	Unit ID	Code	Data	LRC	End
1 char :	2 chars	2 chars	0 up to 2 × 252 char(s)	2 chars	2 chars CR,LF

FIGURE 10.29 ASCII message frame.

The allowable characters transmitted for all other fields are hexadecimal 0–9, A–F (ASCII coded). The devices monitor the bus continuously for the *colon* character. When this character is received, each device decodes the next character until it detects the end of frame.

Intervals of up to one second may elapse between characters within the message. If a greater interval occurs, the receiving device assumes that an error has occurred.

A typical ASCII message frame is shown in Figure 10.29.

Remark: Each data octet needs two characters for encoding. Thus, to ensure compatibility at the Modbus application level between ASCII mode and RTU mode, the maximum data size for the ASCII data field (2 × 252) is twice the maximum data size for the RTU data field (252).

Consequently, the maximum size of a Modbus ASCII frame is 513 characters.

Please consult the *Modbus over serial line* specification at www.Modbus.org for more details on various checks and on the LRC.

The official specification will also provide information about the physical layer.

10.4 Modbus/TCP

10.4.1 General

This section describes Modbus/TCP in general and the client/server encapsulation when the transport layer is TCP/IP. Figure 10.30 shows the Modbus/TCP stack and Table 10.8 shows the positioning with respect to the OSI layers.

The material presented in this section does not contain all the details of the Modbus protocol. The authoritative specifications are publicly available and can be found at the web-based community www.Modbus.org.

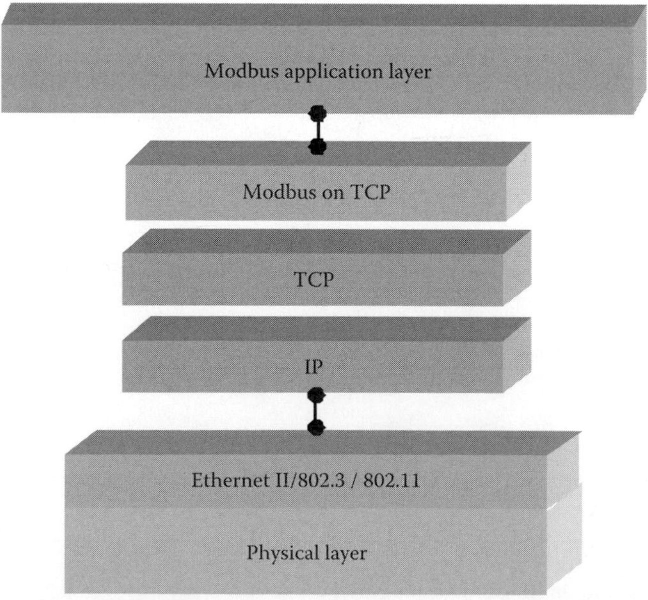

FIGURE 10.30 Modbus/TCP stack.

TABLE 10.8 OSI Layers and Modbus/TCP

Layer	ISO/OSI Model	
7	Application	Modbus application protocol
6	Presentation	Empty
5	Session	Empty
4	Transport	TCP
3	Network	IP
2	Data link	Network access Ethernet II or 802.3 or 802.11
1	Physical	

10.4.2 TCP/IP Encapsulation

The APDU for client/server as described previously is reported for convenience in Figure 10.31.

TCP/IP encapsulation is obtained by adding a header to the APDU. The parameters of the header are described in Table 10.9.

The PDU carried as payload by TCP/IP becomes the one described in Figure 10.32.

10.4.3 Role of the Transaction ID

The set of the servers addressable by a client is determined by the underlying layer. In the case of Modbus/TCP, this set is very large since the address is the combination of the IP address and the unit ID.

Given a client, the transaction ID generated by the client and placed in a request must be unique across all the transaction IDs still pending on the set of its addressable servers. The transaction ID allows the Modbus/TCP stack more flexibility than, for example, the Modbus serial stack.

Unit ID	Code	Data

FIGURE 10.31 APDU format.

TABLE 10.9 Encapsulation Parameters for Client/Server on TCP/IP

Parameters/Fields	Length	Description	Client	Server
Transaction ID	2 octets	Identification of a Modbus request/response transaction	Initialized by the client	Echoed by the server from the received request; returned in the response
Protocol ID	2 octets	0 = Modbus protocol	Initialized by the client	Echoed by the server from the received request; returned in the response
Length	2 octets	Number of following octets in APDU	Initialized by the client based on the request	Initialized by the server based on the response

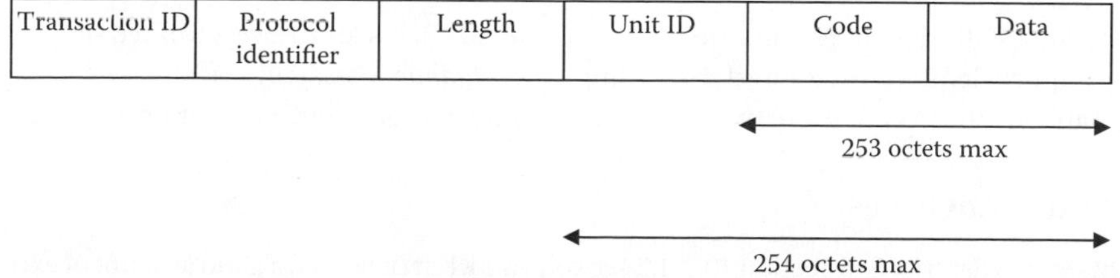

FIGURE 10.32 TCP/IP PDU format.

A client requests a service on the set of its addressable servers with the help of the transaction ID, managed (created and destroyed) by the client. The transaction mechanism is exposed at the application layer due to the client/server possibility of having more than one outstanding request at a time, with the consequent need to properly associate requests and confirmations. It also controls the maximum number of such requests, which could be 1. The capabilities of a client/server application layer client depend on lower layers and on the particular implementation; these factors are captured in the configuration of the transaction mechanism allowing programmatic adaptation.

10.4.4 Assigned Port

Modbus/TCP communicates using port 502, assigned by IANA.

10.4.5 Protocol ID

The protocol ID must be 0.

Note 1: All other protocol IDs are reserved.

10.4.6 Unit ID and Gateways

On TCP/IP, when no gateways or IP colocated application entities are involved, the client and the server are the intended end points of the connection, and they are fully identified using the IP address. In this case, the unit ID may be ignored by the server, and the client should set it to the value of 255.

In case of gateways or IP colocated application entities, the unit ID is used to identify the server connected to the gateway or the server among the IP colocated application entities. In this case, the value of 255 is recommended for addressing the gateway itself or the IP device hosting the application entities.

10.4.7 TCP as a Streaming Protocol

The length field in the TCP envelope is used to identify the transaction payload boundaries, since TCP is a streaming protocol.

The server must be able to handle situations with several outstanding indications in pipelined transaction on the same connection and others in different connections, up to an implementation-dependent number, usually dictated by resource constraints. If such a number is exceeded, the server must respond with exception code = 0x06: server busy.

The aforementioned limit may be per connection, or it may be a shared limit across all the connections on the same server.

The streaming nature of the TCP protocol allows for cases where the server received only a partial transaction, according to a valid length. The server must be able to buffer the partial transaction and wait for the remaining payload. The server may implement mechanisms, for example, via a timer, to reclaim resources if the wait exceeds a configured time.

10.4.8 Connection Establishment and Management

The connection establishment possibilities, with creation of clients and servers limited only by software resources, require dedicated and novel processing. The *Modbus Messaging on TCP/IP Implementation Guide* available at www.Modbus.org provides valuable information about TCP connection management.

10.4.9 State Machines

The state machines described in Section 10.2.4.2 are still valid, but they need a good amount of extrapolation to accommodate the possibility of multiple clients and the added flexibility provided by the transaction ID. Moreover, they need extra mechanisms to handle the streaming protocol nature of TCP.

10.4.10 Modbus/TCP Flowcharts

To go beyond the trivial Modbus/TCP cases where there is only one client on the network and that client cannot have more than one outstanding transaction, it is helpful to explicitly describe flowcharts that cover the more realistic usage/deployment of Modbus/TCP. The flowcharts that follow will not deal with connection establishment and management, which will be considered already in place. The flowcharts describe the behavior considering the request, indication, response, and confirmation between Modbus/TCP clients and servers and between them and the application software above Modbus/TCP, at the user level.

The MBAP referred in the flowcharts is the header that includes transaction ID, protocol ID, and length and unit ID.

The Modbus/TCP client activity is illustrated in Figure 10.33.

The build Modbus request activity of Figure 10.33 is expanded in Figure 10.34.

The process Modbus confirmation activity of Figure 10.33 is expanded in Figure 10.35.

The server activity (indication processing) is illustrated in Figure 10.36.

The Modbus/TCP PDU checking activity of Figure 10.36 is expanded in Figure 10.37.

The Modbus/TCP service processing activity of Figure 10.36 is expanded in Figure 10.38.

10.4.11 [Informative] TCP Interfaces and Parameterization

The Berkeley Software Distribution (BSD) socket interface is often used to communicate using TCP (see, e.g., *TCP/IP Illustrated, Volume 2, Gary R. Wright and W. Richard Stevens*).

A socket is an end point of communication. After the establishment of the TCP connection, the data can be transferred. The `send()` and `recv()` functions are designed specifically to be used with sockets that are already connected.

The `setsockopt()` function allows a socket's creator to associate options with a socket. These options modify the behavior of the socket. The description of these options and recommended settings will follow.

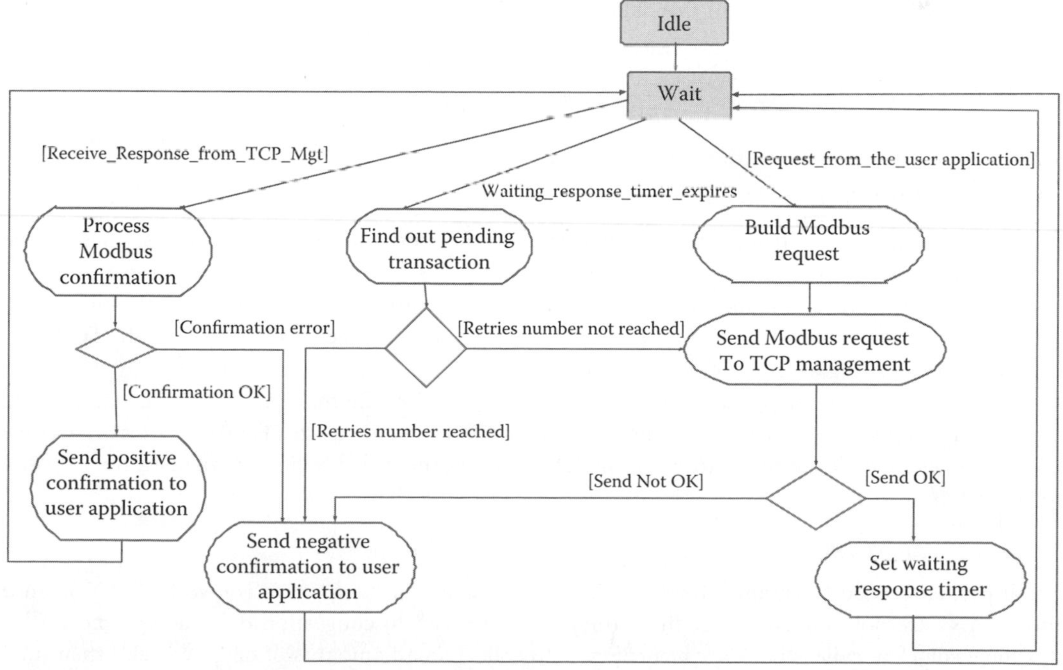

FIGURE 10.33 Modbus/TCP client activity flowchart.

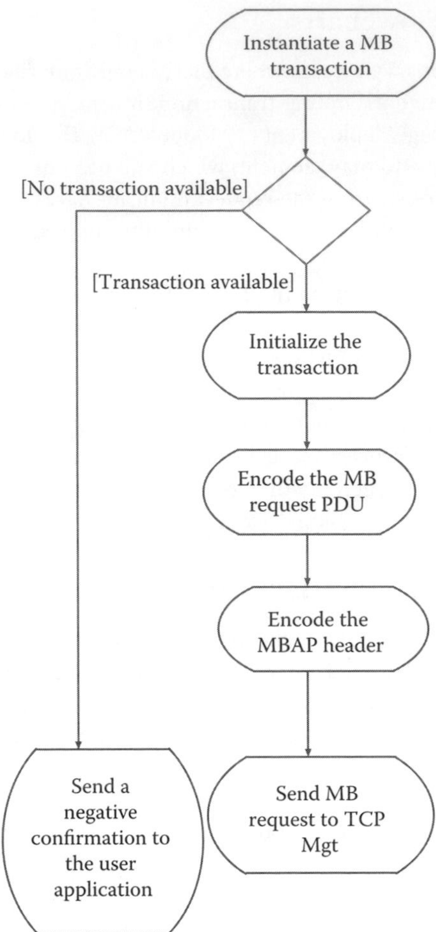

FIGURE 10.34 Modbus/TCP request building (in client) flowchart.

10.4.11.1 Connection Parameters

10.4.11.1.1 SO-RCVBUF, SO-SNDBUF

These parameters allow setting the high watermark for the send and the receive sockets. They can be adjusted for flow control management. The size of the receive buffer is the maximum size advertised window for that connection. Socket buffer sizes must be increased in order to increase performances. Nevertheless, these values must be smaller than internal driver resources in order to close the TCP window before exhausting internal driver resources.

The receive buffer size depends on the TCP Windows size, the TCP maximum segment size, and the time needed to absorb the incoming frames. With a maximum segment size of 300 octets (easily accommodating a client/server request), to accommodate 3 frames, the socket buffer size value can be adjusted to 900 octets.

10.4.11.1.2 TCP-NODELAY

Small packets (called tinygrams) are normally not a problem on Local Area Networks (LANs), since most LANs are not congested, but these tinygrams can lead to congestion on wide-area networks. A simple solution, called the *Nagle algorithm*, is to collect small amounts of data and send them in a single segment when the TCP acknowledgments of a previous packet arrive.

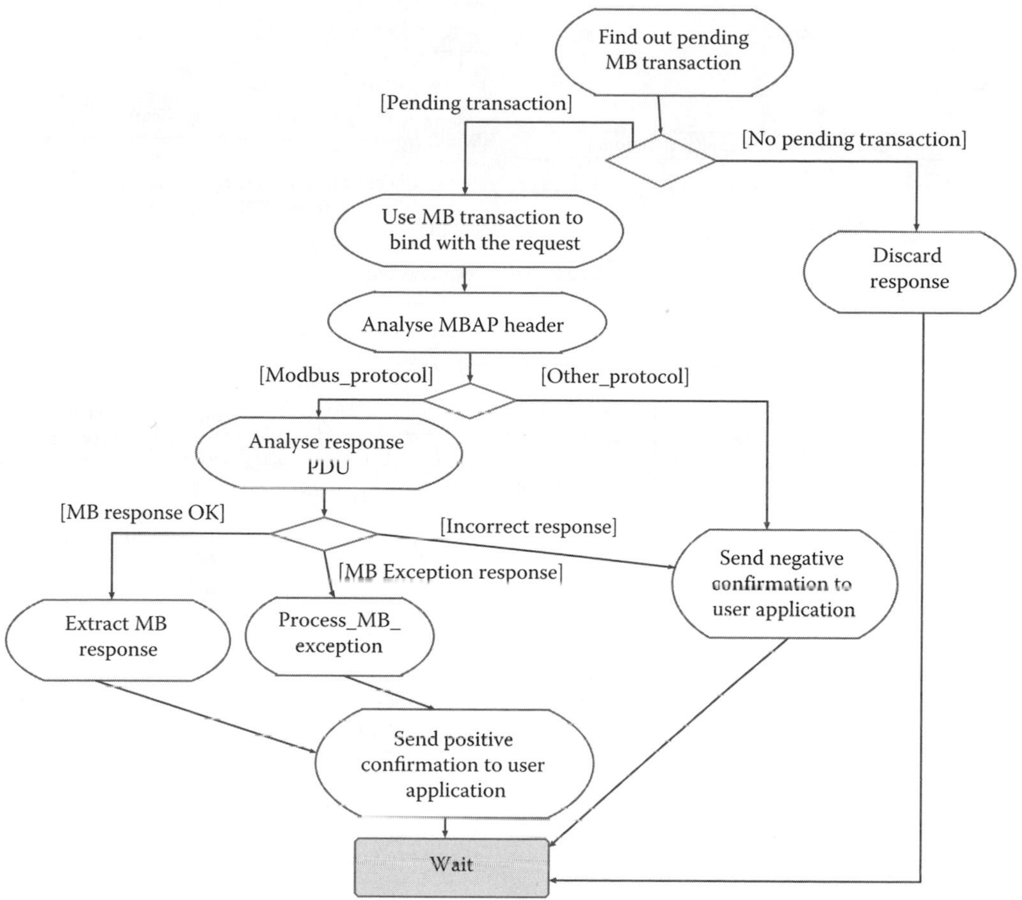

FIGURE 10.35 Process Modbus confirmation (in client) flowchart.

In order to have better behavior, it is recommended to send small amounts of data directly without trying to gather them in a single segment. That is why it is recommended to force the TCP-NODELAY option that disables the *Nagle algorithm* on client and server connections.

10.4.11.1.3 SO-REUSEADDR

When a server closes a TCP connection initialized by a remote client, the local port number used for this connection cannot be reused for a new opening while that connection stays in the *time-wait* state (during two maximum segment lifetime [MSL]).

It is recommended to specify the SO-REUSEADDR option for each client and server connection to bypass this restriction. This option allows the process to assign itself a port number that is part of a connection that is in the two MSL wait for client and listening socket.

10.4.11.1.4 SO-KEEPALIVE

By default on the TCP/IP protocol, there are no data sent across an idle TCP connection. Therefore, if no process at the ends of a TCP connection is sending data to the other, nothing is exchanged.

Under the assumption that either the client application or the server application uses timers to detect inactivity in order to close a connection, it is recommended to enable the KEEPALIVE option on both client and server connections in order to poll the other end to know its status.

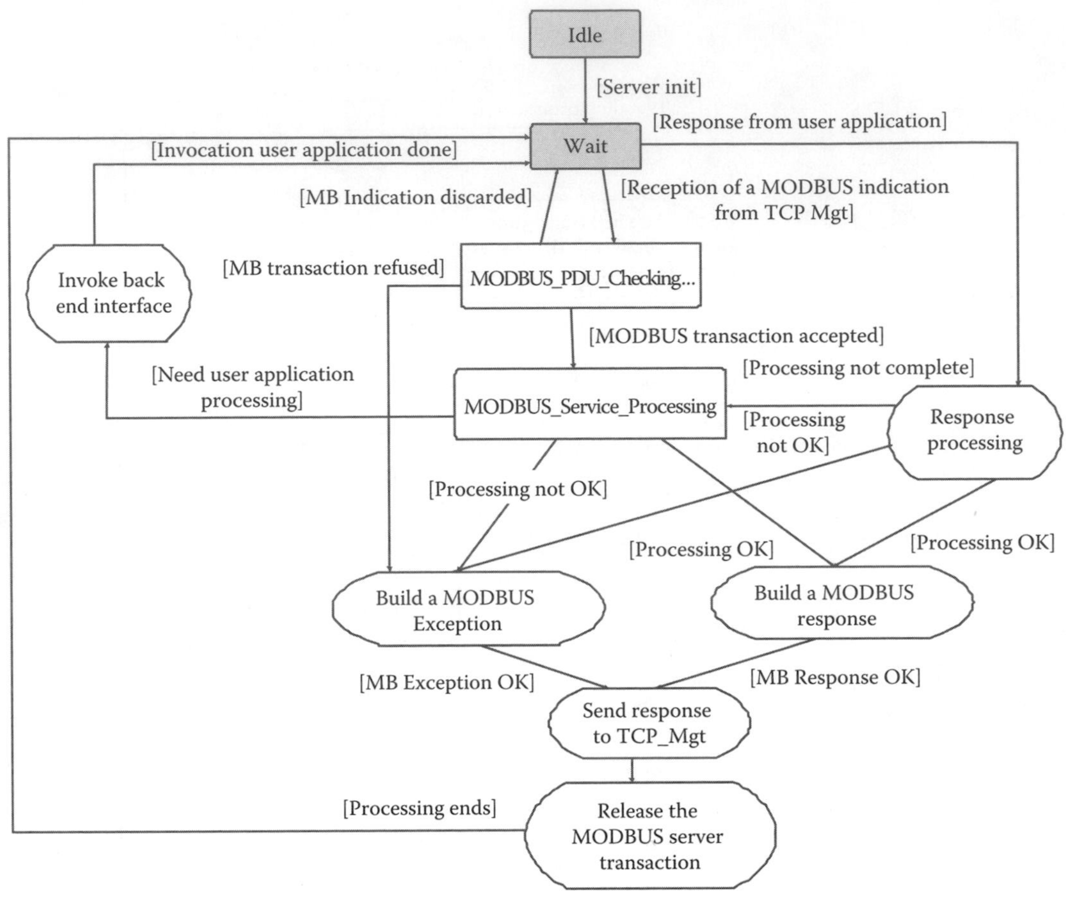

FIGURE 10.36 Modbus/TCP server activity (indication processing) flowchart.

Nevertheless, it must be considered that enabling KEEPALIVE can cause perfectly good connections to be dropped during transient failures and that it consumes unnecessary bandwidth if the keep-alive timer is too short.

10.4.11.2 TCP Layer Parameters

10.4.11.2.1 Time-Out on Establishing a TCP Connection

Most Berkeley-derived systems set a time limit of 75 s on the establishment of a new connection; this default value should be adapted to the constraint of the application.

10.4.11.2.2 Keep-Alive Parameters

The default idle time for a connection is 2 h. Idle times in excess of this value trigger a keep-alive probe. After the first keep-alive probe, a probe is sent every 75 s for a maximum number of times unless a probe response is received.

The maximum number of keep-alive probes sent out on an idle connection is 8. If no probe response is received after sending out the maximum number of keep-alive probes, TCP signals an error to the application that can decide to close the connection.

10.4.11.2.3 Time-Out and Retransmission Parameters

A TCP packet is retransmitted if its loss has been detected. One way to detect the loss is to manage a retransmission time-out (RTO) that expires if no acknowledgment has been received from the remote side.

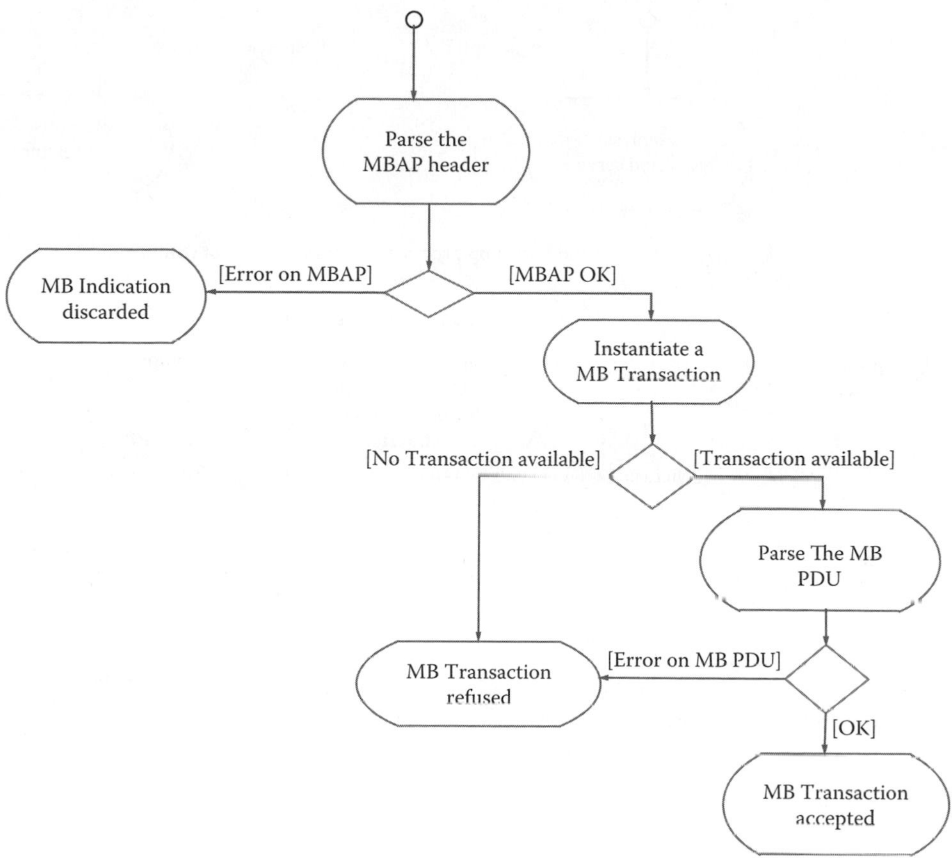

FIGURE 10.37 Modbus/TCP PDU checking (in server) flowchart.

TCP manages a dynamic estimation of the RTO. For that purpose, a round-trip time (RTT) is measured after the sending of every packet that is not a retransmission. The RTT is the time taken for a packet to reach the remote device and to get back an acknowledgment to the sending device. The RTT of a connection is calculated dynamically; nevertheless, if TCP cannot get an estimate within 3 s, the default value of the RTT is set to 3 s.

If the RTO has been estimated, it applies to the sending of the next packet. If the acknowledgment of the next packet is not received before the estimated RTO expiration, the *exponential backoff* (detailed below) is activated. A maximum number of retransmissions of the same packet are allowed during a certain amount of time. After that if no acknowledgment has been received, the connection is aborted.

Some TCP/IP stacks allow the setup of the maximum number of retransmissions and the maximum amount of time before the abort of the connection.

Some retransmission algorithms are defined in TCP IETF RFCs and other papers:

Jacobson's RTO estimation algorithm is used to estimate the RTO.

Karn's algorithm says that the RTO estimation should not be done on a retransmitted segment.

The *exponential backoff* defines that the RTO is doubled for each retransmission with an upper limit of 64 s.

The *fast retransmission algorithm* allows retransmitting after the reception of three duplicate acknowledgments. This algorithm is advised because on a LAN it may lead to a quicker detection of the loss of a packet than waiting for the RTO expiration.

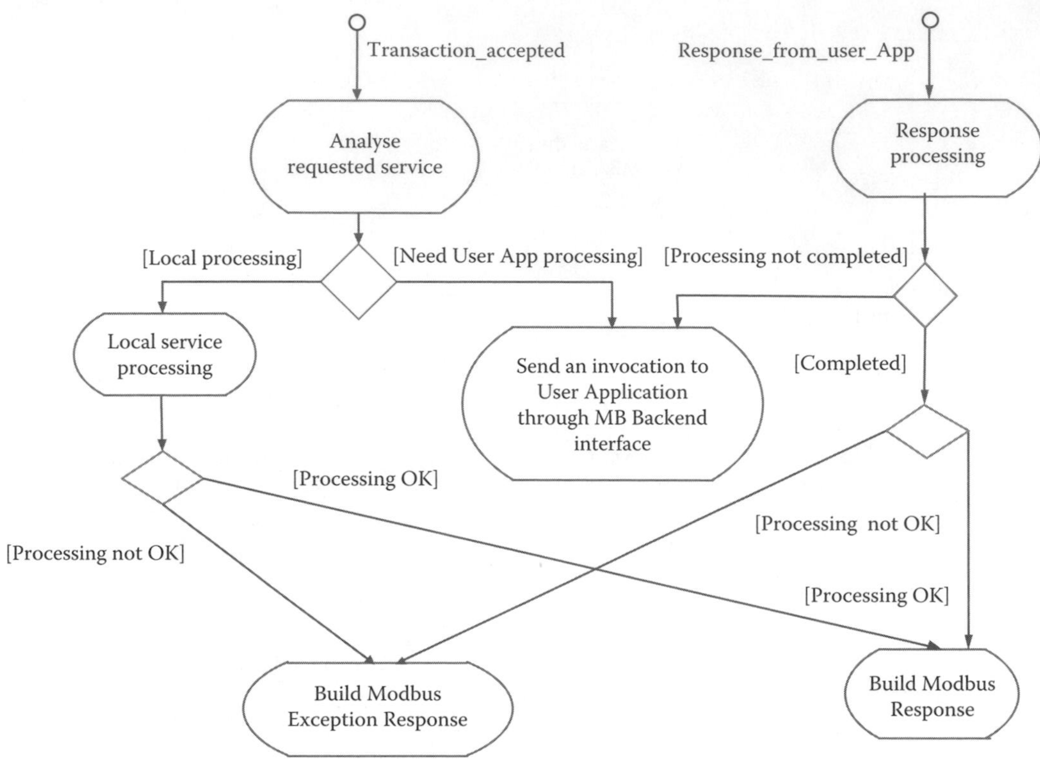

FIGURE 10.38 Modbus/TCP service processing (in server) flowchart.

The use of these algorithms is recommended; they are described in *TCP/IP Illustrated, Volume 2, Gary R. Wright and W. Richard Stevens*, which also points to the original sources.

10.5 Gateways and Similar Devices

10.5.1 General

Gateways involving Modbus have been around for a very long time, and new ones keep being developed, with Modbus being the common second protocol available side by side other protocols, or providing the gateways between different Modbus stacks.

The gateway between Modbus serial and Modbus/TCP, a very successful one, enabled many legacy products to ride the TCP/IP wave without having to be replaced or left behind.

Often gateways perform the duty of proxy clients, for instance, the aforementioned gateway between Modbus serial and Modbus/TCP allows multiple clients on the Modbus/TCP side to access quasi-concurrently servers on the Modbus serial side, by buffering and maintaining separate queues, with no need for user synchronization.

Modbus has been used to access other protocols in two major ways: by interpretation and by mapping. It has also been used with other protocols for tunneling and bridging.

The major reason Modbus is a big player in these protocol activities is that it makes no assumptions about the application semantics and it has an excellent performance/resource ratio on generic services instead.

10.5.2 Interpretation

When using this method, the gateway is knowledgeable about Modbus and the other protocol, or about Modbus on different stacks, and manages services and activities on both protocols, essentially by mapping activities. An example is the aforementioned Modbus/TCP to Modbus serial gateway or various BACnet to Modbus gateways.

10.5.3 Mappings

When using this method, two protocols share a memory mapping, where both can read and write or communicate activities/commands. Examples are the Sunspec Alliance to Modbus register mapping and the Wireless Cooperation Team (WCT) WirelessHART to Modbus register mapping. Once the mapping is agreed, it is very easy to write a Modbus client that can access the other protocol's information with no need to know anything about the other protocol.

10.5.4 Tunneling and Bridging

While the terminology is by no means agreed, in a Modbus tunnel, another protocol carries Modbus as a payload, while a Modbus bridge uses Modbus to carry another protocol in its payload. The market has examples with both.

10.6 Modbus as Part of the CIP Stack, in ODVA

Schneider Electric joined the ODVA organization (see www.odva.org) as a principal member in 2007. The ODVA maintains a suite of industrial protocols called the Common Industrial Protocols (CIPs™). This suite consists of four protocols: DeviceNet™, EtherNet/IP™, ControlNet™, and CompoNet™. One of the goals of this membership was to integrate Modbus into the CIP in order to provide a bridge between CIP networks and Modbus networks. The Modbus translator provides this bridge. A layered architecture view of the CIP protocols and the placement of the Modbus translator is shown in Figure 10.39.

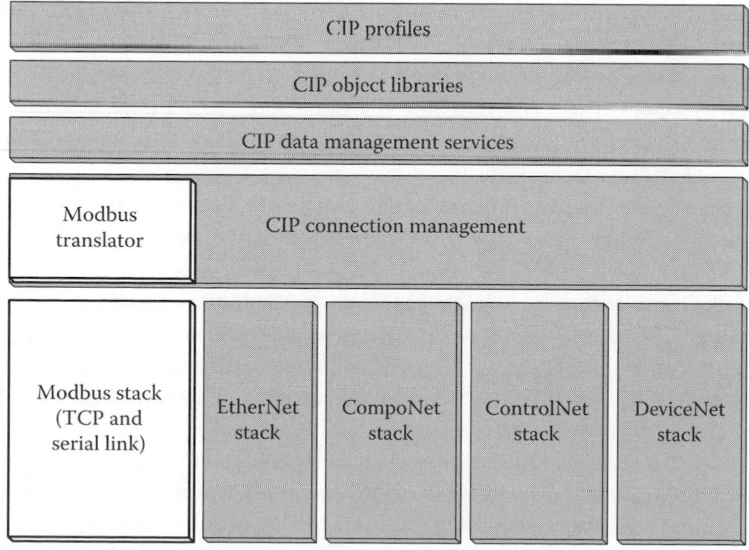

FIGURE 10.39 Modbus integration into CIP through the Modbus translator.

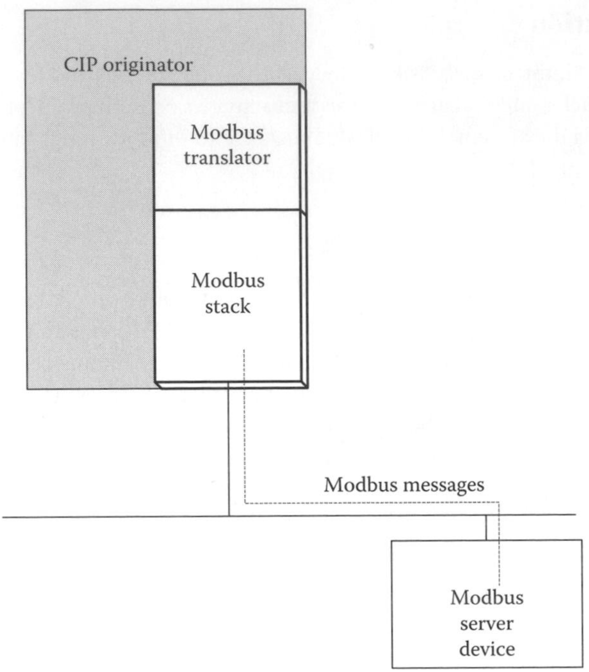

FIGURE 10.40 A Modbus translator located in a CIP originator.

The details of the Modbus translator are located in Volume 7 of the CIP Networks Library. Volume 7 is titled *Integration of Modbus Devices into the CIP Architecture* and is governed by the ODVA.

A Modbus translator houses the Modbus translation functionality used for the bridging. A Modbus translator can be located in a CIP originator (which is a client device as shown in Figure 10.40) or as a stand-alone device (sometimes referred to as a router as shown in Figure 10.41) or embedded in a device between the device's Modbus functionality and the CIP network (called a CIP device with native Modbus as shown in Figure 10.42).

The CIP originators receive the downstream Modbus data in the CIP format consistent with the CIP communication model. The CIP originator communicates with a CIP device, and the Modbus functionality is hidden from the CIP originator through the translation function. A Modbus translation function for Modbus/TCP in a dual IP stack is shown in Figure 10.43.

The Modbus translation function performs the *translation* between the CIP protocols and the Modbus protocols. An example of a CIP service request that is processed by the Modbus translator is shown in Figure 10.44.

The CIP network views the Modbus devices as CIP devices, while the Modbus devices view the communication as though it is from a Modbus device. The Modbus translation function housed in a Modbus translator that sits between the CIP network and the Modbus network allows this seamless communication. Any existing Modbus serial link server device or Modbus/TCP server device can communicate with a CIP network using the Modbus translator.

Throughout the integration of Modbus into the CIP environment, care was taken to ensure that the impact to the CIP originator devices was minimized. There was no impact to existing CIP target devices. Equally important, there was no impact to existing Modbus devices. Furthermore, existing vendor-specific CIP to Modbus gateway products can continue to work without any changes to these devices.

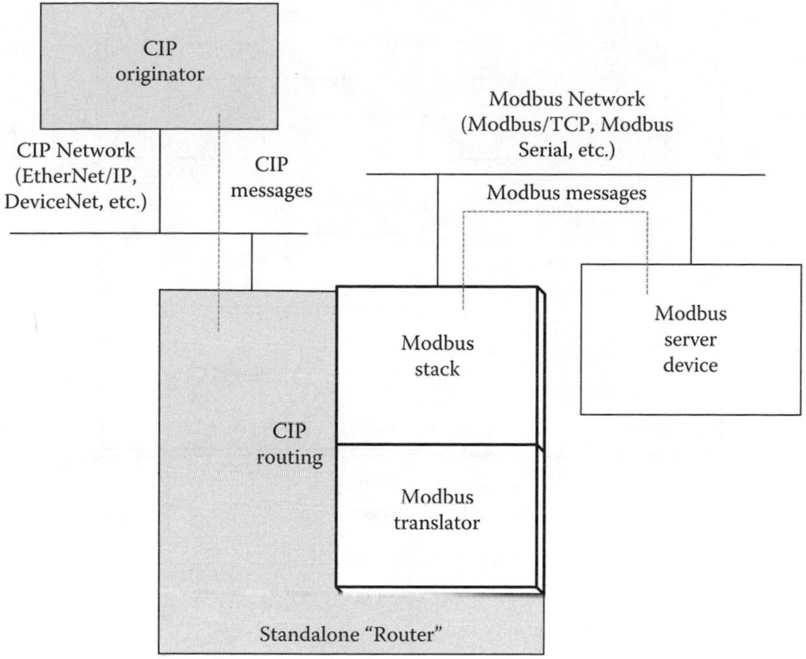

FIGURE 10.41 A Modbus translator as a stand-alone *router* device in a CIP network.

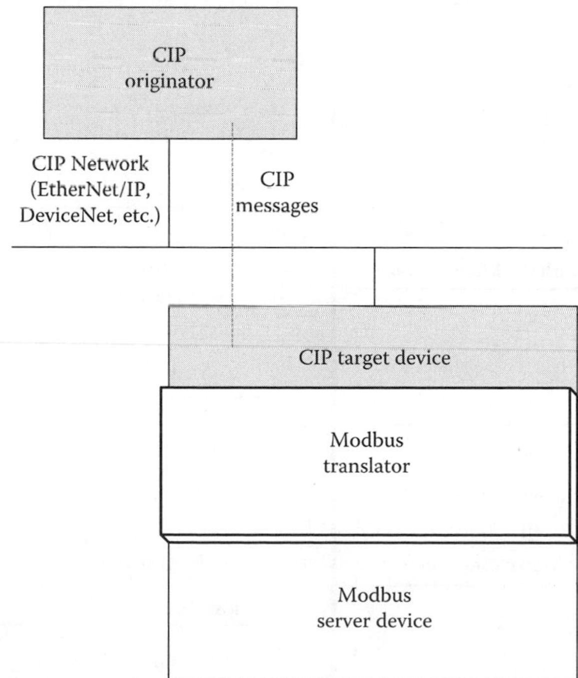

FIGURE 10.42 A Modbus translator embedded in CIP device where the CIP device has native Modbus functionality.

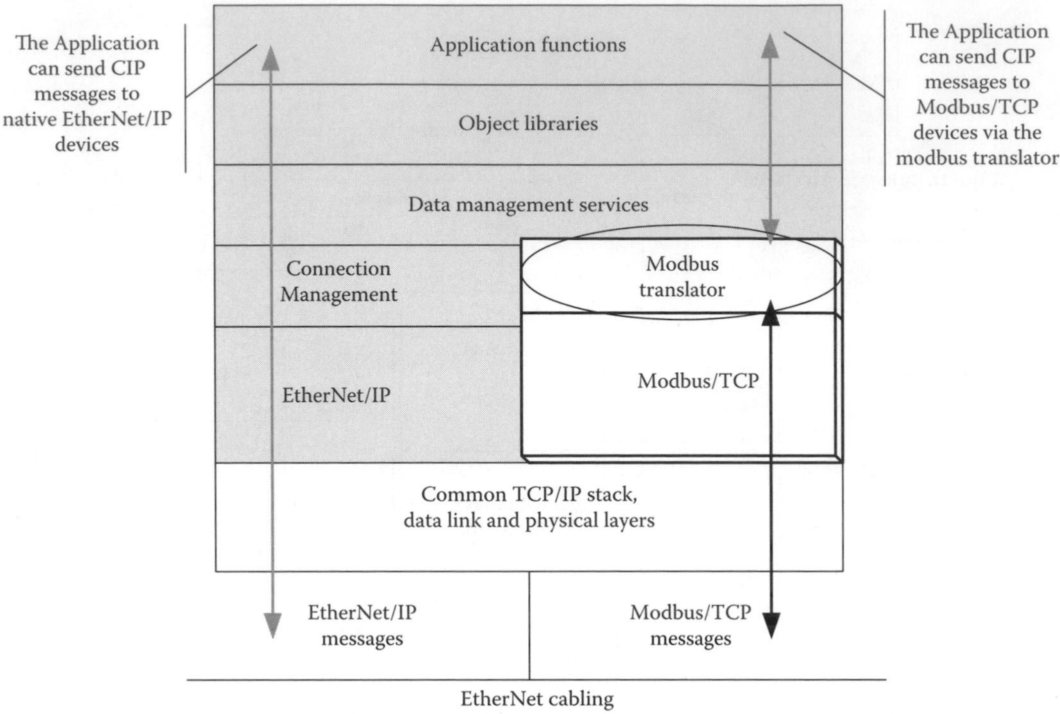

FIGURE 10.43 Diagram of the Modbus translation function in a dual EtherNet/IP and Modbus/TCP stack implementation.

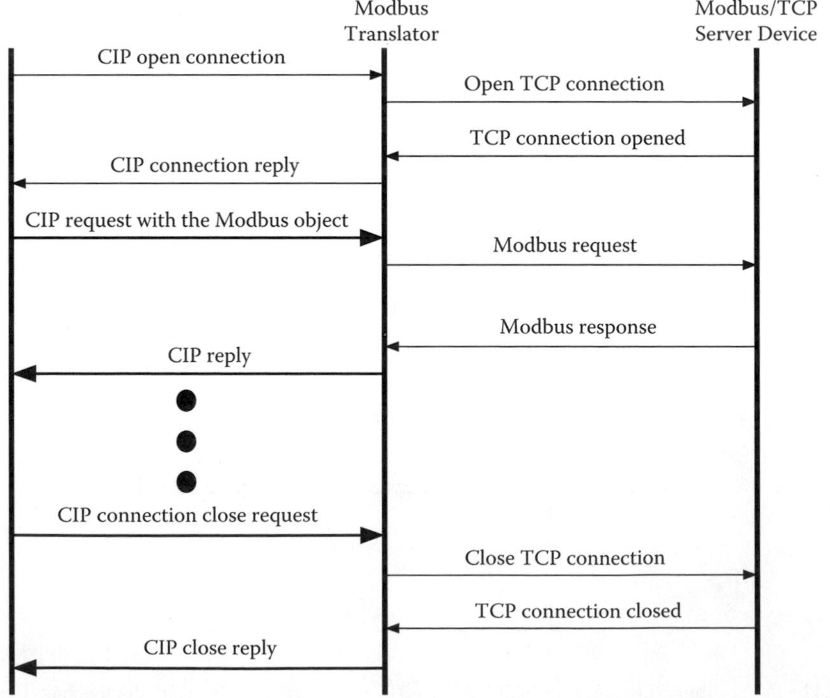

FIGURE 10.44 Sequence diagram of typical communication between a CIP originator and Modbus server (target) device through a Modbus translator.

Simply explained, the Modbus translator translates CIP object service requests into Modbus messages. The Modbus translator can be implemented inside a CIP originator or a CIP router or in a CIP device with native Modbus addressing if desired. This allows for efficient engineering design. CIP object service requests are delivered to the Modbus translator where the translator interprets these service requests, maps the service requests to Modbus function codes, and generates Modbus transactions. For efficient operation, the Modbus object was created inside the CIP library of objects. The Modbus object is structured to better expose CIP services to Modbus functionality and to allow Modbus responses to be easily placed into the CIP object format. For best operation, the CIP originator uses the Modbus object as the basis for communication to the Modbus translator. The translator sends the appropriate function code(s) (reads and writes) to the Modbus devices, which the translator is servicing. CIP communication uses two forms of messaging called explicit messaging and implicit messaging. The Modbus translator can handle both forms of CIP communication, each being turned into Modbus requests to the Modbus devices. The Modbus responses to the translator are formatted into the appropriate CIP object service replies and sent to the CIP originator.

To ease configuration, CIP uses electronic data sheet (EDS) files with CIP configuration tools. Modbus devices can help improve the access by the CIP network to the Modbus device by providing an EDS file for the Modbus device. In order to facilitate the use of EDS files with Modbus devices, a generic Modbus EDS file was created. The generic Modbus EDS file is formatted to walk the Modbus user that is not familiar with CIP through the process of creating a meaningful Modbus device EDS file containing the significant features of the Modbus device. The CIP network manager can use this Modbus EDS file to tailor the CIP objects and services including the Modbus object that provides an interface to the Modbus translator to provide the best and most efficient communication to and from that Modbus device.

Modbus integration into the CIP environment is targeted at developers of CIP originators and at developers of CIP router devices who wish to implement the Modbus translator directly into their CIP originators and routers. The integration is also targeted to Modbus device vendors who wish to understand how their device can be utilized by a Modbus translator. The Modbus vendor through the Modbus translator can access new markets where CIP protocols are the basis for network communication.

This integration does not require a Modbus device to support specific Modbus function codes since Modbus devices may not share a common set of function codes. There are recommended Modbus function codes that a Modbus device supports to allow for the best communication through a Modbus translator. The recommended Modbus function codes are read holding registers (FC 03), write multiple registers (FC 16), read/write multiple registers (FC 23), and read device identification (FC 43/14). Using these four function codes allows for the most efficient communication between CIP originator and a Modbus device. Vendors designing new Modbus products or updating existing Modbus products are encouraged to support the four recommended Modbus function codes for best integration with CIP.

In the Modbus device, it is recommended to minimize address fragmentation when possible. Minimization of address fragmentation in the Modbus device aids communication from CIP network. The work in the Modbus translator is reduced and fewer Modbus transactions are needed to accomplish particular data exchanges.

The CIP and Modbus protocol specifications differ in the architecture, features, and operations they define. For example, the CIP protocol defines time-outs, data refresh/consumption rates, and message sequence numbers, while the Modbus protocol does not. The Modbus translator does not manage these differences between the two protocols. Some of these protocol differences can be mitigated using configuration techniques to tailor the communication system to the application.

For example, the CIP protocol uses a refresh rate to allow for communication timing differences on a CIP network or to accommodate differences in target device performance. Modbus does not define a refresh rate. In order for the CIP device to work best with the Modbus target device, it is important that the CIP refresh rate is not faster than the speed at which the Modbus target device can consume or process a corresponding write. Since the Modbus translator does not manage the refresh rate of CIP or the write processing time of the Modbus target device, it is possible to have multiple pending writes from the CIP network to the corresponding Modbus target device on the Modbus network.

To maintain proper ordering of CIP messages, CIP uses a sequence number on each message. Modbus/TCP uses a similar feature called the Modbus transaction ID. There is no correlation between the CIP sequence number and the Modbus/TCP transaction ID.

The CIP and Modbus protocols transmit multibyte data elements differently on the wire. The Modbus protocol uses *big-endian* encoding and the CIP protocol uses *little-endian* encoding. The Modbus translator will handle the byte swapping between the CIP originator and the Modbus devices.

Changes were made to the CIP protocol to accommodate the Modbus translation function of the Modbus translator. As stated earlier, these changes were kept to a minimum. The changes are as follows:

- Two new port types were added to the CIP port object. The first is the Modbus/TCP port and the second is the Modbus/SL port.
  - A Modbus/TCP port is indicated in the CIP port object as Modbus/TCP and the CIP semantic number in the object is 201.
  - A Modbus serial link port is indicated in the CIP port object as Modbus/SL and has a CIP semantic number of 202.
- The port name attribute of the CIP port object was updated to require that all CIP ports on the same physical network have the same port name.

The Modbus protocol is managed and administered by the Modbus organization. Even though Volume 7 of the CIP Networks Library is administered by the ODVA, the ODVA only governs the Modbus translator and the Modbus translation function. The Modbus protocol is independent of the ODVA, which remains solely the property of Modbus.

10.7 Modbus on Other Stacks

Modbus has been architected and deployed on several stacks, like power line carrier stacks, many 802.15.4 stacks, and cellular.

10.8 Conformance

Modbus.org provides conformance tests; see www.Modbus.org.

11

PROFIBUS

11.1 Basics ... **11**-1
11.2 Transmission Technologies ... **11**-3
EIA/TIA-485 (Former RS485) • Manchester Encoded Bus Powered (MBP, IEC 61158-2) • Fiber Optic
11.3 Communication Protocol ... **11**-9
System Configuration and Device Types • Cyclic and Acyclic Data Communication Protocols
11.4 Application Profiles ... **11**-14
Common Application Profiles • Specific Application Profiles • Host Application Profiles
11.5 Integration Technologies ... **11**-22
GSD • EDD • FDT/DTM • Field Device Integration
11.6 Technical Support ... **11**-25
Quality Assurance • Implementation Support
11.7 Wide Range of Applications ... **11**-28
Abbreviations ... **11**-29
References ... **11**-29

Ulrich Jecht
UJ Process Analytics

Wolfgang Stripf
PROFIBUS and PROFINET International

Peter Wenzel
PROFIBUS and PROFINET International

11.1 Basics

Fieldbuses are industrial communication systems with bit-serial transmission that use a range of media such as copper cable, fiber optics, or radio transmission to connect distributed field devices (sensors, actuators, drives, transducers, analyzers, etc.) to a central control or management system. Fieldbus technology was developed in the 1980s with the aim to save cabling costs by replacing the commonly used central parallel wiring and dominating analog signal transmission (4–20 mA or ±10 V interface) with digital technology. Due to different industry-specific demands, the result of sponsored research and development projects, or the preferred proprietary solutions of large system manufacturers, several bus systems with varying principles and properties were established in the market. The key technologies are now included in the adopted standards IEC 61158 and IEC 61784 [1]. PROFIBUS is an integral part of these standards.

Fieldbuses originally created the basic prerequisite for distributed automation systems in the manufacturing industry. Meanwhile, they evolved to a standard technology also in process automation where they significantly increased productivity and flexibility compared to conventional technology.

PROFIBUS is an open, digital communication system with a wide range of applications, particularly in the fields of factory and process automation, transportation, and power distribution. PROFIBUS is suitable for both fast, time-critical applications and complex communication tasks (Figure 11.1).

The application and engineering aspects are specified in the generally available guidelines of PROFIBUS & PROFINET International (PI) [2]. This fulfills user demand for standardization, manufacturer independence and openness, and ensures communication between the devices of various manufacturers.

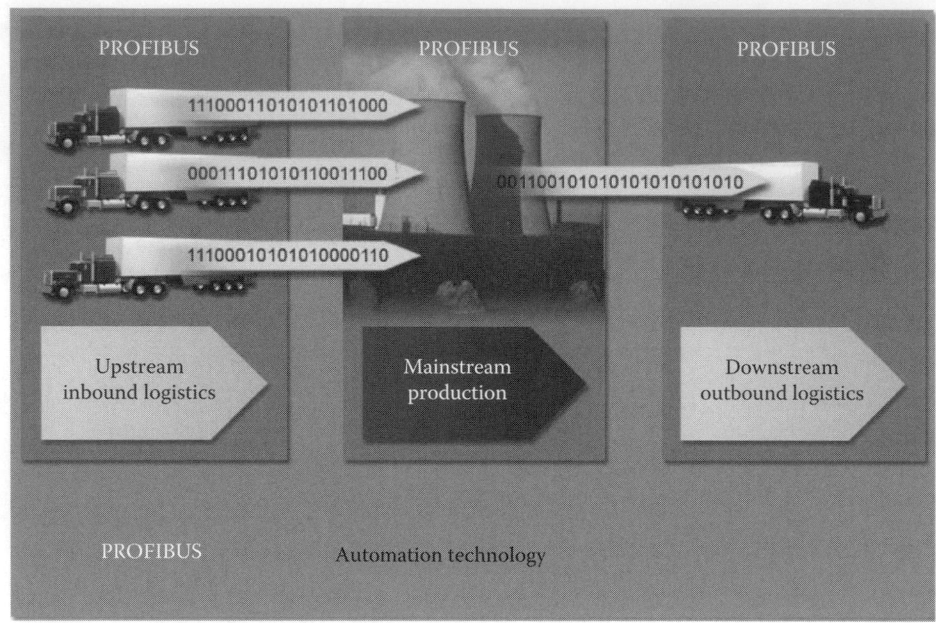

FIGURE 11.1 PROFIBUS suitable for all decentralized applications.

Based on a very efficient and extensible communications protocol combined with the development of numerous application profiles (communication models for device type families) and a fast-growing number of devices and systems, PROFIBUS began its market success initially in factory automation, and then in process automation, from 1995. Today, PROFIBUS is the world leader in the fieldbus market with more than 20% share of the market, over 1 million plants equipped with the PROFIBUS installations, and more than 50 million nodes. Today, there are more than 2500 PROFIBUS products available from a wide range of manufacturers.

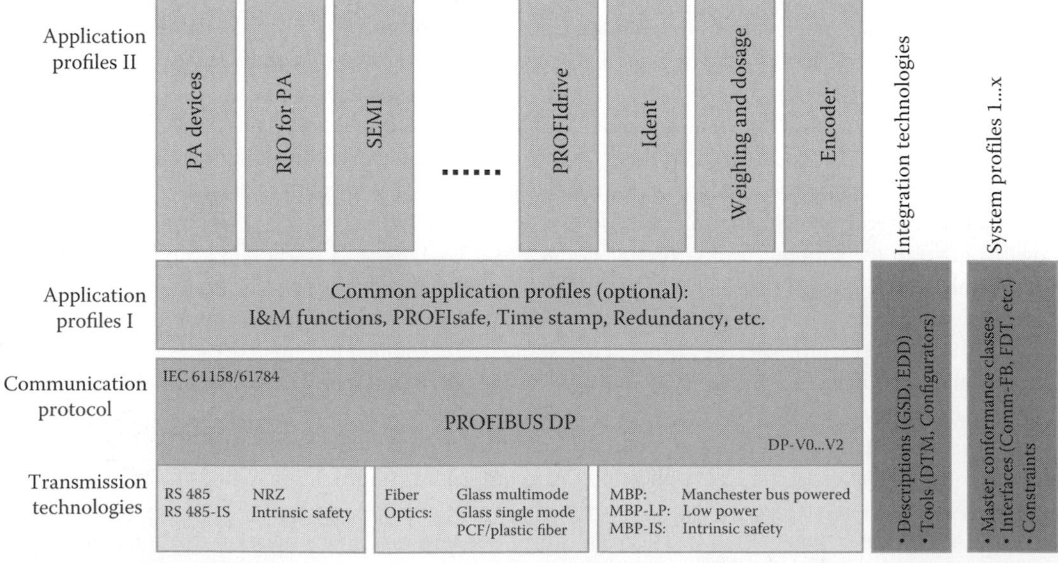

FIGURE 11.2 Structure of PROFIBUS system.

The success of PROFIBUS stems in equal measures from its progressive technology and the strength of its noncommercial PROFIBUS User Organization e.V. (PNO), the trade body of manufacturers and users founded in 1989. Together with other 27 regional associations in countries all around the world, and the international umbrella organization PI founded in 1995, this pan-national organization now totals more than 1400 members worldwide. The objectives are continuing development of the PROFIBUS technology and increase of its worldwide acceptance.

PROFIBUS has a modular structure (PROFIBUS Tool Box) and offers a range of transmission and communication technologies, numerous application and system profiles, as well as device management and integration tools [3]. PROFIBUS covers various and application-specific demands from the field of factory to process automation, from simple to complex applications, by selecting the adequate set of components out of the toolbox (Figure 11.2).

11.2 Transmission Technologies

The physical layer is the first layer of the Open System Interconnection (OSI) model. It deals with bit-level transmission between different devices and supports electrical or mechanical interfaces connecting to the physical medium for synchronized communication. The physical layer is usually a combination of software and hardware programming and may include electromechanical devices. It does not include the physical media as such. PROFIBUS features a number of different transmission technologies, all of which are based on international standards. They all are assigned to PROFIBUS in both IEC 61158 and IEC 61784: RS485/485 IS, MBP/MBP-IS (IS stands for intrinsic safety protection), and fiber optics.

11.2.1 EIA/TIA-485 (Former RS485)

EIA/TIA-485 transmission technology is simple and cost-effective and is primarily used for tasks that require high transmission rates. Shielded, twisted pair copper cable with one conductor pair is used. The bus structure allows the addition or removal of stations, or the step-by-step commissioning of the system without interfering with the operation of other stations. Subsequent expansions (within defined limits) have no effect on stations already in operation.

In 1983, the Electronics Industries Association (EIA) approved a new *balanced transmission standard* called, at that time, Recommended Standard (RS) 485 (RS485). It found widespread acceptance and usage in industrial, medical, and consumer applications. Subsequently, the Telecommunications Industry Association/Electronic Industries Association (TIA/EIA) officially replaced *RS* with *EIA/TIA* to help identify the origin of its standards (TIA/EIA).

EIA/TIA-485 is a standard defining the electrical characteristics of communication in balanced digital multipoint systems. Digital communications networks implementing the EIA/TIA-485 standard can be used effectively over long distances and in electrically noisy environments.

EIA/TIA is an electrical aspect-only standard. In contrast to complete interface standards, which define the functional, mechanical, and electrical specifications, it only defines the electrical characteristics of drivers and receivers that could be used to implement a balanced multipoint transmission line. This standard is intended to be referenced by higher level standards, specifying EIA/TIA-485 as the physical layer standard.

EIA/TIA-485 refers to a physical layer that uses a differential voltage to transmit the data. An electrical circuit in the receiving device measures the voltage between the two incoming lines (Figure 11.3) and looks at the difference between them to determine which stands for a 0 and which for a 1.

- If the difference (A − B) is between −1.5 and −6 V, it is a *logical* 1.
- If the difference is between +1.5 and +6 V, it is a *logical* 0.

As balanced system, EIA/TIA-485 is inherently more immune to noise. If fed via a twisted pair, noise induced into the A line is the same as that induced into the B line. When the difference is calculated, the noise is eliminated and the system will still work.

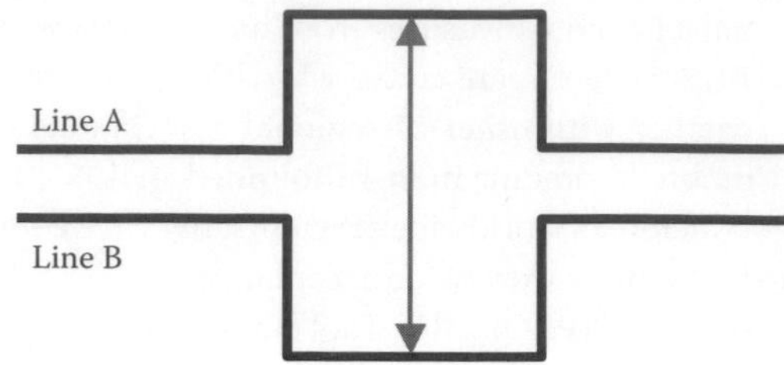

FIGURE 11.3 Logical 1 and 0 determination (EIA/TIA-485).

11.2.1.1 Number of Devices and Type of Cables

EIA/TIA-485 can have as many as 32 devices (masters or slaves) on one segment. A segment is a section of the network, where the same electrical signal flows uninterrupted. The 32 device limit is set by the natural limitations of an electrical circuit. After the signal has been sensed by 31 receivers, the signal degrades to a point where it cannot be read correctly. The number of devices can be expanded by using repeaters.

Different cable types (designated A – D) for different applications are available for connecting devices either to each other or to network elements (segment couplers, links, and repeaters).

The PI recommends that when EIA/TIA-485 is used, cable type A be used.

11.2.1.2 Network Speed and Distances

The PROFIBUS version of EIA/TIA-485 is a modification of the original standard, which could only transmit a maximum rate of 187.5 kbps. The network speed can be set to different transmission rates ranging from 9.6 to 12 Mbps with a typical transmission rate of 1.5 Mbps.

There is a proportional relationship between the transmission rate and the maximum distance of the segment (sum of all cable lengths, including spurs). As the distance increases, the maximum transmission rate decreases (Table 11.1).

11.2.1.3 Intrinsically Safe Operation

EIA/TIA-485-IS, in its IS-variant, supports the use of fast transmission rates within intrinsically safe areas. A PROFIBUS guideline [4] is available for the configuration of intrinsically safe RS485 solutions

TABLE 11.1 Transmission Rates and Cable Lengths

Transmission Rate [Kbit/s]	Transmission Range per Segment [m]	Applies to
9,6 19,2 45,45 93,75	1200	RS485
187.5	1000	RS485
500	400	RS485
1500	200	RS485
3000 6000 12000	100	RS485
31.25	1900	MBP
The values above apply to cable type A with the following properties		
Wave resistance	135 ... 165 Ω	
Capacitance per unit	≤30 pf/m	
Loop resistance	≤110 Ω/km	
Core diameter	>0.64 mm	
Core cross-section	>0,34 mm²	

with simple device interchangeability. The interface specification details the levels for current and voltage that must be adhered to by all stations in order to ensure safe operation during interconnection. An electric circuit limits currents at a specified voltage level. When active sources are connected, the sum of the currents of all stations must not exceed the maximum permissible current. In contrast to the fieldbus intrinsically safe concept (FISCO) model, all stations represent active sources. Up to 32 stations may be connected to the intrinsically safe bus circuit.

11.2.1.4 Conclusion

EIA/TIA-485 is the ideal physical layer for PROFIBUS DP (Decentralized Peripherals) communication:

- It has great noise immunity.
- It can go over long distances.
- It can include up to 32 devices on one segment.
- It can transmit data up to 12 Mbps.

11.2.2 Manchester Encoded Bus Powered (MBP, IEC 61158-2)

MBP (*M*anchester encoded, *B*us *P*owered) transmission technology is a new term that replaces the previously common terms for intrinsically safe transmission such as *physics in accordance with IEC 61158-2*, *1158-2*, etc. The current version of the IEC 61158-2 (physical layer) describes several different transmission technologies, MBP technology being just one of them. Thus, differentiation in naming was necessary.

MBP is a synchronous, Manchester-coded transmission with a defined transmission rate of 31.25 kbps. In its version *MBP-IS*, MBP is frequently used in process automation as it satisfies the key demands of the chemical and petrochemical industries for IS and bus powering using two-wire technology (Figure 11.4).

11.2.2.1 Power and Communication on the Same Cable

The IEC 61158-2 Standard defines the physical layer that uses a special method called MBP to deliver power (direct current) and digital signal communication (alternating current) on the same two-wire

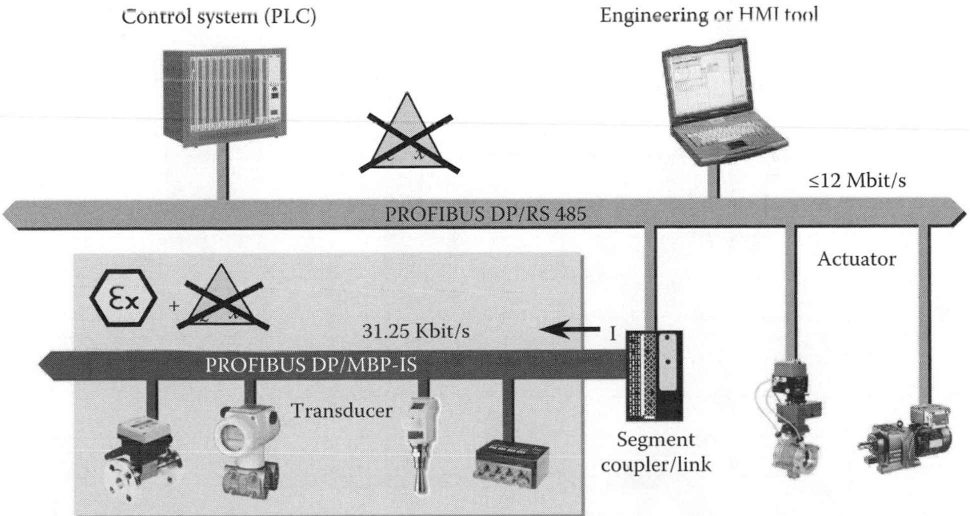

FIGURE 11.4 Intrinsic safety and powering of field devices using MBP-IS.

TABLE 11.2 MBP Characteristics

Fieldbus Standard IEC 61158-2 for MBP Transmission Technology
Up to 32 nodes in one segment
Data transmission rate 31.25 Kbit/s
Per field device: Min. working voltage 9 V DC Min. current consumption 10 mA
Transmission of digital communication signal in zero-mean Manchester II coding (MBP) through ±9 mA amplitude
Signal transmission and remote power supply using twisted-pair cable
Fieldbus cable type A
Connection of field devices via stubs (spur) to a main cable (trunk) for trouble-free disconnection of devices without affecting other nodes
Max. total length of main cable, including all stubs, is 1900 m

cable. Information is transmitted by varying the power draw on the power cable. This standard defines some general conditions for the structure of a fieldbus network. Table 11.2 summarizes the most decisive of these conditions.

11.2.2.2 Manchester Coding

In telecommunication, Manchester coding is a line code in which the encoding of each data bit has at least one transition and takes place in a fixed time period. Manchester code always has a transition in the middle of each bit period and may have, depending on the information to be transmitted, a transition at the start of the period as well. The direction of the mid-bit transition indicates the data, while transitions at the period boundaries do not carry information and are used to place the signal in the correct state to allow the mid-bit transition.

The minimum base current is 10 mA. The physical layer works by having the devices vary the power drawn over time. The transmission of a 1 or 0 depends on whether the power is increased or decreased (Figure 11.5)

- If the power draw at the mid-bit transition goes from high to low, then it is a logical 1.
- If the power goes from a low to a high, then it is a logical 0.

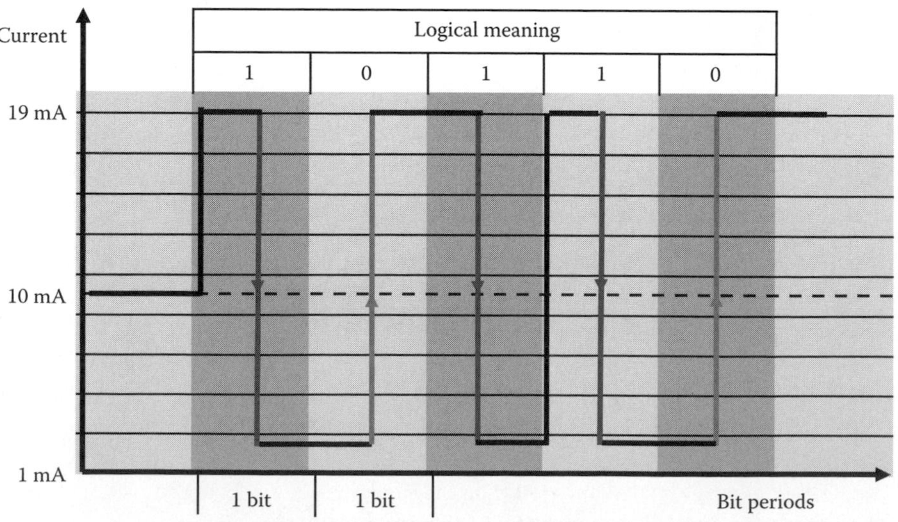

FIGURE 11.5 Logical 1 and 0 determination (MBP).

11.2.2.3 Application and Topology

MBP transmission technology according to IEC 61158-2 type 1 is used with PROFIBUS PA. The same procedure is also used with other fieldbuses, such as the Foundation Fieldbus (FF).

In its version MBP-IS, MBP can be used in potentially explosive environments of the chemical and petrochemical industries for IS and bus powering using two-wire technology (EEx ia/ib IIC).

MBP transmission technology is usually limited to a specific segment (field devices in hazardous areas) of a plant, which is then linked to an RS485 segment via a segment coupler or links (Figure 11.4).

PROFIBUS DP acts as backbone of PROFIBUS PA segments, which are attached through couplers, links, or gateways. Different equipment is available for that from various manufacturers. They provide different solutions, for example, regarding transmission rate on the PROFIBUS DP side (fixed or any transmission rate) or the use of address space (same or different) on PROFIBUS PA side as on PROFIBUS DP side. More information is available in [5]. Tree or line structures (and any combination of the two) are network topologies supported by PROFIBUS with MBP transmission with up to 32 stations per segment and a maximum of 126 per network.

11.2.2.4 FISCO

The internationally recognized FISCO model considerably simplifies the planning, installation, and expansion of PROFIBUS networks in potentially explosive areas. It has been developed by the German Pysikalisch-Technische Bundesanstalt (PTB) [6]. The model is based on the specification that a network is intrinsically safe and requires no individual intrinsic safety calculations when the relevant four bus components (field devices, cables, segment couplers, and bus terminators) fall within predefined limits with regard to voltage, current, output, inductance, and capacity. The corresponding proof can be provided by certification of the components through authorized accreditation agencies, such as PTB (Germany) or Underwriters Laboratories Inc. (UL) and Factory Mutual Global (FM) (the United States) and others.

If FISCO-approved devices are used, not only is it possible to operate more devices on a single line, but the devices can also be replaced during runtime by the devices of other manufacturers, or the line can be expanded—all without the need for time-consuming calculations or system certification. That allows for *plug and play*, even in hazardous areas!

11.2.2.5 High-Power Trunk Concept

IS is the method of choice for instrument connections in hazardous areas, but it does not satisfy completely the needs regarding the cable length and a number of devices, compared with applications outside of hazardous areas.

A new approach to hazardous area applications is based on the *high-power trunk* concept (Figure 11.6), which may be applied to areas where no access is required for maintenance or device replacement during its operation. The high-power trunk concept utilizes a trunk, which is protected using increased safety ignition protection. The trunk is installed in Zone 1/Class I, Div. 1/2 and allows a supply current of up to 1000 mA. Field barriers (known also as field couplers) are connected to the trunk and provide galvanically isolated, intrinsically safe outputs for field device connection. Field devices may be located in Zone 1/Class I, Div. 1/2 or Zone 0/Class I, Div. 1. Up to 40 mA output current is made available for each device, which is sufficient even for high-performance field devices. This allows end users to get the maximum number of devices on a segment while also being able to achieve maximum cable lengths. Typical values for a high-power trunk solution are 30 V at 500 mA.

11.2.2.6 Conclusion

- MBP provides communication and power on the same two wires, saving considerable installation costs because power and communication do not require separate lines.
- MBP works with the current signal on the bus.

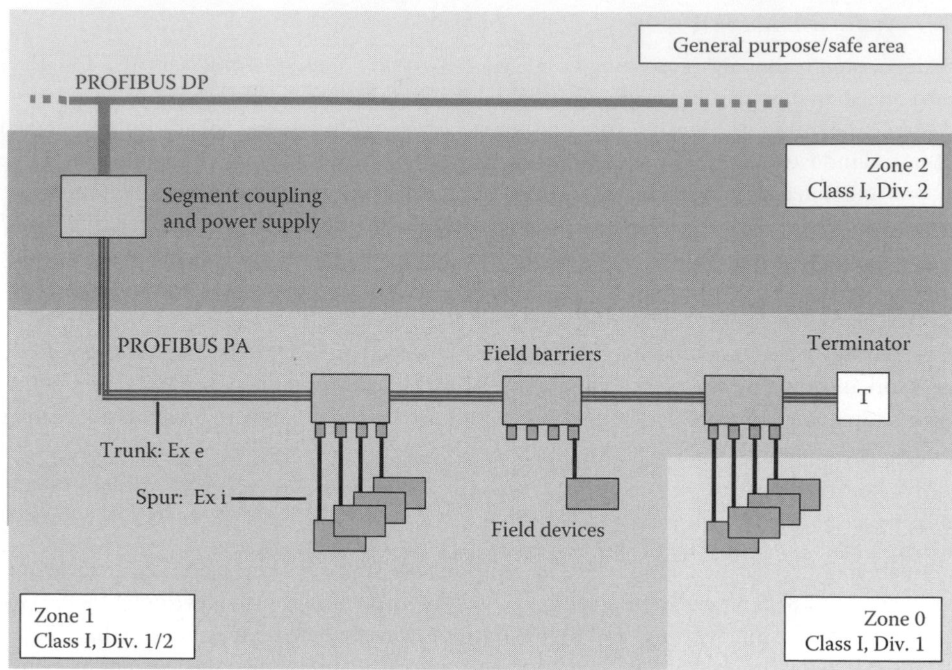

FIGURE 11.6 High-power trunk.

- MBP has considerable electrical noise immunity, so it can be used in industrial settings where electromagnetic noise typically exists.
- A logical "1" is defined as a midpoint transition from low to high, and a "0" is a midpoint transition from high to low.
- MBP segments are created with DP/PA couplers or link modules. Each MBP segment can connect up to 32 stations (including coupler or linking device) with a total distance of up to 1.9 km. With MBP-IS explosion-save installations, these values are reduced.

11.2.3 Fiber Optic

Under certain conditions, wired transmission technology reaches its limits, for example, in an environment with heavy interferences or when bridging long distances. In these cases, optical transmission via fiber optic cables is a good solution. The fiber optical physical layer uses light to transmit data.

Modern fiber optic communication systems generally include an optical transmitter to convert an electrical signal into an optical signal to send into the optical fiber, a cable containing bundles of multiple optical fibers, multiple kinds of amplifiers, and an optical receiver to recover the signal as an electrical signal.

PROFIBUS has a number of different fiber optic solutions specified depending on cost and distances (see Table 11.3). The available technologies for fiber optic transmission include multimode and single-mode glass fiber, plastic fiber, and HCS® fiber.

The implementation of a fiber optic cable network in the simplest case involves the use of electro-optical converters, which are connected to the field device with the RS485 interface and the fiber optic cable on the other side. This also makes it possible to switch between RS485 and fiber optic cable transmission within an automation system, depending on the prevalent conditions. Due to the transmission characteristics, typical topology structures are star and ring, but linear structures are also possible.

TABLE 11.3 Supported Fiber Optic Cable Types

Fiber Type	Core Diameter [μm]	Transmission Range
Multimode glass fiber	62,5/125	2–3 km
Single-mode glass fiber	9/125	>15 km
Plastic fiber	980/1000	Up to 100 m
HCS® fiber	200/230	Approx. 500 m

11.3 Communication Protocol

At the protocol level, PROFIBUS with DP and its versions DP-V0 to DP-V2 offers a broad spectrum of optional services, which enable optimum communication between different applications [3,7]

DP has been designed for a fast data exchange at a field level. Data exchange with distributed devices is primarily cyclic. The communication functions required for this are specified through the DP basic functions (version DP-V0). Geared toward the special demands of the various areas of application, these basic DP functions have been expanded step-by-step with special functions so that DP is now available in three versions, DP-V0, DP-V1, and DP-V2, whereby each version has its own special key features. All versions of DP are specified in detail in the IEC 61158 and IEC 61784.

Version DP-V0 provides the basic functionality of DP, including cyclic data exchange as well as station diagnosis, module diagnosis, and channel-specific diagnosis.

Version DP-V1 contains enhancements geared toward process automation, in particular, acyclic data communication for parameter assignment, operation, visualization, and alarm handling of intelligent field devices, in coexistence to cyclic user data communication. This permits online access to stations using engineering tools. In addition, DP-V1 defines alarms. Examples of different types of alarms are status alarm, update alarm, and a manufacturer-specific alarm.

Version DP-V2 contains further enhancements and is geared primarily toward the demands of the drive technology. Due to additional functionalities, such as isochronous slave mode and slave-to-slave(s) communication (data exchange broadcast [DXB]), etc., the DP-V2 can also be implemented as a drive bus for controlling fast movement sequences of drive axes.

11.3.1 System Configuration and Device Types

DP supports the implementation of both monomaster and multimaster systems. This affords a high degree of flexibility during system configuration. A maximum of 126 devices (masters or slaves) can be connected to a bus segment. In monomaster systems, only *one* master is active on the bus during the operation of the bus system. Figure 11.7 shows the system configuration of a monomaster system. In this case, the master is hosted by a programmable logic controller (PLC).

The PLC is the central control component. The slaves are connected to the PLC via the transmission medium. This system configuration enables the shortest bus cycle times. In multimaster systems, several masters are sharing the same bus. They represent both independent subsystems, comprising masters and their assigned slaves, and additional configuration and diagnostic master devices. The masters coordinate themselves by passing a token from one to the next. Only the master that holds the token can communicate. PROFIBUS DP differentiates three groups of device types on the bus.

DP master class 1 (DPM1) is a central controller that cyclically exchanges information with the distributed stations (slaves) at a specified message cycle. Typical DPM1 devices are PLCs or personal computers (PCs). A DPM1 has active bus access with which it can read measurement data (inputs) of the field devices and write the setpoint values (outputs) of the actuators at fixed times. This continuously repeating cycle is the basis of the automation function.

DP masters class 2 (DPM2) are engineering, configuration, or operating devices. They are put in operation during commissioning and for maintenance and diagnostics in order to configure connected

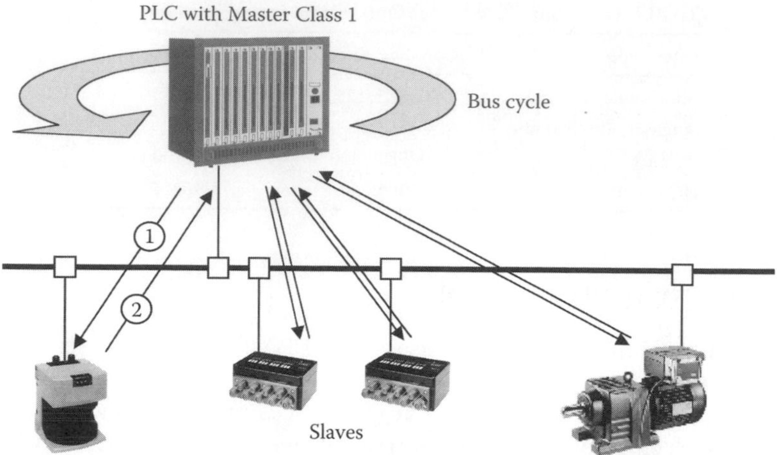

FIGURE 11.7 PROFIBUS DP monomaster system (DP-V0).

devices, evaluate measured values and parameters, and request the device status. A DPM2 does not have to be permanently connected to the bus system. It also has active bus access.

Slaves are peripherals (input/output [I/O] devices, drives, human machine interfaces [HMIs], valves, transducers, and analyzers), which read in process information and/or use output information to intervene in the process. There are also devices that solely provide input information or output information. As far as communication is concerned, slaves are passive devices, and they only respond to direct queries (see Figure 11.7, sequence ① and ②). This behavior is simple and cost-effective to implement. In the case of DP-V0, it is already completely included in the bus application-specific integrated circuit (ASIC).

11.3.2 Cyclic and Acyclic Data Communication Protocols

Cyclic data communication between the DPM1 and its assigned slaves is automatically handled by the DPM1 in a defined, recurring sequence (Figure 11.7). The appropriate services are called cyclic master slave communication services of PROFIBUS DP (MS0). The user defines the assignment of the slave(s) to the DPM1 when configuring the bus system. The user also defines which slaves are to be included/excluded in the cyclic user data communication. DPM1 and the slaves are passing three phases during start-up: parameterization, configuration, and cyclic data exchange (Figure 11.8).

Before entering the cyclic data exchange state, the master first sends information about the transmission rate, the data structures within a protocol data unit (PDU), and other slave-relevant parameters. In a second step, it checks whether the user-defined configuration matches the actual device configuration. Within any state, the master is enabled to request slave diagnosis in order to indicate faults to the user.

An example for the telegram structure for the transmission of information between master and slave is shown in Figure 11.9. The telegram starts with some synchronization bits, the type (SD) and length (LE) of the telegram, source and destination address, and a function code (FC). The FC indicates the type of message or content of the *load* (processing data unit) and serves as a guard to control the state machine of the master. The PDU, which may carry up to 244 bytes, is followed by a safeguard mechanism frame checking sequence (FCS) and a delimiter (see Figure 11.9 for abbreviations).

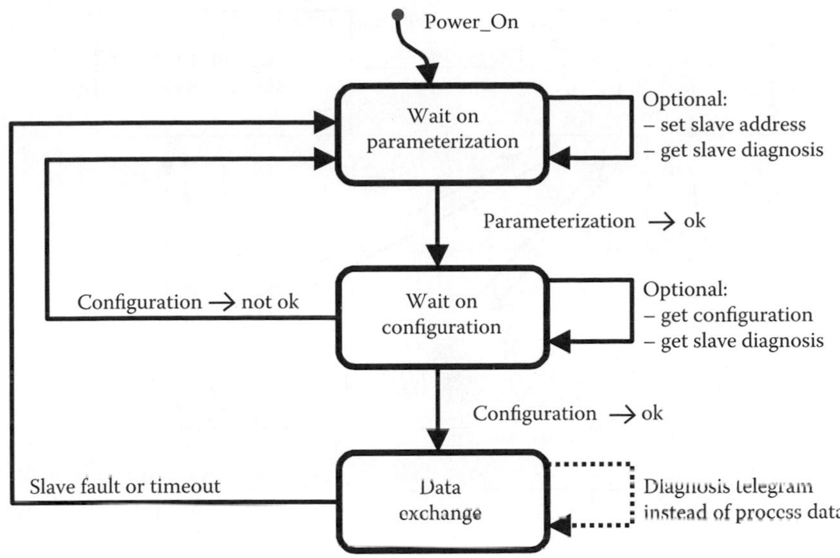

FIGURE 11.8 State machine for slaves.

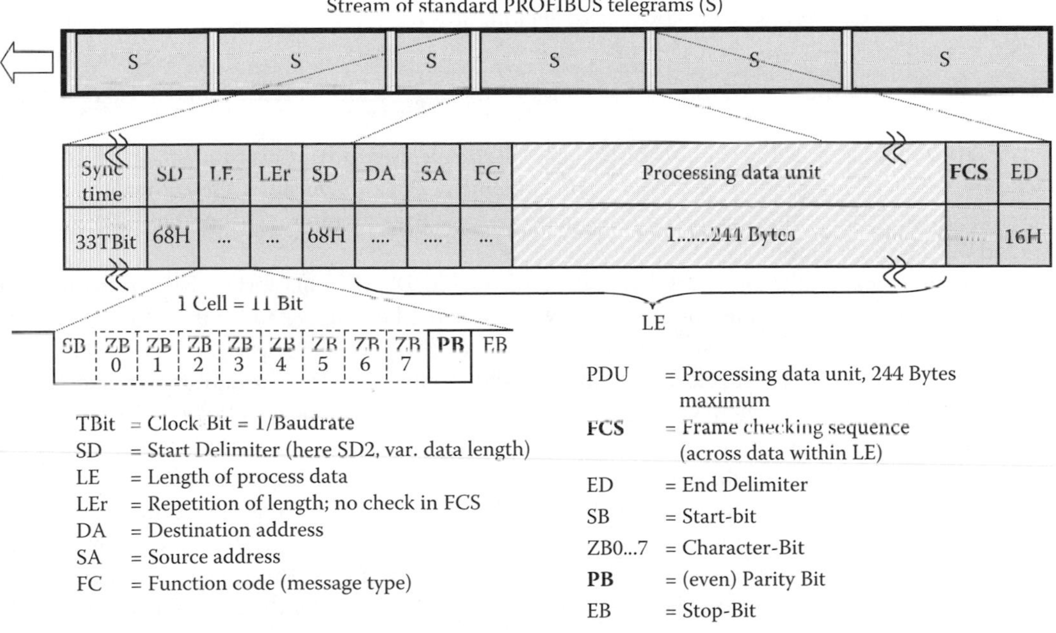

FIGURE 11.9 PROFIBUS DP telegram structure (example).

One example for the usage of the FC is the indication of a fault situation on the slave side. In this case, the master sends a special diagnosis request instead of the normal process data exchange that the slave replies with a diagnosis message. The message comprises six bytes of fixed information and user-definable device and module- or channel-related diagnosis information.

In addition to the single station-related user data communication, which is automatically handled by the DPM1, the master can also send control commands to all slaves or a group of slaves simultaneously.

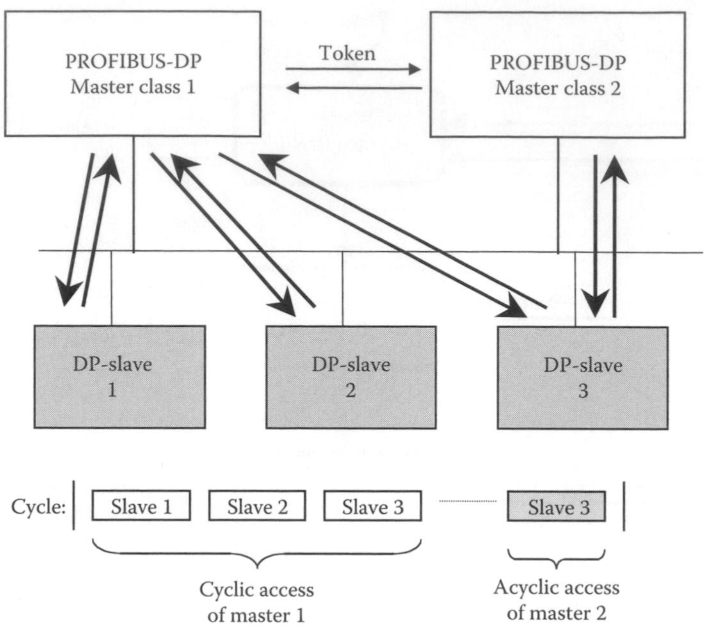

FIGURE 11.10 Cyclic and acyclic data communication with DP-V1.

These control commands are transmitted as multicast messages and enable sync and freeze modes for event-controlled synchronization of the slaves.

For safety reasons, it is necessary to ensure that DP has effective protective functions against incorrect parameterization or failure of transmission equipment. For this purpose, the DP master and the slaves are fitted with monitoring mechanisms in the form of time monitors. The monitoring interval is defined during configuration.

Acyclic data communication is the key feature of version DP-V1. This forms the requirement for parameterization and calibration of the field devices over the bus during runtime and for the introduction of confirmed alarm messages. Transmission of acyclic data is executed parallel to cyclic data communication, but with lower priority. Figure 11.10 shows some sample communication sequences for a master class 2, which is using MS2 services. In using MS1 services, a master class 1 is able to execute acyclic communications also.

Slave-to-slave communications (DP-V2) enable direct and time-saving communication between slaves using broadcast communication without the detour over a master. In this case, the slaves act as *publisher*; that is, the slave response does not go through the coordinating master but directly to other slaves embedded in the sequence, the so-called *subscribers* (Figure 11.11). This enables slaves to directly read data from other slaves and use them as their own input. This opens up the possibility of completely new applications; it also reduces response times on the bus by up to 90%.

Isochronous mode (DP-V2) enables clock synchronous control in masters and slaves, irrespective of the busload.

The function enables highly precise positioning processes with clock deviations of less than one microsecond (<1 μs). All participating device cycles are synchronized to the bus master cycle through a *global control* broadcast message. A special sign of life (consecutive number) allows monitoring of the synchronization.

Clock control (DP-V2) via a new synchronizing mechanism between masters and slaves synchronizes all stations to a system time with a deviation of less than one millisecond (<1 ms). This allows the precise tracking of events. This is particularly useful for the acquisition of timing functions in networks

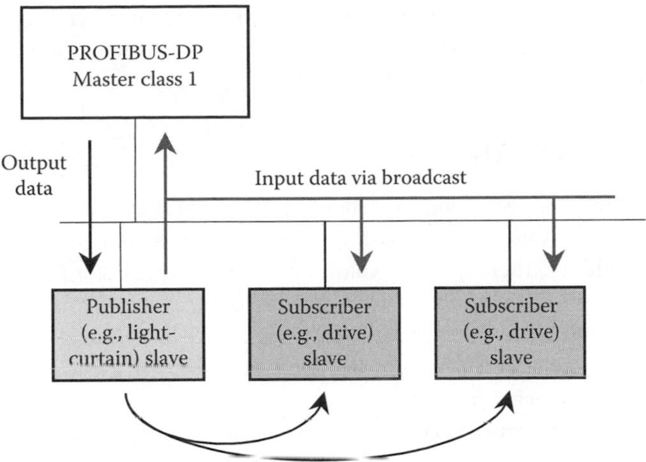

FIGURE 11.11 Slave–to-slave data exchange.

with numerous masters. This facilitates the diagnosis of faults as well as the chronological planning of events.

Upload and download (DP-V2) allows up- and downloading of any amount of data into or from a field device with the help of a few commands. Within IEC 61158, these services are called load region. This enables, for example, programs to be updated or devices replaced without the need for manually loading processes.

Addressing with slot and index is used both for cyclic and acyclic communication services (Figure 11.12).

When addressing data, PROFIBUS assumes that the physical structure of the slaves is modular or can be structured internally in logical functional units, the so-called modules (Figure 11.12). The slot number addresses the module, and the index addresses the data records assigned to a module. Each data record

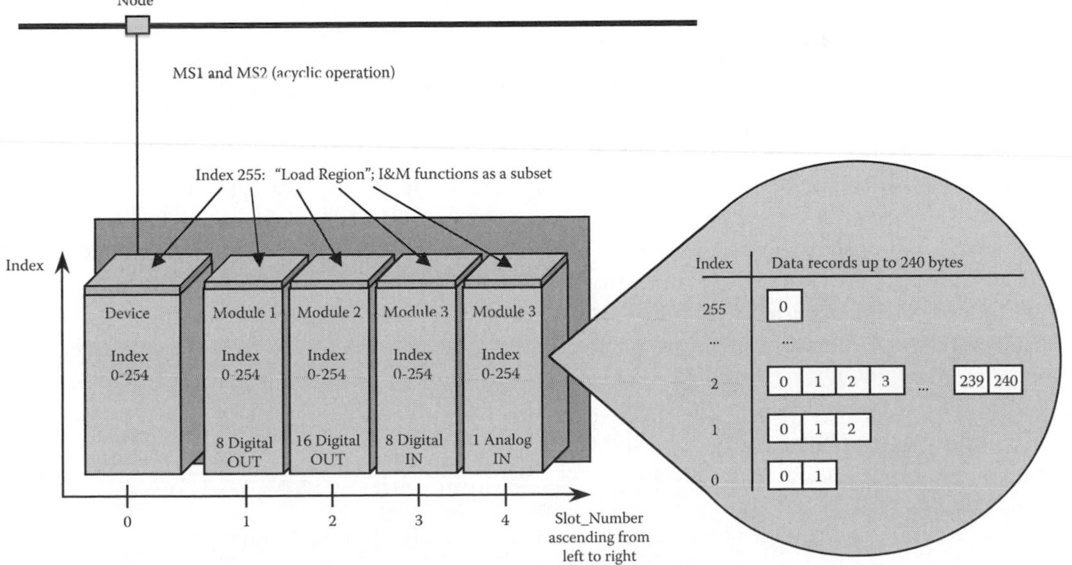

FIGURE 11.12 Slot/index address model of a slave.

can be up to 244 bytes. The modules begin at slot 1 and are numbered in ascending contiguous sequence. The slot number 0 is for the device itself. Compact devices are regarded as a unit of virtual modules. These can also be addressed with slot number and index.

11.4 Application Profiles

Profiles are used in automation technology to define specific properties and behavior for devices, device families, or entire systems. Only devices and systems using a vendor-independent profile provide interoperability on a fieldbus, thereby fully exploiting the advantages of a fieldbus. Profiles take into account application and type-specific special features of field devices, controls, and methods of integration (engineering). The meaning of the term *profile* ranges from just a few specifications for a specific device class through comprehensive specifications for applications in a specific industry. The generic term for all profiles is the application profiles.

A distinction is thus drawn between common application profiles with implementation options for different applications (this includes, e.g., the profiles identification and maintenance [I&M] functions, PROFIsafe, redundancy, and time stamp); specific application profiles, which are developed for a specific application such as PROFIdrive, Ident Systems, or PA Devices; and host profiles, which describe specific system performance that is available to field devices.

PROFIBUS offers a wide range of such application profiles, which allow application-oriented implementation.

11.4.1 Common Application Profiles

I&M functions are mandatory for all PROFIBUS devices with MS1 and/or MS2 services. The main purpose of the I&M functions described hereinafter is to support the end user during various scenarios of a device's life cycle, be it configuration, commissioning, parameterization, diagnosis, repair, firmware update, asset management, audit trailing, and alike. It is kind of a *type plate* or *boiler plate*. The corresponding parameters all are stored at the same address space within the PROFIBUS slot/index address model.

The *call* mechanism of the load region services [1] opens up an additional subindex address space of 65,535 data records. I&M functions are assigned a space between 65,000 and 65,199 for basic, profile-specific, and manufacturer-specific items. Table 11.4 itemizes the individual parameters.

Nowadays, laptops or engineering tools normally have access to the Internet. Thus, it is quite easy to reference a central database (e.g., on the PROFIBUS web server) and retrieve comprehensive and always actual information, even in a desired language (Figure 11.13).

PROFIsafe is a comprehensive, open fieldbus solution for safety-relevant applications without the use of a second relay-based layer or proprietary safety buses. PROFIsafe defines how fail-safe devices (emergency stop pushbuttons, light curtains, level switches, etc.) can communicate over PROFIBUS with fail-safe controllers in such a manner that they can be used for safety-relevant automation tasks up to category 4 compliant with EN954 (ISO 13849) or SIL3 (*S*afety *I*ntegrity *L*evel) according to IEC 61508. It implements safe communications over a profile, that is, over a special PROFIsafe data frame and a special protocol. PROFIsafe is a single-channel software solution, which is implemented in the devices as an additional layer *above* layer 7 (Figure 11.14); the standard PROFIBUS components, such as lines, ASICs, or protocols, remain unchanged. This ensures redundancy mode and retrofit capability. Devices with the PROFIsafe profile can be operated in coexistence with standard devices without restriction on the same bus (cable).

PROFIsafe takes advantage of the acyclic communication (DP-V1) for full maintenance support of the devices and can be used with RS485, fiber optic, or MBP transmission technology. This ensures both fast response times (important for the manufacturing industry) and low power consumption with intrinsically safe operation (important for process automation).

TABLE 11.4 Basic Identification and Maintenance Functions

I&M Function	Data Format	Notes
MANUFACTURERS	2 Octets	I&M functions are using company IDs instead of names. These IDs are harmonized with the list of the HART Foundation and comprise extensions for additional companies.
ORDER_ID	20 Octets	Order number for a particular device type. For virtual modular devices, the root or highest level of the basic device.
SERIAL_NUMBER	16 Octets	Unique identifier for a particular device (counter).
HARDWARE_REVISION	2 Octets	The content of this parameter characterizes the edition of the hardware only.
SOFTWARE_REVISION	4 Octets	The content of this parameter characterizes the edition of the software or firmware of a device or module. The structure supports coarse and detailed differentiation that may be defined by the manufacturer: Vx.y.z.
REV_COUNTER	2 Octets	Indicates unplugging of modules or write access.
PROFILE_ID	2 Octets	Device or module corresponds to a particular PROFIBUS profile.
PROFILE_SPECIFIC_TYPE	2 Octets	This identifier references a device class defined within a PROFIBUS profile.
IM_VERSION	2 Octets	Version of the I&M functions implemented within a device or module.
IM_SUPPORTED	2 Octets	Directory for the subset of I&M functions implemented within a device or module.
TAG_FUNCTION	32 Octets	User-definable information about the *role* of the device within a plant facility.
TAG_LOCATION	22 Octets	User-definable information about the *location coordinates* of the device within a plant facility.
INSTALLATION_DATE	16 Octets	Indicates the date of installation or commissioning of a device or module, e.g., 1995-02-04 16:23.
DESCRIPTOR	54 Octets	User-defined comments.
SIGNATURE	54 Octets	Allows parameterization tools to store a *security* code as a reference for a particular parameterization session and audit trail tools to retrieve the code for integrity checks. Used for safety applications according 21 CFR 11 [6] or hazardous machinery (PROFIsafe).

HART on PROFIBUS DP integrates Highway Addressable Remote Transducer (HART) protocol devices installed in the field, in existing or new PROFIBUS systems. It includes the benefits of the PROFIBUS communication mechanisms without any changes required for the PROFIBUS protocol and services, the PROFIBUS PDUs or the state machines, and functional characteristics. This profile is implemented in the master and slave above layer 7, thus enabling mapping of the HART client-master-server model on PROFIBUS. The cooperation of the HART Foundation on the specification work ensures complete conformity with the HART specifications.

The HART-client application is integrated in a PROFIBUS master, and the HART master in a PROFIBUS slave, whereby the latter serves as a multiplexer and handles communication to the HART devices (Figure 11.15).

The *time stamp* application profile describes mechanisms as to how to supply certain events and actions with a time stamp, which enables precise time assignment. Precondition is a clock control in the slaves through a clock master via special services. An event can be given a precise system time stamp and read out accordingly. A concept of graded messages is used. The message types are summarized under the term *Alerts* and are divided into high-priority *alarms* (these transmit a diagnostics message) and low-priority *events*. In both cases, the master acyclically reads the time-stamped process values and alarm messages from the alarm and event buffer of the field device (Figure 11.16).

The *slave redundancy* application profile provides a slave redundancy mechanism (Figure 11.17): slave devices contain two different PROFIBUS interfaces that are called primary and backup (slave interface). These may be either in a single device or distributed over two devices.

The devices are equipped with two independent protocol stacks with a special redundancy expansion.

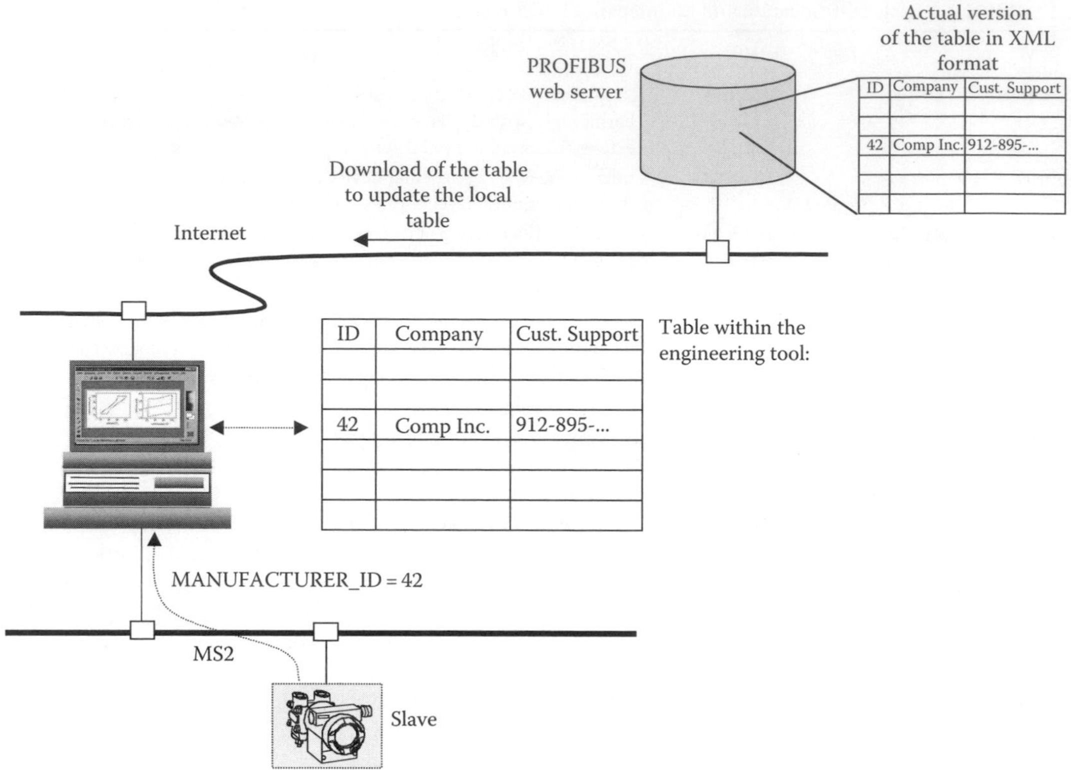

FIGURE 11.13 Referencing IDs via the Internet.

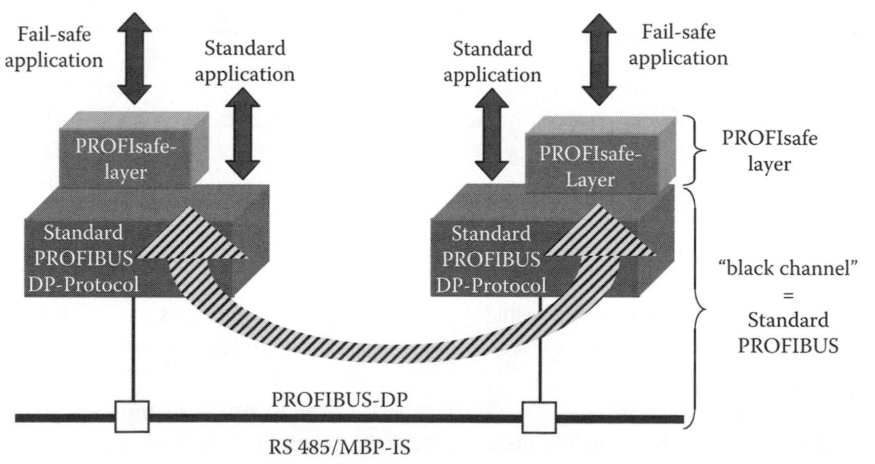

FIGURE 11.14 Safe communication on PROFIBUS DP.

A redundancy communication (RedCom) runs between the protocol stacks, that is, within a device or between two devices. It is independent of PROFIBUS, and its performance capability is largely determined by the redundancy reversing times.

Only one device version is required to implement different redundancy structures, and no additional configuration of the backup slave is necessary. The redundancy of PROFIBUS slave devices provides high availability, short reversing times, no data loss, and ensures fault tolerance.

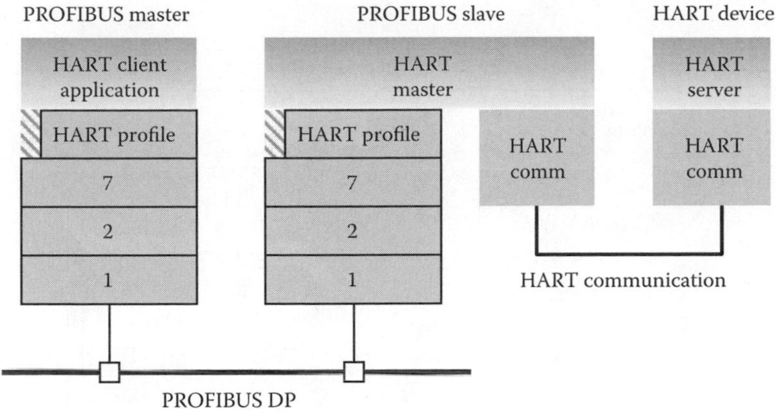

FIGURE 11.15 Operating HART devices over PROFIBUS DP.

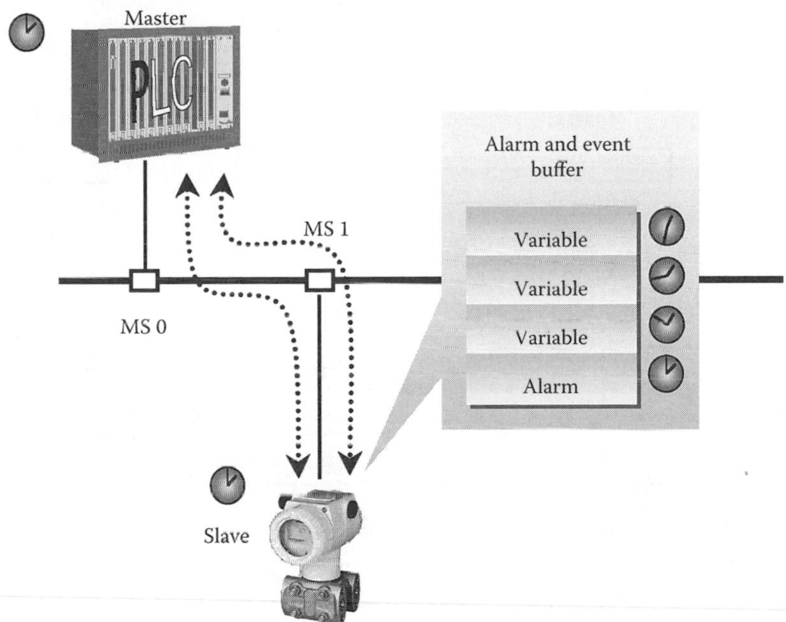

FIGURE 11.16 Time stamping and alarm messages.

11.4.2 Specific Application Profiles

The *PROFIdrive* application profile (standardized in IEC 61800-7) defines device behavior and the access procedure to drive data for electric drives, from simple frequency converters to highly dynamic servo-controls.

The method of integrating drives in automation solutions is highly dependent on the task of the drives (Figure 11.18). The more drives are acting independently from the central host controllers, the more they require slave-to-slave communication capabilities. On the other hand, the more the central host controllers are taking over the computing tasks, the more synchronization of the involved drives is required (Figure 11.19).

For this reason, PROFIdrive defines six classes covering the majority of applications.

With standard drives (class 1), the drive is controlled by means of a main setpoint value (e.g., rotational speed), whereby the speed control is carried out by the drive controller. In case of standard drives

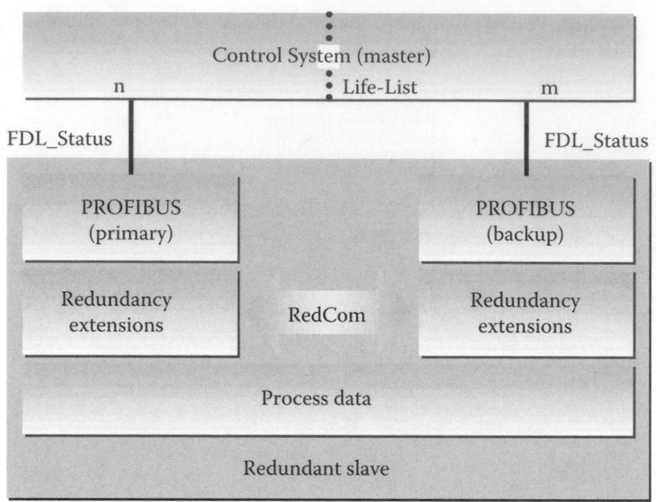

FIGURE 11.17 Slave redundancy in PROFIBUS.

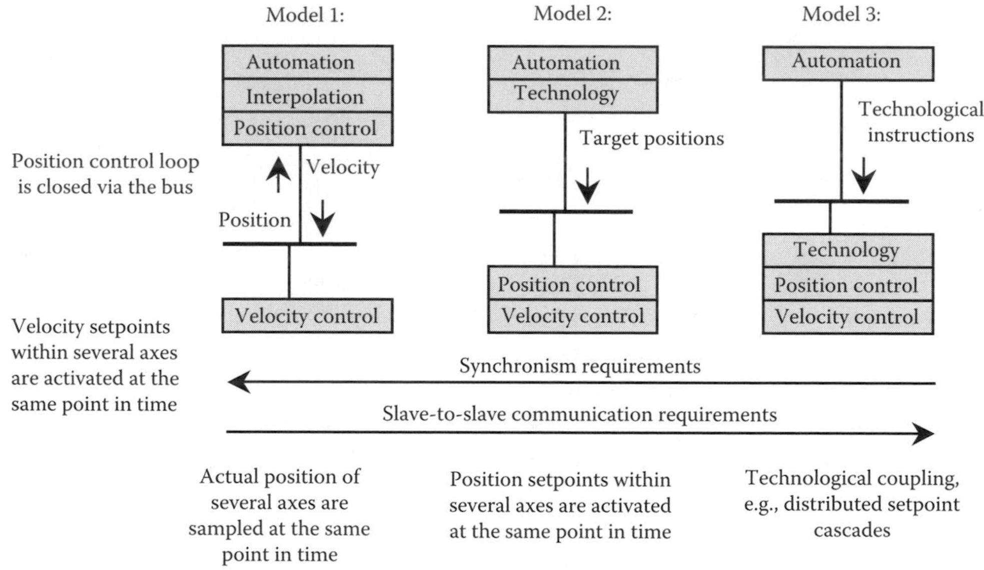

FIGURE 11.18 Different requirements for distributed drive applications.

with a technological function (class 2), the automation process is broken down into several subprocesses, and some of the automation functions are shifted from the central programmable controller to the drive controllers. PROFIBUS serves as the technology interface in this case. Slave-to-slave communication between the individual drive controls is a requirement for this solution.

The positioning drive (class 3) integrates an additional position controller in the drive, thus covering an extremely broad spectrum of applications (e.g., the twisting on and off of bottle tops). The positioning tasks are passed on to the drive controller over PROFIBUS and executed. The central motion control (classes 4 and 5) enables the coordinated motion sequence of multiple drives. The motion is primarily controlled over a Central Numeric Control (CNC). PROFIBUS serves to close the position control loop, as well as synchronize the clock (Figure 11.20).

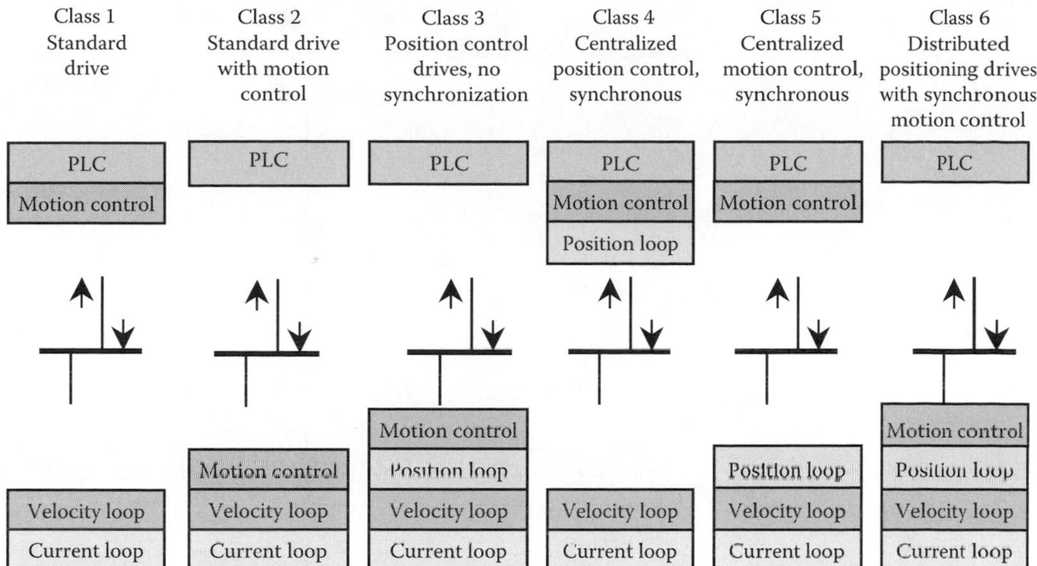

FIGURE 11.19 PROFIdrive defines six application classes.

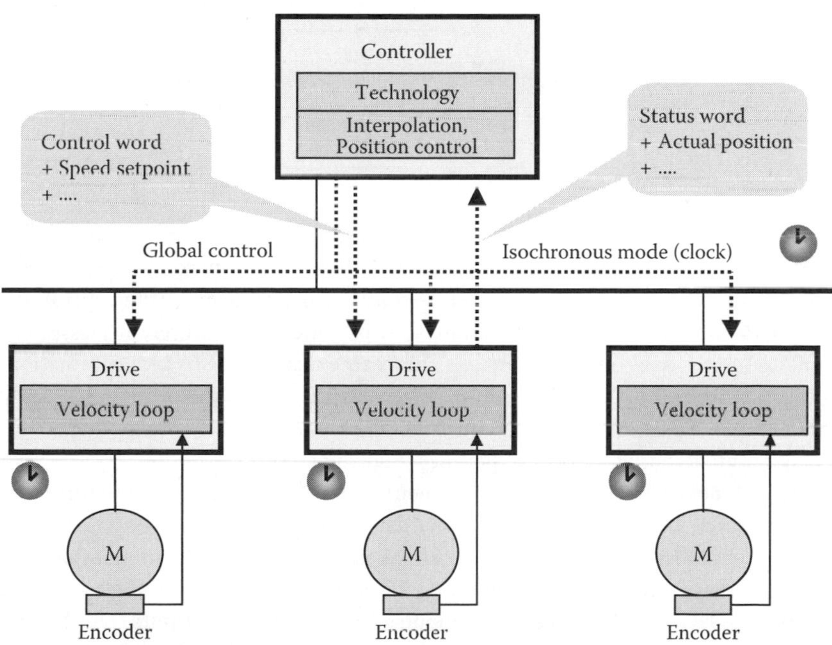

FIGURE 11.20 Positioning with central interpolation and position control.

The position control concept (dynamic servo-control) of this solution also supports extremely sophisticated applications with linear motors. Distributed automation by means of clocked processes and electronic shafts (class 6) can be implemented using slave-to-slave communication and isochronous slaves. Sample applications include *electrical gears*, *curve discs*, and *angular synchronous processes*.

In contrast to other drive profiles, PROFIdrive only defines the access mechanisms to the parameters and a subset of approx. 30 profile parameters, which include fault buffers, drive controllers, device identification, etc.

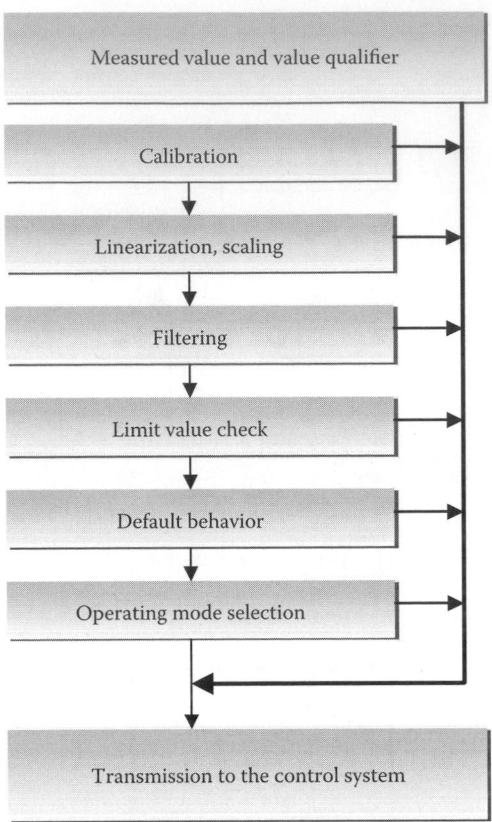

FIGURE 11.21 Signal processing defined in profile PA devices.

All other parameters (which may number more than 1000 in complex devices) are *manufacturer-specific*, which provide drive manufacturers great flexibility when implementing control functions.

The profile for *PA devices* defines all functions and parameters for different classes of devices for process automation with local intelligence [5,8]. They can execute part of the information processing or even take over the overall functionality of the automation systems. The profile includes all steps of a typical signal flow—from process sensor signals to the preprocessed process value that is communicated to the control system together with a value qualifier (Figure 11.21).

The profile for PA devices is documented in a general model description containing the currently valid specifications for all device types and in device data sheets containing the agreed additional specifications for individual device classes. The profile for PA devices includes device data sheets for quantity measurement of pressure and differential pressure; level; temperature; flow rate; and data sheets for valves, actuators, analyzers, analog, and digital inputs (DIs) and digital outputs (DOs).

The current version 3.02 of the profile provides users with powerful possibilities for operating plants more efficiently. The most important of the functions are described in the following.

Field device replacement has been very much simplified and shortened (see Figure 11.22). Once the device has been replaced, the control system detects the device and the device version expected at the bus address based on the message frame of the master. The new device answers with the appropriate Ident Number and is then included in the cyclic communication of the master. The process value is available again immediately after the device start-up. Additional intervention in the control system is not required. User-specific parameters are loaded via the cyclic channel, and the device can be operated as usual by using the abilities of the former profile version. In order to be able to use the new functionality of the new version, the device general station

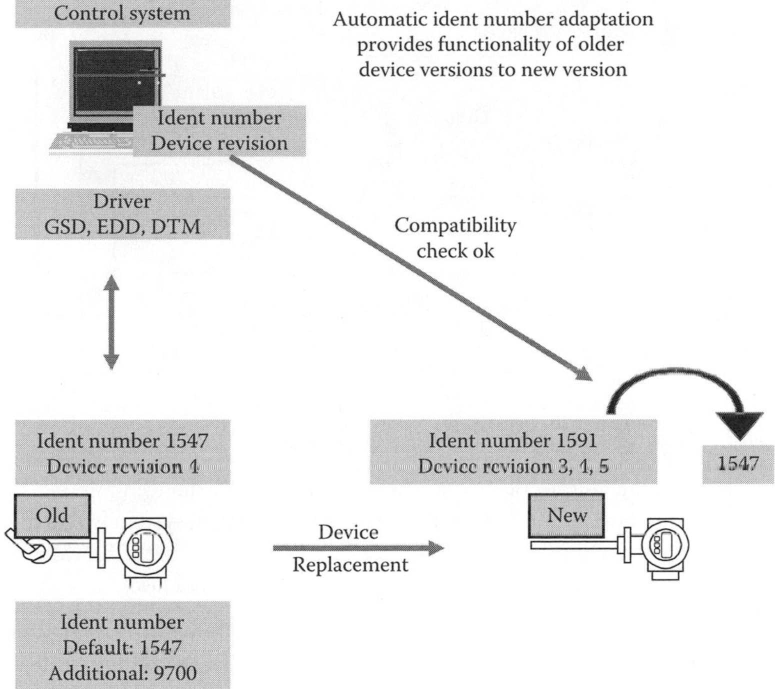

FIGURE 11.22 Device replacement with PA Profile Version 3.02.

description (GSD) and the configuration of the device have to be updated. This can be done during the next scheduled plant shutdown, after which all the new functionality of the device is available.

In order to shorten the time required for uploading or downloading the device parameter assignment, the PA Profile Version 3.02 has introduced a solution: by grouping parameter sets, it is now possible to reduce the time required to transfer parameters by a factor of 5–10, thereby achieving the fastest transfer times for a fieldbus in process automation.

Classification of diagnostic messages has been structured according to NAMUR Recommendation NE 107. The objective is to lessen the burden on the plant operator by providing only the information that is necessary for reliable process operation. The status that is reported to operators is limited to four categories: *failure*, *function check*, *out of specification*, and *maintenance required*.

In process engineering, it is common to use blocks for describing the characteristics and functions of a measuring point, or manipulating point, at a certain control point and to represent an automation application through a combination of these types of blocks. Therefore, the specification for PA devices uses a *function block model* according to IEC 61804 to represent functional sequences as shown in Figure 11.23. The blocks are implemented by the manufacturers as software in the field devices and, taken as a whole, represent the functionality of the device.

The following three block types are used:

1. A Physical Block (PB) contains the characteristic data of a device, such as device name, manufacturer, version and serial number, etc. There can only be one PB in each device.
2. A Transducer Block (TB) contains all the data required for processing an unconditioned signal delivered from a sensor for passing on to a function block. If no processing is required, the TB can be omitted. Multifunctional devices with two or more sensors have a corresponding number of TBs.
3. A Function Block (FB) contains all data for final processing of a measured value prior to transmission to the control system or for adjusting the parameter before it is finally set. Different FBs are available: Analog Input Block (AI), which delivers the measured value from the sensor/TB to the

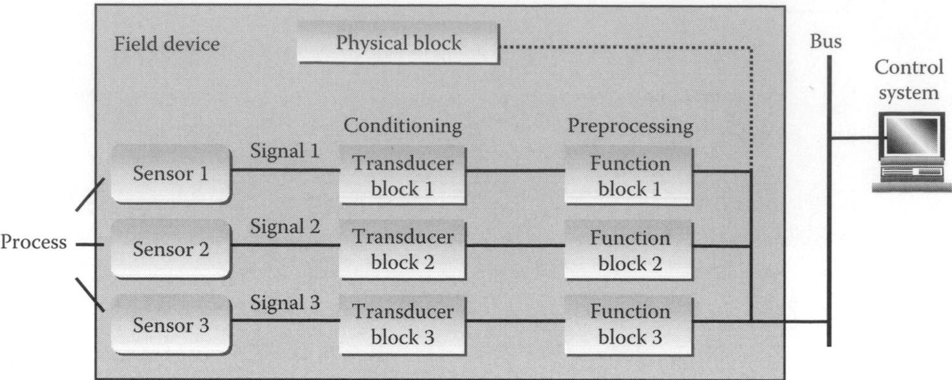

FIGURE 11.23 Block structure of a PA field device.

control system, Analog Output Block (AO), which provides the device with the value specified by the control system, Digital Input (DI), which provides the control system with a digital value from the device, and Digital Output (DO), which provides the device with the value specified by the control system.

The profile for *Ident Systems* defines complete communication and processing models for barcode readers and transponder systems. These are primarily intended for extensive use with the DP-V1 functionality. While the cyclic data transmission channel is used for small data volumes to transfer status/control information, the acyclic channel serves the transmission of large data volumes that result from the information in the barcode reader or transponder. The definition of standard PROXY function blocks [9] according IEC 61131-3 has facilitated the use of these systems and paves the way for the application of open solutions on completion of international standards, such as ISO/IEC 15962 and ISO/IEC18000.

The profile for *weighing and dosage systems* follows similar approaches as the Ident Systems. Communication and processing models are defined for four classes of devices or systems:

1. Simple scale
2. Comfort scale
3. Continuous scale
4. Batch scale

These new types of profiles will dramatically reduce the engineering costs and improve the bidding process during project execution.

11.4.3 Host Application Profiles

Application profiles usually define the functionality from the point of view of field devices. The host application profiles specification document defines the necessary functions of the counterpart of the field device, which is the host with its related engineering system. Typically, the host is a PLC or a Distributed Control System (DCS) controller. The host application profiles specification does not describe any new feature. It rather provides an overview of which PROFIBUS/PROFINET functionality on the host side is required for which application profile.

11.5 Integration Technologies

Modern field devices in both factory and process automation provide a wide range of information and also execute functions that were previously executed in PLCs and control systems. To execute these tasks, the tools for commissioning, maintenance, engineering, and parameterization of these devices

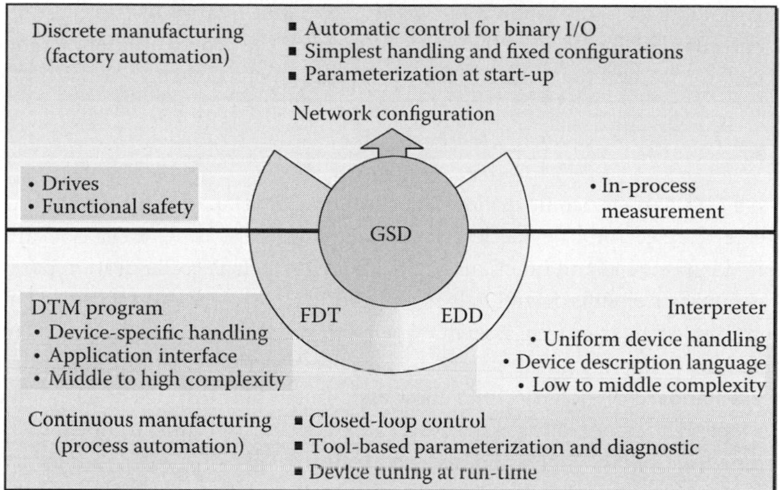

FIGURE 11.24 Integration technologies for PROFIBUS DP.

require an exact and complete description of device data and functions, such as the type of application function, configuration parameters, range of values, units of measurement, default values, limit values, identification, etc. The same applies to the controller/control system, whose device-specific parameters and data formats must also be made known (integrated) to ensure error-free data exchange with the field devices.

PROFIBUS has developed a number of methods and tools (integration technologies) for this type of device description, which enable the standardization of device management. The performance range of these tools is optimized to specific tasks (Figure 11.24), which has given rise to the term *scalable device integration*. General Station Description (GSD) and Electronic Device Description (EDD) are both a sort of *electronic device data sheets*, developed with different languages according to the special scope, whilst a Device Type Manager (DTM) is a software component containing specific field device functions for parameterization, configuration, diagnostics, and maintenance, generated by mapping and to be used together with the universal software interface Field Device Tool (FDT), which is able to implement software components.

11.5.1 GSD

A GSD is an electronically readable ASCII text file and contains both general and device-specific specifications for communication (GSD) and network configuration. Each of the entries describes a feature that is supported by a device. By means of keywords, a configuration tool reads the device identification (ID number), the adjustable parameters, the corresponding data type, and the permitted limit values for the configuration of the device from the GSD. Some of the keywords are mandatory, for example, Vendor_Name. Others are optional, for example, Sync_Mode_supported. A GSD replaces the previously conventional manuals and supports automatic checks for input errors and data consistency, even during the configuration phase.

Distinction is made between a device GSD (for an individual device only) and profile GSD, which may be used for devices that comply exactly with a profile such as PROFIdrive or PA devices.

GSD for compact devices, whose block configuration is already known on delivery, can be created completely by the device manufacturer. GSD for modular devices, whose block configuration is not yet conclusively specified on delivery, must be configured by the user in accordance with the actual module configuration using the configuration tool.

The device manufacturers are responsible for the scope and quality of the GSD of their devices. Submission of a profile GSD (containing the information from the profile of a device family) or an individual device GSD (device-specific) is essential for certification of a device.

11.5.2 EDD

An EDD is, like a GSD, an electronic device data sheet, but developed by using a more powerful and universal language, the electronic device description language (EDDL). An EDD typically describes the application-related parameters and functions of a field device such as configuration parameters, ranges of values, units of measurement, default values, etc. An EDD is a versatile source of information for engineering, commissioning, runtime, asset management, and documentation. It also contains support mechanisms to integrate existing profile descriptions in the device description, allow references to existing objects, access standard dictionaries, and allow assignment of the device description to a device.

An EDD is independent of operating systems and supports the user by its uniform user and operation interface (only one tool, reliable operation, reduced training, and documentation costs) and also the device manufacturer (no specific knowledge required; existing EDDs and libraries can be used).

The EDD concept is suitable for tasks of low to middle complexity.

11.5.3 FDT/DTM

A DTM is software that is generated by mapping the specific functions and dialogs of a field device for parameterization, configuration, diagnostics, and maintenance, complete with user interface, in a software component. This component is called DTM and is integrated in the engineering tool or control system over the FDT interface. A DTM uses the routing function of an engineering system for communicating across the hierarchical levels. It works similar to a printer driver, which the printer supplier includes in delivery and must be installed on the PC by the user. The DTM is generated by the device manufacturer and is included in delivery of the device.

DTM generation may be performed using one of the following options:

- Specific programming in a higher programming language
- Reuse of existing components or tools through their encapsulation as DTM
- Generation from an existing device description using a compiler or interpreter
- Use of the DTM toolkit of MS Visual Basic

With DTMs, it is possible to obtain direct access to all field devices for planning, diagnostics, and maintenance purposes from a central workstation. The FDT/DTM concept is protocol-independent and, with its mapping of device functions in software components, opens up interesting new user options. The DTM/FDT concept is very flexible, resolves nowadays interface and navigation needs, and is suitable for tasks of middle to high complexity.

11.5.4 Field Device Integration

For many years, EDD and FDT/DTM have been the two technologies for managing information from intelligent field devices. Users increasingly asked to form a single, unified solution, which takes account of the various tasks over the entire lifecycle for both simple and complex devices, including configuration, commissioning, diagnosis, and calibration (asset management).

This was how the field device integration (FDI) project was born and supported by the five major automation foundations, including the FDT Group, Fieldbus Foundation, HART Communication Foundation, PI, and OPC Foundation.

The core of FDI technology is the scalable FDI device package (Figure 11.25), which encompasses all features of the electronic description of a field device:

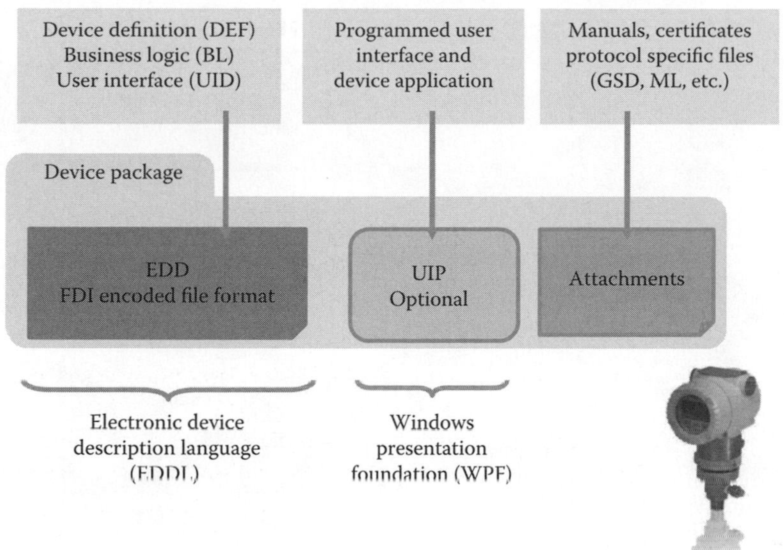

FIGURE 11.25 Conception of the FDI device package.

The Device Definition (DEF), Business Logic (BL), and User Interface Description (UID) are based on EDDL (IEC 61804-3). The optional User Interface Plug-in (UIP) offers the advantages of freely programmable user interfaces familiar from FDT, based on Windows Presentation Foundation (WPF).

Via the FDI device package, the device manufacturers define which data, functions, and user interfaces have to be represented in the information model of the FDI server.

- The DEF describes the field device data and the internal structure (e.g., blocks).
- The BL primarily ensures that the device data remain consistent. The dynamic dependencies, in particular, play a part here.
- UID and UIP define the field device user interfaces.
- Product documentation, protocol-specific files, such as GSD or Common File Format (CFF), etc., will all be added to the FDI package as attachments.

The FDI concept is based on the client-server architecture model. In the architecture of this kind, a server provides services that are accessed by various clients (often distributed).

The FDI architecture is based on the OPC Unified Architecture (OPC UA). The FDI server imports FDI device packages into its internal device catalog (Figure 11.26), which makes version management of FDI packages very easy. The representation of device instances in the FDI server takes place in the information model. The information model maps the communication topology of the automation system by representing the entire communication infrastructure and the field devices as objects. If an FDI client wishes to work with a device, it accesses the information model and, for example, loads the user interface of the device in order to display it on the client side, in a similar manner to a web browser. The FDI server always ensures that the device data remain consistent.

11.6 Technical Support

PI maintains more than 50 approved PI competence centers worldwide. These facilities provide users and manufacturers with all kinds of advice and support. As institutions of PI, they are vendor-neutral service providers and adhere to the mutually agreed-upon rules. The PI competence centers are regularly checked for their suitability as part of an individually tailored approval process. More than 25 PI

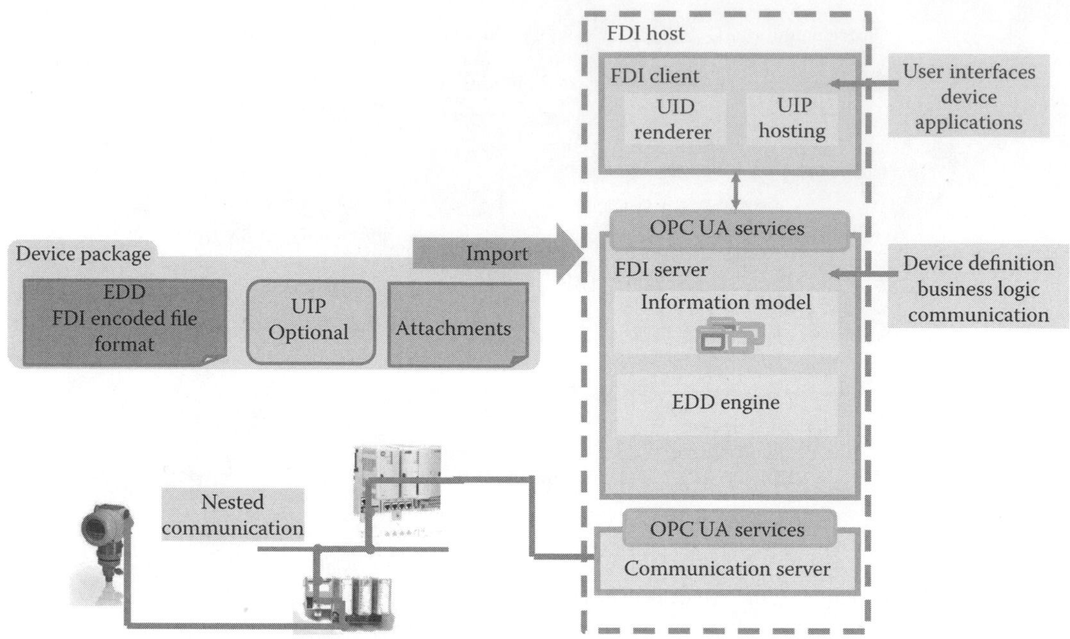

FIGURE 11.26 The architecture of an FDI host.

training centers have been set up to establish a uniform global training standard for engineers and technicians. Approval of the training centers and their experts ensures the quality of the training and, thus, of the engineering and installation services for PROFIBUS. Refer to the website for current addresses.

11.6.1 Quality Assurance

In order for PROFIBUS devices of different types and manufacturers to correctly fulfill tasks in the automation process, it is essential to ensure the error-free exchange of information over the bus. The requirement for this is a standard-compliant implementation of the communications protocol and application profiles by device manufacturers. To ensure that this requirement is fulfilled, the PI has established a quality assurance procedure whereby, on the basis of test reports, certificates are issued to devices that successfully complete the test.

Basis for the certification procedure are the standards ISO 9646 (guides concerning conformance inspection and certification) and EN 45000. The PI has approved manufacturer-independent PI test labs in accordance with the specifications of these standards. Only these test laboratories are authorized to carry out device tests, which form the basis for granting certificates (Figure 11.27).

PI maintains 11 approved PI Test Labs (PITL) worldwide. As institutions of PI, they are vendor-neutral service providers and adhere to the mutually agreed-upon rules. The testing services provided by the PITLs are regularly audited in accordance with a strict accreditation process to ensure that they meet the necessary quality requirements. Refer to the website for current addresses [2].

The test procedure, which is the same for all PI Test Labs, is made up of several parts:

- The GSD/EDD check ensures that the device description files comply with the specification.
- The hardware test tests the electric characteristics of the PROFIBUS interface of the device for compliance with the specifications. This includes terminating resistors, suitability of the implemented drivers and other modules, and the quality of line level.
- The function test examines the bus access and transmission protocol and the functionality of the test device.

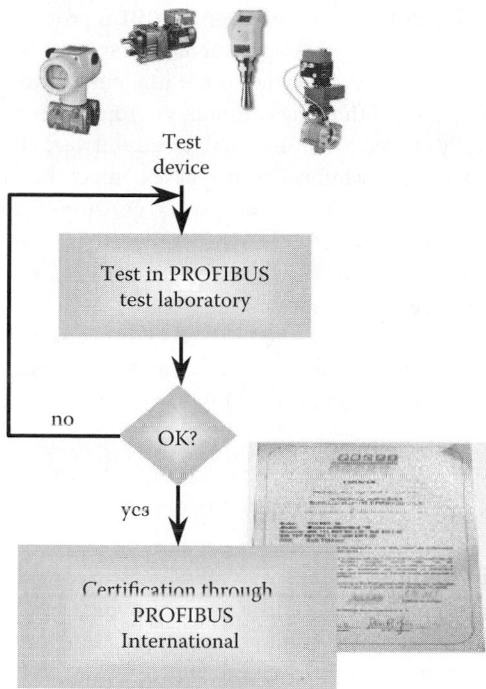

FIGURE 11.27 Certification procedure.

- The conformity test forms the main part of the test. The objective is to test the conformity of the protocol implementation with the standard.
- The interoperability test checks the test device for interoperability with the PROFIBUS devices of other manufacturers in a multivendor plant. This checks that the functionality of the plant is maintained when the test device is added. Operation is also tested with different masters.

Once a device has successfully passed all the tests, the manufacturer can apply for a certificate from PI. Each certified device contains a certification number as a reference.

11.6.2 Implementation Support

For the device development or implementation of the PROFIBUS protocol, a broad spectrum of standard components and development tools (PROFIBUS ASICs, PROFIBUS stacks, monitoring, and commissioning tools) as well as services are available that enable device manufacturers to realize cost-effective development. A corresponding overview is available in the product catalog of PI [2].

PROFIBUS interface modules are ideal for a low/medium volume of devices to be produced. These credit card size modules implement the entire bus protocol. They are fitted on the master board of the device as an additional module.

PROFIBUS protocol chips (single chips, communication chips, and protocol chips) are recommended for an individual implementation in the case of high volume of devices.

The implementation of single-chip ASICs is ideal for simple slaves (I/O devices). All protocol functions already are integrated on the ASIC. No microprocessor or software is required. Only the bus interface driver, the quartz, and the power electronics are required as external components.

For intelligent slaves, parts of the PROFIBUS protocol are implemented on a protocol chip and the remaining protocol parts implemented as software on a microcontroller. In most of the ASICS available on the market, all cyclic protocol parts [4] have been implemented, which are responsible for the transmission of time-critical data.

For complex masters, the time-critical parts of the PROFIBUS protocol are also implemented on a protocol chip and the remaining protocol parts implemented as software on a microcontroller. Various ASICs of different suppliers are currently available for the implementation of complex master devices. They can be operated in combination with many common microprocessors.

An overview for commercially offered PROFIBUS chips and software (PROFIBUS stacks) is available at the PROFIBUS website [2]. For further information, please contact the suppliers directly.

Modem chips are available to realize the (low) power consumption, which is required when implementing a bus-powered field device with MBP transmission technology. Only a feed current of 10–15 mA over the bus cable is available for these devices, which must supply the overall device, including the bus interface and the measuring electronics. These modems take the required operating energy for the overall device from the MBP bus connection and make it available as feed voltage for the other electronic components of the device. At the same time, the digital signals of the connected protocol chip are converted into the bus signal of the MBP connection modulated to the energy supply.

11.7 Wide Range of Applications

PROFIBUS is the world's most popular fieldbus with an unparalleled wide application field in discrete manufacturing and process control as well as in drive technology and safety applications. Profibus offers a fully integrated platform for discrete and process applications, which is a major benefit in the process industries where upstream, mainstream, and downstream processes have to work together (Figure 11.28).

Industries using PROFIBUS range from critical petrochemical operations and high-volume robotic manufacturing plants, right across the spectrum to food and drink, water and waste water, power and glass, pulp and paper, and many others.

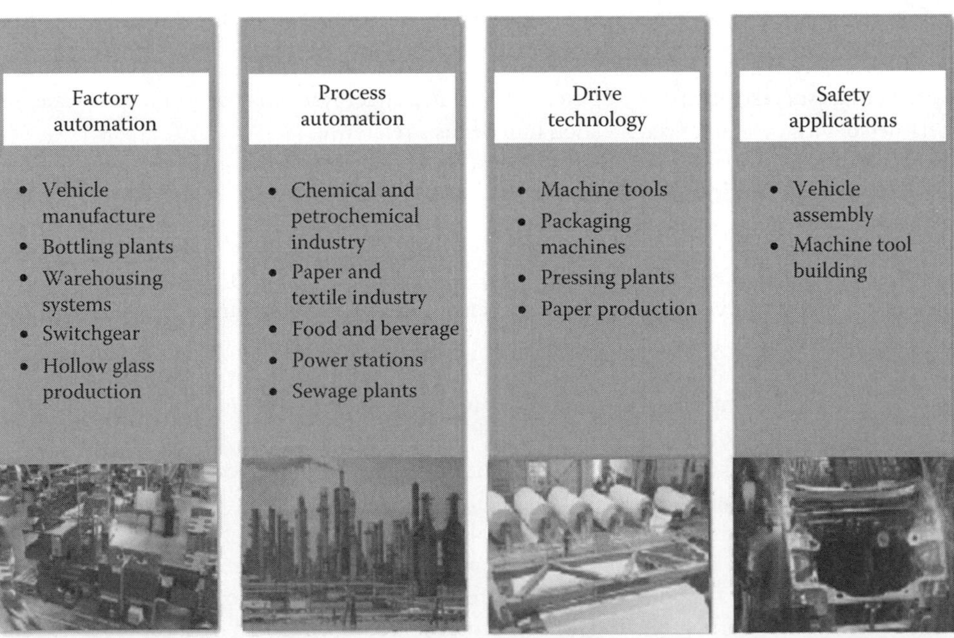

FIGURE 11.28 Application scope for PROFIBUS.

Abbreviations

ASIC	Application-Specific Integrated Circuit
DP	Decentralized Peripherals
DPM1	PROFIBUS DP Master class 1, usually a programmable logic controller
DPM2	PROFIBUS DP Master class 2, usually a laptop or PC
DTM	Device Type Manager
DXB	Data Exchange Broadcast (slave-to-slave(s) communication)
EDD	Electronic Device Description
EIA	Electronics Industries Association
EN, prEN	European standard, preliminary....
FB	Function Block
FDI	Field Device Integration
FDT	Field Device Tool
FF	Foundation Fieldbus
FISCO	Fieldbus Intrinsically Safe Concept
FM	Factory Mutual Global, a commercial and industrial property insurance company with a unique focus on risk management. http://www.fmglobal.com
GSD	General Station Description (electronically readable data sheet)
HART	Highway Addressable Remote Transducer Protocol
HMI	Human Machine Interface
I&M	Identification and Maintenance
IEC	International Electrotechnical Commission
I/O	Input/Output
ISO/OSI	International Standards Organization/Open Systems Interconnection (Reference Model)
MS0	Cyclic Master Slave Communication Services of PROFIBUS DP
MS1/MS2	Acyclic Master Slave Communication Services of PROFIBUS DP
NAMUR	Association of Users of Process Control Technology
PA	Process Automation
PC	Personal Computer
PDU	Protocol Data Unit
PI	PROFIBUS & PROFINET International
PLC	Programmable Logic Controller
PNO	PROFIBUS Nutzerorganisation
PTB	Pysikalisch-Technische Bundesanstalt: national institute of natural and engineering sciences and the highest technical authority for metrology and physical safety engineering of the Federal Republic of Germany. http://www.ptb.de
TIA	Telecommunications Industry Association
UL	Underwriters Laboratories Inc., an independent, not-for-profit product safety testing and certification organization. http://www.ul.com

References

1. IEC 61158/61784 (2010), Digital data communications for measurement and control—Fieldbus for use in industrial control systems.
2. PROFIBUS Website: http://www.profibus.com.
3. PROFIBUS System Description—Technology and Application, November 2010, free download from http://www.profibus.com or Order-No. 4.332.
4. Profibus RS 485-IS User and Installation Guideline, V1.1, June 2003, Order No. 2.262.

5. J. Powell and H. Vanderlinde, On the road with the process fieldbus, 2009, Order-No. 4.192.
6. IEC/TS 60079-27 (2004), Electrical apparatus for explosive gas atmospheres—Part 27: Fieldbus intrinsically safe concept (FISCO) Parts 11, 14, and 25: Constructional and installation requirements.
7. M. Popp, The rapid way to PROFIBUS DP, PROFIBUS (2003), Order No. 4.072.
8. Ch. Diedrich and Th. Bangemann, *Profibus PA, Instrumentation Technology for the Process Industry*, Oldenbourg Industrieverlag, München, Germany, 2007.
9. PROFIBUS Communication and Proxy Function Blocks acc. to IEC 61131-3, V1.2, July 2001, PROFIBUS Order No. 2.182.

12 PROFINET

12.1 Introduction 12-1
Standardization • PROFINET at a Glance • Conformance Classes
12.2 PROFINET Basics 12-5
Device Model of an I/O-Device • Device Descriptions • Communication Relations • Addressing • Configuration and Data Exchange • Diagnostics
12.3 Principles of PROFINET Communication 12-11
Standard Communication with UDP • Real-Time Communication
12.4 Conformance Classes and Their Functions 12-13
Basic Functions of Conformance Class A • Network Diagnostics and Management of Conformance Class B • Isochronous Real-Time with Conformance Class C
12.5 Optional Functions 12-20
Multiple Access to Field Devices • Extended Device Identification • Individual Parameter Server • Configuration in Run • Time Stamping • Fast Restart • High Availability
12.6 Integration of Fieldbus Systems 12-23
12.7 Application Profiles 12-24
PROFIsafe • PROFIdrive • PROFIenergy
12.8 PROFINET for Process Automation 12-25
12.9 Installation Technology for PROFINET 12-25
Network Configuration • Cables for PROFINET • Plug Connectors • Security
12.10 Technical Support 12-29
Technology Development • Tools for Product Development • Certification Test
Abbreviations 12-31
References 12-32

Peter Wenzel
PROFIBUS and PROFINET International

12.1 Introduction

Automation technology is undergoing continuous change due to the ever-shorter innovation cycles for new products. The use of fieldbus technology in the 1990s has represented a significant innovation. It has enabled the migration of automation systems from centralized to decentralized systems. In this regard, PROFIBUS has set the standard as the market leader for more than 20 years. In today's automation technology, not only fieldbuses but also increasingly Ethernet and information technology (IT), with established standards like TCP/IP and XML, are calling the shots. Integrating IT into automation allows significantly better communication options between automation systems, extensive configuration and diagnostic possibilities, and network-wide service functionality.

PROFINET is the open standard for Industrial Ethernet of PROFIBUS & PROFINET International (PI) that covers all automation technology requirements [1]. With PROFINET, real-time (RT) capable solutions can be implemented for discrete manufacturing and continuous process automation, safety applications, and the entire range of drive technology up to and including isochronous motion control applications.

In addition to the RT capability and utilization of IT, protection of investment also plays an important role with PROFINET. PROFINET enables existing fieldbus systems such as PROFIBUS to be integrated without modifications.

The use of open standards, simple handling ability, and integration into existing plant units has defined PROFINET from the start.

12.1.1 Standardization

PROFINET is 100% Ethernet-compatible according to IEEE standards. The standard Ethernet protocol as defined in IEEE 802 was extended only in cases of requirements that could not otherwise be met in a satisfactory manner. PROFINET is standardized in IEC 61158 and IEC 61784 [2]. The mandatory certification test services by authorized PI Test Labs for PROFINET products ensure a high-quality standard.

12.1.2 PROFINET at a Glance

The motivation to create PROFINET came from user requirements and the anticipated cost reduction resulting from manufacturer-independent plant-wide engineering.

The PROFINET architecture supports a modular concept that allows users to choose the functionality they require. The functionality differs mainly in terms of the type of data exchange. This distinction is necessary to satisfy the very stringent requirements for data transmission speed that exist for some applications.

PROFINET defines the following requirements for data communication:

- 100 Mbps data communication with copper or fiber-optic transmission (100 Base TX and 100 Base FX)
- Full duplex transmission
- Switched Ethernet
- Autonegotiation (negotiation of transmission parameters)
- Autocrossover (send and receive cable crossover in the switch)
- Wireless communication based on WLAN and Bluetooth

PROFINET describes an I/O data view of distributed I/O. It includes RT communication and isochronous RT (IRT) communication with distributed I/O. The designations RT and IRT simply refer to the RT properties of communication.

PROFINET uses UDP/IP as the higher-level protocol for demand-oriented data exchange such as parameters, configuration data, and interconnection information. UDP stands for user datagram protocol; it includes the broadcast communication in connection with IP. This satisfies the prerequisites for interfacing the automation levels with other networks (MES and ERP).

In parallel to this UDP/IP communication, cyclic data exchange in PROFINET is based on the scalable RT concept. Based on this concept, PROFINET enables deterministic and isochronous transmission of time-critical process data. Process data and alarms are always transmitted in RT according to the definitions of IEEE and IEC for high-performance data exchange of I/O data. RT communication constitutes the basis for data exchange in PROFINET I/O. RT data are handled with higher priority compared to TCP (UDP)/IP data. This method of data exchange allows bus cycle times in the range of a few hundred milliseconds to be achieved.

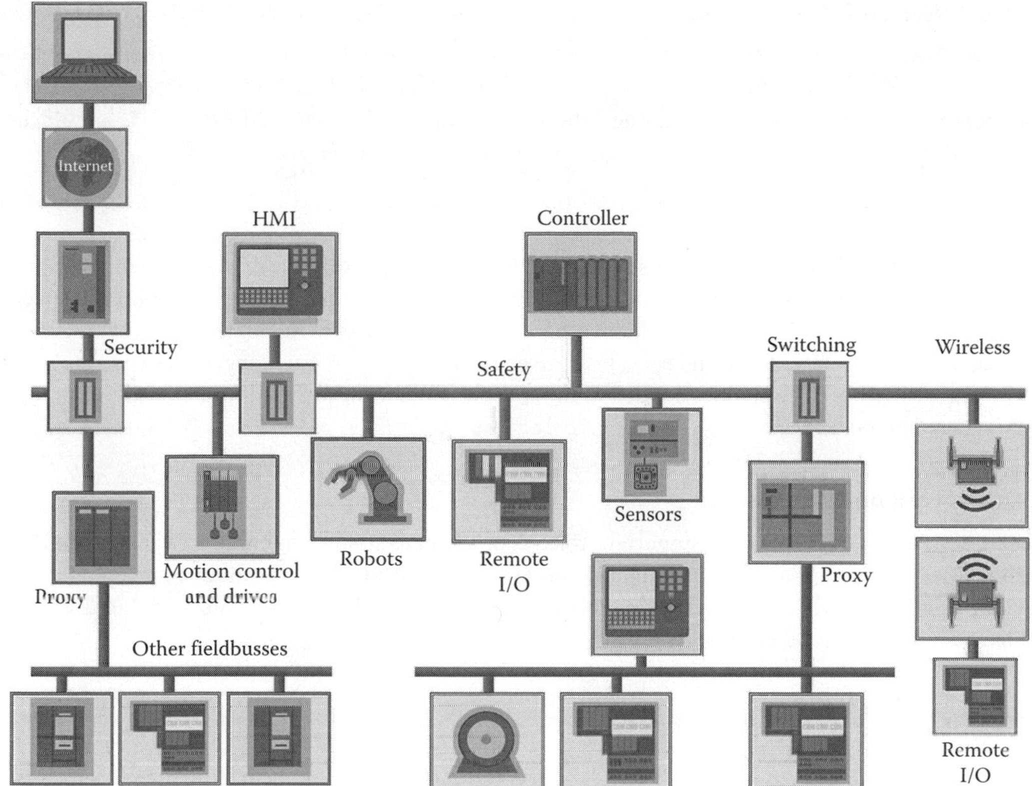

FIGURE 12.1 Example architecture of a PROFINET communication.

Isochronous data exchange with PROFINET is defined in the IRT concept. Here, data exchange cycles are normally in the range of a few tens of microseconds up to 1 ms. IRT communication differs from RT communication mainly in its isochronous behavior, meaning that the bus cycles are started with maximum precision. The start of a bus cycle can deviate by a maximum of 1 μs.

12.1.2.1 Higher-Level Functions

PROFINET provides many application-driven functions for commissioning and operation. These include visualization of plant topology in conjunction with informative diagnostics and support of convenient plant diagnostics through a combination of basic services available by default in the higher-level controller.

For easy replacement of field devices—when a fault occurs—PROFINET has integrated neighborhood detection functionality. This information can be used to represent the plant topology in a very easy-to-understand graphic display.

The proven PROFIsafe safety technology is also available for PROFINET. The ability to use the same cable for standard and safety-related communication saves on devices, engineering, and setup.

PROFINET integrates automatically reacting redundancy solutions and intelligent diagnostic concepts. Acyclic diagnostic data transmission provides important information regarding the status of the network and devices, including a display of the network topology. The defined concepts for media and system redundancy guarantee a smooth changeover from one communication path to another in the event of an error. Both features significantly increase plant availability.

12.1.2.2 Network Installation

PROFINET network installations are oriented toward specific requirements for Ethernet networks in industrial environments. The installation guideline provides plant construction engineers and plant operators with simple rules for installing Ethernet networks and associated cabling. This guideline provides device manufacturers with clear specifications for device interfaces.

12.1.2.3 Web Integration

PROFINET is based on Ethernet and supports TCP/IP. This enables the use of web technologies for accessing an integrated web server in a field device, among other things. Depending on the specific device implementation, diagnostics and other information can be easily called up using a standard web browser, even across network boundaries. Thus, an engineering system is no longer necessary for simple diagnostics. PROFINET itself does not define any specific contents or formats. Rather, it allows an open and free implementation.

12.1.2.4 Fieldbus Integration

In addition, PROFINET plays an important role when it comes to investment protection. PROFINET enables a seamless integration of existing fieldbus systems like PROFIBUS without changes to existing devices. That means that the investments of plant operators, machine and plant manufacturers, and device manufacturers are all protected.

12.1.3 Conformance Classes

The scope of functions supported by PROFINET I/O is clearly divided into conformance classes (CCs). These provide a practical summary of the various minimum properties. There are three CCs that build upon one another and are oriented to typical applications (Figure 12.2).

CC-A provides basic functions for PROFINET I/O with RT communication. All IT services can be used without restriction. Typical applications are found, for example, in building automation. Wireless communication is possible only in this class. CC-B extends the concept to include network diagnostics via IT mechanisms as well as topology information. The system redundancy function important for continuous process automation is contained in an extended version of CC-B named CC-B (PA). CC-C describes the basic functions for devices with hardware-supported bandwidth reservation and synchronization (IRT communication) and is thus the basis for isochronous applications.

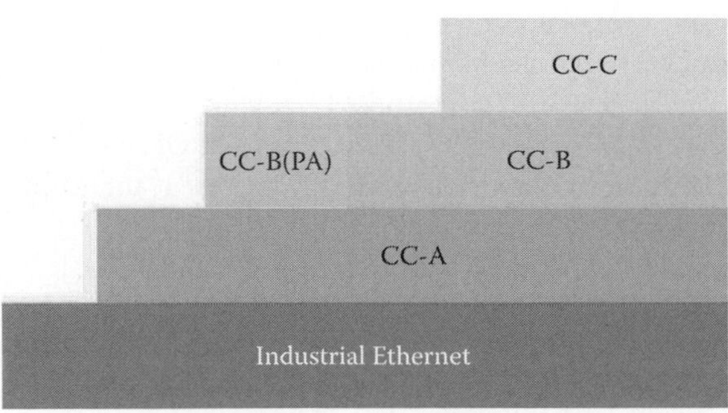

FIGURE 12.2 Structure of conformance classes.

12.2 PROFINET Basics

PROFINET I/O follows the provider/consumer model for data exchange. Configuring a PROFINET I/O system has the same look and feel as in PROFIBUS. The following device classes are defined for PROFINET I/O:

I/O-Controller: This is typically the programmable logic controller (PLC) on which the automation program runs. This is comparable to a class 1 master in PROFIBUS. The I/O-Controller provides output data to the configured I/O-Devices in its role as provider and is the consumer of input data of I/O-Devices.

I/O-Device: An I/O-Device is a distributed I/O field device that is connected to one or more I/O-Controllers via PROFINET I/O. It is comparable to the function of a slave in PROFIBUS. The I/O-Device is the provider of input data and the consumer of output data.

I/O-Supervisor: This can be a programming device, personal computer (PC), or human machine interface (HMI) device for commissioning or diagnostic purposes and corresponds to a class 2 master in PROFIBUS.

A plant unit has at least one I/O-Controller and one or more I/O-Devices. I/O-Supervisors are usually integrated only temporarily for commissioning or troubleshooting purposes.

12.2.1 Device Model of an I/O-Device

The device model describes all field devices in terms of their possible technical and functional features. It is specified by the device access point (DAP) and the defined modules for a particular device family. A DAP is the access point for communication with the Ethernet interface and the processing program. A variety of I/O modules can be assigned to it in order to manage the actual process data communication.

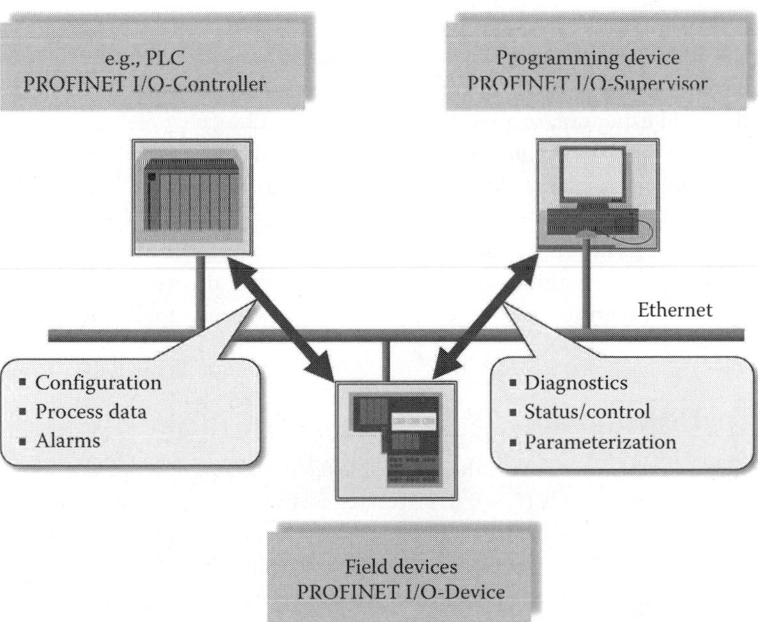

FIGURE 12.3 Communication paths for PROFINET I/O.

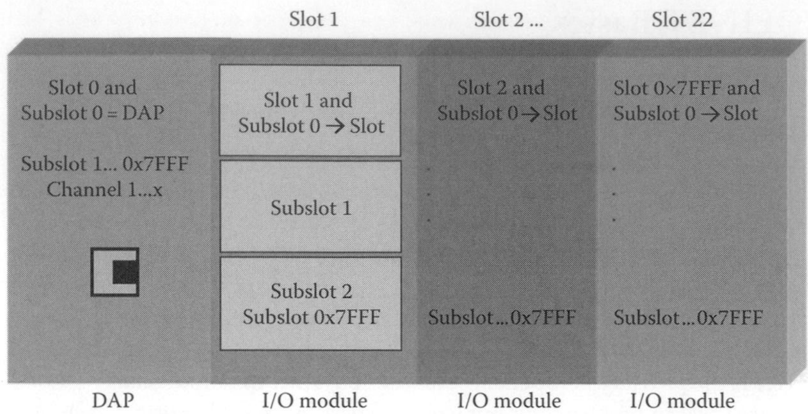

FIGURE 12.4 Addressing of I/O data in PROFINET I/O based on slots and subslots.

The following structures are standardized for an I/O-Device:

- The slot designates the place where an I/O module is inserted in a modular I/O field device. The configured modules containing one or more subslots for data exchange are addressed on the basis of different slots.
- Within a slot, the subslots represent the actual interface to the process (inputs/outputs). The granularity of a subslot (bitwise, bytewise, or wordwise division of I/O data) is determined by the manufacturer. The data content of a subslot is always accompanied by status information, from which the validity of the data can be derived.
- The index specifies the data within a slot/subslot that can be read or written acyclically via read/write services. For example, parameters can be written to a module or manufacturer-specific module data can be read out on the basis of an index.

Cyclic I/O data are addressed by specifying the slot/subslot combination. These can be freely defined by the manufacturer. For acyclic data communication via read/write services, an application can specify the data to be addressed using slot, subslot, and index (Figure 12.4).

To avoid competing accesses in the definition of user profiles (e.g., for PROFIdrive, weighing and dosing, etc.), the application process identifier/instance (API) is defined as an additional addressing level.

PROFINET differentiates between compact field devices, in which the degree of expansion is already specified in the as-delivered condition and cannot be changed by the user, and modular field devices, in which the degree of expansion can be customized for a specific application when the system is configured.

12.2.2 Device Descriptions

All field devices are described in terms of their available technical and functional properties in an XML-based general station description file (GSD) to be created by the field device developer. It contains among other things a representation of the device model that is reproduced by the DAP and the defined modules for a particular device family. The GSD files of the field devices to be configured are required for system engineering. It contains all data relevant for engineering as well as for data exchange with the field device.

The GSD contains all necessary information:

- Properties of the I/O-Device (e.g., communication parameters)
- Insertable modules (number of type)

- Configuration data for individual modules (e.g., 4–20 mA analog input)
- Parameters of modules
- Error texts for diagnostics (e.g., wire break, short circuit)

Because XML is an open, widespread, and accepted standard for describing data, appropriate tools and derived properties are automatically available, including

- Creation and validation through a standard tool
- Foreign language integration
- Hierarchical structuring

The GSD structure corresponds to ISO 15745 and consists of a header, the device description in the application layer (e.g., configuration data and module parameters), and the communication properties description in the transport layer.

12.2.3 Communication Relations

To establish communication between the I/O-Controller and an I/O-Device, the communication paths must be established. These are set up by the I/O-Controller during system start-up based on the configuration data in the engineering system. This specifies the data exchange explicitly.

Every data exchange is embedded into an application relation (AR) (Figure 12.5). Within the AR, communication relations (CRs) specify the data explicitly. As a result, all data for the device modeling, including the general communication parameters, are downloaded to the I/O-Device. An I/O-Device can have multiple ARs established from different I/O-Controllers.

The communication channels for cyclic data exchange (I/O data CR), acyclic data exchange (record data CR), and alarms (alarm CR) are set up simultaneously.

Multiple I/O-Controllers can be used in a PROFINET system (Figure 12.6). If these I/O-Controllers are to be able to access the same data in the I/O Devices, this must be specified when configuring (shared devices, shared inputs).

An I/O-Controller can establish one AR each with multiple I/O-Devices. Within an AR, several 10 Communication Relations (10 CR) and APIs can be used for data exchange. This can be useful, for example, if more than one user profile (PROFIdrive, Encoder, etc.) is involved in the communication and different subslots are required. The specified APIs serve to differentiate the data communication within an I/OCR.

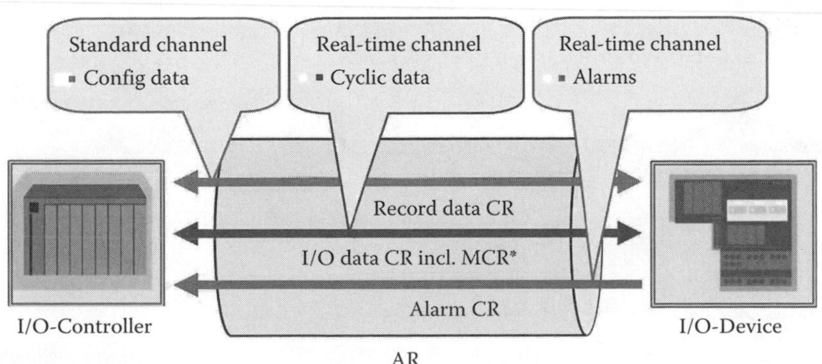

FIGURE 12.5 Application relations and communication relations.

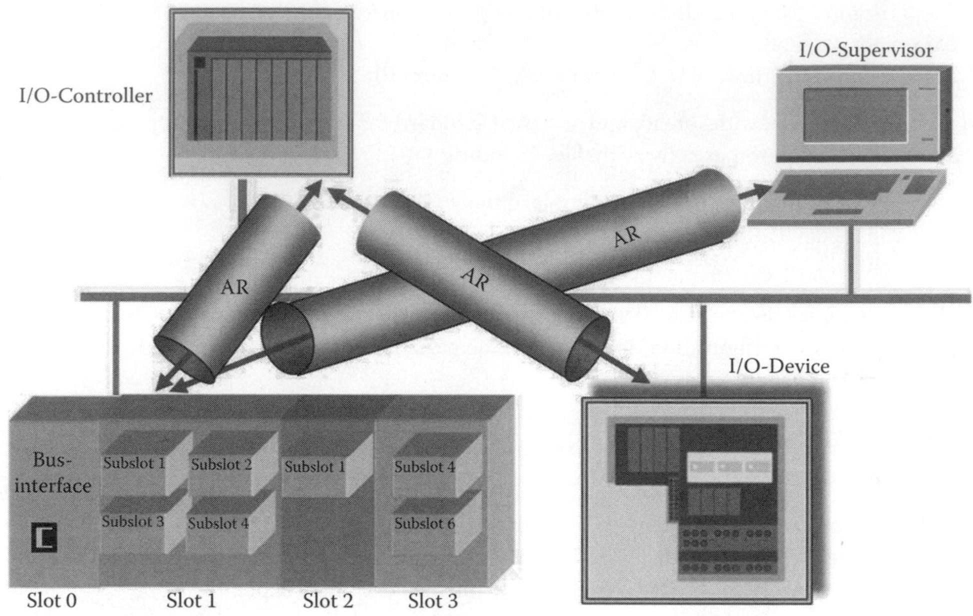

FIGURE 12.6 A field device can be accessed by multiple application relations.

12.2.4 Addressing

In PROFINET I/O, each field device has a symbolic name that uniquely identifies the field device within a PROFINET I/O system. This name is used for assigning the IP address and the MAC address. The dynamic configuration protocol (DCP) integrated in every I/O-Device is used for this purpose. Figure 12.7 shows a network that comprises two subnets. These are represented by the different network IDs (subnet mask).

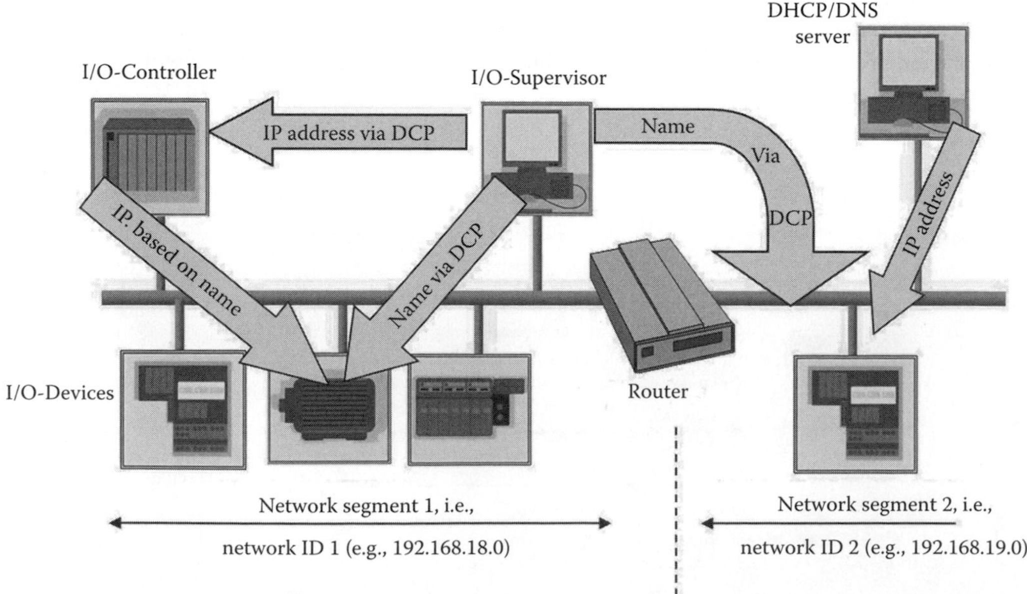

FIGURE 12.7 Example of a PROFINET network consisting of two subnets.

MAC address and OUI (organizationally unique identifier)
Each PROFINET device is addressed based on MAC address. This address is unique worldwide. The company code (bits 47 to 24) can be obtained free of charge from the IEEE Standards Department. This part is called the OUI (organizationally unique identifier).
PI offers MAC addresses to device manufacturers that do not want to apply for their own OUI, in other words, a fixed OUI and the manufacturer-specific portion (bits 23 to 0). This service allows components to acquire MAC addresses from the PI Support Center. The assignment can be completed in 4 K-ranges.
The OUI of PI is 00-0E-CF and is structured as shown in the table. The OUI can be used for up to 16,777,214 products.

Bit significance 47 to 24						Bit significance 23 to 0					
0	0	0	E	C	F	X	X	X	X	X	X
Company code→OUI						Consecutive number					

FIGURE 12.8 OUI for PROFINET.

After the system is configured, the engineering tool loads all information required for data exchange to the I/O-Controller, including the IP addresses of the connected I/O-Devices. Based on the name (and the associated MAC address), an I/O-Controller can recognize the configured field devices and assign them the specified IP addresses using the DCP (address resolution). Alternatively, addressing can be performed via a DHCP server.

Following address resolution, the system powers up, and parameters are transmitted to the I/O-Devices. The system is then available for productive data traffic.

Optionally, the name can also be automatically assigned to the I/O-Device by means of a specified topology based on neighborhood detection.

A PROFINET I/O-Device is addressed for direct data exchange by its MAC address (see Figure 12.8).

PROFINET I/O field devices are always connected via switches as network components (Figure 12.9). This takes the form of a star topology with separate multiport switches or a line topology with switches integrated in the field device (two ports occupied).

PROFINET transmits some message frames (e.g., for synchronization and neighborhood detection) using the MAC address of the respective port instead of the device MAC address. For this reason, each switch port in a field device requires a separate port MAC address. Therefore, a two-port field device has three MAC addresses in the as-delivered condition. However, these port MAC addresses are not visible to users. Because the field devices are connected via switches, PROFINET always sees only point-to-point connections (same as Ethernet). That is, if the connection between two field devices in a line is interrupted, the field devices located after the interruption are no longer accessible. If increased availability is required, provision must be made for redundant communication paths when planning the system, and field devices/switches that support the redundancy concept of PROFINET must be used.

PROFINET-suitable switches must support *autonegotiation* (negotiation of transmission parameters) and *autocrossover* (crossing of send and receive lines in the switch). As a result, communication can be established autonomously.

12.2.5 Configuration and Data Exchange

Each I/O-Controller manufacturer also provides an engineering tool for configuring a PROFINET system.

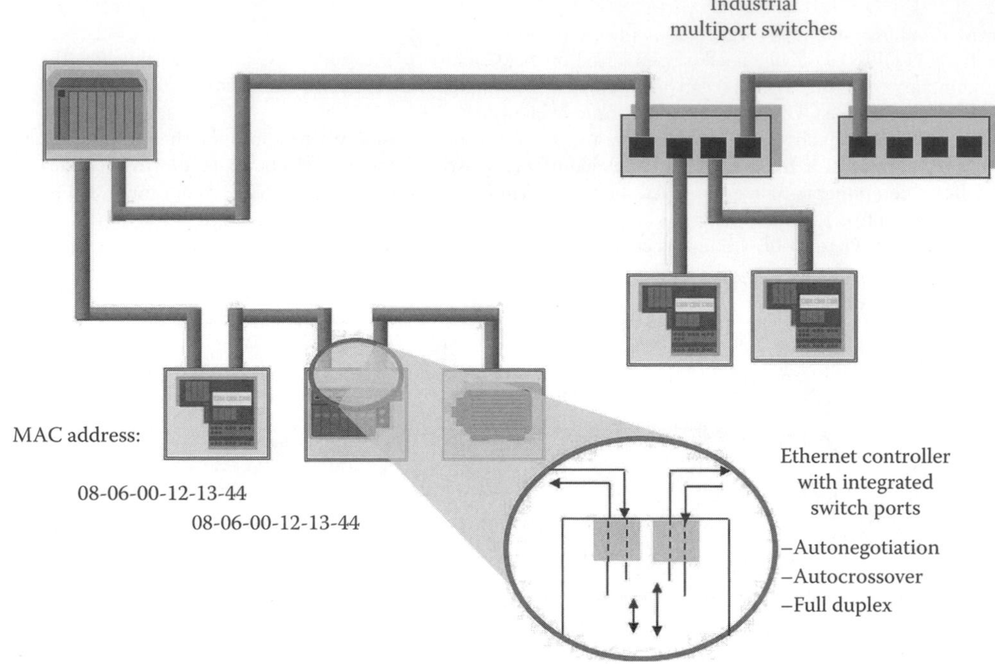

FIGURE 12.9 Usage of switches.

During system engineering, the configuring engineer joins together the modules/submodules of an I/O-Device defined in the GSD file in order to map them to the real system and to assign them to slots/subslots. The configuring engineer configures the real system symbolically in the engineering tool. Figure 12.10 shows the relationship between the GSD definitions, configuration, and real plant view.

After completion of system engineering, the configuring engineer downloads the system data to the I/O-Controller, which also contains the system-specific application. As a result, an I/O-Controller has all the information needed for addressing the I/O-Devices and for data exchange.

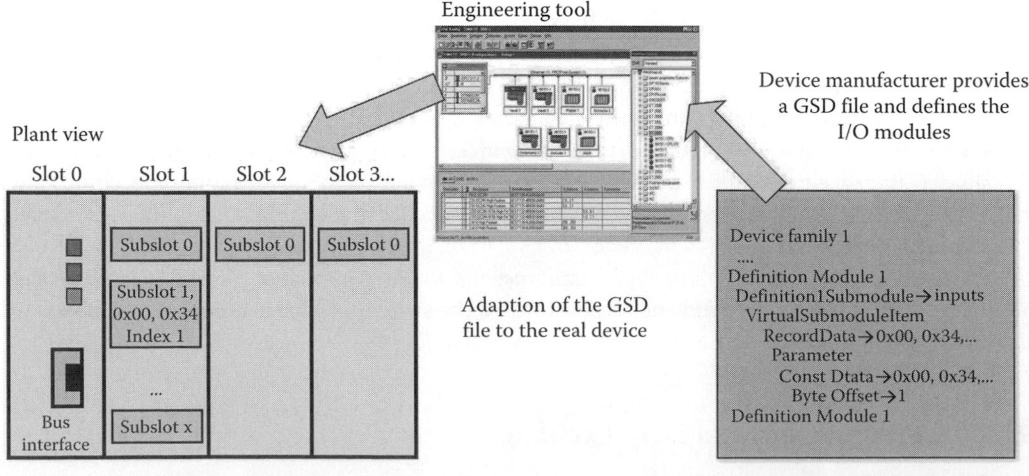

FIGURE 12.10 Assignment of definitions in the GSD file to I/O-Devices when configuring the system.

Before an I/O-Controller can perform data exchange with the I/O-Devices, these must be assigned an IP address based on their configured name. This must take place prior to system power-up. The I/O-Controller performs this automatically using the DCP.

After a start-up/restart, an I/O-Controller always initiates system power-up based on the configuration data without any intervention by the user. During system power-up, an I/O-Controller establishes an explicitly specified CR and AR with an I/O-Device. This specifies the cyclic I/O data, the alarms, the exchange of acyclic read/write services, and the expected modules/submodules. After successful system power-up, the exchange of cyclic process data, alarms, and acyclic data traffic can occur.

12.2.6 Diagnostics

PROFINET I/O supports a multilevel diagnostic concept that enables efficient fault localization and correction. When a fault occurs, the faulty I/O-Device generates a diagnostic alarm to the I/O-Controller. This alarm triggers a call in the PLC program to the appropriate program routine in order to be able to respond to the fault. If a device or module defect requires a complete replacement of the device or module, the I/O-Controller automatically performs a parameter assignment and configuration of the new device or module.

The diagnostic information is structured hierarchically:

- Slot number (module)
- Channel number
- Channel type (input/output)
- Coded fault cause (e.g., wire break, short circuit)
- Additional manufacturer-specific information

When an error occurs in a channel, the I/O-Device generates a diagnostic alarm to the I/O-Controller. This alarm triggers a call in the control program to the appropriate fault routine. After processing of the fault routine, the I/O-Controller acknowledges the fault to the I/O-Device. This acknowledgment mechanism ensures that a sequential fault processing is possible in the I/O-Controller.

12.3 Principles of PROFINET Communication

Standard Ethernet communication via TCP/IP or UDP/IP communication is sufficient for data communication in some cases. In industrial automation, however, requirements regarding time behavior and isochronous operation exist that cannot be fully satisfied using the TCP/IP or UDP/IP channel.

A scalable RT concept is the solution for this. With RT, this concept can be realized with standard network components, such as switches and standard Ethernet controllers. RT communication takes place without TCP/IP information.

The transmission of RT data is based on cyclical data exchange using a provider/consumer model. The communication mechanisms of layer 2 (according to the ISO/OSI model) are sufficient for this. For optimal processing of RT frames within an I/O-Device, the VLAN tag according to IEEE 802.1Q (prioritization of data frames) has been supplemented with a special Ethertype that enables fast channelization of these PROFINET frames in the higher-level software of the field device. Ethertypes are allocated by IEEE and are therefore an unambiguous criterion for differentiation among Ethernet protocols. Ethertype 0x8892 is specified in IEEE and is used for fast data exchange in PROFINET I/O.

PROFINET I/O provides protocol definitions for the following services:

- Address resolution for field devices
- Cyclic transmission of I/O data on RT basis
- Acyclic transmission of alarms to be acknowledged

- Acyclic transmission of data (parameters, detailed diagnostics, I&M data, information functions, etc.) on an as-needed basis
- Redundancy mode for RT frames

The combination of these communication services in the higher-level controller makes it possible to implement convenient system diagnostics, topology detection, and device replacement, among other things.

Many communication services in PROFINET occur in RT.

PROFINET provides scalable Ethernet-based communication. The following three performance levels are differentiated:

1. UDP/IP for data that are not time critical, such as parameter assignment and configuration data
2. RT for time-critical process data in standard application in the area of discrete manufacturing and continuous process automation
3. IRT for particularly challenging application requirements, such as those for motion control

These three PROFINET communication performance levels cover the entire spectrum of automation applications in their overall diversity. The PROFINET communication standard is characterized by the following in particular:

- Coexisting utilization of RT and TCP-based IT communication on one line
- Uniform RT protocol for communication between the controller and the distributed field devices for all applications
- Scalable RT communication from performant to high-performant and isochronous mode

Scalability and a uniform communication basis represent two of the major strengths of PROFINET. They guarantee continuity to the corporate management level and fast response times in the automation process.

12.3.1 Standard Communication with UDP

PROFINET uses Ethernet and UDP with IP as the basis for communication. TCP/UDP with IP is a de facto standard with respect to communication protocols in the IT area. However, for interoperability (i.e., the interaction between applications), establishment of a common communication channel over the field devices, based on TCP/UDP, is insufficient. TCP or UDP represent only the foundation on which Ethernet devices can exchange data via a transport channel in local and distributed networks. Therefore, additional specifications and protocols beyond TCP/UDP—the so-called application protocols—are required. Interoperability is guaranteed only if the same application protocol is used on the devices. Typical application protocols are SMTP (for e-mail), FTP (for file transfer), and HTTP (used on the Internet).

12.3.2 Real-Time Communication

To enable enhanced scaling of communication options and, thus, also of determinism in PROFINET I/O, RT classes have been defined for data exchange. From the user perspective, these classes involve unsynchronized and synchronized communication. The details are managed by the field devices themselves.

RT frames are automatically prioritized in PROFINET compared to UDP/IP frames. This is necessary in order to prioritize the transmission of data in switches to prevent RT frames from being delayed by UDP/IP frames.

RT applications in discrete manufacturing automation require update, or response times, in the range of 5–10 ms. The update time refers to the time that elapses when a variable is generated by an application in a device, then sent to a peer device via the communication system, and then received updated by the application. For devices, the implementation of an RT communication causes only a small load on the processor so that execution of the user program continues to take precedence. From experience, in the case of Fast Ethernet (100 Mb/s Ethernet), the transmission rate on the line in proportion to the execution in the devices can be disregarded. Most of the time is consumed in the application. The time it takes to provide data to the application of the provider is not influenced by the communication. This also applies to processing of data received in the consumer. As a result, noteworthy improvements in update rates and thus the RT performance can be obtained primarily through proper optimization of the communication stack in the provider and the consumer. In order to satisfy RT demands in automation, PROFINET uses an optimized RT communication channel.

Motion control applications require update rates in the range of 1 ms along with a jitter for consecutive update cycles of 1 μs for cases involving up to 100 stations. To satisfy these requirements, PROFINET defines the IRT time slot–controlled transmission process on layer 2 for Fast Ethernet. That means every device knows exactly in which time slot it is allowed to send data over the bus. Through synchronization of the devices involved (network components and PROFINET devices) with the accuracy indicated earlier, a time slot can be specified during which data critical for the automation task are transferred. The communication cycle is split into a deterministic part and an open part. The cyclic RT message frames are dispatched in the deterministic channel, while the RT and UDP/IP message frames are transported in the open channel. The process is comparable to the traffic on a highway where the left lane is reserved for time-critical traffic (IRT traffic) and the remaining traffic elements (RT and UDP/IP traffic) are prevented from switching to this lane. Even if there is traffic jam in the right lane, the time-critical traffic is not impacted.

Isochronous data transmission is realized based on hardware, for example, using application specific integrated circuits (ASICs). Such an ASIC covers the cycle synchronization and time slot reservation functionality for the RT data. Realization in hardware ensures the accuracy requirements to be achieved within the order of magnitude required. Furthermore, the processor in the PROFINET device is relieved from communication tasks. The resulting additional runtime can be made available for automation tasks.

12.4 Conformance Classes and Their Functions

PROFINET I/O is a performance-optimized communication system. It encompasses a broad scope of functions to meet the requirements of all industrial applications. However, the complete scope of functions is not required in every application. For this reason, PROFINET I/O offers scalability with regard to the functionality supported by the particular application.

PI has classified the scope of functions in PROFINET I/O into three CCs. The minimum requirements for these three CCs (CC-A, CC-B, and CC-C) have been defined from the perspective of the plant operator. They enable plant operators to easily select field devices and bus components with explicitly defined minimum properties [3].

12.4.1 Basic Functions of Conformance Class A

The basic functions of CC-A include cyclic exchange of I/O data with RT properties, acyclic data communication for reading and writing of demand-oriented data (parameters and diagnostics), including the identification and maintenance (I&M) function for reading out device information, and a flexible alarm model for signaling device and network errors with three alarm levels (maintenance required, urgent maintenance required, and diagnostics) (Table 12.1).

TABLE 12.1 List of Basic Functions

Requirement	Technical Function/Solution
Cyclic data exchange	PROFINET with RT communication
Acyclic parameter data/device identification (HW/FW)	Read Record/Write Record I&M0
Device/ network diagnostics (alarms)	Diagnostics and maintenance

12.4.1.1 Cyclic Data Exchange

Cyclic I/O data are transmitted via the *I/O Data CR* unacknowledged as RT data between provider and consumer in an assignable time base. The cycle time can be specified individually for connections to the individual devices and are thus adapted to the requirements of the application. Likewise, different cycle times can be selected for the input and output data, within the range of 250 μs to 512 ms. The connection is monitored using a time monitoring setting that is derived from a multiple of the cycle time. During data transmission in the frame, the data of a subslot are followed by a provider status. This status information is evaluated by the respective consumer of the I/O data. As a result, it can evaluate the validity of the data from the cyclic data exchange alone. In addition, the consumer statuses for the opposite direction are transmitted. The data in the message frames are followed by accompanying information that provides information about the data's validity, the redundancy, and the diagnostic status (data status and transfer status). The cycle information (cycle counter) of the provider is also specified so that its update rate can be determined easily. Failure of cyclic data to arrive is monitored by the respective consumer in the CR. If the configured data fail to arrive within the monitoring time, the consumer sends an error message to the application. The cyclic data exchange can be realized with standard network components, such as switches and standard Ethernet controllers and takes place directly on layer 2 with Ethertype 0x8892 and without any UDP/IP information. For optimized processing of cyclic data within a network component, the VLAN tag according to IEEE802.1Q is additionally used with a high priority.

12.4.1.2 Acyclic Parameter Data

Acyclic data exchange can be used for parameter assignment or configuration of I/O-Devices or reading out status information using the *record data CR*. This is accomplished with read/write frames using standard IT services via UDP/IP, in which the different data records are distinguished by their index. In addition to the data records available for use by device manufacturers, the following system data records are also specially defined:

- Diagnostic information on the network and device diagnostics can be read out by the user from any device at any time.
- Error log entries (alarms and error messages), which can be used to determine detailed timing information about events within an I/O-Device.
- I&M information.

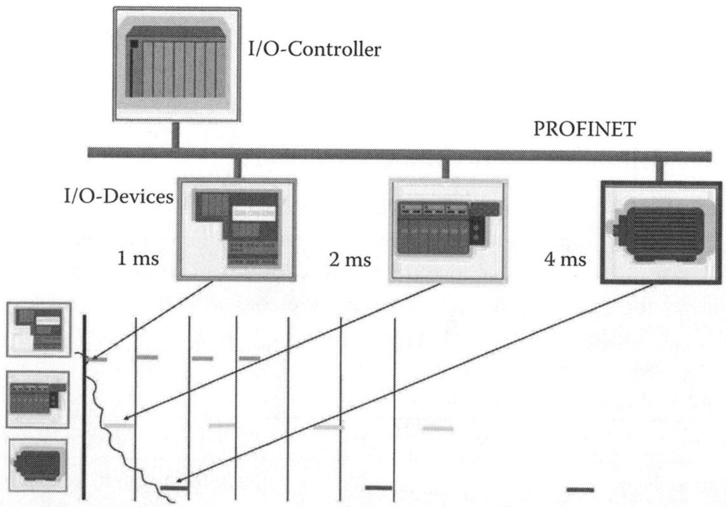

FIGURE 12.11 Real-time communication with cycle time monitoring.

The ability to read out identification information from a field device is very helpful for maintenance purposes. For example, this allows inferences to be drawn in response to incorrect behavior or regarding unsupported functionality in a field device. This information is specified in the I&M data structures. The I&M functions are subdivided into five different blocks (IM0… IM4) and can be addressed separately using their index. Every I/O-Device must support the IM0 function with information about hardware and firmware versions. The I&M specification titled *Identification & Maintenance Functions* can be downloaded from the PI website [4].

12.4.1.3 Device/Network Diagnostics

A status-based maintenance approach is currently gaining relevance for operation and maintenance. It is based on the capability of devices and components to determine their status and to communicate this using agreed mechanisms. A system for reliable signaling of alarms and status messages by the I/O-Device to the I/O-Controller was defined for PROFINET I/O for this purpose. This alarm concept covers both system-defined events (such as removal and insertion of modules) and signaling of faults that were detected in the utilized controller technology (e.g., defective load voltage or wire break). This is based on a state model that defines *good* and *defective* status as well as the *maintenance required* and *maintenance demanded* prewarning levels. A typical example for this is the loss of media redundancy, which signals *maintenance required*, but because of its redundancy, the media is still fully functional.

Diagnostic alarms must be used if the error or event occurs within an I/O-Device or in conjunction with the connected components. They can signal an incoming or outgoing fault status. In addition, the user can define corresponding process alarms for messages from the process, for example, limit temperature exceeded. In this case, the I/O-Device may still be operable. These process alarms can be assigned different priorities than the diagnostic alarms.

12.4.2 Network Diagnostics and Management of Conformance Class B

CC-B expands devices to include functions for additional network diagnostics and for topology detection. PROFINET uses simple network management protocol (SNMP) for this. Portions of management information base 2 (MIB2) and lower link discovery protocol-MIB (LLDP-EXT MIB) are integrated into devices. In parallel to SNMP, all diagnostics and topology information can also be called up from the physical device object (PDEV) using acyclic PROFINET services.

12.4.2.1 Network Management Protocol

In existing networks, SNMP has established itself as the de facto standard for maintenance and monitoring of network components and their functions. SNMP can read access network components, in order to read out statistical data pertaining to the network as well as port-specific data and information for neighborhood detection. In order to monitor PROFINET devices with an established management system, implementation of SNMP is mandatory for devices of CCs B and C.

12.4.2.2 Neighborhood Detection

Automation systems can be configured flexibly in a line, star, or tree structure. To compare the specified and actual topologies, that is, to determine which field devices are connected to which switch port and to identify the respective port neighbor, LLDP according to IEEE 802.1AB was applied in PROFINET I/O. PROFINET field devices exchange existing addressing information with connected neighbor devices via each switch port. The neighbor devices are thereby unambiguously identified, and their physical location is determined (e.g., in Figure 12.12: the delta device is connected to port 003 of switch 1 via port 001).

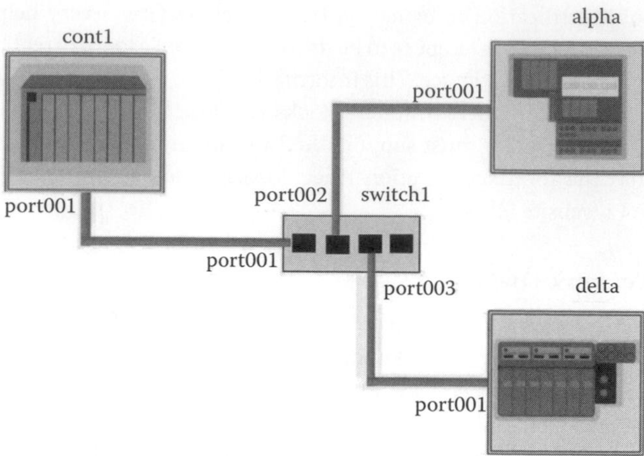

FIGURE 12.12 Example for neighbors.

12.4.2.3 Representation of the Topology

A plant owner can use a suitable tool to graphically display a plant topology and port-granular diagnostics (Figure 12.13). The information found during neighborhood detection is collected using the SNMP protocol. This provides the plant operator a quick overview of the plant status.

12.4.2.4 Device Replacement

If a field device fails in a known topology, it is possible to check whether the replacement device has been reconnected in the proper position. It is even possible to replace devices without the use of an engineering tool: when replaced, a device at a given position in the topology receives the same name and parameters as its predecessor (Figure 12.14).

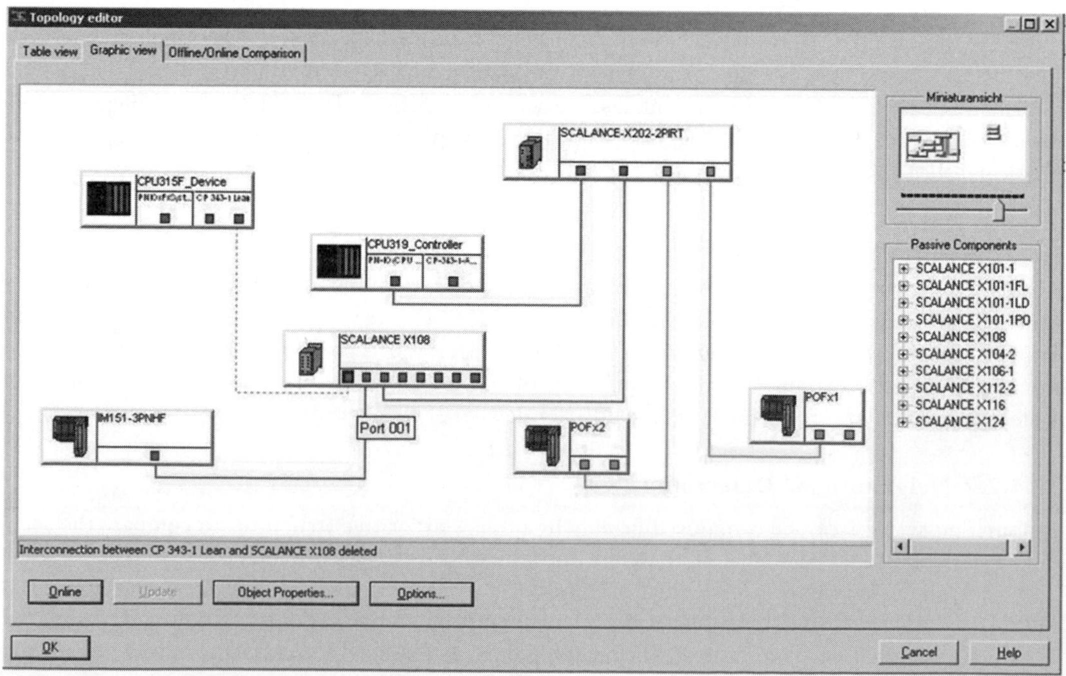

FIGURE 12.13 Example of a topology plan.

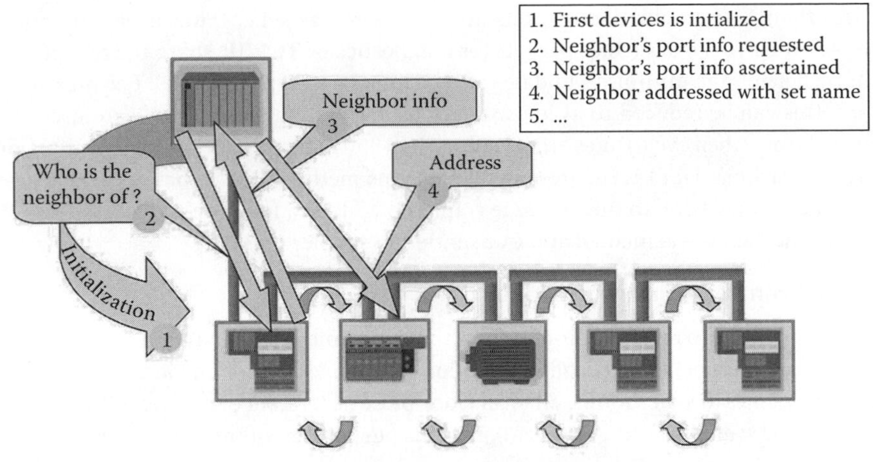

FIGURE 12.14 Device replacement procedure.

12.4.2.5 Integration of Network Diagnostics into the I/O System Diagnostics

A switch must also be configured as a PROFINET I/O-Device and signal the detected network errors of a lower-level Ethernet line directly to the I/O-Controller. Acting as an I/O-Device, this type of switch can signal faults and specific operating modes to its I/O-Controller by transmitting acyclic alarms using the *alarm CR*. In this way, the network diagnostics can be integrated into the I/O system diagnostics (Figure 12.15).

12.4.3 Isochronous Real-Time with Conformance Class C

CC-C includes all necessary network-wide synchronization functions for applications with the most stringent requirements for deterministic behavior. Networks based on CC-C enable applications having

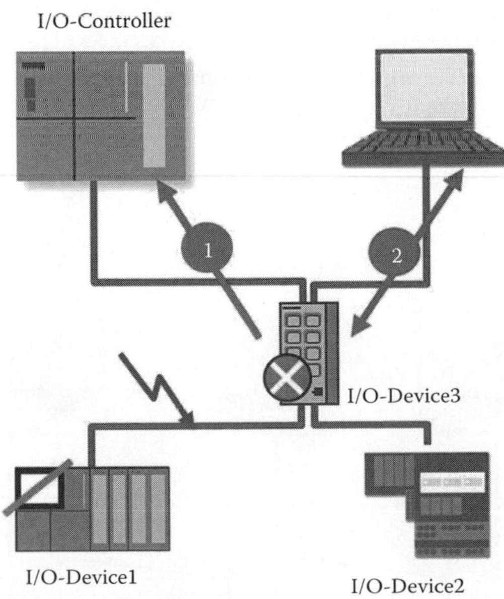

FIGURE 12.15 Integration of network diagnostics into I/O system diagnostics.

a jitter of less than 1 μs. Cyclic data packets are transferred as synchronized packets on a reserved bandwidth. All other packets, such as packets for diagnostics or TCP/IP, share the rest of the Ethernet bandwidth. By default, the minimum update rate is defined at 250 μs in CC-C. For maximum control performance, this can be reduced to as low as 31.25 μs, depending on the hardware used. In order to expand data volumes when cycle times are set at less than 250 μs, a message frame optimization method (dynamic frame packing, DFP) is incorporated. With this method, nodes that are wired together in a line structure are addressed with one message frame. In addition, for cycle times less than 250 μs, the TCP/IP communication is fragmented and transmitted in smaller packets.

12.4.3.1 Synchronized Communication

In order for the bus cycles to run synchronously (at the same time) with a maximum deviation of 1 μs, all devices involved in the synchronous communication must have a common clock. A clock master uses synchronization frames to synchronize all local clock pulse generators of devices within a clock system (IRT domain) to the same clock (Figure 12.16). For this purpose, all of the devices involved in this type of clock system must be connected directly to one another, without crossing through any nonsynchronized devices. Multiple independent clock systems can be defined in one network.

To achieve the desired accuracy for the synchronization and synchronous operation, the runtime on each connecting cable must be measured with defined Ethernet message frames and figured into the synchronization. Special hardware precautions must be taken for implementing this clock synchronization. The bus cycle is divided into different intervals for synchronized communication (Figure 12.17). First, the synchronous data are transmitted in the real-time phase. This phase is protected from delays caused by other data and allows a high level of determinism. In the subsequent open phase, all other data are transmitted according to IEEE 802 and the specified priorities.

The division of the individual intervals can vary. If forwarding of the data before the start of the next reserved interval is not assured, these frames are stored temporarily and sent in the next open phase.

12.4.3.2 Mixed Operation

A combination of synchronous and asynchronous communication within an automation system is possible, if certain preconditions are met. A mixed operation is shown in Figure 12.18. In this example, a synchronizable switch has been integrated in the field device for devices 1–3. This enables the runtime to be determined and precise synchronization of the clock system to be maintained. The other two devices are connected via a standard Ethernet port and thus communicate asynchronously. The switch ensures that this communication occurs only during the open phase.

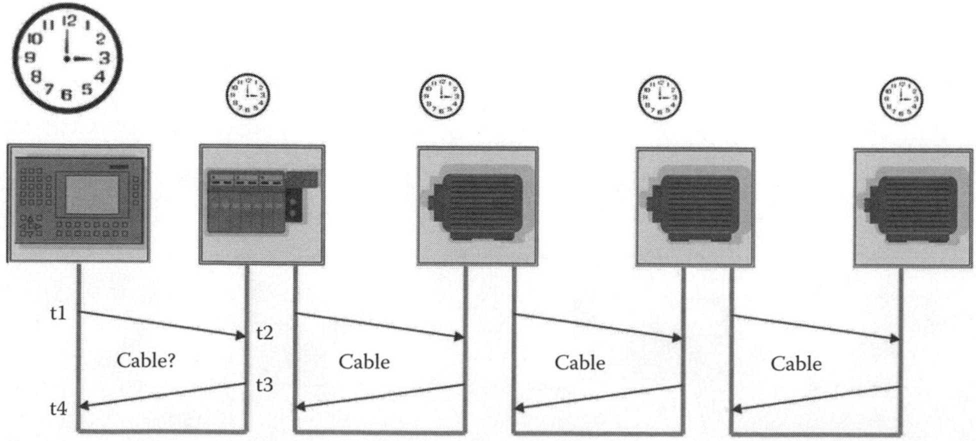

FIGURE 12.16 Synchronization of clock pulse generators within an IRT domain by the clock master.

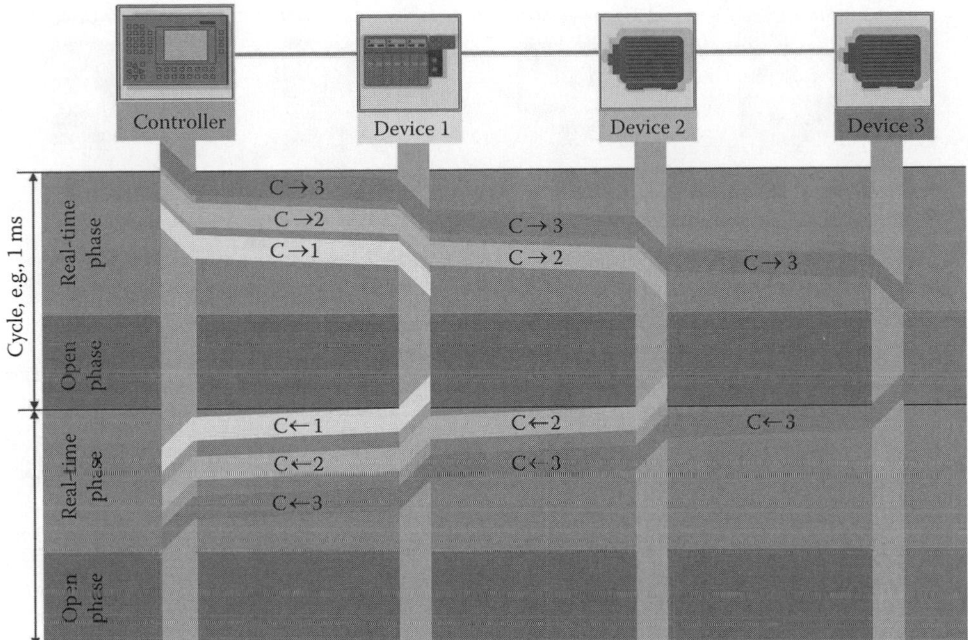

FIGURE 12.17 IRT communication divides the bus cycle into a reserved interval (real-time phase) and an open interval (open phase).

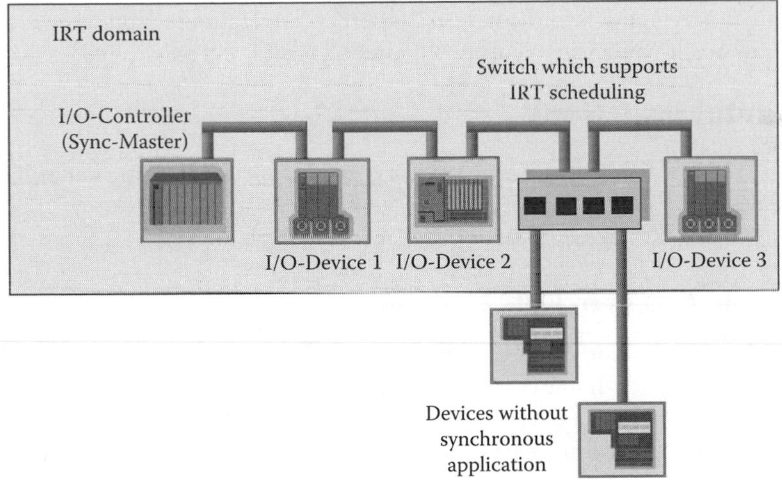

FIGURE 12.18 Mixed operation of synchronized and unsynchronized applications.

12.4.3.3 Optimized IRT Mode

When the time ratios are subject to stringent requirements, the efficiency of the topology-oriented synchronized communication can be optimized using DFP (Figure 12.19). For a line structure, the synchronous data of several devices are optionally combined into one Ethernet frame. The individual cyclic RT data can be extracted for each node. Because the data from the field devices to the controller are also strictly synchronized, these data can be assembled by the switch in a single Ethernet frame. Ideally, only one frame is then transmitted for all the field devices in the real-time phase. This frame is disassembled or assembled in the corresponding switch, if required. This DFP technology is optional for systems with

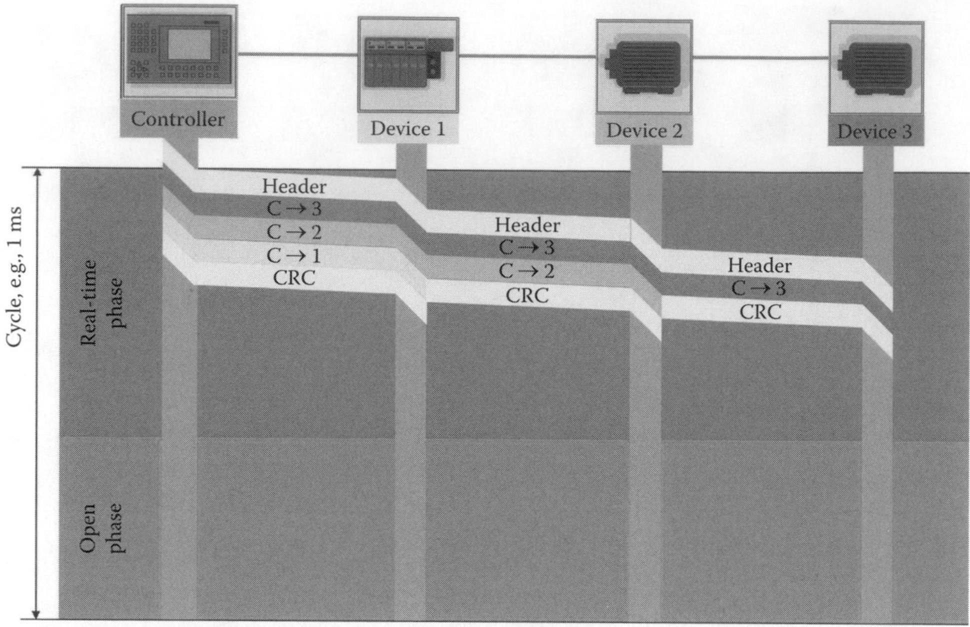

FIGURE 12.19 Packing of individual message frames into a group message frame.

stringent requirements. The functionalities of other intervals are retained; that is, a mixed operation is also possible here. To achieve short cycle times of up to 31.25 μs, however, the open phase must also be sharply reduced. To accomplish this, the standard Ethernet frames for the application are disassembled transparently into smaller fragments, transmitted in small pieces, and reassembled.

12.5 Optional Functions

Additionally, PROFINET provides a large number of optional functions that are not included in devices by default by way of CCs (Table 12.2). If additional functions are to be used, this must be checked on a case-by-case basis using the device properties (data sheet, manuals, and GSD file).

12.5.1 Multiple Access to Field Devices

The innovative starting point for shared devices is the parallel and independent access of two different controllers to the same device. In the case of a shared device, the user configures a fixed assignment of the various I/O modules used in a device to a selected controller. One possible application of a shared device is in fail-safe applications in which a fail-safe CPU controls the safe portion of the device and a

TABLE 12.2 List of Possible Optional Functions

Requirement	Technical Function/Solution
Multiple access to inputs by various controllers	Shared input
Distribution of device functions to various controllers	Shared device
Extended device identification	Identification & Maintenance IM1-4
Automatic parameter assignment of devices using parameter sets	Individual parameter server
Configuration changes during operation	Configuration in Run (CiR)
Time stamping of I/O data	Time sync
Fast restart after voltage recovery	Fast start-up (FSU)
Higher availability through ring redundancy	MRP/MRPD
Call of a device-specific engineering tool	Tool Calling Interface (TCI)

standard controller controls the standard I/O within the same station. In the safety scenario, the F-CPU uses the fail-safe portion to safely switch off the supply voltage of the outputs.

In the case of a shared input, there is parallel access to the same input by two different controllers. Thus, an input signal that must be processed in two different controllers of a plant does not have to be wired twice or transferred via CPU–CPU communication.

12.5.2 Extended Device Identification

Further information for standardized and simplified I&M is defined in additional I&M data records. I&M1–4 (Table 12.3) contain plant-specific information, such as installation location and date, and are created during configuration and written to the device.

12.5.3 Individual Parameter Server

The individual parameter server functionality is available for backing up and reloading of other optional individual parameters of a field device (Figure 12.20). The basic parameter assignment of a field device is carried out using the parameters defined in the GSD file for the field device. A GSD file contains module parameters for I/O modules, among other things. These are stored as static parameters and can be loaded from the I/O-Controller to an I/O-Device during system power-up. For some field devices, it is either impossible or inappropriate to initialize parameters using the GSD approach due to the quantities, the user guidance, or the security requirements involved. Such data for specific devices and technologies are referred to as individual parameters (iPar). Often, they can be specified only during commissioning. If such a field device fails and is replaced, these parameters must also be reloaded to the new field device without the need for an additional tool. The individual parameter server provides plant operators a convenient and uniform solution for this.

TABLE 12.3 Extended Device Identification

IM1	TAG_FUNCTION TAG_LOCATION	Plant designation Location designation
IM2	INSTALLATION_DATE	Installation date
IM3	DESCRIPTOR	Comments
IM4	SIGNATURE	Signature

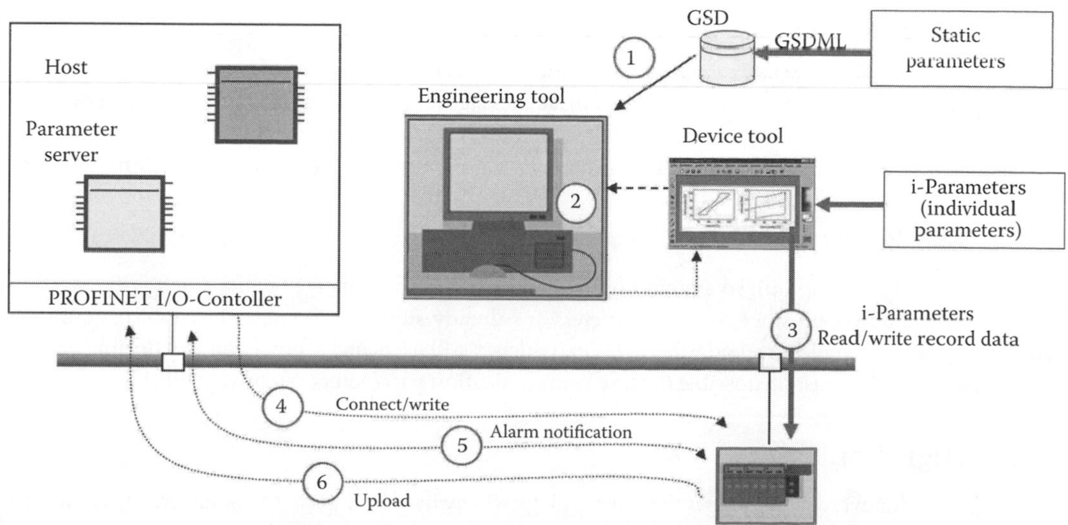

FIGURE 12.20 Automatic reload backed-up data during device replacement using a parameter server.

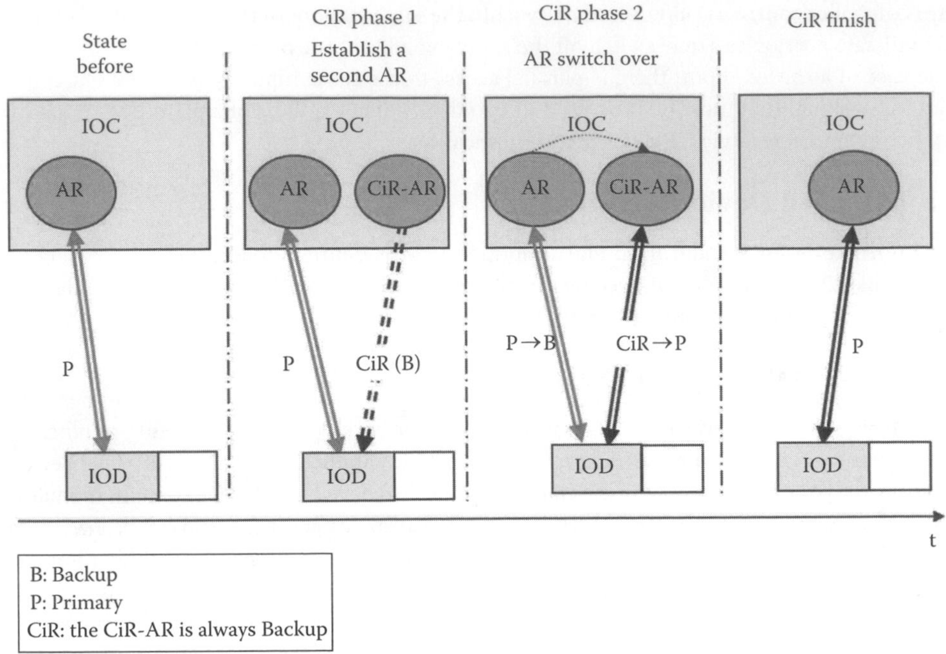

FIGURE 12.21 Configuration changes without plant interruption thanks to redundant connection.

12.5.4 Configuration in Run

Like redundancy, uninterrupted plant operation—including when reconfiguring devices and networks and when inserting, removing, or replacing devices or individual modules—plays an important role in process automation (Figure 12.21). All of these *configuration in run* measures (CiR) are carried out in PROFINET bumpless and without adversely affecting network communication. This ensures that plant repairs, modifications, or expansions can be performed without a plant shutdown in continuous production processes, as well.

12.5.5 Time Stamping

In large plants, the ability to assign alarms and status messages to a sequence of events is often required. For this purpose, an optional time stamping of these messages is possible in PROFINET I/O. In order to time stamp data and alarms, the relevant field devices must have the same time of day. To accomplish this, a master clock and the time synchronization protocol are used to set the clocks to the same time.

12.5.6 Fast Restart

Fast start-up defines an optimized system power-up in which data exchange begins much faster starting with the second power-up since many parameters are already stored in the field devices. This optional path can be used in parallel to standard power-up (which is still used after a power on and during the first power-up or reset). It must be possible to store communication parameters retentively for this.

12.5.7 High Availability

Chaining of multiport switches allows the star topology widely used in Ethernet to be effectively combined with a line structure. This combination is especially well suited for control cabinet connections, that is, line connection between control cabinets and star connection to process-level field device. If the connection between two field devices in a line is interrupted, the field devices located after the interruption are no longer accessible. If increased availability is required, provision must be made for redundant communication paths when planning the system, and field devices/switches that support the redundancy concept of PROFINET must be used. The line can be closed to form a ring to easily provide a redundant communication path. In the event of an error, the connection to all nodes is ensured via the alternative connection (the other direction of the ring line). This achieves a tolerance for one fault. Organizational measures must be taken to ensure that this fault is eliminated before a second error occurs. PROFINET has two mechanisms for setting up ring-shaped media redundancy, depending on the requirements:

12.5.7.1 Media Redundancy Protocol

The media redundancy protocol (MRP) according to IEC 62439 describes PROFINET redundancy with a typical reconfiguring time of <200 ms for communication paths with TCP/IP and RT frames after a fault. Error-free operation of an automation system involves a media redundancy manager and several media redundancy clients arranged in a ring.

The task of a media redundancy manager is to check the functional capability of the configured ring structure. This is done by sending out cyclic test frames. As long as it receives all of its test frames back, the ring structure is intact. As a result of this behavior, a media redundancy manager prevents frames from circulating and converts a ring structure into a line structure. A media redundancy client is a switch that acts only as a *passer* of frames and generally does not assume an active role. It must have two switch ports in order to connect to other media redundancy clients or the media redundancy manager in a single ring.

12.5.7.2 Media Redundancy for Planned Duplication

IEC 61158 describes the redundancy concept media redundancy for planned duplication (MRPD) for topology-optimized IRT communication, which enables smooth switchover from one communication path to another in the event of a fault. During system power-up, the I/O-Controller loads the data of the communication paths for both communication channels (directions) in a communication ring to the individual nodes. Thus, it is immaterial which node fails because the loaded *schedule* for both paths is available in the field devices and is monitored and adhered to without exception. Loading of the *schedule* alone is sufficient to exclude frames from circulating in this variant, because the destination ports are explicitly defined.

12.6 Integration of Fieldbus Systems

PROFINET specifies a model for integrating existing PROFIBUS and other fieldbus systems such as INTERBUS and DeviceNet (Figure 12.22).

This means that any combination of fieldbus and PROFINET-based subsystems can be configured. Thus, a smooth technology transition is possible from fieldbus-based systems to PROFINET. The following requirements are taken into consideration here:

- Plant owners would like the ability to easily integrate existing installations into a newly installed PROFINET system.
- Plant and machine manufacturers would like the ability to use their proven and familiar devices without any modifications for PROFINET automation projects, as well.
- Device manufacturers would like the ability to integrate their existing field devices into PROFINET systems without the need for costly modifications.

Fieldbus solutions can be easily and seamlessly integrated into a PROFINET system using proxies and gateways. The proxy acts as a representative of the fieldbus devices on the Ethernet. It integrates the nodes connected to a lower-level fieldbus system into the higher-level PROFINET system. As a result, the advantages of fieldbuses, such as high dynamic response, pinpoint diagnostics, and automatic system configuration without settings on devices, can be utilized in the PROFINET world, as well. These advantages simplify planning through the use of known sequences. Likewise, commissioning and operation

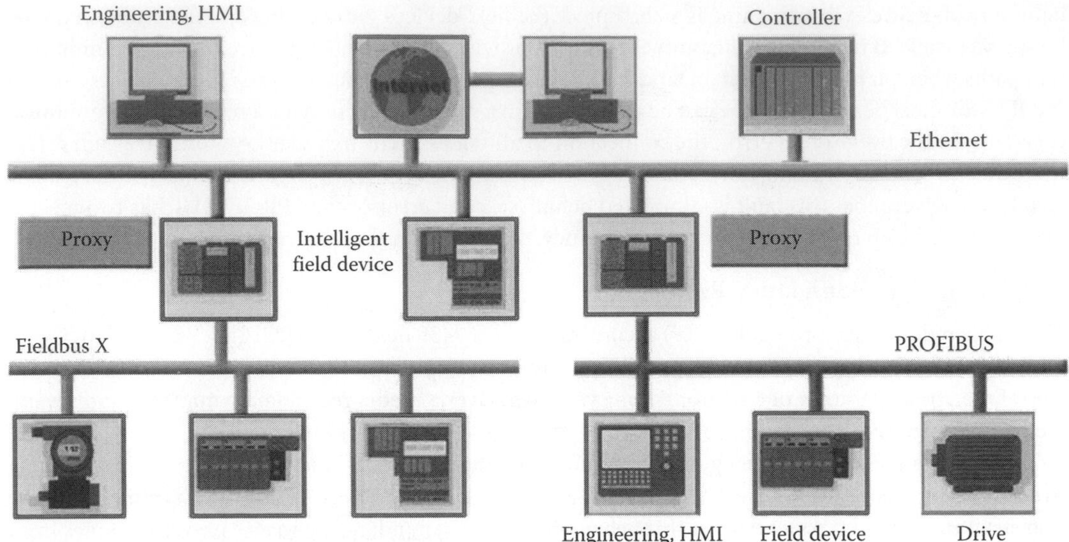

FIGURE 12.22 Integration of fieldbus systems is easy with PROFINET.

are made easier through the comprehensive diagnostics properties of the fieldbus system. Devices and software tools are also supported in the accustomed manner and integrated into the handling of the PROFINET system.

12.7 Application Profiles

By default, PROFINET transfers the specified data transparently. It is up to the user to interpret the sent or received data in the user program of a PC-based solution or PLC. Application profiles are specifications for particular properties, performance characteristics, and behavior of devices and systems that are developed jointly by manufacturers and users. The term *profile* can apply to a few specifications for a particular device class or a comprehensive set of specifications for applications in a particular industry sector. In general, two groups of application profiles are distinguished:

- General application profiles that can be used for different applications (examples of these include the profiles PROFIsafe and PROFIenergy)
- Specific application profiles that are each developed for a specific type of application, such as PROFIdrive or devices for process automation

These application profiles are specified by PI based on market demand and are available on the PI website.

12.7.1 PROFIsafe

The PROFIsafe designation refers to a protocol defined in IEC 61784-3-3 for implementation of functional safety (fail-safe) and approved by IFA and TÜV. PROFIsafe can be used with PROFIBUS and PROFINET alike. The use of PROFIsafe enables elements of a fail-safe controller to be transferred directly to the process control on the same network. The need for additional wiring is eliminated.

12.7.2 PROFIdrive

The PROFIdrive designation refers to the specification of a standardized drive interface for PROFIBUS and PROFINET. This application-oriented profile, which has been standardized in IEC 61800-7, contains standard definitions (syntax and semantics) for communication between drives and automation systems, thus assuring vendor neutrality, interoperability, and investment protection. The PROFIdrive application profile provides the foundation for almost every drive task in the field of industrial automation engineering. It defines the device behavior and the procedure for accessing drive data of electric drives and also optimally integrates the additional PROFIsafe and PROFIenergy profiles.

12.7.3 PROFIenergy

The high cost of energy and compliance with legal obligations are compelling industry to engage in energy conservation. Recent trends toward the use of efficient drives and optimized production processes have been accompanied by significant energy savings. However, in today's plants and production units, it is common for numerous energy-consuming loads to continue running during pauses. PROFIenergy addresses this situation [5]. PROFIenergy enables an active and effective energy management. By purposefully switching off unneeded consumers and/or adapting parameters such as clock rates to the production rate, energy demand and, thus, energy costs can be drastically reduced. In doing so, the power consumption of automation components such as robots and laser cutting machines or other subsystems used in production industries is controlled using PROFIenergy commands. PROFINET nodes in which PROFIenergy functionality is implemented can use the commands to react flexibly to idle times. In this way, individual devices or unneeded portions of a machine can be shut down during short pauses, while a whole plant can be shut down in an orderly manner during long pauses. In addition, PROFIenergy can help optimize a plant's production on the basis of its energy consumption.

12.8 PROFINET for Process Automation

Compared with discrete manufacturing automation, continuous process automation has a few special characteristics that contribute to defining the use of automation to a large extent. Plants can have a service life of many decades. This gives rise to a requirement, on the part of plant operators, for older and newer technologies to coexist in such a way that they are functionally compatible. In addition, requirements for the reliability of process plants, particularly in continuous processes, are often considerably greater. As a result of these two factors, investment decisions regarding new technologies are significantly more conservative in process automation than in discrete manufacturing automation. For optimal use of PROFINET in all sectors of continuous process automation, PI has created a requirements catalog in collaboration with users. In this manner, it is ensured that plant owners can rely on a future-proof system based on PROFIBUS today and can change to PROFINET at any time. The requirements mainly include the functions for cyclic and acyclic data exchange, integration of fieldbuses (PROFIBUS PA, HART, and FF), integration and parameter assignment of devices including configuration in Run, diagnostics and maintenance, redundancy, and time stamping.

The energy-limited bus feed of devices in hazardous areas on Ethernet has not been formulated as a requirement as there is already an ideal, proven solution with PROFINET for PA. In addition, proven, field-tested Ethernet solutions currently do not exist for this.

12.9 Installation Technology for PROFINET

In addition to the described functions, PROFINET defines the passive infrastructure components (cabling, connectors) [6]. Communication may take place on copper or fiber-optic cables. In a CC-A network, communication is also allowed over wireless transmission systems (Bluetooth and WLAN) (Table 12.1).

The cabling guide defines for all CCs, a two-pair cable according to IEC 61784-5-3. The use of four-pair cables is also allowed for transmission systems with Gigabit cabling requirements. For a CA-A network, complete networking with active and passive components according to ISO/IEC-24702 is

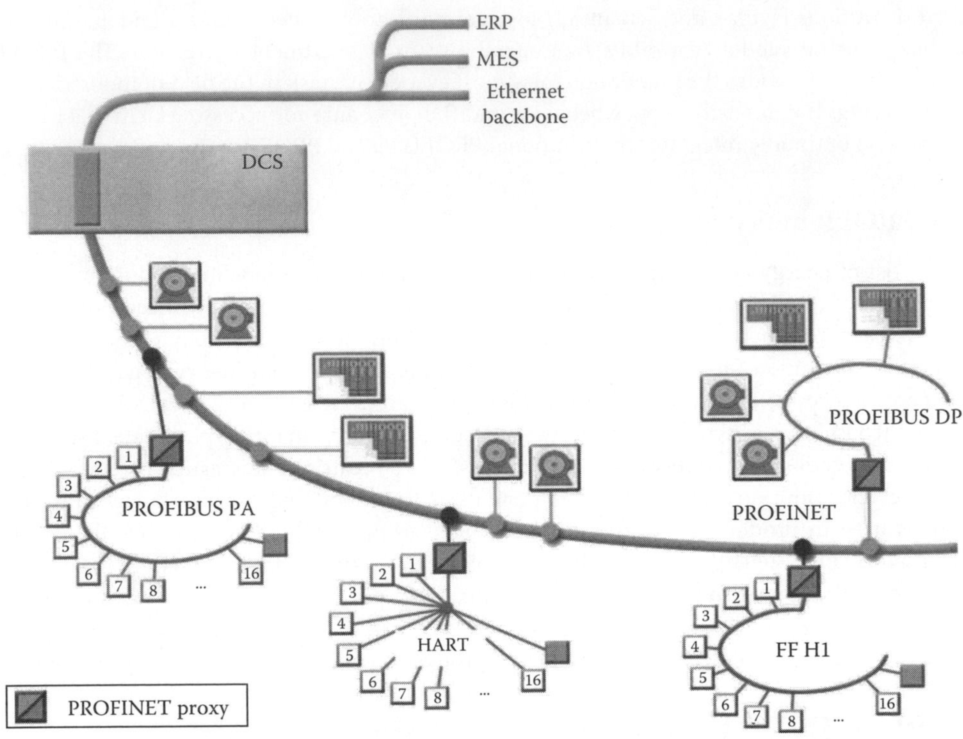

FIGURE 12.23 Example architecture for use of PROFINET in process automation.

TABLE 12.4 Network Installation for Different Conformance Classes

Network Cabling and Infrastructure Components	Solution	Conformance Class
Passive network components (connector, cable)	RJ45, M12	A, B, C
Copper and fiber-optic transmission systems	TX, FX, LX	A, B, C
Wireless connections	WLAN, Bluetooth	A
IT switch	With VLAN tag according to IEEE 802.x	A
Switch with device function	PROFINET with RT	B
Switch with device function and bandwidth reservation	PROFINET with IRT	C

allowed, taking into consideration the CC-A cabling guide. Likewise, active infrastructure components (e.g., switches) according to IEEE 801.x can be used if they support the VLAN tag with prioritization. Easy-to-understand and systematically structured instructions have been prepared to enable problem-free planning, installation, and commissioning of PROFINET I/O. These are available to any interested party on the PI website. These manuals should be consulted for further information.

12.9.1 Network Configuration

The connection of PROFINET I/O field devices occurs exclusively with switches as network components. Switches typically integrated in the field device are used for this (with two ports assigned). PROFINET-suitable switches must support *autonegotiation* (negotiation of transmission parameters)

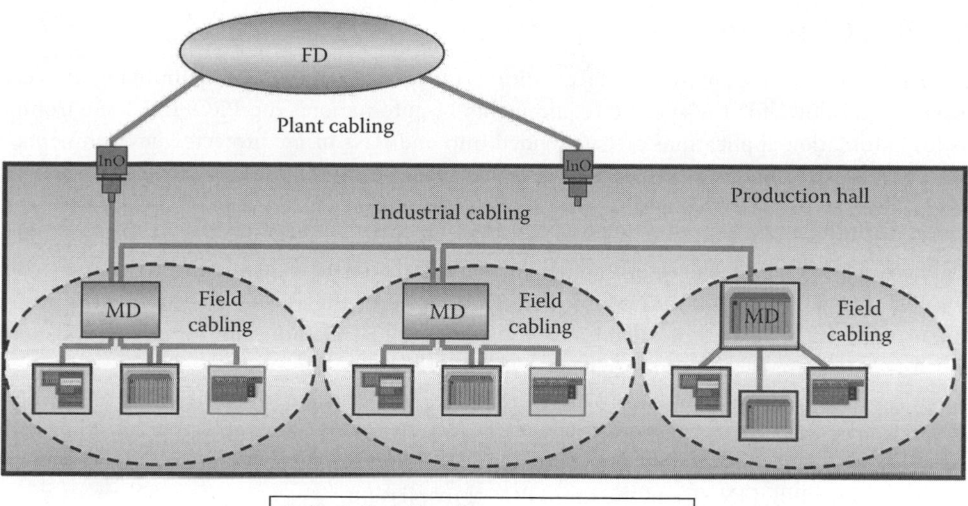

FIGURE 12.24 Ethernet networks in industrial environments often have line topology.

and *autocrossover* (autonomous crossing of send and receive lines). As a result, communication can be established autonomously, and fabrication of the transmission cable is uniform: only 1:1 wired cables can be used.

PROFINET supports the following topologies for Ethernet communication:

- Line topology, which primarily connects terminals with integrated switches in the field (Figure 12.24).
- Star topology, which requires a central switch located preferably in the control cabinet.
- Ring topology, in which a line is closed to form a ring in order to achieve media redundancy.
- Tree topology, in which the topologies indicated earlier are combined.

12.9.2 Cables for PROFINET

The maximum segment length for electrical data transmission with copper cables between two nodes (field devices or switches) is 100 m. The copper cables are designed uniformly in AWG 22. The installation guide defines different cable types, whose range has been optimally adapted to general requirements for industry. Sufficient system reserves allow industry-compatible installation with no limitation on transmission distance. The PROFINET cables conform to the cable types used in industry:

- PROFINET Type A: Standard permanently routed cable, no movement after installation
- PROFINET Type B: Standard flexible cable, occasional movement or vibration
- PROFINET Type C: Special applications: for example, highly flexible, constant movement (trailing cable or torsion)

Due to their electrical isolation, the use of fiber-optic cables for data transmission is especially suitable if equipotential bonding between individual areas of the plant is difficult to establish. Optical fibers also offer advantages over copper in the case of extreme EMC requirements. For fiber-optic transmission, the use of 1 mm polymer optic fibers (POFs) is supported, whose handling conforms optimally to industrial requirements.

12.9.3 Plug Connectors

PROFINET has divided the environmental conditions into just two classes. This eliminates unnecessary complexity and allows for the specific requirements of automation. The PROFINET environmental classes for automation applications are subdivided into one class *inside* protected environments, such as in a control cabinet, and one class *outside* of control cabinets for applications located directly in the field (Figure 17.25).

The selection of suitable PROFINET plug connectors accords with the application. If the emphasis is on a universal network that is to be office-compatible, electrical data transmission is via RJ 45, which is prescribed universally for *inside* environmental conditions. For the *outside* environment, a push–pull plug connector has been developed that is also fitted with the RJ 45 connector for electrical data transmission. The M12 connector is also specified for PROFINET.

For optical data transmission with POFs, the SCRJ plug connector, which is based on the SC plug connector, is specified. The SCRJ is used both in the *inside* environment and in connection with the push–pull housing in the *outside* environment. An optical plug connector is available for the M12 family and can be used for PROFINET and the 1 mm POF transmission.

At the same time, the plug connectors are also specified for the power supply, depending on the topology and the supply voltage. Besides the push–pull plug connector, a 7/8″ plug connector, a hybrid plug connector, or an M12 plug connector can also be used. The difference between these connectors lies in their connectable cross sections and, thus, their maximum amperages.

12.9.4 Security

For networking within a larger production facility or over the Internet, PROFINET relies on a phased security concept. It recommends a security concept optimized for the specific application case, with one or more upstream security zones. On the one hand, this unburdens the PROFINET devices, and on the other, it allows the security concept to be optimized to changing security requirements in a consistent automation engineering solution.

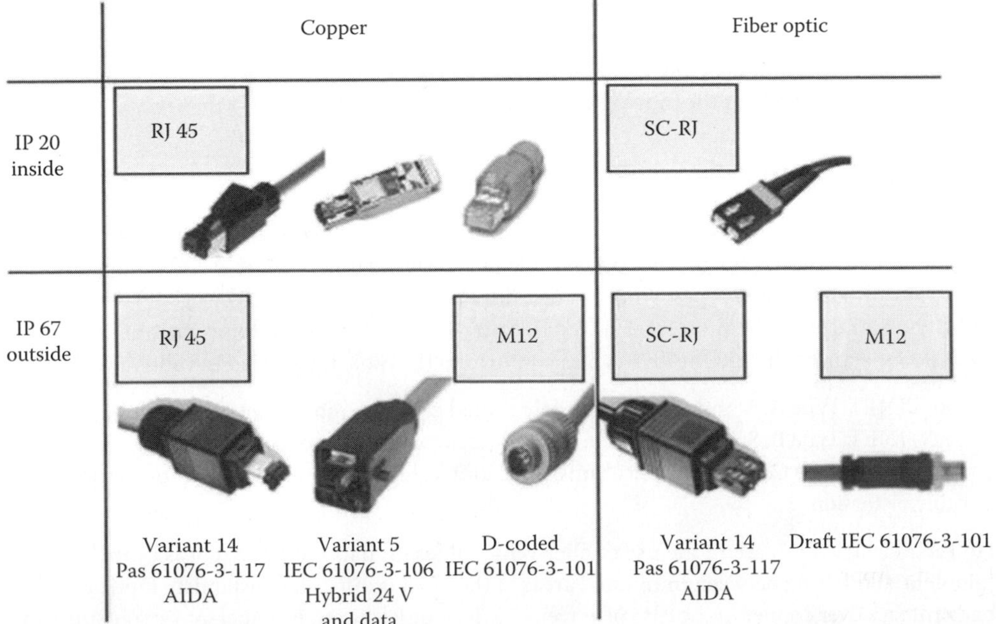

FIGURE 12.25 PROFINET offers a range of industrial connectors.

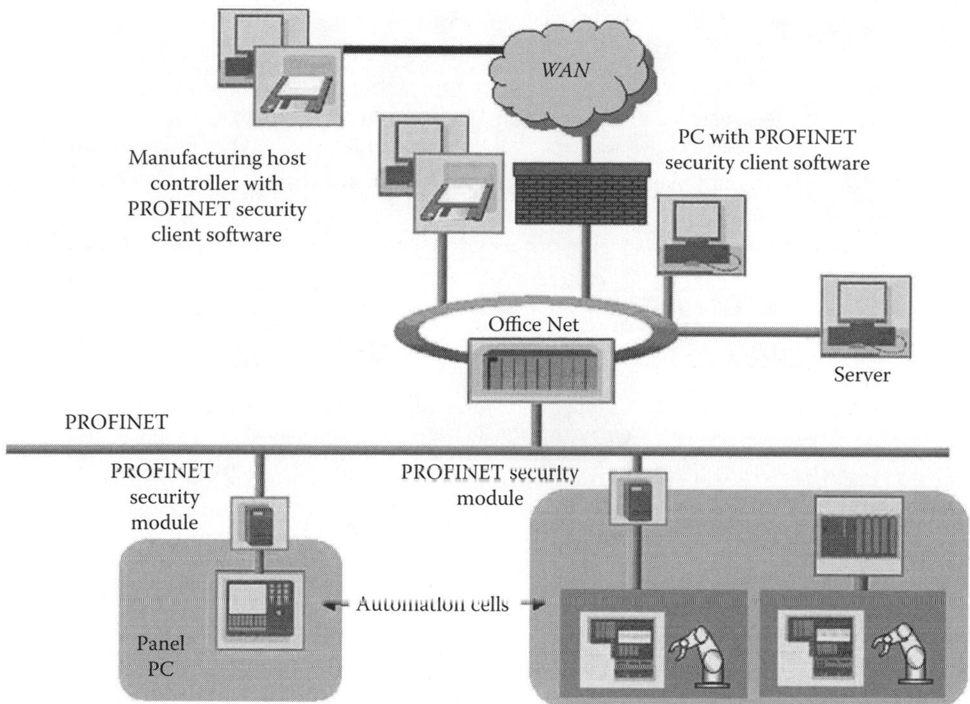

FIGURE 12.26 Segmented automation network.

The security concept provides for protection of both individual devices and whole networks from unauthorized access. In addition, there are security modules that will allow networks to be segmented and, thus, also separated and protected from the safety standpoint. Only uniquely identified and authorized messages will be allowed to reach devices within such segments from outside (Figure 12.26).

12.10 Technical Support

As far as maintenance, ongoing development, and market penetration are concerned, open technologies need a company-independent institution that can serve as a working platform. This is performed for PROFINET (as well as for PROFIBUS) by PI. With its 27 regional PI associations and approximately 1400 members, PI is represented on every continent and is the world's largest interest group for the industrial communications field.

Technology development takes place in the context of approx. 40 working groups with input from more than 500 experts mostly from engineering departments of member companies.

Optimal support by PI is key means for rapid dissemination of PROFINET in the market. In order to guarantee this, a powerful offer of services and products has been established.

PI maintains more than 50 approved PI competence centers worldwide. These facilities provide users and manufacturers with all kinds of advice and support. As institutions of PI, they are vendor-neutral service providers and adhere to the mutually agreed-upon rules. The PI competence centers are regularly checked for their suitability as part of an individually tailored approval process. More than 25 PI training centers have been set up to establish a uniform global training standard for engineers and technicians. Approval of the training centers and their experts ensures the quality of the training and thus of the engineering and installation services for PROFINET.

PI maintains 11 approved PI test labs (PITLs) worldwide. As institutions of PI, they are vendor-neutral service providers and adhere to the mutually agreed-upon rules. The testing services provided by the

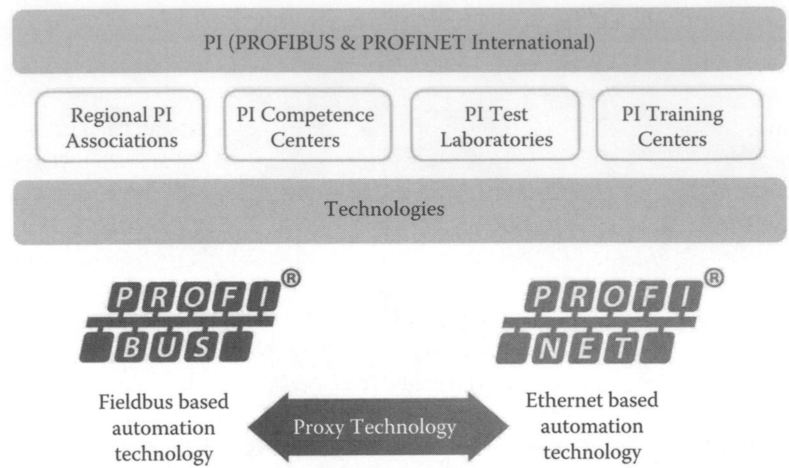

FIGURE 12.27 Structure of PROFIBUS & PROFINET International (PI).

PITLs are regularly audited in accordance with a strict accreditation process to ensure that they meet the necessary quality requirements. Refer to the website for current addresses [2].Current information on PI and the PROFIBUS and PROFINET technologies is available on the PI website www.profibus.com or www.profinet.com. This includes, for example, an online product guide, a glossary, a variety of web-based training content, and the download area containing specifications, profiles, installation guidelines, and other documents.

12.10.1 Technology Development

Device manufacturers that want to develop an interface for PROFINET I/O have the choice of developing field devices based on existing Ethernet controllers. Alternatively, member companies of PI offer many options for efficient implementation of a PROFINET I/O interface. To make the development of a PROFINET I/O interface easier for device manufacturers, the PI competence center and member companies offer PROFINET I/O basic technology (enabling technology). Consulting services and special developer training programs are also available. Before starting a PROFINET I/O development project, device manufacturers should always perform an analysis to determine whether internal development of a PROFINET I/O device is cost-effective or whether the use of a ready-made communication module will satisfy their requirements. More detailed information on this topic can be found in the brochure *PROFINET Technology—The easy way to PROFINET* [7], which can be downloaded from the PI website.

12.10.2 Tools for Product Development

Device manufacturers are assisted by software tools when developing and checking their products. These tools are provided to members of PI at no additional charge. A GSD editor assists the manufacturer when creating the GSD file for its product. This GSD editor can be used to create the proper files and check them. Likewise, PROFINET tester software is available for testing PROFINET functionalities. The current version supports the testing of all CCs as well as IRT functions. The additional security tester allows testing for secure function of a field device, including under load conditions. For detailed analysis, the Wireshark freeware tools can be used for problem-free interpretation of individual PROFINET frames since the decoding of PROFINET is already included in the standard version.

12.10.3 Certification Test

A certification test is a standardized test procedure that is performed by specialists whose knowledge is kept up to date at all times and who are able to interpret the relevant standards unequivocally. The test scope is described in binding terms in a test specification for each laboratory. The tests are implemented as so-called black box tests in which the tester is the first real user. All the defined test cases that are run through in a certification test are field-oriented and reflect industrial requirements. This affords all users the maximum possible security for use of the field device in a system. In very many cases, the dynamic behavior of a system can be simulated in the test laboratory. PI awards the certificate to the manufacturer based on the test report from an accredited test lab. A product must have this certificate in order use the PROFINET designation. For the plant manufacturer/owner, the use of certified products means time savings during commissioning and stable behavior during the entire service life. They therefore require certificates from their suppliers for the field devices used, in accordance with the utilized CC.

PI supports more than 10 accredited PITLs worldwide for the certification of products with a PROFIBUS or PROFINET interface. As institutions of PI, they are independent service providers and adhere to the mutually agreed regulations. The testing services provided by the PITLs are regularly audited in accordance with a strict accreditation process to ensure that they meet the necessary quality requirements. A list of the current addresses can be found on the website.

Abbreviations

AR	Application relation
ASIC	Application specific integrated circuit
CR	Communication relation
DAP	Device access point (This access point allows addressing an I/O-Device as a whole. It is the representative of the device)
GSD	General station description (electronically readable data sheet)
HMI	Human machine interface
I&M	Identification and maintenance
IEC	International Electrotechnical Commission
I/O	Input/output
IP	Internet protocol: the protocol that ensures the transfer of data in the Internet from end node to end node
IRT	Isochronous real time or clocked real time
ISO/OSI	International Standards Organization/Open Systems Interconnection (Reference Model)
LAN	Local area network
LLDP	Lower link discovery protocol
MAC	Media access control
MIB	Management information base
MRP	Media redundancy protocol
MRPD	Media redundancy for planned duplication
OUI	Organizationally unique identifier
PROFIBUS	Process fieldbus (fieldbus developed by PI)
PDEV	Physical device object
PDU	Protocol data unit
PI	PROFIBUS & PROFINET International
PLC	Programmable logic controller
RT	Real time
SNMP	Simple network management protocol

TCP	Transmission control protocol: superimposed protocol of IP to ensure secure data exchange and flow control
UDP	User datagram protocol (unsecured data frame)
VLAN	Virtual local area network
XML	Extensible Markup Language: standardized descriptive language that can be interpreted by almost all parsers
WLAN	Wireless LAN

References

1. PROFINET Website: www.profinet.com or www.profibus.com
2. IEC 61158/61784: Digital data communications for measurement and control—Fieldbus for use in industrial control systems (all parts), 2014.
3. PROFINET I/O Conformance Classes, V1.1 March 2011, Order No. 7.042.
4. Profile Guidelines Part 1: Identification & maintenance functions, V2.0 November 2013 (planned publication date of this version), Order No. 3.502.
5. Common Application Profile PROFIenergy, V1.1 August 2012, Order No. 3.802.
6. PROFINET Cabling and Interconnection Technology, V3.01 October 2011, Order No. 2.252.
7. M. Popp, Industrial communication with PROFINET, 2014, Order No. 4.182.

13

Sercos® Automation Bus*

13.1 Description **13**-2
Introduction • Three Generations of Sercos • Sercos Is an International Standard • How Sercos Works • Sercos Functions • Ideal for Distributed Multiaxis Control Systems • Advantages

13.2 Features and Operation of Sercos III **13**-4
Introduction • Why Ethernet? • Advantages of Sercos III • How Sercos III Communicates • Sercos III Features • Sercos III Profiles • Device Description • Configuration of Sercos III Networks • Implementation of Sercos III • Software Packages • Conformance Testing • Publications • IO-Link Integration • Cooperation between Sercos, ODVA, and OPC Foundation • Performance • Sercos III Applications

13.3 Features and Operation of Sercos II **13**-29
Introduction • How Sercos II Communicates • Error Correction/Diagnostics • System Safety • Sercos II Conformance Testing • Sercos II in Machine Tool Applications

13.4 Future Technical Advancements **13**-43

Acknowledgments **13**-44

References **13**-44

Sources for More Information **13**-45

Scott Hibbard
Bosch Rexroth Corporation

Peter Lutz
Sercos International e.V.

Ronald M. Larsen
Sercos North America

This chapter describes the Sercos automation bus, an international standard (IEC61784/61158/61800) for communication between digital motion controls, drives, I/O, sensors, and ancillary devices. It includes definitions, a description of Sercos communication methodology, an introduction to Sercos hardware, a discussion of speed considerations, information on conformance testing, and information on available development tools. A number of real-world applications are presented, and a list of sources for additional information is provided.

There have been three generations of the Sercos automation bus. The latest generation, industrial Ethernet-based Sercos III is described first, followed by Sercos I/II, the original fiber-optic-based versions of Sercos.

* Much of the text in this chapter is included by permission from www.sercos.com, the website of the Sercos North America Trade Association.

13.1 Description

13.1.1 Introduction

SERCOS is an acronym for *SE*rial *R*ealtime *CO*mmunications *S*ystem, a digital automation bus that interconnects motion controls, drives, I/O, sensors, and ancillary devices such as cameras and weighing stations. It is an open controller-to-intelligent digital device interface, designed for high-speed serial communication of standardized closed-loop data in real time (RT). Sercos I and II transmit data over a noise-immune fiber-optic cable, while Sercos III transmits data over a standard CAT5 Ethernet cable and optionally over a fiber-optic cable.

The Sercos automation bus was created in the 1980s by the German ZVEI and VDW organizations to specify a digital open interface that would ease the transition from analog to digital drive technology. It was originally intended to be a drive interface and, in its beginnings, was mainly used for advanced machine tool applications. It has become a universal machine control interface, accepted worldwide in a myriad of industries. Sercos not only is an RT communication system but also offers more than 700 standardized parameters that describe the interplay of drives, controls, and ancillary devices in terms independent of any manufacturer. It offers advanced motion control and I/O capabilities for electronics, pneumatics, and hydraulics.

Sercos offers short data update times and low communication jitter for any kind of automation applications, including—but not limited to—high-performance multiaxis machine control systems. Like most digital buses, Sercos greatly reduces connectivity problems in control systems. It can connect up to 254 slave devices (drives, I/O, ancillary devices) to a control using one fiber-optic cable ring (Sercos I/II) or up to 511 on a single Ethernet cable (Sercos III), compared to a traditional analog servo system with eight axes of motion that may require over 100 wires between the drive and the control. This reduces system cost, eliminates many types of noise problems, and helps machine designers get motion control systems up and running quickly.

13.1.2 Three Generations of Sercos

There have been three Sercos generations (see Table 13.1). The first two generations utilized an application-specific integrated circuit (ASIC) as a hardware processing platform, fiber-optic transmitters/receivers, and a fiber-optic cable as the transmission medium.

- The first generation (Sercos I, now obsolete) operated at 2 and 4 Mbps, using the Sercon410B ASIC and fiber-optic cabling. The Sercon410B is no longer produced.
- The second generation (Sercos II) operates at 2/4/8/16 Mbps, using the Sercon816 ASIC and fiber-optic cabling.
- The third generation (Sercos III) operates at up to 100 Mbps utilizing a field-programmable gate array (FPGA), an ASIC, or a general-purpose communication controller (GPCC) and is based on standard Ethernet hardware. It links the high-performance of Sercos II with industrial Ethernet, combining the determinism of the original Sercos with the high bandwidth of industrial Ethernet for the best of both worlds. Sercos III maintains backward compatibility with previous versions in regard to profiles, synchronization, and message structures. It retains the set of more than 700 standard parameters that describe all aspects of RT motion and I/O control.

Dozens of manufacturers worldwide offer both Sercos II and Sercos III products, with hundreds of thousands of systems installed. Several manufacturers offer Sercos hardware and software development tools.

13.1.3 Sercos Is an International Standard

The Sercos automation bus is described in a set of standard specifications that may be incorporated into any company's products, with each control and slave device maintaining its own functions and features. Because it is internationally standardized (IEC61784/61158/61800), it allows any manufacturer's

TABLE 13.1 Three Generations of Sercos

	Sercos I	Sercos II	Sercos III
Date implemented	1987	1999	2005
Physical media	Fiber optics	Fiber optics	Ethernet (twisted-pair copper or Fiber optics)
Network topology	Ring	Ring	Line or ring
Transmission speed (Mbps)	2/4	2/4/8/16	100
Cycle time (μs)	Configurable, min. 62.5	Configurable, min. 62.5	Configurable, min. 31.25
Jitter (μs)	<l	<1	<1
Synchronization	Hardware synchronization	Hardware synchronization	Hardware synchronization
Basic protocol	HDLC	HDLC	Ethernet
Real-time protocol	Sercos	Sercos	Sercos
Hardware redundancy	No	No	Yes, with ring topology
Direct cross communication	No	No	Yes
Comm. and synchr. between control systems	No	No	Yes
Service channels	Yes	Yes	Yes
Optional UC channel	No	No	Yes
Hot-plugging	No	No	Yes
Number of masters	1 per ring	1 per ring	1 per ring/line
Number of nodes	254 per ring, multiple rings possible	254 per ring, multiple rings possible	511 per ring/line, multiple rings/lines possible

Sercos-compatible digital control to talk to any other Sercos-compatible digital servo drive, digital spindle drive, pneumatic or hydraulic system, digital I/O, sensors, and ancillary devices over a well-defined fiber-optic link (Sercos I/II) or standard Ethernet technology (Sercos III). Controls and slave devices conforming to the standard comply with a standard medium for transmission, topology, connection techniques, signal levels, message (telegram) structures, timing, and data formats.

13.1.4 How Sercos Works

Sercos II operates at 2/4/8/16 Mbps, using the Sercon816 ASIC. Sercos III operates at up to 100 Mbps, using an FPGA or a GPCC and standard industrial Ethernet hardware.

ASIC-, FPGA-, or GPCC-based Sercos controllers are normally integrated into master motion controls as well as drives, amplifiers, and I/O modules and other devices (e.g., camera systems). They simplify the task of the designer by automatically handling most Sercos communication functions.

A set of over 700 standard software functions (called IDNs) defines standard motion, I/O, and control functions. In addition, the interface allows for manufacturer-specific IDNs, which can be used to define unique functions, not addressed by the standard IDN set.

In a Sercos system, all control loops may be closed in the slave device (e.g., servo loop of a servo drive device). This reduces the computational load on the machine controller, allowing it to control and synchronize more devices (e.g., motion axes and distributed I/Os) than it otherwise could. In addition, closing all the control loops in the slave device reduces the effect of the transport delay between the control system and the slave devices.

13.1.5 Sercos Functions

- Sercos exchanges data between control and slave devices, such as drives, I/Os, and encoders, transmitting command and actual values with extremely short cycle times.

- It guarantees an exact synchronization of all connected Sercos devices for precise data exchange in distributed measurement systems or for precise coordinated moves with as many axes as required.
- It includes mechanisms (service channel and unified communication channel [UCC]) for non-cyclic data transmission, used for the display and input of control internal parameters, data, and diagnostics. Device parameters can be downloaded and uploaded for storage via the service channel and the UCC.
- The bus enables the use of controls, drives, I/Os, and other slave devices from different manufacturers in a system, by standardizing all data, parameters, commands, and feedbacks exchanged between all connected Sercos devices.

13.1.6 Ideal for Distributed Multiaxis Control Systems

Sercos is a foundation for building distributed multiaxis control systems for a myriad of applications. Distributed control improves machine flexibility by moving processing power and decision making from the computer numerical control (CNC) or motion control down into the drives and sensors. These devices then become intelligent building blocks that can easily be added to a machine or production line without major changes in hardware and software.

Sercos is well suited for distributed control because it allows placement of axis-dependent control functions, such as loop closures, interpolation, and registration in the drives, not in the motion controller. Thus, motion controllers can concentrate on motion control profiles and tool paths independent of the axes. In operation, the control issues a position command to the drive, which then closes its own loops and microinterpolates its trajectory, based on previously downloaded parameters. Sercos also integrates input/output functions such as limit switches, pushbuttons, and various sensors.

13.1.7 Advantages

Sercos provides machine manufacturers with the flexibility to configure multivendor control systems with plug and play interoperability. Even more importantly, Sercos has facilitated great advances in machine productivity. A prominent US food manufacturer states as follows: “Our Sercos applications play an important role in our plan for a dynamic enterprise that can flexibly adapt manufacturing for every new product size, shape and packaging configuration. Along with this flexibility, we have realized incredible timesaving advantages – faster delivery, faster installation, faster setup, faster product changeover, and faster production speeds” (Campbell 1999).

Sercos allows manufacturers to create intelligent digital drives with vastly improved capabilities and flexibility. A single drive can be designed to handle multiple motors, such as permanent magnet servomotors, high horsepower induction servomotors (vector drives), and linear motors, with the configuration set up parametrically. Sercos is also used with stepper motors, pneumatic and hydraulic drives, I/O systems, and other devices.

13.2 Features and Operation of Sercos III

13.2.1 Introduction

The third generation of Sercos (Sercos III) is an evolution of Sercos I and II (IEC/EN 61491), based on standard Industrial Ethernet. Work began in 2003, with vendors releasing the first products supporting it in 2005. The well-known and proven Sercos mechanisms (as described in Section 13.3, Features and Operation of *Sercos I/II)* such as motion control profiles, the telegram structure, and the hardware synchronization have been mapped to the Ethernet standard. To ensure hard RT requirements despite the nondeterministic nature of Ethernet, Sercos III uses an RT channel in parallel to a UCC. In the collision-free RT channel, the Sercos-specific frames are transmitted using the EtherType 0x88CD.

The high protocol efficiency in this channel ensures best performance even with a high number of devices connected and rather small amount of data per device. The UCC can be configured in parallel to this RT channel. In it, any standard Ethernet telegrams (EtherType <> 0x88CD) or IP-based protocols like TCP/IP or UDP/IP can be transmitted.

13.2.2 Why Ethernet?

The increasing number of control components, actuators, and sensors in industrial automation adds to the complexity of control networks. The future of industrial communication lies in cost-effective integration of these automation components in low-cost Ethernet networks, which have been engineered to suit the needs of industrial automation.

Specialized fieldbus systems were used initially for simplified networking. Fast Ethernet technology replaces these systems and offers a number of advantages:

- Recognized, future-proof technology
- Data throughput 10–100 times faster than fieldbus solutions
- No need for expensive, proprietary technology
- Use of standard components, for example, CAT5e copper cable with double shielding, connectors, and controllers, such as FPGAs, all produced in large volumes
- Consistent IT implementations stretching from the office to the field level (Figure 13.1)

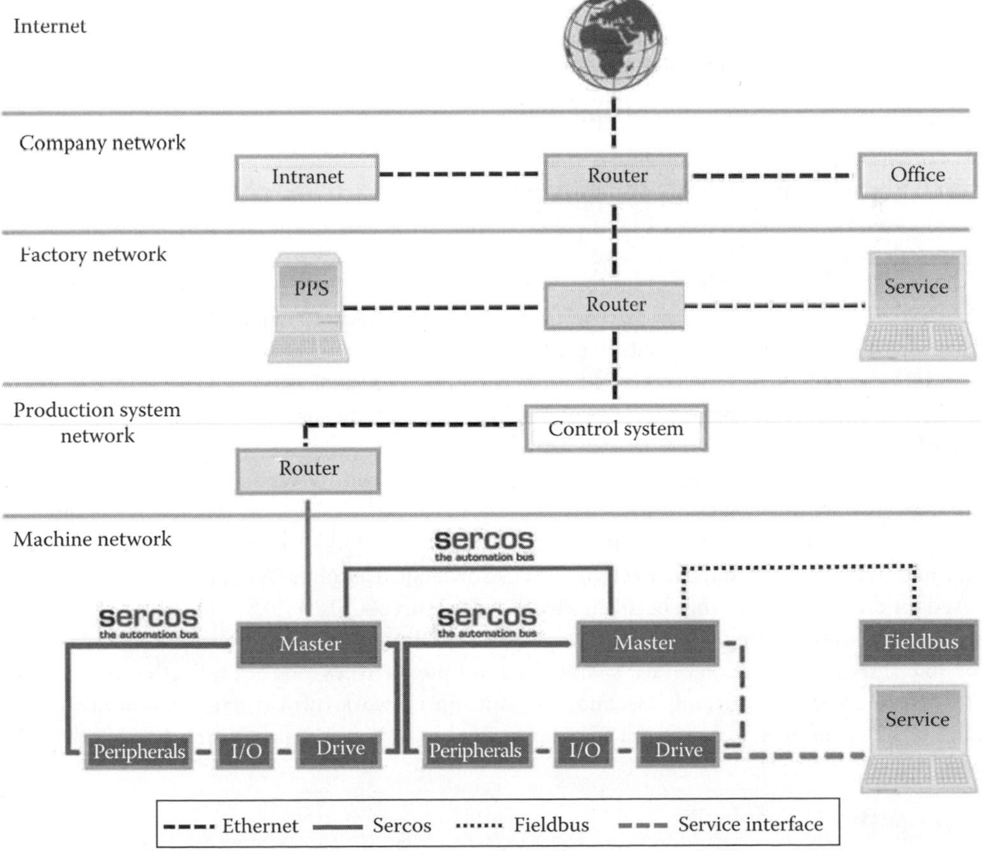

FIGURE 13.1 Ethernet—consistent from the office to the field level.

- Flexible, compatible automation systems based on a global standard
- Supports worldwide networking for diagnostics and maintenance

Ethernet technology combines the peripherals, drives, safety functionality, and office communications in a common medium providing a simple, cost-effective, powerful solution.

13.2.3 Advantages of Sercos III

The evolutionary concept of Sercos III maintains all the proven advantages of previous Sercos versions and at the same time offers a range of new features that allow for a substantially increased number of possible applications. Advantages include the following:

- Protection of investment due to high compatibility with previous Sercos generations (ring topology, profiles, telegram structures, and synchronization)
- Reduction of hardware costs
- Reduced cycle times down to 31.25 μs
- Simultaneous transmission of any Ethernet protocols (e.g., UDP/IP and TCP/IP) over the same cable
- Direct cross communication (CC) between slaves
- Synchronization of several control systems, that is, machine segments
- Transfer of safety-relevant data
- Fault tolerance in case of a cable break

13.2.4 How Sercos III Communicates

13.2.4.1 Physical Layer and Data Link Layer

Sercos III supports standard IEEE 802.3 & ISO/IEC 8802-3 100Base-TX or 100Base-FX (100 Mbit/s baseband) full duplex physical layer (PHY) entities. Standard 802.3–compliant media access controller (MAC) sublayers are used. Autonegotiation must be enabled on each PHY, but only 100 Mbit full duplex is supported. Auto (MAU [media attachment unit]-embedded) crossover is specified between the two physical medium attachment (PMA) units present with a duplex port. These two units are referred to as the primary channel and secondary channel in the Sercos III specification. Dual interfaces are required (two duplex interfaces per device). Within the Sercos III specification, the dual interfaces are referred to as P1 and P2 (Ports 1 and 2). All Sercos III frames use EtherType 0x88CD. For bandwidth optimization, the Sercos III frames are transmitted as broadcast frames that are shared between all Sercos III nodes connected to the network.

The basis of Sercos communication is a time slot protocol with a cyclic transmission of telegrams using a master–slave principle. A central master sends master data telegrams (MDTs) to the subordinate slaves. They can contain command values and timing data. The slaves send process data from the field, for example, actual values, to the master via the acknowledge telegrams (ATs). In Sercos III, the ATs are also used for direct data exchange between slave devices (direct CC).

Various measures ensure that Sercos III uses the 100 Mbps bandwidth of fast Ethernet as efficiently as possible. First, the RT frames are shared by multiple network nodes (collective frames), which reduces the overhead significantly. Second, by omitting network infrastructure components such as switches or hubs and by processing the RT frames *on-the-fly*, the frame runtimes through the network are minimized.

13.2.4.2 Network Topologies

Sercos III defines two basic topologies: ring and line topology. Due to the full-duplex characteristics, the physical ring is a logical double ring structure and the physical line is a logical ring structure (Figure 13.2).

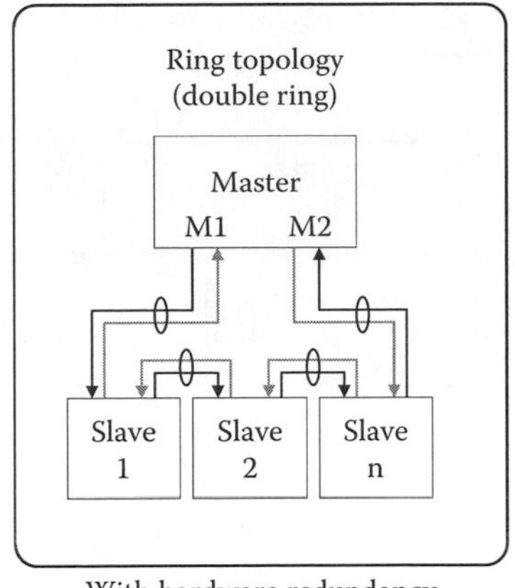

With hardware redundancy

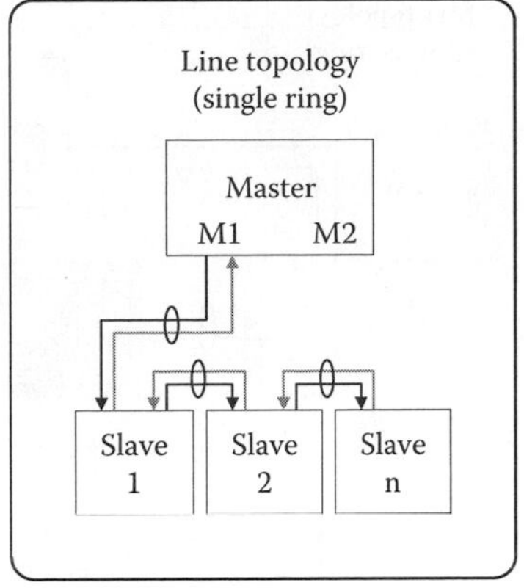

Without hardware redundancy

FIGURE 13.2 Sercos III ring and line topology.

A line topology is the simpler of the two possible arrangements and provides no redundancy. However, this configuration saves the cost of one cable. In it, only one of the two interfaces on the master is used. Telegrams are issued out of the transmit PMA on the master's active port. Either port on the master may be the active one. Sercos III determines this during phase-up (initialization). The first slave receives the telegrams on the connected interface's receive PMA, modifies them as required, and issues them out on the transmit PMA of the second interface. Each cascading slave does likewise until the last slave in the line is reached. That slave, detecting no Sercos III connection on its second port, folds the telegram back on the receiving interface's transmit port. The telegram then makes its way through each slave back to the master. Note the last slave also emits all Sercos III telegrams on its second port, even though no Sercos III connection is detected. This is for snooping and ring closures (see later), as well as hot-plugging. Keep in mind that since the Ethernet destination field in all Sercos III telegrams is the broadcast address of 0xFFFF FFFF FFFF (all 1s), all telegrams issued from this open port will be seen by other devices as broadcast telegrams. This behavior is by design and cannot be disabled. To avoid taxing networks attached to an open Sercos port, an IP switch can be used, or alternately a managed Ethernet switch programmed to block broadcast telegrams received from the Sercos port can be used.

A ring topology simply closes the network by attaching the unused (second) port on the last device in a ring back to the second port on the master. When the Sercos III master senses that a ring exists, it sets up two counter-rotating telegrams. The same data are issued simultaneously out of the transmit PMAs of both ports on the master. From there, both telegrams are managed essentially identically as they make their way through each slave, ending back at the opposite port on the master they were emitted from. Advantages to this topology include tighter synchronization as well as automatic infrastructure redundancy (see later).

No hubs or switches are required for the ring or line topology. This significantly reduces cabling efforts and installation costs.

With both the line or ring structure, Sercos III operates in a *circular* approach. All telegrams leave the master and return there. As with any network that operates in this manner, modified structures can be constructed to appear as a tree or star network, utilizing hardware that manages the branches, but the structure is still circular in nature. The principle of a star topology is shown in Figure 13.3.

13.2.4.3 Network Installation

The installation of a Sercos III network is very simple and does not require any specific network configuration procedures. All nodes are simply connected by patch cables or crossover cables.

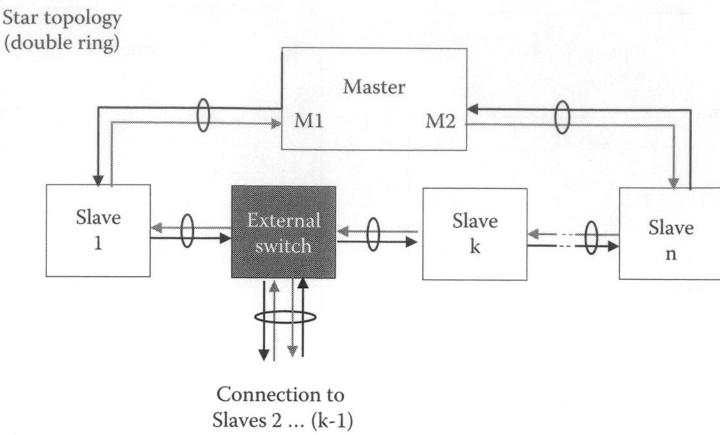

FIGURE 13.3 Sercos III star topology with external switch.

The Ethernet ports of the devices are interchangeable and can even be used to connect standard Ethernet nodes (e.g., laptops) to a Sercos RT network. In this case, full Ethernet and IP connectivity is provided, without affecting the RT behavior of a Sercos III network when in online mode or without requiring active Sercos III communication (off-line mode). In such an off-line mode, standard Ethernet devices, such as a laptop, can still communicate to Sercos devices for diagnostics and up/downloading data.

Sercos III uses a component-based approach for the connection between Sercos III devices. Defined cables and connectors are combined and used to connect the automation devices in a machine or system. The total length of the cabling channel as well as the number of transitions between cable and connectors is considered. The result is transmission lines that fulfill the requirements for the Sercos III cabling with a distinct reserve—without the necessity to perform extensive planning, calculations, and measurements. The Sercos organizations offer a free Installation and Planning Guide to make this process easy and trouble-free.

13.2.4.4 Sercos III Cycle

Just as with Sercos I/II, the communication of Sercos III occurs in strict cyclic intervals. A cycle time is chosen by the user for a given application. Cycle times are 31.25, 62.5, 125, 250 μs, and multiples of 250 μs up to a maximum value of 65 ms.

In Sercos III, a communication cycle is divided into two channels via a time slot technique (Figure 13.4). In the RT channel, the collective telegrams specified by Sercos III are transmitted as a broadcast and processed *on-the-fly* by the slave devices. For this purpose, two types of telegrams are used: MDTs and ATs. After all MDTs and ATs are transmitted, Sercos III nodes allow the remaining time in the cycle to be used as a UCC. In this channel, any other Ethernet frames (e.g., IP protocol frames) can be sent as single telegrams to any device in the network. The cycle times as well as the partitioning of the bandwidth of 100 Mbps Ethernet into the Sercos RT and UCCs can be adjusted to the specific requirements of the application.

The network remains available to UCC traffic until the next cycle begins, at which time the Sercos III nodes close the nodes to UCC traffic again. Sercos III is purposely designed to provide open access at all ports for other protocols between cyclic RT messages. No tunneling is required. This provides the advantage that any Sercos III node is available, whether Sercos III is in cyclic mode or not, to use other protocols, such as TCP/IP, without any additional hardware to process tunneling. Sercos nodes are specified to provide a store-and-forward method of buffering non-Sercos messages should they be received at a node while cyclic communication is active.

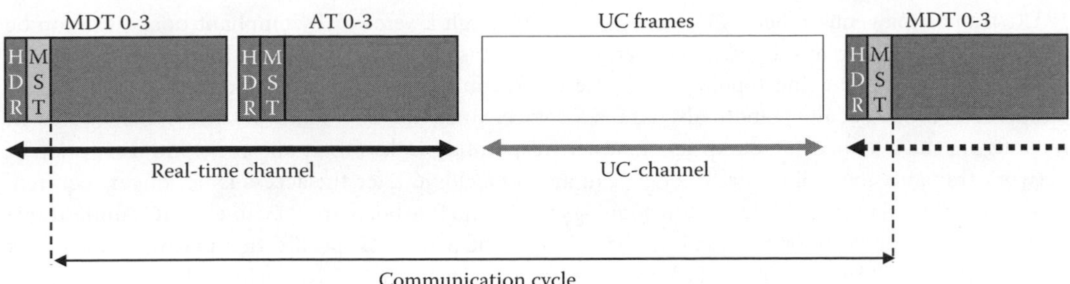

MDT, AT, and UC frames are based on standard Ethernet frames

FIGURE 13.4 Sercos III communication cycle.

13.2.4.5 UC Channel

The time between the end of the transmission of all Sercos III RT cyclic telegrams and the beginning of the next communication cycle is defined as the Sercos III Unified Communication Channel. During this time period, the Sercos network is opened to allow transmission of Ethernet-compliant frames for other services and protocols. For example,

- Web servers can be embedded in Sercos III–compliant devices to respond to standard Hypertext Transfer Protocol messages received via the UCC.
- Frames from other Fieldbus standards that conform to Ethernet frame formatting may be transmitted across a Sercos III network.

Every Sercos III–compliant node must support the passing of UC frames through its Sercos III interface. Whether a Sercos III node actively makes use of the UC feature is determined by the feature set of the product. If, for example, the device has an embedded web server, it could make its IP address available for access by other devices.

A Sercos III network will always pass UC frames, even when cyclic operation has not been initialized. This means that devices always have access to the network for UC messages, as long as the ports are powered.

Sercos III does not define whether a port should operate in cut-through switching or store-and-forward mode when handling UC frames. There are Sercos III products currently on the market that support both modes. Likewise, Sercos III does not define whether a port should intelligently process UC telegrams, such as learn the network topology. The time allotted for UC traffic is dictated by the amount of data transmitted during the RT portion of the cycle. In real-world applications, there is a significant amount of bandwidth available for UC frames. For example, in a typical application with eight axes of motion and a cycle rate of 250 μs, the equivalent of 85 Mbit/s is available for UC use. This amount of time means the UC frames in this example can be as long as the maximum defined for Ethernet (maximum transmission unit [MTU] = 1500). Using the same example of eight axes, but with a cycle time of 62.5 μs, the effective bandwidth available for UC frames would be 40 Mbit/s, and the MTU would be reduced to 325. As with any network where time on the bus is shared, MTU values should be configured to ensure reliable communication. Properly configured Sercos networks will set the Sercos parameter *Requested MTU* (S-0-1027.0.1) to the recommended MTU value, which can then be read by other devices to match their MTU settings. Regardless of the value of this parameter, a Sercos node will allow non-Sercos traffic to pass for the entire UCC time period (i.e., telegrams longer than the MTU setting are not discarded by the Sercos stack). Sercos parameter S-0-1027.0.1 is set by default to 576, the minimum value called out in RFC 791 (request for comment, Internet Protocol).*

* See http://tools.ietf.org/html/rfc791.

UC frames may enter a Sercos III network only through a Sercos III–compliant port. This can be achieved in two different ways. One is to employ the unused Sercos III port at the end of a Sercos III network configured in line topology. In a network configured in ring topology, the ring can be temporarily broken at any point to also attach a device. Since the redundancy feature of Sercos III will reconfigure the network in a bump-less manner (responding in less than one cycle), no disruption of network transmission will occur. The ring can again be closed after the access is no longer required. If access is desired in the middle of a line topology (where no free ports are available), or it is undesirable to break a ring topology for extended periods of time, the Sercos III specification permits the use of a device called an *IP Switch* that can be used to provide access to the UCC anywhere along the network. IP switches supply two Sercos III–compliant ports and one or more ports for UCC access.

Commercially available IP switches block the transmission of Sercos III broadcast telegrams out of their non-Sercos III port(s) to prevent flooding of non-Sercos III networks with Sercos III cyclic data.

13.2.4.6 Telegram Format and Telegram Types

All Sercos III telegrams conform to the IEEE 802.3 & ISO/IEC 8802-3 MAC (media access control) frame format with an EtherType of 0x88CD. Combined frames are used for the optimization of the bandwidth. The MST (synchronization signal) and MDT are combined in a block of a maximum of four Ethernet frames. The ATs are also combined in a block of a maximum of four Ethernet frames. The Sercos III header includes control and status information, which is required for the use of the network and the different devices. Furthermore, the synchronization is derived from the Sercos III header. Subsequent to the Sercos III header, a variable data field follows, which contains the RT data of the Sercos III devices (Figure 13.5).

- Destination address: The destination address for all Sercos III telegrams is always 0xFFFF FFFF FFFF (all 1s), which is defined as a broadcast address for Ethernet telegrams. This is because all telegrams are issued by the master and are intended for all slaves on the network.
- Source address: The source address for all Sercos III telegrams is the MAC address of the master, as it issues all telegrams.
- Ethernet type: A unique EtherType value has been assigned via the IEEE EtherType Field Registration Authority for Sercos III (0x88CD).
- Sercos III header: The beginning of the Ethernet-defined data field always begins with a Sercos III header, which contains control and status information unique to Sercos.
- Sercos III data field: The Sercos III header is followed by the Sercos III data field, which contains a configurable set of variables defined for each device in the network.

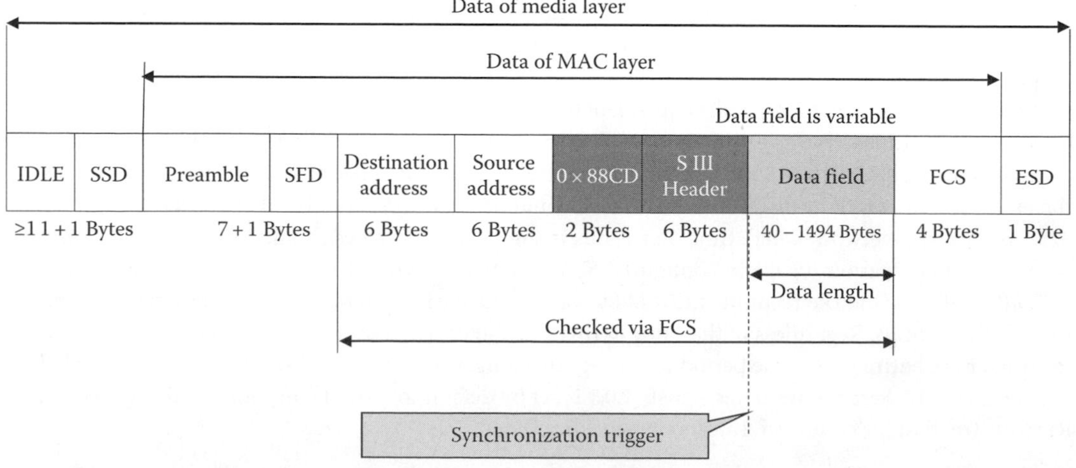

FIGURE 13.5 Sercos III telegram structure.

Two main types of telegrams are used within the Sercos III Cycle: the MDT and the AT. Both telegram types are issued by the master (control). The MDT contains information provided by the master to slaves. It is filled by the master and read by slaves. The AT is issued by the master, but actually populated by each slave with their appropriate response data (feedback values, input states, etc.). More than one slave uses the same AT, filling in its predetermined area in the AT telegram, updating checksums, and then passing the telegram to the next device. This method reduces the impact of the Ethernet frame overhead on the performance of the network without compromising IEEE 802.3 & ISO/IEC 8802-3. The amount of data sent from the master to slaves, as well as the sum of the data returned by the slaves, may exceed the 802.3-specified maximum 1500-byte data field size. To comply with this limit, Sercos III may use more than one MDT telegram in a cycle, as well as more than one AT telegram (up to four in each case). Sercos III specifies that no data are overwritten (destroyed) during a transmission. Every slave on a network may access input and output data for every other slave on the network.

13.2.4.7 Synchronization

There are different ways to achieve synchronicity and simultaneity in distributed systems. Most RT Ethernet systems rely on time synchronization to create a common time basis for the whole network, which is independent of runtimes and runtime fluctuations on the communication medium. The procedure used by Sercos is based on sending a synchronization signal (MST), which is cyclically received and evaluated by all nodes in the network. To ensure optimum synchronization, the signal must be sent and received in a fixed time grid with a minimum of deviation in time. In general, a synchronization accuracy of less than 1 µs is sufficient to realize an adequate machining accuracy for highly dynamic and highly precise production machines and machining systems. It is possible, for example, to achieve a printing accuracy of 10 µm in a shaftless printing press with synchronized single drives and a printing speed of 10 m/s.

In a Sercos III network, the master only has to only measure the ring delay and transmit this value together with a configuration-optimized threshold to all connected devices. Thus, it is possible to prevent a shifting of the synchronization times or the need for reconfiguration of the threshold value when additional devices are added to a Sercos III network in cyclic RT operation (i.e., *hot-plugging*).

The synchronization method ensures that a cyclic and simultaneous synchronization of all connected devices is guaranteed in Sercos III networks, independent of the topology and the number of nodes. Its accuracy remains unchanged even if changes to the topology occur (e.g., redundancy, hot-plugging). The method itself is fast, robust, and easy to apply. On both the master and slave sides, Sercos III controllers can be used, which not only support the fast processing of the RT telegrams and the ring redundancy but also provide an error- and jitter-free synchronization signal. With the described method, Sercos III achieves a synchronization accuracy of less than 20 ns and simultaneity of less than 100 ns. As single Sercos III networks can be combined in interconnected networks on the basis of the controller-to-controller (C2C) profile, it is possible to implement fully synchronized network structures without any restrictions in terms of synchronization performance.

13.2.4.8 Identification Numbers

The identification numbers (IDNs) of Sercos III are based on the IDN concept of Sercos I/II (as described in Section 13.3.2.6).

However, the IDN signification area is extended from 16 bits (Sercos I/II) to 32 bits in Sercos III. This was done in order to expand the parameter area, to allow the logical grouping of IDNs that are used for the same purpose and to enable instances of the same functions with same base data block number (Figure 13.6).

For the 32 bit IDNs in Sercos III, the following symbolic notation of IDNs is used: <IDN>.<SI>.<SE>

- Structure element (SE). The SE is used to address the elements. Consequently, up to 256 elements are possible. The SEs from 1 to 127 are predetermined and the remaining up to 255 are manufacturer specific.

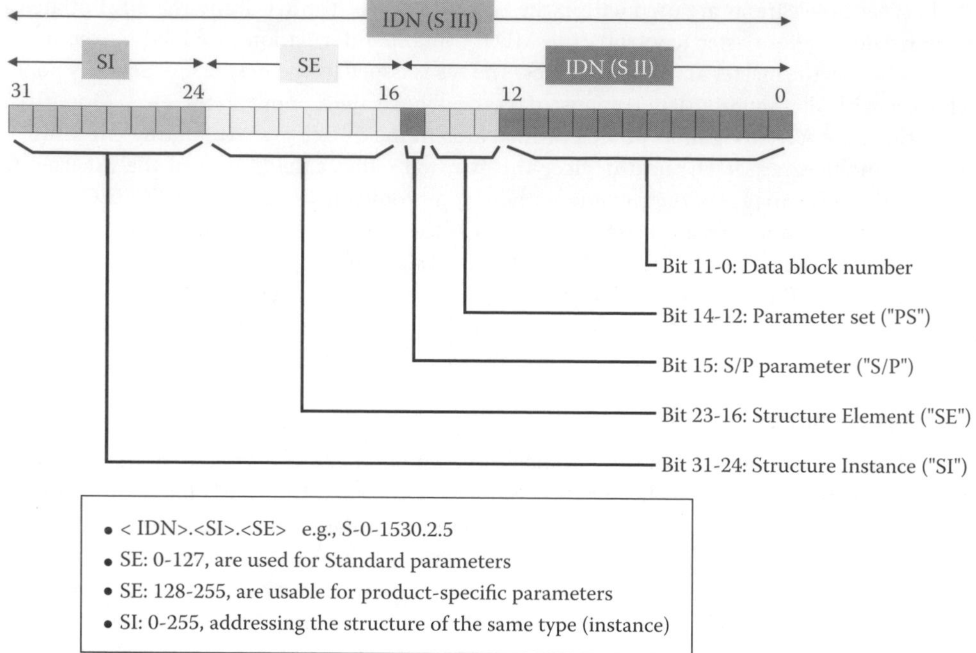

FIGURE 13.6 Structure of 32-bit IDN numbers.

- Structure instance (SI). This serves to address the structure of the same type within one subdevice. So 255 instances of the same structure are possible in one subdevice.

All parameters that were moved from Sercos II are maintained. The upper 16 bits of the IDN are set to zero while the lower 16 bits correspond to the known Sercos II addressing. This has an impact on all IDNs, in which data of type *IDN* are incorporated, because they must use 4 byte values instead of 2 byte values.

An example for the new 32-Bit IDN structure is the electronic label, S-0-1300, which is a group of different IDNs related to the electronic label of a device.

The IDNs S-0-1300.x.x are encompassed by the term *electronic label*. They are used by a master or a configurator device to identify Sercos III devices. Because of this, the two IDNs S-0-1300.x.03 (vendor code) and S-0-1300.x.05 (vendor device ID) are mandatory and must be present in every Sercos III device. All other IDNs in the following list are optional and can be used by a manufacturer to assign additional information to a device.

The SI 0 of all *electronic label* IDNs (e.g., S-0-1300.0.01 for the component name) assigns the physical device that contains the Sercos III Interface itself. The other SIs can be assigned to any component of the device (e.g., modules linked to an internal interface) according to the manufacturer's requirements.

S-0-1300.x.01 component name (optional)
S-0-1300.x.02 vendor name (optional)
S-0-1300.x.03 vendor code (mandatory)
S-0-1300.x.04 device name (optional)
S-0-1300.x.05 vendor device ID (mandatory)
S-0-1300.x.06 connected to subdevice (optional)
S-0-1300.x.07 function revision (optional)
S-0-1300.x.08 hardware revision (optional)
S-0-1300.x.09 software revision (optional)
S-0-1300.x.10 firmware loader revision (optional)

S-0-1300.x.11 order number (optional)
S-0-1300.x.12 serial number (optional)
S-0-1300.x.13 manufacturing date (optional)
S-0-1300.x.14 QA date (optional)
S-0-1300.x.20 operational hours (optional)
S-0-1300.x.21 service date (optional)
S-0-1300.x.22 calibration date (optional)
S-0-1300.x.23 calibration due date (optional)

13.2.4.9 Communication Phases

The Sercos communication is divided into five communication phases (CP0 through CP4) and the NRT (non-RT) state:

- After a Sercos III device has been powered up, internal checks are completed, and it is determined to be error-free, it operates in non-RT state (NRT state).
- Initialization of a Sercos III network always begins with CP0.
- CP0 is used for recognizing the participating slaves.
- CP1 is used to configure the slave devices for noncyclic communication.
- CP2 is used to configure the slave devices for cyclic communication and for parameter setting in the slave via noncyclic communication.
- CP3 is used to further configure the slave devices, the cyclic communication already running but not used.
- In CP4, the initialization process is complete, and the Sercos III network is in normal (cyclic) operation.

It is also possible to enter CP0 from any higher phase. All other phases can be entered only in ascending order when leaving the previous phase.

The master initiates a specific CP by setting the MDT phase in the Sercos III header. The slaves follow accordingly. Only in the case of a communication error do the slaves switch to the NRT state (Figure 13.7).

13.2.4.10 Communication Services

The Sercos III communication connects all participants in a network. Because of this, it is possible that any device can communicate with any other. A connection determines which devices communicate with each other. The Sercos III network supports the application data exchange between the master and all slaves, and between the slaves in both directions (direct CC). Not all devices need to communicate with each other; therefore, the application data exchange between the devices is configured via connections. These logical connections are based on a producer–consumer model. That means, for each transmission of application data, one connection is required (Figure 13.8).

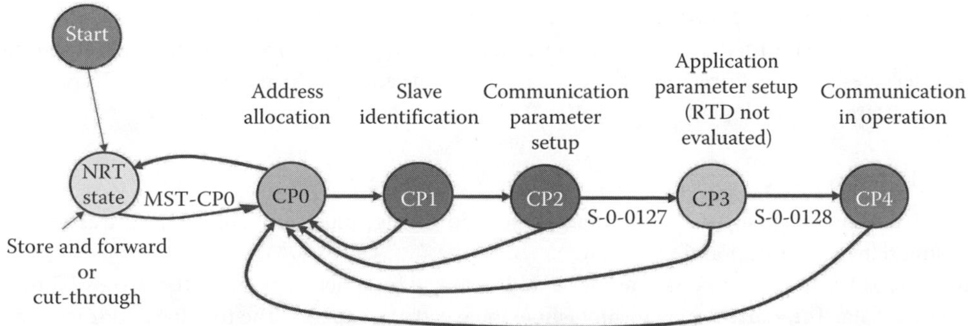

FIGURE 13.7 Communication phases of a Sercos III network.

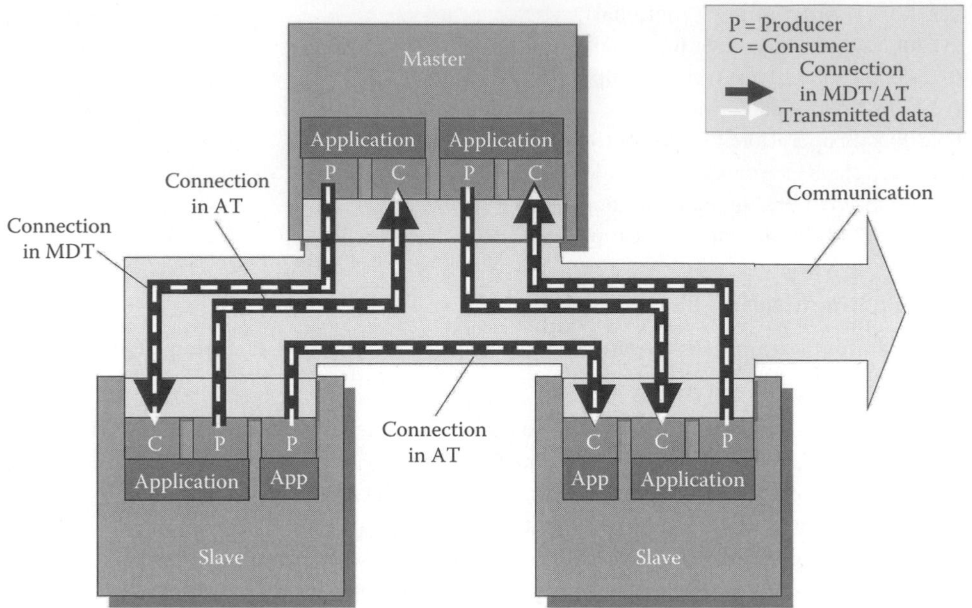

FIGURE 13.8 Sercos III communication model based on connections.

In addition to the cyclic RT data transmission, additional services are available:

The *service channel* allows high level, non-time-critical data transfers without disturbing the synchronous transfer of data. For example, items such as communication settings, parameters, or diagnostic information do not need to be transmitted repeatedly, so it is not necessary to impact loop performance by setting aside bandwidth for them. The service channel is part of the protocol in the RT channel and is identical to the service channel of Sercos I/II.

The *Sercos Messaging Protocol (SMP)* is used to transmit services and data within a cyclically configured data container with a fixed length. Both cyclic and acyclic fragmented data can be transmitted via a cyclic Sercos data container within the RT channel. Data packets with low priority can be interrupted by packets with higher priority. Thus, an efficient multiplexing of several logical communication channels within a single data container is possible.

The *S/IP protocol* enables direct communication as well as uniform data exchange between devices, without a Sercos III master or a subsidiary Sercos III communication being necessary. Even in cyclic RT operation, the S/IP protocol can still be used. In this case, the S/IP telegrams are transferred into the UCC of Sercos III, without having a negative influence on the RT behavior of the network. S/IP provides services to read and write parameters of a device, including the possibility of combining several tasks and sending them to the final device for processing. Furthermore, apart from fixed, preset, standardized protocol services, manufacturer-specific services can also be added, which makes Sercos III particularly flexible for users and developers.

13.2.4.11 Addressing

Sercos III devices must support Ethernet's MAC addressing, plus the Sercos III addressing. Other addressing schemes are optional.

Each Sercos III device contains a numeric address used by other devices on the Sercos III network to exchange data. The address may be any whole integer from 1 to 511. The topology address is new in Sercos III and is necessary for the communication of the master in CP1 and CP2 with all slaves in the ring. By means of this topology address, it is possible to also communicate with slaves that do not have a Sercos address. The topology address also makes it possible to recognize which slaves in the ring use the same Sercos address and, if needed, to change a present Sercos address. The topology address is determined by the slave in CP0 out of the physical topology (cabling). In CP1 and CP2, defined offsets in the frames are assigned to each topology address for the necessary data fields (SVC, C-Dev, and S-Dev). Sercos III does not use an IP address for its own operation. Whether a device contains an IP address or not is dependent on its support of other specifications, either independent (exclusive) of Sercos III operation or via the UCC portion of the cycle.

13.2.5 Sercos III Features

13.2.5.1 Redundancy and Hot-Plugging

When a ring network is employed, Sercos III provides for automatic infrastructure redundancy. If any interconnection point in the ring ceases to function, the associated Sercos III nodes will detect a ring break and loop back the end nodes, effectively operating as two lines rather than one ring. The operation is *bump less*, as the detection and recovery time for such a break is less than 25 μs, which is less than the minimum Sercos III cycle time. Sercos III can also recover from ring breaks and *heal* with no interruption in operation. Since Sercos III telegrams continue to be emitted by transmit PMAs on unconnected ports, and receive PMAs on unconnected ports continue to monitor for incoming data, when a Sercos III port recognizes that a ring has been physically reclosed, it will reactivate the counter rotating telegrams to functionally close the rings again. This operation is also bump-less.

Another feature of Sercos III is hot-plugging and hot-swapping, which is the ability to add devices to an active network or to replace devices. Using the features described for redundancy, a network can detect when a new device is attached to an active network. Processes exist that configure the new device and announce its availability to the master control. After that, the master control can select to make use of the new device based on the application currently running.

13.2.5.2 Direct Cross Communication

Because of the unidirectional communication of the fiber-optic ring, a direct communication between the slaves was not possible with the first and second Sercos generations. However, because of the full duplex characteristic of Ethernet, Sercos III supports this direct CC feature, which is advantageous for several motion control applications (see Figure 13.9). The direct CC supports the data exchange between

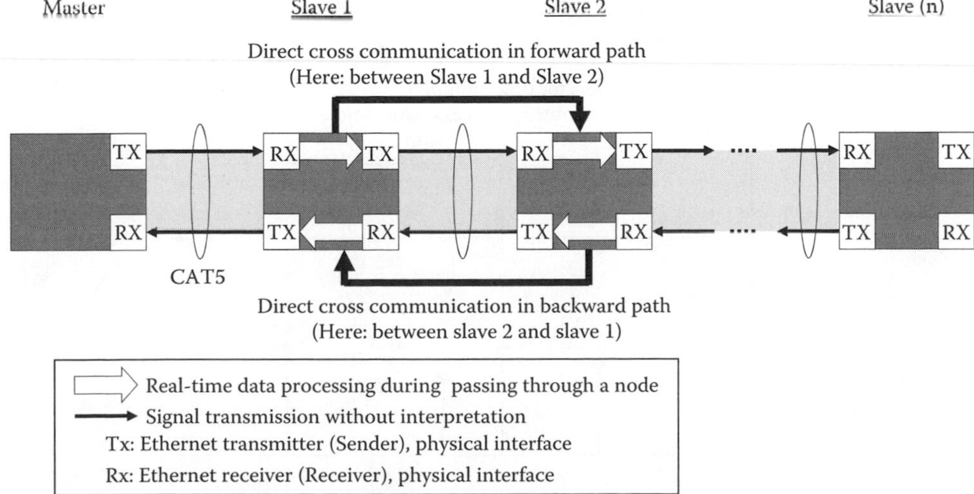

FIGURE 13.9 Direct cross communication between Sercos III slaves.

any two nodes within one communication cycle. This has the advantage that data—even in case of longer cycle times—can always be transferred between slaves within one communication cycle. This results in a minimum delay time. In addition, the central master does not need to be involved to copy and retransmit data in a separate communication cycle. As all devices are synchronized with each other, all target values are activated synchronously and simultaneously, and the actual values across the entire network are collected synchronously and simultaneously (related to the same communication cycle).

Communication and synchronization between multiple motion controls also is supported by Sercos III. Several machine modules in modular and networked machines and systems can be synchronized with each other, and RT data can be exchanged between the controls. For this, a dedicated communication profile (C2C) was developed as an extension to the basic Sercos III communication protocol.

13.2.5.3 Oversampling and Time-Stamping

Sercos III defines two services on the protocol level that are useful for distributed automation applications. The oversampling function makes it possible to transmit more than one target/actual value per clock cycle, while time-stamping allows transmitting event-controlled information independently from the fixed clock cycle (Figure 13.10).

1. An oversampling procedure allows acquisition of equidistant values and activation of equidistant command values—faster than the configured bus or connection cycle. This increases the process control intricacy, for instance, in extremely time-critical laser applications, as it allows for more data to be collected and communicated at a faster speed. Measurement methods were integrated directly in the protocol, thereby opening up the possibility to access these mechanisms across different manufacturers and products.

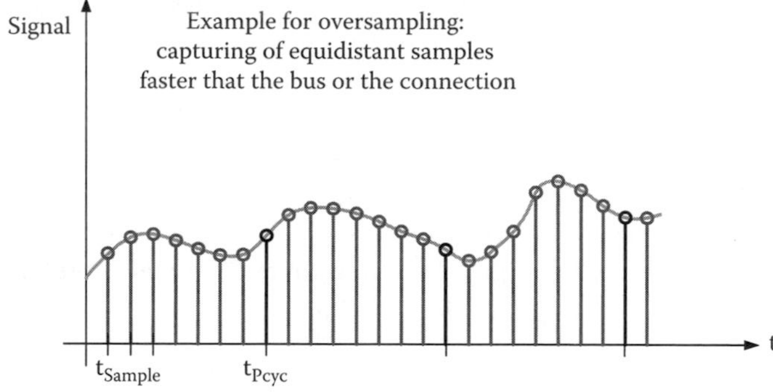

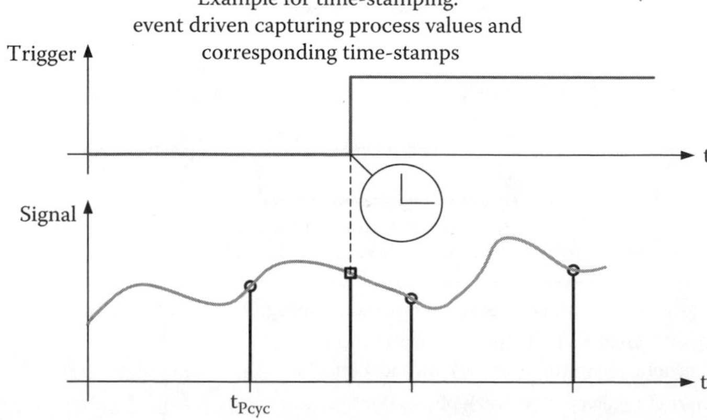

FIGURE 13.10 Oversampling and time-stamping.

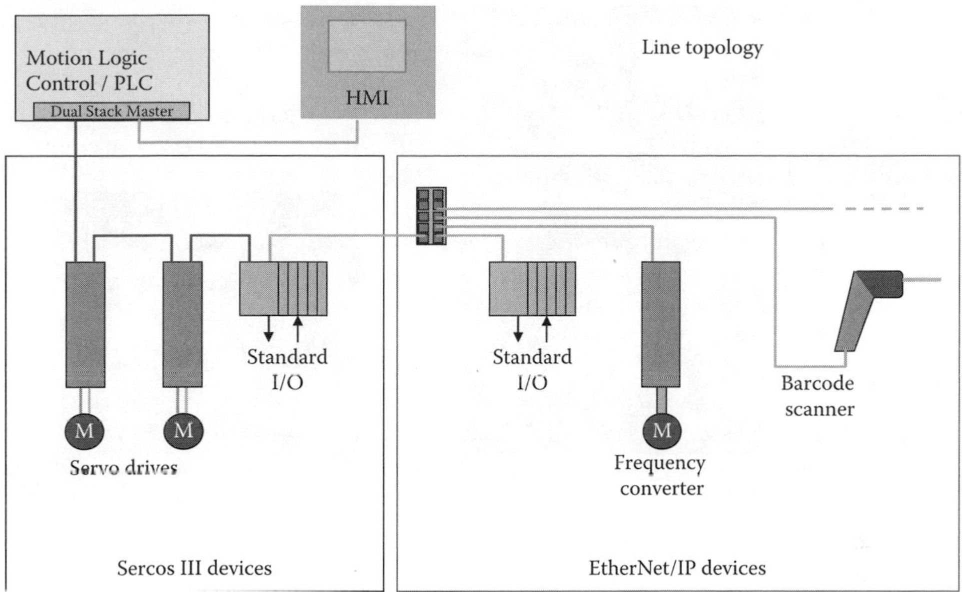

FIGURE 13.11 Blended infrastructure for Sercos III and EtherNet/IP.

2. In addition, a time-stamping procedure for event-triggered acquisition of values with corresponding time stamps, as well as time-triggered activation of command values, has been defined. This function is event controlled, promptly transmitting defined events such as certain measurement values to the controller and switching outputs independently from the clock cycle. This increases process stability, for instance, in complex solutions such as those needed in semiconductor or solar manufacturing.

13.2.5.4 Blended Infrastructure for Combining Sercos III and EtherNet/IP

The blended infrastructure approach was announced in April 2012 by ODVA and Sercos International as one of the first practical results of the machinery initiative in which Sercos International, ODVA, and OPC Foundation collaborate (see Section 13.2.14).

Because the network infrastructure required for EtherNet/IP and Sercos III includes the physics and data link layers of Ethernet, Sercos telegrams, CIP messages, and TCP/IP messages can coexist within a network without requiring additional cables (see Figure 13.11). To keep the cyclic and clocked communication of Sercos III intact, the CIP messages and TCP/IP telegrams are transmitted in the UCC, which is an integral part of the Sercos transmission method. The blended infrastructure approach does not require any modifications to the existing Sercos III and EtherNet/IP specifications, as the respective communication mechanisms have already been integrated in the Sercos III transmission process. Only an implementation guide describing the planning and setup of such multiprotocol networks is required.

13.2.5.5 CIP Safety on Sercos

CIP Safety on Sercos is the protocol used to transfer safety-critical data over the Sercos III network. It was defined in collaboration with ODVA and has been certified for compliance with IEC 61508 up to Safety Integrity Level 3 (SIL3). No additional safety bus is required because the safety information is sent in addition to the standard data on the Sercos network in RT. The combination of drives, peripherals, safety bus, and standard Ethernet in a single network simplifies handling and reduces hardware and installation costs, and it makes it easy to deploy integrated safety controllers and homogeneous safety solutions.

Reliable communication can take place between all network levels including peer-to-peer communication and cross-network communication. The master does not necessarily have to be a safety controller.

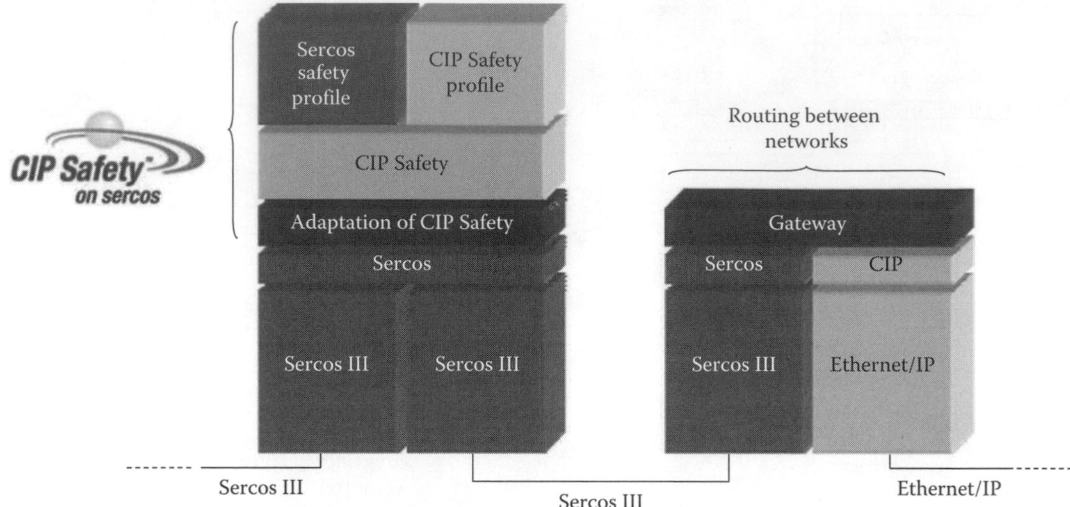

FIGURE 13.12 CIP Safety on Sercos architecture.

It can also route data without being able to interpret it. This makes it possible for users to flexibly configure the safety network architecture for the implementation of safety programmable controllers or peer-to-peer communication between sensors and actuators. In addition, use of a standard CIP network as a backbone network enables seamless communication between safety devices in different subnetworks (see Figure 13.12).

Safety-related data are sent in a safety data container that is placed on the RT data channel (MDT and AT) similar to standard data. A multiplex protocol, the SMP, is used to transmit differently sampled safety data without a loss of bandwidth and despite short cycle times. CIP safety has been certified by TÜV Rheinland for use in applications up to SIL3 and complies with IEC standard 61508 for functional safety (Functional safety of electrical/electronic/programmable electronic safety-related systems).

(For a detailed description of CIP Safety, refer to the ODVA contribution about CIP Safety.)

13.2.6 Sercos III Profiles

13.2.6.1 Introduction

The existing Sercos servo drive profile has evolved into a mature and established device profile during the past 15 years. Due to its broad use in different application fields, it covers an extensive and proven functional range. In the course of adapting the device profile to Sercos III, it was extended and generalized in such a way that not only pure drive, I/O, and control devices are supported, but hybrid devices, which combine different applications in one device, are also covered (see Figure 13.13). Accordingly, new device profiles for additional actuator physics and drive types (closed and open loop drives,…), as well as additional device functions, for example, I/O functions, technology modules, controls, and monitoring systems were defined.

All Sercos-related traffic across a Sercos III network consists of Idents or IDNs (parameters) with attributes (see Section 13.2.4.8). This method was first defined in Sercos I, as an essentially flat set of Idents. They were later grouped into application sets to aid in the selection of pertinent Idents required for a given industry, such as the "Pack Profile" for use with packaging machinery. During the development of the Sercos III specification, this methodology was further refined to group the Idents logically by device class. The definition of the legacy Idents has remained largely untouched; rather their grouping has been reevaluated for a more understandable architecture. This has also enabled the

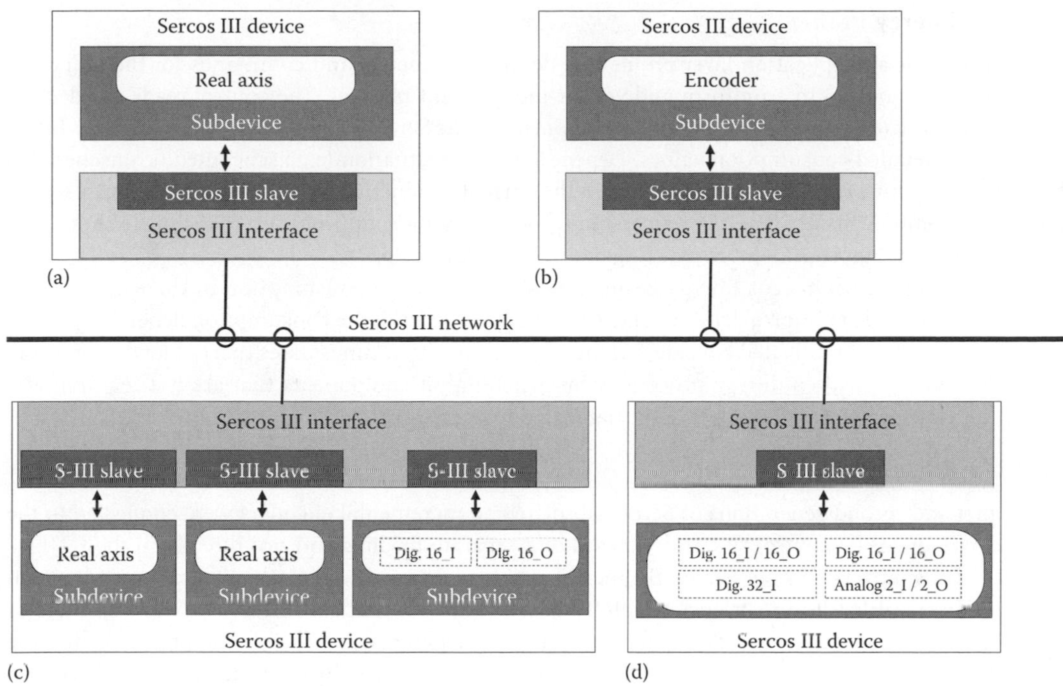

FIGURE 13.13 Profiles of Sercos III devices. (a) servo drive (single axis), (b) encoder, (c) double-axes drive with integrated I/O functions, and (d) modular I/O device.

separation of communication Idents into a logical subset, simplifying migration from Sercos I/II to Sercos III and providing a clear overview to users.

13.2.6.2 Generic Device Profile

The objective of the generic device profile (GDP) is to provide for a view of the subdevice that is independent of the function-specific profile (FSP) that is implemented by the subdevice. The following parts are independent of the FSP:

- Identification
- Administration
- Archiving
- GDP state machines allow a decoupling of communication and the applications state machine

13.2.6.3 I/O Profile

The Sercos I/O profile is a powerful and future-proof device profile for decentralized I/O periphery. The profile can be used for block I/Os as well as for modular I/Os. Furthermore, hybrid devices are supported, which combine miscellaneous functionalities in one single device (e.g., two-axis controller with I/O and master functionality).

For the configuration of corresponding devices, an XML-based device and profile description language was specified. Sercos device description markup language (SDDML) describes which profiles are supported by a certain device. Sercos profile description markup language (SPDML) is applied to specify different profiles on the basis of the parameter model of Sercos. Not only predefined standard parameters can be used, but also manufacturer-specific parameters can be defined. In this context, the Sercos communication profile, the GDP, and the FSP are differentiated. The combination of these profiles forms a so-called application-specific profile.

13.2.6.4 Energy Profile

Sercos Energy is an application-layer profile that defines parameters and commands for the reduction of energy consumption in a uniform and vendor-independent manner. The control reads out defined standard parameters of each Sercos Energy component via the Sercos III network, receiving status information and detailed consumption values. Depending on the situation (e.g., scheduled or unscheduled breaks and machine components not needed in the current production process), the control can switch connected components (drives, I/O, sensors) into energy-saving conditions, up to complete shutdown, in a targeted manner, considerably reducing their energy consumption.

Using this approach, Sercos Energy is able to reduce the energy consumption in three areas. First, the permanent load at motor/machine standstill is reduced; second, the consumption depending on the process is dynamically adjusted considering the target completion times/dates (partial load operation); and third, energy is saved during processing by switching off components that are not required at a particular time or point in the process (partial machine operation).

13.2.6.5 Encoder Profile

In the first and second generations of Sercos, absolute and incremental encoders were connected to the servo drives or to the control system by means of a separate fieldbus and not directly to the RT bus. Because of the universal use of Sercos III, there has been a demand to directly integrate encoders into a Sercos III network. The Sercos encoder profile ensures that the functions of an encoder—absolute or relative—are made available via clearly defined vendor-independent interfaces. The profile defines the functions supported by a device and how these functions may be used by other devices, such as control systems or servo drives.

13.2.7 Device Description

The Sercos device model enables representation of all types of automation devices at the field and control level in the different phases of the life cycle both functionally and logically. For Sercos III, two languages were developed to describe devices and their functions for offline configuration and for a simple display in a generic engineering tool.

SDDML, the device description language for Sercos III devices, is an XML-based description for Sercos III devices and is based on the Sercos III device model. In each case, a device is described that can consist of several subdevices. In addition, more information can be provided for the configuration tool. It is, for example, possible to describe rules, such as which I/O components within a modular device can be combined. The device description also indicates if the range of IDN values deviates from the range defined in the profile, because the complete function is not implemented.

SPDML, the profile description language for Sercos III, offers the opportunity to describe the profile elements. These profile elements are either IDNs or areas in the RT messages of Sercos III. Furthermore, profile elements can be complete profiles. Thus, hierarchical and nested profiles can be described. The description of IDNs and areas in the existing RT messages includes the ability to set value ranges, to impose semantics to a bit sequence, to describe dependencies of profile elements, and to assign plain text in different languages to diagnostic and error codes.

13.2.8 Configuration of Sercos III Networks

With the help of the device and profile description languages (SDDML and SPDML), Sercos III devices can be described uniformly and configured with the appropriate configuration tools using a configuration interface. The configuration interface (SCI) is available in two variants (see Figure 13.14). First, an XML interface is defined based on an XSD schema. Second, there is an optional resource-saving option for systems with limited resources (*embedded systems*) as a binary counterpart. The SCI interface defines the network configuration and defines the slaves that must be present and that may optionally

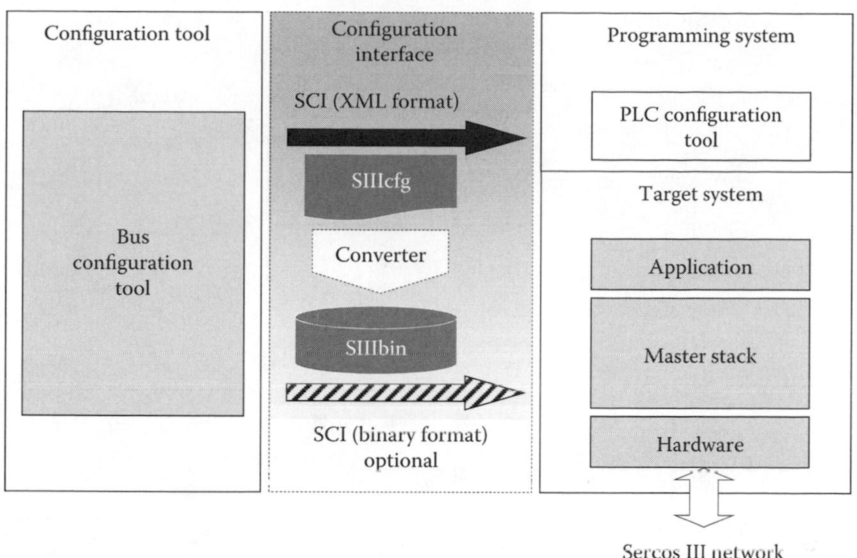

FIGURE 13.14 XML-based configuration interface for Sercos III networks.

be present. The identification of the slave devices is done via criteria described in the file. Furthermore, the configuration file describes the parameters of each device using a generic procedure. In addition, the parameterization of the master can be done.

Sercos III is applying the open and vendor-independent FDT technology in order to standardize the communication interface between field devices and engineering tools. A device manufacturer can deliver a Sercos III product with a device-specific Device Type Manager (DTM). In this case, the DTM can be directly integrated in the FDT framework application. Furthermore, a DTM can be generated from an SDDML-compliant device description using a generic converter. The device description is either already available as a file (offline) or generated from the parameter set of the device that can be accessed and read out via the bus system (online).

13.2.9 Implementation of Sercos III

The first and second generations of Sercos have been supported by a specific communication ASIC (Sercon410B and Sercon816). With Sercos III, a more flexible hardware solution is used (see Figure 13.15). The basis is the development of a Sercos core (Sercos III IP), with which manufacturers of components and systems are able to integrate Sercos III in an FPGA. At present, devices of Xilinx, Altera, and Lattice are supported.

IP core licenses are granted via the Sercos user organization. Different license models are available; member companies of the Sercos organization receive preferential prices.

Furthermore, the Sercos user group cooperates with chip manufacturers and system houses in order to integrate the required hardware functions into ASICs, processors, and universal multiprotocol communication controllers.

Different software packages are made available as open-source software to facilitate the device implementation (see Sections 13.2.10.1 through 13.2.10.3).

A Sercos III master can also be implemented without any special hardware (i.e., SoftMaster). Standard Ethernet controllers are used instead of the specific Sercos III hardware. Sercos-specific hardware functions are moved to the hardware-related and RT capable part of the master driver. Thus, the master can be fully implemented in software. Such a concept is interesting for PC-based control platforms that have an *onboard* Ethernet interface.

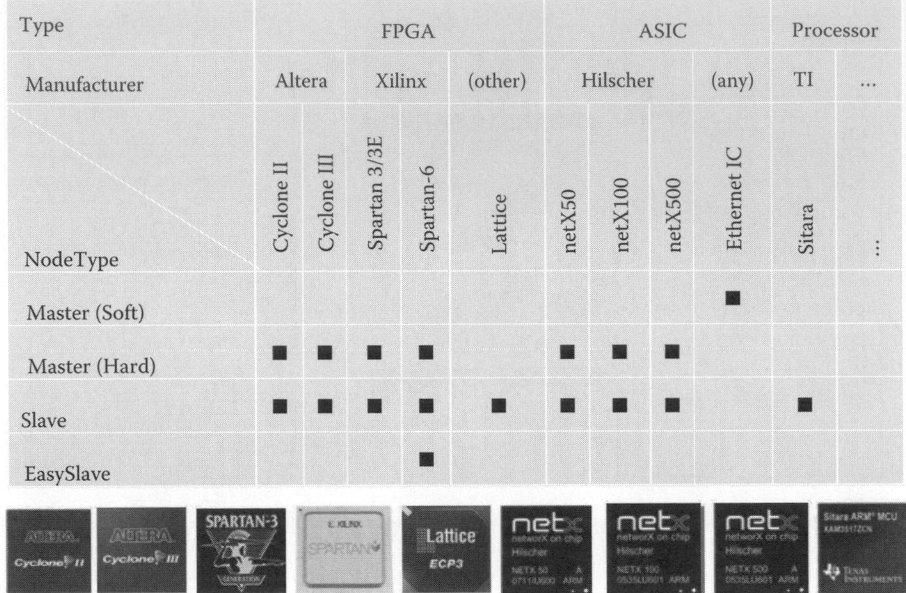

Type	FPGA					ASIC				Processor	
Manufacturer	Altera		Xilinx		(other)	Hilscher			(any)	TI	...
NodeType	Cyclone II	Cyclone III	Spartan 3/3E	Spartan-6	Lattice	netX50	netX100	netX500	Ethernet IC	Sitara	...
Master (Soft)									■		
Master (Hard)	■	■	■	■		■	■	■			
Slave	■	■	■	■	■	■	■	■		■	
EasySlave				■							

FIGURE 13.15 Sercos III controllers for master and slave implementations.

13.2.9.1 EasySlave

The integration of I/O slaves can be easily accomplished with the EasySlave FPGA-based single-chip controller. EasySlave enables the inexpensive and flexible activation of Sercos slave devices, thus making it even easier to connect sensors and actuators. EasySlave is equipped with an RT channel with one input and one output connection and supports cycle times as low as 31.25 µs. The I/O applications are synchronized in the Sercos cycle. Features are the following:

- FPGA-internal RAM for Sercos stack, application code, and data
- RT channel with max. 32 bytes output and 32 bytes input data
- Supported cycle times down to 31.25 µs
- Synchronization of application to Sercos cycle
- Parameterization via service channel
- Support of Sercos hot-plug services
- S/IP services (Nameplate, Identify, SetIp, etc.)
- IP channel for ARP, Ping, and firmware update over TFTP
- Design targeted to Xilinx Spartan-6 FPGA series (XC6SLX16 or higher density) including a MicroBlaze Soft-core Processor

Licenses for applications are granted via Sercos International. Different license models are available; member companies of Sercos International receive preferential prices. A license-free variant of the EasySlave with limited functionality as a bitstream is also available under the product name Sercos EasySlave-IO.

13.2.10 Software Packages

13.2.10.1 Open-Source Sercos III Master Library

The Sercos user organization provides an open-source software driver library (CoSeMa = Common Sercos Master) for a Sercos III master implementation. The source files of this master library are available via the sourceforge.net Online-portal (http://sourceforge.net/projects/cosema/) and can be used by interested users (control manufacturers and software tool suppliers) according to the LGPL license. The library is fully tested software without any functional limitations. All features of Sercos III are supported; for example, the RT channel, the redundancy concept, and hot-plugging.

13.2.10.2 Open-Source S/IP Protocol Driver Library

The Sercos user organization has made available a driver library for the Sercos/IP protocol as open-source software. The driver software is available for download as C# source code via http://sourceforge.net, without any license fees and without any usage limitations (http://sourceforge.net/projects/sercosips/).

The protocol enables direct communication as well as uniform data exchange between devices, without a Sercos master or a running Sercos communication being necessary. Even in cyclic RT operation, the Sercos/IP protocol can still be used. In this case, the Sercos/IP telegrams are transferred into the UCC of Sercos, without having a negative influence on the RT behavior of the network.

With the Sercos/IP protocol, the vertical integration of devices into a production system is improved. The setup, remote maintenance, and diagnosis of Sercos devices can be carried out via Sercos/IP over an intranet or the Internet.

13.2.10.3 Open-Source UCC Ethernet Network Driver

The Sercos user organization also provides a sample code of an Ethernet Network Driver for handling the UCC of a Sercos III network. This driver connects the UCC (CoSeMa API) to, for example, a VxWorks TCP/IP stack (http://sourceforge.net/projects/sercosuccend).

13.2.10.4 CIP Safety Stack Software

This software package contains the protocol stack software for Sercos III and EtherNet/IP, a safety manual, and unit tests. Because all components are made available as ANSI-C code, porting to customer-specific safety platforms is possible with minimum effort. The unit tests provided considerably facilitate the recertification of the CIP Safety protocol software after porting the code to a safe target platform. IXXAT Automation GmbH* provides various services regarding the CIP Safety technology, such as consulting, code introduction, integration workshops, implementation support of the CIP Safety software in customer-specific hardware and software, and assistance in final device certification.

13.2.10.5 Sercos Monitor

The Sercos user organization provides a powerful diagnostic tool for Sercos III networks that can be downloaded free of charge from the Sercos website.† The Sercos Monitor runs under Windows XP and Windows 7 operating systems and is based on the WinPcap interface (see Figure 13.16). The tool allows a comprehensive and detailed analysis of the data traffic in Sercos III networks. It allows both retroactive evaluation of stored network records in the pcap file format and analysis of the network traffic in realtime. A user-friendly interface and overview functions for the typical features of a Sercos III network, such as topology, communications phases, and service channel transmission, enable a fast and targeted launch of the analysis process. Experienced users are additionally provided with comprehensive protocol and analysis functions. Different views and filter mechanisms allow the customized evaluation of Sercos III RT telegrams and other Ethernet telegrams based on the individual requirements of the user. Users can also add additional functions to the Sercos Monitor via user-specific plug-ins, if needed.

---

* See http://www.ixxat.com/cip-safety-originator-target-software-en-html

† See www.sercos.de. Look for Downloads/Tools.

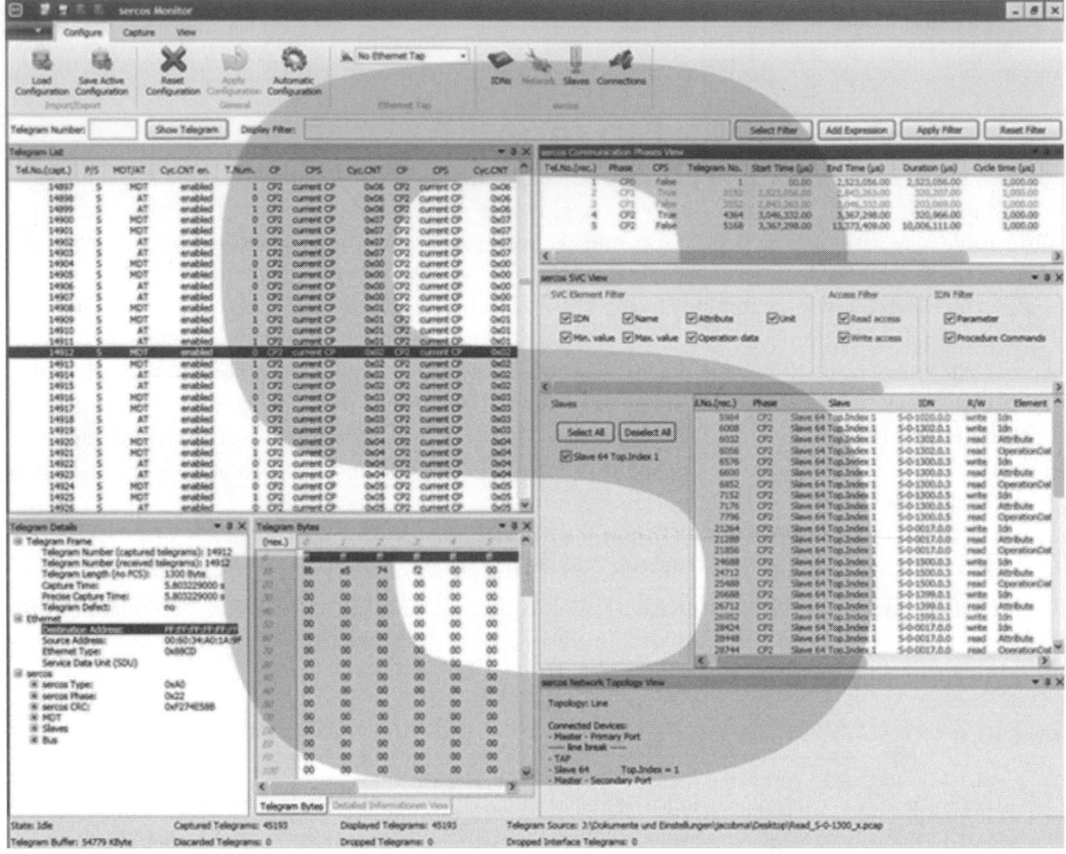

FIGURE 13.16 Diagnostic tool for Sercos III networks.

13.2.10.6 MultiSlave Emulator

The MultiSlave Emulator tool runs under Windows XP and Windows 7 operating systems and is based on an active Sercos PCI board from Automata GmbH & Co. KG.* A user-friendly interface supports the emulation of complete Sercos networks and the connected Sercos slave devices. These devices can be freely configured. By reading out the available parameters from the real device and importing these into the MultiSlave Emulator, the emulation of the individual devices, as well as the complete network, can be executed with a minimum of effort.

13.2.11 Conformance Testing

In order to ensure that Sercos products from various manufacturers are compatible and meet IEC standards, conformance testing and certification of all Sercos products are required. Once a product passes the conformance test, a certificate can be requested, and the product can carry the Sercos trademark.

13.2.11.1 Conformance Test Software

The Sercos III Slave Conformizer is a powerful and independent test system for conformance testing of Sercos III slave devices. This test system, which basically takes the form of a reference master, ensures the compatibility and interoperability of Sercos III slave configurations from different manufacturers.

* See http://www.cannon-automata.com/.

Originally designed for test lab certifications, the software can also be used by companies as a testing tool for product development and quality assurance. This guarantees a high conformity of configurations and allows certification procedures to be carried out quickly and with minimal costs. The Sercos III Slave Conformizer provides a complete testing and development environment for Sercos slave devices. It includes an active PCI Sercos III interface card and test software that runs under Windows operating systems without the need for RT extensions.

With a user-friendly graphic interface based on the Eclipse Framework, individual tests are performed and documented via protocols. The tests themselves are available as readable files using the Ruby script language. Extensive test scripts are included for communication, the GDP and the FSPs for drives and I/Os. Product and manufacturer tests can also be easily integrated.

The Slave Conformizer can be purchased from Sercos International. (see www.serco.de.)

A test procedure is defined for master testing. However, this is a manual test procedure (checklist) instead of an automatic test with a conformance test tool. However, there are plans to extend the MultiSlave Emulator in such a way that master devices can be tested automatically.

13.2.11.2 Conformance Testing Procedure

Sercos III devices are certified by the test lab at the Institute for Control Engineering of Machine Tools and Manufacturing Units (ISW) at the University of Stuttgart, in cooperation with the Steinbeis Transferzentrum für Systemtechnik (TZS) in Esslingen, Germany. Any company interested in conformance testing should contact Sercos International for scheduling and pricing.

13.2.11.3 Plug Fests

Sercos international holds twice-yearly Sercos III Plug Fests at the Sercos Competence Center at ISW, University of Stuttgart. The plug fests are free of charge to any manufacturer of Sercos III control systems, servo drives, I/O systems, and other peripheral devices. The staff of the testing laboratory is present with the Sercos III Slave Conformizer.

Interoperability tests are performed between each present master and slave and between masters and multiple slaves to verify implementation of Sercos III master and slave devices to ensure the best possible compliance of the different products. Various features, such as ring break detection, ring recovery, hot-plugging, UCC connectivity, and S/IP protocol support are tested and verified in different network configurations. This also helps improve the Sercos III specification, consequently limiting the scope of interpretation.

13.2.12 Publications

13.2.12.1 Sercos in IEC Standards

The Sercos technology is part of several IEC standards:

- IEC 61800-7(-1, -204, -304): Sercos Drive Profile
- IEC 61784-1, IEC 61158-2, IEC 61158-3/4/5/6-16: Sercos I/II Communication Profile
- IEC 61784-2, IEC 61158-2, IEC 61158-3/4/5/6-19: Sercos III Communication Profile
- IEC 61784-5-16 Ed 2.0: Sercos Installation Profile
- IEC 61784-3-2: CIP Safety on Sercos

Any of these standards can be obtained via the IEC web store.*

13.2.12.2 Planning and Installation Guide

The Sercos III Planning and Installation Guide is intended to support engineers and electricians during the planning and installation phase of a Sercos III network in order to guarantee trouble-free operation

* See http://webstore.iec.ch/.

of machines and systems. The cabling with copper wire and optical fiber is described, each in consideration of IP20 and IP65/67 protection class. The specification is based on the international standard IEC 61918 (Installation of communication networks in industrial premises), with an added installation profile for Sercos III, which will be standardized in IEC 61784-5.

13.2.12.3 Energy White Paper

The Sercos Energy White Paper describes the increasing importance of energy efficiency in production technology, the application scenarios for Sercos Energy, and the advantages of Sercos Energy. The White Paper is available for download from http://www.Sercos.com/literature/pdf/SERCOS_Energy_Whitepaper_V10.pdf.

13.2.13 IO-Link Integration

The Sercos user organization provides a guide for the standardized mapping of IO-Link devices to Sercos III. IO-Link is a low-level, point-to-point communication interface technology for interconnection of binary and analog sensors and actuators. This user guide helps the automation engineer to uniformly connect and integrate sensors and actuators equipped with an IO-Link interface into a Sercos system. An IO-Link master can be implemented either as part of a modular Sercos slave I/O device or as a stand-alone Sercos III slave. The cyclic data of all inputs and outputs of IO-Link slaves are either transmitted completely within a cyclic IO container or alternatively as separate parameters. The Sercos parameter channel can be used in order to execute acyclic read/write access with regard to the addressed IO-Link slave devices.

13.2.14 Cooperation between Sercos, ODVA, and OPC Foundation

In March 2011, the Sercos user organization established a partnership with the ODVA and the OPC Foundation in order to foster cross-collaboration on topics of mutual interest for machinery applications for the benefit of machine builders and manufacturers (White Paper: Optimization of Machine Integration [OMI]*). The cooperation efforts of the three organizations were set up to work on technical concepts for the following three use cases:

1. *Machine-to-machine communication*: Machine-to-machine communication is the foundation of OMI. Establishing standards will help minimize the time and cost of machine integration and reduce the custom software and special hardware needed to integrate different vendor's equipment. By sharing a unified integration model for key functions needed for machines to interoperate with one another—including command, configure, and control—along with a common data model to communicate key machine attributes, machine builders and manufacturers will find machines from multiple vendors and design cycles to be easier, faster, and cheaper to integrate.
2. *Machine-to-supervisory communication*: Communication between machines and supervisory systems is the accelerant for OMI. By leveraging SCADA and manufacturing execution systems, manufacturers can consolidate and exchange machine information across systems to collect production data, assist in asset management, and report diagnostics and alarms. This machine monitoring and management helps reduce downtime by promoting efficient utilization and multiplies the benefits of machine integration by automating actions that improve overall equipment effectiveness. OMI allows existing automation to incorporate the functionality needed to realize new efficiencies in machinery utilization, thus protecting the user's investment in technology.
3. *Communication connectivity*: Connectivity is the enabler for OMI. Machinery in the industrial ecosystem is characterized by multiple heterogeneous networks connected to multiple physical domains both inside and outside the manufacturer's enterprise. By leveraging common standards

* See http://odva.org/Portals/0/Library/Publications_Numbered/PUB00266R1_ODVA_Optimization_of_Machine_Integration_EN.pdf.

for physical media, network infrastructure, and secure remote access, both users and machine builders will be able to reduce the cost and complexity of machine integration while improving maintainability and safety of machines. As reliability is also essential, the OMI approach will define how communication is handled when machinery is unavailable.

The blended infrastructure approach for combining Sercos III and EtherNet/IP (13.2.5.4) is a first concrete result of this initiative (*communication connectivity* use case, as described earlier).

13.2.15 Performance

Sercos III shows a substantial increase in transmission speed (Figure 13.17) compared to Sercos I/II. For example, more than 70 drives can be provided cyclically with 8 bytes of data with a cycle time of 250 μs. In the following figure, some typical values are listed for different configurations of the UCC.

13.2.16 Sercos III Applications

Sercos III has been used since 2005 in numerous applications in mechanical engineering, for example, printing, food and packaging, machine tools, and robotics. At present, more than 200 Sercos III products from more than 70 manufacturers are offered in the market. They include complete automation solutions (package of controls, drives, and/or I/O peripherals), automation devices (control systems, drives, I/O devices, other devices such as sensors), tools (development, configuration, diagnosis), hardware and software components (controllers, drivers, stacks), and active and passive infrastructure components (switches, cables, plugs). More than a dozen service providers offer consultation and support for device manufacturers who do not wish to develop and integrate a Sercos III interface completely on their own. The range of services covers development support, training, consultation, testing, and even certification of Sercos III devices.

Additional information can be found on www.Sercos.com.

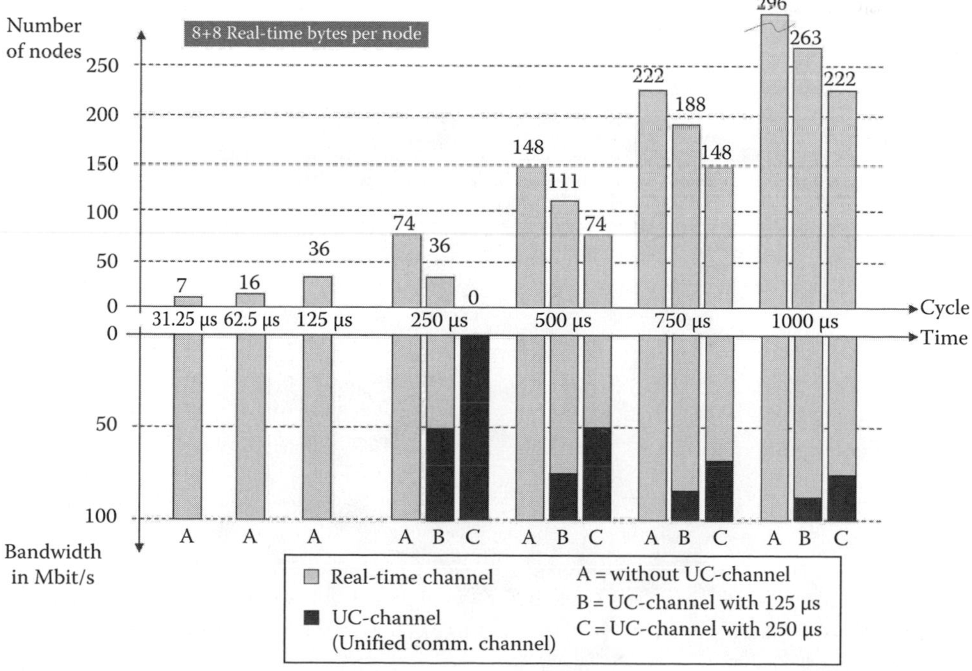

FIGURE 13.17 Sercos III performance.

13.2.16.1 Packaging Machines with Sercos III Communications

Rovema, a German-based packaging machine builder, supplies complete solutions for sophisticated complex packaging applications. Forming, filling, sealing, case packing, and retail packaging for today's market require a holistic approach that goes beyond the individual component level to include the specific customer process. A very powerful and highly flexible control system supports the broad product portfolio and the highly complex packaging operations. Rovema has been using its own proprietary PC-based P@ck-Control system for many years on its packaging machines. The motion control subsystem is based on a centralized control architecture. P@ck-Control takes total control of axis coordination and position control. The interface to the speed-controlled drives had traditionally been a conventional ±10 V interface with incremental encoder feedback. Interbus-S had provided the link between the integral software PLC in the P@ck-Control system and the outside world.

The engineers at Rovema have designed the Ethernet-based Sercos III RT communications system into the new P@ck-Control generation. This new VPL260 tubular bag machine based on Sercos III was displayed for the first time at Interpack 2008. The machine also features continuous volumetric filling. The communication with the speed-controlled drives and the digital I/Os is achieved through the Ethernet-based Sercos III network. With the aid of defined profiles and the open Sercos standard, the design team succeeded in integrating the RT Ethernet network into Rovema's machines within 12 months replacing the analog ±10 V interface (Figure 13.18).

13.2.16.2 Manz Scribing Systems Based on Sercos III: Maximum Thin-Film Module Efficiency

With an installation base of more than 100 machines, Manz Automation AG is a market leader in scribing systems for the thin-film technology. In this technology, very thin layers are applied to a glass surface: two conductive layers as well as the absorber layer. This promises to be a leading technology in large-scale photovoltaic modules. Manz scribing systems are available for various processes: not only mechanical but also laser-based scribing operations can be performed with maximum precision on Manz systems.

FIGURE 13.18 Rovema VPL260 tubular bag machine.

FIGURE 13.19 Manz scribing system based on Sercos III.

Since efficiency of solar modules is key for the shortest return on investment, module manufacturers are strongly looking into process optimization. One approach for increased module efficiency is to minimize the dead area of solar modules. By using the Manz proprietary IPCS (Manz Inline Precision Control System) active tracking system, it is possible to achieve a dead area of <200 μm. This unique feature was introduced to the market in 2009 and has been successfully sold to numerous customers. This is a milestone compared with other systems on the market, where the dead area is approx. 150–300 μm above this figure.

In 2010, the Sercos II motion bus was replaced by the Sercos III automation bus to achieve higher RT bandwidth and optimum integration of machine control, IPCS, and other field devices. Together with its outstanding RT performance, Sercos III allows direct TCP/IP access with engineering and diagnostic tools from the desktop into servo drives, IPCS, and other field devices to achieve optimum service capabilities and minimum downtimes.

Because of the available bandwidth, Sercos III provides the opportunity to integrate all fieldbus-based devices on a single bus and is another contribution to clear system structure and reliability. Sercos III is ready for the fieldbus-independent CIP Safety protocol relevant for safe control of servo drive systems and laser sources. It will reduce total cost of ownership and increase system reliability (Figure 13.19).

13.3 Features and Operation of Sercos II

13.3.1 Introduction

Sercos I and II are the first two generations of Sercos and transmit data over a noise-immune, fiber-optic cable. Sercos I is now obsolete, so this section refers only to Sercos II. The only difference between the two versions is that Sercos I used an ASIC operating at 2 and 4 Mbps, while Sercos II uses an ASIC (Sercon816) that operates at 2/4/8/16 Mbps.

Work began on Sercos I in 1987, first products were introduced in 1990, and a draft international standard was created in 1990 and approved by the IEC in 1995.

ASIC development occurred in parallel with the standardization process. In 1993, SGS Thomson (now STMicroelectronics) developed the Sercon410B ASIC, operating at 2/4 Mbps. This made implementation of Sercos simpler and less expensive.

In 1999, STMicroelectronics developed the Sercon816 ASIC, operating at 2/4/8/16 Mbps, quadrupling available interface speeds. In 2000, Sercos II products from Sercos association members begin shipping with the Sercon816, and product shipments continue strongly today.

13.3.2 How Sercos II Communicates

13.3.2.1 Topology*

Components in a Sercos II–based motion control system are connected via fiber-optic rings using a master/slave configuration. A typical system may include several rings, with up to 254 devices per ring. A Sercos master controls each ring, assigning time slots to ensure deterministic and collision-free access for all slaves.

13.3.2.1.1 Communications Structure

In order to ensure strict synchronization of multiple axes and a predictable update time at each axis, Sercos utilizes a master/slave communication structure, where the machine/motion control acts as a master, communicating with slave devices such as drives. The slaves are only permitted to respond to queries from the control.

13.3.2.2 Fiber Optics

With Sercos II, one fiber-optic ring is used to exchange full 32-bit data between controllers, drives, I/O, and sensors. This includes commands, status, parameters, and diagnostics. Fiber optics provides inherent noise immunity and eliminates the immense requirements for conduit, wiring, and terminations normally required with the analog interface. The ring architecture reduces the number of components required on a motion control. Adding an additional motion axis often requires nothing more than opening the ring and placing the new slave device in the ring.

A Sercos ring is composed of a number of fiber-optic segments as illustrated in Figure 13.20. Each device in a system receives signals via a fiber-optic receiver with an F-SMA connector and transmits the signals to the next device via a fiber-optic transmitter, again with a standard F-SMA connector. Both plastic and glass fiber are specified, with wavelength specified at 650 nm.

Maximum cable lengths are as follows:

- Plastic fiber (POF)—1 mm diameter
  - Node to node—40 m
  - Maximum ring length (254 nodes)—10,000+ m

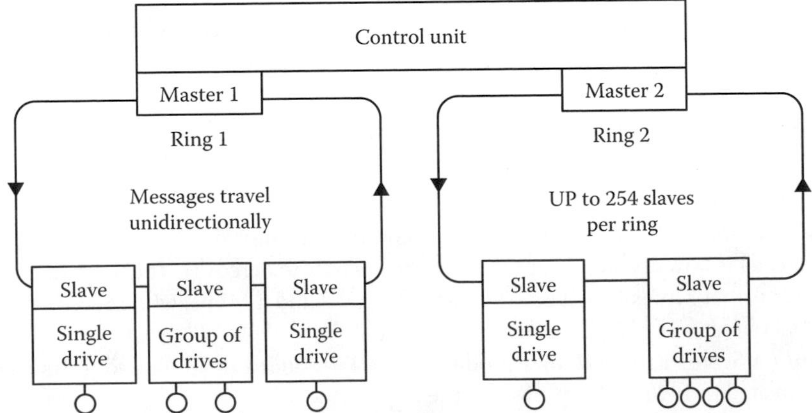

FIGURE 13.20 Sercos generation I and II topology.

* Illustrations and text adapted by permission from the technical paper *Open Drive Interfaces for Advanced Machining Concepts* by Scott Hibbard of Bosch Rexroth, January 1995.

TABLE 13.2 Performance of Sercos Rings: Generations I and II

Cycle Time (ms)	Data Record per Drive (MDT + AT)	Transmission Rate (Mbit/s)	Number of Drives	Data Rate (Noncyclic Data)	Remaining Cycle Time (μs)
2	32 bytes	2	8	8 kbit/s (2 bytes)	390
1	32 bytes	4	8	16 kbit/s (2 bytes)	125
1	36 bytes	8	15	32 kbit/s (4 bytes)	208

- Glass fiber (HCS)—200 μm diameter
  - Node to node—200 m
  - Maximum ring length (254 nodes)—50,000+ m

The maximum number of drops per fiber-optic ring is 254. However, the number of drives that can be serviced per ring depends on three application requirements:

1. The communication cycle time
2. The volume of operational data
3. The communications speed required

Table 13.2 illustrates performance for generations I and II, showing examples of the number of drives per ring at various transmission rates and cycle times.

The number of rings that can be synchronized together is limited only by the controller, which will have a limit to the number of axes it can process. Sercos-compliant controllers can have a much higher axis limit than their predecessors, since Sercos supports distributed processing, which relieves the controller of many time-intensive tasks, which are now handled by the drives or other slave devices.

13.3.2.2.1 Fiber-Optic Transmitters/Receivers

Fiber-optic transmitters and receivers from Agilent Technologies and Honeywell have been tested and approved by the Sercos Technical Working Group (TWG). The Sercos N.A. and Sercos International websites (www.Sercos.com and www.Sercos.de) provide links to downloadable data sheets for these products. Signal levels are defined in the IEC standard.

13.3.2.3 Sercos ASICs

The Sercon816 ASIC is an integrated circuit for Sercos II RT communication systems. A Sercos II communications system consists of one master and several slaves, connected by a fiber-optic ring that starts and ends at the master. While its use is not mandatory, the ASIC contains all the hardware-related functions of Sercos and considerably reduces the hardware costs for a design. Unlike other bus interface devices, the Sercon816 ASIC features the capability to perform all sequencing and synchronization tasks, relieving the host microprocessor of these time-intensive functions. The ASIC is the direct link between the fiber-optic receiver and transmitter and the microprocessor that executes the control algorithms in a Sercos device. The same ASIC is used both for Sercos II masters and slaves.

The Sercon816 is manufactured by STMicroelectronics in the 0.5 μm/5 V HCMOS technology. It operates at 2/4/8/16 Mbps and is downward compatible with the now obsolete Sercon410B (2/4 Mbps) ASIC (Figure 13.21).

Features of this device are as follows*:

- Single-chip controller for Sercos, available in 100-pin flat-pack QFP100 and TQFP100 packages. Physically compatible with the Sercon410B.

* Feature list and Figure 13.21, SERCON816 ASIC Block Diagram, from SERCON816 Sercos Controller reference manual, courtesy of Sercos International e.V., Suessen, Germany, 2001.

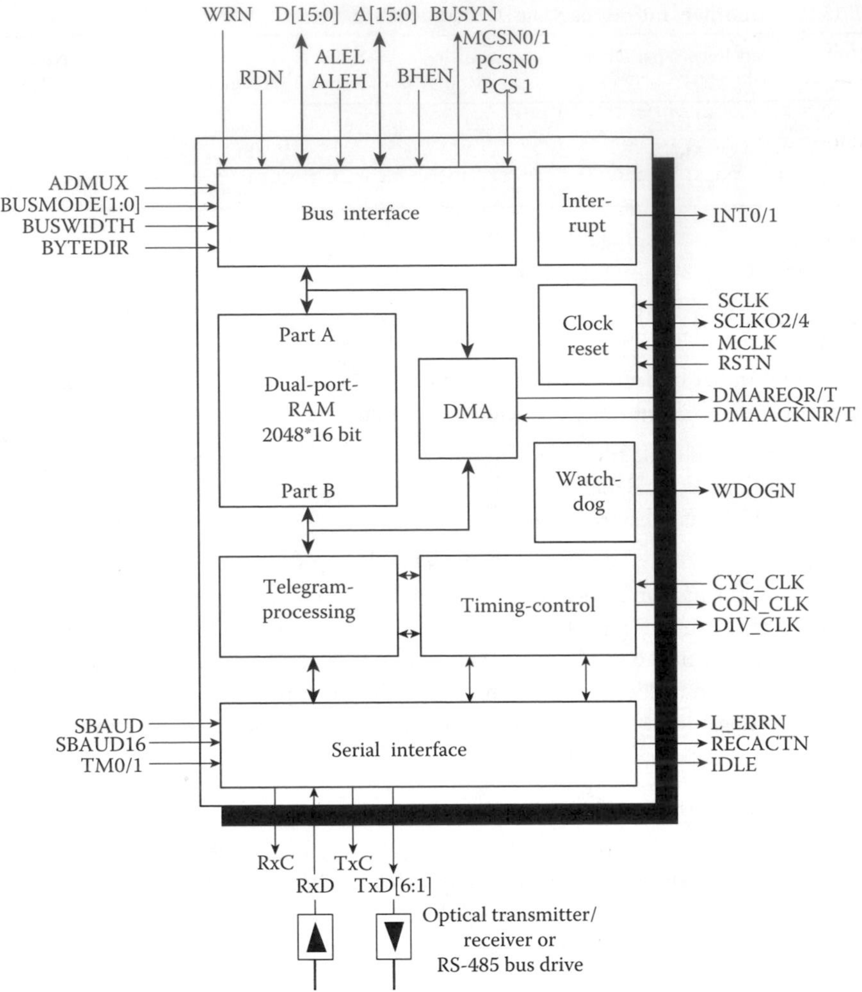

FIGURE 13.21 Sercon816 ASIC block diagram.

- Faster and lower cost than Sercon410B.
- Interface to the microprocessor with a data bus width of 8 or 16 bits and with control lines per Intel or Motorola standards.
- Serial transmission rate of 2/4/8/16 Mbps with an internal clock. The chosen bit rate can be selected by pins or software. The ASIC can be run in Sercon410B compatible mode by tying pin SBAUD16 to a logical 1.
- Data communication is available via fiber-optic rings or RS-485 rings and RS-485 buses. RS-485 is seldom used.
- Dual port RAM with 2048 16-bit words for control and communication data.
- Data and clock regeneration, the repeater for ring topologies, and serial transmitter and receiver are integrated. The signals are monitored, and test signals are generated.
- Full duplex operation.
- Modulation of power of optical transmitter diode.
- Telegram processing for automatic transmission and monitoring of synchronous and data telegrams. The transmission of service channel information over several communication cycles is executed automatically.

- Flexible RAM configuration, communication data stored in RAM (single or double buffer), or transfer via DMA.
- Timing control signals.
- Automatic service channel transmission.
- Watchdog to monitor software and external synchronization signals.
- Reset value for repeater mode of the serial interface is configurable through input pins.

13.3.2.4 Timing

Timing is critical in serial networks because motion controls cannot accurately reconstruct the state of the machine unless everything is precisely synchronized—measurements, transmissions, and replies. Many of the messages that controls and slave devices send each other are for timing and synchronization.

In CNC position loop software, the position error is determined in a software routine that takes time to execute. The system works because the position error updates are designed to occur at predictable points in time. The same logic is applied in the Sercos bus. Methods are specified to keep the jitter on the serial link down to a low level, and then an internal timing sequence is used to ensure that all drives in a loop act upon their command signal at the exact same moment, and all acquire their feedback information at the exact same moment. The result is transparent to the user.

13.3.2.5 Interface Placement

Sercos supports position, velocity, and torque mode control, which maintains technology independence between drives and motion controls. This independence was one of the Sercos design goals, to ensure that the interface restricted neither the design of controls or drives, but allowed both to proceed independently. For this reason, the power stage interface controlling commutation/phase current was ruled out. A power stage interface is highly hardware dependent, requiring the vendor to design motion controls around specific motors, limiting the use of the control to that type of motor. It also involves transmitting high-speed commutation signals over long distances with microsecond update rates. Thus, the demands for speed on the Sercos bus are less than those for competitive interfaces that offer power stage control (Figure 13.22).

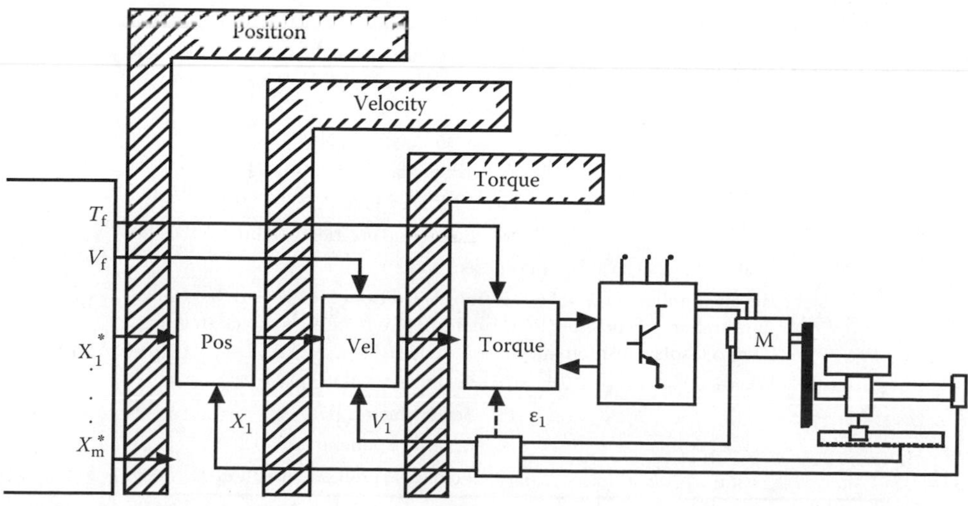

FIGURE 13.22 Interface placement.

13.3.2.6 Sercos IDNs

All communication between master and slaves is performed via a set of Sercos telegrams, each of which includes data records that each has an identification or *Ident* (IDN) number. All parametric data, such as scaling and loop gains, and RT loop closure information are set up this way. This allows Sercos to standardize the most common interface data.

The IEC standard allocates a block of 32,767 IDN numbers for standard commands. Over 700 of these are presently used to define a comprehensive set of motion control and I/O commands. All Sercos products must incorporate a subset of these, but do not necessarily need to include all IDNs.

The format used for a Sercos I/II IDN is shown in Figure 13.23.

13.3.2.6.1 IDNs for Real-Time Data

The standard defines a number of IDNs for RT data. For example, in each communication cycle, the master transmits an MDT that contains a series of IDNs (idents) specifying RT operating data and drive commands for the addressed slaves. The master sends an acknowledge telegram in which each slave places 1–16 IDNs, reporting data such as speed, torque, and position measurements to the master, effectively closing the control loop.

For example, IDN 00047 transfers the specified position command value as a 32-bit value to the drive. IDN 00051 communicates the actual position feedback value 1 (motor feedback) back to the control, as illustrated in Figure 13.24.

13.3.2.6.2 IDNs for Parameters

Figure 13.25 illustrates a Sercos drive system used to control a machine tool axis. It shows some of the control-specific and machine-specific parameter IDNs that would be used for control for such an axis and lists the function of each IDN.

IDN Name (abbreviation)		
Function/description		
Length in bytes	Minimum input value Maximum input value	Scaling/resolution Units

FIGURE 13.23 Standard IDN format

S-00047	Position command value			
	During the position control drive operation mode, the position command values are transferred from the control unit to the drive according to the time pattern of the control cycle.			
	4 (bytes)	$\geq -2^{31}$	Scaling type: IDN 00076	
		Max. $\leq +2^{31}-1$	Scaling factor: IDN 00077	
			Scaling exponent: IDN 00078	
			Rotational position resolution: IDN 00079	
S-00051	Position feedback value 1 (motor feedback)			
	The position feedback value 1 is transferred from the drive to the control unit so that it is possible for the control unit to perform block stepping and to display position information if necessary.			
	4 (bytes)	$\geq -2^{31}$	Scaling type: IDN 00076	
		Max. $\leq +2^{31}-1$	Scaling factor: IDN 00077	
			Scaling exponent: IDN 00078	
			Rotational position resolution: IDN 00079	

FIGURE 13.24 IDNs for position command value and motor feedback value

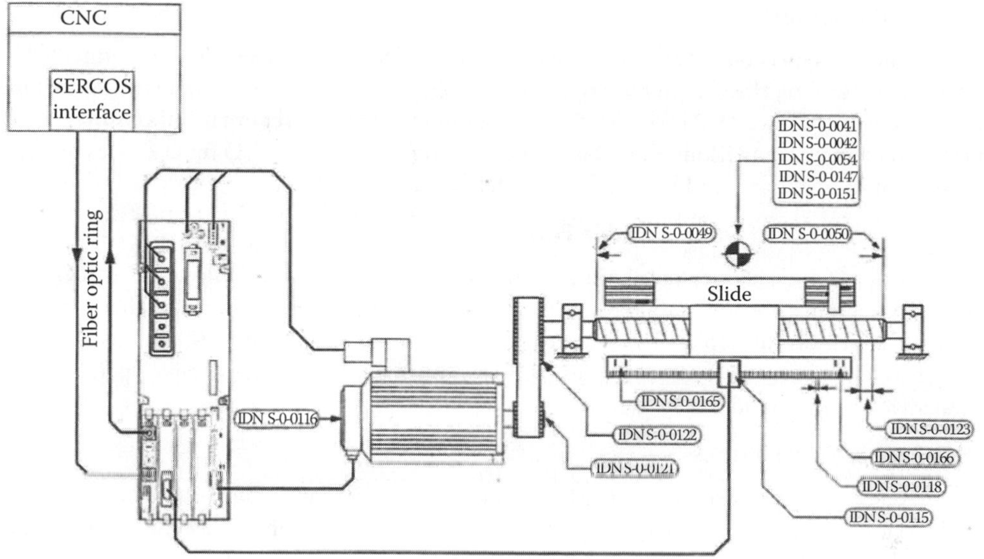

Representative Parameters For Control and Drive Adaptation			
Control-specific applications parameters		Machine-specific application parameters	
IDN	Function	IDN	Function
S-0-0032	Primary mode of operation	S-0-0041	Homing velocity
S-0-0033	Secondary operation mode 1	S-0-0042	Homing acceleration
S-0-0034	Secondary operation mode 2	S-0-0049	Positive position limit value
S-0-0035	Secondary operation mode 3	S-0-0050	Negative position limit value
S-0-0044	Scaling of velocity data	S-0-0054	Actual position feedback 2
S-0-0076	Position data scaling type	S-0-0115	Position feedback type
S-0-0079	Rotational position resolution	S-0-0116	Resolution of rotational feedback 1
S-0-0086	Scaling type for torque force data	S-0-0118	Resolution of linear feedback
S-0-0160	Scaling type for acceleration data	S-0-0121	Load gear input revolutions
		S-0-0122	Load gear output revolutions
		S-0-0123	Feed constant
		S-0-0147	Homing parameter
		S-0-0151	Reference offset 2
		S-0-0165	Distance coded reference dimension 1
		S-0-0166	Distance coded reference dimension 2

FIGURE 13.25 Sercos drive for machine tool axis with representative IDNs. (Courtesy of Sercos Technical Short Description, Sercos International e.V., Sussen, Germany, 1998.)

13.3.2.6.3 Manufacturer-Defined IDNs

In order to avoid restricting drive/control development, the IEC standard defines 32,767 vendor-defined IDNs that can be used by individual vendors to incorporate special features in their products. The standard provides that these vendor-defined IDNs enable *only* additional functionality not already handled by standard functions, thus ensuring interoperability.

One of the values of the vendor-defined IDNs has proven to be their use in incorporating new features into Sercos. A number of vendors have utilized the vendor-defined IDNs to introduce and field-prove some new control/drive feature, such as block mode in which the controller sends the drive a final destination, a velocity, a ramp, and possibly a jerk command, and then the drive operates independently until the destination is reached. Many of these features, including block mode, have been brought to the Sercos TWGs, where they have been incorporated into the standard IDN set for use in a standardized manner by all manufacturers.

13.3.2.7 I/O Functions

A set of I/O functions was defined and accepted as a part of Sercos II, but as Sercos was originally developed as a drive interface, these were not originally included in the formal IEC standard. A document describing these functions and parameters is available from the Sercos user organizations.

On the other hand, Sercos III incorporates a detailed and powerful set of I/O functions, as defined in the IEC standards.

13.3.2.8 System of Units and Variable Format

Wherever possible, tables of acceptable units such as revolutions, inches, and millimeters are defined for standardized telegrams. The format of the byte values is also defined. One of the more valuable results of the standards effort was the application of a predictable system of units for up-to-then unstandardized values, such as loop gains and feedback values. This is another step toward predictable operation of any servo system a machine builder may choose.

13.3.2.9 Sercos Cycle Times

The Sercos cycle time is specified in a flexible format of 62.5, 125, 250, 500 μs, and then multiples of 1 ms. (Sercos III cycle times begin at 31.25 μs.) The amount and type of data contained in a cycle are also variable.

This flexibility permits a designer to vary cycle time, content, and the number of slaves to achieve a particular project's requirements. More data can be sent faster to a smaller number of slaves, such as drives. Slowing the rate down permits a higher density of drives per ring. For example, with Sercos II, four to eight axes are common on a ring in machine tool applications where intense communication is required to control a high-speed tool path. Up to 40 axes may be controlled on a ring in packaging applications, and up to 100 axes may be found in web-fed printing applications, where electronic line shafting is utilized to synchronize a number of print cylinders and auxiliary axes. When necessary, multiple rings can be employed in an application.

13.3.2.10 Cyclic Operation

Sercos devices communicate and stay in sync by sending each other a series of telegrams. A telegram is a rigidly defined bit steam carrying data and timing information. Telegrams are

Master Synchronization Telegram (MST)—establishes the timing for the system
AT—response by the slaves to the MST
MDT—provides data records for all slaves in the loop

Note: In Sercos III, the MDT and MST functions are combined into a single telegram. (See Section 13.2 for details of operation.)

13.3.2.10.1 Telegram Format

All Sercos telegrams consist of five major fields, as illustrated earlier. Telegrams are transmitted in the non-return-to-zero-inverted format. The transmitted signal remains at a 0 or 1 as long as the bit is a 0 or 1. In order to ensure edge synchronization and to prevent the delimiter pattern from reoccurring, the master forces a signal change every six bits by utilizing a *bit stuffing* technique of inserting a zero after five consecutive ones (Figure 13.26).

Administrative		Command and feedback	Administrative	
Delimiter	Address	Data	Error checking	Delimiter
01111110	--------	-------- --------	--------	**01111110**

FIGURE 13.26 Standard telegram format.

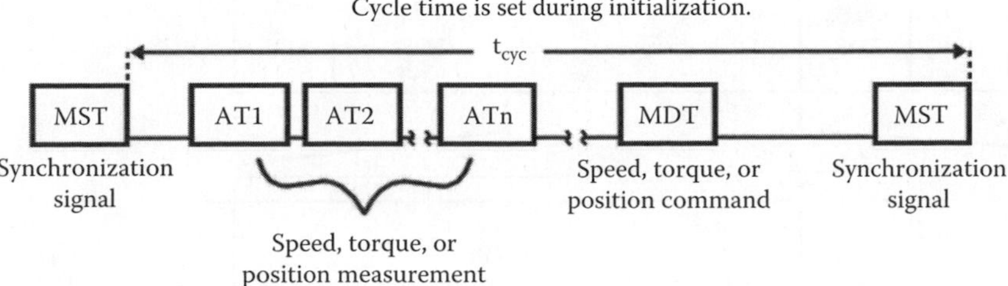

FIGURE 13.27 Timing diagram for cyclic operation.

13.3.2.10.2 Telegram Sequence

A Sercos II cycle involves three different types of telegrams, as illustrated in Figure 13.27.

1. It begins with the transmission of an MST from the master (CNC or motion control). The MST is used as a time mark for all slaves (e.g., drives) to determine when to talk on the bus, when to acquire feedback signals, and so forth.
2. At a predetermined time after the end of the MST, the first slave in the system places its data on the bus in an AT. Each slave follows in turn, all synchronized off the MST. During a Sercos initialization phase, the slaves are instructed when they should transmit their message with respect to the MST
3. After the last slave has placed its data on the bus, the master sends out an MDT. The MDT is one long message with space set aside for each slave in the ring. The slaves have been previously instructed where their data are located within the MDT. As the MDT is received by a slave, it fast-forwards to the start location for its information and retrieves the data.

After the MDT is sent, another MST is transmitted by the master control, signaling the beginning of another cycle, as illustrated earlier.

13.3.2.10.3 Master Synchronization Telegram

The Sercos master begins a communication cycle by transmitting an MST, as detailed in Figure 13.28. The MST sets the timing and synchronization for the network. Each slave device (drive or I/O) synchronizes its clock from the MST and then uses this timing to calculate when to acquire feedback signals and other processing.

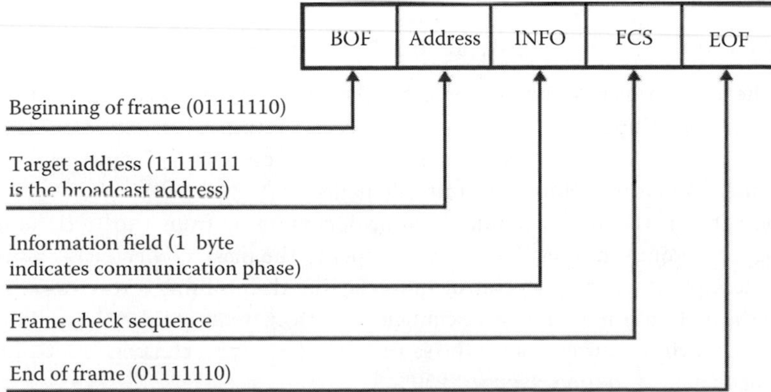

FIGURE 13.28 Master synchronization telegram structure.

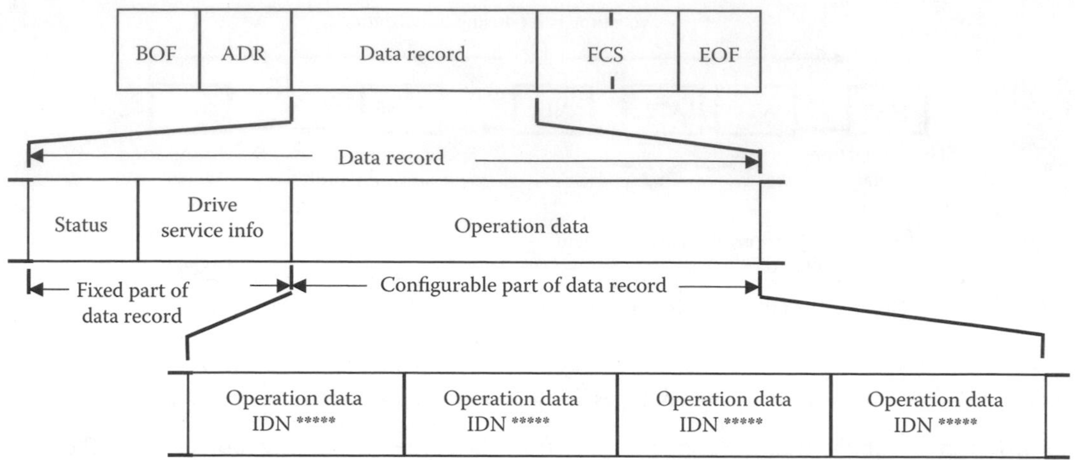

FIGURE 13.29 Acknowledge telegram structure.

The MST data field also contains information. Upon startup, the Sercos loops go through an extensive initialization procedure, where drives are assigned an operating mode (speed, torque, or position) and the exact configuration of their cyclic data. They are also told to synchronize and phase-lock their internal loop-closure frequency, eliminating the risk of harmonics. Cyclic communication doesn't begin until this initialization is complete. The MST data field indicates whether the loop is in one of the initialization phases or in the communication cycle.

13.3.2.10.4 Acknowledge Telegram

Slaves respond to the MSTs via an AT sent by and returned to the master. At a predetermined time after the MST, the first slave in the system places its data in its predetermined slot. Each slave follows in succession, all synchronized off the MST.

An AT is composed of five main fields (Figure 13.29):

1. Beginning of frame
2. Drive address (ADR)
3. Data record
4. Frame check sequence
5. End of frame

The data record is composed of both fixed and variable data in three fields:

1. Status—The 8-bit status field indicates whether the slave (e.g., drive) is ready and verifies that it is in the correct operating mode.
2. Drive service info—This 2-byte service channel field contains non-time-critical data such as torque limits, travel limits, time constants, and gains.
3. Operation data—This is the most important field, containing from 1 to 16 IDNs reporting data such as speed, torque, and position measurements to the master, effectively closing the control loop. For example, if a drive is operating in velocity mode, meaning it receives velocity information from the motion control, it may be configured with current actual velocity in the AT. Because all slaves take their measurements at the same time, the control can construct an instantaneous snapshot of all axes of motion (Figure 13.30).

13.3.2.10.5 Master Data Telegram

The MDT is transmitted by the master after the AT. It consists of one long message, with space allocated for each slave on the ring, as shown in Figure 13.31. All slaves have previously been instructed where

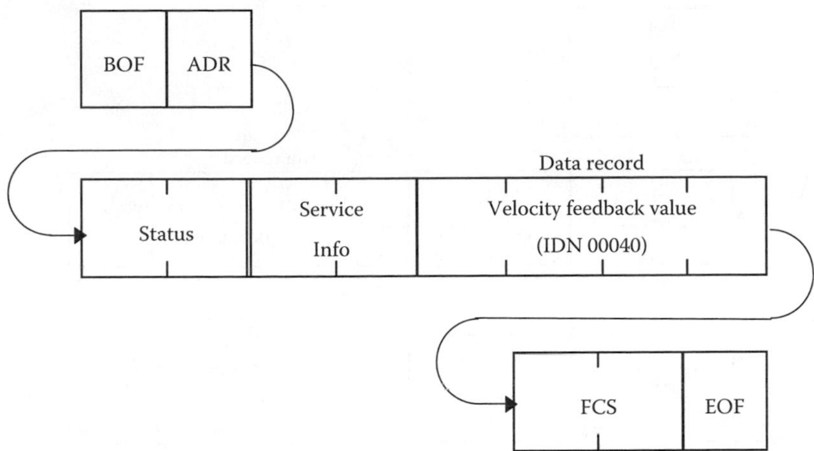

FIGURE 13.30 Acknowledge telegram example.

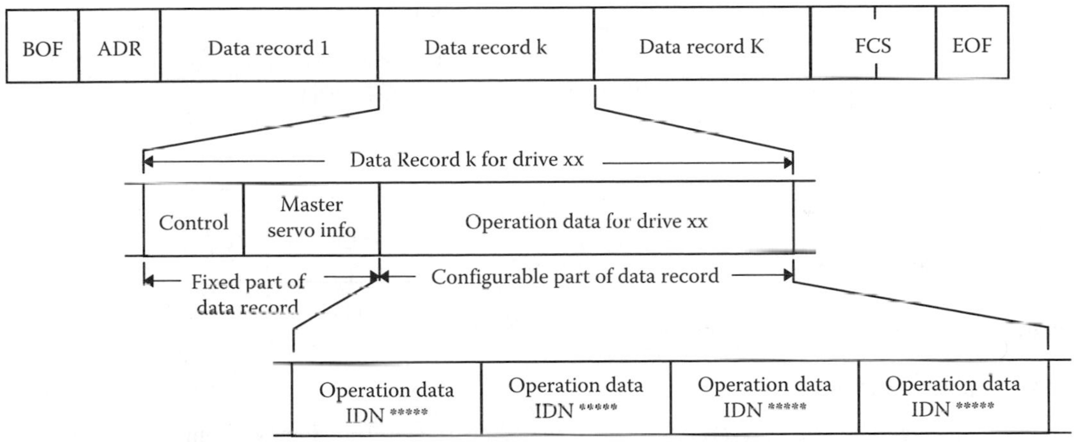

FIGURE 13.31 Structure of the master data telegram.

their data are located within the MDT. As the MDT is received by a slave, it fast-forwards to the starting location for its information and retrieves it.

The MDT is structured much the same as the AT, except that the data field includes a record for each slave on the ring. Like the AT, each data record in the MDT consists of a fixed and variable data portion in three fields:

1. Control field—This 8-bit field enables or disables the slave and configures the service channel.
2. Master servo info—Implements setup parameters and special functions, such as homing, probing, and coordinate offsets.
3. Operation data for slave xx—This field contains a series of IDNs (idents) specifying RT operating data and drive commands for the addressed slave. The chosen idents depend on the application. Using the previous example, in velocity mode, this data record would contain the velocity commands for the drives.

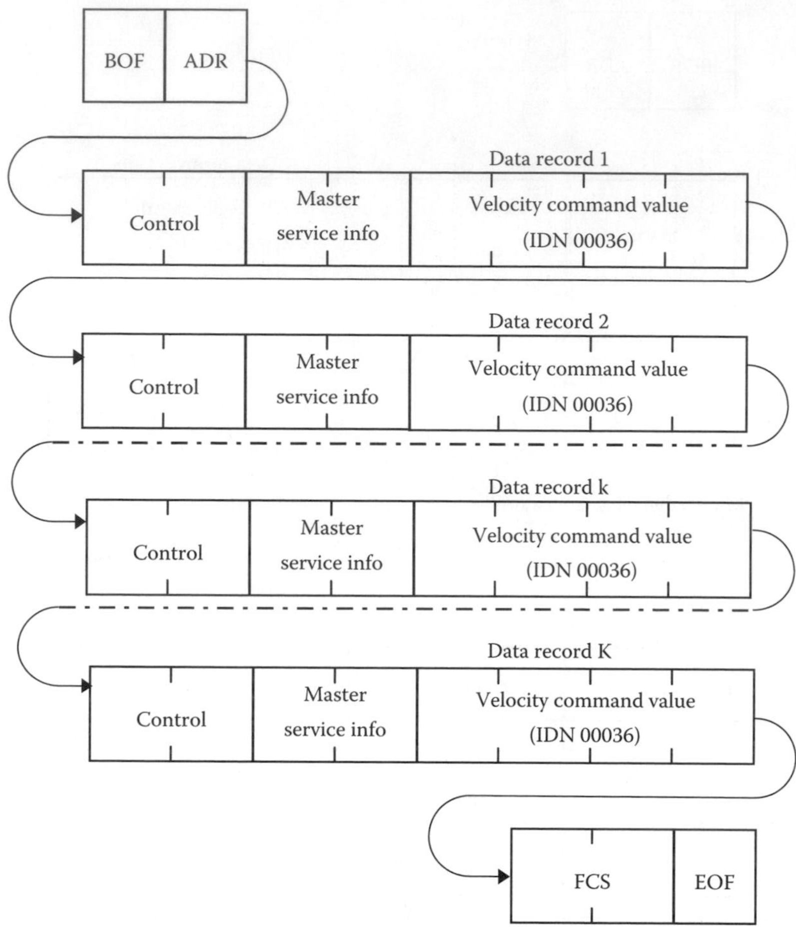

FIGURE 13.32 Example of a master data telegram.

13.3.2.10.6 Service Channel

The service channel allows high-level, non-time-critical data transfers without disturbing the synchronous transfer of data. For example, items such as diagnostics and loop gains do not need to be transmitted repeatedly, so it is not necessary to impact loop performance by setting aside bandwidth for them. Long messages are broken into 2-byte chunks, sent over the service channel in the AT, and reassembled on the other end. Via the service channel, operators can retrieve diagnostic and status information, reprogram or display setup parameters, and change operating data on the fly.

Some manufacturer's drives include a software oscilloscope function that captures data on drive performance, sends the data over the service channel 2 bytes at a time, and formats the information into an oscilloscope-type display on the operating terminal.

13.3.2.10.7 Example of Cyclic Operation: Sercos II

The best way to imagine how a system works is to take a real-world example such as a system with 14 drives and a 500-μs cycle rate using Sercos II. Under these conditions, the following information can be transferred every cycle:

From control to each drive

- 32-bit command value (e.g., velocity or position)
- 16-bit limit value (e.g., torque)

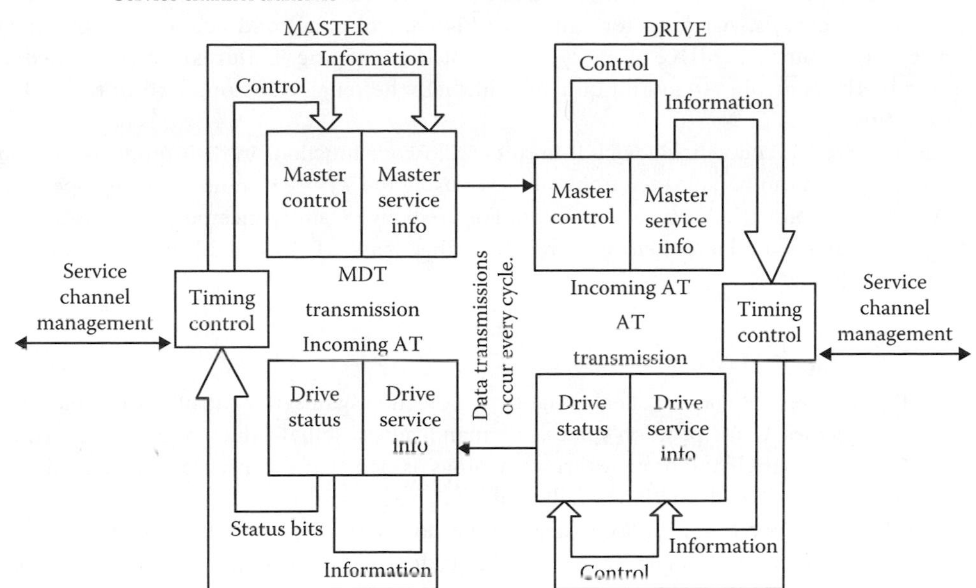

FIGURE 13.33 Service channel operation.

From each drive to control

- 32-bit feedback value (e.g., velocity or position)
- 16-bit feedback value (e.g., torque)

This means that every 500 µs, one Sercos ring can transport a very high-resolution velocity command signal to 14 drives, plus a torque limit command to each of those drives that can be enabled at will. In addition, each cycle is returning two feedback values from each drive. This reduces the customary feedback wiring to the control to nothing.

In addition to all these RT data being exchanged, the service channel provides the equivalent of a 128-kbaud serial port to *each and every drive on the ring*. This can be used to set and archive performance variables, to read diagnostics, etc.

Note that this example does not state that Sercos performance is limited to 500 µs or 14 drives. Neither is the case. Flexibility exists in the update rate (down to 62.5 ms—31.25 ms for Sercos III), the number of drives per ring (up to 511), the amount of data exchanged, and even the number of rings. Performance for Sercos III is even better, as shown in Figure 13.17.

13.3.3 Error Correction/Diagnostics

If a Sercos loop detects an error, it continues to operate for one cycle using previously valid data. However, if a second consecutive error is detected, the slaves are shut down in a predefined, orderly fashion, and a diagnostic message is issued.

Sercos offers extensive diagnostic reporting and produces detailed reports on fault conditions. For example, IDNs 00200 through 00205 report drive and motor temperature warning and shutdown conditions, IDN 00309 indicates spindle synchronization error, IDN 00323 specifies target position outside travel range, and there are many more.

This allows vendors to incorporate detailed context-sensitive troubleshooting procedures in their motion controls. An operator can go to his control panel and acquire instant diagnosis of every axis motor and drive at every section of the machine. Failures can be immediately pinpointed by the control system and quickly and easily be repaired as opposed to hit-and-miss troubleshooting.

A number of Sercos vendors include built-in soft oscilloscope functions in their products to simplify troubleshooting. Data can be uploaded from the slaves using the service channel, allowing operators to display and analyze torque, velocity, and position. For predictive maintenance purposes, a snapshot of operating status can be saved and then compared to another snapshot taken at a later time to determine operation/failure trends.

13.3.4 System Safety

Digital intelligent drives with Sercos offer outstanding protection against uncontrolled drive movements and excessive velocities. Using position values, command values, actual values, and drive parameters, a Sercos drive monitors itself and provides error warnings or error shutdowns, including forced orderly shutdown in the event of a malfunction or failure of the drive processor.

Excessive velocities, overtravel, or axis runaway due to faulty or incorrectly transmitted position commands can be eliminated by logical monitoring by the drive of the values it receives from the control. Safety redundancy can be achieved by the control as it monitors the actual value data fed back from the drive via Sercos. This internal monitoring can ensure safe shutdown in the event of a communications failure.

13.3.5 Sercos II Conformance Testing

Conformance testing of products carrying the Sercos logo is required! Any manufacturer can purchase the IEC standards and design controls and drives according to the standard. However, the Sercos logo is the property of Sercos International e.V. and may be used only on Sercos products that have passed a conformance test. This is important to ensure customers of the compatibility and interoperability of products from various manufacturers.

Conformance testing of Sercos products is done at the Institute for Control Engineering of Machine Tools and Manufacturing Units (ISW) at the University of Stuttgart. The intent of the conformance test is to ensure the compatibility and interoperability of devices from different vendors in multiple-vendor environments.

Manufacturers seeking conformance certification for their products submit their Sercos products to ISW for testing. Once a product has passed the conformance test, the manufacturer can apply to Sercos International for a conformance certificate and can apply the Sercos logo to that product.

13.3.5.1 Conformance Test Environment

The conformance test environment for Sercos II devices consists of Sercos II Conformizer test software running on a standard PC with an enhanced Sercos II card, an optical analyzer, and a digital oscilloscope. There are three facets to the testing:

1. Logical test—Per the script files in the Sercos II Conformizer package
2. Physical test—Measures optical signals
3. Time measurement—Handled by special test hardware

The Sercos II Conformizer is an easy-to-use package that includes a high-level scripting language with control flow statements and over 80 functions, variables, and input/output programming features. It includes a command shell, parameter browser, script editor, function browser, script browser, protocol window, and error code browser. The test results can be directly processed in RTF or HTML format. The Conformizer software runs under Windows NT, 2000, and XP.

The Sercos II Conformizer test software is available for purchase by any Sercos vendor who wishes to pretest his products in his own facilities prior to submitting them for the conformance test. This provides a 90% assurance of passing the formal test.

The Conformizer is sold as a complete test and development environment for Sercos master and slave devices and includes a passive PCI-Sercos card. It operates under the VenturCom RTX system, running Windows NT/2000/XP. A runtime license for VenturCom RTX 5.5 is included with the package.

13.3.6 Sercos II in Machine Tool Applications

Sercos was originally developed to meet the needs of the machine tool industry, thus it incorporates a rich set of parameters for machine tool functions, such as probing, backlash compensation, feed forward, lead screw error compensation, adaptive positioning, move to positive stop, and torque monitoring. The interface provides all parameters necessary for control of machine tool axes, including spindle axis control.

13.3.6.1 Application Note: Contouring Control System*

andron GmbH from Wasserburg, Germany, has been using Sercos as a digital drive interface on its control units since 1992. The andronic 2060 is currently regarded as one of the fastest contouring control systems for use in complex multiaxis contouring interpolation applications.

In contrast to the general rule that speed decreases as the number of axes increases, block change times on this controller remain below 250 μs even with 16 interpolating axes. At a transmission speed of 16 Mbps, the system delivers a guaranteed cycle time of 250 μs regardless of the number of axes present in the ring. This means that all axes that are simultaneously involved in an interpolation receive new position data every 250 μs. In combination with high-performance position controllers in the servo amplifier (position control cycle <250 μs), the system can deliver high path velocities (even at high resolution of, e.g., 1/10 μm) as well as high precision. The stability and predictability of Sercos is indispensable, as any timing error or even a transmission repetition can have fatal consequences at the workpiece.

Tool grinding machines (for complete machining of drills and cutters) can have up to 11 axes. In practice, this normally means simultaneous interpolation of five axes. With two Sercos rings, the cycle time is 500 or 250 μs at machining speeds of <1 m/min, a resolution of <1/10 μm, and a precision of 1 μm.

Similar performance can be achieved during camshaft grinding and grinding of knives and scissors.

For HSC milling applications using Cartesian machines and parallel kinematics machines (Figure 13.34), high path speeds in combination with high contour accuracy and finish quality are required. Processing speeds of 30 m/min at 1/10 μm resolutions and accelerations of up to 8 g ares common.

During gearwheel honing operations, the machines must meet stringent requirements for synchronization of several drives in the start phase and during workpiece machining. This means axis synchronization in the position controller with a cycle time of 500 μs and a position control cycle rate of 250 μs using eight machining axes, five of which are interpolating.

13.4 Future Technical Advancements

Updates to the Sercos specifications' are defined by multicompany Technical Working Groups in both North America and Europe, which ensure that the latest enhancements proposed by member companies are incorporated into the specifications.

* Application note reprinted courtesy of andron GmbH and Sercos International e.V.

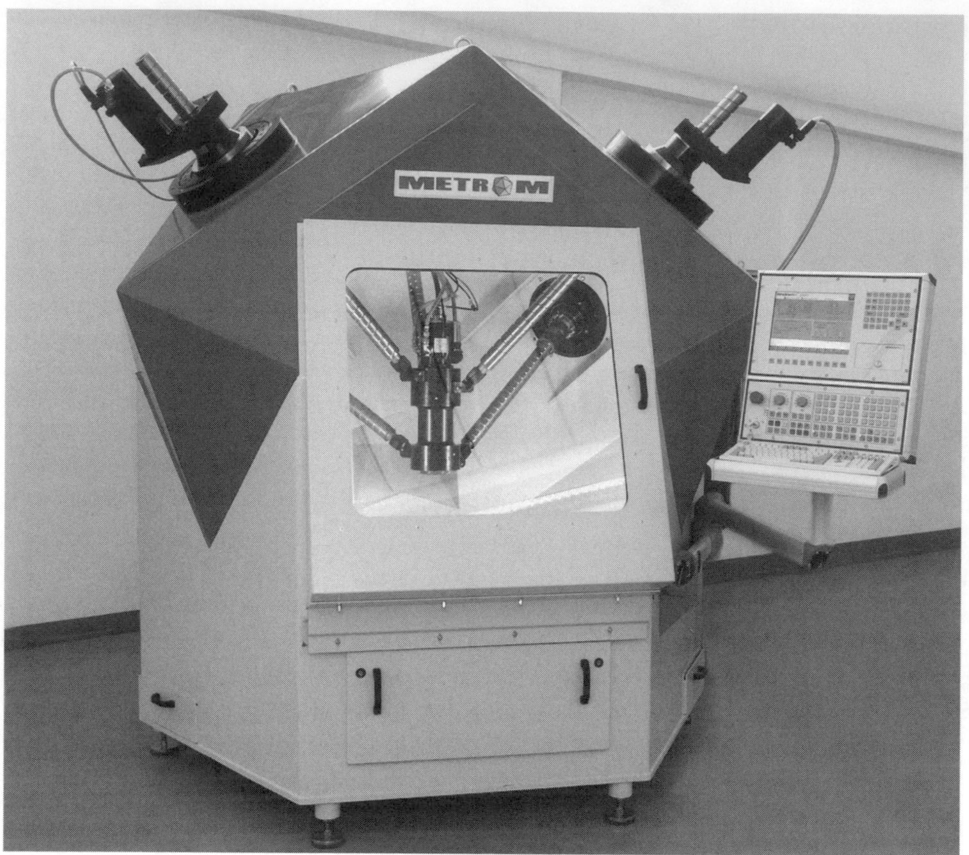

FIGURE 13.34 Parallel Kinematics Machine driven by Andronic 2060 Control.

Acknowledgments

The authors gratefully acknowledge input for this chapter from various members of the Sercos Trade Associations.

This document was prepared using published materials (see footnotes and additional list provided later) and information from the Sercos N.A., Sercos International e.V. and various Sercos vendor websites. Information is believed to be correct, but is subject to change without notice. Please address any comments to info@Sercos.com.

References

Campbell, K., Hershey finds sercos best path to enterprise-wide compatibility, *Instrumentation and Control Systems*, January 1999.

Hibbard, S., Open drive interfaces for advanced machining concepts, Bosch Rexroth, January 1995.

Sources for More Information

Sercos Standards

IEC 61800-7(–1, –204, –304): Sercos Drive Profile
IEC 61784-1, IEC 61158-2, IEC 61158-3/4/5/6-16: Sercos I/II Communication Profile
IEC 61784-2, IEC 61158-2, IEC 61158-3/4/5/6-19: Sercos III Communication Profile
IEC 61784-5-16 Ed 2.0: Sercos Installation Profile
IEC 61784-3-2: CIP Safety on Sercos

- Available from Global Engineering Documents, a source for technical standards—http//global.his.com.
- Available from IEC—International Electrotechnical Commission in Switzerland—www.iec.ch

Support Organizations

A number of support and development tools for the Sercos are available, both from the Sercos trade associations and from various vendors. See the Sercos association websites or contact the organization by phone or e-mail.

Sercos International e.V.—www.sercos.de
Kueblerstrasse 1
73079 Suessen, Germany
Phone: +49 7162-9468-65, Fax: -66
Mobile: +49 171 4041028
e-mail: p.lutz@sercos.de

Sercos North America—www.sercos.com
405 Loblolly Bay Drive
Santa Rosa Beach, FL 32459
Tel: 800/573-7267 or 850/660-1293
Fax: 850/660-1293
e-mail: info@sercos.com

Sercos China—www.sercos.org.cn
Building No. 1 #414, No. 1 Jiao Chang Kou Street,
De Sheng Men Wai, Xi Cheng District,
Beijing 100011, China
Tel: +86 (10) 62015642
Fax: +86 (10) 82015862
e-mail: info@sercos.org.cn

Sercos Japan—www.sercos.org
Lilas Nogizaka Bldg. #901,
Minami Aoyama 1-15-18, Minato-ku
Tokyo, 107-0062, Japan
Tel: +81 3 3470 0640
Fax: +81 3 3478 8648
e-mail: info-japan@sercos.com

14

FOUNDATION Fieldbus

14.1 Introduction 14-1
14.2 Technical Description of FOUNDATION Fieldbus 14-2
H1 FOUNDATION Fieldbus • HSE FOUNDATION Fieldbus • H1 and HSE FOUNDATION Fieldbus User Application Layer
14.3 Wireless Solutions for FOUNDATION Fieldbus 14-15
*Wireless*HART Integration • ISA 100.11a Integration Wireless HSE Backhaul Team • FOUNDATION for Remote Operations Management
14.4 Open Systems Implementation 14-16
Electronic Device Description Language • Field Device Tool • EDDL versus FDT: Field Device Integration
14.5 Conclusions 14-20
References 14-20

Salvatore Cavalieri
University of Catania

14.1 Introduction

FOUNDATION Fieldbus is an all-digital, serial, two-way communication system for industrial applications; its specification has been developed by the not-for-profit Fieldbus Foundation association established in 1994 [1]. Please note that in the following, Fieldbus Foundation will refer to the name of the association while FOUNDATION Fieldbus (FF) will be used for the communication system.

Since the first fieldbus communication systems [2–4] have appeared on the market, the need to achieve just only one fieldbus standard was felt immediately. More than 20 years ago, the International Electrotechnical Commission (IEC) and Instrument Society of America (ISA) embarked on a joint standardization effort identified by two codes: 61158 on the IEC side and SP50 on the ISA one. One of the main aims of the joint standardization committee was the definition of a unique communication system able to merge the main features of the fieldbuses available on the market at that time (e.g., WorldFIP [5] and Profibus [6]). This aim was partially realized by the definition of the IEC/ISA specifications (among which the reader may refer to [7–12]); these specifications do not define a unique standard, but they put under the same umbrella different fieldbus systems, called subtypes in the IEC 61158 specifications.

Among the different fieldbus systems integrated into the common IEC 61158 specifications, there are the two communication specifications produced by the Fieldbus Foundation; they are called H1FOUNDATION Fieldbus (included in the IEC 61158 as type 1, for the physical and data link layer definitions and type 9 for the application layer definition) [8–13] and HSE FOUNDATION Fieldbus (included in the IEC 61158 as type 5 for the application layer definition) [11,12].

H1 FF was the first communication system defined by the Fieldbus Foundation; since its definition, H1 FF has shown two fundamental and novel (at that time, at least) features: an emphasis on the standardization of the description of the fieldbus devices and the adoption of both the distributed and centralized link access mechanisms [4].

During the time, FF has been enriched by new and useful features. The first of them is related to the transmission medium; besides the original IEC/ISA physical medium [7], Fieldbus Foundation decided to adopt also the high-speed Ethernet (HSE) and to integrate FF into a wireless environment. Another important feature is the improvement of interoperability using other standards different from the original one.

This chapter will point out the main (original and novel) technical features of the FF communication system, allowing the reader to clearly understand its key points. In particular, Section 14.1 will describe the main features of the H1 and HSE FFs. Section 14.2 will point out the current wireless extensions; finally, Section 14.3 will describe how implementation of interoperability in FF changed during these last years, pointing out the most recent solutions adopted by Fieldbus Foundation.

14.2 Technical Description of Foundation Fieldbus

As said in the Introduction, FF specifications include two different configurations: H1 and HSE.

H1 FF communication system is mainly devoted to distributed continuous process control; running at 31.25 Kbit/s, it interconnects digital *field* equipment such as sensors, actuators, and I/Os, allowing to integrate existing 4–20 mA devices. H1 FF communication stack is minimal, in order to guarantee maximum speed in the data handling. Below the Fieldbus application layer, H1 FF directly presents the data link layer (DLL) managing access to the communication channel. A physical layer deals with the problem of interfacing with the physical medium. A network and system management layer is also present. Figure 14.1 compares the H1 FF architecture against the ISO OSI reference model.

HSE FF is mainly foreseen for discrete manufacturing applications; it provides integration of controllers (e.g., distributed control systems [DCSs] and programmable logic controllers [PLCs]), H1 subsystems, foreign protocols (e.g., Modbus and Profibus DP [decentralized peripherals]), conventional input/output (I/O) (e.g., digital I/O, 4–20 mA), data servers, and workstations. HSE FF defines an application layer and associated management functions, designed to operate over a standard Transport Control Protocol (TCP)/Unit Datagram Protocol (UDP)/Internet Protocol (IP) stack over twisted-pair or fiber-optic switched Ethernet. Figure 14.2 compares the HSE FF architecture against the ISO OSI reference model.

Looking at Figures 14.1 and 14.2, it becomes evident that the Fieldbus Foundation has specified only one user application layer, for both the H1 and HSE configurations. The H1 and HSE FF user application layer is mainly based on function blocks (FBs), providing a consistent definition of inputs and outputs that allow seamless distribution and integration of functionality from various vendors.

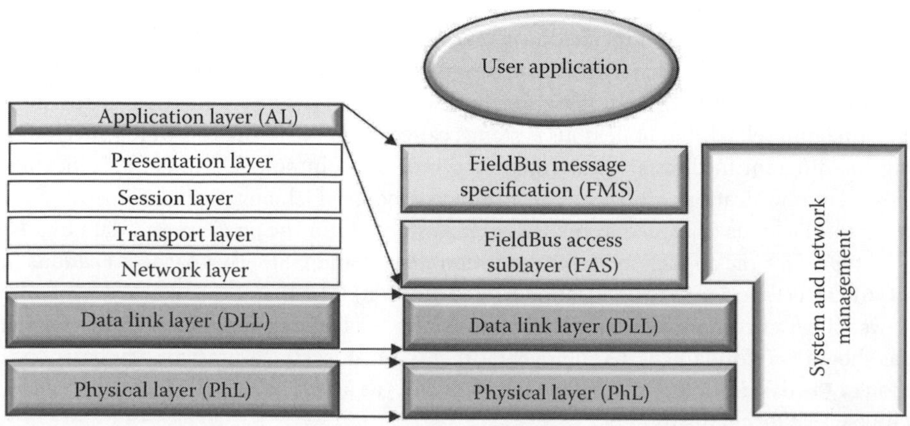

FIGURE 14.1 H1 FF vs. ISO OSI architecture.

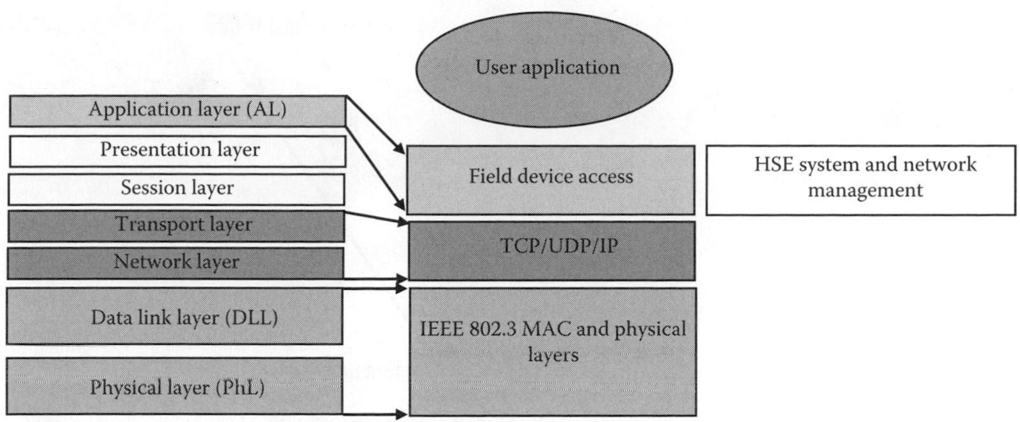

FIGURE 14.2 HSE FF vs. ISO OSI architecture.

14.2.1 H1 FOUNDATION Fieldbus

H1 FF is made up by different layers (as shown in Figure 14.1), whose functionalities will be described in the following.

14.2.1.1 H1 FOUNDATION Fieldbus Physical Layer

H1FF physical layer is the 31.25 KBd version of IEC(type 1)/ISA Fieldbus [7,8,14]. Signals (±10 mA on 50 Ω load) are encoded using the synchronous Manchester Biphase-L technique and can be conveyed on low-cost twisted pair cables. The signal is called "synchronous serial" because the clock information is embedded in the serial data stream. Data are combined with the clock signal while creating the fieldbus signal. The receiver of the fieldbus signal interprets a positive transition in the middle of a bit time as a logical "0" and a negative transition as a logical "1." Special codes are defined for the preamble, start delimiter, and end delimiter. Special N+ and N− characters are used in the start delimiter and end delimiter. Note that the N+ and N− signals do not have a transition in the middle of a bit time.

The transmitting device delivers ±10 mA at 31.25 Kbit/s into a 50 Ω equivalent load to create a 1.0 V peak-to-peak voltage modulated on top of the direct current (DC) supply voltage. The DC supply voltage can range from 9 to 32 V.

H1 FF supports intrinsically safe (I.S.) feature for bus-powered devices. To accomplish this, an I.S. barrier is placed between the power supply in the safe area and the I.S. device in the hazardous area. In the case of I.S. applications, the allowed power supply voltage depends on the barrier rating.

H1 FF wiring is based on trunk cables featuring terminators installed at each end; up to five trunks can be connected in a Fieldbus link by means of repeaters. Spurs may be located anywhere along each trunk, and connected to the trunk by junction box, as shown in Figure 14.3; a single device can be connected by each spur. Existence of spurs allows that a 31.25 Kbit/s devices can operate on wiring previously used for 4–20 mA devices [15,16]. The length of each spur varies from 1 m up to 120 m depending on the number of devices connected to the Fieldbus link.

The maximum number of devices on a Fieldbus link is 32; the actual number depends on factors such as the power consumption of each device, the type of cable, the use of repeaters, etc. In particular, the maximum number of devices is usually 6 for I.S. applications and power delivered through the bus [17], 12 for non-I.S. applications but power still delivered through the bus, and 32 for non-I.S. applications without any power delivered through the bus.

14.2.1.2 H1 FOUNDATION Fieldbus Data Link Layer

H1 FF DLL controls the access to the bus for the transmission of messages; it is based on the idea to merge the two different access paradigms: circulated token and scheduled access [4]. Thus, H1 FF makes available an

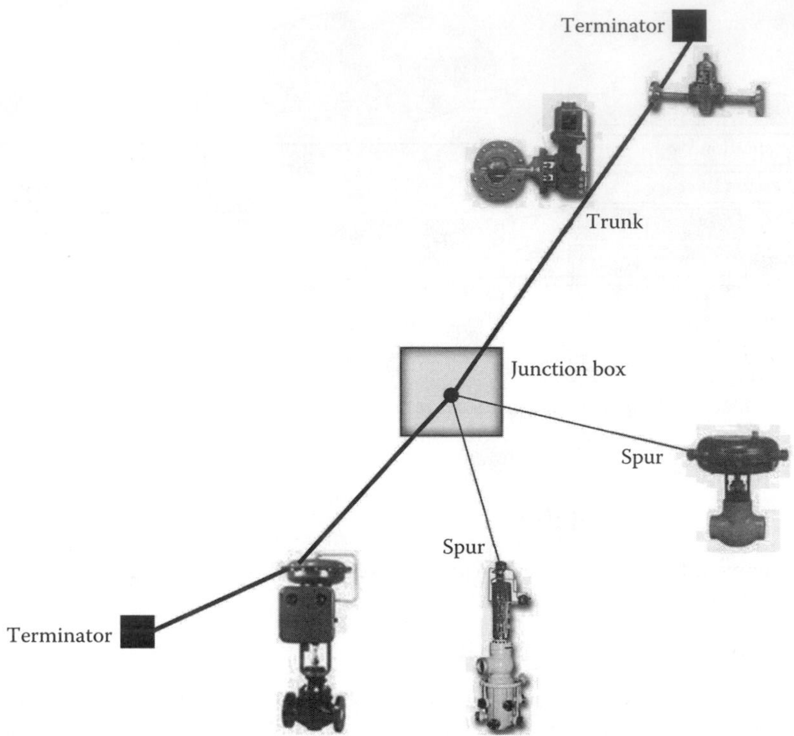

FIGURE 14.3 Trunk, junction box, and spurs in H1 FF.

overall schedule able to guarantee the needed data at the needed time, but also allowing gaps within which a circulated token mechanism can take place while complying with a defined maximum rotation time.

Such a philosophy clearly needs an arbitrator that univocally imposes the transmission of a defined data at a defined time by a defined entity, when so required, but also guarantees a defined minimum amount of free time to each entity. This arbitrator is called link active scheduler (LAS) within IEC(type 1)/ISA DLL and H1 FF DLL [9,10,18,19].

14.2.1.2.1 Scheduled Communication

The way the LAS manages the scheduled access is based on the following mechanism. The LAS has a list of transmitting times for all data buffers in all devices that need to be cyclically transmitted. When it is time for a device to send contents of a buffer, the LAS issues a compel data (CD) message to the device.

Upon the receipt of the CD, the device broadcasts or *publishes* the data item (DT) in the buffer to all devices on the fieldbus. Any device configured to receive the data is called a "subscriber." Figure 14.4 shows this access mechanism.

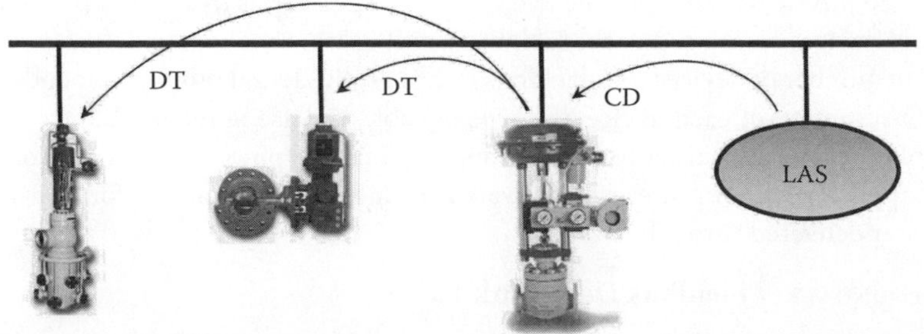

FIGURE 14.4 Scheduled access mechanism.

Scheduled data transfers are typically used for the regular, cyclic transfer of control loop data between devices on the fieldbus.

14.2.1.2.2 Unscheduled Communication

The federal autonomy of each device is given by a bandwidth distribution mechanism based on the use of a circulating token. In unused portions of the bandwidth (i.e., not occupied by the transmission of CDs), the LAS sends a Pass Token (PT) to each device included in a particular list called "live list" (described in the following). Each token is associated with a maximum utilization interval, during which the receiving device can use the available bandwidth to transmit what it needs. On the expiration of the time interval or when the device completes its transmissions, the token is returned to the LAS by using another frame called return token (RT). A target token rotation time (TTRT) defines the interval time desired for each token rotation. The value to be assigned to this parameter is linked to the maximum admissible delay in the transmission of the unscheduled information flow. Figure 14.5 shows the token circulation managed by the LAS.

14.2.1.2.3 Live List Maintenance

The list of all devices that are properly responding to the PT is called the *live list*. New devices may be added to the fieldbus at any time. The LAS periodically sends probe node (PN) messages to all the addresses not yet present in the live list. If a new device appears at an address and receives the PN, it immediately returns a probe response (PR) message. When a device returns a PR, the LAS adds the device to the live list and confirms its addition by sending the device a node activation message.

The LAS is required to probe at least one address after it has completed a cycle of sending PTs to all the devices in the live list.

The device will remain in the live list as long as it responds properly to the PTs sent from the LAS. The LAS will remove a device from the live list if the device does not reply to the relevant PT for three successive tries.

14.2.1.2.4 Data Link Time Synchronization

A DLL time synchronization mechanism is provided so that any device can request the LAS for a scheduled action to be executed at a defined *time* that represents the same absolute instant for all the devices. The LAS periodically broadcasts a time distribution (TD) message on the fieldbus so that all devices have exactly the same data link time. This is important because scheduled communications on the fieldbus and scheduled function block (FB) executions in the user application layer are based on timing derived from these messages.

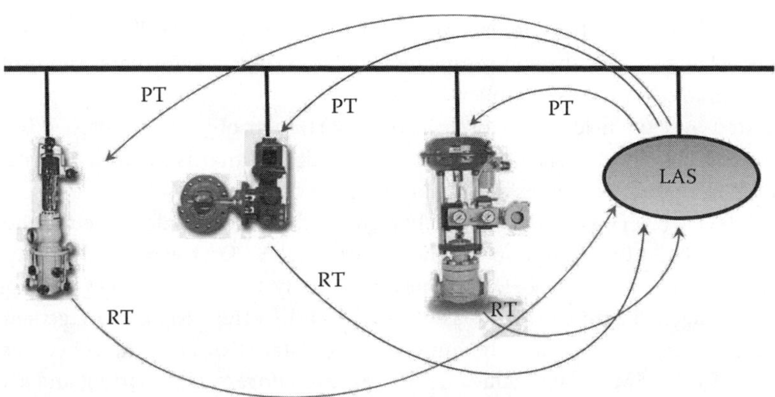

FIGURE 14.5 Token passing mechanism.

14.2.1.3 H1 Foundation Fieldbus Application Layer

The H1 FF application layer includes two sublayers: Fieldbus access sublayer (FAS) and Fieldbus message specification (FMS) [20,21].

14.2.1.3.1 Fieldbus Access Sublayer

FAS uses both the scheduled and unscheduled features of the DLL to provide services for the FMS [20].

The type of each FAS services is described by a virtual communication relationship (VCR). VCR defines the main features of the information exchange between two applications; among them, maybe the number of receivers (one or many) for each transmitter, memory organization (queue or buffer) used to store the messages to be sent/received, and the DLL mechanism used to send the message (PT or CD). The types of VCR defined by Fieldbus Foundation are as follows:

- *Client/server VCR type.* The client/server VCR type is used for queued, unscheduled, user-initiated, one-to-one communications between devices on the fieldbus. "Queued" means that messages are sent and received in the order submitted for transmission, according to their priority, without overwriting previous messages. When a device receives a PT from the LAS, it may send a request message to another device on the fieldbus. The requester is called the client, and the device that receives the request is called the server. The server sends the response when it receives a PT from the LAS. The client/server VCR type is used for operator-initiated requests such as setpoint changes, access to and change of a tuning parameter, alarm acknowledgment, and device upload and download.
- *Report distribution VCR type.* The report distribution VCR type is used for queued, unscheduled, user-initiated, one-to-many communications. When a device, holding an event or a trend report to send, receives a PT from the LAS, it sends its message to a group address defined by its VCR. Devices that are configured to listen to that VCR will receive the report. The report distribution VCR type is normally used by fieldbus devices to send alarm notifications to the operator consoles.
- *Publisher/subscriber VCR type.* The publisher/subscriber VCR type is used for buffered, one-to-many communications. "Buffered" means that only the latest version of the data is maintained within the network. New data completely overwrite previous data. When a device receives CD, the device will publish (broadcast) its message to all devices on the fieldbus. Devices that wish to receive the published message are called subscribers. An attribute of the VCR indicates which method is used. The publisher/subscriber VCR type is normally used by the field devices for cyclic, scheduled publishing of user application FB inputs and outputs.

14.2.1.3.2 Fieldbus Message Specification

FMS services allow user applications to exchange messages across the fieldbus using a standard set of message formats. FMS describes the communication services, message formats, and protocol behavior needed to build messages for the user application [21].

Data transmitted over the fieldbus are described by particular objects maintained by an object dictionary (OD) and called object descriptions. Each object description is identified by its index in OD; index 0, called the OD header, provides a description of the dictionary itself.

A virtual field device (VFD) is used to remotely view local device data described in the OD; the VFD are accessed using VCRs. A typical device will have at least two VFDs: the network and system management VFD and the user application VFD. The network and system management VFD provides access to the network management information base (NMIB) and to the system management information base (SMIB). NMIB data include VCRs, dynamic variables, statistics, and LAS schedules (if the device may assume an LAS role); SMIB data include device tag and address information, and schedules for FB execution (see Section 14.1.3.4). The user application VFD is used to make the device functions visible to the fieldbus communication system.

FMS communication services provide a standard way for user applications to communicate over the fieldbus. Specific FMS communication services are defined for each object type. The list of the communication services available can be found in [21,22]. Detailed description of each service is given in [21].

14.2.1.4 H1 FOUNDATION Fieldbus System Management

Inside H1 FF specification, system management handles important system features ([23]) such as the following:

- FB *scheduling*. FBs on which a user application layer is built (as it will be explained in Section 14.1.3) must often be executed in a particular device at precisely defined intervals and in the proper sequence for correct control system operation. System management of the relevant device synchronizes the execution of the FBs to a common time clock shared by all devices. The execution of the FBs are scheduled inside a so-called macrocycle, which contains the precise sequence of FB execution for each device; a macrocyle is repeated once all the scheduled executions there contained are completed. According to the type of the device, we can have an LAS macrocycle and a device macrocycle. LAS macrocycle allows the system management to synchronize the execution of the FBs across the entire fieldbus link; on the basis of the device macrocycle, instead, the system management can synchronize the execution of FBs inside each device. Further information about FB scheduling will be given inside Section 14.1.3.4.
- *Application clock distribution*. This function allows publication of the time of day to all devices, including automatic switchover to a redundant time publisher. System management has a time publisher, which periodically sends an application clock synchronization message to all fieldbus devices. The data link scheduling time is sampled and sent with the application clock message so that the receiving devices can adjust their local application time. During the intervals between synchronization messages, application clock time is independently maintained within each device relying on its own internal clock.
- *Device address assignment*. Fieldbus devices do not use jumpers or switches to configure addresses; instead, device addresses are set using system management services.
- *Find tag service*. For the convenience of host systems and portable maintenance devices, system management supports a service for finding devices or variables by a tag search. The find tag query message is broadcasted to all fieldbus devices. Upon the receipt of the message, each device searches its VFDs for the requested tag and returns complete path information (if the tag is found), including the network address, VFD number, VCR index, and OD index. Once the path is known, the host or maintenance device can access the data by its tag.

All of the configuration information needed by system management (e.g., the FB schedule) is maintained by object descriptions in the network and system management VFD; this VFD provides access to the SMIB and to the NMIB, as said before.

14.2.1.5 H1 FOUNDATION Fieldbus Network Management

H1 FF network management mainly provides for the configuration management (e.g., initialization and setting of communication parameters), performance management (e.g., capturing data for the description of the performance of device communication), and fault management (e.g., supporting fault detection, fault location, and fault correction) [24]. The following specific tasks are performed by H1 FF network management:

- Loading the list of communication links
- Configuring the communication stack
- Loading the LAS
- Monitoring the performance of device communication
- Detecting communication errors

An H1 FF network has at least one network manager (NMgr), who coordinates the network management tasks of the whole system. Each field device has a network management agent (NMA) who manages the communication stack within the device. An NMA carries out commands issued by the NMgr and can report events and status changes to the NMgr. The NMgr and NMA communicate through the use of the FMS and VCR.

14.2.2 HSE Foundation Fieldbus

HSE FF enhances the H1 applications of FF by providing a high-speed backbone, redundancy, and bridging capabilities for multiple protocols. The main feature of the HSE FF is the use of Internet architecture for high-speed discrete control and, more generally, for interconnecting several H1 segments in order to achieve a plant-wide fieldbus network.

A generic HSE FF field device features the communication stack shown in Figure 14.2. The HSE field device is the HSE counterpart to the H1 field device; instead of an H1 communication stack, it has an HSE communications stack.

There are four typical HSE devices (but several of them are typically combined into a single real device): linking device (LD), Ethernet device (ED), host device (HD), and gateway device (GD). An LD connects one or more H1 networks to an HSE backbone; an ED connects some conventional I/O. A GD interfaces other network protocols (e.g., Modbus, DeviceNet, and Profibus). An HD is a non-HSE device capable of communicating with HSE devices; examples include configurators, operator workstations, and open productivity and connectivity (OPC) servers.

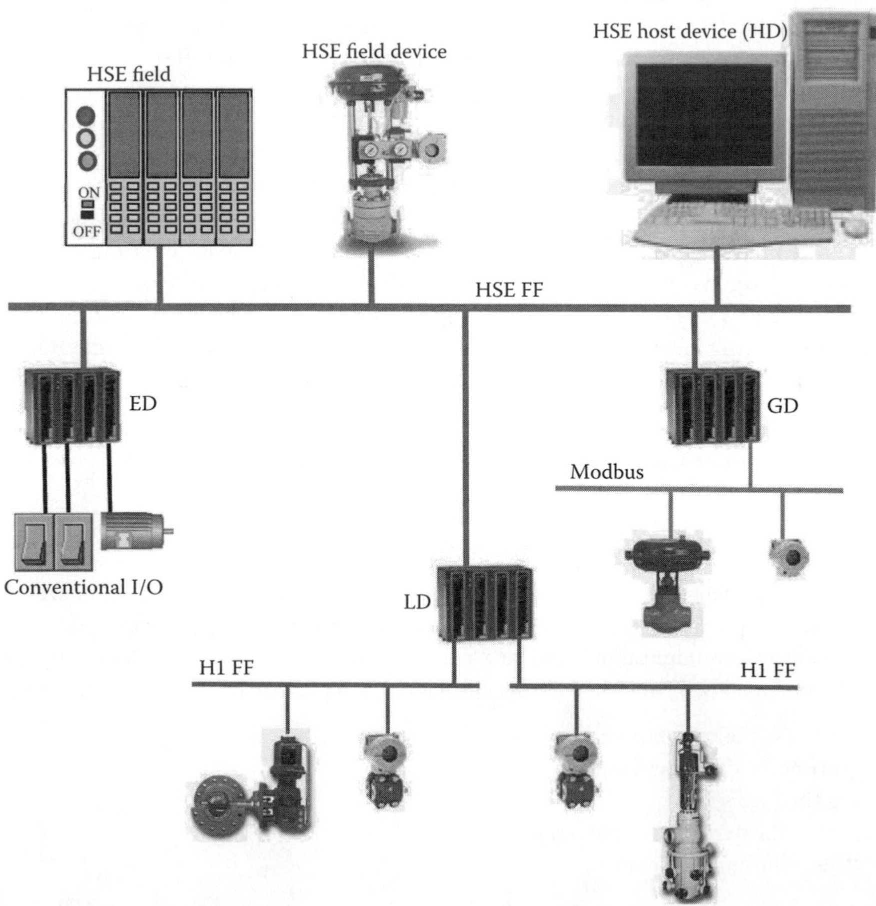

FIGURE 14.6 HSE FF interconnections.

The communication system shown in Figure 14.6 features a realistic example of HSE backbone to which several HSE field devices, an HSE HD, an LD, a GD, and an ED, are connected. The LD allows interconnection with two H1 FF networks; the GD allows interconnection with a Modbus network. Finally, the ED allows the exchange of information with conventional I/Os (e.g., 4–20 mA, digital I/O).

The capabilities of the interconnections allowed by HD, GD, LD, and ED are as follows:

- HSE HD/H1 segment. The HSE HD interacts with an H1 device through an LD. In this situation, the HSE HD is able to configure, diagnose, and publish and subscribe data to or from the H1 device.
- HSE HD/HSE segment. The HSE HD interacts with an HSE field device and is able to configure, diagnose, and publish and subscribe data to or from the HSE field device.
- H1/H1 segment. In this situation, the interaction is between two H1 devices on two distinct H1 bus segments. The segments are connected by LDs. In Figure 14.6, the same LD connects two H1 networks; but interconnections may occur by means of different LDs connected to the HSE backbone. Communications between devices on two H1 segments are functionally equivalent to communications between two H1 devices on the same bus segment. But it is clear that real-time communication between devices belonging to different H1 segments cannot be guaranteed due to the lack of a unique scheduler of the communication (i.e., LAS) among different H1 segments.
- HSE HD/Foreign Protocol. This connection defines the relationship between a foreign device and the FF application environment. The connection is made by GD. The foreign device is seen as a publisher to an HSE resident subscriber; the HSE HD can handle the data stream from the I/O gateway in the same manner as it treats the data streams from devices on H1/HSE segments.

One of the main advantages of using HSE FF is the adoption of commercial, off-the-shelf (COTS) Ethernet equipment; this simplifies the realization of HSE backbone and reduces the relevant installation and maintenance costs. Figure 14.7 shows an example of network interconnections using a commercial Ethernet switch; it allows connections with LD, HD, ED, GD, and with HSE field devices, as clearly depicted by the same figure.

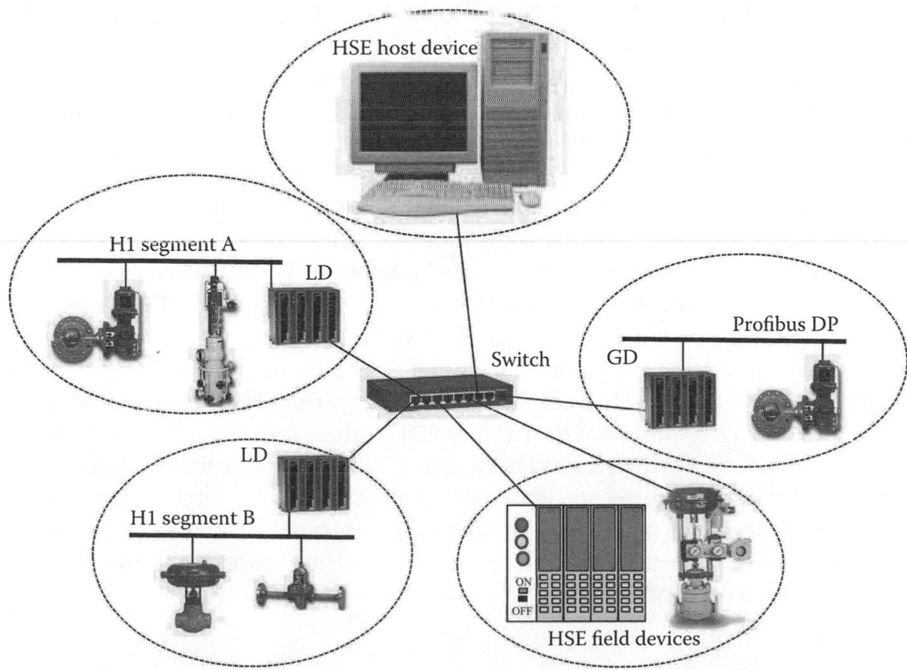

FIGURE 14.7 Interconnection based on COTS Ethernet equipment.

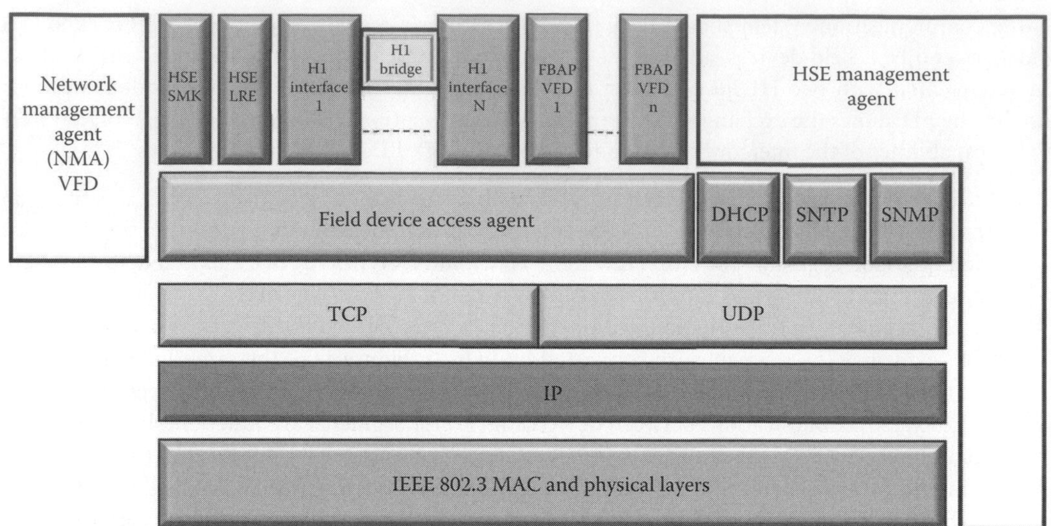

FIGURE 14.8 HSE FF architecture.

14.2.2.1 HSE Foundation Fieldbus Architecture

As said before, Figure 14.2 shows the HSE FF architecture for a generic HSE field device. Figure 14.8 gives a more detailed description of its internal organization.

As shown, the standard IEEE 802.3 physical and media access control (MAC) layers have been adopted at the bottom of the HSE architecture [25,26]. HSE also makes use of the well-established TCP, UDP, and IP [27]. Standard HSE stack components are Distributed Host Configuration Protocol (DHCP), Simple Network Time Protocol (SNTP), and Simple Network Management Protocol (SNMP), which rely on TCP and UDP over IP and over the IEEE 802.3 MAC and physical layers.

Messages sent on Ethernet are bounded by a series of data fields called frames. The combination of a message and frame is called an Ethernet packet. Typically, a packet encoded according to the TCP/IP protocols will be inserted in the message field of the Ethernet packet. HSE FF uses a similar data structure where messages are bounded by addressing and other DTs. What corresponds to a packet in Ethernet is called a protocol data unit (PDU) in HSE FF.

Let us consider a communication between two H1 devices over an interposed HSE backbone, as illustrated in Figures 14.6 and 14.7; the easiest method for LD might be, upon receiving a communication from an H1 device, to simply insert the entire H1 PDU into the message part of the TCP/IP packet. Then LD on the destination H1 segment, upon receiving the Ethernet packet, would merely strip away the Ethernet frame and send the H1 PDU on to the receiving H1 bus segment. This technique is called tunneling and is commonly used in mixed protocol networks.

The solution developed by HSE FF is somewhat more complex, but more efficient than tunneling. HSE FF PDU is inserted into the data field of a TCP/IP message field. However, the Fieldbus address is encoded as a unique TCP/IP address, so the Fieldbus PDU address is used to fill the address field of the TCP/IP packet. The entire TCP/IP packet is then inserted into the message field of the Ethernet packet. Because of the HSE encoding scheme, networks having multiple LDs can locate and transfer messages to the correct destination much more quickly, with far less extraneous bus traffic, as opposed to tunneling. Perhaps even more important, every H1 device (and every HSE device for that matter) has a unique TCP/IP address and can be directly accessed over standard IT and Internet networks.

14.2.2.1.1 Field Device Access Agent

A new technology has been developed and tested to provide a complete high-speed communication and control solution; it is the field device access agent (FDA Agent) [28]. The FDA Agent allows control systems to operate over the HSE and/or through LDs, and it enables remote applications to access field devices of any type across UDP/TCP using a common interface.

In particular, the FDA Agent has the following objectives:

- Conveying H1 system management (SM) services over UDP and H1 FMS services over UDP/TCP. This allows communication between HSE and H1 devices.
- Republishing H1 data from LDs that do not support H1 bridging. This allows LDs to be constructed from multiple standalone H1 interfaces instead of using an H1 bridge (see the next subsection).
- Sending and receiving local area network (LAN) redundancy messages to support the redundancy of HSE interfaces in devices.

14.2.2.1.2 H1 Interfaces and Bridge

Each H1 network attached to an HSE LD requires an H1 interface; an H1 bridge is used to convey H1 network messages directly between other H1 interfaces within the same HSE LD. Figure 14.8 depicts access of several H1 interfaces (numbered from 1 to N) to the FDA Agent; as it can be seen, an H1 bridge may be present in the architecture, performing data forwarding and republishing between H1 interfaces.

14.2.2.1.3 Function Block Application Process Virtual Field Device

The FDA Agent supports the function block application process (FBAP) VFD. The FBAP VFD provides a common means for defining inputs, outputs, algorithms, control variables, and behavior of the automation system. There may be multiple FBAP VFDs (e.g., n FBAP VFDs as shown in Figure 14.8) in a device in order to satisfy the particular needs of an application.

14.2.2.1.4 HSE System Management Kernel

Referring again to Figure 14.8, the HSE system architecture includes an HSE system management kernel (SMK).

The HSE SMK maintains information and a level of coordination that provides an integrated network environment for the execution and interoperation of the FBAP VFD.

HSE SMK provides for routine configuration of certain basic system information prior to device operation. For example at start-up, HSE SMK takes a device through a set of predefined phases for this purpose; during this procedure, a system configuration device recognizes the presence of the device on the network and prepares a basic configuration information, which is sent to the device. Once this happens, the HSE SMK brings the device to an operational state without affecting the operation of other devices on the network. It also enables the HSE FDA Agent for use by other functions in the device.

The HSE SMK also operates to keep a local time and keeps the difference between the local time and a system time provided by a time server within a certain specified value. The local time is used to time stamp events so that event messages from devices may be correlated across the system.

14.2.2.1.5 HSE Local Area Network Redundancy Entity

HSE FF is used at the host level of the control system; the host-level network ties the whole system together linking the various subsystems to the host. Thus, the visibility of hundreds and perhaps thousands of loops depends on the host-level network as does any intra-area control loops. A complete failure could result in heavy losses. High availability for the host-level network is therefore paramount. HSE FF adopts media and device redundancy to reach this aim.

Special integrity-checking diagnostics and redundancy management part of the HSE protocol in each device enables the use of two completely independent networks, redundant communication ports, and also redundant device pairs.

All parts of the network have redundancy, including the hubs (i.e. two independent networks ensuring that communication can continue even if one network fails). This means that the network can sustain multiple faults but still continue to function. All redundant ED pairs and the workstations are connected to both Ethernet buses.

A device may have redundant communication ports, and in this case, the two ports are named "A" and "B." In case of fault, the switchover is totally bumpless and transparent. Every communication port has a unique IP address; the IP address does not change when the communication ports switch.

True device redundancy is implemented using two identical devices, one primary and one secondary; when a primary device fails, the secondary takes over the role of the primary. Primary and secondary devices need to have identical configuration in order to allow for a quick switchover in case of failure. The HSE protocol specifies how these devices communicate with others and how the communication is switched over. However, HSE does not specify how the functionality is switched or how device configuration and data are synchronized. The control application sees either the primary or the secondary ED depending on which one is active, whereas the system diagnostics sees both. Thus, the diagnostics insures that even the inactive devices are fully functional and ready to take over at any moment.

A wide diagnostic coverage is an integral part of the HSE protocol going far beyond mere hardware duplication. Every HSE device and HSE LD contains an HSE local area network redundancy entity (HSE LRE), shown in Figure 14.8. The HSE LRE independently keeps track of the status of the networks and all the devices on it; it periodically sends and receives redundancy diagnostic messages, which represent its view of the network with each other. Each HSE device or LD uses the diagnostic messages to maintain a network status table (NST), which is used for fault detection and transmission port selection. In this way, every device has a complete picture of the network and knows the health of the primary and secondary, as well as communication port A and B, of every other device on the network in order to intelligently select which network, device, and port to communicate with.

Because the redundancy management is distributed to each device, no centralized *redundancy manager* is required [29].

14.2.2.1.6 HSE Management Agent

The following aspects of management are supported in the HSE management agent [30]:

- Acquisition of an IP address for the device, using the DHCP
- Time synchronization with a time server, using the SNTP
- Managing the TCP, UDP, and IP layers, using SNMP

14.2.2.1.7 HSE Network Management Agent VFD

The HSE system architecture includes an NMA VFD for each HSE device and each HSE LD. The NMA VFD provides means for configuring, controlling, and monitoring HSE device and HSE LD operation from the network [31]. The following capabilities are provided by the NMA VFD:

- Configuring the H1 bridge, which performs data forwarding and republishing between H1 interfaces (see Figure 14.8).
- Loading the HSE session list, or single entries in this list, called endpoints. An HSE session endpoint represents a logical communication channel between two or more HSE devices.
- Loading the HSE VCR list, or single entries in this list. An HSE VCR is a communication relationship used for accessing VFDs across HSE.
- Performance monitoring for session endpoints, HSE VCRs, and the H1 bridge.
- Fault detection monitoring.

14.2.3 H1 and HSE Foundation Fieldbus User Application Layer

As shown in Figures 14.1 and 14.2, Fieldbus Foundation has defined a common standard user application layer on both the H1 and HSE FF. User application layer is based on "blocks," which are representations of different types of application functions. The types of blocks used in a user application are resource, function, and transducer. Devices are configured by using resource blocks and transducer blocks; the control strategy is built by using FBs [32–36].

14.2.3.1 Resource Block

The resource block describes the features of the fieldbus device such as the device's name, manufacturer, and serial number. There is only one resource block in a device.

14.2.3.2 Function Block

FBs provide the control system behavior. There can be many FBs in a single user application, both standard and user-defined. The Fieldbus Foundation has defined sets of standard FBs. Ten standard FBs for basic control are defined by the "FBAP" Part 2 specification [33]; they are summarized in Table 14.1. Other, more complex, standard FBs are defined by the Parts 3 and 4 specifications [34,35].

The flexible function block (FFB) is defined by the [36] Part 5; it is a user-defined block. FFB allows a manufacturer or user to define block parameters and algorithms to suit an application that interoperates with standard FBs and host systems.

FBs can be built into fieldbus devices as required in order to achieve the desired (distributed) control. For example, Figure 14.9 shows a complete control loop made up by an analog input, analog output, and a PID control.

In order to realize this control loop in an H1 FF, a simple temperature transmitter may contain an AI FB; a control valve might contain a PID FB as well as the expected AO FB. Thus, as shown in Figure 14.10, the complete control loop shown in Figure 14.9 can be built using only a simple temperature transmitter (FF Device 1) and a control valve (FF Device 2), made up by the earlier-mentioned FBs.

14.2.3.3 Transducer Block

Like the resource block, the transducer blocks are used to configure devices. Transducer blocks decouple FBs from the local input/output functionalities required in order to read sensors or to command actuator's output. They contain information such as calibration date and sensor type [37].

TABLE 14.1 Basic Function Blocks

Function Block Name	Symbol Name
Analog input	AI
Analog output	AO
Bias/gain	BG
Control selector	CS
Discrete input	DI
Discrete output	DO
Manual loader	ML
Proportional/derivative	PD
Proportional/integral/derivative	PID
Ratio	RA

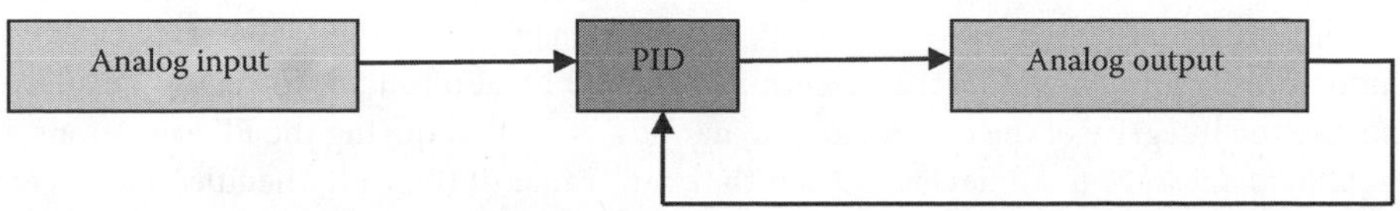

FIGURE 14.9 Example of a PID complete control loop.

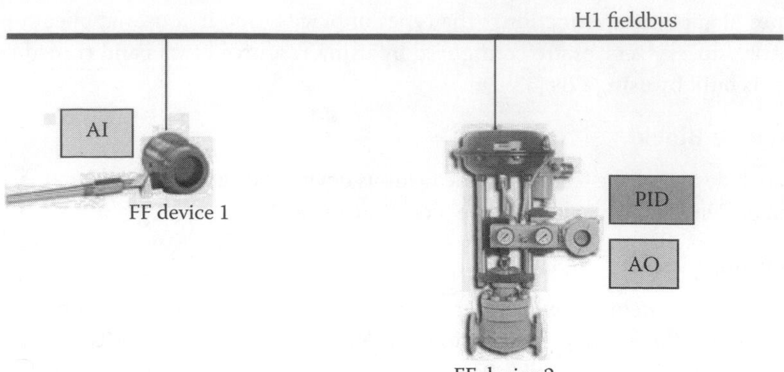

FIGURE 14.10 Realization of a PID complete control loop using FBs in FF devices.

14.2.3.4 Function Block Linking and Scheduling

A control or measurement application consists of FBs connected to each other; as seen, Figure 14.10 shows an example of mapping of a complete control loop on two devices, made up by the AI, AO, and PID basic FBs. These blocks must be connected in order to realize the distributed control loop; for this reason, FF features the link objects, defining the links between FB inputs and outputs internal to the device and across the fieldbus network.

An FB must obtain input parameters prior to the execution of its algorithm; its output parameters must be made available after the execution of the algorithm. Therefore, algorithm execution and the relevant communication must be synchronized even when the FBs are distributed among devices. For this reason, FF allows scheduling of FBs execution in each device. A macrocyle is defined in each device, containing the instants at which each FB must be executed, starting from an absolute start time; the content of each macrocyle may be repeated in each device forever, until a modification in the schedule occurs. The same scheduling may be also realized with communication, involving the LAS if present, in order to realize the exchanges of input/outputs between distributed FBs exactly when they are needed.

As an example, assuming to consider the FBs presented in Figure 14.10 (AI, PID, and AO) and their mapping to real devices shown in the same figure, the following scheduling may be realized in an H1 FF network:

- AI FB in Device 1 is executed at offset 0 (from an absolute start time).
- Communication of AI occurs at offset 20.
- PID FB in Device 2 is executed at offset 30.
- AO in Device 2 is executed at offset 50.
- All the sequence must be repeated in time after a certain offset.

Figure 14.11 shows the macrocyle present in each device of the H1 network, including LAS, which takes into account the previous scheduling. As it can be seen, according to these macrocycles, the previous sequence is repeated in time.

According to the macrocycles, system management in Device 1 will cause the AI FB to be executed at offset 0. At offset 20, the LAS will issue a CD to the AI FB buffer in the transmitter, and data in the buffer will be published on the fieldbus. At offset 30, system management in the valve (Device 2) will cause the execution of PID FB followed by the execution of the AO FB at offset 50. The pattern exactly repeats itself ensuring the integrity of the control loop dynamics. Note that during the FB executions, the LAS is sending the PT message to all devices so that they can transmit their unscheduled messages such as alarm notifications, or operator setpoint changes. For this example, the only time that the fieldbus cannot be used for unscheduled messages is when the AI FB starts to be published on the fieldbus.

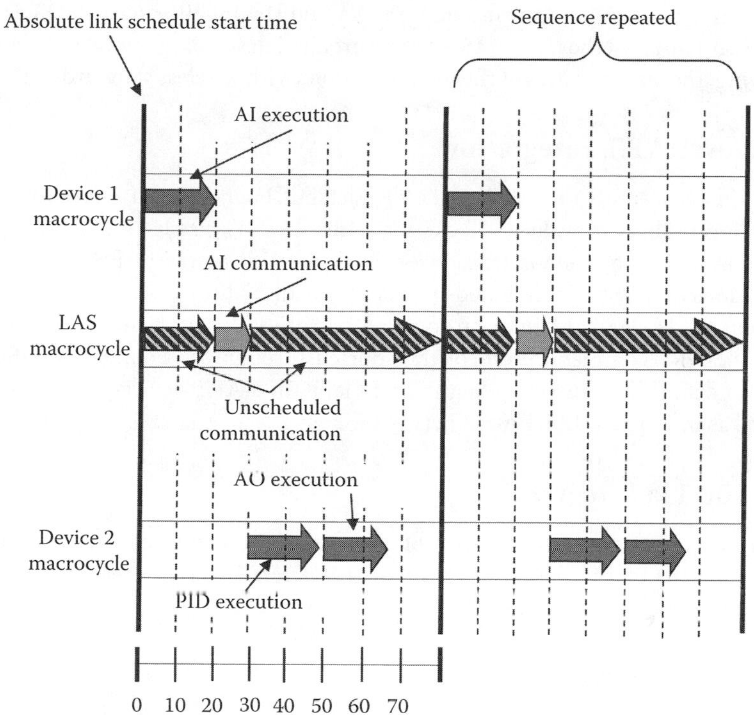

FIGURE 14.11 FB scheduling using macrocyles.

14.3 Wireless Solutions for Foundation Fieldbus

Currently, a large amount of the total cost of a manufacturing plant over its lifetime is spent on installation, set-up, and reconfiguration. If a plant is subjected to changes in its process flow or in its equipment, then the downtime and lifetime costs rise considerably [38].

Some of the main obstacles to fast reconfiguration are the inflexible wired communication infrastructure; this problem can be largely eased by the use of wireless technologies. With wireless technologies, not only the installation costs are much lower, but also the true self-reconfiguration of a system without any rewiring becomes possible as ever did before [38,39]. Together with other technologies like service-oriented architectures, web services, and the use of agent-based technology, wireless at the physical level plays an important role toward flexible and self-reconfiguring systems.

Several wireless solutions to be used in industrial environment are nowadays available.

One of these is the ISA100.11a, "Wireless Systems for Industrial Automation: Process Control and Related Applications"; it is an open wireless standard developed by the ISA SP-100 committee, starting from 2005 [40].

HART Communication Foundation (HCF) [41] is an organization made up by the producers and end users of wireless devices; HCF developed an open standard, the HART 7.0, which extends the HART protocol with wireless features known as *Wireless*HART. In March 2010, the IEC has approved the *Wireless*HART specification as a full international standard (IEC 62591 Ed. 1.0) [42].

In June 2008, ISA100 leaders established a new working group, ISA100.15—Wireless Backhaul Networks Working Group, to develop standards and technical reports to address dedicated or shared wireless backhauls to support technologies running multiple applications. Very recently, ISA published the work as technical report ISA-TR100.15.01, "Backhaul Architecture Model: Secured Connectivity over Untrusted or Trusted Networks," within the ISA100 family of standards [43].

Fieldbus Foundation made a lot of effort to integrate FF specifications with these novel wireless solutions. In order to reach this aim, two project teams have been defined inside Fieldbus Foundation: the Wireless

Sensor Integration Team (for the integration with ISA 100.11a and *Wireless*HART) and the Wireless HSE Backhaul Team (to assist the ISA100.15 working group in developing a wireless backhaul standard).

In the following, the main features of the FF integrations with wireless standards will be pointed out.

14.3.1 *Wireless*HART Integration

About integration between FF and *Wireless*HART, Fieldbus Foundation has recently defined a specification addressing fieldbus transducer blocks for wired and *Wireless*HART devices [44]. This technical specification defines a fieldbus transducer block used to represent HART devices within FF; *Wireless*HART devices may be represented in this block. In addition, the specification describes the expected method for HART configuration tools and asset-managing hosts to access HART devices using the native HART command protocol transported through the Foundation HSE network. The specification also defines structures to identify and maintain HART device status in wired multidrop networks as well as in *Wireless*HART mesh networks.

14.3.2 ISA 100.11a Integration

The ISA 100.11a project has made a considerable effort in the integration of the ISA 100.11a wireless sensor network into the FF infrastructure. Similar to *Wireless*HART devices, ISA 100.11a devices are represented as transducer blocks inside FF.

14.3.3 Wireless HSE Backhaul Team

In late 2008, the Fieldbus Foundation and ISA entered into a cross-licensing agreement allowing the two organizations to collaborate on wireless networks. This agreement will assist the ISA100.15 working group in developing a wireless backhaul standard. Backhaul networks integrated remote locations and applications with central control facilities.

14.3.4 FOUNDATION for Remote Operations Management

In December 2011, the Fieldbus Foundation announced the release of an extension to its technology, called FOUNDATION for Remote Operations Management (ROM) [45]. Integration of FF with existing wireless standard is fully realized into this very novel suite of technologies and additions to the FF specification that provide for both a wireless and wired infrastructure for remote assets and applications. FOUNDATION for ROM provides for direct access to information and diagnostics in wireless and remote I/O devices. Conversely, FOUNDATION for ROM can take the data from those devices and place into the FF environment for data management and quality.

FOUNDATION for ROM provides an open path for the integration of multiple wireless and wired networks, from conventional remote I/O to ISA 100.11a and *Wireless*HART, and enables direct access to device information and diagnostics. It extends the range and capabilities of Foundation Fieldbus to encompass many more devices throughout the plant, regardless of their communications technology.

Figure 14.12 shows the FOUNDATION for ROM architectures; both wireless and wired technologies are integrated inside ROM. Communication with control room occurs through HSE wired and wireless backhaul.

14.4 Open Systems Implementation

One of the main features of FF (in both H1 and HSE configurations) is the availability to build open communication systems. It is clear that this represents a key issue in a communication system as building up a perfect communication stack between two devices is completely useless if those two devices are not able to understand the meaning of each other data, or each other behavior.

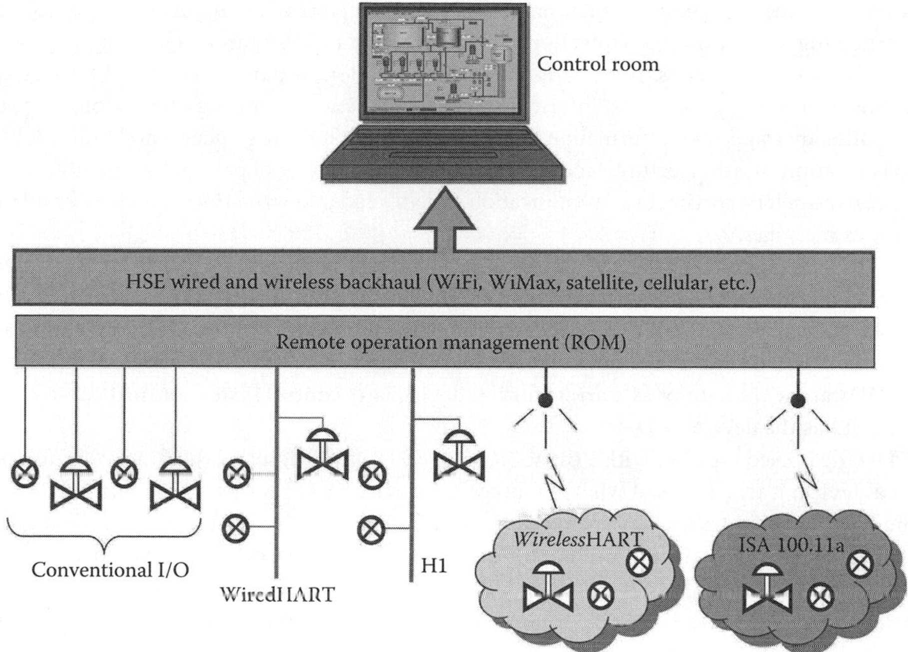

FIGURE 14.12 FOUNDATION for ROM architecture.

Implementation of open systems in FF has been achieved from the beginning through the adoption of a standard way to represent and manage information exchanged by devices: the device description language (DDL).

Very recently, a strong effort has been put by Fieldbus Foundation to integrate FF with the field device tool/device type manager (FDT/DTM) technology. Furthermore, starting from, 2003 Fieldbus Foundation contributes to the FDI, the field device integration initiative, as part of an industry effort to integrate DDL and FDT/DTM.

In the following subsections, a detailed description of both DDL and FDT/DTM will be given, finally, the FDI initiative will be pointed out.

14.4.1 Electronic Device Description Language

The electronic device description language (EDDL) is a language for describing the behavior of the field devices and is defined by the international standard IEC 61804 [46,47].

The IEC 61804 specifies EDDL as a generic language for describing the properties of automation system components, such us device parameters and their dependencies, device functions (e.g., simulation mode and calibration), graphical representations (e.g., menus), interactions with control devices, graphical representations, and persistent data store. EDDL technology was designed to avoid the need for special, proprietary, and operating system-specific host application files. It allows a host system to configure as well as to monitor devices online.

An electronic data description (EDD) is the computer readable file written in EDDL that describes the data in a field device; it is the file that the host application reads in order to learn how to retrieve information from the field device. EDD is a clear and unambiguous structured text description that precisely describes field device data to host systems. EDD is independent from the operating system, DCS platforms, and communication and interface paths. There is no executable code with EDD, which may have an effect on the stability of the operating system; EDD is interpreted and therefore encapsulated.

An EDD contains the following information about the parameters of a device: attributes (e.g., coding, name, engineering unit, and write protection), the arrangement of the parameters in a menu structure; names of menus and submenus, information about the relation of parameters to others, information about help texts and help procedures, information about necessary operating interactions (e.g., calibration), also called methods, and information about visualization tools (e.g., charts and graphs). EDD can describe also communication features: communication orders, sorting of parameters, bit positioning and bit length of parameters, sorting of communication orders, read and write timeouts, and error handling, including user messages.

The IEC 61804 includes the DDL used as an underlying technology by FF to achieve interoperability [48]. A device description (DD) file, written in the DDL, provides an extended description of each object in the VFD; it provides information needed for a control system or a host to understand the meaning of the data in the VFD, including the human interface for functions such as calibration and diagnostics. Thus, the DD can be thought of as a *driver* for the device; any control system or host can operate with the device if it has the device's DD.

In FF, DD file is used together with a capability file (CF) to fully describe the functions and the capabilities of a device; it tells the host what resources the device has in terms of FBs and VCRs, etc. This allows the host to configure the device even if not connected to it. The host can ensure that only functions supported by the device are allocated to it and that other resources are not exceeded.

On account of what is said, a device is always supplied with DD and CFs. New devices are added to the fieldbus by simply connecting the device to the fieldbus wire and providing the control system or host with the DD for the new device.

14.4.1.1 Device Description Services

On the host side, library functions called Device Description Services (DDS) are used to read the device descriptions, but they cannot read operational values [49]. The operational values are read from the fieldbus device over the fieldbus using FMS communication services.

14.4.1.2 Device Description Hierarchy

The Fieldbus Foundation has defined a hierarchy of DDs to make it easier to build devices and perform system configuration. The first level in the hierarchy is referred to as universal parameters, which consist of common attributes such as tag, revision, mode, etc.; all blocks must include the universal parameters. The next level in the hierarchy is FB parameters; at this level, parameters are defined for the standard FBs and resource blocks. The third level is called transducer block parameters; at this level, parameters are defined for the standard transducer blocks. In some cases, the transducer block specification may add parameters to the standard resource block. These first three layers are the standard Fieldbus Foundation DDs. The fourth level of the hierarchy is called manufacturer-specific parameters; at this level, each manufacturer is free to add additional parameters to the FB parameters and transducer block parameters.

14.4.1.3 Device Description Interoperability Test

Before Fieldbus Foundation registers a DD file, the file undergoes a series of tests using FF's Interoperability Test Kit [50]. The first test loads the DD binary into a test system to validate correct identification information. Next the file undergoes multiple test cases to ensure the DD matches the device and follows protocol rules for maintaining interoperability between host and device. Finally, FF logs the DD file information, including name, date, size, and the 32-bit cyclic redundancy checksum value.

14.4.2 Field Device Tool

Field device tool (FDT) technology has been promoted by the not-for-profit organization FDT Group [51]. FDT has been accepted as IEC 62453 standard [52,53]. The FDT standard currently supports several industry network standards, including FF.

According to the IEC 62453 standard, the DTM is the driver supplied by the manufacturer of the device that supports the configuration, operation, diagnostics, maintenance, and calibration of the device. A single DTM may support an entire family or class of devices offered by that manufacturer.

There are two types of DTMs. The most common DTM supports end devices such as pressure and temperature transmitters, I/O blocks, and valve positioners. The second type, called COM (communications) DTM, is provided by the manufacturers of the network infrastructure components such as GDs, SIL barriers, and network power supplies.

The COM DTM has brought a complete revolution to network management; a COM DTM now can monitor the health of the network and its infrastructure such as barriers, network power supplies, and terminators. Gone are the days of having to haul out network analyzers to see if the problem could be related to the network infrastructure. The pulled terminator, the forklift-damaged cable, the colliding packets, the network reaching bandwidth limits, and the failing redundant supply can be proactively detected and reported through the diagnostics of the COM DTM.

The DTM is installed in a host application, which is called the frame. An FDT frame is a higher level software application, such as a DCS, a PLC programming tool, or a maintenance management system, which is compliant with the FDT standard and needs access to the intelligent devices on the fieldbuses or networks associated with the application. The frame gains access to the device intelligence and functionality by installing the appropriate DTMs into the topology tree of the application.

14.4.3 EDDL versus FDT: Field Device Integration

EDDL and FDT are different technologies both aimed at providing easy plug-and-play access to information in smart field devices. For the past few years, many people have seen EDDL and FDT as competing technologies; but the two technologies must be viewed as complementary.

In a growing number of cases, plants are using a combination of EDDL and FDT. There are at least two cases for this. First, the basic integration is done with EDDL, but additional advanced diagnosis functionality is provided by the device vendor at FDT/DTM, which runs in an FDT frame application. In the second case, several device vendors do not provide DTMs for their simple to medium-complex devices; in this instance, those EDDs are run in an EDD interpreter DTM.

Because both EDDL and FDT have their adherents, and both are clearly here for the long run, the industry has started an initiative to integrate the two technologies, mainly because many end users weren't comfortable with these two technologies in tandem.

This effort to integrate EDDL and FDT led to the creation of the Field Device Integration (FDI) initiative, consisting of a combination of representatives from the EDT Group and EDDL Cooperation Team; the simple idea behind FDI was to combine the best features of EDDL and the FDT standards.

In September 2011, a new joint company, FDI Cooperation LLC (a limited liability company under US law), was formed. The new company is headed by a board of managers that includes representatives from the Fieldbus Foundation, FDT Group, HART Communications Foundation, OPC Foundation, and PROFIBUS & PROFInet International.

FDI Cooperation LLC is committed to developing a single technology for the management of information from intelligent devices throughout all areas of the plant. Among its missions, there is the standardization of FDI under the IEC; currently, the FDI standard is under the definition as IEC 62769 [54].

One piece of glue that represents a factor in integrating EDDL and FDT is the IEC 62541, OPC UA (OPC Unified Architecture) [55,56], used to put some standards around the process of getting data from devices.

Basically, the FDI architecture is a client/server structure based on OPC UA. The field device integration is realized by an FDI device package provided by the device supplier. The FDI device package is intended to give information about one or more device type(s) to a system. Typically, an FDI device package describes a single device type; modular devices, such as remote I/O, may be represented as multiple device types.

An FDI device package contains up to four different elements, depending upon the complexities and requirements of the device. It is made up by a mandatory package catalogue and a mandatory EDD for each device type within the FDI package; the FDI device package may also include one or more user interface plug-in (UIP) and attachment. The mandatory EDD acts as the information model of the device and describes the device data and type. Optional UIP is a user-defined software component that defines special device functions/application information and user interfaces to run on the client. Attachments are also optional and include things like product manuals, images, and electronic certifications. EDD is based on EDDL, IEC 61804-3; the optional UIP offers the advantages of freely programmable user interfaces familiar from FDT, based on Windows Presentation Foundation (WPF).

Another important FDI feature is nested communication, that is, the open integration of gateways, and the integration of communication drivers via communication servers. In FDI, all operations and services that are necessary for communication are described and provided in a standardized format as part of an FDI communication package via EDDL code. FDI communication package has a structure similar to the FDI device package and contains the mandatory component EDD. In IEC 62769, two particular profiles have been defined for the description of both FF H1 and HSE communication functionalities inside the EDD component of an FDI communication package [57,58].

14.5 Conclusions

The chapter has given an overview on the main features of the FF. Since its definition, a lot of new technologies have been introduced in the Information and communications technology (ICT), ranging from transmission media to information modeling techniques; updates of the original FF specifications were mandatory in order to integrate these new technologies, when they led to real advantages for the communication system. For this reason, the chapter tried to give an exhaustive overview about all the integrations already done and about all the work that is carried on at this moment to integrate more features.

References

1. www.fieldbus.org.
2. J.D. Decotignie and P. Pleinevaux, Time critical communication networks: Field buses, *IEEE Network*, 2(3): 55–63, 1988.
3. J.D. Decotignie and P. Pleinevaux, A survey on industrial communication networks, *Annales des Telecommunications*, 48: 9–10, 1993.
4. S. Cavalieri, A. Di Stefano, and O. Mirabella, Optimization of acyclic bandwidth allocation exploiting the priority mechanism in the fieldbus data link layer, *IEEE Transactions on Industrial Electronics*, 40(3): 297–306, 1993.
5. CENELEC EN50170-3, General purpose field communication system, WorldFIP, 1996.
6. CENELEC EN50170-2, General purpose field communication system, Profibus, 1996.
7. ISA S50.02, Physical Layer Standard, 1992.
8. IEC 61158, Digital Data Communications for Measurements and Control—Fieldbus for use in Industrial Control Systems—Part 2: Physical Layer Specification, ed.5.0, 2010.
9. IEC 61158, Digital Data Communications for Measurements and Control—Fieldbus for use in Industrial Control Systems—Part 3: Data Link Layer Service Definition, 2007.
10. IEC 61158, Digital Data Communications for Measurements and Control—Fieldbus for use in Industrial Control Systems—Part 4: Data Link Layer Protocol Specification, 2007.
11. IEC 61158, Digital Data Communications for Measurements and Control—Fieldbus for use in Industrial Control Systems—Part 5: Application Layer Service Definition, 2007.
12. IEC 61158, Digital Data Communications for Measurements and Control—Fieldbus for use in Industrial Control Systems—Part 6: Application Layer Protocol Specification, 2007.

13. CENELEC EN50170/A1, General Purpose Field Communication System, Addendum A1: FOUNDATION Fieldbus, 2000.
14. Fieldbus Foundation, 31.25 kbit/s Physical Layer, FF-816, 2011.
15. Fieldbus Foundation, 31.25 kbit/s Wiring and Installation Application Guide, AG-140.
16. Fieldbus Foundation, Fieldbus Installation and Planning Application Guide, AG-165.
17. Fieldbus Foundation, 31.25 kbit/s Intrinsically Safe Systems Application Guide, AG-163.
18. Fieldbus Foundation, H1 Data Link Services, FF-821, 1999.
19. Fieldbus Foundation, H1 Data Link Protocol, FF-822, 2001.
20. Fieldbus Foundation, Fieldbus Access Sublayer, FF-875, 2003.
21. Fieldbus Foundation, Fieldbus Message Specification, FF-870, 2008.
22. S. Cavalieri, FOUNDATION fieldbus: History and features, *The Industrial Communication Technology Handbook*, ed. R. Zurawski, pp. 9–1, 9–17, 2005.
23. Fieldbus Foundation, H1 System Management, FF-880, 2005.
24. Fieldbus Foundation, H1 Network Management, FF-801, 2008.
25. ANSI/IEEE Std 802.3, IEEE Standards for Local Area Networks: CSMA/CD Access Method and Physical Layer Specifications, 1985.
26. ANSI/IEEE Std 802.3u, IEEE Standards for Local Area Networks: Supplement to CSMA/CD Access Method and Physical Layer Specifications. MAC Parameters, Physical Layer, MAUs, and Repeater for 100 Mb/s Operation, Type 100BASE-T, 1995.
27. D.E. Comer, *Internetworking with TCP/IP, Vol. I: Principles, Protocols, and Architecture*, Englewood Cliffs, NJ: Prentice Hall International, 1999.
28. Fieldbus Foundation, Field Device Access Agent, FF-588, 2005.
29. Fieldbus Foundation, HSE Redundancy, FF-593, 2005.
30. Fieldbus Foundation, HSE System Management, FF-589, 2003.
31. Fieldbus Foundation, HSE Network Management, FF-803, 2003.
32. Fieldbus Foundation, Function Block Application Process, Part 1, FF-890, 2012.
33. Fieldbus Foundation, Function Block Application Process, Part 2, FF-891, 2012.
34. Fieldbus Foundation, Function Block Application Process, Part 3, FF-892, 2012.
35. Fieldbus Foundation, Function Block Application Process, Part 4, FF-893, 2008.
36. Fieldbus Foundation, Function Block Application Process, Part 5, FF-894, 2008.
37. Fieldbus Foundation, Transducer Block Common Structure, FF-902, 2012.
38. F. Jammes and S. Harm, Service-oriented paradigms in industrial automation, *IEEE Transactions in Industrial Automation*, 1(1): 62–70, 2005.
39. C. Cardeira, A. Colombo, and R. Schoop, Survey of wireless technologies for automation networking, Pham, D.T., Eldukhri, E.E., and Soroka, A.J., *IPROMS 2006, the 2006 I*PROMS NoE Virtual International Conference on Intelligent Production Machines and Systems*, Elsevier Ltd., July 3–14, 2006.
40. www.isa.org/isa100.
41. www.hartcomm.org.
42. IEC62591, Industrial communication networks—Wireless communication network and communication profiles—WirelessHART™, ed.1, 2010.
43. ISA, ISA-TR100.15.01-2012 Backhaul Architecture Model: Secured Connectivity over Untrusted or Trusted Networks, 2012.
44. Fieldbus Foundation, Transducer Blocks for Wired and Wireless HART® Devices, FF-913 FS 1.0, 2012.
45. http://www.fieldbus.org/images/stories/enduserresources/technicalreferences/documents/foundation_for_remote_operations_management.pdf.
46. IEC 61804-2, Function blocks (FB) for process control—Part 2: Specification of FB concept and Electronic Device Description Language (EDDL), 2004.
47. IEC 61804-3, Function blocks (FB) for process control—Part 3: Electronic Device Description Language (EDDL), 2006.

48. Fieldbus Foundation, Device Description Language, FF-900, 2008.
49. Fieldbus Foundation, DDS User's Guide, FD-110.
50. Fieldbus Foundation, Interoperability Tester User's Guide, FD-210.
51. www.fdtgroup.org.
52. IEC 62453-1, Field device tool (FDT) interface specification—Part 1: Overview and guidance, 2009.
53. IEC 62453-2, Field device tool (FDT) interface specification—Part 2: Concepts and detailed description, 2009.
54. IEC 62769-1/Pre-CDV, Field Device Integration Project, Technical Specification-Part 1: Overview, RC 1.0.0, December 2012.
55. IEC 62541-1, OPC Unified Architecture—Part 1: Overview and Concepts, 2012.
56. www.opcfoundation.org.
57. IEC 62769-101-1/CD, Devices and Integration in Enterprise Systems; Field Device Integration Profiles, Part 101-1: FOUNDATION Fieldbus H1, RC 0.0.13, December 2012.
58. IEC 62769-101-2/CD, Devices and Integration in Enterprise Systems; Field Device Integration Profiles, Part 101-2: FOUNDATION Fieldbus HSE, RC 0.0.12, December 2012.

15

INTERBUS

15.1 Introduction to Field Communication.......... 15-1
15.2 INTERBUS Overview.......... 15-2
15.3 INTERBUS Protocol.......... 15-4
15.4 Diagnostics.......... 15-7
15.5 Functional Safety.......... 15-9
15.6 Interoperability, Certification.......... 15-10
15.7 Connectivity.......... 15-11
15.8 IP over INTERBUS.......... 15-12
15.9 Performance Evaluation.......... 15-13
15.10 Conclusions.......... 15-15
References.......... 15-15

Jürgen Jasperneite
Fraunhofer Institut für Optronik, Systemtechnik und Bildauswertung

15.1 Introduction to Field Communication

The growing degree of automation in machines and systems also increases the amount of cabling required for parallel wiring due to the large number of input/output (I/O) points. This brings with it increased configuration, installation, start-up, and maintenance effort. The cable requirements are often high because, for example, special cables are required for the transmission of analog values. Parallel field wiring thus entails serious cost and time factors. In comparison, the serial networking of components in the field using fieldbus systems is much more cost-effective. The fieldbus replaces the bundle of parallel cables with a single bus cable and connects all levels, from the field to the control level. Regardless of the type of automation device used, for example, programmable logic controllers (PLCs) from various manufacturers or PC-based control systems, the fieldbus transmission medium networks all components. They can be distributed anywhere in the field and are all connected locally. This provides a powerful communication network for today's rationalization concepts. There are numerous advantages to a fieldbus system in comparison with parallel wiring: the reduced amount of cabling saves time during planning and installation, while the cabling, terminal blocks, and the control cabinet dimensions are also reduced. Self-diagnostics minimize downtimes and maintenance times. Open fieldbus systems standardize data transmission and device connection regardless of the manufacturer. The user is therefore independent of any manufacturer-specific standards. The system can be easily extended or modified, offering flexibility as well as investment protection (Figure 15.1).

Fieldbus systems, which are suitable for networking sensors and actuators with control systems, have represented state-of-the-art technology for some time. The main fieldbus systems are combined under the umbrella of IEC 61158 [1]. This also includes INTERBUS as Type 8 of IEC 61158 and CPF6 of IEC 61784, with an installed base of 17 million nodes and more than 1000 device manufacturers.

The requirements of these systems can be grouped according to the following four basic types of arithmetic operations for field communication.

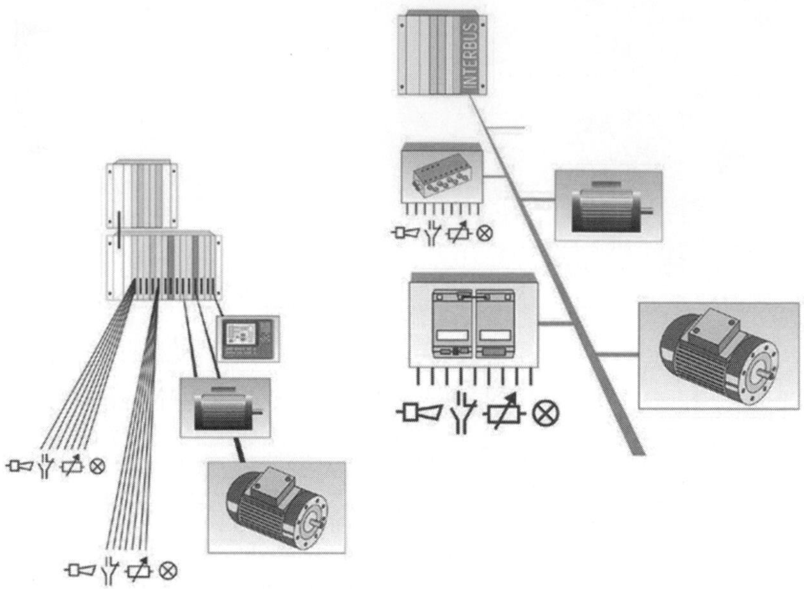

FIGURE 15.1 Serial instead of parallel wiring.

15.2 INTERBUS Overview

INTERBUS has been designed as a fast sensor/actuator bus for transmitting process data in industrial environments. Due to its transmission procedure and ring topology, INTERBUS offers features such as fast, cyclic, and time-equidistant process data transmission, diagnostics to minimize downtime, easy operation and installation, as well as meeting the optimum requirements for fiber optic technology.

In terms of topology, INTERBUS is a ring system, that is, all devices are actively integrated in a closed transmission path (see Figure 15.2). Each device amplifies the incoming signal and forwards it, enabling higher transmission speeds over longer distances.

Unlike other ring systems, the data forward and return lines in the INTERBUS system are led to all devices via a single cable. This means that the general physical appearance of the system is an *open* tree structure. A main line exits the bus master and can be used to form seamless subnetworks up to 16 levels deep. This means that the bus system can be quickly adapted to changing applications.

The INTERBUS master/slave system enables the connection of up to 512 devices, across 16 network levels. The ring is automatically closed by the last device. The point-to-point connection eliminates the need for termination resistors. The system can be adapted flexibly to meet the user's requirements by adding or removing devices. Countless topologies can be created. Branch terminals create branches, which enable the connection and disconnection of devices. The coupling elements between the bus segments enable the connection and disconnection of a subsystem and thus make it possible to work on the subsystem without problems, for example, in the event of an error or when extending the system.

Unlike in other systems where data are assigned by entering a bus address using dual in-line package (DIP) or rotary switches on each individual device, in the INTERBUS system, data are automatically assigned to devices using their physical location in the system. This plug and play function is a great advantage with regard to the installation effort and service-friendliness of the system. The problems and errors, which may occur when manually setting device addresses during installation and servicing, are often underestimated. The ability to assign "easy to understand" software names to the physical addresses enables devices to be added or removed without readdressing.

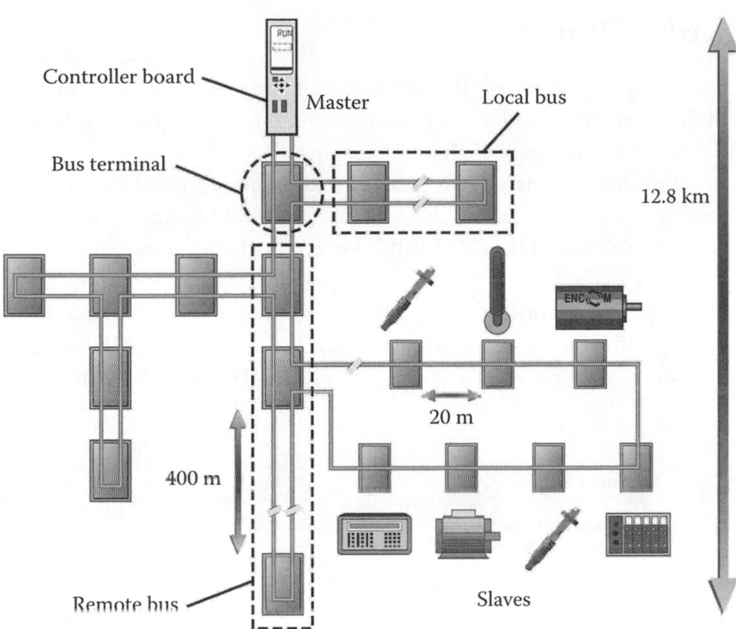

FIGURE 15.2 Topology flexibility.

In order to meet the individual requirements of a system, various basic elements must be used (see Figure 15.2):

1. *Controller Board*. The controller board is the master that controls bus operation. It transfers output data to the corresponding modules, receives input data, and monitors data transfer. In addition, diagnostic messages are displayed, and error messages are transmitted to the host system.
2. *Remote Bus*. The controller board is connected to the remote bus devices via the remote bus. A branch from this connection is referred to as a remote bus branch. Data can be physically transmitted via copper cables (RS-485 standard), fiber optics, optical data links, slip rings, or other media (e.g., wireless). Special bus terminal modules and certain I/O modules or devices such as robots, drives, or operating devices can be used as remote bus devices. Each has a local voltage supply and an electrically isolated outgoing segment. In addition to the data transmission lines, the installation remote bus can also carry the voltage supply for the connected I/O modules and sensors.
3. *Bus Terminal Module*. The bus terminal modules, or devices with embedded bus terminal module functions, are connected to the remote bus. The distributed local buses branch out of the bus terminal module with I/O modules, which establish the connection between INTERBUS and the sensors and actuators. The bus terminal module divides the system into individual segments, thus enabling you to switch individual branches on/off during operation. The bus terminal module amplifies the data signal (repeater function) and electrically isolates the bus segments.
4. *Local Bus*. The local bus branches from the remote bus via a bus coupler and connects the local bus devices. Branches are not allowed at this level. The communications power is supplied by the bus terminal module, while the switching voltage for the outputs is applied separately at the output modules. Local bus devices are typically I/O modules.

15.3 INTERBUS Protocol

INTERBUS recognizes two cycle types: the identification cycle for system configuration and error management and a data transfer cycle for the transmission of user data. Both cycle types are based on a summation frame structure (see Figure 15.3).

The layer 2 summation frame consists of a special 16 bit loopback word (preamble), the user data of all devices, and a terminating 32 bit frame check sequence (FCS). As data can be simultaneously sent and received by the ring structure of the INTERBUS system (full duplex mode), this results in very high protocol efficiency.

The logical method of the operation of an INTERBUS slave can be configured between its incoming and outgoing interface by the register set shown in Figure 15.4. Each INTERBUS slave is part of a large, distributed shift register ring, whose start and end point is the INTERBUS master.

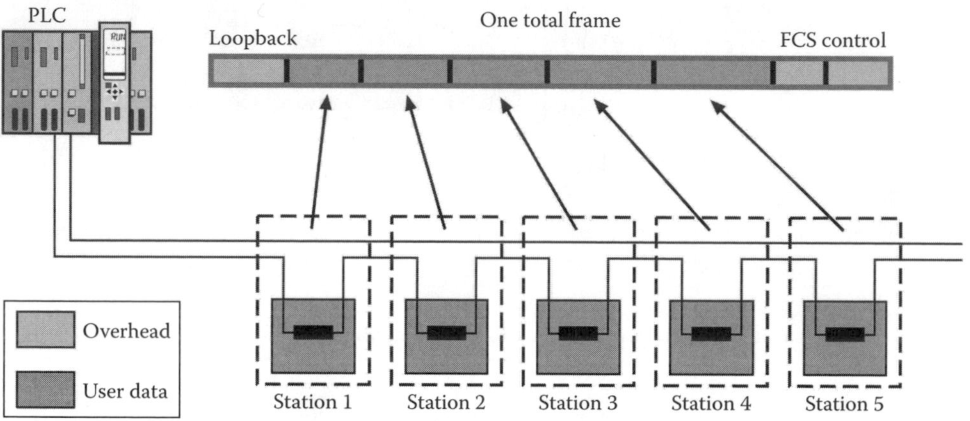

FIGURE 15.3 The layer 2 summation frame structure of INTERBUS.

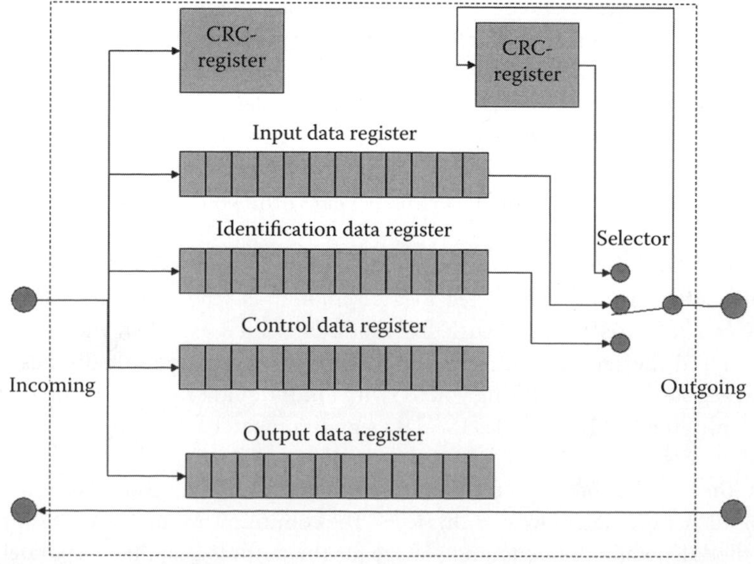

FIGURE 15.4 Basic model of an INTERBUS slave node.

In the input data register, input data, that is, data that are to be transmitted to the master, are loaded during data transfer. The output data registers and the cyclic redundancy check register (cyclic redundancy check) are switched to the input data register in parallel. The Comite Consultatif International Telephonique et Telegraphique (CCITT) polynomial $g(x) = x^{16} + x^{12} + x^5 + 1$ is used for the CRC. After finishing a valid data transfer cycle, output data from the output data register are written to a memory and then accepted by the local application. The CRC registers are used during the FCS to check whether the data have been transmitted correctly. The length of the I/O data registers depends on the number of I/Os of the individual node.

The master needs to know which devices are connected to the bus so that it can assign the right I/O data to the right device.

Once the bus system has been switched on, the master starts a series of identification cycles, which enable it to detect how many and which devices are connected. Each slave has an identification data register, which has a 16 bit ID code. The master can use this ID code to assign a slave node to a defined device class (e.g., digital I/O node and analog I/O node) and detect the length of the I/O data registers in a data cycle. The control data registers are switched in parallel to the identification data registers, whereby the individual devices can be managed by the master. Commands are transmitted, for example, for local resets or outgoing interface shutdown. The identification cycle is also used to find the cause of a transmission error and to check the integrity of the shift register ring.

The individual registers are switched in the different phases of the INTERBUS protocol via the selector in the ring.

On layer 1, INTERBUS uses two telegram formats with start and stop bits similar to universal asynchronous receiver transmitter (UART):

1. The 5 bit status telegram
2. The 13 bit data telegram

The status telegram is used to generate defined activity on the medium during pauses in data transmission. The slave nodes use the status telegram to reset their internal watchdogs, which are used to control a failsafe state. The data telegram is used to transmit a byte of the layer 2 payload. The remaining bits of both telegrams are used to distinguish between data and ID cycles, as well as the phase data transfer and FCS within a cycle. This information is used by the selector shown in Figure 15.4 to switch the relevant register in the ring. INTERBUS uses a physical transmission speed of 500 kbps or 2 Mbps.

The cycle time, that is, the time required for I/O data to be exchanged once with all the connected modules, depends on the amount of user data in an INTERBUS system. Depending on the configuration, INTERBUS can achieve cycle times of just a few milliseconds. The cycle time increases linearly with the number of I/O points, as it depends on the amount of information to be transmitted. For more detailed information, refer to the performance evaluation.

The architecture of the INTERBUS protocol is based on the open systems interconnection (OSI) reference model according to ISO 7498. The protocol architecture of INTERBUS provides the cyclic process data and an acyclic parameter data channel, using the services of the peripheral message specification (PMS) as well as a peripheral network management (PNM) channel (see Figure 15.5). As is typical for fieldbus systems, for reasons of efficiency, ISO layers 3–6 are not explicitly used but are combined in the lower layer interface (LLI) in layer 7. The process data channel enables direct access to the cyclically transmitted process data. It is characterized by its ability to transmit process-relevant data quickly and efficiently. From the application point of view, it acts as a memory interface. The parameter channel enables data to be accessed via a service interface. The data transmitted in the parameter channel have a low dynamic response, and this data transmission occurs relatively infrequently (e.g., updating text in a display).

Network management is used for manufacturer-independent configuration, maintenance, and start-up of the INTERBUS system.

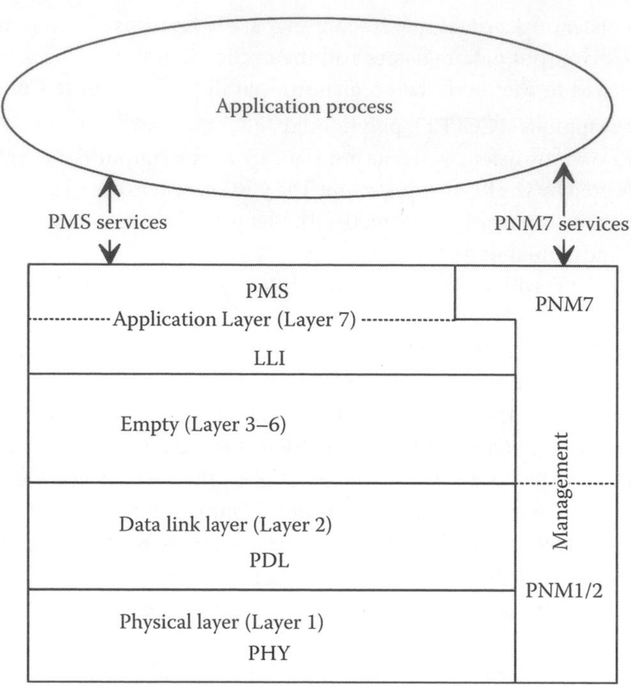

FIGURE 15.5 Protocol architecture of an INTERBUS node.

Network management is used, for example, to start/stop INTERBUS cycles, to execute a system reset, and fault management. Furthermore, logical connections between devices can be established and aborted via the parameter channel in the form of context management, for example.

To transmit parameter data and time-critical process data simultaneously, the data format of the summation frame must be extended by a specific time slot. In several consecutive bus cycles, a different part of the data is inserted in the time slot provided for the addressed devices. The peripherals communication protocol (PCP) performs this task. It inserts a part of the telegram in each summation frame and recombines it at its destination (see Figure 15.6). The parameter channels are activated, if necessary, and they do not affect the transfer of I/O data. The longer transmission time for parameter data that are segmented into several bus cycles is sufficient for the low time requirements that are placed on the transmission of parameter information.

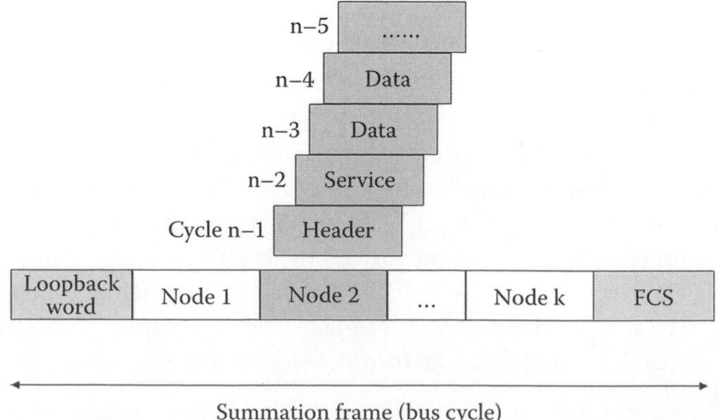

FIGURE 15.6 Transmission of parameter data with a segmentation and recombination mechanism.

INTERBUS uses a master/slave procedure for data transmission. The parameter channel follows the client/server paradigm. It is possible to transmit parameter data between two slaves (peer-to-peer communication). This means that both slaves can adopt both the client and server function. With this function, layer 2 data are not exchanged directly between the two slaves but are implemented by the physical master/slave structure, that is, the data are first transmitted from the client to the master and then forwarded to the server from the master. The server response data are also transmitted via the master. However, this diversion is invisible for slave applications.

The task of a server is described using the model of a virtual field device (VFD).

The VFD model unambiguously represents that part of a real application process, which is visible and accessible through the communication. A real device contains process objects. Process objects include the entire data of an application process (e.g., measured values, programs, or events). The process objects are entered in the object dictionary (OD) as communication objects. The OD is a standardized public list, in which communication objects are entered with their properties. To ensure that data are exchanged smoothly in the network, additional items must be standardized, in addition to the OD, which can be accessed by each device. This includes device features such as the manufacturer name or defined device functions that are manufacturer-independent. These settings are used to achieve a closed and manufacturer-independent representation of a real device from the point of view of the communication system. This kind of modeling is known as a VFD.

15.4 Diagnostics

The system diagnostics play an important role in practical applications. In increasingly complex systems, errors must be located quickly using system diagnostics and clearly indicated to the user. In addition to detecting errors, good error diagnostics include reliable error localization. For message-oriented fieldbus systems with a bus structure, only one telegram is ever transmitted to a device at any one time. An error, which affects the system via a specific device or a device nearby, can even destroy telegrams, which are not themselves directed at the faulty device but may be directed at remote devices. It is therefore virtually impossible to determine the exact error location.

INTERBUS uses the CRC procedure in each device to monitor the transmission paths between two devices and, in the event of CRC errors, can therefore determine in which segment the error occurred.

An important criterion for maintaining data communication is the response of the master in the event of the following errors:

- Cable break
- Failure of a device
- Short circuit on the line
- Diagnostics of temporary interference—electromagnetic interference (EMI)

In all fieldbus systems, in the event of a line interrupt, the devices after the interrupt are no longer reached. The error localization capability depends on the transmission system used. In linear systems, telegrams are still sent to all devices. However, these telegrams are lost because the devices are no longer able to respond. After a certain period, the master detects the data loss. However, it cannot precisely determine the error location because the physical position of the devices is not known. The system diagrams must be consulted so that the service or maintenance personnel can determine the probable error location (see Figure 15.7).

Unlike linear systems, the individual devices in the INTERBUS system are networked so that each one behaves as a separate bus segment. Following a fatal error, the outgoing interfaces of all devices are fed back internally via a bypass switch. In the event of a line interrupt between the devices, the master activates each separate device in turn. To do this, the master opens the outgoing interface, starting from the first device up until the error location, thus clearly identifying the inaccessible device. The controller board can then clearly assign the error location as well as the

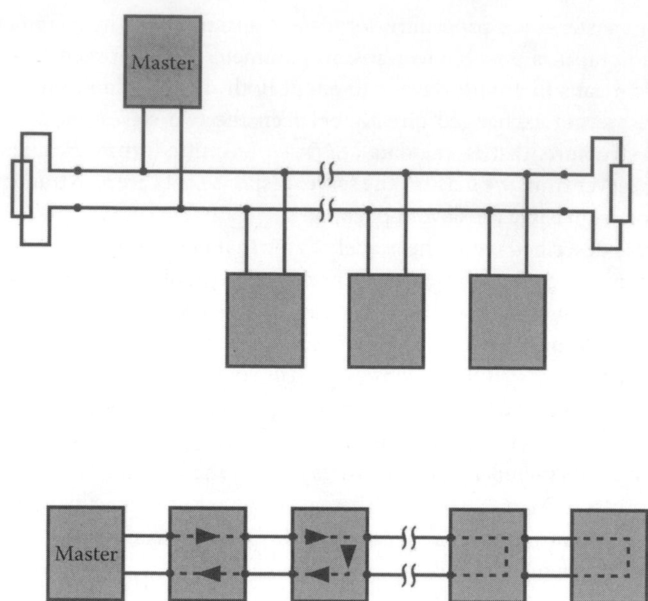

FIGURE 15.7 The behavior of bus systems and ring systems in the event of a cable break.

station or station name and display it in plain text. This is a huge advantage, particularly in large bus structures with numerous devices, where bus systems are typically used.

If a device fails, the fieldbus behaves in the same way as for a line interrupt. However, the functional capability of the remaining stations differs in linear and ring systems. In a linear system, bus operation cannot be maintained because the condition of physical bus termination using a termination resistor is no longer met. This can lead to reflections within the bus configuration. The resulting interference level means that correct operation is not possible. In an INTERBUS ring system, the termination resistor is opened and closed together with a bypass switch, which ensures that the condition of the closed ring is always met. In the event of a line interrupt or device failure, the master can either place the devices in a safe state or start up the remaining bus configuration autonomously.

Short circuits on the line are a major challenge in a bus system. In the event of a direct or indirect (e.g., via ground) short circuit on the line, the transmission path is blocked for the entire section. In linear systems, the transmission line is used equally for all devices, which means that the master cannot reach segment parts either. This considerably reduces further error localization. In the INTERBUS system, the user is aided by the physical separation of the system into different bus segments. As described for the line interrupt, the devices are activated by the master in turn, and the ring is closed prior to the short circuit, which means that subsystems can be started up again. The error location is reported in clear text on the controller board. Linear systems also support a division into different segments. Repeaters, which are placed at specific points, can then perform diagnostic functions. However, a repeater cannot monitor the entire system; it can only cover a defined number of devices per segment. Furthermore, the use of repeaters incurs additional costs, which should not be underestimated, and increased configuration effort.

In summary, the INTERBUS diagnostic features are essentially based on the physical segmentation of the network into numerous point-to-point connections. This feature makes INTERBUS particularly suitable for use with fiber optics, which are used increasingly for data transmission in applications with large drives, welding robots, etc. In linear systems, the use of fiber optics—like bus segmentation—requires expensive repeaters, which simulate a ring structure. Fiber optic path check in the INTERBUS system is another feature, which is not offered by other buses. In this system, a test pattern for the fiber optic cable is transmitted between the interfaces to determine the quality of the connection. If the cable deteriorates due to dirt, loose connections, bending, etc., the transmission power is increased automatically. If a critical value is reached, the system generates a warning message so that the service personnel can intervene before the deterioration leads to expensive downtimes.

Studies by the German association of electrical and electronic manufacturers, ZVEI, and the German Engineering Federation, VDMA, indicate that many bus errors are caused by direct or hidden installation faults. For this reason alone, bus diagnostics simplify start-up and ensure the smooth operation of the system, even in the event of extensions, servicing, and maintenance work. Every bus system should automatically carry out comprehensive diagnostics of all connected bus devices without the need to install and configure additional tools. Additional software tools for system diagnostics often cost several thousand Euro.

In the INTERBUS system, all diagnosed states can be displayed directly on the controller board. If the master has a diagnostic display, various display colors can be used so that serious errors are clearly visible even from a distance. In addition, each master has a diagnostic interface, which can be used to transfer all functions to visualization systems or other software tools.

15.5 Functional Safety

In recent years, safety technology has become increasingly important in machine and system production. This is because complex automation solutions require flexible and cost-effective safety concepts, which offer the same advantages that users have come to appreciate in nonsafe areas. This means that considerable savings can be made, for example, in terms of both cost and time, by changing from parallel to serial wiring.

From the user's point of view, however, various requirements must be taken into consideration. Firstly, safe and nonsafe signals must be separated in order to simplify the programming, operation, and acceptance of safety applications. Secondly, all components used should be operated on a fieldbus, as a standardized installation concept, and standard operation make planning operation, start-up, and maintenance easier. These requirements have led to the safety extension of INTERBUS (Figure 15.8).

As the INTERBUS master, the controller board uses an integrated safe control system.

The INTERBUS controller board with integrated safe controller is the basic unit in the system. It processes all safety-related inputs and confirms them to the standard control system by setting an output or resets the output. This method of operation is similar to existing contact-based safety technology. The enabling of output data is programmed with preapproved blocks such as emergency stop, two-hand control, or electro-sensitive protective equipment in *SafetyProg* Windows software, which is compatible

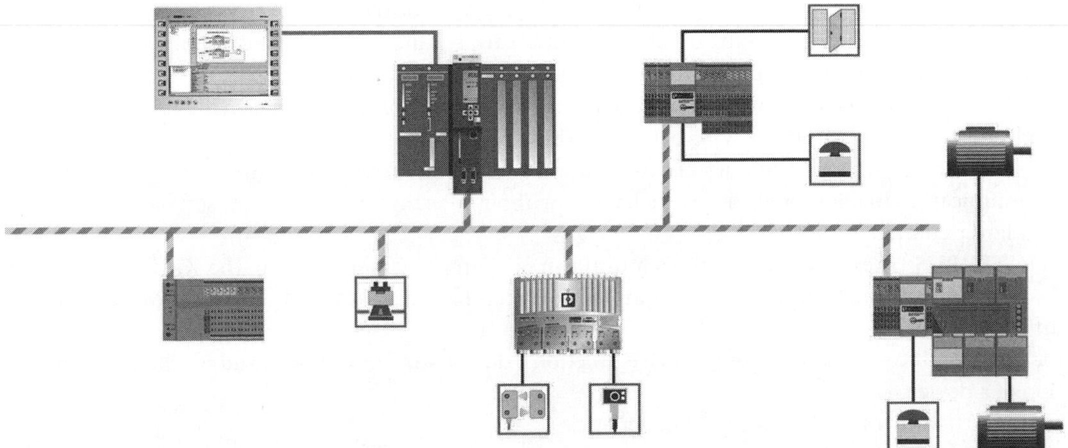

FIGURE 15.8 The safety extension of INTERBUS.

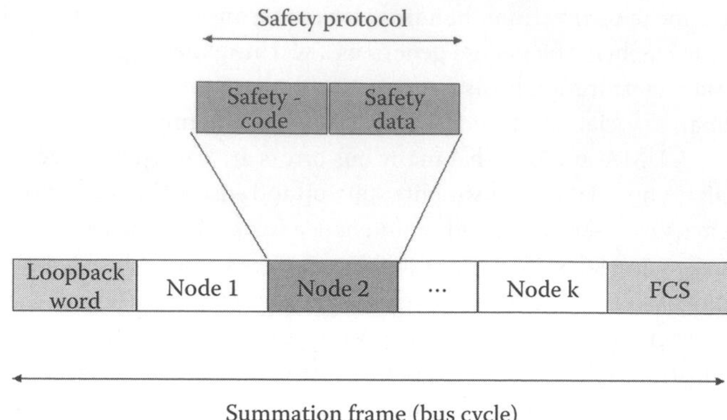

FIGURE 15.9 Safety protocol on top of the INTERBUS summation frame.

with IEC 61131. The amount of programming required is reduced considerably through the use of blocks and the enable principle.

The safe input and output components form the interface to the connected I/Os. They control, for example, contactors or valves and read the input status of the connected safety sensors, including intelligent sensors. The user uses the parameterization function of INTERBUS to select the settings for the I/O components such as clock selection, sensor type, and signal type. The INTERBUS safety system meets safety functions up to Category 4 according to EN 954 [2] and SIL 3 according to IEC 61508 [3]. Depending on the application, the user can choose to use either a *one-cable solution with integrated safety technology* or a *two-cable solution*, where one bus cable is used for standard signals and the other for safety signals.

A safety protocol is used between the safe controller and the connected I/O safety devices. This protocol provides the desired security of data transmission and can only be interpreted by the connected safety devices. The safety data are integrated transparently into the INTERBUS summation frame (see Figure 15.9). This feature ensures the simultaneous operation of standard and safety devices in the bus system.

15.6 Interoperability, Certification

The basic aim of open systems is to enable the exchange of information between application functions implemented on devices made by different manufacturers. This includes fixed application functions, a uniform user interface for communication, and a uniform transmission medium. For the user, this profile definition is a useful supplement to standardized communication and provides a generally valid model for data content and device behavior. These function definitions standardize some essential device parameters. As a result, devices from different manufacturers exhibit the same behavior on the communication medium and can be exchanged without altering the application software when these standard parameters are used.

INTERBUS takes a rigorous approach to the area of interoperability using the XML-based device description Fieldbus Description Configuration Markup Language (FDCML) [4]. FDCML enables the different views of a field device to be described due to the generic device model used. Some examples include identification, connectivity, device functions, diagnostic information, and mechanical description of a device.

FDCML features the following characteristics:

- Can be used for devices of various complexity
- System-independent

- Can be extended by typifying generic FDCML elements
- Supports online changeover to a variety of national languages
- Can be used for devices that support more than one communication system

The FDCML device model is made up of the following basic elements:

- Device identity object
- Device manager object
- Device function object
- Application process object

This electronic device description is used in the configuration software for configuration, start-up, and other engineering aspects. Different applications can use FDCML to evaluate various aspects of a component.

To simplify interoperability and interchangeability, the members of the INTERBUS Club compile a set of standard device profiles in several user groups for common devices such as drives (DRIVECOM Profile 22), human machine interfaces (MMI-COM D1), welding controllers (WELD-COM C0), and I/O devices (Sensor/Actuator Profile 12). These profiles can also be described neutrally with regard to the manufacturer and bus system in FDCML.

The INTERBUS Club and various partner institutes have offered certification for several years to ensure maximum safety when selecting components. Independent test laboratories carry out comprehensive tests on devices as part of this process. The device only receives the *INTERBUS Certified* quality mark, which is increasingly important among users, if the test object passes all the INTERBUS conformance tests. The conformance test comprises tests and examinations, which are carried out by test laboratories using various tools.

The conformance test is divided into the following sections:

- Basic function test (mandatory)
  - General section (valid for all interface types)
  - Fiber optics (for devices with fiber optic interfaces)
  - Burst noise immunity test (mandatory)
- PCP protocol software conformance test (for devices with PCP communication) (dependent)

15.7 Connectivity

As shown in Table 15.1, connectivity is one of the four basic arithmetic operations of field communication. Connectivity is the integration of fieldbus technology in company networks.

However, there are still no standard concepts for *connectivity* solutions. This makes it more difficult to integrate the field level in the company-wide distributed information system and is only possible through increased programming and parameterization.

The Internet protocol (IP) can be used here as an integration tool. IP is increasingly used in automation technology in conjunction with Ethernet and is then frequently referred to as *industrial Ethernet*. In many cases, IP is therefore already well-suited to the field (Figure 15.10).

TABLE 15.1 Four Basic Types of Arithmetic Operations for Field Communication

Signal Acquisition	Quick and Easy Acquisition of Signals from I/O Devices
Functional safety	Transmission of safety-related information (e.g., emergency stop)
Drive synchronization	Quick and precise synchronization of drive functions for distributed closed-loop controllers
Connectivity	Creation of seamless communication between business processes and production

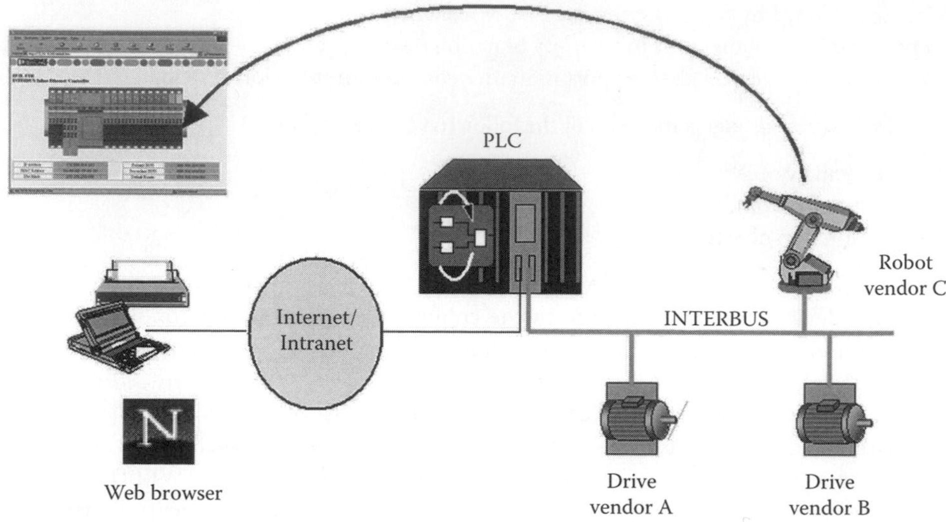

FIGURE 15.10 Connectivity creates new options such as web-based management.

This section deals with transparent IP communication at field level, taking into consideration the real-time requirements. This means that the advantages of fieldbus technology are maintained, and at the same time, the user is provided with a new quality of connectivity. For example, well-known office applications such as browsers can be used to load device descriptions, e-mails can be used to send maintenance messages, or FTP can be used to upload and download files.

Ethernet's role in the future of automation is an important current issue. On the one hand, its specification suggests it could solve all of the communication problems in automation applications and supersede fieldbuses. On the other hand, fieldbuses, with their special characteristics, have arisen because the real world does not consist simply of bits and bytes.

However, Ethernet and INTERBUS can only be fully integrated if transparent communication beyond the boundaries of the system is possible without complex conventional gateway processes. This is achieved by using TCP/IP (Transmission Control Protocol) as the standard communication protocol on both systems. While TCP/IP is now standard on Ethernet, this is by no means the case in the factory environment. Virtually all fieldbus organizations and promoters map their fieldbus protocol to Ethernet TCP/IP in order to protect their existing investments. INTERBUS took a different direction early on and integrated TCP/IP into the hybrid INTERBUS protocol. TCP/IP standard tools and Internet technologies based on TCP/IP can therefore be readily transferred to the factory environment without additional expense.

For example, on INTERBUS, the FTP service can be used to download control programs and other data to a process controller. The use of FTP to upload and download files such as robot programs is just one advantage, since TCP/IP opens up automation to the world of the Internet. Internet browsers will also be the standard user interface of the factory of the future, when all devices have their own integrated web page. Special configuration tools, now supplied for devices by virtually every manufacturer, will no longer be needed, as in the future these devices will be configured through ActiveX controls or Java applets, which are loaded through the network and therefore do not have to be present on the control computer beforehand.

15.8 IP over INTERBUS

Figure 15.11 shows the system architecture for IP tunneling. The known Ethernet TCP/IP structure can be seen on the left, and the extended protocol structure of an INTERBUS device can be seen on the right. An IP router, with the same mechanisms as in the office environment, is used for coupling.

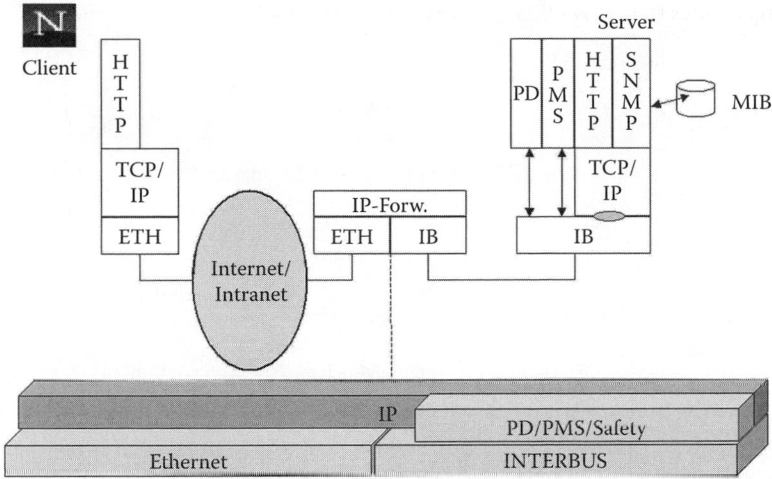

FIGURE 15.11 Basic architecture for IP tunneling.

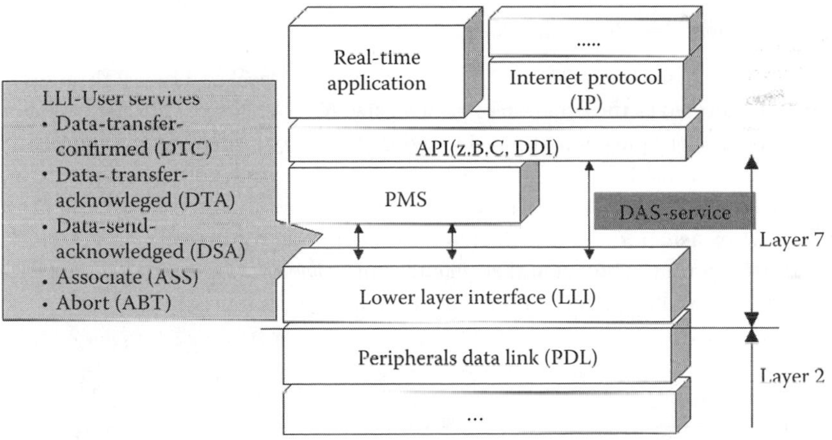

FIGURE 15.12 Protocol architecture of an IP-enabled INTERBUS node.

This function is best performed in the PLC. IP tunneling is performed by introducing a new Data-Send-Acknowledged (DAS) service in the INTERBUS parameter channel (Figure 15.12).

This DAS service enables LLI user protocol data unit (PDU) to be transmitted for unconfirmed, connectionless LLI user services and is used for transparent IP data transmission. These data are transmitted in the same way as the parameter channel peripheral management services (PMS) at the same time as the time-critical process data exchange process data (PD).

15.9 Performance Evaluation

This section considers the performance of the concept in relation to the relevant fieldbus parameters, such as the number of I/O modules and the amount of cyclic I/O data. The achievable data throughput for IP tunneling is a key performance indicator. Due to the determinism of the INTERBUS system, the throughput can be easily calculated.

The following applies to the INTERBUS MAC cycle time T_{IB} [ms]:

$$T_{IB} = 13 \cdot (6 + N) \cdot \frac{1}{Baud\ rate} \cdot 1000 + T_{SW}$$

where

$$N = \sum_{i=1}^{i=k} PL_i$$

The IP throughput (IP_Th) of a device is calculated as follows:

$$IP\_Th = \frac{M-1}{T_{IB}} \cdot 8$$

N	Total payload size: sum of all user data [bytes] of all devices k, where $N \leq 512$ bytes and $k \leq 512$
Baud rate	Physical transmission speed of INTERBUS [bps] {0.5; 2} Mbps
T_{SW}	Software runtime [ms] of the master (0.7 ms, typical, depending on implementation)
PL_i	L2 payload of the ith device [bytes], where $1 \leq i \leq k$, where $k \leq 512$
M	Reserved MAC payload [bytes] for the IP channel of a device (M = 8, typical)
IP_Th	Throughput [kbps] for IP data telegrams

In Figure 15.13, the curve for the INTERBUS cycle time at MAC level and the IP throughput for a baud rate of 2 Mbps is illustrated as the function of payload size N.

For a medium-sized configuration of N = 125 bytes, an IP throughput of 36 kbps at an INTERBUS cycle time of approximately 1.6 ms is achieved. This roughly corresponds to the quality of an analog modem for Internet access. For smaller configurations, even Integrated Services Digital Network (ISDN) quality can be achieved. The calculated values could be confirmed in a practical application. In this configuration, it should be noted that, in addition to the dedicated process data devices, several IP-compatible devices can be operated at the same time with this throughput.

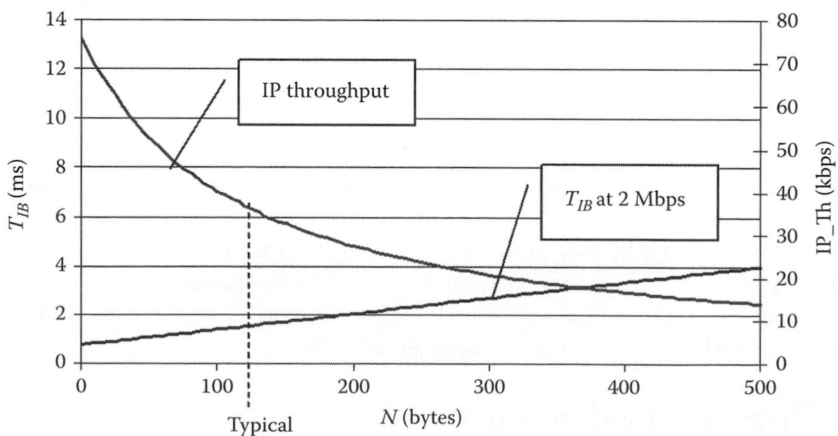

FIGURE 15.13 Performance data.

15.10 Conclusions

The open INTERBUS fieldbus system for modern automation seamlessly connects all the I/O and field devices commonly used in control systems. The serial bus cable can be used to network sensors and actuators, to control machine and system parts, to network production cells, and to connect higher-level systems such as control rooms.

After a comprehensive introduction to INTERBUS, IP tunneling for fieldbuses was described for improved connectivity at field level. An essential requirement is that time-critical process data transport is not affected.

The integration of the IP in INTERBUS creates a seamless communication platform to enable the use of new IP-based applications, which can make it considerably easier to engineer distributed automation solutions. Analysis has shown that the IP throughput can be as high as ISDN quality.

References

1. IEC 61158: Digital data communication for measurement and control—Fieldbus for use in industrial control systems, IEC, Geneva, Switzerland, 2001.
2. DIN EN 954-1: Safety of machinery—Safety-related parts of control systems—Part 1: General principles for design; German version EN 954-1:1996, 1997.
3. IEC 61508: Functional safety of electrical/electronic/programmable electronic safety-related systems—General requirements, 1999.
4. Field Device Configuration Markup Language (FDCML), Specification FDCML 2.0, www.fdcml.org, 2004. (Accessed on March, 2014.)

II

Industrial Ethernet

16 **Switched Ethernet in Automation** *Gunnar Prytz, Per Christian Juel, Rahil Hussain, and Tor Skeie* **16**-1
Introduction • Requirements and Challenges • Network Architectures • Switches • Networking Protocols • Future Outlook • References

17 **Real-Time Ethernet for Automation Applications** *Max Felser* **17**-1
Introduction • Structure of the IEC Standardization • Real-Time Requirements • Practical Realizations • Summary—Conclusion • References

18 **Ethernet for Control Automation Technology** *Gianluca Cena, Stefano Scanzio, Adriano Valenzano, and Claudio Zunino* **18**-1
Introduction • Physical Layer • Data Link Layer • Distributed Clocks • Application Layer • EtherCAT Master Characteristics • References

19 **Ethernet POWERLINK** *Federico Tramarin and Stefano Vitturi* **19**-1
Introduction • Ethernet POWERLINK Protocol • Ethernet POWERLINK Topologies • Ethernet POWERLINK Redundancy • Ring Redundancy • References

20 **IEEE 802.1 Audio/Video Bridging and Time Sensitive Networking** *Wilfried Steiner, Norman Finn, and Matthias Posch* **20**-1
General Overview • Basic Technology • TSN/AVB Protocol Services • Target Domains and Components • References

16

Switched Ethernet in Automation

Gunnar Prytz
ABB Corporate Research

Per Christian Juel
ABB Corporate Research

Rahil Hussain
ABB Corporate Research

Tor Skeie
University of Oslo

16.1 Introduction **16**-1
16.2 Requirements and Challenges **16**-2
Topology • Availability • Performance • Time Synchronization • Security
16.3 Network Architectures **16**-4
16.4 Switches **16**-4
What Is a Switch? • Switch Architecture
16.5 Networking Protocols **16**-7
Redundancy • Quality of Service • Multicast and VLAN Handling • Time Synchronization • Simple Network Time Protocol • Precision Time Protocol
16.6 Future Outlook **16**-14
References **16**-15

16.1 Introduction

Ethernet is a very important communication platform (a physical layer and partly a data link layer) for automation systems. While the fieldbuses are still important in automation with large sales and a huge installed base, there is no doubt that the development of communication technology for automation purposes is now mainly on Ethernet. In addition, wireless communication is being used in some scenarios (mainly wireless sensor networks).

There is a long list of automation protocols developed for various purposes running on top of Ethernet. A key strength of Ethernet is its ability to run many protocols simultaneously on the same network. Ethernet enables flexibility, scalability, and performance in a way not seen before in automation. However, modern networking is a complex domain, which requires investment in competence and product development to be able to succeed. The complexity of automation networks is increasing as new functionality is added to the system. This added complexity is not a negative thing as it both replaces a lot of manual work related to engineering and management and increases the robustness of the systems by the addition of features such as redundancy and accurate synchronization.

A state-of-the-art industrial Ethernet system is therefore not just a communication protocol but includes a carefully designed communication architecture and a set of communication technologies to be used within this architecture. This represents a significant change from fieldbus systems and even early Ethernet-based industrial systems, which utilized Ethernet in a fairly simple way. This chapter will cover some of the most important features of modern switched Ethernet networks.

Before continuing, we need to discuss some basic features of Ethernet and Ethernet-based networks.

1. In the Open Systems Interconnection (OSI—ISO/IEC 7498-1) model, Ethernet covers Layer 1 (the physical layer) and part of Layer 2 (the data link layer) and is defined by IEC 802.3.
2. The term "industrial Ethernet" is often used to describe the combination of Ethernet-based communication protocols targeting industrial automation systems (PROFINET, EtherCAT, IEC 61850, etc.) and a wide variety of fundamental technologies (e.g., switching, routing, and networking protocols).
3. A number of technologies using an enhanced Ethernet data link layer have been developed for the most demanding real-time applications. These solutions (e.g., EtherCAT) support very short communication cycles and deterministic communication requirements. Such technologies typically need some additional hardware support. They are beyond the scope of this chapter.
4. Modern Ethernet networks run at speeds of 100 Mbps and upward. Gigabit Ethernet is now entering into the industrial domain. The development of the Ethernet technology is now going beyond 100 Gbps.
5. Modern Ethernet networks use full duplex communication with no collisions. However, there may be queuing in switches, routers, and end nodes, leading to unpredictable communication delays.

16.2 Requirements and Challenges

16.2.1 Topology

Modern automation systems cover a very wide range of applications, and thus, their physical layout and structure range from simple, small systems with a controller and a few devices to large, very complex systems consisting of very many controllers and devices and IOs in the tens of thousands or even higher. In addition, a single system may consist of various parts responsible for completely different applications such as process control, power distribution, distributed motion, and surveillance. There are also often tough requirements on flexibility and scalability and the coexistence of various protocols on the same physical network [1].

In addition to high numbers of controllers and various types of devices, a system may contain up to hundreds or even thousands of Ethernet switches and tens of routers in addition to a wide range of other nodes. The system may also cover a large geographical area (e.g., in the IEC 61850 standard, it is required that *the communication network within the substation should be capable of covering distances up to 2 km*).

The physical topology of a system can be constructed in many ways, but simple ring topologies and more complex ring-of-rings topologies are common in large installations, for example, within process automation. The line and star topologies are also common, especially in factory automation-related systems. More meshed types of networks are also used, but such networks are typically more complex to handle from a network redundancy perspective (see discussion later in this chapter). This could change with the arrival of new solutions utilizing the potential of meshed networks (e.g., Shortest Path Bridging [SPB] [2]).

16.2.2 Availability

Availability is an important requirement in most automation systems. It is basically a measure of the proportion of time a system is in the operational state. In many automation systems, 24/7 operation is expected and every hour of downtime may have an excessive cost. As an example, there are process automation systems where each hour of downtime may cost millions of dollars in lost production.

The availability of a system depends on a wide range of factors including the quality of the components used, the amount of induced stress, the robustness of the communication, the extent of status and health monitoring of the equipment, the frequency of service, and the system complexity.

The availability of the communication network has a huge impact on the overall availability of the system. The communication system is the backbone of any automation system, and it is crucial that it is working so that the applications running are not negatively affected. The amount of time a system can tolerate that the communication is not working varies from virtually zero to seconds. This time is often expressed as the recovery time. The requirements on recovery time found in the industry can somewhat be simplified and divided into three categories.

In the process automation area (chemical industry, petrochemical, pulp, and paper), the system recovery time requirements are typically from hundreds of milliseconds up to a few seconds. Factory automation applications typically involve movement of machinery such as conveyor belts and may have system recovery time requirements of around 50 ms. The toughest recovery requirements are found in motion control applications involving robots or synchronized motor drives, in systems like drive-by-wire, electrical power distribution systems, and other application areas where maximum recovery times are down to a few milliseconds. The above numbers are just indications but this implies that there are industrial requirements on network recovery times ranging from more or less zero to seconds.

16.2.3 Performance

Communication performance is a term comprised of several factors such as data update rates or cycle times, time synchronization accuracy, jitter, latency, and data throughput.

Automation systems vary widely in their requirements on cycle times (or data update rates), but cycle times ranging from approximately 10 μs up to seconds can be found. Process automation systems typically go down to the milliseconds in cycle times, whereas some power-related applications and discrete automation may have cycle times going down to 10 μs. Some systems use event-based communication instead of cyclic communication, in such systems the fastest update rates may be in the milliseconds range.

Jitter and latency are two important factors. Data needs to get there in time, that is, there is a bound on the tolerable latency. Similarly, the jitter in the communication cannot be higher than a certain limit; otherwise, the process will become unstable or with degraded quality.

In this context, it is relevant to discuss whether a switched network can handle all these requirements or not. A switched network is very efficient and flexible in handling a wide range of protocols and functionality. There are normally many protocols competing about the resources and there are delays in switches related to address lookup. In a store-and-forward switch, a maximum sized Ethernet packet (1500 bytes payload) will be delayed at least 123 μs (one forwarding delay) on 100 Mbps Ethernet (12.3 μs on 1 Gbps Ethernet). For so-called cut-through switches, this time is reduced significantly but this comes together with numerous drawbacks). In addition comes time related to address lookup (a few microseconds) and time spent in waiting for already ongoing transmissions to finish (may range up to several forwarding delays in the worst case). There are Quality of Service (QoS) measures to use to improve this (prioritization, stream reservation functionality, etc.), but there is a limit to what can be achieved with a standard network. For such applications, a set of technologies addressing real-time performance on Ethernet has been developed. EtherCAT is one of these and offers deterministic cycle times down to the 10 μs range with delays through a node of approximately 1 μs (independent of the packet size) [3].

16.2.4 Time Synchronization

Time synchronization is an important requirement in many automation systems and is used in a wide range of scenarios. An automation system typically needs time synchronization for correct time stamping of input data, for control of distributed devices and correct logging of alarms and events.

The components of a distributed automation system thus need to have the same notion of time. Sometimes it is sufficient to use relative time (the nodes share the same time but it is not necessary for the process to relate it to the actual time in the physical world) but often it is needed to use absolute time (e.g., to be able to relate specific events to the actual time). It should be noted that in some rare cases only frequency synchronization is required but this will not be handled here.

In the process automation domain the, typical time synchronization accuracy required for sequence of event timestamping is 1 ms. In power system automation, there is a need for synchronized sampling of voltages and currents and timestamped values for protection functions. The accuracy requirement may be as challenging as 1 μs in these applications. Within discrete automation and motion control, the requirements on time synchronization are ranging from the microseconds level to the hundreds of microseconds, depending on the application.

16.2.5 Security

Security is becoming more and more important in automation systems, and the customers are expecting that the systems are secure against all reasonable threats. This is a huge challenge but the automation industry has been targeting this for years and many solutions are used already. Due to space limitations, the topic of security will not be discussed in this text.

16.3 Network Architectures

The topology of a network refers to the physical layout of the devices. Ethernet-based process automation networks have in recent years favored the ring topology due to its simplicity and the abundance of network redundancy protocols for rings. Star networks are typical for more controller-centric architectures but not very efficient from a networking perspective. Networks may have any layout, so meshed or partially meshed networks are also common; however, it is normally more difficult to achieve determinism and low recovery times in such systems. Recent developments within Ethernet technologies used in other domains may change this and lead to new and flexible topologies (e.g., using technologies related to SPB [2]).

There is no single network topology that is the best in all situations. This depends on the requirements for availability, the physical layout of the system, and other requirements such as cost and real-time performance.

16.4 Switches

16.4.1 What Is a Switch?

Switches are crucial infrastructure components of modern Ethernet-based communication networks. The main function of a switch is to decide for each incoming packet on which port to send the packet out on. The switch may decide to send the packet out on either one, many, all, or no other ports. In this part, we shall see how this is done.

The switch functionality is defined by IEEE 802.1D [4], which describes an architecture and a set of functions distributed throughout this architecture [5]. Ethernet switches are multiport devices designed for efficient communication and utilize various filtering mechanisms to optimize the bandwidth utilization. In addition to the handling of user data communication, they also provide a set of network management–related functionality. Thus, switches are in a completely different league from their predecessors, the hubs, which were basically just signal repeaters, that is, a packet coming in on one port was immediately mirrored out on all other ports.

While a hub acts at the physical layer (Layer 1 in the OSI model), a switch works at the data link layer (Layer 2 in the OSI model). Switches take decisions based on the data link layer destination address of packets, that is, the media access control (MAC) address. Correspondingly, routers work at the transport layer (Layer 3 in the OSI model) and deal with IP addresses.

An actual switch is normally implemented with a combination of switching silicon and a central processing unit (CPU) (sometimes called the management CPU) although some switches have small, inbuilt CPUs and some CPUs have integrated switching capabilities. The communication between the CPU and the switch silicon can occur via various interfaces, including using one of the switch ports.

16.4.2 Switch Architecture

Ethernet switches as defined by IEEE 802.1D in principle consist of three parts as shown in Figure 16.1:

1. Two or more MAC entities
2. A MAC relay entity
3. A number of higher layer entities (HLEs)

These parts interact by exchanging both Ethernet frames and control information.

16.4.2.1 MAC Entity

The MAC entity (sometimes called the MAC port entity) contains the media access part of the switch, including the connectors. A switch with eight physical ports thus has eight MAC entities. The switch MAC entity receives and transmits Ethernet frames (ingress and egress frames, respectively) and validates the Ethernet checksum (it may also calculate the checksum of egress frames). Ingress frames are either sent to the MAC relay entity or to an HLE of the switch if they are not discarded for some reason. Similarly, egress frames come from either the MAC relay entity or from an HLE of the switch. A frame is discarded if the checksum is incorrect or if it does not meet the criteria of a valid frame. Examples of invalid frames are frames with an invalid packet format or an invalid virtual local area network (VLAN) ID. VLAN is a technology where a physical network (local area network [LAN]) can be separated into several virtual networks, LANs. VLANs will be discussed later in this chapter.

A MAC entity should have a unique MAC address; thus, an eight-port switch will need eight unique MAC addresses, one for each port. These addresses should not be used by other functions inside the switching device as that may lead to confusion by other switches in the network.

The MAC entity decides whether an ingress frame is addressed to an HLE of the switch. HLEs are identified by multicast MAC addresses. If there is a match, the frame will be sent to the appropriate HLE. If there is no match, the frame will be checked for VLAN conformance. The MAC entity handles the recognition, interpretation, addition, and removal of VLAN tags; and only frames that match the VLAN criteria will be accepted.

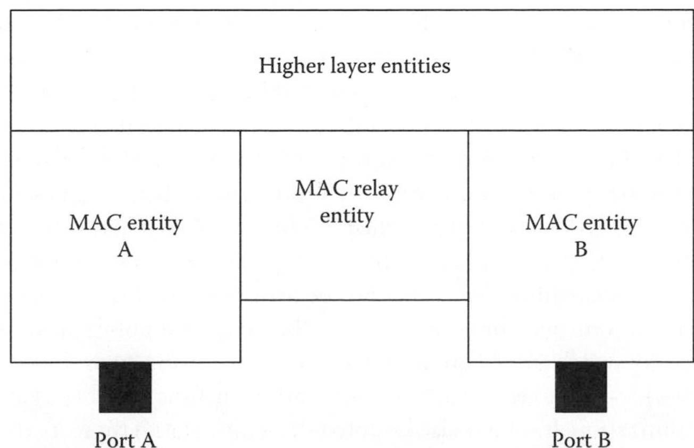

FIGURE 16.1 An Ethernet switch according to IEEE 802.1D.

If it has a VLAN ID, it is checked whether this VLAN ID is valid for the specific port. If so, the frame is forwarded to the MAC relay entity. For egress frames, the MAC entity may add a VLAN tag to the frame.

16.4.2.2 MAC Relay Entity

The MAC relay entity contains the typical switch functions that distribute packets between the switch ports. The main function of a switch is to decide for each incoming packet on which ports to send the packet out again. It does so based on the information found in an address table. This address table contains mappings between MAC addresses and ports; that is, it contains information about which port or ports frames with a certain destination address should be transmitted on. The information in this address table is normally collected by learning from which port traffic with a certain source address is coming in on, but it can also be put there by certain protocols or by manual configuration. The address table contains both unicast and multicast MAC addresses. The address table is more precisely called a filtering database. Based on the information in the filtering database, the switch may decide to send a packet out on one, many, all, or no other ports.

When the MAC relay entity receives a packet with an unknown destination address (i.e., the switch does not know at which port the destination can be found), it normally forwards this frame on all ports. This is similar to how a broadcast frame is handled. Broadcast frames will be forwarded on all ports. When the switch has learned the location of the destination (by receiving a frame from it), an entry in the MAC address table will be made to reflect this. The learned information in the address table will be deleted after the so-called aging time has passed. The aging time is the time passing since a frame from this source has been received. The aging functionality ensures that the address table is not filled up with information that is no longer relevant. The default aging time is 300 s; however, it can be set to as low as 10 s.

The MAC relay entity has (at least) the following functions:

1. Store-and-forward functionality for Ethernet frames
2. Address learning (based on source MAC address)
   a. Unknown destination addresses are normally broadcasted
3. Frame filtering (based on destination MAC addresses)

Furthermore, the MAC relay entity has several address tables:

1. Unicast MAC address table
   a. Port versus MAC address relationships
2. Multicast MAC address table
   a. Port versus MAC address relationships

There are some mandatory entries in the address tables. A certain type of multicast MAC addresses should never be forwarded by the MAC entity. These are multicast MAC addresses in the range 01-80-C2-00-00-00 to 01-80-C2-00-00-0F. These addresses shall be put into a permanent database and shall always be present. Incoming frames with these addresses as destination addresses shall thus never be forwarded by the switch. Such frames are normally associated with HLEs (will be described next) and will be transferred to this part of the switch by the MAC port entity. These frames will only pass into the MAC relay entity if the switch does not support the HLE (protocol) associated with this MAC multicast address. The *never forward* MAC multicast addresses then ensure that such frames are not passed on to any other switch. The logic behind this is that protocols using such MAC multicast addresses as the destination address should only go from one switch to the next (and not do multiple hops) as they are concerned with switch-to-switch information exchange. Transmitting such frames further would lead to confusion by network management tools or degradation in functionality related to, for example, redundancy or synchronization. It should also be noted that frames with these multicast MAC addresses as destination addresses will also pass through so-called blocked ports. Port blocking is discussed later in the availability section of this chapter.

To summarize, the MAC relay entity performs the following functions related to the filtering database:

1. Permanent configuration of reserved addresses
2. Explicit configuration of static filtering information
3. Automatic learning of dynamic filtering information for unicast destination addresses through observation of source addresses of network traffic
4. Aging out of dynamic filtering information that has been learned
5. Automatic addition and removal of dynamic filtering information as a result of Multiple MAC Registration Protocol (MMRP) action

The MAC relay entity then handles all traffic according to the information in the filtering database.

16.4.2.3 Higher Layer Entities

IEEE 802.1D actually requires that at least one HLE be always present, namely, the Rapid Spanning Tree Protocol (RSTP) entity. The RSTP protocol ensures that there are no logical loops in any network as these can lead to forever circulating traffic and network breakdown. RSTP is discussed in more detail later in this chapter.

16.5 Networking Protocols

A modern Ethernet-based network is a combination of basic switching functionality and a set of network management–related protocols. These network management protocols handle crucial functionality in the system such as network redundancy [6], time synchronization, and handling of multicast and VLAN traffic. A network without such network management protocols will only transport date between publishers and receivers in a very simplistic manner and will not be very robust against errors (hardware errors or user errors). There will also be no additional features such as time synchronization. Thus, a modern switched Ethernet network used in automation has a range of protocols in addition to the application-specific protocols (handling the process-related communication). This part describes protocol adding crucial functionality needed in automation systems; however, there are numerous other technologies that are also highly relevant.

16.5.1 Redundancy

As mentioned earlier, availability is an important requirement in many automation systems. The addition of communication redundancy is a common and important means to achieve higher system availability. This section will discuss communication redundancy solutions applicable for automation networks. Redundancy protocols are used to create fault tolerant networks. There are two principal types of network redundancy solutions:

1. Alternate path operation
    - There are parallel paths between devices, but only one path is active.
    - The alternate path is activated if the primary path fails.
    - Some recovery time is needed, typically 10–200 ms.
    - Examples: RSTP and Media Redundancy Protocol (MRP).
2. Parallel path operation
    - There are parallel paths between devices that are used concurrently.
    - Packet duplication.
    - No packet loss in case of a single failure.
    - Requires special HW and/or SW support in the end nodes.
    - Examples: Parallel Redundancy Protocol (PRP) and High-Availability Seamless Redundancy (HSR).

The type of redundancy protocol to use is determined by the requirements of the control application. The main requirement is that a network failure shall not cause loss of the ability to control or protect the process. Industrial requirements on communication are typically expressed in terms of the recovery time: The recovery time equals the time it takes to detect an error after communication has been lost plus the time needed to reconfigure the network to reestablish communication. For solutions based on simultaneous, parallel communication, the recovery time is zero for a single network failure.

The network recovery time is included in the system recovery time, which is the maximum allowable time the system can be down or nonfunctional, either partly or fully. It is important to realize that the network recovery time typically must be significantly lower than the system recovery time for the system to fulfill the availability requirements.

Industrial requirements on recovery time typically range from seconds down to milliseconds. In some special cases, there may be requirements on zero recovery time; that is, the communication has to be functioning at all time. The requirements often assume that only one failure occurs at any given time. This means that a requirement on zero recovery time typically leads to a need for duplicated communication along two independent paths.

It is worth mentioning that network redundancy is only one part of system redundancy measures. When the availability requirements are sufficiently tough, there may be a need to also have duplicated nodes or fully duplicated systems.

16.5.1.1 Spanning Tree Protocol

Before continuing, it is worth taking a short look at the history of network redundancy on Ethernet-based networks. Physical loops were a big problem in the early times of Ethernet usage. Setting up networks with hubs and simple, unmanaged switches could easily lead to physical loops resulting in a situation where broadcasts would circulate forever, rapidly overloading networks into a total crash. The only way to avoid this was through careful manual installation.

It is in practice difficult to avoid loops in medium-to-large-sized networks so the STP was invented to take care of this. STP turns any complex physical network into a logical tree; thus, there is only one available path between any two nodes although several physical paths may exist. STP avoids loops in a network by blocking the traffic on some links (some network management–related traffic still pass through blocked links as mentioned previously). The STP protocol has means to detect a failure of an active path (i.e., a link or a switch) and will, in this case, activate a backup or redundant path so that the nodes can communicate with each other again. With STP, it became possible to design networks with deliberate physical loops to provide communication redundancy.

STP was specified in IEEE 802.1D editions prior to 2004. STP is based on time-outs to discover failures in a network, that is, to check that special test packets arrive within a certain amount of time. These time-outs are fairly large, and in practice, the recovery time using STP is typically around 30 s.

RSTP is an evolution of STP, which significantly decreases the network recovery time. In RSTP, recovery is no longer based on the time-outs found in STP but includes a set of new features to achieve lower recovery times. In RSTP, a failure/recovery detection mechanism is utilized, a handshaking mechanism to rapidly agree on new ports states has been included, and the edge port concept for switch ports connected to end nodes (nodes without switching capabilities) in the network has been introduced. These mechanisms mean that recovery is very much faster than in STP. With RSTP, the recovery time may be approximately 1 s, although considerably faster in some cases.

16.5.1.2 Other Redundancy Solutions for Industrial Networks

Proprietary network redundancy solutions exist in many flavors, but solutions based on the ring topology are popular in industrial networks used in areas such as process automation. Figure 16.2 shows how a ring topology network running a network redundancy protocol typically handles a cable failure.

Typically every switch manufacturer has its own ring redundancy protocol with quite similar working principles, features, and performance. This presents a problem to the industry due to incompatibility

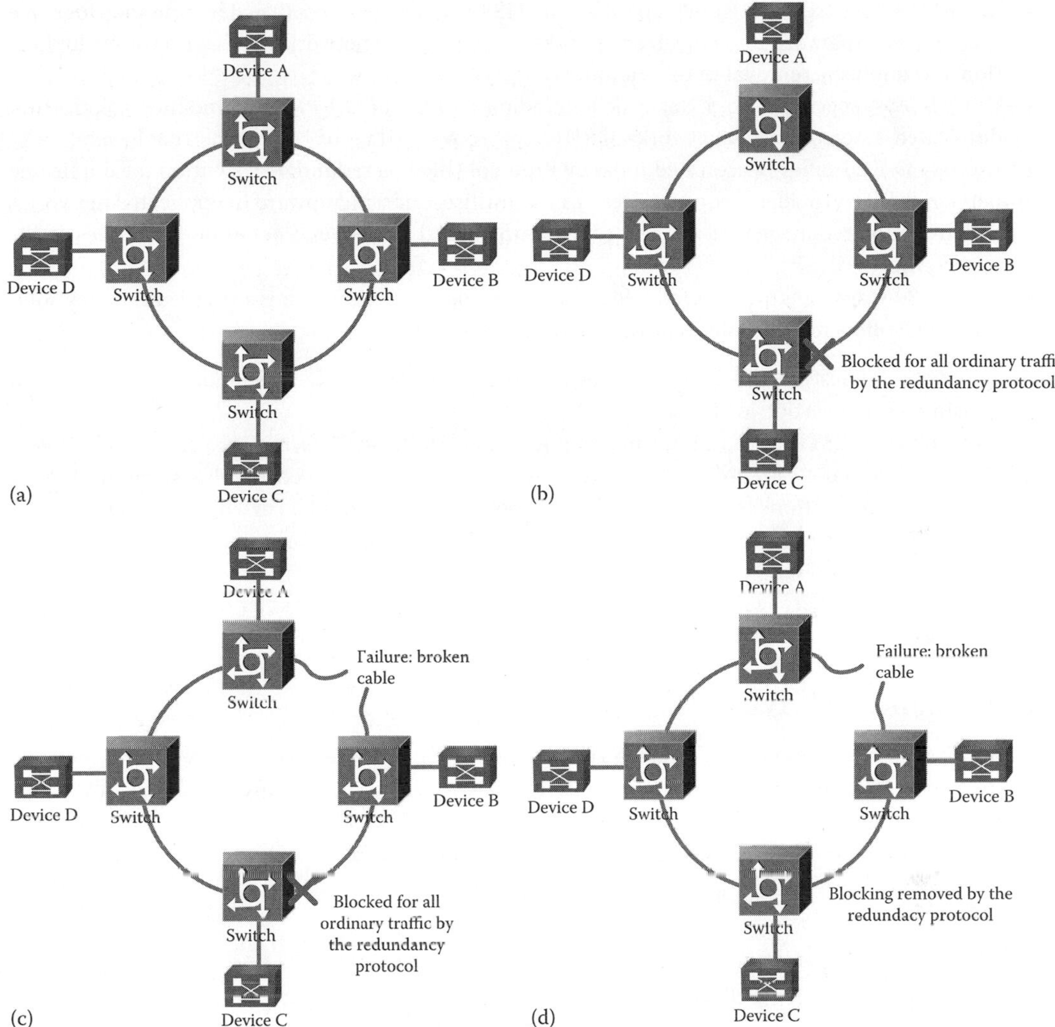

FIGURE 16.2 Ring topology network. (a) Initial network with no blocked ports, (b) the redundancy protocol has blocked a port to avoid a logical loop, but all nodes can still communicate with each other, (c) a failure occurs so that switch B cannot communicate with the other switches, and (d) the redundancy protocol removes the blocking, and all nodes can again communicate with each other.

of switches from different vendors. The ring redundancy protocols are usually, to a great extent, based on the features found in RSTP but with optimizations to get deterministic network recovery times. Ring redundancy protocols provide deterministic recovery times ranging from approximately ten to hundreds of milliseconds.

The IEC 62439 High Availability Automation Networks standard suite has recently been developed to offer standardized network redundancy solutions targeting industrial requirements. This standard currently contains the following variants:

- IEC 62439-1 describes an implementation guideline for RSTP used in ring topologies. This results in typical recovery times of 5 ms multiplied by the number of switches in the ring.
- IEC 62439-2 describes MRP, which is a ring redundancy protocol. MRP provides recovery times of 10, 30, 200, and 500 ms for rings with up to 50 devices (only 14 devices for 10 ms recovery time).

- IEC 62439-3 describes PRP and HSR. PRP and HSR provide seamless (0 ms) redundancy for some specific cases (parallel, unconnected networks or ring-shaped networks, both with frame duplication and duplicate removal in the end nodes).
- IEC 62439-4 specifies Cross-Network Redundancy Protocol (CRP), a redundancy solution for duplicated, interconnected networks. CRP offers recovery times of 1 s or somewhat lower.
- IEC 62439-5 specifies Beacon Redundancy Protocol (BRP), a redundancy solution for duplicated networks. BRP provides recovery times of a few milliseconds and upward in duplicated networks, which are interconnected at one place (and with special beacon end nodes on both sides of the interconnection).
- IEC 62439-6 specifies Distributed Redundancy Protocol (DRP), which is a ring redundancy solution. DRP offers recovery times below 100 ms in most situations.

There are several interesting technologies on the long list of nonstandardized redundancy solutions, but the discussion on these is outside the scope of this document.

When it comes to RSTP, current implementations are typically nondeterministic or at least not very fast. Implementations done according to IEC 62439-1 offer deterministic recovery times, but mixing the two types of implementations leads to unpredictable behavior and should be avoided. The IEC 62439-1 RSTP guideline indicates a recovery time of 5 ms per switch for ring networks, but actual implementations may be faster or slower.

The MRP protocol offers recovery times down to 10 ms for a limited number of devices and higher recovery times for up to 50 devices in a ring.

16.5.2 Quality of Service

QoS on Ethernet is referring to methods to prioritize traffic through the network to reduce latency and jitter for certain types of traffic, that is, to ensure that high priority packets are not delayed by lower-priority packets queued up in a switch [7]. The IEEE 802.1Q [8] Class of Service (CoS) mechanism allows traffic prioritization on Layer 2 Ethernet. There are eight priority levels, 0–7:

- 7—highest priority (network control, e.g., RSTP packets)
- ...
- 1—lowest priority (background)
- 0—default priority (best effort)

A managed switch can assign packets to priority levels by inspecting Layer 2 information in the packet. The value of the priority code point (PCP) field of the VLAN tag in an Ethernet frame maps directly to the CoS priority level. Untagged frames are assigned the port default priority at egress. The priority level can also be set by the end nodes by adding a VLAN tag or a just a priority tag (VLAN ID = 0).

A managed switch usually has four or more priority queues (e.g., critical, high, normal, and background) allowing the scheduling of packets at egress. The mapping of the CoS priorities to the internal priority levels is configurable on the switch. There are different scheduling algorithms applied, for example, strict priority, weighted fair queuing, and weighted round robin. Some managed switches can also use Layer 3 information, the differentiated services code point (DSCP) information in IPv4 and IPv6 headers, to map to the internal priority levels.

16.5.3 Multicast and VLAN Handling

Multicasting means simultaneously sending the same packet to multiple receivers. On an Ethernet network, this is done by using a group or multicast MAC address as the destination address of the Ethernet frame. The network (e.g., the switches) will then distribute the multicast frames to the appropriate devices based on the information in the switch filtering database (a process called multicast filtering).

For unknown multicast MAC addresses (or in switches without support for multicast filtering), this leads to broadcasting of the Ethernet frame on all ports (similar to the handling of unknown unicast MAC addresses).

16.5.3.1 Multicast

Multicasting comes in two flavors, either as MAC (Layer 2) multicasts or as IP (Layer 3) multicasts. MAC multicast frames have a group or multicast MAC address as the destination address of the frame. An IP multicast frame, in addition, has an IP group or multicast address as the destination IP address in the IP header of the frame. For IP multicast frames, the MAC multicast address is calculated from the IP multicast address. However, it should be noted that this is not a one-to-one relationship as there are 32 IP multicast addresses per MAC multicast address.

In the simplest scenario, the multicast filtering is set up manually; that is, the switch MAC address database is configured manually (with multicast address–port relationships). This is called static multicast filtering and may work well in small systems but is difficult to manage in larger networks. In addition to being error prone, manual configurations need to be changed when devices are moved, added, or removed; otherwise, the communication may be disrupted.

In modern networks, most of the multicast filtering is handled by a protocol, a process called dynamic multicast filtering. In this way, a device lets the network switches know what multicast addresses it wants to listen to (i.e., what traffic it wants to receive). This is based on the information the device has about available multicast addresses [9].

For MAC multicasts, the associated protocol is the MMRP which is part of a protocol suite that uses the Multiple Registration Protocol framework. The framework defines various mechanisms (e.g., state machines) for common, proper, and robust parameter registration, whereas MMRP is a protocol that defines the actual parameters and their semantics. The problem is that not all managed switches support the MMRP standard. Industrial switches now start to have support for the MMRP standard.

Most managed switches support a feature called Internet Group Management Protocol (IGMP) snooping. IGMP is a standard specifying how a host can register with a router in order to receive specific IP multicast traffic. IGMP snooping is a feature supported in many switches, facilitating IP multicast filtering for bandwidth optimization. Although IGMP is not a Layer 2 technology, it is still important to have support for it in a switch to configure the filtering behavior of the IP multicast frames; otherwise, IP multicasts may be handled as broadcasts.

It should be noted that IGMP is used only by IPv4 and not by IPv6. In IPv6, another protocol, Multicast Listener Discovery (MLD), which is embedded in ICMPv6, does this task. MLD is used by an IPv6 router to discover the presence of multicast listeners on directly attached links and to discover which multicast addresses are of interest to neighboring nodes.

Layer 2 multicasting is used by protocols like PROFINET, PTP, IEC 61850-8-1 GOOSE, and IEC 61850-9-2 Process Bus, whereas Layer 3 multicasting is used in protocols like FF HSE, Ethernet/IP, and IEEE PTP (PTP may use either Layer 2 or Layer 3 multicasting).

16.5.3.2 Virtual Local Area Network

VLAN technology, specified in IEEE 802.1Q [8], is used to separate logical segments of a network, which operate on the same physical network. VLANs offer complete and secure isolation of the network traffic on Layer 2.

The fundamental principle of VLANs is to limit the broadcast domain. This means that each domain has its own filtering database in the switch, and Ethernet broadcast frames cannot cross domains. The concept of VLANs is realized by tagging each Ethernet frame with a standardized tag. This tag contains the VLAN number a frame belongs to. In a switch, each port is a member of one or more VLANs. When a frame with a specific VLAN tag and a destination address appears, the address is looked up in the filtering database of the VLAN specified by the tag. Note that an end device will work in the same way. The Ethernet port of an end device can also be the member of one or more VLANs, and the frames will be handled differently depending on which VLAN they belong to. An untagged frame will always be assigned to the port default VLAN when entering the switch. The frame will also be assigned the port default priority. The ports used for inter-switch links are normally members of all VLANs. These are often called trunk ports. Frames sent out on a trunk port carry the VLAN tag to specify the VLAN membership.

The VLAN concept increases the complexity of loop-prevention technologies such as RSTP. Since the various VLAN domains are separated, an RSTP instance must run in every domain. Multiple Spanning Tree Protocol (MSTP) has therefore been created to accommodate networks consisting of more than one VLAN.

There are two VLAN configuration techniques. The first technique is to set up VLANs manually; that is, a switch port is assigned to a VLAN. Ingress packets will then normally be tagged with the correct VLAN tag. Only egress packets with the correct VLAN tag will be allowed on the port. This is called static VLAN configuration and is normally used when end devices cannot handle VLANs and VLAN tagging.

In networks where the end devices actively take part in the VLAN configuration and handling, the technique of dynamic VLAN configuration can be used. VLAN configuration is then handled by a protocol called Multiple VLAN Registration Protocol (MVRP) to register and deregister the membership of a VLAN. MVRP is also a protocol using the MRP framework and defines the VLAN-related parameters and their semantics.

16.5.3.3 Multicasting and VLAN Issues in Automation

Flooding of broadcast, unregistered multicast, and unknown unicast messages, causing the devices to receive unwanted messages originating from other protocols, could be a big problem when different protocols coexist on the same physical network if not managed in a proper way.

First of all, this is a concern when applying protocols that use multicast messages

- IEC 61850 GOOSE (L2 multicast)
- FF HSE (L3 multicast)
- EtherNet/IP (L3 multicast)

All devices connected to the LAN would receive all the multicast messages. A typical automation device is usually not designed to handle this kind of traffic, and would potentially not be able to perform its assigned task.

Secondly, an erroneous device might generate a lot of unwanted messages that could disturb all other devices on the network. This can be managed by applying VLAN separation.

16.5.4 Time Synchronization

16.5.4.1 Why Is Time Synchronization Important?

Time synchronization is an important requirement in many automation systems. Time synchronization can be solved by using a dedicated signal line with a common clock signal (e.g., pulse per second [PPS]) between the nodes, but this requires additional cabling and hardware in the nodes and is not very feasible for anything but the simplest of systems. The standard approach is therefore to have synchronized clocks in the nodes and run a protocol on the network that enables the synchronization of these clocks [10].

16.5.5 Simple Network Time Protocol

There are two relevant standard protocols for clock synchronization across an Ethernet-based network—Simple Network Time Protocol (SNTP) and Precision Time Protocol (PTP—IEEE 1588). SNTP is based on Network Time Protocol (NTP) but is, as the name indicates, a simplified version

FIGURE 16.3 The request–reply working principle of SNTP.

TABLE 16.1 Timestamps Used by SNTP

Timestamp Name	ID	When Generated
Originate timestamp	T1	Time request sent by client
Receive timestamp	T2	Time request received by server
Transmit timestamp	T3	Time reply sent by server
Destination timestamp	T4	Time reply received by client

of it. SNTP clients periodically calculate the offset between itself and an SNTP server and adjust their local clocks accordingly. The offset can be calculated with the help of four timestamp values as indicated in Figure 16.3.

The timestamps needed are given in Table 16.1.

The roundtrip delay d and local clock offset t are defined as

$$d = (T4 - T1) - (T2 - T3) \text{ and } t = ([T2 - T1] + [T3 - T4])/2.$$

As can be seen from these equations, an SNTP client assumes that the delays are symmetrical in both directions. This is only valid in very simple situations, and thus, the accuracy that can be achieved is limited. SNTP synchronization can be done every 16 s or slower; thus, the drift of the local clock (i.e., the drift of the clock crystals) may be significant during this time if not properly compensated for. The typical synchronization accuracy that can be achieved with SNTP is in the milliseconds range for moderately sized networks. There are numerous SNTP (NTP) time servers on the Internet, and it is normally possible to synchronize a local clock to such a server to an accuracy in the tens of milliseconds range (SNTP is an IP-based protocol).

16.5.6 Precision Time Protocol

PTP is using a somewhat different method to synchronize local clocks across a network. PTP has several operating modes. The two most important modes are the following:

1. A hierarchical setup with ordinary clocks and boundary clocks. A boundary clock may typically be a switch that is a clock slave of a clock master on one port and a clock master to all connected devices on all other ports. An end node normally has an ordinary clock that can be either a master clock or a slave clock.
2. A setup where network switches have transparent clocks and the end nodes have ordinary clocks (being either a clock master or a clock slave). The transparent clock's primary mission is to add the residence time of the PTP packets inside the device to a data field inside the packets.

PTP clock masters periodically distribute sync messages containing the transmit time. These may be timestamped accurately in hardware at the clock master, or a follow-up packet containing the accurate transmit time from the clock master is sent. These packets (sync and possibly the follow-up) propagate across the network to the slave clock. For accurate synchronization, it is necessary to timestamp the receive time in hardware. Transparent clocks add the accurate residence time so that the slave clocks are able to accurately determine the time that has passed since the master clock timestamp. In this way, it is possible to compensate for the drift between the clock master and the slaves. The offset is similarly compensated for with the addition of a request–response packet exchange initiated from the clock slaves (Delay_request and Delay_response).

The accuracy of a simple, software-only implementation of PTP may be similar to what can be achieved with SNTP, but PTP has some advantages for automation systems. A main advantage is that there are many hardware components (e.g., CPUs, FPGAs, MACs, PHYs, switch silicon) that come with hardware support for accurate timestamping of PTP packets. This means that PTP can accurately determine the amount of time spent for synchronization packets across a network. This feature enables very accurate time synchronization across a network and can also compensate for asymmetric delays. A PTP-compliant network can be synchronized down to an accuracy of better than 1 μs with the assistance of hardware support in all network nodes. Accuracies in the nanoseconds range may even be possible for the most advanced implementations.

It is important to note that while SNTP and PTP can provide the means to synchronize clocks across networks, they do not specify how to update the local clocks. Without taking special care, the result may be that the local clocks of the devices demonstrate significant jumps. As an example, the clock of a device may jump back in time, and the same action may be taken several times. If a clock jumps ahead in time, there is a risk that actions are lost. Thus, it is important to handle the clock update process with care. This is normally handled by implementing some sort of a filtering mechanism so that the clock only adjusts its speed relative to clock master; that is, the clock is only changed in small steps and never back in time.

PTP networks are self-configuring; that is, the devices themselves set up the clock relationships between master clocks and slave clocks. Built-in redundancy ensures that if one clock source fails, then another one will take over. PTP comes as either an Ethernet-based (L2) protocol or an IP-based (L3) protocol. There are many more relevant details of PTP, but these are outside the scope here.

PTP is well established as the time synchronization standard of the future in many domains (both industrial and others) and is already extensively used and fully accepted in industrial networks. Many communication technologies have specified how to use PTP. PROFINET is recommending the usage of IEEE 802.1AS, which is a flavor of PTP. Furthermore, there is a PTP profile called C37.238 developed for usage in IEC 61850 systems.

16.6 Future Outlook

This chapter has described the key functionality of state-of-the-art automation systems. However, there is continuous development ongoing and therefore no trivial thing to foresee what the future will bring. Perhaps the clearest trend is the move from systems using fieldbuses, proprietary communication solutions, and simple Ethernet communication toward automation systems that are advanced Ethernet-based networks with a wide range of functionality as described in this chapter. There is every reason to believe that Ethernet will take over most of the communication in automation systems in the near future. These Ethernet-based networks will be combined with wireless communication for those few applications that need it (e.g., mobility or applications where it is difficult to install cables). Special solutions such as EtherCAT will also be a part of future systems to handle both the most demanding deterministic applications and systems with a large number of IO devices. This will result in a scalability, flexibility, robustness, and richness of functionality far superior to previous solutions. This will also presumably enable completely new applications and functionality.

It is likely that the number of network management protocols in use in automation systems will increase significantly in the years to come. This, in most cases, means that existing technologies used in other domains (e.g., telecom and data centers) will enter the industrial domain.

There are ongoing changes also to many of the network management–related protocols. PTP is currently in the process of being updated, and the communication profiles for automation systems will be revamped to better fit modern communication networks. RSTP (and most other current network redundancy protocols) provides network redundancy, but it is not using the network in a very efficient manner as only one path is used at any given time. The SPB [2] technology is circumventing this by enabling the use of many paths at the same time. This is mainly developed for load balancing in data centers and similar applications but may also offer the opportunity to have parallel communication in automation networks (thus enabling zero recovery time for communication). There is also a somewhat similar approach called Transparent Interconnection of Lots of Links (TRILL).

Security has not been a topic of this chapter, but automation systems will implement an increasing amount of functionality related to secure communication. Similarly, within the area of safe communication, we will see communication networks completely replacing hardwired signals.

Automation systems are clearly in the middle of a paradigm shift with the transition toward fully Ethernet-based networks with advanced networking functionality. Exactly where this development will go remains to be seen, but there are, for sure, both exciting and challenging times ahead. The winners will be the companies that can utilize modern communication technologies in an optimal way to create superior systems.

References

1. P. Ferrari, A. Flammini, S. Rinaldi, and G. Prytz: Mixing real time Ethernet traffic on the IEC 61850 process bus. In *Proceedings of the 9th IEEE Workshop on Factory Communication Systems (WFCS)*, Lemgo/Detmold, Germany, 2012, pp. 153–156.
2. IEEE Std 802.1aq. IEEE standards for local and metropolitan area networks. Media access control (MAC) bridges and virtual bridge local area networks—Amendment 20: Shortest Path Bridging, 2011.
3. G. Prytz: A performance analysis of EtherCAT and PROFINET IRT. In *Proceedings of the 13th IEEE Conference on Emerging Technologies and Factory Automation (ETFA)*, Hamburg, Germany, 2008, pp. 408–415.
4. IEEE 802.1D. IEEE standards for local and metropolitan area networks. Media access control (MAC) bridges, 2004.
5. J. Skaalvik and G. Prytz: Challenges related to automation devices with inbuilt switches. In *Proceedings of the 7th IEEE Workshop on Factory Communication Systems*, Dresden, Germany, 2008, pp. 331–339.
6. G. Prytz: Redundancy in industrial Ethernet networks. In *Proceedings of the 6th IEEE Workshop on Factory Communication Systems*, Torino, Italy, 2006, pp. 380–385.
7. L. Thrybom and G. Prytz: QoS in switched industrial Ethernet. In *Proceedings of the 14th IEEE Conference on Emerging Technologies & Factory Automation (ETFA)*, Palma de Mallorca, Spain, 2009, pp. 1–8.
8. IEEE Std 802.1Q. IEEE standards for local and metropolitan area networks. Media access control (MAC) bridges and virtual bridge local area networks, 2011.
9. L. Thrybom and G. Prytz: Multicast filtering in industrial Ethernet networks. In *Proceedings of the 8th Workshop on Factory Communication Systems (WFCS)*, Nancy, France, 2010, pp. 185–188.
10. T. Skeie, S. Johannessen, and Ø. Holmeide: Highly accurate time synchronization over switched Ethernet. In *Proceedings of 8th IEEE Conference on Emerging Technologies and Factory Automation (ETFA)*, Krakow, Poland, 2001, pp. 195–204.

17

Real-Time Ethernet for Automation Applications

Max Felser
Bern University of Applied Sciences

17.1 Introduction 17-1
17.2 Structure of the IEC Standardization 17-2
17.3 Real-Time Requirements 17-6
User Application Requirements • Performance Indicators of the IEC Standard
17.4 Practical Realizations 17-7
Realization of the "On Top of TCP/IP" Protocols • Realization of the "On Top of Ethernet" • Realizations of the "Modified Ethernet"
17.5 Summary—Conclusion 17-21
References 17-21

17.1 Introduction

International fieldbus standardization has always been a difficult endeavor. After a timely start in 1985 and a few enthusiastic years of development, the quest for the one and only comprehensive international fieldbus gradually became entangled in a web of company politics and marketing interests [1]. What followed was a protracted struggle inside CENELEC* and IEC† committees that finally ended up in the complete abandonment of the original idea. Instead of a single fieldbus, a collection of established systems was standardized. In Europe, CENELEC adopted a series of multivolume standards compiled from the specifications of proven fieldbus systems. On a worldwide scale, IEC defined a matrix of protocol modules, so-called *types* [2], together with guidelines as to how to combine the various modules into actually working fieldbus specifications [3]. With the adoption of the IEC 61158 standard [2] on the memorable date of December 31, 2000, the fieldbus war seemed to be settled just in time for the new millennium.

At the same time, in the office world, we see the penetration of the networks based on Ethernet and Transmission Control Protocol (TCP)/Internet Protocol (IP). The costs of the network infrastructure in the office world are steadily going down, and it is becoming possible to connect almost anything with everything, anywhere, with the help of the Internet technology. But in the field of automation technology, dedicated fieldbuses are used. The only barrier to access devices in the field of the automation world, from the Internet over a network connection, are the fieldbuses. Therefore, the question is why is it not possible to use Ethernet also in the automation technology?

The adoption of the Ethernet technology for industrial communication between controllers, and even for communication with field devices, supports direct Internet capability in the field area, for instance,

* CENELEC = European Committee for Electrotechnical Standardization, see www.cenelec.org.
† IEC = International Electrotechnical Commission, see www.iec.ch.

remote user interfaces via web browser. But it would be unacceptable if the adoption of the Ethernet technology would cause loss of features required in the field area, namely,

- Time deterministic communication
- Time-synchronized actions between field devices such as drives
- Efficient and frequent exchange of very small data records

An implicit but essential requirement is that the office Ethernet communication capability is fully retained so that the entire communication software involved remains usable.

This results in the following requirements:

- Support for migration of the office Ethernet to real-time Ethernet (RTE); the definition is given in the following.
- The use of standard components: bridges, Ethernet controllers, and protocol stacks as far as possible.

To achieve the required higher quality of data transmission with limited jitter and disturbances due to TCP/IP data traffic, it may be necessary to develop further network components. In short, the RTE is a fieldbus specification that uses Ethernet for the lower two layers.

As a matter of fact, neither industrial RTE devices can be as cheap as in the office world (limited by the scale of industrial deployment) nor can plain Ethernet be applied to control applications demanding some sort of hard real-time (RT) behavior; for details of the argument, see [4]. To cope with these limitations, many research projects proposed solutions for the introduction of quality of service, modifications to packet processing in switches, or synchronization between devices.

The IEC/SC65C* committee, in addition to the maintenance of the international fieldbus and its profile, finished a standardization project and defined additional aspects of RTE. And as in the case of the fieldbus, there are several competing solutions and their proponents represented.

This paper will give an outline of this new document and the requirements specified for the RTE standardization. All solutions in this standard able to handle RT requirements will be presented with their key technical features.

17.2 Structure of the IEC Standardization

All industrial protocols are defined in IEC 61158 [2]. This document is structured according to the Open System Interface (OSI) reference model in seven parts, as shown in Table 17.1. Inside parts 2–6, all networks are identified by types. So there exist 24 different types of networks in six different parts.

In the IEC 61784 standard, different sets of profiles are collected as listed in Table 17.2. In IEC 61784-1 [3], the profile sets for continuous and discrete manufacturing relative to fieldbus use in industrial control systems are defined. Inside this first profile, some versions based on Ethernet technology are also defined. In the second standard IEC 61784-2 [5], additional profiles for ISO/IEC 8802-3 (Ethernet)-based communication networks in RT applications are defined. To identify all these profile, a classification with communication profile families (CPF) according to Table 17.3 is introduced. Every CPF is free to define a set of communication profiles (CPs). The complete set of CPs and the related types are listed in Table 17.4.

Additional profiles listed in Table 17.2 cover functional safe communications, secure communications, or installation profiles for communication networks. These profiles are also separated according to the same CPF but are not discussed any further in this chapter.

---

* International Electrotechnical Commission (IEC) is organized in Technical Committees (TC) and Subcommittees (SC), TC65 deals with Industrial-Process Measurement and Control and SC65C with Digital Communication and has the scope to prepare standards on Digital Data Communications subsystems for industrial-process measurement and control as well as on instrumentation systems used for research, development, and testing purposes.

TABLE 17.1 Structure of IEC 61158

IEC 61158-1	Introduction
IEC 61158-2-x	PhL: Physical Layer
IEC 61158-3-x	DLL: Data Link Layer Service
IEC 61158-4-x	DLL: Data Link Layer Protocols
IEC 61158-5-x	AL: Application Layer Services
IEC 61158-6-x	AL: Application Layers Protocol
IEC 61158-7	Network Management

Note: x indicates the related CPF.

TABLE 17.2 Standards Related with Profiles

IEC 61784-1	Profile sets for continuous and discrete manufacturing relative to fieldbus use in industrial control systems
IEC 61784-2	Additional profiles for ISO/IEC 8802-3 based communication networks in RT applications
IEC 61784-3-x	Profiles for functional safe communications in industrial networks
IEC 61784-4-x	Profiles for secure communications in industrial networks
IEC 61784-5-x	Installation profiles for communication networks in industrial control systems

Note: x indicates the related CPF.

TABLE 17.3 List of Communication Profiles Families

CPF1	FOUNDATION™ Fieldbus
CPF2	ControlNet™
CPF3	PROFIBUS
CPF4	P-NET®
CPF5	WorldFIP®
CPF6	INTERBUS®
CPF7	SwiftNet
CPF8	CC-Link
CPF9	HART®
CPF10	VNET/IP
CPF11	TCnet
CPF12	EtherCAT®
CPF13	EPL
CPF14	EPA
CPF15	MODBUS® - RTPS
CPF16	SERCOS
CPF17	RAPIEnet
CPF18	SafetyNET p™
CPF19	MECHATROLINK

TABLE 17.4 Relation between CPF, CP, and Type of Protocol

IEC 61784			IEC 61158 Services and Protocols			
Family	Part 1	Part 2	Phy	DLL	AL	Brand Names
Family 1						*Foundation Fieldbus (FF)*
	Profile 1/1		Type 1	Type 1	Type 9	Foundation—H1
	Profile 1/2		8802-3	TCP/UDP/IP	Type 5	Foundation—HSE
	Profile 1/3		Type 1	Type 1	Type 9	Foundation—H2
Family 2						*CIP*
	Profile 2/1		Type 2	Type 2	Type 2	ControlNet
	Profile 2/2	Profile 2/2	8802-3	TCP/UDP/IP	Type 2	EtherNet/IP
		Profile 2/2.1	8802-3	TCP/UDP/IP	Type 2	EtherNet/IP with time synchronization
	Profile 3/3		Type 2	Type 2	Type 2	DeviceNet
Family 3						*PROFIBUS & PROFINET*
	Profile 3/1		Type 3	Type 3	Type 3	PROFIBUS DP
	Profile 3/2		Type 1	Type 3	Type 3	PROFIBUS PA
	Profile 3/3		8802-3	TCP/IP	Type 10	PROFINET CBA
		Profile 3/4	8802-3	Type 10	Type 10	PROFINET IO Class A
		Profile 3/5	8802-3	Type 10	Type 10	PROFINET IO Class B
		Profile 3/6	8802-3	Type 10	Type 10	PROFINET IO Class C
Family 4						*P-NET*
	Profile 4/1		Type 4	Type 4	Type 4	P-NET RS-485
	Profile 4/2		Type 4	Type 4	Type 4	P-NET RS-232 (removed 2013)
		Profile 4/3				P-NET on IP
Family 5						*WorldFIP*
	Profile 5/1		Type 1	Type 7	Type 7	WorldFIP (MPS,MCS)
	Profile 5/2		Type 1	Type 7	Type 7	WorldFIP (MPS,MCS,SubMMS)
	Profile 5/3		Type 1	Type 7	Type 7	WorldFIP (MPS)
Family 6						*INTERBUS*
	Profile 6/1		Type 8	Type 8	Type 8	INTERBUS
	Profile 6/2		Type 8	Type 8	Type 8	INTERBUS TCP/IP
	Profile 6/3		Type 8	Type 8	Type 8	INTERBUS Subset
		Profile 3/4			Type8/10	Link 3/4 to 6/1
		Profile 3/5			Type8/10	Link 4/5 to 6/1
		Profile 3/6			Type8/10	Link 4/6 to 6/1
Family 7						*Swiftnet* (not in the standard anymore)
Family 8						*CC-Link*
	Profile 8/1		Type 18	Type 18	Type 18	CC-Link/V1
	Profile 8/2		Type 18	Type 18	Type 18	CC-Link/V2

TABLE 17.4 (continued) Relation between CPF, CP, and Type of Protocol

	IEC 61784		IEC 61158 Services and Protocols			
Family	Part 1	Part 2	Phy	DLL	AL	Brand Names
	Profile 8/3		Type 18	Type 18	Type 18	CC-Link/LT (Bus powered—low cost)
		Profile 8/4	8802-3	Type 23	Type 23	CC-Link IE Controller Network
		Profile 8/5	8802-3	Type 23	Type 23	CC-Link IE Field Network
Family 9						*HART*
	Profile 9/1		—	—	Type 20	Universal Command (HART 6)
			—	—	Type 20	WirelessHART (IEC 62591)
Family 10						*Vnet/IP*
		Profile 10/1	8802-3	UDP/IP	Type 17	Vnet/IP
Family 11						*TCnet*
		Profile 11/1	8802-3	Type 11	Type 11	TCnet
		Profile 11/2	8802-3	Type 11	Type 11	TCnet-loop 100
		Profile 11/3	8802-3	Type 11	Type 11	TCnet-loop 1G
Family 12						*EtherCAT*
		Profile 12/1	Type 12	Type 12	Type 12	Simple IO
		Profile 12/2	Type 12	Type 12	Type 12	Mailbox and time synchronization
Family 13						*Ethernet POWERLINK*
		Profile 13/1	8802-3	Type 13	Type 13	POWERLINK
Family 14						*Ethernet for Plant Automation EPA*
		Profile 14/1	8802-3	UDP/TCP/IP	Type 14	EPA Master to Bridge (NRT)
		Profile 14/2	8802-3	Type 14	Type 14	EPA Bridge to Device (RT)
		Profile 14/3	8802-3	Type 14	Type 14	EPA Bridge to Device (FRT)
		Profile 14/4	8802-3	Type 14	Type 14	EPA Bridge to Device (MRT)
Family 15						*MODBUS-RTPS*
		Profile 15/1	8802-3	TCP/IP	Type 15	MODBUS TCP
		Profile 15/2	8802-3	TCP/IP	Type 15	RTPS
Family 16						*SERCOS*
	Profile 16/1		Type 16	Type 16	Type 16	SERCOS I
	Profile 16/2		Type 16	Type 16	Type 16	SERCOS II
		Profile 16/3	8802-3	Type 16	Type 16	SERCOS III
Family 17						*RAPIEnet*
		Profile 17/1	8802-3	Type 21	Type 21	RAPIEnet
Family 18						*SafetyNET p TM*
		Profile 18/1	8802-3	Type 22	Type 22	SafetyNET p RTFL
		Profile 18/2	8802-3	Type 22	Type 22	SafetyNET p RTFN
Family 19						*MECHATROLINK*
	Profile 19/1		Type 24	Type 24	Type 24	MECHATROLINK I
	Profile 19/2		Type 24	Type 24	Type 24	MECHATROLINK II

17.3 Real-Time Requirements

The users of an RTE network have different requirements for different applications. These requirements are defined in [5] as performance indicators (PIs). A list of PIs defines the requirements for a class of applications. Every PI has its limits or ranges, and there exists interdependence between these PIs. Every CP has to define which PIs it fulfills in what conditions.

17.3.1 User Application Requirements

The users of an RTE network have different requirements for different applications. One possible classification structure could be based on the delivery time:

- A low-speed class for *human control* with delivery times around 100 ms. This timing requirement is typical for the case of humans involved in the system observation (10 pictures per second can already be seen as a low-quality movie), for engineering, and for process monitoring. Most processes in process automation and building control fall into this class. This requirement may be fulfilled with a standard system with TCP/IP communication channel without many problems.
- In the second class, for *process control*, the requirement is a delivery time below 10 ms. This is a requirement for most tooling machine control system like PLCs or PC-based control. To reach this timing behavior, special effort has to be taken in the RTE equipment: Powerful and expensive computer resources are needed to handle the TCP/IP protocol in RT, or the protocol stack must be simplified and reduced to get these reaction times on simple and cheap resources.
- The third and most demanding class is imposed by the requirements of *motion control*: To synchronize several axes over a network, a cycle time less than 1 ms is needed with a jitter of not more than 1 μs. This can only be reached with Ethernet network with a minimal bit rate of 100 Mbit/s, if both protocol medium access and hardware structure are modified.

17.3.2 Performance Indicators of the IEC Standard

The following PIs are defined in the CPs for RTE (IEC 61784-2):

- Delivery time
- Number of RTE end-stations
- Basic network topology
- Number of switches between RTE end-stations
- Throughput RTE
- Non-RTE bandwidth
- Time synchronization accuracy
- Non-time-based synchronization accuracy
- Redundancy recovery time

Delivery time is the time needed to convey a service data unit (SDU, message payload) from one node (source) to another node (destination). The delivery time is measured at the application layer interface. The maximum delivery time shall be stated for the two cases of no transmission errors and one lost frame with recovery.

The *number of RTE end-stations* states the maximum number of RTE end-stations supported by a CP.

The *basic network topology* supported by a CP comprises the topologies listed in Table 17.5, or as a combination of these topologies.

The *number of switches between RTE end-stations* supported by a CP defines the possible network layout and is also an important indicator.

The *throughput RTE* is the total amount of application data by octet length on one link received per second.

TABLE 17.5 Possible RTE Topologies

Basic Network Topologie	CP
Hierarchical star	CP m/1
Ring (loop)	CP m/2
Daisy chain	CP m/3

Note: A real topology could be any combination of the three basic topologies.

Non-RTE bandwidth is the percentage of bandwidth, which can be used for non-RTE communication on one link.

Time synchronization accuracy shall indicate the maximum deviation between any two nodes' clocks.

Non-time-based synchronization accuracy indicates the maximum jitter of the cyclic behavior of any two nodes, triggered by periodic events over the network for establishing cyclic behavior.

Redundancy recovery time indicates the maximum time from a single permanent failure to the network becoming fully operational again. In this case of a permanent failure, the delivery time of a message is replaced by the redundancy recovery time.

17.4 Practical Realizations

Standard Ethernet is not able to reach the requirements of the RTE. There exist different propositions to modify the Ethernet technology by the research community [4]. The market has adopted also additional technical solutions. All the solutions included in the standardization are presented here in a short description.

Communication interfaces are structured in different layers. In Figure 17.1, a simplified structure of a communication protocol is described. Common to all Ethernet network is the universal cabling infrastructure. Non-real-time (NRT) applications make use of the Ethernet protocols as defined in ISO 8802-3, and the TCP/User Datagram Protocol (UDP)/IP protocol suite. They use typical IP like, for example, hypertext transfer protocol (HTTP) or file transfer protocol (FTP) for the NRT applications. To build an RTE solution, there are, in principle, three different approaches as shown in Figure 17.1. The first is to keep the TCP/UDP/IP protocols unchanged and concentrate all RT modification in the top layer; here, this solution is called "on top of TCP/IP." In the second approach, the TCP/UDP/IP protocols are bypassed, and the Ethernet functionality is accessed directly ("on top of Ethernet"); in the third approach, the Ethernet mechanism and infrastructure itself are modified to ensure RT performance (modified Ethernet).

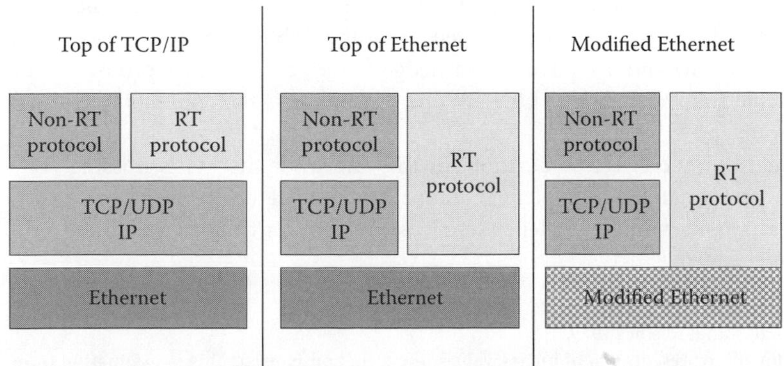

FIGURE 17.1 Possible structures for RT Ethernet.

17.4.1 Realization of the "On Top of TCP/IP" Protocols

Several RTE solutions use the TCP/UDP/IP protocol stack without any modification. With this protocol stack, it is possible to communicate over network boundaries transparently, also through routers. Therefore, it is possible to build automation networks reaching almost every point of the world in the same way as the Internet technology. However, the handling of this communication protocol stack requires reasonable resources in processing power and memory and introduces nondeterministic delays in the communication.

In the international standard IEC 61784-2 [5], all CPs have to list also at least one typical set of PIs as defined in the standard. This allows the end user an easier selection of an appropriate network for his application.

17.4.1.1 Modbus/TCP (Profile 15/1 and 15/2)

Modbus/TCP, defined by Schneider Electric, uses the well-known MODBUS* over a TCP/IP network [7], using port 502, and defined as profile 15/1. This is probably one of the most widely used Ethernet solutions in industrial applications today and fulfills the requirements of the lowest class of applications, which we called human control.

MODBUS is a request/reply protocol (send a request frame and get back a reply frame) and offers services specified by function codes to read or write data objects. These data objects can be discrete inputs, coils,† input registers, or holding registers. In fact, this protocol is very simple, and the actual definition must be extended with service definitions for the integration in international standards.

In addition to the historical MODBUS protocol, new RT extensions have been defined as profile 15/2. These RT extensions use the real-time publisher subscriber (RTPS) protocol [8]. The RTPS protocol provides two main communication models: the publish–subscribe protocol, which transfers data from publishers to subscribers, and the composite state transfer (CST) protocol, which transfers state information from a writer to a reader.

In the CTS protocol, a CTS writer publishes state information as a variable (VAR), which is subscribed by the CTS readers. The user data transmitted in the RTPS protocol from the publisher to one or several subscribers are called an issue. The attributes of the publication service object describe the contents (the topic), the type of the issue, and the quality (e.g., time interval) of the stream of issues that are published on the network. A subscriber defines a minimum separation time between two consecutive issues. It defines the maximum rate at which the subscription is prepared to receive the issues. The persistence indicates how long the issue is valid. The strength is the precedence of the issue sent by the publication. Strength and persistence allow the receiver to arbitrate if issues are received from several matching publications. Publication relation may be best effort (as fast as possible but not faster than the minimum separation), or strict. In the case of the strict publisher–subscriber relation, the timing is ensured with a heartbeat message sent from the publisher to the subscriber (exact timing is middleware-dependent) and a replied acknowledge message. The RTPS protocol is designed to run over an unreliable transport such as UDP/IP, and a message is the contents (payload) of exactly one UDP/IP datagram.

In the standard, any concrete indication for the values for the PIs is missing. They depend very strongly on the performance and implementation of the UDP/IP communication stack. So it is not possible to define an implementation-independent message delivery time, for instance.

---

* Industrial de facto standard since 1979.

† In MODBUS, for the representation of binary values, the term coil is used. This is originating from the ladder-logic where the coil of a relay is used to store binary information.

17.4.1.2 EtherNet/IP (Profile 2/2 and 2/2.1)

EtherNet/IP,* defined by Rockwell and supported by the Open DeviceNet Vendor Association (ODVA)† and ControlNet International,‡ makes use of the common interface protocol (CIP), which is common to the following networks: EtherNet/IP, ControlNet, and DeviceNet [9].

The EtherNet/IP communication technology, standardized in IEC 61784-1 as profile 2/2 (using type 2 specifications in IEC 61158), already provides ISO/IEC 8802-3-based RT communication. In full-duplex switched Ethernet, there is no possibility to get delays due to collisions. But in the switching device, Ethernet frames may be delayed, if an output port is busy with the transmission of an Ethernet frame. This may lead to nondeterministic delays, which are not suitable for RT applications. To reduce these delays, a priority mechanism is defined in IEEE 802.3, which allows the sender of a frame to assign a priority to an Ethernet frame. A VLAN§ tag is added into the Ethernet frame containing a VLAN-ID and a priority level 0–7 of the message. The Ethernet/IP RT messages get the highest priority and are transmitted by the switches before other NRT frames, which results in better accuracy for the RT constraints.

In the CIPsync extensions, the clocks of the devices are synchronized with the IEEE 1588 [10] protocol (accuracy of 0.5 μs). The only problem is that delays may be introduced in the software protocol stack. Based on this time synchronization, the actions in the distributed system are executed based on the planned timing; for example, a device sets its outputs to a defined value not based on the moment a message is received, but on the scheduled time. With this principle, the timing of the application is independent of the delay introduced in the communication network and relies only on the accuracy of the time synchronization. This is defined as profile 2/2.1. When these guidelines are strictly applied, Ethernet/IP is an RT solution usable even for the most demanding classes of applications—compare the range of values in Table 17.6—but it is still not deterministic as a communication network.

CIP defines objects¶ to transport control-oriented data associated with input/output (I/O) devices and other information that are related to the system being controlled, such as configuration parameters and diagnostics. The CIP communication objects and application objects are grouped in classes. Profiles for different types of applications define the objects to be implemented and their relations.

TABLE 17.6 Performance Indicators for Ethernet/IP

Performance Indicator	Profile 2/2	Profile 2/2.1
Delivery time	130 μs–20.4 ms	130–190 μs
Number of end-stations	2–1024	2–90
Number of switches between end-stations	1–1024	1–4
Throughput RTE	0–3.44 M octets/s	0–3.44 M octets/s
Non-RTE bandwidth	0%–100%	0%–100%
Time synchronization accuracy	—	≤1 μs
Non-time-based synchronization accuracy	—	—
Redundancy recovery time	—	—

* EtherNet/IP™ is a trade name of ControlNet International, Ltd. and Open DeviceNet Vendor Association, Inc. IP stands here for Industrial Protocol.

† See www.odva.org.

‡ See www.controlnet.org.

§ VLAN = virtual bridged local area network

¶ An Object in CIP provides an abstract representation of a particular component within a product.

17.4.1.3 P-NET (Profile 4/3)

The *P-NET* on IP specification has been proposed by the Danish national committee and is designed for use in an IP-environment as profile 4/3. P-NET on IP enables the use of P-NET (type 4 in IEC 61158) RT communication wrapped into UDP/IP packages.

P-NET packages can be routed through IP networks in exactly the same way as they can be routed through non-IP networks. Routing can be through any type of P-NET network and in any order.

A P-NET frame has always two P-NET-route elements constructed as a table of destination and source addresses. In the simple case of a fieldbus solution, these two addresses are the node addresses of the fieldbus network. To allow routing over IP-based networks, these P-NET-route tables are now extended to include also IP addresses in the P-NET-route element. For a fieldbus-based P-NET node, these IP addresses are just another format of addresses. This means that any P-NET client can access servers on an IP-network without knowing anything about IP addresses.

In fact, the P-NET on IP specification just defines how the existing P-NET package is tunneled over UDP/IP networks without any special measures to ensure RT behavior on the Ethernet network. The PIs are listed in Table 17.7.

17.4.1.4 Vnet/IP (Profile 10/1)

*Vnet/IP** has been developed by Yokogawa and is included in the IEC document as profile 10/1. The Vnet/IP protocol uses standard TCP/IP protocols for the integration of HTTP or other IPs over the network and special RT extension protocols called Real-time & Reliable Datagram Protocol (RTP).

The Vnet/IP is in fact not an RTE protocol. It just uses the UDP/IP protocol suite to transport the RTP application protocol. No special measures are taken to get a deterministic or even RT behavior. A Vnet/IP network consists of one or more domains connected to each other by routers. The IP unicast and multicast addresses are used as addresses of the data link protocol, and queued communication relations are used.

The minimum cycle time of scheduling of RT traffic is 10 ms, which fulfills the application class of process control. This specification does not cover the limiting of other traffic using the available bandwidth, for example, HTTP or TCP transfer on the same network, which could slow down the RT behavior. The PIs as indicated in the IEC standard are listed in Table 17.8.

At the application layer, different objects like VARs, events, regions, time and network, and the corresponding services are defined. As an example, the VAR object may be accessed over client–server relations with read or write services or publisher–subscriber relations with push or pull mode of operation. In the pull model, the publisher distributes the VAR data periodically by multicasting as requested by a remote subscriber. In the push model, the request is generated locally by the publisher itself.

TABLE 17.7 Performance Indicators for P-NET

Performance Indicator	Profile 4/3
Delivery time	0.564–6.3 ms
Number of end-stations	30
Number of switches between end-stations	4
Throughput RTE	0—3.44 M octets/s
Non-RTE bandwidth	75%
Time synchronization accuracy	—
Non-time-based synchronization accuracy	5.7 ms
Redundancy recovery time	1 s

* Vnet/IP is the trade name of Yokogawa Electric Corporation.

TABLE 17.8 Performance Indicators for Vnet/IP

Performance Indicator	Profile 10/1[a]	Profile 10/1[b]
Delivery time	20 ms	50 ms
Number of end-stations	64	4096
Number of switches between end-stations	7	39
Throughput RTE	10 M octets/s	10 M octets/s
Non-RTE bandwidth	0%–50%	0%–50%
Time synchronization accuracy	<1 ms	<5 ms
Non-time-based synchronization accuracy	—	—
Redundancy recovery time	<200 ms	<600 ms

[a] For two end-stations belonging to the same domain.

[b] For two end-stations belonging to different domains with one lost frame.

17.4.2 Realization of the "On Top of Ethernet"

These RTE realizations do not alter the Ethernet communication hardware in any way but are realized by specifying a special protocol type (Ethertype) in the Ethernet frame. The standard protocol type for IP is Ethertype = 0x0800. These RTE protocols do use, besides the standard IP protocol stack, their own protocol stack identified with their own protocol type. Table 17.9 lists the different values assigned to this Ethertype for these protocols.

17.4.2.1 Ethernet Powerlink (Profile 13/1)

Ethernet Powerlink (EPL) was defined by Bernecker + Rainer (B&R) and is now supported by the Ethernet Powerlink Standardisation Group (EPSG).*

It is based on the principle of using a master–slave scheduling system on a shared Ethernet segment called slot communication network management (SCNM). The master, called managing node (MN), ensures RT access to the cyclic data and lets NRT TCP/IP frame pass through only in time slots reserved for this purpose. All other nodes are called controlled nodes (CN) and are only allowed to send on request by the MN. The MN sends a multicast Start-of-Cycle (SoC) frame to signal the beginning of a cycle. The send and receive time of this frame is the basis for the common timing of all the nodes. It is important to keep the start time of an EPL cycle as exact (jitter-free) as possible. The following time

TABLE 17.9 Protocol Types for Different RTE Profiles Defined in IEC 61784

IEC 61784 Profile	Brand Name	Protocol Type	Owner
Family 3	PROFINET	0x8892	PROFIBUS International
Family 11	TCnet	0x888B	Toshiba Corporation
Family 12	EtherCAT	0x88A4	Beckhoff Industrie Elektronik
Family 13	POWERLINK	0x88AB	ETHERNET Powerlink Standarization Group
Family 14	EPA	0x88CB	Zhejiang University
Family 16	SERCOS	0x88CD	SERCOS International
Family 17	SafetyNET p	0x9C40	Pilz GmbH & Co. KG

* See www.ethernet-powerlink.org.

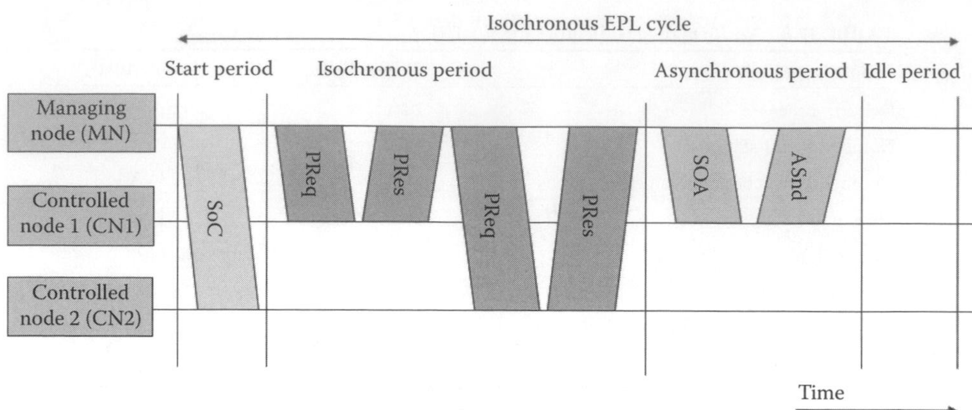

FIGURE 17.2 Ethernet Powerlink timing.

periods exist within one cycle: start period, isochronous* period, asynchronous† period, and an additional idle period. The length of individual periods can vary within the preset period of an EPL cycle. In the isochronous period of the cycle, a Poll-Request (PReq) frame is sent unicast to every configured and active node. The accessed node responds with a multicast Poll-Response (Pres) frame. In the asynchronous period of the cycle, access to the EPL network segment may be granted to one CN or to the MN for the transfer of a single asynchronous message only. The preferred protocol for asynchronous messages is UDP/IP. The Start-of-Asynchronous (SoA) frame is the first frame in the asynchronous period and is a signal for all CNs that all isochronous data have been exchanged during the isochronous period (compare also Figure 17.2). Thus, transmission of isochronous and asynchronous data will never interfere, and precise communication timing is guaranteed.

An EPL network is a *protected Ethernet* defined with one controller acting as the MN and several field devices implemented as CNs. In order to protect the SCNM access mechanism of the MN, non-EPL nodes are not permitted within the *protected Ethernet* itself, as they would corrupt the SCNM access mechanism.

Messages exchanged between the MNs of different *protected Ethernet* segments are synchronized based on distributed clock. With the IEEE 1588 [10] protocol, in every MN, a clock is synchronized, and the messages between the different networks are sent based on the synchronized time in the MNs. The MN includes the routing functionality, including the IP address translation from the network to the outside world. With this synchronization mechanism, RTE communication is also possible among different networks. PIs for a small- and a large-size automation system within a *protected Ethernet* are listed in Table 17.10.

The application layer of the EPL is taken from the CANopen standards provided by the CAN in Automation‡ (CiA) organization [11]. CANopen standards define widely deployed CPs, device profiles, and application profiles. Integration of EPL with CANopen combines profiles, high-performance data exchange, and open, transparent communication with TCP/UDP/IP protocols. These CANopen profiles define process data objects (PDOs) to control the physical process and service data objects (SDOs), which are used to define the behavior of the device as parameters or configuration data. The PDOs are transmitted with the isochronous EPL communication, and the SDOs are transmitted with the UDP/IP protocol. Based on this CP, a variety of CANopen device profiles can be used in an EPL environment without changes.

* From Latin for iso = the same and chronous = time based, so communication at the same time interval.

† Asynchronous is without any synchronization to a reference.

‡ See www.can-cia.org.

TABLE 17.10 Performance Indicators for Ethernet Powerlink

Performance Indicator	Profile 13/1[a]	Profile 13/1[b]
Delivery time	350 μs	2.9 ms
Number of end-stations	4	150
Number of switches between end-stations	0 (1 repeater)	0 (3 repeaters)
Throughput RTE	1.9 M octets/s	4 M octets/s
Non-RTE bandwidth	19.6%	4.4%
Time synchronization accuracy	<1 s	<1 s
Non-time-based synchronization accuracy	<200 ns	<280 ns
Redundancy recovery time	150 μs	2.7 ms

[a] Small-size automation system.
[b] Large-size automation system.

17.4.2.2 TCnet (Profile 11/1)

Time-critical Control Network (TCnet) is a proposal from Toshiba. Like EPL, the TCnet interface goes between the physical and data link layers; the standard MAC* access CSMA/CD† of Ethernet is modified.

In this proposal, there exists a high-speed-transmission period composed of an RT (in TCnet called *time-critical*) cyclic data service and an asynchronous (in TCnet called *sporadic*) message data service. The time-critical cyclic data service is a connection-oriented buffer transfer‡ on preestablished point-to-multipoint connections on the same local link separated by routers, whereas the sporadic message services are unacknowledged messages on an extended link allowed to go through routers.

At the start of the high-speed-transmission period, a special SYN message is broadcasted to all RTE-TCnet nodes. After receiving the SYN-Frame, the node with the number 1 starts sending its data frames as planned during the system configuration. After completion of the transmission of its data frames, it broadcasts a frame called Completed Message (see CMP1 in Figure 17.3). Node *n* upon receiving the CMP (*n* – 1) Completed Message can send out its own data frames. Each node can hold the transmission

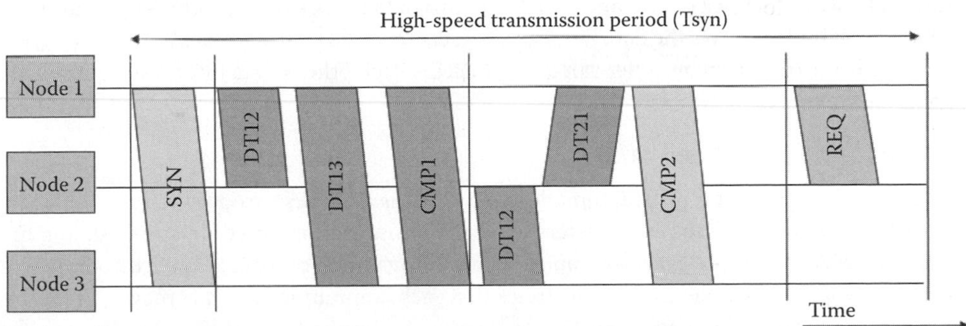

FIGURE 17.3 TCnet timing.

* MAC = media access control.
† CSMA/CD = carrier sense multiple access with collision detection.
‡ In a buffered transfer, a new message overwrites the old value of the previous message in the receiving buffer. This is in contrast to the (standard) queued transfer; where the messages are kept in the receiver in the same order they are sent. Buffered transfer is more suited for control applications than queued: the control application is interested in the actual buffered value and not in the sequence of values.

TABLE 17.11 Performance Indicators for TCnet

Performance Indicator	Profile 11/1[a]	Profile 11/1[b]
Delivery time	2 ms/20 ms/200 ms	2 ms/20 ms/200 ms
Number of end-stations	24	13
Number of switches between end-stations	0 (3 repeaters)	0 (5 repeaters)
Throughput RTE	7.3/6.4/0.9 M octets/s	5.7/5.1/0.6 M octets/s
Non-RTE bandwidth	0%	<20%
Time synchronization accuracy	—	—
Non-time-based synchronization accuracy	<10 μs	<10 μs
Redundancy recovery time	0 s	0 ms

[a] With no non-RTE bandwidth.

[b] With allocated non-RTE bandwidth.

right for a preset time and must transfer the transmission right to the next node within this time. The node holding the transmission right can send cyclic data and sporadic messages. The cyclic data transmission is divided into high-, medium-, and low-speed cyclic data transmission. Each node sends at least the high-speed cyclic data when it receives the transmission right. The other, lower priority, data are sent only depending on the circumstances. Thus, the cycle time for the high-speed cycle is the cycle of the SYN frame, and the cycle time of the medium-speed or low-speed cyclic data is a multiple of the SYN frame cycle.

TCnet is able to handle redundant transmission media. The RTE-TCnet stack manages the selection of two redundant inputs of received frames and two outputs to two redundant transmission media. In the case of collision on one of the mediums, the transmission is continued on the other. The RTE-TCnet accepts the first incoming frame without transmission error from one of the redundant transmission media. This is the reason why in Table 17.11 the recovery time is set to 0.

The RTE-TCnet application layer service defines the common memory system. The common memory is a virtual memory shared over the RTE-TCnet network by the participating application processes running on each node. The common memory is divided into numbers of blocks with different sizes. One node is the publisher of a block of data and broadcasts this data block to all the others by means of cyclic data service. Each node receives the data block as a subscriber and updates its local copy of the common memory. By this means, each controller can quickly access each other's data by accessing its local copy of the common memory.

17.4.2.3 EPA (Profile 14/1 and 14/2)

The *EPA**protocol (Ethernet for plant automation) profile 14 is a Chinese proposal.

It is a distributed approach to realize deterministic communication based on a time slicing mechanism inside the MAC layer. The time to complete a communication procedure is called communication macrocycle and marked as T. Figure 17.4 illustrates that each communication macrocycle (T) is divided into two phases, periodic message transferring phase (Tp) and nonperiodic message transferring phase (Tn). The last part of each device's periodic message contains a nonperiodic message announcement, which indicates whether the device also has a nonperiodic message to transmit or not. Once the Tp is completed, the Tn begins. All devices that announced (during the periodic message transfer phase) that they have a nonperiodic message to send are allowed to transmit their nonperiodic messages in this phase. Two sets of consistent PIs are listed in Table 17.12.

* EPA is the trade name of ZHEJIANG SUPCON CO. LTD.

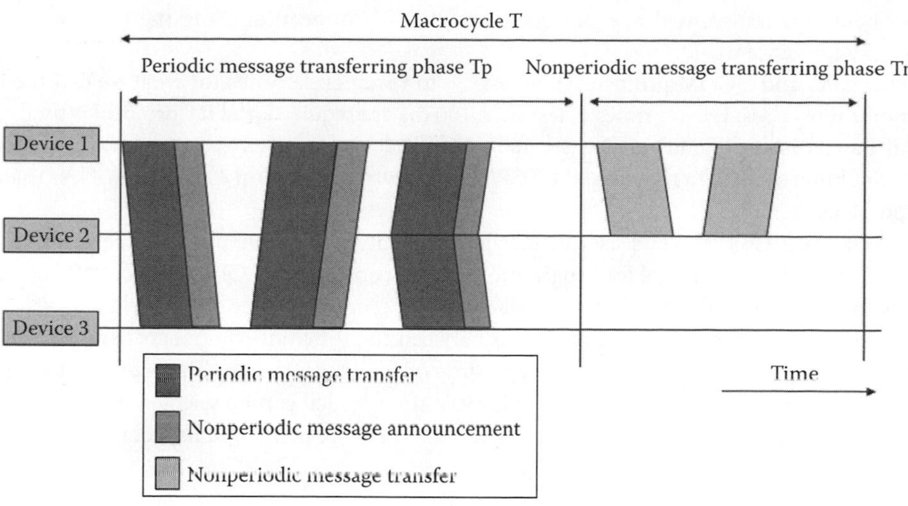

FIGURE 17.4 EPA timing.

TABLE 17.12 Performance Indicators for EPA

Performance Indicator	Profile 14/1	Profile 14/2
Delivery time	5 ms	100 μs
Number of end-stations	32	64
Number of switches between end-stations	4	4
Throughput RTE	1.536 M octets/s	1.536 M octets/s
Non-RTE bandwidth	85%	85%
Time synchronization accuracy	<10 μs	<1 μs
Non-time-based synchronization accuracy	—	—
Redundancy recovery time	<300 ms	<300 ms

In EPA systems, there are two kinds of application processes, EPA function block* application processes and NRT application processes, which may run in parallel in one EPA system. NRT application processes are those based on regular Ethernet and TCP/IP. The interoperation between two function blocks is modeled as connecting the I/O parameters between two function blocks using EPA application services.

17.4.2.4 PROFINET CBA (Profile 3/3)

PROFINET is defined by several manufacturers (including Siemens) and supported by PROFIBUS International† [12]. The first version was based on component-based automation (CBA) and is included in IEC 61784-1 (type 10 in IEC 61158) as profile 3/3.

The mechanical, electrical, and functional elements of an automation device are grouped together into components. Components have inputs and outputs. The values of the input and output VARs of

* A function block is an algorithm with its own associated static memory. Function blocks can be instantiated with another copy of the function block's memory. Function blocks are only accessed via input and output variables.

† See www.profibus.org.

the components are transmitted over the standard TCP/IP connection using the RPC* and DCOM† protocol from the office world.

With this RPC and DCOM protocol, it is possible to reach cycle times for what we call the human control application class. If cycle times of less than 100 ms are required, the RT protocol is used. The RT protocol is based on a special Ethertype (see Table 17.9) and frame prioritization (see explanation in the section about Ethernet/IP). In this case, the TCP/IP stack is bypassed, and cycle times of less than 10 ms become possible.

With PROFINET CBA, the end user defines his automation components with the traditional programming and configuration tool for programming logic controller (PLC). These components are represented by one controller in a machine, a fieldbus network, or any device on the fieldbus itself. For the planning of the installation, logical connections between the different components are defined. These connections specify the data type and the cycle time of the transmission. The supported RT or NRT protocols by the components define the possible cycle time, which can be selected in the planning. As PROFIBUS CBA is defined in the first part of IEC61784, there are no PIs published.

17.4.3 Realizations of the "Modified Ethernet"

Typical cabling topology of Ethernet is the star topology: all devices are connected to a central switching device. With the introduction of the fieldbuses over 10 years ago in the automation applications, this star topology was replaced by bus or ring topologies to reduce the cabling cost. Likewise, the RTE solutions should allow for bus or ring topologies with reduced cabling effort. To permit this daisy-chained bus topology with switched Ethernet, a switch is needed in every connected device.

Most solutions providing hard RT services are based on modifications in the hardware of the device or the network infrastructure (switch or bridge). To allow cabling according to the bus or ring topology and to avoid the star topology, the switching functionality is integrated inside the field device. The modifications are mandatory for all devices inside the RT segment but allow non-RTE traffic to be transmitted without modifications.

17.4.3.1 SERCOS (Profile 16/3)

The former IEC 61491 [13] standard *SERCOS*‡ is well known for its CNC§ control optical ring interface. This standard is now split into an application part and a communication part [5]; the communication part is integrated into the IEC 61158/IEC 61784 set. The SERCOS standard is extended to feature an Ethernet-based solution with the name SERCOS III [14] as profile 16/3.

In a SERCOS system, there is always a master station as a controlling device, and one or up to 254 slave devices as axis controllers each with two Ethernet ports. The basic network topology can be either a daisy chain (line structure) or a ring (ring structure). General use switches are not permitted between any two participants. Only the free port of the last slave in a line structure may be connected to a switch if required by the configuration, for example, for communication with devices via TCP/IP or UDP/UDP.

SERCOS III communication consists of two different logical communication channels: the RT channel and the IP channel.

The communication cycle is initiated by the master and consists of up to four master data telegrams (MDTs) and up to four device telegrams (ATs¶) in the RT channel and the IP channel. MDTs are transmitted by the master and received by each slave (see Figure 17.5). They contain synchronization

* A remote procedure call (RPC) is a protocol that allows a computer program running on one host to cause code to be executed on another host without the programmer needing to explicitly code for this (source: wikipedia.org).

† Distributed Component Object Model (DCOM) is a Microsoft proprietary technology for software components distributed across several networked computers (source: wikipedia.org).

‡ SERCOS stands for SEriell Real time COmmunication System Interface, see also www.sercos.org.

§ CNC stands for computer(ized) numerical(ly) control(led).

¶ Abbreviated from: device (acknowledge) telegram as AT for historical reasons.

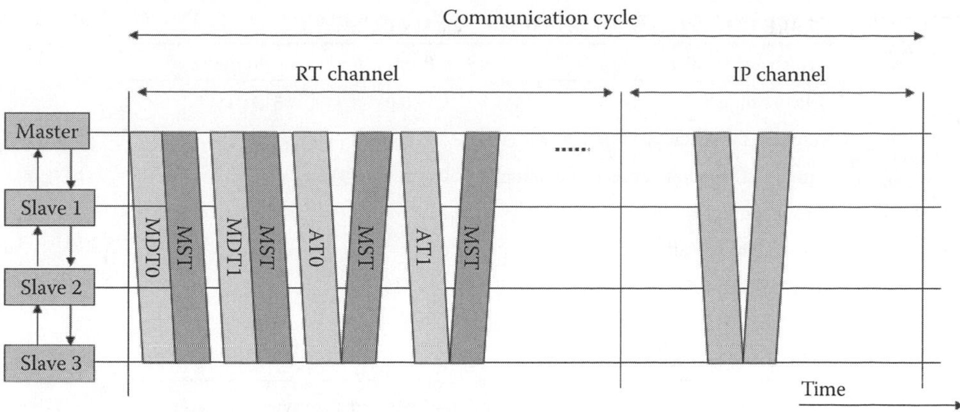

FIGURE 17.5 SERCOS timing.

information and a data record for each slave containing control information, service channel data, and command values sent from the master to the slaves. The ATs are transmitted by the master as an empty frame with predefined fields but without information. Each slave inserts its data into data fields allocated to it in the ATs. Within their data fields in the telegram, the slaves transmit status information, service channel data, and actual values to the master and to other slaves.

The number and the lengths of the RT data telegrams (MDT and AT) are fixed according to a configuration that is also determined during the initialization.

IP telegrams are standard, NRT IP telegrams that can be used for any purpose, and even be omitted. The IP channel length has a fixed duration and determines the maximum number of IP telegrams that can be sent during this duration.

This sequence of transmitting synchronization, RT data telegrams, and IP telegrams is repeated every communication cycle. Defined values for a communication cycle are 31.25, 62.5, 125, 250 μs and integer multiples of 250 μs up to 65,000 μs. The time slots for the RT channel, the IP channel, and the transmission time of the AT are transmitted during initialization and are therefore known to each slave. In every device, a special software, or for a higher performance, an FPGA,* will be needed, which separates the RT channel from the IP channel. PIs for two typical setups are listed in Table 17.13.

The application model of SERCOS is based on the drive model† with a cyclic data exchange. This exchange includes status and actual values transmitted from the drive to the controller, and commands and set-points from the controller to the drive. The functionality of the drive device is determined by setting different parameters in the model.

17.4.3.2 EtherCAT (Profile 12/1 and 12/2)

EtherCAT‡ defined by Beckhoff and supported by the EtherCat Technology Group§ (ETG) uses the Ethernet frames and sends them in a special ring topology [15].

Medium access control employs the master/slave principle, where the master node (typically the control system) sends the Ethernet frames to the slave nodes, which extract data from and insert data into these frames.

From an Ethernet point of view, an EtherCAT segment is a single Ethernet device, which receives and sends standard ISO/IEC 8802-3 Ethernet frames. However, this Ethernet device is not limited to a single

* FPGA = field-programmable gate array, a gate array is a prefabricated circuit, with transistors and standard logic gates.
† A drive model consists of a controller and one or several drives (e.g., motors, servos).
‡ EtherCAT™ is the registered trade name of Beckhoff, Verl.
§ See also www.ethercat.org.

TABLE 17.13 Performance Indicators for SERCOS

Performance Indicator	Profile 16/3[a]	Profile 16/3[b]
Delivery time	<39.8 μs	<513 μs
Number of end-stations	≤9	≤139
Number of switches between end-stations	0	0
Throughput RTE	11.2 M octets/s	≤9.248 M octets/s
Non-RTE bandwidth	0%	25%
Time synchronization accuracy	—	—
Non-time-based synchronization accuracy	<1 μs	<50 μs
Redundancy recovery time	0	0

[a] Scenario with minimum cycle time and high-performance synchronization.

[b] Scenario with non-RTE bandwidth and low-performance synchronization.

Ethernet controller with a downstream microprocessor but may consist of a large number of EtherCAT slave devices. These devices process the incoming frames directly and extract the relevant user data, or insert data and transfer the frame to the next EtherCAT slave device. The last EtherCAT slave device within the segment sends the fully processed frame back so that it is returned by the first slave device to the master as the response frame.

The EtherCAT slave node arrangement represents an open ring bus. The controller is connected to one of the open ends, either directly to the device or via Ethernet switches utilizing the full duplex capabilities of Ethernet; the resulting topology is a physical line (see Figure 17.6). All frames are relayed from the first node to the next ones. The last node returns the telegram back to the first node, via the nodes in between.

In order to achieve maximum performance, the Ethernet frames should be processed "on the fly." This means that the node processes and relays the message to the next node in the line as the message is being received, rather than the other (slower) option of waiting until the message is fully received. If the "on the fly" method of processing is implemented, the slave node recognizes relevant commands and executes them accordingly while the frames are passed on to the next node. To realize such a node, a special ASIC* is needed for medium access, which integrates a two-port switch into the actual device.

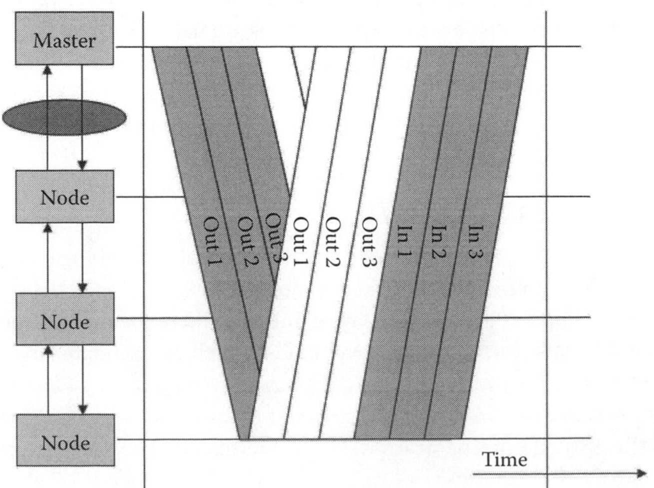

FIGURE 17.6 EtherCAT timing.

* Application-specific integrated circuit.

TABLE 17.14 Performance Indicators for EtherCAT

Performance Indicator	Profile 12/1	Profile 12/2
Delivery time	<150 μs	<519 μs
Number of end-stations	180	650
Number of switches between end-stations	NA	NA
Throughput RTE	10.75 M octets/s	10.5 M octets/s
Non-RTE bandwidth	50.6%	55.9%
Time synchronization accuracy	—	≪1 μs
Non-time-based synchronization accuracy	10 μs	10 μs
Redundancy recovery time	60 μs	200 μs

The nodes have an addressable memory that can be accessed with read or write services, either each node consecutively or several nodes simultaneously. Several EtherCAT telegrams can be embedded within an Ethernet frame, each telegram addressing a data section.* The EtherCAT telegrams are either transported directly in the data area of the Ethernet frame or within the data section of a UDP datagram transported via IP. The first variant is limited to one Ethernet subnet, since associated frames are not relayed by routers. For machine control applications, this usually does not represent a constraint. Multiple EtherCAT segments can be connected to one or several switches. The Ethernet MAC address of the first node within the segment is used for addressing the EtherCAT segment. The second variant via UDP/IP generates a slightly larger overhead (IP and UDP header), but for less time-critical applications, such as building automation, it allows using IP routing. On the master side, any standard UDP/IP implementation can be used on the EtherCAT devices.

For messages, a mailbox mechanism with read and write services is used; for process data output and input, buffered data services are defined.

The performance of the EtherCAT system (when configured to run "on the fly") may reach cycle times of 30 μs if no standard (non-RTE) traffic is added. The maximum transmission unit (MTU) of Ethernet with 1514 bytes corresponding to approximately 125 μs at 100 MBd in the non-RTE phase would enlarge the EtherCAT cycle. Two examples of consistent sets of PIs are shown in Table 17.14. But in EtherCAT, Ethernet telegrams are divided into pieces and reassembled at the destination node, before being relayed as complete Ethernet telegrams to the device connected to the node (see Figure 17.6). This procedure does not restrict the achievable cycle time, since the size of the fragments can be optimized according to the available bandwidth (EtherCAT instead of IP fragmentation). This method permits any EtherCAT device to participate in the normal Ethernet traffic and still have a cycle time for RTE with less than 100 μs.

Similar to EPL, EtherCAT uses the CANopen application layer. The PDOs are mapped to the input and output buffer transfer, which is the same as what is used for EPL. The SDOs, however, are mapped to the mailbox messaging mechanism, rather than the IP protocol, which EPL uses.

17.4.3.3 PROFINET IO (Profile 3/4, 3/5, 3/6)

PROFINET is defined by several manufacturers (including Siemens) and supported by PROFIBUS International† [16]. A second step after the PROFINET CBA definition was the definition of an application model for PROFINET IO based on the well-proven PROFIBUS DP (type 3 of IEC 61158, profile 3/1). The devices are IO controllers to control IO devices with cyclic, buffered data communication. An IO supervisor is used to manage the IO devices and IO controllers in a system.

The exchange of data between the devices may be in different classes of communication service like isochronous RT (IRT), RT, or NRT. NRT traffic is standard TCP/UDP/IP and may also be PROFIBUS

* A data section is a set of memory variables (e.g., inputs or outputs).

† See www.profibus.org.

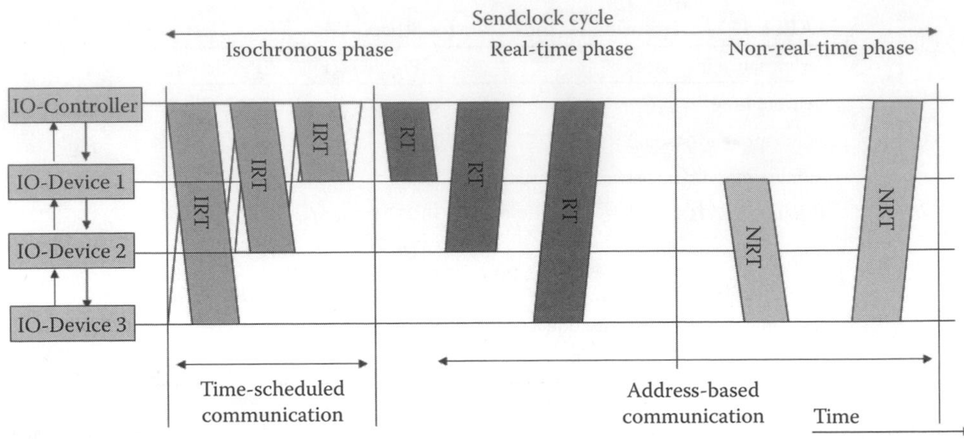

FIGURE 17.7 PROFINET timing.

CBA traffic. In a system with high isochronous cycle requirements, only special PROFINET switching devices are allowed. The Ethernet communication is split into send clock cycles, each with different time phases as presented in Figure 17.7. In the first time phase called isochronous phase, all IRT frames are transmitted. These frames are passed through the switching device without any interpretation of the address information in the Ethernet frame. The switches are set according to a predefined and configured timetable: on every offset time (see Figure 17.7), the planned frame is sent from one port to the other one without interpretation of the address. For even shorter cycle times down to 31.25 µs, these IRT frames are packed in one frame, and every IO device removes its output data and inserts the input data (dynamic frame packing). In the next time phase called RT phase, the switching devices change to address-based communication and behave as standard Ethernet switches. In this addresses-based phases, RT frames are transmitted followed by NRT Ethernet frames (see also Figure 17.7). All PROFINET switching devices are synchronized by means of a modified IEEE 1588 mechanism with "on the fly" stamping [17], to have their cycles and IRT timetables synchronized with 1 µs jitter.

PROFINET CBA and IO do not need any special hardware for RT communication. To ensure good performance, PROFINET IO needs a 100 Mbit/s switched full duplex Ethernet network. For IRT, a special PROFINET-Ethernet switch is needed. It is recommended to integrate this special PROFINET-Ethernet switch in every device to allow all possible Ethernet network topologies as listed in Table 17.15.

The PROFINET specification includes a concept allowing one to integrate existing fieldbuses with proxy devices. A proxy device represents a field device or a fieldbus with several field devices, on the PROFINET network. The user of the PROFINET does not see any difference, if the device is connected to Ethernet or to the fieldbus. This proxy technology is very important to allow for a migration of the

TABLE 17.15 Performance Indicators for PROFINET IO

Performance Indicator	Profile 3/4 and 3/5	Profile 3/6
Delivery time	128 ms	31.25 µs–1 ms
Number of end-stations	60	10–64
Number of switches between end-stations	10	10–63
Throughput RTE	2.324 M octets/s	3.324 M octets/s
Non-RTE bandwidth	23.5%	0%–23.5%
Time synchronization accuracy	<1 ms	<1 µs
Non-time-based synchronization accuracy	—	—
Redundancy recovery time	<200 ms	0 ms

existing fieldbus installations to new Ethernet solutions with PROFINET. Initially, proxies are defined for INTERBUS (type 8 in IEC 61158) and PROFIBUS (type 3 in IEC61158), but today, proxies are also defined for DeviceNet, AS-Interface, and other networks in the field.

17.5 Summary—Conclusion

During the standardization process, there was a long discussion: why is it not possible to reduce the number of technical solutions to three or four RTE profiles? But there was a common understanding from the manufacturers that it is not up to the engineers in the standardization groups to take such decisions. It is agreed that the market should decide which system will be successful in the field. So the problem of selecting a good solution is finally moved to the end user.

This set of standards has been published first in 2007 with 16 different types of networks, was updated in 2010, and will be actualized in 2013 with up to 24 different types of networks. So in fact, the number of different industrial network in the standards is increasing by more than one new network protocol every year. At the moment, there is no upper limit visible for this development.

References

1. M. Felser and T. Sauter: The fieldbus war: History or short break between battles? *IEEE International Workshop on Factory Communication Systems (WFCS)*, August 28–30, 2002, Västerås, Sweden, pp. 73–80.
2. IEC 61158, Digital data communications for measurement and control—Fieldbus for use in industrial control systems, 2013, available at webstore.iec.ch.
3. IEC 61784-1 Ed. 4.0, Digital data communications for measurement and control—Part 1: Profile sets for continuous and discrete manufacturing relative to fieldbus use in industrial control systems, 2013, available at webstore.iec.ch.
4. J.-D. Decotignie: Ethernet-based real-time and industrial communications, *Proceedings of the IEEE*, 93(6), 1102–1117, June 2005.
5. IEC 61784-2 Ed. 3.0: Industrial communication networks—Profiles—Part 2: Additional fieldbus profiles for real-time networks based on ISO/IEC 8802-3, 2013, available at webstore.iec.ch.
6. ISO/IEC 8802-3:2001, Information technology—Telecommunications and information exchange between systems—Local and metropolitan area networks—Specific requirements—Part 3: Carrier sense multiple access with collision detection (CSMA/CD) access method and Physical Layer specifications.
7. Schneider Automation, Modbus messaging on TCP/IP implementation guide, May 2002, available at www.modbus.org.
8. O. Dolejs, P.Smolik, and Z. Hanzalek: On the Ethernet use for real-time publish subscribe based applications, *2004 IEEE International Workshop on Factory Communication Systems*, September 22–24, 2004, Vienna, Austria, pp. 39–44.
9. V. Schiffer: The CIP family of fieldbus protocols and its newest member Ethernet/IP, Emerging Technologies and Factory Automation, 2001. *Proceedings Eighth IEEE International Conference*, Vol. 1, October 15–18, 2001, pp. 377–384.
10. IEC 61588 Ed. 2.0: Precision clock synchronization protocol for networked measurement and control systems, 2009, available at webstore.iec.ch.
11. CiA DS 301: CANopen application layer and communication profile, Version 4.2.0, available at www.can-cia.org.
12. PROFIBUS International: PROFINET: Technology and application, system description, Document number: 4.132, Issue July 2011, available at www.profibus.com.
13. IEC 61491 Ed. 2.0: Electrical equipment of industrial machines—Serial data link for real-time communication between controls and drives, SERCOS, Issue 2002–2010, replaced by TR/IEC 61491 Ed. 1.0 in 2010, available at webstore.iec.ch.

14. E. Schemm: SERCOS to link with Ethernet for its third generation, *Computing and Control Engineering Journal*, 15(2), 30–33, April–May 2004.
15. D. Jansen and H. Buttner: Real-time Ethernet the EtherCAT solution, *Computing and Control Engineering Journal*, 15(1), 16–21, February–March 2004.
16. J. Feld: PROFINET—Scalable factory communication for all applications, *2004 IEEE International Workshop on Factory Communication Systems*, September 22–24, 2004, Vienna, Austria, pp. 33–38.
17. J. Jasperneite, K. Shehab, and K. Weber: Enhancements to the time synchronization standard IEEE-1588 for a system of cascaded bridges, *2004 IEEE International Workshop on Factory Communication Systems*, September 22–24, 2004, Vienna, Austria, pp. 239–244.

18

Ethernet for Control Automation Technology

Gianluca Cena
National Research Council of Italy

Stefano Scanzio
National Research Council of Italy

Adriano Valenzano
National Research Council of Italy

Claudio Zunino
National Research Council of Italy

18.1 Introduction **18**-1
18.2 Physical Layer **18**-2
Communication Support • Network Topology • Device Architecture
18.3 Data Link Layer **18**-4
Frame Format • DLPDU Format • Addressing • Commands • SyncManager
18.4 Distributed Clocks **18**-12
Main Features • DC Mechanism • External Synchronization
18.5 Application Layer **18**-16
Application Protocols
18.6 EtherCAT Master Characteristics **18**-17
Control Loop • Commercial versus Open-Source Implementations • EtherCAT Application Example
References **18**-26

18.1 Introduction

Ethernet for Control Automation Technology (EtherCAT) [ETG12] is a high-performance Ethernet-based fieldbus system. The main reason for its development was the adoption of Ethernet in automation applications, where short cycle times and low communication jitters are required. Nowadays, EtherCAT is a popular solution for enabling access to field devices by control applications in industrial environments, including motion-control systems. Communication equipment and devices are available off-the-shelf, which are based on purposely developed hardware [ET1100].

The EtherCAT protocol is an open standard currently managed by the EtherCAT Technology Group (ETG). Its specification has been integrated into the international fieldbus standards IEC 61158 [IE158] and IEC 61784 [IE784]. EtherCAT is based on a master/slave approach and relies on a ring topology at the physical level. Only one master is allowed in the network, and this is suitable, for instance, to connect a control unit (e.g., a programmable logic controller [PLC]) to decentralized peripherals (sensors, actuators, drives, etc.). By using suitable gateways, EtherCAT can interoperate with both conventional Transmission Control Protocol (TCP)/Internet Protocol (IP)-based networks (intranets) and other real-time Ethernet (RTE) solutions, such as EtherNet/IP and/or PROFINET.

The master node is in complete control of the traffic exchanged over the EtherCAT network. In particular, it is the only device that can take the initiative in the communication; hence, it is responsible for initiating all data exchanges with the slaves. Each slave processes the received frame in order to extract/insert data from/into it. Then, the frame is forwarded to the next slave in the ring.

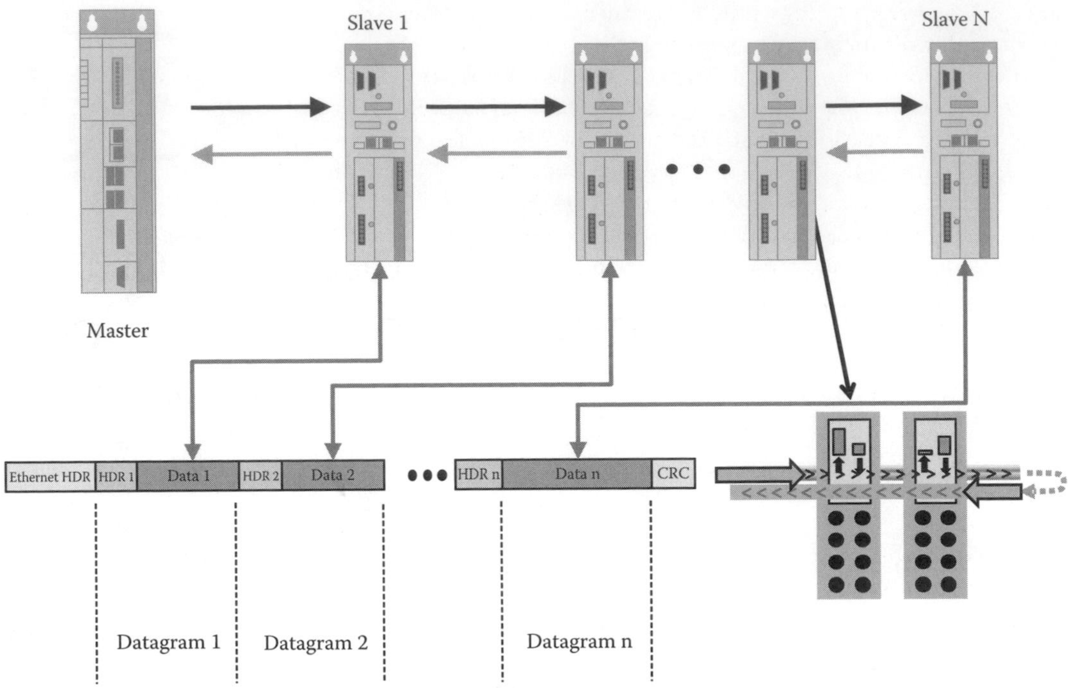

FIGURE 18.1 EtherCAT typical topology, with the *on-the-fly* frame processing.

Unlike Ethernet switches and bridges, slaves do not manage frames according to a conventional store-and-forward approach, which implies receiving the frame, decoding the related protocol control information, and sending the message out. Every frame, instead, is processed on-the-fly by the slave data link layer. In order to ensure high performance, frame processing and relaying take place at the same time so that these operations have to be carried out in hardware. This explains why specialized components are used for slaves, which are known as EtherCAT slave controllers (ESCs) [ET1100] (Figure 18.1).

18.2 Physical Layer

The physical layer of EtherCAT relies on the proven fast Ethernet transmission technology, which enables high data rates that are more than adequate to the communication needs of next-generation industrial plants. A low-cost solution, namely, EBUS, is also foreseen for modular devices, which is able to coexist with conventional Ethernet seamlessly and does not impair performance in any way. Although the use of switches is not recommended to ensure real-time behavior, the EtherCAT network architecture is quite flexible and can also stretch over wide areas.

18.2.1 Communication Support

Though EtherCAT is correctly listed among the industrial Ethernet solutions available today, it actually supports two different types of physical layers, namely, Ethernet and EBUS. The first alternative relies on the conventional 100 Mb/s full-duplex Ethernet technology (either 100BaseTX or 100BaseFX), according to the popular IEEE 802.3 standard [8023]. It is typically used for connecting the master to the network segment to which slaves are attached. Indeed, the whole EtherCAT segment (also known as *Type 12 segment*) is seen as a single, large Ethernet device by the master, which *concurrently* receives and sends Ethernet frames by exploiting full-duplex transmissions. However, this *device* does not consist of a single Ethernet controller but includes a (possibly very large) number of EtherCAT slaves connected so as to form a ring topology. Ethernet is also used to interconnect slave devices, for instance, when they are located in distinct racks or fastened to different mounting rails.

The transmission medium, in this case, consists of a Cat 5 twisted pair (either shielded or unshielded, depending on the amount of electromagnetic interference) although higher category cables are also allowed. Both classic RJ45 (8P8C) and circular M12 D-code connectors can be used.

The Ethernet transceivers (PHYs) needed to interface ESCs to Ethernet have to satisfy a number of requirements in order to be used in EtherCAT. In particular, they must be able to provide either a standard media-independent-interface (MII) or a reduced one (RMII) and have to support the MII management interface, auto-negotiation, and auto-crossover.

Conversely, EBUS can be used only as a backplane bus and is not intended for wire connections. EBUS, in fact, was mainly conceived to interconnect modules in modular devices. It is worth noting that, unlike other fieldbuses that enable a modular design for devices, the sequence of logical bits that EtherCAT transmits over EBUS is exactly the same as for Ethernet. This means that switching from Ethernet to EBUS (and vice versa) can be done quickly, efficiently, and inexpensively (in practice, only transceivers have to be replaced).

EBUS is an inexpensive physical layer that features reduced pass-through delays inside the slaves. Typically, frames experience delays on the order of 120 ÷ 500 ns [PRY08,CEN10] when propagating through EBUS interfaces, whereas longer latencies (about 1 μs) are introduced by Ethernet interfaces. EBUS uses the Manchester (biphase L) encoding scheme and marks the start and end of Ethernet frames with a pair of delimiters, namely, the start of frame (SOF) and end of frame (EOF) identifiers. The beginning of a frame is defined by a Manchester code violation with a positive level following an idle condition. A bit at the logical value "1" follows, which is the first bit of the Ethernet preamble. Then, bits of the whole Ethernet frame are transmitted up to the cyclic redundancy check (CRC). The frame ends with a Manchester violation with negative level, followed by a bit at the logical value "0," which is, in practice, the first bit of the subsequent IDLE phase.

EBUS uses low-voltage differential signaling (LVDS), and the bit rate is selected equal to 100 Mbit/s to match the fast Ethernet specification. The EBUS protocol simply encapsulates Ethernet frames; thus, it can carry any kind of Ethernet message besides EtherCAT.

18.2.2 Network Topology

The topology of a communication system is one of the crucial factors for its successful adoption in industrial environments, since it affects a number of key elements such as cabling, diagnostic features, redundancy options, and hot-plug-and-play features. The star topology, commonly used for switched Ethernet, implies significant cabling and infrastructure costs; hence, line or tree topologies are usually preferable in factory networks. Typically, slave devices in EtherCAT segments are connected in linear structures and exploit a daisy chain wiring scheme. Every slave is provided with (at least) two Ethernet ports, to connect downstream and upstream devices. The last slave in the segment performs a loopback function and returns the frame in the opposite direction to the master. The master is the headend of the structure and requires one Ethernet port only.

Each slave relays all frames it receives to the next device in the EtherCAT segment. Because of the daisy chain connection and loopback, all the slaves in the segment form an open ring. The master transmits frames at one of the ends of this open ring and receives them at the other end, after they have been processed by every slave. This means that on the whole the physical topology of an EtherCAT network is actually a ring.

Thanks to the full-duplex capabilities of Ethernet, which uses two pairs of wires housed in the same cable to carry out communications in both directions simultaneously, the resulting topology resembles, nevertheless, a physical line, as in most legacy fieldbuses. The reduction of wiring complexity helps in making the network deployment easier and lowers the installation costs at the same time.

In principle, branches can be introduced anywhere in an EtherCAT segment, by using devices equipped with three or more ports (EtherCAT couplers). This kind of devices are provided, for example, with two Ethernet ports and one EBUS interface for direct connection of input/output (I/O) modules (also known as EtherCAT terminals) and can be used to enhance the basic line structure setting arbitrarily complex tree networks topologies. It is worth noting that, in this case, every slave device located at the end of a branch has to close the ring on its own using the loopback function.

The maximum number of addressable devices in EtherCAT is quite large, since 2^{16} nodes for each segment are allowed. The limit depends on the data link layer, and in particular on the address field, which is encoded on 16 bits. In the same way, the maximum network extension is usually able to satisfy the requirements of most real applications. The limitation, in this case, is mainly due to the maximum distance allowed between any two adjacent nodes (i.e., the length of the cable), which in turn depends on the underlying transmission support (up to 10 m for EBUS and up to 100 m for Ethernet connections). This means that the whole network size is practically unlimited: in theory, up to 2^{16} devices can be connected in daisy chain using 100 m Ethernet cable segments.

18.2.3 Device Architecture

Many asymmetric communication technologies conceived for industrial environments have different internal architectures for masters and slaves, and this is true for EtherCAT devices too.

EtherCAT masters (EMs) rely on standard communication hardware (full-duplex Ethernet network interface controllers) and dedicated software, even though open-source solutions based on Linux-like operating systems are also available. On the contrary, purposely designed hardware components (ESCs) have to be necessarily used for slaves. In order to reduce the implementation costs, frame processing by ESCs occurs in one direction only, which is known as *processing path*. The reverse direction, known as *forwarding path*, is needed to propagate frames in the ring. From a logical point of view, ESCs exhibit an active behavior only on the processing path (frame modifications on the forwarding path are not allowed), but at the physical layer, they behave as repeaters in both directions. Consequently, they are able to regenerate electrical signals so that network equipment like stand-alone repeaters is no longer necessary. This also reduces connection costs and complexity in large installations.

Most ESCs are internally equipped with two or more ports, depending on device complexity. For example, the Beckhoff ET1100 [ET1100] is provided with four separate ports, which can be individually configured to operate as either EBUS or MII. Port 0 is the *upstream* port whereas the others are used for *downstream* connections and to forward signals. Each port implements two functions, called *auto-forwarder* and *loopback*. The auto-forwarder block performs frame checks, such as CRC error detection, at the physical level and manages the error count. It is also responsible for taking timestamps on frame receptions, a mechanism which is needed, for instance, by the clock synchronization protocol. The loopback function, instead, forwards frames to the next logical port if the related link is not available. In this way, the ring is automatically closed in the case of faults affecting either devices or links.

18.3 Data Link Layer

The data link protocol of EtherCAT was designed to maximize the utilization of the Ethernet bandwidth and to grant a very high communication efficiency. As mentioned earlier, the access mechanism of EtherCAT is based on a master/slave approach, where the master node (typically the control unit, e.g., a PLC) sends Ethernet frames to slave nodes. Slaves, in their turn, either extract data from the frame payload or insert information by overwriting part(s) of the payload itself.

18.3.1 Frame Format

Messages sent over the network are standard Ethernet frames, with EtherCAT frames (also known *Type 12 frames*) encapsulated in the data field. Consequently, they include the conventional fields:

preamble (8 bytes), destination and source MAC addresses (6 bytes each), EtherType (2 bytes, set to 0x88A4 to distinguish them from non-EtherCAT frames), frame check sequence (Fcs, 32 bits), and the interframe gap.

An EtherCAT frame, in turn, contains a frame header (2 bytes) and one or more EtherCAT datagrams, also known as *Type 12 Data Link Protocol Data Units* (DLPDU) according to the data link layer standard specifications. In this way, the large data field made available by conventional Ethernet can be better exploited to increase the communication efficiency. DLPDUs are packed together, one after the other, without intermediate gaps. The payload of the Ethernet frame ends with the last DLPDU, unless its overall size is 63 octets or less. In this case, the frame is padded to 64 octets in length, as required by the Ethernet specifications. The standard Ethernet CRC closes the frame and is used by each device (either master or slave) to check the integrity of the message. Thanks to the EtherType field, EtherCAT can coexist, in theory, with other Ethernet protocols.

Each DLPDU corresponds to a separate EtherCAT command and consists of three sections: header, data, and counter field. Commands are used to perform data exchanges: basically, they are issued by the master for reading or writing specific memory areas in the slave devices.

Ethernet frames, and in particular DLPDUs they embed, are processed in sequence by slaves. Each slave recognizes its commands of interest and executes them while the frames are passing through. Because of the physical ring topology, a frame is returned to the master after being processed by all the slaves. This procedure exploits the full-duplex mode of Ethernet, which means that the two communication directions can work independently.

Several DLPDUs can be embedded in the same Ethernet frame, each one addressing different devices and/or memory areas. As shown in Figure 18.2, DLPDUs are transported either (a) directly in the data field of the Ethernet frame or (b) within the data section of a datagram, by means of the User Datagram Protocol (UDP).

The first variant is limited to a single subnetwork, since Ethernet frames are not relayed by routers. Usually, this is not a limitation at all, for machine control applications. Direct Ethernet encapsulation is by far the most widespread EtherCAT solution at the shop floor of factory automation systems. In theory, multiple EtherCAT segments can be connected to a single master through one or more switches, and the MAC address of the first node in each segment is used for addressing the segment itself. However, this approach can affect the real-time properties of the communication.

On the one hand, the second variant, which relies on UDP and the IP, implies lightly larger overheads (because of the IP and UDP headers) and is also limited by routers, which are typically unable to guarantee a real-time behavior. On the other hand, this solution also enables IP routing; hence, it is suitable for applications having loose timing requirements, such as in process automation. Any standard UDP/IP implementation can be used in this case on the master side.

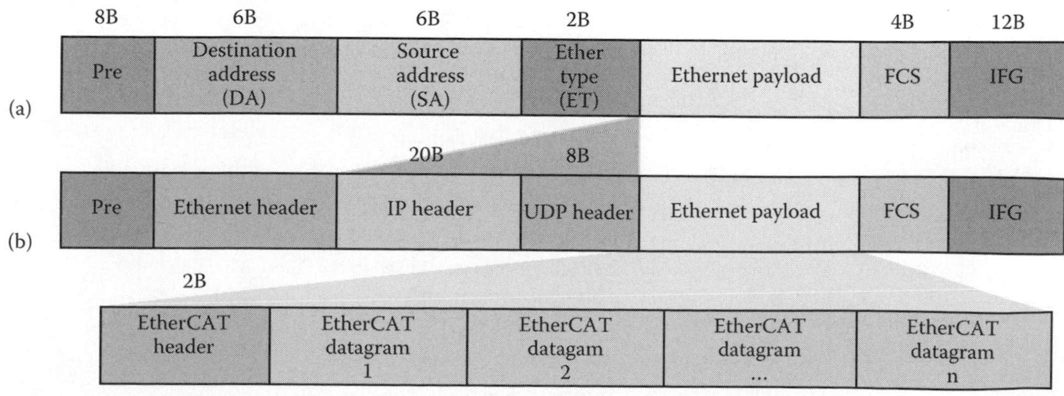

FIGURE 18.2 EtherCAT frame structure.

18.3.2 DLPDU Format

As shown in Figure 18.3, each DLPDU consists of a number of fields. The initial fields (up to IRQ included) can be assumed to belong to the header part, which has a fixed size (10 bytes). The variable-sized data area is placed immediately after the header and includes the information to be exchanged, often referred to as *data link service data unit* (DLSDU). The last field in the frame is the *working counter* (WKC), used mainly for checking whether a command has been successfully executed by the relevant slaves.

Table 18.1 summarizes sizes, formats, and meanings of the DLPDU fields.

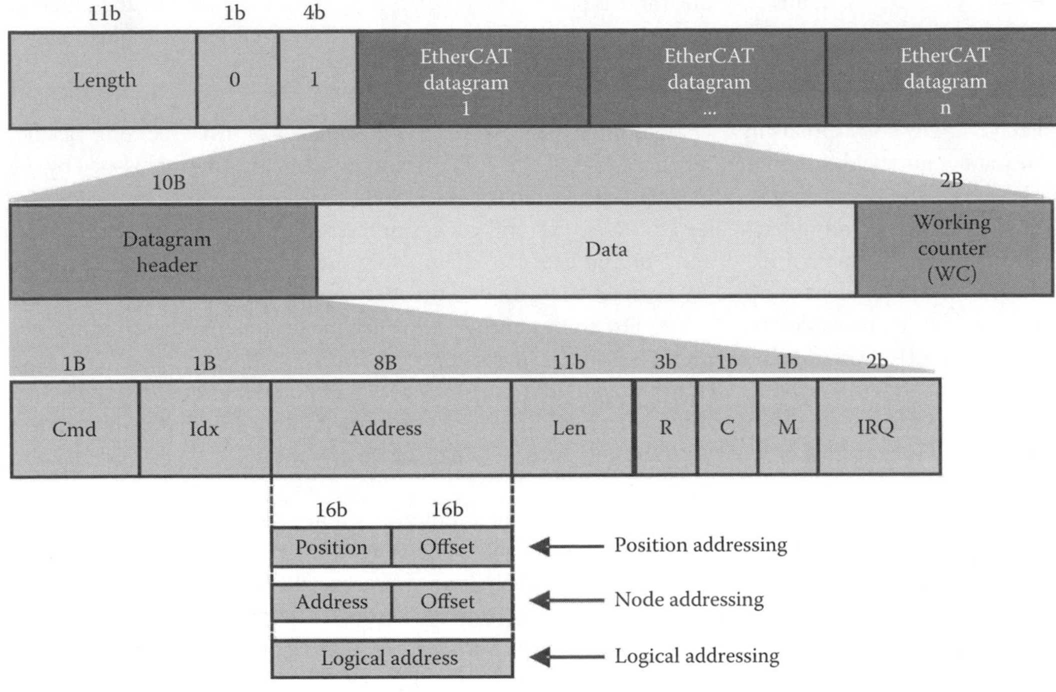

FIGURE 18.3 EtherCAT datagram structure.

TABLE 18.1 Generic Type 12 DLPDU Format

Field	Type	Meaning
CMD	Unsigned8	Service command (0–14, values from 15 to 255 are reserved).
IDX	Unsigned8	Index for identifying duplicate/lost DLPDUs.
Address	DWORD	Address (auto-increment, configured station address, broadcast, or logical address [see below]).
LEN	Unsigned11	Length of the DATA field.
R	Unsigned3	Reserved 0x00.
Circulating frame (C)	Unsigned1	0x00: Frame is not circulating. 0x01: Frame has already circulated once.
NEXT (M)	Unsigned1	0x00: this is the last DLPDU in the EtherCAT frame. 0x01: more DLPDUs follows in the EtherCAT frame.
IRQ	WORD	External Event Request registers of all slaves combined via OR.
DATA	OctetString LEN	Data (DLSDU).
WKC	WORD	Working counter.

The service command (1 byte) is encoded in the CMD parameter. Different types of command exist, which can be used to carry out highly optimized read and write operations on slave devices. Generally speaking, they can be grouped according to the access type:

- *Read* (RD) is used by the master to read memory areas or registers from slave devices.
- *Write* (WR) is used by the master to write to memory areas or registers of slave devices.
- *Read/Write* (RW) is used by the master to carry out both a read and a write operation at the same time; in this case, reading is performed by the slave before writing.
- *Read/Multiple Write* (RMW) is a quite peculiar service, where the addressed slave carries out a read operation while all other slaves are performing a write action.

The IDX field (1 byte) is used by the master to identify a given DLPDU. This is useful, for instance, to pair a frame returned on the forward path with those sent on the processing path, when more frames are allowed to travel along the ring at the same time. This field shall be left unchanged by the slaves.

The address field (4 bytes) can contain different information, encoded as either a single 32 bit double word or two 16 bit words, according to the addressing mode chosen for the DLPDU. A more detailed description is provided in the next section.

The parameter LEN (11 bits) is the size in bytes of the DATA field included in the DLPDU.

The *circulating frame* (C) bit is used to cope with failures of links between slaves. Because of the loop-back function (which forwards Ethernet frames to the next internal port if the current port is neither connected nor available), if a part of the network is cut off from the master, it becomes an insulated ring. If no action is taken, frames that are currently traveling there might continue to circulate. To prevent endless frame circulation, a slave with no active link to the master has to remove frames from the insulated ring according to the following rules:

- If the C bit of the DLPDU is 0, the C bit is set to 1.
- If the C bit of the DLPDU is 1, the frame is not processed but it is destroyed.

The NEXT field (also called Mbit) specifies whether the DLPDU is the last one in the frame or another DLPDU follows.

An important mechanism is implemented through the IRQ field. EtherCAT event requests are used to inform the master of slave events. On each slave, the *External Event* register is combined, using a logical "AND" operation, with a predefined *External Event Mask* register. This mask is used for selecting interrupts that are relevant to the master. The resulting bits are combined with the IRQ field in the incoming DLPDU using a logical "OR" operation, and then written into the IRQ field of the outgoing DLPDU. This implements a quasi-asynchronous event notification mechanism. It is worth noting that, when using the IRQ mechanism, the master is not able to single out what slave(s) originated the interrupt.

Process data that have to be read or written (DLSDU) are placed in the DATA field. When a slave modifies any field in the DLPDU (a quite frequent event), the CRC field has to be recomputed and updated so that it looks correct to the following slaves.

Finally, the WKC field is used as a further error detection mechanism. Basically, it counts the number of devices that were successfully addressed by the DLPDU. Each addressed slave, when reading/writing data to/from the DLPDU, increments the WKC in hardware. In most control applications, each DLPDU has an expected value for WKC, which is known in advance and depends on the number of involved slaves. Therefore, the correctness of datagram processing can be checked by the master by comparing the value of the WKC field in the received DLPDU with the expected value.

WKC is incremented by one in the case at least 1 byte/bit of the data field has been successfully accessed by the slave for read, write, and read/multiple write commands. In the case of read/write commands, instead, WKC is incremented by two for successfully written data and, additionally, by one for a successfully read operation.

18.3.3 Addressing

Different addressing modes are supported for accessing information in the slaves, as shown in Figure 18.3. The DLPDU header contains a 32 bit address field, which is used for both *physical* and *logical* addressing.

18.3.3.1 Physical Addressing

In physical addressing, the address field is split into two 16 bit words: the first one, the address position (ADP) encodes the address of a slave device in the EtherCAT segment, while the second, the address offset (ADO) refers to either a physical memory location or a register in the slave device.

This solution enables up to 2^{16} slave devices in the network, each one associated to a local memory as wide as 2^{16} cells. In other words, every slave is assigned its own 64 kB physical address space. The first 4 kB block (address range from 0x0000 to 0x0fff) is used for *registers* and user memory. The remaining space (from address 0x1000 onwards) is the 60 kB *process memory*. Its actual size, however, depends on the specific device considered. The address space of any ESC is directly addressable by both the EM and the attached microcontroller. In physical addressing, each DLPDU refers to a specific memory area (or register) in either a single slave or all slaves at the same time (broadcast). For this reason, this mechanism is mostly suitable for transferring parameterization information to devices.

Basically, slaves can be physically addressed in two different modes, namely, position and node addressing, which rely on the information included in the ADP field. A third mode is also available for dealing with broadcast DLPDUs. In practice, the following alternatives are possible:

- *Position* (or *auto-increment*) *addressing*: In this case, each slave device is identified via its physical position in the EtherCAT segment. When processing a DLPDU that makes use of this kind of addressing, each slave increments the ADP field by one. When a DLPDU is received with the ADP field set to 0, the device assumes to be the intended destination. As a consequence, the master has to preset ADP to a negative value that corresponds to the absolute position of the target slave (the first slave, which is next to the master, is conventionally assumed to have position 0). For instance, an initial ADP value equal to −2 identifies the third slave in the segment. Because of the mechanism, which updates the address field while the frame is traversing a device, the DLPDU is said to adopt auto-increment addressing. Position addressing should be used only during the start-up phase to scan the EtherCAT segment and can be adopted occasionally to detect new slaves attached to the segment. This is because problems may arise if some network branches are temporarily closed because of link errors. In this case, in fact, a wrong count could lead to address a wrong slave, possibly disrupting the correct system operations.
- *Configured addressing*: In this case, slave devices are identified by means of node addresses (*configured station address* and *configured station alias*) that are assigned at start-up. This choice grants that, even if the segment topology is changed and/or devices are added to/removed from the network at runtime, every slave can nevertheless be safely accessed through the same predefined address. In this working mode, every slave compares the ADP field in the DLPDU with its own configured address(es). If a match is found, the device is the intended destination. Node addressing is typically used for accessing (already configured) devices individually. Each EtherCAT slave can be assigned a pair of station addresses. Assignment can be done by either the EM (configured station address), during the data link start-up phase, or the application running on the slave (configured station alias), for instance, by reading the address value from an internal nonvolatile memory during the device initialization.
- *Broadcast*: In this case, all slaves in the EtherCAT segment are addressed and shall execute the command encoded in the DLPDU. As for position addressing, the ADP field is incremented by one by each slave during the DLPDU propagation so that the master can eventually obtain the number of slaves that are present in the network. It is worth noting, however, that the ADP field is not checked by slaves in this case. Broadcast addressing is used, for instance, to initialize the same memory area in all slaves with the same value at the same time by means of a single command.

Specific areas in the local memory of the target node(s) are then addressed via the ADO field in the DLPDU. In particular, ADO represents an offset into the physical address space of the device and is useful to access a single block of data starting at a given memory location.

18.3.3.2 Logical Addressing

The EtherCAT data link communication services enable the individual addressing of several slaves by means of a single Ethernet frame carrying multiple DLPDUs. The main benefit is a significant improvement of the communication efficiency with respect to conventional solutions where every request (and the related reply) is encoded into a separate frame. The typical Ethernet overhead for frame transmission and protocol control information is quite large, and improvements are surely welcome. However, when dealing with simple devices characterized by very small process data (e.g., for 2 bit digital I/O modules), the overhead introduced by DLPDUs based on physical addressing might be still considered excessive.

This problem can be lessened by adopting the EtherCAT logical addressing mode. In logical addressing, the exchanged information is identified by using the 32 bits in the address field as a whole (ADR). In practice, a logical address space is defined for the EtherCAT segment, which is shared by all devices. When using the logical addressing, the resulting utilization can exceed 90% of the available bandwidth, even when data are exchanged with devices that produce/consume only 2 bits of user data.

Logical addressing requires that slaves include suitable *fieldbus memory management units* (FMMUs), which have to be implemented in hardware in the ESC. An FMMU is similar to memory management unit (MMU) of modern processors and takes care of translating the logical address to a physical address in the device local memory. FMMUs enable individual address mappings for each device.

In contrast to conventional MMUs, which typically map whole memory pages, FMMUs also support bit-wise mapping. This is the reason why, for instance, 2 bits of a specific input device can be mapped individually (and practically anywhere) in the logical address space. If an EtherCAT command is sent to read/write, the logical memory area assigned to the 2 bits (instead of addressing an EtherCAT device explicitly), the FMMU in the related device is able to transfer 2 bits of data from/to the memory and the right position in the payload of the DLPDU. The same happens to other slaves, when they use the logical address within the same DLPDU. Therefore, the combined adoption of FMMUs and logical addressing enables the exchange of data segments that span across several slave devices. As shown in Figure 18.4, a single command is able to address data arbitrarily distributed within several slaves. Clearly, this boosts the communication efficiency significantly.

The access type supported by each FMMU is configurable and can be read, write, or read/write. The configuration of FMMU entities in the slaves is carried out by the master during the data link start-up phase. In particular, the following items are configured for each FMMU to specify how the address translation has to be performed:

- (Bit-oriented) start address in the logical address space
- (Bit-oriented) start address in the device's physical memory
- Size of the translated memory area
- Direction of the mapping (input or output)

When a DLPDU with logical addressing is received, the slave looks for an address match via its FMMU. If any correspondence is found, it transfers data between the local memory position and the data field of the DLPDU.

18.3.4 Commands

Generally speaking, EtherCAT commands are obtained by combining an access type (read, write, read/write, or read/multiple write) and an addressing mode (auto-increment, configured, logical, or broadcast). However, not all possible combinations are actually allowed. The legal EtherCAT commands are listed in Table 18.2.

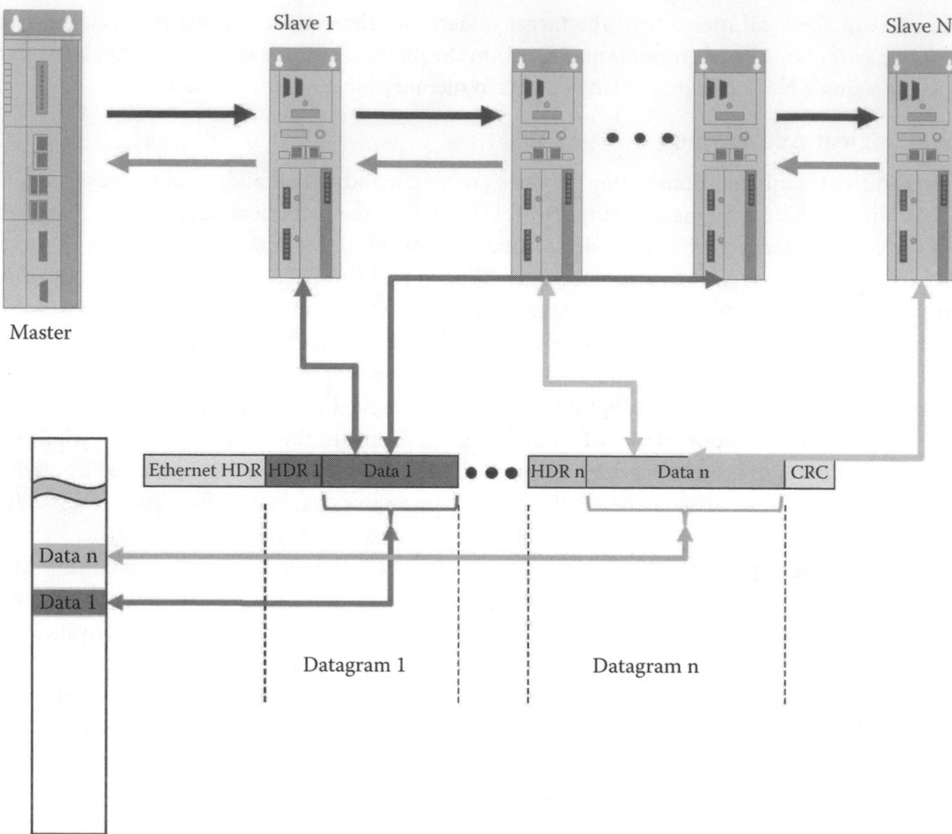

FIGURE 18.4 FMMU mapping example.

TABLE 18.2 EtherCAT Commands

Access Type	Command	CMD	Encoding
Read	Auto-increment physical read	APRD	1
	Configured-address physical read	FPRD	4
	Broadcast read	BRD	7
	Logical read	LRD	10
Write	Auto-increment physical write	APWR	2
	Configured-address physical write	FPWR	5
	Broadcast write	BWR	8
	Logical write	LWR	11
Read/Write	Auto-increment physical read/write	APRW	3
	Configured-address physical read/write	FPRW	6
	Broadcast read/write	BRW	9
	Logical read/write	LRW	12
Read/Multiple Write	Auto-increment physical read/multiple write	ARMW	13
	Configured-address physical read/multiple write	FRMW	14

Read commands (APRD, FPRD) are used to extract information at the specified (ADO) offset in the memory of the slave(s) addressed by ADP and to copy it in the data field of the DLPDU, which will eventually reach the master. Write commands (APWR, FPWR) carry out data exchanges in the opposite direction, while read/write commands (APRW, FPRW) are used for indivisible read and write operations. In this case, in fact, the contents of the DLPDU data field and the addressed memory location are swapped simultaneously.

Logical commands behave almost the same. It is worth remembering that multiple slaves can be involved in the execution of the same command when using logical addressing, depending on the FMMU configuration. As a consequence, the content of the data field in the DLPDU is either obtained by gathering information from the physical memory of several devices (LRD) or scattered over a number of different slaves (LWR). As mentioned before, the logical read/write command (LRW) carries out both operations at the same time.

Broadcast write commands (BWR) are straightforward: as expected, the same value overwrites the memory area addressed by ADO in all slaves. Broadcast read commands (BRD), instead, exploit Boolean operators; while processing an incoming DLPDU, every slave carries out a bitwise "OR" operation between the value found in its memory area addressed by ADO and the data field in the incoming datagram. Then, the result is inserted into the data field of the outgoing DLPDU.

The broadcast read/write command (BRW) combines read and write operations, so that the slave memory content is changed to reflect incoming data. However, according to data sheets [ET1100], this last service is typically not used.

Read/multiple write commands (ARMW and FRMW) complete the set of available services. Only singlecast physical addressing can be used in this case, namely, auto-increment or configured. Read/multiple write operations are quite peculiar and are exploited by EtherCAT in well-defined contexts (i.e., they are not intended for general-purpose data exchange services). In particular, the addressed slave carries out a read operation, whereas all other slaves execute a write operation triggered by the frame reception. If the addressed slave is the first one in the segment, such a mechanism permits a specific value to be read from its local memory and propagated to all the other slaves with a single command. This is profitably exploited, for instance, by the clock synchronization mechanism for distributing the reference time (further details will be discussed in the next sections).

18.3.5 SyncManager

The ESC memory is used for exchanging data between the EM and the application running on the slave. The master can access the memory through the network by using the data link layer services, whereas the local application makes use of the *process data interface* (PDI) provided by the ESC. As a consequence, problems may arise if concurrent accesses are carried out without any restriction. In particular, the consistency of data is not guaranteed by the basic data link communication services, unless a mechanism like semaphores is implemented in software for dealing with data exchanges in a coordinated way. Moreover, both the EM and the application running in the slave have to poll the memory explicitly, in order to determine when it is no longer used by the competing entity.

EtherCAT provides a mechanism for slave memory access control, which is based on *SyncManagers*, and was designed bearing in mind concurrency issues. SyncManagers are implemented in hardware in the ESC and enable consistent and secure data exchanges between the EM and the local application, together with the interrupt generation to notify both sides of changes.

SyncManagers are configured by the EM. The communication direction can be selected, as well as the communication mode. Each SyncManager uses a buffer in the local memory area for exchanging data and transparently controls all accesses to the buffer. The buffer must be accessed beginning with the start address; otherwise, the access is denied. Once access to the start location is granted, the whole buffer can be accessed, either as a whole or in a number of strokes. Accessing the last location also concludes the whole operation. Buffer changes caused by the master are accepted by the SyncManager only if the frame FCS is correct. This also means that such buffer changes take effect immediately after the reception of the end of the frame.

SyncManagers support two communication modes:

1. *Buffered mode*: In this case, the interaction between the producer and the consumer of data is uncorrelated, and each entity can access the buffer at any time. The consumer is always provided with the newest data. In the case data are written into the buffer faster than they are read out, old data are simply discarded. The buffered mode is typically used for cyclic process data. This mechanism is also known as 3-buffer mode, because the SyncManager manages three buffers of identical size (denoted as 0, 1, and 2). One buffer is allocated to the producer (for writing), another buffer to the consumer (for reading), and a third buffer helps as intermediate storage. Reading or writing the last byte of the buffer results in an automatic buffer exchange. It is worth noting that both the EM and the local application must always refer to buffer 0 when accessing memory. It is up to the SyncManager redirecting accesses to the right buffer.
2. *Mailbox mode*: In this case, a handshake mechanism is implemented for data exchanges, which prevents buffer overwriting and ensures that no data will be lost. Just one buffer is allocated for each mailbox; moreover, reading and writing are enabled alternatively. The mechanism implemented by mailboxes is straightforward. At first the producer writes to the mailbox buffer. When done, the SyncManager locks it for writing and enables read access to the consumer. Only when the consumer has finished reading data out of the buffer, the producer is granted write access again. At the same time, the mailbox turns to the locked state for the consumer. The mailbox mode is typically used for application layer (AL) protocols, where the time taken to exchange information typically is not very relevant.

18.4 Distributed Clocks

An important mechanism included in the EtherCAT specification is the distributed clocks (DC) synchronization protocol, which enables all slave devices to share the same system time with high precision and accuracy. Synchronization errors are typically well below 1 μs [CEN12]; in this way, all devices can be synchronized, and consequently, distributed applications are synchronized as well. Possibly, the master can also be synchronized, even though this option requires additional capabilities.

18.4.1 Main Features

The DC mechanism provides a number of features that are very useful for distributed control applications, the most important being

- Synchronization of the slaves (and the master) clocks
- Generation of synchronous output signals (*SyncSignals*)
- Precise timestamping of input events (*LatchSignals*)
- Generation of synchronous interrupts
- Synchronous digital output updates
- Synchronous digital input sampling

DC is placed above the EtherCAT data link protocol, and its implementation is not mandatory. For this reason, both DC-enabled and non-DC-enabled devices can quietly coexist in the same network. It is worth noting that DC is not a general-purpose synchronization protocol, since it relies on specific features of EtherCAT, such as its ring topology, on-the-fly datagram processing, and hardware timestamping capabilities.

To better understand the DC mechanism, the following terms and definitions are needed:

- The *system time* is a 64 bit value representing the time (in nanoseconds) elapsed since 0:00 h of January 1, 2000.
- Typically, the *reference clock* is the slave device with DC capability that is connected closer to the master. Optionally, the reference clock can be adjusted to an extern *global* reference clock, for example, to an IEEE 1588 grandmaster clock. The reference clock provides the system time to all other devices in the EtherCAT segment.
- Each DC-enabled device has a *local clock* that, if not suitably controlled, runs independently of the reference clock.
- The *master clock* is used to initialize the reference clock. Typically, it is the clock of the EM, but it is also possible (and sometimes recommended) to bind it to a global clock reference (IEEE 1588, GPS, etc.), which is made available to the master either directly or through a specific EtherCAT slave.
- The *propagation delay* is the time spent by frames when passing through devices and cables.
- The *offset* is the difference between the local clock of a given slave device and the reference clock. It is due to several elements, such as the propagation delay from the device holding the reference clock to the considered slave, the initial difference of the local clocks resulting from the different instants at which devices were powered on, the skew between oscillator frequencies, and so on. The offset is compensated locally in each slave.
- Because the oscillator periods of local and reference clock sources are subject to small deviations (different quartzes are used), the resulting *drift* has to be compensated regularly.

18.4.2 DC Mechanism

The clock synchronization process consists of the following three main actions:

1. *Propagation delay measurement*: The master sends a synchronization DLPDU at certain time intervals, and each slave stores the time of its local clock; after collecting all timestamps, the master, which is aware of the network topology, computes the propagation delay for each segment.
2. *Offset compensation*: Because the local time of each device is a free-running counter, which typically does not have the same value as the reference clock, the master computes the offset between the reference and local clocks separately for each DC-enabled slave. Then, the offset is written to a specific register of the slave, in order to compensate differences individually. At the end of this step, all devices share the same absolute system time.
3. *Drift compensation*: After propagation delays have been measured and the offsets between clocks compensated, the drift of every local clock is corrected through a time control loop (TCL). This mechanism readjusts the local clock by regularly measuring its difference with the reference clock.

Details on these actions are provided in the following. For the sake of clarity, from now on, DLPDUs will be referred to simply as datagrams.

18.4.2.1 Propagation Delay Measurement

As mentioned before, each slave device introduces small frame processing/forwarding delays, caused by both internal and communication link mechanisms. For this reason, the propagation delay between the reference node and any slave has to be evaluated carefully. This procedure, which is the first action carried out by DC, consists of three main steps. First, the master sends a broadcast message (i.e., a datagram processed by all slaves). Second, each slave stores the value of its local clock when the first bit of that message is received. This operation is performed for each port of the device, on both the processing and forwarding paths (Figure 18.5). Finally, the master collects timestamps and computes the path delays, taking into account the topology of the network.

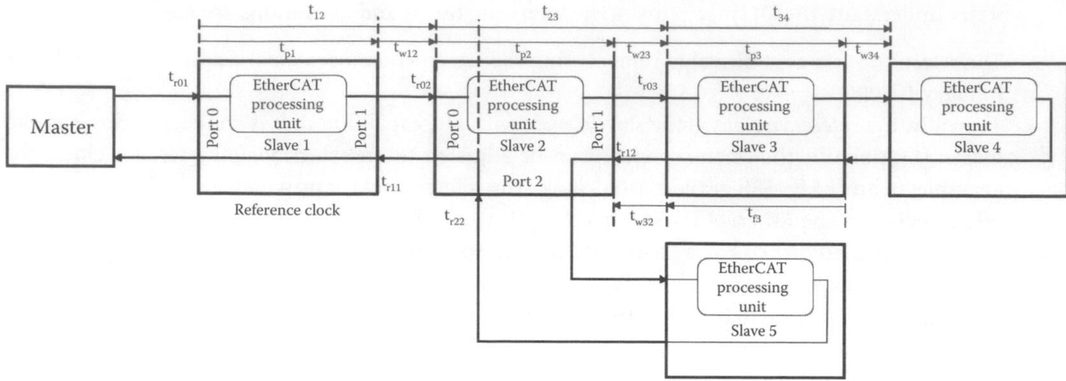

FIGURE 18.5 Time measurement in the DC mechanism.

A synchronization of slave devices is not necessary for the correct operation of the procedure, because only local clock values are used.

The computation of the propagation delay ($t_{PropagationDelay}$) for each slave is based on the differences between receive times at the device ports. Delays introduced by nodes without DC capabilities are taken into account as well. To better understand the propagation delay measurement, a sample EtherCAT network is shown in Figure 18.5. The node that is closer to the master (Slave 1) has been selected as the reference clock. Propagation delay computation makes use of the main elements summarized in Table 18.3.

The algorithm starts when the master sends a set of datagrams to reset the slave DC registers. After this initialization phase, the master transmits a pair of datagrams to each slave: the first message forces the ESC in the device to record timestamps related to the reception of the frame on its different ports, while the second is used to read timestamps back. For simple devices with only two ports, this means that timestamps are taken in both directions; that is, a first timestamp is acquired when the message is received on the processing path, while a second is recorded on the reception of the same frame on the forwarding path.

In sample network in Figure 18.5, Slave 2 records timestamp t_{r02} when the frame reaches its Port 0, t_{r12} when the frame is received back from Slave 3, and, finally, t_{r22} when the message is returned by Slave 5. Since timing information is taken using the same internal clock, differences between each pair of timestamps can be evaluated properly. Differences are then used to evaluate all delays (processing, forwarding, and wire contributions) and to compute the resulting propagation delay for the slave. It should be noted that the forwarding delay for the first slave (Slave 1 in Figure 18.5) cannot be computed in this way, as timestamps are recorded only for incoming messages. A possible solution is to assume that it is the same as the other t_{fx} or, alternatively, it can be measured once and for all (at least in theory) when the device is designed and manufactured. The compensation is then performed directly by the master, likely by using a predefined fixed value.

TABLE 18.3 Definition of DC Timing Parameters

t_{px}	Processing delay of slave x
t_{fx}	Forwarding delay of slave x
t_{xy}	Propagation delay from slave x to slave y
t_{wxy}	Wire propagation delay from slave x to slave y (symmetry assumption)
t_{rky}	Receive time on port k of slave y, recorded on the DC receive time register

18.4.2.2 Offset Compensation

The second main action carried out by DC is the offset compensation. Local clocks ($t_{LocalClock}$), if not synchronized, are free-running counters. At system start-up, they are likely to have different values from the reference clock so that an offset compensation is needed. The technique adopted in EtherCAT to evaluate offsets is based on the same timestamps collected by the master during propagation delay measurement. In fact, after the previous phase has finished, the master is also able to compute the offset of each local clock with respect to the reference time (t_{Offset}). This value is then written into a specific register (*system time offset*) of the slave and then used to adjust the local time. At this point, when the initialization phase is concluded, each DC slave can determine its own copy of the system time autonomously, by using the local time and offset values.

18.4.2.3 Drift Compensation

Finally, DC has to deal with oscillator drifts. The natural drift of every local clock (which depends on variations between the reference and local clock quartz) is corrected by means of a TCL algorithm implemented into each ESC. The master periodically distributes the system time ($t_{ReceivedSystemTime}$, as read from the reference clock) to all slaves in the network. The algorithm in each slave compares the received system time to its local copy. In evaluating the difference Δt, propagation delays have also to be taken into account:

$$\Delta t = (t_{LocalClock} + t_{Offset} - t_{PropagationDelay}) - t_{ReceivedSystemTime}$$

In automation systems, adjusting the clock with abrupt increments (or, even worse, decrements) is typically not tolerable. The clock discipline algorithm defined by DC, namely, TCL, is also able to cope with this requirement. If Δt is positive, the local time is running faster than the system time and has to slow down. On the contrary, when Δt is negative, the local time is too slow and needs to speed up. In this way, the monotonic property of time is assured for each DC-enabled slave. After the initialization of delays and offsets, the master tries to compensate the static clock deviations quickly, that is, by sending a high number of commands (e.g., fifteen thousands) in separate frames. In other words, first the control mechanism takes care of the static deviations to obtain synchronized DCs, then compensation frames are sent cyclically to correct dynamic clock drifts.

18.4.3 External Synchronization

The ability to synchronize to an external time source, so as to define clear timing hierarchies for the whole control system, is quite desirable in current industrial plants. This feature has been recently introduced in EtherCAT so that the time alignment of separate EtherCAT systems is now possible. The basic idea is having the DC time controlled directly by an external device. A possible solution makes use of special devices that provide a common DC reference time to different network segments. In practice, the EBUS is cross-connected to an Ethernet segment, and the user can decide what side (either EBUS or Ethernet) hosts the reference clock with the highest priority. The connecting device is then responsible for sending reference clock correction values to the EM with the lower priority.

Moreover, when the reference time in an EtherCAT system has to be adjusted to a higher-level clock, an external synchronization device can be used that supports synchronization protocols different from DC. For example, some devices are now available that can work with an external Precision Time Protocol (PTP) clock source. In principle, several other popular synchronization protocols and clock sources can be also used, for example, GPS, radio clocks, Network Time Protocol (NTP), and Simple Network Time Protocol (SNTP).

Another remarkable feature has been recently introduced in EtherCAT to cope with systems that demand increased availability; it consists of an optional redundancy mechanism that enables the on-the-fly replacement of devices without having to shut down the network and makes the system resilient to slave failures.

18.5 Application Layer

The AL of EtherCAT implements a state machine, which describes the behavior of a device by means of its states and events that trigger transitions between states. In particular, the state machine is responsible for coordinating master and slave applications during the start-up and operational phases. Depending on the current state, different functions are enabled in the EtherCAT slave. Different commands have also to be sent to the device in each state by the EM, in particular, during the boot sequence of the slave. Commands are acknowledged by the local application after the involved operations have been completed. Unsolicited changes of the local application state are also possible. Moreover, simpler devices, which do not include a microcontroller, can be configured to follow the state machine logic through an emulation mechanism. In this case, any state change has to be accepted and acknowledged.

The state machine is controlled and monitored using some registers included in the slave. The master controls the state transitions by writing to the AL control register. In turn, the slave updates information about its current state by writing in the AL status register, which is also used for error notification by means of suitable error codes written in the register itself.

As Figure 18.6 shows, an EtherCAT slave shall support four basic states and, possibly, one optional state:

- *Init*: EtherCAT slaves enter this state at power-on. In this situation, the master initializes the SyncManager channels for mailbox communications.
- *Preoperational*: Mailbox communications are enabled in the preoperational state, but process data communications are not. The EM initializes the SyncManager channels for process data, the FMMUs and the PDO mapping mechanism, if supported.
- *Safe operational*: In this state, mailbox and process data communications are enabled, but the slave outputs are kept in a safe state, while inputs are updated cyclically.
- *Operational*: In this state, slaves can transfer data between the network and their I/O logic. Mailbox and process data communications are completely enabled. The operational state is the normal working condition for slaves after completing the bootstrap sequence.
- *Bootstrap* (optional): The bootstrap state is mainly aimed at downloading the device firmware. In the bootstrap state, mailboxes are active but restricted to file access via EtherCAT services.

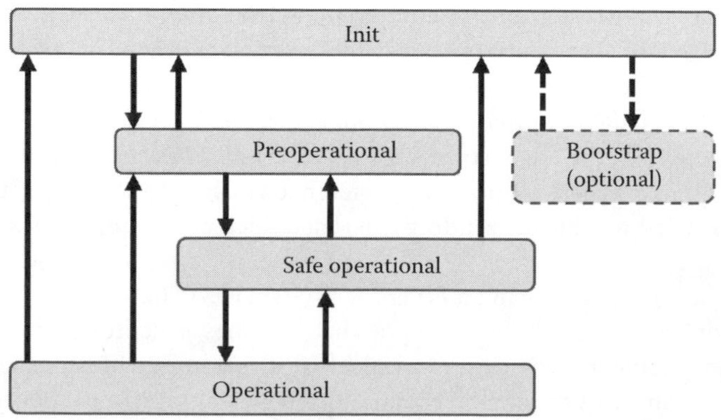

FIGURE 18.6 EtherCAT state machine.

18.5.1 Application Protocols

An important characteristic of EtherCAT is its ability to support multiprotocol higher-level communications using standardized mailboxes. This aspect is particularly appealing when options are offered for popular solutions such as the following:

- *CANopen over EtherCAT* (CoE): This option offers a way to access a CANopen object dictionary (OD) and to exchange CANopen messages according to event-driven mechanisms.
- *Ethernet over EtherCAT* (EoE): This option allows to tunneling standard Ethernet Frames in EtherCAT.
- *File access over EtherCAT* (FoE): This option enables the download/upload of firmware and other files.
- *Servo drive profile over EtherCAT* (SoE): This option is useful to grant access to the device profile of SERCOS.

Supporting popular communication protocols helps in improving compatibility and efficiency of data exchanges between new and old components in automation systems; to this purpose, EtherCAT makes use of well-known and established technologies. For instance, the *CoE* protocol enables the adoption of the complete CANopen profile family in EtherCAT networks. Besides this feature, the service data object (SDO) transport protocol allows the transmission of objects of any size and is equivalent to its CANopen counterpart so that it is possible to reuse existing protocol stacks. Data are organized in process data objects (PDOs), which are transferred using the efficient support of EtherCAT. Moreover, an enhanced mode is defined that overcomes the 8 byte limitation of CAN and enables the readability of the whole object list.

Another appealing feature an industrial Ethernet solution should provide is the support to standard IP-based communication protocols (i.e., TCP/IP and UDP/IP) and all higher-level protocols that rely on them, such as HTTP, FTP, SNMP, etc. To this purpose, the *EoE* feature exploits a mechanism where Ethernet datagrams are tunneled and reassembled in a device, before being relayed as complete Ethernet frames. This procedure has no impact on the achievable cycle time, because the size of fragments can be optimized according to the available bandwidth.

The *FoE* is an EtherCAT service that can be used to download a file from a client to a server or to upload it in the opposite direction. The protocol is similar to the trivial file transfer protocol (TFTP), and both sides are allowed to initiate a read or write request via the corresponding command. This service is typically used to update the device firmware.

Finally, the *Servo drive profile over EtherCAT* (SoE) service enables the use of the SERCOS device profile and is suitable for demanding applications that rely on popular drive technology.

18.6 EtherCAT Master Characteristics

Functionalities of the EM are usually implemented in software. EM coordinates activities of slaves through a control application, mainly by reading and writing data stored in their local memory. In addition, EM manages some network mechanisms (concerning, for instance, clock synchronization) and checks the availability of the system by monitoring the state of slaves (for both detection and active reaction to faults). The master is also responsible for the slave configuration during the network start-up, and it often offers additional features such as the automatic discovery of the network topology and/or support to make the network configuration and reconfiguration easier. Other useful features provided by some EMs are high-level functionalities, such as communications between EMs (carried out at the cell level) that allow interconnecting separate subsystems (at the field level), each one consisting of one EM and a number of connected slaves.

The most critical task carried out by EM is probably dealing with the control loop, which is entered after the configuration of slaves. This is because, in an EtherCAT network, the level of determinism achieved by the master directly affects the real-time performance of the whole control application.

In the following subsections, the concept of control loop is briefly introduced. Then, the main features of two categories of EMs, namely, commercial and open source, as well as their points of strength and weakness, are discussed.

18.6.1 Control Loop

The master enters the cyclic control loop at the end of the configuration phase, after fundamental operations such as the discovery of the network topology and the transmission of configuration data to slaves have been completed. The control loop is a cyclic task with its own predefined period (T_{CYCLE}), which, independently of its complexity, has unavoidably to read and write data from/to the slaves according to the (well-defined) control algorithm. Besides datagrams (i.e., DLPDUs) needed by the control application, additional messages are exchanged over the network under the EM control, to provide essential network services such as synchronization and fault detection.

A distinguishing feature of EtherCAT is its ability to enable very short cycle times T_{CYCLE} (that can be as short as 500 μs or even less), so this technology is also suitable for motion control and other high-precision applications.

Figure 18.7 shows a very basic structure for an EM control loop, where pseudo-code is used to highlight its main phases. The `receive_frame` function is where EtherCAT frames, received through the EM network interface, are made available to the application. This involves a specific sequence of actions: extracting datagrams from the frame, extracting process data from each datagram, and, finally, storing process data into the relevant variables into the process image (collectively represented by the "`data`" variable in the picture). The functional block `execute_control_application` is where the control algorithm is executed, so as to compute new values (`new_data`) for commands and state variables. Data to be sent to actuators are first encoded into suitable datagrams and then collected into a single frame, which may also include other datagrams added by EM for management purposes. At this point, the frame is sent over the EtherCAT network (function `send_frame`). The `wait` function in Figure 18.7 reminds that the execution of the control loop must occur exactly once on every T_{CYCLE}, so some waiting time could be necessary before starting a new iteration.

The timing diagram in Figure 18.8 depicts some iterations of the control loop. In the diagram, t_{CPU} represents the time spent by the CPU in the master to execute the code of the control loop (i.e., the `receive_frame`, `execute_control_application`, and `send_frame` blocks in Figure 18.7). The t_{NET} interval, instead, is the overall time spent by the network interface to send an EtherCAT frame over the network, including the time needed by the frame to traverse all the slaves on the processing path and come back on the forwarding path (propagation delay). To keep the diagram simple, the time needed to store the frame received from the network interface into the EM memory is also included in t_{NET}.

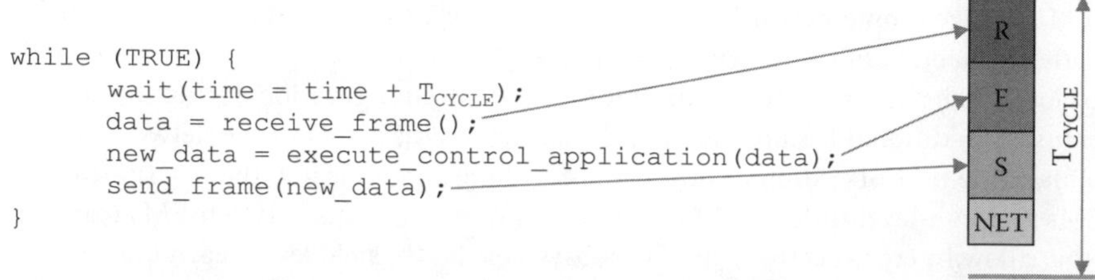
while (TRUE) {
wait(time = time + TCYCLE);
data = receive_frame();
new_data = execute_control_application(data);
send_frame(new_data);
}
R
E
S
NET
TCYCLE

FIGURE 18.7 Pseudo-code of a typical EM control loop.

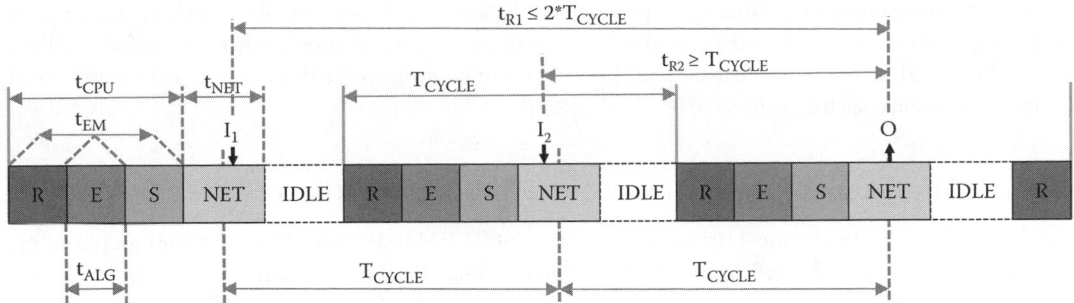

FIGURE 18.8 EtherCAT control loop timing diagram.

Slaves copy sampled input data from their local memory to EtherCAT datagrams in a time interval included in t_{NET}, and the same happens for output data, which have to be transferred from datagrams to slaves' registers.

The *reaction time* t_R is the time required by the control application to react to a variation of sampled input data, by writing commanded output data accordingly. Figure 18.8 shows that t_R is less than or equal to 2^*T_{CYCLE} (the time interval between I_1 and O in Figure 18.8 depicts the worst case). However, when the sampled input signal is made available by the slave in time for being included in the datagram containing the read command, t_R is just a bit larger than T_{CYCLE} (time interval between I_2 and O in Figure 18.8). Summing up, the reaction time t_R is bound by the inequality $T_{CYCLE} < t_R \le 2^*T_{CYCLE}$.

It is worth remarking that the results hold if the input and output operations are performed by the same slave and the execution time of the control algorithm is constant. On the contrary, other factors have to be taken into account, such as the propagation delay on the processing path for the portion of EtherCAT network segment between the involved slaves, and/or the amount of indeterminism introduced by the control algorithm. For instance, when the cycle time is T_{CYCLE} = 500 µs, the maximum reaction time is t_R = 1 ms, but this is true if and only if the inequality $t_{CPU} + t_{NET} \le T_{CYCLE}$ holds.

When DC is used to synchronize the clock of EtherCAT slaves, they can execute a given action at a predefined instant in time. In this case, the maximum reaction time actually increases, because all slaves must receive the related command (included in one or more the datagrams) before the programmed actuation time.

Because of the intrinsic determinism of EtherCAT, the time t_{NET} does not vary across subsequent iterations of the control loop. In practice, t_{NET}, which depends on the frame size and other parameters, can be computed analytically [CEN10b].

Unfortunately, the t_{CPU} time interval is not deterministic. Its duration basically depends on two factors, that is, the time spent by the CPU to execute the control algorithm (t_{ALG}) and delays introduced by EM and its operating system (t_{EM}). t_{ALG} is mostly affected by the code implementing the control algorithm. In most cases, an upper bound on the execution time can be evaluated, either analytically or experimentally. On the contrary, no reasonable bound can be found a priori for t_{EM}. For this reason, the inequality $t_{CPU} + t_{NET} \le T_{CYCLE}$ might not hold in some circumstances.

t_{EM} takes into account the time needed by the protocol stack to send and receive EtherCAT frames and also includes operating system latencies caused by interacting tasks, context switching, and scheduling. To keep t_{EM} bounded, EM implementations usually rely on hard real-time operating systems and protocol stacks. Software drivers for interfacing to the Ethernet controller are often rewritten, in order to ensure that such a requirement is met.

Besides the functionalities a given EM is able to provide, the quality of an EM largely depends on its ability to transfer EtherCAT frames with high determinism. The EM performance analysis presented in [CER11], for instance, shows that jitters on the order of 10 µs can be obtained for a nominal cycle time equal to 1 ms, provided that hard real-time operating systems and protocol stacks are used. EMs based on non real-time operating systems and protocol stacks are usually not suitable for EtherCAT, since in such conditions t_{EM} is not bounded, and delays larger than 1 ms can be commonly experienced. In these cases, the actual determinism featured by EtherCAT can be even worse than non real-time industrial Ethernet solutions, such as EtherNet/IP.

18.6.2 Commercial versus Open-Source Implementations

As already pointed out, the most important requirement for EM is determinism. Even though the determinism achieved by an EtherCAT network is largely enabled by the slave hardware (i.e., by the precision of the DC synchronization mechanism and the ability of slaves to process datagrams on-the-fly), the master plays an important role as well, because all actions summarized in Figure 18.7 have to be completed in a single iteration of the control loop in order to ensure a predetermined and constant cycle time.

In addition, a wide range of other features differentiate EM products today available off-the-shelf. Even though hardware prototype implementations that are based on field programmable gate array (FPGA) boards are described in the scientific literature, EMs are usually implemented as software components. In [MAR12], for example, a mixed hardware-and-software-based architecture was presented that enables the transmission of EtherCAT frames with jitters shorter than 10 ns. The whole system behavior was also sped up significantly, by performing several time-consuming EM operations directly in hardware. Although promising, FPGA-based EM architectures are just prototypes, and, as far as we know, they are not used in real-world EtherCAT applications at the present day. For this reason, the remaining part of this section focuses on software-based EMs.

A number of software EMs are currently available, and designers and systems integrators can choose between commercial and open-source licenses. Commercial products are usually more user-friendly, bug-free, full-featured, and well documented, but open-source products can also be selected, in some situations, for a number of complementary reasons.

On the one hand, Beckhoff TwinCAT [TWIN3] is the most popular commercial EM solution, produced by the company that developed EtherCAT. Other interesting commercial products are NI EM and KPA EM, produced by National Instrument and Koenig-pa GmbH, respectively.

On the other hand, the most popular open-source EM is certainly EtherLab, developed by Ingenieurgemeinschaft IgH and freely downloadable [ELAB]. Another popular open-source solution is the Simple Open EtherCAT Master (SOEM). Unfortunately, SOEM cannot be used in applications demanding high determinism and low T_{CYCLE}, since its implementation is not real-time.

Regardless of the specific implementation, the two EM categories exhibit very different characteristics. In Table 18.4, the most significant differences between commercial and open-source EMs are highlighted, pointing out strengths (+) and weaknesses (–). The absence of weak points for commercial EMs in the table does not mean absence of defects, since open-source EM strengths have to be considered as weaknesses for commercial EMs and vice versa. A more detailed comparison between the two categories can also be found in [SCA12].

TABLE 18.4 Commercial versus Open-Source EMs

Commercial	Open Source
+ Easy-to-use	+ Cost (less expensive)
+ Programming languages compliant with IEC 61131-3	– (Usually) nonstandard programming languages
+ Many libraries	+ Customizable code and (usually) better performance
+ Documentation	– Limited hardware support
+ Integrability	– Some EtherCAT features not implemented

Commercial EMs are usually easy to use, if compared with their open-source counterparts. The network configuration, programming interface, and all configurable features are accessible via a graphical user interface (GUI), which allows reducing the time needed to learn the programming and configuration environment also for not experienced programmers and, consequently, shortens the *time-to-market* of applications.

The biggest advantage of open-source EMs is that they are generally free of charge. For this reason, if a large number of control applications have to be developed, costs for the longer learning time can be compensated by adopting the open licensing scheme.

At the beginning of 2013, the new version of the IEC standard 61131-3 [IEC13] was released, which defines five programming languages to be used with PLCs. This edition replaces the previous version published in 2003. All most important commercial tools for EtherCAT programming (and for the vast majority of other industrial Ethernet protocols as well) are now IEC 61131-3 compliant.

Two textual and three graphical languages are included in the IEC 61131-3 standard:

- *Instruction list* (IL) is a low-level programming language, with a level of abstraction similar to assembler.
- *Structured text* (ST) resembles high-level programming languages like C or C++.
- *Ladder diagram* (LD) allows to represent the program as a sequence of digital circuits, including contacts, coils, and specific blocks like counters, timers, etc.
- *Function block diagram* (FBD) is based on block diagrams, where blocks are connected through oriented arcs. It describes how inputs, through the blocks, are transformed into outputs.
- *Sequential function chart* (SFC) represents the control program as a sequence of states (*steps*) with associated *actions* activated by given conditions (*transitions*).

ST and LD are currently the most used languages. ST has a higher level of abstraction and is suitable for large projects, while LD is well known to technicians working in the industrial automation field. A typical situation found in many projects are programs developed with mixed languages, that is, LD for low-level functions and ST for more complex functions (it also acts as the *glue* for putting different subroutines together).

Unfortunately, open-source EMs seldom support IEC 61131-3-compliant programming languages. This drawback limits the code reusability and portability on other EMs and PLCs and requires, in general, longer training times. However, it may also lead to a better optimization of the produced code.

In most open-source solutions, the programming language used to develop the actual control application is the same as the one adopted for implementing the EM. This means that the EM code, which is freely released in the source form, can be possibly modified to meet specific design requirements. This often results in a more efficient EM (where, for instance, some unused features are disabled at compile time), which sometimes is able to show an improved determinism (the EM can be calibrated with respect to the hardware of the PLC, e.g., by modifying the driver of the network interface card or the operating system scheduler). By contrast, some particular features of EtherCAT may sometimes be unavailable in open-source EMs, and enhanced determinism can be achieved only with the aid of specific hardware components (i.e., selected network interface cards).

The documentation, as well as the availability of a large number of libraries (to ease the implementation of the algorithms used in industrial control systems), is certainly one of the most significant advantages of commercial EMs.

Finally, the ability to integrate a number of subnetworks at the cell level is important in large industrial plants, in order to coordinate the production processes. In this context, the availability of a well-engineered application programming interface (API) for enabling communications between PLCs can make the difference, especially when different industrial Ethernet technologies have to be integrated in the same plant. Open-source EMs often offer communication services that are limited to the field level segment of the network (i.e., in the case of EtherCAT, the segment between EM and its slaves), while no API is provided for communications at the cell level (i.e., between different EMs).

In conclusion, the choice of the master should not be underestimated, since it can strongly constrain the project, and in particular its maintainability and characteristics, in terms of both functionalities and performance of the control application. Since in most situations, applications cannot be easily migrated from one EM to another, requirements about performance, desired features, and costs must be carefully considered during the initial design phase.

18.6.3 EtherCAT Application Example

This section focuses on a simple example of a typical EtherCAT application, so as to offer the reader a flavor of EM configuration and programming. The same application has been implemented with both a commercial (TwinCAT) and an open-source (EtherLab) EM solution. In the case of TwinCAT, two programming languages have been employed, namely, the LD and ST.

The sample application considered here is a frequency divider. Its goal is switching the level of a digital output (DO) each time three rising edges are sampled from a digital input (DI)—for instance, by pressing three times a push button connected to DI. The EtherCAT network involved in this example consists of a single EM (a standard PC, which runs either TwinCAT or EtherLab), one DI slave (Beckhoff EL1202 [EL1202]), and one DO slave (Beckhoff EL2202 [EL2202]).

The EM, independently from its implementation (TwinCAT/EtherLab), must be properly configured before entering the application loop that executes the control algorithm. The configuration takes place by means of a graphical interface in TwinCAT or relies on specific textual commands in the case EtherLab is used. Each EtherCAT slave hosts an OD, containing a number of entries that are addressed through a 16 bit index and an 8 bit subindex. OD entries allow accessing the slave registers that need to be read/written by the program.

In the configuration phase, EM sets up process data in each slave to specify what information is transferred over the network during every iteration of the control loop, in agreement with the application requirements. These PDOs are embedded in EtherCAT datagrams and transferred by means of an EtherCAT frame. In practice, they are the process data that are used by the control application to perform its task.

The final step of the configuration phase is linking process data to the variables in the process image used by the application in the control loop. A detailed analysis of the configuration steps is not reported here: a thorough description of TwinCAT can be found in the public software documentation, while additional details about the configuration phase for both the EMs taken into account are included in [SCA12].

18.6.3.1 Ladder Diagram

A program written in TwinCAT LD consists of two parts: declaration of variables and code. Declaration of variables is performed by means of an IEC 61131-3-compliant textual programming language. For instance, the variables used in the frequency divider example are shown in Figure 18.9.

```
VAR
        (* Process variables *)
        input AT %IX0.0 : BOOL;
        output AT %QX0.0 : BOOL := FALSE;
        (* Internal variables *)
        switch : BOOL := FALSE;
        enable : BOOL := TRUE;
        (* Function block: CTU counter *)
        counter : CTU;
END_VAR
```

FIGURE 18.9 Declaration of variables for the application example.

Process variables `input` and `output` (lines 3 and 4 in the picture) are two Boolean values representing the input DI and output DO; they are linked to the process data in the EtherCAT frame as defined in the configuration phase of the EM. Symbols `%I` and `%Q` are used for input and output process variables, respectively. `X` is one of the possible sizes for process variables (`X`: bit, `B`: byte, `W`: word, and `D`: double word), while the numbers in the format `B.b` specify the position of the variable in the process data memory. `B` and `b` stand for the byte and bit number, respectively (e.g., address `X12.3` selects bit number 3 of byte number 12). Such an addressing method is used to make control application variables largely independent of the underlying communication technology used to interconnect PLCs and remote devices (either a fieldbus or industrial Ethernet).

In Figure 18.9, `switch` and `enable` are two internal variables, while `counter` is a function block, that is, a complex *object* with its own attributes and static variables, which retain their values between successive executions. Variables can be initialized at the same time when they are declared: for instance, `output` is initialized to 0 (`FALSE`), which means that the initial value of the DO slave is reset.

Figure 18.10 shows the LD implementation of the application, as well as the most common symbols used in this language, namely, *contacts* and *coils*. In the LD model, current flows through the circuit from left to right. The basic components are contacts and coils, and each component is associated to a variable. When the variable associated to a contact is TRUE, the current flows through the contact; otherwise, when it is FALSE, the current flow is blocked. When current flows through a coil, the variable associated to that coil is set to TRUE; otherwise its value is FALSE. The symbol "/" over a contact or a coil inverts the behavior of that component; for example, in the case of a contact, the current flow is blocked if the variable associated with the contact is TRUE. An LD program is executed from top to bottom.

Other more complex components are also defined in the LD programming language, which are known as function blocks, the most important being *timers* and *counters*. A count up (CTU) counter is shown in the first row of Figure 18.10. Each time a rising edge is sampled by the CU input of the CTU counter, the CV value of the counter (an unsigned integer variable) is increased by one. When the value of CV is greater than or equal to PV, the output Q is set to TRUE (so that the current flows out from Q). When the current flows through the RESET input, the CTU counter stops counting, and the CV value is set to FALSE. Moreover, the output Q stops emitting the current flow.

The first row of the program in Figure 18.10 means that each time the value of the process variable `input` is TRUE, the current reaches the input CU of the CTU counter, and thus, the CV value is incremented by 1. When the value CV equals 3 (the value of PV), the current exits from output Q and crosses the two coils associated with the `switch` and `enable` variables (both variables are set to TRUE).

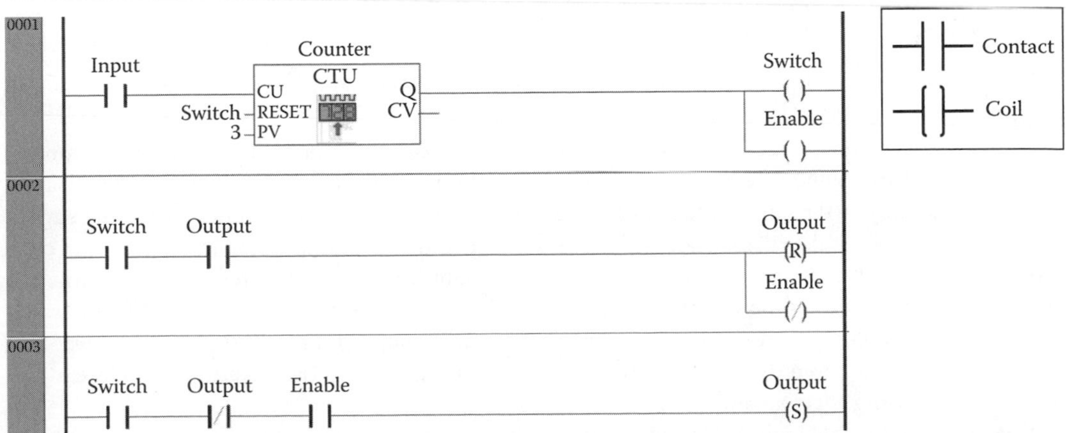

FIGURE 18.10 The example application in ladder diagram.

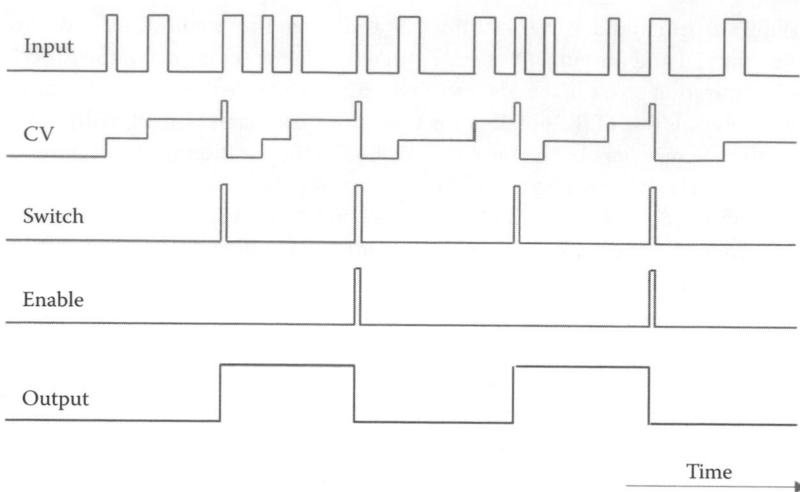

FIGURE 18.11 Timing diagram of the LD program.

The two contacts in row 2 build an AND operator. When both variables `switch` and `output` are TRUE, current flows through both coils associated to the `output` and `enable` variables. The idea behind this scheme is that, when the counter reaches the value of PV (i.e., when `switch` is TRUE), the program switches the `output` variable, and, consequently, also the DO output is switched.

The code in row 2 reverts the output from TRUE to FALSE. The coil with the "R" symbol has a specific behavior: when traversed by a current flow, it resets to FALSE the associated variable. The "S" coil in row 3 is needed to set the associated variable to TRUE. These behaviors are different when compared with classical coils, because actions are performed only in the presence of current flows. In summary, the second row of the LD code sets the variable `output` to FALSE and the `enable` variable to FALSE if the `switch` and `output` variables are both TRUE. The `enable` variable is set to FALSE in order to disable the third row of code in the case a change of output has just been performed in the same execution cycle.

The third row of the code is similar to the second and manages the transition of the `output` variable from FALSE to TRUE. Indeed, if a transition of the output is required (`switch` is TRUE), the `output` variable is FALSE, and the second row has not changed the output yet (`enable` is TRUE), the `output` variable is set to TRUE. In the following cycle, `switch` remains TRUE, the counter is reset, and, since no current exits from the contact Q, the `switch` and `enable` variables are set to FALSE.

A possible example of timing diagram for the variables involved in the example is shown in Figure 18.11.

18.6.3.2 Structured Text

The implementation of the same example in the TwinCAT ST programming language is shown in Figure 18.12.

From a lexical and syntactical point of view, ST resembles the typical high-level programming languages used in computer engineering. The part of the program reserved to variable declaration (lines 2–9) has exactly the same syntax as LD. Code lines 10–14 perform the detection of a rising edge in the `input` variable. If the current sampled value is TRUE and the value sampled in the previous cycle was FALSE (line 11), a rising edge is detected, and the variable `counter` is incremented by 1 (line 12). The `counter` variable has the same meaning as the CT value of the CTU counter in the LD program.

When three rising edges are detected (line 16), the output is switched (lines 17–21), and the `counter` variable is reset to 0. From this point on, the program is put into the initial state, and therefore, it can start counting rising edges again.

This example is useful to understand that, although LD and ST are actually different, their complexity is comparable. Usually, ST is more suitable for large projects since it is more expressive and complete

```
PROGRAM MAIN
VAR
        (* Process variables *)
        input AT %IX0.0 : BOOL;
        output AT %QX0.0 : BOOL := FALSE;
        (* Internal variables *)
        old_input : BOOL := FALSE;
        counter : INT := 0;
END_VAR

(* Detection of a rising edge *)
IF input = TRUE AND old_input = FALSE THEN
        counter := counter + 1;
END_IF
old_input := input;

(* Output switching *)
IF counter = 3 THEN
        IF output = TRUE THEN
                output := FALSE;
        ELSE
                output := TRUE;
        END_IF
        counter := 0;
END_IF
```

FIGURE 18.12 Structured text of the sample application program.

than LD. Moreover, ST can be easily mastered by programmers that are used to classical high-level programming languages, such as C, C++, Java, Ruby, Python, etc.

18.6.3.3 EtherLab C

The same frequency divider has also been implemented by using EtherLab on a Linux-based PC. The EtherLab EM follows a completely different approach with respect to TwinCAT. Indeed, the programming language is C, and the configuration phase is performed directly in the control program by means of C functions and without any graphical interface aid. The configuration phase, not described for reasons of space and because not particularly interesting, consists in a sequence of commands to associate the control application with the EM and activate it and in a mapping between the required entries in the slaves' OD and a memory area in the PC (accessible by the control application).

At the end of the configuration phase, a memory pointer (`domain_pd`) allows the control application to access the memory areas associated to the input of the DI and to the output of the DO terminals. The offsets of these process data with respect to the memory address `domain_pd` are `DI_offset` and `DO_offset` for the input and the output, respectively. The EtherCAT control task must start after the configuration phase. The timing of the control task can be dealt with in EtherLab by using either POSIX system calls, in the case of not demanding applications, or specific services offered by the underlying hard real-time operating systems for increased accuracy.

The control task corresponds to the function `cyclic_task()`, whose body is reported in Figure 18.13. It is periodically executed on every T_{CYCLE}, and it is scheduled by using a timing function similar to `wait` in the pseudo-code in Figure 18.7. Inside the control task, two other functions, `ecrt_master_receive(master)` and `ecrt_domain_process(domain)`, at lines 7 and 8, are used to receive and extract process data from the EtherCAT frame. Data extracted from the datagrams are then stored in the process memory area pointed by the global variable `domain_pd`.

```
1 : int counter = 0;
2 : int old_input = 0;
3 : int input;
4 : int output = 0;

5 : void cyclic_task() {
6 :     /* Receive new process data */
7 :     ecrt_master_receive(master);
8 :     ecrt_domain_process(domain);

9 :     input = EC_READ_BIT(domain_pd+DI_offset, 0);
10:     /* Detection of a rising edge */
11:     if (input==1 && old_input==0) {
12:         counter = counter + 1;
13:     }
14:     old_input = input;

15:     /* Output switching */
16:     if (counter==3) {
17:       if (output==1) {
18:           output = 0;
19:       }else{
20:           output = 1;
21:       }
22:       counter = 0;
23:     }
24:     EC_WRITE_BIT(domain_pd+DO_offset, 0, output);

25:     /* Send new computed process data */
26:     ecrt_domain_queue(domain);
27:     ecrt_master_send(master);
28: }
```

FIGURE 18.13 C code of the sample application program.

In line 9, a bit is read from the process memory at position `domain _ pd+DI _ offset` and stored in the `input` variable. The new value of the variable `input` is then used by the control algorithm (lines 11 to 23) to check if the `output` variable has to be switched again. As for ST in TwinCAT, the `counter` variable keeps track of the number of `input` rising edges. When `counter` reaches the value 3 (line 16), the `output` variable, and thus the value of the DO output, is inverted and the new value is then written in the process memory (line 24). Process data are finally embedded in datagrams (line 26) and sent to the slave devices in the EtherCAT frame (line 27).

References

ETG12 EtherCAT Technology Group, EtherCAT—The Ethernet fieldbus, 2012. [Online]. Available: http://www.ethercat.org/.

ET1100 Beckhoff Automation GmbH, Hardware data sheet ET1100 slave controller, 2013. [Online]. Available: http://www.beckhoff.com/.

IE158 IEC 61158-3/4/5/6-12, Industrial communication networks—Fieldbus specifications—Part 3-12: Data-link layer service definition—Part 4-12: Data-link layer protocol specification—Part 5-12: Application layer service definition—Part 6-12: Application layer protocol specification—Type 12 elements (EtherCAT).

IE784 IEC 61784-2, Industrial communication networks—Profiles—Part 2: Additional fieldbus profiles for real-time networks based on ISO/IEC 8802-3.

8023 IEEE 802.3-2008, IEEE standard for information technology-Specific requirements—Part 3: Carrier sense multiple access with collision detection (CSMA/CD) access method and physical layer specifications.

PRY08 Prytz, G., A performance analysis of EtherCAT and PROFINET IRT, *Proceedings of IEEE International Conference on Emerging Technologies and Factory Automation (ETFA 2008)*, pp. 408–415, September 15–18, 2008.

CEN10 Cena, G.; Bertolotti, I.C.; Scanzio, S.; Valenzano, A.; Zunino, C., On the accuracy of the distributed clock mechanism in EtherCAT, *Proceedings of IEEE International Workshop on Factory Communication Systems (WFCS 2010)*, pp. 43–52, May 18–21, 2010.

CEN12 Cena, G.; Bertolotti, I.C.; Scanzio, S.; Valenzano, A.; Zunino, C., Evaluation of EtherCAT distributed clock performance, *IEEE Transactions on Industrial Informatics*, 8(1), 20–29, February 2012.

CEN10b Cena, G.; Scanzio, S.; Valenzano, A.; Zunino, C., Performance analysis of switched EtherCAT Networks, *Proceedings of IEEE Conference on Emerging Technologies and Factory Automation (ETFA 2010)*, pp. 1–4, September 13–16, 2010.

CER11 Cereia, M.; Cibrario Bertolotti, I.; Scanzio, S., Performance of a real-time EtherCAT master under linux, *IEEE Transactions on Industrial Informatics*, 7(4), 679–687, November 2011.

MAR12 Maruyama, T.; Yamada, T., Hardware acceleration architecture for EtherCAT master controller, *Proceedings of Ninth IEEE International Workshop on Factory Communication Systems (WFCS 2012)*, pp. 223–232, May 21–24, 2012.

TWIN3 Beckhoff Automation GmbH, TwinCAT 3 | eXtended Automation (XA), 2013 [Online]. Available: http://www.beckhoff.com/.

ELAB Ingenieurgemeinschaft IgH, EtherLab—EtherCAT master, 2013. [Online]. Available: http://www.etherlab.org/.

SCA12 Scanzio, S., SoftPLC-based control: A comparison between commercial and open-source EtherCAT technologies, *Handbook of Research on Industrial Informatics and Manufacturing Intelligence: Innovations and Solutions*, IGI Global, pp. 440–463, 2012.

IEC13 International Electrotechnical Commission, Programmable controllers—Part 3: Programming Languages, IEC 61131-3:2013 (Ed. 3.0), 2013.

EL1202 Beckhoff Automation GmbH, EL1202 | 2-channel digital input terminal 24 V DC, TON/TOFF 1 μs, 2013 [Online]. Available: http://www.beckhoff.com/.

EL2202 Beckhoff Automation GmbH, EL2202 | 2-channel digital output terminal 24 V DC, TON/TOFF 1 μs, push-pull outputs, tri-state, 2013 [Online]. Available: http://www.beckhoff.com/.

19

Ethernet POWERLINK

19.1 Introduction 19-1
19.2 Ethernet POWERLINK Protocol 19-1
Low Layers Specifications • Application Layer • Network Operation • Encapsulation of EPL Frames
19.3 Ethernet POWERLINK Topologies 19-5
19.4 Ethernet POWERLINK Redundancy 19-6
Medium Redundancy • Managing Node Redundancy
19.5 Ring Redundancy 19-8
OpenSAFETY Protocol • Basic Performance Analysis
References 19-11

Federico Tramarin
National Research Council of Italy

Stefano Vitturi
National Research Council of Italy

19.1 Introduction

Ethernet POWERLINK (EPL) represents a prominent example of real-time Ethernet (RTE) communication networks. It belongs to the IEC 61784 International Standard [1], where it is referred to as the Communication Profile (CP) #1 within the CP Family #13.

EPL was originally designed by B&R GmbH [2] and the first version was released in 2001. Later on, the Ethernet POWERLINK Standardization Group (EPSG) [3] was in charge of continuing its management and published the first EPSG Draft version of EPL as an open standard in 2003 fostering for a free distribution of its specifications as well as for the open-source distribution of the protocol source code. The current version of this communication standard [4] can actually be downloaded from the EPSG website.

EPL is able to provide the real-time capabilities required by critical processes, control tasks, and management functions typical of the industrial scenario [5]. As a matter of fact, EPL, which is compliant with the Industrial Automation Open Network Alliance (IAONA) real-time class 4 (highest performance) recommendations, is able to cope with communication cycles in the order of hundreds of microseconds, ensuring, at the same time, jitters below 1 μs.

Also, the EPL application layer is based on the popular and settled CANOpen standard [6] (practitioners often refer to EPL as *CANOpen over Ethernet*). This feature ensures, at the high layer of the protocol stack, compatibility with several other industrial communication devices.

19.2 Ethernet POWERLINK Protocol

A schematic representation of the EPL protocol stack is provided in Figure 19.1.

19.2.1 Low Layers Specifications

At the lowest protocol layers, EPL makes use of legacy Ethernet to access the medium, while real-time capabilities are provided by the purposely developed EPL data link layer (EPL DLL) protocol, which is placed on top of the standard Ethernet Medium Access Control (MAC) [7]. Therefore, communications

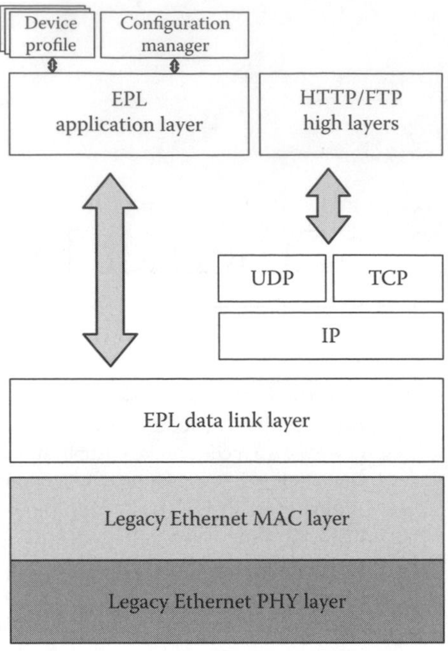

FIGURE 19.1 EPL protocol stack.

on the medium borrow the features provided by Ethernet MAC. Particularly, the transmission medium is accessed exploiting the well-known Carrier Sense Multiple Access with Collision Detection (CSMA/CD) scheme. The EPL reference physical layer is the IEEE 802.3 100BASE-X, used in the half-duplex transmission mode. In practice, the popular 100BASE-TX based on copper wires and the 100BASE-FX fiber-based channel are encompassed, even if the former is the most common choice in real applications. Nevertheless, it is worth noticing that the EPL standard itself foresees that any advancement in the Ethernet specifications (as, e.g., the adoption of Gigabit Ethernet) can be exploited in order to enhance performance.

In the legacy installation case, when copper wires are employed, both RJ-45 and M12 connectors are encompassed and can be used depending on the environment characteristics, following the installation guidelines of IAONA [8]. EPL also provides recommendations on the type of cables to be adopted in network installations. Particularly, AWG26 twisted pair cables with a protection foil shielding (S/UTP, commonly known as Cat5e) shall be used to increase immunity to noise in the potentially harsh industrial environment.

19.2.2 Application Layer

The EPL application layer provides services to the applications by means of *service objects*. From this point of view, the application layer may be seen actually as a collection of objects that, indeed, are listed in a dedicated structure called the *object dictionary*.

Services may be either local or remote. The former type involves only the local objects of a node. These services may be started either by a *request* primitive issued by a specific application or by a local service object, which detects an event and communicates it to the addressed application via the *indication* primitive. Conversely, remote services involve objects that reside on peer nodes connected to the network and may be either nonconfirmed or confirmed. An application that requires a nonconfirmed service issues, at first, a request primitive to a local service object that, in turn, forwards the request to the peer object(s) via an *indication* primitive. On the other hand, confirmed services are peer-to-peer relationships that involve only two service objects, one per each network node.

The EPL application layer ensures data can be exchanged according to several communication models. Particularly, as it will be shown in the following of this chapter, the available relationships are master/slave, client/server, and producer/consumer.

19.2.3 Network Operation

An EPL network comprises two different kinds of nodes, namely, managing node (MN) and controlled nodes (CNs). The former is a device that actually implements both automation and control tasks, taking care, at the same time, of all network management aspects. CNs represent the other entities in the network, which are devices like, for example, sensors and actuators, equipped with communication capabilities. The cyclic exchange of data between MN and CNs is at the basis of the EPL operation.

The free distribution of the EPL specification, as well as the open-source nature of the software codes for both MN and CN, has boosted the implementation of different kinds of these devices, which are hence off-the-shelf available by several vendors, covering the whole range of industrial automation systems, including programmable logic controllers (PLC), several types of sensors and actuators, computer numerical control (CNC) machines, etc.

Each EPL network is managed and supervised by exactly one MN, whereas it can encompass up to 240 CNs deployed in various topologies. To ensure real-time behavior for transmission of time-critical data and, at the same time, to provide mechanisms to allow noncritical communications (traffic based on IP protocol, like UDP, HTTP, or FTP requests) a slot communication network management (SCNM) scheme is defined by EPL. Specifically, operations on the network rely on a fixed-duration periodic cycle, which is mainly split in a first isochronous part, where real-time data are exchanged following a publish/subscribe relationship with CNs, followed by a second asynchronous section. This latter has been conceived for either the transmission of asynchronous real-time data or the seamless integration with general purpose IT applications via the transmission of common TCP-IP (Transmission Control Protocol-Internet Protocol) messages. In practice, a strict time division multiple access (TDMA) technique is at the basis of SCNM, where the MN is in charge of managing the access to the channel. In this way, though legacy CSMA/CD is in use to resolve medium contentions and collisions, traffic will never interfere, also in the case, nonswitched Ethernet configurations are employed.

A schematic representation of the EPL cycle is given in Figure 19.2.

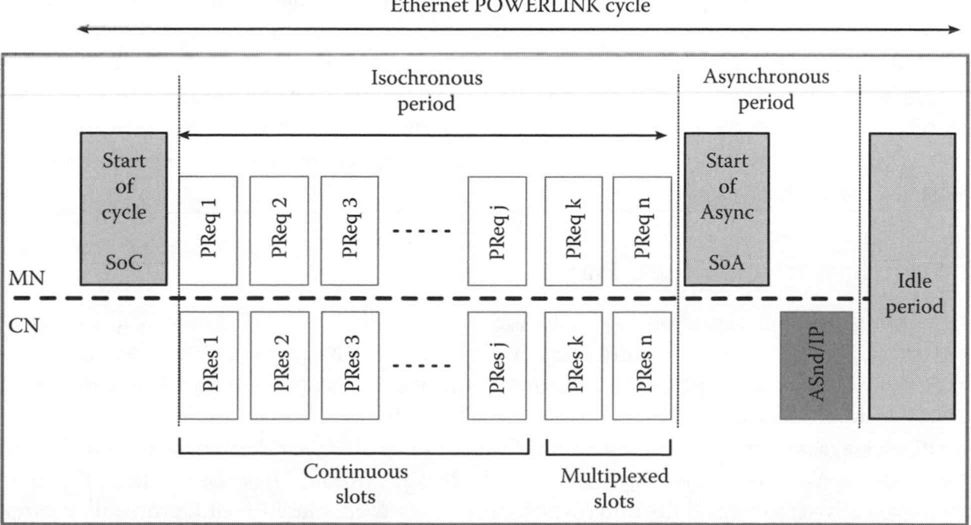

FIGURE 19.2 Ethernet POWERLINK cycle.

The beginning of the EPL cycle is triggered by the MN through the broadcast frame SoC (start of cycle). To maintain a fixed cycle time, the SoC frame is issued on a precise periodic basis, keeping jitter on it as low as possible; this also serves to the purpose of providing time synchronization for all the devices.

After the SoC frame has been issued, the *isochronous period* is entered. In this key part of the cycle, the MN polls each CN by means of the poll request (PReq) frame, which is sent only to the selected CN and can carry output data for it. The accessed CN responds to this query by issuing a poll response (PRes), which is instead a multicast frame carrying input data, made available to all nodes in the network. During this phase, the MN also gathers requests for asynchronous transmissions, which may be solicited by any of the CNs by means of the PRes messages and/or by the MN itself. When the last CN has been polled, the MN has the opportunity to send a PRes frame. According to the protocol specification, such a frame is broadcast, and it is intended to send data to groups of CNs.

Once the isochronous period is concluded, the MN sends a broadcast frame called start of acyclic (SoA), which informs all CNs about the start of the *asynchronous period*. This second phase ensures the transmission of only one asynchronous message by a selected node. The SoA frame is also used to inform which node, among those that previously issued a request, has been selected for the acyclic communication. The choice is based on a priority selection scheme. Acyclic traffic is allowed to serve two different types of transmissions. Specifically, the selected node may transmit either a legacy Ethernet message or an EPL-specific ASyncSend frame. The former type is introduced to allow for general-purpose IT traffic from any source of data, even though the EPL specifications recommend the use of UDP/IP protocol pair since UDP, with respect to the more popular TCP, requires a lower overhead. The second type refers to a particular message purposely defined by the EPL protocol for the exchange of acyclic process data, such as alarms. It is therefore specific of the protocol, and it may be transmitted either to a single node or to a group of nodes.

After both the SoA and acyclic frames have been transmitted, the *idle period* is entered. Here, all the stations wait for the new cycle to start, that is, to receive a new SoC frame from the MN. The purpose of this final period is to ensure that all the message transmissions scheduled during the other periods are concluded before the end of the EPL cycle, and hence, no activity is present on the network during this phase.

The structure of the cycle described earlier represents the baseline scheme defined by the EPL standard. However, a noteworthy feature provided by the protocol is the possibility to poll a subset of the CNs at a predefined multiple of the cycle time. That is, the standard defines two communication classes. The first class is the one just described, referred to as *continuous polling*. CNs belonging to this class are polled at every EPL cycle. The other class enables *multiplexed polling*, where data from a subset of CNs are gathered at a multiple *n* of the cycle time. Both continuous and multiplexed CNs coexist in the isochronous period, as described in Figure 19.2. The multiplexed CNs are hence adequately distributed in slots in such a way their polling period is respected and the duration of the isochronous period is kept as low and uniform as possible.

19.2.4 Encapsulation of EPL Frames

The EPL DLL protocol is placed on top of the MAC layer of the IEEE 802.3 standard. Consequently, any EPL message is encapsulated within a legacy Ethernet II frame, as shown in Figure 19.3. The field Ethernet Type is used to specify that the (Ethernet) frame is actually employed by EPL setting its content to the hexadecimal value 88ABh.

An EPL message is inserted at the beginning of the DATA field of the Ethernet frame and is composed mainly by four consecutive fields. The first octet identifies the *Message type*. In this field, the first bit is reserved (typically set to 0), and the other 7 bits are actually used. The EPL DLL protocol encompasses six main message types, relevant to the messages already mentioned in the quoted protocol description (SoC, SoA, etc.), each one identified by a unique hexadecimal code, as shown in Table 19.1.

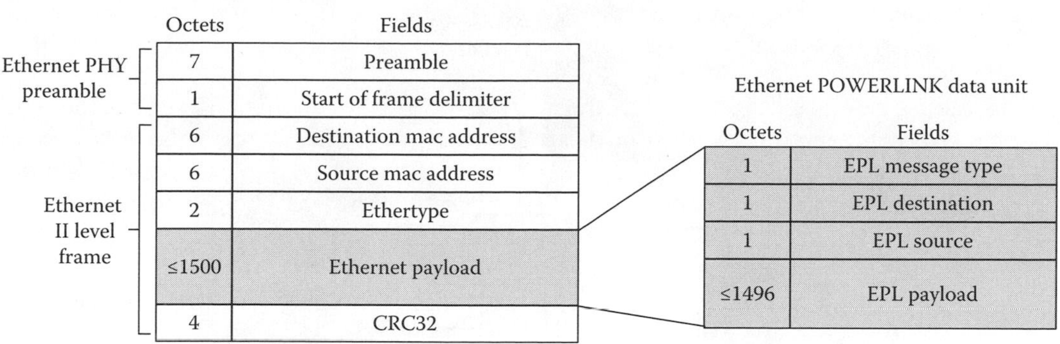

FIGURE 19.3 Ethernet POWERLINK message encapsulation.

TABLE 19.1 Ethernet POWERLINK Message Types

Ethernet POWERLINK Frame	MAC Transfer Type	Hexadecimal Code
Start of cycle	Multicast	01_h
Poll request	Unicast	03_h
Poll response	Multicast	04_h
Start of asynchronous	Multicast	05_h
Asynchronous send	Multicast	06_h

The second and third octets contain, respectively, the *Destination Address* and *Source Address* of the EPL nodes (clearly, they have no relation with the addressing mechanism of Ethernet). Finally, the *Data* field carries the actual message payload, and, on the basis of the message type, it may contain a further internal subdivision to accommodate for message-specific fields.

19.3 Ethernet POWERLINK Topologies

EPL recommends the use of Ethernet hubs in order to cope with the tight timing requirements that are often imposed to the network. However, switches may be employed as well, taking into account their behavior, with particular attention to the additional delays they introduce.

Indeed, hubs are able to provide low path delays (lower than 460 ns) and a very limited jitter (according to the EPL specification, typically no more than 70 ns), while switches are usually characterized by worse performance figures.

The typical fields of application for EPL networks often require the deployment of complex topologies, where several hub levels are introduced. For this reason, hubs may be integrated into the CN network interface card, thus reducing the number of needed components and simplifying the network installation. Specifically, as outlined in Figure 19.4, this allows for the deployment of networks that comprise traditional star topologies combined with linear chains of CNs.

The strict determinism implemented by EPL through the SCNM also ensures the absence of frame collisions on the medium. This allows to deploy even more complex topologies, since the typical Ethernet constraint on the maximum round-trip time being limited to 5120 ns (for the 100BASE-TX physical layer) no longer applies to EPL networks. Thus, for example, linear chains of CNs benefit of this relaxed timing constraint since, in general, they may introduce considerable round-trip times that could exceed the Ethernet upper bound.

Conversely, further timing constraints are introduced by the EPL specifications. Indeed, time-outs are used in the polling mechanism (during the isochronous period, in particular) as a mean to detect either

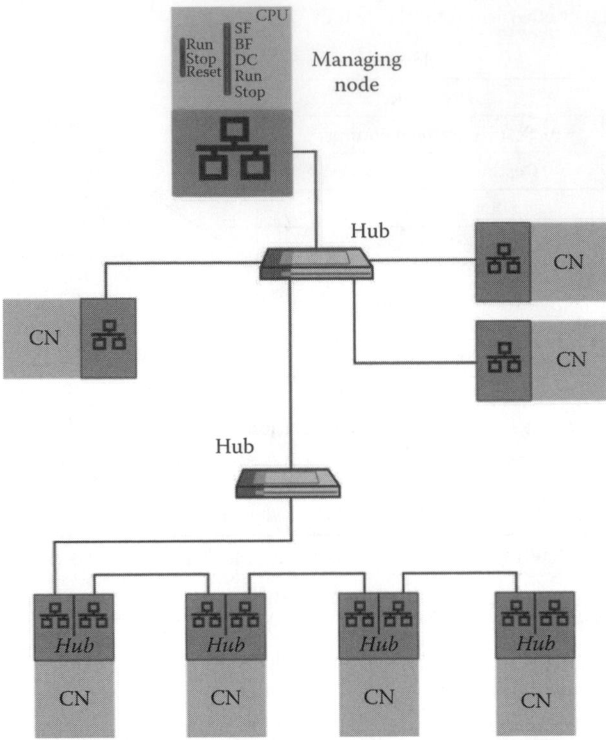

FIGURE 19.4 Example of Ethernet POWERLINK network using combined topologies.

node failures or transmission errors. Particularly, after sending the PReq frame toward a CN, the MN starts a timer whose default value (for the 100 Mbps transmission speed) is set to 25 μs. The corresponding PRes from the addressed CN should be received within this time-out time to avoid misbehavior of the EPL network in the subsequent cycles. Therefore, the latency for a round-trip time between the MN and any of the CNs is upper-bounded by this time-out value. However, since this latency is strictly dependent on both the employed components and the adopted topology, the EPL protocol allows a per-node configuration of the time-out. Also, a global network-wide different time-out selection is possible.

19.4 Ethernet POWERLINK Redundancy

The EPL standard aims at providing a reliable and robust RTE network. To this extent, the POWERLINK High-Availability Service [9], provided as an amendment document to the main specifications, focuses on possible failures of any of the network devices, providing suitable methods to ensure even in those situations a full network functionality. To this purpose, the amendment requires that very fast switchover times be guaranteed in case a failure occurs, including, in particular, a hot standby function. The POWERLINK High-Availability services implement two specific means for achieving the quoted goals, namely, medium redundancy and MN redundancy. Moreover, a further optional technique, referred to as ring redundancy, may be adopted.

19.4.1 Medium Redundancy

A schematic representation of this redundancy technique implementation is given in Figure 19.5.

As can be seen, all devices connected to the EPL network are provided with a second, redundant, cabling. During the normal operation of the network, stations (both MN and CNs) transmit data on

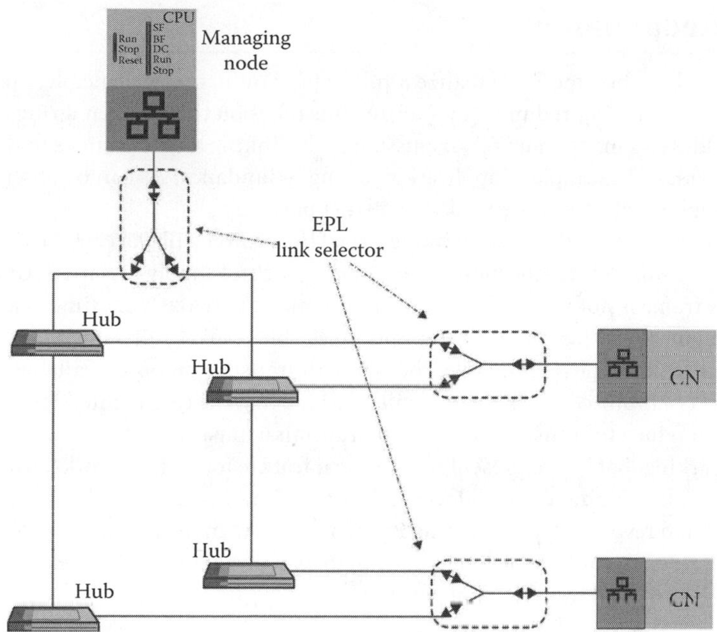

FIGURE 19.5 Example of Ethernet POWERLINK medium redundancy.

both links. A link selector feature should be implemented on each device to decide what is the suitable reception link to be used. The decision relies on a predefined algorithm that can be based on either a single attribute (e.g., the message arrival time) or a set of attributes (e.g., the link quality and the cyclic redundancy codes of the received messages).

Furthermore, the link selector of each CN is required to introduce in the PRes frame the information about the status of both its links. The MN may use the gathered link status information to provide the application it handles with a snapshot of the *quality of the network*. As a final comment, this redundancy feature requires that the operations of the link selector do not introduce a variation in the path delay, measured between any two nodes in the EPL network, larger than 5.76 μs (transmission time of a 64-byte frame on a Fast Ethernet network).

19.4.2 Managing Node Redundancy

This second approach faces the high-availability service from the application layer perspective. Basically, it extends the MN functionality by introducing the redundant managing nodes (RMNs), which are devices with MN capabilities contemporaneously present on the network. This allows defining two new entities, namely, the active managing node (AMN) and the standby managing node (SMN). The former actually represents the legacy MN, and, as such, it implements all its functionalities. Conversely, SMNs are RMNs that are connected to the network and, from the application point of view, behave like legacy CNs. They constantly monitor the network and retrieve information about the current status of all participants. If the current AMN experiences a failure, it is immediately substituted by one of the SMNs, and the takeover shall not require more than an EPL cycle time.

Finally, MN redundancy requires that an election algorithm be implemented to select the SMN that should take over the AMN functionalities, in the case more than one SMN is available. This feature is actually based on priorities assigned to each of the SMNs, which can be either derived from the SMN node address or set through a purposely defined engineering tool.

19.5 Ring Redundancy

Though the earlier-described methods realize a full redundant network principle, optionally, the MN could provide also a basic ring redundancy feature. This relies on the fact that a ring cabling structure intrinsically provides two independent transmission paths linking any two nodes in the network.

Figure 19.6 provides an example of application of ring redundancy. As can be observed, in this case, each device is equipped with two independent Ethernet ports.

Under regular network conditions, each frame issued by the MN will traverse the ring until it reaches the MN again, which will identify the message as *recently sent* and simply discard it. Under faulty conditions, instead, the frame is not received back at the MN, and a previously set time-out time will expire, thus signaling a connectivity loss, due, for example, to a cable break.

As a consequence, in the next EPL cycle, the MN will transmit on both available connections. If it subsequently detects a frame collision, then it will conclude that the ring connectivity has been reestablished so that it can return to transmit on only one transmission path.

It is worth remarking that this represents an optional feature for EPL networks, which should not be confused with the full redundancy introduced with the techniques discussed in the previous sections. However, this method reveals very interesting in the industrial environment since these networks are subjected to harsh stresses, for instance, caused by high temperature variations, extreme humidity conditions, presence of dust, and movement of parts.

19.5.1 OpenSAFETY Protocol

In the context of RTE networks, the main purpose of a safety protocol is to ensure data transfer integrity. Particularly, data duplications, loss of frames, and/or manipulations must be absolutely avoided. Also, a safety protocol should be able to detect excessive delays in data transmission as well as corrupted frames. Typically, only some critical segments of a network are subjected to safety-related control activities, and the implemented safety protocol activates appropriate functions when triggered by a fault situation, leading in extreme cases to the safe shutdown of the plant.

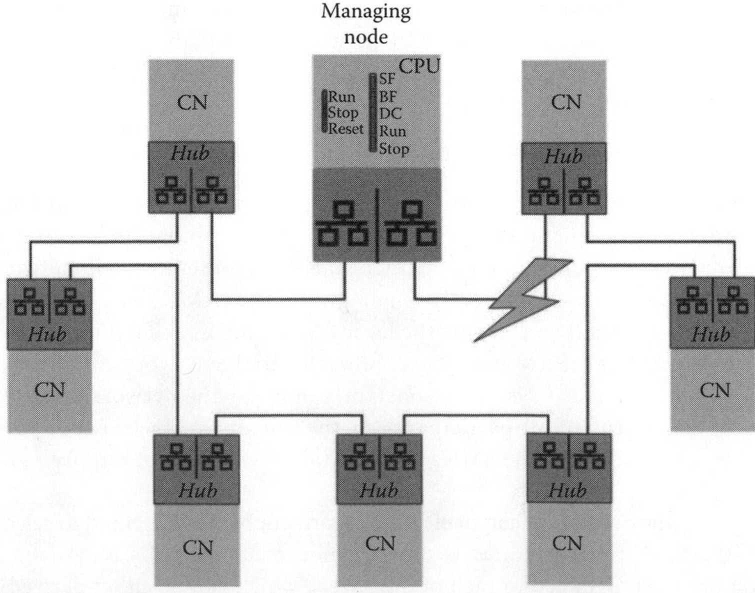

FIGURE 19.6 Example of Ethernet POWERLINK ring redundancy.

To provide an adequate solution to the aforementioned requirements, the EPL Safety Working Group developed the OpenSAFETY protocol, which is now part of the IEC 61784-3 International Standard. OpenSAFETY was originally defined as EPL Safety application layer, released by EPSG in 2007 as an *open* standard. Subsequently, it has been approved by TÜV Rheinland Group in 2008, as being suitable for safety-sensitive applications belonging to the IEC 61508 Safety Integrity Level 3 (SIL 3).

OpenSAFETY, however, has not been exclusively designed for EPL. Indeed, it may be adopted by any other RTE network such as, for example, EtherNet/IP, Sercos, and Modbus, as it is actually independent from the underlying real-time transport protocol, exploiting the *Black channel* principle. This feature derives from the complete implementation of the standard at the application layer, which ensures the so-called OpenSAFETY layer may be, in principle, introduced in the protocol stack of any other RTE network.

A basic method to detect corrupted frames as well as to increase transmission reliability is represented by the use of dedicated frame formats. In particular, user data are duplicated in two adjacent subframes within the same frame. Each one is protected by a checksum calculated differently for both subframes. OpenSAFETY allows a maximum payload data of 354 bytes. At reception, both subframes are independently checked for integrity against their respective cyclic redundancy check (CRC). Further, they are compared to ensure data have been correctly received.

Moreover, as further safety measures, the transmission times of frames and the processing times are taken into account by the protocol, in order to adequately detect delivery failures.

The communication model of the OpenSAFETY protocol relies on the producer/consumer paradigm, in that any safety node (SN, a node part of the safety-controlled network) can be a producer and/or consumer of safety-related information.

In a single OpenSAFETY network, referred to as a safety domain (SD), a unique safe configuration manager (SCM) is responsible for the monitoring of attached SNs, storage of each SN parameters, and node addressing. Indeed, the unique identifier for an SN is provided by the safety address (SADR).

For each SD, there exists only one SCM, to which a maximum of 1023 SNs can be attached. In the case of large OpenSAFETY networks, up to 1023 SDs can be defined, which can extend over several, possibly unrelated and inhomogeneous, networks. In this case, the communication among SDs is ensured by specifically designed OpenSAFETY Domain Gateways (SDGs), which appear as simple SNs from the SD point of view.

19.5.2 Basic Performance Analysis

This section serves the purpose of presenting the major guidelines to be accounted for the performance analysis of communication networks based on the EPL standard, highlighting the existing relationships between protocol design parameters and key performance indicators.

Actually, in the context of RTE networks, performance indicators have been defined by the IEC 61784-2, international standard that introduces some general-purpose metrics as well as specific indicators for each CP.

In the case of EPL, however, some additional complementary indicators may be considered. Namely, cycle time and jitter are indicators of prominent importance for networks like EPL whose operation is based on a periodically repeated cycle. Also, the latency introduced on the delivery of acyclic messages represents a further meaningful metric to be assessed.

19.5.2.1 Cycle Time

From the analytical point of view, the EPL cycle time, which is trivially determined by the time periods shown in Figure 19.2, can be expressed as

$$T_C = T_{SoC} + T_{Isoc} + T_{ASync} + T_{Idle} \tag{19.1}$$

where the four addends represent the durations of, respectively, start, isochronous, asynchronous, and idle periods of the EPL cycle.

The duration of the start period is simply determined by the transmission of the Cycle frame.

The isochronous period is usually the most weighting part of a cycle, and its duration is determined mostly by the time needed to access all the configured CNs. This involves sending of PReq/PRes frames, communication latencies (due, e.g., to network topology), and, finally, component latencies.

The asynchronous period has typically a short duration since, after the SoA frame has been issued by the MN, at most one asynchronous transmission is allowed. The longest duration of such a period can be hence calculated considering that, in any case, the transmitted asynchronous frame cannot exceed a 1500 byte payload.

Finally, the idle period is not related to the transmission of any frame. However, to achieve the best performance, such a period has to be kept as short as possible.

The minimum value that can be set for the cycle time, T_C^{min}, represents a meaningful metric that gives an indication of the achievable performance, helps the design stage of the network, and provides a significant basis for the calculation of other performance indicators. From Equation 19.1, it is clear that, for a given network configuration, the minimum cycle time is achieved when the duration of the idle period is set to zero, and, moreover, the acyclic transmission does not take place. This means that in Equation 19.1, T_{Async} is determined by the transmission of only the SoA frame that, actually, has the same length of the SoC frame. This latter is a minimum size Ethernet frame, and at the transmission rate of 100 Mbps, the time necessary to transmit any of the SoC and SoA frames is $T_{SoC} = T_{SoA} = 5.12 + 0.64\,\mu s = 5.76\,\mu s$, where the second addend (0.64 μs) accounts for the transmission of the Ethernet preamble.

After the transmission of the SoC frame, the MN shall wait for a predefined time, called $t_{SoC-PReq\_MN}$, which represents the minimum delay between the end of SoC transmission and the start of the isochronous period. According to the EPL specification, this time has a minimum value equal to an Ethernet interframe gap (T_{EthIFG}). Hence, $t_{SoC-PReq\_MN} = 960$ ns. The same value of time can be accounted for the gap after the transmission of the SoA frame.

Consequently, the analytical expression of the minimum cycle time, calculated from Equation 19.3 with the aforementioned assumptions, is given by

$$T_C^{min} = 2 \cdot \left(T_{SoC} + T_{EthIFG}\right) + T_{Isoc} \tag{19.2}$$

The duration of the isochronous period can be calculated from the times necessary to poll the CNs, which comprise the transmission time of both the PReq and PRes frames, plus the mandatory gap times to be waited between these frames (which, again, are specified to be equal to an Ethernet interframe gap).

Finally, the calculation of T_C^{min} has to take into consideration both the physical propagation times and the latency introduced by the hubs present in the path between MN and CNs. The propagation delay can be usually approximated by 5 ns per each meter of cable, whereas, the latency introduced by a hub, T_{Hub}, is usually not greater than 460 ns.

Thus, the analytical expression of the minimum EPL cycle time results:

$$T_C^{min} = 2 \cdot \left(T_{SoC} + T_{EthIFG}\right) + \sum \left\{T_{PReq} + T_{PRes} + 2 \cdot \left[T_{EthIFG} + T_{propag} + \left(N_{Hub,n} \cdot T_{Hub}\right)\right]\right\} \tag{19.3}$$

19.5.2.2 Jitter

This indicator is defined as the maximum deviation of the periods of cyclic operations from their nominal values. For example, the polling of a continuous CN is expected to take place with a period equal to the EPL cycle time whereas, in practice, it may be subjected to even minimal fluctuations.

In an EPL network, two main causes of jitter can be identified, namely, the jitter introduced by the MN and that due to the network components.

The former is strongly related to the specific physical device employed as well as to the protocol implementation. On the other hand, the jitter introduced by network components is peculiar of the components themselves (e.g., hubs) being practically independent from the EPL protocol rules.

Limiting the discussion about the jitter introduced by components to the case of hubs, the general expression for the upper bound of the jitter introduced in the path between the MN and a CN is given by

$$J_{TOT} = J_{MN} + N_{Hubs} \cdot J_{Hub} \tag{19.4}$$

where the jitter introduced by the MN is indicated as J_{MN}, whereas the jitter contribution of a single hub is accounted by the term J_{Hub}, under the simplifying assumption that all hubs are homogeneous with respect to their jitter performance. Finally, N_{Hubs} represents the maximum number of hubs in the MN–CN path.

To better highlight the importance of component selection and of topology design in the field of EPL-based RTE, two meaningful (even simple) examples could be introduced, though they remain representative only of a qualitative analysis of this topic. Considering a basic star topology network in which a single hub connects MN and all the CNs. The term J_{MN} accounts for 50 ns, as indicated by the EPL specification, whereas the jitter contribution of a hub is on the order of 70 ns, resulting in an upper bound for the jitter of 120 ns.

In a slightly more complex topology, where two hubs are introduced, the maximum jitter is, trivially, $J_{TOT} = 190$ ns, which however highlights the fundamental impact of the topology on this performance figure.

19.5.2.3 Acyclic Traffic

The main factors affecting the latency of asynchronous messages are both the cycle time and the scheduling algorithm implemented on the MN to select the asynchronous request to serve in the current cycle.

Actually, referring to the worst case, a CN that has asynchronous data to transmit issues its request through the PRes frame when it is polled. The MN receives this request and processes it, using a predefined scheduling mechanism based on assigned priorities (the asynchronous request, however, could be of higher priority for more time-critical events). Simplifying and assuming that all requests have the same priorities, the result is that the latency for asynchronous messages increases linearly with the number of requests. Hence, the EPL capability of providing low cycle times reveals an even more important feature, since it ensures reduced latencies for asynchronous requests.

Finally, the latency of acyclic messages could be reduced adopting techniques requiring slight modifications to the protocol. For example, multiple asynchronous send (ASend) transmissions could be allowed in the same cycle, provided that there is enough bandwidth to accomplish with them. Another possible way of achieving better latencies would consist in reserving part of the EPL asynchronous period exclusively for alarms, as addressed in [10].

References

1. IEC61784 International Standard: Digital data communications for measurement and control. Part 1: Profile sets for continuous and discrete manufacturing relative to fieldbus use in industrial control systems. Part 2: Additional profiles for ISO/IEC8802-3 based communication networks in real-time applications. Part 3: Functional safety fieldbuses—General rules and profile definitions. International Electrotechnical Commission Std., 2010.
2. Bernecker & Rainer Industrie-Elektronik GmbH. [Online]: http://www.br-automation.com. (Accessed March 5, 2014).

3. Ethernet POWERLINK Standardization Group (EPSG). [Online]: http://www.ethernet-powerlink.org. (Accessed March 5, 2014).
4. Ethernet POWERLINK Standardization Group. Ethernet POWERLINK communication profile specification V.1.1.0, EPSG draft std. 301, 2008. [Online]: http://www.ethernet-powerlink.org. (Accessed March 5, 2014).
5. R. Zurawski (Ed.), Industrial communication systems, in *The Industrial Information Technology Handbook*, Boca Raton, FL: CRC Press, 2005, pp. 37.1–47.16.
6. CANopen Application Layer and Communication Profile, CiA/DS301, Version 4.01, CAN In Automation, International Users and Manufacturers Group e.V. Std., June 2000.
7. IEEE 802.3 Standard: Carrier sense multiple access with collision detection (CSMA/CD) access method and physical layer specifications, IEEE Std., October 2000.
8. IOANA. *IAONA Industrial Ethernet Planning and Installation Guide*, [Online]: http://www.iaona-eu.com/home/downloads.php. (Accessed March 5, 2014).
9. Ethernet POWERLINK Standardization Group: Ethernet POWERLINK, Part A: High availability V.1.0.0, EPSG Draft Std. 302-A, 2008. [Online]: http://www.ethernet-powerlink.org. (Accessed March 5, 2014).
10. L. Seno and S. Vitturi, A simulation study of Ethernet POWERLINK networks, *IEEE Conference on Emerging Technologies and Factory Automation*, pp. 740–743, Patras, Greece, September 2007.

20

IEEE 802.1 Audio/Video Bridging and Time-Sensitive Networking

Wilfried Steiner
TTTech Computertechnik AG

Norman Finn
Cisco Systems

Matthias Posch
Vienna University of Technology

20.1 General Overview 20-1
20.2 Basic Technology 20-2
20.3 TSN/AVB Protocol Services 20-5
Stream Reservation Protocol • Clock Synchronization Protocol • Traffic Shaping • Redundancy Management • IEEE 1722 and IEEE 1722.1 (AVBTP)
20.4 Target Domains and Components 20-13
References 20-14

20.1 General Overview

When Ethernet was introduced some 40 years ago, only few would have envisioned the extraordinary success story that would follow. While, at first, Ethernet grew to become the dominant communication standard in the office and IT communication, it quickly entered new application domains, and whenever it did, chances were high that it proved to be a mainstream alternative to niche and potential proprietary protocols. There are a variety of reasons why Ethernet manages over and over again to enter new application domains and become the leading communication technology in the new field. However, one main factor that has contributed to the vast distribution and application of Ethernet is its steady and incremental standards improvements and the industry support that comes hand in hand with that.

The IEEE is the natural home of Ethernet and bridged networks in general; and two main working groups maintain and extend Ethernet's leading position as a communication protocol: IEEE 802.3 and IEEE 802.1. IEEE 802.3 is primarily concerned with standardizing the physical (PHY) and media access control (MAC) layers of Ethernet, while IEEE 802.1 is standardizing general architectures for local area networks (LANs) and metropolitan area architectures (MANs) that use IEEE 802 technology. These general architectures include security mechanisms, interworking of LANs, MANs, and wide area networks (WANs), as well as the overall network management. The protocols and mechanisms that 802.1 develops operate above the MAC layer.

In the IEEE, working groups are split into task groups, and this chapter will mostly focus on the mechanisms and protocols developed by the IEEE 802.1 *time-sensitive networking* (TSN) task group, previously known as the *audio/video bridging* (AVB) task group. As the name suggests, this task group standardizes protocols and mechanisms that improve the real-time behavior in the IEEE 802 family of protocols and technology and, thus, are of immediate interest as industrial communication technologies.

The AVB task group has had its roots in the *residential Ethernet* study group and became later a task group in IEEE 802.1. Today, it is common to refer to the output of the AVB task group as *AVB*. The following four items are considered to be its building blocks:

1. 802.1AS Timing and Synchronization for Time-Sensitive Applications in Bridged Local Area Networks: a protocol and technique to synchronize local clocks in the network to each other
2. 802.1Qat Stream Reservation Protocol (SRP): a protocol that allows applications to dynamically reserve bandwidth in the network
3. 802.1Qav Forwarding and Queuing Enhancements for Time-Sensitive Streams: an enhancement over strict priority-based forwarding and queuing mechanisms that establishes fairness properties for lower priority traffic in the network
4. 802.1BA: list of standards and definition of profiles for AVB systems

IEEE 802.1AS and IEEE 802.1BA are stand-alone IEEE standard documents released in 2011, while 802.1Qat and 802.1Qav are amendments to the IEEE 802.1Q standard, which is considered one of, if not the, central standard documents of IEEE 802.1. These amendments have been incorporated into IEEE 802.1Q-2011 (the most recent version of IEEE 802.1Q at the time of this writing), but for simplicity, the community sometimes continues to refer to the amendment tags.

While AVB equipment is now entering and is about to enter the market for audio and video applications in various domains (e.g., professional theater equipment, in-car information, and entertainment), the task group has broadened the target market for the IEEE 802.1 technology toward industries that demand even more time-sensitive and reliable networks. Therefore, the audio/video bridging task group was renamed to time-sensitive networking in 2012. As of today, the following items are discussed to form the core of TSN:

- 802.1ASbt Timing and Synchronization: Enhancements and Performance Improvements.
- 802.1Qcc Stream Reservation Protocol: Enhancements and Performance Improvements.
- 802.1Qbv Enhancements for Scheduled Traffic: The developed mechanisms will allow a basic form of time-triggered communication in which the outgoing queues from bridges will be drained according to an a priori known schedule. The bridges will execute the schedule synchronously with respect to the IEEE 802.1AS timebase.
- 802.1Qbu Frame Preemption: A mechanism that allows to preempt a frame in transmit to intersperse another frame. While this mechanism was originally intended to improve transmission latency, its primary application will be in combination with scheduled traffic to allow more efficient bandwidth utilization.
- 802.1Qca Path Control and Reservation: Protocols and mechanisms to set up and manage the active communication paths in the network. In particular, it will be possible to set up redundant communication paths between sending and receiving end stations. The concurrent transmission of replicated copies of frames will raise the reliability of frame transmission.
- 802.1CB Frame Replication and Elimination for Reliability: Complementary to the capability of transmitting frames concurrently over redundant communication paths in the network, the TSN task group will also standardize mechanisms to eliminate redundant copies of frames, for example, at the receiver or at bridges directly connecting to a receiving end station.

In this chapter, we will describe the basic IEEE 802.1 technology and elaborate on the AVB and TSN protocols and mechanisms outlined earlier in more detail. We will use the term *TSN/AVB network* to reflect both the new target markets of time-sensitive networks, such as industrial control applications, and the original target markets of audio and video applications.

20.2 Basic Technology

A TSN/AVB network is composed of two types of components: end points and bridges. Frequently, the commercial term *switch* is used synonymously for a bridge. End points will typically be the computational nodes that host and execute applications. As these end points need to exchange information,

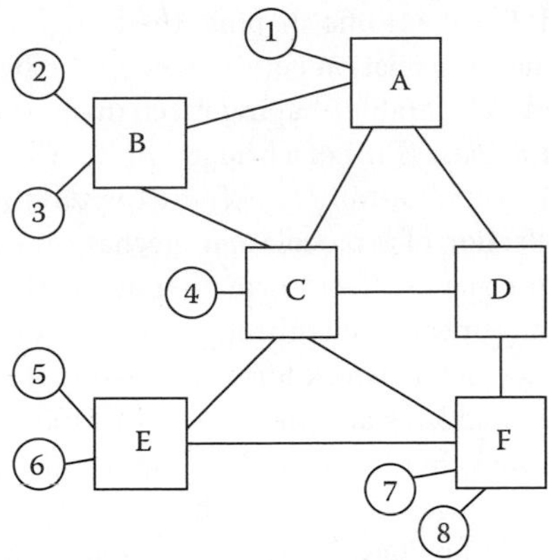

FIGURE 20.1 An example of the physical topology of a TSN/AVB network consisting of eight end stations (labeled 1–8) and six bridges (labeled A–F).

they are connected to each other either directly or indirectly using one or many bridges. An example physical topology of a network consisting of eight end stations (represented by circles) and six bridges (represented by boxes) is depicted in Figure 20.1.

End points and bridges are connected to each other with bidirectional communication links, and for the sake of simplicity, we will assume in the remainder of this chapter that these links are either 100 Mbit/s or 1 Gbit/s Ethernet connections and omit that the 802.1 technology supports also other communication technologies through well-defined interfaces. End points and bridges connect to the links through physical *ports* and communicate solely by the exchange of Ethernet frames. The structure of an Ethernet frame is depicted in Figure 20.2.

End points may send or receive Ethernet frames, or both. Bridges control the forwarding of the frames between end points. In particular, a bridge needs to address two main tasks in frame forwarding:

1. Switching: When a bridge receives an Ethernet frame on a port (the incoming port), it needs to decide to which ports the frame shall be forwarded on to (the outgoing ports).
2. Traffic shaping: When more than one frame is ready for forwarding at an outgoing port, the bridge needs to serialize this set of frames.

Switching and traffic shaping use the information transported in the Ethernet frame: for the switching task of an incoming Ethernet frame, the bridge will take the media access control (MAC) destination and the virtual local area network (VLAN) identifier into account to decide to which outgoing ports

7B	1B	6B	6B	4B	2B	42B–1500B	4B	12B
Preamble	SOF	MAC destination	MAC source	802.1Q "VLAN" Tag	Ethertype/ length	Payload	FCS	IFG

16 bits	3 bits	1 bit	12 bits
Tag protocol identifier	Priority code point	Drop eligible indicator	VLAN identifier

FIGURE 20.2 Structure of an Ethernet frame and detailed representation of the IEEE 802.1Q *VLAN* tag.

the frame needs to be forwarded. For the traffic shaping, the bridge only uses the priority code point (PCP) information in the VLAN tag. The relation between incoming port and outgoing port for a given Ethernet frame as well as the particular traffic-shaping mechanism that needs to be applied is locally stored in the *forwarding information base* (FIB) of a bridge.

It is also common to distinguish the *control plane* from the *data plane* inside a bridge. In simple terms, the control plane is the collection of protocols and mechanisms that populate the FIB, while the data plane refers to the mechanisms that use the information in the FIB to handle the actual data transfers like the buffer and queue management mechanisms.

Sometimes, the physical topology of a network forms a graph that contains loops. As a bridge only uses the Ethernet MAC destination address and the VLAN identifier for the switching task, loops in the graph potentially lead to repeated circulations of Ethernet frames along these loops. To avoid this problem, the bridges use protocols like spanning tree or shortest path bridging that establish a loop-free *active topology* on top of the physical topology. These protocols are considered to be part of the control plane of a bridge. Given the physical topology in Figure 20.1, after the execution of a spanning tree protocol, the active topology may be the one established in Figure 20.3a. If end point 1 needs to communicate with end point 6, it will send its frames to bridge A. Bridge A will learn from the MAC source address in these frames to which port end point 1 is connected to (and update its FIB), but it may not (yet) know the location of end point 6. Hence, bridge A forwards the frame to all ports in the active topology except the one to which end point 1 is connected to. Bridges B, C, D likewise forward the frame and learn the MAC address relation to the incoming port. In the next step, bridges E and F learn the address of end point 1 and their local port on which the frame has been received, and finally, bridge E delivers the frame to end point 6. The active topology avoids circulation of the frame from end point 1, which would happen otherwise if bridge C would forward the frame also to bridge B and bridge B again to bridge A.

Complementary to the dynamic way to establish the active topology and address learning, networks can also be engineered, for example, for active redundancy reasons. Figure 20.3b depicts an engineered ring topology on top of the physical topology in Figure 20.1. Obviously, the ring topology is not loop-free as it only consists of a single loop. Hence, to avoid the problem of circulating frames, great care must be taken in the proper configuration to set up disjoined active paths between senders and receivers. We will continue this discussion in Section 20.3.4.

The bridges in the network will typically be used to transmit frames from many end points and, consequently, become a shared resource. Traffic-shaping mechanisms decide the sequence in which the frames will be forwarded in case several frames are competing for transmission on the same outgoing port.

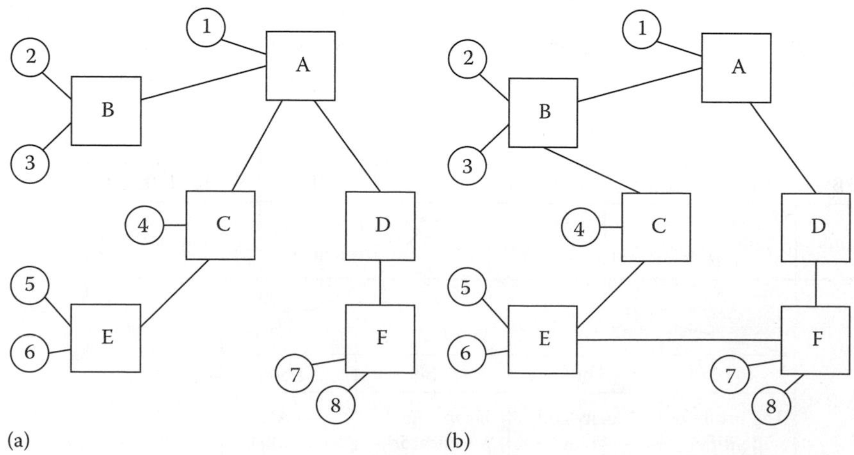

FIGURE 20.3 Possible logical topologies on top of the physical topology in Figure 20.1. (a) Dynamically generated spanning-tree topology and (b) engineered ring topology.

For example, in Figure 20.3a, end points 1 and 4 may need to send a frame each to end point 6. Thus, both frames need to use the connection from bridge C to bridge E. In case both frames are ready to forward at about the same point in time, bridge C needs to decide which frame goes first. This decision can be random or follow a particular traffic-shaping mechanism.

The traffic-shaping mechanisms standardized by IEEE 802.1 operate on classes of frames rather than on individual frames; that is, they operate *per-class* rather than on *per-flow*. The frames are classified based on the PCP in their VLAN tag, and as the PCP has a size of 3 bits, it is common that bridges implement eight outgoing queues (one queue per priority) per outgoing port. The simplest traffic-shaping mechanism is *strict priority*: when multiple frames are ready for forwarding, the bridge will select the frame with the highest priority to transmit next. Should several frames with the same priority be ready for forwarding, the bridge will process them in first-in first-out order. This strict priority traffic-shaping mechanism has several drawbacks, and we will discuss more advanced traffic-shaping mechanisms in Section 20.3.3.

TSN and AVB build on the protocols and mechanisms discussed earlier to enhance the real-time and reliability properties of IEEE 802.1 networks. We continue in the next sections with a more detailed discussion of the specific TSN and AVB protocols and mechanisms.

20.3 TSN/AVB Protocol Services

In this section, we discuss the SRP to dynamically reserve communication bandwidth in a TSN/AVB network, the TSN/AVB clock synchronization protocol, traffic-shaping algorithms, and a current proposal to redundancy management in TSN. Finally, we also give a brief overview of the IEEE 1722 standard, which defines OSI layer 2 protocols for TSN/AVB. A detailed description of how to configure and use AVB in a systems context is standardized in IEEE 802.1BA.

20.3.1 Stream Reservation Protocol

SRP is responsible for reserving bandwidth between two AVB capable devices. SRP itself uses an already existing protocol, the MRP defined in IEEE 802.1Q. Basically, MRP enables to register (and deregister) so-called attributes and distribute them across all ports in a network, or sections of the network confined by a VLAN. This way application-specific configuration information can be distributed, and managed, as registration and deregistration of attributes are handled. Applications of the MRP are the multiple stream reservation protocol (MSRP), the multiple VLAN reservation protocol (MVRP), or the multiple multicast reservation protocol (MMRP). Examples of attributes are MAC addresses, VLAN ID, or stream information in MSRP.

The way MRP works is very simple: every MRP participant periodically sends his registered attributes to the neighboring devices. In particular, a network node that needs to communicate new configuration data declares an attribute (declaration) by sending the information to all its available network ports. When an MRP participant receives information about an attribute, it registers the attribute. The attribute is then, again, declared at the recipient, and therefore, the process propagates from device to device through the network. Using the spanning tree, MRP makes sure the network does not get congested by MRP messages bouncing between neighbors. Figure 20.4 pictures the principle of the MRP operation. In this scenario, end station 4 is the source of the attribute declaration. As depicted, the attribute is registered and declared along the active topology of the network.

AVB uses the MSRP, an application of the MRP to declare streams and propagate the information about streams throughout the network. Using MSRP, the whole set of network nodes is divided into two groups: so-called *talkers* and *listeners*. A talker is a device providing information, like a microphone. A listener is a device consuming the provided information, like a loudspeaker. A talker advertises its stream by registering a *talker advertise* attribute using MRP. So eventually, every participating device will know about the stream and its properties.

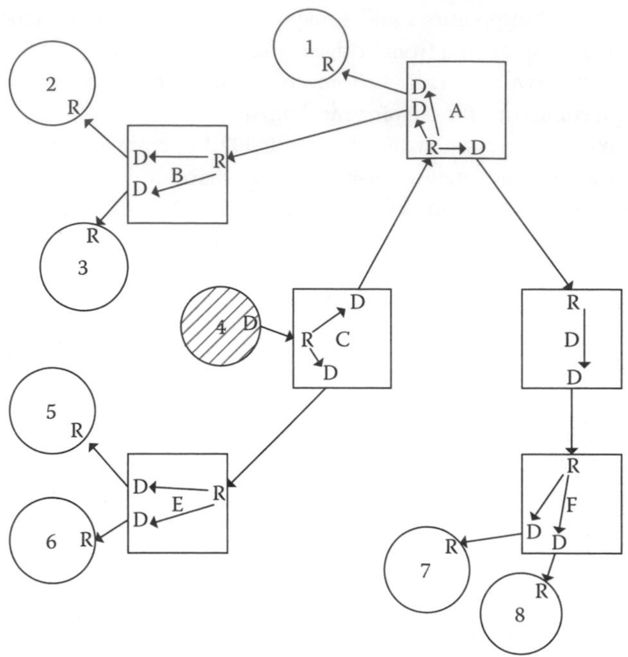

FIGURE 20.4 Example of MRP declaration/registrations as they propagate through the network.

The *talker advertise attribute* contains the following information: StreamID, DataFrameParameters, TSpec, PriorityAndRank, and AccumulatedLatency. StreamID represents a system-wide unique identifier, consisting of the MAC address of the talker and a 16 bit StreamID. DataFrameParameters specifies the destination MAC address StreamDA and the VLAN ID of the stream. TSpec defines the maximum frame size MaxFrameSize of a frame in this stream (excluding media specific overhead) and the maximum number of frames MaxIntervalFrames a talker may send within a *class measurement interval* (CMI). CMI is specified by the *traffic priority class*, where AVB so far defines two classes, A and B, and CMI is 125 μs for class A and 250 μs for class B.

AccumulatedLatency specifies the maximum latency a stream can encounter during the traversal of the network, a value that should not change during the lifetime of the stream reservation; if the value has to be increased (e.g., due to a wireless link), the *talker advertise* attribute will be changed to a *talker failed* with the appropriate error code. PriorityAndRank contains information about the priority of the stream and its rank, a 1 bit field to indicate a critical stream, for example, a speaker system for emergency announcements in a public building.

An example of a *talker advertise* message containing the attributes discussed earlier is depicted in Figure 20.5.

With the information from the *talker advertise* attribute, it is possible to calculate the required bandwidth, BW, for this stream using Equation 20.1:

$$BW = \frac{\text{Tspec}_{\text{MaxIntervalFrames}}}{\text{CMI}} \times 8 \times (\text{Tspec}_{\text{MaxFrameSize}} + \text{Overhead}_{\text{mediaSpecific}}) \tag{20.1}$$

For the captured stream depicted in Figure 20.5, using Equation 20.1 results in $\frac{1}{125\ \mu s} \times 8 \times (224 + 42) =$ 17.024 Mbit/s.

```
▾ Multiple Stream Reservation Protocol
    Protocol Version: 0
  ▾ Message: Talker Advertise (1)
      Attribute Type: Talker Advertise (1)
      Attribute Length: 25
      Attribute List Length: 30
    ▾ Attribute List
      ▾ Vector Attribute
        ▾ Vector Header: 0x0001, Leave All Event: Null
            000. .... .... .... = Leave All Event: Null (0)
            ...0 0000 0000 0001 = Number of Values: 1
        ▾ First Value
            Stream ID: 0x0022970041940000
            Stream DA: MarvellS_ff:07:c4 (01:50:43:ff:07:c4)
            VLAN ID: 0x0002
            TSpec Max Frame Size: 224
            TSpec Max Frame Interval: 1
          ▾ Priority and Rank: 0x70, Priority: Traffic Class A, Rank: Non-emergency, Reserved: Reserved-0
              011. .... = Priority: Traffic Class A (3)
              ...1 .... = Rank: Non-emergency (1)
              .... 0000 = Reserved: Reserved-0 (0)
            Accumulated Latency: 150000
          Attribute Event: JoinMt (3)
        End Mark: 0x0000
  ▸ Message: Domain (4)
```

FIGURE 20.5 Example of a talker advertise message captured with Wireshark.

SRP itself does not specify how the payload of AVB/TSN frames is constructed but relies on other protocols like IEEE 1722 and IEEE 1722.1.

The registration of the talker advertise attributes in an AVB network will typically involve the following steps (Figure 20.6): First, a *talker* declares that it has a stream ready to transmit and registers a *talker advertise* attribute, which in turn gets propagated through the whole network. If a listener is interested in this stream, it can register a *listener ready* attribute (containing the correct stream id). Again, this information will be propagated across the network and eventually reach the talker. The path is constructed on a hop-by-hop basis, as each registers and declares the attribute for the corresponding port. When a bridge device registers a *listener ready* attribute, it knows that a listener exists at this specific port and can check if enough bandwidth is available. This way MSRP can make sure that sufficient bandwidth is available across the whole path of the stream. If an error occurs during this reservation process, a bridge can declare a *listener failed* or *talker failed* attribute and provide an error code to define the cause, so all devices will eventually get informed about the failed stream reservation. As soon as the talker receives a *listener ready* declaration, the talker can be sure that sufficient bandwidth is available along the stream path, and at least, one listener is ready to receive streamed data. The streaming itself is done via higher level protocols like IEEE 1722.

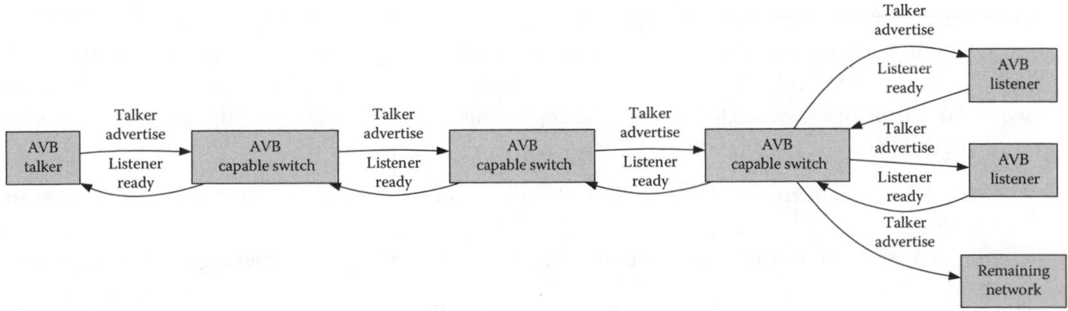

FIGURE 20.6 Example of talker advertise and listener ready propagation through an AVB network.

20.3.2 Clock Synchronization Protocol

The purpose of a clock synchronization protocol is to bring local clocks in the end stations and bridges into close agreement, such that at any point in real-time, two nonfaulty clocks in end points or switches will read approximately the same local computer-time. The clock synchronization protocol of TSN/AVB is standardized as IEEE 802.1AS and is significantly overlapping with the IEEE 1588 standard. We give an overview of IEEE 802.1AS and currently discussed improvements to it in the following. A more detailed introduction to IEEE 802.1AS is given by Garner et al. in [3].

From a literature point of view, IEEE 802.1AS specifies a classical master–slave clock synchronization protocol in which one component (end point or bridge) acts as the master clock toward which all clocks of the other components are synchronized to. In IEEE 802.1AS, the master is called the *grandmaster* and forms the root of a *synchronization spanning tree* that connects all the local clocks of the components. For example, the spanning tree depicted in Figure 20.3a may also be the synchronization spanning tree for the IEEE 802.1AS clock synchronization protocol. In this example, bridge A may act as the grandmaster clock that provides its local time to all other bridges (B–F) and end points (1–8). In summary, IEEE 802.1AS specifies protocols:

- To select a grandmaster out of a set of grandmaster-capable components
- To periodically synchronize to the grandmaster clock
- To measure the forwarding delays in the bridges
- To measure the communication delays of frames on the communication links

For grandmaster selection, IEEE 802.1AS uses the best master clock algorithm (BMCA): components that are capable of acting as grandmaster clocks in the system will announce their presence in so-called "`announce` messages." A receiver of the `announce` messages can directly compare these messages with each other and select the *best* `announce` message using a standardized assessment scheme. Consequently, the component that sent the best `announcement` message will become the grandmaster in the network and the root of the synchronization spanning tree.

The grandmaster will, then, periodically distribute its local time (the `preciseOriginTimestamp`) to the network along the synchronization spanning tree using two messages: the `Sync` and the `Follow_up` messages. The `Sync` message is a time-critical message, and each component precisely timestamps its reception and transmission (in case of the grandmaster and a bridge). The grandmaster and each bridge will then communicate the transmission and, respectively, forwarding delay of the `Sync` message in a second message, the `Follow_up` message. As the `Follow_up` message propagates along the synchronization spanning tree from the grandmaster to the final receivers, each component adds the forwarding delay, that is, the difference between transmission time and reception time of the previous `Sync` message, into the payload of the `Follow_up` message (using the `correctionField` field).

In addition to the forwarding delay, the network periodically also measures the communication delay of the communication links. This measurement is called the *peer delay mechanism* and is done individually on each communication link in the network, for example, on the communication link connecting bridge A to bridge C. We continue to assume that bridge A is the grandmaster clock in the network. Hence, bridge C needs to measure the communication link delay to bridge A. This is done by the exchange of three messages between bridge C and bridge A:

- Bridge C sends a `Pdelay_Req` message to bridge A and locally stores the transmission point in time as t_1.
- Bridge A timestamps the reception point in time of the `Pdelay_Req` message and locally stores the value as t_2.
- At a point in time t_3, bridge A responds to bridge C with a `Pdelay_Resp` message, which contains t_2.

- Bridge C records the point in time (t_4), when it receives the `Pdelay_Resp` message from bridge A.
- To accurately inform bridge C of time t_3, bridge A sends another message, the `Pdelay_Resp_Follow_Up` message, after the `Pdelay_Resp` message, which contains t_3.

Bridge C can then approximate the communication link delay to bridge A, as $[(t_4 - t_1) - (t_3 - t_2)]/2$.

The `Pdelay_Resp` message serves also a second purpose—to measure the relative difference of the clock rates in the components connected by the communication link (e.g., in the example earlier the differences in clock rates between bridge A and bridge C). However, the IEEE 802.1AS standard does not describe a particular mechanism for rate ratio measurement but only requires that the measurement can be done within ±0.1 ppm.

While the resulting synchronization quality of IEEE 802.1AS is quite remarkable and certainly sufficient for a wide range of, for example, automotive and industrial applications, the TSN task group is currently developing reliability enhancements. Some proposed mechanisms discuss the use of redundant grandmasters that operate in hot or warm standby; that is, instead of exactly one grandmaster in the system, at least two are configured to run concurrently, a primary and a secondary grandmaster. In the absence of failures, all components (inclusively the secondary grandmaster) synchronize to the primary grandmaster. Once the primary grandmaster fails, the system continues to use the synchronization information provided from the secondary. The simplicity of a fault tolerance scheme like the hot/warm standby one is certainly attractive and a sufficient reliability increment to broaden the application domain of TSN/AVB significantly. However, there also exist a lot of safety-critical applications that demand more sophisticated fault-tolerant clock synchronization approaches. Hence, to enable TSN/AVB also in these domains, the standardization of interfaces to highly dependable clock synchronization protocols is desirable.

20.3.3 Traffic Shaping

Typically, several applications need to share a single network, and frequently, the messages from several applications will be ready to forward inside a particular bridge to the same communication link, which has then to decide which message should be transmitted next. Furthermore, in several industries, we can observe a trend toward *converged* networks in which more and more applications that used to be physically separated are now being integrated onto the same network. Hence, the question of which message to transmit next becomes more and more complex, as systems become more and more integrated. When the network in addition also has to guarantee end-to-end delays of some critical messages, the serialization problem of the messages in the bridges becomes severely complex.

We have already discussed the simple solution to use a strict priority scheme, but it has several shortcomings: First, high-priority frames only get a significant better service if their number is limited; otherwise, high-priority frames compete regularly for transmission, thereby downgrading the priority effect. Second, if high-priority messages always become precedence over lower-priority messages, then the applications using these lower-priority messages face a starvation problem—they may not be able to continue their execution as an insufficient number of their messages are being serviced by the network.

Hence, more advanced traffic-shaping mechanisms have been developed: AVB standardized the *credit-based shaper*; the TSN task group is currently developing the *time-aware shaper*.

20.3.3.1 Credit-Based Shaper

The credit-based shaper defines two traffic classes, class A and class B, and the classification of Ethernet frames is done based on their PCP in the VLAN tag. Furthermore, the credit-based shaper specifies the *CMI*. Within each CMI, each AVB talker will only send up to a defined bandwidth as granted by the SRP. In praxis, this means that an AVB talker will typically send one Ethernet frame of configurable size per CMI. For class A messages, the CMI is standardized to 125 μs; for class B messages, to 250 μs.

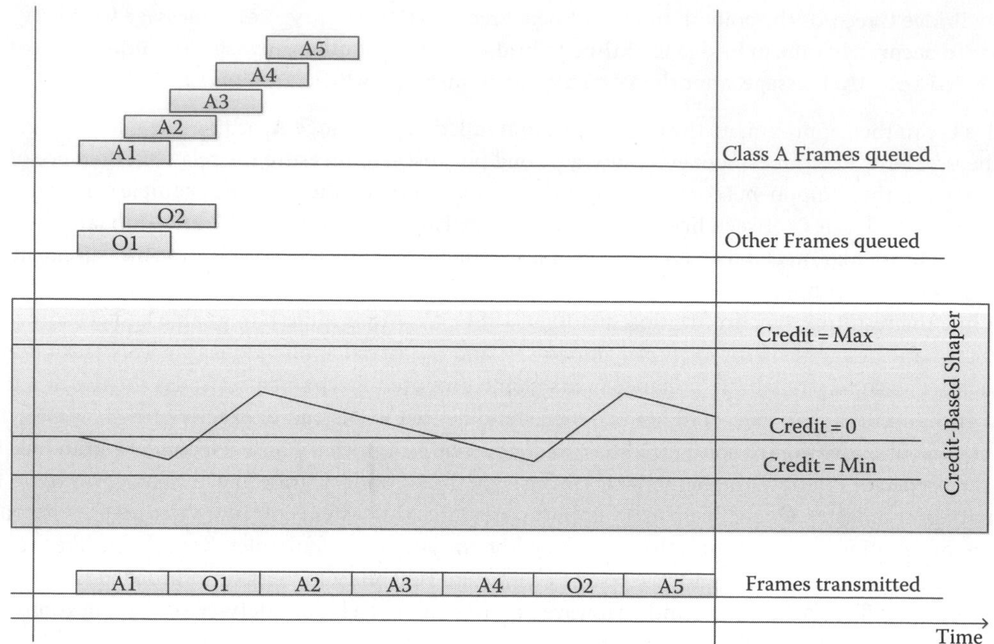

FIGURE 20.7 Example execution trace of traffic shaped according to the credit-based shaper.

The execution of the CMIs is not synchronized between the AVB talkers and the bridges. Hence, class A frames from different AVB talkers can still queue up in the outgoing port of a bridge. An example scenario of a bridge-shaping AVB class A traffic is depicted in Figure 20.7.

Figure 20.7 shows the operation at a single outgoing port of a bridge. In particular, it depicts two outgoing queues of the one outgoing port: a first queue serving AVB class A frames (A1–A5) and a second queue serving other traffic (O1–O2). Class A frames have higher priority than the other frames. Thus, a bridge implementing a strict priority-driven traffic-shaping policy would simply send the AVB frames first followed by the other traffic, thereby causing potential exhaustive delay to the other frames. Hence, in the presented scenario, the credit-based shaper is depicted to show a bridge that will only select an AVB class A frame if the current *credit* of AVB class A frames is nonnegative. Credit is a local variable that changes state with progress in time using the following rules:

- When no AVB class A frame is queued, then credit is set to zero.
- When AVB class A frame is transmitted, then credit is decreased over time with a configurable rate `sendSlope`.
- When at least one AVB class A frame is queued, but not transmitted (because either the credit is negative or another frame is currently in transit), then credit is increased over time with a configurable rate `idleSlope`.

The scenario in Figure 20.7 starts with an initial credit of zero, and AVB class A and other frames become ready for transmit back-to-back. As depicted, the first frame transmitted is A1, rendering the credit negative. Hence, the next frame selected for transmission is O1. As the `sendSlope` is typically set lower than the `idleSlope`, several AVB class A frames are transmitted next (A2–A4) until, again, credit becomes negative. Hence, again, a non-AVB class A frame will be transmitted (O2), and the scenario finally concludes with AVB class A frame A5.

The credit-based shaper operates on AVB class B frames analogously as for class A frames but maintains a separated local variable to keep track of the AVB class B credit.

The original goal of the credit-based shaper was to achieve a latency for highly time-critical messages of 2 ms over network paths consisting of a maximum of six bridges. Although it has been demonstrated that this design goal may not always be satisfied in arbitrary network topologies, the credit-based shaper remains valuable to several industries.

Currently, the static configuration of the CMI of 125 and 250 μs for class A and class B traffic, respectively, is under debate, and it is likely that in future revisions of the standard, other configurations for the CMI will be allowed.

20.3.3.2 Time-Aware Shaper

In order to minimize communication latency and communication jitter, the TSN task group is currently defining another shaper, called the time-aware shaper, which at the time of this writing is currently under standardization as IEEE 802.1Qbv *Enhancements for Scheduled Traffic*. It aims to translate the time-triggered communication paradigm toward the IEEE 802.1 queue-based traffic-shaping model. As in time-triggered communication, the time-aware shaper needs to address two aspects: first, components that use the time-aware shaper require access to a synchronized timebase, and second, the components need to define a communication schedule. The access to the synchronized timebase is inherently present in an AVB/TSN network through the IEEE 802.1AS clock synchronization standard. Hence, all TSN needs to define is how to link this synchronized time to the execution of the communication schedule. The second aspect, how to generate the communication schedule and how to distribute this information in the system, is still in an early phase of technical discussion. However, it is likely that the distribution of the communication schedule will require modifications of the SRP and/or the use of the IS-IS (Intermediate system to Intermediate system) protocol. An example execution trace of what the time-aware shaper is likely to look like is depicted in Figure 20.8.

Again, the scenario depicts two outgoing queues at an outgoing port in a bridge. The higher-priority queue serves time-aware frames (TA1–TA3) while the lower-priority queue serves other nontime-aware frames (O1 and O2). The functionality of the time-aware shaper is shown in the shaded box simply by a TA-enabled/disabled signal: when TA is enabled, the time-aware frames are transmitted; when TA is

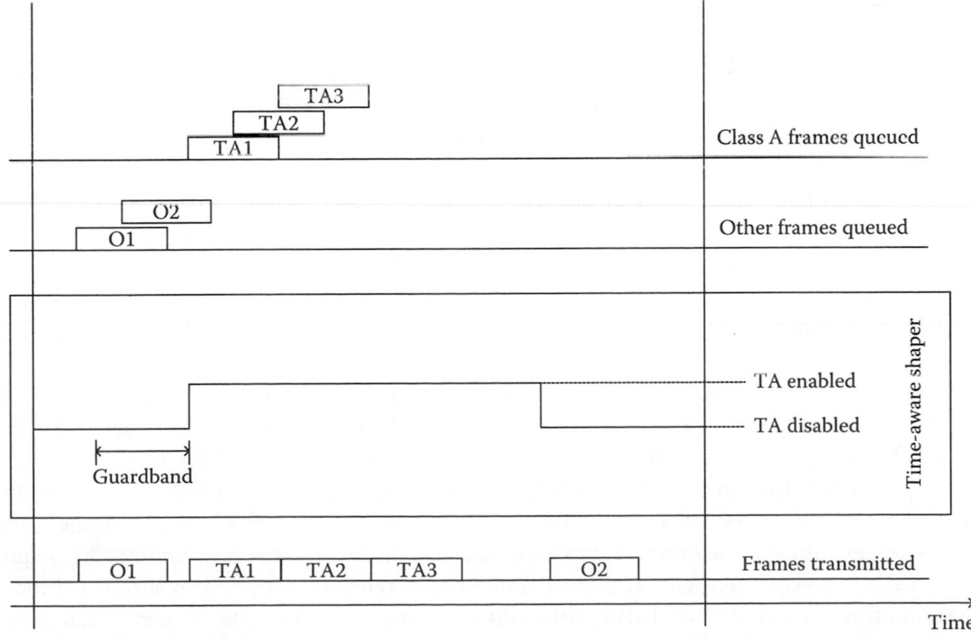

FIGURE 20.8 Example execution trace of traffic shaped according to the time-aware shaper.

disabled, the bridge may forward frames from other queues. As depicted in the scenario, the first frame delivered is O1 followed by the time-aware frames TA1–TA3.

The figure further shows an interval called the *guardband* immediately before the time-aware frames are transmitted. The purpose of this guardband is to ensure that no nontime-aware frame causes a conflict with the time-aware frames once the TA is enabled. Hence, all nontime-aware frames need to finish transmission before the TA is set to be enabled. For example, in the aforementioned scenario, O2 would be ready for transmit before the time-aware frames. However, as the transmission of O2 would not conclude before the TA is enabled, frame O2 cannot be forwarded.

As bandwidth is in general a precious commodity, but even more so in the IEEE 802 community, several ideas have been presented on how to make efficient use of the guardband to avoid idle times. The straightforward solution is to start the transmission of frames during the guardband, but only when it can be guaranteed that the frame transmission will be completed once TA is enabled. However, this means that the closer time advances to the point in time when TA is enabled, the shorter the frames must be that can be selected for transmission. In the IEEE 802.1, it is, furthermore, uncommon, and to a certain degree also prohibited by the standard, to search an outgoing queue for a frame with matching length to transmit next as the order of frames needs to be preserved within certain bounds. Therefore, this straightforward solution does not increase bandwidth efficiency to an optimum. A more sophisticated approach is frame preemption.

IEEE 802.1Qbu investigates a mechanism that allows a bridge to preempt a current frame in transmission and to intersperse another frame. Thereby, the goal is to achieve lower latency for the interspersed frame and to preempt in a way such that the preempted frame can be continued after the communication of the interspersed frame. Preemption as a response to the IEEE 802.1AS time reaching a configured point in time allows the guardband to be used more efficiently than the simple approach discussed earlier as a frame can now be selected for transmission irrespective of its length.

20.3.4 Redundancy Management

Each network may lose frames because of the imperfection of the environment it operates in. For example, electromagnetic emission may cause bit flips on the communication links, rendering the frame check sequence of an Ethernet frame, that is, its cyclic redundancy check (CRC), invalid. A typical office network handles frame losses like this one by implementing higher-layer retransmission protocols, for example, by using TCP/IP. However, in real-time networks, the temporal penalty of retransmissions is often not acceptable, and alternative approaches are required.

In safety-related networks, it is common to install redundant communication paths on which redundant copies of the frames are transmitted in parallel. Hence, this *parallel* form of redundancy increases the reliability of frame transmission in a network as it is more likely that at least one of the redundant copies reaches its destination than in a network without parallel redundancy. At the destination, or at selected bridges in the network, the redundant copies of a frame need to be eliminated such that the information transported in the redundant copies of a frame will be used only once.

In the TSN task group, this form of parallel redundancy is incorporated into the IEEE 802.1 standards. In particular, the setup of redundant communication paths in the network is developed in the 802.1Qca project and is looking into using the IS-IS protocol for this purpose. The 802.1CB project explores how to eliminate the redundant copies at the receiver (or at selected bridges in the network). The current proposals for the elimination use the well-known sequence number mechanism: redundant copies of the same frame are tagged with the same sequence number. This sequence number is then incremented with each new frame of a stream transmitted by a talker. From the sequence number, a listener can distinguish redundant copies of the same frame, which have the same sequence number, from different frames in the stream, which should have sufficiently different sequence numbers.

TABLE 20.1 AVB Maximum Transit Time

Traffic Class	Maximum Transit Time (ms)
A	2
B	50

Source: IEEE Std 1722-2011, IEEE standard for Layer 2 transport protocol for time sensitive applications in a bridged local area network, pp. 1–57, 2011.

20.3.5 IEEE 1722 and IEEE 1722.1 (AVBTP)

IEEE 1722 and IEEE 1722.1, also referred as the AVB Transmission Protocol (AVBTP), are layer 2 protocols that operate on top of the AVB protocols and mechanisms described earlier. IEEE 1722 handles the actual transmission of media data, reusing established protocols (e.g., IEEE 1394). Currently, it is possible to transmit the following formats as can be found in the specification [1]:

- IEC 61883-2:2004: SD-DVCR data transmission
- IEC 61883-4:2004: MPEG2-TS data transmission
- IEC 61883-6:2005: Audio and music data transmission protocol
- IEC 61883-7:2003: Transmission of ITU-R BO.1294 System B
- IEC 61883-8:2008: Transmission of ITU-R BT.601 style Digital Video Data
- IIDC 1394-based Digital Camera Specification [B8]

In addition to standardized frame formats, AVBTP also defines the *presentation time* a timestamp, expressed in nanoseconds referring to the IEEE 802.1AS time when a received data unit of the AVBTP has to be presented by the listener. It represents the time when the sample is sent from the talker plus a constant, the maximum transit time to take the maximum network latency into account. AVBTP defines maximum transit times (i.e., the maximum network latency) per traffic class as shown in Table 20.1.

Therefore, for professional applications, a maximum transit time of 2 ms should not be exceeded. The maximum transit time depends on the number of hops and network speed and can vary. The requirement is not to exceed this limit for a network of seven hops using full-duplex Ethernet for class A, or two wireless plus six Ethernet hops using traffic class B. The talker defines the presentation time *in the future* by adding the maximum network latency to the actual IEEE 802.1AS time, and it is also in the responsibility of the talker to make sure the AVBTP data unit is handled over to the MAC no later than *presentationtime—max.traveltime* to ensure timely delivery. Eventually, the talker needs to further add some time due to the talker's internal implementation, but this is not covered in IEEE 1722 and is up to the application. Any other required timing delays due to format conversion or video compression are also not handled by AVBTP and must be taken into account on an application level. The synchronization of audio and video (also called *lip sync*) has to be done at the application level but can take advantage of the global timebase already established.

20.4 Target Domains and Components

The target market for AVB has been, as the name suggests, audio and video applications. In particular, the AVB technology targets at in-car entertainment, professional audio/video equipment like studio equipment, consumer electronics, and large-scale audio/video installations like in theaters or amusement parks. Large IT networking and semiconductor companies implement AVB in their solutions and have their products operational in the field.

The TSN task group further develops the AVB technology toward more time-sensitive and potentially also time-critical networked applications. Examples of domains targeted by TSN are factory automation and industrial control and automotive driver assistance systems. Indeed, recent academic studies analyze AVB for driver assistance systems [2] and present simulation results. It will be interesting to study the performance improvements taking the TSN-developed mechanisms into account. As the standardization process in IEEE 802.1 usually takes several years, it is common that IT companies and silicon vendors implement prestandard versions of the technology under standardization. Out of the current activities of the TSN task group, the time-aware shaper (IEEE 802.1Qbv) is probably the most stable piece of technology, and it is likely that upcoming products will support this basic form of time-triggered communication.

In general, the benefit of technology becoming part of the IEEE 802.1 family of standards is in its usual broad market acceptance and deployment. Hence, it is fair to expect a constant growth of devices and components that implement the AVB and TSN technologies. The interoperability of these devices is thus a major requirement. Hence, well-known experts like the University of New Hampshire InterOperability Laboratory help to ensure the interoperability of Ethernet devices from different vendors. For AVB (and in the future most likely also for TSN) AVnu Alliance (www.avnu.org) forms the bridge between the IEEE standardization activities and the commercial markets. While TSN/AVB standardizes a baseline technology, particular products will typically implement additional, probably not standardized or prestandard, functionality. For example, bridges typically implement *metering* functions that implement input policing. The significance of the IEEE 802.1 TSN/AVB standardization activities has also influenced other real-time Ethernet products for industrial communication, for example, upcoming TTEthernet products will be compatible with TSN/AVB.

References

1. IEEE Std 1722-2011. IEEE standard for Layer 2 transport protocol for time sensitive applications in a bridged local area network, pp. 1–57, 2011.
2. G. Alderisi, G. Iannizzotto, and L. Lo Bello. Towards IEEE 802.1 Ethernet AVB for advanced driver assistance systems: A preliminary assessment. *IEEE Conference on Emerging Technologies and Factory Automation (ETFA)*, Krakow, Poland, September 2012.
3. G.M. Garner and H. Ryu. Synchronization of audio/video bridging networks using IEEE 802.1as. *IEEE Communications Magazine*, 49(2): 140–147, 2011.

III

Fault-Tolerant Clock Synchronization in Industrial Automation Networks

21 Clock Synchronization in Distributed Systems Using NTP and PTP *Reinhard Exel, Thilo Sauter, Paolo Ferrari, and Stefano Rinaldi* **21**-1
Introduction • Clock Synchronization in a Nutshell • Network Time Protocol • IEEE 1588 Precision Time Protocol • Conclusion • References

21

Clock Synchronization in Distributed Systems Using NTP and PTP

Reinhard Exel
Danube University Krems

Thilo Sauter
Danube University Krems

Paolo Ferrari
University of Brescia

Stefano Rinaldi
University of Brescia

21.1 Introduction **21**-1

21.2 Clock Synchronization in a Nutshell **21**-2
Clocks • Basic Synchronization Principle • Taxonomy of Clock Synchronization Approaches • Synchronization Approaches for Packet-Based Networks

21.3 Network Time Protocol **21**-7
NTP Time Scale, Time Format, and Other Definitions • NTP Implementation Model • On-Wire Protocol • Network Data Packet for NTP • NTP Managing Process • Simple Network Time Protocol

21.4 IEEE 1588 Precision Time Protocol **21**-16
Master Election • Synchronization in PTP • Transparent Clocks • Timestamp Generation • Implementation and Application Issues

21.5 Conclusion **21**-25

References **21**-25

21.1 Introduction

The measurement of time and time intervals has been subject to investigation since the dawn of mankind. With the rise of spatially distributed computers connected by a data network, the need to establish a network-wide timebase has sparked the interest in clock synchronization. One of the first works covering the ordering of events in a distributed network was published by Lamport in 1978 [1]. In the following, a plethora of algorithms and protocols have been designed, not only to enable the ordering of events but also to provide a common notion of time. This degree of synchronization additionally enables to coordinate and schedule actions between multiple computers connected to a common network. Nowadays, clock synchronization is often considered as a basic service of a network technology.

The widespread success of Ethernet and the use of the Internet Protocol (IP) enabled to connect billions of nodes using a universal and flexible communication scheme. Yet, neither Ethernet nor IP have been designed to support real-time capabilities required for numerous applications, such as industrial control, test and measurements networks, or audio and video applications. Common to these applications is that they demand clock synchronization, for example, to synchronize audio to video streams, to fuse sensor data originating from different sensor entities, or to coordinate the access to a shared medium. Fortunately, clock synchronization can be retrofitted to any kind of network even if it has not been designed for it.

This chapter addresses clock synchronization in distributed systems in general and gives particular insight into two state-of-the-art synchronization protocols, namely, the Network Time Protocol (NTP) and the Precision Time Protocol (PTP).

21.2 Clock Synchronization in a Nutshell

This section provides an overview of clocks and their synchronization and briefly describes the most common approaches for synchronizing clocks.

21.2.1 Clocks

A clock can be considered an instrument to measure, keep, and indicate time. Practically, all clocks are built as a two-part device: An oscillating device for defining a reference time interval (e.g., a second or fraction thereof) and a counter device, which accumulates the number of time intervals and provides a time indication. Mechanical oscillators, such as a pendulum or balance wheel, have been widely used for timekeeping until the invention of electronic oscillators. Quartz-crystal oscillators are today the most widely used oscillators, thanks to their low price, ruggedness, and low power consumption. Atomic clocks provide a much better stability than Quartz oscillators or any mechanical oscillator. They exploit the quantum effect, where the transition between two energy levels in an atom requires exactly a certain amount of energy to emit a photon. As the frequency of the emitted photon and its energy are connected by the Planck's constant, the atomic resonance can be exploited to create a well-defined frequency. Due to the stability of the quantum transition, the SI unit second was defined in 1967 as the duration of 9,192,631,770 periods of the radiation of the two hyperfine levels of the ground state of the cesium-133 atom [2].

The second part of a clock is a counter or accumulator, which sums up the oscillations. These oscillators are then divided by a certain number to generate seconds, minutes, hours, and so on. In wristwatches, for example, the 32,768 Hz oscillation of a Quartz is divided by a 15 bit counter (2^{15} = 32,768) to generate a 1 s tick signal. Naturally, a clock must be set up initially to have an offset-free time indication with respect to a given reference clock. An ideal clock, consisting of an oscillator with known and constant frequency and a counter with an unbiased initial offset, would maintain the reference time t infinitely. However, neither the initial offset can be set without error, nor do real oscillators keep a constant and stable oscillation frequency for an infinite amount of time. In particular, the immunity of the oscillation frequency on environmental changes (e.g., temperature, mechanical stress) is critical for the stability of a clock. Due to inevitable offset initialization error and the limited stability of clocks, periodic clock synchronization is required to maintain a time indication with low offset with respect to the reference clock for an extended amount of time. A simple approach to describe the time t_j at clock j by only two parameters a_j and b_j is

$$t_j(t) = a_j t + b_j, \tag{21.1}$$

with a_j being the rate of the clock j and b_j the offset, both with respect to the reference clock t. a_j is commonly termed clock skew as it describes the clock speed ratio with respect to the reference. Synchronization aims for minimizing the clock offset error

$$e = t_j(t) - t \tag{21.2}$$

between the clock under investigation and the reference by an exchange of time information. This information can be shared by various means: It can be transmitted encapsulated as data in the form of a timestamp over a packet network, or implicitly in the form of a pulse signal, which allows the slave clock to synchronize to the reference clock.

21.2.2 Basic Synchronization Principle

Independent of how the time information is shared, the packet-based synchronization of state-of-the-art protocols is either based on a one-way or two-way message exchange. The one-way message exchange mechanism transmits timing information from the time source to the sink. Two-way synchronization extends the one-way exchange by sending packets from the sink back to the source. This allows for measuring the round-trip time and thus compensating for the path delay. Examples for such protocols are the NTP [3], which is commonly used to synchronize PCs over the Internet, and the IEEE 1588 standard defining the PTP [4], which is used in a variety of fields, such as test and measurement, or in industrial applications.

To demonstrate the basic synchronization principle, consider a synchronization scenario with two clocks i and j. Without loss of generality, we assume that clock i is identical to the reference clock, and therefore $t_i = t$ and clock j synchronizes to clock i. We assume the skew $a_j = 1$, which means the clocks differ by a time-invariant constant offset b_j as indicated in Figure 21.1 by the arrow pointing from the time $t_i = 0$ to $t_j = 0$.

The synchronization procedure is initiated by device i at time t_1 by transmitting its current time t_1 in the form of a timestamp to clock j. After an unknown propagation delay T_{ij}, the clock j receives the notification at t_2. The process is reversed by j sending a timestamp at t_3 to i, which receives it after T_{ji} at t_4. The propagation times T_{ij} and T_{ji} can therefore be expressed by

$$T_{ij} = (t_2 + b_j) - t_1, \tag{21.3}$$

$$T_{ji} = t_4 - (t_3 + b_j). \tag{21.4}$$

If (21.3) is substituted into (21.4), the delay b_j can be expressed by

$$b_j = \frac{(t_1 - t_2) - (t_3 - t_4)}{2} + \frac{T_{ij} - T_{ji}}{2}, \tag{21.5}$$

where $(T_{ij} - T_{ji})/2$ represents the asymmetry of the communication channel. If the skew $a_j \neq 1$, the rate of clock j is different than the one of i, and the timestamps between i and j cannot be converted by simply adding the offsets, as the skews scale the time spans,

$$t_j = \frac{a_j}{a_i}(t_i - b_i) + b_j. \tag{21.6}$$

When converting times with respect to the reference clock t with $a = 1$ and $b = 0$ to clock j, (21.6) can be simplified to $t_j = a_j t_i + b_j$. If we stack (21.3) and (21.4) into matrix notation using (21.6), the clock skew

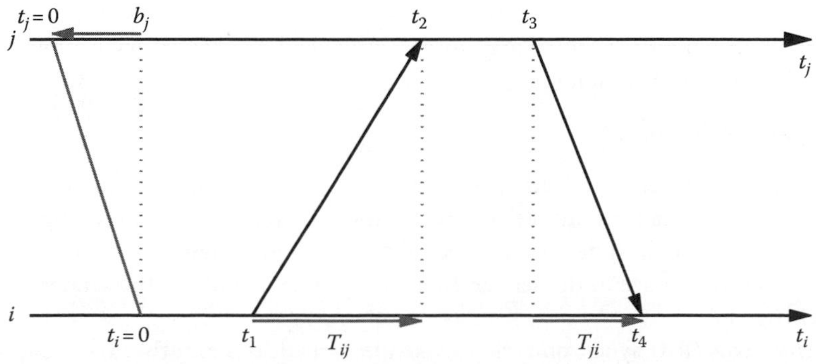

FIGURE 21.1 Generic clock synchronization message flow.

and offsets with respect to the reference clock can be found by solving the equation system using a least-squares solution. The parameter N corresponds to the number of rounds of two-way message exchanges.

$$\begin{bmatrix} t_1^{(1)} \\ \vdots \\ t_1^{(N)} \\ -t_4^{(1)} \\ \vdots \\ -t_4^{(N)} \end{bmatrix} + \begin{bmatrix} T_{ij} \\ \vdots \\ T_{ij} \\ T_{ji} \\ \vdots \\ T_{ji} \end{bmatrix} = \begin{bmatrix} t_2^{(1)} & 1 \\ \vdots & \\ t_2^{(N)} & 1 \\ -t_3^{(1)} & 1 \\ \vdots & \\ -t_{3,}^{(N)} & 1 \end{bmatrix} \begin{bmatrix} a_j \\ b_j \end{bmatrix} \tag{21.7}$$

As visible from (21.5), it is impossible to perform pairwise synchronization on the basis of the timestamps t_1 to t_4. As a result, (21.7) has a rank deficit of one and therefore one degree of freedom, the asymmetry. Even an infinite number of timestamp sets are not sufficient to synchronize two clocks, when T_{ij} and T_{ji} are considered unknown but constant. Yet, even in this general case, the clock skew a_j and round-trip delay $T_{ij} + T_{ji}$ can be calculated just by the timestamps. A common assumption to enable synchronization is to assume constant (or even zero) asymmetry, which renders the last term in (21.5) to a known constant. Hence, the asymmetry needs to be provided by, for example, a previous measurement or calibration. This approach is, for instance, taken by PTP [4].

21.2.3 Taxonomy of Clock Synchronization Approaches

Given the limited stability of real clocks, the aim of clock synchronization is to establish a common notion of time, compensating for offset and rate errors. Clock synchronization approaches can be classified by several criteria as listed in the following. Within each class, the main ideas are only outlined in a representative way and should not be considered comprehensive.

Degrees of synchronization

1. Ordering of events: The protocol ensures that events are ordered in the correct chronological sequence, for instance, by Lamport timestamps [1]. Yet, the exact time of an instant is not estimated. The ordering of events is just a weak notion but might be sufficient for databases or simple monitoring.
2. Relative synchronization: Relative synchronization aims for synchronizing all nodes within a network to a common timebase; it is also referred to as internal synchronization. Relative synchronization is sufficient for many applications such as delay measurements, coordinated actions, or distributed measurements.
3. Absolute synchronization: This type of synchronization aims at synchronizing devices to an absolute (external) reference and is therefore also called external synchronization. It demands access to an absolute time source (e.g., a receiver locked to a global navigation satellite system [GNSS]). It is the strongest notion.

Which devices are synchronized?

1. Sender/receiver (SR) synchronization: A sender submits its time indication in the form of a timestamp message to one (or multiple) receivers. The receivers take a timestamp at the arrival of this message and use the difference between the sender's and their local timestamp, corrected by the offset, to synchronize to the sender. In a steady-state condition, the sender and all receivers are synchronized.
2. Receiver/receiver (RR) synchronization: A sender broadcasts an arbitrary message to multiple receivers. The receivers timestamp the message upon reception and exchange the timestamp information with other receivers that have captured the same message. The difference between

the foreign receivers' timestamps and the own timestamp is used to steer the local clock. Thus, only the group of receivers are synchronized, while the sender remains unsynchronized. RR synchronization does not require any timestamping capabilities in the sender; thus, any device that delivers a sufficient stream of packets to the receivers can be used.

Propagation time correction

1. No offset correction: Clock offsets are not corrected for the propagation time of the message. Thus, in an SR synchronization scheme, the receivers lag behind the sender; in an RR synchronization scheme, the clock offset of two receivers depends on the propagation time difference of the sender with respect to the receivers.
2. Round-trip time-based correction: Clock offsets are corrected by round-trip measurements with the assumption that the individual path delays are exactly half the round-trip time, i.e., no asymmetry. Given this condition is fulfilled, SR synchronization achieves unbiased synchronization.
3. Individual path delay correction: Round trip time based offset correction creates synchronization offsets for asymmetric propagation paths. In this case, the individual propagation delays in each direction (up and down link) need to measured. Note that general purpose synchronization protocols based on timestamps are unable to perform asymmetry measurements. Thus, asymmetry cannot be resolved unless the physical layer provides support for such measurements (e.g., reflectometric cable length measurements).

Timestamping method

1. Software-based timestamping: Timestamps are taken within the device driver or within the application upon the reception of a packet. As the packet processing delay in software depends on the current system and bus load, timestamps are superimposed by nondeterministic jitter.
2. Hardware-based timestamping: Timestamps are taken by dedicated timestamping hardware that removes all jitter impairments from the software. Yet, as hardware timestamps cannot be taken with arbitrary resolution, a small timestamping jitter still remains. Hardware timestamping is used for applications aiming at high synchronization accuracy.

21.2.4 Synchronization Approaches for Packet-Based Networks

Clock synchronization based on the exchange of time information encapsulated in data packets is one of the widely used synchronization techniques due to its seamless integration into existing network and communication techniques. It should be remarked that legacy synchronization technologies exist as well, for example, based on pulses coupling of multiple nodes [5]. Yet, pulse-based synchronization and similar schemes are rather used for niche applications where packet-based message exchange is not available or unsuitable. Messages containing timestamps can be transmitted from the reference clock to the clock to be synchronized, the so-called slave clock, by various means: as wireless signal, for example, from a terrestrial transmitter, satellite, or access point, or using wired communication media, such as Ethernet or an Internet connection. In the following, a number of state-of-the-art synchronization technologies outlining the strengths and limitations of each approach are briefly introduced, yet without providing a comprehensive comparison. Subsequently, the NTP and the PTP are covered in more detail in Sections 21.3 and 21.4.

21.2.4.1 DCF77

Among the first examples of time distribution systems, the DCF77 [6] system should definitely be mentioned. DCF77 sends the time information using long waves with a standard radio station. The primary DCF77 transmitter is located in Mainflingen, Germany, and started radio transmission on January 1, 1959.

The carrier signal at 77.5 kHz is generated from local atomic clocks that are linked with German master clocks. The time provided by the radio transmission is Universal Time Coordinated plus 1 h (UTC + 1), Central Europe Time. The DCF77 signal is modulated using amplitude-shift keying by reducing the carrier amplitude for 0.1 or 0.2 s at the beginning of every second. Every full minute is marked by no carrier power reduction. The same signal is also modulated using phase modulation combined with direct-sequence spread spectrum (DSSS) using a 512 bit long pseudorandom sequence. The synchronization accuracy that can be obtained using the DCF77 signal is limited by wireless propagation conditions and by the technology adopted in the implementation of the receiver. Normal low-cost DCF77 receivers, which typically use only the amplitude-modulated time signal, have a synchronization uncertainty on the order of 0.1 s, enough for radio-controlled low-cost clocks. For more demanding devices, which require an accurate time information, a DCF77 receiver based on the cross-correlation of the phase-modulated time signal can be used to obtain an accurate estimation of the beginning of the second markers. Using this solution, a synchronization accuracy on the order of tens of microseconds can be achieved [7]. Similar radio time signal stations with slightly different modulations and encodings are deployed all over the world, such as the WWVB in Fort Collins, the United States, or the TDF in Allouis, France.

21.2.4.2 Global Positioning System

One of the most commonly used GNSSs is the Navigation Satellite Timing and Ranging System (NAVSTAR GPS). The concept of global positioning system (GPS) was developed for the US Department of Defense (DoD) starting in the 1970s, and the system became fully operational in 1994. Each GPS satellite is equipped with four atomic clocks (two rubidium and two cesium clocks) with a fundamental frequency slightly below 10.23 MHz. This slight frequency skew (−0.4467 ppb) is required to compensate for relativistic effects so that the clock appears as 10.23 MHz for a GPS receiver on the ground [8]. The legacy GPS navigation signals are transmitted on two frequencies, the L1 and L2 band. The L1 band operates with a center frequency of 1575.42 MHz, 154 times the fundamental frequency, and broadcasts the course/acquisition code (C/A code) and the precision code. As the precision code is encrypted, forming the so-called Y-code, it is abbreviated with P(Y). The L2 band at 1227.60 MHz carrier frequency, 120 times the fundamental frequency, transmits the P(Y) code only. The C/A code is free for civilian use, while the encrypted P(Y) code is only accessible for military or special organizations.

The ephemerides of the satellites as well as the clocks are constantly monitored by multiple ground stations. The measurements taken by the ground stations are communicated back to the satellites, which embed this information into the navigation messages sent to the GPS receivers. A GPS receiver tracks the signals of at least four GPS satellites and calculates its four unknown quantities, the three coordinates, and the clock offset to the satellite system. Thus, a GPS receiver can provide an accurate time indication with respect to the synchronized GPS. According to [9], a C/A receiver can provide a median clock offset of below 19 ns with respect to the reference clock network. The versatility and high accuracy of GPS makes clock synchronization using GPS a good choice for synchronization over larger distances. Nevertheless, it should be pointed out that GPS is under the control of the DoD and that deteriorations, such as selective availability, might be reenabled in crisis or wartimes. In addition, the low transmission power of GPS (and in general any GNSS) allows for jamming and spoofing of the signals. Thus, for a safety-critical systems, a synchronization backup solution should be available.

21.2.4.3 Inter-Range Instrumentation Group

The Inter-Range Instrumentation Group (IRIG) standard [10] defines a set of timecodes, each of them characterized by its own specific attributes. The IRIG signal consists of a carrier that is modulated to transfer the timecode. The type of the carrier and its frequency depend on the type of IRIG encoding adopted. The different timecodes are identified by alphabetic designations: A, B, D, E, G, and H. A three-digit numerical suffix defines the type and the frequency of the carrier. The three modulations supported are direct current level shift (DCLS), sine wave carrier (amplitude modulation), and Manchester modulation. The main difference between the codes is the data rate, which ranges between 1 and 10,000 pulses/s. The time information provided by the IRIG signal includes the binary coded decimal (BCD) representation of the day of year, hours, minutes, seconds, and fractions. Additional time information (such as year) can be optionally provided. One of the most common IRIG codes is IRIG-B. This code provides a bit rate of 100 Hz and a frame transmission rate, in BCD coding, of every second. The IRIG code is adopted for transferring time information recovered by a primary time source, like a GPS receiver or an atomic clock, using dedicated network infrastructures. A typical application of an IRIG-B encoded signal is the time distribution to measurement and control devices in substation automation systems (SASs) [11]. Nevertheless, an accurate distribution of the IRIG-B signal requires a dedicated network infrastructure and a periodic calibration of the entire system. Thus, this solution is becoming less and less adopted for the implementation of new systems, for the benefit of network-based synchronization solutions.

21.3 Network Time Protocol

The NTP is defined in RFC 5905 [12]. NTP systems can operate as a primary server, secondary server, or client. The primary server derives its time directly from a UTC reference clock like GPS or Galileo receiver. On the contrary, the client NTP system can be synchronized to one or more upstream servers, but it is not able to synchronize other clients. The last type, the secondary server, lies in between primary servers and clients; for this reason, in large NTP networks, the implementation of secondary servers is critical since they pass synchronization from high layers to end user clients. The NTP protocol can be implemented in three variants: client/server, symmetric, and broadcast:

1. Client/server variant: The client sends request packets to the server that answers, providing synchronization. Generally, a server can deal with several clients, while it does not accept synchronization from any client. Clearly, the server may be directly connected to a UTC standard source (e.g., GPS receiver). Note that in this variant, clients ask for synchronization to the servers
2. Symmetric variant: The NTP implementation is at the same time a server and client. Two possibilities are allowed: symmetric active association or symmetric passive association. The symmetric active association is persistent (i.e., it is created and never destroyed) and operates by sending packets to the correspondent symmetric active peer association. The symmetric passive association is ephemeral (i.e., it is created on packet arrival, and it is destroyed if an error or timeout occurs). Note that in this variant, each peer bidirectionally requests and sends synchronization information. Thus, the peer is also a client.
3. Broadcast variant: The broadcast server associations are both persistent and ephemeral. The broadcast server is persistent and periodically transmits packets that are received by multiple clients. On the contrary, the broadcast client association is ephemeral since it is created when a packet from a broadcast server is received and it remains active until error or timeout occurs. Sometimes, the client starts with the client/server variant and after an initial synchronization (for delay calibration and security setup) becomes a broadcast client. Note that in this variant, the server sends synchronization information to clients or other servers.

NTP is organized as a hierarchical structure of servers to which the clients are connected. The servers' level in the hierarchy is given by the stratum number. Stratum one is the level of the primary servers; all the secondary servers have stratum numbers higher than one: specifically, one greater than the preceding level. The highest stratum number for the current NTP version is 15. Usually, the server accuracy decreases as the stratum number increases, since the network path becomes longer and the cascade of multiple clocks decreases the clock stability. Generally speaking, in NTP networks, the synchronization mean errors are proportional to the server stratum number. For this reason, the NTP topology should minimize the synchronization distance between servers and between servers and clients, although loops should be avoided. A variant of the Bellman-Ford distributed routing algorithm is used to determine the shortest-path spanning tree that originates from the primary servers. A network topology can be rearranged to overcome the results of failures (both physical faults and timing-related errors).

21.3.1 NTP Time Scale, Time Format, and Other Definitions

For NTP implementations, two timescales are considered: UTC time for dating events and the running time, which is the output of the NTP synchronization function. Depending on the context, the timestamp $T(t)$ may represent either the UTC date of an event or the time offset from UTC at a given running time t. If $T(t)$ is the time offset, $R(t)$ is the frequency offset (expressed in seconds-per-second (s/s) or ppm), and $D(t)$ is the aging rate (first derivative of $R(t)$ with respect to t). If $T(t_0)$ is the UTC time offset determined at $t = t_0$, the UTC time offset at time t can be expressed by

$$T(t) = T(t_0) + R(t_0)(t - t_0) + \frac{1}{2}D(t_0)(t - t_0)^2 + e. \tag{21.8}$$

The term e in (21.8) refers to the stochastic error term. The performance of the NTP timekeeping function is computed each time a time information exchange takes place using four statistics. As defined in the standard, these statistics are as follows:

1. The offset θ represents the maximum-likelihood time offset of the server clock relative to the system clock.
2. The delay δ represents the round-trip delay between the client and server.
3. The dispersion ϵ represents the maximum error inherent in the measurement. It increases at a rate equal to the maximum disciplined (hardware) system clock frequency tolerance Φ. Note that Φ is system-dependent, involving an assumption about the system worst-case behavior in case the time source (e.g., the NTP server) has become unreachable.
4. The jitter ψ is defined as the root-mean-square (RMS) average of the most recent offset differences. It represents the nominal error in estimating the offset.

There is a separate group of the four statistics (θ, δ, ϵ, and ψ) for each server the system clock is connected to. However, the strength of the NTP protocol is the way it combines the statistics of many servers in order to increase the overall system clock accuracy. For this reason, the standard also defines the system statistics:

1. The system offset Θ represents the maximum-likelihood offset estimate for the server population.
2. The delta statistic is accumulated at each stratum level from the reference clock to produce the root delay statistic Δ.
3. The epsilon statistic is accumulated at each stratum level from the reference clock to produce the root dispersion statistic E.
4. The system jitter Ψ represents the nominal error in estimating the system offset.
5. The synchronization distance Λ equal to $E + \frac{\Delta}{2}$ represents the maximum error due to all causes.

In the NTP system, there are three time formats, shown in Figure 21.2:

1. The 128 bit date format can be used if sufficient memory space is available in the system. This format has the highest resolution and range, since the 64 bit wide seconds field may reach 584×10^9 years and the 64 bit fraction may resolve 0.5×10^{-18} s. In order to be compatible with other formats, the seconds field is divided into a 32 bit era number field and a 32 bit era offset field. Usually, era numbers are out of the scope of normal NTP implementations.

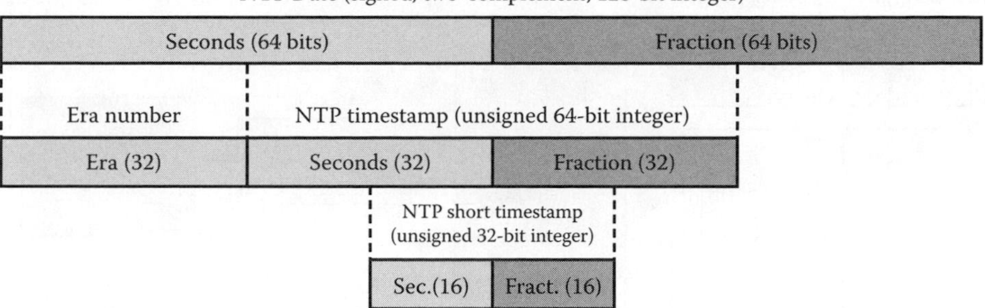

FIGURE 21.2 The three time formats of NTP.

2. The 64 bit time format is commonly used for packet headers and in systems with limited memory space. It is composed of 32 bit unsigned seconds (range of 136 years) and 32 bit fraction (resolution of 232 ps).
3. The 32 bit short format is used only if the range/resolution of the other formats is not required (e.g., for delay and dispersion header fields). It is composed of a 16 bit unsigned seconds field and a 16 bit fraction field.

Every time the 32 bit wide seconds field wraps around, a new era is started; the prime epoch, or base date of era 0, is at 0:00, January 1, 1900, UTC, which corresponds to all bits set to zero. A date value greater than zero represents the time after the prime epoch; conversely, a time value less than zero means a time before the prime epoch. Note that dates are signed values while the timestamps are unsigned values; however, only the subtraction is permitted on dates and timestamps, so both of them can be processed with two's complement arithmetic without problems if the timestamp difference is less than 68 years (which is the half of the 32 bit s field range).

The clock resolution depends on the number of significant bits p in the second fraction field, that is 2^{-p}. Obviously, p can be lower than 32, meaning that some fractional bits are nonsignificant. Note that for minimal bias, the nonsignificant bits should be randomized to unbiased values. In the NTP standard, the clock precision ρ is the larger between the clock resolution and the running time needed for reading the system clock. In other words, the very high resolution cannot be exploited if the access time to the system clock is large.

21.3.2 NTP Implementation Model

The NTP is usually implemented with a multithread structure, since its basic particularity is the ability to exchange time information with several servers at the same time. For each server, two processes exist, one for receiving messages from servers (the peer process) and one for sending requests to servers (the poll process). For each server, a data structure called association is maintained with all statistics described in the previous section, the network addresses, and the protocol ports used by the partner peer. Every time NTP packets are received, the statistics θ, δ, ϵ, and ψ are updated. The time retrieved by each association is fed to the system process, which includes three algorithms (selection, cluster, and combine) to harmonize/mitigate the various servers, determining which are the most accurate and reliable. A block diagram of the NTP system process is shown in Figure 21.3 to illustrate the interactions between the modules.

1. The *selection algorithm* is used to detect Byzantine faults, that is, discarding the apparently incorrect servers (*falsetickers*) and retaining the good servers (*truechimers*). By standard definition, "a truechimer is a clock that maintains timekeeping accuracy to a previously published and trusted standard, while a falseticker is a clock that shows misleading or inconsistent time."

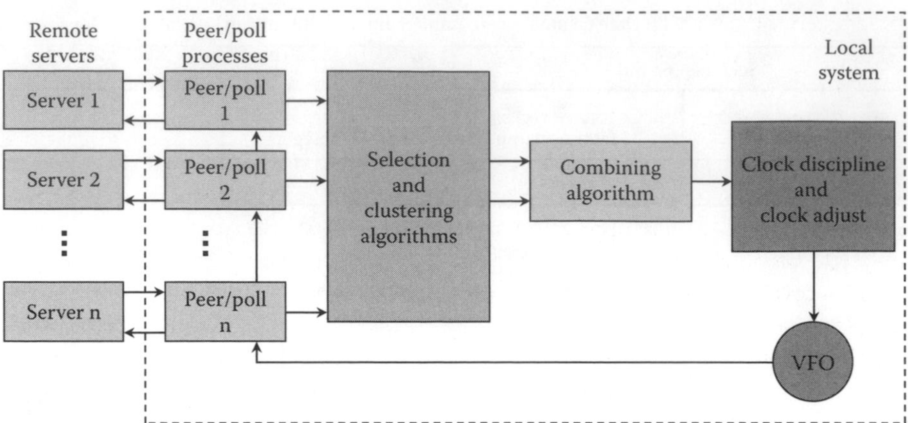

FIGURE 21.3 The block diagram of NTP system process.

2. The *cluster algorithm* is used to classify the truechimers by means of statistical principles, and to identify the most accurate set for further processing.
3. The *combine algorithm* is the last one, and it is used to statistically average the selected truechimers creating the final clock offset.

The obtained offset is passed to the clock discipline process controlling both time and frequency of the system clock, which can be generally represented as a variable frequency oscillator (VFO). The clock discipline process triggers the clock-adjust process every second in order to act on the VFO for time offset and frequency compensation. Note that VFO parameters and drivers depend on both the operating system (OS) and the hardware implementation. The VFO is also the clock source from which all the timestamps are taken, closing the feedback loop of the NTP implementation model.

Note that a client uses the poll process to send messages to the corresponding server using the variable τ for the calculation of the poll interval (2^τ s). In NTP, τ can range from 4 (16 s) to 17 (36 h), and it may vary depending on the clock discipline algorithm that tries to match the *loop–time* constant $T_c = 2^\tau$. NTP provides negotiation means for the peers so that they can agree on a common τ (which is the minimum of both of them).

21.3.3 On-Wire Protocol

The time exchange between peers (servers and/or clients) is the core mechanism of NTP. The on-wire protocol describes how time quantities are exchanged. It is based on four timestamps (t_1, t_2, t_3, t_4) and three state variables (origin timestamp "org," receiving timestamp "rec," and transmitting timestamp "xmt"), as depicted in the most general case in Figure 21.4. The figure shows two peers, *A* and *B*, that autonomously compute the offset and delay with respect to each other. Note that if NTP has direct control over the physical layer, the timestamps refer to the beginning of the first symbol just after the start of frame (SOF). If no physical layer control is available, NTP implementations should attempt to associate the timestamp to the earliest accessible point in the receiving stack. The procedure is the following:

1. The first packet is transmitted by *A* and contains only the origin timestamp t_1. After sending, t_1 is copied to the "xmt" state variable of *A* as T_1.
2. *B* receives the packet at t_2 (the receive timestamp) and copies it to the "rec" state variable of *B* as T_2. The t_1 inside the packet is copied to the "org" state variable of *B* as T_1.

3. Immediately, or some time later than t_2, B sends a packet to A at t_3; the packet contains t_1 and t_2 and the transmit timestamp t_3. The timestamp t_3 is copied to the "xmt" state variable of B as T_3.
4. When A receives the packet at the destination timestamp t_4, it copies t_3 to the "org" state variable of A as T_3 and t_4 to the "rec" state variable of A as T_4.

The algorithm implements two sanity checks: In order to discover duplicated or replayed packets, the transmit timestamp t_3 is compared with the "org" state variable T_3. If they match, the packet is duplicated and should be discarded. In order to discover bogus packets, the origin timestamp t_1 is compared with "xmt" state variable T_1. If they match, the packet is correct (and the "xmt" state variable is set to zero). In case of corruption, the packet is discarded. With the described sanity checks, the protocol can resist the replay of server response packets, but it is still vulnerable to the replay of client packets. In this case, the server replays again with new values of T_2 and T_3. This vulnerability is eliminated, if the "xmt" state variable is set to zero after the offset and delay computation as described in the following. For each peer, the four most recent timestamps (T_1, T_2, T_3, T_4) are used to compute the offset and the round-trip delay of B relative to A as described in the standard:

$$\theta = T_B - T_D = \frac{(T_2 - T_1) + (T_3 - T_4)}{2}, \tag{21.9}$$

$$\delta = T_{ABA} = (T_4 - T_1) - (T_3 - T_2). \tag{21.10}$$

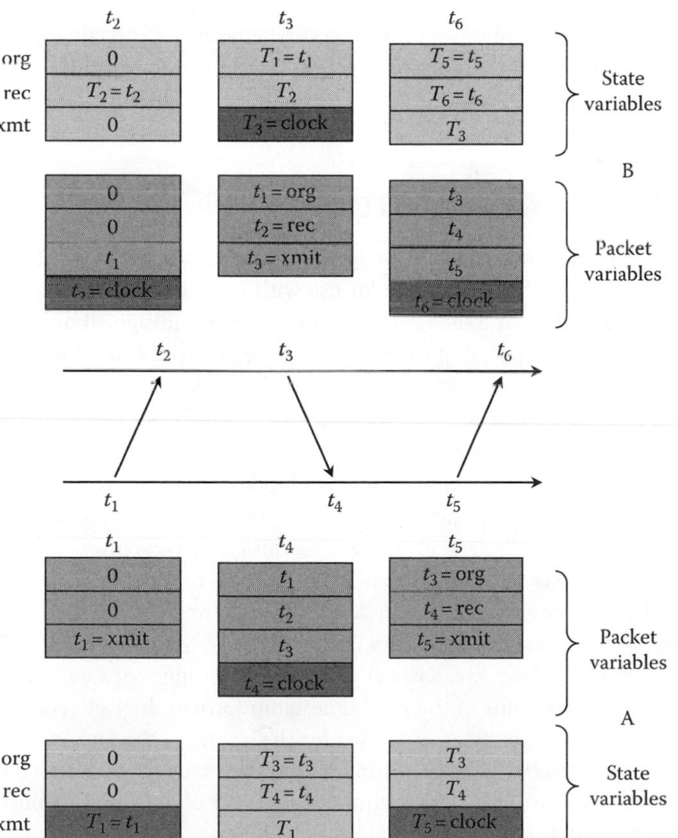

FIGURE 21.4 The NTP time exchange between peers.

These equations match the basic equations of the two-way message exchanges described by (21.3 through 21.5). In addition, the peers need to estimate the dispersion statistic $\epsilon(t)$, which represents the maximum error due to the frequency tolerance and time since the last packet was sent. The initial value of ϵ at t_0 is

$$\epsilon(t_0) = \rho_p + \rho_s + \Phi \cdot (T_4 - T_1), \tag{21.11}$$

where

ρ_p is the precision of the timestamps in the received packet
ρ_s is the system precision

The two terms take into account the uncertainty in reading the system clock for the server and the client, respectively. As a result, the dispersion grows linearly at rate Φ (which is constant, the default value of Φ is 15 ppm or about 1.3 s/day). Note that special care must be taken when these calculations are executed on a real platform in order not to lose resolution. Moreover, known problems may arise if the server and client time has a starting time difference greater than 34 years, or if the initial frequency offset is large and the propagation delay is small. In order to fix these problems, a set of parameters (value clamping and saturation) is provided by the software implementation. Generally speaking (Figure 21.4), the synchronization mechanism is bidirectional: the peer *B* can also estimate the offset of *A* relative to *B*, the next time a packet is exchanged (using timestamps T_3, T_4, T_5, T_6). In case the NTP server *B* is not required to estimate the offset to the client *A*, the server only appends the timestamp T_3 to the timestamps T_1 and T_2 of the client packet, without implementing the "org," "rec," and "xmt" state variables.

In conclusion, the on-wire protocol can be considered resistant to dropped or duplicated packets, without requiring any flow control or retransmission mechanism. Nevertheless, the implementation must pay attention to the fact that the timestamps should be as accurate as possible, meaning that they have to be recalculated in case of retransmission at the link level or in case of security computations during transmission.

21.3.4 Network Data Packet for NTP

The NTP packet is transported inside a UDP datagram. The specific UDP port 123 has been assigned by Internet Assigned Numbers Authority (IANA) for use with NTP. In case of broadcast servers, the IPv4 multicast group address 224.0.1.1 and the IPv6 multicast address ending: 101 are also reserved by IANA for NTP. The NTP packet is composed of three parts: the header, the extension fields, and the message authentication code. Only the header is mandatory since it includes the most relevant variables. The header consists of an integral number of 32 bit words. The optional fields are used for authentication and security mechanisms, like the Autokey public key cryptographic algorithms [13] or the native symmetric key cryptographic authentication algorithm (128 bit MD5) defined in the NTP standard. Authentication and security fields are computed at transmission time, implying additional delays. Such a situation must be properly taken into account when writing the transmission timestamps to the packet header. The packet also includes a 32 bit Reference ID (refid), which is used to identify the time server or reference clock. This 32 bit field can carry a four-character ASCII string defined by IANA in case of servers belonging to stratum 1; otherwise, it encodes the IPv4 address (or the MD5 hash of the IPv6 address) of the server. The primary purpose of the Reference ID is to avoid timing loops across large NTP networks. The header includes four timestamps in the NTP timestamp format: the Reference Timestamp, which is the time when the system clock was set or corrected for the last time; the Origin Timestamp, which corresponds to the "org" state variable (departure time of the client request); the Receive Timestamp, which corresponds to the "rec" state variable (arrival time at the server of the client request); and the Transmit Timestamp, which corresponds to the "xmt" state variable (departure time of the server response). Note that the Destination Timestamp (i.e., the arrival time at the client of the server response) is not transported in the packet, but it is made available in the client receive buffer by the NTP implementation.

21.3.5 NTP Managing Process

The NTP core is focused on filtering and analyzing the time information coming from several servers. In the following, the most important of these processes are described.

21.3.5.1 Clock Filter Algorithm

The clock filter algorithm is used to produce, for each peer, the following variables: offset θ, delay δ, dispersion ϵ, jitter ψ, and time of arrival t. The output of the filter algorithm is fed to the mitigation algorithms, which determines the best control strategy for disciplining the system clock. The clock filter algorithm uses the eight most recent sample tuples (θ, δ, ϵ, t) in the order of packet arrival. When a new packet is received, the least recent tuple is discarded. If a request is not answered by the server, a dummy tuple (0, "maximum_dispersion," "maximum_dispersion," 0) is inserted in the filter stages; suitable controls are implemented to release associations when too many requests are dropped by the server. The operation of the filter algorithm is outlined in the following:

1. The eight tuples in the filter are sorted by increasing δ. Only the valid tuple relative to received packets are considered.
2. The filtered ϵ (i.e. dispersion of the considered peer) is calculated as

$$\epsilon = \sum_{i=0}^{n-1} \frac{\epsilon_i}{2^{(i+1)}},$$

   where
   i is the list order
   ϵ_i is the epsilon variable of the tuple
   n is the number of valid tuples in the filter ($n > 1$)

   Given this formula, the filter produces a dispersion less than 1 s if at least four valid packets were received in the last eight polls.
3. The θ of the first tuple of the ordered list is used as θ_0 for the calculation of the filtered jitter (which is the RMS average of all the valid tuples)

$$\psi = \sqrt{\frac{1}{(n-1)} \sum_{j=1}^{n-1} \left(\theta_0 - \theta_j\right)^2}.$$

   Note that ψ is low-saturated to the system precision for avoiding inconsistency. The jitter is used as indicator of the timekeeping performance.
4. The synchronization distance between server and client is obtained from

$$\lambda = \epsilon + \left(\frac{\delta}{2}\right).$$

   Note that λ is recalculated every time if needed, starting from the current δ and ϵ. It is used by the mitigation algorithms to determine the root synchronization distance of servers and to classify their time quality.

21.3.5.2 Mitigation Algorithm

As already introduced, the mitigation algorithm consists of the selection algorithm, the cluster algorithm, the combine algorithm, and the clock discipline algorithms. The mitigation algorithm is run each time a new tuple (θ, δ, ϵ, ψ, t) is produced by the clock filter algorithm described before. Note that all associations (server–client relationships) are scanned by the mitigation algorithms. The selection algorithm is used to discard the falsetickers and keep the truechimers. With an iterative procedure, the cluster algorithm reduces the number of associations to the specified minimum, discarding the peers that are statistically away from the centroid. The combine algorithm is used to weight the statistics producing the final estimate, which is then passed to the system clock discipline algorithm. Note that the time-related system variables, which are used for propagating the time to other servers and clients and passed to the applications running on the same machine as NTP, are an output of the cluster algorithm; the cluster algorithm selects one of the surviving truechimers as the system peer, and the system inherits its variables (θ, δ, ϵ, ψ, t).

21.3.5.2.1 Selection Algorithm

The selection algorithm tries to find out an intersection interval that contains the majority set of the truechimers. It is focused on the exclusion of Byzantine faults, and it is based on an improved Marzullo algorithm. The selection algorithm is applied only to peers that are usable according to the protocol rules, while others are immediately discarded. For the scope of the algorithm, each server is identified by its midpoint (the offset θ), its upper endpoint $\theta + \lambda$, and its lower endpoint $\theta - \lambda$, where λ is the root synchronization distance. The algorithm assumes initially that there are no falsetickers and tries to find the nonempty intersection interval that contains midpoints of all the truechimers. It may happen that the nonempty interval cannot be found; in this case, the number of falsetickers is increased by one, and the algorithm tries to find the nonempty intersection that contains midpoints of *all minus one* truechimers, and so on. Two possible outcomes are expected: a nonempty interval is found and the number of truechimers is greater than the number of falsetickers; that is, the majority clique has been found. If there are more falsetickers than truechimers, or the number of truechimers is lower than the minimum number of true servers required by the system, no majority clique can be found. In the first (positive) case, the midpoint of each truechimer (θ) represents the candidate for the cluster algorithm. Note that in most NTP implementations, the minimum number of servers required to be truechimers for running the discipline algorithm is just one. Increasing this lower bound results in a more robust system.

21.3.5.2.2 Cluster Algorithm

The following variables are considered by the cluster algorithm for each candidate p belonging to the majority clique: the current offset θ_p, the current jitter ψ_p, and the $QF_p = \text{stratum}_p \cdot \text{MAXDIST} + \lambda$ where stratum_p is the stratum of the considered candidate, MAXDIST is the maximum allowed distance, usually set to 1 s, and λ is the root synchronization distance.

QF_p must be considered a merit factor of candidate p, and it is used to sort the candidates in ascending order. In turn, each candidate is used as the reference, and the selection jitter (RMS average with respect to its θ) is calculated by

$$\psi_s = \sqrt{\frac{1}{(n-1)}\sum_{j=1}^{n-1}\left(\theta_s - \theta_j\right)^2}. \tag{21.12}$$

The algorithm is iterated, and in each round, the candidate with the maximum selection jitter ψ_s is discarded. Two conditions terminate the clustering algorithm: if the ψ_s is less than the minimum peer jitter ψ_p (which means that no improvement is possible by discarding candidates), or the number of survivors is less than a given threshold. The *system peer*, whose time variable will be used to update the system time variable, is the candidate with the minimum ψ_s. On the other hand, the maximum ψ_s value of the survivors is saved as indicator of the system selection jitter Ψ_s.

21.3.5.3 Combine Algorithm

The combine algorithm receives as input the peer offset and jitter statistics of the remaining survivors after the clustering algorithm. Then, the combine algorithm outputs the combined system offset Θ and system peer jitter Ψ_p. The standard recommends weighting the statistics coming from each server j using the weighted average as

$$\Theta = \frac{\sum_{j=0}^{n-1} \frac{\theta_j}{\lambda_j}}{\sum_{j=0}^{n-1} \frac{1}{\lambda_j}}, \tag{21.13}$$

$$\Psi_p = \sqrt{\frac{\sum_{j=0}^{n-1} \frac{(\theta_j - \theta_0)^2}{\lambda_j}}{\sum_{j=0}^{n-1} \frac{1}{\lambda_j}}}, \tag{21.14}$$

where

θ_0 is the offset of the system peer selected by the cluster algorithm
θ_j is the offset of j
λ_j is the root synchronization distance of j

Note that in some implementations, the weighted average can be bypassed, and the variable of a preferred peer (among the survivors) is directly used. Note that Θ is fed to the clock discipline algorithm, while Ψ_p is processed together with the selection jitter Ψ_s in order to obtain the system jitter

$$\Psi = \sqrt{\Psi_s^2 + \Psi_p^2}. \tag{21.15}$$

After this calculation, all the system variables are ready to be used by any applications or program in the local system, so the clock update routine can be called. Resuming, the final system variable set is composed by the system offset Θ (the clock offset relative to the considered synchronization sources), the system jitter Ψ (the estimate of the error in determining the offset), the root delay Δ (the total round-trip delay referred to the primary server), the root dispersion E (the dispersion accumulated from the primary server due to network effect), and finally Λ (the maximum error due to all causes $\Lambda = E + \Delta/2$).

21.3.5.4 Clock Discipline Algorithm

In NTP, the strategy adopted by the clock discipline algorithm is rather complex since it is the combination of two feedback controllers. Basically, there is a phase-locked loop (PLL) controller that periodically corrects the time error with update intervals of microseconds. Besides, a frequency-locked loop (FLL) periodically acts on the frequency error with the update interval μ. The twin approach is used since usually the PLL is better if the network jitter dominates, while the FLL is better if the local oscillator frequency instability is greater than other contributions. The decision on which control action is taken is based on a state machine, meaning that, actually, the control is nonlinear. Such approach allows very fast recovery from unsynchronized situations at the expense of a more complex algorithm. An in-depth description of the controller is available in the standard. The clock adjustment process is executed every second (interrupt-driven) and passes the control values to the OS, usually relying on the OS's functions. The clock adjust process interrupt is also used by the poll process to trigger the transmission of requests to servers.

21.3.6 Simple Network Time Protocol

Servers and clients may comply with a subset of NTP, implementing the RFC4330 Simple Network Time Protocol (SNTP) [14]. The basic difference to NTP is that SNTP does not implement the mitigation algorithms. Generally, the SNTP is used in primary servers that have a single reference clock, or in clients with a single upstream server. Note that NTP and SNTP servers and clients are fully interoperable. The behavior of an SNTP server is indistinguishable from an NTP server from the on-wire protocol point of view. The NTP server has the mitigation algorithms so it is able to manage several upstream time sources, while the SNTP server has a single time reference. When a client request is received, the SNTP server constructs and sends the reply packet only adding the transmit timestamp (no peer state machine is implemented). The SNTP client just implements the on-wire protocol since it has only a single server. The implementation complexity can be scaled, and an arbitrary subset of the on-wire protocol can be implemented. The simplest implementation of the SNTP client just makes use of the transmit timestamp of the server contained in the server response packet; the SNTP client simply updates its system clock to this value.

21.4 IEEE 1588 Precision Time Protocol

The IEEE 1588 standard specifies the PTP enabling highly accurate clock synchronization in packet-based networks. The initial version of PTP, IEEE 1588 version 1, was approved in 2002. In 2008, the second release of IEEE 1588, version 2 or 2008, was published. Apart from the change of the data format, many new features have been introduced and refinements have been made to improve various aspects, such as the synchronization accuracy, fail-over to alternate masters, and management support. As of 2013, a refinement of IEEE 1588 version 2 is considered by the IEEE 1588 committee, which is intended to be backward compatible to version 2. PTP has been designed as a general-purpose synchronization protocol, which can be used over any kind of network and topology. The roots of PTP originate from the test and measurement field where PTP should synchronize multiple distributed measurement instruments. Since the rise of IEEE 1588, the fields of application have extended to power engineering, factory automation systems, real-time communication, telecommunications, and audio–video broadcasting.

PTP is a master–slave synchronization protocol defining mechanisms to find the best master in the network, and to synchronize all slave devices to the master. In the PTP nomenclature, all devices that participate in synchronization are termed *clocks*. Clocks can be categorized into *ordinary clocks, boundary clocks*, *end-to-end transparent clocks*, and *peer-to-peer transparent clocks*. The class of ordinary clocks consists of *master clocks*, which provide the time reference to other nodes, and *slave clocks*, which synchronize to a master clock. Figure 21.5 depicts the model of an ordinary clock. It consists of the local clock, the PTP control engine, the data and port data sets, timestamping functions for event messages, and application-specific functions. The latter are used, for instance, to generate timing signals, to trigger external actions, or to timestamp external signals. The core of an ordinary clock is the local clock, which contains a time indication device, depicted as watch, and a control servo, which performs adjustments in offset and skew as requested by the PTP protocol engine.

A dedicated master clock synchronizes a group of slaves. This master clock can be either a boundary clock, a clock synchronizing itself to another PTP clock, or a PTP *grandmaster*, which is ideally the best synchronized node in a whole system, such as a node connected to a GPS receiver or an atomic clock. The protocol has two phases, the start phase and the synchronization phase. In the start phase, the best master clock (BMC) algorithm is elected by the so-called BMC algorithm, which finds the device with the best accuracy. After the BMC algorithm has decided which node to promote as master clock, the synchronization phase is started where the slave clocks synchronize to the master clock by an exchange of synchronization messages.

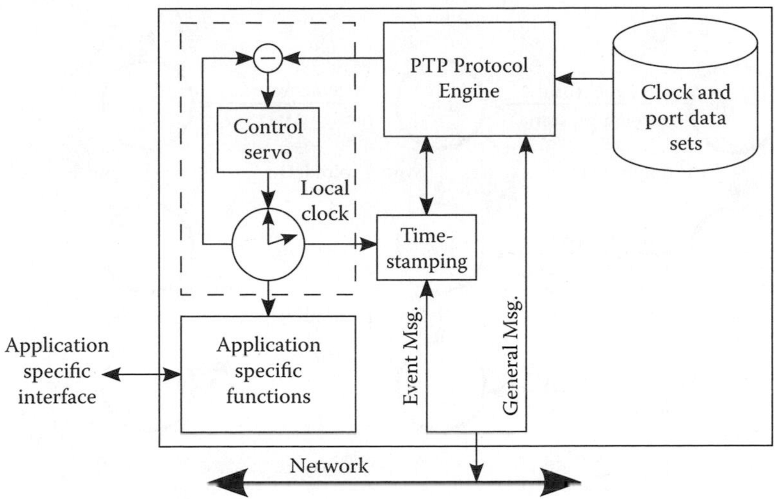

FIGURE 21.5 Model of an ordinary PTP clock.

21.4.1 Master Election

PTP permits two forms of BMC algorithm: the BMC specified in the IEEE 1588 standard and an alternative algorithm specified in a PTP profile. The BMC specifies a way to determine which clock among all contenders is the *best*. The BMC is executed independently on each clock, and every clock makes the master decision on its own; that is, no negotiations between the clocks are done. The BMC is structured into two parts:

1. The data comparison algorithm ranks the clocks and determines which clock is the *best*.
2. An algorithm that computes the recommended state of the synchronization state machine.

A prerequisite for the master election is that each node has knowledge about its data set used for the comparison. This data set includes the priority, clock class, clock accuracy, clock stability, second priority, and clock identity. The data set comparison algorithm first compares the clock identity. If another device has the same identity, the data comparison exits with an error. In Ethernet, the identity corresponds to the MAC address. Thus, this error corresponds to the situation when source and destination address of an announce message are identical, for example, a duplicate MAC address or illegal MAC address modification. If the identity does not match, subsequently the priority, clock class, clock accuracy, clock stability, second priority, and clock identity are compared; and a ranking between the clocks can be made.

Each node in the network starts in the status `INITIALIZED`. The only valid transition from this state is to the state `LISTENING` as denoted by L_{ij} in Figure 21.6. In the `LISTENING` state, the node waits for incoming messages. The nodes broadcast `Announce` messages to inform other nodes about their data set to compete for the master role. If a node receives two messages indicating better clock values than its own clock, it changes into the `SLAVE` state. If no synchronization message is received for 10 synchronization intervals, the node considers itself the master and switches to the `MASTER` state.

In the `MASTER` state, illustrated as `M` in Figure 21.6, a PTP node starts multicasting synchronization messages to all other nodes in the PTP domain. The `MASTER` state can only be left if a synchronization message from another node is received and the analysis shows that the node's clock is less accurate than the clock from which the synchronization message was received. In this case, the master node changes to the `SLAVE` state. A node that is in the slave state (S_i) may change to the `MASTER` state if it has not received a synchronization event for 10 subsequent synchronization periods. This case occurs, for instance, if the data connection to the master is temporarily not available or in case of a master failure.

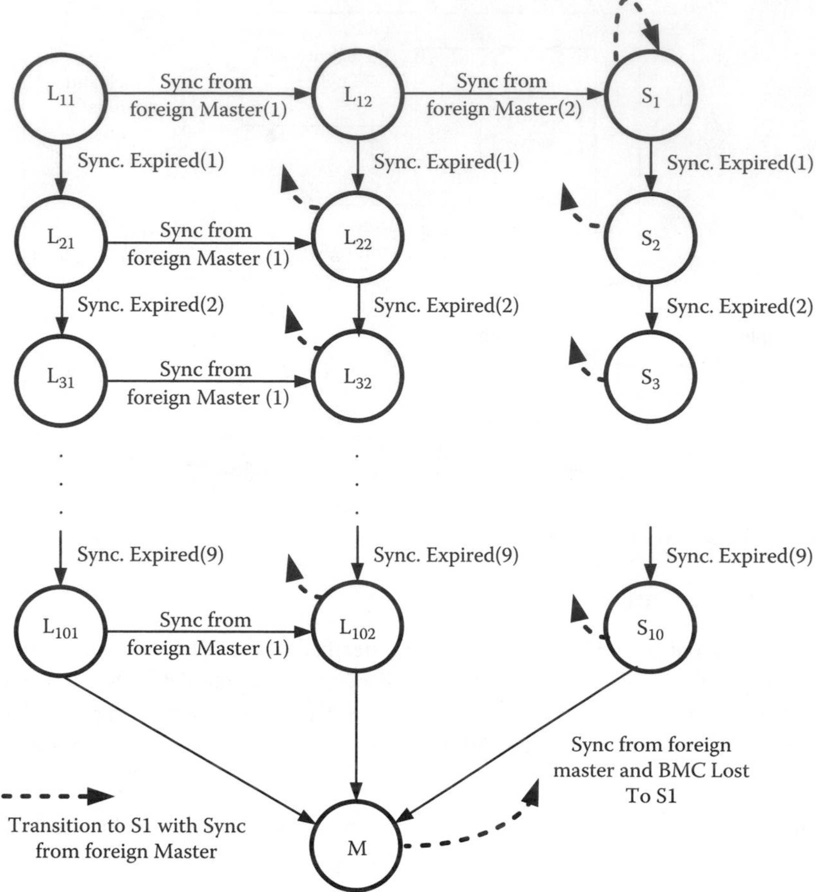

FIGURE 21.6 State transition diagram of the master clock election.

For the latter case, PTP offers the option of alternate master, which allows a fast handover to an alternate master clock. The BMC algorithm also offers the option to prohibit the devices transiting into the SLAVE state by setting the clock class to a value between 1 and 127. One application for this option are grandmaster clocks that are connected to GPS or an atomic clock.

21.4.2 Synchronization in PTP

The synchronization process in PTP is based on timestamps that are taken at the ingress and egress of event messages. IEEE 1588 defines four types of event messages: Sync, Delay_Req, Pdelay_Sync, and Pdelay_Resp. The first two are used for the delay request–response mechanism; the second two are used for the peer delay mechanism. In addition to event messages, general messages are used to transport nontiming critical data between the devices, such as Announce or Follow_Up messages.

21.4.2.1 Offset and Skew Correction

The synchronization in PTP uses the two-way synchronization principle; thus, it compensates for the path delay by round-trip measurements. In the delay request–response mechanism, the master transmits periodic Sync messages to the slaves at time t_1 as shown in Figure 21.7 (cf. [4], p. 34). The slave takes the timestamp t_2 at the arrival of the Sync message. The master may have either included its own egress timestamp t_1 already in the Sync message or it may use the Follow_Up message to transport

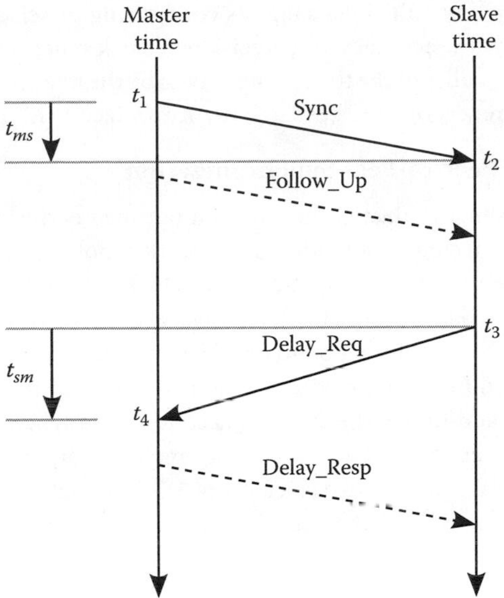

FIGURE 21.7 Synchronization principle of IEEE 1588.

the timestamp to the slave. Devices that are capable of providing the timing information directly within the event message are called one-step clocks. A clock that requires the combination of an event message (e.g., `Sync` message) and a subsequent general message (e.g., `Follow_Up` message) is termed two-step clock.

Similarly, the slave transmits a `Delay_Req` message and takes the egress timestamp t_3. The master receives this message at t_4 and submits t_4 within the `Delay_Resp` message back to the slave. When the slave has received at least two `Sync` messages, it can estimate the clock skew by dividing the timespans elapsed in the master and the slave, and in consequence, it can adjust its rate to the master. When the master and slave clocks have approximately the same clock rate, then only the initial clock offset t_{off} exists due to different bootstrapping times. The offset t_{off} can be exactly determined by the timestamps t_1 to t_4 and the link delays t_{ms} and t_{sm} by

$$t_{off} = \frac{(t_2 - t_1) - (t_4 - t_3)}{2} - \frac{t_{ms} - t_{sm}}{2}. \quad (21.16)$$

21.4.2.2 Asymmetry Correction

The last term in (21.16) is referred to as delayAsymmetry = $(t_{ms} - t_{sm})/2$. Commonly used media, such as switched Ethernet, wireless networks, or wide-area networks, impose different delays for the receive and transmit path and thus create asymmetry. Although asymmetry fields in the protocol have been introduced in PTP version 2, asymmetry is not detectable by PTP. This limitation is not only applicable to PTP but applies to all packet-based time transfer protocols as only round-trip measurements can be performed. Thus, only known asymmetries can be compensated. The delays t_{ms} and t_{sm} are composed of the meanPathDelay = $(t_{ms} + t_{sm})/2$ and the delayAsymmetry as follows:

$$t_{ms} = \text{meanPathDelay} + \text{delayAsymmetry}, \quad (21.17)$$

$$t_{sm} = \text{meanPathDelay} - \text{delayAsymmetry}. \quad (21.18)$$

If the asymmetry is unknown, IEEE 1588 suggests configuring or selecting components that minimize the effects of asymmetry. The accuracy and precision of clock synchronization using PTP therefore depends on three factors: the quality of the timestamps taken by the master and slaves, the stability of the oscillators providing the reference for the clocks to be synchronized, and the knowledge of asymmetry.

21.4.2.3 End-to-End and Peer-to-Peer Synchronization

The delay request–response synchronization mechanism outlined earlier is applicable for ordinary or boundary clocks that are connected by any kind of network topology. Ordinary clocks have a single clock port and may either be master or slave. Boundary clocks are in principle ordinary clocks with multiple ports, where the clock data sets and the local clock are shared to all ports but each port runs its individual protocol state machine. Thus, a boundary clock may serve as slave clock on one port and as master clock on the other port. Boundary clocks are therefore one possible network element to connect multiple timing networks. In addition to the delay request–response synchronization end-to-end mechanism, PTP version 2 introduced the peer-to-peer clock synchronization. The peer-to-peer mechanism is used to measure the link delay between two clock ports. With the gathered information, ordinary and boundary clocks can synchronize using the peer delay information and the information contained in the `Sync` and `Follow_Up` messages.

21.4.3 Transparent Clocks

The accuracy and precision of PTP relies on the timely delivery of event messages; that is, event messages should have constant transmission latency and the latencies in both directions should be identical or known. Real-world equipment often does not offer a constant latency for a number of reasons. For instance, in an Ethernet store-and-forward switch, the entire frame is received, checked, and then forwarded to the destination port queue. Depending on the processing speed of the switching fabric, the queue allocation before the current frame, and medium access delay, the residence time in the switch may vary. This variable delay leads to timestamp jitter and in consequence degrades the synchronization accuracy.

PTP solves this issue by transparent clocks. A transparent clock is a network device (e.g., a switch, router, or gateway) that measures the residence time of PTP event messages within the device and provides this information to clocks receiving these event messages. As a result, the slave clocks still receive event messages with a variable delay, but the magnitude of the variable delay is measured and sent to the slave clocks in order to compensate for the packet delay variation. Figure 21.8 shows the block diagram of a transparent clock with four ports. Each port is connected to a timestamping unit that takes a timestamp for egress and ingress event messages. The timestamping units are supplied by the time indication from the local clock as depicted in the center of Figure 21.8. Messages that have passed the timestamping unit are put into the switch, router, or gateway connecting the ports as indicated with dashed arrows.

In IEEE 1588, two types of transparent clocks are defined: end-to-end transparent clocks and peer-to-peer transparent clocks. An end-to-end transparent clock supports the use of the end-to-end delay measurement mechanism between ordinary and boundary clocks. It timestamps all ingress event messages and forwards them through the switching element to the egress port. The egress port takes another timestamp. The difference between these timestamps is the residence time within the transparent clock. In a one-step, end-to-end transparent clock, the PTP correction field within the event message is directly updated, while a two-step transparent clock measures the residence time of the event message first and then updates the corresponding `Follow_Up` message. One-step transparent clocks have the advantage that the slave clock receives the correct timestamp (the message timestamp plus the correction field) within the event message without any additional latency. Yet, such a transparent clock must be capable of inserting the correction field on the fly. A two-step transparent clock, on the other hand, needs to find the `Follow_Up` frame matching to the event message and modify the correction field. This processing may lead to larger delays and thus reduced synchronization performance as the slave clock cannot adjust its clock before the `Follow_Up` has been received.

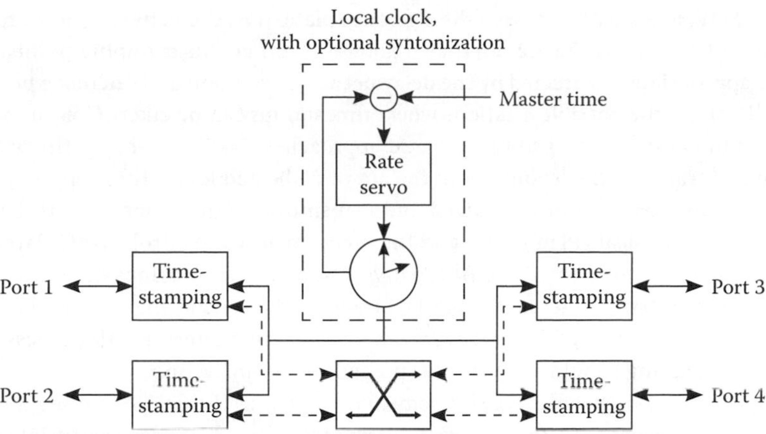

FIGURE 21.8 Block diagram of a PTP transparent clock.

When using end-to-end transparent clocks, the propagation time compensation is done from the master to the slave device without characterizing the intermediate delays. A change in the routing between master and slave changes the path delay between master and slave and therefore temporarily degrades the synchronization until the synchronization has resettled. Thus, end-to-end transparent clocks are particularly suitable for networks with nonvariable routing. Note that end-to-end transparent clocks violate (strictly speaking) most network standards (e.g., IEEE 802.3 for Ethernet) as it is commonly not allowed to modify the contents of a frame without updating the source address to the modifying device. By keeping the source address from the PTP master, transparent clocks appear to be totally transparent to master and slave clocks.

Peer-to-peer transparent clocks additionally provide corrections for the propagation delay of a link that receives PTP event messages. When using the peer-to-peer delay measurement mechanism, propagation delays are corrected on a link basis instead of end to end. Thus, any change in the routing immediately changes the path delay correction, and therefore, the synchronization does not degrade in case of dynamic packet routing.

The residence time measurement in transparent clocks is calculated by the difference of the egress and ingress timestamps generated from the local clock. As any clock offset to the master cancels out in this calculation, transparent clocks do not need to be synchronized to the master. Yet, if the clock rates of the transparent clock and the master clock differ, then the residence times are scaled incorrectly, and an error is created. Depending on the residence time, the rate difference, and the desired synchronization accuracy, the error magnitude may be anything from negligible to unacceptably large (e.g., for large rate differences and/or long residence times). Two solutions for this issue are possible: Master and transparent clocks can be equipped with oscillators that have a very low frequency offset with respect to each other, such as oven controlled oscillators. If standard crystal oscillators are used, frequency offsets of about 50 ppm are typical. The transparent clock may also adjust its local clock frequency to the received upstream synchronization packets from the master clock. If the transparent clock uses a free-running local clock, numerical rate estimates can be used for the rate correction as well. The optional rate synchronization is also depicted in Figure 21.8. To differentiate offset synchronization from rate synchronization, the latter is termed syntonization.

21.4.4 Timestamp Generation

The quality of the timestamps generated for all ingress and egress messages is of paramount importance for PTP. All ingress and egress PTP event messages should be timestamped at a common timestamping point. According to Section 7.3.4.1 of IEEE 1588 [4], the timestamping point is the beginning of the first symbol after the SOF delimiter has passed the reference plane (i.e., the network connector or antenna). Nevertheless, note 1 in Section 7.3.4.2 of [4] also allows for other timestamping points given that the timestamps are appropriately corrected by the delay between detection and reference plane.

Figure 21.9 illustrates the possible locations where timestamps can be taken. Commonly, timestamping is classified into hardware- and software-based approaches. Hardware-based timestamping generates the receive and transmit timestamps with the aid of dedicated logic. These approaches may either estimate the timestamp based on the received and transmitted signal (Signal TS in Figure 21.9), the interface between the physical (PHY) layer and the medium access control (MAC) layer, or the interface between the MAC and the software. Alternatively, software timestamps can be taken within the device driver, when processing interrupt requests, or within the synchronization application. With the increasing number of network layers, the jitter increases due to nondeterministic processing delays, and therefore, the timestamping should be as close as possible to the medium.

The majority of clock synchronization implementations using IEEE 1588 aiming at high accuracy have been based on Ethernet. As Ethernet offers standardized media-independent interfaces between PHY and MAC (e.g., MII, GMII), universal hardware timestamping methods have been developed in the last years able to minimize the timestamp jitter. These Ethernet-based solutions reach accuracies in the nanosecond range or even below [15].

Particular care for the correction to the timestamp reference plane needs to be taken when using software timestamping as illustrated in Figure 21.10. When using the PTP numbering for egress timestamps, t_1 and t_3 (cf. Figure 21.7), let the transmission at master or slave be issued in software at time $\widehat{t_1}$ or $\widehat{t_3}$. After a certain latency (p_1, p_3), the first bit of the message is put on the line. After the propagation delay (t_{ms}, t_{sm}), the first bit of the message arrives at the receiver. Finally, when the entire frame is received, the frame is passed through the network stack, and the payload is presented to the synchronization application (receive complete).

In software timestamping systems, a timestamp at the transmitter is taken when handing a packet over to the network stack of the OS, and at the receiver when retrieving the packet from the OS. This is shown by the egress timestamps $\widehat{t_1}$ and $\widehat{t_3}$, and the ingress timestamps $\widehat{t_2}$ and $\widehat{t_4}$ in Figure 21.10. Thus, p_1 and p_3 are dependent on the implementation of the network stack in the master or slave. As the ingress timestamps $\widehat{t_2}$ and $\widehat{t_4}$ are usually generated after the entire frame is received, p_2 and p_4 contain the transmission delay as well. When the link speed is asymmetric, the transmission delays and therefore the processing delays are different ($p_2 \neq p_4$), leading to a clock offset. Thus, in any case, it is important to correct all timestamps to the reference plane; that is, for software-based timestamping, the frame length, link

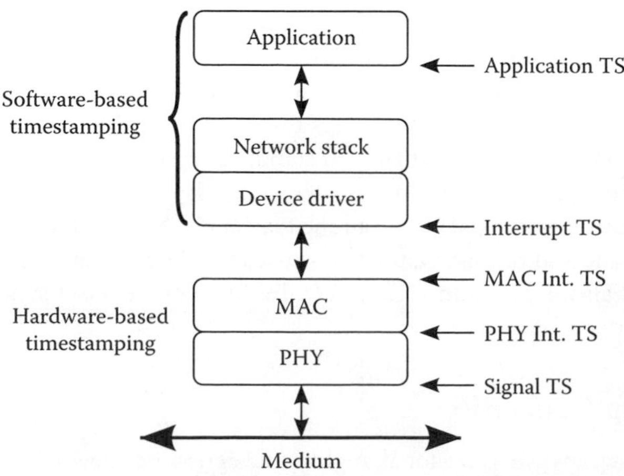

FIGURE 21.9 Locations for timestamping in a clock.

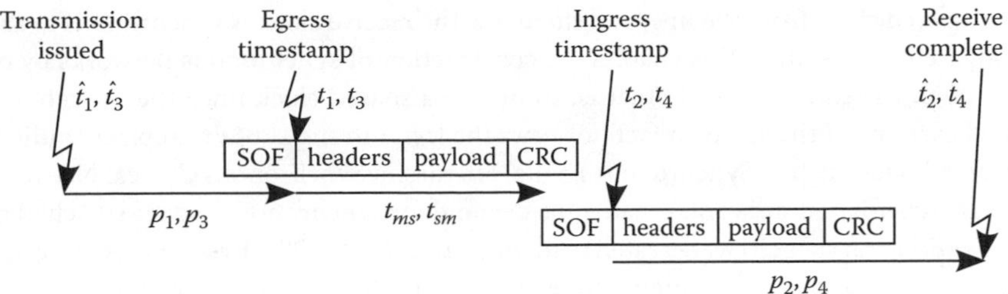

FIGURE 21.10 Software timestamping latencies from master to slave.

speed, and other parameters need to be taken into account. In fact, the issue of correcting timestamps to the reference plane also applies to hardware timestamping. For instance, the DP83848 Ethernet PHY from Texas Instruments [16] has a receive latency of 255 ns and a transmit latency of 50 ns. Thus, a synchronization bias is induced when using PHYs with different latencies ($p_2 - p_3 \neq p_4 - p_1$), for example, different PHY devices.

21.4.5 Implementation and Application Issues

The performance of clock synchronization can be quantified by the statistics of the offset $e = t_s - t_m$ between the slave clock t_s and the master clock t_m. To assess the offset between master and slave, it is common practice to use a one-pulse per second (PPS) signal generated at the master and slave. The PPS signal is asserted every full second, and therefore, the time difference between the rising edges of the PPS signals from the master and slave can be used to determine the clock offset at every full second. Commonly used statistics are the mean clock offset μ and the standard deviation σ, also termed jitter. μ reflects any bias and asymmetry in the synchronization path, while σ is influenced by the quality of the timestamps, the synchronization interval, the stability of the local clocks, as well as the control servos.

It has to be outlined that the clock error measured by the slave, that is the difference between the slave clock and the received timestamp from the master, is not an appropriate measure for the synchronization performance. The clock error is used to steer the slave clock and is filtered by the slave servo to generate the PPS signal. In addition, as the slave is unaware of any asymmetry or bias in the system, it cannot detect clock offsets. Thus, an external unbiased measurement instrument, such as a time interval measurement instrument or an oscilloscope, is required to determine the synchronization performance. For purely software-based synchronization systems, nevertheless, PPS signals might not be available, and the clock error needs to be used for a performance assessment.

21.4.5.1 Synchronous Ethernet and PTP

IEEE 802.3 Ethernet [17] can be considered as the state-of-the-art communication technology for home, office, and even industrial networks. It is used to build local area networks using either fiber or copper cabling with transmission rates of up to 100 Gbit/s. Copper-based Ethernet is widely used and allows for cheap copper cabling and simple rewiring using patch panels. The two most common implementations are 100Base-TX and 1000Base-T. Both 100Base-TX and 1000Base-T transmit the information with a symbol rate of 125 MS/s. 100Base-TX uses a single cable pair in each direction and encodes one bit per symbol using the Multilevel Transmission Encoding-3 levels (MLT-3). In contrast, 1000Base-T uses all four wire pairs of an Ethernet cable simultaneously in both directions and transmits its information using pulse-amplitude modulation (PAM) with five levels.

Besides the modulation-specific differences, 100Base-TX and 1000Base-T differ in the interface to the physical layer transceiver (PHY) and thus require different hardware timestamping logic. For 100Base-TX, the most common interfaces are the media-independent interface (MII) and the reduced media-independent interface (RMII). A particular advantage of MII is that the receive clock of the PHY is simply the recovered clock from the line, and therefore, the receive clock is synchronous to the transmitter on the other end of the line. This enables the construction of synchronous networks by propagating one clock through the entire network, that is, similar to a shared clock line. The concept of exploiting the embedded carrier in Ethernet is in fact not new. The trend in the 2000s to replace traditional circuit switching technologies, such as Synchronous Digital Hierarchy/Synchronous Optical Networking (SDH/SONET), with packet-based networks created the demand for synchronous packet-switched networks.

Synchronous Ethernet (SyncE) was created to fill this gap with the ITU-T recommendation G.8262 [18]. The idea behind SyncE is to provide high frequency stability through syntonization with a clock distribution concept that resembles a cascade of PLLs. The master node, which is connected to an external reference clock signal, embeds the clock into the Ethernet transmit signal. Switches recover the clock, clean it using PLLs, and use it to embed the clock in the transmit signal until it reaches the final destination [19]. Synchronization, if required, can be achieved by protocols on top of SyncE like PTP. The SyncE concept was well accepted by telecom carriers as the core network required no major changes for the introduction of SyncE [20]. When using SyncE or any other form of frequency distributing Ethernet network, the syntonization is maintained through PLLs, and therefore, the clock variance is mainly dependent on the PLLs' phase noise. Nevertheless, the clock bias is unaltered by the introduction of a synchronous network.

A particular issue in both 100Base-TX and 1000Base-T is the asymmetry of the PHY devices and the cabling. The propagation delays of the wire pairs of a category 5E Unshielded Twisted Pair (UTP) cable are permitted to differ by up to 50 ns according to the ANSI/TIA/EIA 568-B specification. This results in an unresolvable synchronization offset of up to 25 ns per link. The asymmetry of real cables is often much lower, such as about ±2 ns for a 60 m cable as measured in [21]. Still, the asymmetry-induced synchronization offset can be considered as one of the main reasons why PTP with hardware timestamping can hardly reach accuracies below 1 ns over copper-based Ethernet. Commercial off-the-shelf PTP equipment using hardware timestamping synchronizes with residual offsets in the range of 10–100 ns depending on the number of transparent clocks, timestamping resolution, synchronization rate, and lastly the asymmetry, which is not detectable by PTP [22]. When using asymmetry determination approaches, such as line swapping [21], or asymmetry calibration, for instance, by the fiber-based WhiteRabbit system [23], the residual clock offsets can be reduced to below 0.5 ns.

PTP can also be used with software timestamping and standard switches instead of PTP transparent or boundary clocks. For synchronization solutions based on software timestamps, the clock offset induced by asymmetric propagation delays may not be relevant due to the large jitter of the software timestamps. Software-based synchronization approaches may reach synchronization accuracies of about 10 μs [24].

21.4.5.2 PTP in Substation Automation Systems

Modern SASs are the core of the smart grid deployment. SASs are based on networked equipment providing the needed protection, remote monitoring, and measurement functions for correct grid operation. The primary and secondary substations are generally structured following the IEC 61850 standard, which describes the communication inside the SAS and sets the base for new interoperating devices in an Ethernet network infrastructure. More recently, the need for clock synchronization among the SAS components for timestamping of critical events (faults and related protection actions) or synchrophasor measurement [25] generated a very high interest in the IEC 61850 community for distributed synchronization protocols like NTP or PTP. Among the two standards, PTP is preferred, since the IEEE C37.238 synchronization profile [26] for power applications requires a synchronization accuracy of end devices below 1 μs. The reason of such constraints is the object-oriented architecture of IEC 61850, where different functions can be separated in different devices inside the substation: the execution of measurement functions like the estimation of synchrophasors is done by acquiring voltage and current signals by means of distributed devices (called Merging Unit—MU); hence, the synchronization performance of the MU became a very critical aspect. Today, new SASs already offer PTP support by means of dedicated hardware, while integration of PTP over existing IEC 61850 SASs (which use NTP, SNTP, or IRIG-B) is basically a trade-off between installation costs and performance, since it may require the replacement of already installed Ethernet infrastructure.

21.5 Conclusion

Distributed systems need clock synchronization as a basic network feature. In packet-based networks, which are the most commonly used network concepts in IT and automation environments, this always boils down to determining the message delays over the network and taking them into account when adjusting the local clock. Historically, NTP and PTP emerged in different application contexts, but their basic operation is very similar as they both essentially rely on the two-way synchronization principle using round-trip delay measurements. Yet, there are many differences in the details, which make the two suitable for different domains. NTP, with its strictly distributed and fault-tolerant approach of finding a good agreement between several time servers, is ideally suited for large-scale networks such as the Internet with unclear and changing topologies. In such environments, resynchronization does not occur often. Evidently, the attainable synchronization accuracy is not comparable with that of PTP, but it is sufficient for most typical IT applications.

PTP, on the other hand, using a simpler master–slave approach, is better suited for smaller-scale networks with clear, well-maintained topology where synchronicity is more essential than in IT environments and thus resynchronization takes place much more frequently. Provisions to achieve a high accuracy are significantly more elaborated such as transparent clocks to measure the residence time of packets in network elements like switches. The extensions included in the recent version of PTP, in particular the boundary clock concept, in principle allow taking PTP beyond network borders and thus widen the spatial extension of network synchronization. Nevertheless, it is to be expected the PTP will remain the synchronization protocol for automation applications and not make NTP obsolete. The two will coexist also in the future, possibly complementing each other in more complex applications.

References

1. L. Lamport, Time, clocks, and the ordering of events in a distributed system, *Communications of the ACM*, 21(7), 558–565, 1978.
2. BIPM, SI brochure, Section 2.1.1.3, July 2013. [Online]. Available: http://www.bipm.org/en/si/si_brochure/chapter2/2-1/second.html (accessed July 24, 2013).
3. D. Mills, Network time protocol (version 2) specification and implementation, RFC 1119 (Standard), Internet Engineering Task Force, Technical report 1119, September 1989. [Online]. Available: http://www.ietf.org/rfc/rfc1119.ps (accessed August 31, 2011).
4. IEEE, IEEE standard for a precision clock synchronization protocol for networked measurement and control systems, IEEE Std, December 2011.
5. Y. Wang, F. Nunez, and F. Doyle, Energy-efficient pulse-coupled synchronization strategy design for wireless sensor networks through reduced idle listening, *IEEE Transactions on Signal Processing*, 60(10), 5293–5306, 2012.
6. P. Hetzel, Die Zeitsignal- und Normalfrequenzaussendungen der PTB über den Sender DCF77: Std 1982, Oldenburg Publishers, pp. 42–57, 1983.
7. D. Engeler, Performance analysis and receiver architectures of DCF77 radio-controlled clocks, *IEEE Transactions on Ultrasonics, Ferroelectrics and Frequency Control*, 59(5), 869–884, 2012.
8. ARINC, Navstar GPS space segment/navigation user interfaces, IS-GPS-200 Rev. D, ARINC Engineering Services, El Segundo, CA, Technical report, March 2006.
9. E. Kaplan and C. Hegarty, *Understanding GPS: Principles and Applications*, Artech House, London, U.K., 2005.

10. IEEE, *IRIG Serial Time Code Formats*, Timing Committee Telecommunications and Timing Group Std, May 1998.
11. C. De Dominicis, P. Ferrari, A. Flammini, S. Rinaldi, and M. Quarantelli, On the use of IEEE 1588 in existing IEC 61850-based SASs: Current behavior and future challenges, *IEEE Transactions on Instrumentation and Measurement*, 60(9), 3070–3081, 2011.
12. D. Mills, J. Martin, J. Burbank, and W. Kasch, Network time protocol version 4: Protocol and algorithms specification, RFC 5905 (Standard), Internet Engineering Task Force, Technical report, June 2010. [Online]. Available: http://www.ietf.org/rfc/rfc5905.txt (accessed August 23, 2013).
13. B. Haberman and D. Mills, Network time protocol version 4: Autokey specification, RFC 5906 (Standard), Internet Engineering Task Force, Technical report, January 2010. [Online]. Available: http://www.ietf.org/rfc/rfc5906.txt (accessed August 21, 2013).
14. D. Mills, Simple network time protocol (SNTP) version 4 for IPv4, IPv6 and OSI, RFC 4330 (Standard), Internet Engineering Task Force, Technical report, January 2006. [Online]. Available: http://www.ietf.org/rfc/rfc4330.txt (accessed August 21, 2013).
15. M. Lipinski, T. Wlostowski, J. Serrano, P. Alvarez, J. Gonzalez Cobas, A. Rubini, and P. Moreira, Performance results of the first white rabbit installation for CNGS time transfer, in *IEEE International Symposium on Precision Clock Synchronization for Measurement, Control and Communication*, IEEE, Piscataway, NJ, September 2012, pp. 1–6.
16. D. Rosselot, DP83848 and DP83849 100MB Data latency, Application Note 1507, August 2006.
17. IEEE, IEEE 802.3 Local and metropolitan area networks specific requirements Part 3: Carrier sense multiple access with collision detection (CSMA/CD) access method and physical layer specifications, IEEE Std, 2002.
18. D. Bui and M. Le Pallec, From ethernet to synchronous ethernet, in *International Conference on Signal Processing and Communication Systems (ICSPCS)*, IEEE, Piscataway, NJ, December 2008, pp. 1–5.
19. J.-L. Ferrant, M. Gilson, S. Jobert, M. Mayer, M. Ouellette, L. Montini, S. Rodrigues, and S. Ruffini, Synchronous ethernet: A method to transport synchronization, *IEEE Communications Magazine*, 46(9), 126–134, September 2008.
20. S. Rodrigues, IEEE-1588 and synchronous ethernet in telecom, in *IEEE International Symposium on Precision Clock Synchronization for Measurement, Control and Communication*, IEEE, Piscataway, NJ, October 2007, pp. 138–142.
21. R. Exel, T. Bigler, and T. Sauter, Asymmetry mitigation in IEEE 802.3 ethernet for high-accuracy clock synchronization, *IEEE Transactions on Instrumentation and Measurement*, 63(3), 729–736, March 2014.
22. K. Han and D.-K. Jeong, Practical considerations in the design and implementation of time synchronization systems using IEEE 1588, *IEEE Communications Magazine*, 47(11), 164–170, November 2009.
23. F. Lamonaca and D. Grimaldi, Trigger realignment by networking synchronized embedded hardware, *IEEE Transactions on Instrumentation and Measurement*, 62(2), 38–49, 2013.
24. J. Ridoux and D. Veitch, Ten microseconds over LAN, for free (extended), *IEEE Transactions on Instrumentation and Measurement*, 58(6), 1841–1848, June 2009.
25. IEEE, IEEE standard C37.118.1-2011 for synchrophasor measurements for power systems (Revision of IEEE Std C37.118-2005), IEEE Std, December 2011.
26. IEEE, IEEE standard profile for use of IEEE 1588 precision time protocol in power system applications, IEEE Standard C37.238-2011, 2011.

IV

Accessing Factory Floor Data

22 Linking Factory Floor and the Internet *Thilo Sauter* **22-1**
Introduction and Historical Background • Interconnection Approaches • Control Data Tunneling • Design Considerations for a Gateway • Gateway Access from the Internet • Role of Industrial Ethernet • Summary • References

23 MTConnect *Dave Edstrom* **23-1**
Classic N2 Problem in Discrete Manufacturing • MTConnect: An Overview • MTConnect Standard • Types of MTConnect-Enabled Devices • Technical Pillars of MTConnect • MTConnect's Powerful Data Dictionary • Your First MTConnect Hello World Application • Summary • Reference

24 Standard Message Specification for Industrial Automation Systems *Karlheinz Schwarz* **24-1**
Introduction • MMS Client/Server Model • Virtual Manufacturing Device • Interfaces • Environment and General Management Services • VMD Support • Domain Management • Program Invocation Management • MMS Variable Model • Resume • References • Websites

25 Extending CEA-709 Control Networks over IP Channels *Dietmar Loy and Stefan Soucek* **25-1**
Introduction and Overview • ANSI/CEA-852 Standard • System Components • Data Communication • Management • Security • Applications • Conclusion • References

26 Web Services for Embedded Devices *Vlado Altmann, Hendrik Bohn, and Frank Golatowski* **26-1**
Introduction • Device-Centric SOAs • DPWS Inside Out • Web Service Orchestration • Software Development Toolkits and Platforms • DPWS in Use • Conclusion • References

22

Linking Factory Floor and the Internet

22.1 Introduction and Historical Background **22-1**
22.2 Interconnection Approaches **22-3**
Protocol Tunneling • Gateways
22.3 Control Data Tunneling **22-6**
Network Parameters for Control Data Tunneling • Performance and Application Classes
22.4 Design Considerations for a Gateway **22-10**
Services • Gateway Structure • Data Representation
22.5 Gateway Access from the Internet **22-14**
Simple Network Management Protocol • LDAP • Structured Query Language • Web-Based Approaches • OPC
22.6 Role of Industrial Ethernet **22-20**
22.7 Summary **22-21**
References **22-22**

Thilo Sauter
Danube University Krems

22.1 Introduction and Historical Background

One of the key issues in automation is the exchange of information, and substantial efforts were devoted in the past to introduce communication systems at all levels of the automation hierarchy. The original goal was geared toward *vertical* data transfer between the levels. Nevertheless, as appropriate networking concepts were still in their infancy or entirely missing at that time, academic and industrial activities focused on the development of dedicated automation networks and thus primarily on *horizontal* integration aspects. Many different networks emerged: fieldbus systems of all kinds, Ethernet-based networks, and wireless concepts tailored to automation purposes [1]. Even though early publications did foresee the use of gateways between the individual networks for protocol conversion [2], true integration was not a primary goal for system developers, and so the lower automation levels remained largely separated from the upper, office-oriented levels in terms of data exchange and application integration [3].

Today, interconnection essentially boils down to a linkage between arbitrary automation systems—which may involve also dedicated automation networks—and networks based on the Internet Protocol (IP). For the sake of simplicity, we will stick with the familiar terms fieldbus and Internet, however keeping in mind that the fieldbus is not a must and that the Internet can as well be an intranet or any other network governed by IP and the protocol suite on top of it. The benefits of such an interconnection are two major, nonetheless interwoven aspects: remote access to automation systems and the promise of easy integration of automation data in a user-friendly environment. In other words, there are two reasons why a fieldbus/Internet connection can be beneficial:

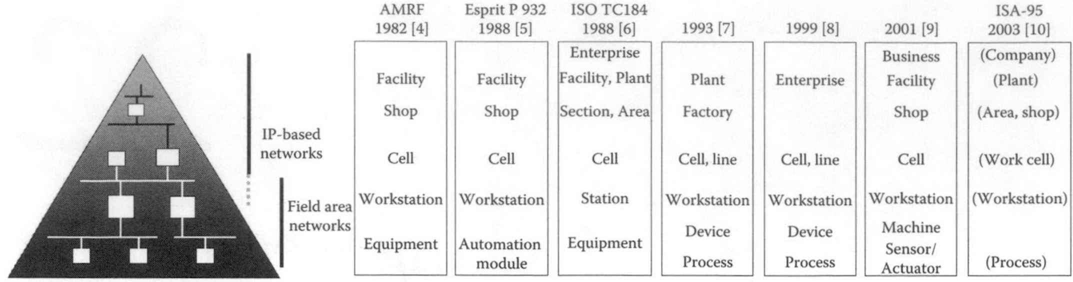

FIGURE 22.1 CIM pyramid and network types used in the various automation levels.

- To extend the physical dimensions of an automation system. Mostly for historical reasons, the extension of a typical automation network is rather limited due to length restrictions in fieldbus segments and the lack of routing capabilities. If an Internet infrastructure is available, it can be used as a kind of backbone to connect distant segments of the installation.
- To provide vertical integration. In the automation domain, this widely used term essentially means bringing information from an automation system into a framework used in the office domain, where it can be used not only for data acquisition but also for strategic operations such as system management or resource planning.

Especially, the second aspect is currently also connected to the growing use of Ethernet in industrial automation. However, it must be stated that the idea of vertical integration is nothing entirely new. The roots date back to the 1970s when the computer-integrated manufacturing (CIM) concept was developed. The hierarchical model was an early attempt to structure the information flow within a company. There are many ways to draw this pyramid, and the names of the individual levels differ according to the application area (Figure 22.1). But the ultimate goal has always been the same: to provide transparent data exchange between the various levels.

The first attempts to implement the CIM concept were more or less futile, for a number of mostly technological reasons. On the one hand, the protocols designed for the communication inside the automation system—essentially the Manufacturing Automation Protocol (MAP) as a full-fledged implementation of the ISO/OSI model—were too complex for implementation. On the other hand, the progress of microelectronics as a technological backbone had not yet been far enough to provide sufficient computing resources at a reasonable price. Therefore, integration of CIM on the field level turned out to be nearly impossible. In addition, fieldbus systems as low-level automation networks were only at a very early stage of their evolution; hence, there was, from a networking point of view, a missing link between the actual source of the process data and the already existing company networks.

Figure 22.2 shows a few milestones in the evolution of both IT and automation networks and demonstrates how the situation changed over the years. Partly driven by the insight that MAP was too clumsy to use (but still following its basic ideas), partly driven by application-specific needs, fieldbus systems as dedicated automation networks emerged. They filled the networking gap at the lowest levels of the automation pyramid, providing an anchor point for subsequent integration efforts. The great leap forward for integration itself came with the success of the Internet and, more precisely, with the invention of the World Wide Web. While in the fieldbus domain, a large variety of approaches still exists even after a lengthy and cumbersome standardization process [1], the office world is dominated by the IP suite and a small number of well-known, widely used applications.

The reason why the old idea of vertical integration was revived in recent years was merely a psychological one. As the timeline in Figure 22.2 shows, the Internet and its basic technologies have been available (and used) for a long time. However, it took the sheer simplicity of the web browser as a ubiquitous tool to boost the acceptance of the Internet and to secure the dominant role of the IP suite in the office

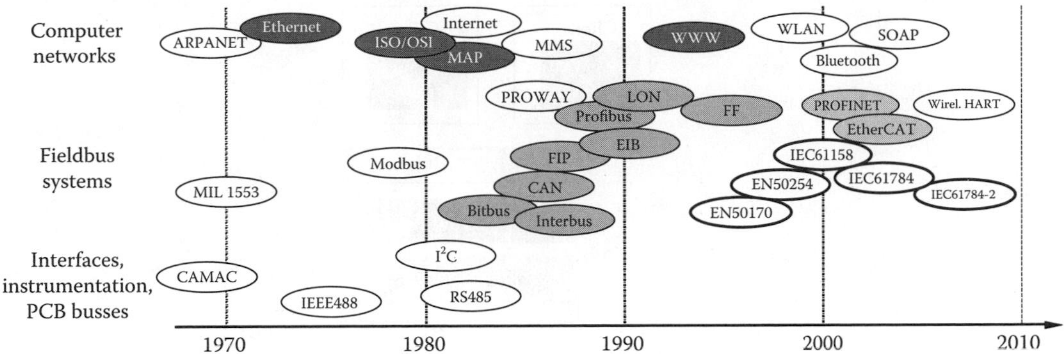

FIGURE 22.2 Evolution of networks in automation and information technology.

world. From the user's point of view, a web browser allows access to distant data in a nearly trivial manner. Hence, it is no wonder that the easy navigating through hypertext documents was soon adopted as a model also for the remote access to automation data. Consequently, many solutions of fieldbus/Internet connectivity rely on web technology and a web browser interface. The great success of IP-based networks also significantly simplified the automation pyramid. Figure 22.1 shows that actually only two network technologies are left in today's automation scenarios: dedicated field-level networks to cope with the special demands of automation and IP-based networks on the upper levels to allow for data transfer over large distances and easy integration into office automation. Still, the impression that *the Internet* naturally entails a uniform way of remote access to automation networks is deceptive. The user interface is only one aspect; the underlying mechanisms and data structures are another. In fact, when it comes to implementing an interconnection between IP-based networks and an automation system fieldbus, there is a surprising variety of possibilities even in the so much *standardized* Internet environment. In particular, the web-based approach is by no means the only one.

22.2 Interconnection Approaches

From an architectural point of view, there are two main possibilities to achieve a fieldbus/Internet interconnection, both of which are being used in practice [11]:

- Tunneling of one protocol over the other
- Providing protocol and service translation via a gateway

As far as the network topology is concerned, both approaches are very similar. In both cases, a central access point between the networks is either required or at least reasonable. What differs is the way the information is processed by this link or, more accurately, how the processing is distributed between the devices involved in the communication.

22.2.1 Protocol Tunneling

In communication networks, tunneling essentially means that data frames of one protocol are simply wrapped into the payload data of another protocol without any modification or translation. The tunneling approach in the automation context falls into two categories. The currently more common solution is to encapsulate IP packets in fieldbus messages, which also opens a channel for the upper layer protocols required for, for example, direct web access to the field devices [12–14], while in most cases not foreseen in the beginning, this possibility has recently been included in several fieldbus systems—a move that has to be seen mainly in connection with the discussion about Ethernet in automation. At first glance, IP tunneling does provide an easy way to achieve vertical integration, as Figure 22.3 shows. The

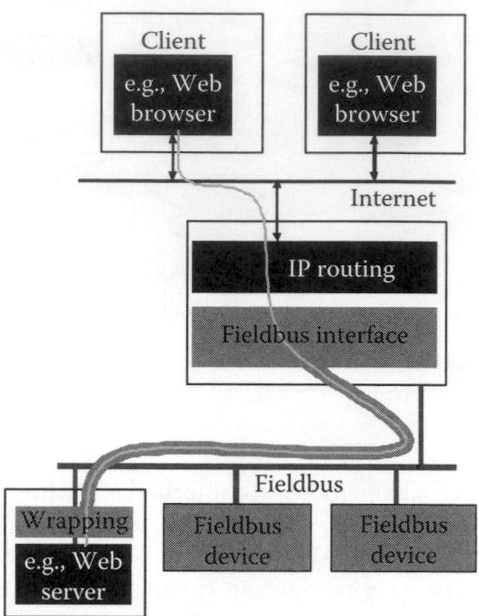

FIGURE 22.3 Structure of IP tunneling over a fieldbus.

field devices run an IP-based service such as a web server providing data for a respective application in the Internet, which can directly access the device or process data.

A second look, however, reveals that the integration is not at all that easy. There are a few critical points to be considered to properly design such a system.

Computing resources at the field device. First of all, it is evident that the field device as the endpoint of a tunnel must be able to run the complete IP stack with all additional protocols required for the particular application. In addition, memory space is required for the actual application program handling the data that are to be accessed via IP. With today's embedded systems, all this may not be overly problematic, as there are very lightweight protocol implementations available (typically at the expense of detailed error handling). At any rate, this often requires additional hardware at the device and might be a cost factor especially for simple devices like sensors or actuators.

Traffic handling at the access point. The node that connects the IP-based network and the fieldbus has to encapsulate the IP packets and forward them to the appropriate fieldbus node, where they are unwrapped and handed over to the IP stack. To this end, the access point has to act as an IP router. More than that, it has to map the IP connections to the respective fieldbus communication channels. This requires an address translation on the fieldbus side of the access point. On the Internet side of the node, an additional translation can become necessary if the IP addresses in the fieldbus are not public ones, but taken from a private address range, which is the usual network configuration. The access point, thus, acts as a simple masquerading firewall with network address translation (NAT). Unlike the case where the IP addresses in the fieldbus are public, the fieldbus as a whole is accessible only via one single IP address, and the distinction between the individual devices (e.g., web servers) hidden behind the firewall can only be done via port forwarding. This means that the services are not reachable through their well-known standard ports, but via dedicated ones that have, of course, to be configured appropriately at the client side in the Internet.

Performance issues. IP packets tend to be rather large, whereas fieldbus data frames are usually optimized for the transmission of only small pieces of data at a time. Squeezing IP packets in the small payload data fields of a fieldbus message, therefore, requires segmentation, which in turn increases the transmission time. An extreme example is Interbus, where IP traffic can

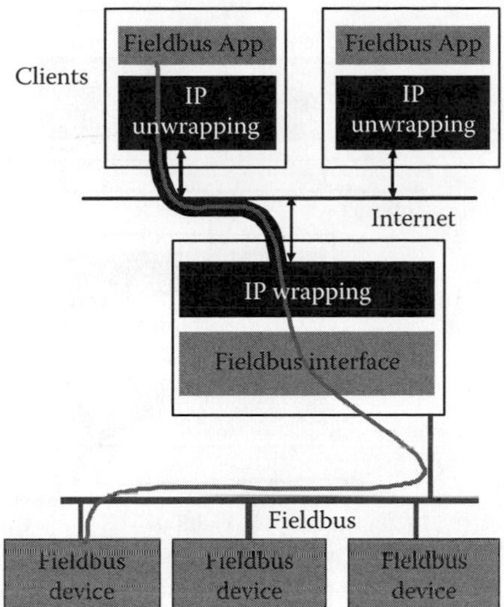

FIGURE 22.4 Structure of a fieldbus tunnel over the Internet.

be transported in the parameter channel [15]. This channel has only a small capacity of, say, 8 bytes per cycle in order not to interfere with the real-time process data, and 1 byte is needed for control purposes. Depending on the size of the network and the selected baud rate, one cycle may take a few milliseconds. As an IP packet usually is of the size of several hundred bytes, it is clear that the performance is rather limited. In other fieldbus systems, the data frames are large enough to avoid excessive segmentation. At any rate, the efficiency of IP over fieldbus tunneling depends strongly on the fieldbus.

The alternative way of tunneling works the other way around. Fieldbus messages are wrapped into IP packets (or those of other protocols such as Transmission Control Protocol [TCP], depending on how the tunnel is set up) and sent to a distant network node (Figure 22.4). From the viewpoint of vertical integration, the disadvantage of this approach is immediately visible. The fieldbus data must be interpreted at the client side, which requires a fieldbus-specific application or at least an experienced user with detailed knowledge about the particular automation system.

This approach does not fulfill the requirements of user-friendliness posed by the idea of vertical integration. In fact, fieldbus tunnels over the Internet are rather used to conveniently connect remote segments of an installation [16]. A wide field of application for this concept are network-based control systems, for instance, in building automation [17,18]. It has to be noted, though, that the interconnection of two fieldbus segments over an Internet tunnel is not fully transparent but has to cope with timing problems introduced by the processing of the data packets and the transmission delays, which has an impact on the design of network-based control systems [19–21]. Therefore, this concept is useful only if the fieldbus protocol provides some routing functionality, where timing requirements can be relaxed.

22.2.2 Gateways

The alternative approach to achieve fieldbus/Internet connectivity is to design the central access point as a gateway. Like with the tunneling approach, the gateway is a full member of the fieldbus on the one side and can be accessed through IP-based mechanisms via the Internet (Figure 22.5). What is different, though, is the way the information exchange is handled. In the IP tunneling approach, the client connects directly to the server running on the field device. The access point just forwards the IP traffic but is not further involved in the data transfer. In the gateway approach, the access point takes the role

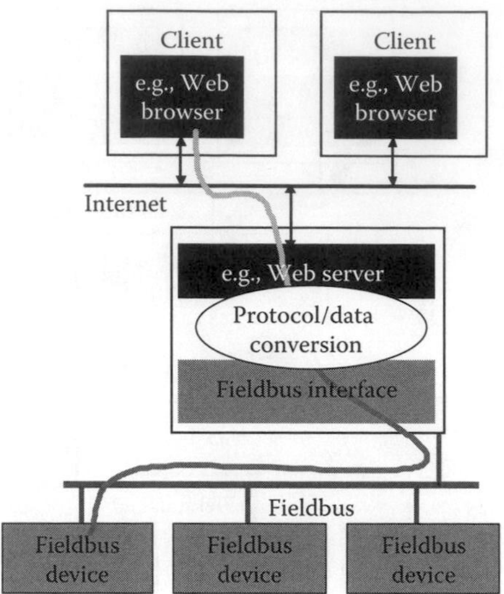

FIGURE 22.5 Network topology for a gateway-based interconnection.

of a proxy representing the fieldbus and its data to the outside world. It fetches the data from the field devices using the usual fieldbus communication methods and is the communication partner addressed by the client.

Unlike the tunneling approach, where each field device has to provide and maintain its data on an individual basis, information processing for the client in the Internet is centralized here. This enables the gateway to set up a comprehensive and consistent view of the automation system, which is undoubtedly a benefit. In addition, the field devices need no special equipment to process IP-based data and services.

For the sake of completeness, it should be noted that a gateway would in principle also allow for the reversal of the communication direction. Therefore, field devices could as well access resources in the Internet, for example, different fieldbus systems connected to the LAN via compatible gateways. Although there have been efforts to foresee this possibility by providing harmonized interfaces (as in the Network Oriented Application Harmonization (NOAH) project [22]), it is irrelevant in practice.

22.3 Control Data Tunneling

A special case of protocol tunneling is the interconnection of two remote fieldbus segments over an IP-based backbone. As stated before, this has become popular in recent years in control engineering. In this type of application, the performance of the backbone is essential. Specifically, the way the *IP channel* connecting the control nodes or networks is set up has a direct influence on the system behavior, and modeling of the network is not a simple task [23].

To implement the tunnel and exchange data between the end points requires a special protocol. Apart from proprietary solutions and those designed in the context of industrial Ethernet, one standard that has been designed for this purpose and enjoys a growing user base is CEA-852 [24]. It describes the system components, the communication protocol, and management requirements to establish an IP channel with control devices and routers. A typical CEA-852 topology is depicted in Figure 22.6. The devices in a CEA-852 system can either be purely IP-based or regular control network (CN) devices connected to the IP network through CN/IP tunneling routers.

CEA-852 control devices communicate directly in a peer-to-peer fashion using a native CN protocol without the need of gateways. The native protocol data units are encapsulated in User Datagram Protocol (UDP) frames and routed to the respective recipients on the IP channel. The choice of UDP transport over TCP has several advantages: First, because of the connection-less nature of CN communication, it

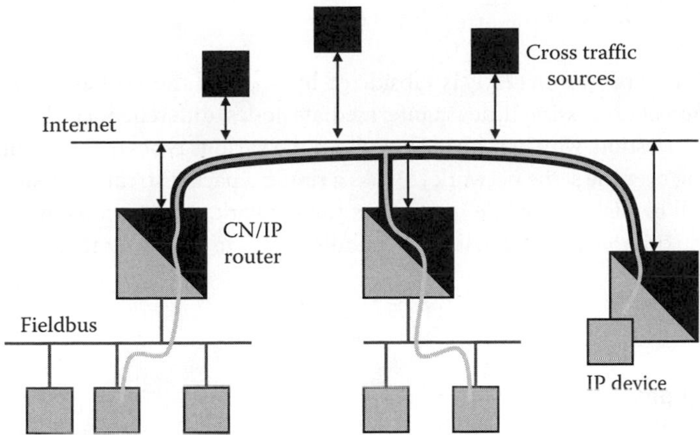

FIGURE 22.6 EIA-852 system configuration.

would incur unnecessary overhead to manage connection setup in TCP. Second, TCP ensures delivery through retransmissions. Because of the (soft) real-time requirements of control applications, however, retransmissions will most likely miss the deadline, which makes them useless. Third, CN protocols typically implement their own retransmission scheme where necessary. So, on top of the actual IP-based network, this tunneling protocol is a viable solution.

22.3.1 Network Parameters for Control Data Tunneling

On the lower layers of the tunnel, that is, regarding the IP-based network, quality of service (QoS) parameters can be used to describe requirements imposed on the network behavior. The origin of QoS definitions was the usage of the Internet to transport multimedia services, specifically the difficulty of using a packet-switched, statistical multiplexing network to transmit stream-based data that normally require directly switched connections. In automation systems, and especially in network-based control, the situation is comparable: Formerly direct connections are being replaced by packet-oriented networks that influence the quality of control (QoC) of the implemented application. QoS parameters are a reasonable means to quantify this influence [25]. In principle, QoS refers to the capability to provide resource assurance and service differentiation in a network. In a narrower sense, timeliness and reliability properties are often referred to as network QoS. Some typical QoS parameters relevant for the interconnection of distributed control systems by IP-based tunnels are discussed in the following.

22.3.1.1 End-to-End Delay

This is an absolute measure for the total delay packets suffer from the emission by the sender until the reception over the network. It is also known as transit delay. A requirement on the end-to-end delay D_i of each packet i can be defined such that the delay D_i of each packet i must not exceed a maximum end-to-end delay D_{max} with a probability greater than or equal to Z_{min},

$$P(D_i \leq D_{max}) \geq Z_{min} \tag{22.1}$$

This statistical bound on the end-to-end delay can be easily converted into a deterministic bound by defining $Z_{min} = 1$.

22.3.1.2 Delay Jitter

The delay jitter J_k is the variance of the end-to-end delay D_k and is a measure for the stochastic part of the delay. It is often defined as

$$D_k = D_{\min} + J_k \tag{22.2}$$

where the deterministic part from (22.1) is subsidized by $D_{\min}$ and the stochastic part by J_k. Delay jitter is a function of different processing times in intermediate nodes, different queue lengths, and typically a result of network congestion, which in turn depends on the various types of data traffic like parallel and cross traffic running over the same network [23]. As a result, a packet stream transmitted in equidistant instants in time will be distorted while it traverses the network. Applications may want to bound the delay jitter J_i to a maximum value $J_{\max}$ with a probability equal to or greater than $U_{\min}$,

$$P(J_i \leq J_{\max}) \geq U_{\min}. \tag{22.3}$$

22.3.1.3 Throughput

The throughput defines how many bits per time unit can be transferred over a given network path. In an end-to-end view, the delay can be seen as a direct function of the instantaneous throughput. The practical notion of throughput, however, implicitly includes a certain time interval. If $A(t)$ denotes the aggregate amount of transferred data up to a time t, the throughput can be defined as

$$\Theta = \frac{A(t+\Delta t) - A(t)}{\Delta t} \tag{22.4}$$

Throughput is especially important for applications transmitting constant data streams. A requirement for the throughput can be written as

$$P(\Theta \geq \Theta_{\min}) \geq \zeta_{\min} \tag{22.5}$$

Since CN applications usually deal with small data transfers only, this requirement will not be as important as for multimedia applications. However, throughput might become an issue over dial-up or other low-speed links.

22.3.1.4 Loss Rate

The drop probabilities for different packets on the individual hops can be subsidized by a global *loss probability* or average *loss rate*. The most common reason for high loss rates is again network congestion. The probability of complement can be defined as the reliability of the network path. The reliability defines how reliable packet transmission should be on the network. A requirement can define at which minimum probability $W_{\min}$ a packet should be received, thus

$$P(\text{packet received}) \geq W_{\min} \tag{22.6}$$

22.3.1.5 Packet Ordering

Some services rely on order preservation of a sequence of packets. Let $\langle p_k \rangle$ be the sequence of transmitted packets and $\langle p'_k \rangle$ be the sequence of received packets, then the network is order-preserving if and only if

$$\langle p_k \rangle = \langle p'_k \rangle, \forall\, k \tag{22.7}$$

Typically, order preservation is a property depending on delay jitter and network topology. If the network provides multiple possible paths connecting any two nodes, control packets may traverse the network in any of those paths. Given that packets suffer different delays in the paths, they may be reordered even if the intermediate nodes obey a strict first-in, first-out (FIFO) policy. More complex queuing policies (e.g., priority queues) implemented in QoS-enabled routers may even lead to reordering on a single path, due to changing packet priorities.

22.3.2 Performance and Application Classes

Another possibility to assess the performance of an IP-based tunnel is from the application point of view. Depending on the application's needs, the network must satisfy certain requirements [17]. With respect to the packet delay as the most important parameter, we can distinguish several *performance classes* and derive characteristic features for the delay density $f(D)$.

Hard real-time. If an application has hard real-time requirements, any given deadline must be met under all circumstances. Missing a deadline will inevitably result in a catastrophic failure of the system. Although the definition of what actually constitutes a catastrophic failure is not always clear, we can at least find a concise requirement for the delay, which must be upper-bounded, that is,

$$f(D) \equiv 0 \text{ for } D \geq D_{max} \tag{22.8}$$

Soft real-time. If missing of a deadline does not lead to a catastrophic failure, the application imposes only soft real-time requirements. For our purpose, this definition is too loose and therefore not useful. In fact, either of the following two classes is a subset of soft real-time.

Guaranteed delivery. For this requirement, the actual delay of a packet is irrelevant. What counts is that the packet is definitely received by the addressed node. There are no particular constraints for the delay density; it may not even be bounded (if it was, we had effectively a hard real-time situation).

Best effort. This scenario is the weakest one. Here, the network only attempts to deliver the packets as fast as possible, without making guarantees concerning deadlines or successful delivery at all. Formally, this comes down to treating the delay density $f(D)$ as conditional distribution,

$$f(D|X) = f(D)\, p(X), \tag{22.9}$$

where X denotes the successful delivery of the packet, which is independent of the delay distribution. For the simulation model, the conditional distribution has the convenient property:

$$\int_{-\infty}^{\infty} f(D|X)dD = p(X = \text{true}) < 1 \tag{22.10}$$

so that the residual probability that a packet is lost in the best effort scenario is implicitly included.

Distributed applications implemented on automation networks typically cope with process data acquisition, processing, and signaling. They can be categorized into several *application classes*, which can be mapped onto the performance classes introduced before (Figure 22.7).

Monitoring/logging. This application class allows to view or record data within a control application. Typically, selected process parameters are sent to a monitoring station where they can be tracked or logged. The QoS requirements can range from hard real-time to no requirements at all (meaning an entirely unreliable network is acceptable as well). Examples are remote power meters, weather sensors, or occupancy detectors.

Application class	RT	GD	BE
Monitoring	+	+	±
Open-loop control	±	++	+
Closed-loop control	+	−	+
File transfer	±	++	−

FIGURE 22.7 Application classes and possible requirements (RT, hard real-time; GD, guaranteed delivery; BE, best effort) imposed on the network.

Open-loop control. Open-loop control appears where the process characteristics and disturbances are well-known so that feedback paths are not required. The network should provide at least reliable data transfer, but the timing requirements can range from nonreal-time to hard real-time. Multidimensional open-loop controls can pose additional timing requirements originating from the time dependencies of the output values. Control data are typically sent periodically, in which case, the overhead for providing guaranteed delivery, for example, by packet retransmission, can be avoided. Examples are light switches, intelligent power outlets, or dot matrix displays.

Closed-loop control. Closed-loop control applications are commonly found in automation tasks where control quality and error compensation are required. The controller does not need to be part of the application; it can be an external system such as a human operator. Timing variations influence the stability and performance of a closed-loop control, so this application class has at least a demand for soft real-time behavior. Again, multidimensional control is possible and can enforce more strict timing requirements. Examples are heating, industrial control, and power management.

File transfer. All distributed automation systems need in one form or the other the transmission of files or at least larger data blocks instead of process data. The typical application is to configure or parameterize field devices either at system start-up or later during runtime. Sometimes, even executable code is transferred, for example, to update programs running on controllers. The minimum requirement for the data transmission is reliable and guaranteed data transfer. In rare cases, when the upload of a device configuration is time-critical, also hard real-time requirements may apply.

22.4 Design Considerations for a Gateway

There are no general guidelines on how to implement a gateway between a fieldbus and the Internet. From the user's perspective, however, there exists a clear-cut requirement for the integration: as the automation data are brought into the office world, the ways and means employed must comply with the standards used there. In particular, this requires some sort of abstraction for both data and functionality. There is a good reason for this. Within the scope of vertical integration, the automation data are accessed from a *strategic* level (rather than control level), so the user wants to have a clear picture of the most relevant data. Too many fieldbus-specific details can obscure this picture and be distractive. After all, the user is typically not a fieldbus expert, let alone the one who set up the automation system.

Another argument speaking in favor of an abstract view to the automation system is the desired user-friendliness of the interface. Typically, the user wants to retrieve data from different systems with only one tool. Therefore, the user interface should be as independent of the underlying system as possible.

It is important to notice that the way data and functionality abstraction can actually be achieved also depends on the protocols and mechanisms used on the Internet side of the gateway. In the sequel, we list and discuss a few design issues. They may vary with the concrete target application. The model for the present discussion is a simple and cost-effective residential gateway for linking a home network to the Internet [26].

22.4.1 Services

The basic function of a gateway evidently is to provide read and write access to process and device data on the fieldbus for monitoring and control operations, but also for network management purposes. However, as the gateway has to accomplish protocol (and maybe data) conversion and therefore needs a certain amount of computational intelligence anyway, additional services can be included. With the gateway acting as a proxy server for the automation data, there are a number of autonomous functions it can (or should) provide [27,28]. Some of these add-on services can in an abstract way be described as *events*. Events are triggered by predefined criteria: if a criterion is fulfilled, an action is taken. Criteria can be the expiration of timers (time-based events) or the crossing of thresholds by the value of a data point (value-based events). The actions to be performed upon the occurrence of an event can be manifold.

Data logging. The storage of data point values based on either time or the recognition of certain trigger conditions enables the gateway to perform stand-alone operation independently of the availability of an online connection to the high-level network. Especially in remote access via unreliable channels (e.g., wireless or heavily congested service providers), the possibility of creating log files (and retrieving them at a later time) may be an essential requirement of the application. A management station can then retrieve logs (fully or incrementally) to perform subsequent analysis or feed specialized software with control system data.

Asynchronous notifications. Contrary to the predominant client/server communication model in the Internet, where the client has to continuously poll the server (the gateway) to get information, the gateway should become active as a response to certain events. The availability of such a mechanism is an important criterion for the choice of a suitable protocol on the Internet side of the gateway. In practical implementations, the lack of native asynchronous notification mechanisms is usually circumvented by the use of services based on some sort of e-mail.

Two other types of services are independent of the fieldbus data and cannot be modeled as events.

Security. Fieldbus systems are traditionally security-unaware [3,29]. When a connection to the Internet is provided, however, the security of protocols that are able to communicate with the gateway is vital for both the field area networks that the gateway is connected to and the gateway itself [11,30]. The gateway has to take care of access control, the minimum level of security being authentication.

Network structure updates. A last point that might be relevant for the gateway depending on the desired application area is some sort of plug-and-play functionality. By this, we understand that the gateway is capable of recognizing changes in the network structure autonomously and adapting its internal data point structure. This does not necessarily eliminate the need for preconfiguration, but it may help to keep the gateway operational. A typical application for this feature are residential gateways that—once installed—have to be as autonomous as possible and must cope on their own with devices joining or leaving the field area network. In this case, the underlying fieldbus must, of course, also support basic plug-and-play features. A key issue here is the ability of the high-level protocol on the Internet side to reflect structural changes in the field area network: nodes in a field area network can be removed, added, or replaced.

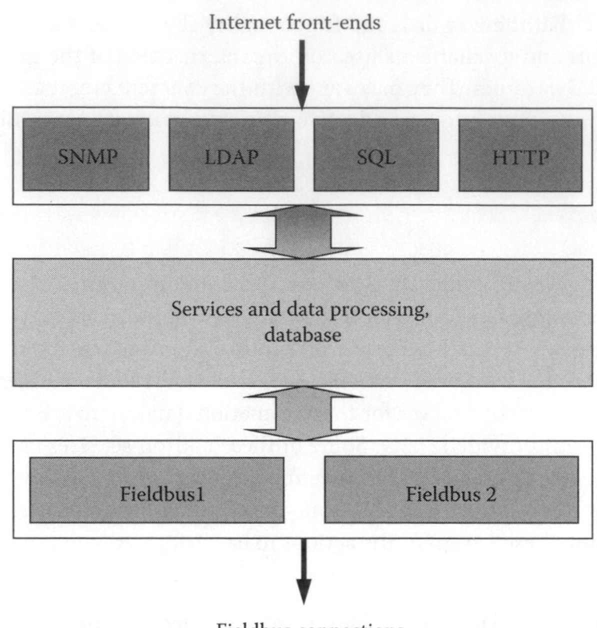

FIGURE 22.8 Modular three-level gateway structure.

This can influence the mapping between protocol addresses and fieldbus addresses. It has to be taken into account how a protocol deals with such changes in the network structure and how objects that existed before a change can be consistently addressed after a change.

22.4.2 Gateway Structure

There is no best architecture for a gateway. Its primary function is to act as an abstract interface to an automation system, independent of both the fieldbus protocol (because it operates on application level) and the fieldbus-specific coding of the data. Therefore, the gateway provides unified access to, maybe, more than one fieldbus simultaneously. The software structure of the gateway implementation can reflect this demand for versatility and exhibit a modular structure [26,31]. Alternatively, the architecture can as well be monolithic and tailored to one specific case. A modular approach will have to provide abstraction as a matter of course; the three-level approach shown in Figure 22.8 is just one possibility.

What has to be considered separately is the low-level access to the fieldbus. One degree of freedom is the strategy used to fetch the data. This can be based on a caching mechanism, where the gateway autonomously retrieves the data with a predefined refresh rate. Upon request from the client, the response is taken from the cache and returned without further delay. This approach provides immediate answers and can reduce the fieldbus traffic, especially in situations where multiple access from different clients needs to be handled. For proper operation, a timestamping should be foreseen so that the client can keep track of the age of the data values.

The alternative is to get the data from the fieldbus only when requested to do so by the client. This avoids cyclic updates as well as timestamps, but the gateway may have to cope with substantial delays of the response from the field device. The connection to the client has to be kept alive in the meantime.

Which strategy can be used depends on both the properties of the fieldbus and the capabilities of the interface used. To avoid excessive bus load in the caching scheme, the gateway can passively listen to the network traffic. Provided the fieldbus interface supports such a monitoring mode, this approach is convenient to track cyclic (in systems like Interbus or Profibus-DP) or event-driven data transfer (in fieldbuses based on the producer/consumer model like Controller Area Network (CAN) and European Installation Bus (EIB)). Unfortunately, many of today's integrated fieldbus controllers perform a low-level data filtering, inhibiting straightforward monitoring. In these cases, the gateway must actively retrieve the data from the bus. The same applies to data that are not normally transferred on a regular basis, such as diagnostic data (e.g., in Profibus-DP). Even worse, the transmission of such specialized information often requires the execution of dedicated commands that cannot be mapped to simple read/write commands. Such fieldbus-specific communication sequences (but also other peculiarities like remote program invocation or special network management functions) must be handled differently. As the high-level access via the Internet most likely will not provide methods for direct command execution, such special functions can be triggered by an ordinary read/write access to specially assigned data points.

Another important restriction may be the medium access control scheme employed by the fieldbus. There are no problems with multimaster systems; here, any node can host the gateway. Single-master systems, on the other hand, imply that the gateway must either be the fieldbus master or cooperate with it, in the sense that there is a dedicated and typically nonstandard communication channel through which the gateway can tell the bus master which data values are to be read or written. Otherwise, writing data points will inevitably lead to inconsistencies, when the gateway sets a value and the bus master overwrites it during a regular update cycle.

22.4.3 Data Representation

A crucial point in the design of a gateway is the way the fieldbus data are represented. A *data point* is the smallest information unit available via the gateway. To achieve fieldbus independence, the data points have to be arranged in a rather general, high-level way. The question how to find a unified view and coding for the sometimes extremely heterogeneous fieldbus objects is an interesting topic on its own and will be left aside here. More relevant for our purpose is the way the complete set of data points is structured inside the gateway. Basically, there are two different possibilities to do so [32]. The first is a function-oriented approach. In this case, emphasis is put on the content of the individual data point, and the underlying network structure is completely irrelevant. The gateway thus offers a simple flat list of data points.

The second approach is a structure-oriented one. Here, the data points are arranged according to the structure of the field area network. The gateway offers a list of nodes, and each node can contain one or several data points. In addition to the simple form of a data point, a data point can also contain a set of values instead of only a single value. This type of data point is referred to as an *aggregate* data point. Data points can also contain multiple values of different types (structures). The advantage of the structure-oriented solution is that properties of nodes (and not only properties of data points) can also be modeled on the gateway. These properties are, for example, a self-identification of the node or a location information that describes the position of a node in a field area network.

Which of the two possibilities is actually used is a matter of taste and also depends on the target application. Furthermore, the choice might be limited by the high-level access protocol and its capabilities of describing and managing data. Every object that is available to a client needs an address that uniquely identifies the object. Protocols typically define their own addressing scheme; the gateway, on the other hand, also defines an addressing scheme that may be similar to or completely abstract from the fieldbus network addressing. These two schemes need to be mapped onto each other to make data points available.

Another point to be observed is how the information of a data point is encoded. This again might be limited by the capabilities of the way the data are transferred from the gateway to the client. Not every protocol may support special data types or encodings used in a particular fieldbus, so there must be some form of data translation and, necessarily, abstraction. This might not be too problematic for scalar data points like temperatures (although there are obviously many possibilities to encode a temperature value), but it will get increasingly difficult for such common things as date and time, not to speak of aggregate data points available only as a structure on the fieldbus level. For all these more complex data points, no one-to-one mapping is possible, and there are basically only two options to cope with the situation: One is to find a compromise in the form of abstract data representations that omit excess information; the other is to collapse the complete data point into a flexible and unspecific data format such as a string—with the disadvantage that the client needs to take the string apart and to interpret it, which again requires fieldbus-specific knowledge.

In addition to the actual value, a data point also may need some attributes to facilitate interaction with the client. Such attributes can be a description of the data point type and encoding (to tell the client how the data are to be handled) and maybe a string providing a textual description of the data point. In some fieldbus systems, such a self-description is foreseen for data objects anyway. Application and device profiles defining functional models for devices are another way to deal with the complexity problem [33,34]. Last not least, there can be a specification of the access mode, that is, what the client is allowed to do with the data point. It should be noted that although the restrictions concerning data representation have been discussed for the gateway approach, they also apply to the IP tunneling solution. In this case, the list of data points has to be maintained by every field device on its own. Typically, the implementation is up to the vendor, and unless there are very stringent rules, this may lead to a substantial inhomogeneity stemming from the many degrees of freedom mentioned before.

22.5 Gateway Access from the Internet

The primary goal of vertical integration is to bring automation data into an IT context, and consequently, there are many ways to incorporate data provided by a gateway into office-level applications. To this end, it is not necessary to employ full-fledged application-level protocols on top of IP/UDP/TCP. A very common approach used in contemporary supervisory control and data acquisition (SCADA) systems is OLE (Object Linking and Embedding) for Process Control (OPC). A more unconventional idea is to treat the gateway as a database and use, for example, SQL to retrieve the data [35].

The protocols discussed in the sequel have been evaluated with respect to their feasibility for home network access, specifically for remote meter reading [32]. The criteria for the evaluation were described in the previous section: a reasonable way of structuring and handling data, a possibility to send asynchronous notifications, and of course security provisions. In addition, the protocol (together with its data representation) should be able to cope with changes in the network structure. This requirement is a direct consequence of the target application in an intended mass-market with a high number of installations, where updates calling for specially trained personnel are not affordable. Likewise, the overall costs of such a gateway should be low, which typically means limited computing power and memory space. Compact implementations on possibly embedded devices are therefore a must. The results of the survey are presented in Table 22.1.

22.5.1 Simple Network Management Protocol

Simple Network Management Protocol (SNMP) is a comparatively old protocol and has been designed for remote administration of networked devices. Data are represented in a tree structure called *management information base* (MIB) where each branch and each leaf have a unique identifier, called *object identifier* (OID). SNMP *managers* (i.e., the client that connects to an SNMP-enabled device) use this MIB to get extended information about the managed *agent* (i.e., the device that allows SNMP access, in the present case the gateway). The MIB is a static representation of all the objects available at the managed agent. As SNMP does not provide mechanisms to download the MIB from the agent, the information which OID stands for which data point needs to be transferred to the manager over mechanisms that are out of scope of SNMP (usually this is done by the administrator of the system during setup). SNMP is based on the connectionless UDP running on top of IP. This is rather efficient because of the low overhead, but unlike TCP, UDP does not guarantee delivery of packets, so packets can be lost on the network without being noticed by the sender.

TABLE 22.1 Features of Several Protocols Applicable for an IP/Fieldbus Application Gateway

Criterion	SNMP	LDAP	SQL	Web Technologies	OPC UA
Transport protocol	UDP	TCP	TCP	HTTP over TCP	TCP; optionally, SOAP/HTTP or UA-specific protocol on top
Data organization and namespace	Tree structure (MIB)	Tree structure (DIT)	Tables	No limitations	Objects organized into a hierarchical or meshed address space
Fieldbus data point mapping	Mapped to OIDs in the MIB	Mapped to attributes of DIT entries	Used as primary keys	Used as identifiers inside HTML/XML pages	Objects
Flexibility of data organization (per installation)	Limited (tables as workaround), MIB must be known by the client	Unproblematic	Unproblematic	Unproblematic	Unproblematic
Dynamic updates of data structure (during runtime)	Can result in corrupted data (nonatomic access to tables)	Possible (atomic data access)	Possible (atomic data access)	Must be handled by the server (consistent access to data within one doc.)	Can be supported by both server and client
Data type support	Available, but unusable inside tables. Limited (e.g., no structural data types)	Data are packed into strings	Available, though limited. Data are packed into ASCII strings. Not usable inside tables	No data types, data are packed into strings	Protocol defines standard data types and allows for manufacturer-specific types
Overhead per request	Low	Low if clients use specific filters in search queries	Low if clients use specific filters in select queries	High, reload of complete page required	High for SOAP/HTTP with XML/text, low for UA TCP with UA Binary
Asynchronous notifications	TRAP and INFORM	Not available	Not available	Not available	Via subscription service
Security	Authentication and encryption in version 3	Kerberos authentication in version 3, can also use TLS	Uses TLS	HTTP over TLS	Integrated encryption, integrity and authentication services
Client support	Good	Weak (mostly *address book* applications)	Programming interfaces and generic applications (like office suites)	Very good (HTML), programming interfaces (SOAP/XML)	Very good, many applications

The major commands provided by SNMP are GET and GETNEXT for reading data (GETNEXT allows the manager to traverse an unknown MIB tree and to retrieve the OIDs) and SET for writing to data points. In contrast to the standard manager-driven communication (in typical client/server fashion), SNMP also defines the unacknowledged TRAP command (from version 2 on also the acknowledged INFORM command), which can be sent by an agent to asynchronously inform a manager about special conditions in the agent. This possibility to reverse the client/server communication is a rare exception among the investigated protocols and together with the good protocol efficiency is a strong argument in favor of SNMP.

One of the big problems of SNMP is the static structure of the MIB together with the lack of native transfer mechanisms between agent and manager, which does not allow dynamic updates of the network (or data point) structure [36]. Therefore, the list of nodes and data points needs to be stored in an SNMP table rather than as a set of separate objects in the MIB. To access individual objects, the manager needs the OID of the table, which is fixed once and for all, and the number of the row containing the object. If the structure of the field area network is changed, the table can be updated by adding or removing the rows that represent the according nodes. This is a critical issue in that SNMP requires the rows in an SNMP table to be numbered continuously; therefore, the insertion or removal of a row changes the numbering of all subsequent rows.

The workaround of using tables to add a somehow dynamic behavior also raises problems if the gateway is accessed by more than one client simultaneously. It is possible that data in a table are modified by one client (e.g., rows are deleted or added) while another client is in the process of reading the table. This results in an inconsistent set of data. A possible solution used in the case study was to introduce a sequence number for every table. This sequence number changes every time data in the table change. The workaround for achieving data consistency is to read the sequence number, read the table, and afterward read the sequence number again to check if it changed [35]. In case it remained unchanged, the table data are consistent; otherwise, the whole process has to be repeated. To this end, for write access, the protocol had to be slightly changed in a way that every SNMP SET command contains a sequence number that uniquely identifies the version of the table to which a data point is written.

Another problem of SNMP is security. The commonly used version of SNMP is version 1, which does not contain any considerable security features. In the most recent version 3, security mechanisms have been introduced that are sufficient (though not widely used at present). Still, difficulties arise from the connectionless UDP used as transport protocol. UDP traffic is usually not permitted to pass through a typical firewall. But even if communication is possible, the underlying UDP does not guarantee packet delivery. This means that an SNMP manager communicating with an agent cannot rely on the last action he has taken on an agent to be successfully finished. Neither can an agent be sure about the status of a communication with a manager. Hence, error monitoring must be implemented within the manager application.

22.5.2 LDAP

The Lightweight Directory Access Protocol (LDAP) is the lightweight counterpart to X.500 directory services. It is oriented toward directories that are organized in a tree, the *directory information tree* (DIT). The DIT is a hierarchically structured organization of data items, the directory entries. Every entry consists of a collection of attributes. Every attribute has an attribute name, an attribute type, and a value associated with it. One special attribute that is mandatory for every tree entry determines the position of the entry in the tree—the *distinguished name* (DN), which is used as a unique *address* for the entry. It specifies the path by listing all parent entries up to the root entry. The directories contain mostly (but not necessarily) small pieces of data, which are mainly read and searched for. Writing of data is intended to occur only rarely, as well as changes in the tree structure.

LDAP offers a set of data querying and manipulating commands for accessing the directory service. The SEARCH command is very powerful and allows specifying search filters that are applied to each entry. It is possible to request just a subset of the available attributes, and in order to narrow a search even more, the scope can be limited to a certain level of the directory tree (search only the current level, one level deeper, or the whole subtree). With the commands ADD, DELETE, and MODIFY, single entries can be added, deleted, or modified, respectively. The COMPARE command is used for comparison of two entries.

Since LDAP uses TCP as transport protocol, there are no problems with firewalls. Also, all security extensions designed for TCP-based protocols such as the *Transport Layer Security* (TLS) can be used. LDAP version 3 also provides appropriate authentication mechanisms in combination with those already defined in TLS.

Dynamic update of the directory information tree is also not particularly problematic, although the changes are not automatically taken over by the client. The DIT as such is not confined to a static structure like the SNMP MIB, which facilitates the adding and deleting of entries even at runtime. Consequently, there is also no problem with the consistency of data, because the address of a data point can be added as an attribute to the directory entries to unambiguously identify the data points. As operations on a DIT entry are always atomic, simultaneous and consistent access to one single data point from multiple clients is inherently handled by the built-in mechanisms of the protocol.

A general difficulty of LDAP is that its original application idea does not exactly fit the purpose of fieldbus gateway access. The main application of LDAP is the access to data that are related to humans, for example, telephone numbers, employee data, or classical white pages services. The client support reflects this field of application, and although clients exist, they are not widely used and very different in their performance with respect to network load. A second point is that the access to white pages directories is typically read-only. Therefore, the powerful search command can provide a convenient means to retrieve fieldbus data [37]. Writing, on the other hand, is more difficult and likely becomes a bottleneck. Hence, LDAP is not so well suited for control tasks where intensive write access is required. Another problem is the lack of an asynchronous notification mechanism, so workarounds (like e-mails or the like) have to be used.

22.5.3 Structured Query Language

To consider SQL as a protocol for remote gateway access may seem a bit exotic, but there are examples for its use as middleware concept in sensor networks [38]. As SQL is the standard language for relational database management systems, it is widely used also in Manufacturing Execution System (MES) and Enterprise Resource Planning (ERP) systems, which are the intended end points for the vertical integration of field-level data. On the other hand, the fieldbus data on the gateway are handled by a database-like mechanism, anyway, so it is natural to investigate the usability of SQL. It should be noted that SQL is no protocol; it only defines the syntax and semantics of a database query. The transmission mechanism depends on the driver (like Open Database Connectivity [ODBC] or Java Database Connectivity [JDBC]); implementations based on TCP are available, which seems to be best from the viewpoint of additional security mechanisms and handling of multiple connections.

The most essential command is SELECT, which allows searching the database by defining a search query. Compared with LDAP, the SELECT statement in SQL provides much more complexity. Write operations on tables are performed with the UPDATE and the INSERT statements, where INSERT creates new rows in a table, while the UPDATE completely or partly overwrites existing rows. The CREATE and DROP statements are used for creating and removing database objects (primarily, tables). The gateway logic must comprise appropriate modules to parse and handle the SQL statements in order to perform the operations on the database.

The way the data are organized in tables is extremely flexible; hence, suitable ways can be found to allow for dynamic updates of the data point structure. In the case study, four tables for the nodes, data points, events, and logs were used according to the services provided by the gateway. The relation between the tables and the physical fieldbus addresses is achieved by using the fieldbus addresses as primary keys in the node and data point tables. Because these addresses are unique and are not changed over the lifetime of a node (or data point), it is not possible to confuse one object with another one. If objects are added between two queries, the response of an SQL query will of course differ before and after the change; however, every query response by itself provides a consistent view of the structure of objects at that time. It is up to the client how it will handle such structural changes; the protocol does not impose any additional constraints here.

A weak point of SQL is client support, and there is no standard *SQL client*. Instead, an additional ODBC driver (C/C++) or JDBC driver (JAVA) needs to be provided for the development of client-side applications. Like LDAP, SQL has no built-in possibility to send notifications from the gateway to the client, and alternative means must be found. Furthermore, due to the organization of data in tables, the use of different data types is limited. A typical but not very convenient way to circumvent this problem is to convert all data to strings and use the native SQL VARCHAR type for all fieldbus values.

22.5.4 Web-Based Approaches

For many solution providers in the area of fieldbus/Internet connectivity, web-based approaches seem to be the only conceivable solution; at least they are strongly favored. However, unlike the protocols described so far, methods based on the WWW are very heterogeneous. They include communication and data representation using web browsers as clients. The common transport protocol is HTTP, which is typically used to transmit HTML documents. HTML documents can contain embedded objects and script languages (e.g. JavaScript) and can be created by multiple technologies on the server side (such as CGI, ASP, JSP). They all have in common that the client receives a predefined view of a set of data. This view can be modified by the user (e.g., by selecting specific objects in a page and filtering out the others), which requires the creation of a new document on the server side. This process is usually not very responsive because of the necessary communication and computation overhead.

The two main commands HTTP supports are GET (for sending requests to a server to retrieve data) and POST (to send data to the server). This, however, does not cover the whole spectrum of possibilities that are based on top of this mechanism. In many cases, however, the user is presented with a view based on a template, which is filled with data by some server mechanism.

A definite strength of web technologies is the availability of many different clients. A lot of Internet browsers for multiple platforms exist, leaving hardly any white space on the map of supported systems. The main problem of pure HTML is the response behavior that is imposed by the necessity to refresh (i.e., completely reload) a page in order to receive current data (this can be circumvented by advanced mechanisms, which only retrieve, for example, the data point value). There is also another performance problem with this approach: As the view on the data depends on HTML documents, typically several data points will be collected on one page to have a better overview. Therefore, even for very small amounts of data (e.g., the value of just one data point), a complete document needs to be created and transmitted. This causes a considerable amount of both computational overhead and network traffic. Likewise, it is not possible to collect data from multiple servers in a single document, which considerably limits the use of this approach. In an IP tunneling scheme, for example, every device providing data needs to maintain a set of HTML pages, and the client needs an instance of the browser for each device. An integrative view of the entire installation, let alone an automatic data acquisition, is not feasible and has to be done manually.

A positive point about the web approach is the comparatively unproblematic security issue. As HTTP is based on TCP, TLS can be used. Furthermore, HTTP is conveniently routable through firewalls [39], which is an important point in an environment becoming increasingly aware of security—although it has to be noted that the simple routing possibility actually opens a way to circumvent security. A drawback is that there is no way for the server to send data to the client without request. Like with all other protocols except for SNMP, other ways have to be found to accomplish such notifications (e.g., special applets or e-mail-based services). Very poor is also the support of different data types. In fact, the data have to be sent as strings and processed appropriately by the client. This lack of ample data type support ideally supports (and also demands) pragmatic, proprietary solutions but is a severe obstacle to standardized, interoperable implementations. A possible way out of this dilemma is the increasing use of XML. If the data are based on (standardized) XML descriptions, they are not only readable for a human operator (a severe drawback of HTML), but also machine parsable, which makes gateway access based on web technologies finally also suited for automatic data acquisition.

A further step in this direction could be the XML-based Simple Object Access Protocol (SOAP), which defines methods for data exchange typically used over HTTP, and on top of it web services allowing for flexible, easy-to-integrate solutions [40–45]. A tremendous benefit is an utmost platform independence as both HTTP and XML are available everywhere. However, as SOAP is a protocol on its own, client applications are still needed, a web browser today is not sufficient. A potential weak point is, as with the somehow related technology Common Object Request Broker Architecture (CORBA), the relatively complex structure that does not lend itself to a lightweight implementation on those resource-limited devices that today make up a significant portion of a typical automation system.

22.5.5 OPC

OPC (originally OLE for Process Control, nowadays the abbreviation no longer has an official meaning) has become a de facto standard for interconnecting fieldbus systems and local area networks. Although the basic intention of OPC is to provide a comprehensive middleware, it is commonly used *only* for direct access to fieldbus data. This also fits the discussion of remote access possibilities to fieldbus gateways: the OPC server itself is the gateway; various OPC modules already implement most of the functionalities desired from a gateway and listed in Section 22.1. Like with all middleware approaches, there are no stand-alone client applications available. The objective of an OPC client is to provide communication and data exchange functionalities to other applications like SCADA or configuration tools [46]. From the viewpoint of vertical integration, it is important to notice that OPC has become a building block for many MES products [47]. In all these applications, OPC serves as a kind of driver to the fieldbus integrated in the application software. Alternatively, there can be an additional layer in the architecture, and OPC can be the fieldbus driver for a gateway, which in turn connects to the actual application via any other protocol, such as HTTP [39].

With regard to the data representation, data in OPC are organized in an arbitrary tree-like namespace consisting of nodes and items (the leaves of the tree) with respective sets of properties like in most other protocols discussed before. Unlike all previous examples, however, the structure relevant for the application is not defined by the OPC server. Rather, the client is free to define its private view on the data. According to the original OPC specification, server presents a flat list of items to the client and lets client organize these items into groups. The newer version OPC XML-DA depreciated the notion of groups and, instead, allowed server to arrange items into a hierarchical *browse space* (somewhat similar to data representation of SNMP and LDAP). Basically, OPC-XML allows the access to process variables only. All specific functions (like fieldbus startup, access to special resources or protocol objects) have to be implemented by the server logic, and access to these functions must be mapped to generic data points on the server.

A major drawback of the original OPC version based on Distributed Component Object Model (DCOM) was a lack of security features, which limited the applicability to closed environments. The technological reason was that DCOM requires a large and a priori unknown number of UDP ports to handle the connections between the distributed objects. For the implementation and administration of a security policy, for example, by means of a firewall, such a situation is highly undesirable and should be avoided. The recent extensions based on XML and SOAP mitigate the problem and permit the use of OPC also over corporate intranets, even over the Internet. Compared with plain web-based approaches, OPC-XML pursues a much more rigorous way of structuring data, which improves interoperability while at the same time exploiting the benefits of web technology.

The most recent OPC UA (Unified Architecture) specification provides further extensions: It drops the reliance on a single communication protocol, or even a single wire format for transmitting messages, and introduces a set of core services that can be made available by a server (gateway) to its clients via a particular *transport* (protocol) in a certain *encoding* (message serialization format). Transport possibilities are *UA TCP* (UA-specific framing on top of TCP with its own session handling, fragmentation, and reassembly capabilities) and SOAP/HTTP (web services, like in the case of XML-DA). The possible encodings are XML/text and UA Binary. The XML encoding is human readable but not very efficient because it requires decent communication bandwidth and sufficient computing resources for XML parsing. With binary encoding, transmitted data are serialized into binary structures for improved efficiency also on resource-limited embedded systems. The standard also leaves room for other protocols and encodings in the future, while maintaining the core set of services [48].

The information model in UA was changed to a new object-oriented concept. Gateway and CN resources are defined in terms of objects that can encapsulate both data variables and methods, that is, gateway-side functions that a client may invoke. Objects are arranged as nodes in the *address space* without any prescribed structure like in earlier versions. Finally, the UA standard addresses issues related to representing CNs and their services. While XML-DA only described data acquisition, UA also describes Alarms and Conditions (asynchronous notifications), Historical Access (logs), and Programs (complex gateway- or fieldbus-side functions that cannot be represented as a single object method). All this makes OPC UA the most suitable protocol out of all considered standards [49].

22.6 Role of Industrial Ethernet

A discussion of fieldbus/Internet connectivity is incomplete without consideration of the growing use of Ethernet in automation. After all, Industrial Ethernet is often praised as *the* solution to vertical integration overcoming all problems encountered in the design of heterogeneous network interconnections. Beyond all well-sounding marketing slogans, it is a matter of fact that the dominant use of Ethernet in local area networks has renewed efforts to employ Ethernet also in the automation domain. These efforts date back to the early days of Ethernet development, but only recent technological advances in the development of full duplex and switching techniques made high data rates possible without the old drawbacks of the Carrier Sense Multiple Access with Collision Detection (CSMA/CA) medium access control. The main benefit is clear: If Ethernet was used as a replacement for the fieldbus, and if IP plus the common transport protocols were used on top of it (which is not necessarily the case in all industrial Ethernet approaches), Internet technologies would be the natural way of accessing automation data. It is thus not astonishing that many contemporary concepts of vertical integration rely strongly on Ethernet as an automation network. For the old CIM pyramid, the penetration of LAN technologies into the classical fieldbus domain has the ultimate consequence that intermediate levels become obsolete. The hierarchy thus gradually turns into a flatter structure with only two, maybe three levels.

How do the facts of Industrial Ethernet compare with the characteristics of vertical integration discussed so far? It is of course shortsighted to postulate that the use of Ethernet alone entails an interoperability with *the office world*. In fact, Ethernet is but a transport medium. Using Ethernet for both office and automation applications may allow for a coexistence on a shared medium (within certain limits posed by, e.g., bandwidth and security considerations) but does not mean integration. What is much more important to this end is the protocol stack employed on top of Ethernet. Some solutions rely on TCP/UDP/IP as a transport layer, and many, especially real-time Ethernet approaches, provide some possibility for a parallel nonreal-time channel that can carry also IP traffic [50,51]. In many cases, however, the higher protocol layers are in fact *fieldbus* application-layer protocols that were retained for compatibility reasons and that are sort of tunneled using TCP/UDP encapsulation methods.

Of course, industrial Ethernet approaches have the benefit of a—thanks to the possibility of using the same medium for office and automation tasks—tight integration from a pure networking point of view. Unless a traffic separation seems reasonable for performance or security reasons, no dedicated connection node is necessary. On the application level, however, integration is by no means so tight. Even though the inclusion of automation data into office applications is simplified by, for example, object-oriented techniques borrowed from the IT world, there is still a substantial amount of fieldbus-specific information to be handled. Consequently, an abstract view on the automation data—one of the requirements for vertical integration—is not a matter of course. Nor is an *automatic* inclusion of industrial Ethernet into the office world easy to achieve. In fact, the effort that needs to be spent on a gateway design has in this case to be put into the development of appropriate nonstandard application-level tools. The only exception where Ethernet together with standard protocols definitely brings an advantage is a web-based human machine interface used, for example, for configuration purposes. But this is only one small aspect of automation.

22.7 Summary

The classical fieldbus systems have been devised as highly specialized communication systems for the factory floor. Their development, though innovative at that time, was aimed at conventional self-contained automation applications. Today's automation tasks, however, are usually no longer stand-alone systems. At least SCADA is required in many cases, with the respective SCADA software running in a typical office environment as regards hardware and software platform or network environment. More advanced demands are remote data access for monitoring or maintenance, or the inclusion of automation data into company-wide management systems. All these functions are in fact not part of the factory floor or the process control level; they belong to a higher, more strategic level of information processing, which is governed by office networks. This circumstance calls for an interconnection between automation and office networks.

As different as the goals of interconnection are the possible realizations. Although the systems on either side are standardized, the link is not—not even in a coarse architectural sense. One reason is the still existing plethora of fieldbus systems. But also on the Internet side, the IP suite is just a least common multiple of an equally large set of possible applications serving as targets for the interconnection. The preceding sections have highlighted the basic system architectures and the many design options to be considered. Evidently, the properties of the networks together with the applications have an important influence on these options and in fact limit the designer's flexibility with respect to the implementation.

Contrary to an initial guess, the architectural approach (i.e., gateway or tunneling) does not seem to affect the overall system complexity very much. As the discussion showed, it rather influences the distribution of the implementation efforts among the involved entities: field devices, clients on the Internet side, and a possible central access node. The particular choice must largely depend on the intended application; there is no golden rule for it. In fact, all approaches have their own strengths and shortcomings. Even though the major part of the article discussed the subject from the viewpoint of a gateway solution, this does not imply that the tunneling approach is inferior. It is simply suited for slightly different types of application, and most of the points described for a gateway are equally valid.

The overall complexity of the interconnection is—according to practical experience—chiefly determined by the desired level of data and functionality abstraction and the degree of vertical integration to be achieved. A direct forwarding of fieldbus data packets *as is*, with the interpretation left up to the client (and, ultimately, the user), is easier to achieve than a reasonable data and protocol translation with the aim of a smooth integration into a nonfieldbus-specific application. Not always is true vertical integration and abstraction really an issue. But if it is, the price to be paid is necessarily a loss of precision, mostly from a functional point of view. The higher the abstraction level, the fewer functions and specific pieces of information can be mapped to the *other side* without problems. Timing precision or real-time behavior should not be a requirement at all because timing relations are typically lost in the node linking the two different networks. Timestamps can help, but only to a certain extent.

The great flexibility of fieldbus/Internet connections sketched in this chapter is a bit deceptive. In practical implementations, one has to live with a number of inconveniences. Timing is just one. Especially if one goes for vertical integration, the high-level target application and the underlying protocol may not support services provided by the automation system. This frequent lack of a one-to-one relation between the two networks requires some creativity from the designer, especially on the Internet side. Good examples are the basic gateway services discussed in Section 22.3. Vital mechanisms such as asynchronous notifications rarely exist as native protocol features and have to be replaced by workarounds outside the actual application. In this respect, the large number of different interconnection approaches is not astonishing but only reflects the fact that the one ideal solution for all occasions does not exist.

References

1. T. Sauter, The three generations of field-level networks—Evolution and compatibility issues, *IEEE Transactions on Industrial Electronics*, 57(11), 2010, 3585–3595.
2. G. G. Wood, Survey of LANs and standards, *Computer Standards & Interfaces*, 6, 1987, 27–36.
3. T. Sauter, The continuing evolution of integration in manufacturing automation, *IEEE Industrial Electronics Magazine*, 1(1), Spring 2007, 10–19.
4. J. N. Daigle, A. Seidmann, J. R. Pimentel, Communications for manufacturing: An overview, *IEEE Network*, 2(3), May 1988, 6–13.
5. G. Doumeingts, B. Vallespir, D. Chen, Methodologies for designing CIM systems: A survey, *Computers in Industry*, 25, 1995, 263–280.
6. L. J. McGuffin, L. O. Reid, S. R. Sparks, MAP/TOP in CIM distributed computing, *IEEE Network*, 2(3), May 1988, 23–31.
7. J.-D. Decotignie, P. Raja, Fulfilling temporal constraints in fieldbus, *Conference on Industrial Electronics, Control, and Instrumentation*, Maui, HI, 1993, pp. 519–524.
8. E. Tovar, F. Vasques, A. Cardoso, Guaranteeing DCCS timing requirements using P-NET fieldbus networks, *IEEE Workshop on Factory Communication Systems* (*WFCS*), Barcelona, Spain, 1999, pp. 831–840.
9. W. Cheng, F. Y. Shun, X. Deyun, Computer integrated manufacturing, in: G. Salvendy (ed.), *Handbook of Industrial Engineering—Technology and Operations Management*, 3rd edn., John Wiley & Sons, New York, 2001, pp. 484–529.
10. IEC Standard 62264-1, Enterprise-control system integration—Part 1: Models and terminology, 2003.
11. T. Sauter, S. Soucek, W. Kastner, D. Dietrich, The evolution of factory and building automation, *IEEE Industrial Electronics Magazine*, 5(3), 2011, 35–48.
12. J. W. Szymanski, Embedded Internet technology in process control devices, *IEEE Workshop on Factory Communication Systems* (*WFCS*), Porto, Portugal, 2000, pp. 301–308.
13. M. Wollschlaeger, Framework for Web integration of factory communication systems, *IEEE International Conference on Emerging Technologies and Factory Automation* (*ETFA*), Antibes Juan-les-Pins, France, October 15–18, 2001, pp. 261–265.
14. G. Moritz, E. Zeeb, S. Prüter, F. Golatowski, D. Timmermann, R. Stoll, Devices profile for Web services and the REST, *IEEE Conference on Industrial Informatics* (*INDIN*), Osaka, Japan, July 2010, pp. 584–591.
15. M. Volz, Quo vadis Layer 7?, *The Industrial Ethernet Book*, Issue 5, Spring 2001, http://www.iebmedia.com/(accessed July 29, 2013).
16. T. Sauter, M. Lobashov, End-to-end communication architecture for smart grids, *IEEE Transactions on Industrial Electronics*, 58(4), April 2011, 1218–1228.
17. S. Soucek, T. Sauter, Quality of service concerns in IP-based control systems, *IEEE Transactions on Industrial Electronics*, 51(6), 2004, 1249–1258.
18. S. R. M. Canovas, C. E. Cugnasca, Implementation of a control loop experiment in a network-based control system with LonWorks technology and IP networks, *IEEE Transactions on Industrial Electronics*, 57(11), 2010, 3857–3867.
19. R. A. Gupta, M.-Y. Chow, Networked control system: Overview and research trends, *IEEE Transactions on Industrial Electronics*, 57(7), July 2010, 2527–2535.

20. K. Natori, R. Oboe, K. Ohnishi, Stability analysis and practical design procedure of time delayed control systems with communication disturbance observer, *IEEE Transactions on Industrial Informatics*, 4(3), August 2008, 185–197.
21. A. Onat, T. Naskali, E. Parlakay, O. Mutluer, Control over imperfect networks: Model-based predictive networked control systems, *IEEE Transactions on Industrial Electronics*, 58(3), March 2011, 905–913.
22. M. Wollschlaeger, C. Diedrich, T. Mangemann, J. Mueller, U. Epple Integration of fieldbus systems into on-line asset management solutions based on fieldbus profile descriptions, *IEEE Workshop on Factory Communication Systems* (*WFCS*), Västeras, Sweden, August 27–30, 2002, pp. 89–96.
23. S. Soucek, T. Sauter, G. Koller, Effect of delay jitter on quality of control in EIA-852-based networks, *29th Annual Conference of the IEEE Industrial Electronics Society* (*IECON*), Roanoke, VA, November 2–6, 2003.
24. ANSI/CEA Standard 852.1, Enhanced protocol for tunneling component network protocols over internet protocol channels, 2010.
25. S. Soucek, T. Sauter, T. Rauscher, A scheme to determine QoS requirements for control network data over IP, *27th Annual Conference of the IEEE Industrial Electronics Society* (*IECON*), Denver, CO, November 29–December 2, 2001, pp. 153–158.
26. G. Pratl, M. Lobachov, T. Sauter, Highly modular gateway architecture for fieldbus/Internet connections, *IFAC International Conference on Fieldbus Systems and their Applications* (*FeT* 2001), Nancy, France, November 15–16, 2001, pp. 267–273.
27. D. Dietrich, D. Bruckner, G. Zucker, P. Palensky, Communication and computation in buildings: A short introduction and overview, *IEEE Transactions on Industrial Electronics*, 57(11), 2010, 3577–3584.
28. M. A. Zamora-Izquierdo, J. Santa, A. F. Gomez-Skarmeta, An integral and networked home automation solution for indoor ambient intelligence, *IEEE Pervasive Computing*, 9(4), 2010, 66–77.
29. W. Granzer, F. Praus, W. Kastner, Security in building automation systems, *IEEE Transactions on Industrial Electronics*, 57(11), 2010, 3622–3630.
30. A. Treytl, T. Sauter, C. Schwaiger, Security measures in automation systems—A practice-oriented approach, *10th IEEE International Conference on Emerging Technologies and Factory Automation* (*ETFA*), Catania, Italy, September 19–25, 2005, Vol. 2, pp. 847–851.
31. I. Kuhl, A. Fay, Reusable and flexible design of communication gateways, *IEEE Conference on Emerging Technologies and Factory Automation* (*ETFA*), Palma de Mallorca, Spain, September 22–25, 2009, pp. 1–9.
32. M. Lobashov, T. Sauter, Vertical communication from the enterprise level to the factory floor—Integrating Fieldbus and IP-based networks, *IEEE Conference on Emerging Technologies and Factory Automation* (*ETFA*), Prague, Czech Republic, September 20–22, 2006, pp. 1214–1221.
33. C. Reinisch, W. Granzer, F. Praus, W. Kastner, Integration of heterogeneous building automation systems using ontologies, *Annual Conference on Industrial Electronics* (*IECON*), Orlando, FL, November 10–13, 2008, pp. 2736–2741.
34. R. Frenzel, M. Wollschlaeger, T. Hadlich, C. Diedrich, Tool support for the development of IEC 62390 compliant fieldbus profiles, *IEEE Conference on Emerging Technologies and Factory Automation* (*ETFA*), Bilbao, Spain, September 13–16, 2010, pp. 1–4.
35. M. Lobashov, G. Pratl, T. Sauter, Applicability of internet protocols for fieldbus access, *IEEE International Workshop on Factory Communication Systems* (*WFCS*), Västerås, Sweden, August 28–30, 2002, pp. 205–213.
36. M. Kunes, T. Sauter, Fieldbus-internet connectivity: The SNMP approach, *IEEE Transactions on Industrial Electronics*, 48(6), 2001, 1248–1256.
37. Y.-G. Ha, Dynamic integration of Zigbee home networks into home gateways using OSGI service registry, *IEEE Transactions on Consumer Electronics*, 55(2), 2009, 470–476.
38. G. Moritz, E. Zeeb, S. Pruter, F. Golatowski, D. Timmermann, R. Stoll, Devices profile for Web services in wireless sensor networks: Adaptations and enhancements, *IEEE Conference on Emerging Technologies and Factory Automation* (*ETFA*), Palma de Mallorca, Spain, September 22–25, 2009, pp. 1–8.

39. V. Kapsalis, K. Charatsis, A. P. Kalogeras, G. Papadopoulos, Web gateway: A platform for industry services over Internet, *IEEE Symposium on Industrial Electronics (ISIE)*, L'Aquila, Italy, July 8–11, 2002, pp. 73–77.
40. T. Cucinotta, A. Mancina, G. F. Anastasi, G. Lipari, L. Mangeruca, R. Checcozzo, F. Rusina, A real-time service-oriented architecture for industrial automation, *IEEE Transactions on Industrial Informatics*, 5(3), 2009, 267–277.
41. E. Zeeb, R. Behnke, C. Hess, D. Timmermann, F. Golatowski, K. Thurow, Generic sensor network gateway architecture for plug and play data management in smart laboratory environments, *IEEE Conference on Emerging Technologies and Factory Automation (ETFA)*, Palma de Mallorca, Spain, September 22–25, 2009, pp. 1–9.
42. H. Cao, J. Chen, Service-oriented transparent interconnection between data-centric WSN and IP networks, *International Conference on Electrical and Control Engineering (ICECE)*, Wuhan, China, June 25–27, 2010, pp. 1884–1887.
43. A. Kalogeras, J. Gialellis, C. Alexacos, M. Georgoudakis, S. Koubias, Vertical integration of enterprise industrial systems utilizing Web services, *IEEE Transactions on Industrial Informatics*, 2(2), 2006, 120–128.
44. M. Velasco, J. M. Fuertes, P. Marti, Process data abstraction/accessibility via Internet, *IEEE Conference on Emerging Technologies and Factory Automation (ETFA)*, Lisbon, Portugal, 2002, Vol. 1, pp. 357–363.
45. J. L. M. Lastra, M. Delamer, Semantic web services in factory automation: Fundamental insights and research roadmap, *IEEE Transactions on Industrial Informatics*, 2(1), February 2006, 1–11.
46. J. Toro, J. Karppinen, J. Hintikka, L. Pohjanheimo, Automatic configuration and diagnostics for fieldbus based automation, *IEEE International Workshop on Factory Communication Systems (WFCS)*, Västerås, Sweden, 2002, pp. 143–148.
47. P. M. Blanco, M. A. Poli, M. R. Pereira Barretto, OPC and CORBA in manufacturing execution systems: A review, *IEEE International Conference on Emerging Technologies and Factory Automation (ETFA)*, Lisbon, Portugal, 2003, Vol. 2, pp. 50–57.
48. C. Diedrich, M. Muhlhause, M. Riedl, T. Bangemann, Mapping of smart field device profiles to web services, *IEEE Workshop on Factory Communication Systems (WFCS)*, Dresden, Germany, May 21–23, 2008, pp. 375–381.
49. T. Sauter, M. Lobashov, How to access factory floor information using internet technologies and gateways, *IEEE Transactions on Industrial Informatics*, 7(4), 2011, 699–712.
50. J.-D. Decotignie, The many faces of industrial Ethernet, *IEEE Industrial Electronics Magazine*, 3(1), 2009, 8–19.
51. J. Jasperneite, J. Imtiaz, M. Schumacher, K. Weber, A proposal for a generic real-time Ethernet system, *IEEE Transactions on Industrial Informatics*, 5(2), 2009, 75–85.

23

MTConnect: Different Devices, Common Connection*

23.1 Classic N2 Problem in Discrete Manufacturing........................ 23-1
Discrete Manufacturing after MTConnect
23.2 MTConnect: An Overview........................ 23-2
What MTConnect Is • What MTConnect Is Not
23.3 MTConnect Standard........................ 23-3
MTConnect Standard Part 1 • MTConnect Standard Part 2 • MTConnect Standard Part 3 • MTConnect Standard Part 4
23.4 Types of MTConnect-Enabled Devices........................ 23-4
23.5 Technical Pillars of MTConnect........................ 23-5
Adapter • Agent • Programmatic Conceptual Terms
23.6 MTConnect's Powerful Data Dictionary........................ 23-9
Application
23.7 Your First MTConnect Hello World Application........................ 23-11
23.8 Summary........................ 23-12
Reference........................ 23-12

Dave Edstrom
Virtual Photons Electrons, LLC

23.1 Classic N2 Problem in Discrete Manufacturing

The huge networking problem with manufacturing was that each machine tool spoke a different proprietary protocol. The problem is depicted in Figure 23.1. As the text in the drawing states, it is a connectivity and integration nightmare.

In Figure 23.2, you will notice that the number of connections is the classic *n*-squared problem of connectivity. This was a huge nightmare for software application developers. Prior to MTConnect, the onus was on software developers to have to write an adapter for each and every piece of manufacturing equipment on the shop floor that the customer was trying to collect data from. MTConnect takes the burden from the application developer and moves it where it should be—closer to the device itself.

23.1.1 Discrete Manufacturing after MTConnect

Figure 23.3 shows how MTConnect addresses this issue. From a developer standpoint, the problem of converting the proprietary formats has been moved from the client side much closer to the actual

* Information and images used in this chapter are from the book, *MTConnect: To Measure Is to Know* [1], and are copyrights of Virtual Photons Electrons, LLC.
MTConnect™ is a registered trademark of AMT—The Association for Manufacturing Technology.

FIGURE 23.1 The manufacturing world prior to MTConnect.

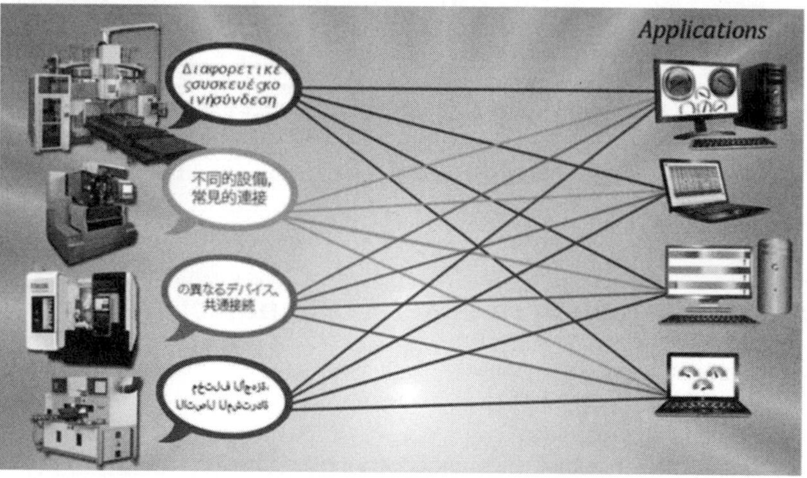

FIGURE 23.2 A version of the classic N^2 problem.

piece of manufacturing equipment. The advantage of this to the application developer is that now with MTConnect, each client application only needs to concern itself with speaking MTConnect.

23.2 MTConnect: An Overview

23.2.1 What MTConnect Is

MTConnect is an open and royalty-free standard for manufacturing that is used for connecting manufacturing equipment with applications by using proven Internet protocols. Think of MTConnect as the Bluetooth for manufacturing. With Bluetooth, both devices must *speak* Bluetooth for anything useful to happen. Just as simply having an OBD-II port on your car does not provide you with any more

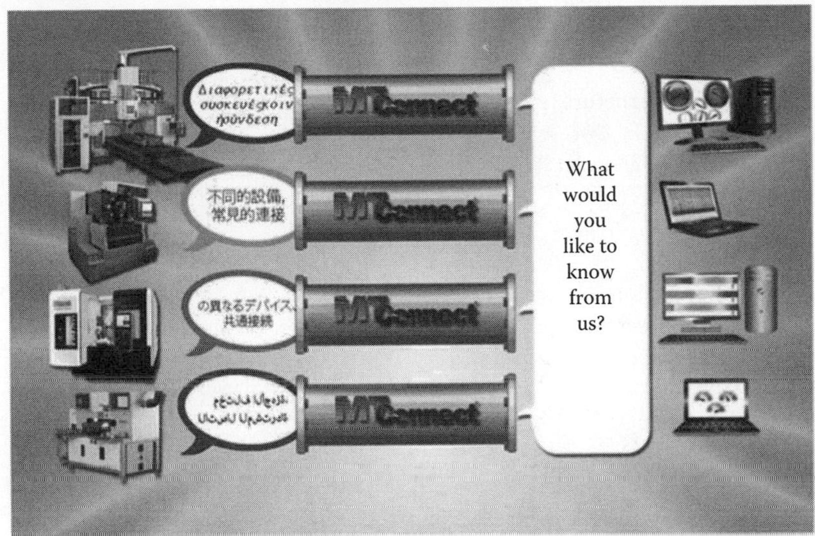

FIGURE 23.3 MTConnect is the Bluetooth for connecting the manufacturing equipment to applications.

data unless you have an OBD-II scanner, the same *pairing* principle applies to MTConnect and software applications. You can have an MTConnect-enabled machine tool or piece of manufacturing equipment, but without the software to read and analyze the data, you do not have both sides of the equation. The OBD-II scanner is really the application or the tool that you use to help you understand what is happening with your engine. In manufacturing, it is software applications, such as shop floor monitoring programs that are the applications, that speak to an MTConnect-enabled piece of manufacturing equipment on the shop floor.

23.2.2 What MTConnect Is Not

MTConnect is not an application. Let me say that again, MTConnect is not an application. MTConnect is not a product. The MTConnect Institute does not sell MTConnect. MTConnect is a standard based on open and royalty-free protocols. This is a common point of confusion with nonsoftware developers. Paul Warndorf likes to tell the story of the individual at IMTS 2010 who pulled out their credit card and wanted to buy MTConnect. It was analogous to a shopper in Best Buy pulling out their credit card and wanting to buy Bluetooth. It is natural to understand where the confusion comes in because MTConnect in and of itself is simply the plug-n-play for making it easy to get data from manufacturing equipment. However, it is what you *do* with that data that turn the MTConnect data into actionable intelligence.

23.3 MTConnect Standard

The MTConnect standard is broken down into four parts. What follows are the four standard sections of MTConnect and its major subsections. Before a developer is able to develop anything more than a very simple application, it will be necessary to read and understand Parts 1, 2, and 3 of the standard. Part 4 is for assets, and this becomes important for developers who are using those typically *mobile* assets of the manufacturing equipment. Mobile assets are those items that are not always connected to a piece of manufacturing equipment. A cutting tool or a part would be good examples. The purpose of listing the sections of each of the four parts is to let the reader know where to go for more specifics on the standard—which is at MTConnect.org

23.3.1 MTConnect Standard Part 1

Overview and Protocol
MTConnect Document Structure
Versioning
HTTP and XML—Brief Reminder
Architecture Overview
Request Structure
Agent Initialization
Application Communication
Agent Data and Agent Asset Storage
Reply XML Document Structure
Devices
Streams
Assets
Error
Protocol (Commands to Agent) Overview
Probe
Sample
Current
Asset
MTConnect Agent and Adapters

23.3.2 MTConnect Standard Part 2

Components
Data Items

23.3.3 MTConnect Standard Part 3

Streams
Events
Samples
Conditions

23.3.4 MTConnect Standard Part 4

Assets

23.4 Types of MTConnect-Enabled Devices

From a manufacturing equipment or device perspective, there are three types of MTConnect-enabled devices:

- Native
- Translation dependent
- Connection dependent

A native device simply means that this device comes with an MTConnect agent already included in the device, and an MTConnect-enabled application could immediately speak to this device with no additional software or hardware that is needed. Native MTConnect devices would be newer devices as MTConnect was first released to the outside world at the end of 2008.

A translation-dependent device is a device, which could be hardware and/or software, that acts as a translator between whatever proprietary hardware and/or software interface that the manufacturing equipment speaks and MTConnect applications.

An example of this might be a small black box that runs Linux and has both an adapter and an agent running. The black box might also have a number of ports on it to speak to a wide variety of protocols as well as have one or more Ethernet ports on it. The adapter speaks to the manufacturing equipment using the physical port and protocol that the manufacturing equipment understands and then that adapter sends information to the agent using the SHDR protocol, which is a simple flat, fast, and low-latency human-readable protocol delimited by pipes (|) that is in ISO 8601 date–time format with optional decimal places. The agent then takes that information and makes it available in an MTConnect-compliant format. A specific example of this might be a black box that has a FANUC FOCAS adapter that is speaking to a FANUC CNC controller as well as the MTConnect agent on the same box that is accepting SHDR from the adapter and then making it available to MTConnect-enabled applications.

The key part of an MTConnect translation-dependent device is that the adapter is typically speaking to an Application Programming Interface (API) to send queries to that device, translating that information into SHDR, and then sending it to the MTConnect agent. There are many examples of this where the CNC controller speaks a specific protocol and you most likely need to license a software development kit or software library to be able to speak to the CNC controller in the language that it knows. There are a variety of adapters at github.com/MTConnect.

A connection-dependent device is a device that could be hardware and/or software that acts as a translator between whatever proprietary hardware and/or software interface that the manufacturing equipment speaks and MTConnect applications. Sometimes there is confusion between a translation unit and a connection unit. Here is the key difference: the translation unit device is speaking to a device that has an API whereas a connection unit has no API. This typically means a connection unit is talking to an older piece of manufacturing equipment and pulling off very low-level signals, such as doing an A/D conversion, which are then transformed and sent to the agent. The simplest example of this might be an old lathe. The only piece of data that can be easily derived is whether the lathe is off, on, or in the process of cutting. A connection unit would have the appropriate ports to determine whether it was off, on, or in the process of cutting, and the adapter would pass the information along to the agent. The agent then would respond to client requests where only those three data items could be requested. While it might be argued that only knowing those three are not that helpful, the truth is that it is much better than a stack light, and even with this low-level example of MTConnect enabling a simple device, there are advantages above and beyond a stack light replacement. For example, with the electrical information, the costs of running that lathe could easily be calculated. Now the manufacturer knows the electrical cost of that device and what percent of the time the device is running and not running. The MTConnect connection unit black boxes are literally plug-n-play for these types of older pieces of manufacturing equipment.

23.5 Technical Pillars of MTConnect

From a software developer's perspective and at an 80,000 ft view, MTConnect has four key pillars:

1. Adapter
2. Agent
3. MTConnect's Data Dictionary aka Namespace or the XML Schema Definition
4. Application

Here are the big picture points to remember:

- Applications speak to an agent.
- The agent speaks to adapters (adapters are optional).
- Adapters or agents speak to manufacturing equipment.

23.5.1 Adapter

The adapter is the optional piece of software that can be thought of as the bridge between the manufacturing equipment and the agent. On one side, the adapter is speaking whatever protocol, as well as providing whatever physical mechanism, that is needed to speak to a piece of manufacturing equipment (could be a machine tool, bar feeder, coordinate measuring machine [CMM], etc.), normalizing that data, putting it into a modified version of the Simple Hierarchical Data Representation (SHDR) format, and sending it to the agent. The adapter is optional because for those pieces of manufacturing equipment where MTConnect is native, there is no need for an adapter.

The following are some key points to remember about MTConnect adapters:

- Do not appear in the MTConnect standard
- Are optional (for nonnative MTConnect devices)
- Speak to devices on one side of the equation
- Constantly stream output to agent(s)
- Typically do not speak to applications directly as this would involve bypassing the agent that makes little logical sense
- Can send data to multiple agents or just one agent—it is a design and implementation choice

The point to remember is that adapters speak to an agent(s) on one side of the equation and manufacturing equipment on the other.

The reference adapter is the adapter that the MTConnect Institute provides at the MTConnect location on github.com. What sometimes is confusing is that the MTConnect Institute provides a reference adapter and a reference agent to make it easy for developers to create these two pieces of software. Keep in mind that *reference* means it is either a binary or the source code that you can use to develop your agent or adapter. You can *refer* to it as you develop your own adapters or agents. This is a standard type of model used in software development. What confuses those outside of software is whether or not the MTConnect Institute mandates the reference agent or adapter—the answer to that is no. A software developer could simply read the MTConnect standard and write their own agents and adapters without ever viewing the reference software. It has been proven that by providing the reference adapters and agents, we have decreased the development time for those who are writing MTConnect agents and adapters.

The difficulty in writing an adapter varies from very easy and quick to very difficult and time consuming. The reason for the range is as follows:

- How much documentation do you have for the device you are trying to connect?
- What is the physical medium that is needed to get information?
- Is there a similar adapter that has been written?
- How much information are you trying to get out?
- Are the costs to get the documentation prohibitive?

Also consider the following examples of reference adapter and agent responsibilities.

1. The term *reference* agent is used loosely in this context because there is no standard for an adapter. This was done by design. While this is true, it is also true that many agent developers have used the SHDR protocol as the mechanism for the adapter to send data to the agent. Again, we give flexibility to the developers.
2. Adapters can provide a socket connection to which a process can connect and receive updates.
3. Adapters send data only when it changes (this is very important) and sends data only after handshake with agent. In other words, the adapter will not just send out data with no known agent.
4. Adapters should make sure all data conforms to a controlled vocabulary, fixed set of values, and only communicate those values. Examples are ControllerMode and Execution.

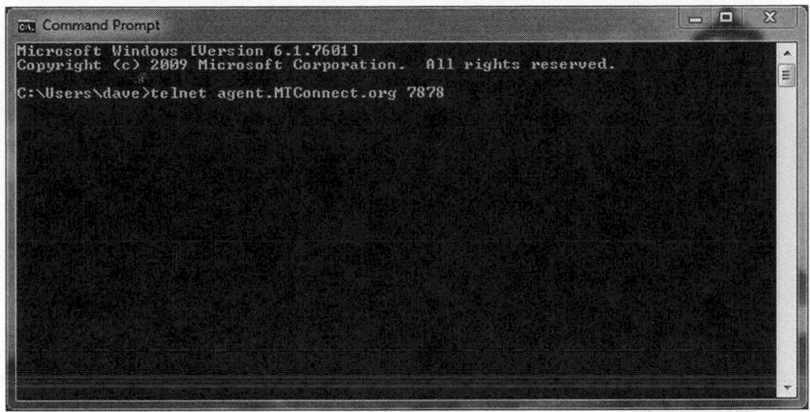

FIGURE 23.4 Talking directly to an adapter.

5. Agents can support multiple connections for testing.
6. Agents send all initial values to new clients when they connect.
7. Adapters format the data according to the SHDR specification.
8. There is a heartbeat between the reference adapter and the agent. For example,
    a. Agent sends a *ping* to the adapter.
    b. Adapter responds with a *pong*.
        i. If the agent does not receive a pong, then first assumption is this is a legacy adapter; if no response after 20 s, then the agent considers the adapter to be dead.
    c. There is a heartbeat between both the adapter and the agent every 10 s, but that can be changed.
    d. If one does not respond to the other's heartbeat for two times the interval, then that (agent/adapter) is considered dead.
9. Adapters perform the connection(s) to the manufacturing equipment interfaces and gather data.

Figure 23.4 shows the simplicity of talking to the adapter by using the standard telnet command. The user can simply telnet to the port defined for your adapter and see output that is going to the agent (Figure 23.5). This is very helpful for troubleshooting when you are developing your adapters and agents or you simply want to verify the adapter is putting out data.

Notice that the reference adapter output is very simple and is delimited by |. The agent is reading the adapter output, puts in a buffer that can then be read from an MTConnect-enabled application or from any application that can speak HTTP and read XML—such as a simple browser.

The following are a few examples of what comes out of the adapter:

2012-08-20T01:34:21.650493|Xact|-1.7883759737|Yact|-0.3661781847|Xcom|-1.7899452213|Ycom|-0.3595617390
2012-08-20T01:34:21.666492|Xact|-1.7894827127|Yact|-0.3615075648|Xcom|-1.7908221776|Ycom|-0.3556611774|path_feedrate|0.3061054894
2012-08-20T01:34:21.682492|Xact|-1.7905533314|Yact|-0.3568308055|Xcom|-1.7916012147|Ycom|-0.3517520805|path_feedrate|0.2972257736

The screen will be racing by with this type of information coming above from the adapter. The adapter determines the sampling rate.

Note: This connection to the machine tool interface and the gathering of data is where all of the hard and time-consuming work is done. Remember, the adapter is *not* part of the standard, and this just provides background on the reference adapter.

FIGURE 23.5 The output of the adapter. Above is SHDR output.

23.5.2 Agent

The agent can be thought of as a very efficient web server that bridges between adapter(s) and application(s). As a developer of client applications, expertise in what the agent is doing and understanding the standard is critical.

The agent *represents* one or more MTConnect-compliant machines. It makes machine data available in uniform MTConnect representations, responds to MTConnect commands, and allows clients to specify/select which data are of interest. The agent can *listen* to multiple adapters. Figure 23.6 is a picture of what the agent is doing.

Think of the agent as having two buckets or buffers that the controller deposits data into:

Samples Events Conditions Buffer default is 131,072, which is also known as 2^{17}.

Note: This must be a factor of two for speed purposes.

- Assets Buffer default is 1024.
- Both buffers can be changed in size.

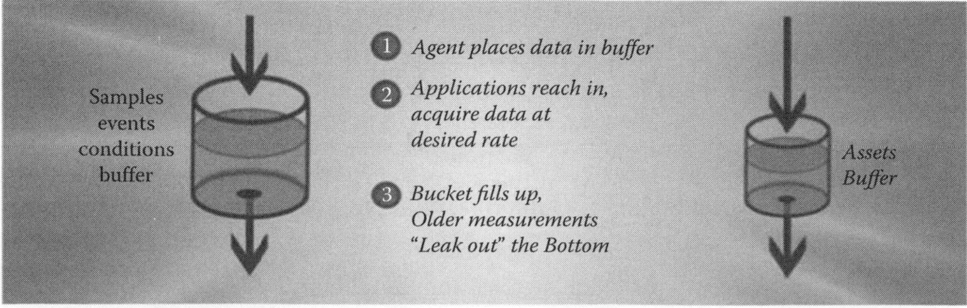

FIGURE 23.6 The agent *bucket* buffer model.

- Both of these can be changed to different sizes based on need. Each application *reaches into the bucket* at the rate it wants. As the bucket fills up, older measurements *leak out* the bottom depending on whether or not any applications have read them and the capacity of bucket, which depends on agent implementation.
- The ability to keep or persist the data can be done either at the application level or by creating a custom agent—whichever is the developer's design choice.

23.5.3 Programmatic Conceptual Terms

The following terms are key conceptual terms that software developers will need to know and understand when developing an application that will speak to the MTConnect agent.

- *Header.* Protocol-related information.
- *Components.* The building blocks of the device.
- *Data Items.* The description of the data available from the device.
- *Streams.* A set of samples or events for components and devices.
- *Samples.* A point-in-time measurement of a data item that is continuously changing.
- *Events.* Unexpected or discrete occurrence in a component. This includes state changes and conditions.
- *Conditions Are Normal, Warning, or Fault.* Warning is trending toward a fault, and fault means the device has stopped and needs intervention to get it working.

23.6 MTConnect's Powerful Data Dictionary

A very powerful component of the MTConnect standard (tons of hard work went into creating this) is the XML schema or the data dictionary. The file http://www.mtconnect.org/schemas/MTConnectDevices_1.2.xsd is over 600 lines long. It is this data dictionary that makes the job of the client application developer much easier. Applications that do not provide a schema place the burden on determining what each value means, as well as the allowable values, back onto the shoulders of the application developer. Figure 23.7 drives home this point.

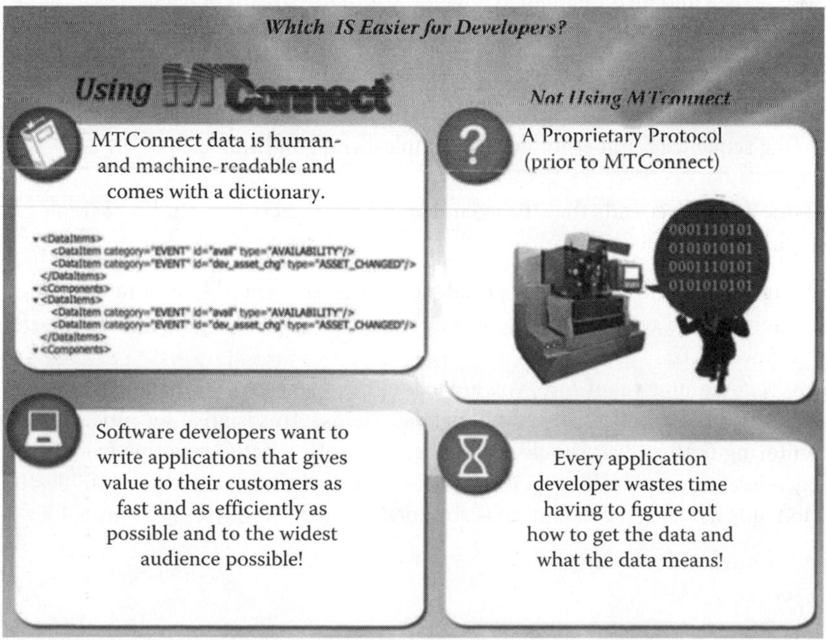

FIGURE 23.7 Using MTConnect vs. not using MTConnect.

In Figure 23.7, notice that AssetCountAttrType is *The number of assets*; it is an integer that has a minimum value of 0 and a maximum value of 4,294,967,295. I doubt 4,294,967,295 will be a limit for assets anytime in my lifetime.

It is impossible to overemphasize the importance and the time that went into creating MTConnect's powerful namespace or data dictionary. It takes years and thousands of man-hours to do it properly. There are standards out there that do not supply a data dictionary and place the entire burden of knowing what bits are coming to the application on the software developer writing client applications.

You may wonder, *How does one making suggestions for additions to the MTConnect standard*? There is a very good white paper at MTConnect.org titled, *Recommending the Addition of Devices, Components, and Data Items to the MTConnect Standard*, which addresses this question.

23.6.1 Application

The application or client is the software that speaks directly to the agent, requesting information and keeping track of the sequence numbers so it can continue to request and process XML information coming from the agent. The most obvious application when connecting to a piece of manufacturing equipment is shop floor monitoring software. A shop floor monitoring program typically requests information from the agent(s), processes that information, and then displays graphs or charts on what the piece of manufacturing equipment is doing.

Applications speak to an MTConnect agent using HTTP and sending simple verbs. The agent returns back well-formed and valid XML.

- Probe—tell me about yourself.
- Sample—information passed with parameters to get specific data back from the agent.
- Current—what is happening right now.
- Asset—to access information on mobile assets or those assets that are not always part of the device such as a cutting tool or a part. That cutting tools can be changed out is why they are mobile.

The sample command retrieves a series of data starting from a position and returns up to the requested number of samples or events. It allows the application to retrieve all data without missing anything. It can stream data as it arrives and be thought of as a window into the stream of data.

The sample request retrieves values for a component's data items.

Path. Xpath expression specifies components and/or data items.
Default is all components in device or devices if no device is specified.
From. Starting sequence number for events, samples, and conditions.
Default is 0.
Interval. Time in milliseconds that the agent should pause between sending samples for events, samples, and conditions.

Sample can be used with an interval in special cases, and this should be done only judiciously (app beware). The interval used to be limited to a minimum of 10 ms, but with the device integration, we allowed it to go down to 0.

When you specify an interval of zero, you are telling the agent that "when a value changes, notify me *immediately* and don't try to collect any additional data items." This is great for processes that need to do real-time monitoring (<10 ms). It should always be used with a path since this will allow only specific events to trigger. Real time is like beauty; it is in the eye of the beholder, so do not think <10 ms is a universal definition; it is not, and the definition is absolutely context sensitive: 1000 ms is 1 s.

Count. Must return next sequence
Default is 100

Run the following MTConnect sample command in a browser—http://agent.mtconnect.org/sample

The following is a snippet from the sample request (with no additional parameters) to show you the streams node; the ComponentStream; the ComponentStream with the name of *VMC-3Axis*; and the various events, conditions, and data items.

```
<Streams>
    <DeviceStream name = "VMC-3Axis" uuid = "000">
        <ComponentStream component = "Rotary" name = "C" componentId = "c1">
    <Samples>
    <ComponentStream component = "Controller" name = "controller"
    componentId = "cn1">
    <Events>
        <EmergencyStop dataItemId = "estop" timestamp =
        "2012-12-31T04:12:51.242511Z" sequence = "2934232704">UNAVAILABLE
        </EmergencyStop>
    </Events>
    <Condition>
    <Unavailable dataItemId = "clp" timestamp = "2012-12-31T04:11:51.227979Z"
    sequence = "2934232644" type = "LOGIC_PROGRAM"/>
    </Condition>
</ComponentStream>
```

The following is a snippet from running the following simple MTConnect command that runs a sample with a count of 10.
http://agent.mtconnect.org/sample?count = 10

```
<Header creationTime = "2012-12-26T13:55:15Z" sender =
"mtconnect"instanceId = "1340212647"version = "1.2.0.10"
bufferSize = "131072"nextSequence = "2933744389"
firstSequence = "2933744379"
lastSequence = "2933875450"/>
<Streams><DeviceStream name = "VMC-3Axis" uuid = "000">
```

Notice these bold XML attributes highlighted in the header. As a developer of applications, the **instanceId** is important. As was just pointed out in a previous section, the **instanceId** was created to facilitate recovery when the agent fails and the application is unaware. It is up to the developer to track sequence numbers. Version is important as there have been changes in the standard. The MTConnect specification is the best place to review these changes.

Please note that if you would like to simply cut-and-paste these commands into a browser, you can go to http://ToMeasureIsToKnow.com, where I have all of these examples listed. It is beyond the scope of this chapter to get into all the specifics of writing an application, so the reader is encouraged to go http://ToMeasureIsToKnow.com to learn more.

What is interesting for software developers is all the ways MTConnect data can be analyzed in the countless other non-shop-floor applications.

23.7 Your First MTConnect Hello World Application

The following is the simplest of Java programs. The program uses Java's basic networking and Input/Output classes to read agent.MTConnect.org with a probe command. Print the XML out until there is nothing left to print. *Corvette* is my prompt on my MacBook Pro in a terminal window. You can go to http://ToMeasureIsToKnow.com if you just want to simply cut and paste the following source code into your editor or IDE.

```
Corvette:more MTConnectHelloWorld.java

import java.net.*;
import java.io.*;
public class MTConnectHelloWorld {
        public static void main(String[] args) throws Exception {
        System.out.println("Your First Hello World Program" + "\n");
        URL MTConnect = new URL("http://agent.MTConnect.org/probe");
        BufferedReader in = new BufferedReader(
        new InputStreamReader(MTConnect.openStream()));
        String inputLine;
        while ((inputLine = in.readLine()) ! = null)
        System.out.println(inputLine);
        in.close();
        System.out.println("\n" + "XML from probe command printed above");
    }
}
```

```
Compile the Java source
  Corvette:javac MTConnectHelloWorld.java
Execute the Java class
  Corvette:java MTConnectHelloWorld
```

At this point, you will see a few pages of XML coming from the simulator that is running as a CNC vertical mill doing a very simple part in an endless loop.

If you are on a *nix (Mac OS X is obviously in this category as well) box, then you can accomplish the same as given earlier from a command line:

```
Corvette:curl http://agent.MTConnect.org/probe
```

If you are a PC user, cURL(Command URL) comes with the PowerShell terminal or you could FTP cURL from SourceForge as well. You could also simply enter the earlier-provided URL from a web browser by just putting in the URL http://agent.MTConnect.org/probe. You can put these in your browser to see the results. Using a browser to enter in MTConnect commands is a great way to learn about MTConnect and do testing as well.

23.8 Summary

MTConnect is not an evolution in manufacturing, but it is a revolution and a true game changer. MTConnect is making possible the dreams and desires of generations of manufacturers, machine tool builders, and manufacturing equipment providers who all want to see the same goal of different devices having a common connection on the plant floor. This is why at the MTConnect Institute, we say, "MTConnect: *Different* Devices, *Common* Connection."

Reference

1. D. Edstrom, *MTConnect: To Measure Is To Know*, Virtual Photons Electrons, Ashburn, VA, 2013.

24

Standard Message Specification for Industrial Automation Systems: ISO 9506 (MMS)

24.1 Introduction 24-1
24.2 MMS Client/Server Model 24-2
24.3 Virtual Manufacturing Device 24-3
General Model • MMS Models and Services • Locality of the VMD
24.4 Interfaces 24-9
24.5 Environment and General Management Services 24-10
24.6 VMD Support 24-11
24.7 Domain Management 24-12
Model Basics • What Is the Domain Scope?
24.8 Program Invocation Management 24-14
Model Basics • Program Invocation Services
24.9 MMS Variable Model 24-16
Model Basics • Access Paths • Objects of the MMS Variable Model • Unnamed Variable • MMS Address of the Unnamed Variable • Services for the Unnamed Variable Object • Explanation of the Type Description • Named Variable • Access to Several Variables
24.10 Resume 24-32
References 24-32
Websites 24-33

Karlheinz Schwarz
Schwarz Consulting Company

24.1 Introduction

The international standard Manufacturing Message Specification (MMS) is an OSI application layer messaging protocol designed for the remote control and monitoring of devices such as remote terminal units, programmable logic controllers (PLCs), numerical controllers, or robot controllers. It provides a set of services allowing the remote manipulation of variables, programs, semaphores, events, journals, etc. MMS offers a wide range of services satisfying both simple and complex applications.

For years, the automation of technical processes has been marked by increasing requirements with regard to flexible functionalities for the transparent control and visualization of any kind of processes. The mere cyclic data exchange will more and more be replaced by systems that join together independent yet coordinated systems—like communication, processing, open and closed loop control, quality protection, monitoring, configuring, and archiving systems—to a whole. These individual systems are interconnected and work together. As a common component, they require a suitable real-time communication system with adequate functions.

The MMS standard defines common functions for distributed automation systems. The expression *manufacturing*, which stands for the first M in MMS, has been badly chosen. The MMS standard does not contain any manufacturing-specific definitions. The application of MMS is as general as the application of a personal computer. MMS offers a platform for a variety of applications.

The first version of MMS documents was published in 1990 by ISO TC 184 (Industrial Automation) as an outcome of the GM initiative *Manufacturing Application Protocols* (MAPs). The current version has been published in 2003:

- *Part 1: ISO 9506-1 Services* (2003): describes the services that are provided to remotely manipulate the MMS objects. For each service, a description is given of the parameters carried by the service primitives. The services are described in an abstract way, which does not imply any particular implementation.
- *Part 2: ISO 9506-2 Protocol* (2003): specifies the MMS protocol in terms of messages. The messages are described with a notation ASN.1, which gives the syntax.

Today, MMS is being implemented on all common communication networks that support a safe transport of data. These can be networks like TCP/IP or ISO/OSI on Ethernet, a fieldbus, or simple point-to-point connections like HDLC, RS 485, or RS 232. MMS is independent of a seven-layer stack. Since MMS was originally developed in the MAP environment, it was generally believed earlier that MMS could be used only in connection with MAP.

MMS is the basis of the international projects IEC 60870-6-TASE.2 (*Inter-control center communication*), IEC 61850 (*Communication networks and systems for power utility automation*), and IEC 61400-25 (*Communications for monitoring and control of wind power plants*).

This chapter introduces the basic concepts of MMS applied in the standards mentioned earlier.

24.2 MMS Client/Server Model

MMS describes the behavior of two communicating devices by the client/server model (see Figure 24.1). The client can, for example, be an operating and monitoring system, a control center, or another intelligent device. The server represents one or several real devices or whole systems. MMS uses an object-oriented modeling method with object classes (Named Variable, Domain, Named Variable List, Journal, etc.), instances from the object classes, and methods (services like read, write, information report, download, and read journal).

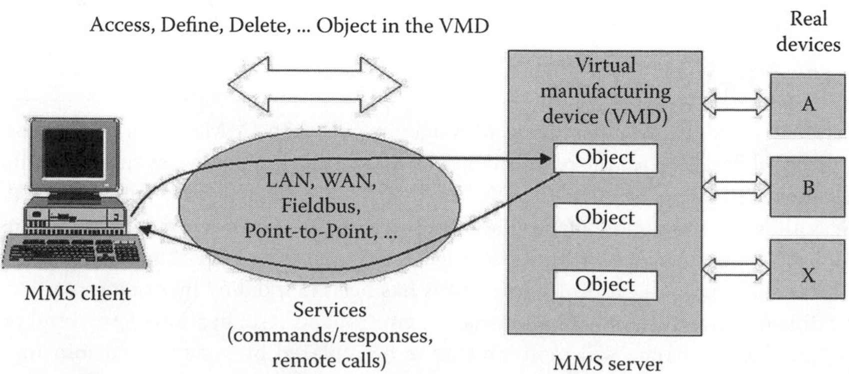

FIGURE 24.1 MMS client/server model.

The standard is comprehensive. This does not at all mean that an MMS implementation must be complex or complicated. If only a simple subset is used, then the implementation can also be simple. Meanwhile, MMS implementations are available in the third generation. They allow the use of MMS both on PC platforms and on embedded controllers.

The MMS server represents the objects that the MMS client can access. The virtual manufacturing device (VMD) object represents the outermost *container* in which all other objects are contained.

Real devices can play both roles (client and server) simultaneously. A server in a control center for its part can be a client with respect to a substation. MMS basically describes the behavior of the server. The server contains the MMS objects, and it also executes services. MMS can be regarded as *server centric.* In principle, in a system, more devices are installed that function as server (e.g., controllers and field devices) than devices that perform the client role (e.g., PC and workstation).

The *calls* that the client sends to the server are described in the part ISO 9506-1 (services). These *calls* are processed and answered by the server. The services can also be referred to as remote calls, commands, or methods. Using these services, the client can access the objects in the server. It can, for example, browse through the server, that is, making visible all available objects and their definitions (configurations). The client can define, delete, change, or access objects via reading and writing.

An MMS server models real data (e.g., temperature measurement, counted measurand, or other data of a device). These real data and their implementation are concealed or hidden by the server. MMS does not define any implementation details of the servers. It is only defined how the objects behave and represent themselves to the outside (from the point of view of the wire) and how a client can access them.

MMS provides the very common classes. The Named Variable, for example, allows structuring any information provided for access by an application. The content (the semantic of the exchanged information) of the Named Variables is outside of the MMS standard. Several other standards define common and domain-specific information models.

IEC 61850 defines the semantic of many *points* in electric substations. For example, *Atlanta26/XCBR3.Pos.stVal* is the position of the third circuit breaker in substation *Atlanta26*. The names *XCBR*, *Pos*, and *stVal* are standardized names.

The standard IEC 61400-25 (communication for wind power plants) defines a comprehensive list of named *points* specific for wind turbines. For example, *Tower24/WROT.RotSpd.mag* is the (deadbanded) measured value of the rotor position of Tower24. *RotSpd.avgVal* is the average value (calculated based on a configuration attribute *avgPer*. These information models are based on common data classes like measured value, three-phase value (delta and Y), single-point status.

24.3 Virtual Manufacturing Device

24.3.1 General Model

According to Figure 24.2, the real data and devices are represented—in the direction of a client—by the VMD). In this regard, the server represents a *standard driver* that maps the real world to a virtual one. The following definition helps to clarify the modeling in the form of a virtual device:

If it's there and you can see it	It's REAL
If it's there and you can't see it	It's TRANSPARENT
If it's not there and you can see it	It's VIRTUAL
If it's not there and you can't see it	It's GONE

Roy Wills

The VMD can represent, for example, a variable *Measurement3* whose value may not permanently exist in reality; only when the variable is being read, a measurand transducer will get started to determinate

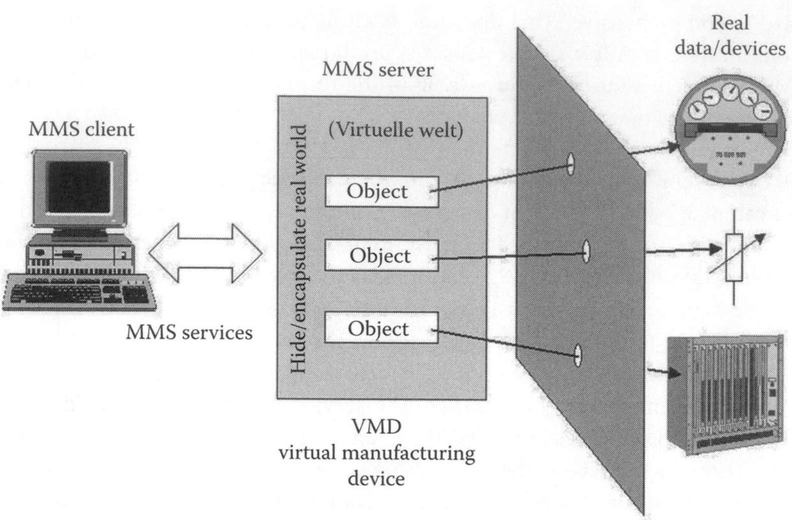

FIGURE 24.2 Hiding real devices in the virtual manufacturing device (VMD).

the value. All objects in a server can already be contained in a device before the delivery of a device. The objects are predefined in this case.

Independent of the implementation of a VMD, data and the access to data are always treated in the same way. This is completely independent of the operating system, the programming language, and memory management. Like printer drivers for a standard operating system hide the various real printers, so a VMD also hides real devices. The server can be understood as a communication driver that hides the specifics of real devices. From the point of view of the client, only the server with its objects and its behavior is visible. The real device is not visible directly.

MMS merely describes the server side of the communication (objects and services) and the messages that are exchanged between client and server.

The VMD describes a VMD completely. This virtual device represents the behavior of a real device as far as it is visible *over the wire*. It contains, for example, an identification of manufacturer, device type, and version. The virtual device contains objects like variables, lists, programs and data areas, semaphores, events, and journals.

The client can read the attributes of the VMD (see Figure 24.3), that is, it can browse through a device. If the client does not have any information about the device, it can view all the objects of the VMD and their attributes by means of the different Get services. With that, the client can perform a first plausibility check on a just installed device by means of a Get(Object-Attribute)-Service. It learns whether the installed device is the ordered device with the right model number (Model Name) and the expected issue number (Revision). All other attributes can also be checked (e.g., variable names and types).

The attributes of all objects represent a self-description of the device. Since they are stored in the device itself, a VMD always has the currently valid and thus consistent configuration information of the respective device. This information can be requested online directly from the device. In this way, the client receives always up-to-date information.

MMS defines some 80 functions:

- Browsing functions about the *contents* of the virtual device: "Which objects are available?"
- Functions for reading, reporting, and writing of arbitrarily structured variable values
- Functions for the transmission of data and programs, the control of programs, and many other functions

The individual groups of the MMS services and objects are shown in Figure 24.4. MMS describes such aspects of the real device that shall be open, that is, standardized. An open device must behave as

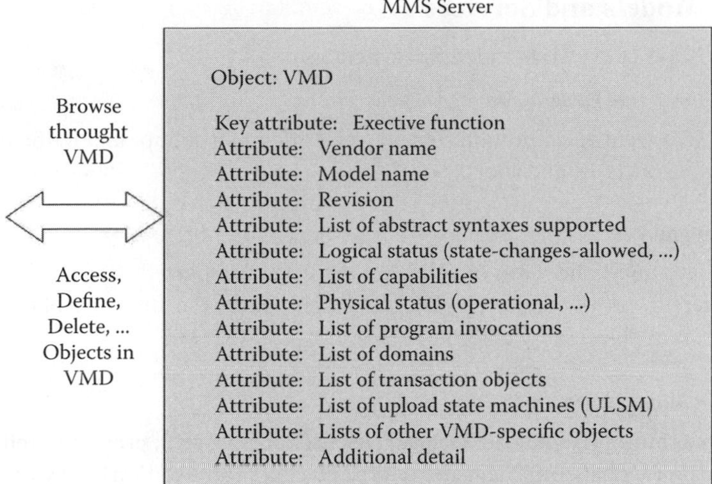

FIGURE 24.3 VMD attributes.

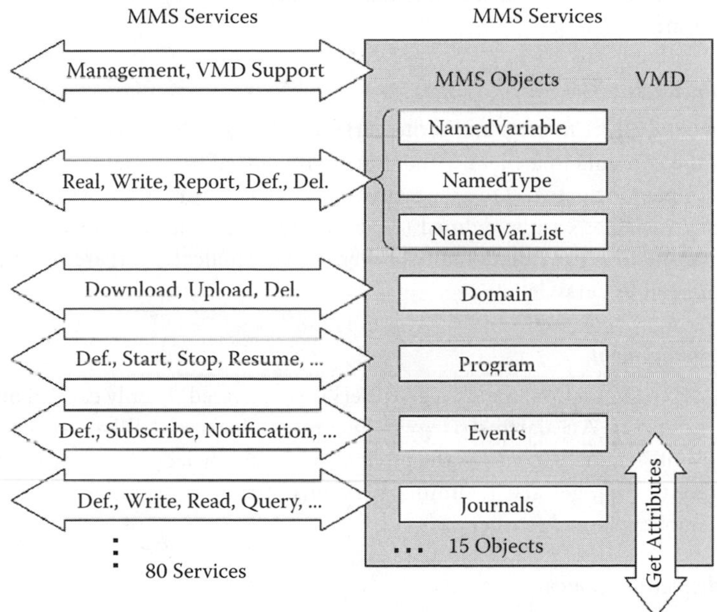

FIGURE 24.4 MMS objects and services.

described by the virtual device. How this behavior is achieved is not visible and also not relevant to the user that accesses the device externally. MMS does not define any local, specific interfaces in the real systems. The interfaces are independent of the functions that shall be used remotely. Interfaces in connection with MMS are always understood in the sense that MMS quasi represents an interface between the devices and not within the devices. This interface could be described as an external interface. Of course, interfaces are also needed for implementations of MMS functions in the individual real devices. These shall not and cannot be defined by a single standard. They are basically dependent on the real systems—and these vary to a great extent.

24.3.2 MMS Models and Services

24.3.2.1 ISO 9506-1 (Part 1): Service Specification

24.3.2.1.1 Environment and General Management Services

Two applications that want to communicate with each other can set up, maintain, and close a logical connection (initiate, conclude, and abort).

24.3.2.1.2 VMD Support

The client can thereby query the status of a VMD or the status is reported (Unsolicited Status); the client can query the different lists of the objects (Get Name List), query the attributes of the VMD (Identify), or change the names of objects (Rename).

24.3.2.1.3 Domain Management

Using a simple flow control (Download, Upload, Delete Domains, etc.), programs and data of arbitrary length can be transmitted between client and server and also a third station (and vice versa). In the case of simple devices, the receiver of the data stream determines the speed of the transmission.

24.3.2.1.4 Program Invocation Management

Services to create, start, stop, and delete modularly structured programs by the client (start, stop, resume, kill, delete, etc.).

24.3.2.1.5 Variable Access

This service allows the client to read and write variables that are defined in the server or a server is enabled to report the contents to a client without being requested (information report). The structures of these data are simple (octet string) to arbitrarily complex (structure of array of …). In addition, data types and arbitrary variables can be defined (Read, Write, information report, define variable, etc.). The variables constitute the core functionality of every MMS application; therefore, the variable access model will be explained in detail later.

24.3.2.1.6 Event Management

It allows an event-driven operation, that is, a given service (e.g., Read) is only carried out if a given event has occurred in the server. An alarm strategy is integrated. Alarms will be reported to one or more clients if certain events occur. These have the possibility to acknowledge the alarms later (define, alter event condition monitoring, get alarm summary, event notification, acknowledge event notification, etc.). This model is not explained further.

24.3.2.1.7 Semaphore Management

The synchronization of several clients and the coordinated access to the resources of real devices is carried out hereby (define semaphore, take/relinquish control, etc.). This model is not explained further.

24.3.2.1.8 Operator Communication

Simple services for the communication with operating consoles integrated in the VMD (input and output). This model is not explained further.

24.3.2.1.9 Journal Management

Several clients can enter data into journals (archives and logbooks) that are defined in the server. Then these data can selectively be retrieved through filters (Write Journal, Read Journal, etc.). This model is not explained further.

24.3.2.2 ISO 9506-2 (Part 2): Protocol Specification

If a client invokes a service, then the server must be informed about the requested type of the service. For a Read service, for example, the name of the variables must be sent to the server. This information that the server needs for the execution is exchanged in so-called protocol data units (PDUs). The set of all the PDUs that can be exchanged between client and server constitutes the MMS protocol.

In other words, the protocol specification—using the standards ISO 8824 (Abstract Syntax Notation One, ASN.1) and 8825 (ASN.1 Basic Encoding Rules, BER)—describes the abstract and concrete syntax of the functions defined in part 1. The syntax is explained later exemplarily.

24.3.3 Locality of the VMD

VMDs are virtual descriptions of real data and devices (e.g., protection devices, measurand transducers, wind turbine, and any other automation device or system). Regarding the implementation of a VMD, there are three very different possibilities where a VMD can be located (see Figure 24.5):

1. In the end device: One or several VMDs are in the real device that is represented by the VMD. The implementations of the VMD have direct access to the data in the device. The modeling can be carried out in such a way that each application field in the device is assigned to its own VMD. The individual VMDs are independent of each other.
2. In the gateway: One or several VMDs are implemented in a separate computer (a so-called gateway or agent). In this case, all MMS objects that describe the access to real data in the devices are at a central location. While being accessed, the data of a VMD can be in the memory of the gateway—or they must be retrieved from the end device only after the request. The modeling can be carried out in such a way that for each device or application, a VMD of its own will be implemented. The VMDs are independent of each other.

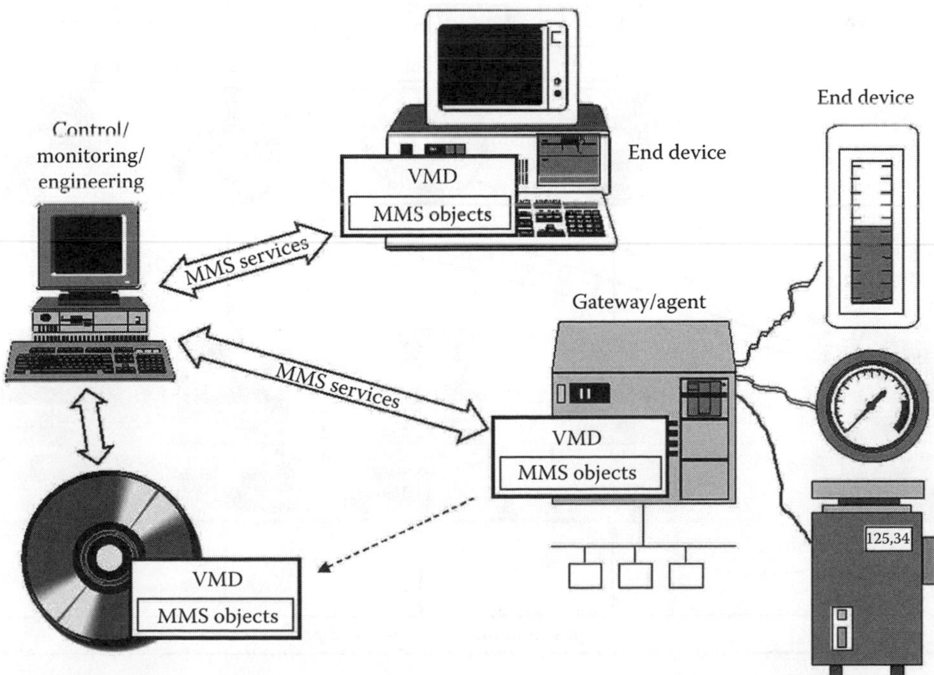

FIGURE 24.5 Location of VMDs.

3. In a file: One or several VMDs are implemented in a database on a computer, on an FTP server, or on a CD ROM (the possibilities under (1) and (2) are also valid here). Thus, all VMDs and all included objects with all their configuration information can be entered directly into engineering systems. Such a CD ROM, which represents the device description, also could be used, for example, to provide a monitoring system with the configuration information: names, data types, object attributes, etc. Before devices are delivered, the engineering tools can already process the accompanying device configuration information (electronic data sheet)! The configuration information can also be read later online from the respective VMD via corresponding MMS requests.

The VMD is independent of the location. This also allows, for example—besides the support during configuration—that several VMDs can be installed for testing purposes on another computer than the final system (see Figure 24.6). Thus, the VMD of several large robots can be tested in the laboratory or office. The VMD will be installed on one or several computers (the computers emulate the real robots). Using a suitable communication (e.g., intranet or also a simple RS 232 connection—available on every PC), the original client (a control system that controls and supervises the robots) can now access and test the VMD in the laboratory. This way, whole systems can be tested beforehand regarding the interaction of individual devices (e.g., monitoring and control system).

If the Internet is used instead of the intranet, global access is possible to any VMD that is connected to the Internet. The author himself tested the access from Germany to a VMD that was implemented on a PLC in the United States. That is to say, through standards like MMS and open transmission systems, it has become possible to set up global communication networks for the real-time process data exchange.

The previous statements about the VMD are valid in full extent also for all standards that are based on MMS.

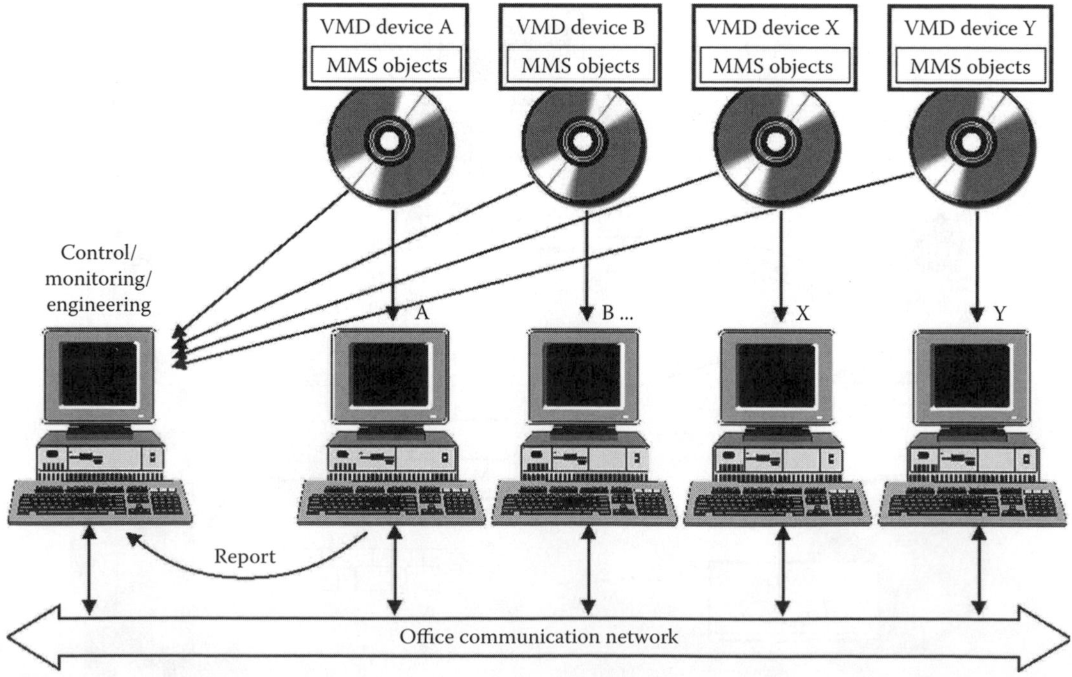

FIGURE 24.6 VMD testing using PC in an office environment.

24.4 Interfaces

The increasing distribution of automation applications and the exploding amount of information require more and more and increasingly more complex interfaces for operation and monitoring. Complex interfaces turn into complicated interfaces very fast. Interfaces *cut* components into two pieces; through this, interactions between the emerged subcomponents—which were hidden in one component before—become visible. An interface discloses which functions are carried out in the individual subcomponents and how they act in combination.

Transmitter and receiver of information must likewise be able to understand these definitions. The request *Read T142* must be formulated *understandably*, transmitted correctly, and understood unambiguously (see Figure 24.7). The semantic (named terms that represent something) of the services and the service parameters are defined in MMS. The content, for example, of Named Variables is defined in domain-specific standards like IE 61850.

Interfaces occur in two forms:

1. Internal program-program interfaces or APIs
2. External interfaces over a network (WAN, LAN, fieldbus, etc.)

Both interfaces affect each other. MMS defines an external interface. The necessity of complex interfaces (complex because of the necessary functionality, not because of an end in itself) is generally known and accepted. To keep the number of complex interfaces as small as possible, they are defined in standards or industry standards—mostly as open interfaces. Open interfaces are in the meantime an integral component of every modern automation. In mid-1997, it was already explained in [17] that the trend in automation engineering obviously leads away from the proprietary solutions to open standardized interfaces—that is, to open systems.

The reason why open interfaces are complicated is not because they were standardized. Proprietary interfaces tend more toward being complicated or even very complicated. The major reasons for the latter observation are found in the permanent *improvement* of the interfaces that expresses itself in the quick changes of version and in the permanent development of new—apparently better—interfaces. Automation systems of one manufacturer often offer—for identical functions—a variety of complicated interfaces that are incompatible to each other.

At first, interfaces can be divided up into two classes (see Figure 24.8): internal interfaces (e.g., in a computer) and external interfaces (over a communication network). The following consideration is strongly simplified because, for example, in reality, both internally and externally several interfaces can lie one above the other. However, it nevertheless shows the differences in principle that must be paid attention to. MMS defines an external interface. Many understand MMS in such a way that it offers—or at least also offers—an internal interface. This notion results in completely false ideas. Therefore, the following consideration is very helpful.

FIGURE 24.7 Sender and receiver of information.

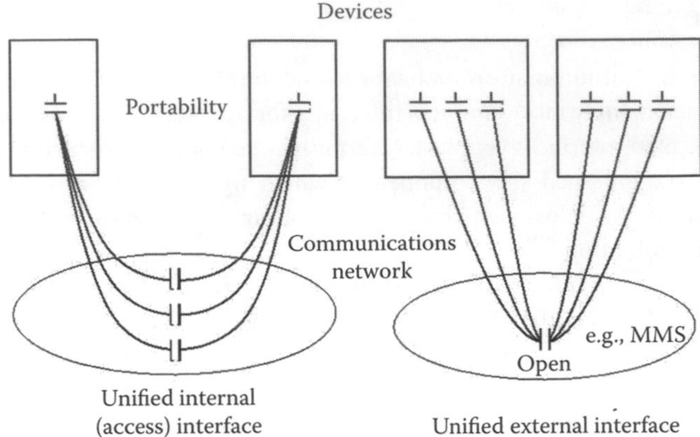

FIGURE 24.8 Internal and external interfaces.

The left-hand side of the figure shows the case with a uniform internal interface and varying external interfaces. This uniform internal interface allows many applications to access the same functions with the same parameters and perhaps the same programming language—independent of the external interface. Uniform internal interfaces basically allow the portability of the application programs over different external interfaces.

The right-hand side of the figure shows the case with the external interface being uniform. The internal interfaces are various (since the programming languages or the operating systems, e.g., are various). The uniform external interface is independent of the internal interface. The consequence of this is, for example, that devices whose local interfaces differ and are implemented in diverse environments can communicate together. Differences can result, for example, from an interface being integrated into the application in a certain device, but being available explicitly in another device. The essential feature of this uniform external interface is the interoperability of different devices. The ISO/OSI reference model is aimed at exactly this feature.

The (internal) MMS interface, for example, in a client (perhaps: $READ [Par. 1, Par. 2, …, Par. N]) depends on the manufacturer, the operating system, and the programming language. MMS implementations are, for example, available for UNIX or Windows NT. On the one hand, this is a disadvantage because applications that want to access an MMS server would have to support—depending on the environment—various real program interfaces. On the other hand, the MMS protocol is completely independent of the—fast changing—operating system platforms. Standardized external interfaces like MMS offer a high degree on stability, because in the first place, the communication can hardly be changed arbitrarily by a manufacturer, and in the second place, several design cycles of devices can survive.

Openness describes in the ISO/OSI world the interface on the *wire*. The protocol of this external interface executes according to defined standardized rules. For an interaction of two components, these rules have to be taken into account on the two sides; otherwise, the two will not understand each other.

24.5 Environment and General Management Services

MMS uses a connection-oriented mode of operation. That is to say, before, for example, a computer can read a value from a PLC for the first time, a connection must be set up between the two.

MMS connections have particular quality features such as the following:

- Exclusive allocation of computer and memory resources to a connection. This is necessary to guarantee that all services (e.g., five Read) that are allowed to be carried out simultaneously find sufficient resources on both sides of the connection.
- Flow control in order to avoid blockages and vain transmissions, if, for example, the receive buffers are full.

- Segmentation of long messages.
- Routing of messages over different networks.
- Supervision of the connection if no communication takes place.
- Acknowledgment of the transmitted data.
- Authentication, access protection (password), and encoding of the messages.

Connections are generally established once and then remain established as long as a device is connected (at least during permanently necessary communication). If, for example, a device is only seldom accessed by a diagnostics system, a connection then does not need to be established permanently (waste of resources). It suffices to establish a connection and, later, to close it to release the needed resources again. The connection can remain established for rare but time-critical transmissions. The subordinate layers supervise the connection permanently. Through this, the interruption of a connection is quickly recognized.

The MMS services for the connection management are

- *Initiate* (connection setup)
- *Conclude* (orderly connection teardown, waiting requests are still being answered)
- *Abort* (abrupt connection teardown, waiting requests are deleted)

Besides these services that are all mapped to the subordinate layers, there are two further services:

- *Cancel*
- *Reject*

After the MMS client has sent a Read request to the MMS server, for example, it may happen that the server leaves the service in its request queue and—for whatever reason—does not process it. Using the service cancel, the client can now delete the request in the server. On the other hand, it may occur that the server shall carry out a service with *forbidden* parameters. Using reject, it rejects the faulty request and reports this back to the client.

Although MMS was originally developed for ISO/OSI networks, a number of implementations are available in the meantime that also use other networks such as the known TCP/IP network. From the point of view of MMS, this is insignificant as long as the necessary quality of the connection is guaranteed.

24.6 VMD Support

The VMD object consists of 12 attributes. The key attribute identifies the *executive function*. The executive function corresponds directly with the entity of a VMD. A VMD is identified by a presentation address):

Object: VMD

Key Attribute:	Executive Function
Attribute:	Vendor Name
Attribute:	Model Name
Attribute:	Revision
Attribute:	Logical Status (STATE-CHANGES-ALLOWED, NO-STATE-CHANGES-ALLOWED, LIMITED-SERVICES-SUPPORTED)
Attribute:	List of Capabilities
Attribute:	Physical Status (OPERATIONAL, PARTIALLY-OPERATIONAL, INOPERABLE NEEDS-COMMISSIONING)
Attribute:	List of Program Invocations
Attribute:	List of Domains
Attribute:	List of Transaction Objects
Attribute:	List of Upload State Machines (ULSM)
Attribute:	List of Other VMD-specific Objects

The attributes Vendor Name, Model Name, and Revision inform about the manufacturer and the device.

The Logical status defines which services may be carried out. The status Limited-Services-Supported allows that only such services may be executed that have read access to the VMD.

The Physical status indicates whether the device works in principle.

The two following services

1. *Unsolicited Status*
2. *Status*

are used to get the status unsolicited (Unsolicited Status) or explicitly requested (Status). Thus, a client can recognize whether a given server—from the point of view of the communication—works at all.

The List of Capabilities offers clients and servers a possibility to define application-specific agreements in the form of features. The available memory of a device, for example, could be a capability. Through the Service Get Capability List, the current value can be queried. The remaining attributes contain the lists of all the MMS objects available in a VMD. The VMD contains therefore an object dictionary in which all objects of a VMD are recorded.

The following three services complete the services of the VMD:

1. *Identify* supplies the VMD attributes Vendor Name, Model Name, and Revision. With that, a plausibility check can be carried out from the side of the client.
2. *Get Name List* returns the names of all MMS objects. It can be selectively determined from which classes of objects (e.g., Named Variable or event condition) the names of the stored objects shall be queried. Let's assume a VMD is not yet known to the client till now (because it is, e.g., a maintenance device); the client can then browse through the VMD and systematically query all names of the objects. Using the Get services, which are defined for every object class (e.g., Get Variable Access Attributes), the client can get detailed knowledge about a given object (e.g., the Named Variable *T142*).
3. *Rename* allows a client to rename the name of an object.

24.7 Domain Management

24.7.1 Model Basics

Domains are to be viewed as containers that represent memory areas. Domain contents can be interchanged between different devices. The object type *domain* with its 12 attributes and 12 direct operations, which create, manipulate, delete a domain, etc., are part of the model.

The abstract structure of the domain object consists of the following attributes:

Object: Domain

Key Attribute:	Domain Name
Attribute:	List of Capabilities
Attribute:	State (LOADING, COMPLETE, INCOMPLETE, READY, IN-USE)
Constraint:	State = (LOADING, COMPLETE, INCOMPLETE)
Attribute:	Assigned Application Association
Attribute:	MMS Deletable
Attribute:	Sharable (TRUE, FALSE)
Attribute:	Domain Content
Attribute:	List of Subordinate Objects
Constraint:	State = (IN-USE)
Attribute:	List of Program Invocation References
Attribute:	Upload In Progress
Attribute:	Additional Detail

The domain name is an identifier of a domain within a VMD.

Domain Content is a dummy for the information that is within a domain. The contents of the data to be transmitted can be coded transparently or according to certain rules agreed upon before. Using the MMS version (2003), the data stream can be coded per default in such a way that a VMD can be transmitted completely—including all MMS object definitions that it contains. This means on the one hand that the contents of a VMD can be loaded from a configuration tool into a device (or saved from a device) and, on the other hand, that the contents can be stored on a disk per default.

Using a visible string, List of Capabilities describes which resources are to be provided—by the real device—for the domain of a VMD.

MMS Deletable indicates whether this domain may be deleted by means of an MMS operation.

Sharable indicates whether a domain may be used by more than one program invocation.

List of Program Invocation lists those program invocation objects that use this domain.

List of Subordinate Objects lists those MMS objects (no domains or program invocations) that are defined within this domain: objects that were (a) created by the domain loading, (b) created dynamically by a program invocation, (c) created dynamically by MMS operations, or (d) created locally.

State describes one of the 10 states in which a domain can be.

Upload in Progress indicates whether the domain content of this domain is being copied to the client at the moment.

MMS defines loading in two directions:

1. Data transmission from the client to the server (download)
2. Data transmission from the server to the client (upload)

Three phases can be distinguished during loading:

1. Open transmission
2. Segmented transmission, controlled by the data sink
3. Close transmission

Transmission during download and upload is initiated by the client, respectively. If the server initiates, then it has the possibility to initiate the transmission indirectly (see Figure 24.9). For this purpose,

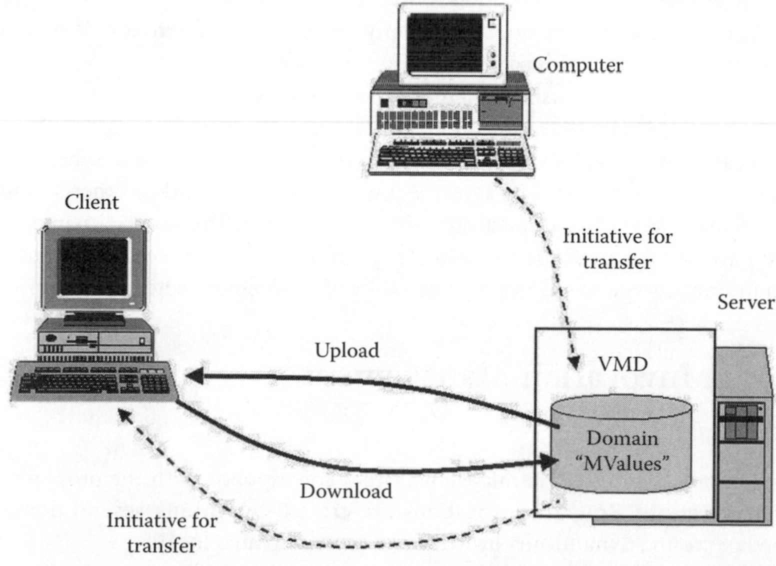

FIGURE 24.9 MMS domain transfer.

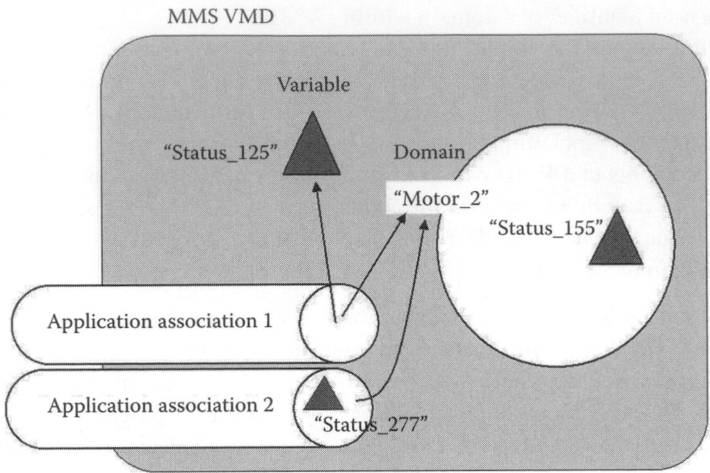

FIGURE 24.10 VMD and domain scope.

the server informs the client that it (the client) shall initiate the loading. Even a third station can initiate the transmission by informing the server, which then informs the client.

24.7.2 What Is the Domain Scope?

Further MMS objects can be defined within a domain: variable objects, event objects, and semaphore objects. That is to say a domain forms a scope (validity range) in which named MMS objects are reversibly unambiguous.

MMS objects can be defined in three different scopes, as shown in Figure 24.10. Objects with VMD-specific scope (e.g., the variable *Status_125*) can be addressed directly through this name by all clients. If an object has domain-specific scope such as the object *Status_155*, then it is identified by two identifiers:

Domain Identifier *Motor_2* and Object Identifier *Status_155*.

A third scope is defined by the Application Association. The object *Status_277* is part of the corresponding connection. This object can be accessed only through this connection. When the connection is closed, all objects are deleted in this scope.

MMS objects can be organized using different scopes. The object names (with or without domain scope) can be compounded from the following character set:

The identifiers can contain 1–32 characters, and they must not start with a number.

The object names can be structured by agreement in a further standard or other specification. Many standards that reference MMS make much use of this possibility. This way, all named variables with the Prefix *RWE_* and similar prefixes, for example, could describe the membership of the data (in a trans-European information network) to a specific utility of an interconnected operation.

24.8 Program Invocation Management

24.8.1 Model Basics

A program invocation object is a dynamic element that corresponds with the program executions in multitasking environments. Program invocations are created by linking several domains. They are either predefined or created dynamically by MMS services or created locally.

A program invocation object is defined by its name, its status (idle, starting, running, stopping, stopped, resuming, unrunnable), the list of the domains to be used, and nine operations.

Object: Program Invocation

Key Attribute:	Program Invocation Name
Attribute:	State (IDLE, STARTING, RUNNING, STOPPING, STOPPED, RESUMING, RESETTING, UNRUNNABLE)
Attribute:	List of Domain References
Attribute:	MMS Deletable (TRUE, FALSE)
Attribute:	Reusable (TRUE, FALSE)
Attribute:	Monitor (TRUE, FALSE)
Constraint:	Monitor = TRUE
Attribute:	Event Condition Reference
Attribute:	Event Action Reference
Attribute:	Event Enrollment Reference
Attribute:	Execution Argument
Attribute:	Additional Detail

Program invocations are structured flatly—though several program invocations can reference the same domains (shared domains). The contents of the individual domains are absolutely transparent both from the point of view of the domain and from the point of view of the program invocations. What is semantically connected with the program invocations is outside the scope of MMS. The user of the MMS objects must therefore define the contents; the semantics result from this context. If a program invocation connects two domains, then the domain contents must define what these domains shall do together—MMS actually only provides a wrapper.

The Program Invocation Name is a clear identifier of a program invocation within a VMD.

State describes the status in which a program invocation can be. Altogether seven states are defined.

List of Domains contains the names of the domains that are combined to a program invocation. This list also includes such domains that are created by the program invocation itself (this can be a domain into which some output is written).

MMS Deletable indicates whether this program invocation may be deleted by means of an MMS operation.

Reusable indicates whether a program invocation can be started again after the program execution. If it cannot be started again, then the program invocation can only be deleted.

Monitor indicates whether the program invocation reports a transition to the client when exiting the running status.

Start Argument contains an application-specific character string that was transferred to a program invocation during the last start operation; this string, for example, could indicate which function has started the program last.

Additional detail allows the companion standards to make application-specific definitions.

24.8.2 Program Invocation Services

Create Program Invocations: This service arranges an operational program, which consists of the indicated domains, in the server. After installation, the program invocation is in the status *idle* from where it can be started. The monitor and the monitor type indicate whether and how the program invocation shall be monitored.

Delete Program Invocation: Deletable program invocations are deleted through this service. Primarily, that is to say that the resources bound to a program invocation are released again.

Start: The start service causes the server to transfer the specified program invocation from the idle into the running state. Further information can be transferred to the VMD through a character string in the start argument. A further parameter (start detail) contains once again additional information that can be defined by companion standards.

Stop: The stop service changes a specified program invocation from the running into the stopped state.

Resume: The resume service changes a specified program invocation from the stopped into the running state.

Reset: The reset service changes a specified program invocation from the running or stopped into the idle state.

Kill: The kill service changes a specified program invocation from arbitrary states into the unrunnable state.

Get Program Invocation Attribute: Through this service, the client can read all attributes of certain program invocations.

24.9 MMS Variable Model

24.9.1 Model Basics

MMS Named Variables are addressed using identifiers made up of the domain name and the Named Variable name within the domain. Components of an MMS Named Variable may also be individually addressed using a scheme called alternate access. The alternate access address of a component consists of the domain name and the Named Variable name, along with a sequence of enclosing component names of the path down to the target component.

The variable access services contain an extensive variable model that offers the user a variety of advanced services for the description and offers the access to arbitrary data of a distributed system.

A wide variety of process data are processed by automation systems. The data, their definition, and representation are usually orientated at the technological requirements and at the available automation equipment. The methods the components employ for the representation of their data and the access to them correspond to the way of thinking of their implementers. This has resulted in a wide variety of data representations and access procedures for one and the same technological datum in different components. If, for example, a certain temperature measurement shall be accessed to in different devices, then a huge quantity of internal details must generally be taken into account for every device (request, parameter, coding of the data, etc.).

As shown in Figure 24.11, the number of the protocols for the access of a client (on the left in the figure) to the data from n servers (S1–Sn) can be reduced to a single protocol (on the right in the figure). Through this, the data rate required for the communication primarily in central devices can be reduced drastically.

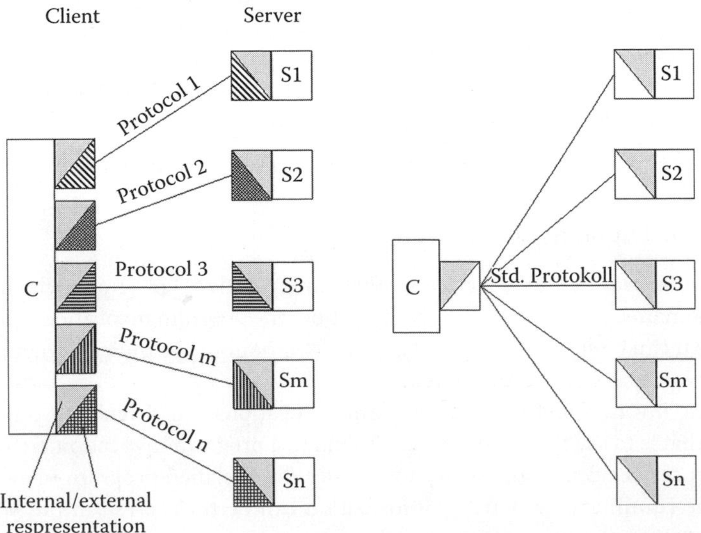

FIGURE 24.11 Unified protocols.

In programs, variables are declared, that is, they get a name, a type, and a value. Described in a simplified way, both the name and the type are converted by the compiler into a memory location and into a reference that is accessible only to the compiled program. Without any further measures, the data of the variable are not identifiable outside the program. It is concealed for the user of the program how a compiler carries out the translation into the representation of a certain real machine.

The data are stored in different ways depending on the processor, that is, primarily that the data are stored in various memory locations. During the runtime of the program, only this representation is available.

These data are not visible from the outside. They must first be made visible for the access from the outside. To enable this, an entity must be provided in the implementation of the application. It is insignificant here whether this entity is separated from or integrated into the program. This entity is acting for all data that shall be accessible from the outside.

This consideration is helpful to the explanation of the MMS variable model: what do protocols through which process data are accessed have in common? Figure 24.12 shows the characteristics in principle. On the right is the memory with the real process data that shall be read. The client must be able to identify the data to be read (gray shade). For this purpose, pointer (start address) and length of the data must be known. By means of this information, the data can be identified in the memory.

Yet, how can a client know the pointer of the data and what the length of the data is? It could have this information somehow and indicate it when reading. Yet, if the data should move some time, then the pointer of the data is not correct any more. There can also be the case that the data do not exist at all at the time of reading but must be calculated first. In this case, there is no pointer. To avoid this, references to the data, which are mapped to the actual pointers (table or algorithm), are used in most cases. In our case, the reference *B* is mapped by table to the corresponding pointer and the length.

The pointer and the length are stored in the type description of the table. The pointer is a system-specific value that generally is not visible on the outside. The length is dependent on the internal representation of the memory and on the type. An individual bit can, for example, be stored as an individual bit in the memory or also as a whole octet. However, this is not relevant from the point of view of the communication.

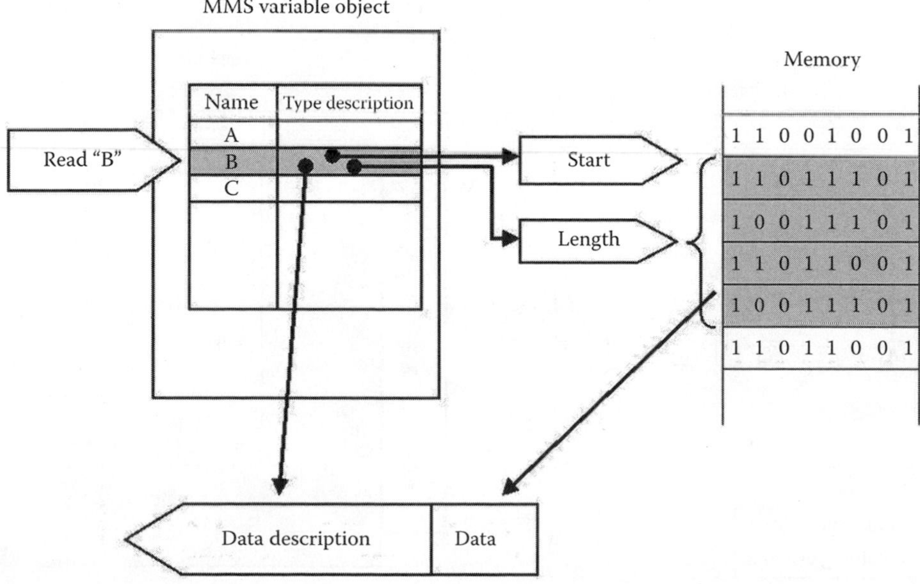

FIGURE 24.12 Data access principle.

The data themselves and their description are important for the message response of the read service. The question of the external representation (is, e.g., an individual bit encoded as a bit or as an octet?) is—unlike the internal representation—of special importance here! The various receivers of the data must be able to interpret the data unambiguously. For this purpose, they need the representation, which is a substantial component of the MMS variable model. The data description is therefore derived from the type description.

For the deeper understanding of the variable model, three aspects have to be explained more exactly:

1. Objects and their functions
2. Services (read, write, etc.) that access the MMS variable objects
3. Data description for the transmission of the data

The object model of an MMS variable object is conceptually different from a variable according to a programming language. The MMS objects describe the access path to structured data. In this sense, they do not have any variable value.

24.9.2 Access Paths

The access path represents an essential feature of the MMS variable model. Starting from a more complex hierarchical structure, we will consider the concept. The abstract and extremely simplified example is deliberately chosen. Here, we are merely concerned about the principle.

Certain data of a machine shall be modeled using MMS methods. The machine has a tool magazine with n similar tools. A tool is represented—according to Figure 24.13—by three components (tool type, number of the blades, and remaining use time).

The machine with its tool magazine *M* is outlined in the figure on the right. The magazine contains three tools *A*, *B*, and C. The appropriate data structure of the magazine is shown on the left. The structure is tree like; the root *M* is drawn as the topmost small circle (node). *M* has three components (branches) *A*, *B*, and *C*, which are also represented as circles. These components in turn also have three components (branches). In this case, the branches end in a leaf (represented in the form of a square). Leaves represent the end-points of the branches. For each tool, three leaves are shown: *T* (tool type), *A* (number of blades), and *R* (remaining use time). The leaves represent the real data. The nodes ordered

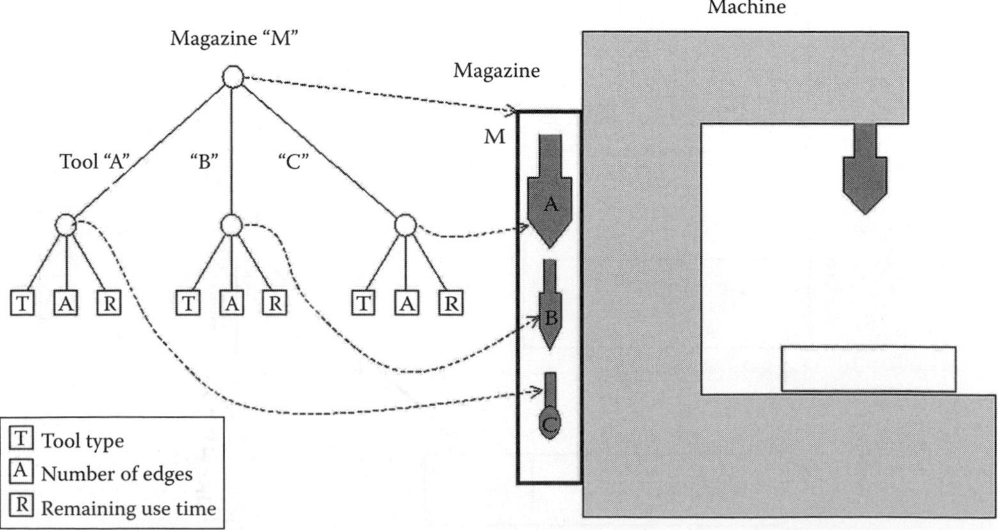

FIGURE 24.13 Data of a machine.

hierarchically are merely introduced for reasons of clustering. Leaves can occur at all nodes. A leaf with the information *magazine full/not full* could, for example, be attached at the topmost node.

With this structure, the MMS features are explained in more detail. The most essential aspect is the definition of the access path. The access to the data (and their use) can be carried out according to various task definitions:

1. Selecting all leaves for reading, writing, etc.
2. Selecting certain leaves for reading, writing, etc.
3. Selecting all leaves as component of a higher-level structure, for example, *machine* with the components magazine *M* and drilling machine
4. Selecting certain leaves as the component of a higher-level structure

Examples of the cases (1) and (2) are shown in Figure 24.14. The case that the complete structure is read is shown in the left top corner (the selected nodes, branches, and leaves are represented in bold lines or squares). All nine data are transmitted as response (3 × (T + A + R)). At first, the representation during transmission is deliberately refrained from here and also in the following examples.

Only a part of the data is read in the top right corner of the figure: respectively, only the leaf *R* of all three components *A*, *B*, and *C*. The notation for the description of the subset M.A.R/.B.R/.C.R is chosen arbitrarily.

The subset M.A.R represents a (access) path that leads from a root to a leaf. It can also be said that one or several paths represent a part of a tree. The read message contains three paths that must be described in the request completely. A path, for example, can also end at *A*. All three components of *A* will be transmitted in this case.

Besides the possibility to describe every conceivable subset, MMS also supports the possibility to read several objects *M1* and *M5* in a read request simultaneously (see lower half of the figure). Of course, every object can only be partly read too (which is, however, not represented).

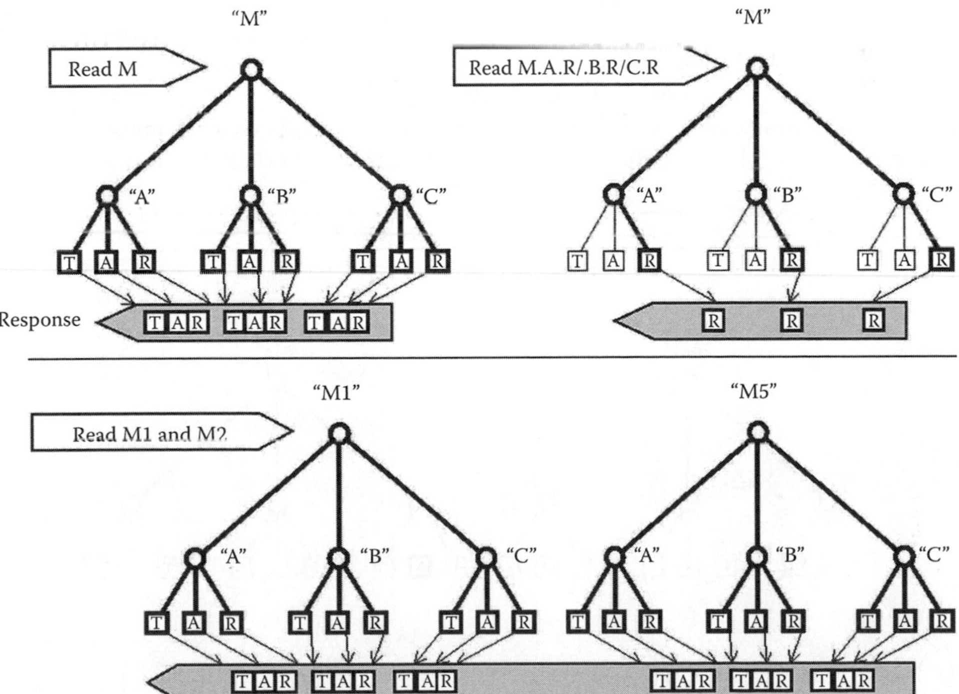

FIGURE 24.14 Access and partial access.

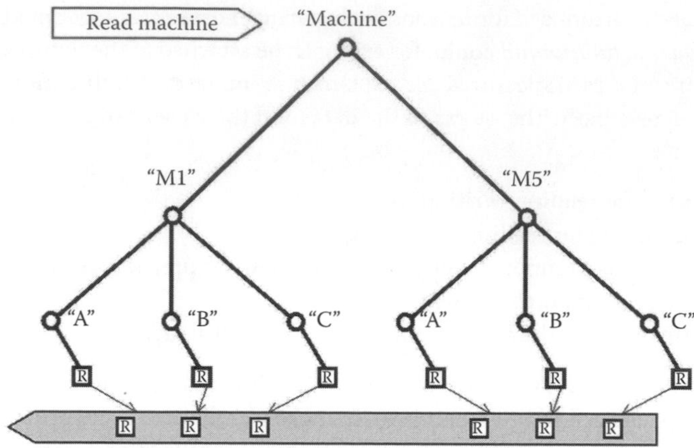

FIGURE 24.15 *Partial* trees.

An example of case 4 is shown in Figure 24.15 (case 4 has to be understood as the generalization of case 3). Here a new structure was defined using two substructures. The object *machine* contains only the *R* component of all six tools of the two magazines *M1* and *M5*. The object *machine* shall not be mixed up with a list (Named Variable List). Read *machine* supplies the six individual *R* values.

The component names, such as *A*, *B*, or *C*, need to be unambiguous only below a node—and only at the next lower level. Thus, *R* can always stand for the remaining life time. The position in the tree indicates the remaining life time of a particular tool. The new structure *machine* has all features that were described in the previous examples also for *M1* and *M5*.

These features of the MMS variable objects can be applied to (1) the definition of new variable objects and (2) the access to existing variable objects. The second case is also interesting. As shown in Figure 24.16, the same result as in Figure 24.15 can be reached by enclosing the description (only the *R* components of the tools shall be read from the two objects *M1* and *M5*) in the read request every time. The results (read answer) are absolutely identical in the two cases.

Every possibility, to assemble hierarchies from other hierarchies (1) or to, for example, read parts of a tree during the access (2), has its useful application. The first case is important in order to avoid enclosing

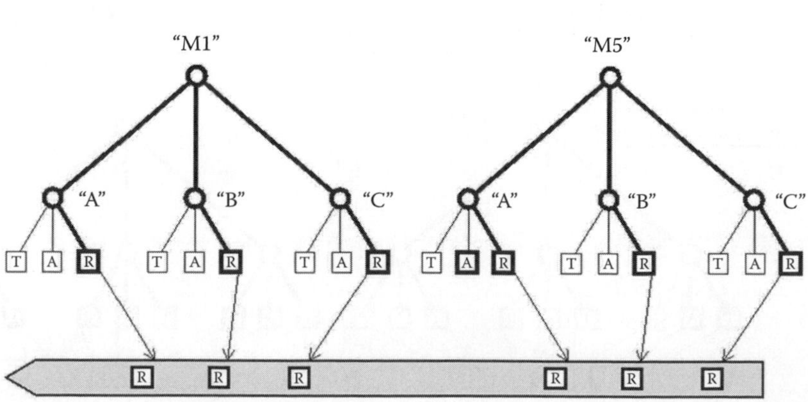

FIGURE 24.16 *Partial* trees used for read requests.

the complete path description every time when reading an extensive part of tree. The second case offers the possibility to construct complex structures based on standardized basic structures (e.g., the structure of *tool data* consisting of the components *T*, *A*, and *R*) and to use them for the definition of new objects.

Summarizing, it can be stated that the access paths accomplish two tasks:

1. Description of a subset of nodes, branches, and leaves of objects during reading, writing, etc.
2. Description of a subset of nodes, branches, and leaves of objects during the definition of new objects

In conclusion, this may again be expressed in the following way: Path descriptions describe the *way* or the *ways* to a single data (leaf) or to several data (leaves). A client can read the structure description (complete tree) through the MMS service Get Variable Access Attributes.

Another aspect is of special importance too. Till now, we have not considered the description of leaves. Every leaf has one of the following MMS basic data types:

- BOOLEAN
- BIT
- INTEGER
- UNSIGNED
- FLOATING POINT
- REAL
- OCTET STRING
- VISIBLE
- GENERALIZED TIME
- BINARY
- BCD

Every node in a tree is either an array or a structure. Arrays have n elements (0 to n − 1) of the same type (this can be a basic type, an array, or a structure). When describing a part of a tree, any number of array elements can be selected (e.g., one element, two adjacent elements, and two arbitrary elements).

Structures consist of one or several components. Every component can be marked by its position and if necessary by an *identifier*—the component name. This component name (e.g., *A*) is used for the access to a component. Parts of trees can describe every subset of the structure.

The path description contains the following three elements:

1. All of the possibilities of the description of the structures (individual and composite paths in the type description of the MMS variables) are defined in the form of an extensive abstract syntax.
2. For every leaf of a structure (these are MMS basic data types), the value range (size) is also defined besides the class. The value range of the class INTEGER can contain one, two, or more octets. The value range four (4) octets (often represented as Int32), for example, indicates that the value cannot exceed these four octets. On the other hand, with ASN.1 BER coding (explained later) and the value range Int32, the decimal value 5 will be transmitted in only one octet (not in four octets!). That is to say, only the length needed for the current value will be transmitted.
3. The aspect of the representation of the data and their structuring during transmission on the line (communication in the original meaning) is dealt with in the following text in the context of the encoding of messages.

The path description is used in a quintuple way (see Figure 24.17):

- During the access to and during the definition of variable objects
- In the type description of variable objects
- In the description of data during reading, for example
- During the definition of Named Type Objects (an object name of its own is assigned to the type description—i.e., to one or several paths—of those objects)
- When reading the attributes of variables and Named Type Objects

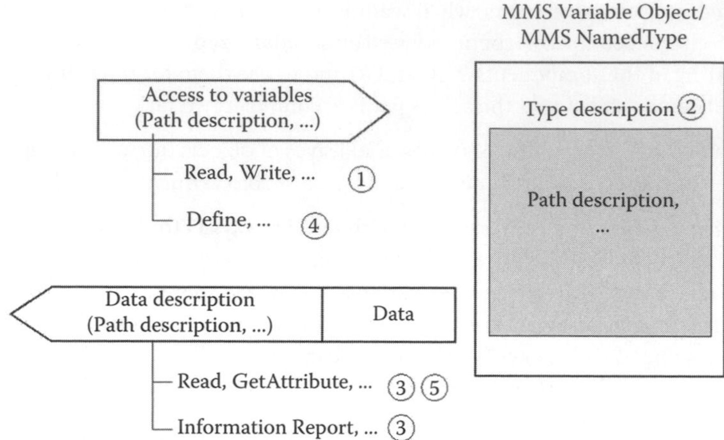

FIGURE 24.17 Application of path descriptions.

24.9.3 Objects of the MMS Variable Model

The five objects of the MMS variable model are as follows:
Description of simple or complex values

- Unnamed Variable
- Named Variable

List of several unnamed variables or named variables

- Named Variable List
- Scattered Access (is not explained here)

Description of the structure by means of a user-defined name

- Named Type

24.9.4 Unnamed Variable

The Unnamed Variable Object describes the assignment of an individual MMS variable to a real variable that is located at a definite address in the device. An Unnamed Variable Object can never be created or deleted. The Unnamed Variable has the following attributes:

Object: Unnamed Variable
Key Attribute: Address
Attribute: MMS Deletable (FALSE)
Attribute: Access Method (PUBLIC)
Attribute: Type Description

Address: Address is used to reference the object. There are three different kinds:

1. Numeric Address (nonnegative integer values)
2. Symbolic Address (character string)
3. Unconstrained Address (implementation-specific format)

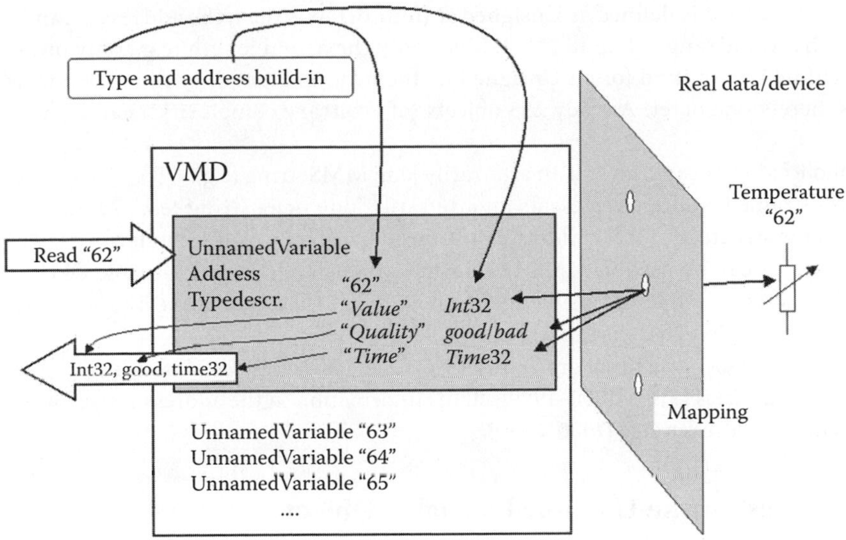

FIGURE 24.18 Unnamed Variable Object.

Even though in (2) the address is represented by character strings, this kind of addressing has absolutely to be distinguished from the Object Name of a Named Variable (see domain scope and explanations later).

MMS Deletable: The attribute is always FALSE here.

Access Method: The attribute is always PUBLIC here.

Type Description: The attribute points to the inherent abstract type of the subordinate real variable as it is seen by MMS. It specifies the class (bitstring, integer, floating-point, etc.), the value range of the values, and the group formation of the real variable (arrays of structures). The attribute type description is completely independent of the addressing.

Figure 24.18 represents the Unnamed Variable roughly sketched. The Unnamed Variable with the address *62* (MMSString) has three components with the names: Value, Quality, and Time. These component names are required only if individual components (specifying the path, e.g., *62/Value*) shall be accessed.

24.9.5 MMS Address of the Unnamed Variable

The MMS Address is a system-specific reference that is used by the system for the internal addressing—it is quasi released for the access via MMS. There the address can assume one of three forms (here the ASN.1 notation is deliberately used for the first time):

```
Address:: = CHOICE {
    NumericAddress          [0] IMPLICIT Unsigned32,
    SymbolicAddress         [1] MMSString,
    UnconstrainedAddress    [2] IMPLICIT OCTET STRING
    }
```

This definition has to be read as follows: Address defines as (:: =) a selection (keyword CHOICE) of three possibilities. The possibilities are numbered here from [0] to [2] to be able to distinguish them. The keyword IMPLICIT is discussed later.

The numeric address is defined as Unsigned32 (four octets). Thus, the addresses can be defined as an index with a value range of up to 2**32. Since only the actual length (e.g., only one octet for the value 65) will be transmitted for an Unsigned32, the minimal length of the index, which can thus be used, is merely one octet. Already 255 objects (of arbitrary complexity!) can be addressed with one octet.

The symbolic address can transmit an arbitrarily long MMSString (e.g., *DB5_DW6*).

The Unconstrained Address represents an arbitrarily long octet string (e.g., 24FE23F2A1hex). The meaning and the structure of these addresses are outside the scope of the standard.

These addresses can be used in MMS Unnamed Variable and Named Variable Objects and in the corresponding services. MMS can neither define nor change these addresses. The address offers a possibility to reference objects by short indexes.

The addresses can be structured arbitrarily. Unnamed Variables could, for example, contain measurements in the address range [1000–1999], status information in the address range [3000–3999], and limit values in the address range [7000–7999].

24.9.6 Services for the Unnamed Variable Object

Read: This service uses the V-Get function to transmit the current value of the real variable, which is described by the Unnamed Variable Object, from a server to a client. Variable Get represents the internal, system-specific function through which an implementation gets the actual data and provides them for the communication.

Write: This service uses the V-Put function to replace the current value of the real variable, which is described by the Unnamed Variable Object, by the enclosed value.

Information Report: As Read, though without prior request by the client. Only the Read.response is sent by the server to the client without being asked. The information report corresponds to a spontaneous message. The application itself determines when the transmission is to be activated.

Get Variable Access Attributes: Through this operation, a client can query the attributes of an Unnamed Variable Object.

24.9.7 Explanation of the Type Description

Features of the structure description of MMS variable objects were explained earlier in principle. For those interested in the details, the formal definition of the MMS type specification is explained according to Figure 24.19.

The description in ASN.1 was also deliberately selected here. The type specification is a CHOICE (selection) of 15 possibilities (Tags [0] to [13] and [15]). Tags are qualifications of the selected possibility. The first possibility is the specification of an Object Name, a Named Type Object. If we remember that one Named Type Object describes one or several paths, then the use is obvious. The path description referenced by the name can be used to define a Named Variable Object. Or, if during reading the path must be specified, it can be referenced by a Named Type Object in the server.

Note that the ASN.1 definitions in MMS are comparable with XML schema. ASN.1 BER (basic encoding rules) provides very efficient message encoding compared to XML documents. The standard IEC 61400-25 provides ASN.1 as well as XML schema for the specification of messages.

The two next possibilities (array and structure) have a common feature. Both refer—through their element type or component type—back to the beginning of the complete definition (type specification). This recursive definition allows the definition of arbitrarily complex structures. Thus, an element of a structure can in turn be a structure or an array.

```
TypeSpecification ::= CHOICE {
    typeName        [0] ObjectName,
    array                   [1] IMPLICIT SEQUENCE {
        packed                      [0] IMPLICIT BOOLEAN DEFAULT FALSE,
        numberOfElements            [1] IMPLICIT Unsigned32,
        elementType                 [2] TypeSpecification },
    structure               [2] IMPLICIT SEQUENCE {
        packed                              [0] IMPLICIT BOOLEAN DEFAULT FALSE,
        components                  [1] IMPLICIT SEQUENCE OF SEQUENCE {
                componentName [0] IMPLICIT Identifier OPTIONAL,
                componentType           [1] TypeSpecification } },

    Simple                  Size                        Class
    boolean                 [3] IMPLICIT NULL,          - BOOLEAN
    bit-string              [4] IMPLICIT Integer32,     - BIT-STRING
    integer                 [5] IMPLICIT Unsigned8,     - INTEGER
    unsigned                [6] IMPLICIT Unsigned8,     - UNSIGNED
    floating-point          [7] IMPLICIT SEQUENCE {
        format-width                Unsigned8,      - # of bits in fraction plus sign
        exponent-width              Unsigned8       size of exponent in bits },
    real                    [8] IMPLICIT SEQUENCE {
        base                        [0] IMPLICIT INTEGER(2|10),
        exponent                    [1] IMPLICIT INTEGER,       - max number of octets
        mantissa                    [2] IMPLICIT INTEGER        - max number of octets },
    octet-string            [9] IMPLICIT Integer32,             - OCTET-STRING
    visible-string          [10] IMPLICIT Integer32,            - VISIBLE-STRING
    generalized-time        [11] IMPLICIT NULL,                 - GENERALIZED-TIME
    binary-time             [12] IMPLICIT BOOLEAN,              - BINARY-TIME
    bcd                     [13] IMPLICIT Unsigned8             - BCD
    objId                   [15] IMPLICIT NULL}
```

FIGURE 24.19 MMS Type specification.

Arrays are defined by three features. Packed defines whether the data are stored optimized. Number-Of Elements indicates the number of the elements of the array of equal type (Element Type).

The data of structures can also be saved as packed. Structures consist of a series of components (components [1] IMPLICIT SEQUENCE OF SEQUENCE). This series is marked by the keyword SEQUENCE OF…, which describes a repetition of the following definition. Next in the list is SEQUENCE {component Name und component Type}, which describes the individual component. Since the SEQUENCE OF… (repetition) can be arbitrarily long, the number of the components at a node is also arbitrary.

Then follow the *simple data types*. They start at the tag [3]. The length of the types is typical for the simple data types. For example, integers of different lengths can be defined. The length (size) is defined as Unsigned8, which allows for an integer with the length of 255 octets.

It should be mentioned here that—in the ASN.1 description of the MMS syntax—expressions like integer (written in small letters) show that they are replaced by another definition (in this case by the Tag [5] with the IMPLICIT-Unsigned8 definition). Capital letters at the beginning indicate that the definition is terminated here—it is not replaced any more. It is here a basic definition.

Figure 24.20 shows an example of an object defined in IEC 61850-7-4. The circuit breaker class is instantiated as *XCBR1*. The hierarchical components of the object are mapped to MMS (according to IEC 61850-8-1). The circuit breaker is defined as a comprehensive MMS Named Variable.

The components of the hierarchical model can be accessed by the description of the Alternate Access: "XCBR1" → component "ST" → component "Pos" → component "stVal." Another possibility of mapping the hierarchy to a flat name is depicted in Figure 24.21. Each path is defined as a character string.

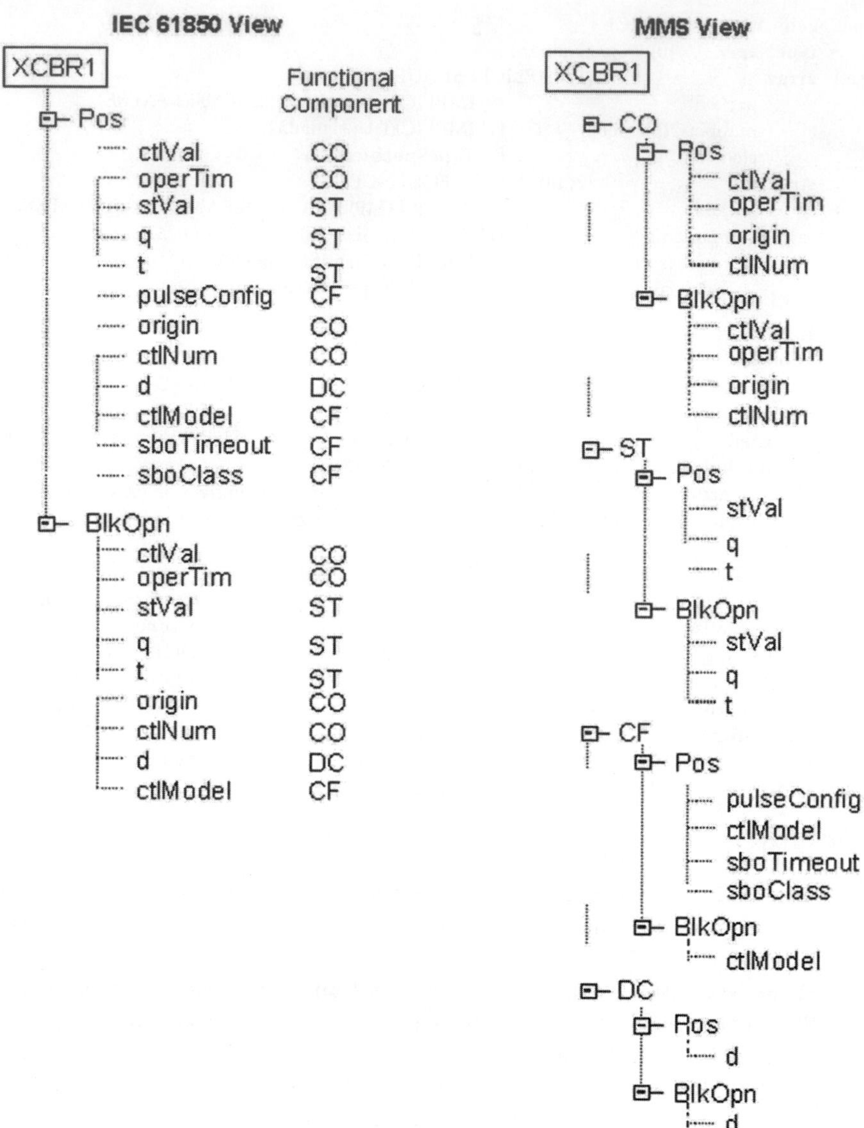

FIGURE 24.20 Example MMS Named Variable.

24.9.8 Named Variable

The Named Variable Object describes the assignment of a named MMS variable to a real variable. Only one Named Variable Object should be assigned to a real variable. The attributes of the object are as follows:

Object: Named Variable

Key Attribute:	Variable Name
Attribute:	MMS Deletable
Attribute:	Type Description
Attribute:	Access Method (PUBLIC,…)
Constraint:	Access Method = PUBLIC
Attribute:	Address

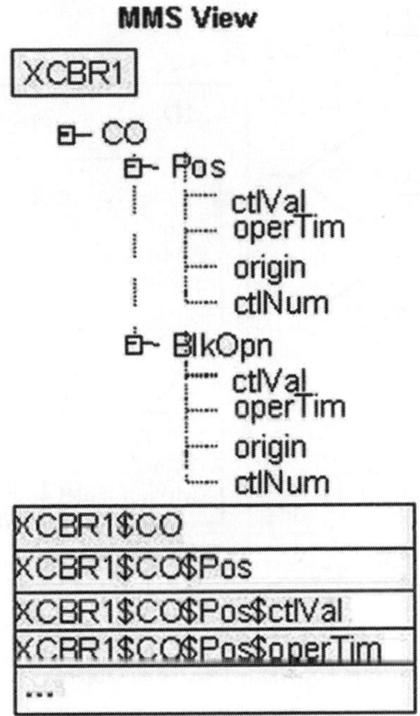

FIGURE 24.21 Example MMS Named Variable.

Variable Name: The Variable Name unambiguously defines the Named Variable Object in a given scope (VMD specific, domain specific, or Application Association specific). The Variable Name can be 32 characters long (plus 32 characters if the object has a domain scope).

MMS Deletable: This attribute shows whether the object may be deleted using a service.

Type Description: This attribute describes the abstract type of the subordinate real variable as it represents itself to the external user. This attribute is not inherent in the system unlike the Unnamed Variable Object; that is, this type description can be defined from the outside.

Access Method: This attribute contains the information that a device needs to identify the real variable. It contains values that are necessary and adequate to find the memory location. The contents lie outside MMS. A special method, the method PUBLIC, is standardized. The attribute Address is also available in the case of PUBLIC. This is the address that identifies an Unnamed Variable Object. Named Variables can thus be addressed by the name and the ADDRESS (see Figure 24.22).

Address: See Unnamed Variable Object. Defining a Named Variable Object does not allocate any memory either, because the real variable must already exist; it is assigned to the Named Variable Object with the corresponding name.

Altogether, six operations are defined on the object:

Read: The service uses the V-Get function to retrieve the current value of the real data that is described by the Object.

Write: The service uses the V-Put function to replace the current value of the real data, which is described by the Object, by the enclosed value.

Define Named Variable: This service creates a new Named Variable Object that is assigned to real data.

Get Variable Access Attribute: Through this operation, a client can query the attributes of a Named Variable Object.

Delete Variable Access: This service deletes a Named Variable Object if Attribute Deletable is (= TRUE).

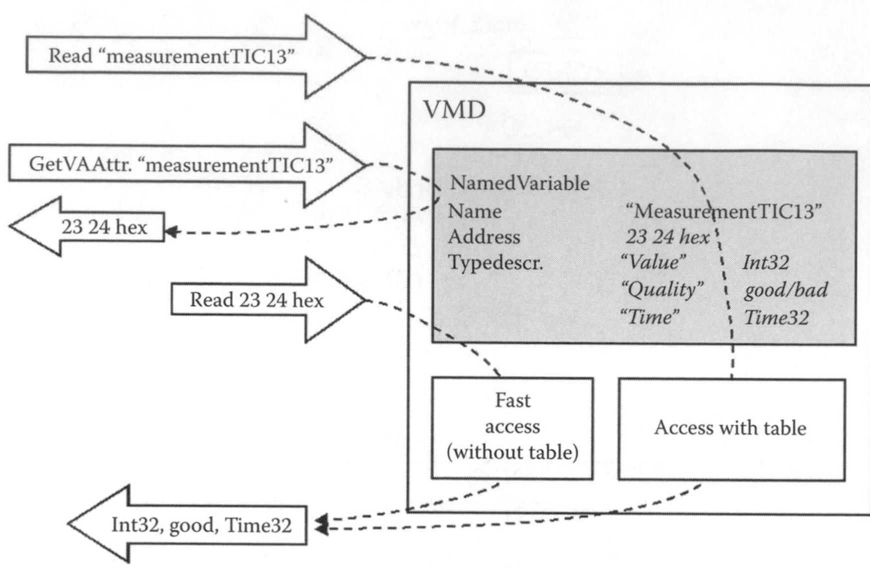

FIGURE 24.22 Address and Variable Name of Named Variable Objects.

Figure 24.23 shows the possibilities to reference a Named Variable Object by names and, if it is required, also by Address (optimal access reference of a given system). For a given name, a client can query the Address by means of the service Get Variable Access Attribute.

This possibility allows the access through technological names (MeasurementTIC13) or with the optimal (index-) address 23 24 hex.

As shown in Figure 24.23, an essential feature of the VMD is the possibility for the client application to define by request Named Variable Objects in the server via the communication. This includes the definition of the name, the type, and the structure. The name by which the client would like to reference the Named Variable later is *TIC42* here. The first component *Value* is of the type Integer32, the second is *Quality* with the values *good* or *bad*, the third is *Time* of the type *Time32*. The type of the Data Value

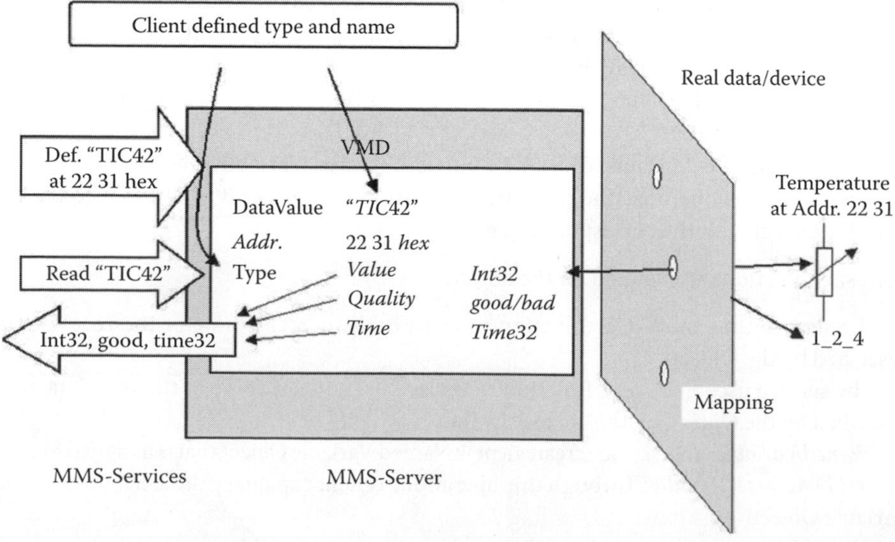

FIGURE 24.23 Client-defined Named Variable Object.

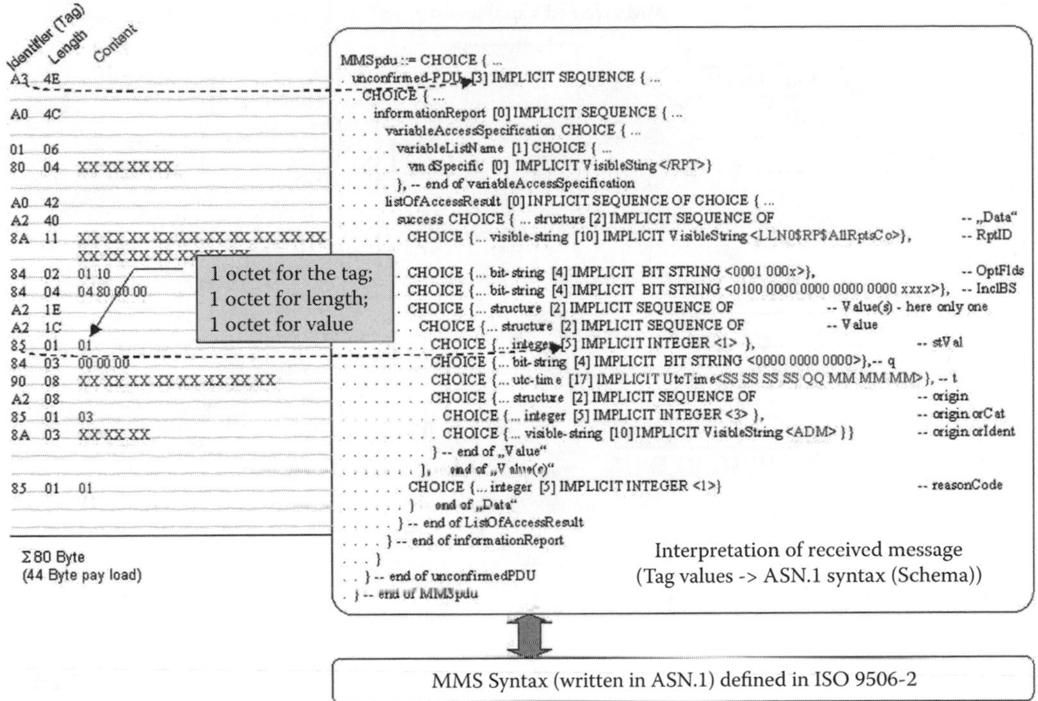

FIGURE 24.24 MMS information report (spontaneous message).

Object can be arbitrarily simple (flat) or complex (hierarchical). As a rule, the Data Value Objects are implicitly created by the local configuring or programming of the server (they are predefined).

The internal assignment of the variable to the real temperature measurement is made by a system-specific optimal reference. This reference whose structure and contents are transparent must be known when defining the Named Variable, though. The reference can, for example, be a relative memory address (e.g., DB5 DW15 of a PLC). So a quick access to the data is allowed.

The Named Variable Object describes how data for the communication are modeled, accessed, encoded, and transmitted. What is transmitted is described independently of the function. From the point of view of the communication, it is not relevant where the data in the server actually come from or where in the client they actually go to and how they are managed—this is deliberately concealed.

Figure 24.24 shows the concrete encoding of the information report message. The message is encoded according to ASN.1 BER (the basic encoding rule for abstract syntax notation number one—ISO 8825).

The encoding using XML would be several times longer as using ASN.1 BER. These octets are packed into further messages that add lower layer-specific control and address information, for example, the TCP header, IP header, and Ethernet frames.

The receiving IED is able to interpret the report message according to the identifier, lengths, names, and other values. The interpretation of the message requires the same stack, that is, knowledge of all layers involved—including the definitions of IEC 61850-7-4, IEC 61850-7-3, IEC 61850-7-2, and IEC 61850-7-1.

24.9.9 Access to Several Variables

24.9.9.1 Named Variable List

The Named Variable List allows to define a single name for a group of references to arbitrary MMS Unnamed Variables and Named Variables. Thus, the Named Variable List offers a grouping for the frequently repeated access to several variables (see Figure 24.25). Although the simultaneous access to

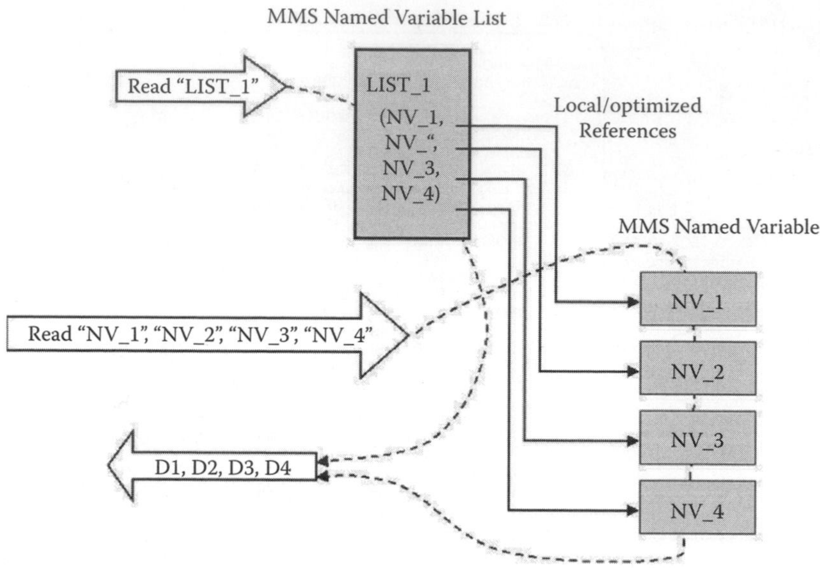

FIGURE 24.25 MMS Named Type and Named Variable.

several MMS variables can be carried out also in a single service (Read or Write), the Named Variable List offers a substantial advantage.

When reading several variables in a Read request, the individual names and the internal access parameters (pointers and lengths), corresponding to the names in the request, must be searched for in a server. This search can last for some time in the case of many names or a low processor performance. By using the Named Variable List Object, the search is not required—except for the search of a single name (the name of the Named Variable List Object)—if the references, for example, have been entered into the Named Variables on the list system specifically and thus optimally. Once the name of the list has been found, the appropriate data can be provided quickly.

Thus, the Named Variable List Object provides optimal access features for the applications. This object class is used in the known applications of MMS very intensively.

The structure of the Named Variable List Object is as follows:

Object: Named Variable List

Key Attribute:	Variable List Name
Attribute:	MMS Deletable (TRUE, FALSE)
Attribute:	List of Variable
Attribute:	Kind of Reference (NAMED, UNNAMED, SCATTERED)
Attribute:	Reference
Attribute:	Access Description

Variable List Name: The Variable List Name unambiguously identifies the Named Variable List Object in a given scope (VMD specific, domain specific, or Application Association specific). See also MMS Object Names provided earlier.

MMS Deletable: This attribute shows whether the object may be deleted.

List of Variable: A list can contain an arbitrary number of objects (Unnamed Variable, Named Variable, or Scattered-Access Object).

Kind of Reference: Lists can refer to three object classes: Named Variables, Unnamed Variables, and Scattered Access; no Named Variable Lists can be included.

Reference: An optimal internal reference to the actual data is assigned to every element of the list. If a referenced object is not (any more) available, the entry into the list will indicate it. When accessing the list, for example, by Read, no data but an error indication will be transmitted to the client for this element.

Access Description: In the same way as for the access of a variable—for example, in the request during Read—could be defined that only parts (part of a tree) of a variable shall be read, can this also be applied to every element of the Named Variable List.

24.9.9.1.1 Services

Read: This service reads the data of all objects being part of the list (Unnamed Variable, Named Variable, and Scattered Access Object). For objects that are not defined, an error is reported in the corresponding place of the list of the returned values.

Write: This service writes the data from the write request into the objects being part of the list (Unnamed Variable, Named Variable, and Scattered Access Object). For objects that are not defined, an error is reported in the corresponding place of the list of the returned values.

Information Report: As the Read service; as if the read data were sent by the server to the client without prior request (Read Request) by the client, that is, as if only a Read Response would be transmitted.

Define Named Variable List: Using this service, a client can create a Named Variable List Object.

Get Named Variable List Attributes: This service queries the attributes of a Named Variable List Object.

Delete Named Variable List: This service deletes the specified Named Variable List Object.

24.9.9.2 Named Type Object

The Named Type Object merely describes structures. The object model is very simple:

Object: Named Type

Key Attribute:	Type Name
Attribute:	MMS Deletable (TRUE, FALSE)
Attribute:	Type Description

Essential attribute is the type description, which was already discussed before for Named and Unnamed Variables. On the one hand, TASE.2 standard data structures can be specified by means of Named Types. This is the most frequent application of the Named Type Objects. On the other hand, Named Types can be used for the access to the server. The read request can refer to a Named Type Object. Or the Named Type Object will be used to define Named Variables.

Figure 24.26 describes the application of the Named Type Objects for the definition of a Named Variable. A variable will be created by the request Define Named Variable.

It shall have the name *TIC42*, the address 22 31 hex, and the type that is defined in the Named Type Object *MWert*. The variable inherits the type from the Named Type Object. The inheritance has the consequence that the variable will only have the type but not the name of the Named Type Object. This inheritance was defined so strictly in order to avoid that through deleting the Named Type, the type of the variable would get undefined or that by subsequent definition of a differently structured Named Type with the old type name *MWert*, the type of the variable would be changed (the Named Type Object and with that the new type description would be referenced by the old name).

Perhaps it is objected now that this strict inheritance has the consequence that also the type would have to be saved for each variable (even though many variables have the same type description). Since these variables can internally be implemented in a system in whatever way the programmer likes it, they can refer through an internal index to a single (!) type description. He must only make sure that this

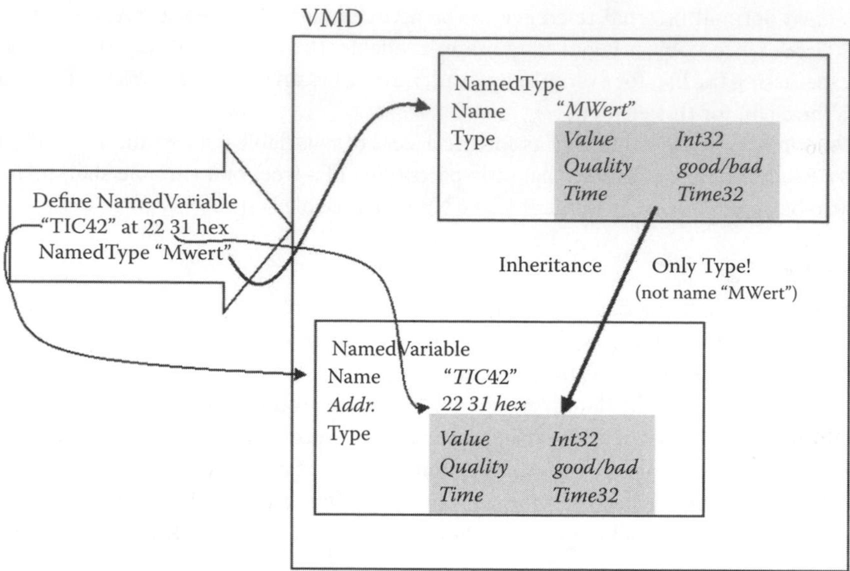

FIGURE 24.26 Inheritance of the Type of the MMS Named Type Objects.

type description is not deleted. If the accompanying Named Type Object gets deleted, then the referenced type description must remain preserved for these many variables.

The disadvantage that the name of the "structure mother," that is, the Named Type, is not known any more as an attribute of the variables has been eliminated in the MMS Revision.

24.9.9.2.1 Services

Define Named Type: This service creates a Named Type Object.

Get Named Type Attribute: This service delivers all attributes of a Named Type Object.

Read, Write, Define Named Variable, Define Scattered Access, Define Named Variable List, and Define Named Type use the type description of the Named Type Object when carrying out their tasks.

24.10 Resume

MMS is a standard messaging specification (comparable to web services), widely implemented by industrial device manufacturers like ABB, Alstom, General Electric, and Siemens. It solves problems of heterogeneity so often found in automation applications. MMS is the lingua franca of industrial devices.

MMS provides much more than TCP/IP, which essentially offers a transfer stream of bytes. MMS transfers commands with parameters between machines.

MMS allows a user to concentrate on the applications, application data to be accessible—and not on communication problems, which are already solved. It provides a basis for the definition of common and domain-specific semantic. Examples are the standards IEC 60870-6 TASE.2, IEC 61850, and IEC 61400-25.

References

1. ISO 9506-1:2003, Industrial automation systems—Manufacturing Message Specification (MMS)—Part 1 Service Definition, ISO, Geneva, Switzerland.
2. ISO 9506-2:2003, Industrial automation systems—Manufacturing Message Specification (MMS)—Part 2 Protocol Definition, ISO, Geneva, Switzerland.

3. ISO 9506-3:1991, Industrial automation systems—Manufacturing Message Specification (MMS)—Part 3 Companion Standard for robotics, ISO, Geneva, Switzerland.
4. ISO 9506-4:1992, Industrial automation systems—Manufacturing Message Specification (MMS)—Part 4 Companion Standard for numerical control , ISO, Geneva, Switzerland.
5. ISO 9506-5:1993, Industrial automation systems—Manufacturing Message Specification (MMS)—Part 5 Companion Standard for Programmable controllers, ISO, Geneva, Switzerland.
6. ISO 9506-6:1993, Industrial automation systems—Manufacturing Message Specification (MMS)—Part 6 Companion Standard for Process control, ISO, Geneva, Switzerland.
7. Preston, K. et al., *MMS: A Communication Language of Manufacturing.*, ESPRIT Consortium CCE-CNMA, Berlin, Germany: Springer-Verlag, 1995.
8. Preston, K. et al., *CCE: An Integration Platform for Distributed Manufacturing Applications*, ESPRIT Consortium CCE-CNMA, Berlin, Germany: Springer-Verlag, 1995.
9. IEC 60870-6-503:1997, Telecontrol equipment and systems—Part 6: Telecontrol protocols compatible with ISO standards and ITU-T recommendations—Section 503: Services and Protocol (ICCP Part 1), IEC, Geneva, Switzerland.
10. IEC 60870-6-802:1997, Telecontrol equipment and systems—Part 6: Telecontrol protocols compatible with ISO standards and ITU-T recommendations—Section 802: Object Models (ICCP Part 4), IEC, Geneva, Switzerland.
11. IEC 61850-6:2009, Communication networks and systems for power utility automation—Part 6. Configuration description language for communication in electrical substations related to IEDs, IEC, Geneva, Switzerland.
12. IEC 61850-7-1:2011, Communication networks and systems in substations—Part 7-1: Basic communication structure—Principles and models, IEC, Geneva, Switzerland.
13. IEC 61850-7-2:2010, Communication networks and systems in substations—Part 7-2: Basic communication structure—Abstract communication service interface (ACSI), IEC, Geneva, Switzerland.
14. IEC 61850-7-3:2010, Communication networks and systems in substations—Part 7-3: Basic communication structure—Common data classes, IEC, Geneva, Switzerland.
15. IEC 61850-7-4:2010, Communication networks and systems in substations—Part 7-4: Basic communication structure—Compatible logical node classes and data classes, IEC, Geneva, Switzerland.
16. IEC 61850-8-1:2011, Communication networks and systems in substations—Part 8-1: Specific communication service mapping (SCSM)—Mappings to MMS (ISO/IEC 9506-1 and ISO/IEC 9506-2) and to ISO/IEC 8802-3.
17. IEC 61400-25:2006, Wind turbines—Part 25: Communications for monitoring and control of wind power plants, IEC, Geneva, Switzerland.

Websites

http://www.sisconet.com.
http://www.systemcorp.com.au
http://www.trianglemicroworks.com/.
http://www.nettedautomation.com.
http://blog.iec61850.com.

25

Extending CEA-709 Control Networks over IP Channels

25.1 Introduction and Overview 25-1
25.2 ANSI/CEA-852 Standard 25-2
25.3 System Components 25-4
25.4 Data Communication 25-5
Encapsulation • Packet Sequencing • Packet Aggregation • Stale Packet Detection • ANSI/CEA-852 Data Packet Routing
25.5 Management 25-11
25.6 Security 25-13
25.7 Applications 25-15
25.8 Conclusion 25-16
References 25-17

Dietmar Loy
LOYTEC electronics GmbH

Stefan Soucek
LOYTEC electronics GmbH

25.1 Introduction and Overview

Control networks (CNs), also known as fieldbus systems, started to emerge rapidly in the early 1990s [1,2]. Back then, CNs were designed to be used in closed systems exchanging data packets on dedicated network channels. These channels included twisted pair media, power-line networks, wireless, and infrared communication. Dedicated network channels have more or less well-defined properties for bit error rates, propagation delays, and average and maximum response times, and they guarantee to maintain the packet order on the network.

With the upcoming hype of the Internet in the late 1990s, requests were made to extend CNs to not only span local communication within relatively small areas but also cover networks that are spread out over cities, countries, and even continents. In building automation, facility managers wanted to connect facilities to a central location for monitoring, logging, alarming, trending, and remote maintenance [3].

This meant for the engineers to either go back to the drawing board and to redesign the CN protocol to make it work in large networks or to use the existing technology like the IP service available on the Internet to transport CN data packets across continents [4]. Of course, there are still restrictions regarding defined maximum response times, proper packet ordering, and defined packet loss—all properties that are easily manageable on dedicated network channels but can be more or less unknown or undefined in Internet Protocol (IP) networks. Nevertheless, all major building automation systems offer IP-based protocols or protocol extensions [5].

When making the transition from dedicated networks to IP channels, the major engineering challenge is to introduce a layer of software, which manages to hide the different behaviors of the IP network and

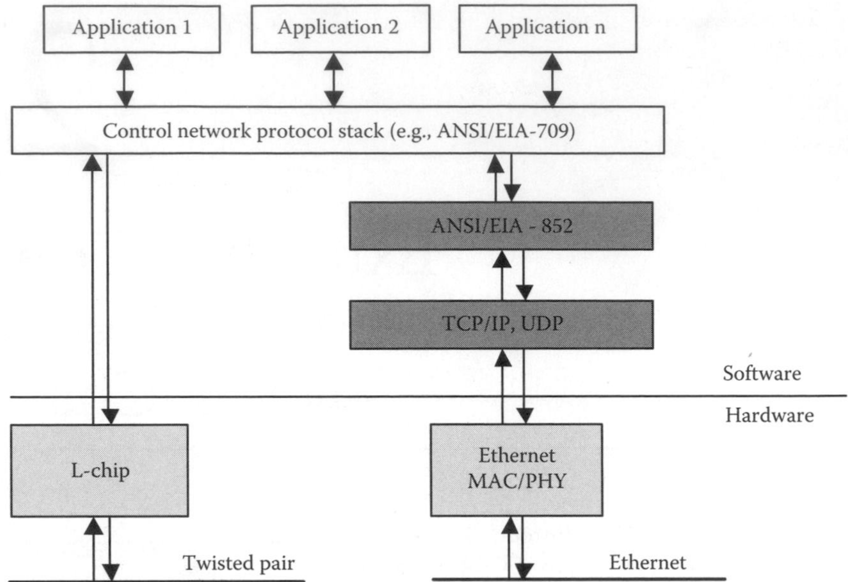

FIGURE 25.1 A dedicated software layer named ANSI/CEA-852 hides limitations of the IP transport service from the control network protocol stack and encapsulates the control network data packet into an IP frame.

ensures that certain CN properties still work by filtering invalid packets (e.g., stale packets). Figure 25.1 shows the software architecture for a typical fieldbus node connected to a dedicated network channel on the left side and a port using the IP transport service on the right side. One can see the software layer called ANSI/CEA-852 [6], which was added between the CN protocol stack and the TCP/IP and UDP socket interface. This layer is responsible to abstract the IP network. It implements the necessary mechanisms to encapsulate CN packets in order to convey them transparently over the IP network. In this chapter, we concentrate on the ANSI/CEA-709 CN protocol [3].

The challenge in tunneling CN data packets through IP networks is to minimize the effect of the unpredictable IP timing parameters on the CN protocol stack. This is a fundamentally different approach than gateways, which try to abstract some application-oriented part of the fieldbus [7,8]. Other aspects to be covered are channel formation and member management, security, and firewall traversal in Internet environments.

25.2 ANSI/CEA-852 Standard

The ANSI/CEA-852 standard [6] was created to ensure interoperability of CN devices, which communicate over an IP network using a native CN protocol. It has since evolved to revision ANSI/CEA-852-B and has been standardized as an ISO standard ISO/IEC14908 [9]. Devices conforming to ANSI/CEA-852 are widely deployed in today's building automation systems. The follow-up standard CEA-852.1 [10] overcomes some of the limitations of ANSI/CEA-852 but achieves this at the cost of dropping backward compatibility. Because ANSI/CEA-852 and CEA-852.1 devices cannot interoperate in the same system, CEA-852.1 devices are still not accepted on the market.

The basic idea is to exchange CN packets over an IP network by embedding them in IP packets. This technique is known as tunneling. In ANSI/CEA-852, there are no application layer transformations involved as in other approaches that use gateways. The tunneling approach is beneficial in systems where a number of CN devices communicate and use the IP network purely as another medium.

One major design criterion of the ANSI/CEA-852 standard is to be generic enough to apply it to different CN networks. Currently, the usage for ANSI/CEA-709 (LonMark IP-852) and EIA-600 (CEBus) is defined. Most ANSI/CEA-852 implementations on the market today are LonWorks based. Therefore, the description of the standard is focused on its application to ANSI/CEA-709.

The network elements defined in the context of the ANSI/CEA-852 standard are CN devices. These CN devices are computing systems (e.g., PCs and embedded systems) equipped with a TCP/IP stack, the ANSI/CEA-852 software, and a LonWorks protocol stack. A number of CN devices are connected over the IP network where they form a logical network known as the *IP channel*. The IP channel functions as a LonWorks communication channel (IP-852 channel), and CN devices exchange data over the IP channel. The CN devices, which are part of the IP channel, are referred to as *channel members*.

It shall be noted that there may exist more than one IP channel on the same IP network. Although the CN devices are connected to the same physical network in this case (e.g., Ethernet), the IP channels are still isolated in the LonWorks domain.

CN devices can have different functions. Depending on their primary function, the ANSI/CEA-852 standard defines the following CN device types:

- CN nodes: They operate solely as nodes in a distributed control application and are members of one IP channel. They have the same functionality as LonWorks nodes on their native communication channels and are connected to the IP channel only.
- CN/IP routers: These CN devices are LonWorks routers that connect one native LonWorks channel with one IP channel. These devices are available on the market as stand-alone LonWorks/IP routers or ANSI/CEA-852 routers, for example, the L-IP by LOYTEC or the i.LON 600 by Echelon Corp.
- IP/IP routers: These CN devices are LonWorks routers that connect two IP channels. These devices are less common on the market today.
- CN proxies: These are CN devices, which, for example, function as LonWorks proxy nodes to establish cross-domain communication.

The ANSI/CEA 852 addressing scheme defines two address types. First, there are IP addresses associated with each CN device on the channel. Second, there are logical CN addresses based on the LonWorks address space. The Unique ID is equivalent to the LonWorks node ID, or Neuron ID, and functions as a unique hardware address. Each CN device is accessible by its unique ID. Logical network addresses include a subnet/node (S/N) scheme. In this addressing scheme, subnet numbers are assigned to individual channels and node addresses to the CN devices on those channels. As a convention, subnets must not span over multiple channels, if the CN routers are used as configured LonWorks routers. The group address identifies a number of nodes independently from subnet addresses. Consequently, groups can span different channels. Finally, those logical addresses are local to a domain.

The mechanisms in the ANSI/CEA-852 standard are designed to take care of typical CN protocol properties, which differentiate them from common IP application protocols such as e-mail, web access, or multimedia streaming. The key characteristics of CN traffic are

- Low throughput
- Small packet sizes
- Higher sensitivity to packet loss and latency

While low throughput allows CN devices to perform more lengthy per-packet calculations, small packet sizes will lead to significant overhead in the encapsulation process. The implications of packet loss and latency on the IP network are twofold in ANSI/CEA-852 systems:

- The control application has to be designed to cope with timing parameters, which differ from native CN channels. This has an impact on the soft real-time properties of some applications. It is important to know the time constants in the application and determine whether the application still operates correctly when run over an IP channel [11,12].

- The ANSI/CEA-852 standard has to ensure the functionality of the CN protocol itself. LonWorks, for instance, makes strict assumptions on the maximum end-to-end delay and packet ordering.

The resulting functional domains defined in the ANSI/CEA-852 standard to implement a CN protocol tunnel over an IP-based network can be summarized as follows:

- Assuring correct and efficient data communication: CN packets must be encapsulated and decapsulated, packet ordering must be ensured, message overhead must be minimized, timeliness must be checked, and optional security measures can be used.
- Routing CN payload to their correct destination on the IP network: The ANSI/CEA-852 standard defines two ways to choose destination CN devices, either by using IP multicasting or by performing selective forwarding.
- Channel formation and managing IP channel members: Configuring the CN devices, distributing information between the CN devices, performing access control on the IP channel, and retrieving statistic information.

25.3 System Components

In addition to assigning logical addresses to the network node as specified in the CN protocol, network nodes on an IP channel must also have an IP address. Thus, when an IP-based node needs to transmit data packets to another node on the network, it must know not only the CN address but also the IP address of the target node. A system component is required that manages the relationship between the CN address and the IP address for a logical IP channel. In the ANSI/CEA-852 standard, this component is the configuration server (CS). Figure 25.2 outlines the system components required to operate and manage an IP channel. The network in Figure 25.2 is partitioned into three parts: two traditional ANSI/CEA-709 networks with subnet numbers 2 and 3 and an IP channel with subnet number 1. The IP channel consists of four CN/IP client devices and a CS. Two IP client devices are routers between the traditional ANSI/CEA-709 network channels and the IP-852 channel (CN/IP router). The other two devices are native IP-based CN nodes. Let us assume that the node with subnet/node address 1/1 on the IP-852 channel wants to send a data packet to the network node with subnet/node address 1/2,

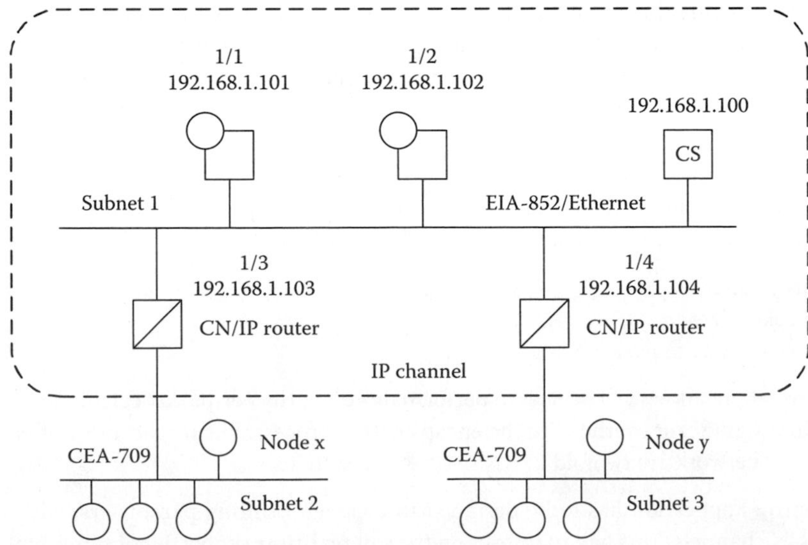

FIGURE 25.2 System components constituting an IP channel including the IP channel, the configuration server (CS), and IP-based nodes acting as configuration clients.

which is also connected to the IP-852 channel. In order for node 1/1 to send out a packet to node 1/2, it must know its peer IP address 192.168.1.102. This information is maintained by the CS hosted on IP node 192.168.1.100 and distributed to all nodes on the IP-852 channel. The node 1/1 queries this information from the CS and therefore knows the mappings between subnet/node addresses and IP addresses for each node on the IP-852 channel. In our example, node 1/1 generates a UDP packet that is sent to node 1/2 at IP address 192.168.1.102. If this packet is sent using acknowledged service, node 1/2 replies with an acknowledgment packet to node 1/1. This is the same behavior as on a traditional ANSI/CEA-709 channel except that node 1/2 must know not only the source subnet/node address to send the reply but also under which IP address this subnet/node address is reachable. The reply packet must be addressed to the IP address of node 1/1, which is 192.168.1.101. To summarize this, the process of selecting an appropriate IP address corresponding to a specific ANSI/CEA-709 destination address is called *channel routing* (CR). Each node that joins an IP channel must send out the information about its assigned ANSI/CEA-709 address information to the CS, which associates this address information with the IP address. This is called CR information. Whenever the CR information changes, a client node on the IP channel must create a new CR packet and send it to the CS, which in turn distributes the new information to all other members on the IP channel.

CN/IP router devices bridge the gap between traditional ANSI/CEA-709 networks and IP channels as shown in Figure 25.2. For example, if node x on Subnet 2 wants to send a message to node y on Subnet 3, then the router with IP address 192.168.1.103 forwards the packet to the router with IP address 192.168.1.104, which forwards the packet to the destination node y on Subnet 3. One can see that even though there is an IP channel on the communication path between node x on Subnet 2 and node y on Subnet 3, the source node x on Subnet 2 does not even notice this IP channel in between the source and the destination node. As a result, it is now possible to connect any two local CNs with IP channels that can span either intranets or the Internet.

The IP channel can be operated on different network media that are offering the IP service. Commonly, 10 BaseT or 100 BaseT Ethernet networks are used for the IP channel. But emerging wireless technologies such as 802.11b or 802.11a and Bluetooth, as well as POTS and ISDN lines, cable networks, and fiber-optic cables are used as well. This opens new application areas to existing CN protocols that could not be satisfied with the traditional technology.

25.4 Data Communication

CN devices exchange CN data packets over the IP channel. This is referred to as data communication to distinguish the traffic from management communication between the CS and the CN devices. There is always a source of a CN data packet and a sink. In the following, we shall refer to a CN device that acts as a source as the sender, to the CN device that acts as the sink as the receiver. It shall be noted that this sender–receiver relation is valid only in the context of a single data packet, or a series of data packets, in the same direction. Generally, each CN device can act as a sender as well as a receiver on the IP channel at the same time.

CN nodes are true sources and sinks of CN data. CN/IP routers are intermediary devices between the actual sources and sinks of the communication. This detail shall be neglected as the CN/IP routers appear on the IP channel as senders and receivers as well.

The diagram in Figure 25.3 shows the functional blocks in the sender and receiver. When a data packet is generated in the CN node or has to be passed on to the IP channel by a CN/IP router, the sender packs the CN packet in an ANSI/CEA-852 data packet, adds a sequence number and a timestamp, routes the ANSI/CEA-852 packet to the appropriate channel members, and finally may collect a bunch of packets before actually sending them on the wire (bunching). The receiver performs the reverse operations: It separates the aggregated packets (debunch), runs them through a sequencer, a stale packet detector, and unpacks the original CN frames. These details of these functions are explained in more detail in the following sections.

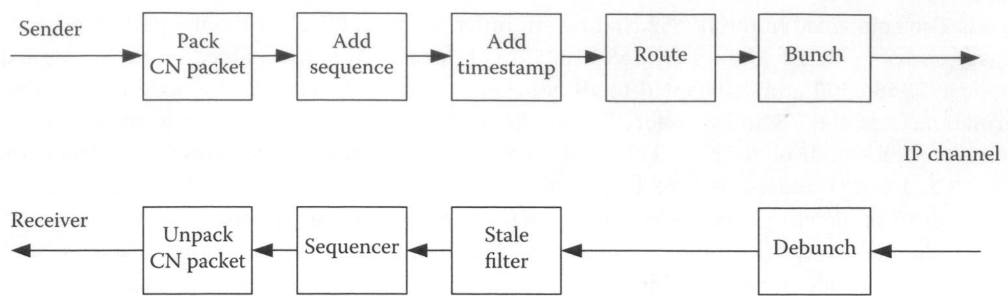

FIGURE 25.3 Functional blocks in ANSI/CEA-852 Data Communication: sender and receiver.

25.4.1 Encapsulation

The process of embedding ANSI/CEA-709 frames in ANSI/CEA-852 data packets is referred to as encapsulating. The native ANSI/CEA-709 packets are encapsulated in UDP frames and routed to the respective CN devices on the IP channel. ANSI/CEA-852 uses the reserved port number 1628 for ANSI/CEA-852 communication on CN devices. The CS uses port 1629. This makes it possible to combine a CS and a CN device on the same system.

The choice of UDP transport over TCP has several advantages: First, because of the connectionless nature of LonWorks communication, it would incur unnecessary overhead to manage connection setup in TCP. Second, TCP ensures delivery through retransmissions. Because of the (soft) real-time requirements of control applications, retransmissions will most likely miss the deadlines. Third, the LonWorks protocol implements its own retransmission scheme where necessary.

Figure 25.4 shows the format of the native LonWorks packet format. The preamble bits and the code violation after the CRC are used for bus arbitration and end-of-frame marking on native ANSI/CEA-709 channels. Not all parts of the native frame are subject to encapsulation. The fields starting with the control field and ending with the CRC information are taken from the native LonWorks frame and encapsulated in the UDP frame. Figure 25.5 shows the ANSI/CEA-852 common header encapsulation format, which

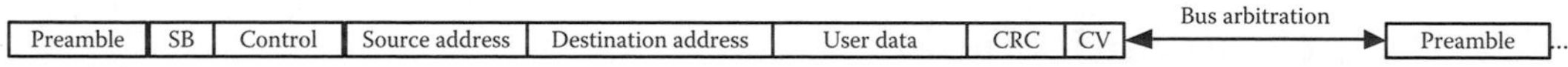

FIGURE 25.4 Native ANSI/CEA-709 frame format.

Byte 0	Byte 1	Byte 2	Byte 3
Data packet length		Version	Packet type
Ext. header size	Protocol flags	Vendor code	
Session ID			
Sequence number			
Timestamp			
: Packet-specific data :			

FIGURE 25.5 ANSI/CEA-852 common header format.

TABLE 25.1 ANSI/CEA-852 Packet Types

Packet Name	Packet Type	Function
Data packet	0x01	Encapsulates the CN frame.
Device configuration request (DC-REQ)	0x63	Request a DC from the CN device.
Device registration (DEVREG)	0x03	Device registers with the CS.
Device configuration (DC)	0x71	DC requested from the device or unsolicited DC sent by the CS.
Channel List Request (REQ-CM)	0x64	Device requests CM from the CS.
Channel List (CM)	0x04	The CS responds with the CM.
Send List Request (REQ-SL)	0x66	Request the send list from the CS.
Send List (SL)	0x06	The CS responds with the send list.
Channel Routing Request (REQ-CR)	0x68	The CN device requests a CR from the CS for a specific member.
Channel Routing (CR) Packet	0x08	The CS responds to a CR request.
Acknowledge (ACK)	0x07	Acknowledgment message from device or CS.
Segment (SEG)	0x7F	Segment of a large data structure.
Statistic Request (REQ-STAT)	0x60	Request statistics from a CN device.
Statistic Response (STAT)	0x70	CN device responds with statistic data.

is the payload of a UDP frame. Data fields in ANSI/CEA-852 packets are defined in network order, transmitting the most significant byte first. The data packet length specifies the complete length of the ANSI/CEA-852 packet including the two length bytes. The version number is currently set to 1. The packet type specifies how the ANSI/CEA-852 payload after the common header is to be interpreted.

An ANSI/CEA-852 data packet, which encapsulates the CN frame, is of type 0x01. For more packet types, refer to Table 25.1. The extended header size can be used to add more fields to the common header. It specifies them in quantities of 4 bytes. This ensures that the common header is always a multiple of 4 bytes. The protocol flags define which CN protocol is to be tunneled (0x00 for LonWorks). The vendor code allows implementers to include their own vendor-specific extensions, which can be transported along with standard ANSI/CEA-852 messages. The session ID is a 32-bit value that a CN device keeps at random. After a CN device reboots or resets, the session ID must be different than the previous one. The sequence number field is valid for ANSI/CEA-852 data packets and is used to detect out-of-sequence packets. The timestamp is also used for ANSI/CEA-852 data messages and is a millisecond value. Therefore, it wraps around every 49 days.

25.4.2 Packet Sequencing

Packet sequencing in ANSI/CEA-852 is based on the pair session ID and sequence number, which are both part of the ANSI/CEA-852 common header. An ANSI/CEA-852 receiver remembers the session ID/sequence number pair for each sending CN device (i.e., for each source IP address). If ANSI/CEA-852 data packets from a given source are received with the same session ID, they are in sequence, provided their sequence numbers are increasing. If data packets get reordered or are dropped on the network, there appear gaps in the sequence numbers. A receiver basically has two choices: (1) continue and drop the missing packets later or (2) wait for the missing packets.

If the receiver waits, it has to keep data packets after such a gap in escrow until a defined escrow timeout (ETO). If the missing packets arrive before the timeout, the packets are passed on in their correct sequence. If they do not arrive in time, they will be dropped later. Without escrowing, all out-of-sequence packets are dropped immediately. This may cause unnecessary packet loss on IP channels where packets may get reordered in a small time window.

If the session ID changes, the receiver has to assume that the source has been reset, and it accepts the first packet with that new session ID as in sequence, clearing the sequence numbers from the previous session. This is a design choice of ANSI/CEA-852 and can cause problems with replay attacks. See the section on security for more detail on this issue.

The ETO is a CN device parameter in most implementations. This time constant has to be considered when designing a control application. It is always a trade-off that has to be made between unnecessary loss of reordered packets and delay of escrowed packets for too long when packets actually are lost on the IP link. If the ETO is big, more packets can be reordered because the wait time for them is longer. For the same reason, if a packet is actually lost on the IP link, the wait time can approach the ETO in the worst case.

Applications that primarily rely on request/response-style communication on LANs should choose a small ETO (e.g., 5 ms) or disable it. Applications that are based on a continuous unidirectional sample stream over a WAN link should choose a large enough ETO. A typical default value on most implementations is 64 ms.

25.4.3 Packet Aggregation

The native message sizes of typical CN traffic are small, around 10–20 bytes in LonWorks. Encapsulating each CN packet into a UDP frame adds the UDP, IP, and Ethernet headers (or other media-specific headers), which account for more than 40 bytes. This incurs unnecessary overhead that may lead to noticeable performance problems on embedded implementations. Those systems typically suffer from an IP transmission bottleneck.

A technique to reduce this overhead is to aggregate more than one ANSI/CEA-852 data packet in a single UDP frame (packet bunching). Clearly, the more packets are bunched, the less per-CN packet overhead is added by the encapsulation. When aggregated, ANSI/CEA-852 data messages are concatenated in the UDP frame. The UDP packet is then transmitted to the destination IP address. Therefore, messages can be aggregated only when they share the same destination. The sender may thus be configured to keep data packets and aggregate them until an aggregation timeout (ATO) before sending the UDP frame on the IP link. The ATO value is a CN device parameter in most implementations.

The effect of packet bunching has to be considered in the design of control applications. There is a trade-off between overhead reduction and extra delay. In the worst case, an ANSI/CEA-852 data packet is delayed for the full ATO until it finally gets on the wire. For highly responsive applications with low packet rates on LANs, packet bunching should be disabled. For applications with high packet rates over WANs, packet bunching should be enabled. A typical default value in most implementations is 16 ms.

25.4.4 Stale Packet Detection

Packet delay on the IP link is a big issue on IP channels. Packet delays can be highly random with a large standard deviation on wide area IP links [13]. There are two problems with this: (1) the control application may miss some of its deadlines, and (2) the CN protocol may break if packets are received after they are delayed too long. For example, the LonWorks duplicate detection mechanism is sensitive to high delay variance on the network. While the first problem must be solved by carefully designing the control application [14], the second problem can be eliminated by discarding packets that are delayed beyond a certain time limit (stale packets). The mechanism is called stale packet detection and is based on one-way packet delay measurements. In order to make such measurements, CN devices must synchronize their clocks. Clock synchronization in ANSI/CEA-852 is performed via SNTP [15]. The sender generates a millisecond timestamp and puts it into the ANSI/CEA-852 common header. If its measured delay exceeds the channel timeout (CTO), the packet is called a stale packet, and the receiver discards it in the stale packet filter.

The CTO is a channel parameter and distributed among the CN devices. Applications on LANs can disable stale packet detection because the delay variance is typically very low and comparable with native LonWorks channels. On WAN links, it is recommended to enable the CTO. Typical values are the average ping delay on the IP channel plus ATO. The lower bound of the CTO is restricted by the resolution of the clock synchronization. On certain systems (e.g., Windows), the resolution may drop to 50 ms. Practical CTO values start at 200 ms. If certain CN devices are Windows implementations, it must be

TABLE 25.2 ANSI/CEA-852 Statistics Data

Data	Explanation
Seconds since cleared	Number of seconds since the statistics have been reset
Date/time of clear (GMT)	Date and time when the statistics have been reset
No. of members	Current number of channel members
CN packets received	CN packets received from LonWorks
CN packets sent	CN packets sent to LonWorks
IP packets sent	CN packets sent onto the IP channel
IP bytes sent	Aggregate number of bytes in CN packets sent onto the IP channel
IP packets received	Number of CN packets received from the IP channel
IP bytes received	Aggregate number of bytes in CN packets received from the IP channel
CN stale packets	Number of packets dropped because they are stale
RFC packets sent	Number of management data packets sent to the CS
RFC packets received	Number of management data packets received from the CS
Avg. aggregation to IP	Average aggregation of transmitted CN packets on the IP channel
Avg. aggregation from IP	Average aggregation of received CN packets from the IP channel
UDP packets sent	Total number of UDP packets sent (including both data and management)
Multicast packets sent	Total number of multicast packets sent (only for send lists)

ensured that the Windows system clock is synchronized. Dropped packets due to stale packet detection can be observed in the ANSI/CEA-852 statistics of stale packets (see Table 25.2). Stale packet filtering also plays an important role in the context of security. Refer to the section on security for more detail.

25.4.5 ANSI/CEA-852 Data Packet Routing

One of the ANSI/CEA-852 CN device tasks is to route ANSI/CEA-852 data packets to the appropriate IP channel members. In this context, it is important to notice that each channel member is associated with an IP address. If a given CN data packet needs to be routed to multiple IP addresses, it is copied into several ANSI/CEA-852 data packets, each sent to one of the selected CN devices on the channel. It shall be noted that sequence numbers are associated per destination address. If CN data packets are routed to different IP addresses, the sequence numbers need not necessarily be the same among all destinations. This stems from the fact that some messages are routed to a specific destination and some are not.

Depending on the available channel data, three variants of routing the CN data packet are possible:

1. Send List (SL) Routing. An SL is an optimized list of IP addresses to reach all channel members. Typically, SLs are used when IP multicast addresses are assigned to groups of channel members. In the best case, all channel members are in the same IP multicast group. In this case, the SL contains this one multicast address. CN devices send ANSI/CEA-852 data packets to all destinations in the SL.
2. Channel Routing: CR is used when only unicast IP addresses are available. The basic idea is to select those destination IP addresses that will accept the packet based on the LonWorks address information. To advertise which LonWorks addresses a CN device accepts, it publishes CR information to the other channel members. Each channel member collects all the available CR and can route outgoing CN data packets in an optimized way. It shall be noted that with CR, all data packets have to be copied multiple times if there are multiple advertised recipients. Especially for large groups and broadcast addresses, SLs appear much more efficient. For subnet/node or NID addressed packets, CR is beneficial because the packets are routed to a single unicast address only not burdening other CN devices with packets that they do not want.
3. Brute Force: Neither SL nor CR information is available. The CN device resorts to the channel list and transmits every data packet to all channel members using unicast addresses. This mode is the least efficient.

The CR algorithm is based on the CR information provided by the CN devices on the channel. The CR information for each channel member contains the following fields:

- The CN device IP address and port
- The CN broadcast flag
- The CN device type (router, node, IPIP router, and proxy)
- The CN router mode (configured, bridge, and repeater)
- A list of node IDs
- A list of subnet/node addresses
- A list of domain items (including subnet and group forwarding flags)

The domain items deserve a more thorough discussion. First, they contain a domain number. Then they specify forwarding flags for both subnets and groups. If the nth subnet flag is set, the CN device accepts all packets destined for subnet n. If the mth group flag is set, the CN device accepts all group packets for group m. CN routers need to specify the subnet and group forwarding flags, while CN nodes only specify the group flags. This is because CN nodes do not forward any packets, but they can be members of multiple groups.

Figure 25.6 summarizes the CR algorithm implemented in most CN/IP router implementations. The CR algorithm iterates over all channel member CR information for each ANSI/CEA-852 data packet. § I checks the CN broadcast flag and routes all packets if the router type is a repeater and it has at least one domain item. The CN broadcast flag is set when the channel member is unconfigured and does not yet have a valid domain. Otherwise, those devices could not be reached over the IP channel by broadcasts.

Leaving the special case of unconfigured CN devices aside, there are two types of devices on a channel: (1) CN nodes, which accept only CN packets addressed to them (e.g., their subnet/node address, their NID, and their group), and (2) CN routers, which forward packets to other channels behind them. The CR algorithm therefore distinguished those two cases.

In § II, the packet is routed by looking at the channel member as a CN router (e.g., configured router and bridge). This step uses information contained in the domain items of the CR information. Therefore, a forwarding device must define at least one domain item in its CR information. § II(i) skips the CR for a certain domain item if the domain does not match. § II(ii) is responsible for forwarding group addressed messages if the corresponding group forwarding flag is set in the domain item. § II(iii) checks the destination subnet part for S/N, NID, and broadcast-addressed messages. If they are domain-wide (i.e., subnet equals "0"), the message is forwarded. Otherwise, those messages are forwarded in § II(iv) only if the corresponding subnet forwarding flag is set in the domain item. As a consequence, CN routers must supply domain items for all domains they need to receive packets in.

I. If (CN broadcast flag is set and address == FMT_BCAST) or (router type == REPEATER and there is a domain item) then route to member.
II. Route to CN router
  (i) If domain is incorrect skip to III.
  (ii) If address == FMT_GROUP and group forward flag is set for GROUP route to member.
  (iii) If SUBNET == 0 route to member.
  (iv) If subnet forward flag is set for SUBNET route to member.
III. Route to CN device
  (i) If address == NID and NID matches route to member.
  (ii) If DOMAIN does not match skip to IV.
  (iii) If address == FMT_SN and subnet/node item matches route to member.
  (iv) If address == FMT_BCAST and SUBNET == 0 route to member
  (v) If address == FMT_BCAST and SUBNET matches subnet/node item route to member.
IV. Go to next member CR information.

FIGURE 25.6 Channel routing algorithm.

In § III, the message is routed by looking at the channel member as a CN node on the IP channel. The IP channel is assigned a separate individual subnet number, and all nodes on that channel get their individual node numbers. Therefore, devices that represent nodes on the IP channel must supply at least one subnet/node, one NID, and one domain entry in their CR information. § III(i) checks if an NID addressed message matches with one of the NID items in the member CR information. If the NID matches, the message is forwarded to that member ignoring any domain information. Therefore, unconfigured nodes on the IP channel can be reached via NID addressed messages. For all other messages, § III(ii) checks if the domain matches with one domain item. If it does not, the algorithm skips to § IV. In § III(iii), subnet/node addressed message are routed. A subnet/node addressed message is forwarded to a specific member if that member has a matching subnet/node entry in its CR information. § III(iv) routes all domain-wide broadcasts to a node on the IP channel. § III(v) routes all subnet broadcasts to the channel member if one of its subnet/node items has a matching subnet number. § IV iterates to the next member in the channel until the CR algorithm has gone through all channel members.

The CR mechanism has the following implications on the CR information, which CN devices must provide:

- Unconfigured channel members must set the CN broadcast flag in order to get configured by LonWorks network management tools (e.g., LonMaker).
- CN routers (configured router, bridges, etc.) must provide corresponding domain items in the CR for all domains they want to exchange packets in.
- CN nodes must provide corresponding subnet/node, NID, and domain items.

25.5 Management

Apart from data communication, a big part of the ANSI/CEA-852 functionality is devoted to the management of IP channels. Management functions include channel formation, which defines the CN devices that are member of a specific IP channel, configuration of individual CN devices, distributing information between CN devices, and retrieving statistics information about CN devices. All relevant configuration data for a CN device is controlled by a device configuration (DC) structure.

The ANSI/CEA-852 management concept is based on a client/server model. A CS is the central part in the system, which is responsible for channel formation. CN devices act as configuration clients and request data structures from the CS. The CS only actively sends the DC to CN devices, if the IP channel has been updated and it needs to announce that change. Therefore, the DC is an unsolicited message from the CS. If no channel changes are pending, the CN devices can operate without a CS. As soon as the IP channel needs to be updated, for example, a CN device is added or recommissioned, the CS must be available again to distribute the change.

The management data structures include the per-device DC, a channel membership (CM) list specifying the members on the IP channel, an SL for SL routing, which is optional, and a per-device CR information. All management data in ANSI/CEA-852 are versioned. The version is a 32-bit value, which increases for newer versions. The version is usually date and time information in NTP second format, which may or may not be synchronized to UTC. Hence, the version is called datetime and is nonzero for valid data structures. Local data structures are checked against published datetime versions and requested if newer data are available. The CM, SL, and per-device DC are versioned by the CS, whereas the CR data are versioned by the individual CN devices.

The CN devices, which constitute the IP channel, are defined in the channel list on the CS. Each CN device is uniquely identified by its IP address in the channel list. An important restriction in ANSI/CEA-852 is that there cannot be two CN devices on a channel, which share the same IP address but have different IP ports.

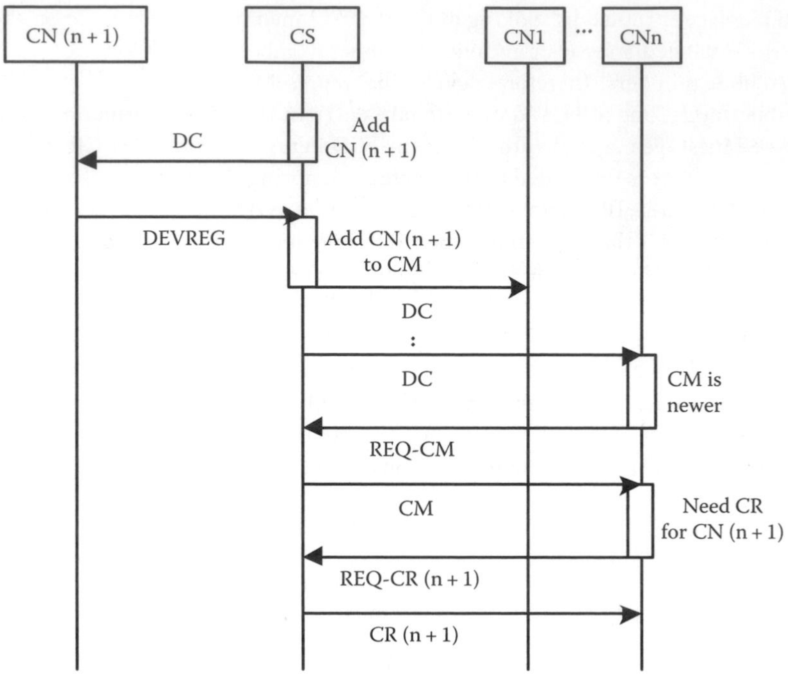

FIGURE 25.7 CS configuration data distribution.

Each CN device in the channel list is configured by a DC. The DC is entered on the CS, for example, through a serial console or a web interface. The following information needs to be supplied in order to add a CN device to an IP channel:

- IP address and port of the CN device
- Name of the CN device (typically 15 characters)

Figure 25.7 depicts the message flow when adding or configuring a CN device. The CS adds a CN device to the channel by sending an updated and unsolicited DC message to the device. Once a CN device is added or has accepted a new configuration, it registers with the CS by sending a device registration (DEVREG). Knowing the channel is updated, the CS sends a new unsolicited DC to all the other members to announce the channel update. The other CN devices in turn request the updated data structures from the CS starting to request the channel list (REQ-CM). If the CN device finds out that the CM advertises new CR information for certain members, it requests them from the CS (REQ-CR) one by one.

If a CN device is not available when it is added on the CS, it is usually marked as *unregistered* in the channel list. In this case, the CS tries to recontact the CN device using an exponential back-off for the retry interval. It starts at 1 s, doubles each time, and is limited by 32 s. As a consequence, it may take up to 32 s for a device to become functional on the IP channel when it was not reachable at the time of adding it. CS implementations usually display CN devices, which have registered successfully as *registered*. If a device has been registered once but is not responding to a channel update, a CS may display it as *not responding*. The operator then can take appropriate action for those devices.

Access control on the CS is also defined through the channel list. If a CN device tries to register with a CS that does not include this particular device in its channel list, the CN device is rejected from the IP channel. Rejected CN devices cannot send or receive CN packets over the IP channel and usually indicate this condition as an error, for example, a red status LED. Some CS implementations collect information about those CN devices in a so-called orphan list. The operator then can select an orphan device and add it manually to a given IP channel.

An important part of ANSI/CEA-852 is the routing of CN packets to the correct CN devices over the IP network. As described in the previous section, there are two possibilities: (1) creating an SL or (2) distributing CR information. The CS generates the SL by looking at the IP address information in the channel list. By appropriately grouping IP multicast addresses and unicast addresses, the SL typically gets much shorter than the channel list. The SL is also versioned and distributed to the CN devices by sending an updated DC to the members.

The other possibility is to distribute CR information. The CR, however, does not originate at the CS, but at the individual CN devices because they are commissioned and only they have knowledge of their CN address information. Each CN device sends a CR update to the CS if its CR changes. The CS in turn generates a new channel list, which reflects the updated CR datetime version of the device. The other devices will request the new channel list and discover that a specific device has a new CR. The other devices then request this CR from the CS as described before. If more than one CR information has changed, the CN devices request all newer CR data from the CS. This is an important detail, because CN devices never request the new CR from the original device directly. Therefore, the CS must be available, if any CN device is recommissioned.

The management part of ANSI/CEA-852 also defines a statistics information structure. This structure can be requested outside of the client/server model. Any CN device or the CS can request the statistic information from a CN device. The statistic data supplied are listed in Table 25.2.

25.6 Security

Security is a growing area of concern for IP networks. This also includes ANSI/CEA-852 traffic. While native LonWorks channels are local (e.g., to a building) and can be accessed only locally, IP channels are established on an open media that may even span across WAN links. Typical security issues are privacy, authenticity, and integrity [16]. Security measures defined in ANSI/CEA-852 refer to the problem of authenticity only. They assure that a received data packet is actually sent by an authentic channel member and not by an attacker.

The principle is based on secure message authentication codes and a shared secret. The secure hash function algorithm MD5 [17] is used to create a unique fingerprint of an ANSI/CEA-852 message. Included in the fingerprint calculation is a secret key, which is not transmitted over the wire. The receiver can perform the same calculation, supplying its own copy of the secret key, and compare the hash codes. If they match, the message is authentic.

The MD5 hash is transmitted right after the ANSI/CEA-852 message. Its fixed length is 128 bits and is not included in the ANSI/CEA-852 packet size field. Instead, channel members are configured for secure operation (MD5 mode) and expect the extra 128 bits after each ANSI/CEA-852 packet. This is especially important for bunched UDP frames. If packets are bunched, the next ANSI/CEA-852 message follows the MD5 hash.

The calculation principle is depicted in Figure 25.8. The MD5 hash following the ANSI/CEA-852 message it set to zero (MBZ). The MD5 algorithm is run over the ANSI/CEA-852 message, the zero hash

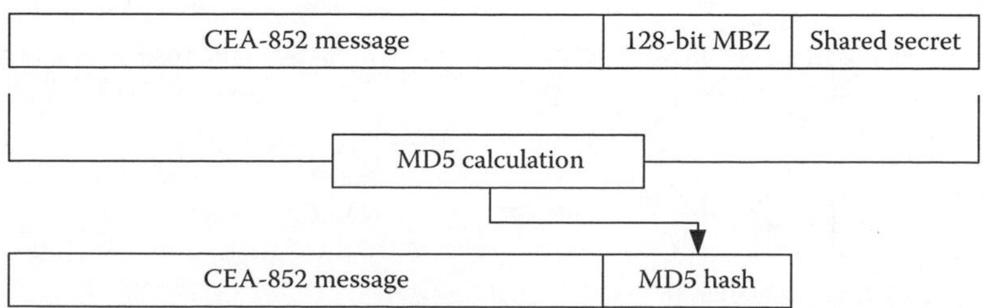

FIGURE 25.8 MD5 authentication of ANSI/CEA-852 packets.

placeholder, and the 128-bit shared secret. The calculated hash value is copied into the placeholder, and the message is transmitted excluding the shared secret. The receiver performs the same calculation. It just saves the MD5 hash and then sets it to zero before calculating the MD5 hash locally. If the saved and the locally computed hash values match, the message contents are authentic.

To prevent replay attacks, stale packet detection must be enabled. The reason for this is the sequencing algorithm in ANSI/CEA-852. An attacker may record a sequence of authenticated messages known to trigger a specific action. These messages will share a common session ID. At a later point in time, the CN device may have chosen a different session ID (e.g., after a power outage). At this moment, the recorded sequence can be replayed successfully because the CN device would accept the old session ID as a new one and start over with counting the sequence numbers. Stale packet detection solves this problem because the ANSI/CEA-852 messages include timestamps in addition. Since the CN devices are synchronized to UTC, a playback at a later time is not possible. The CN devices would drop those packets as stale packets.

Another security aspect in modern networks are firewalls and network address translators (NATs). Firewalls, in principle, only filter traffic. Therefore, CN devices can be operated behind a firewall if the filter rules are set accordingly, namely, allowing UDP traffic on ports 1628 and 1629 to pass.

The case with NAT routers is different. NATs, actually, alter IP addresses and port numbers. This is a problem for ANSI/CEA-852 because channel members on the IP channel are uniquely identified by their IP address and port. If that address or port is changed in the public realm, they could not be part of the IP channel. Some implementations of CN/IP routers therefore allow for a special NAT mode (e.g., the L-IP by LOYTEC). In this mode, a CN device is configured with the public address and port of its NAT router. It represents this public address combination in its DC, and other members will see its public address in the channel list and CR information. The NAT itself must be configured with a port-forwarding rule to route all ANSI/CEA-852 packets from the public interface to the CN device in the private realm. Figure 25.9 illustrates this setup. NAT1 contains a port-forwarding rule to route all packets it receives at IP address 80.41.6.3 on ports 1628 and 1629 to the private IP address 192.168.1.250 on the respective ports.

The drawback of this solution is that only one CN device per IP channel can be operated behind a NAT. If two or more devices would be operated behind the same NAT router, CN devices would need additional information to decide whether the public or private address combination should be used. To overcome this problem, modern devices implement the *extended NAT mode*. This mode uses an extended header format, which contains both private and public addresses of the sender. This makes it possible for senders and receivers to choose the correct address realm. The CS as the central instance for channel formation must be configured with NAT router addresses and private addresses. Since this mode uses an extended header, which is not part of the ANSI/CEA-852 standard, devices that do not support this mode will not be able to join the IP channel. Figure 25.10 shows a possible solution where one IP channel is established over the public realm having two CN devices behind the NAT router 135.23.2.1.

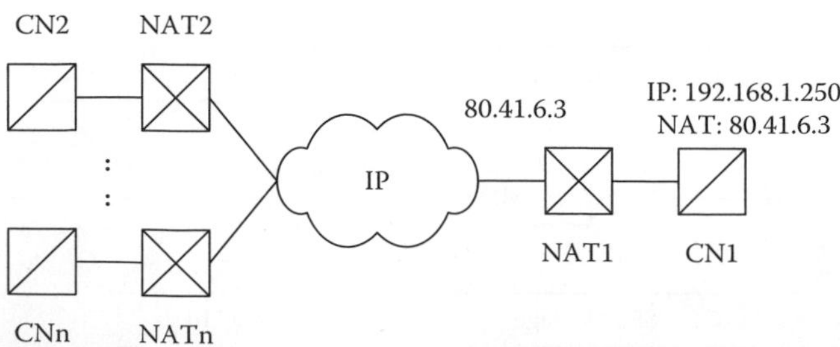

FIGURE 25.9 Simple NAT configuration.

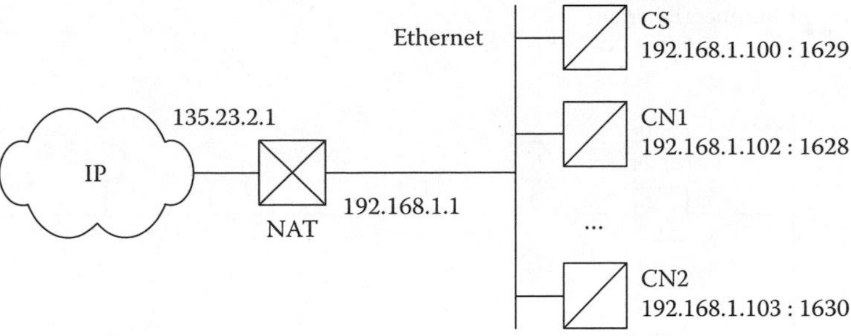

FIGURE 25.10 Complex NAT configuration.

TABLE 25.3 ANSI/CEA-852 Extended NAT Table

Device Name	Local Address	NAT Address
CS	192.168.1.100:1629	135.23.2.1:1629
CN1	192.168.1.100:1628	135.23.2.1:1628
CN2	192.168.1.100:1630	135.23.2.1:1630

Complex NAT configurations require a careful and handcrafted configuration on the NAT routers to make port-forwardings work correctly. Each device needs a separate port forwarding rule in the NAT router. This implies that each device must use a unique client port (e.g., 1628, 1629, and 1630). The port-forwarding rules must be set up so that each port points to one of the ANSI/CEA-852 devices. In the example shown in Figure 25.10, the CS is located at the private address 192.168.1.100. If the CS is located behind a NAT router, the NAT router must have a fixed public IP address. Table 25.3 shows the extended NAT configuration in the CS. For each device, a local and NAT address needs to be specified in order to form the IP channel.

A modern alternative to direct NAT configuration is embedding the IP channel into a virtual private network (VPN). VPNs are standard IT technology and overcome complex NAT configurations. In addition, strong VPN security such as encryption is added to the IP channel communication. However, a VPN solution requires special VPN routers and is not an end-to-end solution. Using a VPN over the Internet is, however, the most secure method available today.

25.7 Applications

So far, we have learned what the new IP tunneling technology following the ANSI/ANSI/CEA-852 standard can do in theory. Here we show the application scenarios that utilize this new technology. We have seen that it is possible to connect CNs via intranets, and this is exactly what is starting to emerge in larger buildings. Typically, office buildings do have a sophisticated IP infrastructure; therefore, it is beneficial to utilize this IP infrastructure for services other than PC networking. As shown in Figure 25.11, CN to IP routers like the L-IP from LOYTEC can utilize the existing IP infrastructure to form a high-speed backbone network in order to connect the CNs on different floors. This setup allows transparent communication between offices on different floors and to a central SCADA system using the IP backbone.

Another steadily growing application area is internetworking between different buildings that are either connected via an intranet or the Internet. Once the building infrastructure is *online*, it is only a small step to remote facility management and remote maintenance. Figure 25.12 shows a typical scenario how building complexes are networked to form one large CN. A management PC connected anywhere to the network allows remote network maintenance, remote trending, alarming, logging, and preventive system repairs.

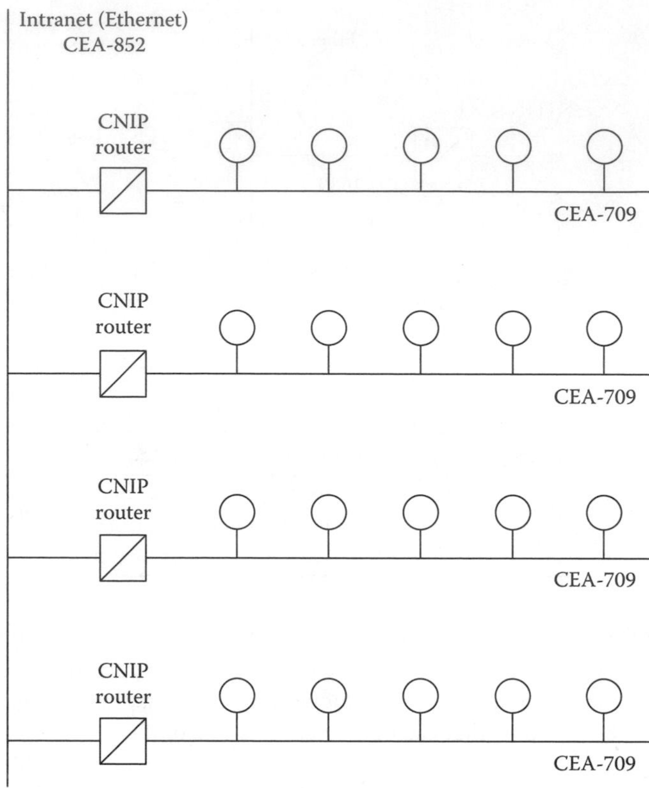

FIGURE 25.11 The IP channel can utilize the existing IP infrastructure in a building to connect several control networks to a large building automation network.

Another advantage, which was initially not quite obvious, is the fact that network management tools and network-troubleshooting tools can be used together with the new IP technology. This means that thousands of installers and system integrators that have been trained over the past 10 years will use this knowledge with a little bit of additional training to install networks that span intranets and the Internet.

Once the CN is able to use the IP service as a transport media, modern wireless technologies can be utilized to create new application areas. One can, for example, walk around a building with a smartphone or tablet and control the building infrastructure such as lighting, heating, AC, blinds, and even carry out network maintenance tasks on the spot by himself or herself. It is now possible to monitor and control remote wastewater pump stations that are connected to the control center via an RF link.

25.8 Conclusion

The world is getting more and more connected, and one of the driving factors is the IP. Connecting PCs to the Internet was the hype of the 1990s; connecting everyday devices to the Internet is the challenge of the new millennium. Modern building automation systems are built on IP-based backbone technologies. With ANSI/CEA-852, this technology is the base of most LonMark systems. NAT support has enabled cross-building automation networks and access ports for remote facility management. Modern installations frequently embed their ANSI/CEA-852 channels in a VPN, which is a widely accepted IT standard today. This method eliminates cumbersome NAT router configuration and provides the enhanced security mechanisms missing in ANSI/CEA-852 such as strong encryption.

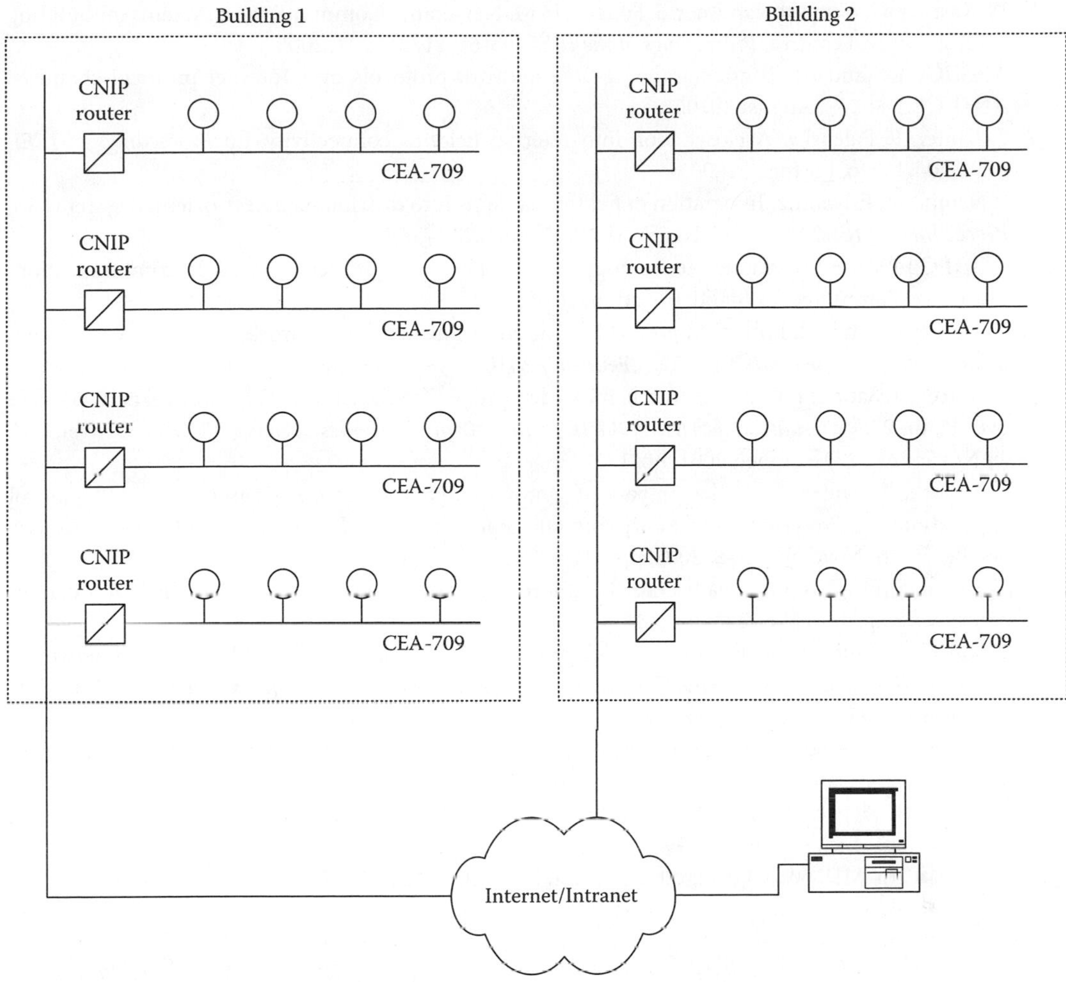

FIGURE 25.12 Connecting different buildings over an intranet or the Internet allows transparent communication of control network data between nodes on either network.

The incompatibility of the enhanced version CEA-852.1 has until now prohibited wide acceptance. Most features such as better authentication and more flexible channel management are currently offset by using VPNs. This is also the only solution to use IPv6 until the enhanced version CEA-852.1 is revised to be natively compatible with IPv6.

References

1. D. Dietrich, P. Neumann, H. Schweinzer. *Fieldbus Technology—System Integration, Networking, and Engineering*, Springer, Wien/New York, 1999.
2. J.-P. Thomesse, M. Leon Chavez. Main paradigms as a basis for current fieldbus concepts, in: *Fieldbus Technology*, Springer, Berlin, Germany, 1999, pp. 2–15.
3. D. Dietrich, D. Loy, H.-J. Schweinzer. *LON-Technologie, Verteilte Systeme in der Anwendung; 2. Auflage*, Hüthig Verlag, Heidelberg, Germany, 1999.
4. A.S. Tanenbaum. *Computer Networks*, 2nd edn., Prentice-Hall Inc., Englewood Cliffs, NJ, 1989.

5. W. Kastner, G. Neugschwandtner, S. Soucek, H.M. Newmann. Communication systems for building automation and control, *Proceedings of the IEEE*, 93(6), 1178–1203, 2005.
6. ANSI/CEA Standard. Tunneling device area network protocols over Internet protocol channels. ANSI/CEA-852-C, August 2010.
7. T. Sauter, P. Palensky. A closer look into Internet-fieldbus connectivity, in: *Proceedings of IEEE WFCS'00*, Porto, Portugal, 2000.
8. P. Neumann, F. Iwanitz. Integration of fieldbus systems into distributed object-oriented systems, in: *Proceedings of IEEE WFCS'97*, New York, 1997, pp. S.247–S.253.
9. ISO/IEC 14908-4. Information technology—Control network protocol—Part 4: IP communication. European Committee for Standardization, 2012.
10. ANSI/CEA Standard. Enhanced protocol for tunneling device area network protocols over internet protocol channels. ANSI/CEA-852.1, February 2010.
11. S. Soucek, T. Sauter, T. Rauscher. A scheme to determine QoS requirements for control network data over IP, in: *27th Annual Conference of the IEEE Industrial Electronics Society* (*IECON*), Denver, CO, 29 November–2 December, 2001, pp. 153–158.
12. S. Soucek, T. Sauter, G. Koller. Impact of QoS parameters on internet-based EIA-709.1 control applications, in: *Proceedings of the 28th Annual Conference of the IEEE Industrial Electronics Society*, Seville, Spain, November 5–8, 2002, pp. 3176–3181.
13. V. Paxon, S. Floyd. Wide-area traffic: The failure of Poisson modeling, in: *ACM/IEEE Transactions on Networking*, 3(3), 226–224, June1995.
14. S. Soucek, T. Sauter, G. Koller. Effect of delay jitter on quality of control in EIA-852-based networks, in: *Proceedings of the 29th Annual Conference of the IEEE Industrial Electronics Society*, Roanoke, VA, November 2003.
15. D. Mills. Simple network time protocol (SNTP) version 4 for IPv4, IPv6 and OSI, RFC 2030, University of Delaware, Newark, DE, October 1996.
16. B. Schneier. *Applied Cryptography: Protocols, Algorithms, and Source Code in C*, 2nd edn., John Wiley & Sons, Inc., New York, 1995.
17. R. Rivest. The MD5 message digest algorithm, RFC 1321, April 1992.

26 Web Services for Embedded Devices

26.1 Introduction 26-1
26.2 Device-Centric SOAs 26-3
SOA Implementations for Devices • Evaluation of Device-Centric SOAs
26.3 DPWS Inside Out 26-5
Interactions between Client, Device, and Service • Messaging • Discovery • Description • Eventing • Security
26.4 Web Service Orchestration 26-22
Web Service Composition • Web Services Business Process Execution Language
26.5 Software Development Toolkits and Platforms 26-24
WS4D, SOA4D • DPWS in Microsoft Windows
26.6 DPWS in Use 26-26
Automation Scenarios
26.7 Conclusion 26-26
References 26-26

Vlado Altmann
University of Rostock

Hendrik Bohn
Nedbank Group Ltd.

Frank Golatowski
University of Rostock

26.1 Introduction

In recent years, an increasing demand for highly automated, process-oriented, and distributed systems consisting of a large number of interconnected heterogeneous hardware and software components is noticeable. These systems—usually composed of components from different manufacturers—require a high interoperability across heterogeneous physical media, platforms, programming languages, and application domains facing several problems:

- Proprietary interfaces. Components have mostly proprietary interfaces weakening compatibility, and developers must know them in order to use such components.
- Limitations in the awareness. Component users are often not aware of other component's functionality.
- Limited discovery capabilities. Component users do not know how to search for a component with certain functionality nor do they know which components are available in a certain scope.
- Proprietary methods for interactions. The interaction mechanisms between components are mostly bound to their application and application domain. The formats for exchanging message as well as the used data formats are mostly proprietary.
- Constrained composition capabilities. Processes/workflows describing the interaction between components are designed for specific applications and domains and seldom applicable for other components and applications.

Hardware components have additional requirements in comparison to software components. Among them are

- Location awareness. In contrast to software components, the location of hardware components is usually important for their application. Offered functionality is often bound to a specific location. For example, the location of a printer is important to its users as it has to be in reachable distance. Therefore, also the printing functionality is bound to the location of the printer, whereas software components can be made available anywhere and where their functionality is required.
- Mobility support. Furthermore, mobile devices can roam between networks possibly resulting in changing addresses of the network protocols.
- Statefulness. From the outside-in view, a software component can be designed in a stateless way by starting a new instance with default values when the component is invoked. Device components mostly have a state that has to be taken care of. For example, a printing device being invoked while printing has to queue the printing job.

These problems are solved for certain proprietary applications and in mostly homogeneous environments, but a solution applicable to a large heterogeneous system is still missing.

The service-oriented architecture (SOA) approach addresses these problems and represents the next step in the endeavor of component-based development and reusability.

SOA [1] is a design paradigm for the interaction between heterogeneous components supporting autonomy and interoperability. The underlying philosophy is that every component offers certain functions that might interest other components or users. SOA provides an outside-in view (orientation) on the functionality of components—whose self-described interfaces to their functionality are called services—rather than deriving the interface from the implementation. This enables developers to build *architectures* from heterogeneous components depending on the needs of their applications.

The SOA paradigm is based on the following foundation and principles: service orientation is based on open *standards* ensuring interoperability. Although *security* is not an attribute of SOA, it is important to foster acceptance. SOA is based on *simplicity* in the sense of flexibility, reusability, and adaptability in the usage of services and the integration of services into heterogeneous service environments. The attribute *distribution* stands for the independence of services, which self-contain their functionality. New applications can be built by designing the interaction of services and the definition of policies (if needed) to formulate interaction constraints. Services are *loosely coupled*, allowing dynamic searching for finding and binding of services. Services are clearly abstracted from their implementation and application environment. The *registry* is a central repository for all available services in a certain environment. SOA implementations without a registry make use of other advanced search mechanisms. *Process orientation* in SOA shows a major advantage of SOAs and moves the technology closer to its application. Processes and workflows can be designed and mapped into an interaction of services by composing the corresponding services.

An SOA defines three roles: service provider, client/user, and optionally a registry. The *service provider* is generally known as *service* and offers its functionality via a service (interface) to the SOA. *Service clients/users* make use of the functionality offered by a service provider. A *service registry* is used by service providers to register themselves and by service clients/users to search for a specific service and to obtain information required to establish a connection to the service provider.

In this chapter, a service provider is simply called service, and a service client/user is called service client or just client, if ambiguity is impossible.

Every service owns a description of its properties and capabilities, which has to be described by the service developer. The format of the description is standardized and has to be machine readable (usually character based in an eXtensible Markup Language [XML] format) and contains all information needed to interact with a service.

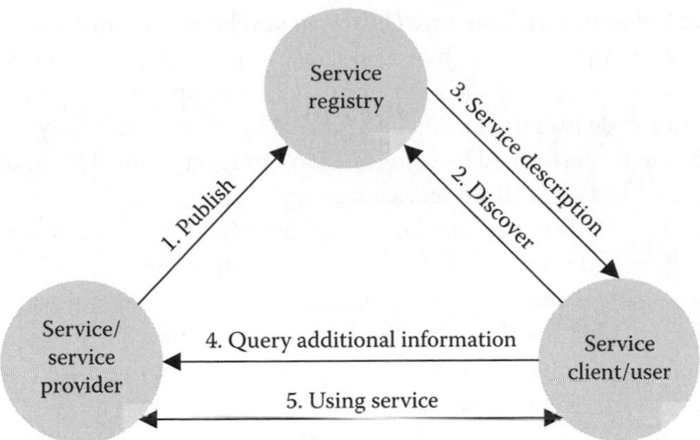

FIGURE 26.1 Roles in an SOA.

A classical SOA implementation involves five stages of service usage as shown in Figure 26.1 [2]:

Stage 1. The service registers a limited description of itself including its general capabilities at the service registry (*publishing*). If no service registry is available, the service announces itself to the network by sending its description including basic capabilities (*announcement*). The limited description fosters scalability by reducing network load as service clients are normally interested only in specific functions of a service and not in all provided functionalities. Further description details are requested in a later step.

Stage 2. The service client sends a description of the desired service to the service registry (*discovery*). In SOAs without a service registry, the discovery message will be sent to the network, or the service client listens for service announcements.

Stage 3. The service registry answers either with a failure response or by sending all service descriptions matching the discovery request (*discovery response*). The service description also includes the network address of the matching service/service provider. In the absence of a service registry, the corresponding service providers will answer directly.

Stage 4. In the next step, service clients will request a detailed description of the capabilities from the matching services to select the most reasonable one (*description query*).

Stage 5. The last step is the usage of the service according to the rules defined between service provider and client (*service usage*). These rules are described in the detailed description of the service provider. If a service provider leaves the network, it should announce its intentions by sending a bye message to the registry and clients, respectively.

26.2 Device-Centric SOAs

There are several implementations following the SOA paradigm. The subsequent paragraphs briefly introduce the implementations being relevant in the context of this chapter.

26.2.1 SOA Implementations for Devices

The Open Service Gateway Initiative (OSGi) specification defines a service platform that serves as a common architecture for service providers, service developers, and software equipment vendors who want to deploy, develop, and manage services [3]. The specification is based on the Java platform and promotes application independence. Thereby, it is enabling the easy integration of existing technologies. Deployed services are called bundles and are plugged into the framework. An OSGi service is a simple Java interface, but the semantics of a service are not clearly specified. The main drawback of OSGi is its reliance on Java.

The Java Intelligent Network Infrastructure (Jini) was developed by Sun Microsystems for spontaneous networking of services and resources based on the Java technology [4]. Services/devices are registered and maintained at a centralized meta-service called Lookup Service but carry the code (proxy) needed to use them. This code is dynamically downloaded by clients when they wish to use the service. Each service access has to be performed by using the lookup service. Jini's main drawbacks are its reliance on Java and the need of a centralized service registry.

Universal Plug-and-Play (UPnP) is a simple, easy-to-use SOA for small networks [5]. It supports ad hoc networking of devices and interaction of services by defining their announcement, discovery, and usage. Programming languages and transmission media are not assumed. Only protocols and interfaces are specified instead. The UPnP specification divides the device life cycle into six phases: Addressing, Discovery, Description (specifying automatic integration of devices and services), Control (operating a remote service/device), Eventing (subscribing to state changes of a remote service/device), and Presentation (URL representation of a service/device specifying its usage). UPnP also specifies a usage profile for distributed audio/video application—the UPnP AV architecture [6]. UPnP supports smaller networks only. With an increasing amount of services/devices, the amount of broadcast messages grows exponentially in a UPnP network. Furthermore, UPnP supports IPv4 only.

Web services (WSs) can be considered as the SOA with the highest market penetration. The WS architecture provides a set of modular protocol building blocks that can be composed in varying ways (called profiles) to meet requirements on the interaction of heterogeneous software components such as interoperability, self-description, and security [7]. *Profiles* define the subset of protocols used for implementing a specific application, required adaptations of the protocols, and the way they should be used in order to achieve interoperability. WSs address networks of any size and provide a set of specifications for service discovery, service description, security, policy, and others. The implementation is entirely hidden from their interfaces and may be exchanged at runtime. WSs use widely accepted standards such as XML and Simple Object Access Protocol (SOAP) for message exchange. For a better adaptation of WSs for embedded environments, Devices Profile for Web Service (DPWS) was developed.

DPWS version 1.0 was announced in August 2004 and revised in May 2005 and February 2006 by a consortium led by Microsoft Corporation [8]. In 2010, DPWS become an Organization for the Advancement of Structured Information Standards (OASIS) standard with version 1.1 [9]. It is a profile identifying a core set of WS protocols that enables dynamic discovery of, and event capabilities for, WSs. The profile arranges several WS specifications such as WS-Addressing, WS-Discovery, WS-Metadata Exchange, and WS-Eventing for devices, particularly. In contrast to UPnP, it supports discovery and interoperability of WSs beyond local networks. Temporarily, DPWSs were considered to be the successor of UPnP and thus subtitled with *A Proposal for UPnP 2.0 Device Architecture* in one of its earlier versions. However, DPWS is not compatible to UPnP, and therefore, it presents an SOA implementation on its own. DPWS debuted in the Microsoft Operating Systems Windows Vista and Windows CE.

It is important to note that the concepts behind two SOA implementations might be entirely different. SOAs are just a paradigm of how future systems could look like and what requirements they should address. The SOA paradigm does neither define design specifications of such system nor address implementation aspects.

26.2.2 Evaluation of Device-Centric SOAs

Table 26.1 is based on previous published research work by Bohn et al. [10] and shows an excerpt of the evaluation results of mentioned SOA implementations with regard to the problems identified in the introduction of this chapter. Although the SOA paradigm theoretically addresses all problems, the individual SOA implementations differ in their support for devices' integration, integration of security concepts, and standards for service composition, plug and play capability, programming language and network media independence, and scalability as shown in Table 26.1.

TABLE 26.1 Evaluation of Service-Oriented Middleware

	OSGi	JINI	UPnP	WS	DPWS
Common interfaces and awareness	✓	✓	✓	✓	✓
Discovery and standardized interaction	✓	✓	✓	✓	✓
Evolved composition standards				✓	
Support for hardware components	✓	✓	✓		✓
Plug and play capability		✓	✓		✓
Programming language independence			✓	✓	✓
Network media independence		✓	✓	✓	✓
Large scalability	✓	✓		✓	✓
Security concepts	✓	✓		✓	✓

Common interfaces and awareness indicated the capability of the standards to provide a common view on their components based on their functionality (called services). A requirement is self-describing interfaces in order that other components are aware of the service's functionality (*awareness*). Furthermore, a *standardized interaction* between clients and services must be ensured including defined protocol, message exchange patterns (MEPs), and data type specification. The capability of searching for certain functionality and the automatic announcement of new services (*discovery capability*) fosters automation in networked environments and applications. Networks of components reveal their full potential if standards that enable designers and nontechnical personnel to design complex applications by simply composing involved components/services (*evolved composition standards*) exist. Although niche solutions exist for other SOAs, only WSs offer a sophisticated set of standards for service composition. The support for hardware components involves addressing their specific requirements such as mobility, location awareness, and publish/subscribe mechanism. OSGi, Jini, UPnP, and DPWS support that in some way. *Plug and Play capabilities* refer to a network that is not dependent on a central service registry. Services and devices announce the entering and leaving of networks, and thereby, clients are informed of their functionality. This is provided by Jini, UPnP, and DPWS. Only UPnP, WS, and DPWS are *programming language independent*, which means clients and services can be developed in any kind of language and executed on any kind of execution environment. *Network media independence* corresponds to message exchanges over any kind of transport protocol. OSGi is somehow out of scope in that way as messages are exchanged over Java. Very complex applications require a *large scalability* supporting a large number of involved services. Furthermore, secure message exchanges based on standardized *security concepts* are often required for such applications. As UPnP mainly focuses on audio/video applications in small home networks, it is not highly scalable and has no security concepts defined yet.

It is apparent that the DPWS is most appropriate to solve the problems stated at the beginning of the introduction. DPWS offers extensive features to support the interoperability between different devices/services. DPWS-compliant components are able to interact with each other forming a highly distributed application. However, each client and service has to be programmed manually. This may result in a huge effort for complex and dynamic applications. Process management capabilities as provided by basic WSs would leverage the easy integration of DPWS clients and services.

26.3 DPWS Inside Out

The DPWS was first announced in 2004 as a possible successor of UPnP, revised in May 2005, February 2006, and July 2009 [9]. It is a WS profile identifying a core set of WS protocols and profile-specific adaptations that enable secure WS messaging, dynamic discovery of, and event capabilities for resource-constrained devices and their services.

Figure 26.2 shows an overview of the WS protocols for DPWS.

Before going into detail, relevant XML basic technologies are briefly introduced.

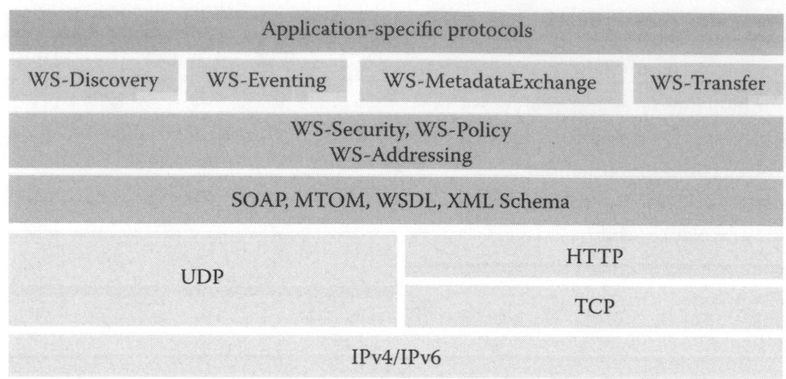

FIGURE 26.2 The Devices Profile for WSs protocol stack.

26.3.1 Interactions between Client, Device, and Service

Three participants are involved in a DPWS interaction: client, device, and the services residing on a device. A DPWS device is also designed as a service called *hosting service*. The hosting service contains information about itself (e.g., manufacturer and model name) and references to its *hosted services*.

Optionally, a DPWS may utilize a central service registry (called *Discovery Proxy* [DP]), which could also be used for device/service discovery spanning several networks.

In order to increase scalability and reduce the network traffic, the invocation of hosted services on devices involve several phases: discovering desired devices/hosting services, obtaining a detailed description about replying device/hosting services and references to the services they host, retrieving further information on the desired hosted services, and finally invoking the desired service. If the client knows the references to desired services already, it may contact them directly. Additionally, DPWS defines constraints on the vertical issues messaging and security mechanisms that are offered for messaging.

The sequence diagram in Figure 26.3 [11] illustrates the phases showing the activities involved in a typical message exchange using DPWS. Details on participating protocols are provided at a later stage.

Message 1: A `Probe` message is sent from the client using UDP multicast to search for a specific device as described in WS-Discovery. The `Probe` message indicates if the client requests security. If a DP is available, multicast suppression is used as specified in WS-Discovery. Clients send their `Probe` messages directly to the DP (UDP unicast).

Message 2: All devices listen for `Probe` messages. In case the desired service matches one of its hosted services, the device responds with a `ProbeMatch` message using unicast UDP. The `ProbeMatch` message contains the device's endpoint reference (EPR), supported transport protocols, and security requirements and capabilities. EPRs are XML structures including a destination address (destination WS is called endpoint) and optional metadata describing the service (e.g., usage requirements) as defined by WS-Addressing. If security is desired, the client sets up a security channel in an additional message.

Messages 3 and 4: In case the EPR does not include a physical address, the client can use a `Resolve` message to retrieve it from the device. The device uses a `ResolveMatch` message for it.

Message 5: The client can directly request more information about the device using a `GetMetadata` message as it is defined in the WS-MetadataExchange (WS-MEX) specification.

Message 6: The desired device will respond either with included metadata or by providing a reference to the metadata. The metadata includes device details and EPR to each hosted service.

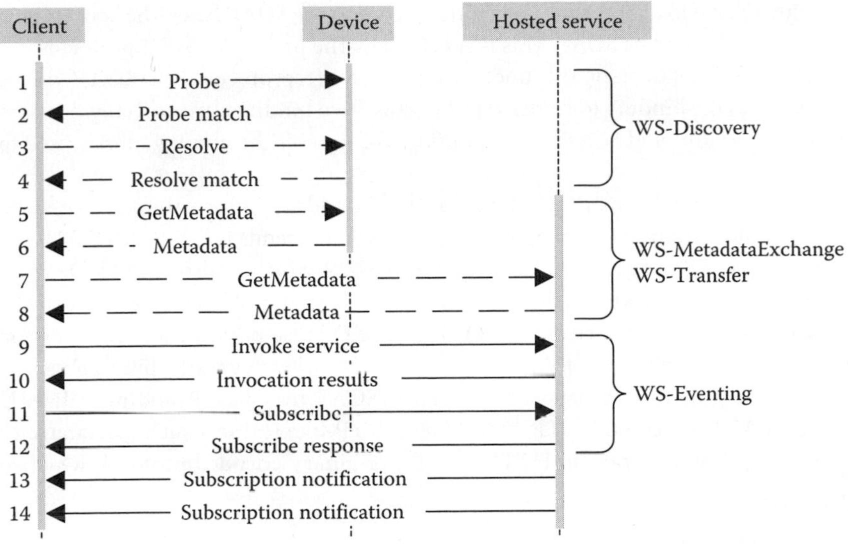

FIGURE 26.3 Typical message exchange using DPWS.

Message 7: Similar to message 3, the client requests more information about the desired hosted service.

Message 8: The desired hosted service sends its metadata to the client including its Web Service Description Language (WSDL) description.

Messages 9 and 10: Hosted services are invoked by receiving an invocation message from the client according to corresponding operation described in the WSDL. The implementation of the invocation of services is not specified in DPWS and left up to the service provider.

Message 11: A client may send a `Subscribe` message to the hosted service to get informed about updates of or the overall status of a service. The publish/subscribe pattern used in DPWS is described by WS-Eventing as well as specifications of subscription renewal and cancellation.

Message 12: The hosted service responds with a `SubscribeResponse` message including a subscription identifier and an expiration time.

Messages 13 and 14: The hosted service sends `Notification` messages to the client whenever a specific event occurs.

This chapter will simply refer to a hosting service as device and to a hosted service as service for easier readability.

26.3.2 Messaging

DPWS makes use of SOAP 1.2, SOAP-over-UDP, HTTP/1.1, WS-Addressing, the URI, and MTOM specifications. These specifications are restricted in such a way that they address the requirements of resource-constraint device implementations by ensuring high WS interoperability.

26.3.2.1 SOAP

SOAP was originally an abbreviation for *Simple Object Access Protocol*, but since version 1.2, it is simply called SOAP [12]. SOAP is a protocol promoting interoperability by specifying an XML-based format for exchanging messages between different applications. SOAP messages consist of a header and a body enclosed by a *SOAP envelope*. The *SOAP header* contains metadata related to delivery and processing of message content. The payload of the message is inserted into the *SOAP body*. The actual content of headers and bodies is not defined by SOAP. This is specified by the particular WS-* protocols.

Since SOAP is independent of the underlying transport protocols, the *SOAP Protocol Binding Framework* describes the binding to transport protocols. Two bindings are officially documented by the W3C: *SOAP Email Binding* and *SOAP HTTP Binding*. The latter is the binding mostly used for historical reasons.

SOAP-over-HTTP/HTTPS has the disadvantages that address information is stored in the HTTP header and only synchronous messaging is supported. Recent trends try to attach SOAP to other binding protocols than HTTP. One approach is *SOAP-over-UDP*, which is also used by DPWS and reduces the network load when no acknowledgment is required (e.g., for WS-Discovery). Unicast as well as multicast transmissions are supported. The SOAP-over-UDP also provides a retransmission algorithm considering the delayed repetition of the same message. Another suggested binding uses Constrained Application Protocol (CoAP) as transport protocol [13]. SOAP-over-CoAP binding utilizes UDP as well; however, since CoAP is based on HTTP, it provides HTTP-like features such as management headers and transport reliability. Contrary to HTTP, CoAP is a binary-encoded protocol, which significantly reduces transport overhead.

26.3.2.2 SOAP Message Transmission Optimization Feature

The conventional way of conveying binary data in SOAP as well as other XML documents is to transform the data into a character-based representation using the *Base64 content-transfer-encoding scheme* as defined by Multipurpose Internet Mail Extensions. Unfortunately, this produces a message overhead of about 33% for large attachments and a possible processing overhead [14].

The *SOAP Message Transmission Optimization Mechanism* (MTOM) defines an *Abstract SOAP Transmission Optimization Feature*, which is encoding parts of the SOAP message while keeping the envelope intact (e.g., SOAP headers). The optimization feature works only if the sender knows or can determine the type of information of the binary element. Conveying binary data using SOAP-over-HTTP is one implementation of the Abstract SOAP Transmission Optimization Feature. This solution is based on *XML-binary Optimized Packaging* conventions, where the binary data are sent as an attachment.

MTOM does not address the general problem of including non-XML content in XML messages.

26.3.2.3 WS-Addressing

The *pure* SOAP has a strong binding to HTTP for historical reasons although meant to be transport protocol independent. This results in the disadvantage that the address of targeted services can only be found in the HTTP header and only synchronous messaging is supported. Message routing depends on HTTP, and publish/subscribe patterns are not supported. WS-Addressing remedies these deficiencies by specifying mechanisms to address WS and messages in a transport-neutral way [15]. SOAP messages can be sent over multiple hops and heterogeneous transport protocols in a consistent manner. Furthermore, it allows sending responses to a third party (WS or WS client). WS-Addressing introduces the concepts of EPRs and message information (MI) header, which are included into the SOAP header.

EPRs release SOAP from the strong binding to HTTP. They are XML structures addressing a message to a WS (destination WS is called *endpoint* including destination address, routing parameters for intermediary nodes, and optional metadata describing the service (e.g., WSDL description for a WS or policies).

MI headers enable asynchronous messaging as well as dispatch possible responses to other endpoints than the source. The mentioned *dispatching of the responses* is realized explicitly defined by assigning specific addresses to source, reply, and fault endpoints, or implicitly by leaving these entries unassigned (responses are sent back to the source). The *asynchronous messaging* is realized by the use of an optional unique message ID. Result messages can be referenced to it. The destination address as well as an *action* URI that identifies the semantics of a message (e.g., fault, subscription, or subscription response message) is mandatory. WS-Addressing defines explicit as well as implicit associations of an action with input, output, and faults elements. Explicit associations are defined in the description of a WS. Implicit association of an action is composed of other message-related information. The optional definitions of relationships in MI headers indicate the nature of the relation to another message (e.g., indication of a response to specific message). It can also be indicated that the message relates to an *unspecified message*.

In case that an endpoint does not have a stable URI, the endpoint can use an *anonymous* URI. Such request must use some other mechanism for delivering replies or faults.

WS-Addressing can also be used to support request-response and solicit-response MEP using SOAP-over-UDP.

26.3.2.4 Adaptations to DPWS

DPWS requires services to meet the *packet length limits* defined in the underlying IP protocol stack. A service retrieving messages exceeding that limit might not be able to process or reject them, which may result in incompatibilities. A hosted service must at least support *SOAP-HTTP-binding* in order to ensure highest possible interoperability as most applications use it. Furthermore, hosted services have to support the flexible HTTP chunked transfer coding. HTTP chunked transfer coding is used to transfer large content of unknown size in chunks. The end of the transmission is indicated by a zero-size chunk. Regarding the MEP, basic WS functionality is achieved by supporting the MEP request-response and one-way (at least).

As devices can also be mobile and roaming across networks, a globally unique UUID (universally unique identifier) for the address property of the EPR is required, which is persistent beyond reinitialization or changes of the device. Hosted services should use an HTTP transport address as the address property in their EPRs. Furthermore, hosted services must assign a reply action to the relationship field in the MI for response or fault messages.

Hosted services must also generate a failure on HTTP Request Message SOAP Envelopes not containing an anonymous reply EPR. This means that DPWS does not support the deferring of reply and fault messages to other endpoints (being different from the sending endpoint). The WS-Addressing 1.0 SOAP Binding specification defines for HTTP Request Message SOAP Envelopes containing an anonymous reply EPR address that their responses must be sent back to the requester in the same HTTP session. This means that DPWS only supports synchronous message exchanges for client–service interactions. The reason for that is the reduced size for the WS implementation for DPWS devices/services and clients. Especially, WS clients cut down on size as they do have to provide a callback interface.

In summary, only the concept of WS-Addressing EPRs and action URIs are used by DPWS. Asynchronous client–service interactions are not supported by DPWS.

A hosted service may support MTOM in order to be able to receive or transmit messages exceeding this packet length limits.

26.3.3 Discovery

DPWS uses WS-Discovery for discovering devices. As illustrated in Figure 26.3, WS-Discovery is not used to search for specific hosted services. Instead, desired services are found by searching for devices and browsing the descriptions of their hosted services.

26.3.3.1 Web Services Dynamic Discovery

WS based on the WS architecture can be used only if their addresses are known to WS clients or if they had registered to a *Universal Description, Discovery and Integration* (UDDI) server. UDDI is a central registry that can be queried for a desired service. The address of a UDDI server can also not be discovered and has to be known to interested WS clients.

The *Web Services Dynamic Discovery* (WS-Discovery) [16] extends the WS discovery mechanisms by concepts for the announcement of services when entering a network, dynamic discovery, and network

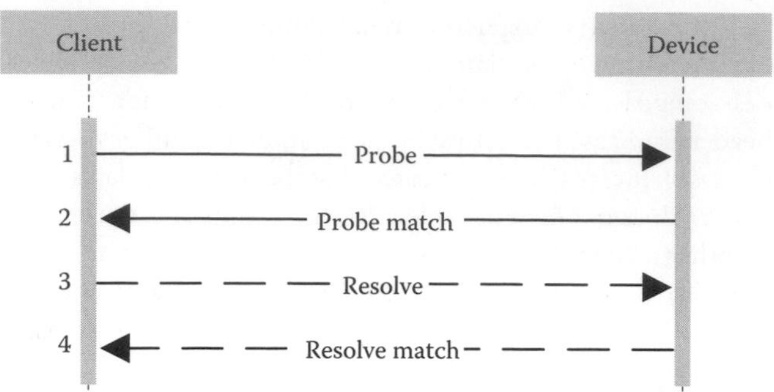

FIGURE 26.4 Interactions in the discovery process defined by WS-Discovery.

spanning discovery of services. It introduces three types of endpoints: Target Service, Client, and DP. Clients are searching for Target Services. DPs enable forwarding of searches to other networks.

The announcement of a service is realized by sending a one-way multicast `Hello` message whenever a Target Service becomes available. This should be used by Clients for implicit discovery through storing the announcement information and thus reducing the amount of discovery messages in the network. When a Target Service intends to leave a network, it should send a one-way multicast `Bye` message.

Dynamic discovery is important for endpoints, which might change their transport addresses during lifetime (e.g., mobile WS traversing different subnets). In such cases, Target Services possess a logical address that can be dynamically resolved into their transport address. Dynamic discovery is realized in a two-stage process (as shown in Figure 26.4): Probe and Resolve. When searching for services, Clients must send a `Probe` message specifying the type of the Target Service and/or its Scope. The specification of the Type and its semantics are left up to the provider. The Scope is a logical group of WSs. The matched Target Services answer to the requester using a unicast `ProbeMatch` message containing the mandatory EPR. Optionally, the `ProbeMatch` can also provide a list of the Target Service addresses (avoiding a following resolve), types, and/or scopes. The next step is to resolve the EPR into a list of Target Service addresses by sending a multicast `Resolve` message containing the EPR. The corresponding Target Service answers with a `ResolveMatch` message including a list of addresses and optionally a list of types and/or scopes.

WS-Discovery also defines a basic set of matching rules for the resolution of URIs, UUIDs, and case-sensitive string.

Network spanning discovery can be realized using DP by relaying discovery messages to other networks. A DP enables the scaling of a large number of endpoints. When a DP receives a multicast `Probe` or `Resolve` message, it announces itself using a `Hello` message to suppress multicast messaging. Clients send their request via unicast to the DP and ignore `Hello` and `Bye` messages from Target Services while the DP is available. Further functionality of the DP is not specified by WS-Discovery.

WS-Discovery uses UDP for messaging. The implementations are mostly bound to SOAP-over-UDP. WS-Addressing is used for the specification of EPRs, action, and fault properties.

26.3.3.2 Adaptations to DPWS

Discovery in DPWS only relates to searching for devices/hosting services. As described at the beginning of this section, hosted services can be found by following the references in the descriptions starting from the device/hosting service description.

A device/hosting service must be a Target Service as defined in WS-Discovery. Furthermore, a device/hosting service is required to support unicast as well as multicast-over-UDP (required by WS-Discovery). In case the HTTP address of a device is known, the discovery client may send a `Probe` message over HTTP (unicast). Due to this, a device must support receiving `Probe` SOAP envelopes as

HTTP requests and be able to send a `ProbeMatch` SOAP envelope in an HTTP response or send an HTTP response with an `HTTP 202 Accepted`.

Devices might not be discoverable if residing in a different subnet than the discovery client. Nevertheless, discovery clients disposed of an EPR and transport address for targeted devices might be able to communicate with the corresponding device. If transport address and EPR of a DP in another subnet are known, this DP can be used to discover devices in that subnet.

Devices must announce the support of DPWS by including the type `wsdp:Device` and may include other types or scopes in `Hello`, `ProbeMatch`, or `ResolveMatch` SOAP envelopes. To be able to match URIs and strings, devices must at least facilitate URI and case-sensitive string comparison defined by the scope matching rules defined by WS-Discovery.

Except for `wsdp:Device`, semantics and classifications of types and scopes are not defined in DPWS. When a device receives a `Probe` message, it uses some internal mechanism to decide if it responds or not. Devices responding to `Probe` messages are not required to publish their supported types and scopes. This might complicate the discovery process if no out-of-band information about responding devices is available. In addition, as messages should not exceed the size of a maximum transmission unit, messages being sent via UDP might be too small to convey supported types and scopes. The support of UDP fragmentation is also not required by DPWS. In such a case, another `Probe` could be sent via HTTP chunked transfer coding as it must be supported. However, this also requires the device to send supported types and scopes in the `ProbeMatch`.

26.3.3.3 Current Research in WS-Discovery

The drawback of the original WS-Discovery standard is the low scalability due to a static application delay—maximum waiting time before sending a response. The rising number of devices leads to an increasing network load and especially to an increasing device processing load as a result of the discovery process, for example, in smart metering or building automation networks. In the worst case, the device can be set out of operation due to buffer overflow or some devices will not be discovered due to packet rejection.

In [17], Altmann et al. provide an approximation formula for the peak data rate estimation of WS-Discovery in Ethernet networks. According to the simulation results, the authors propose optimizations for WS-Discovery to improve its scalability. The proposed approach includes MAC addresses in WS-Discovery requests avoiding large traffic volumes due to the address resolution protocol. Moreover, dynamic adaptation of the application delay in order to reduce the peak data rate on the client is suggested. Furthermore, the suggested additional node aggregation can reduce the data rate independent of the application delay by more than 50%. The optimization of packet repetition allows avoiding the packet interference between original and repeated WS-Discovery responses. For this purpose, the authors propose the division of timing interval into two parts for the first and second responses, respectively. In order to properly set the dynamic values, the knowledge of the network size is required. A new approach for network size discovery is proposed to gain this knowledge. All optimizations allow steady packet processing and significantly relieve the client. The suggested approach can be applied for any network size due to its dynamic nature and consequently provides improved scalability for ad hoc WS-Discovery. The suggested optimization is backward compatible with the standard WS-Discovery specification.

26.3.4 Description

DPWS uses WSDL for describing the functionality and capabilities of hosted services. Devices do not possess a WSDL description as they cannot be invoked by WS clients. The capabilities of a device including the references to its hosted services are defined in a metadata description as defined by WS-MEX. WS-Transfer is used to retrieve such metadata descriptions. Furthermore, hosted services can define their usage requirements by using WS-Policy. All protocols are described in the remainder of this subsection.

26.3.4.1 Web Service Description Language

The WSDL is an XML document format containing definitions used to describe WS as a set of endpoints and its interactions using messages [18]. An endpoint is the location for accessing a service. A WSDL document (or simply referred to as a WSDL) contains an abstract and a concrete part as shown in Figure 26.5 [2].

26.3.4.1.1 Abstract Part

The abstract part of a WSDL document promotes reusability by describing the functionality of the services independent of its technical details such as underlying transport protocols and message format. Firstly, the *data types* that are needed for message exchange are defined. In order to ensure highest interoperability, WSDL refers to XML Schema Definition as the preferred data type scheme. *Messages* are abstract definitions of data being transmitted from and to a WS. Messages consist of one or more *logical parts*. The *port types* definition can be seen as abstract interfaces to services offered by the WSDL. Port types are a set of abstract operations. An *operation* is a set of messages exchanged by following a certain pattern. WSDL supports the following four MEP: one-way, request-response, solicit-response, and notification. They are implicitly identified in the WSDL by the appearing order of input and output messages. Request-response and solicit-response operations may also define one or more fault messages in case a response cannot be sent.

26.3.4.1.2 Concrete Part

The concrete part encloses the technical details of a WS. The *binding* associates a port type, its operations, and messages to specific protocols and data formats. Several bindings to one port type are possible. All port types defined in the abstract part must be reflected here. Concrete fault messages are not defined because they are provided by the protocols. Furthermore, a *service* definition is a collection of ports. *Ports* are network addresses for service endpoints. Each uniquely identified *port* constitutes an endpoint for a binding.

WSDL specifies and recommends the binding to SOAP but is not restricted to it. The WSDL Binding Extension for SOAP defines four *encoding styles*. The style can be either RPC or document. The use can be either literal or SOAP encoded making it to four styles. Each binding has to specify the encoding style it wants to use when translating XML data of a WSDL into a SOAP document. The differences between the encoding styles are as follows:

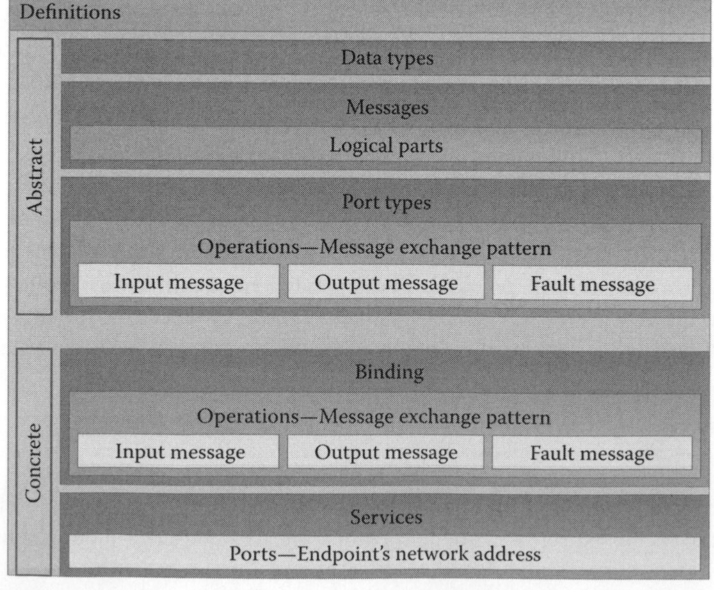

FIGURE 26.5 Major components of a WSDL description.

The *RPC-style* structures the SOAP body into operations embedding a set of parameters (data types and values). Using the *document style*, the SOAP body can be structured in an arbitrary way. The *literal encoding* uses XML Schema to validate the SOAP data. Therefore, the SOAP body must be conforming to specific XML Schema. In contrast, *SOAP encoding* employs a set of rules based on the XML Schema data types to transform the data. A message is not required to comply with a specific schema.

RPC/encoding was the original style but will eventually go away. Document/encoded is rarely supported. Recently, a fifth style was developed—called Document/literal/wrapped, where the parameters are wrapped in an element bearing the operation name. An elaborated comparison of the coding styles can be found in the article by Butek [19]. In most cases, Butek recommends the Document/literal/wrapped encoding style, which is also supported by the .NET framework. The impact of the choice of encoding style on performance, scalability, and reliability of WS interactions is presented in Cohen [20].

26.3.4.2 WS-Policy

WS implementations might rely on certain restrictions although the WS architecture is platform, programming language, and transport protocol independent.

WS-Policy [21] provides a framework to define restrictions, requirements, capabilities, and characteristics of partners in WS interactions as a number of policies. As illustrated in Figure 26.6, a *policy* is an assortment of policy alternatives. In order to comply with a certain policy, an interaction partner can select the best suiting policy alternative from the defined set. If a WS or a client makes use of policies, its interaction partners have to comply with those by choosing one of the offered policy alternatives. A *policy alternative* is defined as a collection of policy assertions. A *policy assertion* is an XML Infoset representing a requirement or capability. A policy assertion can also be optional. Policy assertions can be combined to more complex requirements or capabilities using *policy operators*. Two policy operators are defined by WS-Policy. The *ExactlyOne operator* requires the compliance to one of the stated policy assertions, whereas the *All operator* requires the compliance to all. WS-Policy does not specify how policy assertions are expressed. A way to do so is described in the *Web Services Policy Assertions Language* (WS-PolicyAssertions).

WS-Policy does specify neither the location of policy definitions nor their discovery. This is left to other WS protocols such as WS-PolicyAttachment.

26.3.4.3 WS-PolicyAttachment

WS-PolicyAttachment [22] describes general-purpose mechanisms, how associations between policy expressions and the subject to which they apply can be defined (e.g., associations with WSDL port or binding). Two approaches are offered: policy assertion is part of the subject or externally defined and bound to the subject. WS-PolicyAttachment also describes the mechanisms for WSDL and UDDI.

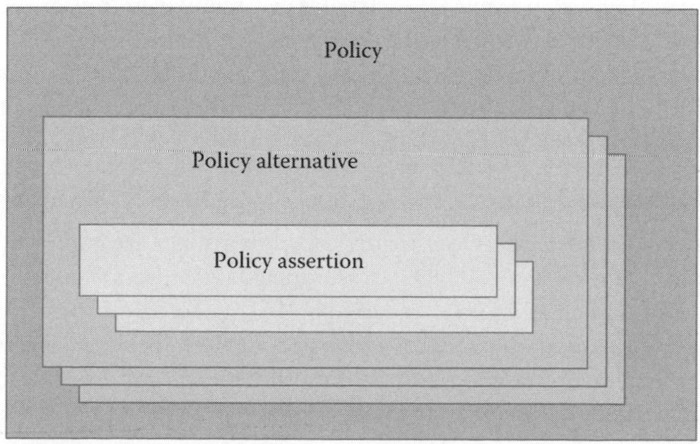

FIGURE 26.6 The structure of a policy as defined by WS-Policy.

These include description of the policy references from WSDL definitions, policy associations with WSDL service instances, and policy associations with UDDI entities.

26.3.4.4 WS-MetadataExchange

The client has received the addresses of targeted services from the device after the discovery process. The next step is to find out more about the Target Services (e.g., their descriptions, requirements, and used data types). This may also be used to find out more about the device itself.

WS-MEX provides a standardized bootstrap mechanism to start the communication with a WS by retrieving the WSDL as well as further information about the WS endpoint such as usage requirements/policies (WS-Policy) and data types (XML Schema) in a well-defined encapsulation format [23]. It defines two request-response operations:

`GetMetadata` is used to retrieve all metadata the WS endpoint provides. It is recommended for a large amount of metadata to provide EPRs (*Metadata references*) or URL (*Metadata Location*) instead of including all metadata in detail in the response. This ensures scalability for service consumers interested only in specific metadata and possibly reduces the network traffic.

`Get` is used to retrieve metadata addressed by a reference. The `Get` operation must be *safe* meaning that the state of requested service must remain unchanged by this operation. It may include a *dialect* characterizing the type of metadata (e.g., metadata defined by WS-Policy), its format, and version. If `Get` is used without a defined dialect and reference for specific metadata, all metadata must be sent in the response.

The default binding of MEX is SOAP 1.1 over HTTP as constrained by the BasicProfile 1.0.

26.3.4.5 WS-Transfer

WS-MEX provides operations to retrieve (metadata) representations of WSs. The resources represented by the WS remain unaffected by that operation. Operations on the WS resources must be defined in a proprietary way and agreed on by both WS interaction partners.

WS-Transfer [24] defines a standardized mechanism supporting *Create, Read, Update, and Delete* (CRUD) operations (called Create, Get, Put, and Delete) on WS resources. A resource is specified as an entity providing an XML representation and being addressable by an EPR. `Get`, `Put`, and `Delete` are operations on resources, while the `Create` operation is performed by resource factories. The WSs capable of creating new resources are called *resource factories*. WSs implementing a resource must provide the `Get` operation and may provide `Put` and `Delete` operations.

The *Get-operation* is used to retrieve a representation of a resource. A *Put-operation* request allows replacing the representation of a resource by the one sent in the request. If a WS accepts a *Delete-operation*, the corresponding resource must be deleted. The *Create-operation*, as the name suggests, is a request to create a new resource according to the representation in the request.

WS-Transfer is a transport-independent implementation similar to the *Representational State Transfer* approach. The persistence of CRUD operations is not guaranteed because the hosting server is responsible for the state maintenance of a resource.

26.3.4.6 Adaptations to DPWS

The metadata of devices and hosted services are described in a structure conform to WS-MEX. Metadata are considered as static details. Changes to the metadata must include the incrementation of their versions (`wsd:MetadataVersion`) reflected in `Hello`, `ProbeMatch`, and `ResolveMatch` SOAP envelopes sent by corresponding devices and hosted services. The `Get` operation of WS-Transfer is used for retrieving metadata in DPWS. Additionally, other means of retrieval can also be used. Only hosted services possess a WSDL description to foster interoperability with other WSs architectures. It is embedded into the metadata of hosted services.

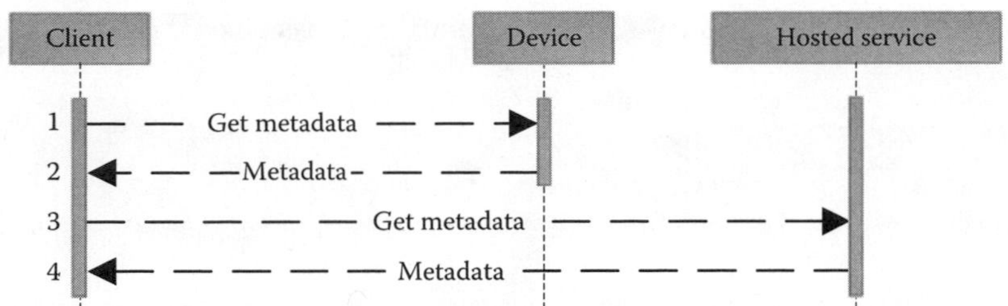

FIGURE 26.7 The metadata exchanges of DPWS.

Figure 26.7 shows the metadata exchanges between client, device, and hosted services. As illustrated by the dashed lines, the metadata exchanges are optional because clients might already be aware of the metadata of devices and hosted service. In such case, the clients may use the hosted services directly as defined in their WSDL descriptions. All metadata must be sent in one wsx:Metadata element in the SOAP Envelope Body.

26.3.4.6.1 Description of Devices

The description of a device contains generic characteristics of the device including EPRs to the hosted services. The `GetResponse` SOAP envelope only includes a `wsx:Metadata` element that contains all metadata for the device. The metadata of a device has the following structure: device class metadata (`ThisModel`), metadata for the specific device (`ThisDevice`), and metadata for the hosted services (Relationship). WS-MEX dialect must be from `http://schemas.xmlsoap.org/ws/2006/02/devprof` for DPWS 1.0 or `http://docs.oasis-open.org/ws-dd/ns/dpws/2009/01` for DPWS 1.1. In the following, only DPWS 1.1 is considered.

ThisModel describes the device class metadata containing information about the manufacturer, model name, and optionally manufacturer URL, model number, model URL as well as a presentation URL (a reference to further information). `ThisModel` is described by the metadata section <dpws:ThisModel ...>, in which the WS-MEX dialect must be equal to http://docs.oasis-open.org/ws-dd/ns/dpws/2009/01/ThisModel.

ThisDevice defines the specifics of a certain device of the class `ThisModel`. It is described by the metadata section <dpws:ThisDevice...> containing a user-friendly name and optionally the version of firmware and serial number. The WS-MEX dialect must be equal to `http://docs.oasis-open.org/ws-dd/ns/dpws/2009/01/ThisDevice`.

26.3.4.6.2 Relationship

The WS-MEX dialect of this metadata section must be equal to `http://docs.oasis-open.org/ws-dd/ns/dpws/2009/01/Relationship`. The `Relationship` definition identifies the nature and content of the relationship between device/hosting service and hosted services indicated by the type `http://docs.oasis-open.org/ws-dd/ns/dpws/2009/01/host`. It includes optional information about the host and the hosted services. The *host section* (<dpws:Host>) contains the EPR of the host, types implemented by the host, and a service ID for the device. The *hosted section* includes the EPR of a hosted service. Therefore, the relationship metadata includes a separate hosted section for each hosted service of the device.

Figure 26.8 presents an excerpt of the SOAP Response message envelope containing the metadata description for a device [8]. The SOAP header elements are specified by WS-Addressing (using the prefix `wsa`). The action (`wsa:Action`) identifies the semantics of the SOAP message—a response message to a Get operation of WS-Transfer. The unique message ID (`wsa:RelatesTo`) relates this response to a certain request message. Furthermore, the message is sent to the address (`wsa:To`) `anonymous` as defined by DPWS for reply messages. As already described in the messaging subsection,

```
<soap:Envelope ...>
 <soap:Header>
  <wsa:Action>http://schemas.xmlsoap.org/ws/2004/09/transfer/GetResponse</wsa:Action>
  <wsa:RelatesTo>urn:uuid:82204a83-52f6-475c-9708-174fa27659ec</wsa:RelatesTo>
  <wsa:To>http://schemas.xmlsoap.org/ws/2004/08/addressing/role/anonymous</wsa:To>
 </soap:Header>
 <soap:Body>
  <wsx:Metadata>
   <wsx:MetadataSection Dialect="http://docs.oasis-open.org/ws-dd/ns/dpws/2009/01/ThisModel">
    <dpws:ThisModel> … </dpws:ThisModel>
   </wsx:MetadataSection>
   <wsx:MetadataSection Dialect="http://docs.oasis-open.org/ws-dd/ns/dpws/2009/01/ThisDevice">
    <dpws:ThisDevice> … </dpws:ThisDevice>
   </wsx:MetadataSection>
   <wsx:MetadataSection Dialect="http://docs.oasis-open.org/ws-dd/ns/dpws/2009/01/Relationship">
    <dpws:Relationship Type="http://docs.oasis-open.org/ws-dd/ns/dpws/2009/01/host">
     <dpws:Host> … </dpws:Host>
     <dpws:Hosted> … </dpws:Hosted>
     <dpws:Hosted> … </dpws:Hosted>
    </dpws:Relationship>
   </wsx:MetadataSection>
  </wsx:Metadata>
 </soap:Body>
</soap:Envelope>
```

FIGURE 26.8 Example for a SOAP Response message envelope.

this reply address indicates that response has to be sent back right after the SOAP Request message (in the same HTTP session).

The SOAP body in Figure 26.8 contains metadata sections for `ThisModel`, `ThisDevice`, and the `Relationship`. The `Relationship` metadata provides details about the device itself and two hosted services.

26.3.4.6.3 Description of Hosted Services

Now that the client has retrieved the metadata for the device including the EPRs of its hosted service, it may request the WS-MEX description of desired hosted services using the `Get` operation of WS-Transfer. The metadata of a hosted service must contain at least one section providing the WSDL description (metadata dialect equal to `http://schemas.xmlsoap.org/wsdl/`). It must be sent in any `GetResponse` message. The WSDL metadata section may include the WSDL description in line or may provide a link referring to it.

26.3.4.6.4 WSDL Constraints of DPWS

Hosted services must at least support the document/literal encoding style and must include the WSDL Binding for SOAP 1.2 for each `portType` at least. If notifications are supported by the hosted service (e.g., for a publish/subscribe mechanism), it must include the notification and/or solicit-response operation in a `portType` specification. Hosted services are not required to include the WSDL services section in a WSDL description since addressing information is included in the `Relationship` section of the device's metadata.

Hosted services must include the policy assertion `dpws:Profile` in their WSDL indicating that they support DPWS. This may be attached to a `port`, should be attached to a `binding`, and must not be attached to a `portType`. The policy assertion can also be stated as being optional indicating that the hosted service supports DPWS but is not restricted to it. Devices have no explicit means to specify the support of DPWS. This must be ascertained by other information.

It should be noted that DPWS does not define a back-reference from hosted services to the device. The only way to find the corresponding device to a certain hosted service is to start a discovery process.

26.3.5 Eventing

Some applications might require to get informed about state changes of a specific WS. DPWS provides a publish/subscribe mechanism based on WS-Eventing.

26.3.5.1 WS-Eventing

WS-Eventing [25] implements a simple publish/subscribe mechanism sending events from one WS to another. WS-Eventing defines four WS roles: the *subscriber* initiates a subscription by subscribing to an event source. The *event source* is publishing (event) notifications whenever an interesting event occurs (e.g., state change). The notifications are sent to an *event sink*, which processes them in some way. In order to decouple the management of subscription from the event source in highly distributed networks, a *subscription manager* may take over that role. However, event sources are also subscription managers and event sinks are also subscribers in many publish/subscribe systems.

Figure 26.9 provides an overview of roles and message exchanges in a publish/subscribe system using WS-Eventing.

Message 1: A subscriber sends a `Subscribe` message to an event source indicating the address for a `SubscriptionEnd` message, address of event sink, the Delivery Mode, subscription expiration time, and subscription filters. The address for a `SubscriptionEnd` message identifies the WS, which receives such message in case of an unexpected subscription cancellation by the event source (e.g., due to an error). The address of subscriber and event sink is equal to the address for the `SubscriptionEnd` message when the subscriber is also the event sink (as in most applications of WS-Eventing).

The *Delivery Mode* specifies the way a notification is retrieved by the event sink. WS-Eventing specifies an abstract delivery mode allowing the use of any kind of delivery mode. One concrete delivery mode is defined as well. The *Push Mode* is used to send individual, unsolicited, and asynchronous notification messages. As an alternative to push delivery, a mechanism

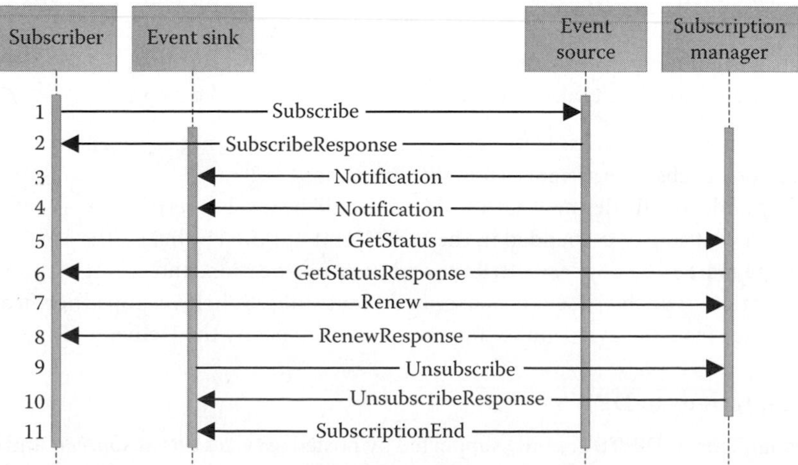

FIGURE 26.9 Roles and message exchanges as defined by WS-Eventing.

for polling notifications could be defined according to the abstract delivery mode (e.g., for applications where the event sink is behind a firewall).

Although subscription can be defined for an unlimited time, WS-Eventing recommends the definition of an absolute expiration time or duration (*subscription expiration time*). This reduces the network load in case of unexpected unavailability of event sinks during a subscription.

Subscription filters are defined to only send notifications the event sink is interested in. A Boolean expression in some *filter dialect* (e.g., string or XML fragment) is used to evaluate the notifications. Notifications are sent only when the expression evaluates to `true`. Any kind of filter dialect can be used (e.g., based on XPath).

Message 2: An event source replies with a `SubscribeResponse` message to the subscriber. When accepting the subscription, the `SubscriptionResponse` contains an EPR of the subscription manager and an expiration time. WS-Addressing MI headers are used to relate a `SubscriptionResponse` to a `subscribe` message. In case the event source is also the subscription manager, the EPR of the subscription manager is one of the event sources.

Messages 3 and 4: *Notification* messages are sent from the event source to the event sink for events defined by the subscription filters. Semantics and syntax of notification messages are not constrained by WS-Eventing as any message can be a notification. However, subscribers may specify `wsa:ReferenceProperties` (WS-Addressing) in the `subscribe` message to request special marked notifications. These reference properties have to be included in each notification message.

Messages 5 and 6: If a subscriber wants to receive information about the expiration time of its subscription, it sends a `GetStatus` message to the subscription manager. The subscription manager includes the expiration time in a `GetStatusResponse`.

Messages 7 and 8: Before the subscription time expires, the subscriber has the chance to renew the subscription by sending a `Renew` message to the subscription manager including the new expiration time. The subscription manager replies with a `RenewResponse` including the new subscription time. If the subscription manager does not want to renew the subscription for any reason, it may reply with a `wse:UnableToRenew` fault in the response message.

Messages 9 and 10: If an event sink wants to terminate the subscription, it should send an `Unsubscribe` message to the subscription manager. If the subscription termination is accepted by the subscription manager, it must reply with an `UnsubscribeResponse` message.

Message 11: If the event source wants to terminate the subscription unexpectedly (e.g., due to an error), it sends a `SubscriptionEnd` message to the event sink. In such case, the event source is required to indicate the error. WS-Eventing specifies the faults `DeliveryFailure` for problems in delivering notifications, `SourceShuttingDown` for indicating a controlled shutdown and for any other fault `SourceCanceling`. Furthermore, the event source may provide an additional textual explanation in the element `wse:Reason`.

Please note that the order the messages exchanged in Figure 26.9 is not representative for all message exchanges in publish/subscribe systems using WS-Eventing.

WS-Eventing additionally defines that WS-MEX should be used to retrieve the event-related information from an event source embedded in the WSDL and optional policies. The services supporting WS-Eventing are marked by a special attribute and their notification, and solicit-response operations generate notifications, which are sent to subscribed event sinks. Solicit-response notifications require a response message from the event sinks. The policy entries specify the Delivery Mode and the filters.

26.3.5.2 Adaptations to DPWS

Eventing mechanisms in DPWS are only supported by hosted services (*event sources*) and clients (*event sinks*). Hosting services/devices do not support eventing. Hosted services supporting WS-Eventing must include a `wse:EventSource = 'true'` in corresponding `wsdl:portType` definitions in their

WSDL description (as defined in the WS-Eventing specification). Furthermore, hosted services must at least support the *Push Delivery Mode* as defined by WS-Eventing.

If a notification cannot be delivered to an event sink, the hosted service may terminate the subscription by sending a `SubscriptionEnd` indicating a `DeliveryFailure`.

As the time synchronization of event source and event sink might require additional resources and protocols being implemented, subscription requests and renewal are not required to be based on an absolute time. However, expressions in duration time must be accepted.

DPWS defines a specific dialect for subscription filtering—action filtering—which allows filtering related to specific actions defined in the MI headers of messages sent by the event source. A filter definition contains a white space–delimited set of URIs identifying the events being subscribed to. The event source evaluates the filter definition using the RFC 2396 prefix matching rules. That means all notifications of a `portType` can be received by subscribing to the action property prefix common to all actions of a `portType`.

26.3.6 Security

Point-to-point security offered by HTTP through authenticating, encrypting, and signing messages is not enough for complex message exchanges using SOAP. SOAP messages may be sent over intermediary nodes traversing different trusted domains, using heterogeneous transport protocols, and security mechanisms. These issues are addressed by WS-Security.

26.3.6.1 Web Services Security: SOAP Message Security 1.0 (WS-Security)

WS-Security [26] provides mechanisms for message integrity and confidentiality to SOAP messaging. It provides a framework to integrate existing security mechanisms into a SOAP message independent of underlying transport technology. WS-Security is based on existing specifications such as XML signatures, XML encryption/decryption, and XML Canonicalization. XML Canonicalization describes the preparation of XML messages for signing and encryption. XML Signature and XML Encryption define mechanisms for signing and encrypting an XML message. Furthermore, WS-Security is based on Kerberos and X.509 for authentication.

WS-Security defines a SOAP header element (`wsse:Security`), which contains security information about used authentication, signature, and encryption mechanisms as well as used security elements (e.g., public encryption keys). WS-Security does not specify which mechanisms should be used but how they are embedded into the security header. The WS-Security header may appear several times in a SOAP message specifying security mechanisms for each intermediary node along the way as well as the ultimate recipient of a SOAP message. A WS-Security header is associated with the corresponding node by using the mandatory `actor` attribute. Two security headers are not allowed to have the same `actor` URI assigned.

In order to sign or encrypt specific parts of a SOAP message, a reference to them is required. Although XPath provides mechanisms to do so, intermediary nodes processing the security headers might not support XPath. Therefore, WS-Security specifies an own mechanism using ID attributes (`wsu:Id`). This makes the processing of security definition easier and faster. An ID has to be unique for the entire SOAP message and is assigned to message elements designated for signing or encrypting. Thereby, it can be simply referred to when defining the parts being encrypted or signed. The ID attribute can also be used to refer to particular security declarations (e.g., public encryption key or signature).

Time often plays an important role in security contexts (e.g., certificates may expire). WS-Security supports this by the attribute `wsu:Timestamp`, which is assigned a security header. It specifies creation time (`wsu:Created`) and/or expiration time (`wsu:Expires`) of a security element.

The attributes `wsu:Id` and `wsu:Timestamp` used to be part of the WS-Security specification. However, they appeared to be useful for a number of WS protocols. Therefore, they have been separately defined in an XML Schema called *Web Services Utility* (indicated by the prefix `wsu`) along with other useful declarations.

26.3.6.1.1 Authentication

WS-Security offers two elements to embed authentication information into a SOAP message. The `UsernameToken` specifies a username and password for user authentication. The `BinarySecurityToken` offers the use of any binary authentication token such as a Kerberos ticket or X.509-certificate. It specifies the type of binary security token (e.g., X.809 certificate or Kerberos ticket) and its XML representation (e.g., Base64 encoded). Figure 26.10 illustrates a common authentication procedure using WS-Security [27]:

Message 1: The user requests a security token from an issuing authority. This request may not be WS based. For example, Kerberos protocols could be used to request a Kerberos service ticket.

Message 2: The issuing authority sends the security token (e.g., ticket, certificate, or encryption key) to the WS client, which embeds it into the SOAP message. The client should also sign and encrypt the message.

Messages 3–5: The client sends the (encrypted and signed) SOAP message to the WS. If necessary, the WS can validate the token by contacting the issuing authority. If the token is valid and the message encrypted, the WS replies in a secure SOAP message.

The concrete message interaction depends on the security mechanisms being used. As Kerberos and X.509 are not in the focus of this chapter, please refer to the corresponding specifications for more detail on it.

Authentication information should be signed and encrypted to ensure that it will not be changed or disclosed.

26.3.6.1.2 Message Integrity

The signing of a message prevents third parties from manipulating its content but does not exclude them from reading it. WS-Security requires the support of XML Signature for signing message content. The whole SOAP message envelope should not be signed as intermediary nodes might include additional security headers for their own security mechanisms. Such message would fail the manipulation check afterward.

Parts of a SOAP message envelope and also security definitions can be signed using the authentication mechanisms described earlier. Thereby, it is guaranteed that the user identified by the X.509 certificate, the `UsernameToken`, or a Kerberos ticket is the one who has signed the message parts.

26.3.6.1.3 Encryption

Encryption prevents third parties from being able to read the content of a message. WS-Security requires the support of XML Encryption. Symmetric or asymmetric encryption mechanism can be used. Symmetric encryption is the use of the same key for encryption and decryption. The key has to be provided to the partners by some other means in order to make sure that the key is not disclosed to untrusted parties. Asymmetric encryption is the use of two keys: a public and a private key. The public key is used for

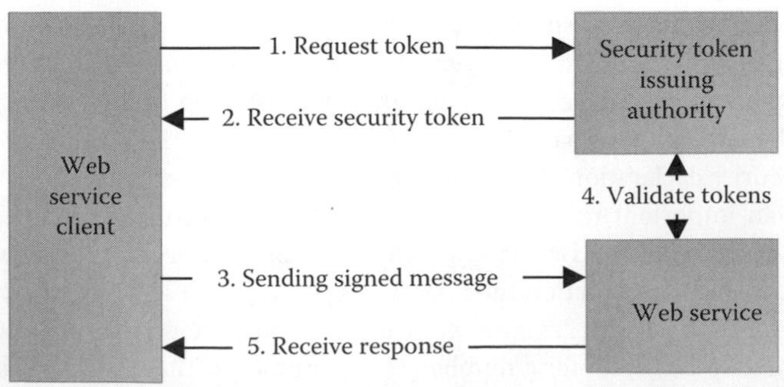

FIGURE 26.10 Message exchanges in an authentication procedure using WS-Security.

encrypting a message. As the public key cannot be used for decryption, it can be sent in any message. The private key is used for decryption. It must only be known and belongs to the partner who has to decrypt a certain message or content. X.509 certificates can be used for that.

In summary, WS-Security provides a container to include security definition and elements into a SOAP message. It provides means to include information about user authentication, digital signatures, and encryption mechanisms. Although any kind of security mechanism can be used, WS-Security provides concrete details on using Kerberos tickets and X.509 certificates as well as an own mechanism to include a username and password.

26.3.6.2 Adaptations to DPWS

WS-Security requires a lot of overhead in supporting and processing different security protocols and credentials in very heterogeneous networks although it offers enhanced security mechanisms for SOAP messaging over multiple hops and heterogeneous transport protocols. DPWS assumes that clients and devices reside in IP-based networks providing HTTP and HTTPS. Additionally, devices might have constrained resources (e.g., processing power and small storage space). Therefore, DPWS restricts the use of WS-Security to the discovery process, and all other messages are sent using a secure channel. The secure channel is established using existing technologies (HTTPS).

Security is an optional feature of DPWS. It can omitted in the administrative trusted domain. DPWS security mechanisms have to be used if messages traverse other administrative domains that cannot be trusted.

Figure 26.11 provides an overview of the security concept defined for DPWS, which is explained later in detail.

26.3.6.2.1 Authentication

The client specifies the level of security, desired security protocols, and credentials in a policy assertion in the `Probe` and `Resolve` message during the discovery. The security policy assertion is included in the metadata section (`wsx:Metadata`) under the SOAP header (`wsa:ReplyTo/wsp:Policy`). The client should also authenticate itself to the device. If several security protocols are proposed by the client, they have to be ordered by decreasing preference. The device selects a protocol for authentication and key establishment from the list and includes it in the corresponding `ProbeMatch` and `ResolveMatch` message. After the security protocols have been negotiated, the client initiates authentication and/or key establishment processes. These processes are out of scope of the DPWS specification and are normally performed using an out-of-band mechanism. If successfully finished, the client establishes a secure channel to the device using Transport Layer Security (TLS)/Secure Socket Layer. The secure channel is also used for communication between client and hosted services.

26.3.6.2.2 Integrity

The integrity of discovery messages of client and devices is ensured by using message-level signatures based on WS-Security. Devices should sign Hello and Bye SOAP message envelopes. If a device has multiple signatures, it must send one message for each of the signatures.

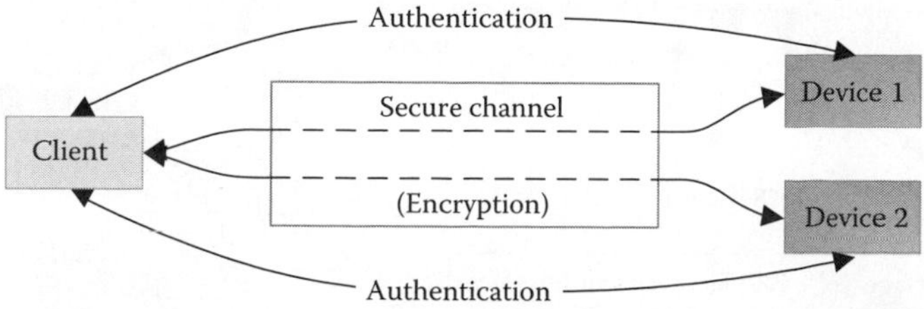

FIGURE 26.11 Security concept defined for DPWS.

Service must protect the following MI header block of their SOAP messages by signatures: Action field (`wsa:Action`), messageID (`wsa:MessageID`), address of recipient (`wsa:To`), address for response (`wsa:ReplyTo`), and the ID of related request message for responses (`wsa:RelatesTo`). Furthermore, the content portion(s) of the SOAP Body associated with those MI headers must also be signed. The requirements for services are automatically fulfilled if using TLS as proposed by DPWS.

26.3.6.2.3 Confidentiality

Discovery messages are not encrypted, whereas all other messages are encrypted using a secure channel. Services must encrypt their SOAP body.

A secure channel should remain active as long as client, device, and hosted services are interacting. The channel is removed when a device or client leaves the network.

26.4 Web Service Orchestration

Although the DPWS is an ideal base to enable connectivity between heterogeneous devices, each client has to be programmed and set up manually for each individual interaction with corresponding devices and services, respectively. This is not a big deal for interactions between one client and several services on devices but involves a huge effort for programming workflows or interaction processes in large networks. WS composition addresses this disadvantage by the definition of process workflows on existing WSs.

26.4.1 Web Service Composition

Service composition is the possibility to combine a number of individual services in such a way that a surplus value is created [2]. Service composition can be divided into two independent aspects: service orchestration and service choreography.

Service orchestration is the execution of a workflow/process of service interactions controlled by a central entity (*service orchestrator*). The central entity plays the role of the initiating client for each individual invocation of services.

Service choreography, by contrast, refers to a decentralized workflow/process that results from each client–service interaction. In other words, each client is aware of its own interactions, but it is not necessarily aware of the interactions between other clients and services. Nevertheless, all client–service interactions form a workflow/process.

Figure 26.12 presents simple examples for service orchestration and choreography. In service orchestration, the service orchestrator (as the central workflow/process coordinator) controls the invocations of three services, their order, and deals with possible failures. In a choreography setup, each participating component is (only) aware of its own interactions. There is no central instance starting and controlling the workflow/processes. The choreography workflow is started by Client 1 when invoking Service 4.

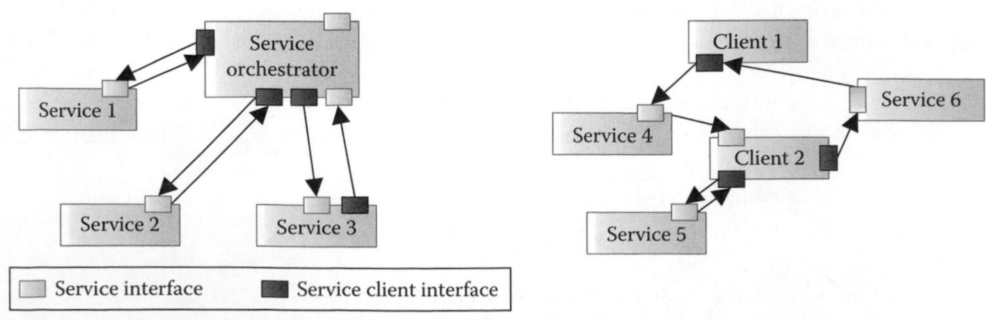

FIGURE 26.12 Service orchestration vs. service choreography.

Another aspect attracting attention while looking at Figure 26.12 is that the individual components do not have clearly assigned roles. Some of them act as both service and service client (illustrated by their interfaces). That is often the case in real-life applications. A strict separation can be done only for a single client–service interaction. For example, the service orchestrator possesses an unconnected service interface. This can be used to start the workflow/process by invoking the service orchestrator. Thereby, services and their functionality can be composed to a single (meta) service. Furthermore, Service 3 might also be a composed service, which takes a long time to finish the execution of its process. Therefore, the service orchestrator provides a callback interface, which will be invoked by Service 3 to deliver the response to its initial invocation (*asynchronous service interaction*). In the meanwhile, the service orchestrator can perform other tasks and does not have to wait for the response of Service 3. Also, the redirection of responses to a third service or client is possible in some SOA implementations as shown in the interaction between Service 4 and Client 2.

It is obvious that the process designer has more influence on aspects such as selection of participating, fault tolerance, compensation of occurring errors, and handling of events by using service orchestration instead of service choreography. Service choreography is useful if some internal aspects of interaction processes are not known (e.g., when involving components from other companies that themselves are interacting somehow).

The orchestration of services to workflow processes falls into the category of business process management (BPM). Although process management originates from the business sector, its concepts can be adapted to the automation of technical processes, which is just the intention of this chapter.

BPM takes care of the planning, modeling, executing, and controlling of processes and workflows, respectively [28]. It bridges the gap between business and IT by offering an outside view that is oriented on the interaction process of components rather than the performance of participating components. Participating components are *black boxes* for the process designer. BPM reacts on the adaptability requirements of constantly changing environments and requires a standardized communication and interface to involved components. Process definitions in the area of IT have to be machine readable to enable automation. Among the WS protocols, the WS-BPEL standard combines all aspects of BPM with the advantages of SOAs represented by WSs.

26.4.2 Web Services Business Process Execution Language

The Web Services Business Process Execution Language (WS-BPEL) is an XML format, which is based on WS standards [29], to describe machine-readable processes. WS-BPEL processes can be designed using graphical tools and can be directly executed by a WS-BPEL engine/orchestrator. WS-BPEL provides several concepts in order to support flexible processes:

- WSs interaction. A WS-BPEL engine provides a WS interface to each WS-BPEL process. Thereby, WS-BPEL processes are started when their corresponding WS interface is invoked. Synchronous and asynchronous interactions with other WSs are supported by WS-BPEL. Synchronous interaction is the immediate response of a WS to its invocation. Asynchronous interaction is the delayed response of a WS by invoking the corresponding callback interface provided by the WS-BPEL process.
- Structuring of processes. WS-BPEL offers validity scopes to distinguish between local and global contexts. Processes can define loops, condition statements, and sequential and parallel execution. Dependencies between process threads can also be defined.
- Manipulation of data. WS-BPEL allows the use of variables for the internal process execution. Data can be extracted from messages, validated, manipulated, and assigned to other variables or messages.
- Correlation. In order to relate a certain message to a specific instance, WS-BPEL introduces the concept of correlation. It allows the identification of a certain context common to all messages of one instance.

- Event handling. Events can be handled in parallel to the process execution. They may be triggered either by the reception of a certain message or by a time-out.
- Fault handling. WS-BPEL also supports mechanisms for exception handling if a fault is thrown.
- Compensation handling. As the execution of a process is not an atomic operation, successfully completed WS interactions have to be possibly undone or compensated in case of a fault. For example, the payment of a flight ticket is not successful, but the seats are already reserved. Then, the reservation has to be compensated, which is facilitated by the WS-BPEL compensation handling.
- Extensibility. WS-BPEL provides means for adding new activities and data assignment operations.

WS-BPEL resides on top of the WS protocols. It supports the Basic Profile for WSs [30], which provides interoperability guidance for the WS core specifications such as WS communication protocols, description, and the service registry. DPWS is not supported by WS-BPEL.

26.4.2.1 WS-BPEL Extension for People (BPEL4 People)

The BPEL4 People enables the involvement of people in the process design (e.g., monitoring by people or realizing manual input to processes) [31]. Process designers can define human interactions and specify their contexts such as individual tasks, roles, and administrative groups of people.

Human tasks can be separated from the process definition. This is specified by Web Services Human Task (WS-HumanTask) [32]. A human task can be exposed as WS and thereby used in a process.

BPEL4 People and WS-HumanTask are OASIS standards since 2010.

26.5 Software Development Toolkits and Platforms

There are currently several initiatives developing software stacks, toolkits, and enhancements to DPWS.

26.5.1 WS4D, SOA4D

The *Web Services for Devices (WS4D) initiative* [33] was started from German partners (University of Rostock, Technical University of Dortmund, and Materna GmbH) of the European ITEAR&D project service infrastructure for real-time embedded networked application (SIRENA). Currently, four DPWS software development kits are available and release as open source. The *Java Multi Edition DPWS stack* is based on the Connected Limited Device Configuration profile enabling DPWS on very resource-constrained devices. The *gSOAP DPWS stack* is based on the gSOAP code generator and SOAP engine and is especially designed for embedded devices. The *Axis2 DPWS stack* is based on the Apache Axis2 WSs engine and focuses on enterprise applications. The uDPWS stack is developed for device with extreme constrained resources such as sensor nodes. All four stacks and their toolkits are available as open-source software.

The *Service-Oriented Architecture for Devices (SOA4D) initiative* [34] originates from the European ITEA R&D project service-oriented device architecture, which is the succeeding project of SIRENA. SOA4D aims at fostering an ecosystem for the development of service-oriented software components adapted to the specific constraints of embedded devices. SOA4D provides two open-source stacks and corresponding toolkits. The *DPWS Core* stack provides an embeddable C WSs stack based on gSOAP. The *DPWS4J Core* stack provides a Java WSs stack for the J2ME CDC platform.

Both initiatives work closely together to ensure the interoperability between their stacks.

26.5.2 DPWS in Microsoft Windows

UPnP and DPWS (here called WSs on Devices WSD) are part of Windows Rally facilitating the integration of network-connected devices in a plug-and-play manner.

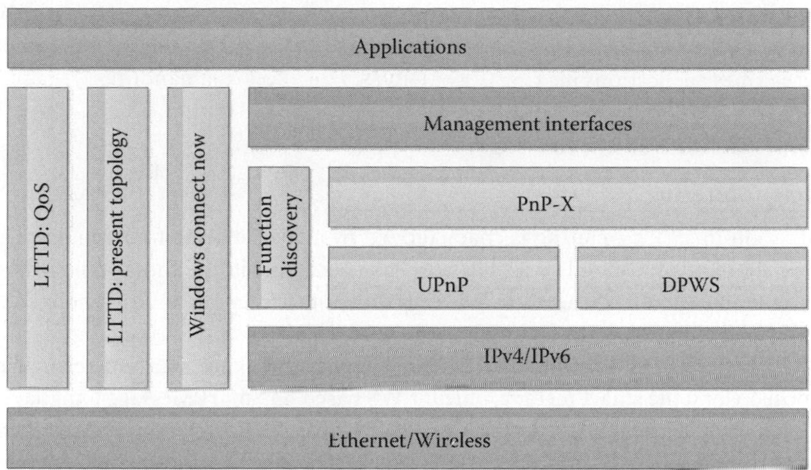

FIGURE 26.13 Windows Rally

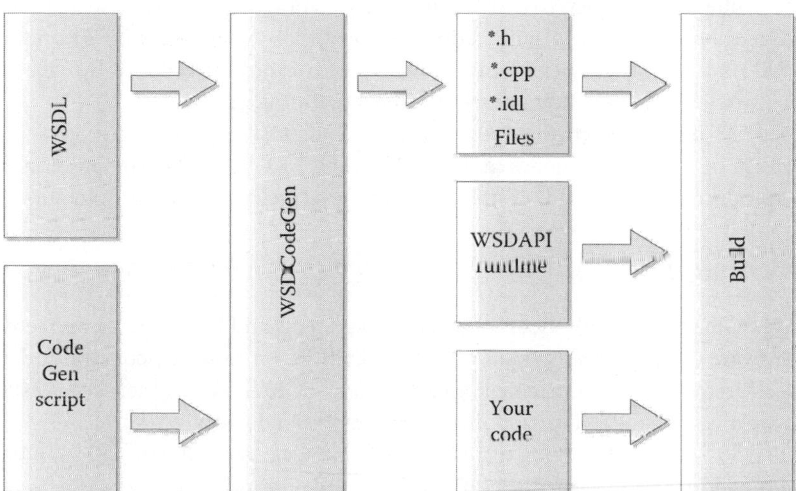

FIGURE 26.14 DPWS Code Generator.

Figure 26.13 presents an overview of Windows Rally and used technologies [35]. The *Link Layer Topology Discovery* enables the discovery of devices using the data-link layer (Layer 2). It can be used to determine the topology of a network, and it provides QoS extensions that enable stream prioritization and quality media streaming experiences. Windows Connect Now is a part of Rally and aims at simpler wireless device configuration. Function discovery is a generic middleware and can be used to execute common discovery enabling the searching for UPnP and DPWS devices.

Web Services on devices API (WSDAPI) is an implementation of the DPWS 1.0 in Microsoft Windows since Windows Vista. The API supports discovery, metadata, control, eventing, and binary attachments to messages. In addition to that, it supports XML schema extensibility. As part of WSDAPI, a code generation tool WSDCodeGen converts WSDLs to COM objects (Figure 26.14) [36].

26.6 DPWS in Use

This section describes applications making use of DPWS.

26.6.1 Automation Scenarios

26.6.1.1 Smart Metering

The actual situation in smart metering is characterized by a coexistence of a large number of proprietary and open standards for wired as well as wireless communication. These standards show low or no interoperability to each other. Therefore, it is very difficult to integrate multi-vendor solutions using one sustainable holistic approach. In [37], proposed approach is to use WS technology as an open widespread Internet standard for the creation of a heterogeneous network for smart metering devices.

In the proposed work, the smart meter profile for WS based on Smart Message Language and DPWS was introduced. It was demonstrated that using DPWS-enabled smart meters, a heterogeneous meter network can be built using existing open standards. From a technical perspective, the suggested WS-based solution is not bound to any physical layer technology so that any possible technology supporting IP can be used. This eases the setup of heterogeneous and vendor-independent meter networks. Due to the sensitivity of meter data, energy profiles, and controlling capabilities, special security requirements have to be satisfied. Authorization, encryption, and signature are common security measures. They are already integrated in DPWS, thereby providing required privacy and security for automated meter reading. In order to reduce communication overhead, CoAP is used for HTTP and Efficient XML Interchange (EXI) for XML binary encoding. Thus, traffic overhead is reduced by over 90%. The suggested approach can be easily applied to nearly all smart metering protocols.

The works in [37,38] break the myth that WSs require high computation power and produce high communication overhead. The uDPWS stack requires ca. 46 KB ROM and 7 KB RAM. The compression through CoAP and EXI do not produce additional CPU or memory load. Nevertheless, the compression rate is over 90%.

26.7 Conclusion

The DPWS solves important requirements of embedded systems and hardware components. It provides an SOA for hardware components by enabling WS capabilities on resource-constraint devices. In particular, DPWS addresses the announcement and discovery of devices and their services, eventing as a publish/subscribe mechanism, and secure connectivity between devices.

Although DPWS is still a quite young technology, it can be expected that it will soon play an important role in building a bridge between device automation and enterprise networks. This results also from the fact that Microsoft and printer manufacturer support it in their products, and a lot of research and development is currently performed with respect to DPWS. A process management standard such as WS-BPEL will reveal the full potential of DPWS and simply the development and controlling of complex automation systems.

References

1. T. Erl. *Service Oriented Architecture: Concepts, Technology, and Design*. Pearson Education Inc., Upper Saddle River, NJ, 2005.
2. W. Dostal, M. Jeckle, I. Melzer, and B. Zengler. *Service-orientierte Architekturen mit Web Services*. Elsevier, Spektrum Akademischer Verlag, München, Germany, 2005.
3. OSGi Alliance. OSGi service platform core specification, Release 5, Version 5.0, March 2012.
4. Sun Microsystems. Jini architecture specification, Version 1.2, December 2001.
5. UPnP Device Architecture 1.1. UPnP forum, October 2008.

6. J. Ritchie, T. Kuehnel, W. van der Beek, and J. Kang. UPnP AV architecture: 2. UPnP forum, December 2010.
7. D. Booth, H. Haas, F. McCabe, E. Newcomer, M. Champion, C. Ferris, and D. Orchard. Web Services Architecture. W3C, February 2004.
8. S. Chan, D. Conti, C. Kaler, T. Kuehnel, A. Regnier, B. Roe, D. Sather et al. Devices profile for Web Services. Microsoft Corporation, February 2006.
9. T. Nixon, A. Regnier, D. Driscoll, and A. Mensch. Devices profile for Web Services Version 1.1. OASIS, July 2009.
10. H. Bohn, A. Bobek, and F. Golatowski. SIRENA—Service infrastructure for real-time embedded networked devices: A service oriented framework for different domains. In *Fifth IEEE International Conference on Networking (ICN'06)*, pp.1–6, Morne, Mauritius, April 2006.
11. J. Schlimmer. A technical introduction to the devices profile for Web Services. *MSDN*, May 2004.
12. N. Mitra. SOAP version 1.2 part 0: Primer. W3C, June 2003.
13. G. Moritz, F. Golatowski, and D. Timmermann. A CoAP based SOAP transport binding. In *The 16th IEEE International Conference on Emerging Technologies & Factory Automation—WIP (ETFA 2011)*, pp. 1–4, Toulouse, France, September 2011.
14. M. Lebold, K. Reichard, C.S. Byington, and R. Orsagh. OSA-CBM architecture development with emphasis on XML implementations. In *The Maintenance and Reliability Conference 2002 (MARCON 2002)*, pp. 1–16, Knoxville, TN, May 2002.
15. M. Gudgin, M. Hadley, and T. Rogers. Web Services Addressing 1.0—Core (WS-Addressing). W3C, May 2006.
16. T. Nixon, A. Regnier, V. Modi, and D. Kemp. Web Services Dynamic Discovery (WS-Discovery) Version 1.1. OASIS, July 2009.
17. V. Altmann, P. Danielis, J. Skodzik, F. Golatowski, and D. Timmermann. Optimization of ad hoc device and service discovery in large scale networks. In *The 18th IEEE Symposium on Computers and Communications (ISCC 2013)*, pp. 1–6, Split, Croatia, July 2013.
18. E. Christensen, F. Curbera, G. Meredith, and S. Weerawarana. Web Service Description Language (WSDL) 1.1. W3C, March 2001.
19. R. Butek. Which style of WSDL should I use? *IBM developerWorks*, May 2005.
20. F. Cohen. Discover SOAP encoding's impact on Web Service performance. *IBM developerWorks*, March 2003.
21. A.S. Vedamuthu, D. Orchard, F. Hirsch, M. Hondo, P. Yendluri, T. Boubez, Ü. Yalçinalp. Web Services Policy 1.5—Framework (WS-Policy). W3C, September 2007.
22. A.S. Vedamuthu, D. Orchard, F. Hirsch, M. Hondo, P. Yendluri, T. Boubez, Ü. Yalçinalp. Web Services Policy 1.5—Attachment (WS-PolicyAttachment). W3C, September 2007.
23. K. Ballinger, D. Box, F. Curbera, D. Ferguson et al. Web Services Metadata Exchange 1.1 (WS-MetadataExchange). W3C, August 2008.
24. J. Alexander, D. Box, L.F. Cabrera, D. Chappell et al. Web Service Transfer (WS-Transfer). W3C, September 2006.
25. D. Box, L.F. Cabrera, C. Critchley, F. Curbera et al. Web Services Eventing (WS-Eventing). W3C, August 2006.
26. K. Lawrence, C. Kaler, A. Nadalin, R. Monzillo, and P. Hallam-Baker. Web Services Security: SOAP message security 1.1 (WS-Security 2004). OASIS Standard, OASIS, February 2006.
27. S. Seely. Understanding WS-Security. MSDN, October 2002.
28. M. Keen, G. Ackerman, I. Azaz, M. Haas, R. Johnson, J.W. Kim, and P. Robertson. *Patterns: SOA Foundation—Business Process Management Scenario*. IBM International Technical Support Organization, Armonk, New York, August 2006.
29. A. Alves, A. Arkin, S. Askary, C. Barreto, B. Bloch, F. Curbera, M. Ford et al. Web Services Business Process Execution Language Version 2.0. OASIS standard, OASIS, April 2007.

30. K. Ballinger, D. Ehnebuske, C. Ferris, M. Gudgin, C.K. Liu, M. Nottingham, and P. Yendluri. Basic profile 1.1. WS-I, April 2006.
31. L. Clement, D. Koenig, V. Mehta, R. Mueller, R. Rangaswamy, M. Rowley, and I. Trickovic. WS-BPEL extension for people (BPEL4People) specification version 1.1. OASIS, August 2010.
32. L. Clement, D. Koenig, V. Mehta, R. Mueller, R. Rangaswamy, M. Rowley, and I. Trickovic. Web Services—Human Task (WS-HumanTask) Version 1.1. OASIS, August 2010.
33. Web Services for Devices (WS4D) initiative, 2013. http://www.ws4d.org/. Accessed July 5, 2013.
34. Service-Oriented Architecture for Devices (SOA4D), 2013. https://forge.soa4d.org/. Accessed July 5, 2013.
35. Microsoft. Windows rally technologies: An overview, Whitepaper, Microsoft Corp., 2006.
36. Microsoft. Web Services on devices. MSDN—Microsoft Developer Network, November 2012.
37. V. Altmann, J. Skodzik, F. Golatowski, and D. Timmermann. Investigation of the use of embedded Web Services in smart metering applications. In *The 38th Annual Conference of the IEEE Industrial Electronics Society* (*IECON 2012*), pp. 6176–6181, Montreal, Canada, October 2012.
38. C. Lerche, N. Laum, G. Moritz, Z. Elmar, F. Golatowski, and D. Timmermann. Implementing powerful Web Services for highly resource-constrained devices. *Pervasive Computing and Communication Workshops* (PerCom Workshops), pp. 332–335, Seattle, WA, March 2011.

V

Safety Technologies in Industrial Networks and Network Security

27 **PROFIsafe** *Wolfgang Stripf and Herbert Barthel* **27**-1
Why Do We Need Functional Safety in Automation? • Dichotomy of Standard and Functional Safety Automation • Motivation and Objectives for PROFIBUS and PROFINET • PROFIsafe, the Solution • Beyond Functional Safety Communication • Abbreviations • References

28 **SafetyNET p Protocol** *Marco Cereia, Jochen Streib, and Reinhard Sperrer* **28**-1
Introduction • SafetyNET p Physical Layer • Communication Model • Precise Clock Synchronization • Safety Communication Model • References

29 **Security in Industrial Communications** *Thilo Sauter and Albert Treytl* **29**-1
Introduction • Basic Security Measures • Security for Automation Networks • Adding Security to Legacy Field-Level Networks • System-Level Approach—Defense in Depth • Conclusions • References

27

PROFIsafe: Functional Safety with PROFIBUS and PROFINET

27.1 Why Do We Need Functional Safety in Automation? 27-1
27.2 Dichotomy of Standard and Functional Safety Automation ... 27-2
27.3 Motivation and Objectives for PROFIBUS and PROFINET ... 27-2
27.4 PROFIsafe, the Solution 27-4
Concept • Black Channel • Possible Transmission Errors and Their Remedies • SIL Monitor • Messages with PROFIsafe PDUs • PROFIsafe Parameters • Structure of PROFIsafe PDUs • Synchronization Means (Finite-State Machines)
27.5 Beyond Functional Safety Communication 27-13
Safety-Related Programmable Control Logic • Commissioning and Repair • Availability • Status of PROFIsafe Related Specifications and Guidelines • Standards Catching Up • Peculiarities for Different Industries • Environmental Conditions • Test and Certification • Development Tools and Support • Products • Prospects
Abbreviations 27-21
References 27-22

Wolfgang Stripf
PROFIBUS and PROFINET International

Herbert Barthel
Siemens AG

27.1 Why Do We Need Functional Safety in Automation?

Any active industrial process is more or less associated with the risk of

- Injuring or killing people
- Destroying nature
- Damaging investments

With most of the processes, it is quite easy to avoid risk without special requirements imposed on automation systems. However, there are typical applications associated with high risk, for example, presses, saws, tooling machines, robots, conveying and packing systems, chemical processes, high-pressure operations, off-shore technology, fire and gas sensing, burners, and cable cars. Those applications need special care and technology.

Over time, the market balances out the reliability and the availability of the *standard* automation technology against a certain economic cost level. That means the failure, or error rate, of the standard automation technology under normal circumstances is acceptable for normal modes of operations, but not sufficient for the safety critical applications, mentioned earlier.

The situation may be compared with the public mail system. While normal letter expedition is expected to be as affordable as possible at certain delivery uncertainty, everybody will use special mail for important messages.

27.2 Dichotomy of Standard and Functional Safety Automation

In the past decades, microcontrollers, software, personal computers, and communication networks have dramatically influenced the automation area, leading to cost reduction, higher flexibility, and availability. In respect to safety, standards and regulations of the day were prohibiting any usage of existing solutions. Safety automation had to be *hard-wired* and based on *relay* technology. This dichotomy, or gap, is quite natural due to the fact that safety relies on trusted technology, trust in experience, and experience accumulated over long time. But adding *classical* safety to modern automation solutions leads to increased cost due to additional wiring and need for engineering, to less flexibility and availability than expected, and to other disadvantages. Recently, the situation has changed significantly. Microcontrollers and software have been in a widespread use in diverse industrial automation applications. The preconditions for their use in safety applications were introduced in the international standard IEC 61508, with its different safety integrity levels, from SIL1 to SIL4 [1].

27.3 Motivation and Objectives for PROFIBUS and PROFINET

Over its lifetime, since 1989, PROFIBUS has emerged to one of the most important fieldbus systems in the world with more than 50 million nodes. PROFINET IO, the Ethernet-based companion, follows the success story since its introduction in 2005. Both are standardized within IEC 61158 and IEC 61784 [2], respectively, and enable decentralized applications in factory and process automation (PA) with a variety of appropriate transmission technologies such as Ethernet, RS485, MBP-IS,* fiber optics, and wireless. It was, thus, merely a matter of time to introduce, in a seamless manner, the necessary means for safety applications on PROFIBUS DP & PROFINET IO and to provide a similar flexibility and availability also for powerful functional safety devices like remote I/O, laser scanners, light curtains, level switches, shutdown valves, drives, robots, and alike.

Back in 1998, when the PROFIBUS organization started its project "functional safe communication across PROFIBUS DP," more than 25 companies, renowned for their involvement in the functional safety area, decided to connect the above-mentioned safety devices to the same transmission line, as the standard devices, and let them communicate with an additional programmable safety-related controller (F-Host in Figure 27.1). The requirement was that neither cables, ASICs, and layer stack software nor other communication devices such as repeaters, links, and couplers should be changed. Configuration, parameterization, programming, and diagnostic means should be as familiar to the user as possible in order to simplify the deployment of the safety option: *what he/she is using, he/she will not be losing.* Fortunately, the working group was able to base its design and development efforts on the new IEC 61508 and on the groundwork of the *railway standard* IEC 62280-1 [3]. PROFIBUS DP and PROFINET IO fall in the category of the so-called *defined transmission systems,* with configured and well-known participants and transmission properties, the preconditions of functionally safe communication. The public phone system or the Internet, for example, falls in the category of the *open transmission systems.*

From the very beginning, it was clear that the transmission of safety messages via the standard PROFIBUS DP (including the PROFIBUS PA), and some years later PROFINET IO, cable would not

* MBP-IS = "Manchester Coded–Bus Powered and Intrinsically safe" replaces the previously used name "IEC 1158-2." Further development has listed additional procedures in the corresponding IEC standard, so that creating an unambiguous designation had become necessary.

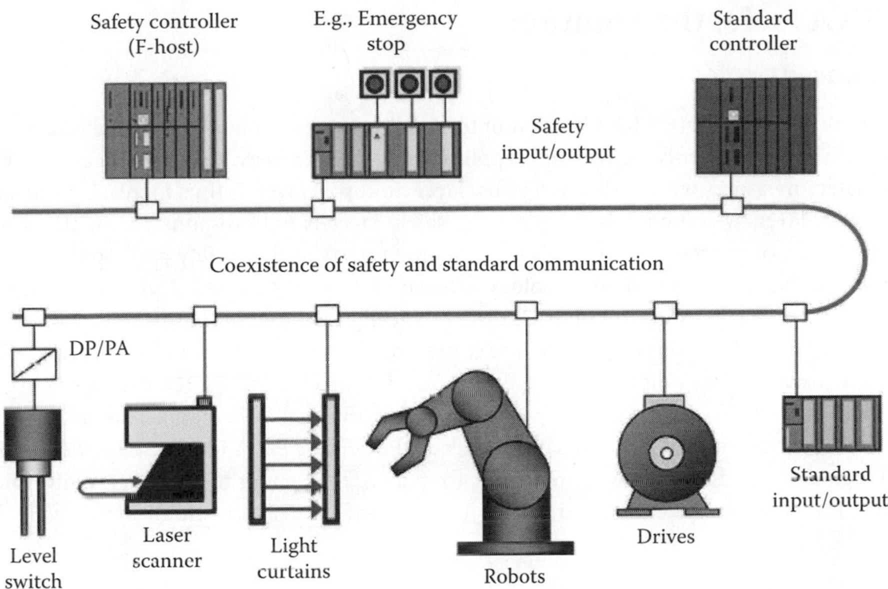

FIGURE 27.1 The PROFIsafe vision.

prove sufficient for the new generation of safety devices that had to be connected to it. What sense would it make, if merely the shutdown signals could be transported via the bus, while parameter value assignment, or diagnostic messages, in the event of a failure still require a time-consuming on-site PC connection via a proprietary interface?

There was a need for the device manufacturers to support rapid device replacement and integration of the device commissioning and diagnostic software with the engineering software of the system developers in order to improve production flexibility (e.g., program-controlled parameter value assignment) and to increase availability of the manufacturing assets (intelligent diagnosis and preventive maintenance).

Safe communication could be achieved by using redundant transmission lines and cross-checking as a basis for safety functions. However, the working group opted for a *single-channel* solution. The redundancy can still be added to the system as an option to provide additional higher availability/reliability for the operational functions (Table 27.1).

TABLE 27.1 Safety and Redundancy Options

	PROFIBUS DP PROFINET IO	PROFIsafe	Redundancy	PROFIsafe and Redundancy
Application	Suitable for all kinds of distributed automation	Factory and process automation:presses, robots, level switches, shutdown valves, burner control, cable cars, etc.	Process automation: chemical or pharmaceutical productions, refineries, offshore, etc.	Process automation: chemical or pharmaceutical productions, refineries, offshore, etc.
High availability	—	—	No downtimes of operational functions at best (fault tolerance)	No downtimes of operational functions at best (fault tolerance)
Functional safety	—	Prevent hazards through safety functions (required by laws or insurances)	Redundancy by itself does not provide functional safety	Prevent hazards through safety functions (required by laws or insurances)

27.4 PROFIsafe, the Solution

27.4.1 Concept

PROFIBUS DP and PROFINET IO, like most of the fieldbus systems, employ only the layers 1, 2, and 7 of the ISO/OSI model [4], only for real-time applications. In order to avoid any changes to any of these layers, the safety measures were added as a safety layer on top of layer 7, thus increasing the size of the OSI application layer by some kilobytes. Since the safety layer is only responsible for the *transport* of safety-related *user*, or process data, it takes the remaining parts of the (safety and) application layer to acquire and process these data. The technology firmware in a safety-related device (e.g., light curtain) can incorporate these layers; in case of SIL3, this is typically a redundant hardware/software structure (Figure 27.18). Similarly to the *non-safety* mode, the process data (signals and/or process values) are packed in the protocol data unit (PDU) of a PROFIBUS DP, or PROFINET IO, message frame. In case of PROFIsafe, the raw process data are supplemented by additional information, discussed in the next sections. The safety-augmented process data are called a *PROFIsafe PDU*. Ideally, a *PROFIsafe PDU* shall be passed as a whole, and completely unmodified, from a (safety) sender to a (safety) receiver no matter what kind of transmission system both are using. Thus, the safety measures are *encapsulated* in the communicating end devices.

27.4.2 Black Channel

Such a transmission system from one safety node to another is called a *black channel*, an analogy to a *black box* (see Figure 27.3). The chosen communication technology does not matter except for a few basic constraints to be discussed later on in this chapter. Another implication is that none of the error detection mechanisms of the chosen communication technology is employed to guarantee the integrity of the transferred process data. Basically, there are no restrictions with respect to transmission rate, number of bus devices, or transmission technology—as long as its parameters are acceptable by the required reaction times of the safety applications. The example in Figure 27.2 demonstrates that PROFIsafe also employs the *black channel* principle for complex PROFIBUS DP and PROFINET IO structures.

Safety-related sensor signals from an F-DI module of a PROFINET IO remote I/O device that can be equipped with other safety-related and non-safety-related modules with input/output channels are routed via the PROFINET IO device node to the PN IO controller. From here, they go via a local backbone bus to the F-Host equipped with a programmable safety-related system (e.g., Safety-PLC). After the safe logic operation, a corresponding output signal is routed via the local bus back to the PN IO controller, and, then, via a PN IO/DP Link to a PROFIBUS DP segment, and from there to its final destination, a drive with integrated safety. In case of explosion-proof areas requiring intrinsic safety, signals are passing a DP-PA Link. The transmission rate is reduced there, and the intrinsically safe transmission technology (MBP-IS) is employed for routing a signal to the safety-related PA slave. At no point on its communication path, the signal has employed a redundant communication path. In other words, this is a *single-channel* transfer.

Up to now, we simply dealt with the communication paths for safety-related messages. Thus, it is still open who is responsible for the transfer and when the transfer takes place. Here, too, PROFINET IO and PROFIBUS DP standard mechanisms are used—the Controller–Device or Master–Slave operation, respectively. Controllers, or Masters, which may be assigned to an F-Host, *cyclically* exchange messages with all their configured Devices, or Slaves. This means that there is always a 1:1 *relationship* between Controller–Master and Device–Slave. The polling operation has the advantage that any failed device will be detected immediately—one of the basic principles of safety technology. Now, we are able to identify the basic constraints for the *black channel* for PROFIsafe operations: the polling principle and the 1:1 relationship.

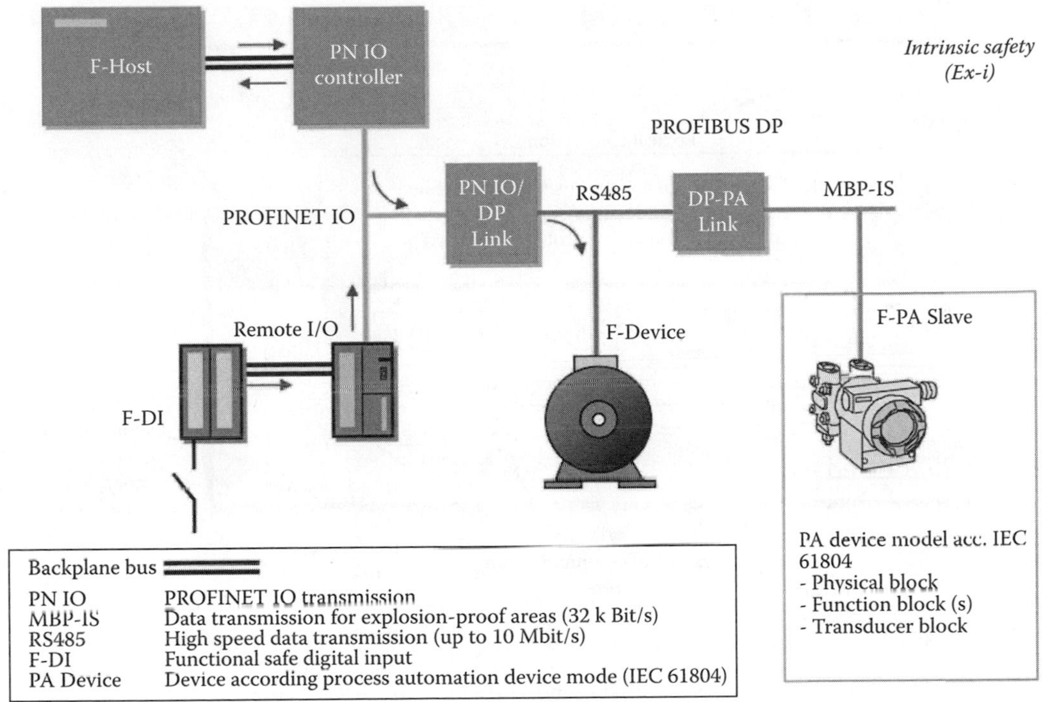

FIGURE 27.2 Complete communication paths for PROFIsafe PDUs.

27.4.3 Possible Transmission Errors and Their Remedies

Various errors may occur when messages are transferred in network topologies of the described complexity, be it due to hardware failures, extraordinary electromagnetic interference (EMI), or other influences. A message of a logical connection between two PROFIsafe safety nodes can be lost, occur repeatedly, or mirrored (loopback), be inserted from somewhere else, appear delayed, in an incorrect sequence, out-of-sequence due to storage elements in the network and/or show corrupted data. In case of safety-related communication, there may also be misrouting due to incorrect addressing, which means that a standard or safety-related message erroneously appears at a safety, or wrong, device and pretends to be an expected safety message. Figure 27.3 shows the black channel for such a logical connection and the possible communication error types.

Out of the numerous remedies known from literature, PROFIsafe focuses on those presented in the matrix shown in Table 27.2.

These include the following:

- The numbering of the PROFIsafe PDUs (*sign-of-life*)
- A time expectation with acknowledgment (*watchdog*)
- A Codename between sender and receiver for each direction (*password*)
- Data integrity checks (cyclic redundancy check [CRC])

Using the numbering, a receiver can see whether it received all of the PROFIsafe PDUs and within the correct sequence. The receiver returns to the sender a PROFIsafe PDU with the correct number as an acknowledgment. Basically, a simple *toggle bit* would have proven sufficient. However, due to possible storing network elements such as routers or switches, a large series of numbers has been selected for PROFIsafe.

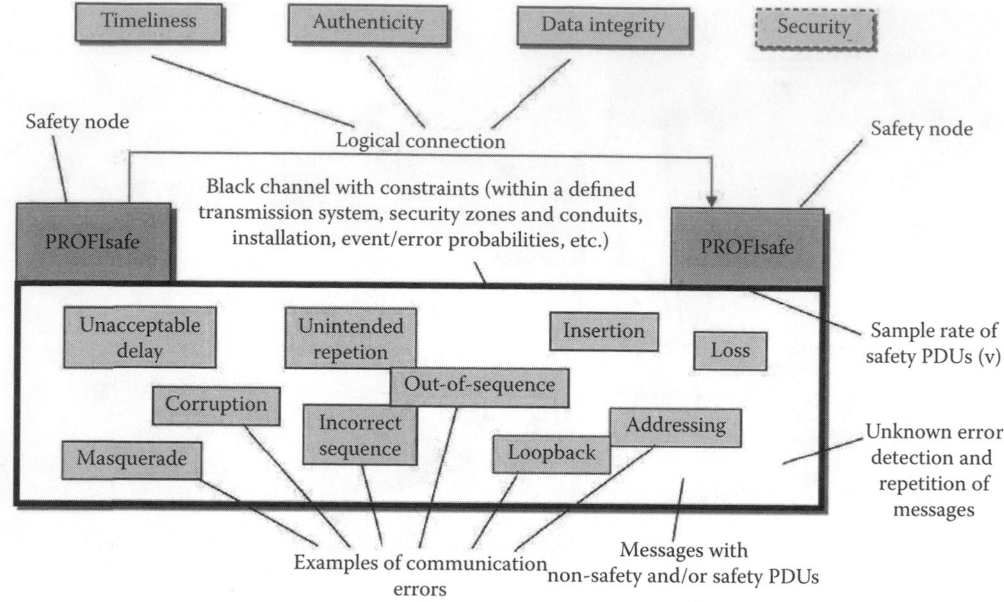

FIGURE 27.3 Black channel and PROFIsafe properties of logical connections.

TABLE 27.2 Black Channel Error Types and Safety Measures of PROFIsafe

	Safety Measures			
Communication Errors	Monitoring Number (Virtual)	Timeout with Receipt	Codename for Sender and Receiver	Data Integrity Check
Corruption				X
Unintended repetition		X		
Incorrect sequence	X			
Loss	X	X		
Unacceptable delay		X		
Insertion	X		X	
Masquerade			X	
Addressing			X	
Out-of-sequence	X			
Loopback of messages			X	

In the functional safety technology, it matters not only that a message transmits the correct process signals or values. The up-to-date values must arrive within a process fault tolerance time, thus enabling the respective device to automatically initiate on-site a safe state, if necessary (e.g., stop of movement). For this purpose, the devices use a watchdog timer that is restarted whenever new PROFIsafe PDUs with incremented numbers have arrived. Numbering and watchdog timer together provide the basis for an important property of functional safety communication, namely, the *Timeliness* (see Figure 27.3). The procedures and sizes of the chosen safety measures for *Timeliness* must meet the quantitative requirements of IEC 61508.

The 1:1 communication relationship (logical connection) between the Controller–Master and a Device–Slave facilitates the detection of misrouted messages. Each and every safety logical connection must have a unique identification (called Codename) inside the defined network, and this Codename can be used to verify the Authenticity of a PROFIsafe PDU. Authenticity is another property of functional safety communication (see Figure 27.3). Again, the procedures and sizes of the chosen safety measures for *Authenticity* must meet the quantitative requirements of IEC 61508.

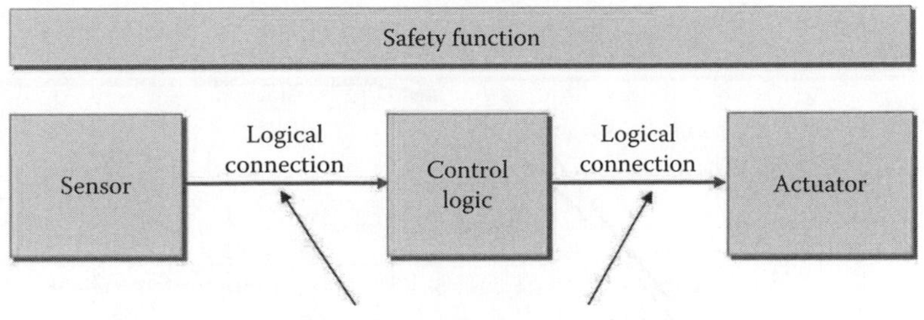
Safety function
Sensor
Logical connection
Control logic
Logical connection
Actuator
The sum of the residual error rates of all logical connections of a safety function shall not exceed 1% of the PFD, PFH of the safety function, e.g., for SIL3 with 10^{-7}/h→ 1× 10^{-9}/h

FIGURE 27.4 Safety function acc. IEC 61508 and the 1% rule.

Data Integrity is the third property of functional safety communication (see Figure 27.3). Detecting corrupted data bits through an additional CRC plays a key role. The procedures and sizes for *Data Integrity* must meet the quantitative requirements of IEC 61508.

The necessary probabilistic examinations for all three properties can benefit from the definitions within the IEC 61508 that considers the probability of failure of entire safety functions. PROFIsafe is following this approach (Figure 27.4).

Accordingly, a safety circuit includes all sensors, actuators, communication elements, and logic processes that are involved in a safety function. IEC 61508 defines overall values for the average probability of failures for different SILs. For SIL3, for example, this is an average probability of failures per hour (PFH) of 10^{-7}/h. For the transmission, PROFIsafe takes up only 1%. This means that the permissible PFH for the sum of the PROFIsafe properties *Timeliness*, *Authenticity*, and *Data Integrity* is 10^{-9}/h.

The major steps of the PROFIsafe protocol development are outlined in the following.

First, suitable CRC generator polynomials had to be determined for a reasonable PROFIsafe PDU length.

The following so-called *proper* polynomials are chosen for the usage with PROFIsafe (r = length of the CRC signature):

r = 16 Bit: 0x**1EAB**
r = 24 Bit: 0x**5D6DCB**
r = 32 Bit: 0x**F4ACFB13**

A polynomial is called *proper* when the residual error probability over an increasing bit error probability never crosses the limit 2^{-r} (e.g., when the curve rises continuously monotonic). Figure 27.5 shows examples of a *proper* and an *improper* polynomial.

The residual error probability (y axis) becomes critical with a high bit error probability. The more bits are corrupted within a PROFIsafe PDU, the higher the probability of a residual error. Usually, EMIs in the form of burst disturbances cause the corruption of a series of adjacent bits within a message. As long as the burst interference is shorter than the bits of the CRC signature, any corruption will be detected. In order to reduce risk further, PROFIsafe uses a patented procedure—the SIL monitor.

27.4.4 SIL Monitor

As mentioned before, PROFIsafe is not relying on the basic data integrity checks of PROFIBUS DP (see Figure 27.7), or PROFINET IO (see Figure 27.8), or other transmission mechanisms. The entire error detection necessary to attain the required SIL (in factory automation usually SIL3 as the maximum) is advantageously only implemented in the superimposed PROFIsafe protocol. Being dependent on the

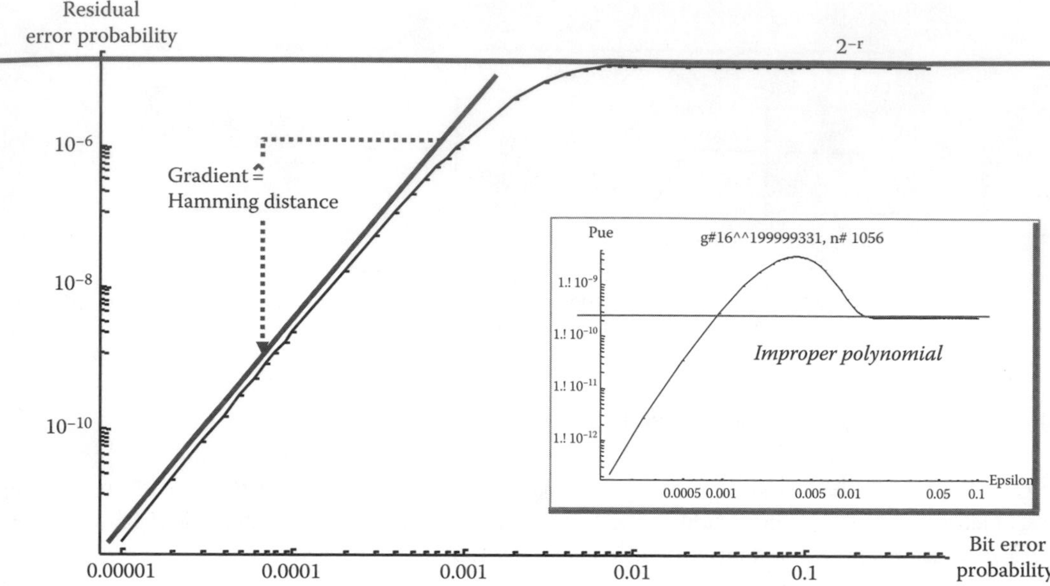

FIGURE 27.5 Examples of proper and improper generator polynomials.

detection mechanisms of the basic transmission systems would require complex verifications for all possible communication configurations (the so-called white channel approach). Thus, a mechanism was created that guarantees the compliance with SIL levels over the service life of a distributed safety-related automation solution, irrespective of the employed components and the configuration (see Figure 27.6).

Here, all influences that may corrupt the PROFIsafe PDUs are looked at as a frequency f_w, whatever the causes of the corruption may be: hardware failures, EMI, etc. For each safety function, the F-Host monitors detected corrupted PROFIsafe PDUs from the associated F-Devices–Slaves via the Status Byte

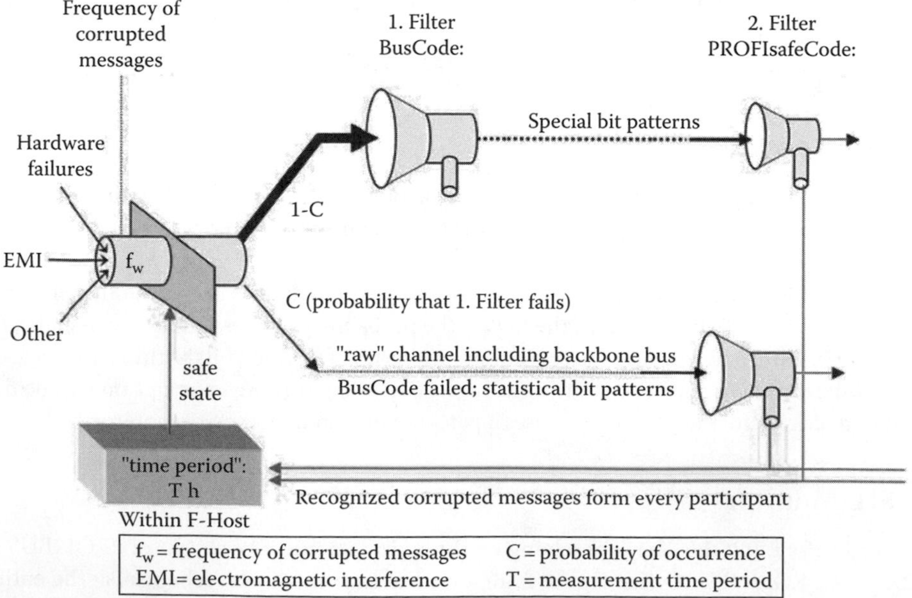

FIGURE 27.6 The SIL monitor.

(see Figure 27.9). It also monitors received detected corrupted PROFIsafe PDUs on its side. Once the frequency of the corrupted PROFIsafe PDUs exceeds a certain limit, the F-Host causes the safe state in the safety function to be activated. The SIL monitor is not a separate component. It is implemented as a part of the PROFIsafe protocol software within the F-Host.

27.4.5 Messages with PROFIsafe PDUs

Up to now, we discussed the mechanisms PROFIsafe employs for safely transporting the PROFIsafe PDUs. This section now deals with the concrete mapping onto the PROFIBUS DP and PROFINET IO communication. Figure 27.7 shows the structure of a PROFIBUS DP message frame.

In this case, PROFIBUS DP achieves data integrity via the parity bit (PB) and the frame checking sequence. Figure 27.8 shows the structure of a PROFINET IO message frame. In this case, PROFINET IO achieves data integrity via a 32-bit CRC signature.

27.4.6 PROFIsafe Parameters

The configuration part of an engineering tool uses the electronic General Station Description (GSD) of a particular F-Device–Slave to arrange for the format of the standard process data to be transferred within the PDU of a message between a PROFINET IO controller–PROFIBUS DP master and its PROFINET IO device–PROFIBUS DP slave. The same happens in case of PROFIsafe PDUs.

There are two groups of safety-related parameters. The *F* parameters (F-Parameter) are associated with the PROFIsafe protocol layer and the *i* (= individual) parameters (iParameter) with the individual safety-related technology firmware of a safety device. The earlier-mentioned CRC signature protection is used not only for cyclically ensuring the data integrity of the process signals and process values but also for ensuring the data integrity of the F-Parameter (such as Codename, watchdog time, and protocol version) and the iParameter stored in the individual slaves. For this purpose, a separate CRC signature value (CRC_FP+) of these parameters is generated and used at start-up and in case of a communication restart after an error.

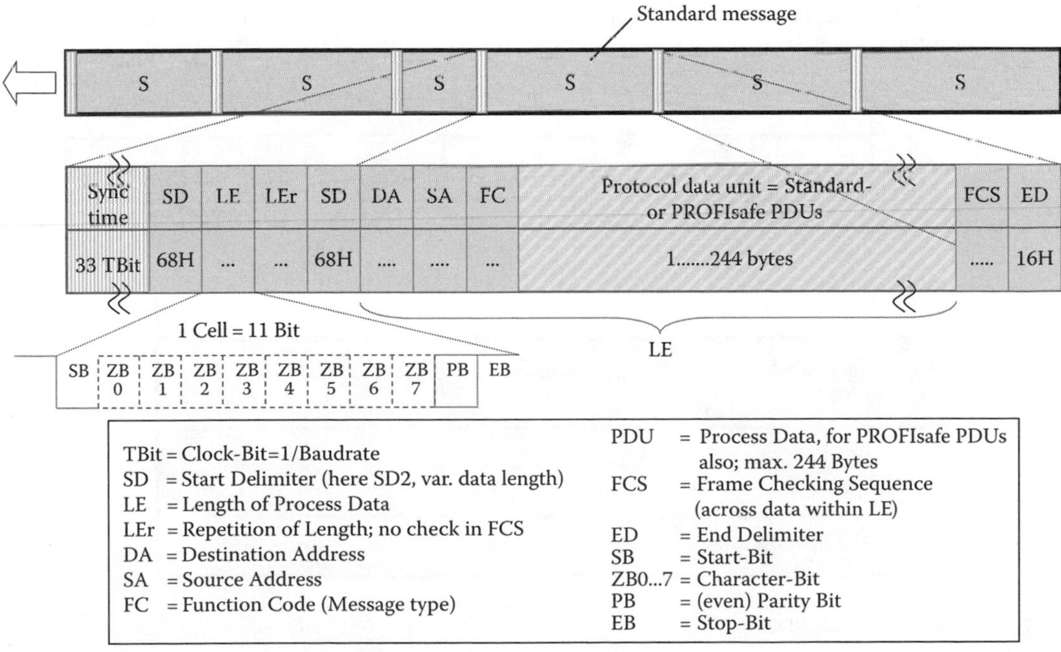

FIGURE 27.7 PROFIBUS DP message structure.

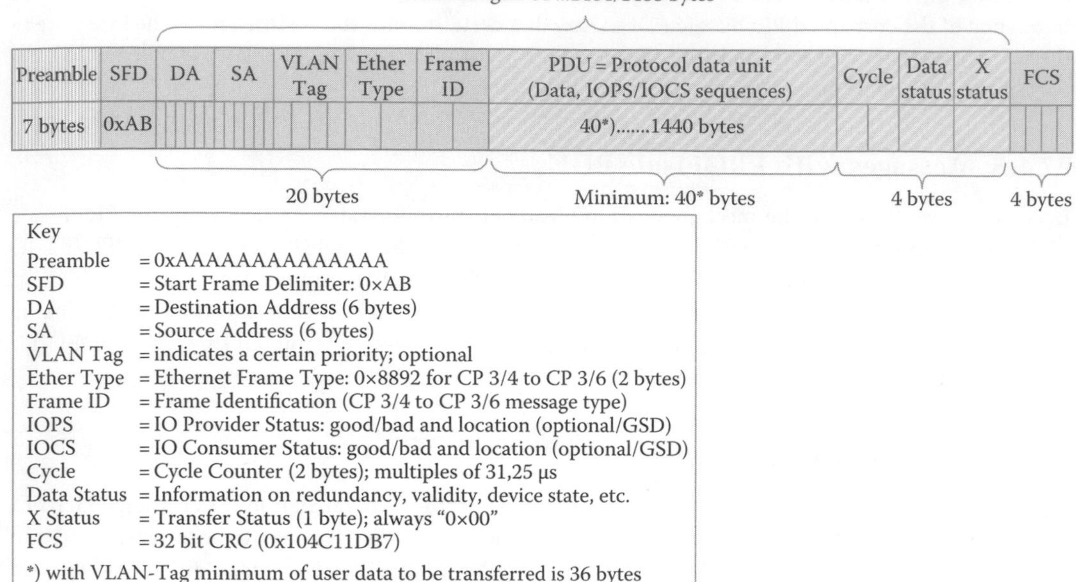

FIGURE 27.8 PROFINET IO message structure.

27.4.7 Structure of PROFIsafe PDUs

The PROFIsafe protocol is located on top of the fieldbus application layer, as shown in Figure 27.3, and realized in the form of state machines on the sender (F-Host*) as well as on the receiver side (F-Device) to cover start-up, regular, and error behavior. In order to synchronize both state machines, the F-Host uses a Control Byte to send commands, and the F-Device a Status Byte to return acknowledgments and additional information for the SIL-Monitor, for example, in case of corrupted PROFIsafe PDUs. The structures of the PROFIsafe PDUs from F-Host and F-Device are

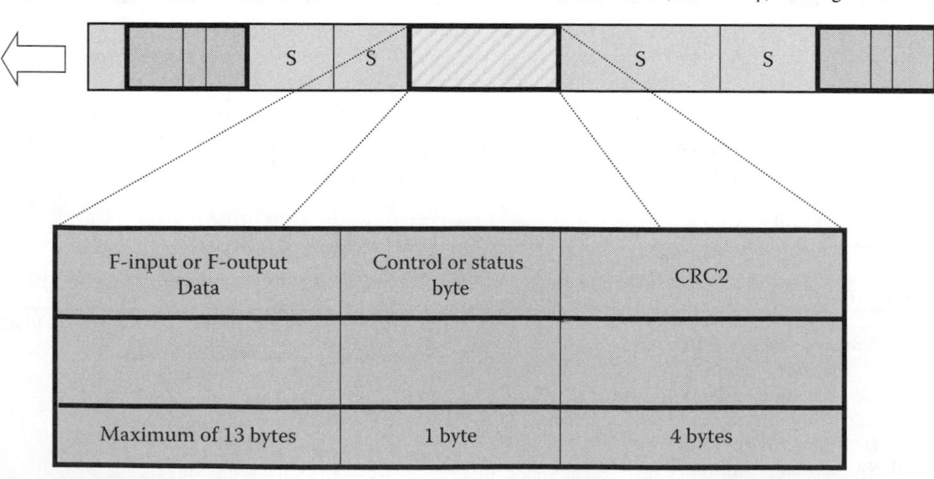

FIGURE 27.9 PROFIsafe PDU structure.

* For simplicity, *F-Host* and *F-Device* are used for both PROFIBUS DP and PROFINET IO.

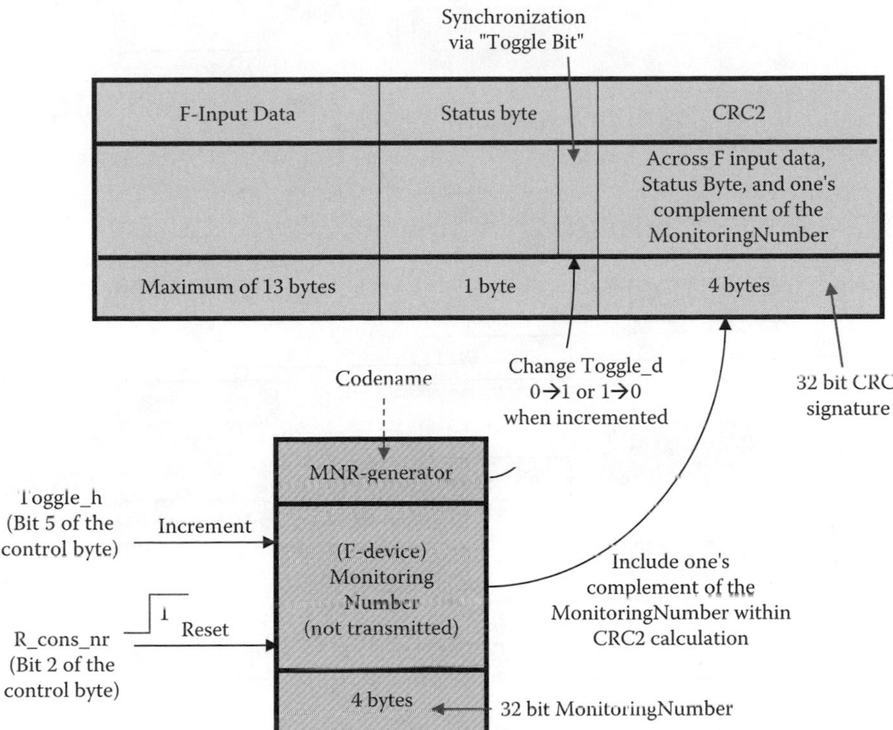

FIGURE 27.10 Implicit method of PROFIsafe via MonitoringNumber (F-Device).

symmetrical as shown in Figure 27.9. The F-Host sends F-Output data (in case of actuators) and the F-Device returns F-Input data (in case of sensors). The maximum length in both cases is 13 bytes. Figure 27.9 also shows 4 bytes of CRC signature to ensure the *Data Integrity*. But obviously, something is missing, isn't it?

As mentioned already, PROFIsafe uses numbering of PROFIsafe PDUs together with a watchdog timer to check *Timeliness* and Codenames to check *Authenticity* of the transmitted safety process data. Researches have shown that both numbering and Codename require each 32 bits. If both would have had to be transmitted in an *explicit* manner, PROFIsafe would have turned out to be very inefficient with a *trailer* of 13 bytes in total for a few F-I/O bits or bytes! Thus, PROFIsafe uses an implicit manner and includes both the numbering and the Codename in the CRC signature (CRC2) calculation. This does not only save bytes, it also prevents this information from being corrupted during transmission. However, since any CRC securing comes with a certain loss of information, a special method is used as shown in Figure 27.10.

A special procedure, called *MNR-Generator*, creates 64-bit-long elements of an exclusive number sequence based on the individual Codename of the particular F-Device and a pseudo random number generator. The initial elements of the sequence are built as follows:

CNNR_64[1] = f (constant, Codename)
CNNR_64[2] = CNNR_64[1] x constant

The cyclically used elements are built as follows (similar to a Fibonacci sequence):

CNNR_64[0] = CNNR_64[1] + CNNR_64[2]
CNNR_64[2] = CNNR_64[1]
CNNR_64[1] = CNNR_64[0]

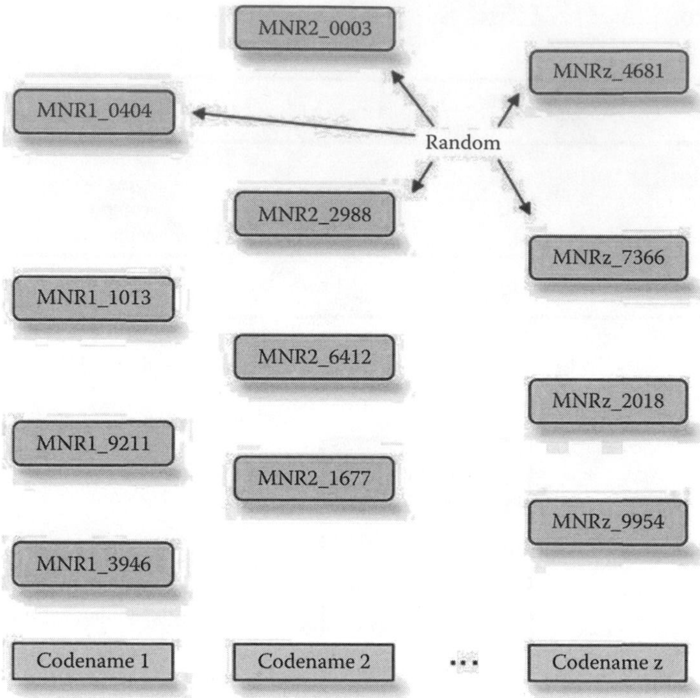

FIGURE 27.11 Principle of the disjoint and random MonitoringNumbers.

The finally used MonitoringNumber (MNR) is built from the 32 most significant bits of CNNR_64[0]. Details of the procedure are specified in [5]. Figure 27.11 shows the principle of the disjoint MNRs.

The MNR enables the receiver to monitor the vitality of sender and communication link, the *Timeliness* of the safety process data, and the *Authenticity* of the sender in one step. Via the acknowledgment mechanism, it is also monitoring the propagation times between sender and receiver (see Figure 27.12).

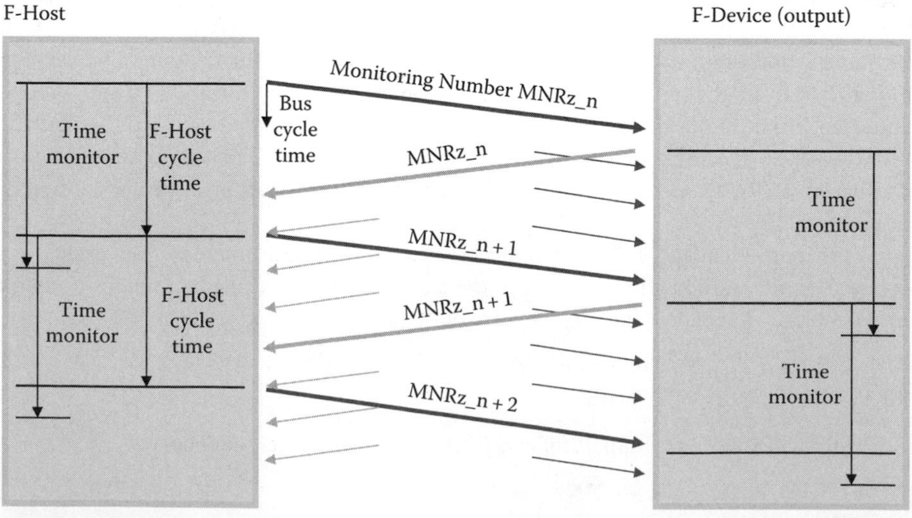

FIGURE 27.12 Mastering propagation times.

27.4.8 Synchronization Means (Finite-State Machines)

In distributed systems, a potential failure of the supply voltage of the devices and their subsequent restart is of particular importance. This hazard potential has been investigated and covered using detailed state and interaction diagrams in the PROFIsafe specification [5].

27.5 Beyond Functional Safety Communication

27.5.1 Safety-Related Programmable Control Logic

PROFIsafe specifies how to connect any type of F-Host to PROFIsafe communications. But it does not dictate how safety-related signals and process values are to be processed within an F-Host. It turned out, however, that programming safety circuits in graphic languages, such as ladder diagram, or function block (FB) diagram according IEC 61131-3, are preferred to any textual language. In the PA world, the representation in the Continuous Function Chart is well accepted. Approved relay-based safety circuits now can be offered as a standard certified software library [6].

There are manufacturers offering the option of retrofitting safety-related user programs for some of their programmable logic controllers (PLCs) with the necessary hardware redundancy (see Figure 27.13).

The functional safety application software has a time diversity structure; it executes in protected areas where it cannot be influenced by the standard application program. Special functional safety I/O modules can supplement the associated standard remote I/O field devices. CNC control systems, too, provide the connection for PROFIsafe field devices.

27.5.2 Commissioning and Repair

In order to be able to *address* the individual signals and/or process values during programming, the fieldbus must be *configured* using an engineering tool. During this configuration, the fieldbus participants arrange for their network address, the input/output signal data structures and types, the transmission rate, etc. In case of F-Devices, additional steps are required supplying the F-Parameters to the PROFIsafe layer: Codename, watchdog time, etc. During start-up of the fieldbus network, the F-Host with its associated PROFINET IO controller or PROFIBUS DP master sequentially provides all devices with the necessary parameters including F-Parameters before it starts cyclic operation [7].

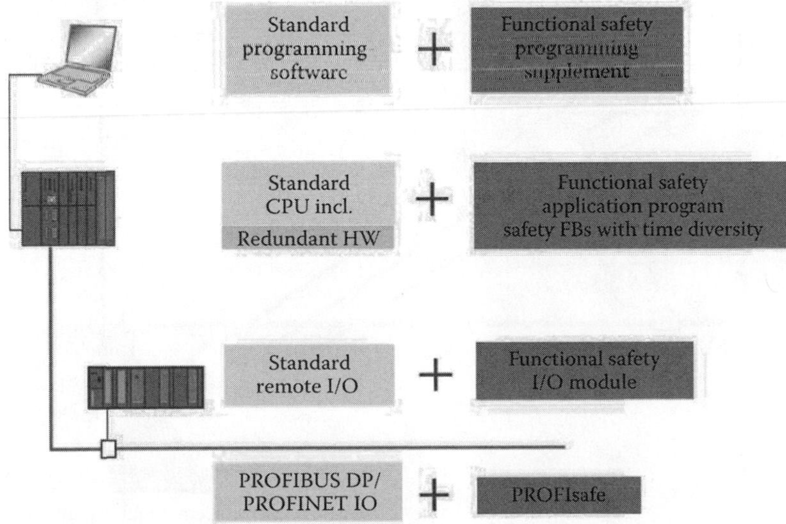

FIGURE 27.13 Possible retrofit options with PROFIsafe.

Caused by the wide acceptance of microcontrollers, various novel devices such as laser scanners and light curtains have turned up in the past even for safety applications that feature a certain complexity and require system support to be able to develop their full functionality. These are, for example, teach-in, rapid device replacement in the event of a malfunction, predictive diagnostics, self-testing and trend analysis, and more.

PROFIsafe has shown ways for mastering these tasks—the so-called three-component model that separates the responsibilities of device manufacturers and system manufacturers by using suitable interfaces. In the simplest case of an F-Device that does not require any technology-related iParameter, a PROFIsafe connection and a corresponding electronic data sheet (GSD file) prove sufficient. A possible example is a level switch. Slightly more effort is required in the case of a safety device with iParameter. It can be connected to a laptop via a separate RS-232, USB, wireless, or other interface. In order to perform diagnostics, it uses a proprietary parameterization and diagnostics program. The program could also be routed to the F-Device using acyclic services of PROFIBUS DP or PROFINET IO that are accessible via the *Tool Calling Interface* [8] or *FDT/DTM* interface technology [9]. Typical representatives of such devices are laser scanner that can be used to protect the entrance to a manufacturing cell. A device manufacturer would have to provide the safety device, a GSD file, and a parameterization and diagnosis program with the appropriate interfaces.

An important customer requirement is the possibility of device replacement without tools in case of faults. PROFIsafe specifies the so-called iPar-Server technology. Figure 27.14 demonstrates the principle steps of the iPar-Server mechanism via an example.

Together with the network configuration and F-Parameterization of an F-Slave/F-Device with the help of the engineering tool, an associated iPar-Server function is instantiated (step 1). The assigned parameterization and diagnosis tool at installation can be launched via an appropriate interface (step 2) from the engineering tool propagating at least the node address of the configured device. Parameterization, commissioning, test, etc., can be executed with the help of the tool (step 3). After finalization, a CRC signature across the iParameter is being calculated and displayed for copy and paste into the entry field of a special F-Parameter ("F_iPar_CRC") within the configuration part of the engineering tool (step 4).

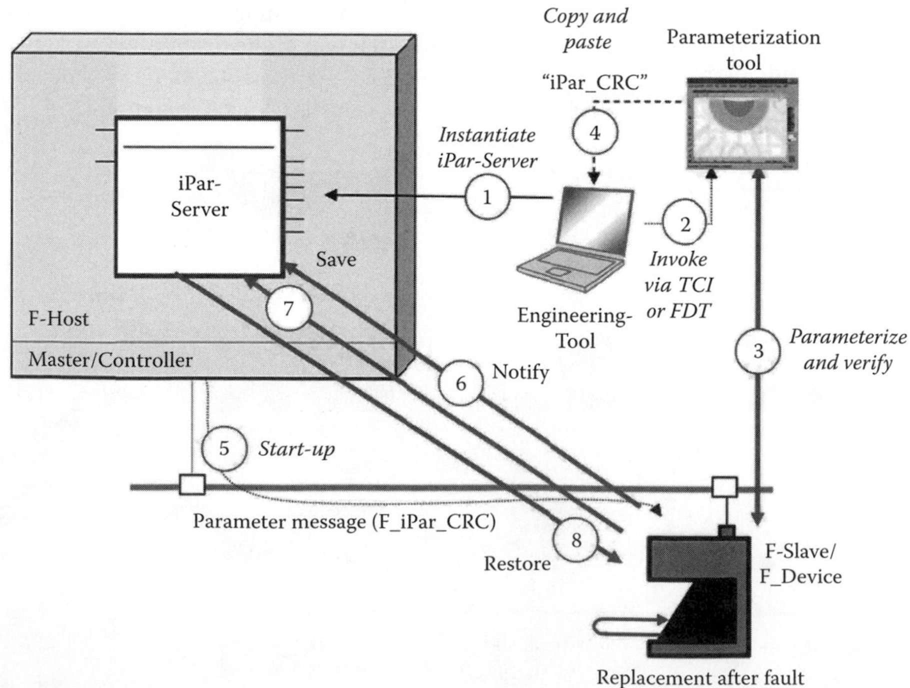

FIGURE 27.14 The iPar-Server for fast device replacement.

A restart of the F-Slave/F-Device is necessary to transfer the "F_iPar_CRC" parameter into the F-Slave/F-Device (step 5). After final verification and release, the F-Slave/F-Device is enabled to initiate an upload notification (step 6) to its iPar-Server instance. The iPar-Server is polling the diagnosis information to interpret the request and to establish the upload process (step 7), which stores the iParameters as instance data within the iPar-Server.

In case of the replacement of a defect F-Slave/F-Device, it receives again its F-Parameters including the "F_iPar_CRC" (same as in step 5) at the start-up. As iParameters normally are missing in a replacement or nonremanent F-Slave/F-Device, it initiates a download notification (step 7) to its iPar-Server instance. The iPar-Server is polling the notification to interpret the request and to establish the download process (step 8). Through this transfer, the F-Slave/F-Device is enabled to provide the original functionality without further software tools.

The most demanding devices are the ones that require user-program-controlled parameter value assignment during production, in addition to the features mentioned earlier. This requires an additional so-called proxy FB in the controller (e.g., F-Host).

27.5.3 Availability

The availability plays a significant role in safety technology. It is a precondition. There is too much of a risk that a frequent—and apparently groundless—*trip* of the safety devices causes the system to shut down (*Bhopal* effect) [10]. There are mainly five areas that influence the availability of a system:

1. Design–assembly of the devices
2. Facility layout and installation
3. Operating conditions
4. Reliability of the components (bathtub curve)
5. Requirements for operator actions

In recent years, most standard automation devices and systems on the market have reached a design and assembly quality standard that ensures high reliability (high meantime between failures). Among the reasons for that are standards, test laboratories, and certifications. However, the availability (up-time in relation to down-time plus up-time) may deteriorate in safety systems that are based on standard systems. Redundant standard components (e.g., two or three controllers) that require synchronization can be the cause. The safety device trips if one component takes longer to execute than the other one due to an external event. This risk exists also in the redundant implementation of sensors whose measured values diverge briefly (discrepancy). PROFIsafe therefore endeavors to support certified (functional safety) F-Devices that take the risk away from the user as much as possible. Standard fieldbus systems, even proven-in-use, without additional mechanisms like PROFIsafe and risk assessments of devices, and the safety functions according to safety standards may not be considered functional safe.

In order to prevent EMI from affecting the operation of a system, the PROFIBUS User Organization have published installation guidelines [11] whose utilization in bidding procedures is recommended, thus creating a clear situation right from the outset. They specify aspects like grounding, shielding, and lightning protection. PROFIsafe makes the satisfaction of these installation guidelines a prerequisite for the operation.

The operating conditions of a system have a strong influence on the reliability–availability. As we all know, a temperature rise of 10 K reduces the service life of an electronic device by 50%. Here, PROFIsafe can merely give recommendations. Devices should be designed for the usage in one of the following zones: office environment, enclosed-type operation within racks, and process participation inside and outside a building.

The means of the *predictive diagnosis* is used in order to avoid unexpected failures toward the end of the service life of a field device. The so-called proxy FBs are provided in the PROFIsafe concept. They can execute, automatically and periodically, test programs in the field device in order to determine its states and the trend (e.g., in electronic shutdown valves that are demanded very seldom). This enables risky units to be replaced at a time when the process is stopped anyway.

The availability of a system is also reduced if the user is unable to handle it properly due to complexity or variety. Solutions with PROFIsafe perfectly fit into the paradigm of fieldbus-based automation systems and thus offer a tremendous risk reduction and increase in flexibility compared to previous solutions. Additional activities on standardization of software user interfaces are complementing the advantages [12].

27.5.4 Status of PROFIsafe-Related Specifications and Guidelines

At Hanover Fair in 1999, the PROFIBUS User Organization had published version 1.0 of the PROFIsafe specification for PROFIBUS DP, followed by version 2.0 in September 2005 incorporating PROFINET IO. Current version 2.6 covers research results on quantitative considerations for the properties of *Timeliness*, *Authenticity*, *Data Integrity*, and CRC behavior in case of the implicit transmission as described in this publication.

All versions have been confirmed by positive reports from the Institute for Occupational Safety and Health of the German Social Accident Insurance (IFA) and the worldwide operating Safety-Assessment Organization (TÜV). Associated guidelines and specifications are described in detail later in this publication. They can be found on the Internet under http://www.profibus.com ([14,16,18,20]).

PROFIsafe was internationally standardized in IEC 61784-3-3.

27.5.5 Standards Catching Up

Figure 27.15 shows an overview of the fieldbus and safety-related standards within the context of PROFIsafe.

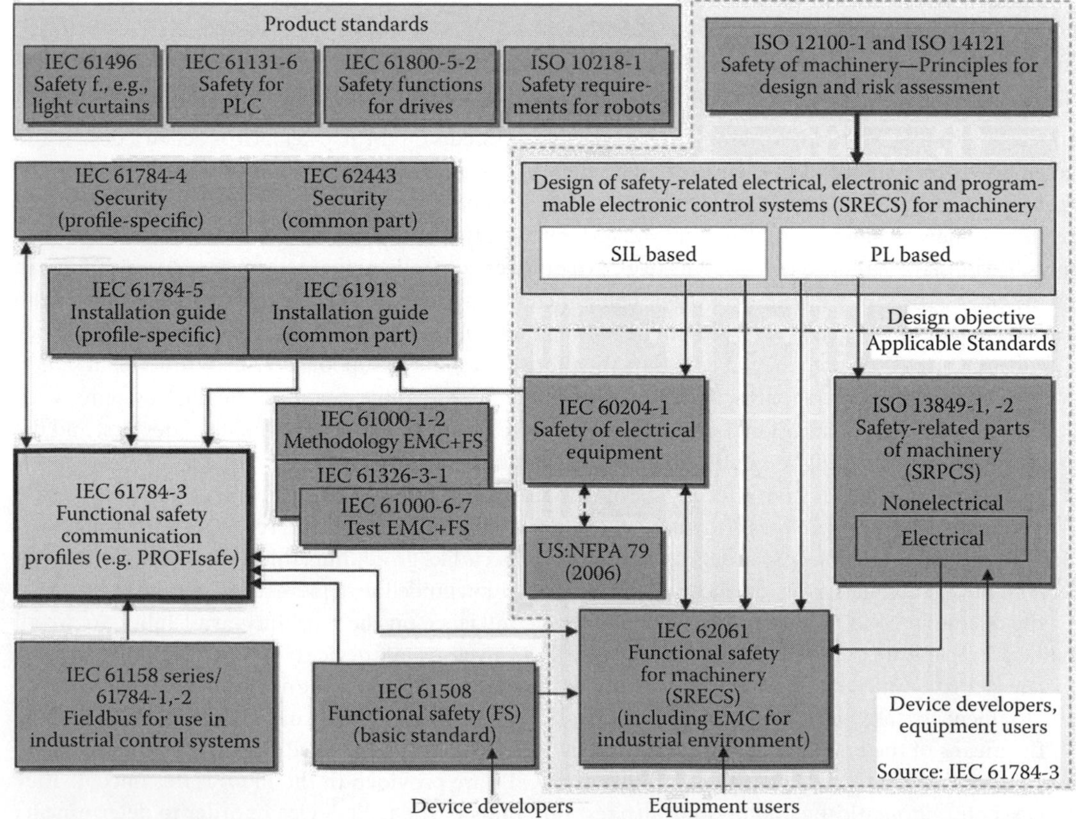

FIGURE 27.15 Relationship between relevant standards for machinery.

The development of PROFIsafe is based on IEC 61508, which focuses on safety functions, on quantitative SILs as opposed to qualitative levels in previous safety standards, and on software development procedures and communication. In order to facilitate the application of the new technology, an IEC 62061 has been created for equipment users, supplemented by IEC 60204-1. The ISO 13849-1 follows the tradition of EN 954. However, it extends its categories by quantitative considerations and introduces performance levels (a–e) instead. It also covers nonelectrical parts of machinery. Recent standards such as IEC 61000-1-2, IEC 61326-3-1, and IEC 61000-6-7 specify the EMC testing procedures and levels. More and more product standards are based on IEC 61508 or ISO 13849-1.

27.5.6 Peculiarities for Different Industries

After 15 years and more, PROFIsafe is very well accepted within factory automation. The discussions about the need of using a separate safety bus in addition to the standard fieldbus for the implementation of distributed safety technology are gone in favor of a single cable solution. Many users are following a migration strategy. First, within existing facilities, the new PROFIsafe technology replaces the relay-based safety technology via a second fieldbus segment. At the next opportunity and after some experience, they switch to the combined solution due to the obvious many advantages such as uniform engineering and diagnosis.

Safety applications in process engineering require a consideration that exceeds the functional safety. Pressure and temperature in a process, for example, cannot always be *monitored* independently of each other. A high availability of sensor functions requires a large degree of design skill and experience. To a large degree, the topic of *proven-in-use* plays a large role.

Up to now, the part of the protection equipment—which mostly represents cases similar to SIL2—has been solved with standard field devices and 4–20-mA transmission technique. Communication faults, such as line interruptions or short circuit, were recognizable, the devices were proven-in use, and safety monitoring was performed in the F-Host system by interpreting redundant signals and voters. In the case of microprocessor-equipped equipment, guidelines like the NAMUR recommendation NE79 defined the necessary prerequisites. These include the proof that the software–firmware has been developed according to generally recognized quality standards. Furthermore, the user expects facilities such as watchdog timer, memory test, and protection against inadvertent parameter changes. The new recommendation NE97, which NAMUR published at the end of 2002, deals with the subject *Fieldbus for Safety Tasks*. Among other things, this recommendation shows two possible solutions that can easily be implemented with PROFIsafe.

A proven-in-use field device, developed according to NE79, connected to PROFIBUS could actually be used for applications up to SIL2, if the possible communication faults that are known in bus operation, such as address corruption, loss, and delay, could be eliminated. As we already know, this is possible, if we employ a PROFIsafe protocol that is implemented in a single-channel manner. In future, this solution permits proven-in-use standard PROFIBUS field devices to be operated in the standard operation mode, or optionally in the safety mode, just by changing a single parameter. The user is at liberty to employ certified products (see Figure 27.16).

In many systems, the installed proven-in-use field devices are retained, and an additional fieldbus wiring is installed via a so-called remote I/O. For this operating mode, too, NE97 provides for a solution.

The PA field devices are usually subject to a strong standardization of device model and device parameters. Parameter value assignment, too, is expected to be as uniform and consistent as possible. To take this into account, PROFIsafe communication has been specified compatibly into the existing *PA device model* [13]. The related specification [14] especially deals with the following topics:

- Switching between standard and safety modes
- Preparation and commissioning phases
- New parameters in the Physical Block of the PA device model
- Safety-related data structures for the cyclic data exchange
- Configuration data

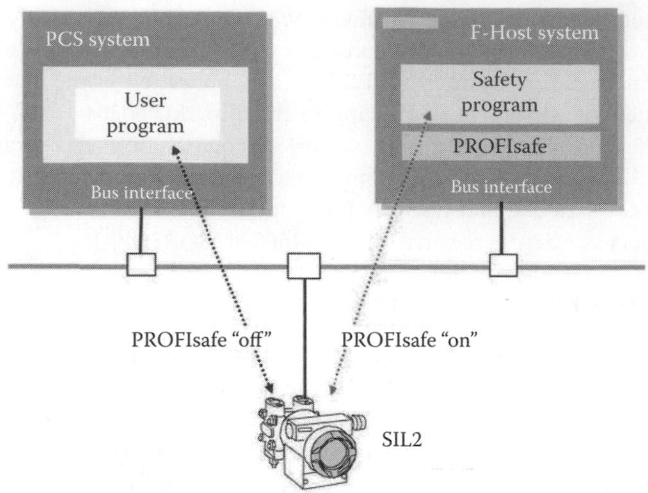

FIGURE 27.16 PROFIsafe in field devices for process automation.

As already mentioned, the field devices should be usable in the standard mode and in the safety mode. It is planned to achieve this switch by corresponding parameter settings in the related GSD file, and thus during the start-up of the field device.

Due to the contingency that parameters have to be edited during operation, particular commissioning phases have been designed that permit existing parameter assignment tools based on Electronic Device Descriptions to be used [15]. In this case, the parameter assignment usually is performed offline, and the device is sealed by soldering or similar procedures.

Additional data formats for the PROFIsafe PDU supplement the standardized data formats for the PA device model for cyclic data exchanges.

27.5.7 Environmental Conditions

In the meantime, further specifications have been published since most safety standards still ignore fieldbus operations. One of these is defining the boundary conditions for safety communications [16]. It deals with the topics of installation guidelines, electrical safety (overvoltage shock protection—SELV–PELV), power supply units, and immunity to EMI. The other is describing a policy for all PROFIBUS member companies and providers of safety-related devices dealing with responsibilities and the quality of products and services [17].

27.5.8 Test and Certification

The PROFIsafe mechanisms are based on finite-state machines. Thus, it was possible via a validation tool for finite-state machines to mathematically prove that PROFIsafe is working correctly even in cases where more than two independent errors or failures may occur. This was systematically achieved by generating all possible cases for *test-to-pass* and *test-to-fail* situations. They have been extracted as test cases for a fully automated PROFIsafe layer tester, which is used to check the PROFIsafe conformance of safety-related devices. It is part of a three-step procedure within the overall certification process (see Figure 27.17) [18].

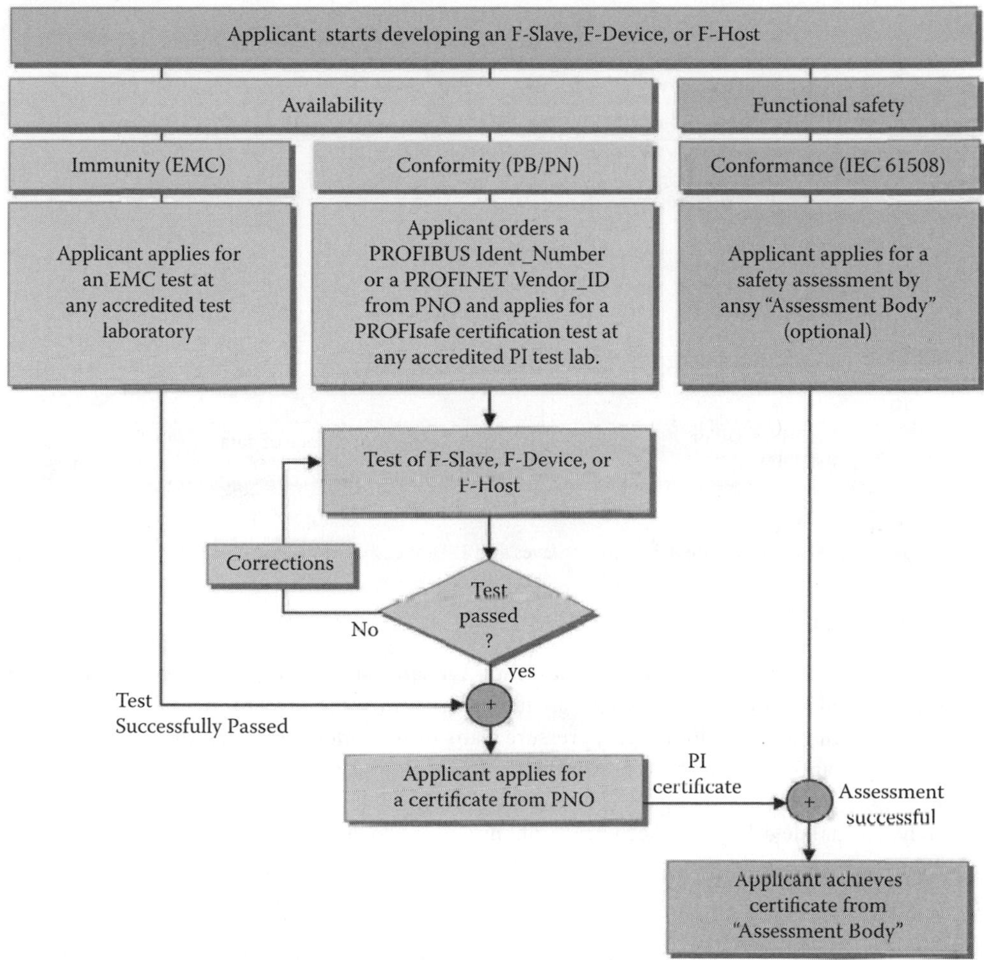

FIGURE 27.17 The PROFIsafe certification procedure.

27.5.9 Development Tools and Support

Using PROFIsafe in factory automation with short reaction times, and—on the other hand—in hazardous areas in PA with lowest power dissipation, places contradictory tasks. Since most modern devices use microprocessors, implementing the PROFIsafe procedure in software is an obvious solution. The operating conditions determine the selection of the microprocessor with respect to high performance or low power dissipation. PROFIsafe, ideally, requires only some kilobytes of free memory: no additional power supply and no additional space in the confined enclosures. PROFIsafe becomes a part or the device-specific safety software.

Immediately after completion of the first specification, the company consortium started a joint development of the *generic PROFIsafe driver software* for F-Devices. Generic in this context means that the driver has been implemented in ANSI-C and according to the encoding rules for safety technology for different microprocessors and C compilers. Although it should preferably be used in dual-channel mode (see Figure 27.18), it may also be used in single-channel mode (see Figure 27.16). A complete development kit (PROFIsafe Starterkit) is available for F-Devices. Upon application for the test at accredited PROFIsafe test laboratories, the F-Device tester software is available for early testing in the development phase.

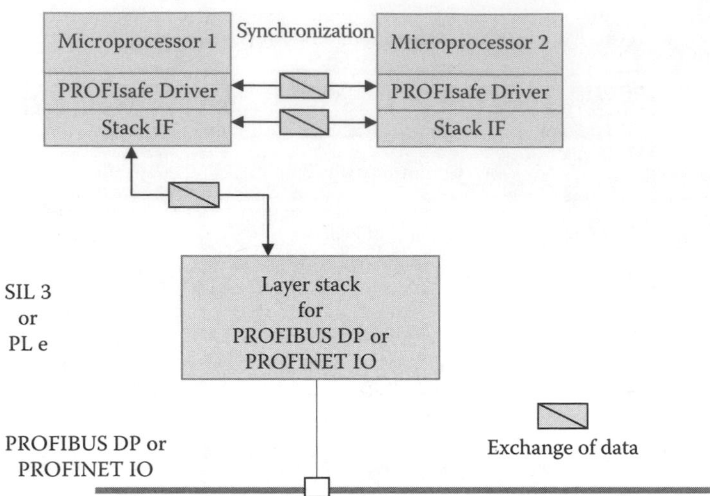

FIGURE 27.18 Generic PROFIsafe driver for F-Slaves and F-Devices.

27.5.10 Products

A large number of PROFIsafe products are currently available such as safety-related PLCs from more than five renowned manufacturers as well as numerical controllers, remote-I/O for IP20 and IP67, laser scanners, light curtains, door locks, pressure transmitters, motor starters, drives with integrated safety, robots, gas and fire sensors, etc. Worldwide, there are more than 2.5 million PROFIsafe nodes installed.

For further actual details, please see www.profibus.com and www.profisafe.net.

27.5.11 Prospects

The sophisticated safety-related devices mentioned in this publication with its many possibilities to use iParameters to adapt to all kinds of safety applications in a flexible manner are causing the requirement for continuous automatic monitoring of security, safety, and data integrity. Thus, regulations like 21 CFR 11 [19] defined "rules for the use of secure, computer-generated, time-stamped audit trails to independently record the date and time of operator entries and actions that create, modify, or delete electronic records. Record changes shall not obscure previously recorded information. Such audit trail documentation shall be retained for a period at least as long as that required for the subject electronic records and shall be available for agency review and copying."

The device-independent Profile Guideline, Part1 "Identification & Maintenance Functions" [20], suggests using asset identification parameters and the parameter SIGNATURE to allow parameterization tools to store a *security* code as a reference for a particular parameterization session. Audit trailing tools can acyclically retrieve the code for integrity checks at any time via the fieldbus (see Figure 27.19). Together with the *asset identification parameters* consisting of MANUFACTURER_ID, ORDER_ID, and SERIAL_NUMBER, the SIGNATURE parameter allows unambiguousness of the session data.

PROFINET IO is gaining recognition on the market. Many car manufacturers are standardizing on PROFINET and PROFIsafe. Due to its *black channel* approach, the PROFIsafe technology can be used on both PROFIBUS DP and PROFINET IO without changes. No matter which one a manufacturer decides to use as the basis, with PROFIsafe, he is on the safe side.

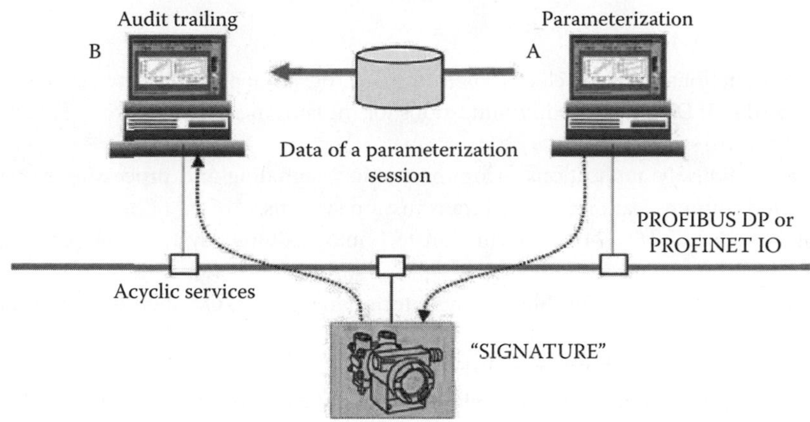

FIGURE 27.19 Audit trailing for secured safety parameters.

Abbreviations

CRC	Cyclic redundancy check
DP	De-centralized peripherals
DTM	Device type manager
EMC	Electromagnetic compatibility
EN, prEN	European standard, preliminary…
F, FS	Failsafe or functional safety
FB	Function block
FDT	Field device tool
GSD	General Station Description (electronically readable data sheet)
HW	Hardware
IEC	International Electrotechnical Commission
IF	Interface
IFA	Institute for Occupational Safety and Health of the German Social Accident Insurance
I/O	Input/output
IP	Ingress protection, e.g., IP20
ISO/OSI	International Standards Organization/Open Systems Interconnection (Reference Model)
NAMUR	Association of users of process control technology
PA	Process automation
PB	PROFIBUS DP or parity bit
PDU	Protocol data unit
PI	PROFIBUS and PROFINET International (user organization)
PL	Performance level (ISO 13849-1)
PLC	Programmable logic controller
PN	PROFINET IO
SIL	Safety integrity level (IEC 61508)
SW	Software
TÜV	Safety Assessment Organization

References

1. IEC 61508: Functional safety of electrical/electronic/programmable El. Safety-related systems.
2. IEC 61158/61784: Digital data communications for measurement and control—Fieldbus for use in industrial control systems.
3. IEC 62288-1: Railway applications—Communication, signaling and processing systems—Part 1: Safety related communication in closed transmission systems.
4. ISO/OSI Model: ISO/IEC 7498 Information technology—Open systems interconnection—Basic reference model.
5. PI specification: PROFIsafe: Profile for safety technology, V2.6, October 2013, PROFIBUS Order No. 3.092.
6. www.plcopen.org (accessed August 8, 2013).
7. PI book: M. Popp: The new rapid way to PROFIBUS DP, PROFIBUS Order No. 4.072.
8. PI specification: Tool Calling Interface, V1.1, October 2008, Order No. 2.602.
9. IEC 62453: Field device tool interface specification.
10. P. Gruhn, H. Cheddie: Safety shutdown systems: Design, analysis and justification, ISA 1998.
11. Installation guidelines for PROFIBUS DP and PROFINET IO: Planning, installation, commissioning, www.profibus.com (accessed August 8, 2013).
12. IEC 62453-61: Field device tool interface specification—Device type manager (DTM) style guide for common object model.
13. PI specification: PROFIBUS PA profile for process control devices, V 3.02, April 2009, PROFIBUS Order No. 3.042.
14. PI specification: Amendment1 to the PROFIBUS profile for process control devices V3.02: PROFIsafe for PA devices, V1.01, March 2009, PROFIBUS Order No. 3.042.
15. Specification for PROFIBUS device description and device integration, Vol. 2: Electronic device description, V2.0, 2013, PROFIBUS Order No. 2.152.
16. PI specification: PROFIsafe—Requirements for installation, immunity, and electrical safety, V2.5, March 2007, PROFIBUS Order No. 2.232.
17. PI guideline: PROFIsafe Policy, V1.5, July 2011, PROFIBUS Order No. 2.282.
18. PI specification: PROFIsafe—Test specification for f-slaves, f-devices, and f-hosts, V2.2, PROFIBUS Order No. 2.242.
19. Title 21 Code of Federal Regulations: U.S. Food and Drug Administration, www.fda.gov (accessed August 8, 2013).
20. PROFIBUS profile guidelines, Part 1: Identification & maintenance functions, V1.9, PROFIBUS Order No. 3.502.

28

SafetyNET p Protocol

Marco Cereia
National Research Council of Italy

Jochen Streib
Safety Network International e.V.

Reinhard Sperrer
Safety Network International e.V.

28.1 Introduction **28**-1
28.2 SafetyNET p Physical Layer **28**-2
RTFL Topology • RTFN Topology • Cables and Connectors
28.3 Communication Model **28**-5
General Frame Structure • Cyclic Data Channel • Message Channel • Communication Management • SafetyNET p Application Layer
28.4 Precise Clock Synchronization **28**-15
28.5 Safety Communication Model **28**-16
Black Channel Approach in SafetyNET p • SafetyNET p Error Detection Mechanisms • Safety Communication Layer • Safety Application Layer
References **28**-24

28.1 Introduction

SafetyNET p [1] is a communication protocol developed by Pilz GmbH & Co. KG, which allows the construction of industrial Ethernet-based networks and is suitable for both real-time communications and safety-oriented applications. The protocol enables the transmission of safety-related data on the same cable used for the transfer of control or process data. In fact, SafetyNET p makes available a logical safe channel that is certified in accordance with the safety integrity level 3 (SIL 3) of IEC 61508 [2] and can be used to transfer safety data without the need of a physical dedicated communication line.

Combining safe and nonsafe communications over the same medium has several advantages: first of all, the reduced amount of cables needed to connect the network elements makes the wiring of components simpler, thus reducing costs for setup and maintenance. Moreover, the ability of connecting safe and nonsafe devices to the same network makes a number of non-safety-related functions (e.g., remote diagnostic, data logging) available for safe devices and at the same time allows standard devices to access safety data for further non-safety-related processing tasks.

SafetyNET p is a multimaster bus system where all devices are given the same rights, and therefore, it does not require a centralized controller. It has been designed to operate according to a publisher–subscriber model, where each device publishes its relevant data and can subscribe, at the same time, the reception of data needed to carry out its tasks, which are produced by other entities in the network.

SafetyNET p implements two communication models, known as real-time frame line (RTFL) and real-time frame network (RTFN).

RTFL has been designed for fast intracell communications, where real-time performance is of primary importance. It is based on a linear topology, and the transport of data is supported through standard Ethernet (OSI layer 2) frames. This communication model enables a minimum bus scan time of 62.5 μs and limits jitters to 100 ns or less, making the protocol suitable for highly demanding control applications. Low jitter is achieved by means of a specialized synchronization mechanism, the so-called precise clock synchronization (PCS), which makes the protocol suitable for distributed processing that requires a high level of simultaneity.

When real-time requirements are not so tight and a cycle time over 1 ms is sufficient, the RTFN communication model can be used to obtain the maximum degree of flexibility and integration with other standard Ethernet-based protocols likely adopted at higher levels of the factory hierarchy. In fact, RTFN can use the standard UDP/IP protocol to transport data, allowing both the routing of frames among different networks and the integration with the existing communication infrastructures. RTFN is usually used for communication from cell to cell and for the connection of cells to the office network by implementing star or tree network topologies. Synchronization of RTFN devices and RTFL networks is made possible by means of the precision time protocol (PTP) [3] with the advantage that it is supported by many standard Ethernet components such as switches.

Regardless of the communication model used to transport data, SafetyNET p supports two distinct logical channels: [4,5] the cyclic data channel (CDC) and the message channel (MSC). When devices are actually in operation, most data are transmitted at regular time intervals or, in other words, their transmission is cyclic. CDC is dedicated to this type of information.

MSC is used to transmit acyclic data such as initialization parameters, diagnostic information, and device programming instructions. The interface of applications to the communication system [6,7] is based on the popular CANopen application layer [8].

Since safety data are cyclic in their nature, the safety application layer [9,10] is built on the top of CDC and relies on the black channel concept: the protocol implements a set of safety-related transmission functions and also monitors the integrity of the safe communication channel, so as to ensure that even the tight requirements of SIL 3 are satisfied. With the exception of the actual safe subscribers, for example, safe sensors, safe PLCs, and safe actuators, all other network components such as switches and cables are regarded as non-safety-related and can be standard devices. Such an idea was introduced in IEC 62280-1 [11,12] and basically originates from the railway signaling technology.

Due to its flexibility and completeness, the protocol is suitable for a wide range of applications that span from factory and safe automation to remote control and protection of railway networks.

28.2 SafetyNET p Physical Layer

As mentioned before, SafetyNET p specifies two communication variants: RTFL for fast intramachine communications and RTFN for the interconnection of different machines and communications with the office networks. A typical structure of a SafetyNET p network is shown in Figure 28.1.

Both SafetyNET p communication variants are based on the well-known Standard Ethernet IEEE 802.3 technology [13]. Since its invention, Ethernet has been evolving to meet the ever-increasing demand for flexibility and transmission bandwidth. According to the IEEE 802.3 specification, standard Ethernet foresees many possible variants. SafetyNET p relies on 100 BASE-TX, which is the most common 100 Mb/s implementation. This allows the use of standard components and makes compatibility easier (one of the most important reasons for Ethernet-based field communications).

28.2.1 RTFL Topology

For highly synchronized communications with fast process data rates, the use of standard Ethernet hardware is not sufficient. SafetyNET p meets these requirements by using special active components that allow on-the-fly processing of 100 BASE-TX frames. Every RTFL device includes one special component and is equipped with two Ethernet ports that can be arranged in a preferred line topology as depicted in Figure 28.2. Each RTFL device is addressed individually by its predecessor in the line using the MAC address, which is modified on the fly while a frame is being processed.

The basic communication is initiated by a special device called root device (RD). At each communication cycle, RD creates an RTFL Ethernet frame and sends it along the line. All the cascading devices,

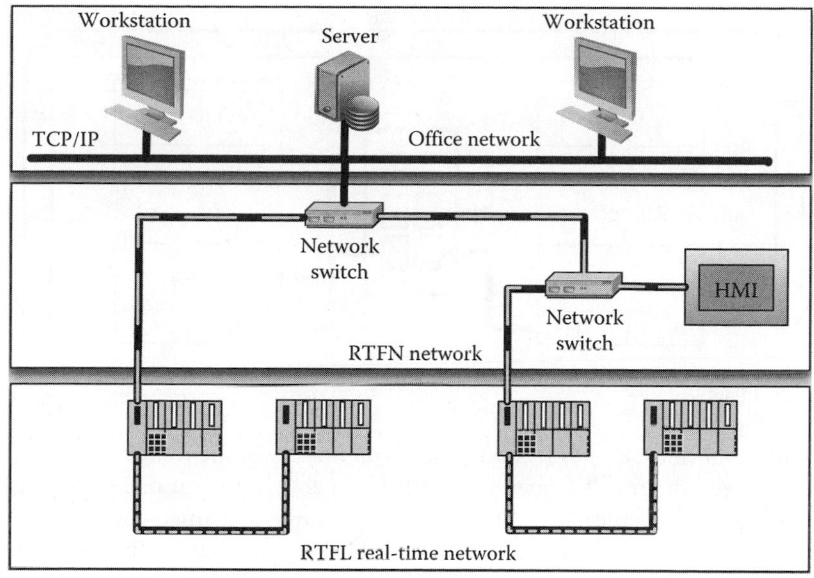

FIGURE 28.1 Typical structure of a SafetyNET p network.

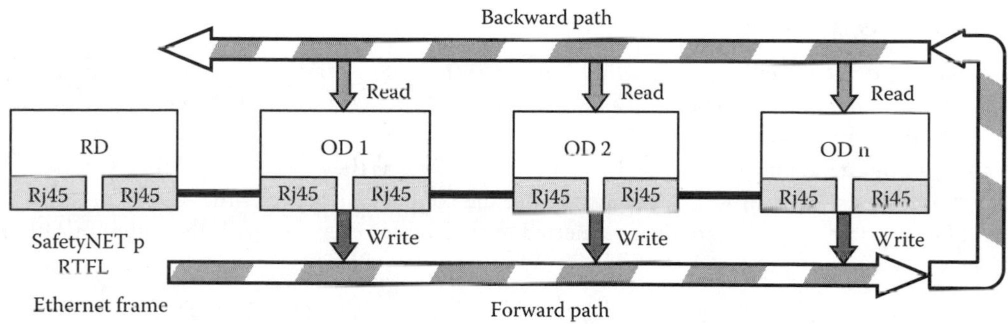

FIGURE 28.2 RTFL physical line.

also known as ordinary devices (ODs), append their data to the frame. When the frame reaches the end of the line (last device), it is forwarded on the way back to RD. While traveling in this direction, relevant data are then read by the different ODs. Each RTFL network requires just one RD.

Ordinary switches store each frame for the time needed to decide which port has to be used for forwarding. This is necessary in order to avoid collisions when frames, which arrive to the switch at the same time, have to be sent out through the same port. In a linear topology, however, this event cannot occur, and cut-through switches can be adopted without causing problems. By contrast, if a star topology is needed for an RTFL system, store-and-forward switches are required, which introduce additional latencies and jitters. It is worth noting that even if the physical topology is a star as depicted in Figure 28.3, the logical topology is a line also in this case. In fact, each device is enabled to forward the frame only to either its logical successor or logical predecessor, depending on whether the frame is traveling on the forward or backward path, respectively.

During start-up, management frames are exchanged between the bus master and the ODs. After entering the steady-state operating cycle, frames are always sent along the (physical) linear structure, with point-to-point transmissions between any two adjacent devices, but virtually emulating broadcast transmissions where all devices are enabled to access the content of the same frame.

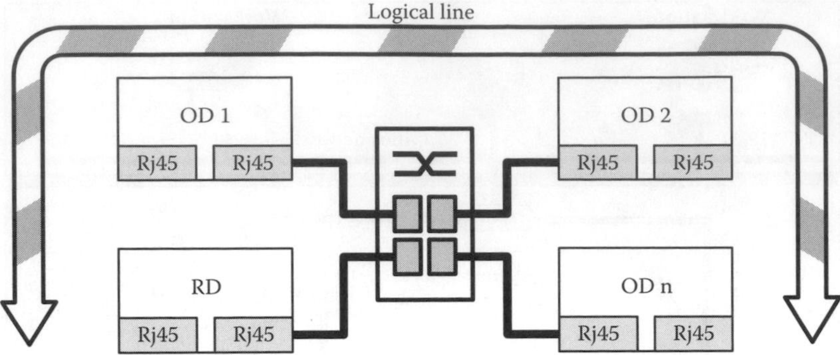

FIGURE 28.3 RTFL logical line.

The communication efficiency of Ethernet depends on the size of the transmitted frames. For small data, the ratio between the frame header and the payload size is not optimal. SafetyNET p RTFL is designed to minimize the overhead by using frames as common containers for data from all devices. This means, for instance, that a single Ethernet frame (which is processed cut-through with only a small delay) can carry all process data needed by devices. Minimizing header overheads is important to reach cycle times as short as 62.5 μs, which are needed especially in motion control applications.

28.2.2 RTFN Topology

RTFN allows SafetyNET p to be used within standard Ethernet networks without the adoption of special hardware. Communication occurs on the top of layers 2 (MAC) and 4 (UDP) and is therefore independent from the physical media.

Standard Ethernet topologies can be used for RTFN as long as they meet the real-time requirements of the application. A typical RTFN use is granting the connection between the factory network and the RTFL subsystems. Devices are then connected to switches or even routers if the communication is UDP-based.

28.2.3 Cables and Connectors

The ability to control the connection technology in an industrial environment is a key criterion in gaining acceptance for an industrial communication system among users. Ethernet connectors commonly used in offices can also be suitable in this case. Essentially, three types of connectors are specified for SafetyNET p, but others may be adopted to fit in well with specific application needs.

28.2.3.1 IP20 Connector

The RJ45 mating interface, which is popular in the office environments (Figure 28.4a), is used in the control cabinets, with a version that has been optimized for the field level. This connector is more robust and easier to install in the field than its office counterpart.

28.2.3.2 IP67 Connector

Two types of connectors are certified for use outside the control cabinet in an IP6x environment, that is,

1. Push–pull connectors with RJ45 mating interface (Figure 28.4b)
2. M12 round connectors, D-coded (Figure 28.4c)

For Gigabit Ethernet also, M12 round connecters X-coded can be used with SafetyNET p.

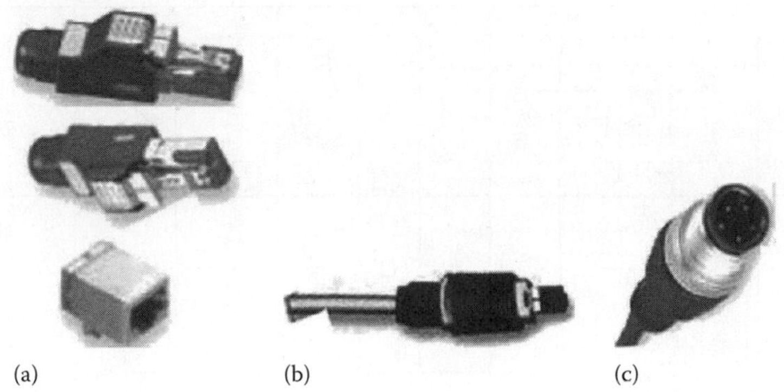

FIGURE 28.4 Connectors for SafetyNET p. (a) RJ45 connector, (b) RJ45 push–pull connector, and (c) M12 D-coded.

28.2.3.3 Cabling

Cabling of SafetyNET p devices is based on the guidelines published in the IAONA "Industrial Ethernet Planning and Installation Guide" document [14]. IAONA created this guide because the Ethernet cabling strategies, originally conceived for building installations, were unsuitable for production environments. SafetyNET p uses Industrial Ethernet cables, which conform to CAT 5 as a minimum, with an AWG22/7 cable cross section. In this way, it is possible to achieve a maximum cable run of 100 m between any two SafetyNET p devices.

28.3 Communication Model

Protocols for RTFL and RTFN have been designed to offer the same set of services for both cyclic data exchanges and acyclic communications [4,5].

For both communication models, two logical channels for data exchange are specified: a CDC to transfer cyclic process data and an MSC to transfer nontime-critical data such as diagnostic information and configuration parameters. Message transmissions on the CDC are not confirmed, while transfers on the MSC can be invoked by means of both confirmed and unconfirmed services.

Logical channels are implemented at the frame level: every SafetyNET p frame is characterized by a 2-byte header field that univocally identifies the logical channel used in the communication. Even if all the frames are transmitted on the same physical line, SafetyNET p devices have the illusion to communicate on two distinct channels.

In addition, SafetyNET p specifies two further acyclic services: a communication management service (CMS) to perform the initialization of a SafetyNET p network and a time synchronization service (TSS) for node synchronization. TSS relies on MSC to carry out its functions and is available only for RTFL-based communications.

Besides differences at the physical layer, as pointed out in Section 28.2, the two SafetyNET p communication techniques mainly differ for both the interaction models adopted in cyclic communications and the way frames are transported.

RTFL is based on a producer–consumer paradigm, where producers make data available to zero or more consumers. In this case, SafetyNET p protocol data units (PDUs) are transported directly in the payload field of standard Ethernet frames as depicted in Figure 28.5. SafetyNET p PDUs carried by Ethernet frames are identified by the value 0x9C40 encoded in the EtherType field. This number, in fact, is the unique identifier that has been assigned to SafetyNET p telegrams by the IEEE EtherType Field Registration Authority [15].

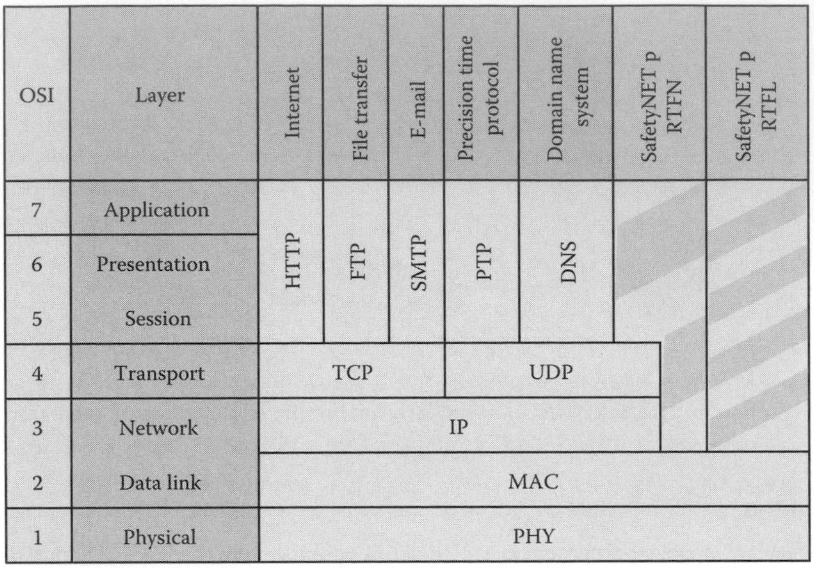

OSI	Layer	Internet	File transfer	E-mail	Precision time protocol	Domain name system	SafetyNET p RTFN	SafetyNET p RTFL
7	Application	HTTP	FTP	SMTP	PTP	DNS		
6	Presentation							
5	Session							
4	Transport	TCP			UDP			
3	Network	IP						
2	Data link	MAC						
1	Physical	PHY						

FIGURE 28.5 Mapping of SafetyNET p on the OSI model.

At every communication cycle, RD initiates the data transmission by generating an RTFL Ethernet frame and by sending it to the first OD in the line. Each OD receives the frame from its predecessor, writes its data into the frame, and forwards the frame itself to the next (downstream) OD. Each device knows the MAC addresses of its predecessor and successor in the line and uses this information to correctly forward the Ethernet frame. The last OD device in the line, after writing its data, reads its relevant data (if any) already inserted in the frame by its predecessors and starts the transmission of the frame in the backward direction. On the backward path, each OD reads the relevant data that have been published by other devices and forwards the frame to its predecessor, until RD is reached. In this way, data can be produced and consumed in the same communication cycle. Each communication cycle consists of one MSC PDU followed by one or more CDC PDUs.

RTFN, instead, adopts a publisher–subscriber interaction model for cyclic data exchange, where each publisher makes data available to a group of one or more subscribers. The model relies on CMS to set up the direct links between publisher and subscribers. Subscribers use the CMS *subscribe request* message to join the publication of data of their interest. Publishers confirm any request with the CMS *subscribe acknowledgment* message. Once the direct links have been created, publishers make cyclic data available to subscribers by sending CDC packets (one for each subscriber).

In RTFN, cyclic PDUs can be transported by either standard Ethernet telegrams or by a higher-level protocol such as UDP/IP as depicted in Figure 28.5. SafetyNET p PDUs carried by UDP frames have to be addressed to port 0x9C40, which is the unique port number assigned to SafetyNET p by the Internet Assigned Numbers Authority [16]. The direct MAC frame transport mechanism reduces frame-processing delays, while the transport mechanism based on the UDP/IP protocol has the advantage of allowing communications among different subnetworks even if segmentation is performed by routers. MSC PDUs are transferred only by means of UDP/IP datagrams.

28.3.1 General Frame Structure

The general structure of SafetyNET p frames consists of a 1-byte header section and a variable-size payload [5]. The header contains a code that identifies either the logical channel (CDC or MSC) used for the communication or a specific message of the CMS. The content of the payload field depends on the header code, and its size can vary from 0 to 1499 bytes.

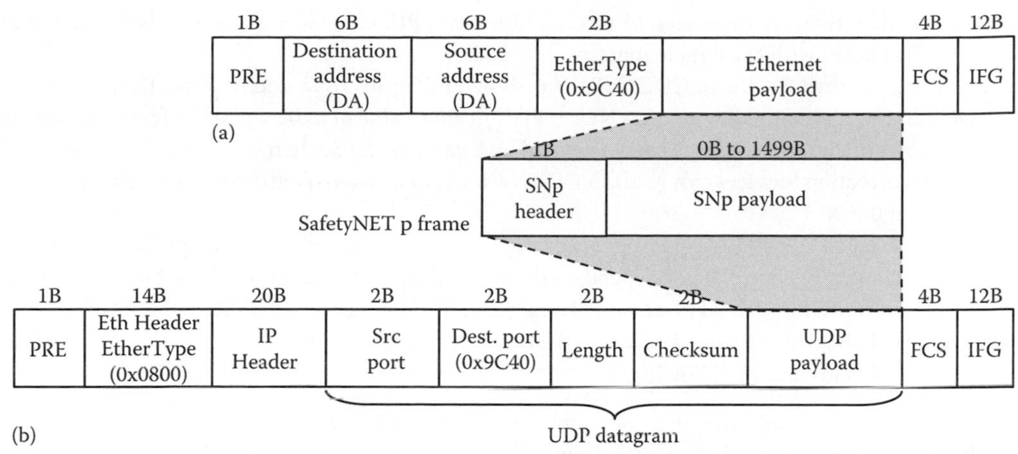

FIGURE 28.6 Mapping of a SafetyNET p frame onto (a) Ethernet frames and (b) UDP datagrams.

In the RTFL communication model, the CDC (CDCL) is assigned the header values 0x02 and 0x03, to write or read the frame to/from the channel (forward/backward direction), respectively. The header values 0x00 and 0x01 identify the MSC (MSCL) for writing and reading messages.

In RTFN, the CDC (CDCN) is identified by the header value 0x60 while the value 0x70 is used for communication on the MSC (MSCN).

The CMS uses a set of messages to perform network initialization and management operations. Each communication model has its own set of services, and each message is identified by its own unique frame header value, since communication management information is transferred with no reference to any logical channel.

SafetyNET p frames can be transported as the payload of either standard Ethernet frames or UDP/IP datagrams, depending on the communication model used to build the network segment. The general structure of a SafetyNET p frame and its mapping onto Ethernet frames and UDP datagrams is shown in Figure 28.6a and b.

28.3.2 Cyclic Data Channel

The CDC is used to transfer cyclic process data. The way the protocol implements the data transfer in the CDC depends on the communication model used in the network segment.

In RTFL, the CDC is a reserved section of the Ethernet frames. In RTFN, instead, the CDC is based on a cyclic sequence of point-to-point communications between couples of devices.

CDCN messages can be transported at either the MAC or UDP level. Each publisher knows the IP or MAC addresses of devices that are subscribed to the reception of its messages, so it is able to create several unidirectional communication links (one for each subscriber) to deliver data consequently. Packets transmitted via the CDCN are unconfirmed, and different cycle times can be used for each link. Only a base RTFN cycle time has to be specified, to set a lower limit on the frequency of CDC message transmissions.

In both communication models, each cyclic data is characterized by a packet identifier (PID), which must be unique in the network segment. If a cyclic datum has to be transferred between different networks, its PID must be unique for all the connected segments.

In RTFL, PIDs allow consumers to identify the sections of the cyclic frame containing data they have to read.

In RTFN, each device knows the set of PIDs that other devices in the network are able to publish. Subscribers use this information to request the setup of unidirectional links with those publishers that provide cyclic data they are interested to receive. Moreover, PIDs enable the correct association of any received cyclic packet with the data it contains.

The structure of the CDCL and CDCN PDUs is shown in Figure 28.7a and b, respectively.

A CDCL frame is identified by a SafetyNET p PDU header value of 0x02 when the frame is transmitted on the forward channel (devices can write) and by a value of 0x03 when the frame is transmitted in the backward direction (devices can read). A CDCN frame, instead, is identified by a SafetyNET p PDU header value of 0x60.

Both CDCL and CDCN PDUs consist of a set of control fields followed by a CDC payload section. The number and meaning of control fields depend on the adopted communication model, while the CDC payload section is identical for both RTFL and RTFN. CDCL PDUs also include an additional status field, which is used to signal if an error has been detected in the FCS field of the Ethernet frame during its transmission.

CDCL control fields are the following:

- *Cycle counter* (2 bytes) identifies the number of the current cycle.
- *Frame counter* (1 byte) identifies the frame number within the actual cycle.
- *Length* (2 bytes) is the length in octets of the *CDC write pointer* field plus the CDC payload section.
- *CDC write pointer* (2 bytes) identifies the write section for cyclic communication.

CDCN PDUs foresee two control fields only:

1. *Version* (1 byte) identifies the version of the CDCN protocol.
2. *Size* (2 bytes) is the size in octets of the CDC payload section.

The CDC payload section is common to both the communication models and consists of a sequence of process data objects (PDOs) as shown in Figure 28.7. Each PDO, in turn, is made up by three fields:

1. *PID* (3 bytes) uniquely identifies the PDO within the network.
2. *Len* (1 byte) is the length of the whole PDO including the *PID* and the *Len* fields.
3. *Data* (Len-4 bytes) are the process data being transmitted.

28.3.3 Message Channel

The MSC is used for acyclic communications to transmit data concerning diagnostics and/or parameter settings, when timing is not a critical issue. Devices write data into MSC packets, and information is

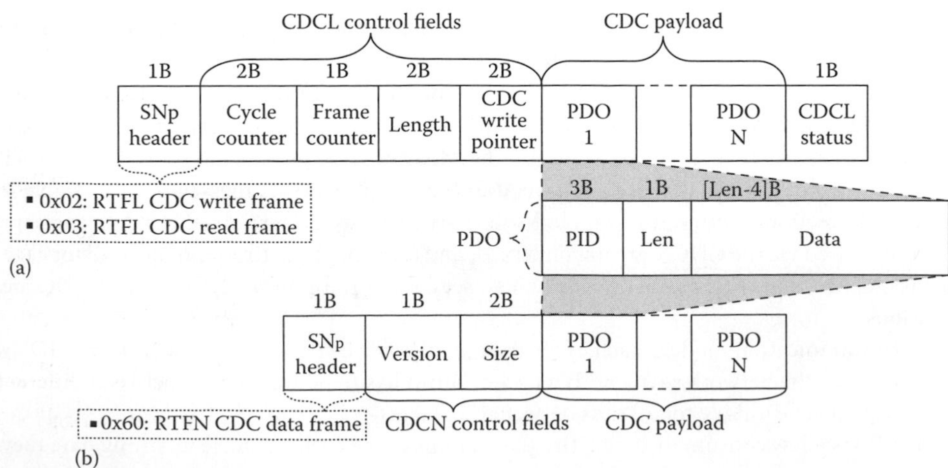

FIGURE 28.7 Structure of CDCL and CDCN PDUs. (a) RTFL CDC frame and (b) RTFN CDC frame.

transferred by means of the MSC message transfer protocol (MSC-MTP). MSC-MTP provides a confirmed service, which implies that the reception of messages is acknowledged of course. MSC-MTP is also in charge of implementing segmented data transmissions in the case the size of the transferred data exceeds a predefined threshold, which is configuration dependent. As for CDC, the implementation of MSC is different for RTFL and RTFN.

When RTFL is used, MSC (MSCL) consists of exactly one frame for each communication cycle. Messages are characterized by a priority that depends on the type of service contained in the message. Three priority levels (from 1 to 3) are foreseen, which are used for bandwidth reservation according to the importance of the message. Priority 1 is the highest level, while 3 is the lowest. The MSCL frame size is configurable, but in the case MSC were not able to accommodate all messages in a single cycle, each OD could reserve additional transfer space for a specific message in one of the following cycles. If more messages require extra transfer space, reservations are made according to the priority of messages.

In RTFN, the implementation of MSC (MSCN) is quite simple. Messages are transferred by means of UDP/IP packets using MSC-MTP without any need of assigning priorities.

MSC-MTP uses six different types of commands to transfer data:

1. *Init frame* is used to initialize a segmented data exchange.
2. *Init fast frame* is used to exchange data without segmentation.
3. *Send frame* is used to transfer a segment of data after the segmented initialization has been performed.
4. *Send last frame* is used to transfer the last segment of data.
5. *Acknowledgment* is used to confirm the reception of data segments.
6. *Abort* is used to interrupt a data exchange.

An MSC-MTP communication session starts with either an *init frame* or *init fast frame* message sent to the destination to propose a handle identifier for the communication session. The communication session is univocally identified by the handle, so that several sessions can be established between the same participants without any risk of ambiguity.

If the size of data to be transmitted is small, the sender invokes the *init_fast frame* to avoid segmentation. In fact, this request also embeds all data to be transmitted. The destination device responds with an *acknowledgment* message, which contains the proposed handle identifier and the maximum segment size (MSS) the device is able to support. If the communication started with an *init_fast frame* and the size of the transmitted data is smaller than MSS, the sender is automatically notified that all data have been successfully received and the communication session is closed without the need of any further message.

Otherwise, if segmentation cannot be avoided, the sender uses *send frame* messages to transfer data segments carrying amounts of data smaller than MSS. Any message also contains the *handle identifier* and *byte counter* fields. The destination device sends *acknowledgment messages* to confirm the reception of segments. An acknowledgment is not transmitted on each segment reception, but it is sent only when the amount of data received since the last acknowledgment is larger than a configurable threshold. However, the initialization and last segment messages are always acknowledged, while this never occurs for broadcast and multicast transmissions.

The last segment of data is transmitted by means of the *send last frame* message. Finally, the *abort* message is used by participants to abruptly interrupt the communication session.

The structure of the MSCL and MSCN frames is shown in Figure 28.8a and b, respectively.

An MSCL frame is identified by a SafetyNET p PDU header value of either 0x00 (forward/write direction) or 0x01 (backward/read direction). A 23-byte section follows the header and contains nine control fields, while the MSC data section has a variable size and includes a sequence of MSC-MTP frames. Finally, a 1-byte status field is used to signal if an error in the Ethernet FCS has been detected.

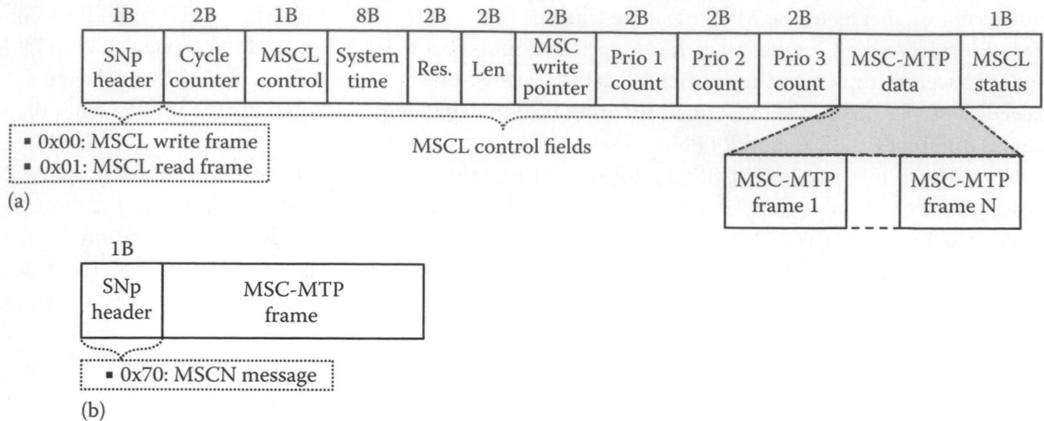

FIGURE 28.8 RTFL and RTFN MSC frame structures. (a) RTFL MSC frame and (b) RTFN MSC frame.

The meaning of the control fields is the following:

- *Cycle counter* (2 bytes) identifies the number of the current cycle.
- *MSCL control* (1 byte) contains control bits to set parameters for the communication. Bit 0 is used to reset previously assigned priorities, bit 1 shows if the line is running in diagnostic mode or not, bit 2 signals whether the PCS clocks are synchronized or a delay measurement is being carried out, and bit 3 is used to know if the local clock is controlled by means of time stamps. The four most significant bits are reserved for future use.
- *System time* (8 bytes) stores the time when the packet was processed by the master clock.
- *Reserved* (2 bytes).
- *Length* (2 bytes) is the length in octets of the *MSC write pointer* header field plus the MSC data section.
- *MSC write pointer* (2 bytes) points to the next write position in the MSC data section.
- *Assigned priority count 1, 2, and 3* (2 bytes each) record how many messages have been assigned to priority levels 1, 2, and 3, respectively.

An MSCN frame is identified by a SafetyNET p PDU unique header value of 0x70 followed by one MSC-MTP frame.

MSC-MTP frames are PDUs used by the MSC message transfer protocol for confirmed acyclic message exchanges. The frame structure is shown in Figure 28.9 and is the same for both RTFL and RTFN. The *address type* is an 8-bit field used to store the addressing mode and priority of the message: bits $0 \div 2$ are reserved and should always be set to 0; bits $3 \div 4$ are used to encode the priority of the message when the communication model is RTFN; bit 5 is used for the version of the IP protocol (0 for IPv4 and 1 for IPv6), while bits $6 \div 7$ are used to distinguish between normal addressing mode (00) or extended addressing mode (01 or 10). In the case of extended addressing, the value is used to correctly identify the information contained in the *IP address* optional field.

The *device DA* field (2 bytes) contains the device address of the destination. Special values 0xFFFF and 0xFFFE are used as broadcast and multicast addresses, respectively. In this case, receiving devices do not have to acknowledge the reception of the frame.

The *message length* field (2 bytes) stores the total size of the MSC-MTP frame in octets.

The *IP address* field is optional, and it is used for the source or destination IP address in the case of extended addressing mode, in accordance with the value of bits $6 \div 7$ in the *address type* field. The size of this field is 4 or 16 bytes, depending on the version of the IP protocol used, as selected by bit 5 of the *address type* field.

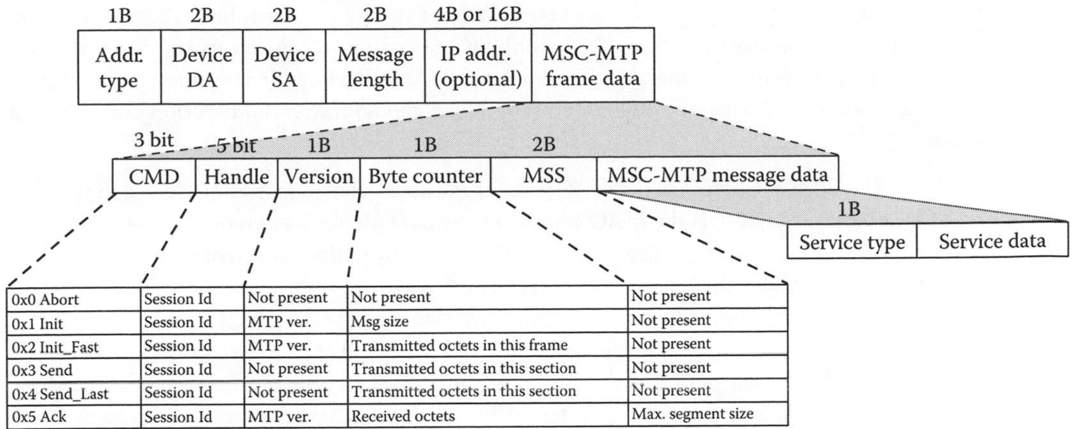

FIGURE 28.9 MSC MTP frame structure.

The last field, that is, *MSC-MTP frame data*, contains a sequence of subfields and the data section. Subfields are used to identify the type of the MSC-MTP message (*command*), the communication session (*handle identifier*), and the number of data octets that have been transferred since the beginning of the communication section (*byte counter*). The data section (MSC-MTP message data) contains data that are being transferred.

28.3.4 Communication Management

The SafetyNET p specification also includes services to initialize and manage the communication. CMSs make use of a specific set of messages that are transferred independently of the protocol logical channels. Each RTF model has its own CMS set of messages, due to the differences in the physical layer and in the way devices transfer data.

In RTFL, CMS are used to set up the communication (physical or logical) double line and to check the device parameter settings. Messages belong to two different classes: network verification and network configuration.

The two most important messages of the network verification class are the *RTFL network verification prepare* and the *RTFL network verification information*. The first message is sent by RD to initialize ODs and to signal that network configuration messages are going to be sent. This message is identified by the value 0x10 in the SafetyNET p frame header and contains a 2-byte sequence number followed by the MAC address of RD.

The *RTFL network verification information* message is used by devices to transmit their status, configuration parameters, and hardware information. This message is identified by a SafetyNET p frame header value of 0x12, followed by a 2-byte sequence number and an *identification data* section. The *identification data* section contains several information fields concerning hardware identification and device status (serial number, vendor ID, product and revision number, status of the physical link, and supported communication models), configuration (symbolic device name, IP configuration parameters, and MAC addresses of both the device and its predecessor), and supported roles (RD, OD, gateway, and switch). The reception of both types of messages mentioned earlier is confirmed by means of a *network verification acknowledgment* message, identified by a SafetyNET p frame header value of 0x13, which contains the sequence number field followed by the same header value as the message that is being confirmed.

The network configuration message class is used by RD to set up the communication line and configure the ODs with functional parameters. Only two messages belong to this class, that is, *RTFL configuration* message and *RTFL configuration acknowledgment.* The former allows RD to configure the ODs, while the latter is used by the ODs to confirm the correct reception of the configuration parameters. The *RTFL configuration* message is identified by the value 0x20 in the SafetyNET p frame header and contains a 2-byte sequence number followed by a *configuration data* section consisting of several fields:

- *Previous MAC address* (6 bytes): is the MAC address of the previous (upstream) device in the line.
- *Next MAC address* (6 Bytes): is the MAC address of the next (downstream) device in the line.
- *Next MAC alternative* (6 bytes) is the MAC address of an alternative successor.
- *Device address* (2 bytes) is the logical address of the device in the line.
- *MSCShortMsgSize* (2 bytes) is the maximum message size for not segmented transfers.
- *Number of frames* (1 byte) is the number of frames for CDC- and MSC-based communications.
- *Cycle time* (4 bytes) sets the period of the communication cycle.
- *RTF timeout* (4 bytes) sets the timeout interval for assuming that not received massages have been lost.
- *Master clock DA* (2 bytes) is the logical address of the device that acts as the master clock.
- *IP configuration section* is a set of fields containing information about the device IP address configuration (IPv4/v6 address, subnet mask, DNS servers, and so on).

The *RTFL configuration acknowledgment* message is identified by a SafetyNET p header value of 0x21 and contains only a 2-byte sequence number.

RTFN CMS adopts two sets of messages: the *RTFN scan network* set, which is used by the protocol to retrieve information about the status of devices, and the *RTFN connection management* set, which is used to initialize and manage links connecting publishers and subscribers.

RTFN scan network consists of two messages. The first one, *RTFN scan network read request,* is broadcast to query the status of the network devices. The structure of the message is very simple: It consists of a SafetyNET p frame with a header value of 0x80 and no additional field. The second message, *RTFN scan network read response,* is the reply of any device in the network to the broadcasted request. The message is a SafetyNET p frame identified by a header value of 0x81 and contains an identification data section with the same structure as the *RTFL Network verification acknowledgment* message.

The *RTFN connection management* set consists of five messages: four of them are needed to set up and destroy communication links between publishers and subscribers, while one message is used by subscribers to notify publishers that the link is still active.

The structure of messages used to set up and destroy communication links is shown in Figure 28.10. The message type depends on the content of the SafetyNET p frame header and the possible meanings are the following:

- *0x40*: the message is a *CDCN subscribe request.* It is used by devices to request the subscription of PDOs published by a specific device.
- *0x41*: the message is a *CDCN subscribe acknowledgment.* It is used by a publisher to confirm that a subscription request has been accepted.
- *0x42*: the message is a *CDCN unsubscribe request.* It is used to unsubscribe the reception of some PDO.
- *0x44*: the message is a *CDCN unpublished.* It is used by a publisher to inform its subscribers that some PDOs will not be published anymore.

The payload of the RTFN connection management frame consists of two control fields and one data section.

The control fields are used to store the version of the CDCN protocol (1 byte) and the number of *ID data fields* that are present in the next frame section (2 bytes).

The *RTFNCM data* section is a sequence of *ID data* elements listing the PDOs to be managed.

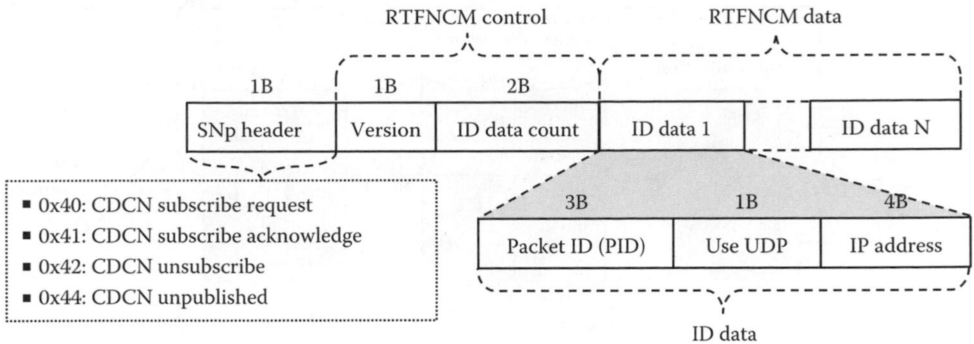

FIGURE 28.10 RTFN connection management frame structure.

Each *ID data* element has a 3-byte *Packet ID* field that contains the PID of the PDO, a 1-byte *Use UDP* field to decide whether pure MAC frames or UDP/IP datagrams are used, and a 4-bytes *IP address* field that contains the address of the subscriber as depicted in Figure 28.10.

To monitor the status of the connection link between a publisher and its subscribers, the *CDCN connection still alive* message is used. The frame header is 0x43, and there are no other data. The message is cyclically sent by subscribers to publishers. The cycle time is configuration dependent and can be different with respect to the period used to transmit process data on CDCN.

28.3.5 SafetyNET p Application Layer

All services provided by any SafetyNET p device are made available to the user via the application layer, which is based on the CANopen specifications [8].

In general, different devices have heterogeneous working states, and this implies different communication functions. Roughly speaking, communication functions can be subdivided into two main classes concerning parameterization, also referred to as service data and process data transfer, respectively.

28.3.5.1 CANopen Application Layer in SafetyNET p

CANopen [8] is a nonproprietary open fieldbus standard, specified and standardized by the CAN in Automation organization, which includes the definition of an application layer for industrial applications.

CANopen includes specifications for both communication, to enable networking of distributed automated field devices and application objects based on device profiles.

Device profiles specify functions and parameters/objects for different application areas and automated device groups. This grants compatibility and interchangeability between devices of different manufacturers.

Since device profiles are not bound to a specific communication medium, they are reusable (and have been reused) in SafetyNET p too. This choice has the twofold advantage of providing the system developer with *application interfaces* equivalent to CANopen and enabling manufacturers to reuse device profile–based applications, projects, and developments for SafetyNET.

The SafetyNET p base element inherited from CANopen is the object dictionary (ODIC), which acts as an interface between the application and the communication subsystems as shown in Figure 28.11. Essentially, ODIC is a technique to group objects in a device and specify uniform communication/device parameters, data, and functions that are then stored and retrieved using objects.

Transmission of process data is taken into account in a device ODIC by means of PDOs. ODIC access from other devices is obtained through service data objects (SDOs) that enable reading/writing information from/to remote device ODICs.

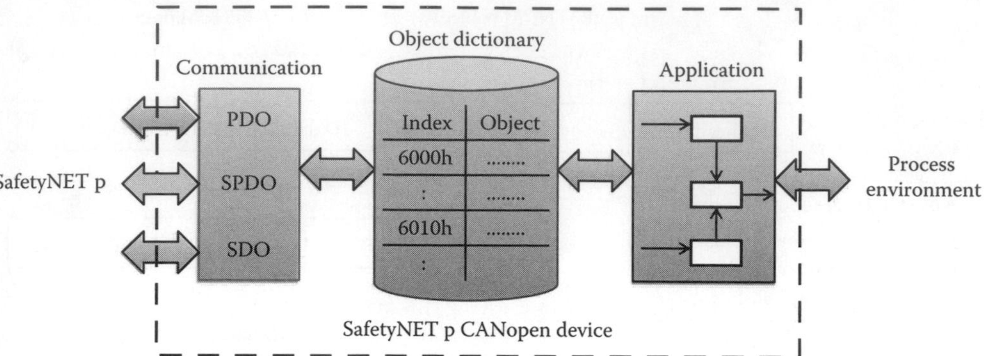

FIGURE 28.11 Interfacing of the communication and application subsystems by means of the object dictionary.

The dictionary is organized in a logical structure as shown in Table 28.1, while a logical referencing scheme is used for accessing the ODIC as depicted in Figure 28.12.

Each ODIC can contain up to 65,536 objects selected with a 16-bit index. The index, in turn, may include an additional 8-bit subindex that allows the definition of complex object structures such as records and arrays. When this feature is not needed, the subindex value is reset to zero, and the whole

TABLE 28.1 Object Dictionary Structure

Index (hex)	Object
0001h—001Fh	Basic data types
0020h—003Fh	Complex data types
0040h—005Fh	Manufacturer-defined data types
0060h—007Fh	Device profile–specific basic data types
0080h—009Fh	Device profile–specific complex data types
00A0h—0FFFh	Reserved
1000h—IFFFh	Communication profile section
2000h—5FFFh	Manufacturer-defined profile section
6000h—9FFFh	Standardized device profile section
A000h—BFFFh	Standardized interface profile section
C000h—C8FFh	SafetyNET RTFN interface profile section
C900h—FFFFh	Reserved

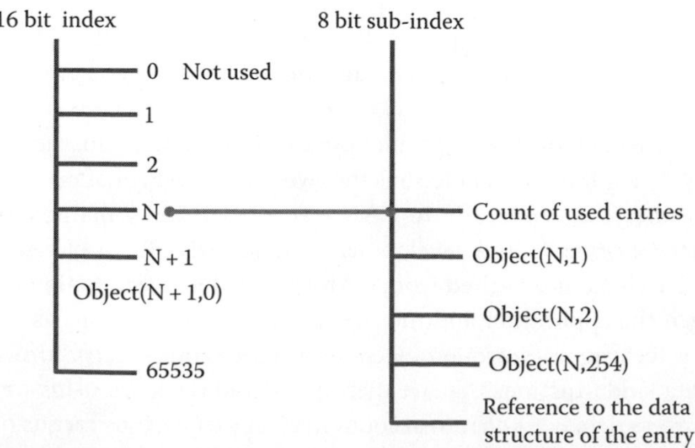

FIGURE 28.12 Object dictionary addressing scheme.

object is managed as a basic type (i.e., integer, Boolean, and so on). Otherwise, the subindex 0 contains the number of elements (subentries) in the object structure. Each subentry, in turn, can be a different basic type.

Process data communication is handled by *PDOs* in SafetyNET p too. PDOs are logical data containers transferred between SafetyNET p devices over either RTFN or RTFL. PDOs encapsulate application objects referenced in the ODIC.

In RTFL, PDO communication is typically cyclic over the CDC logical channel. Event-driven acyclic transfers for PDOs are allowed as well, thus it is possible to transmit PDOs using either CDC or MSC.

The transmission of PDOs always requires a configuration on both the producer (transmit PDO-TxPDO) and consumer (receive PDO—RxPDO) sides. PDO communication parameters define the communication behavior with respect to the communication subsystem. PDO mapping parameters define the content of each PDO and must include one application object at least.

The length of a PDO is application dependent and is stored in a suitable mapping parameter.

As in CANopen, SafetyNET p grants remote access to the ODIC by means of SDOs. SDOs can be used, for instance, to read or write the PDO configurations.

SDO addressing is obtained by specifying the address of the remote device, the index, and the subindex of the object to be accessed. In order to provide concurrent SDO service processing in a device, SafetyNET p extends the CANopen SDO services with a job identification.

SDOs are important, in particular, during the configuration phase, where a SafetyNET p device can be configured completely, depending on what entries are made available in the device's ODIC. SDOs can also be used in normal operation, for instance, for status queries.

The SDO communication is based on the client–server model and is implemented using a point-to-point link between two devices. This is why an SDO is always transferred over MSC in SafetyNET p.

The transmission of an SDO message is a confirmed service, and an abort service is also provided to notify any SDO communication error.

Devices can send diagnostic events using the emergency (EMCY) service, which broadcasts standardized messages using either RTFL or RTFN. Diagnostic messages can also be transmitted directly to a specific device.

The heartbeat service allows to monitor the state of communication partners. A heartbeat producer generates a signal that is received by the heartbeat consumers. Each heartbeat consumer can configure its ODIC to select the heartbeat producers to monitor. When the heartbeat is not received, an event is generated at the heartbeat consumer. The heartbeat packet is cyclically sent using the same PDU as for PDO transfers.

28.4 Precise Clock Synchronization

Synchronous operations of devices across the network (as needed, e.g., in motion control applications) imply synchronized clocks. In RTFN, this can be achieved by implementing standardized protocols as defined, for instance, in the IEEE 1588 (PTP) specifications [3].

In RTFL, because of the cut-through frame-processing mechanism, a dedicated clock synchronization technique is necessary. SafetyNET p provides suitable configuration and control services to this purpose to be used with RTFL. The PCS protocol is based on a master clock that has to be located in the first device of the line. The master clock then delivers the time base for synchronization to the ODs as shown in Figure 28.13.

When the line is started up, the master clock propagates the current system time to all other devices. As the frame is delayed by a constant amount of time due to processing in each device, the overall transmission time increases during the propagation. The delay is measured in each direction by an internal time-stamping unit included in each device. The bus master, which is aware of the line topology, requests the measured delays to all devices and computes the individual offsets accordingly. The devices,

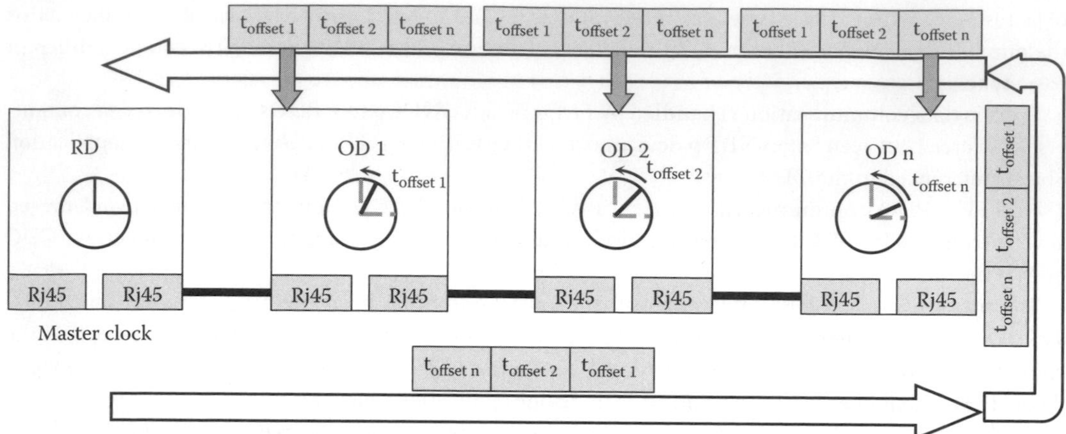

FIGURE 28.13 Precise clock synchronization scheme.

in turn, adjust their internal clocks by using the offsets received from the master. The delay measurement procedure is initiated automatically by the bus master when the line is set up. The PCS protocol uses MSC to transfer messages.

28.5 Safety Communication Model

A safety system is in charge of detecting faults to prevent failures of critical processes that could cause injuries or damages either to people health, environment, or plant components. Any system that has been designed to deal with safety-related functions is a safety system (e.g., emergency stop procedures, and fire and gas detection systems). When the safety system detects a fault, it must bring the plant to a safe state as soon as possible, for instance, by stopping processes in the affected portion of the system and/or starting emergency procedures (e.g., controlled shutdown of machineries, activation of alarms, or fire protection measures). Networks and protocols in safety systems play a key role because they are in charge of transferring information from safety sensors to safety controllers and from controllers to actuators. The safety protocol must ensure a high integrity level of transmitted data as well as provide a set of measures to monitor the reliability of the communication channel.

For example, an emergency stop system can be built as a distributed application: the emergency button could be placed far from the event controller, and actions on several portions of the plant could be necessary in the case the button were pressed. Such a safety system should ensure that, when the button is pressed, the information is transferred to the controller (e.g., a PLC) and then to the actuators that are in charge of stopping the controlled machineries. The key difference between a standard and a safety system is that, in the latter case, the information about the status of the emergency button is regularly transmitted to the controller and then to the actuators. In the case an anomaly is detected by one of these components (communication error or failure of an element in the safety chain), the system must be brought to the safe state, even if the emergency button has not been pressed.

Even if it might appear counterintuitive, neither redundant physical transmission paths nor dedicated communication lines are strictly necessary in order to set up a safety network. The approach commonly adopted to set up safe communications over conventional networks is based on the concept of black channel that originates from the railway signaling technologies and was introduced in IEC 62280-1 [11,12]. In practice, a safety layer is added on top of the conventional communication stack, that is, over the application layer. The safety layer implements the safety-related transmission functions and also monitors the integrity of the communication channel so as to ensure that the requirements of functional safety are satisfied. This technique has two main advantages: first, the conventional and inexpensive transmission technologies used in industrial networks can be adopted, and second, both safety and non-safety-related devices can be connected to the same network. The latter aspect also offers the advantage of a full integration of safety and non-safety-related functions that could be exploited to reduce the risk of failures.

Functional safety is dealt with in the IEC 61508 standard [2]. The standard defines the set of communication errors that a safety function must be able to detect and suggests several measures that can be implemented by the protocol in order to detect communication errors. Moreover, the standard defines four SILs, each one characterized by a maximum failure probability of the safety function that must be guaranteed. Level 4 has the most demanding requirements, and it is applied to extremely critical processes such as nuclear plant controls. SIL 3 is the level commonly adopted for industrial control and process applications. This level requires that the probability of failure in high demand rate (i.e., when the system operates continuously) falls between 10^{-8} and 10^{-7} failures per hour. In low demand rate (i.e., when the system operates on demand), the probability of failure has to be comprised between 10^{-4} and 10^{-3} in order to satisfy SIL 3 requirements [17].

28.5.1 Black Channel Approach in SafetyNET p

SafetyNET p has been designed to support safety-critical communications on the same network infrastructure used to transfer control and process data, thus making the protocol suitable for safety-oriented applications [10]. The protocol makes available a safe communication channel built on top of CDC that depends on neither the used communication model (RTFL/RTFN) nor the transport layer. The safety channel satisfies the strict integrity requirements of SIL 3. The approach used by the protocol to transmit safety-related data on a conventional communication line is based on the concept of black channel, as depicted in Figure 28.14.

The safety layer is built on top of the application layer. Depending on the nature of the information that has to be transferred, the CDC packet section of the SafetyNET p PDU contains either PDOs for standard cyclic communications or safety process data objects (SPDOs) for cyclic safety-related data exchanges. In the same way, acyclic messages contain either SDOs or safety service data objects (SSDOs), for standard and safe transfers on MSC, respectively. The underlying protocol layers are not aware of the kind (standard or safe) of information that is being transmitted and do not need to implement safety-related functions.

The same principle can be applied to all devices used to build the network infrastructure, such as cables, switches, routers, and so on. These devices are only aware that SafetyNET p frames are transmitted (at either the UDP or Ethernet level), but they do not need to know what kind of data a frame is actually transporting, to carry out their interconnection functions as expected. In this

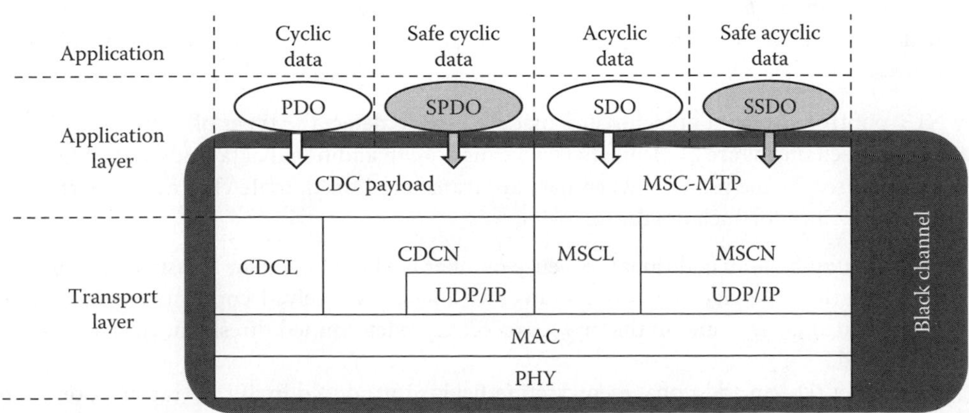

FIGURE 28.14 SafetyNET p black channel implementation.

way, with the exception of the actual safe bus subscribers such as safe sensors, safe PLCs, and safe actuators, all the other network components are considered as non-safety-related devices and are not required to implement any kind of safety function or specific integrity check. The safety layer itself implements a set of detection measures to verify the integrity of the transmission channel and to discover communication errors related to failures in devices used to set up the network infrastructure.

28.5.2 SafetyNET p Error Detection Mechanisms

The IEC 61508 standard defines the communication errors that must be detected by the safety layer as well as a set of detection measures that can be implemented by the protocol to cope with such errors. SafetyNET p is able to cope with the following nine types of communication errors:

- *Corruption*—messages may be corrupted due to either errors on the transmission medium or failures in the hardware of devices (even non-safety-related devices) that participate in the communication.
- *Unintended repetition*—due to errors on the communication channel or failures in devices, old messages are transmitted more than once and out of time.
- *Incorrect sequence*—due to errors on the communication channel or failures in devices, the natural sequence of messages transmitted from a particular source is altered.
- *Loss*—due to errors on the communication channel or failures in devices, either a message or an acknowledgment is not received.
- *Unacceptable delay*—messages must be received within a predefined deadline, but they can be delayed during transmission due to several reasons such as congestion of the communication channel, interferences, processing, and management operations carried out by network devices (e.g., switches, bridges, and routers).
- *Insertion*—because of faults, errors on the communication line, or malicious attacks [18], a message that is not related to any expected sender is injected in the network and received by a safe device.
- *Masquerade*—because of communication errors, faults, or malicious attacks [18], a non-safety-related message is treated as safety relevant by the receiver. For example, when safe and standard data are transmitted on the same communication line, an altered standard message could appear as a safe one.
- *Addressing*—due to faults or errors on the communication line, a safety-related message is sent to a wrong destination that considers the message as valid.
- *Revolving memory failures within switches*—failures in the internal memory of switches can cause the forwarding of altered frames in a way that they are still considered as valid by safety-related devices.

Safety NET p introduces a set of techniques that are used to detect the types of communication errors listed earlier. Mechanisms are implemented by means of both additional data fields and additional specific messages used by the protocol when data are transferred on the safe channel. In particular, the SafetyNET p specification includes the following:

- *Sequence number*—an additional sequence number field is adopted for transmissions of SPDOs.
- *Timeout*—the time elapsing between any two messages received consecutively is checked by safety-related devices, and in the case it exceeds a predetermined threshold, an error occurrence is assumed.
- *Connection ID*—an additional identification field is introduced in SPDO transmissions in order to identify the sender of a frame without any ambiguity.

TABLE 28.2 Communication Errors and Detection Mechanisms

	Safety Measures				
Communication Errors	Sequence Number	Timeout	Connection ID	Data Integrity Assurance	Diff. Data Integrity Assurance
Corruption	—	—	—	X	—
Unintended repetition	X	—	—	—	—
Incorrect sequence	X	—	—	—	—
Loss	X	X	—	—	—
Unacceptable delay	—	X	—	—	—
Insertion	X	—	X	—	—
Masquerade	X	—	X	—	X
Addressing	X	—	X	—	—
Revolving memory failures within switches	X	—	X	X	—

- *Data integrity assurance*—a cyclic redundancy check (CRC) field is introduced to verify the integrity of SPDOs. Moreover, some relevant fields of each SPDO are duplicated to allow cross checking between the two copies of the received data.
- *Different data integrity assurance*—the same CRC field introduced for data integrity assurance is also used as a detection measure. Different data integrity checks are aimed at detecting standard data that, for some reason, are received modified in such a way that could make devices interpret them as safe data. To reduce the probability of this event, a common approach is a diversification of the CRC calculation method (e.g., algorithm and generator polynomials) among the two classes of data. Since SafetyNET p does not use any CRC when standard data are transmitted, different data integrity is ensured by the CRC introduced for normal data integrity.

Table 28.2 shows the relationships between the types of communication errors for safe data transmissions and the detection measures implemented by the SafetyNET p protocol.

28.5.3 Safety Communication Layer

The safe channel is mainly built on top of the standard CDC, since safe data are cyclic in their nature. The MSC is used only to send parameters concerning the configuration of the safety functions by means of SSDOs. Since this kind of communication is not critical, no particular check is needed about the integrity of MSC. The only action taken by SafetyNET p for MSC communications is the protection of SSDOs by means of a CRC field.

The general structure of a cyclical safety PDU is identical to the one used for standard CDC communications. The difference between the standard and safe channels is in the format of the CDC payload section.

In the case of standard communications, in fact, CDC is used only to transfer process data by means of PDOs.

In the case of safe communications, instead, either SPDOs or safety heartbeats (SHBs) are transmitted via CDC. As explained in the previous sections, SPDOs are used to transmit safety application data, while SHBs are used to synchronize the communicating devices and to monitor delays on the communication channel.

28.5.3.1 Safety Process Data Objects

Figure 28.15 shows the typical structures for PDOs, SPDOs, and SHBs. The structure of standard PDOs has already been discussed in Section 28.3.2.

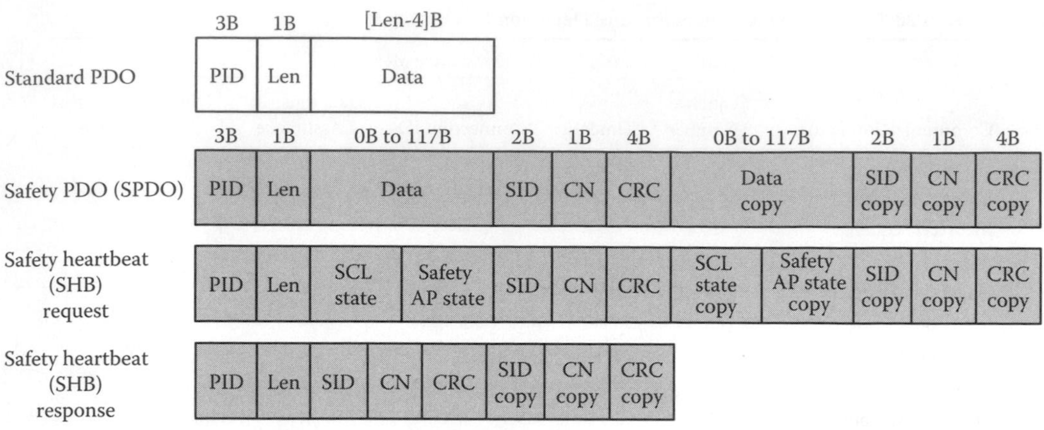

FIGURE 28.15 Comparison of PDO, SPDO, and SHB structures.

The structure of SPDOs is more complex because it includes a set of additional fields that are used for communication error detection according to Table 28.2.

Figure 28.15 shows that an SPDO contains the following:

- A 3-byte *PID* field to store the PID associated to the transmitted safe process data
- A 1-byte *len* field to store the length (in bytes) of the SPDO, including the PID and LEN fields
- A variable-size *safe data* field that contains the safe process data
- A 2-byte *SID* field to store the safe identifier (SID) of the sender
- A 1-byte *consecutive number* field to store a consecutive (sequence) number
- A 4-byte *CRC* field containing a 32-bit CRC that covers the PID, length, safe data, SID, and consecutive number fields
- A copy of the *safe data*, *SID*, *consecutive number*, and CRC fields

The *consecutive number* field implements the *sequence number* safety mechanism, while the couple of PID and SID identifiers is used to build a unique *connection ID*. In particular, SID must be unique across the network, and its value shall not be 0.

The *data integrity assurance* and *different data integrity assurance* are granted by the 32-bit CRC, which is computed using the 0x20044009 polynomial. This polynomial is able to guarantee a residual error probability compatible with the SIL 3 requirements if the overall size of the protected data falls between 0 and 128 octets, including the CRC signature. As a consequence, the maximum size of the *safe data* field is restricted to 117 octets.

The adoption of the redundant *data*, *SID*, *consecutive number*, and *CRC* fields is an additional mechanism used by the safety layer to check the integrity of the communication channel.

When a device receives a CDC message, the safety layer recognizes the type of PDO (standard or safe) in the frame by inspecting the value of the PID field. If the PID identifies an SPDO, the content of the fields following the data section is processed by the safety layer to check the integrity of the received data.

Besides special fields for error detection, the safety layer checks the time elapsed between any two consecutive receptions of SPDOs to monitor the responsiveness of the communication channel. Moreover, since the transmission of safety process data is not a confirmed service, an additional mechanism is required to detect possible failures of devices. In fact, when a device is only a consumer of data, there is no way for the unconfirmed cyclic data transfer service to obtain information about its health status. This check is then carried out by means of the SHB service.

28.5.3.2 Safety Heartbeat Service

The SHB service is used by the protocol to monitor the state of both the safety communication layer (SCL) and safe application and to evaluate delays on the communication channel. SHB cannot rely on other heartbeat mechanisms implemented by the protocol, such as, for instance, the exchange of CDCN *still alive* messages used to monitor the status of the conventional CDC.

SHB is a confirmed service, and messages are exchanged in a cyclic way. The cycle time of the service is independent of other communication cycles and mainly checks the delivery time experienced by messages against a maximum predefined threshold.

Each device involved in a communication relationship is forced to monitor its interlocutor by cyclically sending *SHB request* messages. For example, in the publisher–subscriber model, each publisher must send *SHB requests* to monitor the state of its subscribers, and at the same time, subscribers have to send *SHB requests* to their publishers. Each device that receives an *SHB request* must reply with an *SHB response* message.

SHB PDUs have the same structure of SPDO as shown in Figure 28.15. In SHB request PDUs, the PDO data section used to transport safety-related data is replaced with two fields: *SCL state* and *Safety AP* state. *SCL state* contains information about the state of the SCL. Possible SCL state values are the following:

- *0x00*: SCL is in BOOTUP state
- *0x04*: SCL is in STOPPED state
- *0x05*: SCL is in OPERATIONAL state
- *0x7F*: SCL is in PREOPERATIONAL state

The *Safety AP state* field contains information about the state of the safety application and is implementation specific.

SHB response PDUs do not carry any data and are used to confirm that an *SHB request* has been received.

The SHB service is used by all safety devices to determine the actual delay of the safe communication channel and establish whether the received information can be considered valid or not. Since each *SHB request* is acknowledged by the receiver, the sender can monitor the time elapsed between the request and the reception of the relevant acknowledgment. In the case where the measured delay exceeds the predefined threshold, SCL starts a transition to the *fail safe state* to make the application enter a safe state.

28.5.3.3 Safety Communication Layer Management

The safety communication layer management (SALMT) is a local service conceived to control the behavior of the safety-related components in the device. The service offers a set of commands to trigger the state of the SCL state machine as shown in Figure 28.16.

At the device start-up, the state machine is in the start-up state, and an automatic transition (T1) is enabled to reach the initialization state. In both the start-up and initialization states, the exchange of SPDOs and SHBs is not allowed. If a fault is detected during initialization, a transition (T7) is automatically triggered to move SCL to the initialization failed state. In this state, SCL can start to send and receive SHB messages, but due to the failure condition, the exchange of SPDOs is not allowed. A new initialization must be invoked using the SALMT command *reset node*, which in turn triggers a transition (T8) to bring SCL back to the initialization state.

Another transition (T2) is triggered by the SALMT *enter preoperational* command to drive SCL from the initialization to the preoperational state, where only the exchange of SHB messages is allowed. From the preoperational state, SCL can move back to the initialization state (transition T5 triggered by either the *reset communication* or *reset node* command).

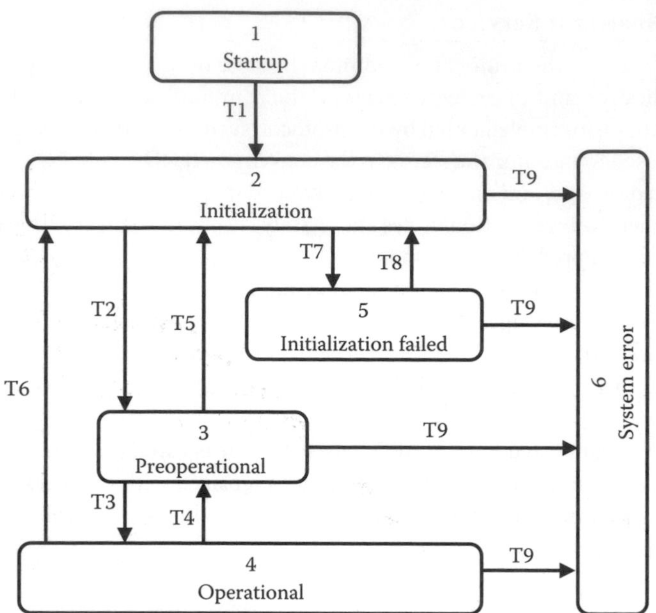

FIGURE 28.16 SALMT state machine.

The command *start remote node* is used to move SCL from the preoperational to the operational state (transition T3). In this state, SCL is fully operative, and the exchange of both SHB messages and SPDOs is allowed. A pair of other transitions (T4 and T6) is foreseen to move SCL back to the preoperational and initialization states, respectively. Transition T4 is triggered by the *stop remote node* command, while T6 is controlled by either *reset communication* or *reset node*.

Finally, in each state with the exception of start-up, the detection of a safety-critical error drives SCL to the system error state (transition T9), where no kind of communication is allowed.

28.5.4 Safety Application Layer

The safety application layer provides applications with the interface to access the SafetyNET p safe services. The basic element in the data model is the safe ODIC, which is an extension to the standard ODIC.

Entries are identified in the same way using data types, indexes, and subindexes. Access to the safety objects is not allowed to nonsafe services such as SDO. Table 28.3 summarizes the objects that are available in the safe ODIC.

The safe address of a SafetyNET p device is its safety ID (SID). SID is one of the mandatory safety objects.

SPDOs are specified as TxSPDO and RxSPDO on the producer and consumer sides, respectively. Communication parameters specify the SPDO transfer mode, while mapping parameters define the SPDO contents by listing the relevant safe application objects.

For SHB producers, the Failsafe Producer Heartbeat is a record similar to the TxSPDO communication parameters. The Failsafe Consumer Heartbeat list contains, in its turn, parameters needed to program the reception of SHBs as well as SHB responses.

Each object description adopts a standardized format. For instance, the structure of the SID is shown in Table 28.4. The table specifies that SID is a simple 16-bit variable that can be accessed only from the SafetyNET p FS part (FSF). It cannot be mapped onto an SPDO, and its value cannot be 0x0000. SID is also a mandatory object with index value equal to 0x1200.

TABLE 28.3 Safe Dictionary Objects

Index	Name	Data Type	Description
0x1000	Device type	Unsigned32	Device classification: Lower 16 bits are *Device Profile Number*, describing the used profile. Upper 16 bits are *Additional Information*.
0x1200	Safety ID	Unsigned16	Unique identifier for the safety device shall be not zero.
0x1216	Safety consumer heartbeat list	Unsigned256	List of remote devices that need to be monitored by the device.
0x1217	Safety producer heartbeat parameter	PDO COM_PAR	Configured in the same way as SPDO transmission. Shall be configured as cyclic transmission.
0x1218	Safety bus cycle time	Unsigned32	Subindex 0: Number of entries. Subindex 1: RTFN Base Cycle Time. Subindex 2: RTFL Base Cycle Time.
0x121E	SPDO Timeout tolerance T_{to}	Unsigned8	Defines how much of an excess of an RxSPDO timeout is acceptable. Unitless number, interpreted as percentage.
0x1C00–0x1CFF	RxSPDO communication parameter	PDO COM_PAR	Parameters that define the communication properties of RxSPDOs.
0x1D00–0x1DFF	RxSPDO mapping parameter	PDO MAPPING	Parameters to define the data mapping of RxSPDOs.
0x1E00 to 0x1EFF	TxSPDO communication parameter	PDO COM_PAR	Parameters that define the communication properties of TxSPDOs.
0x1F00 to 0x1FFF	TxSPDO mapping parameter	PDO MAPPING	Parameters to define the data mapping of TxSPDOs.

TABLE 28.4 Specification of the Safety ID (SID) Object

Attribute	Value
Index	0 × 1200
Name	Safety ID
Description	Unique identifier for device
Object type	VAR
Data type	Unsigned16
Category	Mandatory
Access attribute	FSF
SPDO mapping	No
Value range	0 × 0001 to 0 × FFFF
Value	No

SafetyNET p does not support a master–slave model and delegates the safety configuration management to the system application level. This facilitates the implementation of decentralized and distributed management structures.

The implementation of an ODIC is meaningful only if it is made accessible through the SDO services. Since the safety configuration has to be consistent in the whole system, the access to a single parameter makes little or no sense. In this case, in fact, the ODIC is only a conceptual model description that is used by engineering tools to create the correct SafetyNET p configuration, rather than a physical existing element within a SafetyNET p device.

Note, however, that a safety protocol stack implementation does not necessarily have to provide an ODIC since this is not visible from the outside. A more friendly data model could be used internally instead.

References

1. Safety Network International e.V., Industrial Ethernet communication System. [online]. Available: http://www.safety-network.de/en/technology/safetynet-p (Accessed on March 7, 2014).
2. IEC, IEC 61508-1. Functional safety of electrical/electronic/programmable electronic safety-related systems—Part 1—General requirements, 2010.
3. IEEE, IEEE standard for a precision clock synchronization Protocol for networked measurement and control systems, IEEE Std 1588-2008 (Revision of IEEE Std 1588-2002), vol. 1, pp. 1–269, July 2008.
4. IEC, IEC 61158-3-22, Industrial communication networks—Fieldbus specifications—Part 3-22: Data-link layer service definition—Type 22 elements, 2008.
5. IEC, IEC 61158-4-22, Industrial communication networks—Fieldbus specifications—Part 4-22: Data-link layer protocol specification—Type 22 elements, 2008.
6. IEC, IEC 61158-5-22, Industrial communication networks—Fieldbus specifications—Part 5-22: Application layer service definition—Type 22 elements, 2008.
7. IEC, IEC 61158-6-22, Industrial communication networks—Fieldbus specifications—Part 6-22: Application layer protocol specification—Type 22 elements, 2008.
8. European Committee for Electrotechnical Standardization, Industrial communications subsystem based on ISO 11898 (CAN) for controller-device-interfaces—Part 4: CANopen, EN50325-4:2002, 2002-12-12.
9. IEC, IEC 61784-2-18, Industrial communication networks—Profiles—Part 2-18: Additional field-buses profiles for real-time networks based on ISO/IEC 8802-3—CPF 18—SafetyNET p, 2010.
10. IEC, IEC 61784-3-18, Industrial communication networks—Profiles—Part 3-18: Functional safety fieldbuses—Additional specification for CPF 18—SafetyNET p, 2010.
11. IEC, IEC62280-1, Railways applications—Communication, signaling and processing systems—Part 1—Safety-related communication in closed transmission systems, 2002.
12. IEC, IEC62280-1, Railways applications—Communication, signaling and processing systems—Part 2—Safety-related communication in open transmission systems, 2002.
13. IEEE, IEEE standard for information technology—Telecommunications and information exchange between systems—Local and metropolitan area networks—Specific requirements Part 3: Carrier sense multiple access with collision detection (CSMA/CD) access method and physical layer specifications, Sections 1–5, 2008.
14. IOANA, The IAONA industrial Ethernet planning and installation guide—Release 4.0, 2003. [online]. Available: http://www.iaona-eu.com/home/downloads.php. (Accessed on March 7, 2014).
15. IEEE Standards Association. Use of the IEEE assigned ether type field with IEEE Std 802.3-1998 LAN/MANs. [online]. Available: http://standards.ieee.org/develop/regauth/tut/ethertype.pdf.
16. Internet Assigned Numbers Authority (IANA), Service name and transport protocol port number registry. [online]. Available http://www.iana.org/assignments/service-names-port-numbers/service-names-port-numbers. xhtml (Accessed on March 7, 2014).
17. D.J. Smith, K.G.L. Simpson, *Functional Safety—A Straightforward Guide to Applying IEC 61508 and Related Standards*, Elsevier, Burlington, MA, 2004.
18. G. Cena, M. Cereia, A. Valenzano, Security aspects of safety networks, *16th IEEE International Conference on Emerging Technologies and Factory Automation (ETFA 2011)*, Toulouse, France pp. 1–4, 2011.

29

Security in Industrial Communications

29.1 Introduction 29-1
29.2 Basic Security Measures 29-2
Security System Life Cycle • Common Security Measures
29.3 Security for Automation Networks 29-4
Fieldbus Systems • Industrial Ethernet • Wireless Networks • IP-Based Networks
29.4 Adding Security to Legacy Field-Level Networks 29-9
Network Protocol Aspects • Implementation of a Secure Node • IP Down to the Field Devices?
29.5 System-Level Approach—Defense in Depth 29-16
Three-Zone Security Model • Smart Grid Case Study
29.6 Conclusions 29-19
References 29-20

Thilo Sauter
Danube University Krems

Albert Treytl
Danube University Krems

29.1 Introduction

During the past 30 years, the main focus in the development of field area networks (or fieldbus systems) was on meeting the technical requirements of different application areas, resulting in a wide diversity of systems. Parallel standardization efforts, which still continue, have led to today's widely accepted conclusion that, in order to successfully cope with the requirements found in the application areas of interest, different coexisting solutions are needed [1]. Also conceived in the past two decades were the technologies that make up today's Internet, which, ultimately, led to the enormous surge in Internet usage in the 1990s together with the wide adoption of the Internet protocol (IP) family for the use in local area networks (LANs). This exponential growth of the Internet also sparked new interest in vertical communication flows according to the computer-integrated management (CIM) pyramid.

Vertical integration in the sense of transparent data exchange had been one of the essential goals of the CIM idea already in the 1970s [2]. Nevertheless, automation systems were mostly developed as closed environments without interfaces to the outside, partly also because of missing network solutions. It is thus no wonder that security considerations never played a prominent role in fieldbus systems. Only with the advent of remote access to automation systems and the broad user awareness of security problems in the Internet has security become an issue in automation. Still, it was not before the end of the 1990s that the importance of this topic was highlighted [3–5]. Extending the scope of typical field-level protocols to IP-based networks and interconnecting them even outside the premises by means of well-known IT standards, however, make the need for security obvious. A number of incidents affecting automation systems [6], such as the Stuxnet worm, have clearly demonstrated this. Therefore, solutions have to be found to combat potential security threats well known in the Internet or wireless networks [7–9].

Within the scope of integration in automation, some focus of the discussion is on the connection of field-level networks to *the Internet*. This can be done either via tunneling approaches or through various types of gateways. In this context, the notoriously bad security reputation of the Internet (which in its original form provided no security at all) mandates an awareness of field-level security issues. Still, it is necessary not to limit oneself to studying security issues arising within the context of fieldbus–Internet connections only. A more general approach should be taken instead, one that includes security measures on the field level itself where appropriate. Ultimately, security characteristics should become accepted properties used in the selection of a field-level communication system for a certain application. Security would therefore become as important as other properties such as topology, structure, bus access, and safety features.

29.2 Basic Security Measures

Security measures for IT systems in any organization aim to achieve three basic security goals, namely, confidentiality, integrity, and availability. These measures protect data from unauthorized entities and unauthorized manipulation and ensure that data are accessible when needed. In addition, another security goal that is often desired is nonrepudiation, which binds an entity to its transacted commitments.

Solutions implemented to achieve these goals are typically based on a security policy that explicitly states the following:

- The security objectives and the scope of the policy defining the organization's approach to risk and describing the assets of an organization.
- A definition of business and security goals to avoid conflicts and possible violations of the policy by the users of a system.
- Who is responsible for implementing and maintaining the policy. Security is a continuous process that needs to be updated regularly.
- A description of the implemented security measures and rules through which people are given access to an organization's assets.

Any such security policy must be backed by the organization's management and be communicated to users of the IT system in question. It is noteworthy that 80% of security needs to be invested in organizational measures that allow security technologies to work properly [10].

A good, informal starting point for developing a security policy is the *Site Security Handbook* [11], which is geared toward systems connected to the Internet. Another good introduction is the *IT Baseline Protection Manual* [12], which aids in the establishment of a security policy and additionally offers common best practices for the implementation of security measures similar to British Standard 7799 (later adopted as ISO/IEC 17799 [13]). Unfortunately, no guidelines for developing a security policy for field-level networks exist. Still, for them, the same basic steps need to be obeyed:

- Risk analysis for identification and categorization of risks; threats are manifold and can range from active (modification of data, replay of data, masquerading as an authorized user, denial of service to authorized users) as well as passive (eavesdropping, information disclosure, or traffic analysis) attacks.
- Threat analysis assigning an occurrence probability and extent of the damage to each risk giving them a certain weight in the context of the security policy.
- Analysis of weaknesses tracking vulnerabilities of the system.
- Planning and implementation of security measures.

Next, we discuss security in the context of the life cycle of IT systems and give an overview of the most common technical means to enforce security goals that are also applicable to field area networks (FANs).

29.2.1 Security System Life Cycle

While security activities can be started at any time in the life cycle of a system, it is usually desirable (though often not possible) to integrate them in the whole life cycle [14], where five stages are defined* (Figure 29.1).

During the initiation phase, when the system is designed, a basic security design is established by means of a sensitivity assessment. This assessment results in an estimation of the data to be handled by the system and its impact on security.

In the development stage, the security requirements as well as the methods with which to implement them are selected and included into the overall system specification. Then follows the implementation phase during which security features of the system are activated and tested.

In the operation and maintenance phase, security-related operations such as key exchanges take place. Also in this stage, system audits and monitoring may lead to system changes if new security risks that need to be addressed are found.

Finally, the disposal of information hard- and software needs to be handled. Here, the long-time storage of cryptographic keys is a major task.

Actual approaches such as the German VDI guideline 2182 [16] structure each phase in a closed loop starting with the identification of assets and security objectives, analysis of threats and risks, selection and implementation of measures, and finally ongoing audit that closes the loop. This structured continuous update of the security policy within a life cycle stage reflects the fact that new threats appear, but also that the applications and the configuration of networks continuously change.

29.2.2 Common Security Measures

The notion of security today is most often associated with the Internet. Most security measures presently available—with the exception of cryptography—are in some way tailored to the Internet, although they are also applicable to a wider range of systems. Without going into precise detail, this section introduces the following important security aspects:

- Cryptography
- Authentication and access control
- Firewalls and intrusion detection
- Security evaluations

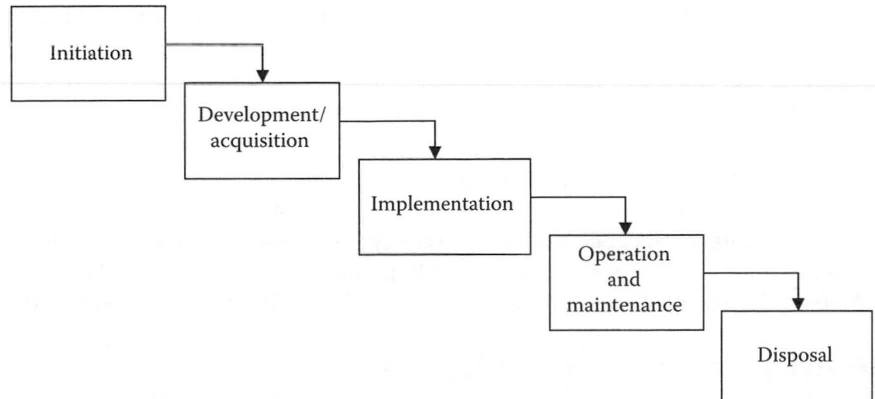

FIGURE 29.1 Basic waterfall life cycle model.

* The activities that follow can also be applied (in a slightly modified way) to other life cycle models like, e.g., the spiral model by Boehm [15].

Cryptography is appropriate for implementing services where confidentiality and integrity are needed. The major building blocks of cryptography are the following:

- Cryptographically secure pseudorandom number generators, used to generate seemingly random entities such as secret keys
- Cryptographic hash functions, either used in conjunction with electronic signatures or to generate message authentication codes (MACs) to protect data integrity
- Symmetric and asymmetric encryption/signature functions used to protect the confidentiality of data or to construct MACs or electronic signatures

Cryptography is discussed in Stinson [17] (with the exception of randomness) and in Menezes et al. [18], which covers all the topics provided earlier and which is a suitable reference for deeper investigation.

Authentication measures are used to establish the identity of an entity, for example, that of a system user. A clear distinction must be made between weak authentication (as found in password-based systems such as Unix or Windows) and strong authentication, which is usually based on challenge–response protocols that rely on either symmetric or asymmetric cryptography like Kerberos [19] and X.509 [20], respectively.

Authenticated entities can be subjected to access control mechanisms that allow the determination of the legality or otherwise of an entity's desired action. Apart from simple models such as access control lists and other discretionary models, mandatory models exist that are mainly found in military systems. Both models can be implemented by the newer role-based access control (RBAC) schemes [21].

Firewalls [22–24] and intrusion detection systems (IDSs) [7,25,26] are complementary technologies usually associated with LAN–fieldbus connections to the Internet, although both are also useful in a pure LAN environment or where a LAN is connected to a field-level network and the LAN is shared by different organizational entities. Firewalls try to prevent illegal access to inbound and outbound connections by, for example, packet filtering while IDSs monitor and analyze network access patterns and try to recognize security breaches and/or attacks, which are then referred to a system administrator (and/or responded to).

The building of a secure system is a complicated task involving a variety of different skills and knowledge, yet might still not succeed even if the necessary care is taken [27]. A prominent example of a failed security system is wired equivalent privacy (WEP), which was designed to protect 802.11 wireless LANs but which failed miserably to do so [28].

Third-party security evaluation is a method with which to build confidence into a system beyond the security statements of the manufacturer. The Common Criteria 2.0 (CC) [29], for example, is an international standard that regulates security evaluations and a positive evaluation of a security target (i.e., the product) against the CC results in an internationally acknowledged certificate. Besides this assurance, which relates to the efforts put into security engineering, the certificate itself contains a description of the security functionality that was assessed during the process.

Additionally, in order to help prepare for certification, there exist so-called protection profiles (PPs) that describe implementation-independent sets of security requirements for specific security areas, for example, firewalls. There are, however, currently no PPs for field-level networks available. Yet, there are initiatives for special application areas such as the German Smart Grid Protection Profile [30].

29.3 Security for Automation Networks

For automation networks, in particular, those used for the field level of the automation hierarchy, many different communication technologies are available today [1]. Classical fieldbus systems were the first networks, followed by a number of developments based on Ethernet. In recent years, wireless technologies are becoming increasingly important. All these different networks are different also with respect to their security capabilities.

29.3.1 Fieldbus Systems

It is a common prejudice that today's fieldbus systems do not exhibit any security measures. In fact, this is not fully true. There are such concepts in communication systems for building automation [31,32], and there are even traces in some industrial automation fieldbuses [33]. However, these approaches are quite limited in scope and capabilities. The fieldbus systems for industrial and process automation currently standardized by the IEC in the standards IEC 61784-1 [34] and the underlying IEC 61158 series [35] are—despite all efforts—a heterogeneous collection of solutions. As far as *security* features are concerned, however, they fall into two categories. Systems like ControlNet, P-Net, and SwiftNet use hardly any appreciable security strategy at all. Foundation Fieldbus, Profibus, WorldFIP, and Interbus share very similar application layer services and protocols modeled after the fieldbus ancestor MMS. The simple access protection mechanism is inspired by the owner/group/world access rights management in, for example, UNIX-based systems. Every object has a password associated with it (8 bits long) as well as a list of access groups (also an 8-bit word, corresponding to eight different groups). Both password and access group are transferred in plain [33]. If they match for a specific request, the additional access rights parameter further details allowed operations. These rights depend on the type of object (like read, write, execute, delete, etc.) and may be activated for the password (equivalent to the owner in UNIX), the groups, and all communication partners without restriction.

Unfortunately, these security means are not mandatory for implementation, which limits their usability. The password itself has no explicit protection, let alone encryption. This concept resembles weak security measures known from, for example, SNMP, where a community string is defined (even mandatory) to protect data access. Alas, this string is transmitted in clear text and hence rarely used at all (the default value *public* is hardly ever changed). A note in the standard reveals the actual intention of the security mechanisms:

> This is not a protection against intentional misuse of the communication facilities of a field device but helps to protect a system for accidental erroneous use of Process Data.

This statement is in fact a generic one and is true for all traditional fieldbus systems in industrial and process automation.

The situation is different in the area of building automation. The two most competing systems on the field level are LonWorks and KNX, both of which have dedicated security features [31]. LonWorks offers an authentication mechanism; KNX offers a password-based access control. However, both approaches are not sufficient from a security viewpoint: The cryptographic algorithm used by LonWorks is based on a four-step challenge–response protocol. The used cryptographic algorithm calculates a 64-bit hash value over the plain message and a random number using a shared secret key. However, the short key length and several protocol flaws render this mechanism insecure [31]. A more advanced one based on MD5 is available for LonWorks/IP. However, since MD5 is not collision resistant, it is also regarded as insecure. KNX provides only a basic access control scheme based on clear text passwords. Up to 255 different access levels can be defined, each of them associated with a different (otherwise unspecified) set of privileges. For each access level, a 4-byte password (key) can be specified. This rudimentary access control mechanism does not, however, provide strong security and is not available for the exchange of process data [36]. The third player in this area, BACnet, is used mainly on the management level, that is, above the others, and is the most recent of the three. The security mechanisms implemented here are the most mature ones and are largely oriented at the current state of the art in information technology, including an authentication service as well as security services that guarantee data confidentiality and integrity. These mechanisms use the symmetric Data Encryption Standard (DES) algorithm and a trusted key server responsible for generating and distributing the secret keys. A more recent security architecture defined in BACnet Addendum g uses AES and HMAC in combination with a message ID and a time stamp. Table 29.1 summarizes the security mechanisms of the various fieldbus systems.

TABLE 29.1 Security Services for Various Fieldbus Systems

Fieldbus	Security Services
Foundation Fieldbus	Eight access groups, 8-bit unencrypted password
ControlNet	Connection auth., unencrypted password
PROFIBUS	Access control for predefined addresses
P-Net	Simple write protection for variables
WorldFIP	Eight access groups, 8-bit unencrypted password
Interbus	Eight access groups, 8-bit unencrypted password
Swift-Net	None
LonWorks	Challenge–response auth., integrity check, MD5 for IP
KNX	Unencrypted password for management access
BACnet	DES-based encryption, integrity check, authentication
BACnet Addendum g	AES for data confidentiality, HMAC for data integrity, time stamp + message ID for data freshness

The fact that the building automation systems care more about security is not astonishing. First, they were developed significantly later than many industrial automation networks. Second and more important, such systems are far more complex because the number of nodes involved in a single building network can exceed by far that in a production plant comprising usually well-identifiable subsystems. The most important aspect, however, is that potential intruders can have nearly unlimited access to the network.

29.3.2 Industrial Ethernet

It has already been stated that security was not a primary concern in the fieldbus development, and that consequently, the mechanisms to protect data and equipment from unauthorized access in contemporary fieldbuses are very limited. By the time the importance of security in automation became obvious [3,4], it was already too late to introduce changes. As far as the field level is concerned, there are in fact not many options to build secure communication channels [37]. With the introduction of Ethernet in automation, a reconsideration of field-level security was possible, and one of the biggest potentials in the development of these new networks would have been to integrate security features on a low level from scratch [38].

The overview of the industrial Ethernet profiles defined in IEC 61784 and their main security features in Table 29.2 shows, however, that unfortunately, this opportunity was not really taken in practice, and the Ethernet-based control networks still today leave security aspects mostly out of their scope. Security mechanisms, if any, focus on special switches to block traffic from network segments or on

TABLE 29.2 Industrial Ethernet Profiles Defined in IEC 61784

IEC 61784 Profile	Security Services
CPF1—Foundation Fieldbus	Simple access protection, optional ISO/IEC 8802-101.
CPF2—Ethernet/IP	No security measures specified.
CPF3—PROFInet	Protection within switches to avoid overloading of segments with real-time traffic.
CPF4—P-NET on IP	No specific security measures specified.
CPF10—VNET/IP	Security should be in data link layer but is not specified.
CPF11—Tcnet	No specific security measures specified.
CPF12—EtherCAT	No specific security measures specified.
CPF13—Ethernet Powerlink	Protection within switches to avoid overloading of segments with real-time traffic.
CPF14—EPA	Firewall-like security bridges between field domains and process monitoring domains or the Internet.
CPF15—MODBUS RTPS	Access control based on IP addresses.
CPF16—SERCOSIII	No specific security measures specified.

proper device design to avoid overloading of the implemented protocol stack. If security is needed that goes beyond simple traffic management aspects between segments (insight-outside view), upper layers of the protocol stack have to be used.

Nevertheless, many industrial Ethernet solutions permit IP traffic in parallel to time-critical process data communication. Therefore, the same technological communication base as in most LANs and the Internet is available at the field level and in principle permits a seamless application of security mechanisms known from the Internet world. Secure versions of the TCP, UDP, and IP stack might offer valuable security measures. This can be on the network layer, for example, by using IPsec [33], on the transport layer (e.g., using TLS/SSL), or on the application layer (e.g., Kerberos). The disadvantages of these protocols are as follows:

- They decrease the real-time properties significantly. Hence they are mainly applicable to secure transport of configuration data and data transmitted (over public networks) to superior management units.
- The complexity of the secure protocol stacks poses heavy requirements on the computational power of the embedded systems found on the field level and sometimes even on the control level. In particular, asymmetric cryptography puts a high load on the field devices.

At any rate, it must be noted that the possible use of secure versions of the IP suite is not genuinely covered by the industrial Ethernet specifications; it remains a workaround for a very limited range of applications, as will also be shown in Section 29.4.

29.3.3 Wireless Networks

Following their success in IT systems and consumer electronics, wireless communications are more and more applied also in automation infrastructures, mostly as hybrid solutions together with wired networks [39–41]. However, industrial wireless networks pose particular security challenges. The typical requirements regarding integrity, authentication, and availability need to be transferred from the traditional wired network domain to wireless networks without impacting the applications' real-time requirements. In comparison to wired installations, which obviously benefit from physical access limitations, for wireless transmission, it is much harder to limit the area of *network* availability because attackers can intrude the system at arbitrary locations. Apart from reliability issues that are still a major obstacle for a straightforward use of wireless technologies, the open medium is evidently prone to eavesdropping and message insertion. Hence the most common wireless network protocols already offer built-in security services on the data link layer, such as data confidentiality, data integrity, and replay protection, though providing different levels of complexity and robustness.

The IEEE 802.15.1 standard supports mutual and one-way authentication and encryption. Four different entities are used for providing this: the device address (48 bits), an authentication key (128 bits), an encryption key (8–128 bits), and a pseudorandom number (128 bits). During the initialization of a connection, the secret keys are derived. The encryption key has varying size and is always based on the authentication key and a pseudorandom number.

IEEE 802.15.4 supports security mechanisms at the data link layer, and the application can choose between eight different security levels depending on its requirements. The security services for no security, only authentication, only encryption, and encryption and authentication are realized using AES for encryption and message authentication applying Cipher Block Chaining (CBC), Counter (CTR), or CCM mode. The message integrity code for data authenticity can have increasing levels. The security classes are using authenticated encryption algorithm CCM*, which is an extension of the generic counter with cipher block chaining MAC (CCM), approach. ZigBee is defined on top of IEEE 802.15.4 and thus utilizes its security mechanisms. However, as important implementation issues are left open in IEEE 802.15.4, ZigBee enhances them with more advanced security services, such as the support of different key types and key management services.

Security services in 802.11 WLANs were initially provided by the WEP algorithm. However, many potential and critical flaws were discovered, and the algorithm was ultimately broken in 2001.

The 802.11i amendment now specifies the robust security network for authentication and encryption. The authentication of a client to an access point consists of a sequence of the mandatory open systems authentication and association, followed by an extensible authentication protocol involving a third entity, the authentication, authorization, and accounting server, in order to derive the 512-bit master session key (MSK) and the pairwise master key (PMK, consisting of the first 256 bits of the MSK). Finally, the PMK is used to derive transient keys for the encryption during the four-way handshake used to set up the secure data exchange.

29.3.4 IP-Based Networks

IP-based networks dominate the management level and also backbone networks interconnecting separated automation systems. Recently, motivated by the low costs for IP cabling and network components, even small microcontrollers are equipped with IP stacks, and IP is gaining a strong momentum at the field level. This IP connectivity down to the field node is also in favor of a real end-to-end communication transparency supporting originally management-level services such as web services or context management.

From the security point of view, this openness and standard conformity makes such networks more prone to security attacks. Many of these vulnerabilities stem from the fact that IP originally had no security measures built in, and that even IPv6, the newest version of the protocol, has its native security mechanisms disabled by default. Threats can be classified into accidental attacks caused by interference of traffic stemming from other applications that now can share the field-level network since gaining (remote) access to the network is easier. Byres and Lowe [10] list such incidents on automation networks. The vision of isolated automation islands is definitely outdated on a protocol level but also often even on a physical level equal to office standards.

The most common IP security measures suitable and used for industrial communication systems are listed in the following:

Internet Protocol Security (IPsec) [42] is the native security extension to the IP v6 implemented on the network layer, which also has been ported to IPv4 commonly used in today's (automation) IP networks. IPsec offers services for data integrity, replay protection, authentication, and confidentiality natively within the IP layer. It is implemented in the two independent services: the Authentication Header (AH) providing authentication and integrity protection of complete IP packets, and the Encapsulating Security Payload (ESP) offering only encryption and integrity verification of the payload [43,44]. In both services, a cryptography integrity check value (ICV), commonly implemented by an HMAC-SHA, is used for message integrity protection and authentication. For encryption, various algorithms such as 3-DES or AES can be selected. For key exchange, the Internet Key Exchange (IKE) protocol is used, which uses asymmetric algorithms like RSA, ECC, or symmetric algorithms with preshared secret keys, alternatively. All security parameters are maintained in a security association (SA) setup between communication partners. Most important for automation networks is the fact that IPsec can operate in two modes: transport mode, which adds security headers (and trailers in ESP) to protect a single packet, and tunnel mode, which, on the other hand, allows securely interconnecting two networks. In this mode, the original IP packets are transmitted as payload of a new security IP packet between two networks. The big overall advantage of IPsec is that its security services are independent from applications and transparent for all traffic. This fact can tremendously ease the security management in multiapplication environments. Additionally, in contrast to layer 2 security services, IPsec is also transparent beyond network borders and avoids possible vulnerabilities caused by a break of end-to-end security at network borders.

Transport Layer Security (TLS) [45] as well as its ancestor Secure Sockets Layer (SSL) is a protocol developed for securing communication between two entities on layer 4 of the OSI stack. TLS can be divided into five sub-protocols: TLS record protocol is responsible for the end-to-end encryption via symmetric ciphers like 3-DES or AES and for message integrity protection and authentication using

MACs like HMAC-SHA. Based on this protocol are the TLS Handshake protocol that is responsible for the initial negotiation of security parameters and the symmetric keys for the TLS record protocol, the TLS change chiper spec protocol changing the ciphers according to the negotiated security parameters, the TLS alert protocol implementing error notification, and finally the TLS application data protocol responsible for fragmentation and management of payload data. The initial negotiation of security parameters is based on asymmetric algorithms and public-key certificates. It serves the proper authentication of sender and receiver and the derivation of the symmetric keys for the TLS record protocol using the Diffie–Hellman key exchange. During the following transmission, payload data are then transmitted fully transparent and not interpreted at all. Both a drawback and an advantage is that TLS only protects traffic between two entities. Hence, there is no possibility to securely interconnect two networks like in IPsec tunnel mode, yet TLS is also very flexible to be retrofitted into existing protocols such as web services, mail, or network management. For example, [46] shows an implementation of a complete secure web server using HTTP and TLS.

Virtual private networks (VPNs) are a technique to interconnect multiple private networks via a public network in a secure way. One can distinguish end-to-site VPNs where nodes are integrated in a private network, end-to-end VPNs creating a completely virtual network, and site-to-site VPNs connecting two compatible private networks. In all cases, the VPN creates a private, closed, and individually configurable virtual network between the connected entities. In contrast to native network security such as IPsec, VPN does not use the security of the public network. IP packets are encapsulated in TCP or UDP packets carrying the VPN packets that are protected by TLS. Most common are end-to-site VPNs for remote access to an automation network fully integrating a computer of, for example, a service engineer into the private automation network via a VPN gateway. Site-to-site VPNs are usually implemented using IPsec and not by a dedicated VPN software. A widely used VPN implementation for IP-based networks is OpenVPN [47]. In OpenVPN, each node opens a secure unicast connection to the VPN gateway, tunneling the whole network traffic to and from the node (end-to-site). The tunneling connection is authenticated by preshared symmetric keys, a username and password, or TLS certificates. The data transmission is then protected using TLS.

It is noteworthy that most of these measures have been designed for different applications and therefore have to be carefully deployed especially if hard real-time and low-bandwidth constraints need to be observed. The next section is addresses such issues.

29.4 Adding Security to Legacy Field-Level Networks

What is currently at hand on the field level in terms of security mechanisms is largely not very promising. Hence, if security is an issue in a given application, appropriate add-ons must be designed. When considering extensions, the common consent is to obey Kerckhoff's principle stating that only well-known and well-tested security mechanisms should be employed, which in turn means using mature approaches known from the IT world. While this is in general the only viable solution, one must observe that the IT and automation environments are considerably different:

- Security protocols for the Internet such as SSL/TLS [45] and IPsec [42,48] operate on top of or at network layers (according to ISO/OSI). These layers are not defined in many fieldbus systems where, for reasons of speed and simplicity, one normally finds only layers 1, 2, and 7.
- Fieldbuses have no built-in provisions for security and are usually designed as lightweight communication systems. Security services added to an existing application area might introduce a large overhead.
- Cryptographic operations such as encryption needed to secure field-level communication might be prohibitive because of the limited processing power of current fieldbus nodes, even more so if real-time needs have to be addressed. Cryptography might thus be limited to maybe configuration or management operations.

- A fieldbus might be installed in *hostile* surroundings where the transmission media and the fieldbus nodes are located at the premises of the adversary. This is the case in building automation; another extreme example is the case of remote metering [49,50].
- The variety of fieldbus systems is too high to allow for unified solutions. Well-known security mechanisms fitting for one fieldbus might be inappropriate for another.
- Proprietary security extensions added to a given fieldbus protocol will inevitably lead to incompatibility with the standard and legacy installations.
- The application of cryptographic operations might be prohibitive in some application areas, for example, where real-time responses are demanded. Execution times on fieldbus controllers with limited processing power in general exceed overall response times or individual channel access limits.
- Security and easy network management are often contradictory. In some cases, it is possible to access configuration and application data of a node without authentication and to reprogram a node by directly writing to its memory [3].

Taking all these into account, the only option left is to use the fieldbus simply as a transport channel and to tunnel secure packets over standard fieldbus protocols and services. Such an application-level approach could easily achieve end-to-end security, which is desired in most applications, anyway. Unfortunately, adding such a security mechanism is likely to cause problems with interoperability unless all nodes in the network adhere to this enhanced standard, that is, a mixture of secure and insecure devices would normally not be feasible. Another problem is the limited message size in some fieldbuses that leaves only little room for efficient security extensions. Especially, integrity services require additional data blocks for, for example, hash codes, thereby reducing the actually available payload per packet and consequently the performance of a secure channel.

29.4.1 Network Protocol Aspects

While security was for a long time no issue in automation networks, things went faster in the world of tightly interconnected computer networks. Security was already discussed in the Internet predecessor Arpanet [51], but without any direct consequences. The security deficiencies in the overall Internet design were recognized only in the 1980s. Finally, in 1989, but still long before the invention of the WWW and the subsequent success of the Internet, the ISO issued a new part 3 of the OSI model covering security issues in open communication systems [52]. This amendment identifies several security services and maps them to the seven layers of the model for possible implementation (Table 29.3). Furthermore, it describes

TABLE 29.3 OSI Security Services in the Different Layers

Security Service	Possible Layer(s)
Peer entity authentication	3, 4, 7
Data origin authentication	3, 4, 7
Access control service	3, 4, 7
Connection confidentiality	1, 2, 3, 4, 6, 7
Connectionless confidentiality	2, 3, 4, 6, 7
Selective field confidentiality	6, 7
Traffic flow confidentiality	1, 3, 7
Connection integrity with recovery	4, 7
Connection integrity without recovery	3, 4, 7
Selective field connection integrity	7
Connectionless integrity	3, 4, 7
Selective field connectionless integrity	7
Nonrepudiation origin	7
Nonrepudiation delivery	7

security mechanisms that may be used to achieve the desired services. While the proposed security architecture is not as widely used and known as the original OSI model, the layered approach is useful, especially for networks that do not implement all seven layers, which is the usual case for fieldbus systems. Security services above the application layer, which are often desirable to achieve end-to-end security, are out of scope and therefore not covered in the standard. The following discussion adheres to this model.

Addressing the dominant security goals mentioned in Section 29.2, we now focus on possible solutions for confidentiality and integrity. Provisions to ensure availability in a field-level environment (apart from the provisions already designed into the fieldbus) will be hard to achieve. In the following discussion, the placement of security functions according to ISO/OSI will be helpful. Attention needs to be paid to the often reduced set of implemented OSI layers. Concerning the placement of confidentiality services in the network stack, the majority of these are placed at layers 1, 2, and 7, although placement above layer 7 is also possible. Integrity services can be placed at layers 3, 4, and 7 of the OSI stack or, again, above layer 7. However, the restriction to layers 1, 2, and 7, often necessary in a fieldbus environment, limits the application of integrity services to layer 7 or above.

Taking into account that on the field level one often finds dedicated microcontrollers designed to handle a specific fieldbus interface and protocol, a security solution that is targeted above layer 7 appears to be a sensible choice in all cases where one tries to secure an established fieldbus system. Moreover, in systems where one finds an abstraction layer above layer 7, an inclusion of security services in this layer would be of special interest. Such a solution would have the special benefit that the user layer of a fieldbus system is typically targeted at the standardization of interoperable services or application profiles. The process of standardizing such services is carried out by a user group (usually comprised of the organizations that are involved in the development of the fieldbus), leading to a broad acceptance in industry. Apart from the expected acceptance, this solution is also of security interest as it provides end-to-end security in its narrowest sense: it secures all communication between two nodes at the application level. In the absence of a widely accepted user layer into which security services could be integrated, or in instances where a specific user group sees no necessity to add such services (as may often presumably be the case), a generic solution above layer 7 can still be devised. The drawback though would be a loss of interoperability with other solutions, resulting in increased development efforts and consequently increased costs.

In both approaches, security services could be provided by a wrapping mechanism. This mechanism takes an engineering value (i.e., a measured input value that has either been sampled with a sensor or first sampled and then preprocessed at the node) and applies this value to either an integrity mechanism or to both an integrity and a confidentiality mechanism. Figure 29.2 shows an example of the second case where the field header contains information about the security transformations applied. The MAC field contains integrity check information.

The integration of both integrity and confidentiality mechanisms into most of today's fieldbus systems could be prohibitive as they require processing power that might be out of range for the typical low-cost 8-bit microprocessors found in typical environments. Another drawback of solutions such as that depicted in Figure 29.2 is that the limited payload of fieldbus messages might not allow for the message

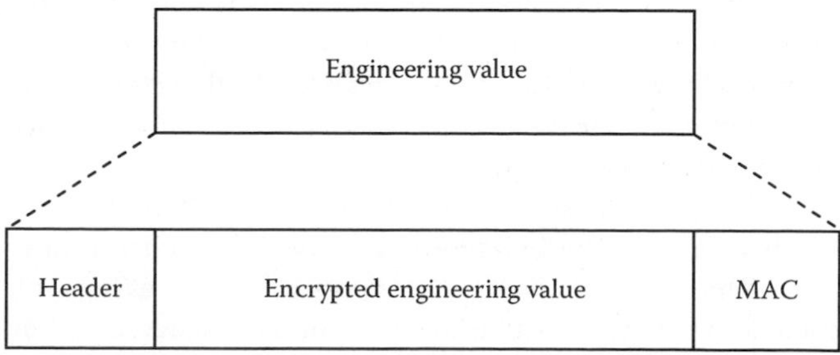

FIGURE 29.2 Wrapping an engineering value.

expansion needed to add confidentiality and/or encryption. This is a problem especially for integrity mechanisms as they always add additional data to the transmitted message. For *confidentiality-only* mode, one could simply encrypt the engineering value and omit the header of the message. In this scenario, the two communicating nodes would need to know in advance whether or not the data to be sent/received was in encrypted form.

If missing processing power in the nodes is a concern, one could also think of applying the necessary security services to the lower layers in the OSI stack. Although a first look at Table 29.3 shows that only confidentiality services can be added at the lower level, it is still possible to add the required services at layer 2 if the necessity arises. This has actually been the case with Ethernet, where IEEE [53] augments the services found in the OSI security model to also include integrity, authentication, and access control. While this could be a feasible solution, one has to keep in mind that the resulting protocol will also need to support nonlayer 2 management functionalities handling, for example, key exchange.

A solution for providing confidentiality and/or integrity would, however, still leave a major security threat unanswered: network management. In today's fieldbuses, this functionality is typically achieved in an unauthenticated manner, an approach that is tantamount to an invitation to an informed attacker who has gained access to the network. Management functionality might even include the possibility to totally reprogram a node by, for example, directly overwriting parts of the node's memory. This would enable an attacker to gain complete knowledge about the programming of the node by reading out configuration and application data from the memory. It is not difficult to imagine situations where such behavior is not acceptable.

The minimum functionality needed to achieve security in this area would be the provision of an access control mechanism, a task that could be supported by integrity mechanisms that allow the authentication of configuration communication messages. The implementation of access control for network management should pose few problems. While the data transmitted might be larger than usual messages and maybe even multipart messages might be needed, such operations are usually not time critical. Restrictions as to the acceptable maximum length of a message and the maximum time a transmission may take are also usually very relaxed.

The situation is even more challenging when the physical access to the field devices as well as to the underlying communication media is controlled by a user (or a group of users) who is not the owner of the system [37]. From a security viewpoint, such users have to be considered to be adversaries. Common examples are a utility company's remote metering device deployed at a customer's site in a smart grid scenario or a fieldbus deployed in a car. Another comparable situation that has been well studied, but where neither security mechanism nor attack results are openly published, is the transmission of secure pay-per-view programs via cable or satellite where only entitled subscribers should be able to unscramble the program.

Such uncontrolled installations seem to be more at risk because the adversary has more possibilities to circumvent possible security measures. As far as ways and means to secure them concerned, one natural possibility are physical measures to secure the different nodes as well as the sensors and actuators connected to them. Depending on the security policy, this can lead to simple solutions where seals are attached to the devices (as is often done today to prevent the manipulation of power meters). In this case, it is the legal implication associated with the removing of the seal that acts as a deterrent. Technologically stronger solutions attempt to construct tamper-proof devices that either notify the system owner if they are tampered with or deactivate themselves. In the latter case, apart from trying to prevent physical access to the nodes, one usually finds special security modules such as smart cards that are useful to protect sensitive data as well as to recognize attempted tampering.

If it is not possible or feasible to physically protect the nodes, their programs, and their memory, then security considerations regarding lower layers of the network stack are in vain: the protection possibly achievable at those layers does nothing to prevent possible attacks at higher levels. Together with physical node protection, lower-layer security is of course of interest again and might help to remedy problems that could arise with the use of integrity services.

Continuing considerations regarding physical security and the automatic deactivation of nodes in case of tampering, the possibility to include intrusion detection mechanisms in a node seems to be an interesting topic of research for the future. Such a mechanism could become feasible in installations that remain unchanged after deployment, meaning that no nodes are added or removed and that the communication patterns between the nodes remain unchanged or only change in a way that can be anticipated. If such an analysis were possible, a node that detects an intrusion could either contact the system owner or deactivate itself. The actual action taken would depend on the system configuration, for example, a failure to contact the system owner after detecting an intrusion would also result in deactivation. It would, however, also be necessary to determine the correct point in time for deactivation as, for example, it would be both inconvenient and dangerous if communication system in a car were to stop working while the car was being driven.

29.4.2 Implementation of a Secure Node

In the absence of native built-in security mechanisms in the communication system, the hardware is also usually not prepared for security extensions. One possibility to cope with both problems is the introduction of dedicated security tokens like smart cards that are widely used in, for example, cellular phones. The benefit of such devices is that they can be used as crypto coprocessor to execute all security algorithms needed for en- or decryption and authentication of data packets. Furthermore, they are a secure container for the keys required for the crypto algorithms. There has always to be one smart card on either side of a communication channel to achieve security. In fact, from the viewpoint of security, the communication takes place directly between the smart cards, and all other network entities see nothing but encrypted and thus incomprehensible data packets.

It must however not be overlooked that the integration of security devices is not at all straightforward. Rigorous security entails the restriction of communication relations, which is usually a problem for all broadcast- or multicast-oriented functions. The performance of fieldbus node processors and the secure serial communication links to smart cards is limited, so that the encryption or authentication may be too time-consuming for real-time process data. Finally, the physical integration in the node hardware is a challenge as well. Fieldbus systems usually have a highly sophisticated network management, which is optimized for efficiency and extensibility and often needs access to at least parts of the controller's internal memory [3], which constitutes a potential security leak. So, if the smart card together with some glue logic is placed between the fieldbus controller and the sensor device, the fieldbus management can no longer access the memory. If both components are integrated behind the fieldbus controller (seen from the fieldbus), a preprocessing of the sensor data in the controller is no longer possible and makes the whole device relatively inflexible.

The following discussion uses LonWorks as an example of how to extend fieldbus nodes by smart cards to implement security services for a secure end-to-end communication. Nevertheless, the main results are applicable to many other fieldbus systems. A schematic of this setup is shown in Figure 29.3.

Since the whole protocol stack of LonWorks is implemented in hard- and firmware within a dedicated microprocessor, security measures have to be implemented above layer 7. In this case, the system still fully complies with the LonWorks standard although interoperability is confined to secured nodes by defining special user-defined structures. If an attacker has appropriate tools and physical access, nodes are not well protected. Properties of a node from simple configuration properties to the whole application and data storage can be read and often changed with common administration tools. Some kind of security token must, therefore, be introduced that securely stores the secret keys and can execute cryptographic operations in a secure manner.

The first step to achieve this goal is the implementation of a smart card interface for the node. Although all interfaces to a smart card are well defined [54], it is generally not possible to integrate any of the ISO 7816 protocols in software, due to limitations in memory and computing power of the node.

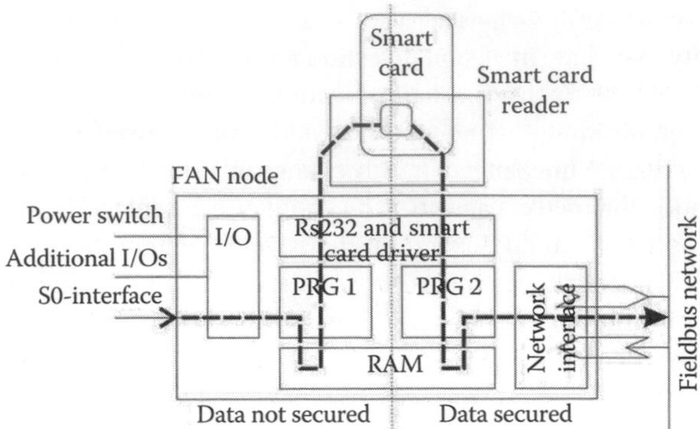

FIGURE 29.3 Schematic of secure fieldbus node.

Hence, an appropriate smart card reader has to be selected, which is a complicated task since only few readers offer a lightweight protocol for low-end microprocessors.

In a second step—during operation of the fieldbus—the raw sensor data (in this example from an S0 power meter input) are preprocessed by the node and transmitted to the smart card that secures the data to be transmitted over the network. The integrity of the data is secured by the HMAC [55] algorithm using SHA-1 as the hash function and encrypted using the 3-DES algorithm. It should be emphasized at this point that the design of a proprietary security algorithm is usually a bad idea and highly error-prone. Instead, one should use algorithms that have been scrutinized by the security community for a substantial amount of time and are considered to be secure. In a third step, the secured data are finally passed to the network protocol stack and transmitted to the recipient's node, where the procedure is executed in reverse.

The selection of the applied security measures is determined not only by the strength of cryptographic algorithms but also by the restrictions of the network. One of the most important restrictions is the limited packet size of fieldbus protocols, because often packet segmentation and reassembly mechanisms (SAR) are not available. In the presented setup, the packet size is limited to 31 ASCII-encoded characters. According to prior research [31], this is no limitation for the processed 2-byte sensor value and should be fine with most other data to be transmitted. However, this limitation could cause problems with asymmetric encryption schemes like RSA [56], which requires a minimum packet size of 128 bytes to achieve adequate security.

Another important issue is the additional delay introduced by transmitting data to and processing the data within the smart card. Whereas this usually is not a big issue for building automation control systems, where delays introduced by the smart card communication can be neglected in normal operation, the impact on networks with real-time requirements such as in industrial automation systems must be analyzed much more carefully. Finally, even in this setup, the node's memory is still unprotected, therefore allowing the possibility of data being read before they are encrypted (Figure 29.3, left side). Solutions to this problem would most likely need additional hardware or the redesign of existing nodes to supply protected areas.

29.4.3 IP Down to the Field Devices?

It is predicted that IP will take on an important role in many control application fields. One of the most prominent fields is the energy sector and the future smart grid communication solutions [57]. Advantages such as improved interoperability and seamless interconnection across network borders in hybrid wired and wireless networks and backbone networks are major benefits seen in having a common protocol independent of the used network technology. But deploying IP and IP security to field-level networks in a straightforward way often violates application requirements such as available bandwidth or latency important on the field level.

In the following, the transport of secure IP over a highly resource-limited field-level network, namely, a last-mile powerline communication (PLC) network, is presented as prototype to efficiently tunnel IP packets. The example is based on IPv6 [58], which offers advantages for the smart grid application due to its much larger address space than IPv4 and its improved packet format designed for efficient packet header processing. Yet, results are also applicable to today's most common IPv4 and other applications, since a versatile IP channel transparent to the field-level network is set up. Using the IPv6 protocol and its security extension, IPsec (see also Section 29.3.4) carries a significant overhead, which must be managed to retain its benefits.

The general approach to reduce the payload penalty of IP is to compress in particular the large address fields of IP (each 16 bytes large) and omit/reconstruct unused flags. Robust header compression (ROHC) [59,60], which is a common protocol that can strip the 40-byte IPv6 header down to 2–4 bytes, was used in this case study. Although ROHCv2 also supports the compression of IPsec IPv6 extension headers, security has two inherent drawbacks for compression: First, encrypted data are random data by nature that cannot be compressed, and second, the combined application of AH and ESP protocols required by the security requirements of many industrial applications produces a large overhead of cryptographic ICVs that are also noncompressible data.

To address the first issue of payload compression, the order of compression is important. In particular, for IPsec tunnel mode, the application protocol data unit data have to be compressed before IPsec protection. For integrity check, the order is not mandatory, yet in the above order, less effort is required for the cryptographic calculation of the ICV, especially interesting for the verification of authenticity at the receiver. Additionally, the IPsec protection building block in Figure 29.4 includes specific compression for the different header and trailer fields of IPsec. In particular, Next Header and Payload Len, Reserved data, Sequence Number fields, and ICV padding can be reduced for transmission under special application-dependent conditions [61].

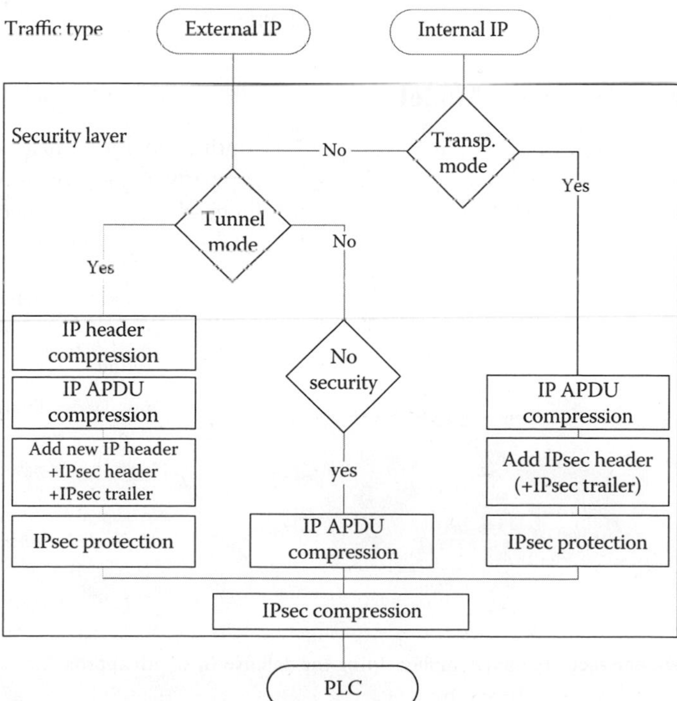

FIGURE 29.4 Security module for efficient transmission of IPsec (internal IP stems from the device hosting the security module, and external IP refers to traffic of an additional network connected to the device).

The second issue stems from the design of IPsec offering two independent security services, namely, AH protecting the integrity of the complete packet including header fields, and ESP only guaranteeing the confidentiality and integrity of the payload, both having their own ICV. Since for automation applications, in general, both services are required,* two 20-byte ICVs and two headers with 30 and 12 bytes have to be transmitted offering no additional security. This is a strong drawback not only concerning packet size but also concerning computational power required to (cryptographically) process these packets. Hirschler and Treytl [62] introduce a new approach called ESP+ to overcome these issues. ESP+ is a modification of the ESP scheme merging the full integrity protection of AH with the ESP ICV calculation offering a constant overhead for all service calculation equal to the smallest IPsec header. Savings up to 50% of the original bandwidth required by standard IPsec can be achieved. Additionally, ESP+ is more energy aware, since one CPU-intensive intense cryptographic operation can be omitted. One of the big advantages of ESP+ is the full standard compatibility. Standard compression such as ROHC can be directly applied, and if required, ESP+ can be used to reversibly compress a combined AH-ESP service if the FAN is just used as a transport network. For smart grids, this is very important, since the backbone networks are not expected to use ESP+, and full IP(sec) transparency down to the nodes is targeted.

29.5 System-Level Approach—Defense in Depth

Given that automation systems today typically consist of a hierarchy of networks, one can also adopt a hierarchical security model following the defense-in-depth approach first suggested in [37]. The idea is to focus on the network levels that are normally separated from each other by dedicated network nodes. These nodes, which link the typical field-level network with the upper network levels, are usually equipped with more powerful hardware and can be used as anchor points for the security strategy. Defense in depth relies on different security layers to protect valuable assets: Using the famous onion analog on, we see that removing the outmost (security) layer reveals another layer and many more remain to be peeled away before the assets become vulnerable.

29.5.1 Three-Zone Security Model

The three-zone security model shown in Figure 29.5 is optimized for industrial communication applications and follows the typical network levels inside a company [63,64]: The inner zone comprises the field-level communication systems located at the shop floor, where security mechanisms might be

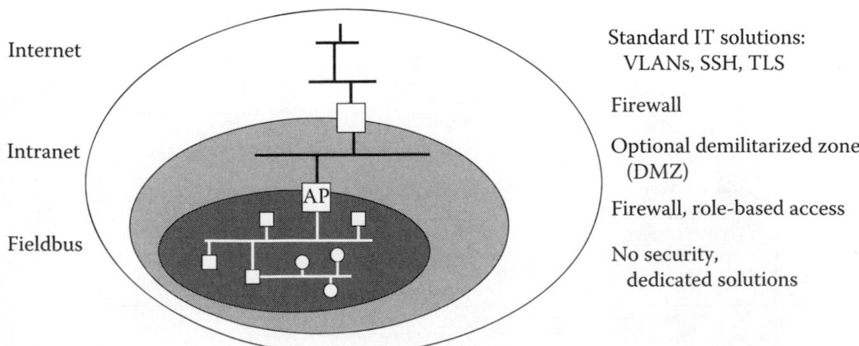

FIGURE 29.5 Three-zone security model implementing the defense-in-depth approach.

* First complete packet authentication is required to identify counterfeit components and devices injecting malicious packets in the network. This is only guaranteed by AH. Second confidentiality is needed not only for privacy protection of consumers but also to prevent disclosure of information between the multiple utilities using the same infrastructure.

limited as discussed in the preceding sections. The intranet zone is often built upon IP-based LANs inside the plant and may also include demilitarized zones (DMZs) and checking domains or inner security structures to strengthen the security. Finally, the Internet zone connects multiple plants, customers, or suppliers horizontally along the value chain. On the IP-based levels of the automation system, the standard security mechanisms known from the IT world such as IPsec or TLS can be used. In the wake of e-commerce, such setups have been considered rather early for the horizontal integration of enterprises, leaving aside the special requirements of automation [65].

The nodes separating different zones always have been natural points to implement security. They are the only part of the automation system where security can easily be applied, since most field-level systems lack security. In the following, this node is called access point due to the fact that it often includes protocol translation and other means to negotiate between the two networks.

The access point itself has so far been the focal point of interest for most researchers. One very common approach in the defense-in-depth model is to combine the access point with a firewall, which is the most widely employed security measure in the design of network interconnections today even in the field of automation [66,67]. For higher security, it is also possible to implement the access point as a DMZ, where the gateway is shielded by two firewalls on each side.

The main purpose of a firewall is to separate a private, restricted intranet from a network that is accessible to a larger community [24,68]. This separation is normally implemented on the IP level by granting access only to computers from a predefined set of IP addresses or domains. Only requests from such addresses are accepted by the firewall, and all others are denied. This requires that the firewall should also work as an IP router connecting the private IP segment and the *open* segment reachable from the Internet. As the entire network traffic is controlled by the firewall, the intranet behind can be completely hidden from the outside world. The other way, from the intranet to the Internet, is less restricted: The firewall directly forwards the request from the internal node. Ideally, the node in the intranet does not even notice the presence of the firewall. Due to this asymmetric operation, the firewall is best tightly integrated into the access point and use, for example, port forwarding to control the traffic [67]. Additionally, a typical scenario is to make the firewall a VPN gateway, where the firewalls route the control network traffic into the VPN. Such implementations are mostly based on IPsec and are especially intriguing as key distribution is built in. A VPN also provides an easy answer for IP-based extensions of clear-text control network protocols.

Apart from providing basic firewall functionality, the access point is also the ideal point to control and manage access to the fieldbus zone. For the implementation of any security policy, end-to-end security is desirable, that is, the client in the Internet establishes a secure channel to the data source in the fieldbus irrespective of the network structure in between. In standard e-business, this can be achieved because only few entities are involved in a secure communication. In automation applications, the number of nodes needing secure communication is potentially very high. In fact, every sensor or actuator is affected, which makes the handling of communication relations and thus security keys extremely complex, even if technical aids like security tokens are employed [37]. A practicable compromise is to intercept the secure channel at the access point and introduce a two-step communication architecture (Figure 29.6).

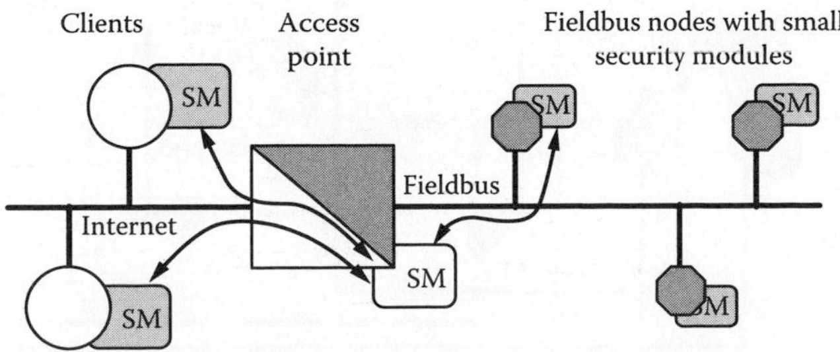

FIGURE 29.6 Two-step security architecture using security modules or security tokens (e.g., smart cards) for improved key manageability.

The access point maintains a list of the authorized connections in the Internet zone and directly controls the access to data objects or communication channels, depending on the access point's operation as tunnel entrance or gateway [68]. The fieldbus nodes, on the other side, only need to connect to *their* access point and can get along with smaller security tokens, if any [3]. This solution makes updates of the user structure easier [69]. A suitable model for handling the access rights in such a scenario is RBAC [21], where all communication partners (users, devices, and tools) are associated with particular roles depending on the context of the data exchange. In the IT world, this concept is widely employed, and it can also be used in an automation context [70,71].

29.5.2 Smart Grid Case Study

Smart grids are among the most novel large-scale industrial networks and transfer sensitive data that must be properly protected. Blackouts, privacy issues, poaching of customers, or even terroristic attacks can be results of missing security [72]. Since smart grids are relatively young, a lot of architectures for the communication systems exist, and multiple communication technologies are deployed. The following overview is based on a case study done in the project DLC+VIT4IP [61]. The structure shown in Figure 29.7 reflects the defense-in-depth approach facilitating a combination of firewalls filtering traffic and various SA endpoints deliberately breaking the communication flow to establish independent security domains.

Most obviously, the web access of customers via the Internet is separated in a single zone. Two firewalls define a DMZ, where only the web server(s) can provide data to the Internet and retrieve data from the intranet. In the intranet (zone), several application servers like billing, SCADA, and demand-side management are located. Pure office PCs are not depicted and usually combined in a distinct zone.

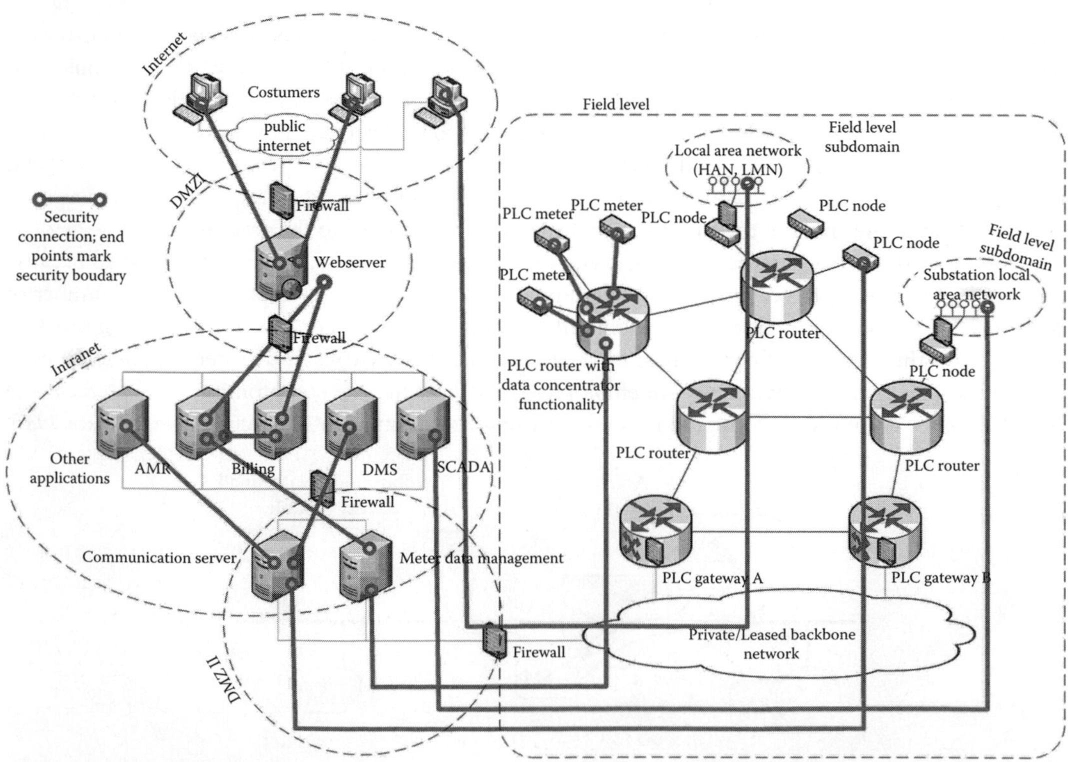

FIGURE 29.7 Communication scenario case study in smart grids.

A second intranet DMZ shields field-level communication. Various communication servers such as meter data management separate the actual field system from the application servers introducing an additional security perimeter. The next security zone is the access network, which is using public lines as well as the utility-owned PLC. The powerline gateway again implements firewall functionality to shield against the public network. Finally, either nodes directly connect to the access network or additionally multiple security zones can exist. In addition to the powerline network, each end node can connect to a LAN. In smart grids, one can distinguish between home automation network (HAN) or local metrology network (LMN) to interconnect home appliances or metering equipment, respectively, on the customer side or LANs for substation automation. The PLC node therefore includes some firewall functionality to separate the powerline access network from the HAN or LMN.

Solid bold lines introduce the security connections inside the network. Often a break of security is desired to either translate data or to deliberately interrupt the connection such as done at the web server providing data to the customers. Also systems such as the AMR are handled by a special server. In this case, the reason is abstraction to hide technical details from the metering system rather than security issues. Domotic applications on the other end can form a complete end-to-end connection from the security point of view. At the access network, communication should be as transparent as possible, although also on this level, devices such as data concentrators doing data preprocessing and aggregation are an interruption of the security chain.

Two design aspects gaining high importance in the implementation are connections transparently concatenating different network technologies and key management. The common trend for transparent communication is the usage of IP. Consequently, also standardized security protocols need to be used to offer end-to-end security and avoid breaking the security at network borders. IPsec and the presented ESP+ can offer such a transparency allowing seamlessly connecting IT equipment (private/leased backbone network) and the powerline access network. This results in a noninterrupted security chain between the desired endpoints independent of network type and also network operator. Key management is the other strong issue. In contrast to the Internet or telecom systems, automation systems require unattended operation and often do not foresee the interaction with an end user who can confirm or deny security requests. Public key cryptography offers the advantage of easier distribution and is therefore favored by several authors. Yet, also symmetric keys have big advantages in terms or execution speed and block size. Discussion is ongoing [73–75].

29.6 Conclusions

Security has become an issue in automation. Though not foreseen at the beginning of the development of automation systems, the ever-increasing degree of integration and interconnection of automation islands has unveiled security deficiencies that need remedy. The general path to security in automation is quite clear: Only proven strategies can be employed, which is tantamount to using security mechanisms developed in and for the much bigger domain of IT applications, where security has been a focal topic for a longer time and where the community devising and testing new security concepts and algorithm is much larger.

Yet, even though the security technologies from the IT world are mature, their application to automation systems is not straightforward. Too different are the requirements in IT and automation applications. Even if the defense-in-depth approach largely solves the technical problems, what remains are the organizational difficulties of defining and enforcing a security strategy. This applies in particular to security management. For instance, methods based on TLS are actually not ideal for data servers in automation applications. TLS requires a separate certificate for each node in the network that provides data. Methods based on plain TLS are therefore mostly found when sites or campuses are interconnected, which possess a well-defined set of access points visible to the Internet.

Especially in building automation or in smart grids, there are numerous access points in the backbone. This makes key distribution during commissioning or later on during operation a logistical challenge and still leaves room for the development of innovative approaches. Even if security tokens are used as easy-to-handle key containers for system installation, key updates during normal operation should rather be done over the network, or if network bandwidth is scarce, distributed key derivation schemes should be devised. Without accounting for such special organizational needs in automation systems, security is difficult to achieve in practice.

References

1. T. Sauter, The three generations of field-level networks—Evolution and compatibility issues, *IEEE Transactions on Industrial Electronics*, 57(11), 2010, 3585–3595.
2. T. Sauter, S. Soucek, W. Kastner, D. Dietrich, The evolution of factory and building automation, *IEEE Industrial Electronics Magazine*, 5(3), 2011, 35–48.
3. P. Palensky, T. Sauter, Security considerations for FAN-Internet connections, *IEEE International Workshop on Factory Communication Systems* (*WFCS*), Porto, Portugal, 2000, pp. 27–35.
4. M. S. DePriest, Network security considerations in TCP/IP-based manufacturing automation, *ISA Transactions*, 36(l), 1997, 37–48.
5. M. Knizak, M. Kunes, M. Manninger, T. Sauter, Modular agent design for fieldbus management, *IEEE International Conference on Emerging Technologies and Factory Automation* (*ETFA*), Barcelona, Spain, 1999, pp. 857–864.
6. E. Byres, J. Lowe, The myths and facts behind cyber security risks for industrial control systems, *VDE Kongress*, Berlin, Germany, 2004, pp. 213–218.
7. S. Shin, T. Kwon, G.-Y. Jo, Y. Park, H. Rhy, An experimental study of hierarchical intrusion detection for wireless industrial sensor networks, *IEEE Transactions on Industrial Informatics*, 6(4), November 2010, 744–757.
8. M. Cheminod, I. C. Bertolotti, L. Durante, P. Maggi, D. Pozza, R. Sisto, A. Valenzano, Detecting chains of vulnerabilities in industrial networks, *IEEE Transactions on Industrial Informatics*, 5(2), May 2009, 181–193.
9. M. Long, C.-H. Wu, J. Y. Hung, Denial of service attacks on network-based control systems: Impact and mitigation, *IEEE Transactions on Industrial Informatics*, 1(2), May 2005, 85–96.
10. E. Byres, J. Lowe, Real world cyber security risks for industrial control systems, *The Online Industrial Ethernet Book*, No. i22, 35, 2005, http://ethernet.industrial-networking.com/origarticles/i22cyber.asp (accessed August 25, 2013).
11. B. Fraser (ed.), *RFC2196 Site Security Handbook*, Network Working Group, 1997, http://www.ietf.org/rfc/rfc2196.txt?number =2196 (accessed August 25, 2013).
12. Bundesamt für Sicherheit in der Informationstechnik, *IT Baseline Protection Manual*, Bundesanzeiger-Verlag, Cologne, Germany, http://www.bsi.de/gshb/english/menue.htm (accessed August 25, 2013).
13. International Standards Organization, ISO/IEC 17799:2000, Information technology—Code of practice for information security management, 2000.
14. National Institute of Standards and Technology, *An Introduction to Computer Security: The NIST Handbook*, NIST Special Publication 800-12, NIST, Gaithersburg, MD, 1995.
15. B. Boehm, A spiral model of software development and enhancement, *IEEE Computer*, 21(5), 1988, 61–72.
16. VDI Guideline 2182, *Information Security in Industrial Automation—A General Process Model*, Beuth Verlag, Berlin, Germany, 2007.
17. D. R. Stinson, *Cryptography, Theory and Practice*, 2nd edn., Chapman & Hall/CRC, Boca Raton, FL, 2002.
18. A. J. Menezes, P. C. van Oorschot, S. A. Vanstone, *Handbook of Applied Cryptography*, CRC Press, Boca Raton, FL, 1996.
19. J. G. Steiner, B. Clifford Neuman, J. I. Schiller, Kerberos: An authentication service for open network systems, *Winter 1988 USENIX Conference*, Dallas, TX, February 1988, pp. 191–202.

20. International Telecommunication Union, ITU-T Recommendation X.509 (1997 E): Information Technology—Open Systems Interconnection—The Directory: Authentication Framework, June 1997.
21. R. Sandhu, E. J. Coyne, H. L. Feinstein, C. E.Youman, Role based access control models, *IEEE Computers*, 29, 1996, 38–47.
22. B. Cheswick, The design of a secure Internet gateway, *USENIX Summer 1990 Technical Conference*, Anaheim, CA, June 1990, pp. 233–238.
23. J. Wack, K. Cutler, J. Pole, *Recommendations of the National Institute of Standards and Technology Guidelines on Firewalls and Firewall Policy*, NIST Special Publication 800-41, NIST, Gaithersburg, MD, 2002.
24. M. Goncalves, *Firewalls Complete*, McGrawHill, New York, 1997.
25. J. von Helden, S. Karsch, BSI-Studie: Intrusion Detection Systeme Grundlagen, Forderungen und Marktübersicht für Intrusion Detection Systeme (IDS) und Intrusion Response Systeme (IRS), 1998, http://www.bsi.de/literat/studien/ids/ids-stud.htm (accessed August 25, 2013).
26. R. Bace, P. Mell, *Intrusion Detection Systems*, NIST Special Publication 800-31, NIST, Gaithersburg, MD, 2001.
27. R. Anderson, *Security Engineering*, Wiley, New York, 2001.
28. N. Borisov, I. Goldberg, D. Wagner, Intercepting mobile communications: The insecurity of 802.11, *Seventh ACM SIGMOBILE Annual International Conference on Mobile Computing and Networking*, Rome, Italy, July 16–21, 2001, pp. 180–189.
29. International Standards Organisation, ISO/IEC 15408-1:1999 Information technology—Security techniques—Evaluation criteria for IT security, 1999.
30. H. Kreutzmann, S. Vollmer, N. Tekampe, A. Abromeit, Protection profile for the gateway of a smart metering system (Gateway PP), Bundesamt für Sicherheit in der Informationstechnik (BSI), Federal Office for Information Security Germany Std., Bonn, Germany, 2011.
31. C. Schwaiger, A. Treytl, Smart card based security for fieldbus systems, *IEEE International Conference on Emerging Technologies and Factory Automation* (*ETFA*), Lisbon, Portugal, 2003, pp. 398–406.
32. W. Granzer, F. Praus, W. Kastner, Security in building automation systems, *IEEE Transactions on Industrial Electronics*, 57(11), 2010, 3622–3630.
33. A. Treytl, T. Sauter, C. Schwaiger, Security measures for industrial fieldbus systems—State of the art and solutions for IP-based approaches, *IEEE International Workshop on Factory Communication Systems* (*WFCS*), Vienna, Austria, 2004, pp. 201–209.
34. International Electrotechnical Commission, Digital data communications for measurement and control—Part 1: Profile sets for continuous and discrete manufacturing relative to fieldbus use in industrial control systems, IEC Standard IEC 61784-1, 2003.
35. International Electrotechnical Commission, Digital data communications for measurement and control—Fieldbus for use in industrial control systems, IEC Standard IEC 61158, 2003.
36. W. Granzer, W. Kastner, G. Neugschwandtner, F. Praus, Security in networked building automation systems, *6th IEEE International Workshop on Factory Communication Systems* (*WFCS*), Torino, Italy, 2006, pp. 283–292.
37. C. Schwaiger, T. Sauter, Security strategies for field area networks, *28th Annual Conference of the IEEE Industrial Electronics Society* (*IECON*), Sevilla, Spain, November 5–8, 2002, pp. 2915–2920.
38. M. Felser, T. Sauter, Standardization of industrial Ethernet—The next battlefield?, *IEEE International Workshop on Factory Communication Systems* (*WFCS*), Vienna, Austria, 2004, pp. 413–421.
39. T. Sauter, J. Jasperneite, L. Lo Bello, Towards new hybrid networks for industrial automation, *IEEE Conference on Emerging Technologies and Factory Automation* (*ETFA*), Palma de Mallorca, Spain, 2009, pp. 1–8.
40. G. Cena, A. Valenzano, S. Vitturi, Hybrid wired/wireless networks for real-time communications, *IEEE Industrial Electronics Magazine*, 2(1), March 2008, 8–20.

41. G. Gaderer, T. Sauter, F. Ring, A. Nagy, A novel, wireless sensor/actuator network for the factory floor, *IEEE Conference on Sensors*, Waikoloa-Big Island, HI, November 1–4, 2010, pp. 940–945.
42. S. Kent, K. Seo, Security architecture for the internet protocol, RFC 4301, 2005.
43. S. Kent, IP authentication header, RFC 4302 (Proposed Standard), Internet Engineering Task Force, December 2005, http://www.ietf.org/rfc/rfc4302.txt (accessed August 25, 2013).
44. S. Kent, IP encapsulating security payload (ESP), RFC 4303 (Proposed Standard), Internet Engineering Task Force, December. 2005, http://www.ietf.org/rfc/rfc4303.txt (accessed August 25, 2013).
45. T. Dierks, K. Resorla, The transport layer security (TLS) protocol version 1.2, RFC 5246, 2008.
46. V. Gupta, M. Wurm, Y. Zhu, M. Millard, S. Fung, N. Gura, H. Eberle, S. C. Shantz, Sizzle: A standards-based end-to-end security architecture for the embedded internet, *Pervasive and Mobile Computing*, 1, 2005, 425–445.
47. OpenVPN, project website, 2009, http://openvpn.net (accessed August 25, 2013).
48. P. Loshin (ed.), *Big Book of IPsec RFCs: Internet Security Architecture*, Morgan Kaufmann, San Diego, CA, 1999.
49. A. Treytl, N. Roberts, G. P. Hancke, Security architecture for power-line metering system, *IEEE International Workshop on Factory Communication Systems* (*WFCS*), Vienna, Austria, 2004, pp. 393–396.
50. A. Treytl, T. Sauter, Security concept for a wide-area low-bandwidth power-line communication system, *International Symposium on Power Line Communications and Its Applications* (*ISPLC*), Vancouver, British Columbia, Canada, 2005, pp. 66–70.
51. P. Baran, On distributed communications: IX security, secrecy, and tamper-free considerations, RAND Memorandum RM-3765-PR, August 1964.
52. International Standardization Organization, Basic reference model for open system interconnection—Part 2: Security architecture, ISO Standard ISO 7498-2, 1989.
53. The Institute of Electrical and Electronics Engineers, IEEE standards for local and metropolitan area networks: Standard for interoperable LAN/MAN security (SILS), IEEE Std 802.10-1998, 1998.
54. International Standardization Organization, Identification cards—Integrated circuit(s) cards with contacts—Part 1-10, ISO Standard 7816, 1996–2002.
55. H. Krawczyk, M. Bellare, R. Canetti, RFC 2104 HMAC: Keyed-hashing for message authentication, February 1997, http://www.ietf.org/rfc/rfc2104.txt (accessed August 25, 2013).
56. R. Rivest, A. Shamir, L. A. Adelman, Method for obtaining digital signatures and public-key cryptosystems, *Communications of the ACM*, 21(2), February 1978, 120–126.
57. F. Baker, D. Meyer, Internet protocols for the smart grid, IETF draft-baker-ietf-core-15, April 2011.
58. S. Deering et al., Internet Protocol, Version 6 (IPv6) Specification, Network Working Group, RFC 2460, December 1998.
59. G. Pelletier, K. Sandlund, RObust Header Compression Version 2 (ROHCv2): Profiles for RTP, UDP, IP, ESP and UDP-Lite, RFC 5225, Internet Engineering Task Force, April 2008, http://www.ietf.org/rfc/rfc5225.txt (accessed August 25, 2013).
60. K. Sandlund, G. Pelletier, L.-E. Jonsson, The RObust Header Compression (ROHC) Framework, RFC 5795, Internet Engineering Task Force, March 2010, http://www.ietf.org/rfc/rfc5795.txt (accessed August 25, 2013).
61. B. Adebisi, A. Treytl, A. Haidine, A. Portnoy, R. U. Shan, D. Lund, H. Pille, B. Honary, IP-centric high rate narrowband PLC for smart grid applications, *IEEE Communications Magazine*, 49(12), 2011, 46–54.
62. B. Hirschler, A. Treytl, Internet protocol security and power line communication, *16th IEEE International Symposium on Power Line Communications and Its Applications* (*ISPLC*), 2012, Beijing, China, pp. 102–107.
63. A. Treytl, T. Sauter, C. Schwaiger, Security measures in automation systems—A practice-oriented approach, *10th IEEE International Conference on Emerging Technologies and Factory Automation* (*ETFA*), Catania, Italy, September 19–22, 2005, vol. 2, pp. 847–854.

64. B. A. Khan, J. Mad, A. Treytl, Security in agent-based automation systems, *IEEE Conference on Emerging Technologies & Factory Automation* (*ETFA*), Patras, Greece, 2007, pp. 768–771.
65. F. Biennier, J. Favrel, Secure collaborative information system for enterprise alliances: A workflow based approach, *IEEE International Conference on Emerging Technologies and Factory Automation* (*ETFA*), Antibes Juan-les-Pins, France, 2001, vol. 2, pp. 33–40.
66. M. Cheminod, A. Pironti, R. Sisto, Formal vulnerability analysis of a security system for remote fieldbus access, *IEEE Transactions on Industrial Informatics*, 7(1), February 2011, 30–40.
67. T. Sauter, C. Schwaiger, Achievement of secure Internet access to fieldbus systems, *Microprocessors and Microsystems*, 26(7), September 2002, 331–339.
68. R. Zalenski, Firewall technologies, *IEEE Potentials*, 21(1), 2002, 24–29.
69. T. Sauter, M. Lobashov, How to access factory floor information using internet technologies and gateways, *IEEE Transactions on Industrial Informatics*, 7(4), 2011, 699–712.
70. V. Kapsalis, K. Charatsis, M. Georgoudakis, G. Papadopoulos, Architecture for web-based services integration, *Annual Conference of the IEEE Industrial Electronics Society* (*IECON*), Roanoke, VA, 2003, vol. 1, pp. 866–871.
71. M. Wollschlaeger, T. Bangemann, Maintenance portals in automation networks—Requirements, structures and model for web-based solutions, *IEEE International Workshop on Factory Communication Systems* (*WFCS*), Vienna, Austria, 2004, pp. 193–199.
72. Y. Yan, Y. Qian, H. Sharif, D. Tipper, A survey on cyber security for smart grid communications, *IEEE Communications Surveys & Tutorials*, 14(4), 2012, 998–1010.
73. J. Xia, Y. Wang, Secure key distribution for the smart grid, *IEEE Transactions on Smart Grid*, 3(3), 2012, 1437–1443.
74. A. R. Metke, R. L. Ekl, Security technology for smart grid networks, *IEEE Transactions on Smart Grid*, 1(1), 2010, 99–107.
75. S. Fuloria, R. Anderson, F. Alvarez, K. McGrath, Key management for substations: Symmetric keys, public keys or no keys?, *2011 IEEE/PES Power Systems Conference and Exposition* (*PSCE*), March 20–23, Phoenix, AZ, 2011, pp. 1–6.

VI

Wireless Industrial Networks

30 Wireless LAN Technology for the Factory Floor *Andreas Willig* **30**-1
Introduction • Wireless Industrial Communications and Wireless Fieldbus: Challenges and Problems • Wireless LAN Technology and Wave Propagation • Physical Layer: Transmission Problems and Solution Approaches • Problems and Solution Approaches on the MAC and Link Layer • Wireless Fieldbus Systems: State of the Art • Wireless Ethernet/IEEE 802.11 • Summary • References

31 WirelessHART *Alessandra Flammini and Emiliano Sisinni* **31**-1
Introduction • Requirements of Wireless Fieldbuses • HART and WirelessHART • WirelessHART Protocol Stack • Message Format • Join Procedure • Security Aspects • Wi-Analys Tool • Designing a New WirelessHART Network • Example of Platform Providers • References

32 ISA100.11a *Stig Petersen and Simon Carlsen* **32**-1
Introduction • System Overview • Communication Protocol • Security • Coexistence • Implementations and Equipment • Conclusion • References

33 Comparison of WirelessHART and ISA100.11a for Wireless Instrumentation *Stig Petersen and Simon Carlsen* **33**-1
Introduction • System Overview • Communication Protocols • Main Differences • Conclusion • References

34 IEC 62601: Wireless Networks for Industrial Automation–Process Automation (WIA-PA) *Ivanovitch Silva and Luiz Affonso Guedes* **34**-1
Introduction • IEC/PAS 62601: WIA-PA • Physical Layer • Data Link Layer • Network Layer • Application Layer • Security • Conclusion and Discussion • References

35 Wireless Extensions of Real-Time Industrial Networks *Gianluca Cena, Adriano Valenzano, and Stefano Vitturi* **35**-1
Introduction • Implementation of Hybrid Networks • IEEE 802.11-Based Extensions • IEEE 802.15.4-Based Extensions • Conclusions • References

36 Wireless Sensor Networks for Automation *Tomas Lennvall, Jan-Erik Frey, and Mikael Gidlund*...**36**-1
Introduction • WSN in Industrial Automation • Development Challenges • Reference Cases • Communication Standards • Low-Power Design • Packaging • Modularity • Power Supply • Acknowledgments • References

37 Design and Implementation of a Truly Wireless Real-Time Sensor/Actuator Interface for Discrete Manufacturing Automation *Guntram Scheible, Dacfey Dzung, Jan Endresen, and Jan-Erik Frey*.. **37**-1
Introduction • Wireless Interface Requirements and System Specifications • Communication Subsystem Design • Communication Subsystem Implementation • Wireless Power Subsystem • Practical Application and Performance Comparison • Summary • References

38 IPv6 over Low-Power Wireless Personal Area Networks (6LoWPAN) and Constrained Application Protocol (CoAP) *Guido Moritz and Frank Golatowski* **38**-1
Introduction • IPv6 over Low-Power Wireless Personal Area Networks—6LoWPAN • Constrained Application Protocol • Related Work • Conclusion • References

30

Wireless LAN Technology for the Factory Floor: Challenges and Approaches

30.1 Introduction 30-1
30.2 Wireless Industrial Communications and Wireless Fieldbus: Challenges and Problems 30-3
System Aspects • Real-Time Transmission over Error-Prone Channels • Integration of Wired and Wireless Stations/Hybrid Systems • Mobility Support • Security Aspects and Coexistence
30.3 Wireless LAN Technology and Wave Propagation 30-5
Wireless LANs • Wave Propagation Effects
30.4 Physical Layer: Transmission Problems and Solution Approaches 30-7
Effects on Transmission • Wireless Transmission Techniques
30.5 Problems and Solution Approaches on the MAC and Link Layer 30-10
Problems for Wireless MAC Protocols • Methods for Combating Channel Errors and Channel Variation
30.6 Wireless Fieldbus Systems: State of the Art 30-13
CAN • FIP/WorldFIP • PROFIBUS • Other Fieldbus Technologies
30.7 Wireless Ethernet/IEEE 802.11 30-15
DCF • QoS Support
30.8 Summary 30-17
References 30-17

Andreas Willig
University of Canterbury

30.1 Introduction

Wireless communication systems have diffused into an ever-increasing number of application areas and have achieved wide popularity. Wireless telephony and wireless mobile Internet access are now an important part of our daily lives, and wireless local area network (WLAN) technologies—for example, WiFi-based—have become more and more the primary way to access business and personal data. Two important benefits of wireless technology are key to this success: the need for cabling is greatly reduced, and computers as well as users can be truly mobile. This saves costs and enables new applications. In factory plants, wireless technology can be used in several interesting ways [37,113,116]:

- WLAN technology can provide the communications services for distributed control applications [42] involving mobile subsystems like autonomous transport vehicles, robots, or turntables.
- Implementation of distributed control systems in explosible areas or in the presence of aggressive chemicals.
- Frequent plant reconfiguration gets easier as fewer cables have to be remounted.
- Mobile plant diagnosis systems and wireless stations for programming and on-site configuration.

However, when adopting WLAN technologies for the factory floor, some problems occur. The first problem is the tension between the hard reliability and timing requirements (hard real-time) pertaining to industrial applications on the one hand and the problem of wireless channels having time-variable and sometimes quite high error rates on the other. A second major source of problems is the desire to *integrate* wireless and wired stations into one single network (henceforth called a *hybrid system* or *hybrid network*). This integration calls for the design of interoperable protocols for the wired and wireless domains. Furthermore, using wireless technology imposes problems not anticipated in the original design of the (wired) fieldbus protocols: security problems, interference, mobility management, and so on.

In this chapter, we survey some issues pertaining to the design and evaluation of protocols and architectures for (integrated) wireless industrial LANs and provide a selective overview of the state of the art. There is an emphasis on aspects influencing the time and reliability behavior of wireless transmission. However, we not only discuss the problems but also present different solution approaches on the physical, medium access control (MAC), or data-link layer. These layers are key to the success of wireless fieldbus systems because they have important responsibilities in fulfilling timing and reliability requirements, and furthermore, they are exposed most directly to the wireless link characteristics. In the second part of this chapter, we focus on specific technologies and the creation of hybrid systems. On the one hand, there are a number of existing fieldbus standards like controller area network (CAN), factory instrumentation protocol (FIP)/WorldFIP, or PROFIBUS. For these systems, we discuss problems and approaches to create hybrid systems. On the other hand, one could start from existing wireless technologies and ask about their capabilities with respect to timeliness and reliability. The most widely deployed WLAN technology is currently the IEEE 802.11 WLAN standard [47], and its suitability for industrial applications is discussed.

This chapter is structured as follows: In Section 30.2, important general considerations and problems of wireless industrial communications and wireless fieldbus systems are presented. In Section 30.3, we discuss some basic aspects of wireless LAN technology and wireless wave propagation. The transmission impairments resulting from certain wave propagation effects and some physical layer approaches to deal with them are presented in Section 30.4. Wireless wave propagation has also some interesting consequences on the operation of the MAC and data-link layer; these are discussed in Section 30.5. The following two sections, Sections 30.6 and 30.7, take a more technology-oriented perspective. Specifically, in Section 30.6, we survey the state of the art regarding wireless industrial communication systems and wireless fieldbus systems. In Section 30.7, we present the important aspects of the IEEE 802.11 WLAN standard with respect to the transmission of real-time data. Finally, in Section 30.8, we provide a brief summary.

The chapter restricts itself to the consideration of protocol-related aspects of wireless transmission; other aspects like signal processing, analog and digital circuitry, and energy aspects are not considered. There are, for example, many introductory and advanced books on wireless networking [1,18,34,50,72,78,87,92,97,98,104]. Several separate topics in wireless communications are treated in [32]. Furthermore, this chapter is not intended to serve as introduction to fieldbus technologies; some background information can be found in [21,69]. Finally, there is also no discussion of wireless industrial sensor networks like, for example, Wireless HART [37,39,40].

30.2 Wireless Industrial Communications and Wireless Fieldbus: Challenges and Problems

In this section, we survey some of the problem areas arising in wireless fieldbus systems.

30.2.1 System Aspects

First of all, wireless fieldbus systems will operate in similar environments as wired ones. Typically, in factory automation applications, a small to moderate number of stations are distributed over geographically small areas with no more than 100 m between any pair of stations [45]. Wired fieldbus systems traditionally offer bitrates ranging from hundreds of kilobits to (tens of) megabits per second (real-time Ethernet systems can achieve higher rates [20]), and wireless fieldbus systems should have comparable bitrates. The wireless transceivers have to meet electromagnetic compatibility requirements, meaning that they not only have to restrict their radiated power and frequencies, but also should be properly shielded from strong magnetic fields and electromagnetic noise emanated by strong motors, high-voltage electrical discharges, and so on. This may pose a serious problem when off-the-shelf wireless transceivers are used (e.g., commercial IEEE 802.11 hardware), since these are typically designed for office environments and have no industrial-strength shielding.

Another problem is that many small fieldbus devices get their energy supply from the same wire as used for data transmission. If the cabling is to be removed from these devices, there is not only the problem of wireless data transmission but also the issue of wireless power transmission [43,95], which requires substantial effort.

For battery-driven devices, the need to conserve energy arises. As is known from the realm of wireless sensor networks, this has important consequences for the design of protocols [26,35,54] but is not discussed any further in this chapter.

30.2.2 Real-Time Transmission over Error-Prone Channels

In industrial applications, often *hard real-time* requirements play a key role. In accordance with [84], we assume the following important characteristics of hard real-time communications: (1) safety-critical messages must be transmitted reliably within an application-dependent deadline, (2) there should be support for message priorities to distinguish between important and unimportant messages, (3) messages with stringent timing constraints typically have a small size, and (4) both periodic and aperiodic/asynchronous traffic are present. The qualifier *hard* stems from the fact that losses or deadline misses of safety-critical packets can damage equipment or can even cost life. Both periodic and aperiodic messages in fieldbus systems can be subject to hard real-time constraints.

Wireless media tend to exhibit time-variable and sometimes high error rates, which creates a problem for fulfilling the hard real-time requirements. As an example, the measurements presented in [115] have shown that in a certain industrial environment, for several seconds, no packet gets through the channel. Therefore, seeking deterministic guarantees regarding timing and reliability is not appropriate. Instead, *stochastic guarantees* become important. The percentage of safety-critical messages which can be transmitted reliably within a prespecified time-bound should be at least 99.x%. Of course, the error behavior limits the application areas of wireless industrial LANs—when a safety-critical application requires deterministic guarantees in the range of 10–100 ms, wireless transmission is essentially ruled out. However, if occasional message loss or deadline misses are tolerable, wireless technologies can offer their potential. The goal is to reduce the frequency of losses and deadline misses.

How transmission reliability can be implemented depends on the communication model. In many fieldbus systems (e.g., PROFIBUS), packets are transmitted from a sender to an *explicitly addressed* receiver station without involving other stations. Reliability can be ensured by several mechanisms, for example, retransmissions, packet duplications, diversity mechanisms, error-correcting codes, or a combination thereof. On the other hand, systems like FIP/WorldFIP [106] and CAN [48] implement the model of a *real-time database* where *data* are identified instead of stations. A piece of data has one *producer* and potentially many *consumers*. The producer broadcasts the data, and all interested consumers copy the data packet into an internal buffer. This broadcast approach inhibits the use of acknowledgments and packet retransmissions, but error-correcting codes and diversity mechanisms can still be used to increase transmission reliability. Often, the data are transmitted periodically, and (repeated) packet losses can be detected by comparing the known period and the time of the last arrival of a data packet. This *freshness* information can be used by the application to react properly.

30.2.3 Integration of Wired and Wireless Stations/Hybrid Systems

There are a huge number of existing and productive fieldbus installations, and it is very attractive to integrate wireless stations into these. Such a network with both wireless stations (stations with a wireless transceiver) and wired stations are called *hybrid systems* [14]. The most important requirements for hybrid systems are as follows:

- *Transparency*: There should be no need to modify the protocol stack of wired stations.
- *Using specifically tailored protocols*: Most fieldbus systems are specified on the layers one (physical), two (MAC and link layer), and seven (application layer) of the OSI reference model. The introduction of a wireless physical layer affects the behavior and performance of both the MAC and link layer. The existing protocols for wired fieldbus systems are not designed for a wireless environment (see [117] for an example) and should be replaced by protocols specifically tailored for the wireless link. However, this comes at the cost of *protocol conversion* between wired and wireless protocols.
- *Portability of higher-layer software*: If the link layer interface is the same for both the wireless and wired protocol stacks, implementations of higher-layer protocols and application software can be used in the same way on both types of stations.

The different approaches to integrate wireless stations into wired fieldbus LANs can be classified according to the layer of the OSI reference model where the integration actually happens [19,112]. Almost all fieldbus systems are restricted to the physical, data-link, and application layers [21]. The classification is as follows:

- *Wireless cable-replacement approach*: All stations are wired stations and thus attached to a cable. A piece of cable can be replaced by a wireless link, and special bridge-like devices translate the framing rules used on the wireless and wired media, respectively. In this approach, no station is aware of the wireless link. A typical application scenario is the wireless interconnection of two fieldbus segments.
- *Wireless MAC-unaware bridging approach*: The network comprises of both wired and wireless stations, but integration happens solely at the physical layer. Again, a bridge-like device translates the framing rules between wired and wireless media. The wireless stations use merely an alternative physical layer (PHY), but the MAC and link layer protocols remain the same as for wired stations.
- *Wireless MAC-aware bridging approach*: The LAN comprises of both wired and wireless stations, and integration happens at the MAC and data-link layer. There are two different MAC and link layer protocol stacks for wired and wireless stations, but both offer the same link layer interface. The wireless MAC and link layer protocols should be (1) specifically tailored to the wireless medium and (2) easily integrable with the wired MAC and link layer protocols. An intelligent bridge-like device is responsible for both translating the different framing rules as well as interoperation of the different MAC protocols.
- *Wireless gateway approach*: In this approach, integration happens at the application layer or even in the application itself. Entirely different protocol stacks can be used on different types of transmission media.
- Some mixture of these approaches.

All of these approaches require special *coupling devices* at the media boundaries. For the wireless cable-replacement and the MAC-unaware bridging approaches, these devices can be relatively simple, whereas the other approaches may require complex and stateful operations. Hence, failures of such devices and proper redundancy need to be addressed.

30.2.4 Mobility Support

The potential station mobility is one of the main attractions of wireless systems. We can assume that wireless fieldbus systems will be mostly infrastructure based (meaning that there are base stations or access points [APs] that facilitate network operation). A *handover* must be performed when a mobile station moves from the range of one AP into the range of another AP [2,28]. Typically, handover processes involve exchange of signaling packets between the mobile station and the APs. Ideally, a station can fulfill timing and reliability requirements even during a handover. The applicability and performance of handover schemes depend on the maximum speed of a mobile station, on the wireless technology, and on other circumstances, for example, how well the trajectory of a mobile station can be predicted. In industrial applications, it is often forklifts, robots, or moving plant subsystems that are mobile, and it is safe to assume that these devices move with velocities of no more than 20 km/h [45].

A simple consequence of mobility is that stations may enter and leave a network at unpredictable times. To support this, a protocol stack at minimum must offer functionalities to make a new station known to the network/the other stations, and sometimes also address assignment is needed. On the other hand, fieldbus systems and their applications often are designed with the assumption that the network is set up once and not changed afterward. Consequently, some fieldbus systems do not easily support some dynamics in their set of stations. Consider, for example, the FIP/WorldFIP fieldbus [106]. This system belongs to the class of real-time database systems, and the producers of data items (called process variables) are polled by a dedicated central station, the *bus arbiter*. The bus arbiter keeps a table of variable identifiers and traverses it cyclically. To include a new station into the system, the arbiter's table has to be modified by a human operator. It is worth noting that the most widely used fieldbus systems do not offer any support for dynamic address assignment.

30.2.5 Security Aspects and Coexistence

Security played no important role in the initial design of the fieldbus standards. This was reasonable, because physical access to a wire is needed to eavesdrop or inject packets. However, the introduction of wireless media allows an attacker to eavesdrop packets at some distance, for example, on the factory's parking lot. Even worse, an attacker could generate interference on the operating frequency of a wireless fieldbus system and distort all transmissions (including the time-critical and important ones). An attacker might also try to inject malicious packets into the network, for example, false valve commands. Therefore, security measures (integrity, authentication, and authorization) have to be added to wireless fieldbus systems [94].

Noise and interference is not only generated deliberately by an attacker, but can also come from collocated wireless systems working in the same frequency band. As an example, both IEEE 802.11 and Bluetooth use the 2.4 GHz ISM band and create mutual interference. This coexistence problem is explored in [16].

30.3 Wireless LAN Technology and Wave Propagation

In this section, we discuss some basic characteristics of WLAN technology and present some of the fundamental wave propagation effects. In Sections 30.4 and 30.5, we discuss physical layer and MAC/link layer approaches to overcome or at least relax some of the problems created by the propagation effects.

30.3.1 Wireless LANs

Wireless LANs are designed for packet-switched communications over short distances (up to a few hundred meters) and with moderate to high bitrates. Nowadays, the dominant wireless LAN technology is IEEE 802.11 [47], also often referred to as WiFi. As an example, the IEEE 802.11a and 11g WLAN standard offers bitrates between 1 and 54 Mbps, whereas the draft 802.11ac standard offers bitrates up to 866.7 Mbit/s, using a system bandwidth of 160 MHz and MIMO transmission. WiFi uses radio frequencies in the 2.4 and 5 GHz frequency bands. More specifically, WiFi uses so-called ISM bands (industrial, scientific, and medical), which are particularly attractive, since they allow for license-free operation, and the only restriction in using this band is a transmit power limit. On the other hand, since anyone can use these bands, several systems have to coexist. Radio waves below 6 GHz propagate through walls and can be reflected on several types of surfaces, depending on both frequency and material. Thus, with radio frequencies, non-line-of-sight communications is possible.

WiFi networks can be operated in two different modes. In the infrastructure mode, centralized facilities like APs or base stations are responsible for tasks like radio resource management, forwarding data to distant stations, mobility management, and so on. In the ad hoc mode, there is no prescribed infrastructure, and the stations have to organize network operation by themselves [102].

Infrastructure-based WLANs offer some advantages for industrial applications. Many industrial communication systems already have an asymmetric structure that can be naturally accommodated by infrastructure-based systems. The often used master–slave communication scheme serves as an example. Furthermore, the opportunity to offload certain protocol processing tasks to the infrastructure allows to keep mobile stations more simple and to make efficient centralized decisions.

As compared to other wireless technologies like cellular systems and cordless telephony, WLAN technologies seem to offer the best compromise between data rate, geographical coverage, and license-free/independent operation.

30.3.2 Wave Propagation Effects

In the wireless channel, waves propagate through free space, which is an unguided medium, that is, propagation is not spatially confined. The wireless channel characteristics are significantly different from those of guided media like cables and fibers, and they create unique challenges for communication protocols.

A transmitted waveform is subjected to phenomena like path loss, attenuation, reflection, diffraction, scattering, adjacent- and co-channel interference, thermal or man-made noise, and finally imperfections in the transmitter and receiver circuitry [12,87].

The path loss characterizes the loss in expected received signal power when increasing the distance between a transmitter T and a receiver R. In general, the average received power level P_{Rx} can be represented as the product of the transmit power P_{Tx} and the path loss PL:

$$P_{Rx} = P_{Tx} \cdot \text{PL}$$

The path loss PL itself is distance dependent. A typical path-loss model for omnidirectional antennas is given by [87, Chapter 4.9]:

$$\text{PL}(d) = C \cdot \left(\frac{d}{d_0} \right)^n$$

where

$d \geq d_0$ is the distance between T and R

PL (d) is the path loss at distance d

d_0 is a reference distance that depends on the antenna technology (it is ≈ 1 m for WiFi systems)

C is a technology- and frequency-dependent scaling factor

n is the so-called path-loss exponent

Typical values for n are in the range between 2 (free-space propagation) and 5 (shadowed urban cellular radio), see also [87, Chapter 4.9].

Reflection occurs when a waveform impinges on a smooth surface with structures significantly larger than the wavelength. Not all signal energy is reflected; some energy is absorbed by the material. The mechanism of diffraction allows a wave to propagate into a shadowed region, provided that some sharp edge exists. Scattering is produced when a wavefront hits a rough surface having structures smaller than the wavelength; it leads to signal diffusion in many directions.

The most important types of interference are co-channel interference and adjacent-channel interference. In co-channel interference, a signal transmitted from T to R on channel c_1 is distorted by a parallel transmission on the same channel. In case of adjacent channel interference, the interferer I transmits on an adjacent channel c_2, but due to imperfect filters, R captures frequency components from c_2. Alternatively, an interferer I transmitting on channel c_2 leaks some signal energy into channel c_1 due to nonperfect transmit circuitry (amplifier).

Noise can be thermal noise or man-made noise. Thermal noise is created in the channel or in transceiver circuitry and can be found in almost any communications channel. Man-made noise in industrial environments can have several sources, for example, remote controls, motors, or microwave ovens. The predominant model for noise is white noise or Gaussian noise, which operates additively on signals. Gaussian noise is characterized by its noise power, which depends on physical factors like the system temperature, the system bandwidth, and others. It is also common to model interference as noise of higher power.

30.4 Physical Layer: Transmission Problems and Solution Approaches

The previously discussed wave propagation effects can lead to channel errors. In general, their impact depends on a multitude of factors, including frequency, modulation scheme, and the current propagation environment. The propagation environment is characterized by the distance between stations, interferers, the number of different paths and their respective loss, and more. These factors can change when stations or parts of the environment move. Consequently, the transmission quality is time variable.

30.4.1 Effects on Transmission

The notion of slow fading refers to significant variations in the mean path loss, as they occur due to significant changes in distance between transmitter T and receiver R, or by moving beyond large obstacles. Slow-fading phenomena usually occur on timescales significantly longer than a packet; they often coincide with human activity like mobility. For short durations in the range of a few seconds, the channel can often be assumed to have constant path loss.

An immediate result of reflection, diffraction, and scattering is that multiple copies of a signal may travel on different paths from T to R. Since these paths usually have different lengths, the copies arrive at different times (delay spread) and with different phase shifts at the receiver, where they are superposed. This has two consequences:

1. Overlapping signals can interfere constructively or destructively. Destructive interference may lead to up to 40 dB loss of received power. Such a situation is often called a deep fade.
2. Delay spread leads to intersymbol interference, since signals belonging to neighbored information symbols overlap at the receiver.

If the stations move relative to each other or to the environment, the number of paths and their phase shifts vary in time. This results in a fast fluctuating signal strength at the receiver (called fast fading or multipath fading). It is important to note that these fluctuations are much faster than those caused by slow fading. Fast fading happens on time scales of milliseconds to tens of milliseconds (which for WiFi can overlap with several packet transmissions), whereas slow fading happens at timescales of seconds or minutes. On the timescale of milliseconds, the mean signal strength is constant. If the delay spread is small relative to the duration of a channel symbol, the channel is called non-frequency-selective or flat; otherwise, it is called frequency selective.

These phenomena translate into bit errors or packet losses. Packet losses occur when the receiver fails to acquire carrier or bit synchronization [115]—a lost packet in this sense is (at the receiver) indistinguishable from a packet never sent. In case of bit errors, synchronization is successfully acquired, but a number of channel symbols are decoded or demodulated incorrectly. The bit error rate can, for example, be reduced by using forward error correction (FEC) techniques [17,67]. The statistical properties of bit errors and packet losses on wireless links were investigated in a number of studies [8,24,76,115,120]. While the results are not immediately comparable, certain trends show up in almost every study:

- Both bit errors and packet losses are *bursty*; they occur in clusters with error-free periods between the clusters. The distributions of the cluster lengths and the lengths of error-free periods often have a large coefficient of variation or even seem to be heavy tailed.
- Bit error rates depend on the modulation scheme; typically, schemes with higher bitrates/symbol rates exhibit higher error rates.
- Wireless channels can be much worse than wired channels; often bit error rates of 10^{-2} to 10^{-5} can be observed. Furthermore, the bit error rate can vary over several orders of magnitudes within minutes.

Some knowledge about error generation patterns and error statistics can be helpful in designing more robust protocols.

30.4.2 Wireless Transmission Techniques

A number of different transmission techniques have been developed to combat the impairments of the wireless channel and to increase the reliability of data transmission.

Many types of WLANs (including the older versions of IEEE 802.11) rely on spread-spectrum techniques [33], where a narrowband information signal is spread to a wideband signal at the transmitter and despread back to a narrowband signal at the receiver. By using a wideband signal, the effects of narrowband noise or narrowband interference are reduced. The two most important spread-spectrum techniques are direct-sequence spread spectrum (DSSS) and frequency-hopping spread spectrum (FHSS).

In DSSS systems, an information bit is multiplied (XORed) with a finite bipolar chip sequence such that transmission takes place at the chip rate instead of the information bitrate. The chip rate is much higher than the information rate and consequently requires more bandwidth; accordingly, the duration of a chip is much smaller than the duration of a user symbol. The chip rate is chosen such that the average delay spread is larger than the chip duration; thus, the channel becomes frequency-selective. Receivers can explore this in different ways. To explain the first one, let us assume that the receiver receives a signal S from an LOS path and a delayed signal S' from another path, such that the delay difference (called lag) between S and S' is more than the duration of a single chip. The chip sequences are designed such that the autocorrelation between the sequence and a shifted version of itself is low for all lags of one chip duration or more. If a coherent matched-filter receiver is synchronized with the direct signal S, the delayed signal S' appears as white noise and produces only a minor distortion. In the RAKE receiver approach, delayed signal copies are not treated as noise but as a useful source of information [97, Section 30.4]. Put briefly, a RAKE receiver tries to acquire the direct signal and the strongest time-delayed copies and to combine them coherently. However, RAKE receivers are much more complex than simple matched-filter DSSS receivers.

In FHSS, the available spectrum is divided into a number of subchannels. The transmitter hops through the subchannels according to a predetermined schedule (often based on a pseudorandom sequence), which is also known to the receiver. The advantage of this scheme is that a subchannel currently subject to transmission errors is used only for a short time before the transmitter hops to the next channel. The hopping frequency is an important parameter of FHSS systems, since high frequencies require fast and accurate synchronization. As an example, the FHSS version of IEEE 802.11 hops with 2.5 Hz, and many packets can be transmitted before the next hop. In Bluetooth, the hopping frequency is 1.6 kHz, and at most one packet can be transmitted before the next hop. Packets are always transmitted without being interrupted by hopping.

In the last decade, there was a strong uptake of OFDM-based transmission technologies [6,66,108], which have been adopted in the LTE [55] technology for 4G mobile broadband, in WiMax (or IEEE 802.16), and also in the newer physical layers adopted for IEEE 802.11 (formerly known as the IEEE 802.11a and g extensions). OFDM stands for orthogonal frequency division multiplexing. In OFDM, the available spectrum is subdivided into a number *N* of small and closely packed subchannels or subcarriers (the 5 GHz OFDM physical layer of IEEE 802.11 uses 52 subcarriers). The bandwidth of each subcarrier is small enough to obtain a frequency-nonselective channel. Data are transmitted over all subcarriers in parallel; the data rate on each subcarrier (using standard modulation schemes like BPSK, QPSK, and so forth) is small enough so that (due to the additional presence of a guard interval) the delay spread is much smaller than the average symbol duration and intersymbol interference can be avoided. The OFDM scheme used in IEEE 802.15.4 additionally applies error-correction coding such that coding is applied *across* subcarriers instead of applying it to each subcarrier separately.

Another large class of transmission techniques are so-called diversity techniques. In general, diversity methods aim to provide the receiver with several independently faded copies of the same signal, which it can then suitably combine [87]. Intuitively, when several *independently faded* signal copies are available at the receiver, the chance that all of them are in a deep fade *simultaneously* (thus leaving the receiver with no chance of proper decoding) shrinks quickly with the number of signal copies. Three basic forms of diversity are time diversity, frequency diversity, and spatial diversity. With time diversity, different copies of the same signal/packet are achieved by repeating the signal at different times on the same channel, such that the spacing between different copies is larger than the so-called channel coherence time, which can be regarded as the time over which the channel stays roughly constant. Even when the coherence time is known, time diversity has unfavorable properties in real-time settings, as generally waiting times need to be inserted before the next copy can be sent. With frequency diversity, the same signal is repeated over different frequencies, where the frequencies are either used sequentially in time (when only one transceiver is available) or in parallel. Finally, in spatial diversity methods, diversity is obtained from placing multiple antennas with a certain spatial separation [23]. For example, in receive diversity, the transmitter uses a single antenna and the receiver uses multiple antennas. This not only effectively increases the received energy per symbol, the receiver can also combine the signals received over different antennas suitably. With transmit diversity, the transmitter has several antennas, but the receiver has only one. The transmitter could use his or her antennas sequentially (sending only on one antenna at a time), or it could transmit on all antennas in parallel. Finally, the case where multiple transmit and multiple receive antennas are present simultaneously is known as Multiple Input Multiple Output (MIMO) transmission [80,81]. MIMO approaches have been adopted recently into the IEEE 802.11 standard; formerly, they were proposed as an extension known as 802.11n. With MIMO systems, it is possible to use the multiple antennas either for increasing transmission rates (in fact, MIMO is the key for driving WiFi rates close to the Gigabit/s range, and theoretical results assert that if both transmitter and receiver have *n* antennas, then the capacity of the MIMO channel scales linearly with *n*), essentially by sending a different symbol over each transmit antenna, or for improving reliability, by sending the same symbol over all transmit antennas. Through usage of so-called space–time codes, it even becomes possible to trade off rate and reliability against each other [64,100].

True MIMO transmission does not only offer increased reliability or increased rates but also presents some problems specifically for energy-constrained (e.g., mobile) devices. One problem is the relatively large form factor of stations that is required to achieve a certain minimum spacing of antennas—antenna separation should be about half a wavelength to obtain uncorrelated signals. The second problem is that MIMO techniques require complex signal processing at the receiver. As an alternative, cooperative communication schemes have been developed recently [58,62,68,77,93,109]. In these schemes, all involved stations require only a single antenna, and virtual antenna arrays are formed by cooperation among them.

30.5 Problems and Solution Approaches on the MAC and Link Layer

The MAC and the link layer are exposed most to the error behavior of wireless channels and should do most of the work needed to improve the channel quality. Specifically for hard real-time communications, the MAC layer is a key component: If the delays on the MAC layer are not properly bounded, the upper layers cannot compensate this. In general, the operation of the MAC protocol is heavily influenced by the properties of the physical layer. Some of the unique problems of wireless media are discussed in this section. For a general discussion of MAC and link layer protocols, we refer the reader to [22,36,59,99,107].

30.5.1 Problems for Wireless MAC Protocols

Several problems arise due to path loss in conjunction with a threshold property—wireless receivers require the signal to have a minimum strength to be detectable. For a given transmit power, this requirement translates into an upper bound for the distance between two stations wishing to communicate*; if the distance between two stations is larger, they cannot hear each other's transmissions. For MAC protocols based on carrier sensing (carrier sense multiple access [CSMA]), this property creates the hidden terminal problem [101] and the exposed terminal problem.

The hidden terminal problem is sketched in Figure 30.1. Consider three stations A, B, and C with transmission radii as indicated by the circles. Stations A and C are in the range of B, but A is not in the range of C and vice versa. If C starts to transmit to B, A cannot detect this by its carrier-sensing mechanism and considers the medium to be free. Consequently, A also starts packet transmission, and a collision occurs at B.

The exposed terminal problem is also a result of false prediction of the channel state at the receiver. An example scenario is shown in Figure 30.2. The four stations A, B, C, and D are placed such that the pairs A/B, B/C, and C/D can hear each other, but all other combinations cannot. Consider the situation where B transmits to A, and one short moment later, C wants to transmit to D. Station C performs

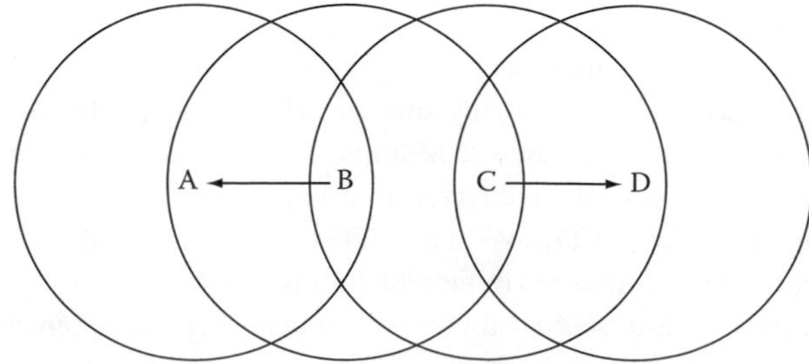

FIGURE 30.1 Hidden terminal scenario.

* To complicate things, wireless links are not necessarily bidirectional: It may well happen that station A can hear station B but not vice versa. This can, for example, happen if the two stations use different transmit powers or if the antennas are directional and both stations' antennas are not well aligned to each other.

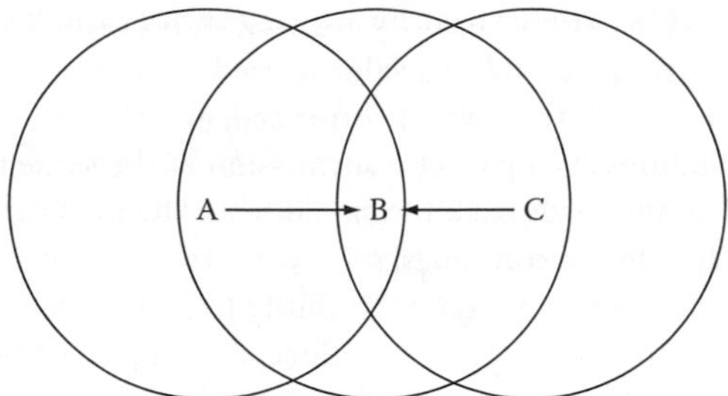

FIGURE 30.2 Exposed terminal scenario.

carrier sensing and senses the medium busy due to B's transmission. Consequently, C postpones its transmission. However, C could safely transmit its packet to D without disturbing B's transmission to A. This leads to a loss of efficiency.

Two approaches to solve these problems are busy tone solutions [101] and the RTS/CTS protocol. In the busy tone solution, two channels are assumed: a data channel and a separate control channel for the busy tone signals. The receiver of a packet transmits a busy tone signal on the control channel during packet reception. If a prospective transmitter wants to perform carrier sensing, it listens on the control channel instead of the data channel. If the control channel is free, the transmitter can start to transmit its packet on the data channel. Under ideal circumstances, this protocol solves the exposed terminal problem. The hidden terminal scenario is also solved, except perhaps for those rare cases where A and C start their transmissions truly simultaneously. However, if the busy tone is transmitted only when the receiver detects a valid packet header, the two colliding stations A and C can abort their transmissions quickly when they perceive the lack of a busy tone. The busy tone solution requires two channels and two transceivers. Furthermore, if the range of the busy tone signal is smaller than the range of the data signal, then collisions and hidden-terminal scenarios can still occur. On the other hand, if the range of the busy tone signal is larger than the range of the data signal, then it might happen that perfectly legal transmissions might be suppressed.

The RTS/CTS protocol attacks the hidden terminal problem using only a single channel. Here we describe the variant used in the IEEE 802.11 WLAN, there are other ones. Consider the case that station A has a data packet for B. After A has obtained channel access, it sends a short request-to-send (RTS) packet to B. This packet includes the time duration needed to finish the whole packet exchange sequence including the final acknowledgment. If B receives the RTS packet properly, it answers with a clear-to-send (CTS) packet, again including the time needed to finish the packet exchange sequence. Station A starts to transmit its data packet immediately after receiving the CTS packet. Any other station C, hearing the RTS and/or the CTS packet, defers its transmissions for the indicated time, this way not disturbing the ongoing packet transmission. It is a conservative choice to defer on any of the RTS or CTS packets, and in fact the exposed terminal problem still exists. One solution could be to let C defer only on reception of a CTS frame and to allow C a packet transmission if it hears an RTS without corresponding CTS frame.* The RTS/CTS protocol described here does not prevent collisions of RTS packets, and it is still susceptible to subtle variants of the hidden terminal problem [86]. Its biggest problem, however, is the large overhead especially for small data packets as they are prevalent in industrial data traffic.

A significant problem of wireless transceivers is that usually they are not able to transmit and receive simultaneously on the same frequency band. Hence, a fast collision detection procedure similar to the CSMA/CD protocol of Ethernet is impossible to implement. Instead, collision detection has to resort to other mechanisms like the busy tone approach described earlier (rarely used) or the use of MAC layer acknowledgments (used frequently). Unfortunately, there are fieldbus systems relying on such a feature,

* Clearly, if C receives a distorted CTS packet, it should defer.

for example, the CAN fieldbus [48] with its priority arbitration protocol. In this class of protocols, each message is tagged with a priority value, and this value is used to deterministically resolve collisions. In the CAN protocol, all stations are tightly time synchronized, and the priority field is always at the start of a packet. All contending stations start packet transmission at the same time. Each contender transmits its priority field bit by bit and reads back the signal from the medium. If the medium state is the same as the transmitted bit, the station continues; otherwise, the station gives up and waits for the next contention cycle. This protocol requires not only the ability to simultaneously listen and receive on the same channel, but also that the channel must also produce meaningful values from overlapping signals. Alternative implementations are sketched in Section 30.6.1.

Even receiver-based collision detection may not work reliably due to the near–far effect. Consider two stations A and B transmitting packets in parallel to a station C. For simplicity, let us assume that both stations use the same transmit power. Station A is very close to C, whereas station B is far away but still in reach of C. Consequently, A's signal at C is much stronger than B's. In this case, it may happen that C successfully decodes a packet sent by A despite B's parallel transmission. This situation is advantageous for the system throughput but disadvantageous for MAC protocols relying on collision detection or collision resolution.

30.5.2 Methods for Combating Channel Errors and Channel Variation

A challenging problem for real-time transmission is the error-prone and time-varying channel. There are many possible control knobs for improving the channel quality, for example, transmit power, bitrate/modulation, coding scheme/redundancy scheme, packet length, choice of retransmission scheme (automatic repeat request [ARQ]), postponing schemes and timing of (re-)transmissions, diversity schemes [79], and adaptation as a meta method [27]. In general, adaptation at the transmitter requires feedback from the receiver. This feedback can be created by using immediate acknowledgment packets after each data packet.

The variation of transmit power and of the bitrate/modulation scheme are both equivalent to varying the energy per bit, which in turn influences the bit error rate [44,85,87]. Roughly, a higher transmit power increases the signal-to-noise ratio and by this reduces the bit error rate and in turn the rate of required retransmissions. However, an increased transmit power might create more interference to other devices sharing the same frequency band (e.g., WiFi and Bluetooth operating in the 2.4 GHz band), which in turn might trigger them to increase their transmit power as well. When adjusting the modulation scheme, the general trade-off is that slower modulation schemes (with fewer levels per time step) have lower bit error rates than faster schemes.

A common way to protect data bits against bit errors is to use redundancy. Example approaches are error detecting and correcting codes (also called FEC) [65] and the transmission of multiple copies of a packet [3]. The latter approach can also be classified as a time-diversity scheme [79]. It is beneficial for the overall throughput to control the amount of redundancy according to the current channel state such that none or only a little redundancy is added when the channel currently shows only few errors [15,24].

Another standard way to deal with transmission errors is retransmissions and suitable ARQ schemes [38]. For channels with deep fades lasting several packet transmission times, it is not useful to retransmit the same packet immediately on the same channel. Specifically, when the mean length of deep fades or error bursts is of the same order or larger than the packet length, both the original packet and its immediate retransmission are likely to be hit by the same error burst. Hence, under these circumstances, an immediate retransmission wastes time and energy.

The transmitter can *postpone* the retransmission for a while and possibly transmit packets to other stations/over other channels in the meantime. If the postponing delay is well chosen, the channel has left the error burst when it is revisited, and the retransmission is successful. Indeed, it has been demonstrated in [5,9,10] that such an approach can reduce the number of wasted packets and increase the throughput significantly. But how to choose the postponing delay? One option is to adopt some fixed value that could be based on measurements or on an a priori knowledge about the channel. Another option is to send occasionally small probing packets [119], which the receiver has to acknowledge.

If the transmitter captures such an acknowledgment, it assumes the channel to be back in good state and continues data transmission. For real-time systems, the postponing decision should consider not only the channel state but also the deadline of a packet. The authors of [25] describe a scheme that takes both the estimated channel state (for postponing decisions) and the packet deadline into account to select one coding scheme from a suite of available schemes.

Retransmissions do not necessarily need to use the same channel as the original packet. It is well known that wireless channels are spatially diverse: A signal transmitted by station A can be in a deep fade at geographical position p_1 and at the same time good enough to be properly received at another position p_2 having the same distance to the transmitter but being separated from p_1 by at least about half a wavelength.* This property is exploited by MIMO and cooperative communication schemes, and it can be explored at the level of ARQ protocols. Assume that station A transmits a packet to station B. The channel from A to B is currently in a deep fade, but another station C successfully captures A's packet. If the channel from C to B is currently in a good state, the packet can be successfully transmitted over this channel. Therefore, station C helps A with its retransmission. This idea has been applied in [112] to the retransmission of data packets as well as to poll packets in a polling-based MAC protocol.

In general, ARQ schemes can be integrated with FEC schemes into hybrid error control schemes [67]. Ideally, for industrial applications, deadlines should be taken into account when designing these schemes. In [4,105], retransmissions and FEC are combined with the concept of deadlines by increasing the coding strength with each retransmitted packet as the packet deadline approaches. This is called deadline dependent coding. Another interesting hybrid error control technique is packet combining [41,52,110]. Put briefly, in these schemes, the receiver tries to take advantage of the partially useful information contained in already received erroneous copies of a packet. For example, if the receiver has received at least three erroneous copies of a packet, it can try to figure out the original packet by applying bit-by-bit majority voting. There are other packet-combining techniques, for example, equal-gain combining or based on intermediate checksums [114].

Sometimes, the packet error probability (and therefore the need for retransmissions) can be reduced by proper tuning of packet sizes. Intuitively, it is clear that larger packets are more likely hit by errors than smaller ones. On the other hand, with smaller packets, the fixed-size packet header becomes more dominant and leads to increased overhead. If the transmitter has estimates of current channel conditions, it can choose the *appropriate* packet size giving the desired trade-off between reliability and efficiency [71,96].

30.6 Wireless Fieldbus Systems: State of the Art

Fieldbus systems are designed to deliver hard real-time services under harsh environmental conditions. A wireless fieldbus [19] should be designed to provide as stringent stochastic timing and reliability guarantees as possible over wireless links. We discuss existing approaches for different popular fieldbus systems.

30.6.1 CAN

As already described in Section 30.5.1, the CAN system [48] uses a priority arbitration protocol on the MAC layer, which cannot be implemented directly on a wireless link. Some approaches have been developed to circumvent this; here we discuss a centralized and two distributed solutions [61].

The distributed WMAC protocol uses the scheme CSMA with collision avoidance (CSMA/CA) with priority-dependent backoffs. A station wishing to transmit a packet uses a carrier sense mechanism to wait for the end of an ongoing packet transmission. After this, the station picks a backoff time depending on the priority value of the current packet. The station listens to the channel during the backoff time. If no other station starts transmission, the station assumes that it has the highest priority and starts transmitting

* More precisely, the two antennas should at minimum have a mutual distance of 40% of the wavelength [87, Chapter 5]. If the system operates in the 2.4 GHz ISM band, this amounts to 5–6 cm.

its own packet. Otherwise, the station defers and starts over after the other packet has been finished. In another distributed scheme, the CAN message priority value is mapped onto the channel using an on–off keying scheme [60]. A station transmits a short burst if the current priority bit is a logical one; otherwise, it switches into receive mode. If the station receives any signal, it gives up; otherwise, it continues with the next bit. The priority bits are considered from the most significant bit to the least significant bit. If the station is still contending after the last bit, it transmits the actual data packet. This approach requires tight synchronization and fast switching between transmit and receive modes of the radio transceiver, which is a problem for certain WLAN technologies. A similar scheme has also been proposed in [82].

The centralized RFMAC protocol leverages the fact that CAN belongs to the class of systems using the real-time database communication model. Data items are identified by unique identifiers. Similar to FIP/WorldFIP, all communications are controlled by a central station broadcasting the variable identifiers and causing the producers of the corresponding data items to transmit the data.

30.6.2 FIP/WorldFIP

The FIP/WorldFIP fieldbus uses a polling table to implement a real-time database [106]. To couple wired and wireless stations, in [74], a wireless-to-wired gateway is introduced, serving as central station for the wireless part. The wireless MAC protocol uses time division multiple access (TDMA), and each TDMA slot is used to transmit one data item (also called process variable).

In the OLCHFA project, a prototype system integrating wired and wireless FIP stations has been developed. This system worked in the 2.4 GHz ISM band using a DSSS physical layer [49]. The available publications put emphasis on the management of configuration data and on distributed algorithms for clock synchronization. The MAC and data-link protocols of FIP were not modified. Since FIP broadcasts the values of process variables periodically, the protocol contains no retransmission scheme for the time-critical data. Instead, the OLCHFA approach is to enhance the FIP process variable model with the so-called time-critical variables, which provide freshness information to the applications. Applications can use this to handle cases of repeated losses.

30.6.3 PROFIBUS

The R-FIELDBUS project (www.rfieldbus.de) evaluated how IEEE 802.11 with DSSS can be used in a PROFIBUS fieldbus system and how such a system can be used for transmission of IP-based multimedia data [45,90]. Two different architectures have been proposed: the single logical ring and the multiple logical ring solution, discussed later. Both solutions run the (almost) unmodified PROFIBUS protocol. The PROFIBUS protocol uses token-passing on top of a broadcast medium. The token is passed between active stations along a logical ring, and much of the protocol's complexity deals with ring maintenance. The token itself is a small control frame.

In the single logical ring solution, all wired and wireless stations are integrated into a single logical token-passing ring. The coupling devices between the wired and wireless domains simply forward all packets. This approach is easy to realize but subjects both data packets and control packets like the token-frame to the errors on wireless links. It is shown in [117] for the PROFIBUS and in [51] for the similar IEEE 802.4. Token Bus protocol that repeated losses of token-frames can create severe problems with the achievable real-time performance. Since there is only a single logical ring, the whole network is affected.

In contrast, in the multiple logical ring solution [30], wireless and wired stations are separated into several logical rings. These rings are coupled by intelligent devices called brouters (a merger from bridge and router). In this solution, transmission problems distort only one ring; the other logical rings remain operational. A second benefit of having multiple rings is traffic segmentation. If the segments are chosen carefully, most of the traffic will be intrasegment, and thus, the overall traffic capacity can be increased. A drawback of the multiple logical ring solution, however, is that intersegment traffic is not natively supported by the PROFIBUS protocol, and extensions are required.

In [111,112], a system following the wireless MAC-aware bridging approach is proposed. On the wireless side, specifically tailored polling-based protocols are used, whereas wired stations run the unmodified PROFIBUS protocol stack. The goal is to avoid token-passing on wireless segments. It is shown that for bursty channel errors, the polling-based protocols achieve substantially better performance in terms of stochastic hard real-time behavior than the PROFIBUS token-passing protocol; for certain kinds of channels, the 99% quantile of the delay needed to successfully transmit a high-priority packet is up to an order of magnitude smaller than for the PROFIBUS protocol. To integrate both protocols, the coupling device between wired and wireless media provides a virtual ring extension [111]. In this scheme, the coupling device acts on the wired side on behalf of the wireless stations. For example, it creates token-frames and executes the ring maintenance mechanisms.

Finally, in [63], a scheme for integration of wireless nodes into a PROFIBUS-DP network (single master, many slaves, no token-passing) is described. An application layer gateway is integrated with a *virtual master* station. The virtual master acts as a proxy for the wireless stations; it polls them using standard IP and IEEE 802.11 distributed coordination function (DCF) protocols.

30.6.4 Other Fieldbus Technologies

For the IEC Fieldbus [46] (which uses a centralized, polling-based access protocol for periodic data and a token-passing protocol for asynchronous data) in [11], an architecture that allows coupling of several fieldbus segments using a wireless backbone based on IEEE 802.11 with point coordination function (PCF) is proposed.

In [75], it is investigated how the MAP/MMS application layer protocol can be enhanced with mobility. In the proposed system, the IEEE 802.11 WLAN with DCF is used; time-critical transmissions and channel errors are not considered. In [73], the same question was investigated with digital European cordless telephone as underlying technology.

30.7 Wireless Ethernet/IEEE 802.11

Instead of developing WLAN technology for the factory floor from scratch, existing technologies can serve as a starting point. A good candidate is the IEEE 802.11 WLAN standard [47], since it is the most widely used WLAN technology. It has also been considered extensively in the context of wireless industrial communications (see, for example, [8,13,53,57,70]).

The IEEE 802.11 standard belongs to the IEEE 802.x family of LAN standards. The standard describes architecture, services, physical layers, and protocols for an Ethernet-like wireless LAN, using a CSMA/CA-based MAC protocol with enhancements for time-bounded services. The protocols run on top of several physical layers, using different transmission schemes ranging from FHSS and DSSS to OFDM. The former 802.11a and g variants of the standard offered physical layer bitrates from 6 to 54 MBit/s, using OFDM and various coding schemes with different code rates. The former 802.11n variant couples MIMO and OFDM to achieve bitrates up to 150 MBit/s.

The standard describes an ad hoc mode and an infrastructure-based mode. In the infrastructure mode, all communications is relayed through fixed APs. An AP constitutes a so-called service set, and mobile stations have to associate with the closest AP. The APs can be connected by a distribution system that allows to forward data packets between mobile stations in different cells. In the ad hoc mode, there are neither APs nor a distribution system, and stations communicate in a peer-to-peer fashion.

30.7.1 DCF

The basic MAC protocol of 802.11 is called the DCF. It is a CSMA/CA protocol using the RTS/CTS scheme described in Section 30.5.1 and different interframe gaps to give control frames (e.g., acknowledgments and CTS frames) priority over data frames. However, data frames cannot be differentiated according to priorities. The IEEE 802.11 MAC provides a connectionless and semireliable best-effort service to its users by performing a bounded number of retransmissions. Furthermore, it is not possible to specify attributes like transmit power, modulation scheme, or the number of retransmissions on a per-packet basis. This control would be desirable for treating different packet types differently. As an example, one could transmit high-priority packets with high transmit power and low bitrate to increase their reliability.

In more detail, the DCF operates as follows. When a station gets a new packet for transmission, it senses the medium (carrier sensing). When the medium is idle for a certain amount of time (called interframe space), the station starts to transmit. Otherwise, the station enters backoff mode. In this mode, it first waits until the medium is idle for at least distributed interframe space (DIFS) time. Then the backoff slots start. If the station was not in backoff mode before, it draws a random number out of the current contention window and stores it in a counter variable. Then the station observes all subsequent timeslots and decrements the counter when the slot was empty. If the counter reaches zero, the station starts to transmit. When a slot is busy, the process of decrementing the counter is suspended and is resumed once the medium becomes idle again for at least DIFS time.

The contention window size changes over time. It is initialized with a preconfigured value of CW_{min}. When a packet transmission fails, the contention window is doubled, until a maximum of CW_{max} is reached. The contention window is set back to CW_{min} after a successful transmission.

30.7.2 QoS Support

The former IEEE 802.11e extension (which has been incorporated into recent versions of the standard) adds quality-of-service (QoS) support to the standard, which provides time-bounded and deterministic services in infrastructure networks. The QoS support is known as the hybrid coordination function and operates on top of the DCF. There are both a distributed scheme and a centralized QoS support scheme; the latter requires a centralized control entity called the hybrid coordinator (HC) that is collocated with an AP. Other stations can send reservation requests to the HC, asking it to schedule so-called transmit opportunities (TXOP). If granted, the requesting station can use the medium exclusively for a limited amount of time to transmit one or more frames including MAC layer acknowledgements to any station in the network. A TXOP can be granted once or periodically. A key function in the HC is the admission control component, which decides whether a new TXOP request can be accommodated without breaking guarantees given to other requesters. The precise admission control algorithm is not prescribed by the standard but left to the implementers.

On this basis, the standard defines two access schemes for QoS support: the enhanced DCF (EDCF) and the hybrid coordination function controlled channel access (HCCA). We describe both of them briefly.

30.7.2.1 EDCF

The EDCF is a distributed scheme that operates similar to the DCF. However, the higher layer labels each packet as belonging to one out of four available QoS classes (also called access classes). To each class corresponds a separate value of the interframe space. For the ith class, the interframe space is called AIFS[i]. The AIFS[i] values can be configured but need to be different. Furthermore, the EDCF defines separate values of the minimum and maximum contention window (CW_{min} and CW_{max}) for each QoS class.

When a station wishes to transmit a packet of priority i, it executes the same protocol as in the DCF, but now using the AIFS[i] as the interframe space and the $CW_{min}[i]$ and $CW_{max}[i]$ values for the contention window. By proper choice of these parameters, it is possible to ensure with high probability that a packet of a better access class wins over a contender having a packet of lower access classes. Hence, a stochastic prioritization is achieved. Another difference between the DCF and the EDCF is that in the DCF, the contention winner is allowed to transmit only one packet, whereas in EDCF, the winner receives a TXOP, which, as explained earlier, may cover several packets of the winning or higher access classes.

A number of performance analyses of the DCF and EDCF are available in the literature [29,56,83,91,57,118].

30.7.2.1.1 HCCA

In the HCCA, the central HC coordinates access to the medium, based on reservation requests from stations. The HC receives requests, performs admission control, and, if granted, is responsible for actual scheduling of TXOPs. However, the HC does not control the channel all the time; scheduled TXOPs alternate with phases where stations contend for the medium using the EDCF. The reservation requests can either be sent as separate packets (using the EDCF) or be piggybacked onto data packets. To gain control over the channel, the HC uses the basic DCF mechanism with an interframe space that is shorter than DIFS and all AIFS[*i*] values. To start a TXOP, the HC sends a poll packet to the owner of the TXOP, which then can transmit one or multiple packets. This approach prevents the system from achieving perfectly periodic services, as the medium might still be busy when the HC tries to gain access to the medium.

The HCCA is rather complex as there are rich interactions with other features of the IEEE 802.11 protocol like, for example, the rate adaptation feature of some of the physical layers. It is similarly complex as the PCF of the original IEEE 802.11 standard, which is still present in the 2007 version and which can be operated jointly with the HCCA. The author is not aware of any existing implementation of the HCCA.

A key research issue is the design of appropriate admission control scheduling policies, which has, for example, been done in [7,88,89]. Its usage in industrial scenarios has been considered in [53,103].

30.8 Summary

This chapter presented some problems and solution approaches to bring WLAN technology to the factory plant and to benefit from reduced cabling and mobility. The basic problem is the tension between the hard timing and reliability requirements of industrial applications on the one hand and the serious error rates and time varying error behavior of wireless channels on the other hand. Many techniques have been developed to improve the reliability and timeliness behavior of lower-layer wireless protocols, but up to now, wireless fieldbus systems have not been deployed on a large scale as the problem of reliable transmission despite channel errors is not solved satisfactorily. It is not clear which combination of mechanisms and technologies has the potential to bound the number of deadline misses under realistic channel conditions.

It seems to be an open question whether just some more engineering is needed to make wireless transmission suitable for fulfilling hard real-time and reliability requirements, or whether there is really a limit of what can be achieved. Fortunately, wireless communications and WLAN technology is a very active field of research and development. New technologies are created, and existing technologies are enhanced. As an example, the IEEE 802.11g and IEEE 802.11e working groups aim to deliver higher bitrates and better quality of service to users. It will be exciting to see how industrial applications can benefit from this.

References

1. L. Ahlin and J. Zander. *Principles of Wireless Communications*. Studentlitteratur, Lund, Sweden, 1998.
2. I. F. Akyildiz, J. Xie, and S. Mohanty. A survey of mobility management in next-generation all-IP-based wireless systems. *IEEE Wireless Communications*, 11(4):16–28, August 2004.
3. A. Annamalai and V. K. Bhargava. Analysis and optimization of adaptive multicopy transmission ARQ protocols for time-varying channels. *IEEE Transactions on Communications*, 46(10):1356–1368, 1998.

4. H. Bengtsson, E. Uhlemann, and P.-A. Wiberg. Protocol for wireless real-time systems. In *Proceedings of 11th Euromicro Conference on Real-Time Systems*, York, U.K., 1999.
5. P. Bhagwat, P. Bhattacharya, A. Krishna, and S. K. Tripathi. Using channel state dependent packet scheduling to improve TCP throughput over wireless LANs. *Wireless Networks*, 3(1):91–102, March 1997.
6. H. Blcskei. MIMO-OFDM wireless systems: Basics, perspectives and challenges. *IEEE Wireless Communications*, 13(4):31–37, August 2006.
7. G. Boggia, P. Camarda, L. A. Grieco, and S. Mascolo. Feedback-based control for providing real-time services with the 802.11e MAC. *IEEE/ACM Transactions on Networking*, 15(2):323–333, April 2007.
8. D. Brevi, D. Mazzocchi, R. Scopigno, A. Bonivento, R. Calcagno, and F. Rusina. A methodology for the analysis of 802.11a links in industrial environments. In *Proceedings of 2006 IEEE International Workshop on Factory Communication Systems (WFCS)*, Torina, Italy, July 2006.
9. R. Cam and C. Leung. Multiplexed ARQ for time-varying channels—Part I: System model and throughput analysis. *IEEE Transactions on Communications*, 46(1):41–51, January 1998.
10. R. Cam and C. Leung. Multiplexed ARQ for time-varying channels—Part II: Postponed retransmission modification and numerical results. *IEEE Transactions on Communications*, 46(3):314–326, March 1998.
11. S. Cavalieri and D. Panno. On the integration of fieldbus traffic within IEEE 802.11 wireless LAN. In *Proceedings of 1997 IEEE International Workshop on Factory Communication Systems (WFCS'97)*, Barcelona, Spain, 1997.
12. J. K. Cavers. *Mobile Channel Characteristics*. Kluwer Academic Publishers, Boston, MA/Dordrecht, the Netherlands, 2000.
13. G. Cena, I. C. Bertolotti, A. Valenzano, and C. Zunino. Evaluation of response times in industrial WLANs. *IEEE Transactions on Industrial Informatics*, 3(3):191–201, August 2007.
14. G. Cena, A. Valenzano, and S. Vitturi. Hybrid wired/wireless networks for real-time communications. *IEEE Industrial Electronics Magazine*, 2(1):8–20, March 2008.
15. R. Chen, K. C. Chua, B. T. Tan, and C. S. Ng. Adaptive error coding using channel prediction. *Wireless Networks*, 5(1):23–32, February 1999.
16. C.-F. Chiasserini and R. R. Rao. Coexistence mechanisms for interference mitigation in the 2.4-GHz ISM band. *IEEE Transactions on Wireless Communications*, 2(5):964–975, September 2003.
17. D. J. Costello, J. Hagenauer, H. Imai, and S. B. Wicker. Applications of error-control coding. *IEEE Transactions on Information Theory*, 44(6):2531–2560, October 1998.
18. K. David and T. Benkner. *Digitale Mobilfunksysteme. Informationstechnik*. B.G. Teubner, Stuttgart, Germany, 1996.
19. J.-D. Decotignie. Wireless fieldbusses—A survey of issues and solutions. In *Proceedings of 15th IFAC World Congress on Automatic Control (IFAC 2002)*, Barcelona, Spain, 2002.
20. J.-D. Decotignie. Ethernet-based real-time and industrial communications. *Proceedings of the IEEE*, 93(6):1102–1117, 2005.
21. J.-D. Decotignie and P. Pleineveaux. A survey on industrial communication networks. *Annals of Telecommunications*, 48(9):435–448, 1993.
22. L. Dellaverson and W. Dellaverson. Distributed channel access on wireless ATM links. *IEEE Communications Magazine*, 35(11):110–113, November 1997.
23. S. N. Diggavi, N. Al-Dhahir, A. Stamoulis, and A. R. Calderbank. Great expectations: The value of spatial diversity in wireless networks. *Proceedings of the IEEE*, 92(2):219–270, February 2004.
24. D. A. Eckhardt and P. Steenkiste. A trace-based evaluation of adaptive error correction for a wireless local area network. *MONET—Mobile Networks and Applications*, 4:273–287, 1999.
25. M. Elaoud and P. Ramanathan. Adaptive use of error-correcting codes for real-time communication in wireless networks. In *Proceedings of IEEE INFOCOM 1998*, San Francisco, CA, March 1998.
26. A. Ephremides. Energy concerns in wireless networks. *IEEE Wireless Communications*, 9(4):48–59, August 2002.

27. A. Farago, A. D. Myers, V. R. Syrotiuk, and G. V. Zaruba. Meta-MAC protocols: Automatic combination of MAC protocols to optimize performance for unknown conditions. *IEEE Journal on Selected Areas in Communications*, 18(9):1670–1681, September 2000.
28. S. Fernandes and A. Karmouch. Vertical mobility management architectures in wireless networks: A comprehensive survey and future directions. *IEEE Communications Surveys and Tutorials*, 14(1):45–63, 2006.
29. P. Ferre, A. Doufexi, A. Nix, and D. Bull. Throughput analysis of IEEE 802.11 and IEEE 802.11e MAC. In *Proceedings of IEEE Wireless Communications and Networking Conference (WCNC), Atlanta, GA, 2004*, pp. 783–788, March 2004.
30. L. Ferreira, M. Alves, and E. Tovar. Hybrid wired/wireless PROFIBUS networks supported by bridges/routers. In *Proceedings of 2002 IEEE Workshop on Factory Communication Systems, WFCS' 2002*, pp. 193–202, Västeras, Sweden, 2002.
31. F.-P. Das. Verbundprojekt Drahtlose Feldbusse im Produktionsumfeld (Funbus)—Abschlußbericht. INTERBUS Club Deutschland e.V., Postf. 1108, 32817 Blomberg, Bestell-Nr: TNR 5121324, October 2000. http://www.softing.de/d/NEWS/Funbusbericht.pdf.
32. J. D. Gibson, ed. *The Communications Handbook*. CRC Press/IEEE Press, Boca Raton, FL, 1996.
33. S. Glisic and B. Vucetic. *Spread Spectrum CDMA Systems for Wireless Communications*. Artech House, Boston, MA, 1997.
34. A. Goldsmith. *Wireless Communications*. Cambridge University Press, Cambridge, U.K., 2005.
35. A. J. Goldsmith and S. B. Wicker. Design challenges for energy constrained ad hoc wireless networks. *IEEE Wireless Communications*, 9(4):8–27, August 2002.
36. A. C. V. Gummalla and J. O. Limb. Wireless medium access control protocols. *IEEE Communications Surveys and Tutorials*, 3(2), 2–15, 2000. http://www.comsoc.org/pubs/surveys.
37. V. C. Gungor and G. P. Hancke. Industrial wireless sensor networks: Challenges, design principles, and technical approaches. *IEEE Transactions on Industrial Electronics*, 56(10):4258–4265, 2009.
38. D. Haccoun and S. Pierre. Automatic repeat request. In J. D. Gibson, ed., *The Communications Handbook*, pp. 181–198. CRC Press/IEEE Press, Boca Raton, FL, 1996.
39. HART Communication Foundation. HART communication protocol specification, HCF SPEC 13 Revision 7.1, 05 June, 2008.
40. HART Communication Foundation. TDMA data link layer specification, HCF SPEC 075 Revision 1.1, 17 May, 2008.
41. B. A. Harvey and S. B. Wicker. Packet combining systems based on the Viterbi decoder. *IEEE Transactions on Communications*, 42(2):1544–1557, February 1994.
42. J. P. Hespanha, P. Naghshtabrizi, and Y. Xu. A survey of recent results in networked control systems. *Proceedings of the IEEE*, 95(1):138–162, January 2007.
43. J. Hirai, T.-W. Kim, and A. Kawamura. Practical study on wireless transmission of power and information for autonomous decentralized manufacturing system. *IEEE Transactions on Industrial Electronics*, 46:349–359, April 1999.
44. G. Holland, N. Vaidya, and P. Bahl. A rate-adaptive MAC protocol for wireless networks. In *Proceedings of Seventh Annual International Conference on Mobile Computing and Networking 2001 (MobiCom)*, Rome, Italy, July 2001.
45. J. Hähniche and L. Rauchhaupt. Radio communication in automation systems: The R-fieldbus approach. In *Proceedings of 2000 IEEE International Workshop on Factory Communication Systems (WFCS 2000)*, pp. 319–326, Porto, Portugal, 2000.
46. IEC—International Electrotechnical Commission. IEC-1158-1, FieldBus specification, Part 1, FieldBus standard for use in industrial control: Functional requirements, 1997.
47. IEEE Computer Society, sponsored by the LAN/MAN Standards Committee. IEEE standard for information technology—Telecommunications and information exchange between systems—Local and metropolitan area networks—Specific Requirements—Part 11: Wireless LAN medium access control (MAC) and physical layer (PHY) specifications, 2012.

48. International Organization for Standardization. ISO standard 11898—Road vehicle—Interchange of digital information—Controller area network (CAN) for high-speed communication. International Organization for Standardization, 1993.
49. I. Izikowitz and M. Solvie. Industrial needs for time-critical wireless communication & wireless data transmission and application layer support for time critical communication. In *Proceedings of Euro-Arch'93*, Munich, 1993. Springer Verlag, Berlin, Germany.
50. W. C. Jakes, ed. *Microwave Mobile Communications*. IEEE Press, Piscataway, NJ, 1993.
51. H. ju Moon, H. S. Park, S. C. Ahn, and W. H. Kwon. Performance degradation of the IEEE 802.4 token bus network in a noisy environment. *Computer Communications*, 21:547–557, 1998.
52. S. Kallel. Analysis of a type-II hybrid ARQ scheme with code combining. *IEEE Transactions on Communications*, 38(8):1133–1137, August 1990.
53. S. P. Karanam, H. Trsek, and J. Jasperneite. Potential of the HCCA scheme defined in IEEE802.11e for QoS enabled industrial wireless networks. In *Proceedings of 2006 IEEE International Workshop on Factory Communication Systems (WFCS)*, Torino, Italy, July 2006.
54. H. Karl and A. Willig. *Protocols and Architectures for Wireless Sensor Networks*. John Wiley & Sons, Chichester, U.K., 2005.
55. F. Khan. *LTE for 4G Mobile Broadband—Air Interface Technologies and Performance*. Cambridge University Press, Cambridge, U.K., 2009.
56. Z.-N. Kong, D. H. K. Tsang, B. Bensaou, and D. Gao. Performance analysis of IEEE 802.11e contention-based channel access. *IEEE Journal on Selected Areas in Communications*, 22(10):2095–2106, December 2004.
57. D. Koscielnik. Simulation study of IEEE 802.11e wireless LAN—Enhancements for real time applications. In *Proceedings of IEEE International Symposium on Industrial Electronics (ISIE)*, Montreal, Quebec, Canada, July 2006.
58. G. Kramer, I. Maric, and R. D. Yates. Cooperative communications. *Foundations and Trends in Networking*, 1(3–4):271–425, 2006.
59. J. F. Kurose, M. Schwartz, and Y. Yemini. Multiple-access protocols and time-constrained communication. *ACM Computing Surveys*, 16:43–70, March 1984.
60. A. Kutlu, H. Ekiz, M. D. Baba, and E. T. Powner. Implementation of "Comb" based wireless access method for control area network. In *Proceedings of 11th International Symposium on Computer and Information Science*, pp. 565–573, Antalaya, Turkey, November 1996.
61. A. Kutlu, H. Ekiz, and E. T. Powner. Performance analysis of MAC protocols for wireless control area network. In *Proceedings of International Symposium on Parallel Architectures, Algorithms and Networks*, pp. 494–499, Beijing, China, June 1996.
62. J. N. Laneman, D. N. C. Tse, and G. W. Wornell. Cooperative diversity in wireless networks: Efficient protocols and outage behaviour. *IEEE Transactions on Information Theory*, 50(12):3062–3080, December 2004.
63. K. C. Lee and S. Lee. Integrated network of PROFIBUS-DP and IEEE 802.11 wireless LAN with hard real-time requirement. In *Proceedings of IEEE 2001 International Symposium on Industrial Electronics*, Korea, 2001.
64. T. H. Liew and L. Hanzo. Space-time codes and concatenated channel codes for wireless communications. *Proceedings of the IEEE*, 90(2):187–219, February 2002.
65. S. Lin and D. J. Costello. *Error Control Coding—Fundamentals and Applications*. Prentice-Hall, Englewood Cliffs, NJ, 1983.
66. H. Liu and G. Li. *OFDM-Based Broadband Wireless Networks*. John Wiley & Sons, Hoboken, NJ, 2005.
67. H. Liu, H. Ma, M. E. Zarki, and S. Gupta. Error control schemes for networks: An overview. *MONET—Mobile Networks and Applications*, 2(2):167–182, 1997.
68. K. J. R. Liu, A. K. Sadek, W. Su, and A. Kwasinski. *Cooperative Communications and Networking*. Cambridge University Press, Cambridge, U.K., 2009.

69. N. P. Mahalik, ed. *Fieldbus Technology—Industrial Network Standards for Real-Time Distributed Control.* Springer, Berlin, Germany, 2003.
70. D. Miorandi and S. Vitturi. Performance analysis of producer/consumer protocols over IEEE802.11 wireless links. In *Proceedings IEEE Workshop on Factory Communication Systems (WFCS)*, Vienna, Austria, September 2004.
71. E. Modiano. An adaptive algorithm for optimizing the packet size used in wireless ARQ protocols. *Wireless Networks*, 5:279–286, 1999.
72. A. F. Molisch. *Wireless Communications.* John Wiley & Sons/IEEE Press, Chichester, U.K., 2005.
73. P. Morel. Mobility in MAP networks using the DECT wireless protocols. In *Proceedings of 1995 IEEE Workshop on Factory Communication Systems, WFCS'95*, Leysin, Switzerland, 1995.
74. P. Morel and A. Croisier. A wireless gateway for fieldbus. In *Proceedings of Sixth International Symposium on Personal, Indoor and Mobile Radio Communications (PIMRC 95)*, Toronto, Canada, 1995.
75. P. Morel and J.-D. Decotignie. Integration of wireless mobile nodes in MAP/MMS. In *Proceedings of 13th IFAC Workshop on Distributed Computer Control Systems DCCS95*, Toulouse, France, 1995.
76. G. T. Nguyen, R. H. Katz, B. Noble, and M. Satyanarayanan. A trace-based approach for modeling wireless channel behavior. In *Proceedings of the Winter Simulation Conference*, Coronado, CA, December 1996.
77. A. Nosratinia, T. E. Hunter, and A. Hedayat. Cooperative communication in wireless networks. *IEEE Communications Magazine*, 42(10):74–80, October 2004.
78. K. Pahlavan and A. H. Levesque. *Wireless Information Networks.* John Wiley & Sons, New York, 1995.
79. A. Paulraj. Diversity techniques. In J. D. Gibson, ed., *The Communications Handbook*, pp. 213–223. CRC Press/IEEE Press, Boca Raton, FL, 1996.
80. A. J. Paulraj, D. A. Gore, R. U. Nabar, and H. Bleskei. An overview of MIMO communications—A key to gigabit wireless. *Proceedings of the IEEE*, 92(2):198–218, February 2004.
81. A. J. Paulraj, R. U. Nabar, and D. A. Gore. *Introduction to Space-Time Wireless Communications.* Cambridge University Press, Cambridge, U.K., 2003.
82. N. Pereira, B. Andersson, and E. Tovar. WiDom: A dominance protocol for wireless medium access. *IEEE Transactions on Industrial Informatics*, 3(2):120–130, May 2007.
83. J. P. Pavon and S. Shankar. Impact of frame size, number of stations and mobility on the throughput performance of IEEE 802.11e. In *Proceedings of IEEE Wireless Communications and Networking Conference (WCNC), 2004*, Atlanta, Georgia, pp. 789–795, March 2004.
84. J. R. Pimentel. *Communication Networks for Manufacturing.* Prentice-Hall International, Upper Saddle River, NJ, 1990.
85. J. G. Proakis. *Digital Communications*, 3rd edn. McGraw-Hill, New York, 1995.
86. C. S. Raghavendra and S. Singh. Pamas—Power aware multi-access protocol with signalling for ad hoc networks. *ACM Computer Communication Review*, 27:5–26, July 1998.
87. T. S. Rappaport. *Wireless Communications—Principles and Practice.* Prentice Hall, Upper Saddle River, NJ, 2002.
88. M. M. Rashid, E. Hossain, and V. K. Bhargava. Queueing analysis of 802.11e HCCA with variable bit rate traffic. In *Proceedings of IEEE International Conference on Communications (ICC)*, Istanbul, Turkey, June 2006.
89. M. M. Rashid, E. Hossain, and V. K. Bhargava. HCCA scheduler design for guaranteed QoS in IEEE 802.11e based WLANs. In *Proceedings of IEEE Wireless Communications and Networking Conference (WCNC)*, Hong Kong, China, March 2007.
90. L. Rauchhaupt. System and device architecture of a radio-based fieldbus—The RFieldbus system. In *Proceedings of Fourth IEEE Workshop on Factory Communication Systems 2002 (WFCS 2002)*, Västeras, Sweden, 2002.
91. J. W. Robinson and T. S. Randhawa. Saturation throughput analysis of IEEE 802.11e enhanced distributed coordination function. *IEEE Journal on Selected Areas in Communications*, 22(5):917–928, 2004.

92. A. Santamaria and F. J. Lopez-Hernandez, eds. *Wireless LAN—Standards and Applications. Mobile Communication Series*. Artech House, Boston, MA/London, U.K., 2001.
93. A. Scaglione, D. L. Goeckel, and J. N. Laneman. Cooperative communications in mobile ad hoc networks. *IEEE Signal Processing Magazine*, 23(5):18–29, September 2006.
94. G. Schäfer. *Security in Fixed and Wireless Networks—An Introduction to Securing Data Communications*. John Wiley & Sons, Chichester, U.K., 2003.
95. G. Scheible, D. Dzung, J. Endresen, and J.-E. Frey. Unplugged but connected—Design and implementation of a truly wireless real-time sensor/actuator interface. *IEEE Industrial Electronics Magazine*, 1(2):25–34, 2007.
96. C. K. Siew and D. J. Goodman. Packet data transmission over mobile radio channels. *IEEE Transactions on Vehicular Technology*, 38(2):95–101, May 1989.
97. W. Stallings. *Wireless Communications and Networks*. Prentice Hall, Upper Saddle River, NJ, 2001.
98. I. Stojmenovic, ed. *Handbook of Wireless Networks and Mobile Computing*. John Wiley & Sons, New York, 2002.
99. A. S. Tanenbaum. *Computernetzwerke*, 3rd edn. Prentice-Hall, Munich, Germany, 1997.
100. V. Tarokh, H. Jafarkhani, and A. R. Calderbank. Space-time block codes from orthogonal designs. *IEEE Transactions on Information Theory*, 45(5):1456–1467, July 1999.
101. F. A. Tobagi and L. Kleinrock. Packet switching in radio channels, Part II: The hidden terminal problem in CSMA and busy-tone solutions. *IEEE Transactions on Communications*, 23(12):1417–1433, 1975.
102. C.-K. Toh. *Ad Hoc Mobile Wireless Networks—Protocols and Systems*. Prentice Hall, Upper Saddle River, NJ, 2002.
103. H. Trsek, J. Jasperneite, and S. P. Karanam. A simulation case study of the new IEEE 802.11e HCCA mechanism in industrial wireless networks. In *Proceedings of 11th IEEE International Conference on Emerging Technologies and Factory Automation (ETFA 2006)*, Prague, Czech Republic, September 2006.
104. D. Tse and P. Viswanath. *Fundamentals of Wireless Communications*. Cambridge University Press, Cambridge, U.K., 2005.
105. E. Uhlemann, P.-A. Wiberg, T. M. Aulin, and L. K. Rasmussen. Deadline-dependent coding—A framework for wireless real-time communication. In *Proceedings of International Conference on Real-Time Computing Systems and Applications*, pp. 135–142, Cheju Island, South Korea, December 2000.
106. Union Technique de l'Electricit'e. General purpose field communication system, EN 50170, Vol. 3: WorldFIP, 1996.
107. H. R. van As. Media access techniques: The evolution towards terabit/s LANs and MANs. *Computer Networks and ISDN Systems*, 26:603–656, 1994.
108. R. van Nee and R. Prasad. *OFDM for Wireless Multimedia Communications*. Artech House, London, U.K., 2000.
109. Various authors. Special issue on models, theory, and codes for relaying and cooperation in communication networks. *IEEE Transactions on Information Theory*, 53(10):3297–3842, October 2007.
110. X. Wang and M. T. Orchard. On reducing the rate of retransmission in time-varying channels. *IEEE Transactions on Communications*, 51(6):900–910, June 2003.
111. A. Willig. An architecture for wireless extension of PROFIBUS. In *Proceedings of IECON 03*, Roanoke, VA, 2003.
112. A. Willig. Polling-based MAC protocols for improving realtime performance in a wireless PROFIBUS. *IEEE Transactions on Industrial Electronics*, 50(4):806–817, August 2003.
113. A. Willig. Recent and emerging topics in wireless industrial communications: A selection. *IEEE Transactions on Industrial Informatics*, 4(2):102–124, May 2008.
114. A. Willig. Memory-efficient segment-based packet-combining schemes in face of deadlines. *IEEE Transactions on Industrial Informatics*, 5(3):338–350, August 2009.
115. A. Willig, M. Kubisch, C. Hoene, and A. Wolisz. Measurements of a wireless link in an industrial environment using an IEEE 802.11-compliant physical layer. *IEEE Transactions on Industrial Electronics*, 49(6):1265–1282, 2002.

116. A. Willig, K. Matheus, and A. Wolisz. Wireless technology in industrial networks. *Proceedings of the IEEE*, 93(6):1130–1151, June 2005.
117. A. Willig and A. Wolisz. Ring stability of the PROFIBUS token passing protocol over error prone links. *IEEE Transactions on Industrial Electronics*, 48(5):1025–1033, October 2001.
118. B. Xiang, M. Yu-Ming, and X. Jun. Performance investigation of IEEE802.11e EDCA under non-saturation condition based on the M/G/1/K model. In *Proceedings of Second IEEE Conference on Industrial Electronics and Applications (ICIEA 2007)*, Harbin, China, May 2007.
119. M. Zorzi and R. R. Rao. Error control and energy consumption in communications for nomadic computing. *IEEE Transactions on Computers*, 46:279–289, March 1997.
120. M. Z. Zamalloa and B. Krishnamachari. An analysis of unreliability and asymmetry in low-power wireless links. *ACM Transactions on Sensor Networks*, 3(2):1–34, June 2007.

31

WirelessHART

31.1 Introduction 31-1
31.2 Requirements of Wireless Fieldbuses 31-2
Physical Layer • Data Link Layer • Network Layer • Application Layer
31.3 HART and WirelessHART 31-3
WirelessHART: The Network Architecture
31.4 WirelessHART Protocol Stack 31-5
WirelessHART Physical Layer • WirelessHART Data Link Layer • WirelessHART Network Layer • WirelessHART Transport Layer • WirelessHART Application Layer
31.5 Message Format 31-13
Physical and Data Link Layer Datagram • Network Layer Datagram • Transport and Application Layer Datagram
31.6 Join Procedure 31-15
31.7 Security Aspects 31-16
31.8 Wi-Analys Tool 31-16
31.9 Designing a New WirelessHART Network 31-16
31.10 Example of Platform Providers 31-19
References 31-19

Alessandra Flammini
University of Brescia

Emiliano Sisinni
University of Brescia

31.1 Introduction

The use of wireless technology for industrial applications is really attractive, as confirmed by the constantly increasing adoption of the *wireless fieldbus* term. Avoiding the use of cables allows to minimize not only costs but also cabling errors, which are the most common reason of communication systems failure in plants (not just in the commissioning stage). As a matter of fact, the adoption of wireless technology in the industrial world is not new, but usually was limited to niche applications due to lack of standard and well-accepted solutions. In particular, peculiar requirements of industrial communications related to determinism, low cost, availability, reliability, and security oblige the development of specific wireless standards, due to the inadequacy of consumer-oriented solutions.

In the following section, for the sake of clarity, the classic International Standards Organization/Open Systems Interconnection (ISO/OSI) model (limited to the physical, data link, network, and application layers) has been used to identify and show the main requirements for each level from the industrial point of view.

31.2 Requirements of Wireless Fieldbuses

31.2.1 Physical Layer

Key issues are the low power consumption and robustness to interference (intentional or unintentional). The wireless nodes must be truly autonomous (i.e., completely untethered), and in most of cases, this need is satisfied using batteries, whose life must be on the order of several years, that is, comparable with wired counterparts. In addition, the communication usually takes place in a license-free industrial, scientific, medical (ISM) band, thus facilitating worldwide market penetration and lowering the cost, but creating coexistence issues with all the other systems operating in the same region of the spectrum.

Less critical aspects are the communication range (which strictly depends on the application, e.g., ranging from less than 100 m for factory automations to several hundreds of meters for process automation) and the data rate, as the typical industrial devices transmit small amounts of data per cycle (e.g., consider the typical transfer rate of commonly adopted wired fieldbuses). What is really important is the end-to-end delay, that is, the throughput at the application level, which mostly depends on upper layers.

31.2.2 Data Link Layer

The industrial communication is normally based on the exchange of periodic events and some sporadic acyclic data such as alarms. Thus, the main requirements are to meet real time and guarantee the determinism.

The technique of media access becomes crucial in this regard. The most suitable technique appears to be the time division multiple access (TDMA, as in wired counterparts), flanked by some form of frequency diversity to improve interference resistance. In turn, this approach implies that the whole network must be somehow synchronized.

31.2.3 Network Layer

A possible solution to ensure high network reliability is the adoption of redundant routes from the source to the destinations. The network level is therefore required to implement some kind of mesh networking through which each device acts not only as a data source or sink but also as a repeater, forwarding the packet until it reaches the desired destination. In such a way, large distance can be covered with low power. The network layer must also ensure high scalability, that is, the capability of the network set to increase the number of active nodes without degradation in performance.

A critical point, which mainly depends on both the data link and the network layer, is the trade-off between the number and density of nodes to be served (affecting the mesh depth) and the end-to-end delay or throughput at the user level. It must be emphasized that this requirement can be very different for factory or process automation, for example, ranging from ms to several seconds or minutes, respectively.

31.2.4 Application Layer

At this level, compatibility and interoperability among devices of different manufacturers must be ensured. This problem is usually solved standardizing not only the application protocol but also different profiles dedicated to the well-defined applications (e.g., clearly specifying how data must be formatted and so on).

In addition, security is a key issue in any communication system and must be ensured across the whole protocol stack. Nowadays, the most common approach adopts encoding and encryption mechanisms to protect the data, trying to mask the data, making them unreadable and complex. Such techniques usually operate across the whole different stack layers, offering increasing security level.

In the following sections, the solutions adopted by WirelessHART (the first open standard available tailored for process control applications) to solve these requirements are briefly resumed.

31.3 HART and WirelessHART

The HART Communications Protocol (an acronym for Highway Addressable Remote Transducer Protocol) is one of the first fieldbuses to appear on the market. It was originally designed to add diagnostic information to process devices still preserving compatibility with legacy 4–20 mA analog instrumentation. In particular, the HART protocol shares the same pair of wires used by the legacy analog system, reducing the cost. Probably, the HART protocol is nowadays one of the most deployed protocols for industrial communications with the largest number of installed nodes.

The protocol was originally proposed by Rosemount Inc. in the mid-1980s for their smart field instruments. It was based on the Bell 202 communications standard, which AT&T developed for serial communication over a modem operating in the audio frequency range. Starting from 1986, it was made an open protocol, paving the way for several improvements, as proved by the numerous revisions to the specification published in the past.

Overall performance has been designed in order to satisfy the needs of process automation. If used in multichannel input/output (I/O) system, it allows for about one transaction per second (if Frequency-Shift Keying (FSK) modulation is adopted), with latencies on the same order of magnitude (depending on the number of nodes). In addition, it is able to work on distances up to 1500 m.

Starting from 2007, with the advent of the HART7 specification, the WirelessHART has been defined as an extension of the traditional wired solution. The target goal was to preserve performance on the same order of magnitude (or even improve them) allowing the implementation of a truly autonomous sensor. In fact, as suggested by its name, it employs a different physical layer that uses radio frequencies (RFs) as the communication medium with a nominal coverage area of hundreds of meters (in line-of-sight [LOS] conditions) and devices that are normally battery powered, but may also exploit power cables or some form of energy harvesting (e.g., solar) or a combination of all these sources.

WirelessHART

- Implements an RF self-healing mesh network
- Allows for network-wide time synchronization
- Enhances the publish/subscribe messaging (i.e., burst mode)
- Adds network and transport layers
- Adds a *fast pipe* for time critical traffic and, last but not least, adds ciphering

A comparison among the traditional HART protocol communication layers, the WirelessHART solution, and the well-known consumer world ZigBee protocol is given in Table 31.1, following the ISO/OSI model.

A resume of main WirelessHART characteristics is provided in the following:

- Low power consumption and low-cost devices (comparable with wired counterparts)
- Raw data rate of 250 kbps per channel in the 2.4 GHz ISM band with 15 different channels worldwide available
- Based on the diffused and well-accepted Institute of Electrical and Electronics Engineers (IEEE) 802.15.4-2006 physical layer
- Based on a proprietary data link layer, based on TDMA and carrier sense multiple access with collision avoidance (CSMA/CA)
- Support of *channel hopping* and *channel blacklisting* mechanisms to improve interference rejection
- Network layer implementing self-healing mesh network, thus introducing redundancy and improving reliability
- Application layer fully compatible with HART

TABLE 31.1 ISO/OSI 7 Layer Model: Comparison among HART, WirelessHART, and ZigBee

Layer \ Standard	HART	WirelessHART	ZigBee
Application	Command oriented, predefined data types and application procedures		Application and security
Presentation	—	—	—
Session	—	—	—
Transport	Auto-segmented transfer of large amount of data, reliable stream transport, and negotiated segment sizes		—
Network	—	Power Optimized redundant path mesh network	Ad-hoc routing, mesh networks
Data link	Token passing master/slave	Time synchronized channel hopping	IEEE 802.15.4-2006
Physical	Simultaneous analog and digital signaling (4–20 mA wire)	IEEE 802.15.4-2006 at 2.4 GHz	IEEE 802.15.4-2006

31.3.1 WirelessHART: The Network Architecture

Each WirelessHART network includes four main elements:

1. The wireless instruments (i.e., the field devices), directly connected to the process equipment in the plant. This kind of devices is further divided into two groups:
   a. The actual WirelessHART process transmitters.
   b. The wireless adapters, that is, devices that provide a wireless interface for legacy analog 4–20 mA transmitters; in addition, if the wired device is compliant with HART, diagnostic information can be extracted.
2. The gateway, which acts as a bridge between the WirelessHART network (reached by means of access points) and the already existing wired infrastructure (e.g., most of commercial solutions exploit an Ethernet or RS485 infrastructure based on Modbus Transmission Control Protocol (TCP) or remote terminal unit (RTU), respectively).
3. The only one network manager, responsible for network configuration, communication among devices, management of routing messages, and monitoring of network conditions. In particular,

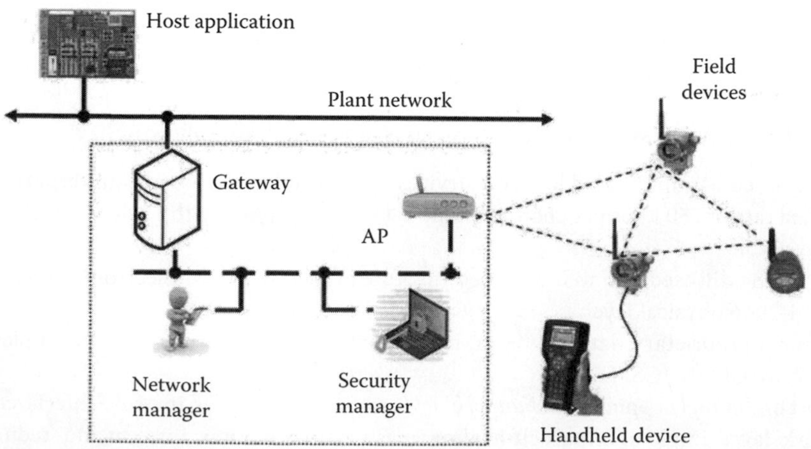

FIGURE 31.1 Simplified diagram of a typical WirelessHART network.

according to the so-called neighbor reports, the network manager dynamically updates the knowledge about the network status adjusting devices and network parameters in order to maximize availability and reliability.

4. The security manager, which deals with the management of security and encryption, setting up and managing the network and session keys, and managing their periodical change.

Moreover, there can be also devices that lack the connection with the process and wireless communication facilities; for instance, they are handheld devices used for commissioning and/or maintenance purposes.

A simplified diagram of a typical WirelessHART network is sketched in Figure 31.1.

31.4 WirelessHART Protocol Stack

31.4.1 WirelessHART Physical Layer

The lowest layer on the OSI model is the physical layer, responsible for sending the *bits* across the network media (as depicted in Figure 31.2). Differently said, it enables the transmission and reception of the protocol data across the physical radio channel. In particular, the WirelessHART physical layer is based on a tailored adoption of the IEEE STD 802.15.4-2006, with the intent of exploiting mass volume production of consumer market devices. This standard is intended to conform to regulations in Europe, Canada, Japan, China, and the United States. The primary goal of the IEEE 802.15.4 is to create a low-cost, low-data-rate, and low-power-consumption wireless communication system, meeting specific needs typically offered by sensors and control devices. In contrast to the IEEE 802.15.1 (Bluetooth) and IEEE 802.11 (WLAN), the IEEE 802.15.4 has been specifically developed for wireless sensor network (WSN) implementations, especially in those applications where many devices sporadically transmit small packets of data towards a sink (acting as the network coordinator and usually mains supplied). Formally, the IEEE 802.15.4 defines the specifications for the physical and the data link layer of the so-called low-rate wireless personal area networks (LR-WPANs) and is not concerned of the higher levels. For instance, the ZigBee solution is an example of a standard protocol defining upper layers.

However, it must be underlined that the use of this protocol (IEEE 802.15.4) without any modification would pose several issues when used in industrial applications. For instance, the use of a fixed channel limits the robustness to interferences, making it coexistent with other colocated wireless solutions operating in the same band unsafe and unreliable. For this reason, only a simplified set of services is inherited and some amendments have been added.

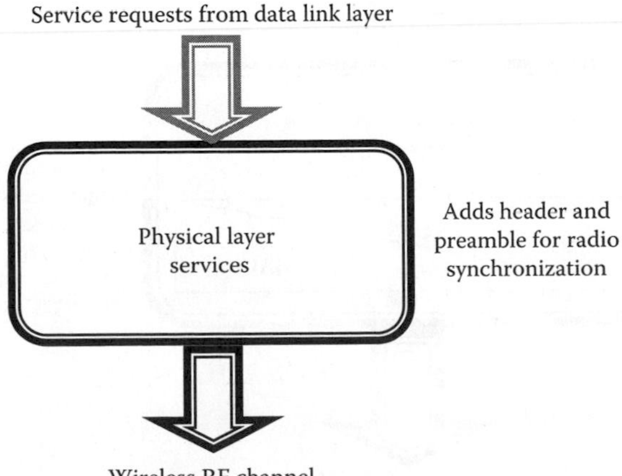

FIGURE 31.2 Schematic representation of the physical layer.

The WirelessHART radio must operate exclusively in the 2.4 GHz band, that is, well above unintentional radiators commonly found in plants. The offset quadrature phase shifting keying (O-QPSK) with direct sequence spread spectrum (DSSS) modulation is used; the raw data rate over the air is 250 kbps. This choice is dictated by a trade-off among the easiness of implementation, the noise rejection, and the communication performance. Sixteen channels (each one 5 MHz apart and with a 3 MHz null-to-null bandwidth) are available; however, channel 26 (centered around 2480 MHz) is not supported since it is not license-free in many countries. Some additional restrictions on clock stability, channel switching time, and power-up sequence have been included to satisfy the WirelessHART medium access strategy and lower the power consumption. The RF transmitting power must be regulated in the range from −10 dBm up to 10 dBm. Some mechanisms to perform the clear channel assessment (CCA) (in particular, the CCA mode 2, carrier sense of IEEE 802.15.4) must be also provided. Additionally, the WirelessHART physical layer defines two different operating states: awake, during normal operation, and sleep, when it is inactive.

As a matter of fact, a WirelessHART node can be implemented using most of IEEE 802.15.4 *commercial off the shelf* available radio transceivers with great benefits to the overall cost.

31.4.2 WirelessHART Data Link Layer

According to the ISO/OSI model, the data link layer provides a reliable means to transfer data between two nodes in the network, detecting and possibly correcting errors that occur in the physical layer due to noise (e.g., retransmitting corrupted data). This level also has the important role of creating and managing data messages. Usually, it is divided into two sublayers (as in Figure 31.3):

1. Medium access control (MAC)
2. Logical link control (LLC)

The LLC sublayer defines the services offered at the next upper level (i.e., the network layer), while the MAC sublayer arbitrates how the common medium must be shared among nodes and tries to avoid or at least to minimize the effect of collisions.

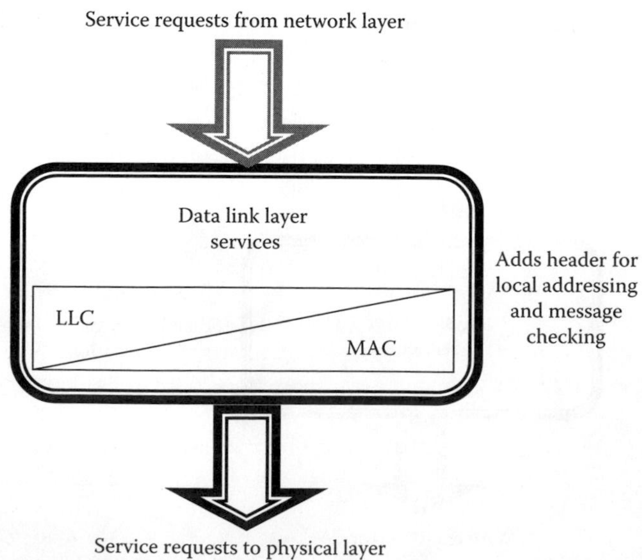

FIGURE 31.3 Schematic representation of the data link layer.

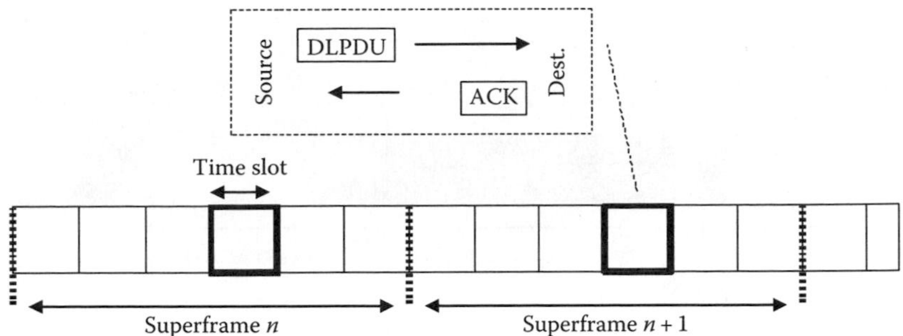

FIGURE 31.4 Basics of WirelessHART TDMA mechanism.

Addresses at the data link layer have only local scope and can be duplicated in other local links. The data link layer is also the lowest information-centric layer.

It must be stressed that the MAC sublayer, directly exploiting services of the physical layer (i.e., controlling the radio itself), has a major impact on the overall energy consumption and, hence, the lifetime of a node.

As previously stated, WirelessHART uses TDMA and channel hopping to control accesses to the network. It is well known that TDMA is a common technique that allows deterministic communications and it is the preferable solution in most of wired fieldbuses. When the TDMA approach is used, communications between devices are organized into time slots, which are grouped together into (continuously repeated) structures called superframes. The actual time slot has a fixed length of 10 ms, enough to contain a message (whose payload is the Data Link Protocol Data Unit [DLPDU]) from the source to the destination and a successive acknowledge (ACK) message. In the ACK, the *response code* specifies if the communication was successful or, conversely, if the request has arrived but was not managed (an *error code* reports the possible causes). A schematization of this mechanism is proposed in Figure 31.4.

Resuming, the main tasks of the WirelessHART data link protocol are as follows:

- Synchronization of the slots
- Identification of the devices that need to access to the medium
- Propagation of received messages at the network layer
- Listening to packet propagated by neighbors

Usually, time slots are univocally dedicated to a couple of devices in what is called a *link* (i.e., an opportunity to communicate) in a time slot of the superframe. However, some of them, called shared slots, may be used by multiple devices using a hybrid approach based on both the TDMA and the CSMA/CA techniques to access the medium. In addition, communication links are specified not only by the time slots they occupy but also by a *channel offset* (used by the channel hopping technique) and by their *direction*. All devices must support multiple links, which are assigned in a centralized way by the network manager. The number of possible links is, typically, equal to the number of channels utilized by a network times the number of slots in the superframe. For example, using 15 channels and 9,000 slots per superframe results in 135,000 possible links.

All devices must support multiple superframes, in order to satisfy different cycle time requirements (e.g., one superframe for data from the field and another one for maintenance).

For effective and efficient TDMA communication, clock synchronization between devices in the network is crucial. Consequently, the tolerances on the local clocks are well defined (oscillator must have ±10 ppm stability), and some synchronization mechanisms have to be implemented.

For sake of clarity, the format of a time slot is shown in Figure 31.5.

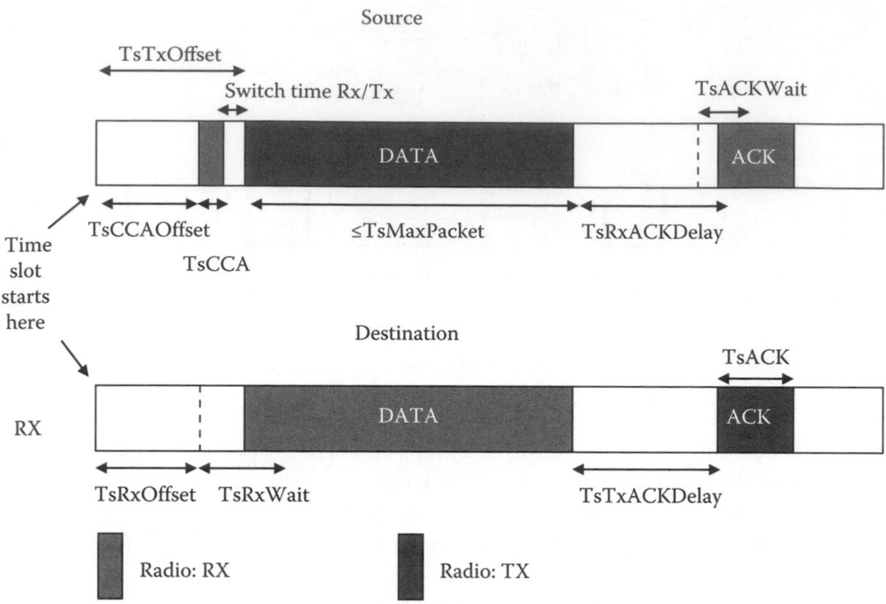

FIGURE 31.5 Packet timing and time slot format. Light gray is for reception state and dark gray is for transmitting state of devices.

In particular, the exchange of a unicast message between a source device (*A*) and a destination device (*B*) is considered. Despite the synchronization mechanism, nodes may have a slightly different sense of time (e.g., due to clock and temperature drift and aging), and for this reason, they may have *different* instants representing the starting time of a slot. According to the standard specifications, the source node has to wait TsTxOffset = 2.12 ms ± 0.1 ms (see Figure 31.5) before starting the actual transmission, whereas the destination node has to wait the interval TsRxOffset = 1.12 ms ± 0.1 ms (as in Figure 31.5) to set up the radio and has to turn on the receiver in the following interval: TsRxWait = 2.2 ms ± 0.1 (as in Figure 31.5). In other words, a guard time of ±1 ms is offered.

In addition, before the source node starts transmitting, it should also perform a *listen before talk* procedure (i.e., CCA), which starts after the TsCCAOffset (=1.8 ± 0.1 ms) and lasts TsCCA (=0.128 ms) in order to improve coexistence with other wireless networks.

The data transaction is completed by the acknowledgment packet (ACK) sent by the destination. The ACK payload specifies the detected synchronization error and notifies possible detected errors. The ACK has a length of 26 bytes (i.e., it lasts TsAck = 0.832 ± 0.1 ms) and the destination node must send the ACK after TsTxAckDelay = 1.0 ± 0.1 ms from the end of the incoming message (such a short time ensures that the cumulated synchronization error is necessarily small).

To enable these services, the WirelessHART device implements and maintains neighbors, superframes, and link tables (the neighbors table is filled each time new neighbor nodes in the radio coverage of the node join the network). All these tables are constantly read by the network manager to maintain an updated snapshot of the network status.

As previously stated, channel hopping (see Figure 31.6) is combined with TDMA in order to enhance reliability. It provides frequency diversity, which can avoid interferers and reduce multipath fading effects. Differently said, each new link will occur on a different channel, thus limiting the effect of interferences that are usually well localized in time and frequency, that is, they have a bursty nature. Each device gets from the network manager an array containing the list of the active (i.e., usable) channels. The current channel to be used is computed by the device as the remainder of the ratio between the sum of the absolute slot number (ASN) (i.e., the time slot sequence number from

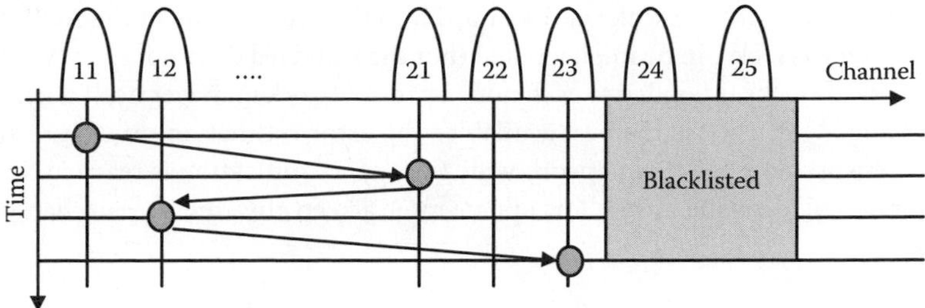

FIGURE 31.6 Frequency diversity mechanism of WirelessHART.

the network creation) plus the channel offset of the current link and the number of available (active) channels, as shown by the following equation:

$$\text{Current Channel} = (\text{Channel Offset} + \text{ASN})\%\ \text{ActiveChannels}$$

This algorithm computes a sort of *logical channel* that moves within the band actually available, and it is identified by the physical channel offset only.

In this way, the same time slot may be used on multiple channels by different nodes, simply changing the channel offsets.

The channel blacklisting mechanism has been also implemented, which allows the network administrator to restrict the active channels usable in the channel hopping procedure, even if its usefulness is not very clear even in a very crowded environment.

31.4.2.1 Link Scheduling Procedure

The main function performed at the data link level by the network manager is the link scheduling, that is, the centralized process by means of which it is decided whether a packet can be sent or received in the current time slot according to its assigned link.

All network details are stored in tables by each node; these tables are initially written by the network manager itself and subsequently appropriately updated according to the evolution of the network.

In particular, each device must be aware of the time slot in which it can communicate (i.e., has the right to occupy the channel) and the channel offset in order to actuate the channel hopping technique; in other words, each device must be aware of the superframe(s) arrangement and the links they are made of. It must be highlighted that this information is handled in a centralized way in order to reduce as much as possible the complexity (and the cost) of field devices. For this reason, only the network manager can dynamically and periodically update superframe arrangement. For instance, depending on the policies adopted by the network manager, the link scheduling decisions can be oriented to the latency reduction, to increase the reliability or to improve load balancing maximizing energy efficiency.

Nevertheless, the planning algorithm should be simple, scalable, and adaptable, for example, to different application requirements. As an example, consider the need to handle messages with different priority requirements.

31.4.2.2 Time Synchronization

The adoption of the TDMA as the basis for the MAC mechanism requires that all the nodes in the network share a common sense of time. Despite the requirement of 10 ppm or better oscillator, local clock skew and drift must be somehow confined in order to allow the node to respect time slot guard limit. Since communications occur on a per packet basis, also timing information must be exchanged using data packets. Devices must be capable to accurately time-stamp the time of arrival of an incoming packet, so that it can be compared with the theoretical (ideal) one, that is, the time slot beginning decided by the network manager. Each destination node of a transaction in a link must estimate this sort of time error (in its local time scale) and forward it to the source node in the ACK payload. In this way, some kind of servo clock can be implemented (but the standard does not specify how). The *quality* (i.e., accuracy) of local clocks is flooded to each node by the network manager itself during the affiliation process, so that nodes can choose the best available. The error estimation procedure can be repeated across multiple network hops, so that network-wide time synchronization is reached. In order to avoid a very long silent period (e.g., due to very low update rate), a keep alive message exchange must occur at least every 30 s.

31.4.3 WirelessHART Network Layer

The network layer is primarily responsible for the routing policies, that is, it routes packets destined for other devices by adding the necessary information (network addresses; see Figure 31.7). In addition, it also provides some security mechanism to end-to-end communications and furnishes some services to the (upper) transport layer. The network layer manages the so-called sessions, that is, the logical confinement of end-to-end data exchanges.

Communication between a pair of network addresses is organized in sessions for security purposes, since each session is ciphered. Four sessions are set up as soon as any device joins the network:

1. Unicast session between the device and network manager—it is used by the network manager to manage the device.
2. Broadcast session between the device and network manager—it is used to globally manage devices. For example, this can be used to roll a new network key out to all network devices. All devices in the network have the same key for this session.
3. Unicast between the device and the gateway—it carries normal communications (e.g., process data) between the gateway and the device.
4. Broadcast session between the device and the gateway—it is used by the gateway to send the identical application data to all devices.

In addition, each device has a join session and it cannot be deleted. The join key is unique; it is the only key that can be written by directly connecting to maintenance port physically implemented on the device itself. It can also be written by the network manager.

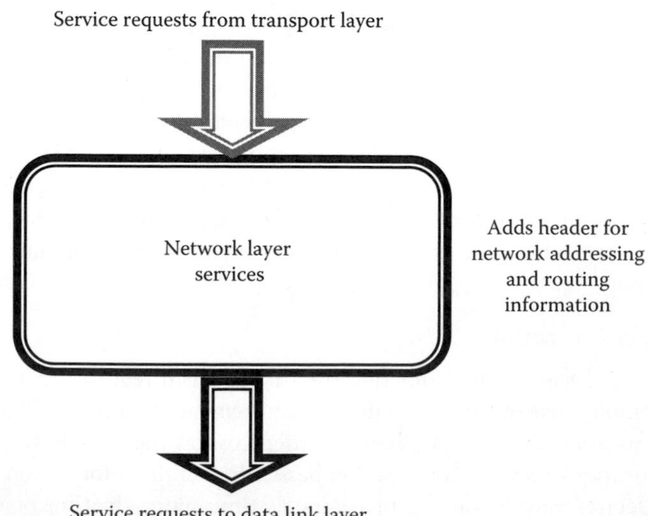

FIGURE 31.7 Schematic representation of the network layer.

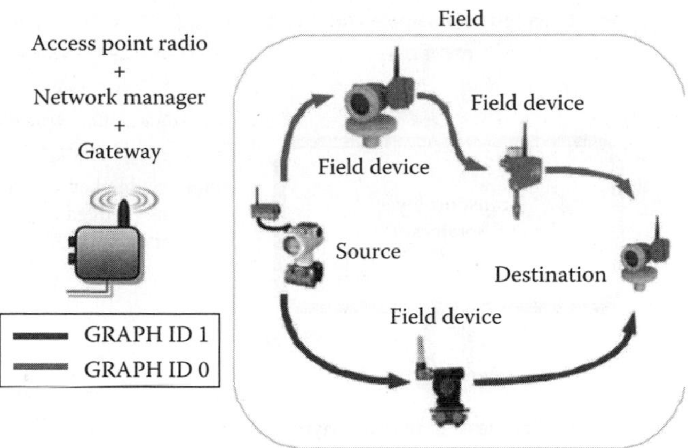

FIGURE 31.8 Graph routing mechanism.

Similarly to the link scheduling described in the data link layer, the routing layer determines in a centralized way where the packet should be sent. This decision is taken by each node according to two possible approaches: the graph routing and the source routing. The graph route is a set of oriented links (*arrows*), each one representing a single hop between one source and one destination device (data link neighbors). Clearly, redundant paths may be arranged to interconnect a couple of nodes, thus greatly improving the overall robustness. On the other hand, in source routing, each message contains the entire list of hops from the source towards the destination; this is not a very efficient routing procedure and it is reserved only for commissioning.

As said, all the computational tasks are performed by the network manager, which is the only device in the network capable to handle a complete list of all the devices and has full knowledge of the network topology. The network manager may, for example, collect all the *neighbor tables*, together with the current status of node batteries, and build the optimal paths (i.e., route graphs). The routing graphs are not unique and can therefore overlap; they are however unidirectional and are identified by a unique graph ID (e.g., refer to the simple example in Figure 31.8). The graph ID is thus the only information that must be propagated by the network layer header in order to determine which path has to be followed. As a rule of thumb, each device should possess at least two neighboring nodes in each graph route to ensure two different forwarding options and, therefore, ensure better reliability.

Routing policies adopted by the network manager may greatly affect overall performance in a number of factors such as energy efficiency, reliability, and latency. The network manager manufacturer has to choose among these (and many possible others) function costs to be minimized (or maximized) by the route scheduling algorithm. Reducing the average hop length decreases energy consumption, but also the redundancy is limited, compromising the reliability of the network; choosing instead longer range will favor direct communication with the gateway (i.e., star topology), reducing latency but increasing the power consumption.

According to the graph routing mechanism previously described, it is clear that WirelessHART generally permits a variety of network topologies, including star, mesh, and a combination of both. This allows taking advantage of the strengths of each: If the focus is on the network's range, reliability, or flexibility, a mesh network will usually be the best choice. If maximum battery life is desirable, a star topology or point-to-point communication is recommended.

31.4.4 WirelessHART Transport Layer

The transport layer ensures the success of end-to-end communication. In other words, it is the transport layer that ensures that packets are successfully delivered to the final destination in a multihop communication (see Figure 31.9).

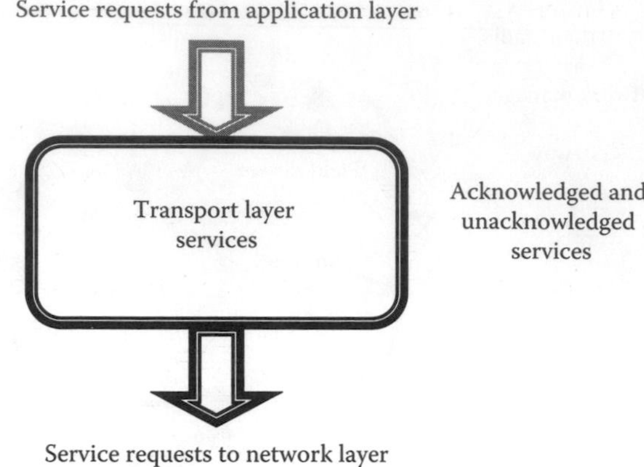

FIGURE 31.9 Schematic representation of the transport layer.

In detail, the transport layer of WirelessHART supports both an acknowledged transfer mechanism (useful for event notification) and an unacknowledged mechanism for the transfer of process data (requiring real-time and lower latencies). The service with acknowledgment provides the use of automatic repeat request (ARQ) mechanism to ensure end-to-end delivery of messages. The default number of retransmission is set to five.

31.4.5 WirelessHART Application Layer

The application layer is the closest to the end user, as shown in Figure 31.10.

It provides the upper host application a way to access to process information traveling on the network. Its functions usually include the identification of originating and sinking data device, the verification of resource availability, and the synchronization of application data.

The WirelessHART application layer follows the well-known HART application layer, being totally compatible and interoperable with it. For these reasons, description of the application layer is out of the scope of this chapter.

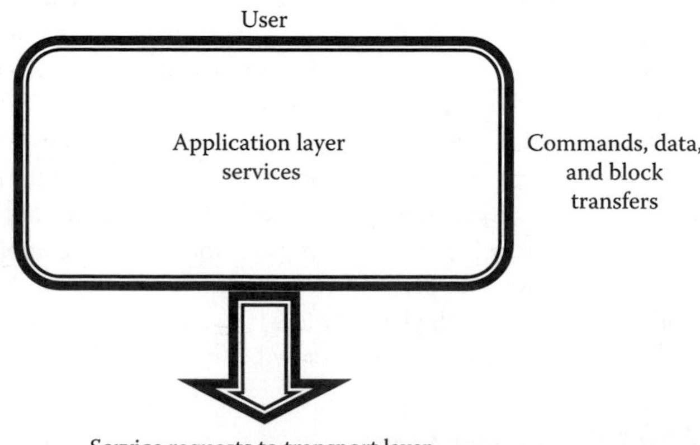

FIGURE 31.10 Schematic representation of the application layer.

As for the wired HART counterpart, all the available commands are divided into classes, which are the following:

- Universal commands, which are commands supported by all HART devices
- Common practice commands, which are supported by a large number of HART devices
- Wireless commands, which are supported by the WirelessHART products
- Device family commands, which are used to set up and configure devices without using device-specific driver
- Device-specific commands, which are defined by the manufacturer for the needs of their own devices

31.5 Message Format

In the following subsections, a brief description of the actual message format is furnished, with some details about the headers added at the different protocol stack levels. Figure 31.11 describes the encapsulation mechanism of different layers according to the WirelessHART specifications, described in the following subsections.

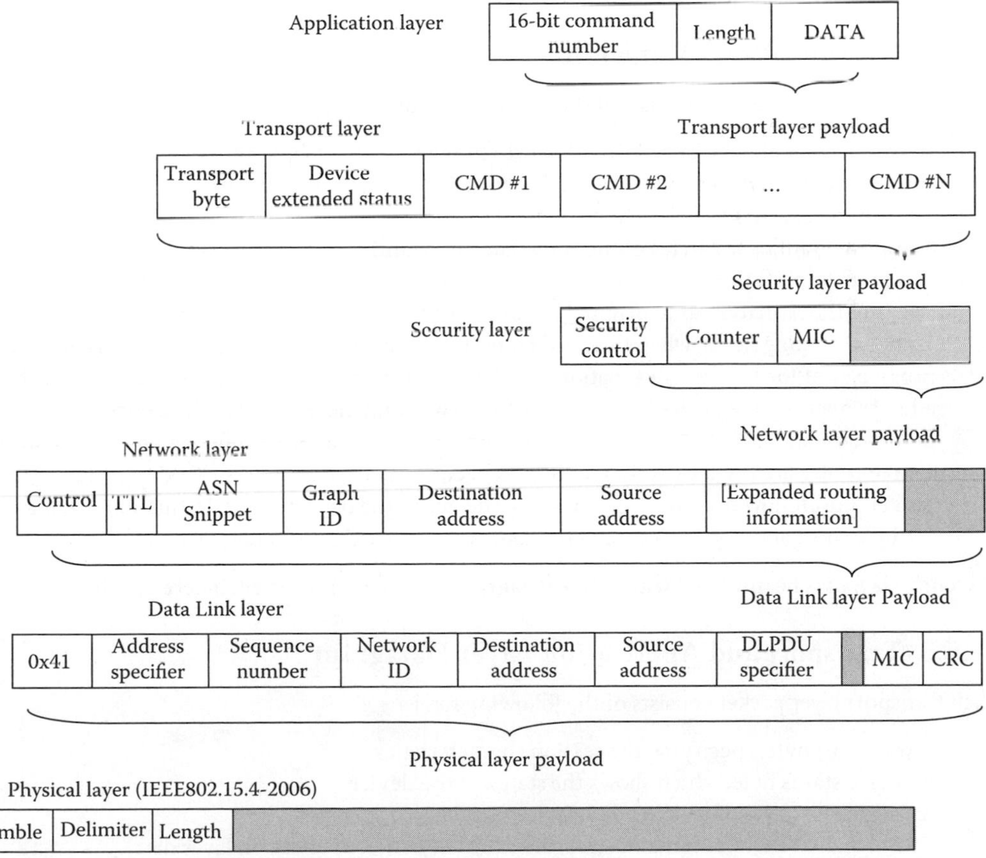

FIGURE 31.11 WirelessHART message format.

31.5.1 Physical and Data Link Layer Datagram

This subsection specifies the format of the data link datagram; each DLPDU consists of the following fields:

- A single byte set to 0x41 (so that WirelessHART messages are recognized as IEEE 802.15.4 data messages)
- A 1-byte address specifier, indicating if long (Extended Unique Identifier [EUI], 64 bits) or short (16 bits) addresses are used
- The 1-byte sequence number
- The 2-byte network ID, which univocally identifies the network
- Destination and source addresses, either of which can be 2 or 8 bytes long
- A 1-byte DLPDU specifier, which allows to distinguish the message priority (*command*, *process* data, *normal*, and *alarm*) and the message type (*ACK*, *advertise*, *keep alive*, *disconnect*, and data)
- The data link layer payload
- A 4-byte keyed message integrity code (MIC)
- A 2-byte cyclic redundancy check (CRC16) (imposed by the IEEE 802.15.4-2006)

In particular, while the data link layer packet is en clair, the MIC is enciphered for link layer authentication of the received datagram; it is generated and confirmed using CCM* mode (combined counter with cipher block chaining-message authentication code (CBC-MAC) [corrected]) in conjunction with the AES-128 block cipher to provide authentication of the originator.

31.5.2 Network Layer Datagram

The network layer datagram consists of the following fields:

- A 1-byte control field, specifying if short (16 bits) or long (EUI, 64 bits) addresses are used for source and destination and if optional routing information are attached.
- The 1-byte time to live (TTL) hop counter.
- The least-significant 2 bytes of the ASN (latency count).
- A 2-byte graph ID.
- The (final) destination and (original) source addresses.
- Optional routing fields; when the third bit of the control byte is set, the 2-byte short address of the proxy parent for the final destination is indicated in the proxy address optional field. If the proxy address matches, the device is responsible for forwarding the message to the device indicated by the destination address. When proxy routing, the destination address must be an EUI-64 address. If source routes are used, each source route field contains four addresses that designate the route the packet must follow from the original source device to the packet's final destination. Each address is a 2-byte short address. Each byte of the addresses not used must be set to 0xFF.

A security layer is encapsulated within this datagram together with the enciphered payload.

31.5.3 Transport and Application Layer Datagram

Each transport layer packet consists of the following fields:

- A transport byte, specifying the session characteristics
- A device status byte, which shows the status of the device
- An additional byte of an extended device status
- One or more HART commands (representing the actual application layer data)

The transport byte indicates if the transfer has to be acknowledged or not; it specifies if the message is a response or not and if the current session is broadcast or unicast; finally, it implements a 5-bit sequence number counter used to sort the data traffic.

Referring to the application level messages, the organization of a single HART command is arranged very similarly to the wired counterpart; the command fields are the following:

- A 2-byte command number, which is used to identify the command
- A 1-byte length, which indicates the number of bytes present in the subsequent data field
- A variable number of bytes containing the request or response code of the command, depending on whether a request or a response was issued, and the actual command-specific data

31.6 Join Procedure

New devices that want to participate to the network exploit information flooded in the *advertise* packets; in particular, these packets allow the device to synchronize the local clock with join superframe, where dedicated join links can be used to start a join request.

The advertise includes basic network information, including the ASN (i.e., the current time slot sequence number, originated by the network manager during the network creation), the position of the join links in the superframe, the security levels supported by the network, and the information necessary for the identification of the channel to be used to transmit. Since advertise packets follow a regular transmission pattern, they can be used by unsynchronized nodes to *follow* their parent clocks.

All the join links are time-shared slots, and the new device that wishes to join the network must communicate only in these time slots until it is authenticated by the network manager.

The join request sent (see Figure 31.12) by the new device is encrypted with the join key (user decided) by means of the AES128 CCM* algorithm; the join request payload is made of the following HART commands in the format of *response*:

- Command 0, containing information on the type and revision of the device and on its unique identification number
- Command 20, containing the long tag of the device
- Command 787, containing information on the health of neighbors

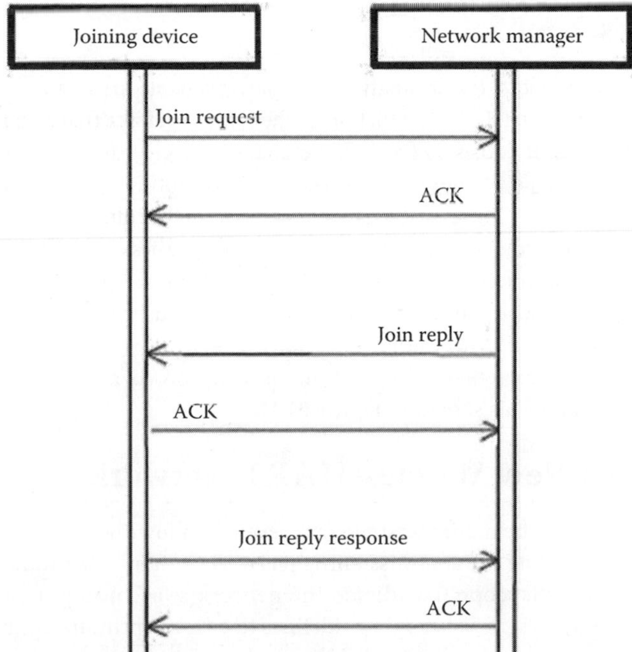

FIGURE 31.12 Sequence diagram of the join phase.

The network manager will first of all send an ACK (see Figure 31.12) to complete the transaction; after a certain amount of time, it will reply with a join reply, also encrypted with the same join key, which contains the following HART commands in the *request* format:

- Command 963, containing the session key to be used
- Command 962, containing the network key to be used
- Command 961, containing the device nickname (short address)

Starting from this point on, the short address will be used as the address of the device in the network, and the payload of all the following messages exchanged with the gateway will be encrypted using the session key, still performing MIC authentication using the network key.

For safety reasons, the session key is periodically changed by the network manager using the command 963. Some details about keys are given in Section 31.7.

31.7 Security Aspects

Security is a very critical aspect in industrial communications solutions. For this reason, WirelessHART has been designed in order to have a security level comparable with wired solutions. Security services are offered not only at the data link layer but also at the network layer. All messages have to be enciphered so that malicious listeners cannot exploit on the air information. Since the ACK packet has to be necessarily sent within the same data time slot, a low computational power mechanism has to be implemented at this level. In particular, the MIC included in the data message is used as a sort of signature just to confirm the identity originating device, without the need of decrypting the message content. The payload of the network layer can be thus decrypted later with relaxed time constraints.

Following the IEEE 802.15.4 approach, the AES-128 algorithm is used with the CCM* mode. Particularly, four different kinds of keys are used in a network: public keys (used at the data link layer by joining devices), network keys (common to all joined devices and used to verify the MIC), join keys (used during the joining process), and session keys (associated to each end-to-end communication).

31.8 Wi-Analys Tool

The Wi-Analys tool is a wireless traffic analyzer (*sniffer*) available from the Hart Communication Foundation (HCF) consortium (HCF_KIT-190), that is, the same task force that also maintains HART and WirelessHART specifications. It consists of a multiple-channel receiver (capable to hear all the 16 IEEE 802.15.4 channels) and is a purposely designed software able to capture packets with a maximum speed of 1000 messages per second (see Figure 31.13). These messages can be filtered according to different stack layer fields and organized and shown in order to easily detect anomalies and errors.

Furthermore, if both the join key and the network ID are known, the Wi-Analys tool is capable to decrypt encrypted packets collected over the air, showing to the user the HART commands actually exchanged and their conformance with the specifications.

Viewing and filtering are implemented by means of a top-down menu whose fields are divided according to the ISO/OSI model, as shown in Figure 31.14.

31.9 Designing a New WirelessHART Network

Designing a new network may be a cumbersome task, even if following some *rule of thumb* allows to minimize troubleshooting during the commissioning phase. As for any other industrial network installation, three stages are needed: scope (i.e., divide the network[s] according to single process unit or subsection of a process unit), design (i.e., apply guidelines to ensure optimum connectivity), and fortify (i.e., troubleshooting to maximize network availability and reliability).

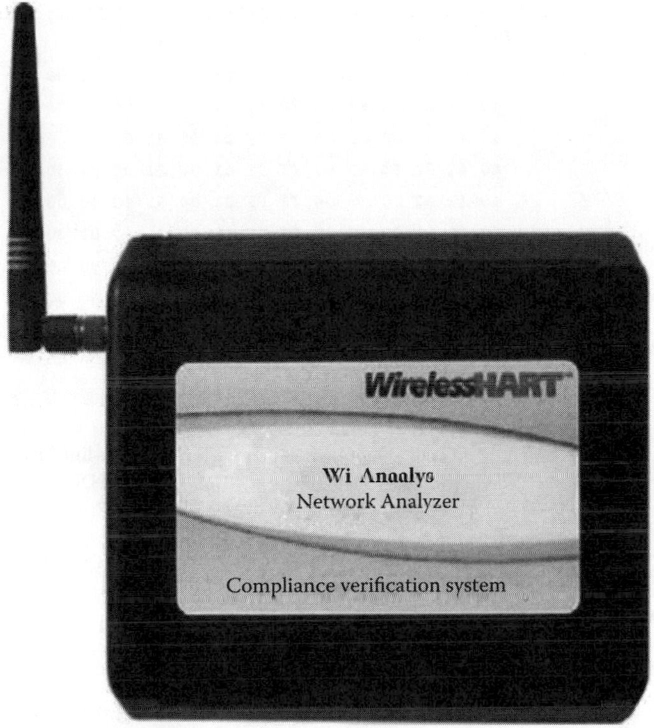

FIGURE 31.13 Wi-Analys network analyzer.

First of all, the number of field devices (including spare parts) must be evaluated, so that the number of gateways can be estimated as a trade-off between the required end-to-end delay and the maximum hop number in the network.

Regarding the design phase, obstructions must be evaluated (in particular, as a function of their distance and their height). Obviously, the LOS scenario must be preferred and if LOS is not possible, additional attenuation, multipath, diffraction, and reflection effects must be taken into account. For the same reasons, moving objects (e.g., people going in and out) have to be considered. It must be also remembered that the antenna is probably the most important component to be considered for correct installation. Since the ground or the floor is somehow conductive (i.e., it may cause reflections), wireless device antenna should be installed at least 1 m above it. Similarly, antennas must be placed as far away as possible from any conductive wall and/or material (e.g., at least 6 cm away from metallic walls running parallel to it). The radiation pattern should be known and the device's antenna has to be aligned accordingly for best results. For instance, do not position WirelessHART devices directly below or above each other since they would be outside each other's antenna range. Performance could be improved connecting a remote antenna to the antenna terminal and mounting the remote antenna as high as possible so that LOS situations are maximized. Also, the use of high-gain (i.e., directional) antennas may improve performance, at the cost of larger dimensions.

In order to exploit intrinsic redundancy of the mesh networking, the number of devices in the effective range of each node must be maximized. As a rule of thumb, each WirelessHART network should have a minimum of five devices in the effective range of the access points (usually placed in the gateway), and every device should have at least three neighbors in its coverage area (thus balancing also the load). Regarding the throughput, it is a good practice to reduce as much as possible the

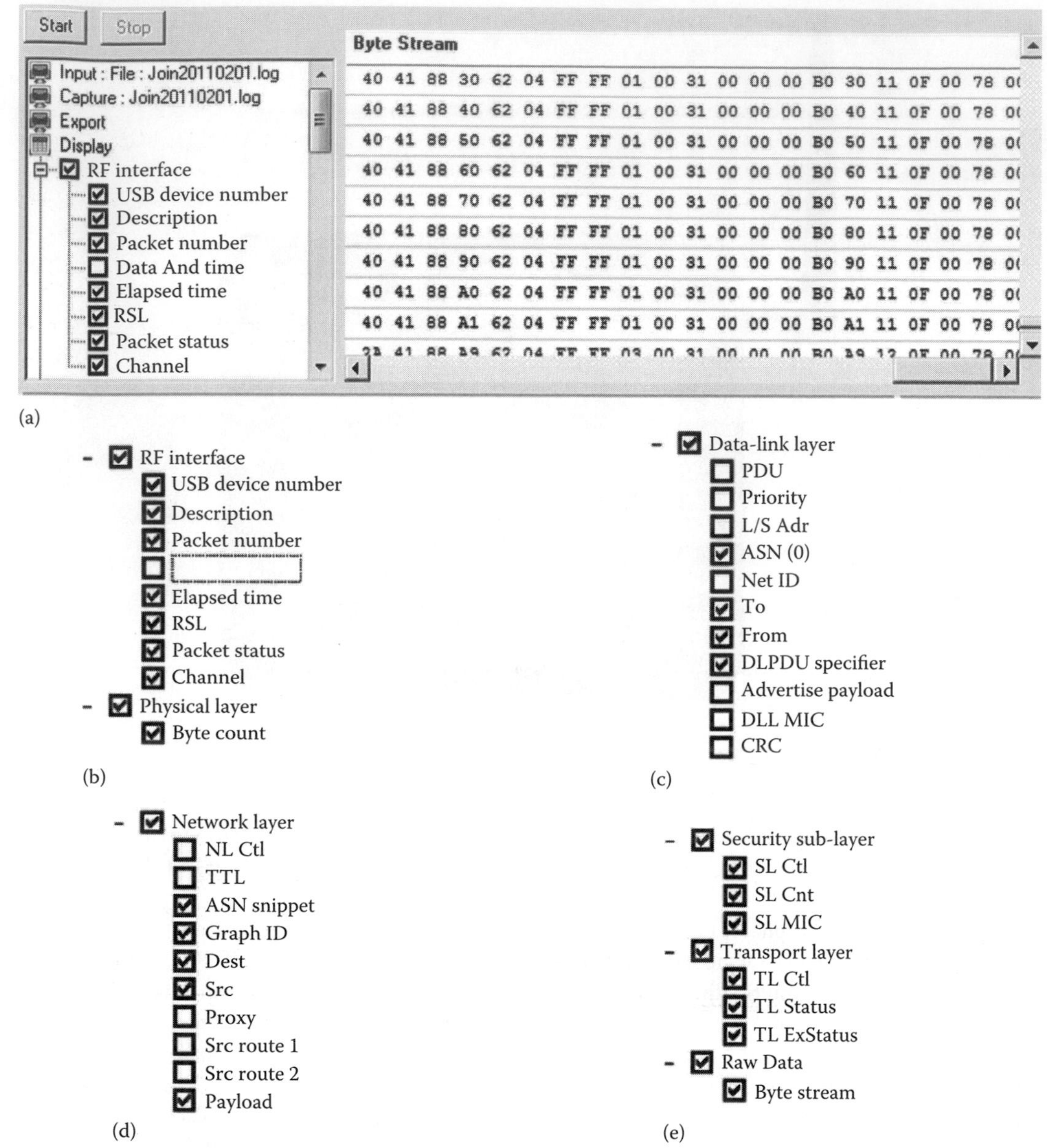

FIGURE 31.14 Graphical user interface of the Wi-Analys software. (a) Example of raw data collection, (b–e) messages filter options.

network depth; in particular, experiments show that at least 25% of the overall devices should be in the effective area of the gateway.

Regarding fixing the troubles, key performance indicators furnished by all commercially available network manager, as the device received signal strength indication (RSSI) or the average packet error rate (PER), should be used to detect anomalies. The simplest (despite not the cheapest) solution to improve network reliability is to increase the node density, that is, add additional routing devices, for example, placing additional instruments not for measuring process variables but specifically to act as routers.

31.10 Example of Platform Providers

Despite the few years since the standardization of the WirelessHART solutions, the market already offers a quite large number of devices, development platforms, and conformance testing tools. In particular, the availability of the previously described *Wi-Analys* tool, together with a Linux box (called Wi-HTest) that automates the execution of device testing scripts, helps in the conformance and interoperability tests. The HCF consortium simplifies testing operations by providing a test specification to developers that clearly explains test requirements and clarifies possible ambiguities in the protocol specifications. These test specifications can be used both in the development phase and in the field device enhancement process. Starting from these test specifications, test scripts for each test case are realized and used as input to feed in the Wi-HTest for establishing the test environments and generating proper test packets. The heart of the Wi-HTest Linux box is the Wi-HTest engine, which is a C++ program executed in the CINT environment, a C/C++ interpreter aimed at processing C/C++ test scripts interactively. A radio module is included to implement the wireless interface with the device under test. On the other hand, the Wi-Analys tool is used to validate the compliance with tests.

The HCF consortium also maintains a list of registered devices on its website (http://www.hart-comm.org/), where almost 50 compliant devices are described. Main manufacturers of gateways (which usually implement gateway, network, and security manager functionalities) and/or transmitters are Rosemount, Emerson Process Management, Aiwa, Dresser, ABB, MACTek, Phoenix Contact, Siemens, Endress+Hauser, Dust Networks, P+F, Softing, and Cooper and Spirax, just to mention some of them.

As an example of multiplatform providers, consider Nivis, a company launched in 1998, which is specialized in sensor networks, supporting an array of standard sensing and control protocols including not only WirelessHART™ but also ISA100.11a and 6LoWPAN. The company supplies wireless sensors, edge routers, a management appliance and application, and a gateway to plant or other enterprise networks.

From the developer perspective, most of the difficulties in realizing an error-free WirelessHART device are, obviously, in the implementation of the protocol stack; to speed up the process, the Wireless Industrial Technology Konsortium (WiTECK) was founded. It was an open, nonprofit membership organization whose mission was to provide a reliable, cost-effective, high-quality portfolio of core enabling system software for industrial wireless sensing applications, under a company- and platform-neutral umbrella. They started working on the realization of a WirelessHART stack, whose development process has now completed by Softing AG, which has made the stack now commercially available.

References

1. P. Ferrari, A. Flammini, D. Marioli, E. Sisinni, and A. Taroni, Wired and wireless sensor networks for industrial applications, *Microelectronics Journal*, 40(9), 1322–1336, September 2009.
2. A. Flammini, D. Marioli, E. Sisinni, and A. Taroni, Design and implementation of a wireless fieldbus for plastic machineries, *IEEE Transactions of Industrial Electronics*, 56(3), 747–755, March 2009.
3. P. Ferrari, A. Flammini, M. Rizzi, and E. Sisinni, Improving simulation of wireless networked control systems based on wireless HART. Computer Standards & Interface 35(6), 605–615, 2013.
4. IEC 62591, Ed1.0 Industrial communication networks—Wireless communication network and communication profiles—WirelessHART, Publication date: April 27, 2010.
5. J. Song, S. Han, A.K. Mok, D. Chen, M. Lucas, M. Nixon, and W. Pratt, WirelessHART: Applying wireless technology in real-time industrial process control, in *Proceedings of 14th IEEE Real-Time and Embedded Technology and Applications Symposium (RTAS)*, St. Louis, Mo, 2007.
6. A.N. Kim, F. Hekland, S. Petersen, and P. Doyle, When HART goes wireless: Understanding and implementing the WH standard, in *Proceeding of ETFA2008*, Hamburg, Germany, pp. 899–907, September 15–18, 2008.

7. J. Åkerberg, M. Gidlund, T. Lennvall, J. Neander, and M. Björkman, Efficient integration of secure and safety critical industrial wireless sensor networks, *EURASIP Journal on Wireless Communications and Networking*, 100, 2011. doi:10.1186/1687-1499-2011-100.
8. J. Song, S. Han, X. Zhu, Al Mok, D. Chen, and M. Nixon, Demonstration of a complete WirelessHART network, in *Proceedings of the Sixth ACM Conference on Embedded Networked Sensor Systems (SenSys 08 Demo)*, Raleigh, NC, 2008, pp. 381–382.
9. I. Muller, J.C. Netto, and C.E. Pereira, WirelessHART field devices, *IEEE Instrumentation & Measurement Magazine*, 14(6), 20–25, December 2011.
10. D. Chen, M. Nixon, and A. Mok, *WirelessHART: Real-Time Mesh Network for Industrial Automation*, Springer, New York, 2010. ISBN 1441960465, 9781441960467.
11. P. Ferrari, A. Flammini, D. Marioli, E. Sisinni, and S. Rinaldi, On the implementation and performance assessment of a WirelessHART distributed packet analyzer, *IEEE Transactions on Instrumentation and Measurement*, 59(5), 1342–1352, May, 2010.
12. C.M. De Dominicis, P. Ferrari, A. Flammini, S. Rinaldi, and E. Sisinni, On the improvement of WirelessHART access points by means of software defined radio, *2010 8th IEEE International Workshop on Factory Communication Systems (WFCS)*, Nancy, France, May 18–21, 2010, pp. 71–74.
13. P. Ferrari, A. Flammini, S. Rinaldi, and E. Sisinni, Performance assessment of a WirelessHART network in a real-world testbed, *2012 IEEE International Instrumentation and Measurement Technology Conference (I2 MTC)*, Graz, Australia, May 13–16, 2012, vol. 35(6), pp. 953–957.
14. S. Han, J. Song, X. Zhu, A. Mok, D. Chen, M.W. Pratt, and V. Gondhalekar, Wi-HTest: Compliance test suite for diagnosing devices in real-time WirelessHART network, *15th IEEE Real-Time and Embedded Technology and Applications Symposium, 2009, RTAS 2009*, San Francisco, CA, April 13–16, 2009, pp. 327–336.

32

ISA100.11a

32.1 Introduction 32-1
32.2 System Overview 32-2
32.3 Communication Protocol 32-3
Physical Layer • Data Link Layer • Network Layer • Transport Layer • Application Layer
32.4 Security 32-8
Payload Encryption and Message Authentication • Security Key Model
32.5 Coexistence 32-10
Spectrum Management
32.6 Implementations and Equipment 32-11
Field Devices • System Manager and Wireless Gateway
32.7 Conclusion 32-14
References 32-14

Stig Petersen
SINTEF ICT

Simon Carlsen
Statoil ASA

32.1 Introduction

The first decade of the new millennium has been the stage for the rapid development of wireless communication technologies for low-cost, low-power wireless solutions capable of robust and reliable communication [1,2]. The Institute of Electrical and Electronics Engineers (IEEE) Std. 802.15.4 for low-rate wireless personal area networks has been the enabling technology for numerous applications within the field of wireless sensor networks (WSNs) [3] and, more recently, wireless instrumentation.

Defined as the merger of WSN technologies and industrial field instruments, wireless instrumentation has become increasingly popular in the process industries since the recent ratifications of the WirelessHART [4] and ISA100.11a specifications [5]. By providing secure and reliable self-configuring and self-healing wireless mesh networks, these standards offer a cost-efficient alternative to traditional wired field instrumentation. Eliminating the need for cables also enables wireless instrumentation to be deployed in remote and hostile areas, as well as provides the opportunity for scalable, flexible, and mobile installations [6].

The ISA100.11a specification is the first standard to emerge from the ISA100 standards committee's work to deliver a family of standards for wireless systems for industrial automation. ISA100.11a was ratified in October 2009, and it is designed to offer secure and robust wireless connectivity for industrial automation applications. It is primarily intended to address noncritical monitoring and control applications, with the goal of targeting control and safety applications in future releases.

32.2 System Overview

In ISA100.11a, devices are defined as the physical embodiment of the behaviors, configuration settings, and capabilities that are necessary to implement and operate a network [5]. Further, a *role* defines a collection of function and capabilities, and all ISA100.11a devices shall implement at least one role. The following roles and their functions are defined [5]:

- *Input/output* (*I/O*): An I/O device provides data to and/or utilizes data from other devices, that is, field sensors and/or actuators. An I/O device implements the minimum characteristics required to participate in an ISA100.11a-compliant network, but it has no message forwarding or routing capabilities.
- *Router*: A router device is capable of forwarding and routing messages from other devices in the network. It also has a clock propagation capability and can act as a proxy.
- *Provisioning*: A provisioning device is able to provision other devices to join the network.
- *Backbone router*: A backbone router is capable of routing messages via the backbone network.
- *Gateway*: The gateway device provides an interface between the wireless instrumentation network and the plant network.
- *System manager*: The system manager is responsible for governing the network, devices, and communications.
- *Security manager*: The security manager is a function that works in conjunction with the system manager to enable secure system operation.
- *System time source*: The system time source is the master time source for the system. The common sense of time is used to manage device operation.

A typical ISA100.11a network configuration with a diverse range of roles and devices is illustrated in Figure 32.1. Since the I/O role is separated from the router role, field instruments conforming to the specification can be defined either as router nodes with routing and forwarding capability or as end nodes with no routing capability. This enables ISA100.11a networks to employ star, star–mesh (as shown in Figure 32.1), or full mesh network topologies depending on the roles of the field devices present in a given network.

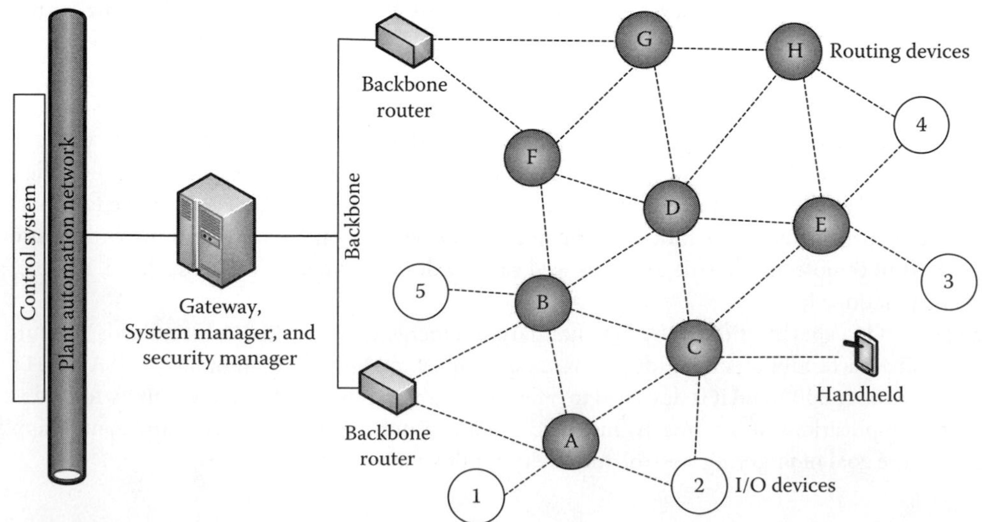

FIGURE 32.1 Typical ISA100.11a network configuration. (From ISA100.11a-2011 Standard, Wireless systems for industrial automation: Process control and related applications, 2011.)

When it comes to the possible number of devices in an ISA100.11a, there are both theoretical and practical limitations. In theory, the protocol addressing space is capable of handling thousands of devices, but for large mesh networks, the network latency and device power consumption will increase to accommodate the communication links in the network. In a growing network, choke points will arise at the devices closest to the backbone router (such as devices A, B, F, and G in Figure 32.1) as these devices have to forward packets from nearly all of the other devices in the network. For large-scale networks, this can lead to a substantial increase in data traffic and, hence, an increase in power consumption. In addition, the maximum achievable sensor sample rate is proportional to the number of devices in the network, as communication bandwidth must be reserved to accommodate the need for sensor data message exchange. Due to these issues with network scalability, currently available ISA100.11a solutions have an upper limit to the number of devices in a network, typically in the range of 50–100.

32.3 Communication Protocol

Most communication protocols use *stacks* to give a layered and abstract description of the network protocol design. A stack consists of different *layers*, where a layer is a collection of related functions needed to perform its defined tasks. Each layer is responsible for providing services to the layer above and, similarly, utilizing services from the layer below. The information and service exchange between layers is performed according to a well-defined message exchange format.

ISA100.11a uses a simplified version of the seven-layered open systems interconnection (OSI) basic reference model [7]. The ISA100.11a stack model is illustrated in Figure 32.2.

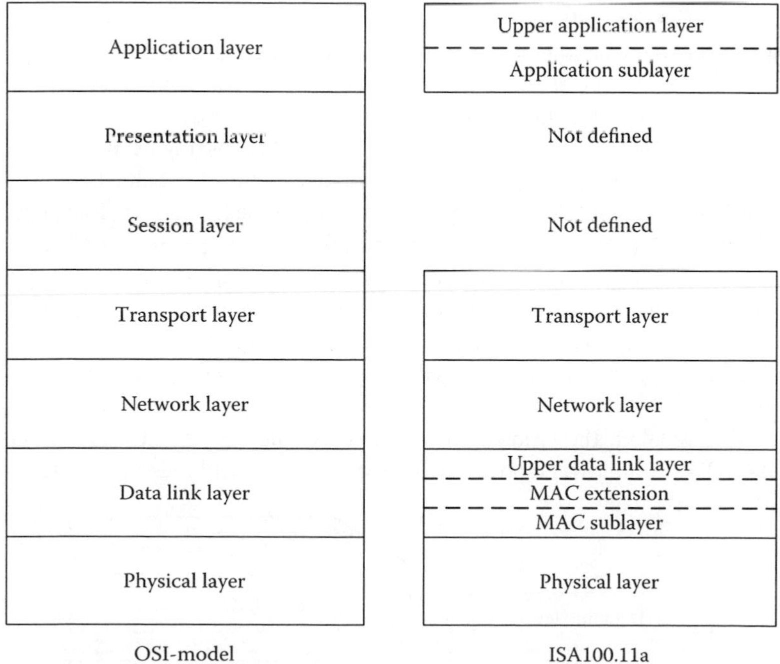

FIGURE 32.2 ISA100.11a stack model. (From ISA100.11a-2011 Standard, Wireless systems for industrial automation: Process control and related applications, 2011; Information Technology, Open systems interconnection—Basic reference model: The basic model, ITU-T X.200 (07/94), 1994.)

32.3.1 Physical Layer

The physical layer (PHY) is responsible for the interface to the physical medium, and it handles the transmission and reception of raw data packets. In addition, it offers mechanisms for selecting operating channels and performing clear channel assessment (CCA) and energy detection. ISA100.11a inherits its PHY from IEEE Std. 802.15.4 [8], with some modifications.

ISA100.11a operation is only defined in the globally available 2.4 GHz band, using channels 11–25 as defined by IEEE Std. 802.15.4. Channel 26 is defined as optional, as it is not available in some countries. Direct sequence spread spectrum (DSSS) is used as the modulation technique, and in combination with offset quadrature phase-shift keying (O-QPSK), this allows for a raw over-the-air data rate of 250 kbps. The maximum transmit power is limited to 10 mW by regulatory authorities, which gives most devices a transmission range of approximately 100 m with direct line of sight, depending on the sensitivity of the radio receiver.

32.3.2 Data Link Layer

The data link layer (DLL) as defined in the OSI model typically encompasses communication between two peer devices in a network. It offers mechanisms for radio channel access, radio synchronization while handling acknowledgement frames, and security control. However, contrary to common definitions, the ISA100.11a DLL also comprises message forwarding and routing within a subnetwork, which is normally defined and handled by the network layer (NL). The ISA100.11a NL on the other hand is responsible for backbone routing.

The ISA100.11a DLL is divided into a media access control (MAC) sublayer, a MAC extension layer, and an upper DLL, as illustrated in Figure 32.3. The MAC sublayer is a subset of the IEEE Std. 802.15.4 MAC, defining transmission and reception of individual data frames. The MAC extension adds extra features not supported by the IEEE Std. 802.15.4 MAC, more specifically changes to the carrier sense multiple access with collision avoidance (CSMA-CA) mechanisms by including spatial, frequency, and time diversity. The upper DLL handles mesh networking above the MAC layer, including routing within a data link (DL) subnet. A DL subnet consists of one or more groups of field devices with a shared system manager (and backbone network). The DL subnet stops at the backbone router, but ISA100.11a messages may be routed into the plant network. The ISA100.11a NL handles routing beyond the backbone router.

Every field device participating in a DL subnet is assigned a short DL subnet address by the system manager. This short address is used for local addressing within the DL subnet. To enable mesh routing, ISA100.11a supports both graph and source routing algorithms. A graph route is a list of paths that connect end points within a network. A single network may have multiple and overlapping graphs, and a device may have multiple graphs going through it. An example illustration of graph routing can be found in Figure 32.4. In the figure, end-to-end communication between device A and device F is enabled by graph 1. To send a message to device F, device A can transmit to either device B or device C. Device B or C will then in turn forward the message towards device F according to their own graph routing configuration. Using graph 1, the following routes between device A and F are possible: A–B–D–F, A–C–D–F, or A–C–E–F. Similarly, graph 2 is used by device A to send messages to device D.

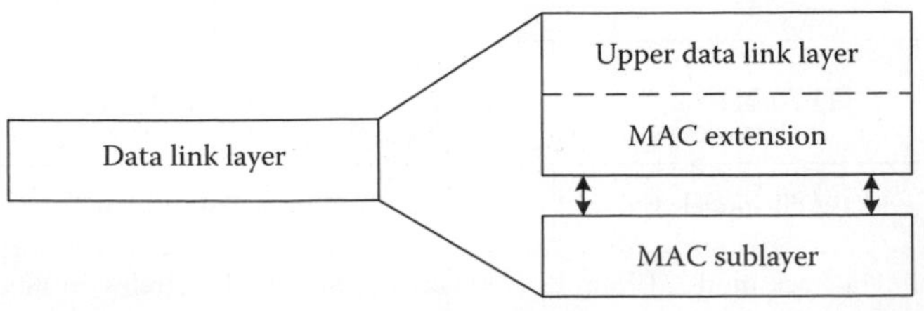

FIGURE 32.3 ISA100.11a DLL structure.

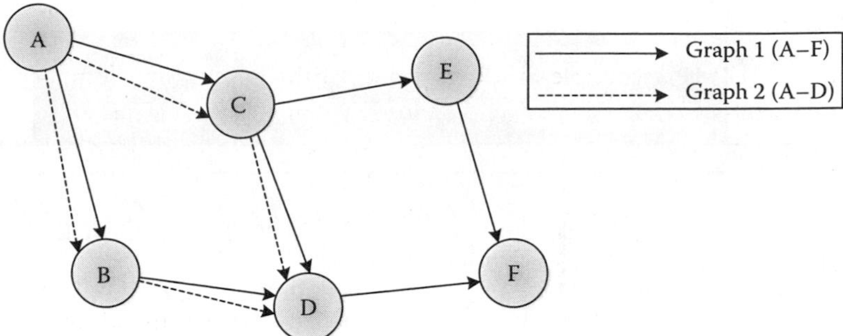

FIGURE 32.4 Illustration of graph routing. (From ISA100.11a-2011 Standard, Wireless systems for industrial automation: Process control and related applications, 2011.)

A source route is a single, unidirectional route between two devices, where one device is defined as the source and the other as the destination. A source route defines the specific path a message must take when traversing from the source to the destination. If a single link in a source route should fail, the message will be lost, while in a graph route, each device will have multiple associated neighbors to which they may transmit messages. Compared to source routing, this enhances both reliability and redundancy. The ISA100.11a DL subnet routes are configured by the system manager based on periodic network health reports from devices, which indicates the quality of the wireless connectivity to their immediate neighbors.

ISA100.11a uses time division multiple access (TDMA) as its main channel access mechanism. Communication is divided into distinct timeslots of configurable duration (typically 10 ms). The timeslot length is set to a specific value by the system manager when a device joins the network. A collection of timeslots forms a superframe that repeats in time throughout the entire lifetime of the network. The term frame is used to separate instances in time of a specific superframe, as illustrated in Figure 32.5. ISA100.11a supports multiple superframes of variable lengths, and one superframe must always be enabled. Superframes can be added and removed by the system manager while the network is operational.

To supervise the communication within a network, the system manager typically assigns two devices to a timeslot, one as a source (transmitter) and the other as the destination (receiver). An exception to this is broadcast messages where multiple devices are assigned as receivers in the same timeslot. Within a timeslot, the source device may transmit a data packet to the destination device. Upon successful reception of a data packet, the destination device will transmit an acknowledgment packet (ACK) to the source device, as depicted in Figure 32.6. If the source device fails to receive an ACK, the data packet will be retransmitted in the next available timeslot. Note that an ACK is not transmitted upon reception of a broadcast message.

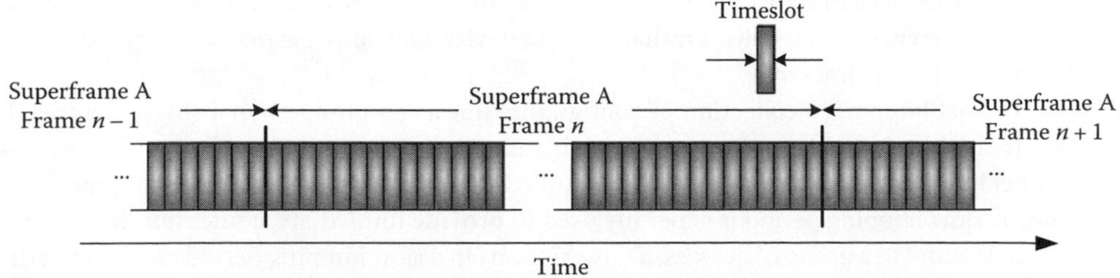

FIGURE 32.5 Structure of TDMA timeslots and superframes.

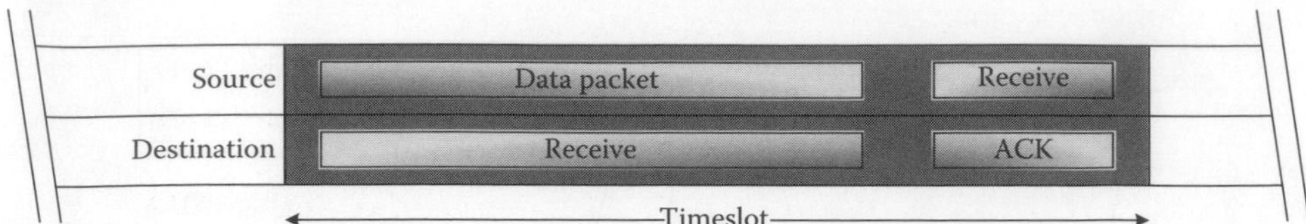

FIGURE 32.6 Data transmission within a timeslot. (From ISA100.11a-2011 Standard, Wireless systems for industrial automation: Process control and related applications, 2011.)

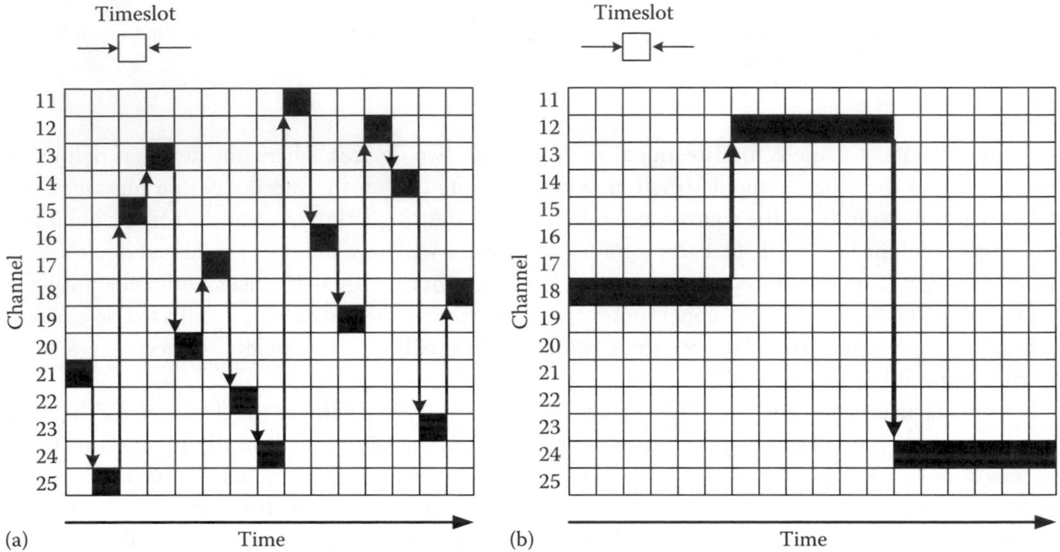

FIGURE 32.7 Frequency hopping patterns: (a) slotted and (b) slow. (From ISA100.11a-2011 Standard, Wireless systems for industrial automation: Process control and related applications, 2011.)

Combined with these TDMA mechanisms, ISA100.11a also employs two types of frequency hopping called *slotted* and *slow* hopping. In slotted hopping, communication is divided into a 2D matrix consisting of timeslot and the 16 available channels. A *link* is thus specified by a superframe, timeslot offset (relative to the first timeslot of the superframe), and channel offset, as illustrated in Figure 32.7a. In consecutive superframes, a link will always have the same timeslot offset, while the communication channel will change according to a pseudorandom hop sequence. As an example, for a given link, communication may occur on channel 19 in timeslot k in frame n of superframe A and on channel 13 in timeslot k in frame $n + 1$ of the same superframe. Combining TDMA and frequency hopping in this manner allows for multiple devices to transmit data at the same time on different channels without generating intranetwork interference. Note, however, that a single device may only participate in communication on one channel (link) per timeslot.

For slow channel hopping, a collection of contiguous timeslot is grouped on a single channel. Each collection is treated as a single slow hopping period, and it will be subjected to channel hopping as for slotted channel hopping, but at a slower rate (see Figure 32.7b). The duration of a slow hopping period is configurable. A slow hopping period is generally used to provide immediate, contention-based channel bandwidth on demand to a group of devices. Transmissions in a slow hopping period are thus not driven by a TDMA scheme; the channel is left open for nondeterministic CSMA-CA-based access. Although the TDMA scheme is not used internally in each slow hopping period, the device must still follow the

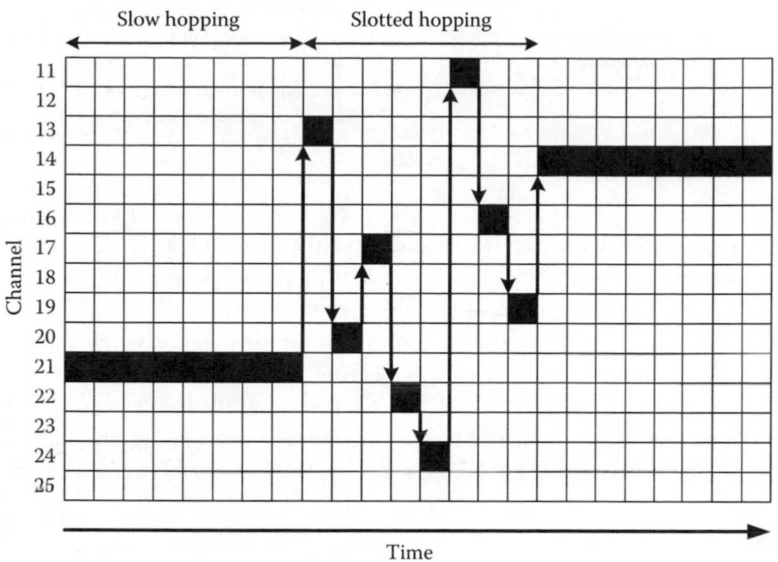

FIGURE 32.8 Hybrid frequency hopping pattern.

overlaying timeslot synchronization and frequency hopping patterns of the network. The slow hopping mechanism enables improved support for event-based and time critical traffic where the occurrence of a given event/alarm may trigger the need for a device to immediately transmit a data packet. With slotted channel hopping, the device would be forced to wait for the next available scheduled timeslot where it is assigned as a transmitter, thereby increasing the communication latency. The drawback of the slow channel hopping method is that the devices designated as receivers in a slow hopping period must continuously listen for incoming traffic, which increases their power consumption compared to slotted channel hopping.

ISA100.11a can also employ a hybrid slotted and slow hopping configuration, where the network will change between periods of slow hopping and periods with slotted hopping as illustrated in Figure 32.8. The order in which slotted and slow hopping are combined is flexible and governed by the system manager.

Five preprogrammed hopping patterns that shall be supported by all field devices (I/O and routers) are defined [5]:

- Pattern 1: 19, 12, 20, 24, 16, 23, 18, 25, 14, 21, 11, 15, 22, 17, 13 (,26)
- Pattern 2: pattern 1 in reverse
- Pattern 3: 15, 20, 25
- Pattern 4: pattern 3 in reverse
- Pattern 5: 11, 12, 13, 14, 15, 16, 17, 18, 19, 20, 21, 22, 23, 24, 25 (,26)

Pattern 3 and 4 are intended for slow hopping channels, while pattern 5 is intended to facilitate coexistence with WirelessHART. The system manager can configure a device to use any of the five patterns for either slotted, slow, or hybrid hopping.

32.3.3 Network Layer

Traditionally, the main responsibilities of the NL are to route packets across the network and to discover and maintain routing tables. However, the mesh-level routing within a DL subnet is handled by the ISA100.11a DLL. The DL subnet stops at the backbone router (see Figure 32.1), and it is the routing

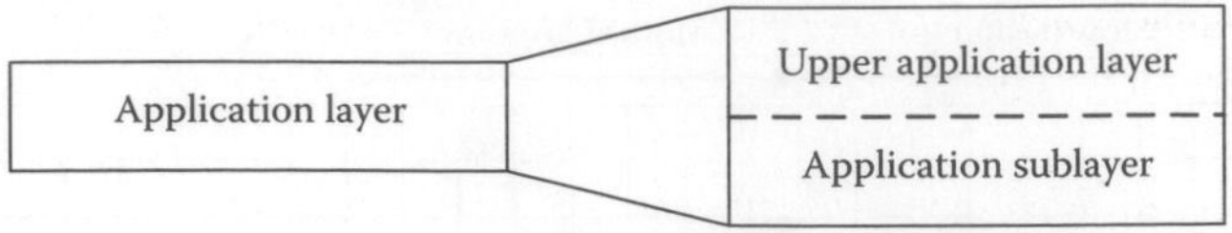

FIGURE 32.9 ISA100.11a AL structure.

beyond the backbone router that is the responsibility of the ISA100.11a NL. The specification of the backbone and plant networks is, on the other hand, outside the scope of ISA100.11a, so the details on how to route traffic over a backbone or plant network are not clearly specified.

The ISA100.11a NL is influenced by the IPv6 over low power wireless wireless personal networks (6LoWPAN) specification [9] of the Internet Engineering Task Force (IETF), with the main goal of facilitating future compatibility with IPv6 [10]. The NL is responsible for determining the appropriate address information, either a 16-bit short address for DL subnets or a 128-bit long address for backbone networks and application end points. The NL must handle translations between the two address types, and all devices shall maintain an address translation table to facilitate these translations. The ISA100.11a NL also fragments and reassembles data packets that exceed the maximum length allowed by the DLL.

32.3.4 Transport Layer

The transport layer (TL) is responsible for end-to-end communication, possibly across several devices, and operates in the communication end points, that is, not on the intermediate routers. The ISA100.11a TL provides connectionless services, which extends the User Datagram Protocol (UDP) [11] over IPv6 with optional compression as defined by the IETF 6LoWPAN specification. The extension to UDP includes better data integrity checks and additional authentication and encryption mechanisms. Acknowledged transactions are not supported by the ISA100.11a TL.

32.3.5 Application Layer

The application layer (AL) provides services to user-defined application processes, and it defines the necessary communication services to enable object-to-object communication between distributed applications. In the ISA100.11a AL, software objects are used to model real-world objects. The AL is divided into two sublayers: the upper AL (UAL) and the application sublayer (ASL), as illustrated in Figure 32.9. The UAL contains the application processes for a device, and it may be used to handle input and/or output hardware, perform a computational function, or support protocol tunneling. The ASL provides the services needed for the UAL to perform its functions, such as object-oriented communication and routing to objects within a user application process across the network.

32.4 Security

When introducing wireless technology in process automation industries, it is important that the communication protocols employ sufficient security mechanisms and algorithms for protection against potential cyber attacks. Data confidentiality, authenticity, and integrity must be maintained, and the networks must be able to prevent the following security issues that wireless systems are susceptible to [12]:

- *Accidental association*: Unintentional access to a wireless network by a foreign computer or device.
- *Malicious association*: Access to a wireless network is obtained by hackers in order to steal user information, passwords, or data or to launch other attacks and install malicious software.
- *Identity theft*: Hacker that is able to impersonate an authorized device or user by listening to credential traffic.

- *Man-in-the-middle attacks*: Hackers gaining access to a network with malicious association and transparently monitor network traffic and/or provide false information and data to other network users.
- *Denial of service*: A target device or gateway is flooded with bogus protocol messages and data in an attempt to reduce or suspend its responsiveness and ability to perform regular functions. Intentional jamming of a wireless communication channel falls under this category.
- *Network injection*: Accessing access points/gateways to introduce bogus network configuration commands that may affect routers, switches, and intelligent hubs. The network devices may crash, shut down, restart, or even require reprogramming.
- *Byzantine attack*: Attack where an intruder reprograms a collection of compromised sensors, whereby they send fictitious sensor readings to the control room.
- *Radio interference*: Interference from other wireless networks operating in the same frequency bands.

ISA100.11a employs payload encryption and message authentication to prevent security breaches, along with a model for distribution and maintenance of encryption keys.

32.4.1 Payload Encryption and Message Authentication

ISA100.11a employs payload encryption and message authentication for both single hop messages and end-to-end messages (multihop). The single hop protection is handled by the DLL, while end-to-end message protection is handled by the TL. The DLL security defends against attackers who are outside the system, while the TL security defends against attackers who may be on the network path between the source and destination devices.

ISA100.11a supports counter with chipper block chaining message authentication code (CCM) mode in conjunction with AES-128 (advanced encryption standard with 128-bit block size) block chipper using symmetric keys for message authentication and encryption. This authenticated encryption algorithm is designed to provide both data authentication and privacy.

32.4.2 Security Key Model

ISA100.11a defines a set of security keys that are used to ensure secure communication in the network. The standard supports both symmetric and asymmetric cryptography.

Symmetric cryptography relies on both communication end points using the same security key when communicating securely. Attackers that do not have access to the security keys cannot modify messages without being detected, and they cannot decrypt the encrypted payload information.

A new device can be provisioned with a join key before it attempts to join a network. The join key is used to authenticate the device for a specific network. Once the device has successfully joined the network, the security manager will provide it with keys for further communication. The use of the join key is optional. A global key, a well-known key with no security guarantees, may also be used in the join process for devices not supporting symmetric keys.

Devices are issued a master key, DL key, and a session key upon joining a network (if the device supports these security features). The master key is used for communication between the security manager and the device, the DL key is used by the DLL to compute the message integrity code (MMIC), and the session key is an optional key used to encrypt and/or authenticate TL messages. The keys are limited in time and need to be periodically updated.

ISA100.11a also supports optional asymmetrical keys. In asymmetric cryptography, different keys are used to encrypt and decrypt a message. Each device has a pair of keys: a public and a private key. The private key is kept secret, while the public key may be freely and openly distributed. Messages encrypted with the public key can be decrypted only with the private key. Unlike symmetric cryptography, this does not require a secure initial exchange of one or more secret keys to the transmitter and receiver. Two asymmetrical keys are defined: CA-root and Cert-A. CA-root is the public key of a certificate authority that signed a device's asymmetric key certificate. It is used to assist in verifying the true identity of the device communicating the certificate. Cert-A is the asymmetric key certificate of device A, used to evidence the true identity of the device during execution of an authenticated asymmetric key establishment protocol.

When joining a network, an ISA100.11a device shall use either symmetric keys, public keys, or no security. The no security option used the global key, and the MIC will then be the equivalent of a cyclic redundancy check (CRC) with no security guarantees. For these devices, no end-to-end security transmissions are allowed.

32.5 Coexistence

ISA100.11a operates in the 2.4 GHz band, one of the very few open frequency bands with near worldwide availability. To ensure a successful adoption in the process industries, it is therefore imperative that ISA100.11a networks are able to peacefully coexist with other wireless networks operating on the same frequency. Examples of such networks are IEEE 802.11-based local area networks (commonly referred to as Wi-Fi), Bluetooth, and cordless telephones. As the widespread adoption of Wi-Fi networks has also reached process and manufacturing plants, it is expected that most ISA100.11a deployments will have to share and contend for the frequency spectrum with nearby Wi-Fi infrastructure.

The 802.11 specifications [13] define 14 overlapping channels in the 2.4 GHz band, each channel being 22 MHz wide and spaced 5 MHz apart. Channel 14 is only available in Japan, while channels 12 and 3 are prohibited in North America and some Central and South American countries. As neighboring channels in Wi-Fi are overlapping, it has become commonplace in industrial deployments to configure Wi-Fi access point to use the nonoverlapping channels 1, 6, and 11 to prevent interference. The frequency distribution of these three channels along with the ISA100.11a channels is illustrated in Figure 32.10. Interference between wireless systems only occurs when they operate at the same frequency, in the same location, and at the same time. As a result, relatively interference-free operation for ISA100.11a can only be achieved in channels 15, 20, 25, and 26 if the network is collocated with a typical industrial Wi-Fi deployment.

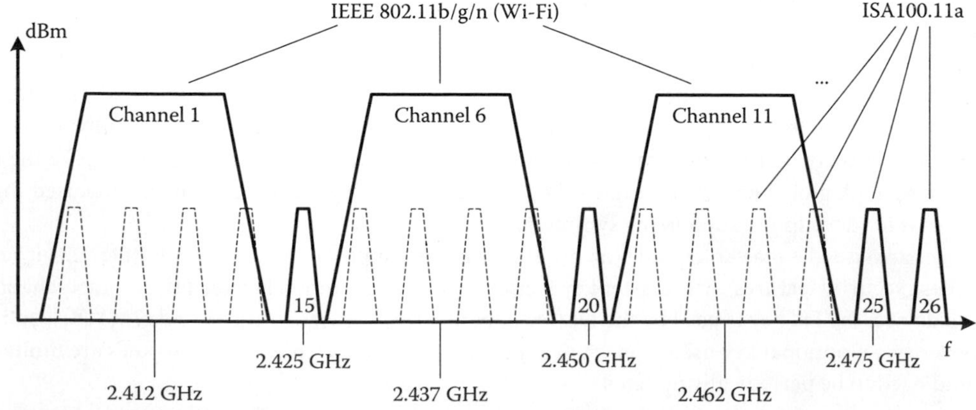

FIGURE 32.10 Wi-Fi and ISA100.11a frequency utilization in the 2.4 GHz band.

32.5.1 Spectrum Management

ISA100.11a employs a set of spectrum management techniques to better cope with coexistence and interference issues. Prior to data transmission, a CCA is performed to ensure that the channel is free to use. Four CCA modes are defined, where modes 1–3 are inherited from the IEEE 802.15.4 PHY [5]:

0—*Disabled*: CCA is not performed prior to data transmission.
1—*Energy above threshold*: CCA reports busy medium upon detecting any energy above a configurable threshold.
2—*Carrier sense only*: CCA reports a busy medium if a signal compliant with IEEE Std. 802.15.4 PHY modulation and spreading characteristics is detected.
3—*Carrier sense with energy above threshold*: CCA reports a busy medium using a logical AND/OR combination of Mode 1 and Mode 2.

To avoid the use of channels with high levels of noise and/or interference, channel blacklisting may be used. If one or more channels are blacklisted, the device will change its hop pattern to not include the blacklisted channel(s). The blacklisting feature is adaptive, giving each device the capability of autonomously blacklisting problematic channels. However, the system manager may disable the adaptive blacklisting mechanisms of any device in the network.

32.6 Implementations and Equipment

32.6.1 Field Devices

A wireless field instrument can be a traditional industrial sensor or actuator equipped with a radio transmitter, antenna, and a power supply (battery), thus eliminating the need for cables. It consists of five main components:

1. *Sensing (or actuator) element*: A sensor element measures a physical phenomenon such as temperature, pressure, or flow, while an actuator element is used for, for example, opening and closing valves.
2. *Microcontroller*: The microcontroller typically runs both the application program for handling and controlling the sensor or actuator, in addition to the software stack for the wireless communication.
3. *Memory/storage*: Temporary and permanent memory and storage elements are used for program code and execution and storing configuration parameters and application data.
4. *Communication*: The communication module consists of an antenna and the radio-frequency (RF) front end and is responsible for transmitting and receiving raw data over the wireless channel.
5. *Power*: To eliminate the need for cables, a wireless sensor typically uses a high-capacity battery as a power source.

The sensor or actuator elements are identical to the wired counterparts, so a wireless field instrument provides the same measurements with regard to characteristics such as precision, range, and dependability. An integrated ISA100.11a field device is illustrated in Figure 32.11.

As described in Section 32.3, stacks are used to aid in the description of network protocol design. The PHY is implemented in hardware, and as ISA100.11a inherits its PHY from the IEEE 802.15.4 specification, any available IEEE 802.15.4 radio module can be used as an ISA100.11a radio. The stack layers above the PHY (from the DLL to the AL) are implemented in software and are commonly referred to as a software stack. This is illustrated in Figure 32.12.

The value chain for ISA100.11a field devices does thus typically consist of three components: IEEE 802.15.4 radio providers, ISA100.11a radio module providers, and ISA100.11a field device providers.

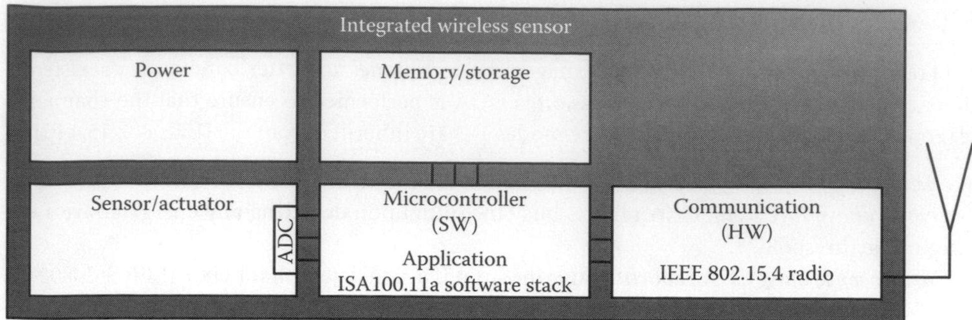

FIGURE 32.11 ISA100.11a field device.

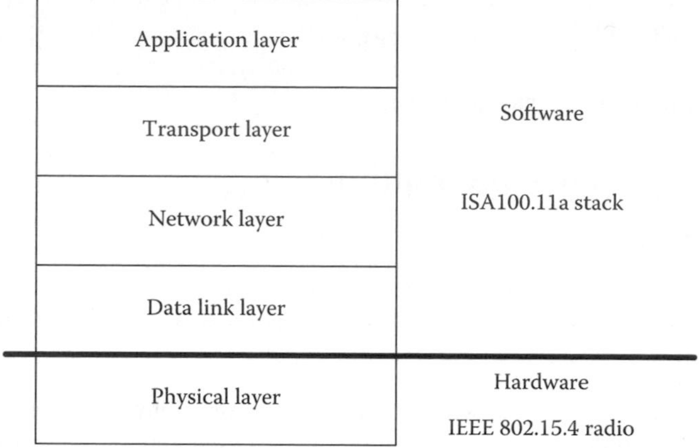

FIGURE 32.12 ISA100.11a radio and software stack structure.

32.6.1.1 IEEE 802.15.4 Radios

An IEEE 802.15.4 radio is an implementation of the IEEE 802.15.4 PHY, which defines frequency, channels, modulation, coding, and transmission power levels. The design and implementation of an IEEE 802.15.4 radio affects the power consumption, transmission range, and receiver sensitivity, and radios from different providers will have (slightly) different performance.

There are numerous providers of IEEE 802.15.4 radios, for example, Atmel, Energy Micro, Freescale, Linear Technologies, and Texas Instruments.

32.6.1.2 ISA100.11a Radio Module Providers and Software Stacks

An ISA100.11a radio module is an integration of an ISA100.11a software stack and IEEE 802.15.4 radio on the same hardware board. All software stacks must adhere strictly to the ISA100.11a specification, and the ISA100.11a Wireless Compliance Institute performs compliance testing to ensure that the behavior of a software stack is according to the specification. For the software stacks, there is little to no room for vendor-specific implementations, with the exception of potential small variations in optional features and parameters.

At the time of writing, Nivis is the only independent vendor for ISA100.11a software stacks and radio modules. Their radio module is based on an IEEE 802.15.4 radio from Freescale. In addition, Honeywell has their own in-house developed ISA100.11a software stack and radio module, running on an IEEE 802.15.4 radio from Texas Instruments. Yokogawa has also developed their own in-house ISA100.11a software stack, which is currently being used for their backbone routers.

32.6.1.3 ISA100.11a Field Device Provider

An ISA100.11a field device is an integrated field instrument with a sensor or actuator element, power, memory/storage, and an ISA100.11a radio module. The wireless performance will, as mentioned in the previous sections, be slightly dependent on the choice of IEEE 802.15.4 radio used with the ISA100.11a radio module.

At the time of writing, the following companies offer certified ISA100.11a field devices (with the ISA100.11a radio module provider in parenthesis): Flowserve (Nivis), GasSecure (Nivis), GE (Nivis), Honeywell (in-house), and Yokogawa (Nivis).

32.6.2 System Manager and Wireless Gateway

The functions of the system manager include security management, address allocation, system performance monitoring, device management, system time services, and communication configuration including contract services and redundancy management [5].

The system manager is responsible for creating and maintaining the routing tables and link distributions (timeslot and channel offsets) for all devices in the network, based on periodic network health reports from the devices. However, the ISA100.11a specification does not define how to actually implement and perform the radio resource management. This means that different implementations of ISA100.11a system managers will have different performances regarding power consumption (of field devices), latency, and reliability.

In the early implementations of ISA100.11a systems, the system manager, security manager, gateway, and backbone router resided encapsulated into the same physical enclosure, which was typically referred to as the network gateway. A schematic of such a network gateway is shown in Figure 32.13. Newer releases of ISA100.11a solutions have enabled the possibility of separating the backbone router from the gateway, system manager, and security manager. This allows for the backbone router to operate in hazardous areas with stricter Atmospheres Explosibles (ATEX) requirements.

At the time of writing, ISA100.11a wireless gateways are available from GasSecure (based on Nivis gateway), Honeywell, Nivis, and Yokogawa.

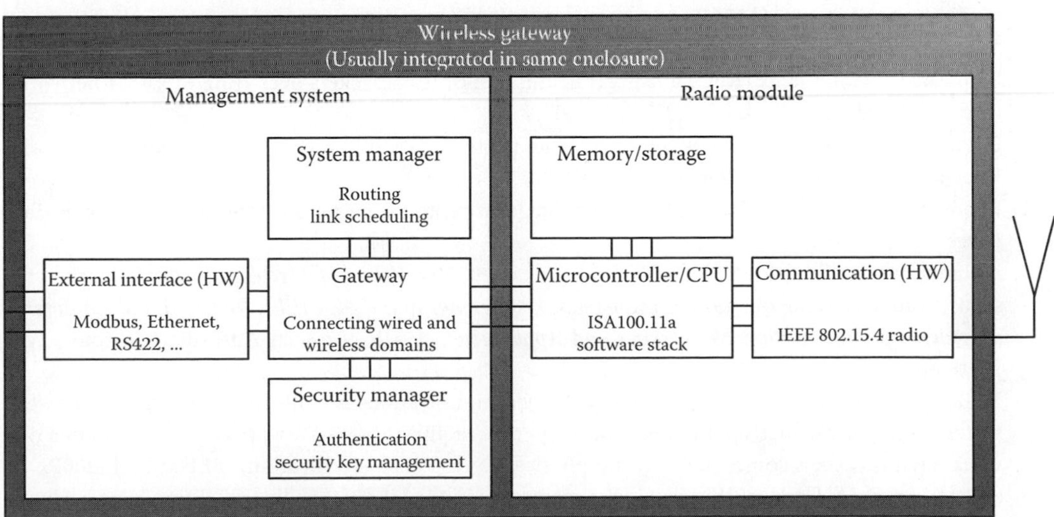

FIGURE 32.13 ISA100.11a wireless gateway.

32.7 Conclusion

This chapter has provided an overview of the ISA100.11a specification and its implementations. Since its release in October 2009, ISA100.11a networks have been successfully deployed in numerous process plants and facilities, providing adequate network performance for noncritical monitoring applications in harsh and hazardous industrial environments [14]. In addition, with the recent development of the wireless hydrocarbon gas detection system from GasSecure, ISA100.11a is also starting to address critical safety applications [14].

As there are currently three competing standards addressing wireless instrumentation for the process automation industries (WirelessHART, ISA100.11a, and WIA-PA), all backed by different groups of automation companies, it will be interesting to follow their *battle* to become the de facto standard for wireless instrumentation in the coming years.

References

1. F. Akyildiz et al., A survey on sensor networks, *IEEE Communications Magazine*, 40(8), Aug. 2002, 102–114.
2. S. Petersen et al., A survey of wireless technology for the oil & gas industry, *SPE Journal on Projects, Facilities & Construction*, 3(4), Dec. 2008, 1–8.
3. Q. Yu, J. Xing, and Y. Zhou, Performance research of the IEEE 802.15.4 protocol in wireless sensor networks, *Proceedings of the IEEE/ASME International Conference on Mechatronic and Embedded Systems and Applications*, Beijing, China, Aug. 2006, pp. 1–4.
4. HART Communication Foundation, HART field communication protocol specification, revision 7.0, Sept. 2007.
5. ISA100.11a-2011 Standard, Wireless systems for industrial automation: Process control and related applications, 2011.
6. S. Petersen et al., Requirements, drivers and analysis of wireless sensor network solutions for the oil & gas industry, *Proceedings of the IEEE ETFA*, Patras, Greece, Sept. 25–28, 2007, pp. 219–226.
7. Information Technology, Open systems interconnection—Basic reference model: The basic model, ITU-T X.200 (07/94), 1994.
8. IEEE Standard for Information Technology, Telecommunications and information exchange between systems—Local and metropolitan networks, Specific requirements—Part 15.4: Wireless medium access control (MAC) and physical layer (PHY) specifications for low rate wireless personal area networks (LR-WPANs), IEEE Computer Society, 2006.
9. Requests for Comments (RFC) 4944—Transmission of IPv6 packets over IEEE 802.15.4 networks, Internet Engineering Task Force (IETF), 2007.
10. Requests for Comments (RFC) 2460—Internet protocol, Version 6 (IPv6) specification, networks, Internet Engineering Task Force (IETF), 1998.
11. Requests for Comments (RFC) 768—User datagram protocol, networks, Internet Engineering Task Force (IETF), 1980.
12. P. Radmand, A. Talevski, S. Petersen, and S. Carlsen, Taxonomy of wireless sensor network cyber security attacks in the oil and gas industries, *Proceedings of the 24th IEEE International Conference on Advanced Information Networking and Applications*, Perth, Western Australia, Australia, Apr. 2010, pp. 949–957.
13. IEEE Standard for Information Technology, Telecommunications and information exchange between systems—Local and metropolitan networks, Specific requirements—Part 11: Wireless local area network medium access control (MAC) and physical layer (PHY) specification, IEEE 802.11, 2007.
14. S. Petersen and S. Carlsen, Wireless instrumentation in the oil & gas industry—From monitoring to control and safety applications, *SPE Intelligent Energy International 2012*, Utrecht, the Netherlands, Mar. 27–29, 2012, pp. 1–14.

33 Comparison of WirelessHART and ISA100.11a for Wireless Instrumentation

Stig Petersen
SINTEF ICT

Simon Carlsen
Statoil ASA

33.1 Introduction 33-1
33.2 System Overview 33-3
33.3 Communication Protocols 33-4
Physical Layer • Data Link Layer • Network Layer • Transport Layer • Application Layer
33.4 Main Differences 33-8
Flexibility • Protocol Support • Coexistence • Security • Quality of Service • Application Potential
33.5 Conclusion 33-13
References 33-14

33.1 Introduction

Wireless instrumentation, defined as the merger of wireless sensor network (WSN) technologies and industrial process automation systems, has become increasingly popular in the process industries in recent years. WSN, the foundation for wireless instrumentation, is a rather new technology, with its origins tracing back to the early 1980s through the distributed sensor networks (DSNs) program at the Defense Advanced Research Project Agency (DARPA) of the US Department of Defense [1]. DSNs were imagined to consist of many spatially distributed, autonomous, and low-cost sensing nodes that collaborated to gather various information about their surroundings. However, in the 1980s, the technology was not quite ready for this application. The sensors were too large and expensive, and the communication was not yet associated with wireless connectivity.

In the late 1990s, advances in computing, communication, and microelectromechanical technologies caused a shift in DSN research, bringing it closer to achieving the original vision. The *second wave* of DSN activities started in 1998, and it attracted large international involvement and attention. New networking techniques and networked information processing suitable for the dynamic ad hoc environments found in sensor networks were the initial focus, with the goal of enabling the required complex applications to run on resource-constrained sensors [1]. The sensors themselves also evolved with new technology, reducing both their cost and size. In addition, advances in wireless technology enabled robust and reliable wireless communication ideally suited for wireless DSNs. DARPA was again the pioneer, leading the efforts of sensor network research. They initiated a research program that provided new insights into ad hoc networking, dynamic querying and tasking, reprogramming, and multi-tasking [1]. At the same time, Institute of Electrical and Electronics Engineers (IEEE) had started to note the potential of WSNs and had begun to work on a specification for low-rate wireless personal area networks (LR-WPAN).

The work of IEEE was finalized in 2003, when the IEEE Std. 802.15.4 specification [2] was ratified, defining the physical layer (PHY) and media access control (MAC) layer for LR-WPAN. The higher layers of the protocol stack are out of scope of the specification. Offering features such as low power, low complexity, and low cost is ideally suited for WSN applications. With a growing number of solutions, both standardized and proprietary, based on the IEEE Std. 802.15.4 appearing in the years since its release, it has become the de facto standard for WSNs.

The ZigBee specification [3], originally released in 2004, was the first full standard to appear based on the IEEE Std. 802.15.4. ZigBee defines the network layer (NL) and application layer on top of the IEEE Std. 802.15.4 PHY and MAC.

Early research and evaluation of the IEEE Std. 802.15.4 identified several potential issues related to information security, in addition to other minor bugs and errors. A new version of the standard was released in 2006, referred to as IEEE Std. 802.15.4-2006 [4], which addressed these shortcomings. The original standard from 2003 is referred to as IEEE Std. 802.15.4-2003, to distinguish the two versions. Shortly after the ratification of IEEE Std. 802.15.4-2006, the ZigBee Alliance released a new version of the ZigBee standard, ZigBee-2006 [5]. The original ZigBee standard is referred to as ZigBee-2004. ZigBee-2006 included improvements for, among other things, addressing issues leading to scalability problems for large networks. It is important to note, however, that ZigBee-2006 was still based on IEEE Std. 802.15.4-2003, and not on the new IEEE Std. 802.15.4-2006. This means that the security issues of IEEE Std. 802.15.4-2003 were still present in ZigBee-2006.

In 2007, the Highway Addressable Remote Transducer (HART) Communication Foundation (HCF) released the HART Field Communication Protocol Specification, Revision 7.0 [6], which included a definition of a wireless interface to field devices, referred to as WirelessHART. WirelessHART was the first specification to be released that was specifically designed for process automation applications. With features such as self-healing and self-configuring multihop mesh networks, WirelessHART offers a viable wireless alternative for the traditionally wired industrial field instrumentation. WirelessHART was approved by the International Electrotechnical Commission (IEC) as international standard IEC 62591 Ed. 1.0 for wireless communication in process automation [7] in March 2010.

The ZigBee specification was initially designed to address applications within home automation and consumer electronics. A ZigBee network operates on the same, user-defined channel throughout its entire lifetime. This makes it susceptible both to interference from other networks operating on the same frequency and to noise from electrical equipment and machinery in the environment. As a result, ZigBee has not been regarded as robust enough for harsh industrial environments [8]. To combat this shortcoming, the ZigBee Alliance released the ZigBee PRO specification [9] in 2007. ZigBee PRO is specifically aimed at the industrial market, having enhanced security features and a frequency agility concept where the entire network may change its operating channel when faced with large amounts of noise and/or interference. Despite these innovations, ZigBee has still not been embraced by the industry.

Parallel to HCF's work on WirelessHART, the International Society of Automation (ISA) initiated work on a family of standards for wireless systems for industrial automation applications. This resulted in the ratification of the ISA100.11a standard in September 2009 [10]. Like WirelessHART, ISA100.11a aims to provide secure and reliable wireless communication for noncritical monitoring and control applications in the process automation industries. A new version of the ISA100.11a was released in 2011 [11], addressing minor faults and errors in the initial specification.

A third specification addressing wireless communication for the process automation industries, WIA-PA, was accepted by the IEC in 2009 as IEC 62601 [12]. WIA-PA was developed by the Chinese Industrial Wireless Alliance (CIWA) under the urgent requirements of process automation. In 2007, CIWA was established by Shenyang Institute of Automation, along with more than 10 universities,

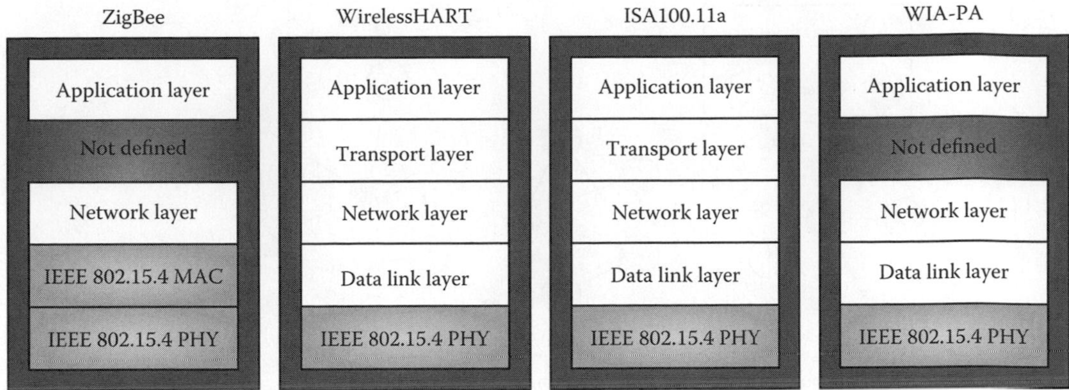

FIGURE 33.1 Wireless instrumentation protocol stacks.

academies, and companies. The scope of WIA-PA is to provide a system architecture and protocol stack for use in industrial monitoring, measurement, and control applications. However, at the time of writing, no products supporting WIA-PA are readily available on the market.

In April 2012, the IEEE Std. 802.15.4e [13] was released as an amendment to the IEEE Std. 802.15.4 specification. It provides additional MAC behavior and frame formats that allow IEEE Std. 802.15.4 devices to support applications within industrial applications such as process control and factory automation. At the time of writing, no devices supporting IEEE Std. 802.15.4e have yet been released.

ZigBee, WirelessHART, ISA100.11a, and WIA-PA are all based on the IEEE Std. 802.15.4 PHY. ZigBee uses the IEEE Std. 802.15.4 MAC as well, while the three industrial standards have made modifications to the IEEE Std. 802.15.4 MAC to allow for frequency hopping and extended security mechanisms (see Figure 33.1). With the current status of ZigBee not being considered robust enough, while WIA-PA products seem to be unavailable, the process industries are now faced with two independent and competing standards specifically designed for wireless instrumentation, each supported by different industry players. The process and automation industry as a whole would benefit more from having one global, wireless standard, but with the current situation, this is unlikely to happen in the near future.

This chapter contains a theoretical and analytical comparison of WirelessHART and ISA100.11a, from a technical and systematical point of view. With two available standards for wireless instrumentation, it is important for the end users to be able to understand the inherent strengths and limitations of the two and how these influence their suitability for different automation applications and usage areas.

Note that two separate chapters in this book are devoted for a more detailed description of WirelessHART and ISA100.11a.

33.2 System Overview

A typical WirelessHART or ISA100.11a installation consists of a collection of components, including physical devices and software modules, each fulfilling one or more defined functions or roles. The following devices and components are defined in WirelessHART [6]:

- Field device
- Adapter
- Handheld
- Gateway
- Network manager
- Security manager

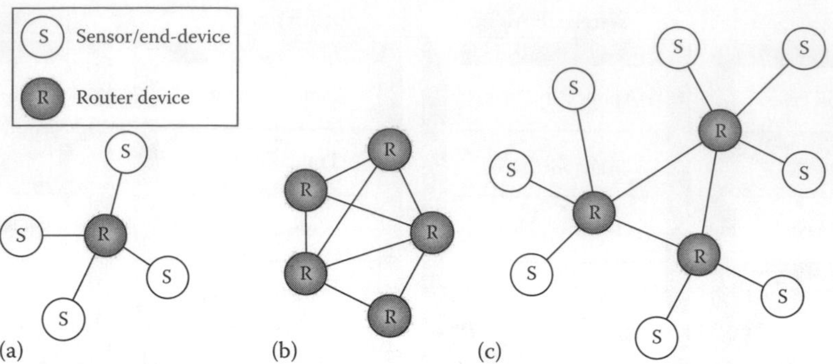

FIGURE 33.2 Network topologies: (a) star, (b) mesh, and (c) hybrid star–mesh.

In ISA100.11a, a set of roles are used to describe the capabilities of a device. The following roles are defined in ISA100.11a [11]:

- Input/output (I/O)
- Router
- Provisioning
- Backbone router
- Gateway
- System manager
- Security manager
- System time source

A more detailed description of the function of each of these components and roles can be found in Chapters 31 and 32.

WirelessHART and ISA100.11a have a fundamental difference on the field device level that influences the possible network topologies for the two standards. In WSNs, the typical network topologies are either star, mesh, or a hybrid combination of the two called star–mesh. These three network topologies are illustrated in Figure 33.2. In WirelessHART, all field devices (and adapters) are required to be routers with the capability of routing packets on behalf of other devices in the networks, enabling a full mesh network topology. In ISA100.11a on the other hand, the sensor and actuator (I/O) role is separated from the router role. An ISA100.11a field device can thus be configured either as end node with no routing capability or as router nodes with routing capacity. As a result, an ISA100.11a network can employ star, mesh, and star–mesh network topologies depending on the defined roles of the devices in a network.

33.3 Communication Protocols

WirelessHART and ISA100.11a use a simplified version of the seven-layered open systems interconnection (OSI) basic reference model, as illustrated in Figure 33.3.

The following section gives a comparison of the different protocol stack layers of WirelessHART and ISA100.11a.

33.3.1 Physical Layer

WirelessHART and ISA100.11a implement the PHY of the IEEE Std. 802.15.4 specification [4], with operational frequency being limited to the 2.4 GHz band. The only difference between the WirelessHART

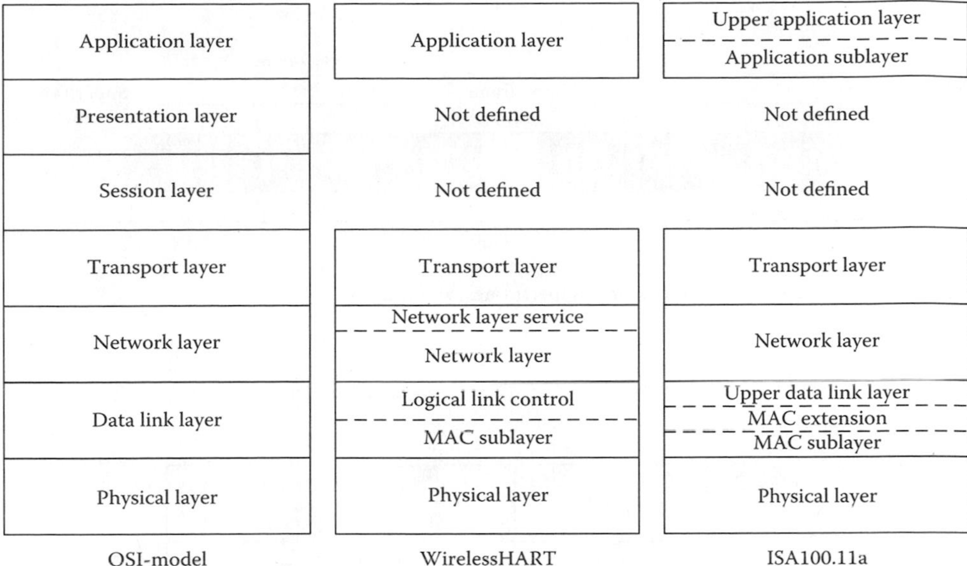

FIGURE 33.3 WirelessHART and ISA100.11a stack structures.

PHY and the ISA100.11a PHY is related to the usage of channel 26, which is not legal to use in some countries. The channel is not included in WirelessHART, while it is defined as optional in ISA100.11a.

33.3.2 Data Link Layer

WirelessHART divides the data link layer (DLL) into a logical link control layer and a MAC (see Figure 33.3). The scope of the WirelessHART DLL is communication between neighboring devices on a one-hop level. Any responsibilities to the network beyond a device's immediate neighbors are allocated to the WirelessHART NL.

ISA100.11a divides the DLL into a MAC sublayer, a MAC extension, and an upper DLL. Unlike WirelessHART and contrary to the DLL definition in the OSI model, the ISA100.11a DLL handles mesh routing within a data link (DL) subnet. An ISA100.11a DL subnet comprises one or more groups of field devices with a shared system manager and backbone network. The DL subnet stops at the backbone router, and network routing beyond the backbone router and into the backbone and plant network is handled by the ISA100.11a NL. More information about the graph and source routing mechanisms employed by ISA100.11a can be found in the separate section on the specification.

Both WirelessHART and ISA100.11a use a combination of time division multiple access (TDMA) and frequency hopping for medium access. Communication is divided into a 2D matrix consisting of timeslots and channels. In WirelessHART, the duration of a timeslot is fixed at 10 ms, while in ISA100.11a, it is configurable (with a typical and default duration of 10 ms). A collection of timeslots is called a superframe, and the superframes repeat continuously throughout the lifetime of the network. The term frame is used to separate different instances in time of a given superframe, as illustrated in Figure 33.4.

The channel used for communication between devices in a given timeslot in a superframe will change in each consecutive frame according to a pseudorandom hop pattern, as illustrated in Figure 33.5. This channel hopping mechanism is called slotted hopping in ISA100.11a, and in addition to slotted hopping, ISA100.11a also supports slow hopping and hybrid combinations of slotted and slow hopping. In slow hopping, a collection of contiguous timeslots are grouped on a single channel, as illustrated

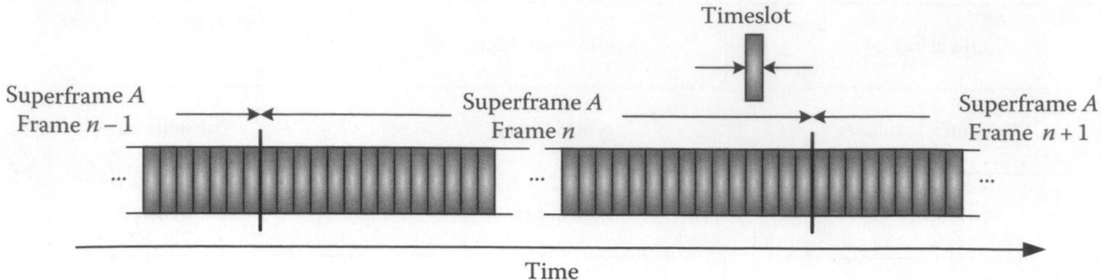

FIGURE 33.4 Structure of TDMA timeslots, superframes, and frames.

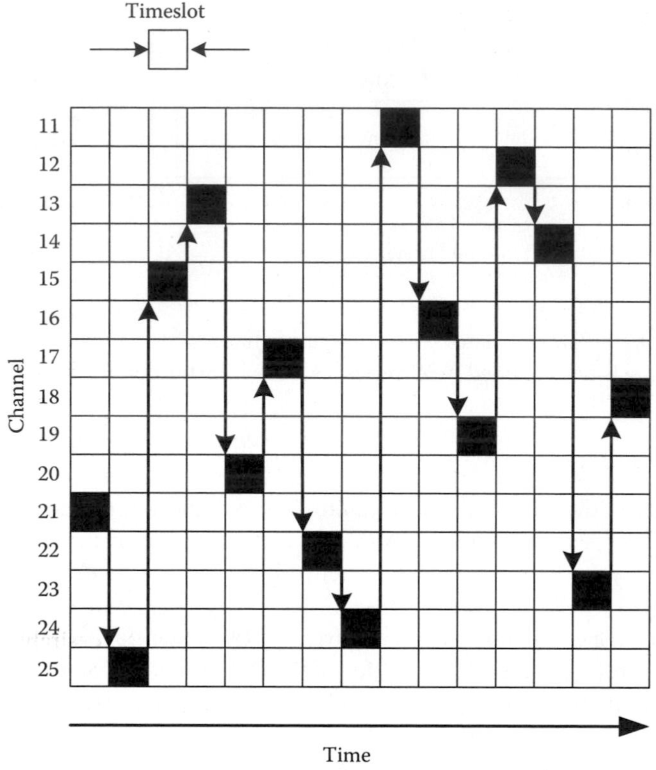

FIGURE 33.5 WirelessHART and ISA100.11a (slotted) frequency hopping.

in Figure 33.6a. The length of a slow hopping period is configurable, and it is intended to be shared by a group of devices that may require immediate and contention-based channel bandwidth on demand. In hybrid slotted and slow hopping configuration, the network will change between periods of slow hopping and periods of slotted hopping. An example of hybrid slotted and slow hopping is shown in Figure 33.6b.

33.3.3 Network Layer

For the HART Field Communication Protocol, the NL is the point of convergence for traditional wired HART token-passing networks and WirelessHART TDMA-based networks. The WirelessHART NL is responsible for routing packets from their initial source to their final destination, similar to the

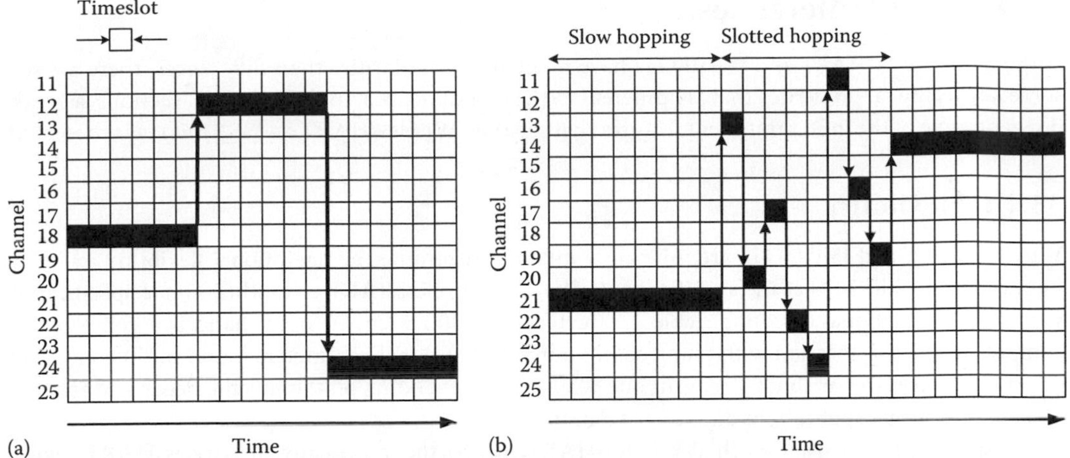

FIGURE 33.6 ISA100.11a frequency hopping mechanisms: (a) slow and (b) hybrid slow and slotted.

ISA100.11a DLL described in the previous section. WirelessHART supports both graph and source routing, and all devices in a WirelessHART network maintain a series of routing tables that control the communications performed by the device. The assignment of routing tables is handled by the network manager.

The ISA100.11a NL is used for backbone routing and has support for IPv6 [14] traffic through inclusion of the IETF 6LoWPAN specification [15]. The ISA100.11a NL is responsible for determining the appropriate address information for data messages, either a 16-bit short address for DL subnets or a 128-bit long address for application end points and backbone networks.

33.3.4 Transport Layer

The WirelessHART transport layer (TL) supports both acknowledged and unacknowledged transactions between devices, allowing the end application to decide whether it is desirable to send packets with or without end-to-end acknowledgment receipts.

The ISA100.11a TL is an extension of the connectionless User Datagram Protocol [16], with optional compression as defined by 6LoWPAN. ISA100.11a does not support acknowledged TL transactions.

33.3.5 Application Layer

WirelessHART inherits its application layer from the wired HART AL. The HART AL defines the commands, responses, data types, and status reporting supported by the HART Field Communication Protocol Specification. All communication between devices on the AL level is through a set of defined commands. The commands are divided into four groups:

1. Universal commands
2. Common practice commands
3. Device families commands
4. Device-specific commands

The ISA100.11a AL is object oriented and has support for handling I/O hardware, protocol tunneling, and execution of computational functions.

33.4 Main Differences

Although WirelessHART and ISA100.11a have many more similarities than differences, there are still some key technical properties that are different in the two standards. In the following sections, a breakdown of some of the most prominent features that separate WirelessHART and ISA100.11a is presented.

33.4.1 Flexibility

WirelessHART and ISA100.11a are inherently different regarding the operational flexibility and configuration possibilities that the specifications allow for. WirelessHART is a rather *simple* specification with very few optional or configurable parameters. ISA100.11a on the other hand is a complex and comprehensive specification with many configurable and optional parameters found in different stack layers. These features are both strengths and weaknesses depending on the specific needs and requirements of the target applications and usage scenarios.

The strict and limited approach of WirelessHART ensures that practically all WirelessHART devices will have identical behavior, regardless of design and implementation choices made by the equipment providers. This should easily facilitate interoperability between multiple vendors, as all products adhering to the standard should be equal. This naturally comes at the cost of a lack of possibility to adapt and tailor the device and network behavior to specific application requirements.

The wide range of available optional and configurable parameters in ISA100.11a allows for great flexibility for adapting network behavior to various application requirements. However, it may lead to interoperability issues if different vendors choose to implement different features of the standard. To combat this, ISA100.11a must define application profiles. A profile is a cross-layer specification that defines which options are mandatory in the different protocol layers. Although profile definitions help with possible interoperability issues, it still requires extensive compliance testing and verification to achieve full vendor flexibility.

33.4.2 Protocol Support

WirelessHART is a wireless extension of the wired HART Field Communication Protocol Specification and is naturally confined to using the command-based HART protocol for message exchange. All information and data in a WirelessHART network must be transmitted in the shape of HART commands.

The ISA100.11a application layer is object oriented and implements tunneling features that allow devices to encapsulate foreign protocols and transport them through the network. Although successful tunneling of protocols depends upon how well ISA100.11a meets the technical requirements of the foreign protocol, it still opens up the possibility of transferring a multitude of wired protocols over an ISA100.11a network.

33.4.3 Coexistence

Since WirelessHART and ISA100.11a operate in the crowded 2.4 GHz band, they are likely to be subjected to interference from other wireless networks operating in the same frequency band. In recent years, IEEE 802.11-based infrastructure has become commonplace in many process plants and facilities, and it is expected that most wireless instrumentation deployments will share the frequency spectrum with IEEE 802.11-based access points and mobile devices. Practical experiments have shown that the performance of IEEE Std. 802.15.4-based networks will be degraded when coexisting with IEEE 802.11 networks [17], and since WirelessHART and ISA100.11a inherit their PHY from IEEE Std. 802.15.4, they will be subjected to such interference as well.

To mitigate the effects of interference, wireless protocols may employ various coexistence mechanisms. In WirelessHART and ISA100.11a, clear channel assessment (CCA) and channel blacklisting are the weapons of choice to combat the degrading influence from other wireless networks. However, the two standards have chosen to implement the two features in slightly different ways. WirelessHART employs manual channel blacklisting, where a network operator must manually configure which channels are available and which channels are blocked. ISA100.11a has an adaptive blacklisting mechanism, where each device in a network may autonomously blacklist channels that suffer from noise and/or interference. Furthermore, ISA100.11a defines four different CCA modes, where modes 1–3 are defined by IEEE Std. 802.15.4:

0—*No CCA*: CCA is disabled and not conducted prior to transmission.
1—*Energy above threshold*: CCA reports a busy medium upon detecting any energy above a configurable threshold.
2—*Carrier sense only*: CCA reports a busy medium if a signal compliant with IEEE Std. 802.15.4 PHY modulation and spreading characteristics is detected.
3—*Carrier sense with energy above threshold*: CCA reports a busy medium using a logical and/or combination of modes 1 and 2.

WirelessHART on the other hand has fixed its CCA mechanism to mode 2.

With the correct configuration, ISA100.11a should be somewhat better equipped to handle coexistence with IEEE 802.11 networks. While WirelessHART only listens to activity from other IEEE Std. 802.15.4 networks, ISA100.11a will by employing CCA modes either 1 or 3 report a busy medium if any energy above a threshold is detected. If there is activity from a nearby IEEE 802.11 access point or client, the ISA100.11a device will back off and delay its transmission to the next available timeslot. This will naturally result in increased latency, but no power is wasted trying to transmit a message that will most likely not be received correctly by the destination device. In addition, the adaptive channel blacklisting mechanism of ISA100.11a can dynamically remove this problem completely by not using channels that show high IEEE 802.11 activity.

33.4.4 Security

Even though WirelessHART and ISA100.11a define a series of security mechanisms to ensure the integrity of the networks, some potential minor security weaknesses have been identified.

In WirelessHART, all security features are mandatory, while ISA100.11a defines many security mechanisms as optional. Considering that security algorithms require additional processing time, memory, and power, making them mandatory means that devices that may not require strict security policies cannot disable them to achieve benefits such as extended battery life. On the other hand, the ISA100.11a concept of having optional security features may be a security threat in itself and also an issue when it comes to interoperability. Vendors might not choose to implement the full security suite, and different vendors might choose to implement different parts of the optional security features.

Both WirelessHART and ISA100.11a rely on a centralized security manager for the authentication of new devices and the generation and management of security keys throughout the lifetime of the network. This means that the loss of the security manager will cause the loss of security mechanisms in the network. New releases of WirelessHART and ISA100.11a are combating this issue by offering redundant network and security manager solutions with automatic and transparent handover from the primary to the secondary system in case of failure.

33.4.5 Quality of Service

Although Quality of Service (QoS) is a term with various meanings and interpretations depending on the context, it can here be accepted as a measure of the service quality that a network offers to applications and/or users [18]. With QoS comes the ability to control the resource sharing of a network by giving different priorities to various applications and data packets depending on their requirements.

Higher performance levels can then be provided to specific applications and data packets through a set of measureable service parameters such as latency, jitter, packet loss, reliability, and availability [19].

Support for QoS in wired networks is generally obtained by overprovisioning and/or traffic engineering [18]. With overprovisioning, extra resources are added to the network so that it is able to provide satisfactory services to all applications. As all users are served at the same service class, overprovisioning may become unpredictable during peak traffic. For resource-constrained WSNs, overprovisioning is not an ideal QoS method as the network often does not have the capacity to provide the required resources. In traffic engineering, users and applications are assigned a different priority through a set of defined service classes. This method is also called service differentiation, and it is a widely adopted scheme for both wired and wireless networks to provide QoS guarantees [19]. For traditional wired computer networks, there are two main models for service differentiation: integrated services (IntServ) [20] and differentiated services (DiffServ) [21]. The IntServ model maintains service on a per-flow basis, while the DiffServ model maintains service on a per-packet basis. For the packet-based nature of WSNs, DiffServ is the best suited mechanism for service differentiation [22]. In the DiffServ model, the source devices know the criticality of the data packets it is sending, and this criticality is translated into predefined priority levels. Other devices in the network also select the appropriate service level for data packets based on their priority.

WirelessHART defines four different priority levels on the DLL [6]:

1. *Command* (highest priority). The command priority is used for packets containing network-related diagnostics, configuration, or control information.
2. *Process data*. Packets containing either process data or network statistics shall be classified as process data priority. Only the control of the network is more important than the delivery of sensor data measurements from field transmitters or set point information to actuators.
3. *Normal*. If a data link protocol data units (DLPDU) do not meet the criteria for any of the other three priority levels (command, process data, or alarm), it shall be classified with normal priority.
4. *Alarm* (lowest priority). Packets containing only network alarm and network event information shall have a priority of alarm.

These priority levels are primarily used for flow control and to mitigate potential network congestion points in the event of either a process upset or noise/interference deteriorating the radio frequency (RF) channel(s). With the aforementioned mechanisms, network management packets have full priority while propagated through the network, allowing the network manager to keep the network operational. Network-induced alarms have a restricted flow through the network, ensuring that alarm floods do not disrupt or hinder the network operation. All other network traffic flows through the network as bandwidth and internal buffer spaces on the devices allow. Unfortunately, there is only one priority level reserved for process data, which means that all sensors and/or actuators in a WirelessHART network share the same priority level, regardless of the requirements and criticality of the application they are serving.

ISA100.11a uses contracts to define the setup and requirement of communication between two devices in a network. A contract is an agreement between the system manager and a device in the network that involves the allocation of network resources by the system manager to support the communication requirements of the device. All contracts are unidirectional, and they are established by the system manager upon reception of a contract request. ISA100.11a supports two priority levels, contract priority and message priority. The contract priority is the base priority for all messages sent using a specific contract. Four contract priorities are supported:

1. *Network control* (highest priority): may be used for critical management of the network by the system manager
2. *Real-time buffer*: may be used for periodic communications in which the message buffer is overwritten whenever a newer message is generated

3. *Real-time sequential*: may be used for applications such as voice or video that needs sequential delivery of messages
4. *Best effort queued* (lowest priority): may be used for client–server communications

The message priority establishes priority within a contract using two message priorities: high and low. The contract priority is specified by the application, during contract establishment time, in its contract request. It may be used by the system manager to establish preferred routes for high-priority contracts and for load balancing the network. The combined contract/message priority is used to resolve contention for scarce resources when these messages are forwarded through the network.

33.4.6 Application Potential

The performance requirements of an industrial field instrument depend upon the nature and criticality of the application it is serving. NAMUR, a user association for automation technologies in the process industries, defines the following three applications classes for wireless instrumentation in their recommendation document NAMUR NE 124 *Wireless Automation Requirements* [23]:

- Application class A—functional safety
- Application class B—process management and control
- Application class C—display and monitoring

Similarly, the ISA has defined six usage classes for wireless instrumentation through their work on the ISA100.11a specification for wireless field devices [11]:

- Application class 0—emergency action
- Application class 1—closed-loop regulatory control
- Application class 2—closed-loop supervisory control
- Application class 3—open-loop control
- Application class 4—alerting and flagging
- Application class 5—logging and downloading/uploading

A mapping between the NAMUR and the ISA application classes is shown in Table 33.1. For simplicity, the three NAMUR application classes will be used in further discussions, referred to as monitoring (C), control (B), and safety (A), respectively.

33.4.6.1 Monitoring

Monitoring applications cover tasks that, by definition, are not of any immediate operational consequence. In addition, they do not affect plant safety in any regard. As a result, the performance

TABLE 33.1 Wireless Instrumentation Usage Classes

Application	NAMUR	ISA
Safety	Class A Functional safety	Class 0 Emergency action
Control	Class B Process management and Control	Class 1 Closed-loop regulatory control
		Class 2 Closed-loop supervisory control
		Class 3 Open-loop control
Monitoring	Class C Display and monitoring	Class 4 Alerting and flagging
		Class 5 Logging and downloading/uploading

requirements for field instruments in this category are somewhat relaxed, but it is still of interest to maintain a certain level of service quality in order for the application to be of any operational value. Monitoring applications typically use continuous data delivery, where sensors measure and transmit data to the control system at periodic intervals.

Most industrial deployments of wireless instrumentation to date have been limited to noncritical monitoring applications, and fulfilling the requirements for this application class has been the initial focus of the WirelessHART and ISA100.11a specifications, as well as the vendor-specific implementations of the specifications. Both WirelessHART and ISA100.11a have been qualified for use by many process industry end users [24]. This is backed by results from extensive laboratory experiments, pilot installations, and deployment in live production environments, which show that both solutions are fully capable of providing adequate application performance for noncritical monitoring applications [24]. For these applications, there should be minimal performance deviations between the two standards.

33.4.6.2 Control

For control applications, sensor data and actuator set points must be provided in a timely and regular manner. Most control algorithms are not designed to handle a periodic reception of sensor data and thus require bounded latency and bounded jitter, where jitter is defined as the variance in latency of consecutive sensor data. Control applications are typically continuous by nature where sensor data and actuator set points are transmitted at periodic intervals. Event-driven control, where sensor data are transmitted only when a measurement exceeds a predefined threshold, is still not very common in the process industries.

Another requirement for many control applications is to have a common timing domain for all components in the system. This means that the clocks of wireless sensors and actuators and the wireless gateway should be synchronized with the clocks of the controllers and control system. Propagating time information through the wireless network should be possible, as a clock accuracy of 1 ms is already required for all wireless devices in order for the TDMA timeslot structure to work properly according to today's wireless standards.

To achieve low latency and jitter, it is recommended to implement proper fieldbus interfaces (e.g., PROFIBUS or Foundation Fieldbus) on the wired side of the wireless gateway and on the instrumentation backbone networks [24]. The definition of these interfaces is beyond the scope of both WirelessHART and ISA100.11a and left open for the implementers to decide what they find most suitable for their systems.

For open-loop control, WirelessHART and ISA100.11a should be able to deliver the low latency and high reliability as required by the applications. However, recent research has shown that latency and determinism issues make WirelessHART less suitable for control applications where bounded and low latency and jitter are the main requirements [25]. Such experiments have at the time of writing not yet been conducted for ISA100.11a, but the results are expected to be quite similar. However, as described in Section 33.4.5, the lack of multiple message priority levels for process data in WirelessHART will most likely hinder future developments in the direction of control applications. ISA100.11a on the other hand has through its contract-based priority levels access to a mechanism for distinguishing between messages of different criticality based on the nature and requirements of the application it is serving.

33.4.6.3 Safety

In safety applications, reliability and timeliness are the main requirements for the communication between sensors and the safety system. As opposed to control loops, rapid update rates are normally not required, but safety communication must have mechanisms that ensure that data packets arrive within a specific deadline. For most safety systems, a query-based data delivery model is used where the safety controller periodically requests data from the sensors.

Safety systems in the process industries are subject to comply with a certain Safety Integrity Levels (SIL). The standard IEC 61508 [26] defines SIL from a set of requirements that accomplish both hardware safety integrity and system safety integrity. There are four SIL levels (1–4), where SIL4 is defined as the most dependable and SIL1 as the least. Neither WirelessHART nor ISA100.11a directly supports the necessary certified SIL safety mechanisms as an integrated part of their specifications. A work-around for this is to use an already established and certified end-to-end communication protocol, such as the SIL2-rated PROFIsafe [27], which is designed to be implemented on top of the PROFINet fieldbus [28].

The recent development of the world's first wireless hydrocarbon gas detection system has proven that it is possible to achieve SIL2 end-to-end communication between a safety controller and a wireless sensor by tunneling PROFIsafe over ISA100.11a [24]. For WirelessHART on the other hand, limitations in currently available HART commands at the application layer make it impossible to implement the tunneling mechanisms needed for full PROFIsafe support. PROFIsafe over WirelessHART will thus not be available before a potential modification and new release of the HART Field Communication Protocol Specification.

33.5 Conclusion

WirelessHART and ISA100.11a are both on the quest to become the de facto standard for wireless instrumentation in the process industries, each being backed by different leading automation providers. The theoretical comparison of the two specifications has shown that although they have some key differences, most features regarding the fundamental wireless communication parameters are the same. They are both able to provide satisfactory network performance in harsh industrial environments, and as a result they have been qualified for use in noncritical monitoring applications in the process industries.

For more advanced applications within industrial control and safety domains, the marginal differences between the two specifications become more apparent. Due to the lacking QoS mechanisms for process data packet priorities, combined with a command-based application layer with limited flexibility, WirelessHART is unlikely to be able to address fast closed-loop control and safety applications with SIL requirement without making alterations to the HART Field Communication Protocol Specification. The more comprehensive ISA100.11a standard, on the other hand, has adequate QoS support through its contract priority levels, and the object-oriented application layer has proven to be suitable for tunneling the SIL2-certified PROFIsafe protocol.

Regardless of these differences and despite ISA100.11a's better suitability for control and safety, it is still nearly impossible to predict which of the two standards, if any, will emerge as the leading global specification for wireless instrumentation. It is expected that nearly all wireless instrumentation installations and deployments in the near future will be within noncritical monitoring applications, so ISA100.11a current advantage in addressing safety applications might not be significant. The ongoing struggle for domination is thus expected to be more of a marketing and political battle than a pure technical one. With a 2-year lead in availability, the current market share is in favor of WirelessHART, and it remains to be seen if ISA100.11a is able to close this gap.

Despite the ongoing struggle to gain market positions, there are some initiatives to merge the two standards. The ISA100 committee has created an ISA100.12 working group that has the stated goal of investigating long-term possibilities for merging ISA100.11a and WirelessHART. This initiative is currently being conducted by leading automation industries, commonly referred to as the Heathrow group. Furthermore, NAMUR, the international user association of companies in the chemical and pharmaceutical process industries, has published recommendations on the convergence of the standards for industrial WSNs [29]. However, these convergence activities are expected to take time, and it is not certain that they will succeed in creating a unified, global standard for wireless instrumentation. So, for the time being, the process automation end users will have to decide between multiple standards when deploying wireless instruments in their plants and facilities, a situation that is far from ideal.

References

1. Chong, C. Y. and Kumar, S. P., Sensor networks: Evolution, opportunities and challenges, *Proceedings of the IEEE*, 91(8), Aug. 2003, 1247–1256.
2. IEEE Standard for Information Technology, Telecommunications and information exchange between systems—Local and metropolitan networks, Specific requirements—Part 15.4: Wireless medium access control (MAC) and physical layer (PHY) specifications for low rate wireless personal area networks (LR-WPANs), IEEE Computer Society, New York, Oct. 2003.
3. ZigBee Alliance, ZigBee Specification Version 1.0, Dec. 2004.
4. IEEE Standard for Information Technology, Telecommunications and information exchange between systems—Local and metropolitan networks, Specific requirements—Part 15.4: Wireless medium access control (MAC) and physical layer (PHY) specifications for low rate wireless personal area networks (LR-WPANs), IEEE Computer Society, New York, 2006.
5. ZigBee Alliance, ZigBee-2006 specification, Dec. 2006.
6. HART Field Communication Protocol Specification, Revision 7.0, HART Communication Foundation, Austin, Tx, Sept. 2007.
7. Industrial Communication Networks, Wireless communication network and communication profiles—WirelessHART, International Electrotechnical Commission (IEC) 62591, 2010.
8. Lennvall, T., Svensson, S., and Hekland, F., A comparison of WirelessHART and ZigBee for industrial applications, *Proceedings of the IEEE International Workshop on Factory Communication Systems*, Nancy, France, May 2008, pp. 85–88.
9. ZigBee Alliance, ZigBee PRO specification, Oct. 2007.
10. ISA100.11a-2009, Wireless systems for industrial automation: Process control and related applications, 2009.
11. ISA100.11a-2011, Wireless systems for industrial automation: Process control and related applications, 2011.
12. Industrial Communication Networks, Fieldbus specifications—WIA-PA communication network and communication profile, International Electrotechnical Commission (IEC) 62601, 2009.
13. IEEE Standard for local and metropolitan area networks— Part 15.4: Low-rat wireless personal area networks (LR-WPANs)—Amendment 1: MAC sublayer, IEEE Computer Society, New York, April 2012.
14. Requests for Comments (RFC) 2460—Internet Protocol, Version 6 (IPv6) Specification, Networks, Internet Engineering Task Force (IETF), 1998.
15. Requests for Comments (RFC) 4944—Transmission of IPv6 Packets over IEEE Std. 802.15.4 Networks, Internet Engineering Task Force (IETF), 2007.
16. Requests for Comments (RFC) 768—User Datagram Protocol, Networks, Internet Engineering Task Force (IETF), 1980.
17. Angrisani, L., Bertocco, M., Fortin, D., and Sona, A., Experimental study of coexistence issues between IEEE 802.11b and IEEE Std. 802.15.4 wireless networks, *IEEE Transactions on Instrumentation and Measurement*, 53(8), Aug. 2008, 1514–1523.
18. Chen, D. and Varshney, P. K., QoS support in wireless sensor networks: A survey, *Communication*, 13244, 2004, 227–233.
19. Yigitel, M. A., Incel, O. D., and Ersoy, C., QoS-aware MAC protocols for wireless sensor networks: A survey, *Computer Networks*, 55, 2011, 1982–2004.
20. Braden, R., Clark, D., and Shenker, S., Integrated services in the internet architecture—An overview, IETF RFC 1663, June 1994.
21. Blake, S. et al., An architecture for differentiated services, IETF RFC 2475, Dec. 1998.
22. Bhatnagar, S., Deb, B., and Nath, B., Service differentiation in sensor networks, *Proceedings of the International Symposium on Wireless Personal Multimedia Communications*, Aalborg, Denmark, Sept. 9–12, 2001.
23. NAMUR NE 124:2010, Wireless automation requirements, 2010.

24. Petersen, S. and Carlsen, S., Wireless instrumentation in the oil & gas industry—From monitoring to control and safety applications, *SPE Intelligent Energy International 2012*, Utrecht, the Netherlands, Mar. 27–29, 2012, pp. 1–14.
25. Akerberg, J., Gidlund, M., Neander, J., Lennvall, T., and Björkman, M., Deterministic downlink transmission in WirelessHART networks enabling wireless control applications, *Proceedings of the 36th Annual Conference on IEEE Industrial Electronics Society*, Glendale, AZ, Nov. 2010, pp. 2120–2125.
26. International Electrotechnical Commission (IEC), IEC 61508—Functional safety of electrical/electronic/programmable electronic safety-related systems—ALL PARTS, 2010.
27. International Electrotechnical Commission (IEC), IEC 61784-3-3—Industrial communication networks—Profiles—Part 3-3: Functional safety fieldbuses, Additional specifications for CPF 3, 2007.
28. International Electrotechnical Commission (IEC), IEC 61158—Industrial communication networks—ieldbus specifications, 2007.
29. NAMUR, NAMUR Recommendation NE 133, Wireless sensor networks: Requirements for the convergence of existing standards, Sept. 2010.

34

IEC 62601: Wireless Networks for Industrial Automation–Process Automation (WIA-PA)

34.1 Introduction 34-1
34.2 IEC/PAS 62601: WIA-PA 34-2
34.3 Physical Layer 34-4
34.4 Data Link Layer 34-5
Medium Access Control • Data Link Sublayer
34.5 Network Layer 34-10
Addressing • Routing, Fragmentation, and Reassembling • Aggregation and Disaggregation Services
34.6 Application Layer 34-12
User Application Process • Application Sublayer
34.7 Security 34-14
Secure Joining Process • Secure Transportation
34.8 Conclusion and Discussion 34-16
References 34-17

Ivanovitch Silva
Federal University of Rio Grande do Norte

Luiz Affonso Guedes
Federal University of Rio Grande do Norte

34.1 Introduction

During the fieldbus war (1986–2002), several technologies appeared as de facto standard for industrial networks (Modbus, Fieldbus Foundation, HART, just to name a few) (Sauter, 2010). All these technologies were based on structured cabling and they had as a goal creating a universal protocol to meet all the industrial demands (Decotignie, 2009). Despite the appearance of wireless technologies, the scenario has not been very different. A variety of technologies coexist and others are yet to be seen. Establishing a conservative scenario, only one wireless solution will hardly be the universal standard for the industry. It is wiser to think that several solutions will coexist, each one having its own advantages and disadvantages and its specific niche applications.

The current solutions for the industrial wireless networks are focused on two distinct groups (Willig, 2008). The first group is related to the WLANs (wireless local area networks), whose technologies are variations of the IEEE 802.11 standard. On the other hand, the second group is represented by WPANs (wireless personal area networks), whose solutions are variations developed by the IEEE 802.15 work group and other standardization commissions such as the ISA (International Society of Automation) and the IEC (International Electrotechnical Commission). Each group has its own peculiarities; however, the WPANs have presented themselves as a better solution when analyzed in industrial environments,

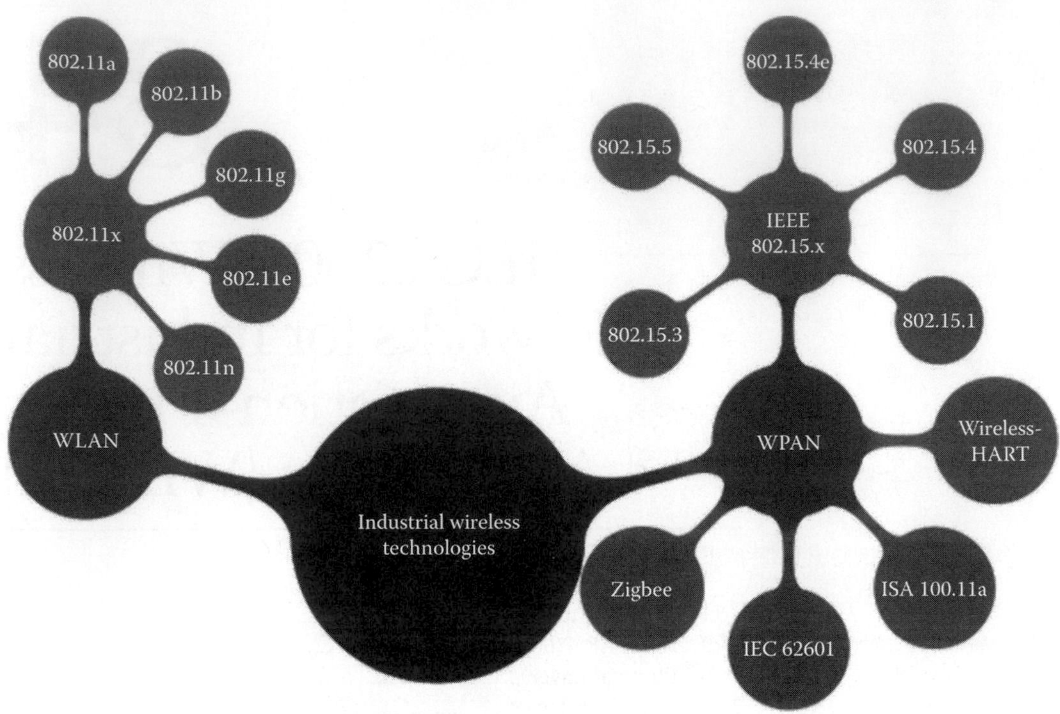

FIGURE 34.1 Wireless communication standards used in the industry.

with special focus on the WirelessHART (IEC-62591, 2010), ISA100.11a (IEC-62734, 2012), and IEC 62601/wireless networks for industrial automation–process automation (WIA-PA) (IEC-62601, 2011) standards, which were created for this kind of environment specifically.

An overview of the main wireless communication standards used in industrial environments is described in Figure 34.1. The chapter mentioned does not have the goal of describing all these standards, shifting the focus to the IEC 62601/WIA-PA standard. In order to have an overall analysis of the standards, it is recommended to read the following references (Silva et al., 2013).

34.2 IEC/PAS 62601: WIA-PA

In 2007, the Chinese Industrial Wireless Alliance (CIWA), under the urgent requirements of process automation, approved a new wireless communication standard, the WIA-PA. In 2011, the IEC approved the WIA-PA standard as a publicly available specification named IEC/PAS 62601 (IEC-62601, 2011). In the same year, the Chinese National Standard approved the WIA-PA as a national specification named GB/T 26790.1-2011. For producing valuable social and economic benefits to China, the GB/T 26790.1-2011 specification won the first prize of the China Standard Innovation Award 2013 (Liang et al., 2013).

The WIA-PA standard follows the trends of industrial network technologies to adopt wireless infrastructure. As its counterparts, the WirelessHART and ISA100.11a standards, the WIA-PA adopts similar physical features (Petersen and Carlsen, 2011). However, there are several differences between them. More details will be described later in this chapter. For the sake of understanding, Figure 34.2 gives an overview about the evolution of industrial wireless technologies.

A typical WIA-PA network defines five types of devices as described in Figure 34.3: host configuration computer, gateway device (GW), router device, field device, and handheld. The host configuration computer is used to provide high-level access to operators. The physical access between the plant automation network and the wireless network is conducted by the gateway. In relation to intranetwork communication,

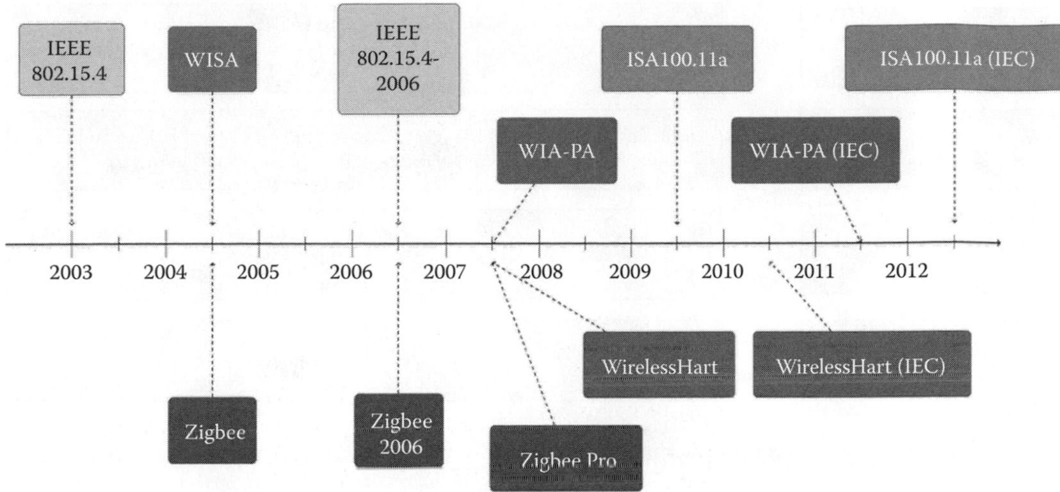

FIGURE 34.2 Evolution of industrial wireless technologies.

routers are responsible for forwarding the messages toward the gateway. The field devices and the hand held devices are considered the most basic devices. The former ones are responsible for the monitoring of process variables, whereas the latter ones are used during the commissioning of network devices.

Additionally, a WIA-PA network supports redundant gateways and redundant routers. The standard suggests a cold standby redundancy approach, where the start-up of a spare device is not later than the primary device. Periodically, the primary device updates its spare device with the information changed

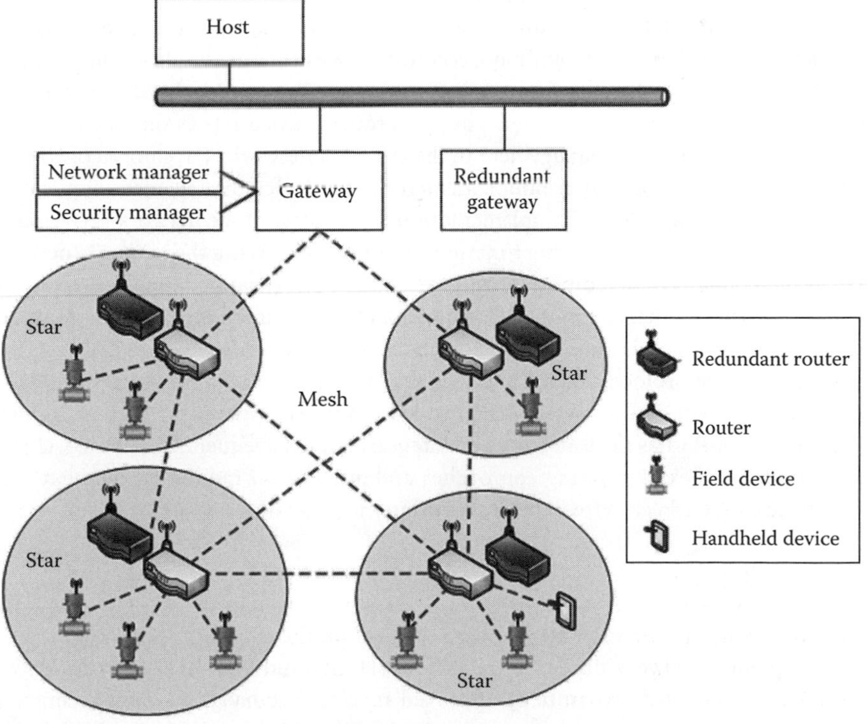

FIGURE 34.3 Typical WIA-PA network.

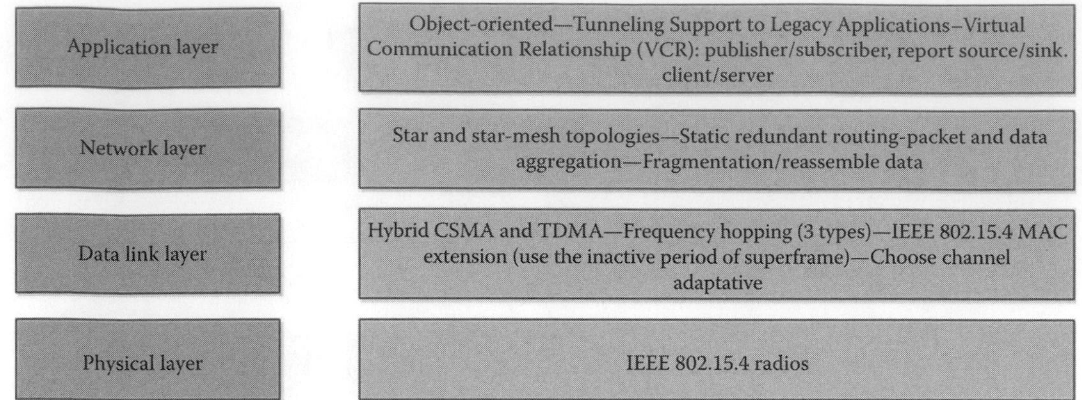

FIGURE 34.4 Overview of protocol stack present in the WIA-PA standard.

more recently. The update must occur in a wired manner; however, it is not specified in the WIA-PA standard. This mechanism allows the reliability of the network to be improved (Liang et al., 2011).

The WIA-PA standard supports two types of topologies: star and mesh star. The communication between routers follows a mesh approach, whereas the communication between a router and field devices follows a star approach, similar to a cluster topology. Thus, from a high-level viewpoint, a field device communicates with the gateway through a mesh-star network.

All communication over the network is centralized by two rules generally implemented inside the gateway. The first rule is called the security manager (SM). Its duties are concentrated on all issues related to security (storage of keys, validation of messages, authentication, etc.). On the other hand, the rule called the network manager (NM) is considered the core of the WIA-PA network. This rule is responsible for controlling all the communications over the network, from the scheduling tables to the choice of communication channels. In addition, secondary rules such as the cluster head and the cluster member are responsible for managing the intracluster communication. A field device and the handheld should only act as the rule of cluster member, whereas a router device acts as the rule of cluster head.

As described earlier, the system management in the WIA-PA network is composed of NM and the SM rules. The functions of management are implemented by the device management application process (DMAP) in each device. In spite of this implementation, the control of the network is centralized on the GW. The DMAP implements the following functions: joining and leaving the network, network address allocation, routing configuration, communication resource configuration, time source configuration, performance monitoring, management information base (MIB) maintenance, and firmware upgrading.

All functionalities of a standard are defined in its protocol stack. Similarly to WirelessHART and ISA100.11a standards, the protocol stack defined in the WIA-PA is based on the OSI model. However, only four layers are defined, as illustrated in Figure 34.4. When compared with other industrial wireless standards, the WIA-PA has the following advantages: adaptive frequency hopping (AFH), data and packet aggregation, and flexible routing approaches and application modes. In the next sections, the main features of each WIA-PA layer will be described in more detail.

34.3 Physical Layer

The physical layer defined in the WIA-PA standard is based on the IEEE 802.15.4-2006 (IEEE-802.15.4, 2006), and consequently, it is very similar to the WirelessHART and ISA100.11a physical layers.

The IEEE 802.15.4 standard was initially approved in 2003; however, its more recent version was approved in 2006. It is considered a de facto standard for the low-rate WPANs (LR-WPANs). These networks were designed to support devices with limited physical features (low energy consumption,

TABLE 34.1 Main Features Present in the IEEE 802.15.4-2006 Physical Layer

Features	IEEE 802.15.4 Physical Layer
Frequency bands	868 MHz (1 channel)
	915 MHz (10 channels)
	2450 MHz (16 channels)
Channel width	2 MHz
Throughput	250, 40, and 20 kbps
Modulation	O-QPSK, BPSK, and ASK
Communication range	100–200 m
Transmission power	0 dBm

a communication range of a few meters, low power processing, and reduced memory), enabling the development of pervasive applications in diverse areas (forest fire and landslide detection, greenhouse and industrial monitoring, passive localization and tracking, etc.). When an LR-WPAN is designed for an industrial environment, it is called an industrial wireless sensor network (IWSN).

The IEEE 802.15.4-2006 physical layer can operate in three distinct frequency bands: the 2450 MHz ISM band (worldwide) with 16 channels, the 915 MHz band (in the United States only) with 10 channels, and the 868 MHz band with only one channel (European and Japanese standards). All of them use the direct-sequence spread spectrum (DSSS) access mode. Signaling in the 2450 MHz band is based on orthogonal quadrature phase shift keying (O-QPSK), while the 868/915 MHz bands rely on binary phase shift keying (BPSK). The throughputs of the bands are 250, 40, and 20 kbps, respectively. In addition to the management of the physical layer, services include the activation and deactivation of the radio transceiver, energy detection (ED) within the current channel, link quality indication (LQI) for received packets, clear channel assessment (CCA) for carrier sense multiple access with collision avoidance (CSMA-CA), channel frequency selection, and data transmission and reception. The main features present in the IEEE 802.15.4-2006 physical layer are summarized in Table 34.1.

Although the physical layer can support three frequency bands, the 2450 MHz ISM band is by far the most frequently used. The use of this band generates a coexistence problem in the sense that other wireless technologies also employ the same frequency band (e.g., IEEE 802.11 and Bluetooth). In relation to coexistence with the IEEE 802.11 devices, it is possible to carry out configuration until there are four nonoverlapping IEEE 802.15.4 channels as described in Figure 34.5. Obviously, as only four channels are available in this scenario, it is expected that the transmissions have a higher probability of suffering interference than a scenario where 16 channels are available (Angrisani et al., 2008). On the other hand, a minor impact is expected when the coexistence with Bluetooth devices is considered, because in the Bluetooth standard, the frequency band is modified 1600 times per second. Thus, it is more likely that the interferences occur in the first transmission attempts. However, an efficient retransmission mechanism can be implemented in the IEEE 802.15.4 to mitigate this interference (Young et al., 2007).

34.4 Data Link Layer

The data link layer (DLL) defined in the WIA-PA standard was designed to provide real-time, secure, and reliable communications. It is based on the IEEE 802.15.4 DLL with the improvement of some properties (frequency hopping, retransmission, TDMA, and CSMA). A message integrity code (MIC) mechanism and encryption are also adopted to guarantee the integrity and confidentiality of the data transmitted, respectively.

According to Figure 34.6, the WIA-PA DLL is composed of two sublayers that include the medium access control (MAC) layer and the data link sublayer (DLSL) and service access points (SAPs). The MAC handles the mechanisms for sending and receiving data frames, whereas the DLSL controls the aspects

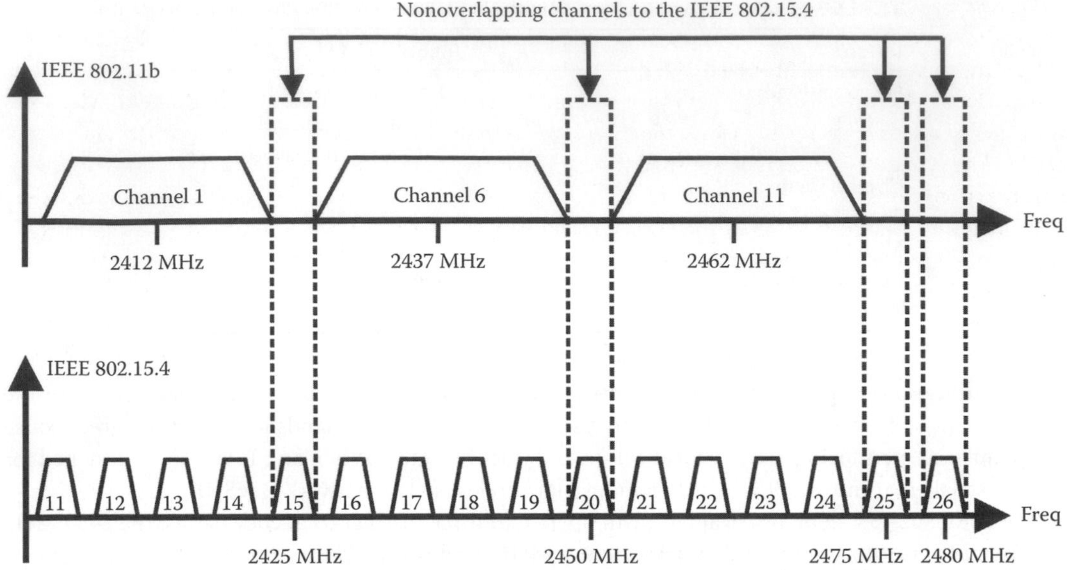

FIGURE 34.5 Coexistence of IEEE 802.11 channels and IEEE 802.15.4 channels.

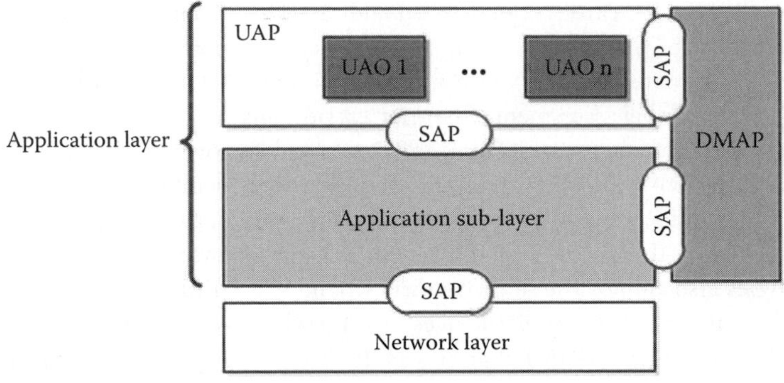

FIGURE 34.6 WIA-PA DLL.

of communication resource. The management of DLL and the exchange of messages between the layers are provided through SAP. The implementation of an interface to the DMAP in the DLL is also mandatory.

34.4.1 Medium Access Control

The MAC layer adopted in the WIA-PA standard is based on the IEEE 802.15.4-2006 MAC layer configured to beacon-enabled mode. The legacy compatibility was necessary to guarantee the following services: generating beacons (only to the gateway and router devices), synchronizing, security and reliability, joining, and maintaining the guaranteed timeslots (GTS).

Another property inherited from the IEEE 802.15.4-2006 MAC layer was the types of devices. According to that standard, the devices are classified as reduced-function devices (RFD) or full-function devices (FFD). In a WIA-PA network, field devices and the handheld act as RFD, whereas router devices and the gateway act as FFD.

One of the most important features implemented in the MAC layer is the synchronization mechanism. In order to guarantee the reliable communication between the devices, it is mandatory that the entire network is synchronized. All devices in a WIA-PA network should synchronize with a time source device. In the intercluster communication, the gateway is time source to the router devices. On the other hand, the router devices are the time sources to the field devices in the intracluster communication. The synchronization is established through beacon (inherited of the IEEE 802.15.4-2006) or synchronization frames. Both use a field of four octets to calibrate the time reference between the devices.

In a similar way to the synchronization mechanism, the devices must implement a mechanism for connection maintenance between neighbor devices. This procedure is implemented through keep-alive messages. Periodically, all devices send keep-alive messages to maintain its connections. The gateway sends keep-alive messages during the intercluster communication, whereas the field devices and the handheld send its keep messages during the intracluster period. Router devices can send keep-alive messages in both periods.

34.4.2 Data Link Sublayer

According to Figure 34.6, the WIA-PA DLSL provides an interface between the network layer (NL) and the MAC layer. Its main duties include the following: avoiding collisions; improving throughput; increasing bandwidth utilization, coexistence, timeslot, and superframe management; frequency hopping; transmission of long cycle data; channel conditional monitoring; security; and other management services.

34.4.2.1 Timeslot, Superframe, and Frequency Hopping

The communication unit is the timeslot. The duration of the timeslot is configured according to the limits imposed by the IEEE 802.15.4-2006. The timeslot length is configured by the NM after a device joins the network. Additionally, a tuple composed of time and frequency (link) is used inside the timeslot for communication. There are three types of links: transmit, receive, and shared. The first two types of links allow only two devices (source and destination) to exchange packets (used for intracluster and intercluster communication of periodic data), whereas the latter one allows more than two devices to use the communication channel (used for transmission of aperiodic data).

A collection of timeslots creates a superframe structure. Each superframe is repeated periodically at intervals (from 15.36 ms up to 251 s) proportional to the number of timeslots used. The WIA-PA superframe is described in Figure 34.7. The IEEE 802.15.4-2006 superframe is included in Figure 34.7 to show that the WIA-PA superframe is an extended version of it.

In a pure beacon-mode IEEE 802.15.4-2006 superframe, a device accesses the medium listening for beacon messages sent by the network coordinator. After receiving a beacon message, the device transmits data by using the superframe structure. As described in Figure 34.7, the IEEE 802.15.4-2006 superframe structure is divided into two parts: the active and inactive periods. During the active period,

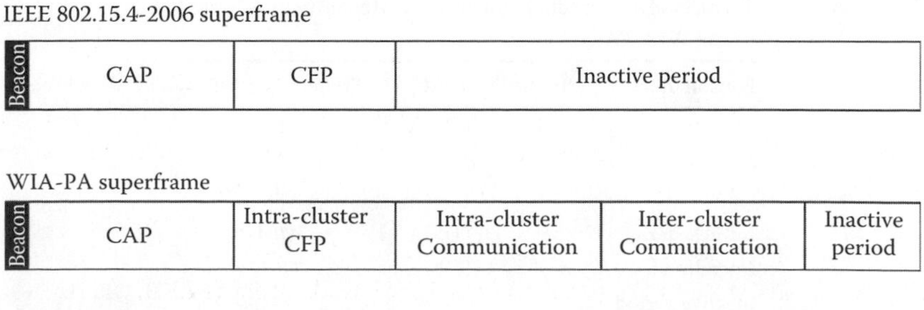

IEEE 802.15.4-2006 superframe

Beacon	CAP	CFP	Inactive period

WIA-PA superframe

Beacon	CAP	Intra-cluster CFP	Intra-cluster Communication	Inter-cluster Communication	Inactive period

FIGURE 34.7 WIA-PA superframe.

the medium can be accessed with or without contention. Devices that wish to communicate during the contention access period (CAP) must compete with other devices using a slotted CSMA-CA mechanism. On the other hand, a contention-free period (CFP) can be used for applications with restrictive requirements. Due to protocol limitations, the CFP has support for only seven timeslots. Finally, the inactive period of the superframe can be used by the devices to configure an operation mode with low energy consumption and consequently increase the network lifetime (assuming that the devices are battery powered).

On the other hand, the WIA-PA superframe added other functions beyond the ones implemented in the IEEE 802.15.4-2006 superframe. In this case, the joining process, intercluster management, and retransmissions occur in the CAP interval of the superframe, where the MAC is based on CSMA-CA. Furthermore, all other communications access the medium using a TDMA approach. During the CFP, mobility devices are authorized to communicate with the cluster head (router). Another important aspect is that part of the inactive period of the IEEE 802.15.4-2006 superframe is used for transmission in the WIA-PA superframe. More specifically, it was designed for intracluster and intercluster communications. The WIA-PA superframe also supports an inactive period when the devices can minimize the energy consumption. However, it has a smaller duration than the IEEE 802.15.4-2006 inactive period.

In order to improve the network reliability, a frequency hopping mechanism was introduced to the WIA-PA network. Basically, three hopping patterns are defined: adaptive frequency switching (AFS), AFH, and timeslot hopping (TH).

The hopping pattern AFS is used during the transmission of beacons and CAP and CFP. In this pattern, all the devices use the same communication channel. If the channel suffers some interference (e.g., the packet loss rate increases), the NM configures the devices to use another channel.

On the other hand, the hopping pattern AFH is used during intracluster communication, where the changes of channels occur irregularly depending on the actual channel condition. A channel is considered bad if the amount of retry times in that channel reaches a threshold. When the sender device identifies that the current channel is bad, it chooses the next channel (immediate) in the channel list. Thus, in the next retry timeslot, the sender notifies the receiver using the same bad channel. If the notification is not received, the receiver device will naturally increment its packet loss rate to that channel. After the threshold of packet loss rate is reached, the receiver device chooses the next channel (immediate) in the channel list. On the other hand, if the receiver device receives the notification (in the next retry timeslot), the channel is changed and an acknowledge packet (ACK) is sent to the sender device. Note that if the number of retry times reaches *macMaxFrameRetries* threshold (range between 0 and 7), the sender discards the current packet and it transmits the next packet using the same bad channel.

Finally, the hopping pattern TH is used during intercluster communication, where the channels are changed per timeslot. This technique is used to combat interference and fading. The sequence used to change the channels is not defined in WIA-PA standard. Table 34.2 summarizes the frequency hopping mechanisms adopted in the WIA-PA network.

TABLE 34.2 Frequency Hopping Mechanism Adopted in the WIA-PA Network

Portion of the Superframe	MAC	Frequency Hopping
Beacon	TDMA	AFS
CAP	CSMA	AFS
CFP	TDMA	AFS
Intracluster	TDMA	AFH
Intercluster	TDMA	TH
Inactive period	—	—

34.4.2.2 Frames Priority and Scheduling Rules

The correct operation of the DLSL is necessary to define priority levels to the different types of frames. According to the WIA-PA standard, four priorities are defined: highest priority (commands), secondary priority (process data), third priority (normal frames), and lowest priority (alarms).

Packets where payload is related to diagnostics, configuration, control information, and emergent alarms must be marked as command (highest priority). On the other hand, any packet where payload contains a process data must be marked as a process data (secondary priority). Nonemergent alarm frames must be marked as alarm priority (lowest priority), whereas a frame that does not meet the previous criteria must be classified as normal priority.

In relation to scheduling rules, the WIA-PA standard suggests the following criteria:

i. Timeslots for beacon frames and active period should be allocated first.
ii. Resources for devices with the fastest update rate should be allocated with a secondary priority.
iii. In a multihop scenario, the packet with the earliest generating time should be allocated first.
iv. Highest-priority packet should be allocated first than a lower-priority packet.

34.4.2.3 Transmission of Long Cycle Data

The long cycle data are scenarios where a device has an update rate greater than the update rate of its cluster head (router device). In a WIA-PA network, the update rate of router devices is configured to the least update rate in cluster. For example, if a cluster has three field devices in which update rates are, respectively, 1s, 4s, and 8s, then the update rate and consequently the superframe duration of cluster head will be 1s. Thus, the field devices with update rates of 4s and 8s will be configured for a long cycle data in that cluster.

The target problem here is to define when a device on a long cycle data will transmit. The WIA-PA standard solves this problem by defining a parameter termed *TransmitFlag* as follows:

$$TransmitFlag = \frac{\text{ASN} - ActiveSlot + 1}{NumberSlots} \% SuperframeMultiple \tag{34.1}$$

where

absolute slot number (ASN) is the number of slots since the beginning of the network
ActiveSlot is the ASN at the moment when the superframe is initialized
NumberSlots is the superframe size
SuperframeMultiple is the ratio between the maximum and the minimum update rate in that superframe

At the beginning of each superframe cycle, beacons are received by the devices and these must decide if they send or not data in this superframe. If a device has a long cycle data, then it must transmit data only if

i. $0 < TransmitFlag < SuperframeMultiple$ and $TransmitFlag = LinkSuperframeNum$
ii. $TransmitFlag = 0$ and $LinkSuperframeNum = SuperframeMultiple$

The parameter *LinkSuperframeNum* is the ratio between the update ratio of a device and the superframe duration (equal to the minimum update rate).

For the sake of understanding, let us describe the example shown in Figure 34.8. We assume only a router device and a field device, in which update rates are, respectively, 4s and 8s (a timeslot has duration of 1s in the example). Thus, the field device has a long cycle data because its update rate is greater than the cluster head. The problem is to identify the slots where the field device can transmit. For this example, the parameters *SuperframeMultiple* and *LinkSuperframeNum* are equals (8/4 = 2). Testing the timeslots where ASN is 8 and 11 and using the Equation 34.1 and the conditions i and ii, then

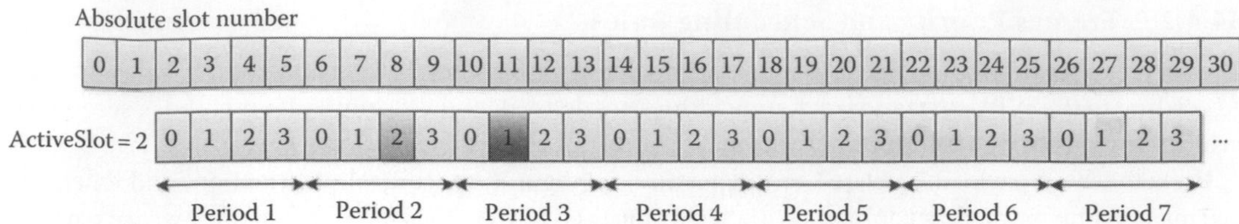

FIGURE 34.8 Transmission of long cycle data assuming an update rate of 8s and a minimum update rate of 4s.

- ASN = 8

  $TransmitFlag = \frac{8-2+1}{4}\%4 = 2$ and $TransmitFlag = LinkSuperframeNum$ (thus, the field device can transmit in this timeslot)
- ASN = 11

  $TransmitFlag = \frac{11-2+1}{4}\%4 = 3$ and $TransmitFlag \neq LinkSuperframeNum$ (thus, the field device cannot transmit in this timeslot)

It is possible to see that the field device can only transmit in any timeslot of an even period.

34.4.2.4 Management Service

The management services provided by the WIA-PA DLSL include network discovery, device joining and leaving, channel condition monitoring, and information about neighbors.

The network discovery service can be used by field devices to search beacons sent by cluster heads or gateways. The device scans the channels for *ScanDuration* until reaching a possible path to the network.

In relation to joining services, there are two moments where this service is required: when a field device joins the star network (through a routing) and when a router joins the mesh network (through the gateway). This service requires that the device has been configured previously with security keys. After the join process, the device receives the confirmation with the information about its address and if it is a time source. Similar procedure is required when a device leaves the network.

The DLSL is also responsible for providing channel condition to the DMAP. The reported data include LQI (in which the range varies between 0 and 255), packet loss rate per channel, and the count of retry per channel. Reports about neighbors are also provided, including information about the number of packets received, time of the last communication, and average level of signals received.

34.5 Network Layer

According to Figure 34.9, the WIA-PA NL provides an interface between the DLL and the application sublayer (ASL). The management of NL and the exchange of messages between the layers are provided through SAP. The implementation of an interface to the DMAP in the NL is also mandatory.

The WIA-PA NL is responsible for the following mechanisms: address assignment, routing, communication resource allocation, management of the life cycle of packets, management for device joining and leaving network, fragmentation/reassembly, and end-to-end network monitoring. All the mechanisms are configured by the NM.

34.5.1 Addressing

Regarding address assignment, each network device has a 64-bit address in the physical layer, whereas in the NL, the address has only 16 bits. These addresses are used in all communications (unicast, broadcast, intracluster, and intercluster).

During the join process, the devices use the 64-bit address to communicate with other devices. After a device has been properly joined to the network, the 16-bit address is used for communication. The 16-bit

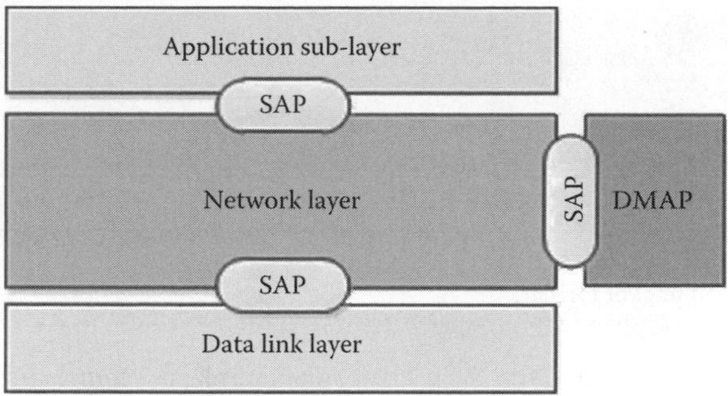

FIGURE 34.9 WIA-PA NL.

address is composed of two 8-bit address numbers, in which the template is X.Y. The most significant number (X) is the cluster address (0–255), whereas the least significant number (Y) is the intracluster address (0–255). The value 255 is used for broadcasting purpose and the routing device address (cluster head) is always 0. Thus, the WIA-PA standard classifies the 16-bit address as follows:

1. Unicast
   - Gateway address is 0.Y (0–254).
   - Router device address is X.0 (1–254).
   - Field device address is X.Y (1–254).
2. Broadcast
   - Intracluster broadcast address is X.255 (0–254).
   - Intercluster broadcast address is 255.0.
   - Global broadcast is 255.255.
   - GW broadcast is 0.255.

34.5.2 Routing, Fragmentation, and Reassembling

The adopted routing approach is based on the network redundancy perspective. The NM generates all routing tables after discovering the complete set of all neighbors from each network router. These tables are used to create a mesh topology where connections between routers and the gateway are made by redundant paths. Each device is configured to periodically transmit information about link failures, battery level, and the status of its neighbors to the gateway. Based on this information, the NM can update the routing table of each device. Thus, the adopted routing approach can guarantee a resilient performance.

Each device maintains its own routing table. This structure is composed of five fields: *RouteID*, *SourceAddress*, *DestinationAddress*, *NextHop*, and *RetryCounter*. The parameter *RouteID* is used to identify a path. The parameters *SourceAddress* and *DestinationAddress* identify the source and destination of path, whereas *NextHop* informs the next hop for the packet. Finally, the parameter *RetryCounter* is used to record the number of retries in the path. In order to ensure redundancy, it is possible to configure a *RouteID* with two *NextHop* values. In this case, the NM should indicate the main and secondary next hop.

Beyond the routing table structure, the information about routing is relayed to the device in the NL packet. As described in Figure 34.10, the routing field is used for this purpose.

Network layer header										Network layer payload
	Routing field									
Control field	Destination address	Source address	RouteID	Time-stamp	Priority	Sequence number	Number of fragment	Fragment sequence number	Payload length	Network layer payload
1 byte	2 bytes	2 bytes	2 bytes	4 bytes	1 byte	1 byte	0/1 byte	0/1 byte	1 byte	Variable length

FIGURE 34.10 Common NL packet format.

	Network layer payload							
		The first aggregated data			...	The nth aggregated data		
Network Layer Header	Aggregated number	Source address 1	Data length 1	Data 1	...	Source address n	Data length n	Data n
14/16 bytes	1 byte	2 bytes	1 byte	Variable	...	2 bytes	1 byte	Variable

FIGURE 34.11 Aggregated packet format.

Other features are also described in the NL packet. For example, *the control field* is used to indicate the type of packet (data, command, or aggregation) or whether the packet is a fragment packet. Fragmentation is used when the NL packet is longer than the DLL payload. In this case, when all fragments reach the receiver, they are reassembled at the NL. The fields *number of fragment* and *fragment sequence number* are used for this purpose. Additionally, the NL records the generation time of a data to limit the packet life cycle. This procedure is controlled by *time-stamp* field. Finally, the fields *priority* and *sequence number* indicate the priority (0–15) and the sequence number of the packet at the NL, respectively.

34.5.3 Aggregation and Disaggregation Services

The WIA-PA standard supports native aggregation services for data and packet. This feature is not present in the WirelessHART and ISA100 standards.

When a field device has many application objects (data) to transmit, it is possible to merge all data into a unique packet. This procedure is known as data aggregation. On the other hand, a router device can receive packets from many cluster members in a given period, being possible to merge all packets for a unique packet. This procedure is known as packet aggregation. Obviously, it is very probable that the merged packet suffers fragmentation because its payload is longer than the standard limit. Data aggregation occurs in the application layer (AL), whereas the packet aggregation occurs in the NL. Both the aggregation procedures try to reduce the number of packet forwarded in order to minimize energy consumption. The drawback of this mechanism is that the longer the packet, the greater is the probability to have interference.

The aggregation service is optional for the field device and the router device. The parameter *AGGSupportFlag* indicates the support for this service. Each device has a device structure table to maintain basic information configured by the NM, including the *AGGSupportFlag* parameter. Another fundamental parameter to the aggregation service is the *AggPeriod*. The NM configures the *AggPeriod* to the minimal update rate of the cluster (if the service is packet aggregation) or the minimal update rate of the field device application object (if the service is data aggregation). The idea is to start a timer right after the first packet or data in order to be received with the aggregation flag active. When the aggregation timer expires, the router device merges all packets previously received, as shown in Figure 34.11 (packet aggregation), or the field device merges all data into a unique packet (data aggregation). The aggregated data format is similar to the one described in Figure 34.10, but in this case, there is not the field *destination address*. The indication about the aggregation service is marked in the *control field* at the NL header.

The process of disaggregation is similar to the aggregation. After receiving a packet, the gateway evaluates whether the HL header indicates an aggregated packet. In an affirmative case, the network payload is disaggregated and right after it is forwarded to the upper layer. If the AL identifies a data aggregation, the payload is fully disaggregated. Nonaggregate packet or data are handled transparently.

34.6 Application Layer

According to Figure 34.12, the WIA-PA AL provides an interface to user application objects (UAOs) and industrial process. It has two sublayers: the user application process (UAP) and the ASL. The former one implements the interaction between the distributed applications of industrial processes (collecting

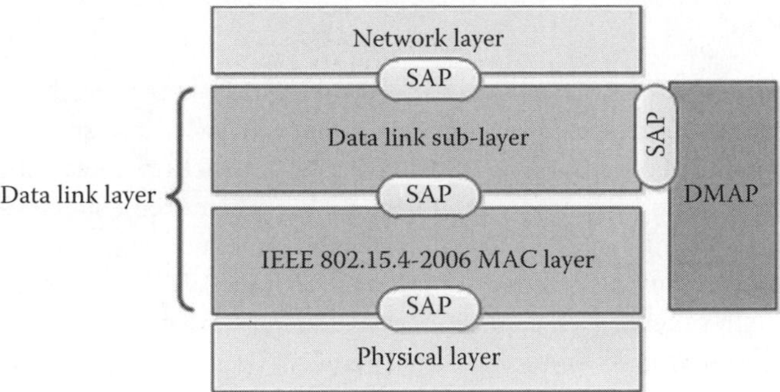

FIGURE 34.12 WIA-PA AL.

process data, data conversion, linearization, generating alarms, etc.), whereas the latter one provides transparent services of data transmission (end-to-end data services, client/server [C/S], publisher/subscriber [P/S], and report source/sink [R/S]). The SAP manages the communication between the AL and the NL. The implementation of a communication interface (SAP) to the DMAP in the AL is also mandatory.

34.6.1 User Application Process

In technical terms, the UAP can be described as an application process defined by network users. According to the WIA-PA standards, the UAP may be defined by either the IEC 61499 or the IEC 61804 standards. Both of them define function block structures that are independent of implementation.

In a typical UAP, several UAOs coexist in the same device. Each UAO may process data of different types of sensors such as temperature, humidity, pressure, level, flow, vacuum, and combustion. The main functions supported by a UAO are the following: range scaling, linearization, compensation (drift), filtering, engineering unit conversion, and time stamp.

34.6.2 Application Sublayer

The ASL provides a transparent interface to communicate with distributed applications (UAO). The communication between two UAOs is defined by a virtual communication relationship (VCR). The ASL defines three VCR application modes:

1. P/S: used for periodic data
2. R/S: used for aperiodic events
3. C/S: used for aperiodic and dynamic unicast messages

In the P/S mode, the publisher (e.g., a field device) sends data periodically according to its update rate. The NM and the cluster head are responsible for ensuring enough resource to the publisher. On the other hand, the subscriber (gateway or a router device) has a prior knowledge about the publisher address and occasionally captures new data. Alternatively, in the R/S mode, a report source (field device or a router device) sends aperiodic alarms or events to the sink (gateway). It is mandatory to send an ACK to the report source. Finally, in the C/S mode, the client (field device, routing, and gateway) sends read or write requests to the server (field device, router, and gateway). The latter one should execute the request and send to the client the respective confirmation. The C/S mode is used for end-to-end retransmission (AL), unicast messages in the intercluster communication (NL), and during the CAP (DL).

34.7 Security

One of the most important advantages in being an open standard is the flexibility to provide interoperability between several manufactures. However, this can be a drawback to security, since an open network standard has potential security risks (invasions, data destruction, replay attacks, etc.). Thus, the implementation of techniques to guarantee the dependability requirements (system availability, data integrity and authenticity, confidentiality, key management) is mandatory.

Additionally, being an industrial communication standard, the WIA-PA must guarantee the support for critical application. In general, such applications have stringent dependability requirements, as a system security failure may result in economic losses, put people in danger, or lead to environmental damages.

Given the context described previously, the WIA-PA implements network security in several levels. The data integrity is provided in the AL and DLSL. MICs are adopted to verify whether the information was compromised. The MIC types are specified in the security header of the protocol data unit (PDU) of AL and DLSL as described in Figure 34.13.

The data confidentiality is also guaranteed in the AL and DLSL. Symmetric keys are used to encrypt and decrypt the payload of packets. MIC and security keys are also adopted to guarantee device authentication, for example, during the join process. The security services in the MAC are provided by the IEEE 802.15.4-2006.

Coordinated by the network security, the management and the maintenance of key are provided in the DMAP module. Each device has an instance of DMAP. According to the WIA-PA standard, the following cryptographic keys are supported:

1. Join key (JK)
   - During the provisioning process, the handheld device configures a temporary key for a device. This key is generated based on the 64-bit-long address of device and it is mandatory for the join process. Eventually, the JK is also necessary to encrypt the key encryption key (KEK) during its first distribution.
2. KEK
   - After the join process, the SM provides for the incoming device a special key to encrypt the transmission of other keys. At the first time, the SM transmits the KEK encrypted by the JK. However, the next transmissions encrypt the new KEK using the old KEK.
3. Data encryption key (DEK)
   - DEK is also distributed (encrypted using the KEK) by the network security after the join process. It is used by the AL and DLSL to protect data during the transmission of packets.

34.7.1 Secure Joining Process

The first step to guarantee the dependability features of a network is maintaining the secure join process. Thus, it is mandatory that the incoming device is authenticated. This procedure is conducted by the SM as described in Figure 34.14.

DLPDU			
Frame control	Security header	Payload	MIC
1 byte	1 byte	Variable	0/4/8/16 bytes

APDU			
ASL Header	Security header	Payload	MIC
5 bytes	1 byte	Variable	0/4 bytes

FIGURE 34.13 Security structure for DLL and ALs.

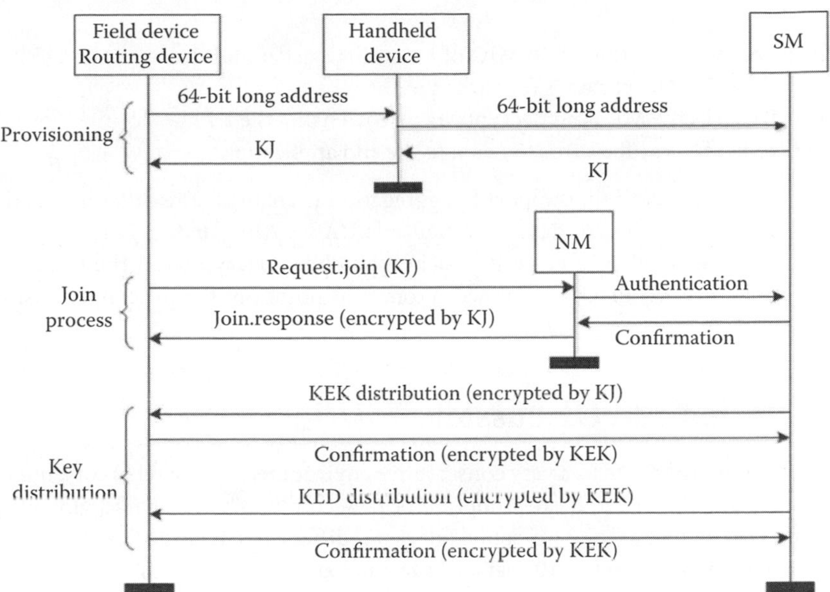

FIGURE 34.14 The secure joining process and the exchange of security keys.

Before joining the network, the incoming device should generate the KJ. The handheld device is responsible for communicating with the SM requesting the generation of the KJ. The parameter of generating function is the 64-bit-long address of incoming device. The join process starts when the incoming device sends the request join (with the JK attached) to the NM. The request is forwarded to the SM where the incoming device is authenticated. A confirmation (encrypted with the JK) is sent to the incoming device acknowledging the secure join.

Right after the join procedure, the SM sends the other security keys to the device. The KEK is the first key to be distributed. Note that the KEK distribution is encrypted with the KJ. Finally, the KED is also sent by the SM to the device; however, the distribution is encrypted with the KEK.

34.7.2 Secure Transportation

For the sake of understanding, in this section, the secure procedure used to transfer data from a field device to the end application is described. The procedure depends on the availability of aggregation service. If the field device and router device do not support the aggregation service, the secure procedure to transfer data is as follows:

- Field device
    - The ASL payload is encrypted using the KED, and after that, it generates the AL MIC. The encrypted material is forwarded to the NL.
    - The DLSL encrypts its payload using the KED, and after that, it gets the DLSL MIC.
    - The packet is sent to the router device.
- Router device
    - The router device receives the packet and checks the integrity with the MIC. Then, the packet is decrypted in the DLSL (using the KED).
    - The next hop is verified in the NL.
    - The router device uses the next-hop DLSL KED to encrypt the DLSL payload, and after that, it generates the MIC.
    - The packet is sent to the gateway.

- Gateway
  - The gateway checks the DLSL MIC of incoming packet and decrypts the DLSL payload using the KED. The packet is forwarded to the NL.
  - The AL checks the MIC and decrypts its payload using the KED.
  - Finally, the gateway forwards the data to the end application.

The procedure is very similar if the support to aggregation is enabled. The difference is that the field device must encrypt the payload of data from multiple UAOs. After that, it generates a unique payload in AL. In the router device, after all the packets had been decrypted in the DLSL, all payloads are aggregated and a new NLPDU is obtained. From that point on, the procedure is similar to the previous one.

34.8 Conclusion and Discussion

Industrial applications are known as a very conservative environment where reliability, interoperability, and security issues are always required. Adoption of new communication paradigms is viewed with great skepticism by plant managers. This context does not differ when considering the emerging industrial wireless technologies. In the last 10 years, this technology has become sufficiently mature to allow its adoption in industrial environments. The main improvements are related to redundant routing protocols, real-time scheduling, and efficient frequency hopping mechanisms. These technologies enable the development of emerging industrial applications where the use of legacy wired technologies is difficult or impracticable, for example, in tank farms, pipelines, distillation columns, wellheads, and rotation equipments (turbines and kilns).

In this chapter, the WIA-PA standard, an emerging Chinese standard for industrial wireless communication, was described in detail. The features described in the chapter can be used to aid the design of applications to develop more resilient networks.

Considering the protocol stack, the WIA-PA supports a physical layer based on the IEEE 802.15.4. However, implementation of the upper layers follows independent approaches. The WIA-PA standard improved the MAC protocol implemented in the IEEE 802.15.4, highlighting the TDMA and frequency hopping approaches with a flexible energy-saving operation. Regarding the routing issues, the WIA-PA network adopts a mesh-star routing approach where the field devices are organized in a cluster. Additionally, there is support to data and packet aggregation. The former one is used by the field devices to aggregate data from multiple processes, whereas the latter one is used by router devices to aggregate multiple packets. Both procedures are adopted to increase the lifetime of devices. Furthermore, the AL provides an interface between the network and the industrial process based on standard application modes (P/S, R/S, and C/S). Finally, security is guaranteed from multiple levels (system availability, data integrity, authenticity, confidentiality, and key management).

From the commercial viewpoint, a wireless battle to establish a communication standard is on the horizon. Currently, various industrial wireless standards coexist. The WirelessHART standard has dominated the marketplace because it was the first standard approved by IEC. However, the ISA100.11a standard has more flexible features than the WirelessHART ones, mainly those related to the tunneling of legacy applications, segregation of network (subnets), and more frequency hopping profiles. On the other hand, in the Chinese market, it is expected that the WIA-PA networks will dominate despite of unannounced Chinese suppliers. To increment the industrial wireless technology competition, the task group IEEE 802.15 (TG4e) is developing a new wireless standard specific to the industrial environment named IEEE 802.15.4e. Another challenge is related to the coexistence of different wireless technologies in the same area. Since the transmission medium is open, there may be situations where different technologies share the same frequency range. Thus, it is important that even when they coexist in the same environment, different technologies can operate without interferences.

References

Angrisani, L., Bertocco, M., Fortin, D., and Sona, A. (2008). Experimental study of coexistence issues between IEEE 802.11b and IEEE 802.15.4 Wireless networks. *IEEE Transactions on Instrumentation and Measurement*, *57*, 1514–1523.

Decotignie, J.-d. (2009). The many faces of industrial ethernet [past and present]. *IEEE Industrial Electronics Magazine*, *3*(1), 8–19.

IEC-62591. (2010). IEC 62591: Industrial communication networks—Wireless communication network and communications profiles—WirelessHART. International Electrotechnical Commission (IEC).

IEC-62601. (2011). IEC/PAS 62601: Industrial communication networks—Fieldbus specifications—WIA-PA communication network and communication profile. International Electrotechnical Commission.

IEC-62734. (2012). IEC 62734: Industrial communication networks—Fieldbus specifications—Wireless systems for industrial automation: Process control and related applications (based on ISA 100.11a). International Electrotechnical Commission.

IEEE-802.15.4. (2006). Part 15.4: Wireless Medium Access Control (MAC) and Physical Layer (PHY) Specifications for Low-Rate Wireless Personal Area Networks (WPANs). Revision of IEEE Std 802.15.4-2003.

Liang, W., Liu, S., Yang, Y., and Li, S. (2013). Research of adaptive frequency hopping technology in WIA-PA industrial wireless network. *Advances in Wireless Sensor Networks*, (pp. 34–37). Berlin, Heidelberg: Springer.

Liang, W., Zhang, X., Xiao, Y., Wang, F., Zeng, P., and Yu, H. (2011). Survey and experiments of WIA-PA specification of industrial wireless network. *Wireless Communications and Mobile Computing*, *11*(8), 1197–1212.

Petersen, S. and Carlsen, S. (2011). WirelessHART Versus ISA100.11a: The format war hits the factory floor. *IEEE Industrial Electronics Magazine*, *5*(4), 23–34.

Sauter, T. (2010). The three generations of field-level networks—Evolution and compatibility issues. *IEEE Transactions on Industrial Electronics*, *57*, 3585–3595.

Silva, I., Portugal, P., and Guedes, L. (2013). Emerging technologies for industrial wireless sensor networks. In: M. Khalgui, O. Mosbahi, and A. Valentini, *Embedded Computing Systems—Applications, Optimization, and Advanced Design* (pp. 343–359). Hershey, PA: IGI Global.

Willig, A. (2008). Recent and emerging topics in wireless industrial communications: A selection. *IEEE Transaction on Industrial Informatics*, *4*(2), 102–124.

Young, S. S., Seong, H. P., Choi, S., and Hyun, W. K. (2007). Packet error rate analysis of ZigBee under WLAN and bluetooth interferences. *IEEE Transactions on Wireless Communications*, *6*, 2825–2830.

35

Wireless Extensions of Real-Time Industrial Networks

Gianluca Cena
National Research Council of Italy

Adriano Valenzano
National Research Council of Italy

Stefano Vitturi
National Research Council of Italy

35.1 Introduction 35-1
35.2 Implementation of Hybrid Networks 35-2
Interconnections at the Physical Layer • Interconnections at the Data-Link Layer • Interconnections at Higher Protocol Layers • Wireless Extensions of Industrial Networks
35.3 IEEE 802.11-Based Extensions 35-7
IEEE 802.11 Extensions of Fieldbuses • IEEE 802.11 Extensions of Real-Time Ethernet Networks
35.4 IEEE 802.15.4-Based Extensions 35-8
IEEE 802.15.4 Extensions of Fieldbuses • IEEE 802.15.4 Extensions of Real-Time Ethernet Networks
35.5 Conclusions 35-9
References 35-10

35.1 Introduction

In the last years, wireless networks have become an interesting opportunity for industrial communication systems [1,2] thanks to the indubitable benefits they can bring. Indeed, introducing wireless communications in industrial environments is particularly appealing since, in principle, it increases system flexibility by reducing setup times and costs. Also, it avoids (or, at least, limits) cabling, which turns out to be cumbersome and/or expensive in many real cases. To this regard, it may be observed, for instance, that reliability as well as availability problems could arise with cables used to connect moving parts, as a consequence of mechanical stress and/or attrition.

Several studies have been carried out to assess the suitability of wireless networks for industrial applications since, as it is well known, such networks may be characterized by problems like fading, multipath propagation, shadowing, and interference, which could have a negative impact on the performance of industrial communication systems often characterized by severe requirements in terms of both timeliness and reliability [3].

The obtained results are interesting, since it has been proved that some very popular, general-purpose wireless solutions can be profitably adopted in several industrial applications, too, including real-time communications at the shop-floor. Also, recently, wireless networks explicitly conceived for industrial communications have been designed, and the number of their deployments is rapidly increasing. It is the case, for example, of both WirelessHART [4] and ISA100.11a [5].

Thus, it may be eventually stated that wireless networks are now definitely available for industrial applications along with other communication systems like both the well-settled fieldbuses [6] and real-time Ethernet (RTE) [7] networks. These three systems, which may be used either in a stand-alone mode or in combination with each other, allow to satisfy the most demanding requirements that may be encountered in factory automation.

As a point on which most experts agree, however, wired solutions will be still massively adopted in the future (to this regard, it is worth observing that the number of RTE installations is progressively increasing [8]). In other words, it is commonly thought that wireless communications are unlikely to replace the wired solutions currently adopted in industrial scenarios, at least in the midterm. Furthermore, in several practical applications, this is not considered as an actual requirement. Conversely, what is often needed in real-world factory automation systems is the ability to connect a few components to an already deployed wired communication system that cannot be reached (easily and/or reliably) with a cable. The most common example is a sensor mounted on a moving/rotating axis that has to exchange data with a controller attached to a wired segment.

This kind of problems may be tackled by providing wireless extensions to the existing wired systems. The resulting configurations are *hybrid* networks in which, typically, wireless segments have limited geographic extension (some tens of meters) and connect only few stations. In such a kind of networks, controller devices (e.g., programmable logic controllers, computer numerical controls, or industrial personal computers) are usually located on the wired segment. This means that wireless networks will have to coexist with more traditional wired communication systems for quite a long time. Consequently, the integration of wireless and wired communications in industrial scenarios is a crucial issue, which has to be dealt with carefully in order to achieve both timely performance and reliability.

This chapter focuses on some of the most relevant aspects concerning the aforementioned hybrid networks. After a general analysis of hybrid wired/wireless configurations and the related issues, the ways in which wireless extensions can be effectively implemented for both fieldbuses and RTE networks are described.

The wireless networks considered as candidate technologies for the extensions are both the IEEE 802.11 *Wireless Local Area Network* (WLAN) [9] and the IEEE 802.15.4 *Low-Rate Wireless Personal Area Network* (LR-WPAN) [10]. WLANs have already been considered in several studies and tested in practical applications that demonstrated their suitability for the use in industrial environment [11–13]. For such a reason, most examples provided in the following are based on that solution. On the other hand, LR-WPAN is a standard for short-range connectivity, mainly conceived for *wireless sensor networks*, that is characterized by long battery lifetime and low cost.

Bluetooth [14] is a further possible candidate for implementing wireless extensions. Such a kind of network, which was originally designed as a mere cable replacement system for mobile equipment, has progressively gained popularity thanks to its features and is now employed in several areas, including industrial automation. However, in spite of its interesting features, Bluetooth is not further considered in this chapter due to page room limits.

35.2 Implementation of Hybrid Networks

From a general point of view, different (and often dissimilar) communication networks can be interconnected through specific devices called *intermediate systems* (ISs) [15]. These devices have different structure, functionality, and complexity, according to the layer of the Open Systems Interconnection (OSI) reference model they refer to. In the following, some practical implementations of ISs are described, along with their features and related issues.

35.2.1 Interconnections at the Physical Layer

The simplest form of ISs are *repeaters*. They operate at the physical layer and are used to interconnect subnetworks that share the same medium access control (MAC) protocol. Repeaters are usually adopted for regenerating physical signals (electrical, optical, etc.) flowing across different network segments.

In Ethernet networks, for example, they operate according to the so-called *3-R principle*: reshaping, retiming, and retransmitting. In this way, it is possible to face the problem of signal attenuation over the cable when the network extension grows larger. Moreover, they enable the number of nodes that can be networked—which is practically limited by the current drained by devices attached to the bus—to be increased and achieve segmented network topologies in addition to the plain linear bus.

Besides simpler two-ported devices, a special kind of repeater exists (called *hub*) that is provided with several ports and can be adopted to set up networks based on star topologies. Because of the increased reliability (a defective cable no longer affects the whole network operation, hence avoiding partitioning faults), hubs are particularly suitable for deploying network infrastructures and for the cabling of buildings. More advanced versions of repeaters may sometimes be employed to interconnect subnetworks that rely on different media (e.g., copper wires and fiber optics) or signaling techniques (e.g., voltage/current levels).

When network segments are connected through repeaters, the MAC protocol that regulates the access to the shared transmission medium is exactly the same on the whole network. Thus, care has to be taken about the increased propagation delays. In many cases, such as, for example, in half-duplex Ethernet, CAN, or Profibus networks, this leads to limitations on the maximum number of repeaters that can be inserted between any two nodes or the maximum cable length for every segment.

Pure repeaters are quite unusual in hybrid wired/wireless networks, since the resulting physical communication support in most practical cases is not seen as a sufficiently uniform medium by the MAC protocol. This is because the signaling schemes and physical transmission techniques adopted by wireless networks are, typically, rather different from wired solutions, which means some form of nontrivial buffering has to be carried out by ISs.

An interesting example of hybrid wired/wireless interconnection carried out at the lowest layer of the protocol stack (i.e., below the MAC) is provided in [16]. In that case, the authors refer to a hybrid architecture between Profibus [17] and Radio Fieldbus [11] (the latter was one of the firstly designed industrial wireless networks that, however, was never implemented, with the exception of some prototypes). The described hybrid network relies on cut-through forwarding devices located between wired segments and the wireless medium. Both networks use the same data-link layer protocol—specifically, the fieldbus data link as defined by Profibus, including its access scheme based on token-passing—but rely on completely different physical layers. Indeed, signal transmission in Profibus is based on the well-known RS 485 standard, whereas Radio Fieldbus is used to adopt a direct sequence spread spectrum technique similar to the one adopted in IEEE 802.11 WLANs.

35.2.2 Interconnections at the Data-Link Layer

An IS operating at the data-link layer is called *bridge*. Its main purpose is transferring frames between systems that are not (or cannot be) treated as a uniform communication support by protocols at the MAC level (e.g., switched LANs that are made up of several separate collision domains). Such devices usually operate according to the store-and-forward principle, that is, when a frame is completely received on one port of the bridge, it is forwarded (possibly, according to a selective strategy based on adaptive message filtering) to the other subnetwork(s) through the related port(s). As an alternative, forwarding may start as soon as a sufficient amount of information is received about the incoming frame (e.g., cut-through bridges). Even though featuring shorter transmission latencies, in this latter case, erroneous frames are not blocked by the bridge, hence reducing the net available bandwidth.

It is worth noting that, thanks to this decoupling mechanism, physical signaling, transmission speeds, medium access techniques, and even the frame formats might differ in the interconnected networks. A bridge, which is equipped with more than two ports, is commonly referred to as a *switch* (this kind of device is currently present in almost every real-life Ethernet network).

Even though the MAC mechanism is not required to be the same for the subnetworks interconnected through bridges/switches, the sets of communication services provided at the data-link layer should be

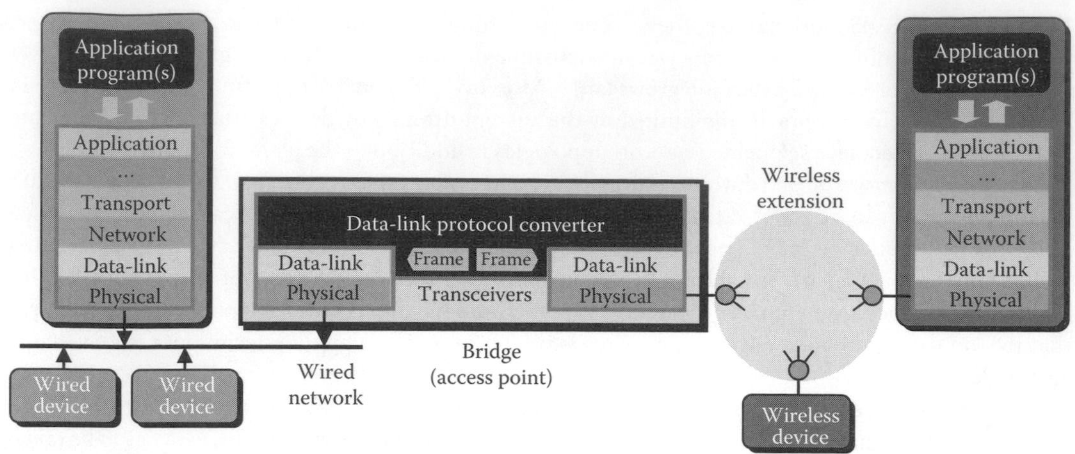

FIGURE 35.1 Interconnection of networks via a bridge.

at least similar. Limitations on the kinds of networks that can be interconnected concern, for example, their addressing scheme (which has to be uniform on the whole network) and the *maximum transfer unit* (MTU) that affects the allowed payload size (this is because fragmentation is not permitted at the data-link level). When subnetworks with different MTUs are interconnected, care has to be taken so that the payload of the exchanged frames never exceeds a threshold equal to the smallest among the supported MTUs.

Figure 35.1 shows an example of interconnection taking place through a bridge for the hybrid networks considered in this chapter. From a practical point of view, the bridge used in hybrid wired/wireless networks is a device equipped with two (or more) transmitting/receiving interfaces (one for each subnetwork) and, optionally, a protocol converter. A frame originated from whatever segment (either the wired network or the wireless extension) is propagated to the other one by the bridge, which receives the frame, converts it into the suitable format, and then retransmits it.

When real-life devices and solutions are considered, *access points* (APs) are a very popular example of bridge. They are used to set up hybrid networks consisting of one (or more) wired IEEE 802.3 segments and an IEEE 802.11 wireless extension. Data-link layer services, in this case, are exactly the same for the two kinds of networks (i.e., those foreseen for the IEEE 802 family of protocols) as well as the 6-byte addressing scheme. On the contrary, frame formats and MTUs differ.

For the sake of truth, it is worth remembering that APs actually have a twofold role: they are used both to enable communication in an infrastructure WiFi network (also known as basic service set or BSS), by relaying frames among wireless stations, and to interconnect the BSS to an existing Ethernet backbone (portal function).

35.2.3 Interconnections at Higher Protocol Layers

For ISs working at the network layer or above, the generic term *gateway* is often adopted. A gateway is responsible for transferring *protocol data units* as well as for converting generic streams of information and application services between application processes executing in the nodes of the interconnected systems, by performing all required protocol translations. From a practical point of view, there is no particular restriction on the kinds of networks that can be interconnected through gateways. On the other hand, gateways are usually complex, expensive devices, often developed ad hoc for the considered systems by means of suitable software modules, and the performance they achieve is to several extents lower than bridges. Generally speaking, noticeably higher latencies have to be expected in this case, which could result in incompatibility with timings requirements of industrial applications.

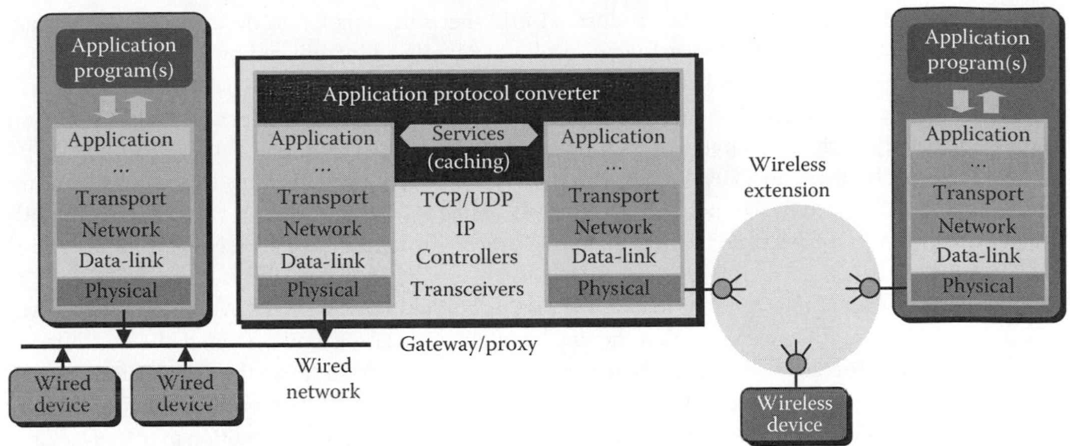

FIGURE 35.2 Interconnection of networks via a gateway.

ISs that operate specifically at the network layer are known as *routers*. At this level, a uniform addressing scheme and protocol are used (i.e., the *Internet protocol* [IP]), which ensure worldwide connectivity irrespective of the physical media, MAC mechanisms, and data-link services of the interconnected subnetworks. This chapter is not concerned with routers, since most industrial networks are deployed as local networks. However, when timing constraints are not very tight, interconnection techniques that rely on routers allow devices, which are natively enabled to communicate over local intranets and/or the Internet, to be directly embedded in the automation system (this is the case of remote monitoring and maintenance tasks).

When the interconnection takes place at the application layer, the IS should take care of the conversion of high-level services. Usually, this implies the gateway has to carry out nontrivial operations for gathering and converting information, which, in turn, often require a number of simpler communication services to be invoked on the interconnected networks. In these cases, the term *proxy* is sometimes adopted. Indeed, a proxy is much more than an application-layer gateway: it usually makes some parts of the network (e.g., a whole subnetwork) look as if they were a single node, possibly hiding the underlying structure for security or performance reasons. Sometimes, some form of data caching is provided as well in order to save bandwidth. For instance, proxies are used in fieldbus/IP configurations to connect legacy fieldbus segments. In this case, they integrate devices connected to the underlying fieldbus into the overlying Ethernet system, maintaining in such a way several advantages of fieldbus networks, like high responsiveness, powerful diagnostic functions, and automatic configuration. At the same time, planning is simplified (because workflows are well known) as well as commissioning and operation (thanks to the diagnostic capabilities offered by the fieldbus).

An example of wireless extension implemented at the application layer is shown in Figure 35.2. The gateway in the figure includes two interface boards that are fully compliant with the overall protocol stack of the relevant subnetworks and a protocol converter, usually implemented as a software module.

35.2.4 Wireless Extensions of Industrial Networks

Although technically feasible, wireless extensions of industrial networks to be used at the shop-floor (either fieldbuses or RTE networks) are not so straightforward. This is basically due to three main reasons: first, the transmission support (wireless medium) is shared among all nodes (possibly including those associated to nearby independent wireless subnetworks). Even operating at the maximum allowable speed, this may result in a low per-station net throughput—compared to that achieved in many wired networks—when the number of connected devices grows higher (this is particularly true when either switched Ethernet or industrial Ethernet solutions based on a combined message are taken into account). Furthermore, the nonnegligible overhead introduced by the communication protocol (larger protocol control fields, acknowledgment and reservation mechanisms, no provision for full-duplex operations, and so on) has to be considered as well. For example, the minimum time needed to reliably exchange 8 bytes of user data over a WLAN—that is, bit rate equal to 54 Mb/s, no RTS/CTS mechanism, no TCP/IP encapsulation, ad hoc network mode, but including interframe spaces as they effectively waste network bandwidth—is about 100 μs. Such a value is only slightly shorter than the worst-case time taken to send the same information over a CAN network operating at 1 Mb/s (i.e., 135 μs).

The second reason is that random medium access techniques (e.g., CSMA/CA) are often employed by wireless networks. This means that, on the one hand, unpredictable—and unbounded—transmission delays might occur (because of the occurrence of repeated collisions) while, on the other hand, network congestions could be experienced, and this is surely the worst aspect. In other words, when the network load generated by devices exceeds a given threshold (even for a limited time), a condition may likely arise—because of the positive feedback due to the increased collision rate—so that the effective net throughput decreases as the load increases. This means that the network may become temporarily unavailable to carry out data exchanges timely, which might not be compatible with proper operation of distributed automation systems. The number of lost messages (discarded by the sender because the retry limit is exceeded) may even grow up to a point that it is no longer acceptable for the control application. Moreover, even in the case of lightly loaded networks, nonnegligible jitter may affect data exchanges and hence the accuracy of the system, unless some form of prioritization scheme is adopted for frame transmission.

Third and last, wireless channels are much more error-prone than wired cabling [18], and this is a serious drawback for industrial communications. Indeed, typical aspects of production environments like the movement of people and/or plant components, along with the interference that could arise from other systems, may increase the bit error rate (BER) and, consequently, compromise the achievement of reliable communications. Besides causing higher latencies and jitter, transmission errors directly affect the network reliability and, hence, the robustness of the overall system. They may also cause consistency problems—for instance, when a multicast message is received only by a part of the addressed devices.

As a consequence, there is a nonnegligible chance that messages sent over the wireless medium never reach the intended destination, even in nonnoisy environments.

Each one of the three drawbacks mentioned earlier can be tackled (at least partially) by means of already existing or soon available technologies. In particular, the throughput problem could be somehow lessened with the adoption of IEEE 802.11n, which exploits the multiple-input multiple-output technology and is theoretically able to provide a big leap (about one order of magnitude) in the network throughput. Indeed, as process data are usually small sized, the same level of improvement cannot be reasonably expected in industrial scenarios, where simple field devices are interconnected. A more viable choice is relying on several (smaller) separate wireless subnetworks (e.g., WLANs), possibly operating on different channels, interconnected by means of a wired backbone in order to limit the packet rate on each one of them.

The second problem (determinism) can be tackled through the adoption of enhanced medium access techniques, such as PCF in the case of IEEE 802.11. Unfortunately, PCF is only seldom implemented in commercially available devices. Variants of PCF, such as the iPCF mechanism introduced by Siemens [19], have the drawback of being proprietary solutions, and hence, compatibility and interoperability with standard equipment can hardly be ensured. Alternatively, the prioritization features offered by both the IEEE 802.11e and IEEE 802.11n standards (already supported by many WLAN adapters currently available off-the-shelf) might be exploited [20]. While not guaranteeing strict determinism, prioritization (mainly developed for multimedia applications) is able to provide tangible improvements, such as, for example, guaranteeing shorter transmission times *on the average* for critical frames, which might be often enough for several industrial automation systems. For example, it is possible to ensure quasi-real-time behavior by assigning higher priorities (i.e., voice and video access classes) to those messages characterized by tight timing constraints.

TDMA techniques may be useful to solve the determinism problem. In such a case, for example, the guaranteed time slots provided by IEEE 802.15.4 represent an interesting opportunity. Also, it is worth observing that both Wireless HART and ISA 100.11a use a combination of TDMA and pseudorandom frequency hopping [21]. Other solutions, as described in literature [22], suggest the adoption of synchronization techniques on WLANs, so as to reduce the likelihood of a collision for data exchanges that take place according to predictable (e.g., periodic) patterns.

The third problem (robustness) can be faced in several ways. As a first measure, the introduction of both spatial and frequency diversity techniques is recommended. Whenever feasible, a proper selection and placement of antennas is envisaged. The use of both APs and adapters with removable antennas is often a good idea, as this permits boosting the strength of signals by replacing the built-in omnidirectional antenna with a high-gain or directional one. Moreover, wireless range extenders can be used to deal with the problem of dead spots (areas with no or weak reception)—even though setting up such devices in a proper way is not always straightforward.

In the case this is not enough, leaky wave antennas [23] can be used to ensure reliable communication over radio links. The drawback, in this case, is the need to deploy the leaky cable infrastructure, which makes it a nonuniversal solution. Usually, such an arrangement is adopted when the equipment includes some parts moving on fixed paths, so as to avoid cable wear and strain. As an alternative, if true mobility is required, devices operating in the 5 GHz band (IEEE 802.11a) can be adopted as a satisfactory choice [21]. In this way, in fact, interferences with other devices operating in the standard (and already jammed) 2.4 GHz ISM band are avoided, hence achieving a better signal-to-noise ratio and, consequently, a reduced error rate on the radio link.

35.3 IEEE 802.11-Based Extensions

35.3.1 IEEE 802.11 Extensions of Fieldbuses

Typically, the transmission protocols adopted by fieldbuses are noticeably different from those employed in wireless networks, at all levels of the communication stack. For example, random access techniques typical of WLANs usually are not employed in any fieldbus network (they are sometimes used in building automation networks, which are characterized by relaxed timing constraints). Other differences are concerned with either the payload size of frames or the address space, both of which are typically quite small in real-world fieldbuses (few tens/hundreds of bytes/nodes, respectively), whereas they are considerable for IEEE 802.11 networks.

As a consequence, these differences discourage the interconnection of fieldbuses and IEEE 802.11 networks at the data-link layer. Thus, wireless extensions of fieldbuses have to be implemented mainly at the application layer, that is, through gateways. However, some remarkable exceptions exist: DeviceNet, for instance, ensures interoperability with all networks based on common industrial protocol (CIP), thanks to its native routing techniques that allow, in principle, the implementation of extensions operating below the application layer (i.e., through direct routing of CIP messages).

35.3.2 IEEE 802.11 Extensions of Real-Time Ethernet Networks

Despite the wide availability of both hardware components and protocols, the implementation of wireless IEEE 802.11-based extensions to RTE networks is not as straightforward as one could think. Indeed, interoperability between Ethernet and WLANs is always possible by means of inexpensive off-the-shelf devices (i.e., APs), along with some commonly available protocols like the well-known IEEE 802.1D bridging protocol [24], which allows for interconnections at the data-link layer. Despite the good efficiency, implementing this kind of solutions in industrial environments might be sometimes not so easy, due to access protocols adopted by RTE networks, which are often not entirely compliant with the original Ethernet specifications. Moreover, the unavoidable uncertainty affecting the IEEE 802.11 communication caused by the random backoff procedure, the possibly high BERs, interference sources, and electromagnetic noise, may seriously impair the determinism of the RTE networks.

An example of IEEE 802.11-based wireless extension is provided in [25] for Ethernet Powerlink [26], a very popular RTE network that adopts an access protocol based on a fast and precise periodic polling carried out on the (wired) nodes; this ensures fast operation (cycle times under 200 μs) and ultra-low jitter (below 1 μs). Results presented in the paper actually confirm the impossibility of maintaining the aforementioned tight timing performance when wireless nodes are introduced. Nonetheless, the adoption of prioritized frames made possible by IEEE 802.11e revealed to be promising, since it allows for a consistent reduction of jitter, especially when a limited number of wireless stations are employed.

A more straightforward interconnection technique, in general, can be obtained through the communication services offered by the TCP/IP protocol suite. Unfortunately, given the complexity of TCP, this solution is not suitable for achieving real-time behaviors. For such a reason, in several cases, the UDP protocol [27] is preferred. UDP, in fact, has lower overheads than TCP and provides simpler communication services, thanks to its unacknowledged transmission scheme (which lacks the sliding window mechanism of TCP). This implies that latencies and jitter become shorter than in TCP, even though the overall behavior cannot be considered as strictly deterministic, yet.

35.4 IEEE 802.15.4-Based Extensions

The IEEE 802.15.4 specification, as all networks defined by the IEEE 802 committee, is only concerned with the two lowest layers of the communication stack and does not specify any type of upper protocol stack. Thus, the access to IEEE 802.15.4 networks by higher layers may be achieved in two different alternative ways, as recommended by the standard, that is,

1. Via logical link control (LLC) Type 1 services [28]
2. Directly, by means of the MAC services

Both techniques are effective, even if it is worth mentioning that IEEE 802.15.4 board manufacturers rarely include LLC services in their products.

Unfortunately, IEEE 802.15.4 has a low transmission speed, and hence, wireless extensions are not able to provide fast communications in this case. Nonetheless, the high efficiency of the protocol makes it particularly appealing for applications where a limited amount of data have to be transmitted. Moreover, the guaranteed time slots available in the beacon-enabled mode allow for deterministic access to the shared medium by wireless nodes, with a consequent reduction of the uncertainty typical of CSMA/CA-based protocols.

35.4.1 IEEE 802.15.4 Extensions of Fieldbuses

The wireless extension of fieldbuses by means of IEEE 802.15.4 suffers from the same problems already discussed for IEEE 802.11. Specifically, differences between the MAC protocols of the two kinds of networks discourage interconnections at the lower layers. Hence, wireless fieldbus extensions based on IEEE 802.15.4 networks can be achieved only through a gateway that maps services foreseen by the fieldbus application layer onto calls to IEEE 802.15.4 services. The gateway may be typically implemented as a station that embeds the interface to the fieldbus and, at the same time, acts as a coordinator for the IEEE 802.15.4 network.

Also in this case, the implementation of the gateway does not impose any restriction on the protocol(s) actually employed in the wireless extension. In particular, the gateway may rely on functions provided by a proprietary protocol explicitly defined for managing data exchanges over the wireless extension.

Alternatively, IEEE 802.15.4 data-link layer services (LLC or MAC) can be directly employed by the gateway to access wireless stations. Once again, it is worth remarking that significant benefits can be obtained from the use of guaranteed time slots. As an example, in this context, the virtual polling algorithm described in [11] for IEEE 802.11 does not need to be implemented for IEEE 802.15.4 since, in practice, such a mechanism is natively provided by the beacon-enabled mode of the access technique.

35.4.2 IEEE 802.15.4 Extensions of Real-Time Ethernet Networks

The availability of the LLC protocol for IEEE 802.15.4 networks could enable, in principle, the implementation of extensions to RTE networks right at the data-link layer, since most networks provide access to such a layer as well. However, the noticeable difference of transmission rate (RTE networks usually work at 100 Mbps, whereas for IEEE 802.15.4, the maximum allowed rate is 250 kbps) practically prevents to include IEEE 802.15.4 devices in the RTE network cycles, since these latter may take values as low as some hundreds of microseconds. Also, the possibly frequent retransmissions that could reveal necessary to eventually deliver a frame would contribute to a further increase in the cycle times. Consequently, also in this case, interconnections are recommended to be implemented at the higher layers via gateways.

As a final remark, it is worth mentioning that ZigBee [29] offers a complete protocol stack based on IEEE 802.15.4, and consequently, it represents an appealing opportunity for implementing wireless extensions of both fieldbuses and RTE networks considered in this chapter. Particularly, ZigBee offers connectivity with whatever type of networks based on the IP protocol, by means of the purposely defined *ZigBee expansion devices*. Consequently, it can be directly interfaced to several RTE networks (practically all those that allow for TCP/IP communications) using a conventional IP router.

35.5 Conclusions

Because of the need to provide flexible, inexpensive, and reliable communications in production plants, it is currently envisaged that hybrid configurations resulting from wireless extensions of conventional (i.e., wired) industrial networks will be adopted more and more in the next future. This is mainly possible thanks to the availability of wireless technologies and devices able to cope with industrial communication requirements in a proper way.

In the same way, both the IEEE 802.11 family of WLAN standards and IEEE 802.15.4 Low Rate WPANs have been taken into account as wireless extensions, since they are some of the most promising technologies for use in industrial automation and control applications, and a lot of devices are already available off-the-shelf at (relatively) low cost.

In general, the extension of almost all fieldbuses can typically take place at the application layer, through the use of suitable proxies. While providing greater flexibility, this may worsen the overall performance, due to delays this kind of devices unavoidably introduce. As an interesting exception, DeviceNet can be interconnected at a lower level via CIP routers. This means that in hybrid networks based on DeviceNet, all devices (both wired and wireless) can be seen and accessed as conventional CIP nodes, whereas in the case a gateway/proxy is adopted, wireless nodes are no longer effectively part of the network itself.

When real-time (industrial) Ethernet networks have to be provided with a wireless extension, more alternatives are possible, in principle, including carrying out interconnection at the data-link layer. However, as these networks often rely on peculiar protocols to ensure predictable communications over conventional Ethernet, the integration with WLANs is not usually straightforward. As a consequence, a loss of performance and determinism in hybrid networks is expected to occur also in this case. Moreover, when IEEE 802.11 is selected for the wireless extension, the native randomness of the DCF medium access technique has to be taken into account carefully in designing/deploying the hybrid network, since it likely leads to additional (possibly critical) delays. Significant improvements can be achieved by using deterministic techniques, such as, for example, PCF in WLANs (which, however, is not usually available in real devices) or guaranteed time slots in IEEE 802.15.4 networks (that, however, suffer from a noticeably lower bit rate).

Finally, an interesting option for achieving quasi-real-time communications over wireless extensions is provided by both the IEEE 802.11e and IEEE 802.11n standards, which support the concept of quality of service. It has been shown that, for example, traffic prioritization could be used to deal with different types of connections (and, hence, of timing requirements) defined and handled by the CIP protocol, so as to transparently enable control applications to meet real-time constraints, even when hybrid network configurations are adopted.

References

1. A. Willig, Recent and emerging topics in wireless industrial communications: A selection, *IEEE Transactions on Industrial Informatics*, 4(2), 102–124, May 2008.
2. A. Willig, K. Matheus, and A. Wolisz, Wireless technology in industrial networks, *Proceedings of the IEEE*, 93(6), 1130–1150, 40–41, June 2005.
3. S. Vitturi, F. Tramarin, and L. Seno, Industrial wireless networks: The significance of timeliness in communication systems, *Industrial Electronics Magazine*, 7(2), 40–51, June 2013.
4. HART Communication Foundation, HART field communication protocol specification, 2007. [Online]. Available at http://www.hartcomm.org/. (accessed March 10, 2014).
5. Wireless systems for industry automation: Process control and related applications, International Society of Automation (ISA) Standard, ISA-100.11a, 2009.
6. J. P. Thomesse, Fieldbus technology in industrial automation, *Proceedings of the IEEE*, 93(6), 1073–1101, June 2005.
7. J. D. Decotignie, Ethernet-based real-time and industrial communications, *Proceedings of the IEEE*, 93(6), 1102–1117, June 2005.
8. *Industrial Ethernet Book*, issue 69/72, April–October 2012. [Online]. Available at http://www.ieb-media.com (accessed March 10, 2014).
9. Institute of Electrical and Electronics Engineers, IEEE standard for information technology—Telecommunications and information exchange between systems—Local and metropolitan area networks—Specific requirements—Part 11: Wireless LAN medium access control (MAC) and physical layer (PHY) specifications, IEEE Std. 802.11–2007, June 12, 2007.
10. Institute of Electrical and Electronics Engineers, IEEE standard for information technology—telecommunications and information exchange between systems—Local and metropolitan area networks—specific requirements—Part 15.4: wireless medium access control (MAC) and physical layer (PHY) specifications for low-rate wireless personal area networks (WPANs), IEEE Std. 802.15.4–2006, September 8, 2006.
11. L. Rauchhaupt, *System and Device Architecture of a Radio Based Fieldbus—The Rfieldbus System*, in *Proceedings of IEEE WFCS 2002*, Vasteras, Sweden, September 2002.
12. S. Lee, K. C. Lee, M. H. Lee, and F. Harashima, Integration of mobile vehicles for automated material handling using PROFIBUS and IEEE 802.11 networks, *IEEE Transaction on Industrial Electronics*, 49(3), 693–701, June 2002.
13. A. Willig, M. Kubisch, C. Hoene, and A. Wolisz, Measurements of a wireless link in an industrial environment using an IEEE 802.11-compliant physical layer, *IEEE Transactions on Industrial Electronics*, 49(6), 1265–1282, December 2002.
14. Institute of Electrical and Electronics Engineers, IEEE standard 802.15.1: Wireless medium access control (MAC) and physical layer (PHY) specifications for personal area networks (WPANs), June 2005.

15. F. Halsall, *Data Communications, Computer Networks and Open Systems*, Addison Wesley Publishing Company, Reading, MA, 1996.
16. C. Koulamas, S. Koubias, and G. Papadopoulos, Using cut-through forwarding to retain the real-time properties of PROFIBUS over hybrid wired/wireless architectures, *IEEE Transactions on Industrial Electronics*, 51(6), 1208–1217, December 2004.
17. Deutsches Institut fuer Normung, *PROFIBUS-DP Standard, Translation of the German National Standard DIN 19245 Parts 1, 2 and 3*, Beuth Verlag GmbH, Berlin, Germany.
18. D. Brevi, D. Mazzocchi, R. Scopigno, A. Bonivento, R. Calcagno, and F. Rusinà, *A Methodology for the Analysis of 802.11a Links for Safety-Critical Industrial Communications*, in *Proceedings of IEEE WFCS 2006*, Torino, Italy, pp. 165–174, June 2006.
19. Siemens SCALANCE W-780 access points family. Available at http://www.automation.siemens.com (accessed March 10, 2014).
20. G. Cena, I. Cibrario Bertolotti, A. Valenzano, and C. Zunino, Evaluation of response times in industrial WLANs, *IEEE Transactions on Industrial Informatics*, 3(3), pp. 191–201, August 2007.
21. P. Popovski, H. Yomo, and R. Prasad, Strategies for adaptive frequency hopping in the unlicensed bands, *IEEE Wireless Communications*, 13(6), 60–67, December 2006.
22. G. Cena, I. Cibrario Bertolotti, A. Valenzano, and C. Zunino, Industrial applications of IEEE 802.11E WLANs, in *Proceedings of WFCS 2006, Sixth IEEE International Workshop on Factory Communication Systems*, Dresden, Germany, pp. 129–138, May 20–23, 2008.
23. G.R. Jackson, C. Calo3, and T. Itoh, Leaky-wave antennas, *Proceedings of the IEEE*, 100(7), 2194–2206, July 2012.
24. Institute of Electrical and Electronics Engineers, IEEE standard for local and metropolitan area networks media access control (MAC) bridges, IEEE Std. 802.1D, June 2004.
25. L. Seno and S. Vitturi, Wireless extension of Ethernet powerlink networks based on the IEEE 802.11 wireless LAN, in *Proceedings of WFCS 2006, Sixth IEEE International Workshop on Factory Communication Systems*, Dresden, Germany, pp. 55–63, May 20–23, 2008.
26. Ethernet Powerlink Standardization Group, Ethernet powerlink communication profile specification, Version 2.0, 2003.
27. Defense Advanced Research Projects Agency, RFC 768, User Datagram Protocol, DARPA Std., August 1980. Available at http://www.ethernet-powerlink.org (accessed\March 10, 2014).
28. Institute of Electrical and Electronics Engineers, IEEE 802-2 standard: Logical link control (with amendments 3, 6 and 7), IEEE Std., 1998.
29. The ZigBee Alliance, ZigBee specification, December 2006. [Online]. Available at http://www.zigbee.org (accessed March 10, 2014).

36

Wireless Sensor Networks for Automation

36.1 Introduction 36-1

36.2 WSN in Industrial Automation 36-2
Classical WSN • (Re)defining WSNs • Industrial Automation Applications

36.3 Development Challenges 36-8
Diverse Requirements • Choosing Communication Technology • Choosing Suppliers and Components • Pulling It All Together

36.4 Reference Cases 36-14
Wireless Vibration Monitoring Application • Communication Standard • Communication Technology • Wireless Process Control Application • Communication Standard • Communication Technology • Challenges for the Future

36.5 Communication Standards 36-20
Standards Landscape • Technology Primer • Why Is ZigBee Not Enough?

36.6 Low-Power Design 36-31
Basic Principle • Sleep Modes • HMI • Energy-Efficient Protocols

36.7 Packaging 36-36
IP Rating • Hazardous Environments (EX) • Other Environmental Considerations • Form Factor

36.8 Modularity 36-40
Levels of Modularity • Interfaces • Subsystem Behavior

36.9 Power Supply 36-44
Requirements • Power Management • Sources and Technology

Acknowledgments 36-50

References 36-51

Tomas Lennvall
ABB AB Corporate Research

Jan-Erik Frey
ABB AB System Automation

Mikael Gidlund
ABB AB Corporate Research

36.1 Introduction

The vision of ubiquitous or pervasive computing prescribes a paradigm shift where the computing power is embedded in our environment rather than concentrated in desktop or laptop machines. This broad vision of the future has fueled a number of narrowly defined research areas, among them wireless sensor networks (WSNs). WSNs originate from military applications such as battlefield surveillance, but are now being deployed in a large number of differing civilian application areas such as environment and habitat monitoring, healthcare applications, home automation, and traffic control. The intrinsic qualities of WSN, such self-powered, self-organizing, and self-healing, make them highly interesting also in the domain of industrial automation. However, industrial automation applications often have very distinct and stringent requirements, which necessitates that we loosen and extend the classical definition of a WSN for it to be more applicable in an industrial setting.

This chapter will provide an overview of typical industrial automation applications, explain how they differ from other domains, and highlight the key challenges in developing WSNs for the industrial automation industry. A reference case will be used throughout the chapter to exemplify these challenges and illustrate practical implications of design choices and requirements trade-offs.

The chapter is organized in to four main parts outlined as follows:

1. *WSN in Industrial Automation*
   We start off with a (re)definition of WSN from an industrial automation point of view and detail the key industry requirements and their impact on the core WSN technology.
2. *Development Challenges*
   Following the description of applications and their requirements, we try to structure a generic development cycle or process that highlights key challenges of developing WSN devices. This section clearly shows how many different aspects are there that need to be considered, how intricate they are, and how intertwined they are with each other.
3. *Reference Case*
   Here, we describe the application of the reference case including the environmental conditions, application requirements, and business motivation. The main elements of the design and implementation are described to provide a basis for comparison in the following sections on development aspects.
4. *Development Aspects*
   The final sections in this chapter elaborate on the key development aspects described earlier such as standards, low-power design considerations, packaging, hardware (HW) and software (SW) modularity, power supply, and component choice.

36.2 WSN in Industrial Automation

36.2.1 Classical WSN

Wikipedia defines a WSN as "a wireless network consisting of spatially distributed autonomous devices using sensors to cooperatively monitor physical or environmental conditions, such as temperature, sound, vibration, pressure, motion or pollutants, at different locations" [1]. The devices, which is the focus of this chapter, are self-contained units comprising a microcontroller, power source (usually, but not always, a battery), radio transceiver, and sensor element (Figure 36.1).

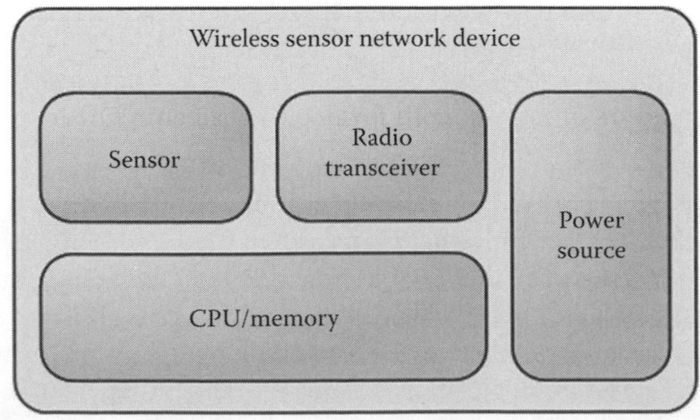

FIGURE 36.1 Components of a WSN device.

WSNs have some intrinsic qualities, which can be derived from the distributed autonomous nature of the devices.

Self-Powered
WSNs are self-contained units with an internal power source (usually, but not always, a battery). Because of the limitations of battery life, nodes are built with power conservation in mind and generally spend large amounts of time in a low-power *sleep* mode.

Self-Organizing
The nodes self-organize their networks in an ad hoc manner rather than having a preprogrammed network topology. For example, in wired network technologies, some designated nodes, usually with custom HW (routers, switches, hubs, and firewalls), perform the task of forwarding data to other nodes in the network. In a WSN, typically each node is willing to forward data for other nodes, resulting in a dynamic network configuration based on the network connectivity.

Self-Healing
WSNs have the ability to self-heal, that is, if one node goes down, the network will find new ways to route the data packets. This way, the network as a whole will survive even if individual nodes lose power or are destroyed. This type of dynamic reorganization can also be triggered by other conditions than node failure, such as maximum node power consumption and communication latency.

36.2.2 (Re)defining WSNs

Industrial applications differ from the definition earlier in a number of respects.

36.2.2.1 Cooperative Sensing Is Not Commonplace in Industrial Automation

First, and maybe most importantly, all sensors are crucial to the operation of the plant. This implies that losing one node is not an option, even if the overall network stays operational. A faulty node will have to be replaced. Increasing the availability by adding redundant sensors would be technically possible but hardly feasible from a cost perspective. Although the cost of radio transceivers is continuously dropping, the majority of the cost of industrial-grade WSN devices is not driven by the cost of the transceiver alone. To this, you need to add the life cycle cost associated with installing and maintaining a device throughout its lifetime (which is a factor of at least two to five times longer than consumer products).

The notion of cooperative sensing prescribed by classical WSN is also not readily applicable to today's automation control systems, which are rather I/O centric. Measurement points are not *dispersed* across the plant but rather installed in very exact locations. The data are collected and fed back to the control system where they are processed and analyzed. Adding a measurement point is associated with manual engineering and application programming before it can be utilized in the system.

The cooperative nature of WSN still adds value though, as setup and installation of the sensors becomes easier and the overall operation more reliable. Since the position of the measurement point might not be optimal from a radio-frequency (RF) perspective, the self-organizing and self-healing nature of WSN help alleviate communication disturbances related to radio propagation in industrial environments. Instead of investing lots of time in frequency planning and network design, performance of the network can be improved by adding router nodes that provide alternate/redundant paths for the wireless sensors.

36.2.2.2 Time Is of Essence

Time is money, also in the industrial automation world. Whereas a data packet in a standard WSN may spend an unknown time from its source to its destination, an industrial application will frequently require hard bounds on the maximum delay allowed. This can put some strict bounds on network topology, for example, prescribing a star configuration rather than a multihop mesh network, and requires a careful trade-off between power conservation and response time. Although there is usually a strict bound on the maximum delay, the requirements can differ quite dramatically between applications. This makes a generic design of the WSN quite difficult, and the *one-size-fits-all* problem quite prevalent in industrial standards groups who seek to minimize the number of overlapping wireless communication standards.

36.2.2.3 Wireless in a Wired World

Although WSNs have made its entry on the industrial automation scene, the majority of sensors and devices have a wired infrastructure. This means that we need to cope with a mix of wireless and wired devices over a foreseeable future. Applications that are built upon a wired infrastructure can afford to *waste* bandwidth and make use of high-performance networks to solve their mission-critical goals. Simply adding a wireless sensor to the equation is often not straightforward.

A quite common mixed scenario on the other hand is to add wireless connectivity to already wired devices. This is done to provide access to diagnostic or asset data in the device in a nonintrusive manner, that is, without modifying or interrupting the wired network.

Since wireless solutions in industry will tend to also have a wired infrastructure, the data will emanate from the sensors and ripple through the network to some wired aggregation point. From here, it will, in general, be transported over a high-speed bus to a controller. Apart from the classical mesh networking topology of WSN, we have two more common topologies in industrial settings (Figure 36.2).

In the star topology, the most prevalent topology today, the wireless nodes communicate with a gateway device that bridges the communication to a wired network. This network configuration will have greater chances of minimizing delays in the network and can thus be used in applications that have strict bounds on the communication delay. Another common intermediate solution of WSN is to have router devices (often mains powered) that communicate with the gateway. The sensors only need to perform point-to-point communication with the routers and can therefore remain simple and low power, while the range and redundancy of the network are improved.

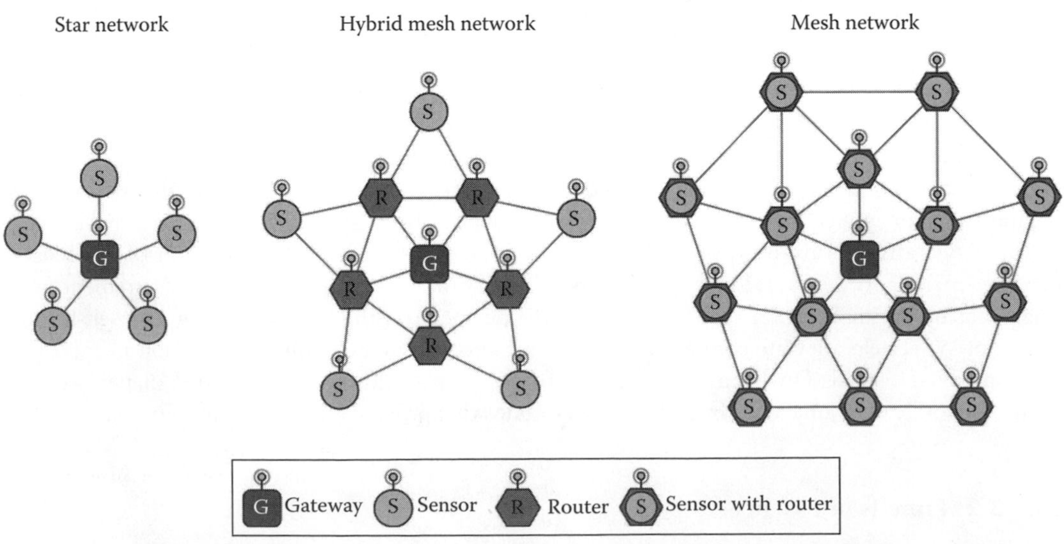

FIGURE 36.2 WSN topologies.

36.2.3 Industrial Automation Applications

36.2.3.1 Industrial Plants, an Unfriendly Setting for WSN

The environmental conditions of industrial automation plants pose some heavy challenges in the development of embedded field devices, in general, and specifically for WSN devices.

Industrial automation environments are characterized by the following:

- *Harsh environments*: Extreme temperature, moisture, vibration, hazardous environments, steel constructions, and (moving) obstructions. Issues facing wireless communication in such environments are heavy multipath fading, fast/slow fading, coverage quality (due to reflection), and local variations in received power.
- Electromagnetic interference (EMI): EMI from electrical activity such as drives and welding, which cause noise in RF bands.
- *Other users*: Disturbance by occupation of frequencies over time (WLAN, Bluetooth, ZigBee, etc).

The following sections will elaborate more on the implications of these severe conditions on the embedded design of WSN devices.

36.2.3.2 Typical WSN Applications

The requirements of any WSN solution will always depend heavily on the particular application in mind. There are many promising applications for wireless technologies in industrial automation, ranging from monitoring of process parameters to mission-critical control applications. ISA (The International Society for Automation [2]) defines six usage classes of wireless communication based on the criticality of the application, which directly govern the importance of the message response time (see Table 36.1).

WSNs in the traditional sense (i.e., low-power, multihop, mesh networks) are today usually deployed in class 4–5 applications. This is logical since most WSN solutions are usually optimized for low-power consumption to increase battery lifetime, and this stands in direct conflict with short and deterministic response times.

However, in real life, the boundaries between these classes are not always that clearly defined/identified. It's not uncommon that from a user's point of view, the application is a mixture of monitoring and control, making the requirements on message delay hard to specify. Defining a suitable wireless communication solution can thus be difficult unless we provide means to influence the network forming and optimization criteria.

TABLE 36.1 ISA-100 Usage Classes

Category	Class	Application	Description	
Safety	0	Emergency action	*(Always critical)*	
Control	1	Closed-loop regulatory control	*(Often critical)*	↑
	2	Closed-loop supervisory control	*(Usually noncritical)*	
	3	Open-loop control	*(Human in the loop)*	
Monitoring	4	Alerting	*Short-term operational consequence (e.g., event-based maintenance)*	Importance of message timeliness increases
	5	Logging and downloading/uploading	*No immediate operational consequence (e.g., history collection, sequence of events, preventive maintenance)*	

TABLE 36.2 Impact of Process Attributes on Wireless Technologies

Process Attribute	Description	Impact on Wireless Technology
Device types	Predominant or common mix of devices, for example, analog and/or discrete devices	The type of data (and consequently also amount) that is sent has a major impact on the design of the communication protocol. This in combination with the timing requirements sets boundaries on the frame/package size utilized.
Device count per unit	Typical number of devices in a manufacturing unit/production cell	The number of devices in a plant and their geographic distribution has a dramatic impact on the communication technology.
Unit physical size [m]	The physical size of a typical manufacturing unit/production cell	The ability to cover a whole factory with wireless communication depends on a number of factors such as • Range of the wireless device/network, which is governed in turn by the reliable operation range, the disturbing range, and *frequency reuse* range • Number of wireless nodes per manufacturing unit • Number of coexisting wireless networks within reliable operation range without performance degradation
Units per plant	Number of manufacturing units/production cells in a typical plant	
Device density (units per 1000 m^2)	Number of devices per 1000 m^2	
Devices per plant	Number of devices in a typical plant	
Production cycle length	Length of a typical production cycle	The production cycle length and start-up times have a direct impact on the setup times for the wireless networks, such as network forming and adding/removing nodes.
Unit start-up time	Start-up time of a manufacturing unit/production cell	
Time		
Fast/subunit	Fast internal communication needs, for example, drives loop	This attribute in combination with the node density has the greatest impact on the technology selection. Depending on the required update rate and equally important the latency (variation in communication delay), very radical differences in design of the wireless communication solution might be required.
Control loop	Time resolution of normal control loops	
Slow/diag./HMI	Time resolution for noncritical data such as diagnostics or HMI	
Field device cost	Typical cost of the field devices	The cost of the device to a large extent puts some upper barriers on the target cost of the wireless communication solution.
Installation cost/device cost	Cost of the installation in comparison to the cost of the device	This metric provides guidance on the target cost of installing and setting up the wireless infrastructure.
Automation technology	Type of control system utilized to automate the process	The use of a DCS versus, for example, a PAC/PLC solution, to some extent, governs the logical network structure. This is also reflected in the device count and unit count measures listed.
Commonly used device networks	Typical wired infrastructure in today's plants	At some point, the wireless network has to interface with the existing network infrastructure. The employed wired network technology and topology have an impact on the way one could/should integrate the wireless system in the plant.
Power availability	Availability of line power/battery size/energy harvesting	The amount of energy that is available has a dramatic impact on the design of the wireless protocol (energy efficiency), as well as the possible duty cycle of the application.

36.2.3.3 Same, Same, but Different

Industrial automation is a broad industry and comprises many segments and application domains. A common way to divide the market is according to the type and characteristics of the process being controlled:

- Process automation (continuous processes)
- Discrete manufacturing (manufacturing of discrete units)
- Hybrid automation (mix of continuous and discrete)

The industry segments can be characterized by a number of process attributes that govern the requirements on wireless communication technologies and consequently also the design and implementation. Table 36.2 provides a description of key process attributes and their impact on the requirements and design of wireless technologies.

Although similar (bad) environmental conditions can be found throughout the various industry segments, there are distinct segment-specific requirements that have a dramatic impact on the wireless technologies. Figure 36.3 provides a comparison of the key process attributes within process automation and discrete manufacturing (note that the scale is logarithmic).

As illustrated in the earlier figure, the two industry segments have quite different characteristics and consequently also pose quite different requirements on the wireless technologies employed.

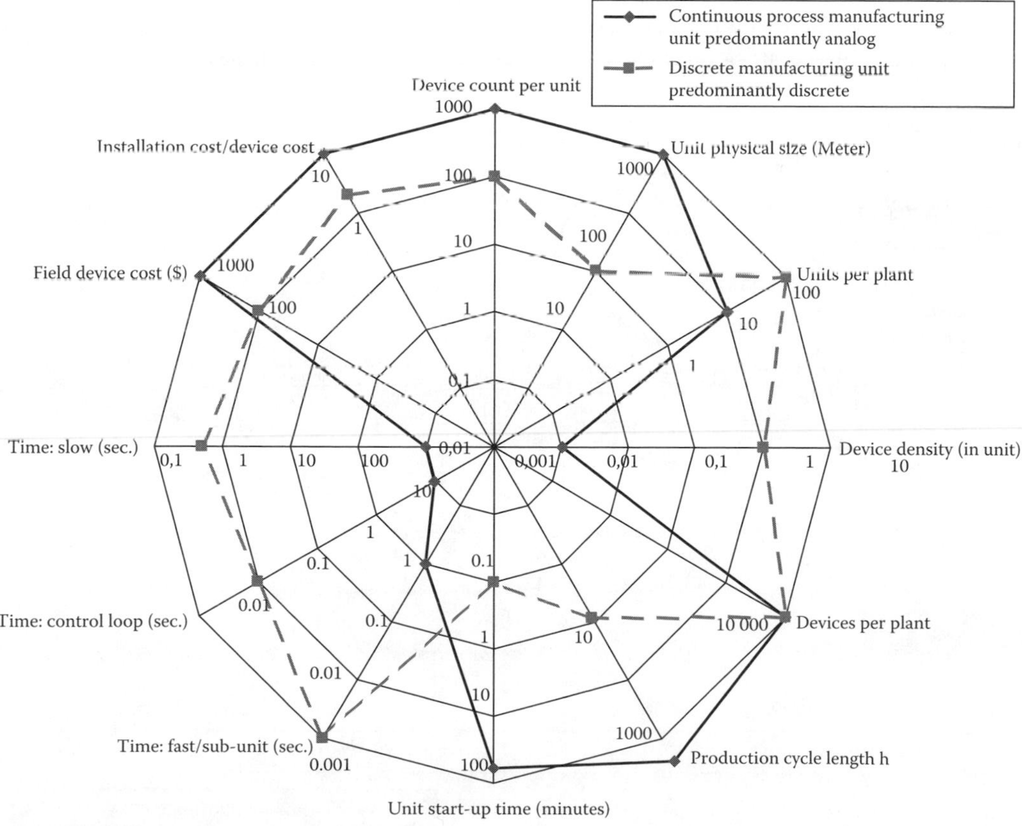

FIGURE 36.3 Process attributes.

Even though industrial automation standardization activities in the wireless domain have started out with a quite holistic view of the processes and target applications, these distinct differences have led to the forming of separate working groups (WGs) targeting process and discrete applications. For an overview of the industrial automation wireless standards landscape, see *Section 36.5*.

36.3 Development Challenges

The previous section presented a definition of the *classical* WSN and contrasted that to the complex set of requirements and applications of industrial automation. These requirements and use cases have a dramatic impact on the development of WSN devices.

Figure 36.4 depicts a generic development process.

Embedded development methodology and development processes in general are complex subject matters that merit a book of its own to do the topic justice. The process depicted previously thus only aims at highlighting some key characteristics of WSN development projects:

1. A thorough requirement analysis is required in order to be able to select the right communication technology.
2. The choice of technology is not driven by the requirements alone. Factors such as standards compliance and legacy systems also have a large impact on the final choice.
3. Component selection to a large extent follows the choice of technology but is many times already *given* by the platform choices of existing legacy systems.
4. Finally, when designing, implementing, and testing the device, we need to pull all the bits together. This usually requires quite an iterative process with many trade-offs and redesigns due to a combination of inaccurate or incomplete requirements and/or component specifications.

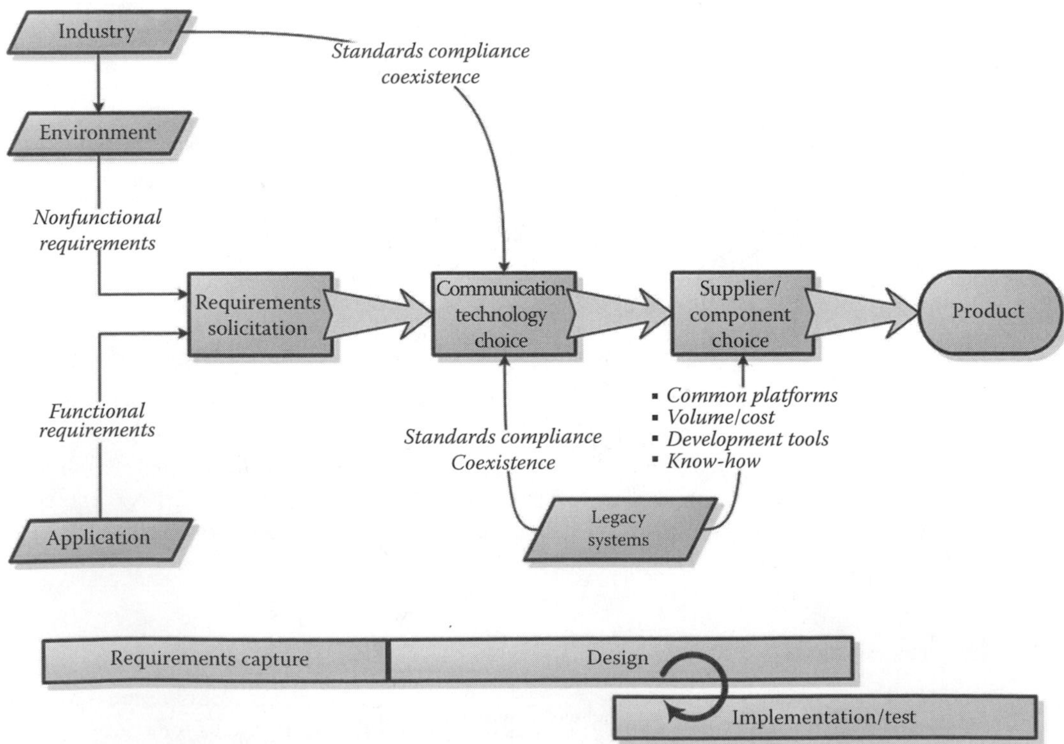

FIGURE 36.4 Principal development process.

36.3.1 Diverse Requirements

As highlighted in the previous section, the target industry and application can pose quite differing requirements on the choice of communication technology as well as impose quite rigid restrictions on the embedded design. The requirements can be grouped into nonfunctional requirements that stem primarily from the target environment and functional requirements given by the target application.

The environmental conditions of the target industry have a major impact on the boundary conditions of the embedded design and choice of communication technology. For example, factors such heat, moisture, and vibration influence the packaging of the device, which in turn imposes additional requirements on the power consumption as well as restrictions on the antenna design and power-supply solution. If the device shall be used in hazardous environments, special consideration must be taken throughout the design process.

The application requirements more or less dictate the duty cycle of the WSN, which has a huge impact on choice of technology and a direct influence on the low-power design and required response time. In addition to the autonomous operation of the device, the human–machine interaction can pose additional requirements on external interfaces that influence both the communication solution as well as the packaging of the device.

36.3.2 Choosing Communication Technology

A clear view of the requirements helps to eliminate obvious communication solutions. However, it's not uncommon that one needs to make a trade-off in the end to be able to address the main requirements that are driving the application.

Many industries, customers, and suppliers have adopted their own communication standards or, when possible, refer to international standards. If a different technology shall be acceptable, as a minimum requirement, it must be able to coexist with other technologies without unduly compromising the performance of installed devices/systems.

36.3.3 Choosing Suppliers and Components

The choice of technology has a direct impact on the availability of suitable SW/HW components and thus also on the total development cost/time. Since industrial devices have a quite long lifetime, legacy systems have a major impact on choice of technology as well as components. Many suppliers have streamlined their development processes and base their product portfolios on standardized platforms. This has great benefits in terms of volume/cost of components and development tools, as well as spreading of best practices and know-how. Introducing a new SW/HW component or platform is often a demanding and time-consuming task and always represents an additional risk. You can never be quite sure that the new component will behave in the same manner as those components that have undergone years of development/refinement and are proven in the field. It might therefore be better to utilize a well-known component and suffer/deal with the performance losses of utilizing a *nonoptimal* component from a WSN perspective.

36.3.4 Pulling It All Together

The development process depicted in Figure 36.4 gives the impression that the development follows a rather linear waterfall-type model. Even if you might start out that way, due to the complexity of the requirements and the boundary conditions imposed by industry and legacy systems, the design/implementation/test phase usually reverts to a highly iterative process. Developing WSN devices for industrial automation applications often comprises a set of *chicken-and-egg* problems, where the order of design choices can result in very different outcomes in the final design. A key success factor is thus to have a very clear view of the driving requirements, that is, the absolute *must-haves*, stick to them and let the other lower priority requirements, the *should-haves*, undergo the inevitable trade-off exercise.

Figure 36.5 illustrates the interdependencies of key development aspects of WSN devices and shows how one design aspect (a row) affects other design aspects (the columns).

The following sections provide a walk-through of the table to demonstrate how design choices in one dimension can, directly or indirectly, affect other dimensions of the design (and vice versa).

36.3.4.1 Communication Standards

Most customers strive toward the use of standards when building their installations. This secures interoperability and availability of suppliers and leads to higher volumes (and ultimately lower cost). The use of a standard is thus often mandatory to be able to compete on the market. One of the key requirements of wireless devices in industry is coexistence. Hence, this needs to be considered when choosing a specific standard.

Once selected, the standard in itself will impose upper/lower bounds for the device in terms of power consumption. Regardless whether you employ low-power electronics and a clever sleep–wake-up scheme, the communication protocol has a vital impact on the final power consumption of the system. Some communication protocols are notoriously inefficient, and no smart embedded programming in the world will help get the consumption down to an acceptable level.

The standard can also impose certain physical interfaces, which will have a direct impact on the design of HW/SW as well as the packaging of the device. For example, the WirelessHART standard [3] requires that you provide a service port on your devices.

The scope and architecture given in the standard indirectly also prescribe the level of integration of HW and SW. This is of course highly linked to the available commercial offering that supports the standard, such as radio transceivers and communication stacks.

So, *simply* by choosing a specific standard, many design choices have been made that will drive the design of the device. In *Section 36.5*, we will dig deeper into the wireless standards jungle and provide some insight to the basic mechanisms of modern WSN solutions targeted toward industrial automation.

36.3.4.2 Low-Power Design

One important aspect of WSN is to keep the power consumption of a node at a minimum while at the same time providing the best possible performance to the system users. The low-power nature of WSN poses serious restrictions on the power consumption of the HW as well as on the SW architecture.

Integrating functionality can be one way to lower the overall power consumption but can put severe limitations on the modularity of the system. This not only affects the HW/SW maintenance of the device but may also limit your options concerning what to power up/down and when, which in turn can affect the overall power consumption as well as performance of the device negatively.

Although WSN devices are autonomous entities, a local human–machine interface (HMI) is often useful or necessary during commissioning and troubleshooting. The specific low-power design directly limits your options concerning the use of a local HMI, for example, liquid crystal display (LCD), light emitting diode (LED) and buttons.

Designing for low consumption involves choosing low-power components. This may seem trivial but is often a complex issue as it usually involves compromises on the performance. Typically, a low-power central processing unit (CPU) runs on reduced clock cycle with fewer on-chip features than its more power hungry counterparts. The trick is to choose elements with just enough performance to do the job. In addition, *overspecifying* components to create margins in your design can lead to cost and maintenance issues as the number of suppliers tends to shrink rapidly.

Impact on

	Communication standards	Low power design	Packaging	Modularity	Power supply	Component selection
Communication standards	✓ Coexistence	✓ Upper bounds ✓ Lower bounds	✓ HW/SW Interfaces	✓ Functional integration		✓ Com stacks ✓ Supported HW architectures
Low-power design			✓ HMI options	✓ Functional integration	✓ Peak power ✓ Average power	✓ Functionality ✓ On/off timing ✓ Second source
Packaging		✓ HMI power consumption ✓ IS adaptations		✓ Integration ✓ Safety barriers ✓ Interfaces	✓ Peak power ✓ Heat buildup ✓ Sparks	✓ Battery ✓ Antenna design ✓ Surface temp
Modularity		✓ Functional integration ✓ Interfaces	✓ Form factor ✓ Interfaces		✓ Make vs buy	✓ Interfaces ✓ Footprint ✓ Debugging
Power supply		✓ Duty cycle	✓ Form factor ✓ Battery change	✓ Make vs buy		✓ Battery technology

FIGURE 36.5 Development aspect matrix.

Finally, the low-power design more or less dictates the required average and peak power that needs to be available in the system, and this may have some serious consequences on the design of the power-supply solution, including the choice of power source.

Less really *is* more in the WSN world, and the more you take away, the more challenging the design task becomes. In *Section 36.6*, we invest some more *energy* in describing the intricacies of low-power design.

36.3.4.3 Packaging

The packaging of the WSN device represents one of the most costly parts of the total product. The environmental conditions and the device rating (IP class, explosive atmosphere [EX] temperature range, etc.) all influence the required ruggedness of the device and prescribe what material to use in the housing and how well it must be sealed off to protect the internal components from dust and moisture.

The use of a local HMI, whether it is an LCD or a single LED, can have a dramatic impact on the overall power consumption of the device and thus needs to be accounted for in the low-power design.

If the device is to be able to operate in a hazardous environment, great care needs to be taken throughout the design of the device. An intrinsically safe (IS) design is commonly required. The basic principles of an IS design are simple and easy to grasp but challenging to achieve in practice. It requires long experience in the field as well as certification by a notified body. Making IS adaptations can have a dramatic impact on the power consumption and the form factor of the device (due to additional protection and power-limiting components). Often, the maximum power that can be provided by the power supply needs to be limited. The major requirement for batteries operating in ATEX applications is for intrinsic safety. Specifically, this calls for a battery design that prevents both excessive heat buildup and the danger of an electric spark. The modularity of the device is governed by the need to define barriers with clear interfaces in the design to contain and transfer energy in a controlled and safe manner. The maximum surface temperature of any component on a circuit board under fault conditions must be considered.

The form factor of the device components is often given indirectly, either through product design guidelines to foster harmonization and common look and feel or by the requirement to utilize an already existing housing. This means that you often have a given space where you need to squeeze in your HW components, which can have a dramatic impact on the required level of component integration (single-chip solutions, double-sided printed circuit boards [PCBs], etc.). It's not uncommon that the size of one specific component can force a complete PCB redesign to fit within the given dimensions. Typical components that drive the PCB design are capacitors and batteries.

Apart from component selection and integration, the form factor also has a direct influence on the antenna design, which is very sensitive to the orientation and proximity of other electronic components.

Appearance is everything! This actually holds true to a large extent when it comes to embedded design of industrial WSN devices. In *Section 36.7*, we find out that what counts is the *inside*, but the appearance of the device (size, shape, material, and weight) does have a major impact on the final design.

36.3.4.4 Modularity

The lifetime of the devices is a key parameter to consider. The office/consumer industry is the main driver for wireless technologies today, with high volume applications. Industrial automation devices, however, tend to have a much longer lifetime than consumer products. This means that special care needs to be taken when integrating wireless components into industrial devices. A modular (HW and SW) design is crucial in enabling effective maintenance of devices—which are built on standard commercial off-the-shelf (COTS) components—throughout their lifetime.

Modular design is necessary in order to reuse elements, but it might seriously limit your control over individual HW/SW functions and thus influence the overall sleep–wake-up scheme of your device. Proper interfaces are thus vital to mediate between reuse and performance, something that can be hard to influence when COTS components are used.

The form factor of commercial components or subassemblies naturally has a direct impact on the packaging of the device, but again, the interfaces are just as important and need to be sufficiently general to allow portability.

One classical example of the separation of modules is the split between the communication protocol and the application SW. The latter is invariably written by the device vendor, but the former is frequently purchased from a third party. Embedding these two components onto the same microcontroller can be difficult. We also run the risk of suboptimizing, that is, the two SW modules are optimized (with respect to power, performance, code size, etc.) individually. This does not necessarily give a globally optimum solution. Even more complex is handling new releases, bug fixes, and documentation when the SW running on the same processor has several sources.

In addition to reuse, modularity is also a means to minimize design risks. For example, to avoid EMC problems in your power-supply design, it is often beneficial to buy predesigned and tested COTS components such as a DC–DC converter. If the device is going to be produced in large volumes, the cost of these COTS components can be prohibitive though, forcing you to make your own low-level design. Component specification can be another limiting factor when utilizing available designs, for example, if you require a large operating range.

The sum of the parts can equal more than the individual parts together, at least from a product maintenance point of view. However, great care must be taken to ensure that sum of the parts also provides adequate performance. In *Section 36.8*, we cut up the WSN device into logical pieces and look how we can put them back together again without unduly compromising the control over the final behavior of the device (power consumption, latency, etc.).

36.3.1.5 Power Supply

The power supply is a central part of the WSN device; it's the *motor* that provides power to the various parts of the system. The challenge is the same as the development team has regarding the final product, to provide the right level, with the right quality, at the right time.

The design of the power supply can put direct bounds on the duty cycle of the application. For example, in an energy-scavenging solution, the harvested energy is usually stored in an intermediate storage. This enables the power supply to provide an even power-supply level in spite of a varying *incoming* power and to handle intermittent peak consumptions. The size of this intermediate storage and the time it takes to charge it directly impact the *maximum* duty cycle.

Although battery technology is progressing, the battery still constitutes a major part of the footprint of the device if we are to reach the 5–10-year lifetime that is typically quoted. The same goes for the energy-scavenging solutions, which usually contain some form of intermediate storage as well.

Typical parameters to consider when choosing battery technology are operating temperature range, energy density, nominal voltage, and Watt hours (for a given cell size). However, there is more than meets the eye here as well. For example, in some types of lithium technologies, passivation may lead to voltage delay, which is the time lag that occurs between the application of a load on the cell and the voltage response. This behavior becomes more pronounced when you have very low duty cycles (since the passivation layer grows thicker).

If battery change in the field is to be supported, special care needs to be taken to allow separation of the battery from the device. This is especially tricky if the device is going to be deployed in a hazardous environment, where any spark must be prevented.

Low-power design is all about powering up/down various parts of the system. In *Section 36.9*, we find out that there is a lot more to consider than ON/OFF.

36.4 Reference Cases

As with other resource-constraint embedded systems, "the devil is in the details." In this section, we present two specific use cases of WSN in industrial automation. These applications will be used as reference cases for the remainder of the chapter to exemplify some of the details that really matter when making design choices.

The first application is *wireless condition monitoring of AC motors using vibration analysis*. In the following sections, all references to this specific use case will be highlighted with a border:

Wireless Vibration Monitoring Case Example

The Wireless Vibration Monitoring application will be used as the main example use-case for reference throughout the rest of the chapter.

36.4.1 Wireless Vibration Monitoring Application

The goal of the vibration monitoring application is to detect early signs of bearing degradation on small AC motors (less than 400 kW), located on offshore oil/gas rigs, in order to be able to schedule maintenance. The motors drive equipment such as pumps, air compressors, and fans, and the load conditions as well as the operating environmental conditions will vary (Figure 36.6).

FIGURE 36.6 Wireless vibration sensor.

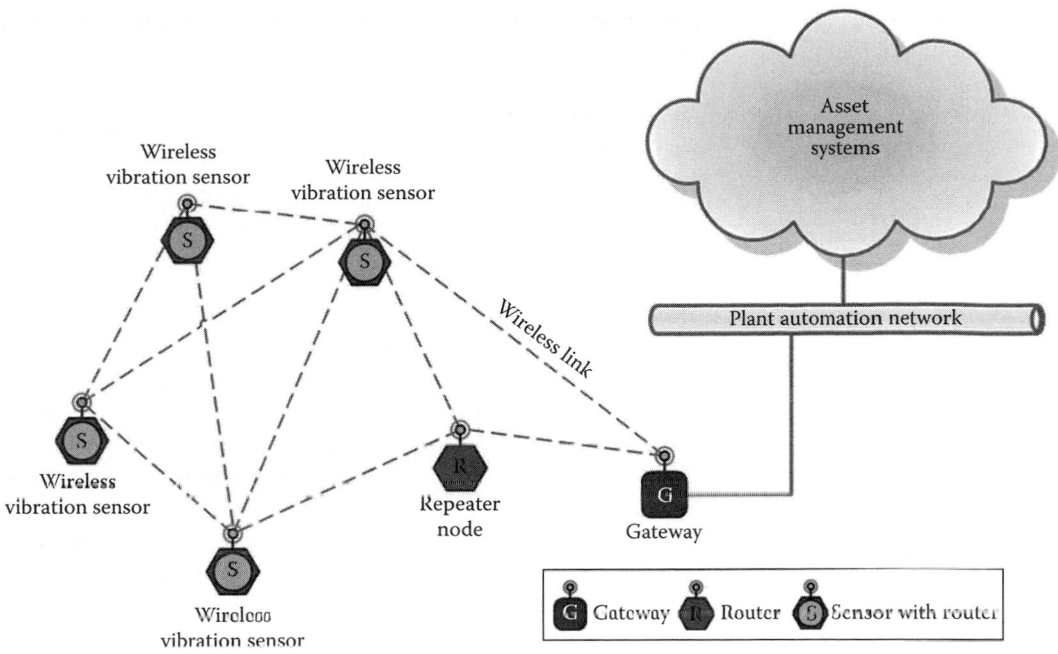

FIGURE 36.7 Wireless vibration monitoring system.

The application consists of several parts as illustrated in Figure 36.7: vibration monitoring sensor nodes, repeater nodes, a gateway, and an asset management system (vibration analysis SW).

For the remainder of the chapter, we will only refer to the WSN part of the application, which consists of the sensor and router nodes. The gateway is actually also part of this but is considered to be part of the infrastructure and often regarded as a COTS component. It is (usually) mains powered and does not face the same design challenges as the resource-constraint sensor nodes. The need for repeater nodes depends on the communication technology and the operating environment of the WSN.

The WSN part of the application has the following design goals and requirements:

- *Low cost*: The sensor/repeater nodes are very restricted in what they are allowed to cost, which has a large impact on component selection and packaging. This also has the effect that the number of nodes that will be deployed in the field should be minimized; that is, the range of the wireless communication should be as long as possible.
- *EX certification*: The environment in oil and gas rigs is hazardous. The sensor and repeater nodes must be certified to operate in an EX Zone 1 area, that is, nodes will intermittently be subjected to flammable material (oil or gas), and must therefore limit the energy of any sparks and surface temperature.
- *Nonintrusive (easy to mount and retrofit)*: The sensor/repeater nodes must be easily mounted on the motors. Some motors have mounting locations prepared, while others don't, and the mounting constructions also differ from case to case. The orientation of the sensor element and location on the motor (preferably close to the bearing) matters from a vibration analysis point of view, which can be in direct conflict with an optimal antenna positioning. All of these must be taken into consideration for the physical design of the sensor nodes.
- *Battery powered and a long battery life*: The requirement is that a sensor node should have a 10-year lifetime without any battery change.
- *Duty cycle*: Each sensor node should perform one vibration measurement per 2 *weeks* and send it to the asset management system for analysis. This is a bit opposed to the *typical* view of a WSN, which sends small amounts of data *fairly often* (minutes, hours, or days).

- *Operating environment*: The temperature can vary from well below zero up to, or even over, 100°C. In addition, the wireless communication solution must be able to cope with the possible interfere caused by the many metal constructions found on an oil rig.
- *Security*: Authenticity (making sure data come from the correct sensor node) and integrity (making sure no one has tampered with the data) are very important. Therefore, it is also important that we only allow authorized nodes to join the network.

Latency is not a critical requirement for the application because the data are not part of a process control loop. Reliability on the other hand is a much more important requirement, because missing data will result in a flawed view of the actual motor condition.

36.4.2 Communication Standard

WirelessHART was the selected communication technology for the vibration monitoring application. It was the first industrial wireless communication standard to become available (released on September 2007). It is explicitly designed with industrial applications and requirements in mind and is the most widespread standard for process automation.

ZigBee is also a currently available wireless communication standard but was not considered sufficient for operation in industrial environments, especially an environment such as the one on an offshore rig. Basically, it lacks effective means with which to alleviate the effects of interference and fading. We provide more details on ZigBee's shortcomings in industrial environments in *Section 36.5.3*.

36.4.3 Communication Technology

As mentioned, WirelessHART is the chosen communication standard for the vibration monitoring application, because it's currently the only available wireless communication standard designed for industrial applications.

However, because it was just recently released, technology suppliers have not yet provided any components supporting the standard. This has led to a decision to use a proprietary technology, as an intermediary solution, until WirelessHART components become available. This leads to another important decision we must make; we must select a proprietary solution that will provide/enable an easy migration to WirelessHART.

In the end, the decision fell on a solution offered by DUST Networks [4] called SmartMesh-XD. It's a system-on-chip (SoC) wireless sensor networking solution called Mote-on-Chip (MoC) (see Figure 36.8). The SmartMesh-XD MoC (DN2140/2040) is a *black-box* solution, which hides all details related to the wireless communication from the rest of the application.

DUST Networks will also provide a WirelessHART-based solution (SmartMesh IA-500), which uses identical HW to the SmartMesh-XD MoC (including the interface), thus only requiring a firmware upgrade in order to become WirelessHART compliant. This will allow for a smooth transition to WirelessHART.

SmartMesh-XD is IEEE 802.15.4 standards compliant and operates in the 900 MHz and 2.4 GHz industrial, scientific and medical (ISM) frequency bands. It is a robust wireless sensor networking technology that combines low-power radio technology with the Time-Synchronized Mesh Protocol (TSMP).* TSMP, which was also the basis for the WirelessHART standard, provides features such as time-division multiple access (TDMA), frequency hopping, and redundant paths (mesh), which makes it suitable for the harsh operating environment of the application. The SmartMesh-XD communication uses a concept called a network frame. The network frame is a fixed-length, cyclically repeating, sequence of time slots (TDMA), in which all nodes have been assigned dedicated slots for message transmission and reception.

* TSMP is a proprietary wireless sensor network protocol developed by DUST Networks.

FIGURE 36.8 DUST Networks SmartMesh MoC solution.

In the following sections, we will describe the impact the decision to use SmartMesh-XD, as well as allowing for future migration to WirelessHART, has on the design of the application.

Wireless Process Control Case Example

Utilizing wireless as an infrastructure for closed loop control presents many advantages such as cost reduction, flexibility and improved performance. However, wireless control also presents many challenges such as deterministic communication, security and battery lifetime decreases.

36.4.4 Wireless Process Control Application

Wireless measurement devices are becoming common features of modern process control systems. They allow plants to quickly and relatively inexpensively add new measurement and control capabilities, especially where it has been too difficult or expensive to use wired instruments. Wireless measurement devices also allow plants to quickly and economically test new control schemes. Today, the most common wireless control applications typically involve noncritical, open-loop control tasks, or closed-loop control applications with longer time constants such as tank level control, temperature control of heated jackets, or control of oil field steam injection for secondary recovery operations.

Utilizing wireless as an infrastructure for closed loop control presents many advantages such as cost reduction, flexibility, and improved performance. However, wireless control also presents many challenges such as deterministic communication, security, and battery lifetime decreases. For the latter, it can be solved by reducing the overall number of signal transmissions, which can significantly extend battery lifetime. Wired devices often *oversample* a process variable, sending measurements to the control 4–10 times more often than the controller performs algorithm executions. Furthermore, many measurement updates that traditional wired devices transmit are essentially identical (or very similar) to previous measurement updates, meaning that it do not serve a practical purpose and only conserve bandwidth.

36.4.4.1 Application: Storage Tank Heating Control

Level monitoring devices are frequently located on top of very large storage and process tanks (Figure 36.9). Using WirelessHART makes it easier and more cost efficient to monitor and control the levels in process and storage tanks by eliminating the need to manually take and record the measurements. As an example application, we consider storage tank heating control where existing manual measurement and control methods were unsatisfactory for maintaining the temperature of fatty nitrites and amines in several tanks where they were stored before shipment to customers. Too much steam was sometimes used to heat materials, and one problem was that the delivered product was too hot. Existing procedures for detecting potential problems required an operator to make regular trips into the field to take snapshot readings from pressure gauges on the tank farm's venting conduits. This is time consuming and fails to provide continuous and immediate information. A typical wired process control system shown in Figure 36.10 was installed before. From the controllers down to the wired instruments, the communication can be achieved by different standards such as HART, Modbus, Profibus, and Profinet.

One possible solution is to install a number of WirelessHART temperature transmitters to control the temperature on a number of tanks (see Figure 36.11 for a conceptual framework) and the controllers are then connected further to the enterprise management system, etc. Measurement data are transmitted every minute to the distributed control system (DCS), which controls a simple ON/OFF

FIGURE 36.9 Storage tank.

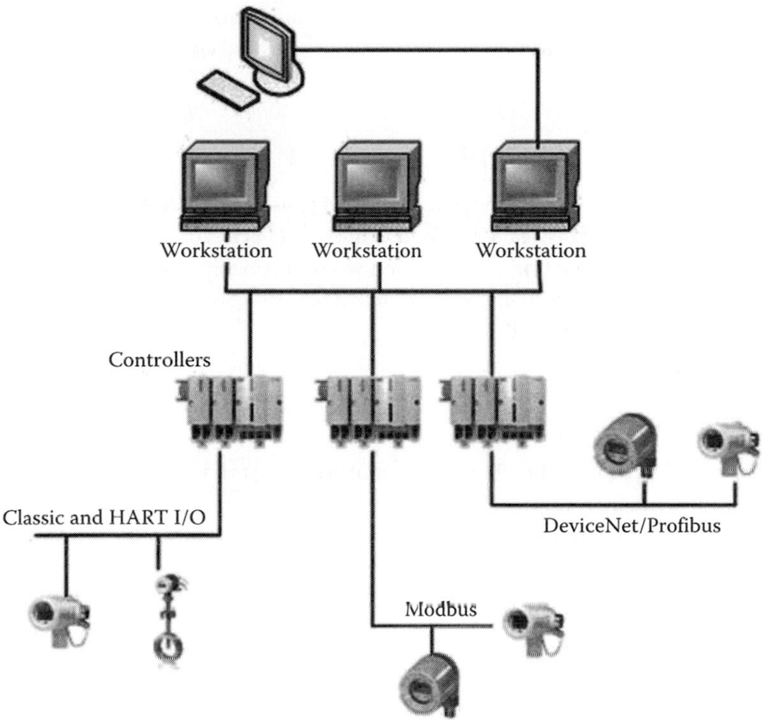

FIGURE 36.10 A wired process control system.

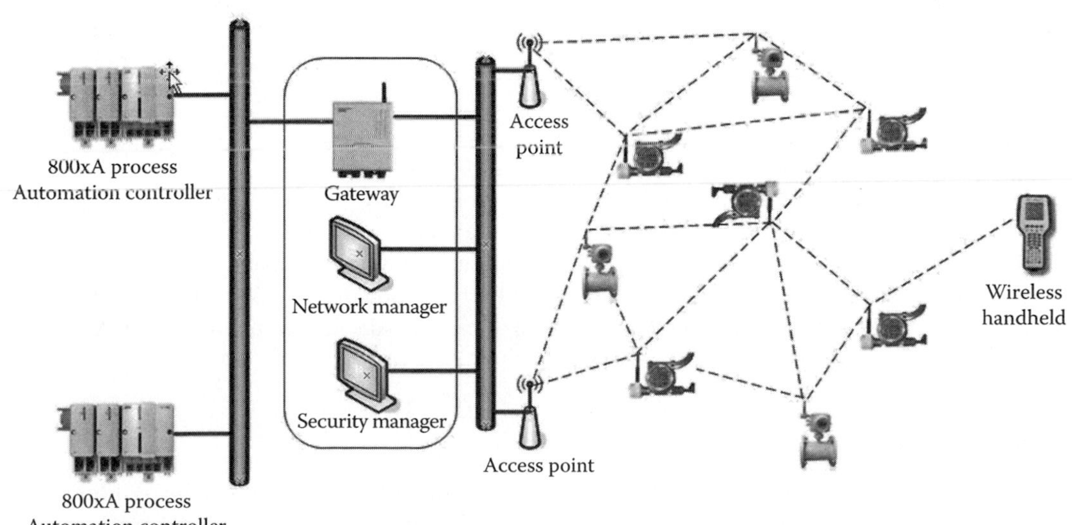

FIGURE 36.11 A wireless process control system.

steam valve. The temperature can now be maintained by using this wireless closed-loop control. This solution leads to reduced steam consumption, lowering operational costs, and ensuring that customers receive the right temperature.

36.4.5 Communication Standard

WirelessHART is the selected communication technology for the wireless control application. The choice of WirelessHART is mainly since it is the most widespread communication standard that currently exists for process automation. The solution would probably also have been working with the ISA 100.11a or *Wireless Networks for Industrial Automation—Process Automation* (WIA-PA) standards, which are very similar to WirelessHART.

36.4.6 Communication Technology

As mentioned, WirelessHART is the chosen communication standard for the storage tank heating control, because it's the most widespread used wireless communication standard for process automation domain. The RF chip is from the same vendor as used in the vibration monitoring application (see Section 36.4.3 for more information).

36.4.7 Challenges for the Future

WirelessHART and the other available standards such as ISA 100.11a, WIA-PA, ZigBee, and *IPv6 over Low-Power Wireless Personal Area Networks* (6LoWPAN) are mainly made for monitoring applications, and the standards lack a proper downlink communication channel. One of the more demanding challenges for control applications in the future is to offer deterministic communication and ensure that packets are delivered correctly within the given deadline. Up to now, neither of the aforementioned standards can guarantee deterministic communication and needs to be revised in order to enable fast wireless control applications, which have refresh rates about milliseconds. It should also be mentioned that 6LoWPAN adds additional overhead to the IEEE 802.15.4 since data/packet aggregation is needed. Furthermore, it will be even worse to guarantee delay constrained applications with the introduction of 6LoWPAN.

36.5 Communication Standards

The advent of WSNs brings many new and exciting technologies into the world of industrial automation. There are currently a number of initiatives underway for standardizing WSN for industrial use. General trend on the market is consolidation of suppliers and migration toward standards, most noticeably the release of WirelessHART and the development of ISA100. Additionally, a major automation network organization, PI (PROFIBUS) and PROFINET International, has decided to use WirelessHART as a base for its wireless process automation solution [7].

36.5.1 Standards Landscape

Table 36.3 provides an overview of the three main WSN standards targeting the industrial (process) automation market. Most fieldbus organizations refer to either one of these standards when they think about how to integrate WSN into their wired bus infrastructure. In addition, there are proprietary solutions out on the market, which could be considered de facto standards in certain niche applications.

TABLE 36.3 Comparison of Key WSN Standards in Industrial Automation

	ZigBee PRO	WirelessHART	ISA100.11a	WIA-PA
Availability	2007	September 7, 2007	September 9, 2009	
PHY layer	IEEE 802.15.4-2006	IEEE 802.15.4-2006	IEEE 802.15.4-2006	IEEE 802.15.4c-2009
Standard specifies	Layers 3 and 7	Layers 2–4 and 7	Layers 2–4 and 7	Layers 2–4 and 7
Frequency	868 MHz 915 MHz 2.4 GHz	2.4 GHz	2.4 GHz	
RF transmission	DSSS (frequency agility)2.4 GHz16 channels868/915:1 + 10 channels	FHSS 2.4 GHz 15 channels	FHSS 2.4 GHz 16 channels	
Media access	CSMA/CA optional GTS	TDMA, optional CSMA/CA in shared slots	TDMA optional CSMA/CA in shared slots	TDMA for CFP and CSMA/CA for CAP
Network topology	Mesh, star, hybrid	Mesh, star, (hybrid possible, but not specified)	Mesh, star, hybrid	Two layers: mesh and star
Bit rate	20 kbps at 868 MHz 40 kbps at 915 MHz 250 kbps at 2.4 GHz	250 kbps	250 kbps	250 kbps
Security	Symmetric keys, AES-128 encryption (enhanced security compared to ZigBee 2007)	Symmetric keys, AES-128 encryption	Symmetric keys, AES-128 encryption; optional use of asymmetric keys	Symmetric keys
Coexistence	CSMA/CA	Optional channel blacklisting, optional CCA, settable transmission power levels	Optional CCA, settable transmission power levels	CCA used in CAP

The following discussion gives some more details on the background of these standards. For a more technical description of the underlying technologies that are employed, see *Section 36.5.2*.

36.5.1.1 *ZigBee*

One of the best-known standards is the ZigBee standard, a low-power, low-cost, low-data-rate wireless specification, which targets home appliances, toys, industrial applications, building automation, and the like. The ZigBee standard is developed, managed, and promoted by the ZigBee Alliance [5]. The scope or focus of the ZigBee Alliance is on defining the network, security, and application SW layers, providing interoperability and conformance testing specifications and promoting the ZigBee brand.

36.5.1.2 WirelessHART

WirelessHART is the only industrial WSN standard currently available. It was released in September 2007 [3] and is a secure and reliable wireless technology for industrial applications ranging from class 2 to 5 (see Table 36.1), that is, from control to process measurement/monitoring. WirelessHART extends the well-known HART standard into the wireless domain, enabling new market areas for the large number of HART users. HART Communication [3] is a bidirectional industrial field communication protocol used to communicate between intelligent field instruments and host systems. HART is the global standard for smart process instrumentation and the majority, more than 25 million of smart field devices installed in plants worldwide, are HART enabled.

36.5.1.3 *ISA100*

Another ongoing initiative is ISA100* (formerly called ISA-SP100) [2]. Originally, the intent of ISA100 was to standardize all elements in the system, instead only to specify the upper levels in the stack (i.e., the open standards interface [OSI] stack), with a number of potential lower-level implementations. This scope has been changed to include all levels of the stack as part of the specification.

The ISA100 workgroup also contained two subgroups that investigated wireless solutions for different use cases: SP100.11a aimed for control use cases, while SP100.14 targeted monitoring use cases.

Compared to WirelessHART, ISA100 takes a more generic path as an industrial WSN technology. It will support the transportation of many different fieldbus formats (where HART is one of them). ISA100 aims to support industrial applications in the range from class 1 to 5, that is, from critical control to monitoring.

The standard was released in 2009.

36.5.1.4 *6LoWPAN*

6LoWPAN is an acronym for *IPv6 over Low-Power Wireless Personal Area Networks*, which is a protocol for transmission of IPv6 packets over IEEE 802.15.4 networks. This technology will enable the extension of the typical WSN structure into an IP-based system, basically allowing any computer in the automation plant to access any WSN device. It also allows reuse of existing IP network tools developed for home and office use, such as commissioning and network debugging support. In 2007, RFC 4919 with the problem statement and RFC 4944 with the protocol specification were published. RFC 4944 has since been extended by RFC 6282 with an improved header compression mechanism in 2011. Neighbor Discovery Protocol for 6LoWPAN technology is described in RFC 6775. There is a specification for applying 6LoWPAN technology to Bluetooth low energy, which is a part of the Bluetooth 4.0 specification, marketed as Bluetooth Smart. There are other Internet Engineering Task Force (IETF) WGs who are involved in 6LoWPAN-related work. The routing over low-power and lossy networks (ROLL) WG has a routing protocol for constrained node networks (RFC 6550). The CoRE WG is in the process of completing an application protocol (CoAP) that does many of the things that HTTP does well, with much less complexity.

36.5.1.5 *WIA-PA*

WIA-PA is an acronym for *Wireless Networks for Industrial Automation—Process Automation*, which is an IEC standard developed by the Shenyang Institute of Automation, Chinese Academy of Sciences. Contrary to WirelessHART and ISA100.11a, WIA-PA retains full compatibility with IEEE 802.15.4, supporting Beacons, Guaranteed Time Slots (GTSs), etc. It also builds a two-tier network where a mesh network forms the backbone and star networks connect the sensor and actuator devices. WIA-PA is targeting the same applications as WirelessHART and ISA100.11a.

36.5.1.6 *Proprietary Solutions*

There are many suppliers of proprietary WSN solutions, who offer a wide range of solutions, from bare technology/components (stack and HW) to complete WSN systems. There are also industrial automation suppliers who offer WSN solutions as part of their automation system.

However, the industry advocates the use of standardized solutions, and there is a strong trend to standardize solutions (like it has been for wired industrial communication technologies).

---

* The International Society for Automation (ISA).

36.5.2 Technology Primer

36.5.2.1 ZigBee

36.5.2.1.1 Architecture

ZigBee is a specification for a cost-effective, low-rate, and low-power wireless communication protocol for home automation, monitoring, and control. It aims to provide short-range wireless networking, which is scalable, self-organizing, and secure, while providing battery life up to 2 years. Although having existed since late 2004, ZigBee has yet to prove its success, at least in the industrial domain where reliability and security are of utmost importance. ZigBee has been more successful in neighboring areas such as building and home automation, as well as for automated meter reading applications.

ZigBee is a specification for the higher protocol layers and builds upon the physical (PHY) and medium-access control (MAC) layers in the IEEE 802.15.4 specification [6], as seen in Figure 36.12.

36.5.2.1.2 Basic Functionality

Mesh networking topology is supported, and routing is achieved through the ad hoc on-demand distance vector (AODV) algorithm. This means that the devices themselves are responsible for route discovery and that peer-to-peer communication is possible. In a ZigBee network, all nodes share the same channel, and frequency agility is minimal. There is no frequency hopping, and the only option is to scan for a channel with the least amount of interference at start-up.

There are two classes of network devices in ZigBee: full-function devices (FFD) and reduced-function devices (RFDs). The former can route messages in mesh networks and act as the network coordinator, whereas the latter can only communicate with one FFD in a star network setup.

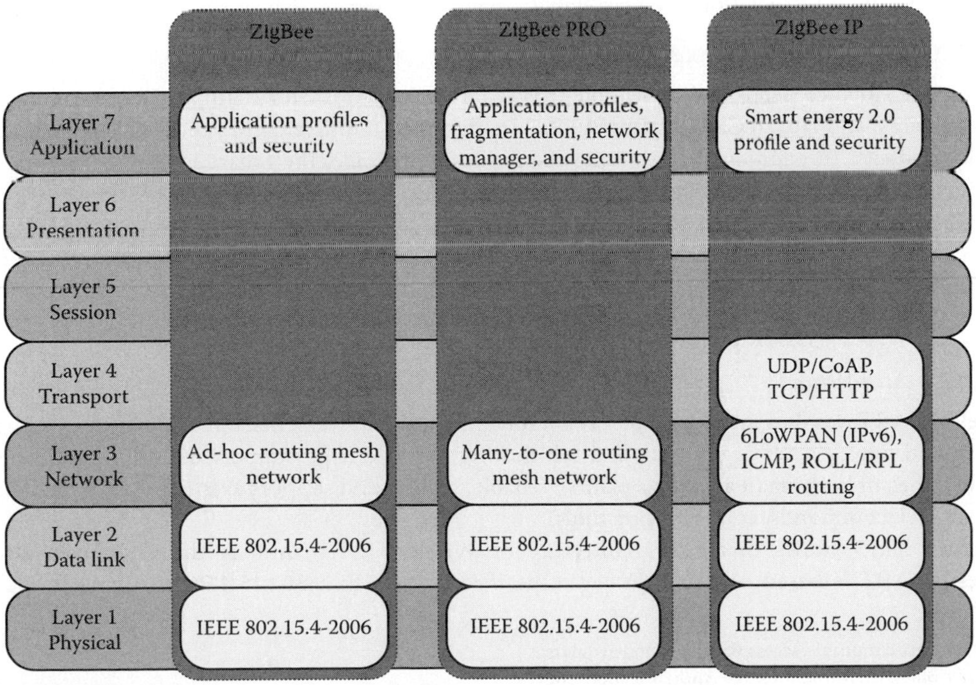

FIGURE 36.12 The different ZigBee protocol stacks.

ZigBee can operate in both beaconed and nonbeaconed modes. In the beaconed mode, the nodes are to some extent synchronized, and the superframe is divided into 16 slots. The slots in the frame are generally contention based, using CSMA/CA.* There is an option to dedicate up to seven of these slots to specific nodes in order to increase determinism, so-called GTS. However, support for this is not mandatory, and use of this feature might break interoperability.

36.5.2.1.3 Security

In the 2006 version of the specification, security is not mandatory. However, support for authentication, integrity, and encryption for both network and application layer is present. MAC layer security available through 802.15.4 is not explicitly addressed in the ZigBee standard, and its use might break interoperability between different vendors' products. Replay attacks are protected against using sequential numbering. ZigBee makes use of the security mechanisms in 802.15.4, Counter with CBC-MAC† (CCM) with AES-128 encryption, but with the option to employ encryption only or integrity only.

Three key types are used: *master key*, *link key*, and *network key*. The master key is necessary to be able to join the network. The link key is used for end-to-end encryption and would by that provide the highest level of security at the price of higher storage requirements. The network key is shared between all devices and thus presents a lower level of security, though with the benefit of reduced storage requirements in the devices. All keys can be set at the factory or be handed out from the trust center (residing in the network coordinator), either over the air or through a physical interface. For commercial grade application, the trust center can control the joining of new devices and periodically refresh the network key.

36.5.2.1.4 ZigBee PRO

ZigBee 2007 is an update of the ZigBee specification, which specifies two profiles for the stack: the first profile, called ZigBee, is compatible with the earlier ZigBee 2006 stack version. The second profile, called ZigBee PRO, provides new features such as multicasting, many-to-one routing, higher security (using Symmetric-Key Key Exchange), frequency agility (possibility to change channel when the interference is too high), and fragmentation and reassembly.

36.5.2.1.5 Application Profiles and ZigBee IP

The ZigBee Alliance publishes application profiles for various areas, for example, Home Automation, Smart Energy, and Health Care. An application profile allows multiple OEM vendors to create interoperable products. In addition to specifying the application profile, the Smart Energy 2.0 profile also defines an IP-based protocol, called ZigBee IP, which replaces the ZigBee PRO stack.

ZigBee IP uses RFC (Request for Comment) specifications developed by the IETF, that is, 6LoWPAN (IPv6 for WPAN), ROLL, RPL (IPv6 Routing Protocol for Low-Power and Lossy Networks), etc., to form the ZigBee IP stack.

36.5.2.2 WirelessHART

36.5.2.2.1 Architecture

WirelessHART was designed based on a set of fundamental requirements: it must be simple (e.g., easy to use and deploy), self-organizing and self-healing, flexible (e.g., support different applications), scalable (i.e., fit both small and large plants), reliable, secure, and support existing HART technology (e.g., HART commands, configuration tools).

Figure 36.13 shows that the architecture of WirelessHART mimics the OSI layer design. WirelessHART is based on the PHY layer specified in the IEEE 802.15.4-2006 standard [6]‡ but

* Carrier-sense multiple access with collision avoidance.
† Cipher Block Chaining Message Authentication Code.
‡ WirelessHART only supports the globally available 2.4 GHz ISM frequency band.

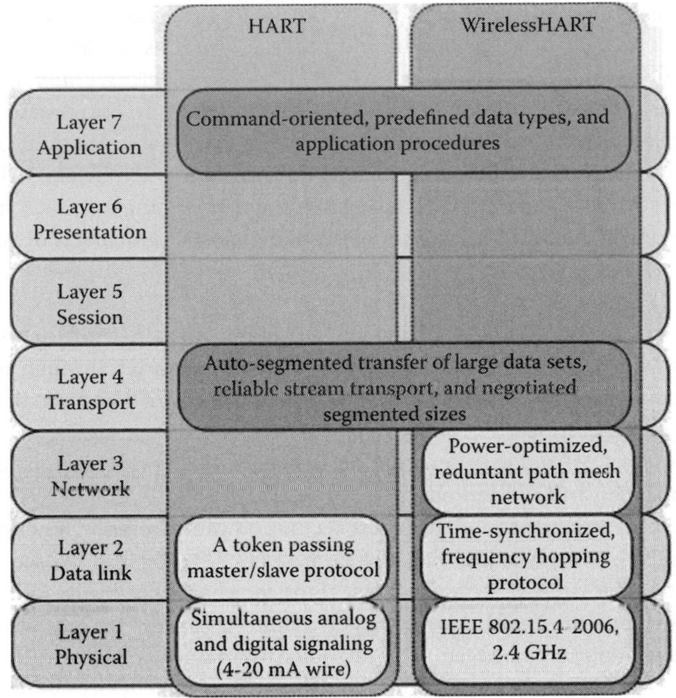

FIGURE 36.13 HART and WirelessHART protocol stacks.

specifies a new data link (including MAC), network, transport, and application layers. The figure also shows the backward compatibility of WirelessHART because it shares the transport and application layer with HART.

36.5.2.2.2 Basic Functionality

WirelessHART is a TDMA-based network. All devices are time synchronized and communicate in prescheduled fixed-length time slots. TDMA minimizes collisions and reduces the power consumption of the devices.

WirelessHART uses several mechanisms in order to successfully coexist in the shared 2.4 GHz ISM band:

- *Frequency-hopping spread spectrum* (FHSS) allows WirelessHART to hop across the 16 channels defined in the IEEE 802.15.4 standard in order to avoid interference.
- *Clear channel assessment* (CCA) is an optional feature that can be performed before transmitting a message. CCA works like CSMA/CA but without using an exponential back-off. This improves coexistence with other neighboring wireless systems.
- The *transmit power level* is configurable at a node level.
- A mechanism to disallow the use of certain channels, called *blacklisting*, is available.

All of these features also ensure WirelessHART does not interfere with other coexisting wireless systems that have real-time constraints.

All WirelessHART devices must have routing capability, that is, there are no RFDs like in ZigBee. Since all devices can be treated equally, in terms of networking capability, installation, formation, and expansion of a WirelessHART network become simple (self-organizing).

WirelessHART forms mesh topology networks (star networks are also possible), providing redundant paths, which allow messages to be routed around physical obstacles, broken links, and interference (self-healing). Two different mechanisms are provided for message routing: *graph routing* uses predetermined paths to route a message from a source to a destination device. To utilize path redundancy, graph routes consist of several different paths between the source and destination devices. Graph routing is the preferred way of routing messages both upstream and downstream in a WirelessHART network. *Source routing* uses ad hoc–created routes for the messages without providing any path redundancy. Source routing is therefore only intended to be used for network diagnostics, not any process-related messages.

36.5.2.2.3 Security

Security is mandatory in WirelessHART; there is no option to turn it off or to scale it up/down. WirelessHART provides end-to-end and hop-to-hop security measures through payload encryption and message authentication on the network and data-link layers (DLLs). However, the security measures are transparent to the application layer.

WirelessHART uses CCM* mode in conjunction with AES-128 block cipher, using symmetric keys, for message authentication and encryption.

A set of different security keys are used to ensure secure communication: A new device is provisioned with a *join key* before it attempts to join the wireless network. The join key is used to authenticate the device for a specific WirelessHART network. Once the device has successfully joined the network, the network manager will provide it with proper session and network keys for further communication.

The actual key generation and management are handled by a *plant-wide security manager*, which is not specified by WirelessHART, but the keys are distributed to the network devices by the network manager. A *session key* is used on the network layer to authenticate the end-to-end communication between two devices (e.g., a field device and the gateway). Different session keys are used for each pairwise communication (e.g., field device to gateway, field device to network manager). The DLL uses a *network key* to authenticate messages on a one-hop basis (e.g., to and from neighbors). A well-known network key is used when a device attempts to join the network, that is, before it has received the proper network key.

Keys are rotated based on the security procedures of the process automation plant.

36.5.2.3 ISA100

The following description of ISA100.11a is a snapshot of what the standard looks like at the time of writing (March 2008). The standard will very likely undergo some (minor) changes before the final version is released.

36.5.2.3.1 Architecture

ISA100.11a is designed to fit the needs of many industrial protocols, such as HART, Profibus, and FOUNDATION Fieldbus. Its scope also includes parts of the plant backbone network.

Figure 36.14 shows an ISA100.11a network that includes a part of the backbone. The leftmost part, from the backbone routers and to the left, is called the DLL subnet, that is, it consists of nonrouting devices, routing devices, and backbone routers. However, in the simplest network scenario, the backbone routers are replaced by one gateway and manager device, which is directly connected to the plant automation network.

Figure 36.15 shows the architecture of ISA100.11a described in terms of the OSI reference. The *physical* layer is based on IEEE 802.15.4-2006 and so is the MAC part of the DLL. The upper part of the DLL handles all communications aspect in the DLL subnet part of an ISA100.11a network. A DLL subnet corresponds to a WirelessHART network, that is, a time-synchronized, frequency-hopping mesh network composed of field devices. The *network* layer is based on a subset of the 6loWPAN (IETF RFC 4944) specification, that is, sending IPv6 packets over 802.15.4 networks. The transport layer is also based on 6loWPAN, which specifies the use of user datagram protocol (UDP) for end-to-end delivery of packets.

---

* Counter with CBC-MAC (corrected).

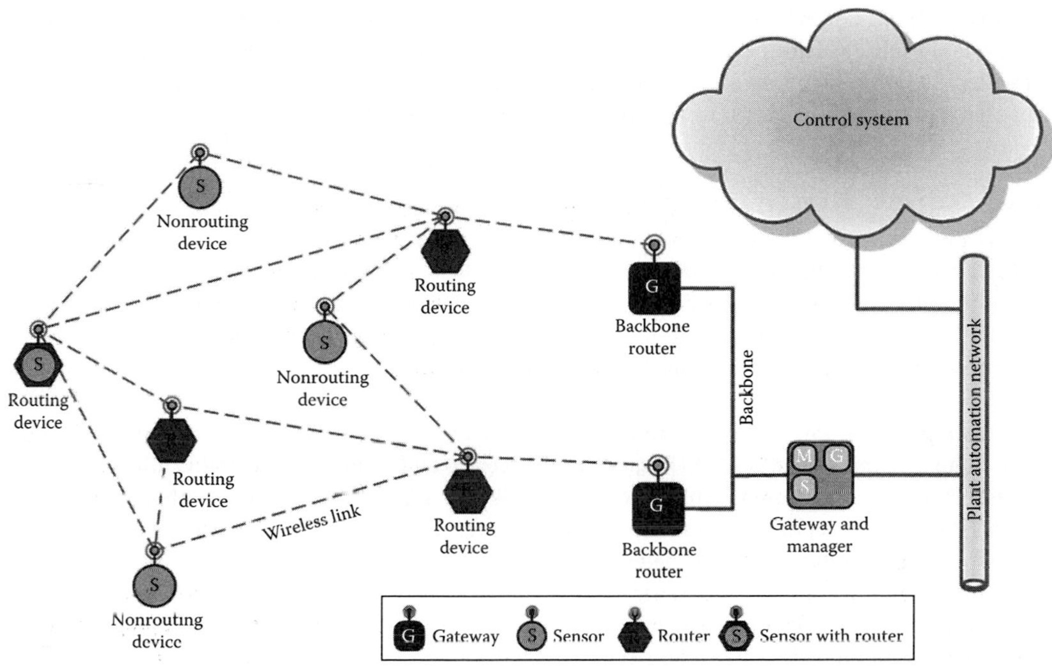

FIGURE 36.14 ISA100.11a network.

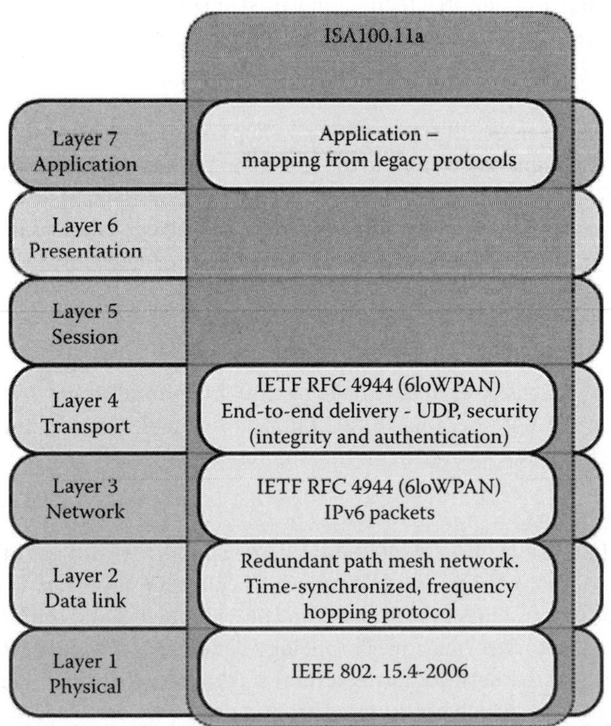

	ISA100.11a
Layer 7 Application	Application – mapping from legacy protocols
Layer 6 Presentation	
Layer 5 Session	
Layer 4 Transport	IETF RFC 4944 (6loWPAN) End-to-end delivery - UDP, security (integrity and authentication)
Layer 3 Network	IETF RFC 4944 (6loWPAN) IPv6 packets
Layer 2 Data link	Redundant path mesh network. Time-synchronized, frequency hopping protocol
Layer 1 Physical	IEEE 802. 15.4-2006

FIGURE 36.15 ISA100.11a protocol stack.

Note that UDP is an unreliable delivery service. Delivery reliability for a DLL subnet is handled in the DLL, whereas reliable delivery on the backbone network is handled by the backbone protocol, which is not within the scope of the standard. The transport layer also contains options for securing the delivery of packets, for example, encryption, authentication, and integrity.

36.5.2.3.2 Basic Functionality

As WirelessHART, ISA100.11a is a TDMA, frequency-hopping, mesh network. In addition, it supports the IPv6 protocol and the UDP transport protocol (through the use of 6loWPAN). Thus, all devices in an ISA100.11a network have an IPv6 address, which makes them globally visible within the plant.

In an ISA100.11a network, there are several different kinds of devices, as can be seen in Figure 36.14. ISA100.11a supports both *routing* and *nonrouting* devices, as well as a special kind of routing devices, *backbone* routers, which utilize the backbone in order to transport messages within, or between, DLL subnets.

The DLL provides time synchronization, frequency hopping, mesh networking, and routing, within a DLL subnet. A *time slot* in ISA100.11a can vary between 10 and 12 ms. This flexibility allows the use of *duocast*, which is a procedure in which each message is sent to two receivers, who both answer with an ACK on successful reception. Even if one transmission fails, the other might succeed, which increases the communication reliability of the network.

Frequency hopping can be either *slotted*, that is, the frequency is changed for every slot, or *slow*, that is, the frequency is only changed when several sequential slots have passed. Slow hopping relaxes the time synchronization requirements for devices, allowing low-cost clocks and crystals to be used.

ISA100.11a DLL subnets can have many topological shapes: mesh, star, or hybrid (a mix of both), and the network as a whole also incorporates the use of a backbone network. The gateway/manager device can be connected either directly to the DLL subnet via a radio or through backbone routers, in which case, it is located somewhere on the backbone network. Routing in DLL subnets is handled through the use of either graph routing (preallocated redundant paths) or source routing (ad hoc–created nonredundant paths), but routing on the backbone network is not within the scope of the standard.

36.5.2.3.3 Security

ISA100.11a provides a two-layer security measure: single-hop and end-to-end security. The single-hop security is handled in the MAC sublayer (DLL), and the end-to-end security is handled in the transport layer.

Both security layers use shared-secret symmetric keys (it is optional to use asymmetric keys) and AES-128 in order to provide encryption, authentication, and integrity checking. Payload encryption is optional in ISA100.11a. The ISA100.11a specification is designed with future *upgrades* of the security specification in mind.

36.5.2.4 WIA-PA

WIA-PA is an industrial WSN standard developed by the Chinese Industrial Wireless Alliance (CIWA) for the Chinese market. WIA-PA became an IEC Publicly Available Specification (PAS) in 2008 (IEC/PAS 62601), and in 2011, it was approved as an IEC standard.

36.5.2.4.1 Architecture

WIA-PA is based on the IEEE 802.15.4-2006 standard and supports two types of network topologies: a hierarchical topology, which combines star and mesh, and a pure star topology. The hierarchical network topology that combines star and mesh (both based on IEEE 802.15.4-2006) is illustrated in Figure 36.16.

The first level of the network forms a mesh topology consisting of the gateway device and routing devices. The second level of the network forms several star topologies, where each star consists of field devices and handheld devices, which are centered around one of the mesh routing devices.

Figure 36.17 shows the OSI layers of the WIA-PA stack. WIA-PA is fully compliant with IEEE 802.15.4c and supports Beacons and GTSs on the PHY and MAC layers. The two lowest layers of the

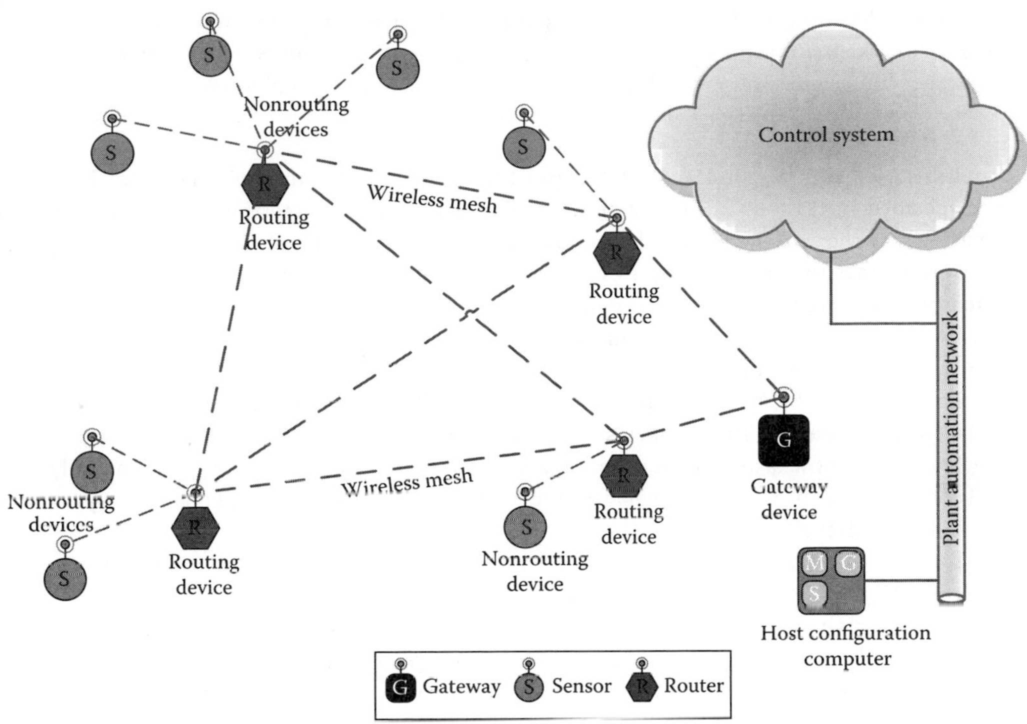

FIGURE 36.16 WIA-PA network.

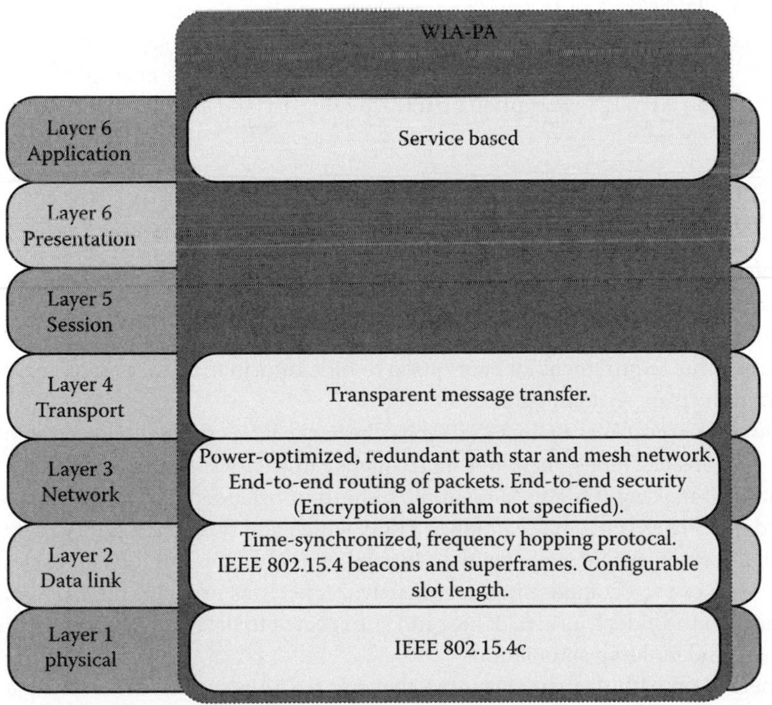

FIGURE 36.17 WIA-PA protocol stack.

stack (PHY and data link) form the star networks in WIA-PAs hybrid topology, while the upper parts of the stack (network and transport layers) form the mesh backbone network. Security is implemented on two levels: per-hop security on the DLL and end-to-end security on the application layer.

36.5.2.4.2 Basic Functionality

As WirelessHART and ISA100.11a, WIA-PA uses time-slotted communication; however, it uses the beacons specified in IEEE 802.15.4. The network manager generates superframes (see for each routing device, where the superframe length is set as the lowest data update rate of all the members in the star network of that router). WIA-PA uses shared time slots for transmission of aperiodic data, and dedicated time slots are used for intra- and intercluster transmission of periodic data.

WIA-PA also supports frequency hopping and uses three different mechanisms for this. Each of the mechanisms is tailored for different types of communication. *Adaptive frequency switching* is used to change the channel for the Beacon, contention access period (CAP), and contention-free period (CFP) for each superframe period. The channel is only changed if the conditions are bad. *Adaptive frequency hopping* is used for intracluster communication and changes the channel per time slot and only if the channel condition is bad. *Time slot hopping* is used for intercluster communication and always changes the channel for each time slot.

WIA-PA uses static routing, which is configured by the network manager based on information on devices' neighbors.

WIA-PA supports both data and packet aggregation. *Data aggregation* is used by the source of the data to combine several data packets into one packet before sending it to the router device. *Packet aggregation* is used by routing devices to combine several data packets from different field devices.

36.5.2.4.3 Security

WIA-PA provides a flexible way to use the security measures defined in the IEEE 802.15.4 standard. Both data integrity (message integrity codes [MICs]) and confidentiality (encryption) are provided on the application and DLLs (i.e., end-to-end and per-hop security). Security is optional in WIA-PA, and it is possible to configure the use of either application or DLL security separately as well as if MIC or encryption-only protection should be used. There is also an option to encrypt the application layer payload. The MIC and AES-128 encryption guarantee integrity and confidentiality for communication on the DLL.

36.5.3 Why Is ZigBee Not Enough?

Since any problems with the equipment translate to economical loss for an industrial user, reliability is a primary concern for these users. Hence, parameters like network robustness, reliable message delivery, authentication, and integrity are all important for industrial use. Moreover, the threat of industrial espionage advances the requirement for encryption to hide information that can reveal anything about the production in the plant to competitors.

One of the loudest arguments against ZigBee has been the lack of industrial-grade robustness. First of all, in older versions of ZigBee, there was no frequency diversity since the entire network shared the same static channel, making it highly susceptible to both unintended and intended interference. This has been remedied in later versions (i.e., ZigBee PRO), which add frequency agility in order to increase the robustness. However, at that time, many industrial efforts to develop robust wireless communication were underway or even completed; unfortunately, ZigBee was probably too late with the additions to have an impact on the industrial area. It has, and is, expected to have a bigger impact on the neighboring areas of home and building automation.

Secondly, there is no path diversity meaning that in case a link is broken, a new path from source to destination must be set up. This increases both delay and overhead and eventually will consume all bandwidth available.

Battery operation for routers with many peers is not realistic, since the CSMA/CA forces it to keep its receiver on for a large part of the frame. Keep in mind that industrial users require several years of battery life, since frequent battery change on thousands of sensors is not an option.

Security in ZigBee can to a great extent be set up to meet the requirements from industrial users, although care must be taken to use equipment from vendors, which support the necessary security mechanisms. A good guideline for ZigBee security is found in [8].

In summary, the harsh operating environment and battery operation and lifetime requirement render ZigBee a nonviable option for the industrial automation application. WirelessHART, ISA100.11a, and WIA-PA on the other hand are very suitable as they are designed to be very robust communications technology in the presence of interference and fading. In addition, these standards have features that allow them to efficiently coexist in a *crowded* frequency band, that is, they are *nice neighbors*.

36.6 Low-Power Design

In this chapter, we dive more deeply into the issues related to low-power design of HW, SW, and the communication protocol.

36.6.1 Basic Principle

The aim is to use the available resources within the acceptable limits of the specification and never have anything powered up, if it doesn't have to be. The task reduces to switching units, such as the sensor, CPU, and transceiver on and off with the right timing. Assume a node needs to wake up at regular intervals and transmit its sensor value if it differs from the last value by more than a predetermined amount. After the value has been sent over the radio channel, the unit waits for an acknowledge message indicating that the packet has been correctly received.

The required behavior of the node is best explained using what is known as a state diagram—a schematic representation of the state the node is in, the events that may cause it to move from one state to another and the actions associated with each state transition.

Figure 36.18 illustrates a typical high-level state diagram for a sensor node, in which the *measure* and *send* states together form an operational mode. The transition from active to sleep is managed by a sleep–wake-up scheme, which has been derived from the measurement of duty cycle requirement.

A natural prerequisite to implement an efficient sleep–wake-up scheme is that the active/sleep state of various system components can be controlled externally. If and how this is accomplished can vary greatly between components and can have quite an impact on the duty cycle of the device.

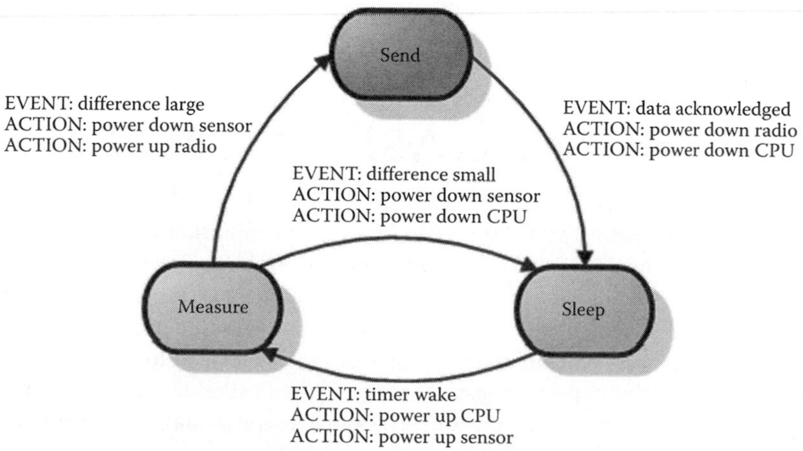

FIGURE 36.18 Sensor node state diagram.

Wireless Vibration Monitoring Case Example

The vibration sensor is build around the following controllable components:

- *CPU*: The microcontroller has several controllable low-power modes.
- External *real-time clock (RTC)*: The clock can be powered on and off as well as controlled using its *normal* SW interface.
- *Radio chip*: The radio is a *black box* that only provides on/off control as well as a reset possibility.
- *Analog filter electronics*: These filters can be powered on and off.

The state diagram shown earlier is rather simplistic as it only shows the *normal* measurement use case. In real life, many additional and alternative use cases need to be considered. As the following sections will show, this simple state chart can become quite complex to design.

36.6.2 Sleep Modes

The first parameter to consider is the power consumption in normal operating mode (i.e., active mode) for the CPU, sensor, radio transceiver, and possibly other elements such as external memory and peripherals. However, since the sensors spend most of their time *sleeping*, it is important that the consumption during sleep mode is low. Often, we can switch power to the sensor and transceiver off completely. However, the CPU will need some form of sleep mode from which it can be woken. The consumption in this sleep mode is absolutely crucial for the overall power budget.

Additionally, one must ensure that all the necessary elements can be controlled by the CPU. It is the master in the system and needs to have complete control over the other functional blocks. However, the possibility to exert this control often varies in the different power modes supported by the CPU. The lowest power-saving mode typically shuts down (stops) the CPU, which will not allow it to control anything.

Wireless Vibration Monitoring Case Example

The CPU, a Texas Instruments MSP430 variant, used in the vibration monitor device supports five low-power modes. The CPU can be operated at voltage levels ranging from 1.8 to 3.6 V. When operating at a 3.0 V supply level, the *lowest* power-saving mode (the CPU is practically turned off) typically requires 0.1 μA; in *standby* mode (next lowest), it requires 1.6 μA; and in *active* mode, it requires 400 μA. Looking at these numbers, it's obviously most desirable to use the *off* mode during sleep time, as it has a 16 × lower power consumption compared to the standby mode. The drawback with the *off* mode is that the CPU can only wake up from external stimuli (i.e., an externally generated interrupt), which is not the case in the standby mode (or any of the other low-power modes).

One aspect that is often overlooked is the time the elements need to turn on and off. For example, the transceiver will need some minimum time for its oscillators to stabilize. While they wait, both the transceiver and the CPU burn power. This therefore needs to be minimized. The same obviously holds true for both the CPU and the sensor.

Wireless Vibration Monitoring Case Example

The sensor node uses analog electronic components to filter the *raw* vibration signal into a desired format. The CPU has control over the activation/deactivation of this analog part and only activates it for the duration of the measurement. The analog part has a fairly long settling time, ~15 s, which forces the node to wait until a usable signal is available for processing (fortunately, the CPU can spend this time in the standby low-power mode).

Another very important settling time is the time it takes for the CPU to become fully operational when waking up from a power-saving mode. For the MSP430 variant used, this time is only 6 μs from both the *off* mode and the standby mode. But the *off* mode has the added benefit of allowing us to further decrease power consumption by shutting off the oscillator signal from the RTC to the CPU (which is used to generate the CPU clock). However, without the oscillator signal, the CPU will default to an inaccurate, internal, RC-generated, lower-frequency clock. Thus, when waking up from the *off* mode, the RTC oscillator signal has to be activated before the CPU will stabilize, which adds to the 6 μs wake-up time.

Timers are an integral part of any sleep–wake-up scheme. Generally, ultralow-power CPUs only have 8- or 16-bit resolution timers, which, when used to generate interrupts at scheduled timeouts, have fairly *short* maximum intervals (in the order of a few seconds). If a longer sleep interval is desired, the timer is forced to wake up (i.e., the CPU becomes active), reset its counter, and go to sleep several times during that interval. Furthermore, in some power-saving modes (as described in the previous example), the CPU shuts down its internal clock, which disallows the use of the internal timers. Thus, if internal timers are the choice for the wake-up scheme, some power-saving modes will not be usable.

Wireless Vibration Monitoring Case Example

The duty cycle of the vibration monitoring device is 2 weeks. The MSP430 CPU used is driven by a 32 kHz crystal, and only provides 16-bit timers, which equates to a maximum interval of only 2 s. This would require the timer interrupt to (1) activate, (2) reset the timer counter, and (3) decrement the total sleeping time, a total of 604,800 times (14 days = 1,209,600 s and a 2 s time-out) during the 2-week interval! In combination with the desire to use the lowest power-saving mode (*off* mode) of the CPU, the solution was to choose an external timer, that is, an RTC, which supports very long time-out intervals (actually years), and allows the CPU to stay in the *off* power-saving mode during the sleep time.

36.6.3 HMI

As noted in previous sections, a local HMI is commonly required to check the status, calibrate, and maybe even instruct the device to perform a measurement. From a low-power-design point of view, this represents additional components that consume power. Depending on the requirements on packaging or the available power budget, several alternatives are used in the industry ranging from simple LEDs and buttons to LCDs. When designing the HMI functionality, great care must be taken how the HMI is used. As battery-powered systems cannot afford to constantly have the HMI active, it must therefore be switched on and off at appropriate times, while maintaining a user-friendly interface (e.g., that responds in a timely manner according to your expectations).

When the HMI is used to asynchronously wake up a device and maybe issue a request to perform a measurement, the normal duty cycle is changed, which will result in a different power consumption of the device. The update rate of an HMI is typically a lot faster than the normal duty cycle of the WSN devices (since it's often used to get a *quick snapshot* of the process). If the HMI is used extensively, not only will the HMI consume power, but the device will also run according to an entirely different, from the planned, duty cycle. This could result in early depletion of batteries or even peak power consumption that exceeds the bounds of the available power.

Wireless Vibration Monitoring Case Example

The vibration monitoring device has an HMI consisting of two LEDs and one button. The LEDs are used to indicate activity, such as powering on/off, status (sleeping or active), and transmitting data. The button is used to activate or deactivate the device, force it to start a measurement, and consequently transmit the data to its destination.

The device uses low-power LEDs, which each have a current consumption of 2–4 mA when lit (LEDs are normally specified for a 4–20 mA consumption); the exact consumption depends on the supply voltage level (2.7–3.6 V). Compared to the current consumption of the CPU in active mode, 400 μA, this is a very high consumption.

To further lower the power consumption of the HMI, it is only activated when a human is interacting with the device. Although it makes no sense to blink a LED if no one is watching, there are situations when this could be necessary, for example, if the sensor is in a hard-to-reach place but can be inspected visually. The button is used to interact with the device, and the response is always indicated by blinking the LEDs in different patterns. This pattern, for example, short blinks and long blinks, indicates the internal state of the device (sleeping, measuring, sending, etc.). In order to even further decrease the power consumption of the LEDS, the digital I/O pins used to control them are set to an inactive state when they are inactive.

It's important to note that an HMI does not necessarily have to be a physical component within the device itself. For example, a handheld unit could communicate (wired or wirelessly) with the device and request data outside of the normal duty cycle. The same goes for a remote terminal connected to the network, which would cause additional traffic routed through the network. This means that in order to make any guarantees in terms of battery lifetime, the topology and usage scenarios need to be clear for the entire network.

36.6.4 Energy-Efficient Protocols

In addition to utilizing low-power electronics and a clever SW architecture (i.e., sleep–wake-up scheme), the communication protocol has a major impact on the final power consumption of the system. The communication protocol prescribes when the system needs to wake up, how it synchronizes, how long it needs to stay on, etc. Relevant protocol design choices from a power consumption point of view include the following:

- Media access scheme: In a contention-free media access scheme, for example, TDMA, collisions are minimized, thus requiring less active time to send the data. The reverse is true for contention-based media access such as CSMA, where the node first has to listen before it can send its data (which degrades rapidly when there is lots of communication).
- Topology: The topology not only affects the timeliness of message delivery but also the power consumption. In a multihop system, a node may have to wake up to forward data from other nodes, which is not the case for a single-hop system. In addition, intermediate nodes don't know when they may be called upon to route packets for others. This is what motivates the hybrid mesh topology with mains-powered router nodes shown in Figure 36.8.
- Data package size: The size of the payload, header, CRC, etc., all have a direct impact on how long the radio needs to be turned on to send its data. This is regardless of what media access scheme is employed. When designing the protocol, a trade-off has to be made between throughput, latency, and power consumption, and at least one of them has to suffer.
- Security: It is important that the data are secured from intentional corruption (i.e., integrity) and that it is sent from a valid device (i.e., authentication). Security adds overhead to the packets and to the computation performed at the devices (i.e., to check the integrity and authenticity). This, in turn, possibly leads to the need of specific HW (i.e., encryption support) in order to achieve the required communication timing requirements.

Details in the communication protocol thus dictate the lower bounds of the consumption. Some communication protocols are notoriously inefficient, and no smart embedded programming in the world will help get the consumption down to an acceptable level. Others are designed to give low consumption without unduly compromising communication performance. The WISA [9] platform is one such low-power protocol. The high performance can be attributed to two reasons: it is single hop and it uses time-division multiplexing (TDM). The former avoids delays in intermediate nodes. The latter guarantees that a node will be alone on the channel, that is, there will be no collisions. WISA is designed for manufacturing control applications, which require high node density and deterministic response times but have rather relaxed requirements on throughput and range.

The recently developed WirelessHART specification with the underlying 802.15.4 protocol is more general but will have lower communication performance. It specifies multihop, where a message can use several radio hops to get to its destination. This introduces uncertainty in the delivery time of a message but extends the range while keeping power consumption acceptable.

In short, the WISA protocol is well adapted to the requirements of discrete manufacturing, while WirelessHART is ideally suited for asset monitoring applications.

Wireless Vibration Monitoring Case Example

In the WSN of the reference case, all nodes have the capability of routing traffic. The network manager continuously changes communication links in the mesh network, in order to cope with fading and interference, as well as trying to evenly spread traffic over all nodes (thus evenly spreading the power consumption between the nodes).

Furthermore, the TSMP solution provided by DUST networks allows the designation of leaf nodes, that is, nodes that will never route any traffic. This provides the possibility to form a mesh *backbone* network consisting of pure routing nodes (i.e., without vibration measurement capability) and to have all sensor nodes configured as leaf nodes, which will further decrease their energy consumption.

36.7 Packaging

36.7.1 IP Rating

The international standard IEC 60529 [10] classifies the degrees of protection of an electric system against solid and/or liquid. The rating consists of the letters IP (International Protection marking) followed by two digits and an optional letter. Certain IP codes require that the device is completely sealed, which might cause problems for the antenna (detuning) and battery design. A completely sealed device does not permit an easy access for battery change. The sealing could also force the use of an internal antenna, that is, inside the housing.

36.7.2 Hazardous Environments (EX)

When designing devices intended for use in potentially explosive atmospheres, special measures need to be taken to prevent or minimize the risk of explosion. This means that you must consider

1. Every possible electrical or nonelectrical source of ignition
2. All potentially hazardous environments where the device could be deployed
3. The different ways the device could be deployed and the technical ability of the person using the device

To design and develop devices for hazardous environment requires long experience in the field as well certification by a notified body. The overview provided in this chapter should thus not be considered a cook book for safe designs.

Conformance to a number of standards is required to achieve EX certification. The IECEx* Scheme [11] was created to facilitate international trade in equipment and services for use in explosive atmospheres, while maintaining the required level of safety. The IECEx Scheme is based on the use of specific international IEC Standards for type of protection of EX equipment. Since July 2003, EX products that are placed on the European market must be certified to the ATEX† directive (ATEX 94/9/EC) [12]. This involves the testing and assessment of such products to the latest ATEX standards.

* International Electrotechnical Commission for Certification to Standards Relating to Equipment for Use in Explosive Atmospheres.

† From the French—Atmospheres Explosibles.

To achieve an EX safe design, a number of principle protection concepts are applied, depending on the applications and target hazardous area (zones):

- Nonsparking devices: Typically applied to transformers, motors, cells, batteries, etc., but not appropriate for equipment or components containing semiconductors. Protection relies upon a dust/watertight enclosure to avoid tracking across live circuits.
- Keep flammable gas out. Encapsulation (enclosing circuit in potting compound), oil immersion, or pressurized enclosures to keep the potentially explosive atmosphere away from the source of ignition. The downside of this approach is that it adds significant cost and weight to your design and sacrifices accessibility and serviceability of the device.
- Limit energy of sparks and surface temperatures. Protection relies upon an electronic design that limits the emitted/transferred power of the device and prevents hot surfaces above the given temperature class. This approach is commonly referred to as *intrinsic safety.*

An IS design has built-in measures to prevent excessive heat buildup and the danger of an electric spark. To achieve this, you need to put limitations on voltage, current, capacitance, and inductance to ensure that the available energy within the device is below the minimum ignition energy of the potentially explosive atmosphere.

A practical implication of this is that you need to limit the maximum amount of energy supplied by the power supply, as well as dividing your design into clearly isolated blocks and limiting the energy transfer between these blocks. The interfaces between the system blocks thus constitute the majority of the design work and also have the most dramatic impact on other design aspects.

Additional components are typically required to limit discharge energy or isolate blocks, for example, resistors, fuses, and zener diodes. These additional components tend to increase the overall power consumption of the device as you will experience losses when traversing through the different blocks of your design.

In addition to providing means for limiting the energy, you also need to provide some guarantees that these protection and power-limiting components will not fail. When designing an IS circuit, all possible scenarios for connections of components need to be considered when calculating the safety circuit parameters. Defining clear interfaces helps to determine worst-case failure mode and devise possible solutions to handle these failures. Typically duplication or triplication of components is required, as well as using the components below their specified ratings and observing sufficient creepage and clearance distances. This further lowers the efficiency of the device and can impact the form factor of the device and require additional redesigns.

Since a component with a high surface temperature could cause an ignition of the gas, the maximum temperature of any component on a circuit board under fault conditions must be considered. Limiting the energy alone does not automatically result in a low component surface temperature, especially when considering small components. This means that you need to take great care when selecting components and ensure that the heat dissipation is satisfactory to avoid an excessive heat buildup. Usually this is solved by choosing sufficiently large components, which again can impact the overall form factor of the electronics and lead to space problems.

Wireless Vibration Monitoring Case Example

A battery with IS approval was selected. The battery is connected to the PCB of the device via an infallible current-limiting resistor. This will ensure that the maximum power loss in any component on the PCB is limited. The temperature rise can then be calculated for each component on the PCB.

In addition, the resistor was carefully chosen with a specified temperature rise that will ensure that the maximum power dissipation in the resistor upon a short will not result in an unacceptable temperature rise in the resistor.

Since the battery voltage level will vary with the surrounding temperature, we might not have enough energy to operate the radio circuit under all conditions (especially during start-up). To solve this, a capacitor bank was connected across the power supply. The capacitor bank stores just enough energy required for the radio transmission. The capacitor bank was carefully designed to comply with relevant safety standards (Figure 36.19).

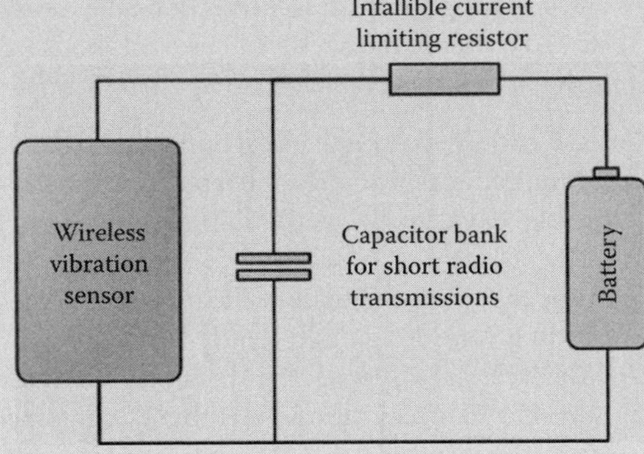

FIGURE 36.19 Principle battery design.

36.7.3 Other Environmental Considerations

In addition to the IP and EX environments described earlier, most industrial environments are also subject to vibration and extreme temperatures.

In an environment where the device is subject to severe continuous or occasional vibrations (shock), there is a risk that internal components will shake loose due to these vibrations. Therefore, as a protection, a device can be filled with a shock-absorbing material. Drawbacks with this solution are that the filling materials usually have a high cost and can force a design solution where the battery is also encased in the material, which make battery changes impossible.

Depending on the specified tolerance of the device components, a high or low temperature can force the device to have a thermally insulating housing or some kind of cooling mechanism to protect its internal components. Thermal insulation housing consists of materials that do not transfer heat quickly, in order to separate and insulate the sensitive internal components from the external temperature. It is also possible to have a thermally insulating barrier, a non-heat-conducting material positioned between the hot/cold part and the critical part of the device.

There are different kinds of cooling mechanisms available: active, such as fans, and passive, such as heat sinks and heat pipes. A problem with active cooling components is that they also wear out and could pose problems if they fail, for example, overheating. This would require maintenance of the cooling mechanism on the device, possibly outside the normal maintenance schedule, which is too costly. Thus, active cooling is not a desirable solution.

36.7.4 Form Factor

The form and shape of an embedded device is very important. It depends on various requirements, such as

- The environment where the device will be located, which usually is very close to the process it will monitor
- Possible reuse of legacy housing
- Mounting requirements
- Design guidelines of the device manufacturer

The housing material of the device is also a very important part of the form factor. As mentioned in the previous subsection, there might be a requirement for the material to be thermally insulating or cooling or just that it should be rugged, which also affects the physical dimensions of the device, for example, weight difference between a metal and a plastic housing. Casting or filling also affects the weight of the device and can also make component replacement or upgrading impossible, for example, battery replacement.

The power source, for example, a battery or capacitor, will affect both the weight and physical shape of the device and thus has a major influence on the design. Batteries (and capacitors) have a correlation between stored power and physical dimensions, that is, larger batteries can store more energy. If the form factor forces a certain maximum physical size for the power source, the device's life length will be effectively limited by that size. Low-power design will push the limits of the device's life length based on the resulting power source.

The resulting physical form can impose limitations and requirements on the design of the PCB and positioning of electronic components, which could have performance and maintenance effects. For example, the form factor could force the antenna to be mounted at a nonoptimal position (i.e., from a radio communication point of view), which could cause degraded communications performance. It is also possible that when selecting certain components, or a combination of components, the whole PCB must be redesigned in order to fit the components onto it in conjunction with having it fit within the device housing.

Wireless Vibration Monitoring Case Example

The physical form and shape of the vibration monitoring device is very much dictated by the dimensions of the battery, which is selected based on the life-length requirements on the device. This in turn restricts the shape of the PCB and thus the area available for mounting electrical components. The result is a cramped area, which has forced a solution that uses both sides of the PCB for component mounting (which has a slightly higher cost).

The weight of the device is also an important requirement. It should be as low as possible because the device weight directly influences the quality of the vibration measurement (a heavy device is harder to move, i.e., vibrate). A related requirement is the size of the mounting area for the device; this should be as small as possible while still be large enough to provide solid mounting, which also affects the quality of the vibration readings. Finally, a more esthetic requirement is that the device should have a similar look and feel as a rugged industrial accelerometer.

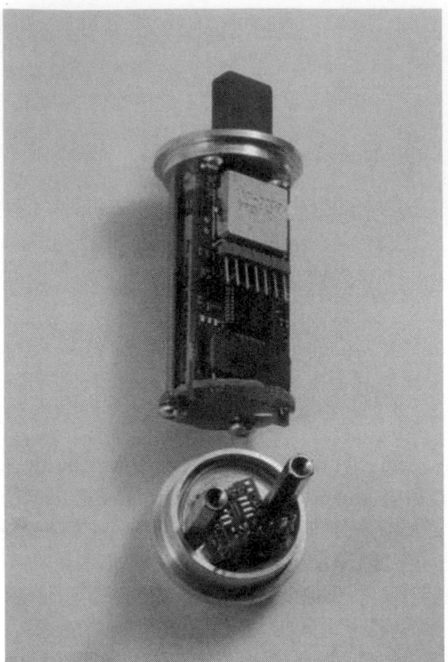

36.8 Modularity

The main driving force behind a modular design is improved maintainability, that is, how easily we can modify the device in order to fix bugs, implement new functionality, and upgrade to another SW/HW version. This is achieved by a clear separation of system features into modules or components that overlap in functionality as little as possible. These HW/SW modules are integrated into the device via clearly defined interfaces, which provide an abstraction of the module, that is, a separation of the external communication with the module from its internal operations. This means that (in theory) you are able to modify the module internally or even replace it with another similar module, without affecting how the rest of the device is operating or interacting with the module.

However, modularity comes at a cost as it places restrictions on the design, and care must be taken to ensure that the interfaces between modules are sufficiently general, while adhering to the nonfunctional requirements (response time, footprint, power consumption, etc). Table 36.4 illustrates some typical aspects or goals of a modular design and their trade-offs.

TABLE 36.4 Modularity Trade-Off Examples

Aspect	Pros	Cons
Reuse elements	Minimized development effort due to reuse of HW/SW design (and code)	Lower performance (footprint, response time, power consumption, etc.) due to increased overhead
Risk reduction	Design based on well-proven and tested components	Integration of several components can result in unclear/untested behavior
Migration	Minimized effort to migrate to other (newer) SW/HW versions	Lower performance due to *wrappers* needed to maintain system interfaces
Cost reduction	Lower cost due to second source and larger volumes (*standardized* components)	Added cost due to additional resources required to support the *standardized* components

As modularity does come at some cost, we have to balance the need for a highly modular design with, on one hand, the nonfunctional requirements and, on the other hand, the products' envisioned life cycle. There is no need to pay the price for a highly modular system if you are going to change it once every 10 years. However, it can be quite tricky to foresee the maintenance need of industrial products very accurately as it depends on large variety of aspects such as new user requirements, deployment in unforeseen applications, and availability of the COTS components that make up the design. Some form of modular design is therefore generally speaking a well-invested development effort.

36.8.1 Levels of Modularity

What does a modular design really mean? Aren't all modern embedded systems modular in some sense since they are built on standardized electronic components that your purchase and integrate on a PCB? The answer is yes and no. Yes, the components are standardized and can be replaced by another brand or type, but this does not automatically equate to a design that enables easy modification, extensions, migration, or reuse of parts of the design.

There is always a value of standardizing on a given set of low-level components, but to achieve a higher level of maintainability, you often need to raise the level of abstraction and divide the device into a set of core system features that overlap in functionality as little as possible.

36.8.1.1 HW versus SW Components

Standardizing on a given (set of) processors (CPU, MCU, DSP) of the same architecture has benefits in terms of reduced training cost of engineers and facilitates porting of SW. Reuse of PCBs, either actual HW or of layout with design modifications, has a direct saving in production cost, testing, and competence buildup. Reuse of mechanical solutions such as the packaging of the device often has large direct cost savings due to the elimination of design costs and cost of tooling. However, you need to balance the cost savings associated with utilizing a common housing; with the design restrictions, this will put on other parts of the design (form factor, antenna placement/design, etc.).

Reuse of SW components is advantageous under the condition that the sum of costs related to purchasing the code (internal or external), competence buildup, modification of components, integration of components, and testing is less than the cost of designing a component from scratch. Experience has shown that this is not easy unless other elements of the development organization are in place such as clear organizational roles, requirement management, and product platform architecture and design.

36.8.1.2 Single versus Multiple Chips

One of the major design choices that must be made is whether to integrate functionality in one HW chip or to have it distributed into different chips. In certain cases, there is no choice; the desired functionality is only supplied as a separate chip, that is, as a COTS component, while in other cases, the component can be designed in house and suitably located. Typical functionality and subsystems available in a WSN device are CPU, radio communication, sensor, signal processing (filters, DSPs, etc.), RTC, external memory (RAM, ROM, flash, etc.), HMI (LEDs, buttons, etc.), and an AD converter. Many of these subsystems could be integrated into one chip, for example, the CPU could contain almost everything.

A typically design choice for a WSN sensor device is whether to separate the radio communication stack and the application into different chips (see Figure 36.20). To conclude on a suitable solution, a trade-off between low-level control, form factor, cost, overhead, and performance has to be made.

Figure 36.20 shows a very clean separation between the application and the communication stack. In reality, this separation is often far from clear. A typical example is what a stack supplier considers being part of the communication stack and what is considered to be the *application* (and thus excludes from the supplied stack offering).

Wireless Vibration Monitoring Case Example

In the reference case, the chosen communication solution, that is, DUST Networks SmartMesh-XD chip, provides a complete SoC solution, where both the HW and SW are hidden within the box.

However, a small portion of the communications interface resides outside the SoC component, possibly in the application. This is functionality that handles packet preparation, such as byte stuffing and FCS* calculation for packets, which will be transmitted, and the corresponding operations for received packets.

This means that we need to implement communication-related functionality in the application, which normally would be considered to be part of the protocol stack.

* Frame check sequence (FCS) calculation.

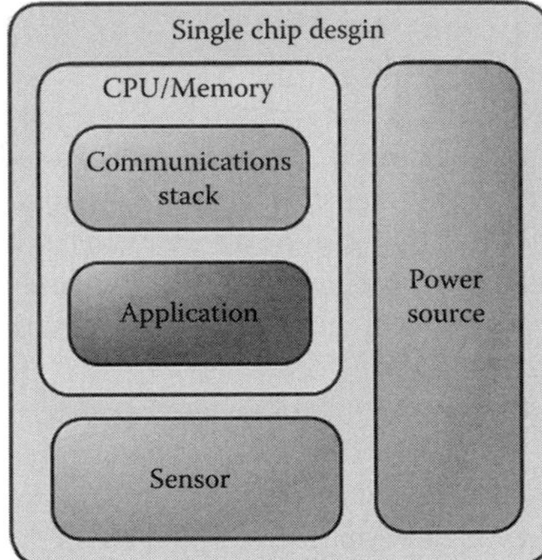

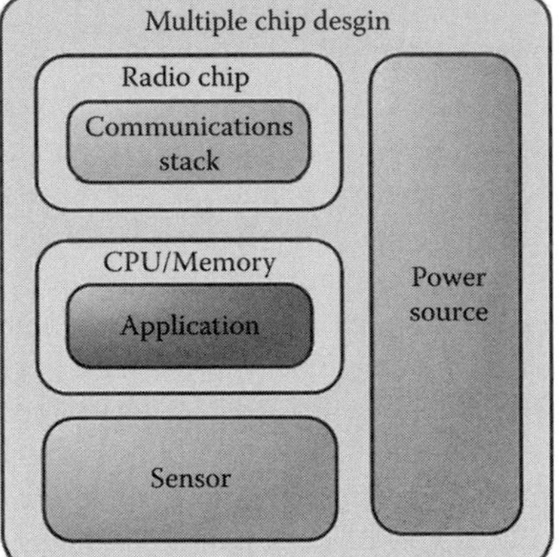

FIGURE 36.20 Example of single- versus multichip solution.

Modularity can also be achieved at a lower level. For instance, the communication protocol can be seen as consisting of several blocks, known as the OSI layers. Given a healthy design procedure, one may be able to exchange a single layer with one from a different source. Obviously, the more the code is split up, the more modular it becomes. At the same time, the risk of suboptimizing increases, that is, the modules are optimized (with respect to power, performance, code size, etc.) individually, but this does not necessarily give a globally optimum solution.

36.8.1.3 Lifetime

Industrial automation devices have a life expectancy that could be measured in 10ths of years. This puts a lot of demand on a modular and flexible design, which is easily maintained. Even if the device functionality would not change over its lifetime (which is highly unlikely), we still need to cope with availability of components. For certain parts of a WSN device, where the technology changes more rapidly, this is especially critical. For example, the sensor part (see Figure 36.7) evolves quite slowly, whereas the radio and power source technologies evolve very quickly. This fact warrants a high degree of modularity around the radio solution, something that is not always advantageous from a performance point of view.

36.8.2 Interfaces

The key to achieving a good modular design, and thus being able to reuse components, are the component interfaces. The interface provides an abstraction of the component, that is, a separation of the external communication with the component from its internal operations. This abstraction enables the internal modification of a component without affecting how the rest of the device is operating or interacting with the component.

A well-defined interface allows for easy maintenance and reuse of the component, but since interfaces are a form of indirection, some additional overhead and decreased performance is incurred versus direct communication. A modular design can thus introduce extra overhead in the sense of protocols needed to communicate between the different components (using the interfaces), which leads to increased power consumption and can result in bottlenecks in the system. The reason is that the system can only operate at the *speed* of its slowest component, at least if that component is part of the central functionality of the system.

Reusing and exchanging integrated circuits (ICs) require that their interfaces are compatible with each other. This is commonly referred to as pin compatibility and comprises the following:

- Functional compatibility implies that two ICs have the same functions (inputs, outputs, power supply, ground, etc.) assigned to the same pins.
- Mechanical compatibility ensures that ICs can be inserted into the same socket or soldered to the same footprint.
- Electrical compatibility implies that the components work with the same supply and signaling voltage levels.

Pin compatibility can play a major role when choosing components as it offers a degree of freedom in terms of migrating from one HW component to another without requiring a complete redesign of the PCB.

36.8.3 Subsystem Behavior

36.8.3.1 Local versus Global Optimum

As discussed previously, the sum of the parts can equal more than the individual parts together, at least from a product maintenance point of view. However, great care must be taken to ensure that sum of the parts also provides adequate performance (speed, power consumption, footprint, etc.). Even if each component has been optimized individually (for speed, power consumption, footprint, etc.), this does not guarantee that the final system behavior is optimal.

The principle operation of a WSN device can be described by the generic state machine depicted in Figure 36.18. This describes the overall behavior of the WSN device. The components that make up your device will in turn have their own state machines. The timing and interaction between the components are thus critical to arrive at an optimal overall system behavior. Care needs to be taken to ensure that a given functionality from one component is available when it is needed (according to the overall device state chart) or that enough memory or power is available to execute the requested operation. This can be difficult to achieve, especially if COTS components are utilized where you have little influence over the internal behavior of the component.

36.8.3.2 Control Freak

When reusing components, either developed in house or COTS, they will provide an interface and level of control that could be nonoptimal from a low-power design perspective. For example, if the communications stack is a *black-box* COTS component, which only provides limited possibility of controlling its internal behavior, it could seriously affect the sleep–wake-up scheme and thus the power consumption of the device. How much control a component allows will thus dictate the low-power design, which we describe in more detail in *Section 36.6*.

The interfaces are thus the Achilles' heel of a modular WSN design. As stated earlier, they need to be sufficiently general to enable efficient integration and porting, while not adding an excessive overhead. In addition, the interfaces also need to provide some control of the internal behavior of the component to be able to mediate between generality and performance. This goes against the basic principle of information hiding employed by component-based development, where internal state information is not shared and interaction is accomplished by exchanging messages carrying data. However, in low-power WSN devices, where the system is a sleep most of the time, this *low-level* control is sometimes crucial to adhere to the sleep–wake-up schedule of the device.

36.8.3.3 Debugging a Black Box

Using *black-box* components has the advantage that the functionality provided by the component is completely hidden. The developer thus does not need to bother about how the component works in detail, nor invest time and effort in testing and validating its functionality. In theory, this is true, but in addition to loss of control as described earlier, there is an additional risk that problems will appear when debugging the total system and the component functionality is part of the debug trace. As the name states, it's a *black box*, and there is no way to peer into the box to see what happens. This increases the difficulty of locating bugs, that is, determine if they are present inside the component or in other parts of the system. This problem is especially pronounced if the COTS component is itself in a *beta stage* (which is very common in early development phases of new technology) and bugs are certain to exist within it.

Wireless Vibration Monitoring Case Example

In the reference case, a *black-box* communication solution was chosen, that is, DUST Networks SmartMesh-XD chip. It's actually a complete SoC solution, where both the HW and SW are hidden within the box. There is very limited control over how the communication is managed, for example, the protocol and wake-up scheme for radio.

A small portion of the communications interface resides outside the SoC component, possibly in the application, which handles packet preparation. This makes the integration between the SoC and the rest of the system very tight.

The advantage is that minimal design effort has to be spent on the communications protocol. But the drawback is that debugging is very difficult because of the limited control and debugging support available through the provided interface.

36.9 Power Supply

The power supply is a central part of the WSN device; it's the *motor* that provides power to the various parts of the system. This motor can take many different shapes and forms and consequently provide varying degree of performance and lifetime. The general requirement is an energy autonomous system that requires little or no manual intervention (charging, maintenance, battery change, etc.) to operate throughout its target lifetime. This requires a thorough requirement analysis of the power demands and a clever design that is able to bridge power outages and handle variations in power consumption.

36.9.1 Requirements

36.9.1.1 Power Demands

Before deciding on a suitable power-supply technology, the power demands of the system need to be clearly defined. The power demand is directly related to the sleep–wake-up scheme of the device as discussed in *Section 36.6*. It can thus be divided into a constant *housekeeping* part corresponding to the losses associated with the operation of sensor and electronics and a time-varying power level that is related to the communication.

36.9.1.2 Energy Buffer

Regardless of the energy source employed, an internal energy storage is commonly used in low-power WSNs. This internal buffer enables the power supply to provide an even supply level in spite of a varying power inflow (e.g., when employing an energy-scavenging solution) and to handle intermittent peak consumptions (e.g., during start-up sequences). If an alternative power supply with a discontinuous power inflow is used, the buffer size will mainly be determined by the time characteristics of the power source. As such, the charging and discharging of the internal buffer need to be accounted for in the sleep–wake-up scheme of the device.

Wireless Vibration Monitoring Case Example

As the wireless vibration sensor is rated for temperatures well below zero, the battery voltage level might decrease too much with the load. is in combination with the current-limiting resistor (IS adaptation) might result in a situation where we do not have enough voltage to operate the radio (especially during start-up). To solve this, a capacitor bank was connected across the power supply, which stores just enough energy required for short radio transmissions (see Figure 36.19 for an overview of the battery design).

36.9.1.3 Lifetime

Typical quoted lifetime values of industrial WSN devices are often closely related to the lifetime of the power supply. This means not only that the components of the power supply should not fail within the specified time frame but also that the available energy is enough to operate the device for at least the specified period.

If the power supply is based on a permanent storage such as a battery, the lifetime is highly dependent on how the device is used in the field, that is, it's dependent on the size of the network, the duty cycle, the amount of data to transmit, and any asynchronous (nonscheduled) data access. Poor radio conditions will result in increased retransmits, which further decrease the actual possible operating time. So unless you specify these conditions, quoting, for example, 5-year lifetime of a battery-operated device really does not say anything about how long you can run your device in the field.

To determine the size (Wh) of the permanent energy storage required to reach the desired lifetime, you also need to account for material aging, leakage, and self-discharge effects.

36.9.1.4 Form Factor

The form factor of the device is a critical parameter when designing WSN devices, and the power supply often has the most dramatic impact on the final form and shape of the device. There must be some volume or area available in the WSN device, which is appropriate to mount the energy storage or energy converter. This volume/area has a direct relation to the required energy density of the storage or power density if an energy converter is used. The often desired small size and weight of WSN devices are thus a clear limiting factor when determining the suitability of available power-supply technologies.

36.9.2 Power Management

Depending on the availability, time behavior, and output characteristics of the source, some kind of power management will be necessary as illustrated in Figure 36.21.

Apart from the primary source of energy converter, a power-supply solution could involve an intermediate storage (energy buffer), a backup supply (e.g., a battery, which would also allow to keep the buffer size within reasonable limits), and a control unit in order to cope with input and output power fluctuations and downtime periods, which are typical for many alternative sources, but also for the power demand of an event-triggered sensor.

Also, a DC/DC converter in the control unit might be necessary to adapt the voltage and/or current level of the source to that of the system and to increase the total efficiency. This is especially important at low available power levels. If, on the other hand, a battery is the main energy source, backup with alternative power can be used to extend its lifetime and reduce the need of service.

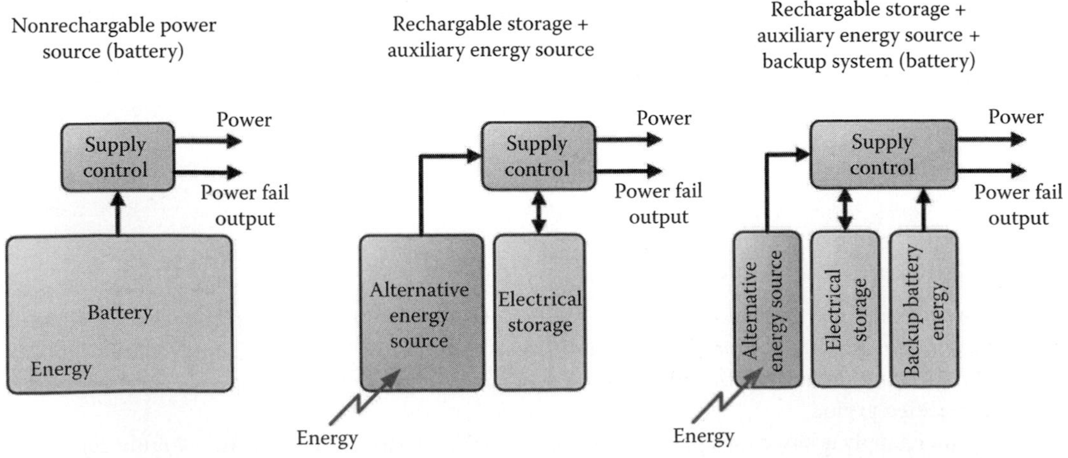

FIGURE 36.21 Power management.

36.9.3 Sources and Technology

The energy required to operate the WSN device can be supplied in three principal manners. It can be stored in the device (energy storage), transmitted from a dedicated external source (energy transmission), or it could be based on converting waste energy from the environment, the monitored/controlled process, or from some user action (energy conversions). The first category is by far the most prevalent method employed in WSN devices, while most research is being conducted in the energy conversion area. The approach of converting available energy, often labeled as energy scavenging or energy harvesting, holds the promise of energy autonomous devices with lifetime limited only by the component lifetime.

36.9.3.1 Energy Storage

If energy is stored in the device, great care needs to be taken when calculating the required energy needed to operate the device for the targeted lifetime (as discussed in *Section 36.9.1*). Energy storages can be divided into permanent storage and electrical buffer storage.

Permanent storages include the following:

- Primary batteries: Primary batteries remain the most common means to power WSN devices today. Although continuous improvement is being made to battery technology in terms of battery capacity and cost, the energy density has not really improved that much over the years. The practical energy density of primary batteries is around 1.2 Wh/cm^3. However, based on which technology you employ, the energy density, open-circuit voltage, self-discharge rates, and properties at high and low temperatures can vary greatly.
- Chemical storage/fuel cells: Combustible materials (chemical fuels) enable the storage of energy with very high densities, which can be converted into electric power by converters like a combination of a machine (motor, turbine) with a generator or a fuel cell. Availability of fuel cells is currently limited to hydrogen, methane, or methanol, with achievable energy densities in the range of a little more than 2 Wh/cm^3.
- Heat storage: Heat storage systems make use of the latent heat involved in melting or evaporation of so-called phase-change materials (PCMs). Latent heat storages are, for example, used for the heating and cooling of buildings. Commercially applicable PCMs are available in a wide temperature range and with melting energies of up to 0.14 Wh/cm^3.

Depending on the required energy and power densities, electrical buffer storages can be used in many cases in order to buffer energy as described in *Section 36.9.2*. Two types of electrical buffer storages are commonly used:

1. Accumulators: Accumulators are based on similar chemical principles as primary batteries, but they are designed for being recharged many times. Although performance is slowly approaching that of primary batteries, lower energy densities, significantly higher self-discharge rates, and the limited lifetime (around 3 years) must be taken into account.
2. Capacitors: Double-layer capacitors (DLCs), also named supercaps, ultracaps, or gold caps, are special capacitors with very large capacitance and low resistance. As no chemical reactions are involved with the charging and discharging processes, rapid changes of the stored charge, that is, high currents, are possible. A major advantage over accumulators is longer lifetime (>10 years) and significantly higher number of load cycles. The technology is rapidly progressing, and commercially available products provide energy densities up to 0.005 Wh/cm^3. Currently, they are often used in photovoltaic systems to bridge the day and night cycle of solar power.

Wireless Vibration Monitoring Case Example

The main requirement when choosing battery technology for the wireless vibration sensor was a high energy density (to comply with size/weight requirements). A primary battery based on a single lithium–thionyl chloride cell was chosen due to its high energy density, high nominal voltage, low self-discharge rate, and overall good properties at both high and low temperatures.

However, this type of primary battery has two main drawbacks. Large batteries may pose safety problems (subject to transportation restrictions) and voltage delay due to passivation.

Due to the size requirements, the chosen cell was small enough not to be subject to any transportation restrictions.

Passivation on the other hand could still be a problem as it might lead to voltage delay. Voltage delay is the time lag that occurs between the application of a load on the cell and the voltage response. As we have an extremely low duty cycle in the wireless vibration application (each sensor node should perform one vibration measurement per 2 weeks), this behavior becomes particularly pronounced (since the passivation layer grows thicker). As described earlier, a capacitor bank was added in order to secure sufficient energy to operate the radio circuit despite a varying operating temperature (see Figure 36.17). This buffer also provides a remedy for potential passivation issues, as the initial energy will be drawn from the capacitor bank, thus eliminating any voltage delays.

36.9.3.2 *Energy Conversion*

Harvesting and converting waste energy from the environment certainly appears as an appealing solution, as the lifetime of the device would (in theory) be limited only by the component lifetime. However, this requires that the *waste* energy is readily available with a constant (or predictable) level. If this is not the case, large internal buffers might be required to guarantee reliable operation (which increases the size/weight/cost of the device). This also means that the applications in which the device can be deployed might be restricted or that special considerations and/or measures need to be taken to guarantee some minimal level of power inflow (which makes device installation more complex and costly).

Energy conversions can be achieved in many ways, but the most common methods include the following:

- Photovoltaic cells: A wide range of photovoltaic systems are available on the market today based on diverse materials and technologies that have reached different maturity levels. Cell types also differ with respect to cost, electrical characteristics, efficiency, topology, thickness, shape, mechanical properties, and adaptation to various lightning conditions. Regardless of the varying energy conversion efficiency (5%–30%), the lightning conditions have by far the greatest impact on the feasibility of photovoltaic cells. In typical environments, the illumination density will vary within at least three orders of magnitude. For example, compare the illumination density of full sunlight (~100 mW/cm^2) with normal office lighting (<1 mW/cm^2). This difference is even more pronounced by the nonlinearity of the cell output power, which results in reduced efficiencies at low illumination intensity.
- Thermoelectric converters (TECs): Utilize temperature gradients to generate electrical energy. If there is waste heat available, which is normally conducted or radiated to the environment, it would be possible to connect a TEC to a hot surface, let part of the heat flow pass through it, and generate electrical energy, which could then be used to power the WSN device. In order to achieve sufficiently high efficiencies, TECs need to be appropriately designed and thermally and electrically matched to the respective application. Especially, in order to increase the heat flow (and with it the temperature

difference across the module), it is important to make very good thermal contacts to the hot surface and to the environment at the same time. The latter requirement usually causes the use of air coolers or other heat spreading elements, which may be even much larger than the module itself.

- Mechanical converters: The most common type of mechanical converters is based on vibrations. In industrial applications, energy-harvesting oscillators need to be adapted to low vibration amplitudes, high forces, and resonance frequencies of typically 50 to several hundred Hz, as this is the range of vibrations in machines. These conditions can be well adapted with piezoelectric converters, which can collect power levels of 10 mW up to several hundred mW dependent on the existing acceleration amplitudes. However, sufficient vibration amplitudes cannot be generally assumed in all applications. Also, the mechanical system must be well tuned to the vibration frequencies (which may change both with the application and with time) in order to achieve high efficiency and power density. Design for broadband response will reduce the available output power amplitudes. One of the first commercial vibration energy harvesters for the supply of small wireless devices like sensors and transmitters was offered by a company called Perpetuum [13]. They offer inductive generators that deliver a few mW and up to 3.3 V depending on the amplitude and frequencies the vibration.

Wireless Temperature Instrument with Energy Harvesting Example

The wireless temperature instrument uses WirelessHART as communication standard since it is widespread and has many proven operational hours. The temperature transmitter has been equipped with a thermal energy-harvesting unit containing two Micropelt Thermogenerators MPG-D651 with a footprint of 6 mm^2 each. With a minimum difference about 30 K between the process and ambient temperatures, the system is able to generate sufficient energy to supply both the measurement and wireless communication electronics. At temperature gradients greater than 30 K, more energy is generated than is needed, which could be used to allow for faster update rates, for example.

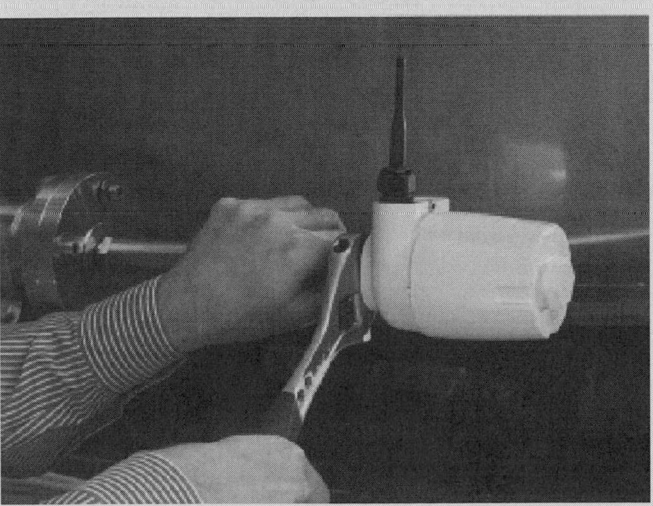

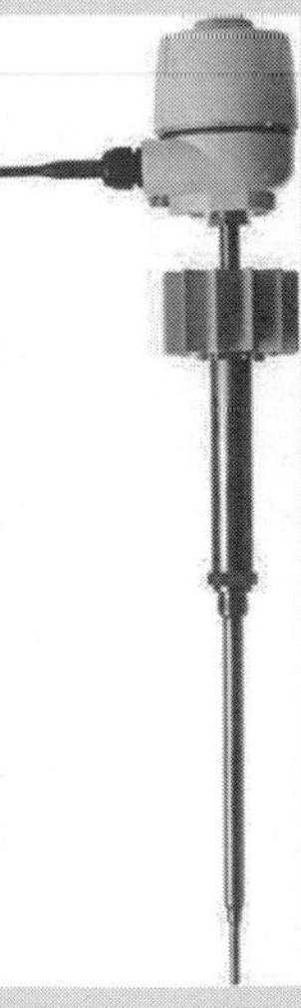

36.9.3.3 Energy Transmission

Another principle method of supplying energy to the WSN device is to transmit the power to the WSN device. This method thus requires some form of external power source that needs to be installed in the plant in *close* proximity of the WSN devices that are going to be supplied. The maximum distance between the power source and the WSN differs depending on the technology employed. Three main technologies can be conceived:

- Inductive transmission: Currently used inductive power transmission systems are different with respect to range (wireless vs. contactless) and operation mode (continuous vs. pulse operation as often used with radio-frequency identification [RFID]). A rather large variety of commercial solutions are available such as contactless rotary power and signal connectors, linear contactless solutions for rail and transport systems, and solutions for the supply of small consumer devices. In addition, RFID and transponder solutions are available from a large number of suppliers. ABB seems to be the only company offering a wireless *long-range* (a couple of meters) inductive power-supply system for sensors and actuators [9]. The system, called WISA-POWER, can deliver 6 mW of continuous power to the wireless sensors even under worst-case conditions and close to 100 mW to concentrator modules with larger receiver coils.
- Capacitive transmission: Power transmission can also be achieved by means of high-frequency electric fields. Contactless solutions for rotary signal transmission are available on the market, but no *wireless* power solutions, as the available power levels are rather limited within the range of allowable field amplitudes.
- Optical transmission: Optical power transmission is possible with fibers, a focused (laser) beam, or with diffuse lighting. In most applications, little would be gained by the replacement of wires by optical fibers. Focused beams are not applicable for the supply of many sensors dispersed throughout the automation facility/cell where there often is no direct line of sight. An alternative could be the simultaneous supply of the sensors in an automation cell with a diffuse artificial lighting. However, reliability, cost-effectiveness, and customer acceptance of such a solution make it a highly questionable approach. Even with such an approach, shadow effects cannot be excluded for general automation applications.

Although this list of power-supply technologies is by far complete, it should provide an idea of the various conceivable power-supply solutions for WSN devices. Noticeable are the varying levels of energy/power density and characteristics of the power-supply solutions, often highly dependent on the conditions under which they are employed. To choose a suitable technology, you thus need to fully understand the static and dynamic power demands of your device, as well as the applications and environmental conditions under which the device will operate. A more comprehensive overview of various solutions is given by [14] and [15].

Acknowledgments

I would like to thank all contributing authors. In particular, authors from industry and industry consortia whose job priorities usually are different from writing chapters in books. I wish to acknowledge the work of Dr. Ronald Schoop of Schneider Electric, who passed away before finishing his contribution. I am grateful to the members of the Handbook's Advisory Board for so generously taking part in the making of this book.

I also wish to thank at CRC Press Nora Konopka, publisher of this book; Michele Smith, Jessica Vakili, and Richard Tressider, the production manager.

References

1. Wikipedia. http://en.wikipedia.org/.
2. ISA. http://www.isa.org/.
3. HART Communication Foundation. http://www.hartcomm2.org.
4. DUST Networks. http://www.dustnetworks.com.
5. ZigBee Alliance. http://www.zigbee.org.
6. Institute of Electrical and Electronics Engineers: Part 15.4: Wireless medium access control (MAC) and physical layer (PHY) specifications for low-rate wireless personal area networks (WPANs), New York, IEEE Computer Society, 2006.
7. International PROFIBUS. http://www.profibus.com/.
8. K. Masica: *Recommended Practices Guide For Securing ZigBee Wireless Networks in Process Control System Environments*, Lawrence Livermore National Laboratory, Livermore, CA, 2007.
9. D. Dzung, G. Scheible, J. Endresen, J-E. Frey: Unplugged but connected: Design and implementation of a truly wireless real-time sensor/actuator interface, *IEEE Industrial Electronics Magazine*, Summer 2007.
10. International Standard IEC 60529, Degrees of protection provided by enclosures (IP Code).
11. International Electrotechnical Commission for Certification to Standards Relating to Equipment for use in Explosive Atmospheres (IECEx Scheme). http://www.iecex.com/index.htm.
12. Guidelines on the application of directive 94/9/EC. http://ec.europa.eu/enterprise/atex/guide/.
13. Perpetuum Ltd. www.perpetuum.com.
14. G. Scheible: Wireless energy autonomous systems: Industrial use? Sensoren und Messysteme VDE/IEEE Conference, Ludwigsburg, Germany, March 11-12 2002.
15. J. A. Paradiso, T. Starner: Energy scavenging for mobile and wireless electronics, *Pervasive Computing*, IEEE CS and IEEE ComSoc, January-March 2005, pp. 18–27.

37

Design and Implementation of a Truly Wireless Real-Time Sensor/Actuator Interface for Discrete Manufacturing Automation

37.1 Introduction 37-1
37.2 Wireless Interface Requirements and System Specifications 37-3
37.3 Communication Subsystem Design 37-5
Medium Access and Retransmission • Frequency Hopping • Antenna Diversity and Switching • Multicell Operation
37.4 Communication Subsystem Implementation 37-9
Coexistence • Performance Measurements • Industrial Electromagnetic Interference • Interference from Other Communication Systems
37.5 Wireless Power Subsystem 37-15
The Magnetic Supply: Wireless-POWER • Resonant, Medium-Frequency Power Supply • Rotating Field • Omnidirectional Receiver Structure • Example Setup
37.6 Practical Application and Performance Comparison 37-22
37.7 Summary 37-26
References 37-27

Guntram Scheible
ABB Automation Products GmbH

Dacfey Dzung
ABB Switzerland Ltd.

Jan Endresen
ABB AS Corporate Research

Jan-Erik Frey
ABB AB System Automation

37.1 Introduction

Wireless communication is increasingly used in many industrial automation applications, which have many differing requirements [1]. The main automation domains are as follows:

- Process Automation, typically dealing with continuous processes (e.g., the continuous or batch processing of liquids)
- Factory Automation, where discrete, stepwise manufacturing dominates (e.g., mounting a product from many different parts)

In discrete manufacturing, again different application levels exist (see Figure 37.1). Consequently, there is a need for different wireless implementations to cover differing applications like mobile operators,

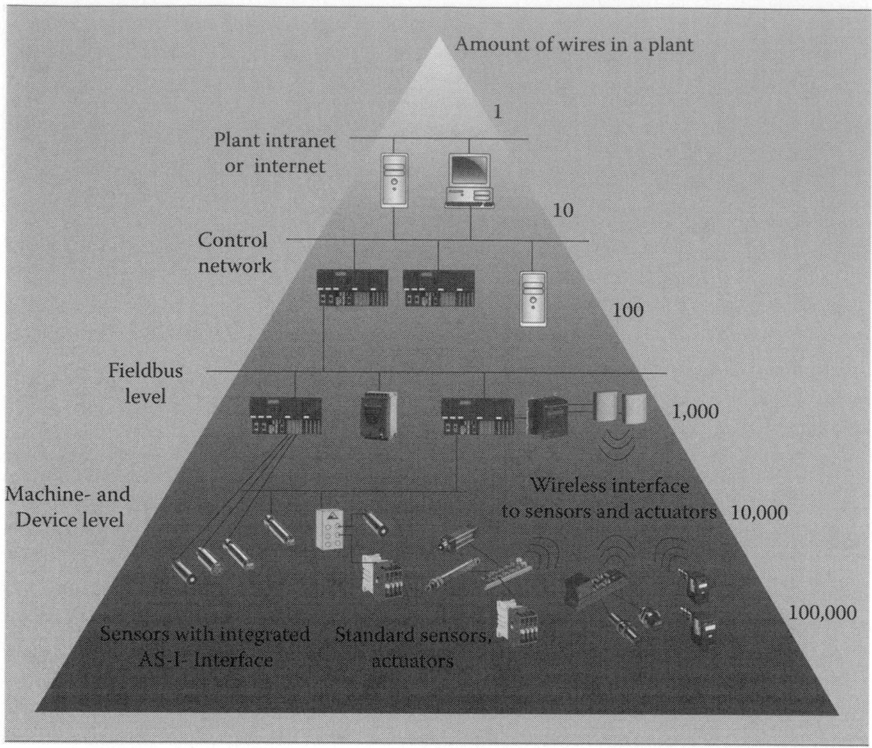

FIGURE 37.1 Automation hierarchy in factory automation, communication levels, and number of possible nodes to be connected wirelessly.

controller-to-controller communication, and the machine and device level. It is obvious that especially in the machine and device level, a large number of wires and connectors are operated in a harsh environment.

Sensors and actuators in the machine and device level operating over wireless links:

- Allow a very flexible installation, a significant advantage in many parts of the machine-building industry, where often a custom or special purpose machine is built just once. The installation and commissioning often starts already in parallel to the ongoing design process.
- Enable fully mobile operation of sensors and actuators.
- Enable new sensor or actuator applications that would not be possible with wired devices.
- Avoid cable *wear and tear* problems.
- Avoid costly and time-consuming integration of mechanical cable tracks/chains, slip-rings, or similar mechanical elements.

In an automotive assembly plant, up to 100,000 input or output points (IO) may be wired, and many of the cables and connectors are subject to repeated movement or reconnection.

Wireless communication systems for the machine and device level in factory automation must guarantee high reliability and low and predictable delay of data transfer. The requirements in the machine level are mission-critical and considerably more stringent than typical requirements for other areas and especially compared to home, building, or office applications.

Requirements on power supply and real-time communication for such applications are highly demanding and cannot be satisfied with today's off-the-shelf wireless systems.

Wireless systems offer advantages in terms of availability, cost, and flexibility also in the device and machine level of factory automation.

To gain most of these advantages, a truly wireless solution, where communication and power supply are wireless, is needed. A novel factory communication system, called Wireless Interface for Sensors and Actuators, has thus been developed, which provides both wireless communication and wireless power supply. The power supply is based on magnetic coupling, while the real-time wireless communication uses a collision-free TDMA protocol, combined with frequency hopping (FH) based on the physical layer of IEEE 802.15.1. The Wireless Interface provides reliable and low-delay transmission for a high number of nodes.

This chapter explains the practical requirements and the system specifications, the design concepts for communication and power supply, and describes practical experience gained especially with respect to reliability and coexistence with other wireless systems. It further derives simple performance profiles for the Wireless Interface and other wireless technologies being discussed for industrial use, based on the requirements in factory automation.

37.2 Wireless Interface Requirements and System Specifications

The basis for system specification of the Wireless Interface has been a thorough analysis of the requirements of typical applications of the targeted machine and device level:

- *Roundtable production machines*: These are used in various forms in many automated discrete production facilities. They can have up to 10 devices per m^3 machine volume (300 per machine). Many of these devices have a relative movement with considerable stress on wires and connectors. Most of them are also triggered at different times in one production cycle. Cycle times may vary from 1 s (one product produced and tested per second) in many steps to 1 min.
- *Production lines*: An automated line-type machine assembling, testing, and packing a standard consumer or household product may have 2000 (1000–3000) IO points.
- *Automotive robot production cells*: These have an IO point density of around two devices per meter cube machine volume (200 per $7 \times 6 \times 2$ m^3). Sometimes, several cells are collocated and work together.

All of the aforementioned applications are typically found several times in a factory hall (5–20), often closely together.

In an *average automotive assembly plant* ($150 \times 150 \times 2$ m^3 machine/application space), around 100,000 IO points can be encountered, leading to an average wireless device density requirement of around 2/m^3. Assuming that only some of the devices are wireless and some are wireless IO concentrators with, for example, eight IO points, an average figure of a suited technology should be at least 0.25 wireless devices/m^3.

There are two related basic wireless requirements in such applications:

- Low additional latency due to the wireless link (e.g., <10 ms) in order to
  - Not unduly increase the cycle time of the application, which otherwise would mean lost product output

    Fifty devices (or IO points) triggered independently and always having to wait 10 ms for each signal to arrive would mean 1 s of additional cycle time already (a 100% increase in a fast roundtable application!)
  - Stay within time limits given by the application (e.g., a critical mechanically controlled movement), which otherwise could mean destruction of the machine, tools in the machine, or at least the produced product
- Very high reliability to get a signal within a defined upper time limit (e.g., in order to avoid the earlier-described destruction)

In contrast to other wireless use cases like mobile operators and service, the very high reliability to get a signal within a defined time limit is an absolute requirement.

TABLE 37.1 Events in a Roundtable Production

	1 production step per 2 s
	min. 1 events per second
100 wireless nodes or 10 pts.[a]	100 telegrams per second
Per hour	360,000 telegrams
Per day (3 shifts)	8 Mio. telegrams
Per year (w/o weekends)	2400 Mio. telegrams

Less than one failure can be tolerated per year, resulting in a needed error rate of $<10^{-9}$.

[a] Digital 10 assumed, concentrating 10 nodes would have the same amount of telegrams, as only rarely 10 points are simultaneously triggered!

Practical values for the error rate of not getting the signal in time depend on the industrial use case assumed. A typical roundtable application can have easily 300 IO points, mainly sensors, of which 100 are here assumed to be wirelessly transmitted in the following Table 37.1. A roundtable is a typical arrangement in a factory, where a stepwise rotating table is used to do many different assembly, testing, or loading/unloading steps in parallel. The substations on the table have additional relative movements, so wireless is an issue due to two-dimensional movements. Typically, most of the IO signals are digital and the events are typically separated in time (unsynchronized), resulting in many telegrams, worst case equaling two times the amount of IO per production cycle (see Table 37.1).

In principle, the distribution of wireless latencies will qualitatively look like in Figure 37.2 in practical implementations. Sketched is a coexistence situation with other frequency users. If the wireless system has a cyclical transmission, for example, as in the Wireless Interface: A, a distribution pattern with steps can be noted (— — curve).

In practical applications, a large difference between the two curves can be noted by comparing the error rate for keeping a certain time limit, like sketched with the vertical bar. A factor of 100 results, which means that in the roundtable application with one wireless system (— — Wireless Interface), there is an error, for example, a production stop or even more severe problem, to be expected less than once a year, and in the other case, nearly once a week: a dramatic difference seen from the user side.

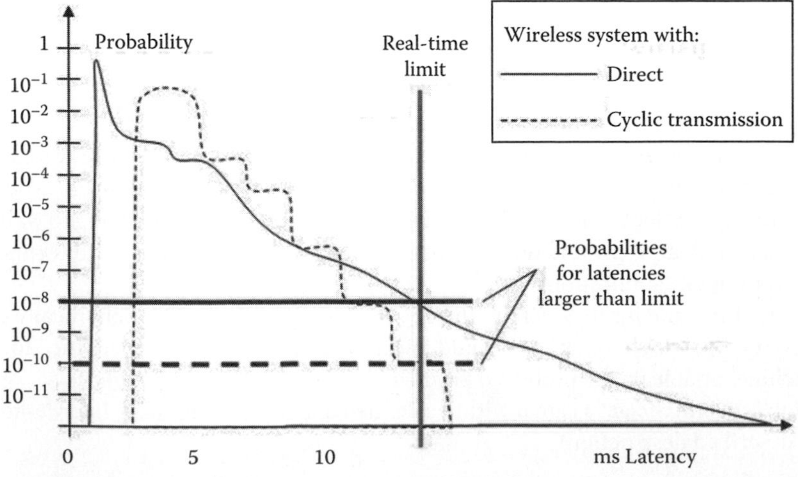

FIGURE 37.2 Typical distribution of latency measurements in wireless automation devices.

The wireless communication subsystem transmits messages between the S/As and a base station. The communication requirements derived from the aforesaid include the following:

- Fast response times (generally less than 15 ms)
- (Practically) guaranteed upper time limit for the response time (e.g., 15 ms on the Wireless Interface with a miss probability of less than 10^{-9} under normal operating conditions)
- Guarantee a high reliability of data transmission even in the unfriendly environment of factories, where radio propagation may be affected by many obstacles and where various sources of interfering signals must be expected, for example, with other users of the same frequency band
- Have low power consumption, which is crucial due to the limited capacity of any wireless power supply
- Serve a large number (up to 120) of sensors/actuators to one base station
- Permit coexistence of multiple cells: Up to 300 devices in a single machine have to be supported
- Have a practically unlimited number of S/As in a large area (>1 device per m^3 volume to cover 100,000 nodes in a factory)
- Permit coexistence with other wireless systems
  - Be robust
  - Efficient use of the valuable frequency band resource

Currently, the most suitable radio frequency band available for high-bandwidth, short range wireless communication is the industrial, scientific, and medical (ISM) frequency band at 2.4 GHz. Radio transceivers operating in this band are readily available, for example, radios adhering to the IEEE 802.15.1 standard (Bluetooth physical layer).

The ISM Band also has a good balance between possible disturbances (more likely at lower frequencies; see Figure 37.9) and a good range. Range is lower at higher frequencies with a given transmit power. Electromagnetic interference is a concern; however, measurements in different environments (see Chapter 4, Industrial Electromagnetic Interference; Section 37.4.3) have shown that typical industrial electromagnetic noise is significantly reduced above 1 GHz, with only arc welding providing broadband noise up to 1.5 GHz.

The protocols of existing standardized wireless systems designed for other purposes do not satisfy the requirements of the sensor and actuator level in machines used in automated factories:

- Wireless local area networks (WLAN) (IEEE 802.11) [2] and Bluetooth (IEEE 802.15.1) [3,4] and also forthcoming systems [5], including ultra wideband (UWB) systems, are all designed for high throughput between a small number of terminals, with less stringent latency, reliability, and power requirements.
- ZigBee (IEEE 802.15.4 based) [6] cannot reliably serve a high number of nodes within the specified cycle time.
- Specialized low energy protocols from building automation like Enocean [5] are not reliable and real-time suited as defined earlier.
- Passive electronic tagging systems (RFID, Radio Frequency Identification tags), as used for electronic article surveillance, do not have sufficient range, speed, and flexibility.

Hence, it was necessary to design a new communication protocol, which however uses much of the available standard hardware.

37.3 Communication Subsystem Design

The radio characteristics of the new Wireless Interface system are those of the IEEE 802.15.1 physical layer specification [7]. The raw data rate is 1 Mb/s and transmit power is 1 mW. This comparatively low transmit power is appropriate and sufficient for a range of 5–10 m, as required for the typical node and base station density.

Not so obvious is that the low transmit power and therefore limited range is also dictated by the requirement of a sufficiently large average node density. A good comparison is human communication, where also a large number of people, for example, at a cocktail party, can communicate reliably and fast, due to fairly low voice/range, which allows others to reuse the transmission medium in a certain short distance. Just a few participants increasing their transmit power/range will disturb many others.

37.3.1 Medium Access and Retransmission

Time division multiple access (TDMA—one fixed time slot per device) with frequency division duplex (FDD—device and base station can send at the same time) and FH is used, as illustrated in Figure 37.3. The downlink transmission (from the base station) is always active, for the purpose of establishing frame and slot synchronization of the S/As. Uplink transmissions from an S/A occur only when it has new data or an acknowledgment to send (event triggered). The Wireless Interface allows up to 120 S/As per cell in different time/frequency slots.

The TDMA frame length, T_{frame}, is 2048 μs. Each S/A may transmit in one out of 30 slots per frame. In order to support up to 120 S/As per cell, each S/A is part of one of four uplink groups {UL_1, UL_2, UL_3, UL_4} with separate frequencies. The five frequencies hop synchronously at each frame boundary.

The slot structure is shown in Figure 37.4. The downlink slot format is a compromise between the efficiency of channel usage and synchronization latency of the downlink.

A device transmits its data packet in its assigned uplink time slot. An acknowledgment is expected in the corresponding downlink slot, which is staggered by 256 μs to allow for TX/RX-turnaround time (see Figure 37.3). If no acknowledgment is received, the S/A will retransmit the data packet in the next frame.

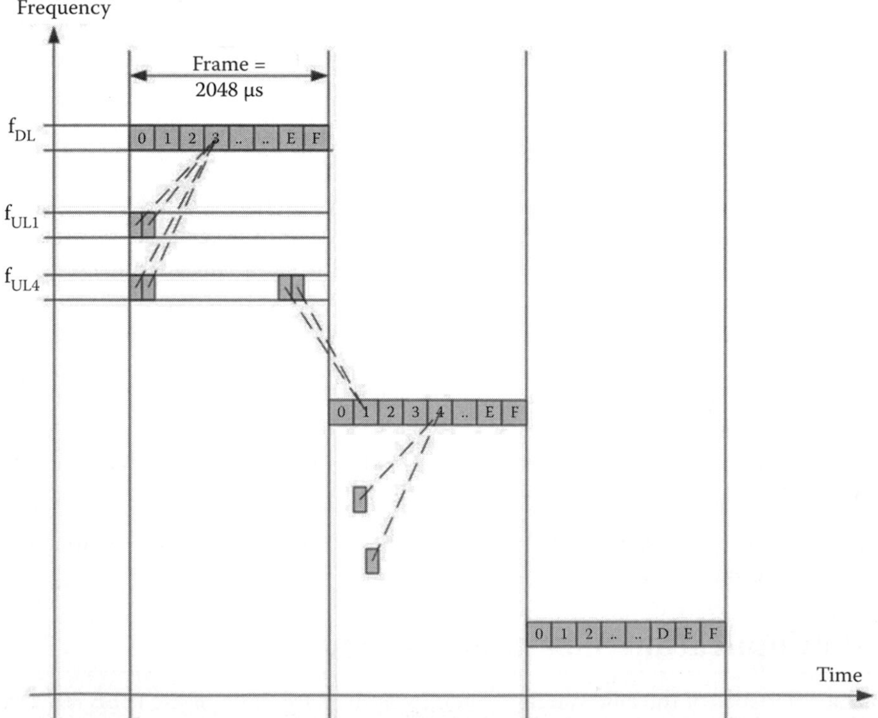

FIGURE 37.3 Wireless Interface TDMA/FDD/FH pattern. *Dotted lines* connect downlink/uplink slots allocated to one sensor/actuator. (TX/RX-turnaround time = 2 double slots = 256 μs; see Figure 37.4).

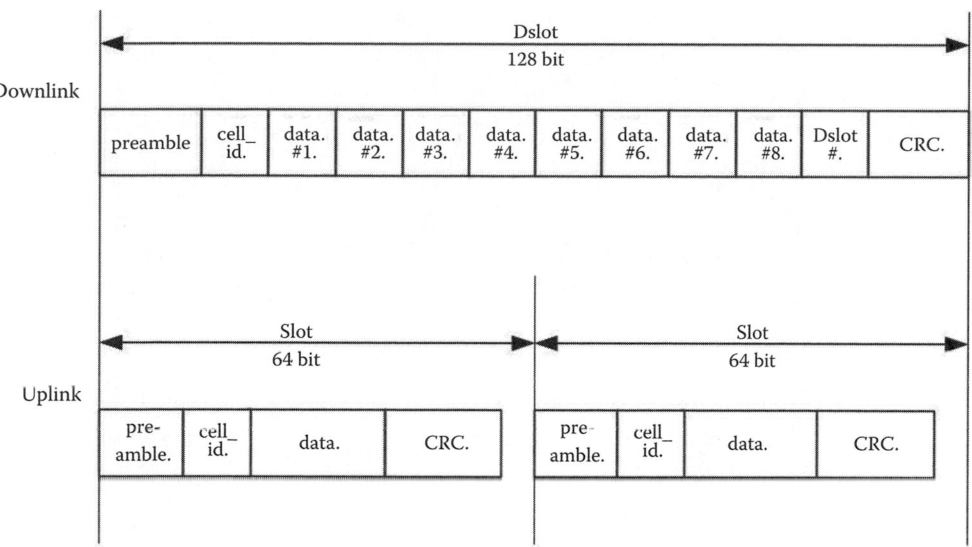

FIGURE 37.4 Wireless Interface slot format: downlink double slot duration = 128 μs.

With frame-by-frame FH, the radio channel used for retransmission will largely be independent of the previous transmission, thus increasing the probability of successful transmission.

37.3.2 Frequency Hopping

Figure 37.5 shows the measured frequency-depending attenuation in the 2.4 GHz ISM band. It can be seen that the coherence bandwidth of the signal may be some tens of megahertz (corresponding to propagation path length differences of 30 m). Wideband FH will thus improve transmission performance. It also reduces the effect of other devices that operate in this band. The GSM [8] and Bluetooth [7] systems are well-known wireless communication standards using TDMA and FH. Their FH sequences are determined in a pseudorandom manner and therefore do not control frequency separation between consecutive hops. Hence, consecutive transmissions may have similar high packet error probability. It is for the Wireless Interface of interest to employ FH sequences with the property that consecutive hops are widely separated in frequency. Its use of the ISM band of 80 MHz is partitioned in seven disjoint subbands, each containing 11 hop frequencies at a given channel spacing of 1 MHz. The subband bandwidth of 11 MHz is normally larger than the typical bandwidth of the fading, and the seven subbands together cover the available band. The method produces 60 different FH sequences with low cross-correlation. The hopping sequences have a period of 7 × 11 = 77 frames. Each hopping frequency is visited exactly once per period. An example of the five FH sequences for a given cell_id is shown in Figure 37.6.

In summary, the constructed FH sequences have a number of desirable properties:

- Consecutive frames are in different subbands.
- The (duplex) spacing between downlink and uplink frequencies is three subbands apart.
- The four concurrent uplinks are in the same subband and hop with a fixed spacing of 3 MHz apart.

37.3.3 Antenna Diversity and Switching

To additionally improve robustness, multiple antennas with separation of at least half a wavelength λ may be used to reduce the effects of multipath fading and shadowing. This effect is exploited by antenna diversity reception. With λ = 12.5 cm, only the base station (BS), for practical reasons, may be fitted with multiple antennas. The simplest way is to switch transmit and receive antennas once per frame in a round robin sequence.

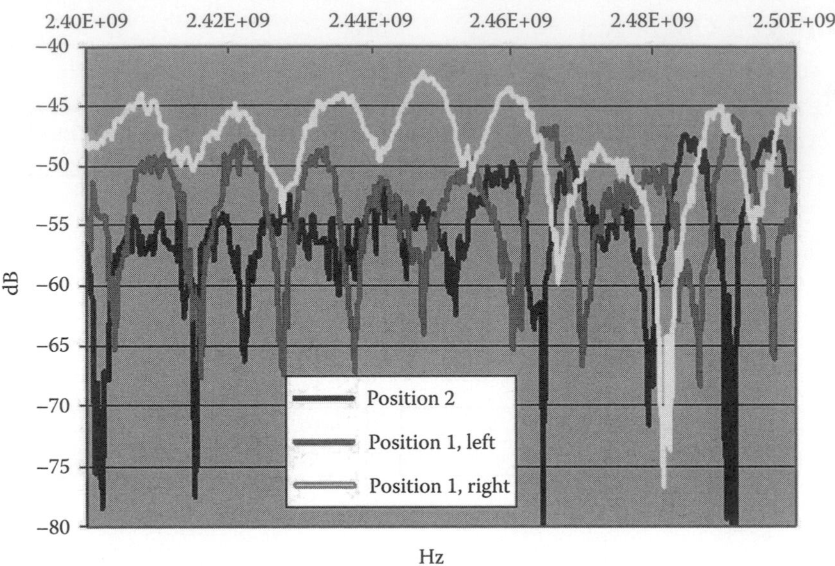

FIGURE 37.5 Frequency selective fading in the 2.4 GHz ISM band: frequency-dependent attenuation from a sensor/actuator to 3 BS antennas in a test installation. *Gray* and *light gray*: pair of base station antennas, separated by 37 cm; *black*: third BS antenna, line of sight path obstructed by a robot arm.

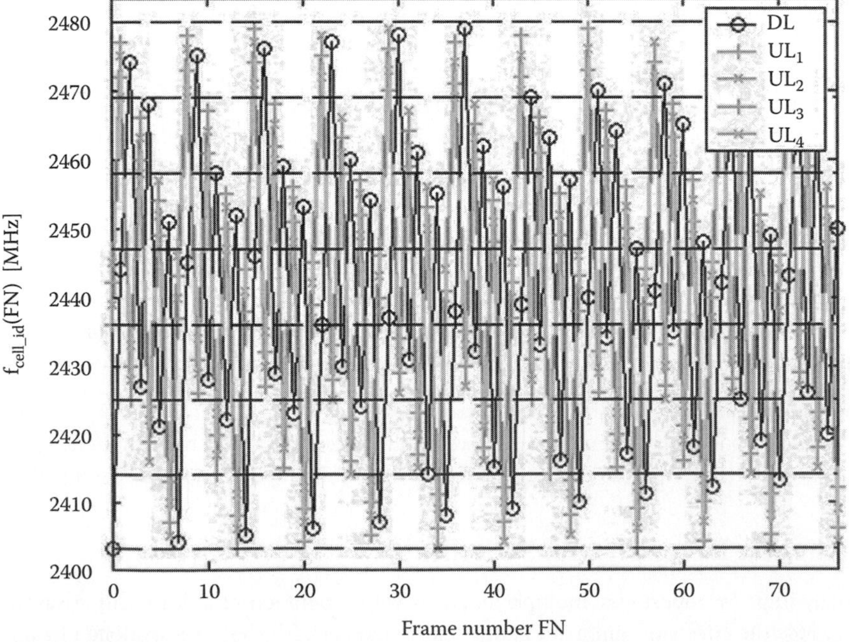

FIGURE 37.6 Example FH sequences (cell_id = 27).

37.3.4 Multicell Operation

The Wireless Interface system is designed such that each cell uses the entire 2.4 GHz ISM band of 80 MHz, without any need for intercell coordination, except that a different cell_id has to be set in each cell. Where cells operate in close proximity, or are even colocated, the interference depends on the correlation properties of the FH sequences and the cell traffic load. Cross-correlation between periodic integer sequences has been studied in [9]. In general, there are a number of interference effects to be considered:

- Co-channel interference from same frequency transmissions becomes relevant if the C/I ratio is below 14 dB [7] (distance effects).
- Adjacent channel interference is relevant if the adjacent channel interference C/I ratio is below –20 dB. This condition may well be true for colocated base stations.
- Multiple interference from any of the five links.
- (DL, UL_1, UL_2, UL_3, UL_4) of an interfering cell within range. While downlink transmissions are continuously active, the uplink load is typically far lower: With a frame rate of 488 frames/sec, and assuming a maximum uplink rate of five transmissions per second per S/A, a slot is only used with a probability of 1.02%.
- Multiple interference from the links of more than one adjacent cell. Links from different cells are mutually asynchronous, causing interference to occur at different times.

To assess these effects, a generalized cross correlation was defined and its properties were analyzed. The results confirm that even in the worst case, the retransmission protocol ensures that messages can be reliably transmitted.

37.4 Communication Subsystem Implementation

Figure 37.7 shows a Wireless Interface base station which is controlled by a microcontroller. Time-critical control of the RF transceiver and baseband signal processing are delegated to an FPGA. The base station maps the wireless links to the addresses of the fieldbus connecting it to the automation controller. Diagnostics and configuration features are supported.

Due to the communication protocol requirements, a standard RF front end cannot be used for the base station implementation. The requirements differ significantly from those found in WLANs and Bluetooth in that the BS must operate in full duplex mode. The BS must simultaneously receive signals from up to four S/As with a dynamic range of more than 60 dB.

In contrast to the BS, the node may use existing RF transceiver hardware (IEEE 802.15.1 based). A sensor module is shown in Figure 37.7 (in this implementation a proximity switch [10]), with the power receiving coil, an IEEE 802.15.1–based transceiver, and a microcontroller and an FPGA for the Wireless Interface protocol. For low power consumption, the transceivers *sleep* until an event occurs. The communication antenna on the module has an omnidirectional radiation characteristic in order not to have to line up the device at installation.

37.4.1 Coexistence

Wireless coexistence is a practical and increasing concern in industrial automation applications. Many different applications may use wireless systems, which may compete for the same frequency band with different technologies. The limited resource are the radio frequencies, and for the Wireless Interface that is in the ISM Band from 2.4 to 2.483 GHz.

The situation may be compared with highways, which have a limited width and many different types of vehicles fulfilling different transport tasks. For each of the transport tasks different requirements apply, the capabilities of the vehicles are quite different also. In addition, certain rules apply to avoid dangerous situations (e.g., speed limits).

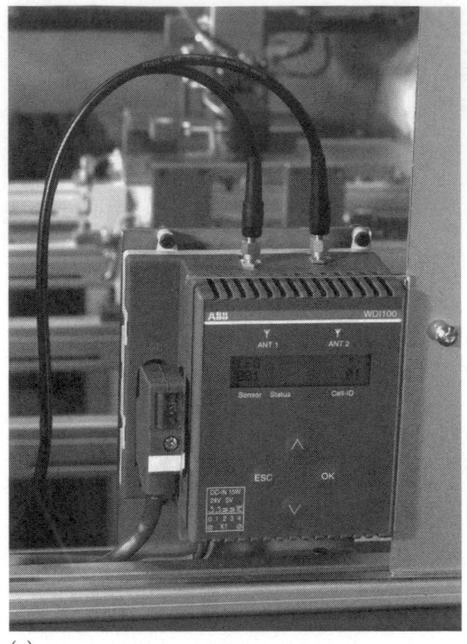

(a)

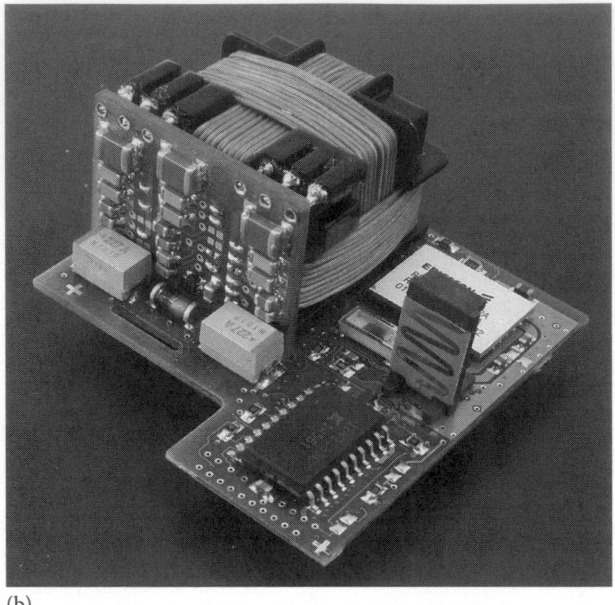

(b)

FIGURE 37.7 Wireless Interface implementation. (a) Base station with antenna connectors and FieldBusPlug (FBP) connection to controller. (b) Sensor module with receiver coil for wireless power supply and communication transceiver, controller, and antenna.

In today's factories, coexistence or frequency management has to be established to coordinate the use, as the frequency bands are a valuable resource. In a larger factory, generally every system has to be announced in advance and released by a person/committee responsible for frequency use allowance [11,12].

Wireless coexistence is an issue only if the same frequency is used at the same time and close by (Figure 37.8).

Separation in location is seldom a free parameter; therefore, time and frequency can be optimized.

If there would be an unlimited bandwidth available, simple frequency separation would be the simplest approach. Nevertheless, the highway example shows that reserving a full lane of the highway just

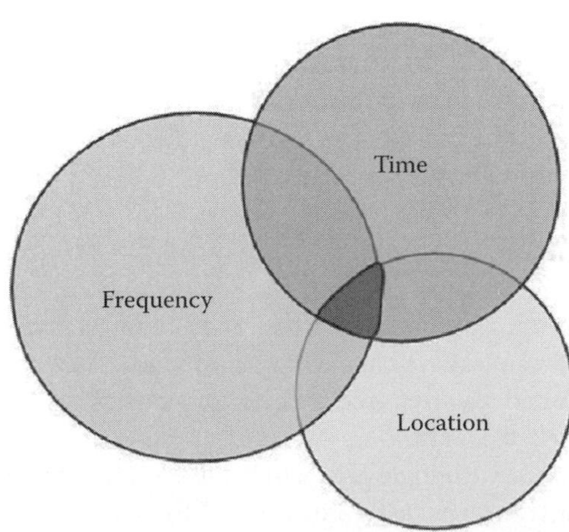

FIGURE 37.8 Coexistence issue: Occurs if there is a *hit* in time *and* frequency *and* location.

for one vehicle type is generally not efficient. Many different vehicles may use the same lanes if they have the ability to change lane and if on average a low duty cycle or low density of vehicles can be guaranteed.

Separation in time, by low duty cycle is exactly the approach being used in many wireless installations by frequency management, thus allowing many users with low duty cycle.

In practical applications, coexistence always has two aspects:

1. Robustness against interference from other wireless users (low susceptibility)
2. Efficient usage of radio frequencies (spectral efficiency and low duty cycle) in order to leave enough free resources for other users

The Wireless Interface was designed to be robust by using adaptive retransmissions. Nevertheless, the continuous use of a significant part of the frequency band by an unfriendly system with a high duty cycle and long telegram lengths (longer than the maximum number of retransmits) will make it impossible for a system to transmit its data within the time limit requirement (e.g., 15 ms).

To discuss coexistence more quantitatively, characteristic numbers are helpful, which are defined here as follows:

Robustness can be measured by the factor between the maximum to the minimum delay values, measured in some defined coexistence test setup.

Robustness (R) = min delay/max delay

R = 4 ms/16 ms = 0.25

In an existing system between two user interfaces under idealistic and realistic but more worse-case conditions, for example, the coexistence with other wireless systems in their typical channel occupation mode. The more close to unity the better is the robustness.

Spectral efficiency can be measured as the ratio of data rate divided by the used bandwidth.

Spectral Efficiency (E) = D/B (D, data rate in b/s;
B, bandwidth in Hz)

E = 1 Mbit/s/1 MHz

This is an implementation- and application-independent value that enables technology comparison.

A more practical performance figure when looking at implemented solutions would be the following:

Performance $(P) = 1/(T_s/n_U \times B)$

(T_s, slot time in ms; n_U, number of parallel uplinks)

= 4 for Wireless Interface,

= 1 for symmetrical systems

$P = 1/(0.064\text{ ms}/4 \times 5\text{ MHz})$

= 12.5 nodes/ms/MHz

The frequency usage can give an indication of the interference potential of a wanted system to other wireless systems. A simple practical value can be derived from multiplying the time/duty cycle and the bandwidth used:

Frequency usage $(F) = T \times B$ (T, fraction of time;
B, bandwidth)

$F \sim 1 \times 1$ MHz

This gives a very practical figure of how much of the available frequency band, here, for example, the ISM band, as a very valuable resource is used for a certain implementation. The frequency usage should, of course, always be as small as possible; nevertheless, the resulting interference depends in practice mainly on the type of the other/victim system. Different victim systems react very different to frequency usage by a wanted system, dependent on if they had been designed for coexistence or not. The Wireless Interface, for example, uses in average not much more than 1 MHz of bandwidth continuously, for example, in an application as defined in Table 37.1.

Another practical aspect of coexistence is stability. The influence of one system on another should be as predictable as possible for automation systems. In practical applications it has a nearly constant frequency usage, which is determined mainly by its downlink, thus making its effect on other systems relatively predictable. The short uplink telegrams generate on average only low traffic load (<10%), even with high number of nodes and events.

Other wireless systems used in industrial applications are as follows:

- IEEE802.11, WLAN:

In factory automation WLAN is applied in many plants for varying applications. There are two typical uses:

- Access of many different applications to the plant/IT network (scanners, order data transfer, production data transfer, network connection with, e.g., a notebook)
- Communication to AGV systems: automated guided vehicles that fulfill transport needs

WLAN has its strength as a widespread standard, allowing a large number of different clients to access the same network. Typically, delay times below 100 ms are not an issue here.

WLAN is very susceptible to disturbance as it has a large receiver bandwidth and its CSMA and CCA features—designed to enable other WLAN devices to access the same system—also react to narrow band frequency user inside the used band. Therefore, the latency of a WLAN system will have a large jitter in practical applications. Measurements have shown values to vary from 1 ms to sometimes over 30 ms in a coexistence situation with other low duty cycle users.

- Bluetooth: Bluetooth is implemented in many mobile phones and notebooks; therefore, it is likely to be present. Bluetooth was made for user device interaction and is a robust protocol. It is suited for HMI purposes in industrial applications. Bluetooth devices do not use CSMA and may thus have an effect on WLAN links (especially in the search/pairing mode).
- IEEE 802.15.4/Zigbee/WirelessHART: Such devices will be used to cover a larger area for condition monitoring or building automation use cases. They will typically have a low duty cycle and are not to be expected in large quantities in factory automation (have therefore not been considered as disturber for the Wireless Interface in more detail). Such devices might be disturbed by a WLAN with a high duty cycle if in close distance and with not enough frequency separation (except Wireless Hart with a FH-option).

In order to estimate distance-dependent interference effects, an empirical model for the path loss (PL) is often used for Bluetooth-type applications (e.g., [13]). In the near field of the transmitting antenna, this attenuation is proportional to d^2, where d is the distance (6dB additional PL if distance is doubled). For larger distances, the empirical formula stipulates attenuation proportional to $d^{3.3}$ (10 dB additional PL if distance is doubled). This is a coarse model; hence, the distance values derived in the following should be considered only as indicative.

37.4.1.1 Interference from IEEE802.11b

Consider factory staff working in the vicinity of a Wireless Interface cell using a laptop computer to upload a large document from an 802.11b base station. The Wireless Interface transmission power is 0 dBm. Using the aforementioned PL model, this results in a carrier reception level C, at the base station at a distance of 5 m, of C = −54 dBm. IEEE 802.15.1 receivers require a co-channel carrier-to-interference ratio of C/I > 14 dB for adequate reception performance. Hence, the interferer level should be I < −54 dBm − 14 dB = −68 dBm. The 802.11b transmitter power may be up to 100 mW in 22 MHz bandwidth, equivalent to 6.5 dBm per Wireless Interface/IEEE802.15.1 receiver bandwidth of 1 MHz. The required PL to ensure safe operation is therefore 6.5 dBm − (−68 dBm) = 74.5 dB. This translates into a minimum safe distance requirement of 25 m for an 802.11b transmitter.

37.4.1.2 Interference from Bluetooth

Assume a factory worker using a Bluetooth phone headset. The typical Bluetooth transmitter power is 0 dBm at 1 MHz bandwidth and the required PL to ensure C/I > 14 dB is therefore 68 dB, which corresponds to a minimum distance requirement of 16 m.

37.4.2 Performance Measurements

The typical Wireless Interface operation is at relatively short distances, with reception signal levels of about −60 dBm, well above the sensitivity level. Hence, performance is in practice not limited by distance or fading effects, but mainly by robustness to external radio interference.

Performance evaluations were done by measuring the message delays. Any loss of message in the uplink (sensor status message) or in the downlink (BS acknowledgment) causes a retransmission in the next frame, resulting in an additional delay of 2048 μs. A measurement setup consisted of 60 nodes set to transmit at a rate of two messages per second over distances of about 70 cm. In the interference-free case, one in about 1000 uplink messages required a retransmission (marked as *reference* in Figure 37.9).

The effect on the transmissions due to interference from other base stations with Wireless Interface was also measured in a realistic scenario. The antennas of the interfering base stations are set to point toward the system under test, thus simulating multicell operation as shown in Figure 37.9 (top).

37.4.3 Industrial Electromagnetic Interference

Investigation of a number of potential industrial sources of interference showed that no interference is to be expected in the 2.4 GHz band from high power welding equipment or frequency converters. Figure 37.9 gives an overview of typical industrial devices, their basic frequency, and indications for their harmonics in the spectrum.

The only interference may come from legal ISM transmitters (Industrial Scientific and Medical). They are not widespread but have to be taken into account if unshielded. They can have very high power levels (e.g., microwave drying applications in the kW range).

A spot welding gun used in induction welding, for example, operates at up to 20 kA and generates very high electromagnetic field strengths. However, the majority of the high-frequency noise decays strongly above 1 GHz and therefore has a minimal impact on the Wireless Interface. The only

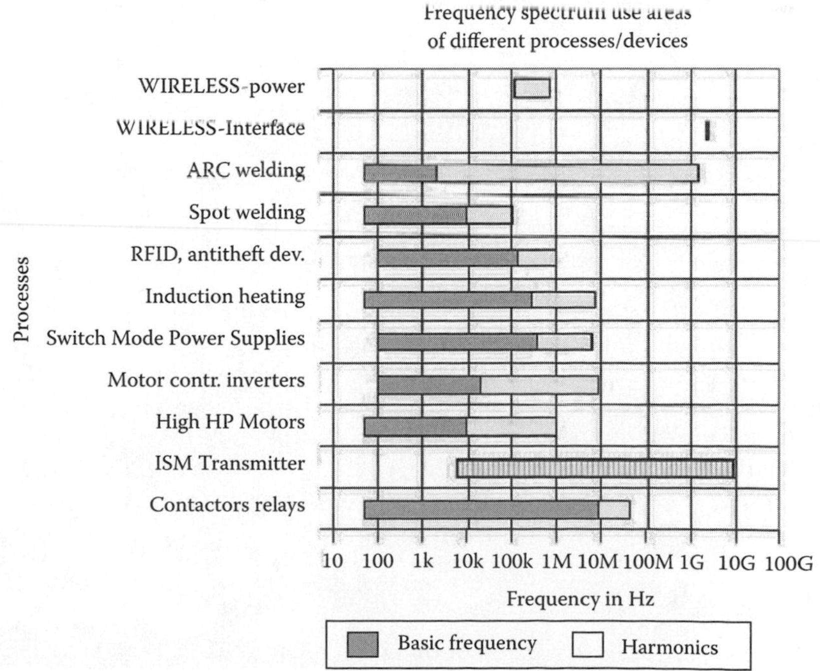

FIGURE 37.9 Electromagnetic interference from typical applications in the industry.

measurable interference with high frequencies (up to 1800 MHz) was identified as being generated by arc welding. Devices with Wireless Interface have been tested with positions only a few centimeters away from the welding spots.

37.4.4 Interference from Other Communication Systems

A number of measurements and tests have been conducted to quantify the effect of interference from other communication systems in the 2.4 GHz band. In a first scenario, interference was caused by a laptop PC with an IEEE 802.11g WLAN communicating to an access point (AP) transmitting at 54 Mb/s on channel 5 at 100 mW (Figure 37.11). Measurements were repeated at several distances (d = 7.5, 4, and 1 m) between the AP and the BS with Wireless Interface, and over 100,000 messages each were taken. Figure 37.11b shows the increased number of retransmissions due to the WLAN interference.

The effective data rate of the WLAN was reduced from 3.1 MB/s in the interference-free case to 2.8 MB/s (d = 7.5 m), 2.25 MB/s (d = 4 m), and 0.86 MB/s (d = 1 m), respectively.

In the second scenario, a Bluetooth-equipped laptop continuously transmits toward a Bluetooth AP at 1 mW and 1 m distance (Figure 37.11c) shows that the effect of Bluetooth interference on the Wireless Interface retransmissions is much smaller than the effect of WLAN interference. In the converse direction, the Bluetooth data rate was reduced from 29.2 to 25.6 kB/s by the interference from the Wireless Interface.

Further work on Wireless Interface is under way in the WSAN (Wireless Sensor Actuator Network) group of the Profibus International group on implementing, for example, WLAN options, which allow a blacklisting of WLAN Channels. As the dominant influence on WLAN is the continuous downlink (compare Figure 37.12, top and middle), just blacklisting the downlink is improving the WLAN statistic significantly (see Figure 37.12, bottom), where the downlink has been in a wire to verify uplink influence. This statistic is already very close to the reference measurement (top).

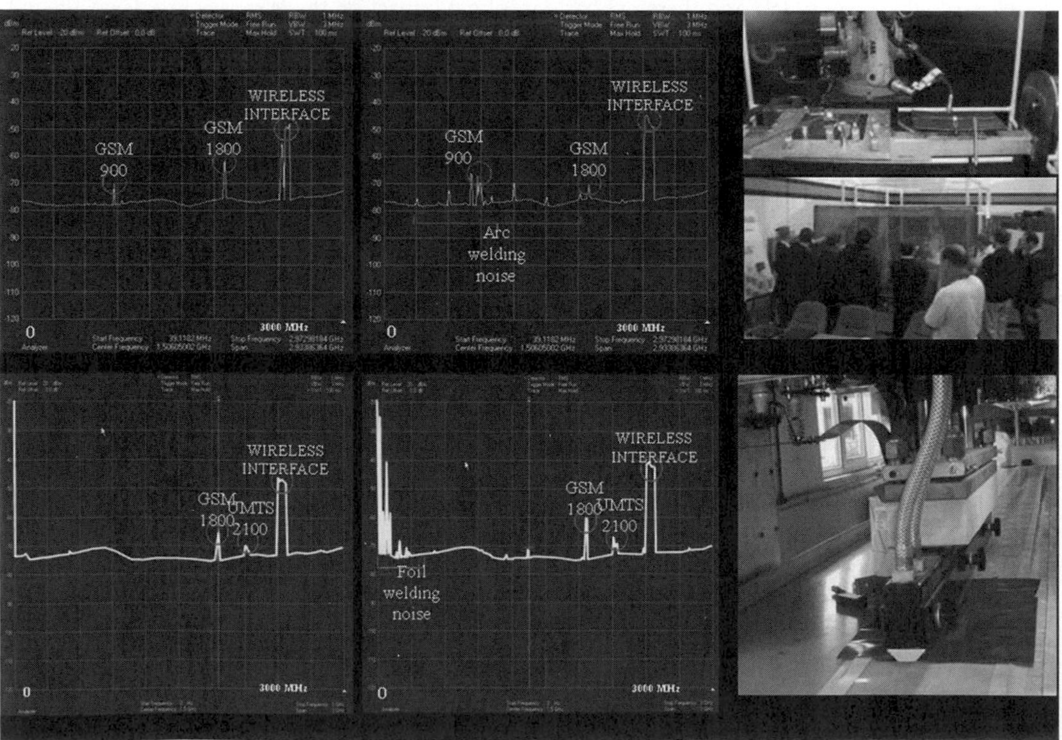

FIGURE 37.10 Frequency spectrum plots of electromagnetic noise of industrial welding equipment.

It has to be noted that generally the disturbance range of a system, outside which, for example, frequency can be reused, is significantly larger than the safe operation range. The disturbance range of a wanted system also varies with respect to a victim system, so it is not a wanted system property only.

37.5 Wireless Power Subsystem

Wireless power in principle can be supplied by different concepts:

- Energy scavenging: Taken from the local environment in the form of light, heat, vibration/motion.
- Energy storage: Included in the system in the form of batteries, fuel cells, etc.
- Energy distribution: Transmitted to the system via optical or radio frequencies, sound, etc.

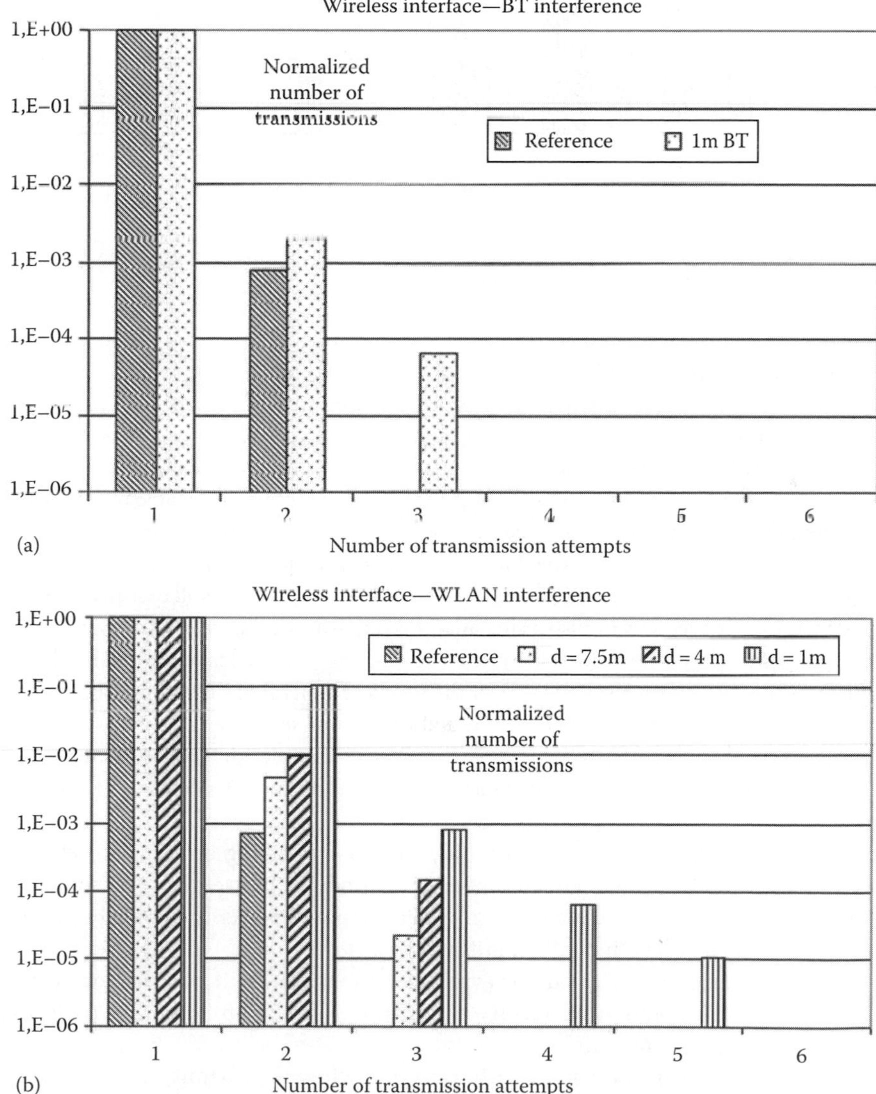

FIGURE 37.11 Measured retransmission statistics for uplink messages due to interference. (1 retransmit = 2048 μs). (a) Wireless Interface/Wireless Interface (1 m distance). (b) Wireless Interface/WLAN.

(continued)

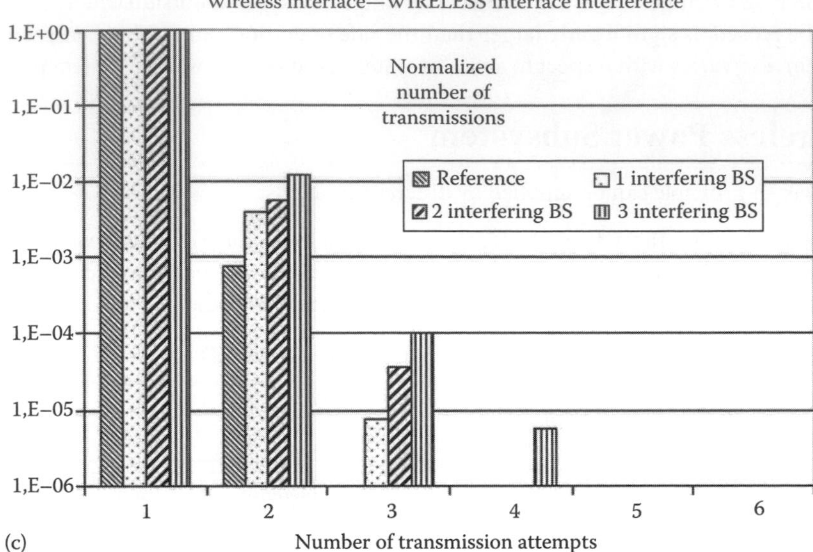

FIGURE 37.11 (continued) Measured retransmission statistics for uplink messages due to interference. (1 retransmit = 2048 μs). (c) Wireless Interface/Bluetooth.

The use of battery power is considered acceptable in the consumer world. However, in general industrial applications, where hundreds of devices require constant, reliable power supply and run day and night, batteries are not an option. Their energy density (normally around 1.2 Wh/cm^3 for primary batteries) is still too low, or the other way round: The consumption of low power radios and sensors is still too high for general use. While for secondary batteries there has been quite some development their volumetric energy density is still significantly lower than for primary batteries, typically by at least a factor of 3–5. For primary batteries, there has not been a noticeable progress regarding energy density.

Fuel cells are potentially somewhat better, but even their realistic potential is little more than 2 Wh/cm^3 (petrol energy density would be around 10 Wh/cm^3) and much development is still required before they could be envisioned to be used in everyday industrial installations [14]. Another fact is to be considered for fuel cells: Like human beings, they simply need air to breathe and evaporate water, a challenge for IP67 housings.

Environmental energy sources also fail to meet the needs of general industry applications, due to their unpredictable nature—both in terms of general usability without special engineering work and reliability. Such solutions would also incur considerable engineering and design costs for every single device. Nevertheless, for some niche applications, there are use cases, for example, vibration or high temperature monitoring (Figure 37.13).

Regarding the energy distribution concepts, only magnetic coupling in the near field can transfer enough power for general use. There are a number of possibilities to use magnetic coupling, depending on the transmission distance, a wide range of applications and power levels can be covered (mW at several meters of air gap up to tenth of kW at millimeters).

After a thorough evaluation of the various available options [14,15], it was concluded that the only viable, generally applicable solution for factory automation is based on long-wave radio frequencies, a form of inductive, magnetic coupling.

In Table 37.2, a comparison for a special use case of a wireless proximity switch is summarized. A figure of merit is defined, which summarizes the characteristic figures:

- Energy amount over 5 years (Wh)
- Energy cost over 5 years (USD)
- Risk of developing such a supply (factor)

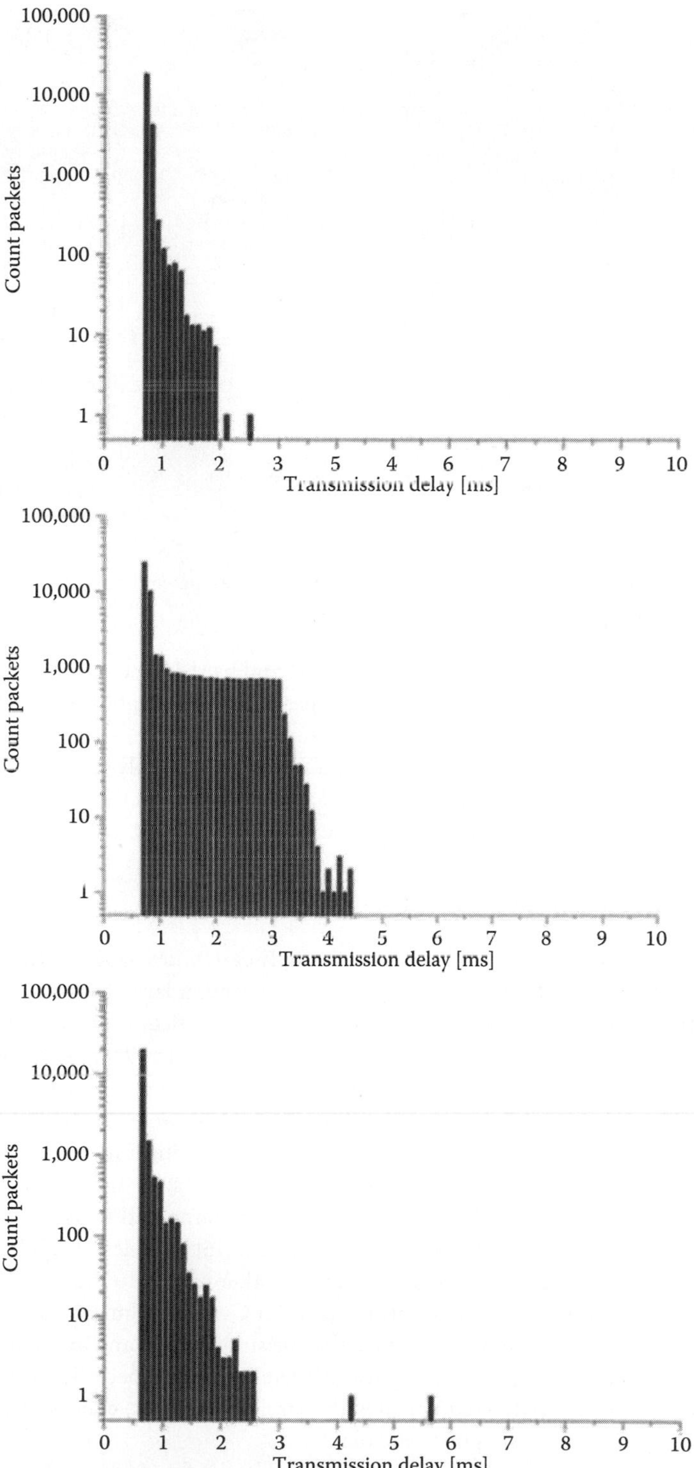

FIGURE 37.12 Measured WLAN-latency statistic (64 byte every 5 ms). Top: Reference measurement (no other frequency user). Middle: With Wireless Interface in parallel, +2 ms downlink influence visible (3 m). Bottom: Same, uplinks only (downlink in cable); 4 consec. Wireless Interface nodes (50 events/s simultaneously).

Concepts	Energy scavenging	Energy storage	Energy Distribution
Technologies ⊙ = Commercial availability	⊙ Photovoltaic ⊙ Temperature gradient O Human power O Wind/air flow O Pressure variations O Vibrations	⊙ Batteries O Micro-batteries O Micro-fuel cells ⊙ Ultra-capacitors O Micro-heat engines O Radioactive power sources	⊙ Electromagnetic coupling ⊙ RF radiation O Wired power grid O Acoustic waves O Optical
Power levels	10 µ–15 m W/cm3	50–3500 Ws/cm^3	µW–kW (m-mm)
Deficiencies	Reliability	Maintenance	Electric Field radiation Power level vs. range
General industrial applications	N\A (only for special single sensors)	Portable/mobile user interfaces	Automation cell power supply many devices

FIGURE 37.13 Overview of wireless power concepts and technologies; power levels, deficiencies, and remark for general application in factory automation.

The figure of merit is computed by dividing energy by cost and by risk according to Table 37.2. Inductive energy transfer at low frequencies of around 100 kHz has clearly the highest figure of merit regarding the intended application.

This is the so-called Wireless-POWER approach. This wireless supply principle, described later in more detail, provides power supply across a distance of a few meters with the help of magnetic fields and is suitable for most sensors and electrical actuators (like pneumatic valves) in discrete factory automation.

37.5.1 The Magnetic Supply: Wireless-POWER

The basic principle of a magnetic field–induced power supply can be described by the well-known transformer principle. A power supply unit feeds a primary winding, a large coil, which can be arranged around a production cell. The secondary side consists of a practically unlimited number of small receiver coils (see Figure 37.7). Each receiver coil is equipped with a ferrite core to increase the amount of flux collected by the coil and to minimize effects of external conductive materials. For this type of *transformer*, magnetic coupling is very low (<0.1%). The receivable power is determined by the amplitude of the magnetic field at the location of the *receiver* (secondary) winding and its size.

Although human operators will rarely work continuously inside such an automated production cell, the strength of the magnetic field at all working positions (including within such a cell) complies with international occupational regulations and recommendations [16]. Wireless-POWER works at the frequency of 120 kHz. The distance to be observed by a pacemaker carrier to this setup would be 2.5 m at the maximum current of the WPU (24 A) according to strict German regulations.

Energy losses in such a wireless power system are surprisingly small and are mainly caused by skin and eddy current effects in the coil itself and in nearby metal objects, especially steel. In many different production cells in factory automation equipped with Wireless-POWER, energy losses have been measured to be around 10–15 W/m^3 of supplied cell volume.

The principle does not depend on the frequency, but the chosen frequency is a good balance between

- Higher amplitudes needed at lower frequencies (higher ohmical losses)
- High-frequency losses in a realistic environment, where always metal will be present to a certain degree (eddy current losses)

TABLE 37.2 Example Power Supply Comparison for a Typical Autonomous Industrial Sensor Application by a Figure of Merit (One-Third of Sensor: 6 cm^2 Usable Surface, 2.5 cm^3 of Usable Volume)

Energy Source	Power	Unit	Energy (per Life of) 5 Years; Wh	Av. Power w/o Service Over Life in mW	Cost (No Service) w/o Encapsul. USD	Figure of Merit Wh per USD/risk	Risks w/o Power 1 = Very Low 4 = Very High.	Risk Descr	Remark
PV 1000 Lux a-Si	0.034	mW/cm^2	9.0	0.21	0.3	15.0	2	Shielding/orient.	One-third surface
PV 100 Lux a-Si[a]	0.0029	mW/cm^2	0.8	0.02	0.3	1.3	2	Shielding/orient.	One-third surface
PV 1000 Lux GaAs[a]	0.102	mW/cm^2	26.9	0.62	3	3.0	3	Rp, costs, shielding/ orient.	One-third surface, 3[a]eta a-si, 1ß[a]Cost a-Si
Thermal (thin film)[a] dT = 10 K	0.0063	mW/cm^2	1.0	0.02	1	0.3	3	Reliability, costs, feasibility?	Mounting area, dT-10 K
HF100 kHz (7 A/m)	1	mW/cm^3	105.7	2,41	1	52.8	2	High Q, E-smog shielding	Simple test setup 0.5 m, 2 A, not yet optimized
HF 13 MHz	0.23	mW/cm^3	24.3	0.55	1.5	8.1	2	High Q, radiation shielding	Ps = 100 mW, antenna 3 cm, distance 1 m
HF 433 MHZ[a]	0.3	mW	13.1	0.30	2	3.3	2	High Q, regulations	Ps = 100 mW; antenna 17 cm, distance 1 m
Battery LiSoCI2	1.1	Wh/cm^3	2.7	0,06	2	1.3	1		One-third volume
Battery AIMn	0.32	Wh/cm^3	0.8	0.02	0.2	3.9	1		One-third volume
Fuel Cell, H2 20 MPa	0.25	Wh/cm^3	0.6	0.01	5	0.03	4	Storage, water	One-third volume, 50% of values of larger systems
Fuel Cell, Hydrid H2	1	Wh/cm^3	2.4	0.06	5	0.16	3	Water, pressure	One-third volume, 50% of values of larger systems
Fuel Cell, Methanol	2.2	Wh/cm^3	5.3	0.12	2	0.66	4	Water, technology/ time	One-third volume, 50% of values of larger systems

[a] Estimation.

- Semiconductor switching losses
- Capacitive dielectric losses
- Component sizes (larger at lower frequencies), which is especially an issue on the receiver side, where you need to have small devices

As the receivers will typically be designed to consume just 5–100 mW (sensors—small actuators) the aforementioned losses will clearly dominate the power consumption. Therefore, the amount of such wireless devices is practically not limited; several hundred or even 1000 can be supplied.

37.5.2 Resonant, Medium-Frequency Power Supply

These unconventional transformers have to be operated in a *resonant* mode, in order to compensate for the large leakage inductances of the transformer. The resonant principle allows the wireless power unit WPU to stimulate the resonant circuit at relatively low voltages (<60 V).

The WPU has to have a control circuit that must also be able to accommodate the following:

- Changes over time in the environment, for example, caused by the movement of large, metal objects such as robots
- Different *load* requirements, caused by differently sized and formed primary coils (inductance values), and losses, caused by factors such as eddy currents in adjacent metal objects
- Other nearby wireless supply systems, which may couple inductively

37.5.3 Rotating Field

Unidirectional magnetic fields can be attenuated or blocked by metal objects, which is the case in applications where the devices have to be placed arbitrarily or are moved, for example, by a robot throughout the application. To minimize the shielding effects, two primary loops can be mounted orthogonally (see Figure 37.14, top). The loops are fed by separate power supplies, whose currents are phase-shifted by 90° with respect to each other. This creates a field vector that rotates at 120 kHz with a constant amplitude, which is sufficient for most practical applications. With a third loop also an undirected field could be created, by rotating the field vector evenly through the volume. Nevertheless, the additional benefit would have been small in the so far realized applications, but a much more complex control approach would be needed.

37.5.4 Omnidirectional Receiver Structure

For sufficient power output, the receiver side also has to be operated in resonant mode. To make the available power independent of the receiver's orientation with respect to the primary field vector, an orthogonal setup of three coils on a common core has been chosen (see Figure 37.7). The available power density for typical *worst-case* shielding conditions in real applications is on the order of 1.2 mW/cm^3. Typically, it is a factor 2–4 higher. The absolute power level can be modified with the secondary coil size and shape (and the distance to the primary coil). With optimized Q-factors of the receiver coils, much higher levels could be transferred, nevertheless only under ideal, nonrealistic environmental conditions.

37.5.5 Example Setup

The resulting field distribution in an installation with rotating field to cover a larger volume is shown in Figure 37.14. The only design quantity needed for the coil placement is the distance D to the next parallel coil. This distance is optimal around two-third of the smallest dimension S of a loop. The smallest dimension S is defining the range of the loop antenna.

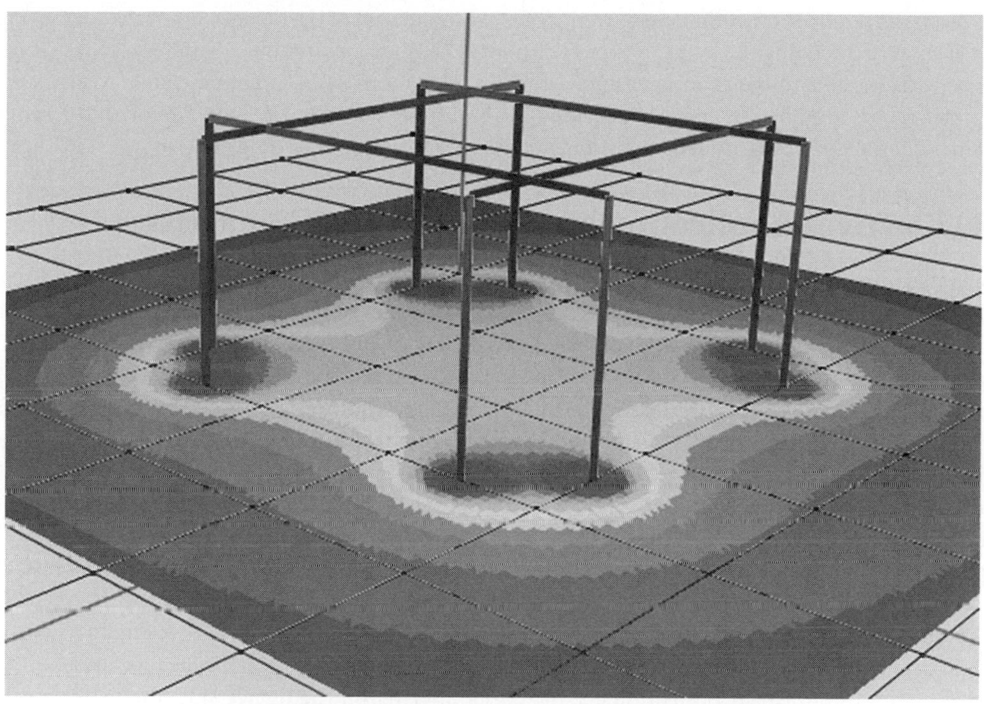

FIGURE 37.14 Top: Simulated field distribution in an example $3 \times 3 \times 3$ m^3 setup. Bottom: Example installation with Wireless-POWER and communication showing some of the integrated power loops and the design parameter distance D, which depends on the smallest length S of the coil.

To reach the required minimum field strength for safe operation of the devices, a certain current has to be set at the WPU feeding the loops. A typical current is 24 A for loop with S = 3 m, respectively, smaller in loops with a smaller dimension. If an application is larger, more loops can be cascaded. Also a number of simpler setups are possible for special cases (line and spot type primary loops) as often the range of movement of the devices is limited. A line type primary loop could be as long as 40 m per WPU unit.

37.6 Practical Application and Performance Comparison

Wireless technologies have led to a range of products. A fully wireless proximity switch and a sensor concentrator are using both wireless technologies and operate with special energy optimized sensor heads. To also allow the integration of other standard sensors and typically much more power hungry actuators, which might be needed for a complete automation application, a sensor/actor distribution box with up to 16 in.-/or outputs is available. For the Wireless communication two profiles (up to 32 bits in both directions) have so far been implemented, to accommodate the different payload data requirements of different devices (Figure 37.15).

Total power delivered even under worst-case conditions to the wireless proximity switch is 6 mW (sufficient for five events per second). The larger receiver in the sensor pad WSP100, which allows the connection and supply of up to eight sensor heads, can worst case provide approximately 50 mW continuously. Both values are two to four times higher in typical working positions.

Cell dimensions that can be covered with the WPU100 power supply devices are $1 \times 1 \times 1\ m^3$ to $8 \times 8 \times 3\ m^3$, respectively, $1\ m^3$ to close to $200\ m^3$ in a single cell arrangement (without vertical conductors inside).

Integrated into industrial IP67 products, including filtering on the device side, the fieldbus timing uncertainty and the handling delays on the base station side of a field bus master, the performance stays within the specified limit.

The base station interface normally is a fieldbus that adds additional latency. The base station also supports over its built-in user interface a special configurable mapping mode, which can map signals, for example, of wireless sensors directly to other devices with outputs. In that mode very simple latency measurements can be done, without being delayed by the bus interfaces.

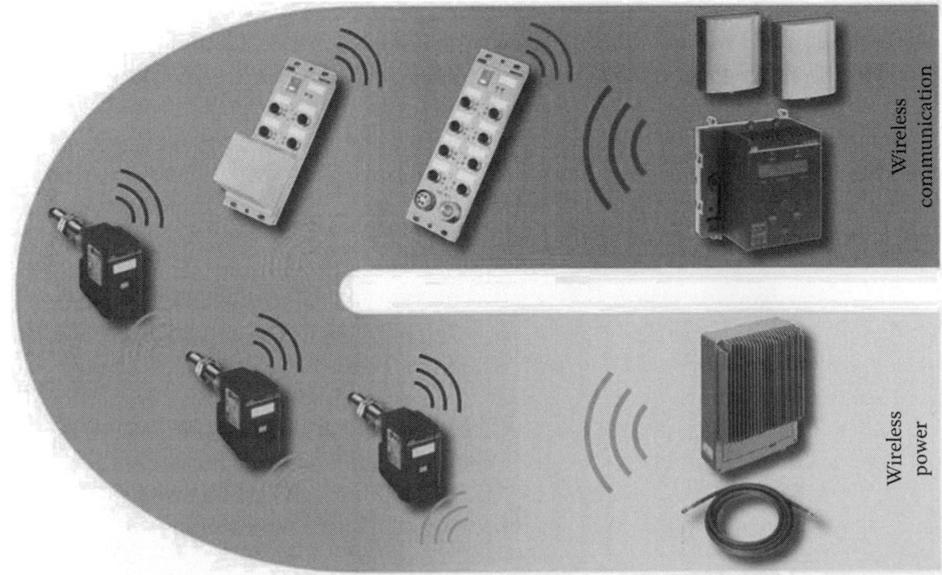

FIGURE 37.15 Industrial automation devices with embedded Wireless-POWER and Wireless Interface-COM technologies (left side) and necessary infrastructure (right side).

Measurements published in Ref. [17] show for the mapping mode a delay statistics with values between 5 and 9 ms increasing in a coexistence setup with two WLAN networks to maximum 11 ms for a few telegrams out of 1 million measured. These latencies measurements were always for two acknowledged telegrams indicating that the Wireless Interface works with latencies of typically 4 ms (2.5–6 ms).

The machine shown in Figure 37.14 is in continuous shift operation since 2003. In total several Wireless Interface applications with several hundred devices have been monitored over several years. Therefore, up to date roughly 4 million device hours have been accumulated, giving a very good feedback of the achieved reliability in production. The main lessons learned were on mechanical issues (usability of housing and connector).

Applications with up to 156 independently rotating nodes in one machine (cable winding machine, Karlskrona Sweden) have been installed in 2005 with up to four colocated base stations. All the sensors are in cable drums, which continuously move in a Ferris wheel–type cage arrangement with multidimensional movement. Several more of such machines are installed or planned.

A further challenging application where the real-time properties are mission-critical is automotive stamping. The delays of the wireless system have to be within the mechanical limits of the press, otherwise drastic failures could occur. The wireless devices are installed in the press tools that are frequently exchanged, so also fast start-up and reconnection times are an issue here. A Wireless Interface device can, after start-up, be transmitting again already after few 100 ms, including internal secondary power supply start-up procedures.

To quantify the achieved performance, another figure of merit was defined (see right column in Table 37.3/I) as number of nodes divided by real-time limit and by power consumption. The empirical quantitative comparison of the technologies mentioned in Chapter 2 was done using the figures as given in Table 37.3. Assumed was the wireless connection of simple proximity switches (1 bit), the most widespread sensor in discrete factory automation.

Data as listed in Table 37.3 are often a subject of controversy and consistent practical data cannot be found in papers. Therefore, typical estimated values have been collected and taken to enable a comparison. Where values cannot be really guaranteed due to the technology, the figures in parentheses have

TABLE 37.3 Empirical Comparison of Wireless Technologies in Discrete Factory Automation for Short Data Packet Applications (Sensors)

I	Nodes/ Network	Networks per Cell[a]	Cycle Time Air Interface in ms	Real Time Limit[b] in ms	Energy/Bit with Retransmits in mWs	Figure o. MERIT (Nodes per ms a. mWs)
WIRELESS Interface	120	3	2	15	1.2	20
BT [11,13]	7	10	8.75	50	10	0.14
WLAN [12]	100[c]	3	(100)	(500)	100	(0.006)
Zigbee [14]	7	16	15	(500)	0.6	(0.37)
Enocean [15]	20	1	(10)[c]	(1000)	0.05	(0.4)

II	Nodes per Cell	Range r in Machine in m	Realizable Avg. Node Density per m3[d]
WIRELESS Interface	360	5	2.29
BT	70	30	0.012
WLAN	300	50	0.019
Zigbee	112	30	0.020
Enocean	20	15	0.014

[a] Without significant performance change.
[b] Air interface (error rate ~10–9, @max node density, and at 5 events per second and device).
[c] Not defined, value assumed for comparison.
[d] Usable volume: Assumed radius of range *r*, height 2 m.

been used to conduct the performance comparison. In Table 37.3, *Enocean* refers to a proprietary technology known for its low power consumption, developed for building automation.

The real-time limit in Table 37.3/I is the time delay, which can be *practically* guaranteed (very small failure probability, goal here at the mentioned number of nodes in operation: 10^{-9}).

As an example, consider a Bluetooth pico-net having up to seven slaves. Ten Bluetooth networks have been reported to operate directly in parallel [3], without a significant decrease in performance. Assuming the typical range in industrial environment, for example, 30 m in for Bluetooth (100 mW class 1 version) and assuming a height of typical applications of 2 m a realizable average node density figure per m^3 can be calculated (Table 37.3/II). This average figure is quite small for Bluetooth due to the longer range, compared with the requirement (0.25–2 wireless devices per m^3 in factory automation applications; see Chapter 2). Bluetooth can operate with a cycle time of ~8.75 ms, but under realistic conditions at high node densities, the maximum latency that can be practically guaranteed is much higher, figures of 50 ms have been reported by various groups. For Bluetooth, a power consumption of 50 mW has been estimated, assuming a certain duty cycle in use.

The resulting figure of merit (last column in Table 37.3/I) for the Wireless Interface is 20 (nodes per ms and mWs), roughly a factor of hundred higher than for any other technology. This relation changes somewhat on the energy per bit scale, with assuming a 32 bit payload, but will still be around a factor of 50 higher.

Looking at Table 37.3/II, it can be clearly seen from the achievable average node density in a factory hall, that only the Wireless Interface can achieve the requirements for a whole factory.

It is worth reminding that generally a smaller range allows a higher node density, due to lower interference between cells. A low range is typically acceptable in factory automation, where a given base station needs only to provide coverage for a manufacturing cell.

For technologies like Enocean, ZigBee, and WLAN, low failure probability is not generally possible due to the impact of, for example, a moving environment in machines within their long range *r* (see Table 37.3/II); therefore, the figures have been put in parentheses. Figure 37.16 visualizes the achieved match of performance of the Wireless Interface compared with the requirements and the discussed competing technologies.

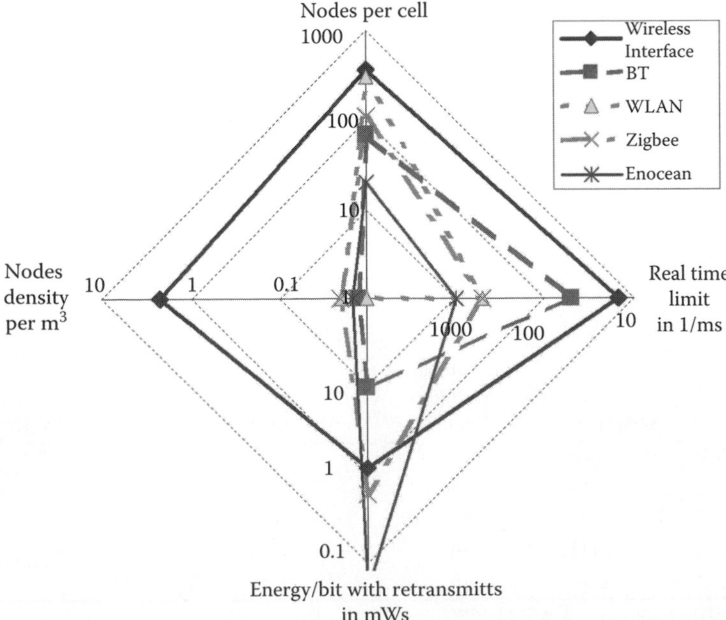

FIGURE 37.16 Visualization of comparison for the most important requirements in discrete factory automation (empirical, mind logarithmical scale).

It should be noted that despite the empirical approach, it is not changing the picture a lot, if a certain value would be differing by a factor of 2, due to the logarithmic scales chosen on each of the axes.

The characteristic figures defined in Chapter 4 coexistence can be used to help practically judge the coexistence behavior of the different in factory automation used systems/technologies. In order to do this, a fictive real-time system with a TDMA approach as implemented in the Wireless Interface on an IEEE802.15.1 physical layer has been assumed implemented with the different physical layers of IEEE802.15.4.-based systems (Zigbee, WirelessHART) and as comparison WLAN (IEEE802.11.g). To calculate a duty cycle for frequency usage, it has been assumed that 100 nodes should be covered as fast as possible, an application cycle time of 20 ms (e.g., PLC) has been assumed, in which all nodes should transmit one telegram correctly.

Table 37.4 shows the assumed realizations and the values used to calculate the characteristic figures. For the 15.4 PHY, it is visible that the resulting TDMA frame of 50 ms would be much larger as the anticipated applications cycle time of 20 ms; therefore, 15.4 would not be practically usable with such a number of wireless nodes. Table 37.5 calculates the characteristic figures defined in Chapter 3 for the three different physical layers.

TABLE 37.4 Assumed Realizations of a TDMA Real-Time System, Based on Different Standard Physical Layers and Resulting TDMA Frame in Which All Nodes Could Transmit

Actual nodes	100	1 byte data		
MS application cycle (PLC)	20			
	Bandwidth MHz	Slot time in ms	Data rate Mbit/s	Resulting TDMA frame time in ms
IEEE 15.4 PHY	3	0.5	0.25	(50[a])
IEEE 15.1 PHY—WIRELESS Interface	5	0.064	5	1.6
IEEE 11.g PHY	22	0.04	54	4

[a] Longer than necessary application cycle: Not realizable.

TABLE 37.5 Characteristic Figures for Three Different Physical Layers in an Assumed TDMA Real-Time Implementation (Empirical, Indicative Only)

Characteristic Figures	Performance		Coexistence	
	E = Spectral efficiency	P = performance	F = Frequency usage (@ max. nodes)	R = Robustness (empirical)
	Bit rate/bandwidth	Nodes/cycle time and per bandwidth	Bandwidth × (nodes × slot time/application cycle time)	Min. latency/ max. latency
		Nodes/(ms × MHz)	(MHz)	Ideal/worst case
IEEE 15.4 PHY	0.083	0.67	7.5	~0[a]
IEEE 15.1 PHY-WIRELESS Interface	1.0	12.5	1.32	0.25[b]
IEEE 11.gPHY	2.45	1.14	4.4	0.02[c]

[a] The worst-case latency for a 15.4 system is very large, as fadings with a large bandwidth can frequently or even statically occur; that is why normally a mesh layer should be used to find alternative routes then.

[b] WIRELESS Interface has been measured, for example, in [4] as *WISA*.

[c] A WLAN latency as, for example, measured in [16] in a very quiet environment can be as fast as 0.7 ms, but in a realistic environment, 100 nodes connected with parallel other WLANS and users, up to 35 ms can be frequently measured.

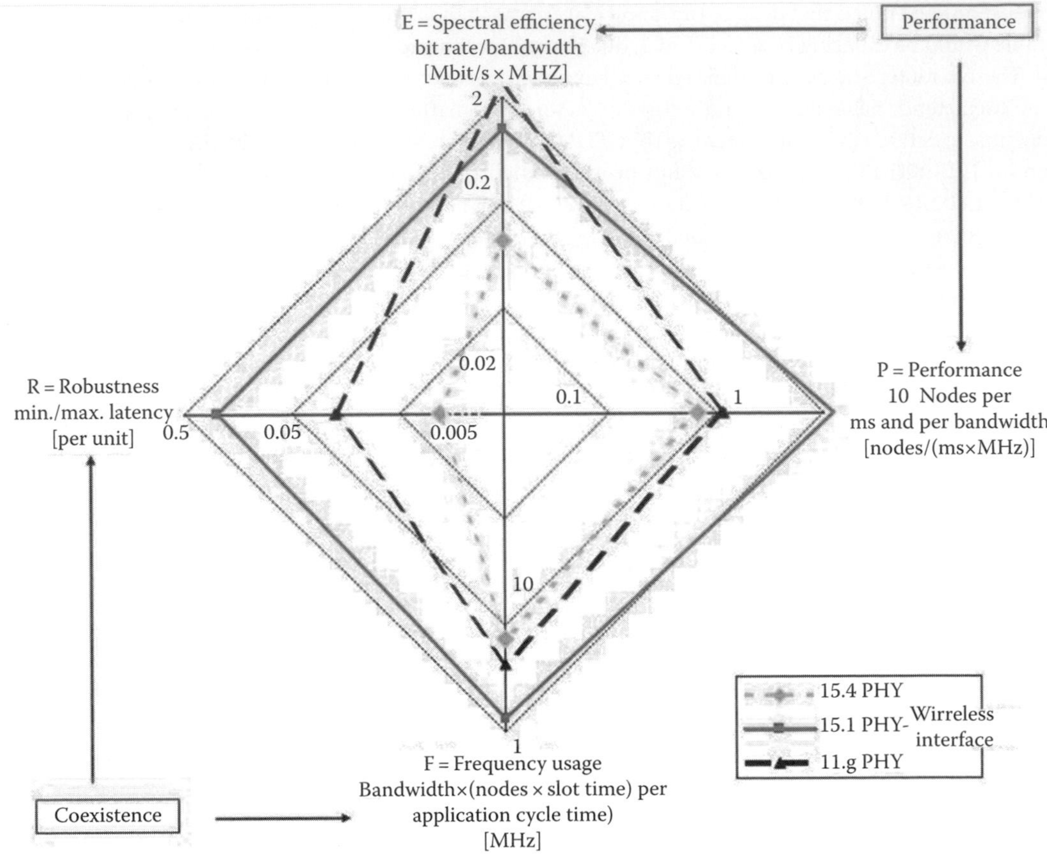

FIGURE 37.17 Characteristic figures for coexistence and performance according to Table 37.5.

The IEEE802.15.1 physical layer–based Wireless Interface implementation has, although it uses five 15.1 frequencies simultaneously, a lower frequency usage F compared with the much higher data rate WLAN, which, if used like that, suffers from its overhead at the assumed small data packet sizes of 1 byte (e.g., a typical simple binary sensor).

The values of Table 37.5 are plotted in Figure 37.17 to visualize again the different profiles of the technologies for short packages.

37.7 Summary

A truly wireless sensor/actuator interface with wireless power and wireless communication for real-time factory has been presented. The Wireless Interface uses IEEE 802.15.1 radio transceivers, but adds an optimized TDMA protocol to support a high number (120) of S/As per base station as well as short cycle times (2048 μs).

It is important to emphasize that Wireless Interface parameters differ significantly from the design goals of existing short-range wireless office systems such as Bluetooth or ZigBee. This difference has been verified graphically and by a figure of merit, which compares the performance to the most important requirements in discrete factory automation applications.

The FH sequences are constructed to guarantee adequate frequency separation between consecutive hops and low correlation (low interference) in the case of multicell operation. In combination with an appropriate ARQ scheme, this provides reliable and low delay data transmission. All cells use the entire available ISM frequency band of 80 MHz at 2.4 GHz, without any requirement for frequency planning.

The effect of interference from other systems likely to operate in the same frequency band has been analyzed, and the Wireless Interface is designed to withstand such interference. For comparison of coexistence and robustness, new characteristic figures have been proposed to quantify discussions.

A wireless power supply system has been designed and verified together with the wireless devices in many practical installations.

The system has been implemented into products and its performance in real factory applications has been verified and more than 4 million node hours have been accumulated. Products based on it are commercially available and are used in a growing number of diverse applications in discrete factory automation [18], some of them already running for more than 5 years.

References

1. A. Willig, K. Matheus, and A. Wolisz, Wireless technology in industrial networks, *Proceedings of the IEEE*, 93(6), 1130–1151, June 2005.
2. S.P. Karanam, H. Trsek, and J. Jasperneite, Potential of the HCCA scheme defined in IEEE802.11e for QoS enabled industrial wireless networks, *Sixth IEEE International Workshop on Factory Communication Systems*, June 28–30, 2006, Torino, Italy.
3. J. Weczerek, Drahtlose Übertragung von Steuersignalen in der Automatisierung mittels Bluetooth, *Wireless Technologies Kongress*, 2004, Mesago Sindelfingen, Germany.
4. L.L. Bello, Communication techniques and architectures for Bluetooth networks in industrial scenarios, *ETFA'2005—10th IEEE International Conference on Emerging Technologies and Factory Automation*, September 19–22, 2005, Catania, Italy.
5. F. Schmidt, Batterielose Funksensoren, *11. ITG/GMA Fachtagung, Sensoren und Mess-Systeme*, März 11. und 12, 2002, Ludwigsburg, Germany.
6. M.L. Mathiesen, R. Indergaard, H. Vefling, and N. Aakvaag, Trial implementation of a wireless human machine interface to field devices, *ETFA 2006—11th IEEE International Conference on Emerging Technologies and Factory Automation*, September 20–22, 2006, Prague, Czech Republic.
7. Bluetooth SIG, Specification of the Bluetooth System, v1.1, February 22, 2001.
8. European Telecommunication Standards Institute, Digital cellular telecommunication system (Phase 2), Multiplexing and multiple access on the radio path (GSM 05.02), ETS 300 574, August 1999.
9. A.A. Shaar and P.A. Davies, A survey of one-coincidence sequences for frequency-hopped spread spectrum systems, *IEE Proceedings*, 131(7), 719–726, December 1984.
10. D. Dzung, Ch. Apneseth, J.-E. Frey, and J. Endresen, Design and implementation of a real-time wireless sensor/actuator communication system, ETFA'2005, *10th IEEE International Conference on Emerging Technologies and Factory Automation*, September 19–22, 2005, Catania, Italy.
11. VDI/VDE Guideline 2185, Radio based communication in industrial automation, Issue 09/2007, VDI/VDE-Gesellschaft Mess- und Automatisierungs-technik.
12. C. Dupler and H. Forbes, Wireless technology for the discrete industries, *Twelfth Annual Orlando Forum-Winning Strategies and Best Practices for Global Manufacturers*, February 4–7, 2008, Orlando, FL.
13. IEEE Standard 802.15.2, IEEE recommended practice for information technology, Part 15.2: Co-existence of wireless personal area networks with other wireless devices operating in unlicensed frequency bands, IEEE Std 802.15.2, 2003.
14. G. Scheible, Wireless energy autonomous systems: Industrial use?, *Sensoren und Messysteme VDE/IEEE Conference*, March 11–12, 2002, Ludwigsburg, Germany.
15. J.A. Paradiso and T. Starner, Energy scavenging for mobile and wireless electronics, *Pervasive Computing*, IEEE CS and IEEE ComSoc, January–March 2005, pp. 18–27.

16. International Commission on Non-Ionizing Radiation Protection (ICNIRP), Guidelines for limiting exposure to time-varying electric, magnetic, and electromagnetic fields (up to 300 GHz), *Health Physics*, 74(4), 494–522, 1998.
17. M. Kraetzig, G. Scheible, and R. Hüppe, So erreicht man Koexistenz—Ergebnisse einer firmenübergreifenden Studie des ZVEI, VDI/GMA Fachtagung Wireless Automation, February 26–28, 2008, Berlin, Germany.
18. J.-E. Frey, G. Scheible, and A. Kreitz, Unplugged but connected, *ABB Review*, 3 and 4, 2005, pp. 65–68, 70–73.

38 IPv6 over Low-Power Wireless Personal Area Networks (6LoWPAN) and Constrained Application Protocol (CoAP)

Guido Moritz
SIV AG

Frank Golatowski
University of Rostock

38.1 Introduction 38-1
38.2 IPv6 over Low-Power Wireless Personal Area Networks—6LoWPAN 38-2
Introducing IEEE 802.15.4 • Introducing 6LoWPAN • Compression • Fragmentation • Neighbor Discovery • Routing • Closing Remarks
38.3 Constrained Application Protocol 38-8
Introducing CoAP • CoAP/HTTP Proxy and Caching • CoAP Enhancements to Binary HTTP
38.4 Related Work 38-11
38.5 Conclusion 38-12
References 38-12

38.1 Introduction

In the past decade, a great deal of research has been conducted in the field of wireless sensor networks (WSNs). One important part of a WSN node is its communication interface. In the early days of WSN, these communication interfaces were not standardized. But today, standardized communication interfaces, which are in line with machine-to-machine (M2M) communication requirements (self-describing interfaces, self-organizing, and self-healing), are required to build the foundation of the Internet of Things (IoT) or Web of Things.

IEEE 802.15.4 defines physical and medium access control layer and is widely used in current WSNs. Recently, this standard was mainly the basis for proprietary protocols, like ZigBee, WirelessHART, and SP100. But today, IEEE 802.15.4 is also used as the basis for Internet Protocol (IP). With the availability and wide acceptance of 6LowPAN—an adaptation of the IP Version 6 for low-power networks with limited resources—the IP functionality is now available for distributed networks with very limited resources.

This chapter describes 6LowPAN—an adaption protocol—which brings IPv6 (IP version 6) protocol to WSNs using IEEE 802.15.4 communication standard. With 6LoWPAN, various challenges have been solved to integrate applications that used to rely on proprietary protocols into the global digital village.

Further, basic principles and features of CoAP, an emerging application protocol initially based on 6LoWPAN, are explained. CoAP forms the basis for RESTful embedded web services, which are often used in IP-based sensor networks for seamless integration.

38.2 IPv6 over Low-Power Wireless Personal Area Networks—6LoWPAN

On low-power wireless devices like WSN nodes and for Smart and Cooperating Objects (SCO), Zigbee over IEEE 802.15.4 is one of the predominating protocols today. Such WSNs lie on the edge of networks, or at the field level in process automation. Today, application layer gateway or proxy solutions are used to integrate WSNs, with their proprietary protocols, into enterprise systems, or higher level services in the IP-based networks. Changes in the application, the protocol data representation, or data encoding, impose high management and computational effort for the application layer gateways. In contrast to gateways, routers of IP-based networks with their end-to-end connectivity just forward packets and do not need to have a deep knowledge about the mission of the specific WSN and applied protocols.

Alternatively to proprietary solutions, for M2M communication and for integration purposes in enterprise systems, IP would be a suitable contributor. IP is the predominating protocol in network infrastructure for information and communication systems, ranging from local area to wide area networks. However, IPv4 is obsolete considering the envisioned trillions of devices, which will appear in the near future. IPv6, the next generation IP, expands address space of IP to 2^{128} addresses and includes miscellaneous conceptual improvements based on experiences with IPv4.

Until recently, IP was not an option for WSNs because there were no IP stacks available that would run on resource-constrained WSN nodes. However, latest developments have born IP stacks, which run on tiny 8-bit microcontrollers like TI's MSP 430—a commonly used microcontroller in WSNs. The footprint of such IP stacks for microcontrollers is around 12 kB (cf. uIPv6 [Dur08]).

Dependent on IP, other higher level protocols, tools, and applications have been established. Protocols like DNS, DHCP, Zeroconf, etc., can be used for naming, addressing, discovery, and configuration. Tools like SNMP, ping, arp, traceroute, etc. are usable for management purposes. Even at the application layer level, matured protocols, data models, and services such as HTTP, HTML, XML, SOAP, REST, etc. are well-established and essential for IT developments. With all these tools and protocols and decades of experience with them, designing and developing WSN becomes easier, more efficient, and thus less expensive.

Summing up, IP is supported by all classes of operating systems ranging from desktop operating systems, embedded and real-time operating systems, to operating systems for WSNs like Contiki [Dun04]. IP protocol stacks are available as an integral part of the operating system, as open source software stacks, and commercial solutions. The uIPv6 [Dur08] stack is an implementation of the 6LoWPAN protocols described in this section. For proprietary protocols, there are only few open source solutions available. They are limited to specific application areas and application niches. Future networks of smart and cooperating objects are not for niche markets only. They will pave the way for very new innovative and broad application areas. That said, IP radically helps to enforce that development, provides the basis for cross-domain interoperability, and will break through technology-specific boundaries.

38.2.1 Introducing IEEE 802.15.4

Based on IEEE 802.15.4 [IEEE154], which defines the physical (PHY) and medium access control (MAC) communication layers as depicted in Figure 38.1, various upper layer protocols have been defined by

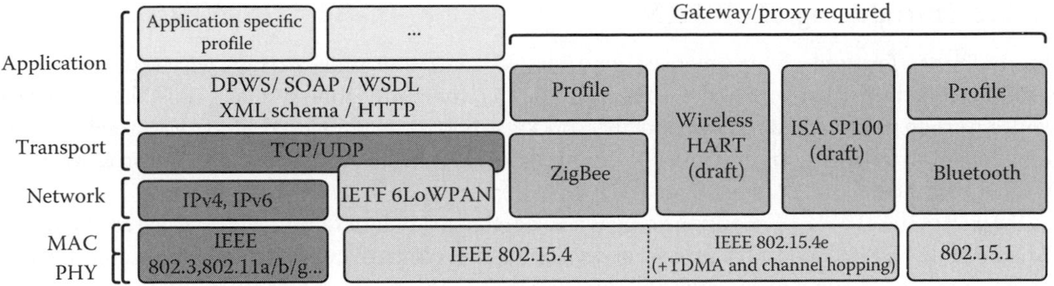

FIGURE 38.1 IEEE 802.15.4 in context.

large industrial consortia; for example, WirelessHART, ISA SP100, and ZigBee. Open standards and proprietary industrial consortia standards will be in a close competition. Key features of all protocols are as follows:

- Long battery life through low-power consumption solutions
- Self-healing networks
- Scalability and support of large networks
- Low cost node
- Low data rate

IEEE 802.15.4 is a standard for wireless personal area networks with low data rates (LR WPAN). The wording "personal area networks" in the 802.15.4 standard is misleading at this point, because 802.15.4 networks are of course not limited to personal use. While technologies like WiFi (802.11) and Bluetooth have been developed for high data throughput, the focus of IEEE 802.15.4 networks was power awareness and low-cost design, which results in low data rates.

The IEEE 802.15.4 standard specifies two physical (PHY) layers—working in the 2.4 GHz and 868/915 MHz band PHY layers. The physical layer depends on local regulation and user preference. The unlicensed 2.4 GHz is valid worldwide and has a higher data rate (240 kbps) compared to 915 MHz band with 40 kbps (US) and 868 MHz (EU) band with 20 kbps data rate. The radio can operate on 16 channels in the 2.4 GHz band, 10 channels in the 915 MHz band, and 1 channel in the 868 MHz band.

IEEE 802.15.4 supports like other low-power link layer several topologies including star and peer-to-peer mesh as depicted in Figure 38.2. Within the topology, the device may form lightweight end points (leaves), which require a minimum of resources in terms of memory, computing power, or energy. With slightly more resources available, the devices may act as full function device (FFD) other endpoints named as reduced function devices (RFD) may connect to. These FFD also provide routing functionalities within the meshed topology. Most resources within the topology are required by the so-called PAN (personal area network) coordinator, which is in charge of managing the network.

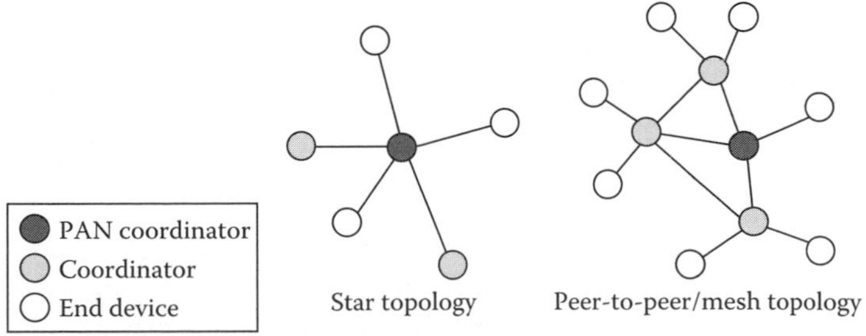

FIGURE 38.2 IEEE 802.15.4 topologies.

38.2.2 Introducing 6LoWPAN

The IETF has foreseen an urgent need for IP-based communication also on wirelessly connected resource-constrained network nodes using IEEE 802.15.4 communication systems. Thus, the 6LoWPAN working group was founded. The name of the working group is currently used widely as acronym for the specific protocol adaptations developed within the 6LoWPAN working group and thus also as name for WSNs using these 6LoWPAN protocols.

The goal of the 6LoWPAN working group was to develop a standard that allows IPv6-based communication over IEEE 802.15.4. In the process of developing the standard, most adaptations became agnostic from the specific 802.15.4 technology and therefore the 6LoWPAN protocols can also be used on top of other low-power links. For example, at the time of writing this section, 6LoWPAN over Bluetooth Low Energy is under development within the IETF.

It is important to notice though, that the 6LoWPAN protocols only describe adaptations and enhancements on the addressing layer. They form the glue between the physical layers and the upper application layer. For higher layers on top of 6LoWPAN, the adaptations and enhancements are transparent and provide identical functionalities as native IPv6, that is, full IPv6 address with global scope if required, end-to-end connectivity, etc. However, when 6LoWPAN nodes communicate with native IPv6 endpoints, the adaptations and enhancements have to be mapped into the corresponding IPv6 pendants. Thus, the 6LoWPAN networks are stub networks at the edge, with dedicated links to other networks and/or the backbone. These links are called border routers and are responsible for forwarding IPv6-only or 6LoWPAN-only communication.

The mentioned adaptations and enhancements that transform native IPv6 to 6LoWPAN and thus to enable IP-based communication in highly resource-constrained networks can be separated in three major parts: (1) compression, (2) fragmentation, and (3) neighbor discovery. All of them are described in the remainder of this section before briefly covering routing.

38.2.3 Compression

Low-power link technologies like IEEE 802.15.4 are characterized by small frame sizes. A typical IEEE 802.15.4 frame is 127 bytes, leaving 81 bytes for upper layers if MAC header and security information are included (worst-case payload = 127–25 bytes MAC header – 21 bytes link layer security = 81 bytes). A native IPv6-header already has a size of 40 bytes, not taking IPv6 extension headers into account. To transport this IPv6 header and, for example, an additional UDP header would result in the remaining 33 bytes (remaining payload = 81–40 bytes IPv6 header – 8 bytes UDP header = 33 bytes).

To transmit IPv6 efficiently over low-power links, the IPv6 headers can be compressed. The objective of this compression is to avoid information carried inline, which is implicitly known or which can be retrieved differently. Examples for the compression of the IPv6 header fields are as follows (c.f. Figure 38.3):

- The protocol version is always IPv6. This field can be omitted.
- The fields Traffic Class and Flow Label are not applicable in meshed low-power networks and can be omitted.
- The field Payload Length is redundant and can be omitted, because this information is already carried in the IEEE 802.15.4 header.
- The Next Header field can be omitted if no next header is carried in the frame.
- The Hop Limit field cannot be compressed and must be carried fully inline.
- The IPv6 address prefixes of the source and the target endpoints may be already known and/or the same, for example, if both endpoints are in the same 6LoWPAN and for the entire 6LoWPAN one prefix is used. In this case, the prefix(es) can be omitted or compressed.

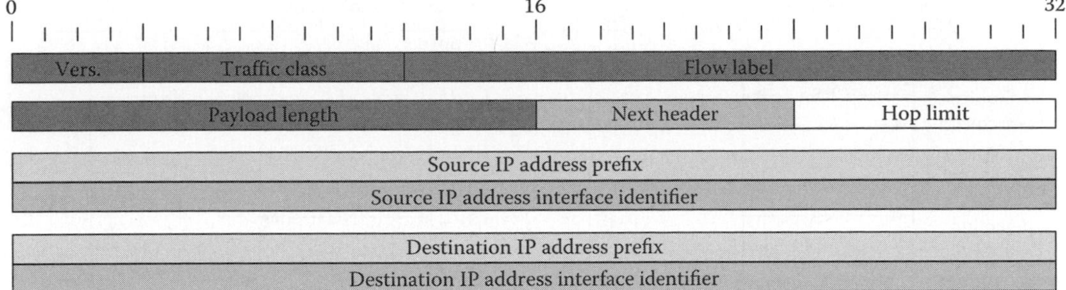

FIGURE 38.3 Native IPv6 header.

- The IPv6 interface identifiers of the source or the target may also be redundant, if the IPv6 interface identifiers are generated from the MAC addresses of the endpoints. For communication within an IEEE 802.15.4, the MAC addresses thus are already carried in the 802.15.4 frames and can be omitted/compressed in the IPv6 headers.

Depending on the specific scenario, native IPv6 headers can be compressed down to a few bytes only. Similar to the IPv6 header compression, the 6LoWPAN protocols also describe a UDP header compression, whereby the UDP header can be compressed from natively 8 bytes down to 4 bytes in the best case.

38.2.4 Fragmentation

The IPv6 standard requires a minimum MTU of 1280 bytes to be supported by the link layer. But low-power link technologies like IEEE 802.15.4 are characterized by small frame sizes, which are typically far below the required MTU of 1280 bytes. Thus, a fragmentation layer is defined for the 6LoWPAN protocols. This layer is located at *layer 2.5*, that is, between MAC and addressing, and thus provides the addressing layer 3 transparently with the required MTU. By using such a dedicated and transparent fragmentation layer, lower layers still can exploit the energy efficiency of small data frames while upper layers can still benefit from the IPv6 compliant packet sizes.

38.2.5 Neighbor Discovery

While the enlargement of the address space is often considered as the major change from IPv4 to IPv6, this modification only affects the format. However, the Neighbor Discovery is the brains behind IPv6. The Neighbor Discovery of IPv6 replaces the mechanisms of IPv4 to retrieve a valid (global scope) IP address and further network configuration parameters such as the address of the default router, etc.

In native IPv6 networks, each endpoint is on the same link as the router that is responsible for the subnetwork. From that router, the endpoint retrieves the address information like network prefix, etc. But in meshed topologies like WSNs/6LoWPANs, the endpoints may not be in direct communication range with the border router that is responsible for the 6LoWPAN subnet. Between the border router and the endpoint, further 6LoWPAN devices acting as forwarding routers may be required. Thus, for 6LoWPAN networks, the Neighbor Discovery protocol has been changed to cope with the topology differences compared to native IPv6 networks (c.f. Figure 38.4).

38.2.6 Routing

Due to the emerging 6LoWPAN protocols, the IETF had to also provide a proper routing solution to be used in such meshed low-power IP networks. Therefore, the ROLL (Routing over Low-Power

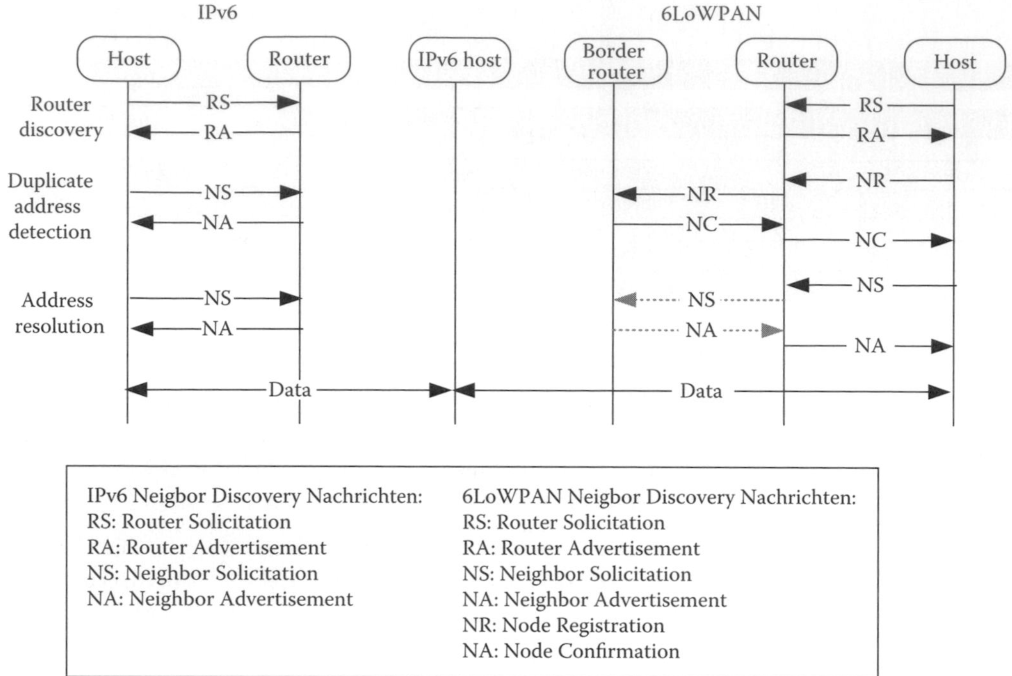

FIGURE 38.4 IPv6 Neighbor Discovery vs. 6LoWPAN Neighbor Discovery.

and Lossy Networks) working group has been formed. Within ROLL, the IPv6 routing protocol RPL (IPv6 Routing Protocol for Low-Power and Lossy Networks; spoken *Ripple*) has been developed. RPL is located on top of IPv6. Thus, it can be used in 6LoWANs as well as in other IPv6-based networks and is agnostic of the link layer. Nevertheless, RPL is optimized for usage in 6LoWPAN networks.

RPL is optimized to pull or push data from the 6LoWPAN into other unconstrained backbone networks. Therefore, RPL establishes a directed, acyclic graph toward a sink. The sink will be the border router of the 6LoWPAN in many use cases, but RPL is not limited to this. The graph (routing tree) can be optimized for several metrics like minimizing energy consumption along the path, minimal hop counts, or spreading traffic load depending on the resources (CPU, RAM, energy) of the routers along the routing path. It is even possible to have more than one such graph in a 6LoWPAN, whereby different routes with different characteristics and metrics may exist from a source to the sink (c.f. Figure 38.5).

Within the routing topology, each endpoint is assigned to one or more routers, whereby one router is the default router. The routers in turn are assigned to routers closer to the sink, or directly to the sink of the routing topology, whereby closer means less costly regarding the used metric and does not necessarily refer to physical distance. Each router informs the sink about its presence and which further endpoints or routers are connected to it. As a result, the sink has knowledge of the whole RPL network. For the communication down the routing topology (i.e., starting from the sink toward the leaves), a source routing is used. In this source routing, all intermediate routers to be traversed are included in the packet by the sink. For communication upward of the topology (i.e., starting in the network and toward the sink) or communication between endpoints in the topology, two different modes exist (c.f. Figure 38.6). To avoid each router to maintain heavyweight and storage-intensive routing tables, the routers may forward all traffic to the sink and from the sink again source routing is used. This mode is called nonstoring mode. To avoid data packets to travel all the way up to the sink and afterward again down, the routers may maintain own routing tables, whereby the packets only have to travel up to the first common ancestor of source and target. This mode is called storing mode.

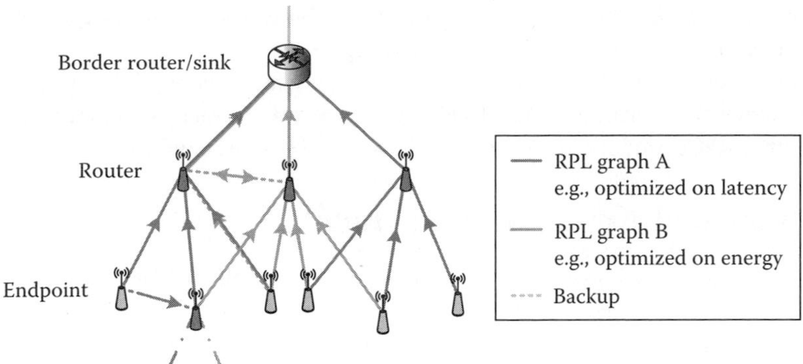

FIGURE 38.5 RPL routing topology.

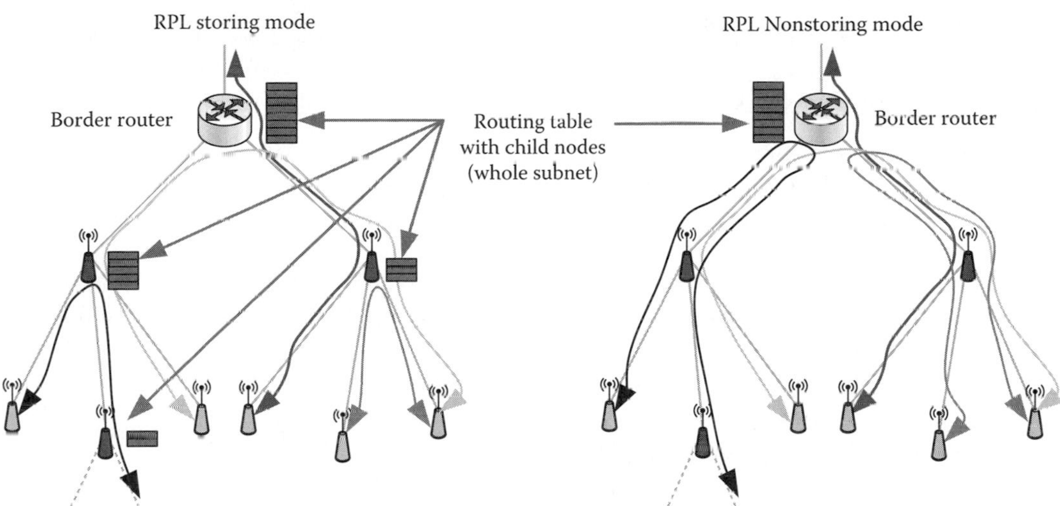

FIGURE 38.6 RPL storing vs. nonstoring mode.

At the time of writing this section, further stateless routing solutions are being developed by the ROLL working group at the IETF. These stateless routing solutions are, for example, based on the well-known Trickle algorithm [RFC2606] and are capable of realizing a so-called controlled flooding behavior. Due to this behavior, such solutions may be more efficient, for example, for IP–Multicast communication in meshed low-power networks.

38.2.7 Closing Remarks

The herein described header compression, fragmentation, Neighbor Discovery, and routing give only a very brief overview about the general concepts. However, the overall solution is more complex. The newest protocols and their different operational modes are not well-established today and not used widely. For example, neither Contiki nor TinyOS, which are the two major operating systems for wireless sensor networks, support the storing mode of RPL.

Thus, many of the protocols and protocol components still have to ensure efficient IP-based communication in WSNs. A good example is the group communication via IP–Multicast, which requires a close cooperation of several layers to be efficient. The IETF copes with this fact by producing documents that will not describe further protocols, but which will have a tutorial characteristic and thus describe how to use the specific protocols. But these documents are still under development, for example, in the Light-Weight Implementation Guidance (LWIG) working group.

A more in-depth overview of the dedicated protocols in general is required only if developers are planning to implement an own 6LoWPAN stack. In this case, readers interested in further details should refer directly to the RFCs produced by the working groups 6LoWPAN and ROLL.

38.3 Constrained Application Protocol

The previous section discussed how IPv6 packets can be transferred efficiently over wireless links with limited bandwidth. Complementary to the efficient transportation, an equally efficient (M2M) application layer protocol is needed that can be used to realize novel and innovative applications. The CoAP protocol can be used to control resource-limited devices via embedded web services. It is an alternative to SOAP web services, which are mainly used on devices with much more resources available as in WSNs. Bear in mind though that research work showing the feasibility of running SOAP web services on wireless sensor nodes does exist.

38.3.1 Introducing CoAP

The Constrained Application Protocol (CoAP) is a comparatively new application layer protocol. CoAP is currently under development by the IETF Constrained RESTful Environments (CoRE) working group, and is designed for the resource-constrained environments such as 6LoWPANs. Nevertheless, CoAP is independent of 6LoWPAN and may also be used, for example, in existing IPv4 networks.

CoAP is designed in its many parts as lightweight alternative to HTTP. Even though there is more to CoAP than merely being binary HTTP, thinking like that makes it easier to understand CoAP. The major common ground for both CoAP and HTTP are the REST design principles described by Fielding in [Fie00]. In a nutshell, REST is about the following:

- Stateless request/response interaction model between application endpoints
- Addressing resources through URIs
- Representation of the resources and resource states through Internet Media Types
- Uniform and well-defined CRUD (Create, Read, Update, Delete) operations on the resources
- Caching and proxying to allow for scalability

It is important to notice at this point that neither HTTP is REST nor REST is HTTP, but HTTP is a protocol that can be mapped on the design principles of REST and so is CoAP. Readers interested in this discussion should refer to Pautasso, who discusses this issue in [Pau08]. Pautasso also pays attention to the *SOA vs. REST* discussion. While REST is often considered to be more lightweight than SOA, mostly the protocols HTTP and SOAP and/or their implementations are compared, but not the architectural styles. It is important not to confuse the design principles, protocols following these principles, and implementations.

To tailor CoAP for bandwidth-limited networks, the protocol headers of CoAP are encoded binary in contrast to the ASCII encoded HTTP headers. Thereby, a typical CoAP header is about 4–20 bytes, while an average HTTP header is about 700–800 bytes in size. Another major difference of CoAP and HTTP is the usage of UDP at transport layer instead of TCP. TCP performs poorly when used for short-lived request/response exchanges, and the SYN/ACK handshake causes a significant overhead when using a low-power link with comparatively long duty cycles. Thus, UDP is the preferred transport protocol for WSNs. Because UDP is not reliable by default, CoAP adds own mechanisms to ensure reliability and for duplication detection and fragmentation by a dedicated CoAP Messaging layer. These features are not implemented in HTTP since TCP takes care of them. An overview about the differences and commonalities of CoAP and HTTP is given in Table 38.1.

TABLE 38.1 Comparison of HTTP and CoAP

	HTTP	CoAP
Role model	Client/server	Client/server
Communication pattern	Request/response	Request/response
Transport	TCP	UDP reliability optional via CoAP messages
Transmission	Unicast, point-to-point	Unicast, multicast, point-to-point, point-to-multipoint
Exchange pattern	Mainly synchronously	Synchronously, asynchronously
Service-/data modeling	Resources	Resources
Service-/data addressing	URI	URI
Methods	GET, PUT, POST, DELETE, TRACE, HEAD, OPTIONS, CONNECT	GET, PUT, POST, DELETE
Media-types	Text, JSON, XML, EXI,...	Text, JSON, XML, EXI,...
Response-codes	Informational 1xx Successful 2xx Redirection 3xx Client Error 4xx Server Error 5xx	Success 2.xx Client Error 4.xx Server Error 5.xx
Header-cording	ASCII	Binary
Proxy und caching	HTTP to HTTP	CoAP to CoAP HTTP to CoAP CoAP to HTTP
Formal description	WADL	WADL
Resource-description	HTTP link header format	CoRE link format
Resource-discovery	None	CoAP, CoAP link format
Eventing (c.f. Push)	Long polling/comet, web sockets, BOSH ...	CoAP observe
Security	SSL/TLS	DTLS

38.3.2 CoAP/HTTP Proxy and Caching

CoAP provides Caching and Proxy features that are similar to HTTP to ensure the required scalability in large-scale networking environments. In that context, caching means that some of the CRUD style interactions with the resources (e.g., GET, PUT, and DELETE) may result in reproducible results or in the delivery of resource representations with a given time of validity and thus a known expiration time. With such information, clients are able to avoid polling of data if it is known that no changes of the resource representations have occurred since the last request. But furthermore, third-party intermediary caches may respond instead of the originally requested endpoints if the result for this request is already known through former interactions with this endpoint. Such intermediary caches are a key enabler for scalability of the Internet applications and are also available for CoAP.

Due to the significant overlapping of the CoAP and HTTP protocols and due to the REST principles both protocols are based on, in major parts, it is even possible to map CoAP seamlessly to HTTP and vice versa. This mapping is already an integral part of the CoAP specifications which is a major difference compared with other binary field bus protocols.

A full deployment, as exemplary depicted in Figure 38.7, may consist of proxies mapping HTTP in to CoAP and vice versa. Thereby, external implementations and deployments are not required to be changed, and native HTTP over TCP still can be used. But in resource-constrained environments like 6LoWPAN networks, the much more tailored CoAP protocol over UDP can be used. Caches can provide the required scalability of the overall architecture for all four possible combinations of endpoints (c.f. Figure 38.8): HTTP/HTTP, CoAP/CoAP, HTTP/CoAP, and CoAP/HTTP. For the CoAP/HTTP and HTTP/CoAP

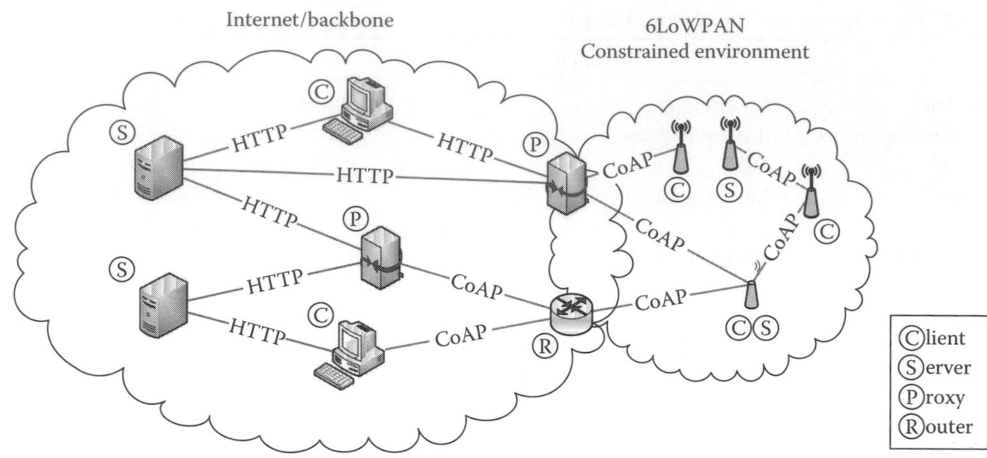

FIGURE 38.7 HTTP + CoAP deployment.

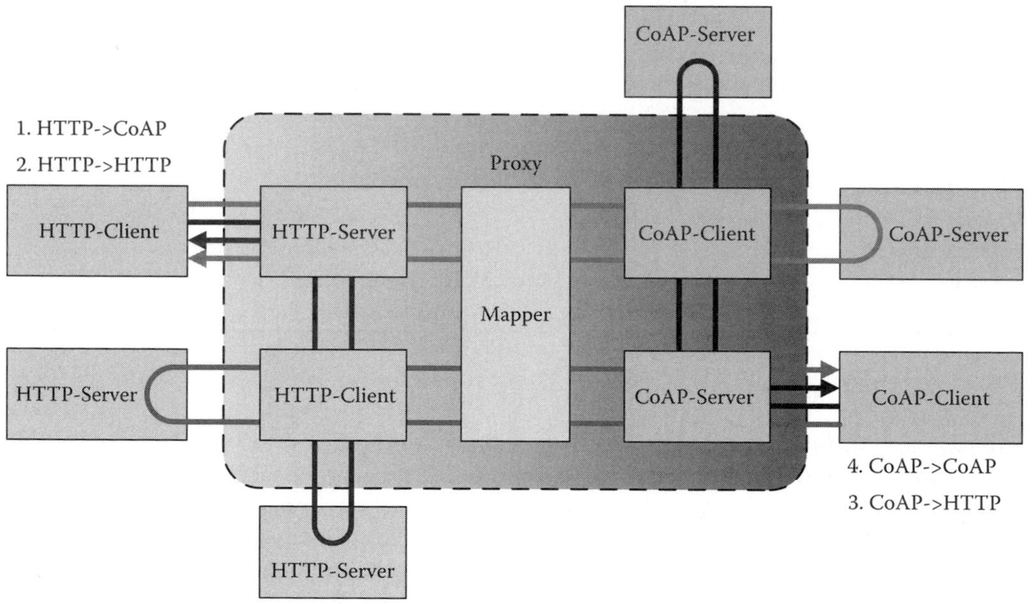

FIGURE 38.8 HTTP/CoAP Proxy.

interactions, the caches may be integrated directly into the proxies but do not have to. Also, even though it appears reasonable to locate the proxies and the caches at the border routers of the 6LoWPANs, the architecture is not bound to such specifics. This is an important aspect and only possible through the usage of the link layer encapsulating IP-based infrastructure.

38.3.3 CoAP Enhancements to Binary HTTP

While HTTP was developed for browser interaction and thus human-to-machine (H2M) communication scenarios, CoAP was designed for M2M applications. Hence, CoAP is not only a binary HTTP but extends HTTP by essential features required in the M2M environments.

While HTTP mainly realizes pull communication, CoAP also features push notifications through a publish/subscribe mechanism. Thereby, clients can *observe* resources for state changes. If subscribed to a specific resource, the server autonomously pushes state changes to the clients, potentially decreasing traffic dramatically since active polling is avoided.

While HTTP uses TCP on transport layer and thus is limited to unicast point-to-point connections, CoAP uses UDP on transport layer. UDP is capable of IP–Multicast communication. Against this background, CoAP interactions with resources are not limited to single nodes, but clients can also interact with a group of devices with one request.

For autonomous M2M environments, a discovery mechanism of endpoints and the provided services/resources is an essential requirement. To provide this feature, CoAP defines two different mechanisms. First, a description format for resources of a device is defined and called CoAP link format. This link format forms a list of available resources at a well-known URI path and uses attributes for each resource to provide information, for example, on the content formats, the expected resource representation size, and further semantic information like the resource type. Second, several mechanisms are defined to retrieve the resource list. Clients may use unicast or multicast requests to retrieve this information from each device in the network. But for CoAP also a central entity called resource directory can be used, where devices register at and upload their resource descriptions. Using such a central entity is a huge advantage in deployments where devices may reside in power-saving sleep states most of the time and wake up only in given time slots for possible client interactions.

38.4 Related Work

Based on the presence of IP in WSNs (through the 6LoWPAN protocols) in research, development, and standardization, further activities have been initiated to benefit from the new infrastructure and possibilities offered.

While HTTP is widely used for H2M communication, and CoAP is designed as a proper M2M alternative for resource-constrained environments, SOAP-based web services are another candidate for M2M applications as they are used widely in the web and in enterprise IT. With the OASIS standard Devices Profile for Web Services (DPWS), a SOAP web services profile dedicated to device-centric communication has been developed. While the WS-* specifications had to be extended to ensure secure M2M communication, DPWS is still based on the WS-Architecture and is thus compliant with existing WS-* specification framework, deployments, and implementations. Nevertheless, SOAP web services are by default too heavy weight and too verbose to be applied to resource-limited systems such as WSNs [Nix09, ABG13].

Moritz et al. [Mor13] have developed a DPWS architecture that pushes DPWS onto WSNs. Therefore, three major extensions were made to the original DPWS: First, DPWS and its mechanisms for discovery and eventing (c.f. push notifications) can be adopted carefully to be a better fit for WSN-typical applications like monitoring temperatures or other environmental parameters. Second, the verbosity of the XML-based message representation had to be removed. Moritz et al. have developed a solution based on the Efficient XML Interchange (EXI) format of the W3C. Third, the compressed SOAP messages need to be transported efficiently. Based on the existing SOAP-over-HTTP transport mechanism and due to the similarities of HTTP and CoAP, Moritz et al. have developed and specified a novel SOAP-over-CoAP transport mechanism to be used in networks such as 6LoWPANs. In contrast to other similar approaches as the one described by Samaras et al. [Sam13], the solution by Moritz et al. is fully compliant with the WS-Architecture and can benefit from several stateless mappings like CoAP in HTTP (for the SOAP transport mechanisms) and EXI in XML (for message encoding).

A completely different application layer protocol was proposed by Hornsby et al. in [Hor09]. The Extensible Messaging and Presence Protocol (XMPP), formerly named Jabber, was originally developed for instant messaging purposes. Nevertheless, over time and through several enhancements to the core protocol, XMPP can also be used for general message exchange and is often described as application layer or XML routing protocol. The reason is the dedicated XMPP infrastructure. Clients connect to servers through long-living TCP connections and the servers are connected to each other. Messages in the form of XML fragments from one client to the other are thus forwarded by the XMPP servers until they reach the final destination. Hornsby et al. have implemented μXMPP, an XMPP client for WSN node class devices.

The earlier mentioned ZigBee protocol was originally developed completely independently of IP and other Internet protocols. The only common basis with 6LoWPAN was the usage of IEEE 802.15.4 as the link layer. Nevertheless, ZigBee does not have a monolithic protocol stack but has a strong separation between the application layer and the lower layers. At the application layer, the so-called ZigBee Application Profiles (ZAP) are defined that consist, for example, of data models including semantics and grouping of data bulks for device types (e.g., lamp and temperature sensor). With the emerging ZigBee Smart Energy 2.0 (SE 2.0), the ZigBee consortium now combines the defined and self-contained ZAP with the IP-based infrastructure. Because ZigBee was designed for low-power and low-cost devices, the 6LoWPAN protocols are of course a major part of the evolving SE 2.0 profile.

38.5 Conclusion

6LoWPAN and CoAP are important protocols used in WSNs. They offer the technological basis for massively scalable networking, and bridge the gap between enterprise networks and the Internet-of-Things (IoT) or Web-of-Objects (WoO).

6LoWPAN ensures applicability of IPv6 to low-power, low-rate wireless radio communication based on the IEEE 802.15.4 standard. It provides protocols for IP header compression, address autoconfiguration, adoption layer to fit IPv6-specific MTU requirements, etc.

CoAP is an application layer protocol that helps integrate wirelessly connected sensors, devices, things, or objects. Beyond that, it can be used in various other scenarios and help downsize huge data sets. By means of those kinds of application layer protocols, system engineers and developers can focus on development of applications and products, instead of putting too much effort on the communication technology and its protocols, as it is today.

Both, 6LoWPAN and CoAP, are completely independent technologies. However, they both are rapidly gaining momentum as they complement each other, making a powerful toolset for integrating formerly separated proprietary networks into our global IP-based infrastructure, and thus enable concepts such as the IoT.

References

ABG13 V. Altmann, H. Bohn, and F. Golatowski, Web Services for embedded devices, in R. Zurawski (ed.), *Handbook Industrial Electronics*, CRC Press, Boca Raton, FL, 2013.

Dun04 A. Dunkels, B. Grönvall, and T. Voigt, Contiki: a lightweight and flexible operating system for tiny networked sensors, *Proceedings of the 29th Annual IEEE International Conference on Local Computer Networks*, pp. 455–462, 2004.

Dur08 M. Durvy, J. Abeillé, P. Wetterwald, C. O'Flynn, B. Leverett, E. Gnoske, M. Vidales et al., Making sensor networks IPv6 ready, *SenSys'08*, Raleigh, NC, 2008.

Fie00 R. T. Fielding, Architectural styles and the design of network-based software architectures, University of California, Irvine, CA, 2000.

Hor09 A. Hornsby and E. Bail, uXMPP: Lightweight implementation for low power operating system Contiki, *International Conference on Ultra Modern Telecommunications Workshops*, pp. 1–5, St. Petersburg, Russia, 2009.

IEEE154 IEEE Computer Society, IEEE Std. 802.15.4–2003, 2003.

Kov11 M. Kovatsch, S. Duquennoy, and A. Dunkels, A low-power CoAP for Contiki. *Proceedings of the Eighth IEEE International Conference on Mobile Ad-hoc and Sensor Systems (MASS 2011)*, Valencia, Spain, October 2011.

Ler11 C. Lerche, N. Laum, G. Moritz, Z. Elmar, F. Golatowski, and D. Timmermann, Implementing powerful Web Services for highly resource-constrained devices, *Pervasive Computing and Communication Workshops (PerCom Workshops)*, pp. 332–335, Seattle, WA, March 2011.

Ler12 C. Lerche, N. Laum, F. Golatowski, C. Niedermeier, and D. Timmermann, Connecting the web with the web of things: Lessons learned from implementing a CoAP-HTTP proxy, *Proceedings of the IoTech Workshop 2012*, Las Vegas, NV, October 2012.

LHK12 C. Lerche, K. Hartke, and M. Kovatsch, Industry adoption of the internet of things: A constrained application protocol survey, *Proceedings of the Seventh International Workshop on Service Oriented Architectures in Converging Networked Environments (SOCNE 2012)*, Kraków, Poland, September 2012.

Mor13 G. Moritz, C. Lerche, F. Golatowski, and D. Timmermann, Beyond 6LoWPAN: Web services in wireless sensor networks, *IEEE Transactions on Industrial Informatics*, 9(4), 1795–1805, 2013.

Nix09 T. Nixon, A. Regnier, D. Driscoll, and A. Mensch. Devices Profile for Web Services Version 1.1., OASIS, July 2009.

Pau08 C. Pautasso, O. Zimmermann, and F. Leymann, Restful web services vs. "big" web services: Making the right architectural decision, *17th International Conference on World Wide Web*, Beijing, China, 2008.

RFC2606 P. Levis, T. Clausen, T. Hui, O. Gnawali, and J. Ko, IETF networking group, The trickle algorithm—RFC2606, March 2011.

Sam13 I. K. Samaras, G. D. Hassapis, and J. V. Gialelis, A modified DPWS protocol stack for 6LoWPAN-based wireless sensor networks, *IEEE Transactions on Industrial Informatics*, 9(1), 209–217, 2013.

VII

Time-Triggered Communication

39 **Concepts of Time-Triggered Communication** *Roman Obermaisser* **39**-1
Time-Triggered and Event-Triggered Communication • Fundamental Services of a Time-Triggered Communication Protocol • Properties of Time-Triggered Communication Systems • References

40 **Time-Triggered Protocol (TTP/C)** *Roman Obermaisser and Michael Paulitsch* **40**-1
Temporal Structuring of Communication • Clock Synchronization • Restart, Reintegration, Integration • Diagnostic Services • Fault Isolation • Configuration Services • References

41 **Time-Triggered CAN** *Roland Kammerer and Roman Obermaisser* **41**-1
Protocol Overview • Protocol Services • Flexible Time-Triggered CAN • Commercial or Prototypical Components • References

42 **Time-Triggered Ethernet** *Wilfried Steiner and Michael Paulitsch* **42**-1
General Overview • TTEthernet Traffic Classes and Traffic Integration • TTEthernet Synchronization Protocols • TTEthernet Example Configurations and Implementations • TTEthernet in Orion • References

39

Concepts of Time-Triggered Communication

39.1 Time-Triggered and Event-Triggered Communication **39**-1
39.2 Fundamental Services of a Time-Triggered Communication Protocol **39**-2
Clock Synchronization • Periodic Exchange of State Messages • Fault Isolation Mechanisms • Diagnostic Services
39.3 Properties of Time-Triggered Communication Systems **39**-4
Composability • Independent Fault Containment Regions • Strict Control on Node Interactions • Replica Determinism • Performance
References **39**-8

Roman Obermaisser
University of Siegen

39.1 Time-Triggered and Event-Triggered Communication

It has been recognized that communication protocols fall into two general categories with corresponding strengths and deficiencies: event-triggered and time-triggered control. Event-triggered protocols (e.g., TCP/IP, CAN, Ethernet, ARINC629) offer flexibility and resource efficiency. Time-triggered protocols (e.g., TTP/C, FlexRay) excel with respect to predictability, composability, error detection, and error containment.

In event-triggered architectures, the system activities, such as sending a message or starting computational activities, are triggered by the occurrence of events in the environment or the computer system. In time-triggered architectures, activities are triggered by the progression of global time. The major contrast between event-triggered and time-triggered approaches lies in the location of control. Time-triggered systems exhibit autonomous control and interact with the environment according to an internal predefined schedule, whereas event-triggered systems are under the control of the environment and must respond to stimuli as they occur.

The time-triggered approach is generally preferred for safety-critical systems [Rus01, Kop95]. For example, in the automotive industry, a time-triggered architecture will provide the ability to handle the communication needs of by-wire cars [Bre01]. In addition to hard real-time performance, time-triggered architectures help in managing the complexity of fault-tolerance and corresponding formal dependability models, as required for the establishment of ultrahigh reliability (failure rates in the order of 10^{-9} failures/h). The predetermined points in time of the periodic message transmissions allow error detection and establishing of membership information. Redundancy can be established transparently to applications, that is, without any modification of the function and timing of application systems. A time-triggered system also supports replica determinism [Pol96], which is essential for establishing fault-tolerance through active redundancy.

Furthermore, time-triggered systems support temporal composability [KO02] via a precise specification of the interfaces between subsystems, both in the value domain and temporal domain. The communication controller in a time-triggered system decides autonomously when a message is transmitted. The communication network interface is a temporal firewall, which isolates the temporal behavior of the host and the rest of the system.

39.2 Fundamental Services of a Time-Triggered Communication Protocol

In the following, four fundamental services of a time-triggered communication protocol are explained, namely, clock synchronization, the periodic exchange of state messages, fault isolation, and diagnostic services.

39.2.1 Clock Synchronization

Due to clock drifts, the clock times in an ensemble of clocks will drift apart, if clocks are not periodically resynchronized. Clock synchronization is concerned with bringing the values of clocks in close relation with respect to each other. A measure for the quality of synchronization is the precision. An ensemble of clocks that are synchronized to each other with a specified precision offers a global time. The global time of the ensemble is represented by the local time of each clock, which serves as an approximation of the global time [KO87].

Two important parameters of a global time base are the granularity and horizon. The granularity determines the minimum interval between two adjacent ticks of the global time (also called macrogranule), that is, the smallest interval that can be measured with the global time. The horizon determines the instant when the time will wrap around.

The reasonableness condition [Kop97] for a global time base ensures that the synchronization error is bounded to less than one macrogranule. However, due to the synchronization and digitalization error, it is impossible to establish the temporal order of occurrences based on their timestamp, if timestamps differ by only a single tick. A solution to this problem is the introduction of a sparse time base [Kop92].

39.2.2 Periodic Exchange of State Messages

A time-triggered communication system is designed for the periodic exchange of messages carrying state information. These messages are called *state messages*.

Information with state semantics contains the absolute value of a real-time entity (e.g., temperature in the environment is 41°C). The self-contained nature and idempotence of state messages eases the establishment of state synchronization, which does not depend on exactly-once processing guarantees. Since applications are often only interested in the most recent value of a real-time object, old state values can be overwritten with newer state values. Hence, a time-triggered communication system does not require message queues.

The periodic transmission of state messages is triggered by the progression of the global time according to a time division multiple access (TDMA) scheme. TDMA statically divides the channel capacity into a number of slots and assigns a unique slot to every node. The communication activities of every node are controlled by a time-triggered communication schedule. The schedule specifies the temporal pattern of messages transmissions, that is, at what points in time nodes send and receive messages. A sequence of sending slots, which allows every node in an ensemble of n nodes to send exactly once, is called a TDMA round. The sequence of the different TDMA rounds forms the cluster cycle and determines the periodicity of the time-triggered communication.

The a priori knowledge about the times of message exchanges enables the communication system to operate autonomously. The temporal control of communication activities is within the sphere of control of the communication system. Hence, the correct temporal behavior of the communication system is independent of the temporal behavior of the application software in the host computer and can be established in isolation.

39.2.3 Fault Isolation Mechanisms

A time-triggered communication protocol and the corresponding system architecture need to provide rules for partitioning a system into independent FCRs. In a system with active redundancy, the dependencies among FCRs (i.e., correlation between failure probability of FCRs) have a significant impact on the system reliability [OKS07]. The independence of FCRs can be compromised by shared physical resources (e.g., hardware, power supply, time-base), external faults (e.g., electromagnetic interference, heat, shock, spatial proximity), and design faults.

In a time-triggered communication protocol, the error containment mechanisms for timing message failures can be enforced transparently to the application. Using the a priori knowledge concerning the global points in time of all intended message sent and receive instants, autonomous guardians can block timing message failures. For this purpose, node-local and centralized guardians have been developed and validated for different time triggered communication protocols (e.g., [BKS03, Gua05]). For example, in TTP, a guardian transforms a message that it judges untimely into a syntactically incorrect message by cutting off its tail [Kop03a].

Error containment for value message failures is generally not part of a time-triggered communication protocol but within the responsibility of the host computers. For example, value failure detection and correction can be performed using N-modular redundancy (NMR). N replicas receive the same requests and provide the same service. The output of all replicas is provided to a voting mechanism, which selects one of the results (e.g., based on majority) or transforms the results to a single one (average voter). The most frequently used N-modular configuration is triple-modular redundancy (TMR). By employing three components and a voter, a single consistent value failure in one of the constituting components can be tolerated.

Although not natively provided by time-triggered communication protocols, N-modular redundancy is enabled by time-triggered communication protocols by supporting replica determinism [Pol96]. Fault-free replicated components exhibit replica determinism, if they deliver identical outputs in an identical order within a specified time interval. Replica determinism simplifies the implementation of fault-tolerance by active redundancy, since failures of components can be detected by carrying out a bit-by-bit comparison of the results of replicas. Replica nondeterminism is introduced either by the interface to the real world or the system's internal behavior.

39.2.4 Diagnostic Services

Diagnostic services are concerned with the identification of failed subsystems. Diagnostic services can trigger the autonomous recovery of a system in case of a transient subsystem failure. In addition, diagnostic services can support the replacement of defective subsystems if a failure is permanent.

An example of diagnostic services that can be found in time-triggered communication protocols is a solution to the membership problem. The membership problem is a fundamental problem in distributed computing, because it allows solutions to other important problems in designing fault-tolerant systems [GP96]. The membership problem is defined as the problem of achieving agreement on the identity of all correctly functioning processes of a process group. A process is correct, if its behavior complies with the specification. Otherwise the process is denoted as faulty.

In the context of integrated architecture, it makes sense to establish membership information for FCRs, since FCRs can be expected to fail independently. Depending on the assumed types of faults, an FCR is either an entire system component or a subsystem within a component (e.g., a task) dedicated to a function.

A service that implements an algorithm for solving the membership problem and offers consistent membership information is called a membership service. A membership service simplifies the provision of many application algorithms, since the architecture offers generic error detection capabilities via this service. Applications can rely on the consistency of the membership information and react to detected failures of FCRs as indicated by the membership service.

A membership service also plays an important role in controlling application-level fault-tolerance mechanisms that deal with failures of functions. If a function fails—since more FCRs have failed than can be tolerated by the given amount of redundancy—all that an integrated architecture can do is to inform other functions about this condition so they can react accordingly by application-level fault-tolerance mechanisms.

In a time-triggered communication system, the periodic message send times are membership points of the sender [KGR91]. Every receiver knows a priori when a message of a sender is supposed to arrive, and interprets the arrival of the message as a life sign at the membership point of the sender. From the arrival of the expected messages at two consecutive membership points, it can be concluded that the sender was operational during the interval delimited by these membership points.

39.3 Properties of Time-Triggered Communication Systems

39.3.1 Composability

In many engineering disciplines, large systems are built from prefabricated components with known and validated properties. Components are connected via stable, understandable, and standardized interfaces. The system engineer has knowledge about the global properties of the components—as they relate to the system functions—and of the detailed specification of the component interfaces. Knowledge about the internal design and implementation of the components is neither needed nor available in many cases. *Composability* deals with all issues that relate to the component-based design of large systems. Composability refers to an architectural framework that supports the smooth integration and reuse of independently developed components in order to increase the level of abstraction in the design process. Architecture instantiations that are derived from a generic platform must support the constructive composition of large systems out of components and subsystems without uncontrolled emerging behavior or side effect. Composability is a concept that relates to the ease of building systems out of subsystems [ART06, p. 24]. An example is the deadlock-freedom of a component-based system.

An architecture must enable the precise specification of the linking interface (LIF) [Kop02] of a node in the domains of value and time. This is a necessary prerequisite for the independent development of nodes on one side and the reuse of existing nodes that is based solely on their LIF specification on the other side. While the operational specification of the value domain of interacting messages is *state-of-the-art* in embedded system design, the temporal properties of these messages are often not considered with the appropriate care. The global time, which is available in time-triggered systems, is essential for the precise specification of the temporal properties of the LIF messages. In the time domain, time-triggered protocols specify the periods and phases of message exchanges with respect to a global time base with a specified precision.

Furthermore, composability requires the stability of prior services upon integration. Services of a node that have been validated in isolation (i.e., prior to the integration of the node into the larger system) remain intact after the integration. This concept is also known as compositionality. The analysis of the component properties can be performed independently for each component. Compositionality avoids exponential analysis complexity. Temporal compositionality is an instantiation of the general notion of compositionality. A communication system supports temporal compositionality, if temporal correctness is not refuted by the system integration [KO02].

Another necessary condition of composability is the support for noninterfering interactions. If there exist two disjoint subgroups of cooperating nodes that share a common communication infrastructure, then the communication activities within one subgroup may not interfere with the communication activities within the other subgroup. If this principle is not satisfied, then the integration within one node-subgroup will depend on the proper behavior of the other (functionally unrelated) node-subgroups. These global interferences compromise the composability of the architecture. Time-triggered communication systems provide predictable message transport latency that is not influenced by the behavior of other, functionally unrelated messages.

Finally, composability requires the preservation of the node abstraction in the case of failures. In a composable architecture, the introduced abstraction of a node must remain intact, even if a node becomes faulty. It must be possible to diagnose and replace a faulty node without any knowledge about the node internals. This requires a certain amount of redundancy for error detection within the architecture. This principle constrains the implementation of a node, because it restricts the implicit sharing of resources among nodes. If a shared resource fails, more than one node can be affected by the failure. In time-triggered communication systems, the communications system contains redundant information about the permitted temporal behavior of nodes and disconnects nodes that violate their temporal specification in order to avoid error propagation from a faulty node (a babbling idiot) into the communication system.

39.3.2 Independent Fault Containment Regions

A fault containment region (FCR) is the boundary of the immediate impact of a fault [LH94]. In conformance with the fault-error-failure chain introduced by Laprie [ALR01], one can distinguish between faults that cause the failure of an FCR (e.g., design of the hardware or software of the FCR, operational fault of the FCR) and faults at the system-level. The latter type of fault is a failure of an FCR, which could propagate to other FCRs through a sent message that deviates from the specification. If the transmission instant of the message violates the specifications, we speak of a *message timing failure*. A *message value failure* means that the data structure contained in a message is incorrect.

Such a failure of an FCR can be tolerated by distributed fault-tolerance mechanisms. The masking of failures of FCRs is denoted as error containment [LH94] because it avoids error propagation by the flow of erroneous messages. The error detection mechanisms must be part of different FCRs than the message sender [Kop03b]. Otherwise, the error detection mechanism may be affected by the same fault that caused the message failure.

For example, a common approach for masking node failures is *N-Modular Redundancy (NMR)* [LA90]. *N* replicas receive the same requests and provide the same service. The output of all replicas is provided to a voting mechanism, which selects one of the results (e.g., based on majority) or transforms the results to a single one (average voter). The most frequently used *N*-modular configuration is triple-modular redundancy (TMR).

We denote three replicas in a TMR configuration a *Fault-Tolerant Unit (FTU)*. In addition, we consider the voter at the input of a node and the node itself as a self-contained unit, which receives the replicated inputs and performs voting by itself without relying on an external voter. We call this behavior *incoming voting*.

Common-mode failures are failures of multiple FCRs, which are correlated and occur due to a common cause. Common-mode failures occur when the assumption of the independence of FCRs is compromised. They can result from replicated design faults or from common operational faults such as a massive transient disturbance. Common-mode failures of the replicas in an FTU must be avoided, because any correlation in the instants of the failures of FCRs significantly decreases the reliability improvements that can be achieved by NMR [HTH97].

The definition of FCRs needs to take into account shared resources that could be affected by a fault. Typically, shared resources include the computing hardware, the power supply, the timing source, the clock source, and the physical space. For example, if two subsystems depend on a single power supply, then these two subsystems are not considered to be independent and therefore belong to the same FCR. Since this definition of independence allows that two FCRs can share the same design, design faults are not part of this fault-model. Diversity is a method to avoid common design faults in FCRs by using different designs that perform the same function.

39.3.3 Strict Control on Node Interactions

A distributed system uses shared networking resources in order to support the interaction between nodes. In order to preserve the independence of FCRs, strict control on the node interactions and the use of these shared resources is required. Based on the specification of the permitted behavior of a node at its linking interface, a time-triggered communication protocol can isolate nodes that violate the linking interface specification. Thereby, the time-triggered communication protocol realizes *temporal partitioning* and *spatial partitioning* [Rus99] at the network-level:

- *Temporal partitioning:* Media access control is concerned with the assignment of time intervals to each node for the transmission of the node's messages. A node that sends untimely messages has the potential of disrupting the communication abilities of the other nodes by delaying their messages. The media access control mechanisms of a time-triggered communication system can be designed to prevent this interference between nodes in the temporal domain. A time-triggered communication system is autonomous and has a priori knowledge of all intended message sent and receive instants. This knowledge can be used by local or central guardians [BKS03, Gua05] in order to block untimely messages.
- *Spatial partitioning:* Spatial partitioning ensures that one node cannot alter the code, private data, or messages of another node. Spatial partitioning must ensure that no messages with corrupted data or wrong addresses are delivered. A message with a wrong address or identification—also called a masquerading failure—could cause correct messages of other nodes to be overwritten at the recipient. Faults occurring at the communication channel can be detected using CRC checks. For faults that occur within the sender, an unforgeable authentication mechanism is necessary. Depending on the failure assumptions, an authentication mechanism can be implemented with simple signature schemes or cryptographic mechanisms [GMR88].

39.3.4 Replica Determinism

Replica determinism has to be supported by the architecture to ensure that the replicas of an FTU produce the same outputs in defined time intervals. A time-triggered communication system addresses key issues of replica determinism. In particular, a time-triggered communication system supports replica determinism by exploiting the sparse global time base in conjunction with preplanned communication and computational schedules. Computational activities are triggered after the last message of a set of input messages has been received by all replicas of an FTU. This instant is a priori known due to the predefined time-triggered schedules. Thus, each replica wakes up at the same global tick, and operates on the same set of input messages. The alignment of communication and computational activities on the sparse global time base ensures temporal predictability and avoids race conditions.

39.3.5 Performance

Important performance attributes in real-time communication networks are the bandwidth, the network delay, and the variability of the network delay (i.e., communication jitter).

The bandwidth is a measure of the available communication resources expressed in bits/second. The bandwidth is an important parameter as it determines the types of functions that can be handled and the number of messages and nodes that can be handled by the communication network.

The network delay denotes the time difference between the production of a message at a sending node and the reception of the last bit of the message at a receiving node. At some instant $t_{request}$, the sending node requests the transmission of a message by invoking a send operation at the node's communication controller. Depending on the communication protocol and the current traffic on the communication channel, the transmission of the message will start after the *access delay* at the send instant.

Depending on whether the communication protocol is time-triggered or event-triggered, the access delays exhibit different characteristics. In a time-triggered system, the send instants of all nodes are periodically recurring instants, which are globally planned in the system and defined with respect to the global time base. The access delay of a message in a time-triggered system is thus locally determined at the sending node. The access delay is independent of the traffic from other nodes and depends solely on the relationship between the request instants and the preplanned send instants. Furthermore, since the next send instant of every node is known a priori, a node can synchronize locally the production of periodic messages with the send instant and thus minimize the access delay of a message.

In addition, the TDMA scheme of a time-triggered communication protocol can contain periodic communication slots for the transport of sporadic and periodic messages. In case of a transmission request, the message is transmitted during the next communication slot in the TDMA scheme. The period of the communication slots can be determined by the maximum permitted network delay, the bandwidth requirements, and knowledge about the message interarrival times (e.g., minimum and average values). Sporadic and aperiodic messages incur a delay depending on the (stochastic) time of the request instant relative to the preplanned send instants of the time-triggered communication protocol. In the worst case, the transmission of a sporadic or aperiodic message is requested immediately after a send instant, thereby incurring a delay equal to the period of the communication slot. At the cost of additional communication resources, this network delay can be reduced by increasing the frequency of the communication slots reserved for sporadic and aperiodic messages. In order to reduce the consumption of communication resources, several time-triggered communication protocols have been extended with means for the event-triggered communication of sporadic and aperiodic messages such as time intervals with dynamic arbitration.

In an event-triggered system, the access delay of a message depends on the state of the communication system at the send instant. If the communication network is idle, the message transmission can start immediately, leading to an access delay close to zero. If the channel is busy, then the access delay depends on the media access strategy implemented in the communication protocol. For example, in the CSMA/CD protocol of Ethernet [IEE00], nodes wait for a random delay before attempting transmission again. In the CSMA/CA protocol of CAN [Int93], the access delay of a message depends on its priority relative to the priorities of other pending messages. Hence, in an event-triggered network the access delay of a message is a global property that depends on the traffic patterns of all nodes.

A bounded network delay with a minimum variability is important in many embedded applications. For example, achievement of control stability in real-time applications depends on the completion of activities (like communicating sensor and control values) in bounded time. Hard real-time systems ensure guaranteed response even in the case of peak load and fault scenarios. Guaranteed response involves assurance of temporal correctness of the design without reference to probabilistic arguments. Guaranteed response requires extensive analysis during the design phase such as an off-line timing and resource analysis [ABG98]. An off-line timing and resource analysis assesses the worst-case behavior of the system in terms of communication delays, computational delays, jitter, end-to-end delays, and temporal interference between different activities.

In hard real-time systems, missed deadlines represent system failures with the potential of consequences as serious as in the case of providing incorrect results. For example, in drive-by-wire applications, the dynamics for steered wheels in closed control loops enforce computer delays of less than 2 ms [HT98]. Taking the vehicle dynamics into account, a transient outage-time of the steering system must not exceed 50 ms [HT98]. In the avionic domain, variable-cycle jet engines can blow up if correct control inputs are not applied every 20–50 ms [LH94].

While control algorithms can be designed to compensate a known delay, delay jitter (i.e., the difference between the maximum and minimum value of delay) brings an additional uncertainty into a control loop that has an adverse effect on the quality of control [Kop97]. Delay jitter represents an uncertainty about the instant a real-time entity was observed and can be expressed as an additional error in the value domain. In case of low jitter or a global time-base with a good precision, state estimation techniques allow to compensate a known delay between the time of observation and the time of use of a real-time image. State estimation uses a model of a real-time entity to compute the probable state of the real-time entity at a future point in time.

References

ABG98 N.C. Audsley, I.J. Bate, and A. Grigg. The role of timing analysis in the certification of IMA systems. *IEE Certification of Ground/Air Systems Seminar (Ref. No. 1998/255)*, February 1998. Department of Computer Science, York University, London, U.K.

ALR01 A. Avizienis, J.C. Laprie, and B. Randell. Fundamental concepts of dependability. Research Report 01-145, LAAS-CNRS, Toulouse, France, April 2001.

ART06 ARTEMIS. Reference designs and architectures—Constraints and requirements. Technical Report Version 13, 2006.

BKS03 G. Bauer, H. Kopetz, and W. Steiner. The central guardian approach to enforce fault isolation in a time-triggered system. In *Proceedings of the Sixth International Symposium on Autonomous Decentralized Systems (ISADS 2003)*, pp. 37–44, Pisa, Italy, April 2003.

Bre01 E. Bretz. By-wire cars turn the corner. *IEEE Spectrum*, 38(4):68–73, April 2001.

GMR88 S. Goldwasser, S. Micali, and R.L. Rivest. A digital signature scheme secure against adaptive chosen-message attacks. *SIAM Journal of Computing*, 17(2):281–308, April 1988.

GP96 A. Galleni and D. Powell. Consensus and membership in synchronous and asynchronous distributed systems. Technical Report 96104, LAAS, Toulouse, France, April 1996.

Gua05 FlexRay Consortium. BMW AG, DaimlerChrysler AG, General Motors Corporation, Freescale GmbH, Philips GmbH, Robert Bosch GmbH, and Volkswagen AG. Node-Local Bus Guardian Specification Version 2.0.9, December 2005.

HT98 G. Heiner and T. Thurner. Time-triggered architecture for safety-related distributed real-time systems in transportation systems. In *Proceedings of the 28th Annual International Symposium on Fault-Tolerant Computing*, Munich, Germany, 17(2):402–407, June 1998.

HTH97 M. Hecht, D. Tang, and H. Hecht. Quantitative reliability and availability assessment for critical systems including software. In *Proceedings of the 12th Annual Conference on Computer Assurance*, Gaithersburg, MD, June 1997.

IEE00 IEEE. IEEE Standard 802.3—Carrier sense multiple access with collision detect (CSMA/CD) access method and physical layer. Technical Report, IEEE, 2000.

Int93 International Standardization Organisation, ISO 11898. *Road Vehicles—Interchange of Digital Information—Controller Area Network (CAN) for High-Speed Communication*, Geneva, Switzerland 1993.

KGR91 H. Kopetz, G. Grünsteidl, and J. Reisinger. Fault-tolerant membership service in a synchronous distributed real-time system. In *Dependable Computing for Critical Applications* (A. Avizienis and J.C. Laprie, eds.), pp. 411–429, Springer-Verlag, Wien, Austria, 1991.

KO02 H. Kopetz and R. Obermaisser. Temporal composability. *Computing & Control Engineering Journal*, 13:156–162, August 2002.

KO87 H. Kopetz and W. Ochsenreiter. Clock synchronization in distributed real-time systems. *IEEE Transactions on Computers*, 36(8):933–940, 1987.

Kop92 H. Kopetz. Sparse time versus dense time in distributed real-time systems. In *Proceedings of 12th International Conference on Distributed Computing Systems*, Yokohama, Japan, June 1992.

Kop95 H. Kopetz. Why time-triggered architectures will succeed in large hard real-time systems. In *Proceedings of the Fifth IEEE Computer Society Workshop on Future Trends of Distributed Computing Systems*, Cheju Island, Korea, August 1995.

Kop97 H. Kopetz. *Real-Time Systems, Design Principles for Distributed Embedded Applications.* Kluwer Academic Publishers, Boston, MA, 1997.

Kop02 H. Kopetz. On the specification of linking interfaces in distributed real-time systems. Research Report 8/2002, Technische Universität Wien, Institut für Technische Informatik, Treitlstr. 1-3/182-1, Vienna, Austria, 2002.

Kop03a H. Kopetz. Fault containment and error detection in the time-triggered architecture. In *Proceedings of the International Symposium on Autonomous Decentralized Systems*, Pisa, Italy, April 2003.

Kop03b H. Kopetz. Fault containment and error detection in the time-triggered architecture. In *Proceedings of the Sixth International Symposium on Autonomous Decentralized Systems*, Pisa, Italy April 2003.

LA90 P.A. Lee and T. Anderson. *Fault Tolerance Principles and Practice*, Vol. 3 of *Dependable Computing and Fault-Tolerant Systems*. Springer-Verlag, Wien, Austria 1990.

LH94 J.H. Lala and R.E. Harper. Architectural principles for safety-critical real-time applications. *Proceedings of the IEEE*, 82:25–40, January 1994.

OKS07 R. Obermaisser, H. Kraut, and C. Salloum. A transient-resilient system-on-a-chip architecture with support for on-chip and off-chip TMR. In *Proceedings of the Seventh European Dependable Computing Conference 2008 (EDCC)*, Kaunas, Lithuania, 2007.

Pol96 S. Poledna. *Fault-Tolerant Real-Time Systems: The Problem of Replica Determinism*. Kluwer Academic Publishers, Boston, Dodrecht, London, 1996.

Rus99 J. Rushby. Partitioning for avionics architectures: Requirements, mechanisms, and assurance. NASA Contractor Report CR-1999-209347, NASA Langley Research Center, Hampton, VA, June 1999.

Rus01 J. Rushby. Bus architectures for safety-critical embedded systems. In *Proceedings of the First Workshop on Embedded Software (EMSOFT 2001)* (T. Henzinger and C. Kirsch, eds.), Vol. 2211 of Lecture Notes in Computer Science, pp. 306–323, Springer-Verlag, Lake Tahoe, CA, October 2001.

40

Time-Triggered Protocol (TTP/C)

Roman Obermaisser
University of Siegen

Michael Paulitsch
Airbus Group Innovations

40.1 Temporal Structuring of Communication 40-4
Timing of a TDMA Slot • Frame Types and States
40.2 Clock Synchronization 40-7
40.3 Restart, Reintegration, Integration 40-7
40.4 Diagnostic Services 40-8
40.5 Fault Isolation 40-9
40.6 Configuration Services 40-10
References 40-11

The Time-Triggered Protocol (TTP) is a communication protocol for distributed fault-tolerant real-time systems. It is designed for applications with stringent requirements concerning safety and availability, such as avionic, automotive, industrial control, and railway systems. TTP was initially named TTP/C and later renamed to TTP. The initial name of the communication protocol originated from the classification of communication protocols of the Society of Automotive Engineers (SAE), which distinguishes four classes of in-vehicle networks based on the performance. TTP/C satisfies the highest performance requirements in this classification of in-vehicle networks and is suitable for network classes C and above.

TTP provides a *consistent distributed computing base* [Kop03] in order to ease the construction of reliable distributed applications. Given the assumptions of the fault hypothesis, TTP guarantees that all correct nodes perceive messages consistently in the value and time domains. In addition, TTP provides consistent information about the operational state of all nodes in the cluster. For example, in the automotive domain, these properties would reduce the efforts for the realization of a safety-critical brake-by-wire application with four braking nodes. Given the consistent information about inputs and node failures, each of the nodes can adjust the braking force to compensate for the failure of other braking nodes. In contrast, the design of distributed algorithms becomes more complex [LSP82], if nodes cannot be certain that every other node works on the same data. In such a case, the agreement problem has to be solved at the application level.

TTP has been integrated into a number of commercial applications. In the railway domain, Thales Rail Signalling Solutions has used TTP for realizing the electronic interlocking system *LockTrac 6131 ELEKTRA* [KK95]. This system has been certified according to CENELEC standards with Safety Integrity Level 4 (SIL4). The ELEKTRA system supports the basic interlocking functions, as well as additional features such as local and remote control, automatic train operation, blocking functionality, and diagnosis capabilities.

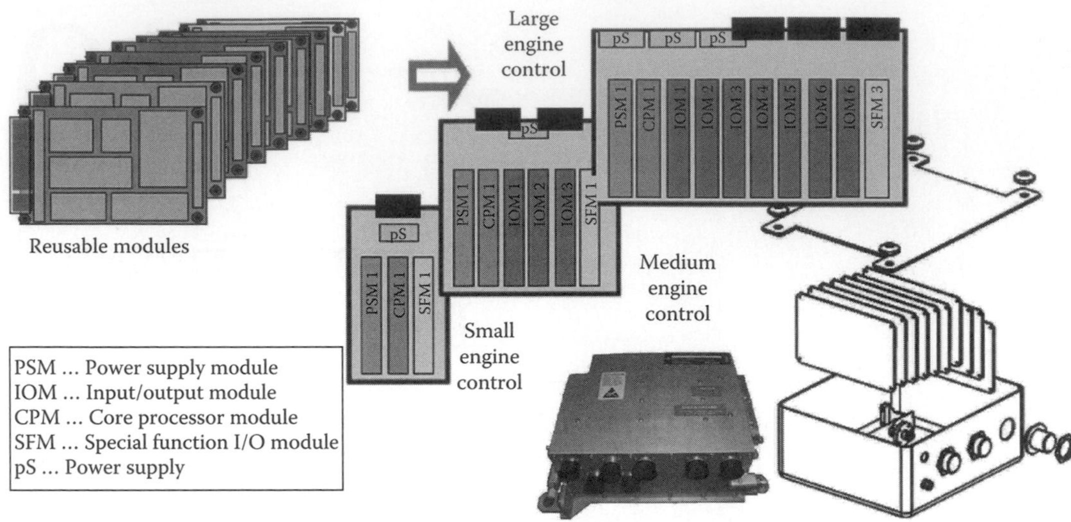

FIGURE 40.1 Concept behind Modular Aerospace Control (MAC) [Tus03]. (Courtesy of Honeywell.)

In the aerospace domain, the Honeywell Modular Aerospace Control (MAC) has been developed using TTP as a backplane databus for intermodule communication. For engine control, MAC facilitates simplified overspeed and uncommanded thrust protection. Honeywell has created the MAC platform having reusable modules in mind. Module functions are scaled to the control needs of multiple applications. Furthermore, reduced development cycle times are possible due to reuse. This reduces nonrecurring cost by reuse of modules. The core architecture comprises multiple modules (i.e., cards) connected by TTP scalable within a wide range of anticipated use cases. The replication level for each module is scalable to the needs of the application for this one module. If one compares this to the most prevalently existing dual-lane engine control dependability paradigm where everything is deployed at the same dual replication needs, this can be more efficient. In addition, this reduces recurring cost. Figure 40.1 depicts the concept behind MAC. Figure 40.2 compares traditional dual-lane and the MAC distributed engine control.

MAC modules are partly prequalified and, hence, decrease certification cost. Furthermore, MAC supports proactive obsolescence management by enabling a building block approach where obsolete modules can easily be redeveloped and exchanged due to well-defined interfaces. Furthermore, TTP should ensure not only proper communication but also partitioning between modules. This supports deterministic integration and keeps efforts low.

MAC is deployed in the FADEC of F124 used, for example, on Alenia Aermacchi's M-346. M-346 is an advanced/lead-in fighter trainer. MAC is also used for the engine control of General Electric's F110 Turbofan deployed, for example, on newer generations F-16 Fighting Falcon.

In the following, protocol services of TTP are described. The *communication services* support the predictable message transport with a small variability of the latency. The *fault-tolerant clock synchronization* maintains a specified precision and accuracy of the global time base, which is initially established by the *restart and start-up services* when transiting from asynchronous to synchronous operation. The *diagnostic services* provide the application with feedback about the operational state of the nodes and the network using a consistent membership vector. The diagnostic services in conjunction with the a priori knowledge about the permitted behavior of nodes are the basis for the *fault isolation services* of TTP. Finally, *configuration services* offer flexibility by switching between predefined modes or programming new communication schedules into the system.

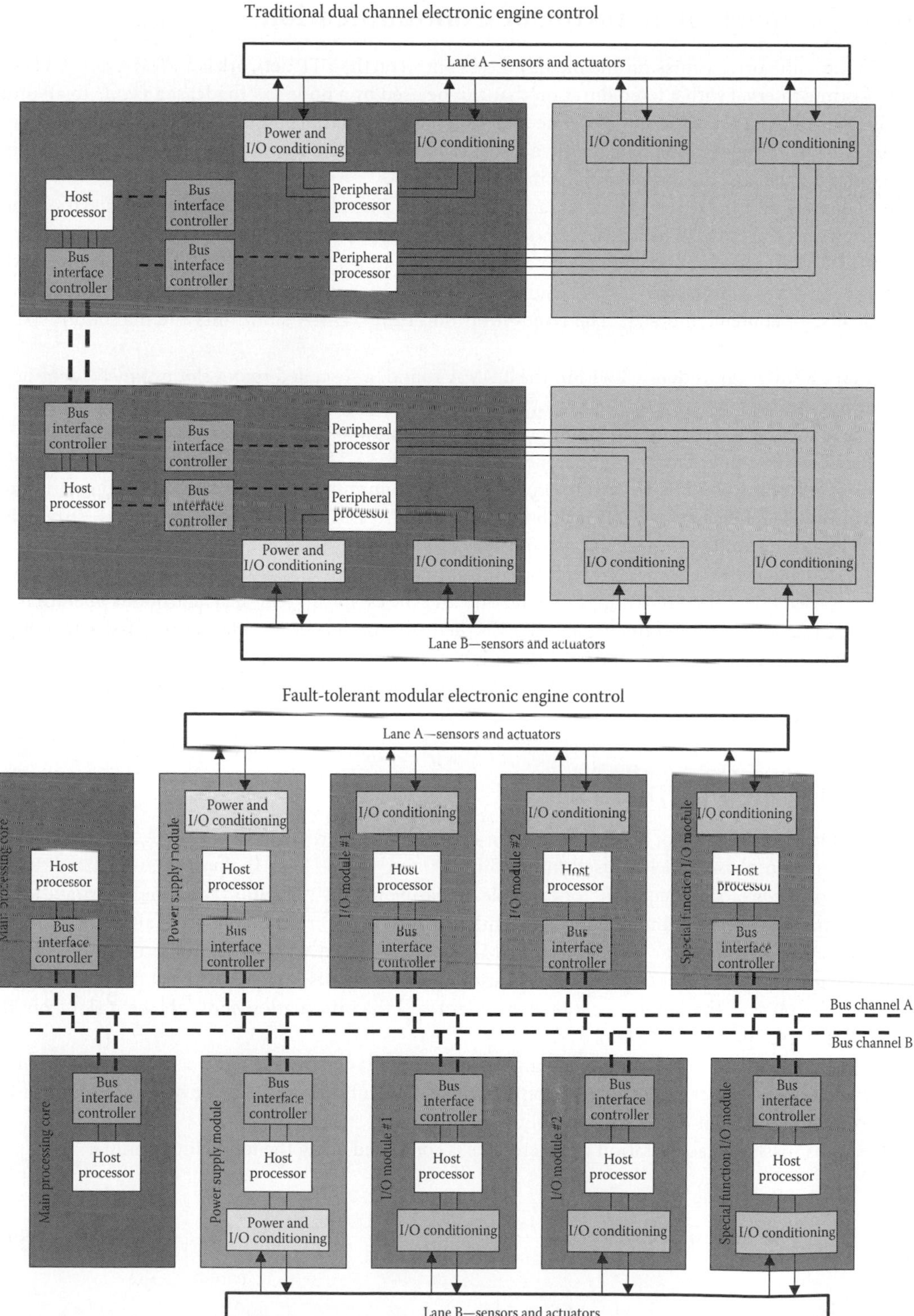

FIGURE 40.2 Engine control architectures—dual-lane versus distributed [Tus03]. (Courtesy of Honeywell.)

40.1 Temporal Structuring of Communication

The smallest unit of transmission and media access control on the TTP network is a *TDMA slot*. A TDMA slot is a time interval with a fixed duration that can be used by a node to broadcast a frame to all other nodes. A frame is a transmission of defined length on a TTP channel containing both application and protocol data. The application data within a frame represents a message, which possesses corresponding semantics for the host (e.g., a speed value in a control loop).

A sequence of TDMA slots is called a *TDMA round*. The *cluster cycle* defines a pattern of periodically recurring TDMA rounds. Although the sequence and the length of the TDMA slots in every TDMA round are equal, the frame communicated in a TDMA slot can differ between TDMA rounds. A TDMA slot in two TDMA rounds can be used for the exchange of frames with different sender nodes, content, and size. The fixed duration of the TDMA slot limits the maximum size of a frame.

While a TDMA slot is defined within the TDMA round, a so-called *round slot* uniquely identifies a slot within the entire cluster cycle. A round slot possesses a unique sender node. In addition, the content and size of the frame communicated in the round slot is known.

If two TDMA rounds have two different sender nodes for a TDMA slot, the nodes are called *multiplexed nodes*. Multiplexed nodes can improve the use of communication bandwidth. If no multiplexed nodes are used, TTP possesses a unique mapping between TDMA slots and sender nodes. Otherwise, the sender is only known at the level of round slots.

Figure 40.3 depicts the structuring of the communication activities on a TTP network. This TDMA scheme is used after the completion of the start-up and the establishment of synchronous operation.

A node that does not send during any slot is called a *passive node*. A node can be passive by design, if it is only required to receive frames (e.g., a bus sniffer for diagnosis). A failure can also lead to a passive node, such as the absence of the timely update of the life-sign or isolation performed by the bus guardian.

40.1.1 Timing of a TDMA Slot

The constituting elements of a TDMA slot are depicted in Figure 40.4. The TDMA slot begins with the *transmission phase*, which is used for sending a frame on the bus. The *postreceive phase* (PRP) is reserved for the protocol execution, for example, updating the membership vector and writing the status information in the Communication Network Interface (CNI). The *idle interval* allows to stretch a TDMA slot in case a specific duration of TDMA slots and TDMA rounds is required in an application. The *presend phase* serves for the preparation of the transmission phase of the next TDMA slot. In this time interval, the communication controller loads the information about the next slot from the Messanger Descriptor List (MEDL).

The durations of the postreceive phase and the presend phase depend on the controller implementation, namely, the number of instructions for the protocol execution, the clock rate of the communication controller, and the memory access times of the CNI memory.

The transmission phase duration in macroticks is computed using the following formula:

$$\text{Transmission phase} = \left\lceil \frac{4\Pi + d_{\text{correction}} + n_{\text{frame}} / f_{\text{bitrate}} + d_{\text{processing}}}{g} \right\rceil + 1 \qquad (40.1)$$

where

Π is the precision of the clocks in the cluster
g is the granularity of the global time base

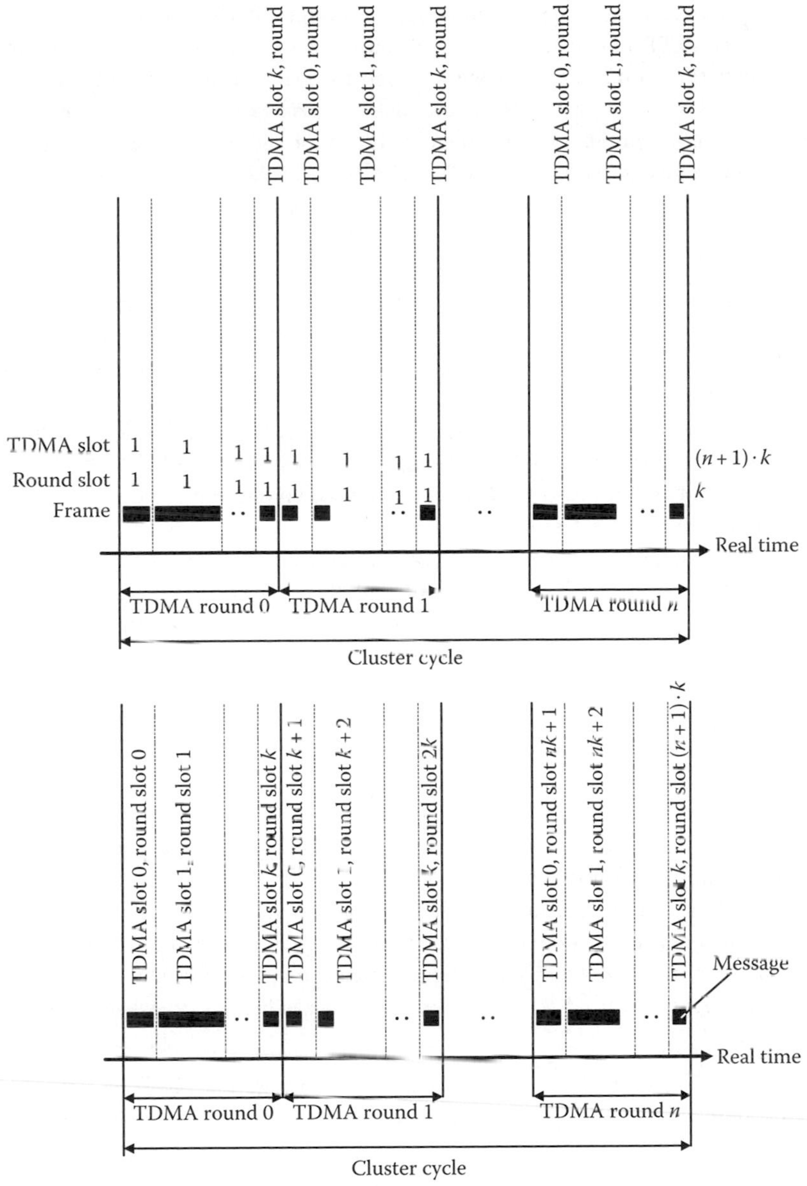

FIGURE 40.3 TDMA scheme.

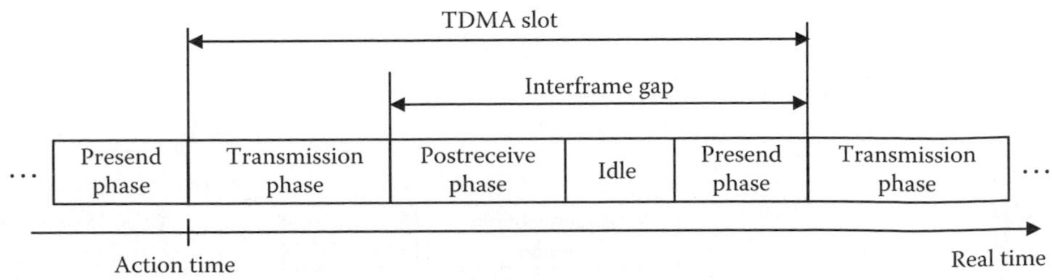

FIGURE 40.4 Timing of a TDMA slot.

The computation of the transmission phase must take into account a maximum deviation of Π between the clocks of different TTP nodes. $d_{correction}$ is the maximum delay correction of all receiving nodes to the sender node of the round slot. The time required for the transmission on the bus is the quotient of the frame length in bits n_{frame} and the bitrate of the TTP bus $f_{bitrate}$. $d_{processing}$ denotes the number of instructions in the communication controller required for the transmission or reception (e.g., 280 instructions at a clock rate of 40 MHz in the C2 controller [TTC02, p. 21]).

40.1.2 Frame Types and States

The TTP protocol distinguishes two main types of frames (cf. Figure 40.5): *cold start frames* are sent during the unsynchronized operation at start-up, while *normal frames* are used in the synchronous operation. In addition, *download frames* can be used to parameterize the TTP protocol or to write an application image into a node.

A cold start frame contains a frame identifier that marks the frame as a cold start frame, the global time of the sender, the current round slot of the sender, and a cyclic redundancy check (CRC). With the knowledge about the current global time and round slot, receiving nodes are enabled to integrate and transit from asynchronous to synchronous operation.

A normal frame consists of a frame type field, information about mode change requests, an optional controller state, application data, and a CRC. If the controller state is included, one speaks of a frame with explicit controller state. Such a frame serves for the integration of nodes during synchronous operation (i.e., no cold start frames are exchanged).

The controller state encompasses internal variables of the TTP communication controller, which are required to be globally consistent among all correct nodes in the TTP cluster. The controller state consists of the global time, the round slot position, the current mode of the cluster, information about pending mode changes, and the membership vector.

Regardless of the inclusion of the controller state in a frame, the controller state is always used in the CRC calculation of the frame. The TTP protocol introduces the notion of *frames with an implicit controller state*, if the state is not contained in a frame. An implicit controller state reduces the protocol overhead and allows a more efficient use of the communication bandwidth at the cost of higher computational complexity.

A divergence of the controller state between receiver and sender is always detected by a CRC error. In particular, this information is used to establish agreement on a consistent view concerning the membership.

After each round slot configured for reception, a node classifies the received frames depending on detected errors:

- *Null frames.* If a receiver observed no transmission activity on a communication channel, the expected frame is called a null frame. A null frame occurs in case of a crash failure of a node.
- *Invalid frames.* A frame is received within the expected time interval of the round slot and the coding rules are satisfied. For example, the Modified Frequency Modulation (MFM) encoding, which is supported by existing TTP communication controllers, imposes limits on the minimum and maximum number of 0 bits that may appear between consecutive 1 bits. A frame that violates these constraints is called an invalid frame.

Normal frame

Frame type	Mode change request	Controller state (optional)	Application data	CRC

Cold start frame

Frame type	Global time	Round slot of sender	CRC

FIGURE 40.5 Frame formats.

- *Incorrect frames.* An incorrect frame is a valid frame with an incorrect CRC. The CRC error indicates a disagreement of the controller states or a transmission error (e.g., bit flip during the transmission).
- *Tentative frames.* As long as the controller state agreement and acknowledgment is in progress, a frame is tentative. In order to perform this agreement, the membership views of successor nodes in the TDMA round are evaluated.
- *Correct frames.* A correct frame is a valid frame, which has passed the CRC check and the acknowledgment.

40.2 Clock Synchronization

The fault-tolerant average (FTA) algorithm [KO87] is used for clock synchronization in TTP. The FTA algorithm computes the convergence function for the clock synchronization within a single TDMA round. It is designed to tolerate k Byzantine faults in a system with N nodes. Therefore, the FTA algorithm bounds the error that can be introduced by arbitrary faulty nodes. These nodes can provide inconsistent information to the other nodes.

Each node collects the $N - 1$ measured time differences between the node's clock and the clocks of the other nodes. The time differences are determined by the difference of the actual arrival time of a frame and the expected arrival time of frames (as defined by the time-triggered communication schedule). These time differences, which indicate the deviations of the local times of sender and receiver, are sorted by size. The k largest and the k smallest time differences are discarded. The average of the remaining time differences is the correction term for the node's clock.

Using the FTA algorithm, a faulty time value is discarded if it is larger or smaller than the other time values. Otherwise, a faulty time value must be within the precision window. As discussed in [KO87], the worst-case scenario occurs if all correct clocks are at opposite ends of the precision window and the Byzantine clock is seen at different corners by two nodes. In this case, each Byzantine clock will cause a difference of $\Pi/(N - 2k)$ in the calculated averages at two different nodes in an ensemble of N clocks. In the worst case, a total of k Byzantine errors will thus cause an error term of $k\Pi/(N - 2k)$.

Considering the jitter ε of the synchronization frames and the drift offset Γ, the convergence function of the FTA algorithm is as follows:

$$(\varepsilon + \Gamma)\frac{N - 2k}{N - 3k} \tag{40.2}$$

In the TTP protocol, the MEDL controls which TDMA slots are used for clock synchronization. The slots at which new correction terms are calculated are marked consistently at all receivers in the MEDL in order to ensure that all nodes correct their clocks at the same time.

In addition to internal clock synchronization, the TTP protocol supports external clock synchronization using a time reference such as GPS. For this purpose, a host acts as a time gateway and possesses a connection to GPS or to another cluster with a different time base. The time gateway periodically computes a correction term and provides this term to its communication controller using the external rate correction field in the CNI. The external rate correction term denotes the number of microticks that need to be corrected in the next synchronization interval.

If the absolute value of the external correction or the absolute value of the total correction term is larger than $\Pi/2$, a node freezes due to a synchronization error.

40.3 Restart, Reintegration, Integration

Cluster start-up is the process of establishing a synchronized cluster after power-on. After power-on, a node listens on the communication channels for a *cold start frame*. A cold start frame consists of a frame type field (identifying the frame as a cold start frame), the global time at the send instant of the frame, a sender round slot position in the MEDL, and a CRC. If the frame is received, then the node sets its controller state accordingly. The node adopts the global time and the round slot position in the MEDL.

If no cold start frame is received within the *listen timeout* and the cold start allowed flag is set in the MEDL of the node, then the node sends a cold start frame itself. The sending of a cold start frame can be repeated until a maximum number of allowed cold starts are reached (as defined in the MEDL). This limit is introduced in order to prevent a node with an incoming link fault to interfere with the synchronization of the other nodes in the cluster.

During start-up, the nodes perform asynchronous access to the communication medium controlled by the *start-up timeout*. This parameter is unique for each node in a cluster. For a given node, it denotes the number of TDMA slots prior to the sending slot of the node. The listen timeout introduced earlier is the sum of the node's controller start-up timeout plus two TDMA rounds. Hence, the duration between two cold starts is always shorter than the listen timeout.

In case of a collision of the cold start frames of two nodes, TTP performs the so-called *big bang mechanism*. In case of large propagation delays, a collision can be perceived inconsistently. Only a subset of the nodes could receive a correct cold start frame, thereby leading to the formation of cliques. All nodes can detect this situation based on the transmission phases of two cold start frames in relation to the sum of the maximum propagation delay and the frame duration. In order to prevent cliques, nodes will not integrate on cold start frames once a big bang scenario has been detected. Thereby, cold starting nodes will not detect traffic and they will restart their start-up timeouts again. Because of the unique start-up timeouts, no second collision will occur between the nodes.

If a node joins a cluster, which is already synchronized, this process is called the *integration* of the node [SP02]. In order to support integration, the MEDL must contain at least one frame with the controller state within the minimum listen timeout (i.e., two TDMA rounds). This constraint ensures that an integrating node does not initiate a cold start.

In order to avoid the integration on faulty frames, an integrating node maintains an *integration counter*. The MEDL contains a parameter called *minimum integration value*, which specifies the number of correct frames that need to be received before a node considers itself integrated and may start to send.

40.4 Diagnostic Services

The TTP protocol detects the crash failure of a host based on a periodic *host life-sign*. In every TDMA round, the host of a node must provide a *life-sign* to the communication controller. More precisely, a host needs to set the life-sign after the start of the node's transmission slot and before the beginning of the presend phase of the node's transmission slot in the next TDMA round. The communication controller verifies whether the host has set the life-sign during the node's transmission slot.

If the life-sign is not set by the host, the communication controller does not send frames and transits into passive mode. Frame transmission is only continued when the host updates the life-sign again. The life-sign is used both during normal operation and at start-up. The set life-sign is also a prerequisite for sending cold start frames to start up the cluster.

In addition to the host life-sign, the updating of the global time and schedule position by the communication controller serves as a *controller life-sign*. Thereby, the host can react to a crash failure of a communication controller in an application-specific way (e.g., enter a safe state).

The TTP protocol informs nodes about the operational state of every other node in the TTP cluster using a membership vector. The membership vector is a vector with a bit for every node, denoting whether the respective node is operational.

A node *A* considers another node *B* as operational, if node *A* has correctly received the frame that was sent by node *B* prior to the membership point. In case redundant communication channels are used, the reception on one of the channels is sufficient in order to consider a sender to be operational.

The delay between the failure of a node and the indication in the membership vector is bounded by the duration of two TDMA rounds. The points in time for establishing the membership information are called *membership points.* In TTP, the PRP of a sending node serves as a membership point.

The *clique detection and avoidance* in TTP [BP00] has the goal of avoiding the partitioning of a cluster into cliques that are not able to communicate with each other. The clique avoidance algorithm selects the largest partition (clique) as a winner, while the nodes of other partitions are shut down by entering the freeze state.

In every TDMA round, the communication controller determines whether it is in agreement with the majority of the other nodes concerning the controller state. For this purpose, the communication controller counts the number of round slots where the frame status is correct, as well as the number of round slots where the frame status is incorrect or invalid. In the presend phase of its own transmission slot, a node checks whether the value of the failed slots counter is larger than the agreed slots counter. In this case, a clique error is detected and the communication controller transits into the freeze state.

40.5 Fault Isolation

The TTP protocol was designed to isolate and tolerate an arbitrary failure of a single node during synchronized operation [KBP01]. After the error detection and the isolation of the node, a consecutive failure can be handled. Given fast error detection and isolation mechanisms, such a single fault hypothesis is considered to be suitable in many safety-critical systems [OP06]. The fault hypothesis assumes an arbitrary failure mode of a single node. TTP does not guarantee to tolerate two independent node failures, that is, a second failure before the detection and isolation of the first failure. Such a scenario is considered very unlikely and addressed by the so-called *Never-Give-Up (NGU) strategy* [Kop04]. If failures outside the fault hypothesis are detected, the communication system informs the application. Depending on the application, a safe-state can be entered (e.g., setting all signals to red in a railway application). Assuming transient faults, a restart of the TTP cluster can be performed in a fail-operational system.

In order to tolerate timing failures, a TTP cluster uses local or central *bus guardians.* In addition, the bus guardian protects the cluster against slightly-off-specification faults [Ade02], which can lead to ambiguous results at the receiver nodes.

A *local bus guardian* is associated with a single TTP node and can be physically implemented as a separate device or within the TTP node (e.g., on the silicon die of the TTP communication controller or as a separate chip). The local bus guardian uses the a priori knowledge about the time-triggered communication schedule in order to ensure fail-silence of the respective node. If the node intends to send outside the preassigned transmission slot in the TDMA scheme, the local bus guardian cuts off the node from the network. In order to avoid common mode failures of the guardian and the node, the TTP protocol suggests the provision of an independent external clock source for the local bus guardian.

The *central bus guardian* is always implemented as a separate device, which protects the TDMA slots of all attached TTP nodes. An advantage compared to the local bus guardians is the higher resilience against spatial proximity faults and the ability to handle slightly-off-specification faults.

In safety-critical systems, a TTP cluster is deployed with two independent bus guardians for the two redundant TTP channels. The failure mode of a central bus guardian is assumed not to be arbitrary. According to the fault hypothesis of TTP, the failure of a central guardian only leads to the transmission of frames that are detectably faulty at the receivers. In the time domain, a failure of a guardian can lead to untimely frames that are perceived at the receiving nodes outside the slots defined by the TDMA scheme. In the value domain, a faulty guardian can produce frames with an invalid CRC. According to the fault hypothesis, the guardian may not generate incorrect frames with valid CRCs and correct timing. The reason for this assumption is that nodes would receive two different frames from the redundant communication channels with correct CRCs and timing. Hence, the receiving nodes would be

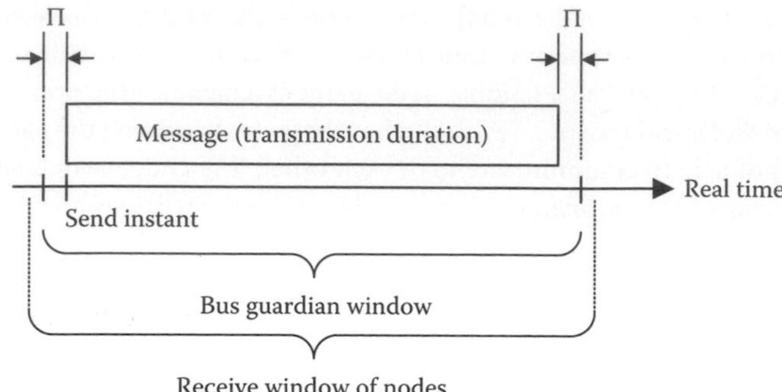

FIGURE 40.6 Bus guardian window and nodes' receive window.

unable to determine which frame is the correct one and should be provided to the application. In order to justify this fault assumption, implementations of the central bus guardian contain no logic for the generation of CRC codes. TTP addresses the replaying of old frames by including the global time in controller state, which is used together with the application data in the CRC calculation.

Both central and local guardians use a *bus guardian window* in order to ensure timely frames (cf. Figure 40.6). The bus guardian window enables access to the communication system for the node at the specified time and for the complete slot duration, but prevents any transmission from the node for the remaining duration of the TDMA round. The start instant and the end instant of the bus guardian window take into account the different views on the global time, which are bounded by the precision of the clock synchronization.

In order to avoid slightly-off-specification failures, the bus guardian uses a bus guardian window that is shorter than the receive windows used by the receiver nodes. This means that the bus guardian is more restrictive concerning the time of a frame transmission than any receiver node. The bus guardian limits the frame transmission in such a way that frames transmitted too early or too late are blocked or truncated, thus resulting in an invalid transmission for all receivers. Thereby, the bus guardian protects the communication system from a node transmitting a correct frame with a temporal deviation close to the precision. Such a transmission can result in an inconsistently perceived failure (i.e., so-called Byzantine fault) when the frame is received correctly by some nodes and incorrectly by other nodes. A detailed discussion of the dimensioning of bus guardian windows can be found in [Rus01].

40.6 Configuration Services

TTP supports the switching between predefined static configurations called *cluster modes*. The rationale for this protocol service is that many applications exhibit mutually exclusive modes of operation. For example, the flight control system of an airplane can support different modes such as on-ground, takeoff, low-altitude, and landing [BM00]. Likewise, cars can exhibit different modes of operation such as a normal mode and a limp-home mode [BP96, TNBM05].

At any time, all nodes of the cluster must be in the same cluster mode. The cluster mode is part of the controller state, thus a divergence of cluster modes is detected by the CRC calculation. Every cluster mode must also possess the same sequence of TDMA slots.

A host can request a mode change using the control area of the CNI. In the next presend phase, the communication controller checks this request against the mode change permissions in the MEDL. If the mode change is permitted by the MEDL, information about the new mode is included in the next sent frame.

Receiving nodes act on this mode change information in the postreceive phase. The communication controller of a receiver node also checks whether the sender is allowed to request the new mode according to the mode change permissions in the MEDL. If the request is permitted, the new mode will become active at the beginning of the next cluster cycle. The ongoing cluster cycle is not preempted.

If another mode change request arrives before the end of the cluster cycle, then the new one overwrites the previous one. Nodes can also cancel a pending mode change request by sending a special value for the mode change request. In this case, the mode of the current cluster cycle remains in place.

References

Ade02 A. Ademaj. Slightly-off-specification failures in the time-triggered architecture. In *Proceedings of the Seventh IEEE International High-Level Design Validation and Test Workshop*, p. 7, Washington, DC, IEEE Computer Society, 2002.

BM00 J.D. Boskovic and R.K. Mehra. Multi-mode switching in flight control. In *Proceedings of the 19th Digital Avionics Systems Conferences (DASC)*, Philadelphia, PA, Vol. 2, pp. 6F2/1–6F2/8, 2000.

BP00 G. Bauer and M. Paulitsch. An investigation of membership and clique avoidance in TTP/C. In *Proceedings of the 19th IEEE Symposium on Reliable Distributed Systems*, Pasadena, CA, pp. 118–124, 2000.

BP96 M.B. Barron and W.F. Powers. The role of electronic controls for future automotive mechatronic systems. *IEEE/ASME Transactions on Mechatronics*, 1(1):80–88, March 1996.

KBP01 H. Kopetz, G. Bauer, and S. Poledna. Tolerating arbitrary node failures in the time-triggered architecture. In *Proceedings of the SAE 2001 World Congress*, Detroit, MI, March 2001.

KK95 H. Kantz and C. Koza. The ELEKTRA railway signalling system: Field experience with an actively replicated system with diversity. In *Proceedings of the 25th International Symposium on Fault-Tolerant Computing (FTCS)*, pp. 453–458, 27–30, 1995.

KO87 H. Kopetz and W. Ochsenreiter. Clock synchronization in distributed real-time systems. *IEEE Transactions on Computers*, 36(8):933–940, 1987.

Kop03 H. Kopetz. Time-triggered real-time computing. *Annual Reviews in Control*, 27(1):3–13, 2003.

Kop04 H. Kopetz. The fault-hypothesis of the time-triggered architecture. In *Proceedings of the 18th Edition of the IFIP World Computer Congress*, Springer, Heidelberg, August 2004.

LSP82 L. Lamport, R. Shostak, and M. Pease. The byzantine generals problem. *ACM Transactions on Programming Languages and Systems (TOPLAS)*, 4(3):382–401, 1982.

OP06 R. Obermaisser and P. Peti. A fault hypothesis for integrated architectures. In *Proceedings of the Fourth International Workshop on Intelligent Solutions in Embedded Systems*, Vienna, Austria, June 2006.

Rus01 J. Rushby. Formal verification of transmission window timing for the time-triggered architecture. Technical report, Computer Science Laboratory, SRI International, Menlo Park, CA, March 2001.

SP02 W. Steiner and M. Paulitsch. The transition from asynchronous to synchronous system operation: An approach for distributed fault-tolerant systems. In *Proceedings of the International Conference on Distributed Computing Systems*, Vienna, Austria, pp. 329–336, 2002.

TNBM05 G. Torrisi, J. Notaro, G. Burlak, and M. Mirowski. Evolution and trends in automotive electrical distribution systems. In *Proceedings of the IEEE Conference on Vehicle Power and Propulsion*, pp. 7–9, Piscataway, NJ, 2005.

TTC02 TTChip. *TTP/C Controller C2: Controller Schedule (MEDL) Structure—Document Protocol Version 2.1*. Vienna, Austria, September 2002.

Tus03 Honeywell Tucson. Design, implementation, and verification of fault-tolerant modular aerospace controls, honeywell ncc-1-377. http://shemesh.larc.nasa.gov/fm/talks/Honeywell—TTTech.ppt, accessed March 20, 2014, April 2003. Aviation Safety Program Single Aircraft Accident Prevention. Coop. Agreement NCC-1-377.

41

Time-Triggered CAN

Roland Kammerer
Vienna University of Technology

Roman Obermaisser
University of Siegen

41.1 Protocol Overview **41**-1
41.2 Protocol Services **41**-2
Communication Services • Clock Synchronization • Sending and Receiving Messages in TTCAN • Restart, Reintegration, Integration • Diagnostic Services • Error Detection and Fault Isolation • Configuration Services
41.3 Flexible Time-Triggered CAN **41**-15
41.4 Commercial or Prototypical Components **41**-16
References **41**-17

41.1 Protocol Overview

Controller area network (CAN) [Int93a] provides an inexpensive and robust network technology, which is widely deployed in many application domains such as automotive, avionic, and industrial control systems. For example, present-day cars contain multiple CAN buses deployed for different domains such as comfort or powertrain subsystems [ES09,HB08]. Properties of CAN that have led to its success include its simplicity, high flexibility, efficiency, and low cost. Adversely, CAN exhibits limitations with respect to reliability, diagnosis, and scalability.

In classic CAN networks, the communication has two main characteristics: the communication is event-triggered, and the shared medium is a bus. Both these properties lead to limitations of the conventional CAN protocol in the context of dependable embedded systems. An example of a hazard to reliability is the missing fault isolation for babbling idiot failures [TH94]. A faulty CAN node can disrupt the communication abilities of all other nodes by continuously transmitting high-priority messages. Besides these limitations, which have their origin in the bus-based topology, conventional CAN does not provide time-triggered concepts like dedicated communication slots for real-time communication, because conventional CAN does not make use of a priori knowledge about communication activities.

Due to the event-triggered nature of CAN, peak loads can occur if nodes in the network try to send messages at the same point in time. The arbitration mechanism of CAN performs sequential transmission of all messages according to their *identifier priority*. This arbitration makes a scheduling analysis of the whole system difficult because worst-case peak loads have to be considered for the analysis. Additionally, the temporal behavior of sending nodes is often unknown or not defined.

Time-Triggered CAN (TTCAN) [Int93b] uses the concept of cyclic communication, divided into slots to implement time-triggered behavior. The standard requires that all activity assigned to one slot (including interrupt handling) be finished until the next slot starts, whereas one message can be assigned to several slots.

It is important to note that TTCAN adds an *additional layer* before the existing standard CAN layers. The physical layer and data link layer of CAN (specified in [Int93c]) are kept unchanged. Mapped to the ISO-OSI model, TTCAN resides on layer 5, the session layer. Within this layer, TTCAN is divided into two different modes of operation. While level 1 and level 2 both support time-triggered behavior, the capabilities of these levels differ in key properties. The most outstanding difference is the notion of a *global time*, which only exists in level 2, and allows a much more fine-grained synchronization of the nodes in a TTCAN network.

The following sections discuss the key properties of the TTCAN protocol, where both levels of TTCAN are taken into consideration.

41.2 Protocol Services

41.2.1 Communication Services

According to the standard [Int93b], TTCAN specifies a serial communication protocol that supports distributed real-time control and multiplexing for use within road vehicles. The underlying CAN-protocol is unchanged; therefore, TTCAN uses the same arbitration and medium access control mechanisms (MAC) as conventional CAN.

TTCAN level 1 provides cyclic message transfer, while level 2 adds the notion of a global time. The key element for the periodic and cyclic communication is the *reference message*, which is sent by a *time master*. TTCAN supports up to eight alternative time masters, where only one is allowed to be the current time master. The reference message is sent by the current time master at the beginning of each *basic cycle*. Figure 41.1 shows the execution of basic cycles over time.

Every basic cycle is divided into windows of communication activity. The sum of all basic cycles forms a *system matrix*. Basic cycles are the rows of the system matrix. Additionally, note that the number of basic cycles in the system matrix shall be a power of 2. The variable *Cycle_Count* refers to the current basic cycle in the system matrix. *Cycle_Count* starts from zero and has a maximum value of *Cycle_Count_Max*.

Figure 41.2 gives an overview about the communication matrix used in TTCAN. A message (e.g., message C in basic cycle 0) can be assigned to one or more time windows. The columns in the system matrix form the so-called *communication columns*, which are important for the periodicity of a message. The two important factors for the periodicity of a message are the number of basic cycles and the number of communication columns.

As shown in Figure 41.2, TTCAN supports different types of communication windows:

- Reference window
- Exclusive window
- Arbitrating window
- Free time window

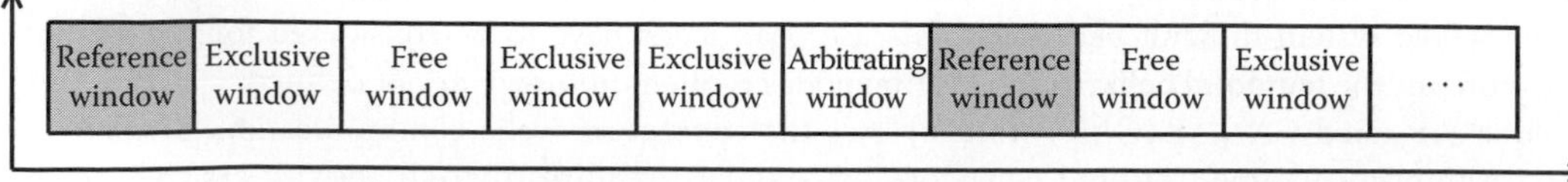

FIGURE 41.1 Basic cycle in TTCAN. (From International Standardization Organisation (ISO), Road vehicles—Controller area network (CAN)—Part 4: Time-triggered communication, ISO 11898-4, 1993.)

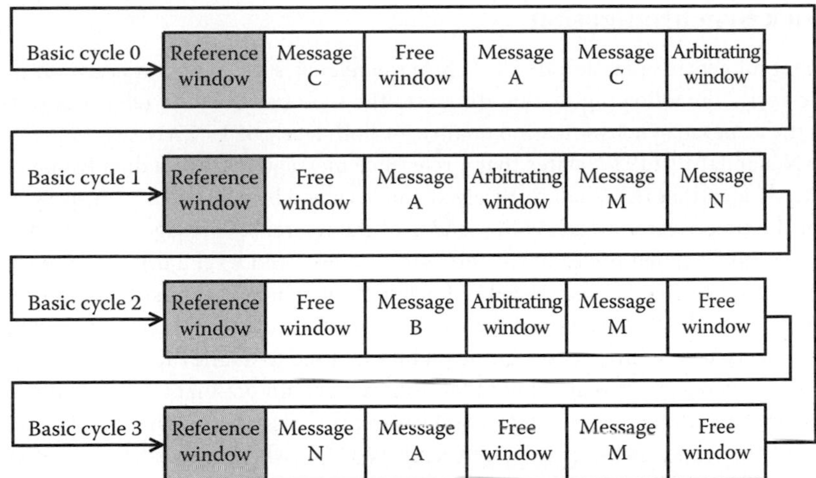

FIGURE 41.2 Communication matrix in TTCAN. (From International Standardization Organisation (ISO), Road vehicles—Controller area network (CAN)—Part 4: Time-triggered communication, ISO 11898-4, 1993.)

Reference windows are used to flag the start of a basic cycle. The format of reference messages and their importance for TTCAN are discussed in Section 41.2.2. *Exclusive windows* are assigned to specific, periodic messages where no other communication is allowed in the same window. There is no competition on the CAN bus within an exclusive window. *Arbitrating windows* are slots where all CAN nodes are allowed to communicate. Conflicts are resolved by the conventional CAN identifier arbitration mechanism (e.g., the message with the highest priority will be transferred). It is important to note that CAN nodes may not start a communication if the bus is not idle. To guarantee that nodes finish their transmission within their assigned window, there is a *time window* (Tx_Enable) at the start of the arbitrating window where nodes are allowed to start their communication. If the attempt to send fails, nodes are not allowed to automatically retransmit their message and have to wait for another arbitrating window. *Free time windows* are reserved slots that will be eventually used in future network extensions.

TTCAN supports two different methods of basic cycle synchronization, namely, time-triggered and event-synchronized. In time-triggered synchronization, the time master sends the reference messages in equidistant time slots. Additionally, TTCAN supports an event-synchronized transmission of reference messages. In this case, the start of a basic cycle is synchronized to a specific event in the current time master. The gap is allowed only between the end of one basic cycle and the start of the next basic cycle (i.e., its reference message). If there is a gap between basic cycle n and basic cycle $n + 1$, the time master has to announce the gap in the reference message of basic cycle n. For this purpose the reference message, which will be described in detail in Section 41.2.2, contains a single-bit flag *Next_is_Gap*. Event-synchronized transmission is shown in Figure 41.3.

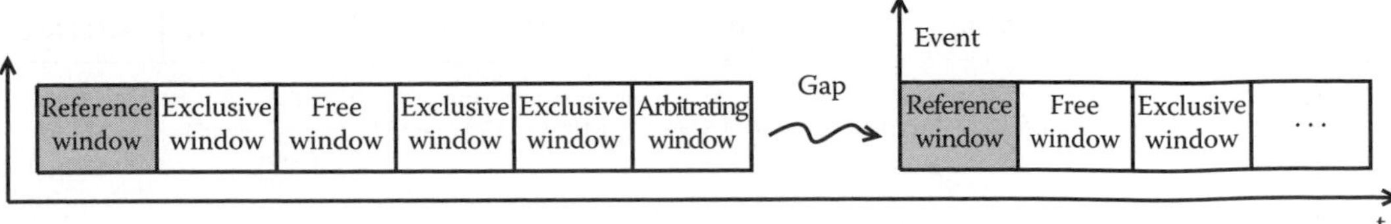

FIGURE 41.3 Event-synchronized basic cycles in TTCAN.

41.2.2 Clock Synchronization

The key element for synchronization in TTCAN is the *reference message* that is sent at the beginning of every basic cycle. The following section discusses the structure and the relevance of the reference message and the achieved synchronization quality for both levels of TTCAN.

The TTCAN standard [Int93b] states that a reference message shall be a data frame characterized by a specific CAN identifier that shall be received and accepted by every FSE* (frame synchronization entity) except the time master, which is the sender of the reference message. For reasons of simplification, FSEs are simply called nodes in the following. Level 1 and level 2 differ in the length of their corresponding reference messages. For level 1, the data length shall be at least 1 byte; for level 2, it shall have a length of at least 4 bytes. All bits of the identifier field, except three, are used to characterize the reference message itself. The three least significant bits are used to define the priority of up to eight (2^3) time masters. The reference messages of both levels have in common, that they shall include the number of the current basic cycle (Cycle_Count) and one bit that signals that there will be a gap between the current and the next reference message (Next_is_Gap), which is used for event-synchronized communication. In general, the reference message is transmitted periodically, with the exception of event-synchronized communication, or if the transmission of a reference message is disturbed. In the latter case, the time master is allowed to retransmit the reference message. In both levels, the reference message may be extended to 8 CAN data bytes. Reserved bits shall be transmitted as logical zero and shall be ignored by the receivers. Additionally, both levels have in common that the most significant bit (bit 7) is transmitted first.

The format of reference messages for level 1 and level 2 is shown in Figures 41.4 and 41.5, respectively. In both figures, grayed boxes represent optional bits in the reference message, which shall be transmitted as logical zero.

All timing in TTCAN, for level 1 and level 2, is controlled by the local clock, where the resolution for time is the network time unit (NTU). For the simpler level 1 the NTU is the nominal CAN bit time. The local time is implemented as a simple incrementing counter that contains 16 bits. In level 1, this counter is increased by one every NTU.

7	6	5	4	3	2	1	0
Next is_Gap	Reserved	Cycle Count (5)	Cycle Count (4)	Cycle Count (3)	Cycle Count (2)	Cycle Count (1)	Cycle Count (0)

FIGURE 41.4 Reference message in TTCAN level 1.

	7	6	5	4	3	2	1	0
Byte 1:	Next is_Gap	Reserved	Cycle Count (5)	Cycle Count (4)	Cycle Count (3)	Cycle Count (2)	Cycle Count (1)	Cycle Count (0)
Byte 2:	NTU Res (6)	NTU Res (5)	NTU Res (4)	NTU Res (3)	NTU Res (2)	NTU Res (1)	NTU Res (0)	Disc Bit
Byte 3:	MRM (7)	MRM (6)	MRM (5)	MRM (4)	MRM (3)	MRM (2)	MRM (1)	MRM (0)
Byte 4:	MRM (15)	MRM (14)	MRM (13)	MRM (12)	MRM (11)	MRM (10)	MRM (9)	MRM (8)

FIGURE 41.5 Reference message in TTCAN level 2.

* Every CAN controller in a time-triggered CAN network has its own FSE.

For level 2 communication, the reference message contains two additional fields, namely, Master_Ref_Mark (MRM), which is a time-stamp measured in global time, and Disc_Bit, which flags that there is a discontinuity in the global time. If the transmission of a reference message is disturbed, the time master shall retransmit the message and the Master_Ref_Mark is updated. The structure of the reference message for TTCAN level 2 is shown in Figure 41.5.

The first byte of a reference message in level 2 contains the same bits as the reference message for a level 1 message. The second byte contains the Disc_Bit, which flags a discontinuity in the global time and the NTU_Res, which is the resolution of the NTU. TTCAN supports four additional bits for a more fine-grained specification of the NTU_Res. If a node does not support these additional bits, they shall be transmitted as logical zero. The third and fourth bytes contain the low and high byte of the Master_Ref_Mark.

In TTCAN, there are three different notions of time [HMFH03], namely, local time, cycle time, and global time. Each of these times and their importance for TTCAN will be discussed in the following.

41.2.2.1 Local Time

As in level 1, the time in level 2 is measured in NTUs. In contrast to level 1, the NTU in level 2 is a fraction of the physical second. This is required to correct the slight differences in the local clock oscillators of the nodes and to synchronize them to the global time. For level 2, the counter for the local time shall contain at least 19 bits, where the 16 most significant bits represent whole NTU ticks. The rest of the counter bits are reserved for the fractional parts of the NTU. Therefore, the counter counts in units of $NTU/2^n$ if NTU_Res contains n bits. If the counter is incremented 2^n times, this is equivalent to 1 NTU. For example if 3 bits are reserved for the fractional part of the NTU, the counter has to increment 2^3 times, to increment one NTU. Every node has a basic time unit (e.g., the frequency of its local clock oscillator). To set this basic time unit in relation to the NTU, there exists the *time unit ratio* (TUR). This usually noninteger value specifies the ratio between the length of an NTU and the length of a basic time unit. The current value of TUR (TUR_Actual) is responsible for the velocity of the NTU counter. The TTCAN standard does not specify the implementation or the data representation of the TUR. The TUR has an essential role for clock synchronization in level 2. Whenever there is a difference between the local time and the global time transmitted by the time master, the TUR gets adjusted. The generation of the *Local_Time* is shown in Figure 41.6

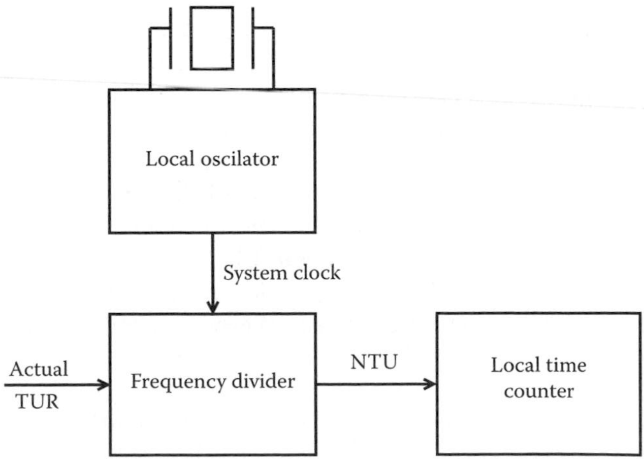

FIGURE 41.6 Local time generation in TTCAN. (From Hartwich, F. et al., Timing in the TTCAN network, Technical report, Robert Bosch GmbH, 2003.)

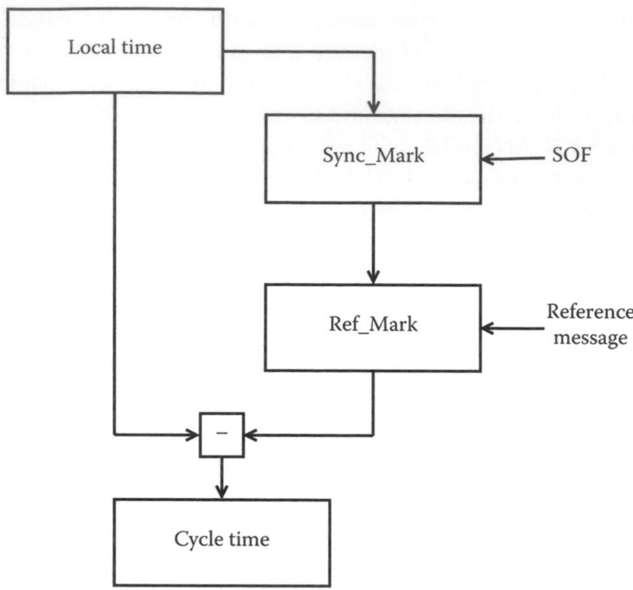

FIGURE 41.7 Cycle time generation in TTCAN. (From Hartwich, F. et al., Timing in the TTCAN network, Technical report, Robert Bosch GmbH, 2003.)

41.2.2.2 Cycle Time

The time within one basic cycle is measured in the *Cycle_Time* parameter. At each start of a data frame (SOF) or remote frame, the node saves its current value of the local time to Sync_Mark. Whenever the node receives a reference message, it saves the current value of Sync_Mark to Ref_Mark. Therefore, the value of Ref_Mark contains the time stamp of that point in time of the start of the last reference message. The Cycle_Time is the difference between the local time of the node and its Ref_Mark. Cycle_Time has no fractional part, therefore only the 16 most significant bits contribute to the Cycle_Time. The generation of the Cycle_Time is shown in Figure 41.7.

41.2.2.3 Global Time

In TTCAN level 1, the common time base is the Cycle_Time, which is restarted at each basic cycle. In this sense, level 1 has a global time with the horizon of one basic cycle. Level 2 adds a more fine-grained concept of global time, which is used to calibrate the local time base of each node in the TTCAN network. To compensate a clock drift in the nodes in level 2, the TUR value is adjusted to the current time master's view of the global time at every reception of a reference message. At a pulse of frame synchronization, a node stores its local view of the global time as its Global_Sync_Mark, which has at least 19 bits. The current time master transmits its Global_Sync_Mark (i.e., its view of the global time) as the bits reserved for Global_Ref_Mark in its reference message. Whenever a node receives a reference message, it calculates a local offset:

$$Local\_Offset = Master\_Ref\_Mark - Ref\_Mark$$

The generation of the global time in TTCAN is shown in Figure 41.8. The current time master's view of Local_Offset remains constant. Every node starts with a Local_Offset of zero. Changes in the value of the

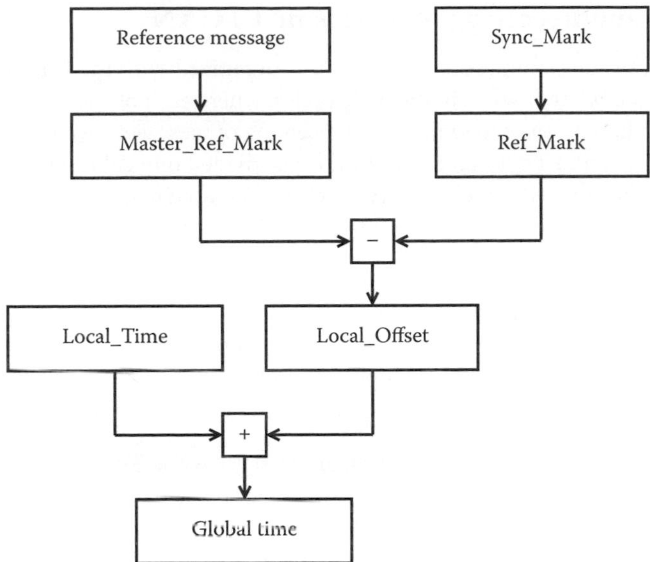

FIGURE 41.8 Global time generation in TTCAN. (From Hartwich, F. et al., Timing in the TTCAN network, Technical report, Robert Bosch GmbH, 2003.)

Local_Offset show that there is drift between the local view of time and the global view, which has to be corrected. This clock rate correction is done by adjusting the NTU value according to the following formula [HMFHO3]:

$$df = \frac{Ref\_Mark - Ref\_Mark_{previous}}{Master\_Ref\_Mark - Master\_Ref\_Mark_{previous}}$$

$$TUR = df * TUR_{previous}$$

TTCAN considers the 16 most significant bits of local time and local offset of the current time master as its *global time*. The local time and local offset of a node are an approximation of the global time.

The time master in TTCAN is allowed to use *external clock synchronization*. The TTCAN standard describes the following three means:

1. *Frequency adjustment*: In this method, the external time period is used as the base for the NTU. To adapt the NTU, the length of the external time period is used as the TUR value. First the time master writes the new value to TUR_Adjust. At the beginning of the next basic cycle, TUR_Actual gets overwritten by TUR_Adjust.
2. *Phase adjustment*: Phase adjustment can be done by continuous frequency adjustment or by inserting a discontinuity in the global time. In the latter case, the time master has to set the Disc_Bit to inform the nodes that a discontinuity has to be expected. At the beginning of next basic cycle, the time master adds the difference between the desired global time and the actual current time to its local offset to influence the global time.
3. *Adjustment of Cycle_Time*: If this synchronization scheme of external clock synchronization is applied, the time master sets the Next_is_Gap flag in its reference message to inform the nodes that the next basic frame will be event-synchronized. After that, the time master waits as long as required by the external clock before it starts sending the next reference message.

41.2.3 Sending and Receiving Messages in TTCAN

As described before, the communication in TTCAN is organized with the help of the system matrix. Within the system matrix the rows form the basic cycles, where each of these basic cycles starts with a reference message. When it comes to sending and receiving of messages, two parameters are of utmost importance: Cycle_Time and Tx_Trigger. A basic cycle is divided into columns that form the time slots when a message shall be sent. Time within a basic cycle is measured in Cycle_Time. *Tx_Trigger* specifies that instant in time when a message corresponding to a time slot shall be sent. A Tx_Trigger contains the following information:

- A reference to a message that shall be sent.
- Activation time mark, which is the column of the system matrix.
- Row of the system matrix measured in Cycle_Count.
- Repeat factor: Position of the same column in which it will be sent next. The value is a power of 2.

A vital parameter for TTCAN and the sending of messages is the *Tx_Enable* parameter. A message has to be sent within a specific time window after the start of its corresponding time slot. If the transmission would start too late, the message would overlap with the start of the next slot. To avoid this erroneous behavior, there exists the Tx_Enable parameter. It specifies a window from the start of the corresponding time slot (Tx_Trigger) within which the transmission of a message is allowed and safe. This window is specified as a number of nominal CAN bit times with a range from 1 to 16. If a node cannot start the transmission of a message within the Tx_Enable window, the transmission may not be started at all. In this case, the node has to wait for the next Tx_Enable to send the message.

The concept of the Rx_Trigger is related to the Tx_Trigger. It contains the same information as a Tx_Trigger, but specifies the point in time when a received message has to be completed and verified.

There are two additional triggers that are important in a TTCAN system: Tx_Ref_Trigger and Watch_Trigger. Tx_Ref_Trigger is a special Tx_Trigger that refers to reference messages and is only used in potential time masters. Whenever this trigger is reached, a potential time master tries to send a reference message. Within a strictly time-triggered TTCAN system (no Next_is_Gap bits) one Tx_Ref_Trigger is enough. If event-synchronized transmission of basic cycles is planed, a second Tx_Ref_Trigger is used. The first trigger is used for the periodic transmission of reference messages, where the second restarts the first one whenever the event that was flagged with the Next_is_Gap bit did not occur or was missed.

Watch_Trigger is also used by nodes that are not potential time masters. These nodes will be called slave nodes in the following. Watch_Trigger is reached if there is no reference message on the bus longer than a specified threshold. This could be the case on a disturbed bus. Like the Tx_Ref_Trigger, there shall be two Watch_Triggers to support event-synchronized communication. When a Watch_Trigger is reached, the application shall be informed about the erroneous behavior in the system. During start-up, the Watch_Trigger is disabled until the first successful transmission or reception of a reference message.

41.2.4 Restart, Reintegration, Integration

TTCAN differentiates between the start-up behavior of a time master and its fault-tolerance capabilities. This section describes the start-up, whereas Section 41.2.6 gives an overview about fault tolerance in TTCAN.

At system start-up (or a reset), all potential time masters try to become the current time master and all of them try to send a reference message. There are two parameters that influence the decision which potential time master becomes the current time master. The potential time master with the highest priority shall use the CAN ID with the highest priority (in relation to the other potential time master). As specified by the TTCAN protocol, CAN IDs of time masters only differ in the three least significant bits.

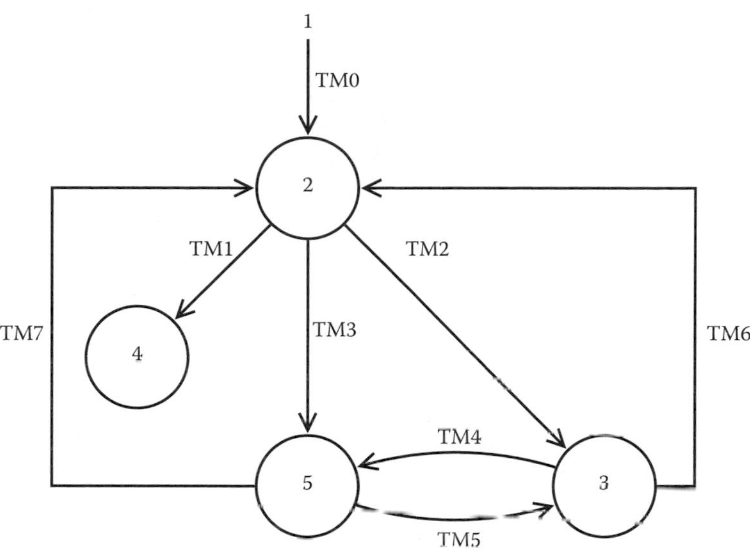

FIGURE 41.9 Master/slave relation in TTCAN. (From International Standardization Organisation (ISO), Road vehicles—Controller area network (CAN)—Part 4: Time-triggered communication, ISO 11898-4, 1993.)

The rest of the bits reserved for the CAN ID are equal for all potential time masters. Additionally, there is a parameter that specifies how long a potential time master is idle before it tries to send a reference message. The potential time master with the highest priority has the shortest delay before it tries to send its initial reference message.

Both of these parameters, the priority of a potential time master and the time a potential time master waits for its initial transmission, influence the decision which potential time master will become the current time master and how long it takes to establish a new time master.

Figure 41.9 shows a state machine of the start-up protocol and the establishment of a new time master. State 1 is the initial state, which is reached for example after a hardware reset. Transition 0 is always taken. State 2 is reached by slaves as well as potential time masters. If the node is a slave, and it receives a reference message on the bus, transition 1 is taken and the node is in state 4, the slave state. If the node is in state 2 and it is a potential time master, two transitions may be taken. First of all, a potential time master will try to send its own reference message. The best case for the potential time master is that it reads its own reference message on the bus. In this case, the potential time master takes transition 3 and becomes the new time master (state 5). On the other hand, a potential time master in state 2 can read a reference message that contains an ID that is not equal to its own. In this case, the potential time master takes transition 2 and becomes a *backup time master* (state 3). From this state a potential time master can become the current time master with transition 4 whenever it reads its own ID in the reference message. This can happen if the prior time master fails and a backup time master can now successfully transmit its own reference message. However, a current time master can also be degraded to a backup time master. This happens if it is the current time master (state 5) and reads a reference message with a higher priority on the bus. In case of an error, the time master as well as backup time masters can be reset to state 2.

After the initialization, but before the synchronization has finished, each node refers to its local time as the global time of the system. The parameter Local_Offset is initially set to zero. The actual time master tries to establish its view of the global time as the global time in the TTCAN system. Therefore, it sends its own Global_Sync_Mark as the Master_Ref_Mark in its reference message. An important fact is that a potential time master that becomes the actual time master (e.g., when the old time master fails) keeps its value for the Local_Offset. This is used to avoid discontinuities of the global time.

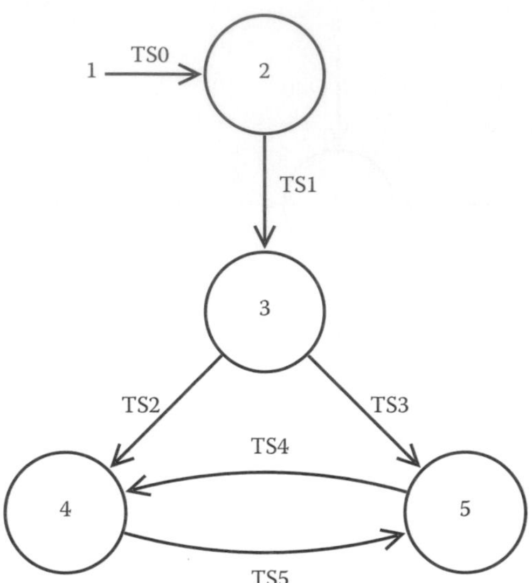

FIGURE 41.10 Synchronization in TTCAN. (From International Standardization Organisation (ISO), Road vehicles—Controller area network (CAN)—Part 4: Time-triggered communication, ISO 11898-4, 1993.)

The synchronization of a node can also be described with the help of a state machine (refer to Figure 41.10). A node is in state 1 after a hardware reset. From this state transition 0 always occurs. In state 2, the node is not yet synchronized (Sync_Off). After leaving an internal configuration mode and setting Cycle_Time to zero, the node is in state 3. From state 3, the node can either take a transition 2 to state 4, where it is In_Gap, or transition 3 to be in schedule (In_Schedule). For both transitions, TS2 and TS3, the node has to receive two successive reference messages where the last of the two did not contain a Disc_Bit. For TS2 the last reference message has to contain a set Next_is_Gap bit, whereas for TS3 Next_is_Gap has to be unset. A node can leave the In_Schedule state with transition TS4 whenever a reference message contains a set Next_is_Gap bit. When a node is in the In_Gap state (state 4), it can reach a synchronized state (state 5) whenever it receives a reference message where the Next_Is_Gap bit is unset.

41.2.5 Diagnostic Services

TTCAN provides simple means for diagnosing the communication on the CAN bus. A *Watch_Trigger* is used to monitor the reception of reference messages. TTCAN uses two different Watch_Triggers, one for the reception of periodic messages and one for event-synchronized messages. If there is no reference message for a specified time, the Watch_Trigger is activated and an error handling procedure is started. Additionally, the application shall be notified. During initialization, the Watch_Trigger is disabled and gets activated upon a successful reception of a reference message.

As stated before, TTCAN provides an additional layer above the conventional CAN protocol. All means of CAN bus error handling support the error handling of TTCAN.

41.2.5.1 Error Counters in CAN

CAN provides a receive error counter as well as a transmit error counter that can be used also in TTCAN. The values of these counters are vital for the CAN node because they define the error state of the node. CAN defines the following three error states for a node:

1. Error active: The node is allowed to send active error frames
2. Error passive: The node sends passive error frames
3. Bus off: The node does not take part in bus communication

According to the state of the error counter of the CAN node, it can send active or passive error frames. When a node is highly disturbed, and therefore has a high error count, it is only allowed to send passive error frames. Within these frames, the node is only allowed to send recessive error flags. Therefore, a node in passive error state cannot disrupt the communication of the bus, because its recessive error flags cannot disturb the communication of working nodes.

When the corresponding error counter has a value between 0 and 96, the node is in *error active state*. If the error counter exceeds 96, it reaches a *warning* limit, where the node sets an error flag and generates an interrupt. Between 97 and 127, the bus is heavily disturbed and the node is still in error active state. Between 128 and 255, the node is in *error passive* state and above 255 the node reaches *bus off* state.

41.2.5.2 Error Frames in CAN

Whenever a node detects an error on the bus, it sends an error frame on the bus. These error frames contain a 6-bit error flag. Additional 0–6 bits are reserved for the superposition of further error flags. After the error flags, an error frame contains 8 bits used for the error delimiter. The delimiter contains 8 recessive bits to allow a restart of the communication after the superposition of error flags. After the delimiter, 3 bits are reserved for an interframe space. The fields of an error frame are shown in Figure 41.11. Superposition of error flags occurs in the following situation: A receiver detects an error on the bus and starts to send an error frame. Within this error frame, the other receivers will detect a violation in the bit stuffing of CAN. Therefore, the other receivers will send error frames themselves, and their first 6 bits will lead to a superposition of error flags in the first error frame. Figure 41.12 shows the superposition of error flags. In this case, receiver 1 detected the error on the bus first and started to send its error frame. After the first 6 bits, receiver 2 also detected an error and started to send its error frame with its error flags. This shows one important mechanism of CAN, the *globalization of local errors*. Whenever a node detects an error on the bus, it will try to establish a global error state on the bus.

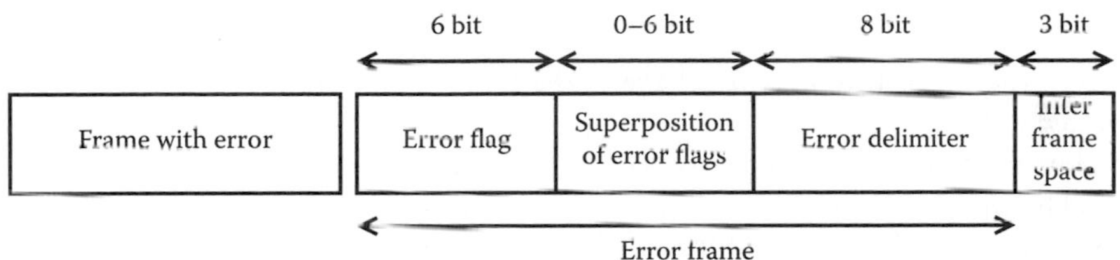

FIGURE 41.11 Error frame in CAN. (www.softing.com. CAN, CANOpen, DeviceNet. Website, August 2010.)

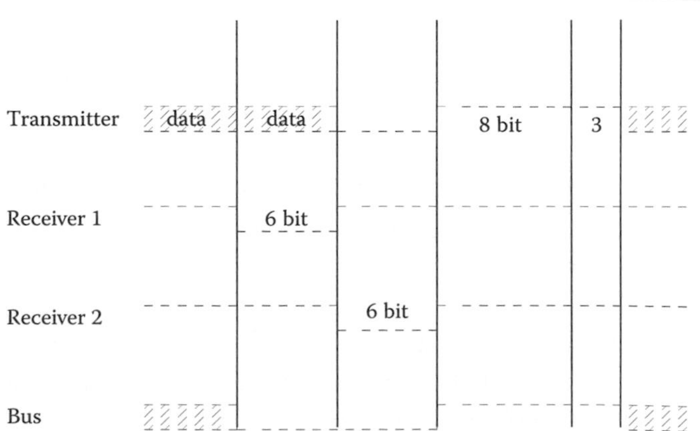

FIGURE 41.12 Error handling in CAN. (www.softing.com. CAN, CANOpen, DeviceNet. Website, August 2010.)

41.2.5.3 Bit Stuffing in CAN

Bit stuffing is used in CAN whenever 5 consecutive bits of the same polarity occur in the CAN bit stream. In this case, one bit of the opposite polarity will be inserted. In CAN, there are fields with a fixed bit pattern, which are not part of the bit stuffing. In CAN, the fields for start of frame (SOF), the arbitration field, the control field, the data field, and the CRC field are part of the bit stuffing area. The bit stuffing property of CAN is used to flag errors on the bus, because whenever 6 consecutive bits of the same polarity are read on the bus, all nodes will interpret this as an error.

41.2.6 Error Detection and Fault Isolation

Fault isolation and tolerance is of utmost importance whenever a network is used in a safety-relevant context. For traditional CAN, it is possible to retransmit a message if an error occurs during the transmission of the message. This simple mechanism cannot be used for TTCAN because it has to follow the schedule for its messages. Therefore, in TTCAN, the automatic retransmission of messages is turned off.

The TTCAN standard itself provides error detection mechanisms and a classification of error severity. It also states that active fault confinement shall be left to a higher layer or to the application. The node shall provide error detection and fail-silence [Int93b].

TTCAN defines the following errors and their severity. Note that the standard does not specify where these bits should be saved (i.e., this could be in registers of a node).

- Scheduling_Error_1 (S1): An error of this kind is flagged whenever the difference between the highest message status count (MSC) and the lowest MSC of all messages is larger than 2. Additionally a scheduling error is flagged if the MSC of an exclusive receive message object is 7. The MSC is used for periodic messages in exclusive time windows. The MSC has a range from 0 to 7. The MSC is updated at the Rx_Trigger event that corresponds to that specific message. If the message is successfully received, the MSC is decremented by 1, else it is incremented by 1.
- Tx_Underflow (S1): This bit is set whenever the count for transmitted messages is lower than its expected value.
- Scheduling_Error_2 (S2): Whenever an MSC of a transmit message object reaches 7, this error bit will be set.
- Tx_Overflow (S2): This flag will be set whenever the expected maximum value (Expected_Tx_Trigger) is reached and there are still Tx_Trigger events.
- Application_Watchdog (S3): The application watchdog gets reset by the application itself via a Host_Alive_Sign. If the application fails to set this sign, an error will be flagged.
- CAN_Bus_Off (S3): This error is flagged whenever the node went off the bus due to CAN-specific errors.
- Config_Error (S3): Slots in TTCAN have a specific time window where the transmission of the data has to start. If the actual point in time exceeds this specified limit, a Config_Error will be flagged.
- Watch_Trigger_Reached (S3): This error flag will be set whenever a reference message is missing.

The TTCAN standard specifies the following severity classes:

- S0: No error.
- S1: Warning—The application will be informed; the reaction is application specific.
- S2: Error—The application will be informed, and all transmissions in exclusive and arbitrating windows will be disabled. Potential time masters are still allowed to transmit reference messages with the maximal allowed offset.
- S3: Severe—The application will be informed, and all bus activities will be stopped. In this case an update of the configuration is required.

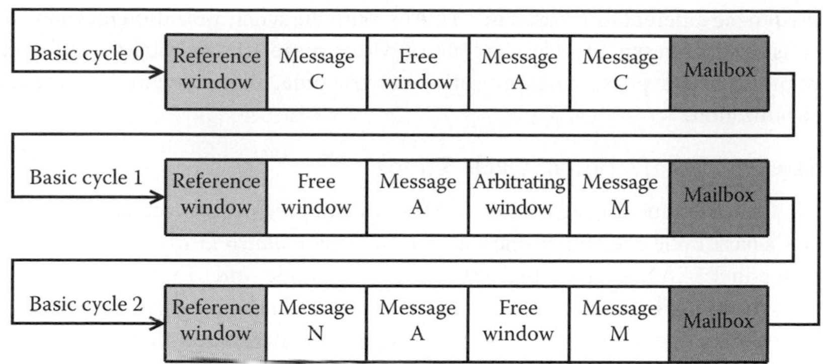

FIGURE 41.13 System matrix with mailbox.

In addition to the fault handling specified by the TTCAN standard itself, there are several approaches as to how to add fault tolerance to TTCAN. One simple approach is described in [BBRN04], where each frame is sent twice.

A different approach is to add a *mailbox system* to TTCAN [MA05]. This is done by reserving an exclusive window, the mailbox, at the end of each basic cycle to retransmit messages that could not be sent in their dedicated slot, as shown in Figure 41.13. The basic idea is to add a software supervisor layer above the TTCAN controller that manages additional send and receive queues. Whenever a message cannot be transmitted in its dedicated slot, this message is scheduled to be transmitted in the mailbox slot at the end of the basic cycle. This solution assumes that bit error rates occur in the magnitude of 10^{-7}. Therefore, a basic cycle would have to be 2×10^7 messages long to corrupt two messages within one basic cycle. This would require a basic cycle of about 2.2 h. The authors conclude that one mailbox is enough to provide sufficient fault tolerance. Whenever an error occurs, the supervisor sends an error frame on the bus. After that, it enqueues the failed message to the send queue and transmits the message in the mailbox slot. At the start of the system matrix (and at every iteration), the send queue gets flushed to prevent queue overflows.

TTCAN provides fault tolerance by combining two (or more) TTCAN buses [MFH+02]. A system of two TTCAN buses is considered a *coupled TTCAN pair* if there is at least one *gateway* node connected to both TTCAN buses. A system of TTCAN buses is considered as coupled if there exists a path from the start bus via coupled nodes to the end bus. More formally [MFH+02]: bus_a and bus_b are TTCAN coupled if there exists a sequence $(bus_1,\ldots,bus_n)$ of TTCAN buses with $bus_1 = bus_a$ and $bus_n = bus_b$ where (bus_i,bus_{i+1}) are coupled TTCAN pairs $\forall i = 1, \ldots, n-1$. Now a fault-tolerant TTCAN network is a system of TTCAN buses where every two of them are TTCAN-coupled. Figure 41.14 shows a coupled TTCAN system where bus_a and bus_c are coupled.

The main problem of providing a coupled, and therefore fault-tolerant, TTCAN system is the synchronization between the different buses. [MFH+02] states three different problems of synchronization,

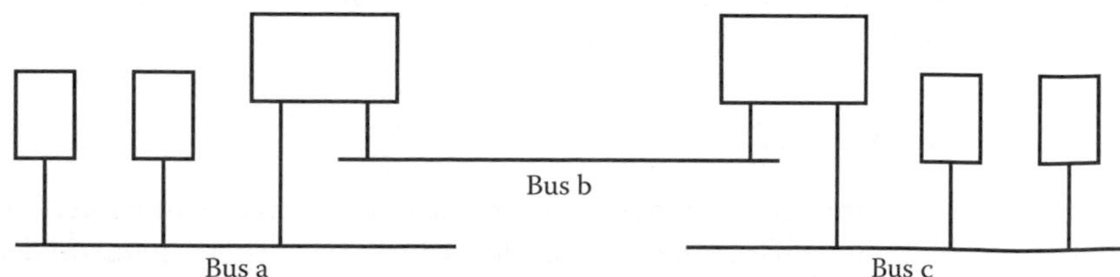

FIGURE 41.14 Coupled TTCAN buses. (From Müller, B. et al., Fault tolerant ttcan networks, Technical report, Robert Bosch GmbH, 2002.)

which are related to the different times used in TTCAN. With the synchronization mechanisms described in the paper, it is possible to synchronize a system of two or more TTCAN buses and therefore provide fault tolerance in the system: phase synchronization of cycle time, phase synchronization of global time, and rate synchronization.

41.2.6.1 Phase Synchronization of Cycle Time

To recapitulate, the cycle time is used to measure the time between two successive reference messages. At every start of a basic cycle a reference mark is saved and *cycle time = local time − local reference mark*. All points in time in TTCAN are specified according to the cycle time. The simplest solution to phase synchronize two (or more) buses is to use one bus as a *master bus* and synchronize the second bus to the first one. The situation can further be simplified if the *gateway* node—the node connected to both buses—is a time master for the second bus. To initiate the desired phase relation, the time master has to set its Next_Is_Gap bit and stall the next reference message until the desired offset. This is the simplest method to solve this issue. If the gateway node is not a time master on the second bus, a simple metaprotocol can be used. In this case, the gateway node transmits the desired phase shift in a TTCAN message to its time master and the time master will take care of the rest. This method of *master buses* can be extended to n TTCAN buses, where the first bus is the master bus and every bus i for $i = 2, \ldots, n$ synchronizes to bus $i - 1$.

41.2.6.2 Phase Synchronization of Global Time

To synchronize the phase of the global time of two time masters on distinct buses, the same methods as for the phase synchronization of the cycle time can be used. The gateway node calculates the desired phase for the slave global time and informs the master node. As well as for the synchronization of the cycle time this can be simplified if the gateway node is the time master of the slave bus. The new global time can then be announced to the slave bus. This is done by setting the Disc_Bit to flag a discontinuity in the global time.

41.2.6.3 Rate Synchronization

To synchronize the rates of two TTCAN buses, first the rates of the two (or more) buses have to be measured. One option is to measure the same physical time interval on both buses. From this measurement the ratio for the slave bus can be calculated. This value will then be sent to the local time master. This time master will then update its TUR value and within the next basic cycle the TTCAN bus will be rate-synchronized.

For details about the different synchronization methods please refer to [MFH+02].

In addition to the presented TTCAN specific methods of fault isolation, classical CAN error detection mechanisms of CRC-checksums can be used in TTCAN networks.

41.2.6.4 CRC in CAN

CAN uses a 15-bit cyclic redundancy check for its messages. The generator polynomial used for CAN is $x^{15} + x^{14} + x^{10} + x^{8} + x^{7} + x^{4} + x^{3} + 1$. The polynomial has a guaranteed minimal hamming distance of 6, which means that up to 5 random bit failures can be detected. Due to bit stuffing, there can be the situation that only two disturbed bits lead to a situation that cannot be detected by the CRC check. For an example refer to Table 41.1.

41.2.7 Configuration Services

The TTCAN standard specifies three main configuration interfaces, where each of these interfaces is divided into subinterfaces that allow the configuration of protocol settings. The TTCAN standard demands that all interfaces shall be lockable against random changes, whereas reading settings is always possible. The following paragraphs provide an overview about the configuration interfaces and their most important

TABLE 41.1 CRC—Hamming Distance Shortfall

Original	1000	0001	1011	1100	0100	0011	01
Stuffed	1000	0010	1101	1110	0010	0001	101
Disturbed	1000	0110	1101	1110	0010	0000	101
De-Stuffed	1000	0110	1101	1110	0010	0000	01
CRC	0000	0111	0110	0010	0110	0011	00

Source: Müller, B. et al., Fault tolerant TTCAN networks, Technical report, Robert Bosch GmbH, 2002.

variables. For a complete list, refer to the corresponding section of the TTCAN standard [Int93b]. Note that the standard does not specify the location of these interfaces or how to access these interfaces. One practical solution is to store the values into registers and provide an API to access these interfaces.

41.2.7.1 General Configuration Interfaces

As the name states, this interface is used to set general protocol parameters for a TTCAN node. Settings include the configuration of the TUR, the operation mode of the node (configuration, CAN communication, time-triggered communication, and event-synchronized time-triggered communication), its role as a slave or a potential master, if external clock synchronization shall be used, which interrupt sources shall be masked, and an 8-bit value for the application watchdog limit.

The main role of the interface is the configuration of the system matrix. With this interface, the system engineer can specify message objects, and the corresponding triggers. The triggers contain a reference to a valid message, a time mark when it should be activated, the position in the transmission column with its first activation, and its repeat factor. The standard states that the configuration can be read and written during the initialization phase, and shall be locked during the time-triggered communication activity.

41.2.7.2 Application Interfaces

The application interfaces provide information about the current state of the node to the application.

Accessible information includes the priority (3 bits) of the current time master, the master state, the current value of the global time in NTUs, the Cycle_Time, the number of the current basic cycle, and the actual value of the TUR (TUR_Actual), which is read-only. Additionally, the interface is used to configure operational and error detection interrupt sources. Operational interrupt sources are, among others, the start of a basic cycle, the start of the system matrix, an interrupt that occurs if the global time wraps, or a change in the local master state. Error detection interrupt sources are, for example, over- and underflows of the Tx-Counter, the application watchdog, or a CAN_Bus_Off interrupt.

41.2.7.3 Optional Interfaces

Optional interfaces are not required to comply with the TTCAN standard. Examples are interfaces that signal that the node is out of synchronization, or that the Disc_Bit should not be used and to prevent a discontinuity in the global time. Additionally, there could be an interface to specify the maximum synchronization deviation.

41.3 Flexible Time-Triggered CAN

Flexible Time-Triggered CAN (FTT-CAN) [PA00] is a CAN-based master/slave protocol for building a predictable time-triggered communication service on top of CAN, while also allowing for slots with conventional event-triggered CAN communication.

FTT-CAN distinguishes two types of traffic, namely, synchronous and asynchronous traffic. The synchronous traffic consists of periodic time-triggered messages. Time windows not used by these periodic messages are available for the asynchronous traffic. Contention of the asynchronous traffic is resolved by the arbitration mechanism of CAN. FTT-CAN structures communication into communication rounds called *elementary cycles*. The master node initiates an elementary cycle by generating a periodic message used to synchronize all other nodes. This periodic message is referred to as *elementary cycle trigger message*. In addition to marking the beginning of a new elementary cycle, the trigger message conveys the identification of synchronous messages that are exchanged within this elementary cycle. Every node participating in synchronous communication is equipped with a local table that contains information about the synchronous messages transmitted and received during the elementary cycle.

FTT-CAN supports two modes. In *controlled mode*, a node may only send an asynchronous message, if the remaining time window before the next synchronous message has a sufficient length for preventing any interference of synchronous and asynchronous messages. The major disadvantage of controlled mode is the introduction of bus idle-time at the end of asynchronous windows, although messages are waiting for transmission. Hence, controlled mode decreases the network utilization. In *uncontrolled mode* nodes can start with the transmission of asynchronous messages at arbitrary points in time. Through assigning higher priorities to synchronous messages, it is ensured that synchronous messages always win in the arbitration process. Nevertheless, the nonpreemptive nature of CAN results in transmission jitter of synchronous messages, in case an asynchronous message is being transmitted when a synchronous message transmission is scheduled.

The advantages of FTT-CAN are as follows:

- *Low jitter mode*: The controlled mode of FTT-CAN ensures predictable communication with low jitter in the absence of faults.
- *Software upgrades for existing systems*: FTT-CAN can be installed in existing systems without involving modifications to hardware. However, in order to reach a fault-tolerant FTT-CAN [FPAF02] communication service, dedicated hardware (bus guardians) are required.
- *Coexistence with conventional CAN*: In uncontrolled mode, synchronous communication can coexist with standard CAN communication. Hence, no changes to the CAN protocol are required in legacy nodes and nodes with application software that does not require deterministic message exchanges.

A major drawback of FTT-CAN is the mutual exclusion of low jitter and support for legacy nodes. The uncontrolled mode of operation offers support for the integration of legacy CAN nodes with an unmodified CAN protocol stack. However, the support for legacy nodes comes at the price of increased transmission jitter due to the inability of CAN to preempt asynchronous messages after the transmission has been started. The controlled mode offers lower jitter, but it restricts the transmission of asynchronous messages to asynchronous windows. At the end of the asynchronous window, transmissions must not be started if an intersection with the synchronous window could take place. These restrictions with respect to the temporal behavior of message transmissions are not part of the CAN protocol specification and exceed the capabilities of standard CAN controllers. Hence, either legacy software has to be modified or CAN communication controllers must be replaced in order to support the controlled mode in existing systems.

41.4 Commercial or Prototypical Components

TTCAN controllers are available both as IP cores and as part of microcontrollers. For example, the TTCAN IP module is available from Bosch [Bos06]. This IP module is described in VHDL and can be integrated as a stand-alone device or as part of an ASIC. The IP module realizes a state machine for the ISO 11898-4 time-triggered communication. The state machine uses the reference messages of the

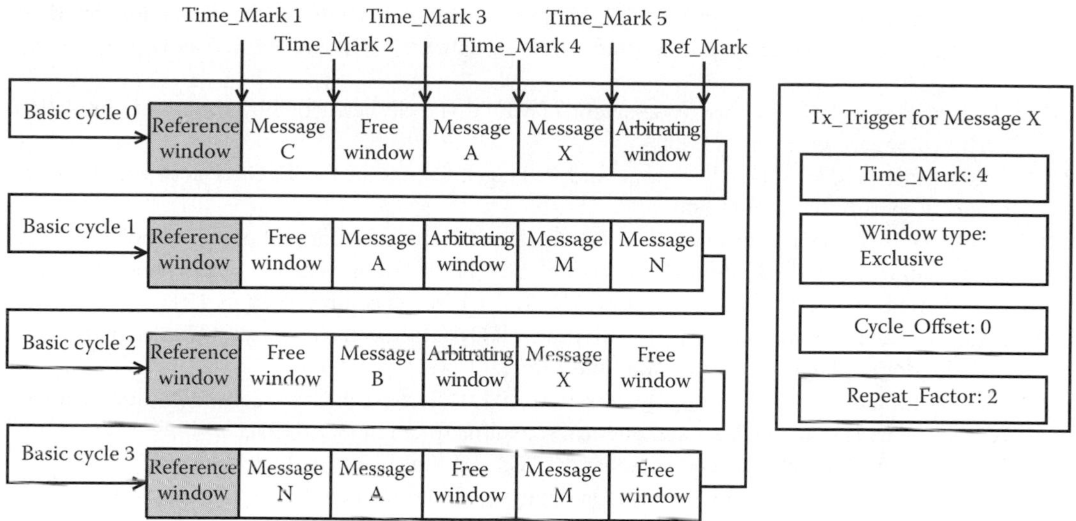

CAN bus as a synchronizing event to establish a global time base and perform message transmissions according to the a priori known communication schedule.

Another IP module supporting TTCAN is the MultiCAN IP of Infineon Technologies [Kel03], which is deployed in several Infineon devices (e.g., TC1130 [Inf03]). Like the TTCAN IP module, the MultiCAN IP performs clock synchronization and offers a scheduler to trigger the predefined message transmissions.

In addition, microcontrollers are available that support the implementation of TTCAN. For example, Atmel provides microcontrollers (e.g., T89C51CC01, T89C51CC02 [Atm08a,Atm08b]) that permit the implementation of TTCAN using higher software layers. In particular, these microcontrollers offer a single-shot transmission mode in order to prevent automatic retransmission upon errors. This transmission mode is required for the implementation of TTCAN.

In academics, the GAST (General Application Development Boards for Safety Critical Time-Triggered Systems) project* provides a controller board that features a TTCAN controller card. The setup was used in a master thesis to evaluate the performance of TTCAN [FU06].

References

Atm08a T89C51CC01—Enhanced 8-bit microcontroller with can controller and flash memory. Technical report, Atmel Corporation, 2008.

Atm08b T89C51CC02—Enhanced 8-bit microcontroller with CAN controller and flash. Technical report, Atmel Corporation, 2008.

BBRN04 I. Broster, A. Burns, and G. Rodriguez-Navas. Comparing real-time communication under electromagnetic interference, 2004.

Bos06 Automotive electronics—Product information TTCAN IP module. Technical report, Bosch, 2006.

CAN, CANOpen, DeviceNet. www10 www.softing.com. Accessed August 2010.

ES09 J. Erjavec and R. Scharff. *Automotive Technology: A Systems Approach*, 5th edition. Delmar Cengage Learning, New York, 2009.

FPAF02 J. Ferreira, P. Pedreiras, L. Almeida, and J. Fonseca. Achieving fault tolerance in FTT-CAN. In *Proceedings of the Fourth IEEE International Workshop on Factory Communication Systems*, Vasteras, Sweden, 2002.

* http://www.chl.chalmers.se/gast/.

FU06 M. Fernström and D. Ungerdahl. TTCAN reference application—An investigation on time-triggered network performance. Master's thesis, Chalmers University of Technology, Göteborg, Sweden, 2006.

HB08 B. Heppner and H. Brauner. Assessment of whole vehicle behaviour by means of simulation. Technical report, Daimler, AG, 2008.

HMFH03 F. Hartwich, B. Müller, T. Führer, and R. Hugel. Timing in the TTCAN network. Technical report, Robert Bosch GmbH, 2003.

Inf03 TC1130—32-bit single-chip microcontroller data sheet. Technical report, Infineon Technologies, 2003.

Int93a International Standardization Organisation, ISO 11898. Road vehicles—Interchange of digital information—Controller area network (CAN) for high-speed communication, 1993.

Int93b International Standardization Organisation (ISO). Road vehicles—Controller area network (CAN)—Part 4: Time-triggered communication, ISO 11898-4, 1993.

Int93c International Standardization Organisation (ISO). Road vehicles—Controller area network (CAN)—Part 1: Data link layer and physical signalling, ISO 11898-1, 1993.

Kel03 U. Kelling. MultiCAN—A step to CAN and TTCAN. Infineon Technologies, 2003.

MA05 S.M. Mahmud and A. Arora. Performance analysis of fault tolerant TTCAN system, 2005.

MFH+02 B. Müller, T. Führer, F. Hartwich, R. Hugel, and H. Weiler. Fault tolerant TTCAN networks. Technical report, Robert Bosch GmbH, 2002.

PA00 P. Pedreiras and L. Almeida. Combining event-triggered and time-triggered traffic in FTT-CAN: Analysis of the asynchronous messaging system. In *Proceedings of Third IEEE International Workshop on Factory Communication Systems*, September 2000.

TH94 K. Tindell and H. Hansson. Babbling idiots, the dual-priority protocol, and smart can controllers. In *Proceedings of the First International CAN Conference*, 1994.

42

Time-Triggered Ethernet

Wilfried Steiner
TTTech Computertechnik AG

Michael Paulitsch
Airbus Group Innovations

42.1 General Overview 42-1
42.2 TTEthernet Traffic Classes and Traffic Integration 42-3
Description of the Traffic Classes • Integration of the Traffic Classes
42.3 TTEthernet Synchronization Protocols 42-6
Start-Up and Integration • Clock Synchronization • Clique Detection and Restart
42.4 TTEthernet Example Configurations and Implementations 42-7
Configurations • Implementations
42.5 TTEthernet in Orion 42-11
References 42-17

42.1 General Overview

Ethernet is the dominant network standard for local area networks (LAN). While originally designed for classic office applications, the growing communication demands in real-time systems led to adapting Ethernet for time-critical applications. Today, we can find Ethernet variants everywhere: in industrial applications (EtherCat, Ethernet Powerlink, ProfiNet, Ethernet IP), in aerospace applications (ARINC 664-p7, ASCB/D), in military naval applications (Gigabit Ethernet Data Multiplex System), in consumer audio/video systems (AVB), as well as in datacenters and cloud computing (DCB). All of these Ethernet variants aim to achieve a certain degree of quality of service (QoS) such that end-to-end transmission guarantees can be ensured. In this multitude of Ethernet-variants, TTEthernet introduces the deterministic time-triggered communication paradigm in an Ethernet flavor, which allows the use of standard Ethernet in safety-critical systems and systems with applications of mixed-criticality.

TTEthernet is the continued industry development of the academic TT-Ethernet research [KAGS05], conducted within a joint research project* between the Vienna University of Technology and TTTech Computertechnik AG. The main objective of the academic project has been the integration of time-triggered with event-triggered messages on a single physical Ethernet network: as event-triggered messages are not synchronized, they typically result in conflicts with time-triggered messages at the outgoing ports in a network switch. The solution proposed within the academic TT-Ethernet project has been a preemptive switch. This TT-Ethernet switch [Ste06] identifies the reception of a time-triggered message based on an identifier within the message and preempts all event-triggered messages under transmission to free the outgoing ports for the time-triggered message. The merit of this solution is two-fold: first, the switch latency for time-triggered messages is constant with negligible error and, second, the switch can be kept almost free from additional configuration data.

The academic TT-Ethernet research has been continued within a joint industrial development between TTTech and Honeywell. Here, the objectives have been extended toward scalable fault tolerance and a finer

* The FIT-IT project TT-Ethernet has been funded by the Austrian ministry for transport, innovation, and technology (BM-VIT) under contract No 808197.

classification of event-triggered messages into rate-constrained and best-effort traffic classes. The main driver for these additional objectives has been a shift in the target application domain toward safety-critical applications, in particular civil aerospace. Fault tolerance is a standard requirement in these applications and the rate-constrained traffic class has been intended as a compatible communication mode to ARINC 664 part 7, which specifies an unsynchronized real-time Ethernet variant for airborne applications. Another key differentiator to the academic version is a nonpreemptive integration mode of event-triggered and time-triggered messages to minimize the number of damaged Ethernet frames on the network, thereby easing diagnosis.

More recently, TTEthernet has been extended toward compatibility with the AVB (audio/video bridging) standard and TSN (time-sensitive networks) prestandard, which are the IEEE 802.1 real-time and robustness evolutions of switched networks, and in particular Ethernet, (see chapter "IEEE 802.1 Audio/Video Bridging and Time-Sensitive Networking" for more details on AVB and TSN). Analogously to ARINC 664 part 7, AVB traffic is available in TTEthernet as a particular flavor of rate-constraint traffic. TSN is currently standardizing a simple form of time-triggered communication (IEEE 802.1Qbv) and upcoming TTEthernet products will also support this TSN version of time-triggered communication. These industrial developments have led to Time-Triggered Ethernet in its current form, which is called TTEthernet and will be discussed in more detail in this chapter.

TTEthernet is intended as cross-industry communication infrastructure: originally designed according to aerospace standards, it is scalable in several directions. As different industrial areas define and use different variants of real-time Ethernet already, the nature of the TTEthernet technology is one of a set of services added to existing standards rather than their replacement. The TTEthernet services have been standardized by the SAE and the SAE AS6802 standard has been published by the SAE in November 2011.

A TTEthernet network consists of end systems and switches and bidirectional communication links to connect these devices to each other. Furthermore, an end system and a switch may be integrated into a single device. Figure 42.1 depicts an example TTEthernet network consisting of eight end systems (1.8) and two redundant communication channels. Channels 1 and 2 consist of the three switches each (channel 1: B,C,E; channel 2: A,D,F). Current TTEthernet implementations allow up to three channels and channels are not required to have the same number of switches each.

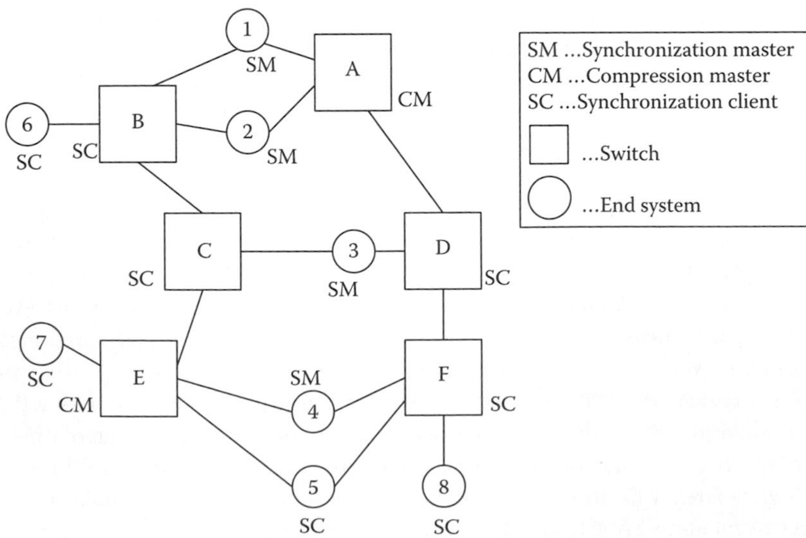

FIGURE 42.1 Example TTEthernet network consisting of two channels each consisting of three switches and eight end systems.

42.2 TTEthernet Traffic Classes and Traffic Integration

TTEthernet is fully compliant with the Ethernet frame format as standardized in IEEE 802.3. An overview of the fields in an Ethernet frame is given in Figure 42.2: the first row lists the number of bytes for a given field, the second row the field's name.

An Ethernet frame starts with a preamble of seven octets followed by a start of frame delimiter (SOF) of one octet. For addressing of the frame, Ethernet specifies the destination address and the source address, each six octets long. Following the address fields, Ethernet IEEE 802.3 specifies two octets that are either used as a type (EtherType) or as a length field. By convention, a value from 64 to 1522 (decimal) of this field indicates its usage as length field, a value of 1536 (decimal) and higher means that the field is used to reflect an EtherType. An example EtherType is 0x0800, which defines the Ethernet frame to carry an Internet Protocol (IPv4) packet. The address fields together with the EtherType/Length field are commonly referred to as the MAC Header. The actual payload of an Ethernet frame has a size of 46–1500 octets. Ethernet specifies a 32-bit CRC called the frame check sequence (FCS), which follows the payload field. The minimum interframe gap (IFG) is 12 octets. As a result, the overall length of an Ethernet frame, as a sum of the fields discussed earlier, is between 84 and 1538 octets.

TTEthernet defines the term *traffic class* to differentiate communication modes. The prime communication mode of TTEthernet is, of course, the time-triggered traffic class (TT). Besides this mandatory traffic class, TTEthernet also supports the optional rate-constrained traffic class (RC) and the optional best-effort traffic class (BE). TTEthernet is an integrative communication protocol capable of communicating frames of these three traffic classes on the same physical network. In the following, we describe the traffic classes in more detail and use Figure 42.3 as running example, in which end systems 1 and 2

7B	1B	6B	6B	4B	2B	42B–1500B	4B	12B
Preamble	SOF	MAC destination	MAC source	802.1Q "VLAN" Tag	Ethertype/ length	Payload	FCS	IFG

FIGURE 42.2 Ethernet frame format.

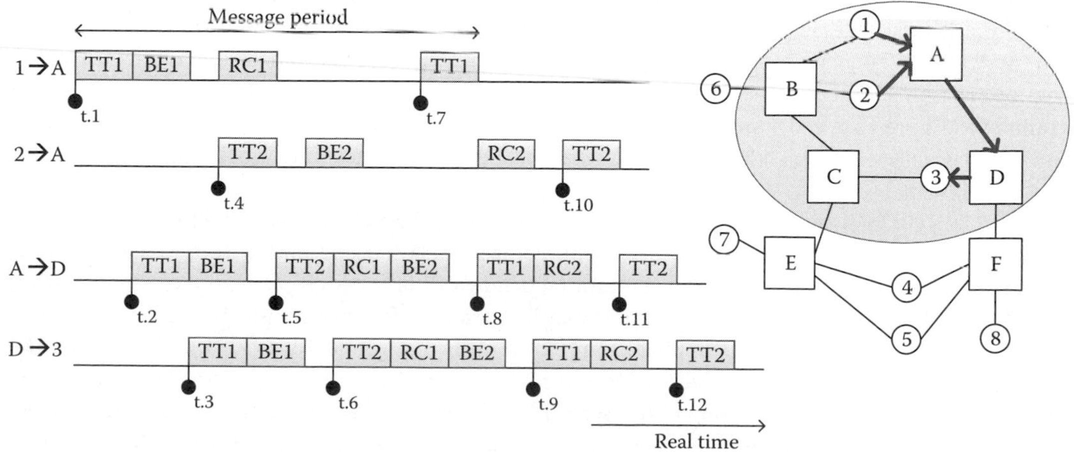

FIGURE 42.3 Example communication scenario in a TTEthernet network using all three traffic classes.

send TT, RC, and BE frames to end system 3 using switches A and D. The TT and RC traffic class also support multicast communication (not depicted).

42.2.1 Description of the Traffic Classes

A TTEthernet end system implements services to synchronize its local clock with the local clocks of other end systems and switches in the system. The end system can then send messages at off-line planned (or online assigned) points in this synchronized global time. These messages are said to be time-triggered. In the example of Figure 42.3, end systems 1 and 2 send TT messages. The points in time of the transmission and forwarding of the TT frames from end systems 1 and 2 to the switch A, from switch A to switch D, and from switch D to end system 3 are depicted by black circles labeled from t.1 to t.12. These points in time are the output of a configuration tool, the (off-line or online) scheduler, which will take care of arranging the TT frames in such a way that they will never be ready for transmission on the same outgoing port of a switch or end system at the same point in time. In this example, end system 1 and end system 2 send one TT message each (TT1 and TT2) with the same message period. However, other than in previous time-triggered communication protocols this is not a requirement in TTEthernet: indeed different TT messages may have different transmission periods.

Time-triggered (TT) messages are used when tight latency, jitter, and determinism are required. All TT messages are sent at predefined times. In cases where an end system decides not to use one of its assigned timed slots, for example, if there are no new data to be sent, the switch recognizes the inactivity of the sender and frees the bandwidth for the other traffic classes. TT messages are optimally suited for communication in distributed real-time systems with regular communication pattern.

Rate-constrained (RC) messages realize a communication paradigm that aims at establishing well-shaped dataflows: successive messages belonging to the same rate-constrained dataflow are guaranteed to be offset by a minimum duration as configured. RC messages are used when determinism and real-time requirements are less strict than provided by time-triggered communication. For RC messages, sufficient bandwidth must be allocated such that delays and temporal deviations have defined limits. In contrast to TT messages, RC messages are not sent with respect to a system-wide synchronized time base. Hence, different communication controllers may send RC messages at the same point in time to the same receiver. Consequently, the RC messages may queue up in the network switches, leading to increased transmission jitter and requiring increased buffer space. As the transmission rate of the RC messages is bound a priori and controlled in the network switches, an upper bound on the transmission latency can be calculated off-line and message loss is avoided. The rate-constrained communication paradigm is used in ARINC 664 part 7 and in similar form in IEEE 802.1 AVB. In the example of Figure 42.3, end systems 1 and 2 send an RC message each (RC1 and RC2). The end systems and switches will transmit and, respectively, forward the RC messages when no TT messages need to be communicated.

Best-Effort (BE) messages implement the classic switched Ethernet approach. There is no guarantee whether or when these messages can be transmitted, what delays occur, and if BE messages arrive at the recipient at all. BE messages use the remaining bandwidth of the network and have lower priority than TT and RC messages. However, BE traffic may be attractive, for example, during maintenance and configuration phases: as during such phases no critical traffic in form of TT or RC may be present, the whole network bandwidth is available for BE traffic without explicitly changing the network mode: RC messages will not be sent and bandwidth reserved for TT messages is automatically reclaimed by the switches. In the example of Figure 42.3, end systems 1 and 2 also send a BE message each (BE1 and BE2). As BE messages have lowest priority, they are communicated once; no TT and no RC message need to be transmitted.

The scheduling of TT messages is often perceived as a complex problem. However, recent studies show that even general purpose tools like SMT-solvers can be used out-of-the-box to schedule a network with about 100 frames. Furthermore, a bit more customized usage of the SMT-solver shows that networks up to 1000 messages can be configured within about half an hour [Ste10]. Other studies show that general purpose model checkers can be used to configure time-triggered systems [VSE09]. Commercial scheduling tools can schedule even bigger networks. The integration of TT messages with RC and BE messages has been studied, for example, in [Ste11,TSPS12].

42.2.2 Integration of the Traffic Classes

While the scheduler guarantees that TT frames are sufficiently scheduled apart from each other, such that no two TT frames will be ready for transmission on the same outgoing port of a switch (or end system), this cannot be guaranteed for mixed TT and RC/BE traffic. For example, in the communication scenario in Figure 42.3, the TT2 frame from end system 2 and the RC1 frame from end system 1 are transmitted from the end systems to switch A in a way such that the two frames become ready for relay by switch A at about the same point in time. For this case, TTEthernet, according the SAE AS6802 standard, defines a priority scheme that recommends using priorities in the following decreasing order: protocol control frames used for synchronization (see next section), TT, RC, BE. Each traffic class may define several priorities. When messages with equal priority become ready for transmit, these messages will be served in FIFO. When a high-priority message (**H**) is being served and a low-priority message (**L**) becomes ready, **L** will be queued. The third case, though, is of particular interest: there are three integration methods to resolve conflicts when a low-priority (**L**) message is already in transmission and a new high-priority (**H**) message becomes ready for transmission [SBH+09]. Such a scenario can occur if end system 1 sends RC1 a tick earlier than depicted in Figure 42.3, and RC1 becomes ready for forwarding in switch A before TT2. In general, the three traffic integration methods for such scenarios are preemption, timely block, and shuffling, and are depicted in Figure 42.4.

Using the preemption method, a switch would stop the ongoing transmission of a low-priority frame and intersperse the high-priority frame. Timely block takes advantage of the planned nature of the TT frames: the switch will know by configuration when to expect TT frames and will not start the transmission of any other traffic before either the expected TT frame has been serviced or the switch can be sure that the TT frame will not arrive (in which case the bandwidth is free for other communication). The last method is called shuffling and simply means that high-priority frames will be in queue and transmitted only after the ongoing transmission of the lower-priority frame has concluded. Current TTEthernet implementations support the timely block and the shuffling integration method, while the early academic TT-Ethernet versions have implemented preemption. Preemption has not been implemented in current TTEthernet implementations for several reasons, for example, preemption systematically generates incorrect Ethernet frames, which makes system health diagnosis difficult.

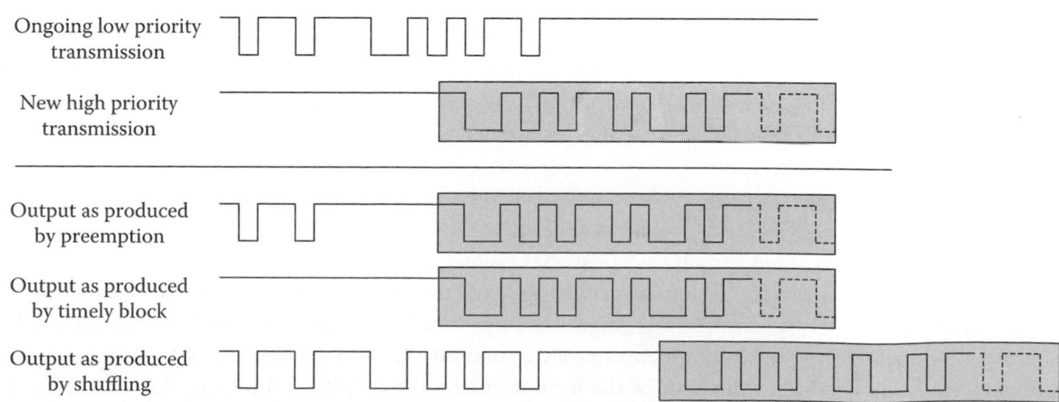

FIGURE 42.4 Integration methods for high-priority and low-priority traffic.

42.3 TTEthernet Synchronization Protocols

TTEthernet specifies and standardizes in SAE AS6802 several synchronization protocols that ensure close synchronization of the local clocks in the end systems and switches during all modes of operation of the network. These protocols have been designed to tolerate the failure of end systems and/or switches in the network. In particular, end systems are allowed to fail either arbitrarily (i.e., faulty end systems may generate sequences of arbitrary messages) or according to an inconsistent-omission failure mode (i.e., may drop messages and/or fail to transmit messages). Switches are always considered to fail according to an inconsistent-omission failure mode. If necessary, the inconsistent-omission failure mode can be guaranteed by dedicated structural design measures such as a self-checking pair design. We give an overview of these protocols in this section.

For all synchronization protocols, the TTEthernet protocol defines three different roles for synchronization: the *Synchronization Master* (SM), the *Synchronization Client* (SC), and the *Compression Master* (CM). The SM role is typically assigned to end systems, while switches act as CMs. As a minimum, one switch per channel is configured as CM. Both, switches and end systems, can be configured as SCs. In the example network in Figure 42.1, the end systems 1, 2, 3, and 4 are configured to be SM, while switches A and E act as CM, and all other devices operate as SC.

Synchronization demands the exchange of information between the devices. In the case of TTEthernet, this information is transported in standard Ethernet frames that are called protocol control frames (PCFs).

42.3.1 Start-Up and Integration

After initial power-on of the network, the start-up protocol ensures that the end systems and switches will synchronize their local clocks within a bounded duration. More precisely, during start-up typically two protocols are executed, the coldstart protocol and the integration protocol.

Coldstart is the actual protocol that establishes synchronized time. However, typically a newly powered-on end system or switch will not know initially, if the system as a whole is currently in start-up mode or already synchronized, and consequently will assume that a subsystem of the system is already synchronized. The component will, thus, aim to synchronize toward these subsystems. Hence, it will try to integrate into an operational synchronized time base. Only if this integration is not successful, then the component will assume that there does not exist an already synchronized subsystem and will undergo the coldstart protocol.

The TTEthernet start-up protocol is only executed between the SMs and the CMs. SCs will synchronize only after a synchronized time base has been established. The start-up protocol involves the exchange of a sequence of PCFs, in particular coldstart and coldstart acknowledgment messages.

Although the start-up protocol is rather small, its real-time behavior, the nature of distributed algorithms, and the behavior of faulty switches and/or end systems make the analysis of protocols like these rather complex. Therefore, model-checking techniques similar to [SRSP04] have been used during the development of the TTEthernet start-up protocol [Ste09].

42.3.2 Clock Synchronization

Once the start-up protocol has initially established the synchronized time base, the clock synchronization protocol is periodically executed to resynchronize the local clocks in the end systems and switches.

The TTEthernet clock synchronization algorithm operates in two steps. In the first step, all SMs concurrently send PCFs to all CMs at their same local time. As the local times of the SMs will be slightly different, for example, due to the drift of the local oscillators, the CMs will receive the PCFs also at slightly different points in time. The CMs will record these deviations and use them to calculate a new point in time at which the CMs will send a PCF (called the *compressed* PCF) back to the SMs and CMs.

The calculation follows a fault-tolerant midpoint selection, which guarantees that the impact of messages from faulty SMs will have only limited impact on the point in time when the CMs send their compressed PCFs. In the second step, all the SMs and SCs use the reception times of the compressed PCFs from the CMs to derive a new reference point in time for synchronization.

Note that in the preceding discussion, we ignored the transmission jitter that may occur at the network, in particular as synchronized messages are integrated with unsynchronized messages. TTEthernet uses the *permanence function* that allows us to do so. The permanence function is a simple means to translate transmission jitter almost entirely into latency. Hence, for the clock synchronization protocol and the TTEthernet synchronization protocols in general we can assume that the communication of PCFs between SMs, SCs, and CMs will be almost constant with known small jitter.

Again, formal methods have been used during the design of the clock synchronization protocol [SD11] as well as to demonstrate the correctness of the permanence function [SD10].

42.3.3 Clique Detection and Restart

The synchronized time base is essential to communicate the TT messages in TTEthernet, while the network is always able to communicate RC and BE messages, whether the components are synchronized or not. However, as the TT messages are likely to be critical messages in a given system TTEthernet incorporates synchronization protocols also for extremely unlikely situations such as system-wide communication blackouts and recovery when one or many components have lost synchronization.

The protocols that detect failures in the synchronization are commonly referred to as clique detection algorithms. TTEthernet implements three clique detection algorithms: the synchronous clique detection algorithm, the asynchronous clique detection algorithm, and the relative clique detection algorithm. Using the synchronous algorithm, a component will continually check as to how many SMs are currently synchronized and will report a clique if this number falls below a configured threshold. Analogously, using the asynchronous algorithm, a component will continually check if there are components in the system it is not synchronized to and will count how many such components there are. When the number of these components grows beyond a given threshold, then, again, a clique will be reported. The relative algorithm uses the information gathered in both, the synchronous and the asynchronous algorithms, and reports a clique if the number of SMs the component is synchronized to falls below the asynchronous components it monitored.

When a component detects a clique, it will restart or reintegrate. The interplay of the clique detection algorithms with the start-up, restart, and reintegration has been studied with model-checking techniques [Ste09].

42.4 TTEthernet Example Configurations and Implementations

TTEthernet has been designed as cross-industry communication infrastructure supporting the implementation of low-cost fieldbus-like networks up to ultrahighly dependable systems. In this section, we discuss two particular configurations of TTEthernet representing the extremes of the possible configuration space: a master-based configuration and a dual-fault-tolerant configuration. Furthermore, we present a system-of-systems configuration.

Current TTEthernet implementations are available in FPGA and in software. We will summarize their key characteristics in this chapter. ASIC implementations of TTEthernet will be developed as well.

42.4.1 Configurations

TTEthernet network configurations may differ with respect to the number of end systems and the number of switches, as well as their connecting topology. In particular for the synchronization algorithms, configurations are determined by the assignment of synchronization roles to TTEthernet devices. As a reminder, we have introduced three roles in the synchronization algorithms of TTEthernet: the synchronization master (SM), the compression master (CM), and the synchronization client (SC). We will discuss different configurations with regard to these aspects. More general characteristics of time-triggered networks include the quality of oscillators used in the TTEthernet devices, the wire speed and associated physical layer, the communication interface, and the activated fault-containment measures.

42.4.1.1 Master-Based Configuration

The most basic TTEthernet configuration consists of an SM and a CM, where both synchronization roles may be integrated into a single device. From this configuration, classic single-master multiple-slave networks as depicted in Figure 42.5 can be built. This example network consists of one SM, one switch configured as CM and six SCs (two switches and four end systems). End systems and switches are connected in a tree topology and there is no redundant channel present.

End system 1 acts as SM and sources PCFs that are relayed by switch A, which acts as CM. As there is only one SM, a compressed PCF generated by switch A will essentially carry the same information as the frame before compression. Hence, a simplified device (integrating SM and CM) that sources PCFs with a fixed period of the integration cycle is sufficient as steady synchronization source for the SCs in the network.

This single-master configuration is attractive for low-cost applications with a requirement of quality-of-service such as real-time performance. This configuration can be enhanced by safety mechanisms like the activation of the central guardian function in the switches or by the high-integrity design of key devices. Hence, given that the appropriate fault-isolation measures are in place, even this basic configuration is appropriate for fail-safe applications (applications that can enter a safe state upon failure in which no protocol operation is required).

Availability, on the other hand, is limited: the failure of an arbitrary number of SCs will not affect the services as provided by TTEthernet. However, as in all single-master-based systems, the loss of the SM or CM means a loss of the synchronization source and so of the TTEthernet protocol services. Availability requires the implementation of a sufficient degree of redundancy. The RC and BE traffic classes are not affected by the failure of the SM and/or CM as they do not require a synchronized time

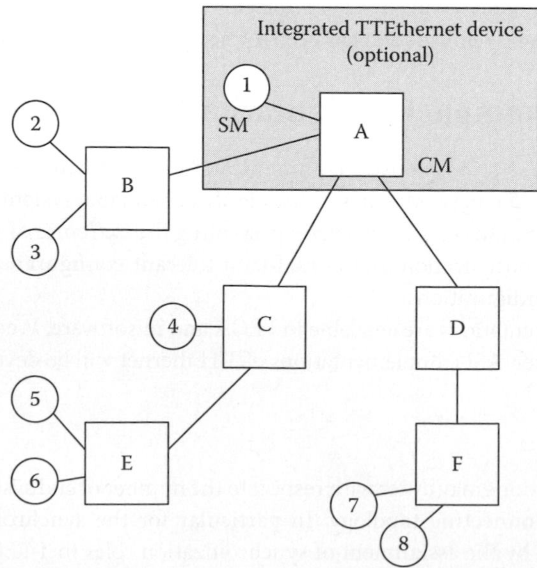

FIGURE 42.5 Master-based configuration with three switches connected in multihop topology.

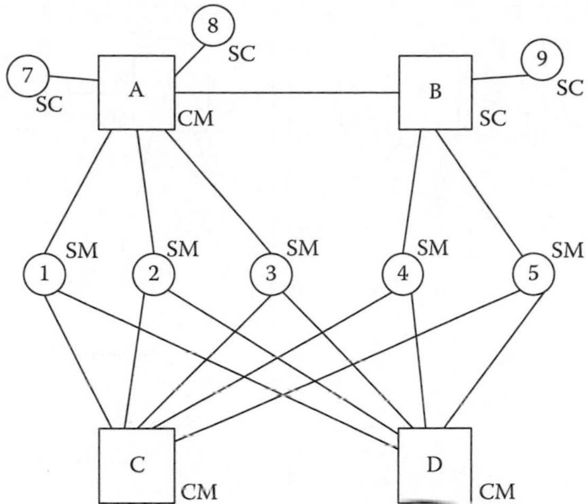

FIGURE 42.6 Dual-fault-tolerant configuration with three redundant channels and five synchronization masters.

base to be operational. Hence, RC and BE traffic can always be communicated as long as there exists a nonfaulty communication path between sender and receiver(s). We discuss a highly dependable and highly available configuration next.

42.4.1.2 Dual-Fault-Tolerant Configuration

Highly dependable systems such as civil airplanes or manned spacecrafts are fail-operational systems. These systems have to remain operational even in the presence of failures. Figure 42.6 shows a redundant TTEthernet network that tolerates two faulty devices without degradation of the TTEthernet services. The network consists of three redundant channels and nine end systems. Two channels are formed by a single switch configured as CM and a third channel is formed by two switches, one of which acts as CM. End systems 1–5 operate as SMs. Furthermore, the SMs and CMs have to be high-integrity devices supporting an inconsistent-omission failure mode. In this configuration, the failure of any two devices is masked by the TTEthernet services without quality degradation.

42.4.1.3 System-of-Systems Configuration

TTEthernet also provides a priority-based mechanism to realize system-of-systems architectures.

Figure 42.7 shows a network architecture consisting of two TTEthernet subnetworks, a high-priority subnetwork and a low-priority subnetwork. The priority of the respective network is stored in the configuration data of each TTEthernet device as well as in the Sync_Priority field in the PCFs. A TTEthernet end system can be configured to automatically synchronize to the highest priority PCF it receives. Alternatively, the change of a TTEthernet current priority to a higher priority can demand host interaction.

This priority mechanism supports the full operation of parts of the network, for example, to realize power-down modes. In the example in Figure 42.7, either one of the subnetworks can be shut down. Once it is powered on again, both subnetworks synchronize either automatically or upon host acknowledgment.

42.4.2 Implementations

TTEthernet specifies a set of services that can be implemented on top of standard Ethernet. End systems realize these services either in the form of a software stack on top of commercial-off-the-shelf (COTS) controllers or in the form of dedicated FPGA solutions. Currently, TTEthernet switches are implemented in FPGA. Furthermore, ASICs for TTEthernet end systems and TTEthernet switches are being developed at the time of this writing.

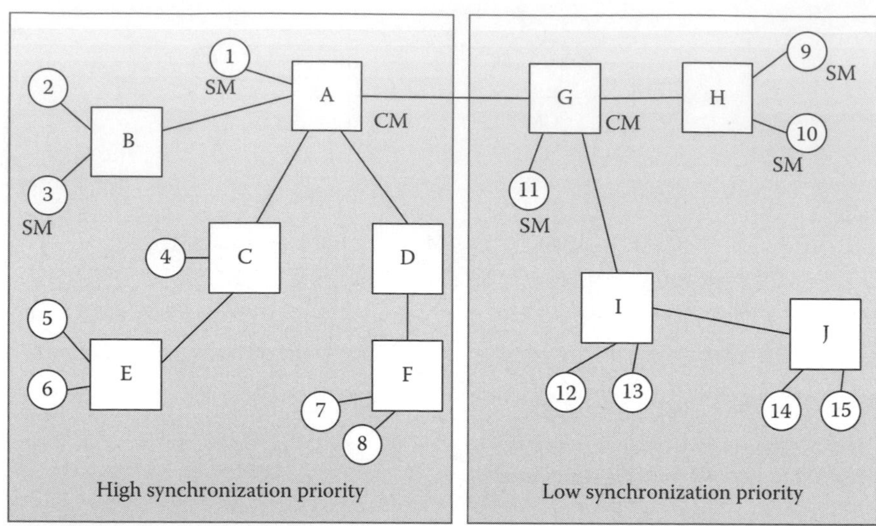

FIGURE 42.7 TTEthernet systems-of-systems configuration consisting of two subnetworks.

The FPGA-based solutions of TTEthernet mainly differ with respect to the communication speed they support. The 100 Mb/s FPGA-based version of the TTEthernet switch is specified as follows:

- Eight 100 Mb/s and one 1 Gb/s uplink port.
- Guaranteed real-time delivery and clock synchronization in the microseconds range.
- Legacy Ethernet devices can synchronize to network time base without knowing about TTEthernet
- Support for legacy and best-effort traffic.
- Standard TCP/IP protocols and applications can be used.
- Flexibility for customer-specific extensions (ALTERA Cyclone III FPGA).
- Digital I/O for triggering measurements.
- Dimensions: 170 × 121 × 55 mm; weight: 800 g; operating temperature: 0°C–70°C; storage: −40°C to +85°C.
- Robust housing.

The 1 Gb/s version of TTEthernet switch has the following characteristics:

- Four 1 Gb/s copper/fiber ports
- Message schedules and routing information stored in internal ROM (loaded by TTE-Load download tool)
- Hardware-based on Altera COTS board and PHY daughter board
- 8 Gb/s full-duplex bandwidth
- Multihop capable
- Single synchronization domain

Also, 12-port and 24-port TTEthernet switches are available. Some TTEthernet switch variants are depicted in Figure 42.8. The TTEthernet switches allow communication of all traffic classes at line-rate communication speeds.

TTEthernet end systems have been realized in FPGA and as a software stack. The FPGA solutions come in different form factors and are depicted in Figure 42.9.

The software stack realizing the TTEthernet services is depicted in Figure 42.10. This stack is designed for a 100 Mb/s software-based TTEthernet end system running a single TTEthernet channel. COTS Ethernet Controllers can be used. In an example setup on an Intel ATOM running at 1.6 GHz with 1 GB of RAM and 0.5 MB cache, the stack was implemented on Standard Linux.

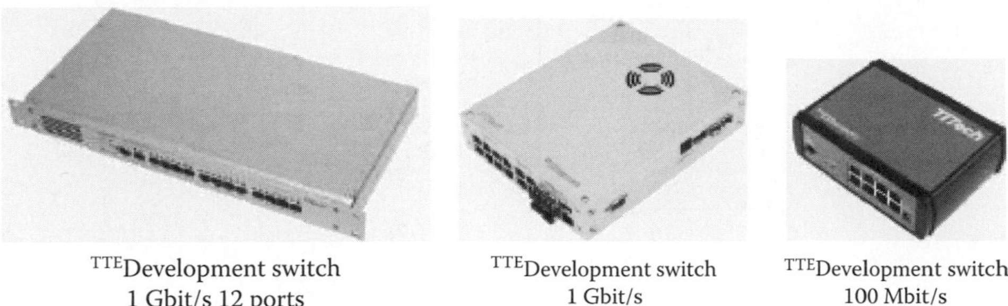

FIGURE 42.8 TTEthernet switches.

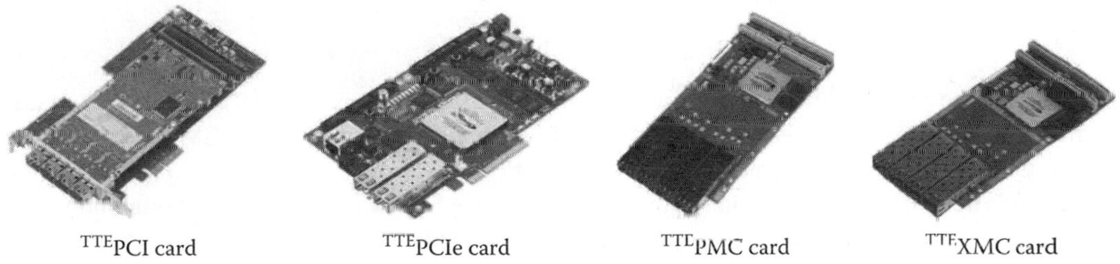

FIGURE 42.9 TTEthernet FPGA-based end systems in different form factors.

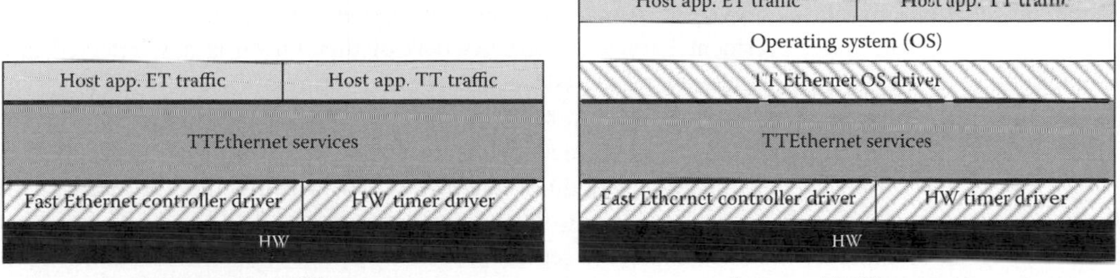

FIGURE 42.10 TTEthernet software stack with and without operating system support.

Using a cluster cycle of 3 ms and communicating 10 time-triggered messages (1500 bytes each), a total network load of 40 Mb/s has been generated. In this configuration, the software stack used about 3% of the ATOM CPU.

Given the cross-industry applicability of TTEthernet, it is currently used from space to new energy applications and is also evaluated for a broad range of use cases including automotive applications. One particular use case is the use of TTEthernet in the Orion space system, which is discussed in the next section.

42.5 TTEthernet in Orion

In the United States under President G.W. Bush, Orion has been planned and started to be developed under the leadership of NASA as the next-generation Crew Exploration Vehicle (CEV) replacing the Space Shuttle. Orion is part of the NASA Constellation program, which comprises several projects, among others the Ares programs, which are the rockets for lifting Orion and other space vehicles into

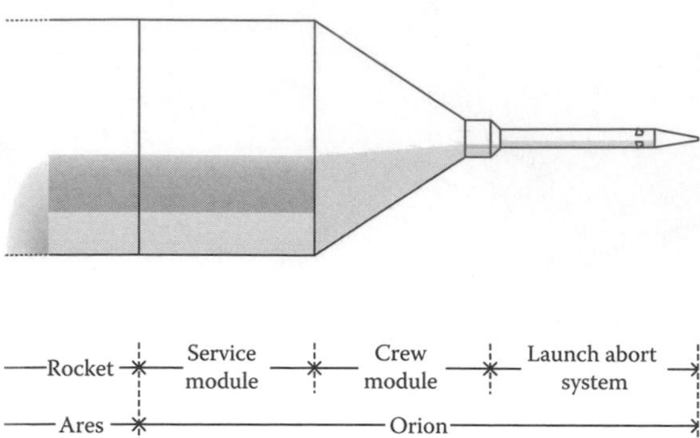

FIGURE 42.11 Sketch of Orion (not to scale).

low earth orbit to go, for example, to the international space station, to our moon, and to Mars—at least this has been the plan. The Orion vehicle consists of two parts, the crew module (CM) and the service module (SM). The crew module provides all the facilities for housing up to six astronauts for about 2 weeks to lower earth orbit in initial configurations and up to 6 months for going to the moon. The crew module is also used for reentry into earth atmosphere to bring the crew back to earth. The service module provides the necessary supplementary equipment for the crew to live in space and is jettisoned before reentry. The concept of operation of the constellation program for space exploration is very similar to the Apollo program. Figure 42.11 sketches the structure of the Orion and Ares I as docked vehicles for the reader to get an idea of the configuration. At the time of writing this text, the future of the Constellation program is slightly rearranged and its development has been slightly changed by the US administration under President Barack Obama. As part of this, Orion is now referred to as Orion Multipurpose Crew Vehicle. However, the basic technical architecture of Orion with TTEthernet remains the same.

Both Ares I and Orion comprise separate avionics architectures due to operational constraints and control aspects. Orion deploys an integrated modular avionics (IMA) architecture approach as this allows an open architecture, a key characteristic to NASA. With an open architecture, the system is easier to integrate and maintain. For example, the central processing unit is standardized, allowing the use of COTS hardware without any single supplier constraint and, hence, reduction of nonrecurring cost for components within the system. Similarly, the use of COTS will decrease the upgrade cost since competition drives supplier upgrades to low or no cost to the project. Additionally, maintenance costs are reduced, since the COTS producers can leverage the development and maintenance tools already available. NASA's systems, such as the human-rated Orion, tend to be deployed for long durations. The space shuttle, for example, has been operational for about 30 years. An IMA architecture used in such systems benefits in more easily supporting midlife upgrades to the system, hence incorporating midlife technology upgrades into the system. Furthermore, with varying mission durations, it is also important that the avionics architecture allows for different safety and criticality protections. IMA architectures support this flexibility [MBV09]. Figure 42.12 shows the Orion/Ares interface and some avionics architecture aspects of Orion and Ares.

Orion has very specific mission requirements. It has to meet a mission requirement of up to 5000 h (about 200 days) of continuous operation at a time. It must be a flexible vehicle that will interface to several other space elements (the International Space Station, the Altair Lunar Lander, Ground Systems, and the Space Network, in addition to the Ares I interface) over the next several decades. Additionally, Orion has strict weight and power restrictions that make it desirable for its avionics system to maintain full protection against erroneous behavior even when operated in a low-power partial system mode, which is a mode when part of the avionics system is shut off to conserve power [MBV09].

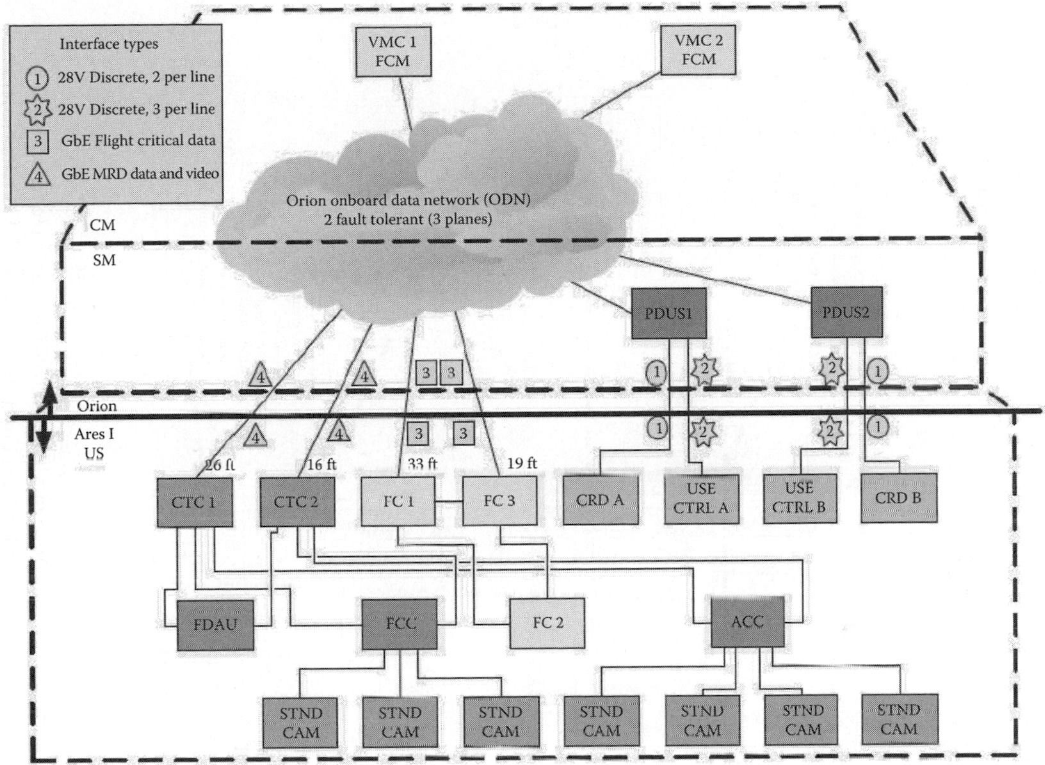

FIGURE 42.12 Orion and Ares I interface. (From McCabe, M. et al., Avionics architecture interface considerations between constellation vehicles, *Proceedings of the 28th Digital Avionics Systems Conference (DASC)*, IEEE/AIAA, pp. 1.E.2-1–1.E.2-10, October 2009.)

Early trade studies that have been conducted favored an IMA approach due to the Orion's requirements [MBV09]:

- The Orion CEV shall meet an ascent reliability allocation of 0.9999.
- The Orion CEV shall meet an overall mission reliability allocation of 0.999.
- The Orion CEV shall be one-fault-tolerant for safe return of the crew with exceptions for design for minimum risk.
- The Orion CEV shall safely recover and return the crew in case of loss of output and erroneous output from the vehicle flight computers due to software common-cause failure.
- The Orion CEV shall allow the crew to manually control, inhibit, and/or override autonomous or ground-controlled critical functions.
- The Orion CEV shall be one-fault-tolerant for mission completion with exceptions for design for minimum risk.

In order to meet requirements like full error detection in low-power partial system mode, a simple high-integrity, self-checking computing approach has been designed building on implementations and knowledge from avionics programs of aircraft like the Boeing 787, Boeing 777, and Boeing 777RS. Figure 42.13 provides a detailed block diagram of a typical self-checking pair computer design, which in a similar manner is deployed on Orion. Orion actually does deploy an IBM PowerPC 750 processor as indicated in Figure 42.13, with self-checking support logic such as an interface to the bus and external memory (BIPM) and backend interfaces. Yet, the actual main interface is a variant of TTEthernet,

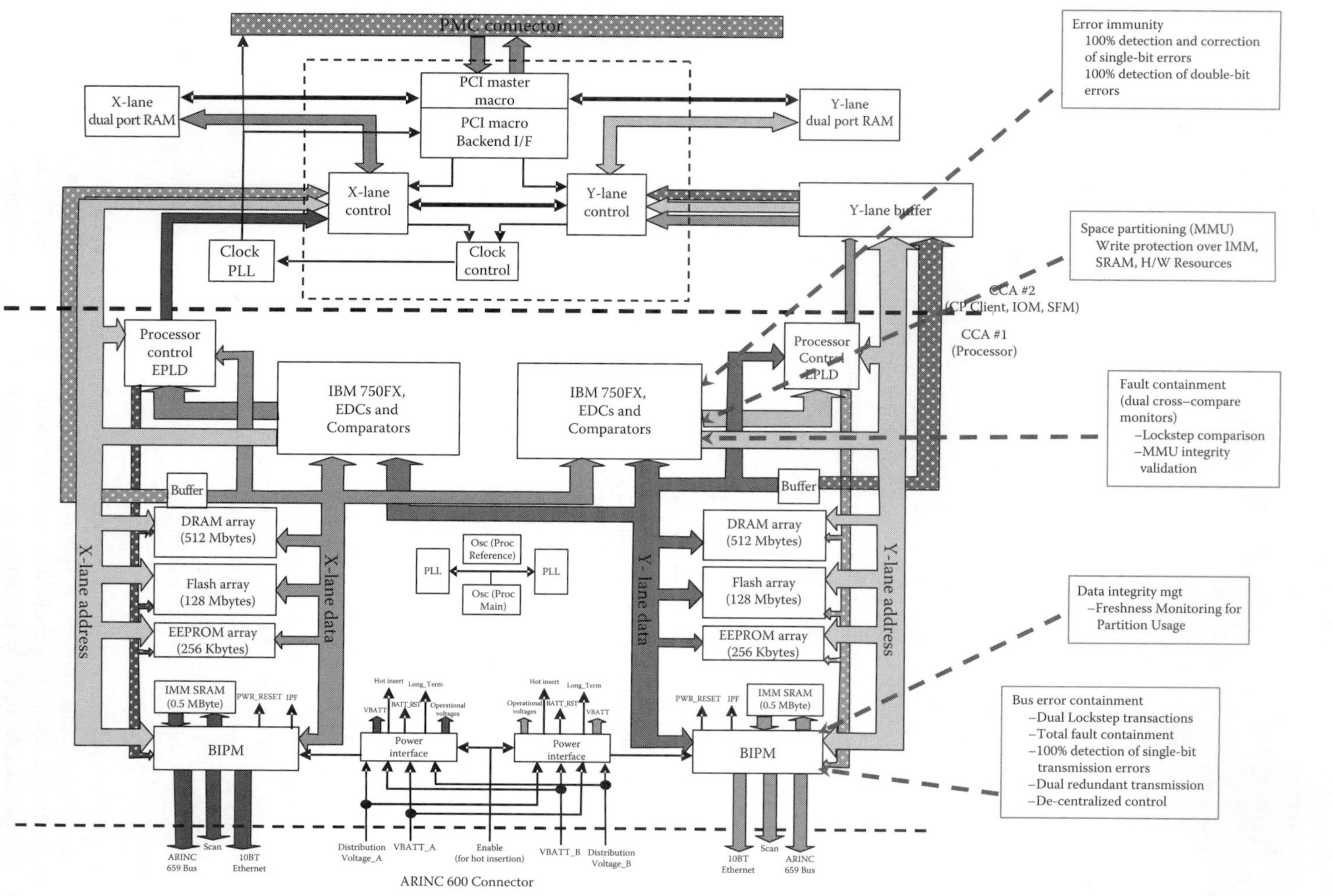

FIGURE 42.13 Self-checking pair computer. (From Fletcher, M., Progression of an open architecture: From Orion to Altair and LSS, Companion to report (Presentation) S65-5000-20-0, Honeywell, *FaultTolerant Spaceborne Computing Employing New Technologies 2009 Conference*, May 2009; Courtesy of Honeywell, Morristown, NJ.)

named TT-GbE, and not ARINC 659 as indicated in Figure 42.13. It shall also be noted that it is typical to perform local power and clock monitoring for a pair of processors. The presented and deployed architecture has the following advantages:

- Good error detection for bit flips in processors as well as in associated memories. In all cases, *no* inadvertent action is performed due to single event upsets with a very high probability (or in other words, the probability of undetected failures is smaller than 10^{-9}).
- No inadvertent violation of space partitioning (memory and I/O) due to features like memory management units (MMUs).

Cross-compare monitors perform constant comparison of computing outputs and validate MMU integrity.

- Data integrity management (freshness monitoring for partition usage).
- Support of bus error containment (more Orion details later).

The system network is an implementation variant of TTEthernet described. The Orion TTEthernet deployment is called *Time-Triggered Gigabit Ethernet* (or *TT-GbE*). As the name implies, it leverages a

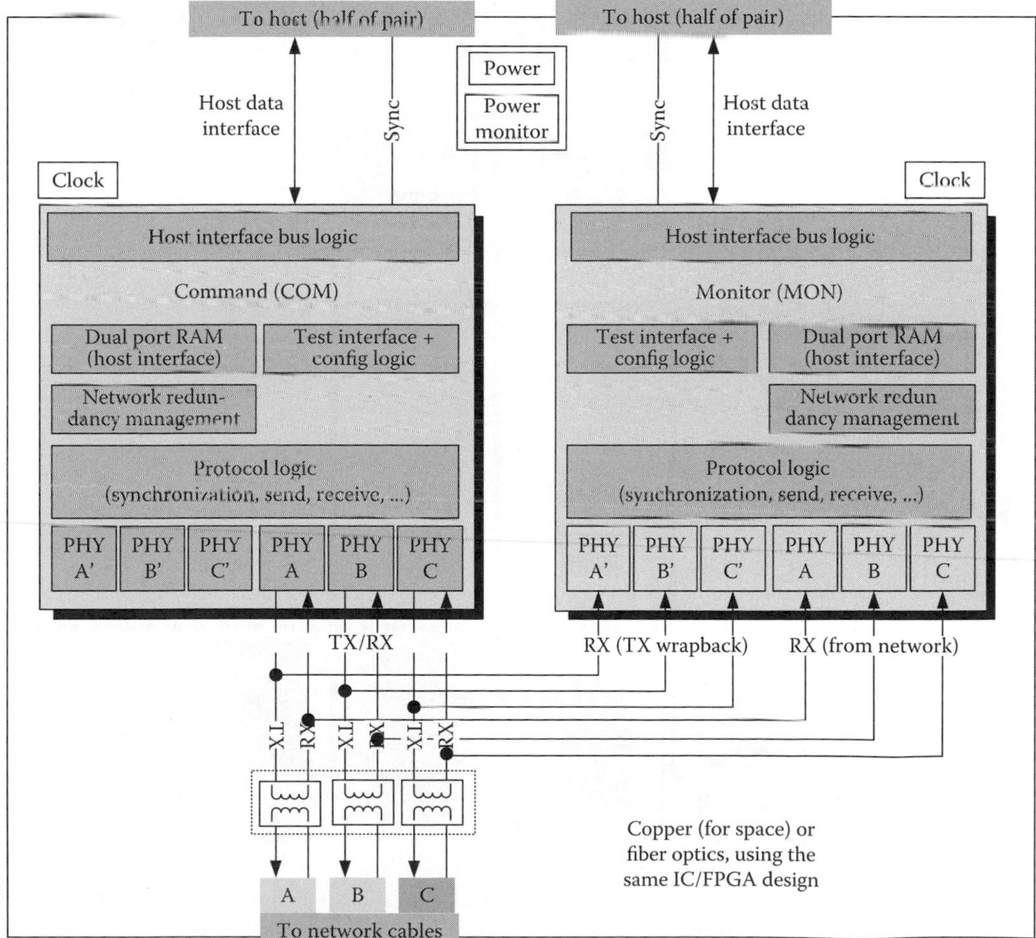

FIGURE 42.14 TTEthernet self-checking pair end system. (From Hall, B. and Driscoll, K., A new aerospace network family, *Presentation to INCOSE*, Honeywell, October 2009; Courtesy of Honeywell, Morristown, NJ.)

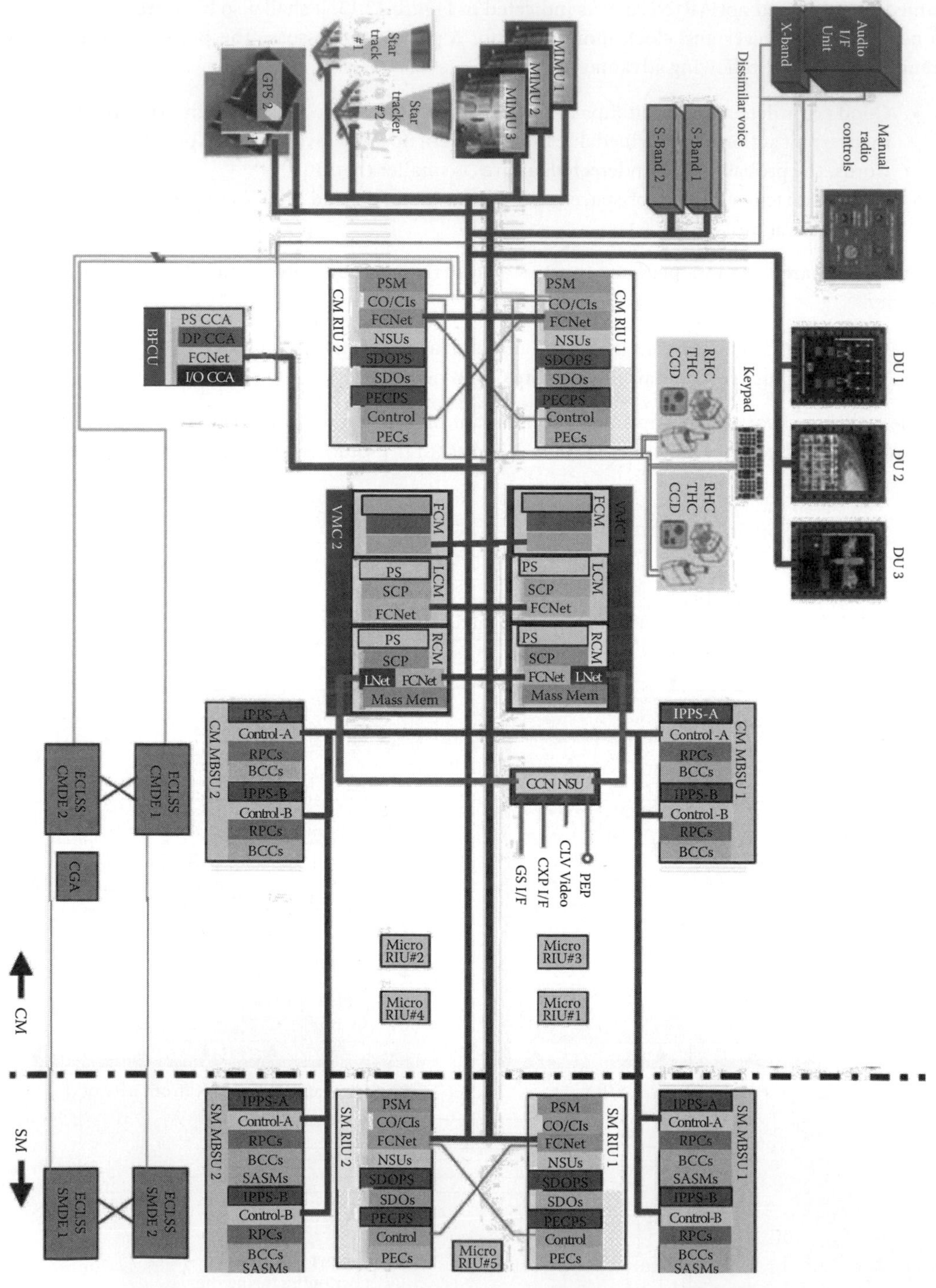

FIGURE 42.15 Orion avionics overview. (From Fletcher, M., Progression of an open architecture: From Orion to Altair and LSS, Companion to report (Presentation) S65-5000-20-0, Honeywell, *FaultTolerant Spaceborne Computing Employing New Technologies 2009 Conference*, May 2009; Courtesy of Honeywell, Morristown, NJ.)

standard Gigabit Ethernet physical layer and the standard TTEthernet IP core including all features and support up to ISO/OSI layer 2 (i.e., not including layer 3 according to ISO/OSI). Given the harsh natural radiation environment in space, the TTEthernet IP is embedded in a space-hardened technology together with a special space-hardened SERDES (serializer/deserializer) physical layer technology, compatible with Ethernet, and other interfaces like PCI. TT-GbE supports normal and high-integrity end systems and high-integrity switches. The high-integrity variant is based on a command/monitor configuration. Figure 42.14 shows the command/monitor configuration for the TTEthernet network for an end system. It is deployed for the switches and some end systems (such as the ones interfacing to the self-checking computers) [HD09,Fle09,MBV09]. Such command/monitor component configurations ensure a validated fail-silence model in that all output of a component is checked bit-for-bit against independent monitor copy and any violation forces a consistently detectable fault output. The command/monitor end system and switch and self-checking compute components enable a reduction of end-to-end integrity augmentation normally deployed in high-integrity end systems. It also helps leaving the high dependability implementation aspects out of the application and at the electronic platform system level, which simplifies platform development.

Figure 42.15 presents a very high-level overview of the Orion avionics architecture at one time during the development. The platform was under development at the time publications [Fle09,MBV09] were written and has slightly changed in the meantime. Orion in general had to adapt to weight requirements due to Ares I rocket design. Adaptation of architectures to a certain extent is a typical system design refinement. On the left side of the dotted line of the diagram labeled CM is the equipment in the crew module. On the right side is the equipment in the SM of Orion. The crew module contains the majority of the different electronic units. The component-connecting lines in Figure 42.15 (in blue in the original publication) show the major onboard data network, which is implemented using TT-GbE and labeled by the FCNet interface cards. The architecture contains—in this instance—two major computer-related cabinets called VMCs (Vehicle Management Computers). VMCs contain three different computer modules (CMs labeled FCM, LCM, and RCM) with associated network cards and power supply modules. Remote Interface Units (RIUs) are cabinets with input/output interfacing cards. Orion has three display units (DUs) implementing modern glass cockpits, interface devices (keypad, RHC, THC, CCD), (external) communication means (e.g., audio I/F Unit, S-Band), absolute and relative navigation units (inertial measurement units (IMUs), star trackers, GPS), environmental control units (ECLSS), and multiple vehicle power management units (MBSUs) among others.

References

Fle09 M. Fletcher. Progression of an open architecture: From Orion to Altair and LSS. Companion to report (Presentation) S65-5000-20-0, Honeywell, *FaultTolerant Spaceborne Computing Employing New Technologies 2009 Conference*, Albuquerque, NM, May 2009.

HD09 B. Hall and K. Driscoll. A new aerospace network family. *Presentation to INCOSE*, Honeywell, , Minneapolis, MN, October 2009.

KAGS05 H. Kopetz, A. Ademaj, P. Grillinger, and K. Steinhammer. The Time-Triggered Ethernet (TTE) design. *Eighth IEEE International Symposium on Object-oriented Real-time distributed Computing (ISORC)*, Seattle, WA, May 2005.

MBV09 M. McCabe, C. Baggerman, and D. Verma. Avionics architecture interface considerations between constellation vehicles. In *Proceedings of the 28th Digital Avionics Systems Conference (DASC)*, Cambridge, MA, pp. 1.E.2-1–1.E.2-10. IEEE/AIAA, October 2009.

SBH+09 W. Steiner, G. Bauer, B. Hall, M. Paulitsch, and S. Varadarajan. TTEthernet dataflow concept. In *NCA*, Washington, DC, pp. 319–322, 2009.

SD10 W. Steiner and B. Dutertre. SMT-based formal verification of a TTEthernet synchronization function. In *FMICS*, Antwerp, Belgium, pp. 148–163, 2010.

SD11 W. Steiner and B. Dutertre. Automated formal verification of the TTEthernet synchronization quality. In *NASA Formal Methods*, Pasadena, CA, pp. 375–390, 2011.

SRSP04 W. Steiner, J.M. Rushby, M. Sorea, and H. Pfeifer. Model checking a fault-tolerant startup algorithm: From design exploration to exhaustive fault simulation. In *DSN*, pp. 189–198, 2004.

Ste06 Klaus Steinhammer. Design of an FPGA-based Time-Triggered Ethernet system. PhD thesis, Technische Universität Wien, Institut für Technische Informatik, Treitlstr. 3/3/182-1, 1040 Vienna, Austria, 2006.

Ste09 W. Steiner. TTEthernet executable formal specification. Research report, 2009. Available at http://www.ttagroup.org/.

Ste10 W. Steiner. An evaluation of SMT-based schedule synthesis for time-triggered multi-hop networks. In *RTSS'10: Proceedings of the 31st IEEE Real-Time Systems Symposium*. IEEE, 2010.

Ste11 W. Steiner. Synthesis of static communication schedules for mixed-criticality systems. In *AMICS 2011: Proceedings of the First International Workshop on Architectures and Applications for Mixed-Criticality Systems*. IEEE, 2011.

TSPS12 D. Tamas-Selicean, P. Pop, and W. Steiner. Synthesis of communication schedules for TTEthernet-based mixed-criticality systems. In *CODES+ISSS*, Tampere, Finland, pp. 473–482, 2012.

VSE09 S. Voss, M. Sorea, and K. Echtle. SAL-based symbolic scheduling in time-triggered networks. In *IFM'09: Proceedings of the Seventh International Conference on Integrated Formal Methods*, Springer-Verlag, Berlin, Germany, pp. 200–214, 2009.

VIII

Avionics and Aerospace

43 **MIL-STD-1553B Digital Time Division Command/Response Multiplex Data Bus** *Chris deLong* **43**-1
Introduction • The Standard • Protocol • Systems-Level Issues • Testing • Further Information

44 **ASCB** *Michael Paulitsch* **44**-1
Protocol Overview • Protocol Services • Protocol Parameterization • Communication Interface • Validation and Verification Efforts • Example Configuration and Implementation: ASCB in Primus Epic • References

45 **ARINC Specification 429 Mark 33 Digital Information Transfer System** *Daniel A. Martinec and Samuel P. Buckwalter* **45**-1
Introduction • ARINC 419 • ARINC 429 • Message and Word Formatting • Timing-Related Elements • Communications Protocols • Applications • ARINC 453

46 **ARINC 629** *Michael Paulitsch* **46**-1
Physical Layer • Data Link Layer • Medium Access Layer • Robustness and Fault Tolerance Features • ARINC 629 in Boeing 777 • References

47 **Commercial Standard Digital Bus** *Lee H. Harrison* **47**-1
Introduction • Bus Architecture • Basic Bus Operation • CSDB Bus Capacity • CSDB Error Detection and Correction • Bus User Monitoring • Integration Considerations • Bus Integration Guidelines • Bus Testing • Aircraft Implementations • Defining Terms • References • Further Information • Bibliography

48 **SAFEbus** *Michael Paulitsch and Kevin R. Driscoll* **48**-1
SAFEbus • Protocol Overview • Protocol Services • Communication Interface • Validation and Verification Efforts • SAFEbus in Boeing 777 • References

49 **Dimensioning of Civilian Avionics Networks** *Jean-Luc Scharbarg and Christian Fraboul* **49**-1
Introduction • Civilian Avionics Systems Evolution and ARINC Context • AFDX End-to-End Delay Analysis • Sure Upper Bounds with Network Calculus • Sure Upper Bounds with Trajectory Approach • Exact Worst-Case Delays with Model Checking • Analysis of a Realistic Configuration • Trends for the Dimensioning of AFDX Network • Conclusion • References

43

MIL-STD-1553B Digital Time Division Command/Response Multiplex Data Bus

43.1 Introduction 43-1
Background • History and Applications
43.2 The Standard.. 43-3
Hardware Elements
43.3 Protocol 43-8
Word Types • Message Formats, Validation, and Timing • Mode Codes
43.4 Systems-Level Issues... 43-21
Subaddress Utilization • Data Wraparound • Data Buffering • Variable Message Blocks • Sample Consistency • Data Validation • Major and Minor Frame Timing • Error Processing
43.5 Testing 43-24
43.6 Further Information.. 43-25

Chris deLong
Honeywell Aerospace

43.1 Introduction

MIL STD-1553 is a standard that defines the electrical and protocol characteristics for a data bus. AS-15531 is the Society of Automotive Engineers (SAE) commercial equivalent to the military standard. A data bus is similar to what the personal computer and office automation industry have dubbed a *local area network* (LAN). In avionics, a data bus is used to provide a medium for the exchange of data and information among various systems and subsystems. MIL-STD-1553 has been deployed on fixed- and rotary-wing aircraft, unmanned aircraft, surface and subsurface ships, ground vehicles, spacecraft and satellites, and the International Space Station. It is undoubtedly the most successful military standard developed.

43.1.1 Background

In the 1950s and 1960s, avionics were simple stand-alone systems. Navigation, communications, flight controls, and displays were analog systems. Often, these systems were composed of multiple boxes interconnected to form a single system. The interconnections between the various boxes were accomplished with point-to-point wiring. The signals mainly consisted of analog voltages, synchro-resolver signals, and relay/switch contacts. The location of these boxes within the aircraft was a function of operator need, available space, and the aircraft weight and balance constraints. As more and more systems were added, the cockpits became crowded due to the number of controls and displays, and the overall weight of the aircraft increased.

By the late 1960s and early 1970s, it was necessary to share information between various systems to reduce the number of black boxes required by each system. A single sensor providing heading and rate information could provide those data to the navigation system, the weapons system, the flight control system, and pilot's display system (see Figure 43.1a). However, the avionics technology was still basically an analog, and while sharing sensors did produce a reduction in the overall number of black boxes, the interconnecting signals became a *rat's nest* of wires and connectors. Moreover, functions or systems that were added later became an integration nightmare as additional connections of a particular signal could have potential system impacts, plus since the system used point-to-point wiring, the system that was the source of the signal typically had to be modified to provide the additional hardware needed to output to the newly added subsystem. As such, intersystem connections were kept to the bare minimums.

By the late 1970s, with the advent of digital technology, digital computers had made their way into avionics systems and subsystems. They offered increased computational capability and easy growth, compared to their analog predecessors. However, the data signals—the inputs and outputs from the sending and receiving systems—were still mainly analog in nature, which led to the configuration of a small number of centralized computers being interfaced to the other systems and subsystems via complex and expensive analog-to-digital and digital-to-analog converters.

As time and technology progressed, the avionics systems became more digital. And with the advent of the microprocessor, things really took off. A benefit of this digital application was the reduction in the number of analog signals, and hence the need for their conversion. Greater sharing of information could be provided by transferring data between users in digital form. An additional side benefit was that digital data could be transferred bidirectionally, whereas analog data were transferred unidirectionally. Serial rather than parallel transmission of the data was used to reduce the number of interconnections within the aircraft and the receiver/driver circuitry required with the black boxes. But this alone was not enough. A data transmission medium that would allow all systems and subsystems to share a single and common set of wires was needed (see Figure 43.1b). By sharing the use of this interconnect, the various subsystems could send data among themselves and to other systems and subsystems, one at a time, and in a defined sequence using a methodology called *time-division multiplexing* (TDM). Enter the 1553 data bus.

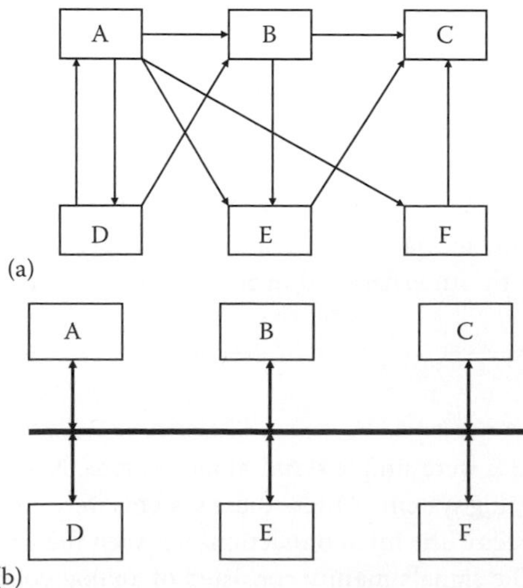

FIGURE 43.1 Systems configurations.

43.1.2 History and Applications

MIL-STD-1553 (USAF) was released in August of 1973. The first user of the standard was the F-16. Further changes and improvements were made and a tri-service version, MIL-STD-1553A, was released in 1975. The first user of the "A" version of the standard was again the Air Force's F-16 and the Army's new attack helicopter, the AH-64A Apache. With some *real-world* experience, it was soon realized that further definitions and additional capabilities were needed. The latest version of the standard, 1553B, was released in 1978.

Today, the 1553 standard is still at the "B" level; however, changes have still been made. In 1980, the Air Force introduced Notice 1. Intended only for Air Force applications, Notice 1 restricted the use of many of the options within the standard. While the Air Force felt this was needed to obtain a common set of avionics systems, many in industry felt that Notice 1 was too restrictive and limited the capabilities in the application of the standard. Released in 1986, the tri-service Notice 2 (which supersedes Notice 1) places tighter definitions upon the options within the standard. And while not restricting an option's use, it tightly defines how an option will be used if implemented. Notice 2, in an effort to obtain a common set of operational characteristics, also places a minimum set of requirements upon the design of the black box. The military standard was converted to its commercial equivalent as SAE AS 15531, as part of the government's effort to increase the use of commercial products.

Other notices have since been released. Notice 3 released in 1993 simply states that the standard has been reviewed and is still valid for use. Notice 4 released in 1996 changed the name of the document from a *military standard* to an *interface standard*. Notice 5 released in 2006 and Notice 6 released in 2007 were an effort by the Air Force to increase the data from 1 megabit per second (Mbps) up to 200 Mbps. This was to be accomplished to providing an additional orthogonal frequency division multiplex signal *on top of* the existing 1553 signal while utilizing the same cables and couplers. These notices and the addition of the higher rate signaling were not immediately accepted by the industry and the Air Force received lots of comments regarding the operation, use, and testing of this addition. Notice 7, released in 2008, cancelled Notices 5 and 6. So the last *meaningful* notice to 1553B is Notice 2. In addition, MIL-STD-1553B has been incorporated into other standards such as MIL-STD-1760 and STANAG 3910.

Since its inception, MIL-STD-1553 has found numerous applications. Notice 2 even removed all references to *aircraft* or *airborne* so as not to limit its applications. The standard has also been accepted and implemented by North Atlantic Treaty Organization (NATO) and many foreign governments. The United Kingdom has issued Def Stan 00-18 (Part 2) and NATO has published STANAG 3838 AVS, both of which are versions of MIL-STD-1553B.

43.2 The Standard

MIL-STD-1553B defines the term TDM as "the transmission of information from several signal sources through one communications system with different signal samples staggered in time to form a composite pulse train." For our example in Figure 43.1b, this means that data can be transferred between multiple avionics units over a single transmission media, with the communications among the different avionics boxes taking place at different moments in time, hence time division. Table 43.1 is a summary of the 1553 data bus characteristics. However, before defining how the data are transferred, it is necessary to understand the data bus hardware.

43.2.1 Hardware Elements

The 1553 standard defines certain aspects regarding the design of the data bus system and the black boxes to which the data bus is connected. The standard defines four hardware elements: transmission media, remote terminals, bus controllers, and bus monitors, each of which is detailed as follows.

TABLE 43.1 Summary of the 1553 Data Bus Characteristics

Data Rate	1 MHz
Word length	20 bits
Data bits per word	16 bits
Message length	Maximum of 32 DWs
Transmission technique	Half-duplex
Operation	Asynchronous
Encoding	Manchester II biphase
Protocol	Command response
Bus control	Single or multiple
Message formats	BC-RT
	RT-BC
	RT-RT
	Broadcast
	System control
Number of remote terminals	Maximum of 31
Terminal types	Remote terminal (RT)
	Bus controller (BC)
	Bus monitor (BM)
Transmission media	Twisted shielded pair cable
Coupling	Transformer or direct

43.2.1.1 Transmission Media

The transmission media, or data bus, is defined as a twisted shielded pair transmission line consisting of the main bus and a number of stubs. There is one stub for each terminal (system) connected to the bus. The main data bus is terminated at each end with a resistance equal to the cable's characteristic impedance. This termination makes the data bus behave electrically like an infinite transmission line. Stubs, which are added to the main bus to connect the terminals, provide *local* loads and produce an impedance mismatch where added. This mismatch, if not properly controlled, produces electrical reflections and degrades the performance of the main bus. Therefore, the characteristics of both the main bus and the stubs are specified within the standard. Table 43.2 is a summary of the transmission media characteristics.

The standard specifies two stub methods: direct and transformer coupled. This refers to the method in which a terminal is connected to the main bus. Figure 43.2 shows the two methods, the primary difference between the two being that the transformer-coupled method utilizes an isolation transformer for connecting the stub cable to the main bus cable. In both methods, two isolation resistors are placed in series with the bus. In the direct-coupled method, the resistors are typically located within the

TABLE 43.2 Summary of Transmission Media Characteristics

Cable Type	Twisted Shielded Pair
Capacitance	30.0 pF/ft max—wire to wire
Characteristic impedance	70.0–85.0 Ω at 1 MHz
Cable attenuation	1.5 dBm/100 ft at 1 MHz
Cable twists	4 twists per foot maximum
Shield coverage	90% minimum
Cable termination	Cable impedance (62%)
Direct-coupled stub length	Maximum of 1 ft
Transformer-coupled stub length	Maximum of 20 ft

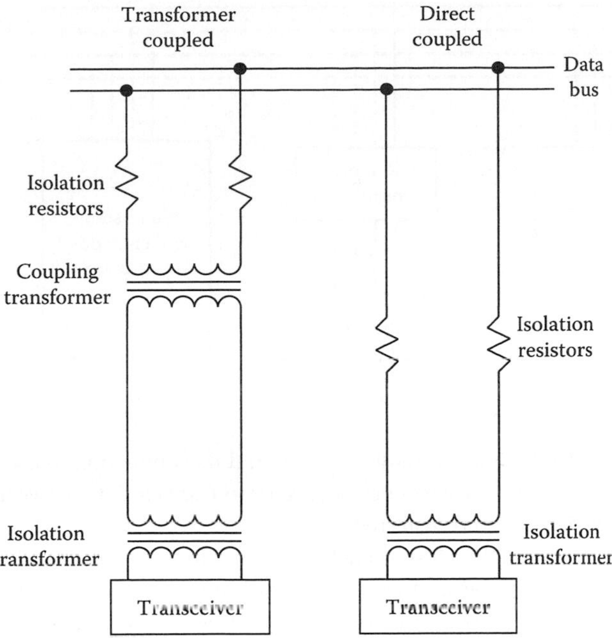

FIGURE 43.2 Terminal connection methods.

terminal, whereas in the transformer-coupled method, the resistors are typically located with the coupling transformer in boxes called data bus couplers. A variety of couplers are available, providing single or multiple stub connections. Bus couplers are available as either boxes using connectors or as *in-line* couplers wherein the stubs are spliced into the main bus cable.

Another difference between the two coupling methods is the length of the stub. For the direct-coupled method, the stub length is limited to a maximum of 1 ft. For the transformer-coupled method, the stub can be up to a maximum length of 20 ft. Therefore, for direct-coupled systems, the data bus must be routed in close proximity to each of the terminals, whereas for a transformer-coupled system, the data bus may be up to 20 ft away from each terminal.

43.2.1.2 Remote Terminal

A remote terminal is defined within the standard as "All terminals not operating as the bus controller or as a bus monitor." Therefore if it is not a controller, monitor, or the main bus or stub, it must be a remote terminal—sort of a *catch all* clause. Basically, the remote terminal is the electronics necessary to transfer data between the data bus and the subsystem. So what is a subsystem? For 1553 applications, the subsystem is the sender or user of the data being transferred.

In the earlier days of 1553, remote terminals were used mainly to convert analog and discrete data to/from a data format compatible with the data bus. The subsystems were still the sensor that provided the data and computer that used the data. As more and more digital avionics became available, the trend became to embed the remote terminal into the sensor and computer. Today, it is common for the subsystem to contain an embedded remote terminal. Figure 43.3 shows the different levels of remote terminals possible.

A remote terminal typically consists of a transceiver, an encoder/decoder, a protocol controller, a buffer or memory, and a subsystem interface. In a modern black box containing a computer or processor, the subsystem interface may consist of the buffers and logic necessary to interface to the computer's address, data, and control buses. For dual redundant systems, two transceivers and two encoders/decoders would be required to meet the requirements of the standard.

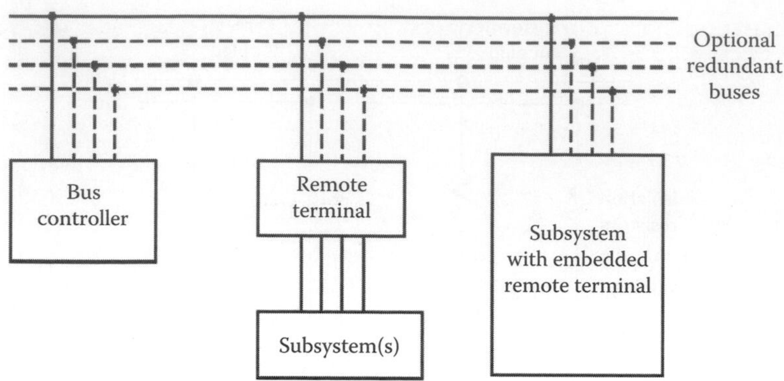

FIGURE 43.3 Simple multiplex architecture.

Figure 43.4 is a block diagram of a remote terminal and its connection to a subsystem. In short, the remote terminal consists of all the electronics necessary to transfer data between the data bus and the user or originator of the data being transferred.

But a remote terminal is more than just a data formatter. It must be capable of receiving and decoding commands from the bus controller and respond accordingly. It must also be capable of buffering a message worth of data, detecting transmission errors and performing validation tests upon the data, and reporting the status of the message transfer. A remote terminal must be capable of performing a few of the bus management commands (referred to as mode commands [MCs]), and for dual redundant applications, it must be capable of listening to and decoding commands on both buses at the same time.

A remote terminal must strictly follow the protocol as defined by the standard. It can only respond to commands received from the bus controller (i.e., it only speaks when spoken to). When it receives a valid

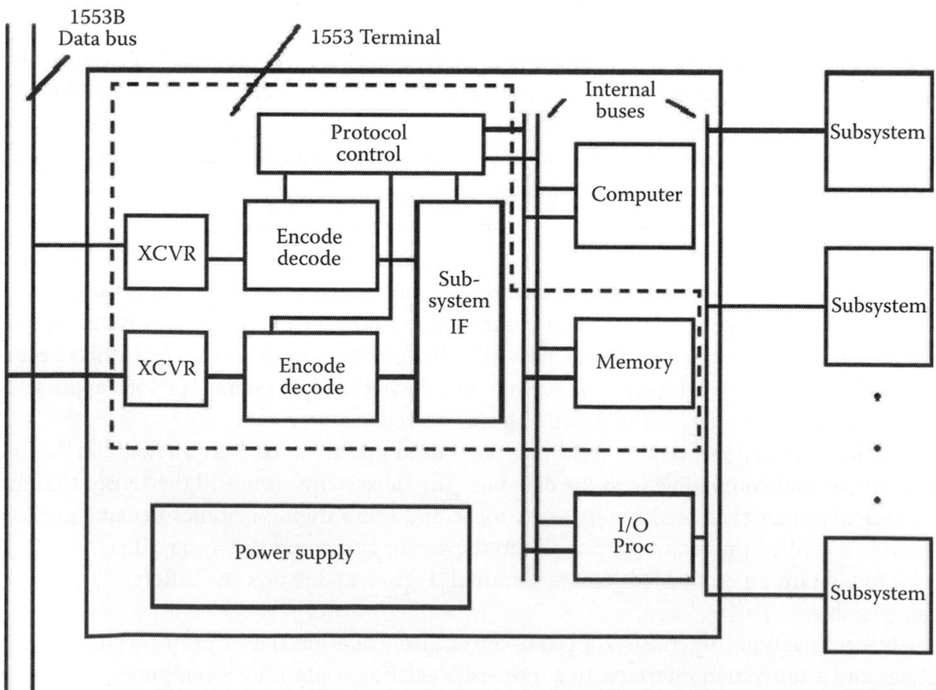

FIGURE 43.4 Terminal definition.

command, it must respond within a defined amount of time. If a message does not meet the validity requirements defined, then the remote terminal must invalidate the entire message and discard the data (not allow the data to be used by the subsystem). In addition to reporting status to the bus controller, most remote terminals today are also capable of providing some level of status information to the subsystem regarding the data received.

43.2.1.3 Bus Controller

The bus controller is responsible for directing the flow of data on the bus. While several terminals may be capable of performing as the bus controller, only one bus controller is allowed to be active at any one time. The bus controller is the only device allowed to issue commands onto the data bus. The commands may be for the transfer of data or the control and management of the bus (referred to as MCs).

Typically, the bus controller is a function that is contained within some other computer, such as a mission computer, a display processor, or a fire control computer. The complexity of the electronics associated with the bus controller is a function of the subsystem interface (the interface to the computer), the amount of error management and processing to be performed, and the architecture of the bus controller. There are three types of bus controller architectures: a word controller, a message controller, and a frame controller.

A *word controller* is the oldest and simplest type. Few word controllers are built today and they are only mentioned herein for completeness. For a word controller, the terminal electronics transfers one word at a time to the subsystem. Message buffering and validation must be performed by the subsystem.

Message controllers output a single message at a time, interfacing with the computer only at the end of the message or perhaps when an error occurs. Some message controllers are capable of performing minor error processing, such as retransmitting once on the alternate data bus, before interrupting the computer. The computer will inform the interface electronics of where the message exists in memory and provide a control word. For each message, the control word typically informs the bus controller electronics of the message type (e.g., a remote terminal to bus controller [RT-BC] or remote terminal to remote terminal [RT-RT] command), which bus to use to transfer the message, where to read or write the data words (DWs) in memory, and what to do if an error occurs. The control words are a function of the hardware design of the electronics and are not standardized among bus controllers.

A *frame controller* is the latest concept in bus controllers. A frame controller is capable of processing multiple messages in a sequence defined by the computer. The frame controller is typically capable of error processing as defined by the message control word. Frame controllers are used to *off-load* the computer as much as possible, interrupting only at the end of a series of messages or when an error it cannot handle is detected.

There is no requirement within the standard as to the internal workings of a bus controller, only that it issue commands onto the bus.

43.2.1.4 Bus Monitor

A bus monitor is just that: a terminal that listens to (monitors) the exchange of information on the data bus. The standard strictly defines what bus monitors may be used for, stating that the information obtained by a bus monitor be used "for offline applications (e.g., flight test recording, maintenance recording, or mission analysis) or to provide a backup bus controller sufficient information to take over as the bus controller." Monitors may collect all the data from the bus or may collect selected data.

The reason for restricting its use is that while a monitor may collect data, it deviates from the command-response protocol of the standard in that a monitor is a passive device that does not transmit a status word (SW), and therefore cannot report on the status of the information transferred. Therefore, bus monitors fall into two categories: a recorder for testing or as a terminal functioning as a backup bus controller.

TABLE 43.3 Terminal Electrical Characteristics

Requirement	Transformer Coupled	Direct Coupled	Condition
	Input characteristics		
Input level	0.86–14.0 V	1.2–20.0 V	p–p, l–l
No response	0.0–0.2 V	0.0–0.28 V	p–p, l–l
Zero crossing stability	±150.0 ns	±150.0 ns	
Rise/fall times	0 ns	0 ns	Sine wave
Noise rejection	140.0 mV WGN[a]	200.0 mV WGN	BER[b] 1 per 10^7
Common mode rejection	±10.0 V peak	±10.0 V peak	line-gnd, DC-2.0 MHz
Input impedance	1000 Ω	2000 Ω	75 kHz–1 MHz
	Output characteristics		
Output level	18.0–27.0 V	6.0–9.0 V	p–p, l–l
Zero crossing stability	25.0 ns	25.0 ns	
Rise/fall times	100–300 ns	100–300 ns	10%–90%
Maximum distortion	±900.0 mV	±300.0 mV	peak, l–l
Maximum output noise	14.0 mV	5.0 mV	rms, l–l
Maximum residual voltage	±250.0 mV	±90.0 mV	peak, l–l

[a] WGN, White Gaussian noise.

[b] BER, Bit error rate.

In collecting data, a monitor must perform the same message validation functions as the remote terminal and, if an error is detected, inform the subsystem of the error (the subsystem may still record the data, but the error should be noted). For monitors that function as recorders for testing, the subsystem is typically a recording device or a telemetry transmitter. For monitors that function as backup bus controllers, the subsystem is the computer.

Today, it is common that bus monitors also contain a remote terminal. When the monitor receives a command addressed to its terminal address (TA), it responds as a remote terminal. For all other commands, it functions as a monitor. The remote terminal portion could be used to provide feedback to the bus controller of the monitor's status, such as the amount of memory or time left, or to reprogram a selective monitor as to what messages to capture.

43.2.1.5 Terminal Hardware

The electronic hardware among a remote terminal, bus controller, and bus monitor does not differ much. Both the remote terminal and bus controller (and bus monitor if it is also a remote terminal) must have the transmitters/receivers and encoders/decoders to format and transfer data. The requirements on the transceivers and the encoders/decoders do not vary between the hardware elements. Table 43.3 lists the electrical characteristics of the terminals.

All three elements have some level of subsystem interface and data buffering. The primary difference lies in the protocol control logic and often this is just a different series of microcoded instructions. For this reason, it is common to find 1553 hardware circuitry that is capable of functioning as all three devices.

There is an abundance of *off-the-shelf* components available today from which to design a terminal. These vary from discrete transceivers, encoders/decoders, and protocol logic devices to a single dual redundant hybrid containing everything but the transformers to design code that can be embedded into gate arrays.

43.3 Protocol

The rules under which the transfers occur are referred to as *protocol*. The control, data flow, status reporting, and management of the bus are provided by three word types.

43.3.1 Word Types

Three distinct word types are defined by the standard. These are command words (CWs), DWs, and SWs. Each word type has a unique format yet all three maintain a common structure. Each word is 20 bits in length. The first three bits are used as a synchronization field, thereby allowing the decode clock to resync at the beginning of each new word. The following 16 bits are the information field and differ among the three word types. The last bit is the parity bit. Parity is based on odd parity for the single word. The three word types are shown in Figure 43.5.

Bit encoding for all words is based on biphase Manchester II format. The Manchester II format provides a self-clocking waveform in which the bit sequence is independent. The positive and negative voltage levels of the Manchester waveform are direct current (DC) balanced (same amount of positive signal as there is negative signal) and as such are well suited for transformer coupling. A transition of the signal occurs at the center of the bit time. A logic "0" is a signal that transitions from a negative level to a positive level. A logic "1" is a signal that transitions from a positive level to a negative level.

The terminal's hardware is responsible for the Manchester encoding and decoding of the word types. The interface that the subsystem sees is the 16 bit information field of all words. The sync and parity fields are not provided directly. However, for received messages, the decoder hardware provides a signal to the protocol logic as to the sync type the word was and as to whether parity was valid or not. For transmitted messages, there is an input to the encoder as to what sync type to place at the beginning of the word, and parity is automatically calculated by the encoder.

43.3.1.1 Sync Fields

The first three bit times of all word types is called the sync field. The sync waveform is in itself an invalid Manchester waveform as the transition only occurs at the middle of the second bit time. The use of this distinct pattern allows the decoder to resync at the beginning of each word received and maintain the overall stability of the transmissions.

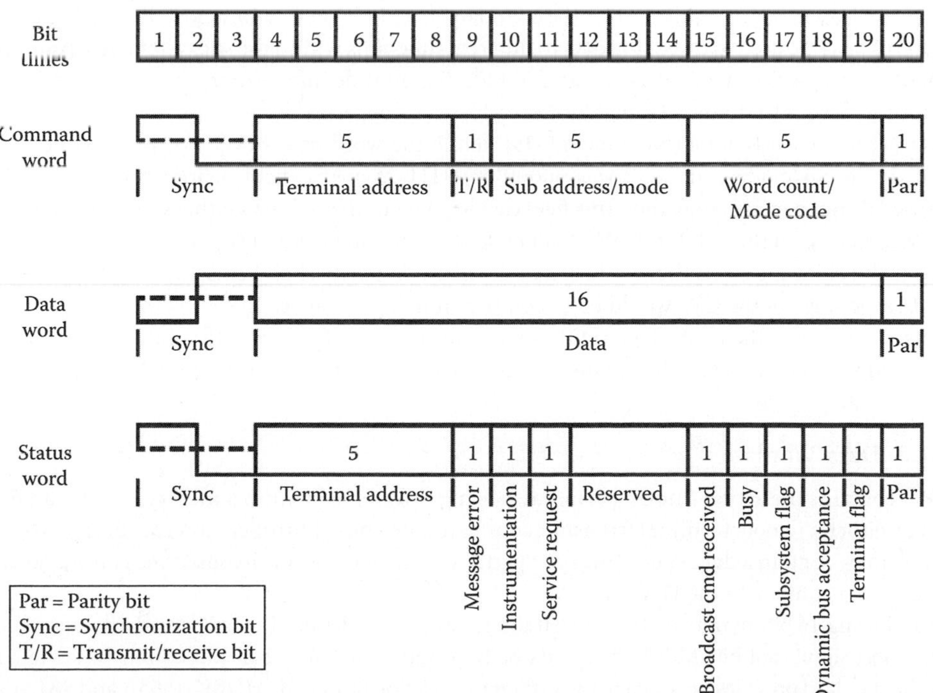

FIGURE 43.5 Word formats.

Two distinct sync patterns are used: the command/status sync and the data sync. The command/status sync has a positive voltage level for the first one and a half bit times, then transitions to a negative voltage level for the second one and a half bit times. The data sync is the opposite—a negative voltage level for the first one and a half bit times, then transitions to a positive voltage level for the second one and a half bit times. The sync patterns are shown in Figure 43.5. As the sync field is the same for both the CW and the SW, the user merely needs to follow the protocol to determine which is which—only a bus controller issues a command sync and only a remote terminal issues status sync.

43.3.1.2 Command Word

The CW specifies the function that a remote terminal(s) is to perform. This word is only transmitted by the active bus controller. The word begins with a command sync in the first three bit times. The following 16-bit information field is as defined in Figure 43.5.

The five-bit TA field (bit times 4–8) states to which unique remote terminal the command is intended (no two terminals may have the same address). Note that an address of 00000 is a valid address and that an address of 11111 is reserved for use as the broadcast address. Also note that there is no requirement that the bus controller be assigned an address; therefore, the maximum number of terminals the data bus can support is 31. Notice 2 to the standard requires that the TA be wire programmable externally to the black box (i.e., an external connector) and that the remote terminal electronics perform a parity test upon the wired TA. The notice basically states that an open circuit on an address line is detected as a logic "1," that connecting an address line to ground is detected as a logic "0," and that odd parity will be used in testing the parity of the wired address field.

The next bit (bit time 9) is the transmit/receive (T/R) bit. This defines the direction of information flow and is always from the point of view of the remote terminal. A transmit command (logic 1) indicates that the remote terminal is to transmit data, while a receive command (logic 0) indicates that the remote terminal is going to receive data. The only exceptions to this rule are associated with MCs.

The following five bits (bit times 10–14) are the subaddress (SA)/MC bits. A logic 00000 or 11111 within this field shall be decoded to indicate that the command is a mode code command. All other logic combinations of this field are used to direct the data to different functions within the subsystem. An example might be that 00001 is position data, 00010 is rate data, 10010 is altitude information, and 10011 is self-test data. The use of the SAs is left to the designer; however, Notice 2 suggests the use of SA 30 for data wraparound.

The next five-bit positions (bit times 15–19) define the word count (WC) or mode code to be performed. If the SA/mode code field was 00000 or 11111, then this field defines the mode code to be performed. If not a mode code, then this field defines the number of DWs either to be received or transmitted depending on the T/R bit. A WC field of 00000 is decoded as 32 DWs.

The last bit (bit time 20) is the word parity bit. Only odd parity shall be used.

The best analogy of the CW would be to compare it to a post office. The TA is the address of the post office itself. The SA is the post-office (PO) box within the post office (directing information to different users/functions) and the WC is the number of letters to be input to or removed from the PO box depending on the T/R bit.

43.3.1.3 Data Word

The DW contains the actual information that is being transferred within a message. DWs can be transmitted by either a remote terminal (transmit command) or a bus controller (receive command). The first three bit times contain a data sync. This sync pattern is the opposite of that used for command and SWs and therefore is unique to the DW type.

The following 16 bits of information are left to the designer to define. The only standard requirements are that the most significant bit (MSB) of the data be transmitted first and that unused bits be set to a logic 0. While the standard provides no guidance as to their use, Section 80 of MIL-HDBK-1553A and SAE AS-15532 provides guidance and lists the formats (i.e., bit patterns, resolutions) of the most commonly used DWs.

The last bit (bit time 20) is the word parity bit. Only odd parity shall be used.

43.3.1.4 Status Word

The SW is only transmitted by a remote terminal in response to a valid message. The SW is used to convey to the bus controller whether a message was properly received or the state of the remote terminal (i.e., service request, busy). The SW is defined in Figure 43.5. Since the SW conveys information to the bus controller, there are two views as to the meaning of each bit—what the setting of the bit means to a remote terminal and what the setting of the bit means to a bus controller. Each field of the SW, and its potential meanings, is examined in the following.

43.3.1.4.1 Resetting the Status Word

The SW, with the exception of the remote TA, is cleared after receipt of a valid CW. The two exceptions to this rule are if the CW received is a transmit SW mode code or a transmit last CW mode code. Conditions that set the individual bits of the word may occur at any time. If after clearing the SW, the conditions for setting the bits still exist, then the bits shall be set again.

Upon detection of an error in the data being received, the message error (ME) bit is set and the transmission of the SW is suppressed. The transmission of the SW is also suppressed upon receipt of a broadcast message. For an illegal message (i.e., an illegal CW), the ME bit is set and the SW is transmitted.

43.3.1.4.2 Status Word Bits

Terminal address: The first five bits (bit times 4–8) of the information field are the TA. These five bits should match the corresponding field within the CW that the terminal received. The remote terminal sets these bits to the address to which it has been programmed. The bus controller should examine these bits to ensure that the terminal responding with its SW was indeed the terminal to which the CW was addressed. In the case of a RT-RT message, the receiving terminal should compare the address of the second CW with that of the received SW. While not required by the standard, it is good design practice to ensure that the data received are from a valid source.

Message error: The next bit (bit time 9) is the ME bit. This bit is set to a logic "1" by the remote terminal upon detection of an error in the message or upon detection of an invalid message (i.e., illegal command) to the terminal. The error may occur in any of the DWs within the message. When the terminal detects an error and sets this bit, none of the data received within the message shall be used. If an error is detected within a message and the ME bit is set, the remote terminal must suppress the transmission of the SW (see Resetting of the Status Word). If the terminal detected an illegal command, the ME bit is set and the SW is transmitted. All remote terminals must implement the ME bit in the SW.

Instrumentation: The instrumentation bit (bit time 10) is provided to differentiate between a CW and an SW (remember, they both have the same sync pattern). The instrumentation bit in the SW is always set to logic "0." If used, the corresponding bit in the CW is set to a logic "1." This bit in the CW is the MSB of the SA field and therefore would limit the SAs used to 10000–11110, hence reducing the number of SAs available from 30 to 15. The instrumentation bit is also the reason why there are two mode code identifiers (00000 and 11111), the latter required when the instrumentation bit is used.

Service request: The service request bit (bit time 11) is such that the remote terminal can inform the bus controller that it needs to be serviced. This bit is set to a logic "1" by the subsystem to indicate that servicing is needed. This bit is typically used when the bus controller is *polling* terminals to determine if they require processing. The bus controller upon receiving this bit set to a logic "1" typically does one of the following. It can take a predetermined action such as issuing a series of messages, or it can request further data from the remote terminal as to its needs. The latter can be accomplished by requesting the terminal to transmit data from a defined SA or by using the transmit vector word mode code.

Reserved: Bit times 12–14 are reserved for future growth of the standard and must be set to a logic "0." The bus controller should declare a message in error if the remote terminal responds with any of these bits set in its SW.

Broadcast command received: The broadcast command received bit (bit time 15) indicates that the remote terminal received a valid broadcast command (e.g., a CW with the TA set to 11111). Upon receipt of a valid broadcast command, the remote terminal sets this bit to logic "1" and suppresses the transmission of its SWs. The bus controller may issue a transmit SW or transmit last CW mode code to determine if the terminal received the message properly.

Busy: The busy bit (bit time 16) is provided as a feedback to the bus controller when the remote terminal is unable to move data between the remote terminal electronics and the subsystem in compliance to a command from the bus controller.

In the earlier days of 1553, the busy bit was required because many of the subsystem interfaces (analogs, synchros, etc.) were much slower compared to the speed of the multiplex data bus. Some terminals were not able to move the data fast enough. So instead of potentially losing data, a terminal was able to set the busy bit, indicating to the bus controller it could not handle new data at that time and for the bus controller to try again later. As new systems have been developed, the need for the use of busy has been reduced. However, there are systems that still need and have a valid use for the busy bit. Examples of these are radios, where the bus controller issues a command to the radio to tune to a certain frequency. It may take the radio several milliseconds to accomplish this, and while it is tuning, it may set the busy bit to inform the bus controller that it is doing as it was told.

When a terminal is busy, it does not need to respond to commands in the *normal* way. For receive commands, the terminal collects the data, but does not have to pass the data to the subsystem. For transmit commands, the terminal transmits its SW only. Therefore, while a terminal is busy, the data it supplies to the rest of the system are not available. This can have an overall effect upon the flow of data within the system and may increase the data latency within time-critical systems (e.g., flight controls).

Some terminals used the busy bit to overcome design problems, setting the busy bit whenever needed. Notice 2 to the standard *strongly discourages* the use of the busy bit. However, as shown in the example earlier, there are valid needs for its use. Therefore, if used, Notice 2 now requires that the busy bit may only be set as the result of a particular command received from the bus controller and not due to an internal periodic or processing function. By following this requirement, the bus controller, with prior knowledge of the remote terminal's characteristics, can determine what will cause a terminal to go busy and minimize the effects on data latency throughout the system.

Subsystem flag: The subsystem flag bit (bit time 17) is used to provide *health* data regarding the subsystems to which the remote terminal is connected. Multiple subsystems may logically *OR* their bits together to form a composite health indicator. This single bit is only to serve as an indicator to the bus controller and user of the data that a fault or failure exists. Further information regarding the nature of the failure must be obtained in some other fashion. Typically, an SA is reserved for built-in-test (BIT) information, with one or two words devoted to subsystem status data.

Dynamic bus control acceptance: The dynamic bus control acceptance bit (bit time 18) is used to inform the bus controller that the remote terminal has received the dynamic bus control mode code and has accepted control of the bus. For the remote terminal, the setting of this bit is controlled by the subsystem and is based upon passing some level of BIT (i.e., a processor passing its power-up and continuous background tests).

The remote terminal upon transmitting its SW becomes the bus controller. The bus controller, upon receipt of the SW from the remote terminal with this bit set, ceases to function as the bus controller and may become a remote terminal or bus monitor.

Terminal flag: The terminal flag bit (bit time 19) is used to inform the bus controller of a fault or failure within the remote terminal circuitry (only the remote terminal). A logic "1" shall indicate a fault condition. This bit is used solely to inform the bus controller of a fault or failure. Further information regarding the nature of the failure must be obtained in some other fashion. Typically, an SA is reserved for BIT information, or the bus controller may issue a transmit BIT word mode code.

Parity: The last bit (bit time 20) is the word parity bit. Only odd parity shall be used.

43.3.2 Message Formats, Validation, and Timing

The primary purpose of the data bus is to provide a common medium for the exchange of data between systems. The exchange of data is based upon message transmissions. The standard defines 10 types of message transmission formats. All of these formats are based upon the three word types just defined. The 10 message formats are shown in Figures 43.6 and 43.7. The message formats have been divided into two groups. These are referred to within the standard as the *information transfer formats* (Figure 43.6) and the *broadcast information transfer formats* (Figure 43.7).

The information transfer formats are based upon the command/response philosophy that all error-free transmissions received by a remote terminal be followed by the transmission of an SW from the terminal to the bus controller. This handshaking principle validates the receipt of the message by the remote terminal.

Broadcast messages are transmitted to multiple remote terminals at the same time. As such, the terminals suppress the transmission of their SWs (not doing so would have multiple boxes trying to talk at the same time and thereby *jam* the bus). In order for the bus controller to determine if a terminal received the message, a polling sequence to each terminal must be initiated to collect the SWs.

Each of the message formats is summarized in the following sections.

43.3.2.1 Bus Controller to Remote Terminal

The bus controller to remote terminal (BC-RT) message is referred to as the receive command since the remote terminal is going to receive data. The bus controller outputs a CW to the terminal defining the SA of the data and the number of DWs it is sending. Immediately (without any gap in the transmission), the number of DWs specified in the CW is sent.

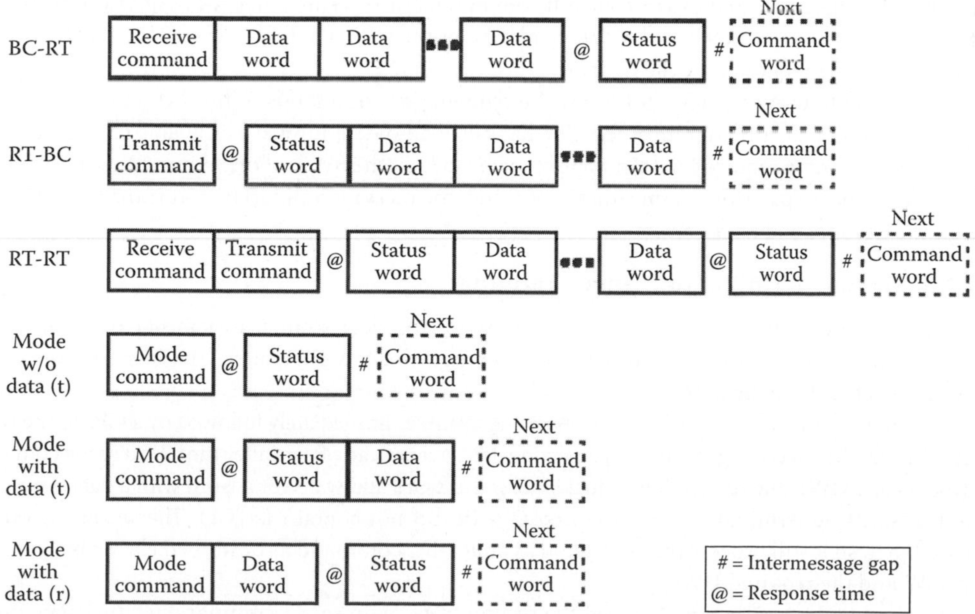

FIGURE 43.6 Information transfer formats.

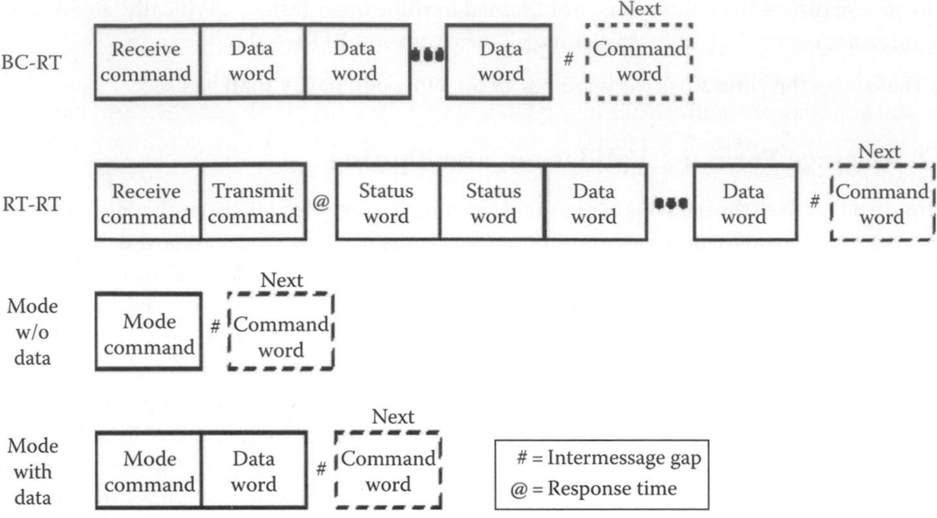

FIGURE 43.7 Broadcast information transfer formats.

The remote terminal upon validating the CW and all of the DWs will issue its SW within the response time requirements (maximum of 12 μs).

The remote terminal must be capable of processing the next command that the bus controller issues. Therefore, the remote terminal has approximately 56 μs (SW response time 12 μs, plus SW transmit time 20 μs, plus intermessage gap minimum 4 μs, plus CW transmit time 20 μs) to either pass the data to the subsystem or buffer the data.

43.3.2.2 Remote Terminal to Bus Controller

The RT-BC message is referred to as a transmit command. The bus controller issues only a transmit CW to the remote terminal. The terminal, upon validation of the CW, will first transmit its SW followed by the number of DWs requested by the CW.

Since the remote terminal does not know the sequence of commands to be sent and does not normally operate upon a command until the CW has been validated, it must be capable of fetching from the subsystem the data required within approximately 28 μs (the SW response time 12 μs, plus the SW transmission time 20 μs, minus some amount of time for message validation and transmission delays through the encoder and transceiver).

43.3.2.3 Remote Terminal to Remote Terminal

The RT-RT command is provided to allow a terminal (the data source) to transfer data directly to another terminal (the data sink) without going through the bus controller. The bus controller may, however, collect and use the data.

The bus controller first issues a CW to the receiving terminal immediately followed by a CW to the transmitting terminal. The receiving terminal is expecting data, but instead of data after the CW, it sees a command sync (the second CW). The receiving terminal ignores this word and waits for a word with a data sync.

The transmitting terminal ignored the first CW (it did not contain its TA). The second word was addressed to it, so it will process the command as an RT-BC command as described earlier by transmitting its SW and the required DWs.

The receiving terminal, having ignored the second CW, again sees a command (status) sync on the next word and waits further. The next word (the first DW sent) now has a data sync and the receiving remote terminal starts collecting data. After receipt of all of the DWs (and validating), the terminal transmits its SW.

43.3.2.3.1 RT-RT Validation

There are several things that the receiving remote terminal of an RT-RT message should do. First, Notice 2 requires that the terminal time is out in 54–60 μs after receipt of the CW. This is required since if the transmitting remote terminal did not validate its CW (and no transmission occurred), then the receiving terminal will not collect data from some new message. This could occur if the next message is either a transmit or receive message, where the terminal ignores all words with a command/status sync and would start collecting DWs beginning with the first data sync. If the same number of DWs were being transferred in the follow-on message and the terminal did not test the command/SW contents, then the potential exists for the terminal to collect erroneous data.

The other function that the receiving terminal should do, but is not required by the standard, is to capture the second CW and the first transmitted DW. The terminal could compare the TA fields of both words to ensure that the terminal doing the transmitting was the one commanded to transmit. This would allow the terminal to provide a level of protection for its data and subsystem.

43.3.2.4 Mode Command Formats

Three MC formats are provided for. This allows for MCs with no DWs and for the MCs with one DW (either transmitted or received). The status/data sequencing is as described for the BC-RT or RT-BC messages except that the data WC is either one or zero. Mode codes and their use are described later.

43.3.2.5 Broadcast Information Transfer Formats

The broadcast information transfer formats, as shown in Figure 43.7, are identical to the nonbroadcast formats described previously with the following two exceptions. First, the bus controller issues commands to TA 31 (11111), which is reserved for this function. And secondly, the remote terminals receiving the messages (those that implement the broadcast option) suppress the transmission of their SW.

The broadcast option can be used with the message formats in which the remote terminal receives data. Obviously, multiple terminals cannot transmit data at the same time, so the RT-BC transfer format and the transmit mode code with data format cannot be used. The broadcast RT-RT allows the bus controller to instruct all remote terminals to receive and then instruct one terminal to transmit, thereby allowing a single subsystem to transfer its data directly to multiple users.

Notice 2 allows the bus controller to only use broadcast commands with mode codes (see Broadcast Mode Codes). Remote terminals are allowed to implement this option for all broadcast message formats. The notice further states that the terminal must differentiate the SAs between broadcast and nonbroadcast messages (see Section 43.4.1).

43.3.2.6 Command and Message Validation

The remote terminal must validate the CW and all DWs received as part of the message. The criteria for a valid CW are that the word begins with a valid command sync and valid TA (matches the assigned address of the terminal or the broadcast address if implemented), all bits are in a valid Manchester code, there are 16 information field bits, and there is a valid parity bit (odd). The criteria for a DW are the same except a valid data sync is required and the TA field is not tested. If a CW fails to meet the criteria, the command is ignored. After the command has been validated, and a DW fails to meet the criteria, then the terminal shall set the ME bit in the SW and suppress the transmission of the SW. Any single error within a message shall invalidate the entire message and the data shall not be used.

43.3.2.7 Illegal Commands

The standard allows remote terminals the option of monitoring for illegal commands. An illegal command is one that meets the valid criteria for a CW but is a command (message) that is not implemented by the terminal. An example is if a terminal is designed to only output 04 DWs to SA 01 and a CW was received by the terminal that requested it to transmit 06 DWs from SA 01, then this command, while still a valid command, could be considered by the terminal as illegal. The standard only states that the bus controller shall not issue illegal or invalid commands.

The standard provides the terminal designer with two options. First, the terminal can respond to all commands as usual (this is referred to as *responding in form*). The data received are typically placed in a series of memory locations that are not accessible by the subsystem or applications programs. This is typically referred to as the *bit bucket*. All invalid commands are placed into the same bit bucket. For invalid transmit commands, the data transmitted are read from the bit bucket. Remember, the bus controller is not supposed to send these invalid commands.

The second option is for the terminal to monitor for illegal commands. For most terminal designs, this is as simple as a lookup table with the T/R bit, SA, and WC fields supplying the address and the output being a single bit that indicates if the command is valid or not. If a terminal implements illegal command detection and an illegal command is received, the terminal sets the ME bit in the SW and responds with the SW.

43.3.2.8 Terminal Response Time

The standard states that a remote terminal, upon validation of a transmit CW or a receive message (CW and all DWs), shall transmit its SW to the bus controller. The response time is the amount of time the terminal has to transmit its SW. To allow for accurate measurements, the time frame is measured from the midcrossing of the parity bit of the CW to the midcrossing of the sync field of the SW. The minimum time is 4.0 μs; the maximum time is 12.0 μs. However, the actual amount of *dead time* on the bus is 2–10 μs since half of the parity bit and sync waveforms are being transmitted during the measured time frame.

The standard also specifies that the bus controller must wait a minimum of 14.0 μs for a SW response before determining that a terminal has failed to respond. In applications where long data buses are used or where other special conditions exist, it may be necessary to extend this time to 20.0 μs or greater.

43.3.2.9 Intermessage Gap

The bus controller must provide for a minimum of 4.0 μs between messages. Again, this time frame is measured from the midcrossing of the parity bit of the last DW or the SW and the midcrossing of the sync field of the next CW. The actual amount of *dead time* on the bus is 2 μs since half of the parity bit and sync waveforms are being transmitted during the measured time frame.

The amount of time required by the bus controller to issue the next command is a function of the controller type (e.g., word, message, or frame). The gap typically associated with word controllers is between 40 and 100 μs. Message controllers typically can issue commands with a gap of 10–30 μs. But frame controllers are capable of issuing commands at the 4 μs rate and often must require a time delay to slow them down.

43.3.2.10 Superseding Commands

A remote terminal must always be capable of receiving a new command. This may occur while operating on a command on bus A and after the minimum intermessage gap, a new command appears or if operating on bus A and a new command appears on bus B. This is referred to as a superseding command. A second valid command (the new command) shall cause the terminal to stop operating on the first command and start on the second. For dual redundant applications, this requirement implies that all terminals must, as a minimum, have two receivers, two decoders, and two sets of CW validation logic.

43.3.3 Mode Codes

Mode codes are defined by the standard to provide the bus controller with data bus management and error handling/recovery capability. The mode codes are divided into two groups: with and without DWs. The DWs that are associated with the mode codes (and only one word per mode code is allowed) contain information pertinent to the control of the bus and do not generally contain information required by

TABLE 43.4 Mode Code

T/R	Mode Code	Function	Data Word	Broadcast
1	00000	Dynamic bus control	No	No
1	00001	Synchronize	No	Yes
1	00010	Transmit SW	No	No
1	00011	Initiate self-test	No	Yes
1	00100	Transmitter shutdown	No	Yes
1	00101	Override transmitter shutdown	No	Yes
1	00110	Inhibit terminal flag bit	No	Yes
1	00111	Override inhibit terminal flag bit	No	Yes
1	01000	Reset	No	Yes
1	01001	Reserved	No	TBD
1	•	•	No	•
1	•	•	No	•
1	01111	Reserved	No	TBD
1	10000	Transmit vector word	Yes	No
0	10001	Synchronize	Yes	Yes
1	10010	Transmit last CW	Yes	No
1	10011	Transmit BIT word	Yes	No
0	10100	Selected transmitter shutdown	Yes	Yes
0	10101	Override selected transmitter shutdown	Yes	Yes
1/0	10110	Reserved	Yes	TBD
	•	•	Yes	•
	•	•	Yes	•
1/0	11111	Reserved	Yes	TBD

the subsystem (the exception may be the synchronize with DW mode code). The mode codes are defined by bit times 15–19 of the CW. The MSB (bit 15) can be used to differentiate between the two mode code groups. When a DW is associated with the mode code, the T/R bit determines if the DW is transmitted or received by the remote terminal. The mode codes are listed in Table 43.4.

43.3.3.1 Mode Code Identifier

The mode code identifier is contained in bits 10–14 of the CW. When this field is either 00000 or 11111, then the contents of bits 15–19 of the CW are to be decoded as a mode code. Two mode code identifiers are provided such that the system can utilize the instrumentation bit if desired. The two mode code identifiers shall not convey different information.

43.3.3.2 Mode Code Functions

The following defines the functionality of each of the mode codes.

Dynamic bus control: The dynamic bus control mode code is used to provide for the passing of the control of the data bus between terminals, thus providing a *round-robin* type of control. Using this methodology, each terminal is responsible for collecting the data it needs from all the other terminals. When it is done collecting, it passes control to the next terminal in line (based on some predefined sequence). This allows the applications program (the end user of the data) to collect the data when it needs it, always ensuring that the data collected are from the latest source sample and have not been sitting around in a buffer waiting to be used.

Notices 1 and 2 to the standard forbid the use of dynamic bus control for Air Force applications. This is due to the problems and concerns of what may occur when a terminal, which has been passed by the control, is unable to perform or does not properly forward the control to the next terminal, thereby forcing the condition of no terminal being in control and having to reestablish control by some terminal. The potential amount of time required to reestablish control could have disastrous effects upon the system (i.e., especially a flight control system).

A remote terminal that is capable of performing as the bus control should be capable of setting the dynamic bus control acceptance bit in the terminal's SW to logic "1" when it receives the mode code command. Typically, the logic associated with the setting of this bit is based on the subsystem's (e.g., computer's) ability to pass some level of confidence test. If the confidence test passes, then the bit is set and the SW is transmitted when the terminal receives the MC, thereby saying that it will assume the role of bus controller.

The bus controller can only issue the dynamic bus control MC to one remote terminal at a time. The command obviously is only issued to terminals that are capable of performing as a bus controller. Upon transmitting the command, the bus controller must check the terminal's SW to determine if the dynamic bus control acceptance bit is set. If set, the bus controller ceases to function as the controller and becomes either a remote terminal or a bus monitor. If the bit in the SW is not set, the remote terminal that was issued the command is not capable of becoming the bus controller; the current controller must either remain the bus controller or attempt to pass the control to some other terminals.

Synchronize: The synchronize mode code is used to establish some form of timing between two or more terminals. This mode code does not use a DW; therefore, the receipt of this command by a terminal must cause some predefined event to occur. Some examples of this event may be the clearing, incrementing, or presetting of a counter; the toggling of an output signal; or the calling of some software routine. Typically, this command is used to time correlate a function such as the sampling of navigation data (i.e., present position, rates) for flight controls or targeting/fire control systems. Other uses have been for the bus controller to *sync* the backup controllers (or monitors) to the beginning of a major/minor frame processing.

When a remote terminal receives the synchronize MC, it should perform its predefined function. For a bus controller, the issuance of the command is all that is needed. The terminal's SW only indicates that the message was received, not that the *sync* function was performed.

Transmit status word: This is one of the two commands that does not cause the remote terminal to reset or clear its SW. Upon receipt of this command, the remote terminal transmits the SW that was associated with the previous message, not the SW of the mode code message.

The bus controller uses this command for control and error management of the data bus. If the remote terminal had detected an error in the message and suppressed its SW, then the bus controller can issue this command to the remote terminal to determine if indeed the nonresponse was due to an error. As this command does not clear the SW from the previous message, a detected error by the remote terminal in a previous message would be indicated by having the ME bit set in the SW.

The bus controller also uses this command when *polling*. If a terminal does not have periodic messages, the RT can indicate when it needs communications by setting the service request bit in the SW. The bus controller, by requesting the terminal to transmit only its SW, can determine if the terminal is in need of servicing and can subsequently issue the necessary commands. This *polling* methodology has the potential of reducing the amount of bus traffic by eliminating the transmission of unnecessary words.

Another use of this command is when broadcast message formats are used. As all of the remote terminals will suppress their SWs, *polling* each terminal for its SW would reveal whether the terminal received the message by having its broadcast command received bit set.

Initiate self-test: This command, when received by the remote terminal, shall cause the remote terminal to enter into its self-test. This command is normally used as a ground-based maintenance function, as part of the system power-on tests, or, in flight, as part of a fault recovery routine. Note that this test is only for the remote terminal, not the subsystem.

In earlier applications, some remote terminals, upon receipt of this command, would enter self-test and go *offline* for long periods of time. Notice 2, in an effort to control the amount of time that a terminal could be *offline*, limited the test time to 100.0 μs following the transmission of the SW by the remote terminal.

While a terminal is performing its self-test, it may respond to a valid command in the following ways: (1) no response on either bus (*offline*), (2) transmit only the SW with the busy bit set, or (3) normal response. The remote terminal may, upon receipt of a valid command received after this mode code, terminate its self-test. As a subsequent command could abort the self-test, the bus controller, after issuing this command, should suspend transmissions to the terminal for the specified amount of time (either a time specified for the remote terminal or the maximum time of 100.0 μs).

Transmitter shutdown: This command is used by the bus controller in the management of the bus. In the event that a terminal's transmitter continuously transmits, this command provides for a mechanism to turn the transmitter off. This command is for dual redundant standby applications only.

Upon receipt of this command, the remote terminal shuts down (i.e., turns off) the transmitter associated with the opposite data bus. That is to say if a terminal's transmitter is babbling on the A bus, the bus controller would send this command to the terminal on the B bus (a command on the A bus would not be received by the terminal).

Override transmitter shutdown: This command is the complement of the previous one in that it provides a mechanism to turn on a transmitter that had previously been turned off. When the remote terminal receives this command, it shall set its control logic such that the transmitter associated with the opposite bus be allowed to transmit when a valid command is received on the opposite bus. The only other command that can enable the transmitter is the reset remote terminal MC.

Inhibit terminal flag: This command provides for the control of the terminal flag bit in a terminal's SW. The terminal flag bit indicates that there is an error within the remote terminal hardware and that the data being transmitted or the data received may be in error. However, the fault within the terminal may not have any effect upon the quality of the data, and the bus controller may elect to continue with the transmissions knowing a fault exists.

The remote terminal receiving this command shall set its terminal flag bit to logic "0" regardless of the true state of this signal. The standard does not state that the BIT that controls this bit be halted, but only that the results be negated to "0".

Override inhibit terminal flag: This command is the complement of the previous one in that it provides a mechanism to turn on the reporting of the terminal flag bit. When the remote terminal receives this command, it shall set its control logic such that the terminal flag bit is properly reported based upon the results of the terminal's BIT functions. The only other command that can enable the response of the terminal flag bit is the reset remote terminal MC.

Reset remote terminal: This command, when received by the remote terminal, shall cause the terminal electronics to reset to its power-up state. This means that if a transmitter had been disabled or the terminal flag bit inhibited, these functions would be reset as if the terminal had just powered up. Again, remember that the reset applies only to the remote terminal electronics and not to the entire box.

Notice 2 restricts the amount of time that a remote terminal can take to reset its electronics. After transmission of its SW, the remote terminal shall reset within 5.0 μs. While a terminal is resetting, it may respond to a valid command in the following ways: (1) no response on either bus (*offline*), (2) transmit only the SW with the busy bit set, or (3) normal response. The remote terminal may, upon receipt of a valid command received after this mode code, terminate its reset function. As a subsequent command could abort the reset, the bus controller, after issuing this command, should suspend transmissions to the terminal for the specified amount of time (either a time specified for the remote terminal or the maximum time of 5.0 μs).

Transmit vector word: This command shall cause the remote terminal to transmit a DW referred to as the vector word. The vector word shall identify to the bus controller service request information relating to the message needs of the remote terminal. While not required, this mode code is often tied to the service request bit in the SW. As indicated, the contents of the DW inform the bus controller of messages that need to be sent.

The bus controller also uses this command when *polling*. Though typically used in conjunction with the service request bit in the SW, wherein the bus controller requests only the SW (transmit SW mode code) and upon seeing the service request bit set would then issue the transmit vector word mode code, the bus controller can always ask for the vector word (always getting the SW anyway) and reduce the amount of time required to respond to the terminal's request.

Synchronize with data word: The purpose of this synchronize command is the same as the synchronize without DW, except this mode code provides a DW to provide additional information to the remote terminal. The contents of the DW are left to the imagination of the user. Examples from *real-world* applications have used this word to provide the remote terminal with a counter or clock value, to provide a backup controller with a frame identification number (minor frame or cycle number), and to provide a terminal with a new base address pointer used in extending the SA capability.

Transmit last CW: This is one of the two commands that does not cause the remote terminal to reset or clear its SW. Upon receipt of this command, the remote terminal transmits the SW that was associated with the previous message and the last CW (valid) that it received.

The bus controller uses this command for control and error management of the data bus. When a remote terminal is not responding properly, then the bus controller can determine the last valid command the terminal received and can reissue subsequent messages as required.

Transmit BIT word: This MC is used to provide detail with regard to the BIT status of the remote terminal. Its contents shall provide information regarding the remote terminal only (remember the definition) and not the subsystem.

While most applications associate this command with the initiate self-test mode code, the standard requires no such association. Typical use is to issue the initiate self-test mode code, allow the required amount of time for the terminal to complete its tests, and then issue the transmit BIT word mode code to collect the results of the test. Other applications have updated the BIT word on a periodic rate based on the results of a continuous background test (e.g., as a data wraparound test performed with every data transmission). This word can then be transmitted to the bus controller, upon request, without having to initiate the test and then wait for the test to be completed. The contents of the DW are left to the terminal designer.

Selected transmitter shutdown: Like the transmitter shutdown mode code, this mode code is used to turn off a babbling transmitter. The difference between the two mode codes is that this mode code has a DW associated with it. The contents of the DW specify which data bus (transmitter) to shutdown. This command is used in systems that provide more than dual redundancy.

Override selected transmitter shutdown: This command is the complement of the previous one in that it provides a mechanism to turn on a transmitter that had previously been turned off. When the remote terminal receives this command, the DW specifies which data bus (transmitter) shall set its control logic such that the transmitter associated with that bus be allowed to transmit when a valid command is received on that bus. The only other command that can enable the selected transmitter is the reset remote terminal MC.

Reserved mode codes: As can be seen from Table 43.4, there are several bit combinations that are set aside as reserved. It was the intent of the standard that these be reserved for future growth. It should also be noticed from the table that certain bit combinations are not listed. The standard allows the remote terminal to respond to these reserved and *undefined* mode codes in the following manner: set the ME bit and respond (see Section 43.3.2.7) or respond in form. The designer of terminal hardware or a multiplex system is forbidden to use the reserved mode codes for any purpose.

43.3.3.3 Required Mode Codes

Notice 2 to the standard requires that all remote terminals implement the following four mode codes: transmit SW, transmitter shutdown, override transmitter shutdown, and reset remote terminal. This requirement was levied so as to provide the multiplex system designer and the bus controller with a minimum set of commands for managing the multiplex system. Note that the aforementioned requirement was placed on the remote terminal. Notice 2 also requires that a bus controller be capable of implementing all of the mode codes; however, for Air Force applications, the dynamic bus control mode code shall never be used.

43.3.3.4 Broadcast Mode Codes

Notice 2 to the standard allows the broadcast of mode codes (see Table 43.4). The use of the broadcast option can be of great assistance in the areas of terminal synchronization. Ground maintenance and troubleshooting can take advantage of broadcast reset remote terminal or initiate self, but these two commands can have disastrous effects if used while in flight. The designer must provide checks to ensure that commands such as these are not issued by the bus controller or operated upon by a remote terminal when certain conditions exists (e.g., in flight).

43.4 Systems-Level Issues

The standard provides very little guidance in how it is applied. Lessons learned from real-world applications have led to design guides, application notes, and handbooks that provide guidance. This section will attempt to answer some of the systems-level questions and identify implied requirements that, while not specifically called out in the standard, are required nonetheless.

43.4.1 Subaddress Utilization

The standard provides no guidance on how to use the SAs. The assignment of SAs and their functions (the data content) is left to the user. Most designers automatically start assigning SAs with 01 and count upwards. If the instrumentation bit is going to be used, then the SAs must start at 16.

The standard also requires that normal SAs be separated from broadcast SAs. If the broadcast option is implemented, then an additional memory block is required to receive broadcast commands.

43.4.1.1 Extended Subaddressing

The number of SAs that a terminal has is limited to 60 (30 transmit and 30 receive). Therefore, the number of unique DWs available to a terminal is 1920 (60×32). For earlier applications, where data being transferred were analog sensor data and switch settings, this was more than sufficient. However, in today's applications, in which digital computers are exchanging data, or for a video sensor passing digitized video data, the number of words is too limited.

Most terminal designs establish a block of memory for use by the 1553 interface circuitry. This block contains an address start pointer and then the memory is offset by the SA number and the WC number to arrive at a particular memory address.

A methodology of extending the range of the SAs has been successfully utilized. This method either uses a dedicated SA and DW or makes use of the synchronize with DW mode code. The DW associated with either of these contains an address pointer that is used to reestablish the starting address of the memory block. The changing of the blocks is controlled by the bus controller and can be done based on numerous functions. Examples are operational modes, wherein one block is used for startup messages, a different block for takeoff and landing, a different block for navigation and cruise, a different block for mission functions (i.e., attack or evade modes), and a different block for maintenance functions.

Another example is that the changing of the start address could also be associated with minor frame cycles. Minor frames could have a separate memory block for each frame. The bus controller could synchronize frames and change memory pointers at the beginning of each new minor frame.

For computers exchanging large amounts of data (e.g., GPS almanac tables) or for computers that receive program loads via the data bus at power-up, the bus controller could set the pointers at the beginning of a message block, send thirty 32-word messages, move the memory pointer to the last location in the remote terminals memory that received data, then send the next block of thirty 32-word messages, continuing this cycle until the memory is loaded. The use is left to the designer.

43.4.2 Data Wraparound

Notice 2 to the standard does require that the terminal is able to perform a data wraparound and SA 30 is suggested for this function. Data wraparound provides the bus controller with a methodology of testing the data bus from its internal circuitry, through the bus media, to the terminal's internal circuitry. This is done by the bus controller sending the remote terminal a message block and then commanding the terminal to send it back. The bus controller can then compare the sent data with that received to determine the state of the data link. There are no special requirements upon the bit patterns of the data being transferred.

The only design requirements are placed upon the remote terminal. These are that the terminal, for the data wraparound function, be capable of sending the number of DWs equal to the largest number of DWs sent for any transmit command. This means that if a terminal maximum data transmission is only four DWs, it need only provide for four DWs in its data wraparound function.

The other requirement is that the remote terminal need only hold the data until the next message. The normal sequence is for the bus controller to send the data, then in the next message, it asks for it back. If another message is received by the remote terminal before the bus controller requests the data, the terminal can discard the data from the wraparound message and operate on the new command.

43.4.3 Data Buffering

The standard specifies that the any error within a message shall invalidate the entire message. This implies that the remote terminal must store the data within a message buffer until the last DW has been received and validated before allowing the subsystem access to the data. To ensure that the subsystem always has the last message of valid data received to work with would require the remote terminal to, as a minimum, double buffer the received data.

There are several methods to accomplish this in hardware. One method is for the terminal electronics to contain a first-in first-out (FIFO) memory that stores the data as it is received. Upon validation of the last DW, the terminal's subsystem interface logic will move the contents of the FIFO into the memory accessible by the subsystem. If an error occurred during the message, the FIFO is reset.

A second method establishes two memory blocks for each message in common memory. The subsystem is directed to read from one block (block A), while the terminal electronics writes to the other (block B). Upon receipt of a valid message, the terminal will switch pointers, indicating that the subsystem is to read from the new memory block (block B) while the terminal will now write to block B. If an error occurs within the message, the memory blocks are not switched.

Some of the *off-the-shelf* components available provide for data buffering. Most provide for double buffering, while some provided for multilevels of buffering.

43.4.4 Variable Message Blocks

Remote terminals should be able to transmit any subset of any message. This means that if a terminal has a transmit message at SA 04 of 30 DWs, it should be capable of transmitting any number of those DWs (01–30) if so commanded by the bus controller. The order in which the subset is transmitted should be the same as if the entire message is being transmitted, that being the contents of DW 01 is the same regardless of the WC.

Terminals that implement illegal command detection should not consider subsets of a message as illegal. That is to say, if in our previous example a command is received for 10 DWs, this should not be illegal. But, if a command is received for 32 DWs, this would be considered as an illegal command.

43.4.5 Sample Consistency

When transmitting data, the remote terminal needs to ensure that each message transmitted is of the same sample set and contains mutually consistent data. Multiple words used to transfer multiple precision parameters or functionally related data must of the same sampling.

If a terminal is transmitting pitch, roll, and yaw rates, and while transmitting the subsystem updates these data in memory, but this occurs after pitch and roll had been read by the terminal's electronics, then the yaw rate transmitted would be of a different sample set. Having data from different sample rates could have undesirable effects on the user of the data.

This implies that the terminal must provide some level of buffering (the reverse of what was described earlier) or some level of control logic to block the subsystem from updating data while being read by the remote terminal.

43.4.6 Data Validation

The standard tightly defines the criteria for the validation of a message. All words must meet certain checks (i.e., valid sync, Manchester encoding, number of bits, odd parity) for each word and each message to be valid. But what about the contents of the DW? MIL-STD-1553 provides the checks to ensure the quality of the data transmission from terminal to terminal, sort of a *data in equals data out*, but is not responsible for the validation tests of the data itself. This is not the responsibility of the 1553 terminal electronics, but of the subsystem. If bad data are sent, then *garbage in equals garbage out*. But the standard does not prevent the user from providing additional levels of protection. The same techniques used in digital computer interfaces (i.e., disk drives, serial interfaces) can be applied to 1553. These techniques include checksums, cyclic redundancy check (CRC) words, and error detection/correction codes. Section 80 of MIL-HDBK-1553A that covers DW formats even offers some examples of these techniques.

But what about using the simple indicators embedded within the standard. Each remote terminal provides an SW—indicating not only the health of the remote terminal's electronics but also that of the subsystem. However, in most designs, the SW is kept within the terminal electronics and not passed to the subsystems. In some *off-the-shelf* components, the SW is not even available to be sent to the subsystem. But two bits from the SW should be made available to the subsystem and the user of the data for further determination as to the validity of the data. These are the subsystem flag and the terminal flag bits.

43.4.7 Major and Minor Frame Timing

The standard specifies the composition of the words (command, data, and status) and the messages (information formats and broadcast formats). It provides a series of management messages (mode codes), but it does not provide any guidance on how to apply these within a system. This is left to the imagination of the user.

Remote terminals, based upon the contents of their data, will typically state how often data are collected and the fastest rate they should be outputted. For input data, the terminal will often state how often it needs certain data to either perform its job or maintain a certain level of accuracy. The rates are referred to as the transmission and update rates. It is the system designer's job to examine the data needs of all of the systems and determine when data are transferred from whom to whom. These data are subdivided into periodic messages, those that must be transferred at some fixed rate, and aperiodic messages, those that are typically either event driven (i.e., the operator pushes a button) or data driven (i.e., a value is now within range).

A major frame is defined such that all periodic messages are transferred at least once. This is therefore defined by the message with the slowest transmission rate. Typical major frame rates used in today's applications vary from 40 to 640 μs. There are some systems that have major frame rates in the 1–5 s range, but these are the exceptions, not the norm. Minor frames are then established to meet the requirements of the higher update rate messages.

The sequence of messages within a minor frame is again left undefined. There are two methodologies that are predominately used. In the first method, the bus controller starts the frame with the transmission of all of the periodic messages (transmit and receive) to be transferred in that minor frame. At the end of the periodic messages, either the bus controller is finished (resulting in dead bus time—no transmissions) until the beginning of the next frame or the bus controller can use this time to transfer aperiodic messages, error handling messages, or transfer data to the backup bus controller(s).

In the second method (typically used in a centralized processing architecture), the bus controller issues all periodic and aperiodic transmit messages (collects the data), then processes the data (possibly using dead time during this processing), and then issues all the receive messages (outputting the results of the processing). Both methods have been used successfully.

43.4.8 Error Processing

The amount and level of error processing is typically left to the systems designer but may be driven by the performance requirements of the system. Error processing is typically only afforded to critical messages, wherein the noncritical messages just await the next normal transmission cycle. If a data bus is 60% loaded and each message received an error, the error processing would exceed 100% of available time and thereby cause problems within the system.

Error processing is again a function of the level of sophistication of the bus controller. Some controllers (typically message or frame controllers) can automatically perform some degree of error processing. This usually is limited to a retransmission of the message either once on the same bus or once on the opposite bus. Should the retried message also fail, the bus controller software is informed of the problem. The message may then be retried at the end of the normal message list for the minor frame.

If the error still persists, then it may be necessary to stop communicating with the terminal, especially if the bus controller is spending a large amount of time performing error processing. Some systems will try to communicate with a terminal for a predefined number of times on each bus. After this, all messages to the terminal are removed from the minor frame lists and substituted with a single transmit SW mode code.

An analysis should be performed on the critical messages to determine the effects upon the system if they are not transmitted or the effects of data latency if they are delayed to the end of the frame.

43.5 Testing

The testing of a MIL-STD-1553 terminal or system is not a trivial task. There are a large number of options available to the designer including message formats, MCs, SW bits, and coupling methodology. In addition, history has shown that different component manufacturers and designers have made different interpretations regarding the standard, thereby introducing products that implement the same function quite differently.

TABLE 43.5 SAE 1553 Test Plans

AS-4111	Remote terminal validation test plan
AS-4112	Remote terminal production test plan
AS-4113	Bus controller validation test plan
AS-4114	Bus controller production test plan
AS-4115	Data bus system test plan
AS-4116	Bus monitor test plan
AS-4117	Bus components test plan

For years, the Air Force provided for the testing of MIL- STD-1553 terminals and components. Today, this testing is the responsibility of the industry. The SAE, in conjunction with the government, has developed a series of test plans for all 1553 elements. These test plans are listed in Table 43.5.

43.6 Further Information

In addition to the SAE test plans listed in Table 43.5, there are other documents that can provide a great deal of insight and assistance in designing with MIL-STD-1553:

- MIL-STD-1553B *Digital Time Division Command/Response Multiplex Data Bus*
- MIL-HDBK-1553A *Multiplex Applications Handbook*
- SAE AS-15531 *Digital Time Division Command/Response Multiplex Data Bus*
- SAE AS-15532 *Standard Data Word Formats*
- SAE AS-12 *Multiplex Systems Integration Handbook*
- SAE AS-19 MIL-STD-1553 *Protocol Reorganized*
- DDC 1553 *Designers Guide*
- UTMC 1553 Handbook

And lastly, there is the SAE 1553 Users Group. This is a collection of industry and military experts in 1553 who provide an open forum for information exchange and provide guidance and interpretations/ clarifications with regard to the standard. This group meets twice a year as part of the SAE Avionics Systems Division conferences.

44

ASCB: Avionics Standard Communications Bus

Michael Paulitsch
Airbus Group Innovations

44.1 Protocol Overview **44**-1
44.2 Protocol Services **44**-1
Communication Services • Clock Synchronization, Restart, Reintegration, and Integration • Diagnostic Services • Fault Isolation • Configuration Services
44.3 Protocol Parameterization **44**-5
44.4 Communication Interface **44**-5
44.5 Validation and Verification Efforts **44**-6
44.6 Example Configuration and Implementation: ASCB in Primus Epic **44**-6
References **44**-10

44.1 Protocol Overview

ASCB stands for *Avionics Standard Communication Bus* and is a proprietary communication protocol for the avionics domain of general, business, and regional aviation aircraft. The general aviation market comprises small multiple-seater aircraft, the business aviation market consists of turbo-engine-powered comfortable small aircraft for business clients, and regional aviation aircraft are single-aisle medium-sized aircraft flying regionally with up to about 100 passengers. Honeywell has developed ASCB for fault-tolerant periodic, real-time communication between avionic modules of its Primus Epic® avionics suit deployed in such aircraft. ASCB is the primary bus in Primus Epic. There are multiple ASCB versions available, some standardized [1], the latest being version D (abbreviated ASCB-D). Detailed information on Primus Epic is available in [2–7].

44.2 Protocol Services

ASCB-D is based on an architecture consisting of four Ethernet buses, two for each side of an airplane. Each of the network interface controllers (NICs) is assigned to one side of the airplane. The NICs on the left side are connected to the left-side primary, the right-side primary, and the left-side secondary (backup) bus. The NICs on the right side are connected to the right-side primary, the left-side primary, and the left-side secondary (backup) bus. Each NIC can listen to and transmit messages on its two buses on its side. Each NIC can also listen to but not transmit on the primary bus of the other side. For example, a left-side NIC can transmit and receive on the left-side primary and backup bus, and listen-only on the right-side primary. It is not connected to the right-side backup bus. From the viewpoint of one NIC, the NICs on the same side are called *onside* NICs; NICs of the other side are called *xside* NICs, where *xside* stands for *cross-side*. Figure 44.1 depicts the structure of the main buses and connections of NICs.

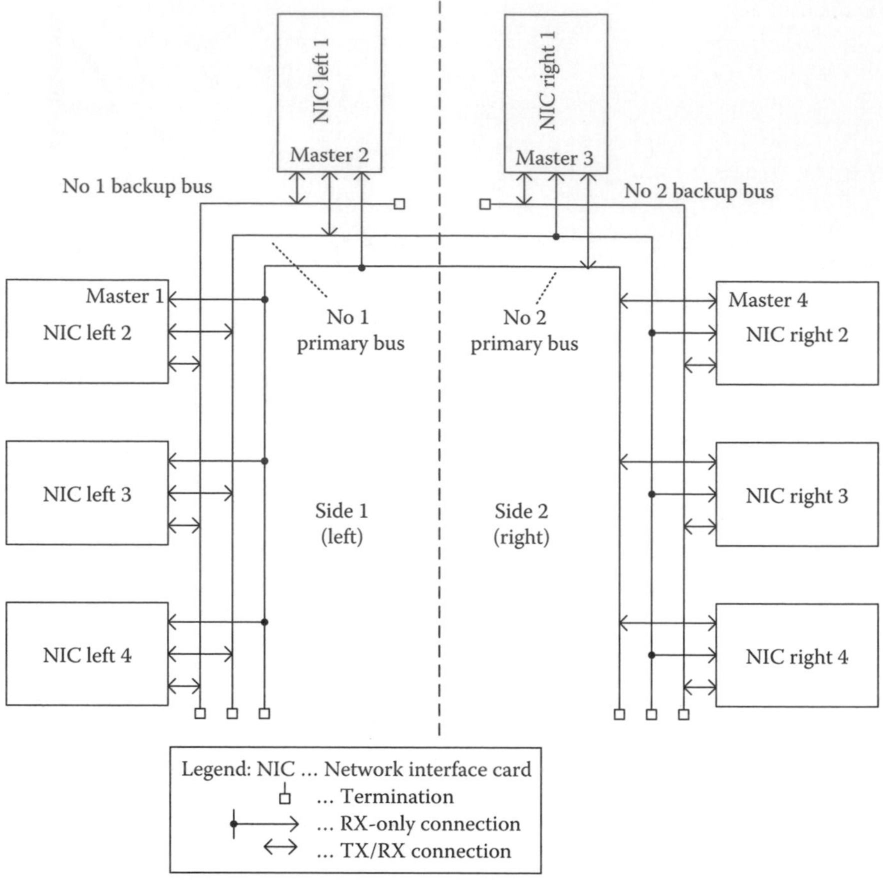

FIGURE 44.1 ASCB architecture.

44.2.1 Communication Services

ASCB-D is a TDMA-based protocol where a minor frame (the fastest period) is running at 80 Hz. Figure 44.2 depicts the bus traffic of one bus. First, master NICs send synchronization pulses (sync pulse) followed by periodic traffic (Ethernet frames). The beginning of each frame is called a *frame tick*. The physical layer of the buses is based on 10 Mbit/s Ethernet using twinax cables and connectors. Figure 44.2 shows the minimum period (called *minor frame*) on the bus. Two sync pulses (special Ethernet frames with synchronization information) are the first frames on the bus. The two time servers that are connected to this bus send these sync pulses. After the sync pulse period, individual NICs (i.e., both the time

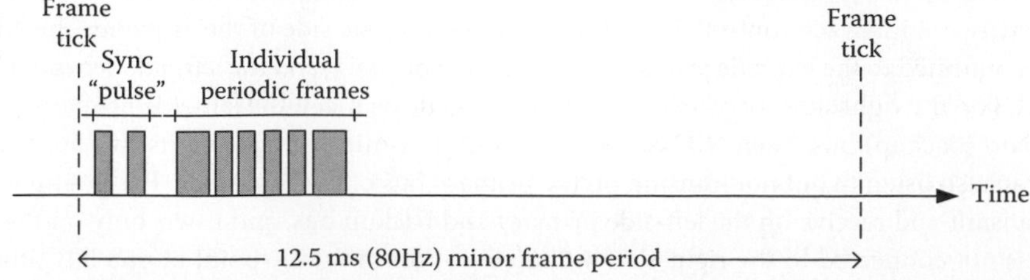

FIGURE 44.2 ASCB minor frame period.

servers and the time slaves) send relative to the frame tick according to a predefined dispatch schedule, where the frame tick is the beginning of a minor frame. The synchronized time of NICs is used for avoidance of collisions on the network. ASCB is part of the Primus Epic architecture, which does not leverage subframe timing (i.e., on actions based on time offsets within the minor period).

44.2.2 Clock Synchronization, Restart, Reintegration, and Integration

Clock synchronization, start-up, or reintegration in Primus Epic is basically the same, as ASCB deploys a master-selection protocol between four cooperating masters called *time servers*. Each of the NICs (including time servers) needs to select one time server. The arbitration on which of the four time servers is selected by other NICs is based on a protocol of cooperating time servers leveraging strike counters and predetermined unique constants in case of conflicts between multiple time servers (multimaster conflict) or no master is active (mastership transition). The time server that first or last counts to its predetermined constant depending on the strike counter communicates its decision of being the timing master, which is then followed by other time servers. Start-up has to complete in 200 ms [4].

In more detail, each time server sends on both of its onside buses and receives the messages on the xside primary bus. A time master receives messages only on one of the onside buses (backup or primary). A time master sends a sync pulse at an offset from its frame tick (i.e., the start of its frame period) based on its ID. Each time server can derive the time difference between another time server's frame tick and its own by measuring the time difference between that server's sync pulse reception time and the expected reception time. Each time server can choose to synchronize to another time server's sync pulse by adjusting its local clock to account for the measured time difference with another time server's sync pulse so that their frame ticks are synchronized.

The protocol provides two strike counters at each time server, called *mastership transition* and *multimaster* strike count, with individual unique thresholds. The *time master* is the currently active time server driving the common time-base for all NICs, hence the names multimaster and mastership transition strike counters. The increment operations of the two strike counters are mutually exclusive. Each time server counts for different periods (using the mastership transition strike counter) in case no time master is detected and sends a sync pulse including indication of mastership (claiming the elected time server is the time master) on the bus in case the mastership transition count reaches the local unique threshold. Once the first one succeeds in sending this, other correct time servers back off and follow the time master. During the period of no time master, time servers run autonomously. If the situation occurs where multiple time servers claim mastership, a time server claiming mastership increases its multimaster strike count. The time server with the largest multimaster threshold wins. A time server gives up mastership and restarts when the multimaster strike count reaches the threshold and multiple time masters are still present. In case of connectivity issues, the synchronization master that does receive multiple synchronization masters does not back off and receives the mastership over other less well-connected masters. This second situation already assumes multiple errors. In case this also fails, the remaining units choose one master and follow its local master. The protocol is not guaranteed to resolve multiple failure scenarios.

The *external timing master test* is performed at certain points in time and is a test that is based upon the receiving sync pulses from other time masters and the relationship of sync pulses to each other. Examples for relationships are tests for periodicity, for temporal gap between successive synchronization signals, for errors in the order of arrival of synchronization signals, and for the absence of particular synchronization signals. The external time master test should prevent any time master from selecting a timing master with unacceptable timing integrity and detect the absence of the timing master. The *self-test* implements a self-monitor and deactivation component in the algorithm. Each time server validates its own timing integrity and deactivates itself when it does not have the required integrity. The time server performs a sync pulse wraparound monitor of the frame counter, a frame tick periodicity monitor, and a local clock integrity monitor. The *break sync action* in Figure 44.3 implies that a master

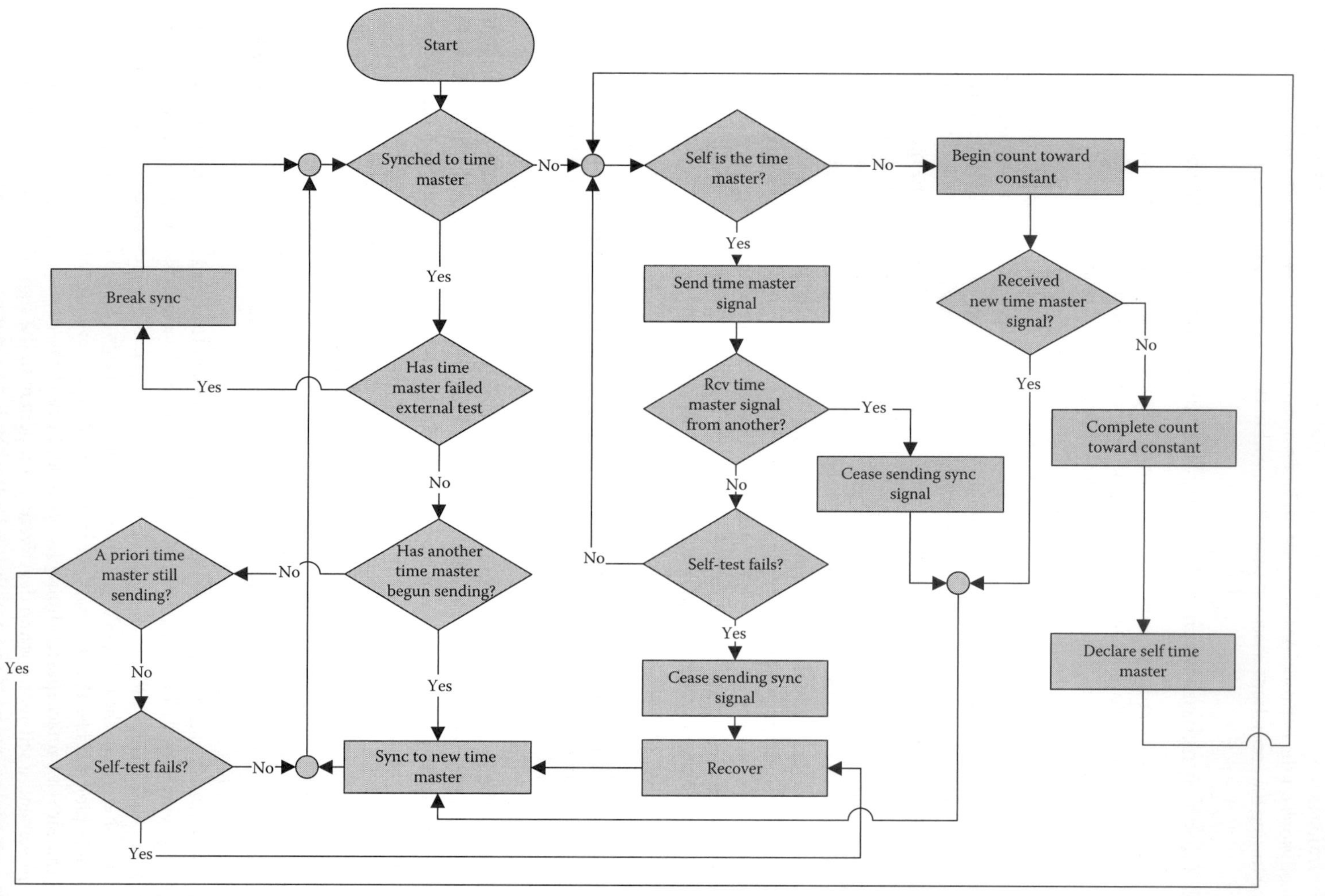

FIGURE 44.3 Flow diagram of ASCB/D synchronization, integration, and start-up algorithms. (From Davidson, D.D. and Chiu, V.Y, Fail-operational global time reference in a redundant synchronous data bus system, Patent Application US 2005/0102586 A1, Honeywell, May 12, 2005.)

that is sure that it is the correct master—based on these locally performed tests (external and self test)—does indicate this to other time masters via sending a synchronization signal claiming mastership or indicating that it wants to take over mastership. Figure 44.3 provides a graphical overview of the ASCB start-up and clock synchronization algorithm.

Please note that the algorithm may have changed slightly due to reported incidents tracing back to synchronization [8].

44.2.3 Diagnostic Services

Examples of diagnostic services are already explained earlier in the synchronization section, where time servers check their correctness based on externally stimulated events. Similarly, NICs perform built-in extensive self-tests at power-up and some continuous self-tests during operation. Heartbeats for power and microcontrollers are deployed and are based, for example, upon the frequency of data transmissions.

44.2.4 Fault Isolation

Fault isolation in ASCB is implemented by the separation of data buses and listen-only connections to primary buses as a first defense. Secondary data buses, which are buses that are limited to one side only, serve as a second level of defense. Transceiver functions for different buses are separated.

44.2.5 Configuration Services

Configuration of ASCB is performed via a special network, a separate local-area connection. This is used to configure local configuration tables of networks and processors. This network can also be used for the monitoring of operation.

44.3 Protocol Parameterization

The communication schedule is composed of multiple minor frames (12.5 ms) in powers of 2 up to 1 s. Exact sending patterns are designed and automatically scheduled by a tool called ESCAPE (Essential System Configuration and Architecture for Primus Epic). Data groups, so-called parameter groups, are handled as a unit, meaning that they are used only if the correctness and completeness checks of all elements of these groups are correct. A correctness check is, for example, whether the error detection codes associated with this data group are correct. A data group is a logical unit, which also implies that a group does not necessarily need to be transmitted in the same Ethernet frame.

Data elements sent on the network are scheduled at frame level (frame is used here in the sense of a frame period). That is, the location of a message (Ethernet frame) within the frame period does not matter for end-to-end latency calculations for applications. The data transmitted during a frame period are not immediately used within this frame period, but only in the next frame period. This allows deploying very simple scheduling algorithms at the expense of delay. Effectively, the frame-period-level scheduling approach decouples any dependencies between the network timing table and operating system (OS) timing tables. Consequently, the delay from one data element in an application is always larger than a minor frame period. Actually, as this principle is deployed based on the frequency of the application, the delay is always larger than the frame period of the application at which the application is run.

44.4 Communication Interface

The communication interfaces in Primus Epic are generally using ping/pong buffers. These are double buffers, where one buffer is used by the application in one period, while the other buffer is used for updates by the network in the same period. This way, network traffic does not overwrite the data that are currently used by the application and vice versa. At a predefined point in time, namely, at the period boundaries, the role of the buffers changes, and the other buffer that has been reserved by the application is now used by the network for sending and receiving.

44.5 Validation and Verification Efforts

ASCB core algorithms, such as clock synchronization, have been verified under the covered fault assumptions and is described in [4,5]. ASCB has been developed using the common applicable aerospace standards for design including DO-178b.

44.6 Example Configuration and Implementation: ASCB in Primus Epic

Primus Epic is one of Honeywell's avionics platforms for general, business, and regional aviation. The general aviation market comprises small multiple-seater aircraft, the business aviation market mainly consists of comfortable small turbo-engine-powered aircraft for business clients, and regional aviation aircraft are single-aisle medium aircraft flying regionally with up to about 100 passengers. Honeywell deploys the Primus Epic avionics suite in such aircrafts. In one instance, Primus Epic is also deployed on a helicopter. The biggest market, however, is business aviation. Specifically, Primus Epic is used on the following certified airframes: Agusta AW139 (helicopter), Cessna Citation Sovereign, Dassault Falcon 2000DX, Dassault Falcon 2000LX, Dassault Falcon 900DX, Dassault Falcon 900EX, Dassault Falcon 900LX, Dassault Falcon 7X, Gulfstream 350, Gulfstream 450, Gulfstream 500, Gulfstream 550, Gulfstream 650, Raytheon Hawker 4000, Embraer 170/175, and Embraer 190/195. Further aircraft using Primus Epic are under development.

Primus Epic is an avionics suite consisting of single or multiple racks and cabinets with integrated circuit cards/modules, which are installed in slots in the cabinets. Each module can contain one or more functions. Each cabinet's configuration can vary by the number of modules installed in each cabinet and the functions loaded into the modules. There can be multiple racks and cabinets installed on the aircraft. The software loaded into the modules determines the functionality of the system modules. The modular avionics unit (MAU) is a hardware cabinet containing field-replaceable modules that represent the *building blocks* of the Primus Epic system. The MAU incorporates input/output (I/O), processing, and database storage modules linked to ASCB and LAN aircraft-wide networks via the NIC modules. Primus Epic deploys a concept called *Virtual Backplane*, meaning that the physical location (e.g., cabinet) of a module is not important. All data generated by any one function within the system are *globally* available to any other function. This makes the MAU flexible and adaptive, allowing for more options in locating and mounting equipment on the aircraft. The integration of processing into a single processing unit (in the meaning of a card) means that it can be shared to perform multiple tasks previously requiring individual processors. This increases the integration results in improved power, weight, reliability, maintainability, and volume. The Primus Epic system redundancy can support multiple redundancy concepts, among others a dual-dual replication arrangement for system redundancy [6].

Summarizing, Honeywell claims that deployment of Primus Epic with its MAU, modular radio cabinets, and flat panel displays provides the following advantages over existing avionics suites mainly due to integration and improved technology deployment:

- Lower total acquisition costs
- Higher reliability and dispatchability
- Improved maintainability
- Lower weight and power
- Lower wire count and installation costs

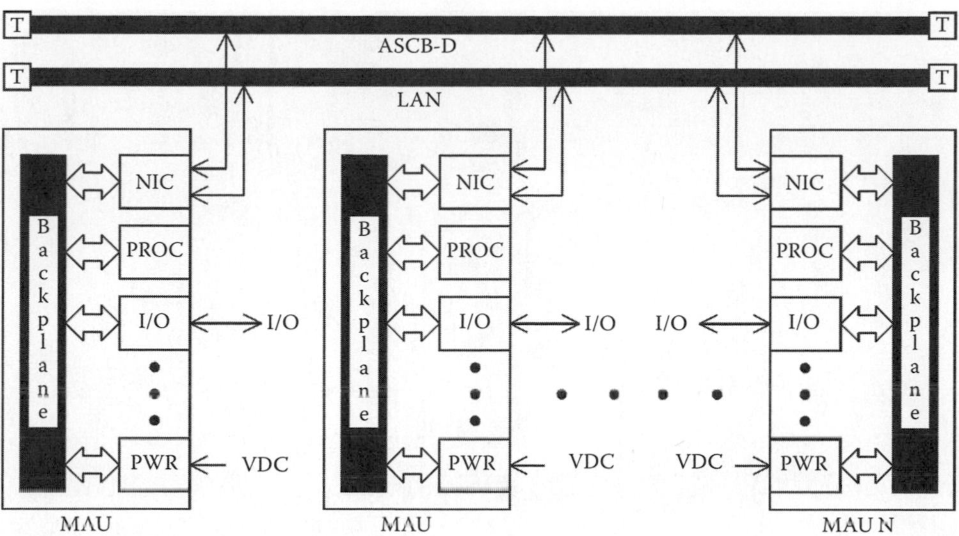

FIGURE 44.4 Primus Epic buses. (Courtesy of Honeywell.)

- Increased functionality
- Large growth capability
- Lower pilot workload and increased safety
- Open architecture by supplying development kits to third-party module suppliers

Figure 44.4 provides a high-level diagram of Primus Epic's compute and on-board communication architecture. Primus Epic in this example consists of multiple MAUs. ASCB-D is depicted as one network in Figure 44.4. The detailed four-bus diagram can be found in Figure 44.6. Furthermore, a LAN is shown, which is used for system maintenance, diagnostics, software and table loading tasks, and a printer. The LAN is an Ethernet 10Base2 network. In each MAU, one or two NICs connect the system-level data sent on ASCB with the backplane databus, a version of Compact PCI (industrial version of PCI derated to 25 MHz being 32-bit-wide). Power modules provide local power to the modules in the cabinet. I/O cards provide interfaces to sensor and actuator data as well as integrate other more federated aircraft systems into the Primus Epic architecture.

Figure 44.5 shows a picture of one Primus Epic cabinet (MAU).

Figure 44.6 depicts an example Primus Epic deployment with two MAUs in the center. Many peripheral, display, utility, sensor, and actuator systems/devices are also shown.

Primus Epic distinguishes itself from other avionics platforms in that it deploys a special OS called DEOS, which stands for *digital engine operating system*. DEOS is a DO-178B level A OS deployed in Federal Aviation Administration–certified environments. It allows multiple software applications/levels to execute on the same processor and provides time and space (memory, I/O) partitioning. Associated with DEOS is a consistent software development method throughout engineering using COTS and custom tools that help in isolating software from hardware. DEOS supports reduction of revalidation and verification efforts of software that changes.

DEOS differs from earlier ARINC 653 [9–11] systems in that it uses preemptive fixed priority (PFP) scheduling, and consequently the time partitioning model includes preemption. The most well-known PFP algorithm is rate-monotonic scheduling [12], where tasks with higher rate have higher priority. Additionally, DEOS supports dynamic thread and dynamic time partition allocation. Upgrades of DEOS slack and time partitioning lead to several improvements. For example, about a threefold increase in communication throughput between a processor and a remote file server and approximately a sevenfold

FIGURE 44.5 Primus Epic cabinet (MAU). (Courtesy of Honeywell.)

reduction in the amount of a priori reserved execution time required for response times of noncritical applications were reported. Also, certain display tasks were able to achieve higher average refresh update rates using slack. DEOS algorithms provide robust time partitioning capabilities, enabling the safe cohosting of COTS software with safety-critical software without decreases in the COTS software performance [13].

The concept of *slack stealing* has been a unique feature in safety-critical OSs deployed in aerospace and in DEOS. It enables many improvements like higher processor utilization and quicker response times. Slack stealing is a preemptive processor scheduling algorithm that delays the execution of high-priority periodic tasks to improve response times of aperiodic tasks provided the periodic tasks will not miss any of their deadlines. When the set of periodic tasks is fixed, there is predictable slack inherent in the execution timeline, known as timeline slack, for threads scheduled using PFP scheduling. Timeline slack can be calculated offline, and table lookups can be used at runtime, in combination with other quantities, to quickly determine the amount of time a periodic task can be delayed and still meet its deadline. Slack stealing also makes available reclaimed slack or equivalently unused worst-case compute time of periodic tasks at the priority at which the compute time was initially reserved [13].

The key for the success of slack stealing in DEOS is the relatively high worst-case execution times compared to actual average execution times. Slack stealing boosts processor utilization efficiency and reduces response times significantly. As can be imagined, slack *moves* the actual execution time of tasks within a frame period. This may be a good illustration for readers to understand why Primus Epic does not deploy subframe-level scheduling. Subframe-level scheduling is a concept denoting the fact that data sent on a network are used by an application with its frame period (i.e., on actions based on time offsets within the frame period). This may tightly couple processor and network execution time. As slack requires some *freedom* of moving actual task execution times within its frame period, Primus Epic designers have chosen *not* to leverage end-to-end worst-case latency improvements obtained by deploying subframe-scheduling, but rather leverage the slack of execution time to boost performance. Each approach has its advantages, but both cannot practically be deployed at the same time.

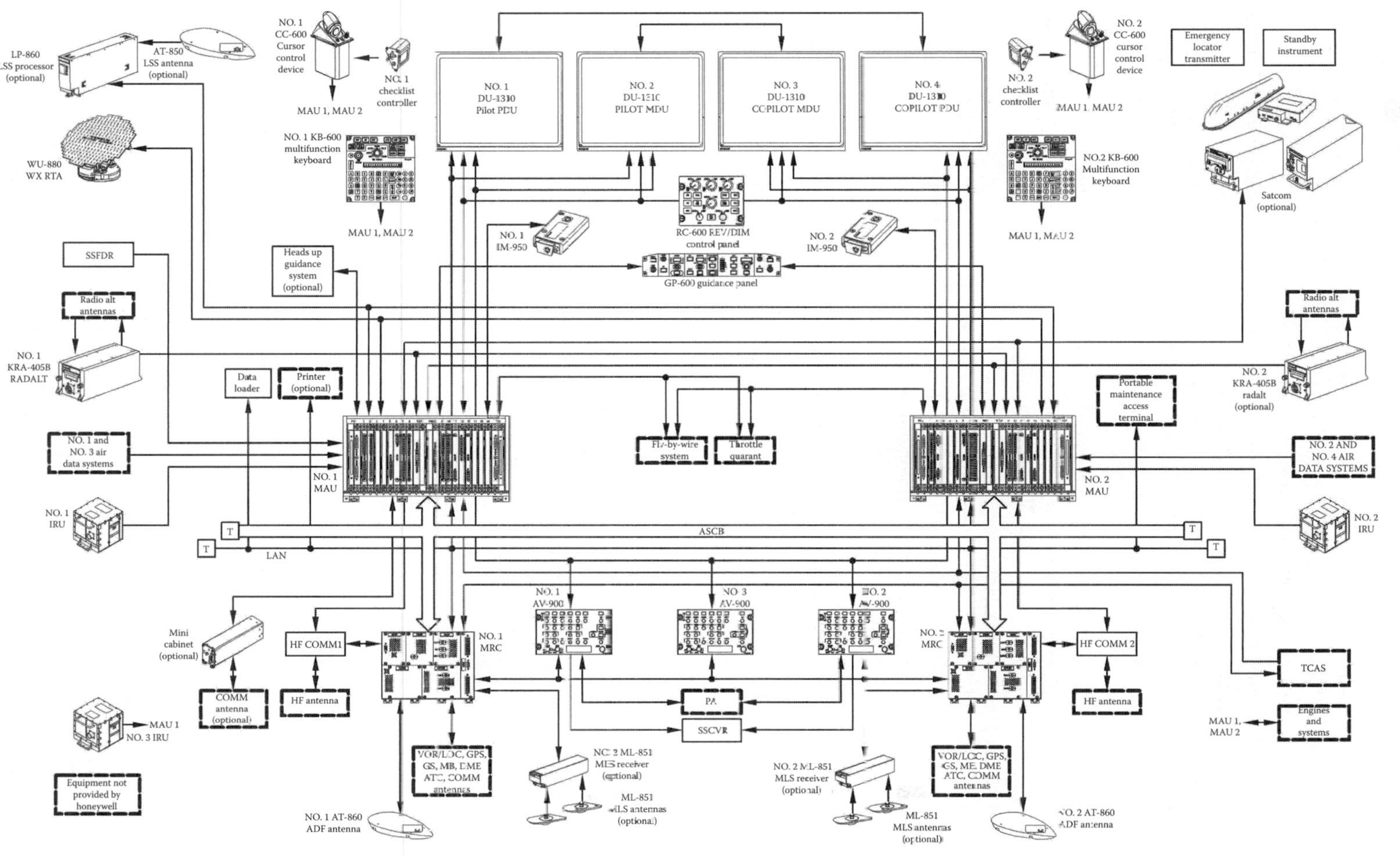

FIGURE 44.6 Primus Epic example system diagram. (Courtesy of Honeywell.)

References

1. GAMA. ASCB: Avionics Standard Communications Bus Version C. General Aviation Manufacturers Association (GAMA), Washington, DC, April 15, 1996.
2. Honeywell. http://www.honeywell.com. Accessed August 2010.
3. D. Michaud. Maintenance Avionique—ATA 100 34 Test Automatique Bus Avionique Langage C. Institut de Maintenance Aronautique, Université Bordeaux I, Talence, France, 2006.
4. N. Weininger and D.D. Cofer. Modeling the ASCB-D synchronization algorithm with SPIN: A case study. In *Proceedings of the Seventh International SPIN Workshop on SPIN Model Checking and Software Verification*, pp. 93–112, Springer-Verlag, London, U.K., 2000.
5. Chick, S.E., Sanchez, P.I., Ferrin, D.M., and Morrice, D.J. (Eds.). *Proceedings of the 35th Winter Simulation Conference: Driving Innovation*; New Orleans, LA, 7–10, December 2003.
6. S. Subbiah and S. Nagaraj. Issues with object orientation in verifying safety-critical systems. In *Sixth IEEE International Symposium on Object-Oriented Real-Time Distributed Computing, 2003*, pp. 99–104, May 14–16, 2003, Hakodate, Japan.
7. D.D. Davidson and V.Y. Chiu. Fail-operational global time reference in a redundant synchronous data bus system. Patent Application US 2005/0102586 A1, Honeywell, May 12, 2005.
8. Federal Aviation Administration (FAA). Airworthiness directives; Dassault model Falcon 2000ex and 900ex series airplanes. *Federal Register* 70, 9853–9856, 2005; FAA, Docket No. FAA-2005-20425; Directorate Identifier 2005-NM-014-AD; Amendment 39-13987; AD 2005-04-15, March 1, 2005.
9. Aeronautical Radio Inc. Avionics application software standard interface Part 1—Required services, ARINC specification 653P1-2 edition, December 2005.
10. Aeronautical Radio Inc. Avionics application software standard interface Part 2—Extended services, ARINC specification 653P2-1 edition, 12 2009.
11. Aeronautical Radio Inc. Avionics application software standard interface Part 3—Conformity test specification, ARINC specification 653P-3 edition, October 2006.
12. C.L. Liu and J.W. Layland. Scheduling algorithms for multiprogramming in a hard real-time environment. *Journal of the ACM*, 20(1):46–61, 1973.
13. P. Binns. A robust high-performance time partitioning algorithm. The digital engine operating system approach. In *Digital Avionics Systems Conference*. AIAA/IEEE, IEEE, 2001.

45

ARINC Specification 429 Mark 33 Digital Information Transfer System

Daniel A. Martinec
ARINC

Samuel P. Buckwalter
ARINC

45.1 Introduction 45-1
45.2 ARINC 419 45-2
45.3 ARINC 429 45-2
General • History • Design Fundamentals
45.4 Message and Word Formatting 45-6
Direction of Information Flow • Information Element • Information Identifier • Source/Destination Identifier • Sign/Status Matrix • Data Standards
45.5 Timing-Related Elements 45-11
Bit Rate • Information Rates • Clocking Method • Word Synchronization • Timing Tolerances
45.6 Communications Protocols 45-12
Development of File Data Transfer • Bit-Oriented Communications Protocol
45.7 Applications 45-15
Initial Implementation • Evolution of Controls • Longevity of ARINC 429
45.8 ARINC 453 45-16

45.1 Introduction

ARINC specifications 419, 429, and 629 and Project Paper 453 are documents prepared by the Airlines Electronic Engineering Committee (AEEC) and published by Aeronautical Radio Inc. (ARINC). These are among over 300 air transport industry avionics standards published since 1949. These documents, commonly referred to as ARINC 419, ARINC 429, ARINC 453, and ARINC 629, describe data communication systems used primarily on commercial transport airplanes, but which are also used by general aviation and military airplanes. The differences between the systems are described in detail in the subsequent sections.

45.2 ARINC 419

ARINC specification 419, "Digital Data Compendium," provides detailed descriptions of the various interfaces used in the ARINC 500 series of avionics standards prior to 1980. ARINC 419 is often incorrectly assumed to be a stand-alone bus standard. It provides a summary of electrical interfaces, protocols, and data standards for avionics built prior to the airlines' selection of a single standard (i.e., ARINC 429) for the distribution of digital information aboard aircraft.

45.3 ARINC 429

45.3.1 General

ARINC specification 429, "Digital Information Transfer System (DITS)," was first published in 1977 and has since become the ARINC standard most widely used by the airlines. The title of this airline standard was chosen so as not to describe it as a "data bus." Although ARINC 429 is a vehicle for data transfer, it does not fit the normal definition of a data bus. A typical data bus provides multidirectional transfer of data between multiple points over a single set of wires. ARINC 429's simplistic one-way flow of data significantly limits this capability, but the associated low cost and the integrity of the installations have provided the airlines with a system exhibiting excellent service for more than two decades. Additional information regarding avionics standards may be found at http://www.aviation-ia.com/aeec.

45.3.2 History

In the early 1970s, the airlines recognized the potential advantage of implementation of digital equipment. Some digital equipment had already been implemented to a certain degree on airplanes existing at that time. However, there were three new transport airplanes on the horizon: the Airbus A-310 and the Boeing B-757 and B-767. The airlines, along with the airframe and equipment manufacturers, established a goal to create an all-new suite of avionics using digital technology.

With digital avionics came the need for an effective means of data communications among the avionics units. The airlines recognized that the military was also in the early stages of development of a data bus that could perform the data transfer functions among military avionics. A joint program to produce a data bus common to the air transport industry and the military suggested a potential for significant economical benefits.

The Society of Automotive Engineers (SAE) took on the early work to develop the military's data bus. Participants in the SAE program came from many parts of the military and private sectors of aviation. Considerable effort went into defining all aspects of the data bus with the goal of meeting the needs of both the military and air transport users. That work culminated in the development of the early version of the data bus identified by Mil-Std 1553 (see Chapter 1).

Early in the process of the Mil-Std 1553 development, representatives from the air transport industry realized that the stringent and wide range of military requirements would cause the Mil-Std 1553 to be overly complex for the commercial user and would not exhibit the flexibility to accommodate the varying applications of transport airplanes. Difficulty in certification also was considered a potential problem. The decision was made to abandon a cooperative data bus development program with the military and pursue work on a data bus to more closely reflect commercial airplane requirements.

The numerous single transmitter and multiple receiver data transfer systems used on airplanes built in the early 1970s proved to be reliable and efficient compared to the more complex data buses of the time. These transfer systems, described in ARINC 419, were considered as candidates for the new digital aircraft.

Although none of the systems addressed in the ARINC specification could adequately perform the task, each exhibited desirable characteristics that could be applied to a new design. The result was the release of a new data transfer system exhibiting a high level of efficiency, extremely good reliability, and ease of certification. ARINC 429 became the industry standard. Subsequent to release of the standard, solid-state component manufacturers produced numerous low-cost integrated circuits. ARINC 429 was used widely by the air transport industry and even found applications in nonaviation commercial and military applications. ARINC 429 has been used as the standard for virtually all ARINC 700-series standards for "digital avionics" used by the air transport industry.

ARINC has maintained and provided the necessary routine updates for new data word assignments and formats. There were no significant changes in the basic design until 1980, when operational experience showed that certain shorted wire conditions would allow the bus to operate in a faulty condition with much reduced noise immunity. This condition also proved to be very difficult to locate during routine maintenance. In response, the airlines suggested that the design be changed to ensure that the bus would not continue to operate when this condition occurred, and a change to the receiver voltage thresholds and impedances solved this problem.

No basic changes to the design have been made since that time. ARINC 429 has remained a reliable system and even today is used extensively in the most modern commercial airplanes.

45.3.3 Design Fundamentals

45.3.3.1 Equipment Interconnection

A single transmitter is connected with up to 20 data receivers via a single twisted and shielded pair of wires. The shields of the wires are grounded at both ends and at any breaks along the length of the cable. The shields' connections to the ground should be kept as short as possible.

45.3.3.2 Modulation

Return-to-Zero (RZ) modulation is used. The voltage levels are used for this modulation scheme.

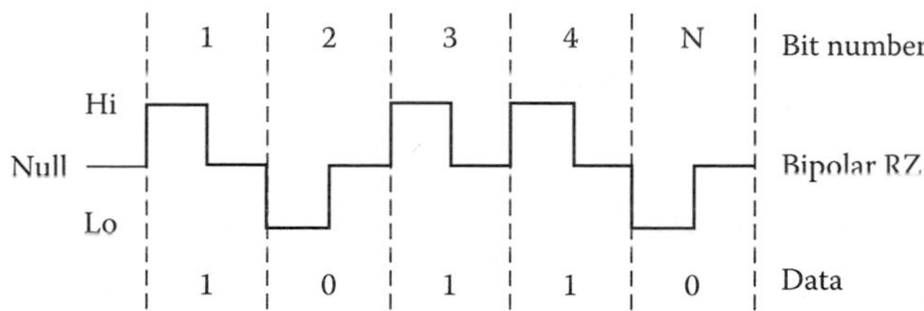

45.3.3.3 Voltage Levels

The differential output voltages across the transmitter output terminal with no load is described in the following table:

	HI (V)	NULL (V)	LO (V)
Line A to line B	$+10 \pm 1.0$	0 ± 0.5	-10 ± 1.0
Line A to ground	5 ± 0.5	0 ± 0.25	-5 ± 0.5
Line B to ground	-5 ± 0.5	0 ± 0.25	5 ± 0.5

The differential voltage seen by the receiver will depend on wire length, loads, stubs, and so on. With no noise present on the signal lines, the nominal voltages at the receiver terminals (A and B) would be the following:

- HI +7.25 V to +11 V
- NULL +0.5 V to −0.5 V
- LO −7.25 V to −11 V

In practical installations impacted by noise, and so on, the following voltage ranges will be typical across the receiver input (A and B):

- HI +6.5 V to +13 V
- NULL +2.5 V to −2.5 V
- LO −6.5 V to −13 V

Line (A or B) to ground voltages are not defined. Receivers are expected to withstand without damage steady-state voltages of 30 volts alternating current (VAC) root mean square (RMS) applied across terminals A and B, or volts direct current (VDC) applied between terminal A or B and the ground.

45.3.3.4 Impedance Levels

45.3.3.4.1 Transmitter Output Impedance

The transmitter output impedance is 70–80 (nominal 75) Ω and is divided equally between lines A and B for all logic states and transitions between those states.

45.3.3.4.2 Receiver Input Impedance

The typical receiver input characteristics are as follows:

- Differential Input Resistance R_I = 12,000 Ω minimum
- Differential Input Capacitance C_I = 50 pF maximum
- Resistance to Ground R_H and $R_G \geq$ 12,000 Ω
- Capacitance to Ground C_H and $C_G \leq$ 50 pF

The total receiver input resistance, including the effects of R_I, R_H, and R_G in parallel, is 8000 Ω minimum (400 Ω minimum for 20 receivers). A maximum of 20 receivers is specified for any one transmitter. See below for the circuit standards.

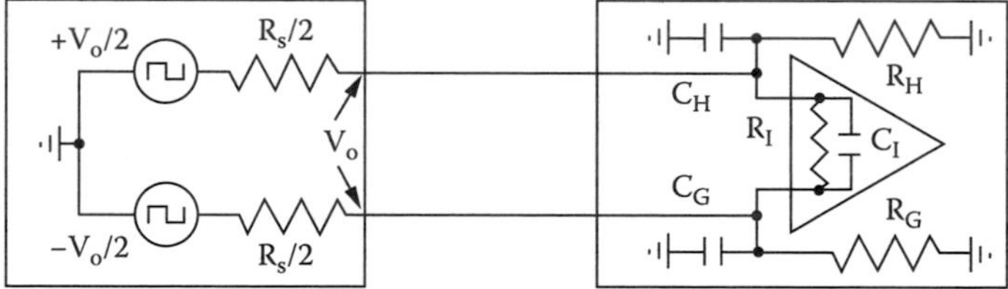

45.3.3.4.3 Cable Impedance

The wire gauges used in the interconnecting cable will typically vary between 20 and 26 depending on desired physical integrity of the cable and weight limitations. Typical characteristic impedances will be in the range of 60–80 ohms. The transmitter output impedance was chosen at 75 ohms nominal to match this range.

45.3.3.5 Fault Tolerance

A generator on each engine provides the electrical power on an airplane. The airplane electrical system is designed to take into account any variation in, for example, engine speeds, phase differentials, and power bus switching. However, it is virtually impossible to ensure that the power source will be perfect at all times. Failures within a system can also cause erratic power levels, and the design of the ARINC 429 components takes power variation into account and is not generally susceptible to either damage or erratic operation when those variations occur. The ranges of those variations are provided in the following sections.

45.3.3.5.1 Transmitter External Fault Voltage

Transmitter failures caused by external fault voltages will not typically cause other transmitters or other circuitry in the unit to function outside of their specification limits or to fail.

45.3.3.5.2 Transmitter External Fault Load Tolerance

Transmitters should indefinitely withstand without sustaining damage a short circuit applied (a) across terminals A and B, (b) from terminal A to ground, (c) from terminal B to ground, or (d) b and c simultaneously.

45.3.3.6 Fault Isolation

45.3.3.6.1 Receiver Fault Isolation

Each receiver incorporates isolation provisions to ensure that the occurrence of any reasonably probable internal line replaceable unit (LRU) or bus receiver failure does not cause any input bus to operate outside its specification limits (both undervoltage and overvoltage).

45.3.3.6.2 Transmitter Fault Isolation

Each transmitter incorporates isolation provisions to ensure that it does not undermine any reasonably probable equipment fault condition providing an output voltage in excess of (a) 30 VAC RMS between terminal A and B, (b) +29 VDC between A and ground, or (c) +29 VDC between B and ground.

45.3.3.7 Logic-Related Elements

This section describes the digital transfer system elements considered to be principally related to the logic aspects of the signal circuit.

45.3.3.7.1 Digital Language

45.3.3.7.1.1 Numeric Data The ARINC 429 accommodates numeric data encoded in two digital languages: BNR expressed in two's complement fractional notation and binary coded decimal (BCD) per the numerical subset of International Standards Organization (ISO) Alphabet No. 5. An information item encoded in both BCD and BNR is assigned unique labels for both BCD and BNR (see Section 45.4.3).

45.3.3.7.1.2 Discrete Data In addition to handling numeric data as specified above, ARINC 429 is also capable of accommodating discrete items of information either in the unused (pad) bits of data words or, when necessary, in dedicated words.

The rule in the assignment of bits in discrete numeric data words is to start with the least significant bit (LSB) of the word and to continue toward the most significant bit available in the word. There are two types of discrete words: general-purpose discrete words and dedicated discrete words. Seven labels (270 XX–276 XX) are assigned to the general-purpose discrete words. These words are assigned in ascending label order (starting with 270 XX), where XX is the equipment identifier.

32	31 30	29 28 27 26 25 24 23 22 21 20 19 18 17 16 15 14 13 12 11	10 9	8 7 6 5 4 3 2 1
P	SSM	Data ⟶ ⟵ Pad ⟵ Discretes	SDI	Label
		MSB LSB		

Generalized BCD word format

P	SSM		BCD CH #1			BCD CH #2				BCD CH #3				BCD CH #4				BCD CH #5				SDI		8 7 6 5 4 3 2 1
			4	2	1	8	4	2	1	8	4	2	1	8	4	2	1	8	4	2	1			
0	0	0	0	1	0	0	1	0	1	0	1	1	1	1	0	0	0	0	1	1	0	0	0	1 0 0 0 0 0 0 1
Example			2			5				7				8				6				DME distance		

BCD word format example (no discretes)

32	31 30 29	28 27 26 25 24 23 22 21 20 19 18 17 16 15 14 13 12 11	10 9	8 7 6 5 4 3 2 1
P	SSM	Data ⟶ ⟵ Pad ⟵ Discretes	SDI	Label
		MSB LSB		

Generalized BCD word format

45.3.3.7.1.3 Maintenance Data (General Purpose) The general-purpose maintenance words are assigned labels in sequential order, as are the labels for the general-purpose discrete words. The lowest octal value label assigned to the maintenance words is used when only one maintenance word is transmitted. When more than one word is transmitted, the lowest octal value label is used first, and the other labels are used sequentially until the message has been completed. The general-purpose maintenance words may contain discrete, BCD, or BNR numeric data. They do not contain ISO Alphabet No. 5 messages. The general-purpose maintenance words are formatted according to the layouts of the corresponding BCD, BNR, and discrete data words shown in the word formats above.

45.4 Message and Word Formatting

45.4.1 Direction of Information Flow

The information output of a system element is transmitted from a designated port (or ports) to which the receiving ports of other system elements in need of that information are connected. In no case does information flow into a port designated for transmission. A separate data bus (twisted and shielded pair of wires) is used for each direction when data are required to flow both ways between two system elements.

45.4.2 Information Element

The basic information element is a digital word containing 32 bits. There are five application groups for such words: BNR data, BCD data, discrete data, general maintenance data and acknowledgment, ISO Alphabet No. 5, and maintenance (ISO Alphabet No. 5) data (AIM). The relevant data-handling rules are set forth in Section 45.4.6. When less than the full data field is needed to accommodate the information conveyed in a word in the desired manner, the unused bit positions are filled with binary zeros or, in the case of BNR and BCD numeric data, valid data bits. If valid data bits are used, the resolution may exceed the accepted standard for an application.

45.4.3 Information Identifier

A six-character label identifies the type of information contained in a word. The first three characters are octal characters coded in binary in the first 8 bits of the word. The 8 bits will identify the information contained within BNR and BCD numeric data words (e.g., DME distance, static air temperature, etc.) and identify the word application for discrete, maintenance, and AIM data.

The last three characters of the six-character label are hexadecimal characters used to identify ARINC 429 bus sources. Each triplet of hexadecimal characters identifies a system element with one or more DITS ports. Each three-character code (and black box) may have up to 255 8-bit labels assigned to it. The code is used administratively to retain distinction between unlike parameters having like label assignments.

Octal label 377 has been assigned for the purpose of electrically identifying the system element. The code appears in the three least significant digits of the 377 word in a BCD word format. Although data encoding is based on the BCD word format, the sign/status matrix (SSM) encoding is per the discrete word criteria to provide enhanced failure warning. The transmission of the equipment identifier word on a bus will permit receivers attached to the bus to recognize the source of the DITS information. Since the transmission of the equipment identifier word is optional, receivers should not depend on that word for correct operation.

45.4.4 Source/Destination Identifier

Bit numbers 9 and 10 of numeric data words are used for a data source and destination identification function. They are not available for this function in alphanumeric (ISO Alphabet No. 5) data words

TABLE 45.1 SDI Encoding

Bit No.		Installation
10	9	See text
0	0	—
0	1	1
1	0	2
1	1	3

Note: In certain specialized applications of the SDI function, the all-call capability may be forfeited so that code "00" is available as an "installation no. 4" identifier.

or when the resolution needed for numeric (BNR/BCD) data necessitates their use for valid data. The source and destination identifier function may be applied when specific words need to be directed to a specific system of a multisystem installation or when the source system of a multisystem installation needs to be recognizable from the word content. When it is used, a source equipment encodes its aircraft installation number in bits 9 and 10 as shown in Table 45.1. A sink equipment will recognize words containing its own installation number code and words containing code "00," the "all-call" code.

Equipment will fall into the categories of source only, sink only, or both source and sink. Use of the SDI bits by equipment functioning only as a source or only as a sink is described above. Both the source and sink texts above are applicable to equipment functioning as both a source and a sink. Such equipment will recognize the SDI bits on the inputs and also encode the SDI bits, as applicable, on the outputs. DME, VOR, ILS (Instrument Landing System), and other sensors are examples of source and sink equipment generally considered to be only source equipment. These are actually sinks for their own control panels. Many other types of equipment are also misconstrued as source only or sink only. If a unit has a 429 input port and a 429 output port, it is a source and sink. With the increase of equipment consolidation (e.g., centralized control panels), the correct use of the SDI bits cannot be overstressed.

When the SDI function is not used, binary zeros or valid data should be transmitted in bits 9 and 10.

45.4.5 Sign/Status Matrix

This section describes the coding of the SSM field. In all cases, the SSM field uses bits 30 and 31; for BNR data words, the SSM field also includes bit 29. The SSM field is used to report hardware equipment condition (fault/normal), operational mode (functional test), or validity of data word content (verified/no computed data). The following definitions apply:

Invalid data: This is defined as any data generated by a source system whose fundamental characteristic is the inability to convey reliable information for the proper performance of a user system. There are two categories of invalid data: (1) no computed data and (2) failure warning.

No computed data: This is a particular case of data invalidity in which the source system is unable to compute reliable data for reasons other than system failure. This inability to compute reliable data is caused exclusively by a definite set of events or conditions whose boundaries are uniquely defined in the system characteristic.

Failure warning: This is a particular case of data invalidity in which the system monitors have detected one or more failures. These failures are uniquely characterized by boundaries defined in the system characteristic.

Displays are normally "flagged invalid" during a "failure warning" condition. When a "no computed data" condition exists, the source system indicates that its outputs are invalid by setting the SSM of the affected words to the "no computed data" code, as defined in the subsections that follow. The system indicators may or may not be flagged depending on system requirements.

While the unit is in the functional test mode, all output data words generated within the unit (i.e., pass-through words are excluded) are coded with "functional test." Pass-through data words are those words received by the unit and retransmitted without alteration.

When the SSM code is used to transmit status and more than one reportable condition exists, the condition with the highest priority is encoded in bits number 30 and 31. The order of condition priorities is shown in the following table:

Failure warning	Priority 1
No computed data	Priority 2
Functional test	Priority 3
Normal operation	Priority 4

Each data word type has its own unique utilization of the SSM field. These various formats are described in the following sections.

45.4.5.1 BCD Numeric

When a failure is detected within a system that would cause one or more of the words normally output by that system to be unreliable, the system stops transmitting the affected word or words on the data bus. Some avionics systems are capable of detecting a fault condition that results in less than normal accuracy. In these systems, when a fault of this nature (for instance, partial sensor loss which results in degraded accuracy) is detected, each unreliable BCD digit is encoded "1111" when transmitted on the data bus. For equipment having a display, the "1111" code should, when received, be recognized as representing an inaccurate digit and a dash or equivalent symbol is normally displayed in place of the inaccurate digit.

The sign (plus/minus, north/south, etc.) of BCD numeric data is encoded in bits 30 and 31 of the word as shown in Table 45.2. Bits 30 and 31 of BCD numeric data words are zero where no sign is needed.

The "no computed data" code is annunciated in the affected BCD numeric data words when a source system is unable to compute reliable data for reasons other than system failure. When the "functional test" code appears in bits 30 and 31 of an instruction input data word, it is interpreted as a command to perform a functional test.

45.4.5.2 BNR Numeric Data Words

The status of the transmitter hardware is encoded in the status matrix field (bit numbers 30 and 31) of BNR numeric data words as shown in Table 45.3.

TABLE 45.2 BCD Numeric Sign/Status Matrix

Bit No.		
31	30	Function
0	0	Plus, north, east, right, to, above
0	1	No computed data
1	0	Functional test
1	1	Minus, south, west, left, from, below

TABLE 45.3 Status Matrix

Bit No.		
31	30	Function
0	0	Failure warning
0	1	No computed data
1	0	Functional test
1	1	Normal operation

A source system annunciates any detected failure that causes one or more of the words normally output by that system to be unreliable by setting bit numbers 30 and 31 in the affected word(s) to the "failure warning" code defined in Table 45.3. Words containing this code continue to be supplied to the data bus during the failure condition.

The "no computed data" code is annunciated in the affected BNR numeric data words when a source system is unable to compute reliable data for reasons other than system failure.

When the "functional test" code appears as a system output, it is interpreted as advice that the data in the word result from the execution of a functional test. A functional test produces indications of one-eighth of positive full-scale values unless indicated otherwise in an ARINC equipment characteristic.

If, during the execution of a functional test, a source system detects a failure that causes one or more of the words normally output by that system to be unreliable, it changes the states of bits 30 and 31 in the affected words such that the "functional test" annunciation is replaced with a "failure warning" annunciation.

The sign (plus, minus, north, south, etc.) of BNR numeric data words are encoded in the sign matrix field (bit 29) as shown in Table 45.4. Bit 29 is zero when no sign is needed.

Some avionics systems are capable of detecting a fault condition that results in less than normal accuracy. In these systems, when a fault of this nature (for instance, partial sensor loss, which results in degraded accuracy) is detected, the equipment will continue to report "normal" for the SSM while indicating the degraded performance by coding bit 11 as shown in Table 45.5.

This implies that degraded accuracy can be coded only in BNR words not exceeding 17 bits of data.

45.4.5.3 Discrete Data Words

A source system annunciates any detected failure that could cause one or more of the words normally output by that system to be unreliable. Three methods are defined. The first method is to set bits 30 and 31 in the affected words to the "failure warning" code defined in Table 45.6. Words containing the "failure warning" code continue to be supplied to the data bus during the failure condition. When using the second method, the equipment may stop transmitting the affected word or words on the data bus. This method

TABLE 45.4 Status Matrix

Bit No. 29	Function
0	Plus, north, east, right, to, above
1	Minus, south, west, left, from, below

TABLE 45.5 Accuracy Status

Bit No.11	Function
0	Nominal accuracy
1	Degraded accuracy

TABLE 45.6 Discrete Data Words

Bit No.		
31	30	Function
0	0	Verified data, normal operation
0	1	No computed data
1	0	Functional test
1	1	Failure warning

is used when the display or utilization of the discrete data by a system is undesirable. The third method applies to data words, which are defined such that they contain failure information within the data field. For these applications, the associated ARINC equipment characteristic specifies the proper SSM reporting. Designers are urged not to mix operational and built-in test equipment (BITE) data in the same word.

The "no computed data" code is annunciated in the affected discrete data words when a source system is unable to compute reliable data for reasons other than system failure. When the "functional test" code appears as a system output, it is interpreted as advice that the data in the discrete data word contents are the result of the execution of a functional test.

45.4.6 Data Standards

The units, ranges, resolutions, refresh rates, number of significant bits, and pad bits for the items of information to be transferred by the Mark 33 DITS are administered by the AEEC and tabulated in ARINC characteristic 429.

ARINC characteristic 429 calls for numeric data to be encoded in BCD and binary, the latter using two's complement fractional notation. In this notation, the most significant bit of the data field represents half of the maximum value chosen for the parameter being defined. Successive bits represent the increments of a binary fraction series. Negative numbers are encoded as the two's complements of positive value and the negative sign is annunciated in the SSM.

In establishing a given parameter's binary data standards, the unit's maximum value and resolution are first determined in that order. The LSB of the word is then given a value equal to the resolution increment, and the number of significant bits is chosen such that the maximum value of the fractional binary series just exceeds the maximum value of the parameter (i.e., it equals the next whole binary number greater than the maximum parameter value less one LSB value). For example, to transfer altitude in units of feet over a range of zero to 100,000 ft with a resolution of 1 ft, the number of significant bits is 17 and the maximum value of the fractional binary series is 131,071 (i.e., 131,072 − 1).

Note that because accuracy is a quality of the measurement process and not the data transfer process, it plays no part in the selection of word characteristics. Obviously, the resolution provided in the data word should equal or exceed the accuracy not to degrade it.

For the binary representation of angular data, the ARINC 429 employs "degrees divided by" as the unit of data transfer and ±1 (semicircle) as the range for two's complement fractional notation encoding (ignoring, for the moment, the subtraction of the LSB value). Thus, the angular range 0 through 359.XXX degrees is encoded as 0 through ±179.XXX degrees, the value of the most significant bit is one half semicircle, and there are no discontinuities in the code.

This may be illustrated as follows. Consider encoding the angular range 0–360 in one degree increments. Per the general encoding rules above, the positive semicircle will cover the range 0–179 (one LSB less than full range). All the bits of the code will be zeros for 0 and ones for 179, and the SSM will indicate the positive sign. The negative semicircle will cover the range 180–359. All the bits will be zeros for 180. The codes for angles between 181 and 359 will be determined by taking the two's complements of the fractional binary series for the result of subtracting each value from 360. Thus, the code for 181 is the two's complement of the code for 179. Throughout the negative semicircle, which includes 180, the SSM contains the negative sign.

45.5 Timing-Related Elements

This section describes the digital data transfer system elements considered to be principally related to the timing aspects of the signal circuit.

45.5.1 Bit Rate

High-speed operation: The bit rate for high-speed operation of the system is 100 kilobits per second (kbps) ±1%.

Low-speed operation: The bit rate for low-speed operation of the system is within the range 12.0–14.5 kbps. The selected rate within this range is maintained within 1%.

45.5.2 Information Rates

The minimum and maximum transmit intervals for each item of information are specific by ARINC 429. Words with like labels but with different SDI codes are treated as unique items of information. Each and every unique item of information is transmitted once during an interval bounded in length by the specified minimum and maximum values. Stated another way, a data word having the same label and four different SDI codes will appear on the bus four times (once for each SDI code) during that time interval.

Discrete bits contained within data words are transferred at the bit rate and repeated at the update rate of the primary data. Words dedicated to discretes should be repeated continuously at specified rates.

45.5.3 Clocking Method

Clocking is inherent in the data transmission. The identification of the bit interval is related to the initiation of either a HI or LO state from a previous NULL state in a bipolar RZ code.

45.5.4 Word Synchronization

The digital word should be synchronized by reference to a gap of four bit times (minimum) between the periods of word transmissions. The beginning of the first transmitted bit following this gap signifies the beginning of the new word.

45.5.5 Timing Tolerances

The waveform timing tolerances are shown below. (Table 45.7)

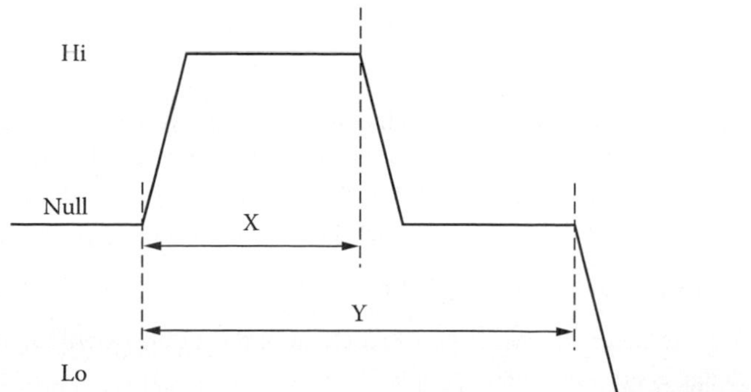

TABLE 45.7 Timing Tolerances

Parameter	High-Speed Operation	Low-Speed Operation
Bit rate	100 kbps ± 1%	12–14.5 kbps
Time Y	10 μs ± 2.5%	Z[a] μs ± 2.5%
Time X	5 μs ± 5%	Y/2 ± 5%
Pulse rise time	1.5 ± 0.5 μs	10 ± 5 μs
Pulse fall time	1.5 ± 0.5 μs	10 ± 5 μs

[a] Z = 1/R, where R = bit rate selected from 12 to 14.5 kbps range.

Note: Pulse rise and fall times are measured between the 10% and 90% voltage amplitude points on the leading and trailing edges of the pulse and include time skew between the transmitter output voltages A-to-ground and B-to-ground.

45.6 Communications Protocols

45.6.1 Development of File Data Transfer

AEEC adopted ARINC 429 in July 1977. It defined a broadcast data bus with general provisions for file data transfer. In October 1989, AEEC updated a file data transfer procedure with a more comprehensive process that will support the transfer of both bit- and character-oriented data. The new protocol became known as the "Williamsburg Protocol."

45.6.1.1 File Data Transfer Techniques

This "file data transfer techniques" specification describes a system in which an LRU may generate binary extended length messages "on demand." Data is sent in the form of link data units (LDUs) organized in 8-bit octets. System address labels (SALs) are used to identify the recipient. Two data bus speeds are supported.

45.6.1.2 Data Transfer

The same principles of the physical layer implementation apply to file data transfer. Any avionics system element having information to transmit does so from a designated output port over a single twisted and shielded pair of wires to all other system elements having need of that information. Unlike the simple broadcast protocol that can deliver data to multiple recipients in a single transmission, the file transfer technique can be used only for point-to-point message delivery.

45.6.1.3 Broadcast Data

The broadcast transmission technique described above can be supported concurrently with file data transfer.

45.6.1.4 Transmission Order

The most significant octet of the file and LSB of each octet should be transmitted first. The label is transmitted ahead of the data in each case. It may be noted that the label field is encoded in reverse order (i.e., the LSB of the word is the most significant bit of the label). This "reversed label" characteristic is a legacy from past systems in which the octal coding of the label field was, apparently, of no significance.

45.6.1.5 Bit-Oriented Protocol Determination

An LRU will require logic to determine which protocol (character- or bit-oriented) and which version to use when prior knowledge is not available.

45.6.2 Bit-Oriented Communications Protocol

This subsection describes Version 1 of the bit-oriented (Williamsburg) protocol and message exchange procedures for file data transfer between units desiring to exchange bit-oriented data assembled in data files. The bit-oriented protocol is designed to accommodate data transfer between sending and receiving units in a form compatible with the Open Systems Interconnect (OSI) model developed by the ISO. This document directs itself to an implementation of the link layer; however, an overview of the first four layers (physical, link, network, and transport) is provided.

Communications will permit the intermixing of bit-oriented file transfer data words (which contain system address labels [SALs]) with conventional data words (which contain label codes). If the sink should receive a conventional data word during the process of accepting a bit-oriented file transfer message, the sink should accept the conventional data word and resume processing of the incoming file transfer message.

The data file and associated protocol control information are encoded into 32-bit words and transmitted over the physical interface. At the link layer, data are transferred using a transparent bit-oriented data file transfer protocol designed to permit the units involved to send and receive information in multiple word frames. It is structured to allow the transmission of any binary data organized into a data file composed of octets.

Physical medium: The physical interface is described above.

Physical layer: The physical layer provides the functions necessary to activate, maintain, and release the physical link that will carry the bit stream of the communication. The interfacing units use the electrical interface, voltage, and timing described earlier. Data words will contain 32 bits; bits 1–8 will contain the SAL and bit 32 will be the parity (odd) bit.

Link layer: The link layer is responsible for transferring information from one logical network entity to another and for enunciating any errors encountered during transmission. The link layer provides a highly reliable virtual channel and some flow control mechanisms.

Network layer: The network layer performs a number of functions to ensure that data packets are properly routed between any two terminals. The network layer expects the link layer to supply data from correctly received frames. The network layer provides for the decoding of information up to the packet level to determine which node (unit) the message should be transferred to. To obtain interoperability, this process, though simple in this application, must be reproduced using the same set of rules throughout all the communications networks (and their subnetworks) on-board the aircraft and on the ground. The bit-oriented data link protocol was designed to operate in a bit-oriented network layer environment. Specifically, ISO 8208 would typically be selected for the subnetwork layer protocol for air and ground subnetworks. There are, however, some applications in which the bit-oriented file transfer protocol will be used under other network layer protocols.

Transport layer: The transport layer controls the transportation of data between a source end-system to a destination end-system. It provides "network independent" data delivery between these processing end-systems. It is the highest order function involved in moving data between systems and it relieves higher layers from any concern specifically with the transportation of information between them.

45.6.2.1 Link Data Units

An LDU contains binary encoded octets. The octets may be set to any possible binary value. The LDU may represent raw data, character data, bit-oriented messages, character-oriented messages, or any string of bits desired. The only restriction is that the bits must be organized into full 8-bit octets. The interpretation of those bits is not a part of the link layer protocol. The LDUs are assembled to make up a data file.

LDUs consist of a set of contiguous ARINC 429 32-bit data words, each containing the SAL (see Section 45.6.2.3) of the sink. The initial data word of each LDU is a start of transmission (SOT). The data described above are contained within the data words that follow. The LDU is concluded with an end of transmission (EOT) data word. No data file should exceed 255 LDUs.

Within the context of this document, LDUs correspond to frames and files correspond to packets.

45.6.2.2 Link Data Unit Size and Word Count

The LDU may vary in size from 3 to 255 ARINC 429 words including the SOT and EOT words. When an LDU is organized for transmission, the total number of ARINC 429 words to be sent (word count) is calculated. The word count is the sum of the SOT word, the data words in the LDU, and the EOT word. In order to obtain maximum system efficiency, the data is typically encoded into the minimum number of LDUs.

The word count field is 8 bits in length. Thus, the maximum number of ARINC 429 words that can be counted in this field is 255. The word count field appears in the request to send (RTS) and clear to send (CTS) data words. The number of LDUs needed to transfer a specific data file will depend upon the method used to encode the data words.

45.6.2.3 System Address Labels

LDUs are sent point-to-point, even though other systems may be connected and listening to the output of a transmitting system. In order to identify the intended recipient of a transmission, the label field (bits 1–8) is used to carry a SAL. Each on-board system is assigned a SAL. When a system sends an LDU to another system, the sending system (the source) addresses each ARINC 429 word to the receiving system (the sink) by setting the label field to the SAL of the sink. When a system receives any data containing its SAL that is not sent through the established conventions of this protocol the data are ignored.

In the data transparent protocol data files are identified by content rather than by an ARINC 429 label. Thus, the label field loses the function of parameter identification available in broadcast communications.

45.6.2.4 Bit Rate and Word Timing

Data transfer may operate at either high speed or low speed. The source introduces a gap between the end of each ARINC 429 word transmitted and the beginning of the next. The gap should be 4 bit times (minimum), and the sink should be capable of receiving the LDU with the minimum word gap of 4 bit times between words. The source should not exceed a maximum average of 64 bit times between data words of an LDU.

The maximum average word gap is intended to compel the source to transmit successive data words of an LDU without excessive delay. This provision prevents a source that is transmitting a short message from using the full available LDU transfer time. The primary value of this provision is realized when assessing a maximum LDU transfer time for short fixed-length LDUs, such as for automatic dependence surveillance (ADS).

If a Williamsburg source device were to synchronously transmit long-length or full LDUs over a single ARINC 429 data bus to several sink devices, the source may not be able to transmit the data words for a given LDU at a rate fast enough to satisfy this requirement because of other bus activity. In aircraft operation, given the asynchronous burst mode nature of Williamsburg LDU transmissions, it is extremely unlikely that a Williamsburg source would synchronously begin sending a long-length or full LDU to more than two Williamsburg sink devices. A failure to meet this requirement will result in either a successful (but slower) LDU transfer or an LDU retransmission due to an LDU transfer time-out.

45.6.2.5 Word Type

The Word Type field occupies bits 31–29 in all bit-oriented LDU words and is used to identify the function of each ARINC 429 data word used by the bit-oriented communication protocol.

45.6.2.6 Protocol Words

The protocol words are identified with a Word Type field of 100 and are used to control the file transfer process.

45.6.2.6.1 Protocol Identifier

The protocol identifier field occupies bits 28–25 of the protocol word and identifies the type of protocol word being transmitted. Protocol words with an invalid protocol identifier field are ignored.

45.6.2.6.2 Destination Code

Some protocol words contain a Destination Code. The Destination Code field (bits 24–17) indicates the final destination of the LDU. If the LDU is intended for the use of the system receiving the message, the destination code may be set to null (hex 00). However, if the LDU is a message intended to be passed on to another on-board system, the Destination Code will indicate the system to which the message is to be passed. The Destination Codes are assigned according to the applications involved and are used in the Destination Code field to indicate the address of the final destination of the LDU.

In an OSI environment, the link layer protocol is not responsible for validating the destination code. It is the responsibility of the higher-level entities to detect invalid destination codes and to initiate error logging and recovery. Within the pre-OSI environment, the Destination Code provides network layer information. In the OSI environment, this field may contain the same information for routing purposes between OSI and non-OSI systems.

45.6.2.6.3 Word Count

Some protocol words contain a Word Count field. The Word Count field (bits 16–9) reflects the number of ARINC 429 words to be transmitted in the subsequent LDU. The maximum word count value is 255 ARINC 429 words and the minimum is 3 ARINC 429 words. An LDU with the minimum word count value of 3 ARINC 429 words would contain a SOT word, a data word, and an EOT word. An LDU with the maximum word count value of 255 ARINC 429 words would contain a SOT word, 253 data words, and an EOT word.

45.7 Applications

45.7.1 Initial Implementation

ARINC 429 was first used in the early 1980s on the Airbus A-310 and Boeing B-757 and B-767 airplanes. Approximately 150 separate buses interconnecting computers, radios, displays, controls, and sensors accommodated virtually all data transfer on these airplanes. Most of these buses operate at the lower speed; the few that operate at the higher speed of 100 kbps are typically connected to critical navigation computers.

45.7.2 Evolution of Controls

The first applications of ARINC 429 for controlling devices were based on the federated avionics approach used on airplanes that comprised mostly analog interfaces. Controllers for tuning communications equipment used an approach defined as two-out-of-five tuning. Each digit of the desired radio frequency was encoded on each set of five wires. Multiple digits dictated the need for multiple sets of wires for each radio receiver.

The introduction of ARINC 429 proved to be a major step toward reduction of wires. A tuning unit needed only one ARINC 429 bus to tune multiple radios of the same type. An entire set of radios and navigation receivers could be tuned with a few control panels, using approximately the same number of wires previously required to tune a single radio.

As cockpit space became more critical, the need to reduce the number of control panels became critical. The industry recognized that a single control panel, properly configured, could replace most of the existing control panels. The multipurpose control/display unit (MCDU), which came from the industry effort, was derived essentially from the control and display approach used by the rather sophisticated controller for the flight management system. For all intents and purposes, the MCDU became the cockpit controller.

A special protocol had to be developed for ARINC 429 to accommodate the capability of addressing different units connected to a single ARINC 429 bus from the MCDU. The protocol employed two-way communications using two pairs of wires between the controlling unit and the controlled device. An addressing scheme provided for selective communications between the controlling unit and any one of the controlled units. Only one output bus from the controller is required to communicate addresses and commands to the receiving units. With the basic ARINC 429 design, up to 20 controlled units could be connected to the output of the controller. Each of the controlled units is addressed by an assigned SAL.

45.7.3 Longevity of ARINC 429

New airplane designs in the twenty-first century continue to employ the ARINC 429 bus for data transmission. The relative simplicity and integrity of the bus, as well as the ease of certification, are characteristics that contribute to the continued selection of the ARINC 429 bus when the required data bandwidth is not critical. The ARINC 629 data bus developed as the replacement for ARINC 429 is used in applications where a large amount of data must be transferred or where many sources and sinks are required on a single bus.

45.8 ARINC 453

ARINC Project Paper 453 was developed by the AEEC in response to an anticipated requirement for data transfer rates higher than achievable with ARINC 429. The original drafts of Project Paper 453 were based on techniques already employed at that time. The electrical characteristics, including the physical medium, voltage thresholds, and modulation techniques, were based on Mil-Std 1553 (see Chapter 1). The data protocols and formats were based on those used in ARINC Specification 429.

During the preparation of the drafts of Project Paper 453, Boeing petitioned AEEC to consider the use of the digital autonomous terminal access communications (DATAC) bus developed by Boeing to accommodate higher data throughput. AEEC accepted Boeing's recommendation for the alternative. ARINC 629 was based on the original version of the Boeing DATAC bus. The work on Project 453 was then curtailed. The latest draft of Project Paper 453 is maintained by ARINC for reference purposes only.

46

ARINC 629

46.1 Physical Layer 46-1
Requirements for Physical Layer • Physical Layer Entities • Physical Layer Signaling
46.2 Data Link Layer 46-4
Message Format • Open Systems Interconnection OSI-Compatible Format • ARINC 629 Non-OSI Data Transfer
46.3 Medium Access Layer 46-7
Basic Protocol • Combined Protocol • Exceptions • Comparison and Summary
46.4 Robustness and Fault Tolerance Features 46-15
Receive Data Monitoring • Transmit Data Monitoring • Bus Protocol Monitoring • Protocol Parameter Monitoring
46.5 ARINC 629 in Boeing 777 46-16
References 46-19

Michael Paulitsch
Airbus Group Innovations

In the late 1980s, there was a push toward integration of multiple functions on a common computing and I/O platform, also referred to as integrated modular avionics (IMA). Combined with this push was a trend to reduce wiring and integrate formerly dedicated buses into system-level communication buses. The advantages are decreased size, cost, and weight; increased reliability; less-frequent maintenance; and more flexibility. The ARINC 629 communication network was the answer in aerospace to the request of integrated system-level buses. The predecessor ARINC 429 [3] had a single sender and multiple receiver topological and logical approach, and ARINC 629 extends the capabilities of the network to support a multiple-sender, multiple-receiver architecture while guaranteeing timely communication. And all this in an environment where high reliability, high availability, and high integrity are required to ensure safety of an aircraft. In addition, ARINC 629 uses current coupling for an electrical physical connection avoiding error-prone physical connections. ARINC 629 is a time division multiplex system. It includes multiple transmitters with broadcast-type, autonomous terminal access.

46.1 Physical Layer

ARINC 629 physical layer specifies the functional, electrical/optical, and mechanical characteristics of the physical layer entities. The physical layer entities and media in conjunction with the terminal controller form a complete communication path. ARINC 629 supports three different physical entity types as part of its physical layers: a current mode bus, a voltage mode bus, and a fiber-optic mode bus.

ARINC 629 deploys 2 Mbit/s serial data transmission rate specified for twisted pair conductors. The standard also supports the use of fiber optics. Within each line-replaceable unit (LRU), a single ARINC 629 terminal and the main computational component interact via shared memory. To reinforce the approach of collision avoidance adopted at the data link level, multiple timers and redundant circuitry are employed within each terminal to prevent single hardware faults causing multiple terminals

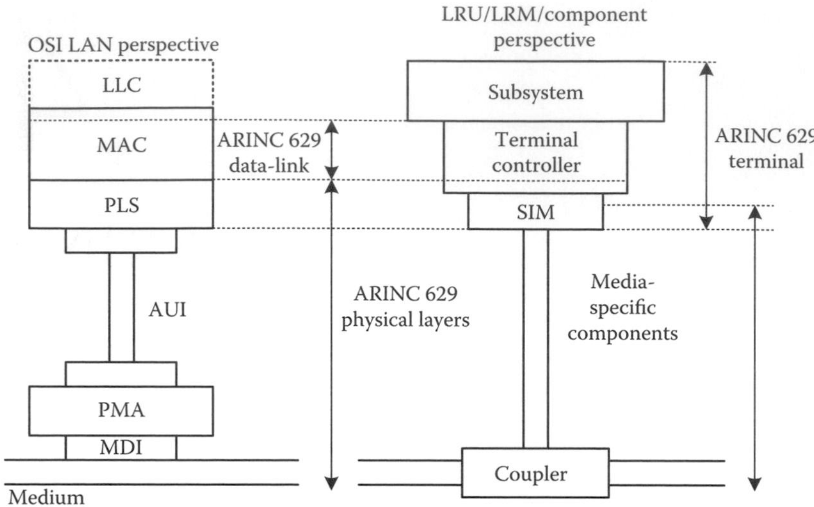

FIGURE 46.1 Scope of ARINC 629 physical layers. LLC, link layer control; MAC, medium access control; PLS, physical layer signaling; AUI, attachment unit interface; PMA, physical medium attachment; MDI, media-dependent interface. (From Airlines Electronic Engineering Committee, Multi-transmitter data bus—Technical description—ARINC specification 629P1-5, Standard 629P1-5, Aeronautical Radio, Inc (ARINC), Annapolis, MD, March 31, 1999.)

to transmit simultaneously. At the physical layer, mechanisms and signals support higher-level data link-level protocols. The maximum number of terminals permissible on the bus is 120. Figure 46.1 illustrates the scope of the physical layer specifications and the mapping of the logical communication layers onto hardware. In the following, we describe the physical layer requirements, entities, and interface essentials for higher layers.

46.1.1 Requirements for Physical Layer

The ARINC 629 system requirements that pertain to the physical layer are as follows [2]:

1. The serial transmission rate should be 2.0 Mbit/s with an accuracy of ±0.01%.
2. The bit error rate should not exceed 1 bit in 10^8 bits. This error rate includes both detected and undetected errors.
3. Each terminal should be capable of receiving the transmissions of every other terminal on the network.
4. The network should support the reliability goals for flight-critical functions, and shortening and opening of a stub at any terminal output should not cause any adverse effect on the bus. A probability of an undetected single point of failure of the terminal affecting the bus should be extremely unlikely. Any terminal should provide an isolation barrier between subsystem and bus, which means that any failure of the subsystem only affects data validity.
5. The installation should provide flexibility in a manner that physical reconfiguration of the network (addition or removal of terminals) should still satisfy the network reliability requirements.
6. The network should meet the RTCA DO-160 [7] standard pertinent to the specific application.
7. The bus medium should enable a receive portion of the transmitting terminal controller to discern simultaneous transmission by other terminal controllers for collision recovery purposes.
8. The bus medium should also enable a receive portion of the transmitting terminal controller to independently monitor the transmitted data directly from the bus for wraparound transmission monitoring.

9. The bus medium should enable detection of *bus quiet (BQ)* and *bus active (BA)* for higher protocol layers and provide signals indicating these states. BQ indicates no sending activity on the bus. BA means a signal is present.

46.1.2 Physical Layer Entities

ARINC 629 specifies three physical layer entity (PLE) types [2]:

Current mode bus: The current mode bus uses the principle of current induction to induce current pulses (so-called doublets) in the conductors of an unshielded twisted pair cable. The advantage of current induction is the nonintrusive coupling for linking terminals to the medium. That is, with current induction, neither cables have to be spliced nor tap connectors are required.

Voltage mode bus: The voltage mode bus implementation is characterized coupling via breaking the cable and insertion of *T* couplers or splicing of cable conductors for a (typically) shielded cable. Conduction is based on application of voltage.

Fiber-optic mode bus: The optical mode bus implementation utilizes electro-optic transceivers to inject pulses of light into an optical fiber either via use of optical coupler or tap and splicing of connectors.

Current mode and fiber-optic bus are defined in detail in the ARINC 629 specification, while a voltage mode bus is not specified.

46.1.2.1 Current Mode Bus

The current mode bus allows up to 120 terminals and supports a physical bus lengths of up to 100 m and stub lengths of up to 40 m.

The serial interface module (SIM) should convert Manchester biphase logic transition transmitter by the terminal controller into analog signals called doublets. These doublets are induced via the stub cable into the bus via the current mode coupler (CMC). Similarly, on the receiving side, signals on the bus are flowing from the bus via the CMC and the stub back to the terminal controller. The SIM provides power to the CMC by conditioning its supply voltage of ±15 V. The SIM also performs fault checking (monitoring) on itself and performs built-in self-test (BIST) when commanded to do so.

Tables 46.1 and 46.2 present an overview of the stub and bus cable requirements. Figure 46.13 also shows a CMC deployed.

46.1.2.2 Fiber-Optic Mode Bus

The optical bus is characterized by passing optical signals through a passive fiber-optic network of either topology (star, linear, or other). Fiber optics has the advantage of immunity to lightning and electromagnetic interference. It also uses passive coupling techniques. Table 46.3 provides an overview of the fiber-optic common characteristics. Details on the transmitter, receiver, and transmission media characteristics can be found in [2].

TABLE 46.1 Stub Cable Requirements

Description	Parameter Value
Insulation rating	600 V RMS
Attenuation	<1.8 dB (CMC to SIM/pair) ($F = 6$ MHz)
Velocity of propagation	>0.75 × c (where $c = 2.9979 \times 10^8$ m/s)
DC resistance	<1 Ω (CMC to SIM/pair)

Source: Airlines Electronic Engineering Committee, Multi-transmitter data bus—Technical description—ARINC specification 629P1-5, Standard 629P1-5, Aeronautical Radio, Inc (ARINC), Annapolis, MD, March 31, 1999.

TABLE 46.2 Current Mode Cable Requirements

Description	Nonshielded Cable Parameter Value	Shielded Cable Parameter Value
Insulation rating	600 V RMS	600 V RMS
Characteristic impedance (per twisted pair)	$Z_O = 130\ \Omega \pm 2\%$	$Z_O = 130\ \Omega \pm 2\%$
Attenuation	<2.8 dB/100 m ($F = 6$ MHz)	<2.6 dB/100 m ($F = 6$ MHz)
Velocity of propagation	>0.78 × c (where $c = 2.9979 \times 10^8$ m/s)	>0.70 × c (where $c = 2.9979 \times 10^8$ m/s)
Conductor size	#20 American wire gauge (AWG) stranded	#20 AWG stranded
Twists	6 ± 0.7 transpositions/ft	6 ± 0.7 transpositions/ft

Source: Airlines Electronic Engineering Committee, Multi-transmitter data bus—Technical description—ARINC specification 629P1-5, Standard 629P1-5, Aeronautical Radio, Inc (ARINC), Annapolis, MD, March 31, 1999.

TABLE 46.3 Common Characteristics of Fiber-Optic Mode Bus

Description	Value	Unit
Data rate	2% ± 0.01%	Mbit/s
Nominal bit rate	500 ± 0.050	ns
Nominal pulse width	62.5	ns
Minimum intermessage gap	3.6875	μs
Upper peak wavelength limit	900	nm
Lower peak wavelength limit	770	nm
Maximum spectral width	80	nm

Source: Airlines Electronic Engineering Committee, Multi-transmitter data bus—Technical description—ARINC specification 629P1-5, Standard 629P1-5, Aeronautical Radio, Inc (ARINC), Annapolis, MD, March 31, 1999.

46.1.3 Physical Layer Signaling

Physical layer signaling specifies the interface between PLE and higher communication layers (or the terminal controller and the SIM when speaking of component perspectives).

The medium access control (MAC) sublayer bus access protocols (see Section 46.3) are reliant on the indication of bus activity from the physical layer for operation. The presence of a signal on the bus is called bus active, while a period of no activity is called bus quiet. Messages are communicated in serial digital pulse code modulation by means of complementary TTL compatible lines, TXO/TXN, and RXI/RXN. The data code used is Manchester II biphase, where a logic one is encoded 1/0 (a half bit time high followed by a half bit time low). A logic zero is encoded 0/1 (a half bit time low followed by a half bit time high). Transitions occur always at the midpoint of each bit time.

MAC sublayer messages are comprised of word strings. A message consists of 1–31 word strings, which are described in detail in the next section. The physical layer should ensure that no gaps are present between successive bit-level encoded words in a word string.

46.2 Data Link Layer

This and the next section describe the two main functions associated with the ARINC 629 data link control procedure to be performed in the medium access sublayer: the media access management and data encapsulation (framing, addressing, and error detection).

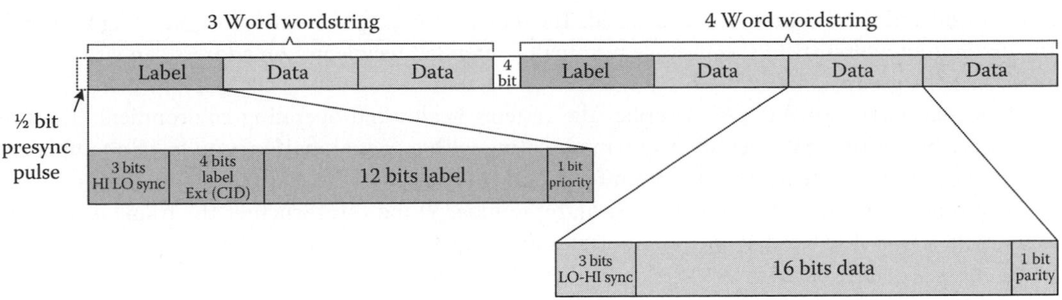

FIGURE 46.2 Example message with two word strings.

46.2.1 Message Format

An ARINC 629 message has variable length up to 31 word strings separated by a 4-bit delimiter. Each word string is made up of one label and several data words. Bus time is decomposed into bus cycles. Each terminal sends one message per bus cycle.

In detail, label words start with a Hi–Lo sync waveform (3 bit times) followed by the channel identification (CID, word bits WB4–WB7), then the label, and finally the parity bit (WB20). Data words start with the Lo–Hi sync waveform (3 bit times) followed by the sign bit (WB4), then the 15-bit data field (word bits WB5–WB19) with the MSB first, and finally the parity bit (WB20).

Data messages consist of word strings with a gap of 4 bit times between word strings and a 3-bit-time sync waveform to initialize each label or data word. The first label word of a message should be preceded by a half-bit-time presync pulse. The beginning of the first transmitted sync waveform following this 4-bit-time gap signifies the beginning of the new word string as indicated in Figure 46.2.

The last bit of each label and data word is a parity bit encoded such that word parity is rendered odd.

46.2.2 Open Systems Interconnection (OSI)-Compatible Format

ARINC 629 supports OSI and non-OSI-compatible formats. Figure 46.3 presents the IEEE Std 802.2 compatible logical link control MAC format. Each MAC should contain two address fields: the source

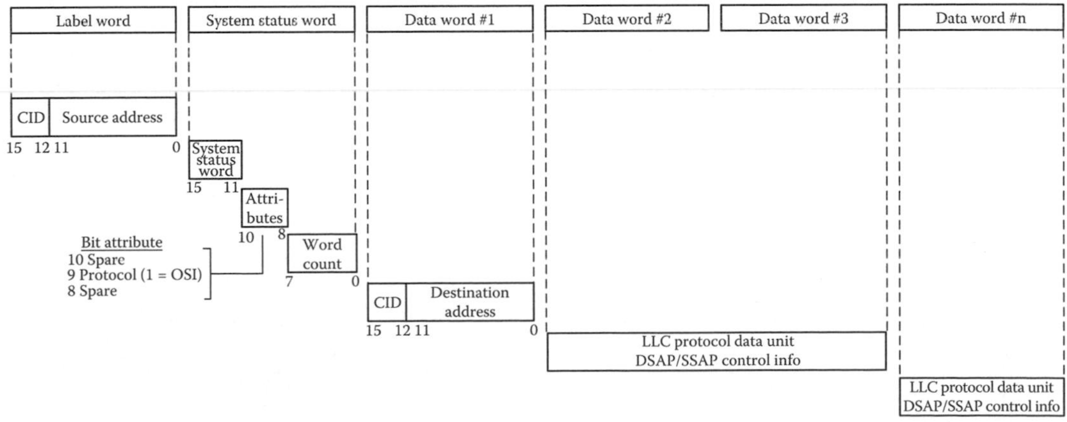

FIGURE 46.3 OSI message format. LLC, logical link control; CID, channel identification; DSAP, destination service access point; SSAP, source service access point. (From Airlines Electronic Engineering Committee, Multi-transmitter data bus—Technical description—ARINC specification 629P1-5, Standard 629P1-5, Aeronautical Radio, Inc (ARINC), Annapolis, MD, March 31, 1999.)

address field and the destination address field. The *source address* field identifies the sending terminal. The *destination address* field specifies the terminal(s) intended recipients. An address can either be an individual or a group address.

The system status word (SSW) describes the systems health and operating environment. The five high-order bits of the SSW identify a system operating with a partial malfunction or in a temporary state that causes the systems data to transmit predefined valid values (i.e., stale data).

The protocol indication field (attribute field, bit number 9) indicates whether the frame is an OSI-compatible frame. The word count field indicates the length of the message. The CRC field contains a frame check sequence based on a 16-bit cyclic redundancy check value.

Receivers of all terminals not transmitting data monitor all labels being transmitted on the bus. For all labels of interest, the receiver uses the memory mapping information associated with that label to store data directly in the user system's memory. Details of mapping into memory and configuration formats can be found in [2].

46.2.3 ARINC 629 Non-OSI Data Transfer

Non-OSI data transfer supports broadcast and directed data. Frames are defined at least 1 digital word containing 20 bits, but only 16 are available to the user, as described in Section 46.2.1. There are multiple predefined word formats for different applications, which are exemplary specified in [2].

Two events are important during normal operation of periodic transmissions on the bus. These are the *minor frame* and *major frame*. Minor frame is a terminal-oriented or bus-oriented event and refers to a specific period on the bus (not to a frame in the sense of a network frame that is physically send or word strings as it is called in ARINC 629). The beginning and end of terminals' minor frame are linked to the transmission of a particular terminal. A bus minor frame is identical to the minor frame of the terminal with the slowest clock for basic protocol (BP) or of the lead terminal in combined protocol (CP), the first terminal to transmit following the longest quiet time [2]. A major frame is a series of unique minor frames as is also signaled. Minor and major frames are related to aircraft-level control loops.

A non-OSI broadcast message includes the following elements: label, CID, word count (if necessary), data status, protocol indication, and information. Figure 46.4 presents the non-OSI broadcast message format.

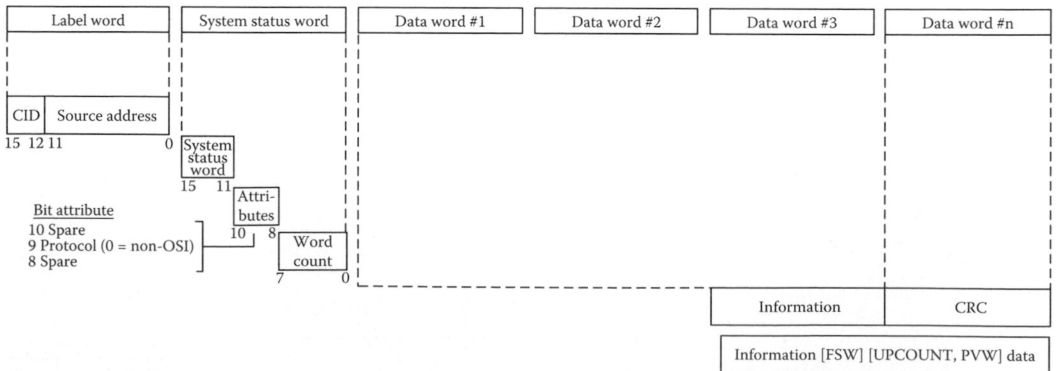

FIGURE 46.4 Non-OSI message broadcast format. LLC, logical link control; CID, channel identification; FSW, function status word; PVW, parameter validity word. (From Airlines Electronic Engineering Committee, Multi-transmitter data bus—Technical description—ARINC specification 629P1-5, Standard 629P1-5, Aeronautical Radio, Inc (ARINC), Annapolis, MD, March 31, 1999.)

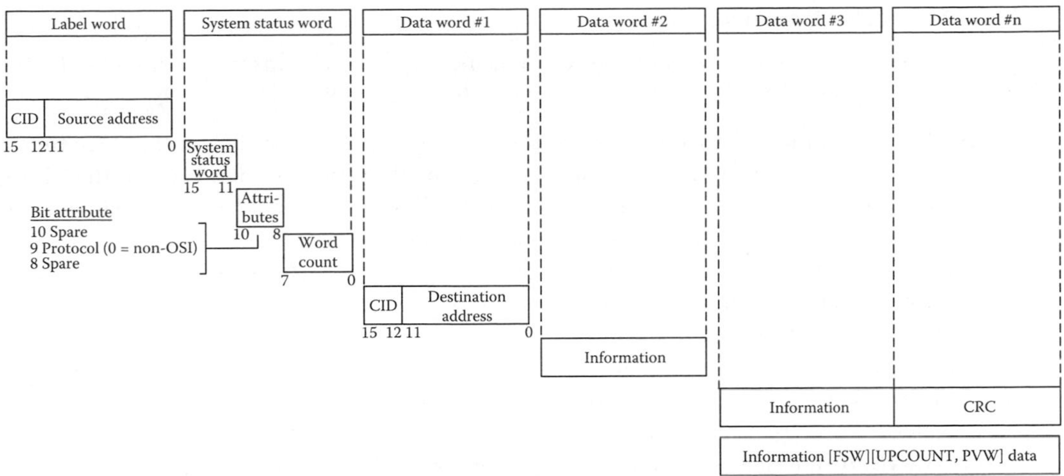

FIGURE 46.5 Non-OSI directed message format. LLC, logical link control; CID, channel identification; FSW, function status word; PVW, parameter validity word. (From Airlines Electronic Engineering Committee, Multi-transmitter data bus—Technical description—ARINC specification 629P1-5, Standard 629P1-5, Aeronautical Radio, Inc (ARINC), Annapolis, MD, March 31, 1999.)

Data messages are composed of word string comprising the label word in the first word followed by 16 data bits in the following words. Each position in a word string contains specific data elements that are specified for an aircraft. These are also sometimes called parameters. Figure 46.5 presents the non-OSI directed message format.

More detailed description of the syntax of non-OSI messages like labels, channel identifiers and its use, data status, word count, protocol indication, information, and frame check sequence can be found in [2].

46.3 Medium Access Layer

The ARINC medium access layer is a carrier-sense multiple access/collision avoidance (CSMA/CA) scheme. Bus access control is distributed among all participating terminals, each of which autonomously determines its transmission sequence. Each terminal achieves access by the use of bus access timers combined status information about the state of the medium. There are two implementations of ARINC 629, which essentially differ at the MAC layer:

1. Basic mode protocol (BP)
2. Combined mode protocol (CP)

The two protocols differ in how stations access the physical medium and especially in the way they handle aperiodic data.

46.3.1 Basic Protocol

The BP is a dual-mode protocol in which the bus can operate in either the periodic mode or the aperiodic mode. The bus system design, that is, factors like the loading factor of the bus, determines which of two modes that the bus will operate in. Buses designed to operate in the periodic mode will automatically transition to the aperiodic mode when they are overloaded. The description in this section is a summary of [2]. It is also slightly more detailed than the description of the CP as it explains also more details about the background and rationale of timers.

46.3.1.1 Timers for Operation

The three timers that are used to access the physical medium are (1) the transmit interval (TI), (2) the synchronization gap (SG), and (3) the terminal gap (TG). In more detail,

1. The TI is identical for all terminals on the same bus. It is the longest time of the three timers and can range from 0.5 to 64.0 ms. It starts every time the terminal starts transmitting. Once a terminal has transmitted, it should wait the length of time specified by the TI before it can transmit again. Hence, it is equal to the minimum possible bus cycle time.

   For a normally periodic bus, the selected value should be such that data are available on the bus at a minimum update rate required of the user system. Hence, TI is a transmission deferral mechanism in periodic mode.

   Lower TI decreases efficiency of the bus due to increasing bus protocol overhead. For aperiodic mode operation, TI should be relatively short, that is, as short as to ensure all users have the opportunity to transmit as often as possible.
2. The SG is also identical for all terminals on the same bus. SG is the second longest timer. The value of SG is selected from the TG-relative binary values (BVs) of 16, 32, 64, and 127 (corresponding to 17.7, 33.7, 65.7, and 127.7 μs for the bit rate of 2 MHz [2,4]). The SG time is effectively dead time on the bus and should be selected to be as small as possible. However, the selection of SG should allow for the use of additional terminals on the bus.

   SG ensures that all terminals are given access and are chosen to be greater than the maximum TG. It starts every time BQ is sensed and may be reset before it has expired if bus activity is sensed. It is restarted the next time the terminal starts to transmit. Allegorically speaking, SG is used to implement a waiting room protocol in aperiodic mode [6]. All stations enter a waiting room by waiting the time of length SG before being serviced in the order of a station's TG with lower TG values first. Hence, for aperiodic mode, SG provides the deferral mechanism for sending.
3. The TG is unique for each terminal on the same bus. It is necessary for each terminal on the bus to have a unique number (N) from 1 to 125 starting at the lowest value. The terminal BV is set to the binary equivalent of the terminal number plus one (i.e., $BV_{10} = N + 1$ [2]). The specific TG values for the bit rate of 2 MHz are between 3.7 and 127.7 μs. The TG timer is reset by the presence of a carrier. TG starts only after SG has expired and the bus is quiet. Unlike the SG, TG is reset if bus activity is sensed. As a consequence, TG and SG should not overlap in time but run consecutively. Also, the value of 1111111_2 or 127_{10} is not used in order that SG be greater than TG.

Figure 46.6 shows the media access flow chart of aperiodic and periodic transmission.

A typical timing diagram for a very simple bus with three terminals operating in periodic mode is shown in Figure 46.7 (attention: diagram is not to scale). The figure shows the terminals in a steady state (some time after start-up and initialization).

The order in which terminals achieve bus access is the same for all bus cycles and is determined by the initialization sequence (if there was no aperiodic mode in between). It does not necessarily (but can) equate to the relative durations of the terminals' TG timers. If the sum of all the TGs, transmission times, and SG is less than TI, then the bus cycle time remains fixed and equal to TI.

If the sum of all the TGs, transmission times, and SG is greater than TI, then the system switches to operation in aperiodic mode as shown in Figure 46.8. For every bus cycle, the terminals transmit in the order of their increasing TG durations. Here, TG2 < TG1 < TG3. Each bus cycle consists of a sequence of transmissions separated by the various TG delays followed by an SG delay that provides synchronization. Referring to the explanation of the aforementioned waiting room protocol, Figure 46.8 shows that SG opens the waiting room door and then each terminal is served in the order of increasing TG values.

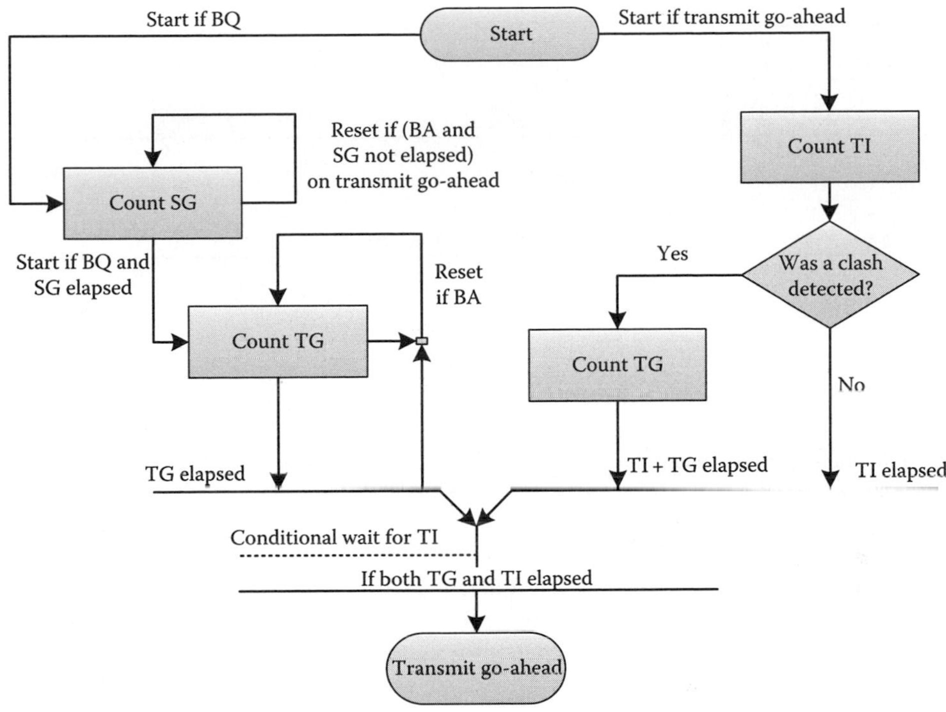

FIGURE 46.6 BP—media access flow chart (not indicating monitoring). BA, bus active; BQ, bus quiet. (From Airlines Electronic Engineering Committee, Multi-transmitter data bus—Technical description—ARINC specification 629P1-5, Standard 629P1-5, Aeronautical Radio, Inc (ARINC), Annapolis, MD, March 31, 1999.)

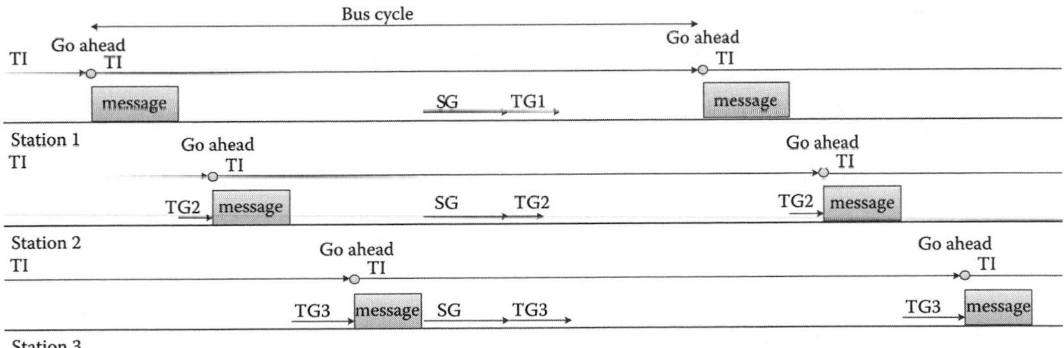

FIGURE 46.7 BP—periodic mode.

46.3.1.2 Bus Cycle Start and Initialization

The initialization phase of bus operation may be relatively uncontrolled depending on the sequence in which local clocks are initialized across multiple terminals. However, as a result of synchronization mechanisms, bus operation will normally settle down to some steady state after a very short time.

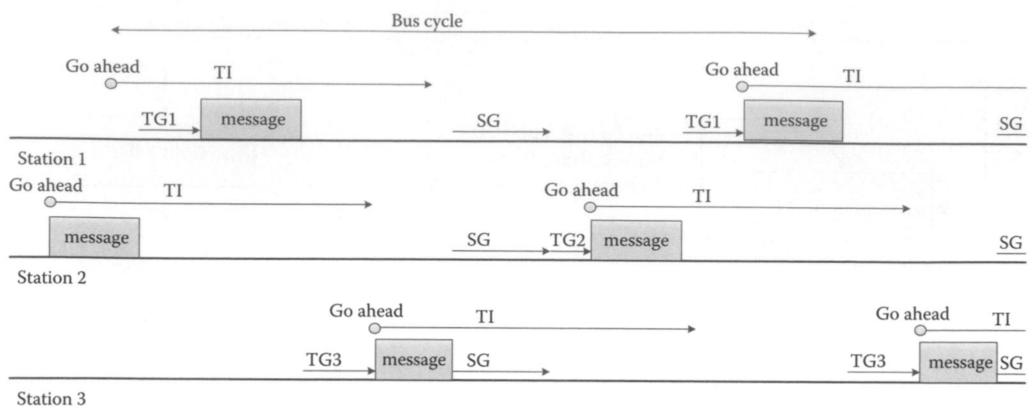

FIGURE 46.8 BP—aperiodic mode.

46.3.2 Combined Protocol

The CP allows sending of periodic and multiple different quality levels of aperiodic messages. The operations of the CP are quite complex. It uses five timers in each station. The combination of the five timers and associated state machines build the kernel of the system network behavior. In the following, we base our description of the ARINC 629 CP closely on an excellent description from Gallon and others provided in [5].

46.3.2.1 Bus Cycle

Media access to the bus is cyclic. A bus cycle is started by one of the ARINC 629 stations—also called the leader in the following. Each bus cycle is divided into four durations also called areas or levels. The first three levels are for periodic traffic (level l), urgent aperiodic traffic (level 2), and nonurgent aperiodic traffic (level 3). Level 3 can be divided into two further sublevels: *level 3 backlog*, for nonurgent aperiodic traffic, which could not be sent in the previous bus cycle, and *level 3 new* for new nonurgent aperiodic traffic. Level 1 traffic has priority and its bus load contribution is evaluated first. Worst-case estimates can be obtained by summing the maximum bus loads. Level 1 traffic does not change significantly between different flight phases. Level 2 and level 3 traffics depend on flight phases. Level 2 traffic is of low periodicity or aperiodic in nature. Level 3 traffic is predominantly aperiodic in nature with some very low-rate periodic data (long messages). Worst-case assessment is not straightforward.

Each station can transmit at most one message per level. Sending of a message is mandatory at level 1, but optional at all other levels. The leader station starts the cycle by sending its first periodic message (i.e., beginning of level 1). The beginning of this first periodic message is the signal for other (nonleading) stations of the beginning of the new bus cycle (this signal is called concatenation event). The fourth area of the bus cycle allows the synchronization of the stations at the end of a cycle in order to define a leader for the new bus cycle (the definition is made by election on local information).

46.3.2.2 Timers for Operation

In the CP, there are two types of timers:

- Type 1 timers—involving three timers—allow the stations to deal with the bus cycle. The behavior of these timers is bound to the transmission activity on the bus as the information BA and BQ are fundamental for the operation of the bus:
  - The timer TG controls the access to the bus at levels 1, 2, and 3. Each station has a different value for TG; we have the following relation: $\forall i, j, TG_j > TG_i + 2\tau$, with τ being the

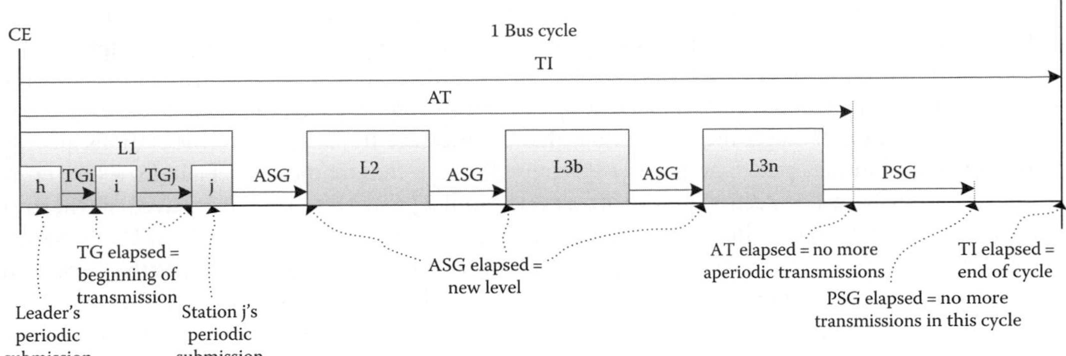

FIGURE 46.9 Role of different timers in CP. (From Gallon, L. et al., Modelling and analysis of the ARINC specification 629 CP MAC layer protocol, in *Proceedings of IEEE International Workshop on Factory Communication Systems*, pp. 91–100, Barcelona, Spain, 1997.)

maximum propagation time on the bus. TG allows specifying a static priority scheme and then implementing a collision avoidance mechanism. As soon as we have the signal BA (beginning of a bus activity), the timer TG is reset, and as soon as we have the signal BQ (end of a bus activity), the timer TG is running. The timer TG is also reset when the station enters a new level of a bus cycle (i.e., when aperiodic synchronization gap [ASG] becomes elapsed). When the timer TG expires, the station can send a message on the bus (if it has not already sent a message at this level).

- The timer ASG controls the level changes. Each station has the same value for ASG and ASG > TG_{max}. As for the timer TG, as soon as a station gets the signal BA (BQ), the timer ASG is reset (running). At the time when the ASG becomes elapsed, the station changes level.
- The timer periodic synchronization gap (PSG) allows all the stations to synchronize themselves at the end of the bus cycle. Each station has the same value for PSG (PSG = 5 ASG). As for the timers TG and ASG, as soon as a station gets the signal BA (BQ), the timer PSG is reset (running). When PSG becomes elapsed, there will be no more transmission in this bus cycle and a station enters a new bus cycle.

- Type 2 timers concern two timers that have the same value in all the stations and which specify the bounds of the bus cycle (i.e., constraints imposed by the stations to the different traffics):
  - Timer TI defines the periodicity of the cycle. It is started at the beginning of the periodic transmission and elapses either by a normal runoff or by a CE interruption (beginning of a new bus cycle). In a permanent normal behavior, the TI of the leader goes to its normal end and then is restarted immediately (new bus cycle). It is then the leader that imposes the periodicity of the cycle (TI).
  - Timer AT (aperiodic access time-out) defines the last instant in a bus cycle for an aperiodic transmission start (i.e., AT = TI − (PSG + ASG + MAL), which MAL being the maximum allowed length for a message).

Figure 46.9 summarizes the definition and the use of the timers.

46.3.2.3 Bus Cycle Start and Initialization

The station that is the leader in one bus cycle normally (i.e., no bus overload, bus initialization, or transient disturbance) stays the leader for the next cycle. This is because the following three local conditions hold at a leader, which are the same the conditions a station satisfies to assert itself as a leader: (1) PSG elapsed (there will be no more transmission in the bus cycle in progress), (2) TI elapsed (the period has finished), and (3) TG elapsed (the station has the right to access to the bus). Normally, the first station powered up will become the leader (considering differences between power-up of stations are greater than differences in TG), but details on the leader election during initialization as described in the next paragraph.

During initialization, which is a transient behavior from the station being switched on to the introduction of the station in a bus cycle (i.e., emitting the first periodic message [at level l]), a station starts off with PSG and wants to get the information that the message transmission of the stations already running is ended in the cycle in progress (PSG elapsed). Next, the station starts TI and tries to introduce itself on the bus as nonleader (waiting for a cycle end [CE] and, then, if CE is received, it will start introducing itself into level 1). In the case of an empty bus, there will not be a CE and then TI will elapse. When TI elapses, TG is restarted and at the end of TG (TG elapsed), the station becomes the leader and starts TI and its periodic transmission in level 1. Using TG when TI becomes elapsed allows to control the phenomenon of collision, which can occur at the initialization if several stations want to join regular bus operation simultaneously (remember the stations have different TG). But we cannot avoid all the collisions with this mechanism (which depend on the dephasing between the stations at the switch on, which can compensate the difference between the TGs of the stations). This protocol avoids collisions in the permanent state, but stations can exhibit collisions in transient phases like the initialization. Whenever a collision occurs, there is a signal Bus Clash (BC), which comes from the physical layer, sent to the MAC entity. The station cannot transmit any more message in this bus cycle, and a reinitialization procedure at the end of the bus cycle (i.e., after the condition PSG elapsed) is started.

46.3.3 Exceptions

Exceptions can be caused in bus overload situations (the duration of level 1 is greater than TI because of two long messages), timer drifts, ghost transmissions (e.g., the lightning-induced noise signals), and failure of a station, which participate to the bus cycle (deaf station and/or dumb station). ARINC 629 implementations deploy significant monitoring mechanisms, but no monitoring is perfect.

46.3.3.1 Architecture of Link Layer

Reference [5] also provides a good overview of the architecture of the link entity, which is represented in Figure 46.10. The MAC is composed of two main modules: (1) the transmitter scheduler and (2) the transmitter. The transmitter scheduler is the core of the implementation of the CSMA/CA mechanism and the cycle bus concept. The signals BA, BQ, and BC are basic signals. The transmitter scheduler implements the mechanisms for the following bus states:

- Initialization of a station on the bus (start-up and reinitialization of the station on the bus after power-up)
- Determining of the leader or nonleader of a station
- Commanding the transmitter to emit messages in levels 1, 2, and 3 of the bus cycle (these messages contain the data from the logical link control)

The architecture of the transmitter scheduler is shown in Figure 46.11, which depicts the transmitter scheduler architecture, states, and dependencies with its two main modules: the *timer module* and the *transmitter control module*. The timer module comprises the timers TG, ASG, PSG, TI, and AT. The transmitter control module includes the submodules' initialization; attribute (leader/nonleader) determination, levels 1, 2, and 3, and a submodule that memorizes the conditions (elapsed, not elapsed) of the timers; and the BC occurrences. Figure 46.11 also shows essential interactions for the behavior of the transmitter scheduler (signals from the physical layer, signals between the timers and the transmitter control).

46.3.3.2 Example Scenario

Figure 46.12 sketches a stable bus cycle with only three stations i, j, and k. Station i is the leader but it does not have the smallest TG. Station j has the smallest TG. Station i transmits in levels 1 and 2; station j

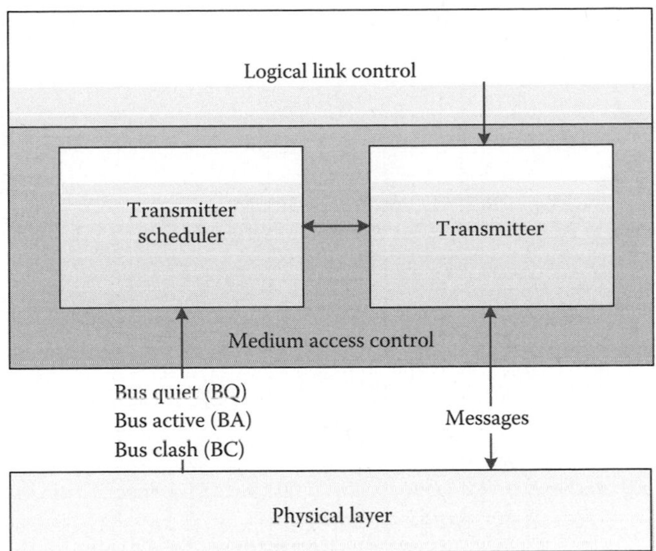

FIGURE 46.10 MAC architecture of CP. (From Gallon, L. et al., Modelling and analysis of the ARINC specification 629 CP MAC layer protocol, in *Proceedings of IEEE International Workshop on Factory Communication Systems*, pp. 91–100, Barcelona, Spain, 1997.)

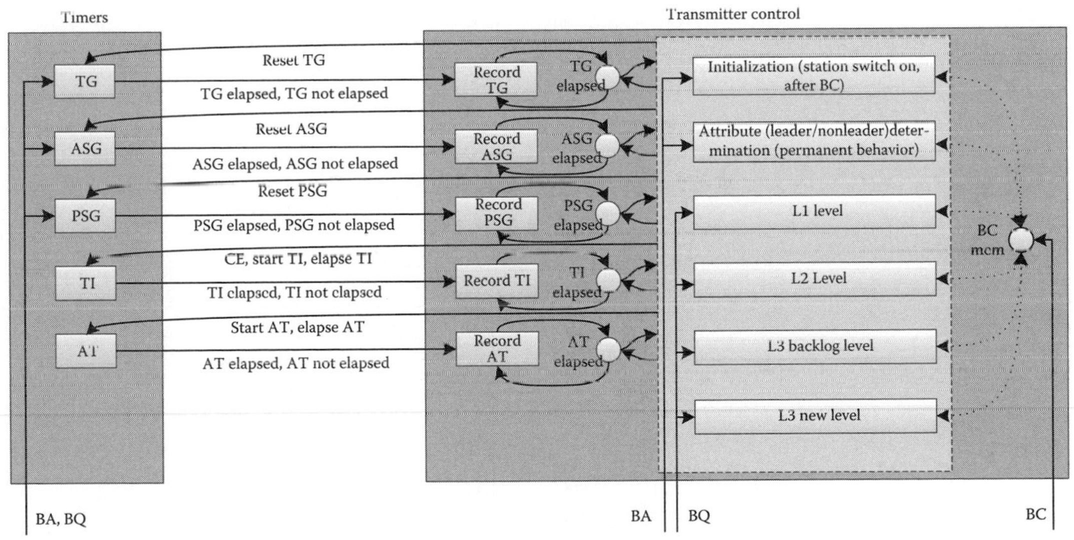

FIGURE 46.11 CP transmitter scheduler architecture. (From Gallon, L. et al., Modelling and analysis of the ARINC specification 629 CP MAC layer protocol, in *Proceedings of IEEE International Workshop on Factory Communication Systems*, pp. 91–100, Barcelona, Spain, 1997.)

transmits in levels 1, 2, and 3 backlog (which means that it had a message to send in level 3 new during the previous bus cycle that could not be sent); station k transmits in levels 1 and 2; the three stations have nothing to send in level 3 new. Figure 46.12 neglects the time propagation τ. In Figure 46.12, circles indicate the actions that are undertaken in the leader and nonleader role. A horizontal piece of line represents a timer. When a line ends with an arrow, it means that the timer expired; otherwise, there is a reset. We can observe the following interesting scenario in line with the aforementioned rules and explanations. Station i starts this bus cycle and the next bus cycle. At level 2, station j sends the first (because it has the smallest TG).

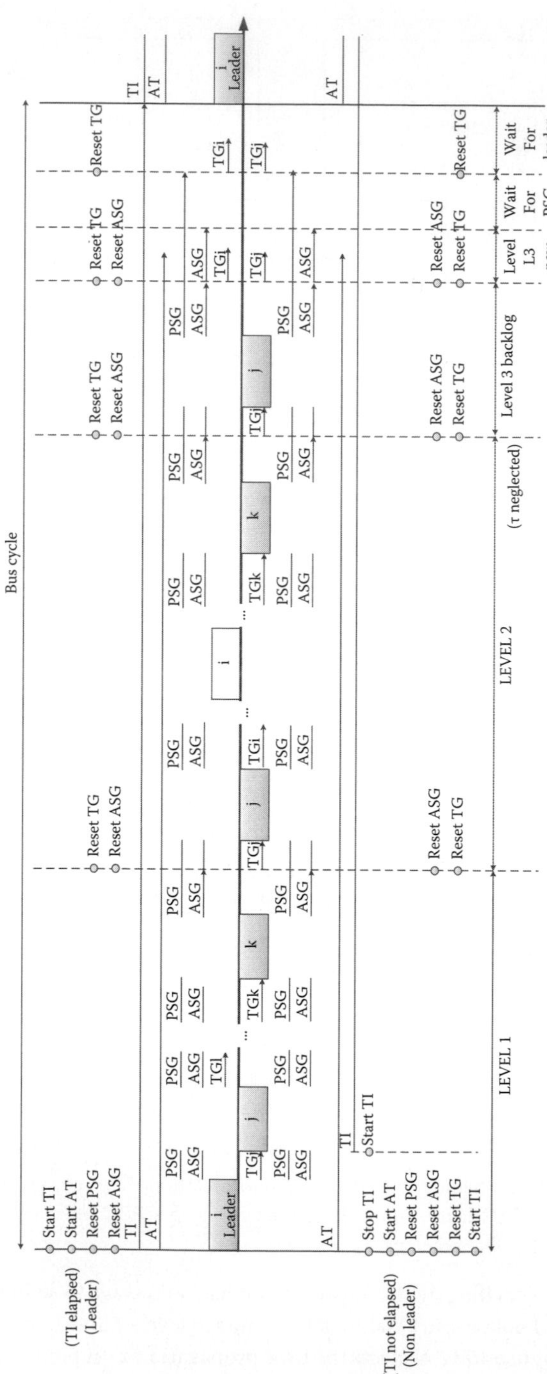

FIGURE 46.12 CP—normal operation.

TABLE 46.4 Main Features of BP and CP

	BP	CP
Operations	Periodic or aperiodic, or aperiodic messages sent in periodic mode	Period and aperiodic (optimization of bandwidth)
	No segregation between periodic and aperiodic messages	Segregation between periodic and aperiodic messages
File transfer	Padding in XPP even if no request; label sent on the bus	Aperiodic word strings sent on request by the host, burst of word strings possible
Determinism for periodic messages	Guaranteed by TI/SG timers	Guaranteed by TI, PSG, and AT timers
Scheduling modes	Block, independent, sync, alternate	Block, independent, sync

Source: Airlines Electronic Engineering Committee, Multi-transmitter data bus—Part 2 applications guide—ARINC specification 629P2-2, Standard 629P2-2, Aeronautical Radio, Inc (ARINC), Annapolis, MD, February 26, 1999.

46.3.4 Comparison and Summary

Table 46.4 presents an overview of main features of the BP and CP.

46.4 Robustness and Fault Tolerance Features

Next to parity for each word string and robustness checks of the physical layer (like amplitude and coding correctness checks), ARINC 629 requires other robustness and fault tolerance features like three types of terminal monitoring: (1) receive data monitoring, (2) transmission monitoring, and (3) protocol check.

46.4.1 Receive Data Monitoring

Receive data monitoring ensures that any received data word that does not have a valid sync pattern, a valid Manchester II biphase modulation pattern, and a valid parity bit is not delivered to higher layers.

The receivers of all terminals that are not transmitting data should monitor all labels being transmitted on the bus. Receivers should check labels against programmed entries in the receive personality Programmable Read-Only Memory (PROM). Labels not destined for this terminal should be discarded.

46.4.2 Transmit Data Monitoring

Transmission monitoring of the terminal uses data a terminal's transmitter has placed on the bus. The goal of the transmission monitoring is detection of events like prevention of impersonation (which happens when a transmitter sends with the label of another transmitter) and babbling (the continuous transmitting of a terminal for an extensive period of time). It detects a label authorized for transmission and ensures a correct CID. The transmit monitor prevents transmission on the bus whenever an invalid sync pattern, modulation, or parity condition exists. Furthermore, the monitor also prevents transmission when the word string length exceeds the specified value, or the message length exceeds the specified value, or an illegal label is being transmitted. In case of a monitoring error, the transmitter should inhibit the remainder of that message. Additionally, a terminal heuristically distinguishes between transient and permanent faults leveraging a transmission error counter. In case of seven successive unsuccessful tries, the terminal is disabled.

46.4.3 Bus Protocol Monitoring

ARINC 629 also specifies monitoring of essential protocol logic in two separate circuits on the same protocol chip. Each of these circuits should operate from a separate crystal oscillator and obtain the essential three controlling parameters from a different source.

These three rules and parameters are as follows:

1. TI has elapsed.
2. The SG—a quiet period on the bus—has existed on the bus.
3. The TG—another quiet period, unique for each terminal—has existed on the bus since the SG occurred.

During execution, the protocol state machine is extended to agreeing on the rules for the go-ahead signal for the transmitter despite expected differences due to crystal oscillator and similar effects. This dual-logic implementation of the protocol machine minimizes the likelihood of a single-point failure causing a terminal to transmit at the wrong time. The protocol logic should initiate a transmit operation and maintain order on the bus.

These three simple rules ensure orderly transmission, without collisions or contentions, for periodic operation, aperiodic operation, or transitional operation. Whether the bus is periodic is not determined by protocol but determined by the loading of the bus and whether variable word strings are allowed.

46.4.4 Protocol Parameter Monitoring

Dual-protocol circuits should control each protocol parameter steered by the LRU program pin strapping. This should be compared to programming in the receive personality PROM, which is a section of memory that contains bus protocol parameters, receive address value, number of word strings in the message, and other key receive message elements.

46.5 ARINC 629 in Boeing 777

In Boeing 777, multiple new electronics technologies have been introduced in order to provide more functionality with higher reliability and easier maintainability, which essentially means higher value for aircraft customers. One of new technologies selected is the new system bus ARINC 629 described in this chapter in more detail.

ARINC 629 is a time division multiplex system. It includes multiple transmitters with broadcast-type, autonomous terminal access. While ARINC 629 specifies that up to 120 LRUs may be connected together, for Boeing 777 a maximum of 44 LRUs connected with couplers are allowed [1, pp. 2–6].

Boeing 777 leverages the current mode coupling physical layer of ARINC 629 (see Section 46.1.2.1) and utilizes the BP (see Section 46.3.1) in periodic transfer mode mainly during normal operation. LRUs communicate to the bus using a coupler and terminal as shown in Figure 46.13.

The terminal controller incorporates the following features [1, p. 2–2]:

- Local transmit and receive message control
- Optional transmit and receive vectored interrupts
- Manchester II biphase coded serial data format
- Onboard direct memory access (DMA) capability
- Automatic power-on initialization
- Periodic and aperiodic transfer modes
- Label identification of data permits global parameters on bus
- Receiver monitoring and shutdown on error of transmitted data

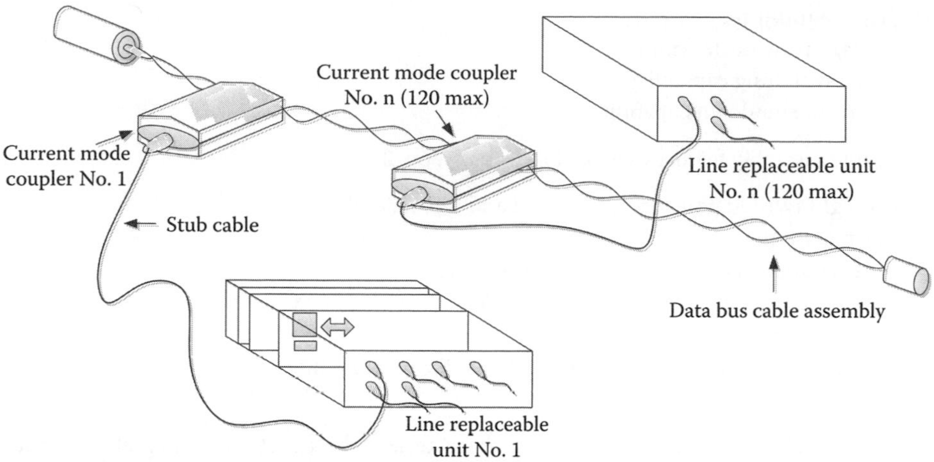

FIGURE 46.13 Interconnect of systems using ARINC 629. (From Yeh, Y.C., Design considerations in Boeing 777 fly-by-wire computers, in *The 3rd IEEE International Symposium on High-Assurance Systems Engineering, HASE'98*, p. 64, Washington, DC, IEEE Computer Society, 1998.)

- Receive parity and format checking
- 2-Mbit continuous transmission collision recovery
- Capable of simultaneous transmission collision recovery
- 180-pin lead ceramic grid array package
- 208-pin lead plastic quad flatpack (PQFP) package

The SIM incorporates the following features [1, p. 2–2]:

- Interface between terminal controller and CMC, which converts serial Manchester encoded data into differential voltage doublets and vice versa
- Checks amplitude and polarity of doublets transmitted to the coupler
- Checks symmetry and polarity of doublets received from coupler
- Notifies terminal controller of SIM/coupler failure symptoms
- Inhibits coupler transmission if both channels are bad
- Shuts down the coupler power if an overcurrent condition is detected
- Administers BIST sequence to determine coupler health
- Prevents propagation of shorts and hot faults into LRU from the stub
- Packages available for through-hole and surface mounting

The following lists the features of the stub cable assembly [1, pp. 2–3]:

- Four conductors, made up of either two twisted pairs, each separately shielded with one overall shield, or quad arrangement with all four wires twisted together
- Circumferential shielding at connectors
- Up to 40 m in length using 22 gauges wires, but only up to 17.3 m in Boeing 777 implementation using 24 gauges wires

The CMC incorporates the following features [1, pp. 2–3]:

- Inductive nonintrusive *clip on* coupling
- Low bus load and reflection
- Low dynamic signal range

- High noise immunity
- Redundant transmitters and receivers
- Channel-switching capability
- Overcurrent shutdown capability

The bus cable assembly features are the following [1, pp. 2–3]:

- Lightweight twisted pair media (shielded and unshielded)
- Low attenuation
- Impedance-matched fault-tolerant terminations
- Capable of supporting up to 120 mode couplers, but only a maximum of 44 couplers in Boeing 777 implementation
- Up to 100 m in length, but only maximum of 59.6 m in Boeing 777 implementation or 71.8 m on stretch airplane version

The avionics architecture in Boeing 777 is characterized by functional integration compared to predecessor aircrafts. Classical avionics at that time had a federated avionics architecture, where a set of functions is implemented in one or more LRUs. In Boeing 777, this has been replaced by a more integrated architecture where multiple sets of functions are combined in fewer LRUs, some in common cabinets instead of many separate boxes.

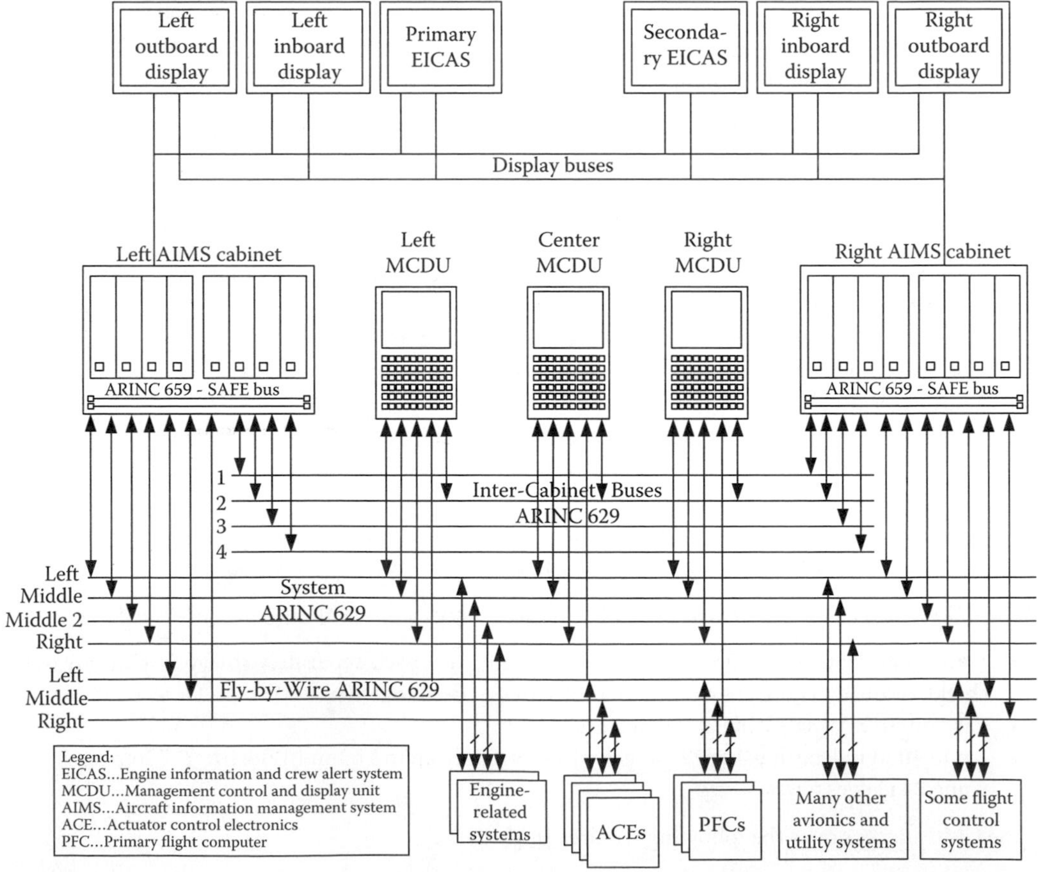

FIGURE 46.14 Boeing B 777 avionics overview. EICAS, engine information and crew alert system; MCDU, management control and display unit; AIMS, aircraft information management system; ACE, actuator control electronics; PFC, primary flight computer. (Courtesy of Honeywell, Morristown, NJ.)

The cabinets and stand-alone LRUs are connected to the airplane interfaces via a combination of ARINC 429, ARINC 629, display buses, and discrete I/O channels. Figure 46.14 shows the most important connected modules; it displays only ARINC 629 and display buses and neither ARINC 429 nor discrete I/O channels. In addition to the cabinets, other flight deck hardware elements that make up the avionics system are six flat panel display units, three control and display units, two electronic flight instrument system (EFIS) display control panels, one display select panel, two cursor control devices, and two display remote light sensors (not all are depicted in Figure 46.14). The system bus in Boeing 777 is mainly ARINC 629 [2]. As can be seen, multiple groups of bus called system ARINC 629 and fly-by-wire ARINC 629 buses are deployed. For the flight control ARINC 629, LRUs receive data on all three ARINC 629 buses but transmit on only one ARINC 629 bus.

References

1. Airlines Electronic Engineering Committee. Multi-transmitter data bus—Part 2 applications guide—ARINC specification 629P2-2. Standard 629P2-2, Aeronautical Radio, Inc. (ARINC), Annapolis, MD, February 26, 1999.
2. Airlines Electronic Engineering Committee. Multi-transmitter data bus—Technical description—ARINC specification 629P1-5. Standard 629P1-5, Aeronautical Radio, Inc (ARINC), Annapolis, MD, March 31, 1999.
3. Airlines Electronic Engineering Committee. Digital information transfer system (DITS)—Part 1—Functional description, electrical interface, label assignments and word formats—ARINC specification 429PART1-16. Standard 629P2-2, Aeronautical Radio, Inc. (ARINC), Annapolis, MD, September 27, 2001.
4. A. Gabillon and L. Gallon. An availability model for avionic databuses. In *Proceedings of the Workshop on Issues in Security and Petri Nets (WISP)*, Vol. 23, Eindhoven, The Netherlands, 2003.
5. L. Gallon, G. Juanole, and I. Blum. Modelling and analysis of the ARINC specification 629 CP MAC layer protocol. In *Proceedings of IEEE International Workshop on Factory Communication Systems*, pp. 91–100, Barcelona, Spain, 1997.
6. H. Kopetz. *Real-Time Systems: Design Principles for Distributed Embedded Applications*. Kluwer Academic Publishers, Norwell, MA, 1st edn., 1997.
7. Special Committee 135 (SC-135). Environmental conditions and test procedures for airborne equipment. Standard DO-160E, RTCA, Inc., Washington, DC, December 9, 2004.
8. Y. C. Yeh. Design considerations in Boeing 777 fly-by-wire computers. In *The 3rd IEEE International Symposium on High-Assurance Systems Engineering, HASE'98*, p. 64, Washington, DC, 1998. IEEE Computer Society.

47

Commercial Standard Digital Bus

47.1 Introduction 47-1
47.2 Bus Architecture 47-2
47.3 Basic Bus Operation 47-2
47.4 CSDB Bus Capacity 47-3
47.5 CSDB Error Detection and Correction 47-3
47.6 Bus User Monitoring 47-4
47.7 Integration Considerations 47-4
Physical Integration • Logical Integration • Software Integration • Functional Integration
47.8 Bus Integration Guidelines 47-6
47.9 Bus Testing 47-6
47.10 Aircraft Implementations 47-7
Defining Terms 47-7
References 47-8
Further Information 47-8
Bibliography 47-8

Lee H. Harrison
Galaxy Scientific Corp.

47.1 Introduction

The Commercial Standard Digital Bus (CSDB) [1–4] is one of several digital serial integration data buses that currently predominate in civilian aircraft. The CSBD finds its primary implementations in the smaller business and private general aviation (GA) aircraft but has also been used in retrofits of some commercial transport aircraft.

Rockwell Collins developed CSDB, a unidirectional data bus. The bus used in a particular aircraft is determined by which company the airframe manufacturer chooses to supply the avionics.

CSDB is an asynchronous linear broadcast bus, specifying the use of a twisted, shielded pair cable for device interconnection. Two bus speeds are defined in the CSDB specification: a low-speed bus operates at 12,500 bits per second (bps) and a high-speed bus operates at 50,000 bps. The bus uses twisted, unterminated, shielded pair cable and has been tested to lengths of 50 m.

The CSDB standard also defines other physical characteristics such as modulation technique, voltage levels, load capacitance, and signal rise and fall times. Fault protection for short circuits of the bus conductors to both 28 volts direct current (VDC) and 115 volts alternating current (VAC) is defined by the standard.

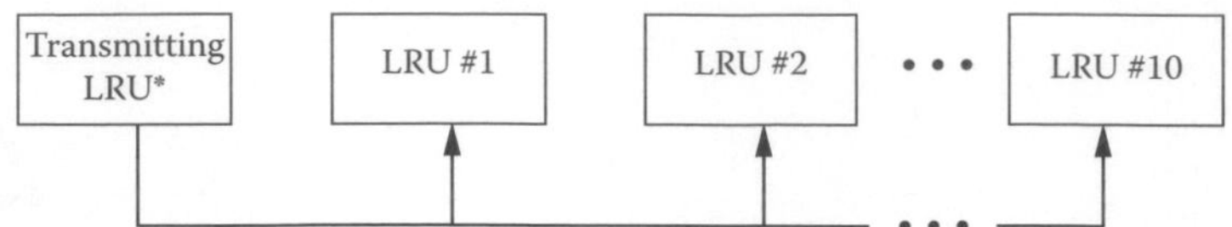

FIGURE 47.1 Unidirectional CSDB communication. *LRU, line replacement unit.

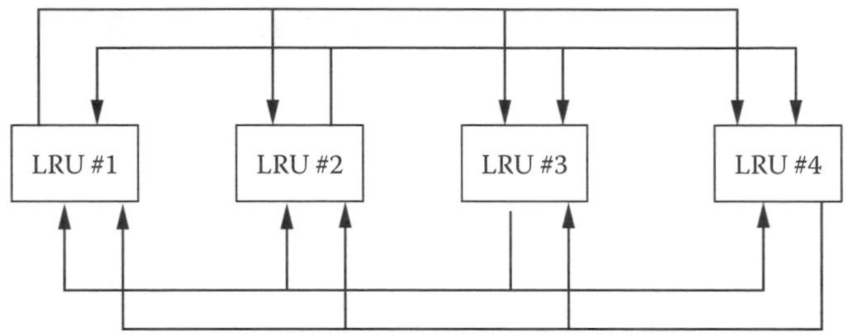

FIGURE 47.2 Bidirectional CSDB communication.

47.2 Bus Architecture

Only one transmitter can be attached to the bus, but it can accommodate up to 10 receivers. Figure 47.1 illustrates the unidirectional bus architecture.

Bidirectional transmission can take place between two bus users. If a receiving bus user is required to send data to any other bus user, a separate bus must be used. Figure 47.2 shows how CSDB may implement bidirectional transmissions between bus users. It can be seen that if each bus user were required to communicate with every other bus user, a significantly greater amount of cabling would be required. In general, total interconnectivity has not been a requirement for CSDB-linked bus users.

It is possible to interface CSDB to other data buses. When this is done, a device known as a gateway interfaces to CSDB and the other bus. If the other bus is ARINC 429 compliant, then messages directed through the gateway from CSDB are converted to the ARINC 429 protocol, and vice versa. The gateway would handle bus timing, error checking, testing, and other necessary functions. The system designers would ensure that data latency introduced by the gateway would not cause a "stale data" problem, resulting in a degradation of system performance. Data are stale when they do not arrive at the destination line replaceable unit (LRU) when required, as specified in the design.

47.3 Basic Bus Operation

In Section 2.1.4 of the CSDB standard [1], three types of transmission are defined: (1) continuous repetition, (2) noncontinuous repetition, and (3) burst transmissions. Continuous repetition transmission refers to the periodic updates of certain bus messages. Some messages on CSDB are transmitted at a greater repetition rate than others. The CSDB standard lists these update rates, along with the message address and message block description. Noncontinuous repetition is used for parameters that are not always valid or available, such as mode or test data. When noncontinuous repetition transmission is in use, it operates the same as continuous repetition. Burst transmission initiates an action (such as radio tuning) or may be used to announce a specific event. Operation in this mode initiates 16 repetitions of the action in each of 16 successive frames, using an update rate of 20 per second.

For CSDB, bytes consist of 11 bits: a start bit, 8 data bits, a parity bit, and a stop bit. The least significant bit (bit 0) follows the start bit. The CSDB standard defines the message block as "a single serial message consisting of a fixed number of bytes transmitted in a fixed sequence" (GAMA CSDB 1986) [1].

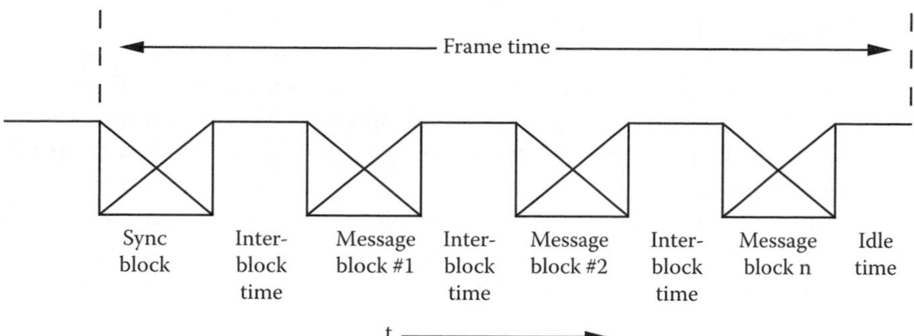

FIGURE 47.3 CSDB data frame structure.

Essentially, a message block consists of a number of bytes concatenated together, with the first byte always being an address byte. A status byte may or may not be included in the message block. When it is, it immediately follows the address byte. The number of data bytes in a message block varies.

Data are sent as frames consisting of a synchronization block followed by a number of message blocks. A particular frame is defined from the start of one synchronization block to the start of the next synchronization block. A "sync" block consists of N bytes of the sync character, which is defined as the hexadecimal character "A5." The sync character is never used as an address. While the data may contain a sync character, it may occur in the data a maximum of N – 1 times. Frames consist of message blocks preceded by a sync block. The start of one sync block to the start of the next sync block is one frame time. Figure 47.3 shows what transpires during a typical frame time.

47.4 CSDB Bus Capacity

The CSDB is similar to ARINC 429 in that it is an asynchronous broadcast bus and operates using character-oriented protocol. Data are sent as frames consisting of a synchronization block followed by a number of message blocks. A particular frame is defined from the start of one synchronization block to the start of the next synchronization block. A message block contains an address byte, a status byte, and a variable number of data bytes. The typical byte consists of one start bit, eight data bits, a parity bit, and a stop bit.

The theoretical bus data rate for a data bus operating at 50,000 bps with an 11-bit data byte is 4545 bytes per second. For CSDB, the update rate is reduced by the address byte and synchronization block overhead required by the standard.

The CSDB interblock and interbyte times also reduce bus throughput. According to the specification, there are no restrictions on these idle times for the data bus. These values, however, are restrained by the defined update rate chosen by the designer. If the update rate needs to be faster, the interblock time and the Interbyte time can be reduced as required, until bus utilization reaches a maximum.

47.5 CSDB Error Detection and Correction

Two methods of error detection are referenced in the standard. They are the use of parity and checksums. A parity bit is appended after each byte of data in a CSDB transmission. The burst transmission makes use of the checksum for error detection. As the General Aviation Manufacturers Association (GAMA) specification states, "It is expected that the receiving unit will accept as a valid message the first message block that contains a verifiable checksum" (GAMA CSDB 1986) [1].

47.6 Bus User Monitoring

Although the CSDB specification defines many parameters, there is no suggestion that receivers should monitor them. The bus frame, consisting of the synchronization block and message block, may be checked for proper format and content. A typical byte, consisting of start, stop, data, and parity bits, may be checked for proper format.

The bus hardware should include the functional capability to monitor these parameters. Parity, frame errors, and buffer overrun errors are typically monitored in the byte format of character-oriented protocols. The message format can be checked and verified by the processor if the hardware does not perform these checks.

47.7 Integration Considerations

The obvious use of a data bus is for integrating various LRUs that need to share data or other resources. The following sections examine various levels of integration considerations for CSDB, including physical, logical, software, and functional considerations.

47.7.1 Physical Integration

The standardization of the bus medium and connectors addresses the physical integration of LRUs connected to the CSDB. These must conform to the Electronic Industries Association (EIA) Recommended Standard (RS)-422-A (1978), "Electrical Characteristics of Balanced Voltage Digital Interface Circuits." The CSDB standard provides for the integration of up to 10 receivers on a single bus, which can be up to 50 m in length. No further constraints or guidelines on the physical layout of the bus are given.

Each LRU connected to a CSDB must satisfy the electrical signals and bit timing that are specified in the EIA RS-422-A. Physical characteristics of the CSDB are given in Table 47.1. Figure 47.4 shows the nonreturn to zero (NRZ) data format used by CSDB LRUs. NRZ codes remain constant throughout a bit interval and either use absolute values of the signal elements or differential encoding in which the polarity of adjacent elements is compared to determine the bit value.

Typical circuit designs for transmitter and receiver interfaces are given in the CSDB standard. Protection against short circuits is also specified for receivers and transmitters. Receiver designs should include

TABLE 47.1 CSDB Physical Characteristics

Modulation Technique	Nonreturn to Zero
Logic sense for logic "0"	Line B positive with respect to line A
Logic sense for logic "1"	Line A positive with respect to line B
Bus receiver	High impedance, differential input
Bus transmitter	Differential line driver
Bus signal rates	Low speed: 12,500 bps
	High speed: 50,000 bps
Signal rise time and fall time	Low speed: 8 μs
Receiver capacitance loading	High speed: 0.8–1.0 μs
	Typical: 600 pF
Transmitter driver capability	Maximum: 1,200 pF
	Maximum: 12,000 pF

Source: *Commercial Standard Digital Bus*, 8th edn., Collins General Aviation Division, Rockwell International Corporation, Cedar Rapids, IA, January 30, 1991.

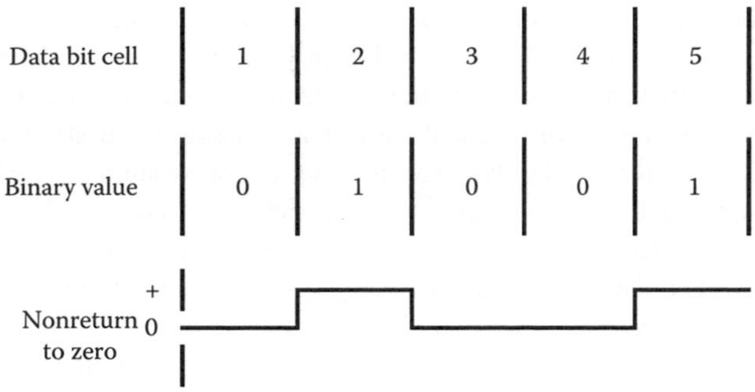

FIGURE 47.4 Nonreturn to zero data example.

protection against bus conductor shorts to 28 VDC and to 115 VAC, and transmitter designs should afford protection against faults propagating to other circuits of the LRU in which the transmitter is located.

To ensure successful integration of CSDB LRUs and avoid potential future integration problems, the electrical load specification must be applied to a fully integrated system, even if the initial design does not include a full complement of receivers. As a result, additional receivers can be integrated at a later time without violating the electrical characteristics of the bus.

47.7.2 Logical Integration

The logical integration of the hardware is controlled by the CSDB standard which establishes the bit patterns that initiate a message block and the start bit, data bits, parity bit, and stop bit pattern that comprises each byte of the message. The system designer, however, must control the number of bytes in each message and ensure that all the messages on a particular bus are of the same length.

47.7.3 Software Integration

Many software integration tasks are left to the system designer for implementation; hence, CSDB does not fully specify software integration. The standard is very thorough in defining the authorized messages and in constraining their signaling and update rates. The synchronization message that begins a new frame of messages is also specified. However, the determination of which messages are sent within a frame for a particular bus is unspecified. Also, there are no guidelines given for choosing the message sequence or frame loading. The frame design is left to the system designer.

In general, the sequencing of the messages does not present an integration problem since receivers recognize messages by the message address, not by the sequence. However, this standard does not disallow an LRU from depending on the message sequence for some other purpose. The system designer must be aware of whether any LRU is depending on the sequence for something other than message recognition, because once the sequence is chosen, it is fixed for every frame.

The bus frame loading is more crucial. There are three types of messages that can occur within a frame: continuous repetition, noncontinuous repetition, and burst transmissions. The system designer must specify which type of transmission to use for each message and ensure that the worst maximum coincidence of the three types within one frame does not exhaust the frame time. The tables of data needed to support this system design are provided, but the system designer must ensure that no parts of the CSDB standard are violated.

47.7.4 Functional Integration

The CSDB standard provides much of the data needed for functional integration. The detailed message block definitions give the interpretation of the address, status byte, and data words for each available message. Given that a particular message is broadcast, the standard completely defines the proper interpretation of the message. The standard even provides a system definition consisting of a suite of predefined buses that satisfy the integration needs of a typical avionics system.

If this predefined system is applicable, most of the system integration questions are already answered. But if there is any variation from the standard, the designer of a subsystem in a CSDB integrated system must inquire to find out which LRUs are generating the messages that the subsystem needs, on which bus each message is transmitted, at what bus speed the messages are transmitted, and the type of transmission. The designer must also ensure that the subsystem provides the messages required by other LRUs. The system designer needs to coordinate this information accurately and comprehensively. The system design must ensure that all the messages on a particular bus are of the same length and must also control the data latencies that may result as data are passed from bus to bus by various LRUs. All testing is left to the system designer.

There are no additional guidelines published for the CSDB. Whatever problems are not addressed by the standard are addressed by Collins during system integration. Furthermore, Collins has not found the need to formalize its integration and testing in internal documents since CSDB-experienced engineers do this work.

47.8 Bus Integration Guidelines

The CSDB, like the ARINC 429 bus, has only one LRU that is capable of transmitting with (usually) multiple LRUs receiving the transmission. Thus, the CSDB has few inherent subsystem integration problems. However, the standard does not address them. The preface to the CSDB standard clearly states its position concerning systems integration: "This specification pertains only to the implementation of CSDB as used in an integrated system. Overall systems design, integration, and certification remain the responsibility of the systems integrator" (GAMA CSDB 1986) [1].

Although this appears to be a problem for the reliability of CSDB-integrated systems, the GA scenario is quite different from the air transport market. The ARINC standards are written to allow any manufacturer to independently produce a compatible LRU. In contrast, the GAMA standard states in its preface that "This specification… is intended to provide the reader with a basic understanding of the data bus and its usage" (CSDB 1986) [1].

The systems integrator for all CSDB installations is the Collins General Aviation Division. That which is not published in the standard is still standardized and controlled because the CSDB is a sole source item.

Deviations from the standard are allowed, however, for cases in which there will be no further interfaces to other subsystem elements. When variations are made, the change first must be approved in a formal design review and the product specification is then updated accordingly. Integration standards and guidelines for CSDB include the CSDB standard and EIA RS-422-A by the Electronic Industries Association.

47.9 Bus Testing

The CSDB connects avionics LRUs point-to-point to provide an asynchronous broadcast method of transmission. Before the bus was used in the avionics environment, it was put through validation tests similar to those used on other commercial data buses. These included the environmental tests presented in RTCA DO-160 and failure analyses. Most environmental tests were done transparently on the bus after it was installed in an aircraft.

As with other avionics data buses, Collins had to develop external tests to show that the bus satisfied specifications in the standard. Test procedures of this nature are not included in the CSDB standard.

TABLE 47.2 Aircraft and Their Use of the CSDB

Boeing 727	Retrofit
Boeing 737	Retrofit
McDonnell-Douglas DC-8	Retrofit
Saab 340, Saab 2000	Primary integration bus
Embraer	Primary integration bus
Short Brothers SD330 and SD360	Primary integration bus
ATR42 and ATR72	Primary integration bus
De Haviland Dash 8	Primary integration bus
Canadair Regional	Primary integration bus

Source: Rockwell Collins, Cedar Rapids, IA.

Internal bus tests that the CSDB standard describes include a checksum test and a parity check. Both of these are used to ensure the integrity of the bus's data. Care should be taken when using these tests because their characteristics do not allow them to be used in systems of all criticality levels.

Simulation is used for development and testing of LRUs with a CSDB interface. To simulate an LRU connection to the bus, manufacturers make black box testers to generate and evaluate messages according to the electrical and logical standards for the bus. These consist of a general-purpose computer connected to bus interface cards. The simplest ones may simulate a single LRU transmitting or receiving. The more complex ones may be able to simulate multiple LRUs simultaneously.

These are not the only external and internal tests that the CSDB manufacturer can perform. Many more characteristics that may require testing are presented in the CSDB specification. Again, it remains the manufacturer's responsibility to prove that exhaustive validation testing of the bus and its related equipment has met all the requirements of the Federal aviation regulations.

47.10 Aircraft Implementations

This section gives a sampling of the aircraft in which the CSDB is installed. Table 47.2 lists some of the commercial transport aircraft and regional airliners using CSDB. CSDB is used both in retrofit installations and as the main integration bus. Additionally, a number of rotorcraft use the CSDB to communicate between the Collins-supplied LRUs.

Defining Terms

Asynchronous: Operating at a speed determined by the circuit functions rather than by timing signals
Checksum: An error detection code produced by performing a binary addition, without carry, of all the words in a message
Frame: A formatted block of data words or bits used to construct messages
Gateway: A bus user that is connected to more than one bus for the purpose of transferring bus messages from one bus to another, where the buses do not follow the same protocol
Linear Bus: A bus where users are connected to the bus medium, one on each end, with the rest connected in between
Parity: An error detection method that adds a bit to a data word based on whether the number of "one" bits is even or odd
Synchronization Block: A special bus pattern, consisting of a certain number of concatenated "sync byte" data words, used to signal the start of a new frame

References

1. GAMA, *Commercial Standard Digital Bus (CSDB)*, General Aviation Manufacturers Association, Washington, DC, June 10, 1986.
2. Eldredge, D. and E. F. Hitt, *Digital System Bus Integrity*, DOT/FAA/CT-86/44, Federal Aviation Administration Technical Center, Atlantic City International Airport, NJ, March 1987.
3. Elwell, D., L. Harrison, J. Hensyl, and N. VanSuetendael, *Avionic Data Bus Integration Technology*, DOT/FAA/CT-91-19, Federal Aviation Administration Technical Center, Atlantic City International Airport, NJ, December 1991.
4. Collins, *Serial Digital Bus Specification*, Part No. 523-0772774, Collins General Aviation Division/ Publications Department, Melbourne, FL.
5. *Commercial Standard Digital Bus*, 8th edn., Collins General Aviation Division, Rockwell International Corporation, Cedar Rapids, IA, January 30, 1991.

Further Information

The most detailed information available for CSDB is the GAMA CSDB Standard, Part Number 523-0772774. It is available from Rockwell Collins, Cedar Rapids, IA.

Bibliography

ARINC Specification 600-7, *Air Transport Avionics Equipment Interfaces*, Aeronautical Radio, Inc., Annapolis, MD, January 1987.

ARINC Specification 607, *Design Guidance for Avionics Equipment*, Aeronautical Radio, Inc., Annapolis, MD, February 17, 1986.

ARINC Specification 607, *Design Guidance for Avionics Equipment*, Supplement 1, Aeronautical Radio, Inc., Annapolis, MD, July 22, 1987.

ARINC Specification 617, *Guidance for Avionics Certification and Configuration Control*, Draft 4, Aeronautical Radio, Inc., Annapolis, MD, December 12, 1990.

Card, M. Ace, Evolution of the digital avionics bus, *Proceedings of the IEEE/AIAA 5th Digital Avionics Systems Conference*, Institute of Electrical and Electronics Engineers, New York, 1983.

Eldredge, D. and E. F. Hitt, *Digital System Bus Integrity*, DOT/FAA/CT-86/44, U.S. Department of Transportation, Federal Aviation Administration, Washington, DC, March 1987.

Eldredge, D. and S. Mangold, *Digital Data Buses for Aviation Applications, Digital Systems Validation Handbook*, Volume II, Chapter 6, DOT/FAA/CT-88/10, U.S. Department of Transportation, Federal Aviation Administration, Leesburg, VA, February 1989.

GAMA, *Commercial Standard Digital Bus* (*CSDB*), General Aviation Manufacturers Association, Washington, DC, June 10, 1986.

Hubacek, P., *The Advanced Avionics Standard Communications Bus, Business and Commuter Aviation Systems Division*, Honeywell, Inc., Phoenix, AZ, July 10, 1990.

Jennings, R. G., Avionics standard communications bus—Its implementation and usage, *Proceedings of the IEEE/AIAA 7th Digital Avionics Systems Conference*, Institute of Electrical and Electronics Engineers, New York, 1986.

RS-422-A, *Electrical Characteristics of Balanced Voltage Digital Interface Circuits*, Electronic Industries Association, Washington, DC, December 1978.

RTCA DO-160C, *Environmental Conditions and Test Procedures for Airborne Equipment*, Radio RTCA DO Technical Commission for Aeronautics, Washington, DC, December 1989.

RTCA DO-178A, *Software Considerations in Airborne Systems and Equipment Certification*, Radio RTCA DO Technical Commission for Aeronautics, Washington, DC, March 1985.

Runo, S. C., Gulfstream IV flight management system, *Proceedings of the 1990 AIAA/FAA Joint Symposium on General Aviation Systems*, DOT/FAA/CT-90/11, U.S. Department of Transportation, Federal Aviation Administration, Atlantic City International Airport, NJ, May 1990.

Spitzer, C. R., Digital avionics architectures—Design and assessment, *Tutorial of the IEEE/AIAA 7th Digital Avionics Systems Conference*, Institute of Electrical and Electronics Engineers, New York, 1986.

Spitzer, C. R., *Digital Avionics Systems*, Prentice Hall, Englewood Cliffs, NJ, 1987.

Thomas, R. E., A standardized digital bus for business aviation, *Proceedings of the IEEE/AIAA 5th Digital Avionics Systems Conference*, Institute of Electrical and Electronics Engineers, New York, 1983.

WD193X Synchronous Data Link Controller, Western Digital Corporation, Irvine, CA, 1983.

WD1931/WD1933 Compatibility Application Notes, Western Digital Corporation, Irvine, CA, 1983.

WD1993 ARINC 429 Receiver/Transmitter and Multi-Character Receiver/Transmitter, Western Digital Corporation, Irvine, CA, 1983.

48

SAFEbus

48.1 SAFEbus 48-1
Background
48.2 Protocol Overview 48-2
48.3 Protocol Services 48-5
Communication Services • Clock Synchronization • Restart, Reintegration, Integration • Diagnostic Services • Fault Isolation • Configuration Services • Protocol Parameterization
48.4 Communication Interface 48-19
48.5 Validation and Verification Efforts 48-21
48.6 SAFEbus in Boeing 777 48-21
References 48-25

Michael Paulitsch
Airbus Group Innovations

Kevin R. Driscoll
Honeywell International Inc.

48.1 SAFEbus

SAFEbus* is the only backplane or local area network standard (ARINC 659) [1] that provides fail-op/fail-safe fault tolerance with near-unity coverage for all of its components—signal lines, terminations, interface electronics, clock sources, and power supplies. This coverage includes tolerating a Byzantine fault. SAFEbus provides a time-based protocol that delivers messages with a precision on the order of 100 ns over a backplane network. The SAFEbus protocol can be implemented in an integrated circuit or FPGA with just 70,000 gates.

48.1.1 Background

In the late 1980s, there was a push toward integration of multiple functions on a common computing and input/output (I/O) platform, also referred to as integrated modular avionics (IMA). This push was due to the advantages of IMA systems over then prevalent federated architectures. The advantages are decreased size, cost, and weight, increased reliability, less-frequent maintenance, and more flexibility. The success of an IMA system hinges on a backplane–bus connecting line-replaceable modules (LRMs). The backplane–bus must be designed to support the requirements of *space and time partitioning*. These requirements are derived from the concept of *robust partitioning* that prevents functions on a common platform from adversely influencing each other, even when some functions may be faulty. Honeywell designed SAFEbus as a backplane for the Aircraft Information Management System (AIMS), which is the IMA part of the avionics for the Boeing B-777 airplane.

Boeing provided some additional design requirements. One requirement was for the number of days the Boeing 777 could be dispatched without maintenance following a failure. The goal was to allow a plane with a failure in an AIMS component to follow its normal schedule, which will eventually bring it to a maintenance base. This requirement meant that individual components of the AIMS system had to be

* SAFEbus is a registered trademark of Honeywell International, Inc.

reliable and that the system as a whole had to be fault tolerant. A second design requirement was that the backplane–bus interface not force complexity on the functions in an LRM. Some LRMs might be high-performance processors, but others might be simple hardwired logic. A third requirement, implicit in the notion of an integrated cabinet, was that the design support a multiprocessor architecture. In particular, the backplane had to provide adequate net throughput for the initial set of functions together with 50% extra capacity to allow for growth. Fourth, the integrity requirements for the avionics system as a whole meant that the backplane–bus had to exhibit total fault containment. There had to be less than one chance in a billion per hour of operation that an error occurring within the backplane system would be passed undetected from or to the application software. Finally, the design had to be one that would support the certification of the system and the recertification of modified functions. In particular, the design could not be one that would force the recertification of all functions when only one function was modified. Honeywell designed SAFEbus because no existing backplane–bus met these requirements. SAFEbus builds on the multiprocessor flight control system (M^2FCS) [2] research done for the US Air Force.

SAFEbus is standardized as ARINC 659 *backplane data bus* [3]. This is in a family of associated standards including ARINC 653 Avionics Application Software Standard Interface, ARINC 651 Design Guide for Integrated Modular Avionics, and ARINC 650 Integrated Modular Avionics Packaging and Interfaces.

48.2 Protocol Overview

The SAFEbus protocol is heavily dependent on its hardware. In order to understand the SAFEbus protocol, one needs to know the basic building blocks of the hardware architecture. The SAFEbus interface logic within each LRM uses paired hardware; each half of the pair consists of a bus interface unit (BIU) ASIC, a table memory, an intermodule memory (IMM), and backplane transceivers. This logic is paired to provide immediate fault detection and containment. The backplane–bus lines are configured in a unique fault-tolerance topology that lies somewhere between quad redundancy and dual–dual redundancy [4]. This topology simultaneously provides high integrity and availability (see Figure 48.1). SAFEbus consists of two self-checking buses (SCBs), A and B, called bus pairs. Each SCB is itself composed of two buses: x and y. Figure 48.2 presents a nomenclature of the different bus-related items. One of the BIUs in an LRM transmits data on one of the buses in each SCB, and its partner BIU transmits on the other bus within each SCB. The data on any two buses, which come from different BIUs, are compared at each receiving LRM. Only bit-for-bit identical data are written into the IMMs. A transmitting LRM checks its transmission using a local loopback. That is, the receiving circuitry in the transmitting LRM also checks what is actually put on the bus for errors. Such self-checking ensures a babbling LRM will be self-detected and will remove itself from SAFEbus. This removal is enforced within an LRM by having each BIU control the other BIU's drivers. If either BIU thinks it should not be transmitting, neither BIU can transmit. Each bus consists of three wires: two for data and one for clock. Thus, the entire SAFEbus set of buses uses a total of 12 wires. The data are transmitted synchronously, 2 bits at a time, at 30 MHz (for throughput of 60 Mb/s). A *bit time* is the time it takes to send 1 bit on one wire. Using multiple wires, 2 bits and 1 dock cycle are sent during 1 bit time. For time, the term *bit time* is preferred to *clock period* because there are multiple clocks within each BIU. SAFEbus uses the *wired-OR*-capable backplane transceiver logic (BTL) defined in the IEEE 1194 standard. The wired-OR drivers and the use of extended clock pulse widths allow SAFEbus to do some out-of-band signaling that supports some features of the protocol.

Table 48.1 defines some SAFEbus terminology that is further described in this paragraph. The bus time is divided into *windows* of design-time-configurable sizes. Each window is sized to contain either one message (from 1 to 256 32-bit words) or one synchronization pulse or some fixed idle time. A set of static cyclic schedules defines the sequence of windows and the size of each window, which LRM(s) may transmit during each window and which LRM(s) may receive a message from the window. Each window

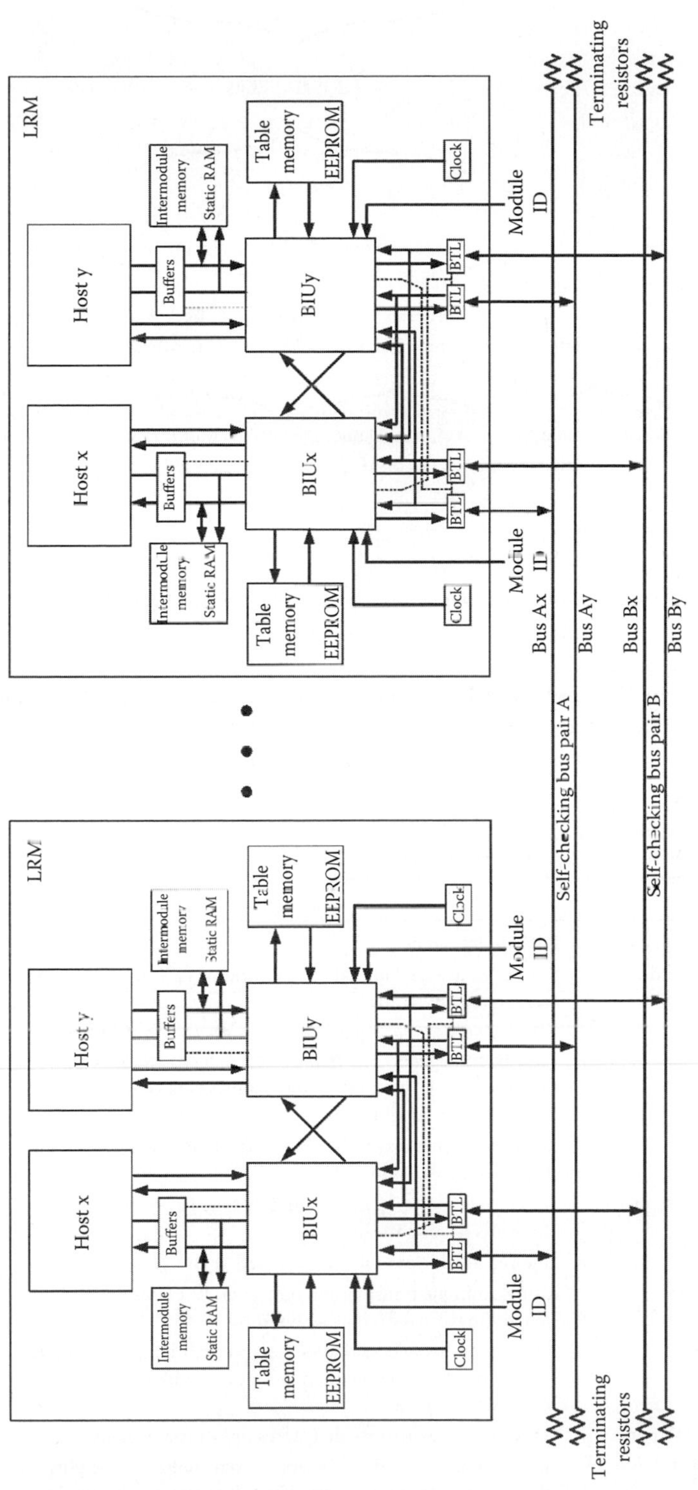

FIGURE 48.1 SAFEbus interface logic.

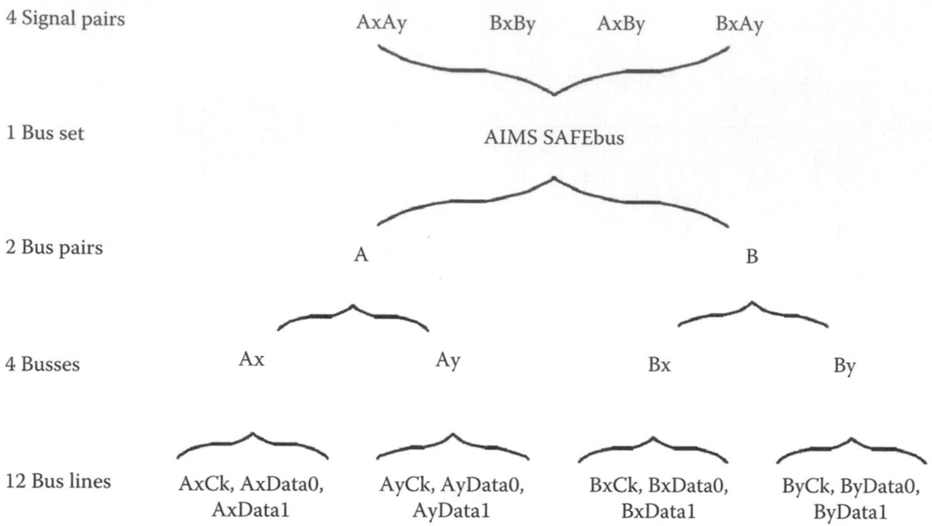

FIGURE 48.2 SAFEbus nomenclature.

ends with a small, fixed intermessage gap time. A typical intermessage gap is 2–4 clock periods (i.e., about 50 ns). Fixed idle time may be inserted to adjust the duration of the full cycle or the time between messages, to a precision of 33 ns.

Messages that are to be transmitted or have been received over the backplane are placed in buffers in IMMs, which are pseudo-dual-port memories shared by the host(s) and the BIUs. This organization permits a simple host interface, because all the hosts can view SAFEbus as a shared multiport memory.

TABLE 48.1 SAFEbus Terminology

Term	Explanation
BIU	Bus interface unit.
Command	Defines what a BIU does during a window (TX, RX, skip, sync, send interrupt to host).
Frame	A cyclical repeating sequence of windows (NB, in SAFEbus and avionics generally); *frame* refers to a repetition of an execution cycle, whereas in nonavionics data communication, *frame* is often roughly synonymous with *message.*
IMM	Intermodule memory (shared message buffer memory between the BIU and its host).
Intermessage gap	The constant reserved minimum idle time between transmissions on the bus.
LRM	Line-replaceable module (a node on SAFEbus).
Message	A single, unique transmission of data on SAFEbus; has a length (in words) fixed at design time.
Table	One or more sequences of commands controlling one or more frames plus their resync jump points and a BIU configuration area.
Table memory	The nonvolatile memory that stores one or more tables.
Window	The bus time reserved for a message, sync pulse, or idle plus the trailing intermessage gap; has a duration (in bit times) fixed at design time.

48.3 Protocol Services

48.3.1 Communication Services

SAFEbus provides data communication with very low jitter (on the order of 100 ns) that is fail-op/fail-safe with near-unity coverage, even with a Byzantine failure.

The data for a message that is to be transmitted over SAFEbus are calculated typically by independent self-checking pair hosts. As with all self-checking pairs, these data are bit-for-bit identical in the fault-free case. The hosts write these data into buffers in the IMMs. Associated with each buffer is a buffer control word (BCW). Whenever a host completes assembling a message buffer in the IMM, it sets a bit in the BCW to say that the buffer is ready for transmission.

The SAFEbus protocol is driven by sequences of commands stored in the BIUs' table memories (see Section 48.3.7.1). Each command corresponds to a single window on the bus. The command indicates whether the BIU should transmit, receive, or ignore the message in that window. The BIUs in every LRM on SAFEbus are synchronized to equivalent points in their respective tables, and mechanisms are provided to quickly attain synchronization if it is ever lost. The tables also contain the local address (in the IMM) of the data to be transmitted or received. The commands in each BIU's Table are organized into multiple frames. Each frame controls a repetitive sequence of windows, and each frame has a fixed total period.

According to a schedule stored in its table memory, each BIU checks its IMM for the buffer assigned to the next time window. If that buffer's BCW says the buffer is ready for transmission, the BIU begins prefetching the message from the IMM. An LRM broadcasts its messages on the four buses when it is scheduled to do so. The two BIUs in an LRM are synced to within 2 bit times of each other via the SAFEbus protocol on the buses (no *backdoor* sync between them). Because a BIU does not know if it is faster or slower than its partner, it turns on its partner's drivers 2 bit times before its first bit transmission and off 2 bit times after its last bit transmission for each message. Off-line scheduling ensures that messages from different sources never collide, taking into account worst-case clock drifts, metastability behavior, resync intervals, read timing errors, LRM positions along the bus (the speed of light does make a difference), etc. Time windows for messages can be set up such that up to four LRMs share a time window in a master/shadow arrangement, using a minislotting scheme to arbitrate for that window.

Receiving BIUs write validated input data into buffers within each of their IMMs. At the completion of a message, the BCWs associated with each buffer are updated with status that includes whether the buffer has valid data and what time it arrived. The fact that the BIUs are synchronized to protocol time allows each of them to independently write identical BCW time stamps into each one of their IMMs without doing an exchange of the received time between the BIUs. Protocol time is measured in bit times from the first window transmitted. The time of each SAFEbus event (e.g., message reception) is the protocol time of the window for which it occurred.

One of the benefits of the table-driven protocol is extremely high efficiency. Control applications typically generate short messages, and most serial protocols perform poorly when messages are short. Efficiencies of between 10% and 30% are typical. In contrast, the SAFEbus protocol is over 89% efficient for a continuous stream of 32-bit messages. Ethernet messages with the same payload would be less than 5% efficient. Because buffer addresses are kept in the tables, they do not need to be transmitted on the bus. The use of transmit and receive commands in the individual tables eliminates the need to send source or destination LRM addresses. Within a message, all clock periods contain data (zero overhead). And because transmissions are scheduled, no transmission time is consumed arbitrating between contending BIUs (with the rare exception of the optional use of master/shadow that typically consumes about 9 bit times per arbitration).

48.3.1.1 Determinism and Partitioning

SAFEbus' determinism and support for robust partitioning warrants more detailed examination, since no other protocol provides these features to this extent. When a system has functions with different levels of criticality, the functions must be partitioned both in space and in time. Space partitioning means that no function can prevent another from obtaining adequate memory space and that the memory space assigned to one function cannot be corrupted by the behavior of another function. Memory management units (MMUs) are usually adequate for simple uniprocessor main memory. But problems arise with multiport memory (including network interfaces) and *memory* that must be shared (CPU registers, cache, I/O registers). Time partitioning means that one function's demand for shared hardware resources will never prevent another function from obtaining a specified level of service and, more importantly, that the timing of a function's access to these resources will not be adversely affected by variable demand or by failure of another function.

Any protocol that includes a destination memory address in a message is a space-partitioning problem. It is extremely difficult to verify correct address usage in a partitioned multiprocessor. To ensure correct usage, the BIU would have to duplicate the typical processor memory management (MMU) function. Then, a difficult protocol would have to be implemented to ensure all BIUs used the same MMU information.

Protocols that use contention arbitration cannot be made strictly time deterministic. Such arbitration is meant to ensure that when two modules contend for the bus, the one with the highest priority request is granted access. But minor jitter in the execution of functions can change which modules contend for the bus on any given bus cycle. As a result, the order in which the modules obtain access can vary from one arbitration to another in ways that cannot be predicted at design time.

SAFEbus achieves both time and space partitioning by placing all message location (IMM) and bus-timing information in its table memories that are frozen at design time. This table information is held in the BIUs' table memories where it cannot be corrupted by any errant software or communications errors. The contents of these memories can be changed only by a very-well-guarded interface (see Section 48.3.6).

To extend SAFEbus' time determinism to include the functions' software, the software execution can be synchronized with the execution of the commands in the bus table. Thus, the software is at the same point during the same bus transmission window in every frame. One benefit is that message latencies can be reduced to insignificance. Results can be scheduled to be transmitted just after they are generated, and input data can be delivered just before they are needed (software never has to ask for input data to be transferred over SAFEbus as data are sent autonomously). A second benefit is that there is less latency jitter on cabinet outputs, which means that a SAFEbus IMA can be used in tighter control loops. A third benefit is that double buffering is rarely necessary because it is possible to schedule the transmission of a data block for a time when it is known that software will not be reading it or modifying it. The elimination of double buffers means the IMMs can be smaller and memory access faster. While the use of software synchronized to the bus and single buffers is the preferred operation, SAFEbus does allow for asynchronous software and double buffering.

48.3.1.2 Data-Message Structure

There are two data-message types: basic and master/shadow. The basic message structure has been chosen to maximize the efficiency of data transmissions. The master/shadow structure supports data transfers by redundant or aperiodic functions.

Basic and master/shadow message structures. Basic messages have a simple structure (see Figure 48.3). Each message consists of a string of 1–256 32-bit data words followed by a programmable intermessage gap of 2–9 bit times. The master/shadow mechanism allows LRMs or applications to be reconfigured or spared without disturbing the traffic pattern on the bus. Master/shadow windows are identified by a field in the associated table command. As many as four transmitters can be assigned to one master/shadow window. Time-slot arbitration determines which of the transmitters actually gets control of the window. If the master is alive and has fresh data to send, it starts transmitting at the beginning of the window. The first shadow begins transmitting *delta* bit times into the window, only if the master did not use its opportunity to transmit. The second shadow begins transmitting two delta bit times into the window, only if the master and the first shadow did not use their opportunities to transmit.

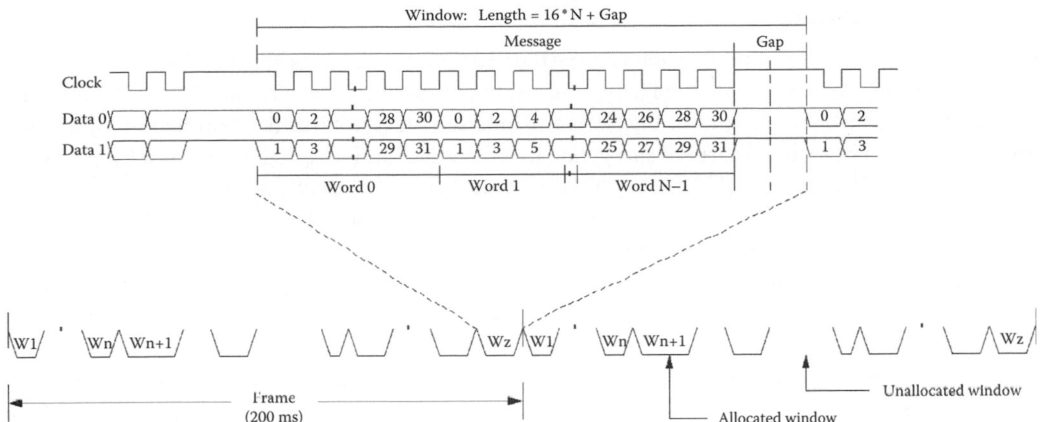

FIGURE 48.3 Basic message structure.

Finally, the third shadow begins transmitting three delta bit times into the window, only if none of the other candidate transmitters use their opportunities to transmit. Delta is a programmable value that is typically set at 1 bit time larger than the selected intermessage gap (values from 3 to 10 bit times may be selected). The selected value depends on the propagation characteristics of the backplane. Examples of the transmission over SAFEbus when the master or third shadow transmits are shown in Figure 48.4.

Time-slot arbitration could reintroduce nondeterminism, but strict measures have been taken to eliminate this danger. First, the total size of a master/shadow window is always the size of its message's data plus three deltas, no matter what happens during arbitration. Thus, the time window remains the same size no matter which transmitter *wins* the arbitration. Second, recipients of a master/shadow message always place the data in the same memory location, no matter which transmitter wins the arbitration. Third, delta can be made large enough to guarantee that the candidate transmitters will never mistake a busy bus for an idle one and begin transmitting in error. Fourth, recipients of a message from this window will be alerted to the presence of this message at exactly the same time after the end of the window, regardless of which LRM wins the arbitration. This allows completely transparent redundancy among the master and the shadows.

The master/shadow mechanism also can be used for sharing bandwidth among asynchronous functions. Of course, partitioning is not maintained in such windows.

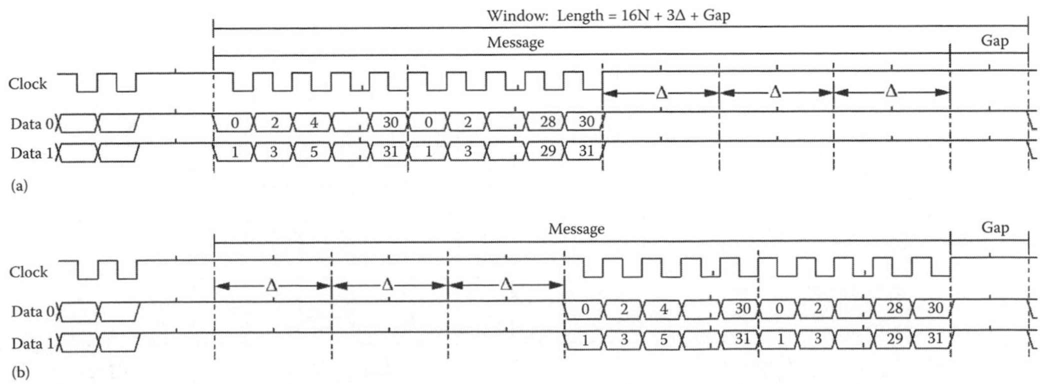

FIGURE 48.4 Master/shadow message structure: (a) master transmits and (b) shadow 3 transmits.

48.3.1.3 Bus Encoding

To improve error-detection coverage, data on the four SAFEbus serial lines are encoded in four different ways. Data on bus Ax have normal polarity. Data on bus Bx are inverted. On bus Ay, every other bit is toggled, starting with the second bit. Bus By is the inverse of bus Ay. This is illustrated in Figure 48.5. This encoding can be seen as a form of *encryption* in which the four buses Ax, Ay, Bx, and By are XORed with the running keys "0000…," "0101…," "1111…," and "1010…," respectively [5].

One type of failure that this encoding catches is bus-to-bus shorts. With identical data on all four buses, a short between the buses could not be detected until another failure occurred such that the failure propagated from the bus with the second fault to the bus with which it was shorted. Having a first fault lay dormant for indefinitely long periods of time and then to make its appearance exactly when a second failure occurs cannot be tolerated. This problem is largely ignored by homogeneous fault-tolerant architectures. One way to mitigate this problem is to do periodic scrubbing of the buses by putting different data on the buses that would ordinarily be the same and see that they actually are received as different. There are a number of problems with this approach. The scrubbing has to disrupt normal communication and it has to disable some of the fault-tolerance features. The latter is dangerous and must be invoked only with a complex set of interlocks. Because of these complications, scrubbing is not continuous and there is an engineering trade-off between the scheduling of scrubbing and exposure time to these latent faults. The use of this bus encoding detects these shorts within 2 bit times of the short onset and without requiring scrubbing.

This encoding scheme can detect unipolar transient upsets that affect several data lines simultaneously. It also allows quick detection of bus collisions caused by a malfunctioning BIU pair (a specific dual-fault scenario that is beyond the basic SAFEbus fault hypothesis). Because bus lines are *wired OR*, if a faulty LRM tries to transmit at the same time as another LRM, illegal encodings appear within 2 bit times of the LRMs starting to transmit differing data.

An additional virtue of this encoding scheme is that power consumption is independent of the data being transmitted. Two bus lines are always high and two are always low (constant average

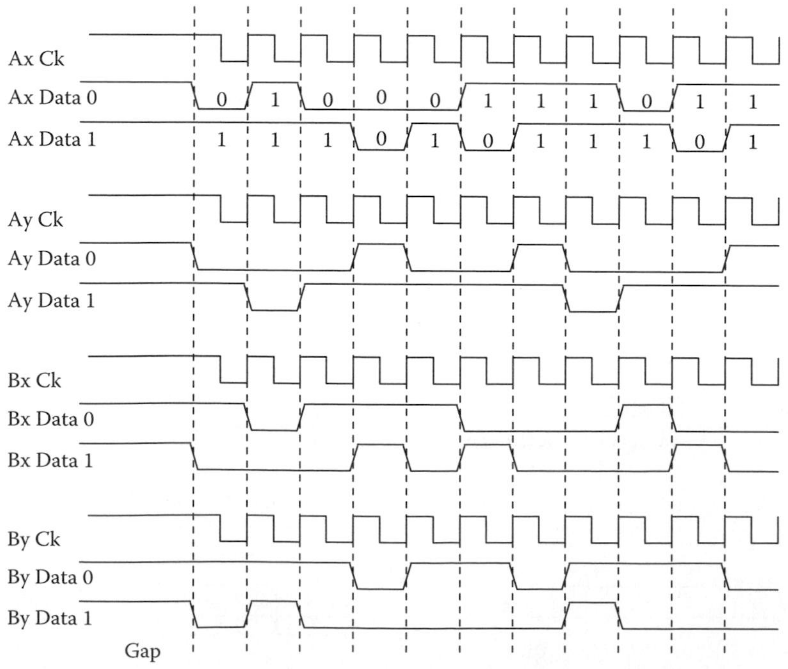

FIGURE 48.5 Bus encoding example.

DC power). When the data change, two of the buses change state and two do not (constant average AC power). Because power consumption is constant, the power supply does not have to be designed for a worst-case data pattern. The fact that bus pairs A and B are inverses of each other provides some of the characteristics of differential signaling, without an additional doubling of the number of signal lines required.

48.3.1.4 Out-of-Band Signaling Pulses

Because SAFEbus messages are pure data (no message-framing overhead) and all data values are used, the only way that protocol-specific information can be conveyed is through the use of out-of-band signaling. One method SAFEbus has for out-of-band signaling uses the clock lines. For data transfer, the clock lines are alternating low for one-half bit time and high for one-half the time (as depicted in Figure 48.3). For out-of-band signaling, the clock lines are driven low for 4 bit times, creating a uniquely identifiable pulse, called the sync pulse. There are four possible variants of this signal depending on the values of the Data0 and Data1 lines during this pulse. Two of these variants are used for synchronization: one of them is used for a debugging mechanism (see Section 48.3.4.1), and one is not used.

The other out-of-band signaling method that the SAFEbus uses drives data lines low, while the clock lines remain high. Of the three variants that are possible with this method, SAFEbus only uses one. It drives Data0 low to indicate initial resync (see Section 48.3.3).

48.3.2 Clock Synchronization

SAFEbus' long resync and the short resync messages both perform precision (sub-bit time level) clock synchronization. The purpose of this synchronization mechanism is to maintain separation of adjacent messages in the presence of oscillator drift and keep the two BIUs in an LRM within 2 bit times of each other. Since the mechanism is the same for both and the short resync message is simpler, only the short resync message will be discussed here. The additional functionality of the long resync message is described in Section 48.3.3. The short resync message is shown in Figure 48.6. It consists of the sync pulse on the clock lines and both data lines being high. To provide availability, all LRMs transmit the sync pulse. The multiple drives are combined into a single pulse by the *wired-OR* action of the open-collector BTL drivers. While all LRMs are scheduled to transmit this pulse at the same time, only one needs to succeed. Because of clock drift, each of the LRMs might turn on its driver at slightly different times. However, these resync pulses happen often enough such that the drift can never be more than the width of the pulse. Thus, this pulse can appear to be from 4 bit times up to eight times wide. When a BIU sees this pulse, it freezes its internal time, and at the end of the pulse, it releases the freeze. It uses an internal effective 4× clock to sample the clock lines. This is shown in Figure 48.7. Because each BIU's time was frozen, its time will have fallen behind real time. The BIUs then enter a catch-up phase where the internal time counters are incremented at a 2× rate until the internal time has caught up with

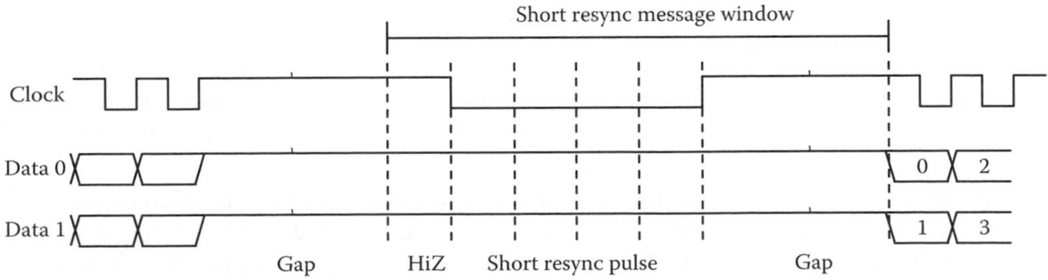

FIGURE 48.6 Short resync message.

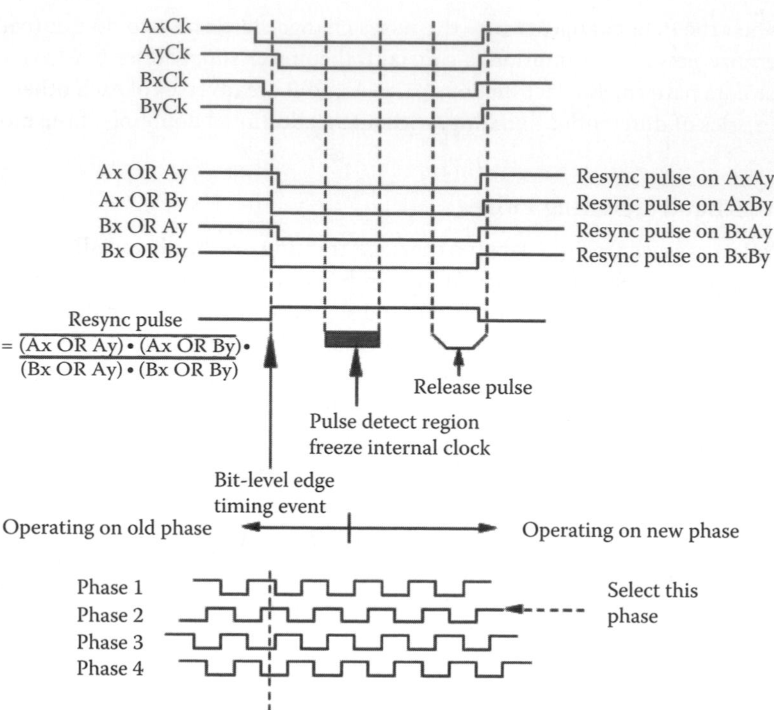

FIGURE 48.7 Resynchronization pulse timing.

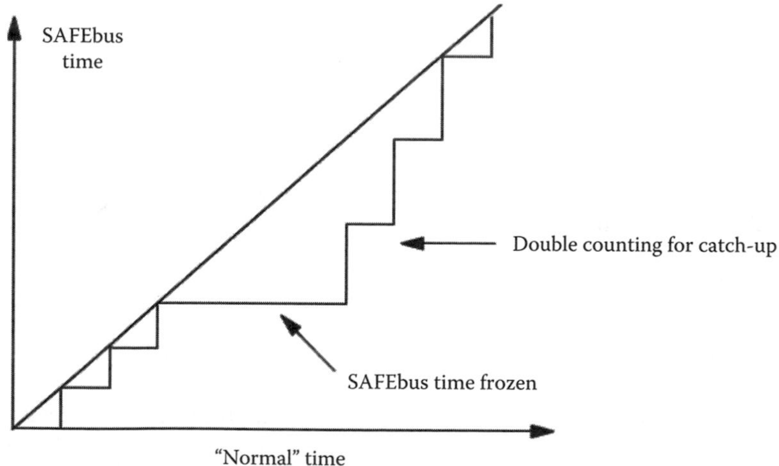

FIGURE 48.8 Time adjustment.

real time (Figure 48.8). The duration of this catch-up phase is equal to the freeze duration. This allows all BIUs to have identical times after each resync event without ever causing time to go backward in any BIU.

The aggregate effect of this resync mechanism is that the slower of the two BIUs in the quickest LRM paces the system time base. Each BIU maintains a counter (called SAFEbus time) driven by its synchronization-corrected oscillator. The synchronization mechanisms make the values in these counters identical in all BIUs with respect to the time that protocol events happen (e.g., the receipt of a message). The time value may be used to time stamp data.

48.3.3 Restart, Reintegration, Integration

Synchronization states. Figure 48.9 shows the major synchronization states of a BIU and their transitions. The major states are as follows:

Initializing. While in this state, the BIU performs such operations as table memory cyclic redundancy checks (CRCs) BIU configuration area loading, and IMM tests. The full-resolution SAFEbus time register value is not valid in this state. IMM access is disabled during much of the initialization process, while the BIU performs IMM pattern testing.

Out_of_Sync. While in this state, the BIU hunts for resynchronization messages transmitted over SAFEbus and attempts to synchronize with them if they are present. If enough time elapses without a synchronization message being seen, the BIU issues an initial sync pulse to start up the backplane. The full-resolution SAFEbus time register value is not valid in this state.

In_Sync. While in this state, the BIU executes command sequences out of a table command sequence area in the table memory. It transmits when executing an appropriate transmit command (and data are fresh) and receives data when executing a receive command. Synchronization is maintained via the transmission and reception of programmed synchronization messages. The full-resolution SAFEbus time register value is valid in this state.

Halted. This is a substate of the In_Sync state. The BIU enters the halted state when it is executing in debug mode and encounters a breakpoint or completes a single-step operation.

Disconnected. While in this state, the BIU suspends all transmission and reception activity on SAFEbus. The host may still read and write memory and BIU registers, but no backplane data will get written into memory. The BIU will enter this state if commanded to by the host or if initialization fails or if it receives a long resync message with mismatching version while in the Out_of_Sync state.

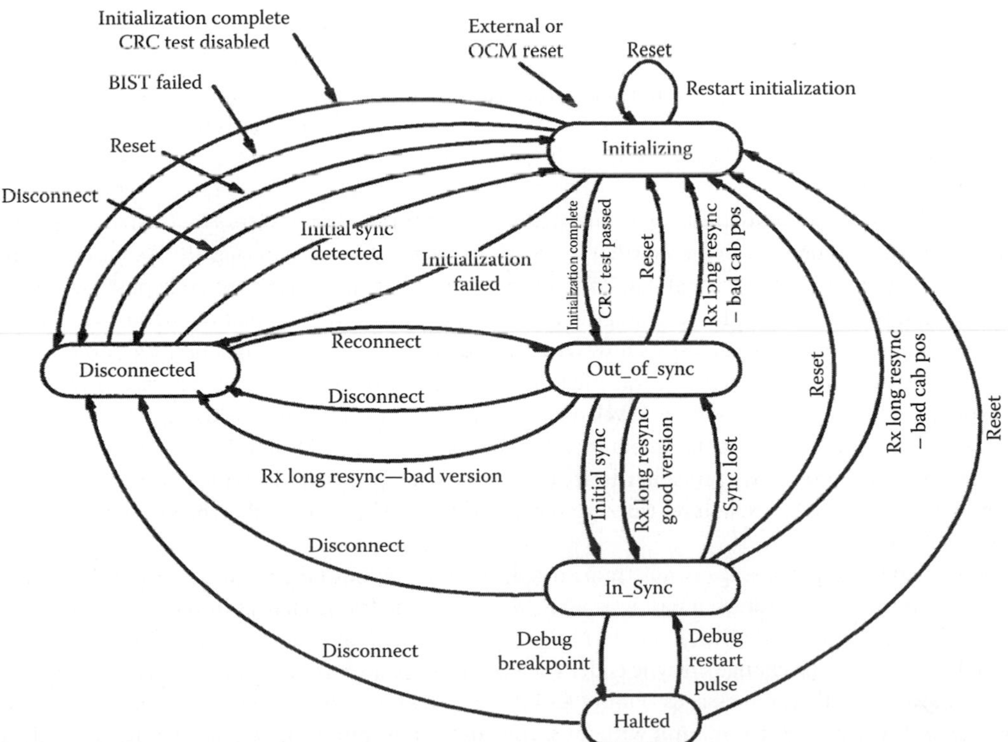

FIGURE 48.9 Synchronization state diagram.

It leaves this state if commanded by the host or if an initial sync pulse is detected on SAFEbus. The full-resolution SAFEbus time register value is not valid in this state.

Synchronization messages. Three uniquely identifiable transmission patterns are provided to support bit-level and frame-level synchronization of the SAFEbus backplane. The initial sync message is used to initialize the SAFEbus after a power-up or in the pathological case of a cabinet-wide loss of synchronization. The short resync message is provided to maintain bit-level synchronization between all BIUs in the cabinet by correcting for oscillator drift between BIUs. The long resync messages are provided to allow lost modules to regain synchronization with an active bus. Long resync messages come in two variants. The entry resync variant is provided simply to allow lost modules to resync to the current frame. The frame change variant is provided to switch between different frame programs in the current table. Both versioned and unversioned forms of the long resync messages exist. Long resync messages also implement the bit-level resynchronization operation (as provided by the short resync message).

Long resync. The structure of the long resync message (Figure 48.10) is separated into two distinct subwindows. The long resync pulse subwindow starts with a unique long resync pulse identified by a low level on the clock lines with all associated Data0 lines low. The pulse is nominally 4 bit times long followed by a maximum gap. The long resync pulse is transmitted by all BIUs that are executing either a transmit or receive long resync command.

The second part of the long resync message is the long resync information subwindow. A single, unique BIU must transmit in the long resync information subwindow. This consists of an 8-bit resync code (a value from 0 to 255), a 1-bit versioned frame indicator, 1 bit of reserved space, a 4-bit cabinet position, 7 bits of reserved space, 43 bits containing SAFEbus time, and a 32-bit table version. The resync code allows the BIU to determine which of 256 locations in the table it should jump to in order to align itself with the other BIUs. In software, this would be called an indirect jump. The code indexes into a resync jump table (see Figure 48.13), which contains addresses that point to the main table command sequence area command that should follow this long resync message in the normal sequence of command execution. The version frame indicator, cabinet position, and table version are used as part of versioning enforcement as described in Section 48.3.7.3. The SAFEbus time field is used to supply this information to LRMs receiving this message while not in sync.

Long resync messages can be sent master/shadow, which combines the properties of both these message types. This benefits the multiway trade-off among fault tolerance, bandwidth used, and resynchronization latency.

Frame change. A variant of the long resync message is the frame change message. The message form is identical. The differences are in the way that it is used. In the long resync message, the resync code points to the next command in the normal sequence. In the frame change message, the code points elsewhere, usually to another frame. The frame change is a conditional jump. An LRM that is scheduled to transmit a frame change cannot do so unless its host writes a code into the BIU that matches the frame's code (a *lock and key* mechanism). A tightly controlled mechanism is provided to switch between frames based on explicit commands generated by Level A operating system (OS) software using the *lock and key*. Normally, *lock and key* fault-containment mechanisms are very weak. However, in the case of SAFEbus, this mechanism is protected against hardware failures by the use of self-checking pair hardware, and it is protected against software failures through the use of an MMU, which limits the lock's access only to the Level A software.

If the frame change message is not transmitted, the *jump* is not taken and the command/window sequence continues with the command, which is next after the frame change in the table. If the frame change message is transmitted, the *jump* is taken and all LRMs fetch their next command from the location pointed to by the frame change's resync code. The use of this feature is described in Section 48.3.6.1.

Initial sync. Initial sync messages (Figure 48.11) are transmitted by an Out_of_Sync BIU that waits longer than the initial sync wait limit without seeing any resync pulses on the backplane. All BIUs that are out of sync at this point will use the initial sync message to synchronize into an unversioned initial frame. The initial sync message starts with the initial sync pulse. It has a unique pattern identified by a

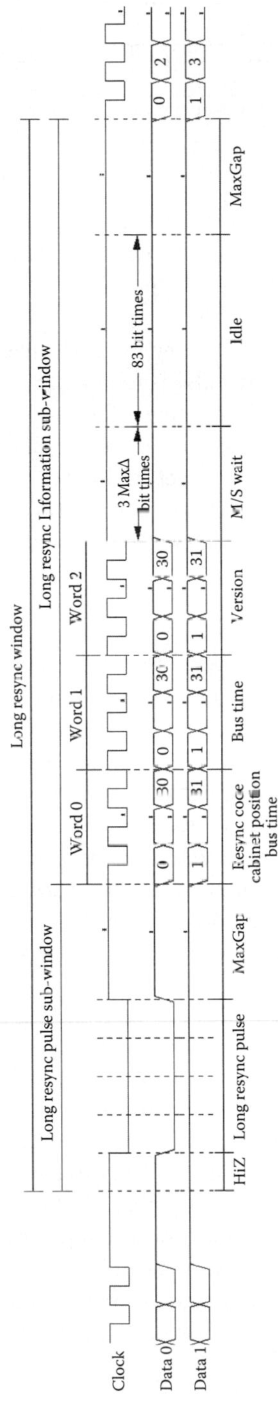

FIGURE 48.10 Long resync message.

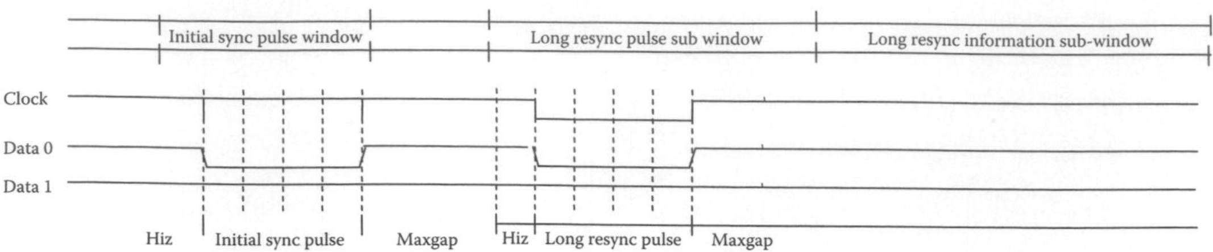

FIGURE 48.11 Initial sync message.

signal pair of buses with a low level on the Data0 line and a high level on the clock line for at least 2 bit times. In addition, for a bus to be considered part of an initial sync pulse signal pair, the Data0 line must have been high at one point during the time while the BIU waited for the initial sync wait limit. The pulse is nominally 4 bit times long.

To more accurately resync the clocks and to allow the BIU time to fetch the command for the first message in the initial frame, the initial sync pulse is followed by a long resync message with no BIU transmitting the information subwindow. The missing information is assumed to be zero. In particular, the resync code is zero and the SAFEbus time is zero. The BIU will enter the In_Sync state at the first bit time of the gap that separates the idle data portion of the long resync message from the first window of the initial frame. The first bit time of the initial frame occurs (106 + 6MaxΔ + 2MaxGap) bit times after the leading edge of the long resync pulse that follows the initial sync pulse.

48.3.4 Diagnostic Services

SAFEbus uses masking fault tolerance, which does not need the complicated diagnostic services required by other protocols, for example, it does not need to keep track of membership. However, it does supply a rich set of diagnostic mechanisms and information for maintenance purposes. This includes full built-in self test (BIST) capability via dual (x and y) IEEE 1149.1 (JTAG) test buses. The x JTAG bus connects only to the x BIU, and the y JTAG (Jaintest action group) bus connects only to the y BIU in order to prevent fault propagation from the two sides of the self-checking pair. For commanded BIST, the BIU goes off-line and scans a set of pseudo randomly generated test vectors into the logic, clocks it through the logic, and accumulates the result of the clocking into the BIST signature result register. The BIST also includes scrubbing of all the critical fault-tolerant circuitry within the BIUs. For scrubbing protection mechanisms, the hardware is invoked to test if the protection mechanism is really invoked when a failure occurs.

48.3.4.1 Debugging Mechanisms

Another unique feature of SAFEbus is its ability to breakpoint and single-step an entire system, including the processors connected to SAFEbus. This is the same as the breakpoint and single-step functionality commonly seen in software debuggers, but on a systemwide basis. The reason this is possible is because SAFEbus acts as the central clock for the system, providing the timing ticks that the OSs of the processors attached to it use for dispatching tasks. If the SAFEbus freezes, so do all of the processors.

A breakpoint can be set for any time within the global SAFEbus timeline. A breakpoint can also be initiated by driving the Ck, Data0, and Data1 lines low simultaneously for at least 4 bit times. To resume from a breakpoint, some LRM transmits a short resync pulse.

Because setting a breakpoint can be dangerous during normal operation, breakpoints are only enabled for special test table versions. This is indicated by the upper 2 bits of the table version being *11*.

48.3.5 Fault Isolation

SAFEbus, with its self-checking approach, provides near-perfect coverage. The checking at the receiving end provides near-perfect error-detection coverage for many faults, including Byzantine faults [6].

It provides better coverage than signature-based error-detection techniques (such as CRCs) [7] while simultaneously not incurring the overhead of these schemes.

Recapping Figure 48.1, SAFEbus consists of four buses (Ax, Ay, Bx, and By) that connect several self-checking pair LRMs. Each of the four buses and each half of an LRM are independent fault-containment zones. If there are N LRMs, there are 2 × N + 4 fault-containment zones. Each of the BTL bus interface parts is in its associated bus' fault-containment zone and gets its power supply from that bus. Thus, the fault-containment zone boundaries are between the BTL parts and the BIU parts. Names ending in x and y denote redundant parts for integrity. Bus names beginning with A and B denote redundant bus pairs for availability. The BTL drivers are connected such that BIUx transmits only on Ax and Bx and BIUy transmits only on Ay and By. Thus, a faulty BIUx can contaminate only buses Ax and Bx, and a faulty BIUy can contaminate only buses Ay and By.

The dashed lines in this diagram are control signals. In particular, BIUx enables BIUy's bus drivers and BIUy enables BIUx's bus drivers. Thus, an LRM cannot transmit unless both BIUx and BIUy agree to do so. The set of buses is fail-op, fail-stop. Each LRM is fail-stop. N redundant LRMs are N-1 fail-op, fail-stop. The major failure scenario that SAFEbus does not cover is two simultaneous active faults in the same LRM that are somehow complementary to escape the bit-for-bit checking and cross-coupled driver enables. This has a probability that has been calculated to be much less than 10^{-10}.

48.3.5.1 Babble Protection

To detect errors, the transmitting LRM checks what it actually puts on the bus. If a BIU sees a miscompare, it stops transmitting and it disables its partner's drivers. The dual nature of this comparison ensures that a babbling module cannot stay on the bus. This is an availability feature rather than an integrity feature. Integrity fault containment is done by the receivers. This availability feature is only applicable to master/shadow windows. In order to prevent loss of resources due to transient failures, LRMs are allowed to restart for a limited time by implementation of a strike counter. This strike counter is incremented for each fault found in a specific interval (say, a maximum of three failures are allowed within 10 min). This ensures that transient faults are dealt with in a constructive manner and permanent or intermittent faults are isolated after the LRM has hit the strike counter limit.

48.3.5.2 Byzantine Protection

SAFEbus has a unique way of tolerating Byzantine faults. Because the transfer of a message from one LRM to another LRM uses four fault zones, it is possible for it to tolerate one Byzantine fault. The BTL receivers are cross-linked to the two BIUs such that each receiving BIU gets a copy of the message from all four buses. This can be seen as the first round of the classical Byzantine exchange. Each BIU creates two 4-bit status vectors, collectively called the *syndrome*, for each 16 bits received within a message.* The first vector has a bit for each bus saying whether anything came in from that bus. The second vector is the result of the comparisons: Ax = Ay, Bx = By, Ax = By, Ay = Bx. The BIUs exchange their syndromes. From these 8 bits, the two BIUs can determine which (if any) of the data bus inputs have arrived error-free. If there is such an error-free source, both BIUs select it as the source data. This can be seen as the second round of the classical Byzantine exchange. This prevents Byzantine failures arriving from outside a pair from confusing a pair into thinking that one of the halves of the pair is faulty. If a message arrives with uncorrectable errors, the BCW associated with its buffer is updated to that status. A self-checking pair host reading a BCW status that indicates an error is not allowed to read the data buffer because differing data in the two IMMs could cause the host pair to split.

While the syndrome exchange prevents a Byzantine fault from splitting a pair, an additional mechanism is needed for Byzantine agreement among pairs. Before SAFEbus, Byzantine algorithms

* The granularity of 16 bits was chosen as an engineering trade-off between the desire for a smaller size to increase availability and the desire for larger size to minimize metastability errors and to make it easier to meet timing constraints on the exchange between the BIUs.

either did a full exchange of an entire message's content or used signatures. The problem with the former method is the large amount of bandwidth it requires. The problem with the latter method is that it does not provide full coverage. SAFEbus introduced a new method: hierarchical Byzantine agreement. In this method, a lower-level agreement prevents Byzantine faults from affecting a pair, as described earlier. An upper-level agreement only needs to send 1 bit of information from every receiving LRM of the message. This bit would indicate whether the LRM rejected the message as being faulty or not. Because the SAFEbus granularity is 32 bits, it can send 32 bits just as well as 1 bit. Using 32 bits allows this upper level of exchange to be implemented with no additional hardware and with no software. All that is done is to have the buffer for this upper-level round of exchange be placed into the IMM such that the data word of this buffer overlaps with the BCW for the original message. Thus, what is exchanged on this upper level is the BCW of the original message. If the original message was rejected, the BCW is zero.

48.3.5.3 Availability versus Integrity Trade-Off

The syndrome exchange mechanism includes an option to select a preference for availability or integrity, for those cases where there is not a generally applicable best choice, for example, if Ax = Ay and Bx = By, but Ax ≠ By and Bx ≠ Ay. This can only happen if there are at least two identical bit errors (after decode). A receiver cannot tell if pair A or pair B is correct. For integrity, both have to be thrown out because neither can be trusted. For availability, either A or B could be arbitrarily chosen.

48.3.5.4 Zombie Module Protection

Each LRM that plugs into SAFEbus must contain a table of commands that is compatible with all other LRMs on that SAFEbus. One mechanism for preventing LRMs with an incompatible table would be to have the LRMs exchange table version information at start-up. However, such a scheme would not cover the scenario where an LRM is dead or comatose at start-up and then *wakes up* in the middle of a critical operation with an incompatible table. This is called the *zombie module* problem. To solve this, SAFEbus requires all LRMs joining a SAFEbus that has active traffic to compare the table version information in a long resync message to its own. It is allowed to join the network only if the versions are compatible. See Section 48.3.7.3 for details on version enforcement.

48.3.6 Configuration Services

The SAFEbus tables are loaded into each BIU's table memory using the same dual IEEE 1149.1 (JTAG) test buses that are used to support BIST. These table memory images can contain multiple tables so that one LRM can play several different roles depending on which slot and in which cabinet the LRM is located. This mechanism can reduce the number of spares required to be held in maintenance facilities. For example, the Boeing 777 AIMS cabinets had one I/O module (IOM) design that was used in eight places. This reduced the number of types of spares required from eight down to one and reduced the total number of spares needed to be held in the logistics pipeline. In theory, this could also allow for onboard sparing and manual reconfiguration. Each slot in every SAFEbus cabinet has five slot ID pins that can allow up to 32 slots per cabinet. To prevent wrong-ID masquerade faults, these pins are protected by parity and cross compare between the two-halves of a pair. Each cabinet also has a cabinet ID. To eliminate the need for cabinet ID pins on every slot, SAFEbus allows some subset of LRMs within the cabinet to have these pins, and then these LRMs broadcast this cabinet ID in their long resync messages' cabinet position field. With the existing 4 bits of cabinet position and 5 bits of four slot IDs, SAFEbus can currently accommodate up to 512 slots systemwide. The long resync message has seven spare bits adjacent to this field that could be used to expand this to a total of 65,536 systemwide slots.

After the tables have been loaded, SAFEbus can be used to download software and other data. To optimize bandwidth usage, special frames can be used for these downloads (as described in Section 48.3.6.1).

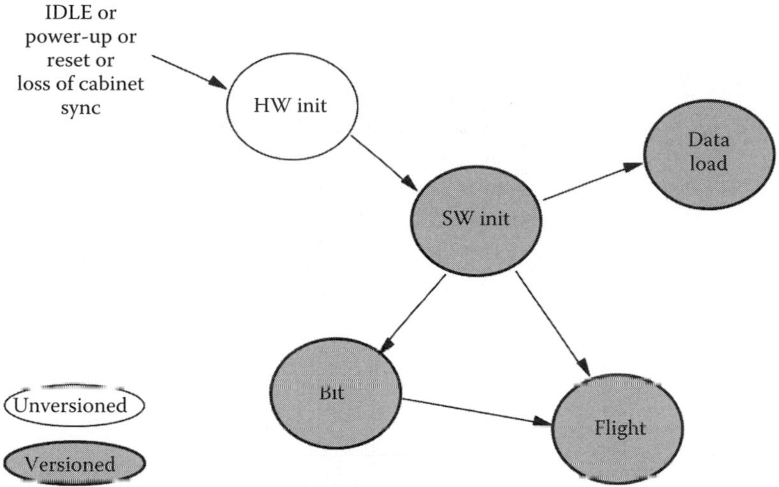

FIGURE 48.12 Example frames and their frame change transitions.

48.3.6.1 Frame Changes

This section provides an example for frames to do hardware initialization (HW Init), software initialization (SW Init), data loading, built-in self-test (BIST), and normal application communication, plus the frame change transitions between them. The following is a brief textual description of each frame:

1. The HW Init frame follows ARINC 659's definition for the initialization frame and, as such, is an unversioned frame. The only data traffic during the HW Init frame is the transfer of version messages from all LRMs. The HW Init frame is approximately 117 μs in duration.
2. The SW Init frame provides time for the processors' environment to be set up before the functional partitions are dispatched in the flight frame.
3. The built-in test (BIT) frame is a versioned frame that provides time to perform power-up BITE.
4. The flight frame is a versioned frame that provides the data transfers for running software partitions.
5. The data load frame is a versioned frame that provides the data load function, using the majority of the SAFEbus bandwidth. While data loading may be performed in the flight frame, a data loading session using the data load frame is much quicker.

Figure 48.12 shows these frames and possible transitions. Note that the frame changes are unidirectional. This means, for example, that once the flight frame starts executing, it cannot be frame changed to any other frame. In order to go from the flight frame to another frame, the SAFEbus must be reset. This is an additional safety precaution against accidental transitions out of flight frame.

48.3.7 Protocol Parameterization

48.3.7.1 Table Memory

As shown in Figure 48.13, a SAFEbus table is divided into three areas: resynchronization jump table, table command sequence area, and BIU configuration area. The resynchronization jump table is used by the BIU to rapidly locate the address in the table memory corresponding to the resync code (see Section 48.3.3). The majority of the table memory is used for the cyclic command sequences that control frames. More than one frame schedule can be present for different system *modes* (such as system initialization, ground checkout, flight operation, software loading), with each mode having a different schedule. The different frames for different modes are contained within a single table command

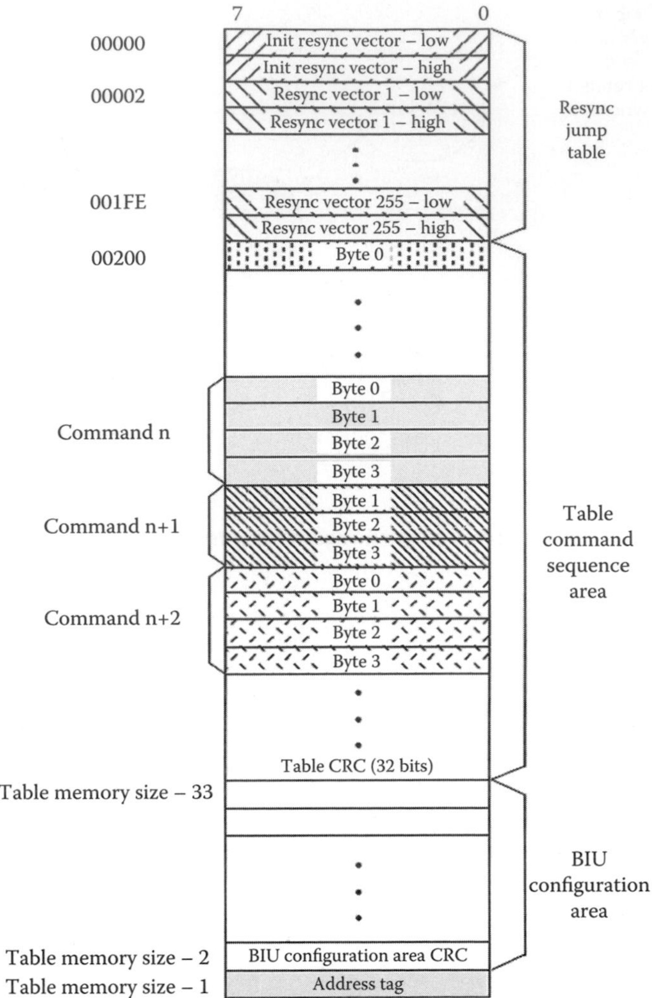

FIGURE 48.13 Table memory structure.

sequence area. Therefore, this selection is in addition to the selection of different tables depending on location (slot and cabinet) roles. The BIU configuration area contains information for BIU customization options such as memory speeds, host interface characteristics, selection of availability or integrity as preferred for those cases where one is not universally preferred over the other, intermessage gap, master/shadow delta, and SAFEbus time increment rate. The contents of the table memories that are associated with the two BIUs on a single module are bit-for-bit identical. However, the table memory contents are different for each module on SAFEbus. This is because the local IMM addresses can be different and the commands are different for TX versus RX, master versus shadow, etc.

48.3.7.2 Frame Description Language

SAFEbus uses a frame description language to define the contents of each frame. This is an intermediary language that can be produced by off-line schedulers. A back-end tool specific to each BIU design translates this FDL into a bit image that can be loaded into table memories. This language has been standardized by ARINC 659 to allow decoupling between vendors of scheduling tools and vendors of BIU hardware. For AIMS, a software tool parses a database of ICD information and generates a schedule that has much more freedom than schedulers for most time-triggered protocols.

48.3.7.3 Table Versioning

The SAFEbus protocol includes a mechanism to ensure that only LRMs with compatible tables are allowed to transmit on the bus. To support this mechanism, SAFEbus uses two types of frames: unversioned and versioned. Unversioned frames allow LRMs of any SAFEbus version to communicate. These frames are used for LRMs to exchange their table version information with each other. Such frames always operate with the maximum programmable gap size between messages in order to be compatible with even the slowest LRM implementations.

After power on, all LRMs enter a standard unversioned initialization frame in which each of the possible 32 LRMs that could exist on one SAFEbus is given a window in which to transmit their table version information. Up to eight of the LRMs are given the authority to do a frame change from the initialization frame to either a versioned frame or another unversioned frame. The decision of which frame change to do is an application-dependent function that uses the gathered table version information as input.

The version field in the frame change command informs all BIUs of the destination frame's version. BIUs with a table version that does not match the destination frame's version cannot follow the frame change and will drop out of sync. BIUs with a table version not matching a currently running frame are prevented from joining the bus traffic because they cannot sync to the versioned long resync messages. Most normal application operations are done in a versioned frame.

48.4 Communication Interface

All data communications between modules on the SAFEbus backplane occur via buffers stored in the IMM address space. Two buffer formats are provided: controlled and noncontrolled.

The controlled buffer memory format consists of a single BCW followed by one or two data sub-buffers (depending on whether the data item requires double buffering) of up to 256 words each. See Figure 48.14.

When a BIU executes a command to transmit from a controlled buffer, it determines the buffer type from the command. The BIU then reads the BCW and checks the TX Fresh bit and the Ping/Pong bit (if the buffer type is double buffered). If the TX Fresh bit is set, the BIU transmits the data and writes the BCW back with the TX Fresh bit cleared to 0. If a transmission failure occurs, the entire BCW is set to 0.

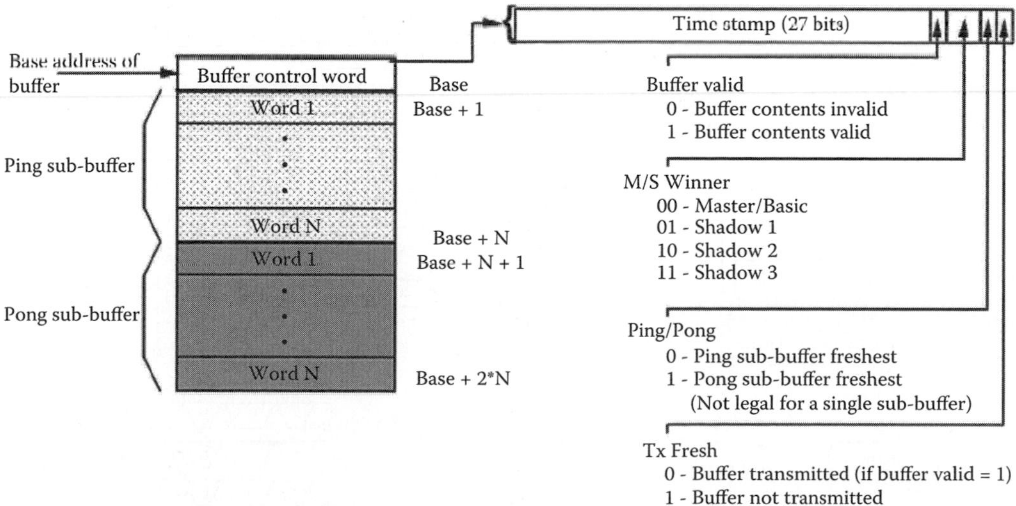

FIGURE 48.14 IMM buffer structure.

When a BIU executes a command to receive into a controlled buffer, it determines the buffer type from the command. If the buffer type is double subbuffered, the BIU reads the BCW and tests the Ping/Pong bit to determine where to place the data. The BIU then places the received and validated data into the stale subbuffer (always Ping for single subbuffers) and then forms a BCW with the current SAFEbus time, buffer valid set, master/shadow winner bits set appropriately, Ping/Pong set to the buffer just written, and TX Fresh bit cleared to 0. If errors occur in reception after the BIU modifies a location in the buffer, the entire BCW is set to 0. When the host reads a controlled receive buffer, it must first read the BCW. If the BCW is 0, a self-checking pair host cannot read any other words in the buffer because erroneous receive data that cause the BCW to be set to zero are not guaranteed to be identical between the x and y sides of the pair. If it is nonzero, the host can read the subbuffer indicated as freshest by the Ping/Pong bit. This bit is valid for both buffer types, since single subbuffered controlled buffers will always have the Ping/Pong bit set to Ping (0). When the host writes a controlled transmit buffer, it must first read the BCW. If it is nonzero, the host can write the subbuffer indicated as not being the freshest by the Ping/Pong bit. The host then sets the TX Fresh bit by requesting the BIU to do a Set_Ping or Set_Pong operation depending upon the subbuffer to be transmitted.

The noncontrolled format consists of 1–256 words without a BCW. No freshness indication or validity information is provided in such a buffer. If such indications are required, they must be provided by the user in the data itself. The BIU considers a noncontrolled buffer to be fresh whenever it attempts to transmit the contents of the buffer.

The BIUs contain special *coincidence* circuitry to detect the case that the host pair is updating the BCW at the same time that the BIUs are trying to read it. Even though these memories are only pseudo-dual port instead of true dual port, the independence of the clocks between the BIUs and between the BIUs and the host pair means that read–write order can be different between the x and the y sides of the pairs. The coincidence circuitry ensures that the BIUs see the same order. If this were not done, it is possible that the two BIUs could disagree on whether a buffer was ready to transmit. A coincidence mechanism similar to that used for the BCW on transmit is also used on receive to prevent the hosts from getting inconsistent data.

Synchronization of the bus schedule and the application software's execution is done by embedding interrupt commands in the SAFEbus tables. On receiving this interrupt, the processor's OS releases an

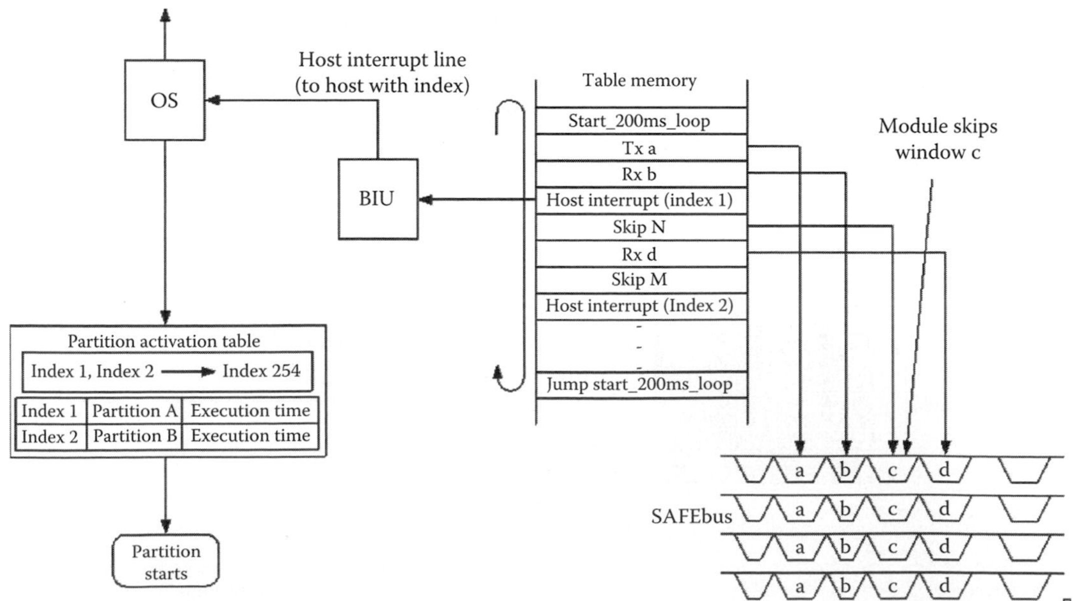

FIGURE 48.15 Data stream time partitioning.

application task (or task set) for execution. This interrupt takes the place of the clock or hardware timer other OSs or real-time executives employ. That is, this SAFEbus interrupt becomes the OS' real-time clock tick. As part of the time and space partitioning, the tasks are grouped into sets of *partitions*. Tasks in one partition are guaranteed not to interfere with tasks of another partition in a mechanism similar to processor virtualization. See Figure 48.15.

48.5 Validation and Verification Efforts

SAFEbus was part of the Boeing 777 airplane certification and passed rigorous testing for design requirements to support Level A applications, the highest level of safety requirements. Newer versions of the Boeing 777 use upgrades to the original SAFEbus BIU design, which also have been certified. Derivatives of SAFEbus have been certified on several other aircraft.

SAFEbus is standardized as ARINC 659 *backplane data bus* [8]. This is in a family of associated standards including ARINC 653 Avionics Application Software Standard Interface, ARINC 651 Design Guide for Integrated Modular Avionics, and ARINC 650 Integrated Modular Avionics Packaging and Interfaces.

48.6 SAFEbus in Boeing 777

The avionics architecture in the Boeing 777 is characterized by functional integration compared to predecessor aircraft. Classical avionics at that time had a federated avionics architecture, where a set of functions is implemented in one or more line-replaceable units (LRUs). In the Boeing 777, this has been replaced by a more integrated architecture where multiple sets of functions are combined in fewer LRUs all in common cabinets instead of many separate boxes. This IMA is formed around the concept of powerful processing modules with a deterministic OS that hosts many applications. This integration offers many benefits, such as lower weight, lower power consumption, increased reliability, less-frequent maintenance, and greater flexibility. Yet because functions share hardware resources, greater care must be taken to ensure they will operate correctly, even if coresident functions fail. Integration increases the risk that unwanted interactions among the functions residing on the shared hardware will lead to unforeseen failures. This is the engineering challenge of integrated architectures [9].

An integrated cabinet typically is larger than a federated LRU, but smaller than the sum of all LRUs that the cabinet replaces. The separate functions in a cabinet share a power supply, general and special I/O units, and processing resources. To increase availability or integrity, functions can be replicated in multiple LRMs or in multiple cabinets. The attraction of the integrated architecture is the economics that can be achieved by sharing resources, like power, mechanical housing, processing, and I/O.

The core avionics of Boeing 777, also called AIMS, attempts to make the execution environment of each function in the cabinet as much like the environment in the discrete LRU as possible. Essentially, all shared resources in the cabinet are rigidly *partitioned* to ensure that one function cannot adversely affect another under any possible operating condition, including the occurrence of faults or design errors in the functions. The best way to ensure adequate partitioning is via strict deterministic control.

Functions must be partitioned both in space (memory and I/O) and in time. Deterministic control over the partitioning of space means that it can be guaranteed that no function can prevent another from obtaining adequate resources like memory space and that the resources are assigned to one function only. In the case where some memory space is assigned to one function, this means that it cannot be corrupted by another function. Preallocated memory areas with hardware-based memory protection such as MMUs prevent contention for memory space. Deterministic control over the partitioning of time means that it can be guaranteed that one function's changing demand for hardware resources will never prevent another function from obtaining a specified minimum level of service and that the timing of a function's access to these resources will not be affected by variable demand or by the failure of another function [10].

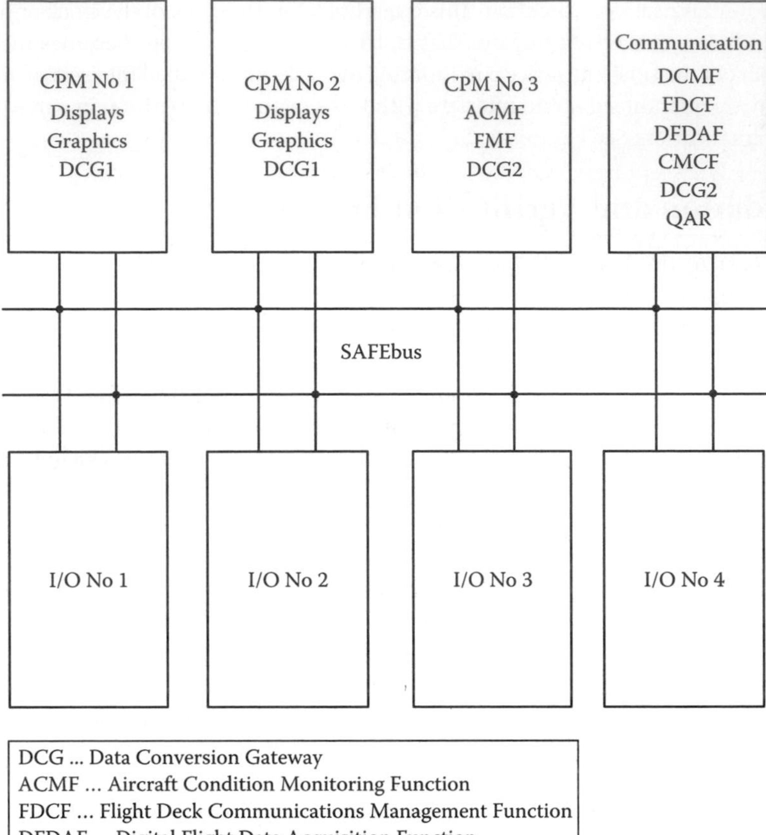

FIGURE 48.16 Overview of one AIMS cabinet module and function allocation (From Witwer. B. *IEEE Transactions on Aerospace and Electronic Systems*, 33, 637, 1997.)

Figure 48.16 provides an overview of one of the Boeing 777 AIMS cabinets. The cabinet comprises three—at that time powerful—core processor modules (CPMs) for computing tasks, running display, and graphic and data conversion gateway applications. One module is dedicated especially to communication (including flight deck communication) and four identical modules to I/O. Other modules like local power modules support all modules located in a cabinet. Furthermore, three slots (for one CPM and two IOMs) are reserved for future growth. The replicated applications hosted on AIMS are as follows, along with the number of redundant copies of each application per shipset in parentheses [12]: displays (4), flight management/thrust management (2), central maintenance (2), data communication management (2), flight deck communication (2), airplane condition monitoring (1), digital flight data acquisition (2), and data conversion gateway (4).

Applications in AIMS use the following shared platform:

1. Common processor, power supply, and mechanical housing
2. Common I/O ports
3. Common backplane–bus (SAFEbus) to move data between CPMs and between CPMs and IOMs
4. Common OS, BIT, and utility software

The CPMs are based on AMD 29k RISC microprocessors using an AMD 29050 for AIMS-1 and HI-29KII for AIMS-2. Each CPM operates two processors in a lockstep configuration, which compares all output of compute applications for consistency similar to SAFEbus. If an error is detected, an error is flagged and all processing is halted ensuring that erroneous data are not propagated to other modules. Once errors are processed, errors are logged, and recovery is attempted. Figure 48.17 depicts the configuration of self-checking processors with SAFEbus.

The predictable time-driven operation of SAFEbus is extended to the processor and OS. The scheduling of the OSs is static and table driven in order to ensure timely partitioning and minimize unwanted interactions between applications running on the same processor. A processor cyclically executes a schedule where different partitions (applications or parts of applications) are assigned different time slots and can use assigned hardware during this period. The time slots of applications (also called partitions) are fixed similar to the time slot of messages on the bus. This means that the dispatcher at partition level is purely time and table driven without any other influence leading to rigid temporal separation. Within the partition time slices, one or more tasks (called processes in [13]) may run, which can be scheduled more flexibly and do share resources. Tasks of one partition may not require separation as they belong to the same partition and, hence, application. The principles of operation and standardized interfaces applied to AIMS have been standardized in ARINC 653 [14–16].

Scheduling of AIMS is described and proposed in [17]. Real-time repetition rates for applications range from 1 to 80 Hz. Synchronization between the AIMS processors is used to control latency and allows the system to *hand over* data at points during the repetition cycle (also called *frame*). When data are shared during cycles and the processors and the network are synchronized, the latency can be minimized. Yet computation and communication are dependent on each other when creating a schedule and can pose a significant challenge for getting a feasible computation and communication schedule that meets all deadlines and ensures data consistency (i.e., neither a process nor the network writes data to a common buffer at the same time). In order to get an idea of the scheduling problem in AIMS, about 3000 unique data items had to be moved with their required frequency in each cabinet. This resulted in 17,000 total messages per period in all cabinets including messages due to oversampling requirements for cabinet-external interactions.

The AIMS cabinets are just one part of the overall avionics. The integrated cabinets are connected to the airplane interfaces via a combination of ARINC 429, ARINC 629, display buses, and discrete I/O channels (see Figure 48.18; note that for clarity, ARINC 429 and discrete channels are not shown). In addition to the cabinet, other flight deck hardware elements that make up the AIMS system are six flat-panel display units, three control and display units, two electronic flight instrument system (EFIS) display control panels, one display select panel, two cursor control devices, and two display remote light sensors. Not all are depicted in Figure 48.18. The system bus in Boeing 777 is mainly ARINC 629 [18], a non-time-triggered, but periodically driven system bus. Boeing 777 avionics is synchronous only within each cabinet (in the domain of ARINC 659), but not at an airplane level.

The synchronization of the bus and software also assists debugging and validation. First, due to the nature of SAFEbus' independent determinism, a function experiences the same system timing whether the cabinet is fully populated or nearly empty. Second, since each processor in a cabinet is synchronized to the bus, the functions running in different processors are implicitly synchronized and timing errors at application level will be more quickly exposed, making the system simpler to debug. In asynchronously scheduled multiprocessor systems, such timing problems show up as intermittent failures, which can be very costly to track down and make it impossible to validate the system. Third, whenever the system is stopped or single-stepped, it passes through a succession of clearly defined states. The clear relationships between the processes, which are defined by the SAFEbus table, make it easier to trace behavior [19].

Many of the elements of Boeing 777 have led to ARINC standards such as the cabinet backplane; SAFEbus has been standardized as ARINC 659 [20] and the system bus as ARINC 629 [21]; and ARINC Report 651 *Design Guidance for Integrated Modular Avionics* is the top-level design guide for IMA.

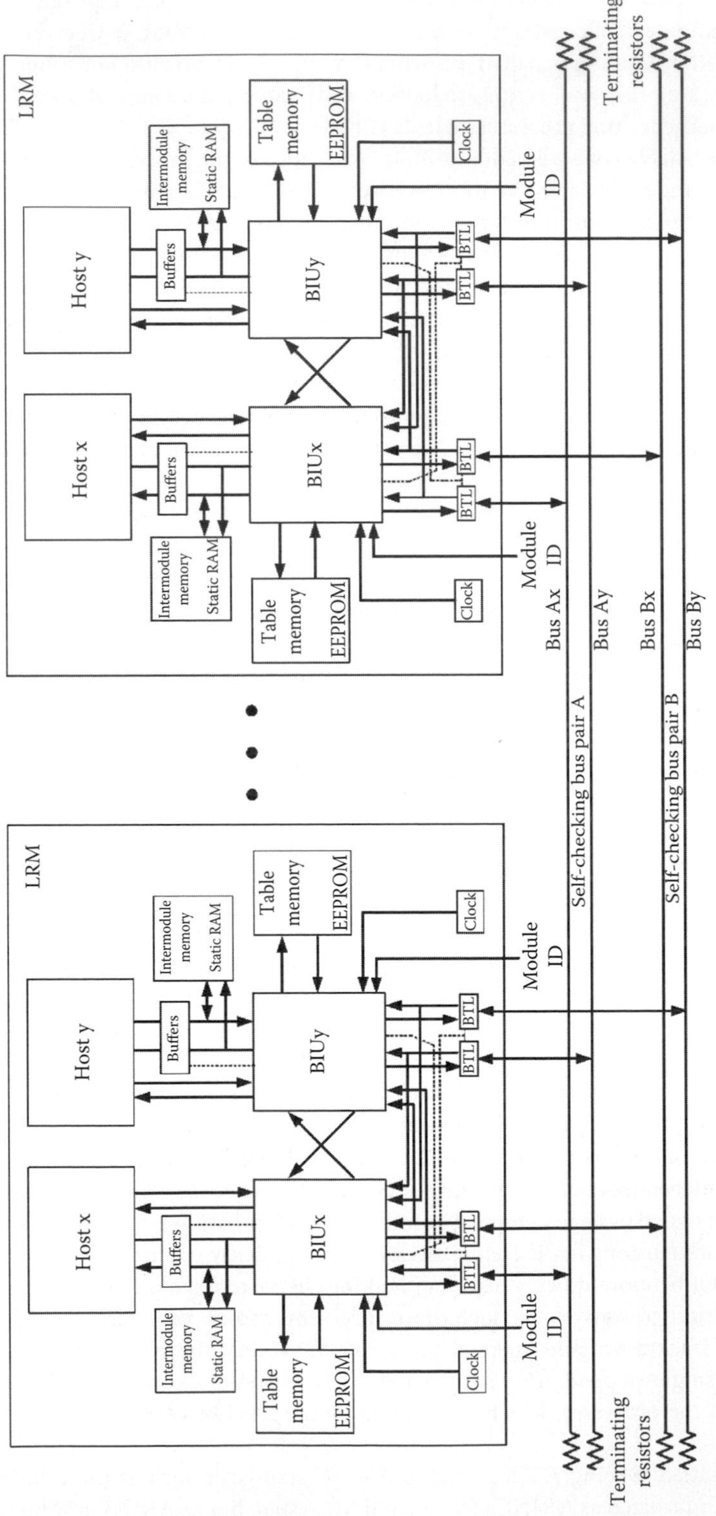

FIGURE 48.17 SAFEbus architecture overview. (Courtesy of Honeywell, Morristown, NJ.)

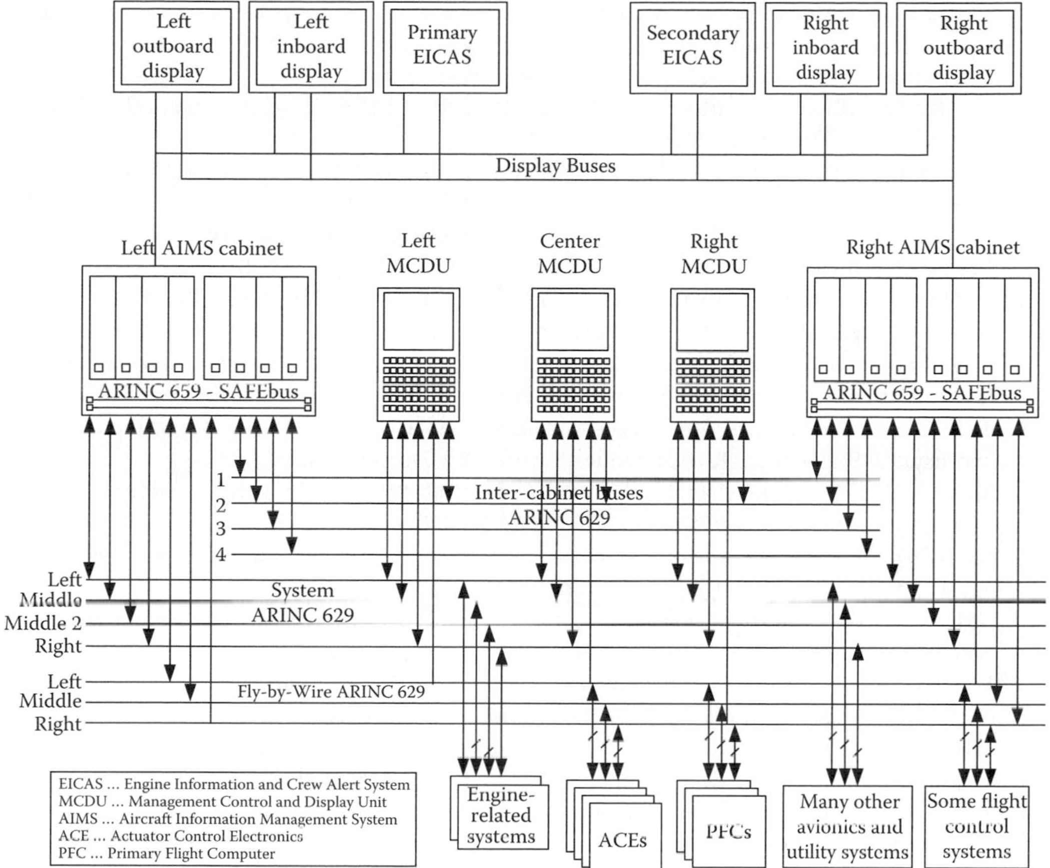

FIGURE 48.18 Boeing B 777 avionics overview. (Courtesy of Honeywell, Morristown, NJ.)

Boeing 777 is said to be the first but not the last plane to leverage IMA extensively. Related to SAFEbus, Versatile Integrated Avionics (VIA) is a derivative of AIMS and also uses SAFEbus on the Boeing 737NG, Boeing 717, McDonnell Douglas MD10, and McDonnell Douglas MD90.

References

1. Aeronautical Radio, Inc. ARINC Specification 659—Backplane data bus. Annapolis, MD, December 27, 1993.
2. K. Driscoll, G.M. Papadoupoulos, S. Nelson, G.L. Hartmann, and G. Ramohalli. M2FCS:1984, Multi-processor flight control system, Honeywell Systems and Research Center, September 1984, AFWAL-TR-84-3076.
3. ARINC659. ARINC, Backplane data bus. ARINC Specification 659, Aeronautical Radio, Inc, Annapolis, MD, December, 1993.
4. K.R. Driscoll. US Patent 5386424, Apparatus and method for transmitting information between dual redundant components utilizing four signal paths, Honeywell, January 31, 1995.
5. K.R. Driscoll. US Patent 5307409, Apparatus and method for fault detection on redundant signal lines via encryption, Honeywell, April 26, 1994.
6. K. Driscoll, B. Hall, M. Paulitsch, P. Zumsteg, and H. Sivencrona. The real byzantine generals, *Proceedings of 23rd Digital Avionics Systems Conference*, Vol. 6.D.4, pp. 61–11, October, 2004.

7. M. Paulitsch, J. Morris, B. Hall, K. Driscoll, E. Latronico, and P. Koopman. Coverage and the use of cyclic redundancy codes in ultra-dependable systems, *Proceedings of the IEEE International Conference on Dependable Systems and Networks (DSN)*, 2005.
8. ARINC659. ARINC, Backplane data bus. ARINC Specification 659, Aeronautical Radio, Inc, Annapolis, MD, December, 1993.
9. K. Hoyme and K. Driscoll. SAFEbus, *IEEE Aerospace and Electronic Systems Magazine*, pp. 34–39, March, 1993.
10. K. Driscoll and K. Hoyme. The airplane information management system: An integrated real-time flight-deck control system, *Real-Time Systems Symposium, 1992*, pp. 267–270, December, 1992.
11. M. Morgan. Boeing B-777. In *The Avionics Handbook*, CRC Press, Boca Raton, FL, 2001.
12. B. Witwer. Developing the 777 airplane information management system (AIMS): A view from program start to one year of service, *IEEE Transactions on Aerospace and Electronic Systems*, 33(2):637–641, April, 1997, doi 10.1109/7.588382.
13. ARINC653P1-2. Avionics application software standard interface Part 1 required services, Aeronautical Radio Inc., ARINC Specification 653P1-2, December, 2005.
14. ARINC653P1-2. Avionics application software standard interface Part 1 required services, Aeronautical Radio Inc., ARINC Specification 653P1-2, December, 2005.
15. ARINC653P3. Avionics application software standard interface Part 3 conformity test specification, Aeronautical Radio Inc., ARINC Specification 653P3, October, 2006.
16. ARINC653P2-1. Avionics application software standard interface Part 2 extended services, Aeronautical Radio Inc., ARINC Specification 653P2-1, December 2009.
17. T. Carpenter, K. Driscoll, K. Hoyme, and J. Carciofini. ARINC 659 scheduling: Problem definition, *Real-Time Systems Symposium, 1994., Proceedings*, IEEE Computer Society, pp. 165–169, December 1994.
18. ARINC. Multi-transmitter data bus: Part 1 technical description. ARINC specification 629p1-5. Technical Report, Aeronautical Radio Inc., Annapolis, MD, March 31, 1999.
19. K. Driscoll and K. Hoyme, The airplane information management system: An integrated real-time flight-deck control system, *Real-Time Systems Symposium, 1992*, pp. 267–270, December 1992.
20. ARINC659. ARINC, Backplane data bus. ARINC Specification 659, Aeronautical Radio, Inc, Annapolis, MD, December 1993.
21. ARINC629. ARINC, Multi-transmitter data bus: Part 1 technical description. ARINC Specification 629P1-5, Aeronautical Radio Inc., Annapolis, MD, March 31, 1999.

49

Dimensioning of Civilian Avionics Networks

49.1 Introduction 49-1
49.2 Civilian Avionics Systems Evolution and ARINC Context 49-2
49.3 AFDX End-to-End Delay Analysis 49-4
49.4 Sure Upper Bounds with Network Calculus 49-5
Basic Network Calculus Approach for the AFDX • Optimizing the Network Calculus Approach for the AFDX
49.5 Sure Upper Bounds with Trajectory Approach 49-9
Main Features of the Trajectory Approach • Illustration on a Sample AFDX Configuration • Trajectory Approach Computation
49.6 Exact Worst-Case Delays with Model Checking 49-13
Overview of Timed Automata • Modeling of an AFDX Network • Computation of the Exact Worst-Case End-to-End Delay
49.7 Analysis of a Realistic Configuration 49-16
49.8 Trends for the Dimensioning of AFDX Network 49-17
More Precise Modeling of Input Flows • Integration of QoS Mechanisms in the Worst-Case Analysis • Pushing the Limits of Model Checking
49.9 Conclusion 49-19
References 49-20

Jean-Luc Scharbarg
University of Toulouse

Christian Fraboul
University of Toulouse

49.1 Introduction

The evolution of civilian aircraft avionics systems is mainly due to an increasing complexity, which is illustrated by a larger number of integrated functions, a larger volume of exchanged data, and a larger number of connections between functions. Consequently, the growth of the number of multipoint communication links cannot be taken into account by classical avionics mono-emitter data–buses (such as ARINC 429 [ARI01]). A first solution proposed for Boeing 777 led to the design of a new multiplexed data–bus based on CSMA-CA medium access control (ARINC 629 standard [ARI99]). The solution adopted by Airbus for the A380 consists in the utilization of a switched Ethernet technology (AFDX: Avionics Full Duplex Switched Ethernet), which allows a reuse of development tools as well as of existing communication components while achieving better performance and which has been standardized in ARINC 664 [ARI02,ARI03]. This new communication standard represents a major step in the deployment of modular avionics architectures (Integrated Modular Avionics: IMA ARINC 651 [ARI91] and 653 [ARI97]). But the main problem is due to the indeterminism of the switched Ethernet, and a network designer must prove that no frame will be lost by the AFDX (no switch queue overflow) and must evaluate the end-to-end transfer delay through the network (guaranteed upper bound and distribution of delays) according to a given avionics applications traffic.

At least three approaches have been proposed in order to compute a worst-case bound for each communication flow of the avionic applications on an AFDX network configuration. They are based on network calculus, trajectories, and model checking. Such a worst-case communication delay analysis allows the comparison between the computed upper bounds and the constraints on the communication delays of each flow. Moreover, it allows the scaling of switch buffers in order to avoid overflow and frame losses.

This chapter summarizes the assumptions of the AFDX network technology and gives an overview of the different approaches that are used for the worst-case analysis of such networks. The approaches are illustrated on a sample configuration, and results on a realistic avionic configuration are shown. Trends on the dimensioning of avionics networks are also listed.

49.2 Civilian Avionics Systems Evolution and ARINC Context

Avionics is the generic name given to the electronic systems installed in an aircraft. It includes the calculators and their software, the sensors and actuators, and all the communication links between these elements. Almost every function of the aircraft is nowadays implemented by an avionic system: command controls, autopilot, navigation, information display, or cabin management. In the 1950s, avionics were very simple stand-alone systems as all functions could be executed by a single calculator. Modern avionics began with the replacement of analog devices by their digital equivalent in the 1960s. Since then, the complexity of avionics has never stopped growing: more and more analog devices are outdated by digital equipments, and new functions are added continuously. A few facts clearly illustrate the growth of the avionics' role in civilian aircrafts: From 1983 (A310) to 1993 (A340), the number of avionic systems onboard has increased by almost 50% (from 77 to 115), while the total processing power has been multiplied by 4, from 60 to 250 MIPS. Consequently, the communication needs increase as the number of embedded systems and functions increases [CC93].

Typically, classical avionics uses pieces of equipment that can be seen as hardware and software black boxes that are responsible for one given function. The physical devices adopt a standardized module form, called line-replaceable unit. New concepts based on a better sharing of execution and communication resources have been proposed in the 1990s. IMA introduces the definition of a generic platform called cabinet composed of standard execution and transmission modules, called line-replaceable modules. The global IMA architecture is thus composed of modules housed by cabinets distributed in the aircraft, leading to new intracabinet and intercabinet communication needs.

The Aeronautical Radio INCorporated (ARINC) standardized the traditional ARINC 429 mono-emitter data–bus [ARI01], which has been widely used in the context of classical avionics architectures. IMA is described in two ARINC standards: ARINC 651 for the modular hardware architecture [ARI91] and ARINC 653 for the modular software architecture [ARI97]. More recently, the ARINC 664 brings an answer to communication increase by multiplexing a huge amount of communication over a full-duplex switched Ethernet network [ARI02,ARI03]. It has become the reference communication technology in the context of civilian aircraft avionics as it provides a backbone network for the avionics platform.

ARINC 429 is the earlier and most used standard for civilian aircraft avionics that has been installed on a majority of Airbus and Boeing aircraft. Messages are transmitted over two twisted pairs by the owner of the bus at a bit rate of either 12.5 or 100 kbps to other systems units. The specification defines the electrical transmission characteristics and protocol data units' features [ARI01]. ARINC 429 protocol implements a mono-emitter broadcast bus with up to 20 receivers. The emitter broadcasts a 32-bit data word (with a given periodicity) on the bus. Filtering is done by the receiver according to the label associated to the data. ARINC 429 data–bus provides high determinism but at the cost of extra wiring and limited data rates. In a system with an increasing number of emitting and receiving units, point-to-point wiring becomes a major problem that ARINC 429 technology cannot face any more.

The IMA concept introduces switched Ethernet technology. All the units are directly connected by a point-to-point link to an Ethernet switch, and cascading switches offer the desired connectivity.

Using full-duplex links eliminates the possibility of transmission collisions on the links: the CSMA-CD medium access control protocol is no more needed. So it eliminates the inherent indeterminism of vintage Ethernet and the collision frame loss, but it shifts in fact the problem to the switch level where various flows enter in competition for the use of the resources of the switches. The switch implements the classical IEEE 802.1d bridging algorithm [IEEE98]: It filters and forwards frames from the incoming ports to the outgoing ports. The store-and-forward bridging mechanism reads the destination address of each received frame and retransmits it according to the port-id stored in the forwarding table. If a temporary congestion on an output port of a switch appears, it can significantly increase end-to-end delays of frames and even lead to frame losses by overflow of buffers. This is why dedicated mechanisms have been added to classical full-duplex Ethernet in order to guarantee determinism.

The AFDX has been standardized as ARINC 664 with the help of many avionics manufacturers: Airbus, Boeing, Rockwell Collins, Honeywell, etc [ARI02,ARI03]. The standard defines the subset of deterministic networks whose main characteristics are static configuration, flow segregation, and controlled traffic (traffic policing mechanisms). Thus, an ARINC 664 deterministic network is based upon Ethernet frame definition and IEEE 802.1d switching protocol, but it includes specific mechanisms in order to guarantee the determinism of avionics communications. Main AFDX-specific assumptions deal with the static definition of avionics flows that have to respect a bandwidth envelope (burst and rate) at network ingress point.

The AFDX network architecture is composed of several interconnected switches. The inputs and outputs of the networks are the end systems. Each end system is connected to exactly one switch port, and each switch port is connected to either an end system or another switch. The links are all full duplex. One first-in first-out (FIFO) buffer is associated to each output port.

The end-to-end traffic flow's characterization is standardized by ARINC 664 in its part 7, where the virtual link (VL) paradigm is presented [ARI03]. The idea behind this concept is to segregate the flows for safety reasons. Thus, the VL concept of virtual communication channels statically defines the flows that enter the network and associates some performance properties to each flow. Each VL can be statically mapped on the network of interconnected AFDX switches. Transmitting an Ethernet frame from one end system to another is based on a VL identifier, which is used for deterministic routing of each VL (the switch-forwarding tables are statically defined after the allocation of all VL on the AFDX network architecture). Each VL defines a logical unidirectional connection from one source end system to one or more destination end systems. For example, Figure 49.1 illustrates different kinds of VL: vx is a unicast VL with path $\{e3-S3-S4-e8\}$, while $v6$ is a multicast VL with paths $\{e1-S1-S2-e7\}$ and $\{e1-S1-S4-e8\}$.

A VL definition includes the bandwidth allocation gap (BAG) value, the minimum frame size (s_{min}), and the maximum frame size (s_{max}). The *BAG* is the minimum delay between two consecutive frames of the associated VL (which actually defines a VL as a sporadic flow). *BAG* and s_{max} values guarantee an

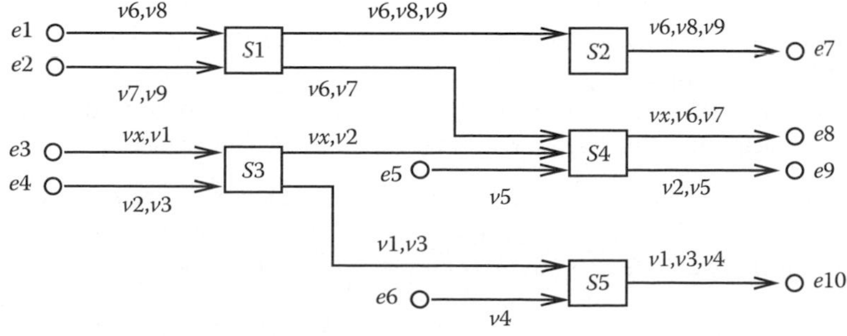

FIGURE 49.1 Illustrative AFDX configuration.

allocated bandwidth for each VL. Moreover, a jitter value is associated to each VL to upper-bound the maximum admissible jitter after multiplexing different regulated VL flows.

From the avionics systems designer point of view, the classical ARINC 429 buses have many interesting features. For example, the single-emitter assumption implies a dedicated link offered to the emitter and thus guaranteed access to the bus, guaranteed bandwidth, and high determinism. Thus, the VL concept is of importance in the definition of the AFDX as it allows direct replacement of ARINC 429 buses, on top of a deterministic ARINC 664 network.

49.3 AFDX End-to-End Delay Analysis

Let us consider a path p_x of a VL v. The end-to-end delay $D(F_v, p_x)$ of a frame F_v transmitted on p_x is defined by

$$D(F_v, p_x) = LD(F_v, p_x) + SD(F_v, p_x) + WD(F_v, p_x)$$

$LD(F_v, p_x)$ is the transmission delay over the links: Thanks to the full-duplex characteristic of the AFDX, there are no collisions on the links. Thus, the transmission delay over a link is $t_{byte} * s(F_v)$, where t_{byte} is the transmission time of one byte and $s(F_v)$ is F_v length. Therefore, considering that all the links have the same bandwidth c (consequently, t_{byte} is the same for all the links),

$$LD(F_v,) = nbl(px) * (t_{byte} * s(F_v))$$

where $nbl(p_x)$ is the number of links in p_x.

$SD(F_v, p_x)$ is the delay in switches between input and output ports: The delay in a switch from an input port to an output port is considered as a constant td, since the only available information about this delay is a guaranteed upper bound of 16 μs. Thus,

$$SD(F_v, p_x) = nbs(px) * td$$

where $nbs(p_x)$ is the number of switches in p_x.

$WD(F_v, p_x)$ is the delay in switches and end system output buffers: This delay highly depends on each output port load at the time where F_v reaches it.

Consequently, $D(F_v, p_x)$ can be divided into a fixed part $LD(F_v, p_x) + SD(F_v, p_x)$ and a variable part $WD(F_v, p_x)$. The fixed part can be statically computed since it depends solely on the path p_x, the length of the frame F_v, and the bandwidth of the links. The variable part depends on dynamic characteristics, such as the sequence of frames that are emitted by each VL (the length of each frame) and the emission instants of the different VL frames.

The end-to-end delay analysis of a path p_x of a VL v has to take into account all the possible scenarios. This analysis should determine the following features of this end-to-end delay.

- The smallest possible value of the end-to-end delay corresponds to the scenarios where the VL v emits a frame with minimal length $s_{min}(v)$, which never waits in output ports. This smallest possible value is denoted $D_{min}(v, p_x)$, and it is computed in the following way:

$$D_{min}(v, p_x) = nbl(p_x) * (t_{byte} * s(F_v)) + nbs(p_x) * td$$

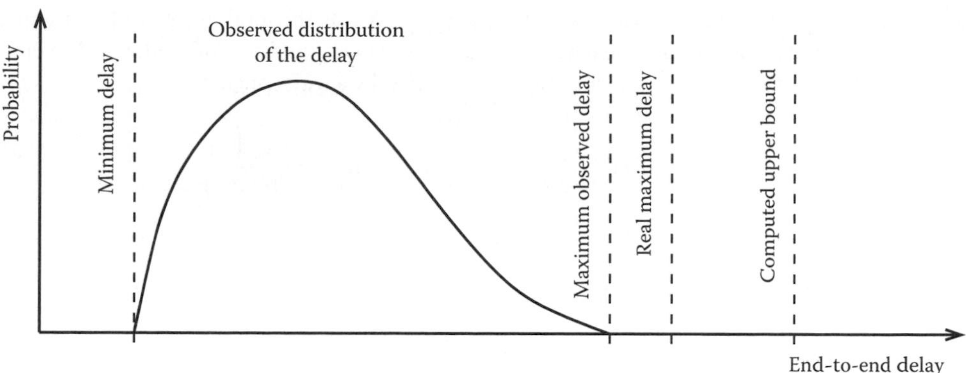

FIGURE 49.2 End-to-end delay features.

- The highest possible value of the end-to-end delay, which corresponds to the worst-case scenario. It is mandatory for the certification of the avionic network. In the general case, finding this worst-case scenario requires an exhaustive analysis of all the possible scenarios. Section 49.6 presents an approach that implements this exhaustive analysis. Such an exhaustive enumeration is impossible for any realistic configuration, since the number of possible scenarios is huge, due to the number of VLs (around 1000). An alternative solution is the computation of a safe upper bound of the end-to-end delay, based on the modeling of the configuration that overestimates the traffic and/or underestimates the service offered by the network (pessimistic assumptions). Sections 49.4 and 49.5 present two approaches that compute a pessimistic safe upper bound of the end-to-end delay of any VL of an industrial AFDX configuration.
- The distribution of the end-to-end delay between its smallest and its highest possible values. This distribution is valuable when prototyping the whole system. This distribution can be obtained thanks to a simulation approach [SRF09]

Figure 49.2 summarizes the characteristics of the end-to-end delay. This delay is always between a minimum and a maximum value. Most of the time, the exact maximum value cannot be computed, and it is lower-bounded by the maximum observed value and upper-bounded by the computed safe upper bound.

49.4 Sure Upper Bounds with Network Calculus

Network calculus [Cha00,LBT01] has been proposed for the computation of an upper bound for the delay and the jitter of a flow transmitted over a network. It can be used on a set of sporadic flows with no assumption concerning the arrival time of packets. It has been applied to AFDX [BSF10], and the certification of the AFDX on board of A380 is based on this approach.

The basic application of network calculus to the AFDX is presented in Section 49.4.1. The improvement of this basic approach in the context of AFDX is described in Section 49.4.2.

49.4.1 Basic Network Calculus Approach for the AFDX

Network calculus is mathematically based on the $(min, +)$ dioid, for which the convolution $\otimes$ and the deconvolution $\oslash$ are defined as follows:

$$(f \otimes g)(t) = inf0 \le u \le t\big(f(t-u)+g(u)\big) \quad \text{and} \quad (f \oslash g)(t) = sup0 \le u\big(f(t+u)-g(u)\big)$$

A flow is represented by its cumulative function R, where $R(t)$ is the total number of bits sent up to time t. A flow R is said to have a function α as arrival curve if α is an upper bound of the traffic generated by the flow. Each VL of an AFDX network (a flow) is modeled by a leaky bucket

$$\gamma_{r,b}(t)=b+r*t \quad \text{with} \quad b=s_{max}(v) \quad \text{and} \quad r=\frac{s_{max}(v)}{BAG(v)}$$

where

The burst b is the capacity of the bucket

The rate r is the leak rate

A server has a service curve β, which is a lower bound of the service offered by the server. Each output port of an AFDX switch offers a service curve:

$$\beta_{R,T}=R[t-T]^{+}$$

T is the maximal technological latency of the switch, that is, 16 μs

R is the servicing rate (100 Mb/s in our context) and $[x]^{+} = \max(0, x)$. Thus, in the context of this chapter, the service curve that is offered by each output port is $\beta_{100,16}$.

The delay experienced by a flow R constrained by an arrival curve α in a node offering a service curve β is bounded by the maximum horizontal difference between curves α and β. This difference is formally defined by

$$h(\alpha,\beta)=sup_{s\geq 0}\left(inf\left\{\left|\tau\geq 0\right|\alpha(s)\leq\beta(s+\tau)\right\}\right\}\right)$$

Figure 49.3 illustrates the delay $h(\alpha, \beta)$ experienced by a flow R constrained by a service curve $\alpha = \gamma_{r,b}$ in an output port of an AFDX switch, provided R is the only flow crossing this output port.

The VLs that compete for a given output port are merged into a single flow by summing their respective arrival curves.

The overall computation is illustrated in the small example in Figure 49.4.

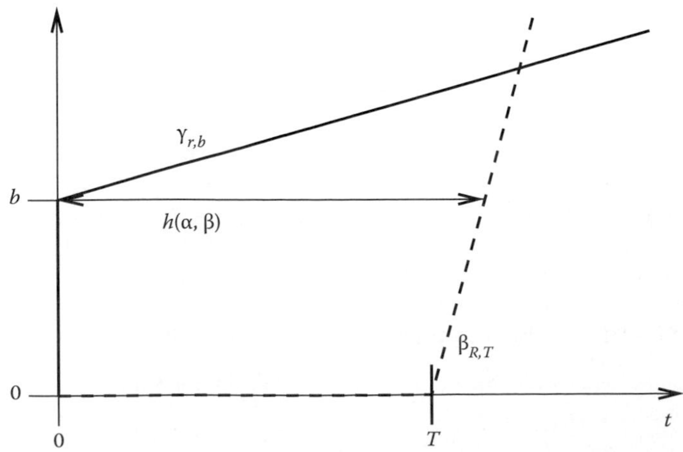

FIGURE 49.3 Maximum delay $h(\alpha, \beta)$.

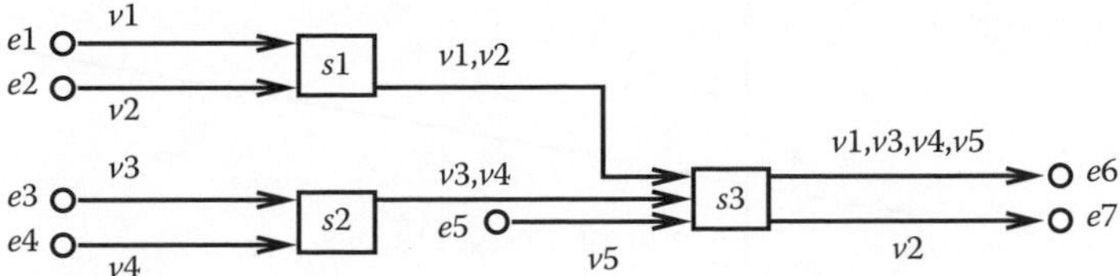

FIGURE 49.4 Sample AFDX configuration.

This configuration includes five unicast VLs *v*1…*v*5. Their parameters are given in the following table:

	BAG (ms)	s_{min}	s_{max}	Path
*v*1	4	300 bytes	500 bytes	*e*1–*S*1–*S*3–*e*6
*v*2	4	300 bytes	500 bytes	*e*2–*S*1–*S*3–*e*7
*v*3	4	300 bytes	500 bytes	*e*3–*S*2–*S*3–*e*6
*v*4	4	300 bytes	500 bytes	*e*4–*S*2–*S*3–*e*6
*v*5	4	300 bytes	500 bytes	*e*5–*S*3–*e*6

Let us consider VL *v*1. Its arrival curve in *S*1 is

$$\alpha 1 = \gamma_{1,4000}$$

since

$$r = \frac{s_{max}(v1)}{BAG(v1)} = \frac{4000}{4000} = 1\frac{\text{Mb}}{s} \quad \text{and} \quad b = s_{max}(v1) = 4000 \text{ bits}$$

*v*1 shares the output port of *S*1 with *v*2, whose arrival curve in *S*1 is

$$\alpha_2 = \gamma_{1,4000}$$

Consequently, the overall arrival curve for the output port of switch *S*1 is

$$\alpha_1 + \alpha_2 = \gamma_{2,8000}$$

As previously mentioned, the service curve of this port is $\beta_{100,16}$. Thus, the delay in this output port is bounded by the maximum horizontal difference between $\gamma_{2,8000}$ and $\beta_{100,16}$, which is 96 μs, as depicted in Figure 49.5.

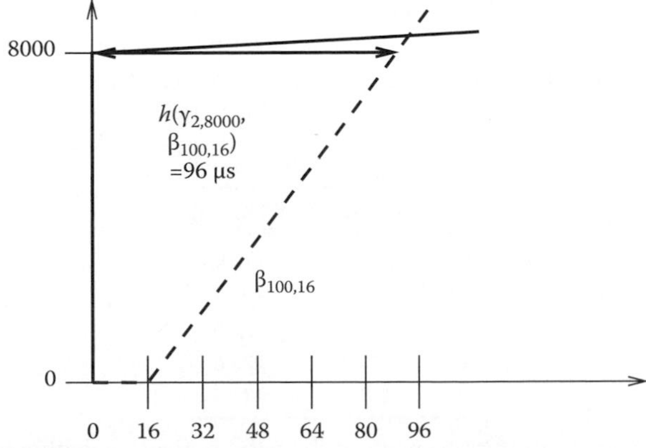

FIGURE 49.5 Output port of switch *S*1.

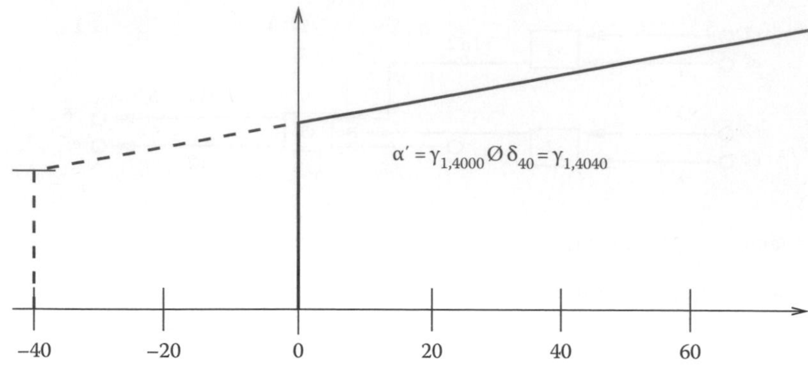

FIGURE 49.6 Output curve after switch *S1* or *S2*.

It includes the technological latency (16 μs), the transmission time (40 μs), and the maximum waiting time in the output buffer (40 μs), since each packet of VL *v1* or *v2* can be delayed by at most one packet of the other VL.

Then, the computation proceeds to switch *S3*, and it needs the input curves of *v1*, *v3*, *v4*, and *v5* in *S3*. These input curves are the output curves of the VLs in their previous crossed output port, that is, *S1* for *v1*, *S2* for *v3* and *v4*, and *e5* for *v5*. In the general case, the output curve α' of the flow is given by

$$\alpha' = \alpha \varnothing \delta_{jitter}$$

where

α is the input curve of the flow in the port
jitter is the maximum jitter encountered by the flow in the port
δ_{jitter} is a guaranteed delay service curve

$$\delta_d(t) = 0 \text{ if } t \leq d, \infty \text{ otherwise}$$

Graphically, it comes down to shift the arrival curve α to the left by the duration of the jitter. The maximum jitter in an output port corresponds to the maximum waiting time in the corresponding buffer. Coming back to *v1*, its maximum jitter when leaving *S1* is 40 μs, that is, the maximum waiting time in the output buffer of *S1*. Then, the input curve α'_1 of *v1* at *S3* is obtained by shifting α_1 by 40 μs to the left:

$$\alpha'_1 = \alpha_1 \varnothing \delta_{40} = \gamma_{1,4040}$$

It is illustrated in Figure 49.6.

The input curves of *v3*, *v4*, and *v5* at *S3* are, respectively, $\gamma_{1,4040}$, $\gamma_{1,4040}$, and $\gamma_{1,4000}$. They lead to an overall arrival curve $\gamma_{4,16120}$ in the output port of *S3*. Then, the maximum delay for *v1* in *S3* is 177.2 μs, leading to a maximum end-to-end delay of 313.2 μs. It is composed of the transmission delay from *e1* to *S1* (40 μs) and the maximum delay computed for *S1* and *S3*, that is, 96 and 177.2 μs.

Column BNC in the following table summarizes the upper bounds computed with this method for the five VLs of Figure 49.4.

	BNC (μs)	ONC (μs)
v1	313.2	273.6
v2	192.4	192.4
v3	313.2	273.6
v4	313.2	273.6
v5	217.2	177.6

The next paragraph presents an improvement of the basic network calculus approach in the context of AFDX.

49.4.2 Optimizing the Network Calculus Approach for the AFDX

The basic network calculus approach presented in the previous section is pessimistic in the context of AFDX. In particular, it does not take into account the property that frames of different flows sharing a link cannot be transmitted at the same time on this link (they are serialized). Consequently, the burst considered in the overall input curves of the basic network calculus approach can never happen, as soon as at least two flows share the same link. This problem is different from the classical *pay burst only once* case described in [LBT01]. Indeed, the objective of *pay burst only once* is to aggregate successive switches in order to give an optimized aggregated service curve. The aggregation of nodes is not possible in our case, since flows can join and leave a path at any switch of the network.

In the example in Figure 49.4, the input curve of the output port of $S3$ shared by $v1$, $v3$, $v4$, and $v5$ is $\gamma_{4,16120}$. The maximum burst (16120 bits) corresponds to the case where four frames—one for each VL—arrive at the same time in the output port. This is definitely impossible, since $v3$ and $v4$ share the same link. The grouping technique integrates this serialization. It proceeds in two steps. First, the overall arrival curve is computed for each link. It is the minimum between, on the one hand the sum of the arrival curves of all the flows sharing the considered link, on the other hand the curve bounding the burst to the maximum burst among the curves of the different flows sharing the link and the rate to the rate of the link. This first step is illustrated in Figure 49.7 for a link shared by two flows with arrival curves $\gamma_{r1,b1}$ and $\gamma_{r2,b2}$.

In the second step, the curves obtained for the different links are added. The gain obtained with this technique is due to the reduction of the maximum burst.

Column *ONC* in the previous table gives the upper bounds computed with this technique in the example in Figure 49.4. Results are clearly improved, compared with the basic network calculus approach.

A more recent approach, based on trajectories, has been proposed for the worst-case delay analysis of distributed systems. The next section shows how this approach can be applied and optimized in the context of the AFDX. The main goal is to compare this new approach with the network calculus one.

49.5 Sure Upper Bounds with Trajectory Approach

The trajectory approach [MM06a,MM06b] has been developed to get deterministic upper bounds on end-to-end response time in distributed systems. This approach considers a set of sporadic flows with no assumption concerning the arrival time of packets. The application of the trajectory approach to the AFDX has been presented in [BSF10]. Main features of the trajectory approach applied to AFDX are summarized and illustrated in Sections 49.5.1 and 49.5.2. The trajectory approach computation is presented in Section 49.5.3.

49.5.1 Main Features of the Trajectory Approach

The approach developed for the analysis of the AFDX considers the results from [MM06a]. A distributed system is composed of a set of interconnected processing nodes. Each flow crossing this system follows

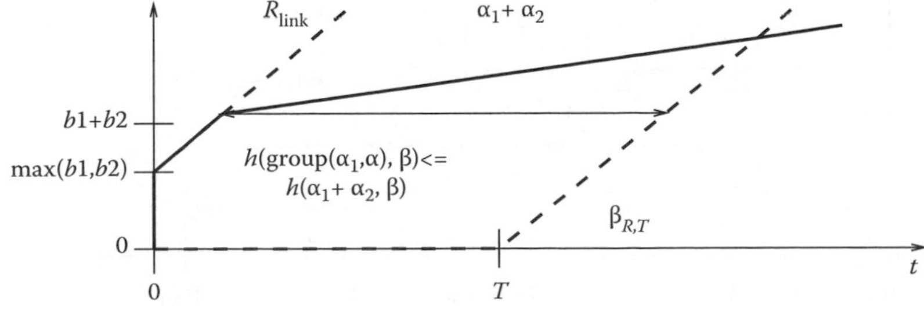

FIGURE 49.7 Grouping of flows.

a static path, which is an ordered sequence of nodes. The trajectory approach assumes, with regard to any flow τ_i following path P_i, that any flow τ_j following path P_j, with $P_j \neq P_i$ and $P_j \cap P_i \neq \emptyset$, never visits a node of path P_i after having left this path.

Flows are scheduled with a FIFO algorithm in every visited node. Each flow τ_i has a minimum interarrival time between two consecutive packets at ingress node, denoted T_i, a maximum release jitter at the ingress node denoted J_i, an end-to-end deadline D_i that is the maximum end-to-end response time acceptable, and a maximum processing time C_i^h on each node N_h, with $N_h \in P_i$.

The transmission time of any packet on any link between nodes has known lower and upper bounds L_{min} and L_{max}, and there are neither collisions nor packet losses on links.

The end-to-end response time of a packet is the sum of the times spent in each crossed node and the transmission delays on links. The transmission delays on links are upper bounded by L_{max}. The time spent by a packet m in a node N_h depends on the pending packets in N_h at the arrival time of m in N_h. The problem is then to upper bound the overall time spent in the visited nodes.

The solution proposed by the trajectory approach is based on the busy period concept. In the FIFO context, a busy period in a node N_h starts at a time t when at least a frame is ready for transmission in the node, and there was previously no ready frame. It ends at a time t' when the packet m under study has been processed.

The trajectory approach considers a packet m from flow τ_i generated at time t. It identifies the busy period and the packets impacting its end-to-end delay on all the nodes visited by m. It enables the computation of the latest starting time of m on its last node. It leads to the worst-case end-to-end response time of the flow τ_i. This computation will be illustrated in the context of AFDX.

The elements of the system considered in the trajectory approach are instantiated in the following way in the context of AFDX. Each node of the system corresponds to an AFDX switch output port, including the output link. Each link of the system corresponds to the switching fabric. Each flow corresponds to a VL path.

The switching fabric delay is upper bounded by a constant value (16 μs), thus $L = L_{min} = L_{max} = 16$ μs. VL parameters match the definition of sporadic flows in the following manner:

$T_i = BAG$, $J_i = 0$, and $C_i^h = C_i = s_{max}/R$ since all the links have the same rate R.

49.5.2 Illustration on a Sample AFDX Configuration

Let us consider the sample AFDX configuration depicted in Figure 49.4. Figure 49.8 shows an arbitrary scheduling of the frames, which are identified by their VL numbers (e.g., frame 3 is a frame from VL $v3$). The scheduling in Figure 49.8 focuses on frame 3.

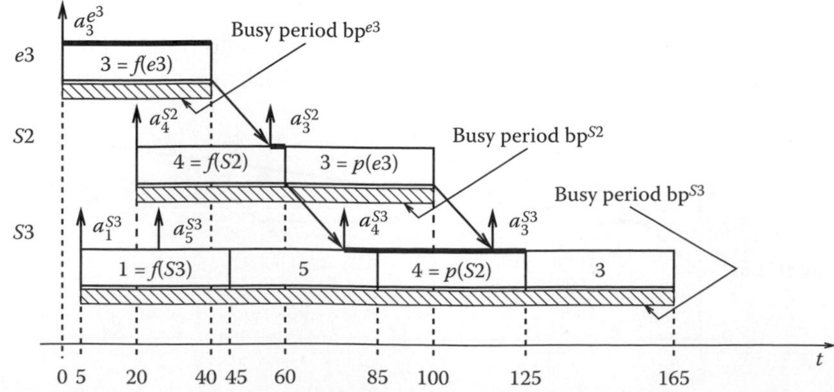

FIGURE 49.8 An arbitrary scheduling of frames.

The arrival time of a frame m in a node N_h is denoted $a_m^{N_h}$. Time origin is arbitrarily chosen as the arrival time of frame 3 in node $e3$. The processing time of a packet in a node is 40 μs. It corresponds to the transmission time of the frame on a link. The delay between the end of the processing of a frame by a node and its arrival in the next node corresponds to the 16 μs switch factory delay. Due to the FIFO policy, frame 3 is delayed by frame 4 in $S2$. In node $S3$, frame 5 is delayed by frame 1 and delays frame 4, which delays frame 3.

Frame 3 from $v3$ crosses three busy periods bp^{e3}, bp^{S2}, and bp^{S3} on its trajectory, corresponding to the three nodes $N_1 = e3$, $N_2 = S2$, and $N_3 = S3$. Let $f(N_i)$ be the first frame that is processed in the busy period bp^{N_i} during which frame 3 is processed. Considering the scheduling in Figure 49.8, we have $f(e3) = 3$, $f(S2) = 4$, and $f(S3) = 1$. As flows do not necessarily follow the same path in the network, it is possible that frame $f(N_i)$ does not come from the same previous node N_{i-1} as frame 3. This case occurs in node $S2$, where frame 4 comes from node $e4$. It also occurs in node $S3$, where frame 1 comes from node $S1$. Therefore, $p(N_{i-1})$ is defined as the first frame, which is processed in bp^{N_i} and comes from node N_{i-1}. Considering the scheduling in Figure 49.8, we have $p(e3) = 3$ and $p(S2) = 4$.

The starting time of frame 3 in node $S3$ is obtained by adding parts of the three busy periods bp^{e3}, bp^{S2}, and bp^{S3} to the delays between the nodes, that is, 2 * 16 μs. From [MM06a], the part of the busy period bp^{N_i} that has to be added is the processing time of frames between $f(N_i)$ and $p(N_i)$ minus the time elapsed between the arrivals of $f(N_i)$ and $p(N_{i-1})$, that is, $\left(a_{p(N_{i-1})}{}^{N_i} - a_{f(N_i)}{}^{N_i}\right)$.

In the example in Figure 49.8, the parts that have to be considered are the transmission of frame 3 in node $e3$, the time elapsed between the arrival of frame 3 and the end of processing of frame 4 in node $S2$, and the time elapsed between the arrival of frame 4 and the end of processing of frame 5 in node $S3$. These parts are shown by thick lines on top of the frames in Figure 49.8. The starting time of frame 3 in node $S3$ in the example in Figure 49.8 is 125 μs.

It has been shown [MM06a] that the latest starting time of a frame m in its last node is reached when

$$\left(a_{p(N_{i-1})}{}^{N_i} - a_{f(N_i)}{}^{N_i}\right) = 0$$

for every node N_i on the path of m. It comes to postpone the arrival time of every frame joining the path of m in the node N_i in order to maximize the waiting time of m in N_i.

The result of this postponing in the example in Figure 49.8 is illustrated in Figure 49.9. The arrival time of frame 4 at node $S2$ is postponed to the arrival time of frame 3 at node $S2$. In node $S3$, frames 1 and 5 have been postponed in order to arrive between frames 4 and 3.

Then, the worst-case end-to-end delay of a frame is obtained by adding its latest starting time on its last visited node and its processing time in this last node. For frame 3 in Figure 49.9, this worst-case end-to-end delay is

$$232 + 40 = 272\ \mu s$$

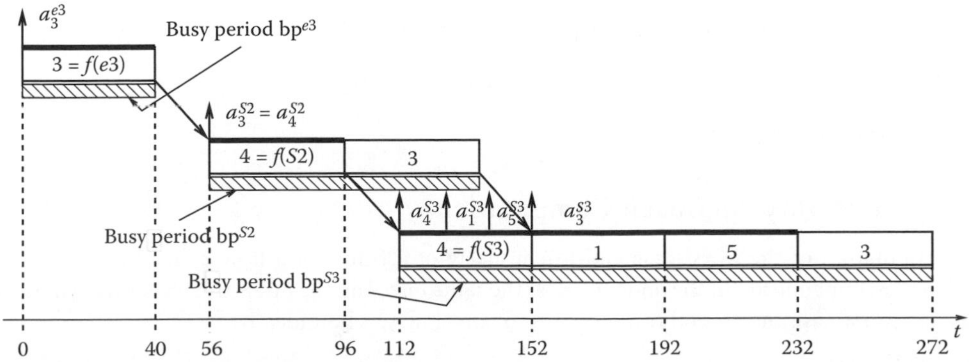

FIGURE 49.9 Latest starting time of frame 3.

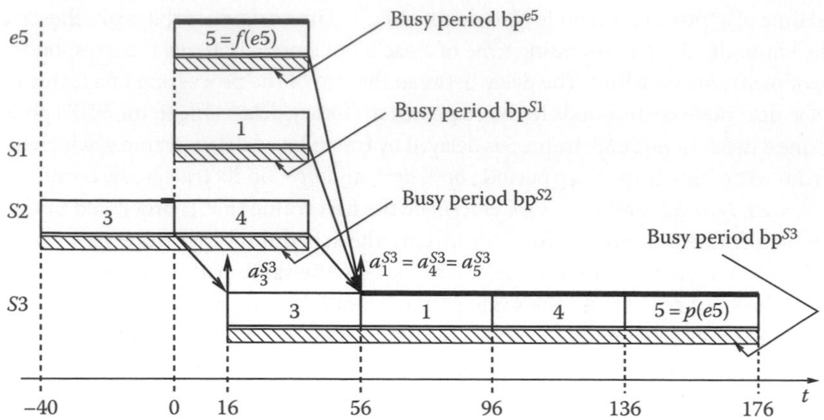

FIGURE 49.10 Latest starting time of 5.

More precisely, this delay includes the transmission times of frame 3 on node *e*3, frame 4 on node *S*2, and frames 4, 1, 5, and 3 on node *S*3. In this example, it can be seen that frames 3 and 4 are counted twice. Actually, it has been shown [MM06b] that exactly one frame has to be counted twice in each node, except the slowest one. In the context of the AFDX, all the nodes work at the same speed. Thus, the slowest node is arbitrarily chosen as the last one. In the example in Figure 49.9, frames 3 and 4 are, respectively, counted twice in nodes *e*3 and *S*2. Frame 3 is the longest one transmitted in nodes *e*3 and *S*2, while frame 4 is the longest one transmitted in nodes *S*2 and *S*3.

In the context of an AFDX network, it is not always possible to find a scheduling that cancels the term $\left(a_{p(N_{i-1})}{}^{N_i} - a_{f(N_i)}{}^{N_i}\right)$ for every node N_i, as proposed in [MM06b]. Let us consider VL *v*5 of the example depicted in Figure 49.4. bp^{S3} is the busy period corresponding to frame 5. In order to maximize the delay of frame 5 in bp^{S3}, the arrival time of frames 3 and 4 in *S*3 has to be as large as possible, but not larger than the arrival time of frame 5 in node *S*3, because of the FIFO scheduling policy:

$$a_3{}^{S3} \le a_5{}^{S3} \text{ and } a_4{}^{S3} \le a_5{}^{S3}$$

Since both frames come from the same link, they are already serialized:

$$|a_3{}^{S3} - a_4{}^{S3}| \ge 40\ \mu s$$

Without loss of generality, let us consider that frame 3 arrives before frame 4. Then, we have

$$a_4{}^{S3} = a_5{}^{S3} \text{ and } a_3{}^{S3} = a_5{}^{S3} - 40\ \mu s$$

The resulting worst-case scheduling is depicted in Figure 49.10. We have

$$(a_{p(e5)}{}^{S3} - a_{f(S3)}{}^{S3}) \ge 40\ \mu s$$

for any possible scheduling.

49.5.3 Trajectory Approach Computation

The computation of the worst-case end-to-end delay of a frame of a flow τ_i has been formalized in [MM06a]. In our context, all the nodes work at the same rate, and the jitter in each emitting node is null. Thus, the worst-case end-to-end response time of any flow τ_i is bounded by

$$R_i = max_{t \ge 0}\left(W_{i,t}{}^{last_i} + C_i - t\right)$$

$last_i$ is the last visited node of flow τ_i

$W_{i,t}^{last_i}$ is a bound on the latest starting time of a frame m generated at time t on its last visited node $W_{i,t}^{last_i}$ is the sum of the following terms:

- $\sum_{j\in[1,n],j\neq i,P_j\cap P_i\neq\emptyset}\left(1+\lfloor(t+A_{ij})/T_j\rfloor\right)*C_j$: It corresponds to the processing time of frames from every flow τ_j crossing the flow τ_i and transmitted in the same busy period as m. $A_{i,j}$ integrates the maximum jitter of frames from τ_i and τ_j on their first shared output port.
- $(1+\lfloor t/T_i\rfloor)*C_i$: It is the processing time, on one node, of the frames of the flow τ_i that are transmitted in the same busy period as m.
- $\sum_{N_h\in P_i,N_h\neq last_i}(max_{j\in[1,n],N_h\in P_j}(C_j))$: It is the processing time of the longest frame for each node of path P_i, except the last one. It represents the frames that have to be counted twice, as explained before.
- $(|P_i|-1)*L_{max}$: It corresponds to the sum of switching delays.
- $\sum_{N_h\in P_i,N_h\neq first_i}(\Delta_{N_h})$: It subtracts for each node N_h in P_i the duration between the beginning of the busy period and the arrival of the first frame coming from the preceding node in P_i, that is, N_{h-1}. This term is null in the context of [M06a]. A lower bound of Δ_{N_h} for each node N_h in P_i has been established in the context of AFDX [BSF10]. At a given node N_h, frames are grouped by input links, and one frame is removed from each group: the smallest one for the input link of the frame under study and the largest one for the other input links. It leads to a set of sequences of frames. If the sequence of the input link of the frame under study is the largest one, Δ_{N_h} is null. Otherwise, Δ_{N_h} is the difference between the largest sequence and the sequence of the input link of the frame under study.
- $-C_i$: It is subtracted, because $W_{i,t}^{last_i}$ is the latest starting time and not the ending time of the frame from τ_i on its last node.

Solving $R_i=max_{t\geq 0}\left(W_{i,t}^{last_i}+C_i-t\right)$ comes to find the maximum vertical deviation between the function $t\,\alpha\,W_{i,t}^{last_i}+C_i$ and the identity function ($t\,\alpha\,t$).

The following table gives the upper bounds computed for the configuration in Figure 49.4. For this configuration, the trajectory approach always gives tighter bounds than the ones obtained by the network calculus approach.

	Traj (μs)
v1	272
v2	192
v3	272
v4	272
v5	176

49.6 Exact Worst-Case Delays with Model Checking

The approach is based on a modeling in timed automata.

49.6.1 Overview of Timed Automata

Timed automata have been first proposed in [AD94] in order to describe systems behavior with time. A timed automaton is a finite automaton with a set of clocks, that is, real and positive variables increasing uniformly with time. Transition labels can be a condition on clock values (a guard), actions, or updates, which assign new value to clocks.

The composition of timed automata is obtained by a synchronous product. Each action a executed by a first timed automaton corresponds to an action with the same name a executed in parallel by a second timed automaton. In other words, a transition that executes the action *a* can be fired only if another transition labeled *a* is possible. The two transitions are performed simultaneously. Thus, communication uses the rendezvous mechanism.

Performing transitions requires no time. Conversely, time can run in nodes. Each node is labeled by an invariant, that is, a Boolean condition on clocks. The node is occupied if the invariant is true.

Several extensions of timed automata have been proposed. One of these extensions is timed automata with shared integer variables. The principle consists in defining a set of integer variables that are shared by different timed automata. The values of these variables can be consulted and updated by the different timed automata [LPY97].

A system modeled with timed automata can be verified using a reachability analysis that is performed by model checking. It consists in encoding each property in terms of the reachability of a given node of one of the automata.

In the general case, reachability analysis is undecidable on timed automata with shared integer variables. In the particular case where the shared variables are represented by nodes of a timed automaton, the reachability analysis is decidable.

The approach considered for the AFDX is based on timed automata with shared integer variables that are represented by nodes of a timed automaton.

49.6.2 Modeling of an AFDX Network

The modeling of an AFDX network considers an automaton for each VL and an automaton for each output port of a switch [CSEF06]. Figure 49.11 depicts the timed automaton of a VL with a BAG equal to *period*. This automaton sends a first message $send_i$ ($send_0$ in the example) delayed by an offset between 0 and *period* and then sends periodically a new message $send_i$.

Figure 49.12 shows an example of an output port of a switch. Each node of the automaton models a location in the FIFO queue associated with the port (3 in the example in Figure 49.12). Each transition

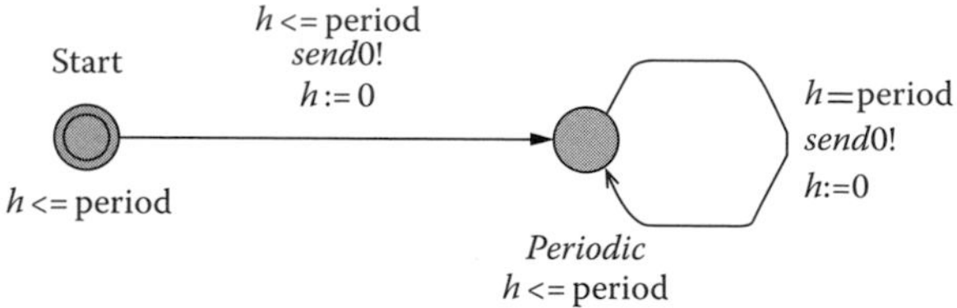

FIGURE 49.11 Automaton of a VL.

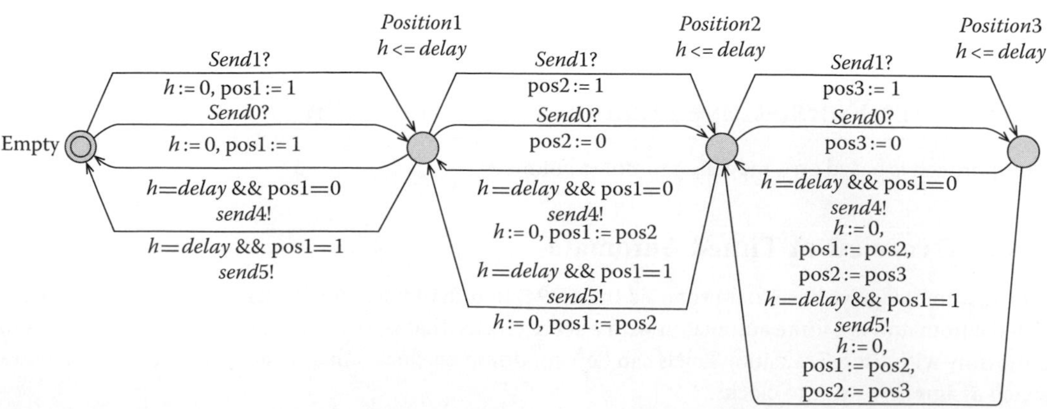

FIGURE 49.12 Automaton of an output port of a switch.

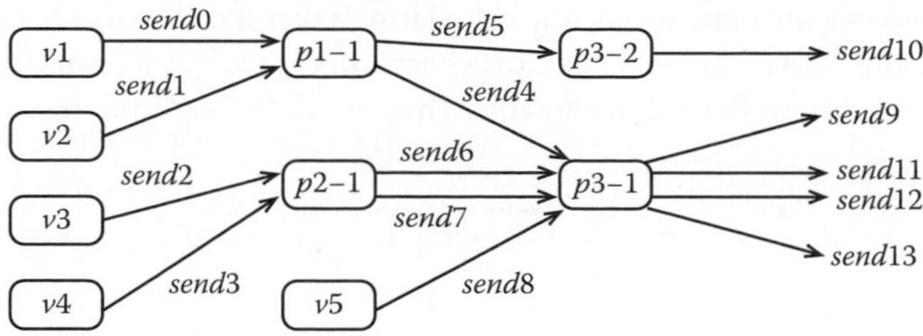

FIGURE 49.13 Global modeled system.

from a node $Position_i$ to a node $Position_{i+1}$ of the automaton models the arrival of one frame at the output port, while each transition from a node $Position_{i+1}$ to a node $Position_i$ models the end of the transmission from this port. The automaton of Figure 49.12 considers two VLs received using signals *send*0 and *send*1 and transmitted using signals *send*4 and *send*5. *delay* is the transmission time of the frame (in the example, all the frames have the same length). pos1, pos2, and pos3 indicate the flows corresponding to the frames waiting in each position of the queue.

The global system is obtained by composing the timed automata of the VLs and the output ports of the switches. For instance, Figure 49.4 is composed of five VLs and four output ports, leading to nine timed automata, as depicted in Figure 49.13. As an example, VL *v*2 is modeled by the timed automaton *v*2, which sends signal *send*1. It is received by timed automaton *p*1 – 1, which models the unique output port of switch *s*1. *v*2 follows the path composed of signals *send*5 and *send*10.

49.6.3 Computation of the Exact Worst-Case End-to-End Delay

Using the test automaton method [BBF+01], the worst-case end-to-end delay of each VL is obtained from the model previously described. The test automaton corresponding to *v*1 is depicted in Figure 49.14.

This automaton models the property *delay of* v1 *is less than bound*. By receiving signal *send*0, it evolves to node *s*2 and waits for *send*9. If this signal is not received within delay *bound*, the automaton evolves to node *reject*. This behavior corresponds to a scenario for which the end-to-end delay of *v*1 is at least *bound*. So, the goal is to find the lowest value of bound for which node *reject* is reached. This value is the maximum end-to-end delay.

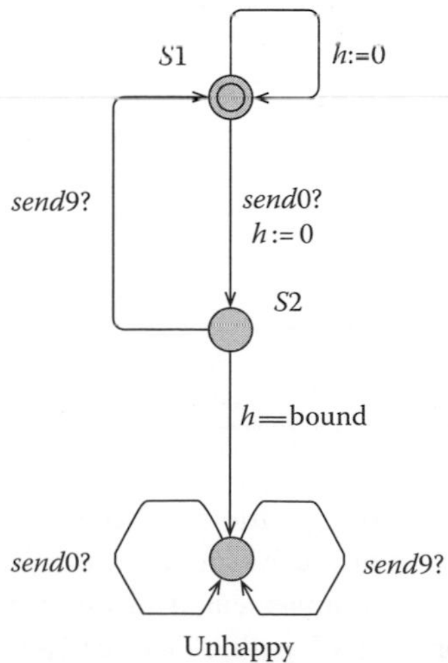

FIGURE 49.14 Test automaton of *v*1.

The approach is implemented in Uppaal. The calculation takes less than 1*s* on a Linux station with a Pentium 4 processor and 2GB of memory size. The exact worst-case end-to-end delays obtained for the VLs in Figure 49.4 are given in the following table. They are identical to the results obtained with the trajectory approach.

	MC (μs)
$v1$	272
$v2$	192
$v3$	272
$v4$	272
$v5$	176

This approach exhibits the exact worst–case, and it is valuable since it helps understand the worst-case behavior of the network. However, it cannot be used for the certification of a realistic network (e.g., the AFDX of the A380), due to the well-known combinatorial explosion problem. Actually, it cannot cope with configurations including more than eight VLs.

49.7 Analysis of a Realistic Configuration

The realistic AFDX network considered in this chapter is composed of two redundant networks [CSEF06]. Each network includes 123 end systems, 8 switches, 964 VLs, and 6412 VL paths (due to VL multicast features). The following table gives the dispatching of VLs among BAGs. It can be seen that BAGs are harmonic between 2 and 128.

BAG (ms)	2	4	8	16	32	64	128
Nb of VLs	20	40	78	142	229	220	255

A second table gives the dispatching of VLs among frame lengths, considering the maximum length s_{max}. Most of the VLs consider short frames.

Frame length (bytes)	0–150	151–300	301–600	601–900	901–1200	1201–1500	>1500
Nb of VLs	561	202	114	57	12	35	3

A third table shows the number of VL paths per length (i.e., the number of crossed switches). Most of the VLs have short paths.

Nb of crossed switches	1	2	3	4
Nb of paths	1797	2787	1537	291

The temporal analysis of this realistic configuration has been conducted. As previously mentioned, the model-checking approach cannot cope with such large-scale configurations. Both the network calculus and the trajectory approaches have been implemented using Python programming language. Upper bounds of the end-to-end delays for each VL path of the realistic configuration have been computed with this tool. This computation takes less than 2 minutes for any approach on a Pentium 4 processor running at 2.8 GHz. The following table gives for both approaches the average upper bounds among all the VLs of the configuration, as well as the minimum and the maximum upper bounds. The obtained upper bounds are slightly tighter with the trajectory approach.

	Network Calculus (ms)	Trajectories (ms)
Minimum upper bound	0.386	0.293
Average upper bound	4.532	4.247
Maximum upper bound	13.194	12.949

Both network calculus and trajectory approaches lead to pessimistic results. Indeed, network calculus is based on arrival curves and service curves that are pessimistic. It has been shown in [Li2011] that the trajectory approach computation can also be pessimistic. Then, the issue is to upper-bound the pessimism of both approaches. This issue has been addressed in [BSF10]. The idea is to build, for each VL, an unfavorable scenario, that is, a scenario that leads to an as-high-as-possible end-to-end delay for the VL. This high end-to-end delay gives an underestimation of the worst-case delay. Thus, the difference between this high end-to-end delay and the computed upper bound gives an overestimation of the pessimism of the computed upper bound.

The overestimation of the pessimism of upper bounds has been conducted on the industrial configuration analyzed in this section [BSF10]. The following table summarizes the obtained results:

	Network Calculus (%)	Trajectories (%)
Minimum pessimism	0.8	0
Average pessimism	13.5	6.5
Maximum pessimism	63	33

The pessimism depends on the flow. The trajectory approach computes the exact worst–case for around 12% of the flows (the minimum pessimism is 0%). This approach is on average at least two times less pessimistic than the network calculus approach. Thus, the trajectory approach significantly reduces the pessimism of the computed upper bounds and the remaining pessimism is very small.

49.8 Trends for the Dimensioning of AFDX Network

Measurements on real AFDX networks show that the end-to-end delays are always much smaller than the computed upper bounds. Thus, the network is overdimensioned. However, it has been shown in the previous section that both worst-case approaches (network calculus and trajectories) provide tight upper bounds on typical industrial configurations. These two observations seem contradictory. Actually, they are not. Indeed, the worst-case delay for a given VL corresponds to very rare scenarios. Moreover, the worst-case approaches consider that all the VLs are independent. This is true for VLs generated by different end systems. This is not true for VLs generated by the same end system. Indeed, each end system implements a scheduling that can generate constraints on the generation times of the VL frames. Some work has been conducted in order to integrate these constraints in the worst-case delay analysis. It is summarized in Section 49.8.1.

A second trend for a better utilization of available resources is to integrate Quality of Service (QoS) mechanisms in the network. Indeed, for future aircraft, the addition of other type of flows (audio, video, best–effort, etc.) on the AFDX network is envisioned. These different flows have different timing constraints and criticality levels. Thus, they have to be differentiated, and the FIFO policy on switch output ports is not suitable. Therefore, it is necessary to design a QoS-aware AFDX switch that implements other service disciplines, such as static priority queueing or weighted fair queueing [PG93]. The issue is to integrate efficiently these QoS mechanisms in the worst-case delay analysis. Work done on this point is summarized in Section 49.8.2.

A third trend concerns the model-checking approach. One valuable feature of this approach is to be able to exhibit a worst-case scenario. Obtaining such a scenario on a medium-sized configuration can bring interesting information on worst-case scenario features. Thus, work has been done in order to mitigate the combinatorial explosion problem. It is summarized in Section 49.8.3.

49.8.1 More Precise Modeling of Input Flows

Input flows generated by a given end system are scheduled. This scheduling leads to minimum duration between frames of different VLs generated by this end system. This fact is illustrated in Figure 49.15.

Two strictly periodic VLs *v*1 and *v*2 (resp. *v*3 and *v*4) with the same BAG are generated by an end system *e*1 (resp. *e*2). Due to the scheduling implemented by *e*1 (resp. *e*2), there is a minimum duration from any frame of *v*1 (resp. *v*3) to any frame of *v*2 (resp. *v*4). Similarly, there is a minimum duration from any frame of *v*2 (resp. *v*4) to any frame of *v*1 (resp. *v*3).

The integration of these minimum duration constraints in the network calculus approach has been proposed in [LSF10]. The idea is to build, for each end system, one arrival curve per group of VLs with minimum duration constraints between them. In the example in Figure 49.15, there is one arrival curve for the group {*v*1,*v*2} and one arrival curve for the group {*v*3,*v*4}. It brings a reduction of the burst generated by each end system. The same principle is also applied at the switch output port level.

Obtained results are promising. Indeed, the integration of these minimum duration constraints on the industrial configuration studied in Section 49.7 can decrease the worst-case upper bounds by 50%, assuming that all the VLs are strictly periodic.

49.8.2 Integration of QoS Mechanisms in the Worst-Case Analysis

The integration of fixed priority (FP) scheduling in the trajectory approach has been proposed in [BSF12]. A priority is associated to each VL. At each output port, the pending frame with the highest priority is transmitted. If two frames have the same priority, the first arrived one is transmitted.

The differences with the trajectory approach for FIFO scheduling are the following:

- One frame *m* can be delayed by a frame with higher priority till their last shared output port, as soon as this higher-priority frame arrives at the output port before *m* leaves. This is taken into account in the computation of competing flow workload, as well as in the computation of the serialization effect factor.
- At each output port, one frame *m* can be delayed by at most one lower-priority frame with maximum size. Thus, a new term is added to take into account this delay.

The trajectory approach for FP has been applied to the industrial configuration of Section 49.7. An overestimation of the pessimism of the obtained upper bounds has been conducted. The average pessimism is 8.2% (it is 6.5% for the trajectory approach for FIFO). It means that an FP scheduling can be efficiently integrated in the worst-case delay analysis.

49.8.3 Pushing the Limits of Model Checking

In order to mitigate the combinatorial explosion problem, the idea is to drastically reduce the number of scenarios to be tested. Therefore, properties have been established in [ASF11,ASEF12]. Any scenario that does not verify these properties cannot lead to the worst case. Thus, it does not have to be tested.

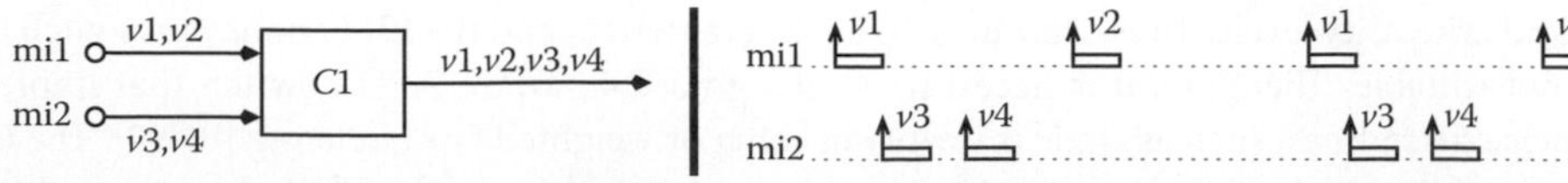

FIGURE 49.15 Minimum duration constraints.

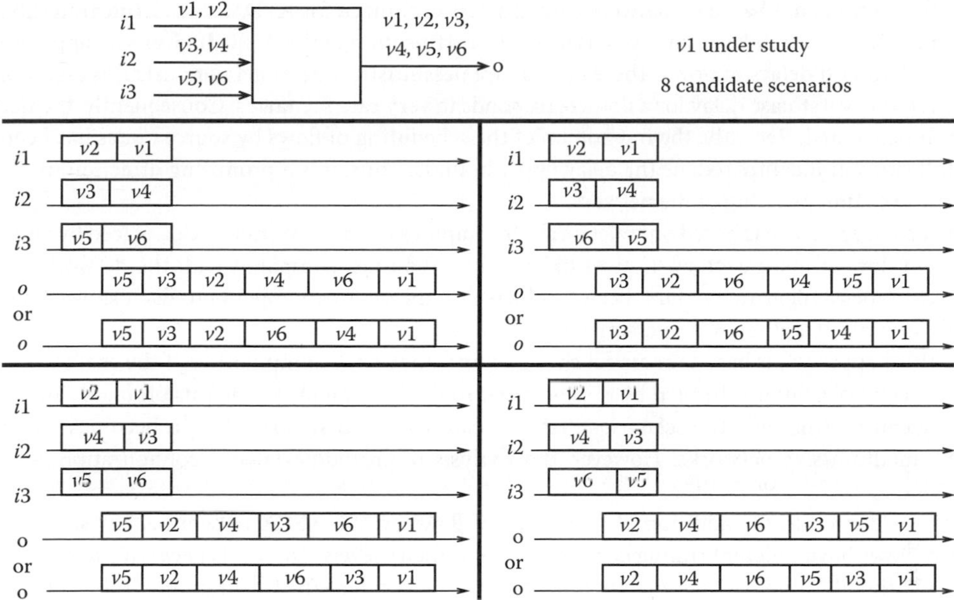

FIGURE 49.16 Worst-case candidate scenarios.

These properties are based on the critical instant concept. Let us consider a frame *m* of a VL *v*. A scenario can lead to the worst case for *m* if it verifies the following properties in each output port crossed by *m*:

- *m* arrives in the output port of a switch at the same instant as one frame from each input port of the switch that has traffic for this output port
- All the frames that can delay *m* arrive at the switch output port as late as possible
- *m* is transmitted last among the frames that have arrived in the output port at the same instant

These properties are illustrated in Figure 49.16. *v*1 is under study. A scenario can lead to the worst-case for *v*1 if

- One frame from *v*3 or *v*4 arrives at the output port at the same time as the frame from *v*1
- One frame from *v*5 or *v*6 arrives at the output port at the same time as the frame from *v*1
- The other frames arrive at the output port as late as possible
- The frame from *v*1 is transmitted last on the output link

It leads to eight candidate scenarios for the worst case for *v*1.

Two approaches taking into account these properties have been proposed. The first one is based on timed automata and Uppaal [ASEF12]. It can analyze configurations with up to 30 VLs. The limitation is mainly due to the fact that the modeling uses arrays that are memory consuming in Uppaal. The second approach is based on a home-made tool in Java [ASF11]. It can cope with configurations including up to 60 VLs.

Such results are promising. The analysis of a medium-sized configuration becomes possible with a model-checking approach. However, further work is needed in order to cope with an industrial size configuration.

49.9 Conclusion

This chapter gives an overview of the temporal analysis of switched Ethernet avionic network (AFDX). Today, three approaches exist for the computation of a safe upper bound of the end-to-end delay of each flow transmitted on the avionic network.

The first approach is based on network calculus. It is optimized for AFDX, since it integrates the serialization of frames on links. It has been used for the certification of the A380. It gives safe upper bounds on the end-to-end delays of flows. These bounds are pessimistic, due to network calculus assumptions. Moreover, the worst-case delay for a flow corresponds to very rare scenarios. Consequently, the network is overdimensioned. Recently, the integration of the scheduling of flows by source nodes has been proposed. It can significantly reduce the delay upper bounds. Thus, it is a promising direction in order to limit the overdimensioning of the network.

The second approach is based on trajectories. It computes the maximum workload faced by any frame of a given flow on its trajectory. It gives tighter upper bounds, compared with the network calculus approach. A pessimism analysis has been conducted on the obtained upper bounds. It shows that these bounds are tight (less than 6.5% pessimism).

The third approach is based on model checking and allows the computation of the exact worst-case delay of each flow, but it is limited by the combinatorial explosion problem. This problem can be mitigated by considering only the scenarios that can lead to the worst case. Thus, it becomes possible to analyze medium-size networks. However, the analysis of an industrial-size configuration is still an open issue.

For future aircraft, the addition of other type of flows on the AFDX network is envisioned. These different flows have different timing constraints and criticity levels. Thus, it is necessary to differentiate them, and the FIFO policy on switch output ports is not suitable. The integration of FP mechanism in the trajectory approach has been proposed, and it has been shown that obtained upper bounds are tight. Further studies are needed in order to allocate these priorities efficiently and to extend the approach to other scheduling policies.

Finally, the sporadic characteristic of avionics flows could be taken into account with the help of a probabilistic modeling, as it has been proposed for the aperiodic traffic in the automotive context [Khan2009]. This leads to a probabilistic analysis of the worst-case delay of flows. Such a preliminary analysis has been proposed in the context of AFDX [SRF09}]. It is based on a stochastic network calculus approach [VLB01,VLB02].

References

AD94 R. Alur and D.L. Dill. Theory of timed automata, *Theoretical Computer Science,* 126(2):183–235, 1994.

ARI91 ARINC 651 Aeronautical Radio Inc., ARINC specification 651: Design guidance for integrated modular avionics, Annapolis, MD, 1991.

ARI97 ARINC 653 Aeronautical Radio Inc., ARINC project 653: Avionics application software standard interface, Annapolis, MD, 1997.

ARI99 ARINC 629 Aeronautical Radio Inc., ARINC specification 629: Multi-transmitter data bus part 1—Technical description, Annapolis, MD, 1999.

ARI01 ARINC 429 Aeronautical Radio Inc., ARINC specification 429-ALL: Mark 33 digital information transfer system (DITS) *Parts* 1, 2, 3, Annapolis, MD, 2001.

ARI02 ARINC 664, Aircraft data network, Part 1: Systems concepts and overview, 2002. ARINC 664, Aircraft data network, Part 2: Ethernet physical and data link layer specification, 2002.

ARI03 ARINC 664, Aircraft data network, Part 7: Deterministic networks, 2003.

ASEF12 M. Adnan, J.L. Scharbarg, J. Ermont, and C. Fraboul. An improved timed automata approach for computing exact worst-case delays of AFDX sporadic flows, in *Proceedings of the ETFA'2012*, Krakow, Poland, September 2012.

ASF11 M. Adnan, J.L. Scharbarg, and C. Fraboul. Minimizing the search space for computing exact worst-case delays of AFDX periodic flows, in *Proceedings of the SIES'2011*, Vasteras, Sweden, June 2011.

BBF+01 B. Bérard, M. Bidoit, A. Finkel, F. Laroussinie, A. Petit, L. Petrucci, and P. Schnoebelen. *Systems and Software Verification. Model-Checking Techniques and Tools*, Springler-Verlag, Berlin, Germany, 2001.

BSF10 H. Bauer, J.L. Scharbarg, and C. Fraboul. Improving the worst-case delay analysis of an AFDX network using an optimized trajectory approach, *IEEE Transactions on Industrial Informatics*, 6(4):521–533, November 2010.

BSF12 H. Bauer, J.L. Scharbarg, and C. Fraboul. Applying trajectory approach with static priority queueing for improving the use of available AFDX resources, *Real Time Systems Journal*, 48(1):101–133, January 2012.

CC93 P. Chanet and V. Cassigneul. How to control the increase in the complexity of civil aircraft on-board systems , *AGARD Meeting on Aerospace Software Engineering for Advanced Systems Architectures*, Paris, France, May 1993.

Cha00 C.S. Chang. *Performance Guarantees in Communication Networks*, Springer-Verlag, London, U.K.

CSEF06 H. Charara, J.L. Scharbarg, J. Ermont, and C. Fraboul. Methods for bounding end-to-end delays on an AFDX network, in *Proceedings of the 18th ECRTS*, Dresden, Germany, July 2006.

IEEE98 IEEE 802.1d. Local and metropolitan area networks: Media access control level bridging, 1998.

IEEE 802.1p: LAN Layer 2 QoS/CoS Protocol for Traffic Prioritization, 1998.

IEEE 802.3, Information technology—Telecommunications and information exchange between systems—Local and metropolitan area networks—CSMA/CD access method and physical layer specifications, 1998.

Khan 2009 D. Khan, N. Navet, B. Bavoux, and J. Migge. A periodic traffic in response time analysis with adjustable safety level, in *Proceedings of the 14th International Conference on Emerging Technologies and Factory Automation*, IEEE, Palma de Mallorca, Spain, September 2009.

LBT01 J.-Y. Le Boudec and P. Thiran. *Network Calculus: A Theory of Deterministic Queuing Systems for the Internet*, volume 2050 of Lecture Notes in Computer Science. Springer-Verlag, Berlin, Germany, 2001.

LPY97 K.G. Larsen, P. Petterson, and W. Yi. Uppaal in a nutshell, *International Journal on Software Tools for Technology Transfer*, 1(1–2):134–152, 1997.

LSF10 X. Li, J.L. Scharbarg, and C. Fraboul. Improving end-to-end delay upper bounds on an AFDX network by integrating offsets in worst-case analysis, in *Proceedings of the ETFA'2010*, Bilbao, Spain, September 2010.

LSF11 X. Li, J.L. Scharbarg, and C. Fraboul. Analysis of the pessimism of the trajectory approach for upper bounding end-to-end delays of sporadic flows sharing a switched Ethernet network, in *Proceedings of the RTNS'2011*, Nantes, France, September 2011.

MM06a S. Martin and P. Minet. Schedulability analysis of flows scheduled with FIFO: Application to the expedited forwarding class, *20th International Parallel and Distributed Processing Symposium*, Rhodes, Greece, April 2006a.

MM06b S. Martin and P. Minet. Worst_case end-to-end response times of flows scheduled with FP/FIFO, *5th International Conference on Networking*, Mauritius, 2006b.

PG93 A. Parekh and R. Gallager. A generalized processor sharing approach to flow control in integrated services networks: The single-node case, *IEEE/ACM Transactions on Networking*, 1(3):344–357, 1993.

SRF09 J.-L. Scharbarg, F. Ridouard, and C. Fraboul. A probabilistic analysis of end-to-end delays on an AFDX avionic network, in *IEEE Transactions on Industrial Informatics*, 5:38–49, 2009.

VLB01 M. Vojnović and J.Y. Le Boudec. Bounds for independent regulated inputs multiplexed in a service curve element, in *Proceedings of the Globecom*, San Antonio, TX, 2001.

VLB02 M. Vojnović and J.Y. Le Boudec. Stochastic analysis of some expedited forwarding networks, in *Proceedings of the Infocom*, New York, June 2002.

IX

Automotive Communication Technologies

50 In-Vehicle Communication Networks *Nicolas Navet and Françoise Simonot-Lion* **50**-1
Automotive Communication Systems: Characteristics and Constraints • In-Car Embedded Networks • Automotive Middleware • Conclusions and Discussion • References

51 Standardized Basic System Software for Automotive Embedded Applications *Thomas M. Galla* **51**-1
Introduction • Hardware Architecture • OSEK/VDX • ISO • ASAM • AUTOSAR • Summary • References

52 Protocols and Services in Controller Area Networks *Gianluca Cena and Adriano Valenzano* **52**-1
Introduction • CAN Protocol Basics • Schedulability of CANs • CAN Advantages and Drawbacks • Time-Triggered CAN • CAN FD • CANopen • Interoperability between Devices • References

53 FlexRay Communication Technology *Roman Nossal-Tueyeni and Dietmar Millinger* **53**-1
Introduction • Automotive Requirements • What Is FlexRay? • System Configuration • Standard Software Components • References

54 The LIN Standard *Antal Rajnák and Anders Kallerdahl* **54**-1
Introduction • Need • History • Some LIN Basics • LIN Physical Layer • LIN Protocol • LIN Conformance Test • Design Process and Workflow • System Definition Process • Debugging • AUTOSAR and LIN • Future • Volcano LIN Tool Chain • LIN Network Architect • LTP • LIN Spector: Test Tool • Summary • Acknowledgments • References • Additional Sources

50

In-Vehicle Communication Networks: A Historical Perspective and Review

Nicolas Navet
University of Luxembourg

Françoise Simonot-Lion
Nancy Université

50.1 Automotive Communication Systems: Characteristics and Constraints 50-1
From Point-to-Point to Multiplexed Communications • Car Domains and Their Evolution • Event Triggered versus Time Triggered • Different Networks for Different Requirements • New Challenges for In-Vehicle Communication Systems
50.2 In-Car Embedded Networks 50-6
CAN Network • Time-Triggered Networks • Low-Cost Automotive Networks • Multimedia and Infotainment Networks • Automotive Ethernet
50.3 Automotive Middleware 50-19
Objectives of an Embedded Middleware • Former Propositions of Middleware • AUTOSAR—A Standard for the Automotive Industry
50.4 Conclusions and Discussion 50-25
References 50-25

50.1 Automotive Communication Systems: Characteristics and Constraints

50.1.1 From Point-to-Point to Multiplexed Communications

Since the 1970s, one observes an exponential increase in the number of electronic systems that have gradually replaced those that are purely mechanical or hydraulic. The growing performance and reliability of hardware components and the possibilities brought by software technologies enabled implementing complex functions that improve the comfort of the vehicle's occupants as well as their safety. In particular, one of the main purposes of electronic systems is to assist the driver to control the vehicle through functions related to the steering, traction (i.e., control of the driving torque), or braking such as the antilock braking system (ABS), electronic stability program (ESP), electric power steering (EPS), active suspensions, or engine control. Another reason for using electronic systems is to control devices in the body of a vehicle such as lights, wipers, doors, windows, as well as entertainment and communication equipment (e.g., radio, DVD, hand-free phones, navigation systems). More recently appeared a large set of advanced driver assistance systems (ADAS), often camera based, such as brake and park assistance, lane departure detection, night vision assistance with pedestrian recognition or suspension proactively scanning the road surface [55]. In the future, vehicles will become entities of large intelligent transportation systems [51], involving cooperation mechanisms [50] through car-to-car and car-to-infrastructure communications.

In the early days of automotive electronics, each new function was implemented as a stand-alone electronic control unit (ECU), which is a subsystem composed of a microcontroller and a set of sensors and actuators. This approach quickly proved to be insufficient with the need for functions to be distributed over several ECUs and the need for information exchanges among functions. For example, the vehicle speed estimated by the engine controller or by wheel rotation sensors has to be known in order to adapt the steering effort, to control the suspension, or simply to choose the right wiping speed. In today's luxury cars, several thousands of signals (i.e., elementary information such as the speed of the vehicle) are exchanged by up to 100 ECUs [1]. Until the beginning of the 1990s, data were exchanged through point-to-point links between ECUs. However, this strategy, which required an amount of communication channels of the order of n^2, where n is the number of ECUs (i.e., if each node is interconnected with all the others, the number of links grows in the square of n), was unable to cope with the increasing use of ECUs due to the problems of weight, cost, complexity, and reliability induced by the wires and the connectors. These issues motivated the use of networks where the communications are multiplexed over a shared medium, which consequently required defining rules—protocols—for managing communications and, in particular, for granting bus access. It was mentioned in a 1998 press release (quoted in Ref. [45]) that the replacement of a "wiring harness with LANs in the four doors of a BMW reduced the weight by 15 kg." In the mid-1980s, the third part supplier Bosch developed controller area network (CAN), which was first integrated in Mercedes production cars in the early 1990s. CAN has become today the most widely used network in automotive systems, and already in 2004, the number of CAN nodes sold per year was estimated [38] to be around 400 millions (all application fields). Other communication networks, providing different services, are now being integrated in automotive applications. A description of the major networks is given in Section 50.2.

50.1.2 Car Domains and Their Evolution

As all the functions embedded in cars do not have the same performance or safety needs, different quality of services (QoS) (e.g., response time, jitter, bandwidth, redundant communication channels for tolerating transmission errors, and efficiency of the error detection mechanisms) are expected from the communication systems. Typically, an in-car embedded system is divided into several functional domains that correspond to different features and constraints. Two of them are concerned specifically with real-time control and safety of the vehicle's behavior: the *powertrain* (i.e., control of engine and transmission) and the *chassis* (i.e., control of suspension, steering, and braking) domains. The third, the *body*, mostly implements comfort functions. The *telematics* (i.e., integration of wireless communications, vehicle monitoring systems, and location devices), *multimedia*, and *human–machine interface* (HMI) domains take advantage of the continuous progress in the field of multimedia and mobile communications. Finally, an emerging domain is concerned with the safety of the occupant and with functions providing driver assistance.

The main function of the powertrain domain is controlling the engine. This domain is increasingly constrained by stringent regulations, such as EURO 5 in force at the time of writing in Europe, for the protection of environment (e.g., emissions of particulate matter) and energy efficiency. The powertrain function is realized through several complex control laws with sampling periods of a magnitude of some milliseconds (due to the rotation speed of the engine) and implemented in microcontrollers with high computing power. In order to cope with the diversity of critical tasks to be treated, multitasking is required and stringent time constraints are imposed on the scheduling of the tasks (see Ref. [58] for typical automotive scheduling solutions). Furthermore, frequent data exchanges with other car domains, such as the chassis (e.g., ESP, ABS) and the body (e.g., dashboard, climate control), are required.

The chassis domain gathers functions such as ABS, ESP, automatic stability control (ASC), and four-wheel drive (4WD), which control the chassis components according to steering/braking solicitations and driving conditions (ground surface, wind, etc.). Communication requirements for this domain are quite similar to those for the powertrain, but because they have a stronger impact on the vehicle's stability, agility, and dynamics, the chassis functions are more critical from a safety standpoint. Furthermore, the *x-by-wire* technology, currently used for avionic systems, is now slowly being introduced to execute steering or braking functions. X-by-wire is a generic term referring to the replacement of mechanical or hydraulic systems by fully electrical and electronic (EE) ones, which led and still leads to new design methods for developing them safely [98] and, in particular, for mastering the interferences between functions [4]. Chassis and powertrain functions operate mainly as closed-loop control systems, and their implementation is moving toward a time-triggered (TT) approach [44,70,74,80], which facilitates composability (i.e., ability to integrate individually developed components) and deterministic real-time behavior of the system.

Dashboard, wipers, lights, doors, windows, seats, mirrors, and climate control are increasingly controlled by software-based systems that make up the *body* domain. This domain is characterized by numerous functions that necessitate many exchanges of small pieces of information among themselves. Not all nodes require a large bandwidth, such as the one offered by CAN; this leads to the design of low-cost networks, such as local interconnect network (LIN; see Section 50.2) or, more recently, Peripheral Sensor Interface 5 (PSI5) and Single-Edge Nibble Transmission (SENT). On LIN, only one node, termed the master, possesses an accurate clock and drives the communication by polling the other nodes—the slaves—periodically. The mixture of different communication needs inside the body domain leads to a hierarchical network architecture where integrated mechatronic subsystems based on low-cost networks are interconnected through a CAN backbone. The activation of body functions is mainly triggered by the driver and passengers' solicitations (e.g., opening a window and locking doors).

Telematics functions are becoming more and more numerous: hand-free phones, car radio, CD, DVD, in-car navigation systems, rear-seat entertainment, remote vehicle diagnostic, stolen vehicle tracking, etc. These functions require a lot of data to be exchanged within the vehicle but also with the external world through the use of wireless technology (see, for instance, Ref. [79]). Here, the emphasis shifts from messages and tasks subject to stringent deadline constraints to multimedia data streams, bandwidth sharing, multimedia QoS where preserving the integrity (i.e., ensuring that information will not be accidentally or maliciously altered) and confidentiality of information is crucial. HMI aims to provide human–machine interfaces that are easy to use and that limit the risk of driver inattention [19].

Electronic-based systems for ensuring the safety of the occupants are increasingly embedded in vehicles. Examples of such systems are impact and roll-over sensors, deployment of air-bags and belt pretensioners, tire pressure monitoring, adaptive cruise control (ACC—the car's speed is adjusted to maintain a safe distance with the car ahead), lane departure warning system, collision avoidance, driver intent and driver drowsiness detection, and night vision assistance. These functions form an emerging domain usually referred to as *active and passive safety*.

50.1.3 Event Triggered versus Time Triggered

One of the main objectives of the design step of an in-vehicle embedded system is to ensure a proper execution of the vehicle functions, with a predefined level of safety, in the normal functioning mode but also when some components fail (e.g., reboot of an ECU) or when the environment of the vehicle creates perturbations (e.g., EMI causing frames to be corrupted). Networks play a central role in maintaining the embedded systems in a *safe* state since most critical functions are now distributed and need to communicate. Thus, the different communication systems have to be analyzed in regard to this objective; in particular, messages transmitted on the bus must meet their real-time constraints, which mainly consist of bounded response times and bounded jitters.

There are two main paradigms for communications in automotive systems: TT and event triggered. Event triggered means that messages are transmitted to signal the occurrence of significant events (e.g., a door has been closed). In this case, the system possesses the ability to take into account, as quickly as possible, any asynchronous events such as an alarm. The communication protocol must define a policy to grant access to the bus in order to avoid collisions; for instance, the strategy used in CAN (see Section 50.2.1.1) is to assign a priority to each frame and to give the bus access to the highest-priority frame. Event-triggered communication is very efficient in terms of bandwidth usage since only necessary messages are transmitted. Furthermore, the evolution of the system without redesigning existing nodes is generally possible, which is important in the automotive industry where incremental design is a usual practice. However, verifying that temporal constraints are met is not obvious, and the detection of node failures is problematic.

When communications are TT, frames are transmitted at predetermined points in time, which is well suited for the periodic transmission of messages as it is required in distributed control loops. Each frame is scheduled for transmission during one predefined interval of time, usually termed a slot, and the schedule repeats itself indefinitely. This medium access strategy is referred to as time division multiple access (TDMA). As the frame scheduling is statically defined, the temporal behavior is fully predictable; thus, it is easy to check whether the timing constraints expressed on data exchanges are met. Another interesting property of TT protocols is that missing messages are immediately identified; this can serve to detect, in a short and bounded amount of time, nodes that are presumably no longer operational. The first downside is the inefficiency in terms of network utilization and response times by comparison with the transmission of aperiodic messages (i.e., messages that are not transmitted in a periodic manner). A second drawback of TT protocols is the lack of flexibility even if different schedules (corresponding to different functioning modes of the application) can be defined and switching from one mode to another is possible at run-time. Finally, the unplanned addition of a new transmitting node on the network induces changes in the message schedule and, thus, may necessitate the update of all other nodes. The Time-Triggered Protocol (TTP, formerly known as TTP/c [95]) is a purely time-triggered network, but there are networks, such as time-triggered CAN (TTCAN) [37], Flexible Time-Triggered CAN FTT-CAN [18], FlexRay [11], or TTEthernet [94], that can support a combination of both time-triggered and event-triggered transmissions. This capability to convey both types of traffic fits in well with the automotive context since data for control loops as well as alarms and events have to be transmitted.

Several comparisons have been done between event-triggered and TT approaches; the reader can refer to [1,18,42] for good starting points.

50.1.4 Different Networks for Different Requirements

The steadily increasing need for bandwidth* and the diversification of performance, costs, and dependability† requirements lead to a diversification of the networks used throughout the car. In 1994, the Society for Automotive Engineers (SAE) defined a classification for automotive communication protocols [13,85,86] based on data transmission speed and functions that are distributed over the network. *Class A* networks have a data rate lower than 10 kbit/s and are used to transmit simple control data with low-cost technology. They are mainly integrated in the *body* domain (seat control, door lock, lighting, trunk release, rain sensor, etc.). Examples of class A networks are LIN [48,77] and TTP/A [27]. *Class B* networks are dedicated to supporting data exchanges between ECUs in order to reduce the number of sensors by sharing information. They operate from 10 to 125 kbit/s. The J1850 [87] and low-speed CAN [33] are the main representatives of this class. Applications that need high-speed real-time communications require *class C* networks (speed of 125 kbit/s to 1 Mbit/s) or *class D* networks‡

* For instance, in [4], the average bandwidth needed for the engine and the chassis control was estimated to reach 1500 kbit/s in 2008, while it was 765 kbit/s in 2004 and 122 kbit/s in 1994.

† Dependability is usually defined as the ability to deliver a service that can justifiably be trusted (see Ref. [3] for more details).

‡ Class D is not formally defined, but it is generally considered that networks over 1 Mb/s belong to class D. This SAE classification is outdated today but was often referred to in the past.

(speed over 1 Mb/s). Class C networks, such as high-speed CAN [35], are used for the powertrain and currently for the chassis domains, while class D networks are devoted to multimedia data (e.g., Media Oriented Systems Transport [MOST] [59]) and safety–critical applications that need predictability and fault tolerance (e.g., TTP/C [93] or FlexRay [11] networks) or serve as gateways between subsystems (see the use of FlexRay at BMW in [83] and [39]).

It is common, in today's vehicles, that the electronic architecture includes four, five, or more different types of networks interconnected by gateways. For example, the first-generation Volvo XC90 [38] embeds up to 40 ECUs interconnected by a LIN bus, a MOST bus, a low-speed CAN, and a high-speed CAN. More recent high-end cars have much more complex architectures, such as the BMW 7 series launched in 2008 (see Ref. [39]), which implements four CAN buses, a FlexRay bus, a MOST bus, several LIN buses, an Ethernet bus, and wireless interfaces. The architecture is organized around a central gateway connecting most of the data buses and providing external connections for diagnostic and large data transfer (code upload, navigational maps, etc.).

50.1.5 New Challenges for In-Vehicle Communication Systems

We discuss here three main evolutions that will probably profoundly shape future automotive communication networks and architectures.

50.1.5.1 Real-Time Processing of Large Data Quantities

The first trend is the requirement for an increasing number of functions to *process, at a high rate, large quantities of data*, for instance, produced by video and infrared cameras. A good example of such functions is a pedestrian detection system with autobrake, which relies on image recognition and machine-learning algorithms, and has to work at the highest possible vehicle speed. In the case of the pedestrian detection system introduced in 2010 in Volvo S60, a collision with a pedestrian can be avoided at up to 35 km/h [96]. This system has to analyze the flow of images and provide, on time, a deterministic verdict about whether there is a pedestrian in front of the car, and then possibly activate braking if the driver fails to respond in time. Such systems, often belonging to the active and passive safety car domain, require powerful computational resources. This is one of the rationales for the introduction of multicore architectures in automotive embedded systems [58]. Furthermore, these functions are often implemented on a distributed architecture and usually require sensor fusion, so that the underlying communication architecture has to provide a strong and guaranteed temporal QoS. Another driver for increasing data streams is of course infotainment systems, consider, for example, a rear-seat video system in HD quality.

50.1.5.2 Reducing Electrical Energy Consumption

Energy efficiency is now a major concern when designing new embedded electronic architectures, while it was for the last 10 years a concern for actuators mainly. A reason is that, by regulation, the average CO_2 emission per passenger car sold in the EU must fall below the 95 g/km threshold from 2020 onward. Similar regulations will come into force in many countries outside EU, for instance, in China or in the United States [100]. At the same time, it is estimated in Ref. [8] that, between 2002 and 2012, the average power requirement for the automotive electronics was multiplied by 4 and reaches now 3 kW.

This leads to the introduction of two main strategies to achieve better energy efficiency: ECU degradation and partial networking. ECU degradation is the ability to switch off or reduce the execution rate of software modules, shut-down some I/O ports, or even put a complete CPU into idle mode. These low-power features are already supported in AUTOSAR. Partial networking means putting temporarily into sleep mode a group of nodes, or some entire network, when they do not need to be operational. For example, a parking assistance system that cannot be used above 20 km/h can be deactivated at higher speeds. Many control modules in the body of the vehicle (seats, mirror, door, etc.) can remain in sleep mode most of the time [8]. Then, the nodes are waken up when needed through specific messages. Audi estimates, for instance, that the potential savings of partial networking would be around 2.6 g CO_2/km [8,95].

Partial networking has been an ongoing work in the automotive industry since 2008 [47], already fully supported in AUTOSAR for CAN. A working group of carmakers and semiconductor manufacturers was set up in 2010 to develop a specification for CAN transceivers with wake-up capability [8,95]. The outcomes of this working group have been submitted for standardization as ISO 11898-6, an extension to the existing CAN standard (ISO 11898). Several components supporting partial networking are commercially available [100], and the first implementation in a car has been announced for 2013 or 2014 [47].

50.1.5.3 Security Issues

The last issue to be dealt with is the security of communication and more generally security of the embedded systems. If safety has always been a matter of concern for automotive embedded systems, security issues have been in our view largely overlooked. The need for more security is driven by the ever-increasing connectivity between the car and the external world, which includes not only telematics services and Internet access but also upcoming vehicle-to-vehicle or vehicle-to-infrastructure communication.

As demonstrated in [43], some existing security mechanisms, for instance, to control ECU reprogramming, are weak, and many current automotive systems usually do not resist to attacks, sometimes even relatively unsophisticated ones according to [43]. With a physical access to the vehicle, through a connection to the standard OBD-II port, the authors were, for instance, able to reflash ECUs, even critical ones controlling the engine and the brakes. The authors were also able to remotely compromise the security of the car, by injecting malicious code, and "ultimately monitor and control the car remotely over the Internet" [43]. Obviously, such security breaches could have catastrophic consequences, especially with ADAS systems and future intelligent transportation systems requiring cooperation between vehicles. One may even imagine a scenario where an infected vehicle would contaminate other vehicles.

The technical answer to this security challenge will probably encompass a spectrum of solutions [43], ranging from cryptography (support for embedding a cryptographic module is available in AUTOSAR since release 4.0 [7]), to message authentication mechanisms, to security breach detection systems, etc. These functionalities will require additional bandwidth, and, for instance, CAN FD (CAN with Flexible Data-Rate, see Section 50.2.1.4) would probably be better suited than normal CAN to fulfill these needs. The use of virtualization technologies [62] will also increase security because it enables to isolate functions or subsystems running on the same ECUs having different security requirements, while enabling them to share information in a controlled manner. Typically, virtualization would help to prevent infotainment applications that are connected to the external world through wireless connections, from accessing the internal communication buses.

50.2 In-Car Embedded Networks

The different performance requirements throughout a vehicle, as well as the competition among companies, have led to the design of a large number of communication networks. The aim of this section is to give a description of the most representative networks for each main domain of utilization.

50.2.1 CAN Network

To ensure at run-time the *freshness** of the exchanged data and the timely delivery of commands to actuators, it is crucial that the medium access control (MAC) protocol is able to ensure bounded response times for frames. An efficient and conceptually simple MAC scheme that possesses this capability is the granting of bus access according to the priority of the messages (the reader can refer to Refs. [14,56,65,91] for how to compute bound on response times for priority buses). To this end, each message is assigned

* The freshness property is verified if data have been produced recently enough to be safely consumed: the difference between the time when the data are used and its production time must be always smaller than a specified value. The delay experienced by a data is made up of the network latency, plus the latencies on the sending and receiving ends, which can be large especially if the application tasks are not synchronized with the transmissions on the network (see Ref. [25]).

an identifier, unique to the whole system. This serves two purposes: giving priority for transmission (the lower the numerical value, the greater the priority) and allowing message filtering upon reception. Besides CAN, two other representatives of such *priority buses* are vehicle area network (VAN) and J1850; they will be briefly presented in Section 50.2.1.3.

50.2.1.1 CAN Protocol

CAN is without a doubt the most widely used in-vehicle network. It was designed by Bosch in the mid-1980s for multiplexing communication between ECUs in vehicles and thus for decreasing the overall wire harness: length of wires and number of dedicated wires (e.g., the number of wires has been reduced by 40%, from 635 to 370, in the Peugeot 307 that embeds two CAN buses with regard to the nonmultiplexed Peugeot 306 [52]). Furthermore, it allows to share sensors among ECUs.

CAN on a twisted pair of copper wires became an ISO standard in 1994 [33,35] and is now a de facto standard in Europe for data transmission in automotive applications, due to its low cost, its robustness, and the bounded communication delays (see Ref. [38]). In today's car, CAN is used as an SAE class C network for real-time control in the powertrain and chassis domains (at 250 or 500 kbit/s), but it also serves as an SAE class B network for the electronics in the body domain, usually at a data rate of 125 kbit/s.

On CAN, data, possibly segmented in several frames, may be transmitted periodically, aperiodically, or on demand (i.e., client–server paradigm). A CAN frame is labeled by an identifier, transmitted within the frame (see Figures 50.1 and 50.2), whose numerical value determines the frame priority. There are two versions of the CAN protocol differing in the size of the identifier: CAN 2.0A (or *standard CAN*) with an 11-bit identifier and CAN 2.0B (or *extended CAN*) with a 29-bit identifier. For in-vehicle communications, only CAN 2.0A is used since it provides a sufficient number of identifiers (i.e., the number of distinct frames exchanged over one CAN network is lower than 2^{11}).

CAN uses non-return-to-zero (NRZ) bit representation with a bit stuffing of length 5. In order not to lose the bit time (i.e., the time between the emission of two successive bits of the same frame), stations need to resynchronize periodically, and this procedure requires edges on the signal. Bit stuffing is an encoding method that enables resynchronization when using NRZ bit representation where the signal level on the bus can remain constant over a longer period of time (e.g., transmission of "000000…"). Edges are inserted

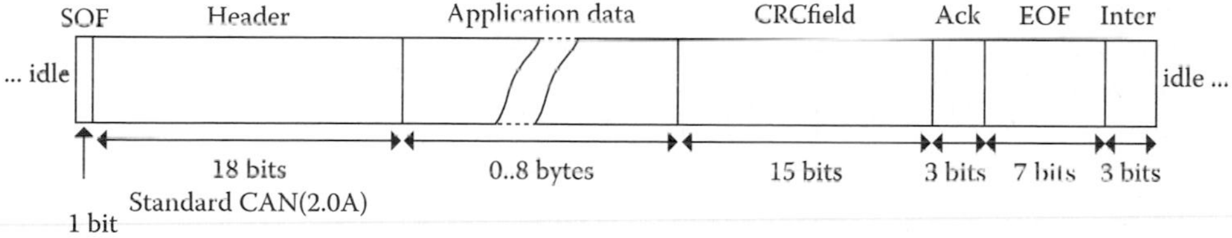

FIGURE 50.1 Format of the CAN 2.0A data frame. SOF, start of frame; EOF, end of frame; Ack, acknowledgment; Inter, intermission.

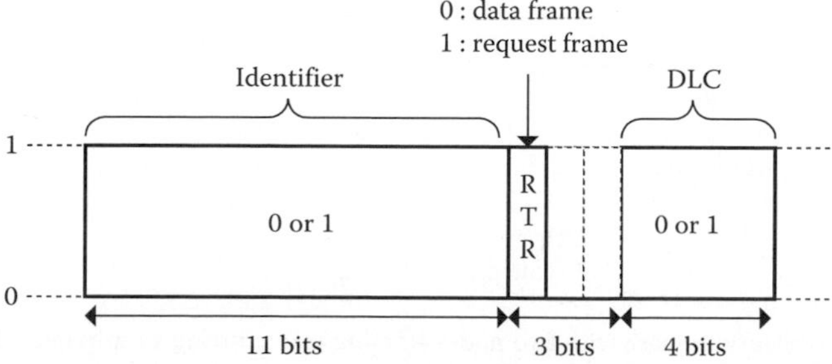

FIGURE 50.2 Format of the header field of the CAN 2.0A data frame.

into the outgoing bit stream in such a way to avoid the transmission of more than a maximum number of consecutive equal-level bits (5 for CAN). The receiver will apply the inverse procedure and destuff the frame. CAN requires the physical layer to implement the logical "and" operator: if at least one node is transmitting the "0"-bit level on the bus, then the bus is in that state regardless if other nodes have transmitted the "1"-bit level. For this reason, "0" is termed the dominant bit value, while "1" is the recessive bit value.

The standard CAN data frame (CAN 2.0A; see Figure 50.1) can contain up to 8 bytes of data for an overall size of, at most, 135 bits, including all the protocol overheads such as the stuff bits. The sections of the frames are as follows:

- The header field (see Figure 50.2), which contains the identifier of the frame, the remote transmission request (RTR) bit that distinguishes between data frame (RTR set to 0) and data request frame (RTR set to 1) and the data length code (DLC) used to inform of the number of bytes of the data field.
- The data field having a maximum length of 8 bytes.
- The 15-bit cyclic redundancy check (CRC) field, which ensures the integrity of the data transmitted.
- The acknowledgment field (ACK). On CAN, the acknowledgment scheme solely enables the sender to know that at least one station, but not necessarily the intended recipient, has received the frame correctly.
- The end-of-frame (EOF) field and the intermission frame space, which is the minimum number of bits separating consecutive messages.

Any CAN node may start a transmission when the bus is idle. Possible conflicts are resolved by a priority-based arbitration process, which is said nondestructive in the sense that, in case of simultaneous transmissions, the highest-priority frame will be sent despite the contention with lower-priority frames. The arbitration is determined by the arbitration fields (identifier plus RTR bit) of the contending nodes. An example illustrating CAN arbitration is shown on Figure 50.3. If one node transmits a recessive bit on the bus while another transmits a dominant bit, the resulting bus level is dominant due to the "and" operator

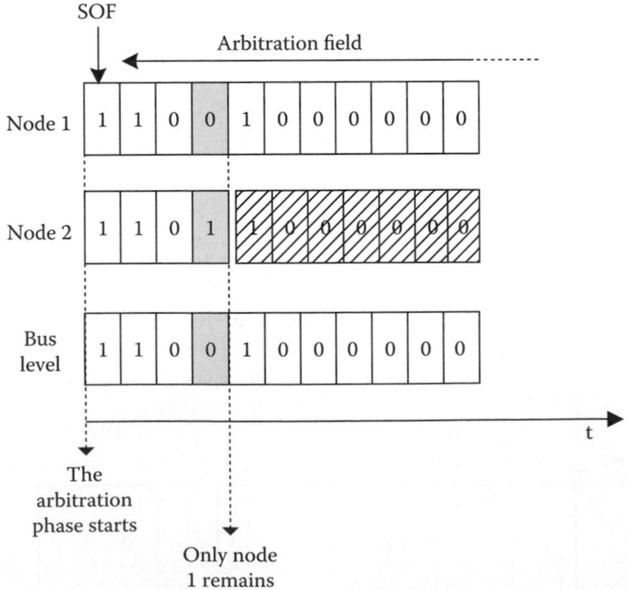

FIGURE 50.3 CAN arbitration phase with two nodes starting transmitting simultaneously. Node 2 detects that a frame with a higher priority than its own is being transmitted when it monitors a level 0 (i.e., dominant level) on the bus while it has sent a bit with a level 1 (i.e., recessive level). Afterward, node 2 immediately stops transmitting.

realized by the physical layer. Therefore, the node transmitting a recessive bit will observe a dominant bit on the bus and then will immediately stop transmitting. Since the identifier is transmitted *most significant bit first*, the node with the numerically lowest identifier field will gain access to the bus. A node that has lost the arbitration will wait until the bus becomes free again before trying to retransmit its frame.

CAN arbitration procedure relies on the fact that a sending node monitors the bus while transmitting. The signal must be able to propagate to the most remote node and return back before the bit value is decided. This requires the bit time to be at least twice as long as the propagation delay, which limits the data rate: for instance, 1 Mbit/s is feasible on a 40 m bus at maximum, while 250 kbit/s can be achieved over 250 m. To alleviate the data rate limit and extend the life-span of CAN further, car manufacturers are starting to optimize the bandwidth usage by implementing *traffic shaping* strategies that are very beneficial in terms of response times (see, for instance, Ref. [24]).

CAN has several mechanisms for error detection. For instance, it is checked that the CRC transmitted in the frame is identical to the CRC computed at the receiver end, that the structure of the frame is valid, and that no bit-stuffing error occurred. Each station that detects an error sends an *error flag*, which is a particular type of frame composed of six consecutive dominant bits that allows all the stations on the bus to be aware of the transmission error. The corrupted frame automatically reenters into the next arbitration phase, which might lead it to miss its deadline due to the additional delay. The error recovery time, defined as the time from detecting an error until the possible start of a new frame, is 17–31-bit times. CAN possesses some fault-confinement mechanisms aimed at identifying permanent failures due to hardware dysfunctioning at the level of the microcontroller, communication controller, or physical layer. The scheme is based on error counters that are increased and decreased according to particular events (e.g., successful reception of a frame, reception of a corrupted frame). The relevance of the algorithms involved is questionable (see Ref. [22]), but the main limitation is that a node has to diagnose itself, which can lead to the nondetection of some critical errors. For instance, a faulty oscillator can cause a node to transmit continuously a dominant bit, which is one manifestation of the *babbling-idiot* fault (see Ref. [72]). Furthermore, other faults such as the partitioning of the network into several subnetworks may prevent all nodes from communicating due to bad signal reflection at the extremities. Without additional fault-tolerance facilities, CAN is not well suited for safety-critical applications. For instance, a single node can perturb the functioning of the whole network by sending messages outside their specification (i.e., length and period of the frames). Many mechanisms were proposed for increasing the dependability of CAN-based networks (see Ref. [72] for an excellent survey), but if each proposal solves a particular problem, they have not necessarily been conceived to be combined. Furthermore, the fault hypotheses used in the design of theses mechanisms are not necessarily the same, and the interactions between them remain to be studied in a formal way.

The CAN standard only defines the physical layer and data link layer (DLL). Several higher-level protocols have been proposed, for instance, for standardizing start-up procedures, implementing data segmentation, or sending periodic messages (see OSEK/VDX and AUTOSAR in Section 50.3). Other higher-level protocols standardize the content of messages in order to ease the interoperability between ECUs. This is the case for J1939, which is widely used, for instance, in Scania's trucks and buses [97].

50.2.1.2 CAN's Use in Today's Automobiles

There is now more than 20 years of experience in automotive CAN applications, and CAN has certainly proven very successful as a robust, cost-effective, and all-around network technology. But the use of CAN in vehicles is evolving, in particular, because of more complex and heterogeneous architectures with FlexRay or Ethernet networks and because of recent needs like hybrid, electric propulsion, or driver assistance that involve more stringent real-time constraints. Besides, there are other new requirements on CAN: more fine-grained ECU mode management for energy savings [8,100], multi-ECU splitted functions, and huge software downloads. In parallel, safety issues request more and more mechanisms to protect against potential failures and provide end-to-end (E2E) integrity (see Section 50.3.3 and Ref. [2]).

50.2.1.2.1 Increased Bandwidth Requirements

The robustness and performance of the CAN technology, as well as the new possibilities brought by distributed software functions, have led engineers to use more and more bandwidth in order to improve existing EE functions and introduce new ones. This trend has never ceased and, along with topology and functional domain constraints, has led to the use of several CAN clusters within a car, sometimes more than four or five [39,63]. Also, the data rates of the CAN buses are now higher (e.g., 250 kbit/s for a body network when it used to be 125 kbit/s), and the bus load level has increased (e.g., >50%).

50.2.1.2.2 More Complex Architectures

At the beginning of CAN, just a few ECUs were connected, while today, there are thousands of signals exchanged by several tens of ECUs, with some signals having timing constraints below 5 ms. Besides, the architectures are becoming complex because of gateways between the CAN buses or between a CAN bus and another networking technology (typically FlexRay). The use of several CAN clusters raises also technical issues regarding, for instance, fault handling, diagnosis timing response, and wake-up and sleep synchronization. And, whatever is done, there is an overlap between the data sent on the buses connected to a gateway, which induces a significant waste of bandwidth. To face the EE architecture complexity and be able to push the limits of CAN, carmakers have established rigorous development processes. Besides, there are now several well-suited COTS tool sets available on the market to help them with the optimization and verification using simulation, schedulability analysis, and trace analysis (see Ref. [63] for how they are used).

50.2.1.2.3 Optimizing CAN Networks

When CAN was introduced, the bus load levels were limited, typically much less than 30% (see Ref. [64] for a typical set of messages of the years 1995–2000). Optimizing CAN networks, which includes reaching higher load levels, has now become an industrial requirement for several reasons:

- It helps to master the complexity of the architectures.
- It reduces the hardware costs, weight, space, consumption, etc.
- It facilitates an incremental design process.
- It may avoid the industrial risk, the costs, and the time to master new technologies such as FlexRay.
- It leads to better communication performances and helps to match the bandwidth needs. Sometimes, a 60%-loaded CAN network can be more efficient than two 40% CAN networks interconnected by a gateway causing delays and high jitters.

The first obvious way to optimize a CAN network is to keep the amount of data transmitted to a minimum, specifically limit the transmission frequency of the frames. This requires a rigorous identification and traceability of the temporal constraints. Given a set of signals or frames, and their associated temporal constraints (freshness, jitters, etc.), they are in addition a few configuration strategies than can be used:

1. Desynchronize the stream of frames by using offsets. The reader may refer to Ref. [26] for comprehensive experiments on the large gains achieved using offsets.
2. Reassign the priorities of the frames, so that the priority order better reflects the timing constraints.
3. Reconsider the frame packing, that is, the allocation of the signals to the frames and choice of the frame periods, so as to minimize the bandwidth usage while meeting timing constraints (see Ref. [81]).
4. Optimize the ECU communication stacks so as to remove all implementation choices that cause a departure from the ideal CAN behavior (see Ref. [63]).

However, because there is less margin for error, using complex CAN-based architectures at high-load levels involves more detailed supplier specifications on the one hand and, on the other hand, spends more time and effort in the integration/validation phase.

50.2.1.3 Other Priority Buses: VAN and J1850

The J1850 [87] is an SAE class B priority bus that was adopted in the United States for communications with nonstringent real-time requirements, such as the control of body electronics or diagnostics. Two variants of the J1850 are defined: a 10.4 kbit/s single-wire version and 41.6 kbit/s two-wire version. For quite a long time, the trend in new designs seems to be the replacement of J1850 by CAN or a low-cost network such as LIN (see Section 50.2.3.1).

In the 1990s, another competing technology was the French VAN (see Ref. [34]) which is very similar to CAN (e.g., frame format, data rate) but possesses some additional or different features that are advantageous from a technical point of view (e.g., no need for bit-stuffing, in-frame response: a node being asked for data answers in the same frame that contained the request). VAN was used for years in production cars by the French carmaker PSA Peugeot Citroen in the body domain (e.g., for the 206 model), but as it was not adopted by the market, it was abandoned in favor of CAN.

50.2.1.4 CAN FD: A High-Speed CAN Network

CAN FD is a new protocol that was presented for the first time by Robert Bosch GmbH at the International CAN Conference in 2012 [30] that combines CAN's core features with a higher data rate and larger data payloads. The data field length can be up to 64 bytes long: from 0 to 8 bytes, then above 8 bytes, the possible values are 12, 16, 20, 24, 32, 48, and 64 bytes. Technically, two bit rates are used successively in the transmission of a CAN FD frame: one lower for the arbitration phase, as required by the CAN bit-wise arbitration (see Section 50.2.2.2), and a higher one used immediately after the arbitration for the transfer of the data payload and related fields (e.g., DLC, CRC). The bit rate for this data part of the frame transmission can be freely chosen (e.g., 10 Mbit/s is mentioned in Ref. [30]), but in practice, it will depend much on the topology of the network and the efficiency of the transceivers. It is not clear at the time of writing what is the kind of data rate and thus the speedup with regard to standard CAN that can be achieved in typical automotive applications, and this will largely determine the acceptance of CAN FD. The target mentioned in Ref. [30] is an average data rate of 2.5 Mbit/s with existing CAN transceivers.

Four main use cases are identified for CAN FD in Ref. [49]: faster software download (end-of-production line or maintenance), avoiding segmentation of long messages (and possibly securing normal CAN messages with message authentication information), offering higher bandwidth to car domains requiring it (e.g., powertrain [49]), and enabling faster communication on long CAN buses (e.g., in trucks and buses). An important advantage of CAN FD is that there is an easy migration path from CAN systems to CAN FD systems since existing CAN application software can basically remain unchanged, the changes taking place in the communication layers and their configuration. Following FPGA's implementations, microcontrollers with CAN FD are already available (e.g., [20]), and the protocol has been submitted as ISO 11898-7 for international standardization.

50.2.2 Time-Triggered Networks

Among communication networks, as discussed previously, one distinguishes TT networks where activities are driven by the progress of time and event triggered once where activities are driven by the occurrence of events. Both types of communication have advantages, but one considers that, in general, dependability is much easier to ensure using a TT bus (refer, for instance, to Ref. [80] for a discussion on this topic). This explains that mainly TT communication systems are being considered for use in the most critical applications such as x-by-wire systems. In this category, multiaccess protocols based on TDMA are particularly well suited; they provide deterministic access to the medium (the order of the transmissions is defined statically at the design time) and thus bounded response times. Moreover, their regular message transmissions can be used as *heartbeats* for detecting station failures. The three TDMA-based networks that could serve as gateways or for supporting safety–critical applications are TTP/C (see Ref. [93]), FlexRay (see Section 50.2.2.1), and TTEthernet (see Section 50.2.5.4). FlexRay, which was backed by the world's automotive industry, is becoming a standard technology in the industry, though without gaining broad acceptance, and is already in use in production cars since 2006 (see Refs. [28,83]). TTCAN (see Section 50.2.2.2) was, before the advent of FlexRay, considered for use and is briefly discussed in this section. In the following, we choose not to discuss further TTP/C, which, to the best of our knowledge, is no more considered for vehicles but is used in aircraft. However, the important experience and know-how gained over the years with TTP/C, in particular regarding its fault-tolerance features (see Ref. [23]) and their formal validation (see Ref. [71]), is certainly beneficial to FlexRay and any Ethernet-based protocols such as TTEthernet (see Section 50.2.5.4) or even Audio/Video Bridging (AVB; see Section 50.2.5.3).

50.2.2.1 FlexRay Protocol

A consortium of major companies from the automotive field developed the FlexRay protocol. The core members of the consortium, now disbanded, were BMW, Bosch, Daimler, General Motors, NXP Semiconductors, Freescale Semiconductor, and Volkswagen. The first publicly available specification of the FlexRay Protocol was released in 2004; the current version of the specification is now available as a set of ISO standards issued in 2013 (ISO 17458 Part 1–5), based on the 3.0.1 release from the FlexRay consortium [11]. FlexRay did not gain very wide acceptance yet, since most car manufacturers still rely on CAN-based architectures, but it has been used since 2006 in some production cars by BMW and Audi. For instance, the latest series 5 from BMW implements up to 17 FlexRay nodes [28].

The FlexRay network is very flexible with regard to topology and transmission support redundancy. It can be configured as a bus, a star, or multistar. It is not mandatory that each station possesses replicated channels nor a bus guardian. At the MAC level, FlexRay defines a communication cycle as the concatenation of a TT (or static) window and an event-triggered (or dynamic) window. In each communication window, size of which is set statically at design time, two distinct protocols are applied. The communication cycles are executed periodically. The TT window uses a TDMA MAC protocol; the main difference with TTP/C is that a station in FlexRay might possess several slots in the TT window, but the size of all the slots is identical (see Figure 50.4). In the event-triggered part of the communication cycle, the protocol is flexible TDMA (FTDMA): the time is divided into so-called minislots, each station possesses* a given number of minislots (not necessarily consecutive), and it can start the transmission of a frame inside each of its own minislots. A minislot remains idle if the station has nothing to transmit, which actually induces a loss of bandwidth (see Ref. [10] for a discussion on that topic). An example of a dynamic window is shown in Figure 50.5: on channel B, frames have been transmitted in minislots n and $n + 2$, while minislot $n + 1$ has not been used. It is noteworthy that frame $n + 4$ is not received simultaneously on channels A and B since, in the dynamic window, transmissions are independent in both channels.

The FlexRay MAC protocol is more flexible than the TTP/C MAC since in the static window nodes are assigned as many slots as necessary (up to 2047 overall) and since the frames are only transmitted if necessary in the dynamic part of the communication cycle. In a similar way as with TTP/C, the structure of the communication cycle is statically stored in the nodes; however, unlike TTP/C, mode changes with a different communication schedule for each mode are not possible.

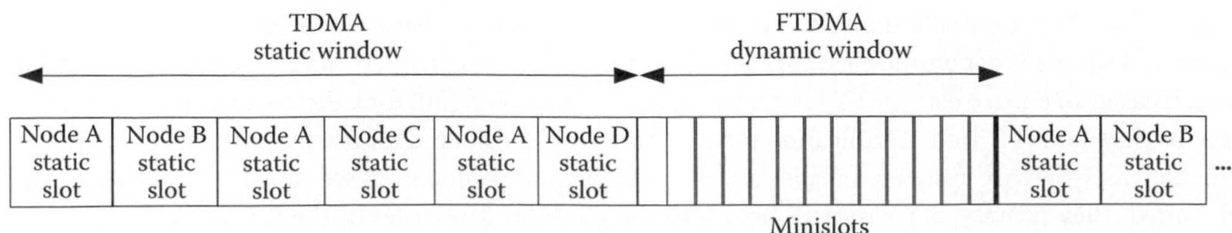

FIGURE 50.4 Example of a FlexRay communication cycle with four nodes: A, B, C, and D.

* Different nodes can send frames in the same slot but in different cycles; this is called *slot multiplexing*, and it is only possible only in the dynamic segment.

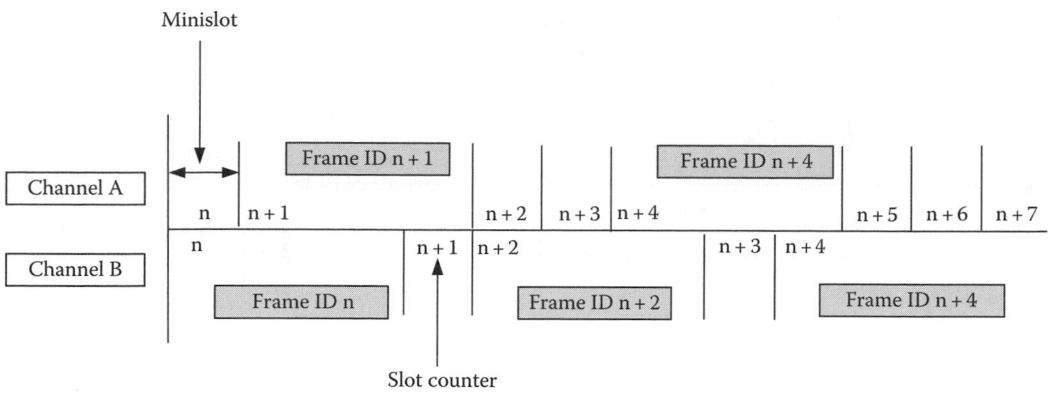

FIGURE 50.5 Example of message scheduling in the dynamic segment of the FlexRay communication cycle.

The FlexRay frame consists of three parts: the header, the payload segment containing up to 254 bytes of data, and the CRC of 24 bits. The header of 5 bytes includes the identifier of the frame and the length of the data payload. The use of identifiers allows to move a software component, which sends a frame *X*, from one ECU to another ECU without changing anything in the nodes that consume frame *X*. It has to be noted that this is no more possible when signals produced by distinct components are packed into the same frame for the purpose of saving bandwidth (i.e., which is referred to as frame packing or protocol data unit [PDU] multiplexing—see Ref. [81] for this problem addressed on CAN).

From the dependability point of view, the FlexRay standard solely specifies a bus guardian (optional), passive and active star coupler (optional), and the clock synchronization algorithms. Other features, such as mode management facilities or a membership service, will have to be implemented in software or hardware layers on top of FlexRay (see, for instance, Ref. [5] for a membership service protocol that could be used along with FlexRay). This will allow to conceive and implement exactly the services that are needed with the drawback that correct and efficient implementations might be more difficult to achieve in a layer above the communication controller.

In the FlexRay specification, it is argued that the protocol provides scalable dependability, that is, the *ability to operate in configurations that provide various degrees of fault tolerance*. Indeed, the protocol allows for mixing links with single- and dual-transmission supports on the same network, subnetworks of nodes without bus guardians or with different fault-tolerance capability with regard to clock synchronization, etc. In the automotive context where critical and noncritical functions coexist and interoperate, this flexibility can prove to be efficient in terms of cost and reuse of existing components. The reader interested in more information about FlexRay can refer to [11,84] and to [25,75,101] for how to configure the communication cycle.

50.2.2.2 TTCAN Protocol

TTCAN (see Ref. [37]) is a communication protocol developed by Robert Bosch GmbH on top of the CAN physical and DLLs. TTCAN uses the CAN standard but, in addition, requires that the controllers have the possibility to disable automatic retransmission of frames upon transmission errors and to provide the upper layers with the point in time at which the first bit of a frame was sent or received. The bus topology of the network, the characteristics of the transmission support, the frame format, as well as the maximum data rate—1 Mbits/s—are imposed by the CAN protocol. Channel redundancy is possible (see Ref. [57] for a proposal) but not standardized, and no bus guardian is implemented in the node. The key idea was to propose, as with FlexRay, a flexible TT/event-triggered protocol. As illustrated in Figure 50.6, TTCAN defines a basic cycle (the equivalent of the FlexRay communication cycle) as the concatenation of one or several TT (or *exclusive*) windows and one event-triggered (or *arbitrating*) window. Exclusive windows are devoted to TT transmissions (i.e., periodic messages), while the arbitrating window is ruled by the standard CAN protocol: transmissions are dynamic and bus access is granted

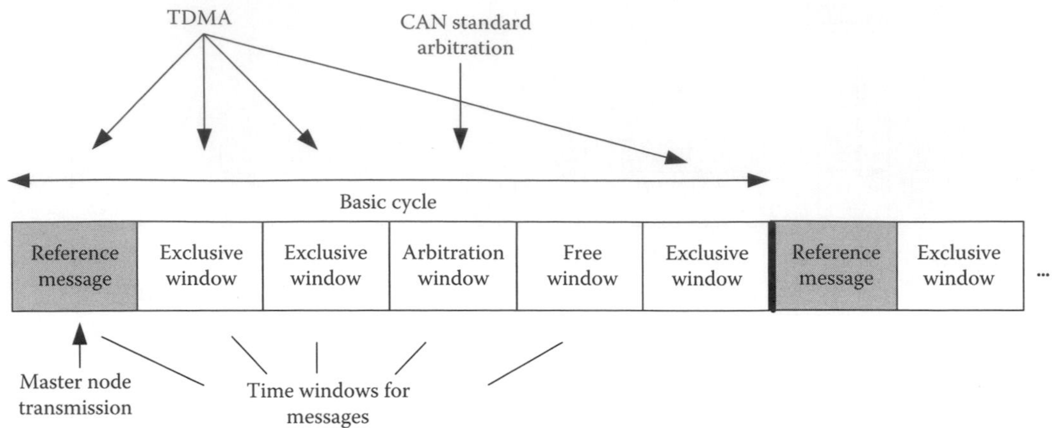

FIGURE 50.6 Example of a TTCAN basic cycle.

according to the priority of the frames. Several basic cycles that differ by their organization in exclusive and arbitrating windows and by the messages sent inside exclusive windows can be defined. The list of successive basic cycles is called the system matrix, which is executed in loops. Interestingly, the protocol enables the master node (i.e., the node that initiates the basic cycle through the transmission of the *reference message*) to stop functioning in TTCAN mode and to resume in standard CAN mode. Later, the master node can switch back to TTCAN mode by sending a reference message.

TTCAN is built on a well-mastered and low-cost technology, CAN, but, as defined by the standard, does not provide important dependability services such as the bus guardian, membership service, and reliable acknowledgment (see Ref. [72]). It is, of course, possible to implement some of these mechanisms at the application or middleware (MW) level but with reduced efficiency. About 10 years ago, it was thought that carmakers could be interested in using TTCAN during a transition period until FlexRay technology is fully mature, but this was not really the case, and TTCAN has not been used in production cars to the best of our knowledge.

50.2.3 Low-Cost Automotive Networks

Several fieldbus networks have been developed to fulfill the need for low-speed/low-cost communication inside mechatronic-based subsystems generally made of an ECU and its set of sensors and actuators. Two representatives of such networks are LIN and TTP/A. The low-cost objective is achieved not only because of the simplicity of the communication controllers but also because the requirements set on the microcontrollers driving the communication are reduced (i.e., low computational power, small amount of memory, low-cost oscillator). Typical applications involving these networks include controlling doors (e.g., door locks, opening/closing windows) or controlling seats (e.g., seat position motors, occupancy control). Besides cost considerations, a hierarchical communication architecture, including a backbone such as CAN and several subnetworks such as LIN, enables reducing the total traffic load on the backbone.

Both LIN and TTP/A are master/slave networks where a single master node, the only node that has to possess a precise and stable time base, coordinates the communication on the bus: a slave is only allowed to send a message when it is polled. More precisely, the dialogue begins with the transmission by the master of a *command frame* that contains the identifier of the message whose transmission is requested. The command frame is then followed by a *data frame* that contains the requested message sent by one of the slaves or by the master itself (i.e., the message can be produced by the master).

This paragraph also presents the SENT and PSI5 networks, which are other more recent low-cost alternatives to CAN.

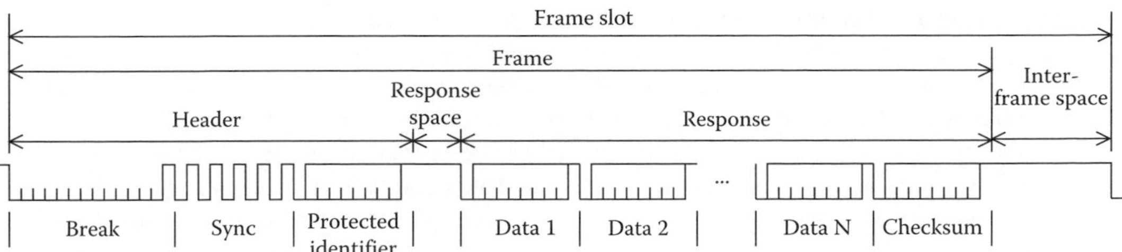

FIGURE 50.7 Format of the LIN frame. A frame is transmitted during its *frame slot*, which corresponds to an entry of the schedule table.

50.2.3.1 LIN Network

LIN (see Refs. [48,77]) is a low-cost serial communication system used as SAE class A network, where the needs in terms of communication do not require the implementation of higher-bandwidth multiplexing networks such as CAN. LIN is developed by a set of major companies from the automotive industry (e.g., Daimler, Volkswagen, BMW, and Volvo) and is widely used in production cars.

The LIN specification package (LIN version 2.2A [48]) includes not only the specification of the transmission protocol (physical layer and DLLs) for master–slave communications but also the specification of a diagnostic protocol on top of the DLL. A language for describing the capability of a node (e.g., bit rates that can be used, characteristics of the frames published and subscribed by the node) and for describing the whole network is provided (e.g., nodes on the network, table of the transmissions' schedule). This description language facilitates the automatic generation of the network configuration by software tools.

A LIN cluster consists of one *master* node and several *slave* nodes connected to a common bus. For achieving a low-cost implementation, the physical layer is defined as a single wire with a data rate limited to 20 kbit/s due to EMI limitations. The master node decides when and which frame shall be transmitted according to the schedule table. The schedule table is a key element in LIN; it contains the list of frames that are to be sent and their associated frame slots, thus ensuring determinism in the transmission order. At the moment a frame is scheduled for transmission, the master sends a header (a kind of transmission request or command frame) inviting a slave node to send its data in response. Any node interested can read a data frame transmitted on the bus. As in CAN, each message has to be identified: 64 distinct message identifiers are available. Figure 50.7 depicts the LIN frame format and the time period, termed a *frame slot*, during which a frame is transmitted.

The header of the frame that contains an identifier is broadcast by the master node, and the slave node that possesses this identifier inserts the data in the response field. The *break* symbol is used to signal the beginning of a frame. It contains at least 13 dominant bits (logical value 0) followed by one recessive bit (logical value 1) as a break delimiter. The rest of the frame is made of byte fields delimited by one start bit (value 0) and one stop bit (value 1), thus resulting in a 10-bit stream per byte. The *sync* byte has a fixed value (which corresponds to a bit stream of alternatively 0 and 1); it allows slave nodes to detect the beginning of a new frame and be synchronized at the start of the identifier field. The so-called protected identifier is composed of two subfields: the first 6 bits is used to encode the identifier, and the last 2 bits, the identifier parity. The data field can contain up to 8 bytes of data. A checksum is calculated over the protected identifier and the data field. Parity bits and checksum enable the receiver of a frame to detect bits that have been inverted during transmission.

LIN defines five different frame types: unconditional, event triggered, sporadic, diagnostic, and user defined. Frames of the latter type are assigned a specific identifier value and are intended to be used in an application-specific way that is not described in the specification. The first three types of frames are used to convey signals. Unconditional frames are the usual type of frames used in the master–slave dialogue and are always sent in their frame slots. Sporadic frames are frames sent by the master, only if at least one signal composing the frame has been updated. Usually, multiple sporadic frames are assigned to the same frame slot, and the higher-priority frame that has an updated signal is transmitted. An event-triggered frame is used by the master willing to obtain a list of several signals from different nodes. A slave will only answer the master if the signals it produces have been updated, thus resulting in bandwidth savings if updates do not take place very often. If more than one slave answers, a collision will occur. The master resolves the collision by requesting all signals in the list one by one. A typical example of the use of the event-triggered transfer given in Ref. [48] is the door knob monitoring in a central locking system. As it is rare that multiple passengers simultaneously press a knob, instead of polling each of the four doors, a single event-triggered frame can be used. Of course, in the rare event when more than one slave responds, a collision will occur. The master will then resolve the collision by sending one by one the individual identifiers of the list during the successive frame slots reserved for polling the list. Finally, diagnostic frames have a fixed size of 8 bytes, fixed value identifiers for both the master's request and the slave answers, and always contain diagnostic or configuration data whose interpretation is defined in the specification.

It is also worth noting that LIN offers services to send nodes into a sleep mode (through a special diagnostic frame termed *go-to-sleep command*) and to wake them up, which is convenient since optimizing energy consumption, especially when the engine is not running, is a real matter of concern in the automotive context.

50.2.3.2 TTP/A Network

As TTP/C, TTP/A [27] was initially invented at the Vienna University of Technology. TTP/A pursues the same aims and shares the main design principles as LIN, and it offers, at the communication controller level, some similar functionalities, in particular, in the areas of plug-and-play capabilities and online diagnostic services. TTP/A implements the classic master–slave dialogue, termed *master–slave round*, where the slave answers the master's request with a data frame having a fixed length data payload of 4 bytes. The *multipartner* rounds enable several slaves to send up to an overall amount of 62 bytes of data after a single command frame. A *broadcast round* is a special master–slave round in which the slaves do not send data; it is, for instance, used to implement sleep/wake-up services. The data rate on a single-wire transmission support is, as for LIN, equal to 20 kbit/s, but other transmission supports enabling higher data rates are possible. To our best knowledge, TTP/A is not currently in use in production cars.

50.2.3.3 PSI5 and SENT Networks

PSI5 (see http://www.psi5.org) is another low-cost bus that was originally developed for communication between ECUs and air-bags but is now considered for use with any kinds of sensors. It provides 125 kbit/s communication and needs only two wires for both data communication and sensor power supply. When used in bus mode with several sensors on the bus, as opposed to the point-to-point mode, the communication is organized in a TT manner according to a TDMA-based MAC protocol, each sensor sending in a predefined time slot.

Another competing technology is SENT, normalized as SAE J2716, which provides one-way communication, but it can be complemented by the SPC protocol (short PWM code, see Ref. [6]), which adds functionalities such as bidirectional communication.

50.2.4 Multimedia and Infotainment Networks

Several protocols have been adapted or specifically conceived for transmitting the large amount of data needed by multimedia and infotainment applications increasingly available in cars. Two prominent protocols in this category are MOST and 1394 Automotive (formerly known as IDB-1394) that are discussed in the following. Another technology, not specifically developed for the automotive domain though that is currently broadly used for infotainment and automotive cameras, is low-voltage differential signaling (LVDS), which enables communication at 655 Mbit/s and above [54] over twisted pair copper cables.

In the near future, it is very likely that Ethernet-based networks, probably compliant with the IEEE 802.1 AVB QoS standard (see Section 50.2.5.3), will be used to transport the high data volumes needed by multimedia and infotainment applications [28,89].

50.2.4.1 MOST Network

MOST (see Ref. [59]) is a multimedia network, development of which was initiated in 1998 by the MOST Cooperation (a consortium of carmakers and component suppliers). MOST provides point-to-point audio and video data transfer with different possible data rates. MOST supports end-user applications like radios, GPS navigation, video displays, and entertainment systems. MOST's physical layer has been so far a polymer optical fiber (POF) transmission support, which provides a better resilience to EMI and higher transmission rates than classical cooper wires, but a new coaxial standard has been introduced recently [41]. It was estimated in 2008 [53] that around 50 model series, for example, from BMW and Daimler, implement a MOST network (e.g., MOST25 at 25 Mbit/s at Daimler, see Ref. [46]). MOST has now become a de facto standard for transporting audio and video streams within vehicles (see Refs. [60,61]).

The third revision of MOST [59] has introduced the support of a channel that can transport standard Ethernet frames and is thus well suited to transmit IP traffic. These features, along with the higher bandwidth provided by MOST150 introduced in 2007, are already taken advantage of in some new vehicle projects [46] to provide access to Internet services and web browsing and independent access to audio and video sources from all seats with a single bus. In Ref. [41], the MOST consortium announces its intention to develop a next generation of the protocol, with a target bandwidth of 5 Gbit/s, that would be suited as well for ADAS requiring the transmission of uncompressed video streams, competing thus with automotive Ethernet solutions.

50.2.4.2 1394 Automotive Network

The 1394 Automotive is an automotive version of IEEE 1394 (FireWire) for in-vehicle multimedia and telematics applications that is developed by the 1394 Trade Association (see http://www.1394ta.org). The system architecture of 1394 Automotive permits existing IEEE 1394 consumer electronics devices to interoperate with embedded automotive grade devices. The 1394 Automotive is advertised to support a data rate of 800 Mbps [82,90] over several physical layers including twisted pair or POF. Thanks to its large bandwidth and the interoperability with existing IEEE 1394 consumer electronic devices, 1394 Automotive was at some time considered a serious competitor for MOST technology, but despite a few early implementations at Renault and Nissan, as far as we know, the protocol did not reach wide acceptance in the automotive market.

50.2.5 Automotive Ethernet

At the time of writing, the question is not if, but when switched Ethernet will become a standard technology in cars [28,94,99]. For instance, Continental expects to start series production of Ethernet-capable control units as soon as 2015 [12]. The introduction of Ethernet will, however, be gradual, and Ethernet is probably going to fulfill different use cases over time. This paragraph discusses these use cases and the main Ethernet technologies that are considered for use.

50.2.5.1 Motivation and Use Cases for Ethernet

The first motivation for Ethernet is that it is a low-cost and mature technology that offers much more bandwidth that what is available today, which is of interest for infotainment and active safety especially [29]. From the user perspective, as well as from an economic point of view, the reuse of nonautomotive-specific networking technologies such as Ethernet can be beneficial. However, automotive-specific requirements must be taken into account: for example, the need for partial networking and power over Ethernet to reduce the wiring, provision of specific QoS, and robustness to severe environmental conditions (e.g., EMI, heat).

In Ref. [29], the authors describe a plausible roadmap for the use of Ethernet. First-generation Ethernet network, based on 100BASE-TX physical layer, will be for diagnostics (using ISO 13400 standard) and code upload (as already done since 2008, see Ref. [39]). The second-generation Ethernet network, from 2015 onward, will support infotainment and camera-based ADAS using Ethernet AVB (see Section 50.2.5.3) and BroadR-Reach physical layer (see Section 50.2.5.2). At the third stage, from 2020 onward, gigabit Ethernet should become the backbone interconnecting most other networks and replacing thus today's gateways (see Section 50.4). This communication architecture will benefit from the ability of Ethernet switches to handle ports having different speeds. The backbone will have to support various kinds of traffic with different QoS requirements and thus require QoS policies that may be offered, according to the authors of Ref. [29], by a second generation of AVB protocols. A new Gb/s physical layer derived from today's 100 Mbit/s BroadR-Reach will be needed, and TT transmissions would be available.

50.2.5.2 BroadR-Reach Physical Layer

BroadR-Reach is a 100 Mbit/s physical layer over low-cost unshielded copper wires that is being developed by the OPEN Alliance SIG consortium (see http://www.opensig.org/). This consortium, which already counts 140 member companies with many of the carmakers along with Bosch and Continental, aims at establishing BroadR-Reach as the standard for the automotive Ethernet physical layer, with Gb/s speed in its roadmap [32]. For that purpose, BroadR-Reach was designed to meet the automotive requirements in terms of cost, power-saving modes, robustness to environmental conditions, etc. One clear advantage of BroadR-Reach over standard 100BASE-TX Fast Ethernet is that it only requires a single pair of copper cables. At the time of writing, the BroadR-Reach specification is close to being final [99].

50.2.5.3 Ethernet AVB

IEEE 802.1 AVB is a set of standards aimed at providing low-latency streaming services on 802.3 Ethernet. The development and adoption of AVB is promoted by the AVnu Alliance consortium (see http://www.avnu.org/), which counts several car manufacturers and automotive suppliers among its members.

IEEE 802.1AS is a synchronization protocol that builds on IEEE 1588 Precision Time Protocol to enable sub-microsecond time synchronization between nodes. IEEE 802.1Qav specifies a leaky-bucket credit-based traffic shaper (CBS) on the output ports of the switches. This shaping mechanism is aimed at enforcing some fairness among traffic streams. 802.1Qat is an E2E bandwidth reservation protocol. Finally, 802.1BA specifies default parameters and profiles that manufacturers of AVB equipment can use. AVB defines three classes of traffic: stream reservation (SR) class A with a guaranteed latency of 2 ms maximum, SR class B with a guaranteed 50 ms maximum latency, and best effort for the rest of the traffic. At most, 75% of the bandwidth can be reserved for both SR class A and SR class B streams.

Experiments in Ref. [89] suggest that AVB would be an excellent option for in-car multimedia streams, with the advantage that it enables to dynamically register data flows and can be used with higher-level protocols for handling audio/video streams such as IEEE 1733. To the best of our knowledge, there is no clear cut answer yet about the suitability of the current AVB standard for the transmission of critical control data. In October 2012, AVnu Alliance has announced the formation of the AVB Gen2 Council with the aims to improve AVB functionality in areas such as time-sensitive transmission and fault tolerance. This new specification is currently in development within the IEEE (AVB Task Group).

50.2.5.4 TTEthernet

TTEthernet [92] is a TT switched Ethernet protocol, promoted by the TTA Group (see http://www.ttagroup.org/), that was standardized as SAE AS6802 in 2011. TTEthernet natively supports mixed-criticality temporal requirements as it defines three types of traffic streams: TT (with off-line configurated transmission schedule), rate constrained (i.e., the maximum workload is bounded), and best effort. Upper bounds on the jitters and latencies can be derived for both the TT and rate-constrained traffic.

The temporal QoS in terms of jitters is better for the TT traffic, but sending and receiving must be done at predefined points in time. Conceptually, the rate-constrained traffic model is similar to virtual links in AFDX and SR classes A and B in AVB.

TTEthernet is designed to be operational in cross-domain critical applications, up to highest criticality level such as DO178 DAL-A in aeronautics, and thus possesses a wide range of fault-tolerant mechanisms. Besides, the TT traffic offers a temporal QoS that cannot be matched by the current version of AVB. These features could be needed for some future automotive applications. Another idea currently investigated [89,99] is the integration of AVB and TTEthernet, which aims to come up with an efficient solution for all common automotive Ethernet use cases.

50.3 Automotive Middleware

50.3.1 Objectives of an Embedded Middleware

The design of automotive electronic systems has to take into account several constraints. First, nowadays, the performance, quality, and safety of a vehicle depend on functions that are mainly implemented in software (e.g., the engine control, body electronics, ADAS) and moreover depend on a tight cooperation between these functions. For example, the control of the engine is done according to requests from the driver (speeding up, slowing down as transmitted by the throttle position sensor or the brake pedal) and requirements from other embedded functions such as climate control, ESP, or brake assist. Second, in-vehicle embedded systems are produced through a complex cooperative multipartner development process shared between OEMs and suppliers. Therefore, in order to increase the efficiency of the production of components and their integration, two important problems have to be solved: (1) the portability of components from one ECU to another one enabling some flexibility in the architecture design and (2) the reuse of components between platforms, which is a key point for car manufacturers. Thus, the cooperative development process raises the problem of interoperability of components. A classic approach for easing the integration of software components is to implement an *MW layer* that provides application programs with common services and a common interface. In particular, the common interface allows the design of an application disregarding the hardware platform and the distribution and therefore enables the designer focusing on the development and the validation of the software components and the software architecture that realize a function.

Among the set of common services usually provided by an MW, those that related to the communication between several application components are crucial. They have to meet several objectives:

- *Hide the distribution* through the availability of services and interfaces that are the same for intra-ECU, inter-ECU, interdomain communications whatever the underlying protocols.
- *Hide the heterogeneity* of the platform by providing an interface independent of the underlying protocols, of the CPU architecture (e.g., 8/16/32 bits, endianness), of the operating systems (OSs), etc.
- *Provide high-level services* in order to shorten the development time and increase quality through the reuse of validated services (e.g., periodic transmission, working mode management such as partial networking). A good example of such a function is the *frame packing* (sometimes also called *signal multiplexing*) that enables application components to exchange *signals* (e.g., the number of revolutions per minute, the speed of the vehicle, the state of a light) while, at run-time, *frames* are transmitted over the network, so the *frame-packing* service of an MW consists in packing the signals into frames and sending the frames at the right points in time for ensuring the deadline constraint on each signal it contains.
- *Ensure QoS properties required by the application*, in particular, it can be necessary to improve the QoS provided by the lower-level protocols, for example, by furnishing an additional CRC (cyclic redundancy code) if the data link layer CRC is not sufficient in regard to the dependability objectives (see for instance in Section 50.3.3.3 the end-to-end communication protection

library of AUTOSAR that protects from message alteration in the communication stacks too). Other examples are the correction of *bugs* in lower-level protocols such as the *inconsistent message duplicate* of CAN (see Ref. [73]), the provision of a reliable acknowledgment service or message authentication mechanism on CAN, the status information on the data consumed by the application components (e.g., data were refreshed since last reading, its freshness constraint was not respected), or filtering mechanisms (e.g., notify the application for each *k* reception or when the data value has changed in a significant way).

Note that a more advanced features would be to come up with adaptive communication services, thanks to algorithms that would modify at run-time the parameters of the communication protocols (e.g., priorities, transmission frequencies) according to the current requirements of the application (e.g., inner-city driving or highway driving) or changing environmental conditions (e.g., EMI level). For the time being, to the best of our knowledge, such features exist in series automotive embedded systems only for low-power modes (see Section 50.1.5 and see also Ref. [102] for a research work in the direction of more adaptive communication systems). By increasing the efficiency and the robustness of the application, such an adaptive strategy would certainly ease the reusability.

50.3.2 Former Propositions of Middleware

Some proprietary MWs were developed by several carmakers in order to support the integration of ECUs and software modules provided by their third-party suppliers. For instance, the TITUS/DBKOM communication stack was a proprietary MW of Daimler that standardizes the cooperation between components according to a client/server model. *Volcano* [9,78] is a commercial product of Mentor Graphics, initially developed in tight cooperation with Volvo. The Volcano Target Package (VTP) consists of a communication layer and a set of off-line configuration tools for application distributed on CAN and LIN. It is aimed at providing the mapping of signals into frames under network bandwidth optimization and ensures a predictable and deterministic real-time communication system, thanks to schedulability analysis techniques (see Refs. [9,91]). To the best of our knowledge, there are no publicly available technically precise descriptions of both TITUS and Volcano.

A first step to define a standard for in-car embedded MW was started by the OSEK/VDX consortium (http://www.osek-vdx.org). In particular, two specifications are of particular interest in the context of this chapter: the *OSEK/VDX communication* layer [68] and the *fault-tolerant communication* layer [66]. The first one specifies a communication layer [68] that defines common software interfaces and common behavior for internal and external communications between application components. How signals are packed into a frame is statically defined off-line, and the *OSEK/VDX communication* layer automatically realizes the packing/unpacking at run-time as well as the handling of queued or unqueued messages at the receiver side. *OSEK/VDX communication* runs on top of a transport layer (e.g., [36]) that takes care mainly of possible segmentation of frames, and it can operate on any OS compliant with *OSEK/VDX OS* services for tasks and events and interrupt management (see Ref. [69]).

OSEK/VDX communication is not intended to be used on top of a TT network, for example, TTP/C or FlexRay. For this purpose, there is *OSEK/VDX FT Communication* specification (fault-tolerant communication, see Ref. [66]) whose main role is to manage the redundancy of data needed to achieve fault tolerance (i.e., the same information can be produced by a set of replicated nodes) by presenting only one copy of data to the receiving application according to the strategy specified by the designer. Two other important services of *OSEK/VDX FT Communication*, already offered by *OSEK/VDX Communication*, are to manage the packing/unpacking of messages [81] and to provide message-filtering mechanisms for passing only *significant* data to the application. *OSEK/VDX FT Communication* was developed to run on top of a TT OS such as OSEK time [67]. OSEK/VDX is no more under active development and has been superseded by AUTOSAR.

50.3.3 AUTOSAR—A Standard for the Automotive Industry

From 2001 to 2004, the ITEA European project EAST-EEA aimed at the specification of an automotive MW for the automotive industry. To the best of our knowledge, this was the first important initiative targeting the specification of both the services to be ensured by the MW and the architecture of the MW itself in terms of components and architecture of components. Similar objectives are guiding the work done in the AUTOSAR consortium (see http://www.autosar.org and [17,21,40]) that gathers most the key players in the automotive industry. The specifications produced by the consortium became quickly de facto standards for the cooperative development of in-vehicle embedded systems (see, for instance, the migration to AUTOSAR at PSA Peugeot Citroen [16]).

50.3.3.1 Reference Model

AUTOSAR specifies the software architecture embedded in an ECU. More precisely, it provides a reference model, which is comprised of three main parts:

1. The application layer
2. The basic software (MW software components)
3. The run-time environment (RTE) that provides standardized software interfaces to the application software

One of AUTOSAR's main objectives is to improve the quality and the reliability of embedded systems. By using a well-suited abstraction, the reference model supports the separation between software and hardware, eases the mastering of the complexity, and allows the portability of application software components and therefore the flexibility for product modification, upgrade and update, as well as the scalability of solutions within and across product lines. Besides, AUTOSAR ensures a smooth integration process of components provided by different companies while protecting the industrial properties of each actor involved. The AUTOSAR reference architecture is schematically illustrated in Figure 50.8. An important issue is the automatic generation of an AUTOSAR MW that has to be

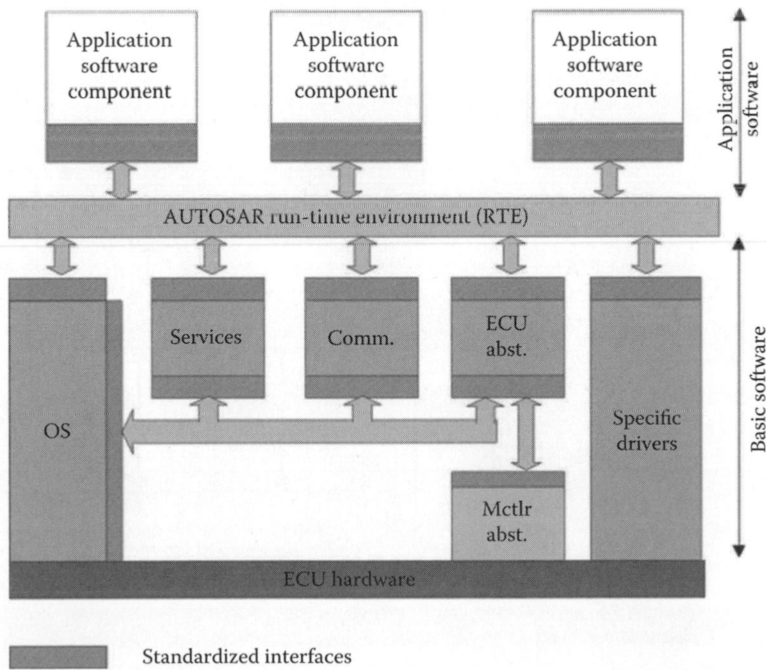

FIGURE 50.8 AUTOSAR reference architecture.

done from the basic software components, generally provided by suppliers, and the specification of the application itself (description of applicative-level tasks, signals sent or received, events, alarms, etc.). The challenge is to realize such a generation so that the deployment of the MW layer can be optimized for each ECU.

50.3.3.2 Communication Services

One of the main objectives of the AUTOSAR MW is to hide the characteristic of the hardware platform as well as the distribution of the application software components. Thus, the inter- or intra-ECU communication services are of major importance and are thoroughly described in the documents provided by the AUTOSAR consortium (see Figure 50.9 for an overview of the different modules). The role of these services is crucial for the behavioral and temporal properties of an embedded and distributed application. So their design, generation, and configuration have to be precisely mastered, and the verification of timing properties becomes an important activity. The problem is complex because, as for the formerly mentioned MW, the objects (e.g., signals, frames, interaction-layer PDU [I-PDU]) that are handled by services at one level are not the same objects that are handled by services at another level. Nevertheless, each object is strongly dependent on one or several objects handled by services belonging to neighboring levels. The AUTOSAR standard proposes two communication models:

1. *Sender–receiver* used for passing information between two application software components (belonging to the same task, to two distinct tasks on the same ECU or to two remote tasks)
2. *Client–server* that supports function invocation

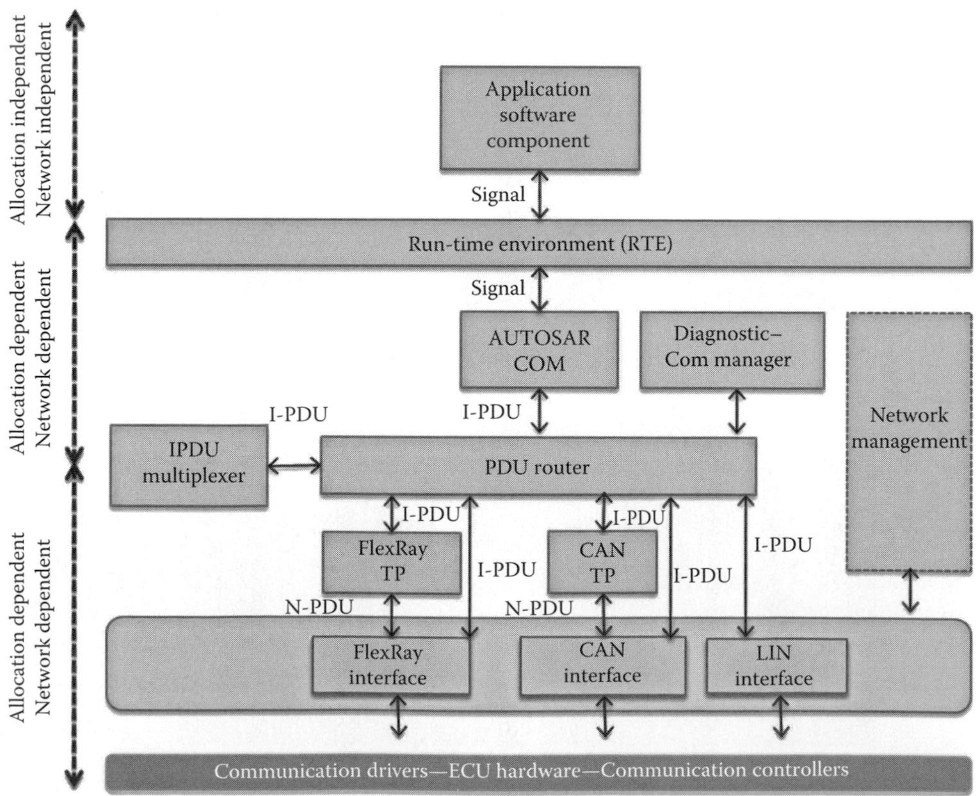

FIGURE 50.9 Communication software components and architecture.

Two communication modes are supported for the *sender–receiver* communication model:

1. The *explicit* mode is specified by a component that makes explicit calls to the AUTOSAR MW for sending or receiving data.
2. The *implicit* mode means that the reading (resp. writing) of data is automatically done by the MW before the invocation (resp. after the end of execution) of a component consuming (resp. producing) the data without any explicit call to AUTOSAR services; this is a way to protect effectively the data between application software components and MW services.

AUTOSAR identifies three main objects regarding the communication: signal exchanged between software components at application level, I-PDU that consists of a group of one or several signals, and the N-PDU (DLL PDU) that will actually be transmitted on the network. Precisely AUTOSAR defines the following:

- *Signals* at application level that are specified by a length and a type. Conceptually, a signal is exchanged between application software components through ports disregarding the distribution of this component. The application needs to indicate a *transfer property* parameter that will impact the behavior of the transmission and whose value can be *triggered* (each time the signal is provided to the MW by the application, it has to be transmitted on the network) or *pending* (the actual transmission of a signal on the network depends only on the emission rule of the frame that contains the signal). Furthermore, when specifying a signal, the designer has to indicate if it is a *data*, an *event*, or a *mode*. For *data* transmission, incoming data are not queued on the receiver side, while for *event* exchanges, signals are queued on the receiver side, and therefore, for each transmission of the signal, a new value will be made available to the application. The handling of buffers or queues is done by the RTE.
- I-PDUs are built by the AUTOSAR COM component. Each I-PDU is made of one or several signals and is passed via the PDU router to the communication interfaces. The maximum length of an I-PDU depends on the maximum length of the L-PDU (i.e., DLL PDU) of the underlying communication interface: for CAN and LIN, the maximum L-PDU length is 8 bytes, while for FlexRay, the maximum L-PDU length is 254 bytes. AUTOSAR COM ensures a local transmission when both components are located on the same ECU, or by building suited objects and triggering the appropriate services of the lower layers when the components are remote. This scheme enables the portability of components and hides their distribution. The transformation from signals to I-PDU and from I-PDU to signals is done according to an off-line generated configuration. Each I-PDU is characterized by a behavioral parameter, termed *Transmission Mode* whose possible value is *direct* (the sending of the I-PDU is done as soon as a *triggered* signal contained in this I-PDU is sent at application layer), *periodic* (the sending of the I-PDU is done only periodically), *mixed* (the rules imposed by the *triggered* signals contained in the I-PDU are taken into account, and additionally the I-PDU is sent periodically if it contains at least one *pending* signal), or *none* (for I-PDUs whose emission rules depend on the underlying network protocol, e.g., FlexRay; no transmission is initiated by AUTOSAR COM in this mode).
- An N-PDU is built by the basic components CAN transport protocol (TP) or FlexRay TP. It consists of the data payload of the frame that will be transmitted on the network and protocol control information. Note that the use of a transport layer is not mandatory and I-PDUs can be transmitted directly to the lower layers (see Figure 50.9).

The RTE implements the AUTOSAR MW interface and the corresponding services. In particular, when using the *sender/receiver* model, the RTE handles the *implicit/explicit* communication modes and the fact that the communication involves *events* (queued) or *data* (unqueued). The *AUTOSAR COM* component is responsible for several functions: on the sender side, it ensures the transmission and notifies the application about its outcome (success or error). In particular, AUTOSAR COM can inform the application if the transmission of an I-PDU did not take place before a specified deadline (i.e., deadline monitoring). On the receiver side, it also notifies the application (success or error of a reception) and supports the filtering mechanism for signals (dispatching each signal of a received I-PDU to the application or to a gateway). Both at the sending and receiving end, the endianness conversion is taken in charge. An important role of the COM component is to pack/unpack *signals* into/from *I-PDUs*. Note that, as the maximal length of an I-PDU depends on the underlying networks, the design of a COM component has to take into account the networks, and therefore, it is not fully independent of the hardware architecture. The COM component has also to determine the points in time where to send the I-PDUs. This is based on the attributes *transmission mode* of an I-PDU and on the attribute *transfer property* of each signal that it contains.

The COM component is generated off-line on the basis of the knowledge of the signals, the I-PDUs, and the allocation of application software components on the ECUs. The *AUTOSAR PDU router* (see Figure 50.9), according to the configuration, dispatches each I-PDU to the right network communication stack. This basic component is statically generated off-line as soon as the allocation of software components and the operational architecture is known. Other basic software components of the communication stack are responsible for the segmenting/reassembling of I-PDU(s) when needed (FlexRay TP, CAN TP) or for providing an interface to the communication drivers (FlexRay interface, CAN interface, LIN interface).

50.3.3.3 End-to-End Communication Protection Library

In order to support safer communications between application software components, the AUTOSAR standard specifies the *E2E communication protection library* [2]. The specified E2E mechanisms aim to ensure that safety–critical data exchanges (up to ASIL D safety level) can be protected against faults that may occur at run-time during the transmission of the data: faults affecting the hardware (e.g., bit flipping due to EMI in the sending communication stack) or software faults due to the incorrect development of a component of the communication stack. In particular, the E2E protection mechanisms allow the following:

- To attach, on the sender side, specific control data to data that are to be sent to the RTE.
- To verify, on the receiver side, through the control data, the correction of the data received from the RTE.
- To report, when it occurs, that received data are faulty. In this case, the fault has to be handled by the receiver software components.

The E2E library is implemented as a set of functions that can be invoked at two levels: by application software components (thanks to a wrapping technique) or from the COM component. These functions are stateless and have to report synchronously a verdict (i.e., the success of the exchange or the detected errors) to their calling component. The E2E communication protection is based on a standardization of mechanisms under the concept of *E2E profile*. Each profile can have several variants. So the instantiation of one profile for a specific information exchange consists in the choice of a variant and therefore the setting of the corresponding configuration options. Each profile offers all or a subset of the following data protection mechanisms:

- Checking the integrity of an exchanged data. This is done by an additional CRC computed by a function belonging to the E2E library.
- Checking the systematic transmission of a data, thanks to a *sequence counter*. The *sequence counter* is incremented upon each transmission of the data, allowing thus the receiver to verify that no instance of the data has been lost.
- Checking, from the receiver side, if a sender is transmitting a data as expected. For this purpose, an *alive counter* is used, it is incremented upon each transmission of the data, and the receiver can verify that the counter has been incremented.
- To check if the received data are the one that are expected. This mechanism is based on the definition of a data ID that is not transmitted with the data but that is used in the computation of the CRC.
- Finally, to verify that the data are sent and/or received on time. Two mechanisms are provided for this purpose: receiver communication time-out and sender acknowledgment time-out.

The CRC and the counters are integrated into an E2E header, which is an additional control field transmitted along with the data payload. When a faulty transmission is detected, the error is then reported to the calling component.

50.4 Conclusions and Discussion

In our view, the development of in-vehicle embedded communications can be schematically subdivided into four main successive periods:

1. Communication through point-to-point links—until the beginning of the 1990s
2. CAN-based architectures as de facto standard—until 2006
3. The advent of AUTOSAR and FlexRay—until 2010
4. The emergence of automotive Ethernet and security concerns since then

One may imagine that the next stage will be when the *car will become an IP node in the World Wide Web* ([28], see also Ref. [31]), integrated into global data infrastructures enabling a more efficient type of *collaborative mobility* [76] relying on interactions between users and infrastructure operators.

Over the years, the technologies needed for the interoperability between applications located on different ECUs and subnetworks have been improved. With AUTOSAR, we are now close to the desirable characteristics listed in Section 50.3. However, if the traditional partitioning of automotive applications into several distinct functional domains with their own characteristics and requirements has proven useful in mastering the complexity, it leads to the development of several independent subsystems with their specific architectures, networks, and software technologies. Some difficulties arise from this partitioning since more and more cross-domain data exchanges are needed. The current practice is to transfer data between different domains through a central gateway (see, e.g., Ref. [39]). This subsystem is recognized as being critical in the vehicle: it constitutes a single point of failure, its design is overly complex, and ensuring a guaranteed QoS through gateways is difficult and requires a careful design (see, for instance, Refs. [15,88]). In the future, the interconnection between vehicle subdomains could be more efficiently ensured by an Ethernet backbone [29], instead of a gateway. This backbone will have to support various traffic with different QoS requirements on the same network, which will require to implement QoS policies such as bandwidth reservation.

A perhaps more difficult challenge ahead of us is security: automotive systems must be designed to be more resilient to security attacks. This entails finding the technical solutions but also the new cooperation schemes between all the automotive stakeholders that will permit to achieve the right trade-off for the automotive domain, between costs/overhead and security level.

These issues will certainly require more research and development work, but the current state of technological maturity in security, network hardware components, communication protocols, and MW engineering is such that in our view, everything is at hand to succeed in building safe, secure, and cost-optimized embedded communication architectures for the next car generations.

References

1. A. Albert. Comparison of event-triggered and time-triggered concepts with regards to distributed control systems. In *Proceedings of Embedded World 2004*, Niirnberg, February 2004.
2. AUTOSAR. Specification of SW-C end-to-end communication protection library, v3.0.0. Available at http://www.autosar.org, 2013 (accessed August 23, 2013).
3. A. Avizienis, J. Laprie, and B. Randell. Fundamental concepts of dependability. In *Proceedings of the 3rd Information Survivability Workshop*, Boston, pp. 7–12, 2000.
4. M. Ayoubi, T. Demmeler, H. Leffler, and P. Köhn. X-by-wire functionality, performance and infrastructure. In *Proceedings of Convergence 2004*, Detroit, MI, 2004.

5. R. Barbosa and J. Karlsson. Formal specification and verification of a protocol for consistent diagnosis in real-time embedded systems. In *Third IEEE International Symposium on Industrial Embedded Systems (SIES'2008)*, La Grande Motte, France, June 2008.
6. L. Beaurenaut. Short PWM Code: A step towards smarter automotive sensors. In *Advanced Microsystems for Automotive Applications 2009*, pp. 383–395. Springer, Switzerland, 2009.
7. S. Bunzel. Autosar architecture expands safety and security applications. EETimes. Available at http://www.eetimes.com/, 2011 (accessed August 23, 2013).
8. C. Butzkamm and D. Bollati. Partial Networking for CAN bus systems: Any saved gram CO2/km is essential to meet stricter EU regulations. In *13th International CAN Conference (iCC2012)*, Hambach, Germany, March 5–6 2012.
9. L. Casparsson, A. Rajnak, K. Tindell, and P. Malmberg. Volcano—A revolution in on-board communications. Technical Report 98-12-10, Volvo, 1999.
10. G. Cena and A. Valenzano. Performance analysis of Byteflight networks. In *Proceedings of the 2004 IEEE Workshop of Factory Communication Systems (WFCS 2004)*, Vienna, Austria, pp. 157–166, September 2004.
11. FlexRay Consortium. FlexRay communications system—Protocol specification—version 3.0.1, December 2010.
12. Continental. Continental shows Ethernet test setup. Press Release, January 2012.
13. Intel Corporation. Introduction to in-vehicle networking. Available at http://support.intel.com/design/auto/autolxbk.htm, 2008 (accessed August 23, 2013).
14. R. Davis, A. Burns, R.J. Bril, and J.J. Lukkien. Controller area network (CAN) schedulability analysis: Refuted, revisited and revised. *Real-Time Systems*, 35(3):239–272, April 2007.
15. R. Davis and N. Navet. Traffic shaping to reduce jitter in controller area network (CAN). *SIGBED Reviews*, 9(4):37–40, November 2012.
16. P.H. Dezaux. Migration strategy of in-house automotive real-time applicative software in AUTOSAR standard. In *Proceedings of the 4th European Congress Embedded Real Time Software (ERTS 2008)*, Toulouse, France, 2008.
17. H. Fennel, S. Bunzel, H. Heinecke, J. Bielefeld, S. Frstand, K.P. Schnelle, W. Grote et al. Achievements and exploitation of the AUTOSAR development partnership. In *Convergence 2006*, Detroit, MI, October 2006.
18. J. Ferreira, P. Pedreiras, L. Almeida, and J.A. Fonseca. The FTT-CAN protocol for flexibility in safety-critical systems. *IEEE Micro, Special Issue on Critical Embedded Automotive Networks*, 22(4):46–55, July–August 2002.
19. Ford Motor Company. Ford to study in-vehicle electronic devices with advanced simulators. Available at http://media.ford.com/article_display.cfm?article_id=7010, 2001 (accessed August 23, 2013).
20. Freescale Semiconductor. New Freescale microcontrollers help streamline automotive body electronics networks and reduce vehicle weight. Press Release, March 18, 2013, Available on Reuters.com.
21. S. Fürst. AUTOSAR for safety-related systems: Objectives, approach and status. In *2nd IEE Conference on Automotive Electronics*, London, U.K., March 2006. IEE.
22. B. Gaujal and N. Navet. Fault confinement mechanisms on CAN: Analysis and improvements. *IEEE Transactions on Vehicular Technology*, 54(3):1103–1113, May 2005.
23. B. Gaujal and N. Navet. Maximizing the robustness of TDMA networks with applications to TTP/C. *Real-Time Systems*, 31(1–3):5–31, December 2005.
24. M. Grenier, L. Havet, and N. Navet. Scheduling frames with offsets in automotive systems: A major performance boost, *The Automotive Embedded Systems Handbook*, pp. 14.1–14.15. CRC Press, Boca Raton, FL, December 2008.
25. M. Grenier, L. Havet, and N. Navet. Configuring the communication on FlexRay: The case of the static segment. In *ERTS Embedded Real Time Software 2008*, Toulouse, France, 2008.

26. M. Grenier, L. Havet, and N. Navet. Pushing the limits of CAN—Scheduling frames with offsets provides a major performance boost. In *Proceedings of the 4th European Congress Embedded Real Time Software (ERTS 2008)*, Toulouse, France, January 29–February 1, 2008.
27. H. Kopetz et al. *Specification of the TTP/A Protocol.* University of Technology, Vienna, Austria, September 2002.
28. C. Hammerschmidt. Beyond FlexRay: BMW airs Ethernet plan. EETimes. Available at http://www.eetimes.com/document.asp?doc_id = 1256700, May 2013 (accessed August 23, 2013).
29. P. Hank, T. Suermann, and S. Mller. Automotive Ethernet, a holistic approach for a next generation in-vehicle networking standard. In *Advanced Microsystems for Automotive Applications 2012*, pp. 79–89. Springer, Switzerland, 2012.
30. F. Hartwich. CAN with flexible data-rate. In *13th International CAN Conference (iCC2012)*, Hambach, Germany, March 5–6, 2012.
31. P. Hoschka. W3C launched work on Web and automotive. In *ERCIM News, ISSN 0926-4981, number 94, Special theme on Intelligent Cars*, p. 5. ERCIM-EEIG, July 2013.
32. C. Humig. Next generation automotive network architecture based on Ethernet. Slides presented at the *3rd International Conference Chassis Electrification*, Darmstadt, Germany, May 2012.
33. International Standard Organization. *ISO 11519-2, Road Vehicles—Low Speed Serial Data Communication—Part 2: Low Speed Controller Area Network.* ISO, Geneva, Switzerland, 1994.
34. International Standard Organization. *ISO 11519-3, Road Vehicles—Low Speed Serial Data Communication—Part 3: Vehicle Area Network (VAN).* ISO, Geneva, Switzerland, 1994.
35. International Standard Organization. *ISO 11898, Road Vehicles—Interchange of Digital Information—Controller Area Network for High-Speed Communication.* ISO, Geneva, Switzerland, 1994.
36. International Standard Organization. *15765-2, Road Vehicles—Diagnostics on CAN—Part 2: Network Layer Services.* ISO, Geneva, Switzerland, 1999.
37. International Standard Organization. *11898-4, Road Vehicles—Controller Area Network (CAN)—Part 4: Time-Triggered Communication.* ISO, Geneva, Switzerland, 2000.
38. K.H. Johansson, M. Trngren, and L. Nielsen. Vehicle applications of controller area network, *Handbook of Networked and Embedded Control Systems*, pp. 741–765. Birkhauser, Boston, MA, 2005.
39. H. Kellermann, G. Nmeth, J. Kostelezky, K. Barbehn, F. El-Dwaik, and L. Hochmuth. Electrical and electronic system architecture. *ATZextra Worldwide*, 13(8):30–37, 2008.
40. F. Kirschke-Biller, S. Frst, S. Lupp, S. Bunzel, S. Schmerler, R. Rimkus, A. Gilberg, K. Nishikawa, and A. Titze. A worldwide standard current developments roll out and outlook. In *15th International VDI Congress Electronic Systems for Vehicles 2011*, Baden-Baden, Germany, 2011.
41. R. Klos. MOST in future automotive connectivity. EETimes. Available at http://www.automotive-eetimes.com/, May 2013 (accessed August 23, 2013).
42. P. Koopman. Critical embedded automotive networks. *IEEE Micro, Special Issue on Critical Embedded Automotive Networks*, 22(4):14–18, July–August 2002.
43. K. Koscher, A. Czeskis, F. Roesner, S. Patel, T. Kohno, S. Checkoway, D. McCoy, B. Kantor, D. Anderson, H. Shacham, and S. Savage. Experimental security analysis of a modern automobile. In *IEEE Symposium on Security and Privacy (SP2010)*, Berkeley, pp. 447–462, 2010.
44. M. Krug and A.V. Schedl. New demands for in-vehicle networks. In *Proceedings of the 23rd EUROMICRO Conference'97*, pp. 601–605, Budapest, Hungary, July 1997.
45. G. Leen and D. Heffernan. Expanding automotive electronic systems. *IEEE Computer*, 35(1):88–93, January 2002.
46. A. Leonhardi, S. Wachter, M. Bsinger, and T. Pech. MOST150's new features in a series project. *EETimes.* Available at http://www.automotive-eetimes.com/, May 2011 (accessed August 23, 2013).
47. T. Liebetrau, U. Kelling, T. Otter, and M. Hel. Energy saving in automotive E/E architectures. Infineon White Paper, December 2012.

48. LIN Consortium. LIN specification package, revision 2.2A. Available at http://www.lin-subbus.org/, December 2010 (accessed August 23, 2013).
49. T. Lindenkreuz. CAN FD—CAN with flexible data rate. Slides presented at *Vector Congress 2012*, May 2012.
50. Z. Lokaj, T. Zelinka, and M. Srotyr. Cooperative systems for car safety improvement. In *ERCIM News*, 94, Special theme on Intelligent Cars, pp. 9–10. ERCIM-EEIG, July 2013.
51. J. Machan and C. Laugier. Intelligent vehicles as an integral part of intelligent transport systems. In *ERCIM News*, 94, Special theme on Intelligent Cars, pp. 6–7. ERCIM-EEIG, July 2013.
52. Y. Martin. *L'argus de l'automobile*, 3969:22–23, March 2005.
53. E. Mayer. Serial bus systems in the automobile—Part 5: MOST for transmission of multimedia data. Summary of Networking Competence, February 2008. Published by Vector Informatik GmbH.
54. K. McCrory. Tech tutorial: LVDS offers efficient data transmission for automotive applications. EETimes. Available at http://www.eetimes.com/, April 2006 (accessed August 23, 2013).
55. Mercedes-Benz. The new Mercedes-Benz S-class. Press Release, May 2013.
56. P. Meumeu-Yomsi, D. Bertrand, N. Navet, and R. Davis. Controller area network (CAN): Response time analysis with offsets. In *9th IEEE International Workshop on Factory Communication System (WFCS 2012)*, Lemgo/Detmold, Germany, May 21–24, 2012.
57. B. Mller, T. Fhrer, F. Hartwich, R. Hugel, and H. Weiler. Fault tolerant TTCAN networks. In *Proceedings of the 8th International CAN Conference (iCC)*, Las Vegas, NV, 2002.
58. A. Monot, N. Navet, B. Bavoux, and F. Simonot-Lion. Multisource software on multicore automotive ECUs—Combining runnable sequencing with task scheduling. *IEEE Transactions on Industrial Electronics*, 59(10):3934–3942, 2012.
59. MOST Cooperation. MOST specification revision 3.0 E2. Available at http://www.mostcooperation.com, July 2010 (accessed August 23, 2013).
60. H. Muyshondt. Consumer and automotive electronics converge: Part 1—Ethernet, USB, and MOST. Available at http://www.automotivedesignline.com/, February 2007 (accessed August 23, 2013).
61. H. Muyshondt. Consumer and automotive electronics converge: Part 2—A MOST implementation. Available at http://www.automotivedesignline.com/, March 2007 (accessed August 23, 2013).
62. N. Navet, B. Delord, and M. Baumeister. Virtualization in automotive embedded systems: An outlook. Slides presented at RTS Embedded Systems 2010 (RTS'2010), March 2010.
63. N. Navet and H. Perrault. CAN in automotive applications: A look forward. In *13th International CAN Conference (iCC2012)*, Hambach, Germany, March 5–6 2012.
64. N. Navet, Y. Song, and F. Simonot. Worst-case deadline failure probability in real-time applications distributed over CAN (Controller Area Network). *Journal of Systems Architecture*, 46(7):607–617, 2000.
65. N. Navet and Y.-Q. Song. Validation of real-time in-vehicle applications. *Computers in Industry*, 46(2):107–122, November 2001.
66. OSEK Consortium. OSEK/VDX Fault-Tolerant Communication, Version 1.0. Available at http://www.osek-vdx.org/, July 2001 (accessed August 23, 2013).
67. OSEK Consortium. OSEK/VDX time triggered operating system, Version 1.0. Available at http://www.osek-vdx.org/, July 2001 (accessed August 23, 2013).
68. OSEK Consortium. OSEK/VDX Communication, Version 3.0.3. Available at http://www.osek-vdx.org/, July 2004 (accessed August 23, 2013).
69. OSEK Consortium. OSEK/VDX Operating System, Version 2.2.3. Available at http://www.osek-vdx.org/, February 2005 (accessed August 23, 2013).
70. M. Peteratzinger, F. Steiner, and R. Schuermans. Use of XCP on FlexRay at BMW. Translated reprint from HANSER Automotive 9/2006. Available at http://www.vector.com, 2006 (accessed August 23, 2013).
71. H. Pfeifer. Formal methods in the automotive domain: The case of TTA, *The Automotive Embedded Systems Handbook*, pp. 15.1–15.27. CRC Press, Boca Raton, FL, December 2008.

72. J. Pimentel, J. Proenza, L. Almeida, G. Rodriguez-Navas, M. Barranco, and J. Ferreira. Dependable automotive CAN networks, *The Automotive Embedded Systems Handbook*, pp. 6.1–6.41. CRC Press/Taylor and Francis, Boca Raton, FL, 2008.
73. L.M. Pinho and F. Vasques. Reliable real-time communication in CAN networks. *IEEE Transactions on Computers*, 52(12):1594–1607, 2003.
74. S. Poledna, W. Ettlmayr, and M. Novak. Communication bus for automotive applications. In *Proceedings of the 27th European Solid-State Circuits Conference*, Villach, Austria, September 2001.
75. T. Pop, P. Pop, P. Eles, and Z. Peng. Bus access optimisation for FlexRay-based distributed embedded systems. In *Proceedings of the Conference on Design, Automation and Test in Europe (DATE'07)*, pp. 51–56, San Jose, CA, 2007.
76. I. Radusch. Collaborative mobility—Beyond communicating vehicles. In *ERCIM News*, 94, Special theme on Intelligent Cars, page 4. ERCIM-EEIG, July 2013.
77. A. Rajnak. The LIN standard, *The Industrial Communication Technology Handbook*, pp. 31.1–31.13. R. Zurawski, editor. CRC Press, Boca Raton, FL, 2005.
78. A. Rajnak and M. Ramnefors. The Volcano communication concept. In *Proceedings of Convergence 2002*, Detroit, MI, 2002.
79. K. Ramaswamy and J. Cooper. Delivering multimedia content to automobiles using wireless networks. In *Proceedings of Convergence 2004*, Detroit, MI, 2004.
80. J. Rushby. A comparison of bus architectures for safety-critical embedded systems. Technical Report NASA/CR-2003-212161, NASA, March 2003.
81. R. Saket and N. Navet. Frame packing algorithms for automotive applications. *Journal of Embedded Computing*, 2:93–102, 2006.
82. W. Saleem. 1394 Automotive network enables powerful, cost-efficient in-vehicle networks for infotainment, navigation, cameras. *EETimes*. Available at http://www.eetimes.com/, May 2009 (accessed August 23, 2013).
83. A. Schedl. Goals and architecture of FlexRay at BMW. Slides presented at *the Vector FlexRay Symposium*, Stuttgart, Germany, March 2007.
84. B. Schtz, C. Khnel, and M. Gonschorek. The FlexRay protocol, *The Automotive Embedded Systems Handbook*, pp. 5.1–5.22. Industrial Information Technology series. CRC Press, Boca Raton, FL, December 2008.
85. Society of Automotive Engineers. J2056/1 class C application requirements classifications. In *SAE Handbook*. SAE, Troy, MI, 1994.
86. Society of Automotive Engineers. J2056/2 survey of known protocols. In *SAE Handbook*, Vol. 2. SAE, Troy, MI, 1994.
87. Society of Automotive Engineers. Class B data communications network interface—SAE J1850 standard—Rev. nov96, 1996.
88 J. Sommer and R. Blind. Optimized resource dimensioning in an embedded CAN-CAN gateway. In *IEEE Second International Symposium on Industrial Embedded Systems (SIES'2007)*, pp. 55–62, July 2007.
89. T. Steinbach, H.-T. Lim, F. Korf, T.C. Schmidt, D. Herrscher, and A. Wolisz. Tomorrow's in-car interconnect? A competitive evaluation of IEEE 802.1 AVB and time-triggered Ethernet (AS6802). In *IEEE Vehicular Technology Conference (VTC Fall)*, Vancouver, Canada, pp. 1–5, 2012.
90. R. Tewell, Z. Freeman, and M. Bassler. The 1394 auto network bus for all high-bandwidth, high-speed communications and infotainment. Slides presented at *SAE 2010 World Congress (Session AE305)*, Detroit, MI, April 2010.
91. K. Tindell, A. Burns, and A.J. Wellings. Calculating controller area network (CAN) message response times. *Control Engineering Practice*, 3(8):1163–1169, 1995.
92. TTTech Computertechnik AG. TTEthernet—A powerful network solution for multiple purpose. Marketing whitepaper, 2013.

93. TTTech Computertechnik GmbH. Time-Triggered Protocol TTP/C, high-level specification document, Protocol Version 1.1. Available at http://www.ttagroup.org, November 2003 (accessed August 23, 2013).
94. S. Tuohy, M. Glavin, E. Jones, M.M. Trivedi, and L. Kilmartin. Next generation wired intra-vehicle networks, a review. In *IEEE Intelligent Vehicles Symposium*, Gold Coast, Australia, June 2013.
95. A. Vollmer. Deutsche OEMs setzen standards. all-electronics.de. Available at http://www.all-electronics.de/texte/anzeigen/42481/Deutsche-OEMs-setzen-Standards, June 2011 (accessed August 23, 2013).
96. Volvo. Volvo unveils innovative safety technology—Pedestrian detection with full auto brake debuts on the all-new Volvo S60. Press Release. Available at https://www.media.volvocars.com/, March 2, 2010 (accessed August 23, 2013).
97. M. Waern. Evaluation of protocols for automotive systems. Master's thesis, KTH Machine Design, Stockholm, Sweden, 2003.
98. C. Wilwert, N. Navet, Y.-Q. Song, and F. Simonot-Lion. Design of automotive x-by-wire systems, *The Industrial Communication Technology Handbook*. R. Zurawski, editor. CRC Press, Boca Raton, FL, 2005.
99. J. Yoshida. NXP: 3 phases of automotive Ethernet. *EETimes*. Available at http://www.eetimes.com/document.asp?doc_id = 1319225, August 2013 (accessed August 23, 2013).
100. H. Zeltwanger. Partial networking reduces CO2 emissions. *CAN Newsletter*, 4/2011, December 2011.
101. H. Zeng, M. Di Natale, A. Ghosal, and A.L. Sangiovanni-Vincentelli. Schedule optimization of time-triggered systems communicating over the FlexRay static segment. *IEEE Transactions in Industrial Informatics*, 7(1):1–17, 2011.
102. T. Ziermann, J. Teich, and Z. Salcic. DynOAA—Dynamic offset adaptation algorithm for improving response times of CAN systems. In *DATE*, pp. 269–272, 2011.

51

Standardized Basic System Software for Automotive Embedded Applications

51.1 Introduction 51-1
51.2 Hardware Architecture 51-2
51.3 OSEK/VDX 51-3
51.4 ISO 51-4
Transport Layer: CAN and FlexRay • Diagnostics: UDS and OBD • Diagnostics over IP
51.5 ASAM 51-8
XCP: The Universal Measurement and Calibration Protocol Family
51.6 AUTOSAR 51-11
Application Software Architecture • System Software Architecture
51.7 Summary 51-30
References 51-30

Thomas M. Galla
Elektrobit Austria GmbH

51.1 Introduction

In the last decade, the percentage of electronic components in today's cars has been ever increasing. Premium cars, for example, use up to 70 electronic control units (ECUs) that are connected via five system networks and realize over 800 different functions [1].

Since 1993, major automotive companies have been striving for the deployment of standard software modules in their applications as the potential benefits of using standard software modules are huge [2]. While the functional software heavily depends on the actual system and is a discriminating factor of competitive importance, this does not apply to the software infrastructure. Furthermore, with continuously shortened development cycles, especially in the electronics area, requirements arise concerning compatibility, reusability, and increased test coverage that can only be fulfilled by setting standards for the various system levels.

This trend has been a key motivation for the formation of several consortia like the OSEK/VDX (Offene Systeme und deren Schnittstellen für die Elektronik im Kraftfahrzeug—Vehicle Distributed Executive)* consortium [3] in 1993/1994, the ASAM (Association for Standardization of Automation and Measuring Systems) initiative [4], the HIS (Herstellerinitiative Software)† group [5], the JasPar (Japan Automotive Software Platform Architecture) consortium [6] in 2004, the EASIS (Electronic Architecture and System Engineering for Integrated Safety Systems) project consortium [7] in 2003, the AUTOSAR (Automotive Open System Architecture) consortium [8] in 2003, and the GENIVI Alliance [9] in 2009.

* Translated into English: *Open Systems and the Corresponding Interfaces for Automotive Electronics*
† Translated into English: *Manufacturers' Software Initiative*

This chapter provides an overview of today's state of the art in the standardization of automotive software infrastructures. The chapter is structured as follows: Section 51.2 provides a short overview of the automotive hardware architecture. Section 51.3 provides information on the software modules specified by the German working groups OSEK/VDX. Section 51.4 illustrates the software modules standardized by the ISO. Section 51.5 targets the software modules defined by the ASAM. Section 51.6 deals with the AUTOSAR initiative. Section 51.7 provides a short summary, concluding the chapter.

51.2 Hardware Architecture

The hardware architecture of automotive systems can be viewed at different levels of abstraction. On the highest level of abstraction, the *system level*, an automotive system consists of a number of networks interconnected via gateways (see Figure 51.1). In general, these networks correspond to the different functional domains that can be found in today's cars (i.e., chassis domain, power train domain, body domain).

The networks themselves comprise a number of ECUs that are interconnected via a communication media (see zoom-in on network A and D in Figure 51.1). The physical topology used for the interconnection is basically arbitrary; however, bus, star, and ring topologies are the most common topologies in today's cars. This *network level* represents the medium level of abstraction.

On the lowest level of abstraction, the *ECU level* (Figure 51.2), the major parts of an ECU are of interest. An ECU comprises one or more microcontroller units (MCUs) as well as one or more communication controllers (CCs). In the past, in most cases, exactly one (single-core) MCU and one CC were used to build up an ECU. Today, however, there is a trend toward multicore MCUs and even ECUs consisting of multiple MCUs for performance and fault-tolerance reasons. In order to be able to control physical processes in the car (e.g., control the injection pump of an engine), the ECU's MCU is connected to actuators via the MCU's analog or digital output ports. To provide means to obtain environmental information, sensors are connected to the MCU's analog or digital input ports. We call this interface the ECU's environmental interface. The CC(s) facilitates the physical connectivity of the ECU to the respective network(s). We call this interface of an ECU the ECU's network interface.

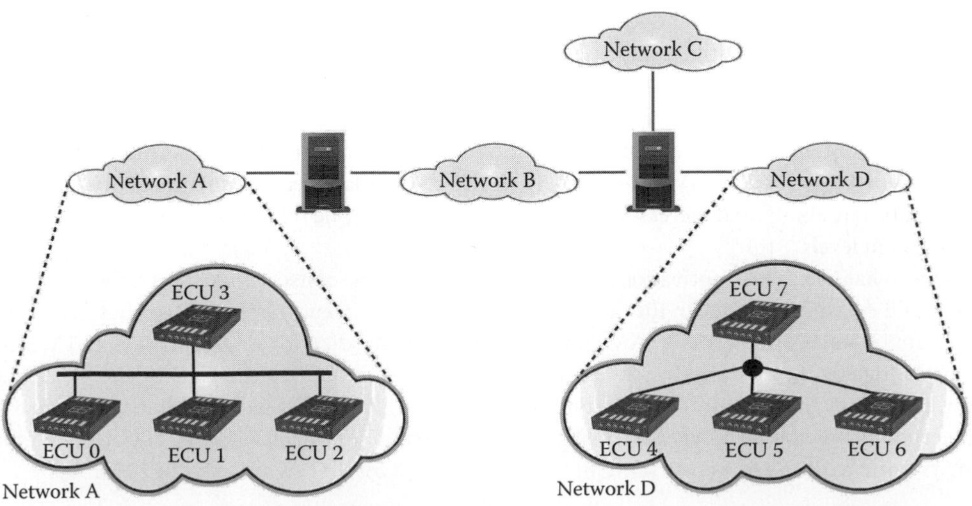

FIGURE 51.1 Hardware architecture—system and network level.

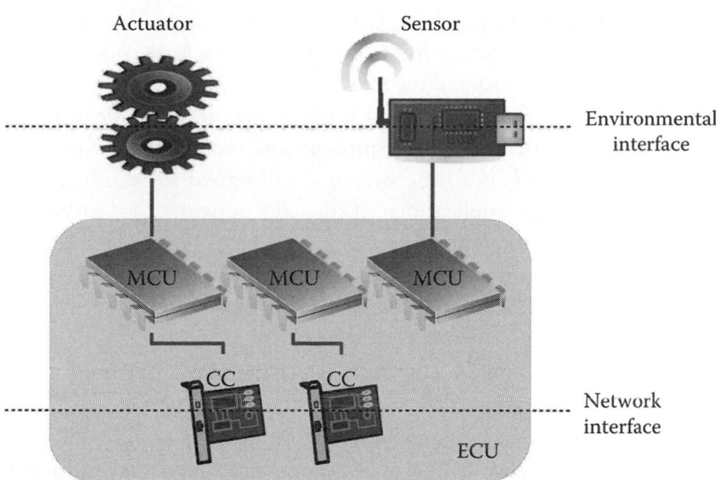

FIGURE 51.2 Hardware architecture—ECU level.

51.3 OSEK/VDX

The OSEK/VDX standard was the result of the endeavors of major German and French car manufacturers and their suppliers to create a standardized software infrastructure for automotive electronics. This standard was initially designed for applications in the area of automotive body electronics or for the power train where autonomous control units build up a loosely coupled network. It comprises the following standardized components:

OSEK OS: The OSEK operating system (OS) [10] is an event-driven OS intended for hard real-time applications. OSEK OS provides services for task management, task activation by means of events and alarms, intertask communication via messages, mutual exclusion by means of resources (implementing the *priority ceiling protocol* [11]), and interrupt handling. Hereby, OSEK OS distinguishes between *basic tasks* (which do not use any blocking interprocess communication [IPC] or synchronization constructs) and *extended tasks* (which are allowed to use blocking IPC or synchronization constructs).

OSEK NM: OSEK network management (NM) [12] provides network management facilities that take care of a *controlled coordinated shutdown of the communication* of multiple ECUs within a network. OSEK NM establishes a *logical ring* among all ECUs participating in OSEK NM. Along this ring, OSEK NM *ring messages* are passed between the participating ECUs. These ring messages contain information on whether the sending ECUs desire to perform a transition into a low-power sleep mode. In case all ECUs along the logical ring agree on this transition (i.e., no ECU objects), a coordinated transition into the sleep mode is performed. In case any ECU objects to this decision because it still requires network communication, a transition into the sleep mode is prevented.

OSEK COM: OSEK communication (CoM) [13] offers services to transfer data between different tasks and/or interrupt service routines (ISRs) residing on the same ECU (internal communication) or possibly being distributed over several ECUs (external communication). OSEK COM supports both *cyclic time-driven communication* and *on-demand event-driven communication*. In case of external communication, the *interaction sublayer* of OSEK COM takes care of the representational issues of signals like byte ordering and alignment. On the sender's side, this layer converts signals from the byte order of the sending ECU into the network byte order. Furthermore, the interaction layer packs multiple signals into a single communication frame in order to reduce communication bandwidth consumption. On the receiver's side, this layer extracts multiple signals from a single communication frame and performs a byte order conversion from the network byte order to the byte order of the receiving ECU.

To support the detection of lost communication frames at the sender and the receiver side, OSEK COM provides mechanisms for transmission and reception time-out monitoring.

OSEKtime OS: OSEKtime OS [14] is a time-driven OS designed for minimal OS footprint and deployment in safety-related applications. Tasks are activated by a dispatcher based on predefined activation times stored in a *dispatcher table* before compile time. The processing of the dispatcher table during runtime is done in a cyclical fashion.

OSEKtime FTCom: OSEKtime FTCom [15] is the fault-tolerant communication layer accompanying OSEKtime OS that only supports *cyclic time-driven communication*. Similar to OSEK COM, OSEKtime FTCom can be used for both external and internal communication.

In case of external communication, the *fault-tolerance sublayer* of OSEKtime FTCom manages all fault-tolerance issues, namely, signal replication and signal reduction. On the sender's side, this layer replicates a single application signal and thus produces multiple signal instances. These signal instances are handed over to the interaction sublayer for byte order conversion and packing. Afterward, the packed signal instances are transmitted via redundant communication paths. Considering redundant transmission paths, temporal redundancy (multiple transmissions on a single communication channel) and spatial redundancy (transmission on multiple communication channels) can be distinguished. On the receiver's side, multiple signal instances are collected from the interaction sublayer and reduced to obtain a single application signal that is handed on to the application layer.

In order to minimize the memory footprint as well as the required execution time, each of these components is configured upon design time via a configuration file in OSEK implementation language (OIL) syntax [16]. Using a generation tool, the appropriate data structures in the used programming language are created based on this configuration file. Thus, none of the components provides services for dynamic resource allocation (like task creation, memory allocation, a.s.o.).

51.4 ISO

Some of the software layers used for automotive networks in today's cars have been standardized by the International Organization for Standardization (ISO). The most prominent of these software layers are the following:

Transport Layer—Controller Area Network (CAN) and FlexRay: This layer provides *segmented data transfer (SDT)* to all higher layers (e.g., the diagnostics layer) and thus facilitates exchange of data amounts larger than the maximum transfer unit (MTU) of the underlying communication system.

Diagnostics Layer—Unified Diagnostic Services (UDS) and Onboard Diagnostics (OBD): The UDS specifications define manufacturer-specific enhanced diagnostic services that *allow an external tester device to control diagnostic functions* in an ECU via a (serial) data link. The OBD specifications define a set of mandatory diagnostic services facilitating the *retrieval of emission-related diagnostic information* from an ECU by an *external tester device*.

51.4.1 Transport Layer: CAN and FlexRay

The ISO transport layer on CAN [17] provides services for *unacknowledged segmented data transfer* (*USDT*) *of known length* to higher software layers (e.g., the ISO diagnostics [see Section 51.4.2]) by facilitating the transmission of messages whose length is greater than the MTU of the underlying communication system. On the sender's side, the transport layer will split such long messages into multiple segments,

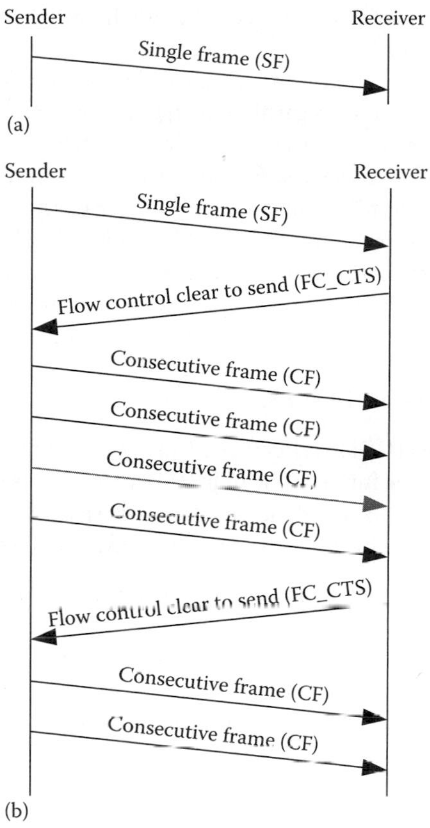

FIGURE 51.3 ISO transport layer on CAN—transmission sequence for unsegmented transmission (a) and segmented transmission (b).

each small enough for the underlying communication system. On the receiver's side, the transport layer reassembles these segments again. As far as the frame format is concerned, the ISO transport layer distinguishes between single frames (SFs), first frames (FFs), consecutive frames (CFs), and flow control frames (FCs), for example, the clear-to-send FC (FC_CTS) depicted in Figure 51.3b. Hereby, SFs are used for the reduction of protocol overhead in case the amount of data to be transmitted does not exceed the MTU. In that case, only an SF is used in order to exchange these data. Otherwise, an FF, which contains the total number of data bytes to be sent as well as the first few data bytes, is transmitted followed by one or more CFs, which then contain the remaining data bytes. For flow control reasons (i.e., to prevent the sender from outpacing the receiver), the receiver is allowed to send FCs at defined points in time. Figure 51.3 illustrates the frame exchange in an unsegmented (Figure 51.3a) and a segmented transmission (Figure 51.3b).

Hereby, the frame format depicted in Figure 51.4 is used by the ISO transport layer on CAN. The *source* and *destination address* fields are used to identify the sender and the receiver(s) of the TP frame. The *address type* field is used to define the semantics of the destination address. Here, the ISO transport protocol (TP) on CAN provides support for single ECU addresses (called *physical addresses* by the ISO TP on CAN specification) and multicast addresses (called *functional addresses* by the ISO TP on CAN specification). In case of multicast transmissions, only unsegmented data transfer is allowed in the ISO TP on CAN.

Src Addr	Dst Addr	Addr Type	PCI	Data

FIGURE 51.4 ISO transport layer on CAN—frame format.

The *protocol control information* (*PCI*) field is used to distinguish between the different frame types (i.e., SF, FF, CF, FC). Furthermore, the PCI field carries the *total number of data bytes* to be transmitted in case of an SF or an FF, a *sequence number* in case of a CF (in order to facilitate early detection of a loss of a CF), and more detailed information regarding the *type of flow control* in case of an FC.

In addition to the service of the ISO transport layer on CAN, the ISO transport layer on FlexRay [18] provides an *acknowledge and retry mechanism* facilitating a positive and a negative acknowledgment by the receiver to indicated successful reception or the request for (early) retransmission to the sender. To this end, a new FC for explicit acknowledgment (FC_ACK) is introduced. Furthermore, the ISO transport layer on FlexRay supports the *transmission of messages with unknown but finite data length* by adding a dedicated last frame (LF) that marks the end of the transmission. By combining these two features, the ISO transport layer on FlexRay facilitates *acknowledged SDT of unknown length*.

51.4.2 Diagnostics: UDS and OBD

The unified diagonistic services (UDS) [19] define client/server diagnostic services that allow a tester device (client) to control diagnostic functions in an ECU (server) via a (serial) data link. Hereby, the provided services, the encoding of the SIDs, and the encoding of the parameters are standardized by the UDS specification. The services provided by the UDS can be grouped into the following six functional units:

Diagnostic Management: This functional unit covers all services realizing diagnostic management functions between the client and the server. Examples for these kinds of services are the services `startDiagnosticSession` for initiating a diagnostic session, `stopDiagnosticSession` for terminating a diagnostic session, `ecuReset` for resetting an ECU, and `readECUIdentification` for reading ECU identification information like the serial number or the programming date of the ECU.

Data Transmission: This functional unit covers all services dealing with data exchange between the client and the server. Prominent examples for these services are the services `readMemoryByAddress` and `writeMemoryByAddress`, which are used to read and write a memory range (defined by start address and the number of bytes to read/write) of the ECU, as well as `readDataByIdentifier` and `writeDataByIdentifier`, which read and write a data item that is uniquely identified via an ID.

Stored Data Transmission: This functional unit covers all services that are used to perform exchange of data that are stored within the ECU (in a nonvolatile [NV] way). Examples for these kinds of services are the service `readDiagnosticTroubleCodes`, which is used to retrieve stored trouble codes from the ECU's error log, as well as the service `clearDiagnosticInformation`, which is used to remove all entries from the ECU's error log.

Input/Output Control: This functional unit covers all services that deal with input and output control functions. An example for a service contained in this group is the service `inputOutputControlByLocalIdentifier`, which can be used by the client to substitute a value for an input signal of the server.

Remote Activation of Routine: This functional unit covers all services that provide remote activation of routines within an ECU. Examples for services in this functional unit are the services `startRoutineByAddress` and `stopRoutineByAddress`, which are used to remotely invoke a routine in the server (ECU) and to terminate an executing routine in the server (ECU). Hereby, the routine to invoke or stop is identified by the routine's address.

Upload/Download: This functional unit covers all services that realize upload/download functionality between the client and the server. Examples for services within this unit are the `requestDownload` service, which gives the client the possibility to request the negotiation of a data transfer from client to server; the `requestUpload` service, which gives the client the possibility to request the negotiation of a data transfer from server to client; and the `transferData` service, which actually takes care of the data transmission between client and server.

In UDS, services are identified by the so-called *service identifier* (*SID*) field that is the first byte in a diagnostic message.* Based on this field, the layout of the remainder of the diagnostic message is completely different. The diagnostic server uses the SID to select the proper service requested by the client and interprets the remainder of the diagnostic message as parameters to the service request according to the SID.

The on board diagnostics (OBD) specifies a set of mandatory diagnostic services focusing on the *retrieval of emission-related information* from an ECU by an *external tester device*. For this purpose, the OBD defines services for the *retrieval of emission-related diagnostic trouble codes (DTCs)*, the *clearing/resetting of emission-related DTCs*, the *controlling of OBD monitoring functionality* like the invocation of defined tests (e.g., a test for leakage in the evaporative system of the vehicle), the *request of results of conducted OBD monitoring tests*, and the *retrieval of vehicle information* like the vehicle identification number (VIN).

Just like in UDS, the OBD encodes the diagnostic service requested by the client in an *SID* located in the first byte in a diagnostic message. Hereby, the SID ranges of ODB and UDS are disjunct to prevent interference between the two standards.

For the exchange of information between the client and the server (and vice versa), both diagnostic protocols use the transport layer described in Section 51.4.1.

51.4.3 Diagnostics over IP

In order to facilitate diagnostic communication between the external tester device and ECUs using the Internet protocol (IP) as well as the transmission control protocol (TCP) and the user datagram protocol (UDP), the diagnostics over IP (DoIP) specification [20] defines features that can be used to *detect a vehicle in a network* and to *enable communication with the vehicle's central diagnostic gateway* as well as with *other ECUs not directly connected to the tester device*.

ECUs that are capable of adhering to the DoIP protocol are termed *DoIP entities* by the DoIP specification. DoIP entities are further distinguished into *DoIP nodes*, which do not forward diagnostic communication requests toward a non-Ethernet in-vehicle network, and *DoIP gateways*, which perform this kind of forwarding and thus relay information between the external tester and non DoIP-capable ECUs.

Vehicle Identification Request and Vehicle Announcement: These two types of messages are required to identify the DoIP entities in a network and to retrieve their IP and logical addresses. The *vehicle identification request* contains either the *VIN* or a *unique hardware ID* (e.g., the medium access control (MAC) address) of the addressed DoIP entity and is broadcasted in the network. The DoIP entity with a matching VIN or hardware ID answers this request with a *vehicle announcement* containing its own IP address (as part of the IP header), its logical address (which is the ECU's physical ISO TP address [see Section 51.4.1]), and port numbers of TCP and UDP ports to be used for subsequent communication with the DoIP entity.

Routing Activation Request and Response: These two message types are required to activate routing in DoIP gateways between the vehicle's external network and the vehicle's internal network in order to facilitate subsequent diagnostic data exchange between the external tester device and non-DoIP-capable ECUs connected to the vehicle's internal network.

Diagnostic Message and Diagnostic Message Acknowledgment: After a successful routing activation, diagnostic message exchange between an external tester device and a non-DoIP-capable ECU connected to the vehicle internal network can take place. The diagnostic message used for the data exchange between the tester and the DoIP gateway hereby contains the ISO TP source address followed by the ISO TP destination address followed by the UDS/OBD message. If the DoIP gateway receives such a diagnostic message, it extracts the two ISO TP addresses and the UDS/OBD message and forwards the UDS/OBD message to the ISO TP of the target network

* This diagnostic message is payload from ISO TP's point of view.

that is selected based on the destination address. Hereby, the ISO TP source address and the ISO TP destination address contained in the DoIP diagnostic message are used as respective source and destination addresses on the target network. A successful or a failed forwarding to the target network is signaled to the external tester device by means of positive or negative diagnostic message acknowledgment.

51.5 ASAM

The ASAM started as an initiative of German car manufacturers with the goal to define standards for data models, interfaces and syntax specifications for testing, evaluation, and simulation applications.

Apart from several data exchange formats like the open diagnostic data exchange format (ODX), the functional specification exchange format (FSX), the metadata exchange format for software module sharing (MDX), and the fieldbus exchange format (FIBEX), the ASAM defines the universal measurement and calibration protocol family (XCP) that is described in the next section.

51.5.1 XCP: The Universal Measurement and Calibration Protocol Family

The universal measurement and calibration protocol family XCP [21] is used for the following main purposes:

- Synchronous data transfer (acquisition and stimulation)
- Online calibration
- Flash programming for development purposes

Prior to describing these main operations of XCP though, we will focus on the protocol's internal structure. XCP itself consists of two main parts, namely, the *XCP protocol layer* and several *XCP transport layers*, one dedicated transport layer for each underlying communication protocol (currently, CAN, FlexRay, universal serial bus (USB) TCP/IP, UDP/IP, and serial peripheral interfaces (SPI) are supported). Figure 51.5 illustrates this XCP protocol stack.

51.5.1.1 XCP Protocol Layer

The XCP protocol layer is the higher layer of the XCP protocol family. This layer implements the main operations of XCP that are described in detail in Sections 51.5.1.3 through 51.5.1.5. The XCP protocol layer itself is independent of a concrete communication protocol (e.g., CAN, FlexRay). Data exchange on the XCP protocol layer level is performed by data objects called *XCP packets*. Figure 51.6 illustrates the structure of an XCP packet.

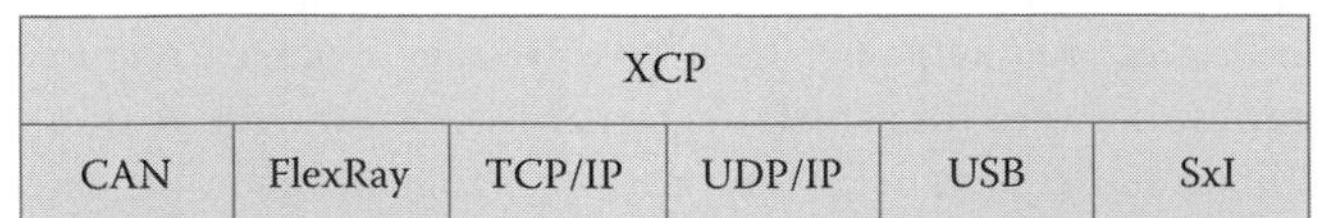

FIGURE 51.5 XCP protocol stack.

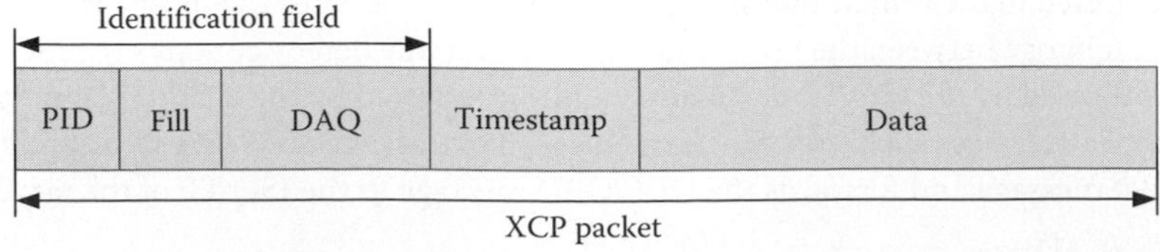

FIGURE 51.6 Structure of an XCP packet.

An XCP packet starts with an *identification field* containing a *packet identifier* (*PID*) that is used to establish a common understanding about the semantics of the packet's data between the XCP master and the XCP slave. The PID is thus used to uniquely identify each of the following two basic packet types and their respective subtypes:

Command Transfer Object (*CTO*) *Packets*: This packet type is used for the transfer of generic control commands from the XCP master to the XCP slave and vice versa. It is used for carrying out *protocol commands* (*CMD*), transferring *command responses* (*RES*), *errors* (*ERR*), *events* (*EV*), and for issuing *service requests* (*SERV*).

Data Transfer Object (*DTO*) *Packets*: This packet type is used for transferring synchronous data between the XCP master and the XCP slave device. *Synchronous data acquisition* (*DAQ*) *data* are transferred from the XCP slave to the XCP master, whereas *synchronous data stimulation* (*STIM*) *data* are transported from the master to the slave.

The *DAQ field* is used to uniquely identify the DAQ list (see Section 51.5.1.3) to be used for data acquisition or stimulation if the XCP packet is of DTO packet type. In order to have the DAQ field aligned to 16-bit boundaries, a fill byte is introduced between the PID field and the DAQ field. In case the packet is a CTO packet, the DAQ field is omitted.

The *time-stamp field* is used in DTO packets to carry a time stamp provided by the XCP slave for the respective data acquisition. The length of the time-stamp field may vary between one to four bytes depending on the configuration. In case the packet is of CTO type, the time-stamp field is omitted.

Command packets (CMD) are explicitly acknowledged on the XCP protocol layer by sending either a command response packet (RES) or an error packet (ERR). Event (EV) packets (i.e., packets informing the XCP master that a specific event has occurred in the XCP slave [e.g., an overload situation has occurred during data acquisition]), service request (SERV) packets (i.e., packets used by the slave to request certain services from the master [e.g., a packet containing text that is to be printed by the master]), and data acquisition packets (DAQ, STIM) are sent asynchronously and unacknowledged at the XCP protocol layer. Therefore, it may not be guaranteed that the master device will receive these packets when using a nonacknowledged transportation link like UDP/IP.

51.5.1.2 XCP Transport Layers

The protocol layer described in the previous sections is independent from the underlying communication protocol. In order to be able to use XCP on top of different communication protocols, the XCP specification defines multiple XCP transport layers that perform the packing of the protocol-independent XCP packets into frames of the respective communication protocol, by adding an XCP header containing a *node address field* used to identify the destination ECU, a *counter field* used for sequence numbering of the XCP packets and a *length field* defining the length of the XCP packet. Depending on the actual communication protocol used, some of these header fields might be missing (e.g., in case TCP/IP is used as communication protocol, the node address is omitted, since IP's addressing scheme is used).

51.5.1.3 Synchronous Data Transfer

The synchronous data transfer feature of XCP allows for a data exchange between the XCP master and the XCP slave that is performed synchronous to the XCP slave's execution. This exchange is carried out by using *DTOs* that are transferred via DTO packets. The memory regions of the XCP slave's memory that are the source or the destination of the transfer are linked to the DTO by so-called object description tables (ODTs). A sequence of one or more ODTs are grouped into a so-called DAQ list (see Figure 51.7).

Hereby, the DAQ–DTO contains a packet identification field (PID), which is used to link the DAQ–DTO to the respective ODT (PID field matches ODT number). For each element within a DAQ–DTO, a corresponding ODT entry is present in the ODT that references a specific part in the ECU's memory by the attribute's address and length. Upon processing of the ODT, the ODT entries are used to transfer

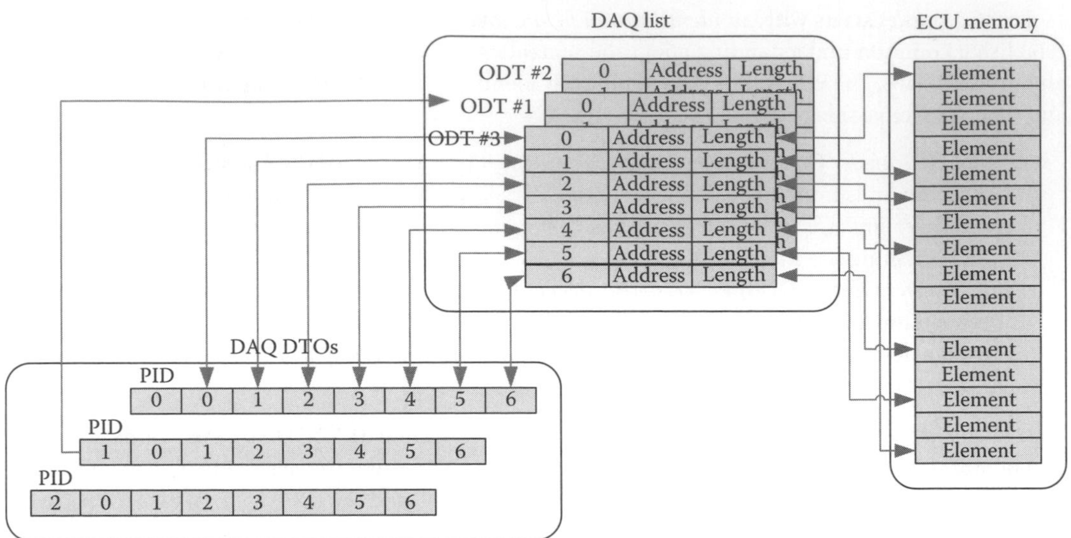

FIGURE 51.7 Structure of a DAQ list.

an element from the ECU's memory into the corresponding element in the DAQ–DTO (in case of data acquisition) and vice versa (in case of data stimulation).

DAQ lists can either be statically stored in an XCP slave ECU's memory or dynamically allocated by the XCP master via special protocol command packets.

51.5.1.4 Online Calibration

For the online calibration feature of XCP, the slave's physical memory is divided into so-called sectors that reflect the memory's size and limit constraints when reprogramming/erasing parts of the memory. This division into sectors thus describes the physical layout of the XCP slave's memory.

The logical layout is described by dividing the memory into *segments*. This division does not need to adhere to the physical limitations of the division into sectors. Each segment can further consist of one or multiple pages where at any given instance in time, only one page of a segment is accessible to the ECU. This page is called the active page for the ECU's application. The same holds true for XCP itself as well (active XCP page). XCP ensures that concurrent access to the same page by the ECU's application and XCP itself is prevented.

Via XCP CMD packets, the XCP master can instruct the XCP slave to switch active pages of the same segment (i.e., to make a page that has previously not been the active one the new active page) and to copy data between pages of the same or of different memory segments.

This way, the calibration data used for the slave ECU's control loops, for example, can be altered upon runtime under the control of the XCP master.

51.5.1.5 Flash Programming

In order to facilitate the exchange of the current image of the application program in the slave ECU's NV memory, XCP defines special commands (exchanged via XCP command packets) for performing flash (re)programming. The following list contains a short description of the main commands used in flash programming:

PROGRAM_START: This indicates the beginning of a programming sequence of the slave ECU's NV memory.

PROGRAM_CLEAR: This clears a part of the slave ECU's NV memory (required before programming new data into this part of memory).

PROGRAM_FORMAT: This is used to specify the format for the data that will be transferred to the slave ECU (e.g., used compression format, used encryption algorithm).

PROGRAM: This command, which is issued in a loop as long as there are still data available that need to be programmed, is used to program a given number of data bytes to a given address within the slave ECU's NV memory.

PROGRAM_VERIFY: This verifies that the programming process has completed successfully by, for example, calculating a checksum over the programmed memory area.

PROGRAM_RESET: This indicates the end of the programming sequence and optionally resets the slave ECU.

By means of these flash programming XCP commands, a flash download process for the ECU's development stage can be implemented. In-system flash programming in series cars, however, is usually performed via the diagnostic protocol (see Section 51.4.2).

51.6 AUTOSAR

The development partnership AUTOSAR is an alliance of car manufacturers, major supplier companies, and tool vendors working together to develop and establish an open industry standard for an automotive electronic architecture together with a development methodology and an XML-based data exchange format that is intended to serve as a basic infrastructure for the management of (soft and hard real-time) functions within automotive applications.

At the time being, AUTOSAR versions 1.0, 2.0, 2.1, 3.0, 3.1, 3.2, 4.0, and quite recently 4.1 as a major update of the standard, which, for example, for the first time included support for Ethernet as an in-vehicle communication network, have been released. Active development of the AUTOSAR standard is limited to the AUTOSAR 3.2 and 4.1 versions though. All other versions are in maintenance mode.

While series projects completed by today* are based on AUTOSAR 2.1 and 3.1, the next-generation vehicles currently under development will mainly be based on AUTOSAR 4.0 [22].

The AUTOSAR software architecture makes a rather strict distinction between application software and basic or system software. While the *basic (or system) software* provides functionality like communication protocol stacks for automotive communication protocols (e.g., FlexRay [23,24]), a real-time OS, and diagnostic modules, the *application software* comprises all application-specific software items (i.e., control loops, interaction with sensor and actuators). This way, the basic or system software provides the foundation the application software is built upon.

The so-called runtime environment (`Rte`) acts as an interface between application software components (SWCs) and the system software modules as well as the infrastructure services that enables communication to occur between application SWCs.

51.6.1 Application Software Architecture

The application software in AUTOSAR is structured into *SWCs* that communicate via *ports* and are interconnected by means of *connectors*. In the following sections, these concepts will be explained in detail.

51.6.1.1 Software Components

The application software in AUTOSAR is composed of reusable *SWCs*. For structuring reasons, AUTOSAR facilitates a hierarchical decomposition of the application software by distinguishing between *atomic SWCs*, which represent the bottom level of this decomposition and thus cannot be divided any further, and *composition SWCs*, which can be divided into multiple atomic SWCs and further composition SWCs.

* The resulting vehicles are already on the road.

The following types of atomic SWCs are defined by AUTOSAR:

Application SWCs: An application SWC implements (part of) the application. Since the application SWC does not directly interact with any particular hardware resources, it is hardware and location independent.

Parameter SWCs: A parameter SWC provides read access to parameter values that can be based either on constant data or on variable data that can be modified by means of calibration (see Section 51.5.1.4). Like application SWCs, the parameter SWCs do not directly interact with any particular hardware resources and are thus hardware and location independent as well.

Service SWCs: The service SWC makes the services of a local (i.e., located on the same ECU) system software module (see Section 51.6.2) available to other SWCs.

Service Proxy SWCs: This type of SWC is responsible for the distribution of mode information throughout the system. Since AUTOSAR does not allow non–service proxy SWCs to directly access to mode information provided by other ECUs, each ECU requiring such remote mode information requires a service proxy SWC.

Complex Driver SWCs: The complex driver SWC facilitates direct access to the MCU's hardware in particular for resource critical applications performing complex sensor evaluation and actuator control.

ECU-Abstraction SWCs: The ECU-abstraction SWC is a specialization of a complex driver SWC that makes the ECU's specific I/O capabilities available to the other SWCs (especially to the sensor–actuator SWCs).

Sensor–Actuator SWCs: The sensor–actuator SWC handles the specifics of a particular sensor and/or actuator. To do so, the sensor–actuator SWC directly interacts with the ECU-abstraction SWCs.

NV Block SWCs: The NV block SWCs provide fine granular (i.e., smaller than the block-based access provided by the NV random access memory [RAM] Manager [`NvM`] [see Section 51.6.2.2]) read-and-write access to NV data to other SWCs by combining these fine granular accesses and mapping them to the blocks provided by the `NvM`.

While parameter SWCs and application SWCs can be relocated among ECUs (deployed on different ECUs), the other types of SWCs are bound to a specific ECU, since they require a particular piece of hardware (e.g., a particular sensor in case of sensor–actuator SWCs and ECU-abstraction SWCs), provide access to the ECU's local NV memory (in case of NV block SWCs), or provide access to the ECU's local system software or complex drivers (in case of service SWCs and complex driver SWCs). Since the sole purpose of service proxy SWCs is to act as a local proxy for the provision of remote mode information, service proxy SWCs are bound to the specific ECU as well.

51.6.1.2 Ports and Interfaces

All SWCs interact with their environment through *ports*. Depending on whether the SWC requires or provides specific services or particular data elements, a distinction between *require ports* (*R ports*) and *provide ports* (*P ports*) is made.

Ports are typed by a specific *interface*, where AUTOSAR defines the following types of interfaces:

Sender–Receiver (S/R) Interfaces: Via a sender–receiver interface, a single *sender* distributes information to one or more *receivers* (1:n communication), or one receiver gets information from one or several senders (n:1 communication). The sender–receiver interface consists of *data elements* that define the data that are exchanged. The type of a data element can either be a simple data type (e.g., an integral value) or a complex data type (e.g., an array or a structure). The transfer of a single data element is hereby performed in an atomic fashion.

Client–Server (C/S) Interfaces: Via a client–server interface, one or more *clients* can invoke a remote operation provided by a *server* (n:1 communication). The client–server interface consists of the *operations* that are provided by the server and can be invoked by the client. This operation invocation can either be *synchronous* (i.e., the client is blocked until the result of the operation is available) or *asynchronous* (the client is not blocked but asynchronously notified as soon as the result is available).

Mode–Switch (M/S) Interfaces: Via a mode–switch interface, a *mode manager* publishes the current mode to one or more *mode users* (1:n communication) in order to have these mode users adjust their behavior according to the current mode or to synchronize activities to mode switches. The mode–switch interface consists of a so-called *mode declaration group* that is the aggregation of all distinct *mode declarations* (i.e., all different modes the mode manager can reside in).

Parameter Interface: Via a parameter interface, a parameter SWC *publishes constant data or calibration data*. Similar to the S/R interface, the parameter interface consists of *data elements* that define the data that are provided.

NV Data Interface: Via an NV data interface, an NV block SWC provides *access to NV data* on a fine-grained level. Similar to the S/R interface, the NV data interface consists of *data elements* that define the NV data that are provided.

Trigger Interface: By means of a trigger interface, an SWC is capable of *directly triggering* the execution of the *runnable entities* (see Section 51.6.1.4) of other SWCs in order to facilitate fast responses to certain events

51.6.1.3 Connectors

The ports of different SWCs can be connected by means of *connectors*. Based on the type of interface of the ports, these connectors represent the exchange of data elements (in case of S/R interfaces), the invocation of a remote operation (in case of C/S interfaces), the information about a performed mode switch (in case of M/S interfaces), the retrieval of constant or calibration data (in case of parameter interfaces), the access to NV data (in case of NV data interfaces), and the direct triggering of the execution of a runnable entity (in case of trigger interfaces). Depending on whether a connector connects two P ports or two R ports or whether the connector connects a P port with an R port, AUTOSAR distinguishes between *delegation connectors* and *assembly connectors*. Whereas the former are used for connection of ports located on *different levels of the SWC hierarchy* (i.e., connection of a port of an SWC with the corresponding port of its enclosing composition SWC), the latter connect ports within the *same level of the SWC hierarchy* (i.e., connection of corresponding ports of SWCs within the same composition SWC).

51.6.1.4 Runnable Entities

The SWCs themselves are composed of one or more *runnable entities* (*REs*) that are executed in the context of OS tasks. Runnable entities are the basic unit of execution in AUTOSAR and basically correspond to a function in common programming languages. Depending on whether or not runnable entities have internal wait points (i.e., they use blocking ICP or synchronization constructs), AUTOSAR distinguishes between *category 1 runnable entities* (without wait points) and *category 2 runnable entities* (with wait points). The former can be assigned to basic tasks of the OS (which are not allowed to invoke any blocking system calls), whereas the latter have to be mapped to extended tasks (which are allowed to use blocking system calls).

Runnable entities are invoked by `Rte` *events*. The following list names the most important `Rte` events that are defined by AUTOSAR:

Timing Event: This kind of event facilitates the periodic activation a runnable entity in an SWC.

Data Received Event: This kind of event is used to activate a runnable entity in a receiver SWC upon the reception of a data element at an S/R port.

Data Send Completed Event: This kind of event is used to activate a runnable entity in a sender SWC upon the successful transmission of a data element at an S/R port.

Operation Invoked Event: This kind of event is used to activate the runnable entity implementing a C/S operation in a server SWC upon the request of the client SWC.

Asynchronous Server Call Returns Event: This kind of event is used to activate a runnable entity in a client SWC as a result of a notification that the server SWC has completed an asynchronous C/S operation.

Mode Switch Event: This kind of event is used to activate a runnable entity in a mode user SWC as a result of a mode switch performed by a mode manager SWC.

Mode Switch Acknowledge Event: This kind of event is used to activate a runnable entity in a mode manager SWC as a result of a notification that all mode users have reacted to the mode switch.

(*Internal/External*) *Trigger Occurred Event*: This kind of event is used to activate a runnable entity in an SWC as a result of an explicit trigger by a runnable entity of the same SWC (internal) or of some other SWC (external).

Init Event: This kind of event is used to activate a runnable entity in an SWC as a result of the initialization of the `Rte`.

51.6.1.5 Communication Scopes

When looking at the different possible scopes of communication, AUTOSAR distinguishes between the following scopes:

Task: Depending on whether or not the communicating runnables are allocated to the same task, a distinction between *intra-task communication* and *inter-task communication* is made.

Partition: Depending on whether or not the tasks of the communicating runnables are allocated to the same memory partition, a distinction between *intra-partition communication* and *inter-partition communication* is made.

Core: Depending on whether or not the tasks of the communicating runnables are allocated to the same MCU core, a distinction between *intra-core communication* and *inter-core communication* is made.

ECU: Finally, depending on whether or not the SWCs of the communicating runnables are deployed to the same ECU, a distinction between *intra-ECU communication* and *inter-ECU communication* is made.

51.6.2 System Software Architecture

In addition to the application SWCs, AUTOSAR also defines a layered architecture of system software modules [25] that provides the basic platform for the execution of the application SWCs. Figure 51.8 provides a coarse-grained overview of the major layers of these system software modules.

Microcontroller Abstraction Layer: This is the lowest software layer of the system software in AUTOSAR. This layer contains software modules that directly access the microcontroller's internal peripheral devices as well as memory-mapped external devices. The task of this layer is to make higher software layers independent of the microcontroller type.

ECU-Abstraction Layer: This layer interfaces with the software modules of the microcontroller abstraction layer and additionally contains drivers for external devices. The layer offers a standardized application programming interface (API) to access the peripheral devices of an ECU regardless of their location (i.e., whether the peripherals are internal or external with respect to the ECU's microcontroller) and their connection (e.g., via port pins, via a serial peripheral interface [SPI], a.s.o.). The task of this layer is to make higher software layers independent of the ECU's hardware layout.

Services Layer: This layer is mainly located on top of the ECU-abstraction layer and provides OS services, vehicle network communication and management services, memory services (e.g., NVRAM management), diagnostic services (e.g., error logger), and state management services of the whole ECU. The task of this layer is the provision of basic services to other

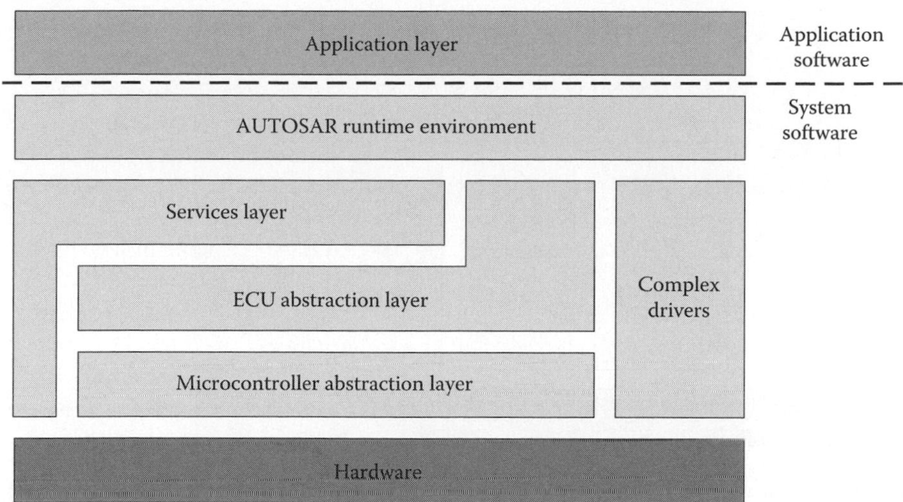

FIGURE 51.8 AUTOSAR—system software layered architecture.

system software modules and the application software. In the latter case, the respective service layer module acts as a service SWC toward the `Rte`.

Complex Drivers: The concept of complex drivers somehow intentionally violates the AUTOSAR layered architecture in order to provide the possibility to deploy legacy device drivers in an AUTOSAR system software module stack. Additionally, the complex device driver module concept has been introduced to facilitate the integration of highly application- and ECU-dependent drivers that require complex sensor evaluation and actuator control with direct access to the microcontroller itself and complex microcontroller peripherals for performance reasons. Complex drivers may provide an interface to (application) SWCs. In that case, the complex drivers act as complex driver SWCs toward the `Rte`.

`Rte`: The AUTOSAR `Rte` provides the interface between application SWCs and the system software modules as well as the infrastructure services that enable communication between application SWCs.

Application Layer: Actually, this layer is not part of the AUTOSAR system software module layered architecture, since this layer contains the AUTOSAR application SWCs described in Section 51.6.1.

Libraries: AUTOSAR libraries are *not* assigned to any particular software layer. Instead, *all* layers (including the SWCs of the application layer) are allowed to make use of these libraries.

In addition to this mainly vertical structuring, AUTOSAR further horizontally subdivides the system software modules into different substacks. This subdivision is depicted in Figure 51.9.

Input/Output Substack: The input/output substack comprises software modules that provide standardized access to sensors, actuators, and ECU onboard peripherals (e.g., D/A or A/D converters).

Memory Substack: The memory substack comprises software modules that facilitate the standardized access to internal and external NV memory for means of persistent data storage.

Communication Substack: The communication substack contains software modules that provide standardized access to vehicle networks (i.e., the local interconnect network [LIN] [26], the CAN [27,28], FlexRay [23,24], and Ethernet).

System Service Substack: Last but not least, the system service substack encompasses all software modules that provide standardized (e.g., OS, timer support, error loggers) and ECU-specific (ECU state management, watchdog management) system services and library functions.

Libraries: Since the libraries are not part of a particular substack, they are depicted separately in Figure 51.9.

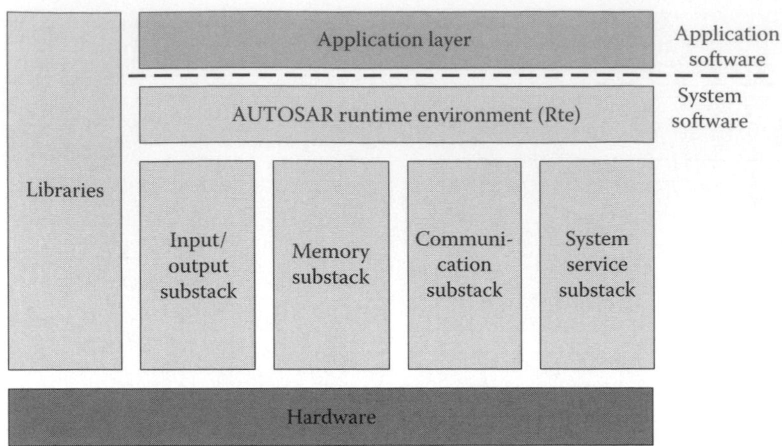

FIGURE 51.9 AUTOSAR—system software substacks overview.

Regardless of the vertical and horizontal structuring, the following classification can be applied to the AUTOSAR system software modules:

Drivers: A driver contains the functionality to *control and access an internal or an external device*. Hereby, internal devices, which are located within the microcontroller, are controlled by internal drivers, whereas external devices, which are located on ECU hardware outside the microcontroller, are controlled by external drivers. Internal drivers can usually be found within the microcontroller abstraction layer, whereas external drivers are situated in the ECU-abstraction layer. Drivers do not change the content of the data handed to them.

Interfaces: An interface contains the functionality to *abstract from the hardware realization of a specific device* and to provide a generic API to access a specific type of device independent of the number of existing devices of that type and independent of the hardware realization of the different devices. Interfaces are generally located within the ECU-abstraction layer. Interfaces do not change the content of the data handed to them.

Handlers: A handler *controls the concurrent, multiple, and asynchronous accesses of one or more clients* to one or more driver or interface modules. Thus, a handler performs buffering, queuing, arbitration, and multiplexing. Handlers do not change the content of the data handed to them.

Managers: A manager offers specific services for multiple clients. Managers are required whenever pure handler functionality is insufficient for accessing and using interface and driver modules. Managers furthermore are allowed to *evaluate and change or adapt the content of the data* handed to them. Managers are usually located in the services layer.

In the following sections, the different substacks of AUTOSAR system software modules are described in detail.

51.6.2.1 Communication Substack

The communication substack contains a group of modules that facilitate communication among the different ECUs in a vehicle via automotive communication protocols (CAN, LIN, FlexRay, and Ethernet). The structure of the communication substack is depicted in Figure 51.10.

Hereby, `<Net>` is used as a placeholder for the respective communication protocol (i.e., CAN, LIN, FlexRay, and Ethernet). Thus, the AUTOSAR communication substack contains communication

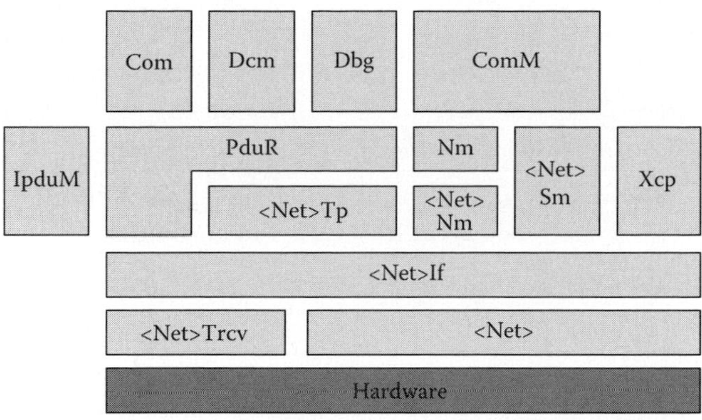

FIGURE 51.10 AUTOSAR—communication substack.

protocol-specific instances of the transport protocol (`Tp`), network management (`Nm`), interface (`If`), state manager (`Sm`), transceiver driver (`Trcv`), and driver (no suffix) modules.

The structure of the communication substack for Ethernet networks and the communication substack for CAN networks for heavy-duty vehicles according to the SAE J1939 standard, however, slightly deviates from the structure depicted in Figure 51.10. For Ethernet networks, the transport protocol functionality is distributed among two modules, namely, the `TcpIp` modules and the `SoAd` module, and two additional modules, namely, the `DoIp` module and the `Sd` module, are introduced. The structure of the communication substack for Ethernet networks is depicted in Figure 51.11.

For CAN networks for heavy-duty vehicles according to the SAE J1939 standard, dedicated modules for transport protocol (`J1939Tp`) and network management (`J1939Nm`) replace the respective modules of the communication substack for standard CAN networks, and an additional `Dcm` module (`J1939Dcm`) and a request management module (`J1939Rm`) are introduced that implement J1939-specific functionality. The structure of the communication substack for CAN networks for heavy-duty vehicles according to the SAE J1939 standard is depicted in Figure 51.12.

In the following, the different modules of the communication substack are described in detail.

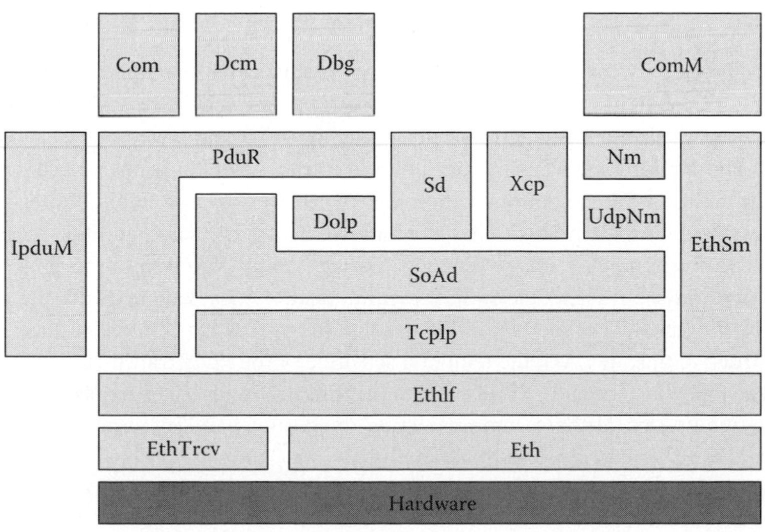

FIGURE 51.11 AUTOSAR—communication substack for Ethernet networks.

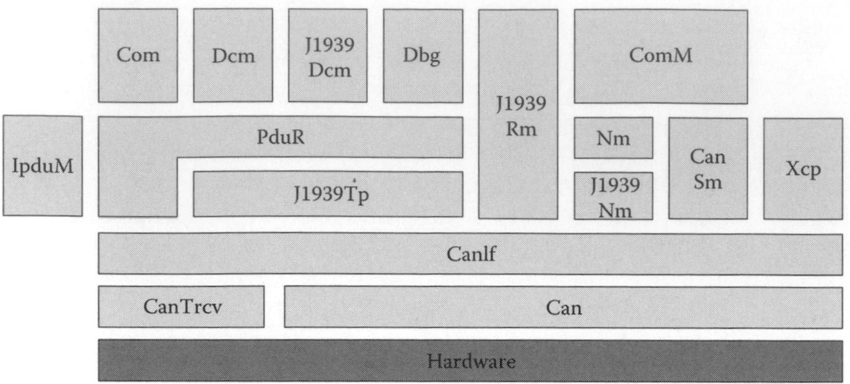

FIGURE 51.12 AUTOSAR—communication substack for CAN networks for heavy-duty vehicles.

51.6.2.1.1 Driver (**<Net>**)

The driver module (`Fr`, `Can`, `Lin`, and `Eth`) provides the basis for the respective interface module by facilitating the *transmission and the reception of frames* via the respective CC. Hereby, the driver is designed to handle multiple CCs of the same type. Thus, if an ECU contains, for example, FlexRay CCs of two different types, two different FlexRay driver modules are required.

In case of CAN, the driver module (`Can`) optionally provides the required support for TTCAN [29] and CAN FD [30] as well.

51.6.2.1.2 Transceiver Driver (**<Net>Trcv**)

The transceiver driver module (i.e., `FrTrcv`, `CanTrcv`, `LinTrcv`, and `EthTrcv`) provides API functions for *controlling the transceiver hardware* (i.e., switching the transceivers into special modes [e.g., listen only or low-power mode]) and for obtaining diagnostic information from the transceiver hardware (e.g., information about short circuits of the different bus lines of CAN or information about wake-up events on the network).

51.6.2.1.3 Interface (**<Net>If**)

Using the frame-based services provided by the driver module, the interface module (`FrIf`, `CanIf`, `LinIf`, and `EthIf`) facilitates *the sending and the reception of protocol data units (PDUs)*. Hereby, multiple PDUs can be packed into a single frame at the sending ECU and have to be extracted again at the receiving ECU.* The point in time when this packing and extracting of PDUs takes place is governed by the temporal scheduling of so-called communication jobs of the FlexRay and the LIN interface. The instant when the frames containing the packed PDUs are handed over to the driver module for transmission or retrieved from the driver module upon reception is triggered by communication jobs of the interface module as well. In FlexRay, the schedule of these communication jobs is aligned with the communication schedule; in LIN, the schedule of the LIN interface module governs the communication schedule on the LIN bus. In contrast to this, in CAN, the temporal schedule of the PDU transmission is governed by the `Com` module (see following texts). In FlexRay, each communication job can consist of one or more *communication operations*, each of these communication operations handling exactly one communication frame (including the PDUs contained in this frame).

In case of CAN, the interface module (`CanIf`) optionally provides the required support for TTCAN [29] and CAN FD [30] as well.

* Currently, only the FlexRay interface module supports the packing of multiple PDUs into a single frame. For the CAN, the LIN, and the Ethernet interface modules there is a 1:1 relationship between PDUs and frames.

In case of Ethernet, the interface module (`EthIf`) additionally provides support for virtual local area networks (VLANs) by taking care of the handling of the VLAN tags (i.e., tag protocol identifier [TPID] and tag control information [TCI]) and thus providing an abstraction from the distinction between VLAN and *normal* LANs to the upper layers.

All interface modules are designed to be able to deal with multiple different drivers for different types of CCs (e.g., freescale's MFR4300 or FlexRay CCs based on the Bosch E-Ray core in the FlexRay case). Furthermore, the interface module wraps the API provided by the transceiver driver module and provides support for multiple different transceiver driver modules (similar to the support for multiple different driver modules).

51.6.2.1.4 Transport Protocol (**`<Net>Tp`**)

The transport protocol is used to perform *segmentation and reassembly of large PDUs*. On CAN (`CanTp`) and FlexRay (`FrTp`), the used protocols are compatible (in certain configuration settings) to their respective ISO TP counterparts (see Section 51.4.1).

For FlexRay networks, AUTOSAR specifies an alternative transport protocol (`FrArTp`) that is backward compatible to the FlexRay transport protocol used in AUTOSAR 3.1 (but incompatible to the ISO TP for FlexRay).

In order to support the SAE J1939 standard and thus make AUTOSAR applicable for the use in heavy-duty vehicles, an alternative transport protocol for CAN (`J1939Tp`) is provided as well that adheres to the respective SAE J1939 specification [31].

A combination of `FrTp` and `FrArTp` or `CanTp` and `J1939Tp` on a single physical network is hereby not supported.

For the LIN communication network, AUTOSAR does not define a separate transport protocol module, but instead integrates the transport protocol functionality (which adheres to the LIN 2.1 specification [26]) into the LIN Interface (`LinIf`) module.

For the Ethernet communication network, AUTOSAR decided to reuse already well-proven protocols, namely, the IP, the Internet control message protocol (ICMP), the address resolution protocol (ARP) for IP address resolution, the UDP for unreliable connectionless communication, the TCP for reliable connection oriented communication, and the dynamic host configuration protocol (DHCP) for automated IP address assignment. The functionality of these protocols is implemented in the `TcpIp` module. The socket adapter (`SoAd`) module, which is located on top of the `TcpIp` module, maps AUTOSAR PDUs (which are identified by a unique PDU identifier) to the network endpoints of TCP connections and/or to the UDP datagrams (identified by a four-tuple of local/remote IP address and port number) by means of static configuration tables. Hereby, the `SoAd` abstracts from TCP specifics like the different methods for local IP address assignment (i.e., stateless address autoconfiguration according to RFC 3927, DHCP according to RFC 2131, or simply by means of static configuration) and details regarding the connection setup and teardown in order to provide an AUTOSAR PDU-based interface to the `PduR`. Similar to the packing of multiple PDUs into a single FlexRay frame in the `FrIf` for efficiency reasons, the `SoAd` allows for packing multiple PDUs into a single UDP datagram or TCP segment in order to achieve a decent header/payload ratio and to reduce the overhead for connection setup and teardown. Especially for the exchange and relaying of diagnostic data between an external tester device, the `DoIp` module that is located above the `SoAd` implements the services specified by the ISO DoIP standard (see Section 51.4.3) and presents itself as conventional AUTOSAR transport protocol (i.e., like `CanTp` or `FrTp`) toward the PduR by providing the respective API.

Just like with ISO TP, the user of the services provided by the various transport protocol modules is the diagnostic layer, called diagnostic communication manager in AUTOSAR (see following texts).

51.6.2.1.5 Network Management (**`<Net>Nm`, `Nm`**)

Similar to OSEK NM (see Section 51.3), the AUTOSAR NM modules provide means for the coordinated transition of the ECUs in a network into and out of a low-power (or even power-down) sleep mode. Hereby, the *network* can be the whole *physical* network with all its connected ECUs or only a *logical* network consisting of a statically defined subset of ECUs of the physical network, a so-called partial network.

The AUTOSAR NM functionality is divided into two modules: a communication protocol–independent module named generic NM (Nm) and communication protocol–dependent modules named FlexRay NM (FrNm), CAN NM (CanNm), LIN NM (LinNm), and UDP NM (UdpNm) for Ethernet networks.

In order to support the SAE J1939 standard and thus make AUTOSAR applicable for the use in heavy-duty vehicles, an alternative network management protocol for CAN (J1939Nm) is provided as well that adheres to the respective SAE J1939 specification [32]. J1939Nm, however, does not take care of a coordinated transition into and out of a low-power sleep mode, but handles the assignment of unique addresses to each ECU instead. A combination of CanNm and J1939Nm on a single physical network is hereby not supported.

51.6.2.1.6 State Manager (**<Net>SM**)

The state manager modules (CanSM, LinSM, FrSM, and EthSM) facilitate the *state management of the respective CCs* with respect to communication system–dependent start-up and shutdown features and provide a common state machine API to the upper layer (i.e., the communication manager [ComM]). This API consists of functions for requesting the communication modes Full, Silent (i.e., listen only), and No Communication.

51.6.2.1.7 XCP

For measurement and calibration purposes via CAN, FlexRay, and Ethernet networks, AUTOSAR includes an Xcp module in the communication stack, which implements the ASAM XCP specification (see Section 51.5.1).

51.6.2.1.8 PDU Router

The PDU router (PduR) module provides two major services. On the one hand, it *dispatches PDUs received via the underlying modules* (i.e., interface and transport layer modules) to the different higher layers (Com, Dcm). On the other hand, the PduR *performs gateway functionalities* between different communication networks by forwarding PDUs from one interface to another of either the same (e.g., FlexRay to FlexRay) or of different type (e.g., CAN to FlexRay). Routing decisions in the PduR are based on a static routing table and on the identifiers of the PDUs.

51.6.2.1.9 PDU Multiplexer

The PDU multiplexer (IpduM) module takes care of *multiplexing parts of a PDU*. Hereby, the value of a dedicated part of the PDU (the *multiplexer switch*) is used to define the semantic content of the remainder of the PDU (just like the tag element in a variant record or a union in programming languages). In the reception case, multiplexed PDUs are forwarded from the PduR to the IpduM for demultiplexing. Once demultiplexed, the IpduM hands the PDUs back to the PduR. In the sending case, the PduR obtains a PDU from Com and hands this PDU to the IpduM for multiplexing. The IpduM returns the multiplexed PDU to the PduR, which routes the multiplexed PDU to its final destination.

51.6.2.1.10 Communication

Similar to OSEK Com (see Section 51.3), the Com module in AUTOSAR provides *signal-based communication* to the upper layer (Rte). The signal-based communication service of Com can be used for intra-ECU communication as well as for inter-ECU communication. In the former case, Com mainly uses shared memory for this intra-ECU communication, whereas for the latter case at the sender side, Com packs multiple signals into a PDU and forwards this PDU to the PduR in order to issue the PDU's transmission via the respective interface module. On the receiver side, Com obtains a PDU from the PduR, extracts the signals contained in the PDU, and forwards the extracted signals to the upper software layer (Rte).

51.6.2.1.11 Diagnostic Communication Manager (Dcm)

The diagnostic communication manager module is a submodule of the AUTOSAR diagnostic module. The Dcm module provides *services that allow a tester device to control diagnostic functions* in an ECU via the communication network (i.e., CAN, LIN, FlexRay). Hereby, the Dcm supports the diagnostic protocols UDS and OBD (see Section 51.4.2 for details).

In order to support the SAE J1939 standard and thus make AUTOSAR applicable for the use in heavy-duty vehicles, an additional Dcm module (J1939Dcm) is provided that implements the diagnostic services mandated by the respective SAE J1939 specification [33]. Since the services provided by the J1939Dcm module are additions to the functionality provided by Dcm module, both modules can be used within a single ECU.

Both the Dcm and the J1939Dcm module act as *service SWCs* (see Section 51.6.1.1) toward the Rte and thus make their services available to (application) SWCs via dedicated ports (see Section 51.6.1.2).

51.6.2.1.12 Communication Manager (ComM)

The communication manager is a resource manager that *encapsulates the control of the underlying communication services*. The ComM collects network communication access requests from communication requesters (e.g., Dcm) and coordinates these requests by interacting with Nm and the respective state manager (<Net>SM) modules. This way the ComM provides a simplified API to the network management where a user of the API does not require any knowledge of the particular physical communication network (or partial network) to use. Via the ComM API, a user simply requests a specific communication mode (i.e., Full, Silent, or No Communication) and the ComM switches (based on a statically configured table mapping users to networks) the communication capability of the corresponding physical or partial networks to On, Silent, or Off.

The ComM module acts as a *service SWC* (see Section 51.6.1.1) toward the Rte and thus makes its services available to (application) SWCs via dedicated ports (see Section 51.6.1.2).

51.6.2.1.13 J1939 Request Manager (J1939Rm)

The J1939 request manager takes care of the SAE J1939-specific feature that the transmission of specific PDUs (termed parameter group number [PGN] in J1939) can explicitly be requested by dedicated request PDU [31]. The J1939Rm module receives this kind of request PDUs and, based on the contained information, triggers the transmission of the requested PDU.

The J1939Rm module acts as a *service SWC* (see Section 51.6.1.1) toward the Rte and thus makes its services available to (application) SWCs via dedicated ports (see Section 51.6.1.2).

51.6.2.1.14 Debugging (Dbg)

The AUTOSAR debugging module supports the remote debugging process of other AUTOSAR modules. For this purpose, the Dbg module collects information like function calls/returns and values/value changes of important variables of other modules during the runtime of the ECU without halting the processor. This, on the one hand, requires proper instrumentation of the debugged modules (e.g., the Dbg module must be notified upon function entry and exit) and, on the other hand, mandates that the relevant variables are accessible to the Dbg module. The collected data are then transmitted (either immediately or upon request) to an external debugging host system for further inspection.

51.6.2.1.15 Service Discovery (Sd)

The service discovery module provides means for a *service requester* (i.e., an ECU that requires a specific service) to detect available services that are offered by *service providers* within the vehicle network. Additionally, the Sd module facilitates to control the send behavior of event information in a way that an *event publisher* (i.e., an ECU sending the event information) only sends this information to the dynamically registered *event subscribers* (i.e., ECUs that want to receive this event information).

51.6.2.2 Memory Substack

The memory substack contains a group of modules that facilitate handling of the ECU's onboard NV memory (i.e., providing API functions to store data in and retrieve data from the ECU's NV memory [e.g., electrically erasable programmable read-only memory (EEPROM) or flash EEPROM]). The structure of the memory substack is depicted in Figure 51.13.

In the following, the different modules of the memory substack are described in detail:

Flash Driver (`Fls`): The flash driver provides services for *reading from, writing to, and erasing parts of the ECU's flash memory.* Furthermore, the flash driver facilitates the setting and resetting of the write/erase protection of the flash EPPROM if such protection is supported by the underlying hardware. In addition to these basic services, the flash driver provides a service for comparing a data block in the flash EEPROM with a data block in the memory (e.g., RAM).

EEPROM Driver (`Eep`): The EEPROM driver provides services for *reading from, writing to, and erasing parts of the ECU's EEPROM.* Additionally, similar to the flash driver, the EEPROM driver provides a service for comparing a data block in the EEPROM with a data block in the memory (e.g., RAM).

Flash EEPROM Emulation (`Fee`): The flash EEPROM emulation module *emulates EEPROM functionality* using the services provided by flash driver module. By making use of multiple flash sectors and smart copying of the data between these sectors, the `Fee` simulates an EEPROM-like behavior, that is, the possibility to perform program/erase operations on subsector granularity.

EEPROM Abstraction (`Ea`): The EEPROM abstraction module *provides uniform mechanisms to access the ECU's internal and external EEPROM devices.* It abstracts from the location of peripheral EEPROM devices (including their connection to the microcontroller), the ECU hardware layout, and the number of EEPROM devices.

Memory Abstraction Interface (`MemIf`): The memory abstraction interface module allows the NV RAM manager module (see following texts) to *access several memory abstraction modules* (`Fee` or `Ea` modules) *in a uniform way.* Hereby, the `MemIf` abstracts from the number of underlying `Fee` or `Ea` modules providing a runtime translation of each block access initiated by the NV RAM manager module to select the corresponding driver functions that are unique for all configured EEPROM or flash EEPROM storage devices.

NV RAM Manager (`NvM`): The NV RAM manager module provides services to *ensure the data storage and maintenance of NV data* according to their individual requirements in an automotive environment, namely, synchronous as well as asynchronous services for the initialization, the reading, the writing, and the control of NV data. The `NvM` operates on a block basis distinguishing the following types of blocks: For *native blocks*, the `NvM` provides a *RAM mirror* that contains a copy of the data stored in the NV memory block. This RAM mirror is initialized with the data from the NV block upon ECU power-up. Upon ECU shutdown, the data from the RAM mirror is flushed to the corresponding NV memory block. Additionally, the `NvM` provides API services that can force the transfer of a memory block from NV memory into

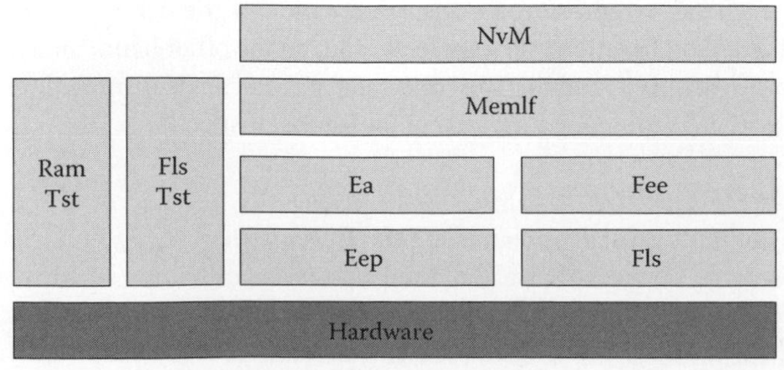

FIGURE 51.13 AUTOSAR—memory substack.

the corresponding RAM mirror and vice versa. In addition to the facilities of native blocks, *redundant blocks* provide enhanced fault tolerance, reliability, and availability. Due to replication of the redundant block in NV memory, the resilience against data corruption is increased.

The NvM module acts as a *service SWC* (see Section 51.6.1.1) toward the `Rte` and thus makes its services available to (application) SWCs via dedicated ports (see Section 51.6.1.2).

RAM Test (`RamTst`): The RAM test module provides means to perform *functional tests of the ECU's internal RAM cells. Complete* tests are performed upon ECU start-up and shutdown as well as on request by special diagnostic commands. During operation, *partial tests* are performed in a periodic manner (e.g., block by block or cell by cell). For both types of tests, several RAM test algorithms, which have been chosen according to the IEC 61508 standard, are available. Depending on the algorithms' diagnostic coverage rate, the algorithms are divided into the following categories: group 1 (low) with a diagnostic coverage rate smaller than 60%, group 2 (medium) exhibiting a diagnostic coverage rate of 60%–90%, and group 3 (high) with a diagnostic coverage rate of 90%–99%.

Flash EEPROM Test (`FlsTst`): The flash EEPROM test module provides means to perform *functional tests on the content of the ECU's nonmutable memory cells (e.g., program/data flash EEPROM content). Foreground tests* are performed on a per-block basis upon request via the module's API functions. *Background tests* are performed in a periodic manner for a set of configured blocks. Both kind of tests compute a configurable signature over the memory block (e.g., 8-, 16-, and 32-bit CRCs or hashes) and compare the computed signature value with a precomputed value stored in the `FlsTst` module's configuration.

51.6.2.3 Input/Output Substack

The input/output substack contains a group of modules that facilitate the handling of the ECU's input/output capabilities. The structure of the input/output substack is depicted in Figure 51.14.

In the following, the different modules of the input/output substack are described in detail:

Port Driver (`Port`): The port driver module provides the service for *initializing the whole port structure* of the microcontroller, allowing for the configuration of different functionalities for each port and port pin (e.g., analog–digital conversion [ADC], digital I/O [DIO]). Hereby, the port pin direction (input/output), the initial level of the port pin, and the fact whether the port pin direction is modifiable during runtime are part of this configuration. Other I/O drivers (e.g., `Icu`, `Pwm`) rely on the configuration performed by the port driver.

Interrupt Capture Unit Driver (`Icu`): The ICU driver is a *module using the ICU hardware* to implement services like signal edge notification, controlling of wake-up interrupts, periodic signal time measurement, edge time stamping (usable for the acquisition of nonperiodic signals), and edge counting. Hereby, the `Icu` module works on pins and ports that have been properly configured by the port driver for this purpose.

Pulse Width Modulation Driver (`Pwm`): The PWM driver module provides functions to *initialize and control the hardware PWM unit* of the microcontroller. The `Pwm` module allows for the generation of pulses with variable pulse width by facilitating the selection of the duty cycle and the signal period time. The `Pwm` module supports multiple PWM channels, where each channel is linked to a hardware PWM unit that belongs to the microcon-

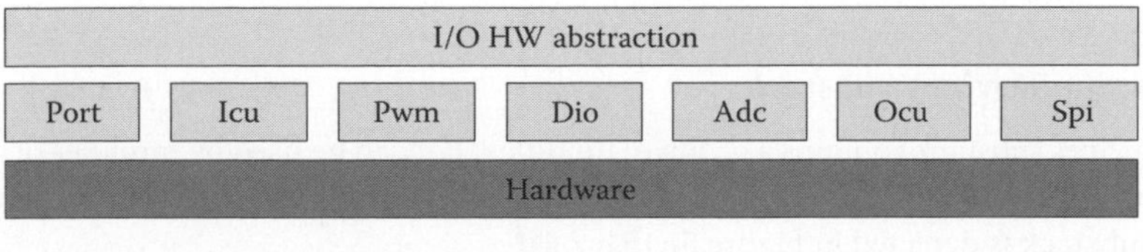

I/O HW abstraction						
Port	Icu	Pwm	Dio	Adc	Ocu	Spi
Hardware						

FIGURE 51.14 AUTOSAR—input/output substack.

troller. Similar to the `Icu` module, the `Pwm` module relies on pins and ports that have been properly configured by the port driver for this purpose.

Digital I/O Driver (`Dio`): The DIO driver provides services for *reading from and writing to DIO channels* (i.e., port pins), DIO ports, and groups of DIO channels. Hereby, the `Dio` module works on pins and ports that have been properly configured by the `Port` driver for this purpose.

Analog/Digital Converter Driver (`Adc`): The ADC driver module *initializes and controls the internal ADC unit(s)* of the microcontroller. The module provides services to start and stop an analog–digital conversion, respectively, to enable and disable the trigger source for a conversion. Furthermore, the module provides services to enable and disable a notification mechanism and routines to query status and result of a conversion. The `Adc` module works on so-called ADC channel groups. An ADC channel group combines an ADC channel (i.e., an analog input pin), the needed ADC circuitry itself, and a conversion result register into an entity that can be individually controlled and accessed via the `Adc` module. The `Adc` module operates on pins and ports that have been properly configured by the port driver for this purpose.

Output Compare Unit Driver (`Ocu`): The OCU driver module provides functions to *initialize and control the hardware OCU unit* of the microcontroller. The `Ocu` module allows comparing and acting automatically (e.g., by calling a notification function) when the value of a free-running counter matches a defined threshold. The module provides services to set the comparison threshold value, to start and stop a comparison, to enable and disable a notification mechanism, and to query the current counter value. The `Ocu` module works on so-called OCU channels. An `OCU` channel group combines a free-running counter and the corresponding threshold value into an entity that can be individually controlled and accessed via the `Ocu` module. The `Ocu` module relies on pins and ports that have been properly configured by the port driver for this purpose.

Standard Peripheral Interface Handle/Driver (`Spi`): The SPI handler/driver module provides services for reading from and writing to peripheral devices connected via SPI buses. These peripheral devices are explicitly addressed by means of a chip select (CS) line. For each peripheral device, a so-called SPI job consisting of multiple *SPI channels* can be configured in the `Spi` module's configuration. SPI channels hereby basically define the atomic units of data transferred between the microcontroller and a peripheral device. SPI jobs are grouped into *SPI sequences* to specify a defined order of data exchange.

Based on these SPI sequences, the `Spi` module provides services to synchronously and asynchronously conduct data transfers and to query the status of such a transfer on SPI sequence and SPI job granularity and notification services upon completion of a particular SPI sequence and/or a particular SPI job.

I/O Hardware Abstraction: The I/O hardware abstraction module provides a signal-based interface to internal and external I/O devices of an ECU. Hereby, the module abstracts from whether a certain I/O device is an MCU internal device or whether a device is externally connected to the MCU, by performing static normalization/inversion of values according to their physical representation at the inputs/outputs of the ECU hardware (i.e., static influences, like voltage division or hardware inversion, on the path between the I/O device and the MCU port pin are compensated).

The I/O hardware abstraction module acts as an *ECU-abstraction SWC* (see Section 51.6.1.1) toward the `Rte` and thus makes its services available to (application) SWCs via dedicated ports (see Section 51.6.1.2).

51.6.2.4 System Services Substack

The system services substack contains a group of modules that can be used by modules of all AUTOSAR layers. Examples are real-time OS, error handler, and watchdog management. The structure of the system services substack is depicted in Figure 51.15.

In the following, the different modules of the system services substack are described in detail.

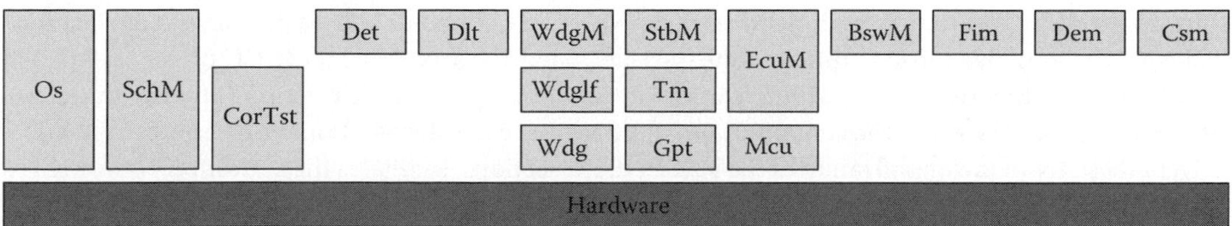

FIGURE 51.15 AUTOSAR—system services substack.

51.6.2.4.1 Operating System (`Os`)

The AUTOSAR OS provides *real-time OS services* to both the other system software modules and the application SWCs of AUTOSAR. The `Os` module is configured and scaled statically, provides a priority-based scheduling policy and protective functions with respect to memory and timing at runtime, and is designed to be hostable on low-end controllers.

Similar to the OSEKtime dispatcher tables (see Section 51.3), AUTOSAR `Os` provides so-called schedule tables consisting of one or more *expiry points*. Hereby, each expiry point is assigned an offset measured in `Os` ticks from the start of the schedule table. Once an expiry point is reached, the action corresponding to the expiry point (e.g., the activation of a task or the setting of an event) is processed. At runtime, the `Os` iterates over the schedule table, processing each expiry point in turn. The iteration is driven by an `Os` counter. In order to facilitate the execution of tasks synchronous to external events (e.g., synchronous to the FlexRay communication schedule), schedule tables can be synchronized to external time sources (e.g., FlexRay's global time).

As far as protection against timing violations is concerned, AUTOSAR `Os` does not provide deadline monitoring (as does OSEKtime `OS`) but provides the facility to track the execution time of each task and ISR and to raise an error in case either exceeds its statically assigned execution time budgets. Regarding memory protection, AUTOSAR `Os` uses the memory protection unit of the MCU to provide coarse-grained memory protection of so-called OS applications, which are a group of related tasks, against each other.

With AUTOSAR 4.0 onward, the AUTOSAR `Os` provides support for multicore MCUs by allowing different OS applications to be statically allocated to the different cores and by supporting data exchange among these cores by means of the inter-OS application communicator (`Ioc`) submodule.

The `Os` module acts as a *service SWC* (see Section 51.6.1.1) toward the `Rte`, thus making its services available to (application) SWCs via dedicated ports (see Section 51.6.1.2).

51.6.2.4.2 Basic Software Scheduler (`SchM`)

The basic software (BSW) scheduler module provides means to embed other AUTOSAR system software module implementations into the context of an AUTOSAR `Os` task or ISR, trigger main processing functions of the system software modules, and apply data consistency mechanisms for these modules. Just like the `Rte` provides the infrastructure for SWCs by embedding runnable entities in a task context, the `SchM` module *provides the infrastructure for the other system software modules* by embedding their main processing functions (termed schedulable entities) in a task context.*

Due to the rather similar functionality between `Rte` and `SchM` and to allow for global scheduling and data consistency optimizations (e.g., automatic elision of data consistency mechanisms in case the involved runnable and schedulable entities cannot preempt one another), the `SchM` module's functionality is integrated into the `Rte` with AUTOSAR 4.0 onward.

51.6.2.4.3 MCU Driver (`Mcu`)

The MCU driver module provides services for *basic microcontroller initialization*, *power-down functionality*, *microcontroller reset*, and microcontroller-specific functions required from other system

* Usually, the main processing functions of multiple system software modules are embedded into a single task in order to keep the number of tasks required for execution of the whole AUTOSAR system software low.

software modules. The initialization services of the MCU driver module allow a flexible and application-related MCU initialization in addition to the start-up code.* The services of the MCU driver include the initialization of the MCU's clock, initialization of the MCU's phase-locked loop (PLL), the initialization of clock prescalers, and the configuration of the MCU's clock distribution. Furthermore, the MCU driver takes care of the initialization of the MCU's RAM sections, facilitates the activation of the MCU's reduced power modes (i.e., putting the MCU into a low-power mode), and provides a service for enforcing a reset of the MCU and a service for obtaining the reset reason from the MCU hardware.

51.6.2.4.4 *ECU State Manager* (`EcuM`)

The ECU state manager module manages all aspects of the ECU related to the `Off`, `Run`, and `Sleep` states of that ECU and the transitions between these states like *start-up and shutdown*. In detail, the ECU state manager is responsible for the initialization and deinitialization of all system software modules including `Os` and `Rte`; cooperates with the `ComM`, and hence indirectly with `Nm`, to shut down the ECU (`Off` state) when needed; manages all wake-up events; and configures the ECU for `Sleep` state when requested. In order to fulfill all these tasks, the `EcuM` makes use of the services provided by the `Mcu` module and implements some important protocols: the *run request protocol*, which is needed to coordinate whether the ECU must be kept alive or is ready to shut down; the *wake-up validation protocol* to distinguish *real* wake-up events from *erratic* ones; and the *time-triggered increased inoperation protocol*, which allows to put the ECU into an increasingly energy-saving `Sleep` state depending on the duration of the ECU's inactivity.

With AUTOSAR 4.0, a distinction between an ECU state manager module with *fixed state machine* (`EcuMfixed`) and an ECU state manager module with *flexible state machine* (`EcuMflex`) has been made. While the former one implements all of the previously mentioned services by itself in a rather fixed way, the latter one focuses on the early start-up and late shutdown phases and initializes only a small number of other system software modules, delegating the rest of the functionality to the basic software mode manager (`BswM`) module.

The `EcuM` module (flex and fixed) acts as a *service SWC* (see Section 51.6.1.1) toward the `Rte` and thus makes its services available to (application) SWCs via dedicated ports (see Section 51.6.1.2).

51.6.2.4.5 *Basic Software Mode Manager* (`BswM`)

The basic software mode manager module arbitrates requests from application SWCs or other system software modules based on statically configurable rules and performs actions based on the arbitration result. Hereby, these configurable rules comprise one or more logical conditions that are combined into a logical expression and a list of actions that are executed in case the logical expression evaluates to true or to false, respectively. The logical conditions can be mode switches performed and announced by other modules and/or SWCs. The performed actions can be mode switches performed by the `BswM` itself that are announced towards other modules and/or SWCs. Thus, the `BswM`, on the one hand, acts as a mode user and, on the other hand, as a mode manager.

Additionally, the `BswM` is in charge of the part of initialization and shutdown process that is not covered by the `EcuMflex` module.

The `BswM` module acts as a *service SWC* (see Section 51.6.1.1) toward the `Rte` and thus makes its services available to (application) SWCs via dedicated ports (see Section 51.6.1.2).

51.6.2.4.6 *Diagnostic Event Manager* (`Dem`)

The diagnostic event manager module realizes part of the diagnostic functionality within AUTOSAR. The `Dem` is responsible for *processing and persistently storing diagnostic events/errors*† and associated data (so-called freeze frame data). To facilitate the persistent storage of these DTCs, the `Dem` makes use

---

* The start-up code itself is not within the scope of AUTOSAR.

† These events can be mapped to UDS/OBD DTCs by means of the `Dem`'s module configuration.

of the services provided by the NvM. Application SWCs as well as other system software modules can raise diagnostic events by means of Dem API calls.

The diagnostic events registered by the Dem serve as triggers for state updates of the Fim and thus might lead to the inhibition of certain runnables. Upon request of the Dcm, the Dem provides an up-to-date list of the currently stored DTCs that are then sent to a tester client by means of the Dcm services.

The Dem module acts as a *service SWC* (see Section 51.6.1.1) toward the Rte and thus makes its services available to (application) SWCs via dedicated ports (see Section 51.6.1.2).

51.6.2.4.7 Function Inhibition Manager (**Fim**)

Like the Dem and the Dcm module, the function inhibition manager realizes part of the diagnostic functionality in AUTOSAR. The Fim is responsible for providing an *execution control mechanism for the runnables* of application SWCs and system software modules. By means of the Fim, these runnables can be inhibited (i.e., deactivated) according to the Fim's static configuration. The functionalities of the runnables are assigned to a unique function identifier (FID) along with an inhibit condition for that particular FID. The functionalities poll for the permission state of their respective FIDs before execution. If an inhibit condition is true for a particular FID, the corresponding functionality is not executed anymore.

The Fim is closely related to the Dem since diagnostic events and their status information can serve as possible inhibit conditions. Hence, functionality that needs to be stopped in case of a failure can be represented by a particular FID. If the failure is detected and the event is reported to the Dem, the Fim then inhibits the FID and therefore the corresponding functionality.

The FiM module acts as a *service SWC* (see Section 51.6.1.1) towards the Rte and thus makes its services available to (application) SWCs via dedicated ports (see Section 51.6.1.2).

51.6.2.4.8 Watchdog Driver (**Wdg**)

This module provides services for initialization, changing of the operation mode (Fast, Slow, Off) and *triggering the ECU's watchdog device.* In case an ECU provides multiple different watchdog devices (e.g., internal and external devices), a dedicated Wdg module has to be present for each of the devices.

51.6.2.4.9 Watchdog Interface (**WdgIf**)

In case of more than one watchdog device and corresponding watchdog driver (e.g., both an internal software watchdog and an external hardware watchdog) is being used on an ECU, the watchdog interface module allows the watchdog manager module (see following texts) to select the correct watchdog driver—and thus the watchdog device—via a device index while retaining the API and functionality of the underlying driver. Thus, the WdgIf module provides *uniform access to services of the underlying watchdog drivers* like mode switching and triggering.

51.6.2.4.10 Watchdog Manager (**WdgM**)

The watchdog manager module is intended to *supervise the program execution* of application SWCs or other system software modules, so-called supervised entities. The WdgM provides services for *alive supervision* for periodic software, *deadline supervision* for aperiodic software, and *logical supervision* monitoring of the program flow of a supervised entity. Hereby, the WdgM monitors each supervised entity autonomously and derives a local status for this supervised entity. The aggregation of the local status of all supervised entities then yields a global status. Based on this global supervision status, the WdgM decides whether or not to trigger the hardware watchdog via the WdgIf's API. Hereby, the set of supervised clients, the desired degree of supervision, and the individual timing constraints are defined by configurable parameters of the WdgM.

The WdgM module acts as a *service SWC* (see Section 51.6.1.1) toward the Rte and thus makes its services available to (application) SWCs via dedicated ports (see Section 51.6.1.2).

51.6.2.4.11 *Development Error Tracer* (Det)

The development error tracer module is the central instance where all other system software modules *report detected development errors* to. The API parameters handed to the Det allow for tracing the source and kind of error, namely, the module and the function in which the error has been detected and the type of the error. The functionality behind the API of the Det is not specified in AUTOSAR. Possible functionalities could be the setting of debugger breakpoints within the error reporting API, the counting of the number of reported errors, the logging of Det calls together with the passed parameters to a RAM buffer for later analysis, and the sending of reported errors via some communication interface (e.g., CanIf) to external logger devices.

The Det module acts as a *service SWC* (see Section 51.6.1.1) towards the Rte and thus makes its services available to (application) SWCs via dedicated ports (see Section 51.6.1.2).

51.6.2.4.12 *Diagnostic Log and Trace* (Dlt)

The diagnostic log and trace module provides services for the logging of errors, warnings, and information messages to SWCs, as well as to the Det and the Dem module. Additionally, the Dlt modules provide trace functionality to the Rte to facilitate the tracing of relevant Rte events (see Section 51.6.1.4). The detail of both the logging and the tracing can hereby be controlled during runtime. The logged/traced data are then transmitted via the PduR or by means of the data transmission services of the Dcm.

The Dlt module acts as a *service SWC* (see Section 51.6.1.1) toward the Rte and thus makes its services available to (application) SWCs via dedicated ports (see Section 51.6.1.2).

51.6.2.4.13 *General-Purpose Timer Driver* (Gpt)

The general-purpose timer (GPT) driver module provides services for starting and stopping a functional timer instance within the hardware GPT module and thus *provides exact and short-term timings* for use in the Os or within other system software modules where an Os alarm service causes too much overhead. Individual time-out periods (*single-shot mode*) as well as repeating time-out periods (*continuous mode*) can be generated via the Gpt module. The user can configure whether a notification shall be invoked when the requested time-out period has expired. These notifications can be enabled and disabled at runtime. Both, the relative time elapsed since the last notification occurred and the time remaining until the next notification will occur, can be queried via API functions of the Gpt module. Additionally, the Gpt module provides a set of free-running timers together with API functions to obtain the current time value of these free-running timers.

51.6.2.4.14 *Time Service* (Tm)

The time service module builds on the free-running timers of the Gpt module and provides a set of *predefined timers with defined tick duration in physical time units* (e.g., microseconds) together with a set of API functions to reset these timers, to (busy) wait for a guaranteed minimum waiting time, to set the timers to a reference time, and to compute the time difference between the current time and this reference time. This way, compatibility of time-based functionality is ensured for all platforms that support the required predefined timers.

51.6.2.4.15 *Synchronized Time Base Manager* (StbM)

The synchronized time base manager *provides a global time* to other system software modules or to the application SWCs (so-called time base users). Hereby, the StbM itself, however, does not provide any facility (e.g., synchronization protocols) for establishing a synchronized time base across multiple ECUs, but relies on the existence of such protocols at so-called time base providers (e.g., the FrIf that provides a synchronized time base for all ECUs connected to the specific FlexRay network). To facilitate a triggering of runnable entities of SWCs or schedulable entities of other system software modules that is synchronous to this global time base, the StbM is capable of synchronizing the schedule tables of the Os to this global time base.

The StbM module acts as a *service SWC* (see Section 51.6.1.1) towards the Rte and thus makes its services available to (application) SWCs via dedicated ports (see Section 51.6.1.2).

51.6.2.4.16 Crypto Service Manager (Csm)

The crypto service manager module provides *cryptographic services* based on a software library or on a dedicated hardware module. These services include means to compute and verify checksums and hash values over a block of data elements; means to generate random numbers; means for symmetric and asymmetric en- and decryption of blocks of data; message authentication and integrity checks by means of digital signatures; derivation of one or more secret keys using a key derivation function; wrapping, serialization, and deserialization of keys for transportation purposes; and generation and secure exchange of shared keys.

The Csm module acts as a *service SWC* (see Section 51.6.1.1) towards the Rte and thus makes its services available to (application) SWCs via dedicated ports (see Section 51.6.1.2).

51.6.2.4.17 Core Test (CorTst)

The core test module provides means to perform *functional tests related to basic functionality of the MCU's core(s)*, namely, tests verifying the correct functionality of the core's registers, the arithmetic logical unit (ALU), the interrupt controller and the exception handling, the memory interface, the memory protection unit (MPU) in case one is available, and the cache controller (w.r.t., cache coherency and consistency). Similar to the RamTst module, the CorTst module supports *foreground tests* that are performed upon request via the module's API functions and *background test* that are performed in a periodic manner.

51.6.2.5 Libraries

In addition to the previously described system software modules, AUTOSAR defines a set of libraries. These libraries are neither assigned to any particular software layer nor to a particular substack. Instead, *all* layers (including the SWCs of the application layer) as well as integration code are allowed to make use of these libraries. To facilitate this flexible use, AUTOSAR defines that the code of libraries

- Is executed synchronously in the context of the caller in the same protection environment
- Is only allowed to call other library code (i.e., library code must not call the code of system software modules or SWCs)
- Is reentrant and does not have any internal state
- Does not require any initialization

As far as the provided functionality is concerned, AUTOSAR specifies libraries for fixed point mathematics, floating point mathematics, interpolation routines for fixed point data, interpolation routines for floating point data, bit handling routines, end-to-end communication protection, CRC calculation, filtering routines, and cryptographical routines.

In the following, a subset of the libraries provided by AUTOSAR is discussed in more detail:

Cyclic Redundancy Check (Crc): The cyclic redundancy check (CRC) library provides bitwise, table-based, and (if available) hardware-assisted computation of CRCs using different 8-, 16-, and 32-bit generator polynomials, namely, the SAE-J1850 CRC8, the CCITT-FALSE CRC16, and the IEEE-802.3 CRC32 used for Ethernet.

End-to-End Protection (E2e): The end-to-end protection library provides means to protect safety-related data exchange at runtime against the effects of faults along the communication path, like random HW faults (e.g., corrupt registers of a CC), interference (e.g., due to EMC), and systematic faults within the communication substack. To this end, additional control fields like CRCs or sequence counters are added to the transmitted data at the sender side. At the receiver side, these control fields are evaluated and checked for correctness. In case this check fails, the received data are marked as corrupt. In order to support the use with different fault hypotheses, multiple profiles that differ in the added control fields (e.g., the strength of the used CRC) are supported.

Crypto Abstraction (`Cal`): From the functional point of view, the crypto abstraction library provides the same services as the `Csm` module. The only difference is the fact that the `Cal` is implemented as a library adhering to all the restrictions AUTOSAR imposes on the implementation of libraries (see earlier text).

51.7 Summary

Since 1993, the major automotive companies are striving for the deployment of standard software modules in their applications to achieve an increased test coverage and higher reliability, requirements that can only be met if standardized modules are used at various system levels.

This chapter provided an overview of today's industry practices in standardized automotive system software. Existing standards proposed by industry partnerships like OSEK/VDX, ASAM, and AUTOSAR and by standardization authorities like ISO have been presented. Of all presented approaches, the AUTOSAR partnership that started off in May 2003 and started putting software modules according to the AUTOSAR standard into production vehicles in 2008 (BMW being first to deploy AUTOSAR in its 7 Series car followed by Daimler, Audi, VW, PSA, and others) turns out to be the most promising one.

This is, on the one hand, due to the fact that several of already approved standards used today (like OSEK OS and COM, ASAM XCP, and ISO transport layer and diagnostics) heavily inspired the corresponding AUTOSAR standards, ensuring that the AUTOSAR standard is built on well-proven technology. On the other hand, the AUTOSAR open standard has massive industrial backup: all AUTOSAR core members including seven of the world's biggest vehicle manufacturers accounting for about 55% of all vehicles produced are strongly committed to the project and released the first revision of the AUTOSAR 4.1 specifications in March 2013, which will serve as a solid basis for future automotive software development.

References

1. S. Fürst. Challenges in the design of automotive software. In *Proceedings of the Conference on Design, Automation and Test in Europe, DATE'10*, Dresden, Germany 2010, pp. 256–258. European Design and Automation Association, Heverlee, Belgium.
2. P. Hansen. AUTOSAR standard software architecture partnership takes shape. *The Hansen Report on Automotive Electronics*, 17(8):1–3, October 2004.
3. OSEK/VDX. OSEK VDX Portal.http://portal.osek-vdx.org/ (Retrieved May, 2013).
4. ASAM—Association for standardization of automation and measuring systems. http://portal.osek-vdx.org/ (Retrieved May, 2013).
5. HIS—Herstellerinitiative software. http://www.automative-his.de/ (Retrieved May, 2013).
6. JASPAR—Japan automotive software platform architecture. http://www.jaspar.jp/english/index-e.php. (Retrieved May 2013).
7. EASIS—Electronic architecture and system engineering for integrated safety systems. http://cordis.europa.eu/projects/507690 (Retrieved May 2013).
8. H. Fennel, S. Bunzel, H. Heinecke, J. Bielefeld, S. Fürst, K.-P. Schnelle, W. Grote et al. Achievements and exploitation of the AUTOSAR development partnership. In *Proceedings of the Convergence 2006*, number SAE 2006-21-0019, Detroit, MI, October 2006.
9. GENIVI Alliance. http://www.genivi.org/ (Retrieved May 2013).
10. T. Wollstadt, W. Kremer, J. Spohr, S. Steinhauer, T. Thurner, K.J. Neumann, H. Kuder et al. OSEK/VDX—Operating system, Version 2.2.3. Technical report, OSEK, February 2005.
11. J. Goodenough and L. Sha. The priority ceiling protocol: A method for minimizing the blocking of high-priority Ada tasks. Technical report SEI-SSR-4, Software Engineering Institute, Pittsburgh, PA, May 1988.

12. C. Hoffmann, J. Minuth, J. Krammer, J. Graf, K.J. Neumann, F. Kaag, A. Maisch et al. OSEK/VDX—Network management—Concept and application programming interface, Version 2.5.3. Technical report, OSEK, July 2004.
13. J. Spohr et al. OSEK/VDX—Communication, Version 3.0.3. Technical report, OSEK, July 2004.
14. V. Barthelmann, A. Schedl, E. Dilger, T. Führer, B. Hedenetz, J. Ruh, M. Kühlewein et al. OSEK/VDX—Time-triggered operating system, Version 1.0. Technical report, OSEK, July 2001.
15. A. Schedl, E. Dilger, T. Führer, B. Hedenetz, J. Ruh, M. Kühlewein, E. Fuchs et al. OSEK/VDX—Fault-tolerant communication, Version 1.0. Technical report, OSEK, July 2001.
16. J. Spohr et al. OSEK/VDX—System generation—OIL: OSEK implementation language, Version 2.5. Technical report, OSEK, July 2004.
17. ISO. Road vehicles—Diagnostics on controller area networks (DoCAN)—Part 2: Transport protocol and network layer services. Technical report ISO 15765-2:2011, ISO (International Organization for Standardization), Geneva, Switzerland, 2011.
18. ISO. Road vehicles—Communication on FlexRay—Part 2: Communication layer services. Technical report ISO 10681-2:2010, ISO (International Organization for Standardization), Geneva, Switzerland, June 2010.
19. ISO. Road vehicles—Unified diagnostic services (UDS)—Part 1: Specification and requirements. Technical report ISO 14229-1:2013, ISO (International Organization for Standardization), Geneva, Switzerland, 2013.
20. ISO. Road vehicles—Diagnostic communication over Internet Protocol (DoIP)—Part 2: Transport protocol and network layer services. Technical report ISO 13400-2:2012, ISO (International Organization for Standardization), Geneva, Switzerland, 2012.
21. R. Schuermans, R. Zaiser, F. Hepperle, H. Schröter, R. Motz, A. Aberfeld, H.-G. Kunz et al. XCP—The universal measurement and calibration protocol family, Version 1.1.0. Technical report, Association for Standardisation of Automation and Measuring Systems (ASAM), Höhenkirchen, Germany, March 2008.
22. L. Lundth. About AUTOSAR achievements phase III, exploitation and rollout. In *Proceedings of the Fifth AUTOSAR Open Conference*, Beijing, China, November 2012.
23. T. Führer, F. Hartwich, R. Hugel, and H. Weiler. FlexRay—The communication system for future control systems in vehicles. In *Proceedings of the SAE 2003 World Congress & Exhibition*, number SAE 2003-01-0110, Detroit, MI, March 2003. Society of Automotive Engineers, Warrendale, PA.
24. R. Mores, G. Hay, R. Belschner, J. Berwanger, C. Ebner, S. Fluhrer, E. Fuchs et al. FlexRay—The communication system for advanced automotive control systems. In *Proceedings of the SAE 2001 World Congress*, number SAE 2006-21-0019, Detroit, MI, March 2001. Society of Automotive Engineers, Warrendale, PA.
25. AUTOSAR Consortium. AUTOSAR—Layered software architecture. Technical report Version 3.3.0, Release 4.1, Rev. 1, AUTOSAR Consortium, Munich, Germany, January 2013.
26. LIN Consortium. LIN specification package. Technical report Version 2.1, LIN Consortium, December 2010.
27. ISO. Road vehicles—Controller area network (CAN)—Part 1: Data link layer and physical signalling. Technical report ISO 11898-1:2003, ISO (International Organization for Standardization), Geneva, Switzerland, 2003.
28. ISO. Road vehicles—Controller area network (CAN)—Part 2: High-speed medium access unit. Technical report ISO 11898-2:2003, ISO (International Organization for Standardization), Geneva, Switzerland, 2003.
29. T. Führer, B. Müller, W. Dieterle, F. Hartwich, R. Hugel, and M. Walther. Time triggered communication on CAN. In *Proceedings of the Seventh International CAN Conference, CAN in Automation*, Amsterdam, the Netherlands, 2000.
30. F. Hartwich. CAN with flexible data-rate. In *Proceedings of the 13th International CAN Conference, CAN in Automation*, Hambach, Germany, 2012.

31. Society of Automotive Engineers (SAE) International. Recommended practice for a serial control and communications vehicle network—Data link layer. Technical report J1939/21, Society of Automotive Engineers (SAE) International, December 2010.
32. Society of Automotive Engineers (SAE) International. Recommended practice for a serial control and communications vehicle network—Network management. Technical report J1939/81, Society of Automotive Engineers (SAE) International, Warrendale, PA, June 2011.
33. Society of Automotive Engineers (SAE) International. Recommended practice for a serial control and communications vehicle network—Application layer—Diagnostics. Technical report J1939/73, Society of Automotive Engineers (SAE) International, Warrendale, PA, February 2010.

52

Protocols and Services in Controller Area Networks

52.1 Introduction 52-1
CAN in Automotive Applications • CAN in Automation and Networked Control Systems
52.2 CAN Protocol Basics 52-3
Physical Layer • Frame Format • Medium Access Technique • Fault Confinement • Logical Link Control
52.3 Schedulability of CANs 52-17
Communication Requirements • Schedulability Analysis • Usage
52.4 CAN Advantages and Drawbacks 52-22
Advantages • Drawbacks and Solutions
52.5 Time-Triggered CAN 52-27
Main Features • Protocol Specification
52.6 CAN FD 52-30
Main Features • Protocol Specification
52.7 CANopen 52-32
Protocol Basics • Device Model • Process Data Objects • Service Data Objects • Other Objects • Network Management
52.8 Interoperability between Devices 52-44
Profiles in CANopen • Device Profile for Generic I/O Modules • Device Behavior
References 52-48

Gianluca Cena
National Research Council of Italy

Adriano Valenzano
National Research Council of Italy

52.1 Introduction

Controller area network (CAN) is one of the leading fieldbuses for networked control systems. Since the history of CAN goes back to 1986, some billion CAN nodes have been sold worldwide. At the beginning of the 1980s, most communication technologies, which were available at that time, were not adequate for in-car communications and unable to fulfill the requirements of a number of automotive applications. For these reasons, Bosch GmbH started a project to develop a new serial bus system suitable for use in passenger cars. The development of CAN also involved academic partners and was partially supported by Intel, which was interested as a potential semiconductor producer.

The new fieldbus called *automotive serial controller area network* (ASAN) appeared officially in 1986 at the Society of Automotive Engineers (SAE) congress in Detroit and included a number of features that were not present in competing fieldbus architectures. ASAN was a real multimaster network, and its access technique to the shared medium resembled the popular carrier sense multiple access (CSMA) approach. The peculiar aspect, however, was the adoption of a new distributed, nondestructive arbitration mechanism, which was able to solve contentions on the bus in real time, by leveraging priorities implicitly assigned to the colliding messages. The robustness and fault tolerance of the whole system were also enhanced by the introduction of several error detection and management mechanisms.

In the following years, controller chips for CAN began to appear and spread on the market. Intel and Philips, in particular, were the main producers, even though they followed two significantly different approaches, whose consequences can be observed in components available off the shelf still today. Actually, the Intel solution (often referred to as FullCAN in the literature) required less host CPU power, since most of the communication and network management functions were carried out directly by the network controller. Philips, instead, relied on a simpler architecture (BasicCAN), which, by contrast, imposes a higher load on the microcontroller used to interface the CAN device.

In the last two decades, all main semiconductor vendors including Siemens, Motorola, and NEC have produced and shipped millions of CAN chips, mainly to car manufacturers such as Mercedes-Benz, Volvo, Saab, Volkswagen, BMW, Renault, and Fiat.

The Bosch specification (CAN version 2.0) was submitted for international standardization at the beginning of the 1990s. The proposal was approved and published as the ISO 11989 standard at the end of 1993 and contained the description of the medium access protocol and the physical layer architecture. In 1995, an addendum to ISO 11898 was approved to describe the extended format for message identifiers (IDs).

The CAN specification was then revised and reorganized, and it was split into several separate parts: Part 1 [ISO1] defines the CAN data link protocol, whereas part 2 [ISO2] and part 3 [ISO3] specify the high-speed (non-fault-tolerant) and low-speed (fault-tolerant) physical layers, respectively. Part 4 [ISO4] specifies the protocol for time-triggered communication (*TTCAN*). More recently, part 5 [ISO5] was added that deals with high-speed medium access units (MAUs) with low-power mode.

Recently, Bosch GmbH has introduced some significant changes to the architecture of CAN, to cope with some limitations of this technology that could not be tackled by simply improving the hardware and software. The new CAN with flexible data rate (CAN FD) specification introduces enhanced mechanisms based on some ideas first published in 1999 [CEN99] that should grant a high level of competitiveness to CAN for many years to come.

52.1.1 CAN in Automotive Applications

The main application area of CAN has been (and still is) the automotive industry. Here, CAN is adopted as an in-vehicle network for engine management and comfort electronics (e.g., doors control, air conditioning, lightning) as well as for some infotainment (information and entertainment) functions. Currently, it is also being used as on-vehicle network for special-purpose cars, such as police cars and taxis.

The different CANs in a vehicle are connected through gateways. In many cases, this functionality is implemented in the dashboard, which may be equipped with a local CAN for connecting displays and instruments.

High-speed CANs (e.g., 500 kbit/s) that comply with the ISO 11898-2 physical layer are used by a vast majority of carmakers in power engine systems. In addition, most passenger cars produced in Europe are equipped with CAN-based multiplex networks, which connect doors and roof control units as well as lighting and seat control units. These networks are often based on either the ISO 11898-3 fault-tolerant physical layer or the ISO 11898-5 high-speed physical layer with low-power functionality.

Sometimes, also infotainment devices are interconnected by means of CAN. Besides proprietary solutions, the SAE defined the Intelligent Transportation Systems (ITS) Data Bus on CAN (IDB-C) application profile in the ITS Data Bus series of specifications (SAE J2366), which is based on high-speed CAN (29-bit IDs) with a bit rate of 250 kbit/s. Finally, a CAN-based diagnostic interface is provided by several passenger cars for connecting diagnostic tools to in-vehicle networks. Several international standards are available in this case, namely, ISO 16765 (diagnostics on CAN including Keyword 2000 over CAN), ISO 14229-1 (Unified Diagnostic Services), and ISO 15031(onboard diagnostic standard).

Electronic control units (ECUs) for vehicles (e.g., passenger cars, trucks, and buses) are in most cases OEM specific. This means they comply with the specifications dictated by the original equipment manufacture (OEM) for what concerns both the physical (e.g., bit timing) and higher-layer protocols. While trucks and buses mostly rely on J1939-based specifications, proprietary higher-layer protocols and profiles are usually adopted in passenger cars. As an exception, CAN-based diagnostic interfaces to scanner tools usually comply with the ISO standards.

52.1.2 CAN in Automation and Networked Control Systems

Even though it was conceived explicitly for vehicle applications, at the beginning of the 1990s, CAN began to be adopted in different scenarios, too. Standard documents provided satisfactory specifications for the lower communication layers, but did not offer guidelines or recommendations for the upper part of the OSI protocol stack, in general, and for the application layer in particular. This is why the earlier applications of CAN outside the automotive scenario (i.e., textile machines, medical systems) adopted ad hoc monolithic solutions. The *CAN in Automation* (CiA) users' group, founded in 1992, was originally concerned with the specification of a *standard* CAN application layer (CAL). This effort led to the development of the general-purpose CAL specification. CAL was intended to fill the gap between the distributed application processes and the underlying communication support, but in practice, it was not successful, the main reason being that, as CAL is really application independent, each user had to develop suitable profiles for his or her specific application fields.

In the same years, Allen-Bradley and Honeywell started a joint distributed control project based on CAN. Although the project was abandoned a few years later, Allen-Bradley and Honeywell continued their works separately and focused on the higher protocol layers. The results of those activities were the Allen-Bradley *DeviceNet* solution and the Honeywell *Smart Distributed System* (SDS). For a number of reasons, SDS remained, in practice, a solution internal to Honeywell Microswitch, while the development and maintenance activities for DeviceNet were soon switched to the *Open DeviceNet Vendor Association* (ODVA). Thereafter, DeviceNet has been widely adopted in the United States in a number of factory automation areas, becoming a serious competitor to widespread solutions such as PROFIBUS-DP and INTERBUS.

Besides DeviceNet and SDS, other significant initiatives were focused on CAN and its application scenarios. *CANopen* was conceived in the framework of the European Esprit project automation and control systems for production units using an installation bus concept (ASPIC) by a consortium led once again by Bosch GmbH. The purpose of CANopen was to define a profile based on CAL, which could support communications inside production cells. The original CANopen specifications were further refined by CiA and released in 1995.

Later on, CANopen, DeviceNet, and SDS became European standards, CANopen as an embedded network for machine distributed control and the other two for factory and process automation. In particular, they are defined in standard EN 50325 [EN1], where part 2 is about DeviceNet [EN2], part 3 deals with SDS [EN3], and part 4 describes CANopen [EN4].

52.2 CAN Protocol Basics

The CAN communication architecture is structured according to the well-known layered approach foreseen by the ISO OSI reference model. However, as in most currently existing networks conceived for use at the field level in automated manufacturing environments, only few layers have been included in its protocol stack. This is to make implementations more efficient and inexpensive. Fewer protocol layers, in fact, imply reduced processing delays when receiving and transmitting messages, as well as simpler communication software (which has typically to run on embedded devices with low computational power). CAN specifications [ISO1,ISO2,ISO3], in particular, include only the *physical* and *data link* layers, as depicted in Figure 52.1.

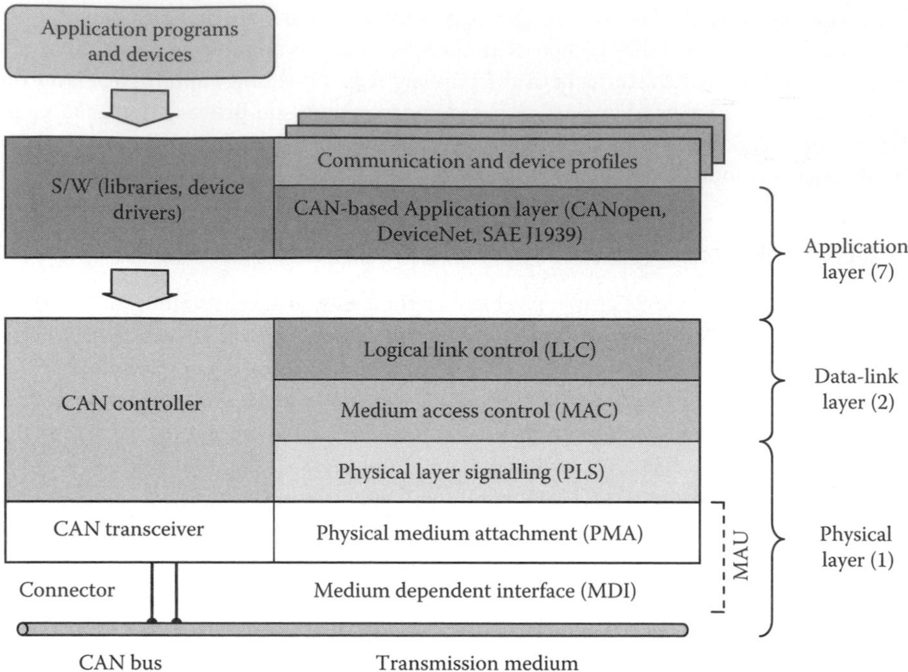

FIGURE 52.1 CAN protocol stack.

The physical layer is aimed at managing the actual transmission of information over the communication medium and is typically concerned with electrical aspects. Bit timing and synchronization, in particular, belong to this layer. The CAN physical layer is conceptually subdivided into three sublayers, namely, the *physical layer signaling* (PLS), the *physical medium attachment* (PMA), and the *medium dependent interface* (MDI). The PLS mostly defines aspects related to bit representation, timing, and synchronization. These features, which are valid for any CAN system, are described in the ISO 11898-1 document [ISO1]. Together, the PMA and the MDI make up the *MAU*, which is the part of the physical layer that actually couples the node to the transmission medium. The PMA covers functional circuitry to deal with transmission and reception over the bus and may also provide means for detecting bus failures. The MDI, instead, is about the mechanical and electrical interfaces between the MAU and the physical medium (i.e., the CAN bus).

The data link layer has to do with the (reliable) exchange of data between network nodes. It is located immediately above the physical layer and exploits the PLS services to carry out the transfer of frames (encoded as sequences of bits). The data link layer is conceptually split into two separate sublayers, namely, the *medium access control* (MAC) and the *logical link control* (LLC). Basically, the main purpose of the MAC is managing the access to the shared transmission support, by providing a mechanism aimed at coordinating the use of the bus, so as to avoid unmanageable collisions. MAC functions include frame encapsulation and decapsulation, arbitration, error checking and signaling. The LLC, instead, offers the user (either application programs or entities belonging to upper protocol layers) a proper interface, which is characterized by a well-defined set of communication services. In addition, it provides the ability to decide whether or not an incoming message is relevant to the node.

It is worth noting that CAN specifications are very flexible concerning both the actual implementation of LLC services and the choice of the physical support, whereas modifications to the behavior of the MAC sublayer are not admissible.

As mentioned before, and unlike most fieldbus networks, CAN specifications do not include any application layer natively. However, a number of such protocols exist that rely on CAN and ease the design and implementation of complex systems.

52.2.1 Physical Layer

PLS functions, as defined in [ISO1], are embedded directly into CAN controllers. Conversely, MAUs are mostly implemented in transceivers and are defined in separate documents. For instance, ISO 11898-2 [ISO2] and ISO 11898-3 [ISO3] deal with high- and low-speed communications, respectively, whereas ISO 11898-5 defines the power modes of high-speed MAUs (for systems requiring low-power consumption while no active communication is taking place on the bus). The definition of the actual interface to the medium (i.e., connectors) is usually covered in other documents: in fact, the ISO specifications mentioned earlier are aimed at standardizing the most important electrical parameters and not to define mechanical and material characteristics.

52.2.1.1 Network Topology and Cabling

CANs are based on a shared bus topology. Buses have to be terminated with resistors at each end (the recommended nominal impedance is 120 Ω), so as to suppress signal reflections. For the same reason, standard documents state that the topology of a CAN should be as close as possible to a single line. Stubs are permitted for connecting devices to the bus, but their length should be kept as short as possible. For example, at the transmission speed of 1 Mbit/s, the maximum allowed length of a stub is 30 cm.

Several kinds of transmission media can be used:

- A two-wire bus enables differential signal transmissions and ensures reliable communications [ISO2]. In this case, shielded twisted pair can be used to further enhance the immunity to electromagnetic interferences. A lower-speed, fault-tolerant version is defined in [ISO3].
- A single-wire bus is a simpler and cheaper solution that features lower immunity to interferences and is mainly suitable for use in comfort electronics in automotive applications [J2411].
- A point-to-point connection over unscreened twisted pair according to [ISO9] can be used, for instance, in the heavy-duty truck industry for communication between the tractor and one or more trailers.
- An optical transmission medium, which ensures complete immunity to electromagnetic noise, can be used in hazardous environments. Fiber optics can be adopted to interconnect different CAN subnetworks through repeaters. This is done to cover plants that spread over large areas.

Several bit rates can be selected for the network, the most commonly adopted being in the range from 50 kbit/s to 1 Mbit/s. Actually, CAN does not mandate any specific value for this parameter, nor does it suggest preferred transmission speeds (which, instead, are typically defined by CAN-based application protocols). One Mbit/s is the maximum allowable bit rate according to the CAN specifications. For reasons explained in the following, it was never increased and is still the same as in the very first version of the CAN specification.

The maximum extension of a CAN depends directly on the bit rate. The exact relation between extension and bit rate also involves parameters such as the latencies introduced by communication hardware and in particular transceivers and optocouplers. When the network speed is not too high, the rule of thumb is that the mathematical product between the length of the bus and the bit rate has to be approximately constant. For example, the maximum extension allowed for a 500 kbit/s network is about 100 m and increases up to about 500 m when the bit rate is 125 kbit/s.

Signal repeaters can be used to take care of signal attenuation, so as to increase the network extension, especially when large plants have to be covered and the bit rate is low to medium. However, since they introduce small additional delays on communication paths, the maximum distance between any two nodes is actually shortened at high bit rates. Using repeaters also enables network topologies different from the bus (e.g., trees or combs). In these cases, a good design may increase the effective area that is covered by the network, by minimizing the worst-case end-to-end delay between any two nodes.

It is worth noting that, unlike other networks conceived for the use in factory and process automation, such as PROFIBUS-PA, there is in general no cost-effective way in CAN to use the very same wire for carrying both signals and power. This is because signals sent over the CAN have a non-null DC component. However, power can be easily provided to devices by including an additional pair of wires in the CAN cable, as foreseen in several commercial solutions such as CANopen and DeviceNet.

Curiously enough, the use of standard connectors is not prescribed in CAN specifications. Instead, several companion or higher-level application standards exist that define their own connectors and pin assignments. CiA, for example, recommends the use of 9-pin D-SUB connectors for general-purpose applications [CiA102], while CANopen [CiA303] and DeviceNet also suggest other kinds of connectors (5-pin ministyle, microstyle, open-style, etc.). In addition, those documents include recommendations for bus lines, cables, and standardized bit rates, which were not included in the original CAN specifications.

52.2.1.2 Bit Encoding

The electrical interface of a node to the CAN bus is conceptually based on a scheme that resembles open-collector circuits. As a consequence, the signal level on the bus can assume two complementary values, which are denoted symbolically as *dominant* and *recessive*. Overall, the behavior of the CAN bus can be described as follows: If the output level of at least one node is dominant, then the level of the whole bus is dominant; conversely, if the output level of every node is recessive, the bus is recessive. The dominant level corresponds to the logical value 0, while the recessive level is the same as the logical value 1.

CAN relies on the *non-return-to-zero* (NRZ) bit encoding, which achieves very good efficiency, for example, noticeably higher than Manchester encoding. Clocking information is not encoded separately from data. Instead, bit synchronization in each node is achieved by means of a *digital phase locked loop* (DPLL), which extracts the timing information directly from the bit stream read on the bus. In particular, edges in the signal are used for synchronizing receiver clocks, so as to compensate unavoidable oscillator tolerances and drifts.

To provide a satisfactory level of synchronization between the sender and receivers, the transmitted bit stream must include a sufficient number of edges. To state it in better terms, the bus is not allowed to stay at the same level for long time intervals when transmission is taking place, so as to avoid clock misalignments and, possibly, bit slips. This is achieved in CAN by means of the so-called *bit stuffing* technique. In practice, whenever five consecutive bits at the same value (either dominant or recessive) appear in the transmitted bit stream (also including those bits that have possibly been added by the bit stuffing mechanism), the transmitting node inserts one additional *stuff bit* at the complementary value, as depicted in Figure 52.2. This is done transparently and automatically by CAN controllers. Stuff bits can be easily and safely removed by receiving nodes, so as to obtain the original stream of bits back.

From a theoretical point of view, the maximum number of stuff bits that may be added when transmitting a CAN message is one every 4 bits in the original frame. For example, in the rightmost part of Figure 52.2, the original bit stream alternates sequences of four consecutive bits at the dominant level followed by 4 bits at the recessive level (actually, the first stuff bit is triggered by a sequence consisting of 5 bits at the same level). This results in a domino effect that leads to the worst case for the number of added stuff bits. As a consequence, the encoding efficiency of the physical layer can be as low as 80%. However,

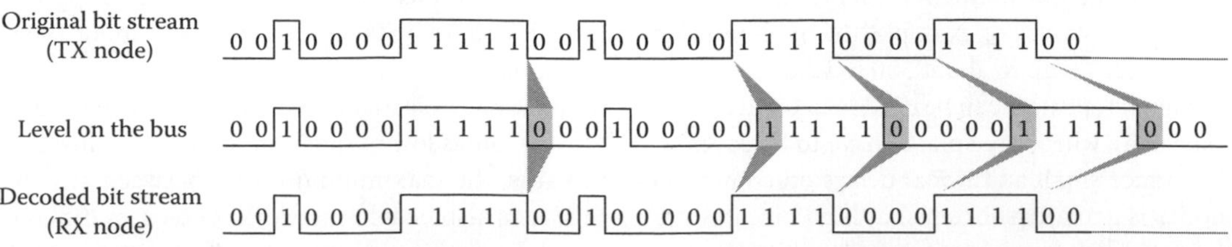

FIGURE 52.2 Bit stuffing technique.

the influence of bit stuffing in real operating conditions is noticeably lower than the theoretical value mentioned earlier. Simulations [CEN12] show that, on the average, only a few stuff bits are effectively added to each frame, depending on the size of the ID and data fields as well as on the generation law for signal values (traffic model).

Despite it is quite efficient, the bit stuffing technique has one main drawback, since the time taken to send a frame over the bus is not fixed. Instead, it depends on the content of the frame itself. This might cause annoying jitters on reception times, which affect the actuation and sampling on remote devices directly. Suitable techniques are available to cope with this undesired effect, which try to reduce the number of stuff bits inserted by the CAN controller by altering the content of the data field before transmission. Either scrambling techniques (which operate by carrying out the exclusive OR (XOR) between the payload and an alternating bit pattern) or true encoding schemes like 8B9B [CEN13] are worth mentioning to this purpose.

Not all the fields in a CAN frame are encoded according to the bit stuffing mechanism, which applies, in particular, only to the initial part of the frames, from the start of frame (SOF) up to the cyclic redundancy check (CRC) sequence (included). The remaining fields (up to the end of the frame) have a fixed form and are not stuffed.

52.2.1.3 DPLL Operation

Operations in the DPLL are timed by the so-called time quanta (T_q). The time quantum is derived from the local oscillator period through a prescaler. As shown in Figure 52.3, the bit time in CAN is divided into four segments, namely, *synchronization segment* (SYNC_SEG), *propagation time segment* (PROP_SEG), and two *phase buffer segments* (PHASE_SEG1 and PHASE_SEG2). Each segment consists of an integral number of time quanta.

Receivers sample the level on the bus on every time quantum. By detecting the edges in the signal read from the bus, which can be inferred from such samples, the DPLL in each receiver is enabled to recover the clocking information of the transmitter. In this way, the actual reading of the current bit is carried out by the CAN controller at the moment in time that maximizes the likelihood of detecting the correct value, which is referred to as *sample point*.

The internal bit timing is adjusted by the DPLL using either the *hard synchronization* or *resynchronization* mechanisms. Whenever a recessive-to-dominant transition occurs while the bus is idle, a hard synchronization takes place that involves all nodes in the network. This transition corresponds to the leading edge of the SOF bit sent by the node that initiated transmission first, and it is located at the very beginning of the frame. The net effect is that, following a hard synchronization, the originating edge is forced to lie within the SYNC_SEG of every node. In all other cases, edges cause a resynchronization. The main difference is that the effect of resynchronization depends on the offset (error) between the actual position of the edge and the point in time when it was expected.

The *phase error* (e) of an edge is measured in time quanta and is defined as follows:

- $e = 0$ when the edge on the bus lies within the intended SYNC_SEG in the receiver node (timely edge).
- $e > 0$ means that the edge on the bus is received late (but before the sample point of the current bit).
- $e < 0$ means that the edge on the bus is received early (but after the sample point of the previous bit).

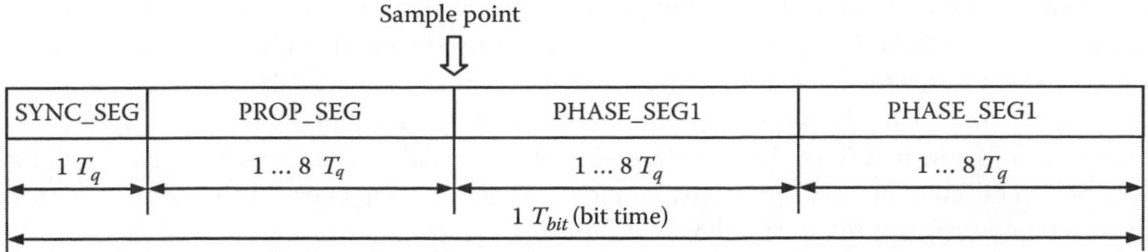

FIGURE 52.3 Segments in a bit time.

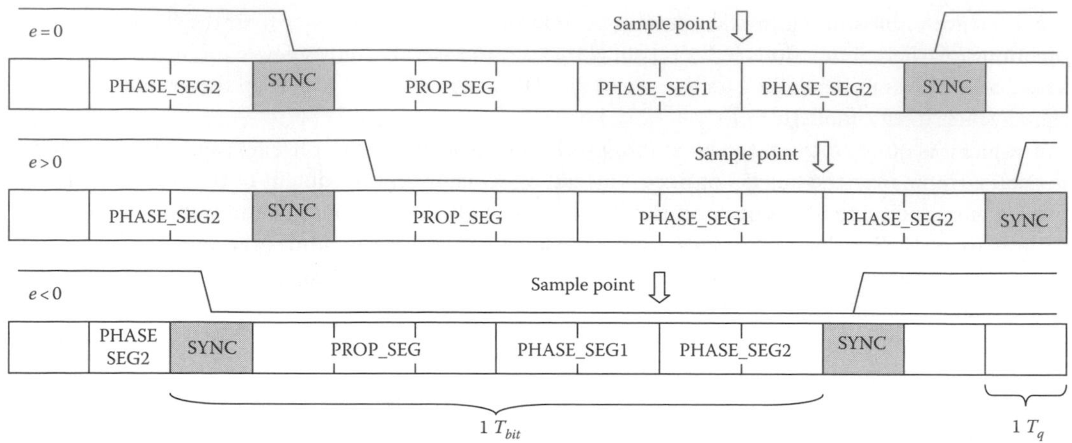

FIGURE 52.4 DPLL synchronization rules.

When the absolute value of the phase error is less than or equal to the *resynchronization jump width* (SJW), a parameter that can be configured in the CAN controller, the effect of resynchronization is just the same as hard synchronization. Otherwise, the amount of correction is bounded by SJW. SJW is programmable in the range between 1 and the minimum between 4 and PHASE_SEG1. Corrections take place by either lengthening PHASE_SEG1 for the current bit when the edge is late or shortening PHASE_SEG2 of the previous bit (during which the level transition was detected) when the edge is early. Such timing rules are sketched in Figure 52.4.

This approach can cope with differences between the actual frequencies of local clocks, as well as drifts, up to about 1.58%. Increasing the duration of the propagation delay segment permits enlarging the network size for any given bit rate. On the other hand, increasing the nominal width of the phase segments allows using cheaper oscillators, which are affected by worse tolerances. The choice on the actual DPLL parameter values in programming the CAN controller is typically a trade-off between system requirements and constraints.

A few additional rules help in reducing the likelihood that glitches on the bus—due, for instance, to short electromagnetic disturbances—may cause synchronization errors. They are basically as follows:

1. Synchronization can occur only once within every bit time; subsequent edges after the first one are ignored and do not restart the local timing.
2. An edge is used for synchronization only if the bus value immediately after the transition differs from the value read at the previous sampling point.

52.2.2 Frame Format

The CAN protocol foresees four kinds of frames only, namely, *data*, *remote*, *error*, and *overload* frames. Data and remote frames are quite similar, and the CAN specification [ISO1] defines both a *base* (or standard) and an *extended* format for them. These formats mainly differ for the size of the ID and for a few bits in the arbitration field. In particular, the base frame format (as defined in the CAN 2.0A specification) adopts 11-bit IDs, which means that up to 2048 different IDs are available to applications running in the same network (several older CAN controllers, however, only support IDs in the range 0–2031). The extended frame format (introduced with CAN 2.0B), instead, assigns 29 bits to the ID, so that up to half a billion different objects could exist, in theory, in the same network. This is a fairly high value, which is virtually sufficient for any kind of application.

Using extended IDs in a network that includes 2.0A compliant CAN controllers, too, usually leads to unmanageable transmission errors, which make the network unstable. Thus, a third category of

CAN controllers was developed, known as 2.0B passive: they can manage the transmission and reception of standard frames in a correct way, while extended frames are simply ignored in order not to hang the network.

It is worth noting that, in most practical cases, the number of different objects allowed by the base frame format is more than adequate. Since standard CAN frames are shorter than the extended ones because of the shorter arbitration field, they enable higher communication efficiency (unless part of the payload is moved into the arbitration field). As a consequence, they are typically adopted in the majority of the existing CAN systems, and also, most of the CAN-based higher-layer protocols, such as CANopen and DeviceNet, basically rely on this format.

Error and overload frames are also very similar. They are not used to convey any explicit information. Rather, they are sent automatically by the CAN controller when some specific condition occurs. In particular, the first kind of frame is used to notify error conditions, whereas the second is used to provide some additional delay between subsequent data (or remote) frames.

The format of all CAN frames, as defined in the current CAN 2.0B specification, is described in detail in the following subsections.

52.2.2.1 Data Frame

Data frames are used to send information over the network. Each CAN data frame begins with an *SOF* bit at the dominant level, as shown in Figure 52.5. Its role is to mark the beginning of the frame as in serial transmissions carried out by means of conventional Universal Asynchronous Receiver-Transmitters (UART). The SOF bit is also used to synchronize the receiving nodes.

Immediately after the SOF bit, the *arbitration* field follows, which includes both the *ID* field and the *remote transmission request* (RTR) bit. The arbitration field as a whole is used by the MAC sublayer to detect and manage the priority of the frame whenever a collision occurs. In particular, the lower the numerical value of the ID, the higher the priority of the frame.

As the name suggests, the ID field also uniquely denotes in the whole network (at the LLC level) the content of the frame that is being exchanged. Its encoding is different for either the base or extended frame format. In the former case, a single 11 bit ID field is used, whereas in the latter, an 11-bit *base ID* is followed by an 18-bit *ID extension*, so as not to change the format too much with respect to the base frame. In both cases, the ID is sent starting from the most significant bit up to the least significant one.

The RTR bit is used to discriminate between data and remote frames. Since a dominant value of RTR denotes a data frame while a recessive value stands for a remote frame, a data frame has a higher priority than a remote frame with the same ID. This helps when they are sent at the same time by different nodes in the network.

The control field comes next to the arbitration field and consists of 6 bits. In the case of base frames, the first two control bits are the identifier extension (IDE) bit (which coincides with the reserved bit r1

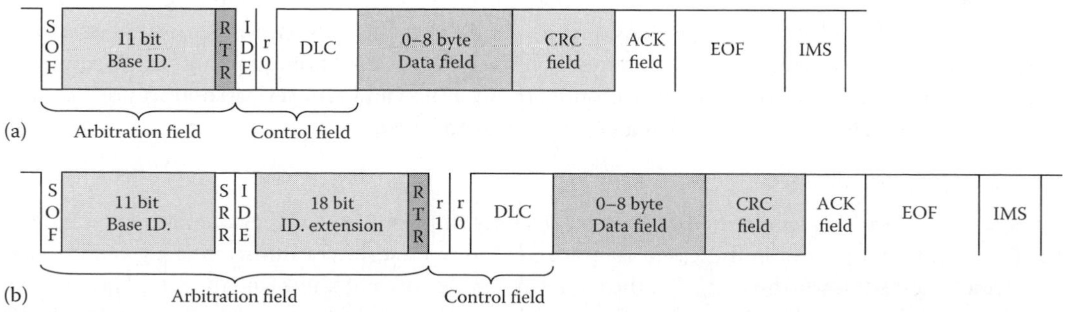

FIGURE 52.5 Format of CAN data frames. (a) Base frame format and (b) extended frame format.

in CAN 2.0A) and the reserved bit r0. The IDE bit discriminates between base and extended frames: in particular, a dominant value identifies the first format, whereas a recessive value is used for the second. Since the IDE bit is dominant in base format frames while it is recessive in the extended ones, standard frames have precedence over their extended counterparts when the same base ID is considered.

It is worth noting that, in extended frames, the IDE bit belongs conceptually to the arbitration field, as well as the *substitute remote request* (SRR) bit—a placeholder that precedes the IDE bit and is sent at recessive value to preserve the structure of the frames. In this case, the IDE bit is followed by the 18-bit IDE and, again, terminates with the RTR bit. The control field follows immediately and begins with the two reserved bits rl and r0. Reserved bits shall be sent by the transmitting node at the dominant value, while receivers simply ignore their values.

The last 4 bits of the control field (after the reserved bit r0) are the *data length code* (DLC), which specifies the length (in number of bytes) of the data field. For the DLC field, values ranging from 0 to 8 are allowed. According to the most recent specifications, higher values (from 9 to 15) can also be used for application-specific purposes. In this case, however, the length of the data field is assumed to be 8 bytes.

The data field is used to store the effective payload of the frame. According to the OSI terminology, it represents the *service data unit* (SDU) of the data link layer also known as link-layer SDU (LSDU). As such, it is not interpreted in any way by the CAN protocol (and by controllers). One of the reasons why the size of the data field is limited to 8 bytes at most is to ensure a high degree of responsiveness, by reducing the maximum transmission time for frames and hence the priority inversion phenomenon.

The CRC and acknowledgment (ACK) fields are placed after the data field. The CRC is made up of a *cyclic redundancy check* sequence encoded on 15 bits, which is followed by a CRC delimiter at the recessive value. The kind of CRC adopted in CAN is particularly suitable to cover short frames (i.e., counting less than 127 bits). The acknowledgment field, instead, is made up of 2 bits, that is to say the ACK slot is followed by the ACK delimiter. Both of them are sent at the recessive level by the transmitter. The ACK slot, however, is overwritten with a dominant value by each node that receives the frame correctly (i.e., no error is detected up to the ACK field). It is worth noting that the ACK slot is actually surrounded by 2 bits at recessive level, namely, the CRC and ACK delimiters. By means of the ACK bit, the transmitting node is enabled to discover whether or not at least one node in the network has received the frame correctly.

The frame is closed by the *end of frame* (EOF) field, made up of seven recessive bits, which notifies all the nodes the end of an error-free transmission. In particular, the transmitting node assumes that the frame has been exchanged correctly if no error is detected until the last bit of the EOF field, while for receivers, the frame is valid if no error is detected until the last but one (i.e., sixth) bit of EOF. When a bus error occurs that affects the very last bit of the frame, this could lead to duplicate message transmissions. In fact, in this case, the sender node retransmits a frame that was already accepted by receivers, which may cause problem when, for instance, incremental or differential values are exchanged.

Subsequent frames are interleaved by the *interframe space*, which is made up of the *intermission* (IMS), possibly followed by a period of time (whose duration is arbitrary) during which no transmission takes place (*bus idle*). The IMS, which consists of three recessive bits, is mandatory and effectively separates any two consecutive frames exchanged on the bus. It is used both to permit the hardware in the CAN controller to prepare for a new transmission/reception and for overload notification. The bus turns in the idle state only if no node has a message ready to be sent.

52.2.2.2 Remote Frames

Remote frames are very similar to data frames. The only difference is that they do not carry data (i.e., the data field is not present in this case). They are used to request the transmission of a given message by a remote node. It is worth noting that the requesting node does not know who the actual producer of the relevant information is. It is up to receivers to discover the one that has to reply, by checking the ID field in the remote frame.

Indeed, the DLC field in remote frames is not used by the CAN protocol. However, it should be set to the same value as in the corresponding data frame, so as to cope with situations where several nodes send remote requests with the same ID at the same time (this is legal in a CAN). In that case, it is necessary for the different request frames to be perfectly identical, so that the related signals on the bus can overlap totally in the case of a collision.

It should be noted that, because of the way the RTR bit is encoded, if a request is made for an object at the same time as the transmission of that object is started by the producer, the contention is resolved in favor of the data frame.

52.2.2.3 Error and Overload Frames

Error frames are used to notify nodes in the network of error occurrences. They consist of two fields, namely, the *error flag* and the *error delimiter*. There are two kinds of error flags: active error flags are made up of six dominant bits, while passive error flags consist of six recessive bits. An active error flag either violates the bit stuffing rules or corrupts the fixed-format parts of the frame that is currently being exchanged. Hence, it enforces an error condition that is detected by all the stations connected to the network. Each node that detects an error condition starts transmitting an error flag on its own. In this way, as a consequence of the transmission of an error flag, a sequence of dominant bits appears on the bus whose duration lies between 6 and 12 bit times.

The error delimiter, instead, is nominally made up of eigth recessive bits. After the transmission of an error flag, each node starts sending recessive bits and, at the same time, it monitors the bus level until a recessive bit is detected. At this point, the node sends seven more recessive bits, hence completing the error delimiter.

Overload frames can be used by slower receivers to slow down operations on the network. This is done by adding extra delays between consecutive data and/or remote frames. The overload frame format is very similar to error frames. In particular, it is made up of an overload flag followed by an overload delimiter. It is worth noting that today's CAN controllers are very fast, so overload frames have become almost useless.

52.2.3 Medium Access Technique

The MAC mechanism, on which CAN relies, is basically the *CSMA* scheme. When no frame is being exchanged, the network is idle; in this condition, the level on the bus is steadily recessive. Before transmitting a frame, a node has to sense the state of the network. If the network is found idle, the frame transmission begins immediately; otherwise, the node must wait for the end of the current frame transmission (including the IMS). Each frame starts with the SOF bit at the dominant level. This bit disrupts the recessive state of the bus and informs all the other nodes that the network is no longer in the idle state.

Although not very frequently, two or more nodes might start sending their frames exactly at the same time. This may actually occur, because both the delay between the beginning of the bit time and the sample point in controllers and propagation delays on the bus, even though very small, are anyway greater than zero. Thus, one node might start its transmission while the SOF bit of another frame is already travelling on the bus. In this case, a collision occurs. In CSMA networks that are based on collision detection, such as nonswitched Ethernet, this unavoidably leads to the corruption of all frames involved, which implies that they have to be retransmitted. The consequence is a waste of time and a net decrease of the available network bandwidth. In high load conditions, this may lead to congestion: when the number of collisions is so high that the net throughput on the Ethernet network falls below the frame arrival rate, the network is stalled.

Unlike Ethernet, the medium access technique of CAN is able to resolve such *contentions* in a deterministic way, so that neither time nor bandwidth is wasted. As a consequence, congestion conditions can no longer occur and all the theoretical system bandwidth is effectively available for communications.

For the sake of truth, it should be said that contentions in CAN occur more often than one may think. In fact, when a node that has a frame ready to be sent finds the bus busy or loses the contention, it waits for the end of the current frame exchange and, immediately after the IMS has elapsed, it starts transmitting. Here, the node may compete with other nodes for which, in the meanwhile, a transmission request was issued. The likelihood that this event takes place is basically related to the number of nodes in the network and their message generation rate. In this case, different nodes synchronize on the falling edge of the first SOF bit that is sensed on the network.

The access technique described earlier implies that, as long as transmission errors are neglected, a CAN could be modeled as a work-conservative system. This means, if there is (at least) one message waiting for transmission in the whole system, it will be sent as soon as the bus becomes idle. CAN effectively behaves like a system-wide distributed priority queue, where messages are selected for transmission according to a priority order.

52.2.3.1 Bus Arbitration

The most distinctive feature of the MAC sublayer of CAN is the ability to resolve in a deterministic way any collision that may occur in the network. In turn, this is made possible by the arbitration mechanism, which is able to find out the most urgent frame each time a contention for the bus takes place between two or more nodes.

The CAN arbitration scheme operates by stopping transmissions of all frames involved except the one that is assigned the highest priority (i.e., the lowest ID). The arbitration technique exploits the peculiar behavior of the physical layer of CAN, which conceptually provides a wired-AND connection among all nodes. In particular, bus level is dominant (i.e., at the 0 level) if at least one node is sending a dominant bit; vice versa, the bus level is recessive (i.e., at the 1 level) only if all nodes are currently sending recessive bits (or, which is the same, if they are not transmitting).

Each node transmits the arbitration field on the bus serially, starting with the most significant bit, immediately after the SOF bit and by means of the so-called *binary countdown* technique. While transmitting, each node compares the level sensed on the bus to the value of the bit it is writing out. If a node is transmitting a recessive bit, whereas the level on the bus is dominant, the node understands it has lost the contention and withdraws immediately. In particular, it ceases transmitting and sets its output port to the recessive level, so as not to interfere with other contending nodes. At the same time, it switches to the receiving mode to be ready to read the incoming (winning) frame.

The binary countdown technique ensures that, in the case of a collision, all nodes that are sending lower-priority frames will abort their transmissions by the end of the arbitration field. Only the node that is sending the frame with the highest priority is allowed to complete the transmission of the arbitration field. It does not even realize that a collision has occurred.

A prerequisite for the correct behavior of arbitration is that, in every CAN, no two nodes can be transmitting frames having the same ID at the same time. If this is not the case, in fact, unmanageable collisions could take place, which in turn cause transmission errors. Because of the automatic retransmission feature of CAN controllers, this would almost certainly lead to a burst of errors on the bus, until the involved stations are disconnected by the fault confinement mechanism. This implies that, in general, only one node can be the producer of each object.

One exception to this rule is given by frames that have no data field, such as remote frames and data frames where the DLC field is set to 0. In this case, should a contention occur involving frames with the same ID, they would overlap perfectly and hence no collision would effectively occur. The same is also true for data frames whose data field is not empty, provided that its content is the same for all the frames sharing the same ID. However, it makes no sense in general to send frames with a fixed data field. Another exception to the rule is when a mechanism is defined by the upper protocol layers that provide nodes the exclusive use of shared IDs at runtime.

All the nodes that lose a contention have to retry the transmission as soon as the exchange of the current (winning) frame ends. They all try to send their frames again, immediately after the IMS is read on the bus. It is worth noting that a new collision may still possibly take place, which could also involve frames sent by nodes for which a transmission request was issued while the bus was busy.

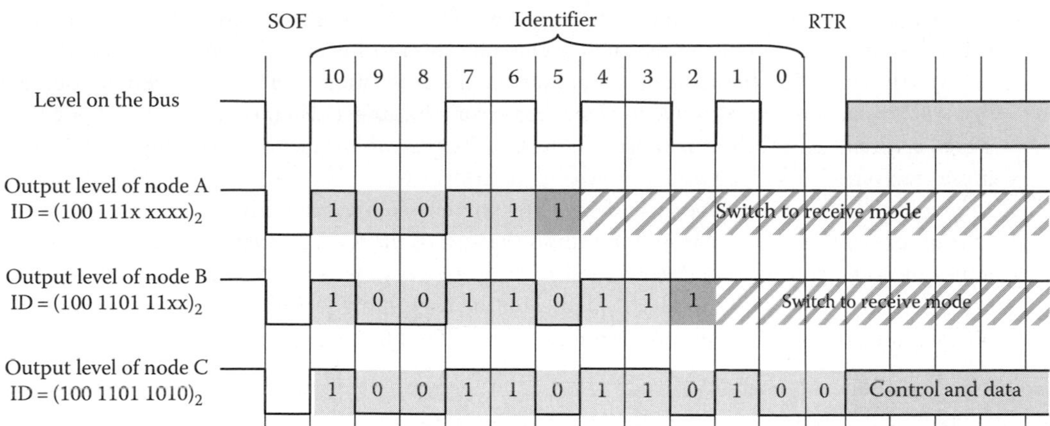

FIGURE 52.6 Arbitration phase in CAN.

An example that shows the detailed behavior of the arbitration phase in CAN is depicted in Figure 52.6. Here, three nodes (which have been indicated symbolically as A, B, and C) start transmitting at the same time—possibly at the end of the IMS following the previous frame exchanged over the bus. The ID of the frame sent by each node is specified in the figure (only the relevant prefix is shown for clarity). As said before, as soon as a node understands it has lost the contention, it switches its output level to the recessive value, so that it no longer interferes with the other transmitting nodes. This event takes place when bit ID_5 is sent for node A, while the same happens at bit ID_2 for node B. Node C manages to send the whole ID field, without incurring in discrepancies between the transmitted bit and the level detected on the bus. Therefore, it is the only winner of the arbitration and keeps on transmitting the remaining part of the frame (which starts with the control field).

52.2.3.2 Error Management

One of the fundamental requirements, which drove the definition of the CAN protocol, was the need to have a robust communication system, that is, one that is able to detect most transmission errors. Hence, particular care was taken in defining the support for error management. The CAN specification foresees five different mechanisms to detect transmission errors:

1. *Cyclic redundancy check*: When transmitting a frame, the originating node appends a 15-bit CRC to the end of the frame, which is calculated on the part of the frame included between the SOF bit and the data field. Receiving nodes reevaluate the CRC for the frame read from the bus and check if it matches the value included in the frame (from a practical viewpoint, the CRC checking is carried out by controllers in a slightly different way). If this is not the case, a CRC error is issued. Unlike other mechanisms, which are able to discover errors as soon as they occur, error detection takes place in this case only after the CRC sequence has been completely received. Generally speaking, the CRC used in CAN is able to discover up to five erroneous bits arbitrarily distributed in the frame and also errors bursts including up to 15 bits.
2. *Frame check*: Fixed-format fields in the received frames can be easily checked against their expected values. For example, the CRC and ACK delimiters as well as the EOF field have to be at the recessive level. If one or more illegal bits are detected, a form error is immediately generated.
3. *Acknowledgment check*: Every node verifies whether the ACK bit in the frame that is being sent has been changed to the dominant value or it is still at the recessive level. In the latter case, an acknowledgment error is immediately issued. Such a kind of check is useful, for example, to discover if there are other active nodes in the network. In particular, it can be used by the transmitting node to determine if its connector (or the related link) is broken.

4. *Bit monitoring*: Each transmitting node compares the level on the bus to the value of the bit that is being written out. Should a mismatch occur, a bit-monitoring error is generated immediately. It is worth noting that the same does not hold for the arbitration field and the acknowledgment slot. Such a check is very effective to detect bus errors that affect also the transmitting node (like global errors). Conversely, it is mostly useless in the case of local errors that only affect one or more receivers.
5. *Bit stuffing*: Each node verifies whether or not the bit stuffing rules have been violated in the portion of the frame from the SOF bit up to the end of the CRC sequence. Whenever 6 bits of identical value are read from the bus, a bit stuffing error is generated immediately. Although this error detection mechanism depends directly on the bit stuffing technique employed in CAN for bit encoding, the two concepts are distinct and should not be confused.

The theoretical residual probability that a corrupted message goes undetected in a CAN—under realistic operating conditions—has been evaluated to be less than 4.7×10^{-11} times the message error rate. Unfortunately, the bit stuffing mechanism may sometimes interfere with the CRC calculation. In particular, a pair of bit flip errors (the former turning a stuff bit into a regular bit and the other doing the opposite) could create error bursts longer that 15 bits, which might trick the CRC mechanism to think no error has taken place. The net result is that the actual residual error probability may be higher than expected, which could be a nonnegligible issue for highly dependable systems.

52.2.4 Fault Confinement

A *fault confinement* mechanism has been included in the CAN specification so as to prevent any node, which is not operating properly, from sending corrupted frames repeatedly and blocking the whole network. The fault confinement unit supervises the correct operation of the MAC sublayer and disconnects the node from the bus if it becomes defective. This mechanism has been conceived so as to discriminate, as long as it is possible, between permanent failures and short disturbances that may cause bursts of errors on the bus. According to this approach, each node can be in one of three states, namely, *error active*, *error passive*, and *bus off*.

The error-active state corresponds to the normal operating mode and is entered automatically upon device start-up or reset. Basically, error-active and error-passive nodes take part in the communication in the same way. The main difference is that, after transmitting a frame, an error-passive node has to wait for eight additional bits after the IMS before starting a new transmission (*suspend transmission*). Moreover, they react to error conditions differently. Active error flags and passive error flags are used alternatively in the two cases. This is because an error-passive node has already experienced several errors, so it should avoid interfering with the network operations. A passive error flag cannot corrupt the ongoing exchange of a frame that is sent by another node.

The fault confinement unit uses two counters to track the behavior of the node with respect to transmission errors: they are known as *transmission error count* (TEC) and *receive error count* (REC), respectively. The rules by which TEC and REC are managed are actually not trivial; however, they can be summarized as follows: Each time an error is detected, the counters are increased by a given amount, whereas successful exchanges decrease them by one. Furthermore, the amount of the increase for the nodes that first detected the error is higher than the nodes that simply replied to the error flag. In this way, counters are likely to increase more quickly in faulty nodes than in nodes that are operating properly, even when sporadic errors, caused, for instance, by electromagnetic noise, are considered.

When one of the error counters exceeds a first threshold (127), the node is switched to the error-passive state, so as to try not to affect the network. It may return to the error-active state as soon as both counters fall below (or equal) the threshold given earlier. When a second threshold (255) is exceeded, instead, the node is switched to the bus-off state. At this point, it can no longer transmit any frame on the network. It can be switched back to the error-active state only after it has been reset and reconfigured.

52.2.5 Logical Link Control

The communication model offered to users of the LLC sublayer is based on the exchange of objects between nodes, where each object is assigned its own ID and has an associated value (LSDU). According to the ISO specification [ISO1], the LLC of CAN provides two communication services only, namely, *L_DATA* and *L_REMOTE*. The former is used to broadcast the value of any object over the network, whereas the latter is used to ask for the value of an object to be broadcast by its (remote) producer. Both services are unacknowledged and mapped on the corresponding frames defined at the MAC level. From a practical point of view, the service primitives are implemented directly in the hardware by all the CAN controllers currently available off the shelf.

52.2.5.1 Model for Information Exchanges

Unlike most network protocols conceived for the use in automated factory environments, which rely on node addressing, CAN adopts an object addressing scheme. In other words, messages are not tagged with the address of the destination (and, possibly, originating) node(s). Instead, each piece of information that is exchanged over the network (often referred to as an *object*) is assigned a unique ID, which denotes unambiguously the meaning of the object itself in the whole system.

This fact has important implications on the way communications are carried out in CAN. In fact, identifying objects according to their meaning rather than to the nodes they are intended for allows implicit multicasting and makes it very easy for the control applications managing interactions among devices according to the producer/consumer paradigm.

In CAN, the exchange of information takes place according to the three phases shown in Figure 52.7:

1. The producer of a given piece of information encodes and sends the related frame on the bus using the MAC transmission services. The arbitration technique will transparently resolve any contention that may occur.
2. Because of the intrinsic broadcast nature of the underlying physical bus, the frame is propagated all over the network, and every node reads its content in a local receive buffer.
3. The *frame acceptance filtering* function in each node determines whether or not information in the frame is relevant to the node itself. In the former case, the frame is passed to the upper communication layers; from a practical point of view, this means that the CAN controller raises an interrupt to the local device logic (e.g., microcontroller), which will then read and use the information in the frame. On the contrary, the frame is simply ignored, which means discarding information.

In the sample data exchange depicted in Figure 52.7, node B is the producer of some kind of information, which is relevant to (i.e., consumed by) nodes A and D. This requires that nodes A and D configure explicitly their filtering function to accept the related frames. Node C, instead, is not interested in such information, which is therefore rejected (this is the default behavior of the filtering function).

52.2.5.2 Model for Device Interaction

The access technique of CAN makes this kind of network particularly suitable for distributed systems where devices interact according to the producer/consumer model. In this case, data frames are used by producer nodes to broadcast new values over the network, and each datum is identified unambiguously by means of its ID. Unlike networks based on the producer/consumer/arbiter model such as the Factory Instrumentation Protocol (FIP), information is sent in CAN as soon as it becomes available from either the control application(s) or the controlled physical system (by means of sensors), without any need for the intervention of a centralized arbiter. This noticeably improves the responsiveness and robustness of the whole system and makes CAN particularly suitable for networked embedded systems.

CANs work equally well when they are used to interconnect devices in systems based on the more conventional master/slave model, which relies on the request–reply paradigm. In this case, the master can

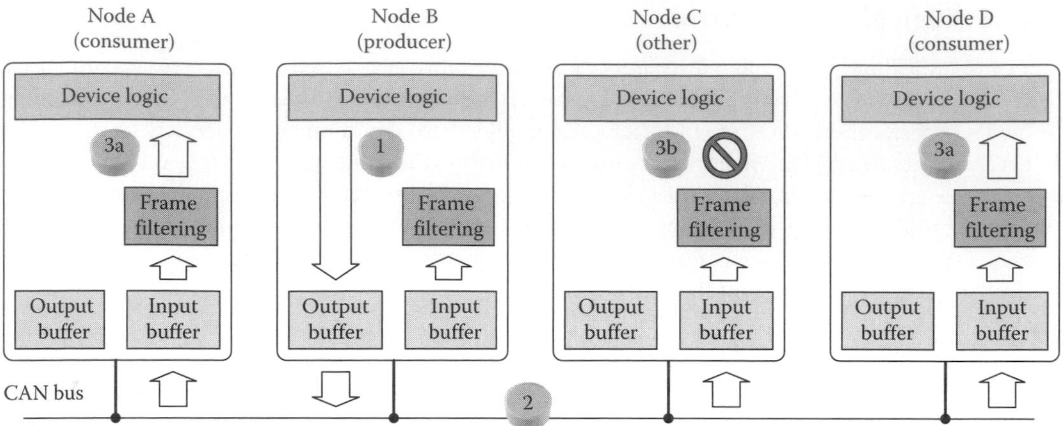

FIGURE 52.7 Producer/consumer model.

use remote frames to solicit some data to be sent on the network. As a consequence, the data producers reply with data frames carrying the relevant values.

It is worth noting that this kind of interaction is implemented in CAN in a way that is fairly more flexible than in conventional master/slave networks, such as PROFIBUS-DP. In CAN, in fact, the reply (data frame) does not need to follow the request (remote frame) immediately. In other words, the network is not kept busy while waiting for the device reply. This allows the system bandwidth to be, in theory at least, fully available to applications. Furthermore, the reply containing the requested value is broadcast on the whole network, and hence it can be read by all the interested nodes (consumers), besides the one that actually issued the remote request.

52.2.5.3 Implementation

According to their internal architecture, CAN controllers can be broadly classified into two categories, namely, *BasicCAN* and *FullCAN*. Differences do not concern the communication protocol, which shall comply to the MAC specification completely, but the interface between the network controller and the host that uses the LLC services.

Conceptually, BasicCAN controllers are provided with one transmit and one receive buffer, as conventional UARTs. The frame filtering function, in this case, is generally left to application programs (i.e., it is under control of the host microcontroller), even though some kind of filtering can typically be carried out by the CAN controller itself. So as to avoid overrun conditions, a double-buffering scheme based on shadow receive buffers is usually provided. This enables the reception of a new frame from the bus while the previous one is being read by the host controller. An example of controller based on the BasicCAN architecture is the PCA82C200 component by Philips. In more recent controllers, such as the Philips SJA 1000, an extended 64-byte receive First In, First Out (FIFO) queue is available.

FullCAN implementations, instead, foresee a number of internal buffers that can be configured to either receive or transmit some particular messages. In this case, the filtering function is implemented directly by the CAN controller. When a new frame, which is of interest for the node, is received from the network, it is stored in the related buffer, where it can be subsequently read by the host controller. Typically, newer values overwrite the older ones. Mechanisms could be also provided in order to avoid overrun conditions. The Intel 82526 and 82527 CAN controllers are based on the FullCAN architecture.

Because of the architectural choices mentioned earlier, FullCAN controllers free the host controller from a number of activities, and hence they are historically considered more powerful than BasicCAN components. It is worth noting, however, that recent CAN controllers are typically packed with many advanced functions and are somehow able to offer the advantages of both architectures. Hence, the earlier claim is no longer true in absolute terms.

52.3 Schedulability of CANs

Since CAN relies on a shared communication support, whose access is regulated through an arbitration phase carried out at runtime, transmissions may unavoidably suffer from delays because of messages that are currently being sent by the other nodes. From a theoretical point of view, the MAC sublayer of CAN can be seen as a (distributed) priority-based nonpreemptive queuing system, where messages are queued for transmission and are served according to their priority. This is why several techniques are available in literature, most of which inspired to fixed-priority scheduling in real-time operating systems, to analyze timings in a CAN. In particular, under well-defined yet typical assumptions, the maximum latency can be evaluated, which a message may experience before it is delivered to the intended destination. This, in turn, enables assessing whether or not a given set of messages, which have to be exchanged over the network as a consequence of the operations of a real-time distributed control application, is actually schedulable (i.e., if the related timing requirements can always be fulfilled).

From a general viewpoint, nothing can be said about communication latencies if messages are generated in a completely arbitrary way. However, if the message generation pattern is known in advance (e.g., either each message is exchanged periodically or a minimum time has to elapse between subsequent instances of the same message), worst-case transmission times can be evaluated. This is also true when the message generation processes on the different nodes are not coordinated in any way, as it happens in CANs. If nodes are not synchronized, the different message transmission patterns are displaced by arbitrary and unknown offsets.

The theory behind schedulability analysis in CAN is not trivial and has been clearly introduced and thoroughly explained in several papers [TIN94,TIN94b,TIN95]. Such an approach may sometimes lead to erroneous conclusions about worst-case transmission times. For this reason, a revised study was published in 2007 [DAV07]. In the following, however, the original approach, which first appeared in 1994, will be described, since it is much simpler and provides nevertheless useful results in many real-world operating conditions, particularly for the networked embedded systems typically found in most automotive applications.

52.3.1 Communication Requirements

The first step to analyze schedulability in CAN is the exact definition of the set of messages exchanged by the distributed control application. In particular, we are interested in message streams, which must not be confused with single message instances even though, for the sake of simplicity, the term message is typically used in the literature to refer to both concepts.

52.3.1.1 Message Set

Let $MS = \{m_1, m_2, \ldots, m_n\}$ be the *message set* (MS), where m_i is the ith message (stream) in MS, characterized by its own unique ID. From the point of view of applications, each message m is completely described by means of the following parameters:

- *Period* (T_m): It is the cycle time for periodically exchanged messages, as specified by the control application. Instead, for messages generated sporadically, T_m specifies the minimum interarrival time, that is, the shortest time interval that may elapse between any two consecutive generations of the same message.
- *Deadline* (D_m): It is the maximum time allowed for completing the reception of message m by the intended destination node(s). If exceeded, the correct behavior of the control application is no longer ensured. In the worse cases, safety problems might occur as well. This is clearly a condition that should never happen in properly operated systems and is what schedulability analysis is aimed to verify.
- *Jitter* (J_m): This parameter (also known as queuing jitter) models all latencies that are not directly caused by the communication protocol. It depends on both hardware overheads (basically

related to the actual implementation of network controllers and, in particular, the host controller interface) and latencies introduced by the communication software (drivers) and task scheduler (operating system). In particular, J_m is the maximum delay that may affect message m, from the instant its transmission is set (at the application level) to the time it is ready to be sent on the bus (i.e., when it is actually available to the protocol machine inside the communication controller).
- *Size* (S_m): It is the size (in bytes) of the message payload. In real control systems, each message may include one or more signals (as defined by the application), each one encoded on several bits. These bit strings are first concatenated and then padded to the next byte boundary.

52.3.1.2 Schedulability of Messages

The parameters given earlier are defined in the design phase of the control system and do not depend on the characteristics of the network used to interconnect devices. Starting from them, several other parameters can be derived, which are explicitly related to the underlying network and are used to assess the message schedulability:

- *Transmission time* (C_m): It is the longest time taken to transmit all the bits of the frame, which embeds message m, serially over the bus (in the case the transmitting node wins the arbitration).
- *Utilization* (U_{MS}): It is the fraction of the available network bandwidth needed to exchange all the messages included in the *MS*.
- *Queuing delay* (w_m), also known as *transmission delay*: It is the maximum delay the transmitting node of message m may experience before it wins the arbitration and starts the successful transmission of the frame on the bus.
- *Worst-case response time* (R_m), also denoted WCRT: It is the maximum end-to-end delay that message m may experience under the specified conditions (namely, for the given *MS* and bit rate R on the network).

The frame transmission time C_m can be computed directly from the message size S_m by taking into account the overhead due to the protocol control information added by CAN. Besides fixed-size fields (ID, DLC, CRC, etc.), a variable number of stuff bits may be inserted into the transmitted frame by the bit stuffing mechanism, which depends on the actual content of the frame. As schedulability has always to be ensured, irrespective of the information carried by messages, the worst case has to be considered (i.e., when the maximum number of stuff bits are added).

Each standard CAN frame includes 44 bits of protocol control information (SOF, arbitration, control, CRC, ACK, and EOF fields), while 3 bits have to be added for the IMS. The bit stuffing mechanism affects 34 of those bits (from the SOF bit up to the CRC field) and the payload (data field) too. Up to one stuff bit could be added, in the worst case every 4 bits of the original frame (an exception is at the very beginning of the frame, where the first stuff bit can be added only after 5 bits). This means that the (worst-case) transmission time C_m for frames with 11-bit IDs is simply given by

$$C_m = \left(34 + 8 \cdot S_m + \left\lfloor \frac{34 + 8 \cdot S_m - 1}{4} \right\rfloor + 13 \right) \cdot T_{bit} = (55 + 10 \cdot S_m) \cdot T_{bit}$$

where

$T_{bit} = 1/R$ is the bit time in the network
R is the transmission speed

In a similar way, the transmission time for frames with extended 29-bit IDs can be found as

$$C_m = (80 + 10 \cdot S_m) \cdot T_{bit}$$

A message m is said to be schedulable if its WCRT is lower than its deadline, that is, $R_m \leq D_m$. The *MS* is said to be schedulable if the response time of every message in the set is always (i.e., in the worst case, too) lower than or equal to its deadline. This means that the condition in the earlier text must hold for every message m in *MS*, that is,

$$R_m \leq D_m, \quad \forall m \in MS$$

Schedulability in a CAN is related to the network utilization U_{MS}. This parameter can be easily computed from the transmission time and period of messages in the set as

$$U_{MS} = \sum_{m \in MS} \frac{C_m}{T_m}$$

Obviously, utilization must be strictly less than 1 in order to have a feasible schedule (this means that the offered load is not allowed to exceed the available bandwidth). It is worth noting, however, that despite this condition being necessary, it is not sufficient to ensure schedulability. Generally speaking, the closer the utilization is to one, the higher the likelihood that the *MS* is not schedulable.

Besides the characteristics of the *MS*, schedulability also depends heavily on the priority assignment. Priorities define an order relation on the *MS*. As a consequence, for every message $m \in MS$, two subsets of *MS* are defined: $hp(m)$ includes all messages whose priority is higher than m, whereas $lp(m)$ includes those having lower priority. For such sets, the property $MS = hp(m) \cup \{m\} \cup lp(m)$ holds. Suitable techniques exist to assign priorities so that some properties hold for the *MS*. Ensuring schedulability is with no doubt the first and most important goal of such techniques.

52.3.1.3 Worst-Case Response Time

The most interesting element in schedulability analysis is certainly the WCRT for messages, which can be defined as follows. Let $t_{sched(m)}$ be the time when the transmission for message m is scheduled in the originating node, $t_{start(m)}$ the time when the transmission of the related frame can be started on the bus (irrespective of whether or not the transmitting node will win the contention), $t_{win(m)}$ the time when the successful transmission is started (i.e., when the node actually wins the contention), and $t_{end(m)}$ the time when the frame is completely and successfully received at the intended destination(s). For simplicity, transmission errors will be neglected in the following.

As shown in Figure 52.8, the WCRT R_m for message m can be evaluated as the sum of the worst-case jitter J_m introduced by the hardware and software in the transmitting node, the queuing delay w_m possibly caused by lost bus contentions, and the time C_m taken to transmit all the bits of the frame serially over the bus, that is,

$$R_m = J_m + w_m + C_m$$

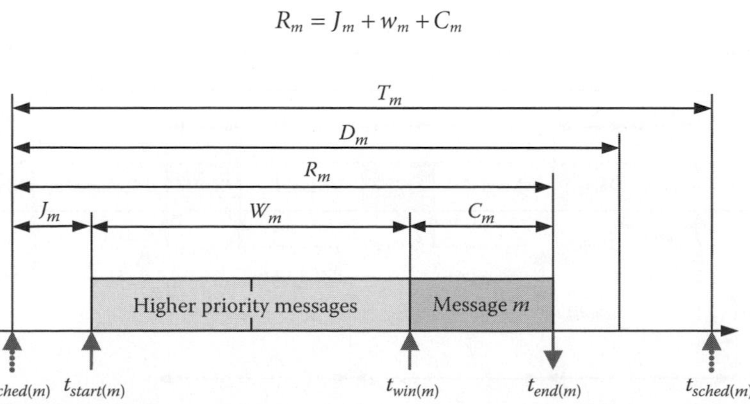

FIGURE 52.8 WCRT in CAN.

52.3.2 Schedulability Analysis

The main problem in schedulability analysis is evaluating the queuing delay w_m for every message in the *MS*. In fact, J_m can be derived from data sheets for what concerns latencies introduced by the communication hardware and code profiling for the software, whereas C_m can be easily computed from network parameters as described earlier.

52.3.2.1 Queuing Delay

The queuing delay w_m basically includes two kinds of contributions. The first one is due to one message at most, with a priority lower than *m*, which may already be in the process of being transmitted at time $t_{start(m)}$. The second concerns every message, with a priority higher than *m*, that might be generated in the time interval when message *m* is queued waiting to be sent. It is worth remembering that in CAN, the (numerically) lower the ID, the higher the priority.

The first kind of contribution gives rise to the so-called blocking time (B_m). It can be computed as the maximum transmission time, evaluated among every message *k* where *k* is (numerically) greater than *m*:

$$B_m = \max_{k \in lp(m)} (C_k)$$

The latter contribution, instead, is known as *interference*. Every message *j*, whose priority is higher than *m*, which is generated (or queued waiting for transmission) in the interval from $t_{start(m)}$ to 1 bit time after $t_{win(m)}$, in fact, certainly wins the contention with message *m* and delays it.

Let $h_j(T)$ be the maximum number of times the higher-priority message $j \in hp(m)$ may interfere with message *m* over a time interval *T*. Basically, $h_j(T)$ depends on the duration *T* of the interval and the generation period of message *j*:

$$h_j(T) = \left\lceil \frac{T + J_j + t_{bit}}{T_j} \right\rceil$$

In the example shown in Figure 52.9, message *j* interferes with message *m* three times, each time making *m* lose the contention. On the contrary, the fourth instance of message *j* no longer causes interference, as the related transmission may only take place after time $t_{sched(j,4)}$, when the SOF bit of message *m* has already been successfully sent (and no other higher-priority message is assumed waiting for transmission at that time).

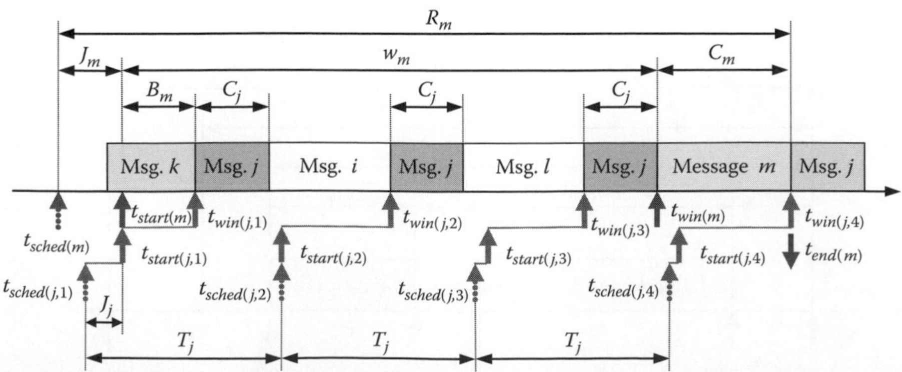

FIGURE 52.9 Blocking time and interference.

It is worth noting that the generation of interfering messages could be affected by jitters as well, because of the overhead caused by network controllers. This is the case of message j in Figure 52.9, whose first occurrence is delayed by J_j.

52.3.2.2 Recurrence Relation

As seen in the earlier text, the interference message m may suffer in a given time interval is related to its duration. This also holds when the considered time interval is the same as the queuing delay w_m. w_m, however, depends on both the blocking time B_m and the interference caused by all the higher-priority messages. The contribution to the overall interference due to each single message j depends in turn on its duration C_j and how many times it may interfere (at most) with m in a time lapse whose width equals w_m (i.e., the h_j parameter introduced earlier). Overall,

$$w_m = B_m + \sum_{j \in hp(m)} h_j(w_m) \cdot C_j$$

This leads to the following recurrence relation, which specifies the queuing delay at step $n + 1$ as a function of w_m at step n:

$$w_m^{n+1} = B_m + \sum_{j \in hp(m)} \left\lceil \frac{w_m^n + J_j + t_{bit}}{T_j} \right\rceil \cdot C_j$$

From a practical point of view, the evaluation of w_m proceeds as follows: the initial value w_m^0 is set to 0, and then the recurrence relation is repeatedly evaluated until either the value of w_m settles (i.e., w_m^{n+1} equals w_m^n) or the deadline D_m is exceeded. Convergence of this iterative approach is ensured, provided that the network utilization is below one, that is, $U_{MS} < 1$. This means that the recurrence relation eventually provides the queuing delay w_m.

The earlier technique enables the evaluation of the queuing delay—and, consequently, of the WCRT—for every single message. By checking them against the related deadlines, schedulability of the whole *MS* can be formally assessed.

In 2007, Davis et al. [DAV07] found that, in some cases, the earlier analysis is optimistic. As a consequence, they developed a new analysis that always provides meaningful results. Despite it corrects the flaws in the original method, the new version of the schedulability test they introduced is more complex and less intuitive.

A sufficient (but not necessary) schedulability test can be obtained by replacing B_m in the recurrence relation given earlier with $\max(B_m, C_m)$. If approximate results are acceptable, an even simpler way to compute the recurrence relation correctly is the following:

$$w_m^{n+1} = B_{MAX} + \sum_{j \in hp(m)} \left\lceil \frac{w_m^n + J_j + t_{bit}}{T_j} \right\rceil \cdot C_j$$

where B_{MAX} is the transmission time of the longest message in the considered network. In its turn, it is upper bounded by the duration of the largest CAN frame (8-byte payload). In this case, the obtained results are pessimistic.

52.3.3 Usage

Schedulability analysis has been (and still is) widely used for designing networked embedded system based on CAN, in order to check whether or not the network is correctly configured and/or dimensioned.

It can be applied to all systems where communications are based on asynchronous data exchanges (either cyclic or sporadic). One of the most interesting and popular real-world examples of this kind of systems is in-vehicle networks used in the automotive domain. In the case of production cars, in fact, the system is made up by several *ECUs*—from about ten up to slightly less than one hundred—each one producing several messages according to either a periodic or a sporadic schedule.

In the earlier analysis, data exchanges have been assumed uncorrelated from the point of view of communications. Indeed, they could be correlated at the application level, as in the case of message exchanges based on the request–response paradigm. Whatever the case, upper bounds are provided on transmission delays, so enabling the verification of whether or not timing constraints are met. More sophisticated approaches are also available if correlations are known, and tighter bounds can be provided in this case.

52.3.3.1 Identifier Assignment

Several aspects have to be considered carefully in the design phase, particularly when assigning IDs to messages that are exchanged in a CAN-based real-time distributed control application. From an intuitive point of view, the most urgent messages (i.e., those characterized by either the tightest deadlines or higher transmission rates) should be assigned the lowest IDs. For example, ID 0 denotes the message that has the highest priority in any CAN.

If the period of cyclic data exchanges (or the minimum interarrival time of acyclic exchanges) is known in advance, a number of techniques exist for assigning IDs, which are based on either the *deadline monotonic* (DM) or *rate monotonic* (RM) approaches [TIN94]. They can be profitably used to find, provided it exists, a suitable assignment of IDs to messages, so that the resulting schedule is feasible (i.e., deadlines of all messages are always met). More sophisticated techniques also exist, which grant optimal assignments according to some specific metric. For instance, the ability to tolerate (a limited amount of) transmission errors without losing schedulability can be a meaningful goal.

Typically, the assignment of priorities is performed statically during the system design so that more sophisticated approaches like the *earliest deadline first* (EDF) are seldom used in practice.

52.4 CAN Advantages and Drawbacks

Each CAN node is enabled to compete for bus ownership directly, so that it can send messages on its own. This means that CAN is a true multimaster network, which can be advantageous for use in systems that rely on the event-driven paradigm. However, it also suffers from several limitations, some of which depend on the adopted arbitration technique and cannot be overcome. Although serious, they are not going to rule out CAN completely, since it is (and will likely remain for several years) the preferred solution to implement cheap and robust networked embedded systems (including most of those used in automotive applications). Advantages and drawbacks of CAN are briefly sketched in the following, along with some solutions conceived to enhance its behavior.

52.4.1 Advantages

CAN is particularly suitable for networked embedded systems. Many inexpensive microcontrollers currently exist, which embed one (or more) CAN controllers at no additional cost and can be readily employed in both existing and new projects. This aspect, together with the fact that the protocol is mature, stable, and very well known, makes CAN a viable solution for many years to come, in spite of the availability of newer real-time solutions whose transmission speed is one or two orders of magnitude higher than CAN.

52.4.1.1 Compared to Fieldbuses

CAN shows several advantages when compared to the traditional networks adopted to exchange real-time process data at the shop floor in factory environments. For instance, it is by far simpler and more robust than access schemes based on token passing (such as PROFIBUS, when used in multimaster configurations). In CAN, there is no need to build or maintain the logical ring nor to manage the circulation of the token around the master stations. In the same way, it is noticeably more flexible than solutions based on the *time division multiple access* (TDMA) or *combined message* approaches—two techniques adopted by SERCOS and INTERBUS, respectively. This is because message exchanges are not required to be known exactly in advance. With respect to elder schemes based on *centralized polling* such as FIP, there is no need to have a node in the network acting as the bus arbiter, which may become a single point of failure for the whole system, and this improves reliability in a noticeable way.

From the point of view of the MAC mechanism, all CAN nodes behave like masters, and hence the notification of asynchronous events, such as alarms or critical error conditions, becomes very simple. In all cases where this aspect is important, CAN is clearly a better option than the other solutions mentioned earlier. Thanks to the arbitration scheme, no message can be delayed by lower-priority exchanges (this phenomenon is known as *priority inversion*). Since CAN, as the vast majority of communication protocols, is not preemptive, a message may still be delayed by a lower-priority transmission if this has already been started. However, as any CAN frame is very small (standard frames are 135-bit long at most, including stuff bits), the blocking time experienced by very urgent messages is quite low (usually well below 1 ms, if medium-to-high transmission speeds are selected). This makes CAN a very responsive network, which explains why it is used in many real-time control systems despite its relatively low bandwidth.

52.4.1.2 Compared to Newer Solutions

When CAN is compared to the communication protocols introduced in the last decade, things are a bit different, but it still keeps some advantages. Although *real-time Ethernet* (RTE) solutions have much higher bit rates, which make them perfectly suitable for industrial plants and factory automation environments, they are not very popular in networked embedded systems. This is because of several reasons, the most relevant being the higher cost. Ethernet controllers, transceivers, magnetics, and cables are typically more expensive than CAN, and often, intermediate network devices like switches are also needed. In some cases, RTE solutions are not completely settled yet or, better, some details may be not as stable as in CAN.

In the same way, CAN proves to be a cheaper and much simpler option than solutions purposely designed for the use in automotive systems (e.g., in-vehicle networks), such as the Time Triggered Protocol-Class C (TTP/C) and, mostly, FlexRay. There is no doubt that, in the future, such technologies will take over CAN in a number of application fields. But, likely, this is not going to happen in a short time. As stated in [FRE02], CAN will remain in use for many years to come. The recent introduction of CAN FD [CANFD], pushed by some leading carmakers, is a further evidence of the intention not to abandon CAN in both automotive applications and networked embedded systems.

52.4.2 Drawbacks and Solutions

CAN is affected by a number of drawbacks, the most important concerning *performance*, *determinism*, and *dependability*. Although they were considered mostly irrelevant when the protocol was introduced at the beginning of the 1990s, with the elapsing of time, they have become more and more limiting in a number of application fields, such as the advanced in-vehicle x-by-wire systems in the automotive industry. Solutions have been consequently proposed, which are aimed at coping with one or more of such limitations.

52.4.2.1 Performance

Even though inherently efficient and quite elegant, the arbitration technique of CAN poses serious limitations on the performance that can be obtained by the network. In fact, in order to ensure the correct behavior of the arbitration mechanism, the signal must propagate from a node located at one end of the bus up to the farthest node (at the other end) and come back to the first node before it samples the level on the bus (this is because all nodes must *see* the same logical value within the same bit time).

Since the sampling point is roughly located near the end of each bit (its position can be programmed through suitable registers in the CAN controller), the end-to-end propagation delay (also including hardware delays of couplers, transceivers, and controllers) must be strictly shorter than about half the bit time, the exact value depending on the bit timing configuration in the DPLL.

Since the propagation speed of signals is fixed (about 200 m/μs on copper wires), this implies that the maximum length allowed for the bus is necessarily limited and depends directly on the bit rate chosen for the network. For example, a 250 kbit/s CAN can span at most 200 m; similarly, the maximum bus length allowed when the bit rate is set equal to 1 Mbit/s is less than 40 m. This, to some degree, explains why the maximum bit rate allowed by the CAN specification [ISO1] has been limited to 1 Mbit/s.

In Figure 52.10, an example is shown where the arbitration no longer works properly because of propagation delays, which in turn disrupts communications. Let us consider a simple CAN, made up by two nodes (A and B), which are located at the opposite ends of the bus. Initially, the bus is assumed to be idle. After some time, a transmission request is issued in node A, which starts transmission (the SOF bit has been grayed for clarity). The signal takes some time to reach node B, due to propagation delays.

Just before the signal arrives in B, let us assume that a transmission request is issued in node B. Since the bus in B is still idle, node B starts transmitting too. The signal generated by B arrives to A after a time (measured from the instant A started transmitting) that is slightly shorter than twice the end-to-end propagation delay. If this happens after node A has sampled the bus to determine the level of the current bit (which is fundamental for arbitration), the value that is read by A is obtained as the (wired) AND between the current bit it is writing out and the previous bit sent by B. In this case, the SOF bit from B, at the dominant level, overwrites the first bit of the ID in the message sent from A, at the recessive level, and tricks node A into thinking it has lost the contention.

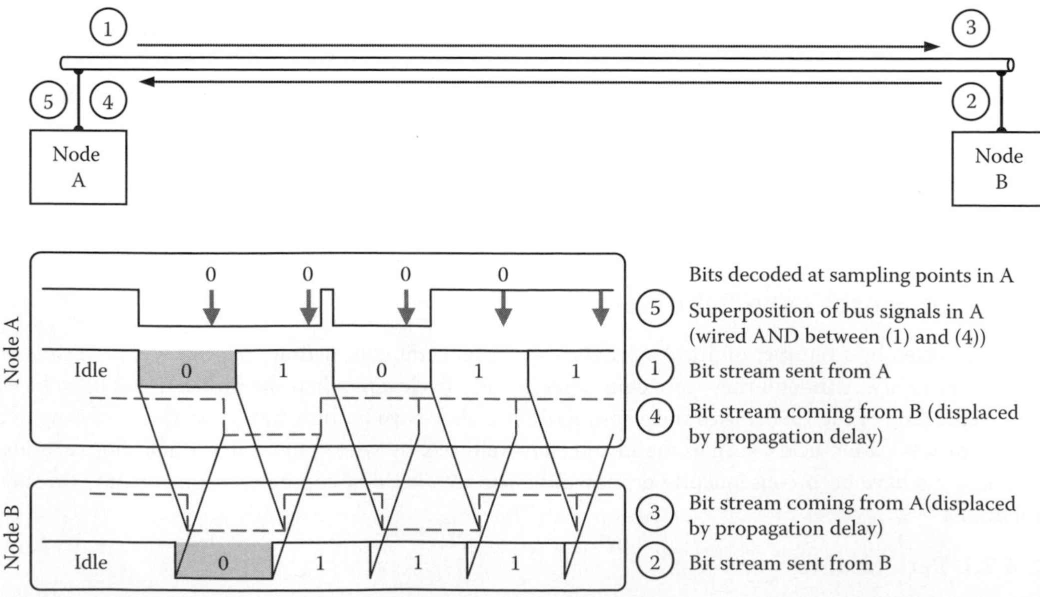

FIGURE 52.10 Effects of propagation delays on arbitration.

Generally speaking, in order for the arbitration to work properly, the following condition must hold:

$$T_{prop\_seg} > 2 \cdot (T_{CC} + T_{TR} + T_{OC} + L_{MAX} \cdot t_p)$$

where

T_{prop_seg} is the duration of the propagation delay segment (as configured in the DPLLs of the nodes)
T_{CC}, T_{TR}, and T_{OC} are the contributions to the end-to-end delay due to the latencies introduced by, respectively, CAN controllers (~50 ns), CAN transceivers (120–250 ns), and, optionally, optocouplers (40–140 ns) on devices at both ends
L_{MAX} is the distance (measured on the bus) between the farthest nodes in the network
t_p is the specific propagation delay of electrical signals over the wire (about 5 ns/m)

From the previous equation, and remembering that T_{prop_seg} is always strictly shorter than the bit time T_{bit}, an upper bound can be found for the maximum bus length allowed for any given bit rate R (inverse of T_{bit}):

$$L_{MAX} < \frac{(1/2R) - (T_{CC} + T_{TR} + T_{OC})}{t_p}$$

Clearly, as long as the bit rate is medium to low, the maximum length can be satisfactorily described as inversely proportional to the bit rate. However, for high bit rates (close to 1 Mbit/s), the contribution of hardware latencies due to controllers and transceivers can no longer be neglected. A detailed analysis of the relationship between bit rate and maximum bus length can be found in [DIN08].

It is worth noting that the earlier limitation depends on physical factors, and hence it cannot be overcome in any way by advances in the technology of transceivers. To make a comparison, it should be mentioned that, at present, several inexpensive communication technologies are available on the market that allow bit rates in the order of tens or hundreds of Mbit/s.

Even though this may appear as a very limiting factor for CAN, it will probably not have any relevant impact in the near future for several application areas—including some automotive and process control applications—where a cheap and well-assessed technology is more important than top performance. However, in other scenarios like factory automation or x-by-wire systems, CAN is suffering from the higher bit rates of its competitors, for example, PROFIBUS-DP (up to 12 Mbit/s), SERCOS (up to 16 Mbit/s), FlexRay (up to 10 Mbit/s), and industrial Ethernet (up to 100 Mbit/s). In fact, such solutions are able to provide noticeably higher data rates, and this is a basic requirement for systems made up of many devices or having very short cycle times (1 ms or less).

It is worth remarking that little can be done to improve CAN performance significantly, unless full backward compatibility is sacrificed. Despite some solutions have appeared in the literature—such as WideCAN [CEN02], which provides higher bit rates while still relying on the conventional CAN arbitration technique—their interest has remained mainly theoretical. Recently, the *CAN FD* protocol [CANFD] has been introduced, which enables a significant increase of the data rate and resembles the original protocol closely at the same time.

52.4.2.2 Determinism

Because of its nondestructive bitwise arbitration scheme, CAN is able to resolve in a deterministic way any collision that might occur on the bus. The MAC mechanism ensures that the bus is always assigned to the node that is sending the message with the highest priority. As shown previously, when dealing with schedulability analysis, it is possible to determine right in the design phase of a CAN system whether or not messages are always delivered before their intended deadlines (or, equivalently, transmission latencies experienced by every message are upper bounded). This means that CAN supports real-time communications.

However, if nodes are allowed to produce asynchronous messages on their own—this is the way event-driven systems usually operate—there is no way to know in advance the exact time when a given message will be sent. This is because the actual number of contentions (i.e., the interference) any message experiences with higher-priority transmissions is not known a priori. This behavior leads to transmission jitters, which in some kinds of applications—including those involved in the automotive field—may affect control algorithms in a negative way and worsen their precision. Moreover, if transmission errors due to electromagnetic noise are taken into account, some messages can even miss their intended deadline.

Another problem related to determinism is that composability is not ensured in CANs. This means that, when several subsystems are developed independently and then interconnected to the same network, the overall system may fail to satisfy the timing constraints, even though each subsystem was tested separately and proved to behave correctly and the overall network bandwidth is theoretically sufficient to carry out all the data exchanges. This severely limits the chance of integrating subsystems made from different vendors and makes the design and test tasks much more difficult.

A possible solution conceived to enhance determinism in CAN is the so-called *time-triggered CAN* (TTCAN) protocol [ISO4], for which commercial network controllers are available even though they are not so popular. By adopting a common clock and a time-triggered approach, TTCAN achieves very-low jitters and provides a fully deterministic behavior.

52.4.2.3 Dependability

The last drawback of CAN concerns dependability. Whenever safety-critical applications are considered, where communication errors may lead to damages to the equipment and even injuries to human beings (as in automotive x-by-wire systems), a highly dependable network has to be adopted.

Reliable error detection should be carried out both in the value and in the time domain. In the former case, conventional techniques such as the use of a suitable CRC are adequate. In the latter case, instead, a time-triggered approach is certainly more appropriate than the event-driven communication scheme provided by CAN. In time-triggered systems, all actions—including sampling of sensors, actuation of commanded values, task activations, as well as message exchanges—are known in advance and must take place at precise points in time. In this context, even the presence/absence of a message at a given instant provides significant information (i.e., it enables the discovery of faults). For example, *masquerading faults* (a node that is sending messages on behalf of another one) can be avoided in time-triggered networks, and a *fail silent* behavior can be easily enforced for nodes by means of *bus guardians*. TTCAN provides some advantages in this respect, by enabling error detection in the time domain.

The so-called *babbling idiot* problem is another issue that may affect CAN systems. In this case, a faulty node that repeatedly keeps transmitting a high-priority message on the bus (usually because of a flaw in the software of the generating task) may slow down the whole network and cause message deadlines to be missed. Sometimes, it can even block the communication completely. Such a kind of failures cannot be detected by the fault confinement unit embedded in CAN chips, as they do not depend on physical faults but rather are due to logical errors (e.g., correct frames that are sent at a higher rate than intended). Although bus guardians can be used, in theory, to cope with these errors in CAN as well, they cannot be implemented easily and efficiently for networks based on the event-driven communication paradigm.

Concerning fault tolerance, several interesting proposals have appeared in the literature to tackle the cases of broken links or faulty transceivers, which can be profitably adopted in real-world systems. For example, in [PRO06], a novel architecture for CAN, which is based on a star topology, is described. Unlike the conventional bus topology, such an arrangement helps avoiding *partitioning faults*, that is, a broken link only affects one node and does not prevent the others from communicating. Such a solution has been subsequently extended in order to deal with redundant media as well.

52.5 Time-Triggered CAN

The TTCAN protocol was introduced by Bosch in 1999 with the aim of making CAN suitable for the new needs of the automotive industry and, in particular, to satisfy, at least in part, the requirements of the upcoming x-by-wire systems. TTCAN can be profitably used in all those applications, characterized by tight timing requirements, that demand for a strictly deterministic behavior. TTCAN, in fact, allows the exact selection of the point in time when messages have to be exchanged, irrespective of the network load.

The TTCAN specification was then standardized by ISO [ISO4]. Its definition was mainly driven by the need to improve determinism of CAN, by providing support for time-triggered communications, as well as to permit higher bus loads while maintaining the highest degree of compatibility with existing devices and development tools. As a consequence, significant savings in communication technology investments were expected. A further advantage has to do with *composability*, which in TTCAN is improved with respect to CAN. In particular, a system can be split into several subsystems in the design phase, which can be developed and tested separately.

Unfortunately, TTCAN failed in establishing itself as the successor of CAN, the main reason being the need for system designers to switch from the event-driven to the time-triggered paradigm. Nevertheless, the availability on the market of CAN controllers that support the TTCAN time-stamping features and time-triggered operations may still be useful in those application contexts where the accuracy of timing is important, in particular when inexpensive implementations are required.

52.5.1 Main Features

One of the most appealing features of TTCAN is that both event-driven and time-triggered operations can coexist in the same network at the same time. In order to ease a graceful migration from CAN, TTCAN foresees two levels of implementation, which are known as level 1 and 2, respectively. Level 1 implements only basic time-triggered communications over CAN. Level 2, which is a proper extension of level 1, also offers a means for maintaining a global system time across the whole network, irrespective of tolerances and drifts of local oscillators. This enables accurate synchronization up to the application level, and hence true time-triggered operations can take place in the system.

The TTCAN protocol is conceptually placed above the unchanged CAN data link layer and allows time-triggered exchanges to take place in a quasi-conventional CAN. For this reason, it suffers from the same performance limitations as the underlying CAN technology. In particular, the transmission speed cannot exceed 1 Mbit/s in practice, and the same limitations concerning the network extension still hold. However, because of the time-triggered paradigm it relies on, TTCAN is able to ensure strictly deterministic communications, which means that, for example, it is deemed suitable for the first generation of drive-by-wire automotive systems—which are provided with hydraulic/mechanical backups. Conversely, TTCAN is likely not adequate for next-generation steer-by-wire applications. In fact, requirements on both fault tolerance and available data rate are noticeably more severe in this case.

52.5.2 Protocol Specification

TTCAN adopts the very same frame format and transmission technology as CAN. In TTCAN, however, nodes are not allowed to send frames at any time. Transmissions, instead, take place according to specific triggers that are driven by time. Accordingly, some means have to be provided by the protocol so that all nodes can synchronize their activities.

52.5.2.1 Time-Triggered Communication

Unlike new-generation automotive protocols (e.g., FlexRay [FR3]), which rely on distributed synchronization mechanisms, TTCAN is based on a centralized approach, where a special node called the *time master* (TM) keeps the whole network synchronized by regularly broadcasting a *reference message* (RM), usually implemented as a high-priority CAN message. Redundant TMs can be envisaged so as to increase reliability. In particular, up to eight *potential TMs* can be deployed in a TTCAN. Their priority is encoded in the three least significant bits of RM. At any time, only one node is allowed to be the actual TM, whereas the others behave as backup substitutes. The procedure for (re)electing the current TM relies on CAN arbitration. It operates in a completely dynamic fashion and ensures that faulty masters are replaced on the fly.

Transmission of data is organized as a sequence of *basic cycles* (BCs). Each BC begins with the reference message, followed by a fixed number of time windows that are configured off-line and can be of three different types:

1. *Exclusive windows*: Each exclusive window is statically reserved to one predefined message, so that collisions cannot occur. Exclusive windows are used for safety-critical data that have to be sent deterministically and without jitters.
2. *Arbitration windows*: These windows are not preallocated to any given message, and thus different messages may possibly be competing for transmission; in this case, any collision is solved thanks to the nondestructive CAN arbitration scheme.
3. *Free windows*: These are placeholders, which are reserved for future expansions of TTCANs.

The beginning of each time window corresponds to a time trigger defined in one (or more) producing nodes. Since boundaries of time windows shall never be exceeded, TTCAN controllers must disable the automatic retransmission feature of CAN (which retransmits the same message whenever the contention is lost or transmission errors are detected). The only exception occurs when several adjacent arbitrating windows are present. In this case, they can be merged together, so as to build a single, larger window, which is able to accommodate a number of asynchronously generated messages in a more flexible way. The only constraint is that the transmission in such a *merged arbitrating window* is allowed only if the complete message can be sent in the time remaining before the end of the window itself.

Despite it seamlessly mixes both synchronous and asynchronous messages (in exclusive and arbitrating windows, respectively), TTCAN is very dependable indeed. Besides the intrinsically deterministic time-driven access to the shared communication support, in fact, the arbitration mechanism of CAN is used to solve the collision whenever a temporary lack of synchronization occurs and more than one node tries to transmit in the same exclusive window.

52.5.2.2 System Matrix

More than one BC can be adopted to increase flexibility. A *system matrix* can be defined, which consists of up to 64 different BCs (the exact number has to be a power of two), which are repeated periodically (see Figure 52.11). Thus, the actual periodicity in TTCAN is given by the so-called matrix cycle. A *cycle*

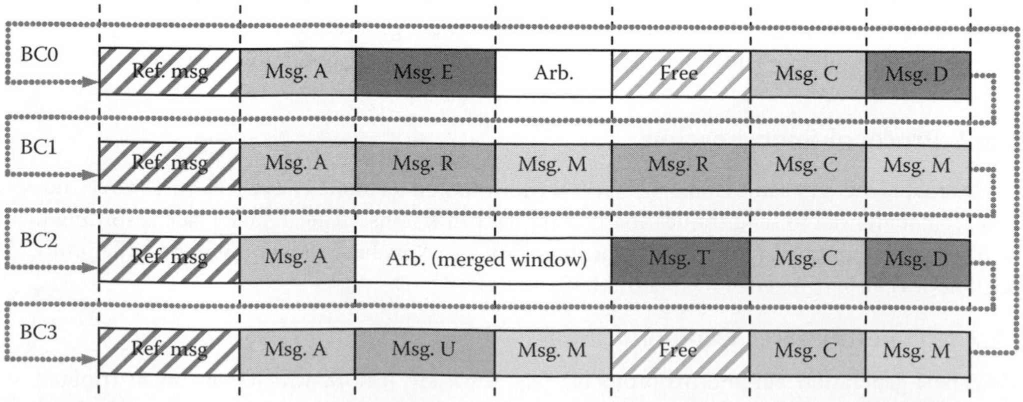

BC0	Ref. msg	Msg. A	Msg. E	Arb.	Free	Msg. C	Msg. D
BC1	Ref. msg	Msg. A	Msg. R	Msg. M	Msg. R	Msg. C	Msg. M
BC2	Ref. msg	Msg. A	Arb. (merged window)		Msg. T	Msg. C	Msg. D
BC3	Ref. msg	Msg. A	Msg. U	Msg. M	Free	Msg. C	Msg. M

FIGURE 52.11 System matrix in TTCAN.

counter—included in the first byte of RM—is used by every node to distinguish the current BC inside the matrix cycle. The TM increases the cycle count by one on every BC, until the maximum value (which is selected on a network-wide basis before operation is started) is reached and the counter is restarted.

The system matrix is highly column oriented. In particular, each BC is made up of the same sequence of time windows, that is, corresponding windows in different BCs do have exactly the same duration. However, they can be used to convey different messages, depending on the cycle counter. In this way, it is possible to have messages in exclusive windows that are repeated once every any given number of BCs. Each message is assigned a *cycle offset* and a *repeat factor* to characterize its transmission schedule (both these parameters are expressed as an integral number of BCs). The cycle offset is the first row in the system matrix when the message shall be sent. It is a displacement to prevent collisions among different messages whose time windows share the same column. The repeat factor, instead, specifies how often a given message has to be sent, and it is expressed as a power of two. This *multiplexing* mechanism enables the optimization of the bus allocation with respect to the timing requirements of different message streams and allows significant savings for the network bandwidth.

More than one exclusive window can be allocated to the same message in the same BC. This is useful for either having redundant transmissions of critical data or refreshing some variables with a rate higher than the BC.

52.5.2.3 Implementation

Each TTCAN controller has its own clock and is provided with a *local time*, which is basically a counter that is increased by one every *network time unit* (NTU). The NTU must be the same (at least, nominally) on the whole network. Each node derives it from the local clock and a local parameter known as *time unit ratio* (TUR). The biggest difference between the level 1 and level 2 variants of the protocol is that the latter defines an adaptive mechanism that is able to keep the NTUs (and, hence, the global time) accurately synchronized by constantly updating the TUR parameters of the different nodes.

Whenever RM is received, each node restarts its *cycle timer*, so that a common view of the elapsing time is ensured across the whole network. In practice, each time the SOF bit of a new message is sampled on the CAN bus, a synchronization event is generated in every network controller, which causes the local time to be copied in a *sync mark* register. After the message has been completely read, a check is made by controllers: if the SOF bit concerns a valid reference message, the sync mark register is copied into the *reference mark* register. At this point, the cycle time can be evaluated simply as the difference between the current local time and the reference mark (see Figure 52.12a).

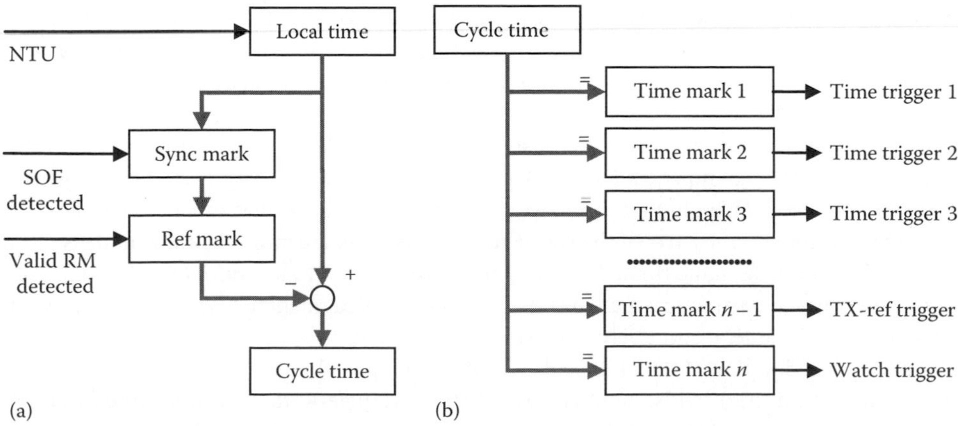

FIGURE 52.12 Time-triggered operation in TTCAN. (a) Synchronization of the cycle time and (b) scheduling driven by the cycle time.

Two kinds of RM are actually foreseen: in level 1 implementations, this message is 1 byte long, whereas level 2 relies on a 4-byte RM that is backward compatible with level 1 (from a practical point of view, three more bytes are appended for distributing the global time as seen by the TM).

In each node, the protocol execution is driven by the progression of the cycle time. In particular, a number of *time marks* are defined in each network controller and work as either transmission or receive triggers to send messages or validate message receptions, as shown in Figure 52.12b. As soon as the cycle time equals a particular time mark, the related trigger is activated. It is worth noting that in TTCAN, nodes are not requested to know the schedule of all messages in the network. Indeed, only the details of those messages the node sends/is interested in are needed.

Two triggers are also foreseen that are not used for exchanging data and are known as *TX-ref trigger* and *watch trigger*, respectively. The former is employed by potential TMs to trigger the generation of RM for the next BC. The latter, instead, is used to detect the absence of reference messages on the network, which implies a faulty condition.

From a general perspective, TTCAN requires slight (and mostly inexpensive) changes to the internal architecture of conventional CAN chips. In particular, transmit and receive triggers plus some counters and registers for managing the cycle time are necessary to ensure time-triggered operations.

Even though level 1 could be implemented completely in software—the ability to support single-shot mode or the cancellation of a transmission request is enough in this case—specialized hardware can reduce the host processor burden for managing time-triggered operations. As level 2 compliant controllers provide support for drift correction and calibration of the local time, they need the hardware to be modified accordingly.

The structure of TTCAN modules resembles the conventional architecture. In particular, only two additional blocks are needed, that is, the *trigger memory* and the *frame synchronization entity*. The former is used for storing the time marks of the system matrix, which are linked to the message buffers held in the controller's memory. The latter, instead, is used to control TTCANs according to the protocol.

At present, a limited number of controllers are available off the shelf that comply with the TTCAN specifications. However, this shows that this solution is not only theoretical, but it can be actually embedded in new projects.

52.6 CAN FD

As pointed out in the previous sections, one of the main bottlenecks of CAN is its limited data throughput. Basically, this is because of two main reasons: The first is the trade-off between network speed and extension, which limits the practical bit rate to 1 Mbit/s (in this case, the bus length is limited to 40 m at best). In order to cover larger areas, the bit rate has to be reduced consequently. The second reason is the limited size of the payload: in fact, in order to ensure high responsiveness, 8 bytes at most can fit in the data field of a frame. This consequently reduces the communication efficiency, which, in the best case, reaches 47% with standard IDs.

Recently, a proposal has been made by Bosch [FLO12] in order to overcome both the drawbacks presented earlier, which is known as CAN FD. CAN FD builds on the basic idea first outlined, more than 10 years ago, in [CEN99]. In that paper, we pointed out that the limitation affecting the maximum bit rate only holds for specific parts of the frame transmission, where more than one node is allowed to transmit on the bus at the same time. These regions were referred to as *multiple-writer-allowed* zones (M-zones) and concern, in practice, the arbitration and acknowledgment fields, placed at the very beginning and end of the frame, respectively. It is worth noting that, in order to denote the end of a frame unambiguously, the duration of the EOF field cannot be changed, too.

In the remaining (inner) fields, denoted in [CEN99] as *single-writer-allowed* zone (S-zone), only one transmitter (i.e., the winner of the arbitration phase) is transmitting over the bus, at least in a properly configured network. In fact, CAN mandates that no two nodes can be enabled to send nonempty data frames having the same ID at the same time. This means that the bit rate can be safely increased in the S-zone (which includes the DLC, DATA, and CRC fields), while it must be left unchanged in the M-zones. This approach was called *overclocking* technique in [CEN99], where the use of *oversized* CAN frames was also envisaged to better exploit the higher bit rate enabled by the payload overclocking, so as to increase the communication efficiency. In fact, the frame header and trailer almost have a fixed duration, so that the communication overhead they introduce becomes proportionally smaller as the payload increases in size. For instance, in [CEN00], the use of an extended length field was assumed that could accommodate up to 80 bytes in the compressed payload.

52.6.1 Main Features

CAN FD brings two main improvements to CAN. First, data frames longer than legacy CAN are foreseen. Up to 64 bytes can be included in the data field of CAN FD frames, which actually increases the maximum amount of data in the frame payload by a factor 8. On the one hand, this improves the communication efficiency, which can now reach 72% in the case of 11-bit IDs. On the other hand, fragmentation is either prevented or significantly reduced, at least, especially when remote device parameterization is carried out.

The second improvement is the increased bit rate for selected portions of the frame. According to the CAN FD specification, two (possibly different) bit rates are defined, which are known as *nominal bit rate* and *data bit rate*. The first one is used during both the arbitration phase and the operation of the ACK mechanism (including the EOF field), whereas the second is used in transmitting the body of the frame (basically the DLC, DATA, and CRC fields). Supporting a higher bit rate, however, requires an alternate set of configuration registers. As a consequence, the duration of the time segments has to be defined separately for the 2 bit rates.

52.6.2 Protocol Specification

From the point of view of the protocol machine, CAN FD basically behaves as conventional CAN. The main differences are the format definition of additional frames and the option to increase the bit rate temporarily for transmitting the inner part of the frame. Other improvements were also brought in order to improve reliability.

52.6.2.1 Frame Format

CAN FD defines two new frame formats [CANFD], namely, CAN FD frames with base (11 bits) and extended (29 bits) IDs. They are sketched in Figure 52.13. As mentioned before, the main goal of CAN FD is increasing the throughput by using either oversized frames or overclocking (or both at the same time). For this reason, no additional format is defined for remote frames; in fact, they do not include any data field and would hardly benefit from new encoding schemes. Conventional remote frames can be used, if needed. As a consequence, the RTR bit in CAN FD frames has been renamed *reserved bit r1* and is set at the dominant value.

As shown in Figure 52.13, a couple of new bits have been added to the control field, which are not found in conventional CAN frames. The *extended data length* (EDL) bit—which takes the place of a reserved (dominant) bit in conventional frames—is sent at the recessive level in CAN FD and allows distinguishing between the two formats.

The *bit rate switch* (BRS) bit, instead, is used to decide whether or not the transmission speed has to be increased from the nominal bit rate (used in the arbitration phase and for the frame trailer) to a higher, alternate bit rate, whose value has to be configured in advance. In particular, a recessive value means a bit rate switching, whereas a dominant value indicates that the same bit rate is used all over the frame. The BRS bit also marks the point in time where the transmission bit rate is possibly switched. The transmission speed is restored to the nominal bit rate during the CRC delimiter.

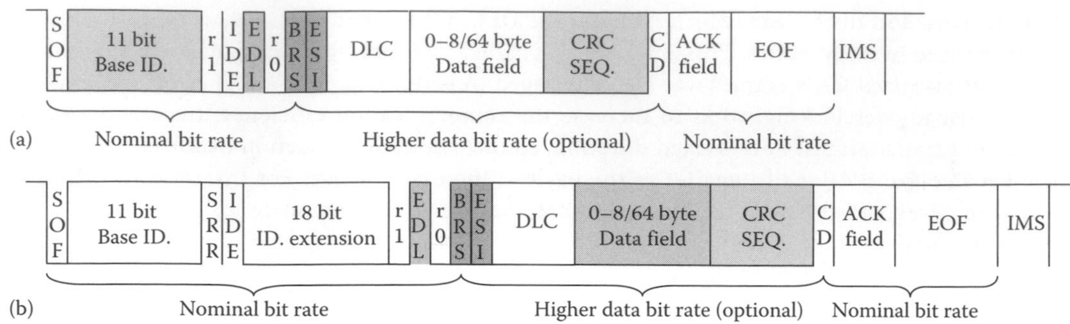

FIGURE 52.13 Format of CAN FD frames. (a) CAN FD base frame format and (b) CAN FD extended frame format.

The third bit defined by CAN FD is the *error state indicator* (ESI) flag. ESI is sent at the dominant value by error-active nodes, whereas passive nodes leave it at the recessive level. For the sake of truth, its purpose has little or nothing to do with enhancing the data rate.

Also the meaning of the DLC field has been extended in CAN FD, in order to cover additional payload sizes besides those originally supported in CAN. In particular, when the value of DLC is greater than 8 (i.e., from 9 to 15), seven additional sizes can be specified for the data field, namely, 12, 16, 20, 24, 32, 48, and 64 bytes. It is worth remarking that the same bit patterns are not illegal in conventional CAN, but they are always interpreted as an 8-byte-long payload.

CAN FD frames can carry a significantly larger amount of information. Three different generator polynomials were then envisaged to maintain the same degree of reliability as conventional CAN; in fact, besides the classical CRC-15, two additional CRCs are defined for frames that comply with the new formats: the first one (CRC-17) is used when the data field is up to 16 bytes in size, whereas the second (CRC-21) is used for larger frames. In this way, a *Hamming distance* (HD) equal to 6 is obtained, irrespective of the actual frame size.

52.6.2.2 Other Differences with CAN

A couple of modifications were also added to the basic CAN protocol. Bit stuffing rules were slightly changed for CAN FD frames. In particular, stuff bits are inserted in fixed positions in the CRC field of such frames. The first stuff bit is inserted just before the first bit of the CRC, while a further stuff bit is added after each fourth bit of the CRC sequence, which is computed as the inverse of the immediately preceding value. Moreover, in the new CAN FD frame formats, stuff bits included in the bit stream are relevant to the CRC calculation, whereas in conventional CAN, they were not considered.

These approaches are intended to improve the error detection capability. In fact, as pointed out in the literature [CHA94], the bit stuffing mechanism may interfere with the CRC calculation and increases the residual error probability (i.e., the probability that a message suffering from a transmission error is considered correct by one or more receivers). In particular, a pair of bit flip errors located more than 15 bits apart lead to an error burst that might go undetected by the conventional CRC mechanism.

In order to decrease implementation costs and improve compatibility with the current controller–host interfaces for CAN, support for all long frame sizes is not mandatory in CAN FD. Nevertheless, every receiver must be able to read frames up to the maximum allowed size without generating any errors. Data bytes exceeding the handling capabilities of the CAN controller can simply be discarded.

52.7 CANopen

Almost all the functions of the CAN data link layer (including services for frame transmission and reception) are implemented directly in the communication hardware. This choice provides a very good degree of efficiency and reliability for data exchanges. Therefore, in many control systems, application processes access the CAN controller directly (possibly through suitable device drivers). Although this solution is acceptable in many scenarios, only basic communication services are offered by the data link layer, and this may be inadequate for other complex applications.

In order to reduce the design and implementation effort of automation and networked systems, a number of higher-level solutions have been defined in the past years, which rely on CAN for message transfers and also define their own *application layer*. The main purpose of the application layer is to provide a usable and well-defined set of service primitives, which can be invoked to interact with devices in a standardized way. The goal of reducing the design and development costs is easily achieved thanks to standardization.

To this purpose, several solutions are currently available off the shelf, such as CANopen [EN4] for networked control systems, DeviceNet [EN2] for factory automation, and J1939 [J1939] for trucks and other vehicles. In the following, the CANopen protocol will be described, focusing in particular on its use in industrial and networked embedded systems.

52.7.1 Protocol Basics

CANopen is a CAN-based application layer protocol with very flexible configuration capabilities that was explicitly defined for automation systems. It unburdens the developer from dealing with CAN-specific details, such as bit timing and implementation-specific functions. Moreover, it defines standardized communication objects (COBs) able to cope with both real-time and configuration data, as well as network management (NMT).

CANopen was initially conceived to rely on the communication services provided by the CAL [CiA20X]. However, the current specifications [CiA301] no longer refer explicitly to CAL. Instead, the relevant services and definitions have been included directly in the CANopen specification. CANopen was subsequently accepted as the international standard EN 50325-4 [EN4]. Specifications include the application layer, the communication profile, as well as several application, device, and interface profiles. These documents are developed and maintained by the CiA organization.

At present, CANopen is used in a very broad range of application fields, such as off-road and rail vehicles, maritime electronics, machine control, medical devices, building and industrial automation, as well as power generation.

52.7.1.1 Communication Objects

In CANopen, information is exchanged by means of *COBs*, which model application layer services. A number of different COBs exist, which are conceived for different functions:

- *Process data objects* (PDOs) are used for real-time data exchanges, such as measurements read from sensors and command values issued to actuators for controlling the physical system.
- *Service data objects* (SDOs) are used for non-real-time communications, for example, device parameterization and diagnostics.
- *Emergency objects* (EMCYs) are used by devices to asynchronously notify the occurrence of error conditions.
- *Synchronization object* (SYNC) is used to achieve synchronized and coordinate operations in the system.
- *Time stamp object* (TIME) is used to broadcast global system time information, which can be used by devices to adjust their local time base.
- *NMT* defines a number of objects for dealing with management services, including mechanisms to check whether devices are alive or not.

Every COB is mapped onto one (or more) CAN frames for transmission over the bus. More details about how such a mapping takes place, as well as the way information is encoded, are provided in the following paragraphs.

52.7.1.2 Communication Protocol

CANs are based on the producer/consumer interaction model and rely on object addressing. However, for a number of application layer services, this arrangement is not the best choice to interact with devices. For this reason, both the master/slave and the client/server protocols are additionally defined in CANopen.

Although every CAN node behaves as a master from the point of view of the MAC mechanism, that is, it can access the bus autonomously whenever it needs to transmit information, the master/slave protocol is a better approach for dealing with NMT. In particular, each device in any CANopen network is identified by means of a unique 7-bit node address, which lies in the range from 1 to 127. In addition, to make the system configuration easier, a *predefined master–slave connection set* must be provided by each device, which consists of a standard allocation scheme of CAN-IDs to COBs and is made available after initialization. Device configuration, instead, is carried out over peer-to-peer connections using the client/server protocol.

52.7.2 Device Model

The behavior of a CANopen device is completely described by means of a number of objects. Each object, in turn, is devoted to some specific aspect concerning the device configuration, the communication on the CAN bus, or the functions for interacting with the controlled physical system (e.g., some objects define the device type, the manufacturer's name, the hardware and software version).

All the objects relevant to a given device are stored in the *object dictionary* (OD) of that node. As shown in Figure 52.14, interactions between frames exchanged on the CAN and local applications that act on the physical system take place through the OD. This choice offers a standardized view of devices and their behavior for control applications, as well as an easy interaction model through the CAN bus.

52.7.2.1 Object Dictionary

The OD can be seen as a table, which stores all information concerning a device. Both parameterization and process data are kept in the OD, which is the only repository of information for the device. The OD can include up to 65,536 different objects, each one accessed through a unique 16-bit address (index 0000_h is not used). Any object, in turn, can be either of simple (Boolean, integer, floating point, string, etc.) or complex (arrays and records) type. In the latter case, it may consist of up to 256 subentries, each one addressed via an 8-bit subindex.

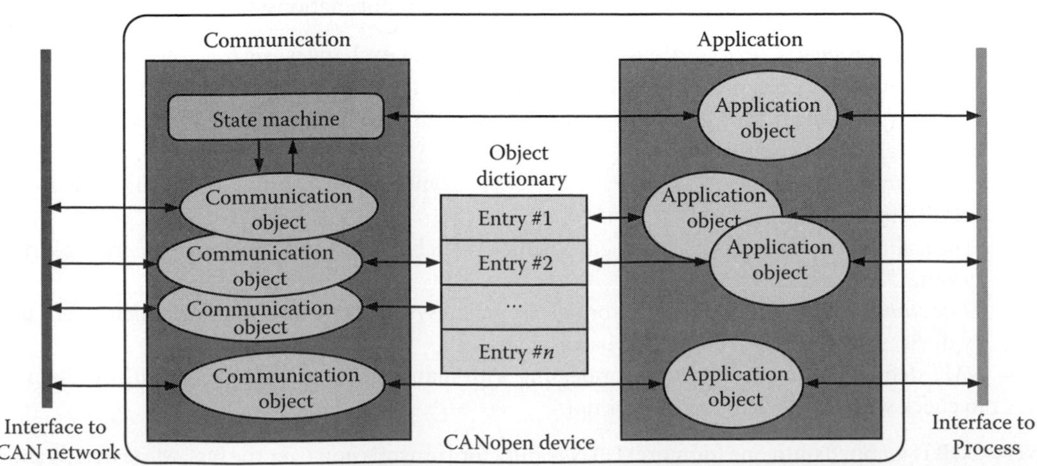

FIGURE 52.14 Model of CANopen devices.

In the case of simple-type entries, subindex 00_h is used to access the information directly. Instead, for complex data types, subindex 00_h encodes the highest subindex supported for the related entry, that is, the number of either array elements or record fields. In this case, subindex FF_h may optionally encode the relevant data type and object type. This means that each single item in the OD can be uniquely identified through a 24-bit multiplexer.

52.7.2.2 OD Structure

From a conceptual point of view, the OD is split into separate areas, according to the indexes of entries.

These areas are then used to store different kinds of information, which concerns, for instance, CANopen communications or device functionalities. The meaning of an entry can be roughly deduced by looking at its index:

- Entries below 1000_h are used to specify *data types* (static, complex, specific to either the manufacturer or the device profile). Most of them (from 0260_h to $0FFF_h$) are reserved for future use.
- Entries from 1000_h to $1FFF_h$ make up the *communication profile area*. They are used to describe communication-specific parameters, that is, to model the interface of the device to the CAN. These objects are common to all CANopen devices.
- Entries from 2000_h to $5FFF_h$ correspond to the *manufacturer-specific profile area*. They can be freely used by manufacturers to extend the basic set of functions of their devices. Their use has to be considered carefully, in that it could make the device no longer interoperable.
- Entries from 6000_h to $9FFF_h$ belong to the *standardized profile area*. They describe all aspects concerning a given class of devices, as defined in the related device profile, in a standardized way. This area is split into up to eight parts, each one corresponding to one logical device. For instance, the entries related to the first logical device are found in the range from 6000_h to $67FF_h$ and those concerning the second one fall between 6800_h and $6FFF_h$.
- The meanings of some entries above $A000_h$ have been defined in the most recent versions of CANopen. Version 4.2.0, in particular, assigns entries from $A000_h$ to $BFFF_h$ to the standardized network and system variable areas. Other indexes (from $C000_h$ on) are either reserved for future use or concern new functionalities that are in the process of being defined at the time of writing.

52.7.3 Process Data Objects

Real-time process data, involved in the physical system control, are exchanged by means of PDOs according to the producer/consumer model. Every PDO is mapped onto exactly one CAN frame and does not introduce any protocol overhead, so that it can be exchanged quickly and reliably. As a direct consequence, the amount of data that can be embedded in one PDO is limited to 8 bytes at most. In the majority of cases, this is more than sufficient to encode one or more signals managed by the control application.

From the point of view of CANopen devices, two kinds of PDO exist: *transmit PDO* (TPDO) and *receive PDO* (RPDO). The former is produced (i.e., sent on the CAN bus) by a device, whereas the latter models consumed objects (which must not be discarded by the frame filtering function). In theory, up to 512 TPDOs and 512 RPDOs can be defined for each device. However, according to the predefined connection set, each node in CANopen is equipped with up to 4 RPDOs and up to 4 TPDOs. RPDOs are used for transferring process data (commands, set points, etc.) from the application master to the device, whereas TPDOs are used to convey process data (e.g., values sampled by sensors) from the device to the application master.

The predefined connection set is often sufficient to set up simple master/slave configurations in small factory automation systems. In the case a larger number of PDOs are needed, the *PDO communication parameters* of the device (stored as entries in the OD) can be used to either define additional messages or just change the existing ones. In particular, the *PDO linking* mechanism can be exploited in those networked embedded systems that need high flexibility and where device-to-device communication or multicasting is required. From a practical point of view, this involves the explicit configuration of matching PDO communication parameters in two or more devices, so that they are enabled to exchange process data directly according to the conventional CAN paradigm.

Finally, a suitable configuration of the *PDO mapping parameters*, if supported by the device, can be exploited to define dynamically what application objects (i.e., signals) have to be included in each PDO. In particular, either a number of OD entries can be collected in the same TPDO for transmission or the information extracted from an RPDO can be scattered over several entries in the OD.

The transmission of PDOs can be triggered by some local event taking place on the node, such as the expiration of a time-out, but it can also be remotely requested from the application master. This gives system designers a high degree of flexibility in choosing the way devices interact and enables a better exploitation of the features offered by intelligent devices.

Process data of a PDO are encoded in the data field of a CAN data frame and, as mentioned before, no additional control information is added by CANopen. As a consequence, PDOs resemble conventional CAN data exchanges closely, and communication efficiency is kept as high as in plain CAN. The meaning of each PDO is determined through the related CAN-ID. As multicasting is allowed for PDOs, their transmission is unconfirmed, that is, the producer has no way to know whether or not a PDO was received by all the intended consumers.

52.7.3.1 Transmission Modes

From a general point of view, two transmission modes are possible in CANopen, namely, *event driven* and *synchronous*. The former resembles the typical interaction model provided by plain CAN, whereas the latter enables synchronous operations for devices.

The transmission of event-driven PDOs can be started by nodes at any time, and hence it is typically asynchronous. The transmission of synchronous PDOs, instead, is driven by the occurrence of a network synchronization event, which is triggered by a special message (SYNC) generated at regular time intervals by a specific node (usually, the node acting as the application master). The transmission period for synchronization messages is referred to as *communication cycle period*. As a consequence, synchronous PDO exchanges take place in periodic cycles.

When synchronous operations are selected for one or more PDOs, commanded values are not actuated immediately by devices as they are received, and the same occurs for the transmission of sampled values. As depicted in Figure 52.15, every time a SYNC message is read from the network, the synchronous PDOs received in the previous communication cycle are actuated by the relevant output devices. At the same time, all sensors sample their input ports and related values are sent as soon as possible in the next cycle. Optionally, a *synchronous windows length* parameter can be defined, which specifies the

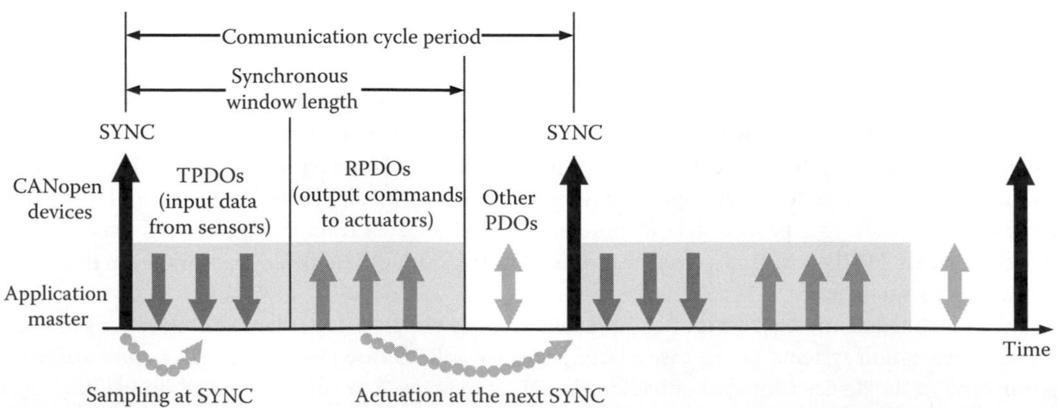

FIGURE 52.15 Synchronous operation.

time when all synchronous values have been exchanged for sure (late PDOs may be discarded). After this time has elapsed, each device can start processing the input/output (I/O) values.

Synchronous operations provide a tangible improvement for lessening the effect of jitters: in this case, in fact, system operations and timings are decoupled from the actual times PDOs are exchanged over the CAN bus.

52.7.3.2 PDO Triggering

Three triggering modes are defined for PDOs in CANopen, which specify in detail the conditions upon which a particular PDO is sent over the network. In particular, they identify the entity that is actually in charge of initiating the data exchange (see Figure 52.16):

- *Event and timer driven*: The decision on when a certain PDO has to be sent is completely up to its producer; the PDO transmission could be triggered by either the occurrence of an application-specific event (whose exact nature depends on the device profile/manufacturer) or the expiration of a local timer because no event has occurred (*event time*).
- *Remotely requested*: The transmission of the PDO has to be requested explicitly by its consumer(s), and it is triggered by a suitable remote frame according to the mechanisms CAN provides at the data link layer.
- *Synchronously triggered*: A third entity in the network (besides the PDO producer and consumers), which is known as SYNC producer, drives the data exchange. In practice, the PDO transmission is triggered by the reception of the SYNC object.

PDO transmission and triggering modes are configured through the *transmission-type* parameter, encoded as an 8-bit unsigned integer. For TPDOs, values in the range from 00_h to $F0_h$ denote *synchronous* transmissions. In particular, the value 01_h means that the PDO has to be sent at every SYNC, whereas values in the range 02_h–$F0_h$ mean that it is sent every *n*th SYNC, where *n* is the numeric

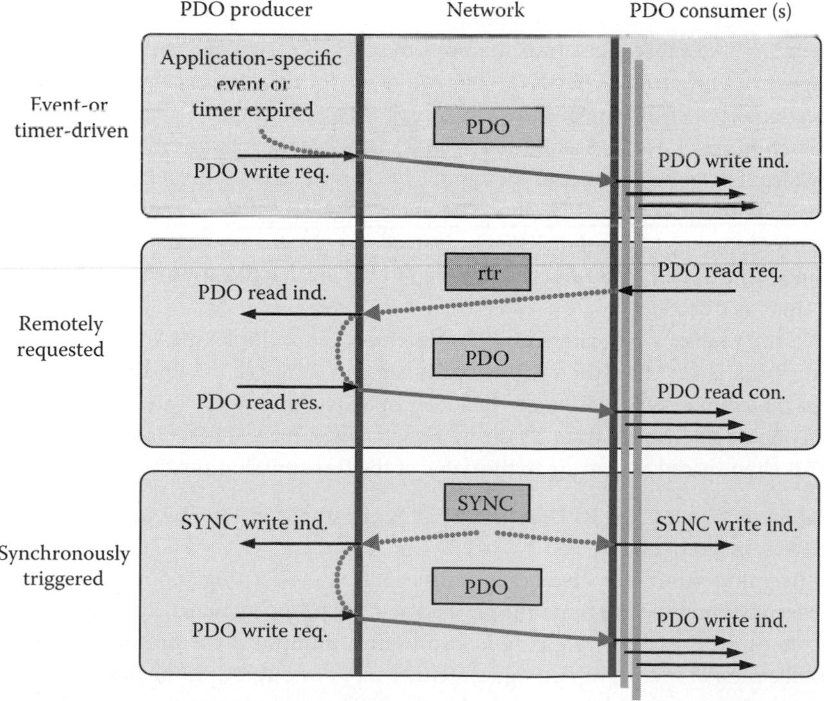

FIGURE 52.16 Triggering modes.

value of the transmission type ($2 \leq n \leq 240$). These values denote *cyclic* transmissions and also specify the PDO transmission period as a multiple of the communication cycle (*n* times the communication cycle period). The value 0, instead, is used when the TPDO generation is *acyclic*. This means that it depends on an internal event, but transmissions have to take place synchronously nevertheless.

Values FC_h and FD_h denote *RTR-only* triggering schemes. In particular, the former (*synchronous RTR only*) means that data are sampled at SYNC, temporarily buffered, and then sent when the related remote frame is received. The latter (*event-driven RTR only*), instead, implies that both sampling and transmission take place (in this order) at the remote frame reception.

Finally, values FE_h and FF_h concern the *event-driven* triggering schemes. Triggering is a pure local matter, which can be either manufacturer specific in the former case or specified in the device/application profile in the latter.

Similar considerations apply to RPDOs, where either event-driven or synchronous transmission can be selected. In the event-driven case (transmission types FE_h and FF_h), data are passed to the application (e.g., actuated) as soon as they are received from the network, whereas in synchronous operations (00_h–$F0_h$), values are first buffered and then actuated at the next SYNC.

52.7.3.3 PDO Configuration

As usual for CANopen, configuration of PDOs takes place via the OD. In particular, every PDO is assigned a number of entries in the OD, the most important being the *PDO communication parameter* and the *PDO mapping parameter*.

Details about the way each PDO is exchanged over the network are defined through the PDO communication parameter. It includes two mandatory items:

1. The ID used for the CAN frame that embeds the PDO (COB-ID)
2. The transmission type, as specified earlier

Optionally, it can also specify the following information:

- An *inhibit time* (expressed as a multiple of 100 μs), which is the shortest time that must elapse between any two subsequent transmissions of the PDO. This is useful to prevent high-priority messages from hogging the network, thus ruling out (or delaying excessively) lower-priority communications. The value 0 means that the inhibit time is disabled.
- An *event timer* (expressed as a multiple of 1 ms), which is the maximum time allowed between any two subsequent transmissions of the same PDO. In the case of TPDOs, when the timer expires, a message is sent automatically. This parameter can be used to enforce a minimum refresh rate for slowly changing values. For RPDOs, instead, the timer may be used for monitoring message deadlines. In this case, time-outs are notified to the local application. The value 0 means that the event timer is disabled.
- A *SYNC start* value, which is used as a displacement for multiplexing synchronous TPDOs whose periodicity is larger than the cycle time. When in the range 1–240, this parameter is checked against the counter field optionally included in SYNC messages. The SYNC message for which equality holds shall be regarded as the first one, on which the TPDO is sent. Following transmissions are then spaced according to the value of the transmission type.

Communication parameters for RPDOs and TPDOs are stored in the OD in the entries 1400_h–$15FF_h$ and 1800_h–$19FF_h$, respectively.

The PDO mapping parameter specifies how information concerning application objects (and stored in the OD) has to be combined to build the payload of a TPDO or, equivalently, how it has to be decoded from an incoming RPDO. PDO mapping is also useful to improve the protocol efficiency: in fact, it enables the collection of several device signals (either input or output) in the payload of the same CAN frame. Obviously, information that is gathered in the PDO must have the same direction (transmission/reception) and triggering mode (including periodicity). The overall size is not allowed to exceed 8 bytes.

Basically, every PDO mapping parameter is made up of a number of 4-byte subentries. In turn, each subentry encodes three elements, namely, the index and subindex of the application object that will be mapped in the PDO (3 bytes) as well as its length in bits (1 byte).

Two kinds of PDO mapping are provided, that is, static and dynamic. Static mapping is built-in in the device and cannot be changed, whereas dynamic mapping can be freely configured by accessing the relevant items in the OD. A procedure is defined to this extent in order to avoid inconsistencies. Mapping parameters for RPDOs and TPDOs are stored in the OD in the entries 1600_h–$17FF_h$ and $1A00_h$–$1BFF_h$, respectively.

52.7.4 Service Data Objects

SDOs provide direct access for reading and writing the entries of the OD remotely. They are used in CANopen for parameterization activities, which usually take place during the configuration phase. Since SDOs are not considered real-time exchanges, they have a lower priority than PDOs. On the other hand, confirmed transmission services are required for them, which ensure a reliable exchange of information between end devices. SDOs rely on a client/server protocol and require a peer-to-peer connection (multicasting is not allowed) where the client is the accessing device, while the server is the device that owns the OD being accessed.

Since information to be exchanged may be of any size (in theory, at least), a fragmentation protocol was defined for SDOs, which derives from the domain transfer services of CAL. This means that information has to be split by the sender in smaller chunks (fragments or *segments* according to the CANopen specification) able to fit in the data field of CAN frames. Fragments must be reassembled on the receiver's side, and this requires that suitable information is included in every segment by the fragmentation protocol, along with the chunk of data. Indeed, the communication efficiency is negatively affected and transmissions are slowed down, but as SDOs are seldom used for the real-time control of the system, this is typically not perceived as a problem.

SDOs are used to access the entries of the OD by applications and/or configuration tools. From a practical point of view, two services are provided, which are used to *upload* and *download* the content of a subentry of the OD, respectively. SDO transfers are always initiated on the client side of the connection, and the protocol involves exchanging one or more pairs of CAN frames with an 8-byte payload. Specifically, one frame in the pair encodes the request from the client to the server, while the other carries the response from the server back to the client, and a separate ID is needed for each direction.

52.7.4.1 SDO Transfer Types

Three kinds of transfers are envisaged for SDOs, namely, *expedited*, *normal* (or *segmented*), and *block*. They basically differ for communication efficiency and implementation complexity. Higher-complexity communication modes are advantageous when large pieces of information have to be transferred:

- Expedited transfer is a particular (simplified) case of SDO that only requires two frame exchanges between the client and the server (i.e., one *SDO download/upload initiate* request and the related response). As a consequence, only up to 4 bytes of data can be exchanged.
- Normal transfer is based on a segmented approach. In this case, following the SDO download/upload initiate message pair, a variable number of segments are sent. Each segment requires a separate *SDO download/upload* request/response pair and can carry up to 7 bytes of data. For this reason, the communication efficiency is quite low.
- Block transfer is an optional mode where information is exchanged as a sequence of blocks, each one consisting of a sequence of up to 127 segments. Every SDO block transfer is required to be both initiated and concluded explicitly (*SDO block download/upload initiate/end* services). Each block is transferred using the *SDO block download/upload subblock* service. A *go-back-n* strategy is adopted in this case to confirm the correct delivery of each block and to improve the efficiency in transferring large amounts of data.

An unconfirmed *SDO abort transfer* service is also provided, which employs only one frame and can be issued on both the client and the server sides, in order to terminate the transfer sequence when some error is detected.

Each CANopen node must provide SDO server functionalities. This means that it has to offer a pair of COB-IDs to support remote access to its OD, one for each transfer direction. Generally speaking, only one SDO client at a time is allowed in the network. In practice, only one SDO client is enabled to connect to any given SDO server, and this is implicitly guaranteed with the static allocation of IDs in the predefined connection set. Optionally, the dynamic establishment of additional SDO connections can be supported by means of a network entity called the *SDO manager.*

52.7.5 Other Objects

Besides PDOs and SDOs, a number of other COBs can be optionally defined in CANopen. They are intended to support specific functions, related to either application (and time) synchronization or error notification.

52.7.5.1 Synchronization Object

The SYNC is periodically broadcast over the network by the *SYNC producer* to enable synchronous operation. The time elapsing between any two subsequent generations of the SYNC message can be configured (as a multiple of 1 μs) through the *communication cycle period.*

Despite its high priority (SYNC is assigned the ID 080_h), the SYNC object may be delayed nevertheless by lower-priority messages and suffer from the blocking time because CAN is not preemptive. The resulting jitters (in the order of hundreds of μs) are much lower than for asynchronous exchanges, but they are not usually negligible. Hence, the synchronization might not be as accurate as high-precision applications demand.

In the initial CANopen specification (as well as in many existing devices), the SYNC object was mapped onto an empty data frame, since it has to provide a sort of *network clock* that ticks for applications, and only the SYNC reception instants are really meaningful. The most recent specification, however, defines an additional format for the SYNC message, which extends the previous one and can now include a 1-byte payload. In this case, the data field contains a counter (initially set to 1), which is increased by 1 on every SYNC transmission. When the counter reaches its maximum value (*synchronous counter overflow*), it is reset (i.e., it is set back to 1). This allows devices to arrange the transmissions of synchronous PDOs with periods greater than the communication cycle, so that the overall traffic is spread evenly over different cycles. As a side effect, this also helps in reducing jitters and latencies.

52.7.5.2 Emergency Object

The EMCY is sent asynchronously by devices as a consequence of internal error conditions. It can be considered as a sort of interrupt, which is raised for notifying urgent alarms so as to provide, for instance, prompt diagnostics. Its implementation is not mandatory. Unlike the error mechanisms provided by CAN, which copes with communication errors, this object is meant to deal with errors at the application level (e.g., a short circuit on an output channel, a low voltage on line power, or an excessive internal temperature).

Only one EMCY shall be sent for each error event. A state machine is then defined that describes the current error condition of the device. As soon as all errors have been repaired, the device returns to the error-free state.

EMCYs are encoded on CAN data frames having an 8-byte data field. Their payload is made up by three elements:

1. The *emergency error code*, encoded on 2 bytes. A number of standardized error code classes are defined in the CANopen communication profile, and others can be added by device profiles.

2. The *error register* (object 1001_h of the OD). This is a 1-byte field that identifies errors that are still pending. Every bit concerns a specific error class, namely, generic, current, voltage, temperature, communication, device profile specific, and manufacturer specific.
3. The *manufacturer-specific error* field, which may include additional information about the error as for the device vendor.

52.7.6 Network Management

CANopen *NMT* relies on a master/slave approach, where one NMT master controls a number of NMT slave devices. In general, all CANopen devices are considered NMT slaves and are identified by means of a *node-ID*, encoded as an integer in the range from 1 to 127. NMT requires that one specific device in the network carries out the function of the NMT master.

Two kinds of services concern NMT, that is, *node control* and *error control*. Node control services, as the name suggests, are used to control the operation of nodes. For example, they can be used to start/stop nodes, to reset them to a predefined state, or to put them in the configuration (preoperational) mode. Error control services, instead, are aimed at verifying if every device is alive and operating properly.

52.7.6.1 Node Control Services

From the point of view of NMT, each device is described through an NMT state machine. At any time, a node can be in one of the following four NMT states, which define its current behavior and allowed operations:

1. *Initialization*: This state actually consists of three substates, whose operations are executed in sequence. The first one (*initializing*), entered autonomously after either power-on or hardware reset, is where basic initialization activities are performed. Parameters of both the manufacturer-specific and the standardized device profile areas are loaded with power-on values in the *reset application* state, whereas in the *reset communication* state, parameters of the communication profile area are set. If the device is equipped with a nonvolatile storage area, power-on values are the last stored parameters; if not, default values are used. After initialization has finished, a boot-up message is sent and the preoperational state is entered automatically.
2. *Preoperational*: This state is used for configuring the device. Communication through SDOs is available and entries of the OD can be remotely accessed (and changed) by the configuration applications and tools. Communication through PDOs, instead, is not allowed yet.
3. *Operational*: In this state, all communications are enabled, and this usually corresponds to a fully functional behavior for the device. Besides PDO communications, which are used to exchange set points and measured values, also SDOs are allowed. However, the access to the OD might be somehow restricted, so as not to alter the system behavior accidentally or leave it in an inconsistent state.
4. *Stopped*: In this state, all communications are completely stopped except for NMT services. Furthermore, a specific behavior (which usually corresponds to a stopped device) can be enforced, as described by the corresponding device profile. Requirements concerning functional safety may also apply.

The resulting behavior of the device can be described by means of the state diagram in Figure 52.17. Transitions are labeled through the following NMT services, whose meaning is self-explanatory:

- Start remote node
- Stop remote node
- Enter preoperational
- Reset node
- Reset communication

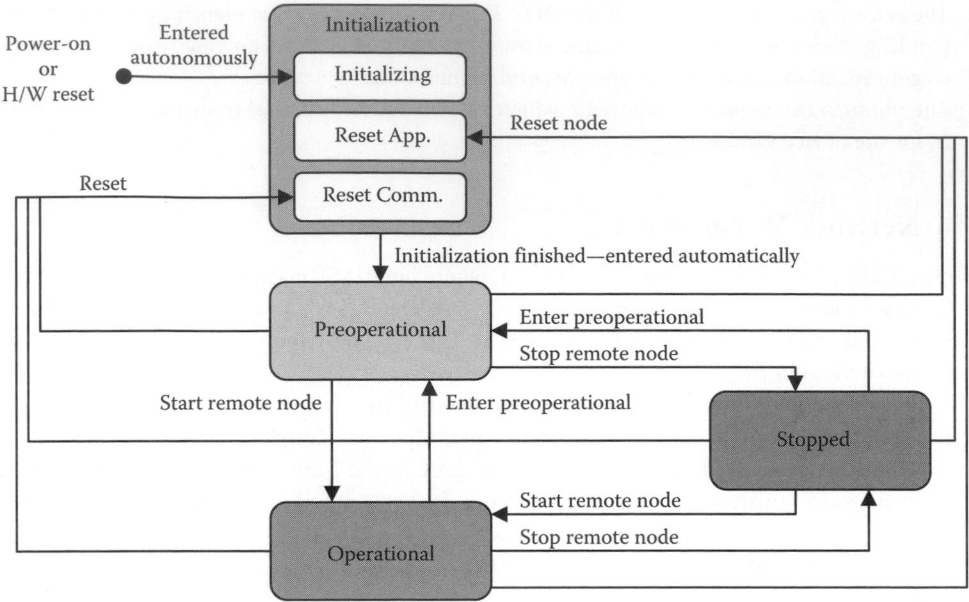

FIGURE 52.17 NMT state machine.

NMT commands are definitely time critical, and hence they use the highest-priority frames available in CAN (i.e., CAN-ID is set to 0). The data field in such frames is 2 bytes long; the first byte (*command specifier*) encodes the service, whereas the second (*node-ID*) identifies the target node(s). The NMT master can invoke these services for either one slave at a time or all of them at the same time. This is useful for starting/halting all devices in the network simultaneously. In the latter case, the node-ID field in NMT messages is set to 00_h.

The way the NMT state machine is actually coupled to the state machine governing the application task(s) depends on the specific device taken into account and falls in the scope of the device profiles.

52.7.6.2 Error Control Services

Error control services are used to detect possible failures by monitoring the correct operation of the network. Two mechanisms are available, that is, *node guarding* and *heartbeat*. In both cases, low-priority messages are periodically exchanged in background, and suitable watchdogs are defined, both in the NMT master and NMT slaves. If a node ceases sending these messages, the NMT entity is made aware of the problem after a short amount of time and can take the appropriate actions.

The main difference between the two mechanisms is that, in the node guarding protocol, the NMT master polls the NMT slaves explicitly at regular intervals (the period is called the *guard time*) by means of suitable requests implemented as CAN remote frames. If a slave does not respond within a predefined time (*lifetime*), whose duration is evaluated as an integer multiple (*lifetime factor*) of the guard time, or the state of the slave has changed unexpectedly, the master application is notified (a *node guarding event* is issued). In this case, *life guarding* is supported as well, that is, slave devices may guard the correct behavior of the NMT master. If a slave is not polled during its lifetime, a *life guarding event* is issued locally.

The heartbeat protocol, instead, foresees that heartbeat messages are produced autonomously and periodically by nodes, which can be monitored by other nodes in the network. If a heartbeat consumer does not hear a heartbeat message from a producer within a given time (*heartbeat consumer time*), the local application is notified. The advantage is that remote frames are not required in this case, so that the network load is reduced and the implementation made easier on some controllers.

Messages sent by these two protocols are very similar. They are encoded in 1-byte data frames and use the very same CAN-IDs. Basically, the NMT state of the transmitting device is included in the seven least significant bits of the payload. In the case of node guarding, a toggle bit is also included as the first bit to improve reliability.

The same kind of messages is also adopted by the *boot-up* protocol, which is used by slave nodes to notify that they have entered the preoperational state after initialization. It is worth noting that the implementation of either node guarding or heartbeat is mandatory.

52.7.6.3 Predefined Connection Set

In order to reduce the efforts involved in setting up a CANopen system, a default allocation scheme can be used to map COBs directly onto CAN-IDs. Such an allocation is made available just after initialization, provided that no modification was stored in the device nonvolatile memory. In many cases, such a scheme is more than adequate and does not need changes, as in the case of simple applications that consist of a single master controlling several slave devices.

According to the *generic predefined connection set*, the CAN-ID of frames used to encode COBs includes two parts: the first one is the *function code* (FC), which takes the four most significant bits of the ID, while the second is the *node-ID* and is encoded in the seven least significant bits, as shown in Figure 52.18. Basically, the FC specifies the type of the COB (SYNC, PDO, SDO, EMCY, NMT, etc.) and is the element that mostly affects the priority of the frame. The inclusion of the node-ID in the CAN-ID, besides providing a very simple allocation scheme, implicitly ensures that no two nodes in the network can send a frame with the same ID (as discussed previously, this could lead to unsolvable collisions on the bus).

Some COBs in the predefined connection set, such as NMT node control, SYNC, and TIME, are intended for transmission to all devices simultaneously (broadcast). The CAN-IDs, in this case, can be easily singled out, as the node-ID field is set to 0. Generally speaking, it is worth noting that the particular format chosen for IDs makes the configuration of the filtering masks easier in CAN controllers.

The scheme for peer-to-peer (and multipeer) communications basically provides each slave device with up to four TPDOs and up to four RPDOs for sending/receiving real-time process data, one EMCY object for asynchronous alarm notification, one SDO to make the OD accessible via the network, and one low-priority message to be used by either the heartbeat or the node guarding error control mechanism. In these cases, the slave device can be either the source or the destination node, depending on the specific COB. The generic predefined connection set is shown in Table 52.1.

CAN-IDs assigned to the SYNC, TIME, and EMCY objects, as well as to any PDO, can be changed through a suitable reconfiguration, if this kind of parameterization is supported by the device.

The generic predefined connection set applies to devices that comply with a certain device profile. Conversely, devices that comply with an application profile must use a *specific predefined connection set*. In this case, the allocation scheme for IDs is defined in the relevant profile specification. COBs concerning NMT and SDO services, however, cannot be redefined.

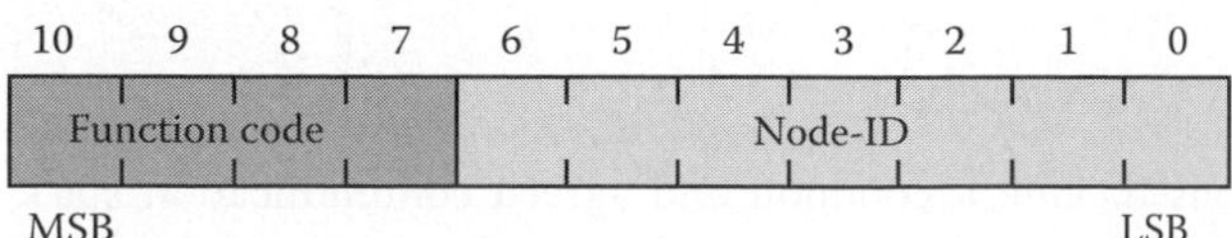

FIGURE 52.18 CAN-ID in the predefined connection set.

TABLE 52.1 COBs Foreseen by the Predefined Connection Set

COB	Address	FC	Node-ID	CAN-ID(s)
NMT node control	Broadcast	0000_b	00_h	000_h
SYNC	Broadcast	0001_b	00_h	080_h
EMCY	Source	0001_b	01_h–$7F_h$	081_h–$0FF_h$
TIME	Broadcast	0010_b	00_h	100_h
TPDO1	Source	0011_b	01_h–$7F_h$	181_h–$1FF_h$
RPDO1	Destination	0100_b	01_h–$7F_h$	201_h–$27F_h$
TPDO2	Source	0101_b	01_h–$7F_h$	281_h–$2FF_h$
RPDO2	Destination	0110_b	01_h–$7F_h$	301_h–$37F_h$
TPDO3	Source	0111_b	01_h–$7F_h$	381_h–$3FF_h$
RPDO3	Destination	1000_b	01_h–$7F_h$	401_h–$47F_h$
TPDO4	Source	1001_b	01_h–$7F_h$	481_h–$4FF_h$
RPDO4	Destination	1010_b	01_h–$7F_h$	501_h–$57F_h$
SDO (tx)	Source	1011_b	01_h–$7F_h$	581_h–$5FF_h$
SDO (rx)	Destination	1100_b	01_h–$7F_h$	601_h–$67F_h$
NMT error control	Source/(RTR)	1110_b	01_h–$7F_h$	701_h–$77F_h$

52.8 Interoperability between Devices

Sharing a common communication protocol is a prerequisite for achieving device interoperability, but it is not a sufficient condition. In fact, besides the rules for exchanging data over the network, also the meaning of data themselves and their impact on the device behavior have to be specified. The vast majority of industrial communication systems introduced in the past two decades (both fieldbuses and RTE solutions) adopt the concept of device profile. Basically, this is an abstract model describing the device behavior, which enables interoperability and interchangeability of equipment coming from different manufacturers. In fact, as long as a device conforms to a given profile, it can be used to replace any other device compliant to the same profile.

CAN-based application protocols are no exception. Popular solutions such as CANopen and DeviceNet define their own device profiles, which are sometimes said to belong to the *user layer* (the user layer is located above the top communication layer in the protocol stack). Though the purpose of profiles is basically the same, irrespective of the different solutions, exact definitions may vary. In particular, DeviceNet relies on a hierarchical model, defined in the *common industrial protocol* (CIP), which is based on the object-oriented paradigm, whereas CANopen is simpler and exploits the OD directly.

It is worth pointing out that device profiles for CAN-based solutions, as well as most part of their communication profiles and application layer services, have been reused in some popular RTE solutions. This is the case, for instance, of both EtherCAT and Ethernet POWERLINK, which rely on CANopen, and EtherNet/IP, which, as DeviceNet, is based on CIP.

52.8.1 Profiles in CANopen

CANopen specifications include a common and agreed communication stack (which includes the physical, data link, and application layers), a standard mechanism to configure the relevant parameters (communication profile area in the OD), and a number of profiles to ensure interoperability among devices.

In particular, *device profiles* can ease the task of system integrators in the design and deployment of CANopen systems, by providing off-the-shelf devices with true plug-and-play capabilities. Each profile concerns the interface of a logical device and is usually described in a separate document. In practice, the common behavior of a particular class of devices is considered. At present, about 20 device profiles have already been defined, including the following:

- The *device profile for generic I/O devices* (CiA 401) deals with both digital and analog I/O devices.
- The *device profile for drives and motion control* (CiA 402), used to describe products for digital motion, such as frequency inverters, stepper motor controllers, and servo controllers.
- The *device profile for measuring devices and closed-loop controllers* (CiA 404) is adopted for measuring and controlling physical variables.
- The *device profile for encoders* (CiA 406) defines the behavior of incremental/absolute linear and rotary encoders for measuring position, velocity, and so on.

Profiles listed earlier have been introduced mainly for general-purpose applications. This means they were not designed to fulfill the requirements of any specific application domain. Compliant devices are typically used, for instance, for assembling automated production lines in factory automation systems.

An additional set of profiles, known as *application profiles*, can be also found in CANopen. They are aimed at fulfilling requirements that are peculiar to a particular application domain and are able to describe a set of virtual device interfaces. This is the case, for example, of the *application profile for special-purpose car add-on devices* (CiA 447). It was defined explicitly for the automotive domain and specifies interfaces for add-on devices to be used in special-purpose cars. Devices for police cars (e.g., the roof bar and digital radio), taxi/cabs (e.g., taximeter), and disabled drivers are included in this profile.

Finally, some *interface profiles*, such as the *interface profile for AS interface gateways* (CiA 446), are provided for devices without application functionality (e.g., nontransparent gateways).

In the following, the CANopen device profile for generic I/O modules (CiA 401) is described in more detail, to give the reader a flavor of the typical profile content and to show how it can help designers to build up a networked system.

52.8.2 Device Profile for Generic I/O Modules

The device profile for generic I/O modules [CiA401] is certainly one of the most popular profiles defined in CANopen. Generic I/O modules provide digital/analog inputs and/or outputs. This means that they are not bound to any application-specific context, such as measuring the temperature of an engine or switching the lights of a car on and off.

Many optional configuration parameters are specified by this profile, as well as a number of PDOs for transmission and reception. For instance, the polarity of each digital I/O port can be defined, or a filtering mask can be applied to selected channels. For analog devices, either raw or converted values can be used (in the latter case, a scaling factor and an offset are applied). In the same way, triggering conditions can be defined when specific thresholds are exceeded. The generic I/O module profile supports different access granularities for digital I/Os and several resolutions (and formats) for analog I/Os.

Input values are sent as TPDOs by sensors over the CAN and their transmission defaults to the asynchronous (i.e., event-driven) triggering. However, also the synchronous or remotely requested triggering schemes can be optionally enabled when supported by the device. In the same way, output values are provided to actuators via asynchronous RPDOs.

I/O values may be accessed by remote applications through SDOs in the OD. However, this is not recommended in control loops, as it leads to reduced communication efficiency and generally higher delays.

52.8.2.1 Profile Predefinitions

Several OD entries are standardized by this profile. One of the most significant objects in the communication profile area is the *device-type* object (entry 1000_h in the OD), which defines both the type of the device and its functionality. In particular, the type is specified through the *device profile number* (which is typically set to the number of the standard document related to the class of device, in this case the value 401), whereas the functionality is encoded in the *additional information* field and reminds what kind of channels (e.g., input vs. output, digital vs. analog) are actually included in the device.

I/O modules compliant to the device profile for generic I/O modules specify OD entries in the standardized device profile area to store process data. Different objects in the OD, concerning digital I/O modules, provide access to process data organized on either 1, 8, 16, or 32 bits. It is worth noting, however, that only the 8-bit access scheme is mandatory, whereas the other lengths are optional. All entries refer to exactly the same value, the only difference being the access granularity. In particular, objects 6000_h and 6200_h are used for 8-bit access to digital input and output modules, respectively.

When analog modules are considered, values encoded as integers on 8, 16, and 32 bits are available. In addition, floating-point and manufacturer-specific formats can be used. Only 16-bit data are mandatory, and objects 6401_h and 6411_h are used for accessing 16-bit analog input and output values, respectively.

The standardized device profile area also includes OD entries that are needed to configure the behavior of the device. A number of default PDOs are defined as well, together with their default mapping:

- *RPDO1* (*digital outputs*): Up to 64 digital outputs can be received from the network through this RPDO, which are actuated immediately.
- *TPDO1* (*digital inputs*): Up to 64 digital inputs can be sent over the network through this TPDO, according to an event-driven scheme. This means that the PDO is transmitted as soon as at least one of the digital inputs changes its value (provided it is not disabled via the interrupt mask).
- *RPDO2* (*analog outputs*): Up to four analog outputs, each one encoded on 16 bits, can be received from the network through this RPDO.
- *TPDO2* (*analog inputs*): Up to four analog inputs, each one encoded on 16 bits, can be sent on the network through this TPDO, according to an event-driven scheme.

Only PDOs concerning the supported functionalities have to be provided by the device. For instance, a digital input module is only required to define TPDOs (and, for simple devices, just TPDO1). However, additional manufacturer-specific PDOs could be included in the module.

52.8.3 Device Behavior

The behavior of devices is formally defined in the profile specification. In the case of generic I/O modules, block diagrams are provided that describe how information is exchanged between the CANopen network and the real world and how sampling and actuation are carried out. Configuration takes place through a set of parameters, which are stored in the OD.

52.8.3.1 Digital Input Devices

In Figure 52.19, the sequence of steps involved in the processing chain of a digital input is shown. Details on the way each operation is carried out can be changed by configuring the appropriate entries in the OD. A filter constant can be activated separately (2) on each single digital input channel (1) by means of object 6003_h. This can be useful, for instance, to suppress undesirable signal fluctuations. Moreover, the

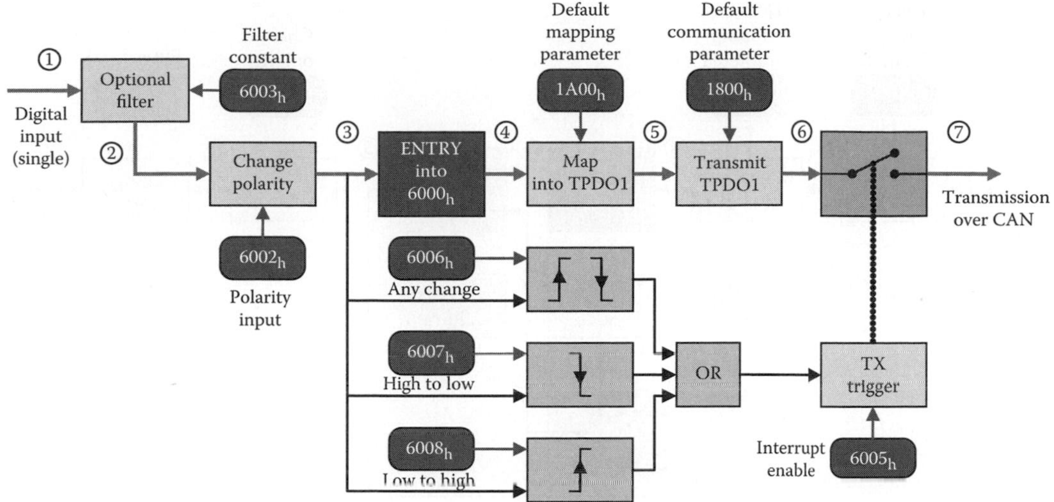

FIGURE 52.19 Abstract model of digital inputs.

channel polarity (normal/inverted) can be optionally changed (3) through object 6002_h. Then, all bits associated to the different input channels are gathered together and stored in the relevant entries of the OD (object 6000_h), from where they will be subsequently extracted and sent over the network (4). In this case, values are first mapped in the payload of a TPDO through the PDO mapping mechanism (5); then the correct CAN-ID is selected (6) by means of the PDO communication parameter.

The actual transmission on the CAN bus (7) is driven by the trigger block. By default, the asynchronous mode is enabled, which triggers the transmission as soon as any of the input channels changes its logical value. However, this behavior can be changed through objects 6006_h, 6007_h, and 6008_h. Moreover, triggering can be globally enabled and disabled through the (Boolean) object 6005_h.

It is clear that, by writing into the relevant entries of the OD, a control application (or a configuration tool) can easily customize the behavior of each single digital input channel, as well as decide how these information are sent over the CAN.

52.8.3.2 Digital Output Devices

Figure 52.20 shows the steps involved in the processing chain affecting the operation of a digital output device. Frames received from the CAN bus (1) are first filtered according to the information stored in the RPDO communication parameter (2). Whenever the ID field in the received frame corresponds to the value configured for RPDO1 in this device, its payload is decoded according to the PDO mapping parameters and the related values are stored (and possibly scattered) (3) through the relevant entries of the OD (object 6200_h). After an optional change of polarity (5), configured through the object 6202_h, each output value is made available on the related output channel (7). A block is also provided (6) through object 6208_h that acts as a filter mask. When disabled, the previous value is retained on the output channel.

The lower part of Figure 52.20 describes the behavior of the actuator in the case a device internal failure (including heartbeat or life guarding events) is detected. In order to improve the functional safety, two objects (6206_h and 6207_h) are provided that specify the value each digital output channel should assume in the case of failure. In particular, each output can be set separately to 0, 1, or its own value before the failure was detected. Whenever a failure is detected, the output is automatically switched (4) to such a value.

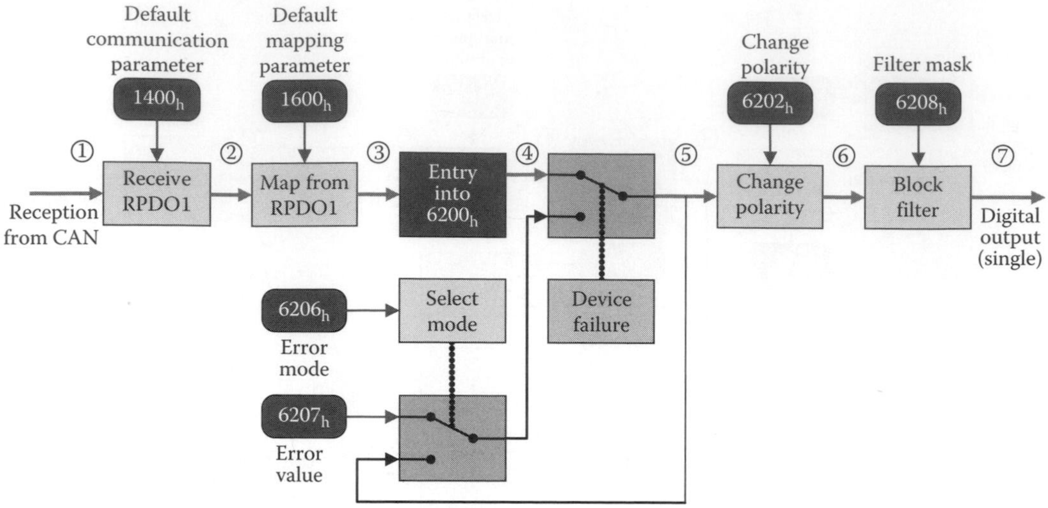

FIGURE 52.20 Abstract model of digital outputs.

References

ISO1 International Standard Organization, Road vehicles—Controller area network—Part 1: Data link layer and physical signalling, ISO 11898-1:2003, 2003-11-19.

ISO2 International Standard Organization, Road vehicles—Controller area network—Part 2: High-speed medium access unit, ISO 11898-2:2003, 2003-11-19.

ISO3 International Standard Organization, Road vehicles—Controller area network—Part 3: Low-speed, fault-tolerant, medium dependent interface, ISO 11898-3:2006, 2006-05-24.

ISO4 International Standard Organization, Road vehicles—Controller area network—Part 4: Time-triggered communication, ISO 11898-4:2004, 2004-08-05.

ISO5 International Standard Organization, Road vehicles—Controller area network—Part 5: High-speed medium access unit with low-power mode, ISO 11898-5:2007, 2007-06-12.

ISO9 International Standard Organization, Road vehicles—Interchange of digital information on electrical connections between towing and towed vehicles—Part 1: Physical and data-link layers, ISO 11992-1:2003.

EN1 European Committee for Electrotechnical Standardisation, Industrial communications subsystem based on ISO 11898 (CAN) for controller-device interfaces—Part 1: General requirements, EN50325-1:2002, 2002-12-12.

EN2 European Committee for Electrotechnical Standardisation, Industrial communications subsystem based on ISO 11898 (CAN) for controller-device interfaces—Part 2: DeviceNet, EN50325-2:2000, 2000-10-30.

EN3 European Committee for Electrotechnical Standardisation, Industrial communications subsystem based on ISO 11898 (CAN) for controller-device interfaces—Part 3: Smart distributed system (SDS), EN50325-3:2001, 2001-04-10.

EN4 European Committee for Electrotechnical Standardisation, Industrial communications subsystem based on ISO 11898 (CAN) for controller-device interfaces—Part 4: CANopen, EN50325-4:2002, 2002-12-12.

CiA102 CAN in Automation International Users and Manufacturers Group e.V., CAN physical layer for industrial applications, CiA 102, Version 3.0, February 5, 2010.

CiA20X CAN in Automation International Users and Manufacturers Group e.V., CAN application layer for industrial applications, CiA DS 201/202/203/204/205/206/207, Version 1.1, February 1996.

CiA301 CAN in Automation International Users and Manufacturers Group e.V., CANopen application layer and communication profile, CiA 301, Version 4.2.0, February 21, 2011.

CiA303 CAN in Automation International Users and Manufacturers Group e.V., CANopen recommendation—Part 1: Cabling and connector pin assignment, CiA 303-1, Version 1.8.0, April 27, 2012.

CiA401 CAN in Automation International Users and Manufacturers Group e.V., CANopen—Device profile for generic I/O modules, CiA 401, Version 3.0.0, June 3, 2008.

CANFD Robert Bosch GmbH, CAN with flexible data-rate (CAN FD), Specification Version 1.0, released April 17, 2012.

FR3 FlexRay Consortium, FlexRay Communications system—Protocol specification, Version 3.0.1, October 2010.

J1939 Society of Automotive Engineers, Serial control and communications heavy duty vehicle network, SAE J1939, 2012-06-01.

J2411 Society of Automotive Engineers, Single wire CAN network for vehicle applications, SAE J2411, 2000-02-14.

CEN99 Cena G., Valenzano A., Overclocking of controller area networks, *IEE Electronics Letters*, 35(22), October 28, 1999, 1923–1925, Institution of Electrical Engineers, London, U.K..

CEN00 Cena G., Valenzano A., A compression-gathering technique for improving the performance of ISO 11898 networks, *Computer Standards & Interfaces*, 22, 2000, 323–335.

CEN02 Cena G., Valenzano A., A multistage hierarchical distributed arbitration technique for priority-based real-time communication systems, *IEEE Transactions on Industrial Electronics*, 49(6), December 2002, 1227–1239, Institute of Electrical and Electronics Engineers, New York.

CEN12 Cena G., Cibrario Bertolotti I., Hu T., Valenzano A., Performance comparison of mechanisms to reduce bit stuffing jitters in controller area networks, in *Proceedings of the 17th IEEE Conference on Emerging Technologies and Factory Automation* (*ETFA 2012*), Cracow, Poland, 2012, pp. 218.1–218.8.

CEN13 Cena G., Cibrario Bertolotti I., Hu T., Valenzano A., Fixed-length payload encoding for low-jitter controller area network communication, *IEEE Transactions on Industrial Informatics*, 9(4), 2013, 2155–2164.

CHA94 Charzinski J., Performance of the error detection mechanisms in CAN, in *Proceedings of the First International CAN Conference* (*iCC 1994*), Mainz, Germany, 1994, pp. 1.20–1.29.

DAV07 Davis R.I., Burns A., Bril R.J., Lukkien J.J., Controller area network (CAN) schedulability analysis: Refuted, revisited and revised, *Real-Time Systems*, 35(3), April 2007, 239–272, Springer, the Netherlands.

DIN08 Di Natale M., Understanding and using the controller area network, October 30, 2008. Available online: http://www-inst.eecs.berkeley.edu/~ee249/fa08/Lectures/handout_canbus2.pdf (accessed on March 10, 2014.)

FLO12 Hartwich F., CAN with flexible data-rate, in *Proceedings of the 13th International CAN Conference* (*iCC 2012*), Hambach Castle, Germany, 2012, pp. 14.1–14.9.

FRE02 Fredriksson L., CAN for critical embedded automotive networks, *IEEE Micro*, 22(4), July–August 2002, 28–35.

PRO06 Barranco M., Proenza J., Rodriguez-Navas G., Almeida L., An active star topology for improving fault confinement in CAN networks, *IEEE Transactions on Industrial Informatics*, 2(2), May 2006, 78–85.

TIN94 Tindell K.W., Burns A., Guaranteeing message latencies on controller area network (CAN), in *Proceedings of the First International CAN Conference* (*iCC 1994*), Mainz, Germany, pp. 1–11.

TIN94b Tindell K.W., Hansson H., Wellings A.J., Analysing real-time communications: Controller area network (CAN), in *Proceedings of the 15th Real-Time Systems Symposium* (*RTSS'94*), San Juan, PR, 1994. IEEE Computer Society Press, Los Alamitos, CA, pp. 259–263.

TIN95 Tindell K.W., Burns A., Wellings A.J., Calculating Controller area network (CAN) messages response times, *Control Engineering Practice*, 3(8), 1995, 1163–1169, Elsevier.

53

FlexRay Communication Technology

Roman Nossal-Tueyeni
Austro Control GmbH

Dietmar Millinger
Technology Consulting

53.1 Introduction 53-1
53.2 Automotive Requirements 53-1
Cutting Costs
53.3 What Is FlexRay? 53-3
Media Access • Clock Synchronization • Start-Up • Bus Guardian • Protocol Services • FlexRay Current State
53.4 System Configuration 53-11
Development Models
53.5 Standard Software Components 53-14
Standardized Interfaces
References 53-15

53.1 Introduction

New electronic technologies have dramatically changed cars and the way we experience driving. Anti-lock braking system (ABS), electronic stability program (ESP), air bags, and many more applications have made cars a lot more convenient, comfortable, and, above all, safer. This trend of the past decade has been rather pleasant for the consumer and a tedious task for the automotive industry. The reasons for this drawback have not to do with the need of higher integration of the involved technologies. The very nature of the deployed communication technologies makes the task of integration itself a lot more complex as well as the design of fault-tolerant systems on top of these communication technologies rather difficult.

These limitations on the one hand as well as requirements and anticipated challenges of future automotive applications on the other hand motivated OEMs and suppliers to join forces. The goal of the 2001 founded FlexRay Consortium [1]: to establish one standard for a high-performance communication technology in the automotive industry.

53.2 Automotive Requirements

Since OEMs and suppliers were the founding members of the FlexRay Consortium, it was clear from the very beginning of the work on the new de facto communication standard that FlexRay has to meet the requirements of the automotive industry. Therefore, two key issues have driven the development work for the communication protocol: the need for a technological basis and solution for future safety-related applications and the need to keep costs down.

53.2.1 Cutting Costs

The cost factor is a key driver for many requirements for the communication system as the push for systematic reuse of existing components in multiple car platforms proves. Due to this approach, subsets of components related to a specific function can be reused in multiple platforms without changes inside the components. This elegant and cost-saving solution, however, is possible only if the communication system offers two decisive qualities:

1. It must be standardized and provide a stable interface to the components.
2. The communication system has to provide a deterministic communication service to the components.

This communication determinism is the solution for the problem of interdependencies between components, which is a major issue and cost factor in today's automotive distributed systems. Since any change in one component can change the behavior of the entire system, integration and testing is of utmost importance in order to ensure the needed system reliability. At the same time, testing is extremely difficult and expensive. A deterministic communication system significantly reduces this integration and test effort as it guarantees that the cross-influence is completely under control of the application and not introduced by the communication system. This property is often referred to as *composability*, meaning that each component of a system can be tested in isolation, and integration of these components does not have any side effects.

53.2.1.1 Migration

A new technology, such as FlexRay, does not make all predecessors obsolete at once. It rather replaces the traditional systems gradually and builds on proven solutions. Therefore, existing components and applications have to be migrated into new systems. In order to create this migration path as smooth and efficient as possible, FlexRay has integrated some key qualities of existing communication technologies, for example, dynamic communication.

53.2.1.2 Scalability

Communication determinism and reuse are also key enablers for scalability, which obviously is yet another cost-driven requirement. Scalability, however, calls for not only communication determinism and reuse but also the support of multiple network topologies and network architectures as well as the applicability of the communication technology in different application domains like power train, chassis control, backbone architectures, or driver assistance systems.

53.2.1.3 Future Proof

Keeping costs down is only one side of the coin. The automotive industry has visions of the future car and applications. The most obvious developments are active safety functions like electronic braking systems, driver convenience functions like active rear steering for parking, or the fast-growing domain of driver assistance systems like active cruise control or the lane departure warning function. These automotive applications demand a high level of reliability and safety from the network infrastructure in the car in order to provide the required level of safety at the system level. Therefore, the communication technology has to meet requirements such as redundant communication channels, deterministic media access scheme, high robustness in case of transient faults, distributed fault-tolerant agreement about the protocol state, and extensive error detection and reporting toward the application. The most stringent particular requirement arises from the deterministic media access scheme. In time division multiple access (TDMA) schemes for networks, all participating communication partners need a common understanding of the time used in order to control the access to the communication medium. Typically, a fault-tolerant distributed mechanism for clock synchronization is required. Additionally, the safety requirement introduces the need to protect individual communication partners from faults of other partners by means of so-called guardians. Otherwise, errors of one partner could cross-influence other partners, thus violating safety demands.

Specific automotive issues complete the broad range of requirements forming the framework for and of FlexRay. These issues range from the use of automotive components like x-tals, automotive electromagnetic compatibility requirements, support for power management to conserve battery power, support for electrical and optical physical layers, to a high bandwidth demand of at least two times 10 Mbps.

53.3 What Is FlexRay?

Before the development of FlexRay was started, a comprehensive evaluation of the existing technologies took place. The results showed that none of the existing communication technologies could fulfill the requirements to a satisfactory degree. Thus, the development of a new technology was started. The resulting communication protocol FlexRay is an open, scalable, deterministic, and high-performance communication technology for automotive applications.

A FlexRay network consists of a set of electronic control units (ECUs) with integrated communication controllers (CCs) (Figure 53.1). Each CC connects the ECU to one or more communication channels via a communication port, which in turn links to a bus driver (BD). The BD connects to the physical layer of the communication channel and can contain a guardian unit that monitors the TDMA access of the controller (the architecture of an ECU is depicted in Figure 53.2). A communication channel can be as simple as a single bus wire but also be as complex as active or passive star configurations.

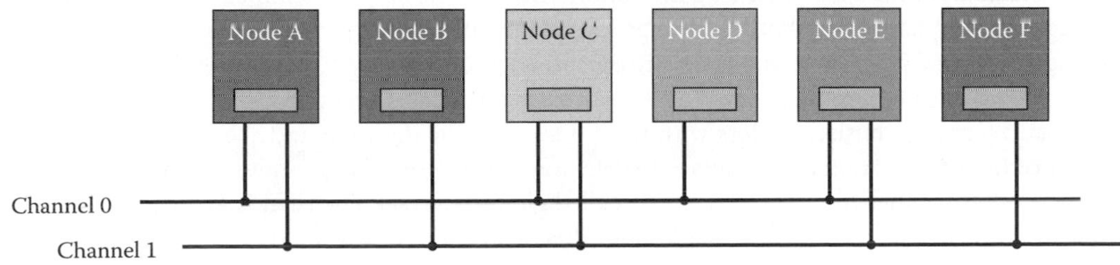

FIGURE 53.1 FlexRay network.

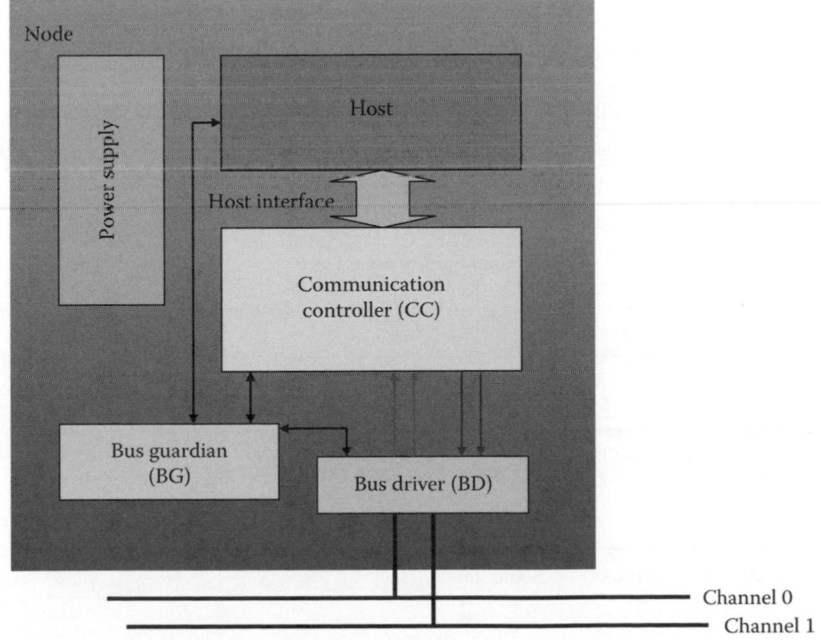

FIGURE 53.2 ECU architecture.

FlexRay supports the operation of a CC with single or redundant communication channels. In case of single communication channel configuration, all controllers are attached to the communication channel via one port. In case of redundant configuration, controllers can be attached to the communication channels via one or two ports. Controllers that are connected to two channels can be configured to transmit data redundantly on two channels at the same time. This redundant transmission allows the masking of a temporary fault of one communication channel and thus constitutes a powerful fault-tolerance feature of the protocol. A second fault-tolerance feature related to transient faults can be constructed by the redundant transmission of data over the same channels with a particular time delay between the redundant transmissions. This delayed transmission allows tolerating transient faults on both channels under particular preconditions.

If the fault-tolerance property of two independent channels is not required for a specific application, the channels can be used to transfer different data, thus effectively doubling the transmission bandwidth.

53.3.1 Media Access

The media access strategy of FlexRay is basically a TDMA scheme with some very specific properties. The basic element of the TDMA scheme is a communication cycle. A communication cycle contains a static segment, a dynamic segment, and two protocol segments called symbol window and network idle time (Figure 53.3). Communication cycles are executed periodically from start-up of the network until shutdown. Two or more communication cycles can form an application cycle.

The static segment consists of slots with fixed duration. The duration and the number of slots are determined by configuration parameters of the FlexRay controllers. These parameters must be identical in all controllers of a network. They form a so-called global contract. Each slot is exclusively owned by one FlexRay CC for the transmission of a frame. This ownership only relates to one channel. On other channels, in the same slot, either the same or another controller can transmit a frame. The identification of the transmitting controllers in one slot is also determined by configuration parameters of the FlexRay controllers. This piece of information is local to the sending controller. The receiving controllers do not possess any knowledge on the transmitter of a frame nor on its content; they are configured solely to receive in a specific slot. Hence, the content of a frame is determined by its positions in the communication cycle.

Alternatively, a message ID in the payload section of the frame can be used to identify a message. The message ID uniquely tells the receivers which information is contained in a frame. By providing a filter mechanism for message IDs, a controller can pick specific data.

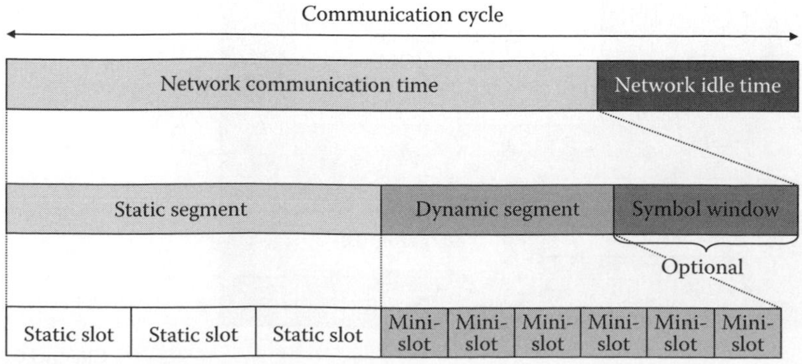

FIGURE 53.3 FlexRay communication cycle.

The static segment provides deterministic communication timing, since it is exactly known when a frame is transmitted on the channel, giving a strong guarantee for the communication latency. This strong guarantee in the static segment comes for the trade-off of fixed bandwidth reservation.

The dynamic segment has fixed duration, which is subdivided into so-called minislots. A minislot has a fixed length that is substantially shorter than that of a static slot. The length of a minislot is not sufficient to accommodate a frame; a minislot only defines a potential start time of a transmission in the dynamic segment. Similar to static slots, each minislot is exclusively owned by one FlexRay controller for the transmission of a frame. During the dynamic segment, all controllers in the network maintain a consistent view about the current minislot. If a controller wants to transmit in a minislot, the controller accesses the medium and starts transmitting the frame. This is detected by all other controllers, which interrupt the counting of minislots. Thus, the minislot is *expanded* to a real slot, which is large enough to accommodate a frame transmission. It is only after the end of the frame transmission that counting of the minislots continues. The expansion of a minislot reduced the number of minislots available in this dynamic segment. The operation of the dynamic segment is illustrated in Figure 53.4: Figure 53.4a shows the situation before minislot 4 occurs. Each of the channels offers 16 minislots for transmission. The owner of minislot 4 on channel 0—in this case controller D—has data to transmit. Hence, the minislot is expanded as shown in Figure 53.4b. The number of available minislots on channel 0 is reduced to 13.

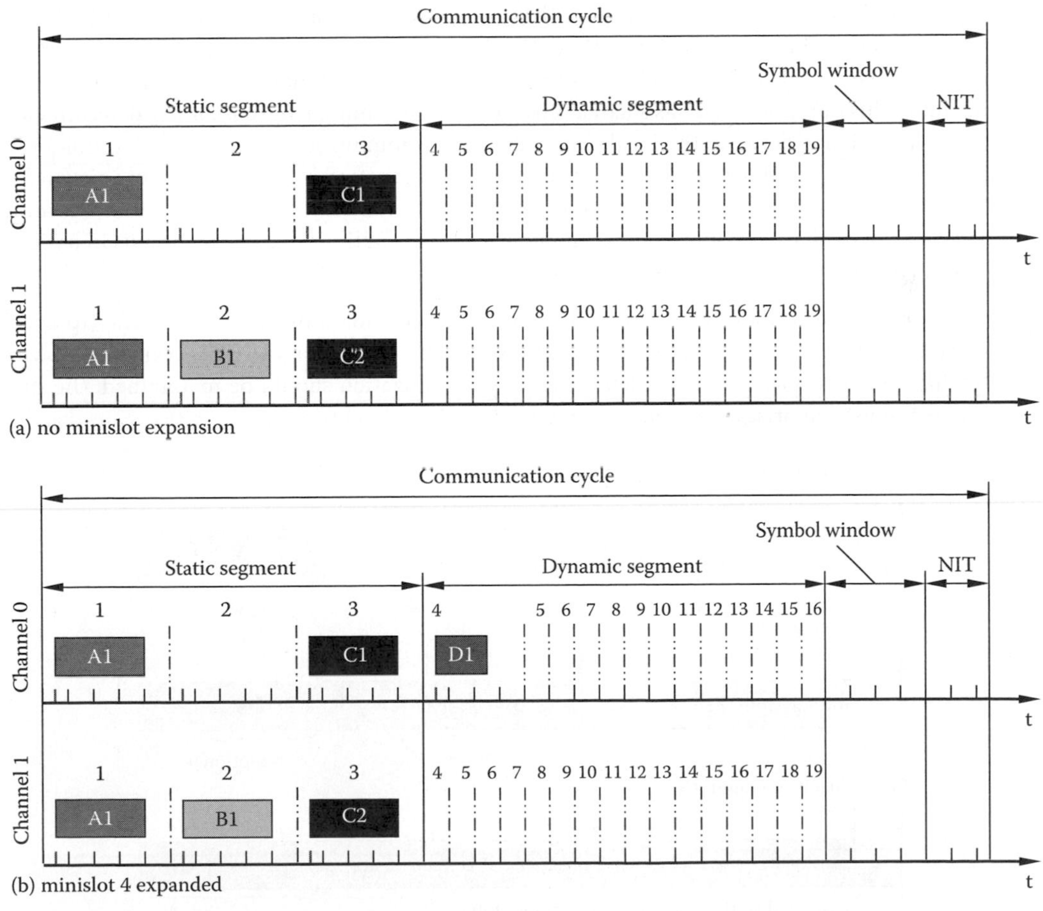

FIGURE 53.4 FlexRay dynamic segment.

If there are no data to transmit by the owner of a minislot, it remains silent. The minislot is not expanded, and slot counting continues with the next minislot. As no minislot expansion occurred, no additional bandwidth beyond the minislot itself is used; hence, other lower-priority minislots that are sequenced later within the dynamic segment have more bandwidth available.

This dynamic media access control scheme produces a priority and demand-driven access pattern that optimally uses the reserved bandwidth for dynamic communication. A controller that owns an *earlier* minislot, that is, a minislot, which has a lower number, has higher priority. The further a minislot is situated in the dynamic segment, the higher is the probability that it will not be in existence in a particular cycle due to the expansion of higher-priority slots. A minislot is only expanded and its bandwidth used if the owning controller has data to transmit. As a consequence, the local controller configuration has to ensure that each minislot is configured only once in a network. The minimum duration of a minislot is mainly determined by physical parameters of the network (delay) and by the maximum deviation of the clock frequency in the controllers. The duration of a minislot and the length of the dynamic segment are global configuration parameters that have to be consistent within all controllers in the network.

The symbol window is a time slot of fixed duration, in which special symbols can be transmitted on the network. Symbols are used for network management purposes.

The network idle time is a protocol-specific time window, in which no traffic is scheduled on the communication channel. The CCs use this time window to execute the clock synchronization algorithm. The offset correction (see the following text) that is done as a consequence of clock synchronization requires that some controllers correct their local view of the time forward and others have to correct backward. The correction is done in the network idle time. Hence, no consistent operations of the media access control can be guaranteed, and thus, silence is required. Since this duration has to be subtracted from net bandwidth, this duration is kept as small as possible. The minimum length is largely determined by the maximum deviation between the local clocks after one communication cycle. The duration of the network idle time is a global parameter that has to be consistent between all controllers in a network.

While at least a minimal static segment and the network idle time are mandatory parts of a communication cycle, the symbol window and the dynamic segment are optional parts. This results in basically three reasonable configurations (see Figure 53.5):

- Pure static configuration, which contains only static slots for transmission. In order to enable clock synchronization, the static segment must consist of at least two slots, which are owned by different controllers. If fault-tolerant clock synchronization should be maintained, the static segment must comprise at least four static slots.

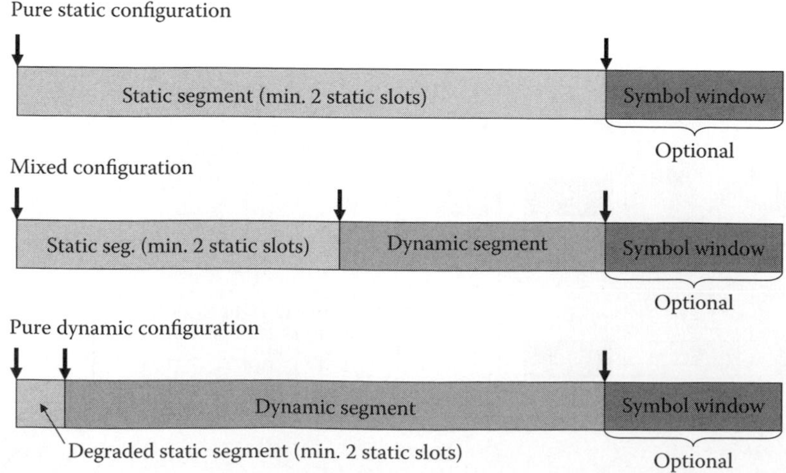

FIGURE 53.5 FlexRay configurations.

- Mixed configuration with a static segment and a dynamic segment, where the ratio between static bandwidth and dynamic bandwidth can vary in a broad range.
- Finally, a pure dynamic configuration with all bandwidth assigned to dynamic communication. This configuration also requires a so-called degraded static segment, which has two static slots.

Considering the most likely application domains, mixed configurations will be dominant. Depending on the actual configuration, FlexRay can achieve a best-case bandwidth utilization of about 70% with the average utilization ranging around 60%.

53.3.2 Clock Synchronization

The media access scheme of FlexRay relies on a common and consistent view about time that is shared between all CCs in the network. It is the task of the clock synchronization service to generate such a time view locally inside each CC. For the detailed description of the clock synchronization, first the representation of time inside a CC is described. The physical basis for each time representation is the tick of the local controller oscillator. This clock signal is divided by an integer to form a clock signal called microtick. This is often done by a frequency divider implemented in hardware. An integer number of microticks form a time unit called macrotick. Minislots and static slots are set up as integer multiple of macroticks. The number of microticks that constitute a macrotick is statically configured. However, for adjustment of the local time, the clock synchronization service can temporarily adjust this ratio in order to accelerate or decelerate the macrotick clock. If, for instance, a clock is too fast, microticks are added to a macrotick in order to slow down the counting.

The clock synchronization service is a distributed control system that produces local macroticks with a defined precision in relation to the local macroticks of the other controllers of a network. The control system takes some globally visible reference events that represent the global time ticks, measures the deviation of the local time ticks to these global ticks, and computes the local adjustments in order to minimize the deviation of the local clock from the global ticks. Due to the distributed nature of the FlexRay system, no explicit global reference event exists. The only event, which is globally observable, is a frame transmission on the communication channel. The start of a transmission is triggered by the local time base of the sending controller. Each controller can collect these reference events to form a virtual global time base by computing a fault-tolerant mean value of the deviation between the local time and the perceived events.

The reasoning behind this approach is based on the assumption that a majority of local clocks in the network is correct. Correctness of a local clock is given, when the local clock does not deviate from every other by more than the precision value. A controller with locally correct clock sees only deviation values within the precision value. By computing the median value of deviations, the actual temporal deviation of the local clock to the virtual reference tick is formed. Next, the local clock is adjusted such that the local deviation is minimized. Since all nodes in the network perform this operation, all local clocks move their ticks toward the tick of the virtual global time. The operation of observation, computing, and correction is performed in every communication cycle.

In case of wrong transmission times of faulty controllers on the communication channel, things get more complicated. Here, a special part of the fault-tolerant median value algorithm takes over. This algorithm uses only the best of the measured deviation values. All other values are discarded. This algorithm ensures that the maximum influence of a faulty controller to the virtual global time is strictly bound. Additionally, the protocol requires the marking of particular synchronization frames that can be used for deviation measurement. The reasoning behind this mechanism is twofold. The first reason is that it is used to pick exactly one frame from a controller in order to avoid monopolization of the global time by one controller with many transmit frames. The second reason is that particular controllers can be excluded from clock synchronization, either because the crystal is not trustworthy or more likely because there are system configurations in which a controller is not available.

In case of a faulty local clock, the faulty controller perceives only deviation values that exceed a particular value. This value is given by the precision value. This condition is checked by the synchronization service, and an error is reported to the application. A specific extension of the clock synchronization services handles the compensation of permanent deviations of one node. In case such a permanent deviation is detected, a permanent correction is applied. The detection and calculation of such permanent deviations is executed less frequently than the correction of temporal deviations.

The error handling of the protocol follows a strategy, which identifies every problem as fast as possible, but keeps the controller alive as long as possible. Problem indicators are frames that are received outside their expected arrival intervals or when the clock synchronization does not receive a sufficient number of synchronization frames. The automatic reaction of the controller is to degrade the operation from a sending mode to a passive mode where reception is still possible. At the same time, the problem is indicated to the application. The application can react in an application-specific manner to the detected problem. This strategy gives the designer of a system a maximum of flexibility for the design of the safety required by the application.

53.3.3 Start-Up

The preceding protocol description handled only the case of an already running system. To reach this state, the start-up service is part of the protocol. Its purpose is to establish a common view on the global time and the position in the communication cycle.

Generally, the start-up service has to handle two different cases. The coldstart case is a start-up of all nodes in the network, while the reintegration case means to integrate a starting controller into an already running set of controllers (see Figure 53.6).

During coldstart, the algorithm has to ensure that really a coldstart situation is given. Otherwise, the starting controller might disturb an already running set of controllers. For this reason, the starting controller has to listen for the so-called listen time-out for traffic on the communication channel. In case no traffic is detected, the controller assumes a coldstart situation and starts to transmit frames for a limited number of rounds. In case another controller responds with frames that fit to the slot counter of the coldstart node, start-up was successful.

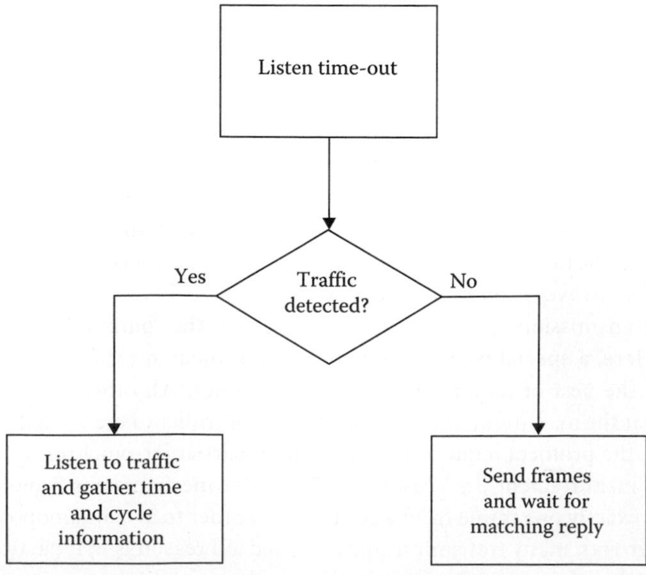

FIGURE 53.6 Start-up process.

In case traffic is detected during the observation period, the controller changes into the reintegration mode. In this mode, the controller has to synchronize the slot counter with the frames seen on the channel. Therefore, the controller receives frames from the channel and sets the slot counter accordingly. For a certain period of time, the controller checks the plausibility of the received frames in relation to the internal slot counter. If there is a match, the controller enters the normal mode, in which active transmission of frames is allowed.

53.3.3.1 Coding and Physical Layer

The frame format for data transmission contains three sections: the header section, the payload section, and the trailer section (see Figure 53.7). The header contains protocol control information like the synchronization frame flag, the frame ID, a null frame indicator and the frame length, and a cycle counter. The payload section contains up to 254 bytes of data. In case the payload does not contain any data, the null frame indicator is set. A null frame is thus a valid frame that does not contain any payload data. It can, however, serve as an alive signal or can be used for clock synchronization purposes.

Optionally, the data section can contain a message ID, which identifies the type of information transported in the frame. The trailer section contains a 24-bit CRC that protects the complete frame.

The existing FlexRay CCs support communication bit rates of up to 10 Mbps on two channels over an electrical physical layer. The physical layer is connected to the controller via a transceiver component.

This physical layer supports bus topologies, star topologies, cascaded star topologies, and bus stubs connected to star couplers as shown in Figure 53.8. This multitude of topologies allows a maximum of scalability and flexibility of electronic architectures in automotive applications.

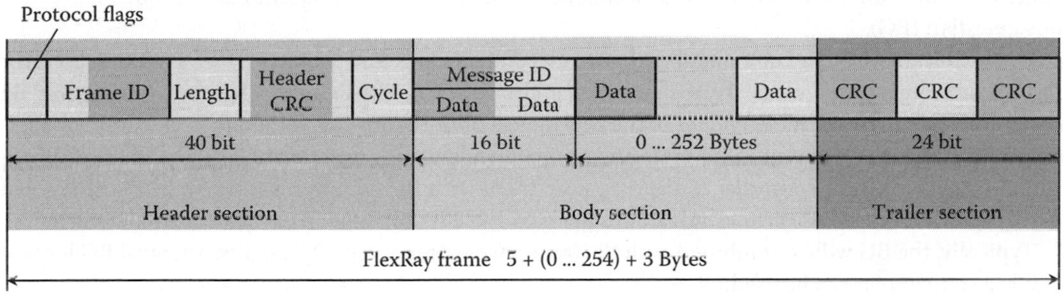

FIGURE 53.7 FlexRay frame format.

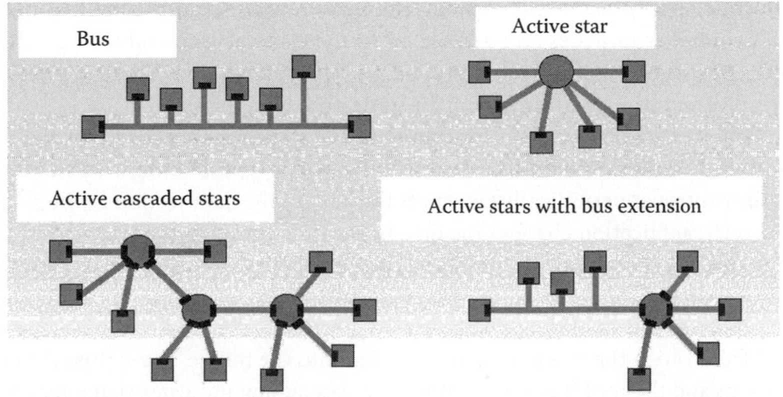

FIGURE 53.8 Topologies.

Besides the transformation of bit streams between the CC and the physical layer, the transceiver component also provides a set of very specific services for an automotive network. The major services are alarm handling and wakeup control. Alarm signals are a very powerful mechanism for diverse information exchange between a sender controller and receiver controllers. A sender transmits an alarm symbol on the bus parallel to alarm information in a frame. A receiver ECU receives the alarm information in the frame like normal data. Additionally, the CC receives the alarm symbol on the physical layer and indicates this symbol to the ECU. Thus, the ECU has two highly independent indicators for an alarm to act on. This scheme can be used for the validation of critical signals like an air-bag fire command.

The second type of service provided by the symbol mechanism is the wakeup function. A wakeup service is required in automotive applications where electronic components have a sleep mode, in which power consumption is extremely reduced. The wakeup service restarts normal operation in all sleeping ECU components. In a network, the wakeup service uses a special signal that is transmitted over the network. In FlexRay, this function relies on the ability of a transceiver component to identify a wakeup symbol and to signal this event to the CC and the remaining components of the ECU to wake these components up.

53.3.4 Bus Guardian

The media access strategy completely relies on the cooperative behavior of every CC in a network. The protocol mechanisms inside a controller ensure this behavior to a considerable high level of confidence. However, for safety-relevant applications, the controller internal mechanisms do not provide a sufficiently high level of safety. An additional and independent component is required to ensure that no controller can disturb the media access mechanism of the network. This additional component is called bus guardian (BG).

In FlexRay, the BG is a component with an independent clock and is constructed such that an error in the controller cannot influence the guardian and vice versa. The BG is configured with its own set of parameters. These are independent of the parameters of the controller, although both parameter sets represent the same communication cycle and slot pattern. During runtime, the BG receives synchronization signals from the controller in order to keep track with the communication cycle. Using its own clock the BG verifies those synchronization signals to avoid being influenced by a faulty controller.

Typically, the BG will be combined with the transceiver component. Optionally, a central BG located inside a star coupler can be used.

53.3.5 Protocol Services

Application information is transmitted by the CC inside of frames. A frame contains one or more application signals. A controller provides an interface for frame transmission and reception that consists of buffers. A buffer consists of a control/status section and the data section. These sections have different semantics for receive and transmit frames and for static and dynamic slots.

The control section of transmit buffers for frames in the static segment contains the slot ID and the channel, in which the frame is transmitted. Once a buffer of this type is configured and the communication is started, the controller periodically transmits the data in the data section in the slot configured in the slot ID. When the application changes the data in the buffer, the subsequent transmission contains the new data. A special control flag allows modifying this behavior such that in case the application does not update the data in the buffer, a null frame is transmitted, signaling the failure to update other controllers. The control section of a receive buffer for frames in the static segment defines the slot ID and the channel from which the frame should be loaded into the buffer. The status section contains the frame receive status and the null frame indicator. One special flag indicates that a new frame has been received. It is important to note that the slot ID and the channel selection for slots in the static segment cannot be changed during operation.

Buffer status and control section for frames in the dynamic segment are similar to the buffers in the static segment. Differences result from the fact that the slot ID and the channel can be changed during normal operation and that multiple buffers can be grouped together to form a FIFO for frame reception from the dynamic section.

The CC provides a set of timers that run clocked by the synchronized time of the network. Several different conditions can be used to generate interrupts based on these timers. These interrupts are efficient means to synchronize the application with the timing on the bus.

53.3.6 FlexRay Current State

By the end of 2006, the FlexRay technology was introduced in a passenger car, proving the readiness of the technology for use in series production. Already in 2008, five silicon vendors provided or announced microcontrollers with FlexRay interfaces for 16- and 32-bit architectures (Freescale, Fujitsu, NEC, TI, and Renesas). Three silicon vendors provide physical layer chips (NXP, AMS, and Elmos). Another important milestone was the introduction of FlexRay in the first mass production car below the premium segment in 2011 (BMW 3 series), thus proving the economic validity of the technology.

The version of the implemented protocol specification is 2.1 rev A. As a result of the experience gained from the first series projects, the FlexRay Consortium continued to improve the FlexRay specification. Those improvements resulted in the FlexRay standard 3.0, which was handed over to the ISO standardization body. In 2013, the standardization process is almost completed (ISO 10681-1,2 and ISO 17458-1,5). However, currently no silicon vendor has FlexRay 3.0 on the roadmap.

Feedback from carmakers and suppliers does not indicate any significant problems with FlexRay observed in the field. Compared to CAN, a high level of robustness was reached by FlexRay much faster. One concern related to FlexRay in automotive applications is the complexity of configuration. The fine-segmented distribution of responsibilities over many parties and suppliers makes the planning process a burden that needs to be overcome. However, the current automotive open system architecture (AUTOSAR) methodology supports this process in many ways.

In the recent years, two alternative technologies got into the focus of the carmakers, Ethernet and CAN FD [2]. Both technologies root in requirements outside the scope of FlexRay, yet they will reduce the scope of application for FlexRay. CAN FD will prolong existing CAN-based architectures without the need to upgrade to FlexRay, and Ethernet can take over areas where the complex configuration of FlexRay is a hindering factor (e.g., backbone networks).

53.4 System Configuration

With the advent of the TDMA communication technologies and especially in the automotive application domain, the off-line configuration of networks gets increasingly important. Off-line configuration means that the configuration parameters of the CCs are not generated during the runtime of the system but are determined throughout the development time of the system. The processes for system development are not only mainly determined by the applied technology, but also driven by industry-specific technical or organizational constraints. In the following section, the background for such a design process is described by first defining a model for the used information (Figure 53.9) and second explaining the information processing.

The functional domain defines entities called functions and the communication relations between them. Functions describe the functionality of the entire system, creating a hierarchy from very abstract high-level functions down to specific tailored functions. A system normally comprises more than one function. In a vehicle, for example, the functional hierarchy would feature *chassis functions* on the top level like *steering functions* and *braking functions*, the latter being even more detailed into *basic braking function*, *antilock brake functions*, and so on. A communication relation between functions or within a function starts at a sender function, connects the sender function with receiver functions, and has an assigned signal.

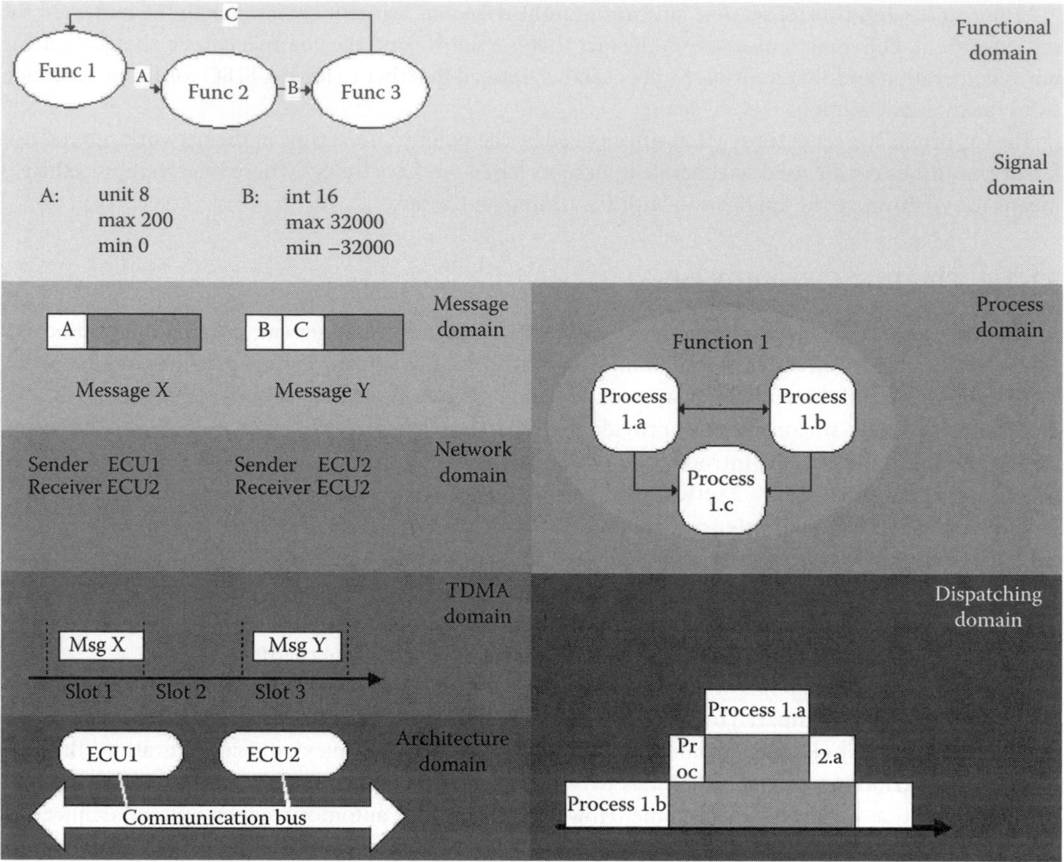

FIGURE 53.9 Information domains.

Signals are defined in the signal domain. A signal definition contains a signal name, the signal semantics, and the value range. Signals are assigned to messages. The message domain defines messages, that is, packages of signals that should be transmitted together. A message has a name and a fixed setup of the signals it contains. The network domain determines which ECU will send a frame containing a message as well as which ECUs should receive this frame. The TDMA domain establishes the exact points in time when frames are transmitted. Finally, the architecture domain defines the physical structure of a system, with all ECUs, communication systems, and the connections of the ECUs to them.

The six information domains provided earlier are of importance for the entire system, that is, for all ECUs of the systems. Hence, they are considered *global information*.

Two additional domains complement the message, network, TDMA, and architecture domains. These two domains are the process domain and the dispatching domain. The process domain describes the software architecture of the system. It lists all processes and their interactions like mutual exclusion, as well as the assignment of processes to functions. Processes are information processing units of an application. They have timing parameters assigned to them that define the period and the time offset of their execution. The dispatching domain is the ECU counterpart of the TDMA domain. It determines the application timing, that is, which process is executed at which point in time. Implicitly, this also defines the preemption of processes, that is, when a process is interrupted by the execution of another process. The latter two information domains feature information that is relevant for only one ECU. Hence, they are considered to be *local information* (local referring to one ECU rather than the entire system).

The categorization of information given by the information domain model leads the way to the development process. It is an organizational constraint in the automotive development processes that the knowledge related to the system to be developed is distributed among the process participants. These participants are typically the automobile manufacturer (OEM) and one or more suppliers. The OEM possesses the information on the intended system functions, the envisioned system architecture, and the allocation of functions to architectural components, that is, to ECUs. Thus, the OEM's knowledge covers three of the six global information domains.

The supplier, on the other hand, is the expert on function implementation and ECU design. This means he provides the knowledge on the process domain, that is, the software architecture underlying the function implementation. Each function of the system or each part thereof is implemented by a set of interacting processes. The functionality of the ECU does rely not only on the software architecture but also on the execution pattern. The supplier has to define the timing of each process executed in his developed ECU. Hence, the dispatching domain is the supplier knowledge as well.

With five of the eight information domains assigned, the open question is whose responsibility are the three remaining information domains. As described in the previous section, these domains—the message, the network, and the TDMA domain—define the communication behavior on the system level. They specify at which point in time, which message is transferred from which sending ECU to which receiving ECUs. These domains obviously affect all ECUs of the system; hence, no single supplier should be able to define these domains. For this reason, the DECOMSYS development process for collaborative system development between an OEM and several suppliers assigns the message, the network, and the TDMA domain to the OEM.

So far, each piece of development information has been assigned to a process participant, which results in a static structure. In the following appropriate dynamic structure, the development process will be described in brief.

The proposed OEM-supplier development process (Figure 53.10) takes a two-phase approach. In the first phase, the OEM has to cover all global aspects; subsequently, supplying companies deal with the local aspects.

The process builds on the functional model of the system, which belongs to the functional domain and the signal domain, and the architectural model belonging to the architecture domain. The functional model describes the functions of the system and their structure consisting of subfunctions. As subfunctions have to exchange information in order to provide the intended function output, the functional

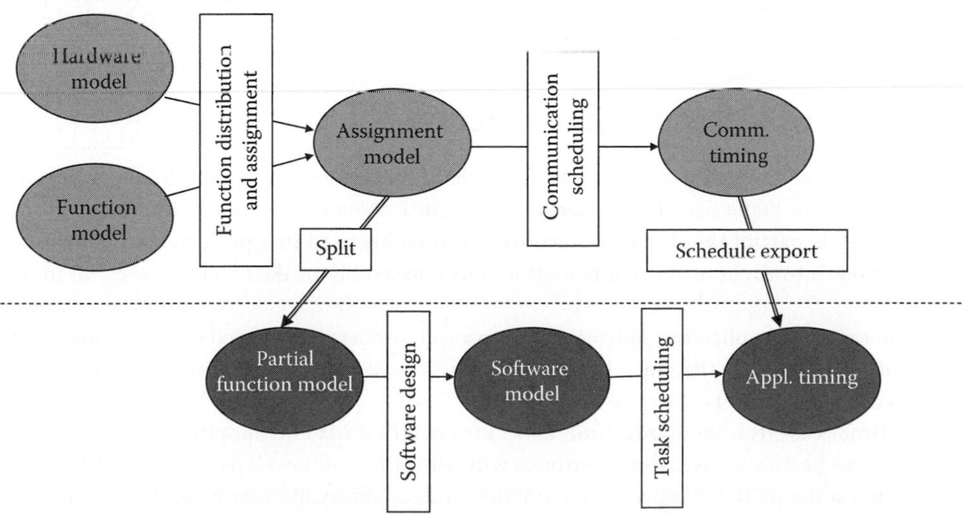

FIGURE 53.10 OEM-supplier development process.

model inherently also defines the signals that are transferred from one subfunction to others. The functional model is complemented by the architectural model, which defines the system topology and the ECUs that are present in the system. The mapping between these two models results in a concrete system architecture for a particular system in a particular vehicle. This is described in the distributed functional model.

Based on the distributed functional model, the OEM performs communication scheduling. During this operation, signals are packed to messages, which in turn are scheduled for transmission at a specific point in time. Communication scheduling concludes the global design steps and thus the OEM's tasks.

The suppliers base their local design steps on the global information given by the OEM. The so-called split of the distributed functional model tells the supplier which functions or subfunctions the ECU has to perform, that is, his responsibility. The supplier conducts software architecture design and creates the software model for each function or subfunction. The resulting list of processes is scheduled for the ECU taking into account local constraints as well as the global constraints defined in the communication schedule.

Note that for performing the local design, the supplier solely requires parts of the global information created by the OEM as well as his own knowledge on the ECU. In principle, the suppliers do not influence each other.

53.4.1 Development Models

The use of development models with a clear purpose and information content is the answer to the challenge of reuse of components for different car lines. Each model focuses on a certain type of information. The full picture of the system consists of these individual models and the mappings between them.

When developing a new car, only those models affected by the differences between the previous version and the new vehicle have to be adapted while the other models remain unchanged. To be more specific, the reuse of system parts calls for the separation of the architectural and the functional models. The functional model, that is, the functions to be executed by the distributed system, is primarily independent of a specific car model and can be reused in different model ranges. The architectural model on the other hand, that is, the concrete number of ECUs and their properties in a certain car, vary between model ranges.

It is decisive for a useful development process to allow the separate development of the functional model and the architectural model. At the same time, the process must support the mapping of the functions to concrete hardware architecture elements.

53.5 Standard Software Components

Standard software components within an ECU deliver a set of services to the actual application. The application itself can make use of these services without implementing any of them. For example, if a transport layer is part of the standard software components used in a project, the application does not need to take into account the segmentation and reassembly of data that exceed the maximum message size.

In order to reuse the application code in another project, the services offered by the standard software in the new project should be the same. If the standard software provides less functionality, the code has to be changed, as missing services have to be added.

In the optimal case, the standardization effort covers all OEMs and suppliers. Only then, reuse of existing code can be guaranteed, thus creating a win–win situation for all participants: The OEMs can purchase tested software that has proven its function and reliability in other projects—suppliers on the other hand have the possibility to sell this software, which they have created with considerable effort, to other OEMs.

53.5.1 Standardized Interfaces

Standard software components are not the only answer to the challenge of reuse. The standard software components and their standardized services must be complemented by standardized interfaces, through which the application software can access these services. Standardizing interfaces for software means to provide one operating system API for the software to access communication as well as other resources like analog-to-digital converters (ADCs). Similarly, standardized network interfaces allow the reuse of entire ECUs in different networks. The hardware of a distributed system can have a standardized interface represented by an abstract description of the network communication.

Standardization of software components as well as interfaces and thus a system architecture is not a competitive issue. Depending on where the interfaces are set, there is ample room for each participating company to use its strengths effectively to achieve its purpose. In our opinion, the real competitive issues are the functions in an ECU or the overall system functionality that is realized by the interaction of ECU functions. The special behavior of an electronic power-steering system as perceived by the car driver is mainly determined by the control algorithms and their application data, rather than by the type of communication interface used for integration.

With respect to standardization efforts, the industry is currently moving in the right direction. Initiatives like the open systems and the corresponding interfaces for automotive electronics (OSEK/VDX) Consortium [3], HIS [4], and many others attempt to standardize certain software components and interfaces. The field bus data exchange format (FIBEX) group that is now part of the Association for standardization of automation and measuring systemsConsortium [5] develops a standardized exchange format between tools based on XML, which is able to hold the complete specification of a distributed system. Many of these initiatives and projects are now united in the AUTOSAR development partnership [6] with the goal to generate an industry-wide standard for the basic software infrastructure.

References

1. FlexRay, Wikipedia, http://en.wikipedia.org/wiki/FlexRay, accessed April 4, 2014.
2. CAN FD, Bosch Semiconductors and Sensors, http://www.bosch-semiconductors.de/en/ubk_semiconductors/safe/ip_modules/can_fd/can.html, accessed April 4, 2014.
3. OSEK-VDX, OSEK/VDX Organization, http://www.osek-vdx.org, accessed April 4, 2014.
4. HIS Hersteller Initiative Software, Daimler AG, http://www.automotive-his.de, accessed April 4, 2014.
5. ASAM, Association for Standardisation of Automation and Measuring Systems, http://www.asam.net, accessed April 4, 2014.
6. AUTOSAR, AUTomotive Open System ARchitecture, http://www.autosar.de, accessed April 4, 2014.

54

The LIN Standard

54.1 Introduction 54-1
54.2 Need 54-2
54.3 History 54-2
54.4 Some LIN Basics 54-3
54.5 LIN Physical Layer 54-3
54.6 LIN Protocol 54-4
54.7 LIN Conformance Test 54-5
54.8 Design Process and Workflow 54-5
54.9 System Definition Process 54-5
54.10 Debugging 54-6
54.11 AUTOSAR and LIN 54-7
54.12 Future 54-7
54.13 Volcano LIN Tool Chain 54-7
54.14 LIN Network Architect 54-7
Requirement Capturing
54.15 LTP 54-9
54.16 LIN Spector: Test Tool 54-10
54.17 Summary 54-13
Acknowledgments 54-14
References 54-14
Additional Sources 54-15

Antal Rajnák
Mentor Graphics Corp.

Anders Kallerdahl
Mocean Laboratories AB

54.1 Introduction

Local interconnect network (LIN) is much more than *just another protocol*! LIN is defining a straightforward design methodology and provides definition of tool interfaces and a signal-based application programming interface (API) in a single package.

LIN is an open communication standard, enabling fast and cost-efficient implementation of low-cost multiplex systems. It supports encapsulation for model-based design and validation, leading to front-loaded development processes that are faster and more cost efficient than traditional development methods.

The LIN standard not only covers the definition of the bus protocol but expands its scope into the domain of application and tool interfaces, reconfiguration mechanisms, and diagnostic services—thus offering a holistic communication solution for automotive, industrial, and consumer applications. In other words, systems engineering at its best enables distributed and parallel development processes.

Availability of dedicated tools to automate the design and system integration process is a key factor for the success of LIN.

54.2 Need

The car industry today is implementing an increasing number of functions in software. Complex electrical architectures using multiple networks with different protocols are the norm in modern high-end cars. The software industry in general is handling software complexity through *best practices* such as the following:

- *Abstraction*: Hiding the unnecessary level of detail.
- *Composability*: Partitioning a solution into a set of separately specified, developed, and validated modules, easily combined into a larger structure inheriting the validity of its components—without the need for revalidation.
- *Parallel processes*: State-of-the-art development processes such as the Rational Unified Process (RUP) are based on parallel and iterative development where the most critical parts are developed and tested in the first iterations.

The automotive industry is under constant pressure to reduce cost and lead time while still providing increasing amount of functionality. This must be managed without sacrificing quality. It is not uncommon today for a car project to spend half a billion US dollars on development and perhaps as much as 150 million on prototypes. By shortening lead time, the carmaker creates benefits in several ways; typically, both development cost and capital costs are reduced. At the same time, an earlier market introduction creates better sales volumes and therefore better profit. One way of reducing lead time is by eliminating traditional prototype loops requiring full-size cars, rather than relying on *virtual development*, replacing traditional development, and testing methods by Computer aided engineering (CAE).

To reduce development time while maintaining quality, a reduction in lead time must occur in a coordinated fashion for all major subsystems of a car such as body, electrical, chassis, and engine. With improved tools and practices for other subsystems and increasing complexity of the electrical system, more focus must be placed on the electrical development process, as it may determine the total lead time and quality of the car. These two challenges—lead time reduction and handling of increased software complexity—will put growing pressure on the industry to handle development of electrical architectures in a more purposeful manner. (Specks and Rajnák [1998, 2000]).

54.3 History

The LIN Consortium started in late 1998 initiated by five car manufacturers Audi, BMW, DaimlerChrysler, Volvo, and Volkswagen, the tool manufacturer Volcano Communications Technologies (VCT), and the semiconductor manufacturer Motorola. The workgroup focused on specification of an open standard for low-cost LINs in vehicles where the bandwidth and versatility of Control Area Network (CAN) are not required. The LIN standard includes the specification of the transmission protocol, the transmission medium, the interface between development tools, and the interfaces for application software programming. LIN promotes scalable architectures and interoperability of network nodes from the viewpoint of hardware and software and a predictable EMC behavior. LIN complements the existing portfolio of automotive multiplex networks. It will be the enabling factor for the implementation of hierarchical vehicle networks, in order to gain further quality enhancement and cost reduction of vehicles. It addresses the needs of increasing complexity, implementation, and maintenance of software in distributed systems by providing a highly automated tool chain.

The main properties of the LIN bus are

- Single master multiple slaves structure
- Low-cost silicon implementation based on common universal asynchronous receiver/transmitter/serial communications Interface (UART/SCI) interface hardware, an equivalent in software, or as pure state machine

- Self-synchronization without a quartz or ceramics resonator in the slave nodes
- Deterministic signal transfer entities, with signal propagation time computable in advance
- Signals-based API

A LIN network comprises one master and one or more slave nodes. The medium access is controlled by a master node—no arbitration or collision management in the slaves is required. Worst-case latency of signal transfer is guaranteed.

The LIN consortium ended its operation in 2010 after 12 years of development of the LIN standard. After 2010, the LIN standard was considered stable for several years and it was difficult to justify continued operation of a consortium maintaining the LIN standard. The last LIN standard released by the LIN consortium by the end of 2010 was the LIN 2.2A. The LIN consortium requested the International Organization for Standardization (ISO) to take over the standard—following the example of the FlexRay specification.

The LIN standard was accepted as an ISO standard and is currently (2013) under continued development. The identification of the specification is ISO/CD 17987, parts 1–7. The diagnostic part is moved to ISO/CD 14229-7.

54.4 Some LIN Basics

LIN is a low-cost, single-wire network. The starting point of the physical layer design was the ISO 9141 standard. In order to meet EMC requirements, the slew rates are controlled. The protocol is a simple master slave protocol based on the common UART format. In order to enable communication between nodes clocked by low-cost RC oscillators, synchronization information is transmitted by the master node on the bus. Slave nodes will synchronize with the master clock, which is regarded to be accurate. The speed of the LIN network is up to 20 kbps, and the transmission is protected by a checksum. The LIN protocol is message identifier based. The identifiers do not address nodes directly, but denote the meaning of the messages. This way any message can have multiple destinations (multicasting). The master sends out the message header consisting of a synchronization break (serving as a unique identifier for the beginning of the frame), a synchronization field carrying the clock information, and the message identifier, which denotes the meaning of the message.

Upon reception of the message identifier, the nodes on the network will know exactly what to do with the message. One of the nodes sends out the message response and the others either listen or do not care. Messages from the master to the slave(s) are carried out in the same manner—in this case, the slave task incorporated into the master node sends the response.

LIN messages are scheduled in a time-triggered fashion. This provides a model for the accurate calculation of latency times—thus supporting fully predictable behavior. Since the master sends out the headers, it is in complete control of the scheduling and is also able to swap between a set of predefined schedule tables, according to the specific requirements/modes of the applications running in the subsystem.

54.5 LIN Physical Layer

The transport medium is a single-line, wired-AND bus being supplied via a termination resistor from the positive battery node (V_{BAT} nominally 12 V). The bus line transceiver is an enhanced ISO 9141 implementation. The bus can take two complementary logical values: the dominant value with an electrical voltage close to ground and representing a logical "0" and the recessive value, with an electrical voltage close to the battery supply and representing a logical "1" (Figure 54.1).

The bus is terminated by a pull-up resistor with a value of 1 kΩ in the master node and 30 kΩ in a slave node. A diode in series with the resistor is required to prevent the electronic control unit (ECU) from being powered by the bus in case of a local loss of battery. The termination capacitance is typically C_{Slave} = 220 pF in the slave nodes, while the capacitance of the master node is higher in order to make the total line capacitance less dependent from the actual number of slave nodes in a particular network.

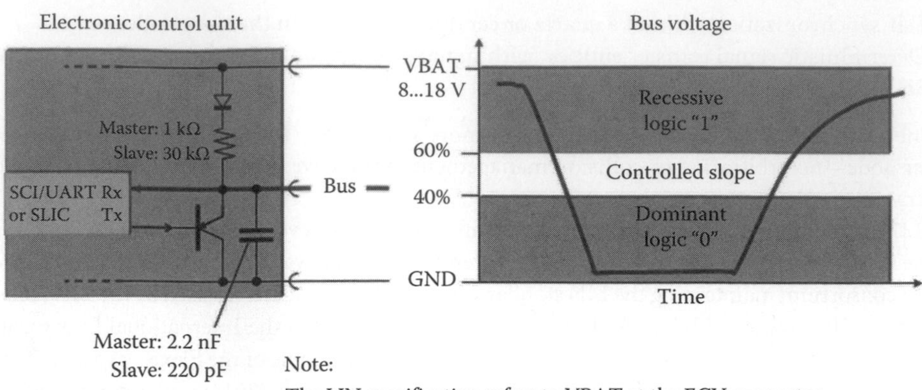

FIGURE 54.1 LIN physical layer.

The maximum signaling rate is limited to 20 kbps. This value is a practical compromise between the conflicting requirements of high slew rates for the purpose of easy synchronization and for slower slew rates for electromagnetic compatibility. The minimum baud rate is 1 kbps—helping to avoid conflicts with the practical implementation of time-out periods.

In addition to the 12 V LIN bus, there is also a specification for a 24 V LIN variant, driven by demand from the truck original equipment manufacturers (OEMs).

54.6 LIN Protocol

The entities that are transferred on the LIN bus are frames. One message frame is formed by the header and the response (data) part. The communication in a LIN network is always initiated by the master task sending out a message header, which comprises the synchronization break, the synchronization byte, and the message identifier. One slave task is activated upon reception and filtering of the identifier and starts the transmission of the message response. The response comprises one to eight data bytes and is protected by one checksum byte.

The time it takes to send a frame is the sum of the time to send each byte, plus the response space, and the interbyte space. The interbyte space is the period between the end of the stop bit of a byte and the start bit of the following byte.

The interframe space is the time from the end of a frame until the start of the next frame. A frame is constructed of a break followed by 4–11 byte fields. The structure of a frame is shown in Figure 54.2.

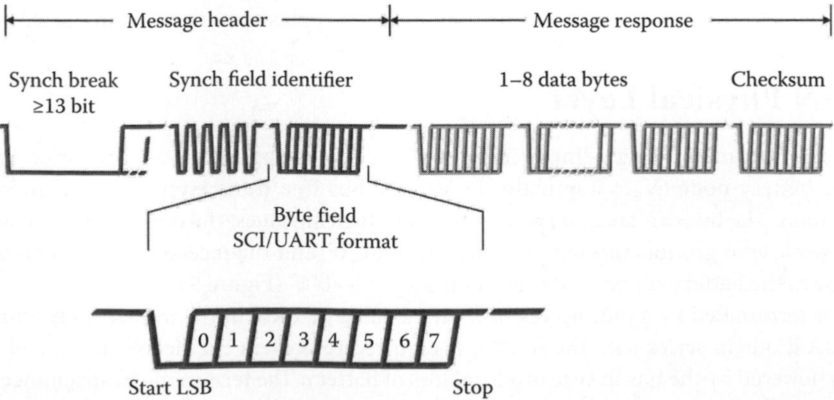

FIGURE 54.2 Frame structure.

In order to allow the detection of signaling errors, the sender of a message is required to monitor the transmission. After transmission of a byte, the subsequent byte may only be transmitted if the received byte was correct. This allows a proper handling of bus collisions and time-outs.

Signals are transported in the data field of a frame. Several signals can be packed into one frame as long as they do not overlap each other. Each signal has exactly one producer, that is, it is always written by the same node in the cluster. Zero, one, or multiple nodes may subscribe to the signal. A key property of the LIN protocol is the use of schedule tables. Schedule table makes it possible to assure that the bus will never be overloaded. They are also the key component to guarantee timely delivery of signals to the subscribing applications. Deterministic behavior is made possible by the fact that all transfers in a LIN cluster are initiated by the master task. It is the responsibility of the master to assure that all frames relevant in a certain mode of operation are given enough time to be transferred.

54.7 LIN Conformance Test

In addition to the LIN specification, the LIN consortium also developed a LIN conformance test specification. The OEMs need an assurance that the LIN nodes that are developed by suppliers do fully conform to the specification. It covers both the physical and the protocol layers. Pre-2010 accredited test houses performed certification of the LIN node implementations. Today, the accredited test houses are typically governed by the OEMs.

54.8 Design Process and Workflow

Regardless of the protocol, a network design process comprises of three major elements:

1. Requirement capturing (signal definitions and timing requirements)
2. Network configuration/design
3. Network validation

The holistic concept of LIN supports the entire development, configuration, and validation of a network by providing definition of all necessary interfaces.

The LIN workflow allows for the implementation of a seamless chain of design and development tools enhancing speed of development and the reliability of the resulting LIN cluster.

The LIN configuration language allows description of a complete LIN network and also contains all information necessary to monitor the network. This information is sufficient to make a limited emulation of one or multiple nodes if it/they are not available. The LIN description file (LDF) can be one component used to generate software for an ECU, which shall be part of the LIN network. An API has been defined by the LIN standard to provide a uniform, abstract way to access the LIN network from applications. The syntax of an LDF is simple and compact enough to be handled manually, but the use of computer-based tools is encouraged. Node capability files (NCFs), as described in LIN node capability language specification, provide one way to (almost) automatically generate LDFs.

54.9 System Definition Process

Defining optimal signals packing and schedule table(s) fulfilling signaling needs in varying modes of operation, with consideration of capabilities of the participating nodes, is called the system definition process. Typically, it will result in generation of the LDF, written by hand for simple systems or generated by high-level network design tools.

However, reusing existing, preconfigured slave nodes to create a cluster of them, starting from scratch, is not that convenient. This is especially true if the defined system contains node address conflicts or frame identifier conflicts. The LIN node capability language, which is a new feature in LIN 2.0, provides

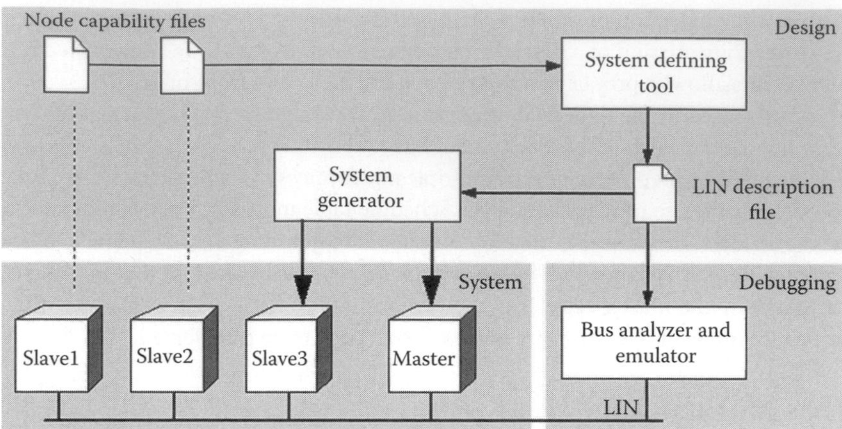

FIGURE 54.3 Workflow.

a standardized syntax for specification of off-the-shelves slave nodes. This will simplify procurement of standard nodes as well as provide possibilities for tools that automate cluster generation. The availability of such nodes is expected to grow rapidly. If accompanied by an NCF, it will be possible to generate both the LIN configuration file and initialization code for the master node. Thus, true plug and play with nodes in a cluster will become a reality.

By receiving an NCF with every existing slave node, the system definition step is automatic: just add the NCF to your project in the system definition tool and it produces the LDF together with C code to configure a conflict free cluster. The configuration C code shall, of course, be run in the master node during start-up of the cluster.

If you want to create new slave nodes as well, the process becomes somewhat more complicated. The steps to perform will depend on the system definition tool being used, which is not part of the LIN specification. A useful tool will allow for entering of additional information before generating the LDF. (It is always possible to write a fictive NCF for the nonexistent slave node, and thus, it will be included.)

An example of the intended workflow is depicted in Figure 54.3.

The slave nodes are connected to the master forming a LIN cluster. The corresponding NCFs are parsed by the system defining tool to generate an LDF in the system definition process. The LDF is parsed by the system generator to automatically generate LIN-related functions in the desired nodes (the Master and Slave3 in the example shown in the picture previously). The LDF is also used by a LIN bus analyzer/emulator tool to allow for cluster debugging.

If the setup and configuration of any LIN cluster is fully automatic, a great step toward plug-and-play development with LIN will be taken. In other words, it will be just as easy to use distributed nodes in a LIN cluster as using a single node with the physical devices connected directly to the node.

It is worth noticing that the generated LDF reflects the configured network; any preexisting conflicts between nodes or frames must have been resolved before activating cluster traffic.

54.10 Debugging

Debugging and node emulation are based on the LDF produced during system definition. Emulation of the master adds the requirement that the cluster must be configured to be conflict free. Hence, the emulator tool must be able to read reconfiguration data produced by the system definition tool.

One example of a comprehensive tool chain built around the open interface definitions of the LIN standard is presented in the following sections of this chapter.

54.11 AUTOSAR and LIN

The Automotive Open System ARchitecture (AUTOSAR) standard partially covers the LIN protocol, covering the master node only. The slave node is not addressed based on the argument that LIN slave nodes do not make use of the AUTOSAR platform. Seen from the bus, the AUTOSAR LIN protocol is identical with the LIN specification. The AUTOSAR LIN software requirements are however different from the original LIN specification. Due to the specific software architecture of AUTOSAR, the LIN interface specifications are not used. Consequently, the configuration API is not used either by AUTOSAR. However, configuration using schedule tables is supported in AUTOSAR.

The relevant AUTOSAR LIN specifications are the following:

LIN interface software specification (*SWS*): Covers the specification of the master node behavior and its interfaces in the AUTOSAR architecture.

LIN network manager: This is today an empty template, since LIN lacks the complex network management defined for CAN.

LIN state manager: Handles the mode change mechanism of the LIN bus.

LIN driver: Handles the μ-controller abstraction to make the other LIN documents μ-controller independent.

LIN transceiver driver: To handle complex LIN transceivers.

54.12 Future

The driving ideas and the resulting technology behind the success of LIN—the structured approach toward the system design process—will most likely migrate to other areas of automotive electronics. LIN itself will find its way to applications outside of the automotive world due to its low cost and versatility. The LIN specification will evolve further to cover upcoming needs. There will be a broad supply of components that are made for LIN. Because of high production volumes, these products can be used cost effectively in many applications, enhancing the functionality of vehicles in a more cost-effective manner.

Fifteen years after the LIN consortium was founded, the number of LIN nodes manufactured has been increasing steadily. Due to the holistic concept that the LIN specifications realize, LIN has gained wide acceptance in the automotive industry.

54.13 Volcano LIN Tool Chain

Process description (Figure 54.4):

1. LIN network requirements are entered into LNA.
2. Automatic frame compilation and schedule table generation by LNA.
3. LDF is generated by LNA.
4. LIN configuration generator tool converts the LDF and private file to target-dependent ".c" and ".h" code.
5. Application code is compiled with target-dependent configuration code and linked to the LIN target package (LTP) library.
6. Analysis and emulation are performed with LIN Spector using the generated LDF.

54.14 LIN Network Architect

LIN Network Architect (LNA) is built for design and management of LIN networks. Starting with the entry of basic data such as signals, encoding types, and nodes, LNA takes the user through all stages of network definition.

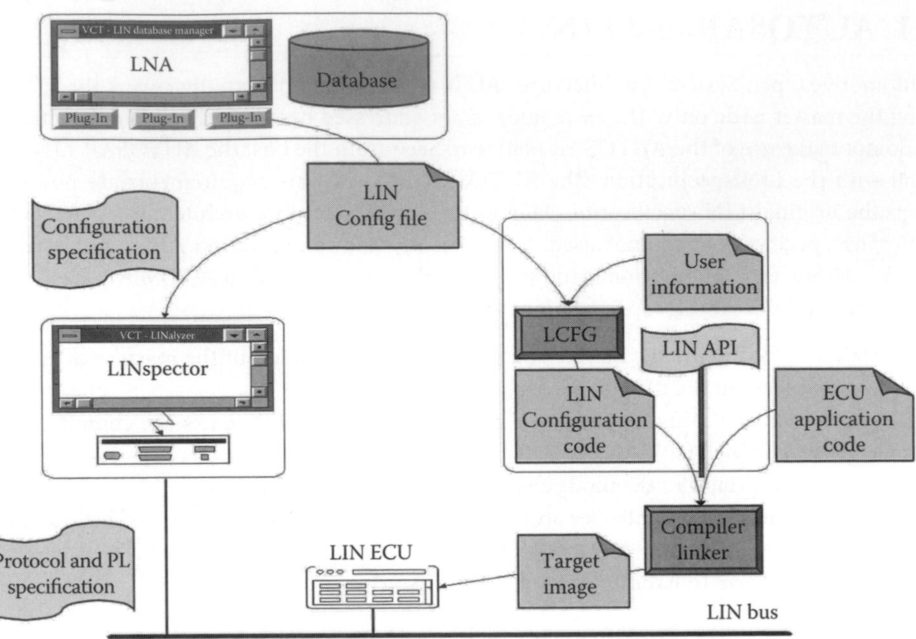

FIGURE 54.4 Volcano LIN tool chain.

54.14.1 Requirement Capturing

There are two types of data administered by LNA:

1. Global objects (signals, encoding types, and nodes)
2. Project-related data (network topology, frames, and schedule tables)

Global objects shall be created first and can then be reused in any number of projects.

They can be defined manually or imported by using a standardized XML input file (based on Fibex rev. 1.0). Future versions of the tool will be able to import data directly from the standardized NCF. Comprehensive version and variant handling is supported (Figure 54.5).

The systems integrator combines subsets of these objects to define networks. Consistency check is being performed continuously during this process. This is followed by automatic packing of signals into frames (Figure 54.6).

The last task to complete is that of generating the schedule table in accordance with the timing requirements captured earlier in the process. The optimization is considering several factors such as bandwidth and memory usage.

Based on the allocation of signals to networks via node interfaces, the tool will automatically identify gateway requirements between subnetworks, regardless whether LIN to LIN or LIN to CAN. The transfer of signals from one subnetwork to another will become completely transparent to the application of the automatically selected gateway node (Figure 54.7).

The tool uses a publish/subscribe model. A signal can only be published by one node but it can be received by any number of other nodes. Different nodes may have different end-to-end timing requirements (Figure 54.8).

The max_age is the most important timing parameter defined in the Volcano timing model. This parameter describes the maximum allowed time between the generation and consumption of a signal involved in a distributed function.

Changes can be introduced in a straightforward manner, with frame definitions and schedule tables automatically recalculated to reflect the changed requirements.

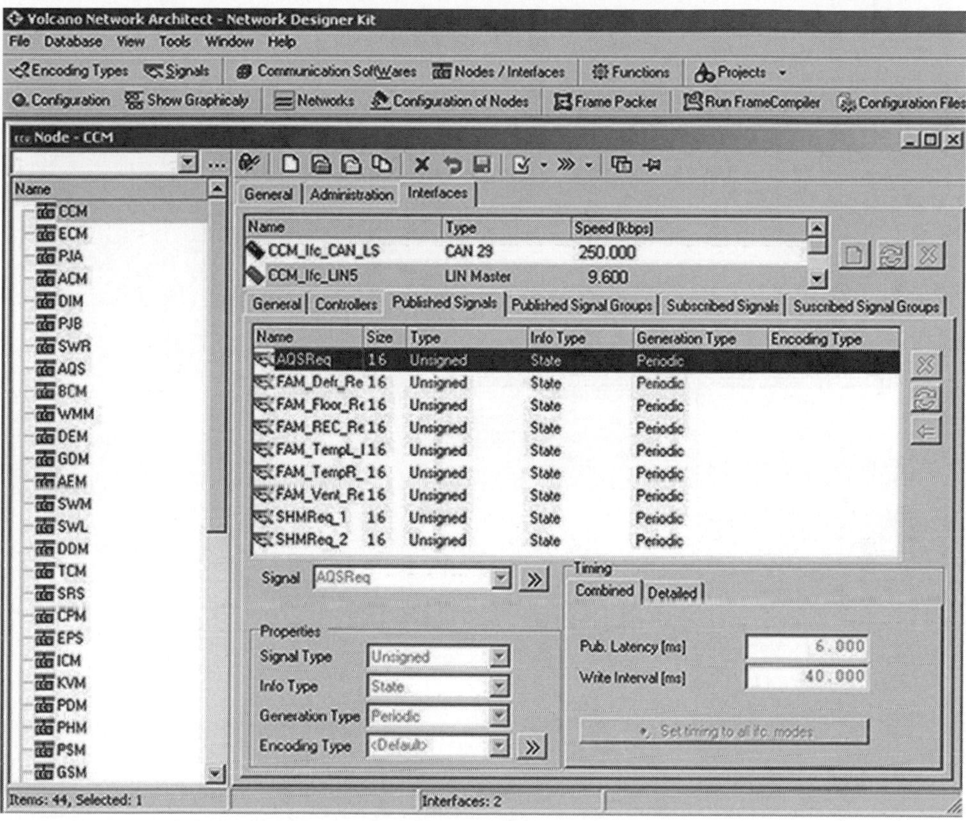

FIGURE 54.5 Definition of global objects in LNA.

When the timing analysis has been performed and the feasibility of the individual subnetworks has been established, LDF will automatically be created for each network (Figure 54.9).

Textual reports can be generated as well to enhance the readability of information for all parties involved in the design, verification, and maintenance process.

54.15 LTP

The LTP represents the embedded software portion of the Volcano tool chain for LIN. The LTP is distributed as a precompiled and fully validated object library also including associated documentation and a command line configuration utility Lin Configuration generator tool (LCFG) with automatic code generation capability, generating the configuration-specific code and set of data structures.

Implementing the LTP with an application program is a simple process. The LDF created by the offline tool contains the communication-related network information. In addition, a target file as an ASCII (American Standard Code for Information Interchange)-based script defines low-level microcontroller information such as memory model, clock, SCI, and other node specifics to the LTP. These two files are run as input through the command line utility LCFG. It converts both files into target-dependent code usable by the microcontroller. The output contains all relevant configuration information formatted into compiler-ready C source code (Figure 54.10).

The target-dependent source code is added to the module build system along with the precompiled object library. After compilation, the LTP gets linked to the application functionality to form the target image that is ready for download.

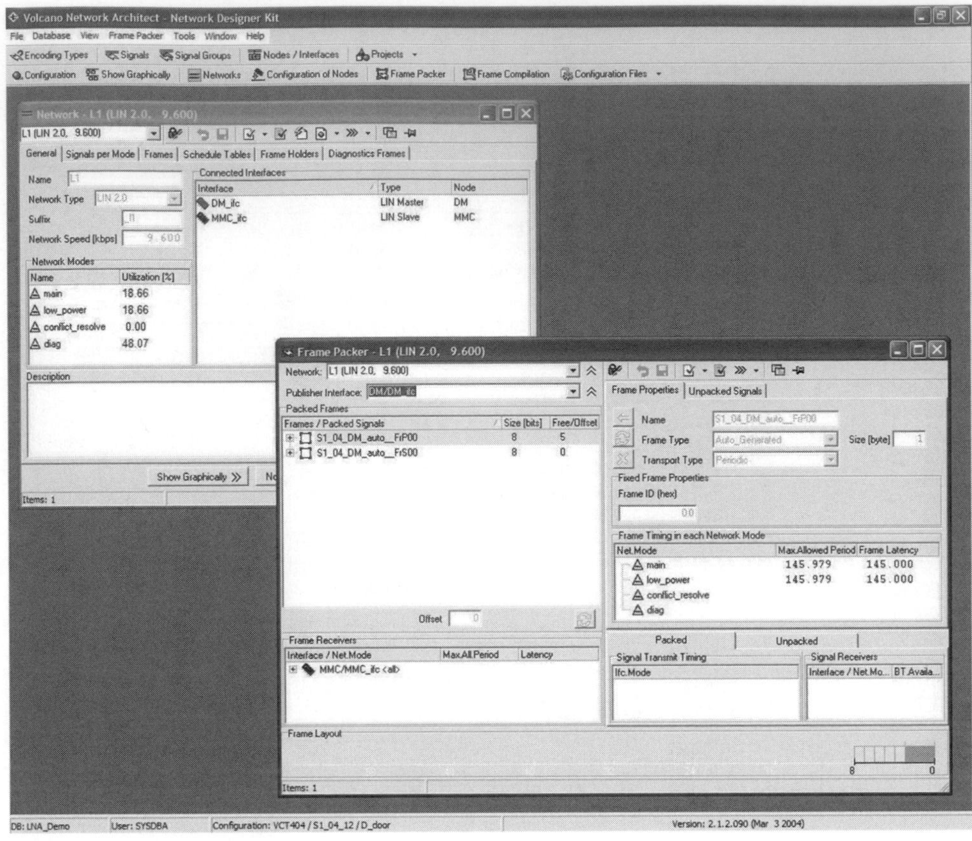

FIGURE 54.6 Network definition and frame packing.

The application programmer can interface to the LTP and therefore to the LIN subnetwork through the standardized LIN API. API calls comprise signal-oriented read and write calls, signal-related flag control but also node-related initialization, interrupt handling, and timed-task management.

The low-level details of communication are hidden to the application programmer using the LTP. Specifics about signal allocation within frames, frame IDs, and others are carried within the LDF so that applications can be reused simply by linking to different configurations described by different LDFs. As long as signal formats do not get changed, a reconfiguration of the network only requires repeating the process described earlier resulting in a new target image without impact to the application. When the node allows for reflashing, the configuration can even be adapted without further supplier involvement allowing for end-of-line programming or after-sales adaptations in case of service.

LTPs are created and built for a specific microcontroller and compiler target platform. A number of ports to popular targets are available and new ports can be made at the customer's request.

54.16 LIN Spector: Test Tool

LIN Spector is a highly flexible analysis and emulation tool used for testing and validation of LIN networks. The tool is divided into two components: an external hardware and PC-based software. Using a 32-bit microcontroller, the hardware portion performs exact low-level real-time bus monitoring (down to 10 μs resolution) while interfacing to the PC via standard RS-232. Other connections are provided for external power and a 9-pin D-Sub for bus and triggering connections. The output trigger is provided for connection to an oscilloscope, allowing the user to externally monitor the bus signalization.

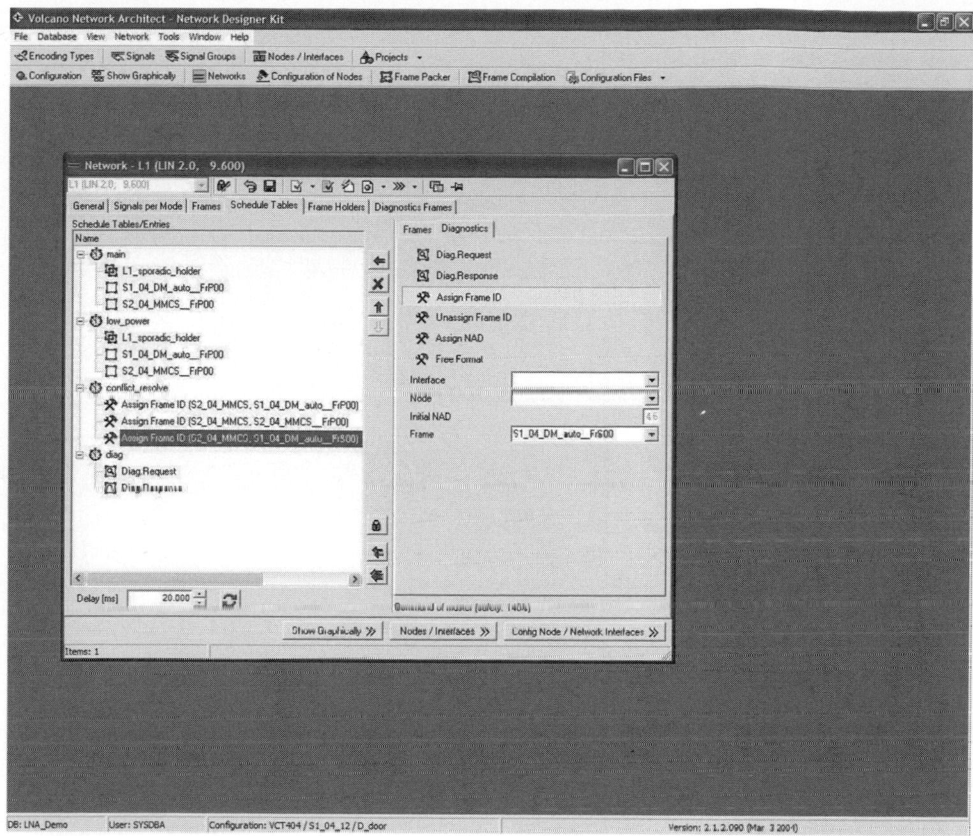

FIGURE 54.7 Manual and/or automatic schedule table generation.

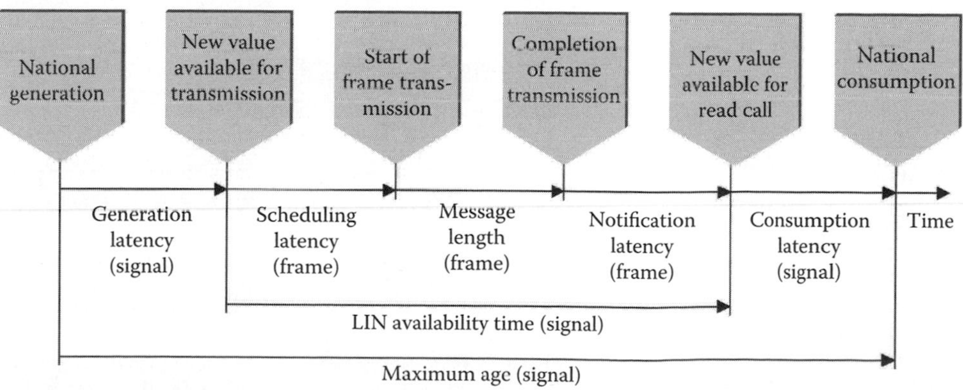

FIGURE 54.8 LIN timing model.

Starting with LDF import, the tool allows for monitoring and display of all network signal data. Advanced analysis is possible with logical name and scaled physical value views. Full emulation of one or many nodes—regardless whether master or slave—is possible using LDF information. Communication logging and replay is possible, including the ability to start a log via logic-based triggers (Figure 54.11).

An optional emulation module enables the user to simulate complete applications or run test cycles when changing emulated signal values and switching schedule tables—all in real time. The functions are specified by the user via LIN Emulation Control (LEC) files created in a C-like programming language.

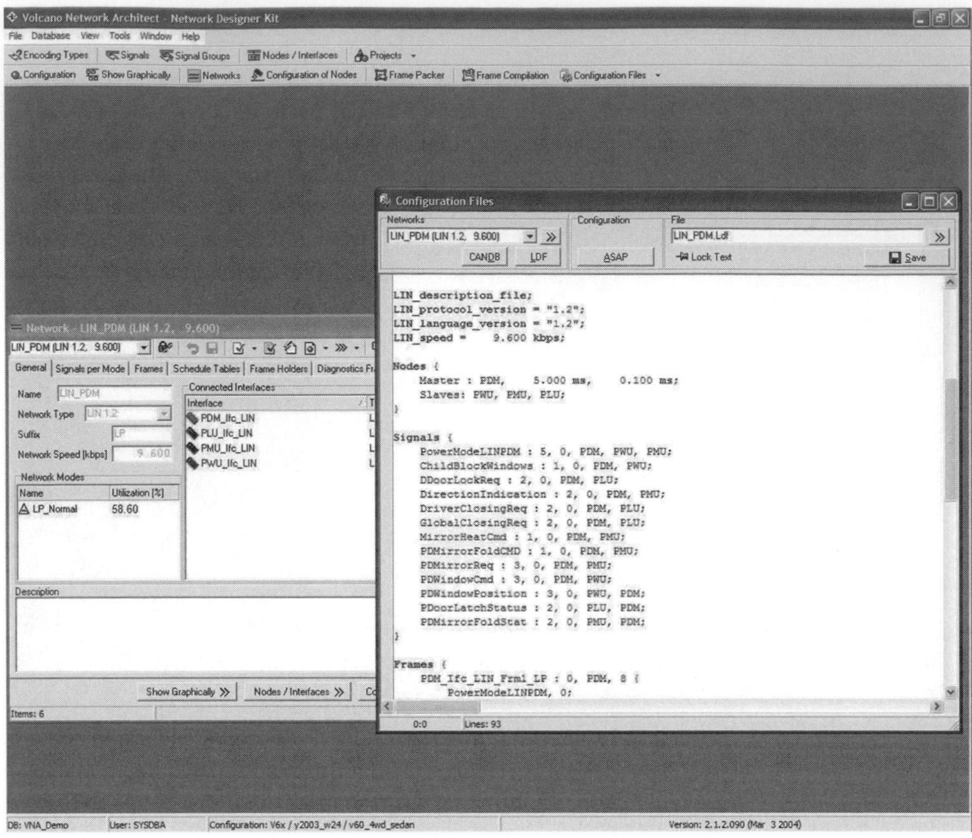

FIGURE 54.9 LDF generation.

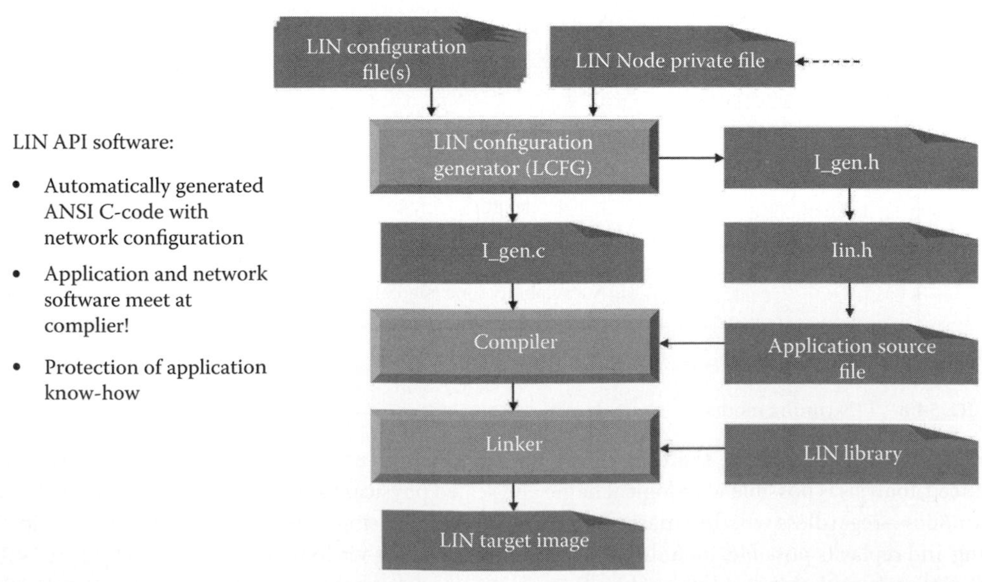

FIGURE 54.10 LTP.

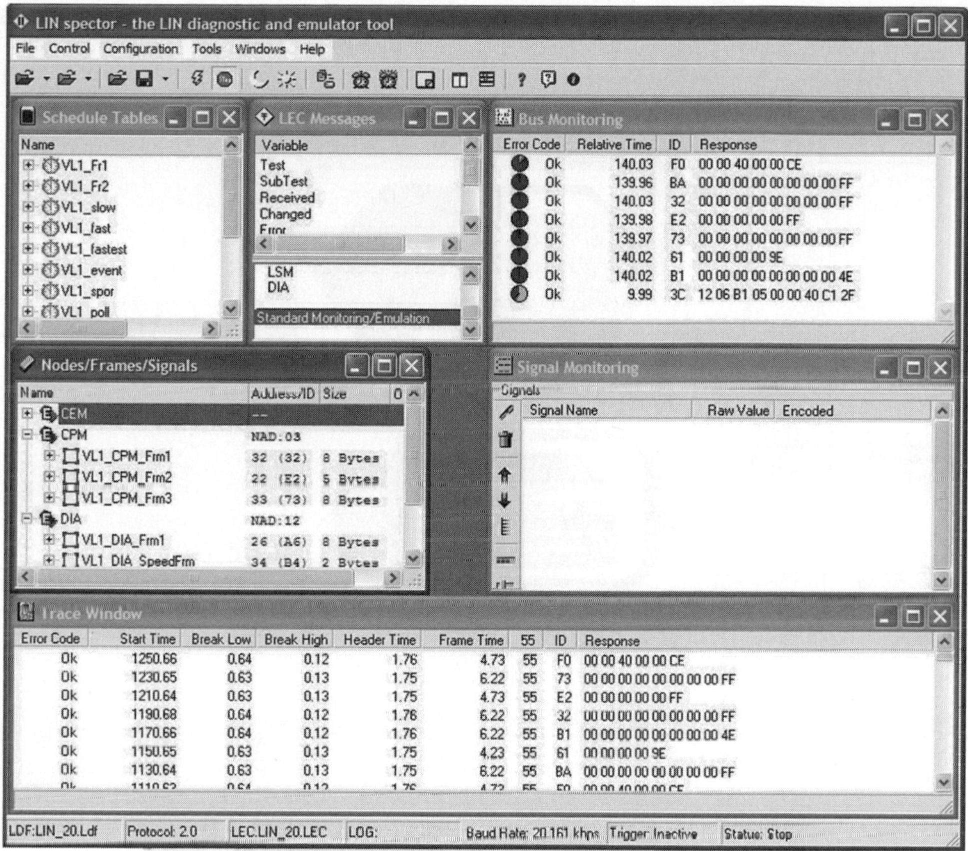

FIGURE 54.11 LIN Spector, diagnostics and emulation tool for LIN.

This can also be used to validate the complete LIN communication in a target module. Test cases are defined stressing bus communication by error injection on bit or protocol timing level.

Sophisticated graphical user interface panels can be created using the LIN Go feature within the test device (Figure 54.12).

These panels interface with the network data defined by the LDF for display and control.

54.17 Summary

LIN is an enabling factor for the implementation of a hierarchical vehicle network to achieve higher quality and reduction of cost for automotive makers. This is enabled by providing best practices of software development to the industry: abstraction and composability. LIN allows for reduction of the many existing low-end multiplex solutions and for cutting the cost of development, production, service, and logistics in vehicle electronics.

The growing number of car lines equipped with LIN and the ambitious plans for the next-generation cars are probably the best proof for the success of LIN. The simplicity and completeness of the LIN specification combined with a holistic networking concept allowing for a high degree of automation has made LIN the perfect complement to CAN as the backbone of in-vehicle communication. Some of the market Potential is related to the opportunity to downsize parts of the vehicle network from CAN to LIN, where limited communication requirements allow for that.

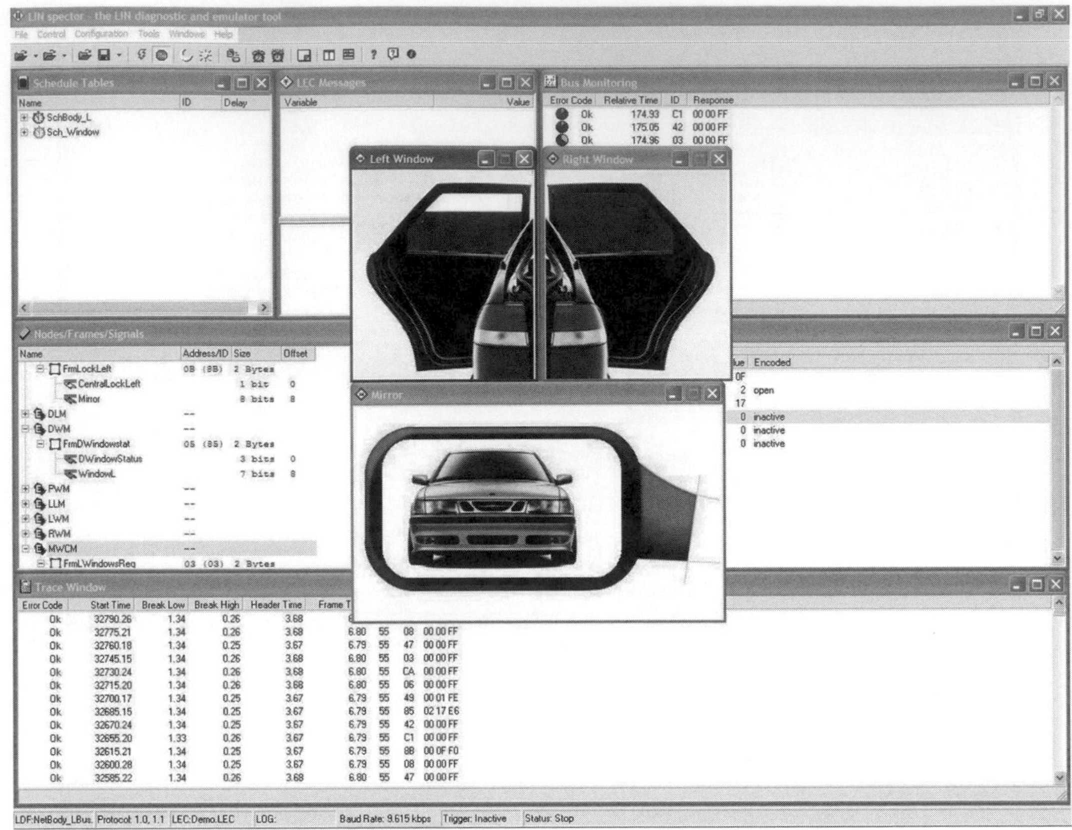

FIGURE 54.12 LIN GO—graphical objects.

The release of LIN 2.0 has further enhanced the reuse of components across car manufacturers and has added a higher degree of automated design capability by introduction of node capability description files and by defining mechanisms for reconfigurability of identical LIN devices in the same network.

VCT is offering the corresponding and highly automated tool chain to guarantee design to correctness. This shortens design cycles and—as a conceptual approach—allows for integration into higher-level tools. LIN solutions provide a means for the automotive industry to drive new technology and functionality in all classes of vehicles.

Acknowledgments

I wish to thank Hans-Christian von der Wense of Freescale Semiconductor Munich and István Horváth and Thomas Engler of Volcano Automotive Group for their contributions to this chapter.

References

Reichart, G., LIN—A subbus standard in an open system architecture, *First International LIN Conference*, Ludwigsburg, Germany, September 19, 2002.

International Standard ISO 9141, Road vehicles—Diagnostics systems —Requirement for interchange of digital information, 1st edn., 1989.

Specks, W. and A. Rajnák, The scaleable network architecture of the Volvo S80, *Eighth International Conference on Electronic Systems for Vehicles*, Baden-Baden, Germany, October 1998, pp. 597–641.

Specks, J.W. and A. Rajnák, LIN—Protocol, development tools, and software interfaces for local interconnect networks in vehicles, *Ninth International Conference on Electronic Systems for Vehicles*, Baden-Baden, Germany, October 5–6, 2000.

Additional Sources

The LIN Specification Package and further background information about LIN and the LIN Consortium is available via the URL: http://www.lin-subbus.org

Since 2010, the maintenance of the LIN standard has been transferred to ISO.

The new document reference is: ISO/CD 17987, parts 1–7.

The diagnostic part was moved to ISO/CD 14229-7.

Information about LIN products referred to in this chapter is available via the URL: http://www.mentor.com

Building Automation

55 **State of the Art in Smart Homes and Buildings** *Wolfgang Kastner, Lukas Krammer, and Andreas Fernbach* **55**-1
Introduction • Functional Model and Reference Architectures • Building Automation Technologies • Integration • Conclusion • References

56 **Fundamentals of LonWorks®/CEA-709 Networks** *Dietmar Loy and Stefan Soucek* **56**-1
Distributed LonWorks® Networks • Node Architecture • Programming Model • Communication Elements • Communication Topologies and Media • ANSI/CEA-709 Protocol Standard (LonTalk) • Network Infrastructure • Interoperability and Profiles • Development and Integration Tools • New Developments for ANSI/CEA-709 Networks • References

57 **BACnet** *Frank Schubert* **57**-1
BACnet as a Standard • BACnet: The Idea of an Open Standard • Basic Idea of BACnet: Interoperability! • BACnet Organizations • Three Major Parts of BACnet • BACnet Network Numbers • Device Address Binding • BACnet Transaction Handling • Encoding BACnet Telegrams • BACnet Procedures • Proprietary Extensions • BACnet Conformance • BACnet Testing, Listing, and Certification • Interoperability Workshops • Character Sets • Network Security • BACnet/IP Specific Functionality • MS/TP Specific Functionality • Vendor IDs • BACnet Mappings • Conclusion

58 **KNX: A Worldwide Standard Protocol for Home and Building Automation** *Michele Ruta, Floriano Scioscia, Giuseppe Loseto, and Eugenio Di Sciascio* **58**-1
Introduction • KNX: Short History • KNX System Specification • Development and Configuration • Research Perspectives • References

59 **Future Trends in Smart Homes and Buildings** *Wolfgang Kastner, Markus Jung, and Lukas Krammer* **59**-1
Introduction • Application Areas • Future Internet of Things Protocols • IoT Protocol Stacks and System Architectures • Conclusion • References

55

State of the Art in Smart Homes and Buildings

Wolfgang Kastner
Vienna University of Technology

Lukas Krammer
Vienna University of Technology

Andreas Fernbach
Vienna University of Technology

55.1 Introduction 55-1
55.2 Functional Model and Reference Architectures 55-2
Automation Hierarchy • Two-Tier Architecture • Bridging the Networks
55.3 Building Automation Technologies 55-4
KNX • LonWorks • BACnet • Modbus • ZigBee • Trade-Specific Technologies in Building Automation • Home Automation Systems
55.4 Integration 55-12
OPC • BACnet/WS • oBIX • DPWS
55.5 Conclusion 55-18
References 55-19

55.1 Introduction

Building automation systems (BASs) are historically grown systems. The core domains of BAS are heating, ventilation, and air conditioning (HVAC) as well as lighting and shading applications. Nowadays, BASs are a well-established way to provide automatic control of indoor conditions in functional buildings [4]. The overall aim is to improve comfort, save energy, and thereby reduce costs arising during the lifetime of a building [1,5]. Relatively new fields of operation where BAS gain importance are security applications like access control. Also, safety-critical applications like fire alarm systems are more and more integrated.

The evolution of communication systems also influences the automation domain which consequently leads to a change of paradigms in automation structures and models. While in the past a centralized control approach was followed, today's automation systems typically adhere to a distributed control approach. Section 55.2 presents an overview of hierarchical models in automation system and their evolution in recent times. For deploying automation systems, many different technologies and products are available. Unfortunately, there exists no technology that can be used for all automation domains at all levels of the control hierarchy. Some technologies are only used for a dedicated hierarchy level (e.g., fieldbus systems), whereas others are also domain specific (e.g., HVAC). Independent of the hierarchy level and the application domain, automation technologies are either based on open standards and specifications or exclusively provided by one manufacturer. Obviously, open standards are the basis for interoperability among different manufacturers. They further guarantee availability of products within the life cycle of a building. Based on these criteria, important representatives of automation systems of different application domains and control levels are presented in Section 55.3. Since no automation system exists that is able to address all domains and levels, interaction among different technologies is crucial for BAS. Besides pure physical interoperability, data are only relevant if denotation and representation is uniquely defined, too. With the gaining importance of the Internet and according network technologies, the demand for integrating automation systems into web-based communication becomes more and more important. Thus, Section 55.4 discusses technologies and standards that allow a manufacturer-independent interaction, control, and visualization whereby also web-based access and interaction are addressed.

55.2 Functional Model and Reference Architectures

55.2.1 Automation Hierarchy

Large-scale BASs are hierarchically structured and typically follow a three-layer approach as depicted in Figure 55.1. At the lowest level, the so-called field-level, interaction with the process under control takes place (i.e., measuring, metering, switching, positioning). Sensors and actuators are connected to a field communication system, which is interfaced by direct digital control (DDC) stations. A DDC system collects and distributes data from sensors and actuators and controls local applications. For controlling complex applications that are distributed over more than one DDC, they cooperate at *automation level*. At the top level of this hierarchy, the *management level* acts as an interface to *building management* (*BM*) and enterprise applications. Centralized access to data points, configuration of the system, visualization, archiving and trending of process data are typical activities at this level. Another task of the management level is to provide interoperability between different systems and technologies used at the lower two tiers of the BAS. The overall application is defined as functions over data points. The relevant world standard [3] divides these functions into three categories: input/output, processing, and management.

All these three levels together form a commonly agreed model of BAS, the automation pyramid. The pyramid shape reflects the hierarchical and centralized control flow from the top level to the bottom level. Data reporting, on the other hand, takes place in the inverse direction. Equipment of an

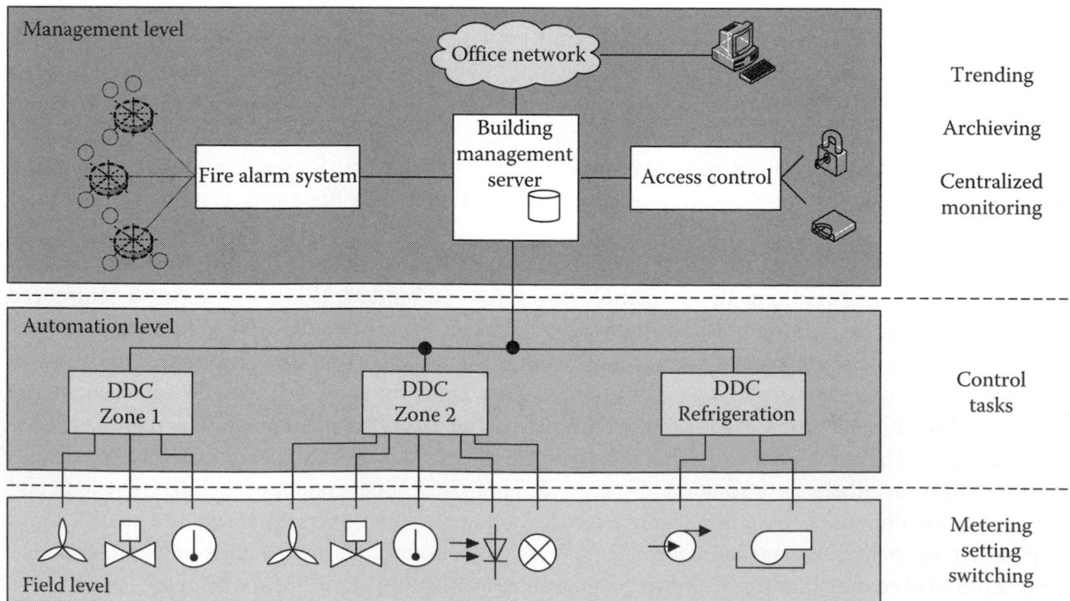

FIGURE 55.1 Three-layer automation pyramid model. (Adapted from Kastner, W. und Neugschwandtner, G., *Datenkommunikation in der verteilten Gebäudeautomation*, Bulletin SEV/VSE, 2006.)

underlying tier provides data to the superior one. Another aspect that is illustrated by the pyramid is the number of devices operating in each level. A high amount of relatively plain and cheap devices can be found at the field level. The number decreases when going upward, whereas the complexity and also costs for a single component rise.

55.2.2 Two-Tier Architecture

In recent times, the previously described three-layer model has been mutated to a two-tier architecture [37]. Due to the increasing processing capabilities and the decreasing costs of embedded systems, sensor and actuator devices are no more restricted to collecting and distributing data. In contrast, they are equipped with powerful hardware and adopt control tasks from formerly used DDCs. Thereby, the level of control changes from the automation level to the so-called control level, which combines the automation and field level. Consequently, the control approach changes from a strictly centralized nature as described in the previous section to a distributed one. At the control level, components act as input/output devices as well as local control instances of a distributed application. They cooperate in so-called control networks. For wide-spanning control applications across different control networks, these are interconnected by *backbone* networks. However, control is still performed at the lower level. The backbone network is further used to provide management communication such as visualization, trend logging, and high-level access.

55.2.3 Bridging the Networks

As mentioned previously, BASs are hierarchically structured and make use of a high diversity of technologies and standards. Besides the use of different technologies in each hierarchy level, different technologies are even used at the same level. In particular at the field level, a plethora of communication and automation protocols are facilitated within one system. To establish a connection between network segments using different technologies, so-called gateways are used. They can establish horizontal connection between two different technologies at the same hierarchy level or likewise connect different technologies used at different automation levels. Gateways can also provide a physical connection among different network media. For example, a gateway can establish a connection between a wireless network and a wired network. The more complex task of a gateway is the semantic cooperation. A gateway has to receive and interpret information from one network and translate and distribute it in another network. Depending on the differences regarding data representations, this causes a lot of processing and engineering effort. Thereby, detailed knowledge of the protocols in use is required. Also, changes or extensions of the protocol influence the behavior of the gateway.

If two network segments that base on the same technology shall be interconnected by an intermediate network, a *tunneling* connection is used. The intermediate network acts as a virtual medium such that so-called tunneling routers forward information in a transparent way. Due to the transparent connection, tunneling routers do not have to interpret or translate information. Thus, tunneling routers do not require as much processing capability as gateways. They are also aware against changes or extensions of the network protocol that has to be tunneled. Additionally, they only forward received data to the addressed tunneling router. In contrast to gateways, tunneling devices do not interpret or translate information, they transparently forward data. Due to the transparent nature of a tunneling connection, even engineering data can be transmitted by the tunneling connection. However, tunneling can cause several communication issues. If two network segments are connected transparently, it needs to be considered that the bus load in the joint network rises to the sum over the loads of each segment since no functional separation of the subnets takes place. To overcome this issue, filtering technologies can be used, which imply the necessity of interpreting data. It has to be further ensured that the intermediate connection is reliable. This essentially includes ordering of data packets and prevention of packet duplication. If more than two subnets are connected, additional filtering and routing concepts have to be considered.

55.3 Building Automation Technologies

After discussing different models of building automation, this section introduces a selection of the most important technologies and standards used in the context of building automation. At first, the big players in building automation are discussed according to the ISO/OSI reference model [9]. These technologies are widely accepted for years or even decades of years. The presented standards are open standards that are applicable in almost all automation domains. Further, well-known representatives of domain-specific technologies are presented. Finally, a number of technologies are listed and briefly described that are mostly used in home environments (*smart homes*). Some of the presented technologies are not based on open standards, but they are mentioned because of their relevance in a dedicated domain.

55.3.1 KNX

KNX is a well-accepted building automation technology, in particular in Europe and German-speaking countries (DACH). It is standardized under IEC 14543-3/EN 50090 [22,23] and maintained by the KNX Association [18]. KNX is typically used in the HVAC domain as well as for lighting and shading applications. However, it can also be used for other applications such as metering and access control. KNX is basically used for field-level communication and control but also specifies IP-based backbone communication mechanisms. It provides a framework for distributed control as well as a flexible communication system. KNX allows manufacturer-independent communication and control. For this purpose, KNX provides a common engineering tool software (ETS) that simplifies the engineering and integration process and allows the configuration and integration of components of different manufacturers. Nevertheless, it is also possible to configure small-scale installations without the use of the ETS by an autoconfiguration mode.

The KNX communication system is designed according to the ISO/OSI reference model [19]. At the physical layer (PL), KNX provides different media for communication, still the most important medium is based on a simple twisted-pair (TP) bus cable (KNX TP1). Due to its low bandwidth, the medium is robust against electromagnetic disturbances and can be installed in nearly any topological structure. A further medium called KNX PL110 facilitates the power line for communication. Thus, no additional communication infrastructure is necessary. However, if PL110 is used, additional filters are required to suppress high-frequency disturbances of the power grid. Repeaters are necessary in three-phase installations. If no wired medium is feasible, KNX supports also a wireless medium called KNX RF. This medium resides in the 868 MHz *industrial, scientific and medical* (ISM) frequency band and can be used without any license fees. Additionally, KNX is able to use the Internet Protocol (IP) as logical medium. Thereby, KNX devices can directly be connected to an IP capable medium.

While the lower part of the data link layer (DLL) depends on the communication medium and thus is responsible for specific medium access control and framing, the upper part of the DLL is independent of the medium and defines the KNX addressing scheme. This scheme foresees unicast, multicast, as well as broadcast addressing. Especially multicast communication allows designing functional and efficient applications. Unicast addresses, so-called individual addresses, allow point-to-point communication among KNX devices, whereas group addresses are used for multicast communication. Additionally, broadcast communication is possible by using a dedicated broadcast address. All KNX addresses have a length of 16 bits and are hierarchically structured. The network layer (NL) is responsible for data exchange among the levels of the communication hierarchy by so-called line couplers and backbone couplers. At the lowest level, a line contains up to 256 devices. Up to 16 lines are connected to a main line and up to 16 main lines are linked to a backbone line. At the transport layer (TL), connection-oriented (unicast) and connectionless communication channels are established. Additionally, the TL is responsible for reliable data transmission. While the session layer (SL) and the presentation layer are not implemented, the application layer (AL) provides an abstract communication interface for implicit communication by using objects and properties. The most important objects are group objects (GOs).

These GOs can be written and read by multiple devices that allow efficient communication. For realizing interoperable applications among devices of different manufacturers, datapoint types (DPTs) are specified. A DPT defines the format and encoding as well as the range and unit of a value. A functional block (FB) is a collection of different DPTs. Besides the input and output of data, an FB specifies a dedicated functional behavior. In an FB, DPTs can be used either as inputs or outputs. An application can be built by interconnecting FBs via their input and output objects of the same DPT, respectively. In most cases, GOs are used for instantiating DPTs, but it is also considerable that other object types like interface objects (IOs) are used.

An extension to KNX allows encapsulating KNX packets into IP frames. This extension, called KNXnet/IP [24], offers the possibility to interconnect KNX devices by a logical IP channel as mentioned earlier. Additionally, KNXnet/IP can be used as tunneling connection to connect two lines or to allow remote configuration and management of a KNX network. KNXnet/IP can also act as backbone line and connect multiple KNX segments by using KNXnet/IP routing. Since KNX uses open and commonly used media (e.g., RF and IP) and conquers security-relevant application domains, security gains importance. To address this, the KNX Association has recently published draft versions of security extensions to KNX [20,21].

55.3.2 LonWorks

Similarly to KNX, LON is an open standard that is used in building automation networks. Initially, LON was specified by the Echelon Cooperation. It was accepted as ANSI standard ANSI/CEA-709.1 [13,14] and became a European norm (EN 14908 [15]). The LON standard is maintained by the LonMark Association [17] and defines the communication as well as data representation and hardware [17]. The communication protocol is known as LonTalk. A special integrated circuit called *Neuron chip* implements the whole communication stack and supports the integration of small user applications. Thanks to the open specifications, second source protocol implementations exist besides the Neuron chip. Moreover, variants with integrated transceivers are available. LON also allows tool-based engineering and commissioning. For this purpose, the so-called LonMaker integration tool can be used. With this tool, networks can be graphically designed, devices can be configured, and applications can be deployed.

The LON standard is specified according to the ISO/OSI reference model, whereby all layers are implemented. At the PL, LON allows different types of media. The most popular medium is TP, which enables flexible wiring in arbitrary physical topologies. Additionally, power line communication can be used, which does not require any further cabling. However, additional hardware is required for phase bridging and filtering. Besides these most important media, communication is further specified via EIA 485, coax cable, and fiber optics. One physical segment is denoted as channel, whereby such channels can be coupled by bridges. The DLL is responsible for media access control as well as frame forming and encoding. LON uses carrier sense multiple access (CSMA) for media access and allows prioritized messages altogether labeled as *predictive p-persistent CSMA*. At the NL, LON specifies addressing and defines multiple network structures. These structures are independent of the physical topology. A LON device can be identified by a globally unique identifier (ID). A LON network is denoted as domain, whereby communication among domains is not foreseen. Within a domain, a device can uniquely be identified by a subnet and a node ID. A device can further be a member (member ID) of multiple groups (group ID), which enables multicast communication. To establish a connection among different subnets, routers are used. These routers are self-learning and also support multicast communication. The TL is responsible for reliable unicast and multicast transmission. It implements an acknowledgment mechanism and is able to detect duplicates and reordered packets. Additionally, it provides message authentication. The SL provides a request/response protocol between client and server nodes, which is the basic communication primitive for upper layers. The presentation and application layer provide implicit communication by so-called network variables (NVs). Besides, services for network management and maintenance are provided.

To guarantee interoperability among devices of different manufacturers, many NVs are specified as standard NV types (SNVTs). A functional profile (FP) contains a number of SNVTs, which either act as inputs or outputs. Besides mandatory SNVTs, an FP contains optional ones and additional configuration properties (some of them are also optional). In addition to the input, output, and configuration parameter, an FP implements a dedicated functional behavior. An FB is the implementation of an FP in a device, whereby a device may consist of different FBs. A distributed application can be realized by connecting the SNVTs of different FBs. This task is also known as *binding*.

LonTalk defines a weak security mechanism based on a 48-bit message authentication code, whereby the authentication algorithm is kept secret. Confidentiality and authorization are not supported. Thus, the security mechanisms are not sufficient for today's applications. For establishing a connection between two subnetworks or enabling remote management, an IP-tunneling standard [16] can be used.

55.3.3 BACnet

Building Automation and Control Network (BACnet) is an open communication standard for BASs. It enables the exchange of data between systems of multiple vendors to provide interworking. Among others, the standard defines services to access data points, to generate alarm and event notification on specific conditions and for network control purposes.

BACnet was developed by the *American Society of Heating, Refrigerating, and Air Conditioning Engineers (ASHRAE)* and was standardized in 1995. Initially designed for the use at the management and automation level of the three-tier automation hierarchy, BACnet has found its use in all kinds of building automation applications, nowadays. The current standard is BACnet 2010 [11]. ISO 16484-5:2012 [12] incorporates BACnet 2010.

BACnet implements the PL, the DLL, and the NL as well as the AL of the ISO/OSI model. The PL and the DL, which are affected by the selection of a so-called network option, are independent protocol standards and therefore not part of the BACnet specification. In the original BACnet standard, five network options describing the physical and the data-link layer are defined: Ethernet, ARCNET, master–slave/token passing (MS/TP), point-to-point (PTP), and LonTalk. Later on, two further network options have been defined, namely, BACnet/IP and ZigBee as a wireless network add-on. The choice of a network option does not influence the upper two protocol layers. It is even possible to use other PL/DLL combinations since BACnet is not limited to these network options. This allows the combination of multiple network technologies in one BACnet network and thus, interworking is provided at the upper protocol layers.

To allow remote devices to access process data, a *network-visible* representation of the stored data has been specified by BACnet. This representation follows an object-oriented approach. Up to now, 50 different *BACnet object types* are defined within the current standard. They differ in the composition of their so-called BACnet properties, which can be seen as data points. Each property has a unique ID referred to as `Property_Identifier`, a designated property data type, and a conformance code attribute. The conformance code defines the access permissions of a property and specifies whether a property is mandatory in a distinct BACnet object or not. Valid values are readable (R), writable (W), and optionally present (O).

Vendors of BACnet devices are free to define their own proprietary object types (referred to as nonstandard object types)—even the definition of proprietary property types is possible. However, there are three mandatory properties that must be defined for each BACnet object: `Object_Identifier`, `Object_Name`, and `Object_Type`. The former two properties must be unique within a BACnet device. Since a BACnet object is always assigned to exactly one device (a BACnet object is never distributed across more than one device), the `Object_Identifier` or `Object_Name` property is used to uniquely identify a BACnet object within a device. The `Object_Type` property is actually an enumerated value, which defines the kind of an object.

The most frequently used BACnet object types are generic ones, for example, the `BACnet Binary Output Object Type` or the `BACnet Analog Value Object Type` representing physical and virtual data points, respectively. However, there exist also more advanced concepts in BACnet, like access to historical data (`Trend Log Object Type`), event scheduling (`Schedule Object Type`) or access control applications (`Access Door Object Type`). It is within the responsibility of the application program to map the control data of a dedicated application to certain BACnet objects. For primitive input or output objects, the `Present_Value` properties represent the signal values of the physical I/O ports of a BACnet controller. Other ones like the `Resolution` or the `Output_Type` properties provide additional meta-information about the physical quantity.

Services in BACnet are based on the client/server communication paradigm. BACnet controllers, which act as servers, hold objects representing physical values of the processes. User control units or management workstations—where values are set and process data are visualized and archived—act as BACnet clients. The complete set of service classes defined in BACnet consists of the following:

- *Object access services* allow manipulation, creation, and destruction of BACnet objects and are also used for getting (`ReadProperty Service`) or setting (`WriteProperty Service`) the value of a distinct property.
- *Alarm and event services* encompass mechanisms for event-based communication. The change of value (COV) subscription (`Confirmed-/UnconfirmedCOVNotification Service`), for example, is used to monitor data points on BACnet controllers by clients like supervisory devices. Other representatives of this class are the `Intrinsic Reporting` services, which allow the communication of fault conditions or the `Algorithmic Change Reporting` services for generating events on a set of conditions.
- *File access services* can access and manipulate files on BACnet devices.
- *Remote device management services* provide mechanisms, for example, network discovery (`Who-Is` and `I-Am Services`), clock synchronization, or device reinitialization.
- *Virtual terminal services* are rarely used routines defined for establishing terminal sessions between two BACnet devices.

55.3.4 Modbus

Modbus is an open standardized communication protocol founded by Modicon in 1979. It was formerly designed for the communication among programmable logic controllers (PLCs) in industrial applications but gained importance in the building automation domain, especially for HVAC systems, DDCs, and room control. Modbus is a lightweight application layer protocol that facilitates different media. Modbus-TCP, a Modbus version based on TCP, is part of the fieldbus standard IEC 61158 [7].

According to ISO/OSI, Modbus focuses on application layer data exchange. However, different media are specified at physical and data-link layer, respectively. Modbus was originally designed for asynchronous serial data exchange via EIA-485 or EIA-232. While EIA-232 specifies only a point-to-point communication, EIA-485 is capable of bus communication. Thereby, the medium access is based on a master/slave approach, where a master requests a response from a dedicated slave identified by its slave address. Additionally, the master can send data to all slaves via broadcast. In this case, no response of the slaves is allowed. For serial communication, two different transmission modes are possible. On the one hand, the remote terminal unit (RTU) mode specifies a binary protocol for data encoding. On the other hand, in ASCII mode, every data byte is encoded by two ASCII signs.

Additionally, Modbus facilitates Ethernet and TCP/IP as additional medium, which is known as Modbus-TCP. Thereby, Modbus packets are encapsulated in TCP frames and IP addresses are used for identifying devices. A Modbus frame contains an additional Modbus application protocol (MBAP) header, which indicates the length of the packet, a transmission ID, and a protocol ID. Additionally, a unit ID is included in the header to address serial Modbus devices behind a gateway.

The application layer protocol is based on a request/response communication scheme between a client and a server. Thereby, a client sends a request to the server. The server may respond if data were requested.

In Modbus-TCP, a unicast connection between two devices is initiated by the client. In serial Modbus protocols, the master acts as client and requests data from the slaves, which act as servers. The medium-independent data frame consists of a function code and variable data field. The function code specifies the action that the server has to perform. For accessing and exchanging data, Modbus defines a simple object model based on network variables that are either readable or writable and accessible either bit-wise or word-wise. Thereby, Modbus distinguishes between four different input/output types. Independent from the access mode, these types can be overlapped in the memory such that read and write access can be granted. In contrast to other automation technologies, Modbus does not specify any data profiles.

Modbus provides no security mechanism at the common application layer. However, the TCP version specifies an access control mechanism that provides authorization. For this purpose, a list of allowed IP addresses is maintained by a server. Only if the remote client IP address is on the list, requests will be accepted. Unfortunately, since no authentication is used, this mechanism can be simply broken by faking the IP address. Since Modbus-TCP uses a dedicated well-known TCP port, firewalls can be used to increase security of the automation network. Extensions to Modbus have been published that propose concepts for securing the communication [38].

55.3.5 ZigBee

ZigBee is an open wireless standard for low-rate and lossy networks [27]. It is mainly used for home and building automation as well as for smart energy applications. The ZigBee standard is specified and maintained by the ZigBee Alliance [26]. ZigBee provides a reliable and flexible communication infrastructure and specifies a high-level information modeling scheme. It natively supports multihop communication and has self-healing capabilities. In contrast to other wireless technologies, ZigBee has less hardware requirements and demands less power. A ZigBee topology contains three different types of devices. A so-called personal area network (PAN) contains exactly one *ZigBee coordinator*. This device initializes the network, assigns addresses to all other participants, and acts as security manager, if required. Furthermore, a PAN contains *ZigBee routers* and *ZigBee end devices* (EDs). ZigBee routers are able to forward packets dynamically, a basic requirement for building mesh networks. ZigBee EDs have no routing capabilities and have to be directly connected to a ZigBee router or the ZigBee coordinator. ZigBee EDs require less memory than other ZigBee devices because they do not need to store routing information. Additionally, they can switch off their receiver for several time intervals, which reduces the energy demand of the device.

ZigBee is specified according to the ISO/OSI reference model whereby the PL and DLL are based on the IEEE 802.15.4 standard [25]. The ZigBee specification only defines the upper levels of the communication stack. At the PL, the IEEE 802.15.4 standard specifies different channels in different frequency bands. Besides the 2.4 GHz band that can be used nearly all over the world, region-dependent sub-GHz bands are specified, too. The DLL is mainly responsible for medium access and addressing. IEEE 802.15.4 provides two medium access mechanisms, whereby one is based on CSMA and the other one follows a slot-based, deterministic approach. However, ZigBee only uses CSMA for medium access. The addressing scheme consists of a unique 64-bit extended address (EUI-64) and a 16-bit short address in combination with a so-called PAN-ID that identifies the network.

At the NL, ZigBee specifies mechanisms for unicast and broadcast communication. For building mesh networks, ZigBee uses a dynamic and efficient routing approach. This increases the reliability and allows overcoming long distances. For point-to-point messages, ZigBee provides end-to-end acknowledgments and uses the ad hoc on-demand distance vector (AODV) routing protocol. To increase the efficiency in pure sensor networks, ZigBee additionally supports a so-called many-to-one routing mechanism. For broadcast transmission, an efficient passive acknowledgment mechanism is used. Multicast communication is realized by defining so-called multicast groups and filtering broadcast messages accordingly.

The AL provides an abstract communication interface and a powerful information modeling framework. ZigBee devices consist of so-called application objects (*AOs*) that specify the functionality of the device.

Every device contains a specific AO, the so-called ZigBee device object (ZDO). This object defines basic properties of the device such as device type and implements the basic functionality. The communication among AOs is realized by *clusters* and *attributes*. An AO contains a set of clusters whereas clusters consist of a set of attributes and commands. For example, the *OnOff* cluster has one attribute that represents the state of an arbitrary device (i.e., on/off) and several commands for turning the device on or off or changing the state somehow. For interoperability reasons, the definition of clusters is specified in the so-called ZigBee cluster library (ZCL) according to their functional domain. Nevertheless, clusters can also be defined application specific. For proving interworking at device or application level, application profiles are defined. These profiles act according to particular application domains such as *home automation*. An AP contains the definition of devices and specifies the interaction between devices. A device consists of dedicated AOs with instantiated clusters. In addition to clusters out of the ZCL, an AP can further specify application specific clusters. For example, an AP defines devices like on/off switches and lighting actuators. In this case, both devices contain an instantiated *OnOff* cluster representing the state of a light.

Since ZigBee uses an open medium for communication, security is a crucial topic. Although IEEE 802.15.4 provides optional security services, ZigBee defines its own, independent ones. In ZigBee, security services are located at two different layers. However, security is optional at both layers. At the NL, basic security mechanisms are located that provide network-wide security. At the AL, the recent version of ZigBee provides an advanced security mechanism including end-to-end security, key establishment, as well as key distribution and management by a so-called security manager. While all these mechanisms are based on the symmetric advanced encryption standard (AES) [39], the *smart energy* AP defines additional asymmetric encryption mechanisms for key distribution.

55.3.6 Trade-Specific Technologies in Building Automation

55.3.6.1 DALI

A major part of building automation concerns about lighting. Besides simple switching, demand-specific and energy-efficient lighting is an important topic in modern buildings. Digital Addressable Lighting Interface (DALI) as ratified in IEC 60929 [6] is an open interface specification that exactly addresses this domain. This interface allows a manufacturer-independent integration of lighting systems. DALI is based on a twisted-pair medium, the so-called DALI loop, which allows flexible network topologies and link-powering sensor devices. The medium access is performed by a master/slave mechanism, whereby at most 64 slaves are possible. Additionally, a multi-master mode is provided. Since DALI covers the whole lighting domain, different types of devices are required. These components are categorized according to their functionality. Sensors can be used for motion detection or brightness measurement. Panels such as light switches or remote controls allow direct control by the user. The category *DALI gear* consists of devices that directly control lights such as dimmers. Devices that establish a connection to analog-controlled actuators or gateways to other BASs (e.g., KNX) are part of the category *interfaces*.

55.3.6.2 SMI

For installing and controlling drives in combination with various shutters and blinds (exterior as well as interior sun protection), a special interface called standard motor interface (SMI) exists [29]. The interface specification includes wiring and switching schemes. Besides the electrical and mechanical properties, a digital communication interface is specified that is based on a master/slave communication scheme. For this purpose, two wires of a five-wire cable are used allowing a bidirectional data transmission with a transmission rate of 2400 bit/s. The remaining wires are used as power supply for the drives. Due to the digital control interface, it is possible to control up to 16 drives by connecting them in parallel. The protocol supports individual addressing, group addressing, and broadcast communication. Besides sending incremental and absolute movement commands and querying of current positions, drive parameters and diagnostic telegrams can be transmitted.

55.3.6.3 EnOcean

EnOcean is a lightweight and ultralow-power wireless technology for building automation applications. It is mainly used for lighting applications but also provides sensors and actuators for the HVAC domain. EnOcean was developed in 2001 by the EnOcean GmbH, a spin-off company of Siemens AG. EnOcean is an emerging technology that still gains importance in the building automation domain. This is mainly caused by the special properties regarding energy demand and simplicity. Due to the ultralow-power consumption, EnOcean is used for devices that draw their energy from the environment, which is also known as *energy harvesting*. For instance, wireless buttons can be powered by the piezoelectric or electromagnetic energy gained from pressing the corresponding button. Other devices use solar cells or thermoelectric effects.

EnOcean was laid down as international standard IEC 14543-3-10 in 2012 [8]. It uses short telegrams of 14 bytes and supports unidirectional as well as bidirectional communication. Thus, instead of an acknowledgment mechanism, data packets are retransmitted at random intervals. Since EnOcean does not support mesh networking due to the limited communication resources, a backbone network such as KNX or ZigBee is required. For this purpose, gateways are used that establish the connection. To guarantee interworking among different manufacturers, EnOcean defines so-called EnOcean equipment profiles. These profiles specify the representation of data and telegram types. Until now, these profiles are not part of the IEC standard.

55.3.6.4 M-Bus

Technologies for metering applications in homes and buildings are important for building smart grids. For this purpose, a specific standard called *Metering Bus* (M-Bus) was ratified as EN 13757 [10] that supports network-based reading of various utility metering systems. Next to the classical wired bus system, a wireless pendant called *wireless M-Bus* (wM-Bus) exists. While the communication in wired standard is based on a master/slave medium access scheme, wM-Bus uses different transmission modes in the 868 MHz ISM band according to IEC 60870-5-2 [41]. Additionally, wM-Bus includes a security mechanism to protect transmitted data over the open medium. Communication is established between two types of devices in a client/server manner. In a wired system, usually the master represents the client and requests data from a slave that can be seen as server. Thereby, the metering device represents the server and the client acts as data collector. The latter displays or possibly forwards metering data (e.g., to a smart grid interface). Nevertheless, it is possible that metering devices actively initiate a communication.

55.3.7 Home Automation Systems

Adopting automation technology is equally attractive in the residential domain. Besides efficiency considerations, increased comfort and peace of mind are key motives here. Also, elderly people can live in their own homes longer (ambient assisted living, smart home care). The importance of this aspect continuously increases as life expectancy does.

Despite having a considerable number of things in common, there is a difference between home automation (residential settings) and building automation (functional buildings). Since mechanical ventilation is seldom present in residential settings (although continuing attention to low energy consumption can be expected to change this situation in the long run), lighting and shading control have a more visible position in home automation than HVAC. Considerable emphasis is placed on the integration of consumer electronics (entertainment systems) and household appliances.

Building automation focuses on the goals of energy efficiency, optimized management, and detailed accounting. While energy efficiency is a growing concern in the residential domain as well, the additional comfort and peace of mind a home automation system promises are still influential decision factors. One obvious difference is the scale of the system: The number of devices involved in a home automation project is by orders of magnitude lower than for large building automation projects.

Nonetheless, the complexity of home automation projects must not be underestimated, not at least since a large number of application domains must be integrated. All mentioned field, automation, and management functions are relevant. Equipment, commissioning and maintenance costs have to be kept very low. Easy configuration is of particular importance, since the disproportionately high setup cost will otherwise reduce the attractiveness of automation. This favors robust *plug-and-play* systems with the capability of communicating via the mains or wireless. Acknowledging these specific challenges, one explicitly speaks of *home automation* or *domotics*. The term *building automation* will remain reserved for functional buildings.

A large number of proprietary, mostly centralized solutions is available. Many of them are application domain specific; a certain focus on multimedia aspects can be noted. Published networking and middleware standards of particular relevance are X10 for simple control via power line (in North America), KNX RF for household appliances (in Europe), ECHONET (in Japan), and UPnP (mainly for multimedia applications). Device profile for web services (DPWS) is intended as a web services (WS) conformant successor to UPnP. Regarding unpublished consortium standards, the OpenTherm protocol has reached importance in Europe for point-to-point communication in home heating systems, while Z-Wave has gained momentum in North America for wireless automation purposes. Dynamic application frameworks—OSGi [36] being the prime, open example—allow providing residential gateways with the flexibility required to have multiple service providers connect from remote to a variety of devices within the home environment. Remote access can be expected to increase in importance as building services equipment, alarm systems, and entertainment systems in homes increase in features and complexity.

In the rest of this section, a collection of selected technologies in the context of *smart homes* is presented. The applications are not restricted to lighting and shading or HVAC applications. Also, in-house network communication and multimedia systems are considered. In contrast to the previously mentioned building automation technologies, the discussed technologies are not necessarily based on open standards. There is also often no need of integration into BM systems.

One technology that is used for network communication by facilitating power line communication is *HomePlug* [32]. This system allows Ethernet communication without having any special network infrastructure. If a network connection is required somewhere, a HomePlug device is simply plugged in a power socket. In contrast to Wi-Fi systems, this technology is still working among different rooms or floors. HomePlug AV further allows the transmission of high definition video and audio signals. Security mechanisms are another feature of HomePlug.

A further technology that facilitates power line communication is called X10. It was developed by General Electric and is mainly used for lighting applications. Due to the limited address space and its unreliable behavior, it is infeasible for large-scale networks or critical domains. CEBus is an open standard that was developed to overcome the limitations of X10 [33]. Similarly to X10, it facilitates the power line medium. It supports different control and monitoring commands that allow simple interaction and even autoconfiguration.

A quite recent home automation system is *digitalSTROM* [31]. It is an open standard that supports different application domains such as lighting and shading, HVAC, audio/video, and security applications. Additionally, it can be used for smart metering applications. In order to support web-based applications, a web server is available that provides RESTful WS, where applications can be based on. A connection to other automation domains or even to the smart grid can be established using this interface as well.

Z-Wave is a wireless home automation technology [30]. It is a proprietary protocol that resides in a sub-GHz frequency band. Z-Wave supports up to 232 network devices and allows mesh networking. Similarly to ZigBee, Z-Wave specifies different device classes. A network contains one *controller* that is aware of the entire topology. Additionally, a network consists of *routing slaves*, which are able to relay data packets along predefined routes and *slaves* that only receive and confirm packets. Besides pure home automation, Z-Wave can also be used for entertainment systems. Since Z-Wave provides secure communication by link encryption, security applications are another field of use.

55.4 Integration

System integration is an everyday reality (and necessity) even for BASs that solely target the HVAC domain. System parts must be combined to form the HVAC control system, and devices must be configured to interact properly. Often, cabling, sensors/actuators (field devices), cabinets, controllers (for those pieces of equipment that do not have their own), and management infrastructure are tendered separately and provided by different specialist contractors (as is, of course, the HVAC equipment itself).

This division is even more pronounced between building disciplines, since they have evolved separately. Consequently, their respective automation systems are as well still entirely separate in most buildings today. However, information exchange between systems of different domains undeniably provides benefits. Actually, such cross-domain integration is indispensable for intelligent buildings [2].

A key area of cross-domain integration is room automation. For example, window blinds have considerable impact on HVAC control strategy, as incident solar radiation causes an increase in air temperature as well as in immediate human thermal sensation. Automatically shutting the blinds on the sunlit side of a building can significantly decrease the energy consumption for cooling purposes. An important guideline defining room control functions independently of the technology (i.e., hardware, software, communications technology) is VDI 3813-2 [42].

A (computer-aided) facility management system that has direct access to data from the BAS becomes an even more efficient tool for, among others, accounting and controlling purposes. For example, take into account cost allocation for climate control and lighting using live metering data.

As another example of BAS integration with IT systems, consider conference rooms to be air-conditioned only (and automatically) when booked. Also, hotel management systems can automatically adjust HVAC operation depending on whether a room is currently rented or vacant.

Yet, in all cases, the benefits reached by tighter integration come with a drawback. In an integrated system, examining subsets of system functionality in an isolated manner becomes more difficult. This introduces additional challenges in fault analysis and debugging as well as functionality assessment. Additionally, if multiple contractors are working on a single integrated system, problems in determining liability may arise.

This assessment problem is of special concern where life safety is involved. For this reason, fire alarm systems traditionally have been kept completely separate from other building control systems. Although a considerable degree of integration has been achieved in some projects, building codes still often disallow BAS to assume the function of life safety systems. This of course does not extend to less critical property safety alarms, such as water leakage detection. Similar considerations apply to security systems.

These issues need to be addressed by carefully selecting the points of interaction between the different systems, with the general goal of making the flow of control traceable. Interfaces have to be defined clearly to ensure that no unwanted influence is possible. Typically, systems stay stand-alone for their core purpose. Only a highly limited number of interaction points are established at the highest system level. However, such *high-level integration* can already yield considerable benefits.

A modern building is not realizable relying on one building automation technology. Such a BAS integrates a variety of different systems taking use of the specific benefits of each single technology. Vendors of automation equipment implement various protocol standards, like BACnet, LonWorks, and KNXnet/IP, as mentioned in the previous sections. Especially at the interface between automation level and management level of BAS, these different technologies join together. Since there is an inherent necessity of interfacing these heterogeneous systems by management level applications, a way must be found to learn them understand the same *language*. In other words, the goal is to achieve *interoperability*.

The classical, but not very convenient, way to reach this goal is to integrate technology-specific interfaces (i.e., drivers communicating with the underlying networks) in the management level applications. The disadvantages of this approach are a lock-in to distinct technologies and vendors. Furthermore, little flexibility for future extensions including new network standards exists.

A more promising way is to define and agree on a *general application model* covering the functionality of these underlying systems. Hereby, a unified interface to these networks is introduced. Since IP-based networks are not only commonly used at the top level of today's BAS but also at lower levels, the most suitable concept of communication is one that is aligned with IT standards. HTTP and HTML certainly are popular in BAS as well but are limited to user interfaces since HTML is ill-suited for machine-to-machine communication. Therefore, by tradition, OPC [28] has fulfilled this role. However, it is in its classic design incarnation restricted to Windows platforms due to its use of object linking and embedding/distributed component object model (OLE/DCOM).

Today, solutions emerge that follow the current mainstream trend of using WS for machine-to-machine interaction over a network. WS are entirely platform independent. They can be implemented using any programming language and run on any hardware platform or operating system. This maximizes interoperability and greatly improves the reusability of the services they provide. Central to WS are the use of XML for data representation and a standardized resource access protocol (typically SOAP, sometimes also plain HTTP). WS follow a modular concept. This allows the use of off-the-shelf standards for transmission, eventing, discovery, security, and many more. However, a drawback is the additional overhead introduced by the XML encoding (especially if small amounts of data are transmitted frequently) and the fact that servers cannot push information to clients (unless the latter implement a server themselves). Given the hardware resources available at the management level, this is typically not an issue, though.

The commitment to use WS technology determines low-level aspects of data representation and transmission. Still, the application domain remains to be modeled. BACnet WS (BACnet/WS) [11] and oBIX [35] have been developed with special regard to building automation and are freely available to the general public. Both provide historical data access and event and alarm management besides a data point abstraction. After a short overview on OPC, a brief introduction to the development of BACnet/WS and oBIX, their information modeling mechanisms as well as the communication concepts are described. Finally, devices profile for WS (DPWS) [40] is outlined, which gained some importance in the field of home automation.

55.4.1 OPC

In the 1990s of the last century, a task force of different vendors of human–machine interface (HMI) and SCADA software (later called *OPC Foundation*) defined a standard that provides services for reading, writing, and monitoring of control data in automation systems. Since it used Microsoft's technology Windows OLE as a unified way to access the network, the name *OLE for process control* (OPC) was chosen. In addition to this first standard release, many other OPC specifications have been defined. Therefore, the initial OPC standard was later renamed to OPC Data Access (OPC DA). Other important standards in this context are OPC Alarms&Events (OPC A&E), OPC Historical Data Access (OPC HDA), OPC Batch, OPC Security, OPC Data Exchange (OPC DX), OPC Compliance Tests, and OPC and XML. Due to the focus-oriented design of OPC, the specification family was very successful. Thanks to the existence of well-defined interfaces, developers of OPC client software do not have to consider communication details of the underlying network technologies.

55.4.1.1 OPC Data Access

OPC DA is the first and the most frequently used kind of the OPC family that allows exchange of real-time data among distributed high-level components. For this purpose, Microsoft's DCOM is used. The communication is based on a client/server relation where an OPC DA server provides data entities to clients in the form of a three-tuple containing the actual value, quality, and timestamp. Typically, OPC DA is used for communication among systems of different manufacturers. It is also used by automation systems to provide data to HMI or monitoring entities.

Due to the unified access method, information can be exchanged independent of the underlying system. However, due to the use of DCOM, systems based on this proprietary interface strongly depend on Microsoft operating systems. Further, dynamic port assignment is applied, which complicates secure data exchange via the Internet.

55.4.1.2 OPC Alarms and Events

In contrast to OPC DA where process data are exchanged periodically, OPC A&E allows event-based communication. Thereby an OPC server offers information of a system behind in form of event notifications. An OPC client can subsequently subscribe on dedicated alarms or events. If an event occurs, the server notifies all subscribed clients. However, events are not triggered by the OPC server but by the system behind the server. OPC A&E defines three different types of events. The `OPCSimpleEvent` specifies the basic properties of an event, whereof the `OPCConditionEvent` and the `OPCTrackingEvent` are adopted. The tracking event extends the simple event in a way that it is able to keep track about the activator of the actual event. A condition event notifies the subscriber about the incidence of a specified condition on a state value. It is further possible that the event requires an acknowledgment by the client. For each subscription, a client can define a filter. Such a filter can be based on different properties of an event such as severity or process area.

55.4.1.3 OPC Historical Data Access

The OPC HDA framework is used by a client to request historical data from a server. While OPC DA is designed to exchange periodic real-time data, OPC HDA is used to make process data accessible from a client, which has been collected by a database for a specified period in time. Thereby, data is only exchanged on request by a client.

55.4.2 BACnet/WS

Due to the evolution of the Internet and IP-based networks, WS are gaining importance within the building automation domain, too. Therefore, in 2006, ASHRAE decided to specify a WS interface for BACnet. This standard extension, also known as BACnet/WS, uses WS as an interface to a generic data model that can be used for integrating control data to existing management applications. Initially defined as BACnet Addendum 135-2004c [34], BACnet/WS is now part of the BACnet standard.

Although the intention of BACnet/WS was to enrich the BACnet protocol with a WS interface, BACnet/WS is not limited to the BACnet protocol. Due to its generic design, BACnet/WS can in principle be used for any kind of technology. Therefore, BACnet/WS can also act as a standardized, WS-oriented way to access data originating from other technologies like KNX.

55.4.2.1 Information Modeling in BACnet/WS

The fundamental elements of the data model in the BACnet/WS are *nodes*. Except the root node, which represents the entry point of the underlying data model, each node has exactly one parent node and may have an arbitrary number of children. Using this way of arranging nodes, a *hierarchy* can be defined, which maps the structure of the underlying data model.

A node represents a data primitive within the data model. The network-visible data of a node are represented as a set of *attributes*. Attributes are used to describe nodes and may themselves have attributes. While it is possible to specify proprietary attributes, BACnet/WS defines a fixed set of standard attributes. Some of these standard attributes are mandatory, while others are optional. However, it can be assumed that the set of available attributes will remain unchanged in normal operation and changes only occur after reconfiguration of the server. There are three types of attributes. Primitive attributes are attributes of data types defined by the XML schema standard (e.g., Boolean, string, or double). Enumerated attributes are enumerations of the XML schema data-type string. The set of allowed values is defined by BACnet/WS. Array attributes are attributes that consist of an array of primitive attributes. In general, neither the definition of proprietary data types for attributes nor an extension of existing data types is possible in BACnet/WS. In the following, the most important attributes are described in more detail.

The `NodeType` attribute is required for all nodes and allows the general classification of nodes. This attribute is an enumeration where the following values are allowed: `Unknown`, `System`, `Network`, `Device`, `Functional`, `Organizational`, `Area`, `Equipment`, `Point`, `Collection`, `Property`, and `Other`. Note that this set of node types cannot be extended. The only way for user-defined node types is to use the `NodeSubtype` attribute that can be set to any arbitrary string.

The `DisplayName` and `Description` attributes are used to provide a human-readable name and a description of the node for user interfaces. The `Value` attribute represents the effective datum of a node (e.g., the actual value of a data point like a temperature value or the actual state of a light). Its type is defined by the `ValueType` attribute, which is mandatory for all nodes. Note that the value attribute is optional. If a node does not represent a real process (but, e.g., a node of type `Area` that models a room), the `ValueType` attribute can be set to `None`. Other important attributes are the `Units` attribute (used to model engineering units), the `Attributes` attribute (lists the available attributes of the node), the `Children` attribute (lists the child nodes), and the `Reference` attribute (used to reference other nodes).

The information model concept of BACnet/WS is very restricted—it is not possible to specify user-defined node types (except using the `NodeSubtype` attribute) or user-defined data types of attributes. It is only possible to define proprietary attributes as well as to use the `NodeSubtype` attribute to define proprietary subtypes of already existing node types. However, defining a type hierarchy and modeling node types in the same way as normal data is not possible in BACnet/WS. Additional semantic can be added to the data model by the hierarchy of the nodes. By linking child nodes to their parent, it is possible to define a structure beneath the data model.

In addition to defining parent–child relationships, it is also possible to specify so-called reference nodes. A reference node is a node that refers to another node (called referent node) within the hierarchy. Except for the attributes `Children`, `Aliases`, `Attributes`, and `Reference`, reading an attribute of the reference node has the same result as reading the attribute of the referent node. Therefore, a client may not be able to distinguish between the reference and its referent node (except for the attributes mentioned before). It is also possible to define multiple reference nodes that point to the same referent node. However, reference loops and self-references are not allowed. Reference nodes are used to model different views and to define multiple hierarchies.

Nodes and attributes are addressed by their path. Paths are character strings that are composed of two parts: the node part followed by the attribute part. The node part reflects the node hierarchy and consists of node IDs separated by forward slash "/" characters. An empty node part refers to the root node of the hierarchy. The attribute part is made up of attribute IDs separated by colon ":"characters. If the attribute part is empty, the `Value` attribute is assumed by default. The complete path form is

[/*nodeid*[/*nodeid*]...][:*attributeid*[:*attributeid*]...]

IDs are case sensitive and have to consist of printable ANSI X3.4 characters except of the following characters:/:;— < > * ?" [] {}. Additionally, node IDs beginning with a period "." character and attribute IDs not beginning with a period "." character are reserved by the BACnet/WS standard. Hence, proprietary node IDs defined by a BACnet/WS server must not begin with a period "." character, whereas proprietary attribute IDs have to.

55.4.2.2 Services and Communication in BACnet/WS

The communication concept of BACnet/WS is based on the client/server model. A BACnet/WS server aims at providing a uniform access to the control data of the underlying BAS. To achieve this, it collects the control data, processes it, and represents the data as BACnet/WS nodes within the server's node hierarchy. Modified values are written back to the originating devices. BACnet/WS clients can access the node hierarchy by reading or writing the individual nodes of the hierarchy.

In BACnet/WS, a set of WS is defined to provide read and write access to the node hierarchy. Currently, 11 different WS are defined, which are mainly used to read and write the attributes of nodes or to retrieve information about the attributes (e.g., the array size). Typical examples are the `getValue()` and `getValues()` services to read the value of one or more attributes as well as the `setValue()` and `setValues()` services that are used to change the value of one or more attribute, respectively. Other services like services to change the data model (e.g., add or delete nodes) or to query the node hierarchy are not supported (browsing the node hierarchy has to be done by reading the `Children` attribute).

The implementation of these services shall conform to the Web Services Interoperability Organization (WS-I) Basic Profile 1.0, which specifies the use of SOAP 1.1 over HTTP 1.1. The data encoding shall be based on XML 1.0 (second edition). In addition, an extension of BACnet/WS that provides the opportunity to use RESTful services has been specified.

55.4.3 oBIX

The open Building Information eXchange (oBIX) specification [38] is maintained by the Organization for the Advancement of Structured Information Standards (OASIS).* The first specification oBIX 1.0 has been published in December 2006 [35]. The main goal of oBIX is to provide an open WS interface for accessing any kind of BASs. As a platform-independent technology, oBIX can be used on top of any existing technology. To exchange data, oBIX defines a small set of WS that can be used over SOAP or HTTP binding. The object model is concise but very flexible due to its support for object-oriented concepts.

The work on oBIX 1.1 started in 2009 and mainly consists of bug fixes as well as some additional features like a binary encoding scheme for the used object model. By 2010, a draft has been released that became a de facto working standard for oBIX 1.1.† Recently, in 2013, the standardization effort on oBIX gained momentum and additional protocol bindings and message encodings were added, which are discussed within Chapter 59, which has a focus on future trends in smart homes and building automation technologies.

55.4.3.1 Information Modeling in oBIX

The basic elements within oBIX are *objects*. Each object is of a certain object type. Currently, 17 standard object types have been specified where each type directly maps to an XML element name. Typical examples are objects that model single data items like `bool`, `int`, `real`, and `str` as well as object types for time and date representations (e.g., `date` and `time`). However, object types that represent references to other oBIX objects (`ref`) and types that can be used for lists (`list`) are also available. Each object has different *attributes* that further describe the object and thus add semantics to it. Some of these predefined sets of attributes are called *facets*, which represent metadata of an object. Typical examples are `displayname`, `min`, `max`, and `units`.

In addition to model normal data (e.g., data points of type `real` or `bool`), oBIX provides the concept of operations (`op`). Operations are methods that can be invoked by a client. They take exactly one oBIX object as parameter (specified by the `in` attribute) and one oBIX object as result (specified by the `out` attribute). If multiple parameters or results are required, nested objects must be used.

The main feature of the object model in oBIX is that objects can have subobjects. Using this nested concept, complex data structures can be defined. Another important feature is that objects and their attributes are represented by XML. Each object (or subobject) is directly mapped to an XML element—attributes and facets are mapped to XML attributes of the corresponding XML element.

An important concept of oBIX is the use of so-called contracts. Contracts can be compared to templates—they are used to define new object types but also provide a possibility to specify default values. Contracts are represented as normal oBIX objects by using the same XML syntax. Thus, there is no difference between a contract and a normal object, and oBIX clients can access type definitions the same

* http://www.oasis-open.org.

† The remainder of this discussion will be based on that draft version of 2010 and focus on the core specification.

way as they can access normal data. To specify that an object implements a contract, the `is` attribute is used. If an object references to a contract using the `is` attribute, the object can be seen as an instance of the contract since it inherits the structure of the contracts as well as the default values. However, to add additional semantics, the object instance can override dedicated subobjects or attributes (e.g., the object name) or extend the structure by adding new subobjects. The `is` attribute can also be a list of contracts and so, multiple inheritance is possible, too. A contract can itself be an implementation of another contract. This means that it is possible to define type hierarchies like it is common in other object-oriented concepts.

Objects are identified by either a name, a hyperreference, or both. Names are used to define the role of an object and used as programmatic IDs. The name of an object is represented by the `name` attribute and should not be used for display purposes—instead, the `displayname` facet shall be used. Names are often used to define semantics of subobjects. The hyperreference of an object is specified by the `href` attribute. In general, uniform resource identifiers (URIs) are chosen to define a hyperreference. For the root object, an absolute URI has to be selected. For all other objects, it is possible to take absolute (e.g., `http://knx/building1/room1/light1`), relative URIs (used to address an object relative to its location within the data structure, e.g., `light1`), or fragmented URIs (used to name objects within the same oBIX document, e.g., `#light1`).

In addition to these basic information modeling concepts, the oBIX specification provides a *core contract library*. Within this library, basic objects like `Units` and `Weekdays` are already defined. Furthermore, oBIX defines the following concepts:

- `Points`: Points provide an abstraction of the datapoint concept within the automation system.
- `Watches`: Watches are used by the client to register to dedicated objects for monitoring object changes.
- `Alarming`: Concepts for querying, watching, and acknowledging of alarms are also provided by oBIX.
- `History`: oBIX also supports mechanisms to retrieve the history of object values.

55.4.3.2 Services and Communication in oBIX

oBIX is also based on a service-oriented client/server architecture where oBIX clients can access the data of an oBIX server using WS. However, compared to other technologies, oBIX only defines three services. The `Read` service is used to read the complete state of an object. The object to be read is specified by the object's URI—the current state of the object is returned as XML document. The `Write` service is used to update the state of an object. As parameters, the object's URI as well as the new object state are passed to the service. If successful, the server responds with the updated state of the object. Finally, the `Invoke` service is used to execute operations. The request contains the input parameter object and the server responds with the output object after having executed the desired operation.

For implementing these three request/response services, two different protocol bindings are specified by the oBIX standard. The HTTP binding simply maps oBIX requests to HTTP 1.1 methods. The second protocol binding maps the services to corresponding SOAP services.

The network messages that are exchanged during service execution have to be encoded. While within oBIX 1.0 only an XML encoding scheme is defined, the new version oBIX 1.1 also defines a binary encoding scheme. The main advantage of this new binary encoding scheme is its efficiency, which makes it suitable for narrow bandwidth networks as well as for small embedded devices.

55.4.4 DPWS

The DPWS was initially created by Microsoft and is meanwhile available as OASIS standard.* It defines a subset of WS-* standards in order to provide interoperability between resource-constrained WS

* http://docs.oasis-open.org/ws-dd/dpws/1.1/os/wsdd-dpws-1.1-spec-os.html.

implementations. The main areas of this specification address messaging, discovery, description, and eventing. The profile intends to constrain the WS protocols and formats to allow implementation on consumer electronics hardware or embedded devices but still be compatible with WS implementations on resource richer devices. The specification addresses the message exchange between clients and services hosted on a device. The specification identifies a device service that contains several hosted service, which are hosted by the device service.

55.4.4.1 Information Modeling in DPWS

DPWS does not offer a metamodel that can be used for information modeling. As the name implies, it is a profile that defines a subset of WS-* standards for the usage on resource constraint devices. For information modeling, DPWS does not come with a custom metamodeling language. For representing devices, process control variables, process data points, or the topology of a BAS XML schema can be used. XML schema is a metamodel that allows to define the structure of XML documents. It uses basic types, which are common in many other information modeling languages like *xs:string*, *xs:decimal*, *xs:integer*, *xs:float*, *xs:float*, *xs:date*, and *xs:time*, and further some XML-schema-specific basic types, which are *QName*, *anyURI*, *language*, *ID*, and *IDREF*. Basic types are neither allowed to contain subelements nor attributes. Furthermore, there are complex types, which may contain nested elements and attributes. It is possible to define value restrictions on XML types and also some requirements regarding the structure.

55.4.4.2 Services and Communication in DPWS

DPWS uses a client/server model for communication and is based on the WS-* standards described earlier. In contrast to other building automation standards using WS, there are no generic or uniform services defined, except the transfer of metadata about the offered services, the services required by the WS-Discovery standard, and several attributes describing the host service and its hosted services. For messaging, the use of SOAP 1.2, SOAP-over-UDP, HTTP 1.1, WS-Addressing, UUIDs (RFC 4122), and Message Transmission Optimization Mechanism (MTOM) are required. WS-Discovery is obligated to be used for discovery. It allows to find WS adhering to a certain port type by using UDP broadcast messages within an IP network. The description of the capabilities of a device and its hosted services is based on WSDL to describe the interfaces and the exchanged messages. The exchange of this metadata about the WS is done using WS-MetaDataExcange, which itself uses WS-Transfer and WS-Addressing. This allows to offer and exchange the metadata in a standardized way. To express the capabilities of a device, DPWS provides several extensible XML complex types, which are *dpws:ThisModel* and *dpws:ThisDevice*. Furthermore, *dpws:Host* and *dpws:Hosted* provide standardized attributes and subelements that describe the services running on a device. This includes, for example, the port types used by WS-Discovery. A client can subscribe to events and receive notifications according to WS-Eventing. DPWS defines a filter dialect that allows to subscribe to one or more notifications exposed by a hosted service. The event types and the type of a notification are not covered by the standard.

55.5 Conclusion

Building automation technologies were primarily used for simple control tasks, but they developed to highly complex systems. Today's BASs are crucial for large buildings. They help to save energy and human resources. Additionally, the comfort of facility users is increased and maintaining safety in buildings is a key asset.

This chapter aimed to give an introduction into state-of-the-art technologies used in building automation. First, structural models of building automation were presented to show the big picture of the topic. Due to the enormous number of different technologies, products, and standards in this domain, a selection of the most important technologies was presented. Many of these technologies were developed years or even decades ago. Nevertheless, they are still widely used today. It is hardly imaginable that these technologies are replaced by any other technology in the near future.

For modern buildings, it is crucial that building automation technologies are able to cooperate with each other. It is common practice that different technologies are used within one system. There are technologies that are used for dedicated applications. Other subsystems are historically grown and therefore use different technologies. For this purpose, high-level approaches were presented that provide a basis for interoperability among this high diversity in technologies. They are further capable for providing an abstract and unified interface for high-level control and visualization.

The recent development toward an Internet of Things (IoT) also influences building automation and technologies. Future automation systems have to be integrated in this new world. This consequently offers new possibilities but also new challenges. Due to the interaction with almost everything and everyone within a building, synergy effects arise that can be exploited by upcoming technologies. Thus, Chapter 59 addresses recent developments and trends in the context of building automation and the IoT.

References

1. W. Kastner and G. Neugschwandtner, Data communications for distributed building automation. In Richard Zurawski, editor, *Embedded Systems Handbook*, 2nd edn, Vol. 2: *Networked Embedded Systems*, Chapter 29, pp. 1–34. CRC Press, Boca Raton, FL, 2009.
2. J.K.W. Wong, H. Li, and S.W. Wang, Intelligent building research: A review, *Automation in Construction*, 14(1), 143–159, 2005.
3. ISO 16484-3, Building automation and control systems (BACS)—Part 3: Functions, 2005.
4. ISO 16484-2, Building automation and control systems (BACS)—Part 2: Hardware, 2004.
5. W. Kastner and G. Neugschwandtner, Datenkommunikation in der verteilten Gebäudeautomation, *Bulletin SEV/VSE*, 17, 9–14, 2006.
6. IEC 60929, AC-supplied electronic ballasts for tubular fluorescent lamps—Performance requirements, 2006.
7. IEC 61158, Industrial communication networks—Fieldbus specifications, 2007.
8. IEC 14543-3-10 Information technology—Home electronic system (HES) architecture—Part 3-10: Wireless short-packet (WSP) protocol optimised for energy harvesting—Architecture and lower layer protocols, 2012.
9. IEC 7498-1 Information technology—Open systems interconnection—Basic reference model: The basic model, 1994.
10. EN 13757, Communication systems for meters and remote reading of meters, 2002.
11. ANSI/ASHRAE Std. 135, BACnet—A data communication protocol for building automation and control networks, 1995–2010.
12. ISO 16484-5, Building automation and control systems (BACS)—Part 5: Data communication protocol, 2012.
13. ANSI/EIA/CEA Std. 709.1, Rev. A, Control network protocol specification, 1999.
14. EIA/CEA Std. 709.1, Rev. B, Control network protocol specification, 2002.
15. EN 14908, Open data communication in building automation, controls and building management—Control network protocol, 2005.
16. ANSI/EIA/CEA Std. 852, Tunneling component network protocols over internet protocol channels, 2002.
17. LonMark International, www.lonmark.org.
18. KNX Association, www.knx.org.
19. KNX Association, KNX specifications, Version 2.0, 2009.
20. KNX Association, Application note 158/13—KNX data security, Version 2 (draft), 2013.
21. KNX Association, Application note 159/13—KNXnet/IP secure, Version 2 (draft), 2013.

22. EN 50090, Home and building electronic systems (HBES), 1994–2007.
23. ISO/IEC 14543, Information technology—Home electronic system (HES) architecture, 2006–2007.
24. EN 13321-2, Open data communication in building automation, controls and building management—Home and building electronic systems—Part 2: KNXnet/IP Communication, 2005.
25. IEEE Std. 802.15.4, Wireless medium access control (MAC) and physical layer (PHY) specifications for low-rate wireless personal area networks (WPANs), 2003–2007.
26. ZigBee Alliance, www.zigbee.org.
27. ZigBee Alliance, ZigBee specification, 2007.
28. OPC Foundation, www.opcfoundation.org.
29. SMI-Arbeitskreis, www.smi-group.com.
30. Z-Wave Alliance, www.z-wavealliance.org.
31. digitalSTROM Alliance, www.digitalstrom.org.
32. HomePlug Alliance, www.homeplug.org.
33. G. Evans, *CEbus Demystified: The ANSI/Eia 600 User's Guide*, McGraw Hill, New York, 2001.
34. American Society of Heating, Refrigerating and Air-Conditioning Engineers, ANSI/ASHRAE Addendum c to ANSI/ASHRAE Standard 135-2004, 2006.
35. OASIS, oBIX 1.0 Committee Specification 01, 2006.
36. OSGi Alliance, www.osgi.org.
37. W. Kastner, G. Neugschwandtner, S. Soucek, and H. M. Newman, Communication systems for building automation and control, *Proceedings of the IEEE*, Vol. 93(6), pp. 1178–1203, 2005.
38. I.N. Favino, A. Carcona, M. Masera, and A. Trombetta, Design and implementation of a secure Modbus protocol, *Critical Infrastructure Protection III*, Vol. 311(6), pp. 83–96, 2009.
39. Federal Information Processing Standards Publication (FIPS) 197, Advanced Encryption Standard (AES), 2001.
40. Organization for the Advancement of Structured Information Standards (OASIS), Devices Profile for Web Services (DPWS), 2009.
41. International Electrotechnical Commission, IEC 60870 Telecontrol equipment and systems—Part 5: Transmission protocols—Section 2: Link transmission procedures, 1992.
42. Verein Deutscher Ingenieure, *VDI 3813-2: Raumautomation—Blatt 2: Funktionen*, Berlin, Germany, 2009.

56

Fundamentals of LonWorks®/CEA-709 Networks

56.1 Distributed LonWorks® Networks 56-1
56.2 Node Architecture 56-2
56.3 Programming Model 56-5
56.4 Communication Elements 56-6
56.5 Communication Topologies and Media 56-6
56.6 ANSI/CEA-709 Protocol Standard (LonTalk) 56-7
56.7 Network Infrastructure 56-11
56.8 Interoperability and Profiles 56-11
56.9 Development and Integration Tools 56-11
56.10 New Developments for ANSI/CEA-709 Networks 56-12
References 56-14

Dietmar Loy
LOYTEC Electronics GmbH

Stefan Soucek
LOYTEC Electronics GmbH

56.1 Distributed LonWorks® Networks

The LonTalk® Communication Protocol defines rules for the communication in LonWorks networks. All nodes connected to a local operating network need to support the LonTalk protocol to such an extent that all communication services are supported. Otherwise, the performance and the openness of the system may not be guaranteed. The LonMark Interoperability Association, founded by significant producers of LonWorks-based devices in 1993, issues guidelines for the development and integration of interoperable LonWorks networks. The LonMark Interoperability Guidelines describe a unique platform for interoperable (manufacturer-independent) LonWorks networks. One task of the LonMark Interoperability Association is conformance testing and certification of LonWorks-based products according to their interoperability guidelines. The protocol has also been advanced to the ISO standard level in ISO/IEC 14908 [1] (Figure 56.1).

Significant application areas for LonWorks-based systems are building and process automation, electrical appliances in homes, restaurants, transportation, and other applications with distributed data acquisition and control. The data exchange between two applications requires reaction times of several milliseconds. This is the time interval in which a message is sent from the application on node A transmitted over the network and received by the application program in node B (end-to-end response time). The LonTalk protocol is only partly suitable for the transmission of audio and video data streams. This characteristic is common to all known control network protocols. It is due to the requirements for short end-to-end response times in real-time control applications, but the overall data throughput is low. Control networks are often used over communication media with heavy disturbances from electromagnetic coupling, signal interference, and mechanical disturbances. Therefore, reliable transmission

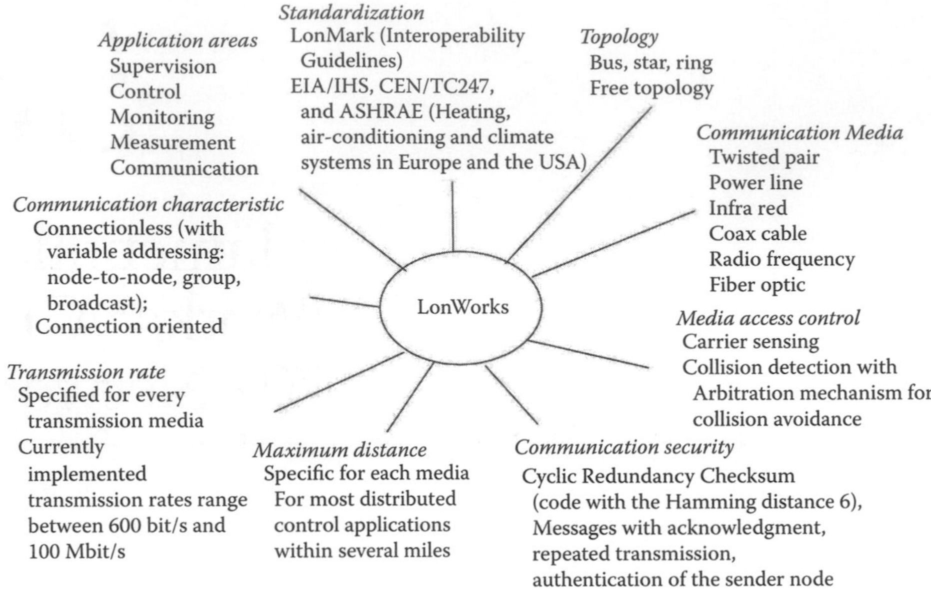

FIGURE 56.1 Features of a LonWorks network.

of control data is often an important criterion for the selection of a control network. The LonTalk protocol defines a 16-bit cycle redundancy check polynomial, which provides the failure recognition with Hamming Distance 6 for each data packet. Furthermore, services for acknowledged end-to-end communication, automatically repeated transmission, and transmitter authentication provide means for secure transmission.

56.2 Node Architecture

For the last 10 years, the Neuron® chip manufactured by Motorola, Toshiba, and Cypress was the only microcontroller that supported the LonTalk protocol. Only in the last few years, with the adoption of the LonTalk protocol as a European and ANSI standard, some platform-independent implementations are available on the market [2]. The traditional LonWorks node is shown in Figure 56.2.

The Neuron chip executes the LonTalk protocol and the application program that interfaces to sensors and actuators through the input/output (I/O) interface. The block diagram of the Neuron chip is shown Figure 56.3.

Two versions of the Neuron chip are available. The 3150 provides an external memory bus interface to expand the internal EEPROM and SRAM with external Flash and SRAM. The 3120 is optimized for low node cost applications and has a built-in 10 kByte ROM, 2/4 kByte EEPROM, and 2/4 kByte SRAM but has no external memory interface [3]. The memory layouts for three versions of the Neuron chip are shown in Figure 56.4.

As shown in Figure 56.3, the Neuron chip has 11 I/O pins available to connect to sensors and actuators. A more detailed view of the I/O interface of the Neuron is shown in Figure 56.5. Each Neuron chip contains two independent timer/counter circuits with a width of 16 bits each. These two timer/counter blocks are used to count pulses from sensors and to generate frequencies or pulse-width-modulated signals to stimulate actuators. These timer/counters as well as the remaining I/O pins can be programmed through software libraries supplied with the development tools for the Neuron chips.

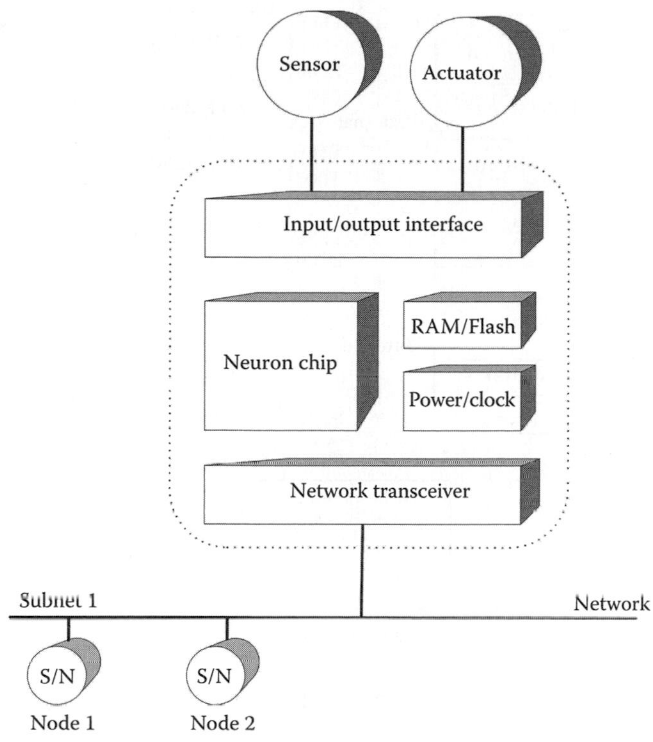

FIGURE 56.2 Typical LonWorks network node architecture.

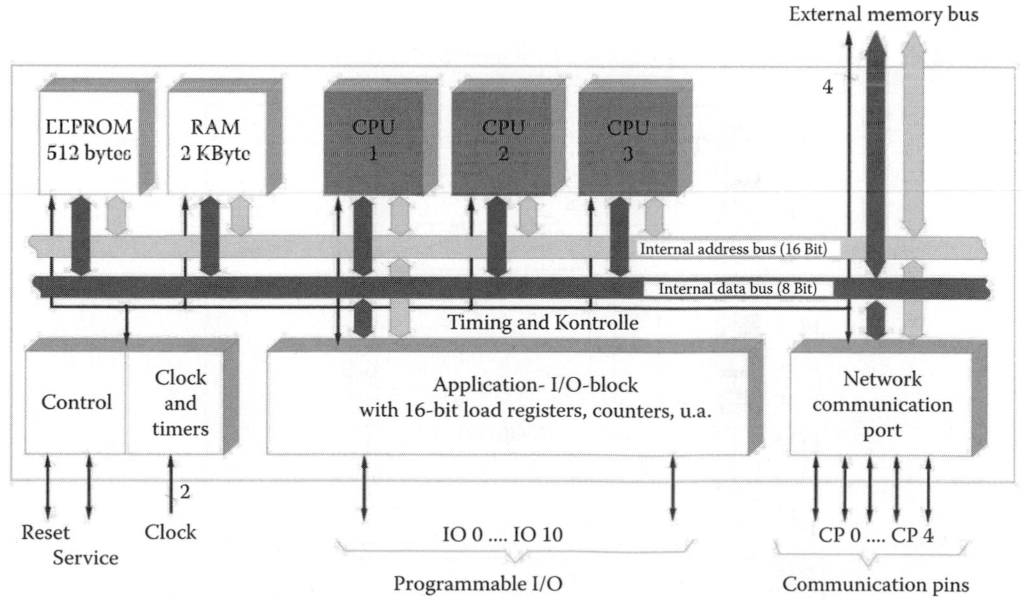

FIGURE 56.3 Neuron chip block diagram for the Neuron 3150.

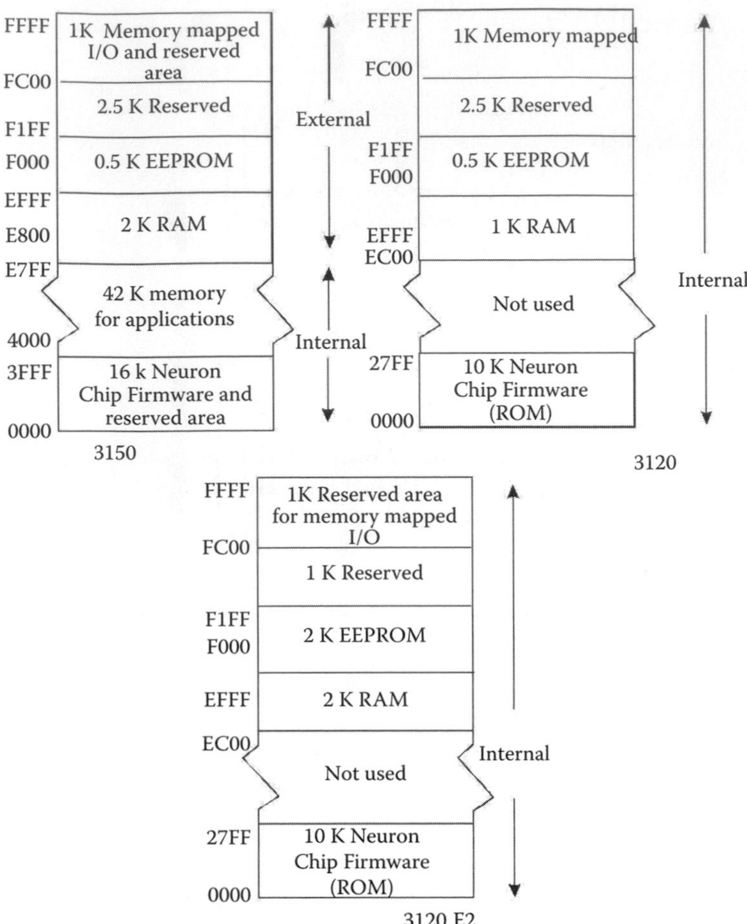

FIGURE 56.4 Memory layout for the 3150, 3120, and 3120E2 Neuron chips.

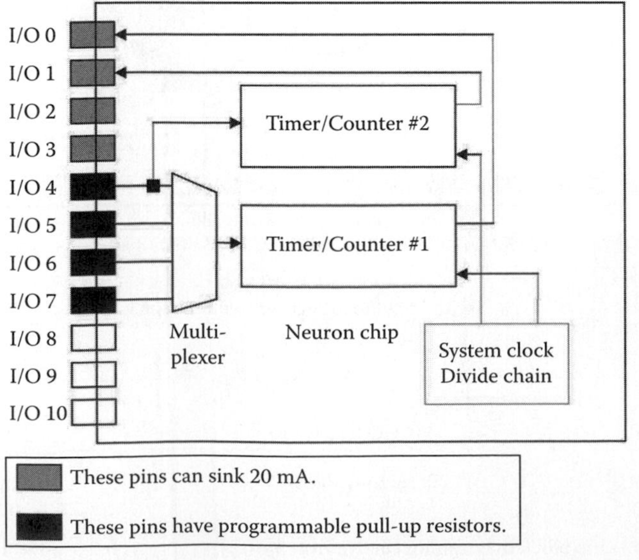

FIGURE 56.5 I/O system of the Neuron chip.

56.3 Programming Model

An ANSI C derivative language called Neuron C is used to program the Neuron chip. Neuron C uses language extensions to schedule application events and to react to incoming data packets (network variables [NVs]) from the network interface. The user does not have to deal with internal details, for instance, how the C compiler produces machine code and how the hardware of the Neuron chip processes this code at run-time. If desired, the compiler generates a comprehensive compilation protocol, which also contains the assembly listing. A number of frequently required functions are available in the firmware library of the Neuron chip, so that there is no need for the user to implement these functions. The C compiler outputs very compact machine code, which efficiently utilizes the memory resources. However, in comparison with other processors, users with assembly language experience will miss possibilities that a direct access to the internal registers on assembly level typically would offer. The lack of software or hardware interrupt capabilities especially in the application programs is sometimes a bit cumbersome. Hence, the development of applications that require fast response times to external events is not trivial. Decentralizing the desired functionality onto multiple Neuron chip–based nodes is, besides using a more powerful platform as explained at the end of this chapter, sometimes the only solution to achieve reasonable response times. The Neuron chip task scheduler that controls the flow of the application program allows event-oriented programming in C. It recognizes external and internal events and executes the associated task (Neuron C *when clause* statement) of the application program. Due to its nonpreemptive mode of operation, it cannot interrupt the active task. Hence, the reaction time to an occurring event greatly depends on the programming style of the software developer.

The operation of the Neuron chip task scheduler is shown in Figure 56.6. The scheduler is initialized after power-on or after a restart of the application program. The scheduler starts with the reset task if

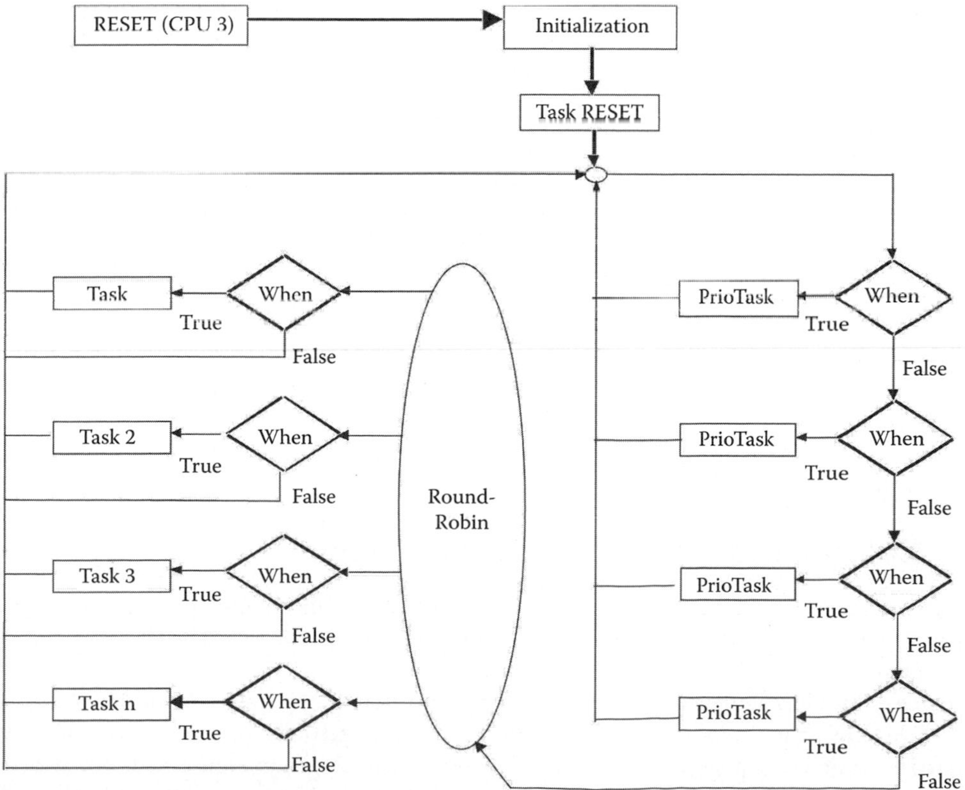

FIGURE 56.6 Neuron C scheduler.

it is present. Inside the reset task, all application-specific initialization can be performed. After that, all the system-specific tasks and priority tasks will be checked. If one of the tasks evaluates true, the task will be executed, and the scheduler returns to the beginning of the scheduler loop. If none of the priority tasks evaluates true, the nonpriority tasks will be checked. These tasks will be selected in a round-robin fashion. Between nonpriority tasks, all system-specific tasks and priority tasks will be checked regardless of whether the task was executed or skipped. The nonpriority tasks will be executed in the order of their definition in the application program. If the last nonpriority task is checked and maybe executed, the scheduler starts again with the first priority task.

56.4 Communication Elements

Communication between network nodes can be done in two ways. The preferred method are so-called NVs. NVs are variables (like variables in a C program) that can be propagated over the network and are defined in the application program. NVs have a type, for example, SNVT_temp to represent a temperature in degree Celsius or SNVT_amp to represent an electric current in Amps or mA. SNVT is the acronym for Standard Network Variable Type. Each SNVT has a valid range defined by the upper and lower boundary, a resolution, and an SI unit. Network nodes communicate with each other by exchanging NVs. Examples for different NV declarations are given as follows:

```
network output SNVT_temp_p Room_Temperature;
network input SNVT_temp_p Outdoor_Temperature;
```

NVs can be input or output NVs. If the application program assigns a new value to an output NV, this value will be automatically transmitted to other nodes on the network. The data packet generation and transmission is transparent for the application program without calling special functions to assemble and to transmit the packet. A simple Neuron C code example for transmitting the value (new_value) for the NV Room_Temperature is shown in the following statement:

```
Room_Temperature = new_value;
```

Another way to communicate between nodes are explicit messages. These messages do not use predefined variable types but can use any combination of bits and bytes assembled in the data packet.

The logical binding between input and output NVs or explicit messages on different nodes can be done differently depending on the system architecture.

- It can be done at the factory if the node is part of a closed system (e.g., copy machine).
- For homogeneous systems, it is possible to implement self-binding strategies.
- Depending on the number of available input pins or application parameters, the node can create the binding itself.
- For complex systems (e.g., large buildings), the binding will be accomplished with an installation tool. This is called system integration.

56.5 Communication Topologies and Media

LonWorks supports a variety of different topologies and communication media. In most cases, the topology of fieldbus systems is a line structure as shown in Figure 56.7. To avoid reflections, nodes are connected to this line by short stub cables.

In home or building automation, pure line topologies are barely applicable. To minimize the wiring cost, stub lengths up to several meters are required. If stubs are getting longer and start to branch out, we go from a pure line topology to a so-called free topology as shown in Figure 56.8. In any case, the network cable must be properly terminated according to the impedance of the network cable, which is typically in the 100 Ω range.

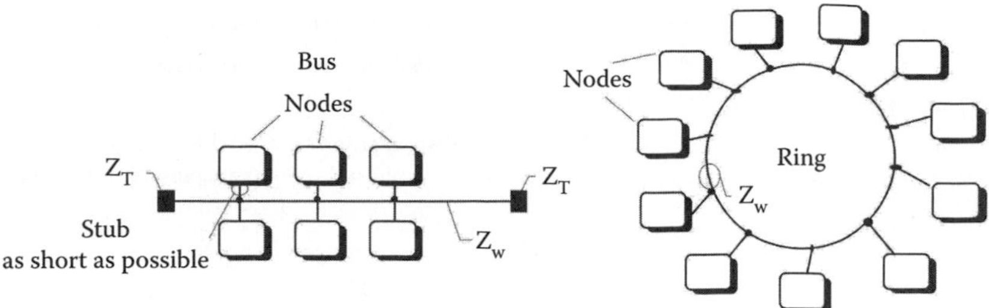

FIGURE 56.7 Line topologies shown as bus and as ring topology.

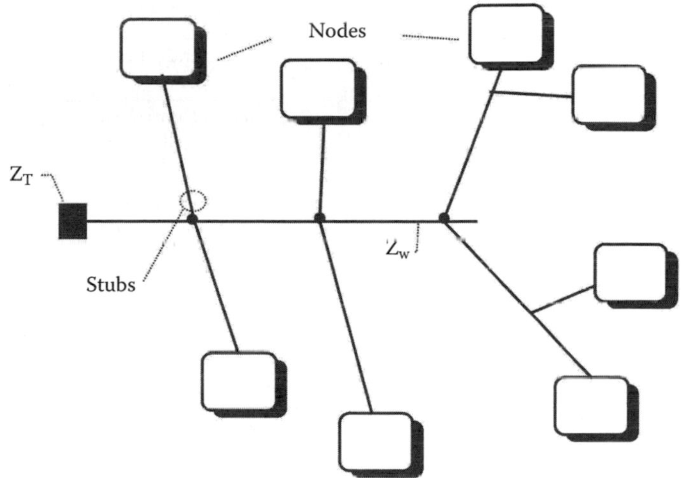

FIGURE 56.8 From long stubs to free topology.

The most popular communication media today are 78 kbps (CEA-709.3) or 1.25 Mbps twisted-pair communication, 4 kbps power line communication (CEA-709.2), a 1.25 Mbps fiber-optics interface, and RS-485 interfaces at various bit rates between a few hundred bps and 1.25 Mbps. A brand-new channel type is the EIA-852 Internet Protocol (IP) channel in order to tunnel CEA-709 data packets through Ethernet IP (Intranet, Internet) networks.

56.6 ANSI/CEA-709 Protocol Standard (LonTalk)

One of the characteristics of LonWorks is its unique seven-layer protocol implementation called LonTalk or LonWorks protocol. Unlike other fieldbus protocols, all seven layers of the ISO/OSI reference model are actually defined and implemented in every single network node. One characteristic is its media-independent OSI Layer 2 that supports various communication media like twisted-pair cables, power line communication, radio frequency channel, infrared connections, fiber-optic channels, as well as IP connections based on the EIA-852 protocol standard. The OSI reference model and the corresponding functionalities of the CEA-709 protocol are shown in Figure 56.9.

The medium access control sublayer provides an arbitration mechanism called predictive p-persistent carrier sense multiple access (CSMA), which avoids collisions on the network. The link layer provides a connectionless service whereby the functionality is restricted to the generation of the frame and the error detection. Detected errors will be indicated to the upper layers to correct them.

The network layer transmits data packets as a connectionless unacknowledged service that neither allows segmentation of messages nor assembling of partial messages to a complete message. Self-learning

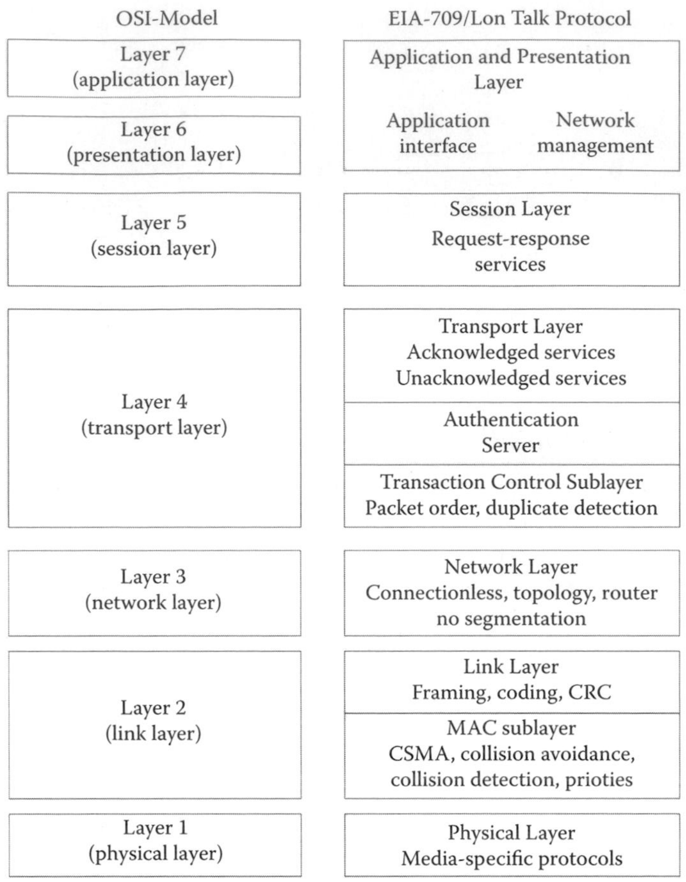

FIGURE 56.9 OSI reference model for the ANSI/CEA-709 protocol.

routers will be supported, but a tree structure of the network will be assumed. In contrast to self-learning routers, configured routers that are using routing tables can be used also in networks with physical loops as long as the logical connections are built in a tree structure.

The main functions of the LonTalk protocol are processed by the transport layer and the session layer. Layer 4 can be split up into (1) transaction control sublayer, (2) authentication server, and (3) transport layer. The duplicate detection in the transaction control sublayer detects duplicated packets to guarantee the correct order of the packets. The purpose of the authentication server is to provide an authenticated data exchange. The (3) transport layer provides a peer-to-peer connection between the source node and one or more destination nodes.

The session layer provides a simple request–response mechanism for the access from clients to distant servers. Based on this functionality, application-specific remote procedure calls can be defined. The network management protocol, for instance, makes use of the request–response mechanism contained in the session layer.

Presentation layer and application layer provide the basic functionality for interoperability. The application layer provides all services required for transmitting and receiving messages. Furthermore, the concept of the NVs is embedded in the application layer. The presentation of data in the presentation layer is independent of the application. This allows simple data exchange between nodes of different manufacturers.

The basic frame layout for a twisted-pair media is sketched in Figure 56.10. The preamble allows for bit synchronization at the receiver, and the start bit indicates the beginning of the byte boundary.

The header field carries information about the importance of the data packet expressed in a priority bit and an estimation of the number of expected data packets on the bus in the near future. This estimation is used to adjust the bus arbitration algorithm. The source address holds the subnet/node address of the transmitting node, and the destination address holds the address of the receiving node(s) followed by the user data that can be between 1 and 228 bytes long. Finally, a 16 bit CRC (CCITT-Polynomial) protects the frame from bit errors (Hamming Distance 4), and the code violation indicates the end of the frame. After the CV, a new round of bus arbitration can take place if nodes have data to be sent over the network [4]. On twisted-pair cables, a p-persistent CSMA bus arbitration scheme is used. For other communication channels, for example, IP over Ethernet, the arbitration scheme defined for this media is used by the CEA-709 protocol stack.

Another characteristic is its elaborated OSI Layer 3 that supports a variety of different addressing schemes and advanced routing capabilities as shown in Figure 56.11. Every node in the network can be identified with a unique 64 bit physical address (Neuron ID) and with a logical address composed of three address elements: the domain identifier, the subnet number, and the node number. Node numbers can be in the range of 1–127, subnet numbers within 1 and 255; hence, a maximum of 127*255 = 32,385 nodes can be addressed within a single domain, and up to 2^{48} domains can be addressed. Domain gateways can be built between logical domains in order to allow communication across domain boundaries. Groups can be formed in order to send a single data packet to a group of nodes using a multicast addressed message.

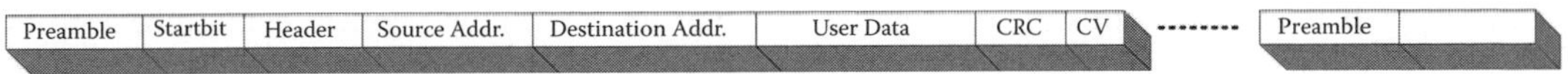

FIGURE 56.10 LonTalk frame layout on a twisted-pair cable.

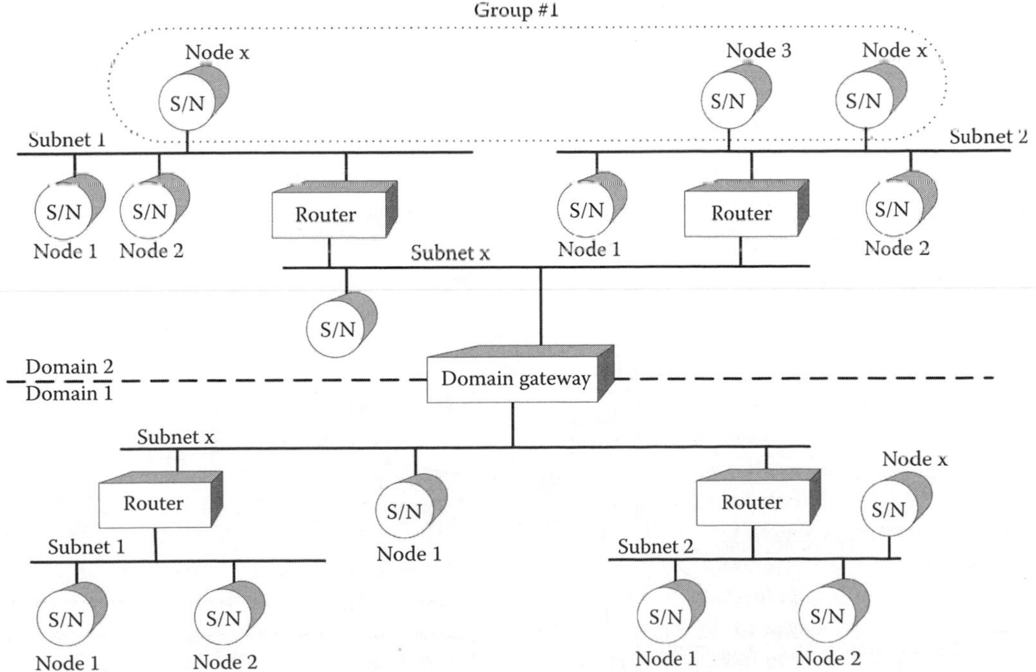

FIGURE 56.11 Addressing elements in CEA-709 networks. Up to 32,385 nodes can be addressed within one domain, and up to 2^{48} domains are possible.

TABLE 56.1 Addressing Schemes

Type	Address Format	Logical Address Components	Destination
#0	Broadcast	Domain, sourceSub-Node, destSubnet = 0	All nodes of a domain
#0	Broadcast	Domain, sourceSub-Node, destSubnet	All nodes of a subnet
#1	Multicast	Domain, sourceSub-Node, destGroup	All nodes of a group
#2a	Unicast	Domain, sourceSub-Node, destSub-Node	One node within the subnet
#2b	Multicast	Domain, sourceSub-Node, destSub-Node, Group, Group Member	One node within the subnet
#3	Unicast	Domain, sourceSub-Node, destSub, Neuron ID	One node with name = Neuron ID

Optionally, routers can be used to keep local traffic within the subnet and only forward packets that are addressed to another subnet as shown in Figure 56.11. A CEA-709 node can send a unicast addressed message to exactly one node using either the physical Neuron ID, or the logical subnet/node (S/N) address, or a multicast addressed message to either a group of nodes (group address) or to all nodes in the subnet or all nodes in the entire domain (broadcast address). All possible addressing schemes are summarized in Table 56.1.

OSI Layer 4 supports four service types. The unacknowledged service simply transmits the data packet from the sender to the receiver and hopes that the message does not get lost or destroyed. The unacknowledged repeated service transmits the same data packet a number of times and assumes that at least one data packet will get through to the receiver. The number of retries is programmable. The acknowledged service transmits the data packet and expects an acknowledgment coming back from the receiver. If the acknowledgment is not received, the transmitter retransmits the same data packet. The number of retries is again programmable. This service is shown in Figure 56.12. Here the original message (transport protocol data unit [TPDU]) is lost on the way to the receiver; hence, the transmitting node is repeating this message after the transmit timer expires. This message is received at the receiving node, but the acknowledgment packet is lost on the way back to the original sender. Hence, the sender again sends out this message after the transmit timer expires. Now the TPDU is received at the receiver, and the acknowledgment is received at the sender, and therefore the transaction is completed and a new transaction can begin.

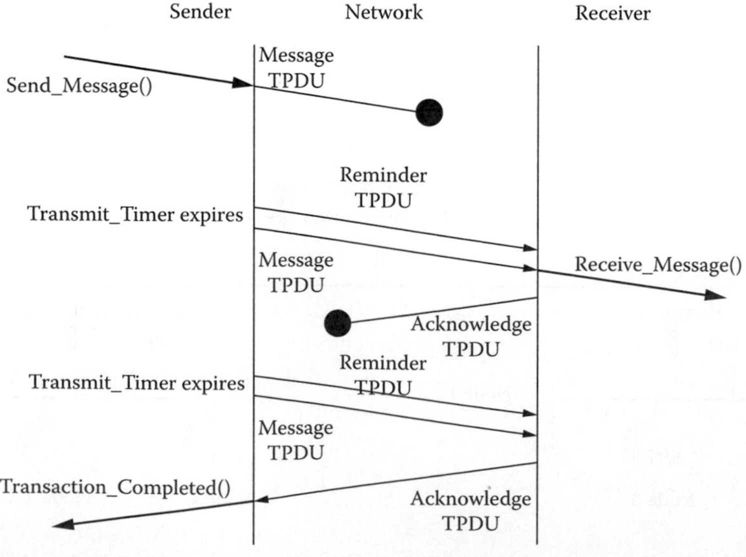

FIGURE 56.12 Acknowledged service with retries.

The request response service sends a request message to the receiver, and the receiver must respond with a response message. This service type can be used to retrieve, for example, statistics information from a network node. Authentication can be used to authenticate the sender at the receiver side.

56.7 Network Infrastructure

Great care must be taken when designing the network infrastructure for a CEA-709 network. There are a number of requirements that must be met in order to guarantee reliable communication between all the nodes in the network. The maximum cable length must be met, the network cables must be properly terminated, the number of nodes on each physical channel must not exceed the specification, the expected network traffic must be estimated, and a network topology chosen to keep local traffic local—only data packets that need to travel across network segments should leave the local segment. The bandwidth utilization must be kept below a certain limit, and the number of CRC errors, due to collisions and noise on the network media, must be measured on each channel and kept within well-defined boundaries. Only a reliable network infrastructure with built-in diagnostics and troubleshooting functionality can guarantee reliable operation of the building or factory.

Depending on the network media and the network transceiver, a variety of network topologies are possible with LonWorks nodes. Traditional bus, ring, star, as well as free topology are supported. Complex networks require networking elements like repeaters, routers, and gateways available in CEA-709 to segment the local traffic from traffic crossing segment boundaries. Such networking elements include Layer 1 repeaters to extend the physical length of a network cable, Layer 2 bridges, and Layer 3 routers to decouple individual segments from a networking backbone and gateways to bridge between the levels of hierarchy in the automation pyramid (see Figure 56.11).

Modern network infrastructure products [2] have built-in network diagnostics capabilities and can monitor the health state of the network 24 h a day 7 days a week. They immediately report network malfunctions or deviations from the normal operation to the system operator. All monitoring and reporting can be done either from local on-site or from remote through an intranet or Internet connection.

56.8 Interoperability and Profiles

Open standards allow individual companies to build single pieces of a bigger puzzle and enable system integrators (SIs) to design complete systems consisting of those building blocks. Node developers must follow certain rules in order to make the *bigger picture* work. These rules are defined in interoperability guidelines and in profiles. Compatible does not necessarily mean interoperable, and not at all plug and play. More elaborated guidelines need to be established for plug-and-play, interchangeable, and interworkable products. Different committees have set guidelines to define those terms and to establish the rules. The LonMark organization, for example, has published interoperability guidelines for nodes that use the LonTalk protocol [5]. It is important to note that interoperability must be guaranteed on all seven OSI layers. Interoperability on all seven OSI layers is still not a guarantee to have interworkable products. A typical node only uses a subset of all the possible protocol features. Profiles define these subsets based on the intended use of the node. Task groups within LonMark define functional profiles for analog input, analog output, temperature sensor, humidity sensor, CO_2 sensor, VAV controller, fan coil unit, chiller, thermostat, damper actuator, etc. As of today, LonMark hosts different task groups for fire, home/utility, petrol station, HVAC, lighting, sun blinds, security, etc., type of applications.

56.9 Development and Integration Tools

In order to program the Neuron chip, the user has two tool options. For node manufacturers, it might be sufficient to buy the NodeBuilder from Echelon; the more advanced and also more expensive tool is the LonBuilder from Echelon. Both tools allow writing Neuron C programs, to compile and link them and download the final application into the target node hardware. The LonBuilder supports simultaneous debugging of multiple nodes whereas the Nodebuilder only supports debugging of one node at the time. The LonBuilder has a built-in protocol analyzer and a network binder to create communication relationships between network nodes.

SIs need tools to design the network and define the communication relationships before the nodes are actually installed in the field. Similar to creating a schematic of resistors, capacitors, ICs, diodes, etc., the SI creates a floor plan of all the nodes in the network. After physically installing the nodes in the network, a network management tool is required to configure (also called commission) the nodes. During commissioning, nodes get a logical address (subnet/node/domain) assigned, and the binding is created. Binding is the name of the process to establish the communication relationships between sensors, actuators, and controller nodes. Protocol analyzers are used to debug communication problems and to gather traffic and error statistics. During the maintenance phase of the network, the network management tool also supports replacing faulty nodes and extending the network with additional nodes and changes in the communication relationships.

56.10 New Developments for ANSI/CEA-709 Networks

As mentioned earlier, until 1999, the Neuron chip was the only vehicle to execute the LonTalk protocol. Due to the limited resources available on the Neuron chip (CPU performance, memory space, I/O capabilities, etc.), many applications could not be realized with the existing technology. Recent hardware platform-independent implementations of the ANSI/EIA709 protocol allow for good scalability of the CPU performance and memory requirements tailored to the specific application. Furthermore, simulation of the network behavior before actual hardware exists is now possible as shown in Figure 56.13.

Different application programs run on top of the ANSI/EIA709 protocol stack that talks to an interface layer (LDI), which communicates with the virtual network devices. As shown on the right-hand

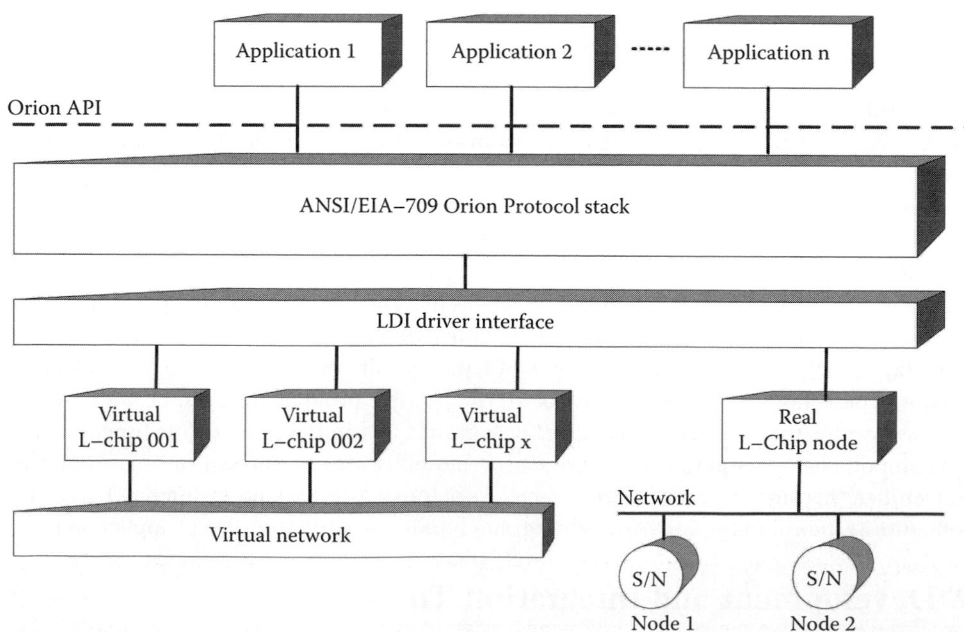

FIGURE 56.13 Platform-independent ANSI/EIA709 implementation. Simulation environment with virtual nodes linked to a physical network.

side in Figure 56.13, the LDI can also talk to a real physical network; hence, existing nodes in a network can be part of the system simulation. The application programs do not see a difference if the node already exists as a physical device or is a virtual device simulated on a PC or workstation [6,7]. This architecture is fully portable between different hardware platforms and different operating systems, and allows cross development, for example, on a Windows platform for an ARM9-based target hardware. Since the application programming interface on top of the CEA-709 protocol stack is the same on all hardware platforms, and under all operating systems, the final application programs can easily be ported between different target systems. This approach allows full hardware–software codesign since the application programs can be developed, while the hardware development of the network node is still going on.

With the new ANSI/CEA-852 standard that defines how CEA-709 packets are tunneled through IP networks, standard hardware platforms with an Ethernet interface like PCs or embedded controllers can be used to develop LonWorks-based network nodes.

Another development that has achieved wide acceptance are multiprotocol processors that can support multiple different protocols. These processors are also very well suited for gateway applications to bridge the gap between various protocol standards. One such development is shown in Figure 56.14. The LC3K chip supports the ANSI/CEA-709 protocol, BACnet, and various IP-based protocols all within a single device and all at the same time [8]. This is especially important in today's converging application domains that formerly have been isolated installations. Modern systems implement cross-domain functions that require gateways or true multiprotocol devices that expose control data into other technologies or provide multitechnology interfaces.

Safety and redundancy are emerging topics in CEA-709 networks. Redundant network topologies offer protection against single failures of cable (e.g., cable break of an FT-10 two-wire installation) and failure of routers (twin-router concept). Products such as the L-IP Redundant enable building network infrastructure that is robust against those single failures and provide an alarming mechanism to notify facility management about failures. This marks one step into the direction of functional safety. Another development is the EU research project SafetyLon. SafetyLon nodes can coexist on the network with regular nodes and embed their control network data into regular CEA-709 frames. That has the advantage of using existing network infrastructure components. The embedded SafetyLon protocol has a much lower residual error probability. In addition, the SafetyLon hardware consists of two tightly coupled CPUs in one device, which monitor each other, so that SIL-3-compliant applications can be built with CEA-709.

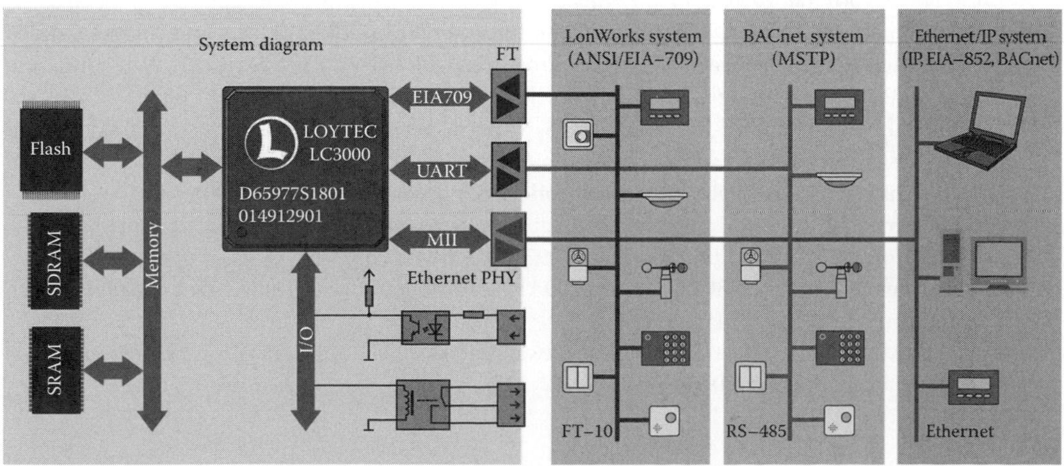

FIGURE 56.14 Multiprotocol processors can support different fieldbus protocols.

References

1. ISO/IEC 14908-1:2012, Information technology—Control network protocol—Part 1: Protocol stack.
2. H. Schweinzer, VENUS—Vienna embedded networking utility suite, Loytec Electronics GmbH, Vienna, Austria, www.loytec.com, October 1999.
3. Motorola Inc., LonWorks technology device data, DL159, Rev 4, Q4/97, 1997.
4. D. Dietrich, D. Loy, and H.-J. Schweinzer, *LON-Technologie, Verteilte Systeme in der Anwendung; 2. Auflage*, Hüthig Verlag, Heidelberg, Germany, 1999.
5. LonMark, Application layer interoperability guidelines, Version 3, LonMark Interoperability Association, San Jose, CA, 1996.
6. A. Bauer, LC-SIM, Loytec Electronics GmbH, Vienna, Austria, www.loytec.com, March 2000.
7. A. Bauer and S. Soucek, Simulation of ANSI/EIA709 Networks, LonWorld, London, U.K., October 2000.
8. D. Loy and H. Schweinzer, LISA—LOYTEC IP Services ASIC, Loytec Electronics GmbH, Vienna, Austria, www.loytec.com, April 2004.
9. D. Dietrich, P. Neumann, and H. Schweinzer, *Fieldbus Technology—System Integration, Networking, and Engineering*, Springer Wien, New York, 1999.
10. A. S. Tanenbaum, *Computer Networks*, 2nd edn., Prentice-Hall Inc., Englewood Cliffs, NJ, 1989.
11. A. S. Tanenbaum, *Distributed Operating Systems*, Prentice-Hall Inc., Upper Saddle River, NJ, 1994.
12. C. J. Koomen, *The Design of Communicating Systems*, Kluwer Academic Publishers, Boston, MA, 1991.
13. H. Kopetz, *Design Principles for Distributed Embedded Applications*, Kluwer Academic Publishers, Boston, MA, 1997.
14. P. K. Sinha, *Distributed Operating Systems*, The Institute of Electrical and Electronics Engineers, Inc., New York, 1997.
15. G. H. Gürtler, Fieldbus standardization, the European approach and experiences, in: *Feldbustechnik in Forschung, Entwicklung und Anwendung*, Springer, Heidelberg, Germany, 1997, pp. 2–11.
16. J.-P. Thomesse and M. Leon Chavez, Main paradigms as a basis for current fieldbus concepts, in: *Fieldbus Technology*, Springer, Berlin, Germany, 1999, pp. 2–15.
17. D. Dietrich and T. Sauter, Evolution potentials for fieldbus systems, in: *Proceedings WFCS'00, IEEE*, Porto, Portugal, 2000.
18. T. Sauter and P. Palensky, A closer look into Internet-fieldbus connectivity, in: *Proceedings WFCS'00, IEEE*, Porto, Portugal, 2000.
19. Th. Bangemann, R. Dübner, and A. Neumann, Integration of fieldbus objects into computer-aided network facility management systems, in: *Proceedings of FeT'99*, D. Dietrich, P. Neumann, and H. Schweinzer (eds.), Springer, Wien, New York, 1999, pp. S.180–S.187.
20. S. Rüping, H. Klugmann, K.-H. Gerdes, and S. Mirbach, Modular OPC-server connecting different fieldbus systems and Internet Java Applets, in: *Proceedings of FeT'99*, D. Dietrich, P. Neumann, and H. Schweinzer (eds.), Springer, Wien, New York, 1999, pp. S.240–S.246.
21. P. Neumann and F. Iwanitz, Integration of fieldbus systems into distributed object-oriented systems, in: *Proceedings WFCS'97, IEEE*, New York, 1997, pp. S.247–S.253.
22. U. Döbrich and P. Noury, ESPRIT Project NOAH—Introduction, in: *Fieldbus Technology*, Springer, Berlin, Germany, 1999, pp. 414–422.
23. P. Palensky, The convergence of intelligent software agents and field area networks, in: *Proceedings of ETFA'99*, Barcelona, Spain, 1999, pp. 917–922.

57

BACnet: Communication Protocol for Building Automation and Control Networks

Frank Schubert
MBS GmbH

57.1 BACnet as a Standard 57-2
57.2 BACnet: The Idea of an Open Standard 57-2
57.3 Basic Idea of BACnet: Interoperability! 57-2
57.4 BACnet Organizations 57-2
57.5 Three Major Parts of BACnet 57-3
Network Media (BACnet Data-Link Layer) • BACnet Object Model • BACnet Services
57.6 BACnet Network Numbers 57-6
57.7 Device Address Binding 57-6
57.8 BACnet Transaction Handling 57-7
57.9 Encoding BACnet Telegrams 57-7
57.10 BACnet Procedures 57-8
Backup/Restore • Command Prioritization • Elapsed Active Time Counting/Status Change Counting • Minimum On or Off Times
57.11 Proprietary Extensions 57-9
57.12 BACnet Conformance 57-9
57.13 BACnet Testing, Listing, and Certification 57-10
57.14 Interoperability Workshops 57-10
57.15 Character Sets 57-11
57.16 Network Security 57-11
57.17 BACnet/IP Specific Functionality 57-11
57.18 MS/TP Specific Functionality 57-12
57.19 Vendor IDs 57-12
57.20 BACnet Mappings 57-13
Web Services • KNX Mapping • OPC UA Mapping
57.21 Conclusion 57-13

57.1 BACnet as a Standard

BACnet (Building automation and control networks), as the name implies, is a communication protocol specifically designed for the purpose of building automation. Since the committee SSPC 135 within the ASHRAE (American Society of Heating, Refrigerating and Air-Conditioning Engineers) started the specification activities in 1987, BACnet was continuously maintained and extended. Right after the first publication in 1995, BACnet became ANSI (American National Standards Institute) standard and has meanwhile become part of the ISO-Norm 16484 (part 5, BACnet Standard, and part 6, Conformance Testing Standard).

57.2 BACnet: The Idea of an Open Standard

The basic idea of BACnet is interoperability between systems of different vendors. In the 1980s, mainly proprietary systems have been sold, and connecting devices from different vendors always required a programmer to develop an interface between two proprietary systems and communication protocols. If a third, or fourth, vendor entered the project, those installations became a nightmare for the facility manager.

Mike H. Newman, responsible for the facility management at Cornell University, started the initiative to specify a communication protocol designed for building automation, later called BACnet.

Meanwhile, many companies changed the communication of their building automation components from proprietary communication to BACnet. Using BACnet in projects does not mean that always multivendor systems must be installed. Today, it may easily happen that BACnet is used by a single company in a project to communicate between devices of the same product family. BACnet, like other open communication protocols, is about to replace proprietary systems and proprietary communication more and more.

57.3 Basic Idea of BACnet: Interoperability!

The basic idea of BACnet is that vendors following the BACnet standard will be able to exchange data on an interoperable basis. As we will see later in this book, BACnet does not stop at the data-exchange level. BACnet specifies highly sophisticated functionality like file access, alarming, scheduling, trend logging, and remote device management.

57.4 BACnet Organizations

The main organization driving BACnet is ASHRAE with their working group SSPC135 maintaining the standard. ASHRAE as the head organization publishes the standard and addenda. SSPC135 over the years has become an international committee with working group members from all over the world.

The ISO organization maintains BACnet as an ISO standard in the international norms ISO 16484-5 and 6.

Several interest groups worldwide promote the BACnet technology; the largest is the BACnet Interest Group Europe (www.big-eu.org).

Last but not least, BACnet International (www.bacnetinternational.org) represents North American vendors (formerly the BACnet Manufacturers Association) and end users (formerly the BACnet Interest Group North America).

57.5 Three Major Parts of BACnet

Like other communication protocols, BACnet specifies three major parts:

1. The network media (so-called data-link layer)
2. Objects to transport the building automation data
3. Services to transport the data between BACnet devices

57.5.1 Network Media (BACnet Data-Link Layer)

BACnet uses a 4-layer collapsed architecture. According to the ISO/OSI (International Organization for Standardization/Open Systems Interconnection) model, layers 1, 2, 3, and 7 are specified by BACnet (layers 4–6 are not required and so omitted) (Figure 57.1).

The network layer 3 provides routing between the different physical media. The most commonly used data-link layers are *BACnet*/Internet Protocol (*IP*) (IPv4 for now, but IPv6 is already specified as well) and master–slave/token passing (*MS/TP*, a layer 2 protocol based upon EIA-485 serial networks specifically designed for BACnet). While BACnet/IP is mainly used for automation stations (programmable direct digital control (DDC) controllers or programmable logic controllers (PLCs)) and operator management stations, devices based on EIA-485 are likely intelligent field devices like variable frequency drives, valves, pumps, VAV boxes, and room controllers.

Point-to-point connections using the BACnet PTP (point to point) protocol may connect different BACnet subnetworks using public phone lines. In the times of the Internet, modem connections likely disappear though.

57.5.2 BACnet Object Model

BACnet uses a decentralized approach to store all information related to the physical or virtual data points. Compared to classical systems where only the present value was requested and all information was added in the management station, BACnet objects provide all information related to the data point in the device processing the data.

As of today, BACnet specifies a total of more than 50 different object types. Object types include analog, binary and multistate input, output, and values. Additionally, primitive value objects represent simple data. In addition to the basic types, objects to represent more specific functionality like load control, structured views, groups, commands, accumulators and pulse converters, trend-log data, schedule programs, calendars, notification classes, lighting output and channels, and counter values are

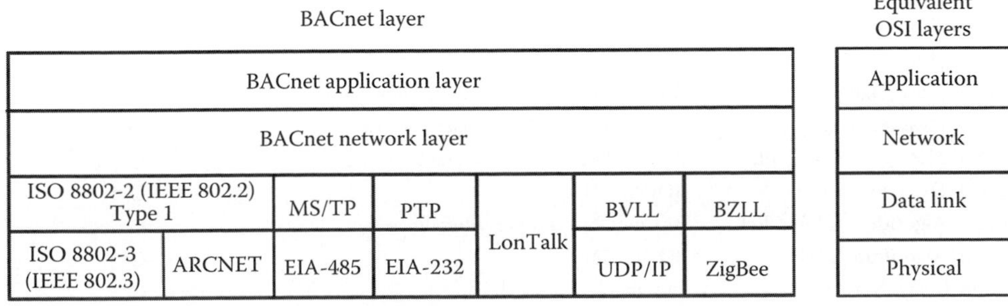

FIGURE 57.1 BACnet data-link layers in the ISO/OSI model. (From the ANSI/ASHRAE 135-2012 standard, Copyright and used with permission from the American Society of Heating, Refrigerating and Air-Conditioning Engineers.)

specified in the standard as well. Devices like fire alarm systems may be represented through the life-safety objects. To represent access-control devices, BACnet specifies a total of seven objects. In a future revision of the standard, BACnet will support representation of elevators and escalators.

All objects consist of properties that represent the physical or virtual information. A property may be the present value, which is likely the most important information. The engineering unit, alarm limits, and even state text information are provided by the properties as well.

A minimum set of properties is specified as being required (conformance code = "R"), others may be implemented as an option ("O"), and few others are required to be writable ("W") (Table 57.1).

TABLE 57.1 Example of an Analog Output Object as Specified in the Standard Model

Property Identifier	Property Datatype	Conformance Code
Object_Identifier	BACnetObjectIdentifier	R
Object_Name	CharacterString	R
Object_Type	BACnetObjectType	R
Present_Value	REAL	W
Description	CharacterString	O
Device_Type	CharacterString	O
Status_Flags	BACnetStatusFlags	R
Event_State	BACnetEventState	R
Reliability	BACnetReliability	O
Out_Of_Service	BOOLEAN	R
Units	BACnetEngineeringUnits	R
Min_Pres_Value	REAL	O
Max_Pres_Value	REAL	O
Resolution	REAL	O
Priority_Array	BACnetPriorityArray	R
Relinquish_Default	REAL	R
COV_Increment	REAL	O[1]
Time_Delay	Unsigned	O[2,4]
Notification_Class	Unsigned	O[2,4]
High_Limit	REAL	O[2,4]
Low_Limit	REAL	O[2,4]
Deadband	REAL	O[2,4]
Limit_Enable	BACnetLimitEnable	O[2,4]
Event_Enable	BACnetEventTransitionBits	O[2,4]
Acked_Transitions	BACnetEventTransitionBits	O[2,4]
Notify_Type	BACnetNotifyType	O[2,4]
Event_Time_Stamps	BACnetARRAY[3] of BACnetTimeStamp	O[2,4]
Event_Message_Texts	BACnetARRAY[3] of CharacterString	O[3,4]
Event_Message_Texts_Config	BACnetARRAY[3] of CharacterString	O[4]
Event_Detection_Enable	BOOLEAN	O[2,4]
Event_Algorithm_Inhibit_Ref	BACnetObjectPropertyReference	O[4]
Event_Algorithm_Inhibit	BOOLEAN	O[4,5]
Time_Delay_Normal	Unsigned	O[4]
Reliability_Evaluation_Inhibit	BOOLEAN	O[6]
Property_List	BACnetARRAY[N] of BACnetPropertyIdentifier	R
Profile_Name	CharacterString	O

Source: The ANSI/ASHRAE 135-2012 standard, Copyright and used with permission from the American Society of Heating, Refrigerating and Air-Conditioning Engineers.

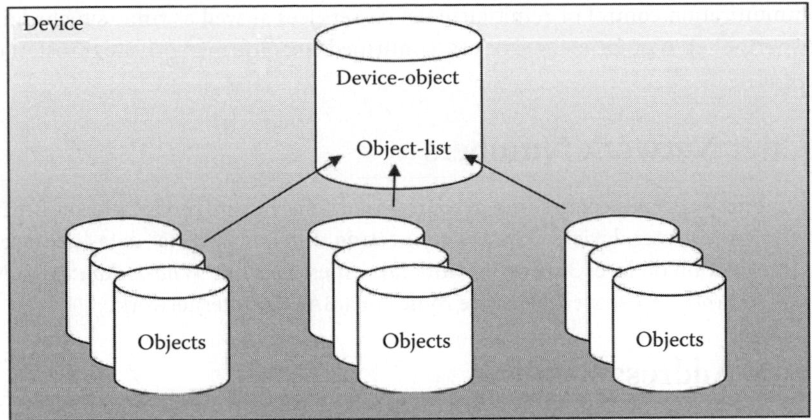

FIGURE 57.2 BACnet objects addressed through the device object.

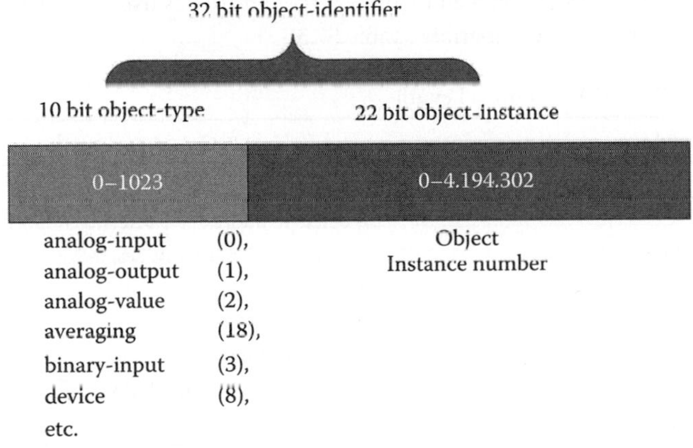

FIGURE 57.3 Object ID.

BACnet objects are represented in the network through the device objects. Every device (and the device object) in the network shall be unique, and every object within the device shall be unique as well.

The object list in the device object presents a list of all objects contained in the device (Figure 57.2).

To uniquely address objects, BACnet uses a 32-bit identifier (ID), which consists of a 10-bit type and a 22-bit instance number. In addition, every object uses a unique object name to allow name resolution.

Object ID 4.194.303 is reserved as a wildcard and shall not be used to address a device (Figure 57.3).

57.5.3 BACnet Services

A total of 35 application services provide access to the devices, objects, and properties.

Another set of 20 services on layer 3 is used to exchange router information and provide security services for authentication.

The BACnet application layer specifies services for data access, alarming, scheduling, trend logging, and device and network management.

Basic services are read or write services that are likely the most commonly used services. More sophisticated services allow file transfer, device discovery in the network, read range from logging objects, and alarm or change-of-value notifications.

Device communication control or reinitialize device services as well as time synchronization, text-message transport, or even proprietary services (confirmed/unconfirmed private transfer) complement the basic services.

57.6 BACnet Network Numbers

Every BACnet subnetwork connected through routers is uniquely identified by a 16-bit BACnet network number known by the router devices. Routers connecting different BACnet data-link layers exchange their routing information on layer 3 using network layer messages like *Who-Is-Router-To-Network* and *I-Am-Router-To-Network* to discover the active routes in a BACnet internetwork.

57.7 Device Address Binding

Beyond the logical addressing using the device-object ID, BACnet physically addresses devices using the so-called BACnet MAC address. The BACnet MAC address consists of the BACnet network number in the range from 1 to 65.535 plus the MAC address assigned to the device (Table 57.2).

Virtual MAC addressing was introduced to allow larger addresses used in Zigbee or IPv6 but to keep the existing network layer header compatible (Table 57.3).

TABLE 57.2 MAC-Address Length

Data-Link Layer	MAC-Address Length
Ethernet	6 octets address of the network card
IPv4	4 octets IP address + 2 octets UDP port
MS/TP (EIA-485)	1 octet 0–127 master nodes, 0–254 slave nodes
Arcnet	1 octet
LonTalk	2–7 octets
Zigbee	3 octets virtual MAC address
IPv6	3 octets virtual MAC address
Virtual networks	Up to 6 octets

TABLE 57.3 Examples of BACnet MAC Addresses

135, 0xC0 A8 01 02 BA C0	Network # 135, address 192.168.1.2, port 47808, BACnet/IP
23, 0x7F	Network # 23, address 127, e.g., MS/TP

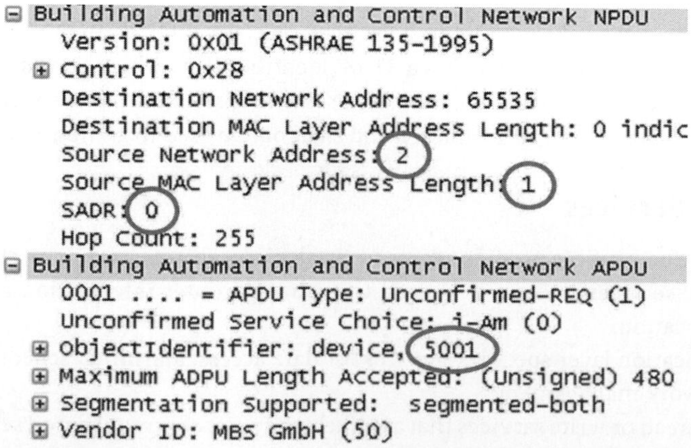

```
⊟ Building Automation and Control Network NPDU
    Version: 0x01 (ASHRAE 135-1995)
  ⊞ Control: 0x28
    Destination Network Address: 65535
    Destination MAC Layer Address Length: 0 indic
    Source Network Address: 2
    Source MAC Layer Address Length: 1
    SADR: 0
    Hop Count: 255
⊟ Building Automation and Control Network APDU
    0001 .... = APDU Type: Unconfirmed-REQ (1)
    Unconfirmed Service Choice: i-Am (0)
  ⊞ ObjectIdentifier: device, 5001
  ⊞ Maximum ADPU Length Accepted: (Unsigned) 480
  ⊞ Segmentation Supported:  segmented-both
  ⊞ Vendor ID: MBS GmbH (50)
```

FIGURE 57.4 Device address binding (physical and logical address).

As an alternative to the optional manual configuration, the services *Who-Is* and *I-Am* provide a method for dynamic device binding. In response to a *Who-Is* request or spontaneously sent (e.g., at the time of the device start), the *I-Am* service provides the logical address (device ID) sent from the physical address. In case the telegram is forwarded by a router, the actual physical address is encoded in the network protocol data unit (NPDU) (network layer) (Figure 57.4).

Dynamic object binding is provided by the *Who-Has* and *I-Have* services. A device may ask for an object name on the network, and other devices containing such an object respond with the respective object ID.

The dynamic device and object binding services are typically used as broadcast messages in the network but may be transported using unicast addressing as well.

57.8 BACnet Transaction Handling

BACnet communication uses a client/server model to exchange requests and responses to or from BACnet devices. Telegrams may be unconfirmed, confirmed, and optionally segmented. Using confirmed services, the requestor may get the expected response (e.g., the current value of the requested property). In case of errors, the responder uses an appropriate error class/error code to inform about the error reason (e.g., Property/Value_out_of_range or Services/File_Access_denied). Additionally, reject-messages and abort-messages demonstrate an invalid communication between two devices on the network.

57.9 Encoding BACnet Telegrams

BACnet messages are encoded into a sequence of tags and values according to the relevant ASN.1 definitions specified in the BACnet standard.

Figures 57.5 through 57.7 demonstrate the encoding of a typical request/response interaction between two devices. The capture was taken using the free open-source protocol analyzer *Wireshark* (www.wireshark.org), which provides an excellent BACnet decoder. For demonstration purposes, the pictures show the telegrams starting on top of the UDP communication and do not show the IP part of the telegram.

Using the ASN.1 syntax, BACnet telegrams are relatively small and compact. The majority of requests in a BACnet network typically have a length between 64 and 256 octets.

```
⊟ BACnet Virtual Link Control
    Type: BACnet/IP (Annex J) (0x81)
    Function: Original-Unicast-NPDU (0x0a)
    BVLC-Length: 4 of 22 bytes BACnet packet length
⊟ Building Automation and Control Network NPDU
    Version: 0x01 (ASHRAE 135-1995)
  ⊞ Control: 0x24
    Destination Network Address: 2
    Destination MAC Layer Address Length: 1
    DADR: 0
    Hop Count: 255
⊟ Building Automation and Control Network APDU
    0000 .... = APDU Type: Confirmed-REQ (0)
  ⊞ .... 0010 = PDU Flags: 0x02
    .101 .... = Max Response Segments accepted: 32 segments (5)
    .... 0101 = Size of Maximum ADPU accepted: Up to 1476 octets (fits in an ISO 8802-3 frame) (5)
    Invoke ID: 1
    Service Choice: readProperty (12)
  ⊞ ObjectIdentifier: device, 5001
  ⊞ Property Identifier: object-name (77)
```

FIGURE 57.5 Example of a ReadProperty request asking for the object name of device, 5001.

```
⊟ BACnet Virtual Link Control
    Type: BACnet/IP (Annex J) (0x81)
    Function: Original-Unicast-NPDU (0x0a)
    BVLC-Length: 4 of 45 bytes BACnet packet length
⊟ Building Automation and Control Network NPDU
    Version: 0x01 (ASHRAE 135-1995)
  ⊞ Control: 0x08
    Source Network Address: 2
    Source MAC Layer Address Length: 1
    SADR: 0
⊟ Building Automation and Control Network APDU
    0011 .... = APDU Type: Complex-ACK (3)
  ⊞ .... 0000 = PDU Flags: 0x00
    Invoke ID: 1
    Service Choice: readProperty (12)
  ⊞ ObjectIdentifier: device, 5001
  ⊞ Property Identifier: object-name (77)
  ⊞ {[3]
  ⊞ object-name: UTF-8 'UBR-01 BACnet router'
  ⊞ }[3]
```

FIGURE 57.6 Example of a ReadProperty response providing the object name of device, 5001: "UBR-01 BACnet router."

```
0000  00 1f f3 8c 20 e4 00 1f  25 00 01 e3 08 00 45 00   .... ... %.....E.
0010  00 49 00 00 40 00 40 11  b7 50 c0 a8 01 02 c0 a8   .I..@.@. .P......
0020  01 01 ba c0 ba c0 00 35  1d 12 81 0a 00 2d 01 08   .......5 .....-..
0030  00 02 01 00 30 01 0c 0c  02 00 13 89 19 4d 3e 75   ....0... .....M>u
0040  15 00 55 42 52 2d 30 31  20 42 41 43 6e 65 74 20   ..UBR-01  BACnet
0050  72 6f 75 74 65 72 3f                               router?
```

FIGURE 57.7 The response telegram in Figure 57.9 in hexadecimal (including all lower layers).

57.10 BACnet Procedures

BACnet is specifically designed to fit the requirements of building automation installations. For this, BACnet specifies procedures for backing up or restoring a device configuration by interoperable services and introduces a schema to realize command prioritization.

57.10.1 Backup/Restore

A client station invokes a backup or restore by calling a special form of the device communication control service, and the device being backed up or restored provides information about the file objects to be written to or read from the device. The content of the files may still be proprietary (some use XML, others binary data—this is up to the vendor), but the general procedure is interoperable between vendors.

57.10.2 Command Prioritization

In building automation systems, multiple entities may control and manipulate an object. There may be a need to arbitrate between a schedule program, manual operation by the facility manager, and access from the service technician. For so-called commandable properties (like the present value of output-object types), BACnet uses a 16-level prioritization schema where 1 is the highest priority (Manual Life Safety) and 16 is the lowest. A default value is taken from the property "Relinquish_Default" in case all slots in the priority array are uninitialized (the data-type "NULL" is used to identify an uninitialized entry = *no value* for either of the 16 slots).

57.10.3 Elapsed Active Time Counting/Status Change Counting

Binary objects may be implemented to support elapsed active time counting and/or change-of-state counting. This functionality is often used in building automation systems to detect necessary maintenance, for example, a filter change.

57.10.4 Minimum On or Off Times

Binary objects may optionally support the tracking of minimum on or off times. For example, chiller devices or gas lamps require a certain on time before they may be shut down. Serious damage will likely appear if those devices are switched off earlier than specified.

The minimum time functionality provided on priority level 6 assures that only services with a higher priority (service technicians, manual life-safety operation, or critical-equipment control) will be allowed to invoke a shutdown.

57.11 Proprietary Extensions

Even though BACnet is designed to be an open and vendor-independent communication protocol, it allows vendors to extend an implementation by using proprietary communication.

Data-link layer may be implemented as long as it is routable (implementing the network layer 3) or may use tunneling. In the past, there have been implementations supporting, for instance, the token ring (by International Business Machines Corporation [IBM]) networks even though this network standard is not specified by BACnet.

Proprietary objects may be designed as long as they support the addressing schema and present at least the three properties: Object_Identifier, Object_Name, and Object_Type. Standard objects may be enhanced by adding proprietary properties.

Services may be defined by using the confirmed private transfer/unconfirmed private transfer services.

57.12 BACnet Conformance

Devices may be specified from implementing a small sensor or actuator device up to freely programmable controller or workstation. This means the level of BACnet functionality implemented varies from device to device.

How to assure interoperability of different devices?

Every vendor may provide a BACnet protocol implementation conformance statement (PICS). This *BACnet data sheet* explains the level of BACnet functionality implemented in the device described in the PICS. The major parts of the PICS are again the supported network layers, the BACnet objects, and the BACnet services.

By comparing the PICS documents, interested parties like project designers may check if the supported functionality matches their requirements in a project and if the level of implementation matches those of other devices intended to interact with the device.

So-called BACnet Interoperability Building Blocks (BIBBs) represent functional blocks of services supported by the device. BIBB type "A" specifies the initiator of BACnet services, while "B" specifies the executor of that services.

Example

A server device provides the capability to allow others to read data from the device:

- Data-Sharing-Read-Property-B (DS-RP-B) demonstrates support for the ReadProperty service executing requests from other devices.
- Data-Sharing-Read-Property-A (DS-RP-A) specifies the capability to initiate a Read Property request to retrieve data from a device providing Read Property B.

If for a specific interoperability area both A and B are present, the functional service specified by these BIBBs may be used to interact between the two.

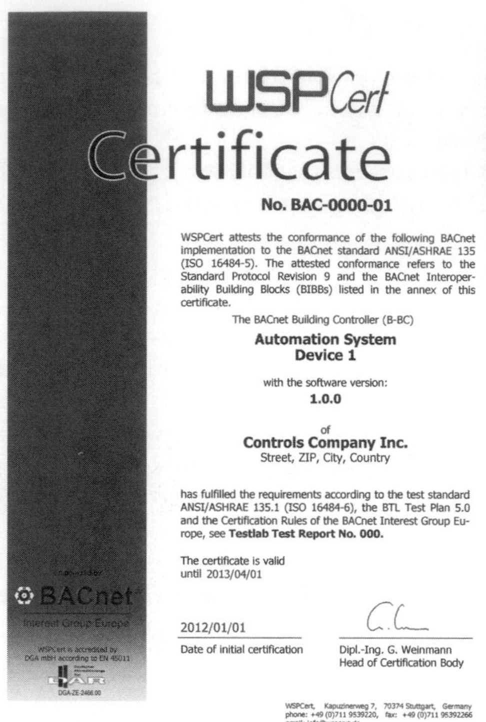

WSPCert

Certificate

No. BAC-0000-01

WSPCert attests the conformance of the following BACnet implementation to the BACnet standard ANSI/ASHRAE 135 (ISO 16484-5). The attested conformance refers to the Standard Protocol Revision 9 and the BACnet Interoperability Building Blocks (BIBBs) listed in the annex of this certificate.

The BACnet Building Controller (B-BC)

Automation System
Device 1

with the software version:

1.0.0

of

Controls Company Inc.
Street, ZIP, City, Country

has fulfilled the requirements according to the test standard ANSI/ASHRAE 135.1 (ISO 16484-6), the BTL Test Plan 5.0 and the Certification Rules of the BACnet Interest Group Europe, see **Testlab Test Report No. 000.**

The certificate is valid
until 2013/04/01

2012/01/01
Date of initial certification

Dipl.-Ing. G. Weinmann
Head of Certification Body

BACnet
Interest Group Europe

WSPCert is accredited by DGA mbH according to EN 45011

DGA-ZE-2466.00

WSPCert, Kapuzinerweg 7, 70374 Stuttgart, Germany
phone: +49 (0)711 9539220, fax: +49 (0)711 95392266
email: info@wspcert.de

(a) Cover page of certificate, general information about the device

WSPCert

Annex to Certificate No. BAC-0000-01

The certification of the BACnet Implementations
Automationssystem, Device 1
of **Controls Company Inc.**

refers to the BACnet Interoperability Building Blocks (BIBBs) listed below. BIBBs other than listed are not part of the certification.

BACnet Interoperability Building Blocks	
DS-RP-A	Data Sharing – Read Property – A
DS-RP-B	Data Sharing – Read Property – B
DS-RPM-A	Data Sharing – Read Property Multiple – A
DS-RPM-B	Data Sharing – Read Property Multiple – B
DS-WP-A	Data Sharing – Write Property – A
DS-WP-B	Data Sharing – Write Property – B
DS-WPM-B	Data Sharing – Write Property Multiple – B
DS-COV-A	Data Sharing – Change Of Value – A
DS-COV-B	Data Sharing – Change Of Value – B
AE-N-I-B	Alarm and Event – Notification Internal – B
AE-ACK-B	Alarm and Event – Acknowledge Alarm – B
AE-ASUM-B	Alarm and Event – Alarm Summary -B
AE-ESUM-B	Alarm and Event – Event-Enrollment Summary-B
AE-INFO-B	Alarm and Event – Information – B
SCHED-I-B	Scheduling – Internal – B
SCHED-E-B	Scheduling – External – B
T-VMT-I-B	Trending – Viewing and Modifying Trends Internal – B
T-ATR-B	Trending – Automated Trend Retrieval – B
DM-DDB-A	Device Management – Dynamic Device Binding – A
DM-DDB-B	Device Management – Dynamic Device Binding – B
DM-DOB-B	Device Management – Dynamic Object Binding – B
DM-DCC-B	Device Management – Device Communication Control – B
DM-TS-B	Device Management – Time Synchronisation – B
DM-UTC-B	Device Management – UTC Time Synchronisation – B
DM-RD-B	Device Management – Reinitialize Device – B
DM-BR-B	Device Management – Backup and Restore – B
DM-OCD-B	Device Management – Object Creation and Deletion – B
DM-LM-B	Device Management – List Manipulation– B

(b) Details page of the certificate, details about the supported BACnet Interoperability Building Blocks (BIBBs)

FIGURE 57.8 Example of a (neutral) BACnet certificate.

57.13 BACnet Testing, Listing, and Certification

The organizations BACnet International and BACnet Interest Group Europe started a global testing, listing, and certification program for BACnet devices. A BACnet Testing Laboratories (BTL)–recognized laboratory is required to be accredited according to ISO 17025 and shall be recognized by the two organizations to perform BACnet testing services. A test report issued by the laboratory may be used to request a listing for the tested product at BACnet International. This includes allowance to use the BTL mark on the device as long as the implementation is kept conform to the standard.

Many project designers and end customers in Europe request a certificate, which is required to fulfill European requirements for energy performance. A test report from one of the recognized laboratories may be used to request a certificate for the device (Figure 57.8).

Round-robin tests performed by all recognized BTL worldwide assure that the laboratories come to comparable results.

The test specifications (BTL test package) are maintained by the international working group BTL-WG of the BACnet International. A major part of the testing specifications is documented in ISO 16484-6 (the testing standard).

57.14 Interoperability Workshops

Vendors wanting to test their implementations against implementations of other vendors may attend interoperability workshops (aka *BACnet Plugfests*). BACnet International and the BACnet Interest Group Europe organize these events (twice a year, in spring at different places in Europe and in fall in the United States) (Figure 57.9).

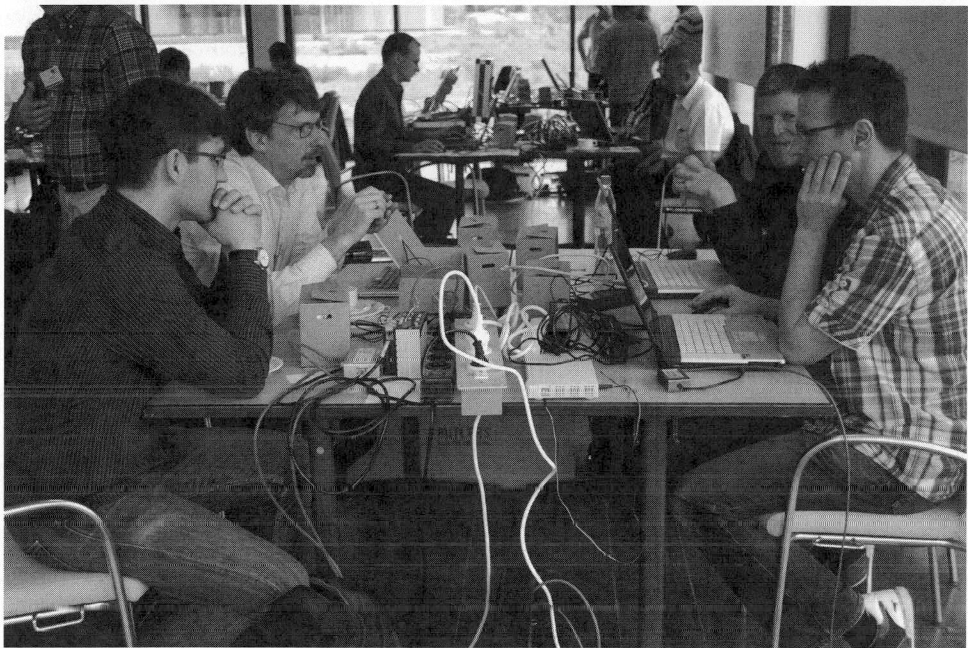

FIGURE 57.9 Picture taken at the European Plugfest 2013 in Berlin, Germany.

57.15 Character Sets

When BACnet started in 1995, the main focus was the North American target market. Character sets supporting special characters have only been rarely used at that time. Mainly the 7-bit character set ANSI X3.4 was used. Over the time, BACnet became an international standard, and BACnet was used in many countries where special characters are required.

To smoothly move from ANSI X3.4 to international character set support in BACnet, 135-2010 UTF-8 (encoded according to ISO 10646) replaced ANSI X3.4—considering this is still upward compatible because the first 7 bits are identical.

The long-term approach is to use international characters only.

57.16 Network Security

Designed as an open communication protocol, BACnet is mainly used openly in projects. Building automation components, as we know today, rarely uses network security yet, even though BACnet provides up-to-date network security features.

There is still space left for improving the network security of such components, especially when connecting them to the Internet. At the moment, the security (like in other communication protocols) in most building automation projects is to be provided by the network infrastructure due to missing security features.

In fire-, security-, or access-control systems, BACnet network security features are most likely implemented while this is often missing in building automation components.

57.17 BACnet/IP Specific Functionality

In IT networks, IP routers connect small subnetworks and most likely block broadcast messages. As we have seen earlier, BACnet uses broadcast messages, for instance, for dynamic device and object binding. In such networks where routers block broadcast messages, every subnetwork may be equipped with a

BBMD (BACnet Broadcast Management Device. A BBMD transports broadcast messages as forwarded unicast messages to other BBMDs, where the broadcast is then locally distributed.

A BACnet-FD (Foreign Device) is a single device connected to a BBMD without the capability to distribute the broadcast to the local subnetwork.

BBMD and FD support is often implemented in building automation controllers and workstations and may be activated where necessary.

Reserved BACnet/IP port: BACnet reserved the port Transmission Control Protocol (TCP) and UDP dec. 47808. This is easy to remember; in hexadecimal, the reserved port is 0xBAC0. BACnet may be transported through all other ports as well.

At the moment, BACnet only uses UDP (the User Datagram Protocol) and does not communicate through TCP (except for other services like a Hypertext Transfer Protocol [HTTP] browser interface). In the context of using the transport layer security, the use of TCP for BACnet is already under discussion in the BACnet community.

57.18 MS/TP Specific Functionality

MS/TP is a serial network where master nodes share the communication on the bus. Slave devices only respond to requests by master nodes. This means only master nodes may be used for dynamic device and object binding (a slave node is not allowed to initiate a telegram on the network).

To allow this feature for slave nodes as well, a master node may support the slave proxy functionality. On behalf of the recognized slaves, the master responds to *Who-Is* and *Who-Has* telegrams as a proxy.

57.19 Vendor IDs

Vendors of BACnet devices may request a unique vendor ID from ASHRAE. The growth of the number of vendor IDs assigned clearly shows the BACnet success over the past years (Figure 57.10).

As of February 19, 2014, a total of 726 vendors worldwide have requested and were assigned a vendor ID.

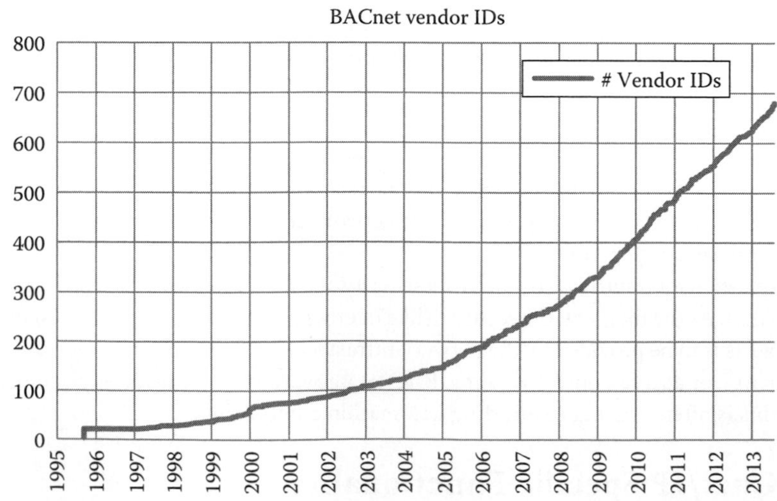

FIGURE 57.10 BACnet vendor ID growth.

57.20 BACnet Mappings

57.20.1 Web Services

To map information contained in objects, BACnet optionally provides web services including a RESTful version. The RESTful version of BACnet/WS extends the logical data model with more complete data modeling, subscriptions, semantic tagging for searching and filtering, and security based on Transport Layer Security (TLS) and OAuth2.0. Users will enjoy greater flexibility of data management as well as increased throughput.

57.20.2 KNX Mapping

The BACnet standard contains a chapter describing the mapping schema from KNX systems into BACnet.

Though gateway mappings in the future will likely not become part of the standard anymore, they are intended to complement the standard by providing separate specifications in addition to the BACnet standard.

57.20.3 OPC UA Mapping

An example of such a mapping profile was presented by the working group OLE for Process Control (OPC) UA (Unified Architecture) and BACnet. The OPC Foundation and BACnet Interest Group founded a working group specifying a mapping of BACnet data (objects and services) to OPC UA as a specification expected to be released by the end of 2013. A second specification will describe how to access OPC UA servers and present the data to BACnet.

57.21 Conclusion

Starting in the late-1980s, BACnet is still up-to-date and is being permanently maintained and extended to encompass new applications. The evolution of BACnet is quite similar to the evolution and success of the IPs. In the 1970s, it was hard to believe in establishing a worldwide video conference just by a simple mouse click (the computer mouse was not even invented at this time). Meanwhile, the IP has become the basis for nearly all computer communication worldwide.

BACnet is on the way to become the same success story as the worldwide standard to connect building automation, life safety, and access-control systems, including specific functions such as lighting control, elevator/escalator representation, or even load management.

International organizations such as BACnet International and the BACnet Interest Group Europe actively support marketing of BACnet, and last but not least, the number of customers requesting BACnet installation (and BACnet providing vendors) is on the rise.

To learn more about BACnet, explore www.bacnet.org.

58

KNX: A Worldwide Standard Protocol for Home and Building Automation: State of the Art and Perspectives

Michele Ruta
Politecnico di Bari

Floriano Scioscia
Politecnico di Bari

Giuseppe Loseto
Politecnico di Bari

Eugenio Di Sciascio
Politecnico di Bari

58.1 Introduction 58-1
58.2 KNX: Short History 58-2
58.3 KNX System Specification 58-3
Transmission Media • Protocol Layers • Network Architecture and Addressing Scheme • Frame Structure • Integration with IP Networks • Application Models
58.4 Development and Configuration 58-10
Interworking Specification • Network Configuration • Development Tools • Basic Installation
58.5 Research Perspectives 58-15
References 58-17

58.1 Introduction

The home and building automation (HBA) is a technological effort aiming to make houses and buildings more controllable, autonomous, and comfortable. Recently, the demand for occupant comfort, flexibility in the management of a building equipment, and the need for efficient energy use have become pressing issues. Safety and comfort requirements, along with energy management, can be achieved only if an intelligent control and monitoring of devices and appliances is performed. A wide array of building automation and control systems (BACS) have been developed and implemented in commercial products, focused on improving energy consumption and minimizing waste and maintenance (reflecting guidelines of the European Standard EN 15232 [1]).

An HBA system requires a large deployment of sensors and actuators in order to detect contextual information and then transfer control data to all building components. A point-to-point interconnecting building device is impractical as it largely increases the amount of wires and consequently the installation complexity and the related costs. A solution is in a deployment of a system using a common, and shared, communication medium by exploiting a well-known and dependable protocol.

BACS are based on bus infrastructures identified as the most effective technological solution for interconnection and communication. Over time, a plethora of protocols have been progressively developed and deployed. Nowadays, the ISO/IEC 14543-3 EIB/KNX [2] (KNX in short) can be recognized as a de facto standard for HBA. It joins and enhances three legacy bus standards, namely, EIB, EHS, and BatiBUS. A KNX medium (twisted pair [TP], radio frequency [RF], power line [PL], or IP/Ethernet) can interconnect heterogeneous bus devices making them able to share and exchange information in a uniform way. Each KNX component can be seen as a sensor detecting an environmental situation (weather stations, lighting sensors, monitoring systems, presence sensors) or an actuator controlling building equipment such as blinds or shutters, safety appliances, energy managers, heating, ventilation and air-conditioning systems, and multimedia devices. Every appliance can be monitored, supervised, and signaled via a shared protocol without the need for single extra control centers.

58.2 KNX: Short History

In 1997, three main European associations were involved in the HBA centered around on-a-bus communication among devices and appliances. Batibus Club International (BCI) was a French no-profit association promoting the BatiBUS medium originally developed by Schneider Electric. European Home Systems Association (EHSA) was a Dutch association promoting the EHS technology, resulting from a European project aiming at the automatic configuration of bus-compatible white (washing machine, cooker) and brown goods (video, hi-fi). European Installation Bus Association (EIBA) was a Belgian cooperative society developing the EIB technology implemented by a consortium of manufacturers headed by Siemens. EIBA consisted of shareholders (*members*) and users of the technology (*licensees*). EIBA developed and marketed a configuration tool called EIB Tool Software.

The fusion of the mentioned associations resulted in a new association called *KNX Association*, and its final aim was the definition of a common standard for a new bus system named *KNX* [2]. It had the following basic features:

- A backward compatibility with EIB
- A simple and effective configuration method essentially based on a PC procedure exploiting ETS but also using central controllers, code wheels, and so on
- The possibility to include heating, ventilation, and air-conditioning (HVAC) systems among supported appliances
- The support for RF and IP media

The KNX standard was approved by the European Committee for Electrotechnical Standardization (CENELEC). In April 1999, the KNX Association was officially founded in Brussels by nine members. Currently, it groups over 340 members in 37 countries (March 2014), including companies that were previously not a member in any of the promoting associations.

In May 2002, the first version of the KNX specification was released among the KNX members starting the KNX Certification Scheme for Products. The intellectual property rights (IPR) clearance was effective from June 2003. As a result, all KNX members may use the KNX technology free of any patent claims of fellow members in KNX-certified products. In 2003, the national committees of CENELEC Technical Committee TC205 (*Home and Building Electronic Systems*) approved a subset of the KNX specification referred to as KNX media and KNX stack, making it a European Standard (as part of the EN 50090 bus family standard). In 2006, the same documentation was approved again by the European CEN for application in BACS. It is registered as EN 13321-1, series EN 50090. The KNX idea to tunnel and route KNX telegrams across IP networks (see later on for details), that is, the KNXnet/IP protocol, was also approved in 2006 by CEN as EN 13321-2, and finally, in the same year, the EN 50090 series achieved the international recognition as ISO/IEC 14543-3. In 2007, the Chinese translation of KNX specification reached the status of national standard as GB/T 20965.

Today, the KNX Association is responsible for all the activities related to the KNX system, and in particular, it oversees the following:

- The specification work conducted by the expert groups
- The technical hotline support to members
- The international standardization of novel features and characteristics but also the KNX certification by incoming organizations
- The logo protection
- The management of the training documentation
- The advertising and communication strategy (e.g., publication of the official association journal twice a year, diffusion of brochures, organization of biannual award for best KNX projects, presence at fairs)
- The management of national groups, that is, groups of member companies in a given country responsible for the local promotion of the KNX system
- The scientific partnership, that is, the involvement of universities, research centers, and technical institutes in scientific and educational initiatives as the annual organization of the KNX Scientific Conference
- The management of the partnership for trained and certified contractors

58.3 KNX System Specification

The KNX specification covers all technical elements of the standard, ranging from the low-level device details and certification rules to testing and application descriptions. In what follows, the fundamentals of the protocol will be outlined according to the KNX specification, version 2.0 [2]. Supported transmission media will be briefly surveyed and both meaning and function of protocol layers defined in the standard will be clarified. Then a description of the reference network architecture and the addressing features will be reported, followed by a section related to the integration of the KNX protocol with IP networks. Finally, details about application-level objects will be given.

58.3.1 Transmission Media

The KNX system offers the possibility for the manufacturers to choose between several transmission media, according to market requirements and specific habits. Moreover, it is also possible to combine them to build multimedia and multivendor network configurations.

The basic TP and PL communication media are accompanied by the RF support in order to make KNX networks flexible and adaptable to multiple application domains and installation situations. Some of the main features of supported media are presented in the following:

- TP TP1 is the basic medium inherited from the EIB protocol. It provides a solution for wired cabling, using a safety extralow-voltage (SELV) network and a supply system. KNX supports both TP1-64 and TP1-256 media, which differ in the number of connectable devices per physical segment (64 or 256). TP1-256 is backward compatible toward TPI-64. The *TP1* transmission rate is 9600 bit/s with a character-oriented asynchronous data transfer mode and a half-duplex bidirectional communication. It also complies with the carrier sense multiple access with collision avoidance (CSMA/CA) protocol. All network topologies (e.g., line or star) may be used and mixed.
- PL PL110 is also inherited from the EIB protocol and enables communication over the main power supply network of a building. It is based on an asynchronous transmission of data packets and a half-duplex bidirectional communication. PL110 uses the central frequency of 110 kHz with a 1200 bit/s data rate; it complies with CSMA and the EN 50065-1 standard [3].
- RF enables a communication via radio in the 868.3 MHz band for short-range devices. It supports the frequency-shift keying, a maximum duty cycle of 1%, and a Manchester data encoding.

KNX allows integrated solutions for IP-enabled media like Ethernet IEEE 802.2 [4], Wireless LAN IEEE 802.11 [5], and FireWire IEEE 1394 [6] exploiting the KNXnet/IP protocol as explained in Section 58.3.5.

The KNX *v*1.1 specification also supported TP0 and PL132 media to ensure a fully backward compatibility with EHS and BatiBUS protocols. Those media are no longer available in the new KNX *v*2.0 protocol. The related documentation was phased out in the last specification and both TP0 and PL132 cannot be used in applications for the KNX certification.

58.3.2 Protocol Layers

The KNX protocol defines a seven-layer stack compliant with the ISO/OSI model [7]. On top of the physical layer, a common kernel model is shared by all devices of the KNX network. The reference structure is shown in Figure 58.1.

The *physical layer* is on top of the communication channel. It mainly consists of a logical unit, a medium attachment unit (MAU), and a medium interface (i.e., the connector). Each medium needs a dedicated MAU and a suitable logical unit. The MAU is used for encoding the logic signals to physical ones and vice versa. In the transmission mode, the logical unit serializes each data octet in a bit sequence; then, it frames the obtained data and transforms them in an asynchronous timed logical signal. In the receiving mode, the logical unit reverts the obtained signal in a data bit stream to be checked, rebuilt, and aggregated into several data octets. Finally, the medium-independent sublayer of the logical unit transfers the received octets to the upper layer.

A general *data link layer*, built above each data link layer specific for each medium, provides a high-level medium access and logical link control. It also takes responsibility for the transmission of single KNX frames between two or more devices on the same subnetwork.

During transmission, the data link layer (i) builds a complete frame starting from data coming from the network layer, (ii) gains access to the medium according to the particular access policy in use, and (iii) transmits the frame to the data link layer of each peer entity, using the services of the physical layer.

When receiving, the data link layer (i) detects possible corruptions in the frame, (ii) transfers the frame to the upper layers, and (iii) issues positive or negative acknowledgments back to the transmitting data link layer entity.

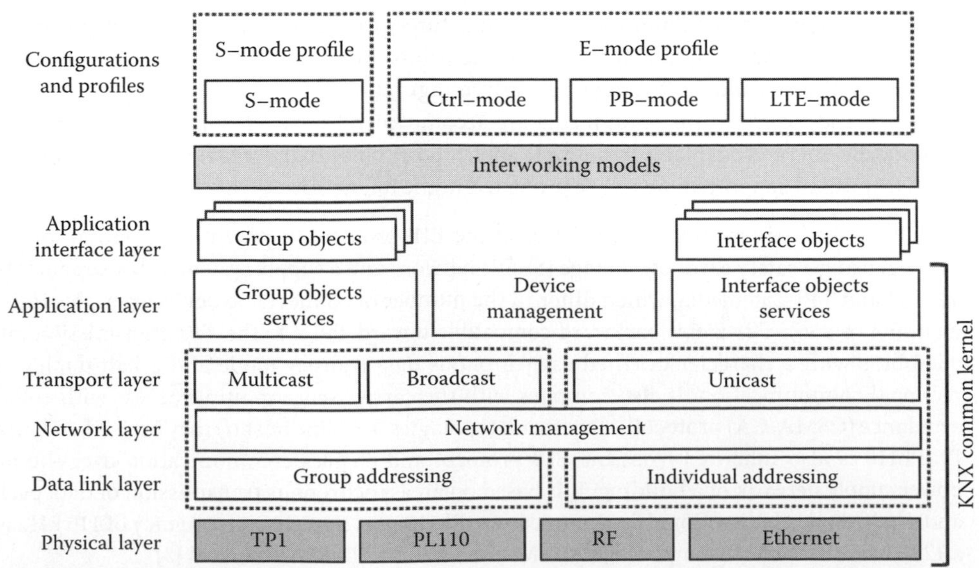

FIGURE 58.1 KNX protocol stack and reference architecture.

The data link layer also manages unicast, multicast, and broadcast communication options and device addressing, as described in Section 58.3.3.

The *network layer* takes care of managing communications across KNX subnetworks. Hence, its role is relevant mainly to the nodes with routing functionalities. As an example, this layer provides a segment-wise acknowledged telegram and controls the hop count of a frame.

The *transport layer* enables the data transmission over different communication modes. Five kinds of communication are defined in the standard:

1. Point-to-multipoint, connectionless (multicast)
2. Point-to-domain, connectionless (broadcast)
3. Point-to-all-points, connectionless (system broadcast)
4. Point-to-point, connectionless (unicast)
5. Point-to-point, connection oriented (unicast)

Finally, *session* and *presentation layers* are unused, while the *application layer* offers several services to application processes. Each different service depends on the communication used at the transport layer. Usually services related to point-to-point communication and broadcast mainly refer to network applications, whereas services related to multicast are used for runtime operations.

Each layer interacts with the layers above and below. In the first case, it acts as a *service provider* making available a set of resources for the above *service user* layer. The interface between both layers defines how the available services can be accessed, how the service parameters must be handled, and how the response should be interpreted. The communication between the interfaces of layers N and N − 1 occurs via *service data units* (SDUs). The KNX specification also defines a rule set for the peer layer communication between devices. In this case, *protocol data units* (PDUs) are used, which mainly consist of user data and layer-specific *protocol control information* (PCI). In addition, four different communication primitives are defined: *request* (req), *indication* (ind), *confirmation* (con), and *response* (res). Services do not always need to use all the primitives and can be classified as in what follows:

- *Locally confirmed* services: require a request, an indication, and a confirmation. In this case, a request and the corresponding PDU are generated at layer N and passed to the lower layers until transmission on the physical medium. On the receiver side, the peer layer N is activated with an indication, the received PDU is progressively decoded, and the data are passed to the layers above. The sender layer N receives a confirmation from the local layer N − 1 indicating if the request has been processed correctly.
- *Confirmed* services: consist of a request, an indication, and a confirmation, but for these services, the peer remote layer generates an acknowledgment after receiving the indication. At the sender side, the received acknowledgment is passed to the local layer N as a confirmation.
- *Answered* services: always need a request, an indication, a response, and a confirmation. As shown in Figure 58.2, the response is generated by the remote service provider as a novel request issued toward the sender that will be received at the local layer N as a confirmation.

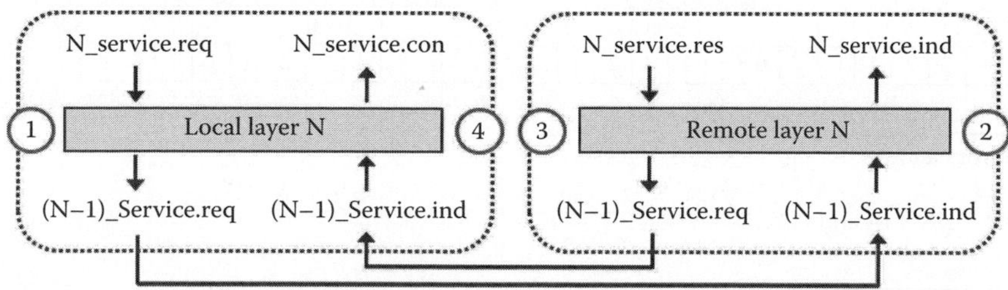

FIGURE 58.2 Answered service communication.

58.3.3 Network Architecture and Addressing Scheme

The KNX specification enables to build a fully distributed network, which accommodates up to 65,536 *devices* by exploiting a 16-bit individual address (IA) space. The logical topology or *subnetwork* structure allows 256 devices for each *line*. As shown in Figure 58.3, lines may be grouped together with a *main line* into an *area*. An overall domain is formed by 15 areas coupled with a *backbone line*. Not considering the addresses reserved for area and line couplers, up to 61,455 end devices may be joined by a KNX network.*

All components, that is, couplers or end devices, are unambiguously identified and subsequently accessed via their IA, reflecting the *area.line.device* topology. As shown in Figure 58.4, the IA consists of a two-octet value: the first byte defines the subnetwork address (SNA), while the last octet specifies the device address (DA) unique within a subnetwork. Routers always have a DA 0; other devices may have a DA with values ranging from 1 to 255. Devices needing no IA for operations may use a default IA (FF_h). This default IA shall consist of the DA for unregistered devices and the medium-dependent default SNA.

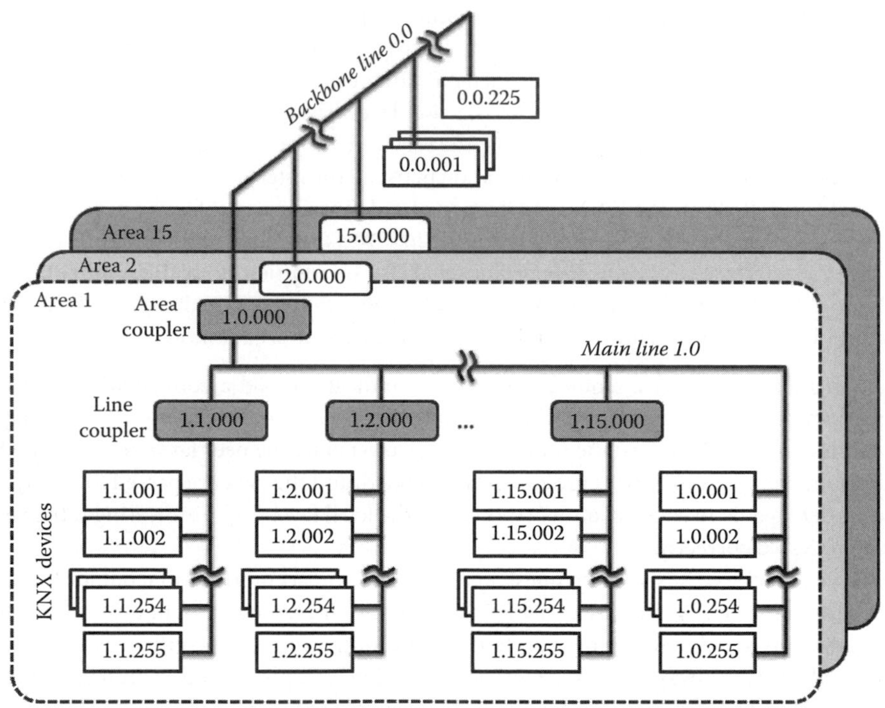

FIGURE 58.3 KNX network architecture.

Individual address															
Octet 0								Octet 1							
7	6	5	4	3	2	1	0	7	6	5	4	3	2	1	0
Area address				Line address				Device address							
Subnetwork address															

Group address															
Octet 0								Octet 1							
7	6	5	4	3	2	1	0	7	6	5	4	3	2	1	0
Main group					Middle group			Little Group							

FIGURE 58.4 KNX IA and GA.

* Installation and product guidelines should be taken into account. Possible restrictions may depend on implementation (e.g., medium, transceiver types, power supply capacity) and environmental factors (e.g., electromagnetic noise).

KNX also supports a full multicast addressing mainly used for the runtime communication, considerably reducing bandwidth requirements. The protocol provides a 16-bit group address (GA) space usually used in the format *main.middle.little* device groups as shown in Figure 58.4. Sometimes GA space can also be adopted with a two-*level* (main group/subgroup) or with a freely defined structure. The GA does not need to be unique. In fact, a device can also have more than one GA. Furthermore, each device is associated by default to the group zero (0/0/0) as request frames addressed to GA 0 will be sent in broadcast. In this way, a device is associated to the sender addresses with the aim of sharing its datapoints (i.e., group communication objects), thus creating a kind of network-wide variables. This communication model will be better described in Section 58.3.6. Since the protocol makes a 16-bit address space available to such GAs, it is possible to define about 2^{16} shared variables each with any number of local device instances.

In addition to IAs and GAs, if the network is built on an open medium (e.g., PL*) shared by devices of different subnetworks, a domain address (DoA) is required as an identifier to logically separate each subnetwork.

58.3.4 Frame Structure

Figure 58.5 shows the structure of a generic KNX frame built above the data link layer.

The control field is used to detect possible *repeated* frames (the flag R identifies a first transmission or already-sent frames), to indicate the frame *priority* (the flag PP can assume a *system*, *high*, *normal*, or *low* value), and to establish the frame *type* (standard or extended†). Then, there is an individual *source address*, in terms of area.line.device, and an individual or group *destination address* (respectively, in case of unicast or multicast communication). The destination address type is determined by a special field (*address type* [AT]) defined at the beginning of the network protocol data unit (NPDU). In addition to AT, the NPDU includes the *network layer protocol control information* (NPCI) field, a hop counter decreased by routers to avoid looping messages. NPCI ensures a frame can appear in seven physical network segments at most in order to avoid packets circulating endlessly due to errors in the network setup; when it becomes zero, the frame is silently discarded. The *length* field specifies the size of the NPDU, while the *transport layer protocol control information* (TPCI) field controls the communication at transport layer level, for example, specifying service codes and sequence numbers used to build up and maintain a point-to-point connection. Analogously, the *application layer protocol control information* (APCI) field contains the application layer service codes. The APCI has a length of 4 or 10 bits, depending on the specific service code. According to both addressing scheme and APCI, the standard frame can carry up to 14 *data* octets. Finally, the frame is closed by the *checksum*, which contains an odd horizontal parity value calculated over all preceding octets. This ensures data consistency and reliable transmissions.

KNX frame						
Octet 0	Octet 1	Octet 2	Octet 3	Octet 4	Octet 5...N−1	Octet N
1 0 R 1 P P 0 0	A A A A L L L L	D D D D D D D D	7 6 5 4 3 2 1 0	7 6 5 4 3 2 1 0	...	7 6 5 4 3 2 1 0
Control field	Source address		Destination address		NPDU	Checksum

Octet 5	Octet 6	Octet 7	Octet 8	...	Octet N−1
7 6 5 4 3 2 1 0	7 6 5 4 3 2 1 0	7 6 5 4 3 2 1 0	7 6 5 4 3 2 1 0		7 6 5 4 3 2 1 0
AT NPCI Length	TPCI	APCI Data/APCI	Data		

FIGURE 58.5 KNX frame structure.

* On a PL, nearby domains are logically separated with a 16-bit DoA.

† Despite the standard frame ensuring a direct backward compatibility toward the EIB protocol, an extended frame (identified by a "00" code in the two most significant bits) can be used to store up to 248 bytes of data.

58.3.5 Integration with IP Networks

The KNX specification describes a compact and flexible Internet protocol (IP) tunneling protocol, named *KNXnet/IP*, used to carry a KNX frame over an IP stretch enabling the communication of KNX implementations (e.g., KNXnet/IP devices) on the top of IP networks. In this way, several KNX subnetworks can be connected via an IP line acting as a fast (if compared to classic KNX transmission speeds) backbone. A widespread deployment of IP-based networks and applications is an opportunity to expand building control and communication beyond a local KNX network. Further benefits include (i) fast interface between local area network (LAN)-based and KNX-based systems and (ii) remote configuration and usage of home and building devices.

A KNXnet/IP system contains at least the following elements:

- A KNX subnetwork (e.g., KNX-TP1, KNX-RF, KNX-PL110) with up to 255 end devices.
- A KNX-to-IP network connection device (named KNXnet/IP *server*). It is a special device having a physical access to a KNX network also implementing the KNXnet/IP protocol to communicate with *clients* or other servers on an IP network channel. A server is by design also a KNX node with a unique IA.
- Additional software for remote operations hosted by client workstations (e.g., configuration tools, building management systems or browsers).

Figure 58.6 shows a basic network configuration where a KNXnet/IP client is connected to multiple KNX subnetworks via IP. The KNXnet/IP client may access one or more KNXnet/IP servers at a time.

When a KNXnet/IP server communicates with another server via IP, it acts also as a KNXnet/IP *router* exploiting a multicast one-to-many communication, where KNX data are simultaneously transferred from a sender device to one or more receivers over IP. A router can also replace a line or an area coupler and directly connect main lines and backbones, allowing the usage of existing cabling (e.g., Ethernet) and ensuring faster transmission times between KNX subnets.

A KNXnet/IP frame is a PDU moving along a non-KNX network. It contains a header, similar to the header of an IP data packet, and a body, if present. As shown in Figure 58.7, the header maintains information about the protocol version, the header length,* the total length of the packet, and the

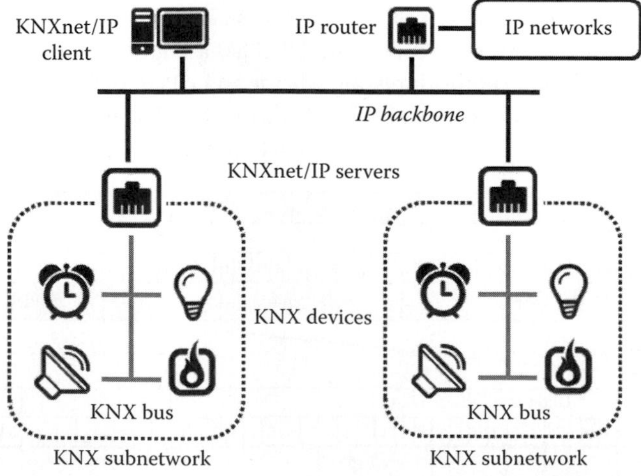

FIGURE 58.6 KNXnet/IP network example.

* Although the length of the header is always fixed, it is possible that it changes with newer versions of the protocol. The header length can be used as an index into the KNXnet/IP frame data to find the beginning of the KNXnet/IP body.

KNXnet/IP header															
7	6	5	4	3	2	1	0	7	6	5	4	3	2	1	0
Header length								Protocol version							
Services type identifier															
Total length															

KNXnet/IP body															
7	6	5	4	3	2	1	0	7	6	5	4	3	2	1	0
Structure length								Communication channel ID							
Sequence counter								Service type specific							
Connection type specific header (variable length, optional)															
cEMI frame (variable length)															

FIGURE 58.7 KNXnet/IP frame structure.

KNXnet/IP service-type identifier. KNXnet/IP services include, but are not limited to, information regarding device discovery, connection management, and KNX data transfer.

The header may be followed by a KNXnet/IP body of variable length, depending on a specific service. An example of a body is shown in Figure 58.7. It embeds the data into a basic KNX frame and starts with data fields that specify additional general information about the connection. Those fields compose the *connection header* whose appearance is fixed, varying only the content according to the connection options. Finally, the IP body includes the so-called cEMI frame consisting of the TP1 telegram structure, excluding the checksum field that is not used in IP communications as the error detection is done by IP.

58.3.6 Application Models

The *application interface layer*, as shown in Figure 58.1, is an additional abstraction level between the application layer and the user applications. It eases the communication tasks by offering a common interface that abstracts from many application layer details. It allows to expose shared objects and use them as local variables. In addition, it takes care of handling network management calls. The application interface layer contains the following objects:

- *Group objects* (GOs): they are accessible via application layer services in the multicast communication mode and provide a communication schema named *shared variable model*. GOs may also refer to one or more interface objects (IOs).
- IOs: they are accessible via application layer services in unicast and broadcast communication modes and are used for a *client/server* interaction. IOs are also classified as *system* IOs (e.g., device objects), relevant to network and device management, and *application* IOs, defined by the user applications.

The GO structure is shown in Figure 58.8. It consists of three parts: *description*, *value*, and *communication flags*. The GO description must at least include the GO *type* and the *transmission priority* (urgent, normal, or low). Optionally, the *config flags* include static information about the GO (e.g., read, write,

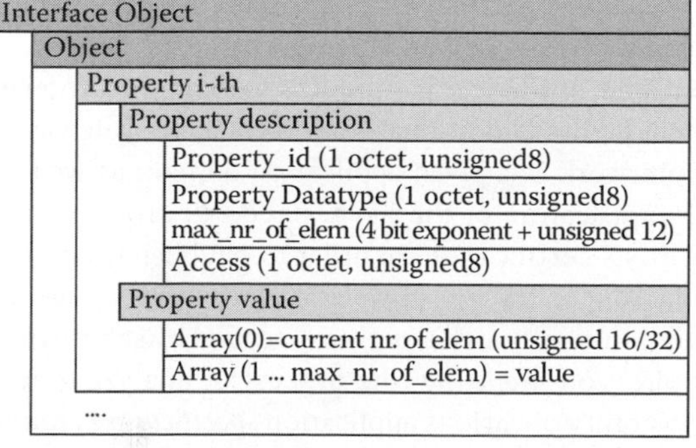

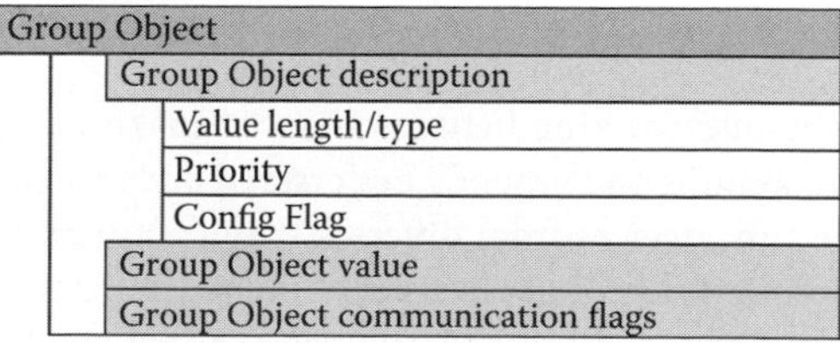

FIGURE 58.8 Interface and GO structure.

or update permissions). According to the GO type, the size of the value field can vary from 1 bit to 14 bytes. Finally, the communication flags show the *state* of a GO. The following states are enabled: *update, read_request, write_request, transmitting, ok, and error.*

The IOs are instances of the general structure shown in Figure 58.8. Starting from this common schema, two object types are derived: *full* IOs and *reduced* IOs.

Full IOs exactly comply with the data structure reported in Figure 58.8. They consist of a number of *object properties* composed by a *description* and a *value* field. In turn, a property description consists of the following fields:

- Property index: it is unique for each IO. The first property has index 0, and further ones are numbered with subsequent values without gaps.
- Property IDentifier (PID): this value is encoded in the *property_id* field and is usually used to identify the specific IO.
- Property data type (PDT): it describes the most appropriate data type for the IO property.
- Maximum number of elements: the value of a property is always an array, so this field specifies the array size: suitable indexes range from 1 to *max_nr_of_elem*, while the element at position 0 contains the current number of valid array elements. This value is automatically updated if an element is written beyond the current last element but always within the maximum allowed number of entries.
- Read/write access levels: this attribute indicates the access level needed to read or write the property value.

Moreover, each property description has also a *write-enable flag* that specifies whether the property value can be written or not. In particular, the property with *property_id* equals to 1, and *index* 0 is named *IO type* (`PID_OBJECT_TYPE`) and contains the description of the IO itself. This property is mandatory for every IO.

As opposed to full IOs, reduced IOs only support a subset of the common IO structure. A property description is only composed of the property_id, while the property value can be a single value or an array.

58.4 Development and Configuration

On the top of KNX networks, several distributed applications can be built, enabling a tight interaction among different devices. This section summarizes some distinctive characteristics of the KNX standard at the application level. Modeling guidelines and interworking models are resumed to give a sketch of a comprehensive and multidomain building communication system. Hence, device configuration modes and development tools are detailed and a basic configuration example of a KNX system is provided.

58.4.1 Interworking Specification

The interworking between devices enables different and multivendor products to send and receive datagrams and properly understand and react on them without additional equipments (e.g., gateways). In this, devices from different manufacturers can interwork in a given application domain as well as across different applications. Interworking is a relevant feature of the KNX standard: products not complying with this specification cannot obtain the KNX certification. In order to establish this communication model, a set of requirements was defined, grouped into the *application interworking specification* (AIS). AIS is defined for various application domains and can be further extended by submitting proposals to the KNX Association. In fact, the application *modeling* (the process of analyzing and defining the model for each application) is the responsibility of various application *specification groups* of the KNX Association.

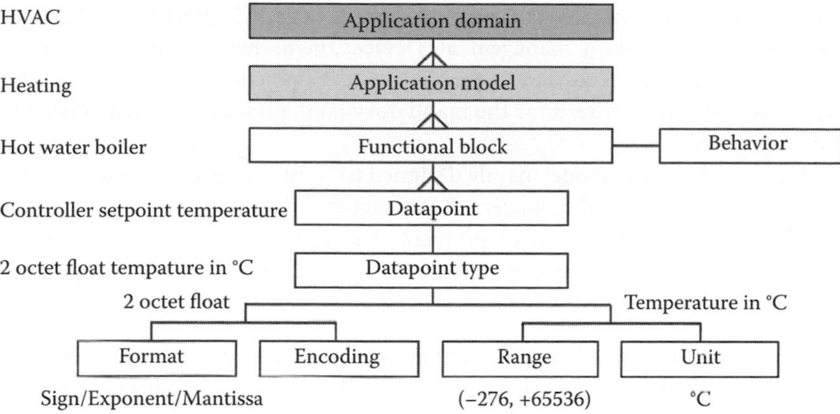

FIGURE 58.9 Example of an application modeling.

As presented in Figure 58.9, each application domain encompasses one or more applications, each of them defined in terms of *functional blocks* (FBs) that communicate with each other. In this way, a *distributed* application is composed of a number of shared FBs implemented in various devices in the network. In addition, a manufacturer may group one or more FBs, of the same, or of different applications, to build a device. An FB is described by the product developer as an object with well-defined behaviors; it uses one or more *datapoints* for interchanging information, which represent inputs, outputs, parameters, or diagnostic data. A datapoint can be seen as a generic interface used to set, receive, or transmit data during the runtime operations.

The information exchanged through datapoints is associated to a *DataPoint Type* (DPT) featuring the data in terms of the following:

- *Format*: describes both sequence and length of the fields composing the DPT.
- *Encoding*: indicates how the data are encoded according to the given format.
- *Range*: describes the limitation of the values that may be encoded in the DPT. This may be a minimum/maximum indication or an explicit list of values.
- *Unit*: specifies the measurement unit of the carried information.

The DPTs are identified by a 16-bit *main number* separated by a dot from a 16-bit *subnumber* (e.g., 7.002). The main number identifies DPTs with the same format and encoding, whereas the subnumber indicates a specific dimension (different range and/or unit).

58.4.2 Network Configuration

The KNX standard allows each manufacturer to select the most proper configuration mode according to different markets, local habits, level of training needed, or application environment. In particular, four different *configuration modes* are available:

1. *System* (S) mode (S-mode): the most versatile and multiusage configuration process. The overall application setup is done by a configuration master. Usually, it is done by means of a set of PC-based tools from the engineering tool software (ETS) family, developed by the KNX Association. With the aid of the ETS tools, described in Section 58.4.3, it is possible to configure the S-mode compliant devices and set them into operation.
2. *Controller* (Ctrl) mode: an external controller device is required to set the IAs and the needed parameters in the devices. The controller is not needed after configuration but may also have additional runtime functionalities.

3. *Push-button* (PB) mode: no tools or external devices (e.g., PC, controller, ETS) are needed for the configuration and linking management. Devices themselves are able to set up the links and assign the IAs and GAs. However, to enable the development and manufacturing of low-cost products, the software overhead for the mandatory configuration and link procedures must be as small as possible.
4. *Logical tag extended* (LTE) mode: mainly designed to cover the specific needs of an easy configuration for HVAC, which needs a longer set of structured data. These data are exchanged via IOs using the extended frame of the KNX protocol. Also in this case, a simple device configuration is possible without using PC tools, manufacturer specification, or the knowledge of device object structures and GAs.

In order to facilitate the network design and configuration and to simplify the device certification, previous configuration modes have been grouped in so-called configuration profiles. The KNX profiles have been designed to cover all needs and habits of the KNX Association community. In particular, a profile specifies how a device should be configured and the minimal device requirements for this process. The KNX specification defines the following profiles:

- *S-mode* profiles: suitable for end devices and system components (i.e., couplers) supporting the S-mode configuration. This type of setting procedure is adopted by well-trained KNX installers in order to realize advanced building control functions enabled via ETS tool using specific product databases provided by device manufacturers. Each database includes all supported functionalities and ETS is also used to set the device parameters as required by the installation. S-mode offers the highest degree of flexibility for building KNX networks.
- Easy mode *(E-mode)* profiles: adopted for devices supporting the controller, PB, and LTE configuration modes. The installation of such components can be also made by users with a basic KNX experience. E-mode compatible products offer limited features compared to S-mode. In this case, components are usually preprogrammed and loaded with a default set of parameters. With a simple configuration software, the device parameter settings and communication links can be partially reconfigured.

58.4.3 Development Tools

In order to plan, design, and commission a KNX installation, a software tool is required to assist designers and electrical installers. For that reason, a manufacturer-independent configuration tool named ETS was developed by the KNX Association. ETS4 [8] is the current release of the tool, replacing since 2010 ETS3. During its development, the user interface has been basically kept very similar to the one of previous releases. On the other hand, both functions and operating philosophy of the ETS4 user interface has been completely reengineered, making them simplified compared to ETS3. ETS is available in three different versions:

1. *Professional*: which enables the development of KNX installations of any size without limits in terms of maximum number of addresses, devices, or project.
2. *Light*: maintaining the same functions as the professional version; this version only allows projects with up to 20 devices. It is intended for training centers and their customers.
3. *Demo*: this version is free of charge but can be used to design networks with three devices at most.

In case of a user with a single PC-dependent license, there is also a *supplementary* version available as an additional license for the full professional version running on a second PC, for example, a laptop used at the construction side. In addition to ETS, the KNX Association also offers the following software tools or libraries:

- *iETS server*: it is a software interface between KNX and IP adopted for connecting iETS clients (such as ETS4) to the KNX network. With iETS, the installer can access the system remotely by means of ETS, enabling remote programming or diagnostics via IP.
- *Falcon Driver Library*: it is a high-performance DCOM (Microsoft Distributed Component Object Model) based library enabling Windows-based PC KNX bus access. Falcon offers an API for sending and receiving telegrams across the KNX network and supports the communication through RS232, USB, and Ethernet.
- *Interworking test tool* (*EITT*): this tool is used by developers and test laboratories for testing, troubleshooting, and monitoring the interworking and system stack compliance of KNX products. EITT is also a powerful tool for the analysis and simulation of the KNX network protocol.

Also KNX members offer a number of KNX-related software packages. In particular, several ETS plug-in modules (named ETS Apps) have been developed, allowing to extend the functionality of ETS, for example, for graphical project designs, data exchange, or efficient project engineering. In fact, any new, or existing, software can be adapted to the ETS App interface and integrated in ETS4 by using the related SDK (software development kit), without the need to completely recompile the software.

Finally, noteworthy is a Java library for KNX access, named *Calimero* [9], offering a clean and lean interface and allowing free client access to KNX networks and a basic KNX server functionality. It is designed to operate also on resource-constrained systems (i.e., mobile and embedded devices) as it requires low memory and CPU. In addition, it is fully compliant with the popular Java Standard Edition (J2SE) environment commonly deployed on home computers, and this ensures a widespread employment of it. It is important to notice that Calimero is an open-source software distributed according to the GNU GPL license.

58.4.4 Basic Installation

In order to illustrate the process of setup and configuration of a KNX network, a simple example is considered. The project design of a KNX system initially does not differ from a typical electrical project. In the preliminary stages, several aspects must be clarified by the planner such as the type and use of the installation, the building system components to use, and their functions and special requirements stipulated by customers. Subsequently, product databases may be downloaded and imported in ETS4 in order to create a new project. Then the network configuration must be defined, for example, building structure, bus topology, and GA levels. The project engineer selects the KNX devices with the corresponding application to insert into the network, assigns them an IA, and sets the parameters according to the requested features. As an example, a network is proposed consisting of a simple TP line (1.1) connected to a backbone IP line (0.0), having six devices connected: a KNX/IP router, a switch and a dimmer actuator, a KNX module with PBs, a shutter controller, and a weather station for outdoor environments (Figure 58.10). The IAs are automatically assigned by ETS4 in ascending order in the respective current line; however, they can be modified at any time by the user. The selected devices are displayed in the tree structure on the left panel. Double-clicking on each device, the properties panel will be displayed containing further information about the product and the parameters that can be modified by the installer. Moreover, selecting a device in the tree view, the related GOs will be shown below it. In the proposed configuration, the weather station (1.1.5) exposes several GOs associated to rain, temperature, wind force, and brightness sensors.

Afterward, GAs can be created. In each project, a different pattern for the definition of GAs can be used. For example, for three-level addresses, the main group can specify floors or rooms, the middle group can specify the functional domains (e.g., switching, dimming), and the subgroup can define articulated operations named scenarios (e.g., turn on/off a specific lamp in the house or open/close the living room according to the weather conditions). The GAs are listed in the GA window, shown in Figure 58.11. In order to ensure that sensor and actuator devices may know which GOs should communicate with each other to realize a given function, the objects of each device are assigned to the GAs.* In the example, the rain sensor of the weather station is connected to the shutter controller by means of the same GA, so when rain is detected, the shutters will be automatically closed.

In this way, all devices are fully configured and the final network topology is defined. In conclusion, the network configuration should be checked before the commissioning of the project. Obviously, it is also possible to deviate from this sequence according to the project requirements. Some steps can be omitted for smaller projects, whereas additional steps could be needed in larger ones.

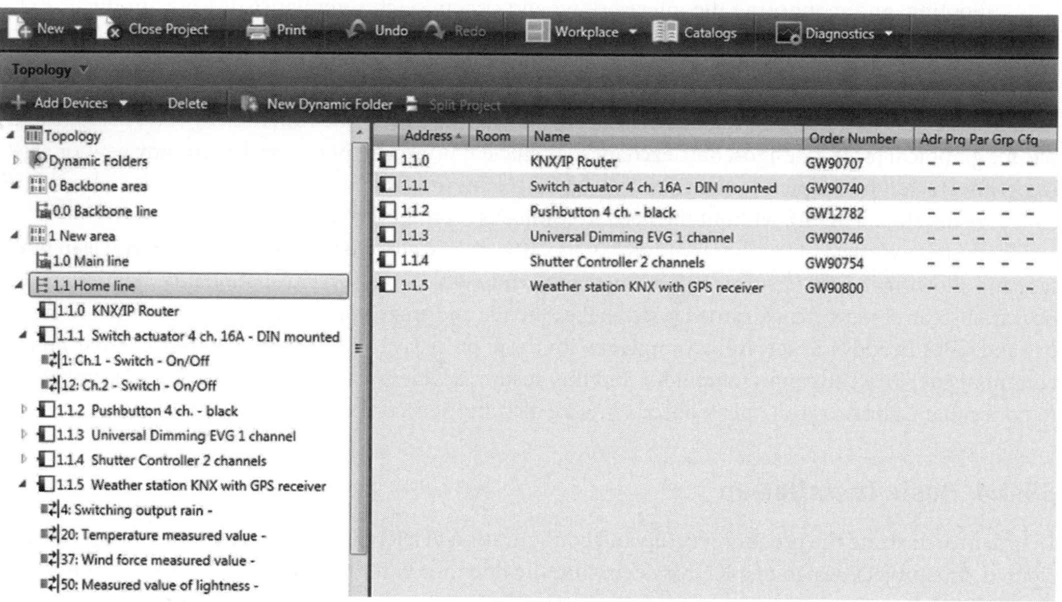

FIGURE 58.10 ETS4 topology panel.

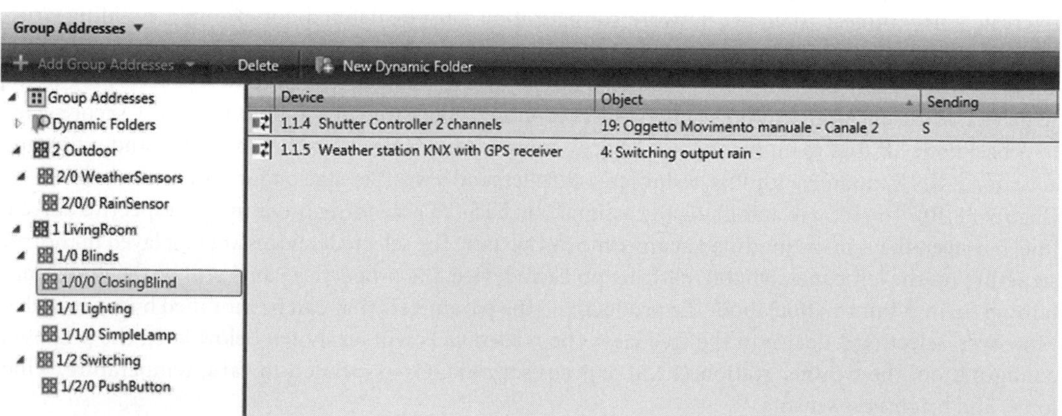

FIGURE 58.11 ETS4 GA panel.

* In a group address, it is only possible to create connections between group objects of the same type (e.g., 1 bit, 1 byte). The group address receives the type of the group objects with the first assignment.

58.5 Research Perspectives

The basic goal of HBA [10] is to improve people's comfort, increase safety and security, reduce energy consumption, and minimize the overall ecological footprint. For this purpose, a more effective management and coordination of building appliances and subsystems is required. In particular, *ambient intelligence* (AmI) [11] proposes a multidisciplinary research vision aiming at defining frameworks and techniques for smart environments, where people are surrounded by intelligent and unobtrusive microcomponents, capable of being sensitive and responsive, recognizing user profiles, and self-adapting behavior accordingly. Devices should communicate and interact autonomously, without the need for direct user intervention, also making decisions based on multiple factors, including contextual information, sensory data patterns, and user presence and preferences. They should be coordinated by intelligent systems acting as supervisors.

At present, EIB/KNX and other current standard HBA systems and technologies are still far from that vision, because they cannot grant such flexibility. As seen in previous sections, they can essentially set static functions and features configured during system setup; explicit interaction with the user is required for all except the most elementary cases. Context-aware dynamic information management and decision-making about users, devices, and services are not yet supported. In order to enable intelligent HBA infrastructures, capable to adapt and control building appliances autonomously, smart environments have to be conceived according to results coming from pervasive and mobile computing, artificial intelligence theory, and agent-based software design [12]. Consequently, AmI research is closely related to studies for effective discovery and coordination in volatile and resource-constrained scenarios.

In latest years, four major research directions were pursued to improve intelligence, effectiveness, and usability of HBA systems:

1. Improve user interaction with HBA systems.
2. Integrate different HBA technologies in a common application layer.
3. Model HBA frameworks as *multiagent systems (MASs)*.
4. Use semantic technologies to enhance information management and decision-making capabilities.

1. Several approaches proposed more advanced and user-friendly device management with respect to current standard technologies. Nevertheless, they still rely only on direct command-based user interaction, through local or network-based interfaces. In particular, [13] introduced a method to control home devices either locally (through an AJAX web application) or remotely (via web services), allowing the user to issue either direct commands to devices or set routines that should be executed periodically. The proposal is not tied to any particular HBA standard. The framework devised in [14] is focused on a low-cost ZigBee home automation system. Also in this case, devices can be controlled directly by the home occupants by means of a ZigBee remote control, while the ZigBee *Home Automation* network exposes a *home gateway* on the Internet, so that the home can also be controlled remotely from a PC.
2. Due to the diversity of HBA technologies, a further aspect widely investigated is the integration of different HBA technologies in a common application layer. Some works achieved limited integration of domotic subnetworks based on different technologies by means of custom gateways. For example, in [15], a gateway is proposed between KNX and ZigBee *Home Automation* application profile. Several proposals [16,17] adopt web service technologies as unified abstraction layer to integrate various communication protocols and mobile service discovery via UDDI (Universal Description, Discovery, and Integration) and UPnP (Universal Plug and Play) in automation contexts. Standard web service description and orchestration languages based on XML are adopted, only allowing syntactic match between users' needs and service/resource attributes, lacking semantic characterization. As a consequence, the improvement in service discovery and composition capabilities with respect to basic HBA protocols is not so relevant to justify the increased architectural complexity.

3. HBA is characterized by the lack of hard real-time constraints and the need for high flexibility and reconfiguration capabilities. As pointed out in [18], such kinds of environments are very well suited to be modeled by means of MASs. Agents can meaningfully represent entities (devices), contexts (rooms, floors), or people, emphasizing interaction capabilities such as communication, concurrency, cooperation, coordination, conflict resolution, and negotiation. Many proposals exist in literature for structuring HBA systems through MAS: while early research focused on agent-based direct control of home appliances, recent MAS proposals concern high-level, complete HBA control systems. In order to achieve greater interoperability, a major goal is to provide a unified view of the home environment and a flexible application model. Particularly, in [19], a platform was proposed to allow an easy integration of multimodal user interfaces in heterogeneous device networks. The MAS was implemented using Java Agent Development Framework (JADE) SDK, whereas IP-based UPnP was exploited as discovery protocol. In [20], a service-oriented smart home architecture was defined, where each component is designed as an agent. When the home is about to perform a service for a user, it compares service requirements with the environment situation to find out rooms whose status and resources are ready for service activation. Similarly, in [21], the use of BDI (belief–desire–intention) agents was proposed to automate service composition tasks, providing transparency from the user standpoint. In latest years, the MAS model has been proposed also for *BEMS* (*Building Energy Management Systems*), due to the growing interest in *smart grid* technologies [22] and in reducing energy consumption via *demand-side management* [23]. An agent-based BEMS typically supports *active load management* techniques [10] and exploits *smart metering* via energy-efficient network protocols [24]. Such approaches *put the human back into the loop* [23], that is, they take into account user profiles and requests in order to maximize both energy efficiency and customer satisfaction. In [25], a *multiagent comfort and energy system* was designed for management and control of both building systems and occupants. It coordinates appliances and users, for example, suggesting changes to occupant meeting schedules. A MAS for energy management in commercial buildings was proposed also in [26]: Agents communicate and negotiate with human occupants to save energy; optimal policies are generated considering multiple criteria—for example, energy cost and personal comfort—as well as uncertainty in occupant preferences.
4. Knowledge representation and reasoning (KRR) technologies allow greater generality and more flexible reuse of models concerning the aforementioned approaches, because *ontologies* provide a conceptual framework to express and share formal and structured descriptions of services and appliances, while general-purpose reasoning procedures can be used for semantic-based service composition in different HBA scenarios. *DomoML* [27] was the first specific proposal of a building automation ontology suite. Reinisch et al. [28] acknowledged the relevance of semantic-enhanced approaches upon current HBA standards, for cost and efficiency motivations; they introduced a theoretical ontology-based framework for the integration of different HBA protocols at the application level. In [21], the use of intelligent agents was proposed to automate service composition tasks. Nevertheless, a very elementary ontology was derived from attribute-based service descriptions in UPnP and Bluetooth service discovery protocols. As a consequence, the approach lacked adequate expressiveness of user, device, and service profiles. In [29], the first self-contained prototype was presented including a reasoning module able to manage and coordinate heterogeneous devices by means of logic rules processing. Classical rule-based inferences are not enough in heterogeneous and dynamic AmI contexts. In order to execute a rule, conditions it imposes must be fully matched by the current system state. Unfortunately, experience shows that full matches are quite unlikely in real-life scenarios, where objects, subjects, and events are featured by different heterogeneous descriptions, often partially in conflict among them. Semantic-based matchmaking frameworks as the one in [30], which exploit standard and nonstandard inference services and allow to match requests and resources based on the meaning of their descriptions (also providing classification and

logic-based ranking), are more effective. Beyond obviously good matches, such as *exact* or *full* ones, they enable so-called *potential or intersection* matches (i.e., those matches where requests and supplied resources have something in common and no conflicting characteristics) and *partial or disjoint* matches (i.e., cases where requests and available services have some conflicting features) that can also be considered useful in scenarios when nothing better exists. In [31], the exploitation of KRR technologies, originally conceived for the semantic web, was proposed to overcome restrictions of common HBA systems. An enhancement to ISO/IEC 14543-3 EIB/KNX standard has been devised in a knowledge-based and context-aware computing framework for building automation, supporting semantic annotation of both user profiles (i.e., needs, moods, features) and device capabilities (i.e., services/resources of home appliances). The integration of a semantic microlayer within the KNX protocol stack enabled novel resource discovery and decision support features in HBA making them autonomous and decentralized while preserving full backward compatibility. Machine-understandable metadata characterize both home environment and user profiles and preferences. Thanks to the integration of semantics at the application layer, each object/subject joining a KNX network can describe itself and advertise managed services/resources. By means of a matchmaking process—based on inference procedures in [30]—the most suitable available services/functionalities for adapting the ambient to a given request or event can be easily detected. Annotations are expressed in ontological formalisms derived from description logics (DLs) [32]: DIG [33], a more compact equivalent of OWL-DL (OWL Web Ontology Language, W3C Recommendation, February 10, 2004, http://www.w3.org/TR/owlfeatures/), has been adopted in particular. In order to reduce the size of semantic annotations referred to both device features and user profiles, an encoding algorithm [34] is exploited for efficiently compacting XML-based ontological languages, generating a header and a body for each encoded document. Furthermore, an IO named *generic profile of the device* (*GPD*) has been introduced to describe general device features, for example, type, manufacturer, or model. A single GPD is associated to a given device. A *specific profile of the device* (*SPD*) object has been also defined to describe individual functionalities and operating modes of a device. Multiple SPDs can be associated to the same device, one for each different service/function it exposes. Furthermore, a user-transparent and device driven interaction is enabled as opposed to current static configuration approaches. At network layer, KNX support for IP communication through KNXnet/IP is leveraged to extend the management of building control beyond the local bus, while IEEE 802.11 and Bluetooth are exploited for wireless communication with the user. The work was extended in [35] with a multiagent approach exploiting semantic-based resource discovery and orchestration in HBA. The semantic annotation of user profiles and device capabilities is used to (1) determine the most suitable home services/functionalities according to implicit and explicit user needs and (2) allow device-driven interaction for autonomous adaptation of the environment to context modification.

References

1. UNI EN 15232, Energy performance of buildings—Impact of building automation, controls and building management, 2007.
2. KNX Association, *KNX Handbook for Home and Building Control*, KNX, Brussels, Belgium, 2009.
3. EN 50065-1, Signalling on low-voltage electrical installations in the frequency range 3 kHz to 148,5 kHz—Part 1: General requirements, frequency bands and electromagnetic disturbances, 2011.
4. ISO/IEC 8802-2, IEEE Standard for Information Technology, Telecommunications and information exchange between systems—Part 2: Logical link control, Washington, DC, 1998.
5. ISO/IEC 8802-11, IEEE Standard for Information Technology, Telecommunications and information exchange between systems—Part 11: Wireless LAN medium access control (MAC) and physical layer (PHY) specifications, 2012.

6. IEEE 1394.1, IEEE standard for high performance serial bus bridges, 2004.
7. ISO/IEC 7498-1, Information technology—Open systems interconnection—Basic reference model: The basic model, 1994.
8. KNX Association, ETS4 (Engineering Tool Software), version 4, 2010. http://www.knx.org/knx-en/software/ets/about/index.php.
9. B. Malinowsky, G. Neugschwandtner, W. Kastner, Calimero: Next generation, in *Proceedings of the KNX Scientific Conference*, Duisburg, Germany, 2007.
10. D. Dietrich, D. Bruckner, G. Zucker, P. Palensky, Communication and computation in buildings: A short introduction and overview, *IEEE Transactions on Industrial Electronics*, 57(11), 3577–3584, IEEE, Washington, DC, 2010.
11. N. Shadbolt, Ambient intelligence, *IEEE Intelligent Systems*, 18, 2–3, IEEE, Washington, DC, 2003.
12. H. Nakashima, H. Aghajan, J.C. Augusto, *Handbook of Ambient Intelligence and Smart Environments*, Springer, New York, 2010.
13. M.E. Pardo, G.E. Strack, D.C. Martinez, A domotic system with remote access based on web services, *Journal of Computer Science & Technology*, 8(2), 91–96, 2008.
14. K. Gill, S. Yang, F. Yao, X. Lu, A ZigBee-based home automation system, *IEEE Transactions on Consumer Electronics*, 55(2), 422–430, IEEE, Washington, DC, 2009.
15. W.S. Lee, S.H. Hong, KNX—ZigBee gateway for home automation, in *IEEE International Conference on Automation Science and Engineering* (*CASE 2008*), Arlington, VA, 2008, pp. 750–755.
16. V. Miori, L. Tarrini, M. Manca, G. Tolomei, An open standard solution for domotic interoperability, *IEEE Transactions on Consumer Electronics*, 52(1), 97–103, 2006.
17. T. Catarci, F. Cincotti, M. Leoni, M. Mecella, G. Santucci, Smart homes for all: Collaborating services in a for-all architecture for domotics, in *5th International Conference on Collaborative Computing: Networking, Applications and Worksharing*, Washington, DC, 2009, pp. 56–69.
18. M. Metzger, G. Polaków, A survey on applications of agent technology in industrial process control, *IEEE Transactions on Industrial Informatics*, 7(4), 570–581, 2011.
19. K.I.-K. Wang, W.H. Abdulla, Z. Salcic, Ambient intelligence platform using multi-agent system and mobile ubiquitous hardware, *Pervasive and Mobile Computing*, 5, 558–573, 2009.
20. C.-L. Wu, L.-C. Fu, Design and realization of a framework for human-system interaction in smart homes, *IEEE Transactions on Systems, Man and Cybernetics, Part A: Systems and Humans*, 42(1), 15–31, IEEE, Washington, DC, 2012.
21. M. Santofimia, F. Moya, F. Villanueva, D. Villa, J. Lopez, Intelligent agents for automatic service composition in ambient intelligence, in *Web Intelligence and Intelligent Agents*, Z. Usmani, ed., InTech, Rijeka, Croatia, pp. 411–428, 2010.
22. V. Gungor, D. Sahin, T. Kocak, S. Ergut, C. Buccella, C. Cecati, G. Hancke, Smart grid technologies: Communications technologies and standards, *IEEE Transactions on Industrial Informatics*, 7(4), 529–539, IEEE, Washington, DC, 2011.
23. P. Palensky, D. Dietrich, Demand side management: Demand response, intelligent energy systems, and smart loads, *IEEE Transactions on Industrial Informatics*, 7(3), 381–388, IEEE, Washington, DC, 2011.
24. M. Aliberti, Green networking in home and building automation systems through power state switching, *IEEE Transactions on Consumer Electronics*, 55(2), 445–452, IEEE, Washington, DC, 2011.
25. L. Klein, J. Kwak, G. Kavulya, F. Jazizadeh, B. Becerik-Gerber, P. Varakantham, M. Tambe, Coordinating occupant behavior for building energy and comfort management using multi-agent systems, *Automation in Construction*, 22, 525–536, 2012.
26. J. Kwak, P. Varakantham, R. Maheswaran, M. Tambe, F. Jazizadeh, G. Kavulya, L. Klein, B. Becerik-Gerber, T. Hayes, W. Wood, SAVES: A sustainable multiagent application to conserve building energy considering occupants, in *11th International Conference on Autonomous Agents and Multiagent Systems* (*AAMAS*), Valencia, Spain, 2012, pp. 21–28.

27. L. Sommaruga, A. Perri, F. Furfari, DomoML-env: An ontology for human home interaction, in *Proceedings of the 2nd Italian Semantic Web Workshop* (*SWAP'05*), *CEUR Workshop*, Trento, Italy, December 2005, pp. 249–255.
28. C. Reinisch, W. Granzer, F. Praus, W. Kastner, Integration of heterogeneous building automation systems using ontologies, in *Proceedings of the 34th Annual Conference on Industrial Electronics* (*IECON'08*), Orlando, FL, 2008, pp. 2736–2741.
29. D. Bonino, E. Castellina, F. Corno, The DOG gateway: Enabling ontology-based intelligent domotic environments, *IEEE Transactions on Consumer Electronics*, 54(4), 1656–1664, IEEE, Washington, DC, November 2008.
30. S. Colucci, T. Di Noia, A. Pinto, A. Ragone, M. Ruta, E. Tinelli, A non-monotonic approach to semantic matchmaking and request refinement in e-marketplaces, *International Journal on Electronic Commerce*, 12(2), 127–154, 2007.
31. M. Ruta, F. Scioscia, E. Di Sciascio, G. Loseto, Semantic-based enhancement of ISO/IEC 14543-3 EIB/KNX standard for building automation, *IEEE Transactions on Industrial Informatics*, 7(4), 731–739, IEEE, Washington, DC, November 2011.
32. F. Baader, D. Calvanese, D. Mc Guinness, D. Nardi, P. Patel-Schneider, *The Description Logic Handbook*, Cambridge University Press, Cambridge, U.K., 2002.
33. S. Bechhofer, R. Möller, P. Crowther, The DIG description logic interface, in *Proceedings of the 16th International Workshop on Description Logics* (*DL'03*), ser. *CEUR Workshop Proceedings*, vol. 81, Rome, Italy, September 2003.
34. F. Scioscia, M. Ruta, Building a semantic web of things: Issues and perspectives in information compression, in *IEEE International Conference on Semantic Computing* (*ICSC'09*), Berkeley, CA, 2009. IEEE, Washington, DC, pp. 589–594.
35. M. Ruta, F. Scioscia, G. Loseto, E. Di Sciascio, Semantic-based resource discovery and orchestration in Home and Building Automation: A multi-agent approach, *IEEE Transactions on Industrial Informatics*, 10(1), 730–741, IEEE, Washington, DC, February 2014.

59

Future Trends in Smart Homes and Buildings

Wolfgang Kastner
Vienna University of Technology

Markus Jung
Vienna University of Technology

Lukas Krammer
Vienna University of Technology

59.1 Introduction 59-1
59.2 Application Areas 59-2
Smart Living • Smart Metering and Energy Management • Smart Grids and Smart Cities
59.3 Future Internet of Things Protocols 59-4
6LoWPAN • ZigBee IP • Constrained RESTful Environments
59.4 IoT Protocol Stacks and System Architectures 59-11
Open Building Information Exchange • OPC UA • MQTT • IPSO Application Framework • ZigBee: Smart Energy Profile 2
59.5 Conclusion 59-18
References 59-18

59.1 Introduction

For home and building automation technologies, two main trends can be observed. The first one arises from the ongoing demand to improve the energy efficiency of buildings. The reason for this trend can be found in decades of continuous increase in the energy prices. Commercial buildings being responsible for about 23% of the worldwide energy consumption are the focus of research on how energy efficiency can be improved and how renewable energy sources can be used to cover the energy demands of buildings [1]. An extensive introduction of the building automation systems in commercial buildings and urban complexes is going to be decisive in achieving the ambitious energy goals that have been set by several regional and national governmental institutions.

Taking into account the continuous urbanization taking place in the Asia-Pacific region that grows annually by the size of a 2 million metropolitan city, the role of buildings including energy consumption is getting increasingly important. In order to improve the energy efficiency of buildings, the various domains of building automation covering heating, ventilation, air-conditioning, cooling, as well as lighting and shading need to cooperate in a cross-domain manner. The use of heterogeneous technologies that are either tailored to specific domains or only addressing one specific layer of a multi-tiered automation system is a commonplace in contemporary buildings (cf. Chapter 55).

Furthermore, building automation services need to be integrated into processes that have a larger scope than the local building. The interaction with smart grids to enable demand-side management or load shifting requires an integration that spans multiple buildings or even a complete region. This requires services that manage multiple buildings and sites and perform real-time data analysis that demands extensive computational resources. Here, cloud computing can provide the required flexibility.

This leads to the second main trend that can be found in building automation systems, which is the adoption of the Internet Protocol (IP) and wireless technologies. Although IP technology is well established at the management tier of automation systems, it is still not treated as a first-class citizen at the lower levels of the automation hierarchy. State-of-the-art technologies like KNX or LonWorks, as introduced in Chapter 55, use IP to tunnel their custom network layer, transport layer, and application layer technologies. IP technologies ease the integration with enterprise and IT systems and can be one enabler for crossing the borders between different automation technologies. However, the adoption of IP technologies does not necessarily require integration with the Internet, although this would allow to fully exploit the potential of having IP in place. Rather, IP can also be operated in closed networks, still leaving the main advantage of using mature technologies with several implementations and tools available.

Also, building management can benefit from the cloud computing. For instance, building maintenance can be taken care of by services and tools offered by a third party. Software-as-a-Service (SaaS) paradigms allow to move the energy management, operation, and maintenance of buildings from the site to a third party.

Although IP technologies ease the integration of the heterogeneous and distributed systems, several issues arise if these technologies are to be used in the field and automation tier. The main problems are the strong resource demands on memory, processing power, and bandwidth that have to be addressed. Nevertheless, the ease of integration through the use of IP technologies for commercial buildings is the main driver for the successful deployment of smart grids and smart cities. Moreover, the application of IP within residential homes is beneficial for coupling together home appliances, consumer electronics, and home energy management systems, ending up in the vision of *smart living*.

Future trends in building automation also tackle the wide-spanning area of the *Internet of Things* (IoT) [2] that was initially characterized by radio-frequency identification (RFID) components and the use of the Electronic Product Code (EPC) to identify physical objects. Since IP does not provide much more than a network protocol to interconnect different devices, the layers mentioned earlier are of utmost importance for building *interoperable* systems feasible for the IoT. The use of Web-based communication is a possible way to provide these upper protocol layers. The term *Web of Things* [3] can be used to summarize these efforts that aim at building interoperable systems that adhere to the paradigms that made the World Wide Web successful.

59.2 Application Areas

In the future, home and building automation systems will be used in novel application areas that require the interconnection of heterogeneous infrastructures and information systems. Selected areas and their related communication protocols are summarized in the following sections.

59.2.1 Smart Living

The term smart living generally denotes a residential dwelling where automation technologies take over the automatic control of (at least) parts of the building services in so-called smart homes. It shares its roots with building automation but, nowadays, needs to be regarded as an independent application and research area [4]. In order to differentiate it from *classic* building automation, it is often referred to as home automation or sometimes domotics. Apart from the core automation domains heating, ventilation, and air-conditioning (HVAC) and lighting/shading, a smart home also interconnects household appliances as well as consumer electronics and entertainment devices. Moreover, security and safety components as well as equipment from the *Ambient Assisted Living* (*AAL*) domain (e.g., sensors monitoring human vital parameters) are integrated into a smart home system.

Research is related to ongoing work in the area of ambient intelligence where systems should be sensitive and responsive to people and the environment. Thereby, computing devices are fully integrated into the physical environment and people are using them without being aware of it. This concept is denoted by the term ubiquitous computing. Unfortunately, smart homes are often regarded as a future vision that seems to be practically available only in some decades of time. In practice, however, the technological and scientific foundations have long been developed so that a realization of smart homes is already practically feasible today. Still, the commercial breakthrough of the smart home is awaited. However, several demonstration projects addressing smart homes have already been put into practice (e.g., PlaceLab, Georgia Tech Broadband Institute Residential Laboratory, T-Com Haus, inHouse, iHomeLab). They illustrate different approaches regarding the concept of a smart home and also provide valuable information on the challenges as well as lessons-learned information for the design of upcoming smart home concepts and systems. Nevertheless, in many cases, existing prototypes cannot fully achieve their promises. Their systems are too complex for a practical use, seldom tailored to pervasive energy savings and often not characterized by real smartness. A smart home system needs to combine many different parts (e.g., novel control strategies that operate on additional knowledge and are executed autonomously) so that the overall solution is more than the mere sum of its parts [5].

For this reason, a smart home cannot be seen as an isolated instance that solely interacts with its inhabitants. It incorporates with different kinds of devices and adopts information of its environment in order to realize *smart* applications. Thereby, the IoT allows direct communication among devices (i.e., *machines*). Since machines do not have the human capabilities of interpreting different types of information, precise foundations of semantics have to be assigned to the objects, their data, and functionalities. Ontology frameworks that emerged within recent research work of the IoT community (e.g., SensorML, DomoML, or the ThinkHome ontology) can provide such a foundation.

59.2.2 Smart Metering and Energy Management

Smart meter devices are the cornerstone of energy management systems and a new energy grid (see next section) and will facilitate the power authorities to deal with real-time power consumption values. Here, the integration of devices resting upon the (Wireless) M-Bus standard defined in EN 13757 will play an important role [6]. However, also other approaches have to be considered, like the International Electrotechnical Commission (IEC) 62056 where the Device Language Message Specification (DLMS) and the Companion Specification for Energy Metering (COSEM) have been published. Also, Smart Message Language (SML) provides a communication protocol for meter devices [7]. Within the North American market, ANSI C12.18 gained importance [8].

While energy management systems usually aimed at optimizing the local energy consumption of a building, in the context of upcoming smart grids, also the external systems are in the scope of this optimization. For commercial buildings, substantial energy savings can be achieved. In the residential area, there is still a strong potential for the deployment of home energy management systems. The key criteria are the global optimization of the user comfort and energy efficiency. For this optimization task, information technology is exploited to host and execute control algorithms that are capable of multi-criteria optimizations. Minimum inputs for these algorithms are goals of the global level, the states of the underlying automation systems and their devices, as well as information on the state of the environment. This mandatory context awareness as well as the optional availability of predictions of future environmental states (e.g., anticipated user presence, predicted weather situations) are foundations of energy management systems. Decisive to the widespread deployment of energy management systems will be the rising costs of energy, regulations mandating energy efficiency, flexible pricing schemes, and the need to protect the environment. Still, strong inhibitors are the low savings potential for residential customers, high initial investment costs, the heterogeneity of technologies, and the risk aversion of utilities that are reluctant to go beyond pilot deployments [9]. Security and privacy issues may impede the deployment.

The IoT and the use of interoperable protocols can be a solution to the problem of the underlying heterogeneous smart (metering) technologies. The key application for the home and building automation systems will be energy demand response, which is the adjustment of the electric energy consumption in reaction to external parameters such as pricing signals or the grid stability. For large commercial customers, demand response is well-known and might even be performed in a manual way. However, specialized protocols for the automated demand response such as Open Automated Demand Response

(OpenADR) [10] make the deployment more attractive. Thus, a strong market growth is to be expected within the next decade for demand response in the commercial and residential area.

59.2.3 Smart Grids and Smart Cities

The term smart grids reflects the need of adding information and communication technologies to the existing power grid in order to successfully integrate variable renewable energy sources and to increase the grid's stability. Traditional power grids adhere to the paradigm that the production has to follow the consumption pattern of energy consumers. Usually, centralized power plants generate the energy that is distributed to the consumer using high-voltage transmission grids as well as medium- and low-voltage distribution grids. Renewable energy sources like photovoltaic and wind power or small- and middle-sized power decentralized power plants lead to revolutionary challenges put on the electricity grid.

Within smart grids, the IEC 61850 [11] standard series will play an important role in the design and operation of power utility automation systems. While the focus of Edition 1 of the standard series concentrated on substation automation systems, most parts of the standard series have been revised in Edition 2 where the focus was broadened to interconnected power utility automation systems in general. IEC 61850 defines technical aspects of automation system communication, such as communication protocols and data objects, but also aspects on the design, configuration, and operation of power utility automation systems are specified. The standard covers communication within a substation, as well as inter-substation communication and communication between control centers and substations. IEC 61850 uses a two-layered approach for defining its communication services: Abstract Communication Service Interface (ACSI) defines all communication services in an object-oriented way, while several Specific Communication Service Mappings (SCSMs) map these abstract services to specific communication protocols. Examples for SCSMs are the client/server protocol Manufacturing Messaging Specification (MMS) and the real-time protocols Generic Object Oriented Substation Events (GOOSE) and Sampled Measured Values (SMV).

Buildings are a fertile ground for the demand-side management. The heating and cooling processes deployed in commercial buildings can be used to provide the required flexibility with respect to the energy consumption. Distributed generation puts stress on the distribution grid since the energy flow can now be reversed. Smart meters and advanced metering infrastructure allow to measure in real-time energy consumption and the power quality. They also allow data analysis and provide feedback to the customers on energy consumption. The IoT and its protocols can be one of the key factors in the successful development and deployment of smart grids [12]. Smart cities—focused on the urbanized area—have a similar goal: to increase the energy efficiency and optimize the usage of renewable energy sources. In addition to the built environment and the energy grid, they include other systems and infrastructures, like telecommunications, traffic management, or the health, safety, and security domains [13].

59.3 Future Internet of Things Protocols

There are several technologies and protocols that are relevant for the future IoT that will interconnect billions of devices equipped with embedded intelligence or so-called smart objects. One way to realize such an infrastructure is to reuse the IP that proofed its reliability and scalability. Besides the use of *plain* IP, Web technologies like *Web services* revolutionized the landscape of information systems and their integration within different applications. Thus, the following subsections provide an overview on these protocols.

59.3.1 6LoWPAN

6LoWPAN is an acronym for IPv6 over Low-power Wireless Personal Area Networks (LoWPAN). Its idea is to deploy IP even on most constrained devices with limits regarding memory and processing power that are wirelessly connected with narrow bandwidth and strong requirements on energy efficiency [14]. 6LoWPAN is not a fully specified protocol stack; it is more a framework of standards that respectively address several issues of the IoT. As stated in [15], several problems have to be covered if IPv6 is used on top of Institute of Electrical and Electronics Engineers (IEEE) 802.15.4 networks [45]:

- IPv6 connectivity: IPv6 connectivity is desirable since LoWPANs will consist of many devices. IPv6 provides the required address space and mechanisms for stateless auto-configuration. Furthermore, this allows the seamless connectivity with the Internet.
- Topologies: Different topologies needed to be supported within a LoWPAN including mesh and star topologies. Routing may happen within the local mesh at the link layer or on a more global level between different IP networks that may use other links. Furthermore, the resource demand of routing protocols needs to be taken into account with a routing overhead balanced regarding topology changes and power conservation.
- Packet size: IPv6 demands the maximum transmission unit (MTU) to be at least 1280 bytes. However, for IEEE 802.15.4, the standard packet size is 127 bytes. Further, the overhead of the header and, optionally, some link layer security features needed to be considered, leaving a remaining size of about 81 bytes that can be used by upper layers (i.e., payload).

  IPv6 provides its own fragmentation mechanism for packets that are larger than the minimum MTU of 1280 bytes but relies on the link layer to be able to transmit packets of this size. 6LoWPAN takes care of this gap and provides a fragmentation format. Packets are split into multiple fragments that can be reassembled at the receiving node. It defines an 11-bit `datagram_size` field that is transmitted in every fragment and reflects the size of the resulting reassembled packet limited to 2047 bytes. A 16-bit `datagram_tag` field together with the sender and destination link layer addresses and datagram size is used to distinguish between different packets. An 8-bit `datagram_offset` provides the position of the fragment within the packet and counts in 8 byte units.

  In general, protocols should be designed to allow the exchange of data or control commands within one IEEE 802.15.4 frame to have optimal efficiency.
- Limited configuration and management: Since devices within a LoWPAN have limited resources, usually no user interfaces are available. Thus, it is required to keep the configuration and management efforts as low as possible. In the best case, the protocols support bootstrap mechanisms without manual configuration.
- Service discovery: Services delivered by nodes within a LoWPAN need to be discoverable. Efficient protocols are required to supply such mechanisms to advertise and to search for certain service types. 6LoWPAN by itself does not define any service discovery protocol. However, for basic node discovery mechanisms, IPv6 neighbor discovery or the optimized version for LoWPANs can be used [16]. For application-layer-specific discovery requests, the multicast domain name system (mDNS) or the domain name system service discovery protocol can be applied.
- Security: Security is a key issue if sensors and actuators are connected wirelessly by IP. Using IEEE 802.15.4 as link layer option, basic security is provided by using the Advanced Encryption Standard (AES). However, within 6LoWPAN, no details on the key exchange and key management are specified. The bootstrapping process is not described and applications need to take care about it by themselves.

6LoWPAN provides an adaption layer that makes it possible to use IPv6 on constrained links. A popular link for 6LoWPAN is, for example, provided by IEEE 802.15.4 networks. Additionally, efficient mechanisms for multicast and unicast routing are specified, which facilitate a special neighbor discovery approach.

59.3.1.1 6LoWPAN Frame Format

Since IEEE 802.15.4 does not define a field for identifying the carried protocol, 6LoWPAN provides this information by using a dispatch byte as first element of the protocol header. The first 2 bits are used to group the possible dispatch values. All values starting with 00 are reserved for non-6LoWPAN frames. A value of 01 identifies a regular 6LoWPAN frame, 10 is used to identify a mesh header, and 11 is taken for fragmentation headers.

For addressing within 6LoWPAN, a relation between link layer addresses and network layer addresses is necessary. While IEEE 802.15.4 uses either 64-bit EUI addresses or 16-bit short addresses in conjunction with 16-bit network IDs, 6LoWPAN uses IPv6 addresses. Furthermore, a lot of information sent within the IPv6 header is redundant for subsequent packets that are exchanged between two communicating nodes. The full address space of 128 bits of IPv6 is usually not needed if the packets are transmitted within a 6LoWPAN network. 6LoWPAN addresses this issue by using header compression. Header compression can work either *end-to-end* or *hop-by-hop*. Since the full IP header is required to perform routing, *hop-by-hop* is a more attractive solution since it keeps the header compression as a local aspect between two adjacent nodes. 6LoWPAN aims at *stateless* compression, which has the advantage of not having any synchronization problem and simple algorithms. The 16-byte IPv6 addresses are the largest part in the IPv6 header and 6LoWPAN provides a stateless header compression for link-local addresses. Unfortunately, for compressing global routable addresses, at least some state information is required; therefore, a context-based header compression is also available.

59.3.1.2 Stateless Header Compression

For stateless header compression, two schemes are specified in [17]. HC1 is used to compress IPv6 headers and HC2 to narrow down User Datagram Protocol (UDP) headers.

The HC1 header contains two flags identifying the source address encoding (SAE) and the destination address encoding (DAE). These flags contain one bit for the prefix and the interface identifier (ID). If set to 0, the according part of the address is transmitted; if set to 1, the part is elided. For the prefix, this can only be used for link-local addresses (FE80::/64). For the interface ID, a 1 indicates that the address can be derived from the layer 2 address. Another optimization is to leave away the traffic class and the flow label of the IPv6 header that are in the most common cases set to 0. The C flag indicates this optimization. The two-bit next header (NH) flag is used to indicate the NH, which is limited to the options sent in-line, UDP, Internet Control Message Protocol (ICMP), or Transmission Control Protocol (TCP). The final bit indicates whether an HC2 header is present or not. Furthermore, the IPv6 version flag is never transmitted since this is an obvious IP version 6 in 6LoWPAN. The payload length can be inferred and is also elided.

The HC2 header only consists of three-bit flags. The first two bits are used to control the compression of the source and destination port that is limited to the port range between 61616 and 61631 (0xFBn), which allows to compress the port to use only the last four bits. The L flag is most of the time set and allows to leave away the payload length that can be derived.

The best case header compression that can be achieved with stateless compression is impressive, although it has to be mentioned that this case only applies for rather limited local interaction scenarios.

59.3.1.3 Stateful Header Compression

In cases where globally distributed nodes need to communicate, the optimizations used in stateless header compression cannot be applied. In order to allow compression of the global routable address information, some kind of *context* is required, that is, kept synchronized between a sender and receiver. The stateful header compression as stated in [17] is again separated into two compression headers. The first is the compression for the IP header (LOWPAN_IPHC) and the second one is the optional compression scheme for the NH (LOWPAN_NHC).

59.3.1.4 Neighbor Discovery

IPv6 neighbor discovery (IPv6 ND) is an essential part of IPv6 as it provides the functionality to discover routers, perform address resolution, and duplicate address detection and redirection of messages. It can be further used to discover the used network prefix and other parameters. IPv6 ND relies on IPv6 links that usually form a single broadcast domain and provide transitivity. In this context, transitivity means that if node A can communicate using the link with node B and node B can communicate to node C, then node A can also communicate with node C using the link without any intermediaries. 6LoWPANs provide a nontransitive link; therefore, the IPv6 ND—heavily using multicasting—does not operate well, as the realization of efficient multicasting is still an ongoing topic. To address this issue, the 6LoWPAN working group proposed an optimization for IPv6 ND that finally resulted in Requests for Comments (RFC) 6775 Neighbor Discovery Optimization for 6LoWPANs [16]. It lists several optimizations and extensions to IPv6 ND:

- Unsolicited router advertisements are avoided by a new host-initiated refresh of router advertisement information.
- For EUI-64-based IPv6 addresses, there is no duplicate address detection; for DHCPv6 assigned addresses, it is optional.
- A new address registration mechanism between hosts and routers avoids the use of multicast neighbor solicitation messages.
- A new router advertisement option supplies context information used for header compression.
- New messages for duplicate address request and duplicate address confirmation are used in features for performing duplicate address detection.
- A new mechanism to distribute prefixes and context information is available.
- New and optimized protocol constants are introduced.

59.3.1.5 Routing

Routing is a crucial aspect of IP-based network communication. For LoWPANs, approaches for unicast and multicast routing in IPv6-based wireless networks are defined.

For unicast (i.e., point-to-point [P2P]) communication, two different routing approaches are applicable. On the one hand, routing can be performed at the data link layer by the underlying mesh network (i.e., *mesh-under*). In this case, the adaption layer ensures that a pairwise communication is possible. However, it is not specified which routing mechanism is applied. On the other hand, a routing protocol at the network layer can be taken that addresses the special demands of wireless sensor and actuator networks (i.e., *route-over*).

For this purpose, the Routing Protocol for Low power and Lossy Networks (RPLs) as specified in RFC 6550 [18] is used. The routing mechanism is based on destination-oriented directed acyclic graphs (DODAG). Every graph has exactly one destination node (root), where every path in the graph finally leads to. The graph is stored in a distributed way, whereas the building process is initiated by the root node. For the computation and maintenance of the graph, ICMPv6 messages are used. Generally, messages are sent upward in the graph, but it is also possible that the root node sends messages to its children. RPL operates either in a non-storing mode or in a storing mode. In the case of P2P traffic, all downward traffic in the non-storing mode has first to reach the DODAG root. For the storing mode, a routing node processes the Destination Advertisement Objects (DAO) messages and constructs downward routes. This improves the P2P traffic by allowing a packet to be transmitted via a common ancestor between the source and the destination before reaching the DODAG root.

Immediate communication among nodes in different paths of the tree is not possible. Thus, some routes are far more expensive in terms of the used metric than the optimal one. If such links are frequently used, the overall performance drops. In order to overcome this issue, multiple instances of RPL can be defined. Each instance constructs its own DODAG using a different metric or a different root node; however, an RPL node has to belong to at most one DODAG within an RPL instance.

The storing mode can be operated with multicast or without multicast support. With multicast support, DAOs are used to relay group registrations up toward the DODAG root. In contrast to unicast traffic in which a packet is forwarded only to a single child node, multicast traffic is forwarded to all nodes that have registered to a multicast group. Multicasting routing states are therefore installed on each router between the listeners and the DODAG root.

A different approach is provided by Multicast Protocol for Low power and Lossy Networks (MPLs) [19] that is at the time being still in draft state. MPL avoids the creation and maintenance of a multicast forwarding topology. It uses the Trickle algorithm [20] for controlling packet and control information transmissions. Trickle optimizes the propagation of information efficiently and mainly specifies when messages need to be transmitted, without defining a concrete frame format or even the overall purpose the Trickle algorithm is used for. The basic idea is that when two neighbors are consistent (i.e., share the same state information), the message exchange rate between them is slowed down exponentially. For this purpose, the so-called trickle timer is used. If a node receives consistent information, it increases this timer. If an inconsistency is detected, the trickle timer is reset. This dynamic behavior prevents from *wasting* messages and allows to propagate changes within milliseconds.

MPL forwarders subscribe to multicast groups and are responsible for the dissemination of MPL multicast packets. An MPL domain consists of a set of connected forwarders that define the degree of the multicast distribution process. The behavior of the forwarders regarding which packets to buffer and which ones to transmit is defined by MPL. For the distribution, a proactive propagation mode and a reactive propagation mode are defined. In the proactive mode, the forwarder transmits the message based on the Trickle algorithm without discovering that a neighboring forwarder has not received the message. In the reactive mode, control messages are used by neighboring forwarders and buffered multicast packets are only sent, if it is discovered that they have not yet been received by other forwarders.

The main difference between RPL- and MPL-based multicasting is that in the case of RPL, a multicasting topology is maintained, while in case of MPL, only a per-packet state is preserved. This leads to several advantages and disadvantages of the mechanisms regarding scalability, performance, and complexity. A thorough evaluation is provided within [21]. Since MPL is in a draft state, the protocol can be expected to be improved.

59.3.2 ZigBee IP

ZigBee IP (ZIP) [22] is an upcoming technology in the context of Wireless Sensor and Actuator Networks (WSAN). It is specified by the ZigBee Alliance [23] and addresses recent developments in Web-based technologies. ZIP completely differs from the *original* ZigBee version [47]. In contrast to ZigBee that uses IEEE 802.15.4 addressing, ZIP is based on IPv6. However, ZigBee as well as ZIP are based on the IEEE 802.15.4 physical layer (PHY) and media access control (MAC). ZIP is quite similar to 6LoWPAN. While 6LoWPAN is a collection of possible IoT standards, ZIP strictly specifies communication modes and referees to dedicated standards. Thus, the ZigBee Alliance states that ZIP is the *first open standard for IP-based wireless mesh networks.*

59.3.2.1 ZigBee IP Protocol Stack

ZIP is based on IEEE 802.15.4-2006 [24] that defines the PHY and data link layer and supports communication in different frequency bands. While the 2.4 GHz frequency band can be used globally, there are two sub-GHz bands that are restricted to specific regions. Immediately above, the *6LoWPAN adaption layer* resides in the ZIP communication stack. As stated in Section 59.3.1, IPv6 requires an MTU of at least 1280 bytes, but IEEE 802.15.4 allows only packet sizes of 127 bytes. Thus, the adaption layer is responsible for compressing and decompressing IPv6 headers as well as fragmenting and reassembling IPv6 packets. For packet fragmentation, ZIP refers to RFC 4944 [25] with the additional requirement that fragments of an IPv6 packet are transmitted and received in order. Since already one IPv6 address is as long as an IEEE 802.15.4 packet, the header has to be compressed and decompressed according to RFC 6282 [17]. Thereby, a ZIP device must support all defined compression modes and shall use the most effective one. IPv6 headers can be compressed in a stateless as well as in a stateful way as described in Sections 59.3.1.2 and 59.3.1.3, respectively. The adaption layer further handles neighbor discovery as specified in RFC 6775 [16]. This mechanism allows a node to determine its immediate neighbors. In contrast to conventional networks, wireless links have special properties that imply requirements concerning neighbor discovery. Thus, ZIP refers to RFC 6775 [16] that optimizes the common neighbor discovery mechanism of IPv6, whereby ZIP requires that even optional mechanisms of the standard are supported.

The network layer implements the IPv6 protocol suite with special optimizations for low-rate wireless networks. Since ZIP is based on IEEE 802.15.4, two different types of addresses are possible. On the one hand, a device has to be uniquely identified by a EUI-64 address [26], while on the other hand a device can be identified by a combination of a 16-bit short address and a 16-bit personal area network id (PAN-ID). To derive an IPv6 interface ID from a EUI-64 address, link-local unicast addresses (cf. [27]) with a prefix of 64 bits are used. Similarly, a link-local unicast address with a prefix of 64 bits can be further derived from a concatenation of the PAN-ID, 16 zero bits followed by the 16-bit short address. Both address types have to be compressed in a stateless way by the *adaption layer*. Furthermore, a global unicast address can also be derived from the short address. Thereby, a unique prefix has to be defined and the address has to be checked for uniqueness. ZIP also facilitates IPv6 multicast communication. For this reason, *subnet-local* addresses are used (i.e., scope value 3, cf. RFC 4291). The network layer is further responsible for routing. In contrast to 6LoWPAN where routing can alternatively be performed at the adaption layer, ZIP uses the RPL as specified in RFC 6550 (i.e., *route-over*) [18]. For multicast routing, a special multicast forwarding mechanism called MPLs is used in ZIP according to an Internet Engineering Task Force (IETF) draft standard [28]. These basic properties of the mechanisms are described in Section 59.3.1.5.

59.3.2.2 Security

The transport layer of ZIP supports connectionless UDP as well as connection-oriented TCP services. Besides network encryption according to IEEE 802.15.4, the main security mechanism resides at this layer. The security suite of ZIP is based on *Protocol for Carrying Authentication for Network Access (PANA)* as described in RFC 5191 [29], which acts as carrier for other security mechanisms and additional security-related material. For example, security information for configuring IEEE 802.15.4 security mechanisms are distributed by the authentication server. PANA is based on UDP and is therefore capable of multihop transmissions. Thus, a joining device not necessarily is directly reachable by the authentication server. Mutual authentication between a joining node and the authentication server is established by the *Extensible Authentication Protocol (EAP)*. At highest level, Transport Layer Security (TLS) authentication is used, which is why EAP-TLS must be used for carrying TLS data. TLS provides two different types of authentication: one is based on preshared keys and the other one is based on elliptic curve cryptography in conjunction with an AES-based mode of operation.

59.3.2.3 Topology

A ZIP network consists of devices of different types. Initially, a network contains exactly one ZIP coordinator that acts as PAN coordinator according to IEEE 802.15.4. Additionally, a ZIP router and ZIP hosts can join the network depending on their hardware resources and capabilities. While a ZIP router is able to forward packets in the mesh network, ZIP hosts are associated with a ZIP router only and are not capable of routing.

59.3.3 Constrained RESTful Environments

Web services, especially RESTful Web services [30], are a key technology for integration in enterprise technologies. The IETF working group on constrained RESTful environments* (CoRE) aims at a framework for resource-oriented applications that brings down the advantages of REST to constrained IP networks. These networks are constrained in the sense of providing limited bandwidth and reliability and having nodes with limited computing capacities and strong energy constraints. Moreover, these devices tend to be not online all the time and may have limited wake up times. LoWPANs are one

* https://datatracker.ietf.org/wg/core/.

example of such network types. The framework is tailored to limited classes of applications that are working on simple resources like sensors and actuators found in a typical automation system setup and the application domains range, for example, from home and building automation systems to energy management systems.

Within the CoRE framework, several standardization drafts emerged. The Constrained Application Protocol (CoAP) [31] is a messaging protocol to interact with resources on constrained devices and is the most mature standardization draft of CoRE that will soon become an IETF standard. CoAP defines a similar request/response interaction protocol like Hypertext Transfer Protocol (HTTP), but due to the high resource demand of HTTP and TCP, it provides several optimizations for the protocol header and uses UDP instead of TCP. This opens the possibility of further message exchange patterns like asynchronous communication or group communication but forces CoAP also to deal with the reliability issues when UDP is used as transport layer protocol. Complementary to the messaging protocol, there are standard drafts for observing resources and block-wise transfer, and finally a link format that allows to describe the resources that are provided by a CoAP device.

The architecture of CoRE targets a seamless integration with the existing Internet and the Web. Figure 59.1 provides an overview of the architecture of CoRE.

59.3.3.1 CoAP Architecture

CoAP uses a client/server interaction model where nodes may act at the same time as server and client. At transport layer, UDP is used. Since UDP is packet oriented and furthermore provides no functionalities to ensure reliability, several extensions to CoAP compared to HTTP are required. CoAP defines four message types, which are confirmable, non-confirmable, acknowledgment, and reset. These messages are provided in a sublayer and used for a request/response interaction semantics to the application layer, which does not have to deal with the messaging. All requests are sent to a uniform resource identifier (URI) that uniquely identifies the target resource.

For requests, four methods are defined: `GET` to retrieve the representation of the identified resource, `POST` to process the request by the identified resource or to update the identified resource, `PUT` to create or update the identified resource, and `DELETE` to delete the resource identified.

CoAP uses a 4-byte binary header followed by binary options and the payload. Each message contains a message ID in order to detect duplicate messages and for optional reliability mechanism. CoAP marks messages as confirmable if an acknowledgment should be provided by the server. In this case, the server responds with an acknowledgment. If the acknowledgment is not received, the client retransmits

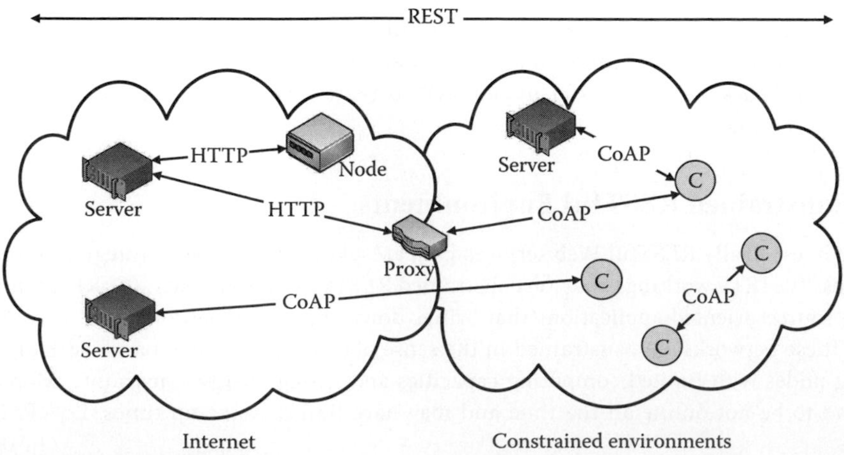

FIGURE 59.1 CoRE architecture.

the message for several times until an acknowledgment is returned. In order to avoid congestion, an exponential back-off mechanism is used. For some use cases, reliable messaging is not required; in this case, only one message is passed from the client to the server, which makes the communication more efficient but unreliable.

The request/response semantics of CoAP are similar to HTTP. The request contains the method or response code and the optional information such as the resource URI and the payload media types. A token is used to match responses to requests and can use up to 8 bytes. The token should be generated with strong randomization mechanisms in order to prevent easy spoofing of response messages. Requests can be marked confirmable or non-confirmable, and if possible, response messages can be piggybacked in the acknowledgment message.

Due to the use of UDP, CoAP further provides asynchronous communication abilities allowing a client to subscribe to a resource and to be notified about changes without polling permanently for updates. In this case, the observe option can be specified for the GET request. The client then continuously receives updates until a reset request is sent.

59.3.3.2 Message Format

A CoAP message starts with a fixed-size 4-byte header, consisting of a 2-bit flag indicating the CoAP version number, followed by a type ID (T) that maps to the different message types that are possible. An 8-bit field is used for specifying the message code. All possible values are maintained within a code registry. Finally, the 16-bit message ID is part of the fixed header. The message ID is used for message exchange purpose and reliability as described earlier. Afterward, a variable length token with the size between 0 and 8 bytes is used for mapping requests and responses. Subsequently, 0 or more options are closed by a 1-byte payload marker. Each option is identified through a number and has a variable length and an option value. The mechanism for specifying options is rather sophisticated due to optimizations but is also quite flexible. The option number is in the most basic case expressed through a 4-bit delta value that is accumulated while the header is processed and a 4-bit length field that provides the length in bytes. Reserved values for the option number delta and length allow to use up to 16-bit fields for identifying the option number and the length. One example for an option is the URI path that identifies a resource.

59.4 IoT Protocol Stacks and System Architectures

The previous section presented several upcoming standards of relevant communication protocols for the IoT. Typically, these protocols are tailored to a very specific functionality at a certain layer of the ISO/OSI reference model. For providing an alternative to existing home and building automation systems, a full system stack is required that allows two nodes to communicate with each other or, in a more advanced case, even a complete system architecture. The following subsections overview selected standards and technologies that address this issue.

59.4.1 Open Building Information Exchange

The Open Building Information Exchange (oBIX) as already presented in Chapter 55 can provide missing application layer services and semantics for the IoT, especially when it comes to building automation systems [32]. This can be seen in the recent standardization activities around oBIX that gained substantial momentum.

By 2010, the current working draft of the oBIX specification proposed a binary message encoding for the usage within 6LoWPAN networks. However, the performance issue of using HTTP and TCP remained. In the latest oBIX 1.1 committee specification that was released for public review, this shortcoming was addressed by new protocol bindings and encodings that make oBIX more suitable for the IoT and a deployment in constrained environments.

59.4.1.1 IoT Protocol Bindings and Message Encodings

In the latest oBIX 1.1 specification [50], the protocol bindings and encodings moved to separate specification documents and include a new REST protocol binding to CoAP. This allows to take advantage of the communication features of CoAP. Alternatively to the former oBIX watch concept, objects can be observed using the CoAP observe feature. In that way, a client does not need to poll for updates. Instead, they are pushed by the server that makes this kind of conversion more efficient. Furthermore, new message encodings for JavaScript Object Notation (JSON) and the Efficient XML Interchange (EXI) format have been defined. The JSON encoding has become recently quite popular in the IoT community since it is slightly more lightweight than Extensible Markup Language (XML) and easy to process in Web browsers using JavaScript. EXI is an attractive alternative to a proprietary binary encoding as it is as efficient regarding message size and also based on an open standard. If schema-informed EXI is used, the message encoding efficiency compares to the oBIX binary encoding and is therefore well suited for the use within LoWPANs. With these improvements, oBIX can be directly deployed on constrained sensors and actuators and directly integrated in the IoT through technologies like 6LoWPAN and CoAP.

59.4.1.2 oBIX 2.0 Outlook

The current standardization activities within oBIX are mainly driven through the requirements identified by the IoT. Also, oBIX has been recognized as a relevant technology providing interoperability within smart grids by the National Institute of Standards and Technology (NIST) [33]. The work plan for oBIX 2.0 is yet to be defined but several goals are already determined.* oBIX 2.0 should be a main technology in the IoT, and therefore peer-to-peer communication services should enrich the optimized protocol bindings. Besides this, the focus is put on so-called enterprise services. Specific contracts for the energy domain should enrich the core contract sets and allow the easy integration with Energy Interoperation standard that subsumes OpenADR 2.0. Furthermore, the requirements of the Energy Market Information Exchange standard should be addressed. Reporting and advanced history services are also on the roadmap. Additionally, it is planned to enrich current alarming services by a powerful standardized way to specify alarm logic. Building information models will play an essential role. Moreover, the integration toward enterprise scheduling will be addressed to schedule interactions within building systems. Finally, oBIX 1.1 does not support a custom security framework. This is going to be solved by oBIX 2.0 with a special attention on how access control can be achieved and aligned with enterprise services.

59.4.2 OPC UA

OPC as introduced in Chapter 55 was originally designed for reading, writing, and monitoring of control data. Therefore, different kinds of access schemes were defined such as OPC Data Access or OPC Alarms and Events. However, using classic OPC in modern automation systems leads to several disadvantages. First, OPC is based on Microsoft's technology (Distributed) Component Object Model (COM/DCOM), which limits its use to Windows-based systems. Furthermore, COM/DCOM has additional drawbacks like limited support for routing and security as well as incompatibility between different Windows versions. Another major drawback is its rudimentary data model where only very restricted information models can be created. Complex data structures, for example, are not provided by these OPC standards.

To overcome the drawbacks of these so-called classical OPC specifications, the OPC Foundation [48] decided to define a new specification that is referred to as *OPC Unified Architecture* (*OPC UA*) [34]. It is also standardized as IEC 62541 [35]. To eliminate the dependency on Windows, OPC UA is based on a service-oriented architecture (SOA). To achieve this, OPC UA uses Web services or a TCP-based protocol for data exchange that both enable full platform independence. In addition to this new communication

* http://www.newdaedalus.com/articles/2013/3/26/work-plan-for-obix-20.html.

concept, another key feature within OPC UA is the support of a powerful and extensible information model that can be used to model any kind of information and data.

59.4.2.1 Information Modeling in OPC UA

The comprehensive information modeling capabilities of OPC UA can be used to provide a standardized, well-defined way to model the data that are originating from different technologies. In order to model information in OPC UA, a so-called address space model is introduced that provides the following features [36]:

- The address space model supports object-oriented mechanisms allowing the definition of type hierarchies and inheritance.
- The modeling concept allows a definition of models that consist of full-meshed nodes.
- Information models are defined on server side, which entails that clients do not need to be aware of the models since they can be retrieved from the server.
- Type information and type hierarchies can be accessed by clients since they are exposed like normal data.
- A key concept of OPC UA information modeling is that the base information model can be extended by defining own models on top of the existing one.

Information modeling in OPC UA is based on defining *nodes* and *references* between them. Nodes are used to model any kind of information. This includes the representation of data instances or the definition of data types. Each node belongs to exactly one of the following *node classes*:

- `Base NodeClass`: This node class represents the base class from which all other node classes are derived.
- `Object NodeClass`: Objects are used to represent real-world entities like system components, hardware and software components, or even whole systems.
- `ObjectType NodeClass`: This node class holds type definitions for objects.
- `Variable NodeClass`: Variables are used to model values of the system. OPC UA distinguishes between data variables (e.g., data points that represent physical values of the process under control) and properties (e.g., metadata that further describes nodes).
- `VariableType NodeClass`: This node class is used to provide type definitions for variables.
- `DataType NodeClass`: The `DataType NodeClass` is used to provide type definitions for the values of variables.
- `ReferenceType NodeClass`: This kind of nodes is used to define different reference types.
- `Method NodeClass`: Methods are used to model callable functions that initiate actions within a server.
- `View NodeClass`: A view summarizes a subset of the nodes of the address space. Views are used to make only parts of the address space visible to clients.

Depending on the node class, a node has several attributes that characterize its properties. Some attributes are common to all nodes like the `NodeId` (uniquely identifying a node within the address space), the `NodeClass` (defining the node class), the `DisplayName` (localized text that can be used by a client for displaying the name), and the `BrowseName` (identifies a node when browsing the address space). Other attributes are only available for certain node classes. For example, variables have the attribute `Value` that represents the value of the variable and the attribute `DataType` that indicates the data type of the `Value` attribute.

The list of attributes cannot be extended by the user. Adding additional semantics to nodes has to be done by introducing a type hierarchy and by adding references to other nodes. A key concept within OPC UA is the modeling of *complex type definitions* for objects and variables. Such a complex type is an object or variable that embeds other nodes as subcomponents. The assignment of these subnodes to their parent node is done by using references.

Another powerful modeling mechanism is the use of *references*. References are always defined from one node to another—references are therefore always one-to-one relations. References can be defined in a symmetric or asymmetric way and further be divided into hierarchical or non-hierarchical. Hierarchical references can be used to introduce topologies and hierarchies within the model. Typical examples of standard OPC UA hierarchical references are `Organizes` (used to introduce a general hierarchy of nodes) and `HasComponent` (used to reference a complex type to its subnodes). An example of a standard OPC UA non-hierarchical reference is `HasTypeDefinition` that allows referencing an object or variable to its type definition. OPC Foundation [34] provides further details about modeling mechanisms in OPC UA and a full list of all attributes.

One of the most important features of OPC UA is that it already defines a set of standardized information models. The OPC UA base information model provides basic object, variable, reference, and data types as well as a set of predefined objects and variables. On top of this basic model, four additional information models are defined:

1. OPC UA Data Access (Part 8): This model specifies basic concepts to model data points of automation systems.
2. OPC UA Alarm and Conditions (Part 9): Within this part, concepts for alarm handling and defining alarm conditions are available.
3. OPC UA Programs (Part 10): OPC UA methods are only useful for invoking operations that immediately finish after execution. For operations that require a longer execution time, OPC UA programs shall be used.
4. OPC UA Historical Access (Part 11): To store and access historical data within the information model, the concepts covered within this part can be used.

59.4.2.2 Services and Communication in OPC UA

OPC UA is completely based on the client/server model—the communication is performed in *sessions*. To gain access to data, an OPC client establishes a connection to one or more OPC servers. In OPC UA, devices can choose one of two different transport protocols for communication. The first one is based on WS-I Basic Profile 1.1 [37] where Simple Object Access Protocol (SOAP) and HTTP are used. The second one is a binary TCP-based protocol that has been developed by the OPC Foundation.

Since OPC devices are mainly located at the management level where the network may be shared with other application domains, a security model has been introduced in OPC UA. To secure the communication, a *secured channel* is set up during session establishment. This secured channel is provided by the OPC communication layer that is located between the transport layer and the application layer. The secured channel uses cryptographic techniques to guarantee data confidentiality, freshness, and data origin authentication. Entity authentication is provided during session establishment with the help of certificates. Authorization and access control to the control data and services, is left up to the application.

Based on these secure transport protocols, OPC UA defines different services that are used by clients to communicate with the servers and vice versa. These services are grouped into so-called service sets. Typical service sets are the `Attribute Service Set` that contains services to read and write attributes of nodes as well as the `Session Service Set` that consists of services to open and close a session. For a complete list of all services specified by OPC UA, see OPC Foundation [34].

59.4.3 MQTT

Message Queuing Telemetry Transport (MQTT) follows a publish/subscribe integration approach. It aims to be lightweight, simple, and open. MQTT is developed by International Business Machines Corporation (IBM) and is now moving toward becoming an Organization for the Advancement of Structured Information Standards (OASIS) standard.

The main features of the protocol are [38]:

- One-to-many message distribution and decoupling of information sources and applications
- Payload agnostic message transport
- TCP/IP for network connectivity
- Three qualities of service for message delivery
- Small transport overhead
- Notification mechanism for abnormal client disconnection

MQTT follows a data-centric design paradigm. Information systems that need to interact with each other do not use a direct communication; instead, a broker is used as middleware and topics identify the information of interest. MQTT clients can then subscribe to topics and publish new messages to topics. In that way, a one-to-many interaction is possible and the communicating systems are decoupled through the broker. A drawback of this architecture is that the broker becomes a single point of failure and a bottleneck for scalability.

59.4.3.1 Message Format

The MQTT message format uses a fixed 2-byte header. There is a 4-bit field identifying the message type, where 14 message types are defined. The message types range from connection-related messages like connect, disconnect, ping, and acknowledgment messages over to messaging related types like publishing and subscription management. The remaining flags are used for the duplicate delivery (DUP), quality of service (QoS), and retain flags. The DUP flag is used to indicate the retransmission of messages, the QoS sets the level of assurance (at most once, at least once, exactly once), and the retain bit holds semantics if a message is persistently stored at the broker and also transmitted to subscribers that enroll on the topic after the message has been published. A second byte represents the remaining length of the payload. There, 7 bits are used to encode the length in bytes and 1 bit is used as continuation bit to extend the remaining length representation for a variable number of further bytes.

Depending on the message type, several variable headers are defined and come between the fixed header and the message payload. For example, the topic name is presented as variable header in a publish message and can take up to 32.767 characters. Within the subscribe message, multiple topics can be placed in the payload.

59.4.3.2 MQTT for Sensor Networks

MQTT for sensor networks (MQTT-S) [39] specifies optimizations that allow to use the MQTT protocol within sensor networks. It is adapted to the requirements on constrained networks regarding computing resources and limited bandwidth and device energy. The main adaptions are the following:

- Multiple connect messages combined with other message types to reduce the number of messages.
- Short 2-byte-long topic ID instead of long character-based identification. A registration mechanism allows to register names for topic IDs and to query by name at the broker.
- Predefined topic IDs and short topic names.
- Discovery mechanism for brokers and gateways.
- Keep alive mechanism for sleeping clients.

The overall architecture of MQTT-S is outlined in Figure 59.2. The architecture consists of four components. The original MQTT broker is left unchanged by MQTT-S, but three new components are added which are clients, forwarders, and gateways. The clients use the optimized MQTT-S protocol and the forwarders simply relay the protocol unchanged, which might happen at the border of a LoWPAN to a network with more bandwidth and reliability. Finally, the gateway does the translation between MQTT-S and MQTT.

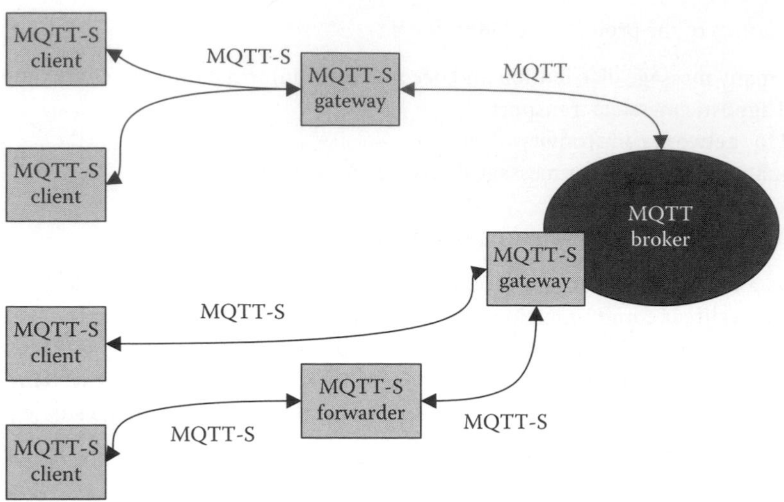

FIGURE 59.2 MQTT-S architecture [39].

59.4.4 IPSO Application Framework

The Internet Protocol for Smart Objects (IPSO) application framework defines the required application layer to provide interoperability among sensors and actuators using CoAP for message exchange. The IPSO Alliance is an organization that promotes the use of IP directly on smart objects ranging from home over building automation systems to smart appliances and other M2M applications, for example, for energy management systems.

The application framework defines the interfaces and the resource representation of smart objects equipped with IP. The framework does not aim to be a standard and intends to complement existing application-layer-specific standards like oBIX or ZigBee Smart Energy Profile 2.

Certain resource types are grouped into function sets. A function set has a resource type parameter that can be used for discovery and comes with a recommended root path. The subresources are organized below the root path. The root path is a recommendation but resources might be represented also using different addressing schemes.

Path templates are defined that contain index parameters like {#}, which provide means for identification within multiple instances of one resource type. The resource type is described through a namespace and a type ID. This information is also used for the `rt` field if the CoRE link format description is requested.

The framework currently supports two namespaces, namely, `ipso` and `ucum`. The `ipso` namespace is used for namespaces that are defined with the application framework. `ucum` refers to the Unified Code for Units of Measure (UCUM) specification, which is used for referencing standard unit types. The interface specification is based on REST interfaces and references the basic core resource types of Shelby and Vial [40].

The available data types are the basic XML schema data types or the data types defined by Sensor Markup Language (SenML) [41]. For the message encoding, it is possible to use either `text/plain` encoding or `application/senml+json`. However, an encoding using XML or more efficiently EXI is also possible.

As interaction model, a classic client/server communication model is used, based on either HTTP or CoAP. For discovery, only a kind of *lobby* mechanism is provided that describes the resources using the CoRE link format and returns links to those resources when querying the path `/ .well-known/core`; additionally, a device may register its resources at a resource directory [42].

59.4.4.1 Function Sets

Table 59.1 provides an overview of the function sets defined by the application framework. Each function set covers a certain domain or specific functionality and comes with a root path and a base

TABLE 59.1 IPSO Application Framework Function Sets

Function Set	Root Path	Resource Type
Device	/dev	ipso.dev
General-purpose IO	/gpio	ipso.gpio
Power	/pwr	ipso.pwr
Load control	/load	ipso.load
Sensors	/sen	ipso.sen
Light control	/lt	ipso.lt
Message	/msg	ipso.msg
Location	/loc	ipso.loc
Configuration	/cfg	ipso.cfg

resource type. A function type can be seen as a complex type containing further subresource types. These resource types are more equivalent to the concept of a data point, since for those types, a certain interface, data type, and (in case of numerical values) a unit are defined.

The device, configuration, and location function sets cover operational and meta-information of the smart object like the manufacturer, serial number, and name or the location and allow to modify the configuration. The General purpose input-output (GPIO) function set covers generic digital and analog inputs and outputs. The other function sets like power, load control, sensors, and light control are more domain-specific and come with specific types and resource representations. For example, within the sensor function set, a motion type is defined for motion sensors, or within the light control function set, a subtype for dimming is specified. Finally, the message function set can be used to express status, alarming, and display messages.

59.4.4.2 Data Formats

The data types used by the application framework to specify the subtypes within the function set are illustrated in Table 59.2; also, the mapping to the plain-text encoding and to SenML fields are described. If no additional metadata is required, the plain-text encoding provides an efficient way to exchange information. In this encoding, the data fields are represented using ASCII and the according XML schema type definitions. SenML provides a simple information model for representing sensor devices and corresponding measurements. Each message represents one single object with attributes that hold an array of entries. An entry consists of attributes like a unique ID within the sensor, the measurement time, and the value. One goal of SenML is to keep metadata as much as possible out of communication and to rely on out-of-band exchange for the required metadata.

59.4.5 ZigBee: Smart Energy Profile 2

With the recent development of ZIP, the ZigBee Alliance also revised the Smart Energy Profile (SEP) used with the *original* ZigBee. In cooperation with the HomePlug Alliance [43], the ZigBee Alliance [23] introduced the Smart Energy Profile 2 (SEP2) [44]. Both versions are application layer information

TABLE 59.2 IPSO Application Framework Data Formats

Data Type	Text/Plain Format	SenML Field
b (boolean)	xsd:boolean	Boolean value (v)
s (string)	xsd:string	String value (sv)
e (enum)	xsd:integer	Value (v)
i (integer)	xsd:integer	Value (v)
d (decimal)	xsd:decimal	Value (v)

representation and interaction models. While SEP is based on the ZigBee information modeling framework as introduced in Chapter 55, SEP2 implements a RESTful architecture based on HTTP. The communication interface is defined using Web Application Description Language (WADL), a machine-reachable description language. The application payload is encoded in either XML or EXI. Although SEP2 is designed to be used in conjunction with ZIP, the use of standard technologies (i.e., HTTP, XML) allows of using different IP-related transmission technologies such as WiFi or Ethernet. Since different—possibly unsecured—transmission media can be used, SEP2 specifies additional application layer security mechanisms. Besides device authentication, SEP2 provides methods for authorization with fine-grained access control for specific resources.

SEP2 defines services in form of function sets and resources for domain-specific applications. Thereby, sets and resources for metering, pricing, and billing are specified. Further domains are demand response and load control as well as prepayment. An important domain deals with distributed energy resources. Additionally, a group of common resources and function sets is specified for timing network management and configuration.

59.5 Conclusion

Home and building automation technologies are the target of a rapidly proceeding technology revolution driven by the recent research activities around smart grids, smart cities, and the IoT. Besides new protocols to be deployed, integration and interoperability remain key challenges to be solved. A clear trend toward the use of IP and Web technologies can be observed that will finally lead to a convergence of control and data network infrastructures. New protocols are available, but the definition of a complete system is to be defined, yet. There is ongoing work on different approaches and system architectures that have a different level of maturity. Several standards are still in a draft state and open issues exist. Nevertheless, vendors are already starting to produce devices equipped with such communication stacks, but it will take more time until a broad industry acceptance can be seen with large-scale deployments that go beyond research or pilot installations.

References

1. D. Emmerich and E. Bloom, *Commercial Building Automation Systems—Security and Access, HVAC Controls, Fire and Life Safety, Building Management Systems and Lighting Controls: Global Market Analysis and Forecasts*, Pike Research, Boulder, CO, 2012.
2. L. Atzori, A. Ieara, and G. Morabito, The Internet of things: A survey, *Computer Networks*, 54(15), 2787–2805, 2010, Elsevier.
3. D. Guinard and V. Trifa, Towards the Web of Things: Web mashups for embedded devices, in *Proceedings of Workshop on Mashups, Enterprise Mashups and Lightweight Composition on the Web (MEM 2009)*, Madrid, Spain, 2009.
4. F. Aldrich, Smart homes: Past, present and future, in R. Harper (ed.), *Inside the Smart Home*, Springer, London, U.K., 2003.
5. C. Reinisch, A framework for distributed intelligence for energy efficient operation of smart homes, PhD thesis, TU Vienna, Vienna, Austria, 2012.
6. EN 13757-x, Communication system for and remote reading of meters, 2012.
7. DE Verband der Elektrotechnik Elektronik Informationstechnik e.V., Sym 2 Alliance, www.vde.com/de/fnn/arbeitsgebiete/messwesen/Sym2.
8. American National Standards Institute, Protocol specification for ANSI Type 2 optical port C12.18, 2006.
9. N. Strother and B. Gohn, *Home Energy Management—In-Home Displays, Networked HEM Systems, Standalone HEM Systems, Web Portals, and Paper Bill HEM Reports: Market Analysis and Forecasts*, Pike Research, Boulder, CO, www.pikeresearch.com, 2012.

10. openADR Alliance, www.openadr.org, 2012.
11. IEC Std. 61850-x, Communication networks and systems in substations, 2003.
12. O. Hersent, D. Boswarthick, and O. Elloumi, *The Internet of Things: Key Applications and Protocols*, Wiley, Chichester, U.K., 2011.
13. M. Naphade, G. Banavar, C. Harrison, J. Parszczak, and R. Morris, Smarter cities and their innovation challenges, *Computer*, 44(6), 32–39, 2010, IEEE.
14. G. Mulligan, The 6LoWPAN architecture, in *Proceedings of the Fourth ACM Workshop on Embedded Networked Sensors*, Cork, Ireland, 2007, pp. 78–82.
15. IETF RFC 4919, IPv6 over low-power wireless personal area networks (6LoWPANs): Overview, assumptions, problem statement, and goals, 2007.
16. IETF RFC 6775, Neighbor discovery optimization for IPv6 over low-power wireless personal area networks (6LoWPANs), 2012.
17. IETF RFC 6282, Compression format for IPv6 datagrams over IEEE 802.15.4-based networks, 2011.
18. IETF RFC 6550, RPL: IPv6 routing protocol for low-power and Lossy networks, 2012.
19. J. Hui and R. Kelsey, Multicast protocol for low power and lossy networks (MPL), IETF Active Internet-Draft, 2013.
20. IETF RFC 6206, The Trickle algorithm, 2011.
21. G. Oikonomou and I. Philips, Stateless multicast forwarding with RPL in 6LoWPAN sensor networks, in *Proceedings of the Eighth IEEE International Workshop on Sensor Networks and Systems for Pervasive Computing*, Lugano, Switzerland, 2012.
22. ZigBee Alliance, ZigBee IP specification, 2013.
23. ZigBee Alliance, www.zigbee.org.
24. IEEE Std. 802.15.4-2006, Wireless medium access control (MAC) and physical layer (PHY) specifications for low-rate wireless personal area networks (WPANs), 2006.
25. IETF RFC 4944, Transmission of IPv6 packets over IEEE 802.15.4 networks, 2007.
26. IEEE, Guidelines for 64-bit global identifier (EUI-64) registration authority, 1997.
27. IETF RFC 4291, IP version 6 addressing architecture, 2006.
28. IETF RFC, Multicast protocol for low power and lossy networks (MPL) (Internet-Draft, Version 04), 2013.
29. IETF RFC 5191, Protocol for carrying authentication for network access (PANA), 2008.
30. R.T. Fielding, Architectural styles and the design of network-based software architectures, PhD dissertation, University of California, Irvine, CA, 2000.
31. Z. Shelby, K. Hartke, and C. Bormann, Constrained application protocol, Active Internet Draft, 2013.
32. Z. Shelby and C. Bormann, *6LoWPAN: The Wireless Embedded Internet*, John Wiley & Sons, 2011.
33. NIST, NIST framework and roadmap for smart grid interoperability standards, Release 2.0, 2012.
34. OPC Foundation, OPC UA specification, 2013.
35. IEC 62541, OPC unified architecture, 2012.
36. W. Mahnke, S.-H. Leitner, and M. Damm, *OPC Unified Architecture*, Springer, Heidelberg, Germany, 2009.
37. Web Services-Interoperability Organization, Basic profile version 1.1, 2004.
38. Dave Locke, MQTT V3.1 protocol specification, International Business Machines Corporation (IBM), 2010.
39. A. Stanford-Clark and H.L. Trong, MQTT for sensor networks protocol specification version 1.2, International Business Machines Corporation (IBM), 2013.
40. Z. Shelby and M.V. Vial, CoRE interfaces, IETF Active Internet-Draft, 2013.
41. Open Geospatial Consortium, Sensor Model Language (SensorML), 2007.
42. Z. Shelby, S. Krco, and C. Bormann, CoRE resource directory, Active Internet-Draft, 2013.
43. HomePlug Alliance, www.homeplug.org.
44. ZigBee Alliance, HomePlug Alliance smart energy profile 2—Application protocol standard, 2013.
45. IEEE Std. 802.15.4, Wireless medium access control (MAC) and physical layer (PHY) specifications for low-rate wireless personal area networks (WPANs), 2003–2007.

46. ZigBee Alliance, ZigBee specification, 2007.
47. OPC Foundation, www.opcfoundation.org.
48. American Society of Heating, Refrigerating and Air-Conditioning Engineers, ANSI/ASHRAE Addendum c to ANSI/ASHRAE Standard 135-2004, 2006.
49. OASIS, oBIX 1.1 committee specification, 2006.
50. International Alliance for Interoperability, www.iai-international.org.
51. B. Graf, M. Hans, and R.D. Schraft, Care-O-bot II—Development of a next generation robotic home assistant, *Autonomous Robots*, 16(2), 193–205, 2004, Springer.
52. IEC Std. 62056-x, Electricity metering—Data exchange for meter reading, Tariff and load control, 2002.

XI

Energy and Power Systems

60 **Protocols for Automatic Meter Reading** *Klaas De Craemer, Geert Deconinck, and Matthias Stifter* **60**-1
Introduction • Available Standards • Suitability in the SG Architecture • Conclusions • Bibliography

61 **Communication Protocols for Power System Automation** *Peter Palensky, Friederich Kupzog, Thomas Strasser, Matthias Stifter, and Thomas Leber* **61**-1
Introduction • IEC Seamless Integration Architecture • Common Information Model • Telecontrol Communication • Intercontrol Center Communication • Distributed Network Protocol • Other SCADA Communication Protocols • Data and Communication Security • Modbus • OpenADR • EMIX, Energy Interoperability • Open Platform Communications • Building Automation • References

62 **IEC 61850: A Single Standard for the Engineering, Protection, Automation, and Supervision of Power Systems on All Voltage Levels** *Karlheinz Schwarz* **62**-1
Introduction • Overview • Information Modeling and Information Models • Parts and Application Domains of the Standard • Functional Description of Applications • Information Exchange Models • System Configuration Language • Overview of Implementation Layering • Related International Projects and Application Example • UCA International Users Group (UCA IUG) and Quality Assurance • Third-Party Solutions for IEC 61850 • References

63 **Fundamentals of the IEC 61400-25 Standard** *Federico Pérez, Elisabet Estévez, Isidro Calvo, Darío Orive, and Marga Marcos* **63**-1
Introduction • Overall Description of the IEC 61400-25 Series • Wind Power Plant Information Model • Wind Power Plant Information Exchange Model • Mapping to Communication Protocols • Condition Monitoring • Conclusions • References • Additional Web Information

60

Protocols for Automatic Meter Reading

Klaas De Craemer
Katholieke Universiteit Leuven

Geert Deconinck
Katholieke Universiteit Leuven

Matthias Stifter
Austrian Institute of Technology

60.1 Introduction **60**-1
Smart Grid Communication Architecture
60.2 Available Standards **60**-2
IEC 61334-5 PLC • IEEE P1901.2/PRIME PLC/PLC G3 • ANSI C12 Protocol Suite • IEC 62056-21/IEC 61107 • SITRED/Telegestore/Meters and More • SML (SyM² Project) • EN 13757/M-Bus • DLMS/COSEM and IEC 62056 • IEC 61968 Part 9: System Interfaces for Meter Reading and Control • IEC 62056-31 *Euridis* • Other Standards with Relevance to Smart Metering
60.3 Suitability in the SG Architecture **60**-27
60.4 Conclusions **60**-27
Bibliography **60**-27

60.1 Introduction

Automatic meter reading (AMR) can be considered as the predecessor of AMI (advanced metering infrastructure), and it allows utilities to perform some basic readout functions of a customer's meter. It does not allow control of the meter itself (e.g., disconnect or the uploading of new tariff tables), let alone enable demand-side management. Together with new generation challenges (distributed energy resources [DER] and intermittent), the next logical step will be the emergence of AMI systems under distributed control: the smart grid.

The rise of the smart grid is shaping up to follow a gentle evolutionary trajectory. New wind farms, solar installations, and substations will be equipped with distributed control and monitoring features, enabling utility automation. At the consumer side, the installation of smart meters will trigger a (slow) adoption of *energy-aware* devices that can communicate with it.

60.1.1 Smart Grid Communication Architecture

Figure 60.1 shows a generic smart metering system outlining the basic system components. It is very similar to the *system components and interfaces* defined in the Open Meter project and the *metering installation description* from the DSMR document. A short overview of the main components:

Electricity meter and communication hub: This device functions as the gateway to the smart grid. Most of the time, the house's main electricity meter will be integrated with it, as well as an externally controllable breaker switch (e.g., for contract control). Being also a communication hub, the device will communicate with other metering devices and home appliances connected to a home area/automation network (HAN).

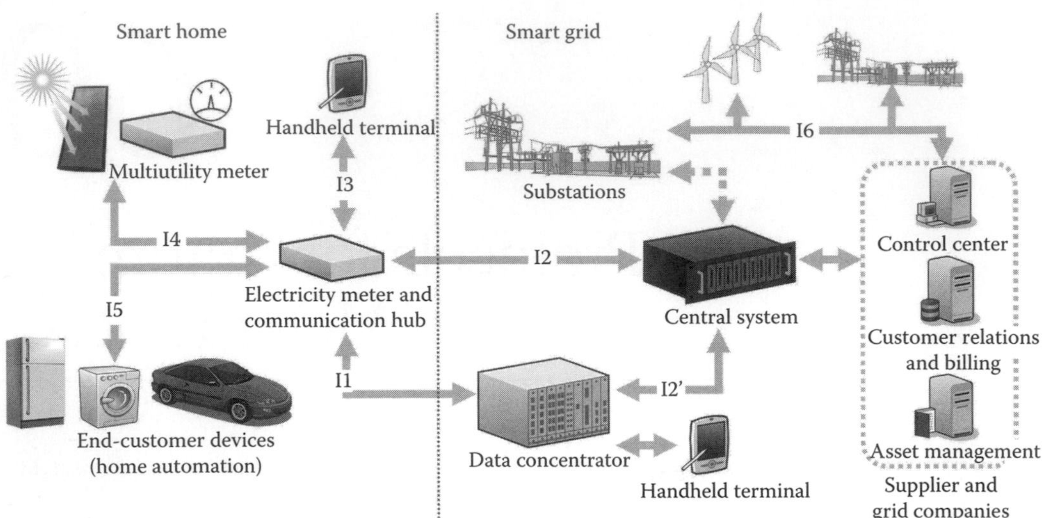

FIGURE 60.1 Smart grid system architecture.

Multiutility meter: A meter measuring, for example, gas, water, or heat, or an electricity submeter such as for a residential PV installation.

End-customer device: Appliances that communicate with the hub using a HAN. Examples include heating and ventilation air conditioning (HVAC) units, lighting and major appliances, and also plug-in (hybrid) electric vehicles (P(H)EVs).

Data concentrator: In between the smart meter and the central system, an intermediate data concentrator can be present. Its main purpose is to manage the connected meters, aggregate the data, and act as a relay for the central system.

Central system: The task of the central system is to provide the core communication functions of the smart grid. It interfaces with the supplier and grid companies, which can query customer meters for billing purposes or issue tariff commands and demand response alarms. It is possible that the central system will be operated by new service providers, which then grant access to the various market players. Eventually, the control center of the transmission system operator and distribution system operator is still responsible for the management of the legacy grid infrastructure as well as its centralized and distributed energy sources.

Handheld units (HHUs): An extra port on the meter/hub and concentrator allows service technicians to locally read and change parameters during installation or maintenance. It is also useful in case problems arise that cannot be remedied remotely.

The relevant communication channels and relations that arise are marked interface I1–I6. The protocols introduced in this chapter are dealing with the interactions I1–I5 (see Section 60.3).

60.2 Available Standards

In this section, a short introduction to each analyzed standard is presented. The relevant criterion for selection was the *openness* of the standard. Because of the smart grid context, interoperability and extendibility must be guaranteed across different devices and manufacturers. Therefore, only open standards have been considered in this overview.

A rough referencing of the protocol to the OSI layer position is shown in Figure 60.2. For example, if it mostly concerns hardware aspects (such as IEC 61334-5 or IEC 62056-21) or does it also specify a data model (e.g., Distribution Line Message Specification [DLMS]/COmpanion Specification for Energy Metering [COSEM]) or both (such as KNX and LonWorks).

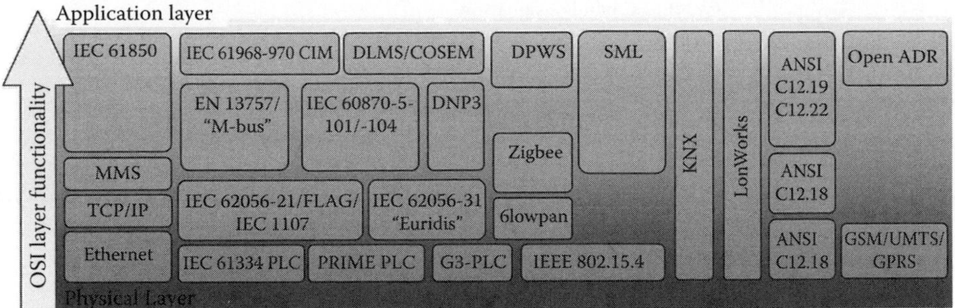

FIGURE 60.2 OSI layer placement of the analyzed standards.

60.2.1 IEC 61334-5 PLC

This IEC suite of standards was developed at the end of the 1990s by IEC working group 57 and defines several narrowband power line communication (NB-PLC) systems. The full name is "Distribution automation using distribution line carrier systems."

While IEC 61334-5 is mostly used with S-FSK, the standard contains a suite of different lower layers (numbered 61334-5 Part 1–5):

- Part 1: Spread frequency shift keying (S-FSK) in the CENELEC A-band (9–95 kHz). This allows for a few hundreds of kbps using about 100 mW of transmit power.
- Part 2: Frequency shift keying (FSK).
- Part 3: Spread-spectrum adaptive wideband (SS-AW).
- Part 4: Multicarrier modulation (MCM).
- Part 5: Spread spectrum—fast frequency hopping (SS-FFH).

The upper OSI layers (network layer, application layer, and DLMS protocol) are also specified (in IEC 61334-4), but their use is optional.

A typical narrowband PLC deployment consists of a concentrator close to the MV/LV transformer, where it is coupled to a backbone network (Figure 60.3). All communication is initiated by the concentrator or gateway, which means that meters down the line can access the medium only when requested by the concentrator. Once the data are in the concentrator, it can be forwarded to the management center.

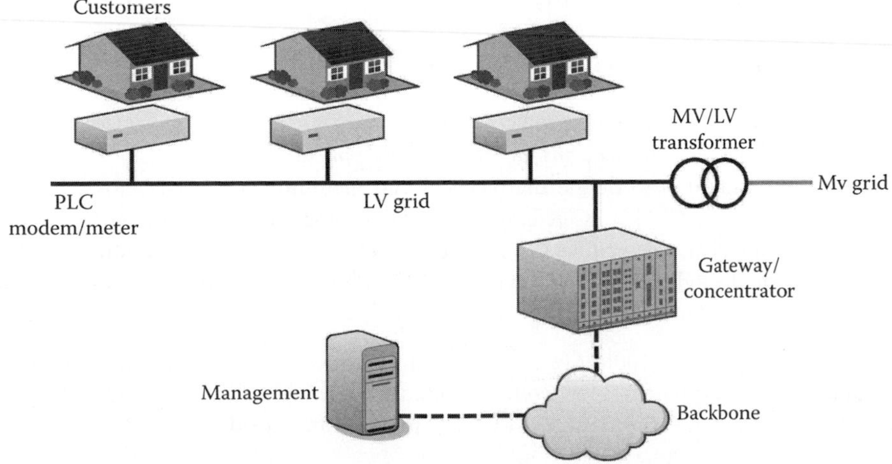

FIGURE 60.3 Overview of typical narrowband PLC deployment.

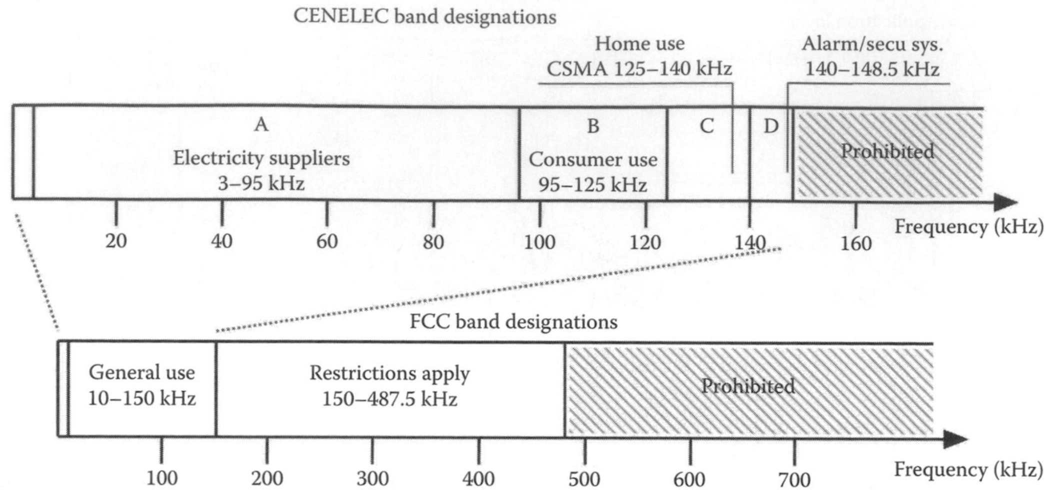

FIGURE 60.4 Comparison of CENELEC and FCC EMC regulations regarding the use of PLC.

Because of the allowed frequency range (3–148.5 kHz in Europe) and limited transmission power, the available bandwidth is quite low (up to a few kbit/s). Also, the small packet size makes IEC 61334-5 not very suitable for, for example, TCP/IP communication as this combination would result in unacceptable fragmentation and delays. However, this standard is robust, mature, and widely used.

In the United States and Asia, the regulation is different. The FCC and ARIB allow PLC devices to work in the frequency band from 3 to 500 kHz. A comparison of the CENELEC and FCC regulation is depicted in Figure 60.4.

Recent narrowband PLC technologies such as PRIME (2009) and PLC G3 (2011) are based on more sophisticated orthogonal frequency-division multiplexing (OFDM) techniques to provide higher data rates. Both of these technologies are under consideration for the physical and medium access layers of the upcoming IEEE 1901.2 standard.

Also worth mentioning is the installation of filters that isolate the LV loads from the outside grid or the different feeders in the LV grid to improve SNR ratios and the use of multiple gateways per line segment.

Despite the difficulties, PLC technologies are at an advantage when viewed from the perspective of the utility company. They already have full access to the network infrastructure, which means more control and less dependence from third parties.

60.2.2 IEEE P1901.2/PRIME PLC/PLC G3

In March 2010, the IEEE authorized a project *Draft Standard for Low Frequency (less than 500 kHz) Narrow Band Power Line Communications for Smart Grid Applications.*

The goal of this standard is to define the physical and medium access layers for a low-frequency (below 500 kHz and inside CENELEC A-band) NB-PLC protocol operating on AC and DC electric power lines. Typical applications include smart meters, electric vehicle charge stations, intrahome networking, and additionally lighting and photovoltaic converter communications.

IEEE P1901.2 is intended to be usable

- On the low-voltage lines <1000 V (e.g., between the transformer and consumer's meter)
- Through the transformer from low-voltage to medium-voltage grid (1000 V–72 kV)
- Through the transformer from medium-voltage to low-voltage grid

In North America, the low-voltage segment between the house and utility may extend 3–4 m. In such scenarios, putting a concentrator before the MV/LV transformers would not be cost-effective, making

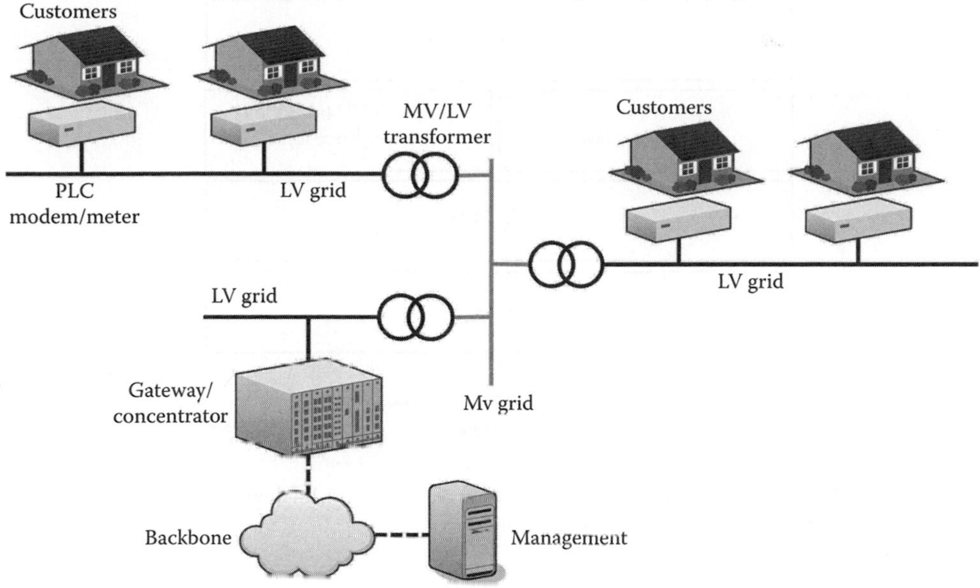

FIGURE 60.5 Overview of a typical IEEE P1902.1 PLC deployment.

transformer crossing an interesting capability. It enables the concentrator to be placed where it can aggregate data from more locations (Figure 60.5).

Multikilometer rural communications and/or data rates up to 500 kbit/s should be possible depending on the application. The standard ensures coexistence with broadband power line (also part of IEEE 1901) devices by minimizing emissions in bands >500 kHz.

Two technologies are under consideration for the physical and medium access layers of the upcoming IEEE 1901.2 standard, PRIME and PLC G3. Both drafts are based upon the use of OFDM, whereas IEC 61334 is based on the use of FSK on a single carrier frequency.

60.2.2.1 PRIME

PRIME stands for power line–related intelligent metering evolution and defines the lower layers of a PLC narrowband system that operates within the CENELEC A-band, between 42 and 89 kHz. A maximum data rate of 128.6 kbit/s can be attained under good circumstances or 21.4 kbit/s using its most robust encoding scheme. An IPv4 convergence layer allows the transfer of TCP/IP traffic, and a security profile based on 128-bit Advanced Encryption Standard (AES) allows secure authentication and data encryption. Simplicity, low cost, and flexibility are the main goals in the design of the PRIME system. An overview of the layers in the PRIME system is shown in Figure 60.6.

The main coordinator behind this technology is the Spanish utility Iberdrola, which started a pilot project back in 2009. As of May 2013, the PRIME Alliance reported that over 2M meters using this technology have been installed.

Work is progressing on a PRIME extension that operates with frequencies up to 500 kHz (for use outside Europe). PRIME FCC bands will span from 40 to 490 kHz, allowing for maximum data rates of 1 Mbps. The relevant PHY parameters are to be defined by end of April 2013.

Additionally, a COSEM Profile Task Force has been formed with the scope of defining an optional, common DLMS/COSEM profile for PRIME.

60.2.2.2 G3-PLC

In 2010, ERDF announced its plans to develop and implement an automatic metering management (AMM) system to manage the complete electricity supply chain, from electricity suppliers all the way

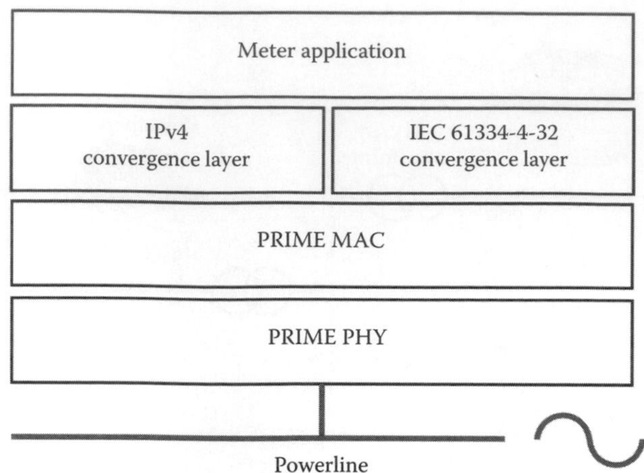

FIGURE 60.6 PRIME layers. (Derived from PRIME Project, PRIME technology whitepaper: PHY, MAC and convergence layers, 2008, Available at http://www.iberdrola.com/webibd/gc/prod/en/doc/MAC_Spec_white_paper_1_0_080721.pdf.)

to the consumers. The system provides two-way communication using OFDM-PLC between meters installed at the customers' premises and the concentrator, communicating in a master–slave setup.

During 2010, after a successful experimental field test, over 250,000 *Linky* meters supporting PLC-3G were installed by ERDF. In 2011, the G3-PLC alliance was founded to promote the technology. The alliance includes meter vendors such as Landis+Gyr and chip manufacturers Maxim, STMicroelectronics, and Texas Instruments.

G3-PLC supports the portion between 35.9 and 90.6 kHz of the CELENEC-A band. Using the *normal mode* modulation schemes, data rates of up to 33.4 kbit/s are supported. In *robust mode*, additional error detection and correction is added, reducing throughput to about a fourth of that. There is a potential to use this standard to support communication in frequencies up to 180 kHz, in the FCC band, using an extension proposed by Maxim (G3-FCC). Data rates up to 400 kbit/s are then possible.

G3-PLC coexists with S-FSK and other legacy PLC technologies by notch filtering on the S-FSK carrier frequencies and seamlessly supports DLMS/COSEM (IEC 62065). Data security is provided through 128-bit AES and support for IPv6 enables PLC-G3 to converge IPv4 and IPv6 devices and networks in an efficient manner.

60.2.3 ANSI C12 Protocol Suite

Back in 1996, a number of partners from industry and utility companies developed two ANSI standards for the readout of electricity meters. In ANSI C12.19, standard meter data and tables are described, together with the basic requirements for reading and writing these tables (referred to as services), independent of the transport protocol. In ANSI C12.18, a low-overhead protocol for the ANSI Type 2 Optical Port (ANSI C12.13) for on-site meter reading is described (Figure 60.7). It can be compared to IEC 1107-FLAG in Europe and is basically the first protocol for the transport of C12.19 table data.

In 1999, C12.18 was adapted for remote access over telephone modems, as C12.21, making it one of the early solutions for AMI projects. Both are essentially session-based serial point-to-point protocols and known as "Protocol Specification for Electricity Metering" (PSEM). Communication takes place in a request/response manner, in which both sides can initiate request commands.

C12.21 also adds authentication enhancements due to the fact that the person reading the meter is not necessarily on-site. Subsequent data reads and writes are done unencrypted though, making C12.21 not well-suited for use over the Internet.

FIGURE 60.7 An Elster A3 Alpha Meter showing the D-shaped ANSI Type 2 Optical port. (Copyright Elster Metering, Elster A3 Alpha Meter, [Online] 2013, Available at http://www.elstersolutions.com/en/product-details-na/73/en/A3_ALPHA_meter (Accessed on August 2, 2013).)

In both C12.18 and C12.21, specific media (optical and telephone) are specified. The goal of C12.22, added in 2007, is to describe transport of C12.19 meter tables using PSEM messages, using already existing communication networks such as TCP/IP over Ethernet, SMS/GSM, and so on. Typically, C12.22 uses TCP as its transport protocol. The commonly used TCP port for C12.22 traffic is 1153.

C12.22 makes it possible to chain request commands and support communications over shared media. It also provides for both session and session-less communications, which reduces the amount of packets and complexity (e.g., in the case when using SMS over GSM). AES encryption is also specified as an option.

Some countries such as Australia and South Africa have adopted both IEC 1107 and ANSI standards for metering, and a variant called open smart grid protocol is used by Echelon in their AMI solutions.

60.2.4 IEC 62056-21/IEC 61107

Similar in purpose to ANSI C12.18 and 19, this standard is aimed at meter reading and programming using a directly connected handheld device, but is also used over modem connections. In 2002, IEC 1107 was incorporated into IEC 62056 as *IEC 62056-21 Electrical metering—Data exchange for meter reading, tariff and load control—Part 21: direct local data exchange, 2002–05*. It was originally part of IEC 1107:1990.

Together with Euridis, IEC 1107 is one of the physical layer standards provided by IEC 62056. Part 21 contains the third edition and describes a bidirectional optical interface along a specification with data rates, character format, and transmission protocol. This standard is mainly used in Europe.

Optical probes used for IEC 1107 are sometimes called FLAG probes, after an earlier interface by Ferranti and Landis+Gyr. While the protocols are the same, FLAG is a subset of IEC 1107 and is more specific regarding the media. The hardware of the interface is identical though, and IEC 1107 optical probes will work on FLAG meters.

It is also possible to use the IEC 1107 protocol over other media than optical, such as a two-wire system similar to EIA-485, inductively coupled loops, modems, or even Internet protocols. Multiple devices can be connected to a shared communications bus as well, for example, in an apartment building. Different operating modes are specified, labeled from A to D, all based on ASCII but differing in baud rate, directionality, and security. A special mode E allows the use of DLMS/COSEM via High-Level Data Link Control (HDLC) (see IEC 62056-46).

The mode is negotiated between the readout unit and the meter. More than one mode can be supported: nowadays, support for Mode C and Mode E with DLMS/COSEM is common practice.

As one of the first meter data exchange standards, IEC 62056-21 is widely used today. However, it does not use a data model or uniform memory mapping. Therefore, meter communication requires manufacturer-specific information, limiting interchangeability, unless DLMS/COSEM is used in Mode E.

Every manufacturer has their own three-letter identification code, whose list is maintained by the FLAG Association and the DLMS User Association.

60.2.5 SITRED/Telegestore/Meters and More

At the beginning of the 1990s, Italy's largest distribution company, ENEL, developed a transmission system called SITRED to read and manage meters remotely, using distribution line communication. A large test comprising 80,000 installations proved that remote management via the LV network was technically viable, but the use of Ferraris meters equipped with electronics was not cost-effective enough.

At the end of the 1990s, technology had advanced enough, and ENEL concluded that changing all the low-voltage meters for electronic ones would soon be profitable, and in October 1999, the Telegestore project was started.

60.2.5.1 Architecture

At the consumer side, fully electronic meters communicate with a concentrator situated close to a transformer via PLC (Figure 60.8). The concentrator communicates with the acquisition center through an access server using GSM, PSTN, or satellite. Local access is also provided through an RS-232 and IEC 1107 Mode C optical port.

The central management system does all the customer relationship management, including remote contract management (disconnection, tariff plans, etc.), monthly billing, and demand-side management.

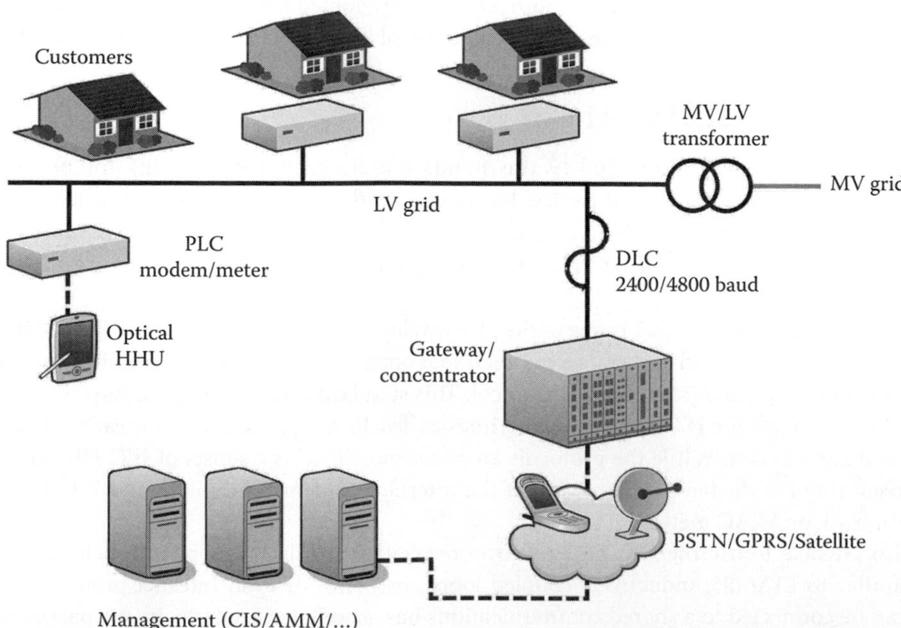

FIGURE 60.8 Overview of the Telegestore architecture. (Derived from Botte, B. et al., *18th International Conference and Exhibition on Electricity Distribution*, 2005, pp. 1–4.)

FIGURE 60.9 Enel Single Phase Smart Meters as installed during the Telegestore project. (Copyright Enel, Smart Metering Diffusion, [Online] 2013, Available at http://www.enel.com/en-GB/innovation/smart_grids/smart_metering/spread.aspx (Accessed on August 2, 2013).)

For the communication between concentrator and meters, two protocols (using narrowband PLC) are used:

- ENEL's own SITRED protocol, based on a simple narrowband FSK-based solution (similar to IEC 61334-5-2, in CENELEC band A). It uses a carrier frequency of 82 kHz and is relatively reliable throughout their whole diverse grid. The attainable speed is only about 2400–4800 baud, but the cost of transceivers and coupling devices is limited.
- An enhanced version of LonTalk PLC, using two carriers at 86 and 75 kHz. Depending on the noise level, the system can switch between the two.

LonTalk and SITRED differ in the PHY and MAC layers, but the same proprietary application layer is used on top of both, ensuring transparency for the acquisition center. The details on this application layer are sparse, but apparently, it allows autodiscovery of meters and network autoconfiguration.

During the peak of deployment, more than 40,000 m were installed a day by about 4,000 workers (Figure 60.9). Installation and initialization takes 15–30 min using an HHU after which the control center commissions the meter via the data concentrator. The latter detects the communication path and the phase to which the meter is connected without topology information, in order to assign the meter's real address. The control center reaches 97%–99% of the meters on the first connection attempt. This is not very high with such an installed user base (30 million meters and 350,000 data concentrators).

By the end of 2006, about 29.8 million meters had been installed. Eventually, ENEL announced to make its system open to the market and started the *Meters and More* alliance together with Spanish utility Endesa and IBM to further promote the technology.

With Meters and More, focus is being put on making energy data available in the home, via the smart meter. This would enable new applications, such as in-home energy displays and appliances that react to the current or future tariff. In 2009, Enel performed a field test in 1000 households with in-home displays to analyze the impact on customer behavior.

60.2.6 SML (SyM^2 Project)

The Smart Message Language (SML) is a communication protocol for data acquisition and parameterization developed by the German utility companies RWE, EON, and EnBW. The main idea

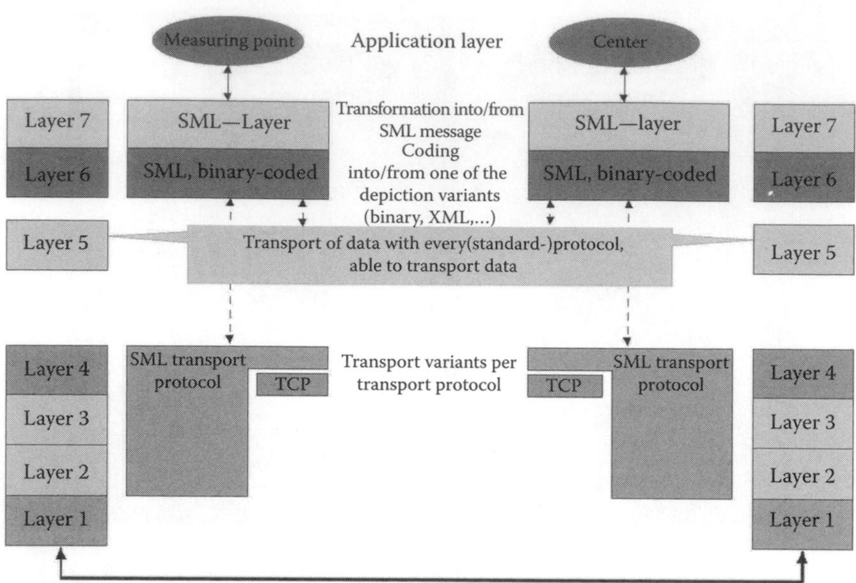

FIGURE 60.10 SML communication model overview, from the SML 1.02 specification document.

is to have a simple structure that is usable in low-power embedded devices (Figure 60.10). Therefore, SML message can be transported over stateless, secure communication paths, and both push and pull operations are supported (not limited to a fixed client–server model).

SML is tailored toward electricity metering and has to be viewed alongside the SyM² project. This is a joint venture between RWE, EON, and Landis+Gyr to develop an industry standard specification for load profile meters in Germany.

60.2.6.1 Application Layer

The application layer defines a file and document structure to carry a useful payload between the measuring point and, for example, the data collection center.

SML messages can be grouped together in self-contained information units called *files* that are independent of the transport technology used. An SML file is always structured as a chain of SML messages and can be segmented to reduce the size. An SML message must be either a Request or Response message and contains tasks and assigned attributes (including a transaction ID and a group number to process messages in a particular sequence or simultaneously).

The SML documentation does not refer to encryption, because it only provides a framework for the transmission of signatures. The encryption method itself is determined by the selected application. This is typically defined in a so-called companion specification, such as Sym² or Multiutility Communications Controller.

60.2.6.2 Presentation Layer

In the presentation layer, SML provides two options: SML binary coding and XML encoding. The binary coding is optimized to be data efficient, where the XML format is human readable but takes up much more data volume. The SML binary encoding is detailed in Section 6 of the specification. It is based on a classic type–length–value structure. The type and length are grouped together in one single byte for most cases.

60.2.6.3 Transport Layer and Below

In typical metering applications, SML messages are transported using TCP/UDP transport layers over IP networks. However, for (unsecured) serial links such as GSM/PSTN or direct readout links, the SML transport protocol is available.

First of all, SML is focused and tailored specifically toward electricity metering and has to be viewed alongside the SyM² project. SYM² is a set of specifications for a modular smart meter, with a strong focus on clock synchronized and signed meter data, and interchangeable communication modules.

SML only defines a message structure but does not specify an object model, nor does it specify what functions should be supported. There are also no interface classes such as in COSEM. SML does however support the Object Identification System (OBIS) codes as maintained by the DLMS User Association. However, SML may specify additional functionality for which no OBIS codes exist. Thus, the use of companion specifications is necessary (such as SyM²).

SML is under consideration as messaging method for DLMS/COSEM, as IEC 62056-5-3-8. At the moment, SML is not used outside of Germany. In the rest of Europe, DLMS/COSEM has received wider industry support and is already an IEC standard.

Additionally, it was decided (BSI-TR-03109-1) that SML has to be supported by the Smart Meter Gateway with the smart meters in Germany, but it is only mentioned for the communication with the local meters (LMN), not in the protocol stack for the WAN communication.

60.2.7 EN 13757/M-Bus

EN 13757 (Meter bus) is a European standard for the remote interaction with utility meters (originally electricity and gas, called *meters* in the specification) and various sensors and actuators (*other*), which was developed by the University of Paderborn and Texas Instruments.

M-Bus is primarily focused at the special requirements of simple remotely or battery-powered systems. When interrogated, the meters on the bus deliver the data they have collected to a common master, such as a utility's handheld computer. Alternatively, data can also be collected by transmitting them over a PSTN modem line.

Several physical media are supported including twisted pair and wireless M-Bus (in the ISM band). The twisted pair medium typically operates at 2400–9600 baud, and a single segment can be up to a few kilometers long. Noteworthy is the support for DLMS/COSEM in the lower layers.

The DSMR and German Technische Richtlinie BSI TR-03109-1 both specify wired and/or wireless M-Bus as the means of communication between a metering installation and other (gas, water, etc.) meters, albeit with improved security (AES instead of DES). As this standard is already widely used in meters and reasonably future proof, it is almost indispensable for local data exchange in the smart grid architecture.

60.2.7.1 Structure

M-Bus uses a reduced OSI three-layer structure, as shown in Table 60.1, consisting of the following.

60.2.7.1.1 Physical Layer

Several physical media are covered, such as twisted pair baseband signaling (EN 13757-2) and *local bus* accordingly (EN 13757-6).

Using the twisted pair system (Figure 60.11), signaling from master to slave takes place by voltage-level changes (up to ±42 V). The slaves thereby act as constant current sinks. Range is up to a few kilometers, and typical data rate is 2400–9600 baud, but up to 38,400 baud is supported.

Meters that are connected with the twisted pair media can also be powered through these, plus the wires can be interchanged without damage to the connected equipment.

The wireless version (EN 13757-4) operates in the ISM band of 869–980 MHz for short-range devices. This allows data exchange up to about 15 m, extendable by using relaying methods (specified in EN 13757-5).

Several types of communication are defined: Mode S (stationary mode), Mode T (frequent transmit mode, where the meter periodically transmits its data), and Mode R (frequent receive mode). They differ

TABLE 60.1 Parts of the EN 13757 Standard

ISO Layer	Type of Layer	Functions	Standard Part
Application layer	Application-oriented layers	Data structures, data types, actions	EN 13757-3
Presentation layer		Empty	
Session layer		Empty	
Transport layer	Transport-oriented layers	Empty	
Network layer		Routing (optional)	
Data link layer		Transmission parameters, telegram formats, addressing, data integrity	EN 13757-2
Physical layer		Cable type, bit representation, bus extensions, topology, electrical specification	EN 13757-2 EN 13757-4

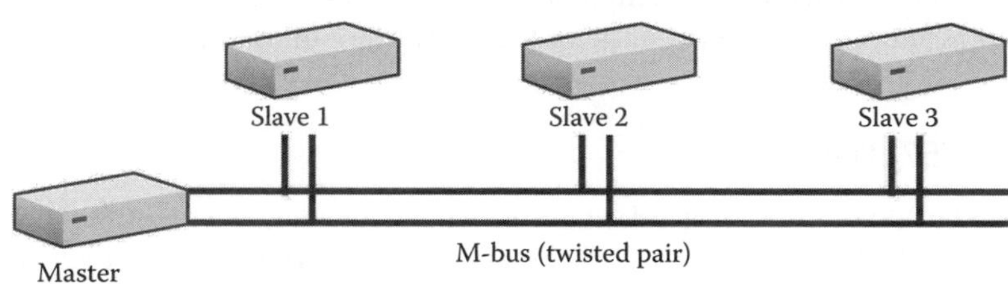

FIGURE 60.11 Master–slave M-Bus twisted pair system.

by transmission speed (4.8 to 100 kchip/s*), coding, bandwidth, and some other technical aspects. The physical layer and data link layer of KNX-RF were defined jointly with the wireless MBUS standard, and S-mode corresponds with KNX-RF.

Recently, new modes have been added to Wireless M-Bus: C-mode (*compact mode*; derived from the T-mode but reduced complexity) and N-mode (at 169 MHz and high power, for long-range meter reading +1 km). There are also extensions following the meter standardization work in the Netherlands (NTA 8130), Denmark, Germany (Open Metering Specification), and France.

60.2.7.1.2 Link Layer

A link layer based on the format class FT1.2 (defined in EN 60870-5-1) and a telegram structure based on EN 60870-5-2. This standard, also known as IEC 870-5, defines the transmission protocols for telecontrol equipment and systems (also used in SCADA communication). The M-Bus protocol derives from this standard but does not use all the IEC functions.

60.2.7.1.3 Application Layer

The dedicated M-Bus application layer as defined in EN 13757-3 can have an either fixed or variable data structure and can transport either the dedicated layer specified in EN 13757-3 or DLMS/COSEM type of data. The maximum length of a telegram is recommended to be 255 bytes, as longer messages could lead to problems in modem communication. The upper limit for characters in the variable data blocks is therefore limited to 240 bytes.

Encryption is mandatory. EN 13757-3 (Section 60.5.10) defines DES encryption for the M-Bus telegrams, which is nowadays considered as insecure. Authentication is not supported. The Dutch DSMR

* In systems using spread-spectrum signals, a single data bit is transmitted using a sequence of symbols or chips. The chip rate thus refers to the symbol rate of the physically transmitted signal and is generally lower than the effective data bit rate.

project deviates from the standard and specifies that the preferred encryption is AES (Advanced Encryption Standard; Federal Information Processing Standard [FIPS] 197).

DLMS/COSEM devices such as electricity meters can be set up as M-Bus masters (using M-Bus client objects). M-Bus protocol parameters are mapped to attributes, and M-Bus functions are mapped to methods. Then, data can be exchanged with M-Bus slaves, such as gas, heat, or water meters, and stored in COSEM objects. The objects are then suitable to be collected by the DCS (Data Collection System). The clients can also send commands to the M-Bus devices. It should be noted that this is not the same as the DLMS/COSEM support in the M-Bus protocol itself. The extensions resulting from the DSMR specifications are likely to be included in the next revision of EN 13757.

60.2.7.2 Conclusion

M-Bus is a specification mainly targeted at local in-house data exchange. It can be implemented with a twisted pair cabling system or using a wireless system in the ISM band. M-Bus is optimized to be able to work on simple battery-powered devices with modest requirements. The lower layers of M-Bus support DLMS/COSEM and DLMS/COSEM devices can be set up as M-Bus masters, making the two very compatible.

60.2.8 DLMS/COSEM and IEC 62056

DLMS originally stands for the Distribution Line Message Specification and is an application layer specification, aimed at messaging between (energy) distribution devices. It is an international standard established by IEC TC 57 and published as IEC 61334-4-41.

Later on, this standard became the Device Language Message Specification and integrated into the IEC 62056 series of standards. It specifies general concepts for the modeling of object-related services, communication entities, and protocols for energy metering.

COSEM stands for COmpanion Specification for Energy Metering and provides an object-oriented interface model of communicating metering equipment, based on OBIS (Object Identification System) codes.

Finally, xDLMS is an extension to the original DLMS and describes how to access attributes and methods of COSEM objects. The original DLMS standard was IEC 61334-4-41 and did not provide a meter data model or a naming system. When combined with COSEM, as DLMS/COSEM, only a part of the original DLMS standard is used, but a few extensions were added to make more efficient use of the COSEM data model. This is what constitutes xDLMS.

The full specification of DLMS/COSEM is divided over four *colored books* that are maintained by the DLMS User Association:

1. The Blue book details the data model, comprising the COSEM interface classes and the OBIS codes for the various energy types. This specification is standardized by the IEC and CEN.
2. The Green book specifies the protocols that can support DLMS and refer to various ISO, IEC, NIST, and Internet standards.
3. The Yellow book specifies the conformance and certification test plans for the COSEM object model and the communication layers.
4. The White book contains a glossary of terms.

60.2.8.1 COSEM Data Modeling

In order to structure the information and functionality of some physical metering equipment and to control access to it, COSEM standardized building blocks can be combined to model a metering device in a hierarchical structure. There are four groups of COSEM interface classes, relating to storage, access control, time and scheduling, and communication.

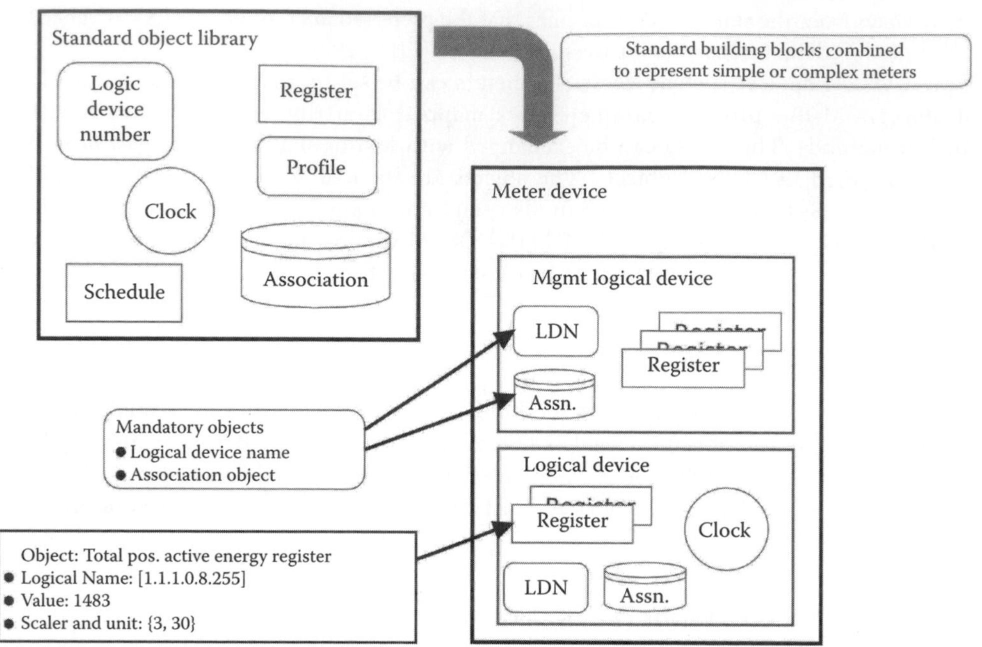

FIGURE 60.12 Combining COSEM standard building blocks to represent meter functionality.

For example, a meter is modeled as a physical device that contains one or more logical devices (LDs). Each LD has a number of interface objects, holding the data in the form of attributes and presenting the functionality of the LD as methods. Methods can be used to examine or modify attributes.

In a physical device, at least a *Management LD* is mandatory. In turn, each LD should at least contain an *LD Name*, a unique identifier for each and every LD worldwide, and an *Association object* to control the access to the objects. Thus, by combining standard objects, complex metering systems can be built. A residential meter will contain on order of a 10 few objects, while an industrial meter could contain more than 100 objects.

Using this system, a wide variety of interoperable products from different manufacturers, from simple to very complex metering equipment, can be supported. An example is shown in Figure 60.12.

A more complex example would be a concentrator handling data for a number of meters. Sticking to the DLMS/COSEM meter model, each set of metering data is handled by an LD implemented in a physical device.

60.2.8.1.1 Management LD

The management LD is the LD, which must always be present in the physical device and for which the upper HDLC address 0x01 is reserved. It holds a list of LDs that are inside the physical device and should at least be accessible for a public client using the lowest security level. This ensures that a device with an unknown structure can be *discovered*.

60.2.8.1.2 LD Name

The *LD name* is used for a worldwide unique identification of each LD and is an octet string of up to 16 octets. The first three octets uniquely identify the manufacturer of the device. It is assigned by the DLMS UA in cooperation with the FLAG Association. The manufacturer is responsible for guaranteeing the uniqueness of the octets that follow (up to 13 octets).

60.2.8.1.3 Association Objects

Access to the objects in a device is controlled by association objects, which provides the list of objects available, authentication rules, and access rights. Associations are established between a client and an LD. They are identified by the pair of client and LD addresses.

60.2.8.1.4 Profile Objects

To efficiently store and process tables of data, such as tariff profiles or logged voltage events, COSEM provides profile objects. Such profiles can be freely parameterized, and the objects have attributes and methods to describe the contents and ways to deliver the data.

60.2.8.2 Messaging and Referencing

The role of the messaging system is to transform COSEM object information into messages that can be transported by the lower layers. This transformation is performed by the DLMS services, according to a client–server concept. Herein, the DCS acts as a client to the meter and requests certain abstract services (read/write/action) while the LDs in the meter act as servers and respond to those requests.

To access any information modeled with COSEM, such as attributes and methods of objects, they have to be referenced. In the metering equipment (server), the COSEM application layer provides two mechanisms to access attributes: Short Name (SN) referencing and Logical Name (LN) referencing (Figure 60.13). SN referencing has been inherited from the original DLMS standard. The DCS (client) application layer can only use LN referencing.

60.2.8.2.1 Short Name Referencing

With SN referencing, attributes and methods of COSEM objects are mapped to DLMS named variables during the design of the meter. Each named variable is identified with an SN, which is a 16-bit unsigned integer. Attribute 1, the LN of the object, is mapped to a DLMS named variable identified by a base name. There are no general rules defined for assigning base names.

The map with named variables can be retrieved from the meter after which access using SN referencing is provided by the DLMS read/write services.

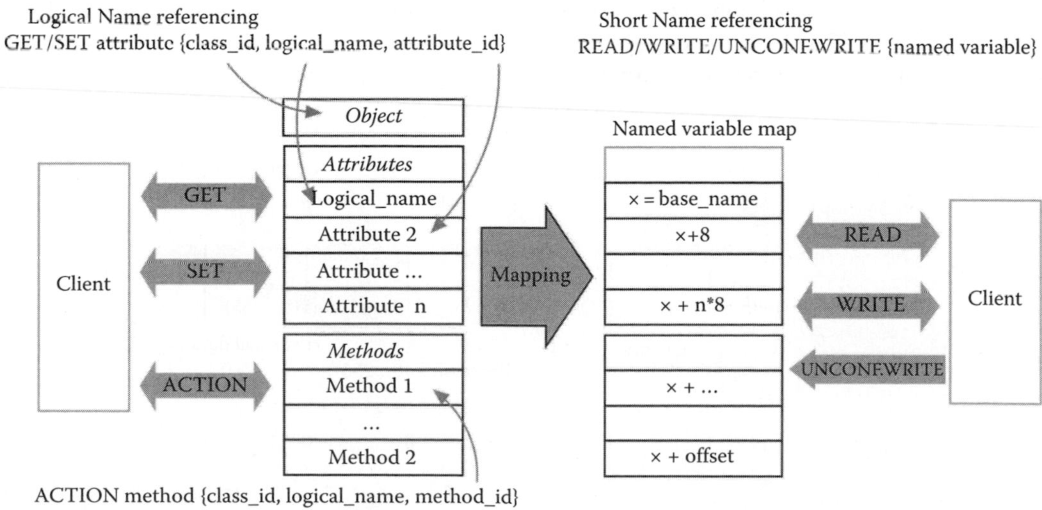

FIGURE 60.13 Accessing objects/attributes/methods using Logical Name referencing (left) or Short Name referencing (right).

60.2.8.2.2 LN Referencing

With LN referencing, attributes and methods are accessed via the LN of the object, by specifying the indexes of the attributes or methods. LNs are defined according to OBIS.

When a meter is read using LN referencing, the necessary attributes of certain objects are accessed using an xDLMS service and transformed into a series of bytes, called Application Protocol Data Units (APDUs). OBIS naming is used to identify COSEM objects to make them self-describing. Standard codes for electricity, cold water, hot water, gas, heating, and cooling are defined in the DLMS UA Blue book. A full list of standard OBIS codes and valid combinations of standard values in each group is maintained by the DLMS User Association.

To support future functionality, specific elements such as new OBIS codes, attributes, methods, and interface classes are allowed. However, the information on such elements has to be made available by the manufacturer. Obviously, for standard features, manufacturer-specific *hacks* are not allowed.

60.2.8.3 OBIS Codes

Names of objects are specified by OBIS codes. The list of codes follows a hierarchical structure, using six value groups, in which the meaningful combinations are standardized (Figure 60.14). The OBIS system has its origins in the EDIS system.

The value group A defines the media (energy type) that the metering is related to (with A = 0 for abstract values). Value group B identifies the measuring channel. Value group C defines the physical quantity measured, while value group D identifies processing methods and country-specific codes. Value group E is typically used for identifying rates. Value group F is used for identifying historical values. Groups B–D have code space for manufacturer-specific identifiers. Some value groups may be suppressed if they are not relevant to an application, for example, by setting group F value to 255.

For example, the LN of a register object holding the total positive active energy could be 1.1.1.8.0.255, while the one holding total positive reactive energy could be 1.1.3.8.0.255.

Billing data are modeled as physical quantities and processed as energy (integration) and demand (averaging) quantities. Energy and demand registers can be allocated to tariff schemes and stored over several billing periods.

60.2.8.4 Communication Profiles

In reality, data communication takes place through a communication protocol stack and via the communication media. As in the OSI system, the communication process is divided into independent layers, and various lower layers can be used below the application layer. A combination of the

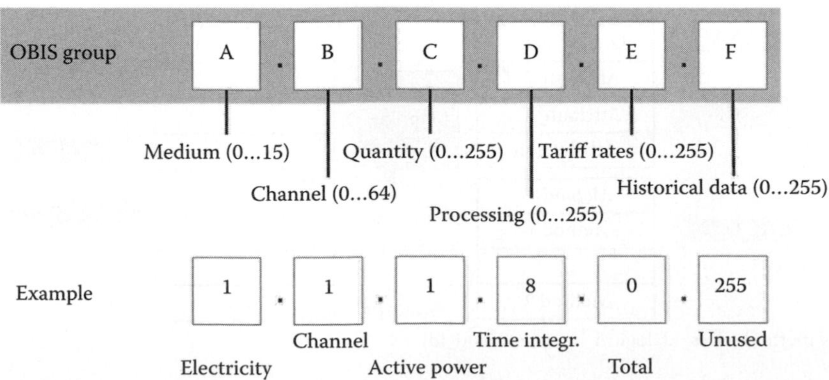

FIGURE 60.14 Structure of the OBIS naming system.

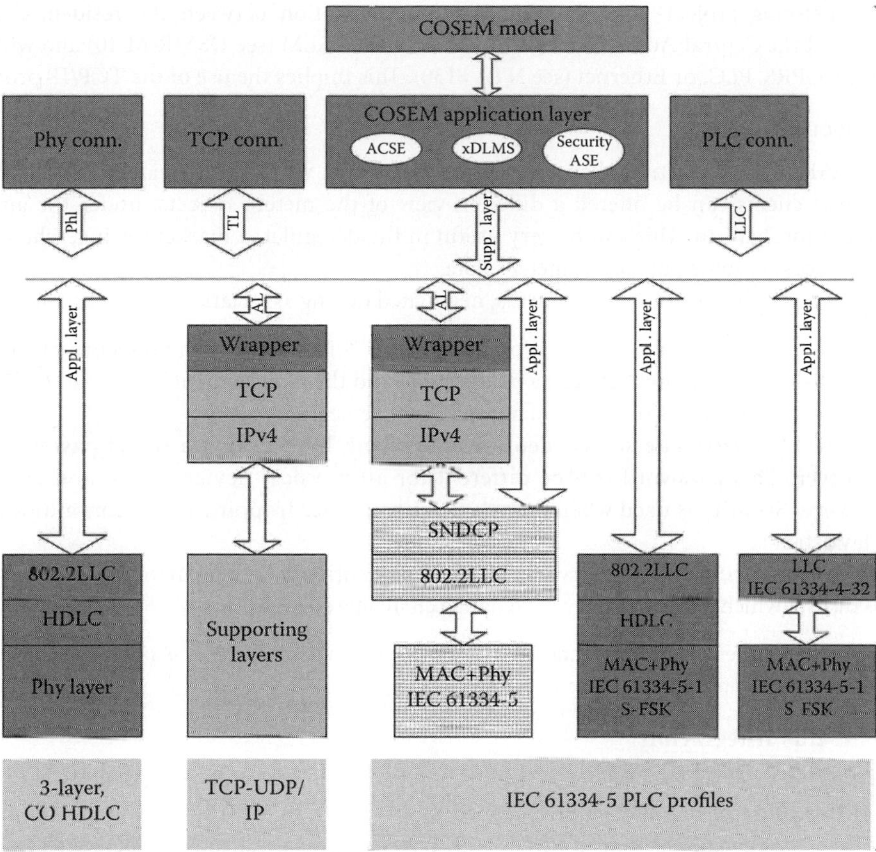

FIGURE 60.15 DLMS/COSEM communication profiles. (From the DLMS/COSEM Architecture and Protocols document, DLMS User Association, Companion specification for energy metering: Architecture and protocols, 2009.)

layers, with always the same application layer on top working directly with the application process, is called a communication profile.

As the communication protocol stack (called a profile) is completely independent of the application layer, the servers and clients may independently support one or more communication profiles to communicate over various media. The COSEM model, used for modeling the application process and the application layer, remains the same.

The DLMS services are defined in a verbal form and are formally using the ASN.1 standard. The information is then turned into bytes using the A-XDR standard (a performance-optimized version of BER). The resulting APDUs are transported by the lower protocol layer over the selected media.

DLMS/COSEM specifies a number of possible communication profiles, some of which are shown in Figure 60.15:

- The three-layer HDLC-based connection-oriented profile. HDLC is a bit-oriented synchronous data link layer protocol. The profile supports data exchange over certain local ports such as RS232, optical links, current loops, and GSM/PSTN connections. Also see IEC 62056-21 *Mode E*.
- A TCP/IP-based communication profile, to support data exchange over the Internet and GPRS connections.
- The PLC profile, as specified in IEC 61334.
- An M-Bus/EN 13757 profile.
- The Euridis/IEC 62056-31 profile, using twisted pair cables.

The Dutch metering project* specifies that the communication between the residential metering installation and the Central Access Server will use DLMS/COSEM (see DSMR-M 40) and will transfer data either by GPRS, PLC, or Ethernet (see NTA 8130). This implies the use of the TCP/IP profile.

60.2.8.5 Security

DLMS/COSEM provides controlled access to the meter's data with the association objects. Based on these, different clients can be offered a different view of the meter's objects, under the appropriate authentication mechanism. This can be very useful in the deregulated market setting, where multiple parties need access to some part of the meter's data.

There are three authentication mechanisms, negotiated during association establishment:

1. *Lowest-Level Security*: No identification takes place. This is used when no security is required or to establish an association between a public client and the management LD of the server, in order to be able to retrieve the resources of an unknown device.
2. *Low-Level Security*: The server identifies the client, by checking a secret password held by the server. This password can be different for all physical devices, LDs, and associations. Low-Level Security is used when there is no risk of eavesdropping of the communication and replay attacks.
3. *High-Level Security*: Both the server and the client identify the other partner. It is a four-step process during which client and server exchange challenge strings.

In addition to the authentication mechanisms, encryption can be used. This is provided by the xDLMS service element of the application layer.

60.2.8.6 M-Bus Integration

DLMS/COSEM devices such as electricity meters can be set up as M-Bus masters (using M-Bus client objects). M-Bus protocol parameters are mapped to attributes, and M-Bus functions are mapped to methods. Then, data can be exchanged with M-Bus slaves, such as gas, heat, or water meters, and stored in COSEM objects. The objects are then suitable to be collected by the DCS. The clients can also send commands to the M-Bus devices.

60.2.8.7 Conclusion

DLMS/COSEM is positioning itself as the all-round contender for smart grid communication. The support for DLMS/COSEM in a lot of other standards (such as M-Bus, IEC 62056-21, -31, and recently ZigBee), projects (Dutch DSMR), and existing meters illustrates this.

60.2.9 IEC 61968 Part 9: System Interfaces for Meter Reading and Control

The Common Information Model specified by the standards IEC 61970 and IEC 61968 is a semantic model that describes the components of an electrical system and their interrelations. It consists of many parts that specify format, for example, exchange of power system network models, and describe processes like asset management, work scheduling, and customer billing. Basically, it consists of the definition of an *information model* (UML-based canonical model), *contextual* profiles (context-specific subsets), and their implementation schemas (XML/RDF).

Part 9 specifies the exchange of information between the AMR metering and other systems and business functions within the Enterprise Resource Planning (ERP) of the utility enterprise. Specific communication protocols (e.g., TCP/IP) are not defined by this standard. It can be used also for

* The document "Dutch smart meter requirements" (DSMR) is an elaboration of the NTA 8130, commissioned by the Dutch grid companies (ENBIN) and aimed at meter interoperability. Also, requirements have been added, mainly with respect to installation and maintenance, quality, and performance).

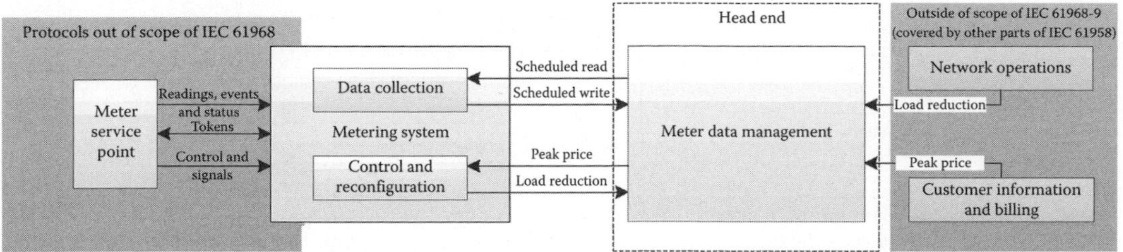

FIGURE 60.16 Simplified IEC 61968-9 reference model with relation to other parts of the standard.

communicating metered quantities of gas or water infrastructure. Information is modeled by a set of messages that support reading and control functions of smart meters. Typical uses are

- Meter reading and control
- Meter events
- Customer Data Sync and Switching

From the common reference model point of view, the smart meter is an *end device*, with a *unique id*. It can be considered as a *physical asset*, which receives *control requests* and *collects* and *reports* measured values. The smart meter is part of a *business process* including customer and respective contracts, energy market, aggregators, and other ancillary service providers.

The reference model specifies the following functionalities for meter reading and control:

- Metering system includes data collection, control, and reconfiguration.
- Meter data management.
- Maintenance.
- Load management including load control and load analysis.
- Meter asset management.
- Meter administrative functions.

In detail, the relations of the reference model are depicted in Figure 60.16.

60.2.9.1 Message Structure

The message profile according to IEC 61968 (Figure 60.17) is basically structured in the header, containing meta-information and the payload. It can also contain data defining either a request or a reply.

The message types specified include messages for

- CustomerMeterDataSet
- MeterAsset
- MeterAssetReading
- EndDeviceControls
- EndDeviceEvents
- MeterReadings
- MeterReadSchedule
- MeterServiceRequest
- MeterSystemEvents
- EndDeviceFirmware

The EndDeviceEvent is shown in Figure 60.18 as an example for a message type. The reason of the event is also specified and includes, for instance, a sustained outage, a high- or low-voltage threshold, a distortion meter, and a revenue.

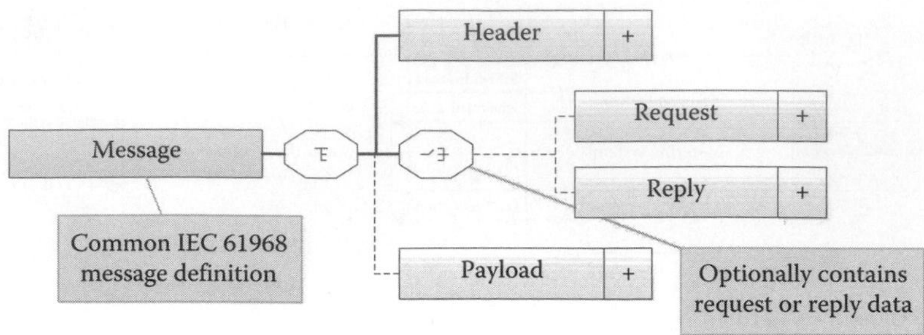

FIGURE 60.17 Common message definition by IEC 61968.

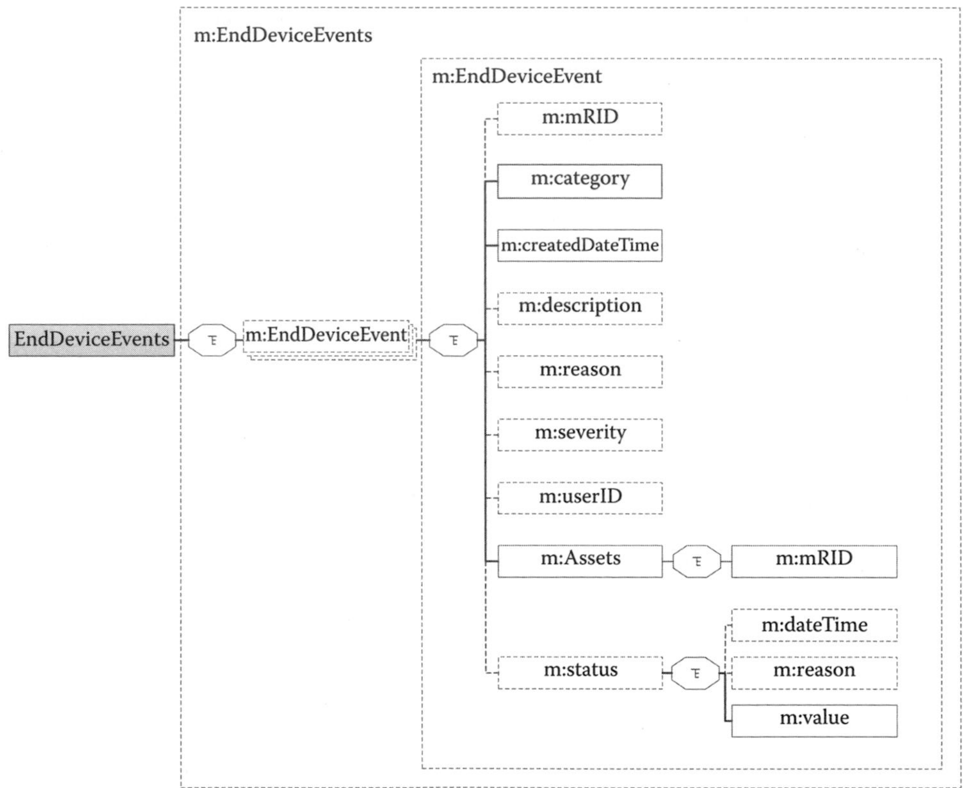

FIGURE 60.18 Example message "EndDeviceEvents."

60.2.9.2 IEC 61968-9 and DSLM/COSEM

An extension of CIM has been proposed to include the DSLM/COSEM object model (e.g., physical device and LD) in CIM. IEC 61968 specifies the communication within the ERP to the head-end of the metering system, while the meters can be compliant to DSLM/COSEM. This means, in case of a request from the ERP, that the head-end has a DSLM/COSEM protocol session with the meters, where the IEC 61968 message can provide all the details for addressing the meter and the objects and attributes of interest.

60.2.10 IEC 62056-31 *Euridis*

Euridis is a standard (IEC 1142) for remote and local meter reading introduced at the beginning of the 1990s. In 1999, it was integrated into IEC 62056 as part 31 (*IEC 62056-31 Ed. 1.0:1999 Electricity metering—Data exchange for meter reading, tariff, and load control—Part 31: Using local area networks on twisted pair with carrier signalling*). Euridis is mostly used in francophone countries, and it is estimated that about 10 million meters communicate using Euridis.

Euridis uses a twisted pair cabling system, the local bus, onto which all meters in a building can be linked. It uses OOK (On-Off Keying) amplitude modulation on a 50 kHz carrier. A magnetic coupler then allows connecting an HHU for readout or programming, in a master–slave configuration (referred to as primary and secondary stations). The bus can be up to 500 m or 100 devices and allows a data rate of 1200 baud half-duplex. Functions are provided to perform a complete automatic identification of the stations on the bus.

EURIDIS also provides authentication features upon connection and simple encryption scheme for the data exchanges.

The original Euridis specification does not include a data model. However, at the beginning of 2009, the Euridis+protocol stack was announced, supporting DLMS and allowing data rates of 9600 baud, while retaining backward compatibility.

60.2.11 Other Standards with Relevance to Smart Metering

60.2.11.1 KNX

KNX is the result of the joint effort of three European consortia working on home and building control, namely, BatiBUS, EIB, and EHS. At the beginning of 2002, a first specification of KNX was made public, which was primarily based on EIB, extended with new communication media and configuration mechanisms originally developed for BatiBUS and EHS. KNX was made into standard ISO/IEC 14543-3-x in November 2006.

KNX defines a hierarchically structured bus (implemented using twisted pair, radio frequency, power line, or IP/Ethernet, or a combination) to which all devices are connected. Devices are sensors and actuators needed for the control of building management equipment such as lighting, blinds, security/monitoring systems, energy management, HVAC systems, metering, audio/video control, and white goods. All their functions can be controlled, monitored, and signaled via a uniform system without the need for any extra control centers.

The KNX specification encompasses the following concepts (Figure 60.19):

- Interworking and Distributed Application Models for the various tasks of home and building automation
- Schemes for configuration and management (configuration modes, E, S, and A), typically with the PC-based ETS tool
- A communication system, with a set of physical media, a message protocol, and communication stack models
- Device models, summarized in profiles

The KNX specifications are spread over 10 volumes, with volume 3 *System Specification* handling the bulk of the hardware and software implementation. The other volumes cover general architecture, information on specification tests and certification, conformance tests, device profile definitions, etc.

60.2.11.1.1 Application and Interworking Models

A central idea in the KNX Device Network is the data point. Data points can be inputs, outputs, parameters, etc., and implement standardized data point types. They represent the process and control variables of the system.

When a local application in a device (e.g., such as a sensor) writes a value to a *sending data point*, the communication system sends a write message with the corresponding address and the new value.

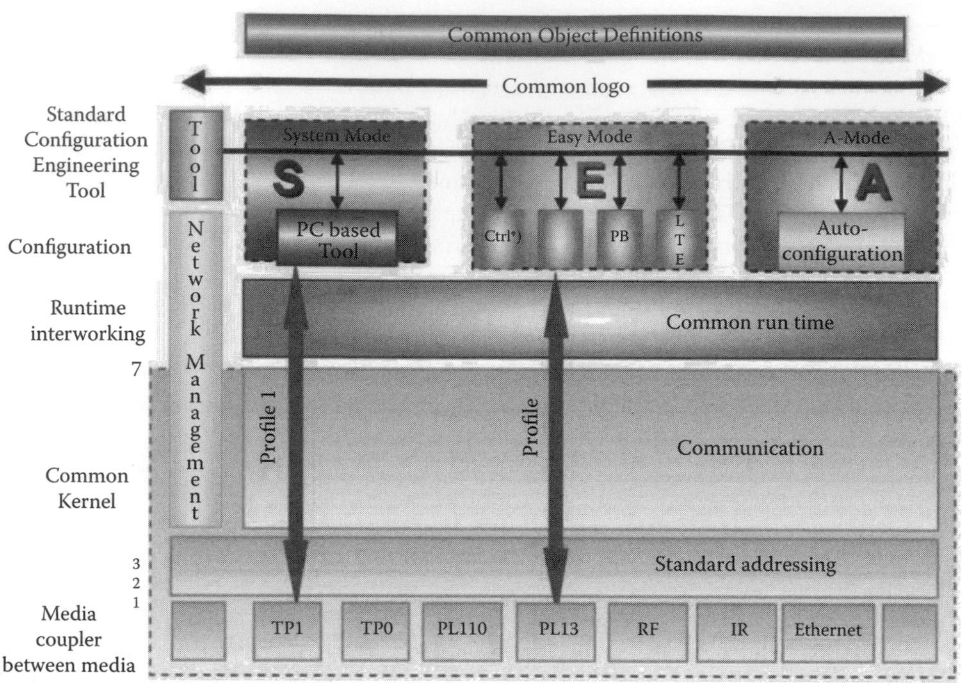

FIGURE 60.19 The KNX model. (From Konnex Association, KNX system architecture V3.0, 2009, Available at http://www.knx.org/fileadmin/downloads/03-KNXStandard/KNXStandardPublicDocuments/03_01_01Architecturev3.0.zip.)

A *receiving data point* with this same address will inform the local application of the new value, which in turn can then take an action. This action could be an internal state change, updating one of its own *sending data points* or modifying some output in an actuator.

The principal way to realize data points is with group objects. Group objects are accessed via multicast group addressing. Thus, one or more sensors emit a value, and many actuators listen on a given group address. In this way, local applications in a number of devices, with linked data points, combine to form a distributed application.

KNX is not limited to grouping devices, as each device can also publish several of its own data points that can be grouped independently into network-wide shared variables.

60.2.11.1.2 Communication System

The logical topology of a KNX installation is structured into a backbone line, main lines, and lines (Figure 60.20). During commissioning, each KNX device is assigned a unique 16-bit *individual address* that reflects the position in this topology. A KNX network contains a single backbone line, to which up to 15 main lines are attached, which in turn allow up to 15 lines to be connected. The connections between the lines are done by couplers.

A subnetwork is any part of this topology with the same main line number and the same line number. The backbone line, any main line, and any line are subnetworks.

The KNX end devices can be connected anywhere in this topology, and up to 255 KNX end devices can be addressed in any subnetwork. This results in a maximum 65,536 devices in a backbone line. While different physical layers can be mixed within a KNX network, a subnetwork is the smallest part of the topology that can be implemented with one single physical layer type; it cannot be split.

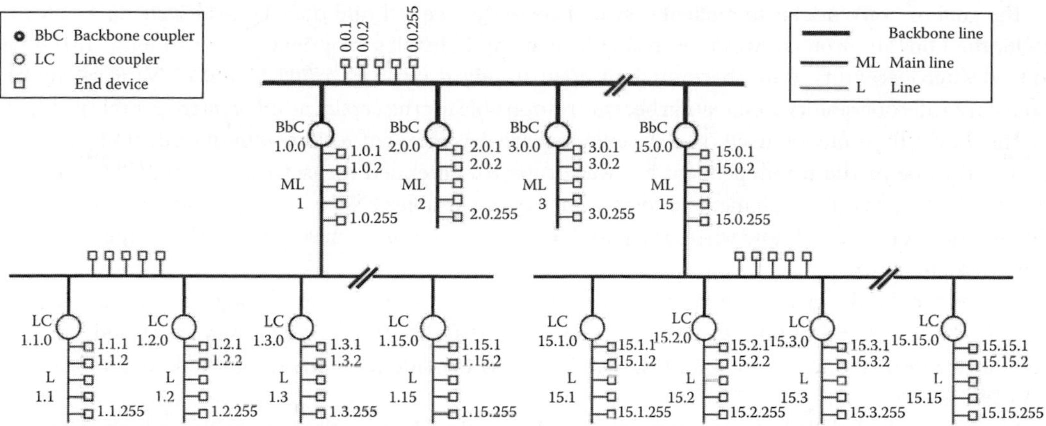

FIGURE 60.20 Logical topology of a KNX system. (From Konnex Association, KNX logical topology FAQ, 2005, Available at http://www.knx.org/fileadmin/downloads/03-KNXStandard/KNXStandardPublicDocuments/FAQLogicalTopology.pdf.)

60.2.11.1.3 Physical Layers

As mentioned, KNX can be used on top of a number of media:

- *Twisted pair*: There are two twisted pair media specified; TP0, with a bit rate of 4800 bits/s, has been taken over from BatiBUS, and TP1, with an arbitrate of 9600 bits/s, has been taken over from EIB. Devices can be powered over the bus.
- *Power line*: Again, two variants are specified—PL110 (operating at 110 kHz, from EIB) and PL132 (at 132 kHz, from EHS).
- *Wireless*: KNX-RF operates in the 838 MHz ISM band. Separate wiring is needed for power, making KNX-RF not widely used. The physical layer and the data link layer of KNX-RF were defined jointly with the wireless MBUS standard, and S-mode complies with the KNX-RF standard. KNX-RF and wireless M-Bus S-mode are equal on the PHY and MAC layer but differ on the higher layers.
- *KNX over IP*: Using UDP (unicast and multicast) allows the use of Ethernet, Bluetooth, Wi-Fi, etc. Usable for high-bandwidth applications.

60.2.11.1.4 Configuration and Management

For the configuration, KNX specifies three modes: system (S), easy (E), and automatic (A). System mode allows sophisticated building setups but needs a separate configuration master (such as the association's own ETS software) and trained installers. Easy mode is meant is meant for installers with basic KNX knowledge, and such devices can only be partly reconfigured. Automatic mode is suitable for end-user installation. Some devices support more than one configuration mode. The KNX standard also has predefined sets of features for many common applications, called profiles.

KNX aims to provide a complete solution for home and building automation and is backed by a lot of manufacturers worldwide. It must be noted that most KNX success stories about reduced energy consumption involve a complex interaction of KNX-enabled boilers, lighting, etc., making the installation costs very high, especially for retrofitting.

60.2.11.2 LonWorks/LonTalk

The heart of the LonWorks technology (Local Operating Network) is the proprietary LonTalk protocol, developed by Echelon. Development of the LonWorks platform began in 1988, and the technology was heavily marketed to the utility industry.

The goal of LonWorks is to make it easy and cost-effective to build open control systems, but before 1996, the LonTalk protocol was only available by using Echelon's proprietary Neuron chip. This guaranteed interoperability. After becoming an open standard (with EIA 709.1 *Control Networking*), the LonMark Interoperability Association became responsible for the certification of interoperable products.

The LonTalk protocol itself is a layered, packet-based, peer-to-peer communications protocol designed to be medium independent but was originally intended for twisted pair and PLC medium. Just as KNX, LonTalk implements a form of network variables (NVs). The application program in a device does not need to know where input and output NVs come from or go to as this is the task of the LonWorks firmware.

Altogether, LonWorks is similar to KNX but is used in a much wider range of applications, well outside the home and building space (e.g., train controls). In 1999, LonTalk was submitted for ANSI standardization, and at the end of 2008, the ISO and IEC made the LonTalk technology into standard ISO/IEC 14908-1 to -4.

More recently, Echelon started marketing its Networked Energy Services (NES), which is an integrated (smart meters, data concentrators, and system software) smart metering solution using the power lines as communication medium, based on LonWorks. The AMM system deployed in Sweden by Vattenfall and E.ON is using Echelon's NES.

60.2.11.3 BACnet

BACnet stands for Building and Automation Control Networking and is developed by the BACnet Committee since 1987. In 1995, it was adopted by the American Society of Heating Refrigeration and Air-Conditioning Engineers and became ISO standard 16484-1 to -5 in 2003.

BACnet is an entirely nonproprietary system, with typical applications in the HVAC, lighting, and security domains. A number of network technologies can be used, including Ethernet, LonTalk, ARCnet, ZigBee networks, and BACnet/IP. The latter allows the use of BACnet over virtually any medium.

60.2.11.3.1 Concept

All data in a BACnet system are represented using an object-oriented approach, resulting in objects, properties, and services.

An object contains information relating to physical inputs and outputs, or nonphysical concepts like software, or a group of set points. Objects provide a set of properties that are used to get information from the object or give information and commands to an object. The BACnet specification has a list of standard objects, with predefined properties and behavior.

For example, a temperature sensor might be represented as a BACnet Analog Input object. This object would contain a *Present_Value* property that contains the current reading, and possibly Min and Max values.

BACnet specifies 123 properties of objects, of which *name*, *type*, and *identifier* have to be present in every object. Some properties can accept writes, while others can only be read.

The act of reading or writing a property is called a *service*, which, following a client–server model, is carried out by the server on behalf of the client. There are 32 standard services in the specification.

BACnet has a Smart Grid Working Group (SG-WG) focused on enabling buildings to interact in the grid. Standard BACnet objects such as the Load Control Object can already be used to track consumption and execute preprogrammed actions accordingly. The BACnet/web services (WS) specification allows external applications to interact with a building automation system and is used in the OpenADR project. The group has also developed definitions for a standard meter object as well as energy profiles (for load, generation, and storage status).

The BACnet Committee has also added ZigBee as a possible data link layer, and the ZigBee Commercial Building Automation Task Group is defining a way to tunnel BACnet traffic through ZigBee networks.

LonWorks and BACnet have overlapping scopes, but the latter has become the first choice at the system management level.

60.2.11.4 ZigBee (Smart Energy Profile)

ZigBee is a low-power wireless communications technology designed for monitoring and control of devices and is maintained and published by the ZigBee Alliance. Home automation is one of the major market areas. ZigBee works on top of the IEEE 802.15.4 standard, in the unlicensed 2.4 GHz or 915/868 MHz bands.

An important feature of ZigBee is the possibility to handle mesh-networking, thereby extending the range and making a ZigBee network self-healing.

The ZigBee Smart Energy Profile (numbered 0x0109) was defined in cooperation with the Homeplug Alliance in order to further enhance earlier HAN specifications. The profile defines device descriptions for meter reading, demand response, meter prepayment, etc.

A collaborative effort between the ZigBee Alliance and the DLMS UA was announced to define a method to tunnel standard DLMS/COSEM messages with metering data through ZigBee Smart Energy networks. In 2011, the smart energy profile was updated to version 1.1, with additions related to dynamic pricing and tariff blocks, prepayment, and enhancements to allow tunneling of other protocols (such as DLMS).

In April 2013, the Smart Energy Profile 2 was published. This profile focuses on IP-based (6LoWPAN) energy management and is not directly compatible with the 1.1 profile. Noteworthy is the support for very large size multidwelling units (where direct wireless connections between meters in a basement and the HAN device in an apartment may not be possible) and features to control PHEVs.

Considering the low power requirements, robustness, availability of cheap ZigBee *kits*, and the specific profile for metering applications, ZigBee has a lot of potential in HANs.

60.2.11.5 Homeplug (Command and Control)

The Homeplug 1.0 standard was published in 2001 by the Homeplug Power line Alliance and allows communication over power lines at 14 Mbps half-duplex. In 2005, it was succeeded by Homeplug AV, allowing over 100 Mbps and meant for HD multimedia applications.

In 2007, version 1.0 of Homeplug Command and Control was announced, providing a PHY and MAC specification for low-speed (up to 5 kbps), low-cost PLC usable in house-control applications (lighting, HVAC, security, and metering) and meter reading. Work on network, transport, and session layers is still reported ongoing. Device profiles will provide a description language to define supported services and actions. A transfer rate of about 5 kbps when using 128 byte packets (at application level) is possible.

Currently, no Homeplug C&C products have hit the market, and it is uncertain whether this technology will be further developed.

60.2.11.6 6LoWPAN

The 6LoWPAN is a standard from the IETF designed from the ground up to be used in small sensor networks, on top of low-power wireless (mesh) networks, specifically IEEE 802.15.4 (thus directly competing with ZigBee). Implementations of 6LoWPAN should easily fit into a few kBs of memory. Highlights include support for the Zero-Conf and Neighbor Discovery capabilities of IPv6 and stateless header compression that allows the packets to be as small as 4 bytes. 6LoWPAN could realize the main concept of the *Internet of Things* by making it feasible to assign an IP address to the smallest of devices, sensors, and actuators.

60.2.11.7 DPWS

DPWS stands for Devices Profile for Web Services, and its goal is to integrate devices with Internet WS. DPWS 1.1 was approved as an OASIS Standard in June 2009. The full protocol stack is composed of several web standards, such as WSDL, XML, SOAP, and a host of WS standards. DPWS is similar to Universal Plug and Play but puts more focus on WS technology. DPWS enables secure WS messaging, discovery, description, and eventing on embedded, resource-constrained devices.

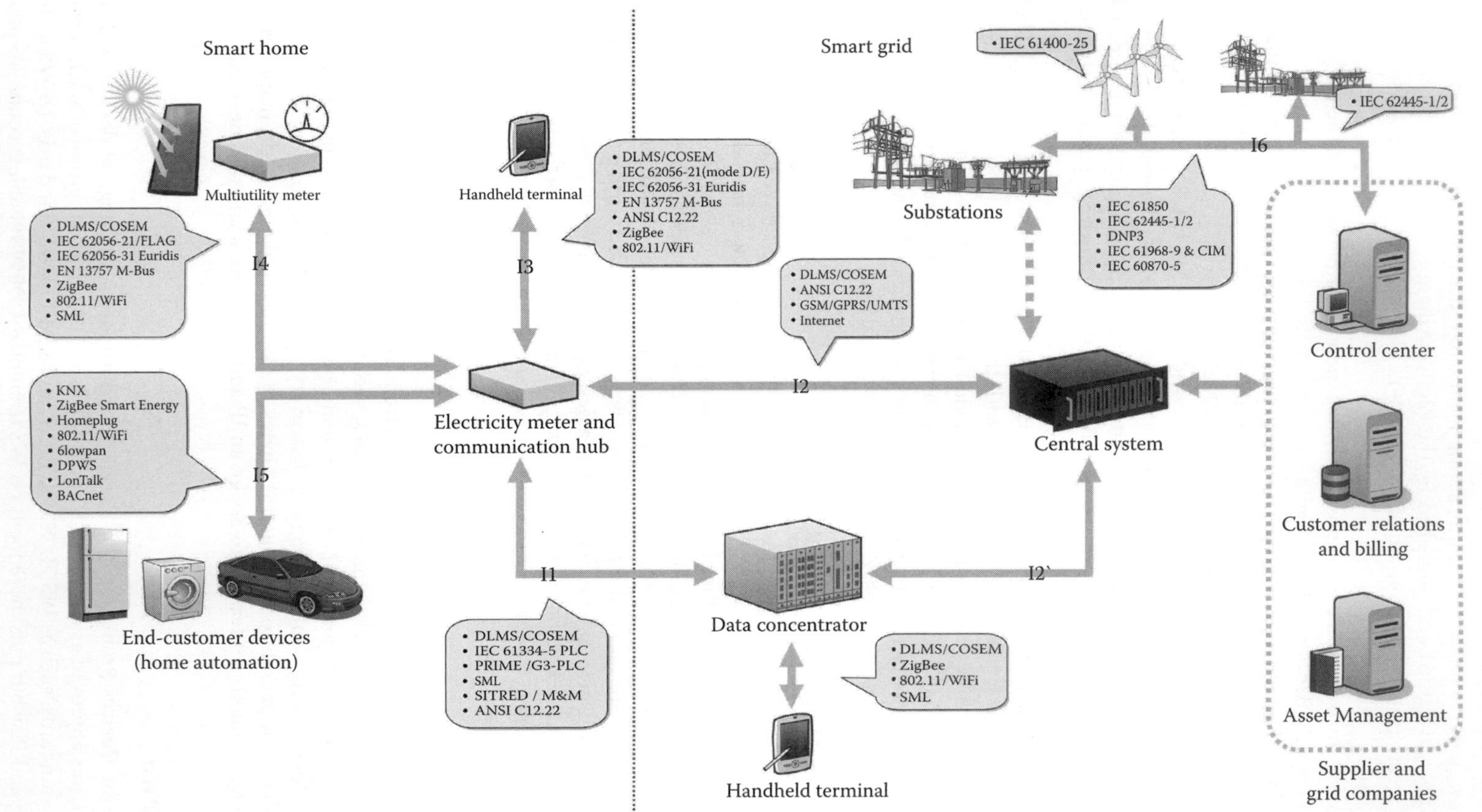

FIGURE 60.21 Placement of the analyzed standards in the smart grid architecture.

60.3 Suitability in the SG Architecture

In Figure 60.21, most of the legacy systems are situated within the dotted area marked as *supplier and grid companies*, and the transmission and substation systems pictured above them. On the lower levels of the net, command-and-control systems are less common. In a typical system for smart meter, we find the following interfaces:

- Interface I1: The communication between a data concentrator and a consumer's electricity meter. Data concentrators are used when a direct connection from the central server to the meter is not possible (e.g., PLC systems) or practical. An HHU can be connected for maintenance.
- Interface I2: Direct communication between the meter and the central system. This will typically be based on GPRS/UMTS/LTE or an already available broadband Internet connection.
- Interface I3: Connection between the meter and a local terminal (for installation and configuration). It is comparable to the case of the data concentrator, with wireless functionality making cost effective *drive-by meter reading* possible.
- Interface I4: Communication between the central meter and secondary meters (e.g., domestic solar panel arrays) or multiutility meters for gas, water, or heat.
- Interface I5: Communication between the meter and a HAN for home automation and domotics, enabling advanced demand response and load-shedding functionality. In-home displays and controllers will communicate with the meter via this interface.
- Interface I6: Communication for command and control of DER and control centers. These protocols were not covered in this chapter.

60.4 Conclusions

The rollout of smart meters that are currently under progress in many countries necessitates the need for interoperable and future-proof solutions that are available today. In this chapter, an overview was given of communication standards relevant to AMR and smart meters. Of course, many more technologies are out there, but most of them lack wide acceptance and flexibility or are still nascent or vendor controlled. The latter is especially true in connection with home and building automation space.

When talking strictly about meter data exchange itself, the leading standard would be DLMS/COSEM. As an application layer protocol, it can be used over virtually any medium, and more importantly, commercial solutions are available today.

Bibliography

Berger, L.T., Schwager, A., and Escudero-Garzás, J.J. (2013). Power line communications for smart grid applications. *Journal of Electrical and Computer Engineering*, 1–16.

Beroset, E. Overview of ANSI C12. http://www.electricenergyonline.com/?page=show_article&mag=18&article=138 (accessed July 25, 2013).

Botte, B., Cannatelli, V., and Rogai, S. (2005). The Telegestore project in ENEL's metering system. *18th International Conference and Exhibition on Electricity Distribution,* Turin, Italy, (pp. 1–4).

BSI. (2013). Technische Richtlinie BSI TR-03109-1. https://www.bsi.bund.de/SharedDocs/Downloads/DE/BSI/Publikationen/TechnischeRichtlinien/TR03109/TR03109-1.pdf.

Caleno, F. (2009). The Enel Smart Info—A first smart grids step to addressing in-home energy efficiency. *The 20th International Conference and Exhibition on Electricity Distribution (CIRED), Prague, Czech Republic.*

CENELEC. EN 13757: Communication system for meters and remote reading of meters, 2003.

CENELEC. (2005). EN 13757-4: Communication systems for meters and remote reading of meters—Part 4: Wireless meter readout (radio meter reading for operation in the 868 MHz to 870 MHz SRD Band).

CIM User Group. cimug.ucaiug.org.

Comellini, E., Gargiuli, R., Mirra, C., Mirandola, P., and Pioli, M. (1989). ENEL standardised telecontrol system for MV distribution network automation. *10th International Conference on Electricity Distribution (CIRED)*, Brighton, UK.

DLMS User Association. (2009). Companion specification for energy metering: Architecture and protocols.

DLMS User Association. FLAG application for ID. http://www.dlms.com/organization/flagapplicationforid/.

DLMS User Association. http://www.dlms.com (accessed July 25, 2013).

Echelon. The open smart grid protocol. http://www.echelon.com/technology/osgp/.

Echelon Corporation. Echelon's smart metering solution. http://www.echelon.com/applications/smart-metering/ (accessed July 15, 2013).

EDN. (2013). How a standard is born: IEEE P1901.2 for narrowband OFDM PLC. http://www.edn.com/design/wireless-networking/4415005/How-a-standard-is-born—IEEE-P1901-2-for-narrowband-OFDM-PLC (accessed July 22, 2013).

Endesa. Endesasmartmetering. http://www.endesa.com/en/aboutEndesa/businessLines/principalesproyectos/Telegestion (accessed July 16, 2013).

EnelSpA, IBM and Enel will team to offer automated metering solution worldwide, March 18, 2004. http://www.enel.it/eWCM/salastampa/comunicati_eng/1595207-1_PDF-1.pdf.

ERDF. (2009). Performance of open-standard PLC technologies on ERDF distribution network. http://www.maximintegrated.com/products/powerline/pdfs/plc-erdf.pdf.

ERDF Linky. http://www.erdfdistribution.fr/Linky/ (accessed July 16, 2013).

Euridis Association. http://www.euridis.org (accessed July 14, 2013).

Ferreira, H.C., Grove, H.M., Hooijen, O., and Han Vinck, A.J. (n.d.). Power line communications: An overview. *Proceedings of IEEE. AFRICON'96* Stellenbosch, South Africa (Vol. 2, pp. 558–563). IEEE. doi: 10.1109/AFRCON.1996.562949.

Feuerhahn, S., Zillgith, M., Wittwer, C., and Wietfeld, C. (2011). Comparison of the communication protocols DLMS/COSEM, SML and IEC 61850 for smart metering applications. *2011 IEEE International Conference on Smart Grid Communications (SmartGridComm)*, Brussels, Belgium, (pp. 410–415). IEEE. doi: 10.1109/SmartGridComm.2011.6102357.

G3-PLC Alliance. http://www.g3-plc.com/ (accessed July 14, 2013).

Gebhardt, M., Weinmann, F., and Dostert, K. (2003). Physical and regulatory constraints for communication over the power supply grid. *IEEE Communications Magazine*, 41(5), 84–90.

GyozoKmethy, *IEC 62056 DLMS/COSEM Workshop, Metering—Billing/CRM Asia*, Bangkok, Thailand, May 13–14, 2009.

Hersent, O., Boswarthick, D., and Elloumi, O. (2011). The ANSI C12 Suite. *The Internet of Things: Key Applications and Protocols*. Chichester, U.K.: John Wiley & Sons.

Hoch, M. (2011). Comparison of PLC G3 and PRIME. *2011 IEEE International Symposium on Power Line Communications and Its Applications* Udine, Italy (pp. 165–169). IEEE. doi: 10.1109/ISPLC.2011.5764384.

Holmberg, D.G. and Bushby, S.T. (November 2009). BACnet and the smart grid: BACnet today. *Supplement to ASHRAE Journal*, 51(11), B8–B12. Available at: http://www.bacnet.org/Bibliography/BACnet-Today-09/Bushby_2009.pdf.

Iberdrola Smart Metering. http://www.iberdrola.es/webibd/corporativa/iberdrola?IDPAG=ENWEBPROVEEBASDOCCONT&codCache=12632064318298617 (accessed July 14, 2013).

IEC. (2012). ISO/IEC 14908-1 ed1.0—Information technology—Control network protocol—Part 1: Protocol stack.

IEC. (1998). IEC/TS 61334-5-2 ed1.0—Distribution automation using distribution line carrier systems—Part 5-2: Lower layer profiles—Frequency shift keying (FSK) profile.

IEC. (2002). IEC 62056-21 ed1.0—Electricity metering—Data exchange for meter reading, tariff and load control—Part 21: Direct local data exchange.

IEC. (1999). IEC 62056-31 ed1.0—Electricity metering—Data exchange for meter reading, tariff and load control—Part 31: Use of local area networks on twisted pair with carrier signalling.

IEC. (2006). ISO/IEC 14543-3-1 ed1.0—Information technology—Home electronic system (HES) architecture—Part 3-1: Communication layers—Application layer for network based control of HES Class 1.

IEC. (2002). IEC 62056-42 ed1.0—Electricity metering—Data exchange for meter reading, tariff and load control—Part 42: Physical layer services and procedures for connection-oriented asynchronous data exchange.

IEC. (2006). IEC 62056-53 ed2.0—Electricity metering—Data exchange for meter reading, tariff and load control—Part 53: COSEM application layer.

IEC 61968-9. Application integration at electric utilities—System interfaces for distribution management; Part 9: Interfaces for meter reading and control.

IEC 61968-9. Message profiles for DLMS/COSEM. Study Report by D. Taylor and G. Kmethy, June 14, 2010.

IEEE P1901.2. IEEE draft standard for low frequency (less than 500 kHz) narrow band power line communications for smart grid applications. http://standards.ieee.org/develop/project/1901.2.html.

ISO. (2012). ISO 16484-5:2012—Building automation and control systems (BACS)—Part 1: Project specification and implementation.

iTrona GmbH. (2011). Beschreibung SML Datenprotokoll für SMART METER. Retrieved from http://itrona.ch/stuff/F2-2_PJM_5_BeschreibungSMLDatenprotokollV1.0_28.02.2011.pdf.

KEMA Consulting. (2010). Dutch smart meter requirements 3.0: main document. http://www.netbeheernederland.nl/DecosDocument/Download/?fileName=a-n579CEVhC-PRRFqtN5bZIjj5OT0osfD6_rJVUBnU68t2AawYRWyWBDAldRRBLWrKVQpjwrWTzy4aqBYn-Tm3XabvqaDw5G8_M0dpTDWOc&name=DSMR3.0-final-Main.

Konnex Association. (2005). KNX logical topology FAQ. http://www.knx.org/fileadmin/downloads/03-KNXStandard/KNXStandardPublicDocuments/FAQLogicalTopology.pdf.

Konnex Association. (2009). KNX system architecture V3.0. http://www.knx.org/fileadmin/downloads/03-KNXStandard/KNXStandardPublicDocuments/03_01_01Architecturev3.0.zip.

Landis+Gyr. (2012). Landis+Gyr E750 electricity meter. http://www.landisgyr.com/webfoo/wp-content/uploads/2012/12/Landis+Gyr-E750_Brochure-English.pdf.

LonMark International. http://www.lonmark.org/ (accessed July 16, 2013).

Luc Henderieckx. (2010). Distributor power line communication system (Patent nb 20100204850).

M-Bus original website (EN 1434). http://www.m-bus.com/ (accessed July 14, 2013).

Maxim. PLC G3 physical layer specification. http://www.maximintegrated.com/products/powerline/g3-plc/ (accessed July 22, 2013).

Meters & More Alliance. http://www.metersandmore.com/ (accessed July 16, 2013).

NEMA. (2004). ANSI C12.18-2006: American national standard for protocol specification for ANSI Type 2 Optical Port. http://www.nema.org/stds/c12-18.cfm.

NEMA. (2008). ANSI C12.22-2008: American national standard for protocol specification for interfacing to data communication networks. http://www.nema.org/stds/c12-22.cfm.

OPEN Meter Project. (2011). http://www.openmeter.com/.

PRIME Project. (2008). PRIME technology whitepaper: PHY, MAC and convergence layers. http://www.iberdrola.com/webibd/gc/prod/en/doc/MAC_Spec_white_paper_1_0_080721.pdf.

Sikora, A. and Lill, D. (2012). Design, simulation, and verification techniques for highly portable and flexible wireless M-bus protocol stacks. *International Journal of Smart Grid and Clean Energy*, 1(1), 97–102.

Snyder, A.F. and Stuber, M.T.G. (2007). The ANSI C12 protocol suite—Updated and now with network capabilities. *2007 Power Systems Conference: Advanced Metering, Protection, Control, Communication, and Distributed Resources* Clemson, SC, (pp. 117–122). IEEE. doi: 10.1109/PSAMP.2007.4740906.

The Flag Protocol. http://www.theflagprotocol.com (accessed July 14, 2013).

tLZ Project Group. (2009). SyM² clocked load profile meter general specification. http://www.vde.com/en/fnn/extras/sym2/Infomaterial/Pages/SyM2-The-clocked-load-profile-meter.aspx.

Tonello, A.M., D'Alessandro, S., Versolatto, F., and Tornelli, C. (2011). Comparison of narrow-band OFDM PLC solutions and I-UWB modulation over distribution grids. *2011 IEEE International Conference on Smart Grid Communications (SmartGridComm)* (pp. 149–154). IEEE. Brussels, Belgium doi: 10.1109/SmartGridComm.2011.6102308.

Willig, A. (2008). Recent and emerging topics in wireless industrial communications: A selection. *IEEE Transactions on Industrial Informatics*, 4(2), 102–124.

Wisy. (2008). Smart Message Language Version 1.02. http://www.vde.com/EN/FNN/EXTRAS/SYM2/INFOMATERIAL/Pages/SML-specification.aspx.

ZigBee Alliance. (2012). ZigBee smart energy profile overview. http://www.zigbee.org/Standards/ZigBeeSmartEnergy/Overview.aspx (accessed July 23, 2013).

61

Communication Protocols for Power System Automation

Peter Palensky
Austrian Institute of Technology

Friederich Kupzog
Austrian Institute of Technology

Thomas Strasser
Austrian Institute of Technology

Matthias Stifter
Austrian Institute of Technology

Thomas Leber
Vienna University of Technology

61.1 Introduction **61**-1
61.2 IEC Seamless Integration Architecture **61**-2
61.3 Common Information Model **61**-4
IEC CIM • CIM Data Model • Component Interface Specification • CIM Profiles • XML-Based Message Exchange
61.4 Telecontrol Communication **61**-7
61.5 Intercontrol Center Communication **61**-8
61.6 Distributed Network Protocol **61**-9
61.7 Other SCADA Communication Protocols **61**-10
61.8 Data and Communication Security **61**-11
61.9 Modbus **61**-12
61.10 OpenADR **61**-13
61.11 EMIX, Energy Interoperability **61**-15
61.12 Open Platform Communications **61**-15
61.13 Building Automation **61**-16
LonWorks • KNX • BACnet
References **61**-20

61.1 Introduction

The area of power system automation has developed over more than 100 years. Compared to other automation application areas, the automation pyramid is rather large, spanning from field-level controls such as voltage and frequency control at generation units up to wide-area monitoring systems with phasor measurement units and cross-nation SCADA systems for transmission networks. However, due to stability of the architecture and operation principles of power grid over many decades, the job to be done by power grid automation and its underlying communication protocols did not change a lot over long periods. This left room for improvements that became available with technology innovations. The introduction of high-bandwidth wide-area communications was one of these technical improvements.

In recent years, requirements have changed dramatically. The massive integration of renewable energies into the existing infrastructure has changed operation paradigms of transmission and distribution power grids. Automation infrastructure is needed more urgently than ever before. This results in a paradigm shift, where automation systems, which in the past were usually considered as secondary infrastructure, become a central role in the electricity system. Typical tasks for these systems are remote access of field devices, substations, etc., or metering data for billing purposes on the enterprise side. New protocols have developed, dealing, for example, with large-scale meter readings, communication with electric cars, or management of demand response. Compared to the well-established automation protocols, there are manifold approaches and competing solutions with very different levels of standardization.

The typical power system automation architecture starts from operation centers, connecting the user interfaces of the SCADA systems via a wide-area network (WAN) connection typically isolated from other communication networks such as enterprise local-area network (LAN) or Internet with substations and associated field devices (sensors, switches, and protection devices). A wide range of protocols are used within this architecture: IEC 60870-5 within the WAN, Modbus, DNP3, IEC 61850 within substations, RS485, RS232, wired connections, field buses on the field level, and many others. In the following, a light will put on the most widely used candidates among these protocols. There is, however, usually just a very simple paradigm on the application layer: each data point is connected from its physical position like a virtual connection to the SCADA system.

In parallel with such substation automation systems, dedicated infrastructure for remote metering and wide-area sensor networks (e.g., phasor measurement units) are in operation. These systems are usually physically separated from each other for security and dependability reasons. Further, at the control center level, many centralized IT systems exist for applications like billing or energy trading.

A number of communication-related challenges have to be resolved in future smart grids, such as interoperability (see, e.g., the common information model [CIM]), security, and infrastructure costs. Most of these challenges are not caused by the centralized parts, but happen at the interface between centralized and distributed controls. First of all, the smart grid automation system is far from being homogeneous. The IEEE Guide for Smart Grid Interoperability describes the smart grid as a system of systems [26]. Many of the relevant subsystems already exist today. Others are currently emerging and driven by new smart grid use cases (such as electric mobility operators or load aggregators). From the functionality side, there is a strong trend toward interconnection of currently separated systems (e.g., metering infrastructure and grid operation networks) in order to use meters as field sensors. Also from an economic viewpoint, setting up dedicated infrastructure for different applications is not favorable. Currently, the rollout of smart grid technologies is being held back by significant infrastructure costs. Multiple-use and modular extension capabilities are key factors to reduce investment costs and to avoid stranded costs. However, from a security and dependability viewpoint, this convergence of infrastructure is not always ideal. Most discussions and contributions in the area of power grid communication systems are currently driven by this conflict.

61.2 IEC Seamless Integration Architecture

As already mentioned in the introduction, the area of power system automation is characterized by the usage of various approaches and communication protocols. The interoperability and data exchange between the different solutions, automation systems, and control devices is a highly important requirement.

The International Electrotechnical Commission (IEC) describes in the technical report (TR) IEC TR 62357 "Power system control and associated communications: Reference architecture for object models, services and protocols" [25] of the Technical Commission (TC) TC 57 "Power Systems Management and Associated Information Exchange" different IEC standards families and proposals, which provide the basis for the implementation of a Seamless Integration Architecture (SIA) for the power generation, transmission, and distribution domain. Moreover, this IEC TR should also help to identify inconsistencies in the usage of these standards and to identify open gaps, which should be filled by the development of new IEC standards in the power and energy domain. An overview of the IEC SIA is provided in Figure 61.1.

IEC TR 62357 describes all existing object models, services, and protocols and their interrelation, which have been suggested by the IEC TC 57 for the domain of power generation, transmission, and

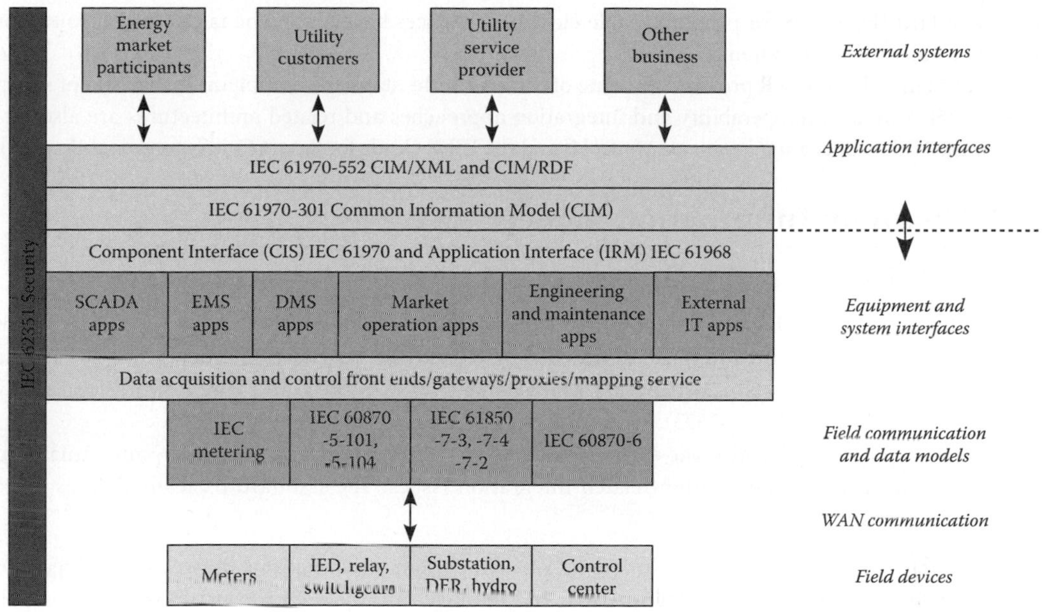

FIGURE 61.1 IEC TR 62357: TC 57 Seamless Integration Architecture. (From Appelrath, H.-J. et al., *Future Energy Grid*, Springer DE, 2012; IEC TC 57, IEC/TR 62357: Power system control and associated communications—Reference architecture for object models, services and protocols, Technical Report IEC/TR 62357, International Electrotechnical Commission, Geneva, Switzerland, 2003.)

distribution. It covers interfaces, data models, exchange formats, and communication protocols from the field up to the business process layer. This includes the following core IEC standard families for the realization of smart grids:

- IEC 61970/61968: The CIM provides a kind of domain ontology for power transmission and distribution systems represented in UML.* In addition, it defines also an exchange format and protocol for network model data using XML† and RDF.‡ Further details of this important energy systems approach are given in Section 61.3.
- IEC 61970: Deals with the information exchange of energy management system (EMS) used for the automation of power transmission grids. It defines especially an application programming interface (API) for EMS.
- IEC 61968: Similar to IEC 61970, this standards family covers the information exchange between distribution management system (DMS) and also defines a corresponding API for these kinds of automation systems.
- IEC 61850: Covers the interoperability of substation and power utility automation systems by the definition of an object-oriented modeling approach and high-level communication protocols. Moreover, it defines an XML-based description language—called Substation Configuration Language (SCL)—for the modeling of the automation equipment and corresponding devices.
- IEC 62351: Defines security concepts and mechanisms for protecting critical infrastructure in the power and energy domain against cyber attacks.

In addition, the IEC 60870-5 remote control approach (for details, see Section 61.4) and the IEC 60870-6 for intercontrol center communication (details see Section 61.5) as well as the well-known functional

* Unified Modeling Language.
† eXtensible Markup Language.
‡ Resource Description Framework.

safety standard IEC 61508 for programmable electronic devices have also to be taken into account for the domain of smart grid systems.

Summarizing, this IEC TR provides the state of the art for the standard-compliant realization of smart grids [10,25]. Similar interoperability and integration approaches and related architectures are also targeted by the NIST *Conceptual Reference Model* [4] and the IEEE Guide for Smart Grid Interoperability [26].

61.3 Common Information Model

61.3.1 IEC CIM

CIM is defined by a series of documents summarized under the three standards for "Energy Management System Application Program Interfaces EMS-API" (IEC 61970), "Application Integration at Electric Utilities—System Interfaces for Distribution Management" (IEC 61968), and "Framework for Energy Market Communication" (IEC 62325).

A typical application of CIM is the exchange of a power system model data between companies and applications, or other interapplication-related integration issues. The standard basically defines three fundamental parts:

1. *Information model*: A canonical model that defines semantic relations between the objects of the power system. It is modeled in UML and managed in SPARX Enterprise Architect® UML modeling tool.
2. *Contextual profiles*: Specific subset of the CIM, for example, Common Power System Model (CPSM).
3. *Implementation models*: Rules for coding CIM in XML/RDF schemas, for example, schema for power system model exchange or schema for messages.

A semantic information model defines the basic vocabulary or terms of the objects. It defines an ontology that represents the knowledge, business concepts, relationships, and a set of rules. The knowledge can be organized in a discrete layer for independent use by other systems and thus is more scalable, maintainable, and manageable.

The context-specific use is restricted by the profile. It restricts the information model and thus is standardized to provide interoperability.

The message syntax is needed to implement the format of the exchanged information. It has to obtain the rules of the format and is covered in different part of the standard.

Besides the data model, the standard series specify interface architectures, general guidelines, implementation profiles, and common services for, for example, data access and exchange, as well as examples.

61.3.2 CIM Data Model

The CIM core data model is specified by parts of three standards:

1. IEC 61970-301 describes a semantic model or domain ontology of all power system components of an electrical system and their relationship. Physical aspects and complex relations of EMS can be modeled with the classes defined in the packages depicted in Figure 61.2.
2. IEC 61968-11 extends it with other aspects of power system information such as asset management, workforce scheduling, or customer billing. Information exchange requirements for distribution systems, which extend the base model, are modeled with the classed defined by the packages depicted in Figure 61.3.
3. IEC 62325-301 describes the framework for energy market communications. Dealing with the market transaction for energy exchange and market operation (both for the US and EU markets) can be modeled with the packages depicted in Figure 61.4.

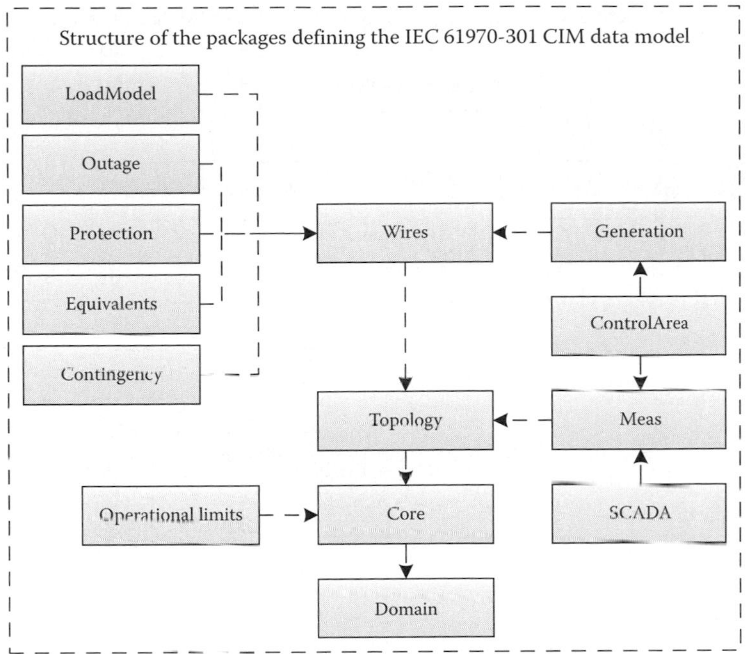

FIGURE 61.2 IEC 61970-301 CIM base data model. (From IEC 61970, Common Information Model (CIM)/ Energy Management, Part 1–405, 2013, www.iec.ch.)

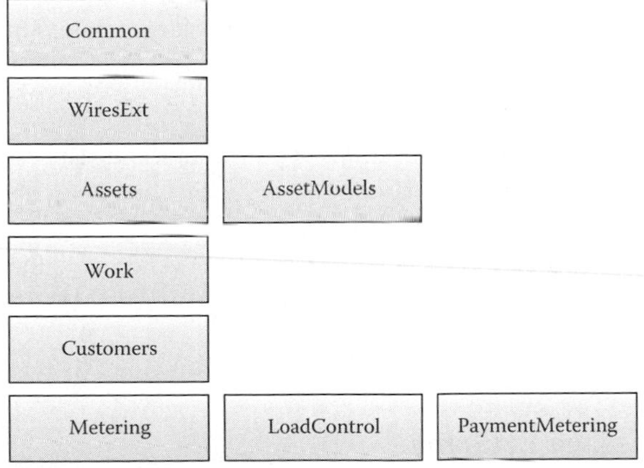

FIGURE 61.3 IEC 61968-11 CIM base data model. (From IEC 61968, Common Information Model (CIM)/ Distribution Management, Part 1–13, 2013, www.iec.ch.)

61.3.3 Component Interface Specification

This part of the standard series deals mainly with interoperability: system integration issues and their implementation. From application-independent services and interface definition language (IDL), application-specific information exchange model to the mapping to specific technologies (e.g., web services [WS] and XML) are defined in the part 400 documents [46].

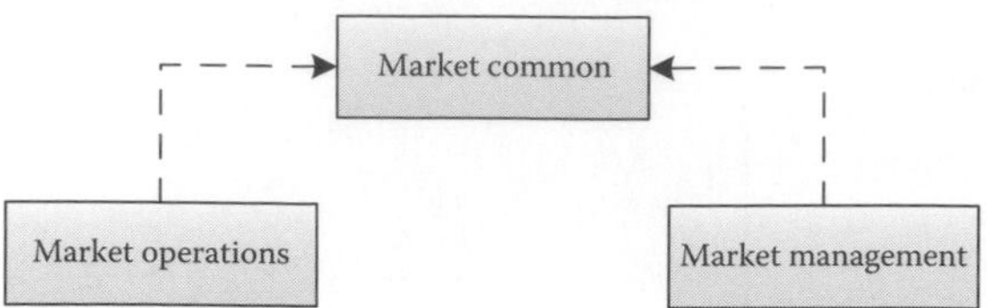

FIGURE 61.4 IEC 62325-301 CIM base data model. (From IEC 62325, Framework for Energy Market Communications, Part 1–502, 2013, www.iec.ch.)

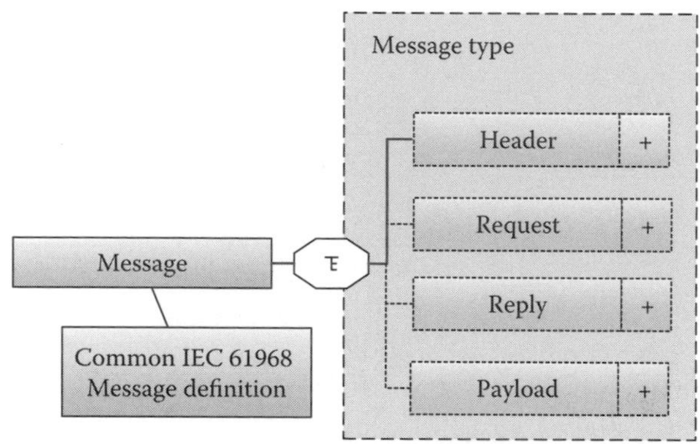

FIGURE 61.5 IEC 61968-11 Generic message structure. (From IEC 61968, Common Information Model (CIM)/Distribution Management, Part 1–13, 2013, www.iec.ch.)

61.3.4 CIM Profiles

Profiles represent a subset of the CIM model. Since CIM is a very large model, these subsets limit the scope to an application-specific scenario, contextual or domain model [46].

Examples of profiles are as follows:

- CPSM (IEC 61970-452). Execute EMS applications of power flow and state estimation.
- Planning Profile Model for transmission planning power flow. Used to exchange info about contingency, load forecast, scheduling, and operation limits.
- Dynamic stability analysis. Handling of dynamic parameters of vendor dynamic stability packages.
- Common Distribution Power System Model (IEC 61968-13) (CDPSM). Used to execute DMS applications.

61.3.5 XML-Based Message Exchange

Due to the loosely coupled heterogeneous infrastructure of many different power system applications and protocols, the purpose of using CIM is to achieve interoperability and format independency of the communicating subsystems. The proposed middleware architectures based on message technologies like WS and Enterprise Service Buses using common integration technologies. The message structure is defined starting from a generic level 5 down to application-specific definitions for, for example, metering data exchange [46] (Figure 61.5):

- Header: basically required and using a common structure of the service interfaces
- Request: optional with qualifier, for example, *get, close, cancel*

- Reply: only required for response messages with status indicators
- Payload: only required when transporting message information specified in the header

61.4 Telecontrol Communication

The IEC 60870-5 standards family describes an open communication approach for industrial automation, which is being applied in (critical) electricity infrastructures such as substations and power grids, but it can generally be also applied to other domains. With IEC 60870-5, it is possible to describe the functions of telecontrol equipment, which is used in geographically widespread processes controlled by SCADA systems. The goal of this specification is to have a standardized approach for transmitting telemetry control and information data in such an automation infrastructure [1,23,36,43].

This protocol specification is based on the enhanced performance architecture (EPA) [21] definition, which describes a three-layer architecture. Originally, IEC 60870-5 was mainly designed for low-bandwidth serial communication, but over the time, the specification has been adopted to be also transmitted over IP networks using TCP. An overview of the IEC 60870-5 protocol specification is provided in Figure 61.6.

IEC 60870-5 has been defined for different purposes, which are described in several parts of this standards family. The following list provides a brief overview of the application areas and the corresponding specifications:

- IEC 60870-5-101: This part of the IEC 60870-5 standards family describes general transmission protocols that have been developed for basic telecontrol tasks. It is mainly used for power system monitoring, control, and the associated communication for control and protection functions in electric power systems and its corresponding components.
- IEC 60870-5-102: It defines a protocol for the transmission of meter data in power systems. This part of the IEC 60870-5 standards family is currently not widely used.

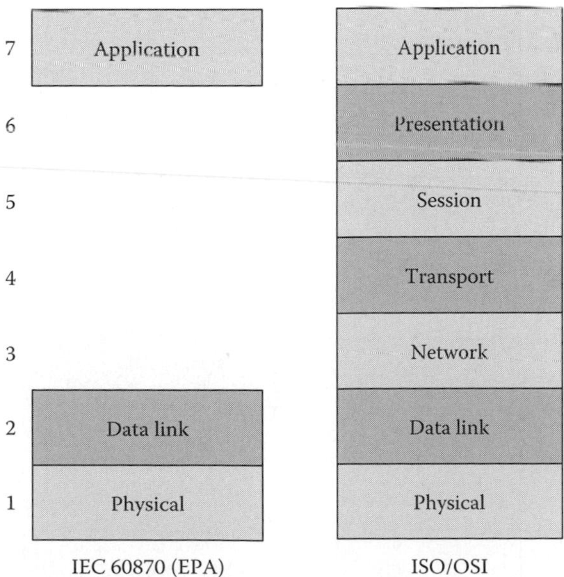

FIGURE 61.6 IEC 60870-5 three-layer architecture based on EPA specification.

- IEC 60870-5-103: It specifies a protocol for the interoperability of protection equipment/devices and the automation system in substations.
- IEC 60870-5-104: This part of the standard describes the extension of the IEC 60870-5-101 protocol specification for its usage in IP networks. The data transmission is accomplished by encapsulating IEC 60870-5-104 messages in TCP/IP packages.

61.5 Intercontrol Center Communication

Intercontrol Center Protocol (ICCP)—also known as Telecontrol Application Service Element 2 (TASE.2)—is an application-layer protocol specifications, which is used for the data exchange between control centers of electric utilities in WAN and LAN. It was developed by a consortium lead by the Electric Power Research Institute (EPRI) [5]. Today, ICCP/TASE.2 is standardized by IEC in the corresponding IEC 60870-6 series [25].

One very interesting point of this specification is the fact that it is independent of an execution platform/operating systems as well as from the underlying communication layer. In other words, ICCP/TASE.2 is independent of the transport protocol. The protocol specification is based on the ISO 9506 Manufacturing Messaging Specification [2], which is also used by the IEC 61850 interoperability approach for power utility automation [25] and the Association Control Service Element. Originally, ICCP/TASE.2 was developed using only ISO-compliant transport protocol standards, but today, also implementations using packet-oriented approaches like TCP/IP and UDP/IP are very common. An overview of ICCP/TASE.2 in the ISO/OSI model is provided in Figure 61.7.

The ICCP/TASE.2 functions for the intercontrol center data exchange are defined in nine groups called *Conformance Blocks*, whereas the implementation of the first group (i.e., basic services like DataValue, DataSet, and DataSet Transfer objects) is mandatory. The other groups providing more advanced functions and services are related to condition monitoring, messages/alarms, program control, etc.

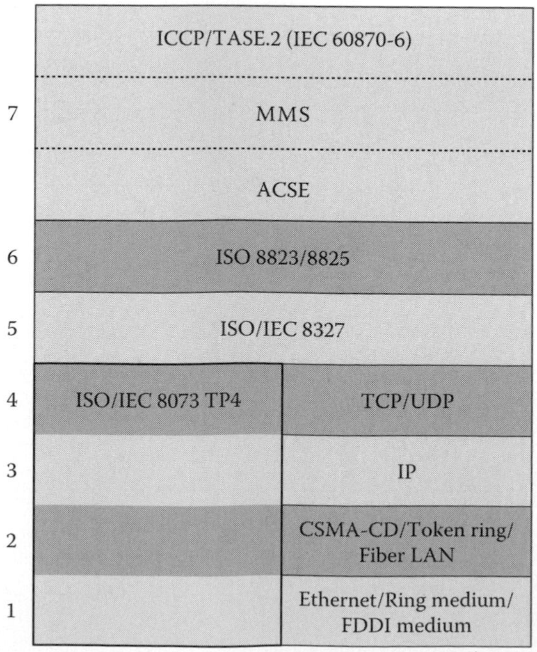

FIGURE 61.7 Overview of the ICCP/TASE2 specification using the ISO/OSI model.

61.6 Distributed Network Protocol

The Distributed Network Protocol Version 3 (DNP3) is a communication approach that has been developed for the information and data exchange in process automation (PA) systems. It is especially being used by electric and water utilities managing their distribution networks. The goal of DNP3 is to connect various devices and equipment used in SCADA systems covering control centers, RTUs, and intelligent electronic devices (IEDs) [14,22].

Originally developed by Westronic (now GE Harris), DNP3 is managed today by the DNP Users Group.* This SCADA communication approach is based on an early version of the IEC 60870-5 telecontrol specification (details see Section 61.4), but it is not really compatible with it anymore. Since 2012, the DNP3 specification is available as IEEE 1815-2012 standard [6].

Like other SCADA protocols, DNP3 is also based on the EPA specification. Compared with the ISO/OSI model, EPA is a reduced specification covering only the application layer, the data link layer, and the physical layer (normally based on serial communication using RS 232 and RS 485). As a main difference to the EPA and the IEC 60870-5 specification, an additional pseudotransport layer is introduced in DNP3. Figure 61.8 provides an overview of the DNP3 four-layer architecture based on the EPA specification.

The physical layer is not really specified in the DNP3 approach, but general guidelines and rules for networked applications are provided. Moreover, for the avoidance of collisions due to several sending devices, corresponding concepts are defined. Normally, RS 232 or RS 485 serial communication is used by various DNP3 implementations. It is, in principle, also possible to tunnel DNP3 messages over packet-oriented Ethernet networks using TCP/IP or UDP/IP.

Asynchronous and synchronous communication is supported with the DNP3 data link layer. Since RTUs or IEDs support advanced functions and messages that are larger than the maximum data link frame length, a pseudotransport layer has been introduced in the DNP3 specification as mentioned earlier. This additional layer is responsible for message assembly and disassembly and provides corresponding services. With the application layer, various operations on data objects provided by the devices (e.g., RTU and IED) are accomplished. This includes reading and writing values.

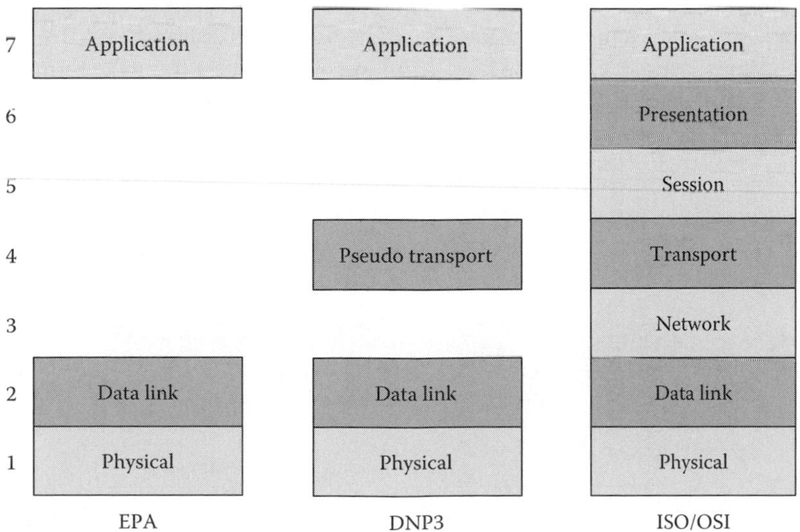

FIGURE 61.8 DNP3 four-layer architecture based on EPA specification.

* http://www.dnp.org.

In general, the following three simple communication modes are supported by the DNP3 specification [16]:

1. *Unicast transaction*: A master (e.g., control center) sends out a request message to a specific device (e.g., RTU and IED), whereas this device usually responds with a corresponding message.
2. *Broadcast transaction*: In difference to the unicast transaction, a master sends out a request message to all field devices, whereas the outstation devices are not responding to this request.
3. *Unsolicited responses*: They are sent by the field devices in order to provide periodic updates or alerts (e.g., a measurement value exceeds a threshold).

61.7 Other SCADA Communication Protocols

Besides the commonly used and standardized SCADA system communication protocols like IEC 60870-5-101/-104 (for details, see Section 61.4), IEC 60870-6 (details are described in Section 61.5), IEC 61850, and DNP3 (for details, see Section 61.6) also Utility Communications Architecture (UCA) as well as further legacy approaches like Modbus RTU/TCP (for details, see Section 61.9, RP-570), Profibus, and OPC-DA (for details, see Section 61.12) is still applied in industrial systems [12,14,28]. A short summary of those legacy communication protocols is given as follows:

- UCA was an attempt by EPRI and IEEE to define a standardized communication protocol for electric utilities. The development of this approach was started in the beginning of the 1990s. Initially, the focus was on control center communication as well as substation to control center communication (Ed. 1.0). The EPRI/IEEE consortium focuses with version 2.0 on the definition of a common station bus. At the same time, IEC TC 57 started with a similar goal and came out with the well-known IEC 61850 interoperability approach for substation (Ed. 1.0) and power utility automation (Ed. 2.0). In 1997, both groups (i.e., EPRI/IEEE and IEC) joined their forces together resulting in today's IEC 61850 specification. Therefore, UCA 2.0 can be considered as a subset of IEC 61850 [12,15].
- RP-570 (i.e., RTU* protocol based on IEC 57 part 5-1† version 0 or 1) was originally developed in the 1990s by ABB as industrial communication protocol for the data exchange between a front-end computer and substations. Therefore, it can be considered as legacy SCADA communication protocol, and it is based on the low-level definition of IEC 60870 using the format class 1.2 [27]. Figure 61.9 shows the RP-570 protocol specification.

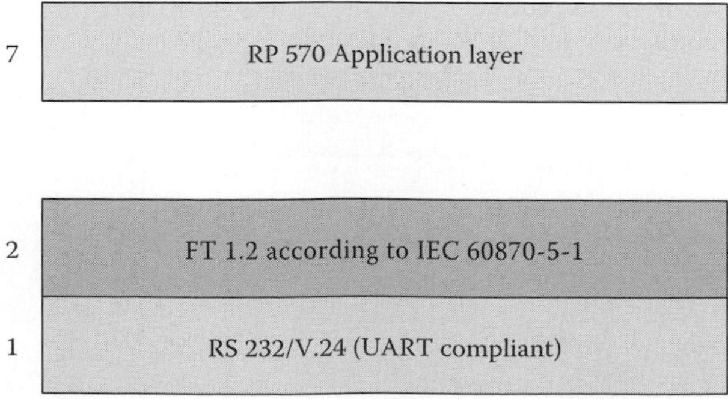

FIGURE 61.9 RP-570 protocol according to the OSI reference model.

* Remote Terminal Unit.
† Present IEC 60870.

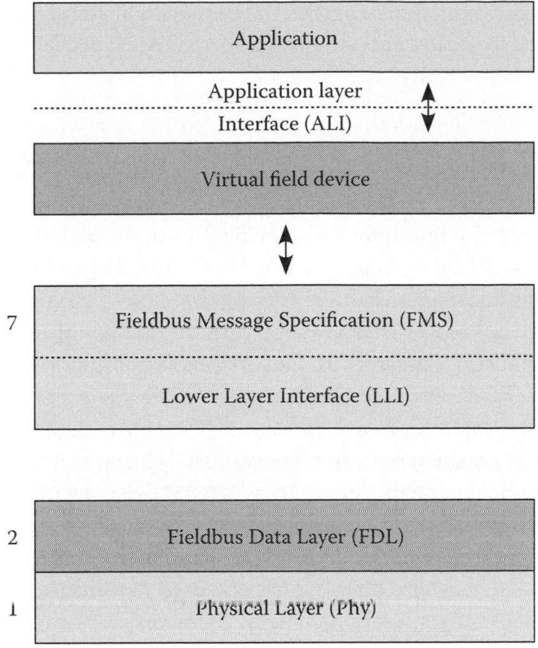

FIGURE 61.10 PROFIBUS protocol according to the OSI reference model.

- Process Field Bus (PROFIBUS) PA originally developed by Siemens is a communication protocol for the data exchange between measurement and control devices (e.g., RTU, PLC*) and the SCADA system. It uses the same protocol specification as PROFIBUS decentralized peripherals, which is applied as process field bus for connecting devices on the field layer [19]. An overview of the PROFIBUS protocol specification is provided in Figure 61.10.

In the near future, WS-based communication approaches become widely used [35]. A very promising candidate that is based on this technology is OPC Unified Architecture (OPC-UA) [33]. This upcoming industry standard for machine-to-machine communication uses a service-oriented architecture and is specified in a platform-independent way. This allows implementing it in various embedded devices, RTUs, PLCs, IEDs, as well as in industry PCs. In the IEC 62541 standards family, this important communication approach for the automation industry is defined. Also the mapping of the CIM and IEC 61850 specifications to the OPC-UA object model are currently ongoing (research) activities in the power and energy domain [32,42].

61.8 Data and Communication Security

A very important issue in SCADA communication is the protection of the data and information exchange between the power system components, devices, and the control center against cyber attacks [17,24,47]. Therefore, the IEC introduced the IEC 62351 series dealing with information security for power system automation and control applications (e.g., EMS, DMS, distribution automation, smart homes, and automatic meter readings) [3]. It mainly addresses secure communication for the IEC TC 57 developments IEC 60870-5, IEC 60870-6, IEC 61850, IEC 61970, and IEC 61968.

With this important IEC standard, various security mechanisms and objects are introduced. This includes the authentication of data and information transfer using digital signatures. Only authenticated

* Programmable Logic Controller.

access allows controlling the power system and its connected components. Moreover, the prevention of eavesdropping as well as playback and spoofing is supported. Also mechanisms for intrusion detection are provided by this IEC cyber security specification.

61.9 Modbus

Modbus is a flexible and widespread protocol for supervision and control of automation equipment. It is flexible in terms of its data access philosophy, and also in terms of the underlying network technology and topology. In short, Modbus is able to operate over serial lines (EIA 232 and EIA 485) as well as over TCP/IP networks. Modbus communication is client–server–oriented (with multimaster capability) and implements a read/write model of bits or words in a Modbus-specific address space. Modbus/TCP is published in the form of a de-factor industry standard with the Modbus Organization hosting the current specification of the protocol.

Modbus is intended as an application-layer protocol. It supports different types of underlying networks. Most commonly used are the serial variants over EIA 232 and EIA 485, where a Modbus connection can factually be established over a minimum of two (four) twisted wires, as well as the Modbus over TCP/IP version, with Ethernet as underlying physical layer (see Figure 61.11). For high-speed data transfer, the Modbus specification proposes Modbus plus, which is a token-passing network. Since Modbus stays on application level, other physical links can also be used to transmit Modbus data packets.

The Modbus message structure is straightforward; it consists of a function code and an optional data block. The message length can vary. Its maximum is inherited from the first physical layer implementation that was used with Modbus, which was EIA 485 with its limitation to 256 bytes for each protocol data unit (PDU). In Modbus, the pair of function code and data block is referred to as PDU. For communication over a specific network technology, the PDU is extended by a leading protocol-specific addressing block and a trialing error-correction code. The PDU together with these additions is called application data unit.

Modbus has a broad range of applications; these can be found mainly in the field level of power grid automation systems. Modbus is generally used to connect sensors and actuators to a local station's automation bus and thus make them available to the local station SCADA system or even a remote SCADA application. In some cases, the station bus itself is realized as Modbus/TCP bus. In other cases, Modbus is only used for a subset of station equipment.

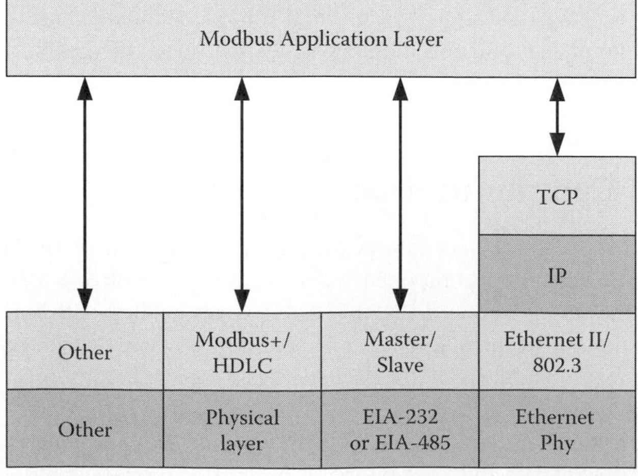

FIGURE 61.11 Modbus protocol stack varieties. (From The Modbus Organisation, Modbus application protocol specification v1.1b3, Technical report, The Modbus Organisation, 2012.)

As an application example, the case of an on-load tap changer transformer (transformer with selectable winding ratio) shall be considered. State-of-the-art tap changer controllers often feature Modbus interfaces. The transformer manufacturer defines a number of function-specific Modbus registers. The current tap position as well as the minimum and maximum tap setting can, for example, be held in individual read-only registers. In order to change the tap position, one could choose a solution where just a single register with the destination position is written. For save operation, it is however useful to use two registers, where the first holds the desired position and the second enables the position change. A typical tap changer controller would also feature many more sensor registers, with line currents, nominal values, etc., as well as different mode registers with delay times, etc. By reading and writing the register set via Modbus, all functionalities of the tap changer transformer can easily be accessed.

61.10 OpenADR

In recent years, one can observe a worldwide trend of utilizing the flexibilities in electrical demand for maintaining the demand–supply balance and management of power grid congestions. Traditionally, it is the task of a few but large bulk generation units to fulfill these functions, but more and more involvement of the demand side in the electric power grid can be observed. These measures are referred to as demand-side management or demand response [30,39]. The drivers for this development are manifold and range from aging infrastructure, restricted generation capacities up to partial loss of controllability on the generation side due to a growing share of renewable energies in the generation mix. While the first demand response schemes were based on telephone calls, where customers were asked in advance to change their load, the first automated demand response (ADR) experiments in the United States were conducted in the 1990s [34].

In order to implement schemes in which electric consumers can offer their flexibility as a service, an appropriate communication system needs to be in place. This is exactly where OpenADR comes into the game. OpenADR is an application-level standard for communication between entities that are required to support demand response, especially utilities, aggregators, and commercial buildings. The standard has been worked out by Demand Response Research Center (DRRC), which itself is managed by US Lawrence Berkeley National Laboratory. It defines a domain-specific data model and different options for message exchange sequences based on this data model. The protocol uses Internet protocols and WS.

In contrast to other ADR protocols, OpenADR's data model does not reference any specific devices. It rather defines a flexible and device-independent set of methods to issue demand response. This can be a concrete request to reduce load by x kW in 2 h as well as the communication of a time-varying energy tariff. The energy customer does always have the possibility for an opt-out, that is, in critical situations, electrical appliances can remain operative [34]. OpenADR does not specify the way utilities interact with their customers. It rather provides a standardized tool set, from which the utility can choose the elements that support the way the specific demand response scheme is designed.

This philosophy is reflected in the way OpenADR communication is specified. The basic entity for the communication between the demand response automation server and the clients is the *EventState*. An event state contains specific information for a demand response event. Since the design of the demand response program can vary significantly depending on the ideas and requirements of the utility company that defines the program, the data items in an EventState are very generic. It usually has a specific time period associated to it in which the event is active. The demand response event itself can be a special energy tariff for the time period or a direct request to reduce load (in categories *moderate*, *high*, and *special* [41], or some other means to specify a demand response event). Using this EventState, events can be scheduled as well as activated immediately.

The data fields of an EventState are the following [41]:

- EventState ID
- Associated demand response program

- Identifier for the event
- Notification time
- Start time
- End time
- Event info instances, etc.

Figure 61.12 depicts a straightforward OpenADR application scenario. The Demand Response Automation Server organizes the EventStates and sends them out to the associated sites, where demand can be remotely managed. The server can use a push or pull model for communication with the sites and therefore is not always a server in the strict sense of the client–server model [41]. The reception of the event state is confirmed by the sites. The on-site OpenADR client can be either an embedded system with simple logic or a PC-based solution with more computation power. The OpenADR specification therefore distinguishes simple and smart clients. The EventState format does differ for the two client types, so that embedded clients do not have to perform extensive parsing and event scheduling [41].

OpenADR does also allow realizing load aggregators. A load aggregator aggregates a number of (flexible) customers and represents them as one large customer to the utility. In this case, the aggregator's OpenADR software is a client toward the utility and a demand response automation server toward the aggregator's contractors.

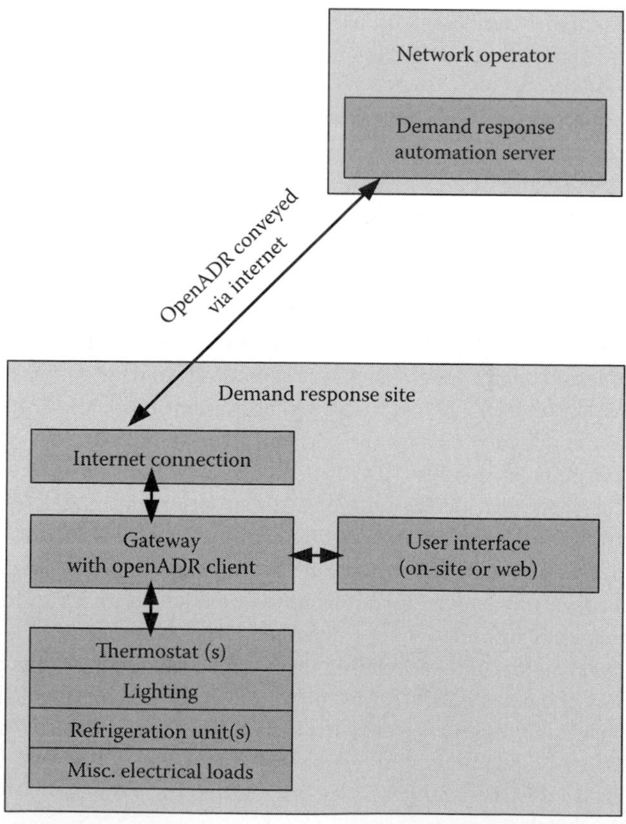

FIGURE 61.12 Typical OpenADR application architecture (From Page, J., Kiliccote, S., Dudley, J.H., Piette, M.A., Chiu, A.K., Kellow, B., Koch, E., and Lipkin, P. Automated demand response technology demonstration project for small and medium commercial buildings Ernest Orlando Lawrence Berkeley National Laboratory, Berkeley, CA, 2011.).

61.11 EMIX, Energy Interoperability

The Organization for the Advancement of Structured Information Standards (OASIS) has formed its Energy Market Information Exchange Technical Committee (EMIX-TC) in order to develop a standard for electricity prices, associated to energy services like power quality or a certain schedule. The outcome is EMIX Version 1.0, available via OASIS. The purpose of this standard is to provide an XML-based language to express [37]:

- Price information
- Bid information
- Time for use or availability
- Units and quantity to be traded
- Characteristics of what is to be traded
- Deal, bid, and acceptance confirmations

If, for instance, a building or a factory participates in a load-shed bidding process, it can use EMIX-based communication to administer the process. EMIX is strongly linked to OASIS WS-Calendar for specifying time aspects.

As EMIX is as generic as possible, it offers extensions to adapt it to particular energy markets like the power market. With this, it is possible to describe electricity-specific terms like reactive power, transport service, or spinning reserve.

One of the profiles defined by EMIX is the Profile for Transactive Power (TeMIX) that defines power delivery in blocks of constant duration (e.g., hourly blocks). Using this, electric vehicles or buildings can express their capabilities and energy market needs on the fly. Also power quality and ancillary services are covered by EMIX.

61.12 Open Platform Communications

The OPC* standard was originally invented to achieve interoperability between common and also exotic process control system objects and interfaces. Also the integration of data sources in a plant into the plants business logic was a major goal for the OPC Foundation. A normal plant topology consists of a field bus, which interconnects the actor and sensor devices with the control unit and maybe with direct access for the operator console. In a second layer, the plant network interconnects several process management systems and access terminals. In addition, there is a third layer, which interconnects the business logic with the plant network (see Figure 61.13). Each of the field devices and process control system needs its own driver to access the process data. OPC tries to solve this problem by accessing the data sources and databases by an OPC server. The OPC server itself provides a standardized interface for client applications to capture the needed process data. The first standard, OPC-DA (Data Access), is based on Microsoft's COM, DCOM, and OLE and is now at revision 3. Additionally, there are several other specifications to OPC-DA, carried out by the OPC Foundation. The major ones are OPC-A/E, Alarm and Events, and OPC-HDA (Historical Data Access). OPC-DA transports real-time data within the system. The OPC server is not capable of storing historic data. This gap fills OPC-HDA, where capable devices can be accessed to provide historical process data. OPC-A/E is the specification for transporting alarms and events between the devices. It is also based on DCOM and is event driven. Over the years, it has emerged that the relationship of OPC and DCOM is the weak point. DCOM is not capable to communicate between domains and over firewalls, without a loss of flexibility and privacy. The most used way to solve this problem is using specialized programs, which convert the communication from OPC into TCP and vice versa.

* Open Platform Communications.

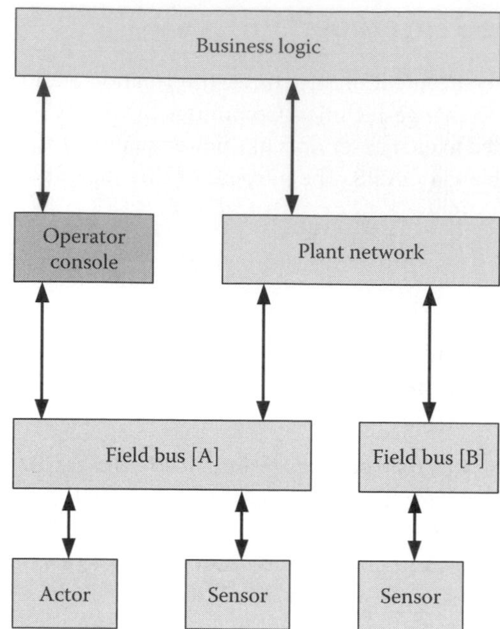

FIGURE 61.13 Typical plant architecture.

61.13 Building Automation

Buildings are responsible for 40% of the energy demand in Western countries. ICT can help to assess, understand, and reduce this demand, both in its dynamics (important for supporting renewable energy sources) and in its total amount (i.e., energy efficiency). One element in implementing the infrastructure for smart buildings is Building Automation and Control Systems (BACSs [11]). BACSs, previously proprietary technologies, dominated by big manufacturers, are now mainly based on public standards and supplied by a large number of small and medium enterprises as well.

Most BACS technologies are application agnostic, but there are still several aspects that expose their relevance for energy-related issues. Applications like demand response [39] and grid-friendly operations [48] are implemented by exploiting the degrees of freedom a BACS offers: shiftable loads, load-shedding, model-based controls [40], smart inverters for local generation, intelligent electric loads [31], etc. The various energy-relevant elements in an automated building are integrated via a BACS network that must fulfill a number of requirements:

- Control cycles/latency of 200 ms (lighting) to 1 min (heating)
- Bandwidth of some kbit/s

These requirements are relaxed and typically easy to achieve with any network technology, wired and wireless. If networks get larger, and if network segments are connected via a backbone, these requirements naturally scale up and might become a challenge. Other requirements that are not so easy to satisfy are as follows:

- Information security: Especially in the case of remote usage and Internet connections.
- Low cost for nodes and communication infrastructure: Some large buildings have more than 100,000 data points.
- Robustness: Self-healing communication channels, reliable transport mechanisms, immunity to electromagnetic interference, etc.
- Interoperability: Multivendor installations in buildings are the standard case, so frictions and errors need to be avoided in a systematic fashion.

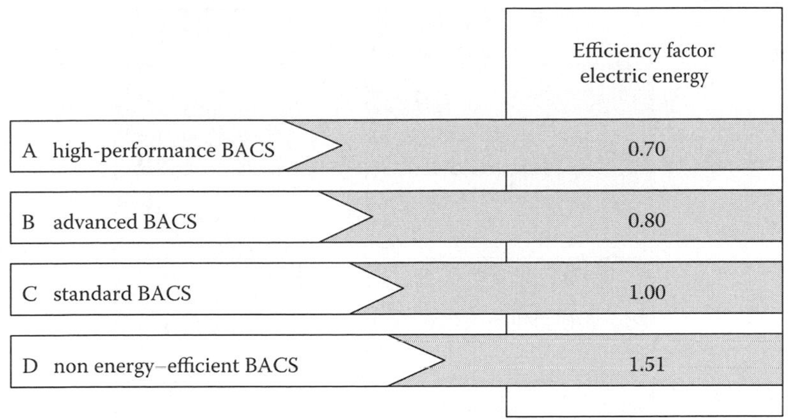

FIGURE 61.14 The impact of BACS onto the energy consumption of an office, based on EN 15232.

Costs are a major factor: Often, energy is simply too cheap to justify expensive investment in ICT [44]. Information security was long neglected, relying on the fact that BACS networks are isolated from hostile public networks and that technologies are proprietary and unknown to the average hacker.

In future, BACS networks will have to incorporate more scalability and flexibility [38], to keep up with the fast pace of IT and Internet business.

Figure 61.14 gives an impression on how BACS can influence the energy balance of buildings. The EN 15232 standard offers evidence-based figures for hotels, schools, and other buildings, depending on how sophisticated their BACS is.

One of the most important aspects of BACS networks in the energy domain is interoperability [29]. The typical two steps to achieve interoperability are to extend the plain protocol specification by standardized data types and standardized functional profiles. They might be called differently in the various BACS technologies (e.g., objects) but are necessary to bring devices of different origin together. Ideally, interoperability is tested and certified [20] in order to allow for clear customer communication.

The following sections will cover the most important BACS network technologies and their relevant aspects for energy applications. Three international standards are LonWorks (ISO/IEC 14908), KNX (ISO/IEC 14543-3), and BACnet (ISO/IEC 16484-5).

61.13.1 LonWorks

LonWorks [13] is a general-purpose control network that has a high popularity in building automation applications. While LonWorks describes the entire technology family, parts of it are sometimes used in other products and technologies. BACnet (see later) for instance can use a LonTalk (the protocol, defined by LonWorks) channel to communicate. Parts of the LonWorks standard were adopted in the Open Smart Grid Protocol (OSGP) [18], an ETSI standard that covers remote metering, demand control, billing, and other utility-relevant functions. OSGP's networking layer is based on part 1 (protocol stack) of EN14908, enhanced by various IT security features implemented on the application layer. The classical usage is, however, to use all LonTalk layers plus the interoperability rules, developed by LonMark, a nonprofit organization for the promotion and improvement of LonWorks.

LonWorks implements all seven ISO/ISO protocol layers in its LonTalk protocol stack (Figure 61.15). Most remarkable is a sophisticated Layer 2 with collision avoidance and transport channels with configurable retries (if packets are not acknowledged by the recipient). It also offers simple information security via an authentication mechanism.

LonMark offers a number of functional profiles for energy use. Examples are Utility Meter (profile #2201) for meter reading, Automatic Transfer Switch (profile #13120) for selecting a power source

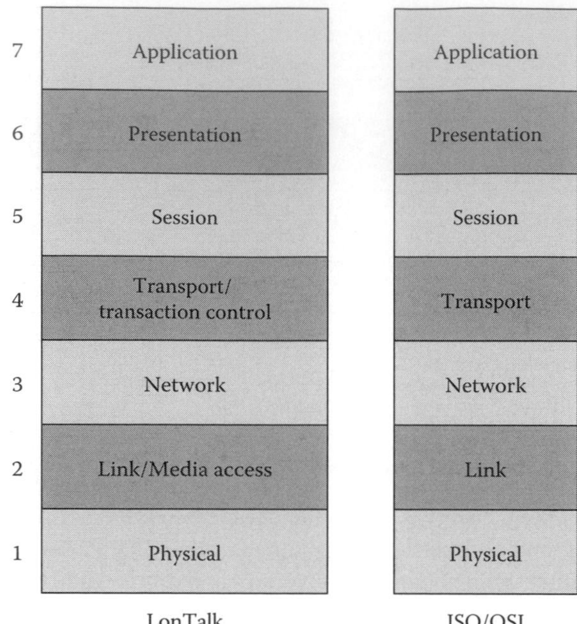

FIGURE 61.15 The LonWorks Protocol Stack.

in case of failure, and the Generator Set (profile #13110) for power sources. Other profiles are under development, like Wind Turbine Generator, Demand-Response Load Actuator, and the Photovoltaic Converter profile.

As Figure 61.16 shows, a functional profile consists of a selection of network variables, based on standard network variable types. The given example shows a meter profile where current and historic meter values can be retrieved.

The physical layer of LonWorks systems can be twisted pair, power line, radio frequency, and much more. Large buildings typically have an IP or Ethernet-based backbone for its LonWorks network.

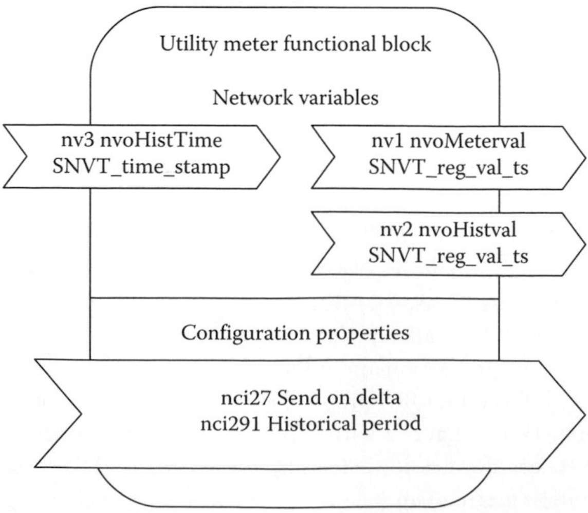

FIGURE 61.16 The LonMark Utility Meter profile (only the mandatory part).

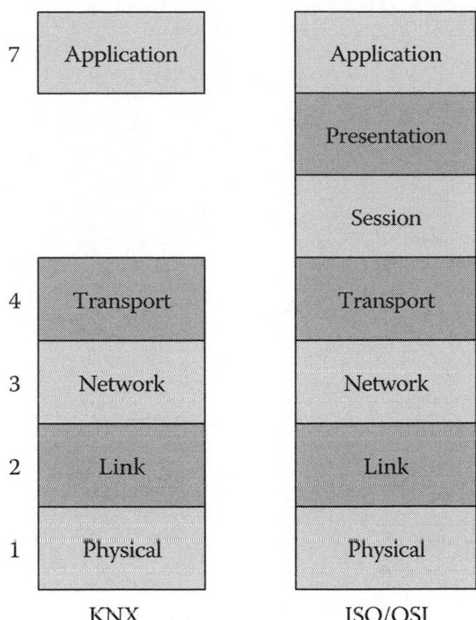

FIGURE 61.17 KNX Stack.

61.13.2 KNX

KNX, the fusion of three earlier technologies (EIB, EHS, and BatiBUS), is specialized on home and building automation [29]. Its original topology is tree-based, new media like Ethernet, however, allow for other topologies as well. KNX supports several physical media; the most popular are twisted pair and Ethernet. Data types for load management are under development, but there are many other data types that are of relevance for energy applications like *HVAC Operating Mode* or *Boiler Controller Demand Signal*. There are products out for meter reading, M-Bus data collection, and other energy-relevant services.

The biggest benefit of KNX is its reduced cost. It is, however, designed for small to medium buildings. A large complex might bring it to its limits. KNX is specialized for building applications, which is why it only implements the most necessary features. It, as can be seen in Figure 61.17, has, for instance, omitted to offer connection-oriented transport, which is practically never needed in BACS applications.

61.13.3 BACnet

BACnet was designed top down, in contrast to its two aforementioned competitors. The applications were the starting point, and transport and wires were done later or not at all: BACnet simply specifies some existing communication standards as selected, one of them even being LonWorks. Building functions were projected into functional blocks, called objects. By doing so, the BACnet designers worked mainly on interoperability questions and (sometimes complex) functions.

One example, relevant for energy automation, is the BACnet load control object (Figure 61.18, Addendum 135-2004e-1). It describes a simple state machine that can represent a device (e.g., a chiller) or an entire building. Its strength is its simplicity: a clear interface that can be incorporated in many energy-relevant processes.

A so-called Protocol Implementation and Conformance Statement (PICS) describes which functions a BACnet device implements. With its PICS, a device can tell that it is a BACnet Building Controller (B-BC), that it supports BACnet/IP as transport, that it contains a schedule object, and many things more, important to compare the device with others. One important part is the BIBBs

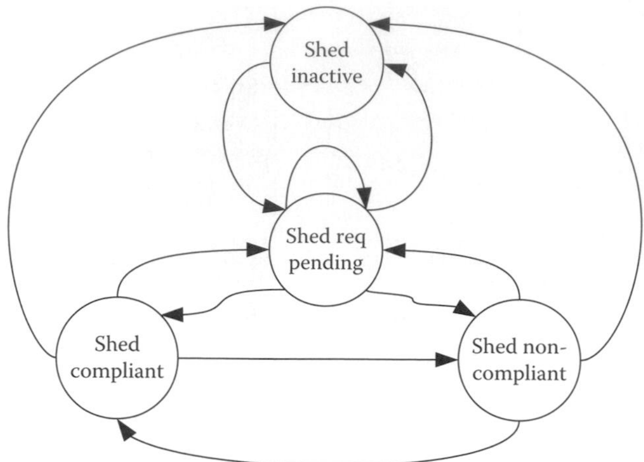

FIGURE 61.18 The BACnet Load Control Object (simplified).

(BACnet Interoperability Building Blocks) that tell which interoperability functions are available. BIBBs are one of six categories (e.g., network management, alarm management, and data sharing), and a device can either act as source of data (client, initiating a function) or as sink of data (server, executing a function).

References

1. IEC 60870-5 Telecontrol equipment and systems—Part 5: Transmission protocols. Technical report, International Electrotechnical Commission, Switzerland, 1990.
2. ISO 9506 Industrial automation systems—Manufacturing message specification. Technical report, International Organization for Standardization, Switzerland, 2003.
3. IEC 62351 Power systems management and associated information exchange—Data and communications security. Technical report, International Electrotechnical Commission, Switzerland, 2007.
4. NIST Framework and roadmap for smart grid interoperability standards. Technical report, NIST Publication 1108, National Institute of Standards and Technology—U.S. Department of Commerce, 2010.
5. Standards and technology adoption case study—Inter-control center protocol (ICCP/TASE.2). Technical report, Electric Power Research Institute, 2012.
6. IEEE standard for electric power systems communications—Distributed network protocol (DNP3). IEEE Std 1815-2012 (Revision of IEEE Std 1815-2010), pp. 1–821, 2012. doi: 10.1109/IEEESTD.2012.6327578.
7. IEC 61968: Common information model (CIM)/Distribution management. Part 1–13, 2013. www.iec.ch.
8. IEC 61970: Common information model (CIM)/Energy management. Part 1–405, 2013. www.iec.ch.
9. IEC 62325: Framework for energy market communications. Part 1–502, 2013. www.iec.ch.
10. H.-J. Appelrath, H. Kagermann, and C. Mayer. *Future Energy Grid*. Springer DE, Heidelberg, Germany, 2012.
11. D. Bruckner, D. Dietrich, G. Zucker, and P. Palensky. Guest editorial building automation, control and management. *IEEE Transactions on Industrial Electronics*, 57(11):2010.
12. K. Caird. Integrating substation automation. *Spectrum, IEEE*, 34(8):64–69, 1997.
13. CEN. En14908: Open data communication in building automation, controls and building management control network protocol, 2010.
14. G. Clarke, D. Reynders, and E. Wright. Practical modern SCADA protocols. *IDC Technologies*, 2004.

15. K.N. Clinard. Comparison of IEC 61850 and UCA 2.0 data models. In *Power Engineering Society Summer Meeting, 2002 IEEE*, vol. 1, pp. 289–290. IEEE, Chicago, IL, 2002.
16. S. East, J. Butts, M. Papa, and S. Shenoi. A taxonomy of attacks on the dnp3 protocol. In *Critical Infrastructure Protection III*, pp. 67–81. Springer, Hanover, New Hampshire, 2009.
17. G.N. Ericsson. Cyber security and power system communication—Essential parts of a smart grid infrastructure. *Power Delivery, IEEE Transactions on*, 25(3):1501–1507, 2010.
18. ETSI. Etsi gs osg 001 v1.1.1 open smart grid protocol (osgp), 2012.
19. M. Felser. The fieldbus standards: History and structures. *Technology Leadership Day*, 2002.
20. P. Fischer and P. Palensky. The importance of being certified—The role of conformance testing and certification of communication systems in building automation and control devices. In *Proceedings of 2004 IEEE Africon, 7th Africon Conference in Africa, Technology Innovation*, vol. 2, pp. 1223–1227. IEEE, Gaborone, Botswana, 2004. doi: 10.1109/AFRICON.2004.1406886.
21. GM MAP Task Force. The MAP enhanced performance architecture (EPA), 1987.
22. DNP Users Group. DNP3 introduction, 2002.
23. C. Hoga. New Ethernet technologies for substation automation. In *Power Tech, 2007 IEEE Lausanne*, Lausanne, Switzerland, pp. 707–712, 2007. doi: 10.1109/PCT.2007.4538402.
24. S. Hong and M. Lee. Challenges and direction toward secure communication in the SCADA system. In *Communication Networks and Services Research Conference (CNSR), 2010 Eighth Annual*, pp. 381–386, 2010. doi: 10.1109/CNSR.2010.52.
25. IEC TC 57. IEC/TR 62357: Power system control and associated communications—Reference architecture for object models, services and protocols. Technical Report IEC/TR 62357, International Electrotechnical Commission, Geneva, Switzerland, 2003.
26. IEEE. IEEE guide for smart grid interoperability of energy technology and information technology operation with the electric power system (EPS), end-use applications, and loads, 10, 2011.
27. S. Jaloudi, E. Ortjohann, A. Schmelter, P. Wirasanti, and D. Morton. Communication strategy for grid control and monitoring of distributed generators in smart grids using IEC and IEEE standards. In *Innovative Smart Grid Technologies (ISGT Europe), 2011 2nd IEEE PES International Conference and Exhibition on*, Manchester, UK, pp. 1–6, 2011. doi: 10.1109/ISGTEurope.2011.6162689.
28. R. Kalapatapu. SCADA protocols and communication trends. *ISA EXPO*, Houstan, TX, 2004.
29. W. Kastner, P. Palensky, T. Rauscher, and C. Roesener. A closer look on today's home and building networks. In *Proceedings of 2004 IEEE Africon, 7th Africon Conference in Africa, Technology Innovation*, vol. 2, pp. 1239–1245. IEEE, Gaborone, Botswana, 2004. doi: 10.1109/AFRICON.2004.1406889.
30. F. Kupzog and C. Roesener. A closer look on load management. In *Industrial Informatics, 2007 5th IEEE International Conference on*, vol. 2, pp. 1151–1156, June 2007. doi: 10.1109/INDIN.2007.4384893.
31. Y.P. Landaburu, P. Palensky, and M. Lobashov. Requirements and prospects for consumers of electrical energy regarding demand side management. In *IEWT 2003*, Vienna, Austria, *12.02.200314.02.2003, in: Die Zukunft der Energiewirtschaft im liberalisierten Markt-Kurzfassungsband (2003)*, pp. 101–102, 2003.
32. S. Lehnhoff, W. Mahnke, S. Rohjans, and M. Uslar. IEC 61850 based OPC UA Communication—The future of smart grid automation. In *17th Power Systems Computation Conference (PSCC 2011)*, Stockholm, Sweden, 2011.
33. W. Mahnke, S.-H. Leitner, and M. Damm. *OPC Unified Architecture*. Springer, Berlin, Heidelberg, Germany, 2009.
34. C. McParland. OpenADR open source toolkit: Developing open source software for the smart grid. In *Power and Energy Society General Meeting, 2011 IEEE*, pp. 1–7, 2011. doi: 10.1109/PES.2011.6039816.
35. A. Mercurio, A. Di Giorgio, and P. Cioci. Open-source implementation of monitoring and controlling services for EMS/SCADA systems by means of web services—IEC 61850 and IEC 61970 standards. *IEEE Transactions on Power Delivery*, 24(3):1148–1153, 2009.

36. S. Mohagheghi, J. Stoupis, and Z. Wang. Communication protocols and networks for power systems—Current status and future trends. In *Power Systems Conference and Exposition, 2009. PSCE'09. IEEE/PES*, Seattle, WA, pp. 1–9, 2009. doi: 10.1109/PSCE.2009.4840174.
37. OASIS. Energy market information exchange (emix) version 1.0, 2012.
38. P. Palensky. Requirements for the next generation of building networks. In *Proceedings of International Conference on Cybernetics and Information Technologies, Systems and Applications (ISAS CITSA 2004)*, Orlando, Fl, pp. 225–230, 2004.
39. P. Palensky and D. Dietrich. Demand side management: Demand response, intelligent energy systems, and smart loads. *IEEE Transactions on Industrial Informatics*, 7(3):381–388, 2011.
40. P. Palensky, G. Zucker, F. Judex, R. Braun, F. Kupzog, T. Gamauf, and J. Haase. Demand response with functional buildings using simplified process models. In *Proceedings of the 37th IEEE Conference on Industrial Electronics*, Melbourne, Australia, 2011.
41. M.A. Piette. Open automated demand response communications specification (version 1.0). Technical Report CEC-500-2009-063, Lawrence Berkeley National Laboratory, Akuacom, 2009.
42. S. Rohjans, M. Uslar, and H.J. Appelrath. OPC UA and CIM: Semantics for the smart grid. In *Transmission and Distribution Conference and Exposition, 2010 IEEE PES*, New Orleans, LA, pp. 1–8, 2010. doi: 10.1109/TDC.2010.5484299.
43. G. Sanchez, I. Gomez, J. Luque, J. Benjumea, and O. Rivera. Using Internet protocols to implement IEC 60870-5 telecontrol functions. *IEEE Transactions on Power Delivery*, 25(1):407–416, 2010.
44. M. Stadler, F. Kupzog, and P. Palensky. Distributed energy resource allocation and dispatch: An economic and technological perception. *International Journal of Electronic Business Management*, 5(4):59–73, 2007.
45. The Modbus Organisation. Modbus application protocol specification v1.1b3. Technical report, The Modbus Organisation, 2012.
46. M. Uslar, M. Specht, S. Rohjans, J. Trefke, and J.M.V. Gonzalez. *The Common Information Model CIM: IEC 61968/61970 and 62325—A Practical Introduction to the CIM (Power Systems)*. Springer, Berlin, Heidelberg, Germany, 2012.
47. Y. Wang, D. Ruan, D. Gu, J. Gao, D. Liu, J. Xu, F. Chen, F. Dai, and J. Yang. Analysis of smart grid security standards. In *Computer Science and Automation Engineering (CSAE), 2011 IEEE International Conference on*, vol. 4, pp. 697–701, 2011.
48. G. Zucker, P. Palensky, F. Judex, C. Hettfleisch, R.-R. Schmidt, and D. Basciotti. Energy aware building automation enables smart grid-friendly buildings. *Elektrotechnik & Informationstechnik*, 129(4):271–277, 2012.
49. Page, J., Kiliccote, S., Dudley, J.H., Piette, M.A., Chiu, A.K., Kellow, B., Koch, E., and Lipkin, P. Automated demand response technology demonstration project small and medium commercial buildings. Ernest Orlando Lawrence Berkeley National Laboratory, Berkeley, CA, 2011.

62

IEC 61850: A Single Standard for the Engineering, Protection, Automation, and Supervision of Power Systems on All Voltage Levels

62.1 Introduction 62-1
62.2 Overview 62-2
62.3 Information Modeling and Information Models 62-7
62.4 Parts and Application Domains of the Standard 62-14
62.5 Functional Description of Applications 62-19
62.6 Information Exchange Models 62-21
62.7 System Configuration Language 62-30
62.8 Overview of Implementation Layering 62-32
62.9 Related International Projects and Application Example 62-37
Gateway for Smart Grid Applications
62.10 UCA International Users Group (UCA IUG) and Quality Assurance 62-40
62.11 Third-Party Solutions for IEC 61850 62-41
References 62-42

Karlheinz Schwarz
Schwarz Consulting Company

62.1 Introduction

Automation, information, and communication systems in power systems are widely used. Usually, they are based on an immense number of proprietary specifications or (de facto) standards. To reduce this variety and to meet future requirements, the standard IEC 61850 (Communication networks and systems for power utility automation) [1] has been developed by users, vendors, and system integrators since 1995.

The standard IEC 61850 provides the basis for many application domains outside of electrical side substations. One of the biggest success stories of IEC 61850 in power delivery systems is the deployment of the standard for wind power systems. Utilities and operators of wind turbines are struggling with the fast-growing diversity of communication systems for wind turbines. Utilities have initiated the standardization of a communication system in 2001: IEC 61400-25 (Communications for monitoring and control of wind power plants) [2]. IEC 61400-25 is based on IEC 61850.

The standards referenced in IEC 61850, for example, Ethernet, TCP/IP, MMS (ISO 9506), and XML, make use of advanced IT solutions, the reduced bandwidth costs, and increased processor capabilities in the end devices. IEC 61850 focuses on process data and metadata. More than 3000 standardized objects with concrete names and type information have been defined. The standard overcomes the huge point lists to be created for each installation in case traditional communication systems are applied.

The focus of standard series IEC 61850 was on substation operational aspects (mainly protection and control) during the first couple of years. Various groups have identified that IEC 61850 could be used as the basis of further applications, for example, monitoring of functions, processes, primary equipment, and the communication infrastructure in substations and many other power system application domains. The scope and title changed in 2010 from "Communication networks and systems for substations" to "Communication networks and systems for power utility automation." This reflects the many new application domains that use the standard.

Revised and new parts provide additional definitions for critical measurements like temperatures, oil levels, gas densities, maximum number of connections exceeded, etc. Such extensions cover the monitoring of equipment like switchgear, transformers, on-load tap changers, automatic voltage regulation devices, gas compartments, and lines; generators, gearboxes, and towers in wind turbines; and communication infrastructure like Ethernet switches and routers. Myriads of sensors are needed to monitor the condition of the wind power foundation, tower, rotors, gearboxes, and generators to name just a few.

IEC 61850 is the standard in power and energy delivery systems.

62.2 Overview

IEC 61850 mainly provides solutions for the following:

- The *information model* of the real world: a circuit breaker model, protection functions, measurement units, nameplate information, condition monitoring information, wave forms, disturbance records, limit violations, etc. Many hundred *signals* are already defined in various standards or draft standards (for substation automation, asset management and monitoring, wind power plants, hydropower plants, and other domain). The models are defined as data object that are contained in logical nodes (LNs)—comprising all typical information that is to be exposed from a real function.
- The *communication services* like get, set, reporting events, logging events, control devices, exchanging GOOSE messages for real-time information, exchanging sampled of current and voltage sensors, recording, transfer of COMTRADE (and other) files, retrieve self-description of the devices *content*, and other services.
- The *mapping* of the information models and communication services to concrete communication protocols like Manufacturing Message Specification (MMS, ISO 9506).
- The *communication networks* (TCP/IP, Ethernet, switches, and routers).
- The *configuration* of the information models and information flow of the whole substation (IEC 61850-6, system configuration language [SCL]) and mainly which devices provide the information to be sent to whom comprising the following:
  - Single-line diagram
  - Binding of the information models to the single-line diagram (lines, current and voltage sensors, circuit breakers, etc.)
  - Binding of the information model to the automation functions and primary equipment
  - Engineering of all communication details: relations between intelligent electronic devices (IEDs) and the message flow (traffic engineering)

The crucial features and characteristics of IEC 61850 are

- Information modeling method (LogicalDevice/LogicalNode.DataObject)
- Information models of real-world objects (LD/LN.DO)
- SCL based on XML
- Information exchange services (abstract definitions that can be mapped to multiple services and protocols)
- Communication services (concrete messages applying Ethernet, TCP/IP, MMS, etc.)
- Applying native standards
- Supporting interoperability of IEDs in small and huge multivendor systems
- Long-term stability of definitions (like AC power system)
- Information models for many application domains (growing quite fast)

The information exchange mechanisms access well-defined information models. IEC 61850 uses the approach of modeling the common information found in real devices as shown in Figure 62.1. All common information made available to be exchanged with other devices is defined. The standard applies an advanced name space concept. The name space allows defining private extensions for the information that is required by an application but not yet standardized.

IEC 61850 defines the information and information exchange independent of a concrete implementation (i.e., it uses abstract models). The standard also uses the concept of virtualization. Virtualization provides a view of those aspects of a real device that are of interest for the information exchange with other devices. Only those details that are required to provide interoperability of devices are defined in IEC 61850.

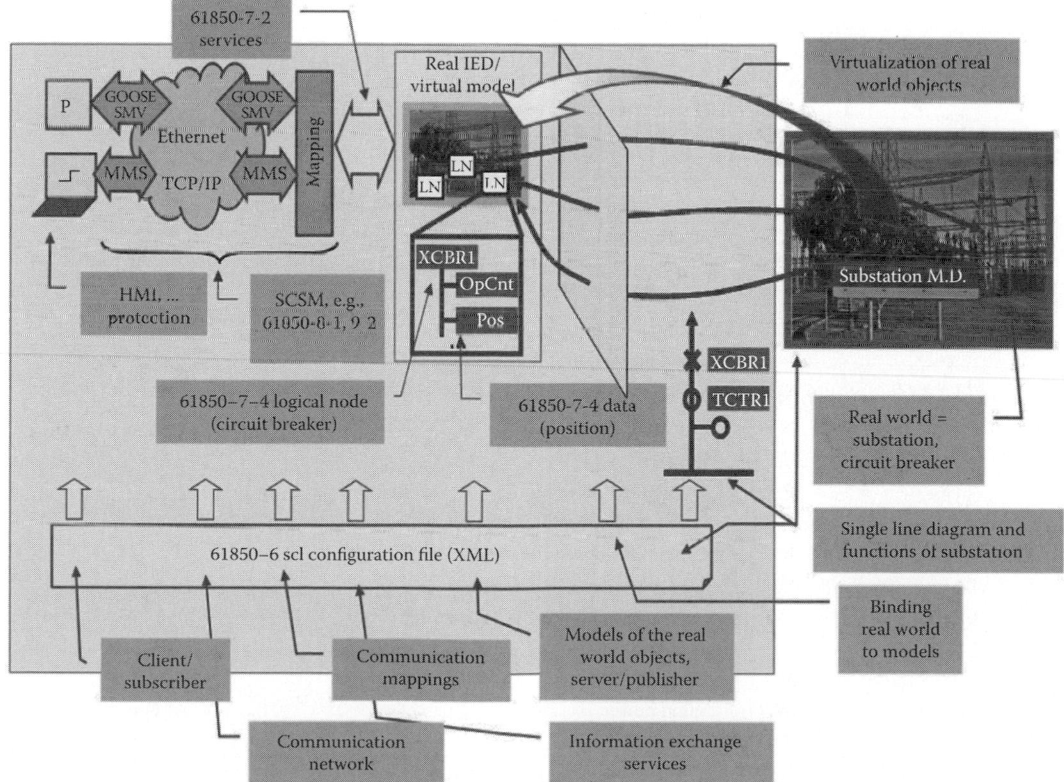

FIGURE 62.1 Topology of IEC 61850.

The approach of the standard is to decompose the application functions into the smallest entities, which are used to exchange information. The granularity is given by a reasonable distributed allocation of these entities to dedicated devices. These entities are called logical nodes (e.g., a virtual representation of a circuit breaker class, with the standardized class name XCBR). Several LNs build a logical device (e.g., a representation of a bay unit). A logical device is always implemented in one IED (intelligent electronic device).

Real devices of an air-insulated substation on the right-hand side of Figure 62.1 are modeled as a virtual model in the middle of the figure. The LNs defined in the logical device (e.g., bay) correspond to well-known functions in the real devices. In this example, the LN XCBR1 represents a specific circuit breaker.

Based on their functionality, an LN contains a list of data object (e.g., Pos for position) with dedicated data attributes. The data objects have a structure and a well-defined semantic (meaning in the context of substation automation systems [SASs]). The semantic is partly very common (status or measurement). Other information is substation specific (directional mode information of a distance protection function). The information represented by the data and their attributes is accessed by various services. The services are implemented by specific and concrete communication means (e.g., using MMS, TCP/IP, and Ethernet among others). The LNs and the data contained in the LNs are crucial for the description and information exchange of SAS to reach interoperability (Table 62.1).

TABLE 62.1 Logical Node Groups

Id	Logical Node Groups
A	Automatic control
B	Reserved
C	Supervisory control
D	Distributed energy resources
E	Enthalpy (steam and gas power plants)
F	Functional blocks
G	Generic function references
H	Hydropower
I	Interfacing and archiving
J	Reserved
K	Mechanical and nonelectrical primary equipment
L	System logical nodes
M	Metering and measurement
N	Reserved
O	Reserved
P	Protection functions
Q	Power quality events detection related
R	Protection-related functions
S	Supervision and monitoring
T	Instrument transformer and sensors
U	Reserved
V	Reserved
W	Wind power
X	Switchgear
Y	Power transformer and related functions
Z	Further (power system) equipment

The logical devices, the LNs, and the data they contain need to be configured. The main reason for the configuration is to select the appropriate LNs and data from the standard and to assign the instance-specific values, for example, concrete references between instances of the LNs (their data) and the exchange mechanisms, and initial values for process data.

Another task of the configuration language is the binding between the model and the real world. It must be known which real circuit breaker in the switch yard is opened or closed when a control command is sent to the model instance XCBR1. The binding is contained in an SCL file for a whole substation.

An example of a change of the circuit breaker position and how this change can be communicated is shown in Figure 62.2. The real circuit breaker that is represented by the XCBR1 model opens at time t_n. This change of the status represented by the model XCBR1.Pos.stVal is an event that could cause sending a spontaneous report message over TCP/IP to a supervision device. Report messages are understood to be non-time-critical messages; the timeliness is in the range of 1 s.

The supervision device may also just read (poll) the current status of the circuit breaker. Another option is to store (log) the change in a local log of the device that implements the object model (e.g., a bay controller). The log (historical storage) could be queried at any time.

The fourth possibility is to distribute the state change in real time to all devices in the substation that need to know continuously what the status of the switchgears are at any time. GOOSE messages (using Ethertype and multicast) are carrying this kind of information within a few milliseconds to all devices that need this information. All devices that have information to be shared with other devices are configured by an SCL file. In case of GOOSE messages, it is required that a device that wants to receive the status change of XCBR1.Pos.stVal is configured to listen to that specific GOOSE messages. The GOOSE messages are just sent to the network not knowing which devices are interested in receiving the values carried; the system SCL document specifies the publisher and which subscriber listens to which publisher.

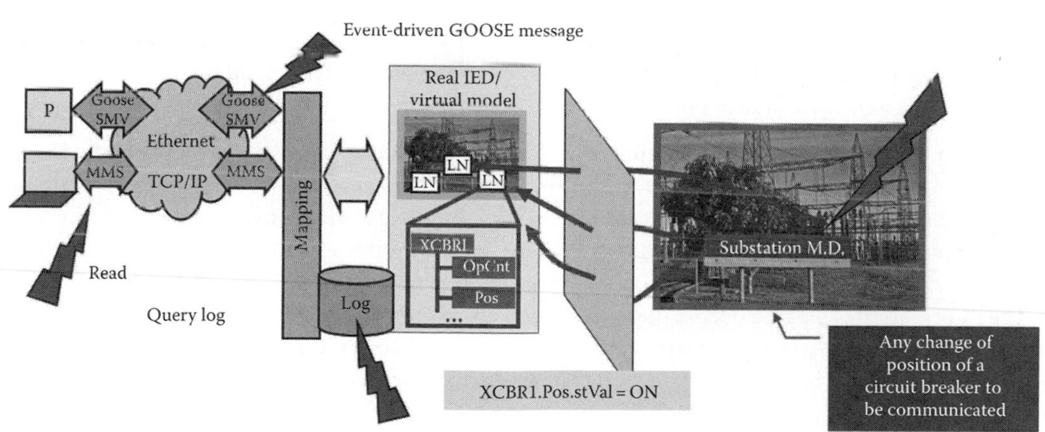

Event-driven or spontaneous reports/GOOSE from server/publisher to client/subscriber (or store entries into IED log or read value)

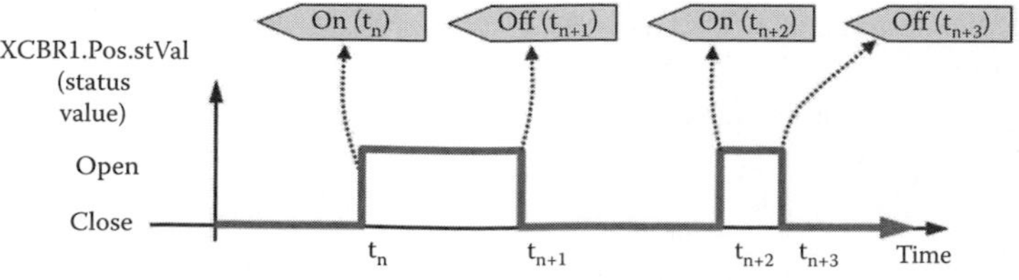

FIGURE 62.2 Communication example.

The four main building blocks of IEC 61850 for substations are (see Figure 62.1) as follows:

Information models

IEC 61850-7-4 defines substation-specific and common information models for substation automation functions (e.g., breaker with the status of breaker position, settings for a protection function, etc.)—what is modeled and could be accessed for information exchange. The models could be used to derive an electronic data sheet embedded in the IED for self-identification and self-description.

IEC 61850-7-3 defines a list of commonly used information templates (e.g., double-point control, three-phase electrical measured value, etc.); these are the most common information building blocks.

Information exchange methods

IEC 61850-7-2 provides the services to exchange information for the different kinds of functions (e.g., control, report, log, and get and set); it defines how to exchange information. The information to be accessed comprises also the IED's self-identification and self-description.

Mapping to concrete communication protocols

IEC 61850-8-1 defines the concrete means to communicate the information between IEDs (e.g., the application layer and the encoding)—how to serialize the information in corresponding messages.

Configuration of a substation (system configuration language [SCL])

IEC 61850-6 offers the formal configuration description of a substation and all devices including the description of the relations with other devices and with the power process (single-line diagram)—how to describe the configuration. This offline data sheet completely describes the devices information provided by all devices and how it could be exchanged. The SCL file (XML document) eliminates the need to manually input these data when configuring a device. This not only reduces system configuration time, but also increases the general integrity and reliability of systems by reducing human error. And it can be used as comprehensive system documentation.

These four building blocks are independent of each other to a high degree. The information models can easily be extended by definition of new LNs and new data according to specific and flexible rules—as required by other application domains (e.g., wind power plants). In the same way, additional communication stacks may be used following the advancement in communication technology.

The information is separated from the presentation and from the information exchange services. The information exchange services are separated from the concrete communication profiles.

The standard series has an impact on many phases during the life of an SAS. The most crucial aspects are as follows:

- Design and development of devices and systems
- Procurement process
- Engineering and configuration of devices and systems
- System simulation
- Commissioning and testing of systems
- Network engineering and management
- Operation
- Error diagnostics
- Maintenance
- Asset management

- Condition monitoring
- Organization
- Remote control and monitoring of SS from control centers
- Integration in company networks
- Integration of control and monitoring of renewable and decentralized Energy

The standard series IEC 61850 is not just another communication protocol; it is a series of documents that cover many aspects to get interoperable multivendor systems for the protection, control, supervision, and automation of the power delivery system.

For that reason, it is recommended for users and vendors to create teams that have members with different background, like power system protection and power engineering, network infrastructure, communication protocols, applications, SCADA, security issues, and other people.

For IEC 61850, it is really true what has been said: Teamwork makes the Dream work.

Most important is still the fact that the main task still is to keep the power flowing and the meter running. The business case of utilities is power—not only communication.

62.3 Information Modeling and Information Models

Information models represent measurements and status information taken at the process level and other kinds of processed information like rms measurements and metering information and settings. The information models are independent of any communication protocol and network solution. They are intended to have a *long life*—a Phase C Voltage in a 50 Hz system is a Phase C Voltage today, tomorrow, in 20 years, in Karlsruhe and in Reykjavik.

Figure 62.3 shows the different levels of standard definitions: from *long life* at the top to *short life* at the bottom.

The models are organized in LNs containing data objects. An LN is, for example, the *MMXU*: the measurements and calculated values of a three-phase electrical system. Figure 62.4 depicts the application of the standard LN *MMXU* for different voltage levels.

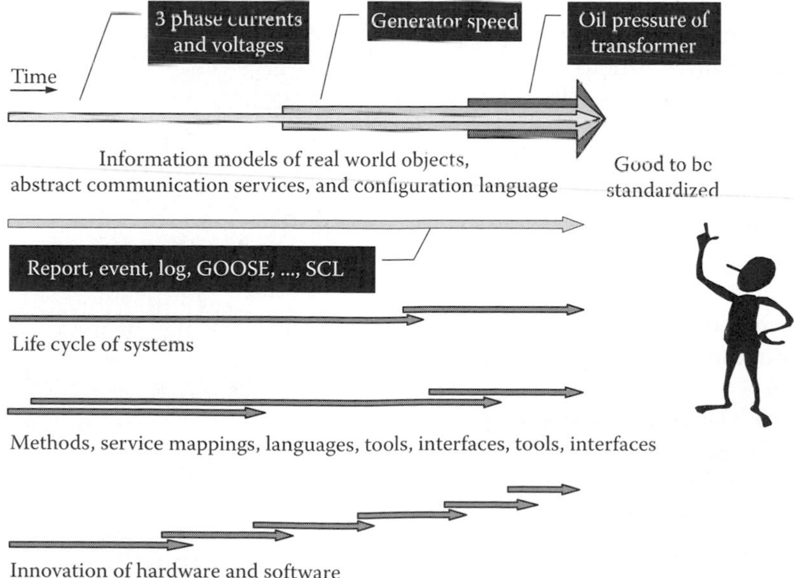

FIGURE 62.3 Information models and implementation issues.

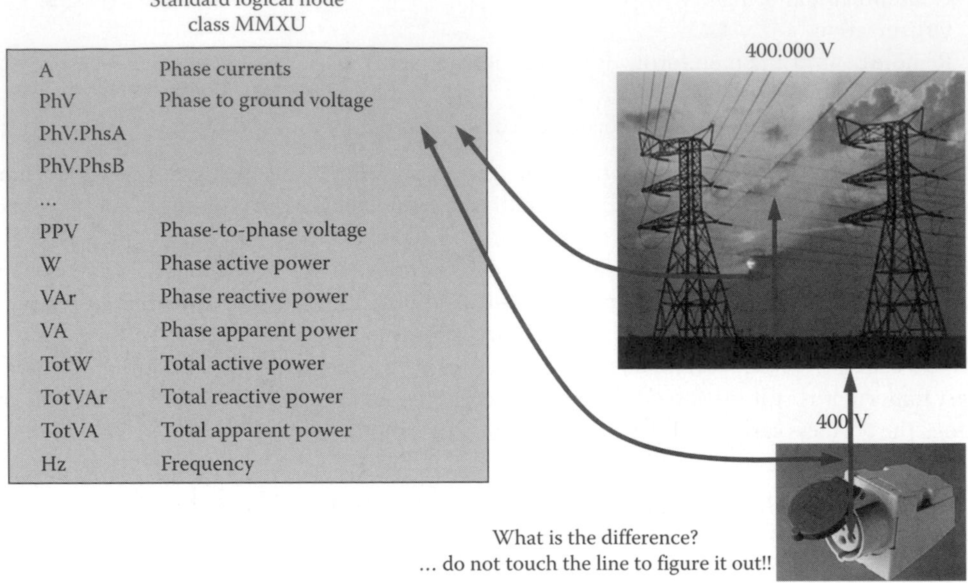

FIGURE 62.4 Information model of electrical values.

IEC 61850 and related standards define thousands of *signals* in data objects organized in LN; more than 280 standardized LNs are already defined.

Information models are one of the key elements of the standard series IEC 61850 and related standards.

Several LNs covering the most common applications in electric power delivery systems are defined. The applications inside and beyond substations are

- System information
- Protection functions
- Protection-related functions
- Supervisory control
- Generic references
- Interfacing and archiving
- Automatic control
- Metering and measurement
- Sensors and monitoring
- Switchgear
- Instrument transformer
- Power transformer
- Further power system equipment
- Decentralized energy resources (DER)
- Hydropower plants
- Wind power turbines
- Distribution automation
- Power quality
- Electric vehicle charging station and others

Most LNs provide information that can be categorized as depicted in Figure 62.5. The semantic of an LN is represented by data and data attributes. LNs may provide a few or up to 30 or more data objects.

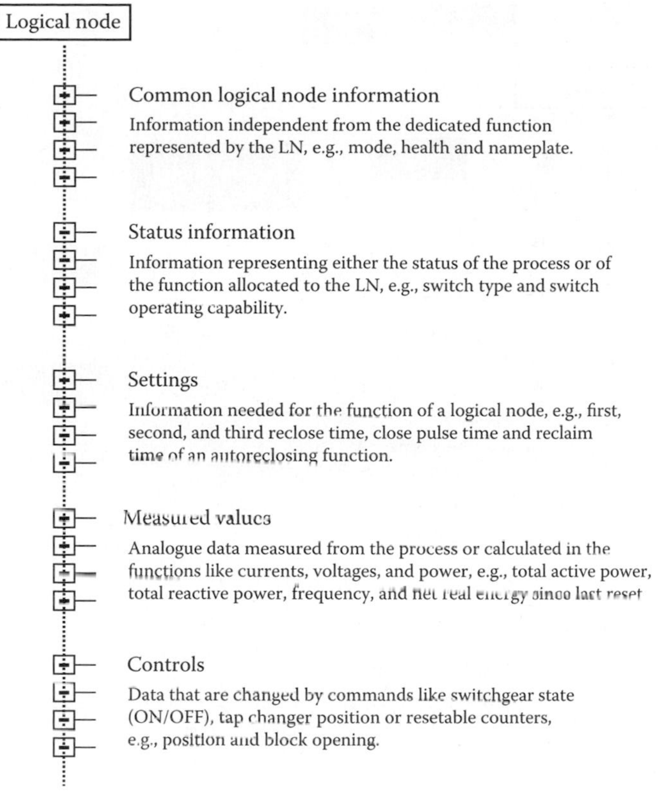

Logical node
Common logical node information
Information independent from the dedicated function represented by the LN, e.g., mode, health and nameplate.
Status information
Information representing either the status of the process or of the function allocated to the LN, e.g., switch type and switch operating capability.
Settings
Information needed for the function of a logical node, e.g., first, second, and third reclose time, close pulse time and reclaim time of an autoreclosing function.
Measured values
Analogue data measured from the process or calculated in the functions like currents, voltages, and power, e.g., total active power, total reactive power, frequency, and net real energy since last reset
Controls
Data that are changed by commands like switchgear state (ON/OFF), tap changer position or resetable counters, e.g., position and block opening.

FIGURE 62.5 Logical node information categories.

Data objects may contain a few or even more than 20 data attributes. LNs may contain more than 100 individual information (points) organized in a hierarchical structure.

The first letter in the name of the LN class indicates the application domain. The following list contains the assigned first letters (2012):

Some examples of LNs of the group measurements and metering information are provided in the following text:

Environmental information	MENV
Flicker measurement name	MFLK
Flow measurements	MFLW
Fuel characteristics	MFUL
Harmonics or interharmonics	MHAI
Heat measured values	MHET
Hydrological information	MHYD
DC measurement	MMDC
Measurement (electrical three-phase system)	MMXU
Meteorological information	MMET
Metering (three phase)	MMTR
Metering (single phase)	MMTN
Metering statistics	MSTA
Non-phase-related harmonics or interharmonics	MHAN
Non-phase-related measurement	MMXN
Pressure measurements	MPRS
Sequence and imbalance	MSQI

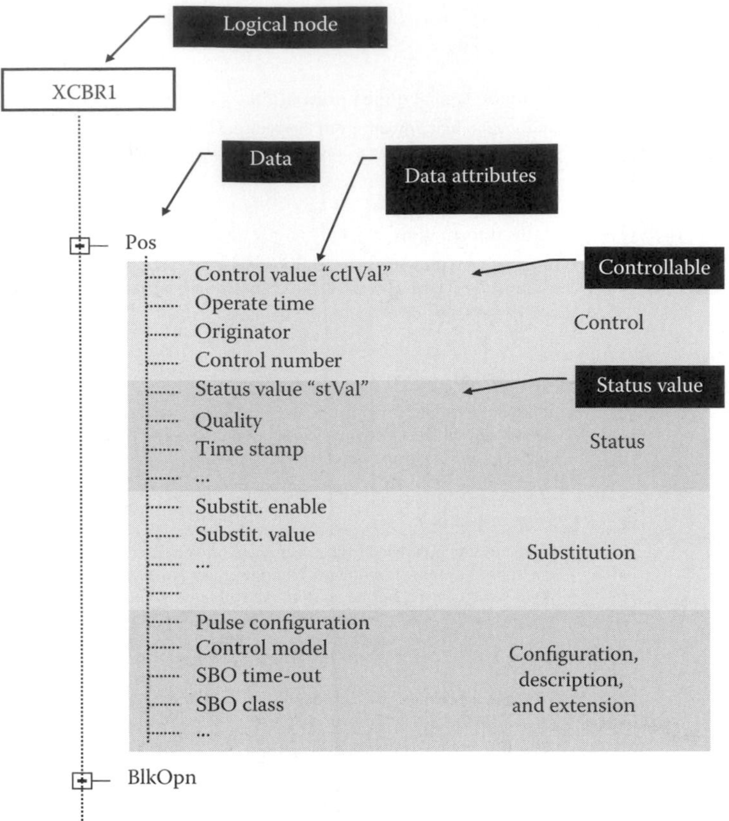

FIGURE 62.6 Position information depicted as a tree (conceptual).

The length of the LN class name is always four characters. The mean number of data objects provided by LNs defined in IEC 61850-7-4 is approximately 20. Each of the data (e.g., position of a circuit breaker) comprises several details (the data attributes). The position (named *Pos*) of a circuit breaker is defined in the LN *XCBR* (see Figure 62.6). The position is defined as data object.

The position *Pos* is more than just a simple *point* in the sense of traditional RTU protocols. It contains several data attributes. The data attributes are categorized as follows:

- Information to control a circuit breaker
- Status or measured/metered value
- Substitution
- Configuration, description, and extension

The example data object *Pos* has approximately 20 data attributes; some are mandatory while others are optional. The data attribute "Pos.ctlVal" represents the controllable information (can be used to operate a switch to open or close). The data attribute "Pos.stVal" represents the position of the real breaker (could be in intermediate state, off, on, or bad state).

The position also has optional information about when to process the control command (Operate time), the originator that issued the command, and the control number (given by the originator in the request). The quality and time stamp information indicates the current validity of the status value and the time of the last change of the status value.

The current values for "stVal," "q" (quality), and "t" (time stamp) can be read, reported, or logged in a buffer of the IED or exchanged by GOOSE.

The values for "stVal" and "q" can be remotely substituted. The substituted values take effect immediately after enabling substitution.

Common data classes (CDCs, templates) are used to define data attributes of data. The crucial CDCs defined in IEC 61850-7-3 are as follows:

CDCs for status information

- Single-point status (SPS)
- Double-point status (DPS)
- Integer status
- Binary counter reading

CDCs for measurand information

- Measured value (MV)
- Complex measured value
- Sampled value (SAV)
- Phase-to-ground-related measured values of a three-phase system (WYE)
- Phase-to-phase-related measured values of a three-phase system (DEL)
- Harmonic value
- Harmonic value for WYE
- Harmonic value for DEL

CDCs for controllable status information

- Controllable single point
- Controllable double point (DPC)
- Controllable integer status
- Binary controlled step position information
- Integer controlled step position information

CDCs for description information

- Device nameplate
- LN nameplate

The minimum subset of data attributes of the CDC MV is as follows:

"mag" Deadbanded value. Value based on a deadband calculation. The value of mag is updated when the value has changed according to the configuration parameter db specifying a percentage change of the value (deadband—online changeable)

"q" Quality of the attribute(s) representing the value of the data

"t" Time stamp of the last change in one of the attribute(s) representing the value of the data or in the q attribute (UTC-based)

These three data attributes can be used for immediate reading, reporting, or logging. Almost all RTU protocols like IEC 60870-5 or DNP3 communicate usually these three data attributes. These protocols have been developed to exchange status information, measurements, metered values, and control commands.

The XCBR LN Model in IEC 61850-7-4 comprises a list of generic information of the circuit breaker as shown in Table 62.2. This LN class contains some 20 data objects that expose specific information of a real circuit breaker. Most of the data objects are optional (O). The most common data attributes are mandatory (M). The data object Pos (Position) of the circuit breaker is used to control (open or close) it or to provide the status of the position.

TABLE 62.2 Circuit Breaker Model

XCBR Class				
Data Object Name	Common Data Class	Explanation	T	M/O/C
LNName		The name shall be composed of the class name, the LN-Prefix and LN-Instance ID according to IEC 61850-7-2, Clause 22.		
Data objects				
Descriptions				
EEName	DPL	External equipment nameplate		O
Status Information				
EEHealth	ENS	External equipment health		O
LocKey	SPS	Local or remote key (local means without substation automation communication, hardwired direct control)		O
Loc	SPS	Local control behavior		M
OpCnt	INS	Operation counter		M
CBOpCap	ENS	Circuit breaker operating capability		O
POWCap	ENS	Point on wave switching capability		O
MaxOpCap	INS	Circuit breaker operating capability when fully charged		O
Dsc	SPS	Discrepancy		O
Measured and metered values				
SumSwARs	BCR	Sum of switched amperes, resettable		O
Controls				
LocSta	SPC	Switching authority at station level		O
Pos	DPC	Switch position		M
BlkOpn	SPC	Block opening		M
BlkCls	SPC	Block closing		M
ChaMotEna	SPC	Charger motor enabled		O
Settings				
CBTmms	ING	Closing time of breaker		O

The Pos is of type DPC (CDC DPC). This type defines the details of the object (see Table 62.3). It contains several switchgear-specific attributes.

The currently published documents use a simple table notation as shown in Table 62.2. Due to the fact that the number of LNs is growing very fast, it was decided to convert the models into UML models using the tool Enterprise Architect. This UML notation is used for maintenance and extensions of the various object models. The circuit breaker model in UML notation is shown in Figure 62.7 as an example.

The content is the same in both diagrams. The benefit of using UML is this: It requires specifying the detail more precisely. The consistency is easier to check. This model allows also to automatically generating Word-, PDF-, and html-formatted documents. The models can be exported also in an XML format.

For people that are used to read models in the form of UML notation, it is now easier to study the models.

The data object Pos of the circuit breaker position has the CDC DPC. Just six data attributes of the DPC CDC are mandatory; the others are optional or conditional (indicated by the term "constraints"). The abstract class Substitution CDC (the right box in Table 62.3) has four attributes that are dependent on the constraints {PICS_SUBST}—if substitution is supported, then all four substitution-specific attributes are mandatory.

TABLE 62.3 Double-Point Common Data Class (DPC) in Table Notation

DPC class					
Data Attribute Name	Type	FC	TrgOp	Value/Value Range	M/O/C
DataName	Inherited from GenDataObject Class or from GenSubDataObject Class (see IEC 61850-7-2)				
DataAttribute					
Status and control mirror					
origin	Originator	ST			AC_CO_O
ctlNum	INT8U	ST		0..255	AC_CO_O
stVal	CODED ENUM	ST	dchg	intermediate-state | off | on | bad-state	M
q	Quality	ST	qchg		M
t	TimeStamp	ST			M
stSeld	BOOLEAN	ST	dchg		O
opRcvd	BOOLEAN	OR	dchg		O
opOk	BOOLEAN	OR	dchg		O
tOpOk	TimeStamp	OR			O
Substitution and blocked					
subEna	BOOLEAN	SV			PICS_SUBST
subVal	CODED ENUM	SV		intermediate-state | off | on | bad-state	PICS_SUBST
subQ	Quality	SV			PICS_SUBST
subID	VISIBLE STRING64	SV			PICS_SUBST
blkEna	BOOLEAN	BL			O
Configuration, description, and extension					
pulseConfig	PulseConfig	CF	dchg		AC_CO_O
ctlModel	CtlModels	CF	dchg		M
sboTimeout	INT32U	CF	dchg		AC_CO_O
sboClass	SboClasses	CF	dchg		AC_CO_O
operTimeout	INT32U	CF	dchg		AC_CO_O
d	VISIBLE STRING255	DC		Text	O
dU	UNICODE STRING255	DC			O
cdcNs	VISIBLE STRING255	EX			AC_DLNDA_M
cdcName	VISIBLE STRING255	EX			AC_DLNDA_M
dataNs	VISIBLE STRING255	EX			AC_DLN_M

The most crucial data attributes are shown in the lower box DPC that indicates that the five gray marked data attributes are mandatory:

- ctlVal (open or close)
- ctlModel (selects different control features: direct control, select before operate, etc.)
- q (quality of status signal)
- t (time of last change)
- stVal (DPS)
- The ctlVal is represented by a control service parameter and then mapped to the data attribute ctlVal in the mapping IEC 61850-8-1

The CDC for DPC is depicted in Table 62.3.

These are the minimum data attributes that are required to implement a simple control function to open or close a circuit breaker or other switchgear.

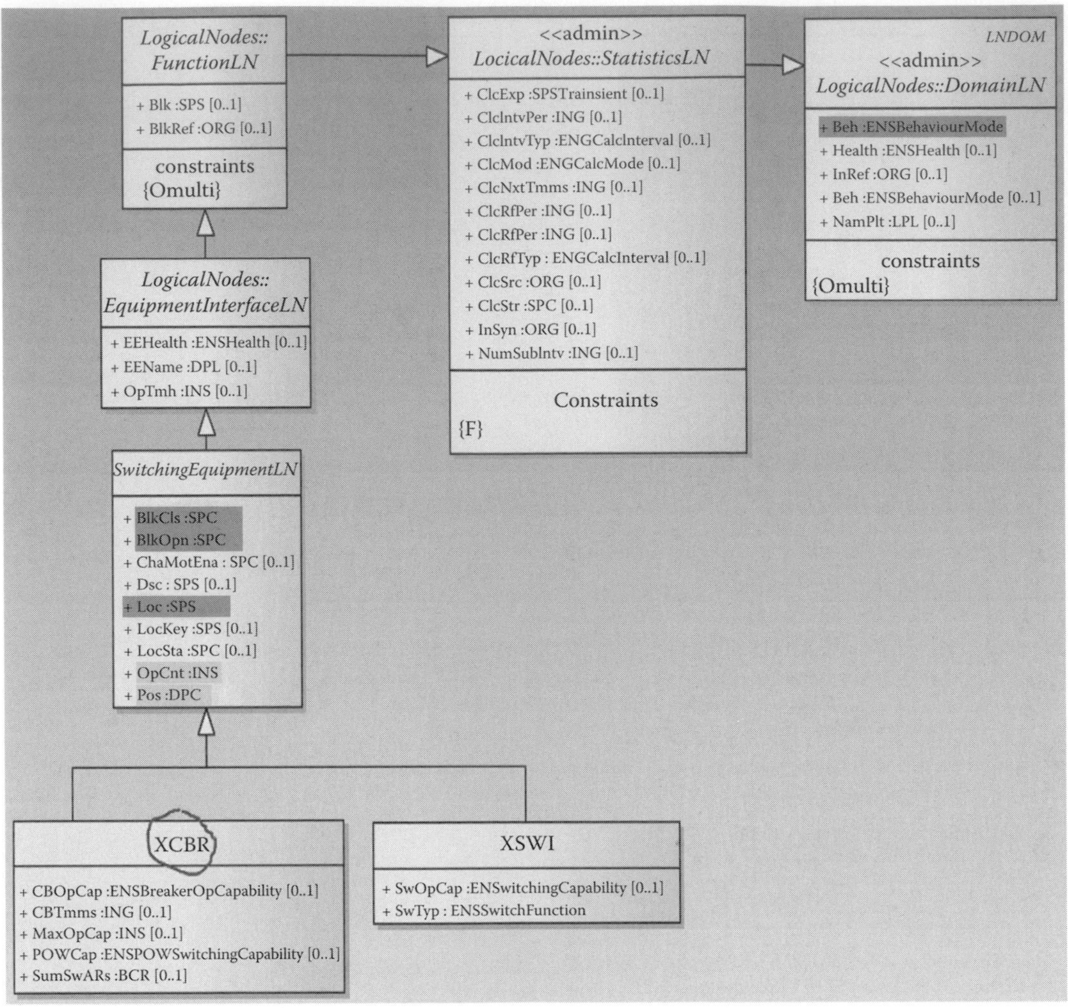

FIGURE 62.7 Circuit breaker model in UML notation.

The various parts of IEC 61850 define some 300 LNs. The number of information models is growing quite fast.

The UML models are working group internal tools. The main objective is to keep the information models consistent and at a high quality level.

62.4 Parts and Application Domains of the Standard

A summary of the coverage of the many parts of the IEC 61850 standard series is sketched in Figure 62.8. There are five main areas:

1. Information modeling method
2. Information models
3. Configuration language (SCL)
4. Abstract services
5. Communication protocols
    - TCP/IP-based SCADA (polling, spontaneous messages (events, reporting), logging, control, etc.)

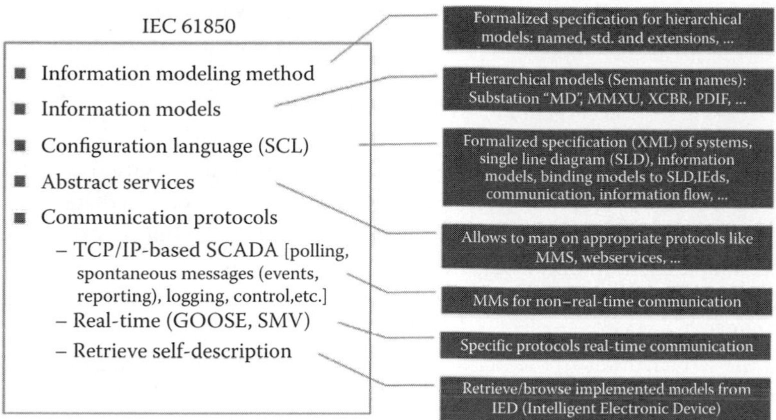

FIGURE 62.8 Summary of modeling.

- Real time (GOOSE, SMV)
- Retrieve self-description

Figure 62.9 shows simple examples for the five areas. The single-line diagram shows a bay with a bus bar (line at bottom), a circuit breaker (cross), current and voltage sensors (circles), and four LNs: XCBR1 (circuit breaker), TCTR1 (current sensor), TVTR1 (voltage sensor), and MMXU1 (electrical measurements in rms). The line between the sensors and the MMXU1 specifies that the voltage and

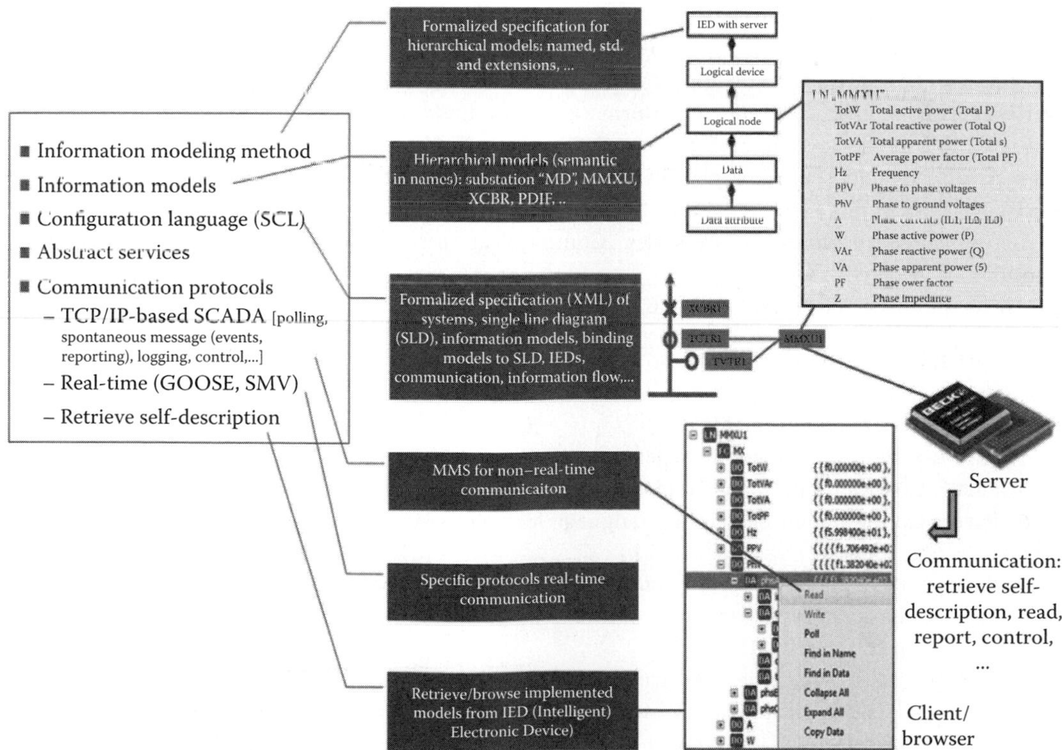

FIGURE 62.9 Modeling example.

current samples from the two sensors are communicated to the MMXU1. The MMXU1 represents the calculated rms values of voltages, currents, frequency, active power, reactive power, and other data that are bound to the exact position in the electrical drawing (single line). The line between the MMXU and the embedded controller to the right means that the MMXU1 is implemented on that platform.

The non-real-time communication that retrieves, for example, the information model from the server device is used to get all information directly from the device and allows immediate access to the model of the device without any further configuration at the client side. The self-description leads to a self-configuration of simple client (monitoring and control) applications.

The various aspects of IEC 61850 go far beyond the definitions in fieldbus standards, telecontrol standards like IEC 60870-5, or DNP.

The scope of the standard was in the early days focusing on high-voltage substations. The market for high-voltage substation automation and protection is quite small compared to the industrial automation for factories and petrochemical plants. This niche market was one of the crucial reasons why the utilities and vendors decided to start the standardization project IEC 61850 in order to prevent proliferation of solutions compared to the fieldbus standardization project with the many solution defined in one standard.

The experience with the MAP project, fieldbus standardization, and the telecontrol applications was taken into account when the IEC TC 57 Working Groups 10, 11, and 12 started to work on a single standard covering all aspects of information modeling, information models, communication, and configuration of the whole system as well as the devices that implement protection, control, and monitoring of power delivery systems.

Due to the fact that the power automation in high-voltage systems was a very special application domain, there was little interest for the vendors of industrial automation systems to enter into that market niche. The work on IEC 61850 was done by protection engineers, power engineers, SCADA engineers, and communication specialists. Several communication experts had long-term experience in the many power- and non-power-specific application domains.

Since the first projects with IEC 61850 were commissioned and in operation, there was an ever-increasing interest using existing information model and defining further domain-specific models. Figure 62.10 lists the most interesting extensions. All extensions except those listed over the right cylinder are already published as official IEC documents. The interest in industrial applications is growing these days.

The published documents of the series "Communication networks and systems for power utility automation" are as follows:

Already published documents (2012-07)

1. Part 1: Introduction and overview
2. Part 2: Glossary
3. Part 3: General requirements
4. Part 4: System and project management
5. Part 5: Communication requirements for functions and device models
6. Part 6: Configuration description language for communication in electrical substations related to IEDs
7. Part 7-1: Basic communication structure—Principles and models
8. Part 7-2: Basic information and communication structure—Abstract communication service interface (ACSI)
9. Part 7-3: Basic communication structure—Common data classes
10. Part 7-4: Basic communication structure—Compatible logical node classes and data object classes
11. Part 7-410: Hydroelectric power plants—Communication for monitoring and control

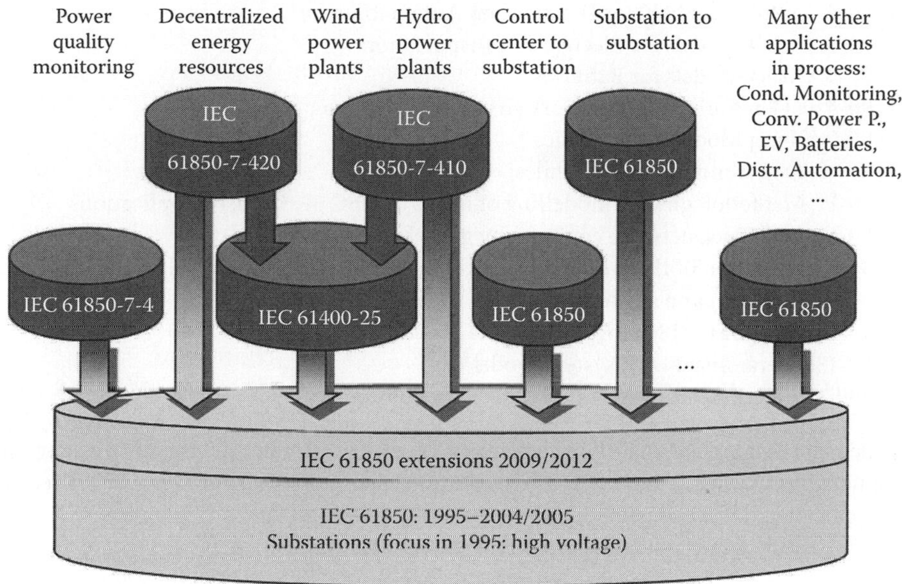

FIGURE 62.10 Application domains of IEC 61850.

12. Part 7-420: Basic communication structure—Distributed energy resources logical nodes
13. Part 7-510: Basic communication structure—Hydroelectric power plants—Modelling concepts and guidelines
14. Part 8-1: Specific communication service mapping (SCSM)—Mappings to MMS (ISO 9506-1 and -2) and to ISO/IEC 8802-3
15. Part 9-2: Specific communication service mapping (SCSM)—Sampled values over ISO/IEC 8802-3
16. Part 10: Conformance testing
17. Part 80-1: Guideline to exchanging information from a CDC-based data model using IEC 60870-5-101 or IEC 60870-5-104
18. Part 90-1: Use of IEC 61850 for the communication between substations
19. Part 90-5: Use of IEC 61850 to transmit synchrophasor information according to IEEE C37.118
20. Part 90-7: Object Models for PV, Storage … inverters

Documents in *bold* are often called the IEC 61850 Core Parts. These documents contain basic definitions for the modeling approach and for the models in other domains. A CDC DPC or MV can be used in any other domain. The reusability of the models is very high—the term "common" in CDC is truly very generic and can be applied not only in the electrical world but also in any other application domain. A temperature LN STMP (see details further down in this clause) could expose temperature measurements in building automation, computer farms, and power plants.

The following documents are in preparation (2012-07):

21. Part 7-5: Usage of information models SAS (substation automation system)
22. Part 7-500: Use of LN to model functions (SAS)
23. Part 7-520: Use of LN (DER)
24. Part 8-2: Web service mapping
25. Part 10-2: Interoperability test for hydro equipment
26. Part 90-2: Using IEC 61850 for SS-CC communication
27. Part 90-3: Using IEC 61850 for Condition Monitoring
28. Part 90-4: Network Engineering Guidelines

29. Part 90-6: Use of IEC 61850 for Distribution Automation
30. Part 90-8: Object Models for Electrical Transportation
31. Part 90-9: Object Models for Batteries
32. Part 90-410: Communication network structures in hydro power plants
33. Part 90-9: Object Models for Batteries
34. Part 90-10: Object models for schedules
35. Part 90-11: Methodologies for modelling of logics for IEC 61850 based applications
36. Part 90-12: Wide area network engineering guidelines
37. Part 90-13: Extension of IEC 61850 information models to include logical nodes and data models for steam and gas turbines
38. Part 90-14: Using IEC 61850 for FACTS data modelling
39. Part 90-15: Hierarchical DER system model
40. Part 100-1: Functional testing of IEC 61850 based systems

The published documents are mainly independent of each other (except the core documents). For specific application domains, it is helpful to give some guidance on which documents are relevant for a particular application domain.

The security requirements are defined in a separate series: IEC 62351

IEC 62351-1: Security for TC57 standards and end-to-end security
IEC 62351-2: Glossary
IEC 62351-3: Security using transport layer security (TLS)
IEC 62351-4: Security for manufacturing messaging specification (MMS)
IEC 62351-5: Security for IEC 60870-5 and DNP3
IEC 62351-6: Security for IEC 61850
IEC 62351-7: Network and system management objects
IEC 62351-8: Role-based access control (RBAC)
IEC 62351-9: Cyber security key management for power system equipment
IEC 62351-10: Security architecture guidelines for TC 57 systems
IEC 62351-11: Security for XML files

From a model point of view, a device can comprise any LN model from any of the 11 parts that define object models (LNs, data, and CDCs). Applications that need models that are not (yet) standardized can be defined as extended models. The extensions are marked as "Functional Constraint" FC = "EX" (extended) in the corresponding name space attribute (for LNs, data, and CDCs).

The name space concept is appreciated very well, because very often devices and applications have more data to be communicated than what the standards provide currently. From the communication software viewpoint, the handling of the models in the communication software is the same if the model is extended or not.

The comprehensive information models are of interest by several other groups. Note that the IEC 61850 information models are independent of the communication system that carries the data values. The following groups use subsets of the IEC 61850 models and put them into their own context:

- IETF EMAN (Energy Management Group) that used many models and translated them into the Management Information Base. This would allow accessing the models by Simple Network Management Protocol [3].
- ISO/TC 205/WG 3 (Building Automation and Control System Design) has published recently a new work proposal on power system information models [4].
- SunSpec Alliance uses many information models and maps them to Modbus [5].

62.5 Functional Description of Applications

Most of the models defined in the standards listed are static in the sense that they mainly provide input and output signals and settings for well-known functions in the different application domains (see Figure 62.11). The functions are expected to be known by the domain experts. The meaning of the setting "MinOpTmms = Minimum operate time" of several protection LNs (e.g., PTOV = Time over current protection) is understood by protection engineers (only). This setting can have a value that is set by the application (thus not configurable or online writable); the setting may also be configured in the configuration file for that device, or it may be additionally writable remotely. These possibilities could be specified in the SCL file.

The standardization work was mainly done by protection engineers at the beginning. These engineers have not defined the details of the protection functions in IEC 61850, because they understood the details anyway. Other groups define the protection functions, for example, IEC TC 95 (Measuring relays and protection equipment). There are also general LNs that define data objects that are self-explanatory, that is, their names indicate uniquely what the function behind the data object is all about. Example: The temperature measurement model STMP (Supervision temperature) has a data object Tmp; this is of course the object that exposes the measured temperature. The data object TmpAlmSpt (Set point) represents the value of the Alarm level or limit. The data object Alm exposes the case that the Tmp value reached the alarm level.

How and by whom the inputs and output are used is also outside the standard. In the real world, there is often a need to specify and to know how LNs interoperate—which signals (data attributes) are needed when and from which other LN(s). This is usually the task of the system engineering. The engineering of the signal flow and how these signals are used internally to implement specific functions are outside the standard. The need for functional specifications is known by various standardization groups. One crucial reason why the functional specification is very important is the need of standardized definitions for implementing functional tests of devices and the whole system.

Examples of LNs that define functions can be found in IEC 61850-90-7 "IEC 61850 object models for photovoltaic, storage, and other DER inverters." The need for a precise definition on how hundreds of photovoltaic inverter could connect to the electrical grid and how to monitor and control them is a very crucial issue in electrical networks with high penetration of PV into the grid. Figure 62.12 shows the PV performance on 2013-07-27 of 20.7 GW—with an average total consumption of some 60 GW [6].

The installation of PV systems and connecting them to the grid is still growing. This situation requires that (at least larger) PV systems have to be monitored and controlled by grid operators by means of remote access. IEC 61850-90-7 is more often the reference solution. The Italian power industry has

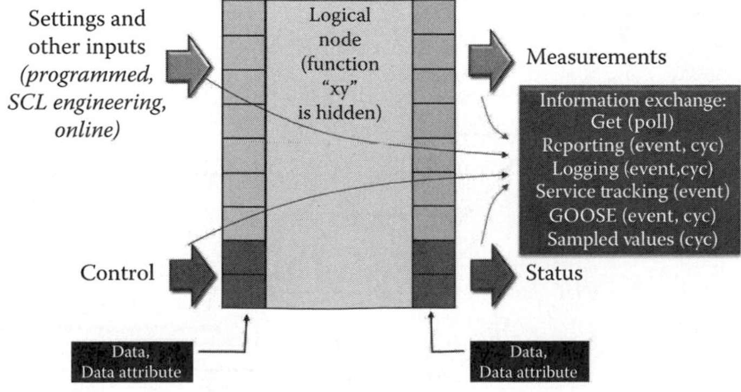

FIGURE 62.11 Function block view of IEC 61850 models.

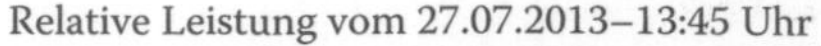

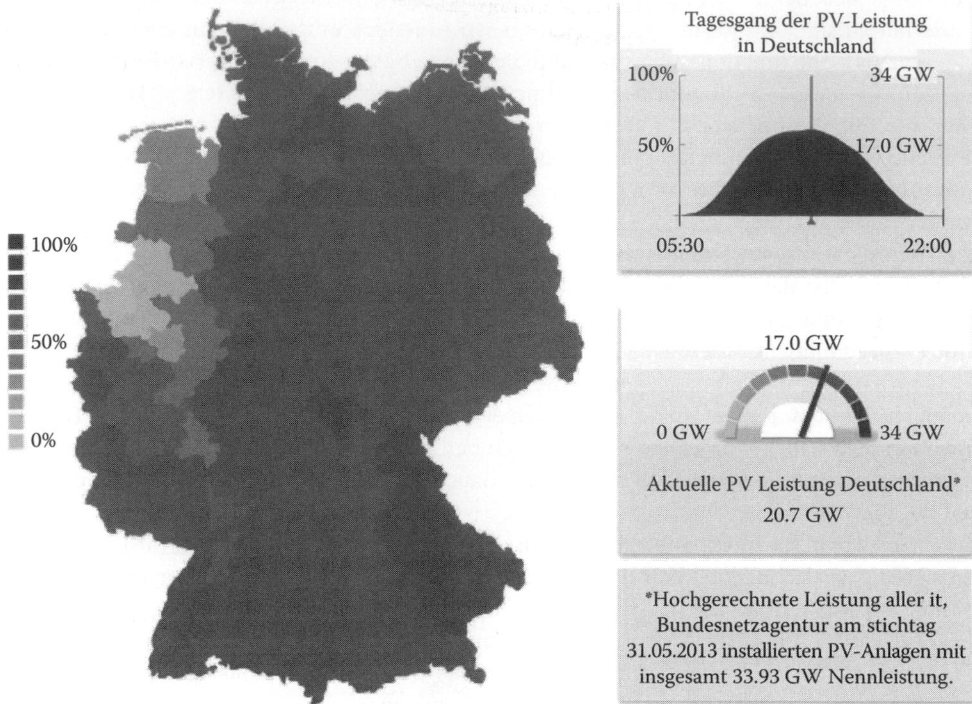

FIGURE 62.12 PV performance into German grids.

published two norms (CEI 0-16 and CEI 0-21) to use IEC 61850 for connecting control centers with PV systems at high- and low-voltage levels. First pilot systems are in operation since summer 2013.

IEC 61850-90-7 defines exactly how various control functions are defined and how they can be configured and used. Figure 62.13 shows the diagram for one of the many different electrical behaviors of the PV inverter at the electrical connection point. The reactive power production

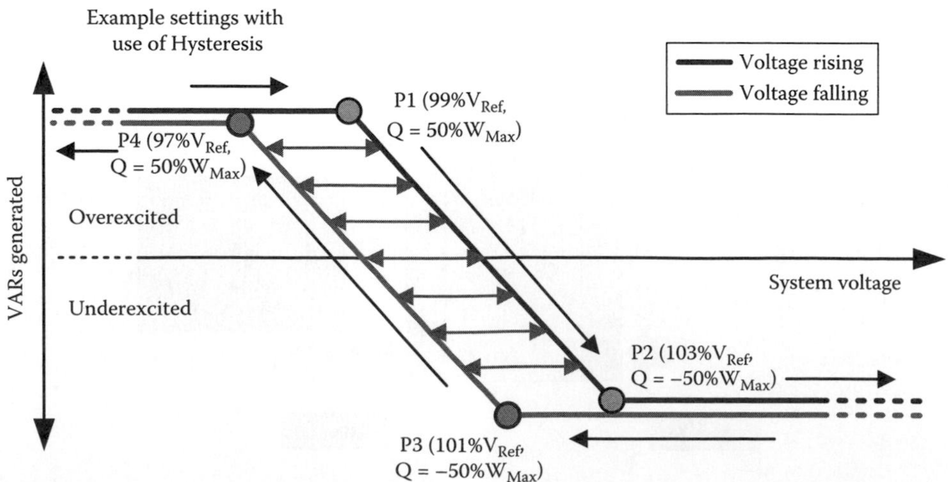

FIGURE 62.13 Example of hysteresis in volt-var curves.

(over- or underexcited) has to follow a defined curve that is configured based on the data objects in the LN models defined in IEC 61850-90-7.

These models allow grid operators to manage and manipulate the electrical grid with a single international standard. Applying IEC 61850-90-7 reduces the proliferation of grid codes for connecting PV and other inverters to the electric power grid. The benefit for grid operators is this: There is only one specification and requirement documentation needed per utility. The vendors of inverters do not need to implement a variety of grid codes—one per utility of region.

The standard could also be used to build closed loop control mechanisms that could be implemented by horizontal communication between an energy manager next to the power transformer and the loads and generation units along the feeder line.

The definition of functions is also crucial for testing an inverter. The function and the communication with the inverter can now be tested together. The electrical output of the inverter can be tested against the settings of the LN model. The behavior of the electrical output must be the same as defined by the configured curve as shown in Figure 62.13. The certification of the communication and the inverter function by one test lab would build confidence that the inverter does what it is intended for: managing the electrical connecting point with an open information and information exchange solution.

The inverter functions modeled in IEC 61850-90-7 range from simple to complex control operations. Most inverter functions are based on settings or curves that allow inverters to respond autonomously to local electrical conditions, while some require direct control commands. The following list describes briefly some crucial functions that are defined in IEC 61850-90-7:

Immediate control functions for inverters

- Function INV1: Connect/disconnect from grid
- Function INV2: Adjust maximum generation level up/down
- Function INV3: Adjust power factor
- Function INV4: Request active power (charge or discharge storage)
- Function INV5: Request action through a pricing signal

Volt-var management modes

- Volt-var mode VV11: Available var mode with no impact on watts
- Volt-var mode VV12: Maximum var support mode based on maximum watts
- Volt-var mode VV13: Static inverter mode based on settings
- Volt-var mode VV14: Passive mode with no var support

Frequency-watt management modes

- Frequency-watt mode FW21: High frequency reduces active power or low frequency reduces charging
- Frequency-watt mode FW22: Constraining generating/charging by frequency
- Frequency-watt mode FW23: Watt generation/absorption counteractions to frequency deviations

Dynamic reactive current support during abnormally high- or low-voltage levels

- Dynamic reactive current support TV31: Volt-var support during abnormally high- or low-voltage levels

62.6 Information Exchange Models

Information exchange models in IEC 61850 are first defined in an abstract way: the ACSI (IEC 61850-7-2). The service models comprise

- Association
- Information model
    - Directory services to retrieve the information (logical device, LN, dada object, and data attributes) and communication models (data sets and control blocks)

    - Read and write data or data sets (polling)
    - Create/delete data sets
    - Substitute data (freeze)
- Setting groups for device settings contained in LNs
    - Edit setting groups
    - Activate setting groups
- Service tracking (useful for security solutions)
    - Specialized reporting and logging
    - Track service parameters of message exchange with a device
- Spontaneous or event-driven data transmission (reporting)
    - Configuring reports
    - Transmit buffered or unbuffered reports (very efficient communication)
- Storing sequence of events (SOE; logging)
    - Configuring logs
    - Read logged information
- Real-time transmission of events and other signals
    - Multicasting of events (Generic Object–Oriented Event [GOOSE])
- Transmission of SVs
    - Multicasting of samples (current, voltage, vibration, etc.)
- Control model
    - Select and operate
    - Time-activated operation
- Time synchronization
    - SNTP (for SCADA)
    - IEEE 1588 or extra fiber for SVs
- File transfer
    - Read and write files
    - Delete files

Most service models defined in IEC 61850-7-2 are based on a client/server model. The client and the server are roles that can be implemented by real devices. The server exposes information (models) that can be accessed through TC/IP connections by one or several clients (see Figure 62.14). This client/server concept maps directly to the MMS approach. Real devices may implement both roles independently.

The second concept is the multicast-based communication for GOOSE and SV exchange. The publisher (role) sends Ethernet (Ethertype) multicast messages on the network. All devices that need to receive multicast messages take the messages they wants to receive from the network. The publisher role usually is implemented in a device that also implements a server. The server model is needed in case the information objects, data sets, and control blocks are required to be exposed by client/server services.

All models (application information models and communication models) have to be configured. Especially this comprises the LNs, data objects, data sets, and control blocks. An SCL document can contain all these details of the sending devices (server and publisher). In addition, the SCL document can also contain clients and subscriber information. The subscriber, for example, can be configured to listen to a specific GOOSE message and to use a specific subset of the payload for a specific function. The function has inputs and outputs. These are modeled as LNs as well (see Figure 62.15).

At configuration level, there is another feature defined in IEC 61850-6: the reference to the physical or logical input and outputs (sAddr—short address for inputs; and intAdr—internal address). This internal configuration is not visible to the online communication. The concept of SCL comprises the compete signal from a source to the model, through online exchange (client/server or publisher/subscriber), to an application or output at the receiving device. The IEC 61850-6 allows specifying the complete traffic in a system with a few or hundreds of devices.

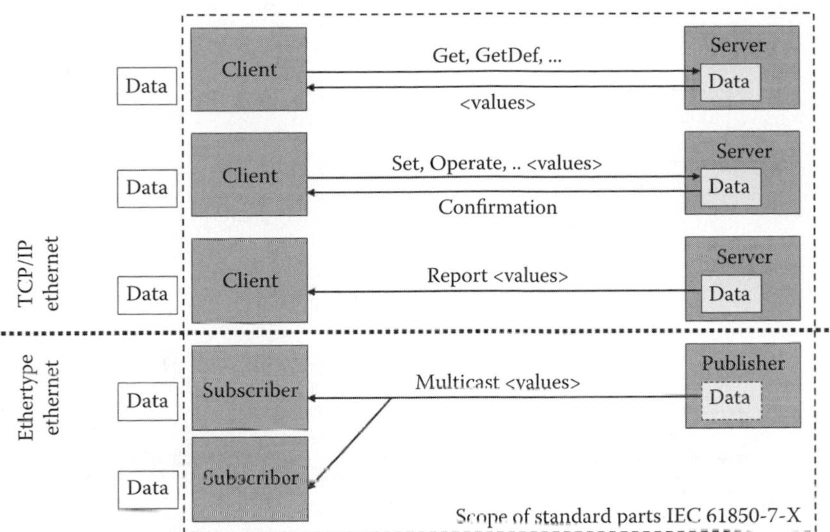

FIGURE 62.14 Service concepts in IEC 61850-7-2.

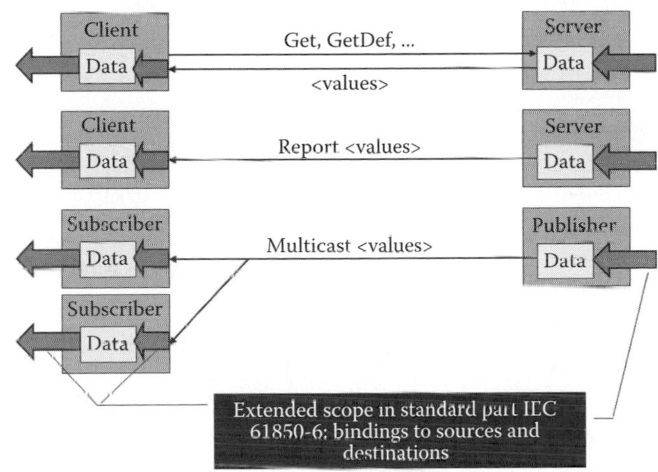

FIGURE 62.15 Services used and extended in the configuration.

An excerpt of the information exchange services for accessing information models in a device (implementing a server and subscriber role) is displayed in Figure 62.16. The circles with the numbers refer to the bulleted list provided later.

The operate service manipulates the control-specific data attributes of a circuit breaker position (open or close the circuit breaker). The report services spontaneously inform another device that the position of the circuit breaker has been changed. The substitute service forces a specific data attribute to be set to a specific value; this may be required for test purposes or when the position sensors are defective.

The categories of services (defined in IEC 61850-7-2) are as follows:

- Control devices (operate service or by multicast trip signals) [1]
- Fast and reliable peer-to-peer exchange of status information (tripping or blocking of functions or devices) [2]
- Reporting of any set of data (data attributes), SOE—cyclic and event triggered [7]
- Logging and retrieving of any set of data (data attributes)—cyclic and event-triggered [8]

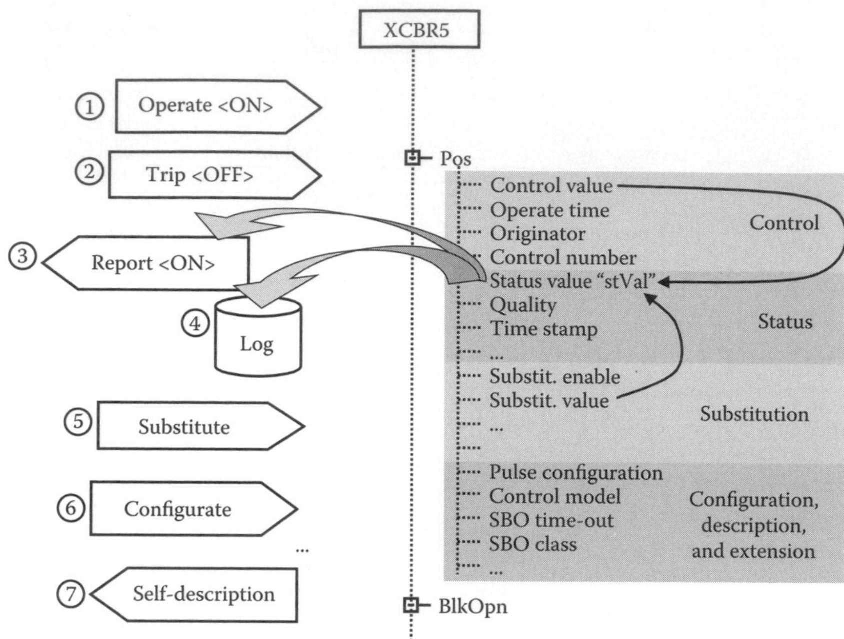

FIGURE 62.16 Service excerpt.

- Substitution [9]
- Handling and setting of parameter setting groups
- Transmission of SVs from sensors
- Time synchronization
- File transfer
- Online configuration [10]
- Retrieving the self-description of a device [3]

The information exchange models defined in IEC 61850-7-2 are listed in Table 62.4.

The information exchange supports various methods to access process data. The polling is the most common method. The use is restricted to applications that do not rely on real-time information and SOE. Data values may be lost because the value may have been overwritten by the application between two GetDataValue requests.

The buffered and unbuffered reporting starts with the configuration of the report control blocks. The basic buffered reporting mechanism is shown in Figure 62.17. The reporting starts with setting the enable buffer attribute to TRUE; setting to FALSE stops the reporting.

The specific characteristic of the buffered report control block is that it continues buffering the event data as they occur according to the enabled trigger options in case of, for example, a communication loss. The reporting process continues as soon as the communication is available again. The buffered report control block guarantees the SOE up to some practical limits (e.g., buffer size and maximum interruption time).

The basic architecture of reporting, logging, GOOSE, and SVs is the existence of an information model, data sets (that refers to data attributes), and control blocks that refer to a data set (see Figure 62.18). A specific data set can be used by any of the four control block types (reporting, logging, GOOSE, and SVs). And a specific data set can be used by multiple control blocks at the same time. This makes the information model and how the values are exchanged independent of each other.

TABLE 62.4 Information Exchange Models

Information Exchange Model	Description	Services
Server	Represents the external visible behavior of a device. All other ACSI models are part of the server.	ServerDirectory
Application association	Provision of how two or more devices can be connected. Provides different views to a device: restricted access to the server's information and functions.	Associate Abort Release
Logical device	Represents a group of functions; each function is defined as a logical node.	LogicalDeviceDirectory GetAllDataValues
Logical node	Represents a specific function of the substation system, for example, overvoltage protection.	LogicalNodeDirectory
Data	Provides a means to specify typed information, for example, position of a switch with quality information, and time stamp.	GetDataValues SetDataValues GetDataDefinition GetDataDirectory
Data set	Allow to group various data together.	GetDataSetValue SetDataSetValue CreateDataSet DeleteDataSet GetDataSetDirectory
Service tracking	Specialized reporting and logging services that allow to report and log (track) the service parameters of received service indications and service responses. Special common data classes for service tracking are defined in IEC 61850-7-2.	Services of reporting and logging are applied.
Substitution	The client can request the server to replace a process value by a value set by the client, for example, in the case of an invalid measurement value.	SetDataValues
Setting group control	Defines how to switch from one set of setting values to another one and how to edit setting groups.	SelectActiveSG SelectEditSG SetSGValues ConfirmEditSGValues GetSGValues GetSGCBValues
Reporting and logging	Describes the conditions for generating reports and logs based on parameters set by the client. Reports may be triggered by changes of process data values (e.g., state change or deadband) or by quality changes. Logs can be queried for later retrieval. Reports may be sent immediately or deferred (buffered). Reports provide change-of-state and sequence-of-events information exchange.	**Buffered RCB:** Report GetBRCBValues SetBRCBValues **Unbuffered RCB:** Report GetURCBValues SetURCBValues **Log CB:** GetLCBValues SetLCBValues **Log:** QueryLogByTime QueryLogAfter GetLogStatusValues
Generic substation events (GSE)	Provides fast and reliable system-wide distribution of data; peer-to-peer exchange of IED binary status information. GOOSE means Generic Object–Oriented Substation Event and supports the exchange of a wide range of possible common data organized by a data set.	**GOOSE CB:** SendGOOSEMessage GetGoReference GetGOOSEElementNumber GetGoCBValues SetGoCBValues

(*continued*)

TABLE 62.4 (continued) Information Exchange Models

Information Exchange Model	Description	Services
Transmission of sampled values	Fast and cyclic transfer of samples, for example, of instrument transformers.	**Multicast SVC:** SendMSVMessage GetMSVCBValues SetMSVCBValues **Unicast SVC:** SendUSVMessage GetUSVCBValues SetUSVCBValues
Control	Describes the services to control, for example, devices or parameter setting groups.	Select SelectWithValue Cancel Operate CommandTermination TimeActivatedOperate
Time and time synchronization	Provides the time base for the device and system.	services in SCSM
File transfer	Defines the exchange of huge data blocks such as programs.	GetFile SetFile DeleteFile GetFileAttributeValues

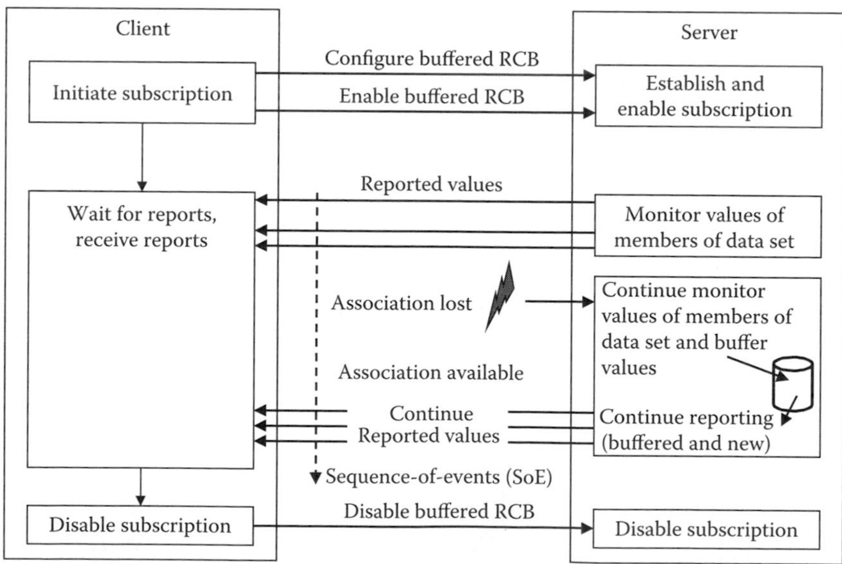

FIGURE 62.17 Buffered reporting (conceptual).

The reporting model supports two kinds: unbuffered and buffered (see Figures 62.19 and 62.20). In the case of unbuffered reporting, the events that issue a report are communicated only when communication is available. In case of a communication interrupt, the buffered reporting will store (buffer) the events during the interrupt and send the reports when communication is available again.

The buffered reporting is used to get an SOE. The SOE is crucial to a user for providing all changes that happened in order to react on changes.

The data attributes (like the position of a switchgear, XSWI.Pos) of several LNs and data objects are listed in a data set as shown in Figure 62.21. The members of the data set are references to the models

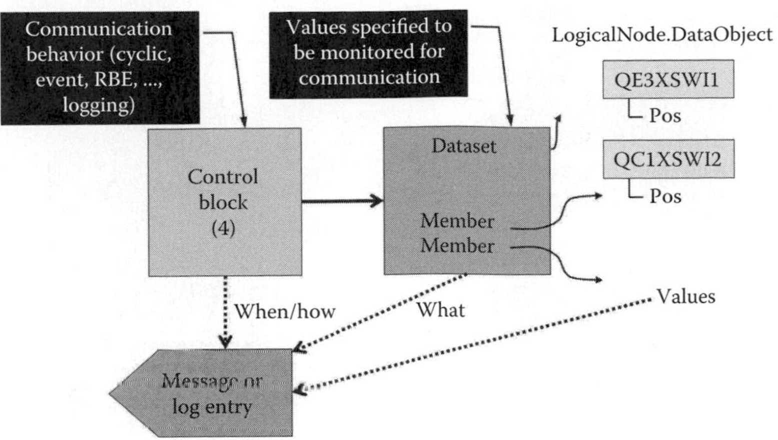

FIGURE 62.18 Control blocks and data set.

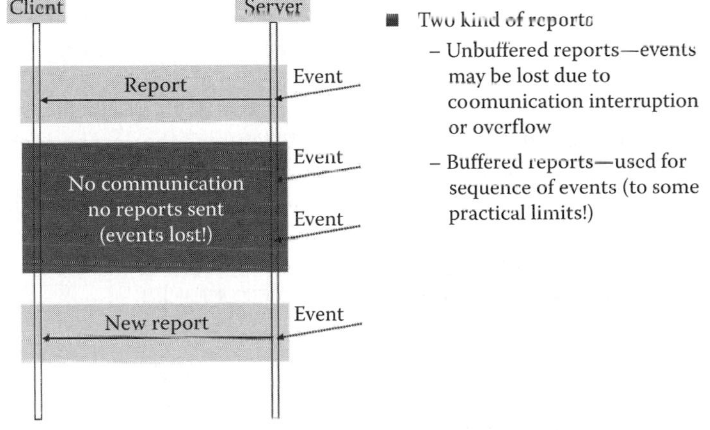

FIGURE 62.19 Unbuffered event reporting mechanism.

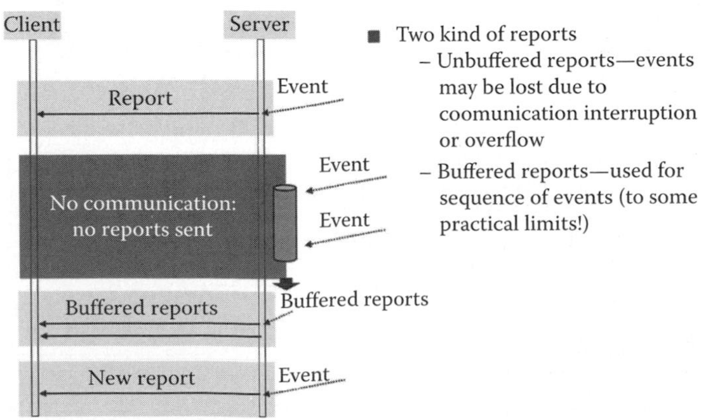

FIGURE 62.20 Buffered event reporting mechanism.

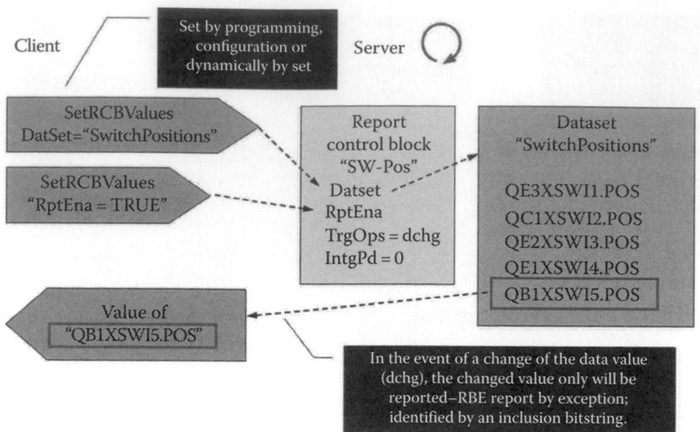

FIGURE 62.21 Report by exception.

that expose the real status values of the switch yard. If a position of one of the members of the data set changes, then only the value of that object will be reported according to the configuration of the report control block, for example, the value of QB1XSWI5.Pos. This behavior is called report by exception (RBE). This reduces the number of octets to be communicated to the minimum.

In our example, the report message just communicates one value—the fifth member of the data set. The report message included a so-called inclusion bitstring that indicates to the receiving side which object has issued the report—in this case number 5. The structure of the data set is quite important and should not be changed once the device is running.

Note that the report control block model is more comprehensive than shown in the earlier example.

Figure 62.22 shows an example of a log and three log control blocks. The first step is to configure and enable log control blocks. After enabling, the association with that server may be closed. The log entries are stored into the log as they arrive for inclusion into the log. The logs are stored in time sequence order. This allows the retrieval of an SOE list.

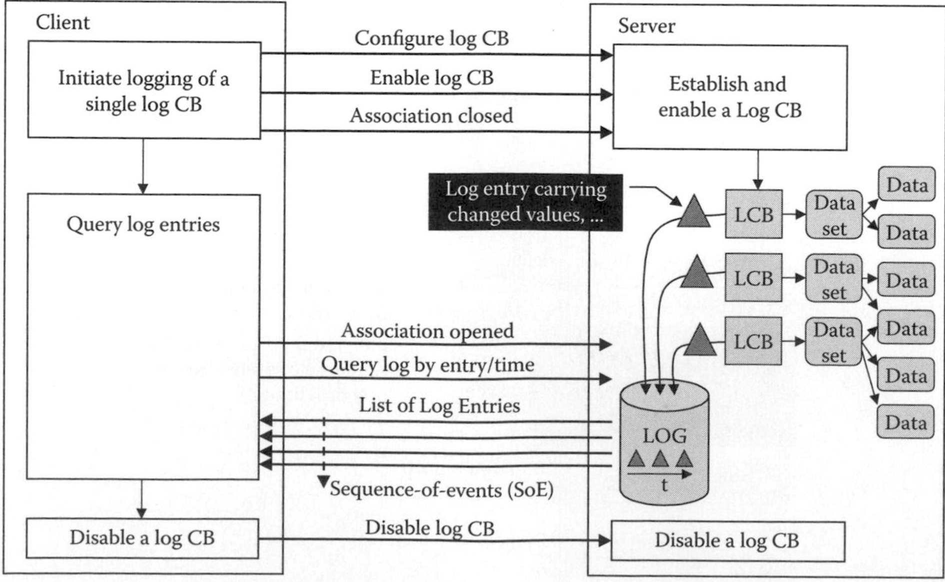

FIGURE 62.22 Log control block (conceptual).

The different log control blocks allow storage of information from different data sets into a log instance. Each log control block is independent of the other control blocks.

Two models are dedicated to the transmission of information with high priority in real time:

GOOSE Event is used to model the transmission of high-priority information like trip commands or interlocking information. The model is based on cyclic and high-priority transmission of status information. Information like a trip command is transmitted spontaneously and then cyclically at increasing intervals.

SV is used to model the exchange of the sampled MVs from current and voltage transducers to any IED that need the samples. The model is based on an unconfirmed transmission of a set of SVs. A counter is added to time-correlate samples from different sources and to detect the loss of a set of samples.

GOOSE and SV service models are defined in IEC 61850-7-2. The mapping of the SMV model to a concrete communication system is specified in IEC 61850-9-1.

A summary of all services that allow exchanging information between two devices could be found in Table 62.5. The application of one of the various possibilities depends on the needs of an application.

The two main approaches (TCP/IP-based reporting and Ethernet-based communication) are compared in Table 62.6. The main differences is the maximum length of the messages: The use of TCP/IP for the message routing respective using Ethernet directly without routing capabilities.

The resulting timeliness is quite different in both solutions. With TCP/IP, reaction times in the range of 1 s could be expected. This seems to be sufficient for most SCADA applications. The multicast messaging could be applied for real-time applications in the power transmission and distribution domain. Reaction times in the range of a few milliseconds could be implemented.

TABLE 62.5 Information Exchange Mechanisms

Get a value of single data object (GetDataValues – Client initiated)
Get a list of values of data objects (GetDataValues with list sent in each request – Client initiated)
Get the complete list values of data objects using a dataset object (GetDataSetValues – Client initiated)
Get the complete list of values of data objects (of a dataset) using reporting (reporting, General Interrogation – Client initiated)
Get the complete list of values of data objects (of a dataset) using reporting (reporting, Integrity period – Server initiated)
Get one (ButTm=0) or more (ButTm>0) value(s) of data objects (of a dataset) using reporting on data and quality change and data update – Server initiated
GOOSE and Sampled Values ... exchanges complete list of values of data objects of a dataset (events: application specific – Server initiated)
Get sequence of value(s) of data objects (of a dataset) using logging (on data and quality change and data update – Client initiate)

TABLE 62.6 Comparison of Event Reporting Mechanism

	Reporting	GOOSE/SMV
Max. message size	… 64.000 … octets	< 1.500 octets
Segmentation	Yes (option in report)	No
Flow control	Yes (TCP)	No
Detection of loss	Yes (TCP)	Yes (var. counters)
Reliable transmission	Yes (TCP)	Yes (re-transm./cyc.)
Network routing	Yes (IP)	No (yes: 90-5 UDP/IP)
Relation – 1:1	Yes	No/yes
Relation – 1:n	Yes (multiple CBs)	Yes (layer 2 multicast)
Timeliness	… 1 sec …	… few msec …
General interrogation	Yes	No
Layers	1 … 7, application	1,2,7, application
Documents	IEC 61850-7-2, 8-1	IEC 61850-7-2, 8-1/9-2

The multicast messages are according to IEC 61850-8-1 (GOOSE) and IEC 61850-9-2 (SVs) not routable through a network. The requirement for wide-area awareness of the situation of the electrical network has led to a new part IEC 61850-90-5 applied for wide-area phasor measurement exchange. IEC 61850-90-5 defines that GOOSE and SVs can also be exchanged using UDP/IP. The timeliness is less stringent; it is typically for crucial applications in the range of several seconds.

The timeliness of the communication according to the IEC 61850 standard series depends mainly on the implementation at the senders and receivers of messages. The communication system comprised of Ethernet switches and routers is usually less critical.

62.7 System Configuration Language

The use of IEC 61850 requires to be very precise on how and when to communicate which information to whom. All this configuration information is formally defined by applying a huge XML schema according to IEC 61850-6. Vendors have to use these templates to configure their devices. Users can use the same template independent of the devices' manufacturer.

IEC 61850-6 also describes the electrical single-line drawing in the form of an XML document derived from the part 6 XML schemas (Figure 62.23). The textual description is on the left. The single-line diagram depicts also a couple of LNs that indicate that there will be a distance protection function (PDIS) at the top and a differential protection at the transformer. The current sensor (TCTR) is below the transformer. Next to the circuit breaker, there are three LNs that describe the circuit breaker information (XCBR) and how it can be controlled (CSWI and CILO). The CSWI is exposing the information for the control function, while CILO represents the interlocking information, that is, closing or opening of the circuit breaker allowed at the moment or not.

The formal representation of this diagram (only the topology and not the graphical representation) is contained in an XML document (see Figure 62.24).

This document is the result of a system design process. The document will be used as input for the system engineering process to map the information and communication needed to real devices. The protection function may be implemented in a single protection device and the control functions in another. The current sensor may be implemented in a third device (usually called merging unit).

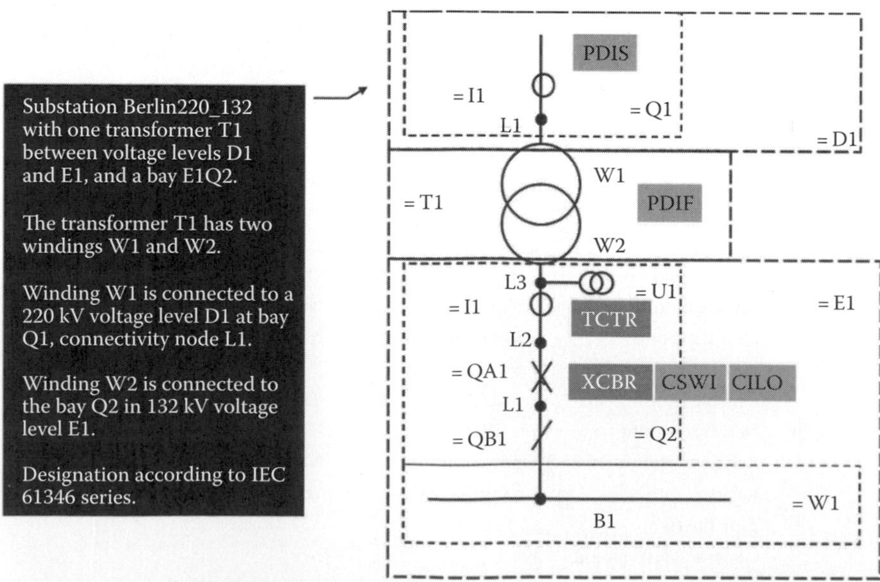

FIGURE 62.23 Single-line diagram with logical nodes assigned.

```
<?xml version="1.0"?>
<SCL xmlns="http://www.iec.ch/61850/2003/SCL"
xmlns:xsi="http://www.w3.org/2001/XMLSchema-instance" version="2007" revision="A">
    <Header id="SSD Example"/>
    <Substation name="Berlin220_132">
      <PowerTransformer name="T1" type="PTR">
          <LNode lnInst="1" lnClass="PDIF" ldInst="F1"/>
          <LNode lnInst="1" lnClass="TCTR" ldInst="C1"/>
          <TransformerWinding name="W1" type="PTW">
              <Terminal connectivityNode="baden220_132/D1/Q1/L1"
              substationName="baden220_132" voltageLevelName="D1"
              bayName="Q1" cNodeName="L1"/>
          </TransformerWinding>
          <TransformerWinding name="W2" type="PTW">
              ...
      </PowerTransformer>
      <VoltageLevel name="D1">
          <Voltage multiplier="k" unit="V">220</Voltage>
          <Bay name="Q1">
              <LNode lnInst="1" lnClass="PDIS" ldInst="F1"/>
              <ConductingEquipment name="I1" type="CTR">
              <Terminal connectivityNode="baden220_132/D1/Q1/L1"
              substationName="baden220_132" voltageLevelName="D1" bayName="Q1"
              cNodeName="L1"/>
              </ConductingEquipment>
              <ConnectivityNode name="L1" pathName="berlin220_132/D1/Q1/L1"/>
          </Bay>
      ...
```

FIGURE 62.24 Single-line diagram according to SCL schema.

The merging unit typically may send 80 samples per nominal period or 4.000 per second at 50 Hz systems. This means that for protection functions, the merging unit sends 4.000 messages per second over the network—continuously.

The substation specification description as partly shown in Figure 62.24 contains the complete topology of the substation from an electrical and information point of view. The information represents all information needed and provided by the underlying function like CILO for interlocking.

There are various tools available that support the design of the single-line diagram.

The next step in substation description is to map the information indicated by the shown LNs to real devices like merging units for the current and voltage sampling and SV exchange.

The third step in the engineering process is the configuration of the needed devices:

- Instantiate devices
- Define unique names for devices and logical devices
- Map logical in the single-line diagram to LNs in the instantiated devices
- Design communication network
- Design data flow

The design of the data flow (traffic engineering) is one of the most crucial steps to reach a stable communication system at the end of the day. The reason is quite simple: The traffic in systems like substation protection and automation will be characterized by cyclical and spontaneous information exchange. The transmission of SVs generates a constant flow, for example, 4.000 messages with samples of voltages and currents for protection devices. There is a lot of traffic that is issued by events (state changes, limit violations, quality changes, etc.) and alarms. The number of messages is increased when the monitoring process is checking for rather small changes (change of measurement by 1%) instead of changes in the range of 10%. IEC 61850 can also send messages cyclically—configured by a setting.

The settings for monitoring values and cyclical reports could end up in a situation that the many interworking devices exchange many more messages than required by the application. There could be showers or even avalanches of messages occur that congest the communication system to some degree, causing inacceptable delays.

```
<?xml version="1.0" encoding="UTF-8"?>
- <SCL xmlns="http://www.iec.ch/61850/2003/SCL" xmlns:xsd="http://www.w3.org/2001/XMLSchema">
      <!-- ICD file for Demo on Beck IPC com.tom WEB-PLC - generated for demo on 2013-07-15; 2013-07-14 19:30-->
      <!-- Field 1: digital=0/analog=1; Field 2: ind=0/ctrl=1; Field3: DO Index; Field4: value=0, q=1, t=2, d=3 -->
   <Header version="3" id=""/>
 + <Communication>
 - <IED name="ComTom" type="WEBPLC" configVersion="1.0" manufacturer="NettedAutomation GmbH">
   + <Services>
   - <AccessPoint name="SubstationRing1">
     - <Server timeout="30">
         <Authentication/>
       - <LDevice desc="" inst="LDevice1">
         - <LN0 inst="" lnType="LLN0_0" lnClass="LLN0">
           - <DataSet name="Indicate_DataSet">
               <FCDA ldInst="LDevice1" lnClass="GGIO" fc="ST" daName="stVal" doName="Ind1" lnInst="1" prefix="WEBPLC_"/>
               <FCDA ldInst="LDevice1" lnClass="GGIO" fc="ST" daName="stVal" doName="Ind2" lnInst="1" prefix="WEBPLC_"/>
               <FCDA ldInst="LDevice1" lnClass="GGIO" fc="ST" daName="stVal" doName="Ind3" lnInst="1" prefix="WEBPLC_"/>
               <FCDA ldInst="LDevice1" lnClass="GGIO" fc="ST" daName="stVal" doName="Ind4" lnInst="1" prefix="WEBPLC_"/>
             </DataSet>
           - <ReportControl name="UNBUFFERED_RCB" desc="Unbuf RCB" datSet="Indicate_DataSet" intgPd="5000" confRev="1"
             rptID="MyRepURCB_ID">
               <TrgOps period="true" dupd="true" qchg="true" dchg="true"/>
               <OptFields reasonCode="true" dataSet="true" timeStamp="true" seqNum="true"/>
             - <RptEnabled max="1">
                 <ClientLN iedName="MyClient" ldInst="none" lnClass="IHMI" lnInst="1" prefix=""/>
               </RptEnabled>
             </ReportControl>
           + <DOI name="NamPlt">
           + <DOI name="Mod">
           </LN0>
         + <LN inst="0" lnType="LPHD_0" lnClass="LPHD" prefix="">
         + <LN inst="1" lnType="GGIO_7" lnClass="GGIO" prefix="WEBPLC_">
         </LDevice>
       </Server>
     </AccessPoint>
   </IED>
 - <DataTypeTemplates>
   - <LNodeType id="LLN0_0" lnClass="LLN0">
       <DO name="Mod" type="INC_1"/>
       <DO name="Beh" type="INS_2"/>
       <DO name="Health" type="INS_3"/>
       <DO name="NamPlt" type="LPL_0"/>
     </LNodeType>
```


FIGURE 62.25 SCL document for a simple device.

For that reason, a good understanding of the needs is very crucial during the traffic engineering. Based on the needs, the following steps have to be processed:

- Configure reports and required data sets
- Preconfigure a client for the report
- Configure data flow for publish/subscriber services
- Configure GOOSE and SMV controls and required data sets
- Optionally configure subscribing IEDs
- Configure inputs to LNs
- Data required as inputs can be transmitted by GOOSE/SMV or as reports

An example of an SCL document of a simple device is shown in Figure 62.25.

The figure shows a device (IED) with a logical device LDevicel and logical nodes LN0, LPHD, and WEBPLC_GGIO1. The logical node LN0 comprises a data set and a report control block.

SCL documents of real devices are usually quite comprehensive exposing hundreds or thousands of signals modeled in umpteen LNs. The document usually is used to be imported by a device-specific configuration to generate a device-specific file that is loaded onto the device to configure the device. There are also devices that can interpret SCL files and configure the information and communication models during the start-up of the device. An example will be introduced further down in the clause.

62.8 Overview of Implementation Layering

One of the challenges with IEC 61850 is that there are multiple layers involved from models all the way down to messages exchanged over the network. The standard IEC 61850 is more than just a protocol. Therefore, the various levels have to be mapped to the next underlying layer and provide in the end messages that carry meaningful information—information that can be encoded and decoded in a multivendor environment to create interoperable systems.

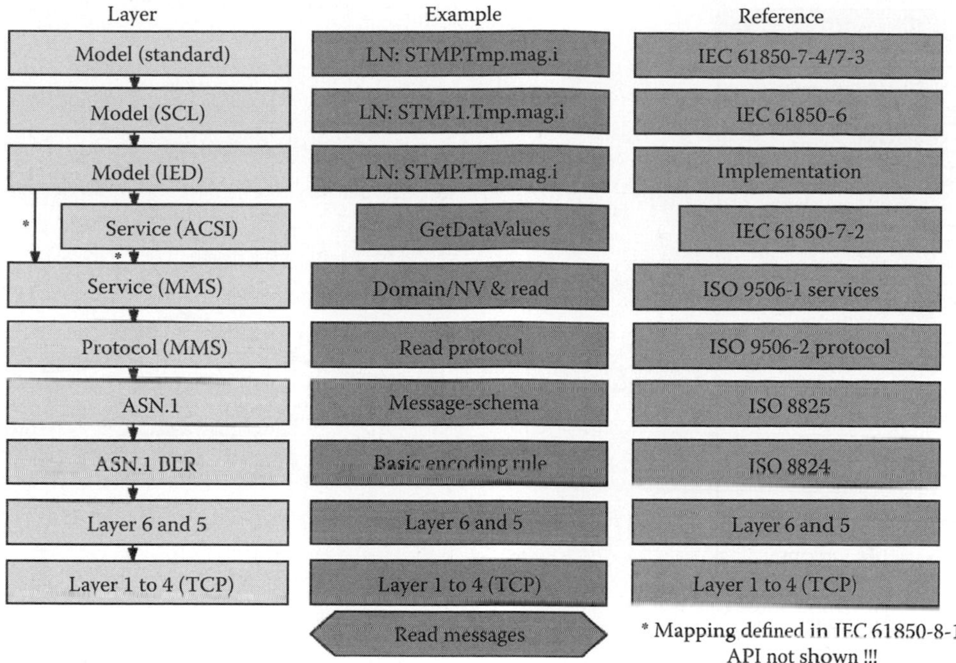

FIGURE 62.26 Layers from models to messages.

Figure 62.26 shows how the 10 layers involved from models down to messages. The left column defines the layer, the middle column shows an example, and the right contains the documents that define the corresponding layer.

The top layer shows the model of a temperature supervision LN STMP (see Figure 62.27).The model comprises data objects that represent information of specific monitoring functions. The attribute

LN: STMP.Tmp.mag.i
LN: STMP.Tmp.mag.i
LN: STMP.Tmp.mag.i
GetDataValues
Domain/NV & read
Read protocol
Message-schema
Basic encoding
Layer 6 and 5
Layer 1 to 4 (TCP)
Read messages

7-4 STMP class

Data object name	Common data class	Expla
LNname		
Data objects		
Descriptions		
EEname		
Status information		
EEhealth	ENS	External equipment health
Alm	SPS	Temperature alarm level reached
Trip	SPS	Temperature trip level reached
Measured and metered values		
Tmp	MV	Temperature
Controls		
OpCntRs	INC	Resettable operation counter
Settings		
TmpAlmSpt	ASG	Temperature alarm level set-point
TmpTripSpt	ASG	Temperature trip level set-point

7-3

instMag	AnalogueValue	MX	
mag	AnalogueValue	MX	dchg
range	Enumerated	MX	dchg
q	Quality	MX	qchg
t	TimeStamp	MX	

FIGURE 62.27 Model standard.

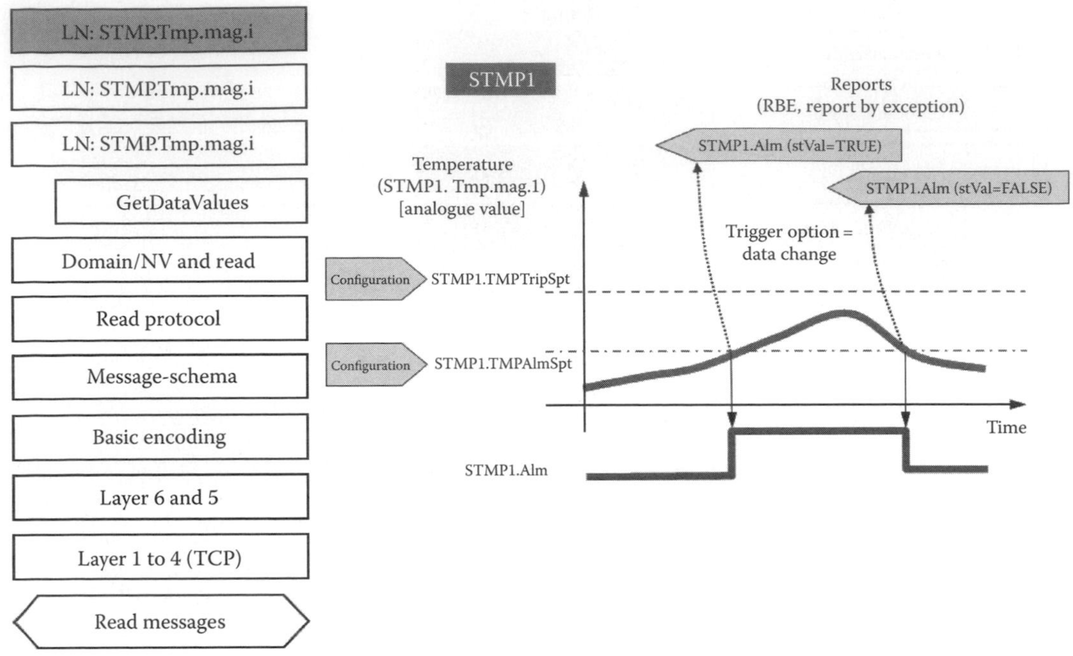

FIGURE 62.28 STMP supervision function.

STMP.Tmp.mag.i exposes the temperature value of a temperature sensor. The sensor itself is modeled with the LN TTMP.

The STMP LN is composed of a list of data objects that are used to define input and output values for the function. The main data object is the object Tmp. The temperature is an instance of the CDC MV. An MV has several data attributes like instMag (the current rms value), mag (a deadband filtered value), q (quality), and t (time stamp of the last change of the mag attribute). These attributes are all characterized as measurement (FC = MX). The last column in the MV CDC is defining the trigger option (dchg = data value change, qchg = quality change). The trigger option is used to control the reporting and logging procedures. If the value of mag changes (based on a percentage value change, e.g., value changes by 5%), then a report message with the new value, q and t, could be issued automatically.

Further data objects are TmpAlmSpt (alarm set point) and Alm (alarm). The diagram in Figure 62.28 shows the overview of the supervision function.

In this case, the supervision function is more or less trivial. The monitoring of temperatures is quite straightforward. Further monitoring features like supervising the change rate of the temperature and others are not (yet) standardized. The standard IEC 61850 is open to add more supervision functions in these basic supervision functions.

The second layer (see Figure 62.29) is the description of the model as an instance of the SCL schema. The example shown here is the LN element that has attributes lnClass (= STMP), an instance number 1, no prefix, and lnType (= STMP_0). The lnClass is the definition of the standard model; the lnType is a subset of the standard model and possibly including some preconfigured values like scaling factor or offset of an integer typed analog value. The underlying elements like DOI contain the data object instances—here the Alm, Tmp, and TmpAlmSpt.

The next level (see Figure 62.30) describes an example of the implementation of the model in a real IEC 61850-8-1 communication stack (using MMS) and API. The stack and application software use the private binding element (five values in five fields) to bind the application memory to the IEC 61850 information model (signal) and vice versa. These five fields are exchanged between the application and stack. This API is vendor specific.

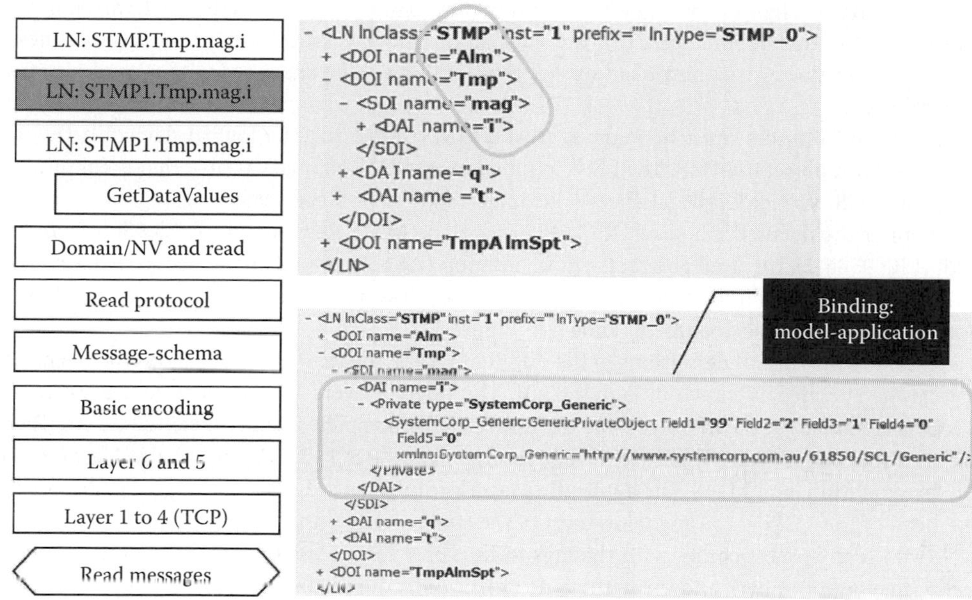

FIGURE 62.29 SCL document of example.

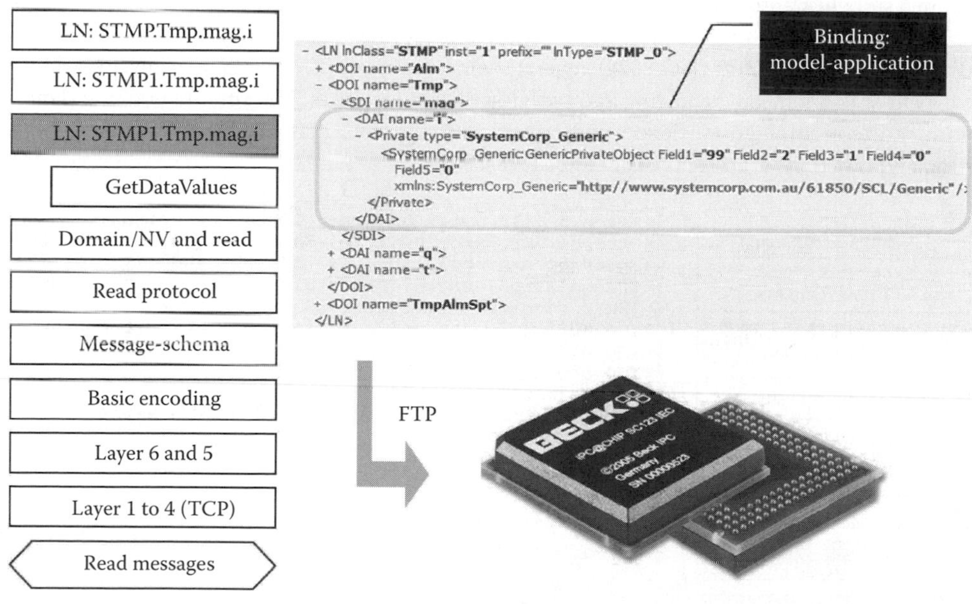

FIGURE 62.30 Model implementation with real stack and API.

The SCL document with that LN and binding is transferred to the embedded controller. This controller runs the stack, API, and application. When the application starts, it reads the SCL file, interprets, and implements the model. The stack instantiates all data sets and control blocks for reporting, logging, GOOSE, and SVs. There is no further configuration for IEC 61850 functions required. Note that the LNs are just interfaces to the functions. The real values of the data objects of the STMP LN are calculated and processed by the application program; it is outside of IEC 61850.

The application that figured out a new value of the STMP.Tmp.mag informs the stack through the API of a new value (API update function call). The stack autonomously issues a report message, or the client can read the new value by issuing a Read service request that will be answered with a Read response by the server device.

The model and services must be mapped to the MMS standard. The logical device in IEC 61850 maps to a Domain object in MMS. The LN is mapped to an MMS NamedVariable that is contained in a Domain. The ACSI service GetDataObjectValue maps to an MMS Read service.

The layers of the detailed mapping (IEC 61850-8-1), the MMS Read service (ISO 9506-1) and Read protocol (ISO 9506-2), the used abstract syntax notation (ASN.1) as well as the concrete basic message encoding (ASN.1 BER), and the other layers (Presentation, Session, Transport, Network, Data Link, and Physical) are defined in the documents shown in Figure 62.26.

As a result of applying all definitions of the different layers, the message that carries the Read service request can be encoded as shown in Figure 62.31. The various layers are indicated in the figure (using the Wireshark to analyze the message). The top indicates the complete frame of size 123 octets. Then the details like Ethernet, IP, TCP, TPKT (RFC 1006 for the mapping of ISO Transport Layer class 0 to TCP), session, presentation, and MMS. The application layer protocol encoding of messages is not important; the overhead of the TCP/IP routing takes most of the time.

The MMS Read request comes with the tags to interpret this message as a MMS Read request, the invokeID (message number), and the path name for the object to read: STMP1.Tmp.mag.i. The message in the mapping to MMS adds the functional constraint FC = MX into the name and replaces the "." with "$" due to restrictions in MMS.

It is important to note that all layers and the mapping to the corresponding underlying layer are defined in a suite of standards.

The message encoding of MMS has been criticized since the very beginning in the late 1980s. The encoding rules of ASN.1 BER (ISO 8825) generate some overhead. This could be neglected when using event-driven or spontaneous reporting and applying RBE mechanism. The amount of bits and messages to be transmitted with high-efficient encoding and applying polling is much higher than less efficient encoding and just a few crucial messages to be transmitted.

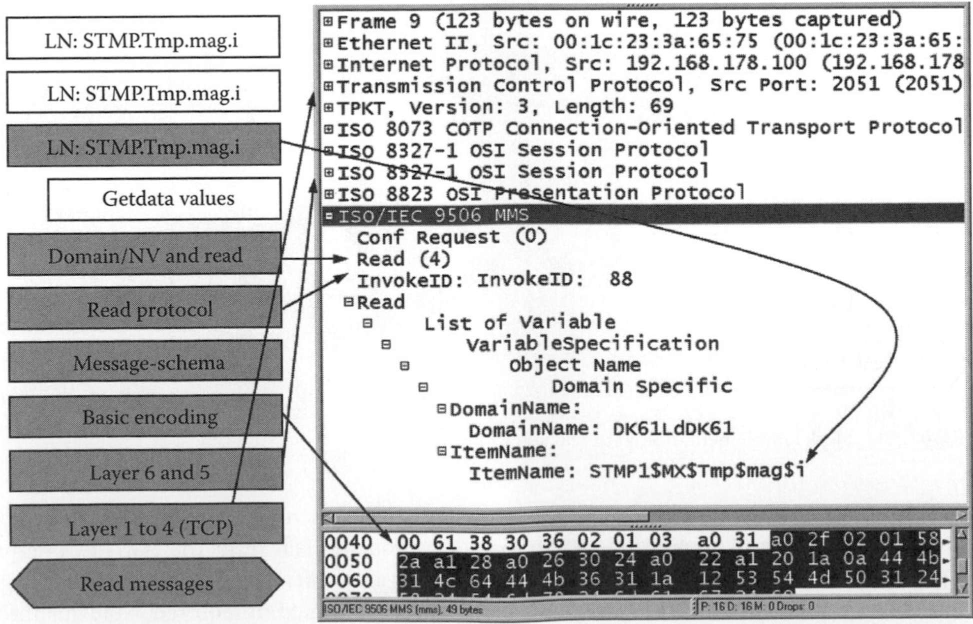

FIGURE 62.31 MMS Read request message example.

In addition to event-driven messages, it is also important to understand the process requirements and configure a system such that only the severe signals are transmitted. The overhead of a single message is one issue; another is the optimization of event and alarm generation and how these are packed into one or several messages. In case there is a data set comprising a lot of status signals to be monitored for changes, it is crucial to configure the system in a way that a minimum of messages are issued and transmitted.

IEC 61850 can collect several status changes that occur in a short time, for example, 1 s. All status changes that happen within 1 s after the first change of a signal in that data set will be sent together in one message after 1 s expired. If five changes happened in that second, there is a tremendous saving of bits and messages on the wire.

The optimum is mainly dependent on the configuration of the signal flow (traffic engineering). The configuration language of IEC 61850-6 carries the configuration values set by human beings that should know the applications needs very well. Garbage in–garbage out! IEC 61850 is just a tool that can be used by humans.

It is highly recommended to spend appropriate time in analyzing the application needs before any device is configured for flow control. This is true for any other communication system as well—even for communication that is based on cyclic polling. Polling does not scale if there are many signals that need to be communicated with short reaction time, especially in the case of very few events and alarms over time. Spontaneous messages require powerful receivers. In case of event showers these devices have to handle the amount of messages that arrive in a short time. The number of messages is more critical than the length of messages.

For real-time information exchange like GOOSE messages (repetitions in the range of milliseconds) and SV exchange (cyclic transmissions up to several thousand messages per second), the encoding is critical. IEC 61850 encodes the payload for these messages as an octet string with fixed-length values. The encoding and decoding are very efficient; the flexibility is of course reduced, and a change in the structure is not easy to figure out.

62.9 Related International Projects and Application Example

The requirement for secure and reliable information exchange is one of the most crucial requirements in the future of the power delivery systems. It is no surprise that many activities refer to the list of standards published by IEC and especially by the Technical Committee 57.

The following list shows some of the international projects that recommend applying international standard for information exchange and communication; most documents refer especially to IEC 61850:

- US NIST IOP Roadmap (Revision 2, 2012-02)
- www.nist.gov/smartgrid/upload/NIST_Framework_Release_2-0_corr.pdf
- IEC Smart Grid Standardization Roadmap
- http://www.iec.ch/zone/smartgrid/pdf/sg3_roadmap.pdf
- Hydro Quebec: Distribution System Automation Roadmap—2005–2020
- http://grouper.ieee.org/groups/td/dist/da/doc/2006-08-31HQDA_roadmap_final.pdf
- European Electricity Grid Initiative (EEGI) http://ec.europa.eu/energy/technology/initiatives/doc/grid_implementation_plan_final.pdf
- German BDI Internet of Energy—ICT for Energy Markets of the Future
- http://www.bdi.eu/bdi_english/download_content/Marketing/Brochure_Internet_of_Energy.pdf
- BDEW Roadmap—Realistische Schritte zur Umsetzung von Smart Grids in Deutschland
- http://www.bdew.de/internet.nsf/id/F8B8CCE3B35FF53BC1257B0F0038D328/$file/BDEW-Roadmap_Smart_Grids.pdf
- Microsoft SERA (Smart Energy Reference Architecture)

- http://download.microsoft.com/download/0/C/2/0C2F64B1-241D-4433-9665-5F802E7510D6/Microsoft%20Smart%20Energy%20Reference%20Architecture.pdf
- European Union: Mandate CEN/CENELEC M/490
- http://blog.iec61850.com/2012/12/europe-smart-grid-standards-are-on-way.html
- CEN-CENELEC-ETSI Smart Grid Coordination Group—Sustainable Processes
- http://ec.europa.eu/energy/gas_electricity/smartgrids/doc/xpert_group1_sustainable_processes.pdf
- CEN-CENELEC-ETSI Smart Grid Coordination Group—Smart Grid Reference Architecture
- http://ec.europa.eu/energy/gas_electricity/smartgrids/doc/xpert_group1_reference_architecture.pdf
- Smart Grids in Deutschland—Handlungsfelder für Verteilnetzbetreiber auf dem Weg zu intelligenten Netzen
- http://www.bdew.de/internet.nsf/res/86B8189509AE3126C12579CE0035F374/$file/120327%20BDEW%20ZVEI%20Smart-Grid-Broschuere%20final.pdf
- German Roadmap E-Energy/Smart Grid 2.0
- https://www.vde.com/en/dke/std/KoEn/Documents/DKE_Normungsroadmap_ENG-20%20-%20gV.pdf
- State Grid Cooperation China (SGCC) "Framework and Roadmap for Strong and Smart Grid Standards"
- http://collaborate.nist.gov/twiki-sggrid/pub/SmartGrid/SGIPDocumentsAndReferencesSGAC/China_State_Grid_Framework_and_Roadmap_for_SG_Standards.pdf
- Vattenfalls VHPREADY—Virtual Heat and Power
- http://www.vattenfall.de/de/vhp-ready.htm?WT.ac=search_success
- OGEMA—Open Energy Management Gateway
- http://ogema.org/

One of the key questions in the years to come is this: How can gateways for smart grid applications implemented that allow using the common RTU protocols and giving a migration path to IEC 61850? The following example is used to demonstrate that IEC 61850 could be implemented in small embedded devices.

62.9.1 Gateway for Smart Grid Applications

First implementations of IEC 61850 were focused on high-voltage substations. The devices and software for IEC 61850 were quite expensive, complex, and with restricted services. There were more of less just two vendors of stack software; that was acceptable for the beginning of the new approach. Many R&D projects all over interested in IEC 61850 lab or pilot projects failed to purchase or port the source code on their platforms.

The OGEMA project purchased a source code—which was at the end of the day not ported due to restricted resources. The need for simple stack and API for small devices in power generation, distribution automation, and smart grids was more than due.

In order to focus on the many challenges of the automation and management of the power system, it was quite important to hide details of MMS, IEC 61850 software, and their configuration—and to expose only those features that are very crucial. This would have allowed a fast start with low cost and low risk.

Due to the many challenges with the IEC 61850 implementation, many projects implemented appropriate LNs and data objects and used proprietary web services to exchange values with IEC 61850 models. One of the involved institutes (ISE—Fraunhofer Institute for Solar Systems in Freiburg, Germany) implemented MMS in Java and C as open-source projects. The open-source code implements a small subset of IEC 61850 models and services.

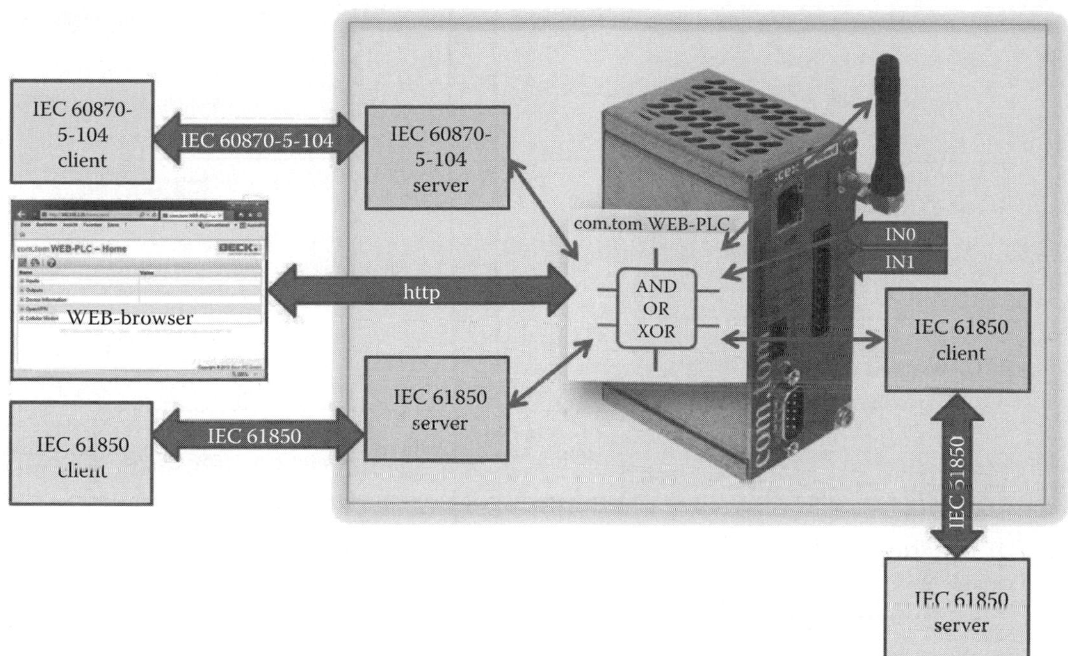

FIGURE 62.32 WEB-PLC and multiprotocol gateway.

Finally, for the OGEMA gateway, the IEC 61850 stack and API from SystemCorp was integrated into the OGEMA framework. This allows applying almost all services of IEC 61850-7-2 and IEC 61850-8-1 including reporting and GOOSE.

IEC 61850 has been successfully integrated as well into the embedded controller com.tom provided by Beck IPC (see Figure 62.32). This small platform uses a real-time operating system that has a lot of components like serial links, I/Os, wireless, GPRS, and openVPN already integrated in the platform. It supports all crucial protocols like IEC 61850, IEC 60870-5-104, DNP3, Modbus, CAN bus, and Profibus. The programming languages C/C#, IEC 61158-3 (PLC Programming languages), and a very simple web-based logic programming (WEB-PLC) are implemented.

The WEB-PLC could be used as core component to map signals from one communication system to another. All major protocols used in the energy world are implemented and could be used without a need for a vendor-specific tool—it requires just a standard web browser once connected to the server (see Figure 62.33).

Once the application (logic) diagram is drawn, the diagram can be loaded to the platform (com.tom) and started by pushing a button on the web browser. No vendor-specific engineering tool is required.

The WEB-PLC objects "IEC 61850 Indication1-from-DK61" and "IEC 61850 Indication2-from-DK61" are representing two binary signals from an underlying device with an IEC 61850 server. The two signals are reported from the server device at any time when they change. The two signals are used as input to the AND gate. The output of the AND gate is then communicated internally to the WEB-PLC output object IEC 60870-5-104 server Indication0-from-61850-Client. In case the two signals from the IEC 61850 server are both becoming true, then the AND output will change to true as well, and the change to true will be communicated to a master station via an IEC 60870-5-104 message.

Any control center, HMI or SCADA system, that could receive messages according to IEC 60870-5-104 would be able to receive the AND output signal. The signals from the IEC 61850 server could of course directly map 1:1 to an IEC 60870-5-104 or DNP3 message.

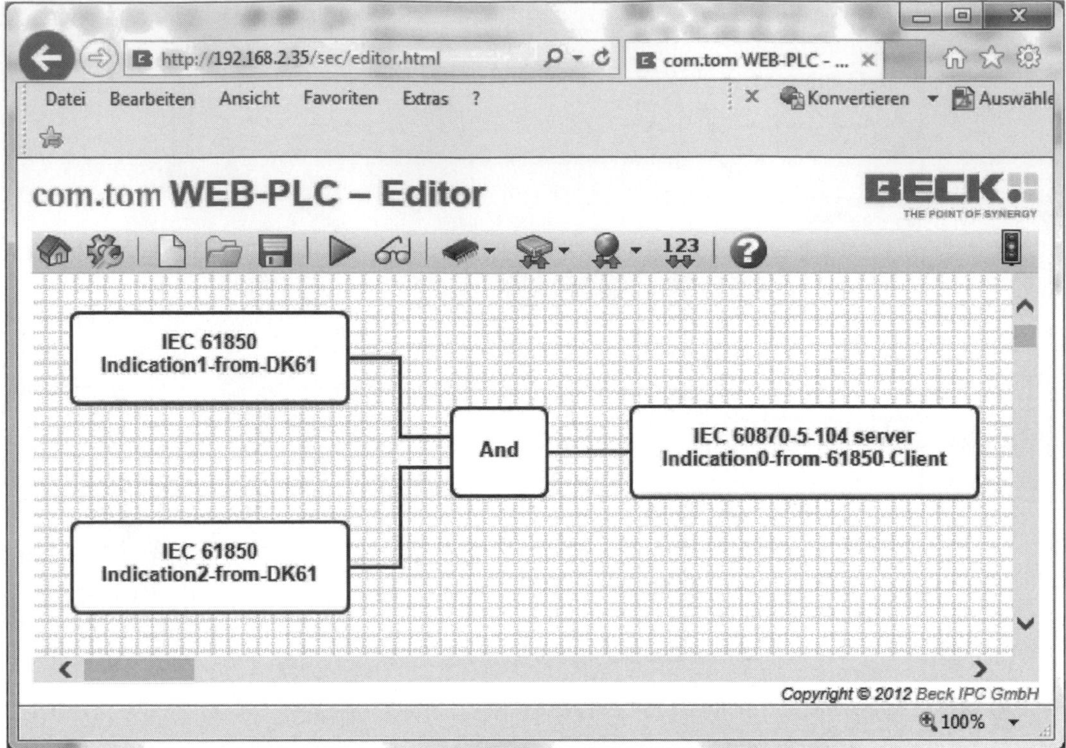

FIGURE 62.33 WEB-PLC to bind inputs to outputs.

In order to program a function all input and output objects (shown as rectangle icons of the WEB-PLC) on both sides (uplink with IEC 60870-5-104 and downlink with IEC 61850) need to be bound to the incoming respective the outgoing signals.

62.10 UCA International Users Group (UCA IUG) and Quality Assurance

The UCA International Users Group [11] is a not-for-profit corporation focused on assisting users and vendors in the deployment of standards for real-time applications for several industries with related requirements. The Users Group does not write standards; however, it works closely with those bodies that have primary responsibility for the completion of standards (notably IEC TC 57: Power Systems Management and Associated Information Exchange).

The UCAIug as well as its member groups (CIMug, Open Smart Grid, and IEC61850) draws its membership from utility user and supplier companies. The mission of the UCA International Users Group is to enable integration through the deployment of open standards by providing a forum in which the various stakeholders in the energy and utility industry can work cooperatively together as members of a common organization to do the following:

- Influence, select, and/or endorse open and public standards appropriate to the energy and utility market based upon the needs of the membership.
- Specify, develop, and/or accredit product/system-testing programs that facilitate the field interoperability of products and systems based upon these standards.

- Implement educational and promotional activities that increase awareness and deployment of these standards in the energy and utility industry.
- Influence and promote the adoption of standards and technologies specific to the ever-increasing smart grid initiatives worldwide.

The USAIUG is responsible for the definition of requirements and procedures for IEC 61850 conformance tests and accrediting and auditing test laboratories globally. The test procedures cover server tests, client tests, and tests of merging units.

Five third-party test labs and five company internal test labs are accredited. As of May 2013, the UCAIUG has certified 24 client devices, 344 server devices, and 9 merging units.

To collect and discuss technical errors and document the resulting correction, the standardization experts of the various groups that are defining, maintaining, and extending the standard definitions maintain the so-called tissue data base (technical issues) [12].

The tissue data base is a very helpful tool to keep track of the known and reported issues like errors, missing text, inconsistencies.

TÜV SÜD (Munich Germany) is one of the five third-party testers [13]. They are also deeply involved and support interoperability tests. It is understood that interoperability tests are very crucial for the acceptance and widespread use of the standard. TÜV SÜD acts also as a competence center for IEC 61850.

62.11 Third-Party Solutions for IEC 61850

The standard series IEC 61850 and IEC 61400-25 are quite comprehensive and abstract. One lesson learned since the mid-1980 (when MMS was defined and published) is this: Talking about benefits of a standard is needed, *but* it is most crucial that vendors are developing standard complaint devices. The implementation of MMS and the underlying layer services and protocols—as a core component in the message exchange between all devices in a system—was understood as calling for an unparalleled and huge effort and amount of resources.

Until 2010, there were mainly two source code solutions available. These two had been developed in the late 1980 and early 1990. Most of the early supporters and vendors of IEC 61850-compliant devices used one or the other of the two. Very few vendors implemented the complete stack and API on their own. If they did, it took years before they finished. Even with the source code, very often it took a year or two before the software was ported and integrated into the applications.
Stack and API especially for embedded systems

- SystemCorp, Beckenham, Western Australia, 6107 Australia
- http://www.systemcorp.com.au

Stack for general purpose

- Triangle Micro Works, Raleigh, North Carolina 27612 USA
- http://trianglemicroworks.com/

MMS for PC and other application

- SISCO, Sterling Heights, Michigan 48314-1408 USA
  http://sisconet.com/
- Monfox, Cumming, GA 30041 USA
  http://www.monfox.com/dosi/java-mms-sdk.html
- Open source (C and Java), OpenMUC,
  http://libiec61850.com/libiec61850/
  http://www.openmuc.org/index.php?id=24

References

1. IEC 61850—Communication networks and systems for power utility automation

Part 1	Introduction and overview
Part 2	Glossary
Part 3	General requirements
Part 4	System and project management
Part 5	Communication requirements for functions and device models
Part 6	Configuration description language for communication in electrical substations related to IEDs
Part 7-1	Basic communication structure—Principles and models
Part 7-2	Basic information and communication structure—Abstract communication service interface (ACSI)
Part 7-3	Basic communication structure—Common data classes
Part 7-4	Basic communication structure—Compatible logical node classes and data object classes
Part 7-410	Hydroelectric power plants—Communication for monitoring and control
Part 7-420	Basic communication structure—Distributed energy resources logical nodes
Part 7-5	Usage of information models SAS (substation automation system)
Part 7-500	Use of LN to model functions (SAS)
Part 7-510	Basic communication structure—Hydroelectric power plants—Modeling concepts, and guidelines
Part 7-520	Use of LN (DER)
Part 8-1	Specific communication service mapping (SCSM)—Mappings to MMS (ISO 9506-1 and -2) and to ISO/IEC 8802-3
Part 8-2	Web service mapping
Part 9-2	Specific communication service mapping (SCSM)—Sampled values over ISO/IEC 8802-3
Part 10	Conformance testing
Part 10-2	Interoperability test for hydro equipment
Part 80-1	Guideline to exchanging information from a CDC-based data model using IEC 60870-5-101 or IEC 60870-5-104
Part 90-1	Use of IEC 61850 for the communication between substations
Part 90-2	Using IEC 61850 for SS-CC communication
Part 90-3	Using IEC 61850 for Condition Monitoring
Part 90-4	Network Engineering Guidelines
Part 90-410	Communication network structures in hydropower plants
Part 90-5	Use of IEC 61850 to transmit synchrophasor information according to IEEE C37.118
Part 90-6	Use of IEC 61850 for Distribution Automation
Part 90-7	Object Models for PV, Storage … inverters
Part 90-8	Object Models for Electrical Transportation
Part 90-9	Object Models for Batteries
Part 90-10	Object models for schedules
Part 90-11	Methodologies for modeling of logics for IEC 61850-based applications
Part 90-12	Wide-area network engineering guidelines
Part 90-13	Extension of IEC 61850 information models to include logical nodes and data models for steam and gas turbines
Part 90-14	Using IEC 61850 for FACTS data modeling
Part 90-15	Hierarchical DER system model
Part 100-1	Functional testing of IEC 61850-based systems

2. IEC 61400-25—Wind turbines—Communications for monitoring and control of wind power plants
   Part 1 Overall description of principles and models
   Part 2 Information models
   Part 3 Information exchange models
   Part 4 Mapping to communication profiles
   Part 5 Conformance testing
   Part 6 Logical node classes and data classes for condition monitoring
3. IETF EMAN (Energy Management Group), http://tools.ietf.org/wg/eman/.
4. ISO/TC 205/WG 3 (Building Automation and Control System (BACS) Design) [ISO/TC 205/SC N 410].
5. SunSpec Alliance, http://www.sunspec.org/about/.
6. http://www.sma.de/en/company/pv-electricity-produced-in-germany.html.
7. IEC 61158—Industrial communication networks—Fieldbus specifications (93 parts).
8. IEC 61784—Industrial communication networks—Profiles (21 parts).
9. IEC 60870 6 TASE.2: Telecontrol application service element 2 (ICCP—Intercontrol-center communication protocol).
10. IEEE Technical Report 1550 (1999): Utility Communications Architecture, UCA; http://www.nettedautomation.com/standardization/IEEE_SCC36_UCA.
11. UCA International Users Group, http://www.iec61850.ucaiug.org.
12. Technical Issues, IEC 61850 http://tissue.iec61850.com.
13. TÜV SÜD testing, http://www.tuev-sued.de/home-en/services-by-industry/cross-sector-services/embedded-systems/iec-61850/iec-61850-conformance-test.

More information on the IPC@CHIP can be found in English and German:
http://www.beck-ipc.com.
General information, trends, and news on IEC 61850:
http://blog.iec61850.com.

63

Fundamentals of the IEC 61400-25 Standard: Communications for Monitoring and Control of Wind Power Plants

Federico Pérez
University of the Basque Country

Elisabet Estévez
University of Jaén

Isidro Calvo
University of the Basque Country

Darío Orive
University of the Basque Country

Marga Marcos
University of the Basque Country

63.1 Introduction 63-1
63.2 Overall Description of the IEC 61400-25 Series 63-2
Wind Power Plants • Requirements on Communications • Communication Model
63.3 Wind Power Plant Information Model 63-5
Logical Device • Logical Node • Data Class
63.4 Wind Power Plant Information Exchange Model 63-11
Communication Functions • Abstract Communication Service Interface • Service Modeling Convention
63.5 Mapping to Communication Protocols 63-15
Mapping to SOAP-Based WSs • Mapping to OPC XML DA • Mapping to MMS • Mapping to IEC 60870-5-104 • Mapping to DNP3
63.6 Condition Monitoring 63-24
63.7 Conclusions 63-24
References 63-25
Additional Web Information 63-25

63.1 Introduction

Building, managing, and monitoring systems for renewable energy plants are challenging tasks since they require the integration of big amount of data of diverse nature produced by different equipment from several vendors. Commonly, manufacturers provide proprietary software packages for specific data access. This introduces a higher level of complexity to the operators, administrators, and system integrators as they must deal with equipment from different vendors offering specific data access and formats.

In this scenario, the International Electrotechnical Commission (IEC) (Technical Committee TC88/PT25) proposed a new communications standard for the wind power industry aiming at providing a common communication approach for wind power plant (WPP) monitoring and control. This standard makes the integration of heterogeneous devices in large and complex WPPs easy, and at present, it is the only available standard for wind turbine (WT) communications.

The IEC 61400-25 standard may be regarded as an extension of the IEC 61850 standard [1,13], which deals with communication networks and systems in substations. More specifically, the IEC 61850 series cover the integration of the so-called intelligent electronic devices (IEDs), that is, any device incorporating one or more processors with communication capabilities within substation automated systems (SASs). IEC 61400-25 reuses the information and exchange models defined in IEC 61850 for electric power systems but adapted to WPPs. Thus, it defines the relevant information specific to WPPs as well as manufacturer-independent mechanisms for information exchange.

The communication approach is based on three different areas: (1) the *information model* (IM), which defines the hierarchy of relevant information in WPPs using standardized names and semantics; (2) the *information exchange model* (IEM), which defines the access to the content and structure of the WPP IMs; and (3) the *mapping* of these two models to *standard protocol stacks*, such as OPC [2], MMS [3], or DNP3 [4]. It is important to remark that feeders and substation equipment are not covered in IEC 61400-25 as they are included in IEC 61850.

As a result, the IEC 61400-25 series allows applications such as SCADA (Supervisory Control and Data Acquisition) systems, local real-time build-in control systems, or energy dispatch centers, to access WPP information data from different vendor equipment while providing scalability, connectivity, and interoperability.

In 2008, the USE 61400-25 (IEC 61400-25 users group) [5] was created to facilitate the adoption of the standard in the wind power generation industry. Major WT vendors, system integrators, and users are actively involved in this initiative.

63.2 Overall Description of the IEC 61400-25 Series

The main objective of the IEC 61400-25 series is to define a standard basis for manufacturer-independent communications for monitoring and control of WPPs. They define what vendors (manufacturer and suppliers) of WPP's components should implement in their systems and devices to be compliant to the standard and consequently to achieve vendor-independent interoperability between the WPP and external actors (applications).

The IEC 61400-25 standard is divided into six different parts:

1. IEC 61400-25-1 Overall description of principles and models: Part 1 gives a general overview of the standard, analyzing the major communication requirements, establishing the terminology used through all the parts, and giving a brief description of the information and exchange models.
2. IEC 61400-25-2 Information models: Part 2 defines the elements of the models representing the WPP's components from the communication point of view (data objects) and the naming convention for both the elements and the data (communication content).
3. IEC 61400-25-3 Information exchange models: Part 3 is dedicated to the description of the data exchange mechanisms, such as authentication, control commands, subscription to data publication, or device description.
4. IEC 61400-25-4 Mapping to communication profiles: Part 4 proposes several communication profiles for mapping the models described in previous sections to standard communication protocols.
5. IEC 61400-25-5 Conformance testing: Part 5 defines methods and abstract test cases for conformance testing of devices used in WPPs.
6. IEC 61400-25-6 Logical node (LN) classes and data classes (DCs) for condition monitoring: Finally, part 6 defines additional IMs to be used in condition monitoring systems.

The IEC 61400-25 series defines models, but recommendations for implementing the communication interface and the application program interface are not included.

63.2.1 Wind Power Plants

According to the IEC 61400-25 series, WPPs may be regarded as a particular case of SAS of the IEC 61850 series. WPPs are complex systems composed of the following components:

- *WT* is the main component of a WPP, and it is responsible for generating energy by converting kinetic wind energy into electric energy. The main subcomponents within a WT are the rotor, the transmission system, the generator, the converter nacelle, the yaw system, the tower, and the alarm system.
- *Meteorological system* is in charge of monitoring the ambient conditions. Ambient variables such as the wind speed and direction, the pressure, or the temperature are measured and provided for different purposes such as correlating the meteorological data to the electrical energy output.
- *WPP management system* is responsible for ensuring that the complete system adapts itself to the static and dynamic conditions as well as to the requirements of the electrical power connection (i.e., interoperation of the WTs with substations and other power network–related devices).
- *Electrical system* is responsible for collecting and transmitting the energy produced in the WTs.

63.2.2 Requirements on Communications

Monitoring and control operations over WPPs are typically executed by SCADA systems, local real-time build-in control systems, or energy dispatch centers. In this scenario, communication technologies play an important role since they enable the communication between these systems and the whole WPP or any of its components, that is, the individual devices. WTs include local controllers, which are primarily responsible for the monitoring and control of the internal processes within the WT. Besides, these local controllers must provide the mechanisms to interact with external systems or applications performing monitoring and control operations of the whole plant.

The communication requirements referred to both the content (type and formats) and the functions to get or set such content, that is, the communication functions. The following sections briefly summarize the requirements on communication defined in the IEC 61400-25 series.

63.2.2.1 Communication Content

The IEC 61400-25 series differentiates five types of information: (1) *process information* that contains information about the behavior and the current state of systems and components; (2) *statistical information* that allows evaluating the operation of WPPs; (3) *historical information* in the style of logs and reports, useful to track operational trends; (4) *control information* intended to perform control operations over the WPP, such as access profiles, set points, parameters setting, and commands; and (5) *descriptive information* that is related to the type and accuracy of the information.

63.2.2.2 Communication Functions

External systems and applications that perform the control and monitoring of WPPs need to be provided with specific functions to operate and manage WPPs. They are divided into two categories:

1. *Operational functions*: They are used by the external systems to get information from/send control commands to WPPs, as Table 63.1 illustrates.
2. *Management functions*: They are used by the external system to manage users and access to WPP data. Management functions are illustrated in Table 63.2.

TABLE 63.1 Operational Functions

Operational Functions	Range of Use
Monitoring	Local or remote observation of a system or process in order to track any changes that may happen over time. It may include single data values or groups of data values
Control	Changing, modifying, intervening, switching, controlling, parameterizing, and optimizing WPPs
Data retrieval	Collecting of WPP data
Logging	Recording of data and events in chronological order
Reporting	Data transfer initiated from a server triggered by event or condition

TABLE 63.2 Management Functions

Management Functions	Range of Use
User/access management	Setting up, modifying and deleting users, managing access rights, monitoring access
Time synchronization	Synchronization of devices within a communication system
Diagnostics (self-monitoring)	Self-monitoring of the communication system
System setup functions	Definition of the way the information exchange takes place: setting, changing, and retrieval of system setup data

63.2.3 Communication Model

The IEC 61400-25 series defines an abstract communication model for monitoring and controlling the operation of WPPs. As in the IEC 61850 series, the communication model follows the client–server paradigm. The approach is based on the definition of *abstract classes* and *services* independent of specific implementations of the protocol stack as well as of the underlying operating system.

The communication model, depicted in Figure 63.1, uses three different views:

1. The *IM* standardizes object-oriented abstract models that represent the components of a WPP (devices) and the information to be exchanged between the devices within WPPs and the external world. The modeling approach represents the physical devices of the WPP as a set of reusable DCs associated to devices through logical entities. DCs can be instanced in objects that contain information such as analog values, binary status, commands, or set points. The IM is described in IEC 61400-25-2.
2. The *IEM* describes an abstract representation of the services available at the servers through which the clients can access and modify the IM data. This model is described in IEC 61400-25-3.
3. *Mapping to standard communication profiles*: The information exchange between a client and a server requires a uniform communication protocol stack at both ends in order to transport the data and implement the services defined by the IM and IEM, respectively. The mapping is structured in layers according to the OSI reference model [14], and it specifies the communication protocol stacks used in the IEC 61400-25 series. It is described in IEC 61400-25-4.

Although the IEC 61400-25 standard defines the model of information data, the model of exchange services, and the mapping to standard communication protocols, it does not give any recommendation on implementation issues. The distribution of objects among the servers available or the implementation of the modeling approach on the servers is completely open.

As commented earlier, the IEC 61400-25 series can be considered as an extension of the IEC 61850 series, and thus, it uses the communication structure described in IEC 61850-7 [1]. In particular, IEC 61400-25-3 describes the communication services described in IEC 61850-7-2 (Abstract Communication

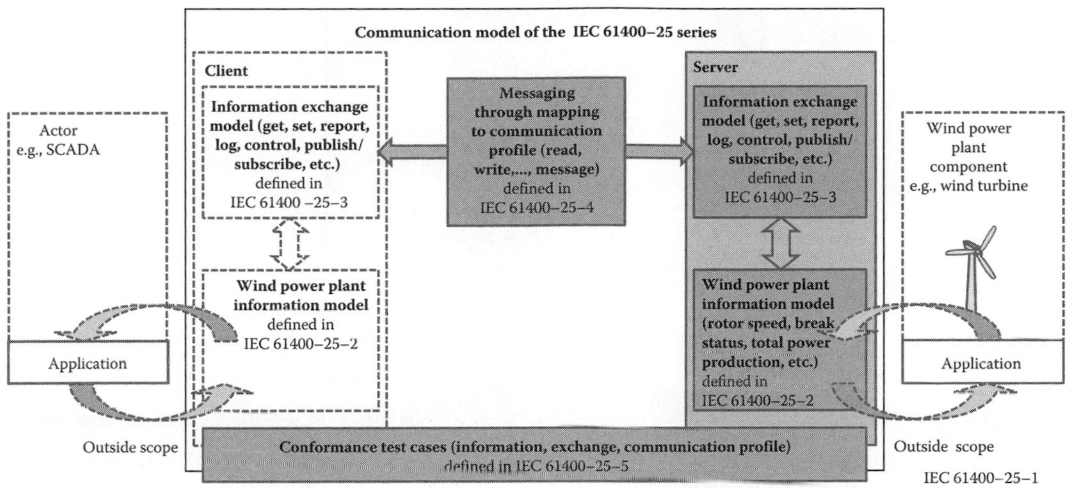

FIGURE 63.1 Conceptual communication model of the IEC 61400-25 series.

Service Interface [ACSI]) [6]. The IM proposed in IEC 61400-25-2 conforms to IEC 61850-7-3 (common data classes [CDC]) [7] and IEC 61850-7-4 (compatible logical node classes and data classes) [8]. Neither IEC 61850 nor IEC 61400-25 standards define new process data, but they assign data and data types to physical entities. The mapping described in IEC 61400-25-4 follows the same methodology as the one described in IEC 61850-8 (specific communication service mapping [SCSM]) [9], and it uses the device description approach specified in IEC 61850-6 (configuration description language for communication) [10]. As a consequence, the WPP models can be accessed from tools and communication protocols conforming to the IEC 61850 standard.

63.3 Wind Power Plant Information Model

The IM of WPPs defines the relevant aspects of the real system from the point of view of monitoring and control, abstracting the nonessential details. All the information to be exchanged between devices of the WPP or between a device and the external world is defined in the model. The target is to have a standard frame of interpretation via which the server may process all data into relevant and semantically standardized information and may grant the client access to these data in a component-oriented view.

The conceptual modeling approach is depicted in Figure 63.2. It abstracts the physical devices of the real world in logical devices (LDs) to which relevant and semantically standardized information is associated. The server processes this information and makes it available to clients.

The central part of Figure 63.2 illustrates the hierarchy of the WPP IM. It is a model of the real world consisting of physical devices. In the context of the IEC 61400-25 modeling approach, a physical device is a device connected to the network (a server). They are typically defined by their network address. A physical device may have associated one or more LD. An example of LD is a WT. LDs are decomposed into LNs representing the physical device's components with given functionality, for example, an LD representing a WT can be composed of a rotor (with a standardized name WROT), a transmission (WRTM), a generator (WGEN), a nacelle (WNAC), and a tower (WTOW) among others. LNs are the basic objects containing a collection of data objects, instances of predefined DCs, depending on its functionality. Each data class inherits a collection of properties, as defined by the corresponding CDC. A CDC consists of a collection of data records (*attributes*). DCs are standard templates for representing the exchangeable information through the services defined by the IEM. An example of a DC instance is represented in Figure 63.2, corresponding to an instance of the DC RotSpd that contains the speed of the rotor of the WT.

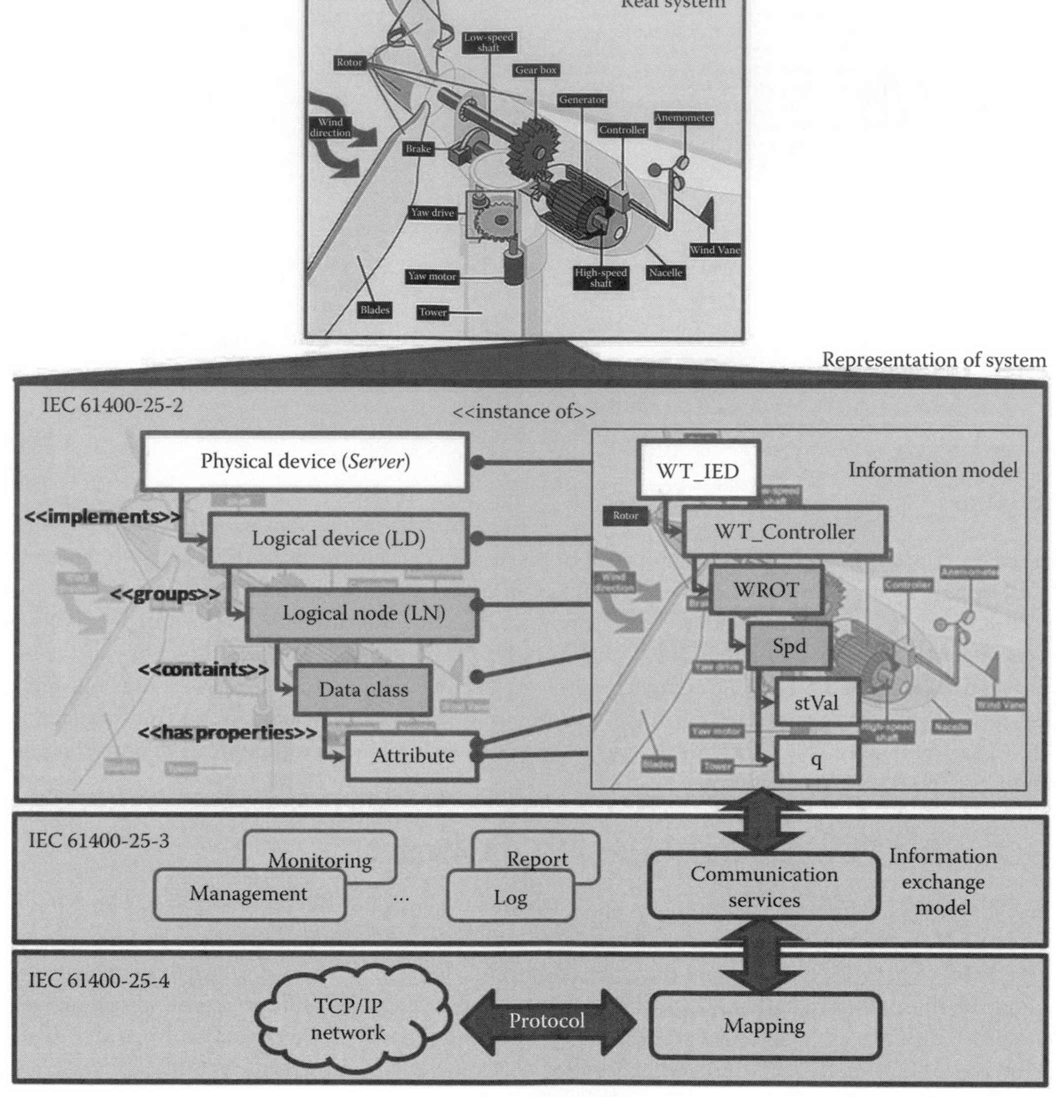

FIGURE 63.2 Modeling approach of the IEC 61400-25 standard.

The IM of a concrete WPP is then an instance of the metamodel depicted in Figure 63.3. The following sections detail the basic concepts of the metamodel, that is, the concept of LD, LN, and DC.

63.3.1 Logical Device

As commented earlier, an LD encapsulates all the needed LNs that contain the data corresponding to a WPP device. An LD residing on a server is composed of a collection of specific LNs that will further contain data instances reflecting the actual physical state of the real WPP device.

Besides the specific LNs within a WPP device, an LD also contains mandatory data related to the physical device (IED or server) that hosts it and WPP-independent information. This common information is contained in specific LNs serving only for this purpose: the logical node physical device (LPHD)

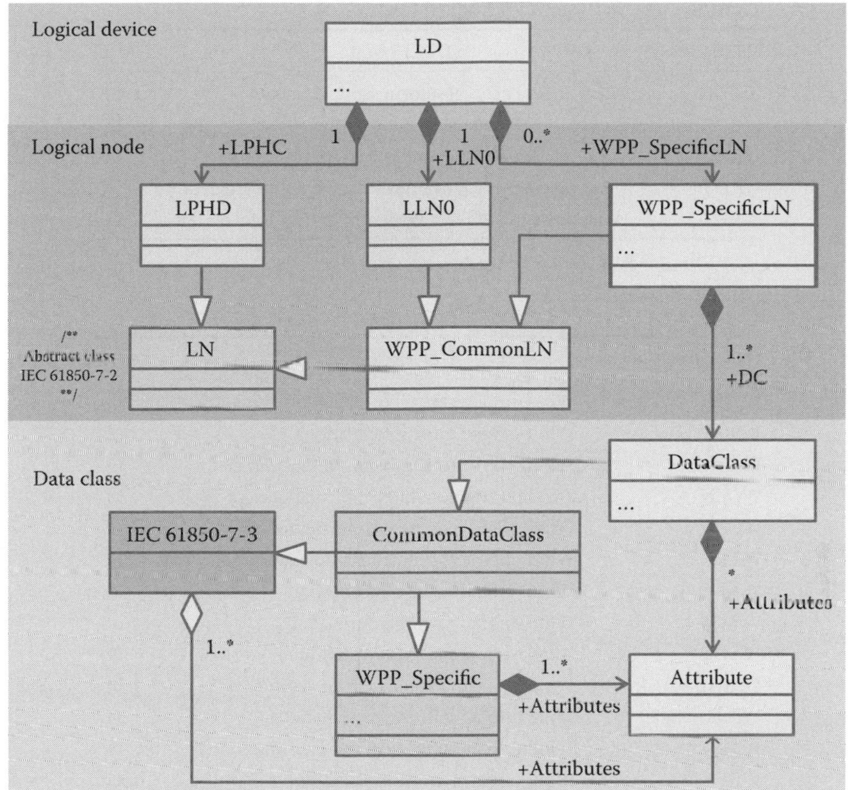

FIGURE 63.3 Metamodel of the WPP information model.

represents common data of the physical device hosting the LD. The logical node zero (LLN0) represents information specific to the LD, such as the LD name space. They are called *system-specific* LNs.

As any other LN, LPHD and LLN0 must belong to an LD. If the WPP only has one LD associated, then it will also contain LPHD and LLN0. However, if there is more than one LD, LPHD can be associated to the specially named LD0, together with an LLN0 and the common data for all LNs (communication-related data sets and control blocks). This allows dedicating the rest of LDs exclusively for functional modeling. These *system-specific* LNs are described in Clause 8 of IEC 61850-7-1.

63.3.2 Logical Node

All functions performed in WPPs have been split into the smallest entities that communicate with each other and with the external world. These entities or objects are called *WPP_Specific Logical Nodes* (WPP_SpecificLN in Figure 63.3). They contain all exchangeable function-related data and their attributes.

All LN classes defined in IEC 61400-25-2 series inherit their structure from the LN abstract class (see Figure 63.3) defined in 9.1.1 of IEC 61850-7-2. Apart from the LPHD LN class, LLN0 and WPP-specific LNs inherit at least the mandatory information of the WPP common LN (WPP_CommonLN in Figure 63.3). The features of LNs are detailed in 9.1.1 of IEC 61850 standard.

The standard defines a set of LNs specific to WPPs (WPP_SpecificLN in Figure 63.3). They have a standardized mnemonic name of four letters, beginning with "W" followed by three capitals representing the content. Instances of these LNs may be implemented single or multiple in any server, but their name must be unique. Commonly, LNs for common applications are grouped into the same LD. As commented earlier, the common device properties are described in the LPHD.

TABLE 63.3 WPP-Specific Logical Nodes

LN Classes	Description	M/O
WTUR	Wind turbine general information	M
WALM	Wind power plant alarm information	O
WMET	Wind power plant meteorological information	O
WAPC	Wind power plant active power control information	O
WRPC	Wind power plant reactive power control information	O

TABLE 63.4 Wind Turbine–Specific Logical Nodes

LN Classes	Description	M/O
WTUR	Wind turbine general information	M
WROT	Wind turbine rotor information	M
WTRM	Wind turbine transmission information	O
WGEN	Wind turbine generator information	M
WCNV	Wind turbine converter information	O
WTRF	Wind turbine transformer information	O
WNAC	Wind turbine nacelle information	M
WYAW	Wind turbine yawing information	M
WTOW	Wind turbine tower information	O
WALM	Wind power plant alarm information	O
WSLG	Wind turbine state log information	O
WALG	Wind turbine analog log information	O
WREP	Wind turbine report information	O

The LN classes can be instantiated several times on the device, implementing its functionality. If not mentioned especially, this document always means instances when talking about LNs in the context of modeling.

LNs consist of a collection of related data, defined by means of the corresponding DCs. All the information in an LN is contained in instances of DCs. The structure of all LNs is similar, and it is possible to build different types of LNs by combining different optional DCs.

Clause 5 of IEC 61400-25-2 defines five categories of WPP-Specific LNs, as Table 63.3 illustrates, being the WT general information (WTUR) the only mandatory LN.

A WT model is composed of a set of turbine-specific LNs. Table 63.4 lists the WT-specific LNs highlighting the mandatory ones.

As shown in Tables 63.3 and 63.4, the information is mainly modeled by a set of LN classes that correspond to the physical turbine physical components. The only exception is the information related to alarms that are represented in a separate LN. There are also specific LNs for logged events (status, alarms, commands, event counters, and state timers) and logged analog time series (long period, demands, and transient recording).

Figure 63.4 illustrates an example of real WT that instances different LNs.

63.3.3 Data Class

As commented in Section 63.2.2.1, the information related to WPPs is organized into three main categories: (1) *process information* that stands for manufacturer-independent features, state status, and analog behavior information; (2) *control information* that is related to commands, set points, and configuration parameters; and (3) *derived information* that consists of statistical, logging, and report information.

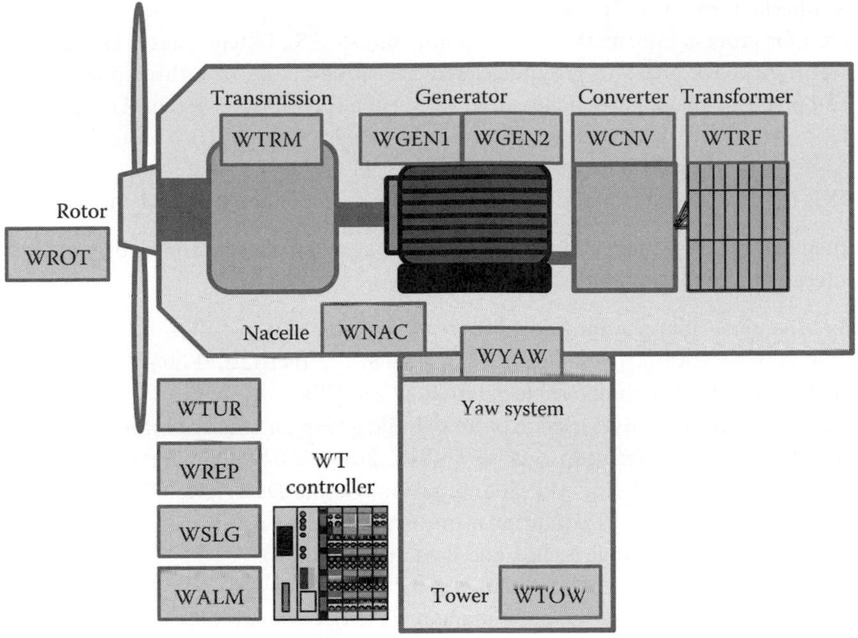

FIGURE 63.4 LN instances in a real WT.

Clause 6.2 of the IEC 61400-25-2 series defines the DCs corresponding to LNs of a WPP. Figure 63.5 illustrates the general format of DCs.

Based on WPP information requirements, a set of specific CDC have been specified. The name of all of them is unique and is formed by three capital letters. They can be grouped into two main groups (see Data Class elements in the metamodel in Figure 63.3):

A. WPP-specific CDCs (WPP_Specific)
   1. WPP_Specific DC for Process Information: Set Point Value (SPV) and STatus Value (STV)
   2. WPP_Specific DC for Control Information: ALarM (ALM) and CoMmanD (CMD)
   3. WPP_Specific DC for Derived Information CDCs: Event counting (CTE), State TiMing (TMS), and Alarm Set Status (ASS)

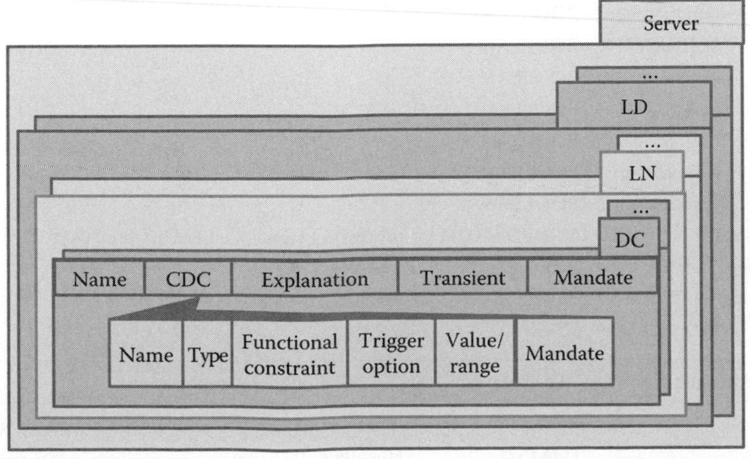

FIGURE 63.5 Identification of DC attributes.

B. CDCs inherited from IEC 61850-7-3
   1. DCs for process information: single point status (SPS), INteger status (INS), binary counter reading (BCR), phase-to-ground-related measured values of a three-phase system (WYE), and phase-to-phase-related measured values of a three-phase system (DEL)
   2. DCs for control information: controllable single point (SPC), controllable integer status (INC), and controllable analog process value (APC)
   3. DCs for derived information: Logical name plate (LPL) and device name plate (DPL)

As commented earlier, all of them collect a set of information attributes. The fields required for a complete characterization of the attributes are as follows:

1. *Name*: Mnemonic abbreviation of the attribute (e.g., clockFailure).
2. *Type*: It can be BASIC (e.g., BOOLEAN, INT8 …, INT32U, FLOAT32, CODED ENUM, ENUMERATED, …) or composed (e.g. PACKAGE LIST).
3. *Functional Constraint*: This is used in order to build group for efficient information exchange. The possible list of values is defined in clause 2 of IEC 61850-7. Examples are: status (ST), Measurand (MX), Control (CO), Set Point (SP), Configuration (CF), or Description (DC).
4. *Trigger Option*: It is used as conditional notification when a value or state has changed. Data change (dchg), quality change (qchg), and data update (dupd).
5. *Explanation/Range*: Description and range of an attribute record.
6. *Mandate*: It indicates if the attribute is mandatory (M), optional (O), or conditional (C).

As an illustrative example of CDCs, Table 63.5 presents the attributes that Characterize Quality and Analog CDCs. Quality CDC is defined by five attributes. Validity and source are *enumerated*; detailQual is a complex attribute constituted by other eight basic *Boolean* fields. Finally, test and operatorBlocked attributes are *Boolean*. Among those attributes, only the former is mandatory and the rest are

TABLE 63.5 Examples of Characterization of Common Data Class Attributes

Quality			
Attribute Name	Attribute Type	Value/Range/Explanation	M/O
	PACKED LIST		
validity	CODED ENUM	good|invalid|reserved|questionable	M
detailQual	PACKED LIST		O
overflow	BOOLEAN		O
outOfRange	BOOLEAN		O
badReference	BOOLEAN		O
oscillatory	BOOLEAN		O
Failure	BOOLEAN		O
oldData	BOOLEAN		O
inconsistent	BOOLEAN		O
Inaccurate	BOOLEAN		O
source	CODED ENUM	process|substituted, DEFAULT process	O
test	BOOLEAN	DEFAULT FALSE	O
operatorBlocked	BOOLEAN	DEFAULT FALSE	O

Analogue Value			
Attribute Name	Attribute Type	Value/Range/Explanation	M/O
i	INT32	Integer value	(*)
f	FOAT32	Floating point value	(*)

*, At least one of the attributes must appear in the instance.

TABLE 63.6 CDC: Status Value (STV)

STV—Status Value					
Attribute Name	Attribute Type	FC	TrgOp	Explanation and Value/Range	M/O
Data Name	Inherited from DC (see IEC 61850-7-2)				
Data					
Status information					
actSt	INS			Actual status	M
stVal	CtxInt	ST	dchg		M
q	Quality	ST	qchg		M
t	TimeStamp	ST			M
oldSt	INS			Previous status	O
stVal	CtxInt	ST	dchg		M
q	Quality	ST	qchg		M
t	TimeStamp	ST			M
Statistical information					
stTm	TMS			Time duration of active status	O
stCt	CTE			Number of changes to active status	O

Note: Configurations, Description, and Extension Information Services (detailed in Section 63.4).

optional. The analog CDC is composed of two attributes: one for integer values (*i*) and the other for float values, being mandatory to have one of them in an instance.

Following the same philosophy, Table 63.6 illustrates the characterization of the Status Value WPP-specific CDC. STV data class contains attributes that represent the status information of a status data. In this case, a complex attribute for actual status (actSt) is mandatory. Optionally, it could be information about the previous status using the same complex attribute (oldSt).

63.4 Wind Power Plant Information Exchange Model

The IEC 61400-25 standard provides a platform-independent interface between a client and a server through the WPP IM and the IEM. In Section 63.3, the IM, which provides a uniform, component-oriented view of the WPP data, has been described. This section focuses on the IEM, which defines an ACSI with a complete collection of services that represent the whole functionality of the server. More specifically, these services include the following operations: (1) data access and retrieval; (2) device control; (3) event reporting and logging; (4) publish/subscribe of data; (5) self-description of devices; and (6) retrieving the WPP IM.

It is important to remark that the IEC 61400-25 standard defines a hierarchical model for representing real devices (see Figure 63.2). Following the object orientation paradigm, the data associated to real devices are contained in LNs, and the server grants access to authorized clients through a set of specified services. However, the access to real physical devices is the responsibility of the device that hosts the server. The standard only defines the information exchange through the model.

The IEC 61400-25 series excludes a definition of how and where to implement the communication interface, the application program interface, and implementation recommendations. However, the objective is that the information associated with a single WPP component (such as a WT) be accessible through a corresponding LD. Clause 6 of IEC 61400-25-3 gives an overview of the IEM for operational functions and management functions. Clause 7 introduces the IEMs for operational functions: authorization, control, monitoring, reporting and logging. Clause 8 gives an overview of the IEMs for management functions. Clause 9 provides the details of the services for the service model classes. Annex A provides examples of the reporting and logging services required. Annex B provides relationship between ACSI services and functional constraints, and Annex C provides relationship between ACSIs defined in IEC 61850-7-2 and IEC 61400-25-3.

The following sections summarize the most important concepts of the IEM.

63.4.1 Communication Functions

The IEM defines services for accessing the IM data. The services may be classified in two major categories: those that provide *operational functions* and those that are dedicated to *management functions*. All these functions generate data traffic of different nature, for example, periodic/aperiodic and synchronous/asynchronous. As a consequence, the IEM must support different transfer principles and provide the required abstract communication services. Every instance of the WPP IM defined in the previous section, such as LD, LN, data (DC), data attributes (ATTR), and control block (CB) objects, will be accessed by instances of the IEM described later. Figure 63.6 summarizes the communication as well as the management functions required from the clients.

63.4.1.1 Operational Functions

Operational functions may be classified in the following categories:

1. *Authorization and association*: Aiming at providing a secure information exchange via an association between a client and a server. These functions provide client authentication and control the access to server functions. They provide authentication, authorization and access control, integrity against unauthorized modification or destruction, confidentiality, nonrepudiation, and prevention of denial of service.
2. *Control*: Aiming at defining the IEM for sending commands and controlling groups of set points in an operational device. The operate commands set the values of controllable data, which are derived from controllable CDCs.

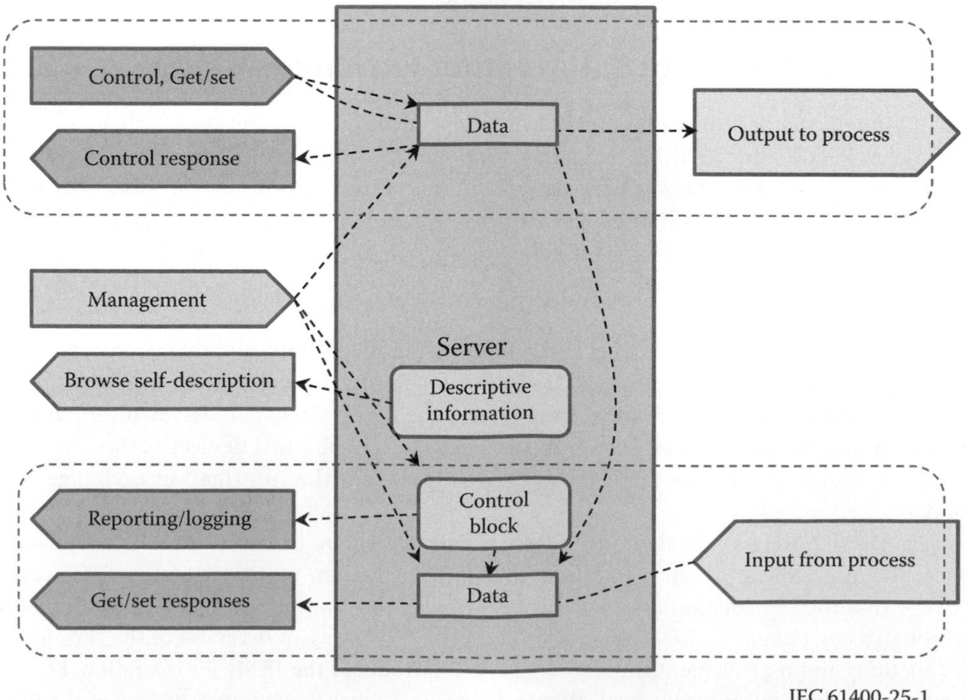

FIGURE 63.6 IEM services and interactions.

3. *Monitoring, reporting, and logging*: Aiming at retrieving information from one server. Three independent information retrieval methods are included: (1) *Retrieval on demand*: It includes operations such as Get or Read in which the response must be transmitted immediately. (2) *Publish/Subscribe*: Servers can be configured to transmit values periodically or triggered by changes or events. They may buffer values if necessary. Clients will initiate and finalize subscriptions. Finally, (3) *Logging*: Allow delivering values logged at the server. These values may be grouped by means of Data Sets. The client can query for all logged entries or for the entries between two time stamps.

63.4.1.2 Management Functions

These functions are used to configure or maintain a system. More specifically, they allow setting and changing configuration data and retrieving configuration information from the system. They also include services for the management of users and access security, configuration, time synchronization, and diagnosis operations.

63.4.2 Abstract Communication Service Interface

The set of basic services used to accomplish the information exchange between the clients and the virtual components that represent real-world devices are referred to as the Abstract Communication Service Interface. Since WPP automation requirements are similar, but not exactly equal, to those required by Substation Automation Systems, ACSI models defined in IEC 61400-25-3 are similar to those defined in IEC 61850-7-2 [6]. However, some entities defined in IEC 61850-7-2 are not used in WPP, whereas the IEC 61400-25-3 solves some restrictions for WPPs and adds new operations to some entities.
Figure 63.7 depicts the basic IEMs, illustrating the various components of the ACSI services.

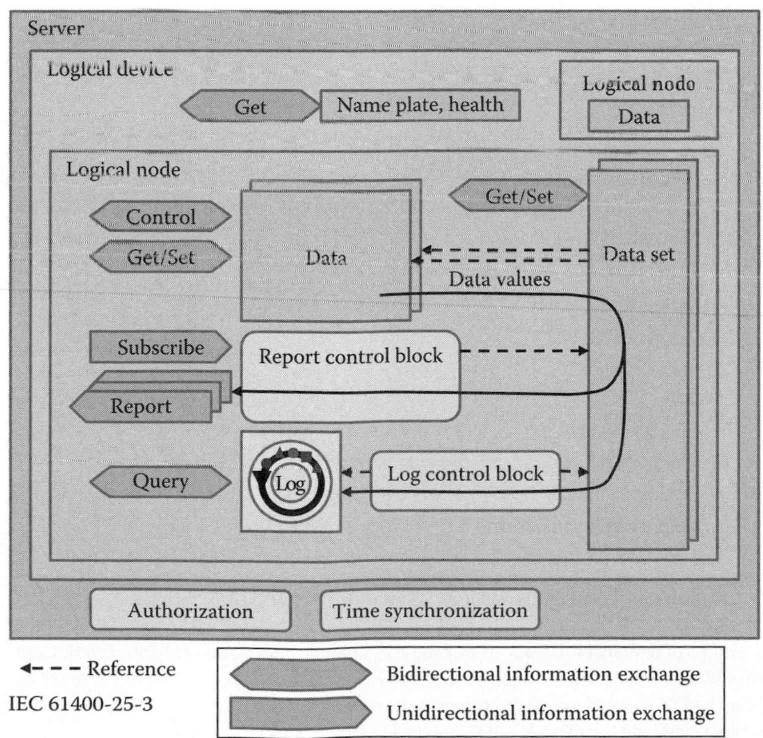

FIGURE 63.7 Conceptual information exchange model for a WPP.

LNs contain data that can be written or read individually or in groups (Data Sets), they respond to control inputs, provide solicited and unsolicited reports, and contain logs that can be queried. Get/Set and Control services are provided to read (get) and write (set) the data in LNs. The rules for logging and reporting are defined in the Log Control Block (LCB) and Report Control Block (RCB), respectively. Table 63.7 represents the IEC 61400-25 ACSI services associated to the corresponding IM entity.

TABLE 63.7 IEC 61400-25 ACSI Services

IEC 61400-25 ACSI Services			
IM Entity	Service	Explanation	M/O
Server	Associate	Used for two-party application association	M
	Abort		O
	Release		O
	GetServerDirectory	Retrieve the list of names of all LDs	O
Logical Device	GetLogicalDeviceDirectory	Retrieve the list of names of all LNs	O
Logical Node	GetLogicalNodeDirectory	Retrieve the list of names of all Data	O
Data	GetDataValues	Apply to CF, DC, ST, MX, SP, EX[a]	M
	SetDataValues	Apply to CF, SP[a]	M
	GetDataDirectory	Apply to CF, DC, ST, MX, SP, CO, EX[a]	O
	GetDataDefinition	Apply to CF, DC, ST, MX, SP, CO, EX[a]	O
DataSet[b]	GetDataSetValues		M
	SetDataSetValues		O
	CreateDataSet		O
	DeleteDataSet		O
	GetDataSetDirectory		O
ReportControl[c]	Report	A report can include any elements of any	O
	GetBRCBValues	Only applies to Buffered Report Control Blocks	O
	SetBRCBValues		O
	GetURCBValues	Only applies to Unbuffered Report Control Blocks	O
	SetURCBValues		O
	AddSubscription	Applies to Buffered/Unbuffered Report Control Blocks	O
	RemoveSubscription		O
LogControlBlock[d]	GetLCBValues		O
	SetLCBValues		O
Log[e]	GetLogStatusValues		O
	QueryLogByTime		O
	QueryLogAfter		O
Control[f]	Select		O
	SelectWithValue		O
	Cancel		O
	Operate		M
	CommandTermination		O
	TimeActivatedOperate		O

[a] Functional constraints: CF, Configuration; DC, Description; ST, Status; MX, Measure; SP, Set Point; CO, Control; EX, Extended Definition.

[b] DataSets are groups of references to Data. They can be used for Report Control Blocks or Log Control Blocks.

[c] Report Control Blocks provide the mechanism to report data values on specific criteria. There are two types: Buffered Report Control Block (BRCB) and Unbuffered Report Control Block (URCB).

[d] Log Control Blocks provide the mechanism to log data values on specific criteria.

[e] Log provides the query service to receive data values stored between two time stamps.

[f] Control provides the mechanism to control functions and real devices represented by a server.

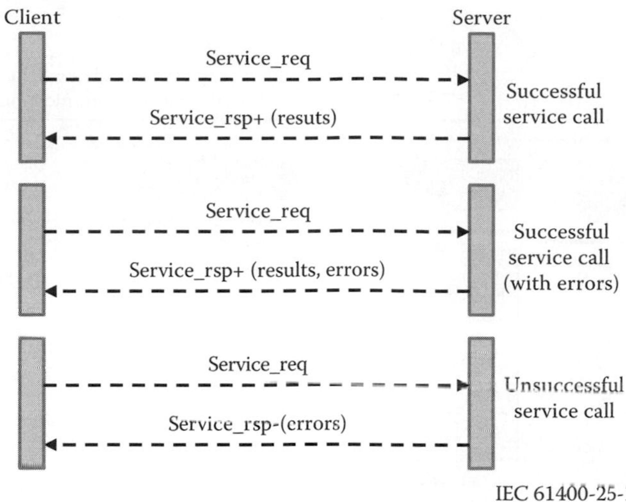

FIGURE 63.8 Sequence diagram.

63.4.3 Service Modeling Convention

ACSI services are generally defined by the following:

1. A set of rules for the definition of messages so that receivers can unambiguously understand messages sent from a peer
2. The service request parameters as well as results and errors that may be returned to the service caller
3. An agreed-on action to be executed by the service

All services are based on three message primitives: request, positive response, and negative response, as illustrated in Figure 63.8 by means of a sequence diagram. A message primitive may have a number of parameters, called results and errors in case of response primitives.

63.5 Mapping to Communication Protocols

Both IMs and IEMs need to be mapped to appropriate protocols. Even though the IEC 61400-25 series follow a more independent format, they are still based on IEC 61850. Consequently, IEC 61400-25 inherits and references many of the IEC 61850 rules and definitions keeping the compatibility between the two standards.

IEC 61400-25-4 focuses on the mapping of services, objects, and parameters defined in the standard to specific communication protocols. Independent mappings may be implemented using the IEC 61850 *SCSM*, aimed to provide independence from a specific communication stack, as illustrated in Figure 63.9.

The SCSM defines how the services and the server objects (LD, LN, data, data sets, report controls, log controls, or setting group objects) are mapped to the application layer by means of specific communication stacks compatible with the OSI reference model. These mappings and the application layer define the syntax for the data exchanged over the network.

The mappings may have different complexities depending on the communication network technology used. Thus, some of the ACSI services defined in the previous section may not be supported in all mappings. However, when a service is provided in a mapping, that service is equivalent to the same service in the benchmark mapping. It is also possible that the application layer combine one or more stacks simultaneously.

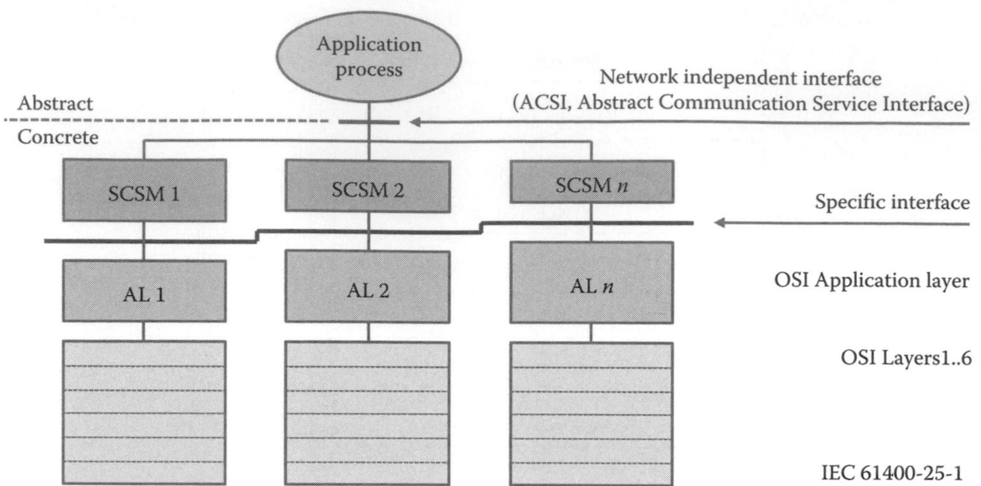

FIGURE 63.9 ACSI mapping to communication stacks/profiles.

IEC 61400-25-4 maps both, the IM and the IEM, within a client–server relation application using different, optional, communication profiles. The general mapping architecture consists of three parts:

1. Mapping of the IM, including the mapping of the DCs
2. Mapping of the information exchange services
3. Definition of the communication stacks

The IM and the IEM define the interface specification between a client and a server. The IM provides a logical view of the WPP data that are used by the server to offer a uniform and component-oriented view of the WPP data to the client.

Typically, TCP/IP protocols are used at the transport/network layers in most communication profiles (see Figure 63.10). Specifications for the data link and the physical layers are implementation specific and beyond the scope of the IEC 61400-25 series.

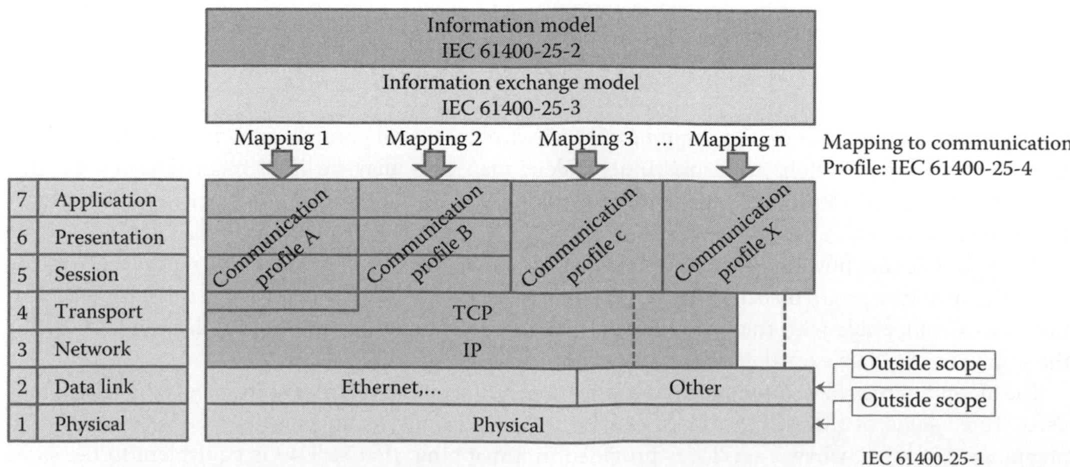

FIGURE 63.10 Communication profiles.

According to the market relevance, the IEC 61400-25-4 standard specifies mappings to five communication protocol stacks:

1. Mapping to SOAP-based web services (WSs)
2. Mapping to OPC/XML-DA
3. Mapping to IEC 61850-8-1 MMS
4. Mapping to IEC 60870-5-104
5. Mapping to DNP3

Due to the proximity between the IEC 61400-25 and IEC 61850 standards, the IEC 61850-8-1 MMS [9] mapping is a natural choice. IEC 60870-5-104 [11] and DNP3 [4] mappings are included due to their current use in power system automation applications. Also, market trends back the use of WSs and OPC [2] in electrical and automation applications.

All mappings are optional, but at least one mapping must be implemented in order to be compliant with IEC 61400-25-4. These mappings must provide the abstract services defined in IEC 61400-25-3 fully or in part. However, some services are only implemented in some mappings, for example, the IEM services aiming at retrieving the WPP IM of a device (self-description) are only defined in the MMS and in the WS mappings.

The use of a specific mapping may influence the overall performance of the applications. For example, the IEC 61850-8-1 mapping, which uses binary encoded messages, provides a higher performance than the WS mapping, based on ASCII characters.

63.5.1 Mapping to SOAP-Based WSs

The mapping to WSs is based on using XML to represent the IM defined in IEC 61400-25-2 for the WPP components and SOAP to implement the IEM services according to IEC 61400-25-3 (see Figure 63.11).

The WPP IM defined in IEC 61400-25-2 is mapped to a hierarchical structure that is preserved when mapped to WSs. This means the following:

- The server represents its internal data according to a hierarchical structure that follows the IEC 61400-25-2 standard. These data can be retrieved by a full set of services.

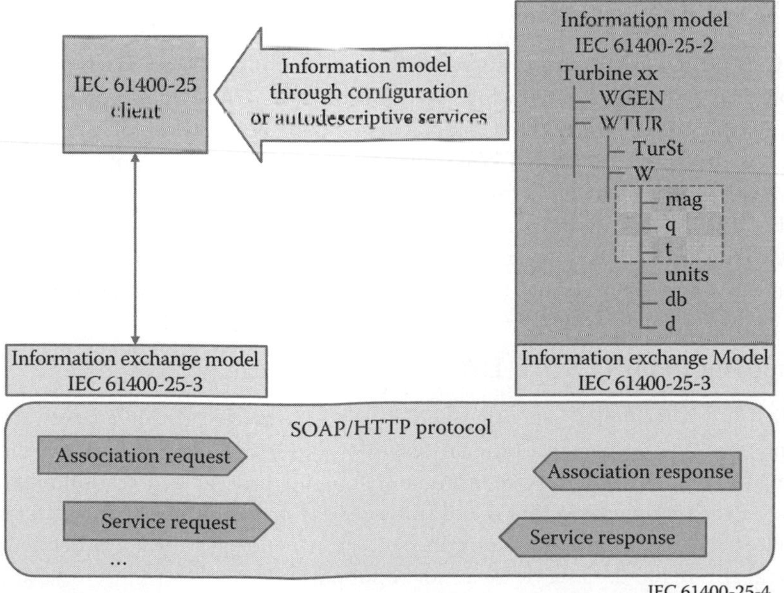

FIGURE 63.11 Conceptual Web Services mapping architecture.

TABLE 63.8 Web Services Protocol Stack for IEC 61400-25 Communication Profile

OSI Layer	Specification: Name	Specification: Service/Protocol Specification	M/O
Application	Web Services	Web Services Architecture, http://www.w3.org/TR/2002/WD-wsarch-20021114/	M
	SOAP	SOAP ver. 1.1	M
	HTTP ver. 1.1	RFC 2616, RFC 2817	M
	SSL/TLS ver. 1.0	RFC 2246	O
Presentation	—	—	
Session	—	—	
Transport	TCP	RFC 793	M
Network	ICMP	RFC 792	M
	IP	RFC 791	M
	ARP	RFC 826	M
	Broadcasting Internet Datagrams	RFC 922	M
		RFC 922	
		RFC 919	
	Host Extensions for IP Multicasting	RFC 1112	M
Data Link Layer	Implementation specific and beyond the scope of IEC 61400-25		
Physical Layer	Implementation specific and beyond the scope of IEC 61400-25		

- The client is configured with the hierarchical structure representing the server's IM or may retrieve it from the server by means of self-description services.
- The client is capable of accessing the server's hierarchy through the services provided by WSs according to the IEC 61400-25-3.

As in other SCSMs, the TCP/IP protocol stack is the basic communication stack provided for the WS mapping. Table 63.8 describes in detail which are the protocols used in this mapping.

The WSDL (Web Services Description Language (WSDL) specification for the mapping to WSs is defined in the standard and is given as a WSDL text file that includes all definitions and mappings.

This mapping provides a complete communication solution for all data objects and services for implementing the IEM of a WPP as defined in the standard. However, since this mapping requires the translation of the IM to XML format and the IEM as SOAP messages, and consequently into ASCII text, the need of bandwidth increases, and problems may arise when big amounts of information have to be exchanged.

63.5.2 Mapping to OPC XML DA

OPC XML DA [2] is a technology specified by OPC Foundation to provide a hardware and software independent platform to exchange real-time process information by means of XML messages.

The mapping to OPC XML-DA services aims at exchanging process information required for operational purposes. Even though the WPP IM and IEM are intended to be preserved when mapped to OPC XML-DA services, the mapping of some complex data sets may not be available as defined in IEC 61400-25-2, and several IEC 61400-25-3 services cannot be supported.

In this mapping, the IEC 61400-25 server is implemented as an OPC XML-DA web server. Every OPC XML-DA WS is executed inside the web server according to the rules defined in the OPC XML-DA

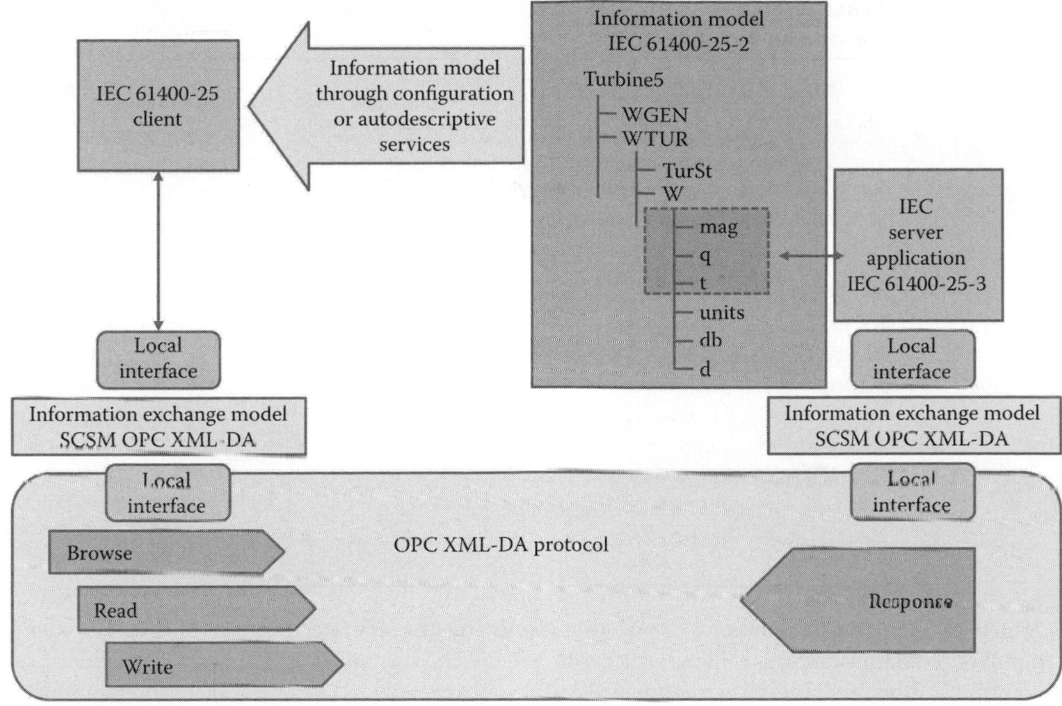

FIGURE 63.12 Conceptual OPC XML-DA mapping architecture.

specification. WPP LDs, LNs, and data objects are implemented as branches in the OPC hierarchy (see Figure 63.12). Primary data attributes are implemented as OPC items.

The WPP IM defined in IEC 61400-25-2 is mapped to a hierarchical structure. Again, the WPP IM must be preserved when mapped to WSs. This means the following:

- The server implements the hierarchical WPP IM according to IEC 61400-25-2. These IMs may be retrieved by means of IEC 61400-25-3 services.
- The client implements the WPP IM by configuration, or it can be retrieved by using self-description services.
- The client application accesses the hierarchical WPP IM through the services provided by OPC XML-DA mapping to WPP IEM.

Table 63.9 details the protocols used in this mapping.

A wide set of IEC 61400-25 services cannot be supported. For example, the management of complex data sets is the responsibility of the client. Also, due to the OPC XML-DA behavior, this mapping does not define how the authentication of the client must be implemented, requiring that client should authenticate at every request. However, the availability of commercial solutions that use this accepted and well-proven technology provides a strong advantage to the OPC XML DA approach when compared with the WS mapping.

63.5.3 Mapping to MMS

IEC 61850-8 describes the SCSM within the IEC 61850 series. More specifically, IEC 61850-8-1 is the part that defines the mappings to MMS (ISO/IEC 9506—Ed.2) [3].

TABLE 63.9 OPC XML-DA Protocol Stack for IEC 61400-25 Communication Profile

	Specification		
OSI Layer	Name	Service/Protocol Specification	M/O
Application	OPC XML-DA	OPC XMLDA 1.01	M
	SOAP	SOAP ver. 1.1	M
	HTTP ver. 1.1	RFC 2616	M
	TLS ver. 1.0	RFC 2246	O
Presentation	—	—	
Session	—	—	
Transport	TCP	RFC 793	M
Network	IP	RFC 791	M
	ICMP	RFC 792	M
	ARP	RFC 826	M
Data Link Layer	Implementation specific and beyond the scope of IEC 61400-25		
Physical Layer	Implementation specific and beyond the scope of IEC 61400-25		

ISO 9506, Industrial automation systems–Manufacturing Message Specification (MMS), is an application layer communication specification according to the OSI communication model. It provides a set of communication services between automated equipment and systems used for control and monitoring purposes, and it follows the client–server paradigm. MMS is suitable over network technologies that support full-duplex, reliable communication. It is divided into two major documents: ISO 9506-1:2003, which provides the Service Definitions, and ISO 9506-2:2003, which provides the Protocol specification of the messages exchanged between client and server.

Most service models defined in IEC 61400-25-3 are mapped to ISO 9506 following the same approach defined in IEC 61850-8-1.

The WPP IM is mapped to a hierarchical structure and must be preserved when mapped to IEC 61850-8-1 MMS services. The mapping is illustrated in Figure 63.13. This especially means the following:

- The server implements the hierarchical WPP IM of IEC 61400-25-2 that can be retrieved by the services.
- The client needs to interpret the WPP IM.
- The client application accesses the hierarchical WPP IM through the services provided by IEC 61850-8-1 and ISO 9506.

This mapping requires the use of OSI protocols as defined on ISO MMS (see Table 63.10). These protocols cannot directly be implemented over the TCP/IP stack, so the RFC 1006 protocol specification ("ISO Transport over TCP") must be added over the transport layer.

The IEC 61850-8-1 mapping is mainly intended to achieve real-time exchange of information between clients and servers. Clients need to receive real-time information from servers in order to monitor and control the power generation process and the related equipment. However, IEC 61850 service models such as Generic Object-Oriented System Event (GOOSE) or Sampled Value (SV) are not referenced in IEC 61400-25 since they require high-priority real-time communication protocols that are out of the scope of the IEC 61400-25 series.

The major advantages of using MMS are the reduction of bandwidth needs due to its binary data encoding, the fact that it is, up to date, the only tested protocol in real systems, and the possibility of making use of the full set of abstract communication services.

However, MMS uses the full OSI stack, so the processing requirements at the devices are relatively high. Also, the software complexity of the full stack makes the development of the applications difficult.

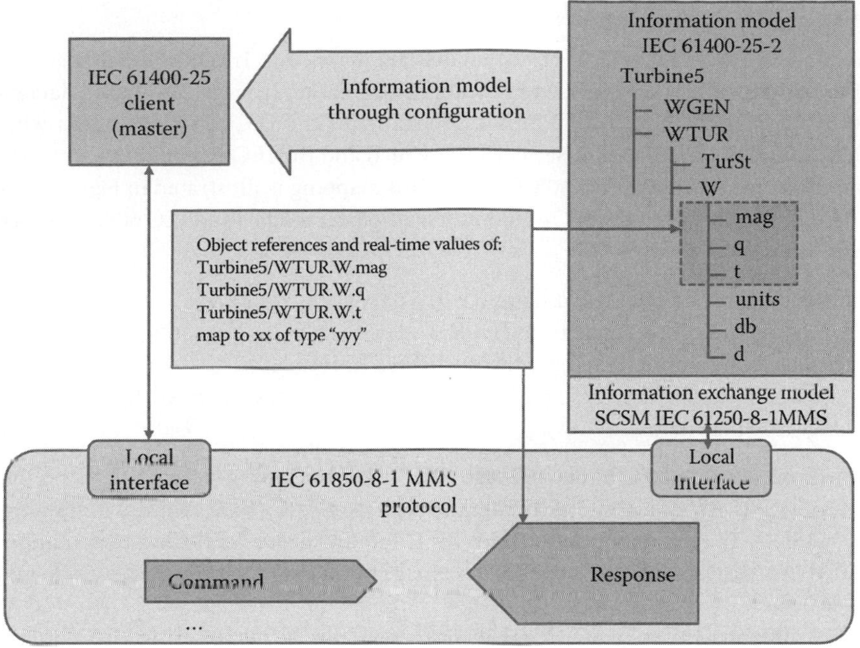

FIGURE 63.13 Conceptual MMS mapping architecture.

TABLE 63.10 MMS Protocol Stack for IEC 61400-25 Communication Profile

	Specification		
OSI Layer	Name	Service/Protocol Specification	M/O
Application	Manufacturing Message Specification	ISO 9506-1:2003 ISO 9506-2:2003	M
	Association Control Service Element	ISO/IEC 8649:1996 ISO IEC 8650-1:1996	M
Presentation	Connection-Oriented Presentation	ISO/IEC 8822:1994 ISO/IEC 8823-1:1994	M
	Abstract Syntax	ISO/IEC 8824-1:2002 ISO/IEC 8825-1:2002	M
Session	Connection-Oriented Session	ISO/IEC 8326:1996 ISO/IEC 8327-1:1996	M
Transport	ISO transport on top of TCP	RFC 1006	M
	TCP	RFC 793	M
Network	ICMP	RFC 792	M
	IP	RFC 791	M
	ARP	RFC 826	M
Data Link Layer	Implementation specific and beyond the scope of IEC 61400-25		
Physical Layer	Implementation specific and beyond the scope of IEC 61400-25		

63.5.4 Mapping to IEC 60870-5-104

In this mapping, the IEC 61400-25-3 service models are mapped to IEC 60870-5-104 according to the IEC/TS 61850-80-1 standard (Guideline to exchange information from a CDC-based data model using IEC 60870-5-101 or IEC 60870-5-104). This approach maps the IEC 61400-25-2 IMs with the attributes of CDC to ASDUs (Application Service Data Units) and the IEC 61400-25-3 services to the Basic Application functions defined in IEC 60870-5-104. This mapping is illustrated in Figure 63.14.

Again, the mapping between the IEC 61400-25-based object models and IEC 60870-5-104 defines the following:

- Both the server and client are configured with the specific WPP IM.
- The configuration can be done in different ways, either by online configuration or by using the Substation Configuration Language (SCL), defined in IEC 61850-6.
- The client application accesses the IM through the services provided by IEC 60870-5-104 to exchange real-time data.

The configured data is defined as Process Image (PI). The IM mapping is based on using the Common Address of ASDU (CASDU) and Information Object Address (IOA) within a standardized IEC 60870-5-104 ASDU. This accommodates the WPP IM and transfers real-time information data. It is recommended to use a Data Engineering tool to manage the use of CASDU and IOA following a unique numbering scheme.

The IEC/TS 61850-80-1 uses an SCL file that covers not only the mapping but also a broader scope of IEC 61850-based models. Selected parts of IEC/TS 61850-80-1 can be used for WPPs.

The WPP IES specified in the IEC 61400-25-3 IEM must be mapped to the services defined by IEC 60870-5-104. However, not all IEC 61400-25-3 services could be mapped due to the nature of the IEC 60870-5-104 standard.

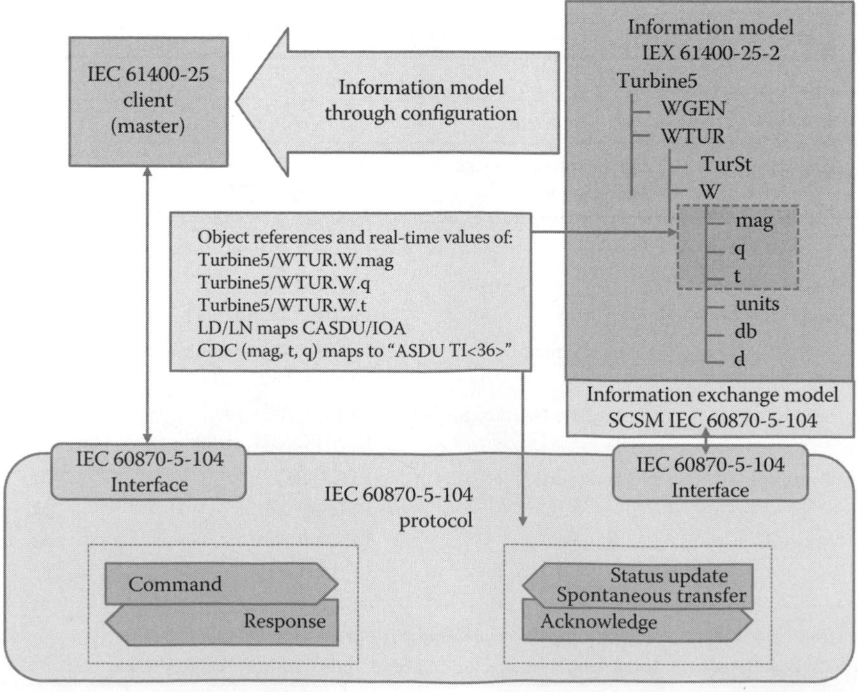

FIGURE 63.14 Conceptual IEC 60870-5-104 mapping architecture.

IEC 60870-5-104 is one of the most frequently used standards to control remote stations from control centers. However, as in other mappings, a wide set of IEC 61400-25 services cannot be supported, and management of data sets is not available.

63.5.5 Mapping to DNP3

Distributed Network Protocol 3 (DNP3) is an open and public protocol aiming at applications of telecontrol of substations. DNP3 is a derived branch of the IEC 60860-5 protocol adapted to the North American application. DNP3 is administered by an independent organization called the DNP Users Group, which is a consortium of device vendors and users.

DNP3 uses data set prototypes to specify the structure and ordering of data within an entire data set or a portion of a data set. It uses name spaces to assign unique meaning to common names. The DNP3 Users Group accepts voluntary registration of name spaces submitted by any vendor or user. Once a name space is registered, the vendor or user is free to define, redefine, or delete any data set prototype(s) within that name space. Each data set prototype is assigned a UUID (Universally Unique Identifier) that identifies it.

In DNP3, any IEC 61400-25-2 entity is mapped to a specific DNP3 data set prototype. A DNP3 UUID is reserved for each CDC in IEC 61400-25-2. The DNP3 name space assigned to the IM is "IEC 61400-25-2."

This mapping, illustrated in Figure 63.15, only provides services for process data exchange. Therefore, the following IEC 61400-25-3 services should be mapped to services defined in DNP3:

- Data model (GetDataValue, SetDataValue)
- Control model (Select, SelectWithValue, Operate, CommandTermination)

The specified mapping for IEC 61400-25 DCs can be based on either one of these mechanisms:

1. *Mandatory*: DNP3 XML schema mapping of DNP3 data points to IEC 61850 object models. This schema can be used to describe a mapping between DNP3 data points and the data attributes of the object models described in IEC 61400-25-2.
2. *Optional*: DNP3 Data Sets can transport the IM of the WPP components. If DNP3 Data Sets are used for the mapping of DCs, then the IEC 61400-25-2 WPP IM must be mapped to DNP3 Data Set descriptors with DNP3 objects.

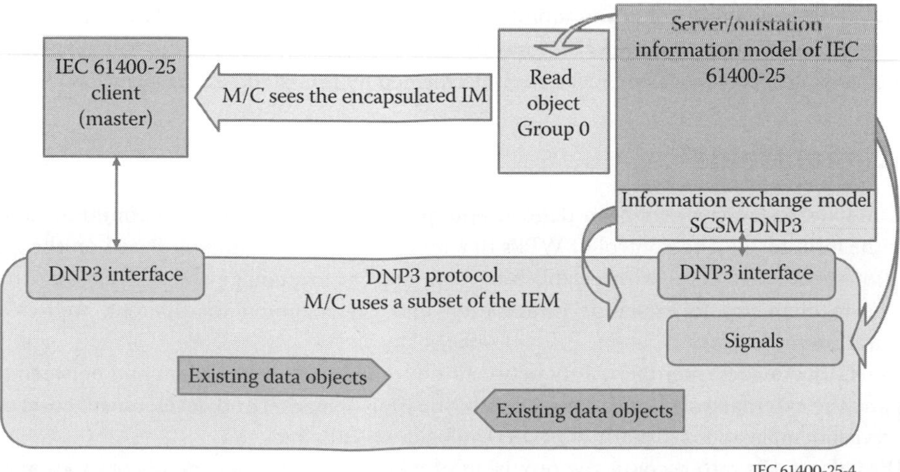

FIGURE 63.15 Conceptual DNP3 mapping architecture.

When the optional DNP3 Data Set mapping procedure is used, the DNP3 XML mapping procedure may be left unused.

Where DNP3 Data Sets are used for the mapping,

- The server encapsulates the WPP IM that can be read by the services by reading DNP3 Object Group 0
- The master/client station accesses the Data Attributes from the WPP IM through the services provided by DNP3

63.6 Condition Monitoring

Condition monitoring focuses on observing components or structures for a period of time in order to evaluate the state and changes at these entities and to detect early indications of failure. Condition monitoring is most frequently used as a predictive or condition-based maintenance (CBM) technique. In WPPs, condition monitoring includes tasks as vibration measurements and analysis, oil debris analysis, temperature measurement, or strain gauge measurement.

The information used by condition monitoring may be located at different physical devices. For example, some information could be located in a turbine controller device (TCD) while other information could be located in additional specialized condition monitoring devices (CMDs).

The purpose of IEC 61400-25-6 is to define the IM related to condition monitoring for WPPs and the information exchange of data values related to these models. It is based on the models and mappings defined in previous parts of the IEC 61400-25 standard, but it extends these models for condition monitoring purposes. For instance, LNs like WTUR (Wind TURbine general information) or WALM (Wind power plant ALarM information) are extended and new LNs are introduced as, for instance, WCON (Wind CONdition monitoring information). The condition monitoring data are defined in specific CDCs having different related categories and data types.

Many CDC attribute types such as analog value, time stamp or vector, and point are inherited from IEC 61400-25-2 or IEC 61400-25-3. However, additional specific CDCs are needed for condition monitoring such as SV or Data File (DAT).

The data types are divided into two groups of data types: one group is mandatory and another group is optional or vendor specific. The data types specified for the condition monitoring system for a particular part of the turbine are assigned a unique number. Then, this data type can be used in the data model as a reference when referring to measurements of a certain data type.

Due to the big amount of condition monitoring information over the WCON LN, it is necessary to relate many measurements to the specific location of the sensors on the WT. For this purpose, a methodology to specify specific measure locations is performed by IEC 61400-25-6.

63.7 Conclusions

The IEC 61400-25 standard has been defined aiming at providing a uniform communication backbone for the monitoring and control of WPPs in a multivendor environment. Based on the IEC 61850 standard, it specifically deals with establishing data objects to manage the relevant information of WPPs, the mechanisms to exchange information and the mapping to different, well-established, communication protocols.

The standard provides interoperability between the components of the plant and between the components and the external world, and, consequently, the time to market and development costs of control and supervision applications (usually SCADA) can be drastically reduced.

The IEM also takes into account the provision of mechanisms to manage history logs and reports containing a great amount of information that is very common in WPPs (the order is typically thousands).

On the other hand, in order to cover different communication requirements, it specifies mappings to very well-known protocols. Hence, the mapping to OPCX XML-DA allows integrating commercially available SCADAs. MMS covers the management of large WPPs. WS provides a universal access to small, delocalized WPPs. Finally, IEC 60870-5-104 and DNP3 cover the need of remote control in both European and American systems.

Far from being a standard with an unclear future, IEC 61400-25 appears to be a safe bet in the development of monitoring and control systems for WPPs. Besides, there are companies with a proven experience in the development of monitoring and control systems in the sector of management and distribution of electrical energy that are applying their experience with the IEC 61850 in the sector of WPPs. Besides that, companies that develop data acquisition and control systems applications are introducing IEC 61400-25 clients as an access mechanism to process data looking at enhancing their application environment.

References

1. IEC 61850-7-1:2003, Communication networks and systems in substations—Part 7-1: Basic communication structure for substation and feeder equipment—Principles and models.
2. OPC XML-DA 1.01 Specification:2003, http://www.opcfoundation.org/.
3. ISO 9506-1:2003, Industrial automation systems—Manufacturing message specification.
4. DNP3 Specification, Volume 2–8:2007.
5. IEC User Group: http://use61400-25.com/.
6. IEC 61850-7-2:2003, Communication networks and systems in substations—Part 7-2: Basic communication structure for substation and feeder equipment—Abstract communication service interface (ACSI).
7. IEC 61850-7-3:2003, Communication networks and systems in substations—Part 7-3: Basic communication structure for substation and feeder equipment—Common data classes.
8. IEC 61850-7-4:2003, Communication networks and systems in substations—Part 7-4: Basic communication structure for substation and feeder equipment—Compatible logical node classes and data classes.
9. IEC 61850-8-1:2004, Communication networks and systems in substations—Part 8-1: Specific Communication Service Mapping (SCSM)—Mappings to MMS (ISO 9506-1 and ISO 9506-2) and to ISO/IEC 8802-3.
10. IEC 61850-6:2004, Communication networks and systems in substations—Part 6: Configuration description language for communication in electrical substations related to IEDs.
11. IEC 60870-5-104:2006, Telecontrol equipment and systems—Part 5-104: Transmission protocols—Network access for IEC 60870-5-101 using standard transport profiles.
12. IEC 61400-25 (all parts), Wind turbines—Part 25: Communications for monitoring and control of wind power plants.
13. ISO 7498-1:1994, Information technology—Open systems interconnection—Basic reference model: The basic model.

Additional Web Information

1. News on IEC 61850 and related Standards: http://blog.iec61850.com/.
2. LinkedIn Group for IEC 61400-25: http://www.linkedin.com/groups/USE61400-25-4552476.
3. LinkedIn Group for IEC 61850: http://www.linkedin.com/groups/61850-119621.

XII

Communication Networks and Services in Railway Applications

64 Communication in Train Control *Heinz Kantz, Stefan Resch, and Christoph Scherrer* **64**-1
Introduction • General Aspects of Railway Systems • RAMS by Construction • TAS Control Platform • One Channel Safe Concept • Operations Management Center • From Human and System Action to Signals and Switches • References

65 Ethernet Extensions for Train Communication Network *Stephan Rupp and Reiner Gruebmeyer* **65**-1
Ethernet in TCN • Native Ethernet Technology • Ethernet in Embedded Environments • Specific Requirements in TCN • Acronyms • References

64

Communication in Train Control

64.1 Introduction 64-1
64.2 General Aspects of Railway Systems 64-2
Standards and Certification • Safety • Fail-Safe • Reliability and Availability • Maintainability • Real Time
64.3 RAMS by Construction 64-3
Basic Concepts • Computation • Communication
64.4 TAS Control Platform 64-7
Structure and Interfaces • Application Properties
64.5 One Channel Safe Concept 64-8
Message Creation • OCS and Availability
64.6 Operations Management Center 64-9
Overall Architecture • Common Interface for Substations • Versatile Communication Interfaces
64.7 From Human and System Action to Signals and Switches 64-12
From Planner to Operator • Operator to Switch • ILP to Substation via Datacryptors • Mobile Shunting Device to ILP • Ensuring ETCS Conform Trains on ETCS Tracks
References 64-15

Heinz Kantz
Thales Austria GmbH

Stefan Resch
Thales Austria GmbH

Christoph Scherrer
Thales Austria GmbH

64.1 Introduction

The main goal of a railway network operator is to provide a timely and safe train service to the customer. Automation and computer-driven train control are essential elements toward achieving such a service for large railway systems. Historically, interlocking systems were locally manned. Dispatchers controlled local train and shunting movements and exchanged the relevant information, following strict protocols, via landline.

Their tasks include route control and train supervision, as well as work gang protection and customer information. Today, some railway lines are still controlled this way. A major drawback of this approach is the limited support for automatic operations and dispatching. Automation of local interlocking systems eased and gradually replaced the work performed by local dispatchers and enabled train management on a large scale. With the interconnection and remote control of these automated interlocking systems, it is possible to supervise the train service for a large part of the railway network from one operational center.

The Austrian main line network, operated by Österreichische Bundesbahnen (ÖBB—Austrian Federal Railways), is supervised with one Network Management Center (NMC) and five Operation Management Centers (OMCs) as illustrated in Figure 64.1. Within the NMC, all information about the traffic situation on the whole network is available, and traffic is coordinated on a large scale. The OMCs perform regional disposition, control and management of traffic, as well as customer information and

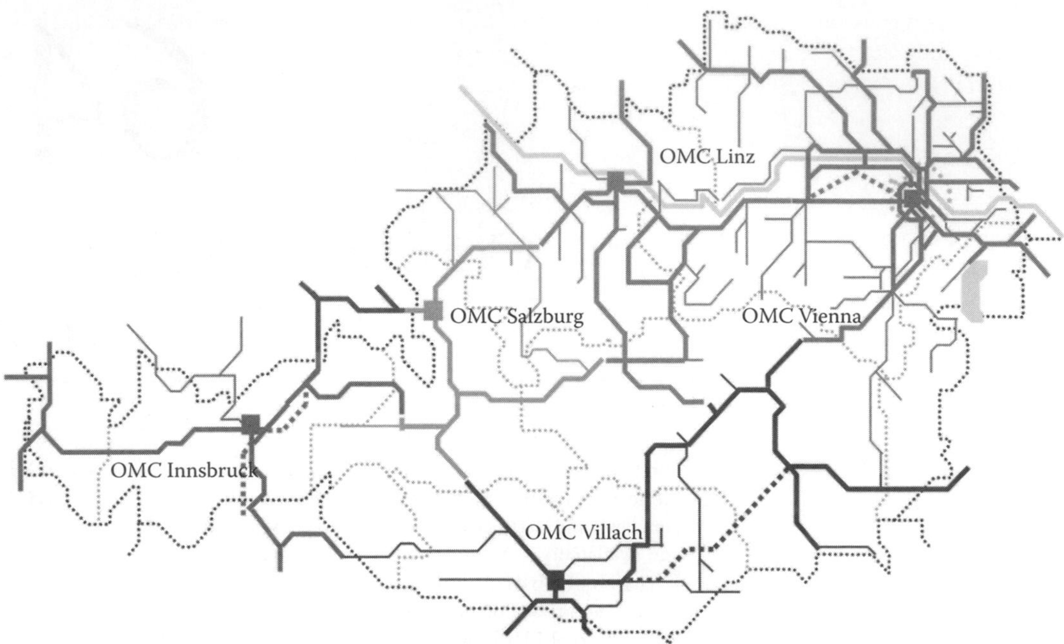

FIGURE 64.1 Operation management centers in Austria.

function as technical service centers. Each OMC connects numerous people and systems to fulfill its task. In this chapter, we will describe communication paths in these systems and their properties. These paths take us from operators, the automatic train routing system, and work gangs, through the OMC, down to the signals and switches of the railway network.

In the next section, we discuss general aspects of railway systems, before describing the Reliability, Availability, Maintainability, and Safety (RAMS) by construction approach in Section 64.3. This approach is the foundation for the TAS Control Platform and OCS concepts presented in Sections 64.4 and 64.5. The OMC architecture and its interfaces are outlined in Section 64.6, which is the basis for Section 64.7. There we follow example communication paths through the railway control systems and have a detailed look on the requirements and properties along these paths.

64.2 General Aspects of Railway Systems

Railway systems share common aspects regarding their general properties, which define their architecture. In the following, the most important aspects will be briefly discussed.

64.2.1 Standards and Certification

To ensure safe and reliable functionality of railway systems, they are subject to certification. Systems targeted for the European market are certified according to the corresponding CENELEC standards. Basis is the EN 50126 standard [1], which specifies RAMS requirements applicable for railway applications. EN 50128 [2] and EN 50129 [3] specify requirements for software development and electronic applications in general. These standards are the basis for railway system development and operation.

64.2.2 Safety

Less than 10^{-9} safety-critical failures per hour are required by EN 50129, for systems where a failure may result in a catastrophic event. Within the context of an application, such safety requirements can be reached using various strategies and methods for implementation. In the course of this chapter, we discuss different technical and conceptual methods for achieving safety in communication and computation.

64.2.3 Fail-Safe

Unlike other domains such as avionics, most railway applications can enter a safe state upon the detection of anomalies. An example for such a safe state is when all signals are set to red, that is, all trains stop. Here the difference of safe and available is evident. Setting all signals to red is safe, but the train service is not available. Please note that, in the context of rail, safe communication does not necessarily imply high availability.

64.2.4 Reliability and Availability

Apart from safety, high reliability and availability are essential for providing a quality railway service. This is equally true for applications, as well as for safe and nonsafe communication. As an example, the Austrian Federal Railway Authority requires less than one service interruption in 10 years.

64.2.5 Maintainability

Railway products typically have a long lifetime in the order of 20–30 years. This requires concepts for

- Hardware maintenance
- Software maintenance
- Function upgrades
- Delivery and replacement of faulty components

They are essential to achieve such long lifetimes. Again, this applies to the applications, as well as the communication systems.

64.2.6 Real Time

Since railway applications are fail-safe, they can be viewed as soft real-time systems. The most important time constraint is on the detection of anomalies to trigger an appropriate safety reaction. The time it takes from occurrence of anomalies to the safety reaction is influenced by the application architecture, as well as its communication. Here, supervised time-outs are key elements to ensure safety. These time-outs have to be carefully chosen by the application designer to guarantee fast failure detection on the one hand and avoid unnecessary failure reactions on the other hand. These time-outs implicitly define the tolerance of the system with respect to communication jitter.

64.3 RAMS by Construction

Conformance to the RAMS criteria is reflected by the architecture and behavior of a railway application and its components. With the use of basic concepts, generic architectures can be created for computation and communication. They in turn enable the modular construction of complex railway applications fulfilling the RAMS criteria.

64.3.1 Basic Concepts

The basic concepts presented here can be utilized for safety, reliability, and/or availability. Furthermore, breaking down the system architecture in such entities is a natural support, but no guarantee, for maintainability.

64.3.1.1 Fault Detection

With fault detection, a system can avert the sending or processing of possibly incorrect input and output caused by a fault. This is a technique suitable for fail-safe system, where upon the detection of a fault, the system is put into a safe state, and the service interrupted. Continuation of service is possible when either the system has verified that the fault no longer persists or appropriate maintenance actions have been taken to remove the fault.

64.3.1.2 Fault Tolerance

Fault tolerance enables continuous operation in the presence of a single or even several faults. For communication, fault tolerance can be achieved with, for example, redundant networking equipment, retransmission of messages, or error-correcting codes.* Example architectures for computation are triple modular redundancy (TMR) and active replication for availability.

The correctness of the output generated by a TMR system is guaranteed by the fault hypothesis for the TMR architecture and the subsequent voting on the output. Output generated by an active replicated system for availability has to be valid for each replica on its own.

64.3.1.3 Diversity

Diversity is aimed at the detection of failures. Two computation or communication channels apply different methods to calculate or transmit output, using the same input. Based on the assumption that a failure occurring in one or both channels affects each method differently, the correctness of the output can be verified by voting. The diversity of these methods can be achieved in various ways, such as by using, for example,

- Different programming languages
- Different development teams (N-Version programming [4])
- Different hardware or software components
- Different designs, for example, compute and validate
- Different specifications
- Monitoring techniques
- Automatic transformation of one method to another

Another aspect of diversity is the possibility to protect against software bugs introduced during the development phase. Here, it is crucial that the bug was either introduced into only one channel or, if found in both channels, results in different output. With the consequent application of diversity methods, the existence of an undetectable software bug in the compound software system is extremely unlikely even when executed on the same hardware.

Apart from the explicit introduction of diversity, similar systems may expose diversity inherently. This is caused by having slightly different initial states or ordering of input and output events [5].

64.3.1.4 Redundancy Concepts

Redundancy can be used for availability, as well as for fault detection. Different redundancy concepts and architectures target various properties. Replication [6], diversity [4], and clustering are strategies for redundancy used within software railway systems. Figure 64.2 illustrates the different redundancy concepts.

Redundancy models for continuation of service in case of failure of one entity are active and semiactive replication (Figure 64.2a and b). The most prominent representative of active replication is TMR. Master–slave replication where both replicas actively compute the result is an example for semiactive replication.

---

* Note that EN 50159 discourages the use of error-correction techniques and requires the use of a safety code for safe communication.

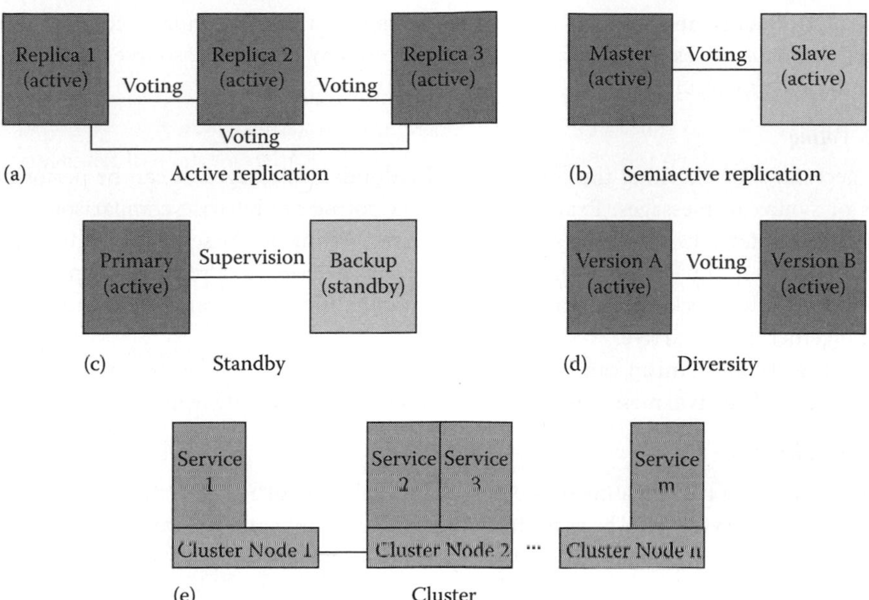

FIGURE 64.2 Redundancy concepts for safety and/or availability.

Replication with cold or warm standby (Figure 64.2c) is suitable if a short interruption of the provided service is tolerable. This interruption occurs since the service has to be started on passive replica in case of failure of the active one.

Execution of services or replicas on top of a cluster middleware (Figure 64.2e) introduces a further layer of hardware abstraction. This enables a fast restart of such services or replicas, in case their executing computing channel fails. Whether or not the service was interrupted due to that failure depends on the replication mechanism of the individual service or replica. With the hardware abstraction of the cluster service also the scalability of a computing system can be increased. New computing channels may be simply added to the cluster for more performance. However, the application executed within the cluster has to support this concept through its design.

All redundancy models without active replication are purely applied for availability and cannot provide support for safety. They have no option to check the validity of a generated output.* Diversity (Figure 64.2d), on the other hand, is a redundancy model for safety and not availability, which we discussed above.

Independent of the type of redundancy, all systems have to implement mechanism to ensure real-time properties and detect anomalies. Additionally, redundancy concepts can be combined for reaching both—availability and safety.

64.3.1.5 Active Replication Concepts

Active replicated systems require their involved components to coordinately interact with their environment. The key here is to ensure the replica-deterministic execution of the replicated programs. Time base, voting, and membership services enable this coordination and are implemented either implicitly or explicitly in any active replicated system.

64.3.1.5.1 Time Base

For fault-tolerant redundant systems, a consistent view of the environment is necessary to interact correctly with the latter. A common time base allows the ordering of observed events and

* Note that also regular output may be interpreted as "heartbeat."

triggering of time-outs and implicitly monitors progress of the redundant computing channels. Replica deterministically executed programs must not use any other time source, as not to break the determinism in their execution.

64.3.1.5.2 Voting

Voting is necessary for checking the correctness of redundant messages. It can be performed on the semantics or syntax of messages. Examples for syntax voting are bit-wise comparison and majority voting for failure detection. For systems utilizing diversity, voting on the semantics of the output might be necessary. However, equality of the datum is not the only criterion, as also the timeliness of the output has to be fulfilled. To check for timeliness, a voter might define time frames within which messages belonging together have to arrive.

Additional safety mechanism can be included in the voting process, for example, the green signal may be activated only by two messages, whereas one message may be sufficient for setting the red signal.

64.3.1.5.3 Membership

For identifying replicas participating in the redundant execution of a service, the internal membership service is used. This service is the basis for including and excluding replicas from execution and reflects the health state of the redundant entities. The timeliness of membership information is crucial for an accurate view of the system.

Please note that also semiactive and standby redundant systems have to provide some kind of membership service. They too have to identify the failure or absence of the master replica and designate a new one.

64.3.1.5.4 Reintegration of Components

Fault-tolerant systems require the reintegration of components, after they have been excluded due to a transient fault. This is crucial for providing an interruption-free service. Without this reintegration capability, also called recovery, a TMR system would degrade to a redundant system with two replicas after the first fault and interrupt its service at the second. Two prerequisites are required to provide recovery in the first place. First, enough communication and computation resources have to be available to satisfy the additional load during recovery. Second, the recovery algorithm has to exactly know which parts of an application to recover.

64.3.2 Computation

Safe computation can be realized with a wide variety of architectures and concepts. Different architectures and safety concepts enable the achievement of objectives with respect to maintenance and costs.

64.3.2.1 Health Monitoring

Computing channels usually monitor their health by executing a set of hardware tests as well as incorporating information retrieved by the membership service. Health monitors are particularly targeted against accumulated stochastic faults, which are not immediately detected by the functional implementation of the service.

64.3.2.2 1-out-of-1 (1oo1)

A 1-out-of-1 system achieves safe program executing on a single computing channel. The common approach for ensuring safety on such a system is software diversity. Naturally, fault tolerance is not possible with one 1oo1 system alone.

64.3.2.3 2-out-of-2 (2oo2)

Two computing channels are used for a 2oo2 system. In this setup fault detection of replica-deterministic applications, but no tolerance is possible. For fault tolerance, a 2oo2 system would need to ensure that each of the two entities functions as 1oo1 on their own.

64.3.2.4 2-out-of-3 (2oo3)

2oo3 is the classical TMR architecture. Replica-deterministic execution of the safety-critical software components is essential in this architecture. In case of failure of one entity, the system switches into a 2oo2 configuration. When the excluded replica is eventually tested successfully for being free of persistent failures, it is recovered, and operation continued in the 2oo3 configuration.

64.3.3 Communication

As with computation, communication in railway systems is heterogeneous and incorporates safe and nonsafe communication networks next to each other. An example for nonsafe communication is a command requesting the setting of a route from the automatic train routing system to the interlocking system. This is not safety critical, since the interlocking system guarantees the safe setting of routes. However, a command from the operator to the interlocking system changing the state of a switch marked as occupied is safety critical. This command overrules the route protection of the interlocking system, thus, the safe execution has to be guaranteed.

Safe communication is subject to EN 50159 [7]. This standard defines requirements for communication between safety-related railway equipment. The key properties for safe communication are

- Authenticity
- Integrity
- Timeliness
- Sequence

The standard discriminates network architectures in three categories:

Category 1 applies to closed transmission systems with a fixed number of participants, negligible risk of unauthorized access, and static physical characteristics of the transmission system during its life cycle.

Categories 2 and 3 concern open transmission systems, which may have

- A changing set of participants
- Possibly unknown participants, which are not part of the railway application and may generate arbitrary communication loads
- Changing properties of the transmission media
- Changing message routes through the system

Only in category 3, the open transmission system may also be subject to unauthorized, malicious access.

Based on this categorization, the standard identifies possible threats and lists measures and methods that can protect the safety-related communication against these threats. The appropriate measures have to be implemented in an independent layer above the transmission system according to EN 50128 and EN 50129. However, safety is not the only criterion for communication systems, as all RAMS requirements have to be taken into account.

64.4 TAS Control Platform

The idea of TAS Control Platform [8] is to build railway applications on top of a generic computing platform and, with this approach, support the fulfillment of the overall RAMS requirements. This separation reduces the direct dependency of long-lived railway applications on short-lived hardware. TAS Control Platform includes all necessary generic elements for constructing replica-deterministic fault-detecting 2oo2 or fault-tolerant 2oo3 architectures. A basic set of services is also provided for 1oo1 architectures. The services include global time base, voting, membership, recovery, and fault management services, as well as a health monitor. The use of POSIX as generic interface for applications enables the transparent integration of fault-detection mechanisms. Applications can build upon the generic TAS Control Platform safety concept and safety case to reach their targeted safety goals.

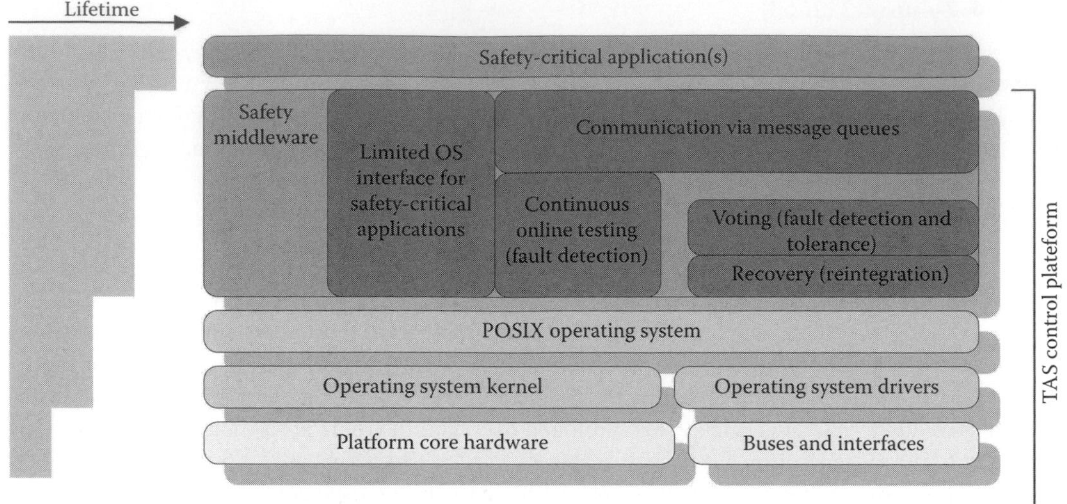

FIGURE 64.3 TAS Control Platform layer structure.

64.4.1 Structure and Interfaces

The TAS Control Platform itself is structured in layers as depicted in Figure 64.3. The safety middleware layer enables safe application execution on top of a Component off-the-shelf POSIX operating system with its kernel and drivers. Within the safety layer, all safety-relevant platform services are executed to ensure detection and/or tolerance of faults occurring in the layers below. The figure also shows the relative lifetimes of components in the individual layers. This illustrates how the safety and operating system layers are used to achieve a long lifetime for the interface to the application, as well as the application itself.

The common time base, membership, and voting are implemented within the voting service of the safety layer. They are provided to the application via a message queue interface. Continuous online testing is performed by the health monitor service, and recovery is implemented as a separate service.

64.4.2 Application Properties

TAS Control Platform Applications are structured in sets of closely related processes called *task sets* (TSs). They communicate by message passing through message queues. These message queues are used as transparent and generic interface to the communication system of TAS Control Platform. Each TS is assigned a redundancy model (1oo1 to 2oo3), and messages are distributed and voted accordingly. Replica-deterministic execution is enforced through the delivery strategy of messages and time. Thus, the applications do not need to implement these mechanisms themselves. However, time-outs on the interarrival and processing times of messages still have to be defined and evaluated in the application context. This ensures timeliness of the services and possible failure reactions.

For communication between applications, the concept presented in the next section is suitable.

64.5 One Channel Safe Concept

The fundamental idea of the *one channel safe* (OCS) communication concept is to establish a safe communication cannel via one physical link. This channel is established without assumptions on the properties of this link, other than the failure rate and failure behavior. The application can be certain that a datum received via an OCS link is correct in the value and time domain (Figure 64.4).

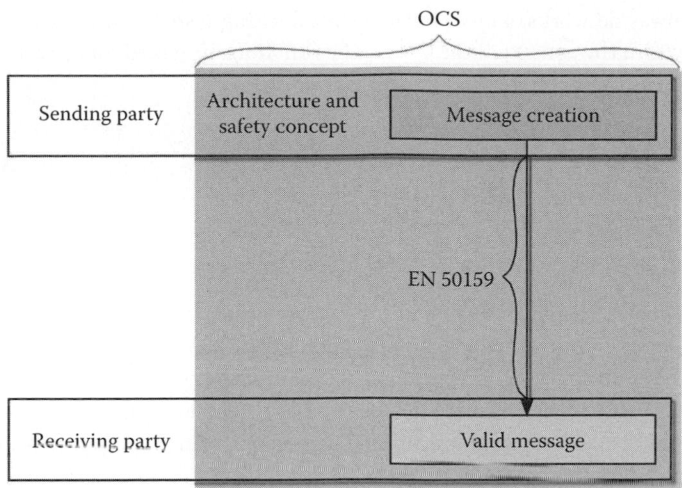

FIGURE 64.4 Components of OCS-based communication.

To reach this level of assurance, the OCS communication must, first, implement the required measures of EN 50159 in accordance with the nature of the underlying link. Note that this is where properties described in Section 64.3.3 of the underlying channel are taken into account. Second, the construction of these messages by the sending party has to be consistent with the sending parties' architecture and safety concept. This enables to interconnect applications having different architectures and safety concepts. However, guarantees with respect to timeliness are still achieved by the applications and OCS together and must be treated as such in each application's safety concepts.

64.5.1 Message Creation

Depending on the used architecture and safety concept, different approaches may be used for safe message creation. It is important to note that voting and safety code creation before sending the message is not sufficient, as the voter might fail. One approach for replica-deterministic 2oo2 or 2oo3 configurations is to create different parts of the safety code on different replicas and combine them for the message sent to the receiving party. That way, no replica on its own can create a valid safety code, and the detection of faults is ensured.

64.5.2 OCS and Availability

The OCS concept can be extended with availability by applying redundancy. Due to the nature of OCS, this is possible either by introducing redundancy transparently on the underlying link or by adding another link, which is handled by OCS itself. In the latter approach, the messages can be sent via one or both links, depending on the replication model. Due to the safety property of OCS messages, duplicate messages received on different links can be simply ignored.

64.6 Operations Management Center

After presenting the RAMS requirements and basic concepts for railway systems, we will now illustrate how these are used within the system for operative train control by the example of the OMC [9] and its communication structure. The OMC integrates all necessary systems to handle the operative train services for large railway networks of up to 100 local interlocking systems. These interlocking systems are called substations within the context of the OMC. The functions provided for operative train control are

- Shunting
- Automatic train traffic control
- Train tracking
- Monitoring of train traffic
- Manual train control
- Dispositive planning

These functions are complemented with a recording service and a central administration and diagnosis service. The recording service documents critical events and human actions, whereas the administration and diagnosis service is used for the technical supervision of the computerized systems. Furthermore, each OMC workplace provides an interface to the automatic time-tabling service ARAMIS.

Within the OMC, all aspects of communication in train control are found: safe systems communicate with nonsafe systems; some systems have to be highly available, whereas others are not as crucial. All of these systems communicate using many different protocols and interfaces. Still, the RAMS criteria have to be met by the overall OMC to provide a satisfactory train service to the railway customers.

64.6.1 Overall Architecture

A key element in the OMC architecture is the breakdown of functionality in smallest replaceable units (SRUs). This enables continuous service for all components that are not directly affected by the failure of such an SRU. Furthermore, a clean separation of safe and nonsafe components and services ensures that the safe part of the system is not influenced by failures within the nonsafe part. This includes the separation of safe and nonsafe interfaces for human planners or operators. Such a separation also provides a scope for certification and subsequently enforces the underlying structure of the system. Another requirement affecting the architecture of the OMC is the support of the stepwise integration of existing local interlocking systems.

Following this philosophy, the OMC architecture is divided in safe and nonsafe components. Figure 64.5 gives an overview of the individual components of the OMC. Nonsafe components are illustrated with the dark background, whereas safe ones on the right-hand side have a brighter color. The human machine interface (HMI), the Interlocking Proxy Clusters, and their Interlocking Proxies (ILPs), as well as the interconnection of these components are safety-critical and strictly separated from all other components within the OMC.

From each HMI, that is, workplace of the OMC, operators can issue commands to, and view the network state of, all substations connected with the OMC. Within each OMC, several of these workplaces are installed. Their safety concept is based on a 1oo1 diverse architecture, as presented earlier, and a method for safe human interaction. The HMI workplaces exchange data with the ILPs, which safely and timely provide the state of the railway network elements at their corresponding substation. Only with the use of these ILPs it is possible to remotely control substations, since communication links to the substations are restricted in bandwidth and have a nonnegligible transmission delay. Furthermore, they provide first feasibility checks of safety-critical commands before forwarding these to the substation. ILPs and substations are organized in cells, where each cell is controlled with a dedicated ILP cluster. This increases the availability of individual ILPs and enables the stepwise integration of substations. The connections between OMC workplaces and ILPs are independent from each other, such that the failure of one workplace or ILP affects only this very part of the system.

The ILP also provides the elements' states to the nonsafe interlocking cache (ILC), which fulfills a similar function as the ILP for all nonsafe components of the operational functions:

- Automatic train routing service (ATRS), which is the interface to ARAMIS
- Mobile Shunting Service (MOVUS), providing a direct shunting interface to shunting personnel

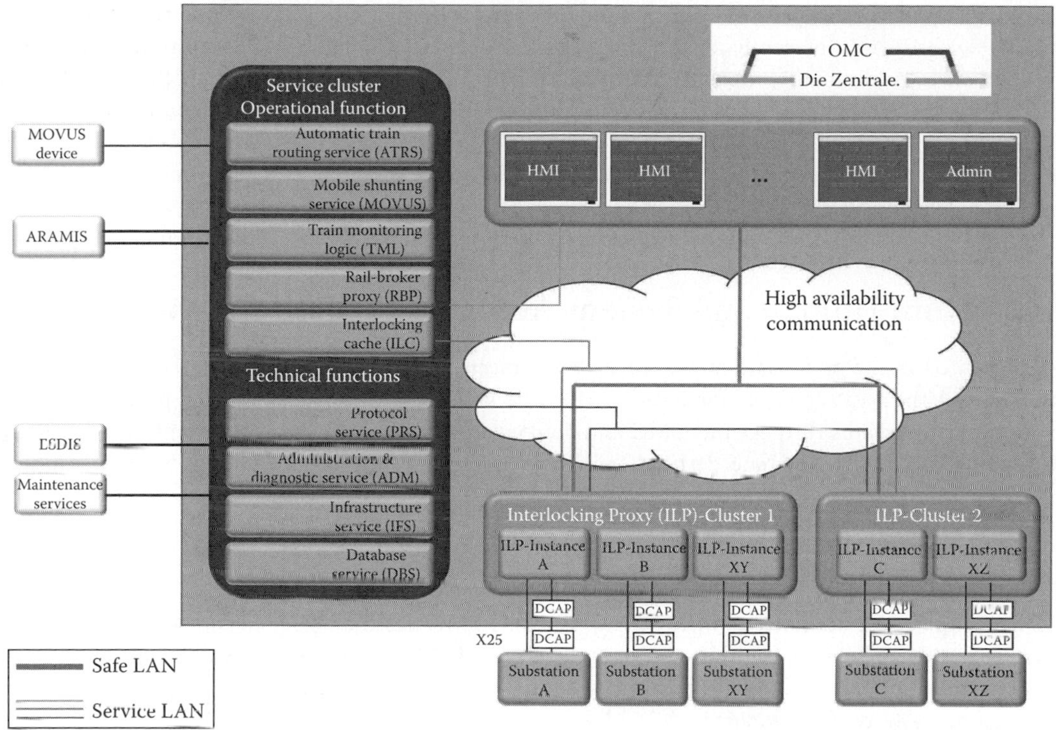

FIGURE 64.5 Architectural components and interfaces of the OMC.

- Rail-broker proxy, which is used by the ATRS for finding free train routes
- Train Monitoring Logic (TML), a service tracking train-specific information like the number of passenger wagons

They all exchange commands and states for the ILPs via the ILC and send information necessary for the notification of the operator to the HMI.

All nonsafe components are restricted to issue commands, which will later be checked by safe components for their validity and freedom of harm. For availability and the ability to increase computational power, the components of the operational functions are executed within a cluster.

The high-availability communication system within the OMC is realized with a redundant network of switches and cables based on Ethernet.

64.6.2 Common Interface for Substations

A prerequisite for the integration of a substation into the OMC is a suitable interface for the OMC at the substation. The X25 protocol has been defined to provide a generic protocol for railway automation and is an application layer protocol for the Austrian railway network. It is used for all communication between OMCs and substations. This requires that substations either already provide this interface or are extended with it.

64.6.3 Versatile Communication Interfaces

Many different systems using numerous interfaces are integrated through and within the OMC. ARAMIS is connected via an XML-based interface over TCP/IP, as well as through the X25 protocol. This protocol is also used for connections of substations to the OMS, as already outlined. The mobile shunting devices exchange their information with the OMC via GSM-R. The OMC components use TCP/IP with VLANs internally for the high-availability network and also different protocols within this network. Additionally, firewalls are placed at the TCP/IP interface to ARAMIS, as well as to the error-reporting system of the railway network operator and remote maintenance service for the OMC.

Again, these interfaces can be divided into safe and nonsafe ones, but security is still an issue here, since the communication channels may be connected via open networks according to EN 50159. All these properties are discussed in the next section, where we follow different paths through the system.

64.7 From Human and System Action to Signals and Switches

As presented in the previous section, the OMC connects and provides various systems and services by applying the basic concepts for RAMS. After the architectural aspects, we now present example communication paths through the OMC and connected systems, which are representative for many different command sources and destinations. They exhibit not only various communication requirements in railway but also the organizational structure and safety philosophy.

64.7.1 From Planner to Operator

Each workplace for planning or operation within the OMC has the layout as shown in Figure 64.6. The separation of safe and nonsafe interfaces is not explicitly visible to its user, but the underlying systems are completely separated. The interface for time-tabling and general planning information of the regional railway network, that is, ARAMIS, has no direct connection to the HMI of the OMC. Their interfaces are physically close at the workplace, but they are two completely different systems. ARAMIS is connected to the OMC only via the nonsafe part of the OMC. Here the strict separation of nonsafe and safe functions, that is, planning and operation, is apparent.

The HMI is controlled through method safe operation, that is, safety-relevant information is exchanged twofold and has to be acknowledged by the user. For (nonsafe) planning, no such procedure is required. Within the OMC, different roles are assigned to the personnel. These roles define the

FIGURE 64.6 OMC workplace.

capabilities available at the workplace and allowed actions. In our example, a planner changes the schedule for a specific train, and we follow the path of communication to the operator observing a different arrival time of the train.

The planner issues his change request in the planning interface. For these changes, it is guaranteed that they will be executed on time, provided they are issued at least 5 min before their schedule. Via a standby redundant IP connection, the change request is sent to ARAMIS. This communication is nonsafe and uses time-outs to detect a broken IP connection. Furthermore, a closed IP network is provided for this traffic. When the train route is scheduled, the ARAMIS system forwards the route to the OMC via the same network structure and protocol. The nonsafe ATRS then sends the information in XML format via TCP/IP to the nonsafe ILC. The ILC forwards the command to set a route to the corresponding ILP, again through TCP/IP. The ILP and substation check the validity of the set route command. However, the details of the communication path for setting the route are explained in the next section.

The ILP is implemented in a diverse manner on a 1oo1 TAS Control Platform, with the concepts for RAMS as presented earlier. It forwards the changes performed by setting a route from the substation to the HMI. It does so by using safe message construction according to OCS and uses the SAHARA protocol for EN 50159–conform communication. Availability within the OMC is achieved through the transparent high-availability network. The SAHARA protocol implementation has to conform to the open network classification without malicious attacks according to EN 50159. This is required since the internal network is logically separated via VLANs, but uses the same physical infrastructure as nonsafe VLANs.

Finally, the HMI will display the set train route via its interface on the OMC workplace.

64.7.2 Operator to Switch

An operator altering the position of a switch marked as occupied, that is, a switch that is already used within a train route, is an example of a communication path involving only safe components within the OMC. This is an override situation, where the operator has to guarantee the safety of his operation. The OMC provides a checklist of safe preconditions as support, but does not check the validity of the operation. However, the system has to ensure the correct execution of the command.

There are no explicit time restrictions on when the execution of this command is executed by the switch. The operator himself is provided with a timely view of the railway network elements, which only depicts the actual state and no pending commands. Should a command fail, for example, a switch's motor is broken, the operator will observe it by the absence of a reaction.

The operator method-safely issues the switch change command to the HMI. The HMI forwards this command via the SAHARA protocol to the ILP. Through one of two X25 connections, the command is then sent from the ILP to the substation. Either these X25 connections can be directly used in closed networks or datacryptors can be added on both ends for use in any open network. The redundancy model here is standby, and only heartbeat messages are sent on the passive channel to check its availability. A loss of the active connection is detected with a time-out. In such a case, the channels are switched, and the ILP requests the state of all connected elements of the substation. Until it has new information, the ILP notifies the HMIs that no valid state of the network elements is currently available.

The substations themselves are implemented with different architectures and safety concepts. These architectures may be one of 1oo1, 2oo2, or 2oo3. Also NX interlocking systems, which are based on relay logic, can be connected when extended with a suitable electronic interface. This is one of the main challenges for the introduction of the OMCs: the integration of substations with largely differing technology and age. This is solved with the X25 protocol as described earlier. The substation forwards the state change command to the switch. There the switch motor will change its state. On completion, the substation sends the new state of the switch back to the ILP. The ILP in turn notifies the HMI of the new switch state and, by this, the human operator.

64.7.3 ILP to Substation via Datacryptors

A common scenario in Austria is the connection of the ILP to the substation via two X25 channels, where one channel is routed through a closed network and the other via an open one [9]. In the latter one, measures against malicious attacks have to be provided according to the classification of EN 50159.

The X25 protocol itself does not include any security measures suitable for open network communication. This functionality is added with the use of a special device, the datacryptor. This device can encrypt and decrypt the communication received via its interfaces. It is used at both ends of the X25 communication, that is, the ILP and the substation. Its interfaces to the ILP and OMC behave like interfaces for data communication equipments and toward the open network like data terminal equipment. An example datacryptor is DCAP-LE, which is mounted in the substation and OMC on a regular 19 in. rack as shown in Figure 64.7. Defined by its configuration, it either uses the advanced encryption standard or data encryption standard method for encryption and decryption.

With the use of datacryptors, the X25 connection can also be routed through these open networks.

64.7.4 Mobile Shunting Device to ILP

Shunting personnel working at the tracks can use a mobile shunting device for interaction with the OMC to set and monitor shunting train routes. This eases shunting tremendously, when compared to the method of phoning the operator and requesting the same while following a strict protocol on the landline. The mobile device sends the request for a shunting train route by GSM-R to MOVUS in the OMC. MOVUS then requests the route at the ILC using an XML-based protocol via TCP/IP. All these components are nonsafe, and again the border between safe and nonsafe components is crossed between ILC and ILP. If the requested route cannot be set, an error is reported to MOVUS. MOVUS forwards this error to the mobile shunting device, as well as the HMI.

64.7.5 Ensuring ETCS Conform Trains on ETCS Tracks

The European Train Control System (ETCS) is a standard for systems designed to provide signaling information and other services through on-board equipment of the train to its driver. ETCS classifies

FIGURE 64.7 Datacryptors.

signaling systems in levels. With the use of the ETCS system, all segments of a railway network conforming to ETCS Level 2 or 3 can omit signals along their tracks. Such tracks must be used only by trains that are equipped with the corresponding ETCS on-board devices.

Within the OMC, a supervision is implemented to prevent the inadvertent entering of insufficiently equipped trains into such segments. The ARAMIS system provides the train ID and other train-related data. This includes the availability of an ETCS on-board device, to the OMC via its TCP/IP interface. Within the OMC, the TML collects these data and provides the train information to the automated train routing service on request via XML-formatted messages over TCP/IP. This service then denies the setting of a train route, when no ETCS is available on board. It, furthermore, informs the HMI, again through XML and TCP/IP, of the failed attempt. The operator in front of the OMC workplace must then resolve this conflict, for example, by setting a different route. The task of the ETCS check can be performed within the nonsafe part of the OMC, since an additional system is installed at the end points of the ETCS segments. This system stops the train if it fails a further ETCS on-board device check.

References

1. CENELEC, E.N. 50126 Railway applications: The specification and demonstration of Reliability, Availability, Maintainability and Safety (RAMS). European Committee for Electrotechnical Standardization, Brussels, September 1999.
2. CENELEC, E.N. 50128 Railway applications: Software for railway control and protection systems. European Committee for Electrotechnical Standardization, Brussels, March 2011.
3. CENELEC, E.N. 50129 Railway applications: Safety related electronic systems for signaling. European Committee for Electrotechnical Standardization, Brussels, May 2002.
4. Avizienis, A. The N-version approach to fault-tolerant software. *IEEE Transactions on Software Engineering*, 12, 1491–1501, 1985.
5. Gaiswinkler, G. and A. Gerstinger. Automated software diversity for hardware fault detection. In *IEEE Conference on Emerging Technologies & Factory Automation, 2009 (ETFA 2009)*, Palma de Mallorca, Spain, 2009, IEEE, pp. 1–7.
6. Poledna, S. Replica determinism in distributed real-time systems: A brief survey. *Real-Time Systems*, 6(3), 289–316, 1994.
7. CENELEC, E.N. 50159 Railway applications—Communication, signalling and processing systems—Safety-related communication in transmission systems. European Committee for Electrotechnical Standardization, Brussels, September 2010.
8. Gerstinger, A., H. Kantz, and C. Scherrer. TAS control platform: A platform for safety-critical railway applications. *ECRIM NEWS* 2008(75), 49–50, 2008.
9. Messauer, C. and G. Grünsteidi. Die Betriebsführungszentrale Salzburg in Betrieb. Signal und Draht, Hamburg, 102, 6–12, 2010.

65

Ethernet Extensions for Train Communication Network

Stephan Rupp
Duale Hochschule Baden-Württemberg Stuttgart

Reiner Gruebmeyer
Kontron AG

65.1 Ethernet in TCN 65-1
65.2 Native Ethernet Technology 65-2
65.3 Ethernet in Embedded Environments 65-5
65.4 Specific Requirements in TCN 65-8
Acronyms 65-13
References 65-13

65.1 Ethernet in TCN

The Train Communication Network (TCN) represents a fieldbus, which is used in railway vehicles. This established fieldbus has been standardized as an IEC standard by an international alliance of railway manufacturers (IEC 61375-1, see [1]). The TCN according to the IEC standard is structured in a hierarchical way with the following components: (1) the interconnection of train segments (so-called consists) is handled by a Wire Train Bus (WTB) and (2) the interconnection of devices within each train segment is controlled by a Multifunction Vehicle Bus (MVB).*

Like in other industrial applications and in avionics, there is also an Ethernet-based version for the TCN on the way. It is also specified by an international IEC working group and currently in progress (for the current state, please refer to the dashboard of the TC-9 at Ref. [1]). The Ethernet-based version of TCN comprises the specifications IEC 61375-3-4 (Ethernet Consist Network [ECN], representing the MVB) as well as 61375-2-5 (Ethernet Train Backbone [ETB], representing the WTB). Thus, it follows the same principles as IEC 61375. Figure 65.1 shows the overall architecture.

The figure shows that the Ethernet-based TCN maintains the hierarchical architecture of the IEC 61375-1. Within a train segment, devices are controlled via an ECN, which interconnects end devices (EDs) by consist switches (CSs). As one train may contain multiple train segments (cars, consists), the segments (which are, respectively, consist networks) are interconnected via Ethernet Train Backbone Nodes (ETBNs).

The network topology within a Consist Network (ECN) is not restricted by the standard. For instance, ring topologies can be applied in combination with a protocol providing ring redundancy.

* For additional information on TCN, see *The IEC/IEEE Train Communication Network*, Hubert Kirrmann and Pierre A. Zuber, Industrial Communication Technology, CRC Press, 2005.

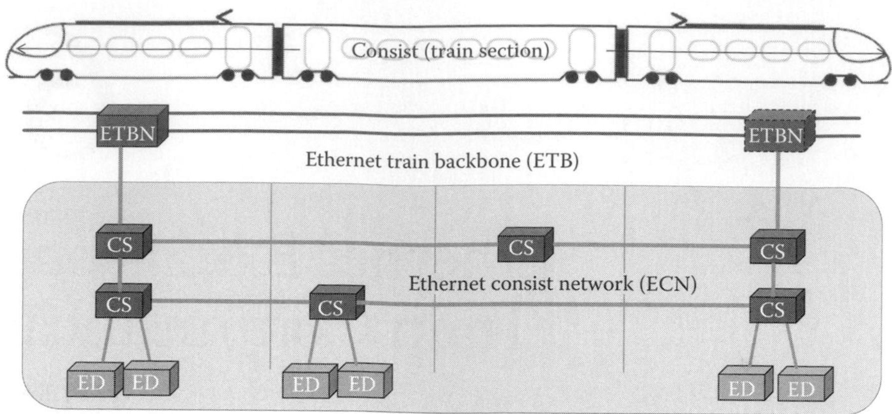

FIGURE 65.1 Ethernet-based TCN architecture.

Apart from the ring configuration indicated in the figure, there are other typical requirements in the vehicle network, which need to be met by the Ethernet-based TCN:

- Real-time (RT) processing
- Fault protection at ECN and ETB level
- Redundant ETBN configuration
- Separation of vital process data from other traffic (measurements, announcements, passenger information, surveillance cameras, etc.)
- Network organization at Ethernet and Internet Protocol (IP) level
- Safety protocols

One specific requirement on a train is the dynamic organization of the network: Unlike a road vehicle or an airplane, a train may need to be reconfigured while stopping at a railway station. One reason may be that it needs to drive in the opposite direction at a head station, which changes the driver's position and the head of the train. Another reason is that a train segment may be added or dropped. Such changes affect the network topology and thus need to be supported by the TCN.

While driving, the network topology is stable. Still the TCN represents a complete Layer 3 network including routing functions and administration of IP addresses. Thus, the TCN does not represent a simple Ethernet-based subnetwork (respectively, Layer 2 network).

Moving from any proven technology to a new generation is not a trivial step. In the case of TCN, the fundamental conditions are

- The need for a full functional equivalent to WTB and MVB on the new network technology
- The need for embedded network products, which suit the specific requirements in railway systems

Native Ethernet technology did not have any of such applications in mind. However, Ethernet has been following an evolutionary approach concerning its specifications and thus has been capable to adjust to a huge variety of applications.

65.2 Native Ethernet Technology

Ethernet originated some 30 years ago in computer networks in the office environment. Since the 1980s, Ethernet has been spreading fast as a universal medium to interconnect computers and all sorts of peripheral devices. One reason for this success is that the Ethernet specifications take an evolutionary approach.

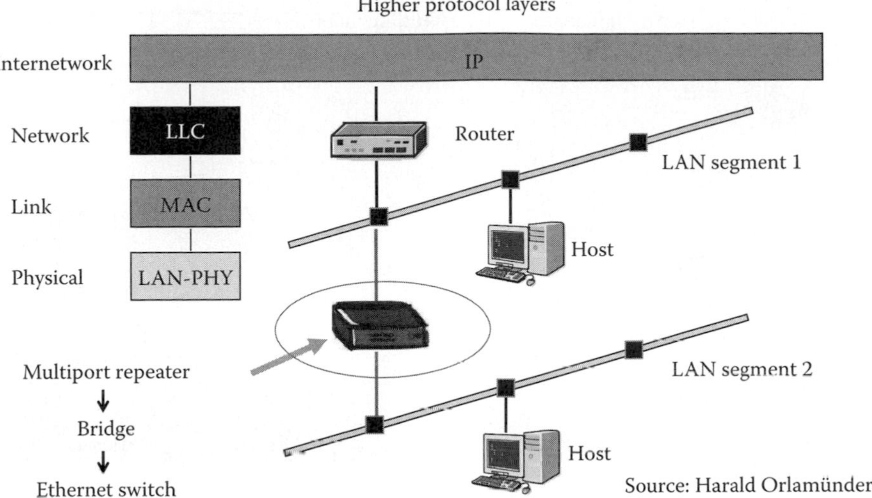

FIGURE 65.2 The origin and structure of Ethernet.

Ethernet protocols are structured in a layered and modular way (Figure 65.2). The basic specification of Ethernet contains the Layer 2 protocols: MAC (medium access control, IEEE 802.3) and LLC (logical link control, IEEE 802.2). This specification could be extended by further control protocols, such as IEEE 802.1 that—among others—contains the spanning tree protocols, virtual local area networks (VLANs), and port-based access control, as well as application-specific extensions (IEEE 802.4 and higher).

At the lowest layer (the physical layer), Ethernet protocols accommodate different media, such as Ethernet cables at a variety of speeds, Wi-Fi (wireless LAN), or optical fibers. This way, the elementary protocols for MAC and LLC could grow beyond their original physical medium.

The function of an Ethernet switch is based on the Ethernet message format (Ethernet/bridging protocols), that is, the Layer 2 protocols. This layer contains local network addresses (MAC addresses). Elementary Ethernet switches (hubs) have been simple multiport repeaters, that is, a message received on one port is repeated to all other ports. Devices connected to such a network receive all messages and respond only to messages, which are directed to their own MAC address.

With the growth of networks, bridges have been introduced, which only transmit traffic intended for the specific network segments connected to the bridge. In order to filter the traffic, bridges had to learn the MAC addresses of devices connected to their ports. In a subsequent step, the traffic capacity of the Ethernet bridges was increased: Ethernet switches allow all ports to operate at their nominal speed.

Layer 2 protocols have been evolving and today typically include link aggregation (802.3ad), VLANs (802.1Q), spanning tree (802.1D, 802.1w), quality of service (QoS) (802.1p), flow control (802.3x), as well as GVRP (dynamic VLAN registration) and GMRP (dynamic L2 multicast registration).

With larger networks and for wide area connectivity, it made sense to break down the network infrastructure into segments: local area networks (LANs), which are organized by the Layer 2 Ethernet and MAC addresses, and a higher network layer—wide area networks (WANs), which are organized by Layer 3 protocols, that is, the IP and IP addresses.

From the perspective of the higher network layer, the local networks are operated as IP subnetworks. This allows the use of private IP addresses for hosts and devices in home and office networks. Also, the network can be structured flexibly at the Layer 3. In such a network, IP routing protocols may be applied. More sophisticated Ethernet switches also support Layer 3 protocols such as OSPFv2, RIPv2, Virtual Router Redundancy Protocol (VRRP), IGMP snooping, IPv4 forwarding, DiffServ, ARP, ICMP, as well as DCHP client/server to receive or distribute IP addresses.

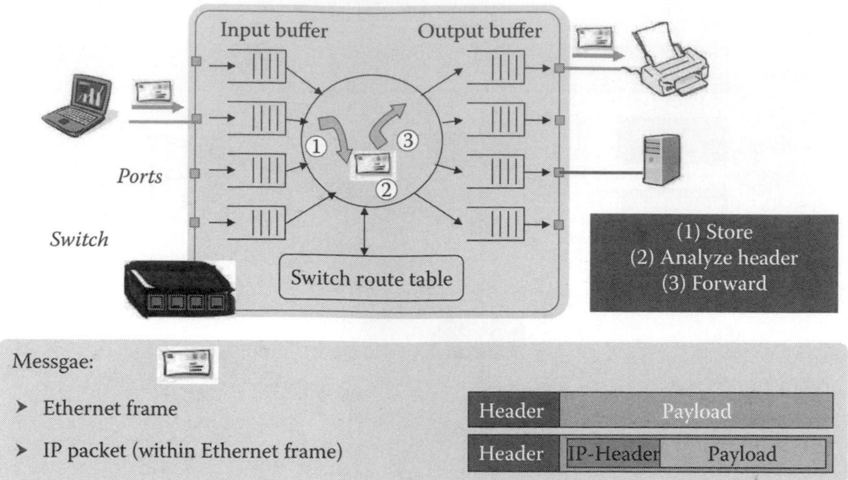

FIGURE 65.3 Ethernet switches.

Because of their many configuration options, advanced Ethernet switches need a user interface, which allows operation via a command-line interface (CLI), remote terminal (Telnet), or browser (web server), as well as a management interface (SNMP). Devices with these options are known as *managed switches*. Such switches also receive an IP address because they need to be addressed as hosts for configuration.

Irrespective of protocol diversity, every Ethernet switch operates in basically the same manner. Acting in a similar way as a post office, a switch receives messages at one of its ports (message in), analyzes the destination address of the message, and forwards it to a suitable port (message out). The source address and destination address (MAC addresses) of each message (Ethernet frame) are contained in the frame header.

The switch matches ports with addresses from the source addresses that are received at the input port. Addresses and matching ports are maintained in a table (switch route table), which can be looked up each time a message is received. If a destination port is known, the message is forwarded to this matching port only. Otherwise, the message is repeated to every port. Figure 65.3 shows the storage and forwarding of messages in an Ethernet switch.

As shown in the Figure 65.3, an Ethernet switch receives messages at a port, analyzes the destination address, and forwards it to a suitable port. The MAC address of each message is contained in the frame header. The header of an Ethernet message (IEEE 802.3 Frame) may contain further information beyond source and destination addresses, such as VLAN tags, which qualify different traffic classes. This information classifies a category or group a message belongs to and may also decide which way to forward the message. Depending on its classification, a message can be placed in a priority queue on a port. If a message is part of a multicast group, it will be copied and forwarded to all members (destination addresses) of the group.

In the same way, IP header information, which is part of an Ethernet frame's payload, can be accessed by the switch and used for the classification and handling of messages. This, for instance, allows mapping IP multicasts into MAC multicast over Ethernet ports. This so-called IGMP snooping may be used in to reduce unicast streams. Ethernet switches, which support Layer 3, sometimes provide such features.

Figure 65.4 shows a message forwarded to a group of receivers. To do so, the sender sends a message to a multicast address (i.e., an address representing the group). The Ethernet switch resolves the multicast address into individual addresses and repeats the message to all respective ports. This is much faster and more efficient because it avoids generation of many unicast messages by the sender.

The use of VLAN tags within an Ethernet frame's header allows bigger LANs to be broken into segments or a VLAN. Figure 65.5 illustrates the principle by using the colors as VLAN tags on Ethernet frames and switch ports. In this case, the Ethernet switch is configured so that the ports are allocated to specific VLANs. The switch marks all incoming messages to such a port with the corresponding VLAN tag and

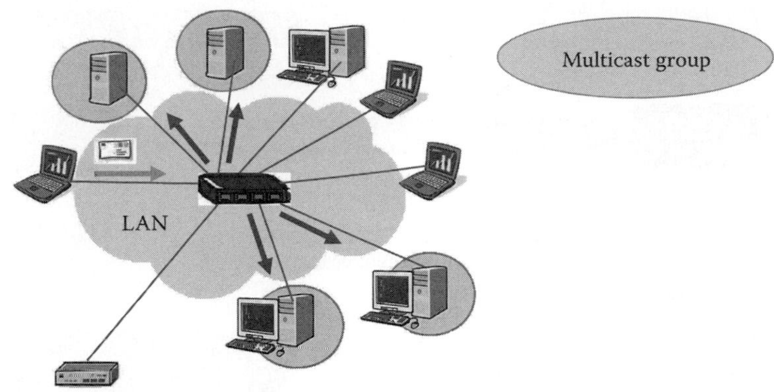

FIGURE 65.4 Multicast transmission.

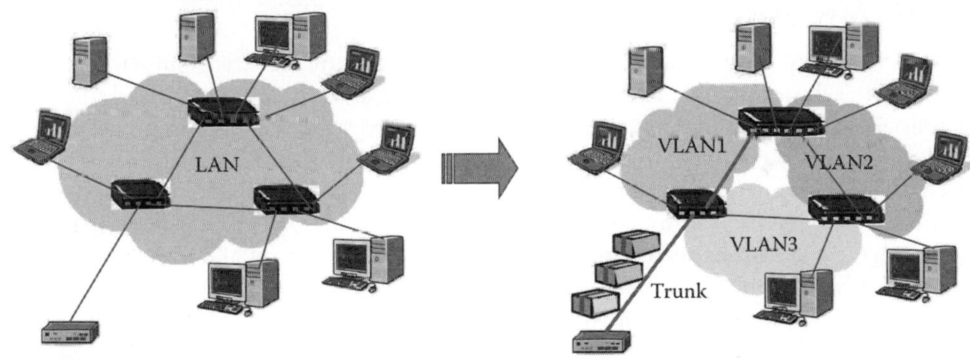

FIGURE 65.5 VLANs.

forwards packets received with a VLAN tag only to those ports that match the corresponding VLAN. For traffic from multiple LAN segments, ports can be configured to provide such interconnection: the trunks.

Multiple LAN segments can also be implemented physically using enough Ethernet switches. In VLANs, the segmentation is achieved by configuring the switches accordingly. Layer 2 features like multicast, VLAN, or priority queues for traffic classes are entirely based on information contained in individual frames, the port configuration, and address tables. They do not need any state-based information and can be performed at wire speed by Layer 2 switches.

Layer 3–based functions, such as the organization of networks with IP addresses (such as assigning IP addresses via DHCP or the routing of IP packets), the handling of sessions (which needs state-based information beyond the context of the individual Ethernet frame), or the processing of sequence numbers (which needs to keep the sequence in memory), are usually handled by routers or hosts in the network.

65.3 Ethernet in Embedded Environments

Ethernet has been also adopted in the industrial applications. Controllers and drives in industrial automation are increasingly being interconnected and controlled via Industrial Ethernet (see [2–5]). Ethernet is also used as fieldbus in aircrafts (see [6]). Such applications demand RT performance and safety levels beyond the scope of native Ethernet. Embedded network products also need to meet the specific nonfunctional requirements (such as environmental conditions, EMC, shock and vibration, materials used). So far, Ethernet has been growing and meeting new demands.

Ethernet switches for embedded systems are implemented to match specific environmental conditions such as an extended temperature range (between −40°C and +70°C). Embedded network products are

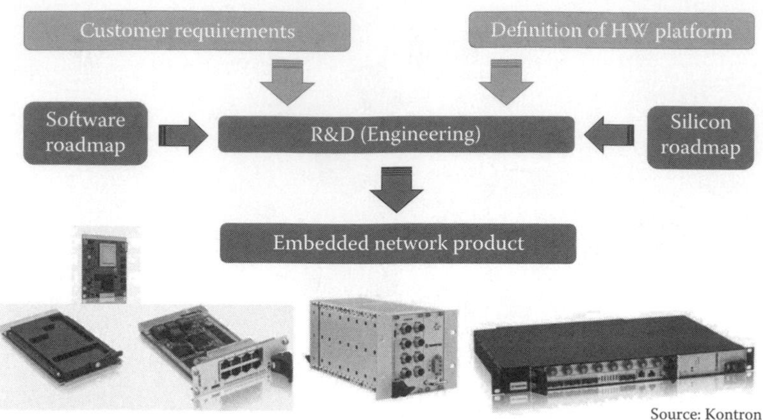

FIGURE 65.6 Embedded network platforms.

hardened to withstand greater electromagnetic irradiation and support strict limits for its emission. Switches used in vehicles, aircraft, and industrial plant must also withstand severe shock and vibration and be resistant to dust, humidity, and many other hazardous substances. Moreover, embedded systems are frequently adapted to individual customer requirements and must match specific needs in terms of product availability, maintainability, service, and repair.

The functional composition of an embedded switch, or router, follows the process shown in Figure 65.6. Depending on the application- and customer-specific requirements, the set of feature is defined and an appropriate hardware platform is chosen. Embedded hardware offers a choice of standardized form factors and modular components to support specific application areas and customer-specific setups.

If Ethernet is used as a fieldbus to control drives or vehicles, the network must respond within a certain period of time. An industrial environment demands RT responsiveness within a specified interval. Figure 65.7 illustrates the environment and RT requirements. As shown on the right, a fieldbus transmits control messages from a controller to an actor (e.g., a drive) and from a sensor to the controller.

Message transfer times between devices and the controller are subject to variable delays called jitter. While a constant delay can easily be compensated for in an industrial control with a high level of accuracy (see [7]), delay variations cannot. The challenge is to keep the variations of delay (jitter) low with respect to the response times, which vary according to the application. Controlling drives needs a response time way below 1 ms. For controlling a process, response times of below 10 ms may be sufficient. Machines or vehicles operated by a user terminal typically need response times in the same order of magnitude as the human response time: about 100 ms.

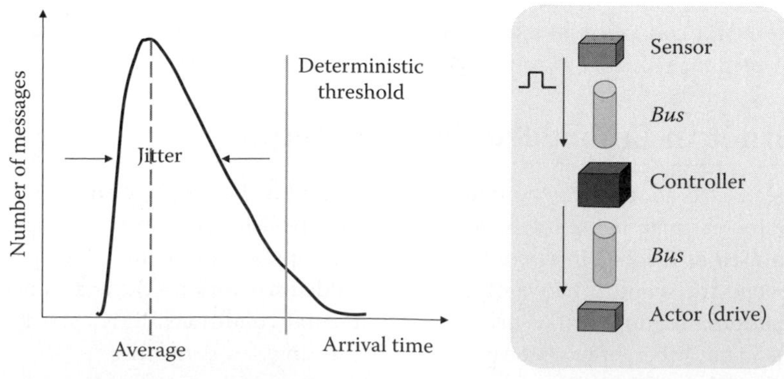

FIGURE 65.7 RT requirements on the fieldbus.

RT performance typically demands deterministic response times. Sensors and devices do not change their state within a deterministic threshold. But when knowledge of their state is requested in a cyclic interval, they do respond. The response times that an Ethernet-based fieldbus can achieve depends upon network size and topology, transmission speed, and message size in bytes. One reason is the storage and forwarding of messages within each switch. Ethernet cannot ensure deterministic response times.

Figure 65.7 shows the arrival of messages over a fieldbus as a histogram (number of messages arriving at a variety of delay times). The variations of delay (jitter) depend on the traffic conditions on the network, in just the same as for road traffic. If roads are crowded with vehicles and trucks, arrival times become hard to predict.

In an industrial environment, a fieldbus also will need to handle different traffic classes, such as process data relevant to control, diagnostics data, and status messages, as well as any other traffic. Response times matter for control, which depends on process data. Other traffic is less critical or not critical at all. Such traffic may travel in a lower class with a lower priority.

Different approaches may be used to meet specific requirements. Some of these approaches are compatible with IEEE 802 and some are not. The handling of traffic classes, for instance, is part of regular Ethernet and IP standards as applied to Internet-based telecommunication networks. Under QoS, suitable methods for handling different classes of traffic are described both for Layer 2 (IEEE 802.1p) and for Layer 3 (DiffServ, corresponding to RFC 2474 and 2475, as well as some supplementary request for comments [RFC]).

The principle is based on the classification of messages (Ethernet frames or IP packets) in a specified field in the respective message header. At each node (Ethernet switch or IP router), messages are handled according to their traffic class. Higher-class messages are handled with priority. The procedure corresponds to an airline check-in: for each flight (port), there is a counter and a queue for each traffic class. The highest-class passenger (message) is dispatched with priority and forwarded to the gate first. This procedure rearranges the order of messages to a specific destination according to their priority (class of service) at each node (switch or router).

This procedure cannot guarantee a specific response time, which depends upon the total traffic volume. Regarding the highest class of service, the amount of process data that needs to be communicated in this class is known, so the traffic situation within the class can be regulated using suitable cycles and the corresponding volume of data and messages.

Still there is interference with traffic from the lower classes. For instance, while a second class message (*Business* class in Figure 65.8) is in transmission over the destination port, the first-class message (*Senator* class in Figure 65.8) needs to wait. The delay depends on the length of the lower-class message. The length of Ethernet frames can vary significantly between 64 and 1518 bytes (9000 bytes for *Jumbo* frames). IP packets may extend up to 64 kB.

By not discarding Jumbo frames, the length of *vehicles* (data frames) on the data highway varies by a factor of more than 1000. At 100 Mbps transmission rate (fast Ethernet), a Jumbo frame takes 0.7 ms to transmit. The transmission of a 64-byte control message would have taken just 5 μs. Depending on the number of nodes between sender and receiver, delays and jitter accumulate.

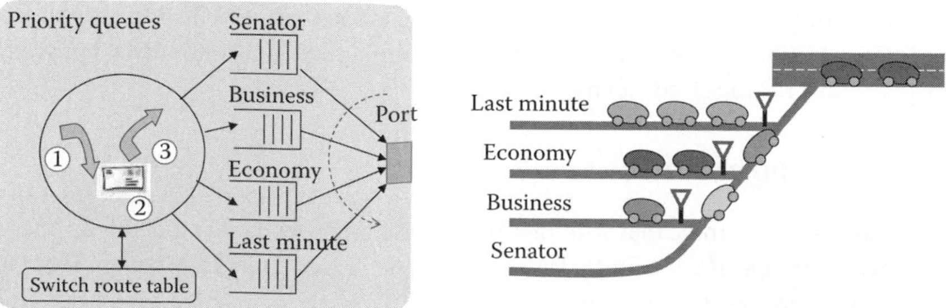

FIGURE 65.8 Priority for process data.

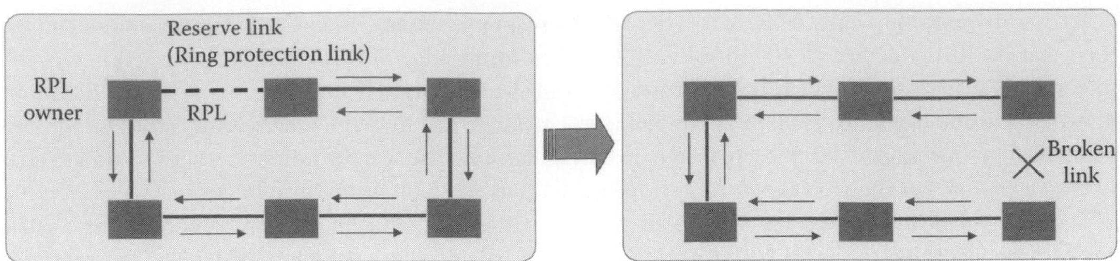

FIGURE 65.9 Redundant ring configuration.

There are some simple measures to limit delays and jitter: (1) limiting the maximum frame size and (2) limiting the size of the network. For instance, a maximum frame size of 512 bytes/message corresponds to a 40 μs transmission time. Total delay and jitter then depend on the number of network nodes. In a control application, the network topology and number of control segment nodes can be designed accordingly.

Not allowing different classes of traffic to use the same network could avoid interference between process data and any other traffic altogether. This does not necessarily mean that physically separated networks (LANs) need to be used. The same effect can be achieved by configuring VLANs in a suitable way (see Section 65.2). A VLAN can assign a *private* port to process data, which avoids the conflict.

For operation in an industrial environment, some redundancy is desirable allowing protection against failures. A ring topology achieves such protection in a very simple and cost-effective way. It provides fault protection for single failures. Figure 65.9 shows the configuration of Ethernet rings. In general, network administrators do not like loops in the network, because they lead to pointless traffic situations. As the figure shows, the ring only exists on physical layer.

The nodes (Ethernet switches) are cabled such that the linear topology cabling becomes a ring. The nodes still operate as in a linear topology with one link representing a reserve link—the so-called ring protection link (RPL)—which is not used to transmit regular messages. If one link of the network fails, the reserve link is activated. After recovery, all nodes are still connected to each other and form a new linear topology. Failover to the new topology needs to be fast enough to meet the demands of the industrial environment concerned—typical failover times are below 500 ms.

Detecting failures, activating the reserve link, and failing over to a new topology mean extra tasks that need a suitable control protocol and controller. One of the switches takes charge of this task and becomes owner of the reserve link (RPL owner). To check the ring for potential failures, the RPL owner sends control messages in both directions across the ring (including the reserve link, which does not carry conventional traffic).

If the RPL owner receives messages from both directions, the ring is operational. If a broken link occurs, the RPL owner activates the reserve link and changes the topology. Because messages now need to be passed along different ways, switches need to change their MAC address tables accordingly. To allow the interoperation of switches from different manufacturers, a standardized procedure for ring supervision and failover is required (such as in [8]).

65.4 Specific Requirements in TCN

Most requirements in TCN can be met with methods, which are specific to embedded Ethernet applications but not necessarily specific for TCN. Thus, an Ethernet-based TCN may benefit from mainstream embedded Ethernet applications such as in industrial automation, energy provision, and avionics. In general, the requirements on RT processing and fault protection for rail vehicles are comparable to those for industrial automation.

Concerning the RT requirements, a controller for drives has much harder time constraints than railway vehicles, which are controlled by a human via an human machine interface (HMI) (see response times in Section 65.3). Concerning redundancy, requirements in transportation are higher than in industrial automation but still within the range of the more demanding applications there. At the level of ECN, the same type of ring topologies can be applied as in industrial control in combination with a ring redundancy protocol: like MRP and HSR; see [8] and [9], respectively.

The ETBN can be implemented in a redundant way. The connections between the train segments via ETB can also be implemented by using dual Ethernet links. If fault tolerance and recovery time allow, protocols such as Ethernet link aggregation could be used on the ETB connections. Operating redundant ETBNs requires extra attention at Layer 3. At IP level, there can be only one active ETBN per consist in a redundant configuration. This situation may be covered by native Layer 3 redundancy protocols, such as Virtual Router Redundancy Protocol (VRRP) (IETF RFC 5798).

The CSs represent regular Layer 2 switches. In order to separate process data from all other traffic (measurements, announcements of camera images, etc.), they should support different classes of services (QoS) according to the native procedures of Ethernet and IP. The use of VLANs in a ring or line topology would require an extra port and extra cabling for process data. Whether such an effort can be justified depends on the specific situation.

As with other fieldbuses, the concept and data models of the proceeding fieldbus may be kept and implemented at application layer. This way, the process data may be migrated from the WTB/MVB type of TCN. As in other traditional fieldbuses, WTB/MVB devices operate on a shared medium (much like the original yellow cable Ethernet). Figure 65.10 shows the traditional fieldbus in operation.

When operating at a transmission rate of 1 Mbit/s, the transmission of a single bit takes 1 μs. This time corresponds to a length of 200 m at a propagation rate of 200×10^6 m/s over the bus cable. The extension of MVB corresponds to the length of one car or one consist, which is much lower, so the bus operates in a quasistationary mode concerning signal propagation.

The lower part of Figure 65.10 shows the bus cycle. The bus master arranges communication over the shared medium, that is, by sending cyclic requests to each device on the bus. The devices respond in a specified way, that is, following reception of a request. The duration of each bus cycle depends on the number of devices and the size of the messages. For instance, with 10 devices and a message size of 100 bits (12 bytes), the minimum bus cycle takes 2 ms. This number results from 100 μs transmission time per message, 10 devices on the bus, and 2 messages per device.

Concerning signals and response times, Figure 65.11 shows a sample scenario. In this scenario, a sensor is placed as device number 2, the controller as device number 8. While the bus master operates the bus cycle, messages from devices are transmitted over the bus in their corresponding time slot. As soon as the state of a sensor changes, its signal is buffered in its I/O device and transmitted in a message in the appropriate time slot.

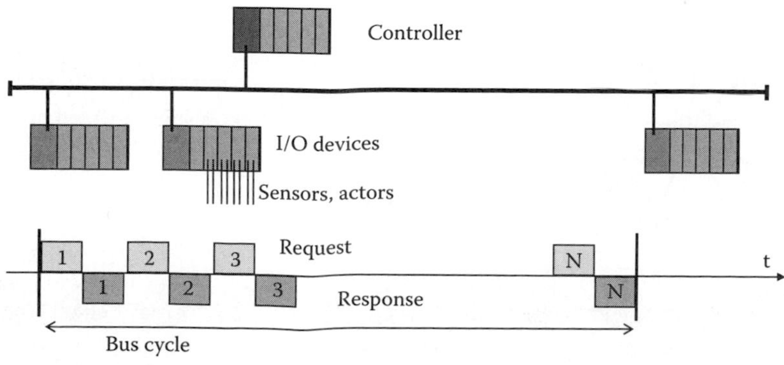

FIGURE 65.10 Traditional fieldbus (WTB/MVB).

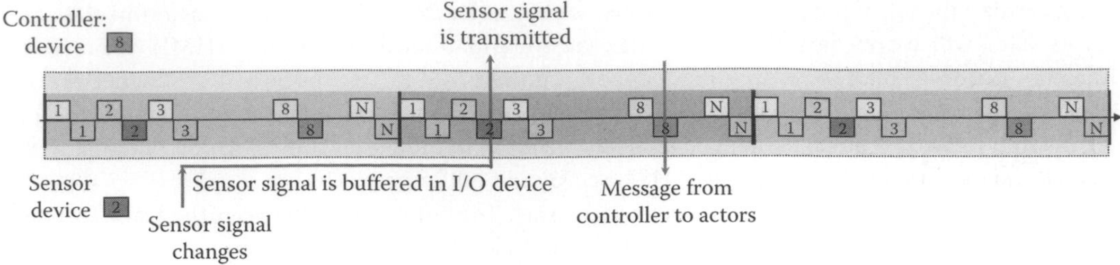

FIGURE 65.11 Signals and response times on the bus.

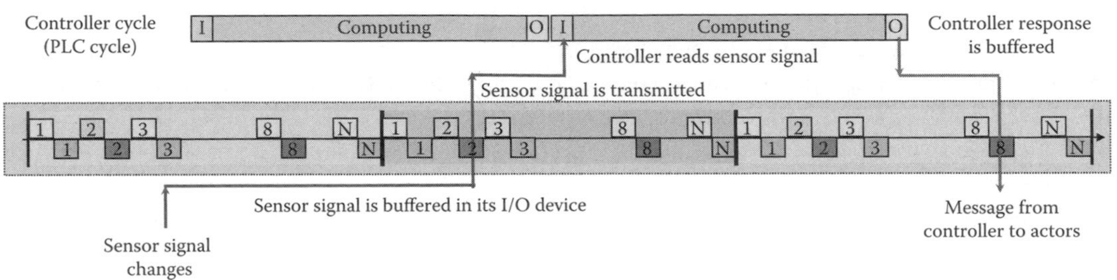

FIGURE 65.12 Response times including controller.

Depending on the time it takes the controller to compute an instruction based in the message, the controller response could be transmitted within the same or the following cycle. If the controller is operating in its own cycle, the response is transmitted as indicated in Figure 65.12.

How does the traditional fieldbus operation translate to Ethernet-based TCN? In the simplest case, the bus cycle is maintained and placed upon the Ethernet-based bus. The response times depend on the chosen bus topology, as well as number of devices, message size, and transmission rate. Ethernet can be operated in a variety of topologies, such as a star topology, tree topology, or linear topology (which includes the redundant ring topology). Figure 65.13 shows a star and a ring in combination with the bus cycle, with the controller at the center, respectively in front. The linear topology corresponds to both ECN and ETB.

Most Ethernet fieldbuses today operate at transmission rates of 100 Mbit/s (fast Ethernet). The minimum message size is 64 bytes (including header information). Hence, the transmission of one message takes about 5 μs. With 10 devices and 2 messages per device, this results in a minimum bus cycle of 100 μs. Thus, Ethernet-based TCN would be significantly faster than MVB (minimum of 2 ms for the same number of devices). This figure is correct when using a star topology for the bus.

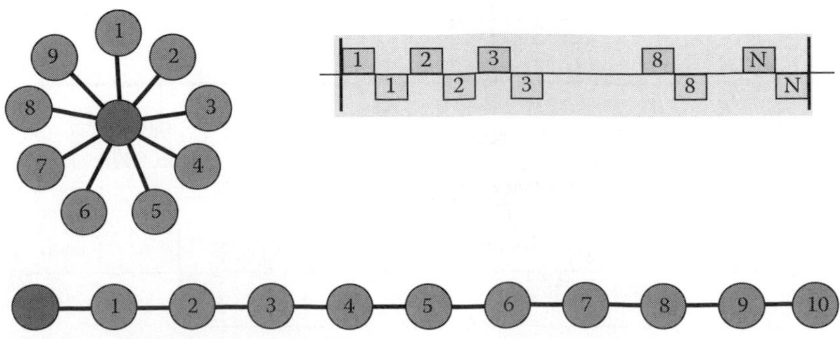

FIGURE 65.13 Ethernet bus topologies.

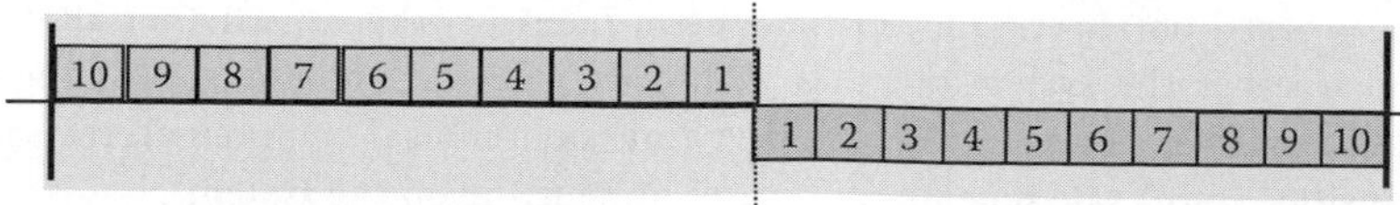

FIGURE 65.14 Optimized request pattern.

While the linear topology looks close to the traditional configuration, it does not use a shared medium but connects devices over a chain of Ethernet switches. At each stage, the messages are stored and forwarded, which consume about 5 μs for the collecting and processing of header information. In a chain of 10 devices, it takes 50 μs for one message to travel from the front to the end. When addressing devices on the bus in a randomly chosen cycle, requests need to be placed at intervals of 100 μs to allow responses traveling back the chain. Hence, the minimum bus cycle for 10 devices results in 1 ms (which still is slightly lower than the MVB cycle).

Still, the same cycle time (100 μs) as the star configuration may be achieved in a linear topology with a specific request pattern, which starts requesting the devices at the end of the chain, as shown in Figure 65.14. This way, it will take half a bus cycle for a request to travel to the most distant device and another half bus cycle for the response to travel back. The overall bus cycle with 10 devices then takes 100 μs.

In industrial automation, there are other mechanisms in place to optimize linear topologies, such as discarding the processing of MAC addresses to pass messages between the switches in a chain. In this case, the switches take information for their connected devices from the message, as it is passed by (see [2]). In TCN, there is currently no need for such optimization, because response times of 10–100 ms are perfectly acceptable.

In TCN, a dedicated role is assigned to the ETBN. While the train is in driving mode, it simply operates as an ordinary router, which connects train segments at Layer 3. While driving (moving), it allows the translation of local IP addresses, for example, by Network Address Translation. Before entering the driving state, there is an initial phase, at which IP addresses are allocated dynamically via DHCP, with the SHCP server implemented on the ETBN.

On the train, a segment may be added or removed, and the direction of driving may be changed. At such changes, the ETBN discovers the topology and enumerates all train sections correctly. This is important, so as to, for example, confirm that when the driver clears doors for opening at a station, doors will only open on the platform side of the train.

At higher protocol layers, an Ethernet-based TCN may handle safety issues in much the same way as it is applied in industrial automation. In this approach, a safety application is used to check the integrity of the system, as shown in Figure 65.15. The safety application, its supporting libraries, as well as a safety protocol need to be certified by an authorized institution.

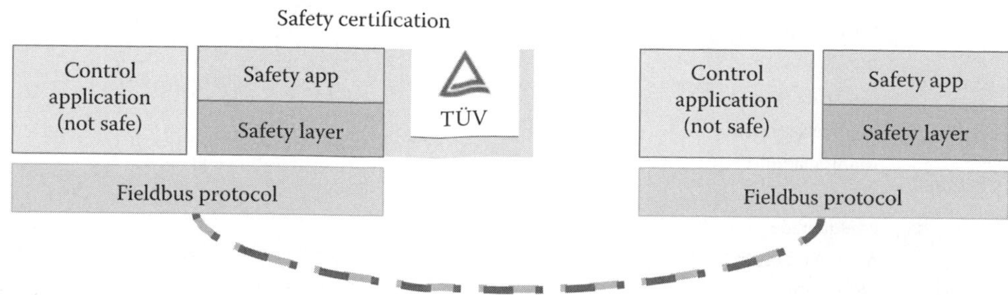

FIGURE 65.15 Safety layer.

This concept is based on surveillance of the system integrity, that is, whether all components are operational and whether the system operates within the specified boundaries. If the system leaves the operational boundaries, for example, by losing a mission-critical component, the safety processor initiates the transition into a safe state (in most cases by shutting off the system).

For the communication between devices over a fieldbus, the safety layer introduces a safety protocol. Messages between safety controllers and safety devices are communicated just as any other messages (i.e., Ethernet frames) and may be mixed with other TCN traffic. Examples for a safety controller and an associated safety device are a controller and an associated HMI (e.g., communicated the actual speed for display at the HMI or sending a command from the HMI to the controller), respectively, the controller and actors like drives, breaks, and doors.

A safe communication means that the integrity of messages on the bus is maintained. In any communication network, loss of messages cannot be prevented, neither accidental messages nor misplaced messages. Apart from error conditions, the communication between mission critical devices should also be protected against manipulation. Table 65.1 shows a summary of fault conditions and corrective measures, which are applied for a safe communication protocol. An "×" in Table 65.1 indicates, that a specific fault condition may be covered by the corrective measure in the respective column.

The unintentional repetition of a message should not repeat the action invoked by this message. The safety protocol may simply detect a duplicated message by using sequence numbers with each message. A duplicated message will have the same sequence number as the original one and can be discarded by the safety protocol.

A measure to protect against message loss takes a watchdog timer, which is set to the expected response time. Should the message not have arrived within this period, it is considered as lost. The safety protocol will repeat the message. Other indicators for lost messages is a missing sequence number, as well as a checksum, which indicates that the message has been corrupted and thus also has to be discarded and is lost.

A simple way of manipulation is to record and later replay a message. In an unprotected system, this could invoke an action. The way of protection is simple: a recorded message would have an unacceptable sequence number and thus may be discarded. A more clever kind of attack, which manipulates the content of a message (masquerade), could easily time out, respectively not have a matching checksum.

The figure shows that just three corrective measures (sequence numbers, watchdog timers, and checksums) allow covering many fault conditions. Another corrective measure against misplaced messages is the use of connection IDs or session IDs. Such an identifier classifies all messages belonging to the same connection, for example, a dialog between a controller and a specific actor.

TABLE 65.1 Fault Conditions and Safety Measures at Protocol Level

	Corrective Measures			
Fault Condition	Sequence Number	Watchdog Timer	Connection ID	CRC (Checksum)
Unintentional repetition	×			×
Message loss	×	×		×
Inserted message	×			×
Incorrect sequence	×			×
Manipulation				×
Unacceptable delay		×		
Masquerade		×		×
Memory fault in switches	×			×
Incorrect transmission between segments			×	

Acronyms

ECN Ethernet Consist Network
ETB Ethernet Train Backbone
ETBN Ethernet Train Backbone Node
HSR High-availability seamless redundancy
IP Internet Protocol
LAN Local area network
MVB Mobile Vehicle Bus (IEC 61375-1)
MRP Media Redundancy Protocol
PLC Programmable logic controller
RPL Ring protection link
RT Real time
TCN Train Communication Network
WAN Wide area network
WTB Wire Train Bus (IEC 61375-1)

References

1. TCN: Train Communication Network, IEC 61375-1, 2012, www.iec.ch.
2. EtherCAT Technology Group, www.ethercat.org.
3. Profinet and Profibus Association, www.profibus.com.
4. Ethernet/IP Consortium, www.odva.org, www.ethernetip.de.
5. Ethernet Powerlink, www.ethernet-powerlink.org.
6. AFDX: Avionics full duplex switched Ethernet, ARINC Standard 664, www.aeec-amcfsemc.com/standards/index.html.
7. IEEE 1588. IEEE standard for a precision clock synchronisation protocol for networked measurement and control systems. IEEE Std, 1588, 2002, www.ieee1588.com.
8. MRP: Media redundancy protocol for Ethernet ring topologies, IEC 62439, www.iec.ch.
9. HSR: High-availability seamless redundancy protocol, IEC 62439-3, www.iec.ch.

XIII

Semiconductor Equipment and Materials International

66 **Semiconductor Equipment and Materials International Interface and Communication Standards** *Kiah Moh Goh, Lay Siong Goh, Geok Hong Phua, Aik Meng Fong, Yan Guan Lim, Ke Yi, and Oo Tin* **66**-1
Introduction to SEMI • SEMI Equipment Communication Standard • Survey of SECS/GEM-Compliant Toolkits • Conclusion • Acknowledgment • References

66

Semiconductor Equipment and Materials International Interface and Communication Standards: An Overview with Case Studies

Kiah Mok Goh
Singapore Institute of Manufacturing Technology

Lay Siong Goh
Singapore Institute of Manufacturing Technology

Geok Hong Phua
Singapore Institute of Manufacturing Technology

Aik Meng Fong
Singapore Institute of Manufacturing Technology

Yan Guan Lim
Singapore Institute of Manufacturing Technology

Ke Yi
Singapore Institute of Manufacturing Technology

Oo Tin
Singapore Institute of Manufacturing Technology

66.1 Introduction to SEMI **66**-2

66.2 SEMI Equipment Communication Standard **66**-2
Four Main SEMI Communications/Control Standards • SECS-I • High-Speed SECS Message Services Standard • SEMI Communication Standard (SECS-II) • Generic Equipment Model • Limitations of SECS/GEM • Emergence and Evolving New Standards

66.3 Survey of SECS/GEM-Compliant Toolkits **66**-18
Overview of Equipment Integration Software • SEMI Equipment Communication Software • Analysis of Current Available SECS/GEM Solutions

66.4 Conclusion **66**-36

Acknowledgment **66**-36

References **66**-37

66.1 Introduction to SEMI

Semiconductor Equipment and Materials International (SEMI), founded in the United States in 1970, is a global industry association with a worldwide membership of companies, representing semiconductor-related industries. SEMI has 26 subcommittees that include, among others, the subcommittees for Automated Test Equipment, Environment, Health and Safety, and Information and Control.

This chapter will give an overview of the fundamentals of the SEMI Equipment Communication Standards, commonly referred to as the *SECS*, its interpretation, the available software tools, and case study applications.

66.2 SEMI Equipment Communication Standard

The SECS defines the computer-to-computer communication protocols intended to assist in the automation of electronics-manufacturing facilities. SECS enables the automation, communications, and control of semiconductor process equipment, from a variety of vendors, using a set of standards and reliable protocols. These standards have the capacity to resolve/define the physical layers, signals, block semantics, etc., thus ensuring repeatable, nondeadlocked communications from simple to extended-hierarchy networks.

In its basic form, SECS provides a one-to-one link between a particular unit of equipment and a host computer. The equipment can be in the form of wafer processing equipment, test and metrology equipment, die/package assembly equipment, surface mount equipment, and others.

Similarly, the host computer can range from minicomputers, small-dedicated desktop PCs, and laptops to a network of embedded systems. In many cases, the host computer system manages and coordinates the equipment on the factory floor via cell controllers, line controllers, and material controllers. These must be compatible with current *manufacturing execution systems and manufacturing legacy systems* (*MLS*).

66.2.1 Four Main SEMI Communications/Control Standards

SEMI Equipment Communications are based on four main standards, namely, the SECS-I, SECS-II, High-Speed SECS Message Services Standard (HSMS), and Generic Equipment Model (GEM) Standard as shown in Table 66.1. Both SECS-I and HSMS define the message transport layer of the protocol. SECS-II defines the message format and GEM defines the equipment behavior.

66.2.1.1 Components for Executing the SECS Protocol

Host systems are linked to different equipments that have to be SECS compliant (i.e., based on SECS standards). The connectivity to the host can be through serial (RS-232) or TCP/IP ports. GEM is incorporated in some of the newer equipments. Older equipments are generally not GEM compliant, having only serial connectivity. An example of *SECS Implementation Architecture* is shown in Figure 66.1.

Host system: The main function of the host system is to manage the connectivity to a group of equipments via a configuration file, recipe management, equipment load balancing, equipment monitoring, and connection to enterprise-level applications. The host systems that manage the same type of equipment are referred to as cell controllers. Similarly, the host systems that manage and coordinate different equipments are referred to as line controllers.

TABLE 66.1 SEMI Equipment Communication Standard

Protocol	SECS-I	SECS-II	GEM	HSMS
Date of launch	1978 →	1982 →	1992 →	1994 →
Message definition	Communication protocol and physical definition.	Message format.	Additional SECS-II sequences define document for communication.	Supersede SECS-I for TCP/IP compatibility.

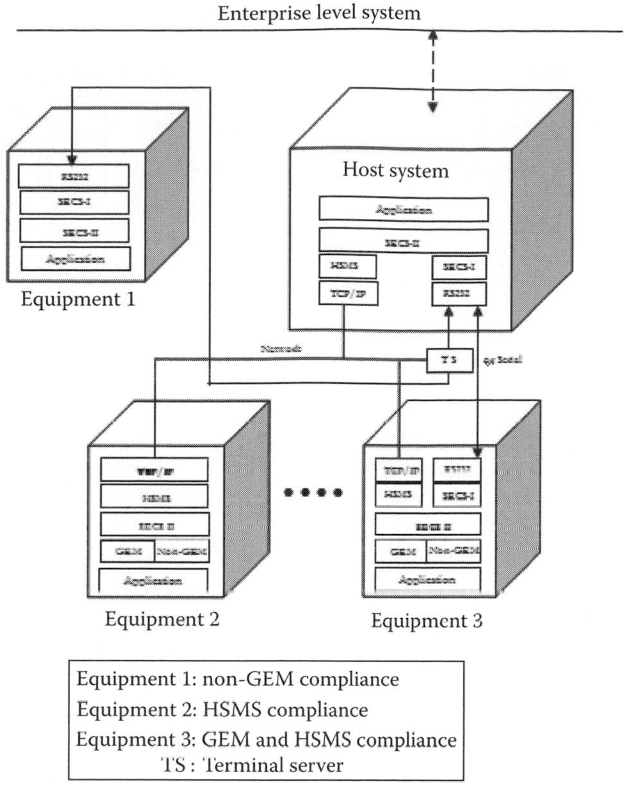

FIGURE 66.1 The components for the execution of SECS protocol.

Terminal server: It is a serial hub (also sometimes called a *serial device server*) that allows equipment with serial port connections to connect to this hub and provides information sharing among other equipments through Ethernet-ready TCP/IP connections.

Equipment 1 (*Non-GEM compliant*): This is normally the older semiconductor equipment that is not GEM compliant but supports SECS-I and SECS-II messages. Most of the older equipment will only provide serial connections to the host system.

Equipment 2 (*HSMS compliant*): This is typically the newest semiconductor equipment that *has* point-to-point communication protocol for TCP/IP Ethernet communications. Some of the equipments are already GEM compliant for host/equipment communications.

Equipment 3 (*GEM and HSMS compliant*): These are typically the up-to-date models of semiconductor equipment that are both GEM and HSMS compliant, having both serial and TCP/IP connections. Nevertheless, some of these equipments are expected to be *simultaneously serial port and TCP/IP capable*. Usually there is a problem of *connectivity to MLS*, which we cannot easily ignore.

66.2.2 SECS-I

SECS-I is also called SEMI E4 standard. It was launched in 1978 and released in 1980. It describes a transmission interface for passing messages between the equipment and a host over the RS-232 communication as shown in Figure 66.2. This standard defines point-to-point communication of data, utilizing a subset of the international standard EIA RS-232-C for the connector and voltage levels. The communication is bidirectional and asynchronous but is half duplex. The communication speed normally varies from 9,600 to 19,200 baud rate. SECS-I protocol establishes multiblock transfers based on blocks of 256 bytes.

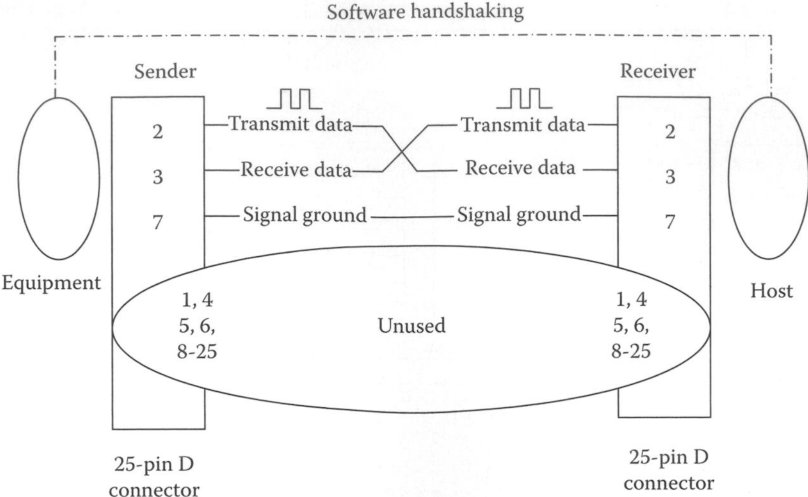

FIGURE 66.2 Three-wire serial data configuration for SECS-I (with software handshake).

TABLE 66.2 Layer Definition of the SECS-I Standard

SECS-I Layers	Definitions
Transaction layer	Primary message
Message layer	Block header
Block layer	Block format, block protocol, packet format, packet protocol
Physical layer	Physical connector, voltage, bit coding, baud rate

SECS-I includes specifications for the electrical connection between two communicating entities. It defines the message header structure, the block transfer protocol, and the message transfer protocol. The block transfer protocol is the procedure used by the serial line to establish the direction of communication and to provide the environment for passing messages. The message transfer protocol details how the block transfer protocol is used to send and receive messages.

Table 66.2 shows the layer definition of SECS-I standard. The physical layer defines the physical connectors, electrical characteristics, bit and character coding, and baud rate. The block layer defines the block or packet format and protocols that are used to transfer a single block. The message layer defines the block header. The transaction layer defines how primary messages and its corresponding reply messages are related.

66.2.2.1 SECS-I Message Structure

Figure 66.3 shows the structure of the SECS-I message. This message structure allows multiblock messages with one stream and function. The alphabet in Figure 66.3 can be described as follows:

a. The first byte represents the length of the message.
b. Byte 0–1 of the header represents the device ID. The device ID is to identify the equipment in the factory. There is no device ID for the host system. Bit 8 of first byte is a reserved bit, or R-bit, that is used for indicating the direction of the message, that is, host to equipment or equipment to host. If the R-bit is 0, it means that the message is from the host to the equipment. If the R-bit is 1, it means that the message is from the equipment to the host.
c. Byte 2 of the header represents the stream number for the message. The stream represents the category of the messages that are defined in SECS-II.
d. Byte 3 of the header represents the function numbers within the stream.

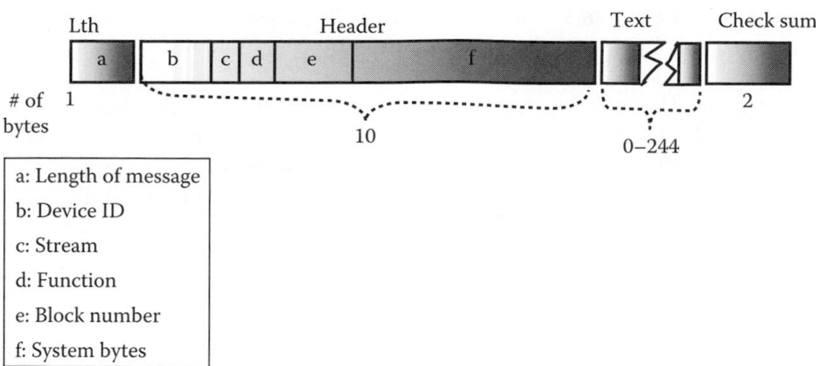

FIGURE 66.3 Structure of SECS-I message.

e. Bytes 4–5 of the header represent the block number of the message. If the message is longer than 244 bytes, the message will be broken into multiple blocks. The system will use these 2 bytes to identify the sequence of the block number.
f. Bytes 6–9 of the header represent the system bytes. The purpose of system bytes is to avoid duplication of message.

66.2.2.2 Synchronization Mechanism

SECS-I uses software-based time-outs to handle communication synchronization between the host system and the equipment. As is shown in Figure 66.4, time-outs (T1 to T4) are used in SECS-I for managing the transmission of blocks between the sender and receiver. Each time-out has a special function, as explained in the following:

T1 defines intercharacter time-out in the block transfer protocol. It measures the duration of characters transferred within the same block. T1 time-out will be activated if the host system does not receive the block of characters within the defined time.

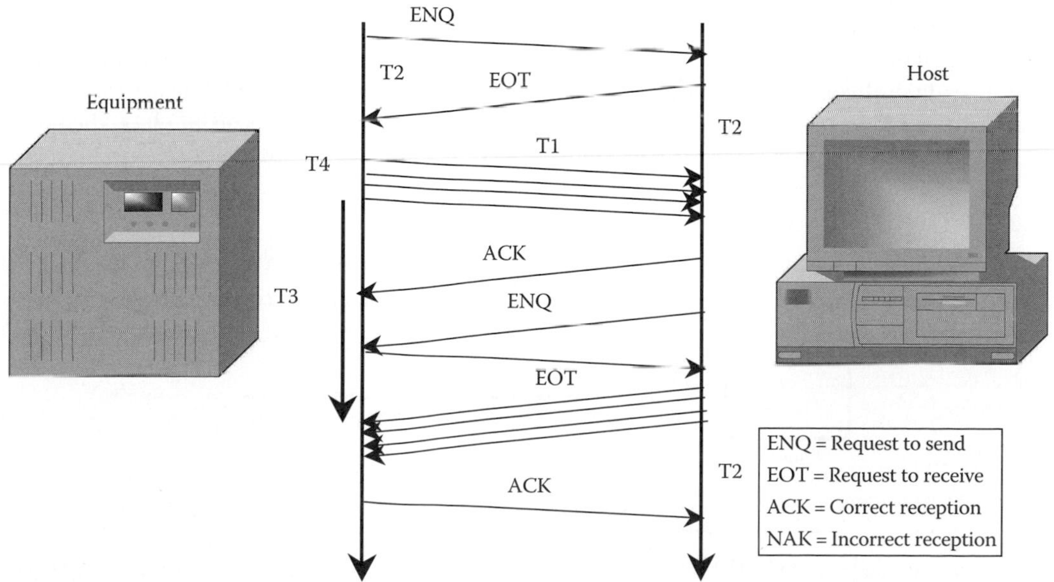

FIGURE 66.4 Establishing synchronization via time-outs (T1 to T4).

TABLE 66.3 Time-Out (T1 to T4) Definition

Time-Out	Description
T1	Intercharacter time-out
T2	Protocol waiting time time-out
T3	Time-out for reply message
T4	Interblock communication time-out

T2 defines the protocol time-out. It limits the time required to communicate among the following messages:

- Request to send (ENQ) and request to receive (EOT, end-of-transmission)
- EOT and the receipt of the block of characters
- Sending block of characters and correct reception (ACK)

T3 defines the time-out for the reply message. It sets the maximum waiting time for multiple block message communications. Within a specified time, if the last block is not received, T3 time-out will be triggered.

T4 defines the time-out for interblock communications. It sets the time limit to receive the next block of message within the multiblock communications. If the next block of message is not received within the specific time, T4 will be activated.

SECS-I provides a mechanism for synchronization and software-deadlock resolution via 4 time-outs (T1 to T4) as shown in Table 66.3.

66.2.3 High-Speed SECS Message Services Standard

With the increasing trend toward network communication for high-performance components and hardware, HSMS was introduced in 1995. HSMS (also referred to as E31 standard) defines a communication interface suitable for the exchange of messages between computers via TCP/IP. HSMS uses TCP/IP streams to provide reliable two-way simultaneous transmission of streams of adjoining bytes. It can be used as a replacement for SECS-I communication as well as other more advanced communication environments.

It was intended as an alternative for applications where higher-speed communication is needed or when a simple point-to-point topology is insufficient. Most of the new SECS/GEM implementations use this standard for the message transport layer. HSMS was designed with the intention that existing SECS-II and GEM-compliant applications could be easily converted to use HSMS instead of SECS-I. HSMS also provides the means for independent equipment manufacturers to produce their own SECS-II/GEM implementations while still maintaining interequipment compatibility.

66.2.3.1 HSMS State Model

The *state model* is a methodology developed by David Harel and adapted by SEMI to describe the expected functionalities in the standard. As shown in Figure 66.5, a state is a static condition of an operation.

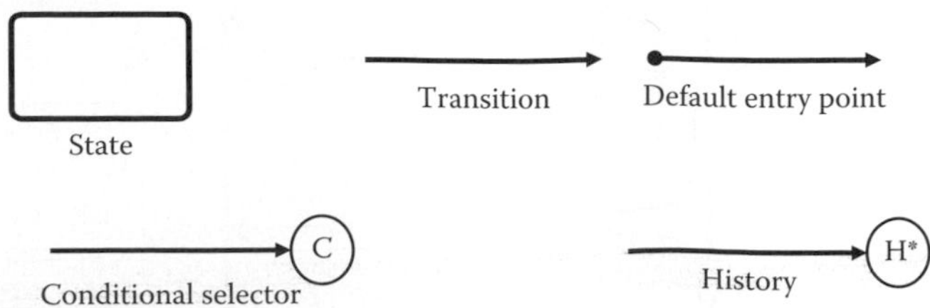

FIGURE 66.5 Harel's notation of state models.

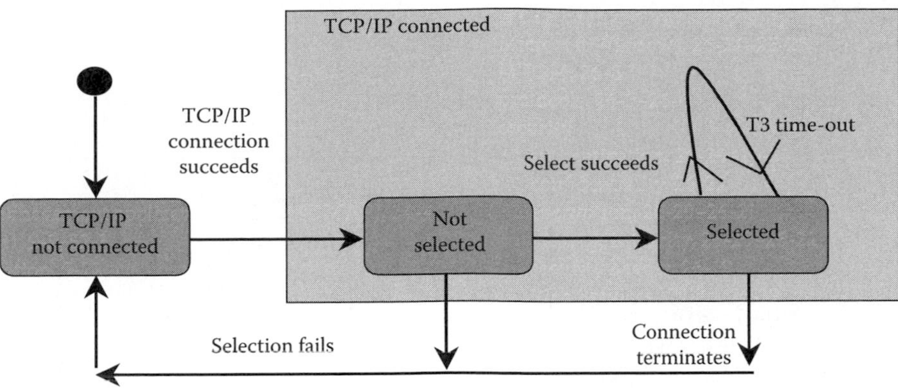

FIGURE 66.6 State model of HSMS.

A state could have multiple substates. The arrows represent the transitions from one state to another state. In Figure 66.6, the round black dot is the start of the process. When HSMS first starts, it will be in *TCP/IP not connected* state. If the TCP/IP connection is successful, then the process will move to *TCP/IP connected* state. In *TCP/IP connected* state, there are two substates, namely, *not selected* and *selected*. The system could be *connected on TCP/IP but HSMS has not been selected*. When HSMS is selected, it will go into a substate called *selected* and stays there until a T3 time-out occurs. When *connection terminated or selection fails* occur, the state will go back to *TCP/IP not connected*.

66.2.3.2 HSMS Message Structure

Figure 66.7 shows the structure of the HSMS message. The message structure is similar to SECS-I, but it is for TCP/IP network communication. The alphabets in Figure 66.7 describe as follows:

a. Bytes 1–4 represent the length of the message. HSMS can handle about 8 Mb of data. The length defines the total length of the header and the actual message.
b. Bytes 0 and one of the headers represent session ID. Session ID is used to identify whether the communication message is a control message or subsequent data information.
c. Bytes 2–3 of the header represent the stream and function number or status code of the message.
d. Byte 4 of the header represents the presentation type (or P-type) of how the message is encoded. A value 0 means that it will be used in SECS-II. Value 1–127 is reserved for *subsidiary standards*. Value 128–255 is reserved but is not used.

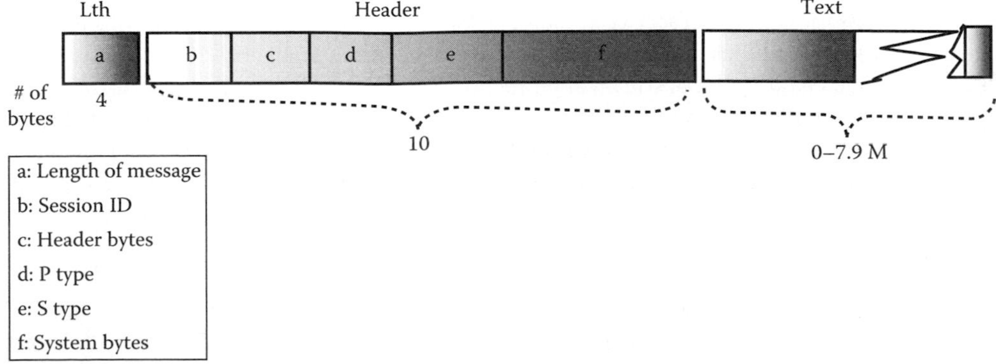

FIGURE 66.7 Structure of HSMS message.

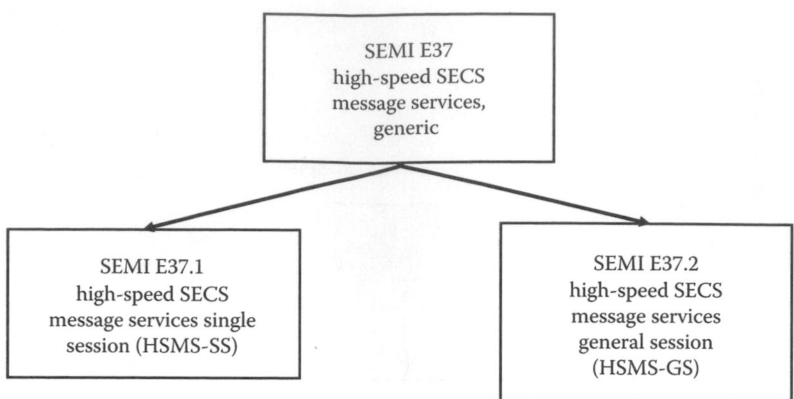

FIGURE 66.8 Subclassification of HSMS.

e. Byte 5 of the header represents the session type or S-type. Value 0 means that the session is sending data information. Other values indicate that the message is *control information.*
f. Bytes 6–9 of the header represent the system bytes. The purpose of system bytes is to avoid duplication of message.

66.2.3.3 HSMS Subclassification

SEMI has subclassified HSMS into single session (HSMS-SS) and general session (HSMS-GS) as shown in Figure 66.8. HSMS-SS defines single session equipment connection so that the implementation does not have to consider other equipment or knowledge of other systems. It will be similar to SECS-I connection, which is a one-to-one connection. The HSMS generic definition and structure will still apply. HSMS-GS is used for complex systems that have many independent equipments as subsystems. HSMS-GS defines the convention for identifying independent system entities.

66.2.4 SEMI Communication Standard (SECS-II)

This SECS-II standard was released in 1982. SECS-II (also referred to as the E5 standard) defines the message content passing between the equipment and the host. SECS-II specifies the format of the text and the minimal unit of semantics, but it does not specify the exact action that a recipient machine should take. With the definition of SECS-II, the host/equipment software may be constructed with minimal knowledge of each other. In addition, SECS-II allows the creation of *user-specific* messages.

SECS-II defines the method of conveying information between the equipment and the host in the form of messages. Every SECS-II message is identified by a stream and a function number. Each of the streams represents the category of service available as shown in Table 66.4. The stream will contain specific messages called functions. It also defines the *request for information* and the corresponding data transmission. Each combination of stream and function represents a distinct *message classification.* In every stream, function 0 is reserved for aborting transactions and the others are constructed in pairs. For example, Table 66.5 shows that each pair odd number function code is reserved for the primary message and the even number function code is reserved for the reply or the secondary message. If the primary message does not require a reply message, the even number message is discarded.

66.2.4.1 Data Structure

The messages that are sent through the physical layer through serial (RS232) or TCP/IP (Ethernet network) are broken down by SECS-I or HSMS definitions, yielding the length, header, and data blocks. SECS-II defines the structure of messages into entities called items and *lists of items.* This structure allows for a self-describing data format to guarantee proper interpretation of the message. Interchange

TABLE 66.4 Streams Defined in SECS-II

Stream	Description
1	Equipment status
2	Equipment control and diagnostics
3	Material status
4	Material control
5	Exception handling
6	Data collection
7	Process program management
8	Control program transfer
9	System errors
10	Terminal services
11	Host file services
12	Wafer mapping
13	Data set transfer
14	Object services
15	Recipe management
16	Processing management
17	Equipment control and diagnostics

TABLE 66.5 Odd/Even Message Numbering of SECS-II Functions

Stream	Function	Description
1	0	Abort transaction (S1 F0)
1	1	"Are you there" request
1	2	Online data
1	3	"Select equipment status" request
1	4	"Select equipment status" data
1	5	"Formatted status" request
1	6	"Formatted status" data
...	...	...

of messages is governed by a set of rules for handling messages called the transaction protocol. SECS-II further defines the data into multiple blocks structure, yielding the format code, length, and data item. All messages are packed into multiple data blocks with headers and checksums as shown in Figure 66.9.

66.2.4.2 Streams and Functions in SECS-II

Table 66.4 shows the category of various SECS-II messages that are classified under the different streams. As an example, all messages in stream 1 will be part of the equipment status category. This category will help the user and the integrator in understanding, designing, and developing their SECS solutions.

Each stream can specify multiple functions, as shown in Table 66.5, where an example for stream 1 is explained. Function 0 is reserved for *abort* purposes. The remaining functions are operated upon in pairs of requests (odd function or primary message) and reply/acknowledgments (even function or secondary message). The primary message is the *request message* sent by either the host or the equipment. All the odd number functions, for example, S1 F1 and S2 F1, are examples of primary messages. The user can indicate if a reply is needed for the primary message using the wait bit (W-bit). The W-bit is the first bit from the header byte, which is used to indicate that the sender of a primary message expects a reply. The *secondary message* is the reply message. All the even number functions, for example, S1F2

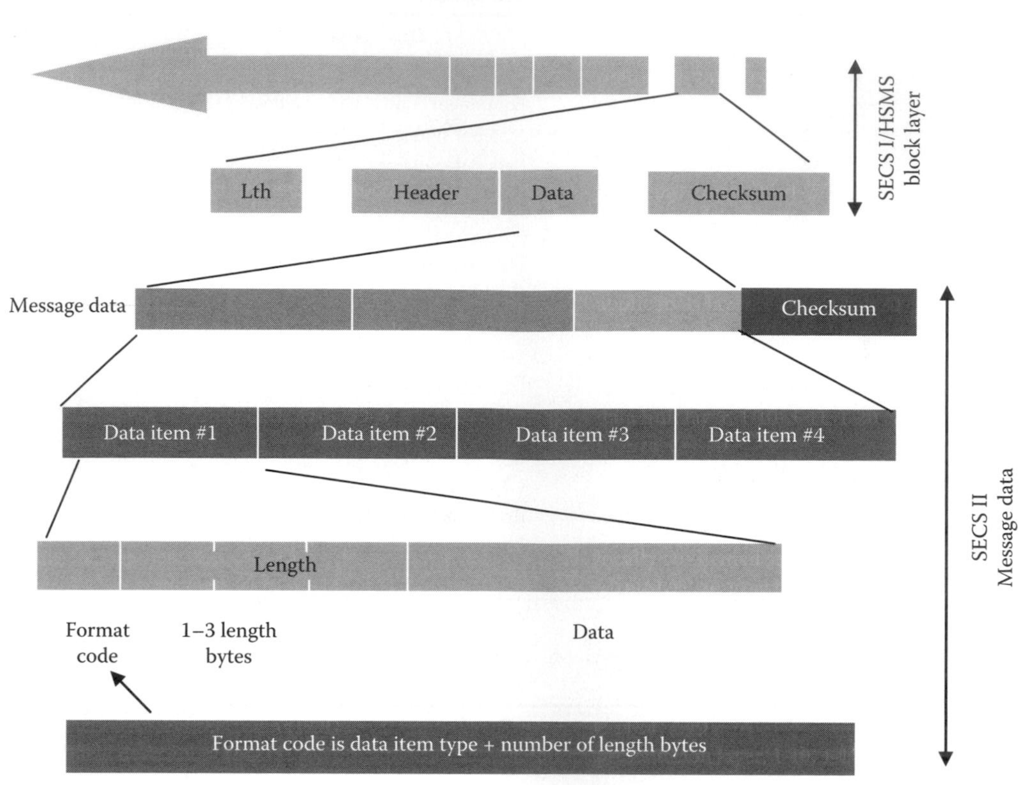

FIGURE 66.9 Data structures of SECS-I/-II/HSMS.

and S2F2, are examples of secondary messages. For any messages that cannot be processed by the equipment, the appropriate error message on stream 9 will be used. Upon detection of a transaction time-out, the equipment sends S9 F9 to the host. Upon receipt of function 0 as a reply to a primary message, the related transaction is terminated. No error message should be sent to the host by the equipment.

66.2.4.3 Conversation Protocols

A *conversation* is a series of one or more related SECS-II messages used to complete a specific task. A conversation should include all transactions necessary to accomplish the task and leave both the originator and interpreter free of resource commitments at its conclusion. There are a few types of conversations, as follows:

66.2.4.3.1 Primary Message with No Reply

Once the W-bit is set to zero, the primary message will not require a reply. It is the simplest conversation. The primary message will send a single block of information without reply. This conversation is used where the originator can do nothing if the message is rejected.

66.2.4.3.2 Primary Message with Data Return (Request/Data Conversation)

Originator will need to get information from interpreter. The W-bit will have to be set to one and data is returned as secondary reply message.

66.2.4.3.3 Send/Acknowledge Conversation

Originator sends data in a single block and expects an acknowledgment.

66.2.4.3.4 *Inquire, Grant, Send, and Acknowledge Conversation*

Originator must ask for permission from the interpreter. The interpreter will grant permission as well as set *time-out* for getting the information. If the wait time is longer, then *time-out* time, S9 F13, will be sent out. Once the information is received, the acknowledgment will be sent. This type of conversation is commonly used in equipment-to-host conversations and only the equipment should send error messages to the host.

66.2.4.3.5 *Unformatted Data Set Conversation*

Stream 13 is used for unformatted data conversation. Unformatted data could be in the form of a bitmap file or a binary file.

66.2.4.3.6 *Material-Handling Conversation*

Stream 4 is used for material control and handling of material between equipments.

66.2.4.3.7 *Conversation with Delay*

The originator may request information from the interpreter, which requires some time to obtain. The interpreter could collect the information immediately and return. The interpreter could also indicate if the information has be obtained or returned in the subsequent transaction.

Messages can also be transferred as single-block messages or multiblock messages. The SECS-I maximum message length for a single block is 244 bytes. The maximum message length for HSMS is 7.9 megabytes. For longer messages, SECS-II will have to use multiblock messages. For example, if messages that are longer than 244 bytes are needed, they have to be sent by multiple blocks or packets. Each packet will have a block number so that the receiver will be able to join all the blocks together.

66.2.4.4 Sample of SECS-II Message

66.2.4.4.1 *Stream 1 Function 1*

Figure 66.10 shows the S1 F1 and S1 F2 message structure and description.

S1, F1 Are There Request (R)[a] *S,H <-> E, Reply*[b]

Description[c]: *To establish if the equipment is online. A function 0 response to this message means the communication is inoperative. In the equipment, a function 0 is equivalent to a time-out of the receive timer after issuing S1, F1 to the host.*

Structure: Header only[d]

The above example is form the SEMI standard E5

S1. F2 On line data(D) S,H <-> E

Description: Data signifying that the equipment is alive.

Structure[e]: L, 2

1. <MDLN>
2. <SOFTREV>

FIGURE 66.10 S1 F1 and F1 F2 message structure.

TABLE 66.6 Status of Streams and Functions (Reserved/Unreserved)

Stream	Status
1–14	Currently defined in the standard
15–63	Reserved for standard messages
64–127	Reserved for user-defined messages
Function	*Status*
1–63	Reserved for standard messages
64–255	Reserved for user-defined messages

Exception: The host sends a zero-length list to the equipment:

a. Shows the stream (S), function (F), and title of the function (R).
b. Shows the type of message. S represents single-block message. Multiblock message will be represented by M. H ↔ E means the message could be sent from either the equipment (E) or the host (H). Other alternatives are E → H, which means *the equipment sends to the host*, and H → E, which is from the host to the equipment. The message requires *reply*. Other alternatives are *optional to reply* or *blank*, which means no reply is needed.
c. The description of the message.
d. Shows the structure of the message. In this case, it is just the header of the message and there are no data.
e. Shows the structure of S1 F2. L, 2 means that it has two listed items. The first listed item is <MDLN>, which represents the equipment model type with maximum 6-byte length. The second item is <SOFTREV>, which represents software revision with maximum of 6 bytes as well.

As shown in Table 66.6, SECS-II only uses streams 1–14 and functions 1–63. In order to provide for expansion of features, streams 15–63 are reserved for future SEMI standards. Streams 64–127 and functions are available for custom messages.

While SECS-II makes room for vendor-independent compatibility, problems do arise when vendor applies different semantics for the same message pairs. This creates intervendor incompatibility. GEM was created to define *generic equipment* features so that intervendor compatibility can be established.

66.2.5 Generic Equipment Model

The GEM standard (also referred to as the E30 standard) was published in 1992. It defines which SECS-II messages should be used, in which situation, and what the resulting activity should be.

GEM is intended to specify the following:

- A behavior model of semiconductor equipment in a SECS-II communication environment
- A description of information and control functions needed in a semiconductor-manufacturing environment
- A definition of the basic SECS-II communication capabilities
- A single consistent means of accomplishing an action when SECS-II provides multiple possible methods
- Standard message dialogs that is necessary to achieve useful communication

The GEM standard contains two types of requirements:

1. Fundamental (or foundation) requirements
2. Requirement for additional GEM capabilities

As shown in Figure 66.11, the fundamental GEM requirements form the foundation of the GEM standard, which are basic requirements for all types of semiconductor equipment. The additional GEM

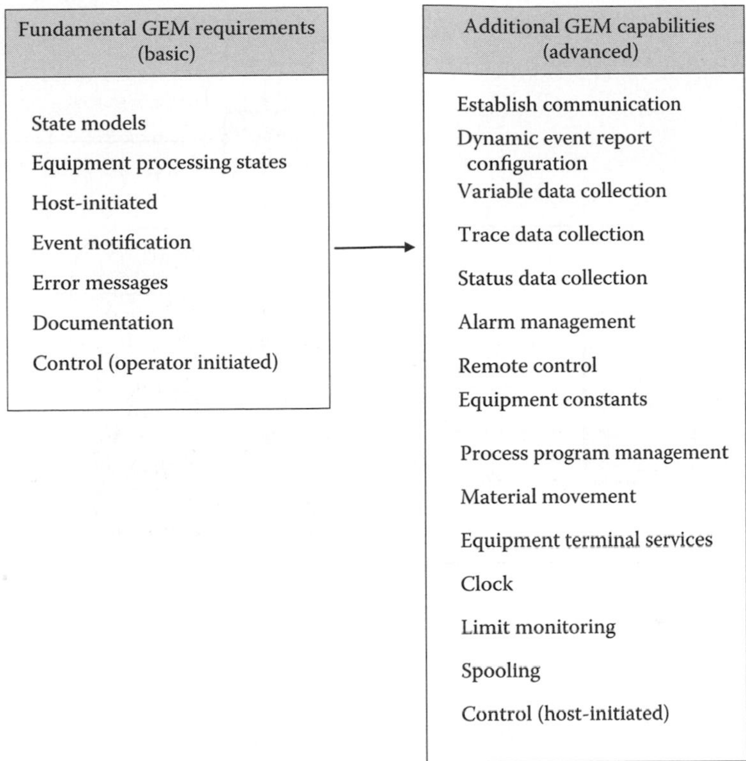

FIGURE 66.11 GEM requirements.

capabilities provide the functionality required for some instances of factory automation or functionalities that are applicable to specific types of equipment. Capabilities are operations performed by semiconductor-manufacturing equipment. Each of the capability consists of a statement of purpose, pertinent definitions, detailed description, requirements, and various scenarios. A *scenario* is a group of SECS-II messages arranged in a sequence to perform a function. These *scenarios* handle communications that involve more than one pair of messages.

66.2.5.1 GEM Fundamental Requirements

66.2.5.1.1 State Models

State models describe the behavior of the equipment from a host perspective. The same model can be used in different equipment for identical functions. The fundamental GEM specifies the state model of process (equipment specific), communication, alarm, control, material movement, and spooling.

Figure 66.12 shows a sample state model for equipment processing. There are also step numbers defined in the diagram. Here is how the state model can be interpreted:

Step 1: The process will be initialized (INIT). After being initialized, the process could be in "Idle" state.

Step 2: The next state is "Processing Active." It will be in the substate of "Process" doing an operation of "Setup." The setup operation could be semimanual, as the operator needs to log on and fine-tune the equipment.

Step 3: After setup, the equipment will be in "Ready" state for operation.

Step 4: In normal processes, the equipment should be either in the "Executing" state, producing some part/component, or in Step 5, which is "Idle" waiting for the next part/component.

Step 5: The equipment is idle and waiting for the next part.

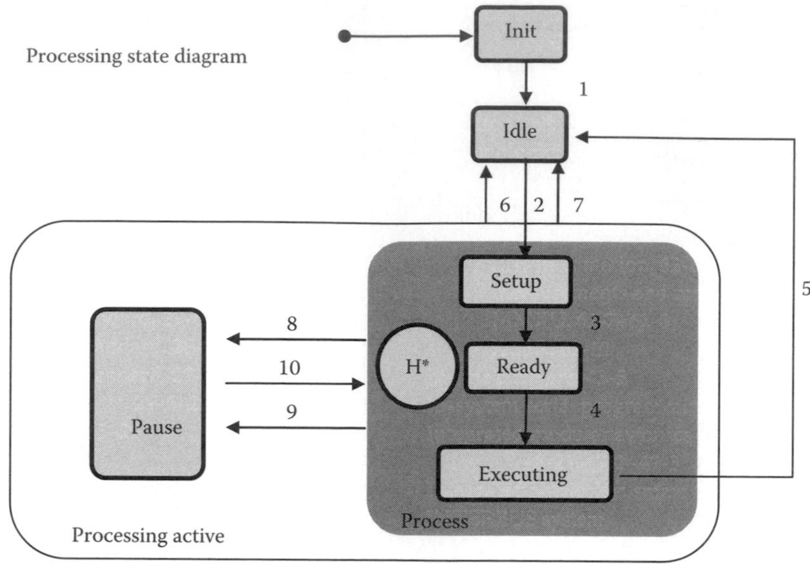

FIGURE 66.12 Equipment processing states.

Step 6: The equipment could be in "Processing Active" and receive a "STOP" command from the operator or the host system. This will put the equipment back to "Idle" state.

Step 7: The equipment could be in "Processing Active" and receive an "ABORT" command from the operator or the host system. This will put the equipment back to "Idle" state.

Step 8: The equipment may receive an alarm and it will then be in "Pause" state.

Step 9: The equipment may receive a "Pause" command by the operator or the host system.

Step 10: The equipment may receive a "Resume" command by the operator or the host system.

An example of the GEM transactions for event-reporting setup is illustrated in Table 66.7.

Step 1: The host defines the report and request for reports:

- The host initiates a report definition by sending S2 F33. Stream 2 is the equipment control and diagnostics stream and function 33 is to define report. This message allows the host system to define and to request the group of reports available from the equipment. A unique ID identifies each report.
- The equipment receives and acknowledges by replying with S2 F34. If the primary message has errors or the report ID is not found, the primary message will be rejected.

Step 2: The host sets an ID for the required reports:

- The host initiates an *event and report* link definition by sending S2 F35. Function 35 will only link the reports that are requested by the host, with an ID. The link will be disabled until the equipment is ready to send.
- The equipment receives and acknowledges by replying S2 F36. Again if there is any error condition, the whole message will be rejected.

Step 3: The host enables sending of the reports:

- The host sends the event-enabling definition by sending with S2 F37. At this stage, the equipment knows what reports the host requires and the host will enable the sending process.
- The equipment receives and acknowledges by replying with S2 F38. After the equipment acknowledges the message from the host system, it will then proceed to send the reports to the host system.

TABLE 66.7 Example of GEM Transactions for Event-Reporting Setup

Step	Host	Direction	Equipment
1	Send report definitions S2, F33.	⟶	
		⟵	Send acknowledgment S2, F34.
2	Send event and report link definitions S2, F35.	⟶	
		⟵	Send acknowledgment S2, F36.
3	Send event-enabling information S2, F37.	⟶	
		⟵	Send acknowledgment S2, F38.

66.2.5.1.2 *Host Initiates S1 F13, S1 F14*

It is a fundamental requirement in GEM that the required equipment is able to accept a connect request from the host. The host can request connections by sending S1 F13 to the equipment. Upon receipt of the message, the equipment will send the S1 F14 "Connect Acknowledge" reply.

66.2.5.1.3 *Event Notification*

This capability provides data to the host at specified points during the equipment operation. The equipment will send S6 F5 to get permission from the host. The host will reply with S6 F6 to grant permission. Once the permission is granted, the equipment will send the report using S6 F11. The host will acknowledge the report using F6 F12.

66.2.5.1.4 *Online Identification*

Online identification is the most basic SEMI requirement. The originator (which could be the host or the equipment) will send S1 F1 to ask if the interpreter (which could be the equipment or the host) is there. The interpreter will then reply using S1 F2, which has the information of the equipment type and revision.

66.2.5.1.5 *Error Messages*

Fundamental GEM requires the system to be able to detect communication link errors in SECS-I and detect SECS-II and GEM format errors. It will also need to support all stream 9 messages and provide appropriate error-handling routines.

66.2.5.1.6 *Documentation*

Fundamental GEM requires that the equipment should be able to provide a detailed specification document for the SECS/GEM interface on the equipment.

66.2.5.1.7 *Operator-Initiated Control*

Fundamental GEM requires that the equipment should have operator-initiated control-related capabilities to allow the configuration and manipulation of the control state model. In this way, the host and/or user may modify the equipment's control-related behavior. For example, in equipment report event scenario, the operator at the equipment will send S6 F5 to the host. The host will send S6 F6 to grant online link to the equipment. The equipment will send control state local event report using S6 F11. The host will acknowledge the report with S6 F12.

66.2.5.2 Additional GEM Requirement

Additional GEM requires that the system should be able to support communication state model, operator communication state display, operator enable/disable commands, enabled or disabled power-up states, and establishment of communications timer equipment constants.

66.2.5.2.1 Establishment of Communications

Establishment of communications capability provides a means of formally establishing communications upon system initialization or upon any loss of communication between communication partners. Notification of the period of noncommunication is then possible. The host system initiates communication by sending S1 F13 and the equipment will reply using S1 F14.

66.2.5.2.2 Event Notification

This capability provides data to the host at specified points during equipment operation. The equipment will ask the host for permission to send by using S6 F5 message. The host will give permission using the S6 F6 message. Once the permission is given, the equipment will send the event report using S6 F11, and the host will acknowledge the report using S6 F12 message.

66.2.5.2.3 Dynamic Event Report Configuration

This capability provides the data-reporting flexibility required in some manufacturing environments. The host will send S2 F39 to request permission from the equipment and will get the permission from the equipment using S2 F40. Having received the permission, the host will send S2 F33 to define the report requirements. The equipment will acknowledge the report requirements by S2 F34. The host will send S2 F35 to collect the report from the equipment and the equipment will then send the report using S2 F36.

66.2.5.2.4 Variable Data Collection

This capability allows the host to query for the equipment data variables and is useful during initialization and synchronization. The host will send S2 F19 to ask for the variable data and the equipment will send the data using S2 F20.

66.2.5.2.5 Trace Data Collection

Trace data collection provides a method of sampling data on a periodic basis. The host will initiate the trace of certain data using S2 F23. The equipment will initialize the trace data and acknowledge using S2 F24.

66.2.5.2.6 Status Data Collection

This capability allows the host to query the equipment for selected information and is useful in synchronizing with the equipment status. The host will send S1 F11 to request for the variable equipment status and the equipment will reply the variable status using S1 F12.

66.2.5.2.7 Alarm Management

This capability provides for host notification and management of alarm conditions occurring on the equipment. The host will ask for the alarm data and text using S5 F5. The equipment will send the alarm data to the host using S5 F6.

66.2.5.2.8 Remote Control

This capability provides the host with a level of control over equipment operations. The host will send the remote command using S2 F41. The equipment will acknowledge using S2 F42 and send the data report using F6 F11. The host will acknowledge the report using S6 F12.

66.2.5.2.9 Equipment Constants

This capability provides a method for the host to read and change the value of selected equipment constants on the equipment. These equipment constants include nonvolatile storage, validation/verification of equipment constants, host equipment constants, and name-list request messages. The host will send the new equipment constant using S2 F15 and the equipment will set the equipment constant using S2 F16. The host will then send S2 F13 to request for the equipment constants and the equipment will send the equipment constants using S2 F14.

66.2.5.2.10 *Process Program Management*

This capability provides a means to transfer process programs and share the management of those process programs between the host and the equipment. The host will ask for the equipment process program using S7 F19 and the equipment will send the process program data using S7 F20. To upload the process program, the host will send S7 F5 and the equipment will send the process program using S7 F6. To download the process program, the host will send S7 F1 and the equipment will grant permission using S7 F2. Once permission is received, the host will send the process program using S7 F3 and the equipment will acknowledge with S7 F4.

66.2.5.2.11 *Material Movement*

This capability includes the physical transfer of material among equipment, buffers, and storage facilities. The transfer of material can be performed by the operator, automated guided vehicle (AGV) robots, tracks, or dedicated fixed material-handling equipment. The equipment will send the collection of event using S6 F11 and the host will acknowledge with S6 F12.

66.2.5.2.12 *Equipment Terminal Services*

This capability allows the host to display information on the equipment's display device or the operator of the equipment to send information to the host. The host will send the textual information to the equipment using S10 F3. The equipment will display the information and acknowledge with S10 F4.

66.2.5.2.13 *Clock*

The clock capability enables the host management of time-related activities and occurrences to associate with the equipment and across multiple pieces of the equipment. The equipment will send S2 F17 to request time information from the host and the host will respond with time value using S2 F18.

66.2.5.2.14 *Limit Monitoring*

This capability relates to the monitoring of the selected equipment variables. The host will get the equipment's current limit–monitoring status using S2 F47 and the equipment will send the status using S2 F48.

66.2.5.2.15 *Pooling*

Spooling as a capability whereby the equipment can query message intended for the host during times of communication failure and subsequently deliver these messages when communication is restored. The host will define messages that need spooling in case of communication failure using S2 F43. The equipment will reset the spooling message and acknowledge with S2 F44.

66.2.5.2.16 *Host-Initiated Control*

The control-related capabilities allow for configuration and manipulation of the control state model. In this way, the host and/or user may modify the equipment's control-related behavior. The host will request to be offline using S1 F15. The equipment uses S1 F0 to abort transaction, sends S1 F16 to acknowledge, and uses S6 F11 to send event report. The host will acknowledge with S6 F12.

66.2.6 Limitations of SECS/GEM

The limitations faced by currently defined SECS/GEM standards are as follows:

1. *Manual integration that cannot be totally avoided.* The factory automation system cannot query SECS/GEM software to determine its full capabilities because different vendors have customized the same message pairs for different applications. Therefore, SECS standard requires significant manual intervention/customization efforts for each application function.

TABLE 66.8 Trends of the SEMI Standard

First Generation	Second Generation	Third Generation	Next Generation
Characteristics	*Characteristics*	*Characteristics*	*Characteristics*
Simple text-based reporting	Some recipe management	Full recipe management	Intelligent control
	Some equipment performance matrix	Equipment performance matrix	Remote diagnostics/ monitoring
	Low standardization	SEMI standardization	Mainstream: distributed computing standards
	Standards	*Standards*	*Standards*
	SECS-I	SECS-I/-II	Object-Based Standards
	De facto	GEM	
	Proprietary	HSMS	
	Technology	*Technology*	*Technology*
	Serial	Serial	Middleware
	TCP/IP	TCP/IP	
	Issues	*Issues*	*Issues*
	Limited solution opportunities	Highest integration cost	Nonsemiconductor-specific PC-to-PC communication
	Highest integration cost	Long lead time	Lower integration cost
	Longest development	Semiconductor-specific interfaces	Object-oriented approach

2. *Weak security mechanism*. SECS is weak in its *security mechanism*. There is no concept of client authentication or access permission in SECS. Issues of security are becoming important operational considerations because of the need for remote equipment engineering and remote diagnostics.
3. *Single-client architecture*. The implementation of SECS is based on a point-to-point link, and there is only one software process (i.e., one point) on the factory side. This is a critical limitation because the newly developed factory applications such as advanced process control (APC), factory automation, and advanced planning and scheduling (APS) systems require a steady stream of accurate real-time data, even in the form of peer-to-peer communications to facilitate a decision-support process.

66.2.7 Emergence and Evolving New Standards

The industry associations such as SEMI and SEMATECH as well as universities have tried to develop a new standard that can handle the various applications needed by today's semiconductor-manufacturing environment. SEMI communication and interface standards have adopted some of the popular object-oriented programming like the XML and distributed computing techniques, but these are still not very commonly used. Table 66.8 shows the trend in the SEMI communication standards.

66.3 Survey of SECS/GEM-Compliant Toolkits

SECS/GEM-compliant software comes in different layers. In this section, a survey of different toolkits that are available in the market is presented together with their different features. The different features that are provided by each vendor are summarized in one table to highlight the capabilities among them.

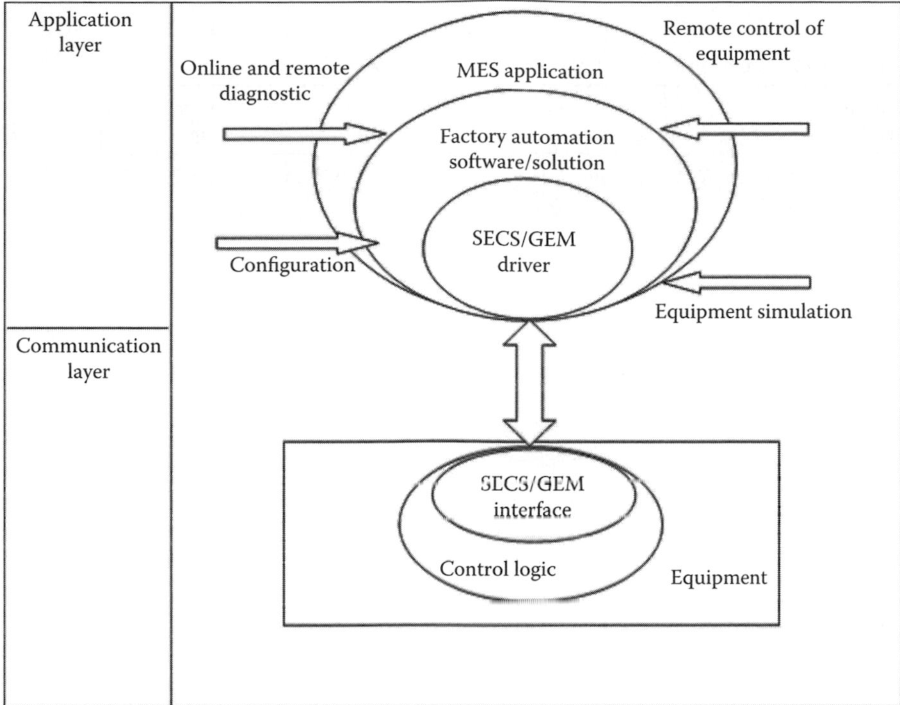

FIGURE 66.13 Application layer and communication layer of factory automation software.

66.3.1 Overview of Equipment Integration Software

The equipment integration software is an important part in semiconductor manufacturing or wafer fabrication. The factory automation software must accommodate various factory applications such as equipment configuration, online and remote diagnostics, remote control, and equipment simulation. These applications are the essential functionalities in today's semiconductor wafer fabrication. Most of the software vendors would like to provide solutions across the span of *process control, system automation, production data acquisition, and communication systems.*

The factory automation software can be divided into two levels: the application level and the communication-enabling level. The function of the application level is to carry out the various applications such as remote diagnostic, data collection, tracking of materials, and testing and simulation. The function of the communication enabler is to interface the factory networks or host with the equipment, which may be SECS/GEM-compatible or non-SECS/GEM-compatible equipment such as programmable logic controller (PLC)–controlled machines. Figure 66.13 gives an overview of the factory automation software.

Software toolkit comes with different features and functionalities. They can be conveniently classified as toolkits for MES application, factory automation software/solution, and SECS/GEM drivers.

66.3.2 SEMI Equipment Communication Software

Some of the commercially available factory automation software are described in this section. These products are obtained from company-provided information. The survey is based on several important features listed in Table 66.9. A few software vendors provide the complete solution, but most of them provide solutions for the application level. They generally use the communication-enabler component from prominent players in the market such as Cimetrix in conjunction with their application package.

TABLE 66.9 Comparative Analysis of the Available Software

Software Vendor	Yokogawa			Cimetrix	Peer Group		Hume
SDK Software Development Kit	EQ Brain*	EQ Brain 300*	Cell Brain*	CIM Connect	GWGEM	SECS4Hosts	Datahub SDK
SECS-I	✓	✓	✓	✓	✓	✓	✓
SECS-II	✓	✓	✓	✓	✓	✓	✓
HSMS				✓	✓	✓	✓
GEM	✓	✓	✓	✓	✓	✓	✓
OBEM							
O/S	Win32	Win32	Win32	Win32	Win7 (32 bit, 64 bit), WinServer2008R2, Linux, Solanis	Win7 (32 bit, 64 bit), Linux, WinServer2008R2	Win7, Win Vista, WinXP, Linux, Solaris
Language	VC++, VB.NET	VC++, VB	VC++, VB	VC++, C#, VB.NET	C#, VC++, VB6, Java	C#, VC++, VB6, Java	C#, Java, VB, C, VC++, Tcl
API	✓	✓	✓	✓	✓		
Recipe management				✓			
ARAM				✓			
Testing tools				✓	✓	✓	✓
Middleware				ActiveX/ COM	COM	ActiveX	✓
Good user interface				✓		✓	
GEM to XML				✓		✓	✓

Insphere Technology		ABAKUS	Kinesys		Ergo Tech		SDI	Quest Adaptations		Savigent Software
Secs To Host. Net	Secs To Tool.Net	AHEAD active X.GEM interface, SCI Spy	GEMBox	ALPS	Tran SECS	JAVA SECS	sdiRelayer, EDA Gateway, sdiStation Host Developer, SMS Developer. Reseller/ Integrator of Cimetrix TEST Connect, CLMConnect, CIM300	QLink4Tool	QGEMSerive	SECS/ GEM Agent
✓	✓	✓	✓	✓	✓	✓	✓	✓	✓	✓
✓	✓	✓	✓	✓	✓	✓	✓	✓	✓	✓
✓	✓	✓	✓	✓	✓	✓	✓	✓	✓	✓
✓	✓	✓	✓	✓	✓	✓	✓	✓	✓	✓
✓	✓			✓	✓	✓		✓	✓	✓
Win32	Win32	Win	Win2K, WinNT	Win2K, WinNT	Win32, Linux	Win32, Linux	MS-Windows, Unix, Linux	Win32	Win32	Win32
C#, VC++, VB6	J#, C++, Phyton		VC++		Java	Java	C++, C, VB	VC++, VB6	VC++, VB6	C#, VB.Net
✓	✓				✓	✓	✓	✓	✓	✓
✓	✓				✓	✓		✓	✓	✓
✓	✓							✓		✓
			✓	✓	✓		✓	✓	✓	✓
COM	COM			DCOM	JAVABean	JAVABean	CORBA, TIBCO, SOAP, XML, ACE, TAO	ActiveX/ COM	ActiveX/ COM	ActiveX
			✓	✓	✓	✓	✓			
✓	✓			✓	✓	✓	✓	✓	✓	✓

66.3.3 Analysis of Current Available SECS/GEM Solutions

An analysis of the SECS/GEM solutions is shown later, based on the features of the software and the intended application.

66.3.3.1 Analysis Based on the Features of the Software

The analysis on the current available solutions is consolidated from public information provided by the companies. Feedback from vendors is also used to compile this analysis.

66.3.3.2 Analysis Based on the Intended Application

Due to the fact that SECS/GEM communication software has the propensity of being tailored to a variety of applications, it is necessary to map such applications to varying communication levels as shown in Figure 66.14 and Table 66.10. For example, it can be used by the equipment manufacturer for developing the host interface just as it can also be used by the semiconductor manufacturer to communicate with the equipment, to test the equipment for its interface, etc. Figure 66.8 shows one example that highlights the application of the SECS/GEM software. Further, Table 66.10 compares the application intent of the software.

66.3.3.3 Case Study I: Design and Implementation of SECS/GEM for a New Semiconductor Back-End Machine (Implementation of SECS/GEM Compliance on a Virgin Machine)

This section presents a case study in the design and implementation of SECS and GEM for a case where a back-end semiconductor machine is to be built with SECS/GEM compliance. We shall see how facile it

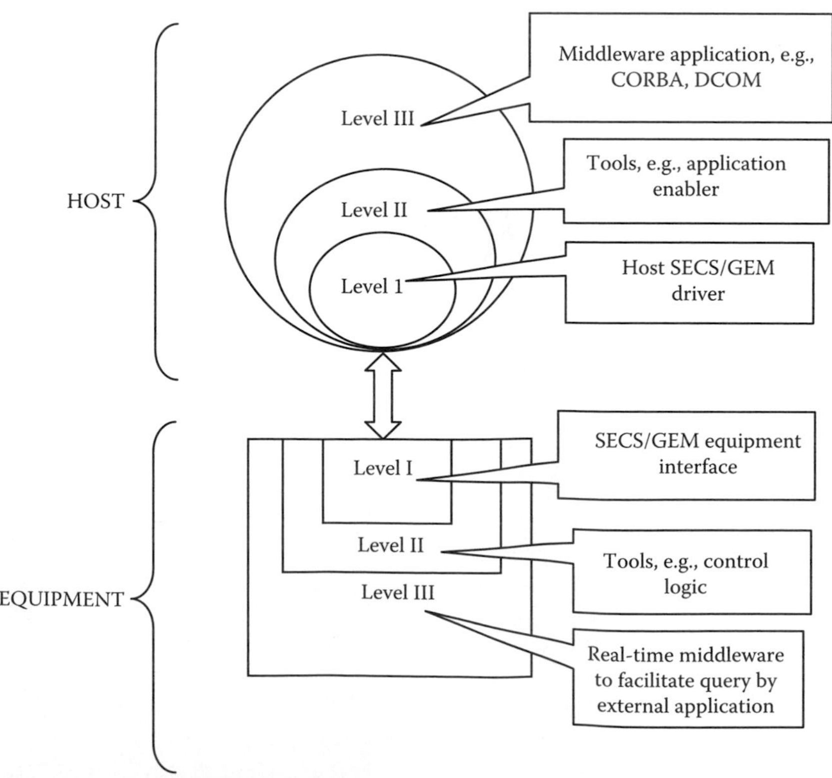

FIGURE 66.14 The functionality of SECS/GEM software.

TABLE 66.10 Analysis of the Software Based on Communication Levels

Software Vendor	Yokogawa			Cimetrix	Peer Group		Hume	Insphere Technology		ABAKUS	Kinesys		Ergo Tech		SDI	Quest Adaptations		Savigent Software
Software	EQ Brain*	EQ Brain 300*	Cell Brain*	CIM Connect	GWGEM	SECS4 Hosts	Datahub SDK	Secs To Host.Net	Secs To Tool.Net	AHEAD ActiveX GEM interface, SCI Spy	GEM Box	ALPS	Tran SECS	JAVA SECS	sdi Relayer, EDA Gateway, sdi Station Host Developer, SMS Developer Reseller/ Integrator of Cimetrix TESTConnect, CIMConnect, CIM300	Qlink4Tool	QGEM Service	SECS/ GEM Agent
Level III (host)			✓	✓					✓			✓			✓			
Level II (host)			✓	✓	✓	✓			✓	✓		✓						✓
Level I (host)			✓	✓	✓	✓			✓	✓		✓		✓				✓
Level I equipment	✓	✓		✓	✓		✓	✓		✓	✓		✓	✓		✓	✓	✓
Level II equipment	✓	✓		✓	✓		✓	✓		✓	✓		✓			✓	✓	✓
Level III equipment	✓	✓		✓			✓	✓										

is to implement SECS-II and GEM for a PC-based equipment controller by leveraging on commercially available software. We have chosen a package from GW Inc. We will show the steps toward the implementation of a SECS/GEM, in the case where the equipment initiates the communication with the host system:

a. A statement of the specifications and requirements
   The equipment maker is required to develop a GEM-compliant equipment. This equipment is required to be directly integrated into existing SECS-II-enabled host (cell control) system. The detailed requirements and capabilities are as follows, being able to
   1. Connect the network to the host system with both SECS-II and HSMS protocols
   2. Upload and download process programs for equipment configuration
   3. React to the host's requests, based on the process state of the equipment
   4. Send event or alarm reports to the host system in the event of predefined events or *alarms occur*
   5. Time-synchronize with the host system on equipment power-up
   6. Implement GEM fundamental requirements
b. System architecture and software tools
   Figure 66.15 shows the recommended system architecture and communication interfaces. The SECS/GEM software package from GW Inc. is utilized in this example.
   The package includes

   - GEM configuration data (GCD) file
   - GWGEM daemon
   - GWGEM extension task
   - GWGEM primary message handler task
   - GWGEM application programming interface (API) in C++

   As we mentioned in the earlier section, in order to run GEM application that is built on the earlier four major GWGEM components, the SDR must be installed and run as a message driver (HSMS) to communicate with the host system.

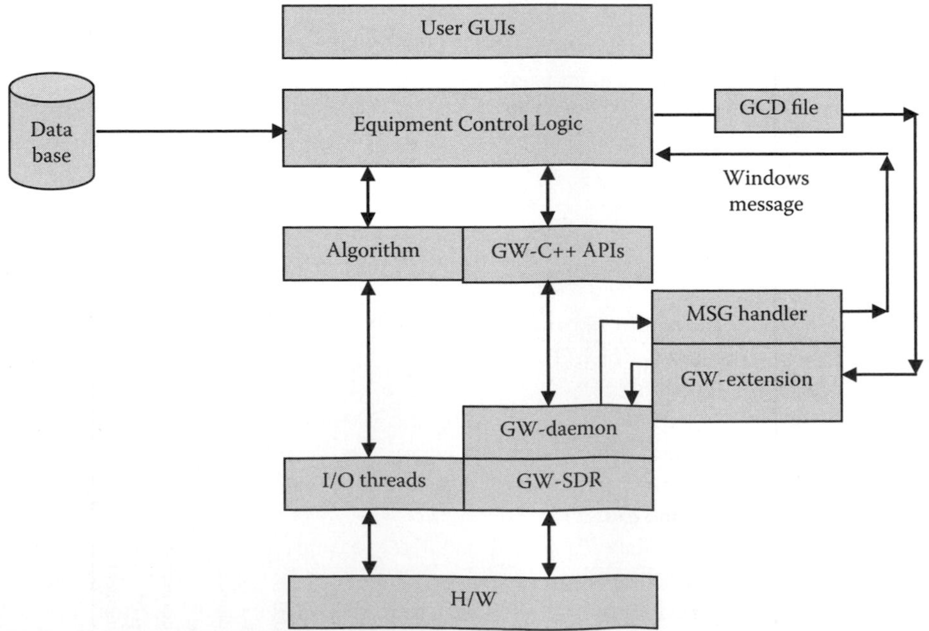

FIGURE 66.15 System architecture for SEMI interface creation.

c. Functional mapping of modules
   The GWGEM daemon process coordinates all the communications between user applications, GWGEM extension, and message handler tasks. The GCD file defines all SECS variables such as status variables, equipment constants, and data variables. The extension task is a process that contains various GWGEM extension routines that are called by the GWGEM daemon process from time to time. Extension routines are user-written application programs that provide application-specific or equipment-specific handling of GEM messages.
d. Message partition
   A primary message handler task is used to process a particular type of incoming SECS primary message from the host system. It must provide for the following conditions:

   - Messages that are beyond the GEM standard set
   - Messages that vary from one equipment type to another
   - The ability to override GWGEM built-in handling of a particular incoming primary message
   - The ability to scan (peek at) an incoming SECS-II message without actually processing it

e. Levels of implementation
   In this project, the following GEM fundamental requirements are implemented:

   1. State model
   2. Equipment processing status
   3. Host-initiated S1 F13/F14 scenario
   4. Event notification
   5. Online identification
   6. Error message
   7. Documentation
   8. Control (operator-initiated)

   By implementing the GEM fundamental requirements, the equipment is GEM compliant, which means that this equipment is able to communicate with any GEM-compliant equipment no matter which hardware vendor it comes from. In addition to the aforementioned, some additional GEM capabilities are also implemented. These include

   1. Initialization and establishment of communications
   2. Alarm management
   3. Equipment terminal services
   4. Process program management
   5. Clock

   GWGEM C++ encapsulates the GEM requirements into several classes. The developer treats the GEM functionalities as some objects. GWGEM C++ for Windows packs as Windows dynamic link library (DLL). It takes advantage of window programming for WIN32-based applications such as Microsoft Foundation Class (MFC). GWGEM C++ uses Windows event notification and message service to notify or trigger WIN32 application to handle incoming SECS-II primary messages.
f. Integrate GWGEM C++ with an MFC-based equipment controller
   Normally, the development team of GEM-compliant semiconductor equipment is made up of mechanical engineers, control engineers, and SECS/GEM consultants. The first step for the consultant is to work with the mechanical and control engineer to collect equipment data, status data, event data, and alarm data. The consultant needs to translate these data related to the equipment's operation into variables defined in the GCD file; see GCD file sample.
   Once the control engineer completes code writing including GUI portions and I/O control logic, it is time to integrate GWGEM C++ with the control program. To improve the real-time performance, GWGEM C++ maps events such as equipment state change, process state

Code samples

```
/* GWGEMCPP Base-line GCD Source File */
Constant fixup GemOfflineSubstate = <U1 3>      /* data type is 1 byte unsigned
                                                   integer */
       vid = 43                                 /* variable ID is 43 */
       name = "OFFLINESUBSTATE"                 /* 1: offline/equipment offline */
       units = ""                               /* no unit */
       min = <U1 1>                             /* minimum value is 1*/
       max = <U1 3>                             /* maximum value is 3*/
       default = <U1 3>                         /* default value is 1 */
event ProcessRunningEvent                       /* event name */
       ceid = 122                               /* constant even ID is 122 */
event = ProcessAbortEventceid = 123
status AramsStateVariable = <A [0..5]"">        /* character array with null value */
       vid = 500                                /* variable ID is 500 */
       name = "AramsStateVariable"              /* status name */
       units = ""
```

(a)

By using GCD compiler, these equipment constants in GCD file are converted to standard C++ header file, which is recognized by standard C++ compilers.

```
#include "gwgemcpptemplate.h";                  /*for template class */
#include "gemcppbase.h";                        /* base class definition */
#include "GemVirtImp.h";                        /* virtual classes for application-
                                                   specific overwrite */
#include "GemMessages.h"                        /* SECS II message header
structures, etc. */
#include "gwcontrol.h"                          /* equipment control state */
```

(b)

Note that these header files are provided with GWGEM C++ package for Windows platform.

```
/* This is a sample program to be used in conjunction with GWGEM class library to
demonstrate how to create, initialize, access and remove GWGEM object in MFC code */
CGWGemCPP* pGem = new CGWGemCPP;                /*create a GWGemCPP instance */
pGlobalGem = pGem;                              /* setup global pointer */
```

(c)

```
/* Create a control state message, set initial condition states and call the
  message handler. */
pGlobalGem -> GWControl().GetControlState(ControlState); /*get current control
state */
pGemConfControlStateChange pControlStateChange = new GemConfControlStateChange;
pControlStateChange->NewControlState = ControlState; /*set state */
pControlStateChange->ControlStateName = getStrControlState(ControlState);
OnGemConfControlStateChange(0, (LPARAM) pControlStateChange); /*notify state
  change */
```

(d)

```
pGlobalGem -> GWLink().GetLinkState(LinkState);   /* get initial communications state */
SetDlgItemText(IDS_LINKSTATE, getStrLinkState(LinkState));  /* display on GUI */
```

(e)

```
pGlobalGem -> GWSpool().GetSpoolState(SpoolState); */       /* get the spool state
SetDlgItemText(IDS_SPOOLSTATE, getStrSpoolState(SpoolState));   /*display */
```

(f)

FIGURE 66.16 (a) Sample GCD file. (b) C++ sample, header files. (c) C++ sample, create GWGEM object. (d) C++ sample, setup control state. (e) C++ sample, setup communication state. (f) C++ sample, setup spooling state.

```
delete pGEM; /* remove GWGEM object */
```
(g)

```
int status = pGlobalGem->GWEvent().Send(EventID); /*send event with ID = EventID*/
```
(h)

```
pGlobalGem->GWLink().Disable();
```
(i)

```
pGlobalGem->GWLink().Disable();
```
(j)

FIGURE 66.16 (continued) (g) C++ sample, remove GWGEM object. (h) C++ sample, fire an event. (i) C++ sample, disable the communication link. (j) C++ sample, send S1F13 to the host.

change, or the arrival of SECS-II primary messages with window's message handler so that these events are handled concurrently without performance trade-offs. Due to space limit, the code samples in Figure 66.16a through j only show the necessary steps for declarations: initialization and deletion of GEGEM objects in an MFC-based application.

By using GWGEM C++ for Windows, there is no need for the developer to code the low-level communication details as shown in Figure 66.17. GW daemon and its supporting components such as GWGEM extension tasks do the process of incoming SECS-II messages. As a result, it significantly shortens the development cycle and reduces software bugs.

Typically, after 2 months of development and testing, the equipment successfully passes the GEM tests and is declared fit for delivery.

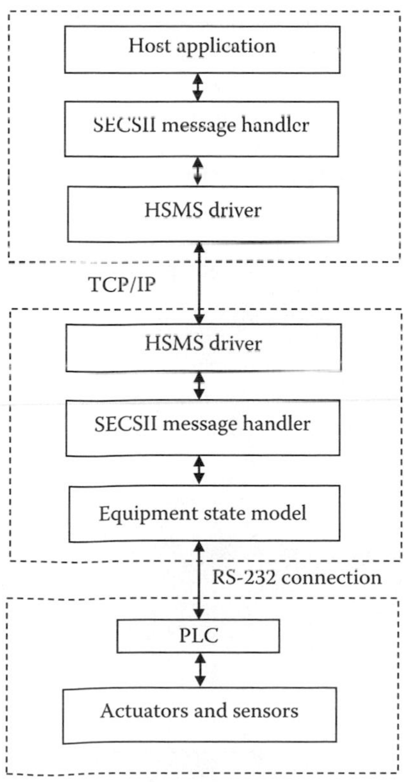

FIGURE 66.17 Design of intercommunication process.

66.3.3.4 Case Study II: Design and Implementation of SECS-I, SECS-II, and GEM Driver

This section presents a case study in the design and implementation of SECS-I (serial port using RS232), SECS-II, and GEM driver for a case where a new machine is to be integrated with a computer-integrated manufacturing (CIM) system.

a. Software design

The driver is developed using Microsoft C#.Net version 4.0, and it provides communication between host computer system and the equipment using SECS-I serial communication (RS232) with specifies SECS-II message format content. Figure 66.18 shows the software architecture of the driver in the form of DLL and is statically linked to the machine application.

The SECS-I and SECS-II define communication interface standards between the semiconductor equipment and the host CIM system. The SECS-I standards describe details for physical connection, signal size, data rate, and protocol logic, and SECS-II standards describe message (information) exchange between the equipment and the host. All the messages are transferred based on the SECS protocol, and each of the message transfer is processed as one transaction. Combinations of those transactions control and manage the semiconductor fabricating process. Table 66.11 shows the SECS-II messages transferred between the equipment and the host CIM system. In this *equipment-driven scenario*, the equipment sends an S1 F13 representing communication establishment to the host in order to start the semiconductor fabricating process. The host transfers an acknowledgment message S1 F14 to the equipment in order to respond to the message from the equipment. The equipment then sends an S1 F1 to the host, which is a query inquiring whether the host exists and receives an acknowledgment message of S1 F2 from the host. Subsequently, the equipment transfers S6 F11, which represents a mode change from offline to online of the host and receives an acknowledgment message S6 F12 from the host. The operator for the equipment puts a product on the worktable prior to starting the semiconductor fabricating process.

After a product is loaded to the worktable, the equipment sends an S6 F11 representing the material that has been set up at the worktable to the host and then receives an acknowledgment message of S6 F12 from the host. The equipment sends an S7 F5 to request for process recipe for this product to the host and then receives S7 F6, the recipe data from the host. Upon receiving

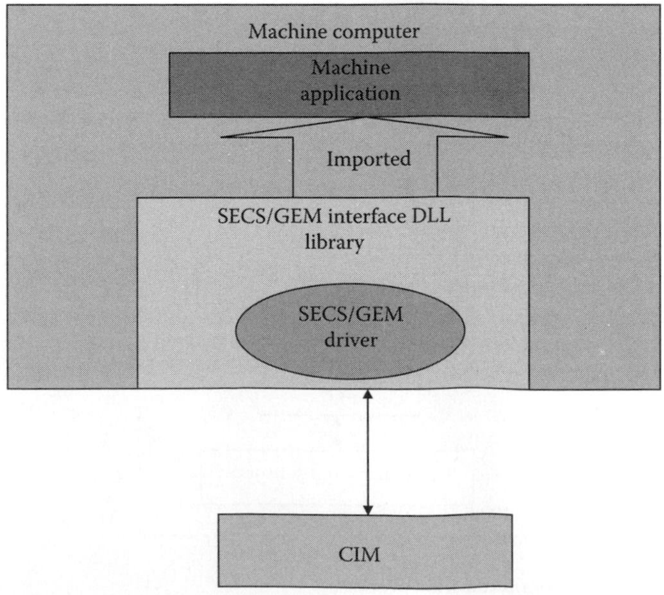

FIGURE 66.18 Software architecture.

TABLE 66.11 Equipment-Driven Scenario

Host	SECS Message	Equipment
	←S1F13S1F14→	Establish COMM request.
	←S1F1S1F2→	Online request.
	←S6F11S6F12→	Online report event.
		Operator puts a product on the worktable prior to starting the semiconductor fabricating process (product loading).
	←S6F11S6F12→	Material setup event.
	←S7F5S7F6→	Request recipe program.
	←S6F11S6F12→	Process start event.
	←S6F11S6F12→	Process completed event.

Sn,Fm	Function	Transfer direction	Response required
S6 F11	Event report send (ERS)	Host <- equipment	Yes

L, 3

<DATAID>
<CEID>
L, 2

<RPTID>

L, 3

<PORT>

<LOTID>

<PPID>

FIGURE 66.19 Example of SECS-II message.

the recipe data, the equipment starts the process fabrication procedure according to the data defined in the recipe program. After the fabrication process is completed, the equipment sends an S6 F11 representing the completion in fabrication to the host and then receives an acknowledgment message of S6 F12 from the host.

Figure 66.19 shows an example of a SECS-II message transferred between the equipment and the host. Each SECS-II message presents a unique expression thereof such as S6 F11, a combination of one of stream number (S1–S128) and one of function numbers (F1–F128). S6 F11 is message for Event Report Send (ERS) from the equipment to the host, and it is necessary for the host to respond to the message toward the equipment. The structure of the SECS data in S6 F11 message is "L,3," which means three data lines: the first line contains <DATAID> representing a series of "data ID," the second line contains <CEID> representing "collective event ID," and the third line contains "L,2" composed of two lines. The first line of "L,2" contains <RPTID> representing "report ID," and the second line of L,2 is composed of another three lines "L,3," which contain <PORT> representing "port number," <LOTID> representing "lot or product ID," and <PPID> representing "process program ID."

A SECS-II message comprises of data variables that have unique data format. The data format is defined as the data type with "," as separator, followed by data length and data value inside

TABLE 66.12 Data Format

Data Format	Description
A, (1 − n)	ASCII char array, array max size n = 65,535
B, (1 − n)	Binary number array, array max size n = 65,535
11, (1 − n)	1-byte signed integer, array max size n = 65,535
12, (1 − n)	2-byte signed integer, array max size n = 35,535
14, (1 − n)	4-byte signed integer, array max size n = 65,535
18, (1 − n)	8-byte signed integer array max size n = 65,535
U1, (1 − n)	1-byte unsigned integer array max size n = 65,535
U2, (1 − n)	2-byte unsigned integer array max size n = 65,535
U4, (1 − n)	4-byte unsigned integer array max size n = 65,535
U8, (1 − n)	8-byte unsigned integer array max size n = 65,535
F4, (1 − n)	4-byte float number array max size n = 65,535
F8, (1 − n)	8-byte float number, array max size n = 65,535
Bool, (1 − n)	1-byte Boolean number array max size n = 65,535
L, (1 − n)	List number, max size n = 65,535

the "<>." Data type comprises of various numeric numbers and string characters, with length in single (1) or array type (1<). Table 66.12 shows the various data types used in SECS message.

SECS-II message is formed by both header and data blocks. Header block stores the system information, and data block stores the information of data variables. The message is constructed with the structure and format as defined in Figure 66.20 where header block stores system information such as byte length, device ID, stream ID, function ID, block number, system byte, and checksum value. Data block stores data variables, which are grouped within a list element. A list allows the grouping of related data items, which can have different formats. A list can contain sublist elements. Figure 66.20 also showed the value before and after conversion between text and binary format of SECS-II message.

Data format: data type, data length <data value>
String data: A, 18 <This is an example.>
Unsigned array data: U1, 5 <1 2 3 4 5>

b. Communication method

The communication method between the host and the machine is by RS232 connection with the following master and slave relationship. The master is referring to the party that initiates the communication, and the slave responds to the requests. The master is usually referred to the host, and the slave to the machine. In Figure 66.21, a handshake ENQ/EOT is required before the sending of the message, and followed by a handshake to acknowledge message, ACK/NAK is successfully passed or not. Upon message contention as shown in Figure 66.22, where both parties request to send messages at the same time, in this situation, the machine will back off and allow the host to send first. The sequence of the communication is the same as the earlier communication when the direction is opposite.

A single message can only send max 256 bytes. In multiblock message, E-bit of block number in Figure 66.23 is used to track the block number of the multiblock message. A handshake ENQ/EOT is required before the sending of the message, and followed by a handshake to acknowledge message, ACK is successfully passed. E-bit incremented by 1 and followed by the sending of the subsequent message and so on. If there is any negative acknowledge NAK from the receiving party, the message transmission is abruptly terminated.

c. GUI for SECS-I interface (RS232)

As shown in Figure 66.24, the GUI comprises of various panels, namely, "control and data table" panel, "message to host" panel, "message from host" panel, and "history data log" panel:

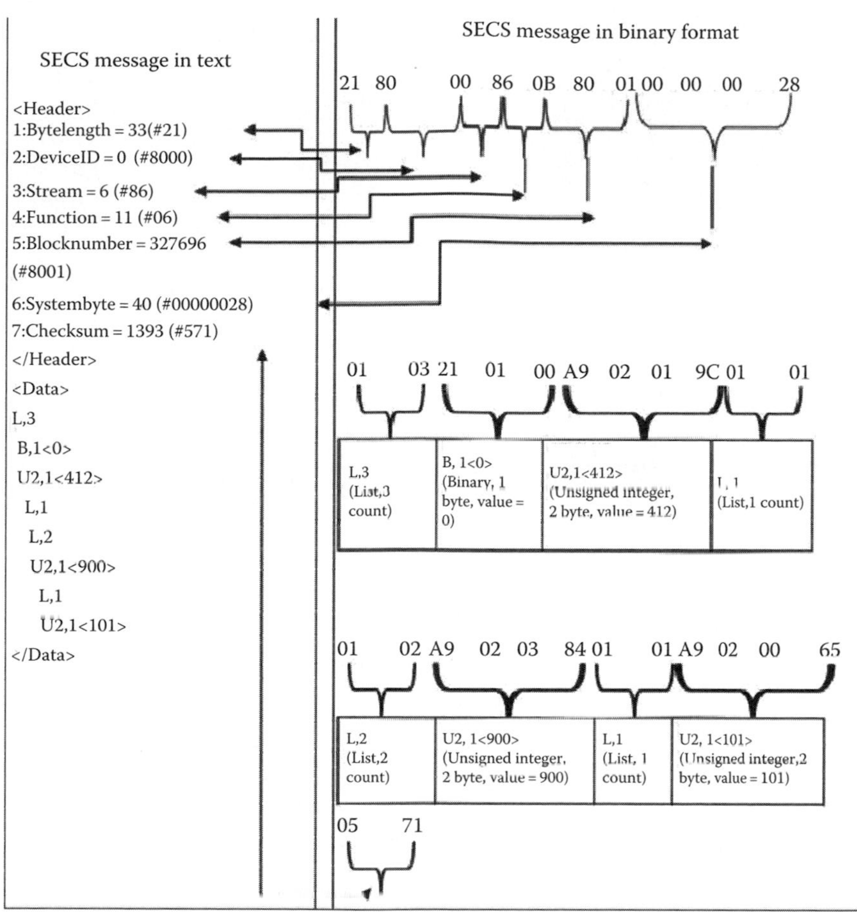

FIGURE 66.20 Format and structure of SECS-II message.

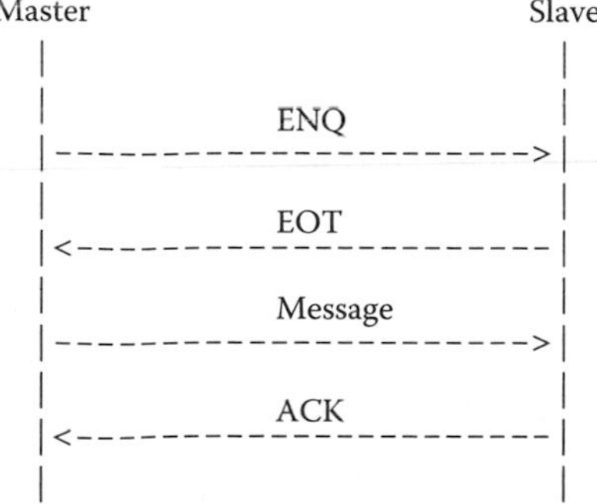

FIGURE 66.21 Single-block message.

1. "Control and data table" panel consists of user interfaces that allow users to control and monitor the communication state between the equipment and the host. Establishment of communication mode can be either operator- or host-initiated. There are altogether 6 timers (T1 to T6) in place to monitor the health of the communication link; time-out value can be remotely changed by the host or locally changed by the operator. The communication port number is selectable and checked on the button above to open port. The interface also

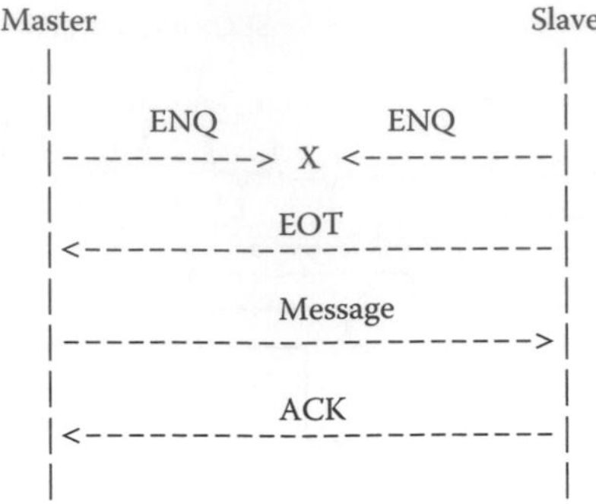

FIGURE 66.22 In case of contention.

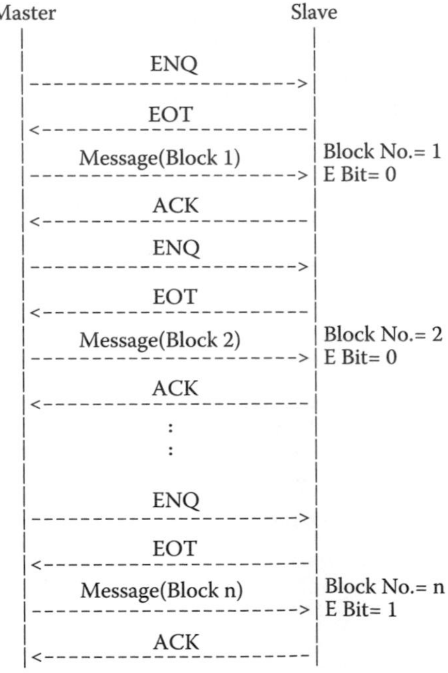

FIGURE 66.23 Multiblock message.

provides an auto or manual selection for user to run the driver in two different modes. The start button is for starting the driver's process, and the initialize button is for resetting the driver's process to its initial state. The update button is for updating SECS data value to the data tables of the software driver.

Figure 66.25 shows the detail "data table" panel box. Data table stored various tables for SECS parameters and data. These are equipment constant ID (ECID), collective event ID (CEID), remote command ID (RCMD), report ID (REPID), process program ID (PPID), data variable (DV), status variable (SV), alarm ID (ALID), etc. These variables are the key data elements for the data exchanges between the host and the machine. For stream and function panel, predefined streams and function's messages are defined here. Each individual function panel consists of a textbox in which SECS-II message and variables are formulated.

2. As shown in Figure 66.26, the "message to host" panel consists of panel boxes where SECS binary data are computed. Information, such as device ID, Reply expected, and

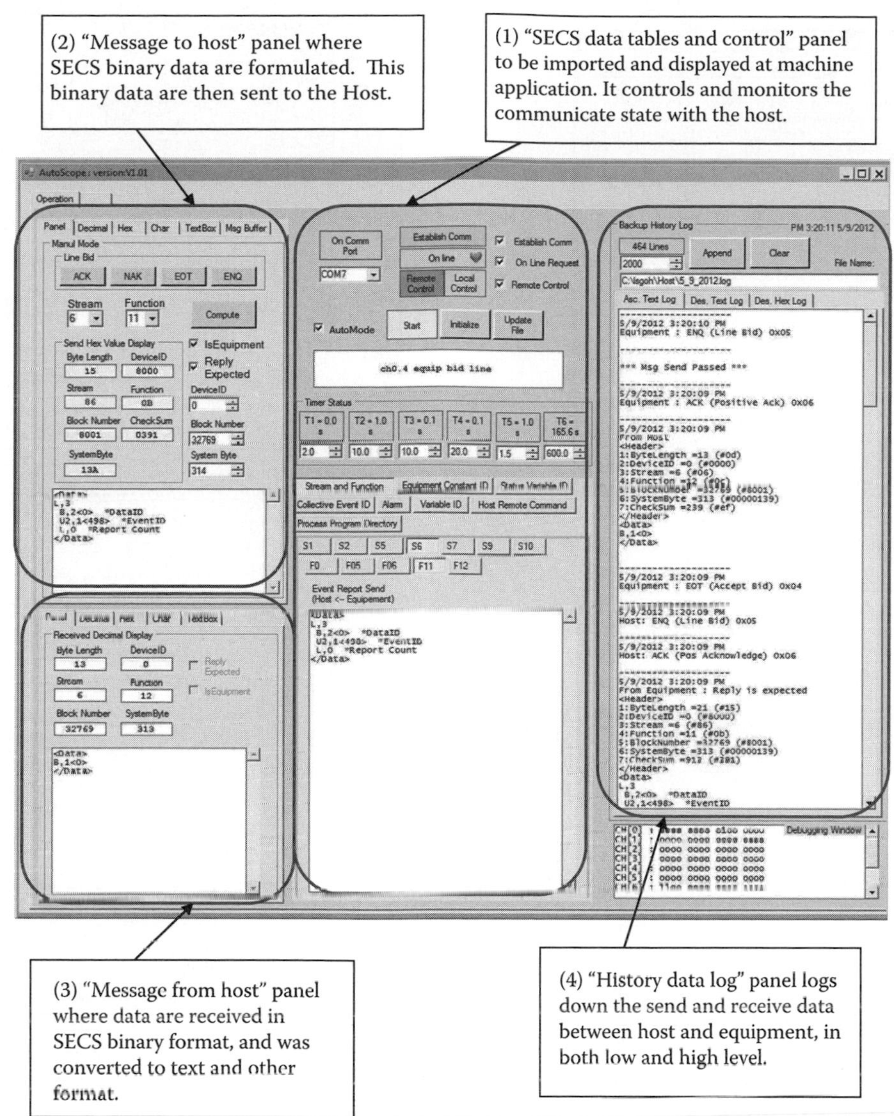

FIGURE 66.24 GUI for SECS-I interface driver.

Is-Equipment status, forms the header block, and the data at the panel's textbox formed the data block. The message that is ready to be sent to the host can be viewed in various formats, namely, hexadecimal, char, decimal, and text as shown on the right side of Figure 66.26. Buttons such as ENQ, EOT, ACK, and NAK are used for communication's handshake with the host. The "Compute" button is used to start text to SECS binary format conversion.

3. "Message from host" panel consists of panel boxes where SECS binary messages from the host are received and displayed. The binary message is decoded to text format, and header and data block information are then extracted and archived to the respective data tables. The received SECS message is displayed in various formats, namely, hexadecimal, char, decimal, and text as shown on the right side of Figure 66.27.

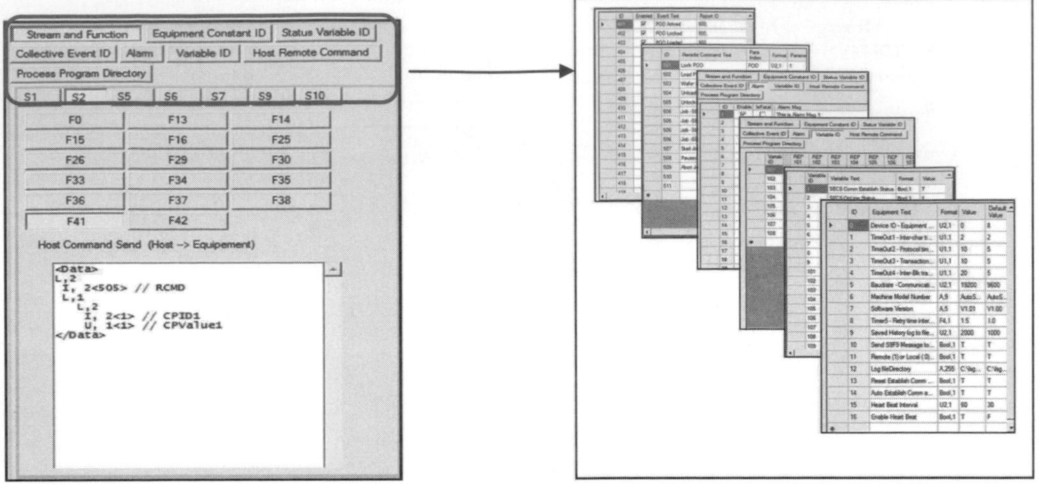

FIGURE 66.25 "Data table" panel box.

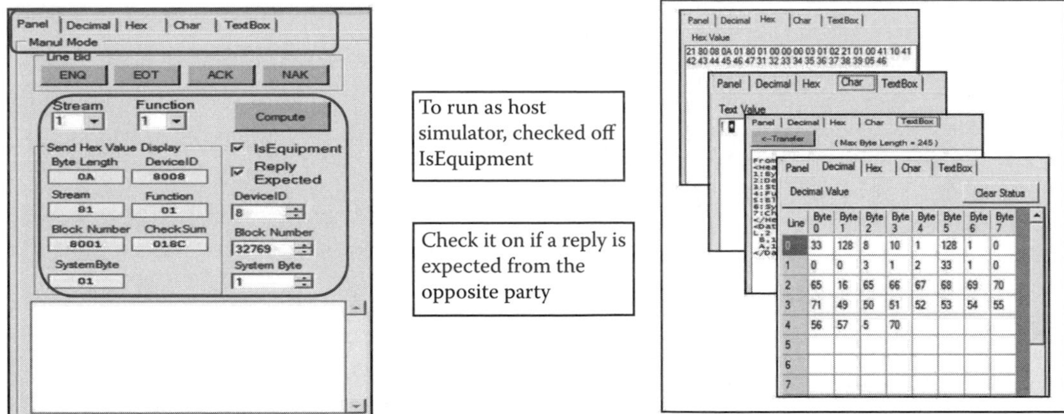

FIGURE 66.26 "Message to host" panel box.

4. As shown in Figure 66.28, the "history data log" panel boxes contain transmitted and received log data between the host and the equipment, in both SECS binary and text format with date and time stamp. Handshakes and communication status, for example, timer status, are saved in the data log too. When the data lines are greater than the preset count, the log is automatically saved, with the "today" filename at the user's predefined directory.

d. Software integration method

The integration of SECS driver with machine application software is direct and simple. Basically, it consists of the following three parts. The first and second parts are mandatory, while the third part is optional.

The first part is the "Message In" that is from machine application to SECS driver. These messages are, for example, machine's event and alarm, to be sent from the machine to the host through the SECS driver. The message is pushed into the input message storage by machine application and is retrieved in first-in–first-out (FIFO) order by the driver. In the event of communication breakdown at the RS232 link, the message from the machine is spooled in the buffer and can be retrieved by the driver and send to the host when communication is up again.

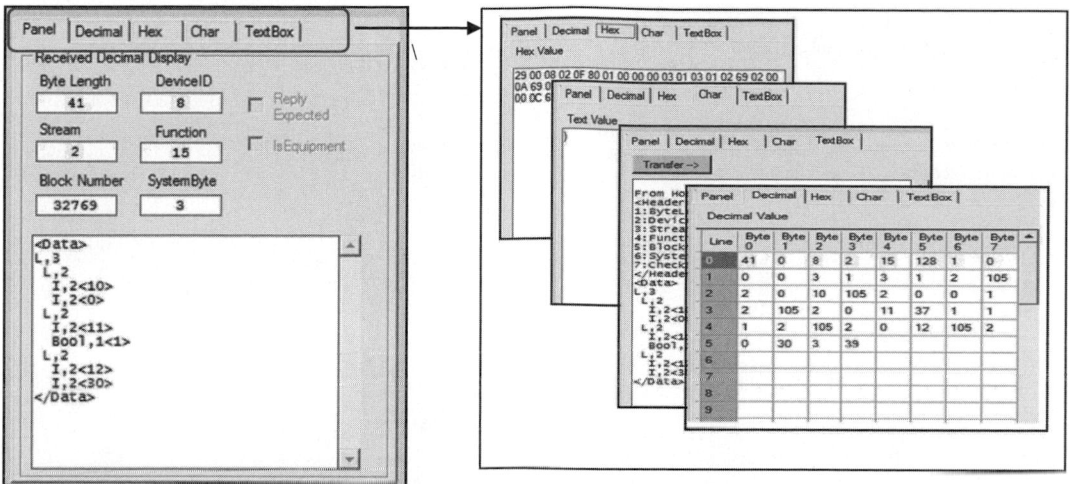

FIGURE 66.27 "Message from host" panel box.

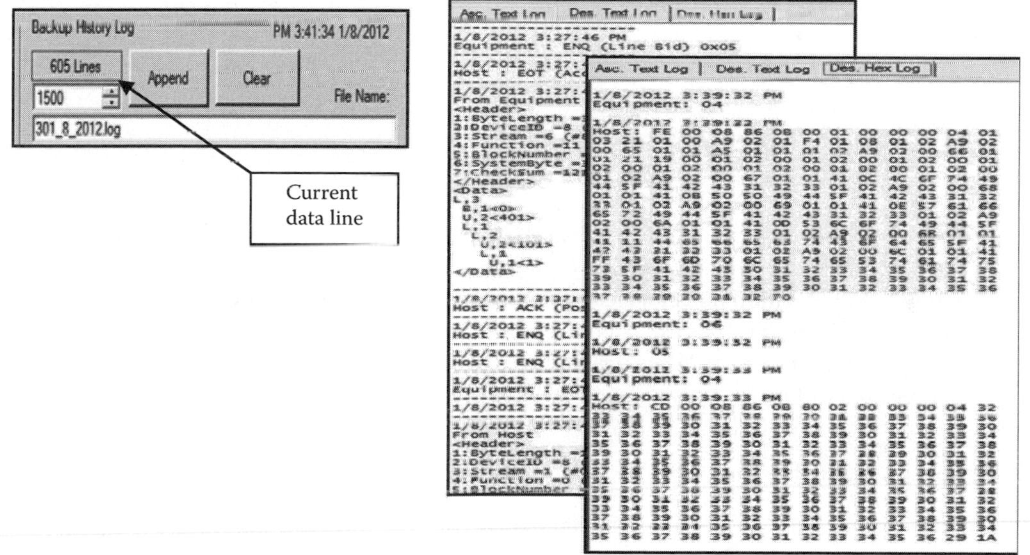

FIGURE 66.28 Data logging.

As shown in Figure 66.29, the second part is the "Message Out" that is from the SECS driver to machine application. These messages could be the process's result of the driver or message from the host. These messages are passed to the caller, which is the machine application in the form of *event callback*. With these two methods of conversation, it formed a closed loop communication system between the machine application, the SECS driver, and the host.

The final part is an optional feature that allows the data tables to be directly accessible by the machine application. User can manually or by software program method view, edit, and update the data table that resides at the SECS driver side.

Figure 66.30 shows the imports of "data tables and control" panel from the driver to the machine application. A software example is also given to illustrate the "Message In" and "Message Out" or call-back methods. In the example, an alarm "Message In" is sent from the machine to the driver's message queue. Upon the detection of message available at the

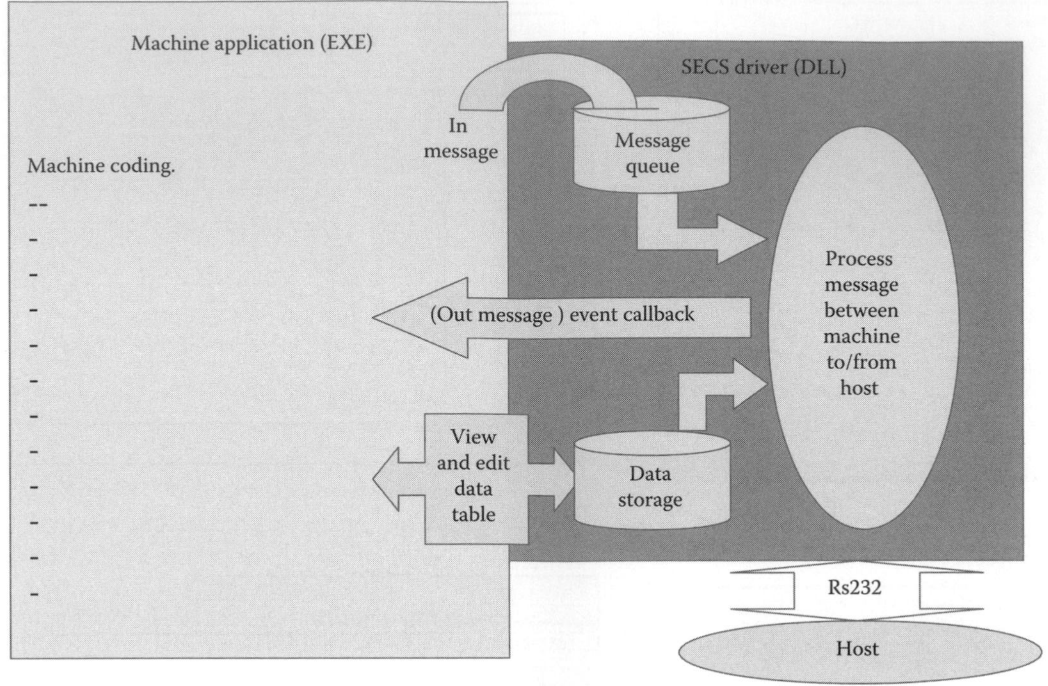

FIGURE 66.29 Integration diagram.

queue, the driver popped out the queued Message In FIFO order and searched through the data table to look for data item related to the message's ID. Once the item is found, the message's ID (ALID) with data is formed to text message, converted to SECS-II binary format, and sent over through the RS232 connection to the host computer. After the message is sent successfully or not, a "Message Out" event callback is triggered and sent back to the machine application.

66.4 Conclusion

This chapter has given a brief overview of the standards of SEMI SECS-I/-II/GEM. While keeping within the special (and at time strange) nomenclature and conventions adopted by SEMI, the authors have provided ample comments and interpretations on the pertinent points to enable the reader to understand the complexity and integrity of SECS equipment. A survey of existing SECS-compliant software is also conducted.

Two case studies are also introduced to illustrate how SECS/GEM compliance can be implemented on virgin machines and interfaced to host system.

Presently, there are no systematic alternatives to SEMI standards. Although the SEMI communication standard is still evolving, it continues to be critical as a means of sustaining maximum functionality/availability in the wake of new semiconductor processes and new semiconductor-manufacturing automation.

Acknowledgment

The authors would like to express their sincere thanks to Dr. Zeng Xianting, Director of Knowledge Transfer Office of SIMTech, for his support leading toward the completion of this chapter.

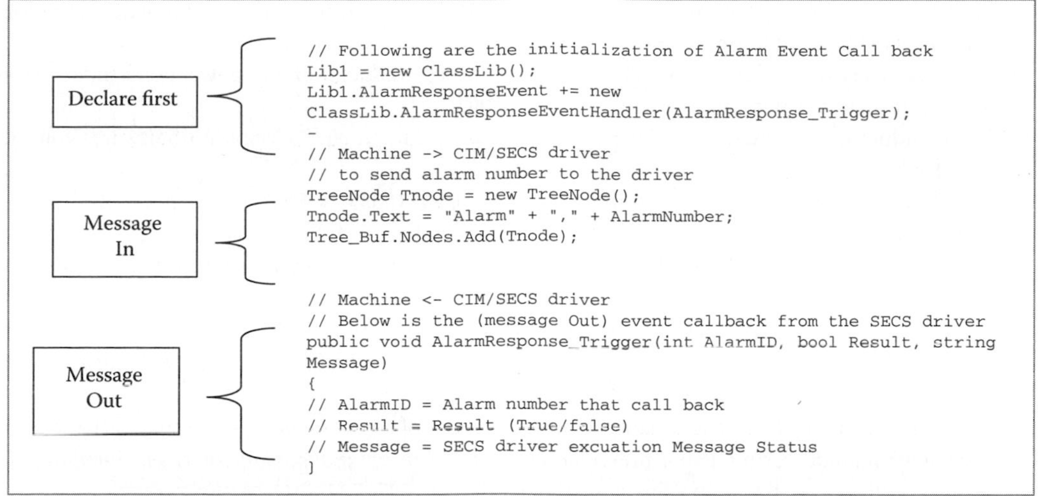

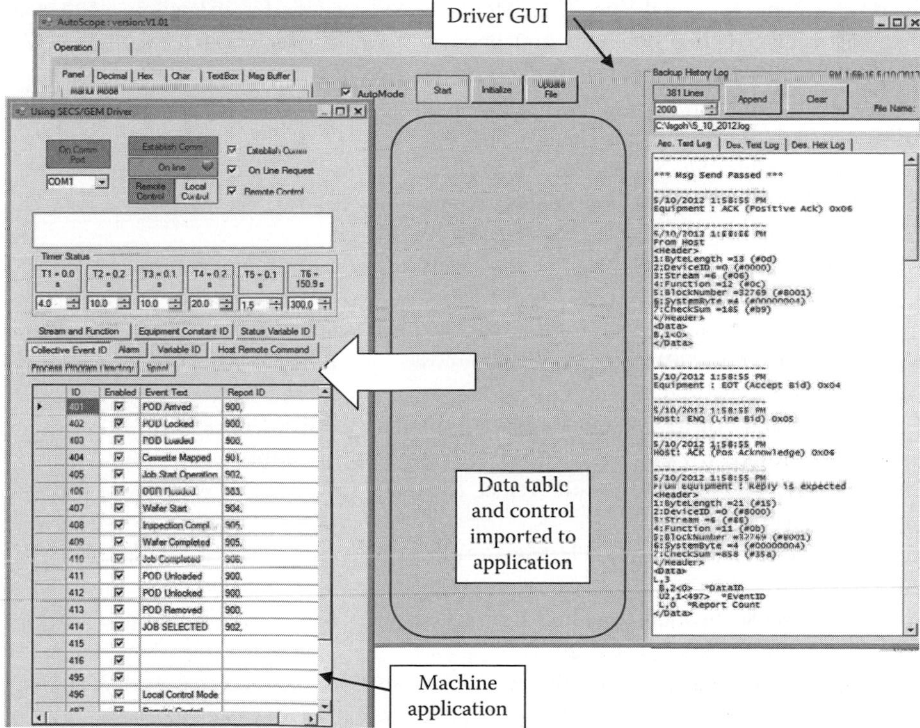

FIGURE 66.30 From driver to machine application.

References

1. Tin, O., Competitive analysis and conceptual design of SEMI equipment communication standards and middleware technology, Master of Science (Computer Integrated Manufacturing) Dissertation, Nanyang Technological University, Singapore, 2003.
2. SEMATECH, Generic equipment model (GEM) specification manual: The GEM specification as viewed from host, Technology transfer 97093366A-XFR, pp. 4–39.

3. SEMATECH, High speed message services (HSMS), Technical education report, Technology transfer 95092974A-TR, pp. 11–34.
4. GW Associates, Inc., Solutions for SECS communications, Product training (power point slides), 1999.
5. SEMI International Standards CD-ROM, SEMI, 2003.
6. Semiconductor Equipment and Materials International Equipment Automation/Software, Volumes 1–2, 1995.
7. SEMATECH, *CIM Framework Architecture Guide 1.0*, #97103379A-ENG 1997, pp. 1–31.
8. SEMATECH, *CIM Framework Architecture Guide 2.0*, 1998, pp. 1–24.
9. SEMI, *Standard for the Object-Based Equipment Model*, SEMI Draft Document #2748, 1998, pp. 1–52.
10. SEMI E98, Provisional standard for the object-based equipment model.
11. Weiss, M., *Increasing Productivity in Existing Fabs by Simplified Tool Interconnection*, Semiconductor Fabtech, 12th edition, 2001, pp. 21–24.
12. Yang, H.-C., Cheng, F.-T., and Huang, D., Development of a generic equipment manager for semiconductor manufacturing, Paper presented at *7th IEEE International Conference on Emerging and Factory Automation*, Barcelona, Spain, October 1996, pp. 727–732.
13. Feng, C., Cheng, F.-T., and Kuo, T.-L., Modeling and analysis for an equipment manager of the manufacturing execution system in semiconductor packaging factories, 1998, pp. 469–474.
14. ControlPRo™, *Developer Guide*, Realtime Performance, Inc., 1996.
15. Kaufmann, T., *The Paradigm Shift for Manufacturing Execution Systems in European Projects and SEMI Activities*, Semiconductor Fabtech, 8th edition, pp. 17–25.
16. GW Associates, Inc., *SECSIMPro GEM Compliance Scripts User's Guide*, 2001.
17. GW Associates, Inc., *SECSIMPro, SSL Reference Guide*, 2001.
18. GW Associates, Inc., *SECSIMPro, Users Guide*, 2001.
19. SEMATECH, SEMASPEC GEM purchasing guidelines 2.0, Technology transfer, #93031573B-STD 1994, pp. 10–30.
20. Cimetrix Software, Factory automation, manufacturing control software solutions, www.cimetrix.com (accessed April 3, 2014).
21. Abakus Software, www.abakus-soft.de (accessed April 3, 2014).
22. Kinesys Software—The GEM box, www.kinesysinc.com (accessed April 3, 2014).
23. Ergo Tech Software, www.secsandgem.com (accessed April 3, 2014).
24. SDI Software, TN3270 Plus secure Telnet terminal emulation for Windows, www.sdiusa.com (accessed April 3, 2014).
25. Yokogawa Software, Yokogawa Electric Corporation, www.yokogawa.com.sg (accessed April 3, 2014).
26. Asyst Software, Semiconductor automation software for fabs, assembly and equipment suppliers, www.asyst.com (accessed April 3, 2014).
27. SI Automation Software, SiAutomation, www.siautomation.com (accessed April 3, 2014).
28. VECC Product, AIS Automation Dresden GmbH, www.ais-dresden.de (accessed April 3, 2014).
29. Agilent Software, Agilent Technology, www.agilent.com (accessed April 3, 2014).

XIV

Emerging Protocols and Technologies

67 **Brain** *Michael Paulitsch, Brendan Hall, and Kevin R. Driscoll* **67**-1
Protocol Overview • Protocol Mechanisms and Services • Fault Isolation • Diagnostic and Agreement Services • Validation and Verification Efforts • Example Configurations, Implementations, and Deployment Considerations • References

67 BRAIN

Michael Paulitsch
Airbus Group Innovations

Brendan Hall
Honeywell International Inc.

Kevin R. Driscoll
Honeywell International Inc.

67.1 Protocol Overview 67-1
Development History and Design Goals • Minimal Overhead Replication and Input Agreement
67.2 Protocol Mechanisms and Services 67-4
High-Integrity Data Propagation • Clock Synchronization, Start-Up, and Clique Resolution
67.3 Fault Isolation 67-14
Time-Triggered Sequenced Guardian Roles • Asynchronous Guardian Roles
67.4 Diagnostic and Agreement Services 67-18
Host Task Set Agreement
67.5 Validation and Verification Efforts 67-18
67.6 Example Configurations, Implementations, and Deployment Considerations 67-18
References 67-20

67.1 Protocol Overview

The BRAIN (Braided-Ring Availability Integrity Network) is a novel communication architecture supporting fault-tolerant time-triggered communication. As the name suggests, the BRAIN is built upon a braided-ring topology. This topology augments the standard ring topology with increased connectivity. In addition to the *direct link* connections between a node and its immediate neighboring nodes (as is used in simple rings), a braided-ring node also is connected to its neighbor's neighbor via a link called the braid or skip link (see Figure 67.1). The BRAIN utilizes the additional connectivity to achieve both high-coverage integrity and availability concurrently. This is in contrast to previous braided rings, which use these additional links only for availability. The BRAIN can use almost any existing local area network (LAN) technology to implement its communication links, including any of the IEEE 802.3 Ethernet variants.

The BRAIN uses the least amount of hardware to achieve single fault tolerance (including Byzantine failure) of any known data network. The BRAIN also can tolerate two benign faults with no additional redundancy. The BRAIN topology enables adjacent nodes to collaboratively form self-checking pairs (SCPs). This allows standard simplex computational hardware to be run-time configured into high-integrity fail-silent computational platforms, which provides the high fault coverage for processing that one would find in architectures supported by SAFEbus (see Chapter 48) but without requiring any special SCP hardware for the processors.

The BRAIN's benefits derive from its time-triggered data flow and its use of high-coverage fault tolerance.

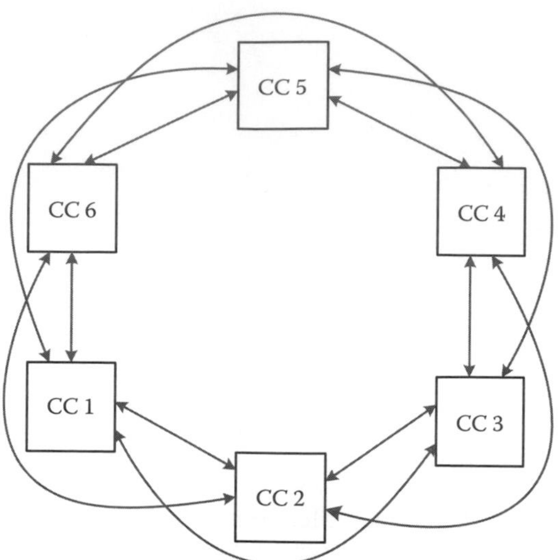

FIGURE 67.1 BRAIN's braided-ring basic architecture.

67.1.1 Development History and Design Goals

The BRAIN was originally conceived as a field-bus-type protocol, targeting low-bandwidth applications. Acknowledging the simplicity of high-coverage self-checking architectures, such as SAFEbus (see Chapter 48), the BRAIN was conceived to apply these techniques to a field-bus protocol. The intent was to deliver SAFEbus levels of integrity and availability within a field-bus environment. Exploiting lessons learned from the application of other field-bus protocols in safety-relevant applications, the BRAIN has been designed with two principles in mind: (1) limited reliance on in-line error detection mechanisms such as cyclic redundancy codes (CRCs) for error detection needs and (2) the avoidance of dedicated guardian hardware commonly used in other field-bus protocols targeting safety-relevant deployment [1]. The BRAIN has evolved to become a more general-purpose architecture, and it continues to evolve.

Limited reliance on coding-based techniques for error detection. It can be argued that the error detection properties of code-based in-line error detection like CRCs are not strong enough to cover the failure modes of active interstages. Active interstages are defined as relaying stations with active circuits (e.g., integrated circuits). These typically are a network switch or hub in a communication architecture [2]. Codes like CRCs have been designed for good error detection coverage of typical failure modes of wires, such as occasional bit flips and burst errors. The failure modes of interstages and the potential for correlated failures are unlikely to fit these assumptions. Hence, the goal of the BRAIN is to not rely on coverage of CRCs or similar error detection schemes for interstages. This is described in more detail in [2].

Avoidance of dedicated guardian hardware. The requirement for local or central guardian hardware constitutes an intolerable overhead in cost-constrained applications. They also introduce complexities when the guardian action needs to be verified within a deployed system. Similarly, each fault-tolerance feature needs scrubbing logic—logic that tests whether the fault-tolerance mechanism is still operational. Such scrubbing logic needs to detect latent faults of the logic implementing the fault tolerance, in order to ensure the logic is available when needed. More allegorically explained, scrubbing ensures that the guard is awake and well when really needed. Distributed, built-in guardian logic as

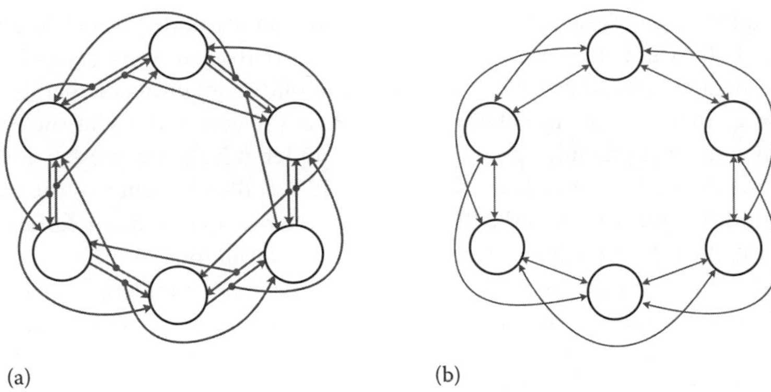

FIGURE 67.2 The two BRAIN connection duplex configurations. (a) Full-duplex BRAIN configuration and (b) half-duplex BRAIN configuration.

deployed in the BRAIN can be more easily scrubbed for latent faults during operation than central or local guardians. Central guardians may additionally introduce architectural constraints, for example, by restricting the order of component power-up. Therefore, a core goal of the BRAIN was to remove the need for any dedicated guardian components.

The BRAIN was conceived as a broadcast flooding network targeting network bandwidth on the order of 5–20 Mbit/s. At these speeds, the intranode propagation delay is minimal, comprising only a few bits of elasticity delay encountered at each link as the message floods around the segmented medium. A global, time-triggered, a priori agreed schedule coordinates the transmission sequencing, in a manner conceptually similar to the media access of a typical time-triggered bus, that is, time-division multiple access (TDMA). For targeting very-low-end applications, the BRAIN can use half-duplex instead of full-duplex BRAIN links. The half-duplex BRAIN sends data in both directions on each link medium, while the full-duplex BRAIN sends data through unidirectional link media. Half-duplex operation is described in [3]. The half-duplex configuration uses less physical layer hardware, at the expense of cutting effective bandwidth by about half (Figure 67.2).

Targeting medium- to high-performance applications, one recent development in the evolution of the BRAIN uses store-and-forward propagation to be able to more efficiently use emerging high-speed serial protocols, such as the gigabit per second (Gbs) Ethernet. However, the algorithms for this higher-speed variant of the BRAIN were still undergoing formal analysis at the time of this publication. Therefore, the details of this development are not discussed here.

67.1.1.1 Fault Tolerance and Fault Hypothesis

With point-to-point links, the BRAIN topology implicitly addresses the spatial and other physical layer damage issues that impact bus topologies. The BRAIN topology is able to tolerate complete loss of communications at any single geographic location on the ring. Similarly, any single node may fail and drop out from the BRAIN without adversely impacting the system-level communication (availability and integrity) guarantees. In addition to these passive failures, the BRAIN protocol also has been designed to tolerate an active malicious fault. The BRAIN mitigates such faults by leveraging bit-for-bit comparisons between the independent data paths that exist between the skip and direct links and/or the two opposing directions of traversing the ring (clockwise and counterclockwise) (see Section 67.2.1). In addition, the BRAIN can employ a *brother's keeper* guardian action, where each node may act as a guardian for its adjacent neighbors. Note that this is consistent with the original design goals, since it supports the deployment of guardian actions without incurring additional component overheads.

The BRAIN originally was designed to tolerate at least one faulty node.* A node's failure mode may be either (1) passive (i.e., fail stop) on any combination of its ingress or egress links or (2) actively malicious, where the node acts actively to corrupt data and/or disrupt protocol operation, for example, by masquerading as another node or babbling. In practice, the connectivity of the BRAIN provides tolerance to many more benign faults, providing service under at least two passive node faults or link faults. However, to tolerate the active fault, the initial field-bus BRAIN requires the full connectivity of the nonfaulty nodes. A discussion of the BRAIN's sensitivity to node and link failures together with a comparison of the BRAIN's policies with other protocols is given in [4].

Note that, similar to SAFEbus, the BRAIN has not been designed to tolerate dual active colluding faults, that is, where two nodes act in a coordinated fashion to corrupt data flow and/or disrupt the system. This fault hypothesis is backed by two observations. First, statistical analysis shows that it is highly unlikely that two faults manifest in correlated active failure modes in two independent† devices within the very tight time window of self-checking action. Second, the combination of the self-checking forwarding action, the periodically scrubbed guardian function, and the potential reconfiguration actions of Section 67.3 are sufficient to tolerate an active fault for the time of exposure between scrubs.

67.1.2 Minimal Overhead Replication and Input Agreement

In Byzantine-tolerant real-time systems, the bandwidth and scheduling overheads associated with the Byzantine data exchange and agreement often constitute significant network, system, and software burdens. This observation created another design goal of the BRAIN, which was to provide a low-overhead framework to support task replication and data agreement. To this end, the BRAIN introduces additional mechanisms to contain Byzantine failures, for example, preventing a node from supplying different data on each of its outgoing directions. With such fault containment in place, an inconsistent omission fault model may be assumed for those messages needing agreement. This reduces the expenses of higher-level agreement exchanges required for such messages typically used in other Byzantine-tolerant protocols. In place of the fault message exchange and voting typically used in Byzantine agreement, a simpler agreement on reception status may be implemented. This is a hierarchical Byzantine agreement mechanism, similar to the one developed for SAFEbus. The BRAIN also encompasses hardware assistance mechanisms for this exchange that does not need software interaction (see Section 67.4.1).

67.2 Protocol Mechanisms and Services

67.2.1 High-Integrity Data Propagation

The core service of the BRAIN is the fault-tolerant, high-integrity, high-availability broadcast distribution of data. This mechanism provides the foundation for all of the higher-level protocol services. Abstractly, the BRAIN data-propagation scheme is very simple and can be viewed independently of the higher-level protocol services. These techniques may also be applicable to other time-triggered protocols such as FlexRay (see Chapter 6 in KT), Time-triggered protocol (TTP) (see Chapter 5 in KT), or time-triggered Ethernet (see Chapter 42).

A conceptual representation of the BRAIN's high-integrity data propagation is illustrated in Figure 67.3. At a high level, the BRAIN is best viewed as offering two concurrent modes of data propagation, with both modes collaborating to maximize data integrity and data availability. The two modes are self-checking data relay and independent path data integrity reconstitution.

* Later variants are targeting tolerance to two faults.

† This independence can be made arbitrarily large—separate boxes, separate power supplies, etc.

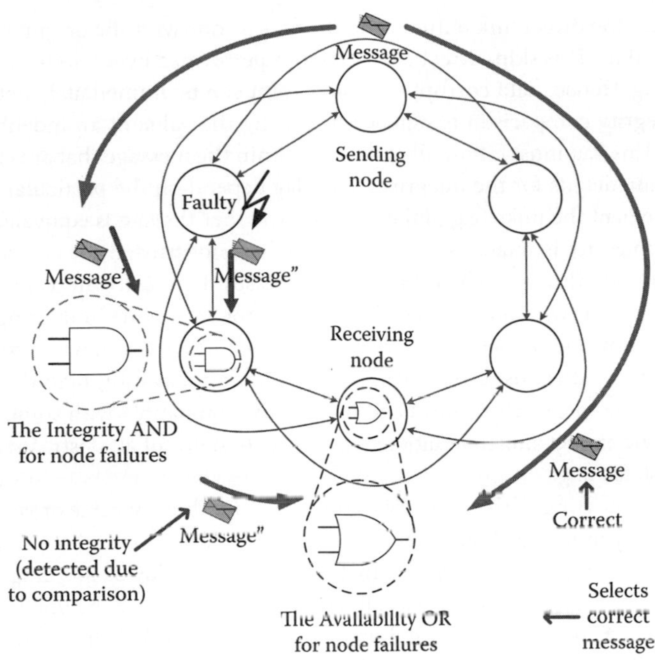

FIGURE 67.3 Conceptual BRAIN operation for tolerating one arbitrary fault.

67.2.1.1 Self-Checking Data Relay

The BRAIN's first propagation mode focuses on in-line integrity failure detection, that is, the possible corruption of data during a node's data relaying action and availability in the event of a single fault. As illustrated by the long outer arrows in Figure 67.3, a sending node transmits its message in both directions around the ring (*clockwise* and *counterclockwise*). Broadcasting a message in both directions provides availability. A message will be delivered successfully if either one of the directions is intact (the *Availability OR* in the figure). The independence of these two paths ensures that there will always be one success path available from any arbitrary sending node to any arbitrary receiving node, given a single fault.

If a receiving node gets two copies of a message (one from each direction) that are different, it must decide which one is good. This decision is supported by the use of message integrity status provided by an integrity scheme employed within each direction. This integrity scheme prevents a message from being undetectably corrupted in transit. For every potentially faulty component within the data-propagation path, there exists completely independent hardware, which checks that this component has not caused any corruption. If corruption is detected, the message is indelibly flagged as suspect (i.e., has lost its integrity guarantee). One instance of the mechanism supporting this scheme is shown in Figure 67.3 as the *Integrity AND* function for the counterclockwise direction. This is replicated for every node in both directions.

As a message traverses the ring, it tries to pass through every node using the direct links. Each node is also bypassed by a skip link. For the general case, in a particular direction, a node receives two inputs, one from the direct link and one from the skip link. Each node compares the data it receives from its direct link with the data it receives from its neighbor's neighbor via the skip link. This bit-for-bit comparison of all data transmitted ensures that any data corruption injected by a faulty node is immediately detected. For example, consider the node marked *faulty* of Figure 67.3 corrupting the message transmission from the sending node as it is being relayed. The erroneous data from the faulty node are immediately detected by the Integrity AND comparison function performed at the next downstream node, which compares the direct link output from the faulty node with the original data (sending node) available on the skip link. This skip–direct comparison is performed by all nodes as the data are propagated around the ring. Hence, data corruption at any point can be immediately detected. The status of the propagation integrity comparison is signaled by setting the value of an indelible status flag at the end of the message. This flag must follow all of the bits within the message that are checked for integrity.

The implementation details for the integrity status flag depend on the particular physical layer technology used to implement the links (e.g., Ethernet) and whether the flag is equivalent to a single binary value (i.e., message integrity is intact vs. message integrity is not trusted) or is a vector of such binary values, one for each node through which the message passes. For vector implementations, the vector either can be a fixed length (with one element of the vector for each node in the ring) or can be variable length (growing in size by one element for every node through which the message passes). Vector implementations enable precise diagnostics to isolate where in the propagation path the errors are induced. However, it increases the message overhead and may not be compatible with some physical layer technologies. For the single binary implementations, an aggregate status of all upstream comparisons can be carried by a shared status flag.

It is essential that the setting of the integrity status flag (single binary value or vector) by all repeating nodes is strictly one way indelible. That is, repeating nodes are limited to either propagating the current integrity status value or setting it to *not trusted*, that is, they are not allowed to change a nontrusted status back to one that says the message has integrity. To enforce this one-way behavior, the value of the integrity status is also checked in the direct–skip link comparison described earlier. If either link says that the message integrity is no longer trusted, the relayed message will have its integrity status flag say that the integrity is no longer trusted. In framed formats, such as Ethernet, the loss-of-integrity status can be signaled by truncating the frame (e.g., deleting the end-of-frame marker), forcing a detectable error at the end of the message (e.g., using the wrong disparity in 8B/10B encoding) or forcing a frame check sequence to be invalid (e.g., inverting Ethernet's CRC field) that causes the frame to be detectability erroneous by the physical technology used to implement the links and consumes no bandwidth for this flag.

67.2.1.2 Independent Path Data Integrity Reconstitution

The BRAIN's second propagation mode focuses on tolerating a second benign (fail-stop or omission) failure. The BRAIN is able to tolerate a benign second fault without any increase in redundancy. This provides the BRAIN an additional degree of fault tolerance. However, the first version of the BRAIN (described here) could not tolerate an active fault and an arbitrary benign fault at the same time. The first propagation mode provides for fail-op/fail-stop operation. The second propagation mode adds fail-op/fail-op/fail-stop operation for benign faults. This is worst case. The BRAIN can tolerate many more faults for most cases. For example, the BRAIN can tolerate any number of benign node failures, as long as no two or more failed nodes are adjacent when three or more nodes have failed.

This second propagation mode reverses the locations of the *Integrity AND* and *Availability OR* mechanisms. See Figure 67.4 and compare it to Figure 67.3. In this mode, the skip and direct links provide increased channel availability, while the Integrity AND is performed by comparing the data a node receives from each direction. If the data received from both directions match bit for bit and the data from neither direction violate the link data encoding nor framing checks, the integrity of the data received may be *reconstituted*. This allows the data to be used, despite having the integrity flags from both directions showing that integrity is suspect. The situation of getting a correct high-integrity message via integrity reconstitution may happen only if there are multiple benign node failures (fail-omission or broken links). By leveraging the reconstitution of integrity, a system may lose multiple skip and direct link connections yet still be able to deliver high-integrity data to the application.

In the second propagation mode, the Availability OR is implemented by each node selecting the data from either link to be forwarded, with the loss-of-integrity status signaled if the two incoming links do

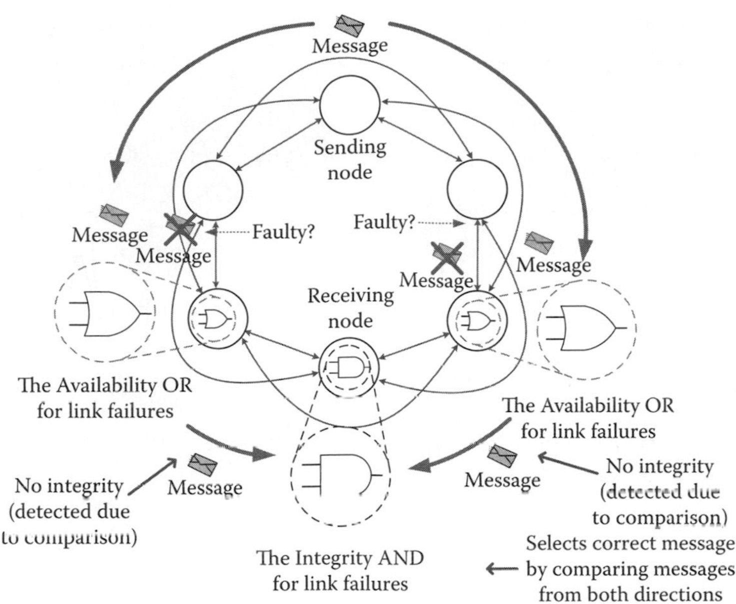

FIGURE 67.4 Conceptual BRAIN operation for a second benign failure.

not match bit for bit. This selection can be arbitrary in most cases. The exception rules for selection are discussed with the rules for the time sequenced guardian roles in Section 67.3.1.2.

It is emphasized that both propagation modes of the BRAIN coexist simultaneously and there are not any specific mode selections nor additional bandwidth required to support them both. Therefore, there are three success propagations that may constitute a high-integrity data reception, either clockwise or counterclockwise with the in-line integrity status confirmed or the reconstituted path from the comparison of both directions.

67.2.1.3 Self-Checking Processor Pair Broadcast

The connectivity of the BRAIN that enables the data relaying policies of the BRAIN, as described earlier, also can be used to compare the output of two adjacent nodes. This allows for adjacent nodes to be configured into high-integrity *message-based SCPs*. Implementing paired transmissions is as simple as configuring the communication schedule to make the two halves of a pair transmit in a shared slot (the time allocated on the media to transmit one message). The synchronous nature of the BRAIN and the high-integrity forwarding mechanism ensure that the receiving nodes receive a single high-integrity message when the data sent from the two halves of the pair are identical. Such a configuration is depicted in Figure 67.5, where the copies of message *msg a* appears as the copies of a single message at all receivers.

Using the time-triggered network communication primitives to implement the high-integrity computational comparison presents significant advantages when contrasted with other self-checking approaches. Traditionally, high-integrity computational hardware platforms have required specialized design to *clock step* or *lock step* the processing platform. In modern processors, such lock-step comparisons can often introduce a significant run-time performance degradation. In addition, with the emergence of multicore processing engines, the complexity associated with such comparisons may impact the very feasibility of the lock-step approach. The loosely coupled time-triggered message-stepping comparison* of the BRAIN does not suffer from performance degradation, since the comparison is made at the communication line

* Integrity comparisons are done on a message-by-message basis rather than on every clock tick of the processor.

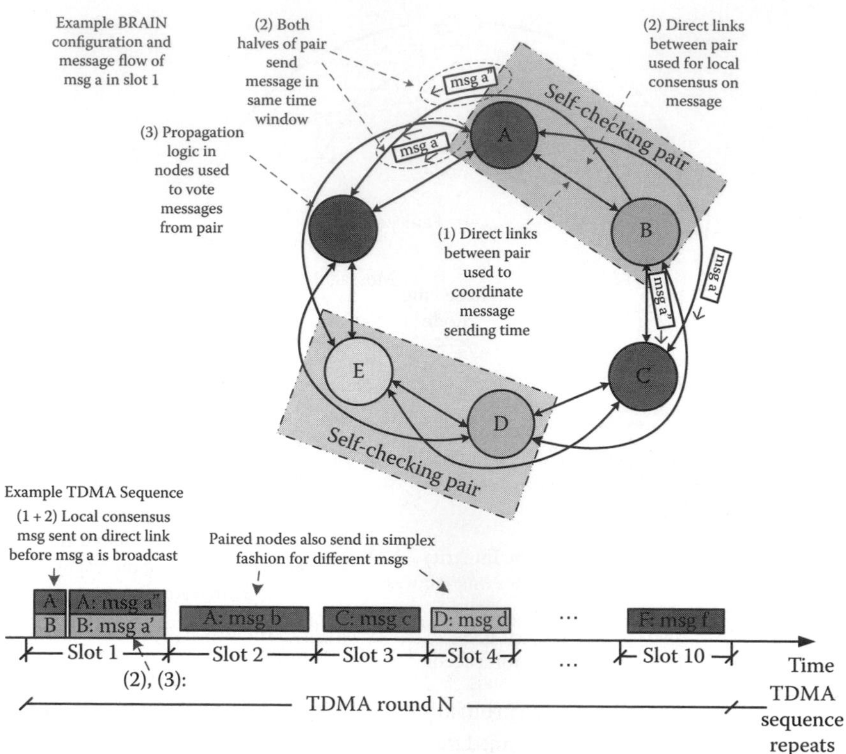

FIGURE 67.5 SCPs.

rate as a by-product of the BRAIN's normal message propagation mechanisms. Additionally, the communications-based comparison approach may arguably offer superior fault coverage, since it detects failures that occur within the communication hardware itself. It should also be noted that the BRAIN message-stepping scheme enables nodes to selectively perform self-checking or non-self-checking message transmissions. As illustrated in Figure 67.5, nodes can operate as message-stepping pairs, during one TDMA slot (see nodes A and B in slot 1), and the same nodes can make independent simplex transmissions in other slots (node A in slot 2). This scheme may provide improved processing hardware utilization, since it provides the freedom to replicate only safety-relevant software tasks and does not require the full replication of all computation on the self-checking host.

There are advantages in reducing the tight redundancy coupling required for processor integrity mechanisms. For example, in missions and applications where spare resources are scarce or the logistics pipeline is expensive, system designers often strive for *generic sparing* strategies that enable common reconfigurable hardware to be used in more than one application or mission role. Hence, the ability to configure standard COTS processing hardware into high-integrity fail-silent computational pairs via a simple change of the TDMA schedule is very attractive. Additionally, each half of the pair may use dissimilar hardware, providing a path for generic design failure mitigation. A triple-modular redundant variant of the self-checking scheme is also possible, as presented in Figure 67.6. Using such a configuration, it may be possible to guarantee fail-operational behavior with a generic processing fault while using minimal hardware.

To implement the message-stepping pairs, input agreement is required between each half of the pair to ensure that a fault entering the two halves of the pair cannot force each half of the pair to diverge or disagree. To implement such agreement, the direct connections between the paired nodes may be used. Section 67.4.1 outlines the use of agreement hardware.

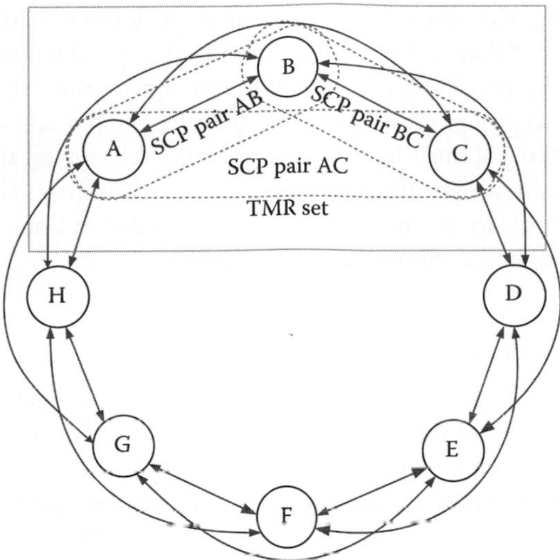

FIGURE 67.6 Triple-modular replication (TMR) deployed on BRAIN.

67.2.2 Clock Synchronization, Start-Up, and Clique Resolution

The self-checking data-propagation mechanism described earlier is extensively utilized by the BRAIN's clock synchronization and start-up protocols. Within a BRAIN network, adjacent nodes act as SCPs to send protocol synchronization messages. Unlike other protocols that utilize distributed algorithms to vote and converge output from multiple *peer* nodes, the BRAIN adopts a hierarchical protocol strategy. In place of a fault-tolerant convergence function, the BRAIN employs a fault-tolerant priority-based selection function to select a master-clock pair. The selection of a pair is performed once each synchronization interval. It is, therefore, tolerant to dynamic system membership changes and will reselect an alternative driving pair when the highest-priority selected master fails. Similarly, when the higher-priority pair recovers, the BRAIN protocol incorporates mechanisms to ensure that the recovering pair will assume the running timeline before it resumes its mastership. As is generally true for the creation of BRAIN SCPs, clock synchronization SCPs are created simply by including this information in the schedule tables. Nodes that are not configured to be in the set of potential SCP clock masters are called *slaves*. The protocol messages sourced from the SCPs are simple, comprising two fields as noted in Table 67.1. In the BRAIN, the same message format is used for synchronization and integration; hence, the messages are termed Synchronization–INTegration (SINT) messages.

Within the SINT message, the TDMA POSITION field enables nodes to agree and integrate on the TDMA schedule phase. The PRIORITY field is used to arbitrate among clock-pair masters, in the event that the pairs disagree on timing and/or schedule phase. The PRIORITY is statically allocated to pairs at design time and the pair with the highest priority wins. Following any detected contention, the highest-priority pair will continue running; the lower-priority pairs will yield and resynchronize according to the content of the higher-priority pair's SINT message. The timing of the highest-priority SINT messages also is used for global synchronization, with lower-priority pairs and slave nodes

TABLE 67.1 SINT Message Fields

Field	Description
TDMA position	The TDMA slot counter
Priority	The priority of the sending SCP

correcting their local timing with respect to the highest-priority master's SINT message arrival. There is a relationship between precision and drift of synchronized time. The more frequently an algorithm synchronizes, the tighter the precision, but frequent resynchronization may lead to a larger drift of the global time (affecting accuracy). The BRAIN master-/slave-based approach performs better than classical distributed algorithms because it maximizes accuracy (eliminates the integration of *read error* suffered by peer synchronization algorithms) and minimizes jitter impact. The time between scheduled SINT messages (the synchronization interval) is determined at design time as an engineering trade-off between better clock precision and minimizing bandwidth overhead, which can be different for every application.

67.2.2.1 Self-Checking Master Coordination

The accuracy and validity of the BRAIN's leader-elect master/slave synchronization approach is dependent on the fault coverage of the master clocks. For this reason, the BRAIN introduces special qualification logic to cross-check a pair's health prior to them initiating any protocol traffic. In the BRAIN, all SCPs are formed from adjacent nodes. Hence, each half of a master-clock pair is connected by the direct link (private link) that connects the nodes. The pairs use this link to initiate a pairing rendezvous. This rendezvous is performed by each pair once per synchronization interval and is scheduled to occur just prior to the sending of the associated pair's SINT frames. This ensures that the clocks are checked for accuracy before each half of the pair initiates SINT transmission. To achieve this, as part of a pair's rendezvousing action, each half of the pair monitors the timing of its other half. Subsequently, one half of a pair will not support a SINT transmission if the timing of its partner's clock fails beyond a configured tolerance.

To ensure clock correctness at start-up, the rendezvous procedure is also performed as part of the protocol start-up. On power-up, the nodes of each SCP perform a rendezvous prior to entering a listen–time-out period, that is, the period during which the nodes listen and wait to detect running TDMA traffic. If the SCP nodes complete the listen–time-out without detecting traffic, the nodes rendezvous again on completion of the listen period. Thus, the duration of the listen period constitutes the period over which the start-up clock frequency monitoring occurs. To ensure that the clocks are operating within the configured tolerance, both halves of the pair monitor and cross-check the duration of the listen period as indicated by its partner.

Should the listen activity result in a change of schedule phase, that is, following the reception of a SINT frame from another master, each half of the pair inhibits SINT transmission for the first synchronization interval following integration. This ensures that a full synchronization interval duration is

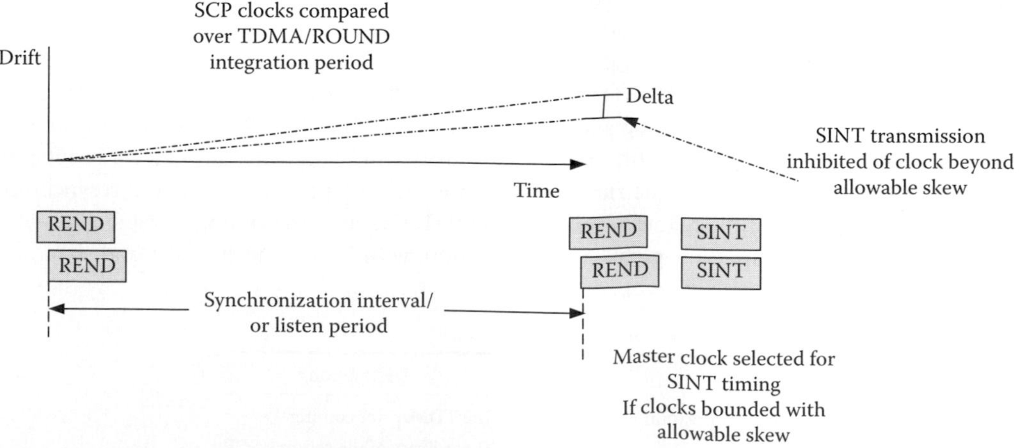

FIGURE 67.7 An SCP clock monitoring and rendezvous.

used for clock cross-checking prior to the pair sourcing of any coordinated SINT messages. Similarly, if one half of a pair has successfully integrated into running TDMA traffic, it will not support its partner in the establishment of a disjoint-phase clique. Figure 67.7 illustrates the rendezvous procedure, showing the sequence of SCP rendezvous (REND) messages used to align their respective timelines.

To maximize clock accuracy, the two halves of a pair do not perform mutual clock averaging. Instead, one of the pair's clocks is used as the master clock, and all protocol timing is derived from this single clock. The clock of the other half of the pair is used only for monitoring, to ensure that the master clock remains within the expected accuracy. This improves the clock accuracy.

67.2.2.2 Connectivity Building and Clique Aggregation

The self-checking leader-elect synchronization strategy described earlier is conceptually very simple. However, it is not sufficient, by itself, to address the multihop and segmented-media topology of the BRAIN architecture. Without mitigation, the segmentation of the BRAIN may cause start-up clique formation. This vulnerability is illustrated in Figure 67.8, which depicts a hypothetical scenario of how two cliques could be formed. If the skip links around nodes A and E have failed and these two nodes are powered on later than the other nodes, disjoint cliques can form on each side (because no messages can get past the locations of nodes A and E until they power up). Once either of these nodes power up, a communication path is created between the two cliques and clique resolution is required. As described in [5], existing clique resolution protocols (which are based on the properties of a globally perceived broadcast medium) cannot resolve such clique scenarios in all cases. This is because the disjoint phases of the nonaligned TDMA schedules can cause *collisions* that prevent nodes from one TDMA timeline from hearing messages sent by nodes on another TDMA timeline. For example, in Figure 67.8 the power-up timing of nodes A and E are such that node A joins the right-hand clique (H, G, and F) and node E joins the left-hand clique (B, C, and D). The nodes A, B, E, and F are called frontier nodes because they are the *outermost* nodes on their respective cliques. Frontier nodes belong to one clique, but they may hear messages from another clique. These messages are rejected because they do not meet the timeline constraints of the clique to which the frontier node belongs.

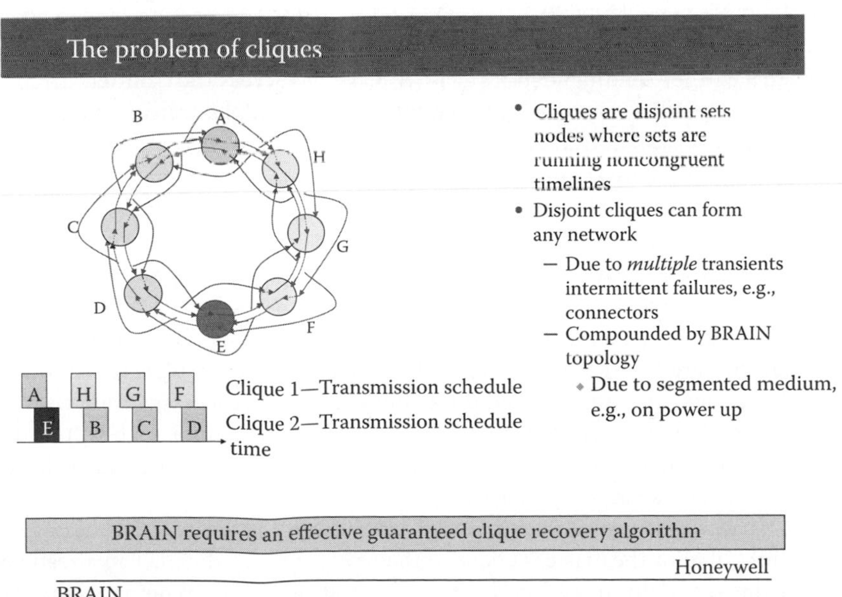

FIGURE 67.8 A BRAIN clique scenario.

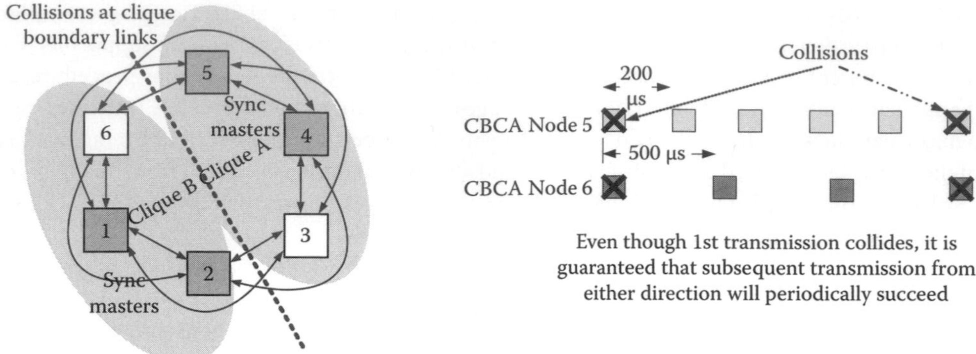

FIGURE 67.9 CBCA link arbitration resolves asynchronous TMDA clique boundaries.

To address these issues, the BRAIN approaches the start-up problem differently from previous clique resolution protocols. In place of a monolithic protocol that targets both synchronization and clique resolution, the BRAIN splits the protocol action into two layers. The BRAIN introduces a novel foundation layer in the form of a constructive connectivity-building and clique-aggregation (CBCA) algorithm that is used to build network connectivity and clique allegiance between adjacent frontier nodes. As such, the CBCA algorithm is conceived to run underneath the normal TMDA protocol activity. This means that it is unable to use, and does not need to use, the TDMA-coordinated activity to implement its signaling and/or link arbitration. For this reason, the foundation of the CBCA algorithm is a simple asynchronous medium arbitration mechanism. For half-duplex rings, nodes that share an adjacent link are configured with relatively prime periods of transmission, as illustrated in Figure 67.9. By using relatively prime periods, for example, 200 and 500 µs as shown, it can be assured that should a collision occur during one transmission, the next transmission will be collision-free (in the fault-free case). Therefore, periodically successful transmissions will occur from each direction. For full-duplex rings, the *collisions* occur within frontier nodes, which receive messages that are out of sync with its TDMA timeline. This form of collision prevents these out-of-sync messages from being forwarded into the clique.

Similar to SINT messages, the CBCA messages leverage SCP protection, where adjacent nodes collaborate to signal the CBCA messages. Each frontier node and its nearest neighbor within the same clique form a pair for sending identical CBCA messages across the frontier.

The transmission of CBCA messages is timed relative to the SINT transmission schedule of the initiating clique. The decision to transmit CBCA messages is made by each node determining if it is on a frontier. The full procedure is illustrated in Figure 67.10.

In a simple start-up scenario (to illustrate the CBCA process), one master-clock SCP of nodes happens to start up before the other nodes, for example, nodes 4 and 5 in Figure 67.10. At the beginning of this scenario, these nodes are the master pair and the frontier nodes. The CBCA process is initiated from this pair, following the transmission of its SINT message. Because the nodes adjacent to the pair have not sent or relayed any messages indicating that they are in the same clique as the master pair, the master pair assumes that they are frontier nodes and initiate the coordinated CBCA transmissions, sending repetitive CBCA messages on both skip and direct links. The CBCA message comprises two fields as illustrated in Table 67.2. The TIME_TO_NEXT_SINT field is updated for each CBCA transmission, relative to the sourcing-clique's clock master's SINT timing interval, with the values decrementing as the time of the next SINT transition approaches.

When a node adjacent to the master pair powers up and sees identical CBCA messages on both its direct and skip links, it joins the master's clique. If the newly connected node had already been part of a different clique (instead of this power-up scenario), the node would have compared the priority of the sending clique (as sent in its CBCA message's MASTER-PRIORITY field) with the priority of the clique to which it was already synchronized. If the CBCA message had a higher priority, the node would join

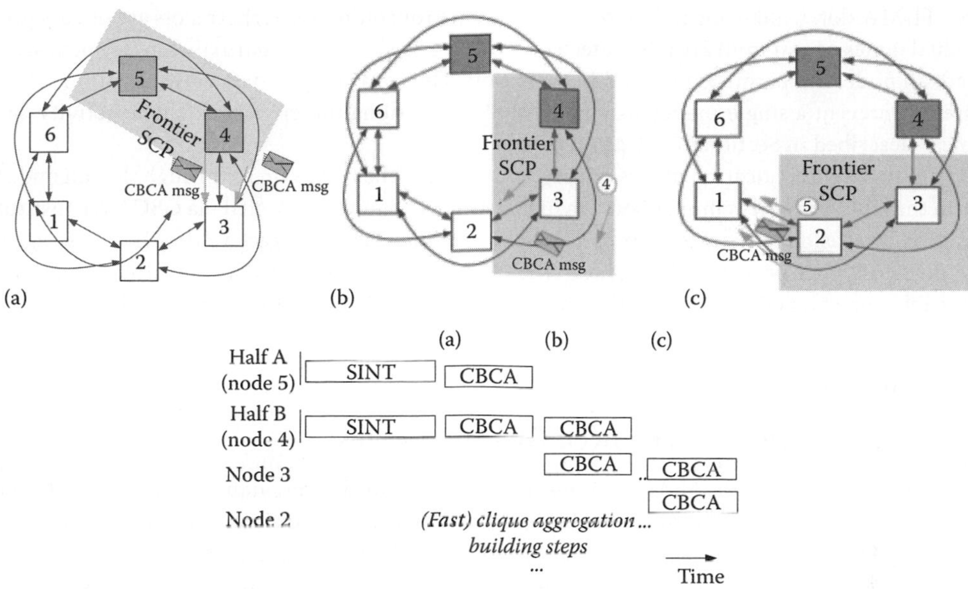

FIGURE 67.10 CBCA.

the higher-priority clique and synchronize to its schedule, that is, defecting from their previous clique/schedule. Following a node's defection or integration, it extends the self-checking frontier action with its immediate neighbor not already in the higher-priority clique. This propagates the clique status and associated priority to its neighbors, which expands the frontier out by one node. This CBCA action continues for a duration proportional to the value of the TIME TO NEXT SINT field received in the CBCA messages. All CBCA action is timed to cease before the next scheduled SINT message. During the CBCA propagation in frontier expansions, no other traffic is allowed on the newly converted links. Thus, prior to the next SINT message transmission, the media among all newly joined nodes is reserved to enable clean SINT message propagation through all of them. Note that, as the CBCA mechanism executes, each node records the status of which neighboring nodes have joined its clique. Then the nodes cease CBCA activity on those links that have indicated membership in its clique, following the next SINT transmission. Therefore, normal TDMA activity can commence between all nodes confirmed to be in the same clique, immediately following the next SINT message transmission.

67.2.2.3 Synchronous Mode Clique-Aggregation Breakthrough

The CBCA mechanism earlier is sufficient to mitigate all cliques given a BRAIN with a single node fault. However, to improve robustness, additional mechanisms have been added to resolve dual-link and dual-benign-node failure scenarios. In such scenarios, it is necessary to utilize the integrity reconstitution mechanisms of the BRAIN, to qualify integrity of the CBCA status information by using information received from the two different ring directions. To facilitate CBCA integrity reconstitution, special CBCA *breakthrough* slots are added to the normal TDMA schedule. In a breakthrough slot, each node at a frontier of the clique sends the received CBCA priority of the neighboring clique,

TABLE 67.2 CBCA Fields

Field	Description
Master_priority	The priority of clock master to which the frontier node is synchronized
Time_to_next_SINT	A countdown field that denotes time or number of remaining CBCA transmissions that may occur before the next scheduled SINT message

via this TDMA slot. Unlike normal TDMA data or synchronization slots, these slots are not assigned to individual nodes or pairs and are not protected by the Brother's Keeper Guardianship (see next section). Instead, all nodes are permitted to send in these slots. Hence, a suitable data authentication scheme is required to prevent a single maliciously faulty node from signaling erroneous clique activity. Such a scheme is described in Section 67.3.2.2.

Given suitable authentication of messages received during the CBCA breakthrough slot, all nodes can perform the CBCA priority comparison. If a higher-priority clique is detected via CBCA reconstitution, nodes yield by ceasing to execute their current schedule and prepare to receive the SINT message from the confirmed higher-priority SINT source. Therefore, by using the CBCA breakthrough mechanism, it is possible to resolve cliques even in the presence of two link faults or two benign node faults.

67.3 Fault Isolation

67.3.1 Time-Triggered Sequenced Guardian Roles

The basic high-integrity data-propagation mechanisms described in previous sections are sufficient to protect data during its transmission on the BRAIN fabric. However, additional mechanisms are required to qualify data as it enters the BRAIN, to ensure that the BRAIN's data integrity is consistent for all member nodes. These additional mechanisms can be viewed as guardian roles that cross-check and police data as it enters the BRAIN. The specific roles are selected in accordance with the TDMA schedule and are performed by the active transmitting nodes' immediate neighbors (direct links) and neighbors' neighbors (skip links). Hence, it is called Brother's Keeper Guardianship. In synchronous operation, the nodes adjacent to the currently scheduled transmitter implement guardian enforcement actions. Thus, guardianship can be pictured as moving around the ring as the TDMA communication sequence progresses, with Brother's Keeper Guardians only standing on either side of the currently transmitting node. Note that the guardian, being an independent neighboring node, ensures that guardian action is fully independent of the transmitter it is guarding, even if it is embodied in the communication controller hardware. This gives all the benefits of fully independent redundant guardian hardware, without requiring the addition of any redundant hardware components (such as central guardians or dedicated hardware components at each transmitter). Note that using the Brother's Keeper Guardian strategy, the early limitations restricting slot order and slot size for bus topology protocols such as TTP and FlexRay can be removed.

The generic guardian role implements the TDMA selection of the schedule transmitted, granting its neighbor access to initiate a transmission, during its assigned transmission slot. This mechanism prevents a node, disturbing traffic outside its schedule slot. Additional guardian roles, described later, may also be selected according to a node's relative placement to the active sender and/or self-checking senders and the data consistency requirements for the schedule slot.

67.3.1.1 Directional Integrity Exchange

The directional integrity mechanism is illustrated in Figure 67.11. It is performed by nodes adjacent to an active simplex sender to ensure that the sender sends consistent data in each direction, that is, the copies of the data sent in each direction are identical. For a TDMA slot in which a simplex node is scheduled to transmit, each node on either side of the sender receives data from the sender via their direct links and immediately relays it to the partner guardian on the other side of the sending node via the skip links that connect them. The guardian nodes also perform an integrity comparison of the data transmitted directly from the sending node with the data reflected via the other guardian. Data integrity is signaled using the BRAIN's main data integrity mechanism, that is, by setting the integrity status trailing the transmission. This guardian action ensures that the transmitting node sends consistently in both directions. If the data sent in the two different directions are not identical, both copies of the transmitted data are marked with integrity loss.

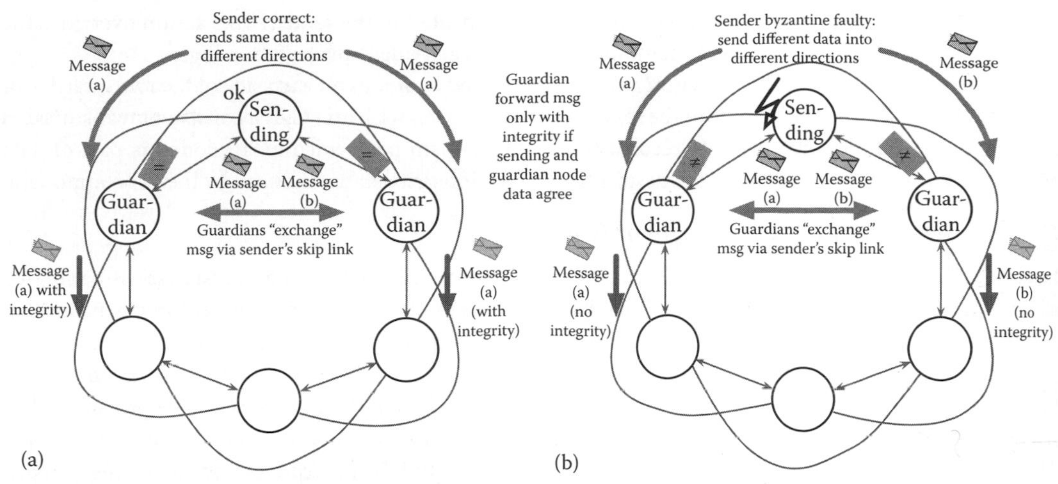

FIGURE 67.11 Consistent data guardian exchange. (a) Sender correct and (b) sender Byzantine faulty.

67.3.1.2 Skip Guardian Link Forwarding

The directional integrity mechanism described is sufficient to guarantee that data broadcast in both directions around the ring is the same. However, it has the disadvantage that it requires a guardian influence on the data transmissions in both directions of the BRAIN. Therefore, under guardian failure, the directional integrity mechanism may impact communication availability (but integrity *is* assured). For example, consider the faulty guardian (FG) in Figure 67.12. This guardian corrupts the data in both the clockwise and counterclockwise directions, that is, the data it is forwarding directly from the sender and the data it is reflecting to the adjacent guardian located on the other side of the sender. In this case, the integrity is flagged as lost in both directions. However, given that data integrity can be reconstituted, this situation can be mitigated by the nodes adjacent to the guardians. These nodes bias their selection of the data they forward such that in the event of data conflict/miscompare, the data from the skip link

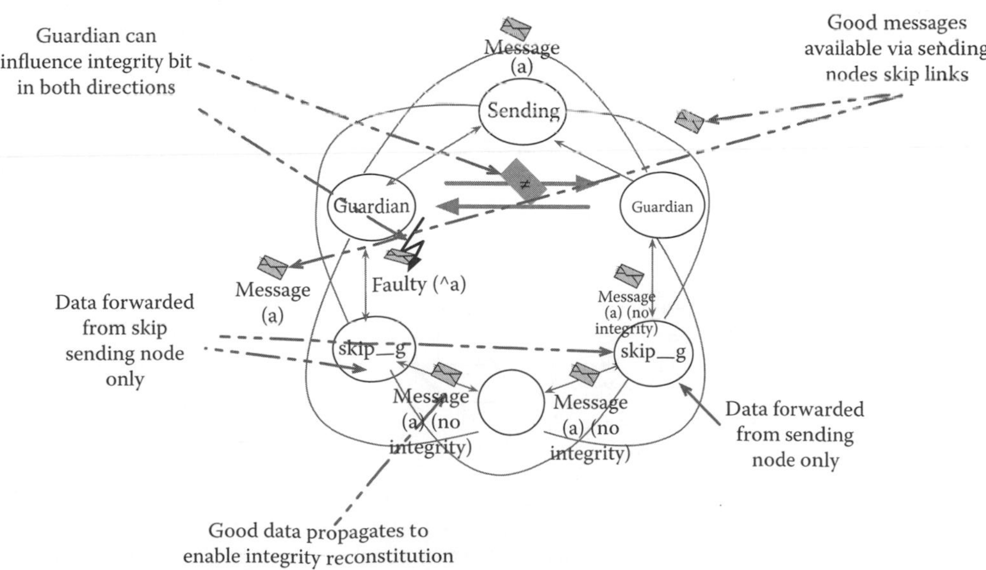

FIGURE 67.12 Skip guardian action mitigate.

are forwarded, that is, the data from the link directly connected to the sender. This action overrides the normal Availability OR selection of the standard high-integrity data-propagation logic.

As illustrated in Figure 67.12, this mitigation is performed by nodes on each side of the two guardians. Using this mechanism, the data from the sender propagate, even with an FG. If a consistent transmission is made in both directions, the data reconstitution mechanism performed by all nodes as part of data reception will accept the data as good, despite the loss-of-integrity status induced by the FG component.

67.3.1.3 Self-Checking Pair Neighbor Guardian

For the BRAIN's SCP assumptions to be valid, the output from each half of the pair must be cross-compared as part of any data validity evaluation. Therefore, the BRAIN needs to implement enforcement mechanisms to ensure that such an evaluation is always done. The required roles of SCP neighbors are illustrated in Figure 67.13. To ensure that data from each half of the pair can be utilized by the integrity reconstitution logic, the data flow from the pair is restricted such that the nodes adjacent to the SCP only propagate data from the direct links, with the data received from the skip links used solely for data integrity comparison. This data flow restriction exists even in the event of the direct link being *dead*, that is, indicating no data. As illustrated in Figure 67.13, this restriction prevents integrity reconstitution from evaluating the data from a single node (which would be a violation of the integrity separation guarantees for SCPs).

67.3.2 Asynchronous Guardian Roles

67.3.2.1 Start-Up Enforcement

Prior to entering synchronous operation, the time-triggered guardian roles earlier are not applicable. Hence, the BRAIN deploys alternative enforcement actions to provide the protection during protocol start-up. Following power-up, the start-up guardian action is performed by all nodes. Before protocol start-up, the SINT message is the only message expected to flow around the BRAIN. As outlined earlier, the SINT message is a very small message, comprising the frame ID and clock master priority (as well as information to avoid a single node imitating a SCP, which is discussed later. Therefore, the guardian is only required to limit the size and rate of SINT message activity and ensure that no erroneous nodes send messages larger than the SINT message. The behavior is illustrated in Figure 67.14.

Initially, a node performs a race-based arbitration on skip and direct links in each direction. Once a link (i.e., direct or skip) is identified as the winner, the guardian enables data propagation for a window

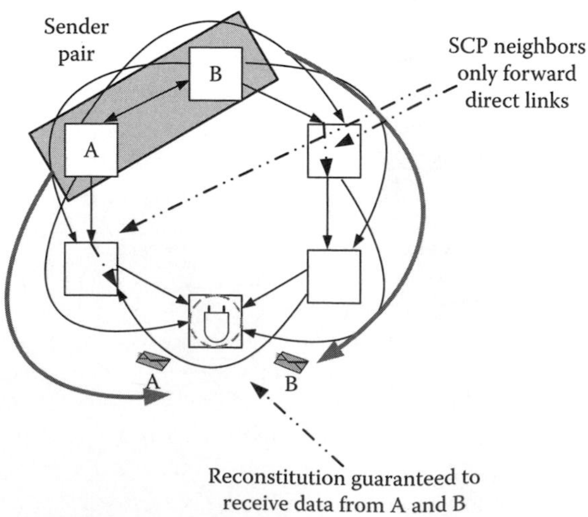

FIGURE 67.13 SCP guardian action.

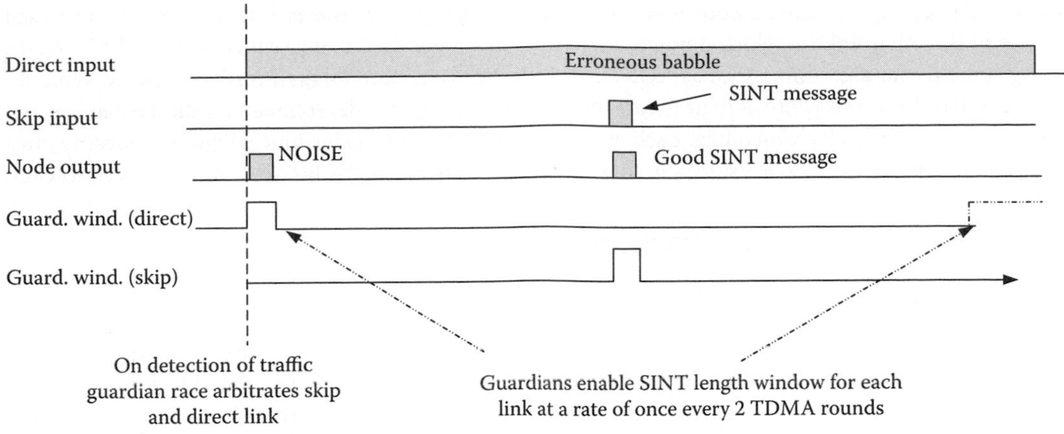

FIGURE 67.14 Start-up guardian action.

slightly larger than the SINT message. Once the SINT message window has expired, the guardian blocks the link, truncating any activity that is still in process. The winning link is then blocked for a period that is larger than two rounds of the minimum TDMA synchronization cycle. This blocking action can be viewed as a toggle-based enforcement where, during the blocking period, the guardian allows only the nonwinning link to take part in arbitration. Once the blocking time-out expires, the winning link is reenabled to take part in arbitration action.

67.3.2.2 Source Authentication

In the asynchronous start-up mode and/or in time-triggered slots where guardian action is disabled (e.g., the CBCA breakthrough slots of Section 67.2.2.3), an additional source authentication mechanism is required to prevent a node from masquerading as two independent sources. Without authentication, a faulty node would be able to unduly influence the system and potentially disrupt protocol operation. Many authentication schemes are possible. The key requirement for all schemes is to ensure that data from a single node are not used for both inputs in cross-checking. The initial authentication scheme of the BRAIN utilized a simple *port stamping* scheme. This mechanism required frontier nodes to write their ID into the message, together with the port of entry (i.e., skip or direct), upon reception. The topology of the BRAIN ensures that the port stamps are independently applied from each direction. Since all data used for integrity reconstitution are a function of two directions, this simple port stamping scheme is single-fault tolerant, requiring two nodes to agree before the data are validated. Alternative authentication algorithms in the BRAIN can be based on a hop count [1]. It should be noted that in all cases, only the port stamp or hop count data are used to implement the authentication, and no application data entering the BRAIN have influence on the authentication decision.

67.3.2.3 Additional Guardian Fault Containment Behavior

67.3.2.3.1 Short-Circuit Detection

One vulnerability flow through guardian mechanisms, such as those used by the BRAIN, is undetected short circuits between links or within nodes, which may render the node's respective guardian ineffective (e.g., by the short circuit bypassing the blocking or flag-setting actions of a guardian). Although such failures may be mitigated by the active fault scrubbing of the guardian hardware, the overhead and system complications of in situ guardian testing is often nontrivial, potentially introducing significant software and system complications. The period of guardian scrubbing can also become a limit of system reliability. Therefore, the BRAIN implements continuous scrubbing. This scrubbing uses dissimilar link encoding, which is similar to the bus encoding of SAFEbus [6]. Each link that might short to another link in a way that could bypass a guardian function is given a trivial *cryptographic key* that is different from the keys used by any of the other links to which it might short. These keys are used to *encrypt* each link's data streams using some trivial algorithms such as XOR. If the physical layer technology uses block-coded symbols, such as 8B/10B, the *encryption* is done before the block encoding, in order to maintain the desired spectral characteristics of the block encoding. Each encrypted signal that meets in some geographic locality must have a different key. Six keys are sufficient for most physical layouts of the BRAIN. This means a 3-bit key.

67.4 Diagnostic and Agreement Services

67.4.1 Host Task Set Agreement

The message-stepping SCP configuration of the BRAIN is an efficient scheme for software replication and output comparison. However, for such a scheme to be effective, the replicated task sets on both halves of the SCP processor need to be replica determinate. It is also important to prevent a fault from entering the pair and upsetting the pair's consistency (agreement of state between the two halves of the pair). Therefore, the BRAIN utilizes the direct link that connects configured self-checking nodes to implement input state agreement. Using this private link, both halves of the pair agree on the data they received. Given the inconsistent omission fault model of the BRAIN data exchange, this agreement mechanism is similar to the SAFEbus syndrome exchange, where the nodes simply agree on the received status. A full value exchange is not required. This mechanism ensures that any data not received consistently by both nodes are dropped by both nodes. Hence, data entering inconsistently are prevented from causing state divergence within the pair.

67.5 Validation and Verification Efforts

The braided-ring topology of the BRAIN is arguably optimal with respect to hardware and bandwidth usage for fault-tolerant systems. The reliability of braided-ring topologies has been investigated in the literature and research on this related to the BRAIN is described in [1]. Summarizing, the additional links provide significant additional reliability, especially for long mission times. In [4], the reliability of the braided ring is compared to dual-star topologies, evaluated in the context of extended dispatch scenarios very similar to the time-limited dispatch in engine control systems as described in ARP5107 [7]. The results are that braided rings significantly outperform dual stars in typical architectures and deployment scenarios with respect to reliability while using less hardware.

Certain aspects of the BRAIN, such as the BRAIN high-integrity forwarding logic, have been model checked using SRI's Symbolic Analysis Laboratory (SAL).

67.6 Example Configurations, Implementations, and Deployment Considerations

The BRAIN is still an evolving concept. It has a wide variety of possible implementations. It can use almost any physical layer technology to create its links. It can be half duplex or full duplex. The links within a single BRAIN can use multiple types of physical layer technologies, as long as the net speed and the type of duplexing are the same.

As discussed earlier, high-speed variants of the BRAIN are also under development. Targeting higher-speed protocols, such as 100 Mbs or Gbs Ethernet, these variants incorporate both store-and-forward and cut-through messaging. They are also targeted to tolerate multifault scenarios, that is, to be tolerant to multiple simultaneous active and benign node failures. Although many of the core protocol mechanisms of these BRAIN variants are the same as those discussed earlier, these new variants of the BRAIN replace the sequenced guardian roles of Section 67.3.1 with a more aggressive source authentication scheme. Such a strategy enables greater availability to be leveraged from the same BRAIN connectivity and amount of hardware components while simultaneously reducing protocol complexity. Targeting full-duplex connectivity, these variants also remove the CBCA algorithm, since the possibility of store-and-forward messaging enables messages to propagate without regard for TDMA phase alignment.

A potential physical-layer BRAIN deployment option is an optimization that is possible with respect to BRAIN media routing. Often, the conceptual diagrams of the BRAIN lead readers to consider that it introduces a large cabling overhead. However, the cabling of the BRAIN does *not* have to be routed separately, and the BRAIN does not suffer any loss of dependability when cable layout is optimized. For example, consider the physical wiring of Figure 67.15. Here the physical routing of skip links passes through the same shield as the direct links and the skip connections are routed through the neighboring nodes. This results in simple physical ring topology with respect to wiring and connections. It should also be noted that the number of connections for a BRAIN is equivalent to a dual-star despite the BRAIN's greater dependability.

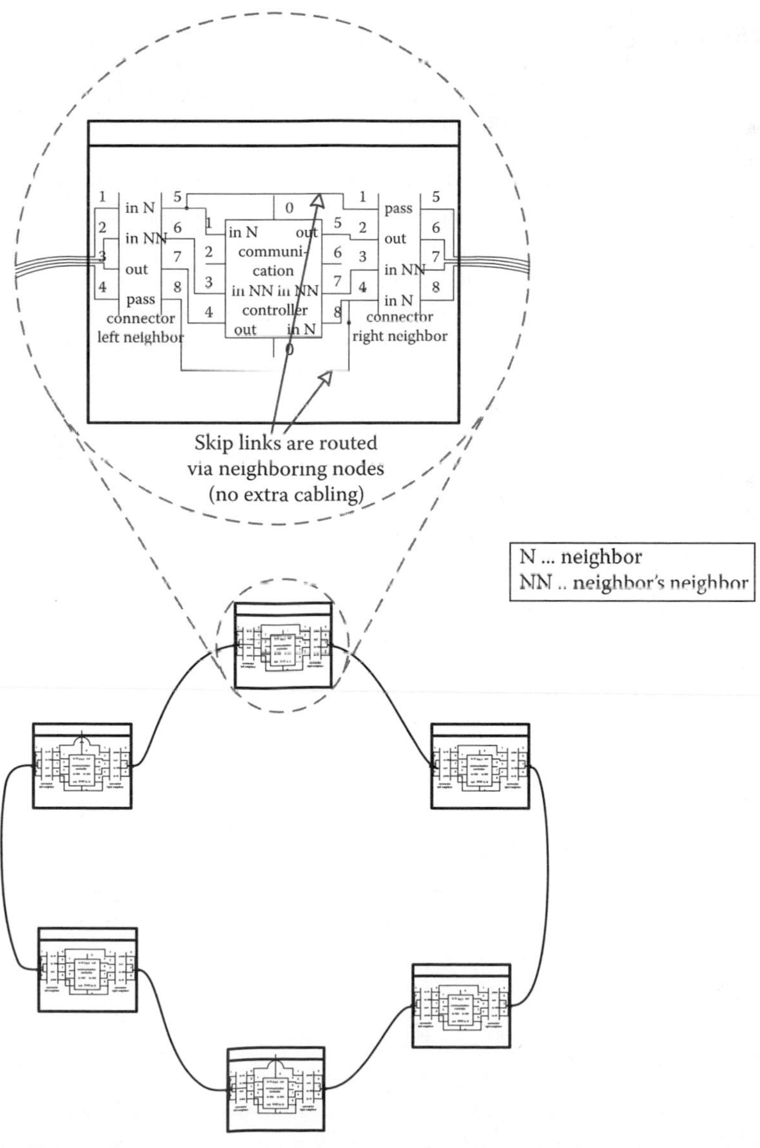

FIGURE 67.15 Cabling of BRAIN routed within neighboring nodes.

The BRAIN has yet to be deployed in series production or completely prototyped within a full-up system. However, the efficiencies offered by the BRAIN approach is beneficial to high-integrity control architectures. In addition, as Ethernet and time-triggered Ethernets become more pervasive, hybrid approaches of mixed BRAIN and COTS technology present advantages, for example, by allowing a high-dependability by-wire infrastructure to coexist with noncritical Ethernet. This would allow generic Ethernet to be used for system maintenance and loading, allowing for the use of COTS and standard laptop- or desktop-derived test equipment while still providing very strong assurance and fault tolerance guarantees for the by-wire backbones. In addition, the BRAIN's ability to support message-stepping SCPs, together with its guaranteed inconsistent omission fault handling for data exchange, also allows for significant reduction of software complexity and associated overheads. With its various flavors and forms, a version of the BRAIN can be chosen to fit a wide variety of cost and performance targets while providing a guaranteed level of fault tolerance with the least amount of hardware possible.

References

1. B. Hall, K.R. Driscoll, M.Paulitsch, and S. Dajani-Brown. Ringing out fault tolerance. A new ring network for superior low-cost dependability. *International Conference on Dependable Systems and Networks*, pp. 928–307, 2005.
2. M. Paulitsch, J. Morris, B. Hall, K.R. Driscoll, E. Latronico, and P. Koopman. Coverage and the use of cyclic redundancy codes in ultra-dependable systems. *International Conference on Dependable Systems and Networks*, Yokohama, Japan, pp. 346–355, 2005.
3. B. Hall, M. Paulitsch, and K.R. Driscoll. FlexRay BRAIN fusion: A FlexRay-based braided ring availability integrity network. *SAE World Congress, Paper No 2007-01-1492*, 2007.
4. M. Paulitsch and B. Hall. Insights into the sensitivity of the BRAIN (braided ring availability integrity network)-on platform robustness in extended operation. *International Conference on Dependable Systems and Networks*, pp. 154–163, 2007.
5. W. Steiner, M. Paulitsch, and H. Kopetz. The TTA's approach to resilience after transient upsets. *Real-Time Systems*, 32(3):213–233, 2006.
6. K.R. Driscoll. Apparatus and method for fault detection on redundant signal lines via encryption. Patent U.S. 5307409, Honeywell, April 26th 1994.
7. SAE. ARP 5107 (aerospace recommended practice), guidelines for time-limited-dispatch analysis for electronic engine control systems. Technical Report Rev. B, Society of Automotive Engineers, November 2006.

Index

A

Abstract communication service interface (ACSI)
services, 63-13–63-15
Active managing node (AMN), 19-7
Actuator/sensor interface, *see* AS-Interface (AS-i)
Address Conflict Detection (ACD) mechanism, 9-52
Advanced encryption standard (AES), 60-5
Advanced metering infrastructure (AMI), 60-2
Aeronautical Radio INCorporated (ARINC) standard
full-duplex link, 49-2–49-3
function, 49-2
line-replaceable modules, 49-2
mono-emitter broadcast bus, 49-2
virtual link, 49-3–49-4
AES, *see* Advanced encryption standard (AES)
AES128 CCM* algorithm, 31-15
AFDX, *see* Avionics Full Duplex Switched
Ethernet (AFDX)
Aircraft Information Management System (AIMS)
applications, 48-22
cabinet module and function allocation, 48-22
CPM, 48-22–48-23
design requirements, 48-1–48-2
scheduling, 48-23
space and time partitioning, 48-1, 48-21
VIA, 48-25
American Standard Code for Information Interchange
(ASCII), 10-16, 10-18–10-20
AMI, *see* Advanced metering infrastructure (AMI)
AMM system, *see* Automatic metering management
(AMM) system
AMR, *see* Automatic meter reading (AMR)
ANSI C derivative language, 56-5–56-6
ANSI/CEA-709 protocol
addressing elements, 56-9
addressing schemes, 56-10
frame layout, 56-8–56-9
OSI reference model, 56-7–56-8
TPDU, 56-10–56-11
ANSI/CEA-709 standard, *see* Control networks (CNs)
ANSI/CEA-852 standard, *see* Control networks (CNs)
ANSI/ISA-88.01-1995 hierarchical model, 3-14
Application protocol data units (APDUs), 10-2, 60-16
code field, 10-6–10-7
data field, 10-7
structure, 10-5–10-6
unit ID field, 10-6
Application-specific integrated circuit (ASIC), 11-10,
11-27, 13-2, 13-31–13-33, 27-2
ARINC 419 standard, 45-2
ARINC 429 standard
Airbus A-310, 45-15
bit-oriented communications protocol
bit rate and word timing, 45-14
file transfer data words, 45-13
link data unit (LDU), 45-13–45-14
physical, link, network, and transport
layers, 45-13
protocol words, 45-15
system address labels, 45-14
word count, 45-14
word type, 45-15
Boeing B-757 and B-767 airplanes, 45-15
controllers, 45-15–45-16
equipment interconnection, 45-3
fault isolation, 45-5
fault tolerance, 45-4–45-5
file data transfer, 45-12–45-13
history, 45-2–45-3
impedance levels, 45-4
logic-related elements, 45-5–45-6
longevity, 45-16
message and word formatting
data standards, 45-10–45-11
information element, 45-6
information flow direction, 45-6
information identifier, 45-6
source/destination identifier, 45-6–45-7
SSM (*see* Sign/status matrix (SSM))
modulation, 45-3
one-way data flow, 45-2
timing-related elements
bit rate, 45-11
clocking method, 45-11
information rates, 45-11

timing tolerances, 45-12
word synchronization, 45-11
voltage levels, 45-3–45-4
ARINC 453 standard, 45-16
ARINC 629 standard
Boeing 777
avionics overview, 46-18–46-19
bus cable assembly features, 46-18
CMC features, 46-17–46-18
SIM features, 46-17
stub cable assembly features, 46-17
system interconnects, 46-16–46-17
terminal controller features, 46-16–46-17
data link layer
message format, 46-4–46-5
non-OSI data transfer, 46-6–46-7
OSI-compatible format, 46-5–46-6
medium access layer (*see* Medium access layer)
physical layer
current mode bus, 46-1, 46-3–46-4
fiber optic mode bus, 46-1, 46-3–46-4
LRU, 46-1
requirements, 46-2–46-3
signaling, 46-4
voltage mode bus, 46-1, 46-3
robustness and fault tolerance
bus protocol monitoring, 46-16
protocol parameter monitoring, 46-16
receive data monitoring, 46-15
transmit data monitoring, 46-15
ASAN, *see* Automotive serial controller area network (ASAN)
ASCB, *see* Avionics Standard Communications Bus (ASCB)
ASCII, *see* American Standard Code for Information Interchange (ASCII)
ASIC, *see* Application-specific integrated circuit (ASIC)
AS-Interface (AS-i)
access procedure, 6-8–6-9
addressing device, 6-30
addressing interfaces, 6-29–6-30
automatic addressing, 6-30–6-31
balancing, 6-3
cable, 6-4–6-5
combined transaction types (*see* Combined transaction types)
data-decoupling network, 6-3
data safety, 6-12–6-14
device profiles (*see* Slave profiles)
electromechanics, 6-8–6-9
external AS-i, 6-32
functional safety
code table, 6-53
guidelines, 6-47
safe inputs, 6-52–6-53
safe outputs, 6-53
safety at work, 6-47–6-48
safety monitor, 6-49–6-51
transmission principle, 6-48–6-50
hardware components, 6-14–6-15
implementation, 6-1
integrated AS-i, 6-31–6-32
interoperability and certification, 6-53–6-54
master, 6-15
addressing call, 6-40
analog part, 6-42–6-44
broadcast, 6-41–6-42
clear operating address, 6-40–6-41
coupler, 6-44–6-47
data call, 6-40
function sequence, 6-36–6-38
higher-level control system, 6-36
logical structure, 6-38
parameter call, 6-40
PROFIBUS/AS-i gateway, 6-44–6-45
read ID code, 6-40
read I/O configuration, 6-39
read status, 6-41
reset slave, 6-40
supported functions, 6-44–6-45
message, 6-9–6-11
modulation procedure, 6-6–6-7
Power24 technology, 6-4
power supply, 6-3
requirements, 6-2
safety, 6-3
sequence control, 6-41–6-42
technical data, 6-2
transmission layer, 6-16–6-17
transmission phases, 6-17–6-19
transmission system, 6-15
Audio/video bridging (AVB), *see* Time-sensitive networking (TSN)
Audio/video bridging transmission protocol (AVBTP), 20-13
Automatic metering management (AMM) system, 60-5–60-6
Automatic meter reading (AMR)
ANSI C12 Protocol Suite, 60-6–60-7
BACnet, 60-24
DLMS/COSEM (*see* Device Language Message Specification/Companion Specification for Energy Metering (DLMS/COSEM))
DPWS, 60-25
EN 13757/M-Bus
application layer, 60-12–60-13
link layer, 60-12
OSI three-layer structure, 60-11–60-12
physical layer, 60-11–60-12
Homeplug Command and Control, 60-25

IEC 61334-5 PLC
CENELEC and FCC regulation, 60-4
layers, 60-3
narrowband deployment, 60-3
IEC 61968-9
Common Information Model, 60-18
EndDeviceEvent, 60-19–60-20
message definition, 60-19–60-20
reference model, 60-19
IEC 62056, 60-13
IEC 62056-21/IEC 61107, 60-7
IEC 62056-31 Euridis, 60-20–60-21
IEEE P1902.1
G3-PLC, 60-5–60-6
goal, 60-4
PRIME layers, 60-5–60-6
KNX
application and interworking models, 60-22
configuration and management, 60-23
logical topology, 60-22–60-23
physical layers, 60-23
specification, 60-22
LonWorks/LonTalk, 60-23–60-24
OSI layer, 60-2–60-3
SITRED, 60-7–60-9
6LoWPAN, 60-25
smart grid system architecture
central system, 60-2
data concentrator, 60-2
electricity meter and communication hub, 60-1
end-customer device, 60-2
HHUs, 60-2
interfaces, 60-27
multiutility meter, 60-2
standards, 60-26–60-27
SML, 60-10–60-11
Telegestore, 60-7–60-9
ZigBee Smart Energy Profile, 60-25
Automaton model
global system, 49-15
output port, 49-14–49-15
timed automata, 49-13–49-14
VL automaton, 49-14
worst-case end-to-end delay, 49-7, 49-15–49-16
1394 automotive network, 50-16–50-17
Automotive open system architecture (AUTOSAR)
application software
communication scopes, 51-14
connectors, 51-13
ports and interfaces, 51-12–51-13
runnable entities (REs), 51-13–51-14
software components, 51-11–51-12
communication services, 50-21–50-23
communication substack
communication manager (ComM), 51-21
debugging, 51-21
diagnostic communication manager, 51-21
driver module, 51-18
Ethernet network, 51-17
heavy-duty vehicles, 51-17–51-18
interface, 51-18–51-19
J1939 request manager, 51-21
network management, 51-19–51-20
PDU multiplexer, 51-20
service discovery, 51-21
signal-based communication, 51-20
state manager module, 51-20
structure, 51-16–51-17
transceiver driver module, 51-18
transport protocol, 51-19
XCP module, 51-20
E2E communication protection library, 50-24
input/output substack, 51-23–51-24
library, 51-29–51-30
memory substack, 51-22–51-23
reference model, 50-21
software layered architecture, 51-14–51-15
system services substack
BswM, 51-26
BSW scheduler, 51-25
core test, 51-29
crypto service manager, 51-29
development error tracer, 51-28
diagnostic event manager, 51-26–51-27
diagnostic log and trace, 51-28
ECU state manager, 51-26
function inhibition manager, 51-27
general-purpose timer, 51-28
MCU driver, 51-25–51-26
operating system, 51-25
structure, 51-24–51-25
synchronized time base manager, 51-28–51-29
time service, 51-28
watchdog driver, 51-27
watchdog manager, 51-27
Automotive serial controller area network (ASAN), 52-1
Avionics Full Duplex Switched Ethernet (AFDX)
ARINC standard
full-duplex link, 49-2–49-3
function, 49-2
line-replaceable modules, 49-2
mono-emitter broadcast bus, 49-2
virtual link, 49-3–49-4
automaton model
global system, 49-15
output port, 49-14–49-15
timed automata, 49-13–49-14
VL automaton, 49-14
worst-case end-to-end delay, 49-7, 49-15–49-16
end-to-end delay analysis, 49-4–49-5
fixed priority scheduling, 49-18
minimum duration constraints, 49-17–49-18
model-checking approach, 49-18–49-19

network calculus
arrival curve, 49-6–49-8
BNC, 49-8
convolution and deconvolution, 49-5
leaky bucket, 49-6
maximum delay, 49-7
optimization, 49-7, 49-9
output port, switch, 49-7–49-8
parameters, 49-7
sample configuration, 49-6–49-7
service curve, 49-6–49-8
realistic configuration
temporal analysis, 49-16
upper bound, 49-16–49-17
trajectory approach
features, 49-9–49-10
sample configuration, 49-7, 49-10–49-12
worst-case end-to-end delay, 49-7, 49-12–49-13
Avionics Standard Communications Bus (ASCB)
architecture, 44-1–44-2
communication interfaces, 44-5–44-6
configuration, 44-5
diagnostic services, 44-5
Ethernet buses, 44-1
fault isolation, 44-5
frame tick, 44-2–44-3
minor frame period, 44-2–44-3
NICs, 44-1
overview, 44-1
parameterization, 44-5
Primus Epic®
cabinet (MAU), 44-6–44-8
certified airframes, 44-6
DEOS, 44-7–44-8
four-bus diagram, 44-7, 44-9
high-level diagram, 44-7
slack stealing, 44-8
subframe-level scheduling, 44-8
synchronization, start-up, and reintegration
break sync action, 44-3
flow diagram, 44-3–44-4
mastership transition, 44-3
multimaster strike, 44-3
time master, 44-3, 44-5
time server, 44-3
validation and verification efforts, 44-6

B

Babbling idiot problem, 52-26
Backhaul networks, 14-16
BACnet, *see* Building Automation and Control Networks (BACnet)
BACnet Broadcast Management Device (BBMD), 57-12
BACnet Interoperability Building Blocks (BIBBs), 57-9, 61-19–61-20
Baseline Protection Manual, 29-2
Basic protocol (BP)
aperiodic mode, 46-7–46-8, 46-10
bus cycle start and initialization, 46-9
features, 46-15
flow chart, 46-8–46-9
periodic mode, 46-7–46-9
synchronization gap (SG), 46-8
terminal gap (TG), 46-8
transmit interval (TI), 46-8
Basic software mode manager (BswM), 51-26
Basic software (BSW) scheduler, 51-25
BASs, *see* Building automation systems (BASs)
BIBBs, *see* BACnet Interoperability Building Blocks (BIBBs)
Bidirectional CSDB communication, 47-2
Binary countdown technique, 52-12
Bit bucket, 43-16
Bit monitoring, 52-14
Bit-oriented communication protocol
bit rate and word timing, 45-14
file transfer data words, 45-13
link data unit (LDU), 45-13–45-14
physical, link, network, and transport layers, 45-13
protocol words, 45-15
system address labels, 45-14
word count, 45-14
word type, 45-15
Bit rate switch (BRS) bit, 52-31
Bit stuffing, 52-14
Bluetooth, 2-22–2-25, 34-14, 34-16, 37-5, 37-12
Boeing 777
ARINC 629
avionics overview, 46-18–46-19
bus cable assembly features, 46-18
CMC features, 46-17–46-18
SIM features, 46-17
stub cable assembly features, 46-17
system interconnects, 46-16–46-17
terminal controller features, 46-16–46-17
SAFEbus
AIMS, 48-21–48-23
avionics architecture, 48-20
avionics overview, 48-23, 48-25
design requirements, 48-1–48-2
integrated architecture, 48-21
VIA, 48-25
Braided-ring availability integrity network (BRAIN)
advantages, 67-1
architecture, 67-1–67-2
asynchronous guardian roles
short-circuit detection, 67-17–67-18
source authentication, 67-17
start-up enforcement, 67-16–67-17
CBCA
asynchronous mode, 67-12–67-13
clique scenario, 67-11

synchronous mode, 67-13–67-14
TIME_TO_NEXT_SINT, 67-12–67-13
CRCs, 67-2
definition, 67-1
error detection, 67-2
fault tolerance/hypothesis, 67-3–67-4
full-duplex configuration, 67-3
guardian hardware, 67-2–67-3
half-duplex configuration, 67-3
high-integrity data propagation
one arbitrary fault, 67-4–67-5
SCPs, 67-7–67-8
second benign failure, 67-6–67-7
self-checking data relay, 67-5–67-6
TMR, 67-8–67-9
host task set agreement, 67-18
implementations, 67-18–67-19
ring topology, 67-19
self-checking master coordination, 67-10–67-11
SINT messages, 67-9–67-10
time-triggered guardian roles
Brother's Keeper Guardianship, 67-14
directional integrity mechanism, 67-14–67-15
SCPs, 67-16
skip link forwarding, 67-15–67-16
validation and verification, 67-18
Broadcast write commands (BWR), 18-9–18-11
Buffer control word (BCW), 48-5, 48-15–48-16, 48-19–48-20
Building Automation and Control Networks (BACnet), 55-6–55-7, 60-24
backup/restore, 57-8
BBMD, 57-12
BIBBs, 57-9
certification, 57-10
character sets, 57-11
command prioritization, 57-8
device address binding, 57-6–57-7
foreign device, 57-12
interoperability, 57-2, 57-10–57-11
KNX mapping, 57-13
network media, 57-3
network numbers, 57-6
network security, 57-11
object model
analog, 57-3–57-4
application services, 57-5–57-6
binary, 57-3
32-bit identifier (ID), 57-5
device objects, 57-4–57-5
multistate, 57-3
on/off times, 57-9
OPC UA mapping, 57-13
organizations, 57-2
PICS, 57-9
proprietary objects, 57-9
slave devices, 57-12
standards, 57-2
status change counting, 57-8
telegram, 57-7–57-8
testing, 57-10
transaction handling, 57-7
vendor IDs, 57-12
web services, 57-13
WS, 55-14–55-16
Building automation systems (BASs)
automation domain, 55-1
automation pyramid model, 55-2–55-3
BACnet, 55-6–55-7
bridging networks, 55-3
DALI, 55-9
definition, 55-1
EnOcean, 55-10
home automation systems, 55-10–55-11
KNX, 55-4–55-5
LonWorks, 55-5–55-6
M-Bus, 55-10
Modbus, 55-7–55-8
SMI, 55-9
system integration
BACnet/WS, 55-14–55-16
DPWS, 55-17–55-18
facility management system, 55-12
oBIX, 55-16–55-17
OPC, 55-13–55-14
two-tier architecture, 55-3
Building energy management systems (BEMS), 58-16
Burst transmission, 47-2, 47-3, 47-5
Bus controller, 43-7
Bus interface unit (BIU), 48-2, 48-5
Bus monitor, 43-7–43-8

C

Cable impedance, 45-4
CAN, *see* Controller area network (CAN)
CAN in Automation (CiA), 52-3
CANopen
COBs, 52-33
communication protocol, 52-34
device model, 52-34
digital input devices, 52-46–52-47
digital output devices, 52-47–52-48
emergency object, 52-40–52-41
error control services, 52-42–52-43
generic I/O modules device profile, 52-45–52-46
NMT state machine, 52-41–52-42
object dictionary (OD), 52-34–52-35
PDOs
configuration, 52-38–52-39
event-driven transmission, 52-36
RPDO, 52-35–52-36
SDOs, 52-39–52-40
synchronous operations, 52-36–52-37

TPDO, 52-35–52-36
triggering modes, 52-37–52-38
predefined connection set, 52-43–52-44
profiles, 52-44–52-45
synchronization object, 52-40
Capacitive transmission, 36-49
Carrier-sense multiple access/collision avoidance (CSMA/CA), 46-7
CBCA, *see* Connectivity-building and clique-aggregation (CBCA)
CCs, *see* Conformance classes (CCs)
CDC, *see* Common data class; Cyclic data channel
Channel routing (CR), 25-9
address selection, 25-5
algorithm, 25-10
client node, 25-5
function, 25-7
mechanism, 25-11
Chinese Industrial Wireless Alliance (CIWA), 33-2–33-3, 34-2, 36-28
Chinese National Standard, 34-2
CiA, *see* CAN in Automation (CiA)
CIM, *see* Common information model (CIM); Computer-integrated manufacturing (CIM)
CIP™, *see* Common Industrial Protocol (CIP™)
Civilian Avionics System, *see* Avionics Full Duplex Switched Ethernet (AFDX)
CIWA, *see* Chinese Industrial Wireless Alliance (CIWA)
Client–server model, 1-31–1-32
Clock synchronization
clock offset error, 21-2
degrees of synchronization, 21-4
FlexRay technology, 53-7–53-8
NTP
client/server, symmetric, and broadcast variant, 21-7
clock discipline algorithm, 21-15
clock filter algorithm, 21-13
combine algorithm, 21-15
implementation model, 21-9–21-10
mitigation algorithm, 21-13–21-14
network data packet, 21-12
on-wire protocol, 21-10–21-12
server stratum number, 21-7–21-8
simple network time protocol, 21-16
time format, 21-8–21-9
time scale, 21-8–21-9
packet-based network
DCF77, 21-5–21-6
GPS, 21-6
IRIG, 21-6–21-7
principle, 21-3–21-4
propagation time correction, 21-5
PTP
asymmetry correction, 21-19–21-20
end-to-end synchronization, 21-20
master clock election, 21-17–21-18
model, 21-16–21-17
offset and skew correction, 21-18–21-19
peer-to-peer synchronization, 21-20
SAS, 21-24–21-25
synchronous Ethernet, 21-23–21-24
timestamp generation, 21-21–21-23
transparent clock, 21-20–21-21
receiver/receiver (RR) synchronization, 21-4–21-5
sender/receiver (SR) synchronization, 21-4
timestamping method, 21-5
time-triggered communication, 39-2
TTCAN
cycle time, 41-6
global time, 41-6–41-7
local time, 41-5
reference message, 41-4–41-5
TTEthernet, 42-6–42-7
TTP/C, 40-7
Clock synchronization protocol, 20-8–20-9
Clock synchronization, SAFEbus
disconnected state, 48-11
frame change message, 48-12
halted state, 48-11
initializing state, 48-11
initial sync message, 48-12, 48-14
In_Sync state, 48-11
long resync message, 48-9, 48-12–48-13
Out_of_Sync state, 48-11
resynchronization pulse timing, 48-9–48-10
short resync message, 48-9
time adjustment, 48-10
Cluster mode, 40-10–40-11
CMS, *see* Communication management service (CMS)
CoAP, *see* Constrained Application Protocol (CoAP)
COBs, *see* Communication objects (COBs)
Combined protocol (CP)
bus cycle, 46-10–46-12
features, 46-15
MAC architecture, 46-12–46-13
normal operation, 46-13–46-14
timers, 46-10–46-11
transmitter scheduler architecture, 46-12–46-13
Combined transaction types (CTT)
CTT1, 6-24–6-26
CTT2, 6-26–6-27
CTT3, 6-27
CTT4, 6-28
CTT5, 6-28–6-29
transaction, 6-23–6-24
Command transfer object (CTO) packet, 51-9
Command word (CW), 43-10
Commercial Standard Digital Bus (CSDB)
aircraft implementations, 47-7
bidirectional transmissions, 47-2
burst transmission, 47-2
capacity, 47-3

continuous repetition, 47-2
data frame structure, 47-3
error detection and correction, 47-3
high-speed bus, 47-1
integration considerations
functional integration, 47-5–47-6
logical integration, 47-5
physical integration, 47-4–47-5
software integration, 47-5
integration guidelines, 47-6
low-speed bus, 47-1
noncontinuous repetition, 47-2
physical characteristics, 47-1
testing, 47-6–47-7
unidirectional bus architecture, 47-2
user monitoring, 47-4
Common data class (CDC), 62-11
attributes characterization, 63-10–63-11
IEC 61850-7-3, 63-10
Status Value (STV), 63-11
WPP specific, 63-9
Common Industrial Protocol (CIP™)
advantages, 9-70
big-endian encoding, 10-34
CompoNet (*see* CompoNet)
conformance testing, 9-93–9-94
ControlNet (*see* ControlNet)
data throughput, 9-71
deterministic system, 9-71
DeviceNet (*see* DeviceNet)
distributed motion control, 9-3
energy application, 9-3–9-4
Energy Objects
Base Energy Application Object, 9-90–9-92
Electrical Energy Object, 9-91–9-92
Nonelectrical Energy Object, 9-90, 9-92
Power Management Object, 9-90, 9-92–9-93
EtherNet/IP (*see* EtherNet/IP)
functional safety, 9-3
IEEE 1588 clock synchronization, 9-72
jitter, 9-71
layered architecture view, 10-29
little-endian encoding, 10-34
Modbus function codes, 10-33
Modbus translation function, 10-30, 10-32
Motion, 9-74–9-75
native Modbus functionality, 10-30–10-31
Networks Library
auxiliary power distribution system, 9-5, 9-22
client, 9-5
configuration, 9-18
data management, 9-21
device profiles, 9-17–9-18
EDS, 9-18–9-20
explicit messaging, 9-6, 9-10
features, 9-5
implicit (I/O) message, 9-6, 9-10
maintenance, 9-22
messaging protocol, 9-8–9-9
object library (*see* Object library)
object modeling, 9-6–9-8
producer/consumer model, 9-5–9-6
routing, 9-20
server, 9-5
service codes, 9-7–9-8
SIG, 9-22
volumes, 9-4–9-5
non-CIP networks, 9-87–9-90
originators, 10-30
port types, 10-34
reaction time, 9-71
real-time behavior, 9-71
safety
application layer functionality, 9-76
configuration implementation, 9-84
connection establishment, 9-82, 9-84
CRC, 9-80
data routers, 9-77
device configuration, 9-81–9-83
device objects, 9-84–9-85
diverse measures, 9-80
error detection measures, 9-78–9-79
German Safety Bus committee, 9-76
implementation, 9-77–9-78
message packet sections, 9-81
Multicast Connections, 9-80–9-83
network routing, 9-77
PID, 9-80
Safety Supervisor Object, 9-84
Safety Validator Object, 9-84
Sercos, 9-85–9-87
time expectation, 9-78–9-79
Unicast Connections, 9-80–9-81
sequence diagram of communication, 10-30, 10-32
stand-alone *router* device, 10-30–10-31
sync
applications, 9-74
communication, 9-72, 9-74
message prioritization, 9-72–9-73
Time Sync Object, 9-72
synchronization, 9-3, 9-71
Common Information Model (CIM)
component interface specification, 61-5–61-6
data model
IEC 61968-11, 61-4–61-5
IEC 61970-301, 61-4–61-5
IEC 62325-301, 61-4, 61-6
IEC, 61-4
profiles, 61-6
XML-based message exchange, 61-6–61-7
Communication management service (CMS), 28-5
Communication objects (COBs), 52-33
Communication profile (CP), 17-2, 17-4–17-5
Communication profile families (CPF), 17-2–17-5

Communication stack
  communication layers, 8-15–8-16
  wired protocol
    application layer, 8-20
    data link layer (DLL), 8-17–8-19
    multimaster setup, 8-15–8-16
    physical layer, 8-16–8-17
    transport layer, 8-19
  wireless protocol
    application layer, 8-21
    data link layer (DLL), 8-20
    network layer, 8-20
    physical layers, 8-20
    transport layer, 8-21
Communication standards, 36-10, 36-20–36-21
  6LoWPAN, 36-19–36-20, 36-22
  ISA100
    architecture, 36-26–36-27
    functionality, 36-27
    security, 36-28
    standards landscape, 36-21
  proprietary solutions, 36-22
  WIA-PA
    architecture, 36-28–36-29
    functionality, 36-29
    security, 36-29
    standards landscape, 36-22
  WirelessHART
    architecture, 36-24–36-25
    functionality, 36-25
    security, 36-25–36-26
    standards landscape, 36-20–36-21
  wireless process control application, 36-18–36-19
  wireless vibration monitoring application, 36-16
  ZigBee
    application profiles, 36-24
    architecture, 36-22–36-23
    battery operation, 36-30
    frequency diversity, 36-30
    functionality, 36-22–36-23
    path diversity, 36-30
    security, 36-23, 36-30
    standards landscape, 36-20
    ZigBee IP, 36-24
    ZigBee PRO, 36-24
Communication subsystem
  Bluetooth interference, 34-12, 34-14, 34-16
  design
    antenna diversity and switching, 37-7–37-8
    frequency hopping, 37-7
    IEEE 802.15.1 physical layer, 37-5
    medium access and retransmission, 37-6–37-7
    multicell operation, 37-9
  industrial electromagnetic interference, 37-13–37-14
  performance measurements, 37-13
  wireless coexistence
    frequency usage, 37-11
    IEEE802.11b, 37-12
    ISM Band radio frequencies, 37-9
    issue, 37-10
    location separation, 37-10
    performance, 37-11
    robustness, 37-11
    spectral efficiency, 37-11
    stability, 37-11–37-12
    time separation, 37-11
  WLAN interference, 37-12, 37-14–37-15, 37-17
  WSAN, 37-14
Communication technology
  wireless process control application, 36-19
  wireless vibration monitoring application, 36-16–36-17
CompoNet, 9-3
  adaptation, 9-62
  Bit Slave/Word Slave, 9-68
  Byte Size Differences, 9-68–9-69
  CompoNet Link Object, 9-65
  CompoNet Repeater Object, 9-65
  configuration, 9-69
  development factors, 9-69
  explicit messaging, 9-67
  features, 9-62–9-63
  frame structure, 9-64–9-65
  indicators, 9-65
  I/O connection establishment, 9-68
  I/O messaging, 9-68
  network access, 9-66–9-67
  network schedule, 9-65–9-66
  physical layer, 9-63–9-64
  protocol adaptation, 9-65
  switches, 9-65
Composite state transfer (CST) protocol, 17-8
Computer-integrated manufacturing (CIM) system, 1-4
  communication method, 66-30–66-32
  data format, 66-30
  equipment-driven scenario, 66-28–66-29
  GUI interface, 66-30, 66-32–66-34
  SECS-II message, 66-29–66-31
  software architecture, 66-28
  software integration method, 66-34–66-36
Conformance blocks, 61-8
Conformance classes (CCs)
  CC-A
    acyclic data exchange, 12-13–12-15
    cyclic data exchange, 12-13–12-14
    device/network diagnostics, 12-13, 12-15
  CC-B
    device replacement procedure, 12-16–12-17
    neighborhood detection, 12-15–12-16
    network diagnostics integration, 12-17

network management protocol, 12-15
topology plan, 12-16
CC-C
mixed operation, 12-18–12-19
optimized IRT mode, 12-19–12-20
synchronized communication, 12-18–12-19
update rate, 12-18
network installation, 12-25–12-26
types, 12-4
Connectivity-building and clique-aggregation (CBCA)
asynchronous mode, 67-12–67-13
clique scenario, 67-11
synchronous mode, 67-13–67-14
TIME_TO_NEXT_SINT, 67-12–67-13
Constrained Application Protocol (CoAP), 3-13, 59-10
enhancements, 38-10–38-11
vs. HTTP, 38-8–38-9
proxy and caching, 38-9–38-10
REST design principles, 38-8
Constrained RESTful environments (CoRE)
architecture, 59-10–59-11
CoAP, 59-10
message format, 59-11
Continuous repetition, 47-2
Control data tunneling
application classes, 22-9–22-10
delay jitter, 22-7–22-8
EIA-852 system configuration, 22-6–22-7
end-to-end delay, 22-7
loss probability/loss rate, 22-8
packet ordering, 22-8–22-9
performance classes, 22-9
throughput, 22-8
Controller area network (CAN), 30-13–30-14
advantages, 52-22–52-23
arbitration phase, 50-8–50-9
ASAN, 52-1
automotive applications, 52-2–52-3
bandwidth requirement, 50-9–50-10
Bosch specification, 52-2
CAN 2.0A data frame, 50-7–50-8
CAN FD, 50-11
bit stuffing rules, 52-32
features, 52-31
frame formats, 52-31–52-32
M-zone, 52-30–52-31
S-zone, 52-30–52-31
CANopen (*see* CANopen)
CiA and networked control systems, 52-3
complex architecture, 50-9–50-10
data frames, 52-9–52-10
data link layer, 50-9
DeviceNet, 2-13–2-14, 9-22
digital input devices, 52-46–52-47
disadvantage, 2-14
drawbacks and solutions
dependability, 52-26
determinism, 52-25–52-26
performance, 52-23–52-25
error and overload frames, 52-11
error detection, 50-9
fault confinement, 52-14
flexible data rate (FD)
bit stuffing rules, 52-32
features, 52-31
frame formats, 52-31–52-32
M-zone, 52-30–52-31
S-zone, 52-30–52-31
industrial requirement, 50-9–50-10
J1850, 50-11
LLC
BasicCAN implementation, 52-16
FullCAN implementation, 52-16
information exchanges, 52-15
producer/consumer model, 52-15–52-16
MAC
arbitration phase, 52-12–52-13
CSMA networks, 52-11
error management, 52-13–52-14
non-return-to-zero, 50-7–50-8
physical layer, 50-9
bit stuffing technique, 52-6–52-7
DPLL (*see* Digital phase locked loop (DPLL))
network topology and cabling, 52-5–52-6
protocol stack, 52-4
remote frames, 52-10–52-11
schedulability analysis
blocking time and interference, 52-20–52-21
message schedulability, 52-18–52-19
message set (MS), 52-17–52-18
recurrence relation, 52-21
usage, 52-21–52-22
WCRT, 52-19
temporal constraints, 50-9–50-10
traffic shaping strategy, 50-9
transport layer, 51-4–51-6
TTCAN
features, 52-27
implementation, 52-29–52-30
system matrix, 52-28–52-29
time-triggered communication, 52-27–52-28
VAN, 50-11
ControlNet, 9-3
configuration, 9-46–9-47
configuration tools, 9-47
conformance testing, 9-47
connection establishment, 9-46
ControlNet Object, 9-43
development factors, 9-47–9-48
device classes, 9-46
device profiles, 9-46
explicit messaging, 9-45–9-46

features, 9-40
frame structure, 9-41–9-43
I/O messaging, 9-46
ISO standard, 9-39
Keeper Object, 9-43
monitoring tools, 9-47
network access, 9-44–9-45
network start-up, 9-45
network status indicators, 9-43
physical layer, 9-40–9-41
physical layer tools, 9-47
protocol adaptation, 9-43
Scheduling Object, 9-43
switches, 9-43
time-based message scheduling mechanism, 9-39
Control networks (CNs)
address types, 25-3
application, 25-15–25-17
channel members, 25-3
data communication
data packet routing, 25-9–25-11
encapsulation, 25-6–25-7
functional blocks, 25-5–25-6
packet aggregation, 25-8
packet sequencing, 25-7–25-8
stale packet detection, 25-8–25-9
functional domain, 25-4
low throughput, 25-3
management function, 25-11–25-13
packet loss and latency, 25-3–25-4
security
calculation principle, 25-13–25-14
NAT configuration, 25-14–25-15
replay attack, 25-14
VPN, 25-15
small packet size, 25-3
software layer, 25-1–25-2
system components, 25-4–25-5
tunneling approach, 25-2–25-3
CoRE, *see* Constrained RESTful environments (CoRE)
Core processor modules (CPMs), 48-22–48-23
CP, *see* Combined protocol (CP)
CPF, *see* Communication profile families (CPF)
CR, *see* Channel routing (CR)
CRC, *see* Cyclic redundancy check; Cyclic redundancy code
Cryptography, 29-3–29-4
CSDB, *see* Commercial Standard Digital Bus (CSDB)
CST protocol, *see* Composite state transfer (CST) protocol
CTT, *see* Combined transaction types (CTT)
Cyclic and acyclic data communication
clock control, 11-12–11-13
communication sequences, 11-12
configuration, 11-10–11-11
cyclic data exchange, 11-10–11-11
data up-and downloading, 11-13
isochronous mode, 11-12
monitoring interval, 11-12
parameterization, 11-10–11-11
slave-to-slave communications, 11-12–11-13
slot/index address model, 11-13
telegram structure, 11-10–11-11
Cyclic data channel (CDC), 28-2, 28-7–28-8
Cyclic redundancy check (CRC), 15-5, 52-10, 52-13
Cyclic redundancy code (CRC), 9-80, 67-2

D

DALI, *see* Digital addressable lighting interface (DALI)
DARPA, *see* Defense Advanced Research Project Agency (DARPA)
Data buffering, 43-22
Data communication
data packet routing, 25-9–25-11
encapsulation, 25-6–25-7
functional blocks, 25-5–25-6
packet aggregation, 25-8
packet sequencing, 25-7–25-8
stale packet detection, 25-8–25-9
Data-decoupling network, 6-3
Data length code (DLC), 52-10
Data Link Protocol Data Units (DLPDU) format, 18-6–18-7
Datapoint types (DPTs), 55-5
Data-Send-Acknowledged (DAS) service, 15-13
Data transfer object (DTO) packet, 51-9–51-10
Data validation, 43-23
Data word (DW), 43-10
Data wraparound, 43-22
DCF, *see* Distributed coordination function (DCF)
DCP, *see* Dynamic configuration protocol (DCP)
DDC, *see* Direct digital control (DDC)
Defense Advanced Research Project Agency (DARPA), 33-1–33-2
Delay and jitter
pre-and postprocessing times, 2-6–2-7
timing diagram, 2-6–2-7
waiting time (T_{wait}), 2-7–2-9
Demand response automation server, 61-14
Demand Response Research Center (DRRC), 61-13
DEOS, *see* Digital engine operating system (DEOS)
Destination Code field, 45-15
Device Language Message Specification/Companion Specification for Energy Metering (DLMS/COSEM)
communication profiles, 60-16–60-18
data modeling
association objects, 60-15
building blocks, 60-13–60-14
LD name, 60-14
management LD, 60-14
profile objects, 60-15

IEC 61968-9, 60-20
M-Bus integration, 60-18
messaging and referencing, 60-15–60-16
OBIS naming system, 60-16
security, 60-18
specification, colored books, 60-13
xDLMS, 60-13
Device management application process (DMAP), 34-4
DeviceNet, 9-2, 52-3
adaptation, 9-22
CAN Identifier, 9-29–9-30
CAN protocol, 9-22
connection establishment, 9-30
DeviceNet messages, 9-26
DeviceNet Object, 9-25
explicit messaging, 9-26–9-27
features, 9-23–9-24
frame structure, 9-24
indicators, 9-25
I/O messaging, 9-27–9-28
network access, 9-25
Offline Connection Set, 9-26
online, 9-26
physical layer, 9-24
Predefined Master/Slave Connection Set
1:n communication structure, 9-31
Allocate Message, 9-32
Allocation Choice Byte, 9-32
BasicCAN controllers, 9-37
Bit-Strobe I/O Connection, 9-34–9-35
CAN Identifier (CID), 9-30
Configuration Assembly, 9-38
configuration tools, 9-38
conformance testing, 9-38
connection IDs, 9-31–9-32
COS/cyclic connection, 9-32
COS/cyclic I/O connection, 9-34–9-36
development factors, 9-38–9-39
device profiles, 9-38
Group 2 Only Unconnected Explicit Request Port, 9-31
I/O Data Sharing, 9-35
I/O Scanner, 9-36–9-37
monitoring tools, 9-38
Multicast Polled I/O Connection, 9-35
physical layer tools, 9-38
Polled I/O Connection, 9-33–9-34
proxy function, 9-37
quick connect connection establishment, 9-37
UCMM, 9-31
protocol adaptations, 9-24–9-25
switches, 9-25
Devices Profile for Web Services (DPWS), 3-12, 38-11
AMR, 60-25
BASs, 55-17–55-18
client, device, and service, 26-6–26-7
hardware and software components, 26-1–26-2
HTTP chunked transfer coding, 26-9
Microsoft Windows, 26-24–26-25
MTOM, 26-8
protocols, 26-5–26-6
security
authentication, 26-20–26-21
confidentiality, 26-22
encryption, 26-20–26-21
message integrity, 26-20–26-22
security header, 26-19
Web Services Utility, 26-19
XML Canonicalization, 26-19
service composition, 26-22–26-23
smart metering, 26-25–26-26
SOA
evaluation, 26-4–26-5
Jini, 26-4
OSGi specification, 26-3
profiles, 26-4
services, 26-2–26-3
UPnP, 26-4
SOA4D, 26-24
SOAP, 26-7–26-8
WS4D, 26-24
WS-Addressing, 26-8–26-9
WS-BPEL, 26-23–26-24
WS-Discovery
low scalability, 26-11
optimization, 26-11
ProbeMatch, 26-10–26-11
UDDI server, 26-9–26-10
WSDL
abstract part, 26-12
concrete part, 26-12–26-13
WS-Eventing, 26-17–26-19
WS-MEX
Get operation, 26-14
hosted service, 26-16
policy assertion, 26-16–26-17
Relationship metadata, 26-15–25-16
ThisDevice, 26-15–25-16
ThisModel, 26-15–25-16
WS-Policy, 26-13
WS-PolicyAttachment, 26-13–26-14
WS-Transfer, 26-14
Device type manager (DTM), 3-10, 11-19, 11-24
Diagnostics over IP (DoIP) specification, 51-7–51-8
Digital addressable lighting interface (DALI), 55-9
Digital Data Compendium, *see* ARINC 419
Digital engine operating system (DEOS), 44-7–44-8
Digital information transfer system (DITS), *see* ARINC 429
Digital language, 45-5
Digital phase locked loop (DPLL)
bit encoding, 52-6
bit time segments, 52-7

synchronization rules, 52-8
time quanta (T_q), 52-7
DigitalSTROM, 55-11
Direct-coupled stub method, 43-4–43-5
Direct digital control (DDC), 55-2–55-3
Direct-sequence spread spectrum (DSSS), 21-6, 30-8, 30-14–30-16, 32-4–32-5
Discrete manufacturing automation
application and performance
automotive stamping, 37-22
characteristic figures, 37-24, 37-26
empirical quantitative comparison, 37-22–37-24
latency measurements, 37-22
TDMA real-time system, 37-24–37-25
wireless communication profiles, 37-21–37-22
hierarchy, 37-1–37-2
machine and device level, 37-2
wireless interface
automotive robot production cells, 37-2
basic requirements, 37-2
communication requirements, 37-5
communication subsystem (*see* Communication subsystem)
latency measurements, 37-3
production lines, 37-2
roundtable production machines, 37-2–37-3
wireless power subsystem
energy density, 37-16
energy distribution, 37-15–37-16
energy scavenging, 37-15
energy storage, 37-15
figure of merit, 37-16, 37-18–37-19
fuel cells, 37-16
magnetic supply, 37-18, 37-20
omnidirectional receiver structure, 37-20
resonant, medium-frequency power supply, 37-20
rotating field, 37-20–37-21
Distributed clocks (DC)
drift compensation, 18-15
external synchronization, 18-15–18-16
features, 18-12–18-13
offset compensation, 18-15
propagation delay measurement, 18-13–18-14
Distributed coordination function (DCF), 30-15–30-16
Distributed Network Protocol Version 3 (DNP3), 61-9–61-10, 63-23–63-24
Distributed sensor networks (DSNs), 33-1
Distributed system, *see* Clock synchronization
DLC, *see* Data length code (DLC)
DLMS/COSEM, *see* Device Language Message Specification/Companion Specification for Energy Metering
DMAP, *see* Device management application process (DMAP)
DNP3, *see* Distributed Network Protocol Version 3 (DNP3)
Domain management
domain content, 24-13
domain scope, 24-14
structure, 24-12–24-13
transmission, 24-13–24-14
Domain-specific language (DSL), 8-23
Double-point common data class (DPC), 62-13
DPLL, *see* Digital phase locked loop (DPLL)
DPTs, *see* Datapoint types (DPTs)
DPWS, *see* Devices Profile for Web Services (DPWS)
DRRC, *see* Demand Response Research Center (DRRC)
DSL, *see* Domain-specific language (DSL)
DSNs, *see* Distributed sensor networks (DSNs)
DSSS, *see* Direct-sequence spread spectrum (DSSS)
DTM, *see* Device type manager (DTM)
DUST networks, 36-16–36-17
Dutch metering project, 60-18
Dynamic configuration protocol (DCP), 12-8

E

EDD, *see* Electronic device description (EDD)
EDDL, *see* Electronic device description language (EDDL)
EDS, *see* Electronic data sheets (EDS)
Efficient XML interchange (EXI), 3-12–3-13
EIA/TIA-485 transmission technology, 11-3–11-5
Electromagnetic interference (EMI), 36-5
Electronic data sheets (EDS), 9-18–9-20, 10-33
Electronic device description (EDD), 11-24
Electronic device description language (EDDL), 3-10, 8-7–8-8, 14-17–14-20
Electronic product code (EPC), 59-2
EN 13757/M-Bus
application layer, 60-12–60-13
link layer, 60-12
OSI three-layer structure, 60-11–60-12
physical layer, 60-11–60-12
Encapsulating Security Payload (ESP), 29-8, 29-15–29-16, 50-1
Energy Market Information Exchange Technical Committee (EMIX), 61-15
Engineering tool software (ETS), 55-4, 58-11–58-13
Enhanced performance architecture (EPA), 61-7
EnOcean, 37-23–37-24, 55-10
EPL, *see* Ethernet POWERLINK (EPL)
Equipment communication software, 66-19–66-21
Equipment integration software, 66-19
Error processing, 43-24
Error state indicator (ESI), 52-32
ESP, *see* Encapsulating Security Payload (ESP)
EtherCAT, *see* Ethernet for Control Automation Technology (EtherCAT)
Ethernet; *see also* Train communication network (TCN)
automotive-specific requirements, 50-17–50-18
AVB, 50-18

BroadR-Reach, 50-18
de facto standard, 2-14
Ethernet/IP, 1-39
high-speed and time-critical applications
precise clock synchronization, 2-19
real-time capabilities, 2-18
real-time Ethernet, 2-18
high-speed Ethernet, 1-39
hub-based Ethernet, 2-14–2-15
industrial Ethernet, 1-40–1-41, 2-16–2-17
Modbus/TCP, 1-39
nonautomotive-specific requirements, 50-17–50-18
OPC, 2-20
physical and data link layer, 1-39
plausible roadmap, 50-18
P-NET, 1-39
protocol architecture, 1-40
RTE, 1-39–1-40
structures, 1-38
switched Ethernet, 2-15–2-16
traffic smoothing techniques, 1-38
TTEthernet, 50-18–50-19
Ethernet for Control Automation Technology (EtherCAT)
application layer, 18-16–18-17
commercial *vs.* open-source implementation, 18-20–18-22
control loop, 18-18–18-20
data link layer
BWR, 18-9–18-11
DLPDU format, 18-6–18-7
frame format, 18-4–18-5
logical addressing, 18-9–18-10
logical commands, 18-9–18-11
physical addressing, 18-8–18-9
read commands, 18-9–18-11
read/multiple write commands, 18-9–18-11
SyncManager, 18-11–18-12
distributed clocks (DC)
drift compensation, 18-15
external synchronization, 18-15–18-16
features, 18-12–18-13
offset compensation, 18-15
propagation delay measurement, 18-13–18-14
EtherLab C, 18-25–18-26
frame processing, 18-2
ladder diagram, 18-22–18-24
mailbox mechanism, 17-19
medium access control, 17-17
open ring bus, 17-18
performance indicator, 17-19
physical layer
communication support, 18-2–18-3
device architecture, 18-4
network topology, 18-3–18-4
slave device, 17-17–17-18
structured text, 18-24–18-25
Ethernet for plant automation (EPA) protocol, 17-14–17-15
EtherNet/IP, 9-2
ACD mechanism, 9-52
Base Switch Object, 9-51
commissioning tools, 9-61
configuration, 9-56
configuration tools, 9-61
conformance testing, 9-56
development factors, 9-61–9-62
device classes, 9-56
Device Level Ring Object, 9-51
device profiles, 9-56
encapsulation mechanism
explicit messaging, 9-48, 9-54
I/O Message encapsulation, 9-55
UCMM request encapsulation, 9-54
Ethernet-based protocols, 9-57
Ethernet Link Object, 9-51
features, 9-49–9-50
frame structure, 9-50
indicators, 9-50
infrastructure, 9-58
IP address assignment, 9-51–9-52
linear topology, 9-58
message prioritization, 9-72–9-73
monitoring tools, 9-61
multiple Ethernet ports, 9-58
Parallel Redundancy Protocol (PRP) Object, 9-51
physical layer, 9-50
physical layer tools, 9-61
Power Management Object, 9-51
protocol adaptation, 9-50
PRP Node Table Object, 9-51
Quality of Service (QoS) Object, 9-51
quick connect connection establishment, 9-55–9-56
ring topology, 9-58–9-60
RSTP, 9-51, 9-60
Simple Network Management Object, 9-51
switches, 9-50
TCP/IP Interface Object, 9-51
TCP/IP Support, 9-56–9-57
Ethernet POWERLINK (EPL), 17-11–17-13
application layer, 19-2–19-3
encapsulation, 19-4–19-5
low layers specification, 19-1–19-2
managing node redundancy, 19-7
medium redundancy, 19-6–19-7
network operation, 19-3–19-4
ring redundancy
acyclic traffic, 19-11
cycle time, 19-9–19-10
jitter, 19-10–19-11
OpenSAFETY protocol, 19-8–19-9
topology, 19-5–19-6
Ethernet train backbone node (ETBN), 65-9
ETS, *see* Engineering tool software (ETS)

European train control system (ETCS), 64-14–64-15
EventState, 61-13–61-14
Event-triggered communication, 39-1
EX certification, 36-36
EXI, *see* Efficient XML interchange (EXI)
Extended data length (EDL) bit, 52-31
Extended subaddress, 43-21–43-22
Extensible Messaging and Presence Protocol
(XMPP), 38-11–38-12

F

Factory floor, *see* Wireless local area network
(WLAN)
FAN, *see* Field area networks (FAN)
Fault containment region (FCR), 39-5–39-6
Fault isolation, SAFEbus
availability *vs.* integrity trade-off, 48-16
babble protection, 48-15
Byzantine protection, 48-15–48-16
fault-containment zone, 48-15
near-perfect error-detection, 48-14
zombie module protection, 48-16
Fault-tolerant average (FTA) algorithm, 40-7
FCR, *see* Fault containment region (FCR)
FDA agent, *see* Field device access (FDA) agent
FDI, *see* Field device integration (FDI)
FDT, *see* Field device tool (FDT)
FF, *see* Foundation fieldbus (FF)
FHSS, *see* Frequency-hopping spread spectrum
(FHSS)
Fiber optic communication, 11-8–11-9
Field area networks (FAN), 1-5
Fieldbus
advantages, 1-3
application-related management, 1-36
automation pyramid, 1-4
automotive and aircraft fieldbusses, 1-43
building and home automation, 1-45
CAMAC standard, 1-6
CIM, 1-4
client–server model, 1-31–1-32
communication-related management, 1-36
definition, 1-1–1-2
Ethernet (*see* Ethernet)
evolution, 1-7–1-9
FAN, 1-5
features, 1-2
field multiplexers, 1-3
future evolution, 1-41–1-42
GAN, 1-4
GPIB standard, 1-4
hierarchical network levels, 1-4
high-level network management protocol, 1-37
historical roots, 1-6
industrial and process automation, 1-44
LAN, 1-4–1-5
medium access control, 1-23
master–slave approach, 1-22
multimaster approach, 1-22
polling, 1-23–1-25
random access, 1-28–1-30
TDMA strategy, 1-23
time-slot-based methods, 1-27–1-28
token passing, 1-25–1-27
network topologies, 1-20–1-22
OSI model
answered service, 1-15
application layer, 1-10, 1-12, 1-20
BACnet, 1-19
confirmed service, 1-14–1-15
ControlNet, 1-19
data link layer, 1-10–1-11, 1-20
interface, 1-10
interface data unit (IDU), 1-13
interoperability, 1-34–1-36
layer structure, 1-10
locally confirmed service, 1-14
MiniMAP approach, 1-18
multiport masters, 1-19
network layer, 1-10–1-11, 1-20
open system, 1-9
physical layer, 1-10–1-11, 1-20
P-NET, 1-19
presentation layer, 1-10, 1-12, 1-20
protocol, 1-9
SAP, 1-12
service primitives, 1-13
service provider, 1-10
service user, 1-10
session layer, 1-10, 1-12, 1-20
three-layer structure, 1-18–1-19
transport layer, 1-10–1-12, 1-20
PCB-level busses, 1-43
publisher–subscriber model, 1-31–1-34
remote I/O systems, 1-3
serial data transmission, 1-6
traffic
aperiodic, acyclic, or spontaneous traffic, 1-17
classes and properties, 1-16
cyclic/identified traffic, 1-17
event-based approach, 1-16
implications, 1-18
parallel traffic, 1-17–1-18
parameterization/configuration data, 1-17
state-based approach, 1-16
WAN, 1-4
Fieldbus Description Configuration Markup Language
(FDCML), 15-10–15-11
Fieldbus memory management units (FMMUs), 18-9
Fieldbus system, *see* Control networks (CNs)
Field device access (FDA) agent, 14-10–14-11
Field device integration (FDI), 11-24–11-25
Field device tool (FDT), 3-10, 11-24, 14-18–14-20

Field terminal assembly (FTA), 8-22–8-23
FLAG probes, 60-7
Flexible Time-Triggered CAN (FTT-CAN), 41-15–41-16
FlexRay protocol, 50-11–50-13
FlexRay technology
  bus guardian, 53-10
  carmakers and suppliers, 53-11
  clock synchronization, 53-7–53-8
  communication controller, 53-3–53-4
  cost factor
    composability, 53-2
    future car and application, 53-2–53-3
    migration path, 53-2
    quality, 53-2
    scalability, 53-2
  ECU architecture, 53-3
  media access
    communication cycle, 53-4
    configuration, 53-6–53-7
    dynamic segment, 53-4–53-5
    minislot, 53-6
    static segment, 53-4–53-5
    symbol window, 53-6
  protocol services, 53-10–53-11
  redundant transmission, 53-4
  standardized automotive system software, 51 6
  standard software components, 53-14–53-15
  start-up service
    case traffic, 53-9
    coding and physical layer, 53-9–53-10
    coldstart case, 53-8
  system configuration
    development model, 53-14
    functional domain, 53-11
    information domain, 53-12–53-13
    OEM-supplier development process, 53-13–53-14
    signal domain, 53-12
Foundation fieldbus (FF)
  EDDL, 14-17–14-20
  FDT, 14-18–14-20
  features, 14-16
  H1 FF
    data link layer (DLL), 14-3–14-5
    definition, 14-1
    fieldbus access sublayer, 14-6
    fieldbus message specification, 14-6–14-7
    vs. ISO OSI architecture, 14-2
    network management, 14-7–14-8
    physical layer, 14-3
    system management, 14-7
    user application layer (*see* User application layer)
  HSE FF
    advantages, 14-9
    Ethernet device (ED), 14-8–14-9
    FBAP VFD, 14-11
    FDA agent, 14-10–14-11
    frames, 14-10
    gateway device (GD), 14-8–14-9
    H1 interfaces and bridge, 14-11
    host device (HD), 14-8–14-9
    interconnections, 14-8
    internal organization, 14-10
    vs. ISO OSI architecture, 14-2–14-3
    linking device (LD), 14-8–14-9
    local area network redundancy entity, 14-11–14-12
    management agent, 14-12
    network management, 14-12
    SMK, 14-11
    user application layer (*see* User application layer)
  wireless solutions, 14-15–14-17
Frame controller, 43-7
Frequency-hopping spread spectrum (FHSS), 30-8–30-9, 30-25
Frequency shift keying (FSK) principle, 7-2, 8-1
FTA algorithm, *see* Fault-tolerant average (FTA) algorithm
FTT-CAN, *see* Flexible Time-Triggered CAN (FTT-CAN)
Full function devices (FFD), 36-23, 38-3
Function block application process (FBAP) virtual field device (VFD), 14-11

G

GAN, *see* Global area networks (GAN)
Generic Equipment Model (GEM)
  alarm management, 66-16
  clock, 66-17
  communications, 66-16
  documentation, 66-15
  dynamic event report configuration, 66-16
  equipment constants, 66-16
  equipment terminal services, 66-17
  error messages, 66-15
  event notification, 66-15–66-16
  host-initiated control, 66-15, 66-17
  limitations, 66-17–66-18
  limit monitoring, 66-17
  material movement, 66-17
  online identification, 66-15
  operator-initiated control, 66-15
  pooling, 66-17
  process program management, 66-17
  remote control, 66-16
  vs. SECS-II, 66-12
  state models, 66-13–66-15
  status data collection, 66-16
  trace data collection, 66-16
  variable data collection, 66-16

Global area networks (GAN), 1-4
Global positioning system (GPS), 21-6–21-8, 40-7
GOOSE messages, 62-5

H

Handheld units (HHUs), 60-2, 60-9
HART device, *see* Highway addressable remote transducer (HART) device
Hello World application, 23-11–23-12
High-Level Data Link Control (HDLC), 60-7, 60-17
High-Speed SECS Message Services Standard (HSMS)
  message structure, 66-7–66-8
  state model, 66-6–66-7
  subclassification, 66-8
Highway addressable remote transducer (HART) device
  analog and digital communication, 7-2–7-3, 8-1–8-2
  analog signaling, 7-11–7-12
  applications, 8-4
  bioreactor, 8-31–8-32
  burst-mode communications, 7-7
  cable capacitance, 7-11
  checkout, 8-4–8-5
  commands, 8-13–8-15
  communication channels, 7-2
  communication modes
    block data transfer, 8-11–8-12
    burst mode, 8-9–8-11
    event notification, 8-11
    request/response messages, 8-9
  communication stack (*see* Communication stack)
  compatible control systems, 8-22
  configuration, 8-4–8-5
  control applications, 8-5
  data link layer (DLL)
    device display, 8-23, 8-25
    device layout, 8-23–8-24
    DSL, 8-23
    item declarations, 8-23, 8-25
  evolution, 7-1–7-2, 8-1–8-2
  field devices
    common practice commands, 7-4
    device variables *vs.* dynamic variables, 7-4
    loop current calibration, 7-6
    process variable (PV) calibration, 7-5–7-6
    process variable (PV) scaling, 7-6
  four-wire connection, 7-10, 7-12–7-14
  FSK principle, 7-2, 8-1
  FTA, 8-22–8-23
  installation, 8-4–8-5
  intrinsically safe circuits, 7-14
  latency, 7-6–7-8
  modulated analog signal, 7-8
  monitoring, 8-5
  multidrop connections, 7-7
  multidrop interface, 8-21–8-22
  multidrop mode of operation, 7-3–7-4, 8-13
  multidropped slave devices, 7-10
  network guidelines, 7-10
  pass-through messages, 8-22
  physical layer, 7-8
  plant operations, 8-4
  point-to-point interface, 8-21
  point-to-point mode of operation, 8-12–8-13
  primary and secondary master devices, 7-10
  safety, 8-5
  setup, 7-2–7-3, 8-2–8-3
  theory of operation
    data access, 8-8
    data writing, 8-8
    device identification, 8-7
    device-specific commands, 8-8–8-9
    EDDL, 8-7–8-8
    message-oriented communications, 8-8
    specifications, 8-8
    universal commands, 8-8
  throughput, 7-6–7-8
  two-wire connection, 7-9, 7-12–7-13
  Wired HART, 8-28–8-29
  WirelessHART (*see* WirelessHART)
Home and building automation (HBA) systems, 58-1–58-2
  CoRE
    architecture, 59-10–59-11
    CoAP, 59-10
    message format, 59-11
  energy management systems, 59-3–59-4
  EPC, 59-2
  IoT protocol stacks (*see* IoT protocol stacks)
  IPSO application framework
    data formats, 59-17
    function sets, 59-16–59-17
    UCUM, 59-16
  MQTT
    features, 59-15
    message format, 59-15–59-16
    sensor networks, 59-15
  RFID, 59-2
  SEP, 59-17–59-18
  6LoWPAN
    configuration and management, 59-5
    connectivity, 59-5
    frame format, 59-5–59-6
    neighbor discovery, 59-6–59-7
    packet size, 59-5
    routing, 59-7–59-8
    security, 59-5
    stateful header compression, 59-6
    stateless header compression, 59-6
    topologies, 59-5
  smart cities, 59-4
  smart grids, 59-4
  smart living, 59-2–59-3

- smart meter devices, 59-3
- ZigBee IP
  - IP protocol stack, 59-8–59-9
  - MAC, 59-8
  - physical layer, 59-8
  - security, 59-9
  - topology, 59-9
  - WSAN, 59-8

Homeplug Command and Control, 60-25
HTTP
- binary, 38-10–38-11
- *vs.* CoAP, 38-8–38-9
- proxy and caching, 38-9–38-10
- REST design principles, 38-8

Human–machine interface (HMI), 36-33–36-34
Human-to-machine (H2M) communication, 38-10–38-11
Hybrid networks
- higher protocol layers, 35-4–35-5
- intermediate systems, 35-2
- physical layer, 35-2–35-3

Hybrid systems, 30-2, 30-4–30-5

I

IEC 60870-5, 61-3–61-4, 61-7–61-8
IEC 61400-25 standard
- communication contents, 63-3
- communication model, 63-4–63-5
- description, 63-2
- IEDs, 63-2
- management functions, 63-3–63-4
- operational functions, 63-3–63-4
- WPPs
  - components, 63-3
  - IEM (*see* Wind power plant information exchange model (WPP IEM))
  - IM (*see* Wind power plant information model (WPP IM))

IEC 61850 standard
- application domains, 62-16–62-18
- circuit breaker, 62-5
- communication networks, 62-2
- communication services, 62-2
- concrete communication protocols, 62-6
- control functions, 62-21
- features and characteristics, 62-3
- frequency-watt management modes, 62-21
- function block view, 62-19
- GOOSE messages, 62-5
- high-voltage substations, 62-16
- implementation layering
  - cyclic polling, 62-37
  - LN STMP, 62-33
  - MMS Read request, 62-36
  - models to messages, 62-33
  - real stack and API, 62-34–62-35
  - SCL document, 62-34–62-35
  - STMP supervision function, 62-34
- information exchange models (*see* Information exchange models, IEC 61850)
- information models (*see* Information models, IEC 61850)
- inverter functions, 62-21
- LNs, 62-19
- logical nodes, 62-4–62-5
- MMXU1, 62-16
- non-real-time communication, 62-16
- OGEMA gateway, 62-38
- power delivery systems, 62-1
- projects, 62-37–62-38
- PV performance, 62-19–62-20
- SCL, 62-6, 62-30–62-32
- standard series impacts, 62-6–62-7
- third-party solutions, 62-41
- topology, 62-3
- UCA IUG, 62-40–62-41
- virtualization, 62-3
- volt-var curves, 62-20–62-21
- volt-var management modes, 62-21
- WEB-PLC, 62-39–62-40
- XCBR1 model, 62-5

IEC 61968-9 standard
- Common Information Model, 60-18
- DSLM/COSEM, 60-20
- EndDeviceEvent, 60-19–60-20
- message definition, 60-19–60-20
- reference model, 60-19

IEC 62056-31 Euridis, 60-20–60-21
IEC/PAS 62601, *see* Wireless networks for industrial automation–process automation (WIA-PA)
IEC Seamless Integration Architecture (IEC SIA), 61-2–61-4
IEEE 802.1 technology, *see* Time-sensitive networking (TSN)
IEEE 802.11 standard, *see* Wireless local area network (WLAN)
IEEE 802.11i standard, 29-8
IEEE 802.15.1 standard, 29-7, 37-5, 37-12
IEEE 802.15.4 standard, 29-7, 37-5, 37-12, 38-2–38-3
IEEE 1451 standards
- advantages, 4-11, 4-13–4-14
- application, 4-16
- family of standards, 4-5–4-6
- goals, 4-6
- IEEE 1451.0 standard, 4-6–4-8
- IEEE 1451.2 standard, 4-8–4-9
- IEEE 1451.3 distributed multidrop systems, 4-9–4-10
- IEEE 1451.4 MMI, 4-10
- IEEE 1451.5 wireless transducer interface, 4-10–4-12
- IEEE 1451.7 transducers, 4-11, 4-13
- IEEE P1451.1, 4-8

network interfaces, 4-6, 4-8
reference model, 4-5
TEDS formats, 4-4
IEEE 1451.2 standard, 3-10, 3-12, 4-14–4-15
IFS, *see* Interface file system (IFS)
Inductive transmission, 36-49
Industrial communications
access control mechanism, 29-12
automation networks
ESP, 29-8
fieldbus systems, 29-5–29-6
IEEE 802.11i standard, 29-8
IEEE 802.15.1 standard, 29-7
IEEE 802.15.4 standard, 29-7
Industrial Ethernet, 29-6–29-7
IPsec, 29-8, 29-15–29-16
TLS, 29-8–29-9
VPN, 29-9
end-to-end security, 29-11
integrity and confidentiality mechanisms, 29-11–29-12
microcontrollers, 29-11
OSI security services, 29-10
PLC network, 29-15
ROHC, 29-15
secure node implementation, 29-13–29-14
security measures
authentication measures, 29-4
basic steps, 29-2
cryptography, 29-3–29-4
firewalls, 29-4
security goals, 29-2
third-party security evaluation, 29-4
waterfall life cycle model, 29-3
smart grids, 29-18–29-19
three-zone security model, 29-16–29-18
wrapping mechanism, 29-11
Industrial wireless technologies evolution, 34-2–34-3, 34-16
Information exchange models, IEC 61850, 62-6
accessing information models, 62-23–62-24
buffered reporting, 62-24, 62-26–62-27
client–server model, 62-22–62-23
configuration level, 62-22–62-23
control blocks and data set, 62-24, 62-27
event reporting mechanism, 62-29
GOOSE Event, 62-29
inclusion bitstring, 62-28
log control blocks, 62-28–62-29
multicast-based communication, 62-22
RBE, 62-28
SCL document, 62-22
service excerpt, 62-23–62-24
service models, 62-21–62-22
SVs, 62-29–62-30
unbuffered event reporting, 62-26–62-27
Information models, IEC 61850, 62-2, 62-6, 62-14–62-15
CDCs, 62-11
circuit breaker model, 62-11–62-12, 62-14
data attributes, 62-10–62-11
double-point common data class (DPC), 62-13
implementation issues, 62-7
logical nodes (LNs)
applications, 62-8
LNs information, 62-8–62-9
measurements and metering information, 62-9
MMXU, 62-7–62-8
XCBR, 62-10
In-line couplers, 43-5
In-system serial programming approach, 3-18–3-19
Integrated modular avionics (IMA), 42-12–42-13, 48-1, 48-21, 49-2–49-3
Integrated networked smart transducer, 4-3
INTERBUS system
certification, 15-11
connectivity, 15-11–15-12
CRC register, 15-5
data transfer cycle, 15-4–15-5
features, 15-2
identification cycle, 15-4–15-5
interoperability, 15-10–15-11
IP tunneling, 15-12–15-13
network management, 15-5
parallel wiring, 15-1–15-2
parameter data transmission, 15-6–15-7
performance evaluation, 15-13–15-14
protocol architecture, 15-5–15-6
safety technology, 15-9–15-10
slave node, 15-4
system diagnostics, 15-7–15-9
topology flexibility, 15-2–15-3
VFD model, 15-7
Intercontrol Center Protocol/Telecontrol Application Service Element 2 (ICCP/TASE.2), 61-8
Interface file system (IFS), 3-5–3-6
Intermediate systems (ISs)
bridge, 35-3–35-4
definition, 35-2
gateway, 35-4–35-5
repeaters, 35-2–35-3
Intermodule memory (IMM), 48-5
International Organization for Standardization (ISO)
CAN, 51-4–51-6
DoIP specification, 51-7–51-8
FlexRay, 51-6
OBD, 51-7
UDS, 51-6–51-7
International Society of Automation (ISA), 14-1, 33-2, 33-11, 36-5
Internet
application area, 22-2–22-3
control data tunneling

application classes, 22-9–22-10
delay jitter, 22-7–22-8
EIA-852 system configuration, 22-6–22-7
end-to-end delay, 22-7
loss probability/loss rate, 22-8
packet ordering, 22-8–22-9
performance classes, 22-9
throughput, 22-8
evolution, 22-3–22-4
gateway
data point, 22-13–22-14
features, 22-14–22-15
LDAP, 22-16–22-17
OPC, 22-19–22-20
services, 22-11–22-12
SNMP, 22-14, 22-16
SQL, 22-17–22-18
structure, 22-12–22-13
tunneling approach, 22-5–22-6
user interface, 22-10–22-11
web-based approach, 22-18–22-19
industrial Ethernet, 22-20–22-21
protocol tunneling, 22-3–22-5
Internet Group Management Protocol (IGMP), 9-55, 16-11, 65-4
Internet Protocol Security (IPsec), 29-8, 29-15–29-16
Internet protocol (IP) tunneling, 15-12–15-13, 22-3–22-4, 22-18
Inter-Range Instrumentation Group (IRIG), 21-6–21-7
In-vehicle communication network
CAN
arbitration phase, 50-8–50-9
bandwidth requirement, 50-9–50-10
CAN 2.0A data frame, 50-7–50-8
CAN FD, 50-11
complex architecture, 50-9–50-10
data link layer, 50-9
error detection, 50-9
industrial requirement, 50-9–50-10
J1850, 50-11
non-return-to-zero, 50-7–50-8
physical layer, 50-9
temporal constraints, 50-9–50-10
traffic shaping strategy, 50-9
VAN, 50-11
car domain, 50-2–50-3
class A network, 50-4
class B network, 50-4
class C network, 50-4–50-5
class D network, 50-4–50-5
energy efficiency, 50-5–50-6
Ethernet
automotive-specific requirements, 50-17–50-18
AVB, 50-18
BroadR-Reach, 50-18
nonautomotive-specific requirements, 50-17–50-18
plausible roadmap, 50-18
TTEthernet, 50-18–50-19
event triggered *vs.* time triggered, 50-3–50-4
low-cost automotive network
application, 50-14
LIN, 50-14–50-16
PSI5 network, 50-16
SENT network, 50-16
TTP/A, 50-16
middleware (MW) layer
application components, 50-19–50-20
AUTOSAR (*see* Automotive Open System Architecture (AUTOSAR))
design, 50-19
features, 50-20
OSEK/VDX communication, 50-20
multimedia and infotainment network, 50-16–50-17
multiplexed communication, 50-1–50-2
point-to-point link, 50-1–50-2
real-time processing, 50-5
security, 50-6
time-triggered network
FlexRay protocol, 50-11–50-13
TTCAN protocol, 50-13–50-14
IO-Link technology
automation hierarchy, 5-2–5-3
basic concept, 5-2
communication features
application layer (AL), 5-5–5-6
data link layer (DL), 5-5
generic device model, 5-3–5-4
Master and Device, 5-4–5-7
M-sequence, 5-4
configuration, 5-6
connectors, 5-3
engineering support, 5-7
functional safety, 5-8
implementation, 5-7
power, 5-3
profiles, 5-7
specification, 5-7
standardization, 5-8
test and certification, 5-7
wiring, 5-3
IoT protocol stacks
OPC UA
drawbacks, 59-12
information modeling, 59-13–59-14
services and communication, 59-14
open building information exchange
oBIX 2.0 outlook, 59-12
protocol bindings and message encodings, 59-12
IPsec, *see* Internet Protocol Security (IPsec)
IPv4, 38-2, 38-5, 42-3

IPv6 over Low-Power Wireless Personal Area Networks (6LoWPAN), 36-22, 60-25
adaptations and enhancements, 38-4
compression, 38-4–38-5
fragmentation, 38-5
IEEE 802.15.4, 38-2–38-3
monitoring applications, 36-19–36-20
neighbor discovery, 38-5–38-6
resource-constrained networks, 38-4
routing
ROLL, 38-5–38-6
RPL, 38-6–38-7
ISA, *see* International Society of Automation (ISA)
ISA100
architecture, 36-26–36-27
functionality, 36-27
security, 36-28
standards landscape, 36-21
ISA100.11a
application layer, 32-8
data link layer
frequency hopping patterns, 32-6–32-7
graph routing, 32-4–32-5
MAC extension layer, 32-4
MAC sublayer, 32-4
message forwarding and routing, 32-4
source routing algorithms, 32-4–32-5
structure, 32-4
TDMA, 32-5
upper DLL, 32-4
field devices, 32-11–32-13
frequency utilization, 32-10
message authentication, 32-9
network configuration, 32-2
network layer, 32-7–32-8
payload encryption, 32-9
physical layer, 32-4
roles and functions, 32-2
security issues, 32-8–32-9
security key model, 32-9–32-10
specifications, 32-10
spectrum management, 32-11
stack model, 32-3
system manager, 32-13
transport layer, 32-8
wireless gateway, 32-13
WirelessHART
application layer, 33-7
coexistence mechanisms, 33-8–33-9
data link layer, 33-5–33-7
flexibility, 33-8
installation, 33-3–33-4
ISA, 33-11–33-13
NAMUR, 33-11–33-13
network layer, 33-6–33-7
network topologies, 33-4
physical layer, 33-4–33-5
protocol stacks, 33-3
protocol support, 33-8
QoS, 33-9–33-11
security mechanisms, 33-9
stack structures, 33-4–33-5
transport layer, 33-7
ISs, *see* Intermediate systems (ISs)

J

Java Intelligent Network Infrastructure (Jini), 26-4

K

Kerckhoff's principle, 29-9
Knowledge representation and reasoning (KRR) technologies, 58-16–58-17
KNX system
addressing scheme, 58-6–58-7
AMR
application and interworking models, 60-22
configuration and management, 60-23
logical topology, 60-22–60-23
physical layers, 60-23
specification, 60-22
answered service communication, 58-5
application layer, 58-4–58-5
BACS, 58-1–58-2
BEMS, 58-16
data link layer, 58-4–58-5
development tools, 58-12–58-13
frame structure, 58-7
group objects, 58-9–58-10
HBA, 58-1–58-2
history, 58-2–58-3
installation, 58-13–58-14
interface object, 58-9–58-10
interworking specification, 58-10–58-11
KNXnet/IP network, 55-5, 58-8–58-9
KNX PL110, 55-4
KNX RF, 55-4
KRR technologies, 58-16–58-17
network architecture, 58-6–58-7
network configuration
Ctrl mode, 58-11
LTE mode, 58-12
PB mode, 58-12
S-mode, 58-11
network layer, 58-4–58-5
PCI, 58-5
PDUs, 58-5
physical layers, 58-4
SDUs, 58-5
transmission media, 58-3–58-4
transport layer, 58-4–58-5
ZigBee home automation system, 58-15

L

LAN, *see* Local area network (LAN)
LAS, *see* Link active scheduler (LAS)
Lightweight Directory Access Protocol (LDAP), 22-16–22-17
LIN, *see* Local interconnect network (LIN)
LIN description file (LDF), 54-5–54-7, 54-11–54-12
Line-replaceable modules (LRMs), 48-1–48-2, 48-5
Line-replaceable unit (LRU), 45-5, 45-12–45-13, 46-1, 46-16, 47-5
Link active scheduler (LAS), 6-46, 14-4–14-7, 14-14
Link data unit (LDU), 45-13–45-15
LIN Network Architect (LNA), 54-7
  global objects, 54-8–54-9
  LDF generation, 54-9, 54-12
  manual and automatic schedule table generation, 54-8–54-11
  network definition and frame packing, 54-8, 54-10
  timing model, 54-8–54-11
LIN target package (LTP), 54-9–54-10, 54-12
LLC, *see* Logical link control (LLC)
LNA, *see* LIN Network Architect (LNA)
Local area network (LAN), 1-4–1-5, 29-4
Local interconnect network (LIN), 50-14–50-16
  automotive industry, 54-2
  AUTOSAR, 54-7
  conformance test, 54-5
  debugging, 54-6
  design process and workflow, 54-5
  frame structure, 54-4–54-5
  LIN consortium, 54-2–54-3
  LIN Spector
    diagnostics and emulation tool, 54-10–54-11, 54-13
    LIN GO, graphical objects, 54-13–54-14
  LNA
    global objects, 54-8–54-9
    LDF generation, 54-9, 54-12
    manual and automatic schedule table generation, 54-8–54-11
    network definition and frame packing, 54-8, 54-10
    timing model, 54-8–54-11
  LTP, 54-9–54-10, 54-12
  message identifier, 54-3
  physical layer, 54-3–54-4
  properties, 54-2
  software industry, 54-2
  system definition process, 54-5–54-6
  Volcano LIN tool chain, 54-7–54-8
LockTrac 6131 ELEKTRA system, 40-1
Logical device (LD), 62-4, 63-6–63-7
Logical link control (LLC)
  BasicCAN implementation, 52-16
  FullCAN implementation, 52-16
  information exchanges, 52-15
  producer/consumer model, 52-15–52-16
Logical Name (LN) referencing, 60-15–60-16
Logical tag extended (LTE) mode, 58-12
LonBuilder, 56-11–56-12
LonTalk protocol, 60-23–60-24; *see also* ANSI/CEA-709 protocol
LonWorks networks, 60-23–60-24
  ANSI/CEA-709 protocol (*see* ANSI/CEA-709 protocol)
  bus topology, 56-6–56-7
  communication media, 56-6–56-7
  explicit messages, 56-6
  features, 56-1–56-2
  free topology, 56-6–56-7
  interoperability and profiles, 56-11
  LonBuilder, 56-11–56-12
  LonTalk protocol, 56-1–56-2
  multiprotocol processors, 56-14
  network infrastructure, 56-11
  Neuron C derivative language, 56-5–56-6
  Neuron chip
    block diagram, 56-2–56-3
    I/O system, 56-2, 56-4
    memory layout, 56-2, 56-4
    node architecture, 56-2–56-3
  NodeBuilder, 56-11–56-12
  NVs, 56-6
  ring topology, 56-6–56-7
  SNVT, 56-6
LTP, *see* LIN target package (LTP)

M

MAC, *see* Modular Aerospace Control (MAC)
Maintenance data, 45-6
Major and minor frame, 43-23–43-24
Manchester Coding and Bus Powered (MBP) technology
  applications, 11-7
  bus powering, 11-5–11-6
  characteristics, 11-5–11-6
  current version, 11-5
  FISCO model, 11-7
  high-power trunk concept, 11-7–11-8
  logical 1 and 0 determination, 11-5–11-6
  mid-bit transition, 11-5
  network topologies, 11-7
  process automation, 11-5–11-6
Manufacturing Message Specification (MMS), 63-19–63-21
  client/server model, 24-2–24-3
  connection management, 24-10–24-11
  domain management
    domain content, 24-13
    domain scope, 24-14

structure, 24-12–24-13
transmission, 24-13–24-14
interfaces, 24-9–24-10
ISO 9506-1 services, 24-6
ISO 9506-2 protocol, 24-7
program invocation, 24-14–24-16
variable model (*see* Variable model)
VMD
attributes, 24-4–24-5
executive function, 24-11–24-12
functions, 24-4
location, 24-7–24-8
logical status, 24-12
objects and services, 24-4–24-5
physical status, 24-12
Service Get Capability List, 24-12
standard driver, 24-3–24-4
Manz scribing systems, 13-28–13-29
Master data telegram (MDT), 13-38–13-40, 17-16–17-17
Master synchronization telegram (MST), 13-37–13-38
MAUs, *see* Medium access units (MAUs)
MBAP, *see* Modbus application protocol (MBAP)
MBP technology, *see* Manchester Coding and Bus Powered (MBP) technology
MDI, *see* Medium dependent interface (MDI)
MDT, *see* Master data telegram (MDT)
Media redundancy for planned duplication (MRPD), 12-20, 12-23
Media redundancy protocol (MRP), 12-20, 12-23, 20-5–20-6
Medium access control (MAC) protocol, 46-4
arbitration phase, 50-8–50-9
CAN 2.0A data frame, 50-7–50-8
data link layer, 50-9
error detection, 50-9
non-return-to-zero, 50-7–50-8
physical layer, 50-9
traffic shaping strategy, 50-9
Medium access layer
basic protocol
aperiodic mode, 46-7–46-8, 46-10
bus cycle start and initialization, 46-9
features, 46-15
flow chart, 46-8–46-9
periodic mode, 46-7–46-9
synchronization gap (SG), 46-8
terminal gap (TG), 46-8
transmit interval (TI), 46-8
combined protocol
bus cycle, 46-10–46-12
features, 46-15
MAC architecture, 46-12–46-13
normal operation, 46-13–46-14
timers, 46-10–46-11
transmitter scheduler architecture, 46-12–46-13
CSMA/CA, 46-7
Medium access units (MAUs), 44-6–44-8, 52-4–52-5, 58-4
Medium dependent interface (MDI), 52-4
MEDL, *see* Message Descriptor List (MEDL)
Memory management unit (MMU) integrity, 42-15, 48-6
Message block, 47-2–47-3, 47-5–47-6
Message channel (MSC), 28-2
acknowledgment messages, 28-9
acyclic communications, 28-8
control fields, 28-10
data transfer commands, 28-9
init_fast frame, 28-9
init frame, 28-9
message transfer protocol (MTP), 28-9–28-11
MSCL frame, 28-9–28-10
MSCN frame, 28-9–28-10
priority levels, 28-9
send frame messages, 28-9
Message controller, 43-7
Message Descriptor List (MEDL)
configuration services, 40-10–40-11
minimum integration value, 40-8
TDMA slot, 40-4–40-6
Message error (ME), 43-11
Message Queuing Telemetry Transport (MQTT)
features, 59-15
message format, 59-15–59-16
sensor networks, 59-15
Message Transmission Optimization Mechanism (MTOM), 26-8
Metering Bus (M-Bus), 55-9–55-10, 60-18
Microcontroller unit (MCU) driver, 51-25–51-26
MIL-STD-1553, 45-2
analog avionic systems, 43-1–43-2
data bus characteristics, 43-3–43-4
digital avionic systems, 43-2
hardware elements
bus controller, 43-7
bus monitor, 43-7–43-8
remote terminal, 43-5–43-7
terminal hardware, 43-7–43-8
transmission media, 43-4–43-4
history and applications, 43-3
message transmission formats
broadcast information transfer formats, 43-13–43-15
bus controller to remote terminal, 43-13–43-14
command and message validation, 43-15
illegal commands, 43-15–43-16
information transfer formats, 43-13
intermessage gap, 43-16
mode command formats, 43-15
remote terminal to bus controller, 43-14
remote terminal to remote terminal, 43-14–43-15

superseding commands, 43-16
terminal response time, 43-16
mode codes (*see* Mode codes)
systems configurations, 43-2
systems-level issues
data buffering, 43-22
data validation, 43-23
data wraparound, 43-22
error processing, 43-24
major and minor frame, 43-23–43-24
sample consistency, 43-23
subaddress utilization, 43-21–43-22
variable message blocks, 43-23
TDM, 43-2–43-3
testing, 43-24–43-25
word types
bit encoding, 43-9
command word, 43-10
data word, 43-10
status word, 43-11–43-13
sync fields, 43-9–43-10
word formats, 43-9
word length, 43-9
MIMO transmission, *see* Multiple input multiple output (MIMO) transmission
Mixed-mode transducer interface (MMI), 4-10
MMS, *see* Manufacturing Message Specification (MMS)
Modbus application protocol (MBAP), 55-7
Modbus™ protocol, 55-7–55-8, 61-12–61-13
in abstract, 10-4
APDUs, 10-2
code field, 10-6–10-7
data field, 10-7
structure, 10-5–10-6
unit ID field, 10-6
CIP, 10-29–10-34
client–server
broadcast interaction, 10-7–10-8
colocations, 10-4–10-5
function codes, 10-12–10-13
I/O device, 10-4
messaging, 10-2, 10-5
scanner, 10-5
state machines, 10-9–10-10
system response time, 10-4
unicast interaction, 10-7–10-8
communication stacks, 10-2
conformance tests, 10-34
de facto standard, 10-1
distinct table interpretation, 10-10–10-11
exception codes, 10-14
features, 10-3–10-4
gateways, 10-28–10-29
network example, 10-2–10-3
overlapping table interpretation, 10-10–10-11
primitive data types, 10-10
serial line
ASCII, 10-16, 10-18–10-20
broadcast interaction, 10-15
error checking field, 10-16
OSI layers, 10-12, 10-14
PDU, 10-16
RS232 network, 10-13, 10-15
RS485 network, 10-13, 10-15
RTU transmission mode, 10-16–10-18
unicast interaction, 10-15
TCP
assigned port, 10-22
BSD socket interface, 10-23
client activity flowchart, 10-23
confirmation activity flowchart, 10-23, 10-25
connection establishment, 10-22
gateways, 10-22
and IP encapsulation, 10-21
layer parameters, 10-26–10-28
OSI layers, 10-20
PDU checking flowchart, 10-23, 10-27
protocol ID, 10-22
request activity flowchart, 10-23–10-24
server activity flowchart, 10-23, 10-26
service processing activity flowchart, 10-23, 10-28
SO-KEEPALIVE parameters, 10-25–10-26
SO-RCVBUF parameters, 10-24
SO-REUSEADDR parameters, 10-25
SO-SNDBUF parameters, 10-24
stack, 10-20
state machines, 10-22
streaming protocol, 10-22
TCP-NODELAY parameters, 10-24–10-25
transaction ID, 10-21–10-22
unit ID, 10-22
transmission order, 10-12
web-based community, 10-2
web services
application level, 10-35
business agility, 10-37
infrastructure services, 10-37
plug&play mechanism, 10-36
service broker, 10-35
service provider, 10-35
SOA (*see* Service-oriented architecture (SOA))
system-level services, 10-37
Mode codes, 43-17
broadcast, 43-21
data bus management, 43-16
error handling/recovery, 43-16
functions
dynamic bus control, 43-17–43-18
inhibit terminal flag, 43-19
initiate self-test, 43-18–43-19
override inhibit terminal flag, 43-19
override selected transmitter shutdown, 43-20
override transmitter shutdown, 43-19

reserved mode codes, 43-20–43-21
reset remote terminal, 43-19
selected transmitter shutdown, 43-20
synchronize, 43-18, 43-20
transmit BIT word, 43-20
transmit last CW, 43-20
transmit status word, 43-18
transmitter shutdown, 43-19
transmit vector word, 43-20
identifier, 43-17
required mode codes, 43-21
Modular Aerospace Control (MAC), 40-2–40-3
Modular avionics unit (MAU), 44-6–44-8
MOST network, 50-16–50-17
MQTT, *see* Message Queuing Telemetry Transport (MQTT)
MRP, *see* Media redundancy protocol (MRP)
MRPD, *see* Media redundancy for planned duplication (MRPD)
MSC, *see* Message channel (MSC)
MTConnect
adapter, 23-6–23-8
agent, 23-8–23-9
classic N^2 problem, discrete manufacturing, 23-1–23-3
connection-dependent device, 23-5
data dictionary, 23-9–23-11
Hello World application, 23-11–23-12
native device, 23-4
OBD-II scanner, 23-2–23-3
programmatic terms, 23-9
standards, 23-3–23-4
translation-dependent device, 23-5
Multifunction vehicle bus (MVB), 65-1
Multiple input multiple output (MIMO) transmission, 30-8–30-9
Multiple-writer-allowed zones (M-zones), 52-30–52-31

N

Named Type Object, 24-31–24-32
Named Variable List Object, 24-20, 24-29–24-31
Named Variable Object
attributes, 24-26–24-28
client application, 24-28–24-29
operations, 24-27–24-28
report message, 24-29
system-specific optimal reference, 24-29
NCFs, *see* Node capability files (NCFs)
Network address translator (NAT), 22-4, 25-14–25-15
Networked control systems (NCS)
analytical perspective, 2-30–2-31
control network operation, 2-2
diagnostics, 2-32
differentiation
CAN, 2-13–2-14
Ethernet (*see* Ethernet)
MAC sublayer protocol, 2-12
TDM, 2-12–2-13
wireless networks (*see* Wireless networks)
event-based control, 2-31
experimental perspective, 2-29
factory-floor control, 2-32
intermediate level, 2-32
lower-level networked control, 2-31
network protocols, 2-3
network speed, 2-3
parameterization
collision avoidance (CA), 2-4
delay and jitter (*see* Delay and jitter)
QoS, 2-10–2-11
random access (RA), 2-4
speed and bandwidth, 2-5–2-6
TDM, 2-4
real-time control, 2-31
RFT, 2-33–2-36
safety, 2-2, 2-32–2-33
SCADA systems, 2-2
theoretical information perspective, 2-28
Networked embedded devices
application development
abstract application models, 3-15
ANSI/ISA-88.01-1995 hierarchical model, 3-14
block diagram, 3-14
cluster description meta-information, 3-16
cluster node information, 3-16
communication configuration information, 3-16
communication properties, 3-15
event-triggered communication, 3-16
function blocks mapping, 3-15
multicycle polling, 3-16
object dictionary (OD), 3-15
time-triggered communication model, 3-16
two-level design approach, 3-14
complexity management, 3-2
configuration and management, 3-4–3-5
constrained application protocol, 3-13
distributed sensing, 3-2
DPWS, 3-12
DTM, 3-10
EDDL, 3-10
EXI, 3-12–3-13
FDT, 3-10
hardware configuration, 3-17
h-state, 3-3
in-system serial programming approach, 3-18–3-19
interface file system, 3-11–3-12
interface separation
configuration and planning, 3-5
diagnostic and management, 3-5
IFS, 3-5–3-6
real-time service, 3-5
resource-oriented architectures, 3-7–3-8
service-oriented device architectures, 3-7

i-state, 3-3
legacy systems, 3-2
maintenance, 3-20–3-22
management interfaces, 3-19–3-20
plug and participate, 3-3, 3-17
plug and play, 3-3
profiles, 3-9–3-10
service discovery mechanisms, 3-18
smart/intelligent device, 3-3
STD, 3-11–3-12
TEDS, 3-10–3-11
UPnP standard, 3-12
Network media, 57-3
Network time protocol (NTP)
client/server, symmetric, and broadcast variant, 21-7
clock discipline algorithm, 21-15
clock filter algorithm, 21-13
combine algorithm, 21-15
implementation model, 21-9–21-10
mitigation algorithm, 21-13–21-14
network data packet, 21-12
on-wire protocol, 21-10–21-12
server stratum number, 21-7–21-8
simple network time protocol, 21-16
time format, 21-8–21-9
time scale, 21-8–21-9
Network time unit (NTU), 41-4–41-5, 52-29
Network variables (NVs), 55-5, 56-6, 60-24
Neuron C derivative language, 56-5–56-6
Neuron chip
block diagram, 56-2–56-3
I/O system, 56-2, 56-4
memory layout, 56-2, 56-4
node architecture, 56-2–56-3
Never-Give-Up (NGU) strategy, 40-9
N-modular redundancy (NMR), 39-3, 39-5
NodeBuilder, 56-11–56-12
Node capability files (NCFs), 54-5–54-6
Noncontinuous repetition, 47-2
Non-real-time (NRT) application, 17-7
Non-return-to-zero (NRZ) bit encoding, 52-6
Non-return to zero (NRZ) data format, 47-4–47-5
Nonstoring mode, 38-6–38-7
NTU, *see* Network time unit (NTU)

O

OASIS, *see* Organization for the Advancement of Structured Information Standards (OASIS)
OBIS naming system, *see* Object Identification System (OBIS) naming system
OBIX, *see* Open building information exchange (OBIX)
Object description tables (ODTs), 51-9–51-10
Object dictionary (OD), 3-15, 14-6, 15-7, 18-17, 19-2, 52-34–52-35
Object Identification System (OBIS) naming system, 60-16
Object library
application specific, 9-11
Assembly Objects, 9-15–9-17
general-use objects, 9-11
Identity Object, 9-12–9-15
network specific, 9-12
object model, 9-12–9-13
parameter object, 9-15
Object linking and embedding for process control (OPC), 13-26–13-27, 22-19–22-20, 55-13–55-16
OD, *see* Object dictionary (OD)
Onboard diagnostics (OBD), 51-7
One channel safe (OCS) communication
availability, 64-9
components, 64-8–64-9
message creation, 64-9
OPC, *see* Open platform communications (OPC)
OPC Alarms and Events (OPC A&E), 55-14
OPC Data Access (OPC DA), 55-13–55-14
OPC Historical Data Access (OPC HDA), 55-14
OPC Unified Architecture (OPC-UA), 59-12–59-14, 61-11
OpenADR, 61-13–61-14
Open building information exchange (OBIX), 55-16–55-17
Open platform communications (OPC), 2-20, 61-15–61-16
OpenSAFETY protocol, 19-8–19-9
Open Service Gateway Initiative (OSGi) specification, 26-3
Open Smart Grid Protocol (OSGP), 61-17
Operations management center (OMC), 64-1–64-2
architecture, 64-10–64-11
communication interfaces, 64-11–64-12
functions, 64-10
interface, 64-11
interlocking systems/substations, 64-9–64-10
Optical transmission, 36-49
Organization for the Advancement of Structured Information Standards (OASIS), 61-15
Orion space system
TTEthernet
Ares I interface, 42-12–42-13
avionics architecture interface, 42-16–42-17
command/monitor configuration, 42-15, 42-17
crew and service module, 42-12
IMA architecture approach, 42-12–42-13
MMU integrity, 42-15
requirements, 42-12–42-13
self-checking pair computer, 42-13–42-15
structure, 42-12
TT-GbE, 42-15, 42-17
VMC, 42-17
OSGP, *see* Open Smart Grid Protocol (OSGP)

P

PAN, *see* Personal area network (PAN)
PCS, *see* Precise clock synchronization (PCS)
PDOs, *see* Process data objects (PDOs)
PDU, *see* Protocol data units (PDUs)
Peer delay mechanism, 20-8–20-9
Personal area network (PAN), 55-8
Physical layer signaling (PLS), 52-4
Physical medium attachment (PMA), 13-6–13-7, 52-4
PICS, *see* Protocol implementation conformance statement (PICS)
PID, *see* Production Identifier (PID)
PLC network, *see* Powerline communication (PLC) network
PLS, *see* Physical layer signaling (PLS)
PMA, *see* Physical medium attachment (PMA)
Power24 technology, 6-4
Powerline communication (PLC) network, 29-15
Power line–related intelligent metering evolution (PRIME), 60-5
Power system automation
  building automation and control systems (BACSs)
    BACnet, 61-19–61-20
    energy consumption, 61-17
    KNX Stack, 61-19
    LonWorks, 61-17–61-18
    requirements, 61-16
  CIM (*see* Common Information Model (CIM))
  data and communication security, 61-11–61-12
  DNP3, 61-9–61-10
  EMIX, 61-15
  ICCP/TASE.2, 61-8
  IEC 60870-5, 61-7–61-8
  IEC SIA, 61-2–61-4
  Modbus, 61-12–61-13
  OPC, 61-15–61-16
  OpenADR, 61-13–61-14
  PROFIBUS protocol, 61-11
  RP-570 protocol, 61-10
  UCA, 61-10
Precise clock synchronization (PCS), 28-2
Precision time protocol (PTP), 16-13–16-14
  asymmetry correction, 21-19–21-20
  end-to-end synchronization, 21-20
  master clock election, 21-17–21-18
  model, 21-16–21-17
  offset and skew correction, 21-18–21-19
  peer-to-peer synchronization, 21-20
  SAS, 21-24–21-25
  synchronous Ethernet, 21-23–21-24
  timestamp generation, 21-21–21-23
  transparent clock, 21-20–21-21
Preemptive fixed priority (PFP), 44-7–44-8
PRIME, *see* Power line–related intelligent metering evolution (PRIME)
Primus Epic®
  cabinet (MAU), 44-6–44-8
  certified airframes, 44-6
  DEOS, 44-7–44-8
  four-bus diagram, 44-7, 44-9
  high-level diagram, 44-7
  slack stealing, 44-8
  subframe-level scheduling, 44-8
Process data objects (PDOs)
  configuration, 52-38–52-39
  event-driven transmission, 52-36
  RPDO, 52-35–52-36
  SDOs, 52-39–52-40
  synchronous operations, 52-36–52-37
  TPDO, 52-35–52-36
  triggering modes, 52-37–52-38
Process field bus (PROFIBUS)
  application profiles
    HART, 11-15, 11-17
    host application profiles, 11-22
    identification and maintenance (I&M) functions, 11-14–11-15
    ID referencing, 11-14, 11-16
    PROFIdrive application profile, 11-17–11-19
    profile PA devices, 11-20–11-22
    redundancy communication (RedCom), 11-16
    slave redundancy application profile, 11-15, 11-18
    time stamp application profile, 11-15, 11-17
  applications, 11-28
  decentralized applications, 11-1–11-2
  decentralized peripherals (DP)
    cyclic and acyclic data communication (*see* Cyclic and acyclic data communication)
    DP master class 1 (DPM1), 11-9
    DP master class 2 (DPM2), 11-9–11-10
    DP-V0 version, 11-9
    DP-V1 version, 11-9
    DP-V2 version, 11-9
    monomaster system, 11-9–11-10
    slaves, 11-10
  EIA/TIA-485 transmission technology, 11-3–11-5
  fiber optic communication, 11-8–11-9
  fieldbus system, 30-14–30-15
  functional safety (*see* PROFIsafe)
  implementation support, 11-27–11-28
  integration technologies
    EDD, 11-24
    FDI, 11-24–11-26
    FDT/DTM, 11-24
    GSD, 11-23–11-24
    performance range, 11-23
  MBP, IEC 61158-2
    applications, 11-7
    bus powering, 11-5–11-6
    characteristics, 11-5–11-6
    current version, 11-5

FISCO model, 11-7
high-power trunk concept, 11-7–11-8
logical 1 and 0 determination, 11-5–11-6
mid-bit transition, 11-5
network topologies, 11-7
process automation, 11-5–11-6
protocol, 61-11
quality assurance, 11-26–11-27
structure, 11-2–11-3
Production Identifier (PID), 9-80
PROFIBUS, *see* Process field bus (PROFIBUS)
PROFIdrive, 12-24–12-25
PROFIenergy, 12-25
PROFINET
architecture, 12-2–12-3
autocrossover, 12-9
automatically reacting redundancy solutions, 12-3
autonegotiation, 12-9
certification test, 12-31
communication paths, 12-5
communication principles
Ethernet based communication, 12-12
Ethertype 0x8892, 12-11
protocol definitions, 12-11–12-12
real-time communication, 12-12–12-13
standard communication, 12-12
VLAN tag, 12-11
configuration in run (CiR) measures, 12-20, 12-22
conformance classes (*see* Conformance classes (CCs))
extended device identification, 12-20–12-21
fast start-up, 12-20, 12-22
fieldbus systems integration, 12-4, 12-23–12-24
general station description file (GSD), 12-6–12-7
individual parameter server functionality, 12-20–12-21
installation technology
cables, 12-27
network configuration, 12-26–12-27
plug connectors, 12-28
security concept, 12-28–12-29
IO-Controller
application relations, 12-7–12-8
communication relations, 12-7–12-8
data exchange, 12-9, 12-11
diagnostics, 12-11
engineering tool, 12-9
programmable logic controller (PLC), 12-5
IO-Device
application relations, 12-7
communication relations, 12-7
data exchange, 12-11
DCP, 12-8
device model, 12-5
diagnostics, 12-11
I/O data addressing, 12-6
name assigning, 12-9
OUI, 12-9
structures, 12-6
switches, 12-9–12-10
IO-Supervisors, 12-5
MRP, 12-20, 12-23
MRPD, 12-20, 12-23
multiple access to field devices, 12-20–12-21
neighborhood detection functionality, 12-3
network installations, 12-4
PI test labs (PITLs), 12-29
process automation, 12-25–12-26
product development, 12-30
PROFIdrive, 12-24–12-25
PROFIenergy, 12-25
PROFIsafe (*see* PROFIsafe)
real-time capable solutions, 12-2
requirements, 12-2
standardization, 12-2
structure, 12-29–12-30
subnets, 12-8
technology development, 12-30
time stamping, 12-20, 12-22
UDP/IP, 12-2
web integration, 12-4
PROFIsafe, 11-14, 11-16, 12-3, 12-24
audit trailing, 27-21
authenticity, 27-6
availability, 27-15–27-16
black channel, 27-4–27-6
codename, 27-6
1:1 communication relationship, 27-6
data integrity, 27-7
defined transmission systems, 27-2
development tools, 27-19–27-20
environmental conditions, 27-18
finite state machines, 27-13
F-Parameters, 27-13
functional safety automation, 27-1–27-2
guidelines, 27-16
IEC 61508, 27-2, 27-6–27-7
IEC 62280-1 railway standard, 27-2
iParameters, 27-20
iPar-Server mechanism, 27-14–27-15
objectives, 27-2
open transmission systems, 27-2
parameters, 27-9
products, 27-20
PROFIsafe PDU
CRC securing, 27-11
definition, 27-4
F-Device, 27-10–27-11
F-Host monitor, 27-8–27-10
message frame structure, 27-9–27-10
MNR-generator, 27-11–27-12
propagation times monitoring, 27-12
programmable logic controllers (PLCs), 27-13
protection equipment, 27-17

protocol development, 27-7–27-8
proven-in-use field device, 27-17
regulations, 27-20
remote I/O, 27-17
safety and redundancy options, 27-3
safety layer, 27-4
SIL monitor
configuration, 27-8
error detection, 27-7
PROFIBUS DP message structure, 27-9
PROFINET IO message structure, 27-10
white channel approach, 27-8
specification, 27-17
standards, 27-16–27-17
test and certification, 27-18–27-19
three-component model, 27-14
timeliness, 27-6–27-7
vision, 27-2–27-3
watchdog timer, 27-6
Program invocation management, 24-14–24-16
proprietary solutions, 36-22
Protocol control information (PCI), 51-6, 58-5
Protocol data units (PDUs), 51-20, 58-5, 61-12
Protocol identifier, 45-15
Protocol implementation conformance statement (PICS), 57-9, 61-19
Protocol Specification for Electricity Metering (PSEM), 60-6
Protocol words, 45-15
Publisher–subscriber model, 1-31–1-34
Push-button (PB) mode, 58-12

Q

Queuing delay, 52-20–52-21

R

Radio-frequency identification (RFID), 59-2
Rapid Spanning Tree Protocol (RSTP), 9-51, 9-60
quality of service, 16-10
ring topology network, 16-9
STP, 16-8
VLAN, 16-12
RBE, *see* Report by exception (RBE)
Real-time Ethernet (RTE), 1-39–1-40
communication interfaces, 17-7
CP, 17-2, 17-4–17-5
CPF, 17-2–17-5
EPA protocol, 17-14–17-15
EPL, 17-11–17-13
EtherCAT
mailbox mechanism, 17-19
medium access control, 17-17
open ring bus, 17-18
performance indicator, 17-19
slave device, 17-17–17-18
NRT application, 17-7
OSI reference model, 17-2–17-3
performance indicator (PI), 17-6–17-7
PROFINET CBA, 17-15–17-16
PROFINET IO, 17-19–17-21
SERCOS, 17-16–17-17
TCnet, 17-13–17-14
TCP/IP protocol
EtherNet/IP, 17-9
Modbus, 17-8
P-NET package, 17-10
Vnet/IP, 17-10–17-11
user application requirements, 17-6
Real-time publisher subscriber (RTPS) protocol, 17-8
Real-time service, 3-5
Receive error count (REC), 52-14
Receive process data objects (RPDO), 52-35–52-36
Receiver fault isolation, 45-5
Receiver input impedance, 45-4
Reconfigurable factory testbed (RFT), 2-33–2-36
Reduced-function devices (RFDs), 36-23
Redundant managing nodes (RMNs), 19-7
Reliability, availability, maintainability, and safety (RAMS), 64-2–64-3
communication, 64-7
computation, 64-6–64-7
diversity, 64-4
fault detection, 64-4
fault tolerance, 64-4
redundancy concepts, 64-4–64-5
replication concepts, 64-4–64-5
Remote Operations Management (ROM), 14-16–14-17
Remote terminal, 43-5–43-7
Remote transmission request (RTR) bit, 52-9
Report by exception (RBE), 62-28
Residual error probability, 6-13
Resource-oriented architectures, 3-7–3-8
Return-to-zero (RZ) modulation, 45-3
RFT, *see* Reconfigurable factory testbed (RFT)
Ring protection link (RPL), 65-8
Robust header compression (ROHC), 29-15
ROM, *see* Remote Operations Management (ROM)
Routing over low-power and lossy networks (ROLL), 38-5–38-6
Routing protocol for low-power and lossy networks (RPL), 38-6–38-7
RP-570 protocol, 61-10
RPDO, *see* Receive process data objects (RPDO)
RS-485 standard, 11-3–11-5
RSTP, *see* Rapid Spanning Tree Protocol (RSTP)
RTE, *see* Real-time Ethernet (RTE)

S

SAFEbus
- architecture overview, 48-23–48-24
- Boeing 777
  - AIMS, 48-21–48-23
  - avionics architecture, 48-20
  - avionics overview, 48-23, 48-25
  - design requirements, 48-1–48-2
  - integrated architecture, 48-21
  - VIA, 48-25
- bus time, 48-2
- communication interface
  - controlled buffer memory format, 48-19–48-20
  - data stream time partitioning, 48-20–48-21
  - noncontrolled buffer memory format, 48-20
- communication services
  - BCW, 48-5
  - bus encoding, 48-8–48-9
  - clock synchronization (*see* Clock synchronization, SAFEbus)
  - data-message structure, 48-6–48-7
  - determinism and partitioning, 48-5–48-6
  - out-of-band signaling pulses, 48-9
  - table-driven protocol, 48-5
- configuration services
  - frame changes, 48-17
  - sparing and manual reconfiguration, 48-16
- diagnostic services
  - availability *vs.* integrity trade-off, 48-16
  - babble protection, 48-15
  - BIST, 48-14
  - Byzantine protection, 48-15–48-16
  - debugging mechanisms, 48-14
  - fault-containment zone, 48-15
  - near-perfect error-detection, 48-14
  - protection mechanism, 48-14
  - zombie module protection, 48-16
- hardware architecture, 48-2
- IMA system, 48-1
- interface logic, 48-2–48-3
- LRM, 48-1–48-2
- nomenclature, 48-2, 48-4
- protocol parameterization
  - frame description language, 48-18
  - table memory, 48-17–48-18
  - table versioning, 48-19
- terminology, 48-2, 48-4
- validation and verification efforts, 48-20

Safety communication layer management (SALMT), 28-21–28-22

Safety heartbeat service (SHB), 28-21

SafetyNET p protocol
- advantages, 28-16–28-17
- application layer, 28-22–28-23
- black channel approach, 28-17–28-18
- cabling, 28-5
- CANopen application layer, 28-13–28-15
- communication layer
  - process data objects, 28-19–28-20
  - SALMT, 28-21–28-22
  - SHB, 28-20–28-21
- emergency stop system, 28-16
- error detection mechanisms, 28-18–28-19
- general frame structure, 28-7
- IP20 connector, 28-4–28-5
- IP67 connector, 28-4–28-5
- overview, 28-1
- precise clock synchronization, 28-15–28-16
- RTFL
  - CDC, 28-2, 28-7–28-8
  - CMS, 28-5
  - configuration acknowledgment message, 28-12
  - cut-through switches, 28-3
  - Ethernet frame, 28-6
  - intracell communications, 28-1
  - logical line, 28-3–28-4
  - MAC address, 28-2
  - management frames, 28-3
  - MSC (*see* Message channel (MSC))
  - network verification information, 28-11–28-12
  - PCS, 28-2
  - PDUs, 28-5–28-6
  - physical line, 28-2–28-3
  - producer–consumer paradigm, 28-5
  - root device (RD), 28-2–28-3
  - store-and-forward switches, 28-3
  - TSS, 28-5
- RTFN
  - CDC, 28-2, 28-7–28-8
  - cell to cell communication, 28-2
  - CMS, 28-5
  - connection management frame, 28-12–28-13
  - connection management set, 28-12
  - cyclic PDUs, 28-6
  - Ethernet topologies, 28-4
  - MSC (*see* Message channel (MSC))
  - publisher–subscriber interaction model, 28-6
  - scan network read request, 28-12
  - scan network read response, 28-12
  - TSS, 28-5
  - UDP/IP protocol, 28-2
- structure, 28-2–28-3

SafetyProg Windows software, 15-9–15-10

SALMT, *see* Safety communication layer management (SALMT)

Sample consistency, 43-23

SAP, *see* Service access point (SAP)

SCADA systems, *see* Supervisory control and data acquisition (SCADA) systems

Scheduled access mechanism, 14-4

SCL, *see* Substation configuration language (SCL)

SDOs, *see* Service data objects (SDOs)

SDS, *see* Smart distributed system (SDS)

Self-checking buses (SCBs), 48-2
Self-checking pairs (SCPs), 67-7–67-8
SEMI Equipment Communication Standard (SECS), 66-12
  back-end semiconductor machine
    functional mapping, 66-25
    implementation levels, 66-25
    integration, 66-25–66-27
    intercommunication process, 66-27
    message partition, 66-25
    software tools, 66-24
    specifications and requirements, 66-24
    system architecture, 66-24
  CIM system (*see* Computer-integrated manufacturing (CIM) system)
  components, 66-2–66-3
  definition, 66-2
  equipment communication software, 66-19–66-21
  equipment integration software, 66-19
  GEM
    alarm management, 66-16
    clock, 66-17
    communications, 66-16
    documentation, 66-15
    dynamic event report configuration, 66-16
    equipment constants, 66-16
    equipment terminal services, 66-17
    error messages, 66-15
    event notification, 66-15–66-16
    host-initiated control, 66-15, 66-17
    limitations, 66-17–66-18
    limit monitoring, 66-17
    material movement, 66-17
    online identification, 66-15
    operator-initiated control, 66-15
    pooling, 66-17
    process program management, 66-17
    remote control, 66-16
    state models, 66-13–66-15
    status data collection, 66-16
    trace data collection, 66-16
    variable data collection, 66-16
  HSMS
    message structure, 66-7–66-8
    state model, 66-6–66-7
    subclassification, 66-8
  SECS/GEM solutions, 66-22–66-23
  SECS-I
    block transfer protocol, 66-4
    layer definition, 66-4
    message structure, 66-4–66-5
    point-to-point communication, 66-3
    synchronization mechanism, 66-5–66-6
  SECS-II
    conversation protocols, 66-10–66-12
    data structure, 66-8–66-9
    limitations, 66-17–66-18
    streams and functions, 66-9–66-10
Sercos® automation bus
  advantages, 13-4
  distributed multiaxis control systems, 13-4
  functions, 13-3–13-4
  international standard, 13-2–13-3
  overview, 13-2
  Sercos I, 13-2–13-3
  Sercos II, 13-2–13-3
    acknowledge telegram structure, 13-38–13-39
    ASIC, 13-31–13-33
    conformance testing, 13-42–13-43
    cycle times, 13-36
    cyclic operation, 13-40–13-41
    error correction/diagnostics, 13-41–13-42
    fiber optics, 13-30–13-31
    FPGA-based Sercos controller, 13-3
    GPCC-based Sercos controller, 13-3
    identification numbers (IDNs), 13-34–13-35
    interface placement, 13-33
    I/O functions, 13-36
    machine tool applications, 13-43–13-44
    MDT, 13-38–13-40
    MST, 13-37–13-38
    safety redundancy, 13-42
    Sercon816 ASIC, 13-3
    service channel, 13-40–13-41
    system of units, 13-36
    telegram format and sequence, 13-36–13-37
    timing, 13-33
    topology, 13-30
  Sercos III, 13-2–13-3
    addressing, 13-14–13-15
    advantages, 13-5
    applications, 13-27–13-29
    blended infrastructure approach, 13-17
    CIP safety, 13-17–13-18
    communication connectivity, 13-26–13-27
    communication cycle, 13-8–13-9
    communication phases, 13-13
    configuration, 13-20–13-21
    conformance testing, 13-24–13-25
    data link layer, 13-5–13-6
    device description language, 13-20
    direct cross communication, 13-15–13-16
    encoder profile, 13-20
    energy profile, 13-20
    Energy White Paper, 13-26
    Ethernet technology, 13-5–13-6
    generic device profile (GDP), 13-19
    hot-plugging, 13-15
    hot-swapping, 13-15
    identification numbers (IDNs), 13-11–13-13
    IEC standards, 13-25
    IO-Link integration, 13-26
    I/O profile, 13-19

line topology, 9-85–9-86, 13-7
machine-to-machine communication, 13-26
machine-to-supervisory communication, 13-26
master and slave implementations, 13-21–13-22
network installation, 13-7–13-8
oversampling, 13-16
performance, 13-27
physical layer, 13-5–13-6
planning and installation guide, 13-25–13-26
producer–consumer model, 13-13–13-14
redundancy, 13-15
ring topology, 9-85–9-86, 13-7
software packages, 13-22–13-24
star topology, 13-7–13-8
synchronization, 13-11
telegrams, 13-10–13-11
time-stamping, 13-16–13-17
unified communication channel, 13-9–13-10
Serial real-time communications system, *see* Sercos® automation bus
Service access point (SAP), 1-12
Service data objects (SDOs), 52-39–52-40
Service data units (SDUs), 58-5
Service discovery mechanisms, 3-18
Service Get Capability List, 24-12
Service-oriented architecture (SOA)
contract-first design, 10-40–10-41
evaluation, 26-4–26-5
features, 10-36
interface contracts, 10-37
interoperability, 10-39
Jini, 26-4
loose coupling, 10-37–10-38
nonfunctional requirements, 10-39
OSGi specification, 26-3
principles, 10-36
profiles, 26-4
qualities of a service, 10-39
Schneider SOA recommendations, 10-38
security, 10-39
services, 26-2–26-3
strong cohesion, 10-38
UPnP, 26-4
wire format, 10-38
Service-Oriented Architecture for Devices (SOA4D), 3-7, 26-24
SG-WG, *see* Smart Grid Working Group (SG-WG)
SHB, *see* Safety heartbeat service (SHB)
Short Name (SN) referencing, 60-15
Sign/status matrix (SSM)
BCD numeric, 45-8–45-9
discrete data words, 45-9–45-10
failure warning, 45-7–45-8
invalid data, 45-7
no computed data, 45-7–45-8
Simple network management protocol (SNMP), 22-14, 22-15
Simple network time protocol (SNTP), 16-12–16-13
Simple Object Access Protocol (SOAP), 26-7–26-8
Single-drop digital communication system, *see* IO-Link technology
Single-writer-allowed zone (S-zone), 52-30–52-31
Site Security Handbook, 29-2
6LoWPAN, *see* IPv6 over Low-Power Wireless Personal Area Networks (6LoWPAN)
Slave profiles, 6-15
actual ID code, 6-23
address register, 6-34
block diagram, 6-33
data exchange blocked flag, 6-34
data output register, 6-34
extended ID code 1, 6-23
extended ID code 2, 6-23
function request, 6-19–6-20
identification register, 6-34
I/O configuration, 6-22–6-23
jabber inhibit logic, 6-33
overview, 6-21–6-22
parameter output register, 6-34
pin assignment, 6-20–6-21
properties, 6-20
receive register, 6-34
send register, 6-34
status diagram, 6-20–6-21
status display, 6-35
status register, 6-34
synchronization register, 6-34
watchdog, 6-35
Slow-fading phenomena, 30-7
Smart distributed system (SDS), 1-8, 1-20, 52-3
Smart energy profile (SEP), 59-17–59-18
Smart Grid Working Group (SG-WG), 60-24
SmartMesh-XD, 36-16–36-17
SmartMesh-XD Mote-on-Chip (MoC), 36-16–36-17
Smart Message Language (SML), 60-10–60-11
Smart Transducer Descriptions (STD), 3-11–3-12
Smart transducer interface module (STIM), 4-3
Smart transducers
advantages, 4-4
block diagram, 4-2
definition, 4-2
features, 4-4
functional partitioning, 4-3
integrated networked smart transducer, 4-3
standards (*see* IEEE 1451.2 standard; IEEE 1451 standards)
STIM, 4-3
SMI, *see* Standard motor interface (SMI)
SMK, *see* System management kernel (SMK)
SML, *see* Smart Message Language (SML)
SNMP, *see* Simple network management protocol (SNMP)
SNVTs, *see* Standard network variables types (SNVTs)
SOA, *see* Service-oriented architecture (SOA)

SOA4D, *see* Service-Oriented Architecture for Devices (SOA4D)
SOAP, *see* Simple Object Access Protocol (SOAP)
SOAP-based web services, 38-11
Spanning Tree Protocol (STP), 16-8
Special Interest Group (SIG), 9-22
SSM, *see* Sign/status matrix (SSM)
Standardized automotive system software
  ASAM (*see* XCP)
  AUTOSAR (*see* Automotive open system architecture (AUTOSAR))
  hardware architecture, 51-2–51-3
  ISO
    CAN, 51-4–51-6
    DoIP specification, 51-7–51-8
    FlexRay, 51-6
    OBD, 51-7
    UDS, 51-6–51-7
  OSEK/VDX standard, 51-3–51-4
Standard motor interface (SMI), 55-9
Standard network variables types (SNVTs), 55-6, 56-6
Standby managing node (SMN), 19-7
Start-of-cycle (SoC) frame, 17-11–17-12, 19-4, 19-10
Status word (SW)
  bits
    broadcast command received, 43-12
    busy, 43-12
    dynamic bus control acceptance, 43-12
    instrumentation, 43-11
    message error, 43-11
    parity, 43-13
    reserved, 43-11
    service request, 43-11
    subsystem flag, 43-12
    terminal address, 43-11
    terminal flag, 43-12–43-13
  reset, 43-11
STD, *see* Smart transducer descriptions (STD)
STIM, *see* Smart transducer interface module (STIM)
STMicroelectronics, 13-29
Storing mode, 38-6–38-7
Stream Reservation Protocol (SRP)
  listener ready attribute, 20-7
  MRP, 20-5–20-6
  talker advertise attribute, 20-6–20-7
Structured Query Language (SQL), 22-17–22-18
Subaddress (SAs), 43-21–43-22
Substation automation system (SAS), 21-24–21-25
Substation configuration language (SCL), 62-6, 62-30–62-32
Substitute remote request (SRR) bit, 52-10
Supervisory control and data acquisition (SCADA) systems, 2-2
Switched Ethernet
  availability, 16-2–16-3
  communication performance, 16-3
  definition, 16-4–16-5
  EtherCAT, 16-14
  features, 16-2
  HLE, 16-7
  MAC entity, 16-5–16-7
  multicasting, 16-10–16-12
  network architecture, 16-4
  PTP, 16-13–16-14
  quality of service (QoS), 16-10–16-11
  redundancy protocol
    IEC 62439 standard, 16-9–16-10
    industrial requirements, 16-8
    principles, 16-7
    ring topology network, 16-8–16-9
    STP, 16-8
  security, 16-4
  SNTP, 16-12–16-13
  time synchronization, 16-3–16-4, 16-12
  topology, 16-2
  VLAN, 16-11–16-12
SyM² Project, *see* Smart Message Language (SML)
Synchronization block, 47-3
Synchronization-INTegration (SINT) messages, 67-9–67-10
Synchronous data acquisition (DAQ) data, 51-9–51-10
System management kernel (SMK), 14-11
S-zone, *see* Single-writer-allowed zone (S-zone)

T

TDM, *see* Time-division multiplexing (TDM)
TDMA strategy, *see* Time division multiple access (TDMA)
TEC, *see* Transmission error count (TEC)
TEDS formats, *see* Transducer electronic data sheet (TEDS) formats
TeMIX, 61-15
Terminal address (TA), 43-11
Terminal electrical characteristics, 43-7–43-8
Third-party security evaluation, 29-4
Three-zone security model, 29-16–29-18
Time-critical Control Network (TCnet), 17-13–17-14
Time division multiple access (TDMA), 1-23
  communication medium, 53-2
  communication service, 40-4–40-6
  ECU architecture, 53-3
  information domain, 53-1
  signal domain, 53-12
  symbol window, 53-4
Time-division multiplexing (TDM), 2-4, 2-12–2-13, 43-2–43-3
Timer module, 46-12
Time-sensitive networking (TSN)
  clock synchronization protocol, 20-8–20-9
  IEEE 1722 and IEEE 1722.1, 20-13
  physical topology, 20-2–20-4
  redundancy management, 20-12

SRP
listener ready attributes, 20-7
MRP, 20-5–20-6
talker advertise attributes, 20-6–20-7
switching, 20-3–20-4
target domain and components, 20-13–20-14
traffic shaping mechanism
active topology, 20-4–20-5
credit-based shaper, 20-9–20-11
engineered ring topology, 20-4
time-aware shaper, 20-11–20-12
VLAN identifier, 20-3–20-4
Time synchronization service (TSS), 28-5
Time-triggered communication, 3-16
access delay, 39-7
bandwidth, 39-7
clock synchronization, 39-2
composability, 39-4–39-5
delay jitter, 39-8
diagnostic services, 39-3–39-4
event-triggered communication, 39-1–39-2
fault isolation mechanism, 39-3
FCR, 39-5–39-6
network delay, 39-7–39-8
replica determinism, 39-6
spatial partitioning, 39-6
state messages, 39-2–39-3
temporal partitioning, 39-6
time division multiple access, 39-2, 39-7
Time-triggered controller area network (TTCAN)
Bosch IP module, 41-16–41-17
clock synchronization
cycle time, 41-6
global time, 41-6–41-7
local time, 41-5
reference message, 41-4–41-5
communication services, 41-2–41-3
configuration interface, 41-14–41-15
coupled system, 41-13–41-14
cyclic redundancy check, 41-14–41-15
diagnostic services
bit stuffing, 41-12
error counters, 41-10–41-11
error frames, 41-11
error detection mechanism, 41-12
error severity, 41-12
features, 52-27
FTT-CAN, 41-15–41-16
implementation, 52-29–52-30
mailbox system, 41-13
MultiCAN IP module, 41-17
protocol, 50-13–50-14
Rx_Trigger, 41-8
start-up protocol, 41-8–41-10
synchronization problem, 41-14
system matrix, 52-28–52-29
time-triggered communication, 52-27–52-28
Tx_Trigger, 41-8
Time-triggered Ethernet (TTEthernet)
compatibility, 42-2
cross-industry communication, 42-2
event-triggered messages, 41-1–42-2
FPGA-based system, 42-9–42-11
frame format, 42-3
implementation, 42-2
network configuration
dual-fault-tolerant, 42-9
master-based, 42-8–42-9
system-of-systems, 42-9–42-10
network topology, 42-2
Orion space system (*see* Orion space system)
software stack, 42-10–42-11
switches, 42-10–42-11
synchronization protocol
clique detection algorithm, 42-7
clock synchronization, 42-6–42-7
start-up protocol, 42-6
traffic class
best-effort messages, 42-4
communication modes, 42-3
preemption method, 42-5
priority scheme, 42-5
rate-constrained messages, 42-4
time-triggered messages, 42-4
Time-Triggered Gigabit Ethernet (TT-GbE), 42-15, 45-17
Time-Triggered Protocol (TTP/C)
big bang mechanism, 40-8
bus guardian, 40-9–40-10
clique avoidance algorithm, 40-9
clock synchronization, 40-7
cluster start-up process, 40-7–40-8
communication service
frame format, 40-6–40-7
TDMA slot, 40-4–40-6
configuration service, 40-10–40-11
consistent distributed computing, 40-1
controller life-sign, 40-8
integration process, 40-8
LockTrac 6131 ELEKTRA system, 40-1
MAC, 40-2–40-3
Message Descriptor List, 40-8
NGU strategy, 40-9
railway domain, 40-1
Time unit ratio (TUR), 41-6–41-8, 52-29
Tissue data base, 62-41
TLS, *see* Transport Layer Security (TLS)
Token passing mechanism, 14-5
TPDO, *see* Transmit process data objects (TPDO)
Train communication network (TCN)
architecture, 65-1–65-2

embedded environment
data processing, 65-7
jitters, 65-6–65-7
network platforms, 65-6
real time requirements, 65-6
ring configuration, 65-8
RPL, 65-8
ETBN, 65-9
Ethernet bus topology, 65-10
Ethernet switches, 65-3–65-4
fault conditions and safety measures, 65-12
field bus, 65-9
Layer 2 protocols, 65-3
Layer 3 protocols, 65-3
multicast transmission, 65-4–65-5
MVB, 65-1
requirements, 65-2
safety layer, 65-11
signals and response times, 65-9–65-10
VLANs, 65-4–65-5
WTB, 65-1
Train control
dispatchers, 64-1
fail-safe system, 64-3
NMC, 64-1
OCS communication
availability, 64-9
components, 64-8–64-9
message creation, 64-9
OMC, 64-1–64-2
architecture, 64-10–64-11
communication interfaces, 64-11–64-12
functions, 64-10
interface, 64-11
interlocking systems/substations, 64-9–64-10
RAMS (*see* Reliability, availability, maintainability, and safety (RAMS))
real-time systems, 64-3
signals and switches
datacryptors, 64-14
ETCS, 64-14–64-15
mobile shunting device, 64-14
operator to switch, 64-13
planner to operator, 64-12–64-13
standards and certification, 64-2
TAS control platform, 64-7–64-8
Transducer electronic data sheet (TEDS) formats, 3-10–3-11, 4-4, 4-6–4-8
Transformer coupled stub method, 43-4–43-5
Transmission Control Protocol (TCP)/Internet Protocol (IP)
EtherNet/IP, 17-9
Modbus, 17-8
P-NET package, 17-10
Vnet/IP, 17-10–17-11
Transmission error count (TEC), 52-14
Transmission media, 43-4–43-4
Transmit process data objects (TPDO), 52-35–52-36
Transmitter control module, 46-12
Transmitter external fault load tolerance, 45-5
Transmitter external fault voltage, 45-4
Transmitter fault isolation, 45-5
Transmitter output impedance, 45-4
Transport Layer Security (TLS), 29-8–29-9
Transport protocol data unit (TPDU), 56-10–56-11
Triple-modular replication (TMR), 67-8–67-9
TSS, *see* Time synchronization service (TSS)
TTCAN, *see* Time-triggered controller area network (TTCAN)
TTEthernet, *see* Time-triggered Ethernet (TTEthernet)
TTP/C, *see* Time-Triggered Protocol (TTP/C)
TUR, *see* Time unit ratio (TUR)

U

UCA International Users Group (UCA IUG), 62-40–62-41
Unconnected Message Manager (UCMM), 9-31
Unidirectional CSDB communication, 47-2
Unified Diagnostic Services (UDS), 51-6–51-7
Universal Plug-and-Play (UPnP) standard, 3-12–3-13, 26-4
Unnamed Variable Object
address, 24-23–24-24
attributes, 24-22
services, 24-24
value, quality, and time, 24-23
User application layer
function block, 14-13
function block linking and scheduling, 14-14–14-15
resource block, 14-13
transducer block, 14-13

V

Variable message blocks, 43-23
Variable model
access path
data types, 24-21
partial trees, 24-19–24-21
path description, 24-21–24-22
task definition, 24-19
tool magazine, 24-18–24-19
data access, 24-17
Named Type Object, 24-31–24-32
Named Variable List Object, 24-29–24-31
Named Variable Object
attributes, 24-26–24-28
client application, 24-28–24-29
operations, 24-27–24-28
report message, 24-29
system-specific optimal reference, 24-29
services, 24-32

type description
arrays, 24-25
character string, 24-25, 24-27
circuit breaker, 24-25–24-26
type specification, 24-24–24-25
unified protocol, 24-16
Unnamed Variable Object
address, 24-23–24-24
attributes, 24-22
services, 24-24
value, quality, and time, 24-23
Vehicle area network (VAN), 50-11
Vehicle distributed executive (VDX) standard, 51-3–51-4
Vehicle management computer (VMC), 42-17
Versatile Integrated Avionics (VIA), 48-25
Virtual field device (VFD) model, 15-7
Virtual local area network (VLAN), 16-11–16-12, 20-3–20-4
Virtual manufacturing device (VMD)
attributes, 24-4–24-5
executive function, 24-11–24-12
functions, 24-4
location, 24-7–24-8
logical status, 24-12
objects and services, 24-4–24-5
physical status, 24-12
Service Get Capability List, 24-12
standard driver, 24-3–24-4
Virtual private networks (VPNs), 25-15, 29-9
Volcano LIN tool chain, 54-7–54-8

W

WCRT, *see* Worst-case response time (WCRT)
Web Service Description Language (WSDL)
abstract part, 26-12
concrete part, 26-12–26-13
Web Service-Policy (WS-Policy), 26-13
Web Services Business Process Execution Language (WS-BPEL), 26-23–26-24
Web Services Dynamic Discovery (WS-Discovery)
low scalability, 26-11
optimization, 26-11
ProbeMatch, 26-10–26-11
UDDI server, 26-9–26-10
Web Services for Devices (WS4D), 26-24
Web Service Transfer (WS-Transfer), 26-14
Wi-Analys tool, 8-26–8-28, 31-16–31-17
Wide area networks (WAN), 1-4
WiFi networks, 30-6
Wi-HTest tool, 8-24, 8-26, 31-19
Wind power plant information exchange model (WPP IEM)
ACSI services, 63-13–63-15
definition, 63-11
management functions, 63-13
mapping with IM
ACSI services, 63-15–63-16
communication profiles, 63-16
condition monitoring, 63-24
DNP3, 63-23–63-24
IEC 60870-5-104, 63-22–63-23
MMS, 63-19–63-21
SCSM, 63-15
SOAP-based web services, 63-17–63-18
operational functions, 63-12–63-13
services and interactions, 63-12
Wind power plant information model (WPP IM)
CDC
attributes characterization, 63-10–63-11
IEC 61850-7-3, 63-10
Status Value (STV), 63-11
WPP-specific, 63-9
data class attributes, 63-9
LD, 63-6–63-7
mapping with IEM
ACSI services, 63-15–63-16
communication profiles, 63-16
condition monitoring, 63-24
DNP3, 63-23–63-24
IEC 60870-5-104, 63-22–63-23
MMS, 63-19–63-21
SCSM, 63-15
SOAP-based web services, 63-17–63-18
metamodel, 63-6–63-7
modeling approach, 63-5–63-6
real WT, 63-8–63-9
wind turbine–specific LNs, 63-7–63-8
WPP-Specific LNs, 63-7–63-8
Wind power plants (WPPs), 63-3
Wind turbine–specific LNs, 63-7–63-8
Wired HART, 8-28–8-29
Wireless cable-replacement approach, 30-4
Wireless communication standards, 34-2
Wireless extensions
determinism, 35-6–35-7
IEEE 802.11 extensions, 35-7–35-8
IEEE 802.15.4 extensions, 35-8–35-9
random medium access techniques, 35-6
robustness, 35-7
transmission support, 35-5–35-6
wireless channels, 35-6
Wireless gateway approach, 30-4
WirelessHART, 8-3, 37-12
AES-128 algorithm, 31-16
application layer, 31-2, 31-12–31-13
architecture, 36-24–36-25
characteristics, 31-3
commands, 8-14–8-15
data link layer, 31-2
channel blacklisting mechanism, 31-9
channel hopping, 31-7–31-8
frequency diversity mechanism, 31-8–31-9

function, 31-7
link scheduling procedure, 31-9
logical link control (LLC) sublayer, 31-6
medium access control (MAC) layer, 31-6–31-7
neighbors table, 31-8
packet timing, 31-7–31-8
schematic representation, 31-6
TDMA, 31-7
time slot format, 31-7–31-8
time synchronization, 31-9–31-10
design, 31-16
fortify, 31-16
functionality, 36-25
goal, 31-3
HCF consortium, 31-19
installation and maintenance cost, 8-6
intrinsic redundancy, 31-17
ISA100.11a
application layer, 33-7
coexistence mechanisms, 33-8–33-9
data link layer, 33-5–33-7
flexibility, 33-8
installation, 33-3–33-4
ISA, 33-11–33-13
NAMUR, 33-11–33-13
network layer, 33-6–33-7
network topologies, 33-4
physical layer, 33-4–33-5
protocol stacks, 33-3
protocol support, 33-8
QoS, 33-9–33-11
security mechanisms, 33-9
stack structures, 33-4–33-5
transport layer, 33-7
ISO/OSI 7 layer model, 31-3–31-4
join procedure, 31-15–31-16
key factors, 8-4
LOS scenario, 31-17
message format, 31-13–31-15
network architecture, 31-4–31-5
network health reports, 8-6
network layer, 31-2, 31-10–31-11
network management, 8-6
physical layer, 31-2, 31-5–31-6
planning and installation, 8-29–8-31
scope, 31-16
security, 8-6, 36-25–36-26
standards landscape, 36-20–36-21
storage tank heating control, 36-19
topology, 8-13
transport layer, 31-11–31-12
vibration monitoring application, 36-16
Wi-Analys tool, 8-26–8-28, 31-16–31-18
Wi-HTest, 8-24, 8-26, 31-19
wireless control application, 36-19
wireless handheld, 8-27–8-28
Wireless local area network (WLAN), 37-5, 37-12
ARQ schemes, 30-13
bitrate/modulation scheme, 30-12
CAN, 30-13–30-14
channel quality, 30-12
distributed control systems, 30-2
FEC schemes, 30-13
FIP/WorldFIP fieldbus, 30-14
frequent plant reconfiguration, 30-2
IEEE 802.11 standard
DCF, 30-15–30-16
QoS, 30-16–30-17
infrastructure-based WLANs, 30-6
latency statistics, 37-14, 37-17
MAC protocols, 30-10–30-12
mobile plant diagnosis systems, 30-2
mobile subsystems, 30-2
packet error probability, 30-13
packet-switched communications, 30-6
physical layer
bit errors, 30-8
diversity techniques, 30-8
DSSS, 30-8
fast-fading phenomena, 30-8
FHSS, 30-8–30-9
frequency selective, 30-8
MIMO transmission, 30-8–30-9
OFDM, 30-8
packet losses, 30-8
slow-fading phenomena, 30-7
PROFIBUS fieldbus system, 30-14–30-15
redundancy, 30-12
retransmissions, 30-13, 37-14–37-15
wave propagation effects, 30-6–30-7
WiFi networks, 30-6
wireless fieldbus systems
energy supply, 30-3
hybrid systems, 30-4–30-5
mobility, 30-5
noise and interference, 30-5
real-time transmission, 30-3–30-4
transceivers, 30-3
Wireless MAC-aware bridging approach, 30-4
Wireless MAC-unaware bridging approach, 30-4
Wireless networks
active radio interference, 2-26–2-27
Bluetooth, 2-22–2-25
channel access errors, 2-28
passive channel errors, 2-27
radio propagation path loss, 2-25–2-26
trade-offs, 2-20–2-21
wireless Ethernet, 2-21–2-23
Wireless networks for industrial automation–process automation (WIA-PA)
application layer, 34-12–34-13
architecture, 36-28–36-29
Chinese National Standard, 34-2

cold standby redundancy approach, 34-3
cryptographic keys, 34-14
data link layer, 34-5–34-6
data link sublayer
frames priority, 34-9
frequency hopping mechanism, 34-8
long cycle data, 34-9–34-10
management services, 34-10
scheduling rules, 34-9
superframe, 34-7–34-8
timeslot, 34-7
DMAP, 34-4
field device, 34-2–34-3
functionality, 36-29
gateway device, 34-2–34-3
handheld device, 34-2–34-3
host configuration computer, 34-2–34-3
MAC layer, 34-6–34-7
mesh star topology, 34-4
network layer
address assignment, 34-10–34-11
adopted routing approach, 34-11–34-12
aggregation services, 34-12
disaggregation services, 34-12
fragmentation, 34-12
reassembling, 34-12
network manager (NM) rule, 34-4
physical layer, 34-4–34-6
protocol stack, 34-4
router device, 34-2–34-3
secure joining process, 34-14–34-15
secure transportation, 34-15–34-16
security, 36-29
security manager (SM) rule, 34-4
standards landscape, 36-22
star topology, 34-4
Wireless process control application
6LoWPAN, 36-19–36-20
communication standard, 36-18–36-19
communication technology, 36-19
deterministic communication, 36-19
storage tank heating control, 36-17–36-18
Wireless sensor actuator network (WSAN), 37-14, 59-8
Wireless sensor network (WSN), 38-2, 38-11
applications, 36-5
characteristics, 36-6
communication standards (*see* Communication standards)
communication technology selection, 36-8–36-9
components, 36-2
cooperative sensing, 36-3
definition, 36-2
development aspect matrix, 36-10–36-11
development process, 36-6, 36-9
discrete manufacturing, 36-6
diverse requirements, 36-6, 36-8
EMI, 36-5
harsh environments, 36-5
hybrid automation, 36-6
low-power design, 36-12
energy-efficient protocols, 36-34–36-35
functionality integration, 36-10
HMI, 36-33–36-34
low-power components, 36-10
principle, 36-30–36-31
sleep modes, 36-31–36-33
maximum delay, 36-4
modularity
black-box components, 36-43
control freak, 36-43
COTS components, 36-12–36-13, 36-43
HW *vs.* SW components, 36-40
interfaces, 36-42
lifetime, 36-12, 36-42
local *vs.* global optimum, 36-42–36-43
maintainability, 36-39
single *vs.* multiple chips, 36-41–36-42
trade-offs, 36-40
network topology, 36-4
packaging, 36-12
form factor, 36-38–36-39
hazardous environments, 36-36–36-37
IP rating, 36-35
vibration and extreme temperatures, 36-38
power supply, 36-13
energy buffer, 36-44
energy conversion, 36-47–36-48
energy storage, 36-46–36-47
energy transmission, 36-49
form factor, 36-45
lifetime, 36-44–36-45
power demands, 36-44
power management, 36-45
process attributes, 36-6–36-8
process automation, 36-6
self-healing, 36-3
self-organizing, 36-3
self-powered, 36-3
suppliers and components selection, 36-9
wireless process control application (*see* Wireless process control application)
wireless vibration monitoring application (*see* Wireless vibration monitoring application)
Wireless vibration monitoring application
bearing degradation signs, 36-14
communication standard, 36-16
communication technology, 36-16–36-17
design goals and requirements, 36-15–36-16
gateway, 36-15
repeater nodes, 36-14–36-15
wireless vibration sensor, 36-14
Wire train bus (WTB), 65-1
WLAN, *see* Wireless local area network (WLAN)
Word controller, 43-7

Word count, 45-15
Worst-case response time (WCRT), 52-18–52-19, 52-19
WPP IEM, *see* Wind power plant information exchange model (WPP IEM)
WPP IM, *see* Wind power plant Information Model (WPP IM)
WPPs, *see* Wind power plants (WPPs)
WPP-specific logical nodes (WPP-specific LNs), 63-7–63-8
WS-Discovery, *see* Web Services Dynamic Discovery (WS-Discovery)
WSDL, *see* Web Service Description Language (WSDL)
WS-MetadataExchange (WS-MEX)
    Get operation, 26-14
    hosted service, 26-16
    policy assertion, 26-16–26-17
    relationship metadata, 26-15–25-16
    ThisDevice, 26-15–25-16
    ThisModel, 26-15–25-16
WSN, *see* Wireless sensor network (WSN)
WS-PolicyAttachment, 26-13–26-14

X

XCP
    flash programming, 51-10–51-11
    online calibration, 51-10
    protocol layer, 51-8–51-9
    synchronous data transfer, 51-9–51-10
    transport layer, 51-9
XMPP, *see* Extensible Messaging and Presence Protocol (XMPP)

Z

ZigBee
    addressing scheme, 55-8
    application objects, 55-8–55-9
    application profiles, 36-24
    architecture, 36-22–36-23
    battery operation, 36-30
    coordinator, 55-8
    CSMA, 55-8
    end devices, 55-8
    frequency diversity, 36-30
    functionality, 36-22–36-23
    path diversity, 36-30
    routers, 55-8
    security, 36-23, 36-30
    Smart Energy Profile, 60-25
    standard, 33-2
    standards landscape, 36-20
    ZAP, 38-12
    ZigBee IP, 36-24
    ZigBee PRO, 36-24
ZigBee Application Profiles (ZAP), 38-12
ZigBee cluster library (ZCL), 55-9
ZigBee device object (ZDO), 55-9
Z-Wave, 55-11